World Scientific Proceedings Series on
Computer Engineering and Information Science 11

Data Science and Knowledge Engineering for Sensing Decision Support

Proceedings of the 13th International FLINS Conference

World Scientific Proceedings Series on Computer Engineering and Information Science

Series Founding Editor: Da Ruan
Series Editor: Jie Lu, University of Technology Sydney

Vol. 4 Computational Intelligence: Foundations and Applications
edited by Da Ruan, Tianrui Li, Yang Xu, Guoqing Chen and Etienne E. Kerre

Vol. 5 Quantitative Logic and Soft Computing
edited by Guojun Wang, Bin Zhao and Yongming Li

Vol. 6 Methods for Decision Making in an Uncertain Environment
edited by Jaime Gil-Aluja and Antonio Terceño

Vol. 7 Uncertainty Modeling in Knowledge Engineering and Decision Making
edited by Cengiz Kahraman, Etienne E Kerre and Faik Tunc Bozbura

Vol. 8 Decision Making Systems in Business Administration
edited by Anna M. Gil Lafuente, Luciano Barcellos de Paula, José M. Merigó Lindahl, Fernando Augusto Silva-Marins and Antonio Carlos de Azevedo-Ritto

Vol. 9 Decision Making and Soft Computing
edited by Ronei Marcos de Moraes, Etienne E Kerre, Liliane dos Santos Machado and Jie Lu

Vol. 10 Uncertainty Modelling in Knowledge Engineering and Decision Making
edited by Xianyi Zeng, Jie Lu, Etienne E. Kerre, Luis Martinez and Ludovic Koehl

Vol. 11 Data Science and Knowledge Engineering for Sensing Decision Support
edited by Jun Liu, Jie Lu, Yang Xu, Luis Martinez and Etienne E. Kerre

*For the complete list of titles in this series, please visit
http://www.worldscientific.com/series/wspsceis

World Scientific Proceedings Series on
Computer Engineering and Information Science 11

Data Science and Knowledge Engineering for Sensing Decision Support

Proceedings of the 13th International FLINS Conference

Belfast, Northern Ireland, UK   21–24 August 2018

editors

Jun Liu
Ulster University, UK

Jie Lu
University of Technology Sydney, Australia

Yang Xu
Southwest Jiaotong University, China

Luis Martinez
University of Jaén, Spain

Etienne E Kerre
University of Ghent, Belgium

NEW JERSEY • LONDON • SINGAPORE • BEIJING • SHANGHAI • HONG KONG • TAIPEI • CHENNAI

Published by

World Scientific Publishing Co. Pte. Ltd.
5 Toh Tuck Link, Singapore 596224
USA office: 27 Warren Street, Suite 401-402, Hackensack, NJ 07601
UK office: 57 Shelton Street, Covent Garden, London WC2H 9HE

British Library Cataloguing-in-Publication Data
A catalogue record for this book is available from the British Library.

World Scientific Proceedings Series on Computer Engineering and Information Science — Vol. 11
DATA SCIENCE AND KNOWLEDGE ENGINEERING FOR SENSING DECISION SUPPORT
Proceedings of the 13th International FLINS Conference (FLINS 2018)

Copyright © 2018 by World Scientific Publishing Co. Pte. Ltd.

All rights reserved. This book, or parts thereof, may not be reproduced in any form or by any means, electronic or mechanical, including photocopying, recording or any information storage and retrieval system now known or to be invented, without written permission from the publisher.

For photocopying of material in this volume, please pay a copying fee through the Copyright Clearance Center, Inc., 222 Rosewood Drive, Danvers, MA 01923, USA. In this case permission to photocopy is not required from the publisher.

ISBN 978-981-3273-22-1

For any available supplementary material, please visit
https://www.worldscientific.com/worldscibooks/10.1142/11069#t=suppl

Printed in Singapore

Preface

FLINS, an acronym originally for Fuzzy Logic and Intelligent Technologies in Nuclear Science, was launched in 1994 by an internationally reputed researcher Prof. Da Ruan in the Belgian Nuclear Research Center (SCK·CEN) aiming to give researchers the opportunity to carry out future-oriented research in fuzzy logic and its applications. For more than twenty years FLINS has been extended to a well-established worldwide and multidisciplinary research forum to advance the foundations and applications of computational intelligence for applied research in general and for complex engineering and decision support systems.

The 2018 International FLINS Conference (FLINS 2018) is the thirteenth in the successful conferences series: FLINS 1994 and FLINS 1996 in Mol, FLINS 1998 in Antwerp, FLINS 2000 in Bruges, FLINS 2002 in Ghent, FLINS 2004 in Blankenberge, FLINS 2006 in Genova, FLINS 2008 in Madrid, FLINS 2010 in Chengdu, FLINS 2012 in Istanbul, FLINS 2014 in Joao Pessoa, and FLINS 2016 in Roubaix.

The principal mission of FLINS is bridging the gap between machine intelligence and real complex systems via joint research between universities and international research institutions, encouraging interdisciplinary research and bringing multidisciplinary researchers together. After Prof. Da Ruan's sudden passing away in 2011 at the age of 50, his successors, including the organizers of FLINS 2012, FLINS 2014, FLINS 2016 and FLINS 2018, have followed his spirit of opening, compatibility and innovation, and continued to consolidate and enlarge this research consortium by developing and presenting new theories and applications in related areas.

Supervised by the FLINS Steering Committee, FLINS 2018 is held in Belfast (Northern Ireland, UK), is co-organized by the Ulster University and financially co-sponsored by Belfast City Council and Tourism NI, as well as the IEEE UK&I Section and IEEE System, Man and Cybernetics Society (IEEE-SMC). It is technically co-sponsored by Southwest Jiaotong University (China), National Association of Non-Classical Logic and Computation, Chinese Society of Logic (NANCLC, CSL), IEEE-SMC Ireland Chapter, European Society for Fuzzy Logic and Technology (EUSFLAT), and International Fuzzy System Association (IFSA).

Within the key theme of FLINS on computational intelligence and its applications, FLINS 2018 aims at providing an international forum that brings together researchers in mathematics, artificial intelligence, information science, and engineering, actively involved in areas of interest to Data Science and Knowledge Engineering with their applications in decision making problems under uncertainty to support sensible decision making in different areas, to report their latest innovations and developments, summarize the state-of-the-art, and exchange their ideas and progress, specially the contributions within the following four research fields, which are theoretical formalisms, methodologies; algorithms investigation, survey or practical applications:

1) Contributions on development of models, algorithms and systems to handle multiple and heterogeneous information for sensing decision making purpose. The approaches are either numerical or symbolic, or both; based on fuzzy/possibility theory, neural network, genetic algorithm, rough set theory, different varieties of machine learning/data mining methods, probability theory and belief function theory, and different varieties of logic based approaches (classical logics or non-classical logics);

2) Contributions on investigating, reviewing and assessing the principles, explanation, and strategies on how humans represent and use incomplete and uncertain data and knowledge from a cognitive science perspective;

3) Contributions on investigating, reviewing and assessing the principles, algorithms, methodologies and tools on how artificial intelligence can help in data science, data mining, big data, machine learning, and predictive analytics;

4) Written contributions representing advanced theories and innovative applications in computational intelligence as well as the integration of both quantitative and qualitative formalisms and modelling approaches in science, engineering, business and education.

Established also by Prof. Da Ruan and launched in 2006, the 13th Symposium of ISKE 2018: Intelligent Systems and Knowledge Engineering has also been organized in the frame of FLINS 2018. All the papers in both conferences will be published in this proceedings. ISKE 2018 emphasizes current practice, experience and promising new ideas in the broad area of intelligent systems and knowledge engineering, and its connection with FLINS 2018 is consistent and mutually beneficial.

The FLINS 2018 proceedings consists of four invited keynotes' abstracts and 192 conference papers from 30 countries. The four keynotes' talks include: 1) *From Data to Information Granules: A Data Science Perspective*, by Witold Pedrycz from University of Alberta, Canada; 2) *State-of-the-Art Evolutionary Algorithms for Many Objective Optimization*, by Gary G. Yen from Oklahoma State University, USA; 3) *AI 2.0: Augmented Intelligence,* by Paulo Lisboa from Liverpool John Moores University, UK; 4) *Data Driven Decision Making in Manufacturing — How did we get there and where next?*, by Adrian Johnston from Seagate Technology, Northern Ireland.

Moreover, the contributions of the 13 special sessions are playing an important role in FLINS 2018. These special sessions include the topics on data-driven decision making with uncertainty, intuitionistic fuzzy decision making, understandable and interpretable machine learning, logic based automated reasoning with applications, soft computing approaches in data mining and text retrieval, and soft computing methods in image and dynamic data processing, and so on.

The regular papers of the conference proceedings are organized as follows. Parts 1–6 (110 papers) mainly focus on the new theoretical contributions on different topics, such as theoretical foundation of computational intelligence, (fuzzy) multi-criteria decision analysis: advanced theory and methods, decision analytics and modeling under uncertainty, data mining and text retrieval, data analytics for classification, and logic and automated reasoning. Parts 7–13 (82 papers) present application-oriented contributions, including applied computing intelligence, (fuzzy) multi-criteria decision analysis and intelligent support system for marketing decisions, qualitative and quantitative decision making in management and social science, decision making under uncertainty in health care systems, decision model and intelligent system for risk and security analysis, soft computing methods in image and dynamic data processing, and soft computing and data analytic in fashion design and textile production.

We wish to express our special gratitude to all authors, reviewers, special sessions organizers, session chairs, and members of the International Committee of FLINS 2018 for their kind cooperation and contributions to the success of the Conference. We also thank Yolande Koh and Kim Tan (of World Scientific Publishing) for their kind advice and help to produce this book.

Acknowledgments

Special thanks go to the Steering Committee, International Scientific Committee, Organizing Committee and Organizers of Special Sessions as well as Reviewers for their invaluable contributions to the success of FLINS2018 Conference and ensuring the book is scientifically sound.

Honorary Chair
L.A. Zadeh (1921–09/2017) (USA)

Founding Chair
Da Ruan (1960–2011) (Belgium)

Steering Committee
Guoqing Chen (China)
Cengiz Kahraman (Turkey)
Etienne Kerre (Chair, Belgium)
Jun Liu (UK)
Jie Lu (Australia)
Luis Martinez (Spain)
Javier Montero (Spain)
Ronei Marcos de Moraes (Brazil)
Yang Xu (China)
Xianyi Zeng (France)

Program Chairs
Jun Liu (UK)
Jie Lu (Australia)

Program Co-Chairs
Javier Montero (Spain)
Yang Xu (China)

Conference Chair
Jun Liu (UK)

Organizing Chair
Liam Maguire (UK)

Organizing Co-Chairs
Chris Nugent (UK)
James Uhomoibhi (UK)
Hui Wang (UK)

Special Session Chairs
Juan Carlos Leyva Lopez (Mexico)
Shuwei Chen (China)

Poster Sessions Chairs
Guangquan Zhang (Australia)
Li Zou (China)

Tutorials Chairs
Guoqing Chen (China)
Cengiz Kahraman (Turkey)

Award Chairs
Etienne Kerre (Belgium)
Luis Martinez (Spain)

Publicity Chairs
Idri Ali (Morocco)
Peijun Guo (Japan)
Han-Xiong Li (Hong Kong)
Tianrui Li (China)
Wenjiang Li (China)
Xiaohong Liu (China)
Ronei Marcos de Moraes (Brazil)
Rosa Rodriguez (Spain)
Zhengming Song (China)

Wen-June Wang (Taiwan)
Dong-Ling Xu (UK)
Jian-Bo Yang (UK)
Xianyi Zeng (France)

Local Organizing Committee
Stuart Blair
Fiona Browne
Isaac Essel
David Glass
Glenn Hawe
Zhiwei Lin
Orla McHugh
Xinran Ning
Claire Orr
Joseph Rafferty
Boomadevi Sekar
Niloofer Shanavas
Colin Shewell
Roy Sterritt
Xin Tian
Huan Wan
Xin Wei
Chris Wilson
Chunlin Xu

Web Chair
Jesus Jaime Solano Noriega
(Mexico)

Demo Chair
Jonathan Synnott (UK)

International Program Committee
Jose Alonso (Spain)
Benjamin R. C. Bedregal (Brazil)
Vahid Behbood (Australia)
Gloria Bordogna (Italy)
Yves Breitmoser (Germany)
Fiona Browne (UK)
Humberto Bustince (Spain)
Jinli Cao (Australia)
Ivan Palomares Carrascosa (UK)
Pavel Anselmo Álvarez Carrillo (Mexico)
Oscar Castillo (Mexico)
Sondes Chaabane (France)
Guoqing Chen (China)
Shuwei Chen (China)
Yuwang Chen (UK)
Francisco Chiclana (UK)
Nicolai Christov (France)
Chris Cornelis (Belgium)
Cécile Coulon-Leroy (France)
Bernard De Baets (Belgium)
Martine De Cock (Belgium)
François Delmotte (France)
Ashok Deshpande (India)
Didier Dubois (France)
Rafael Espin (Mexico)
Macarena Espinilla (Spain)
Pasi Gabriella (Italy)
Bogdan Gabrys (Australia)
Alexander Gegov (UK)
Patrick Glauner (Luxembourg)
Fernando Gomide (Brazil)
Peijun Guo (Japan)
Wolfgang Halang (Germany)
Slim Hammadi (France)
Glenn Hawe (UK)
Francisco Herrera (Spain)
Farookh K. Hussain (Australia)
Van Nam Huynh (Japan)
Atsushi Inoue (USA)
Imed Kacem (France)
Cengiz Kahraman (Turkey)
Etienne Kerre (Belgium)
Ludovic Koehl (France)
Grzegorz Kolaczek (Poland)

Kan Li (China)
Tianrui Li (China)
Xiaohong Li (China)
Wenjiang Li (China)
Zhong Li (Germany)
Alan Wee-Chung Liew (Australia)
Chin-Teng Lin (Taiwan)
Yi Liu (China)
Juan Carlos Leyva Lopez (Mexico)
Victoria López (Spain)
Helen Lu (Australia)
Jie Lu (Australia)
Chuan Luo (China)
Jun Ma (Australia)
Luis Magdalena (Spain)
Enrico Marchioni (France)
Luis Martinez (Spain)
Patricia Melin (Mexico)
Dan Meng (China)
Javier Montero (Spain)
Jacky Montmain (France)
Ronei Marcos de Moraes (Brazil)
Oussalah Mourad (Finland)
Ngoc-Thanh Nguyen (Poland)
Vilem Novak (Czech Republic)
Marie-Pierre Pacaux-Lemoine (France)
Matilde Santos Penas (Spain)
Irina Perfilieva (Czech Republic)
Henri Prade (France)
Xiaoping Qiu (China)
Dragan Radojevic (Serbia)
Rita A. Ribeiro (Portugal)
Rosa Rodriguez (Spain)
Sandra Sandri (Brazil)
Regivan H. N. Santiago (Brazil)
Yongjun Shen (Belgium)
Jaime Solano (Mexico)
Qinglin Sun (China)
Peter Sussner (Brazil)
Jonathan Tan (UK)
Isis Truck (France)
Frédéric Vanderhagegen (France)
Kaburlasos Vassilis (Greece)
Dong Wang (China)
Jin Wang (UK)
Yingming Wang (China)
Junzo Watada (Japan)
Dianshuang Wu (Australia)
Fatos Xhafa (Spain)
Dongling Xu (UK)
Yang Xu (China)
Ronald R. Yager (USA)
Farouk Yalaoui (France)
Jian-Bo Yang (UK)
Yan Yang (China)
Xianyi Zeng (France)
Guangquan Zhang (Australia)
Pei Zheng (China)
Yue Zheng (China)
Zheng Zheng (China)
Hans J. Zimmermann (Germany)
Li Zou (China)

Organizers of Special Sessions
Antonio Javier Barragán Piña (Spain)
Humberto Bustince (Spain)
Shuwei Chen (China)
Yu-Wang Chen (UK)
Milena Cukic (Spain)
Ronei Marcos de Moraes (Brazil)
Jozo Dujmovic (USA)
Javier Fernández (Spain)
Yolanda García (Spain)
Patrick Glauner (Luxembourg)
Eloy Irigogyen Gordo (Spain)
Xingxing He (China)
Juan Carlos Leyva Lopez (Mexico)
Chee Peng Lim (Australia)

Xiaohong Liu (China)
Victoria López (Spain)
Jie Lu (Australia)
Marcin Jaroszewski (Poland)
Ulf Johansson (Sweden)
Cengiz Kahraman (Turkey)
Grzegorz Kołaczek (Poland)
Sezi Cevik Onar (Turkey)
Mourad Oussalah (Finland)
Basar Oztaysi (Turkey)
Jun Ma (Australia)
Luis Martinez (Spain)
Guadalupe Miñana (Spain)
Javier Montero (Spain)
Lie Meng Pang (Malaysia)
Pascal Perez (Australia)
Irina Perfilieva (Czech Republic)
José Carlos R. Alcantud (Spain)

Rosa Rodriguez (Spain)
Matilde Santos (Spain)
Liliane dos Santos Machado (Brazil)
Jesus Jaime Solano Noriega (Mexico)
Radu State (Luxembourg)
Kai Meng Tay (Malaysia)
Kewen Wang (Australia)
Yisong Wang (China)
Ying Wang (China)
Dianshuang Wu (Australia)
Zheng Yan (Australia)
Jian-Bo Yang (UK)
Xianyi Zeng (France)
Guanquan Zhang (Australia)
Li Zhang (UK)
Liangjie Zhao (China)
Ningning Zhou (China)

Contents

Preface   v

Acknowledgments   ix

Invited Lectures

From data to information granules: A data science perspective   3
Witold Pedrycz

State-of-the-art evolutionary algorithms for dynamic multiobjective optimization   7
Gary G. Yen

AI 2.0: Augmented Intelligence   10
Paulo J.G. Lisboa

Data driven decision making in manufacturing — How did we get there and where next?   12
Adrian Johnston

Part 1: Theoretical Foundation of Computational Intelligence

A novel OWA weighs determination based on ordinal dispersion   17
Nuria Martinez, Daniel Gómez, Pablo Olaso, Javier Montero and Karina Rojas

Hesitant fuzzy soft prefilters and filters over EQ-algebras   24
Jia-Yin Peng and Yu-Ting Jiang

Quantum controlled teleportation of five-qubit state   33
Zhi-wen Mo, Lan Shu, Xue Yang and Yu-ting Jiang

Analysis of numerical method for semilinear stochastic differential equation   38
Luoping Chen

New definition of the indefinite integral of fuzzy valued function linearly generated by structural elements 46
Tian-jun Shu and Zhi-wen Mo

Robustness of the interval-valued fuzzy inference method based on logic metrics 57
Yajing Wang, Minxia Luo and Huarong Zhang

Novel intuitionistic 2-tuple linguistic representation model 65
Yi Liu, Jun Liu and Ya Qin

Generalized bidirectional quantum operation teleportation via multiqubit cluster states 73
Si-qi Zhou, Ming-qiang Bai and Zhi-wen Mo

Bidirectional remote teleportation of single-qubit operations via cluster states 81
Si-qi Zhou, Ming-qiang Bai and Zhi-wen Mo

Similarity measure between fuzzy sets based on axiomatic definition 89
Xiaodong Pan and Yang Xu

Development of the automated system for design solutions' analysis at CAD KOMPAS-3D 95
Alexander Nikilaevich Afanasyev, Sergey Igorevich Brigadnov, Nikolay Nikolaevich Voit and Tatyana Vasilievna Afanasyeva

Solution of a fuzzy differential equation with interactivity via fuzzy Laplace transform 103
Silvio Salgado, Laécio C. Barros and Estevão Esmi

Robustness of the interval-valued fuzzy inference algorithms 109
Minxia Luo, Lixian Wu and Li Fu

Energy optimization in cloud servers using static threshold VM consolidation technique (STVMC) 117
Bilal Ahmad, Sally McClean, Darryl Charles and Gerard Parr

The properties of soft product topological space 129
Fu Li

Fuzzy TOPSIS: Violation of basic axioms   137
Boris Yatsalo, Alexander Korobov and Luis Martínez

s-grouping functions: A generalization of bivariate grouping functions by   145
t-conorms
Hugo Zapata, Graçaliz Pereira Dimuro, Javier Fernández and Humberto Bustince

Upper and lower general aggregation operators based on a strong fuzzy   154
metric
Pavels Orlovs and Svetlana Asmuss

Part 2: (Fuzzy) Multi-Criteria Decision Analysis: Advanced Theory and Methods

Interval-valued Pythagorean fuzzy interaction aggregation operators and   165
their application in multiple attribute decision making
Wei Yang and Yongfeng Pang

Approach for multi-attribute group decision making based on   173
linguistic-valued aggregation operator
Li Yao, Yiming Cao, Pengsen Liu, Yingying Xu and Li Zou

Linguistic preference relation and its application in group decision   182
making
Xinzi Li, Hongyue Diao, Li Yao and Li Zou

Managing consensus with an established goal in the DG model with an   191
undirected graph: An approach based on minimizing adjusted opinions
Xia Chen, Yucheng Dong and Zhaogang Ding

A consensus model for large scale using hesitant information   198
Rosa M. Rodríguez, Luis Martínez and Guy De Tré

An optimization-based approach to aggregating multi-granular hesitant   205
fuzzy linguistic term sets
Zhen Zhang and Wenyu Yu

The effect of opinion evolution in consensus reaching in group decision making 213
Quanbo Zha and Yucheng Dong

Improved score based decision making method by using fuzzy soft sets 220
Yaya Liu, Rosa M. Rodríguez, Keyun Qin and Luis Martínez

On group recommendation supported by a minimum cost consensus model 227
Raciel Yera, Álvaro Labella, Jorge Castro and Luis Martínez

A new visualization for preferences evolution in group decision making 235
Álvaro Labella and Luis Martínez

An alternative core information of hesitant fuzzy linguistic term sets for linguistic decision making 243
Fangling Ren, Ying Qiao, Mingming Kong and Zheng Pei

An unbalanced linguistic terms transformation method for linguistic decision making 252
Jing Liu, Ying Qiao and Zheng Pei

Absolute and relative preferences in AHP-like matrices 260
David Koloseni, Tove Helldin and Vicenç Torra

Generalized hesitant fuzzy linguistic term sets for linguistic decision making 268
Li Yan, Mingming Kong and Zheng Pei

Type-2 intuitionistic fuzzy (IFS2) WASPAS 276
Cengiz Kahraman, Sezi Çevik Onar, Basar Oztaysi and Esra İlbahar

Part 3: Decision Analytics and Modeling under Uncertainty

An evidential-reasoning based model for probabilistic inference with uncertain data acquired from different data sources 287
Huaying Zhu, Jian-Bo Yang, Dong-Ling Xu and Cong Xu

	xvii
A further discussion on matching value for cooperative games and its application in profit allocation decision-making *Fei Guan, Yuxin Cui and Hu Zhang*	294
Regret theory-based decision-making with evidential reasoning *Liuqian Jin and Xin Fang*	302
Decision roughness measurement and application research based on data effect *Fachao Li, Xiao Zhang and Chenxia Jin*	308
Semi-supervised transfer learning in Takagi-Sugeno fuzzy models *Hua Zuo, Guangquan Zhang and Jie Lu*	316
RNN-based traffic flow prediction for dynamic reversible lane control decision *Yang Song and Jie Lu*	323
Road traffic flow prediction using deep transfer learning *Bin Wang, Zheng Yan, Jie Lu, Guangquan Zhang and Tianrui Li*	331
Incremental possibilistic decision trees in non-specificity approach *Mohamed Sofien Boutaib and Zied Elouedi*	339
Maintaining case knowledge vocabulary using a new evidential attribute clustering method *Safa Ben Ayed, Zied Elouedi and Eric Lefevre*	347
A workforce health insurance plan recommender system *Dianshuang Wu, Jie Lu, Farookh Hussain, Craig Doumouras and Guangquan Zhang*	355
New product development using disjunctive belief rule base *Jianguo Xu, Mengjun Li, Leilei Chang, Jiang Jiang, Yuwang Chen and Longhao Yong*	363
Restricted multi-pruning of decision trees *Mohammad Azad, Igor Chikalov, Mikhail Moshkov and Shahid Hussain*	371

Pipeline leak alarm decision via dynamic evidence fusion 379
Xiaobin Xu, Haiyang Xu, Guo Li, Pingzhi Hou, Darong Huang and Weifeng Liu

A belief rule-based inference model for prediction of R&D project success 389
Ying Yang, Yu-Xiao Li and Yu-Wang Chen

Causal inference with Gaussian processes for support of terminating or maintaining an existing program 397
Adi Lin, Junyu Xuan, Guangquan Zhang and Jie Lu

Social influenced decision making: A belief-based model analysis 405
Lei Ni, Yu-Wang Chen and Oscar de Bruijn

Pareto-smoothed inverse propensity weighing for causal inference 413
Fujin Zhu, Adi Lin, Guangquan Zhang, Jie Lu and Donghua Zhu

Feature selection and weighing for case-based reasoning system using random forests 421
Booma Devi Sekar and Hui Wang

Comparative analysis on extended belief rule-based system for activity recognition 430
Long-Hao Yang, Jun Liu, Ying-Ming Wang and Luis Martínez

Part 4: Data Mining and Text Retrieval

On the reduction of biases in big data sets for the detection of irregular power usage 439
Patrick Glauner, Radu State, Petko Valtchev and Diogo Duarte

Dynamic neighborhood-based decision-theoretic rough set 446
Xin Yang, Tianrui Li, Dun Liu and Qianqian Huang

Incremental updating of approximations in composite ordered decision systems under attribute generalization 453
Qianqian Huang, Tianrui Li, Yanyong Huang and Xin Yang

Data-driven competitor identification using text mining and word network analysis *Xueyan Zhong, Guoqing Chen, Leilei Sun and Lan Liu*	461
Parallel attribute reduction algorithms based on CUDA *Yunmeng Hu, Tianrui Li, Jie Hu and Hongmei Chen*	469
A multi-kernel spectral clustering algorithm based on incomplete views *Wei Zhang, Yan Yang and Jie Hu*	477
Attention-based bidirectional LSTM for Chinese punctuation prediction *Jinliang Li, Chengfeng Yin, Zhen Jia, Tianrui Li and Min Tang*	485
Clutter reduction of parallel coordinates based on an approximate measure of line crossing *Yunlong Li, Tianrui Li, Shengdong Du and Xun Gong*	492
A novel preprocessing approach for imbalanced learning in software defect prediction *Kamal Bashir, Tianrui Li, Chubato Wondaferaw Yohannese, Mahama Yahaya and Tayseer Ali*	500
A new tool for static and dynamic Android malware analysis *Alejandro Martín, Raúl Lara-Cabrera and David Camacho*	509
A super-hyper network model based on matrix operation *Shengjiu Liu, Tianrui Li and Yunlong Li*	517
Visualization methods for tracking the dynamics of coronal dimmings *Yuhang Yang, Bo Peng, Tianrui Li and Yunmeng Hu*	526
DP_DETECTION: An outlier detection algorithm based on density of big data *Xiaodi Li, Ping Deng, Ming Huang, Dingcheng Li and Hongjun Wang*	534
An iterative multi-criteria optimization of product snippets enhanced by feature extraction from online reviews *Yao Mu, Qiang Wei, Guoqing Chen and Xunhua Guo*	545

Using fuzzy representation in educational data mining and learning analytics — 553
Jun Ma, Jie Yang, Sarah K. Howard, Carlos Gonzalez and Dany Lopez

Compressive sensing based feature selection: A case study for commuter behaviour modelling — 560
Jie Yang, Jun Ma and Xiangqian Wang

ALES: An Arabic Legal query Expansion System — 568
Imen Bouaziz Mezghanni and Faiez Gargouri

Extracting a diverse information subset by considering information coverage and redundancy simultaneously — 576
Baojun Ma, Qiang Wei, Guoqing Chen and Qiongwei Ye

Term frequency occurrences on web pages for textual information retrieval — 585
Karthika Sivapathasundaram, Xiaochun Cheng and Miltos Petridis

Community finding in dynamic networks using a genetic algorithm improved via a hybrid immigrants scheme — 591
Angel Panizo Lledot, Gema Bello-Orgaz, Alfonso Ortega and David Camacho

Using graph-theoretic methods for text classification — 599
Niloofer Shanavas, Hui Wang, Zhiwei Lin and Glenn Hawe

Argumentation system for intelligent assistants using fuzzy-based reasoning — 608
T. Koivuaho, M. Ibrahim, F. Ummul and M. Oussalah

A fuzzy based approach for wordsense disambiguation using morphological transformation and domain link knowledge — 617
F. Farooghian, M. Oussalah and E. Gilian

A novel path planning approach for unmanned ships based on deep reinforcement learning — 626
Chen Chen, Feng Ma, Jia-Lun Liu, Xin-Ping Yan and Xian-Qiao Chen

Part 5: Data Analytics for Classification

Naive Bayes clusterer — 637
Mujiexin Liu, Hongjun Wang, Tian Rui Li and Ping Deng

Semi-supervised cluster ensemble based on density peaks — 645
Kadhim Mustafa, Hongjun Wang, Yuan Zhou and Jian Song

Ensemble Evidential Editing k-NNs through rough set reducts — 652
Asma Trabelsi, Zied Elouedi and Eric Lefèvre

A new evidential collaborative filtering: A hybrid memory- and model-based approach — 660
Raoua Abdelkhalek, Imen Boukhris and Zied Elouedi

Improving the activities of a robotic guide by taking into account human reactions — 668
J. Javier Rainer, Fernado López Hernández and Ramón Galán

Hierarchical classification learning: A novel two-layer framework for multiclass classification — 674
Yuanyuan Liu, Xiaoshuang Qiao, Gongde Guo and Hui Wang

A rough set based hybrid approach for classification — 683
Ahmed Saad Hussein, Tianrui Li, Noora Sabah Jaber and Chubato Wondaferaw Yohannese

A Fuzzy Gamma Naive Bayes classifier — 691
Ronei Marcos de Moraes, Elaine Anita de Melo Gomes Soares and Liliane dos Santos Machado

Spectral data classification using locally weighted partial least squares classifier — 700
Weiran Song, Hui Wang, Paul Maguire and Omar Nibouche

Attention-based bidirectional LSTM for Chinese punctuation prediction — 708
Jinliang Li, Chengfeng Yin, Zhen Jia, Tianrui Li and Min Tang

A new map for combination of spatial clustering methods 715
Danielly Cristina De Souza Costa Holmes,
Rodrigo De Pinheiro De Toledo Vianna and Ronei Marcos De Moraes

Part 6: Logic and Automated Reasoning

Distinctive features of the contradiction separation based dynamic automated deduction 725
Yang Xu, Shuwei Chen, Jun Liu, Xiaomei Zhong and Xingxing He

The empirical study of imported genetic algorithm combined with ant colony algorithm based on 3-SAT problems 733
Huimin Fu, Yang Xu, Xinran Ning and Wuyang Zhang

Verifying Deutsch-Schorr-Waite algorithm in first-order logic with arithmetic 740
Bo Yang, Sheng Liang, Ying Zhang and Mingyi Zhang

Look-ahead clause selection strategy for contradiction separation based automated deduction 750
Shuwei Chen, Yang Xu, Jun Liu and Feng Cao

α-Generalized resolution method based on linguistic truth-valued first-order logic system 758
Weitao Xu and Yang Xu

Deductive control strategies based on contradiction separation rule 766
Feng Cao, Yang Xu, Xinran Ning and Xuecheng Wang

Preliminary framework to combine contradiction separation based automated deduction with superposition and given-clause algorithm 774
Xinran Ning, Yang Xu, Feng Cao and Jun Liu

Chaotic mappings in symbol space 782
Inese Bula

A new strategy for preventing repeated and redundant clauses in contradiction separation based automated deduction 790
Xingxing He, Yang Xu, Jun Liu, Xiaoping Qiu and Yingfang Li

Logical difference of propositional theories *Yisong Wang, Hong Liu, Ying Zhang, Mingyi Zhang and Danning Li*	798
Reducing Answer Set Programs under partial assignments *Jianmin Ji and Hai Wan*	806
Improvements of categorical propositions on quantification and systemization *Yinsheng Zhang*	814
Non-prioritised belief revision for DL-Lite TBoxes *Quan Yu, Zhiqiang Zhuang, Zhe Wang and Kewen Wang*	824
Strongest necessary and weakest sufficient conditions in S5 *Renyan Feng, Yisong Wang, Panfeng Chen and Jincheng Zhou*	832
The number of minimal preference contraction operation *Maonian Wu, Shaojun Zhu, Bo Zheng, Weihua Yang and Mingyi Zhang*	840
Protection zone designation of railway radio environment based on TD-LTE system *Peng Xu, Meirong Chen, Lifang Feng, Fangli Ma, Guanfeng Wu and Yang Xu*	848
Study on transition *Long Hong and Jianyang Zhao*	855
A scale-free model for random ASP programs *Lian Wen, Kaile Su and Zhe Wang*	862
Procedural extensions for executable ontologies in conventional software development *Selena S. Baset and Kilian Stoffel*	870
Graphic deduction based on set (I): Venn graphic representation of judgment *J** *Xia He, Guoping Du and Long Hong*	878
Graphic deduction based on set (II): The graphic argument of *J** *Xia He, Guoping Du and Long Hong*	884

Parallel Monte Carlo for calculating π value — 891
Hui Zong, Jianyang Zhao and Long Hong

Learnt clause deletion based on clause deepness — 897
Zhonghe Du and Zhenming Song

Part 7: Applied Computing Intelligence

Compound trajectory optimization methodology for parafoil delivery system based on quantum genetic algorithm — 907
Hao Sun, Qinglin Sun, Shuzhen Luo, Zengqiang Chen, Wannan Wu and Jin Tao

Multiple μ-stability of complex-valued Cohen-Grossberg neural networks with unbounded time-varying delays — 914
Yunfeng Liu, Manchun Tan and Desheng Xu

SIFT-based textile defects detection by using adaptive neural-fuzzy inference system — 922
Xueqing Zhao, Xinjuan Zhu, Tao Xue, Kaixuan Liu and Yongmei Deng

Fuzzy explicit simplified MPC with adjustment parameter — 929
Juan Manuel Escaño, Kritchai Witheephanich, Samira Roshany-Yamchi and Carlos Bordons

A control to soybean aphid via fuzzy linear programming — 937
Magda S. Peixoto, Silvia Carvalho, Laécio C. Barros, Rodney Bassanezi, Estevão Laureano and Weldon Lodwick

An approach to recognize coral species on the coast of Brazil using image analysis and fuzzy associative memories based on equivalent measures — 943
Estevão Esmi, João Batista Florindo, Flávio Pérez and Marcos Barbeitos

Tasks scheduling in computational grids: A proposal considering an uncertainty regime — 951
Bruno M. P. Moura, Guilherme B. Schneider, Adenauer C. Yamin, Mauricio L. Pilla and Renata H. S. Reiser

Safety helmet recognition based on deep convolution neural networks — 959
Ningning Zhou, Guofang Huang and Shaodong Shi

Research on optimal trajectory curve of intelligent vehicles based on neighborhood system — 968
Xing Wang, Hailiang Zhao and Zhigang Wang

Goodwin model via *p*-fuzzy system — 977
Daniel Sanchez, Laecio Barros, Estevao Esmi and Alessandro Miebach

Part 8: (Fuzzy) Multi-Criteria Decision Analysis and Intelligent Support System for Marketing Decisions

The new product design problem using novel preference approach — 987
Juan Carlos Leyva Lopez, Leon-Santiesteban Martín, Omar Ahumada-Valenzuela and Jesus Jaime Solano-Noriega

An intelligent decision support system for the design of new products — 995
Juan Francisco Figueroa Pérez, Juan Carlos Leyva López, Luis Carlos Santillán Hernández and Edgar Omar Pérez Contreras

A multicriteria and multiobjective approach for the market segmentation problem — 1003
Diego Alonso Gastelum-Chavira, Juan Carlos Leyva Lopez and Elsa Veronica Larreta-Ramirez

A new disaggregation preference method for new products design — 1010
Pavel Anselmo Alvarez, Juan Leyva Lopez and Pavel Lopez Parra

A choice model for the product design problem based on the outranking approach — 1018
Juan Carlos Leyva-Lopez, Leon-Santiesteban Martín, Omar Ahumada-Valenzuela and Alma Montserrat Romero-Serrano

Interval valued neutrosophic CODAS method for renewable energy selection — 1026
Eda Bolturk and Ali Karasan

Natural gas technology selection using Pythagorean fuzzy CODAS — 1034
Eda Bolturk and Cengiz Kahraman

A multicriteria model for launching a new product for a group of decision-makers ... 1042
Pedro Flores Leal, Pavel Anselmo Alvarez, Diego Alonso Gastelum Chavira and Juan Carlos Leyva Lopez

Ground handling services firm evaluation based on neutrosophic MULTIMOORA method ... 1050
Serhat Aydin and Mehmet Yörükoğlu

Creating alternatives for the eggplant waste problem in a horticultural company ... 1058
Elsa Veronica Larreta-Ramirez, Diego Alonso Gastelum-Chavira, Juan Carlos Leyva-Lopez and Octavio Valdez-Lafarga

Interval-valued intuitionistic fuzzy MULTIMOORA approach for new product development ... 1066
Orhan Feyzioğlu, Fethullah Gocer and Gulcin Buyukozkan

Analysis of companies' digital maturity by hesitant fuzzy linguistic MCDM methods ... 1074
Merve Güler, Esin Mukul and Gülçin Büyükozkan

A multi-criteria decision support system for multi-UAV mission planning ... 1083
Cristian Ramírez-Atencia, Víctor Rodríguez-Fernández and David Camacho

The use of decision support systems for new product design: A review ... 1091
Juan Francisco Figueroa Pérez, Juan Carlos Leyva López, Luis Carlos Santillán Hernández and Edgar Omar Pérez Contreras

Prioritization of the requirements for collaborative feedback platform for course contents using Pythagorean fuzzy sets ... 1099
Basar Oztaysi, Sezi Cevik Onar and Cengiz Kahraman

Solar energy project selection by using hesitant Pythagorean fuzzy TOPSIS ... 1107
Veysel Çoban, Sezi Çevik Onar, Basar Oztaysi and Cengiz Kahraman

Interval-valued intuitionistic fuzzy based QFD application for smart hospital design 1115
Deniz Uztürk, Gülçin Büyükozkan, Ahmet Fahri Negüs and M. Yaman Öztek

Part 9: Qualitative and Quantitative Decision Making in Management and Social Science

Causality and inference identification method of individual poverty 1127
Xiaohong Liu and Xianyi Zeng

The influence of ecological protection in poverty-stricken areas of western China 1135
Haozhen Liu and Bin Luo

An interval valued intuitionistic fuzzy location based recommendation system utilizing social platforms 1143
S. Ceren Öner, Başar Öztayşi and Mahir Öner

The effects of CEO's facial trustworthiness on the investment decision 1152
Jinping Gao, Si Long, Chang Yuan, Jing Fan and Yan Wan

The application of evaluation method based on prospect theory in stock investment 1159
Chang Yuan, Yujia Sui, Jinping Gao, Yu Pan and Li Gao

Sustainable supplier selection using two-phase DFD and TOPSIS within PD-HFLTS context 1168
Zhen-Song Chen

Neutrosophic AHP and prioritization of legal service outsourcing firms/law offices 1176
Cengiz Kahraman, Başar Öztayşi, Sezi Çevik Onar and Eda Boltürk

Evaluation of smart cities with integrated hesitant fuzzy linguistic AHP–COPRAS method 1184
Esin Mukul, Merve Güler and Gülçin Büyüközkan

Weighting performance indicators of law offices by using interval valued intuitionistic fuzzy AHP 1192
Basar Oztaysi, Sezi Cevik Onar and Cengiz Kahraman

Multi-criteria evaluation of law firms by using dynamic intuitionistic fuzzy sets 1199
Sezi Çevik Onar, Basar Oztaysi and Cengiz Kahraman

Analysis on the effectiveness of education in institutions of higher learning based on cultural confidence 1208
Yanyan Ding and Xiaohong Liu

2-tuple combined group decision making methodology for climate change strategy selection 1216
Gülçin Büyüközkan and Deniz Uztürk

Analysis of balance between reform of judicial system and supervision to power in China 1225
Cheng-Gao Liu, Yun Yuan and Zhao-Yi Zhang

Influence factors of enterprises' social responsibility performance in China 1233
Yun Yuan, Cheng-Gao Liu and Yin-Ye Liao

Study on management decision of China's dry cleaning enterprises based on the current situation of industry development 1245
Ming Jiang and Caijuan Zhang

A study on the use and development strategy of official Weibo of Minzu universities in China 1254
Siyuan Song and Ming Jiang

Part 10: Decision Making Under Uncertainty in Health Care Systems

Frequency domain analysis of telephone helpline call data 1267
Alexander Grigorash, Raymond R. Bond, Maurice D. Mulvenna, Siobhan O'Neill, Cherie Armour and Colette Ramsey

Predicting assistive technology adoption for people with Parkinson's 1273
disease using mobile data from a smartphone
Jonathan Greer, Ian Cleland and Sally McClean

Machine learning using synthetic and real data: Similarity of evaluation 1281
metrics for different healthcare datasets and for different algorithms
Rachel Heyburn, Raymond R. Bond, Michaela Black, Maurice Mulvenna,
Jonathan Wallace, Deborah Rankin and Brian Cleland

Fuzzy framework for activity recognition in a multi-occupant smart 1292
environment based on wearable devices and proximity beacons
Macarena Espinilla, Javier Medina Qeuro, Naomi Irvine, Ian Cleland
and Chris Nugent

The impact of dataset quality on the performance of data-driven 1300
approaches for human activity recognition
Naomi Irvine, Chris Nugent, Shuai Zhang, Hui Wang, W. Y. Ng Wing,
Ian Cleland and Macarena Espinilla

Features selection and improving for trauma outcomes prediction models 1309
Fatima Almaghrabi, Prof. Dong-Ling Xu and Prof. Jian-Bo Yang

Computer aided diagnostic tool for prostate cancer with rule extraction 1315
from support vector machines
Guanjin Wang, Jie Lu, Jeremy Yuen-Chun Teoh and Kup-Sze Choi

Map-based medical practice behavior analysis: Methodology and a case 1323
study on Australia's medical practices
Yi Zhang, Wei Wang, Junyu Xuan, Jie Lu, Guangquan Zhang and
Hua Lin

Part 11: Decision Model and Intelligent System for Risk and Security Analysis

Analysis of sentencing reasoning on the traffic accident crime in China 1333
Caijuan Zhang and Ming Jiang

Ranking road safety risk factors using preference structures and fuzzy 1341
preference structures
Yongjun Shen, Elke Hermans and Qiong Bao

Chaos in hydrology: A case study in Konya Basin, Turkey 1349
Didem Odabasi Cingi, Ergun Eray Akkaya and Dilek Eren Akyuz

Intuitionistic fuzzy decision making: Multi-drug resistant tuberculosis 1357
risk assessment
Elif Dogu and Y. Esra Albayrak

Software fault prediction using data reduction approaches 1364
*Chubato Wondaferaw Yohannese, Tianrui Li, Kamal Bashir,
Macmillan Simfukwe and Ahmed Saad Hussein*

The sensitive data leakage detection model based on Bayesian 1373
convolution neural network
Chunliang Zhou, Zhengqiu Lu and Yangguang Liu

Digital supply chain risk analysis with intuitionistic fuzzy cognitive map 1385
Gulcin Buyukozkan and Fethullah Gocer

Development of the approach to check the correctness of workflows 1392
Alexander Afanasyev, Nikolay Voit, Maria Ukhanova and Irina Ionova

Estimation of fuzzy reliability: A case study for flash vessel in ammonia 1400
storage tank
Satish Salunkhe and Ashok Deshpande

A fuzzy telematics data-driven approach for vehicle insurance 1407
policyholder risk assessment
Mohammad Siami, Anahita Namvar, Mohsen Naderpour and Jie Lu

Part 12: Soft Computing Methods in Image and Dynamic Data Processing

Detection of structural breaks and perceptionally important points in time 1417
series
Vilém Novák

Post-processing in edge detection based on segments 1425
P. A. Flores-Vidal, Nuria Martínez and Daniel Gómez

Noise influence in FzT+JPEG image compression   1433
Petr Hurtik and Irina Perfilieva

Synthetic dataset for compositional learning   1440
Vojtech Molek and Jan Hula

Differentiation of organic and non-organic apples using image processing — a cost-effective approach   1446
Jiang Nan Feng, Wang Hui and Guo Gong De

Analysis of senile dementia from the brain magnetic resonance imaging data with clustering   1454
Xiaobo Zhang, Yan Yang, Hongjun Wang and Ping Deng

Cross-domain image description generation using transfer learning   1462
Philip Kinghorn and Li Zhang

Heart sound de-noising using wavelet and empirical mode decomposition based thresholding methods   1470
Shaocan Fan, Booma Devi Sekar, Peng Un Mak, Sio Hang Pun and Mang I Vai

Weak boundary value problem: Fuzzy partition in Galerkin method   1478
Linh Nguyen, Irina Perfilieva and Michal Holčapek

Part 13: Soft Computing and Data Analytic in Fashion Design and Textile Production

Intelligent application of data fusion in garment manufacturing under the thinking of "Internet plus"   1489
Liu Cui, Hong Dai and Kai Xuan Liu

Garment fit evaluation using artificial intelligence technology   1498
Kaixuan Liu, Hong Dai, Yanbo Ji, Yongmei Deng, Yongchi Xu, Yue Wang, Xueqing Zhao and Tao Xue

A two staged forecasting scheme considering the constraints of sales forecasting in the fashion industry   1504
Rohan Maleku Shrestha, Giuseppe Craparotta, Sébastien Thomassey and Ronald Moore

A collaborative platform with negotiation mechanism for make-to-order textile supply chain: A study based on multi-agent simulation 1512
Ke Ma, Sébastien Thomassey and Xianyi Zeng

A fuzzy signal processing approach integrated into an intelligent garment for online fetal movement monitoring 1520
Xin Zhao, Xianyi Zeng, Guillaume Tartare, Ludovic Koehl and Julien De Jonckheere

Optimization of the body part recognition method of 3D human mesh based on propagation algorithm 1528
Yu Chen, Xianyi Zeng and Zhebin Xue

The application of process modeling in denim manufacturing 1536
Zhenglei He, Sebastien Thomassey, Xianyi Zeng, Danying Zuo and Changhai Yi

A proposal of a rapid evaluation and analysis system of fashion design using expression analyzer and eye tracker 1544
Yu Chen, Xianyi Zeng, Xie Hong and Zhebin Xue

Explorative multi-objective optimization of marketing campaigns for the fashion retail industry 1551
Håkan Sundell, Tuwe Löfström and Ulf Johansson

A data-driven approach to online fitting services 1559
Tuwe Löfström, Ulf Johansson, Jenny Balkow and Håkan Sundell

Analysis of consumer emotions about fashion brands: An exploratory study 1567
Chandadevi Giri, Nitin Harale, Sebatien Thomassey and Xianyi Zeng

An intelligent approach to study suiting fabric formability 1575
Zhebin Xue, Lei Shen, Jianli Liu and Yu Chen

Author Index 1585

Invited Lectures

From data to information granules:
A data science perspective

Witold Pedrycz

Department of Electrical & Computer Engineering, University of Alberta
Edmonton AB T6R 2V4, Canada

The apparent reliance on data and experimental evidence in system modeling, decision-making, pattern recognition, and control engineering entails their centrality and a paramount role of data science. To capture the essence of data and avoid various artifacts (noise, outliers, incompleteness) as well as facilitate building essential descriptors and revealing key relationships, we advocate a need of transforming data into information granules. Information granules are regarded as conceptually sound knowledge tidbits over which a number of various models are developed.

1. Introductory Notes

The paradigm shift implied by the engagement of information granules becomes manifested in several tangible ways including: (i) stronger dependence on data when building structure-free and versatile models spanned over selected representatives of experimental data, (ii) emergence of models at various levels of abstraction being delivered by the specificity/generality of information granules, and (iii) building a collection of individual local models and supporting their efficient aggregation.

A framework of Granular Computing along with a diversity of its formal settings offers a critically needed conceptual and algorithmic environment. A suitable perspective built with the aid of information granules is advantageous in realizing a suitable level of abstraction and becomes instrumental when forming sound and pragmatic problem-oriented tradeoffs among precision of results, their easiness of interpretation, value, and stability.

Those aspects emphasize the importance of *actionability* and *interestingness* of the produced findings considered e.g., in decision-making. Granular models built on a basis of available numeric models offer a comprehensive view at real world systems. More specifically, granular spaces, viz. spaces of granular parameters of the models and granular input and output spaces play a pivotal role in making the original *numeric* constructs more realistic. The granular results may also emerge as a direct outcome of the aggregation of locally constructed models.

2. The Functional Scheme

The functional scheme advocated in this study is portrayed in the following way:

data → numeric prototypes → granular prototypes → granular models

The formation of numeric prototypes occurs through invoking clustering algorithms, which as a result yield a partition matrix and a collection of the prototypes. The typical representatives of such clustering methods are the *K*-Means algorithm and its fuzzy set-based counterpart of the Fuzzy C-Means (*FCM*). It is worth noting that the clustering realizes a certain process of abstraction producing a small number of the prototypes out of a large number of numeric data. Clustering can be also completed in the feature space returning a small collection of abstracted features (groups of features) that might be referred to as meta-features. The prototypes are numeric and this format of the produced results could be limited and somewhat counterintuitive anticipating that any generalization (abstraction) may give rise to more abstract (viz. non-numeric) representatives. This motivates the buildup of *granular* prototypes, which arise as a direct consequence to realize a more comprehensive representation of the data. In the sequence, any model (say, a predictor, classifier, associator) is constructed in the presence of information granules. The main advantage stems from the fact that the data set of information granules is far smaller than the original numeric data. This directly contributes to the reduced development (learning) overhead; it is worth stressing that typically $c<<N$ where c stands for the number of clusters and N is the number of data. The limitation, however, resides with the results of lower specificity. This entails that the formation of information granules, especially their level of abstraction, has to be prudently selected to achieve the required quality of the granular model.

3. Evaluation and Design

At each phase of the overall scheme presented above, there is a mechanism supporting a comprehensive assessment of the quality of the developed artifacts. In light of the granular format of the results present at several phases, their evaluation has to take this facet into consideration. Two performance indexes are considered:

(i) *coverage*. The measure expresses a validity of a certain information granule visa-a-vis experimental evidence (usually of numeric character). on a basis of which information granule or a collection of information granules have been developed. The detailed formulation is dependent upon the formalism of information granules being used. For instance, for interval information granule, the coverage is the count of the number of data included in the granule A, viz. $cov(A)$ = card $(x|\ A(x)=1\}$. In case of A being realized as a fuzzy set, the coverage is a so-called σ-count, viz. a sum of the membership degrees.

(ii) *specificity*. The measure of specificity $sp(.)$ expresses how detailed (specific) a given information granule A is. The boundary conditions pertain to a case when A is composed of a single element and here specificity attains 1 and the information granule being the entire universe (space), in which case the specificity is set to zero.

The criteria of coverage and specificity are in conflict: an increase of one of them yields the decrease of the second one. The optimization of the product of the coverage and specificity can be sought as a viable way of striking a sound compromise. The definitions presented above can be generalized to situations of granular experimental evidence.

Elaborating the performance evaluation processes, they are associated with the consecutive phases of the scheme.

Evaluation of representation capabilities of numeric prototypes. This assessment is quantified by means of a so-called reconstruction error, which is taken as a sum of distances between the original data and the reconstructed ones where the reconstruction is accomplished by means of the prototypes. The reconstruction is carried out by running a sequence of the two steps, namely a granulation phase followed by the degranulation phase. The granulation G results in an internal representation of input data x through the use of the prototypes; in case of the *FCM* algorithm, this results in a family of membership grades u, namely $u = G(x;\ v_1,\ v_2,\ ..,\ v_c)$. The degranulation G^{-1} is regarded as a inverse procedure of building (reconstructing) x based on the internal representation and the prototypes, namely $\hat{x} = G^{-1}(x;\ v_1,\ v_2,\ ..,\ v_c)$. Usually the reconstruction result is not ideal, i.e., $x \neq \hat{x}$. The higher the reconstruction error (with the sum taken over the data for which granulation process was applied), the weaker the representation capabilities of the developed information granules become.

Evaluation of representation capabilities of granular prototypes. Starting from numeric prototypes, one constructs their granular counterparts whose parameters are optimized wit the use of the principle of justifiable granularity where the product of coverage and specificity is maximized through the adjustments of the parameters of the granule. The result is V expressed as a pair $(v,\ \rho)$ with ρ

optimized via this principle. The representation capabilities of granular prototypes $\{V_1, V_2, ..., V_c\}$ are again evaluated by the coverage and specificity criteria. For any original datum, one develops an internal representation in terms of the membership grades $u_1, u_2,..., u_c$ and forms an information granule $\hat{V} = (\hat{v}, \hat{\rho})$ where $\hat{v} = \sum_{i=1}^{c} u_i^2 v_i / \sum_{i=1}^{c} u_i^2$ and $\hat{\rho} = \sum_{i=1}^{c} u_i^2 \rho_i / \sum_{i=1}^{c} u_i^2$ The coverage and specificity is determined for the granular result evaluated with regard to x.

Evaluation of the quality of the granular model. The model is constructed in the presence of granular data and given their format of the pairs $\{(v_i,\rho_i), (w_i,\tau_i)\}$, $i =$ 1, 2,..., c, the design (learning) is carried out in a standard manner by noting, however, that both w_i and τ_i have to be approximated by the outputs of the model. The minimized performance index is expressed as $\sum_{i=1}^{c} \| GM(v_i, \rho_i) - z_i \|^2 = \sum_{i=1}^{c} \| \hat{z}_i - z_i \|^2$ where $\hat{z}_i = [\hat{w}_i \ \hat{\tau}_i] = GM(v_i, \rho_i)$ is the vector of the outputs produced by the granular model $GM(.)$ while $z_i = [w_i \ \rho_i]$. As the results of modeling are information granules (instead of numeric outputs), the quality of the model is evaluated by engaging both the coverage and specificity measures. It has been noted here that the nature of granular data (say, their specificity) needs to be adjusted to achieve the required characteristics of the model.

State-of-the-art evolutionary algorithms for dynamic multiobjective optimization

Gary G. Yen
School of Electrical and Computer Engineering, Oklahoma State University
Oklahoma, USA

Evolutionary computation is the study of biologically motivated computational paradigms which exert novel ideas and inspiration from natural evolution and adaptation. The applications of population-based meta-heuristics in solving multiobjective optimization problems have been receiving a growing attention. To search for a family of Pareto optimal solutions based on nature-inspiring metaphors, *Evolutionary Multiobjective Optimization Algorithms* have been successfully exploited to solve optimization problems in which the fitness measures and even constraints are uncertain and changed over time. This is generally regarded as the most fundamental characteristics of *Dynamic Multiobjective Optimization Problems* (DMOPs) [1]. This underlying problem characteristic bears significant implications for real-world applications, such as deregulated electricity markets in which the operations of different power stations are controlled and coordinated to maximize profit while minimizing risk [2]. Solving the DMOPs efficiently and effectively has become an important research issue in evolutionary computation community [3].

In recent years, a great deal of progress has been made and different types of algorithms have been proposed. In all of these methods, one class of approaches, the *prediction based*, has gained much interest. This class of approaches allows evolutionary algorithm and machine learning to be seamlessly integrated. After deriving a prediction model via machine learning techniques, the evolutionary algorithms can sustain the needed performance even if the environment changes over time. For example, in [4], the authors proposed a memory-based EA which introduced two types of prediction models. The first one used the linear/nonlinear regression model to predict when the environment would change, while the second model was based on Markov chains which was used to forecast changes. In [5], the authors suggested integrating motion information into an evolutionary algorithm, such that the algorithm can track a time-changing optimum. In [6], the authors proposed a Kalman-extended genetic algorithm and this algorithm was designed to determine when to re-evaluate an existing individual, when to produce a new individual, and which individual to re-evaluate. The basic idea of these methods is "keeping track of good (partial) solutions in order to reuse them under periodically changing environment". If we consider this view from a statistical point of view, this idea implies that the solutions of a dynamic optimization problem obey an identical distribution. In other words, the solutions which are used to construct the prediction model and the solutions forecasted by the prediction model meet the Independent Identical Distribution hypothesis to certain degree. This assumption undoubtedly simplifies the complexity of the problem, however we still have to understand there is an appreciable difference between the good, but out-of-date solutions and the proper and newly generated solutions, especially under a dynamic environment. That is to say, the changing Pareto optimal front may lead to the different distributions of the training samples and the predicted samples, and this problem is very difficult for the traditional machine learning methods.

The findings from machine leaning community already proven that a prediction model built by traditional machine learning methods leaves much room to be improved when the training samples and the predicted samples fail to meet the IID hypothesis. Transfer learning [7] allows the distribution of data used in training and testing to be different and it is becoming a useful tool to overcome this difficulty. Therefore, the dynamic multiobjective optimization algorithms based on traditional machine learning methods, especially the prediction based algorithms, can also have significant performance improvements by overcoming the limitation caused by the IID, and transfer learning approach is a powerful instrument we can use to improve performance of evolutionary algorithms for DMOPs. In this talk, we argue that integrating transfer leaning approaches into an evolutionary algorithm can offer significant benefits to performance and robustness for designing better Dynamic Multiobjective Evolutionary Algorithms (DMOEAs). We adopt a domain adaptation method, called transfer component analysis, to construct a prediction model. This model uses the gained knowledge of finding Pareto optimal solutions, but not the population, to generate an initial population pool for the optimization function at the next instance. Based on this initial population pool, the optima of the changed environment can be found more efficiently and effectively. The proposed domain adaptation learning approach can be easily incorporated into any evolutionary-based multiobjective optimization algorithms. The contribution of this research is the integration between transfer learning and classical evolutionary multiobjective optimization algorithms. This combination provides two benefits. First, the advantages of the evolutionary algorithms are preserved in the improved design for DMOPs. Secondly, the proposed design can significantly improve the search efficiency via reusing past experience which is critical for solving the DMOPs.

References

[1] M. Farina, K. Deb, and P. Amato, "Dynamic multiobjective optimization problems: test cases approximations, and applications," *IEEE Transactions on Evolutionary Computation*, vol. 8, no. 5, pp. 425–442, 2004.

[2] C. Cruz, J. R. Gonzalez, and D. A. Pelta, "Optimization in dynamic environments: a survey on problems methods and measures," Soft Computing, vol. 15, no. 7, pp. 1427–1448, 2010.

[3] M. Daneshyari and G. G. Yen, "Cultural-based particle swarm for dynamic optimization problems," *International Journal of Systems Science*, vol. 43, no. 7, pp. 1284–1304, 2012.

[4] A. Simoes and E. Costa, "Prediction in evolutionary algorithms for dynamic environments," *Soft Computing*, vol. 18, no. 8, pp. 1471–1497, 2013.

[5] C. Rossi, M. Abderrahim, and J. C. Diaz, "Tracking moving optima using Kalman-based predictions," *Evolutionary Computation*, vol. 16, no. 1, pp. 1–30, 2008.

[6] P. Stroud, "Kalman-extended genetic algorithm for search in nonstationary environments with noisy fitness evaluations," *IEEE Transactions on Evolutionary Computation*, vol. 5, no. 1, pp. 66–77, 2001.

[7] S. J. Pan and Q. Yang, "A survey on transfer learning," *IEEE Transactions on Knowledge and Data Engineering*, vol. 22, no. 10, pp. 1345–1359, 2010.

AI 2.0: Augmented Intelligence

Paulo J.G. Lisboa

Department of Applied Mathematics, Liverpool John Moores University
Byrom Street, Liverpool L3 3F, UK

Computational intelligence (CI) models are often evaluated solely on their predictive performance, lacking appropriate consideration of other aspects which might make a claim to the intelligence of the model and which can be critical for their use by a subject expert who is not a CI expert. Yet appearances can be deceiving, especially with summary performance measures, e.g. AUROC. This is especially the case for non-linear models given their ability to exploit any weaknesses in the data, for instance structural artefacts that can add confounding effects. In addition, many applied CI models work well for well-classified cases but cannot explain predictions for borderline cases. In other words, they confirm to expert users what they already know but do not add insights to the data in the difficult cases for which CI is most needed.

There is a drive for the use of CI to complement rather than automate decision-making. This is fundamental to make CI useful in practice and has been termed Augmented Intelligence, or AI 2.0.

The talk will illustrate some of the pitfalls in the design and validation of databased models. It will then describe how rules can be efficiently derived from neural networks so opening the black-box.

1. Context

A new paradigm is gaining ground in AI in the form of explainable machine learning, by rendering generic non-linear models interpretable by domain experts. AI 2.0, augmented intelligence, complements rather than competes with human decision-making. It formally recognises that applied AI needs to be human rather than machine centred and promotes the level of transparency that is essential for achieving the most critical formal requirement of software applications, validation against the initial specification. This is the most fundamental criterion for functional software testing. It is what builds trust in the machine.

Moreover, it is also a requirement to demonstrate compliance with the 'right to explanation' believed to be enshrined in the General Data Protection Regulation (GDPR) taking effect in Europe from May 2018. In the case of automated decision-making, Article 13 mandates in the case of automated decision-making, to provide "meaningful information about the logic involved, as well as the significance and the envisaged consequences of such processing for the data subject". This raises yet unanswered questions, the more so as increasingly complex models such as deep learning, are coming to prominence [1].

The principles of interpretable models in machine learning and directions for practical application are not new [2,3]. It is even possible to improve on the performance of large models by simplifying them, for instance by "distilling the knowledge" of large neural networks using much smaller ones to replicate their functional response [4].

2. Review of Interpretable Methods

An alternative and popular way of presenting and using complex models e.g. to clinicians, is the use of nomograms. It will be shown how to derive them to explain SVMs [6]. Further, case-based reasoning will be explored using information geometry to calculate similarity metrics directly to identify patients-like-mine with reference to specific clinical queries e.g. diagnosis or prognosis. This defines a statistically principled intelligent query system that enables subject experts to identify relevant cases for a new diagnosis and so apply clinical judgement to validate the inference of the decision support algorithm by ascertaining whether the new case is associated with the correct reference cases. This provides a direct route to interpretation and a way for domain experts to access generic non-linear models by using them for intelligent data retrieval, through delivering the analytical models in the form of an inferred data structure.

These methods make some headway towards the goal of a rigorous science of interpretable machine learning [7]. This is important because, ultimately, machine explanation remains the best, perhaps the only guarantee of plausible generalisation.

References

1. B. Goodman and S. Flaxman, *arXive preprint atXiv:16006.08813* (2016).
2. P.J.G. Lisboa, in F. Masulli, G. Pasi, and R. Yager (Eds.): *WILF 2013*, LNAI 8256, pp. 15–21. Springer International Publishing, Switzerland (2013).
3. A. Vellido, J.D. Martín-Guerrero, and P.J.G. Lisboa, *European Symposium on Artificial Neural Networks, Computational Intelligence and Machine Learning (ESANN)*, 163 (2012).
4. G. Hinton, O. Vinyals and J. Dean, *arXive preprint atXiv:1503.02531v1* (2015).
6. V. Van Belle, V. B. Van Calster, S. Van Huffel, S. J.A.K. Suykens, and P.J.G. Lisboa, *PLOS ONE*, **11**, 01 (2016).
7. F. Doshi-Velez and B. Kim, *arXive preprint atXiv:1702.08608v2* (2017).

Data driven decision making in manufacturing — How did we get there and where next?

Adrian Johnston

Seagate Technology, 1 Disc Drive, Springtown, N. Ireland, BT48 0BF, UK

While all the current focus is on Manufacturing 4.0 it should be acknowledged that many initiatives have been implemented across industry sectors over the past 15 years. Initiatives such as the Six Sigma business [1] and Lean Manufacturing business improvement strategies helped lay the foundations and innovative thinking that would facilitate more advanced analysis and control systems to in manufacturing facilities [2]. In particular, the semiconductor manufacturing industry has been at the forefront of data system innovation for two decades. Through continuous improvement programs there was a clear understanding that real value lay in the data collated from processing steps, manufacturing equipment, manufacturing execution systems and measurement systems. This paper will demonstrate how data driven decision making is not a new concept, more an evolving one that has gained significant traction as new and emerging technology and data systems have been developed. A short introduction will provide an overview of how simple business improvement techniques identified solutions that would replace some simple computational tasks. The second section of this paper will highlight how industrial practice evolved to capture subject matter expertise within rule based control systems. This knowledge was then augmented with other automated factory systems to make real time process control a reality. The final section of the paper will demonstrate how the dawn of new advances in machine learning and computational systems has enabled a rapid growth of advanced analytics within current manufacturing environments. Covering descriptive and predictive analytics, the final section will review how further advances in cognitive analytics is being implemented within current manufacturing control systems to move beyond simple tasks to more complex automated solutions.

1. Introduction

1.1. *Business improvement strategies*

Six Sigma business improvement strategies [3] where first introduced in 1986 but became an integral part of manufacturing systems from the mid 1990s. A blend of data analysis and business tools, it provided a structured roadmap to identify and eliminate unwanted and costly process variability. A key to the success of these programs is the "Control" strategy required to ensure sustainability of improvements made. In essence this requires automating some of the engineering knowledge uncovered during a BI project within systems.

This in turn requires knowledge based system development and corresponding data architecture.

1.2. Advanced analytics and systems

Figure 1 illustrates the path that Business Improvements has aided the progression of analytics from descriptive to diagnostic.

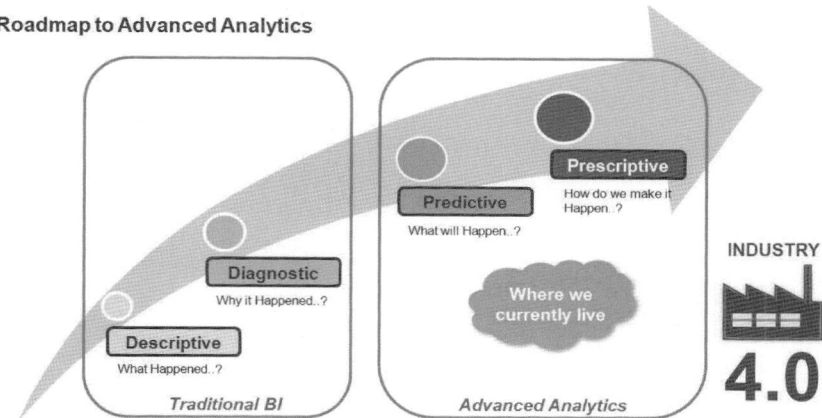

Figure 1. Roadmap to Advanced Analytics.

This subsequently led to the deployment of predictive and prescriptive analytics capability with the rise in Machine Learning capabilities [4]. These brought significant advantages to businesses in the areas such as Predictive Maintenance, Virtual Metrology and Automated Process Control. However, these were somewhat limited by computation constraints of real time analytics coupled with a lack of appropriate data and infrastructure. These constraints have been the focus of recent Big Data developments with a significant capability improvement in data and compute infrastructure. Complex computation tasks have now become a reality and with it the ability to model human knowledge base and decision making. This will see the implementation of complex decision support systems across many business sectors with Cognitive Analytics capability [5].

References

1. F.W. Breyfogle III, *Implementing Six Sigma: Smarter Solutions Using Statistical Methods*. Wiley (1999).
2. A.B. Johnston, L.P. Maguire and T.M. McGinnity, Using business improvement techniques to inform the optimisation of production cycle time: An industrial case

study, *Proceedings of the IEEE SMC UK-RI Chapter Conference 2004 on Intelligent Cybernetic Systems.* September 7–8 (2004).
3. P.S. Pande, R.P. Neumann and R.R. Cavanagh, The six sigma way. An implementation guide for process improvement teams. McGraw-Hill (2002).
4. P.K.S. Prakash, B. Honari, A.B. Johnston and S.F. McLoone, Optimal wafer site selection using Forward Selection Component Analysis, IEEE/SEMI Advanced Semiconductor Manufacturing Conference (2012).
5. MIT Technology Review, https://www.technologyreview.com/lists/technologies/2017/ (2017).

Part 1

Theoretical Foundation of Computational Intelligence

A novel OWA weights determination based on ordinal dispersion

N. Martinez,* D. Gómez, P. Olaso, J. Montero and K. Rojas

Complutense University of Madrid
Madrid, Spain
*dagomez@estad.ucm.es

One of the most common techniques to find the adequate weights in OWA operators is based on the orness concept. The weights are determined by maximizing the entropy (variation) for a fixed orness value. The entropy can be viewed as a dispersion measure for nominal variables. Taking into account that the weights in OWA operators can be viewed as an ordinal measure (instead of a nominal measure), in this paper we propose a novel way to determine these weights based on ordinal dispersion measures instead of the entropy measure. We find an explicit formula for the weights in this case and we show the differences in some multicriteria decision making examples.

Keywords: Aggregation operators; OWA; ordinal dispersion; multicriteria decision making problems.

1. Introduction

An aggregation operator is usually defined as a real function $A : [0,1]^n \to [0,1]$, such that for n items in $[0,1]$, yields an aggregation value in the same interval.[1,10–12,15] OWA aggregation operators introduced by Yager[16] have been discussed and studied in a large number of papers. Many notable operators functions such as the max, arithmetic average, median and min can be viewed as a particular case of OWA aggregation operator.

One important issue in the theory of Ordered Weighted Averaging (OWA) operators is the determination of the associated weights. There exist several approaches[13] to determine the weights in OWA operators. In[4] it is suggested that the weight vector should be the one that maximizes the operator entropy while some predetermined orness is guaranteed. The same problem was studied in[8] where the maximum entropy model is transformed into a polynomial equation, which can be solved analytically. Also in this paper the entropy deviation measure is changed by the classical variance for a fixed level of orness.

Finally in [14] the following question is presented. Let us suppose that we have an aggregation problem and after a discussion with the decision-maker we want to use a disjunctive operator. Obviously, the maximum operator could be used but it would imply losing too much information in the aggregation process. Now let us suppose that the variability of the aggregation operators is associated with the risk and that it is fixed, then it could be interesting to maximize its orness.

When the OWA operators are used to aggregate the different membership degrees of a sequence of items, the weights could be viewed as an ordinal variable since the i-th weight is associated with the i-th item once the membership degrees have been ordered.

Although OWA aggregation operator is a symmetric function (i.e $OWA(\mu(x)) = OWA(\mu_1, \ldots, \mu_n) = OWA(\mu_{\pi(1)}, \ldots, \mu_{\pi(n)})$), their weights are clearly not. Taking this into account, the the dispersion measure should be sensitive to changes in the order of the weights. Otherwise vectors such as $w = (0.5, 0, 0, 0.5)$ and $w' = (0, 0.5, 0.5, 0)$ will have the same dispersion as it happens with the classical entropy commonly used.

In this work we propose to solve the weights determination problem in OWA aggregation operators by using an ordinal dispersion measure instead of nominal dispersion measure like the entropy (MEOWA weights).

The paper is organized as follows: in Section 2 we present some preliminaries that are used in the work. Section 3 contains the novel way to determine the weights based on ordinal dispersion and an explicit expression is also presented. In this section we present a small application to multicriteria decision making problems. Finally in Section 4 some remarks are drawn.

2. Preliminaries

In this section, we recall some concepts and properties of OWA aggregation operators, the orness concept and the concept of ordinal measure that we will use in this work.

2.1. OWA aggregation Operators

An OWA aggregation operator of dimension n can be defined as a mapping $F : [0,1]^n \longrightarrow [0,1]$ that has an associated weight vector $w = (w_1, \ldots, w_n)$

having the properties

$$\sum_{i=1}^{i=n} w_i = 1; \quad 0 \le w_i \le 1 \quad \forall i = 1, \ldots n$$

and such that

$$F(x_1, \ldots, x_n) = \sum_{i=1}^{i=n} w_i x_{(i)}.$$

2.2. OWA weights determination based on entropy and orness

First of all, let us define the orness associated with an OWA operator. Orness measure represent in some way the degree in which an operator is disjunctive. Formally, given an OWA aggregation operator, the orness is defined as

$$orness(W) = \sum_{i=1}^{i=n} \frac{n-i}{n-1} w_i.$$

Let us observe that the orness measure for $W_{max} = (1, \ldots, 0)$ is 1 (totally disjunctive), $W_{min} = (0, \ldots, 1)$ is 0 (totally conjuctive) and for any other weights vector it takes a value between 0 and 1, in particular for $W_{average} = (\frac{1}{n}, \ldots, \frac{1}{n})$ it is 0.5.

In [4] the weights are determined maximizing the entropy of the weights vector for a given level of orness. Formally the mathematical programming problems is as follows:

$$\begin{array}{l} maximize\ dispersion(w) = -\sum_{i=1}^{i=n} w_i Ln(w_i) \\ s.t. \quad \frac{1}{n-1} \sum_{i=1}^{i=n} (n-i) w_i = \alpha \\ \quad \sum_{i=1}^{i=n} w_i = 1 \\ \quad 0 \le w_i \le 1,\ i = 1, \ldots, n. \end{array} \quad (1)$$

Using the method of Lagrange multipliers it is possible to transform this problem to a polynomial equation which has to be solved to determine the optimal weights. Nevertheless, these polynomial equations are not trivial and present some computational problems.[4,5,7,8] As a consequence, there is not an explicit formula for its calculation and due to its complexity there are some authors (see for example [7] among others) that used heuristic approaches for a simple calculation.

2.3. *Ordinal and nominal dispersion measures*

Variability, dispersion, variance among other topics are key concepts in Statistics. Although these concepts can be extended to random variables and can be used in different frameworks, in this paper we focus in the concept of dispersion of a statistical variable for a given sample. In statistics, dispersion measure is usually associated with continuous statistical variables. When the dispersion has to be measured in ordinal variables (like for example a Likert-scale) the common approach is to convert the ordinal estimation into a numerical one by assigning numerical values on each ordinal variable category. And after it is possible to used a classical dispersion measure. Some authors have stated that [2,6,9] this procedure is undesirable because it can lead to misunderstanding and misinterpretation of the measurement results.

Taking into account the above consideration, there have been defined some ordinal dispersion measures [2,6,9] that work with ordinal statistical variables better than other classical measures such as entropy, standard deviation, variance or quasivariance.

Although other measures could be used for the approach we develop here, in this paper we will center on the ordinal dispersion measure defined in 1992 by Berry and Mielke[3] usually called as IOV. Given and ordinal variable with values $X = \{1, \ldots, n\}$ and relative frequency $f = (f_1, \ldots, f_n$ the ordinal dispersion measure IOV is defined as

$$IOV = \sum_{i=1}^{i=n-1} \sum_{j=i+1}^{j=n} f_i f_j (j-i). \qquad (2)$$

3. A New Way to Determine the Weights for OWA Operators

The main difference between entropy and an ordinal solution is that entropy reaches its maximum value when all the weights have the same value, or if this is not a feasible solution, they are close to this idea. On the other hand, an ordinal solution reaches the maximum value in the polar case, i.e. when we have the maximum value in the extreme cases.

Formally, the mathematical programming problem is as follows:

$$\begin{array}{l} maximize\ dispersion(w) = \sum_{i=1}^{i=n-1} \sum_{j=i}^{i=n}(j-i)w_i w_j \\ s.a. \qquad \frac{1}{n-1}\sum_{i=1}^{i=n}(n-i)w_i = \alpha \\ \qquad \sum_{i=1}^{i=n} w_i = 1 \\ \qquad 0 \leq w_i \leq 1,\ i=1,\ldots,n. \end{array} \qquad (3)$$

With a similar analysis using Lagrange Multipliers it is possible to obtain the following result.

Given an α value in the previous optimization problem, it is possible to prove that the vector weight $w = (\alpha, 0, \ldots, 0, 1 - \alpha)$ is the optimal weight vector that maximize the ordinal dispersion. We will refer to this weight as MOOWA vector.

Example 3.1. Let us suppose that we have 10 candidates that apply for a job in a company. After some considerations this firm has decided to take into account 3 criteria $C_1, \ldots C_3$ to evaluate those 10 candidates. In Figure 1, we show the degree to which each solicitant reaches the maximum qualification to each criteria. In order to rank the participants, we decide to use an OWA aggregation operators. We show below the final aggregation for a given orness (α) value using the MEOWA weights and MOOWA (ordinal weights).

	c1	c2	c3	MEOWA α=0.3	MEOWA α=0.4	MEOWA α=0.5	MEOWA α=0.6	MOOWA α=0.3	MOOWA α=0.4	MOOWA α=0.5	MOOWA α=0.6
p1	1	1	0	2	1	1	1	6-7	3	3-4	3-4
p2	0,8	0,2	0	10	10	7-10	7	8	8	7-8	7
p3	0,6	0,4	0	9	9	7-10	8	9	9	9	9
p4	0,6	0,2	0,2	6	7	7-10	10	5	7	7-8	8
p5	0,5	0,5	0	7	8	7-10	9	10	10	10	10
p6	0,5	0,5	0,5	1	2	4	5	2	2	3-4	5-6
p7	1	0,5	0,1	4	4	3	3	3	4	2	2
p8	1	0,2	0	8	6	6	4	6-7	5-6	5	3-4
p9	1	0,4	0,3	3	3	2	2	1	1	1	1
p10	0,7	0,3	0,2	5	5	6	6	4	5-6	6	5-6

Fig. 1. Ranking in terms of orness level with MEOWA and MOOWA weights.

Note that regardless of the level of exigency (orness: $\alpha = 0.3$, 0.4, 0.5, 0.6), for both aggregations the candidates who occupy the firsts positions are very similar. Now let's look at the first candidate, the C_1 and C_2 criteria are very successful, while C_3 presents the worst score. If we assume that all criteria are important and focus on the most demanding context ($\alpha = .3$), it does not seem reasonable that p_1 is placed in second place considering that there are other candidates who have a better evaluation in C_3 and half of the other criteria. However, MEOWA assigns this good ranking position to this candidate because of the high score in the other criteria, being affected by the classic trade off problem. This does not happen with

MOOWA, since it penalizes the low value in C_3, placing the candidate in the sixth position, and positioning those that surpass the minimum value in this criterion in a better ranking (p_9, p_7, p_{10} and p_4). Therefore, and considering all relevant criteria, it seems less accurate to place a candidate with zero achievement in a better position in one criterion despite having another criterion with a maximum score (p_1) than a candidate with average skill in both criteria (p_9).

Therefore, it seems less accurate to place better a candidate with zero achievement in one criterion, (despite the fact of having maximum scores on other criteria (p_1)) than a candidate with average skills in all criteria (p_9).

4. Final Remarks

In this paper we propose an alternative way to determine the weights based on ordinal dispersion for a fixed level or orness that we have called MOOWA weights. Weights in OWA aggregation operators can be viewed as an ordinal measure in which the order is relevant. Classical entropy dispersion measure does not take into account the order of the vector and thus vectors as $w = (0.5, 0.5, 0, 0, 0)$ (that aggregates the information considering the two most greater values) has the same dispersion as $w = (0.5, 0, 0, 0, 0.5)$ that is a linear combination of the maximal and minimal values. There exist some situations in which MEOWA weights produce non desirable compensations (trade off) between criteria due to the fact that entropy tries to spread the weights. With our approach we avoid this problem.

Although we think that classical MEOWA are very useful to aggregate the information in some practical situations, they present two important inconvenients in some real cases. The first one is the complexity of its calculations and the second one is the use of entropy. In this paper we have avoided these two important issues and we have sketched a practical example showing the differences between both approaches.

References

1. T. Calvo, A. Kolesarova, M. Komornikova, R. Mesiar. Aggregation operators, properties, classes and construction methods. In Aggregation Operators New Trends and Applications (Physica-Verlag, Heidelberg, 3-104, 2002).
2. J. Blair, M.G. Lacy. Measures of variation for ordinal data as functions of the cumulative distribution Perceptual and Motor Skills, 82 (1996), pp. 411-418.
3. K.J. Berry, P.W. Mielke Jr. Assessment of variation in ordinal data Perceptual and Motor Skills, 74 (1992), pp. 63-66.

4. M. O'Hagan. Aggregating template or rule antecedents in real-time expert systems with fuzzy set logic, Proc. 22nd Annu. IEEE Asilomar Conf. on Signals, Systems, Computers, Pacific Grove, CA, 1988, pp. 81-689.
5. Filev, D., Yager, R. R. On the issue of obtaining OWA operator weights. Fuzzy sets and systems, 94(2), (1998) 157-169.
6. F. Franceschini, D. Romano. Control chart for linguistic variables: A method based on the use of linguistic quantifiers International Journal of Production Research, 37(16) (1999), pp. 3791-3801.
7. Fuller, R., Majlender, P. (2001). An analytic approach for obtaining maximal entropy OWA operator weights. Fuzzy Sets and Systems, 124(1), 53-57.
8. Fuller, R., Majlender, P. (2003). On obtaining minimal variability OWA operator weights. Fuzzy Sets and Systems, 136(2), 203-215.
9. Gadrich, T., Bashkansky, E. (2012). ORDANOVA: analysis of ordinal variation. Journal of Statistical Planning and Inference, 142(12), 3174-3188.
10. D. Gómez, J. Montero: A discussion of aggregation functions. Kybernetika **40**, 107-120 (2004).
11. D. Gomez, K. Rojas, J. Montero, and J. T. Rodriguez in *Consistency and Stability in Aggregation Operators: An Application to Missing Data Problems*, ed. by H. Bustince et al. Aggregation Functions in Theory and in Practise, Advances in Intelligent Systems and Computing, $DOI: 10.1007/978-3-642-39165-1_48$ (Springer-Verlag Berlin Heidelberg 2012), pp. 507-518.
12. M. Grabisch, J. Marichal, R. Mesiar and E. Pap. Aggregation Functions. Encyclopedia of Mathematics and its Applications (2009).
13. Liu, X. (2011). A review of the OWA determination methods: Classification and some extensions. Recent Developments in the Ordered Weighted Averaging Operators: Theory and Practice, 49-90.
14. Marchant, T. (2006). Maximal orness weights with a fixed variability for OWA operators. International Journal of Uncertainty, Fuzziness and Knowledge-Based Systems, 14(03), 271-276.
15. K. Rojas, D. Gómez, J.T. Rodriguez and J. Montero. Strict Stability in Aggregation Operators. Fuzzy sets and Systems **228**, 44-63 (2013).
16. R.R. Yager. On ordered weighted averaging aggregation operators in multicriteria decisionmaking. IEEE Transactions on systems, Man, and Cybernetics, 18(1), 183-190 (1988).

Hesitant fuzzy soft prefilters and filters over EQ-algebras

Jia-Yin Peng

School of Mathematics and Information Science, Neijiang Normal University
Neijiang 641199, China
pengjiayin62226@163.com

Yu-Ting Jiang

College of Mathematics and Software Science, Sichuan Normal University
Chengdu, 610066, China
250326748@qq.com

Application of hesitant fuzzy soft sets to EQ-algebras is considered. The concepts of hesitant fuzzy soft prefilters (filters) are proposed, and some properties are studied. Characterizations of hesitant fuzzy soft prefilters (filters) are discussed.

Keywords: EQ-algebra; hesitant fuzzy soft set; hesitant fuzzy soft prefilter (filter).

1. Introduction

In recent years, algebras including EQ-algebras have played a vital role in real lives and scientific experiments, and have its momentous applications in a lot of aspects such as in dynamical systems and genetic code of biology [1-4]. Starting from the four DNA bases order in the Boolean lattice, Sáanchez et al. [5] proposed a novel Lie Algebra of the genetic code which shows strong connections among algebraic relationship, codon assignments and physicochemical properties of amino acids. In 2008, an important algebra called evolution algebra was proposed by Tian [6] and has been applied in many aspects including Non-Mendelian inheritance. In 2009, a new class of algebras, called EQ-algebras, were introduced by V. Novák and B. De Baets [7], which are a natural algebra presented as an algebra of truth values on the basis of which the fuzzy type theory (a higher-order fuzzy logic) should be developed.

The classical soft sets are not appropriate to handle imprecise and fuzzy parameters, even though soft set theory [8] is newly useful mathematical tool to deal with uncertainty information. In order to overcome this situation, the notion of fuzzy soft sets as a generalization of the standard soft sets was introduced by Maji et al. and applied to a decision making problem.

In this paper, we apply the concept of hesitant fuzzy soft set to EQ-algebras. We introduce the notions of hesitant fuzzy soft prefilters (filters) of EQ-algebras, and investigate related properties.

2. Preliminaries

Firstly, we display basic definition and properties of EQ-algebras that will be used in this paper. For more details of EQ-algebras, we refer the reader to [8].

Definition 2.1. An algebra $\mathcal{L} := (L, \wedge, \otimes, \sim, 1)$ of type (2,2,2,0) is called an EQ-algebra if it satisfies the following axioms:

(L1) $(L, \wedge, 1)$ is a commutative idempotent monoid (i.e. $\wedge$-semilattice with top element 1),

(L2) $(L, \otimes, 1)$ is a monoid and $\otimes$ is isotone with respect to $\leq$ (with $x \leq y$ defined as $x \wedge y = x$),

(L3) $x \sim x = 1$,

(L4) $((x \wedge y) \sim z) \otimes (a \sim x) \leq z \sim (a \wedge y)$,

(L5) $(x \sim y) \otimes (a \sim b) \leq (x \sim a) \sim (y \sim b)$,

(L6) $(x \wedge y \wedge z) \sim x \leq (x \wedge y) \sim x$,

(L7) $x \otimes y \leq x \sim y$

for all $x, y, z, a, b \in L$.

The operation "$\wedge$" is called meet and "$\otimes$" is called multiplication. If the multiplication is commutative in an EQ-algebra $\mathcal{L}$, then we say that $\mathcal{L}$ is a commutative EQ-algebra.

Let $\mathcal{L}$ be an EQ-algebra. For all $x \in L$, we put $\tilde{x} = x \sim 1$. Define the implication operation $\rightarrow$ as follow:

$$(\forall x, y \in L)(x \rightarrow y = (x \wedge y) \sim x).$$

An EQ-algebra $\mathcal{L}$ is said to be residuated if $(x \otimes y) \wedge z = x \otimes y$ if and only if $x \wedge ((y \wedge z) \sim y) = x$ for all $x, y, z \in L$.

Proposition 2.1. Every (commutative) EQ-algebra $\mathcal{L}$ satisfies the following conditions for all $a, b, c, d \in L$:

(1) If $a \leq b$, then $a \rightarrow b = 1$, $a \sim b = b \sim a$, $\tilde{a} \leq \tilde{b}$, $c \rightarrow a \leq c \rightarrow b$ and $b \rightarrow c \leq a \rightarrow c$.

(2) $(a \to b) \otimes (b \to a) \leq a \sim b \leq (a \to b) \wedge (b \to a)$.
(3) $a \otimes b \leq a \wedge b \leq a, b$ and $b \otimes a \leq a \wedge b \leq a, b$.
(4) $a \to b = a \to (a \wedge b)$.
(5) $a \sim d \leq (a \sim c) \sim (d \sim c)$.
(6) $a \sim b = b \sim a$.
(7) $a \sim d \leq (a \wedge b) \sim (d \wedge b)$.
(8) $(a \to b) \otimes (b \to c) \leq a \to c$.
(9) $a \sim b \leq a \to b$ and $a \to a = 1$.
(10) $a \to b \leq (a \wedge c) \to (b \wedge c)$.
(11) $a \to b \leq (b \to c) \to (a \to c)$.

Definition 2.2. A subset f of an EQ-algebra $\mathcal{L}$ is called a prefilter of $\mathcal{L}$ if it satisfies the following conditions:
(i) $1 \in f$,
(ii) $(\forall x, y \in L)(x \to y \in f, x \in f \Rightarrow y \in f)$.

Definition 2.3. A subset f of an EQ-algebra $\mathcal{L}$ is called a filter of $\mathcal{L}$ if it is a prefilter of $\mathcal{L}$ with the following additional condition:
(iii) $(\forall x, y, z \in L)(x \to y \in f \Rightarrow (x \otimes z) \to (y \otimes z) \in f, (z \otimes x) \to (z \otimes y) \in f)$.

Definition 2.4. A prefilter (resp. filter) f of an EQ-algebra $\mathcal{L}$ is said to be positive implicative if the following assertion is valid:
(iv) $(\forall x, y, z \in L)(x \to (y \to z) \in f, x \to y \in f \Rightarrow x \to z \in f)$.

Now, we cite below some notations and definitions which will be needed in the sequel.

3. Hesitant Fuzzy Soft Prefilters (filters)

In what follows let E be a set of parameters and take a commutative EQ-algebra $\mathcal{L}$ as a reference set unless otherwise specified.

Definition 3.1. For a subset A of E, a hesitant fuzzy soft set $(\tilde{F}, A)$ over $\mathcal{L}$ is called a hesitant fuzzy soft prefilter based on $e \in A$ (berifly, e-hesitant fuzzy soft prefilter) over $\mathcal{L}$ if the hesitant fuzzy set $\tilde{F}[e] := \{(x, h_{\tilde{F}[e]}(x)) | x \in L\}$ satisfies the following conditions:

(F1) $(\forall x \in L)(h_{\tilde{F}[e]}(1) \supseteq h_{\tilde{F}[e]}(x))$,
(F2) $(\forall x, y \in L)(h_{\tilde{F}[e]}(y) \supseteq h_{\tilde{F}[e]}(x \to y) \cap h_{\tilde{F}[e]}(x))$.

If $(\tilde{F}, A)$ is an e-hesitant fuzzy soft prefilter over $\mathcal{L}$ for all $e \in A$, we say that $(\tilde{F}, A)$ is a hesitant fuzzy soft prefilter over $\mathcal{L}$.

Example 3.1. Let $L = \{0, a, b, 1\}$ be a chain. We define two binary operations '$\otimes$' and '$\sim$' by Table 1 and Table 2, respectively.

Table 1. Cayley table for the binary operation $\otimes$ in X.

$\otimes$	0	a	b	1
0	0	0	0	0
a	0	a	a	a
b	0	a	b	b
1	0	a	b	1

Table 2. Cayley table for the binary operation $\sim$ in X.

$\sim$	0	a	b	1
0	1	0	0	0
a	0	1	a	a
b	0	a	1	1
1	0	a	1	1

Then $\mathcal{L} := (L, \wedge, \otimes, \sim, 1)$ is an EQ-algebra [7]. The implication "$\to$" is described by Table 3. Consider a set of parameters $E := \{e_1, e_2, e_3, e_4, e_5\}$.

Table 3. Cayley table for the binary operation $\to$ in X.

$\to$	0	a	b	1
0	1	1	1	1
a	0	1	1	1
b	0	a	1	1
1	0	a	1	1

(1) Let $(\tilde{F}, A)$ be a hesitant fuzzy soft set over $\tilde{L}$ where $A = \{e_1, e_2, e_3\}$ which is given in Table 4. It is routine to verify that for all $i \in \{1, 2, 3\}$, $(\tilde{F}, A)$ is an e_i-hesitant fuzzy soft prefilter over $\mathcal{L}$, that is, $(\tilde{F}, A)$ is a hesitant fuzzy soft prefilter over $\tilde{L}$.

Table 4. Tabular representation of the hesitant fuzzy soft $(\tilde{F}, A)$ over $\mathcal{L}$.

$\tilde{F}$	0	a	b	1
e_1	$[0.4, 0.6]$	$[0.4, 0.6]$	$(0.3, 0.7]$	$(0.3, 0.7]$
e_2	$\{0.5, 0.6\}$	$(0.4, 0.7)$	$[0.2, , 0.8)$	$[0.2, 0.8)$
e_3	$\{0.3\}$	$(0.2, 0.4) \cup (0, 5, 0.6]$	$[0.2, 0.6]$	$[0.2, 0.6]$

(2) Let $(\tilde{G}, B)$ be a hesitant fuzzy soft set over $\tilde{L}$ where $B = \{e_4, e_5\}$ which is defined in Table 5. It is easy that $(\tilde{G}, B)$ is an e_4-hesitant fuzzy soft prefilter over $\mathcal{L}$. But it is not an e_5-hesitant fuzzy soft prefilter over $\mathcal{L}$ since $h_{\tilde{G}[e_5]}(a \to b) \cap h_{\tilde{G}[e_5]}(a) = h_{\tilde{G}[e_5]}(1) \cap h_{\tilde{G}[e_5]}(a) = \{0.2, 0.4\} \nsubseteq \{0.1, 0.5\} = h_{\tilde{G}[e_5]}(b)$. Therefore, $(\tilde{G}, B)$ is not a hesitant fuzzy soft prefilter over $\mathcal{L}$.

Table 5. Tabular representation of the hesitant fuzzy soft $(\tilde{G}, B)$ over $\mathcal{L}$.

$\tilde{G}$	0	a	b	1
e_4	$[0.6, 0.7]$	$[0.5, 0.7]$	$[0.4, 0.9)$	$[0.4, 0.9)$
e_5	$(0.1, 0.3)$	$\{0.2, 0.4\}$	$\{0.1, 0.5\}$	$[0.1, 0.5]$

Definition 3.2. For a subset A of E, a hesitant fuzzy soft set $(\tilde{F}, A)$ over $\mathcal{L}$ is called a hesitant fuzzy soft filter over $\mathcal{L}$ if it is a hesitant fuzzy soft prefilter over $\mathcal{L}$ that satisfies the additional condition:

(F3) $(\forall x, y, z \in L)(\forall e \in A)(h_{\tilde{F}[e]}(x \to y) \subseteq h_{\tilde{F}[e]}((x \otimes z) \to (y \otimes z))$.

Example 3.2. In Example 3.1(1), $(\tilde{F}, A)$ is a hesitant fuzzy soft filter over $\mathcal{L}$.

Theorem 3.1. Let $(\tilde{F}, A)$ is a hesitant fuzzy soft prefilter (reps. filter) over $\mathcal{L}$. If B is a subset of A, then $(\tilde{F}|_B, B)$ is a hesitant fuzzy soft prefilter (reps. filter) over $\mathcal{L}$.

Theorem 3.2. A hesitant fuzzy soft set $(\tilde{F}, A)$ over $\mathcal{L}$ is a hesitant fuzzy soft prefilter over $\mathcal{L}$ if and only if the set

$$(\tilde{F}[e])_\tau := \{x \in E | \tau \subseteq h_{\tilde{F}[e]}(x)\}$$

is a prefilter of $\mathcal{L}$ for all $e \in A$ and for all $\tau \in \mathcal{P}([0, 1])$ with $(\tilde{F}[e])_\tau \neq \emptyset$.

Proof. Suppose that $(\tilde{F}, A)$ is a hesitant fuzzy soft prefilter over $\mathcal{L}$. Let $x, y \in L$, $e \in A$ and $\tau \in \mathcal{P}([0,1])$ be such that $x \in (\tilde{F}[e])_\tau$ and $x \to y \in (\tilde{F}[e])_\tau$. Then $\tau \subseteq h_{\tilde{F}[e]}(x)$ and $\tau \subseteq h_{\tilde{F}[e]}(x \to y)$. It follows from (F1) and (F2) that $\tau \subseteq h_{\tilde{F}[e]}(x) \cap h_{\tilde{F}[e]}(x \to y) \subseteq h_{\tilde{F}[e]}(y) \subseteq h_{\tilde{F}[e]}(1)$. Hence, $y \in (\tilde{F}[e])_\tau$ and $1 \in (\tilde{F}[e])_\tau$. Therefore, $(\tilde{F}[e])_\tau$ is a prefilter of $\mathcal{L}$ for all $e \in A$ and for all $\tau \in \mathcal{P}([0,1])$ with $(\tilde{F}[e])_\tau \neq \emptyset$.

Conversely, assume that $(\tilde{F}[e])_\tau$ is a prefilter of $\mathcal{L}$ for all $e \in A$ and for all $\tau \in \mathcal{P}([0,1])$ with $(\tilde{F}[e])_\tau \neq \emptyset$. For any $x \in L$ and for any $e \in A$, let $h_{\tilde{F}[e]}(x) = \gamma$, then $x \in (\tilde{F}[e])_\gamma$. Since $(\tilde{F}[e])_\gamma$ is a prefilter of $\mathcal{L}$, we have $1 \in (\tilde{F}[e])_\gamma$ and so $h_{\tilde{F}[e]}(x) = \gamma \subseteq h_{\tilde{F}[e]}(1)$ for all $x \in L$ and for all $e \in A$. For any $x, y \in L$ and for any $e \in A$, let $\varepsilon = h_{\tilde{F}[e]}(x) \cap h_{\tilde{F}[e]}(x \to y)$. Then $h_{\tilde{F}[e]}(x) \supseteq \varepsilon$ and $h_{\tilde{F}[e]}(x \to y) \supseteq \varepsilon$, that is, $x \in (\tilde{F}[e])_\varepsilon$ and $x \to y \in (\tilde{F}[e])_\varepsilon$. It follows from Definition 2.2 (ii) that $y \in (\tilde{F}[e])_\varepsilon$ and $h_{\tilde{F}[e]}(y) \supseteq \varepsilon = h_{\tilde{F}[e]}(x) \cap h_{\tilde{F}[e]}(x \to y)$ for any $x, y \in L$ and for any $e \in A$. This completes the proof.

Theorem 3.3. A hesitant fuzzy soft set $(\tilde{F}, A)$ over $\mathcal{L}$ is a hesitant fuzzy soft filter over $\mathcal{L}$ if and only if the set $(\tilde{F}[e])_\tau$ is a filter of $\mathcal{L}$ for all $e \in A$ and for all $\tau \in \mathcal{P}([0,1])$ with $(\tilde{F}[e])_\tau \neq \emptyset$.

Proposition 3.1. For every hesitant fuzzy soft prefilter $(\tilde{F}, A)$ over $\mathcal{L}$, we have the following properties: for all $a, b, x, y, z \in L$ and for all $e \in A$,

(1) If $x \leq y$, then $h_{\tilde{F}[e]}(x) \subseteq h_{\tilde{F}[e]}(y)$.
(2) $h_{\tilde{F}[e]}(x) \cap h_{\tilde{F}[e]}(x \sim y) \subseteq h_{\tilde{F}[e]}(y)$.
(3) $h_{\tilde{F}[e]}(x) \cap h_{\tilde{F}[e]}(y) \subseteq h_{\tilde{F}[e]}(x \wedge y)$.
(4) $h_{\tilde{F}[e]}(x \sim y) \cap h_{\tilde{F}[e]}(y \sim z) \subseteq h_{\tilde{F}[e]}(x \sim z)$.
(5) $h_{\tilde{F}[e]}((x \to y) \otimes (y \to x)) \subseteq h_{\tilde{F}[e]}(x \sim y) \subseteq h_{\tilde{F}[e]}((x \to y) \wedge (y \to x))$.
(6) $h_{\tilde{F}[e]}(1 \sim x) = h_{\tilde{F}[e]}(x)$.
(7) $h_{\tilde{F}[e]}(y) \subseteq h_{\tilde{F}[e]}(x \to y)$.
(8) $h_{\tilde{F}[e]}((x \to (a \wedge b)) \otimes (a \sim y)) \subseteq h_{\tilde{F}[e]}(x \to (y \wedge b))$.
(9) $h_{\tilde{F}[e]}((x \to y) \otimes (y \sim z)) \subseteq h_{\tilde{F}[e]}(x \to z)$.

Proof. (1) Let $x, y \in L$ be such that $x \leq y$. Then $x \to y = 1$ by Proposition 2.1(1). It follows (F1) and (F2) that $h_{\tilde{F}[e]}(x) = h_{\tilde{F}[e]}(x) \cap h_{\tilde{F}[e]}(1) = h_{\tilde{F}[e]}(x) \cap h_{\tilde{F}[e]}(x \to y) \subseteq h_{\tilde{F}[e]}(y)$ for all $e \in A$.

(2) Let $x, y \in L$. Note that $x \sim y \leq x \to y$ by (9) of Proposition 2.1. Using (F2) and item (1) implies that $h_{\tilde{F}[e]}(x) \cap h_{\tilde{F}[e]}(x \sim y) \subseteq h_{\tilde{F}[e]}(x) \cap h_{\tilde{F}[e]}(x \to y) \subseteq h_{\tilde{F}[e]}(y)$ for all $e \in A$.

(3) Note that $y \leq x \to y = x \to (x \wedge y)$ for all $x, y \in L$. It follows from (F2) and item (1) that $h_{\tilde{F}[e]}(x) \cap h_{\tilde{F}[e]}(y) \subseteq h_{\tilde{F}[e]}(x) \cap h_{\tilde{F}[e]}(x \to (x \wedge y)) \subseteq h_{\tilde{F}[e]}(x \wedge y)$ for all $x, y \in L$ and for all $e \in A$.

(4) Since $x \sim y = y \sim x \leq (y \sim z) \sim (x \sim z)$ by (6) and (5) of Proposition 2.1, it follows from item (1) and (2) that $h_{\tilde{F}[e]}(x \sim y) \cap h_{\tilde{F}[e]}(y \sim z) \subseteq h_{\tilde{F}[e]}(y \to z) \cap h_{\tilde{F}[e]}((y \sim z) \to (x \to z)) \subseteq h_{\tilde{F}[e]}(x \sim z)$ for all $x, y, z \in L$ and $e \in A$.

(5) follows from Proposition 2.1(2) and item (1).

(6) Using item (2) and (F1), we have $h_{\tilde{F}[e]}(x) \supseteq h_{\tilde{F}[e]}(1) \cap h_{\tilde{F}[e]}(1 \sim x) = h_{\tilde{F}[e]}(1 \sim x)$ for all $x \in L$ and for all $e \in A$. Since $x \leq 1 \sim x$ for all $x \in L$, it follows from item (1) and Proposition 2.1(6) that $h_{\tilde{F}[e]}(x) \subseteq h_{\tilde{F}[e]}(1 \sim x) = h_{\tilde{F}[e]}(x \sim 1)$ for all $x \in L$ and for all $e \in A$. Therefore (6) holds.

(7) For any $x, y \in L$ and for any $e \in A$, we have $h_{\tilde{F}[e]}(y) \subseteq h_{\tilde{F}[e]}(1 \sim y) = h_{\tilde{F}[e]}((1 \wedge y) \sim 1) \subseteq h_{\tilde{F}[e]}(x \wedge y) \sim x) = h_{\tilde{F}[e]}(x \to y)$.

(8) Using the commutativity and associativity of $\wedge$, (L4), the definition of implication ($\to$) and item (1), we have $h_{\tilde{F}[e]}((x \to (a \wedge b)) \otimes (a \sim y)) = h_{\tilde{F}[e]}(((x \wedge (a \wedge b)) \sim x) \otimes (a \sim y)) = h_{\tilde{F}[e]}(((a \wedge (b \wedge x)) \sim x) \otimes (y \sim a)) \subseteq h_{\tilde{F}[e]}(x \sim (y \wedge (b \wedge x))) = h_{\tilde{F}[e]}((x \wedge (y \wedge b)) \sim x) = h_{\tilde{F}[e]}(x \to (y \wedge b))$ for all $x, y, a, b \in L$ and for all $e \in A$.

(9) Using item (8), (L6) and the definition of implication ($\to$), we get $h_{\tilde{F}[e]}((x \to y) \otimes (y \sim z)) = h_{\tilde{F}[e]}((x \to (y \wedge y)) \otimes (y \sim z)) \subseteq h_{\tilde{F}[e]}(x \to (z \wedge y)) = h_{\tilde{F}[e]}((x \wedge (z \wedge y)) \sim x) \subseteq h_{\tilde{F}[e]}((x \wedge z) \sim x) = h_{\tilde{F}[e]}((z \wedge x) \sim x) = h_{\tilde{F}[e]}(x \to z)$ for all $x, y, z \in L$ and for all $e \in A$.

Proposition 3.2. Every Hesitant fuzzy prefilter $(\tilde{F}, A)$ over $\mathcal{L}$ satisfies the following assertions.

(1) $h_{\tilde{F}[e]}(x \sim y) \cap h_{\tilde{F}[e]}(a \sim b) \subseteq h_{\tilde{F}[e]}((a \wedge x) \sim (b \wedge y)) \cap h_{\tilde{F}[e]}((a \sim x) \sim (b \sim y)) \cap h_{\tilde{F}[e]}((a \to x) \sim (b \to y))$,

(2) $h_{\tilde{F}[e]}(x \otimes y) \subseteq h_{\tilde{F}[e]}(x) \cap h_{\tilde{F}[e]}(y)$,

(3) If $x \leq y$, then $h_{\tilde{F}[e]}(\tilde{x}) \subseteq h_{\tilde{F}[e]}(\tilde{y})$ and $h_{\tilde{F}[e]}(a \to x) \subseteq h_{\tilde{F}[e]}(a \to y)$,

for all $x, y, a, b \in L$ and for all $e \in A$.

Theorem 3.4. For a hesitant fuzzy soft set $(\tilde{F}, A)$ over $\mathcal{L}$, the following are equivalent.

(1) $(\tilde{F}, A)$ is a hesitant fuzzy soft prefilter over $\mathcal{L}$.

(2) $(\forall x, y, z \in L)(\forall e \in A)(x \leq y \to z \Rightarrow h_{\tilde{F}[e]}(x) \cap h_{\tilde{F}[e]}(y) \subseteq h_{\tilde{F}[e]}(z))$.

(3) $(\forall x, y, z \in L)(\forall e \in A)(x \to (y \to z) = 1 \Rightarrow h_{\tilde{F}[e]}(x) \cap h_{\tilde{F}[e]}(y) \subseteq h_{\tilde{F}[e]}(z))$.

Proof. (1)⇒(2) Let $x, y, z \in L$ be such that $x \leq y \to z$. Then $h_{\tilde{F}[e]}(x) \subseteq h_{\tilde{F}[e]}(y \to z)$ for all $e \in A$ by Proposition 3.1(1). Using (F2), we have $h_{\tilde{F}[e]}(x) \cap h_{\tilde{F}[e]}(y) \subseteq h_{\tilde{F}[e]}(y \to z) \cap h_{\tilde{F}[e]}(y) \subseteq h_{\tilde{F}[e]}(z)$ for all $e \in A$.

(2)⇒(3) Let $x, y, z \in L$ be such that $x \to (y \to z) = 1$. Then $x \leq 1 = x \to (y \to z)$, and so $h_{\tilde{F}[e]}(x) \subseteq h_{\tilde{F}[e]}(y \to z)$ for all $e \in A$ by item (2). Since $y \to z \leq y \to z$, it follows from item (2) that $h_{\tilde{F}[e]}(x) \cap h_{\tilde{F}[e]}(y) \subseteq h_{\tilde{F}[e]}(y \to z) \cap h_{\tilde{F}[e]}(y) \subseteq h_{\tilde{F}[e]}(z)$ for all $e \in A$.

(3)⇒(1) Since $x \to (x \to 1) = 1$ for all $x \in L$, it follows from item (3) that $h_{\tilde{F}[e]}(x) \subseteq h_{\tilde{F}[e]}(1)$ for all $e \in A$ and for all $x \in L$. Note that $(x \to y) \to (x \to y) = 1$ for all $x, y \in L$, thus $h_{\tilde{F}[e]}(x \to y) \cap h_{\tilde{F}[e]}(x) \subseteq h_{\tilde{F}[e]}(y)$ for all $x, y \in L$ and for all $e \in A$ by item (3). Therefore $(\tilde{F}, A)$ is a hesitant fuzzy soft prefilter over $\mathcal{L}$.

Proposition 3.3. For any hesitant fuzzy soft filter $(\tilde{F}, A)$ over $\mathcal{L}$, the following assertions are valid.
(1) $h_{\tilde{F}[e]}(x \otimes y) = h_{\tilde{F}[e]}(x) \cap h_{\tilde{F}[e]}(y)$,
(2) $h_{\tilde{F}[e]}(x \to y) \cap h_{\tilde{F}[e]}(y \to z) \subseteq h_{\tilde{F}[e]}(x \to z)$,
(3) $h_{\tilde{F}[e]}(x \sim y) \subseteq h_{\tilde{F}[e]}((x \otimes z) \sim (y \otimes z))$,
for all $x, y, z \in L$ and for all $e \in A$.

4. Conclusions

Soft sets, initiated by Molodtsov [8], are a newly emerging mathematical tool for dealing with uncertainties. As a generalization of the the standard soft sets, its applications in a decision making problem have been presented. On the other hand, filter theory plays an important role in the study of algebraic structures and the completeness of the corresponding non-classical logics. From a logical point of view, various filters correspond to various sets of provable formulas. In this paper, we develop the hesitant fuzzy soft filter theory over EQ-algebra. Mainly, we introduce the concepts of hesitant fuzzy soft prefilters (filters), and investigate their basic properties. We give some Characterizations of hesitant fuzzy soft prefilters (filters) over EQ-algebra. Meanwhile, we hope that it will be of great use to provide theoretical foundation to design intelligent information processing systems.

Acknowledgments

This work was supported by National Natural Science Foundation of P.R. China (Grant no. 11071178), the Speciality Comprehensive Reform of Mathematics and Applied Mathematics of Ministry of Education of

P.R. China (ZG064), the Speciality Comprehensive Reform of Mathematics and Applied Mathematics of Ministry of Education of Sichuan Province of China (01249).

References

1. M. K. Kinyou, A. A. Sagle. Quadratic dynamical systems and algebras. *J. Differential Equations*, 117 (1995) 67–126.
2. J. D. Bashford, P. D. Jarvis. The genetic code as a priodic table: Algebriaic aspects. *BioSystems*, 57 (2000) 147–161.
3. L. Frappat, A. Sciarrino, P. Sorba. Crystalizing the genetic code. *J. Biological Physics*, 27 (2001) 1–34.
4. J. J. Tian, B. L. Li. Coalgebraic structure of genetics inheritance. *Mathematical Biosciences and Engineering*, 1 (2004) 243–266.
5. R. Sáanchez, R. Grau, E. Morgado. A novel Lie algebra of the genetic code over the Galois field of four DNA bases. *Mathematical Biosciences*, 202 (2006) 156–174.
6. J. J. Tian. *Evolution Algebras and Their Applications* (Springer-Verlag, Berlin, Heidelberg, 2008).
7. V. Novák, B. De Baets. EQ-algebras. *Fuzzy Sets and Systems*, 160 (2009) 2956–2978.
8. D. Molodtsov. Soft set theory - first results. *Computers & Mathematics with Applications*, 37(4–5) (1999) 19–31.

Quantum controlled teleportation of five-qubit state

Zhi-wen Mo

College of Mathematics and Software Science, Sichuan Normal University
Chengdu, 610066, China
mozhiwen@sicnu.edu.cn

Lan Shu

College of Mathematics, University of Electronic Science and Technology of China
Chengdu, 610054, China
shul@uestc.edu.cn

Xue Yang and Yu-ting Jiang

College of Mathematics and Software Science, Sichuan Normal University
Chengdu, 610066, China
yx12290552@163.com

Quantum teleportation can transmit an unknown quantum state from a sender to a receiver via a quantum channel with the help of local operations and classical communications. We propose a quantum controlled teleportation for certain class of five-qubit state with a seven-qubit cluster state as quantum channel. In our scheme, controller Charlie needs to perform a Bell-basis joint measurement on his particles. Alice carries out a measurement in the orthonormal basis, then the original state with deterministic probability can be reconstructed by the receiver Bob.

Keywords: Quantum controlled teleportation; cluster state; five-qubit state.

1. Introduction

Quantum entanglement is one of the most interesting features in quantum mechanics. It provides promising and wide applications in quantum information processing, for example, teleportation[1-4], dense coding[5,6], quantum information sharing[7-10], quantum information concentration[11-13] and so on. Quantum teleportation[14] is one of the most important applications of quantum entanglement. It can transmit an unknown quantum state from a sender to a receiver via a quantum channel with the help of local operations and classical communications. The first controlled teleportation(CT) scheme, proposed in 1998, uses a three-particle

Greenberger-Horne-Zeilinger(GHZ) state[15]. In this scheme, the teleportation procedure is controlled by a controller, such that the arbitrary quantum state can be teleported from sender to receiver only with the participation of the controller[16].

Entanglement in multi-particle case is more complicated than in three-particle case. Briegel and Raussendorf introduced a novel kind of n-particle entangled state, cluster state[17]. In this paper, Briegel and Raussendorf showed that cluster states have some particular characters in the case of $N > 3$. For instance, the cluster states have the properties of both the GHZ-class and the W-class entangled states, and they are harder to be destroyed by local operations than GHZ-class states[18]. In 2017, Li and Zhao propose a novel quantum protocol for five-qubit state[19]. Jiang and Mo use control power to describe the controlled teleportation schemes via four-particle cluster state[20].

In this paper, we propose a quantum controlled teleportation with a special five-qubit state by using a seven-qubit cluster state. Controller Charlie performs a Bell-basis joint measurement on his particles. Charlie always informs Alice and Bob of his measurement result by the classical channel. The successful possibility of our scheme is 1.

2. Quantum Controlled Teleportation

The sender Alice wants to teleport an unknown five-state to the receiver Bob with the controller Charlie's help. This five-state can be written as

$$|\chi\rangle_{abcde} = (\alpha|00000\rangle + \beta|00011\rangle + \gamma|11100\rangle + \delta|11111\rangle)_{abcde}, \quad (1)$$

where $|\alpha|^2 + |\beta|^2 + |\gamma|^2 + |\delta|^2 = 1$.

The sender Alice, receiver Bob and controlled Charlie share a seven-qubit cluster state as quantum channel, the seven-qubit state can be written in the standard form

$$|\Phi\rangle_{1234567} = \frac{1}{2}(|0000000\rangle + |0001101\rangle + |1110010\rangle + |1111111\rangle)_{1234567}, \quad (2)$$

the qubits a,b,c,d and e belong to Alice, qubits 1,2,3,4 and 5 belong to Bob, qubits 6 and 7 belong to Charlie, respectively. The joint state of the five-qubit state and the quantum channel is given by,

$$|\tau\rangle = |\chi\rangle_{abcde} \otimes |\Phi\rangle_{1234567}$$
$$= \frac{\sqrt{2}}{4}(|\phi^+\rangle_{67}|\varphi^1\rangle_{abcde12345} + |\phi^-\rangle_{67}|\varphi^2\rangle_{abcde12345} \quad (3)$$
$$+ |\psi^+\rangle_{67}|\varphi^3\rangle_{abcde12345} + |\psi^-\rangle_{67}|\varphi^3\rangle_{abcde12345}),$$

where $|\varphi^i\rangle_{abcde12345}(i=1,2,3,4)$ are

$$\begin{aligned}|\varphi^1\rangle = & \alpha|00000\rangle|00000\rangle + \alpha|00000\rangle|11111\rangle + \beta|00011\rangle|00000\rangle \\ & + \beta|00011\rangle|11111\rangle + \gamma|11100\rangle|00000\rangle + \gamma|11100\rangle|11111\rangle \\ & + \delta|11111\rangle|00000\rangle + \delta|11111\rangle|11111\rangle,\end{aligned} \quad (4)$$

$$\begin{aligned}|\varphi^2\rangle = & \alpha|00000\rangle|00000\rangle - \alpha|00000\rangle|11111\rangle + \beta|00011\rangle|00000\rangle \\ & - \beta|00011\rangle|11111\rangle + \gamma|11100\rangle|00000\rangle - \gamma|11100\rangle|11111\rangle \\ & + \delta|11111\rangle|00000\rangle - \delta|11111\rangle|11111\rangle,\end{aligned} \quad (5)$$

$$\begin{aligned}|\varphi^3\rangle = & \alpha|00000\rangle|00011\rangle + \alpha|00000\rangle|11100\rangle + \beta|00011\rangle|00011\rangle \\ & + \beta|00011\rangle|11100\rangle + \gamma|11100\rangle|00011\rangle + \gamma|11100\rangle|11100\rangle \\ & + \delta|11111\rangle|00011\rangle + \delta|11111\rangle|11100\rangle,\end{aligned} \quad (6)$$

$$\begin{aligned}|\varphi^4\rangle = & \alpha|00000\rangle|00011\rangle - \alpha|00000\rangle|11100\rangle + \beta|00011\rangle|00011\rangle \\ & - \beta|00011\rangle|11100\rangle + \gamma|11100\rangle|00011\rangle - \gamma|11100\rangle|11100\rangle \\ & + \delta|11111\rangle|00011\rangle - \delta|11111\rangle|11100\rangle.\end{aligned} \quad (7)$$

Charlie performs a Bell-basis joint measurement on his particles 6 and 7, where the Bell-basis is

$$\begin{aligned}|\phi^\pm\rangle &= \frac{1}{\sqrt{2}}(|00\rangle \pm |11\rangle), \\ |\psi^\pm\rangle &= \frac{1}{\sqrt{2}}(|01\rangle \pm |10\rangle).\end{aligned} \quad (8)$$

After the measurement, Charlie informs his result to Alice and Bob through classical channel. Then Bob implements some unitary operations. If Charlie's measurement result is $|\phi^+\rangle$ or $|\psi^+\rangle$, Bob needs to perform $I = |0\rangle\langle 0| + |1\rangle\langle 1|$ on his particles. If Charlie's measurement result is $|\phi^-\rangle$ or $|\psi^-\rangle$, Bob needs to perform $\sigma_z = |0\rangle\langle 0| - |1\rangle\langle 1|$ on particle 5 to reconstruct the original state. Then the state of particles is transformed into $|\varphi^1\rangle$ and $|\varphi^3\rangle$.

For Alice, If Charlie's measurement result is $|\phi^\pm\rangle$, Alice carries out a measurement in the orthonormal basis of $\{\frac{1}{\sqrt{2}}|00000\rangle, \frac{1}{\sqrt{2}}|11111\rangle\}$ on her particles.

$$|\varphi^1\rangle = |00000\rangle(\alpha|00000\rangle + \beta|00011\rangle + \gamma|11100\rangle + \delta|11111\rangle) \quad (9)$$
$$+|11111\rangle(\alpha|00000\rangle + \beta|00011\rangle + \gamma|11100\rangle + \delta|11111\rangle). \quad (10)$$

If Charlie's measurement result is $|\psi^{\pm}\rangle$, Alice carries out a measurement in the orthonormal basis of $\{\frac{1}{\sqrt{2}}|000011\rangle, \frac{1}{\sqrt{2}}|11100\rangle\}$ on her particles.

$$|\varphi^1\rangle = |00011\rangle(\alpha|00000\rangle + \beta|00011\rangle + \gamma|11100\rangle + \delta|11111\rangle) \quad (11)$$
$$+|11100\rangle(\alpha|00000\rangle + \beta|00011\rangle + \gamma|11100\rangle + \delta|11111\rangle). \quad (12)$$

Then Bob can acquire the original entangled state.

Table 1. Measurement Results and Unitary Operations.

Charlie's measurement results	Bob's unitary operations	Alice's measurement results
$|\phi^+\rangle$	I	$\{\frac{1}{\sqrt{2}}|000000\rangle, \frac{1}{\sqrt{2}}|11111\rangle\}$
$|\phi^-\rangle$	σ_z	$\{\frac{1}{\sqrt{2}}|000000\rangle, \frac{1}{\sqrt{2}}|11111\rangle\}$
$|\psi^+\rangle$	I	$\{\frac{1}{\sqrt{2}}|000011\rangle, \frac{1}{\sqrt{2}}|11100\rangle\}$
$|\psi^-\rangle$	σ_z	$\{\frac{1}{\sqrt{2}}|000011\rangle, \frac{1}{\sqrt{2}}|11100\rangle\}$

3. Conclusion

In this paper, we display a quantum controlled teleportation with a seven-qubit cluster state. We propose a quantum controlled teleportation of a special five-qubit state. In our scheme, Controller Charlie performs a Bell-basis joint measurement on his particles. And Bob makes a local single operation on his particles. We can recover the original entangled state and our scheme is a deterministic. The successful possibility of our scheme is 1. We also expect to further realize the quantum state transmission of an arbitrary five-qubit state.

Acknowledgments

This work is supported by the National Natural Science Foundation of China (Grant No. 11671284), Specialized Research Fund for the Doctoral Program of Higher Education (Grant No. 20135134110003), Sichuan Provincial Natural Science Foundation of China (Grant No. 2015JY0002), the Research Foundation of the Education Department of Sichuan Province (Grant No. 15ZA0032) and Sichuan Provincial Natural Science Foundation of China (Grant No. 2017JY0197).

References

1. Bai M. Q., Peng J. Y., Mo Z. W. Transformation from Probabilistic Channel to Deterministic Channel Based on Eight-Qubit Quantum Channel[J]. Modern Physics Letters B, 2013, 27(04): 1350030.

2. Hou P. Y., Huang Y. Y., Yuan X. X. et al. Quantum teleportation from light beams to vibrational states of a macroscopic diamond[J]. Nature communications, 2016, 7.
3. Jia-Yin P., Zhi-Wen M. Several teleportation schemes of an arbitrary unknown multi-particle state via different quantum channels[J]. Chinese Physics B, 2013, 22(5): 050310.
4. Santos A. C., Silva R. D., Sarandy M. S. Shortcut to adiabatic gate teleportation[J]. Physical Review A, 2016, 93(1): 012311.
5. Liu J., Mo Z., Sun S. Controlled Dense Coding Using the Maximal Slice States[J]. International Journal of Theoretical Physics, 2016, 55(4): 2182-2188.
6. Kögler R. A., Neves L. Optimal probabilistic dense coding schemes[J]. Quantum Information Processing, 2017, 16(4): 92.
7. Xiang Y., Mo Z. W. Quantum secret sharing protocol based on four-dimensional three-particle entangled states[J]. Modern Physics Letters B, 2016, 30(02): 1550267.
8. Peng J. Y., Bai M., Mo Z. W. Bidirectional quantum states sharing[J]. International Journal of Theoretical Physics, 2016, 55(5): 2481-2489.
9. Kogias I., Xiang Y., He Q. et al. Unconditional security of entanglement-based continuous-variable quantum secret sharing[J]. Physical Review A, 2017, 95(1): 012315.
10. Bai M. Q., Mo Z. W. Hierarchical quantum information splitting with eight-qubit cluster states[J]. Quantum information processing, 2013: 1-12.
11. Bai M., Peng J. Y., Mo Z. W. Three schemes of remote information concentration based on ancilla-free phase-covariant telecloning[J]. Quantum information processing, 2014, 13(5): 1067-1083.
12. Wang J., Shu L., Mo Z. Controlled Remote Information Concentration via Non-Maximally Entangled GHZ-Type States[J]. International Journal of Theoretical Physics, 2016, 55(2): 746-753.
13. Peng J. Y., Bai M., Mo Z. W. Remote information concentration via W state: reverse of ancilla-free phase-covariant telecloning[J]. Quantum information processing, 2013, 12(11): 3511-3525.
14. Bennett C. H., Brassard G., Crpeau C. et al. Teleporting an unknown quantum state via dual classical and Einstein-Podolsky-Rosen channels[J]. Physical review letters, 1993, 70(13): 1895.
15. Karlsson A., Bourennane M. Quantum teleportation using three-particle entanglement[J]. Physical Review A, 1998, 58(6): 4394.
16. Hillery M., Bužek V., Berthiaume A. Quantum secret sharing[J]. Physical Review A, 1999, 59(3): 1829.
17. Briegel H. J., Raussendorf R. Persistent entanglement in arrays of interacting particles[J]. Physical Review Letters, 2001, 86(5): 910.
18. Dong P., Xue Z. Y., Yang M. et al. Generation of cluster states[J]. Physical Review A, 2006, 73(3): 033818.
19. Li M., Zhao N., Chen N. et al. Quantum Teleportation of Five-qubit State[J]. International Journal of Theoretical Physics, 2017: 1-6.
20. Jiang Y. T., Mo Z. W. Comparison and Analysis of the Control Power Between Two Different Perfect Controlled Teleportation Schemes Using Four-particle Cluster State[J]. International Journal of Theoretical Physics, 2017, 56(10): 3084-3091.

Analysis of numerical method for semilinear stochastic differential equation

Luoping Chen

School of Mathematics, Southwest Jiaotong University
Chengdu, 611756, P. R. China
cherrychen@home.swjtu.edu.cn

In this work, we will investigate the numerical method we proposed in[1] and study the general convergence property of the method. We prove that for the model problem we studied, when its solution is analytic to each random variable in a bounded complex domain, the proposed method is a convergent and efficient scheme for solving the model problem. Some numerical examples are implemented.

Keywords: Stochastic models; numerical solutions; convergence analysis.

1. Introduction

Stochastic differential equations (SDEs), especially nonlinear SDEs, provide mathematical models for the quantification of uncertainties in many complex physical, engineering and financial applications. For instance, flow in heterogeneous porous media,[2] thermo-fluid processes,[3] optimal portfolio problems and prices of options on open market[4–6] and so on.

In this work, we will study the numerical methods proposed in our previous work[1] and prove the theoretical convergence property of the method in a bounded random domain Γ. In order to prove the convergence of the efficient numerical method, we need to analyze the interpolation error of the solutions in $L^4(\Gamma)$ norm. In our previous work, to guarantee the convergence property theoretically, we need to restrict the solutions of the model problem to be sufficient smooth functions and have bounded derivatives of any order to each random variable. However, this condition can not be satisfied for lots of problems in applications. In this study, we will prove that for any solution that is analytic to each random variable in a bounded complex domain, the numerical method proposed in[1] is convergent.

2. Model Problem and Approximate Method

The stochastic differential equation we investigate is

$$\begin{cases} -\nabla \cdot (a_N(\omega,x)\nabla u(\omega,x)) + f(\omega,x,u(\omega,x)) = 0, & x \in D, \\ u(\omega,x) = 0, & x \in \partial D, \end{cases} \quad (1)$$

where $a_N(\omega,x) = a_N(Y_1(\omega), Y_2(\omega), \cdots, Y_N(\omega), x)$ is a truncated Karhunen-Loève expansion of real-valued random diffusion coefficient, $D \subset \mathbb{R}^d$ is a bounded domain and ∂D is a smooth boundary. Suppose a_N is uniformly bounded from below and $f(\omega,x,u(\omega,x))$ is smooth enough. For brevity, we shall drop the dependence of ω, x in $f(\omega,x,u(\omega,x))$ in the following exposition.

The stochastic differential model (1) is a prototype stationary reaction-diffusion problems that can be founded in many applications. For example, it appears in the semi-discretization in time of the nonlinear reaction-diffusion problem modeling the conversion of starch into sugars in growing apples.[7] The main contents of this work is to provide theoretical analysis for the efficient numerical method we proposed for this model problem.

Let $(\cdot,\cdot)$ be the inner product of $\mathcal{L}^2(D)$ and $\mathcal{W}^{m,q}(D)$ be Sobolev space with norm given by $\|v\|_{m,q}^q = \sum_{|\alpha| \le m} \|\frac{\partial^\alpha v}{\partial x^\alpha}\|_{\mathcal{L}^q}^q$. When $q = 2$, we denote $\mathcal{H}^m(D) = \mathcal{W}^{m,2}(D)$. Let $\mathcal{H}_0^1(D)$ be the subspace of $\mathcal{H}^1(D)$ with vanishing trace on ∂D. $\|\cdot\|_m = \|\cdot\|_{m,2}$ and $\|\cdot\| = \|\cdot\|_{0,2}$. Let "$\lesssim$" be "$\le C$" for some positive constant C.

We denote $\{\rho_n\}_{n=1}^N$ as the probability density functions of $\{Y_n\}_{n=1}^N$. The expectation of $\mu(y)$ is $E(\mu(y)) = \int_\Gamma \mu(y)\rho(y)dy$ with Γ being a bounded domain. For a.e. (almost everywhere) $y \in \Gamma$, we suppose that problem (1) has at least one solution in $\mathcal{H}_0^1(D) \cap \mathcal{H}^2(D)$ and the linearized operator $L_v := -\nabla \cdot (a_N \nabla) + f'(v)$ is nonsingular.

Let $(f,g)_{\mathcal{L}_\rho^2}$, tensor product spaces $\mathcal{L}_\rho^2(\Gamma) \otimes V$, $\mathcal{L}_\rho^q(\Gamma) \otimes V$ and their corresponding norms be defined as in paper.[1] $C^0(\Gamma; V)$ is a space with function $v(\cdot, x)$ being continuous to random variable y and $v(y, \cdot) \in V$.

The variational equation of problem (1) is given by

$$(a_N(y)\nabla u(y), \nabla w) + (f(u(y)), w) = 0, \quad \forall w \in \mathcal{H}_0^1(D), \quad \text{a.e. in } \Gamma \quad (2)$$

and the weak formulation for the linearized problem of (1) can be represented as: for some $v \in \mathcal{L}_\rho^2(\Gamma) \otimes \mathcal{W}^{1,p}(D)$ and a.e. $y \in \Gamma$,

$$(a_N \nabla \bar{u}, \nabla w) + (f'(v)\bar{u}, w) = (-f(v) + f'(v)v, w), \quad \forall w \in \mathcal{H}_0^1(D). \quad (3)$$

The finite dimensional approximate space is subspace of $\mathcal{L}_\rho^2(\Gamma) \otimes \mathcal{H}_0^1(D)$, which is denoted by $\mathcal{P}_p(\Gamma) \otimes \mathcal{X}_h(D)(\subset \mathcal{L}_\rho^2(\Gamma) \otimes \mathcal{H}_0^1(D))$ and defined as in[1] with polynomial degree p and mesh size h of triangulation $\mathcal{T}_h$.

Then, the semi-discrete approximation u_h of (2) is derived by: for a.e. $y \in \Gamma$,

$$(a_N(y)\nabla u_h(y), \nabla w) + (f(u_h(y)), w) = 0, \quad \forall w \in \mathcal{X}_h(D). \tag{4}$$

Similarly, the semi-discrete approximate u^h of (3) satisfying

$$(a_N \nabla u^h, \nabla w) + (f'(v)u^h, w) = (-f(v) + f'(v)v, w), \quad \forall w \in \mathcal{X}_h(D). \tag{5}$$

The fully discrete solution $u_{h,p}$ is derived by collocating equation (4) on the roots of Legendre polynomials, i.e. $u_{h,p} \in \mathcal{P}_{\boldsymbol{p}}(\Gamma) \otimes \mathcal{X}_h(D)$,

$$u_{h,\boldsymbol{p}}(y,x) = \sum_{k=1}^{N_{\boldsymbol{p}}} u_h(\hat{y}_k, x)\psi_k(y) = \mathcal{I}_{\boldsymbol{p}} u_h,$$

where $\{\hat{y}_k\}_{k=1}^{N_{\boldsymbol{p}}}$ are collocation points and $\{\psi_k(y)\}_{k=1}^{N_{\boldsymbol{p}}}$ are Lagrange basis to $\{\hat{y}_k\}_{k=1}^{N_{\boldsymbol{p}}}$ and $\mathcal{I}_{\boldsymbol{p}} : \mathcal{C}^0(\Gamma; \mathcal{H}_0^1(D)) \to \mathcal{P}_{\boldsymbol{p}}(\Gamma) \otimes \mathcal{H}_0^1(D)$ is interpolation operator.

Two-level numerical approximation method:

Step 1: on the coarse mesh $\mathcal{T}_H$, solve the semilinear equation on a small number of collocation points, i.e., for $k = 1, 2, \cdots, N_{\boldsymbol{P}}$, find $u_H(\hat{y}_k^c, \cdot)$ on the coarse mesh such that

$$(a_N(\hat{y}_k^c, \cdot)\nabla u_H(\hat{y}_k^c, \cdot), \nabla w) + (f(u_H(\hat{y}_k^c, \cdot)), w) = 0, \quad \forall w \in \mathcal{X}_H(D), \tag{6}$$

where $\{\hat{y}_k^c\}_{k=1}^{N_{\boldsymbol{P}}}$ are collocation points of polynomial space $\mathcal{P}_{\boldsymbol{P}}(\Gamma)$.

The fully approximated solution of (2) in $\mathcal{P}_{\boldsymbol{P}}(\Gamma) \otimes \mathcal{X}_H(D)$ is given by

$$u_{H,\boldsymbol{P}}(y,x) = (\mathcal{I}_{\boldsymbol{P}} u_H)(y) = \sum_{k=1}^{N_{\boldsymbol{P}}} u_H(\hat{y}_k^c, x)\psi_k^{\boldsymbol{P}}(y),$$

with $\{\psi_k^{\boldsymbol{P}}\}_{k=1}^{N_{\boldsymbol{P}}}$ being Lagrange basis of $\mathcal{P}_{\boldsymbol{P}}(\Gamma)$. For each collocation point, the semilinear equation (6) is solved by Newton iteration methods.

Step 2: Solve the linearized problem on a larger set of collocation points. That is, find $u^h(\hat{y}_k, \cdot)$ on the fine mesh $\mathcal{T}_h$ to satisfy

$$\begin{aligned}(a_N(\hat{y}_k, \cdot)\nabla u^h(\hat{y}_k, \cdot), \nabla w) + (f'(u_{H,\boldsymbol{P}}(\hat{y}_k, \cdot))u^h(\hat{y}_k, \cdot), w) \\ = (-f(u_{H,\boldsymbol{P}}(\hat{y}_k, \cdot)) + f'(u_{H,\boldsymbol{P}}(\hat{y}_k, \cdot))u_{H,\boldsymbol{P}}(\hat{y}_k, \cdot), w), \quad \forall w \in \mathcal{X}_h(D),\end{aligned} \tag{7}$$

with $\{\hat{y}_k\}_{k=1}^{N_{\boldsymbol{p}}}$ being the set of collocation points to $\mathcal{P}_{\boldsymbol{p}}(\Gamma)$ and the fully approximate solution $u^{h,\boldsymbol{p}}$ of (3) with $v = u_{H,\boldsymbol{P}}$ is

$$u^{h,\boldsymbol{p}} = (\mathcal{I}_{\boldsymbol{p}} u^h)(y) = \sum_{k=1}^{N_{\boldsymbol{p}}} u^h(\hat{y}_k, x)\psi_k^{\boldsymbol{p}}(y), \tag{8}$$

where $\{\psi_k^{\boldsymbol{p}}\}_{k=1}^{N_{\boldsymbol{p}}}$ are the Lagrange basis functions to $\{\hat{y}_k\}_{k=1}^{N_{\boldsymbol{p}}}$.

3. Convergence Analysis

Let $u^{h,p}$ be the numerical solution of (8), then, the approximate error can be divided into three parts, i.e.

$$u - u^{h,p} = (u-u_h)+(u_h-u^h)+(u^h-u^{h,p}) = (u-u_h)+(u_h-u^h)+(u^h-\mathcal{I}_p(u^h)),$$

where u^h is the semi-discrete solution satisfying the equation (5) with v being chosen as $u_{H,P}$ and $u^h - \mathcal{I}_p(u^h)$ is the Lagrange interpolation error. First, we have the following result for semi-discrete error.

Lemma 3.1.[1] *Let $u_h : \Gamma \to \mathcal{X}_h(D)$ be the semi-discrete solution satisfying (4) with bounded coefficient $a_N(y,x)$, then, for $2 \leq p < \infty$, $2 \leq q < \infty$,*

$$\|u-u_h\|_{\mathcal{L}_\rho^q(\Gamma)\otimes\mathcal{L}^p(D)} + h\|u-u_h\|_{\mathcal{L}_\rho^q(\Gamma)\otimes W^{1,p}(D)} \lesssim h^2\|u\|_{\mathcal{L}_\rho^q(\Gamma)\otimes W^{2,p}(D)}.$$

Then, we will analyze the convergence property of the interpolation operator $\mathcal{I}_p : \mathcal{C}^0(\Gamma;V) \to \mathcal{L}_\rho^q(\Gamma;V)$ with $q = 2, 4$. When Γ is bounded, the corresponding collocation points are the cartesian products determined by the roots of Legendre polynomials. We first give the results for $q = 2$.

Lemma 3.2.[8] *There exist positive constants $r_n, n = 1, 2, \cdots, N$, independent of p, such that for any $v \in \mathcal{C}^0(\Gamma;\mathcal{H}_0^1(D))$,*

$$\|v - \mathcal{I}_p v\|_{\mathcal{L}_\rho^2(\Gamma)\otimes\mathcal{H}_0^1(D)} \lesssim \sum_{n=1}^N e^{-r_n p_n},$$

where $r_n = \log\left[\frac{2\tau_n}{|\Gamma_n|}\left(1+\sqrt{1+\frac{|\Gamma_n|^2}{4(\tau_n)^2}}\right)\right]$ and τ_n is smaller than the distance between Γ_n and the nearest singularity in the complex plane.

Let $I_p v$ be a Lagrange interpolation polynomial related to the zeros of the $(p+1)$-th Legendre orthogonal polynomial. We denote $D_\varrho \subset \mathbb{C}$ with $\varrho > 1$ as an elliptic disc with foci at ± 1, where ϱ is the sum of its semi-axis. Then, we have the following result for $I_p : \mathcal{C}^0(\Gamma;V) \to \mathcal{L}_\rho^4(\Gamma;V)$.

Lemma 3.3. *If $v \in \mathcal{C}^0(\Gamma;V)$ is analytic in D_ϱ and continuous on $\bar{D}_\varrho$, then*

$$\|v - I_p v\|_{\mathcal{L}_\rho^4(\Gamma;V)} \lesssim \frac{\sqrt{p+1}}{\varrho^{p+1}}\|v\|_{\mathcal{C}^0(\Gamma;V)}.$$

Proof. Using Theorem 1 and Corollary 2 in,[9] we can prove

$$\|v - I_p v\|_{\mathcal{C}^0(\Gamma;V)} \lesssim \frac{\sqrt{p+1}}{\varrho^{p+1}}\|v\|_{\mathcal{C}^0(\Gamma;V)},$$

since $\mathcal{C}^0(\Gamma;V) \hookrightarrow \mathcal{L}_\rho^4(\Gamma;V)$, which results in the following relation

$$\|v - I_p v\|_{\mathcal{L}_\rho^4(\Gamma;V)} \lesssim \|v - I_p v\|_{\mathcal{C}^0(\Gamma;V)} \lesssim \frac{\sqrt{p+1}}{\varrho^{p+1}}\|v\|_{\mathcal{C}^0(\Gamma;V)}.$$

□

By the one dimensional argument in[8] and result of Lemma 3.3, we have

$$\|v - \mathcal{I}_{\boldsymbol{p}}v\|_{\mathcal{L}_\rho^4(\Gamma;V)} \lesssim \sum_{n=1}^{N} \sqrt{p_n+1} e^{-(p_n+1)\log \varrho_n} \|v\|_{\mathcal{C}^0(\Gamma;V)}. \tag{9}$$

For the term $\|u_h - u^h\|_{\mathcal{L}_\rho^2(\Gamma) \otimes \mathcal{H}_0^1(D)}$, we can follow the same procedure of Theorem 4.10 in[1] and estimate error of $\|u_h - u^h\|_{\mathcal{L}_\rho^2(\Gamma) \otimes \mathcal{H}_0^1(D)}$ by semi-discrete error on coarse mesh and interpolation error with small number of collocation points, which is represented in detail as follows.

Lemma 3.4. *Let u^h be the semi-discrete solution satisfying (5) with $v = u_{H,\boldsymbol{P}}$, u_h and u_H be the semi-discrete solutions of equation (4) on fine and coarse meshes respectively and u_H satisfy condition in Lemma 3.3, then*

$$\|u_h - u^h\|_{\mathcal{L}_\rho^2(\Gamma) \otimes \mathcal{H}_0^1(D)} \lesssim \|u - u_H\|_{\mathcal{L}_\rho^4(\Gamma) \otimes \mathcal{L}^p(D)}^2 + \|u_H - \mathcal{I}_{\boldsymbol{P}}u_H\|_{\mathcal{L}_\rho^4(\Gamma) \otimes \mathcal{L}^p(D)}^2$$

$$\lesssim H^4 + \sum_{n=1}^{N} \beta_n^2(P_n) e^{-2r_n^H P_n},$$

where $\beta_n = \mathcal{O}(\sqrt{P_n})$, $r_n^H = \log\left[\frac{2\tau_n^H}{|\Gamma_n|}\left(1 + \sqrt{1 + \frac{|\Gamma_n|^2}{4(\tau_n^H)^2}}\right)\right]$ with τ_n^H are defined as in Lemma 3.2 and $4 \leq p \leq \infty$ when $d = 2$.

By using Lemma 3.1, 3.4, inequality (9) and Lemma 3.2, we finally get the following estimation for the numerical approximate solution.

Theorem 3.1. *Let $u^{h,\boldsymbol{p}}$ be the fully approximate solution derived from (8) and u be the exact solution of (1). When conditions of Lemma 3.4 are satisfied, we have*

$$\|u - u^{h,\boldsymbol{p}}\|_{\mathcal{L}_\rho^2(\Gamma) \otimes \mathcal{H}_0^1(D)} \lesssim \|u - u_h\|_{\mathcal{L}_\rho^2(\Gamma) \otimes \mathcal{H}_0^1(D)} + \|u - u_H\|_{\mathcal{L}_\rho^4(\Gamma) \otimes \mathcal{L}^p(D)}^2$$

$$+ \|u_H - \mathcal{I}_{\boldsymbol{P}} u_H\|_{\mathcal{L}_\rho^4(\Gamma) \otimes \mathcal{L}^p(D)}^2 + \|u^h - \mathcal{I}_{\boldsymbol{p}} u^h\|_{\mathcal{L}_\rho^2(\Gamma) \otimes \mathcal{H}_0^1(D)}$$

$$\lesssim h + H^4 + \sum_{n=1}^{N} \beta_n^2(P_n) e^{-2\bar{r}_n P_n} + \sum_{n=1}^{N} e^{-\bar{r}_n p_n},$$

$$\|u - u^{h,\boldsymbol{p}}\|_{\mathcal{L}_\rho^2(\Gamma) \otimes \mathcal{L}^2(D)} \lesssim \|u - u_h\|_{\mathcal{L}_\rho^2(\Gamma) \otimes \mathcal{L}^2(D)} + \|u - u_H\|_{\mathcal{L}_\rho^4(\Gamma) \otimes \mathcal{L}^p(D)}^2$$

$$+ \|u_H - \mathcal{I}_{\boldsymbol{P}} u_H\|_{\mathcal{L}_\rho^4(\Gamma) \otimes \mathcal{L}^p(D)}^2 + \|u^h - \mathcal{I}_{\boldsymbol{p}} u^h\|_{\mathcal{L}_\rho^2(\Gamma) \otimes \mathcal{L}^2(D)}$$

$$\lesssim h^2 + H^4 + \sum_{n=1}^{N} \beta_n^2(P_n) e^{-2\bar{r}_n P_n} + \sum_{n=1}^{N} e^{-\bar{r}_n p_n}.$$

where β_n are defined as in Lemma 3.4 and $\bar{r}_n = \max\{r_n^H, r_n\}$.

Remark 3.1. In reference,[1] we have estimated the convergence property for special functions in

$$\tilde{\mathcal{C}}^\infty(\Gamma; V) = \left\{\phi \left| \frac{|\Gamma|^{m+1}}{(m+1)!} \left\|\frac{\partial^{m+1}\phi}{\partial y^{m+1}}\right\|_{\mathcal{C}^0(\Gamma;V)} \leq \|\phi\|_{\mathcal{C}^0(\Gamma;V)}, \forall m \in \mathbb{N} \right.\right\},$$

which implies the solutions should maintain bounded derivatives of any order. Theorem 3.1 reduces the condition and shows that for any function that is analytic to each random variable in a bounded complex domain, the approximate method is convergent and the numerical solutions have the same convergence order when the parameters $H, h, \boldsymbol{P}, \boldsymbol{p}$ are chosen appropriately.

4. Numerical Experiments

In the numerical tests, we choose $f(u) = u^3 + g$. We first show the result when u maintains different smooth property of Y. For simplicity, we only study cases with one random variable that satisfying uniform distribution and choose $a(Y) = 3 + Y$. Functions g are chosen such that the exact solutions have the following expressions

Ex1/2/3/4: $u(Y, x) = Y^8 / \frac{1}{1+Y^2} / e^{-Y^{-2}} / |Y|^3 \cdot \sin(\pi x_1) \sin(\pi x_2)$.

These four examples represent different smooth properties to Y. The first is smooth and analytic in the complex plane; the second is analytic in a bounded complex domain; the third one is smooth but not analytic and the last one only has a third derivative of bounded variation.

We also test a 2D example to see the efficiency of the numerical method by comparing with general Newton iteration method and Monte Carlo method. We choose $g(\omega, x)$ such that the exact solution is

Ex5: $u(Y_1(\omega), Y_2(\omega), x_1, x_2) = a^{-1}(Y_1, Y_2, x_1, x_2) \sin \pi x_1 \sin \pi x_2$, $a(Y_1(\omega), Y_2(\omega), x_1, x_2) = 3 + Y_1(\omega) + Y_2(\omega)$ with $Y_n(\omega)$, $(n = 1, 2)$ are independent and identically random variables satisfying uniform distribution.

We choose $\Gamma = [-1, 1]$ or $[-1, 1]^2$ and $D = [-1, 1]^2$. We use piecewise linear interpolation for the spatial approximation. The stopping criterion for Newton iteration is chosen to be the relative error between two adjacent iterates less than 10^{-3} in the tests. Algebraic multigrid method with tolerance 10^{-9} is used for solving the linear system.

To see the convergence in random field, we choose mesh sizes in physical space as $(H, h) = (1/64, 1/4096)$ with error approximately equals 10^{-8} and test the degree pairs $(P, p) = (1, 2); (2, 4); (3, 6); (4, 8)$. To see the convergence in physical space, we choose degrees of polynomial in random spaces as $(P, p) = (8, 16)$ with error approximately equals to 10^{-7} and test physical mesh pairs $(H, h) = (1/4, 1/16); (1/8, 1/64); (1/16, 1/256)$.

Numerical results presented in Table 1 show that even though u has different smooth property to Y, the method presented in Section 2 is convergent and it converges faster when the solution has better smooth

Table 1. Errors of $(\boldsymbol{P}, \boldsymbol{p}) = (1,2); (2,4); (3,6); (4,8)$ for **Ex1/2/3/4**.

Error $(\boldsymbol{P}, \boldsymbol{p})$	$(1,2)$	$(2,4)$	$(3,6)$	$(4,8)$
$E(\|u - u^{h,\boldsymbol{p}}\|)$-**Ex1**	$7.19E-3$	$1.39E-3$	$2.14E-4$	$3.92E-7$
$E(\|u - u^{h,\boldsymbol{p}}\|)$-**Ex2**	$1.74E-3$	$2.51E-4$	$4.33E-5$	$8.78E-6$
$E(\|u - u^{h,\boldsymbol{p}}\|)$-**Ex3**	$1.35E-3$	$6.87E-4$	$1.84E-4$	$1.52E-4$
$E(\|u - u^{h,\boldsymbol{p}}\|)$-**Ex4**	$6.60E-3$	$3.84E-3$	$2.38E-3$	$1.34E-3$

Table 2. Errors of $(H, h) = (1/4, 1/16); (1/8, 1/64); (1/16, 1/256)$ for **Ex1/2/3/4**.

Error (H, h)	$(1/4, 1/16)$	$(1/8, 1/64)$	$(1/16, 1/256)$
$E(\|u - u^{h,\boldsymbol{p}}\|)$-**Ex1**	$2.75E-2$	$6.90E-3$	$4.73E-4$
$E(\|u - u^{h,\boldsymbol{p}}\|)$-**Ex2**	$9.28E-2$	$2.42E-2$	$1.60E-3$
$E(\|u - u^{h,\boldsymbol{p}}\|)$-**Ex3**	$1.37E-2$	$4.10E-3$	$2.58E-4$
$E(\|u - u^{h,\boldsymbol{p}}\|)$-**Ex4**	$6.68E-2$	$1.74E-2$	$2.30E-3$

property to variable Y. We can also observe the expected convergence order in physical domain from Table 2.

We also compare the approximate error and computing time of Monte Carlo method (with number of collocation points equals to 100, 1000, 3000 and $h = \frac{1}{64}$), Newton iteration method (Newton) (with $h = \frac{1}{64}$, $\boldsymbol{p} = 8$) and two-level (T-L) numerical approximate method (with $\boldsymbol{P} = 4$, $\boldsymbol{p} = 8$; $H = \frac{1}{8}$, $h = \frac{1}{64}$) for **Ex5** in Table 3. From the data in Table 3, we can see the two-level method we proposed is efficient than Monte Carlo and Newton iteration methods for this kind of model problem.

Table 3. Errors and computing time (in seconds) with $h = 1/64$ for Monte Carlo method (MC), Newton iteration (Newton) and two-level numerical approximate method (T-L).

Methods	$\mathcal{L}^2_\rho(\Gamma) \otimes \mathcal{L}^2(D)$-norm	$\mathcal{L}^2_\rho(\Gamma) \otimes \mathcal{H}^1_0(D)$-norm	Compute time(s)
MC(100)	0.0163	0.0406	111.77
MC(1000)	0.0017	0.0397	1116.1
MC(4000)	$1.9741E-4$	0.0412	4495.4
Newton($\boldsymbol{p} = 8$)	$2.6276E-4$	0.0418	81.361
T-L($\boldsymbol{P} = 4, \boldsymbol{p} = 8$)	$2.5832E-4$	0.0418	29.552

5. Conclusion

In this work, we study the numerical approximate method proposed in [1] for semilinear stochastic differential model and prove the theoretical convergence of the numerical method. We claimed that for the model problem (1), when its solution is analytic to each random variable in a bounded complex domain, the numerical approximate method is a convergent and efficient scheme and achieves approximately the same convergence property as Newton iteration method when $h, H, \boldsymbol{p}, \boldsymbol{P}$ are chosen appropriately.

Acknowledgments

This work is supported by the National Natural Science Foundation of China under Grant No. 11501473, 11426189 and the Fundamental Research Funds for the Central Universities of China (2682016CX108).

References

1. L. Chen, B. Zheng and G. Lin et al., *J. Comput. Appl. Math.* **315**, 195(2017).
2. X. Ma and N. Zabaras, *J. Comput. Phys.* **230**, 4696(2011).
3. O. M. Knio, H. N. Najm, and R. G. Ghanem, *J. Comput. Phys.* **173**, 481(2001).
4. G. Adomian, *Nonlinear Stochastic Systems: Theory and Application to Physics*, volume 46. (Springer, Netherlands, 1989).
5. L. Arnold, *B. Lond. Math. Soc.* **8**, 326(1976).
6. N. Bellomo and R. Riganti, *Nonlinear Stochastic Systems in Physics and Mechanics* (World Scientific, Singapore, 1987).
7. E. Rosseel, N. Scheerlinck and S. Vandewalle, *Numer. Math-Theroy M. E.* **5**, 62(2012).
8. I. Babuška, F. Nobile and R. Tempone, *SIAM Rev.* **52**, 317(2010).
9. P. Zhang, *Chinese Quarterly Journal of Mathematics (Chinese Edition)* **4**, 67(1989).

New definition of the indefinite integral of fuzzy valued function linearly generated by structural elements

Tian-jun Shu* and Zhi-wen Mo[†]

Institute of Intelligent Information and Quantum Information
Sichuan Normal University, Chengdu, 610066, China
*605519161@qq.com
[†]mozhiwen@263.net

This paper proposes a new definition of the indefinite integral of fuzzy valued function linearly generated by structural elements using a kind of fuzzy distance. Then the new definition is used to study the basic properties of the indefinite integral of fuzzy valued function linearly generated by structural elements. They are the existence of the original function, addition together with multiplication, integration by substitution and integration by parts. Subsequently, by giving the Cauchy convergence criterion for the anomalous integral of fuzzy value function linearly generated by structural elements, the integrable conditions of the anomalous integral of fuzzy value function linearly generated by structural elements are discussed. They are absolute convergence, Dirichlet test and Abel test.

Keywords: Structural element; fuzzy valued function; fuzzy distance; indefinite integral; integrable condition.

1. Introduction

The indefinite integral of fuzzy value function has different properties due to the different fuzzy distance in the definition. The notion of fuzzy valued function for linear generation of structural elements is proposed by Guo S. Z. in literature [8]. For the definite integral of fuzzy valued function linearly generated by structural elements, it has the different form of expression because of the different form of fuzzy distance. In this paper, the definite integral of fuzzy valued function linearly generated by structural elements is defined by this fuzzy distance which was given in literature [10]. Particularly, this fuzzy distance needs to satisfy the convergence of horizontal condition. Then the related properties about the indefinite integral of fuzzy valued functions linearly generated by structural elements are investigated. Soon afterwards, we discuss the integrable conditions of the anomalous integral of fuzzy valued function linearly generated by structural elements.

2. Brief Introduction of the Fuzzy Valued Function of Linear Construction Theory

Definition 2.1.[7] E is the fuzzy structural element over the field R of real numbers. If its membership function E(x)($x \in R$) has following:

(1) E(0)=1, and E(1+0)=E(-1-0)=0.

(2) If $x \in [-1, 0)$, then E(x)is increasingly monotonic function being right continuous, and if $x \in (0, 1]$, then E(x)is decreasing monotonic function being left continuous.

(3) If $x \in (-\infty, -1) \cup (1, +\infty)$, then E(x)=0.

Obviously, E is a regularly convex fuzzy set over the field R, and it is a boundedly closed fuzzy number.

If the membership function of E is a symmetric function, then E is called symmetric structure element, it is recorded as E'.

Definition 2.2.[8] $\widetilde{A}$ is finite fuzzy number. If there is a fuzzy structural element E, and finite real number $a \in R$, r$\in R^+$, such that $\widetilde{A} = a + rE($ $r \to 0^+)$, then $\widetilde{A}$ is said to be a fuzzy number linearly generated by E. All fuzzy numbers linearly generated by E is denoted as the symbol $\varepsilon(E)$, and write $\varepsilon(E) = \{\widetilde{A} \mid \widetilde{A} = a + rE, \forall a \in R, r \in R^+\}$.

All in this paper, there must be $\widetilde{A} \in \varepsilon(E)$, on account of the decomposition theorem of fuzzy set, $\widetilde{A} = \bigcup_{\lambda \in [0,1]} \lambda \widetilde{A}_\lambda = \bigcup_{\lambda \in [0,1]} \lambda [a + rE_\lambda^-, a + rE_\lambda^+]$.

Definition 2.3.[9] If X and Y are two real number sets. $\widetilde{N}(f)$ is a set consisted of all fuzzy numbers on Y, and $\widetilde{f}$ is a mapping from X to $\widetilde{N}(f)$. In other words, for arbitrary $x \in X$, there exist only $\widetilde{y} \in \widetilde{N}(f)$ with it correspondence. It is recorded as $\widetilde{y} = \widetilde{f}(x)$, then $\widetilde{f}(x)$ is said to be a fuzzy value function defined on X. If E is a regular fuzzy structural element on $\widetilde{N}(f)$, such that $\widetilde{f}(x) = h(x) + \omega(x)E$ ($x \in X$, h(x) and $\omega(x)$ are bounded function on X, even $\omega(x) > 0$), then $\widetilde{f}(x)$ is said to be a fuzzy valued function linearly generated by E. We use the symbol $\widetilde{N}(E_f)$ to denote all of fuzzy valued function linearly generated by E, and write $\widetilde{N}(E_f) = \{\widetilde{f}(x) \mid \widetilde{f}(x) = h(x) + \omega(x)E, \forall x \in X, \omega(x) > 0\}$.

All in this paper, there must be $\widetilde{f}(x) \in \widetilde{N}(E_f)$, on account of the decomposition theorem of fuzzy set, $\widetilde{f}(x) = \bigcup_{\lambda \in [0,1]} \lambda \widetilde{f}_\lambda(x) = \bigcup_{\lambda \in [0,1]} (h(x) + \omega(x)E_\lambda) = \bigcup_{\lambda \in [0,1]} [h(x) + \omega(x)E_\lambda^-, h(x) + \omega(x)E_\lambda^+]$.

Definition 2.4.[10] $\widetilde{a}$ and $\widetilde{b}$ are arbitrary fuzzy numbers. The fuzzy distance of $\widetilde{a}$ and $\widetilde{b}$ is defined as $\widetilde{d}(\widetilde{a},\widetilde{b}) = \bigcup_{\lambda \in [0,1]} \lambda[\sup_{\lambda \leq \mu \leq 1} |\widetilde{a}_\mu^- - \widetilde{b}_\mu^-|, \sup_{0 \leq \lambda \leq \mu} (|\widetilde{a}_\mu^- - \widetilde{b}_\mu^-| \vee |\widetilde{a}_\mu^+ - \widetilde{b}_\mu^+|)]$.

If $\widetilde{a}$, $\widetilde{b}$ and $\widetilde{c}$ are the fuzzy numbers, then the distance obviously follows:
(1) $\widetilde{d}(\widetilde{a},\widetilde{b}) \geq 0$, $\widetilde{a} = \widetilde{b}$ if and only if $\widetilde{d}(\widetilde{a},\widetilde{b}) = 0$.
(2) $\widetilde{d}(\widetilde{a},\widetilde{b}) = \widetilde{d}(\widetilde{b},\widetilde{a})$.
(3) $\widetilde{d}(\widetilde{a},\widetilde{b}) \leq \widetilde{d}(\widetilde{a},\widetilde{c}) + \widetilde{d}(\widetilde{c},\widetilde{b})$.

3. A New Definition of the Indefinite Integral of $\widetilde{f}(x)$ Belonging to $\widetilde{N}(E_f)$

Definition 3.1. Let's assume $\widetilde{f}(x)$ be defined on $U^0(x_0;\delta')$, $\lambda \in (0,1]$, the cut set of $\widetilde{f}(x)$ is $\widetilde{f}_\lambda(x) = [\widetilde{f}_\lambda^-(x), \widetilde{f}_\lambda^+(x)]$. For functions $\widetilde{f}_\lambda^-(x)$ and $\widetilde{f}_\lambda^+(x)$, if for any $\varepsilon > 0$, there exists $\delta > 0 (\delta < \delta')$, when $x', x'' \in U^0(x_0;\delta')$, such that $|\widetilde{f}_\lambda^-(x') - \widetilde{f}_\lambda^-(x'')| < \varepsilon$, and $|\widetilde{f}_\lambda^+(x') - \widetilde{f}_\lambda^+(x'')| < \varepsilon$, then the limit of $\widetilde{f}(x)$ exists. Furthermore, whenever $x_0 \in U^0(x';x'')$, such that $\widetilde{d}(\widetilde{f}(x), \widetilde{f}(x_0)) < \varepsilon$. Then $\widetilde{A}$ is the limit of $\widetilde{f}(x)$ when x tends to x_0, and it is recorded as $\lim_{x \to x_0} \widetilde{f}(x) = \widetilde{f}(x_0)$.

Definition 3.2. ($\varepsilon - \delta$ definition of Definition 3.1) Let $\widetilde{f}(x)$ be defined on $U^0(x_0;\delta')$, $\widetilde{A} \in \varepsilon(E)$. If for any real number $\varepsilon > 0$, there exists positive $\delta(< \delta')$, whenever $0 < |x - x_0| < \delta$, such that $\widetilde{d}(\widetilde{f}(x), \widetilde{f}(x_0)) < \varepsilon$, then $\widetilde{f}(x_0)$ is the limit of $\widetilde{f}(x)$ when x tends to x_0, and write $\lim_{x \to x_0} \widetilde{f}(x) = \widetilde{f}(x_0)$.

Definition 3.3. Let $\widetilde{f}(x)$ be defined on I. If for any real number $\varepsilon > 0$, there exists positive $\delta(< \delta')$, whenever $0 < |x - x_0| < \delta$, such that $\widetilde{d}(\frac{\widetilde{f}(x) - \widetilde{f}(x_0)}{x - x_0}, \widetilde{f}'(x_0)) < \varepsilon$, then $\widetilde{f}'(x_0)$ is called the derivative of $\widetilde{f}(x)$ at x_0, and write $\widetilde{f}'(x_0) = \lim_{x \to x_0} \frac{\widetilde{f}(x) - \widetilde{f}(x_0)}{x - x_0}$.

Definition 3.4. If $\widetilde{F}'(x) = \widetilde{f}(x)$, $x \in I$, then $\widetilde{F}(x)$ is named the primitive function of $\widetilde{f}(x)$ on interval I. All primitive functions of $\widetilde{f}(x)$ on interval I are named indefinite integral of $\widetilde{f}(x)$ on the interval I, it is recorded as $\widetilde{A} = \int \widetilde{f}(x) dx$. If $\widetilde{F}(x)$ is the primitive functions of $\widetilde{f}(x)$, then $\widetilde{A}(x) = \widetilde{F}(x) + \widetilde{C}$.

3.1. *The properties of the indefinite integral of $\widetilde{f}(x)$ belonging to $\widetilde{N}(E_f)$*

Theorem 3.1.1. If $\widetilde{F}(x)$ is primitive function of $\widetilde{f}(x)$ on interval I, then

(1) $\widetilde{F}(x)+\widetilde{C}$ is primitive function of $\widetilde{f}(x)$ on interval I, $\widetilde{C}$ is an arbitrary constant fuzzy value function.

(2) On the interval I, there is only one constant fuzzy value function between two primitive functions of $\widetilde{f}(x)$.

Proof. (1) By definition 3.4, $[\widetilde{F}(x) + \widetilde{C}]' = \widetilde{F}'(x) + \widetilde{C}' = \widetilde{f}(x)$.

(2) Let $\widetilde{F_1}(x)$ and $\widetilde{F_2}(x)$ are primitive functions of $\widetilde{f}(x)$ on interval I. There is $[\widetilde{F_1}(x) - \widetilde{F_2}(x)]' = \widetilde{F_1}'(x) - \widetilde{F_2}'(x) = \widetilde{f}(x) - \widetilde{f}(x) = 0$. And $\widetilde{F_1}(x) - \widetilde{F_2}(x)$ is satisfied the Lagrangian central theorem on I, hence $\widetilde{F_1}(x) - \widetilde{F_2}(x) = \widetilde{C}$.

Definition 3.1.1. For the closed interval $[a, b]$, there are n-1 points, it follows $a = x_0 < x_1 < x_2 < \cdots < x_{n-1} < x_n = b$. Sub $[a, b]$ into n intervals as $\Delta_i = [x_{i-1}, x_i]$, $i = 1, 2 \cdots n$. These closed intervals constitute a segmentation on [a, b], which is denoted as $T = \{\Delta_1, \Delta_2 \cdots \Delta_n\}$. The length of the interval Δ_i is $\Delta x_i = x_i - x_{i-1}$. We have $\|T\| = \max_{1 \leq i \leq n} \{\Delta x_i\}$ as the modulus of segmentation T. Arbitrarily pick point $\xi_i \in \Delta_i$, $i = 1, 2 \ldots n$, there the formula $\sum_{i=1}^{n} \widetilde{f}(\xi_i) \cdot \Delta x_i$ is named an integral sum of $\widetilde{f}(x)$ on [a, b]. If $\forall \varepsilon > 0$, there exists positive δ, related to segmentation T on[a, b] and the set of points $\{\xi_i\}(\xi_i \in \Delta_i)$ selected in it. Whenever $\|T\| < \delta$, such that $\widetilde{d}(\sum_{i=1}^{n} \widetilde{f}(\xi_i) \cdot \Delta x_i, \widetilde{A}) < \varepsilon$, then $\widetilde{f}(x)$ is fuzzy integrable on [a, b], and $\widetilde{A}$ is the fuzzy definite integral of $\widetilde{f}(x)$ on [a, b]. It is recorded as $\widetilde{A} = \int_a^b \widetilde{f}(x)dx$.

Definition 3.1.2. Let $\widetilde{f}(x)$ be defined on $[a, +\infty)$, and be integrable on any finite interval $[a, u]$. If there is $\lim_{u \to +\infty} \int_a^u \widetilde{f}(x)dx = \widetilde{A}$, then $\widetilde{A}$ is called an infinite improper integral, and it is recorded as $\widetilde{A} = \int_a^{+\infty} \widetilde{f}(x)dx$.

Similarly, the infinite improper integral of $\widetilde{f}(x)$ on $(-\infty, b]$ can be defined.

Definition 3.1.3. (Variable limit integral) If $\widetilde{f}(x)$ is integrable on interval [a, b], for $x \in [a, b]$, $\widetilde{f}(x)$ is also integrable on interval [a, x], then $\widetilde{\Phi}(x) = \int_a^x \widetilde{f}(t)dt$, $x \in [a, b]$. This integral is called the fuzzy definite integral of the variable limit. Similarly, the fuzzy definite integral of the lower limit can be defined, write $\widetilde{\Psi}(x) = \int_x^b \widetilde{f}(t)dt$, $x \in [a, b]$. They are collectively referred to as the variable limits.

Theorem 3.1.2. If both $\widetilde{f}(x_1)$ and $\widetilde{f}(x_2)$ have primitive functions on the interval I, $k \in R$, then

(1) $\widetilde{f}(x_1) + \widetilde{f}(x_2)$ have primitive functions on the interval I, and $\int \widetilde{f}(x_1) + \widetilde{f}(x_1)dx = \int \widetilde{f}(x_1)dx + \int \widetilde{f}(x_2)dx$.

(2) $\widetilde{f}(x_1) - \widetilde{f}(x_2)$ have primitive functions on the interval I, and $\int \widetilde{f}(x_1) - \widetilde{f}(x_2)dx = \int \widetilde{f}(x_1)dx - \int \widetilde{f}(x_2)dx$.

(3) $k\widetilde{f}_1(x)$ have primitive functions on the interval I, and $\int k\widetilde{f}_1(x)dx = k\int \widetilde{f}(x)dx$.

Proof. (1) Because $\widetilde{f}_1(x)$ and $\widetilde{f}_2(x)$ are integrable functions on the interval I, by Theorem 3.1.1. Let $\int \widetilde{f}(x_1)dx = \int_a^x \widetilde{f}_1(x)dx + \widetilde{C}_1$, and $\int \widetilde{f}(x_2)dx = \int_a^x \widetilde{f}_2(x)dx + \widetilde{C}_2$. On account of $[\int_a^x \widetilde{f}_1(x)dx + \int_a^x \widetilde{f}_2(x)dx]' = [\int_a^x \widetilde{f}_1(x)dx]' + [\int_a^x \widetilde{f}_2(x)dx]' = \widetilde{f}_1(x) + \widetilde{f}_2(x)$. In other words, $\int \widetilde{f}_1(x) + \widetilde{f}_2(x)dx = \int_a^x \widetilde{f}_1(x)dx + \int_a^x \widetilde{f}_2(x)dx + \widetilde{C} = \int_a^x \widetilde{f}_1(x)dx + \widetilde{C}_1 + \int_a^x \widetilde{f}_2(x)dx + \widetilde{C}_2 = \int \widetilde{f}(x_1)dx + \int \widetilde{f}(x_2)dx$.

(2) The proof is similar to (1).

(3) $\widetilde{f}_1(x)$ is integrable functions on the interval I. Let $\int \widetilde{f}(x_1)dx = \int_a^x \widetilde{f}_1(x)dx + \widetilde{C}_1$. Because of $[k\int_a^x \widetilde{f}_1(x)dx]' = k[\int_a^x \widetilde{f}_1(x)dx]' = k\widetilde{f}_1(x)$. $\int k\widetilde{f}_1(x)dx = [\int_a^x k\widetilde{f}_1(x)dx + \widetilde{C}] = \int_a^x k\widetilde{f}_1(x)dx + k\widetilde{C}_1 = k[\int_a^x \widetilde{f}_1(x)dx + \widetilde{C}_1] = k\int \widetilde{f}_1(x)dx$.

Theorem 3.1.3 (Integration by substitution). Let $\widetilde{g}(u)$ be define on $[\alpha, \beta]$, $u = \varphi(x)$ is derivable on interval $[a, b]$, and $\alpha \leq \varphi(x) \leq \beta$, $x \in [a, b]$, there is $\widetilde{f}(x) = \widetilde{g}(\varphi(x)) \cdot \varphi'(x), x \in [a, b]$. If $\widetilde{G}(u)$ is primitive function of $\widetilde{g}(u)$ on interval $[\alpha, \beta]$, then $\widetilde{F}(x)$ is primitive function of $\widetilde{f}(x)$ on interval $[a, b]$, and $\widetilde{F}(x) = \widetilde{G}(\varphi(x)) + \widetilde{C}$, that is $\int \widetilde{f}(x)dx = \int \widetilde{g}(\varphi(x))\psi'(x)dx = \int \widetilde{g}(\varphi(x))d\varphi(x) = \widetilde{G}(u) + \widetilde{C} = \widetilde{G}(\varphi(x)) + \widetilde{C}$.

Proof. Owing to derivation rule of the compound functions, $[\widetilde{G}(\varphi(x))]' = \widetilde{G}'(\varphi(x))\varphi'(x) = \widetilde{g}(\varphi(x))\varphi'(x) = \widetilde{f}(x)$, $\widetilde{G}(\varphi(x))$ is an primitive function of $\widetilde{f}(x)$. There is $\int \widetilde{f}(x)dx = \widetilde{G}(\varphi(x)) + \widetilde{C}$. That is to say equality of the theorem is established.

Theorem 3.1.4 (Integration by parts). Let's assume $\widetilde{u}(x), \widetilde{v}(x) \in \widetilde{N}(E'_f)$. If indefinite integral $\int \widetilde{v}'(x) \cdot \widetilde{u}(x))dx$ and $\int \widetilde{v}(x) \cdot \widetilde{u}'(x)dx$ are present, then $\int \widetilde{v}'(x) \cdot \widetilde{u}(x)dx = \widetilde{u}(x) \cdot \widetilde{v}(x) - \int \widetilde{u}'(x)\widetilde{v}(x)dx$.

Proof. Owing to Definition 3.3,

$$[\widetilde{u}(x_0)\widetilde{v}(x_0)]' = \lim_{x \to x_0} \frac{\widetilde{u}(x)\widetilde{v}(x) - \widetilde{u}(x_0)\widetilde{v}(x_0)}{x - x_0} =$$

$$\lim_{x \to x_0} \frac{1}{x - x_0}[\widetilde{u}(x)\widetilde{v}(x) - \widetilde{u}(x)\widetilde{v}(x_0) + \widetilde{u}(x)\widetilde{v}(x_0) - \widetilde{u}(x_0)\widetilde{v}(x_0)] =$$

$$1\widetilde{u}(x)\lim_{x \to x_0}\frac{\widetilde{v}(x) - \widetilde{v}(x_0)}{x - x_0} + \widetilde{v}(x) \cdot \lim_{x \to x_0}\frac{\widetilde{u}(x) - \widetilde{u}(x_0)}{x - x_0} = \widetilde{u}(x_0)\widetilde{v}'(x_0) + \widetilde{u}'(x_0)\widetilde{v}(x_0).$$

then $\widetilde{v}(x)'\widetilde{u}(x) = [\widetilde{u}(x)\widetilde{v}(x)]' - \widetilde{v}(x)'\widetilde{u}(x)$. Equivalence on both sides of the equation, we can obtain $\int \widetilde{v}(x)' \cdot \widetilde{u}(x)dx = \widetilde{u}(x) \cdot \widetilde{v}(x) - \int \widetilde{u}'(x)\widetilde{v}(x)dx$.

3.2. Integrable conditions of anomalous integral of $\widetilde{f}(x)$ belonging to $\widetilde{N}(E_f)$

Theorem 3.2.1 (Anomalous integral Cauchy convergence condition). $\int_a^{+\infty} \widetilde{f}(x)dx$ is convergent if and only if for any real number $\varepsilon > 0$, there exists $M \geq a$, whenever $u_1, u_2 > M$, such that $\widetilde{d}(\int_a^{u_2} \widetilde{f}(x)dx, \int_{u_1}^a \widetilde{f}(x)dx) = |\int_{u_1}^{u_2} \widetilde{f}(x)dx| < \varepsilon$.

Proof. Necessity Owing to $\int_a^{+\infty} \widetilde{f}(x)dx$ is convergent, by Theorem 3.1, $\int_a^{+\infty} \widetilde{f}(x)dx = \int_a^x \widetilde{f}(t)dt + \widetilde{C}$. Then $\int_a^{+\infty} \widetilde{f}(x)dx$ is convergent if as only if $\int_a^x \widetilde{f}(t)dt$ is convergent. For any real number $\varepsilon > 0$, there exists $M \geq a$, whenever $u_1, u_2 > M$, in the closed interval $[a, u_1]$, there is $\widetilde{d}(\sum_{i=1}^{n} \widetilde{f}(\xi_i) \cdot \Delta x_i, \widetilde{A}) < \frac{\varepsilon}{2}$, and in the closed interval $[a, u_2]$, there is $\widetilde{d}(\sum_{i=1}^{n} \widetilde{f}(\eta_i) \cdot \Delta x_i, \widetilde{A}) < \frac{\varepsilon}{2}$. Then in the closed interval $[u_1, u_2]$, $\widetilde{d}(\int_{u_1}^{u_2} \widetilde{f}(x)dx, 0) = \widetilde{d}(\sum_{i=1}^{n} \widetilde{f}(\eta_i) \cdot \Delta x_i - \sum_{i=1}^{n} \widetilde{f}(\xi_i) \cdot \Delta x_i, 0) = \widetilde{d}(\sum_{i=1}^{n} \widetilde{f}(\eta_i) \cdot \Delta x_i, \sum_{i=1}^{n} \widetilde{f}(\xi_i) \cdot \Delta x_i) \leq \widetilde{d}(\sum_{i=1}^{n} \widetilde{f}(\eta_i) \cdot \Delta x_i, \widetilde{A}) + \widetilde{d}(\sum_{i=1}^{n} \widetilde{f}(\xi_i) \cdot \Delta x_i, \widetilde{A}) < \frac{\varepsilon}{2} + \frac{\varepsilon}{2} = \varepsilon$. That proved $\widetilde{d}(\int_a^{u_2} \widetilde{f}(x)dx, \int_a^{u_1} \widetilde{f}(x)dx) = |\int_{u_1}^{u_2} \widetilde{f}(x)dx| < \varepsilon$.

Sufficiency. $\int_a^{+\infty} \widetilde{f}(x)dx = \int_a^x \widetilde{f}(t)dt + \widetilde{C}$. To prove $\int_a^{+\infty} \widetilde{f}(x)dx$ is convergent, only need to prove $\int_a^x \widetilde{f}(t)dt$ is convergent. Let sequence $\{x_n\}$ subset $[u_1, u_2]$ and $\lim_{n \to \infty} x_n = x_0$. Follow the assumptions, for any real number $\varepsilon > 0$, there exists $M \geq a$, whenever $u_1, u_2 > M$, such that $\widetilde{d}(\int_a^{u_2} \widetilde{f}(x)dx, \int_a^{u_1} \widetilde{f}(x)dx) < \varepsilon$. Then for any real number $\varepsilon > 0$, there exists

$\delta > 0$, take a positive integer N, when $p, q > N$, there is $x_p, x_q \in u^0(x_0, \delta)$, such that $\widetilde{d}(\int_a^{x_p} \widetilde{f}(t)dt, \int_a^{x_q} \widetilde{f}(t)dt) < \varepsilon$. Convergence criteria by the series of Cauchy, there is $\lim_{n\to\infty} \int_a^{x_n} \widetilde{f}(t)dt$, write $\int_a^{x_0} \widetilde{f}(t)dt = \lim_{n\to\infty} \int_a^{x_n} \widetilde{f}(t)dt$. Suppose another sequence $\{y_n\} \subset U^0(x_0, \delta')$ and $\lim_{n\to\infty} y_n = x_0$. In summary, there is $\lim_{n\to\infty} \int_a^{y_n} \widetilde{f}(t)dt$, write $\int_a^{y_0} \widetilde{f}(t)dt = \lim_{n\to\infty} \int_a^{y_n} \widetilde{f}(t)dt$. Easy to prove $\int_a^{x_0} \widetilde{f}(t)dt = \int_a^{y_0} \widetilde{f}(t)dt$. Particularly, introduce the sequence $\{z_n\}$: $x_1, y_1, x_2, y_2, ..., x_n, y_n, ...$. Easy to see $\{z_n\} \subset U^0(x_0, \delta')$. Because $\lim_{n\to\infty} x_n = \lim_{n\to\infty} y_n = x_0$, we can notice for any $\delta' > 0$, Items $\{\int_a^{x_n} \widetilde{f}(t)dt\}$ and $\{\int_a^{y_n} \widetilde{f}(t)dt\}$ fell outside of $U^0(x_0, \delta')$ have at least a limited term, then item $\{z_n\}$ fell outside of $U^0(x_0, \delta')$ has at least a limited term. Hence $\lim_{n\to\infty} z_n = x_0$. Similarly, $\{\int_a^{z_n} \widetilde{f}(t)dt\}$ is also convergent. Therefore, $\lim_{n\to\infty} \int_a^{x_n} \widetilde{f}(t)dt$ and $\lim_{n\to\infty} \int_a^{y_n} \widetilde{f}(t)dt$ are equal. That is $\int_a^{x_0} \widetilde{f}(t)dt = \int_a^{y_0} \widetilde{f}(t)dt$. By the principle of attribution, $\widetilde{d}(\int_a^x \widetilde{f}(x)dx, \int_a^{x_0} \widetilde{f}(x)dx) < \varepsilon$. In other words, $\int_a^x \widetilde{f}(x)dx$ is convergent.

Theorem 3.2.2. If $\widetilde{f}(x)$ is integrable on any finite interval $[a, u]$, $a < b$, then $\int_a^{+\infty} \widetilde{f}(x)dx$ is convergent if and only if $\int_b^{+\infty} \widetilde{f}(x)dx$ is convergent, and $\int_a^{+\infty} \widetilde{f}(x)dx = \int_a^b \widetilde{f}(x)dx + \int_b^{+\infty} \widetilde{f}(x)dx$.

Proof. Sufficiency. By the known conditions, $\int_b^{+\infty} \widetilde{f}(x)dx$ is convergent, and $\widetilde{f}(x)$ is integrable on any finite interval $[a, u]$, $a < b$. let $\int_b^{+\infty} \widetilde{f}(x)dx = \lim_{u \to +\infty} \int_b^u \widetilde{f}(x)dx = \widetilde{A}_1$, and $\int_a^b \widetilde{f}(x)dx = \widetilde{A}_2$. For each $\varepsilon > 0$, the segmentations of [a, b] and [b,u] are $\|T_1\|$ and $\|T_2\|$ respectively. Whenever $\|T_1\| < \delta_1$, such that $\widetilde{d}(\sum_{i=1}^n \widetilde{f}_1(\xi_i) \cdot \Delta x_i, \widetilde{A}_1) = \bigcup_{\lambda \in [0,1]} \lambda \sum_{i=1}^n [\sup_{\lambda \le \mu \le 1} |h_1(\xi_i) \cdot \Delta x_i + \omega_1(\xi_i) \cdot \Delta x_i E_\mu^- - a_1 - r_1 E_\mu^-|, \sup_{0 \le \lambda \le \mu} (|h_1(\xi_i) \cdot \Delta x_i + \omega_1(\xi_i) \cdot \Delta x_i E_\mu^- - a_1 - r_1 E_\mu^-| \vee |h_1(\xi_i) \cdot \Delta x_i + \omega_1(\xi_i) \cdot \Delta x_i E_\mu^+ - a_1 - r_1 E_\mu^+|)] < \frac{\varepsilon}{2}$. Whenever $\|T_2\| < \delta_2$, such that $\widetilde{d}(\sum_{i=1}^n \widetilde{f}_2(\xi_i) \cdot \Delta x_i, \widetilde{A}_2) = \bigcup_{\lambda \in [0,1]} \lambda \sum_{i=1}^n [\sup_{\lambda \le \mu \le 1} |h_2(\xi_i) \cdot \Delta x_i + \omega_2(\xi_i) \cdot \Delta x_i E_\mu^- - a_2 - r_2 E_\mu^-|, \sup_{0 \le \lambda \le \mu} (|h_2(\xi_i) \cdot \Delta x_i + \omega_2(\xi_i) \cdot \Delta x_i E_\mu^- - a_2 - r_2 E_\mu^-| \vee |h_2(\xi_i) \cdot \Delta x_i + \omega_2(\xi_i) \cdot \Delta x_i E_\mu^+ - a_2 - r_2 E_\mu^+|)] < \frac{\varepsilon}{2}$. Take $\delta = \min\{\delta_1, \delta_2\}$, whenever $\|T\| < \delta$, we have $\widetilde{d}(\sum_{i=1}^n \widetilde{f}_1(\xi_i) \cdot \Delta x_i + \sum_{i=1}^n \widetilde{f}_2(\xi_i) \cdot \Delta x_i, \widetilde{A}_1 + \widetilde{A}_2) = \bigcup_{\lambda \in [0,1]} \lambda \sum_{i=1}^n [\sup_{\lambda \le \mu \le 1} |(h_1(\xi_i) \cdot \Delta x_i + \omega_1(\xi_i) \cdot \Delta x_i E_\mu^- + h_2(\xi_i) \cdot \Delta x_i + \omega_2(\xi_i) \cdot$

$\Delta x_i E_\mu^-) - (a_1 + r_1 E_\mu^- + a_2 + r_2 E_\mu^-)|, \sup_{0 \le \lambda \le \mu} (|(h_1(\xi_i) \cdot \Delta x_i + \omega_1(\xi_i) \cdot \Delta x_i E_\mu^- + h_2(\xi_i) \cdot \Delta x_i + \omega_2(\xi_i) \cdot \Delta x_i E_\mu^-) - (a_1 + r_1 E_\mu^- + a_2 + r_2 E_\mu^-)| \vee |(h_1(\xi_i) \cdot \Delta x_i + \omega_1(\xi_i) \cdot \Delta x_i E_\mu^+ + h_2(\xi_i) \cdot \Delta x_i + \omega_2(\xi_i) \cdot \Delta x_i E_\mu^+) - (a_1 + r_1 E_\mu^+ + a_2 + r_2 E_\mu^+)|)] =$
$\bigcup_{\lambda \in [0,1]} \lambda \sum_{i=1}^{n} [\sup_{\lambda \le \mu \le 1} |(h_1(\xi_i) \cdot \Delta x_i + \omega_1(\xi_i) \cdot \Delta x_i E_\mu^- - a_1 - r_1 E_\mu^-) + (h_2(\xi_i) \cdot \Delta x_i + \omega_2(\xi_i) \cdot \Delta x_i E_\mu^- - a_2 - r_2 E_\mu^-)|, \sup_{0 \le \lambda \le \mu} (|(h_1(\xi_i) \cdot \Delta x_i + \omega_1(\xi_i) \cdot \Delta x_i E_\mu^- - a_1 - r_1 E_\mu^-) + (h_2(\xi_i) \cdot \Delta x_i + \omega_2(\xi_i) \cdot \Delta x_i E_\mu^- - a_2 - r_2 E_\mu^-)| \vee |(h_1(\xi_i) \cdot \Delta x_i + \omega_1(\xi_i) \cdot \Delta x_i E_\lambda^+ - a_1 - r_1 E_\lambda^+) + (h_2(\xi_i) \cdot \Delta x_i + \omega_2(\xi_i) \cdot \Delta x_i E_\lambda^+ - a_2 - r_2 E_\lambda^+)|)] \le \bigcup_{\lambda \in [0,1]} \lambda \sum_{i=1}^{n} [\sup_{\lambda \le \mu \le 1} |h_1(\xi_i) \cdot \Delta x_i + \omega_1(\xi_i) \cdot \Delta x_i E_\mu^- - a_1 - r_1 E_\mu^-| + |h_2(\xi_i) \cdot \Delta x_i + \omega_2(\xi_i) \cdot \Delta x_i \cdot E_\mu^- - a_2 - r_2 E_\mu^-|, \sup_{0 \le \lambda \le \mu} (|h_1(\xi_i) \cdot \Delta x_i + \omega_1(\xi_i) \cdot \Delta x_i E_\mu^- - a_1 - r_1 E_\mu^-| + |h_2(\xi_i) \cdot \Delta x_i + \omega_2(\xi_i) \cdot \Delta x_i E_\mu^- - a_2 - r_2 E_\mu^-| \vee |h_1(\xi_i) \cdot \Delta x_i + \omega_1(\xi_i) \cdot \Delta x_i E_\mu^+ - a_1 - r_1 E_\mu^+| + |h_2(\xi_i) \cdot \Delta x_i + \omega_2(\xi_i) \cdot \Delta x_i E_\mu^+ - a_2 - r_2 E_\mu^+|)] = \bigcup_{\lambda \in [0,1]} \lambda \sum_{i=1}^{n} [\sup_{\lambda \le \mu \le 1} |h_1(\xi_i) \cdot \Delta x_i + \omega_1(\xi_i) \cdot \Delta x_i E_\mu^- - a_1 - r_1 E_\mu^-|, \sup_{0 \le \lambda \le \mu} (|h_1(\xi_i) \cdot \Delta x_i + \omega_1(\xi_i) \cdot \Delta x_i E_\mu^- - a_1 - r_1 E_\mu^-| \vee |h_1(\xi_i) \cdot \Delta x_i + \omega_1(\xi_i) \cdot \Delta x_i E_\mu^+ - a_1 - r_1 E_\mu^+|)] + \bigcup_{\lambda \in [0,1]} \lambda \sum_{i=1}^{n} [\sup_{\lambda \le \mu \le 1} |h_2(\xi_i) \cdot \Delta x_i + \omega_2(\xi_i) \cdot \Delta x_i E_\mu^- - a_2 - r_2 E_\mu^-|, \sup_{0 \le \lambda \le \mu} (|h_2(\xi_i) \cdot \Delta x_i + \omega_2(\xi_i) \cdot \Delta x_i E_\mu^- - a_2 - r_2 E_\mu^-| \vee |h_2(\xi_i) \cdot \Delta x_i + \omega_2(\xi_i) \cdot \Delta x_i E_\mu^+ - a_2 - r_2 E_\mu^+|)] < \frac{\varepsilon}{2} + \frac{\varepsilon}{2} = \varepsilon.$ It has been proved that $\int_a^u \widetilde{f}(x) dx = \widetilde{A}_1 + \widetilde{A}_2$, then $\int_a^u \widetilde{f}(x) dx$ is convergent, that is $\int_a^{+\infty} \widetilde{f}(x) dx$ is convergent.

Necessity. Because $\int_a^{+\infty} \widetilde{f}(x) dx$ is convergent, for any u > a, there is a segmentation T on [a, u]. For each $\varepsilon > 0$, whenever $\|T\| < \delta$, such that $\widetilde{d}(\sum_{i=1}^{n} \widetilde{f}(\xi_i) \cdot \Delta x_i, \widetilde{A}) < \varepsilon$. Add a point u on T to get a new segmentation T^*, and $\widetilde{d}(\sum_{i=1}^{n} \widetilde{f}(\xi_i^*) \cdot \Delta x_i^*, \widetilde{A}^*) \le \widetilde{d}(\sum_{i=1}^{n} \widetilde{f}(\xi_i) \cdot \Delta x_i, \widetilde{A}) < \varepsilon$. T^* exists in parts of [a, b] and [b, u], they are denoted as T' and T''. Whenever $\|T'\| < \delta$, such that $\widetilde{d}(\sum_{i=1}^{n} \widetilde{f}(\xi_i') \cdot \Delta x_i', \widetilde{A}') \le \widetilde{d}(\sum_{i=1}^{n} \widetilde{f}(\xi_i^*) \cdot \Delta x_i, \widetilde{A}) < \varepsilon$; and whenever $\|T''\| < \delta$, such that $\widetilde{d}(\sum_{i=1}^{n} \widetilde{f}(\xi_i)'' \cdot \Delta x_i'', \widetilde{A}'') \le \widetilde{d}(\sum_{i=1}^{n} \widetilde{f}(\xi_i^*) \cdot \Delta x_i, \widetilde{A}) < \varepsilon$. Then $\int_a^u \widetilde{f}(x) dx$ is convergent.

The segmentation on [a, u] is T, the constant point b is one of the points.

The segmentation T^* exists on parts of $[a,b]$ and $[b,u]$, Because $\sum_{i=1}^{n} \widetilde{f}(\xi_i) \cdot \Delta x_i = \sum_{i=1}^{n} \widetilde{f}(\xi_i') \cdot \Delta x_i' + \sum_{i=1}^{n} \widetilde{f}(\xi_i'') \cdot \Delta x_i''$. Whenever $\|T\|<\delta$, such that $\|T'\|<\delta$ and $\|T''\|<\delta$. Namely, $\int_a^u \widetilde{f}(x)dx = \int_a^b \widetilde{f}(x)dx + \int_b^u \widetilde{f}(x)dx$. This theorem is fully proved.

Corollary 3.2.1. $\int_a^{+\infty} \widetilde{f}(x)dx$ is convergent if and only if for any real number ε, there exists $M \geq a$, whenever $u > M$, such that $|\int_u^{+\infty} \widetilde{f}(x)dx| < \varepsilon$.

Theorem 3.2.3 (Absolute convergence). If $\widetilde{f}(x)$ is integrable on any finite interval $[a, +\infty)$, and $\int_a^{+\infty} |\widetilde{f}(x)|dx$ is convergent, then $\int_a^{+\infty} \widetilde{f}(x)dx$ is also convergent, and $|\int_a^{+\infty} \widetilde{f}(x)dx| \leq \int_a^{+\infty} |\widetilde{f}(x)|dx$.

Proof. Because $\int_a^{+\infty} |\widetilde{f}(x)|dx$ is integrable, by Theorem 3.2.1, for any real number $\varepsilon > 0$, there exists $M \geq a$, whenever $u_1, u_2 > M$, such that $\int_{u_1}^{u_2} |\widetilde{f}(x)|dx < \varepsilon$. Owing to the properties of absolute value inequality, $|\int_{u_1}^{u_2} \widetilde{f}(x)dx| \leq \int_{u_1}^{u_2} |\widetilde{f}(x)|dx < \varepsilon$. It is proved that $\int_a^{+\infty} \widetilde{f}(x)dx$ is also convergent, and $|\int_a^{+\infty} \widetilde{f}(x)dx| \leq \int_a^{+\infty} |\widetilde{f}(x)|dx$.

Corollary 3.2.2. Let's assume $\widetilde{f}(x), \widetilde{g}(x) \in \widetilde{N}(E_f')$, and be integrable on any finite interval $[a, u]$, $\widetilde{g}(x) > 0$. If for any real number $\varepsilon > 0$, there exists positive integer M, whenever $x > M$, such that $\widetilde{d}(|\widetilde{f}(x)|, \widetilde{g}(x) \cdot c) < \varepsilon$, then when $0 \leq c < +\infty$, $\int_a^{+\infty} |\widetilde{f}(x)|dx$ is convergent if and only if $\int_a^{+\infty} \widetilde{g}(x)dx$ is convergent.

Theorem 3.2.4 (Dirichlet test). Let's assume $\widetilde{f}(x), \widetilde{g}(x) \in \widetilde{N}(E_f')$, and be integrable on any finite interval $[a, u]$. If $\widetilde{F}(x) = \int_a^u \widetilde{f}(x)dx$ is bounded on $[a, +\infty)$, $\widetilde{g}(x)$ tends to be zero when $x \to o$, $x \in [a, +\infty)$, then $\int_a^{+\infty} \widetilde{f}(x) \cdot \widetilde{g}(x)dx$ is convergent.

Proof. According to $\widetilde{F}(x) = \int_a^u \widetilde{f}(x)dx$ is bounded on $[a, +\infty)$, we assumes $|\int_a^u \widetilde{f}(x)dx| < M$, $u \in [a, +\infty)$. For any real number $\varepsilon > 0$, owing to $\lim_{x \to \infty} \widetilde{g}(x) = 0$, there exists $G \geq a$, when $x > G$, there is $|g(x)| < \frac{\varepsilon}{4M}$. Further because $\widetilde{g}(x)$ is monotonous function, use integral second median theorem. For any $u_2 > u_1 > G$, there exists $\xi \in [u_1, u_2]$, such that $\int_{u_1}^{u_2} \widetilde{g}(x)dx = g(u_1) \int_{u_1}^{\varepsilon} \widetilde{f}(x)dx + g(u_2) \int_{\xi}^{u_2} \widetilde{f}(x)dx$.

Hence $|\int_{u_1}^{u_2} \widetilde{f}(x) dx| \leq |\widetilde{g}(u_1)| \cdot |\int_{u_1}^{\xi} \widetilde{f}(x) dx + |\widetilde{g}(u_2)| \cdot |\int_{\xi}^{u_2} \widetilde{f}(x) dx = |\widetilde{g}(u_1)| \cdot |\int_{a}^{\xi} \widetilde{f}(x) dx - \int_{a}^{u_1} \widetilde{f}(x) dx| + |\widetilde{g}(u_2)| \cdot |\int_{a}^{u_2} \widetilde{f}(x) \cdot dx - \int_{a}^{\xi} \widetilde{f}(x) \cdot dx| < \frac{\varepsilon}{4M} \cdot 2M + \frac{\varepsilon}{4M} \cdot 2M = \varepsilon$.

Theorem 3.2.5 (Abel test). Let's assume $\widetilde{f}(x), \widetilde{g}(x) \in \widetilde{N}(E'_f)$, and be integrable on any finite interval $[a, u]$. If $\int_{a}^{+\infty} \widetilde{f}(x) dx$ is convergent, and $\widetilde{g}(x)$ is monotonous and bounded on $[a, +\infty)$, then $\int_{a}^{+\infty} \widetilde{f}(x) \cdot \widetilde{g}(x) dx$ is convergent.

Proof. Due to boundedness of $\widetilde{g}(x)$ on $[a, +\infty)$, there exists $M > 0$, for any $x \in [a, +\infty)$, there is $|\widetilde{g}(x)| \leq M$. Additionally $\int_{a}^{+\infty} \widetilde{f}(x) dx$ is convergent, owing to cauchy by the convergence of the criteria, for any real number $\varepsilon > 0$, there exists real number $M' > a$, when $u_2 > u_1 > M'$, for any $x \in [a, +\infty)$, such that $|\int_{u_1}^{u_2} \widetilde{f}(x) \cdot dx| < \frac{\varepsilon}{2(M+1)}$. And for $x \in [a, +\infty)$, $\widetilde{g}(x)$ is monotonous. According to integral second median theorem, for any $x \in [a, +\infty)$, there exists real number $\eta \in [u_1, u_2]$, such that $\int_{u_1}^{u_2} \widetilde{g}(x) dx = g(u_1) \int_{u_1}^{\eta} \widetilde{f}(x) dx + g(u_2) \int_{\eta}^{u_2} \widetilde{f}(x) dx$. Then $\int_{u_1}^{u_2} \widetilde{f}(x) \cdot \widetilde{g}(x) dx \leq M \cdot \frac{\varepsilon}{2(M+1)} + M \cdot \frac{\varepsilon}{2(M+1)} = \frac{M}{M+1} \cdot \varepsilon < \varepsilon$. Hence $\int_{a}^{+\infty} \widetilde{f}(x) \cdot \widetilde{g}(x) dx$ is convergent.

References

1. L. I. An-Gui, H. W. Jin, Z. H. Zhang and W. B. Hua, A new definition of fuzzy limit. Journal of Liaoning Technical University, 2004, 23(6), 845–847.
2. Z. T. Gong and Y. W. Guo, Choquet integrals of interval-valued functions and fuzzy number-valued functions[J]. Journal of Lanzhou University, 2009, 45(4), 112–117.
3. C. Wu and H. Liu, On RSu integral of interval-valued functions and fuzzy-valued functions [J]. Fuzzy Sets Systems, 1993, 55(1), 93–106.
4. A. Bayoumi, Mean-value theorem for definite integrals of vector-valued functions of p-Banach spaces[J]. Algebras Groups Geometries, 2005, 22(3), 353–363.
5. S. J. Bi and X. D. Zhang, Convergence and continuity of fuzzy-valued functions. Journal of Heilongjiang Commercial College, 2002, 18(3), 330–333.
6. A. R. Didonato, Recurrence Relations for the Indefinite Integrals of the Associated Legendre Functions[J]. Mathematics of Computation, 1982, 38(158), 547–551.
7. S. Z. Guo, Brief introduction to fuzzy-valued function analysis based on the fuzzy structured element method(I). Mathematics in Practice and Theory, 2002, 21(5), 87–93.
8. S. Z. Guo, Brief introduction of fuzzy-valued function analytics base on fuzzy structured element method(II). Mathematics in Practice and Theory, 2008, 38(2), 73–79.
9. F. Yin and P. F. Wang, A New Definition of Fuzzy-Valued Function Limit(Continuity) and Its Derivability. Journal of North University of China, 2011, 32(6), 662–666.

10. S. Z. Guo, Principle of fuzzy mathematical analysis based on structural element theory. ShenYang: Northeastern University Press, 2004, 97–113.
11. L. Tran and L. Duckstein, Comparison of fuzzy numbers using a fuzzy distance measure. Fuzzy Sets and Systems, 2002, 130(3), 331–341.
12. R. Goetschel and W. Voxman, Elementary fuzzy calculus. Fuzzy Sets and Systems, 2005, 19(1), 82–86.

Robustness of the interval-valued fuzzy inference method based on logic metrics

Y.J. Wang, M.X. Luo* and H.R. Zhang

*Department of Information and Computing Science, China Jiliang University
Hangzhou 310018, PR China
minxialuo@163.com

In this paper, a new distance metric between interval-valued fuzzy sets is proposed. The four special interval-valued fuzzy metric spaces are studied, which are induced by four well-known interval-valued residual implication operations. It proved that the interval-valued fuzzy metric spaces induced by Łukasiewicz implication and Goguen implication are more suitable for interval-valued fuzzy reasoning. Moreover, based on these two interval-valued fuzzy metric spaces, we discuss the robustness of interval-valued fuzzy reasoning triple I methods.

Keywords: Interval-valued fuzzy sets; distance metric; logic metric spaces; triple I methods; robustness.

1. Introduction

In order to solve the most fundamental forms of fuzzy reasoning problem, i.e. fuzzy modus ponens (FMP) and fuzzy modus tollens (FMT), Zadeh [19,20] proposed the compositional rule of inference (CRI for short). However, CRI method don't have reducibility, and it has some arbitrariness and lacks solid logical basis, so a new method called the full implication triple I method, or simply called the triple I method, was proposed by Wang.[16]

It is well known that fuzzy connectives determine the internal structure of a fuzzy logic system. Nevertheless, the construction of distance metrics does not involve fuzzy connectives in previous works.[11,12,18] According to this point, Dai et al.[3] and Jin et al.[8] proposed the notion of logic similarity degree between fuzzy sets based on fuzzy connectives, and discuss robustness of fuzzy reasoning. Wang et al.[17] proposed a new distance metric based on residuated implication and conjunction connective, and discussed the robustness of full implication triple I inference method. Then Duan[5] studied the structures of four specific logic metrics.

However, when we solve some problems with uncertain, lots of fuzzy information are lost. So Zadeh[21] introduced interval-valued fuzzy sets (IVFSs for short), which can solve the problem of vagueness and uncertainty, and it can effectively reduce the loss of fuzzy information. Li et al.[10] extended CRI method to the case of IVFSs and discussed the robustness of interval-valued CRI method. Luo et al.[13,14] extended triple I method and reverse triple I method to the case of IVFSs and studied their robustness. Luo and Zhou[15] proposed interval-valued quintuple implication principle of fuzzy reasoning. These researches of robustness of interval-valued fuzzy reasoning methods are based on moore metric.

But moore metric has its drawback. For example, suppose $SI(X)$ is the class of all interval-valued fuzzy subsets of non-empty sets X. Let $X = \{x_1, x_2, \cdots, x_{10}\}$, A,B and $C \in SI(X)$, $A(x_1) = [1,1]$, $A(x_i) = [0,0](2 \leq i \leq 10)$, $B(x_i) = [0,0](1 \leq i \leq 10)$, and $C(x_i) = [1,1](1 \leq i \leq 10)$. If we use the moore metric, then we obtain $d(A,B) = \max\{\max_i\{|A_l(x_i) - B_l(x_i)|, |A_r(x_i) - B_r(x_i)|\}\} = 1$ and $d(A,C) = \max\{\max_i\{|A_l(x_i) - C_l(x_i)|, |A_r(x_i) - C_r(x_i)|\}\} = 1$. Obviously, this result is unreasonable. And because the behavior of a fuzzy logic system is mainly determined by its fuzzy connectives and fuzzy implication operators, based on these points, we propose a new distance metric of interval-valued fuzzy sets based on residual implication induced by left-continuous interval-valued t-representable t-norm.

2. Preliminaries

Let $SI = \{[x,y]|x \leq y, x, y \in [0,1]\}$. An order on SI as $[a,b] \leq [c,d]$ if $a \leq c$ and $b \leq d$ is called component-wise order or Kulisch-Miranker order.[4] Furthermore, $[a,b] \wedge [c,d] = [a \wedge c, b \wedge d]$, $[a,b] \vee [c,d] = [a \vee c, b \vee d]$. In this paper, let $X = \{x_1, x_2, \cdots, x_n\}$ be a non-empty set, $SI(X)$ denote interval-valued fuzzy subsets of X, for $1 \leq i \leq n$, $A(x_i) \in SI(X)$ ($A(x_i)$ denoted by $[A_l(x_i), A_r(x_i)]$). A^c is complement of interval-valued A, where $A^c = [1 - A_r, 1 - A_l]$.

Definition 2.1 (Klement, 2000, Fodor, 1994). *A function T:* $[0,1]^2 \to [0,1]$ *is called a triangular norm (t-norm) if it commutative, associative, non-decreasing in each argument and* $T(x,1) = x$ *for all* $a \in [0,1]$. *The associated residuated implication induced by left-continuous t-norm is defined by* $R(a,b) = \sup \{x \in [0,1] | T(a,x) \leq b\}$.

Definition 2.2 (Jenei, 1997). *Let* T *be a t-norm defined on the interval* $[0,1]$, *the associated interval-valued t-norm on SI is defined as follows:*
$\mathcal{T} : SI \times SI \to SI$, *where* $\mathcal{T}([a,b],[c,d]) = [T(a,c), T(b,d)]$.

The associated t-norm $\mathcal{T}$ on SI is called left-continuous, if T is left-continuous t-norm on the interval $[0, 1]$.

Note. The associated t-norm is also called interval-valued t-representable t-norm.[13]

Definition 2.3 (Alcalde, 2005).
For every $[a, b], [c, d] \in SI$, an interval-valued $\mathcal{R}$-implication is defined by:
$\mathcal{R}([a, b], [c, d]) = \bigvee \{[x, y] \in SI | \mathcal{T}([a, b], [x, y]) \leq [c, d]\}$.

Lemma 2.1 (Alcalde, 2005). *Given $\mathcal{T}$ the associated t-norm on SI, residuated implication $\mathcal{R}$ induced by the associated t-norm $\mathcal{T}$ has the form:*
$\mathcal{R}([a, b], [c, d]) = [R_T(a, c) \wedge R_T(b, d), R_T(b, d)]$.

Definition 2.4. *The interval-valued biresiduation is defined by*

$$\wp([a, b], [c, d]) = \mathcal{R}([a, b], [c, d]) \wedge \mathcal{R}([c, d], [a, b])$$

where $\mathcal{R}$ is residuated implication induced by left-continuous interval-valued t-representable t-norm on SI.

3. A New Distance Metric Between Interval-valued Fuzzy Sets

Definition 3.1. *Let $X = \{x_1, x_2, \cdots, x_n\}$, $A, B \in SI(X)$. Suppose $\mathcal{R}$ is interval-valued residuated implication. We denote*

$$d(A, B) = 1 - \bigwedge_{x_i \in X} (\wp(A(x_i), B(x_i)))_l$$

where $\wp(A(x_i), B(x_i)) = \mathcal{R}(A(x_i), B(x_i)) \wedge \mathcal{R}(B(x_i), A(x_i))$. Then d is called a distance metric on $SI(X)$ and $(SI(X), d)$ is called a logic metric space.

Remark 3.1. *If interval-valued fuzzy set A and B are reduced to fuzzy sets, then we have $d(A, B) = 1 - \bigwedge_{x_i \in X} (R(A(x_i), B(x_i)) \wedge R(B(x_i), A(x_i)))$, $(i = 1, 2, \cdots, n)$, which is the distance metric in Ref. 5.*

Lemma 3.1. *Let $\mathcal{T}$ be a left-continuous interval-valued t-norm on SI. Then the interval-valued biresiduation $\wp$ has the following property:*
$\mathcal{T}(\wp(A, B), \wp([B, C)) \leq \wp(A, C), A, B, C \in SI$.

Theorem 3.1. *d defined by Definition 3.1 is a distance metric on $SI(X)$.*

Proof. Obviously, $(D1)$, $(D2)$ are established.

$(D3)$: According to Lemma 3.1, we have
$\mathcal{T}(\wp(A(x_i), B(x_i)), \wp(B(x_i), C(x_i))) \leq \wp(A(x_i), C(x_i))$.
Thus, $\wp_l(A(x_i), B(x_i)) \leq R(\wp_l(B(x_i), C(x_i)), \wp_l(A(x_i), C(x_i)))$. Then we have $\wp_l(A(x_i), B(x_i)) + \wp_l(B(x_i), C(x_i)) \leq \wp_l(B(x_i), C(x_i)) + R(\wp_l(B(x_i), C(x_i)), \wp_l(A(x_i), C(x_i)))$.

Now we prove $\wp_l(B(x_i), C(x_i)) + R(\wp_l(B(x_i), C(x_i)), \wp_l(A(x_i), C(x_i))) \leq 1 + \wp_l(A(x_i), C(x_i))$, i.e. $a + R(a, b) \leq 1 + b$, for all $a, b \in [0, 1]$. In fact, if R is one of the four specific implications, we can easily prove that $a + R(a, b) \leq 1 + b$, for all $a, b \in [0, 1]$. So, we have $\wp_l(A(x_i), B(x_i)) + \wp_l(B(x_i), C(x_i)) \leq 1 + \wp_l(A(x_i), C(x_i))$.

Therefore, we can get $d(A(x), B(x)) + d(B(x), C(x)) \geq d(A(x), C(x))$. Therefore, d is a distance metric on $SI(X)$.

4. Analysis of Four Interval-valued Fuzzy Metric Spaces Structures

In a metric space (Y, d), let $y \in Y$. For arbitrary $0 < \varepsilon < 1$, if we could always find another point $y' \in Y$ such that $d(y, y') < \varepsilon$, then we say that y is a condensation point of Y. Otherwise, if exists $\delta > 0$ such that for every $y' \in Y$, $d(y, y') \geq \delta$, then we say that y is an isolated point. In this section, we analyze condensation points and isolated points in these four metric spaces $(SI(X), d_G)$, $(SI(X), d_0)$, $(SI(X), d_{Go})$ and $(SI(X), d_L)$ respectively. Let $X = \{x_1, \cdots, x_n\}$, $A \in SI(X)$, if there exists $x_k \in X$, such that $A(x_k) = [1, 1]$, then A is a normal interval-valued fuzzy set.

Theorem 4.1. *Let $X = \{x_1, \cdots, x_n\}$ and $A \in SI(X)$. Then A is condensation point of $(SI(X), d_G)$ if and only if A is normal interval-valued fuzzy set on X.*

Theorem 4.2. *Let $X = \{x_1, \cdots, x_n\}$ and $A \in SI(X)$. Then A is condensation point of $(SI(X), d_0)$ if and only if A or A^c is normal interval-valued fuzzy set on X.*

Theorem 4.3. *Let $X = \{x_1, \cdots, x_n\}$ and $A \in SI(X)$. $A(x_i) = [0, 0](1 \leq i \leq n)$ is the only isolated point of $(SI(X), d_{Go})$.*

Proof. Let $A(x_i) = [0, 0], i = 1, \cdots, n$. Then for arbitrary $B \in SI(X)$ satisfying $B \neq A$, for arbitrary $x_i \in X$, $B(x_i) \neq [0, 0]$, then we have $\bigwedge_{x_i \in X} (\mathcal{R}_{Go}(A(x_i), B(x_i)) \wedge \mathcal{R}_{Go}(B(x_i), A(x_i)))_l = 0$. Hence, $d_{Go}(A, B) =$

$1 - \bigwedge_{x_i \in X} (\mathcal{R}_{Go}(A(x_i), B(x_i)) \wedge \mathcal{R}_{Go}(B(x_i), A(x_i)))_l = 1 > \varepsilon$. Therefore, A is isolated point of $(SI(X), d_{Go})$.

If for arbitrary $x_i \in X$, then $A(x_i) \neq [0,0]$, i.e. $A_l(x_i) \neq 0$ and $A_r(x_i) \neq 0$, then exists $x_k \in X$, such that $A(x_k) = [c_1, c_2](c_1 \neq 0)$. Given $0 < \varepsilon < 1$ and given $m_1 > \frac{1}{\varepsilon}$ and $m_2 > \frac{1}{\varepsilon}$, let $b_1 = (1 - \frac{1}{m_1})c_1$ and $b_2 = (1 - \frac{1}{m_2})c_2$. Let $B \in SI(X)$, such that $B(x_k) = [b_1, b_2]$, $B(x_j) = A(x_j)$ $(j \neq k)$, thus $d_{Go}(A, B) = 1 - (1 - \frac{1}{m_1}) \wedge (1 - \frac{1}{m_2}) = \max\{\frac{1}{m_1}, \frac{1}{m_2}\} < \varepsilon$. Hence, A is condensation point of $(SI(X), d_{Go})$.

Theorem 4.4. *Let $X = \{x_1, \cdots, x_n\}$ and $A \in SI(X)$. Then there is no isolated point in the logic metric space $(SI(X), d_L)$.*

Proof. $d_L(A, B) = \bigvee_{x_i \in X} (|A_l(x_i) - B_l(x_i)| \vee |A_r(x_i) - B_r(x_i)|)$

For arbitrary $\varepsilon > 0$, we can find $B \in SI$ such that $|A_l(x_i) - B_l(x_i)| < \frac{\varepsilon}{2}$ and $|A_r(x_i) - B_r(x_i)| < \frac{\varepsilon}{2}$ $(i = 1, \cdots, n)$. Then $d_L(A, B) < \varepsilon$, i.e., A is condensation point of $(SI(X), d_L)$.

We can see that there is no isolated point in $(SI(X), d_L)$, and there is only one isolated point in $(SI(X), d_{Go})$. However, there are too many isolated points in $(SI(X), d_G)$ and $(SI(X), d_0)$. In order to discuss the robustness of interval-valued fuzzy reasoning better, we don't expect there are too many isolated points in interval-valued fuzzy metric spaces. Therefore, interval-valued fuzzy metric spaces $(SI(X), d_L)$ and $(SI(X), d_{Go})$ are more suitable for interval-valued fuzzy reasoning.

5. Robustness of Interval-valued Fuzzy Connectives and Interval-valued Triple I Methods

Proposition 5.1. *Suppose that $X = \{x_1, x_2, \cdots, x_n\}$, A, A', B and $B' \in SI(X)$. If $d(A, A') \leq \varepsilon_1$ and $d(B, B') \leq \varepsilon_2$, then $d(f(A, B), f(A', B')) \leq \varepsilon_1 + \varepsilon_2$, where $f \in \{\mathcal{T}_L, \mathcal{R}_L\}$.*

Proof. (1) Suppose that $f = \mathcal{T}_L$. Let $\mathcal{T}_L(A(x_i), B(x_i)) = t_i$ and $\mathcal{T}_L(A'(x_i), B'(x_i)) = t'_i$. Then $(\mathcal{R}(t_i, t'_i))_l = 1 - \max\{(A_l(x_i) - A'_l(x_i)) + (B_l(x_i) - B'_l(x_i)), (A_r(x_i) - A'_r(x_i)) + (B_r(x_i) - B'_r(x_i))\}$, thus $d(\mathcal{T}_L(A, B), \mathcal{T}_L(A', B')) \leq d(A, A') + d(B, B')$.

(2) Suppose that $f = \mathcal{R}_L$. Let $\mathcal{R}_L(A(x_i), B(x_i)) = t_i$ and $\mathcal{R}_L(A'(x_i), B'(x_i)) = t'_i$. Then $(\mathcal{R}(t_i, t'_i))_l \geq (\mathcal{R}(A(x_i), A'(x_i)))_l + (\mathcal{R}(B(x_i), B'(x_i)))_l - 1$. Thus $d(\mathcal{R}_L(A, B), \mathcal{R}_L(A', B')) \leq d(A, A') + d(B, B')$.

In conclusion, we can obtain: $d(f(A,B), f(A^{'}, B^{'})) \leq d(A, A^{'}) + d(B, B^{'}) \leq \varepsilon_1 + \varepsilon_2$.

Proposition 5.2. *Suppose that $X = \{x_1, x_2, \cdots, x_n\}$, $A, A^{'}, B$ and $B^{'} \in SI(X)$. If $d(A, A^{'}) \leq \varepsilon_1$ and $d(B, B^{'}) \leq \varepsilon_2$, then $d(f(A,B), f(A^{'}, B^{'})) \leq \varepsilon_1 + \varepsilon_2 - \varepsilon_1 \varepsilon_2$, where $f \in \{\mathcal{T}_{Go}, \mathcal{R}_{Go}\}$.*

Let $A, A^* \in SI(X)$ and $B, B^* \in SI(Y)$. Then the interval-valued $\mathcal{R}$-type triple I solution B^* of interval-valued fuzzy modus ponens ($IFMP$) and interval-valued fuzzy modus tollens ($IFMT$) can be respectively expressed as following two theorems.

Theorem 5.1 (Luo, Zhang, 2015). *Suppose that $\mathcal{R}$ is an interval-valued residuated implication. Then the interval-valued $\mathcal{R}$-type triple I solution B^* of $IFMP$ is given by the following formula:*

$$B^*(y) = \bigvee_{x \in X} \mathcal{T}(\mathcal{R}(A(x), B(y)), A^*(x)), \ y \in Y.$$

Theorem 5.2 (Luo, Zhang, 2015). *Suppose that $\mathcal{R}$ is an interval-valued residuated implication. Then the interval-valued $\mathcal{R}$-type triple I solution A^* of $IFMT$ is given by the following formula:*

$$A^*(x) = \bigwedge_{y \in Y} \mathcal{R}(\mathcal{R}(A(x), B(y)), B^*(y)), \ x \in X.$$

Theorem 5.3. *Let $A, A^{'}, A^*$ and $A^{'*}$ be interval-valued fuzzy sets on X, B and $B^{'}$ be interval-valued fuzzy sets on Y. B^* and $B^{'*}$ are the interval-valued $\mathcal{R}$-type triple I solutions of $IFMP(A, B, A^*)$ and $IFMP(A^{'}, B^{'}, A^{'*})$ given by Theorem 5.1 respectively. Suppose $d(A, A^{'}) \leq \varepsilon_1$, $d(B, B^{'}) \leq \varepsilon_2$ and $d(A^*, A^{'*}) \leq \varepsilon_3$, then*

$$d(B^*, B^{'*}) \leq \begin{cases} \varepsilon_1 + \varepsilon_2 + \varepsilon_3, & \text{if } \mathcal{R} = \mathcal{R}_L, \\ 1 - (1 - \varepsilon_1)(1 - \varepsilon_2)(1 - \varepsilon_3), & \text{if } \mathcal{R} = \mathcal{R}_{Go}. \end{cases}$$

Proof. We only proof that the conclusion is true when $\mathcal{R} = \mathcal{R}_{Go}$.

Let $\mathcal{R}(A(x_i), B(y_j)) = r_{ij}$, $\mathcal{R}^{'}(A(x_i), B(y_j)) = r^{'}_{ij}$, $1 \leq i \leq n$, $1 \leq j \leq m$. For interval-valued Goguen implication, from Proposition 5.2, we have $d(\mathcal{R}_{Go}(A,B), \mathcal{R}_{Go}(A^{'}, B^{'})) \leq \varepsilon_1 + \varepsilon_2 - \varepsilon_1 \varepsilon_2$. Then $(\mathcal{R}(B^*(y_j), B^{'*}(y_j)))_l \geq \min_{1 \leq i \leq n}(\mathcal{R}(r_{ij}, r^{'}_{ij}))_l \cdot (\mathcal{R}(A^*(x_i), A^{'*}(x_i)))_l$.

Thus we have $d(B^*, B^{'*}) \leq 1 - \min_{1 \leq i \leq n}(\mathcal{R}(r_{ij}, r^{'}_{ij}))_l \cdot (\mathcal{R}(A^*(x_i), A^{'*}(x_i)))_l$ $\wedge \min_{1 \leq i \leq n}(\mathcal{R}(r^{'}_{ij}, r_{ij}))_l \cdot (\mathcal{R}(A^{'*}(x_i), A^*(x_i)))_l \leq 1 - (1 - \varepsilon_1)(1 - \varepsilon_2)(1 - \varepsilon_3)$.

Therefore, if $\mathcal{R} = \mathcal{R}_{Go}$, we have $d(B^*, B^{'*}) \leq 1 - (1 - \varepsilon_1)(1 - \varepsilon_2)(1 - \varepsilon_3)$.

Theorem 5.4. Let A and A' be interval-valued fuzzy sets on X, B, B', B^* and B'^* be interval-valued fuzzy sets on Y. A^* and A'^* are the interval-valued $\mathcal{R}$-type triple I solutions of $IFMT(A, B, B^*)$ and $IFMT(A', B', B'^*)$ given by Theorem 5.2 respectively. Suppose $d(A, A') \leq \varepsilon_1$, $d(B, B') \leq \varepsilon_2$ and $d(B^*, B'^*) \leq \varepsilon_3$, then

$$d(A^*, A'^*) \leq \begin{cases} \varepsilon_1 + \varepsilon_2 + \varepsilon_3, & \text{if } \mathcal{R} = \mathcal{R}_L, \\ 1 - (1 - \varepsilon_1)(1 - \varepsilon_2)(1 - \varepsilon_3), & \text{if } \mathcal{R} = \mathcal{R}_{Go}. \end{cases}$$

Example 5.1. Suppose that $X = \{x_1, x_2, x_3, x_4, x_5\}$, $Y = \{y_1, y_2, y_3, y_4, y_5\}$, $A, A', A^*, A'^* \in SI(X)$, $B, B' \in SI(Y)$. And the corresponding membership of interval-valued fuzzy sets are given in the following table:

Item	x_1	x_2	x_3	x_4	x_5
A	[0.2,0.2]	[0.4,0.4]	[0.8,0.8]	[0.1,0.6]	[0.3,0.8]
A'	[0.2,0.2]	[0.4,0.4]	[0.0,0.0]	[0.1,0.6]	[0.3,0.8]
A^*	[0.0,0.0]	[0.2,0.3]	[0.1,0.7]	[0.0,0.0]	[0.0,0.0]
A'^*	[0.1,0.1]	[0.2,0.3]	[0.1,0.6]	[0.0,0.0]	[0.0,0.0]
Item	y_1	y_2	y_3	y_4	y_5
B	[0.4,0.6]	[0.4,0.8]	[0.8,0.8]	[0.0,0.0]	[0.2,0.2]
B'	[0.4,0.6]	[0.4,0.7]	[0.0,0.8]	[0.9,0.9]	[0.4,0.4]

Let B^* and B'^* are interval-valued triple I solutions based on Łukasiewicz implication for $IFMP(A, B, A^*)$ and $IFMP(A', B', A'^*)$, then we can obtain:
$B^*(y_1) = [0.2, 0.5], B^*(y_2) = [0.2, 0.7], B^*(y_3) = [0.2, 0.7], B^*(y_4) = [0.0, 0.0], B^*(y_5) = [0.0, 0.1], B'^*(y_1) = [0.2, 0.6], B'^*(y_2) = [0.2, 0.6], B'^*(y_3) = [0.1, 0.6], B'^*(y_4) = [0.2, 0.6], B'^*(y_5) = [0.2, 0.6]$.

By using the distance metric d_L, we have: $d_L(A, A') = 0.8$, $d_L(A^*, A'^*) = 0.1$, $d_L(B, B') = 0.9$, $d_L(B^*, B'^*) = 0.6 < 0.8 + 0.1 + 0.9$. This result satisfies Theorem 5.3.

6. Conclusions and Future Scope

In this paper, a new distance metric between interval-valued fuzzy sets based on the residuated implication induced by the left-continuous interval-valued t-representable t-norm is proposed and four interval-valued fuzzy metric spaces structures are studied. We found that interval-valued fuzzy metric spaces $(SI(X), d_{Go})$ and $(SI(X), d_L)$ fit for researching the issues of interval-valued fuzzy reasoning. Furthermore, we prove the robustness about interval-valued fuzzy reasoning algorithms based on Goguen implication and Łukasiewicz implication.

The distance metrics between the interval-valued fuzzy sets is studied, we can study the distance metrics between the neutrosophic sets[2] in the future, which can solve the problems of uncertainty information better.

Acknowledgments

This work is supported by the National Natural Science Foundation of China (Grant Nos. 61773019 and 11701540).

References

1. C. Alcalde, A. Burusco, R. Fuentes-Gonzalez, *Fuzzy Sets and Systems*, **153**, 211(2005).
2. S. Broumi, F. Smarandache, *Neutrosophic Sets and Systems*, **2**, 9(2014).
3. S.S. Dai, D.W. Pei, D.H. Guo, *International Journal of Approximate Reasoning*, **54**, 653(2013).
4. B.A. Davey, H.A. Priestley, *Cambridge University Press*, Cambridge, 1990.
5. J.Y. Duan, Y.M. Li, *International Journal of Approximate Reasoning*, **61**, 33(2015).
6. J.C. Fodor, M. Roubens, *Fuzzy Preference Modelling and Multicriteria Decision Support*, **14**, (1994).
7. S. Jenei, *Fuzzy Sets and Systems*, **90**, 25(1997).
8. J.H. Jin, Y.M. Li, C.Q. Li, *Information Sciences*, **177**, 5103(2007).
9. E. Klement, R. Mesiar, E. Pap, *Triangular Norms*, Kluwer Academic Publishers, Dordrecht, 2000.
10. D.Ch. Li, Y.M. Li, Y.J. Xie, *Information Sciences*, **181**, 4754(2011).
11. Y.M. Li, *Analysis of Fuzzy System*, Science Press, Beijing, 2005.
12. Y.M. Li, D.Ch. Li, P. Witold, J.J. Wu, *International Journal of Intelligent Systems*, **20**, 393(2005).
13. M.X. Luo, K. Zhang, *International Journal of Approximate Reasoning*, **62**, 61(2015).
14. M.X. Luo, X.L. Zhou, *International Journal of Approximate Reasoning*, **66**, 16(2015).
15. M.X. Luo, X.L. Zhou, *International Journal of Approximate Reasoning*, **84**, 23(2017).
16. G.J. Wang, *Sciences in China (Series E)*, **29**, 43(1999).
17. G.J. Wang, J.Y. Duan, *International Journal of Approximate Reasoning*, **55**, 787(2014).
18. M.S. Ying, *Perturbation of fuzzy reasoning*, IEEE Transactions on Fuzzy Sets and Systems, **7**, 625(1999).
19. L.A. Zadeh, *Information Sciences*, **9**, 43(1975).
20. L.A. Zadeh, *IEEE Transactions on Systems Man and Cybernetics*, **3**, 28(1973).
21. L.A. Zadeh, *Theory of approximate reasoning*, in: J.E. Hayes, D. Michie, L.I. Mikulich (Eds.), Machine Intelligence, Ellis Horwood, Chichester, 149(1970).

Novel intuitionistic 2-tuple linguistic representation model

Yi Liu

Data Recovery Key Laboratory of Sichuan Province, Neijiang Normal University
Neijiang, Sichuan 641112, P. R. China
liuyiyl@163.com

Jun Liu

School of Computing, Ulster University
Jordanstown Campus, Northern Ireland BT37 0QB, UK
j.liu@ulster.ac.uk

Ya Qin

School of Mathematics and Information Sciences, Neijiang Normal University
Neijiang, Sichuan 641112, P. R. China
qinyaqy@126.com

Intuitionistic fuzzy sets (IFS) were proposed to manage situations in which experts have some membership and nonmembership degree to assess an alternative, and linguistic values was used to assess an alternative and variable in qualitative settings. In this paper, on the basis of Beg's work,[15] we introduce a new intuitionistic 2-tuple linguistic model which can provide a linguistic and computational basis to manage the situations where experts assess an alternative in possible and impossible linguistic variable and their translation parameter.

Keywords: Intuitionistic fuzzy set; 2-tuple; linguistic term set; representation model.

1. Introduction

As an important extension of fuzzy set, intuitionistic fuzzy set (IFS)[1] is characterized by three parameters, namely, a membership degree, a nonmembership degree and an indeterminacy degree are adopted at the same time. Therefore, IFS is considered to be more appropriate to represent and deal with imprecise, uncertain and vague information in some decision making problems. In last few years, some fuzzy multi-attribute decision making methods based on IFS have been proposed. There are many complicated or ill-defined problems are not to be amenable for expressions in conventional quantitative ways in the real world, so it is not always adequate to represent such problems by only numerical based modelling. Therefore, the decision makers (DMs) utilize linguistic descriptors to express their assessments on the uncertain knowledge when they encounter such problems. Many studies on using the linguistic variables to model the problems have been carried out and have applied successfully in different fields. In multi-attribute decision making (MADM) problems, the linguistic decision information needs to be aggregated by some proper methods in order to rank the given decision alternatives and then to get the best one. On basis of the concept of symbolic translation, Herrera et al.[2,3] proposed 2-tuple linguistic representation model which was characterized by a linguistic term and a numeric value. It has exact characteristic in linguistic information processing and can effectively avoid information distortion and losing which occur formerly in the linguistic information processing. The 2-tuple linguistic model has received more and more attention since its appearance. Some extensions of 2-tuple linguistic model have been developed, e.g. hesitant 2-tuple linguistic information model,[4-11,14] intuitionistic 2-tuple linguistic information model.[15] Whilst, a variety of decision making methods based on 2-tuple linguistic model are also developed.

On basis of 2-tuple linguistic model[2,3] and intuitionistic 2-tuple linguistic information model,[15] we propose a novel intuitionistic 2-tuple linguistic model and it's representation model in this paper in order to provide a linguistic and computational basis to manage the situations in which experts assess an alternative in possible and impossible linguistic variable and their translation parameter.

2. Preliminaries

In this section, firstly some basic concepts related to intuitionistic fuzzy set and 2-tuple linguistic model are reviewed, which are the basis of the present work.

2.1. *Intuitionistic fuzzy set*

Definition 1.[1] Let $X = \{x_1, x_2, \cdots, x_n\}$ be a finite universe of discourse, an intuitionistic fuzzy set (IFS) A in X characterized by a membership function $\mu_A : X \to [0, 1]$ and a non-membership function $\nu_A : X \to [0, 1]$, which satisfies the condition $0 \leq \mu_A(x) + \nu_A(x) \leq 1$. An IFS A can be expressed as

$$A = \{\langle x, (\mu_A(x), \nu_A(x))\rangle | x \in X\}.$$

$\pi_A(x) = 1 - \mu_A(x) - \nu_A(x)$ is called the degree of indeterminacy, $\pi_A(x)$ represents the degree of hesitance of x to A and is also called intuitionistic index. For convenience, called $(\mu_A(x), \nu_A(x))$ is an intuitionistic fuzzy number (IFN) and denoted by (μ_A, ν_A).

2.2. 2-tuple linguistic representation model

Let $S = \{s_i | i = 0, 1, \cdots, g\}$ be a linguistic term set with odd cardinality, for any label s_i, which represents a possible values for a linguistic variable and satisfy the following characteristics:[2]

(1) $s_i > s_j$ if and only if $i > j$;
(2) if $s_i \geq s_j$, then $max(s_i, s_j) = s_i$;
(3) if $s_i \geq s_j$, then $min(s_i, s_j) = s_j$;
(4) $Neg(s_i) = s_j$, such that $j = g - i$.

To compute with words without loss of information, the 2-tuple linguistic model based on the concept of symbolic translation was proposed in.[2,3] The model uses a 2-tuple (s_k, α) to represent linguistic information, where $s_k \in S$, α denotes the value of symbolic translation and $\alpha \in [-0.5, 0.5)$. The specific definition of 2-tuple linguistic model is given as follows.

Let $S = \{s_0, s_1, \cdots, s_g\}$ be a linguistic term set and $\beta \in [0, g]$ be a value representing the result of a symbolic aggregation operation, then the 2-tuple that expresses the equivalent information to β is obtained with the following function:[2]

$$\Delta : [0, g] \to S \times [-0.5, 0.5)$$
$$\Delta(\beta) = (s_i, \alpha), with \begin{cases} s_i, & i = round(\beta) \\ \alpha = \beta - i, & \alpha \in [-0.5, 0.5) \end{cases}$$

where $round(\beta)$ is the usual round operation, s_i has the closest index label to β and α is the value of symbolic translation.

Let $S = \{s_0, s_1, \cdots, s_g\}$ be a linguistic term set and (s_i, α) be a 2-tuple, there is a function Δ^{-1}, which can transform a 2-tuple into its equivalent numerical value $\beta \in [0, g]$. The transformation function[2] can be defined as

$$\Delta^{-1} : S \times [-0.5, 0.5) \to [0, g]$$
$$\Delta^{-1}(s_i, \alpha) = i + \alpha = \beta.$$

It easily follows from above definitions that a linguistic term can be considered as a linguistic 2-tuple by adding a value 0 to it as symbolic translation, i.e. $\Delta(s_i) = (s_i, 0)$.

Herrera et al.[3] proposed 2-tuple linguistic model, in which the linguistic term s_i in a 2-tuple (s_i, α) has the closest index label to the symbolic aggregation value β, and symbolic translation $\alpha \in [-0.5, 0.5)$ represents the deviation value of i and β. So the aggregation result represented by a 2-tuple has a clear implication. However, Wei[12] pointed out that the meaning of symbolic translation α in a 2-tuple (s_i, α) defined by Chen[13] is not clear. Thus, in order to deal with the linguistic information and describe the aggregation result, Wei[12] modified the translation functions as follows:

Definition 2.[12] Let $S = \{s_0, s_1, \cdots, s_g\}$ be a linguistic term set and $\beta \in [0, g]$ be a value representing the result of a symbolic aggregation operation, then the 2-tuple that expresses the equivalent information to β is obtained with the following function:

$$\Delta : [0, 1] \to S \times [-0.5, 0.5)$$
$$\Delta(\beta) = (s_i, \alpha), \text{with} \begin{cases} s_i, & i = round(\beta g), \\ \alpha = \beta g - i, & \alpha \in [-0.5, 0.5) \end{cases}$$

where $round(\beta g)$ is the usual round operation, s_i has the closest index label to β and α is the value of symbolic translation.

Definition 3. Let $S = \{s_0, s_1, \cdots, s_g\}$ be a linguistic term set and (s_i, α) be a 2-tuple, there is a function Δ^{-1}, which can transform a 2-tuple into its equivalent numerical value $\beta \in [0, 1]$. The transformation function can be defined as

$$\Delta^{-1} : S \times [-0.5, 0.5) \to [0, 1]$$
$$\Delta^{-1}(s_i, \alpha) = (i + \alpha)/g = \beta.$$

3. Novel Intuitionistic 2-Tuple Linguistic Representation Model

Motivated by Def. 2 and Def. 3, we put forward the a new intuitionistic 2-tuple linguistic representation model based on Beg's[15] work.

Definition 4.[15] Let X be a universe of discourse, and $S = \{s_0, s_1, \cdots, s_g\}$ be a linguistic term set, an intuitionistic linguistic term set (ILTS) on X is an expression $A = \{(x, \mu(x), \nu(x)) | x \in X\}$, where $\mu(x), \nu(x)$ are linguistic terms in S such that $\mu(x) = s_i$ and $\nu(x) = s_j$ with the condition $i + j \in [0, g]$ for any $x \in X$.

Definition 5. Let (s_i, s_j) be a member of ILTS in X, $((s_i, \alpha), (s_j, \beta))$ is an intuitionisitic 2-tuple linguistic model if it satisfies:
  (1) s_i, s_j are linguistic terms;
  (2) $\alpha, \beta \in [-0.5, 0.5)$ are numeric values representing the symbolic translation and satisfy the following conditions:
    (a) when $0 \leq i + j < g$, $\alpha + \beta \in [-1, 1)$;
    (b) when $i + j = g$, $\alpha + \beta \in [-1, 0]$.

Definition 6. Let $S = \{s_0, s_1, \cdots, s_g\}$ be a linguistic term set and $t = g + 1$ be the cardinality of S. An intuitionistic 2-tuple is composed of two linguistic terms and two crisp numbers, denoted by $(s_i, \alpha_1), (s_j, \alpha_2)$. s_i, s_j represent the linguistic label of the linguistic term set S and α_1, α_2 represent the symbol translation. The intuitionistic 2-tuple that express the equivalent information to an intuitionistic fuzzy value $\beta_1, \beta_2 \in [0, 1]$ is derived by the following function

$$\Delta : [0, 1] \times [0, 1] \to (S \times [-0.5, 0.5)) \times (S \times [-0.5, 0.5))$$

such that

$$\Delta(\beta_1, \beta_2) = ((s_i, \alpha_1), (s_j, \alpha_2)), \text{with} \begin{cases} s_i, & i = round(\beta_1 g), \\ s_j, & j = round(\beta_2 g), \\ \alpha_1 = \beta_1 g - i, & \alpha_1 \in [-0.5, 0.5), \\ \alpha_2 = \beta_2 g - j, & \alpha_2 \in [-0.5, 0.5). \end{cases} \quad (1)$$

Conversely, there exist a function Δ^{-1} such that intuitionistic 2-tuple can be translated into an intuitionistic fuzzy value $(\beta_1, \beta_2), \beta_1, \beta_2 \in [0, 1]$ as follows:

$$\Delta^{-1} : (S \times [-0.5, 0.5)) \times (S \times [-0.5, 0.5)) \to [0, 1] \times [0, 1]$$

such that

$$\Delta^{-1}((s_i, \alpha_1), (s_j, \alpha_2)) = ((\alpha_1 + i)/g, (\alpha_2 + j)/g) = (\beta_1, \beta_2). \quad (2)$$

We can see from Eq. (1) and Eq. (2) that the translation functions Δ and Δ^{-1} can help us to aggregate the multigranularity intuitionistic linguistic information. Since $i + j \in [0, g]$ and $\alpha_1, \alpha_2 \in [-0.5, 0.5)$, therefore it follows from Def. 2 that $\beta_1 + \beta_2 \in [0, 1]$, that is, for an intuitionistic linguistic 2-tuple, we can obtain an intuitionistic fuzzy value by the translation function Δ^{-1}.

Example 1. Let $S = \{s_0, s_1, \cdots, s_6\}$ be a linguistic term set and $((s_4, 0.2), (s_2, -0.2))$ be an intuitionistic 2-tuple on linguistic term set on

S. According Eq. (1), we have
$$\Delta^{-1}((s_4, 0.2), (s_2, -0.2)) = \left(\frac{4+0.2}{6}, \frac{2-0.2}{6}\right) = (0.7, 0.3).$$
Conversely, assume an intuitionistic value $(0.6, 0.4)$, according to Eq. (1),
$$\Delta(0.6, 0.4) = ((s_4, -0.4), (s_2, 0.4)).$$
From above definitions of Δ, Δ^{-1}, we have the
$$\Delta(\Delta^{-1}((s_i, \alpha_1), (s_j, \alpha_2))) = (s_i, \alpha_1), (s_j, \alpha_2)$$
for any intuitionistic 2-tuple linguistic $((s_i, \alpha_1), (s_j, \alpha_2))$ on linguistic term set S.

4. Score Function and Accuracy Function

Let $S = \{s_0, s_1, \cdots, s_\tau\}$ be a linguistic term set with granularity $g = \tau + 1$. For an intuitionistic 2-tuple $A = [(s_i, \alpha_1), (s_j, \alpha_2)]$ on the linguistic term set S, the score function of A is defined as follows:
$$S(A) = \frac{1}{2}(\Delta^{-1}(s_i, \alpha_1) + \Delta^{-1}(s_j, \alpha_2)). \tag{3}$$
The accuracy function of A is defined as follows:
$$H(A) = \Delta^{-1}(s_j, \alpha_2) - \Delta^{-1}(s_i, \alpha_1)). \tag{4}$$
It is obvious that $S(A) \in [0, 1]$ and $H(A) \in [0, 1]$. Now, the compare rule of two intuitionistic 2-tuple is listed as follows:

Let S be a linguistic term set with granularity g. And A, B are two intuitionistic 2-tuples on S.
If $S(A) > S(B)$, then $A > B$.
If $S(A) = S(B)$, then:
(1) $H(A) > H(B)$, then $A > B$;
(2) $H(A) < H(B)$, then $A < B$;
(3) $H(A) = H(B)$, then $A = B$.

Example 2. let $A = ((s_2, 0.1), (s_4, -0.2))$ and $B = ((s_3, 0.2), (s_2, -0.1))$ be two intuitionistic 2-tuples on linguistic term set $S = \{s_0, s_1, \cdots, s_6\}$. Since
$$S(A) = \frac{1}{2}(\Delta^{-1}(s_2, 0.1) + \Delta^{-1}(s_4, -0.2)) = 0.4915;$$
$$S(B) = \frac{1}{2}(\Delta^{-1}(s_3, 0.2) + \Delta^{-1}(s_2, -0.1)) = 0.425,$$
we have $A > B$.

Conclusions

On basis of 2-tuple linguistic model[2,3] and intuitionistic 2-tuple linguistic information model,[15] we investigated novel intuitionistic 2-tuple linguistic model along with it's representation model in this paper in order to provide a linguistic and computational basis to manage the situations in which experts assess an alternative in possible and impossible linguistic variable and their translation parameter. In our future work, we will work for some information measures, aggregation operators and some decision making method based on intuitionistic 2-tuple linguistic information model.

Acknowledgments

This work is supported by National Natural Science Foundation of P.R. China (Grant no. 61673320); The Application Basic Research Plan Project of Sichuan Province (No. 2015JY0120); The Scientific Research Project of Department of Education of Sichuan Province (15TD0027, 15ZB0270); Natural Science Foundation of Guangdong Province (2016A030310003).

References

1. K. T. Atanassov, Intuitionistic fuzzy sets, Fuzzy Sets Syst. 20 (1) (1986) 87–96.
2. F. Herrera, L. Martnez, A 2-Tuple Fuzzy Linguistic Representation Model for Computing with Words, Ieee Transactions on Fuzzy Systems, 8 (2000) 746–752.
3. F. Herrera, L. Martnez, A Model Based on Linguistic 2-Tuples for Dealing with Multigranular Hierarchical Linguistic Contexts in Multi-Expert Decision-Making, Ieee transactions on systems, man, and cybernetics part b: cybernetics, 31 (2001) 227–234.
4. I. Beg, T. Rashid, Hesitant 2-tuple linguistic information in multiple attributes group decision making, Journal of Intelligent and Fuzzy Systems, 30 (2015) 109–116.
5. Y. Dong, C. C. Li, F. Herrera, Connecting the linguistic hierarchy and the numerical scale for the 2-tuple linguistic model and its use to deal with hesitant unbalanced linguistic information, Information Sciences, 367–368 (2016) 259–278.
6. W. Li, X. Q. Zhou, G. Q. Guo, Hesitant fuzzy Maclaurin symmetric mean operators and their application in multiple attribute decision making, J. Comput. Anal. Appl. 20 (2016): 459–469.
7. J. D. Qin, X. W. Liu, W. Pedrycz, Hesitant fuzzy Maclaurin symmetric mean operators and its application to multiple attribute decision making, Int. J. Fuzzy Syst. 17 (2015) 509–520.
8. R.M. Rodriguez, L. Martnez, F. Herrera, A Linguistic 2-Tuple Multicriteria Decision Making Model dealing with Hesitant Linguistic Information, 2015 IEEE International Conference on Fuzzy Systems (FUZZ-IEEE), (2015).
9. C. Tan, Y. Jia, X. Chen, 2-Tuple Linguistic Hesitant Fuzzy Aggregation Operators and Its Application to Multi-Attribute Decision Making, Informatica-Lithuan, 28 (2017) 329–358.

10. I. Truck, M.-A. Abchir, Toward a Classification of Hesitant Operators in the 2-Tuple Linguistic Model, International Journal of Intelligent Systems, 29 (2014) 560–578.
11. J. Wang, J.Q. Wang, H.Y. Zhang, X.H. Chen, Multi-criteria Group Decision-Making Approach Based on 2-Tuple Linguistic Aggregation Operators with Multi-hesitant Fuzzy Linguistic Information, Int J Fuzzy Syst, 18 (2016) 81–97.
12. C. Wei, H. Liao, A Multigranularity Linguistic Group Decision-Making Method Based on Hesitant 2-Tuple Sets, International Journal of Intelligent Systems, 31 (2016) 612–634.
13. C.T. Chen, W. S. Tai, Measuring the intellectual capital performance based on 2-tuple fuzzy linguistic information, in: Proceedings of the 10th Annual Meeting of Asia Pacific Region of Decision Sciences Institute, Taiwan, 2005.
14. Y. Xu, H. Wang, A group consensus decision support model for hesitant 2-tuple fuzzy linguistic preference relations with additive consistency, Journal of Intelligent and Fuzzy Systems, 33 (2017) 41–54.
15. I. Beg, T. Rashid, An Intuitionistic 2-Tuple Linguistic Information Model and Aggregation Operators, International Journal of Intelligent Systems, 31 (2016) 569–592.

Generalized bidirectional quantum operation teleportation via multiqubit cluster states

Si Qi Zhou,* Ming Qiang Bai[†] and Zhi Wen Mo

School of Mathematics, Sichuan Normal University
Chengdu, 610066, China
Institute of Intelligent Information & Quantum Information,
Sichuan Normal University, Chengdu, 610066, China
*siqi_zhou@163.com
[†]baimq@sicnu.edu.cn

A protocol of bidirectional quantum operation teleportation (BQOT) is proposed. The basic idea of BQOT is essentially a combination of the ideas of quantum operation teleportation (QOT) and bidirectional quantum teleportation (QT). We put forward a protocol utilizing a product state of two multiqubit cluster states as a quantum channel. Moreover, some concrete discussions are made to study its important features from five aspects, i.e., quantum resource consumption, operation complexity, classical resource consumption, success probability and efficiency.

Keywords: Bidirectional teleportation; quantum operation teleportation; cluster state; multiqubit.

1. Introduction

Quantum operation teleportation (QOT) was initially proposed by Huelga et al. [1] in 2001. They examine the issue of teleportation, not of an unknown quantum state, but rather of an unknown quantum operation on a qubit. Obviously, QOT as a useful form of quantum remote control has been attracting much attention [2-9]. For instance, Wang [2] extended the original QOT protocol which deal with single-qubit operations to those involving multiple qubits. Zhao and Wang [3] present a protocol for local and remote implementation of nonlocal operations with block diagonal forms. Although much attention was focused on QOT in both theoretical and experimental aspects, bidirectional quantum operation teleportation (BQOT) is not seriously considered up to now.

According to this bidirectional idea, we present a protocol for bidirectional quantum operation teleportation (BQOT) with two multiqubit cluster states by integrating the ideas of quantum operation teleportation

(QOT) and bidirectional quantum teleportation (QT). The basic idea of our protocol is as follows. Based on the quantum channel shared in advance, Alice wants to execute a unitary operation U_A on an arbitrary qubit state which is in the possession of Bob; and at the same time Bob can perform a unitary operation U_B on the target qubit in Alice's site. We put forward a generalized protocol state of $2m$- and $2n$-qubit cluster states as a quantum channel, where n, m are integer and $n, m \geq 2$.

Cluster state was first introduced by Briegel and Raussendorf [10]. Owing to their distinct advantages, e.g., their robustness against decoherence [11], they have been applied for many quantum information processing. Using four- and five-partite cluster states, Muralidharan and Panigrahi [12] provided various schemes for splitting arbitrary single-qubit or two-qubit quantum information into two parts. Paul et al. [13] showed that six-photon cluster states can be used as shared entanglements for splitting arbitrary two-qubit states between two parties. Moreover, Muralidharan et al. [14] further provided a number of quantum state sharing schemes for splitting any two-qubit state among k parties using a N-qubit linear cluster state as a quantum channel.

The rest of this paper is organized as follows. Section 2 describes a BQOT with two multiqubit cluster states. Then, in Section 3, we discuss our scheme. Finally, we make a concise summary in Section 4.

2. Generalized BQOT Protocol via Multiqubit Cluster States

In this section, let us amply depict our generalized BQOT scheme by using multiqubit cluster states.

In our scheme, there are two legitimate participants, say, Alice and Bob. Alice wants to perform a unitary operation U_A on the target qubit state which is in the possession of Bob; at the same time, Bob performs a unitary operation U_B on the other target qubit in Alice's location. Actually, Alice and Bob may not hear of the concerned operations U_A and U_B before. Without loss of generality, at the beginning we suppose the target state $|\varphi\rangle_A$ ($|\varphi\rangle_B$) in Alice's (Bob's) location reads as follows

$$|\varphi\rangle_A = (\alpha_0|0\rangle + \alpha_1|1\rangle)_A, \quad |\varphi\rangle_B = (\beta_0|0\rangle + \beta_1|1\rangle)_B, \tag{1}$$

where α_0, α_1, β_0 and β_1 are complex and satisfy $|\alpha_0|^2 + |\alpha_1|^2 = 1$, $|\beta_0|^2 + |\beta_1|^2 = 1$.

To start with, the form of the N-qubit cluster state $|C_N\rangle$ was proposed by Briegel and Raussendorf [10] is $|C_N\rangle = \frac{1}{2^{N/2}} \bigotimes_{k=1}^{N}(|0\rangle_k \sigma_{k+1}^{(1,1)} + |1\rangle_k)$, with $\sigma_{k+1}^{(1,1)} = 1$. And in this paper, the quantum channel linking the two legitimate users is a product state of two multiqubit cluster states, i.e.,

$$|\Omega\rangle_{s_1,\cdots,s_m,s_{m+1},\cdots,s_{2m},t_1,\cdots,t_n,t_{n+1},\cdots,t_{2n}}$$
$$= |C_{2m}\rangle_{s_1,\cdots,s_m,s_{m+1},\cdots,s_{2m}} \otimes |C_{2n}\rangle_{t_1,\cdots,t_n,t_{n+1},\cdots,t_{2n}}, \qquad (2)$$

where

$$|C_{2m}\rangle_{s_1,\cdots,s_m,s_{m+1},\cdots,s_{2m}} = \frac{1}{2}(|\underbrace{0\cdots0}_{m}\underbrace{0\cdots0}_{m}\rangle + |\underbrace{0\cdots0}_{m}\underbrace{1\cdots1}_{m}\rangle$$
$$+ |\underbrace{1\cdots1}_{m}\underbrace{0\cdots0}_{m}\rangle - |\underbrace{1\cdots1}_{m}\underbrace{1\cdots1}_{m}\rangle), \qquad (3)$$

$$|C_{2n}\rangle_{t_1,\cdots t_n, t_{n+1}\cdots t_{2n}} = \frac{1}{2}(|\underbrace{0\cdots0}_{n}\underbrace{0\cdots0}_{n}\rangle + |\underbrace{0\cdots0}_{n}\underbrace{1\cdots1}_{n}\rangle$$
$$+ |\underbrace{1\cdots1}_{n}\underbrace{0\cdots0}_{n}\rangle - |\underbrace{1\cdots1}_{n}\underbrace{1\cdots1}_{n}\rangle), \qquad (4)$$

and n, m are integer and $n, m \geq 2$.

We let Alice possess qubit pair $(s_1, t_2, \cdots, t_{2n})$, the qubit pair $(t_1, s_2, \cdots, s_{2m})$ are possessed to Bob. As a result, the quantum channel can be reexpressed as

$$|\Omega'\rangle_{a_1,a_2,\cdots,a_{2n},b_1,b_2,\cdots,b_{2m}} = |\Omega\rangle_{s_1,t_2,\cdots,t_{2n},t_1,s_2,\cdots,s_{2m}}. \qquad (5)$$

Here, the state of the whole quantum system can be expressed as

$$|\Psi\rangle = |\varphi\rangle_A \otimes |\varphi\rangle_B \otimes |\Omega'\rangle_{a_1,a_2,\cdots,a_{2n},b_1,b_2,\cdots,b_{2m}}. \qquad (6)$$

Our scheme is designed to own six steps, and the details are as follows.

Step 1. Alice and Bob execute two-qubit unitary operations $P_{a_n,a_{n+1}}^C$ and $P_{b_m,b_{m+1}}^C$ on qubit pairs (a_n, a_{n+1}) and (b_m, b_{m+1}), respectively.

Here, the unitary operation $P^C = |0\rangle\langle 0| \otimes \sigma^{(0,0)} + |1\rangle\langle 1| \otimes \sigma^{(1,1)}$.

Step 2. Alice and Bob carry out Bell-state measurements on qubit pairs (A, a_{2n}) and (B, b_{2m}), respectively.

In this paper, the Bell states are defined as: $|\mathcal{B}_{0,0}\rangle = \frac{1}{\sqrt{2}}(|00\rangle + |11\rangle)$, $|\mathcal{B}_{0,1}\rangle = \frac{1}{\sqrt{2}}(|01\rangle + |10\rangle)$, $|\mathcal{B}_{1,0}\rangle = \frac{1}{\sqrt{2}}(|01\rangle - |10\rangle)$, $|\mathcal{B}_{1,1}\rangle = \frac{1}{\sqrt{2}}(|00\rangle - |11\rangle)$. Therefore, the outcomes of these measurements and the corresponding Pauli operations are shown in Tab.1, where $|\gamma\rangle = \frac{1}{8}(|00\cdots000\cdots0\rangle + |01\cdots110\cdots0\rangle + |10\cdots001\cdots1\rangle + |11\cdots111\cdots1\rangle)_{a_1,a_2,\cdots,a_n,b_1,b_2,\cdots,b_m}$.

Tab.1. The outcome of measurements are performed by Alice and Bob, and the corresponding Alice and Bob's operation.

Alice and Bob's measurement	The state obtained and their joint operation
$\|\mathcal{B}_{0,0}\rangle_{A,a_{2n}}\|\mathcal{B}_{0,0}\rangle_{B,b_{2m}}$	$\|\gamma\rangle \sum_{i,j=0}^{1} \alpha_i\beta_j \|i\rangle_{a_{n+1},\cdots,a_{2n-1}}^{\otimes(n-1)} \otimes \|j\rangle_{b_{m+1},\cdots,b_{2m-1}}^{\otimes(m-1)}$ $\sigma_{a_{n+1}}^{(0,0)} \otimes \cdots \otimes \sigma_{a_{2n-1}}^{(0,0)} \otimes \sigma_{b_{m+1}}^{(0,0)} \otimes \cdots \otimes \sigma_{b_{2m-1}}^{(0,0)}$
$\|\mathcal{B}_{0,0}\rangle_{A,a_{2n}}\|\mathcal{B}_{0,1}\rangle_{B,b_{2m}}$	$\|\gamma\rangle \sum_{i,j=0}^{1} \alpha_i\beta_j \|i\rangle_{a_{n+1},\cdots,a_{2n-1}}^{\otimes(n-1)} \otimes \|j\oplus 1\rangle_{b_{m+1},\cdots,b_{2m-1}}^{\otimes(m-1)}$ $\sigma_{a_{n+1}}^{(0,0)} \otimes \cdots \otimes \sigma_{a_{2n-1}}^{(0,0)} \otimes \sigma_{b_{m+1}}^{(0,1)} \otimes \cdots \otimes \sigma_{b_{2m-1}}^{(0,1)}$
$\|\mathcal{B}_{0,0}\rangle_{A,a_{2n}}\|\mathcal{B}_{1,0}\rangle_{B,b_{2m}}$	$\|\gamma\rangle \sum_{i,j=0}^{1} (-1)^j \alpha_i\beta_j \|i\rangle_{a_{n+1},\cdots,a_{2n-1}}^{\otimes(n-1)} \otimes \|j\oplus 1\rangle_{b_{m+1},\cdots,b_{2m-1}}^{\otimes(m-1)}$ $\sigma_{a_{n+1}}^{(0,0)} \otimes \cdots \otimes \sigma_{a_{2n-1}}^{(0,0)} \otimes \sigma_{b_{m+1}}^{(1,0)} \otimes \cdots \otimes \sigma_{b_{2m-1}}^{(1,0)}$
$\|\mathcal{B}_{0,0}\rangle_{A,a_{2n}}\|\mathcal{B}_{1,1}\rangle_{B,b_{2m}}$	$\|\gamma\rangle \sum_{i,j=0}^{1} (-1)^j \alpha_i\beta_j \|i\rangle_{a_{n+1},\cdots,a_{2n-1}}^{\otimes(n-1)} \otimes \|j\rangle_{b_{m+1},\cdots,b_{2m-1}}^{\otimes(m-1)}$ $\sigma_{a_{n+1}}^{(0,0)} \otimes \cdots \otimes \sigma_{a_{2n-1}}^{(0,0)} \otimes \sigma_{b_{m+1}}^{(1,1)} \otimes \cdots \otimes \sigma_{b_{2m-1}}^{(1,1)}$
$\|\mathcal{B}_{0,1}\rangle_{A,a_{2n}}\|\mathcal{B}_{0,0}\rangle_{B,b_{2m}}$	$\|\gamma\rangle \sum_{i,j=0}^{1} \alpha_i\beta_j \|i\oplus 1\rangle_{a_{n+1},\cdots,a_{2n-1}}^{\otimes(n-1)} \otimes \|j\rangle_{b_{m+1},\cdots,b_{2m-1}}^{\otimes(m-1)}$ $\sigma_{a_{n+1}}^{(0,1)} \otimes \cdots \otimes \sigma_{a_{2n-1}}^{(0,1)} \otimes \sigma_{b_{m+1}}^{(0,0)} \otimes \cdots \otimes \sigma_{b_{2m-1}}^{(0,0)}$
$\|\mathcal{B}_{0,1}\rangle_{A,a_{2n}}\|\mathcal{B}_{0,1}\rangle_{B,b_{2m}}$	$\|\gamma\rangle \sum_{i,j=0}^{1} \alpha_i\beta_j \|i\oplus 1\rangle_{a_{n+1},\cdots,a_{2n-1}}^{\otimes(n-1)} \otimes \|j\oplus 1\rangle_{b_{m+1},\cdots,b_{2m-1}}^{\otimes(m-1)}$ $\sigma_{a_{n+1}}^{(0,1)} \otimes \cdots \otimes \sigma_{a_{2n-1}}^{(0,1)} \otimes \sigma_{b_{m+1}}^{(0,1)} \otimes \cdots \otimes \sigma_{b_{2m-1}}^{(0,1)}$
$\|\mathcal{B}_{0,1}\rangle_{A,a_{2n}}\|\mathcal{B}_{1,0}\rangle_{B,b_{2m}}$	$\|\gamma\rangle \sum_{i,j=0}^{1} (-1)^j \alpha_i\beta_j \|i\oplus 1\rangle_{a_{n+1},\cdots,a_{2n-1}}^{\otimes(n-1)} \otimes \|j\oplus 1\rangle_{b_{m+1},\cdots,b_{2m-1}}^{\otimes(m-1)}$ $\sigma_{a_{n+1}}^{(0,1)} \otimes \cdots \otimes \sigma_{a_{2n-1}}^{(0,1)} \otimes \sigma_{b_{m+1}}^{(1,0)} \otimes \cdots \otimes \sigma_{b_{2m-1}}^{(1,0)}$
$\|\mathcal{B}_{0,1}\rangle_{A,a_{2n}}\|\mathcal{B}_{1,1}\rangle_{B,b_{2m}}$	$\|\gamma\rangle \sum_{i,j=0}^{1} (-1)^j \alpha_i\beta_j \|i\oplus 1\rangle_{a_{n+1},\cdots,a_{2n-1}}^{\otimes(n-1)} \otimes \|j\rangle_{b_{m+1},\cdots,b_{2m-1}}^{\otimes(m-1)}$ $\sigma_{a_{n+1}}^{(0,1)} \otimes \cdots \otimes \sigma_{a_{2n-1}}^{(0,1)} \otimes \sigma_{b_{m+1}}^{(1,1)} \otimes \cdots \otimes \sigma_{b_{2m-1}}^{(1,1)}$
$\|\mathcal{B}_{1,0}\rangle_{A,a_{2n}}\|\mathcal{B}_{0,0}\rangle_{B,b_{2m}}$	$\|\gamma\rangle \sum_{i,j=0}^{1} (-1)^i \alpha_i\beta_j \|i\oplus 1\rangle_{a_{n+1},\cdots,a_{2n-1}}^{\otimes(n-1)} \otimes \|j\rangle_{b_{m+1},\cdots,b_{2m-1}}^{\otimes(m-1)}$ $\sigma_{a_{n+1}}^{(1,0)} \otimes \cdots \otimes \sigma_{a_{2n-1}}^{(1,0)} \otimes \sigma_{b_{m+1}}^{(0,0)} \otimes \cdots \otimes \sigma_{b_{2m-1}}^{(0,0)}$
$\|\mathcal{B}_{1,0}\rangle_{A,a_{2n}}\|\mathcal{B}_{0,1}\rangle_{B,b_{2m}}$	$\|\gamma\rangle \sum_{i,j=0}^{1} (-1)^i \alpha_i\beta_j \|i\oplus 1\rangle_{a_{n+1},\cdots,a_{2n-1}}^{\otimes(n-1)} \otimes \|j\oplus 1\rangle_{b_{m+1},\cdots,b_{2m-1}}^{\otimes(m-1)}$ $\sigma_{a_{n+1}}^{(1,0)} \otimes \cdots \otimes \sigma_{a_{2n-1}}^{(1,0)} \otimes \sigma_{b_{m+1}}^{(0,1)} \otimes \cdots \otimes \sigma_{b_{2m-1}}^{(0,1)}$
$\|\mathcal{B}_{1,0}\rangle_{A,a_{2n}}\|\mathcal{B}_{1,0}\rangle_{B,b_{2m}}$	$\|\gamma\rangle \sum_{i,j=0}^{1} (-1)^{i+j} \alpha_i\beta_j \|i\oplus 1\rangle_{a_{n+1},\cdots,a_{2n-1}}^{\otimes(n-1)} \otimes \|j\oplus 1\rangle_{b_{m+1},\cdots,b_{2m-1}}^{\otimes(m-1)}$ $\sigma_{a_{n+1}}^{(1,0)} \otimes \cdots \otimes \sigma_{a_{2n-1}}^{(1,0)} \otimes \sigma_{b_{m+1}}^{(1,0)} \otimes \cdots \otimes \sigma_{b_{2m-1}}^{(1,0)}$
$\|\mathcal{B}_{1,0}\rangle_{A,a_{2n}}\|\mathcal{B}_{1,1}\rangle_{B,b_{2m}}$	$\|\gamma\rangle \sum_{i,j=0}^{1} (-1)^{i+j} \alpha_i\beta_j \|i\oplus 1\rangle_{a_{n+1},\cdots,a_{2n-1}}^{\otimes(n-1)} \otimes \|j\rangle_{b_{m+1},\cdots,b_{2m-1}}^{\otimes(m-1)}$ $\sigma_{a_{n+1}}^{(1,0)} \otimes \cdots \otimes \sigma_{a_{2n-1}}^{(1,0)} \otimes \sigma_{b_{m+1}}^{(1,1)} \otimes \cdots \otimes \sigma_{b_{2m-1}}^{(1,1)}$
$\|\mathcal{B}_{1,1}\rangle_{A,a_{2n}}\|\mathcal{B}_{0,0}\rangle_{B,b_{2m}}$	$\|\gamma\rangle \sum_{i,j=0}^{1} (-1)^i \alpha_i\beta_j \|i\rangle_{a_{n+1},\cdots,a_{2n-1}}^{\otimes(n-1)} \otimes \|j\rangle_{b_{m+1},\cdots,b_{2m-1}}^{\otimes(m-1)}$ $\sigma_{a_{n+1}}^{(1,1)} \otimes \cdots \otimes \sigma_{a_{2n-1}}^{(1,1)} \otimes \sigma_{b_{m+1}}^{(0,0)} \otimes \cdots \otimes \sigma_{b_{2m-1}}^{(0,0)}$
$\|\mathcal{B}_{1,1}\rangle_{A,a_{2n}}\|\mathcal{B}_{0,1}\rangle_{B,b_{2m}}$	$\|\gamma\rangle \sum_{i,j=0}^{1} (-1)^i \alpha_i\beta_j \|i\rangle_{a_{n+1},\cdots,a_{2n-1}}^{\otimes(n-1)} \otimes \|j\oplus 1\rangle_{b_{m+1},\cdots,b_{2m-1}}^{\otimes(m-1)}$ $\sigma_{a_{n+1}}^{(1,1)} \otimes \cdots \otimes \sigma_{a_{2n-1}}^{(1,1)} \otimes \sigma_{b_{m+1}}^{(0,1)} \otimes \cdots \otimes \sigma_{b_{2m-1}}^{(0,1)}$
$\|\mathcal{B}_{1,1}\rangle_{A,a_{2n}}\|\mathcal{B}_{1,0}\rangle_{B,b_{2m}}$	$\|\gamma\rangle \sum_{i,j=0}^{1} (-1)^{i+j} \alpha_i\beta_j \|i\rangle_{a_{n+1},\cdots,a_{2n-1}}^{\otimes(n-1)} \otimes \|j\oplus 1\rangle_{b_{m+1},\cdots,b_{2m-1}}^{\otimes(m-1)}$ $\sigma_{a_{n+1}}^{(1,1)} \otimes \cdots \otimes \sigma_{a_{2n-1}}^{(1,1)} \otimes \sigma_{b_{m+1}}^{(1,0)} \otimes \cdots \otimes \sigma_{b_{2m-1}}^{(1,0)}$
$\|\mathcal{B}_{1,1}\rangle_{A,a_{2n}}\|\mathcal{B}_{1,1}\rangle_{B,b_{2m}}$	$\|\gamma\rangle \sum_{i,j=0}^{1} (-1)^{i+j} \alpha_i\beta_j \|i\rangle_{a_{n+1},\cdots,a_{2n-1}}^{\otimes(n-1)} \otimes \|j\rangle_{b_{m+1},\cdots,b_{2m-1}}^{\otimes(m-1)}$ $\sigma_{a_{n+1}}^{(1,1)} \otimes \cdots \otimes \sigma_{a_{2n-1}}^{(1,1)} \otimes \sigma_{b_{m+1}}^{(1,1)} \otimes \cdots \otimes \sigma_{b_{2m-1}}^{(1,1)}$

From Tab.1, we can see that whatever Alice and Bob's measurements are, the corresponding collapsed state can be transformed to

$$|\Psi^1\rangle_{a_1,a_2,\cdots,a_n,a_{n+1},\cdots,a_{2n-1},b_1,b_2,\cdots,b_m,b_{m+1},\cdots,b_{2m-1}}$$
$$= |\gamma\rangle \sum_{i,j=0}^{1} \alpha_i\beta_j |i\rangle_{a_{n+1},\cdots,a_{2n-1}}^{\otimes(n-1)} \otimes |j\rangle_{b_{m+1},\cdots,b_{2m-1}}^{\otimes(m-1)}, \qquad (7)$$

by performing the corresponding operations which are shown in Tab.1.

Step 3. Alice and Bob execute composite unitary operations $\mathcal{W}_1$ and $\mathcal{W}_2$, respectively. Here,

$$\mathcal{W}_1 = \mathcal{N}^c_{a_{2n-1},a_{n+1}} \otimes \cdots \otimes \mathcal{N}^c_{a_{2n-1},a_{2n-2}},$$
$$\mathcal{W}_2 = \mathcal{N}^c_{b_{2m-1},b_{m+1}} \otimes \cdots \otimes \mathcal{N}^c_{b_{2m-1},b_{2m-2}},$$

and the unitary operation $\mathcal{N}^c$ is actually a two-qubit controlled-NOT (CNOT) gate. So, the operations $\mathcal{W}_1$ and $\mathcal{W}_2$ transform the state $|\Psi^1\rangle$ in Eq. (7) into

$$|\Psi^2\rangle_{a_1,a_2,\cdots,a_n,a_{n+1},\cdots,a_{2n-2},b_1,b_2,\cdots,b_m,b_{m+1},\cdots,b_{2m-2}}$$
$$= \frac{1}{8}(\alpha_0|0\rangle + \alpha_1|1\rangle)_{a_{2n-1}} \otimes (\beta_0|0\rangle + \beta_1|1\rangle)_{b_{2m-1}}$$
$$\otimes (|00\cdots00\cdots000\cdots00\cdots0\rangle + |01\cdots10\cdots010\cdots00\cdots0\rangle$$
$$+ |10\cdots00\cdots001\cdots10\cdots0\rangle + |11\cdots10\cdots011\cdots10\cdots0\rangle). \quad (8)$$

Remark: It is easy to see that when $n = 2$, we only need Bob excutes the composite unitary operation $\mathcal{W}_2$; when $m = 2$, Alice performs the composite unitary operation $\mathcal{W}_1$ only; and when $n = m = 2$, we can skip this step and straightforward to step 4.

Step 4. Alice and Bob carry out the concerned operations U_A and U_B on their qubits a_{2n-1} and b_{2m-1}, respectively, i.e.,

$$U_A|\varphi\rangle_{a_{2n-1}} = (\alpha'_0|0\rangle + \alpha'_1|1\rangle)_{a_{2n-1}}, \quad U_B|\varphi\rangle_{b_{2m-1}} = (\beta'_0|0\rangle + \beta'_1|1\rangle)_{b_{2m-1}}.$$

This indicates the two concerned operations U_A and U_B have been performed on the target states $|\varphi\rangle_{a_{2n-1}}$ and $|\varphi\rangle_{b_{2m-1}}$ at the same time. So, the total joint state can be written as

$$|\Psi^3\rangle_{a_{2n-1},a_1,a_2,\cdots,a_n,a_{n+1},\cdots,a_{2n-2},b_{2m-1},b_1,b_2,\cdots,b_m,b_{m+1},\cdots,b_{2m-2}}$$
$$= \frac{1}{8}(\alpha'_0|0\rangle + \alpha'_1|1\rangle)_{a_{2n-1}} \otimes (\beta'_0|0\rangle + \beta'_1|1\rangle)_{b_{2m-1}}$$
$$\otimes (|00\cdots00\cdots000\cdots00\cdots0\rangle + |01\cdots10\cdots010\cdots00\cdots0\rangle$$
$$+ |10\cdots00\cdots001\cdots10\cdots0\rangle + |11\cdots10\cdots011\cdots10\cdots0\rangle). \quad (9)$$

Step 5. Alice and Bob measure their qubits $a_3, \cdots, a_n, a_{n+1}, \cdots, a_{2n-2}, b_3, \cdots, b_m, b_{m+1}, \cdots, b_{2m-2}$ using the othornormal bases $\{|+\rangle, |-\rangle\}$, respectively.

Here, the othornormal bases $\{|+\rangle, |-\rangle\}$ can be defined as: $|+\rangle = \frac{1}{\sqrt{2}}(|0\rangle + |1\rangle)$, $|-\rangle = \frac{1}{\sqrt{2}}(|0\rangle - |1\rangle)$. Then they announce the results to each other via classical channel. Therefore, with the measure base $|-\rangle$, we may perform the Pauli operation $\sigma^{(1,1)}$ corresponding to those particles of $|1\rangle$. That is, Alice uses $[2+(-1)^{(m-2)}]$ (mod 3) times $\sigma_{a_1}^{(1,1)}$ and Bob performs $[2+(-1)^{(n-2)}]$ (mod 3) times $\sigma_{b_1}^{(1,1)}$. So, the state $|\Psi^3\rangle$ in Eq. (9) can be transformed to

$$|\Psi^4\rangle_{a_{2n-1},b_{2m-1},a_1,a_2,b_1,b_2}$$
$$=(\frac{1}{2})^{n+m-1}[\alpha_0'\beta_0'(|000000\rangle+|000110\rangle+|001001\rangle+|001111\rangle)$$
$$+\alpha_0'\beta_1'(|010000\rangle+|010110\rangle+|011001\rangle+|011111\rangle)$$
$$+\alpha_1'\beta_0'(|100000\rangle+|100110\rangle+|101001\rangle+|101111\rangle)$$
$$+\alpha_1'\beta_1'(|110000\rangle+|110110\rangle+|111001\rangle+|111111\rangle)]. \qquad (10)$$

Step 6. Alice and Bob carry out Bell-state measurements on qubit pairs (a_{2n-1}, a_1) and (b_{2m-1}, b_1), respectively. The outcomes of these measurements and the corresponding operations are shown in Tab.2.

Tab.2. The outcome of measurements are performed by Alice and Bob, and the corresponding Alice and Bob's operations.

Alice and Bob's measurement	The state obtained	Alice and Bob's joint operation
$\|\mathcal{B}_{0,0}\rangle_{a_{2n-1},a_1}\|\mathcal{B}_{0,0}\rangle_{b_{2m-1},b_1}$	$(\frac{1}{2})^{n+m}(\alpha_0'\|0\rangle+\alpha_1'\|1\rangle)_{b_2}\otimes(\beta_0'\|0\rangle+\beta_1'\|1\rangle)_{a_2}$	$\sigma_{b_2}^{(0,0)}\otimes\sigma_{a_2}^{(0,0)}$
$\|\mathcal{B}_{0,0}\rangle_{a_{2n-1},a_1}\|\mathcal{B}_{0,1}\rangle_{b_{2m-1},b_1}$	$(\frac{1}{2})^{n+m}(\alpha_0'\|1\rangle+\alpha_1'\|0\rangle)_{b_2}\otimes(\beta_0'\|0\rangle+\beta_1'\|1\rangle)_{a_2}$	$\sigma_{b_2}^{(0,1)}\otimes\sigma_{a_2}^{(0,0)}$
$\|\mathcal{B}_{0,0}\rangle_{a_{2n-1},a_1}\|\mathcal{B}_{1,0}\rangle_{b_{2m-1},b_1}$	$(\frac{1}{2})^{n+m}(\alpha_0'\|1\rangle-\alpha_1'\|0\rangle)_{b_2}\otimes(\beta_0'\|0\rangle+\beta_1'\|1\rangle)_{a_2}$	$\sigma_{b_2}^{(1,0)}\otimes\sigma_{a_2}^{(0,0)}$
$\|\mathcal{B}_{0,0}\rangle_{a_{2n-1},a_1}\|\mathcal{B}_{1,1}\rangle_{b_{2m-1},b_1}$	$(\frac{1}{2})^{n+m}(\alpha_0'\|0\rangle-\alpha_1'\|1\rangle)_{b_2}\otimes(\beta_0'\|0\rangle+\beta_1'\|1\rangle)_{a_2}$	$\sigma_{b_2}^{(1,1)}\otimes\sigma_{a_2}^{(0,0)}$
$\|\mathcal{B}_{0,1}\rangle_{a_{2n-1},a_1}\|\mathcal{B}_{0,0}\rangle_{b_{2m-1},b_1}$	$(\frac{1}{2})^{n+m}(\alpha_0'\|0\rangle+\alpha_1'\|1\rangle)_{b_2}\otimes(\beta_0'\|1\rangle+\beta_1'\|0\rangle)_{a_2}$	$\sigma_{b_2}^{(0,0)}\otimes\sigma_{a_2}^{(0,1)}$
$\|\mathcal{B}_{0,1}\rangle_{a_{2n-1},a_1}\|\mathcal{B}_{0,1}\rangle_{b_{2m-1},b_1}$	$(\frac{1}{2})^{n+m}(\alpha_0'\|1\rangle+\alpha_1'\|0\rangle)_{b_2}\otimes(\beta_0'\|1\rangle+\beta_1'\|0\rangle)_{a_2}$	$\sigma_{b_2}^{(0,1)}\otimes\sigma_{a_2}^{(0,1)}$
$\|\mathcal{B}_{0,1}\rangle_{a_{2n-1},a_1}\|\mathcal{B}_{1,0}\rangle_{b_{2m-1},b_1}$	$(\frac{1}{2})^{n+m}(\alpha_0'\|1\rangle-\alpha_1'\|0\rangle)_{b_2}\otimes(\beta_0'\|1\rangle+\beta_1'\|0\rangle)_{a_2}$	$\sigma_{b_2}^{(1,0)}\otimes\sigma_{a_2}^{(0,1)}$
$\|\mathcal{B}_{0,1}\rangle_{a_{2n-1},a_1}\|\mathcal{B}_{1,1}\rangle_{b_{2m-1},b_1}$	$(\frac{1}{2})^{n+m}(\alpha_0'\|0\rangle-\alpha_1'\|1\rangle)_{b_2}\otimes(\beta_0'\|1\rangle+\beta_1'\|0\rangle)_{a_2}$	$\sigma_{b_2}^{(1,1)}\otimes\sigma_{a_2}^{(0,1)}$
$\|\mathcal{B}_{1,0}\rangle_{a_{2n-1},a_1}\|\mathcal{B}_{0,0}\rangle_{b_{2m-1},b_1}$	$(\frac{1}{2})^{n+m}(\alpha_0'\|0\rangle+\alpha_1'\|1\rangle)_{b_2}\otimes(\beta_0'\|1\rangle-\beta_1'\|0\rangle)_{a_2}$	$\sigma_{b_2}^{(0,0)}\otimes\sigma_{a_2}^{(1,0)}$
$\|\mathcal{B}_{1,0}\rangle_{a_{2n-1},a_1}\|\mathcal{B}_{0,1}\rangle_{b_{2m-1},b_1}$	$(\frac{1}{2})^{n+m}(\alpha_0'\|1\rangle+\alpha_1'\|0\rangle)_{b_2}\otimes(\beta_0'\|1\rangle-\beta_1'\|0\rangle)_{a_2}$	$\sigma_{b_2}^{(0,1)}\otimes\sigma_{a_2}^{(1,0)}$
$\|\mathcal{B}_{1,0}\rangle_{a_{2n-1},a_1}\|\mathcal{B}_{1,0}\rangle_{b_{2m-1},b_1}$	$(\frac{1}{2})^{n+m}(\alpha_0'\|1\rangle-\alpha_1'\|0\rangle)_{b_2}\otimes(\beta_0'\|1\rangle-\beta_1'\|0\rangle)_{a_2}$	$\sigma_{b_2}^{(1,0)}\otimes\sigma_{a_2}^{(1,0)}$
$\|\mathcal{B}_{1,0}\rangle_{a_{2n-1},a_1}\|\mathcal{B}_{1,1}\rangle_{b_{2m-1},b_1}$	$(\frac{1}{2})^{n+m}(\alpha_0'\|0\rangle-\alpha_1'\|1\rangle)_{b_2}\otimes(\beta_0'\|1\rangle-\beta_1'\|0\rangle)_{a_2}$	$\sigma_{b_2}^{(1,1)}\otimes\sigma_{a_2}^{(1,0)}$
$\|\mathcal{B}_{1,1}\rangle_{a_{2n-1},a_1}\|\mathcal{B}_{0,0}\rangle_{b_{2m-1},b_1}$	$(\frac{1}{2})^{n+m}(\alpha_0'\|0\rangle+\alpha_1'\|1\rangle)_{b_2}\otimes(\beta_0'\|0\rangle-\beta_1'\|1\rangle)_{a_2}$	$\sigma_{b_2}^{(0,0)}\otimes\sigma_{a_2}^{(1,1)}$
$\|\mathcal{B}_{1,1}\rangle_{a_{2n-1},a_1}\|\mathcal{B}_{0,1}\rangle_{b_{2m-1},b_1}$	$(\frac{1}{2})^{n+m}(\alpha_0'\|1\rangle+\alpha_1'\|0\rangle)_{b_2}\otimes(\beta_0'\|0\rangle-\beta_1'\|1\rangle)_{a_2}$	$\sigma_{b_2}^{(0,1)}\otimes\sigma_{a_2}^{(1,1)}$
$\|\mathcal{B}_{1,1}\rangle_{a_{2n-1},a_1}\|\mathcal{B}_{1,0}\rangle_{b_{2m-1},b_1}$	$(\frac{1}{2})^{n+m}(\alpha_0'\|1\rangle-\alpha_1'\|0\rangle)_{b_2}\otimes(\beta_0'\|0\rangle-\beta_1'\|1\rangle)_{a_2}$	$\sigma_{b_2}^{(1,0)}\otimes\sigma_{a_2}^{(1,1)}$
$\|\mathcal{B}_{1,1}\rangle_{a_{2n-1},a_1}\|\mathcal{B}_{1,1}\rangle_{b_{2m-1},b_1}$	$(\frac{1}{2})^{n+m}(\alpha_0'\|0\rangle-\alpha_1'\|1\rangle)_{b_2}\otimes(\beta_0'\|0\rangle-\beta_1'\|1\rangle)_{a_2}$	$\sigma_{b_2}^{(1,1)}\otimes\sigma_{a_2}^{(1,1)}$

Therefore, one is readily to see that after performing the Pauli operations, Alice and Bob's measurements induce the following collapses

$$|\Psi^5\rangle=(\frac{1}{2})^{n+m}(U_A|\varphi\rangle_{b_2})\otimes(U_B|\varphi\rangle_{a_2}). \qquad (11)$$

Thus, the generalized BQOT protocol with multiqubit cluster states is successfully finished.

3. Discussion

Let us consider our scheme from the following five aspects: quantum resource consumption, classical resource consumption, operation complexity, success probability and intrinsic efficiency. We have already summarized these in Tab.3. Similar to [15], the intrinsic efficiency of the BQOT scheme is defined as $\eta = P/(q+t)$, where q is the number of the qubits which are used as quantum channels (except for those chose for security checking), t is the classical bits transmitted, and P is the final success probability.

Tab.3. Discussion of our scheme.

QRC	NO	CRC	P	η
2CS	4BM$_s$, $(2n+2m-8)$SM$_s$, $(n+m+2)$SO$_s$, $(n+m-2)$TO$_s$	$(2n+2m)$ cbits	1	$1/(4m+4n)$

QRC: quantum resource consumption. NO: necessary operations.
CRC: classical resource consumption. BM: Bell state measurement.
CS: multiqubit cluster state. SM: single-qubit state measurement.
SO: single-qubit unitary operation. TO: two-qubit unitary operation.

It is important to consider the quantum resource consumption of the scheme. In the scheme, to achieve the goal of BQOT, two multiqubit cluster states need to be consumed as the quantum channel. To achieve the teleportation, it is also indispensable to transmit some classical information between the initial owner and the final receiver. This means the use of classical resource is necessary. In this case, it is obviously interesting to ask whether the use can be economic. From Tab.3, we can see the classical resource consumption of the scheme is $(2n+2m)$ cbits. As for operation complexity, in any quantum information processing, measurement and unitary operations are indispensable. In this case, it is intriguing to ask whether the difficulty or intensity of the necessary operations can be degraded for sharing given quantum information. Obviously the scheme is using $(n+m-2)$ two-qubit unitary operations and $(n+m+2)$ single-qubit unitary operations. The success probabilities in the scheme is completely determined by the entanglement structures of the employed states. From the table, one can very easily find that the scheme is deterministic, that is, the BQOT task can be achieved with unit probability. Similarly, the intrinsic efficiencies are also determined by the entanglement structures, as can be seen from the Tab.3. The efficiency of the scheme is $1/(4m+4n)$.

4. Summary

To summarize, by integrating the ideas of QOT and bidirectional QT, we have presented a scheme for BQOT with two multiqubit cluster states. Namely, based on the quantum channel shared in advance, Alice wants to execute a unitary operation U_A on an arbitrary qubit state which is in the possession of Bob; and at the same time, Bob can perform a unitary operation U_B on the target qubit in Alice's site. Moreover, we made a discussion of our scheme.

Acknowledgments

This work is supported by the National Natural Science Foundation of China (Grant No. 11671284), Sichuan Provincial Natural Science Foundation of China (Grant No. 2015JY0002, 2017JY0197) and the Research Foundation of the Education Department of Sichuan Province (Grant No. 15ZA0032).

References

1. S. F. Huelga, J. A. Vaccaro, A. Chefles and M. B. Plenio, *Phys. Rev. A* **63**, 042303 (2001).
2. A. M. Wang, *Phys. Rev. A* **74**, 032317 (2006).
3. N. B. Zhao and A. M. Wang, *Phys. Rev. A* **78**, 014305 (2008).
4. Z. J. Zhang and C. Y. Cheung, *J. Phys. B* **44**, 165508 (2011).
5. A. Garon, S. J. Glaser and D. Sugny, *Phys. Rev. A* **88**, 3544–3549 (2013).
6. Y. W. Wang, X. H. Wei and Z. H. Zhu, *Acta Physica Sinica* **62**, 581–586 (2013).
7. J. Y. Peng, M. Q. Bai and Z. W. Mo, *Quantum Inf. Process.* **12**, 3511–3525 (2016).
8. J. Y. Peng, M. Q. Bai and Z. W. Mo, *Int. J. Theor. Phys.* **55**, 2481–2489 (2016).
9. J. H. Tian, J. Z. Zhang and Y. P. Li, *Int. J. Theor. Phys.* **55**, 2303–2310 (2016).
10. H. J. Briegel and R. Raussendorf, *Phys. Rev. Lett.* **86**, 910 (2001).
11. M. Hein, W. Dr and H. J. Briegel, *Phys. Rev. A* **71**, 032350 (2005).
12. S. Muralidharan and P. K. Panigrahi, *Phys. Rev. A* **78**, 062333 (2008).
13. N. Paul, J. V. Menon, S. Karumanchi, S. Muralidharan and P. K. Panigrahi, *Quantum Inf. Process.* **10**, 619–632 (2011).
14. S. Muralidharan, S. Jain and P. K. Panigrahi, *Optics Communications* **4**, 284 (2011).
15. H. Yuan, Y. M. Liu, W. Zhang and Z. J. Zhang, *J. Phys. B* **41**, 145506 (2008).

Bidirectional remote teleportation of single-qubit operations via cluster states

Si Qi Zhou,* Ming Qiang Bai[†] and Zhi Wen Mo

School of Mathematics, Sichuan Normal University
Chengdu, 610066, China
Institute of Intelligent Information & Quantum Information,
Sichuan Normal University, Chengdu, 610066, China
* siqi_zhou@163.com
[†] baimq@sicnu.edu.cn

By integrating the ideas of quantum operation teleportation (QOT) and bidirectional quantum teleportation (QT), we present a protocol for bidirectional quantum operation teleportation (BQOT) with two cluster states. This means that Alice can transmit an unknown single-qubit operation U_A on the remotely Bob's quantum system; and at the same time, Bob can also transmit an arbitrary single-qubit operation U_B on Alice's quantum system. Furthermore, some concrete discussions are made to study its important features, including the determinacy and security of the scheme, the nowaday's experimental feasibility as well as the intrinsic efficiency.

Keywords: Bidirectional teleportation; quantum operation teleportation; cluster state.

1. Introduction

It is well known that quantum entanglement as a crucial quantum resource has been widely utilized in various quantum information processing tasks [1-5]. In 1993, Bennett et al. [1] first put forward the creative protocol of quantum teleportation (QT), where an unknown quantum state can be transmitted to a distant site without sending any physical particles. After Bennett et al.'s pioneering work, much attention was focused on QT in both theoretical and experimental aspects [2-5]. Since then, in 2001 Huelga and his coworkers [6] first introduced quantum operation teleportation (QOT), the manipulated object in QOT is quantum operation. Therefore, QOT as a newly developing branch in the field of quantum information processing has been attracting much attention [7-10].

Recently, some researchers have devoted much interest to the study of bidirectional quantum communication [11-15], but bidirectional quantum operation teleportation is not seriously considered up to now. In this paper, we propose a scheme of BQOT which is essentially a union of the ideas of QOT and bidirectional QT. We will utilize a product state of two cluster state as a quantum channel. After we present our proposal, we discuss its important features, including the scheme determinacy, the scheme security, the nowaday's experimental feasibility as well as the intrinsic efficiency.

The rest of this paper is organized as follows. Section 2 describes the scheme of BQOT with two cluster states. And in Section 3, we describe some features of the scheme, including determinacy, security, current experimental feasibility and efficiency. Finally, we make a concise summary about this work in Section 4.

2. BQOT Protocol with Two Cluster States

Now let us amply depict our protocol. There are two legitimate users, say, Alice and Bob. Alice wants to perform a unitary operation U_A on the target qubit state which is in the possession of Bob; at the same time, Bob performs a unitary operation U_B on the other target qubit in Alice's location. Actually, Alice and Bob may not hear of the concerned operations U_A and U_B before. Without loss of generality, at the beginning we suppose the target state $|\varphi\rangle_A(|\varphi\rangle_B)$ in Alice's(Bob's) location read as follows

$$|\varphi\rangle_A = (\alpha_0|0\rangle + \alpha_1|1\rangle)_A, \quad |\varphi\rangle_B = (\beta_0|0\rangle + \beta_1|1\rangle)_B, \qquad (1)$$

where α_0, α_1, β_0 and β_1 are complex and satisfy $|\alpha_0|^2 + |\alpha_1|^2 = 1$, $|\beta_0|^2 + |\beta_1|^2 = 1$.

The shared entanglements are four-qubit cluster state and six-qubit cluster state:

$$|C_4\rangle = \frac{1}{2}(|0000\rangle + |0011\rangle + |1100\rangle - |1111\rangle)_{a_1 b_2 b_3 b_4}, \qquad (2)$$

and

$$|C_6\rangle = \frac{1}{2}(|000000\rangle + |000111\rangle + |111000\rangle - |111111\rangle)_{b_1 a_2 a_3 a_4 a_5 a_6}, \qquad (3)$$

where the subscripts denote different qubits. In advance, these qubits are assumed to be safely distributed among the two legitimate users, i.e., Alice has the the qubit pair $(a_1, a_2, a_3, a_4, a_5, a_6)$ and Bob the qubit pair (b_1, b_2, b_3, b_4).

So, the quantum state of composed system with twelve particles can be expressed as

$$|\Psi\rangle_{AB a_1 b_2 b_3 b_4 b_1 a_2 a_3 a_4 a_5 a_6} = |\varphi\rangle_A \otimes |\varphi\rangle_B \otimes |C_4\rangle \otimes |C_6\rangle. \qquad (4)$$

Our scheme is designed to own six steps, and the details are as follows.

Step 1. Alice and Bob carry out two-qubit unitary operations $\mathcal{P}^c_{a_3 a_4}$ and $\mathcal{P}^c_{b_2 b_3}$, respectively.

Here, the unitary operation $\mathcal{P}^C = |0\rangle\langle 0| \otimes \sigma^{(0,0)} + |1\rangle\langle 1| \otimes \sigma^{(1,1)}$, and the action of this operation can be described as following equations

$$|00\rangle \mapsto |00\rangle, \ |01\rangle \mapsto |01\rangle, \ |10\rangle \mapsto |10\rangle, \ |11\rangle \mapsto -|11\rangle.$$

Step 2. Alice measures her qubit pair (A, a_6) with Bell-state measurements $\{|\mathcal{B}_{i,j}\rangle_{A,a_6} : i, j = 0, 1\}$; at the same time, Bob measures his qubit pair (B, b_4) with Bell-state measurements $\{|\mathcal{B}_{i',j'}\rangle_{B,b_4} : i', j' = 0, 1\}$.

In this paper, the Bell-state bases can be defined as

$$|\mathcal{B}_{0,0}\rangle = \frac{1}{\sqrt{2}}(|00\rangle + |11\rangle), \ |\mathcal{B}_{0,1}\rangle = \frac{1}{\sqrt{2}}(|01\rangle + |10\rangle),$$

$$|\mathcal{B}_{1,0}\rangle = \frac{1}{\sqrt{2}}(|01\rangle - |10\rangle), \ |\mathcal{B}_{1,1}\rangle = \frac{1}{\sqrt{2}}(|00\rangle - |11\rangle).$$

Then they announce the results to each other via classical channels. To be specific, the state $|\mathcal{B}_{i,j}\rangle_{A,a_6}$, $|\mathcal{B}_{i',j'}\rangle_{B,b_4}$ corresponds to the classical-bit pair (i, j) and (i', j'), respectively. After receiving the results, Alice and Bob can perform Pauli operations $\sigma^{(i,j)}_{a_4} \otimes \sigma^{(i,j)}_{a_5}$ and $\sigma^{(i',j')}_{b_3}$, respectively. In this paper, Pauli operations are $\sigma^{(0,0)} = |0\rangle\langle 0| + |1\rangle\langle 1|$, $\sigma^{(0,1)} = |0\rangle\langle 1| + |1\rangle\langle 0|$, $\sigma^{(1,0)} = |0\rangle\langle 1| - |1\rangle\langle 0|$, $\sigma^{(1,1)} = |0\rangle\langle 0| - |1\rangle\langle 1|$.

It is easy to see that whatever Alice and Bob's measurements are, the corresponding collapsed state can be transformed to

$$\begin{aligned}|\Psi^1\rangle = \frac{1}{8}[\alpha_0\beta_0(&|00000000\rangle + |01100100\rangle + |10000010\rangle \\ &+ |11100110\rangle) + \alpha_0\beta_1(|00000001\rangle + |01100101\rangle \\ &+ |10000011\rangle + |11100111\rangle) + \alpha_1\beta_0(|00011000\rangle \\ &+ |01111100\rangle + |10011010\rangle + |11111110\rangle) \\ &+ \alpha_1\beta_1(|00011001\rangle + |01111101\rangle + |10011011\rangle \\ &+ |11111111\rangle)]_{a_1 a_2 a_3 a_4 a_5 b_1 b_2 b_3}.\end{aligned} \quad (5)$$

Step 3. Alice executes the two-qubit unitary operation $\mathcal{N}^c_{a_5 a_4}$.

The unitary operation $\mathcal{N}^c$ is actually a two-qubit controlled-NOT (CNOT) gate, which is defined as $\mathcal{N}^c = |0\rangle\langle 0| \otimes \sigma^{(0,0)} + |1\rangle\langle 1| \otimes \sigma^{(0,1)}$. In CNOT gate, the state of target particle is changed if and only if the state of control particle is $|1\rangle$. That is, when the state of the control particle is $|1\rangle$, the state of the target particle is $|1\rangle$ ($|0\rangle$) is changed into $|0\rangle$ ($|1\rangle$).

But the state of target particle is not changed when the state of control particle is $|0\rangle$.

Thus, after performing the operation $\mathcal{N}^c_{a_5 a_4}$, the state $|\Psi^1\rangle$ in Eq. (5) is changed into

$$|\Psi^2\rangle = \frac{1}{8}[\alpha_0\beta_0(|00000000\rangle + |01100100\rangle + |10000010\rangle$$
$$+ |11100110\rangle) + \alpha_0\beta_1(|00000001\rangle + |01100101\rangle$$
$$+ |10000011\rangle + |11100111\rangle) + \alpha_1\beta_0(|00001000\rangle$$
$$+ |01101100\rangle + |10001010\rangle + |11101110\rangle)$$
$$+ \alpha_1\beta_1(|00001001\rangle + |01101101\rangle + |10001011\rangle$$
$$+ |11101111\rangle)]_{a_1 a_2 a_3 a_4 a_5 b_1 b_2 b_3}$$
$$= \frac{1}{8}(\alpha_0|0\rangle + \alpha_1|1\rangle)_{a_5} \otimes (\beta_0|0\rangle + \beta_1|1\rangle)_{b_3}$$
$$\otimes (|000000\rangle + |011010\rangle + |100001\rangle + |111011\rangle)_{a_1 a_2 a_3 a_4 b_1 b_2}$$
$$= \frac{1}{8}|\varphi\rangle_{a_5} \otimes |\varphi\rangle_{b_3} \otimes (|000000\rangle + |011010\rangle$$
$$+ |100001\rangle + |111011\rangle)_{a_1 a_2 a_3 a_4 b_1 b_2}. \qquad (6)$$

Step 4. Alice and Bob carry out the concerned operations U_A and U_B on their qubits a_5 and b_3, respectively, i.e.,

$$U_A|\varphi\rangle_{a_5} = (\alpha'_0|0\rangle + \alpha'_1|1\rangle)_{a_5},$$

$$U_B|\varphi\rangle_{b_3} = (\beta'_0|0\rangle + \beta'_1|1\rangle)_{b_3}.$$

This indicates the two concerned operations U_A and U_B have been performed on the target states $|\varphi\rangle_{a_5}$ and $|\varphi\rangle_{b_3}$ at the same time.

So, the total joint state can be written as

$$|\Psi^3\rangle = \frac{1}{8}(\alpha'_0|0\rangle + \alpha'_1|1\rangle)_{a_5} \otimes (\beta'_0|0\rangle + \beta'_1|1\rangle)_{b_3}$$
$$\otimes (|000000\rangle + |011010\rangle + |100001\rangle + |111011\rangle)_{a_1 a_2 a_3 a_4 b_1 b_2}$$
$$= \frac{1}{8}[\alpha'_0\beta'_0(|00000000\rangle + |00011010\rangle + |00100001\rangle + |00111011\rangle)$$
$$+ \alpha'_0\beta'_1(|01000000\rangle + |01011010\rangle + |01100001\rangle + |01111011\rangle)$$
$$+ \alpha'_1\beta'_0(|10000000\rangle + |10011010\rangle + |10100001\rangle + |10111011\rangle)$$
$$+ \alpha'_1\beta'_1(|11000000\rangle + |11011010\rangle + |11100001\rangle + |11111011\rangle)] \qquad (7)$$

Step 5. Alice measures her qubits a_3 and a_4 using the othornormal bases $\{|+\rangle, |-\rangle\}$.

Here, the othornormal bases $\{|+\rangle, |-\rangle\}$ can be defined as

$$|+\rangle = \frac{1}{\sqrt{2}}(|0\rangle + |1\rangle), \quad |-\rangle = \frac{1}{\sqrt{2}}(|0\rangle - |1\rangle).$$

And then she announces the results to Bob via classical channel. To be specific, the states $|+\rangle_{a_4}$ and $|-\rangle_{a_4}$ correspond to the classical-bit 0 and 1, respectively. After performing a unitary operation as shown in Tab.1, the corresponding collapsed state can be transformed to

$$\begin{aligned}|\Psi^4\rangle &= \frac{1}{16}\sum_{i,j=0,1}\alpha'_i\beta'_j|i,j\rangle_{a_5b_3}(|0000\rangle + |0110\rangle + |1001\rangle + |1111\rangle)_{a_1a_2b_1b_2} \\ &= \frac{1}{16}[\alpha'_0\beta'_0(|000000\rangle + |000110\rangle + |001001\rangle + |001111\rangle) \\ &+ \alpha'_0\beta'_1(|010000\rangle + |010110\rangle + |011001\rangle + |011111\rangle) \\ &+ \alpha'_1\beta'_0(|100000\rangle + |100110\rangle + |101001\rangle + |101111\rangle) \\ &+ \alpha'_1\beta'_1(|110000\rangle + |110110\rangle + |111001\rangle + |111111\rangle)]_{a_5b_3a_1a_2b_1b_2} \quad (8)\end{aligned}$$

Tab.1. The outcome of measurements and the corresponding operations are performed by Alice.

Alice's measurements	The state obtained	Alice's operations
$|+\rangle_{a_3}|+\rangle_{a_4}$	$\frac{1}{16}\sum_{i,j=0,1}\alpha'_i\beta'_j|i,j\rangle_{a_5b_3}(|0000\rangle + |0110\rangle + |1001\rangle + |1111\rangle)_{a_1a_2b_1b_2}$	$\sigma_{a_2}^{(0,0)}$
$|-\rangle_{a_3}|+\rangle_{a_4}$		
$|+\rangle_{a_3}|-\rangle_{a_4}$	$\frac{1}{16}\sum_{i,j=0,1}\alpha'_i\beta'_j|i,j\rangle_{a_5b_3}(|0000\rangle - |0110\rangle + |1001\rangle - |1111\rangle)_{a_1a_2b_1b_2}$	$\sigma_{a_2}^{(1,1)}$
$|-\rangle_{a_3}|-\rangle_{a_4}$		

Step 6. Alice performs a Bell-state measurement $|\mathcal{B}_{m,n}\rangle_{a_5a_1}$ on her qubit pair (a_5, a_1), and at the same time, Bob measures his qubit pair (b_3, b_1) with the Bell-state $|\mathcal{B}_{m',n'}\rangle_{b_3b_1}$.

And then they announce the results to each other via classical channels. To be specific, the state $|\mathcal{B}_{m,n}\rangle_{A6}$, $|\mathcal{B}_{m',n'}\rangle_{B2}$ corresponds to the classical-bit pair (m, n) and (m', n'), respectively. And then, according to the classical information they have received, Alice and Bob perform the unitary Pauli operations $\sigma^{(m,n)}$ on qubit a_2 and $\sigma^{(m',n')}$ on particle b_2, respectively.

Thus, it is readily to see that the corresponding collapsed state can be transformed to

$$\begin{aligned}|\Psi^5\rangle &= \frac{1}{32} \sum_{i,j=0,1} \alpha'_i \beta'_j |j,i\rangle_{a_2 b_2} \\ &= \frac{1}{32}(\alpha'_0 \beta'_0 |00\rangle + \alpha'_0 \beta'_1 |10\rangle + \alpha'_1 \beta'_0 |01\rangle + \alpha'_1 \beta'_1 |11\rangle)_{a_2 b_2} \\ &= \frac{1}{32}(U_A|\varphi\rangle_{b_2}) \otimes (U_B|\varphi\rangle_{a_2}). \end{aligned} \quad (9)$$

Obviously, Alice's operation U_A has conclusively and successfully been carried out on the state $|\varphi\rangle_{b_2}$ in Bob's site, and at the same time, Bob's operation U_B has been successfully performed on the state $|\varphi\rangle_{a_2}$ in Alice's location.

Thus, the BQOT scheme which using a product state $|C_4\rangle \otimes |C_6\rangle$ as quantum channel is successfully realized.

3. Analysis and Discussions

Now, we will briefly discuss our scheme. Let's consider the features of the scheme. From the scheme descriptions, one can readily see that our scheme is deterministic, that is, the bidirectional teleportation is conclusively fulfilled with unit probability.

Let us analyze the security of our scheme. As a teleportation scheme, its security should be assured anytime. The aim of our scheme is to transmit two arbitrary single-qubit operations from two directions at the same time. Hence, it is necessary to consider the issue of the quantum channel security before the legitimate users' action. By virtue of the same matured check strategies in treating other similar quantum task [16-18], then any evil outsider's attack can be easily detected. That is, the four-qubit and six-qubit in the cluster states are not attacked or disturbed by any evil outsider's attack. In general, BQOT is the combination of quantum operation teleportation as well as bidirectional quantum teleportation, quantum teleportation is used for the reconstruction of target state $|\varphi\rangle$, similarly, quantum operation teleportation is employed to recover the achievable state $U|\varphi\rangle$, in the latter circumstances in this paper, the process is a bidirectional and simultaneous one. In Refs. [16-18], similar security analysis has been described in detail, so we don't repeat any more.

Let us further discuss the feasibility of experimental implementation about our scheme. It is obvious that all the unitary operations in our scheme are local. They are either single-qubit operations or two-qubit control gates, and single-qubit measurements as well as two-qubit measurements (i.e., Bell-state measurements). Incidentally, it is reported that the Bell-state measurements, two-qubit control gates and the single-qubit unitary operation have already been realized in various quantum system [19-21]. In addition, cluster state preparation and the local operations used in our scheme is completely accessible. This means that our BQOT scheme is thoroughly feasible with respect to the current experimental technologies.

Finally, we simply analyze the efficiency of our scheme. Similar to [22], the intrinsic efficiency of the QOT scheme is defined as $\eta = P/(Q + C)$, where Q is the number of the qubits which are used as quantum channels, C is the classical bits transmitted, and P is the final success probability. As mentioned before, our BQOT scheme can be achieved with unit probability. From our scheme, one can see that the Q is 10 bits, the C is 10 cbits and η is 1/20.

4. Summary

To summarize, in this paper, we have put forward a BQOT scheme with two cluster states. The scheme is applicable to deterministically transmit two arbitrary single-qubit operations bidirectionally and simultaneously. Furthermore, we have revealed some important features of our scheme, such as its determinacy and security are like those in other similar schemes. Moreover, we have exhibited the experimental feasibility in term of current techniques and analyzed the efficiency of our scheme.

Acknowledgments

This work is supported by the National Natural Science Foundation of China (Grant No. 11671284), Sichuan Provincial Natural Science Foundation of China (Grant No. 2015JY0002, 2017JY0197) and the Research Foundation of the Education Department of Sichuan Province (Grant No. 15ZA0032).

References

1. C. H. Bennett, G. Brassard, C. Crepeau, R. Jozsa, A. Peres and W. K. Wootters, *Phys. Rev. Lett.* **70**, 1895 (1993).
2. C. Y. Cheung and Z. J. Zhang, *Phys. Rev. A* **80**, 022327 (2009).
3. M. Q. Bai and Z. W. Mo, *Quantum Inf. Process.* **12**, 1053–1064 (2013).
4. E. O. Kiktenko, A. A. Popov and A. K. Fedorov, *Phys. Rev. A* **93**, 062305 (2016).
5. F. Raphael and R. Gustavo, *Phys. Rev. A* **93**, 062330 (2016).
6. S. F. Huelga, J. A. Vaccaro and A. Chefles, *Phys. Rev. A* **63**, 042303 (2001).

7. X. B. Zou, K. Pahlke and W. Mathis, *Phys. Rev. A* **65**, 064305 (2002).
8. W. Dur, G. Vidal and J. I. Cirac, *Phys. Rev. Lett.* **89**, 057901 (2002).
9. Y. S. Zhang, M. Y. Ye and G. C. Guo, *Phys. Rev. A* **71**, 062331 (2005).
10. N. B. Zhao and A. M. Wang, *Phys. Rev. A* **78**, 014305 (2008).
11. Y. H. Li and L. P. Nie, *Int. J. Theor. Phys.* **52**, 1630–1634 (2013).
12. Y. H. Li, X. L. Li, M. H. Sang, Y. Y. Nie and Z. S. Wang, *Quantum Inf. Process.* **12**, 3835–3844 (2013).
13. Y. H. Li and X. M. Jin, *Quantum Inf. Process.* **15**, 929–945 (2016).
14. J. Y. Peng, M. Q. Bai and Z. W. Mo, *Quantum Inf. Process.* **12**, 3511–3525 (2016).
15. J. Y. Peng, M. Q. Bai and Z. W. Mo, *Int. J. Theor. Phys.* **55**, 2481–2489 (2016).
16. Z. J. Zhang and Z. X. Man, *Phys. Rev. A* **72**, 022303 (2005).
17. Z. J. Zhang, Y. Li and Z. X. Man, *Phys. Rev. A* **71**, 044301 (2005).
18. F. L. Yan and T. Gao, *Phys. Rev. A* **72**, 012304 (2005).
19. S. B. Zheng, *Phys. Rev. A* **69**, 064302 (2004).
20. E. Solano and C. L. Cesar, *Eur. Phys. J. D* **13**, 121–128 (2001).
21. M. Riebe and C. F. Roos, *Nature* **429**, 734–737 (2004).
22. H. Yuan, Y. M. Liu, W. Zhang and Z. J. Zhang, *J. Phys. B* **41**, 145506 (2008).

Similarity measure between fuzzy sets based on axiomatic definition

Xiaodong Pan* and Yang Xu

School of Mathematics, Southwest Jiaotong University
West Section, High-tech Zone, Chengdu, Sichuan, 611756, P.R. China
*xdpan1@163.com

The axiomatic definition of fuzzy set, which is based on the concept of vague partition which is used to model the essence and key features of vague phenomena, has been introduced by Xiaodong Pan and Yang Xu. Inspired by the idea of the angle of two vectors in linear algebra, this paper redefines the concept of similarity measure between fuzzy sets based on the axiomatic definition by introducing the concepts of inner product and angle, and discussed its some properties.

Keywords: Fuzzy set; axiomatic definition; similarity measure; inner product.

1. Introduction

Fuzzy sets [14] are designed to model the extensions of vague concepts by a numerical (or formalized) way, although some important results have been obtained in fuzzy set and fuzzy logic by Zadeh and others (Refs. [1-6], [8-10], [13-15]), however, there are some important fundamental concepts and operations in fuzzy set theory and fuzzy logic that still require further explanation, and still need further to be more normalized, explicitly and strictly. Among them, what exactly fuzzy phenomena are? And what is its essence? In [12], we presented the following point of view:

Vagueness arises in the process of classifying objects, it is a kind of manifestation of the continuity and gradualness existing in the process of development and evolution of objects, and is the result of combined effect by the continuity and gradualness existing in the objects themselves and human subjective cognitive style. Vague phenomena are the external manifestations of vagueness. An usual way to discretize a continual and evolutionary process is to delineate the process by using some natural linguistic terms (words or phrases, i.e., vague predicates), a natural linguistic term labels a stage in the process, these natural linguistic terms denote the main characteristic of each corresponding stage respectively. In other words, natural language provides us with a very useful tool to describe a continuous process in a discrete way. Hence, vagueness usually arises together with the use of natural language.

Based on the above analysis, the occurrence of vagueness is not due to the lack of human cognitive ability and technological means, and is also not due to the complexity of objective things themselves, but is due to a specific cognitive style (discretizing the continual and evolutionary process of objects) for objective things based on a certain cognitive aim (classifying objects). In this sense, vagueness is a bridge to link discretization and continuity.

In what follow, unless otherwise stated, the symbols $\mathbb{N}^+$ denotes the set of nonzero natural numbers $\mathbb{N} \setminus \{0\}$. For any $n \in \mathbb{N}^+$, the set $\{1, 2, \cdots, n\}$ be denoted by $\overline{n}$.

In [11, 12], we introduced the concept of vague partition as follow:

Definition 1.1. Let $U = [a, b] \subset \mathbb{R}$. A vague partition of U is an object having the following form

$$\widetilde{U} = \{\mu_{A_1}(x), \cdots, \mu_{A_n}(x)\}, n \in \mathbb{N}^+,$$

where the functions $\mu_{A_i} : U \to [0, 1]$ $(i = 1, \cdots, n)$ define the degrees of memberships of the element $x \in U$ to the class A_i, respectively, and satisfy the following conditions:

(1) for any $x \in U$, there is at least one $i \in \overline{n}$ such that $\mu_{A_i}(x) > 0$;
(2) for any $i \in \overline{n}$, $\mu_{A_i}(x)$ is continuous on U;
(3) for any $i \in \overline{n}$, there is at least one $x_0 \in U$ such that $\mu_{A_i}(x_0) = 1$;
(4) for any $i \in \overline{n}$, if $\mu_{A_i}(x_0) = 1$ for $x_0 \in U$, then $\mu_{A_i}(x)$ is non-decreasing on $[a, x_0]$, and is non-increasing on $[x_0, b]$;
(5) $0 < \mu_{A_1}(x) + \cdots + \mu_{A_n}(x) \leq 1$ holds for any $x \in U$.

If $\mu_{A_1}(x) + \cdots + \mu_{A_n}(x) = 1$ in (5), then $\widetilde{U}$ is said to be a regular vague partition.

Based on the essence of vagueness and vague partition, we redefined the concept of fuzzy set in [12, 13] from the perspective of axiomatization:

Definition 1.2. Let $U = [a, b] \subset \mathbb{R}$ and $\widetilde{U} = \{\mu_{A_1}(x), \cdots, \mu_{A_n}(x)\}$, $n \in \mathbb{N}^+$, a vague partition of U. The set $\mathcal{F}(\widetilde{U})$ of fuzzy sets in U with respect to $\widetilde{U}$ consists of the following elements:

(1) if there exists $i \in \overline{n}$ such that $\mu_A(x) = \mu_{A_i}(x)$ for all $x \in U$, then $A = \{(x, \mu_A(x)) \mid x \in U\} \in \mathcal{F}(\widetilde{U})$;

(2) if $\mu_A(x) = \overline{\mu}(x) = 1$ for all $x \in U$, then $A = \{(x, \mu_A(x)) \mid x \in U\} \in \mathcal{F}(\widetilde{U})$;

(3) if $\mu_A(x) = \underline{\mu}(x) = 0$ for all $x \in U$, then $A = \{(x, \mu_A(x)) \mid x \in U\} \in \mathcal{F}(\widetilde{U})$;

(4) if $A = \{(x, \mu_A(x)) \mid x \in U\} \in \mathcal{F}(\widetilde{U})$ and $r \in \mathbb{Q}^+$, then $A^r = \{(x, (\mu_A(x))^r) \mid x \in U\} \in \mathcal{F}(\widetilde{U})$;

(5) if $A = \{(x, \mu_A(x)) \mid x \in U\} \in \mathcal{F}(\widetilde{U})$, and N is a strong negation on $[0, 1]$, then $A^N = \{(x, (\mu_A(x))^N) \mid x \in U\} \in \mathcal{F}(\widetilde{U})$;

(6) if $A = \{(x, \mu_A(x)) \mid x \in U\}, B = \{(x, \mu_B(x)) \mid x \in U\} \in \mathcal{F}(\widetilde{U})$, and $\otimes$ is a triangular norm, then $A \cap_\otimes B = \{(x, \mu_A(x) \otimes \mu_B(x)) \mid x \in U\} \in \mathcal{F}(\widetilde{U})$;

(7) if $A = \{(x, \mu_A(x)) \mid x \in U\}, B = \{(x, \mu_B(x)) \mid x \in U\} \in \mathcal{F}(\widetilde{U})$, and $\oplus$ is a triangular conorm, then $A \cup_\oplus B = \{(x, \mu_A(x) \oplus \mu_B(x)) \mid x \in U\} \in \mathcal{F}(\widetilde{U})$;

(8) $\mathcal{F}(\widetilde{U})$ not include other elements.

This paper will focus on similarity measures between fuzzy sets based on the above axiomatic definition, and our discussion will be limited only to fuzzy sets under the same vague partition. It is our hope that the proposed method in this paper could serve as the theoretical foundation of fuzzy set theory and its application in fuzzy pattern recognition, medical diagnosis, multiple criteria decision making, etc.

2. Similarity Measure Between Fuzzy Sets Based on Axiomatic Definition

Inspired by the idea of the angle of two vectors in linear algebra, in order to establish the concept of similarity measure between fuzzy sets, we first introduce the concept of inner product of two fuzzy sets, which generalizes the vector inner product in linear algebra.

For convenience's sake, unless otherwise stated, we will always assume in the following that $U = [a, b] \subset \mathbb{R}$ and $\widetilde{U} = \{\mu_{A_1}(x), \cdots, \mu_{A_n}(x)\}$, $n \in \mathbb{N}^+$, a vague partition of U.

Definition 2.1. Let $A = \{(x, \mu_A(x)) \mid x \in U\}, B = \{(x, \mu_B(x)) \mid x \in U\} \in \mathcal{F}(\widetilde{U})$, the inner product between A and B is the scalar quantity in $\mathbb{R}$,

$$\langle A, B \rangle = \int_a^b \mu_A(x) \mu_B(x) dx.$$

From Definition 2.1, the inner product between fuzzy sets is a mapping from $\mathcal{F}(\widetilde{U}) \times \mathcal{F}(\widetilde{U})$ to $\mathbb{R}$, and for any $A, B \in \mathcal{F}(\widetilde{U})$, $\langle A, B \rangle \geq 0$. We call

$\|A\| = \sqrt{\langle A, A\rangle} = \sqrt{\int_a^b (\mu_A(x))^2 dx}$ the norm of A, obviously, $\|A\| = 0$ if and only if $\mu_A(x)) = 0$ for any $x \in U$.

Note that $\mu_A(x)\mu_B(x) \le \min\{\mu_A(x), \mu_B(x)\}$ for any $x \in U$, the following result can be obtained from properties of definite integral.

Proposition 2.1. *For any $A = \{(x, \mu_A(x)) \mid x \in U\}, B = \{(x, \mu_B(x)) \mid x \in U\} \in \mathcal{F}(\widetilde{U})$, we have*

$$\langle A, B\rangle \le \int_a^b \min\{\mu_A(x), \mu_B(x)\}dx.$$

Similar to the angle between vectors, we can define the angle between fuzzy sets by the inner product between fuzzy sets.

Definition 2.2. Let $A = \{(x, \mu_A(x)) \mid x \in U\}, B = \{(x, \mu_B(x)) \mid x \in U\} \in \mathcal{F}(\widetilde{U})$ and $\mu_A(x) \ne 0$ for some $x_1 \in U$, $\mu_B(x) \ne 0$ for some $x_2 \in U$ the angle θ between A and B is given by

$$\theta = \cos^{-1}\left(\frac{\langle A, B\rangle}{\|A\| \cdot \|B\|}\right) = \cos^{-1}\left(\frac{\int_a^b \mu_A(x)\mu_B(x)dx}{\sqrt{\int_a^b (\mu_A(x))^2 dx}\sqrt{\int_a^b (\mu_B(x))^2 dx}}\right).$$

According to Cauchy-Schwarz inequality in integral form, we have $\int_a^b \mu_A(x)\mu_B(x)dx \le \sqrt{\int_a^b (\mu_A(x))^2 dx}\sqrt{\int_a^b (\mu_B(x))^2 dx}$, hence, the above definition is well-defined. Usually, the angle θ between A and B is also written as **Agl**(A, B).

Definition 2.3. The similarity measure S between fuzzy sets based on axiomatic definition is a real function $S : \mathcal{F}(\widetilde{U}) \times \mathcal{F}(\widetilde{U}) \to [0, 1]$ such that $S(A, B) = \cos\theta$, where $A = \{(x, \mu_A(x)) \mid x \in U\}, B = \{(x, \mu_B(x)) \mid x \in U\} \in \mathcal{F}(\widetilde{U})$ and θ is the angle between A and B. That is,

$$S(A, B) = \frac{\langle A, B\rangle}{\|A\| \cdot \|B\|} = \frac{\int_a^b \mu_A(x)\mu_B(x)dx}{\sqrt{\int_a^b (\mu_A(x))^2 dx}\sqrt{\int_a^b (\mu_B(x))^2 dx}}.$$

From Definitions 2.2 and 2.3, the following proposition is obvious.

Proposition 2.2. *For any $A = \{(x, \mu_A(x)) \mid x \in U\}, B = \{(x, \mu_B(x)) \mid x \in U\} \in \mathcal{F}(\widetilde{U})$, we have*

(1) $0 \le S(A, B) \le 1$;
(2) $S(A, B) = 1$ if and only if $\mu_A(x) = k \cdot \mu_B(x)$, $k \ne 0$;
(3) $S(A, B) = S(B, A)$.

Remark 2.1. This similarity measure proposed in this paper is based on the axiomatic definition of fuzzy sets, we hope to characterize the similarity between fuzzy sets from the essence of vagueness and the corresponding axioms proposed in [12], in this sense, our method is different from known similarity measures [7].

3. Conclusion

In this paper, we present a similarity measure between fuzzy sets based on axiomatic definition. Compared with other similarity measures between fuzzy sets in Zadeh's sense which have been proposed in the literature, the similarity measure in this paper has a different domain, i.e., the fuzzy sets based on axiomatic definition.

Acknowledgments

The work was partially supported by the National Natural Science Foundation of China (Grant No. 61673320, 61771016).

References

1. B. Bede, *Mathematics of fuzzy sets and fuzzy logic*, Springer-Verlag Berlin Heidelberg, 2013.
2. P. Cintula, P. Hájek, C. Noguera (eds.), *Handbook of Mathematical Fuzzy Logic*, Studies in Logic, Mathematical Logic and Foundations, vol. 37 and 38. College Publications, London, 2011.
3. D. Dubois, H. Prade, *The legacy of 50 years of fuzzy sets: A discussion*, Fuzzy Sets and Systems 281, 21–31, 2015.
4. A. Dvořák, V. Novák, *Formal theories and linguistic descriptions*, Fuzzy Sets and Systems 143, 169–188, 2004.
5. D. Dubois, H. Prade, E. P. Klement (eds.), *Fuzzy sets, logics and reasoning about knowledge*, Kluwer Academic Publishers, 1999.
6. P. Hájek, *Metamathematics of fuzzy logic*, Dordrecht: Kluwer, 1998.
7. H.B. Mitchell, *On the Dengfeng-Chuntian similarity measure and its application to pattern recognition*, Pattern Recogn. Lett. 24, 3101–3104, 2003.
8. V. Novák, I. Perfilieva, J. Mockor, *Mathematical principles of fuzzy logic*, Boston: Kluwer, 1999.
9. V. Novák, *Which logic is the real fuzzy logic?* Fuzzy Sets and Systems 157, 635–641, 2006.
10. V. Novák, *Reasoning about mathematical fuzzy logic and its future*, Fuzzy Sets and Systems 192, 25–44, 2012.
11. X.D. Pan, *Vague partition*, Proceedings of 2015 International Conference on Intelligent Systems and Knowledge Engineering (ISKE 2015), 24–27 November 2015, Taipei, Taiwan, 83–88.
12. X.D. Pan, Yang Xu, *Redefinition of the concept of fuzzy set based on vague partition from the perspective of axiomatization*, Soft Computing 22, 1777–1789, 2018.

13. X.D. Pan, Yang Xu, *Redefinition of the concept of fuzzy set based on vague partition from the perspective of axiomatization*, Soft Computing 22, 2079, 2018.
14. J. Pavelka, *On fuzzy logic I: Many-valued rules of inference, II: Enriched residuated lattices and semantics of propositional calculi, III: Semantical completeness of some many-valued propositional calculi*. Zeitschr. F. Math. Logik und Grundlagend. Math. 25, 45–52, 119–134, 447–464, 1979.
15. L. A. Zadeh, *Fuzzy sets*, Infirm. and Control 8, 338–353, 1965.
16. L. A. Zadeh, *Fuzzy logic-a personal perspective*, Fuzzy Sets and Systems 281, 4–20, 2015.

Development of the automated system for design solutions' analysis at CAD KOMPAS-3D

Alexander Nikolaevich Afanasyev

Ulyanovsk State Technical University
32, Severny Venets, Ulyanovsk, 432027, Russia

Sergey Igorevich Brigadnov

Ulyanovsk State Technical University
32, Severny Venets, Ulyanovsk, 432027, Russia

Nikolay Nikolaevich Voit

Computing Technique Department, Ulyanovsk State Technical University
32, Severny Venets, Ulyanovsk, 432027, Russia

Tatyana Vasilievna Afanasyeva

Information Systems Department, Ulyanovsk State Technical University
32, Severny Venets, Ulyanovsk, 432027, Russia

This paper presents an automated system for CAD KOMPAS-3D-based design solutions' analysis and its main operation modes. The method of a structural-parametric analysis for CAD KOMPAS-3D-based design solutions is proposed. An algorithm of forming a sequence of optimal design operations is developed. An example of the design solution's analysis based on a specific detail performed at CAD KOMPAS-3D is given.

1. Introduction

There are various analysis subsystems in the computer-aided design of engineering objects, for example: a strength analysis. It consists of static calculation, stability calculation, calculation of natural frequencies and modes of natural oscillations, calculation of stationary thermal conductivity and thermoelasticity; the analysis of the dynamic behavior of machinery and mechanisms; subsystems for checking compliance with the design standards (the distance between dimension lines, a text arrangement, intersections in the dimension line, line styles and ticks, etc.), compliance with the company's restrictive lists (valid roughness values, quality grade, threads, etc.), compliance with the rules on working in CAD (manual dimensions input, linking the position designation to the specification, the use of an axial object not lines with the axial style, etc.); calculation of dimensional chains and springs; optimization of gearing; selection of electric motors, reducers and coupling. At the same time, there is no analysis of designers' actions in CAD.

In practice, the project activities for developing 3D model engineering objects often encounter operations executed by designers, which lead to suboptimal design solution (in terms of the quantity and sequence of design operations). As a result, the design solutions' tree and the automated programs for CNC machines become complicated (unnecessary operations make these programs more complex).

CAD (KOMPAS, INDORCAD, Autodesk Inventor [1], CATIA, SolidWorks, T-FLEX CAD) and design solutions' analysis tools (KOMPAS-Expert, INDORCAD 10 [2], ANSYS [3]) are widely used in the production. There are no functions for determining suboptimal sequences of design operations. It is not possible to automatically rebuild a 3D model product based on the design solutions' analysis.

Thus, the topical task in the field of computer-aided design of engineering objects is the structural-parametric analysis of a design solution in order to identify suboptimal sequences of design operations, automatically restructure a model, and generate appropriate recommendations for a designer.

The paper deals with a new developed method of a structural- parametric analysis for CAD KOMPAS-3D based design solutions, the algorithm for forming recommendations to the designer. We have evaluated the developed method efficiency.

2. Problem

The research work describes the problem concerned with the quality improvement of engineering objects' design solutions performed at CAD.

The main objectives for developing an automated system for design solutions' analysis are.
1. Improve the quality of design solutions and simplify design solutions.
2. Increase CAD productivity by reducing the number of objects obtained in a model tree after design solutions' analysis.
3. Upgrade designers' skills and performance.

In order to achieve the above functional requirements, the following tasks have been completed.
1. A method of structural-parametric analysis for design solutions has been developed on basis of the design operations' sequence performed at CAD KOMPAS-3D, including the development of:

- models of the design operations' sequence, initial data for computer-aided rebuilding of an object 3D modeling, 3D model details, 3D model variables and parameters;
- algorithm of forming a design operations' sequence based on the design solutions' analysis [4];
- algorithm of searching for suboptimally executed design operations and their replacement by operations with fewer actions;
- rules for searching for suboptimal design operations and their replacement by operations with fewer actions.
2. The proposed models, method and algorithms are implemented as a software package.

3. Related Works

In [5], the authors propose to use Siemens PLM Software's NX. It is a flagship product for CAD/CAM/CAE which offers integrated, automated design validation capabilities. Cimatron CAD/CAM software [6] designs complex details based on a mesh generation system for finite elements. In [7] describes the use of parametric control at both part and assembly levels and the ability to include technological information at all stages of product development in Creo Parametric. The end-to-end object-oriented system CADdy developed by the German company Ziegler (DataSolid GmbH) [8] provides standardization and unification of design solutions based on parameterization. In [9], the CATIA system is represented as a system that can effectively perform all product design engineering tasks: from conceptual design to drawings, specifications, wiring diagrams and control programs for CNC machines.

A common disadvantage of the above research works is the lack of search function of suboptimal sequence of design operations and automatic rebuilding a three-dimensional model of a product.

4. Solution

The method of structural- parametric analysis of design solutions is to search for design operations performed suboptimally by a designer. It differs from known methods in that the model tree of a design solution and 3D model objects' operations performed at CAD KOMPAS-3D are analyzed. The method allows designers to reconstruct the design solution's tree and to classify the products of engineering objects.

The model tree of the design solution displays a part (assembly) as a list of objects in accordance with the order in which they were designed. The History of Construction mode shows the assembly in the model tree of the design solution. This mode is used to represent the sequence of design operations, and it is used to correct operations in which the result of previous designer's actions influences on next ones. Each element of a model tree has certain properties and parameters: external parameters, coverage, material of manufacture, etc.

The sequence of design operations' analysis of 3D modeling objects executed in the CAD KOMPAS-3D environment is carried out on the basis of rules. The rule for the design operations' analysis consists of the following components: an operation type, a text description of the rule, a condition for triggering the rule. If the rule is found for the project operations sequence, a corresponding recommendation is given to a designer.

The initial data for CAD KOMPAS-based design solutions' analysis are the sequence of design operations performed by the designer. The model is given as:

$$PPrOperations = (Operations, TypesOperation, ParamsOp),$$

where $Operations = \{op_i | i = 1..k\}$ — a set of design operations,
$TypesOperation = \{o3d_i | i = 0..159\}$ — many types of operations at CAD KOMPAS (e.g., $o3d\_fillet = 34$ — the "Round" operation; $o3d\_chamfer = 33$ — the "Chamfer" operation),
$ParamsOp = \{pr_i | i = 1..PR\}$ — is a set of parameters for operations with a value.

The operation model is given as:

$$Operation = (number, type, params),$$

where $number$ — is the number of an operation in the operations' sequence,
$type \in TypesOperation$ — is a type of an operation,
$params \in ParamsOp$ — a list of operation's parameters with a value.

The model of the initial data for computer-aided rebuilding of a 3D model object is given as:

$$RebuildModel = (Details, Operations, Rules),$$

where $Details = \{dt_i | i = 1..k\}$ — a set of details included in the three-dimensional model of a CAD KOMPAS-based product,
$Operations = \{op \in Operation\}$ — a set of design operations,
$Rules = \{r_i | i = 1..k\}$ is a set of rules for searching for suboptimal design operations and their replacement by operations with fewer actions.

The model of details in a 3D model of CAD KOMPAS-based product is as follows:

$$Details = (id, class, attribute, material),$$

where id — a unique identifier of a detail,
$class$ — a class of a detail (for example, "Ring", "Bush", "Flange", etc.).

attribute — *a* set of variables and parameters of a three-dimensional model,
material — material for the detail production (for example, "Steel 10 GOST 1050-88" (All-Union State Standard)).

The model of variables and parameters of a three-dimensional model is given as:

$$attribute = (name, value, note),$$

where *name* — a name of a variable or parameter of a three-dimensional model,
value — a value of a variable or parameter of a three-dimensional model,
note — a description of a variable or parameter of the three-dimensional model (e.g., "d = 24 — a diameter of a frame mounting surface")

The rule model is given as:

$$Rules = (template, result),$$

where $template = \{tpl_i | i = 1..k\}$ is a first-order logic formula of searching for suboptimal operations in a design operations' sequence,
$result = \{res_i | i = 1..n\}$, $res = (type, params)$ — (type, params) is a set of optimal design operations (operation type, operation parameter with value), where the operation type is a constant, and the operation parameter with a value is a first-order logic formula.

The rule template in general terms has a following structure:

$$TPL = (id, type, txt, action),$$

where *id* — *an* identifier of a rule,
type $\in$ *TypesOperation* — a type of an operation,
txt — a description of a rule,
action — a rule's trigger condition.

Improving the quality of design solutions performed at CAD KOMPAS as well as increasing the designer's performance can be achieved through the search for suboptimally executed design operations and their replacement by operations with fewer actions.

The initial data for searching and analyzing the rules of an operations' sequence is ABC KOMPAS.

The algorithm of forming a sequence of optimal design operations consists of the following steps.
1. Start the designer's work with the project.
2. Generate operations based on an existing project (the initial data is the XML description of the assembly).
3. Add an operation to a sequence of operations.
4. Search for a rule that corresponds to an operation's sequence.

5. Form the optimal set of operations.
6. Generate recommendations for replacing suboptimal operations based on sets of optimal and suboptimal operations.
7. Add a recommendation to the individual designer's list of recommendations.
8. Replacing the set of suboptimal operations by a sequence with fewer actions.
9. Rebuild the design solution based on the sequence from step 8.
10. Save the design solution and display it in CAD KOMPAS.

5. Examples

Let us consider the rule of searching for suboptimal operations using the example of the "Round" operation. This rule has the following description:

"Do not use the" Round "operation for each edge individually. If it is possible, specify as many edges as possible, which parameters are the same."

Several "Round" operations with the same parameters are a condition of triggering the rule.

For example, the detail of "Casing" was considered, its model tree, prior to the design solution analysis, consists of 92 operations, 13 of which are forming operations, 72 — construction of geometric objects, and 7 — construction of sketches. The total number of actions taken in building is 395. The model tree contains 6 "Round" operations with the same parameters.

For this detail, the result of the design solutions' analysis is the following recommendation:

«You have 6 identical operations ("Round: 5", "Round: 6", "Round: 7", "Round: 8", "Round: 9", "Round: 10"). Do not use the "Round" operation for each edge individually. If it is possible, specify as many edges as possible, which parameters are the same. This will result to 63% reduction in the number of actions.

If recommendations are fulfilled, the total number of actions will decrease from 395 to 380 or by 4%».

6. Experiment

As a measure of the designer's work effectiveness [10] we choose the number of actions (design operations) performed in CAD KOMPAS-3D when building a 3D part.

In order to assess the method of a structural-parametric analysis of design solutions, we analyzed assemblies developed in CAD KOMPAS-3D. The results are presented in the summary Table 1.

Table 1. Results of the analysis of product assemblies.

Assembly	Number of operations	Number of actions	Number of recommendations	Relative value of reducing the number of actions, %
Pump	830	3578	16	3,0%
Reducer	220	913	2	1,1%
Block	318	1330	7	4,0%
Suspension	937	3967	6	1,1%
Hud	162	742	4	1,0%
Assembly 1	46	212	3	20,8%
Assembly 2	123	542	1	3,1%
Assembly 3	58	287	1	5,9%
Assembly 4	119	552	3	3,4%
Assembly 5	1425	6075	6	0,9%
TOTAL	4238	18198	49	4,4%

The developed method of a structural-parametric analysis allows us to reduce the actions number performed by the designer on average by 4.4%.

7. Conclusion

The models of a design operations' sequence, initial data for computer-aided rebuilding of a 3D model object, details, variables and parameters of a 3D model are proposed.

Using the proposed models, we have developed a method for a structural-parametric analysis of design solutions based on the designer's action flow at CAD. It allows designers to increase their performance, as well as to improve the quality and simplify CAD-based design solutions.

Acknowledgments

The reported research was funded by the Ministry of Education of the Russian Federation, Grant № 2.1615.2017/4.6 and by Russian Foundation of Basic Research and the Government of the Ulyanovsk region, Grant № 16-47-732152.

References

1. T. Kishore, Getting Started with Autodesk Inventor 2018, *In book: Learn Autodesk Inventor 2018 Basics*.
2. V. Boykov. N. Mirza, D. Petrenko, A. Skvortsov IndorCAD 10 as BIM-tool for project analysis and conflicts detection. doi: 10.17273/CADGIS.2015.2.16
3. M. K. Thompson, J. Thompson, Interacting with ANSYS, *In book: ANSYS Mechanical APDL for Finite Element Analysis*.
4. A. Afanasyev, N. Voit, D. Kanev, Development of expert systems for evaluating user's actions in training systems and virtual laboratories, in *Proc. of The 11th IEEE International Conference on Application of Information and Communication Technologies (AICT-2017)*, (Moscow, Russia, 2017).

5. *NX for Design: Advantages and Benefits* (2018). https://www.plm.automation.siemens.com/en/products/nx/for-design/advantages-benefits.shtml.
6. *3D Systems Software* (2018). http://www.3dsystemssoftware.it/ShowProduct.aspx?ID=Cimatron.
7. *3D CAD Software Creo PTC* (2018). https://www.ptc.com/en/products/cad/creo.
8. *CADdy++ mechanical professional* (2018). http://english.datasolid.com/products/caddy-mechanical-design/caddy-mechanical-design-professional?showall=&start=1.
9. *CATIA Engineering* (2018). https://www.3ds.com/products-services/catia/products/v6/portfolio/d/digital-product-experience/s/catia-engineering/.
10. A. Afanasyev, N. Voit, I. Ionova, M. Ukhanova, V. Yepifanov, Development of the Intelligent System of Engineering Education for Corporate Use in the University and Enterprises. *In book: Teaching and Learning in a Digital World* (2018). doi: 10.1007/978-3-319-73210-7_84

Solution of a fuzzy differential equation with interactivity via fuzzy Laplace transform

S. A. B. Salgado

*Institute of Applied Social Sciences, Federal University of Alfenas
Varginha, Minas Gerais, Brazil, 37130-001
silvio.salgado@unifal-mg.edu.br*

L. C. Barros* and E. Esmi[†]

*Department of Applied Mathematics, University of Campinas
Campinas, São Paulo, Brazil, 13083-859
*laeciocb@ime.unicamp.br
[†] eelaureano@ime.unicamp.br*

This presentation considers a type of fuzzy harmonic oscillator described by differential equation of the form $x'' + a^2 x = 0$ with initial conditions given by fuzzy numbers, where $x''(t)$ and $x(t)$ are linearly correlated fuzzy numbers for each t. We obtain a solution for this problem using a fuzzy Laplace transform method.

Keywords: Interactive fuzzy numbers; fuzzy differential equation; fuzzy Laplace transform; fuzzy harmonic oscillator.

1. Introduction

Differential equations appear in many disciplines, such as engineering, physics, demography, economics, etc. For instance, Sandoval Jr. and França modelled a period of crisis in the stock market using an equation of the form $x'' + a^2 x = 0$.[1] Since phenomena of these types usually involve uncertainties, it is reasonable to deal with these problems using fuzzy set theory. Thus, in the case where $x(t)$ and $x''(t)$ are fuzzy numbers for each t in above equation, we are faced with a fuzzy differential equation given by an addition of two fuzzy numbers whose sum is zero. According to Barros and Santo Pedro,[2] this sum only makes sense when one uses the arithmetic of linearly correlated fuzzy numbers.

Several authors have studied different techniques for solving fuzzy initial value problems (FIVP). Allahviranloo and Ahmadi[3] introduced the concept of a fuzzy Laplace transform using the notion of strongly generalized differential. This operator can be used to obtain solutions of FIVPs of first and second orders.[3,4] Conditions for the existence of the fuzzy Laplace transform have been established.[4]

Here, we study the solution of the differential equation $x'' + a^2 x = 0$, $a \neq 0$ with uncertain initial conditions given by fuzzy triangular numbers, using the arithmetic of linearly correlated fuzzy numbers and the fuzzy Laplace transform method.

2. Basic Concepts

Let U be a topological space. A fuzzy subset A of U is characterized by a membership function $\mu_A : U \longrightarrow [0,1]$, where $\mu_A(x)$ denotes the membership degree that element x belongs to the set A. For notation convenience, we may use the symbol $A(x)$ instead of $\mu_A(x)$. If A is a classical subset of U, its membership function is given by the characteristic function $\chi_A : U \to \{0,1\}$ where $\chi_A(x) = 1 \Leftrightarrow x \in A$. The α-level of the fuzzy subset A is defined by $[A]^\alpha = \{x \in U; A(x) \geq \alpha\}$ for $\alpha \in (0,1]$ and $[A]^0 = cl\{x \in U; A(x) > 0\}$, that is, the closure of the support of A.

A fuzzy set A of $\mathbb{R}$ is said to be a fuzzy number if its α-level are non-empty closed nested intervals of $\mathbb{R}$[5]. Thus, the α-level of a fuzzy number A is given by $[A]^\alpha = [a_-^\alpha, a_+^\alpha]$ for each $\alpha \in [0,1]$. We use the symbol $\mathbb{R}_\mathcal{F}$ to denote the set of all fuzzy numbers.

The addition and scalar multiplication operations of two fuzzy numbers can be defined for all $A, B \in \mathbb{R}_\mathcal{F}$ as follows:

$$(A \oplus B)(z) = \sup_{z=x+y} \min\{A(x), B(y)\}, \text{ and } (k \odot A)(z) = \begin{cases} A\left(\frac{z}{k}\right), k \neq 0 \\ \chi_{\{0\}}(x), k = 0 \end{cases}$$

for all $z \in \mathbb{R}$. The Hukuhara difference (H-difference), $B \ominus A = C$ is the fuzzy number C, such that $B = A \oplus C$, if it exists.[6] The set $\mathbb{R}_\mathcal{F}$ forms a complete metric space with the Hausdorff-Pompeiu distance D_∞ given by

$$D_\infty(A, B) = \sup_{0 \leq \alpha \leq 1} \max\{|a_-^\alpha - b_-^\alpha|, |a_+^\alpha - b_+^\alpha|\}. \tag{1}$$

Definition 2.1.[7] Let $f : (a,b) \longrightarrow \mathbb{R}_\mathcal{F}$ and $t_0 \in (a,b)$. We say that f is strongly generalized differentiable at t_0 if there exists an element $f'(t_0) \in \mathbb{R}_\mathcal{F}$, such that for all $h > 0$ sufficiently small, one of these conditions holds true
(i)

$$\lim_{h \searrow 0} \frac{f(t_0 + h) \ominus f(t_0)}{h} = \lim_{h \searrow 0} \frac{f(t_0) \ominus f(t_0 - h)}{h} = f'(t_0) \tag{2}$$

(ii)
$$\lim_{h \searrow 0} \frac{f(t_0) \ominus f(t_0 + h)}{-h} = \lim_{h \searrow 0} \frac{f(t_0 - h) \ominus f(t_0)}{-h} = f'(t_0) \qquad (3)$$

(iii)
$$\lim_{h \searrow 0} \frac{f(t_0 + h) \ominus f(t_0)}{h} = \lim_{h \searrow 0} \frac{f(t_0 - h) \ominus f(t_0)}{-h} = f'(t_0) \qquad (4)$$

(iv)
$$\lim_{h \searrow 0} \frac{f(t_0) \ominus f(t_0 + h)}{-h} = \lim_{h \searrow 0} \frac{f(t_0) \ominus f(t_0 - h)}{h} = f'(t_0) \qquad (5)$$

where all limits are taken in the metric D_∞.

Remark 2.1. Other types of arithmetics can be defined by means of the notion of interactivity between fuzzy number.[2,8] For example, two fuzzy numbers A and B are said to be linearly correlated if there exists $q, r \in \mathbb{R}$ such that $[B]^\alpha = q[A]^\alpha + r$ for each $\alpha \in [0, 1]$. In this case, it is written $B = qA + r$.[2,8] The addition and subtraction of A and B are given respectively by $[B +_L A]^\alpha = (q+1)[A]^\alpha + r$ and $[B -_L A]^\alpha = (q-1)[A]^\alpha + r$, for each $\alpha \in [0, 1]$. Note that if $q = -1$ and $r = 0$ then $B +_L A = 0$.

The fuzzy Laplace transform method introduced by Allahviranloo and Ahmadi[3] can be used to solve the linear fuzzy initial value problems.

Definition 2.2. Let $f : [0, \infty[\longrightarrow \mathbb{R}_\mathcal{F}$. If $f(t) \odot e^{-st}$ is improper fuzzy Riemann integrable[3] on $[0, \infty[$ for each $s > 0$, then
$$\mathcal{L}\{f(t)\} = \int_0^\infty f(t) \odot e^{-st} dt$$
is called a fuzzy Laplace transform.

In Definition 2.2, if f is a continuous function and $[f(t)]^\alpha = [f_-^\alpha(t), f_+^\alpha(t)]$ for $\alpha \in [0, 1]$ and $t \geq 0$, then we have that
$$[\mathcal{L}\{f(t)\}]^\alpha = [L\{f_-^\alpha(t)\}, L\{f_+^\alpha(t)\}]. \qquad (6)$$
for all $\alpha \in [0, 1]$ and $s > 0$, where L denotes the usual Laplace transform, that is, $L\{g(t)\} = \int_0^\infty g(t) e^{-st} dt$.[4] The next propositions establish other properties of the fuzzy Laplace transform operator.

Proposition 2.1.[3] *Let f, g be continuous fuzzy-number-valued functions. If $c_1, c_2 \in \mathbb{R}$ are constants, then $\mathcal{L}\{(c_1 \odot f(t)) \oplus (c_2 \odot g(t))\} = (c_1 \odot \mathcal{L}\{f(t)\}) \oplus (c_2 \odot \mathcal{L}\{g(t)\}).*

Proposition 2.2.[3] *Let f be an integrable fuzzy-number-valued function. If f is the primitive of f' on $[0, \infty[$, then*

1) $\mathcal{L}\left\{f'(t)\right\} = (s \odot \mathcal{L}\{f(t)\}) \ominus f(0)$ *if f is (i)-differentiable;*

2) $\mathcal{L}\left\{f'(t)\right\} = ((-1) \odot f(0)) \ominus (-s \odot \mathcal{L}\{f(t)\})$ *if f is (ii)-differentiable.*

Proposition 2.3.[4] *Under the conditions of Proposition 2.2, if f' is a continuous functions on $[0, \infty[$ such that its fuzzy Laplace transform exists, then*

1) $\mathcal{L}\left\{f''(t)\right\} = (s^2 \odot \mathcal{L}\{f(t)\}) \ominus (s \odot f(0)) \ominus f'(0)$ *if f and f' are differentiable in the sense (i);*

2) $\mathcal{L}\left\{f''(t)\right\} = \left((-1) \odot f'(0)\right) \ominus (-s^2 \odot \mathcal{L}\{f(t)\}) \oplus (-s \odot f(0))$ *is differentiable in the sense (i) and f' is differentiable in the sense (ii);*

3) $\mathcal{L}\left\{f''(t)\right\} = (-s \odot f(0)) \ominus (-s^2 \odot \mathcal{L}\{f(t)\}) \ominus \left(f'(0)\right)$ *if f is differentiable in the sense (ii) and f' is differentiable in the sense (i);*

4) $\mathcal{L}\left\{f''(t)\right\} = (s^2 \odot \mathcal{L}\{f(t)\}) \ominus (s \odot f(0)) \oplus \left((-1) \odot f'(0)\right)$ *if f and f' are differentiable in the sense (ii).*

3. A Solution for a Fuzzy Harmonic Oscillator

We consider in this section a FIVP of the form

$$\begin{cases} x'' + a^2 x = 0 \\ x(0) = A \in \mathbb{R}_{\mathcal{F}} \\ x'(0) = B \in \mathbb{R}_{\mathcal{F}} \end{cases} \tag{7}$$

where $x, x'' : [0, \infty[\longrightarrow \mathbb{R}_{\mathcal{F}}$, $[A]^\alpha = [a_-^\alpha, a_+^\alpha]$ and $[B]^\alpha = [b_-^\alpha, b_+^\alpha]$ for all $\alpha \in [0,1]$, and $a > 0$. Note that, for each $t \in [0, \infty[$, $x(t)$ and $x''(t)$ are fuzzy numbers whose the sum is zero. This sum makes sense if $x''(t)$ and $a^2 x(t)$ is linearly correlated with $q = -1$ and $r = 0$ from Remark 2.1. This last observation implies that $x''(t) = -a^2 x(t)$ for each $t \in [0, \infty[$. Thus, the FPVI (7) can be rewritten as follows:

$$\begin{cases} x'' = -a^2 x \\ x(0) = A \in \mathbb{R}_{\mathcal{F}} \\ x'(0) = B \in \mathbb{R}_{\mathcal{F}} \end{cases} \tag{8}$$

We obtain a fuzzy solution, $x(\cdot)$, with $[x(t)]^\alpha = [x_-^\alpha(t), x_+^\alpha(t)]$ for (8) by applying the fuzzy Laplace transform in both sides of the differential equation $x'' = -a^2 x$, supposing that the solution obeys Proposition 2.3.

Thus, we have $[\mathcal{L}\{x(t)\}]^\alpha = [L\{x_-^\alpha(t)\}, L\{x_+^\alpha(t)\}]$ where

$$\begin{cases} L\{x_-^\alpha(t)\} = -\frac{s^3}{a^4-s^4}a_-^\alpha - \frac{s^2}{a^4-s^4}b_-^\alpha + \frac{a^2 s}{a^4-s^4}a_+^\alpha + \frac{a^2}{a^4-s^4}b_+^\alpha \\ L\{x_+^\alpha(t)\} = -\frac{s^3}{a^4-s^4}a_+^\alpha - \frac{s^2}{a^4-s^4}b_+^\alpha + \frac{a^2 s}{a^4-s^4}a_-^\alpha + \frac{a^2}{a^4-s^4}b_-^\alpha \end{cases} \quad (9)$$

Finally, from Equation (9), we can apply the usual inverse Laplace transform to obtain $[x(t)]^\alpha = [x_-^\alpha(t), x_+^\alpha(t)]$ where

$$\begin{cases} x_-^\alpha(t) = \frac{1}{2}\left(\cosh at + \cos at\right) a_-^\alpha + \frac{1}{2a}\left(\sinh at + \sin at\right) b_-^\alpha \\ \qquad + \frac{1}{2}\left(\cos at - \cosh at\right) a_+^\alpha + \frac{1}{2a}\left(\sin at - \sinh at\right) b_+^\alpha \\ x_+^\alpha(t) = \frac{1}{2}\left(\cosh at + \cos at\right) a_+^\alpha + \frac{1}{2a}\left(\sinh at + \sin at\right) b_+^\alpha \\ \qquad + \frac{1}{2}\left(\cos at - \cosh at\right) a_-^\alpha + \frac{1}{2a}\left(\sin at - \sinh at\right) b_-^\alpha \end{cases} \quad (10)$$

It is easy to see that if the initial conditions are real numbers, that is, $a_-^\alpha = a_+^\alpha = \bar{a}$ and $b_-^\alpha = b_+^\alpha = \bar{b}$, for all $\alpha \in [0,1]$, then Equation (10) becomes the classical solution of harmonic oscillator

$$x(t) = \bar{a}\cos(at) + \bar{b}\sin(at). \quad (11)$$

Moreover, this solution is preferred in the sense that, for each $t > 0$, it belongs to fuzzy solution (10) with membership degree equal to 1. Deterministic solution (11) is depicted in Figure 1(a), for $\bar{a} = 1$, $\bar{b} = 0$ and $a = 0.1$. The fuzzy solution $x(\cdot)$ given by Equation (10) with the triangular fuzzy initial conditions $A = (0.9; 1; 1.1)$ and $B = (-0.05; 0; 0)$ and the parameter $a = 0.1$ is depicted in Figure 1(b).

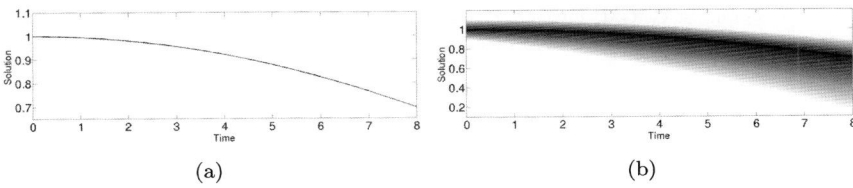

Fig. 1. (a) Deterministic solution $x(t)$ given by (10), for $\bar{a} = 1$, $\bar{b} = 0$ and $a = 0.1$. (b) The top figure of the fuzzy solution $x(t)$ (10), with initial conditions given by triangular fuzzy numbers $A = (0.9; 1; 1.1)$ and $B = (-0.05; 0; 0)$ and parameter $a = 0.1$. The endpoints of the α−level for α varying from 0 to 1 are represented respectively by the gray-scale lines varying from white to black.

4. Conclusion

It is common to use differential equations to describe the temporal evolution of certain phenomena such as in demography, stock pricing, and so on, even those with oscillatory behavior. In this work we studied the differential equation $x'' + a^2x = 0$, $a > 0$, which models small oscillations of the so-called simple harmonic motion. Assuming that the phenomena present uncertainties in its variable state, we treated $x(t)$ by means of the fuzzy set theory. Thus, $x'' + a^2x = 0$ is, typically, a fuzzy differential equation. However, in fuzzy arithmetic, the $x'' + a^2x = 0$ makes sense if this sum is a linearly correlated fuzzy numbers. That is, $x'' + a^2x = 0$ is equivalent $x'' = -a^2x$. We treated this last fuzzy differential equation by means of the strongly generalized differentiability and we used the fuzzy Laplace transform method to find the fuzzy solution.

Acknowledgments

This research was partially supported CNPq under grant no. 306546/2017-5 and FAPESP under grant no. 2016/26040-7.

References

1. L. S. Jr. and I. D. França, *Physica A* **391**, 187 (2012).
2. L. C. Barros and F. S. Pedro, *Fuzzy Sets and Systems* **309**, 64 (2017).
3. T. Allhviranloo and M. B. Ahmadi, *Soft Comput-Springer* **14**, 235 (2010).
4. T. Allhviranloo and S. Salahshour, *Soft Comput-Springer* **17**, 145 (2013).
5. L. C. Barros, R. C. Bassanezi and W. A. Lodwick, *A First Course in Fuzzy Logic, Fuzzy Dynamical Systems, and Biomathematics* (Springer-Verlag Berlin Heidelberg, 2017).
6. M. L. Puri and D. A. Ralescu, *Journal of Mathematical Analysis and Applications* **91**, 552 (1983).
7. B. Bede and S. G. Gal, *Fuzzy Sets and Systems* **147**, 385 (2004).
8. C. Carlsson, R. Fuller and P. Majlender, *Proc. IEEE Inter. Conference* **1**, 535 (2004).

Robustness of the interval-valued fuzzy inference algorithms

M. X. Luo* and L. X. Wu

Department of Information and Computing Science,
China Jiliang University,
Hangzhou, 310018, PR China
*minxialuo@163.com

L. Fu

School of Mathematics and Statistics,
Qinghai Nationalities University,
Xining, 810007, PR China
fl0971@163.com

A new distance metric between interval-valued fuzzy sets is proposed. And based on this metric, we analyze robustness of the interval-valued fuzzy reasoning algorithms based on two well-known interval-valued residual implication operations. It was proved that the interval-valued fuzzy reasoning algorithm based the Łukasiewicz interval-valued implication and Goguen interval-valued implication are robust.

Keywords: Interval-valued fuzzy sets; distance metric; logic metric spaces; triple I methods; robustness.

1. Introduction

Let M be an inference mechanism, and A, B the input and the output of the inference mechanism M respectively. If small disturbance of input A without causing large changes of output B in inference process, then we say that the inference mechanism M has a good robustness. Researchers in different areas have different answers on the concepts of disturbance. Therefore, various related notions such as the largest perturbation, the average perturbation, δ-equalities, δ-sensitivity, largest δ-sensitivity, average δ-sensitivity and so on were proposed. The largest perturbation and the average perturbation of fuzzy sets were proposed and the perturbation of several fuzzy reasoning systems were discussed.[21] Cai discussed the robustness of fuzzy inference using δ-equalities of connectives and fuzzy implication operators.[3] Literature[11,13] proposed concepts of δ-sensitivity, largest

δ-sensitivity and average δ-sensitivity of fuzzy connectives, and discussed the robustness of fuzzy reasoning. By comparing and analyzing these concepts, we notice that these concepts are based on different distance metrics. It is well known that fuzzy connectives determine the internal structure of a fuzzy logic system. However, the construction of distance metrics does not involve fuzzy connectives in previous works. Taking this into account, Dai et al.[5] and Jin et al.[8] proposed the notion of logic similarity degree between fuzzy sets based on fuzzy connectives, and discuss robustness of fuzzy reasoning. Wang et al.[19] proposed a new distance metric based on residuated implication and conjunction connective, and discussed the robustness of full implication triple I inference method. Then Duan studied the structures of four specific logic metric.[9]

However, there are some limitations when we deal with imprecise information using fuzzy sets. Therefore, interval-valued fuzzy set was introduced by Zadeh,[22] which can not only effectively reduce the loss of fuzzy information but also reflect the vagueness and uncertainty in information processing. And then many researchers have studied this topic and extended approximate inference to the case of interval-valued fuzzy sets. Li et al.[12] extended CRI method to the case of interval-valued fuzzy set and discussed the robustness of interval-valued CRI method. Luo et al.[15,16] extended triple I method and reverse triple I method to the case of interval-valued fuzzy sets and studied their robustness respectively. These researches of robustness of interval-valued fuzzy reasoning methods are based on moore metric.

However, it is easy to lose information when we use moore metric to study interval-valued fuzzy inference. For example, suppose $SI(X)$ is the class of all interval-valued fuzzy subsets of non-empty sets X. Let $X = \{x_1, x_2, \cdots, x_{10}\}$, A, B and $C \in SI(X)$, $A(x_i) = [1,1](1 \leq i \leq 10)$, $B(x_i) = [1,1](1 \leq i \leq 9)$, $B(x_{10}) = [0,0]$ and $C(x_i) = [0,0](1 \leq i \leq 10)$. If we use the moore metric, then we obtain $d(A, B) = \max\{\max_i\{|A_l(x_i) - B_l(x_i)|, |A_r(x_i) - B_r(x_i)|\}\} = 1$ and $d(A, C) = \max\{\max_i\{|A_l(x_i) - C_l(x_i)|, |A_r(x_i) - C_r(x_i)|\}\} = 1$. Obviously, this result is not reasonable. And because the behavior of a fuzzy logic system is mainly determined by its fuzzy connectives and fuzzy implication operators, based on this point, we propose a new distance metric of interval-valued fuzzy sets based on left-continuous t-norm and its residual implication. We analyze the robustness of the interval-valued fuzzy inference full implication method.

The remaining part is organized as the follows. Section 2 review some concepts needed for the paper. In Section 3, a new distance between interval-valued fuzzy sets based on left-continuous t-norm and its residuated implication is proposed. In Section 4, the robustness of interval-valued fuzzy connectives and interval-valued fuzzy inference triple I method. The final Section includes our conclusions.

2. Preliminaries

In this section, we review some concepts that will be required for our following work. Let $SI = \{[x,y] | x \leq y, x, y \in [0,1]\}$. An ordering on SI as $[a,b] \leq [c,d]$ if $a \leq c$ and $b \leq d$ is called component-wise order or Kulisch-Miranker order.[4] It is easy to verify that the ordering just defined is a partially ordering on SI. Furthermore, take $[a,b] \wedge [c,d] = [a,b]$ iff $[a,b] \leq [c,d]$ and $[a,b] \vee [c,d] = [c,d]$ iff $[a,b] \leq [c,d]$. In this paper, let $X = \{x_1, x_2, \cdots, x_n\}$ be non-empty sets, $SI(X)$ denote interval-valued fuzzy subsets of non-empty sets X, for $1 \leq i \leq n$, $A(x_i) \in SI(X)$ ($A(x_i)$ denoted by $[A_l(x_i), A_r(x_i)]$). A^c is complement of interval-valued A, where $A^c = [1 - A_r, 1 - A_l]$.

Definition 2.1 (Klement, 2000). *A function $T: [0,1]^2 \to [0,1]$ is called a triangular norm (t-norm) if it is commutative, associative, non-decreasing in each argument and $T(a, 1) = a$ for all $a \in [0, 1]$.*

Definition 2.2 (Wang, 2000). *Let L be a bounded lattice and T be a t-norm on L. If there exists another operator R_T: $L^2 \to L$ such that $T(a,b) \leq c$ if and only if $a \leq R_T(b,c)$, for all $a, b, c \in L$, then R_T is called the residuum of T, (T, R_T) is called a residuated pair on L.*

Many other properties of residuated implication could be found:[2,20]

(R1) $R_T(a,b) = 1$ if and only if $a \leq b$ for all $a,b \in [0,1]$,
(R2) $R_T(a, \wedge_{j \in J} b_j) = \wedge_{j \in J} R_T(a, b_j)$,
(R3) $R_T(\vee_{j \in J} a_j, b) = \wedge_{j \in J} R_T(a_j, b)$.

Example 2.1 (Wang, 2009). *Defined on the unit interval $[0,1]$ that:*
(1) Product t-norm and its residuum, Goguen implication:

$$T_{G_o}(a,b) = ab, \quad R_{G_o}(a,b) = \begin{cases} 1, & \text{if } a \leq b, \\ \frac{b}{a}, & \text{if } a > b. \end{cases}$$

(2) Lukasiewicz t-norm and its residuum, Lukasiewicz implication:

$$T_L(a,b) = 0 \vee (a+b-1), \quad R_L(a,b) = 1 \wedge (1-a+b).$$

Definition 2.3 (Jenei, 1997). *Let T be a t-norm defined on the interval $[0,1]$, the associated t-norm on SI is defined as: $\mathcal{T}: SI \times SI \to SI$ where $\mathcal{T}([a,b],[c,e]) = [T(a,c), T(b,e)]$. The associated t-norm $\mathcal{T}$ on SI is called left continuous, if T is left continuous t-norm on the interval $[0,1]$.*

Definition 2.4 (Alcalde, 2005). *For every $[a,b], [c,e] \in SI$, an interval-valued $\mathcal{R}$-implication is defined by:*

$$\mathcal{R}([a,b],[c,e]) = \bigvee\{[x,y] \in SI | \mathcal{T}([a,b],[x,y]) \leq [c,e]\},$$

where $\mathcal{T}$ is a t-norm on SI.

Definition 2.5 (Liu, 1992). *A real function $d: F(X) \times F(X) \to [0,1]$ is called a distance, if d satisfied the following properties:*
(D1): $d(A,B) = d(B,A)$ for all $A, B \in F(X)$;
(D2): $d(A, A^c) = 1 \iff A \in \mathcal{P}(X)$;
(D3): $d(A,B) = 0$ if and only if $A = B$, for all $A, B \in F(X)$;
(D4): for all $A, B, C \in F(X)$, if $A \leq B \leq C$, then $d(A,C) \geq d(A,B)$ and $d(A,C) \geq d(B,C)$.
Where $F(X)$ is fuzzy subsets of all non-empty set X and $\mathcal{P}(X)$ stands for the sets of all crisp sets in X.

3. A New Distance Metric Between Interval-valued Fuzzy Sets

Definition 3.1. *Let $X = \{x_1, x_2, \cdots, x_n\}$, $A, B \in SI(X)$. Suppose R_T is residuated implication introduced by the left-continuous t-norm T. We denote $d(A,B) = 1 - \frac{1}{n}\sum_{i=1}^{n} T(\varrho(A(x_i), B(x_i)), \varrho(B(x_i), A(x_i)))$ where $\varrho(A(x_i), B(x_i)) = \min\{R_T(A_l(x_i), B_l(x_i)), R_T(A_r(x_i), B_r(x_i))\}$. Then d is called a distance metric on $SI(X)$ and $(SI(X), d)$ is called a logic metric space.*

Remark 3.1. *If $A_l = A_r$ and $B_l = B_r$, then we have $d(A,B) = 1 - \frac{1}{n}\sum_{i=1}^{n} \rho(A(x_i), B(x_i))$, which is the distance metric in literature.*[19]

Theorem 3.1. *d defined by Definition 3.1 is a distance metric on $SI(X)$.*

Proposition 3.1. *Let $X = \{x_1, x_2, \cdots, x_n\}$, $A, B \in SI(X)$.*
(1) *If R is Goguen implication and T is corresponding t-norm, then $d_{Go}(A,B) = 1 - \frac{1}{n}\sum_{i=1}^{n}\{\varrho_{Go}(A(x_i), B(x_i)) \cdot \varrho_{Go}(B(x_i), A(x_i))\} = 1 - \frac{1}{n}\sum_{i=1}^{n}\{(\frac{B_l(x_i)}{A_l(x_i)} \wedge \frac{B_r(x_i)}{A_r(x_i)} \wedge 1) \cdot (\frac{A_l(x_i)}{B_l(x_i)} \wedge \frac{A_r(x_i)}{B_r(x_i)} \wedge 1)\}$.*
(2) *If R is Łukasiewicz implication and T is corresponding t-norm, then $d_L(A,B) = 1 - \frac{1}{n}\sum_{i=1}^{n}\{(\varrho_L(A(x_i), B(x_i)) + \varrho_L(B(x_i), A(x_i)) - 1) \vee 0\}$.*

By above two interval-valued distance metrics, we can construct two interval-valued fuzzy metric spaces $(SI(X), d_{Go})$ and $(SI(X), d_L)$.

4. Robustness of Interval-valued Fuzzy Reasoning Triple I Methods

Proposition 4.1. *Suppose that* $X = \{x_1, x_2, \cdots, x_n\}$, A, A', B *and* $B' \in SI(X)$. *If* $d(A, A') \leq \varepsilon_1$ *and* $d(B, B') \leq \varepsilon_2$, *then* $d(f(A, B), f(A', B')) \leq \varepsilon_1 + \varepsilon_2$, *where* $f \in \{\mathcal{T}_L, \mathcal{R}_L\}$.

Proposition 4.2. *Suppose that* $X = \{x_1, x_2, \cdots, x_n\}$, A, A', B *and* $B' \in SI(X)$. *If* $d(A, A') \leq \varepsilon_1$ *and* $d(B, B') \leq \varepsilon_2$, *then* $d(f(A, B), f(A', B')) \leq \varepsilon_1 + \varepsilon_2 - \varepsilon_1\varepsilon_2$, *where* $f \in \{\mathcal{T}_{Go}, \mathcal{R}_{Go}\}$.

Let $A, A^* \in SI(X)$ and $B, B^* \in SI(Y)$. Then the interval-valued $\mathcal{R}$-type I solution B^* of interval-valued fuzzy modus ponens ($IFMP$) and interval-valued fuzzy modus tollens ($IFMT$) can be respectively expressed as following two theorems.

Theorem 4.1 (Luo, Zhang, 2015). *Suppose that* $\mathcal{R}$ *is an interval-valued residuated implication induced by a left continuous t-norm* $\mathcal{T}$ *on* SI. *Then the interval-valued* $\mathcal{R}$-*type triple I solution* B^* *of* $IFMP$ *is given by the following formula:* $B^*(y) = \bigvee_{x \in X} \mathcal{T}(\mathcal{R}(A(x), B(y)), A^*(x))$, $y \in Y$.

Theorem 4.2 (Luo, Zhang, 2015). *Suppose that* $\mathcal{R}$ *is an interval-valued residuated implication induced by a left continuous t-norm* $\mathcal{T}$ *on* SI. *Then the interval-valued* $\mathcal{R}$-*type triple I solution* A^* *of* $IFMT$ *is given by the following formula:* $A^*(x) = \bigwedge_{y \in Y} \mathcal{R}(\mathcal{R}(A(x), B(y)), B^*(y))$, $x \in X$.

Theorem 4.3. *Let* A, A', A^* *and* A'^* *be interval-valued fuzzy sets on* X, B *and* B' *be interval-valued fuzzy sets on* Y. B^* *and* B'^* *are the interval-valued* $\mathcal{R}$-*type triple I solutions of* $IFMP(A, B, A^*)$ *and* $IFMP(A', B', A'^*)$ *given by Theorem 4.1 respectively. Suppose* $d(A, A') \leq \varepsilon_1$, $d(B, B') \leq \varepsilon_2$ *and* $d(A^*, A'^*) \leq \varepsilon_3$, *then*

$$d(B^*, B'^*) \leq \begin{cases} \varepsilon_1 + \varepsilon_2 + \varepsilon_3, & \text{if } \mathcal{R} = \mathcal{R}_L, \\ 1 - (1-\varepsilon_1)(1-\varepsilon_2)(1-\varepsilon_3), & \text{if } \mathcal{R} = \mathcal{R}_{Go}. \end{cases}$$

Proof. Let $\mathcal{R}(A(x_i), B(y_j)) = r_{ij}$, $\mathcal{R}'(A(x_i), B(y_j)) = r'_{ij}$, $1 \leq i \leq n$, $1 \leq j \leq m$.

(1) For interval-valued Łukasiewicz implication and corresponding t-norm. From Proposition 4.1, we have $d(\mathcal{R}_L(A, B), \mathcal{R}_L(A', B')) \leq \varepsilon_1 + \varepsilon_2$.

Then we have

$$\varrho(B^*(y_j), B^{'*}(y_j))$$
$$\geq 1 - \max\{\max_{1\leq j\leq m}|(r_{ij})_l - (r'_{ij})_l| + \max_{1\leq j\leq m}|A_l^*(y_j) - A_l^{'*}(y_j)|,$$
$$\max_{1\leq j\leq m}|(r_{ij})_r - (r'_{ij})_r| + \max_{1\leq j\leq m}|A_r^*(y_j) - A_r^{'*}(y_j)|\}$$

Similarly, $\varrho(B^{'*}(y_j), B^*(y_j)) \geq 1 - \max\{\max_{1\leq j\leq m}|(r_{ij})_l - (r'_{ij})_l| + \max_{1\leq j\leq m}|B_l^*(y_j) - B_l^{'*}(y_j)|, \max_{1\leq j\leq m}|(r_{ij})_r - (r'_{ij})_r| + \max_{1\leq j\leq m}|B_r^*(y_j) - B_r^{'*}(y_j)|\}$.

Thus we have

$$d(B^*, B^{'*}) \leq 1 - \frac{1}{m}\sum_{j=1}^m (2\varrho(B^*(y_j), B^{'*}(y_j)) - 1)$$

$$\leq 2 \cdot \frac{1}{m}\sum_{j=1}^m \{\max\{\max_{1\leq i\leq n}|(r_{ij})_l - (r'_{ij})_l|, \max_{1\leq i\leq n}|(r_{ij})_r - (r'_{ij})_r|\}$$
$$+ \max\{\max_{1\leq i\leq n}|A_l^*(x_i) - A_l^{'*}(x_i)|, \max_{1\leq i\leq n}|A_r^*(x_i) - A_r^{'*}(x_i)|\}\}$$

$$\leq 2 \cdot \frac{1}{m}\sum_{j=1}^m \{\max_{1\leq i\leq n}\{|(r_{ij})_l - (r'_{ij})_l|, |(r_{ij})_r - (r'_{ij})_r|\} +$$
$$\max_{1\leq i\leq n}\{|A_l^*(x_i) - A_l^{'*}(x_i)|, |A_r^*(x_i) - A_r^{'*}(x_i)|\}\}$$

$$\leq \varepsilon_1 + \varepsilon_2 + \varepsilon_3$$

(2) For interval-valued Goguen implication and corresponding t-norm. From Proposition 4.2, we have $d(\mathcal{R}_{Go}(A, B), \mathcal{R}_{Go}(A', B')) \leq \varepsilon_1 + \varepsilon_2 - \varepsilon_1 \cdot \varepsilon_2$. Then $\varrho(B^*(y_j), B^{'*}(y_j)) \geq \min_{1\leq i\leq n} \varrho(r_{ij}, r'_{ij}) \cdot \varrho(A^*(x_i), A^{'*}(x_i))$. Similarly, we have $\varrho(B^{'*}(y_j), B^*(y_j)) \geq \min_{1\leq i\leq n} \varrho(r'_{ij}, r_{ij}) \cdot \varrho(A^{'*}(x_i), A^*(x_i))$. Thus we have

$$d(B^*, B^{'*}) = 1 - \frac{1}{m}\sum_{j=1}^m T_{Go}(\varrho(B^*(y_j), B^{'*}(y_j)), \varrho(B^{'*}(y_j), B^*(y_j)))$$

$$\leq 1 - \frac{1}{m}\sum_{j=1}^m \min_{1\leq i\leq n}(1 - d(r_{ij}, r'_{ij})) \cdot (1 - d(A^*(x_i), A^{'*}(x_i)))$$

$$\leq 1 - (1 - \varepsilon_1)(1 - \varepsilon_2)(1 - \varepsilon_3).$$

□

Theorem 4.4. *Let A and A' be interval-valued fuzzy sets on X, B, B', B^* and B'^* be interval-valued fuzzy sets on Y. A^* and A'^**

are the interval-valued $\mathcal{R}$-type triple I solutions of $IFMT(A, B, B^*)$ and $IFMT(A', B', B'^*)$ given by Theorem 4.2 respectively. Suppose $d(A, A') \leq \varepsilon_1$, $d(B, B') \leq \varepsilon_2$ and $d(B^*, B'^*) \leq \varepsilon_3$, then

$$d(A^*, A'^*) \leq \begin{cases} \varepsilon_1 + \varepsilon_2 + \varepsilon_3, & \text{if } \mathcal{R} = \mathcal{R}_L, \\ 1 - (1-\varepsilon_1)(1-\varepsilon_2)(1-\varepsilon_3), & \text{if } \mathcal{R} = \mathcal{R}_{Go}. \end{cases}$$

Example 4.1. Suppose that $X = \{x_1, x_2, x_3, x_4, x_5\}$, $Y = \{y_1, y_2, y_3, y_4, y_5\}$, $A, A', A^*, A'^* \in SI(X)$, $B, B' \in SI(Y)$.

Item	x_1	x_2	x_3	x_4	x_5
A	[0.2,0.2]	[0.4,0.4]	[0.8,0.8]	[0.1,0.6]	[0.3,0.8]
A'	[0.2,0.2]	[0.4,0.4]	[0.0,0.0]	[0.1,0.6]	[0.3,0.8]
A^*	[0.0,0.0]	[0.2,0.3]	[0.1,0.7]	[0.0,0.0]	[0.0,0.0]
A'^*	[0.1,0.1]	[0.2,0.3]	[0.1,0.6]	[0.0,0.0]	[0.0,0.0]
Item	y_1	y_2	y_3	y_4	y_5
B	[0.4,0.6]	[0.4,0.8]	[0.8,0.8]	[0.0,0.0]	[0.2,0.2]
B'	[0.4,0.6]	[0.4,0.7]	[0.0,0.8]	[0.9,0.9]	[0.4,0.4]

Let B^* and B'^* are interval-valued triple I solutions based on Łukasiewicz implication for $IFMP(A, B, A^*)$ and $IFMP(A', B', A'^*)$, then we can obtain by Theorem 4.1

Item	y_1	y_2	y_3	y_4	y_5
B^*	[0.2,0.5]	[0.2,0.7]	[0.2,0.7]	[0.0,0.0]	[0.0,0.1]
B'^*	[0.2,0.6]	[0.2,0.6]	[0.1,0.6]	[0.2,0.6]	[0.2,0.6]

By using the distance metric d_L in Proposition 3.1, we have: $d_L(A, A') = 0.16$, $d_L(A^*, A'^*) = 0.04$, $d_L(B, B') = 0.2$, $d_L(B^*, B'^*) = 0.28 < 0.16 + 0.04 + 0.2$. This result satisfies Theorem 4.3.

5. Conclusions

In this paper, a new distance metric between interval-valued fuzzy sets is proposed, which is induced by left-continuous t-norm and corresponding residuated implication. we analysis robustness of the interval-valued fuzzy reasoning algorithms based on two well-known interval-valued residual implication operations. It proved that the interval-valued fuzzy reasoning algorithm based the Łukasiewicz interval-valued implication and Goguen interval-valued implication are robust.

Acknowledgments

This work is supported by the National Natural Science Foundation of China (Grant Nos. 61773019 and 61273018).

References

1. C. Alcalde, A. Burusco, R. Fuentes-Gonzalez, *Fuzzy Sets and Systems*, **153**, 211(2005).
2. H. Bustince, P. Burillo, F. Soria, Automorphisms, *Fuzzy Sets and Systems*, **134**, 209(2003).
3. K.Y. Cai, *IEEE Trans Fuzzy Syst*, **9**, 738(2001).
4. B.A. Davey, H.A. Priestley, *Introduction to Lattices and Order*, Cambridge University Press, Cambridge, 1990.
5. S.S. Dai, D.W. Pei, D.H. Guo, *International Journal of Approximate Reasoning*, **54**, 653(2013).
6. S. Jenei, *Fuzzy Sets and Systems*, **90**, 25(1997).
7. D.H. Hong, S.Y. Hwang, *Fuzzy Sets and Systems*, **66**, 383(1994).
8. J.H. Jin, Y.M. Li, C.Q. Li, *Information Sciences*, **177**, 5103(2007).
9. J.Y. Duan, Y.M. Li, *International Journal of Approximate Reasoning*, **61**, 33(2015).
10. E. Klement, R. Mesiar, E. Pap, *Triangular Norms*, Kluwer Academic Publishers, Dordrecht, 2000.
11. Y.M. Li, *Analysis of Fuzzy System*, Science Press, Beijing, 2005.
12. D.Ch. Li, Y.M. Li, Y.J. Xie, *Information Sciences*, **181**, 4754(2011).
13. Y.M. Li, D.Ch. Li, P. Witold, J.J. Wu, *International Journal of Intelligent Systems*, **20**, 393(2005).
14. X.C. Liu, Entropy, *Fuzzy Sets and Systems*, **52**, 305(1992).
15. M.X. Luo, K. Zhang, *International Journal of Approximate Reasoning*, **62**, 61(2015).
16. M.X. Luo, X.L. Zhou, *International Journal of Approximate Reasoning*, **66**, 16(2015).
17. G.J. Wang, *Non-classical Mathematical Logic and Approximate Reasoning*, Science Press, Beijing, 2000(in Chinese).
18. G.J. Wang, H.J. Zhou, *Introduction to Mathematical Logic and Resolution Principle*, Science Press/U.K. Alpha Science International Limited, Beijing/Oxford, 2009.
19. G.J. Wang, J.Y. Duan, *International Journal of Approximate Reasoning*, **55**, 787(2014).
20. R.R. Yager, *Fuzzy Sets and Systems*, **106**, 3(1999).
21. M.S. Ying, *Perturbation of fuzzy reasoning*, IEEE Trans Fuzzy Syst, **7**, 625(1999).
22. L.A. Zadeh, *Theory of approximate reasoning*, in: J.E. Hayes, D. Michie, L.I. Mikulich (Eds.), Machine Intelligence, Ellis Horwood, Chichester, 1970, pp. 149–194.

Energy optimisation in cloud servers using a static threshold VM consolidation technique (STVMC)

Bilal Ahmad,[†] Sally McClean,[†] Darryl Charles[†] and Gerard Parr[*]

[†]*School of Computing and Information Engineering, Ulster University*
Coleraine, UK
[*]*School of Computing Sciences, University of East Anglia*
Norwich Research Park, Norwich, UK
[†]*(ahmad-b, si.mcclean, dk.charles)@ulster.ac.uk,* [*]*g.parr@uea.ac.uk*

The IT industry has improved in the past decade from grid to distributed computing. Further advancements are making cloud computing a useful approach for users placed globally. It enables users to choose pay as you go service using the internet as a communication medium. Energy consumption has attracted a lot of focus from the research community. It is dependent upon a number of factors e.g. virtual machine allocation, workload type, service level agreement, optimisation policies. This paper carries out experimentation relating loaded hosts and defines various algorithms and policies which can be used to control the energy consumption. At the same time, it shows how quality of service parameters could be achieved by having minimum service level agreement violations for gaming workloads. The novelty includes the use of static threshold detection for under/over loaded hosts (STVMC) concept along with minimum migration time, maximum correlation, and minimum utilisation VM selection policies for multi-players. Results show how quality of service (QoS) can be maintained along with minimum energy consumption in heterogeneous data centers.

1. Introduction

Advancements in technology have changed the computing industry tremendously. Nowadays, many services are available as pay as you go over the internet. This leverages the IT industry to outsource their services which are then made available to the user at their door-step by the service providers. The challenge is to meet quality of service along with good quality of experience. Service level agreements are thus required to be kept in mind by service providers. This also puts constraints on how quality of service is affected while maintaining energy consumption requirements.

All kinds of cloud services (gaming, Big Data, internet of things, etc.) are maintained by large servers that are located globally. It can be seen that these servers commonly only use 15%–50% of their full performance while remaining idle at other times. Lots of energy is wasted in operation of such servers 24/7 while not being loaded to 100% utilisation. Therefore, if virtualisation is used the server's energy could be saved and used in better ways for maintaining quality of service. Hosts that have a lower load can be moved from one place to another, using the virtualisation concept to save energy [1].

However, a number of factors must to be kept in mind by service providers in the cloud environment, e.g. energy consumption, quality of service, service level agreements and quality of experience [2]. Virtual machine migration might result in a reduction of energy consumption, but it can also degrade the service. Therefore, a trade-off exists between quality of service and service level agreements which is required to be followed for acceptable quality of experience by users. This paper proposes a mechanism in which energy consumption is reduced using virtualisation techniques. It reduces active idle hosts by means of threshold virtualisation technique while meeting the minimum service level agreements between the user and service provider. The proposed work is carried out using the CloudSim simulation toolkit and a gaming workload is used for test purposes. The rest of the paper is organised as follows: Section 2 (related research work), Section 3 (cloud gaming model and its working), Section 4 (proposed experimental environment), Section 5 (results and discussion) and Section 6 (conclusion and future work).

2. Related Work

Researchers around the world have been addressing energy related issues in the field of cloud computing. This also requires maintenance of quality of service and service level agreements for better user experience. The authors of [2] use the concept of virtualising central processing unit and graphical processing unit which are present in cloud servers. Their results are used for improvement of different latency factors which are present in cloud servers in the form of game frames. The paper shows that quality of service could be maintained by using trade-off between game latency, buffering, scalability and data redundancy.

The authors in [3], presented an idea that energy consumption can be reduced using dynamic frequency scaling. It works by using the load on the central processing unit and on the basis of workload frequency scaling is performed. Work shows that energy can be saved in big gaming servers that are located globally by using this technique. Experiments have been performed using the CloudSim simulator for validation of energy aware functionality. The author in [4], have implemented the concept of virtual machine migration. The virtual machines that are idle in the server are migrated from one place to another for load balancing. This also helps in maintenance of quality of service while reducing significant amount of energy. The drawback in this work is that it used live migration techniques that sometimes caused a problem for system performance.

The authors in [1], performs experiments using virtual machine consolidation technique using the virtualisation concept and implements testing using LrMmt algorithm. The workload is used from Planet Lab. Whereas, work show that energy reduction can be performed to some extent using this technique along with maintenance of quality of service.

In [5] the authors use the Amazon Elastic Cloud Computing (EC2) servers to monitor network performance. They have carried a measurement study for monitoring the virtualisation effect on the network latency and performed tests using packet loss tolerance, with TCP/UDP throughput and for measuring performance of virtual machines located on Amazon EC2 servers. The authors recommend that virtualisation can cause adverse effects on the network if it is not handled properly depending upon several factors.

3. Cloud Gaming Model and Its Working

In cloud gaming a lot of data is transferred from the server to the users located globally. This user is connected to the cloud server through the internet. The user performs an action in the game on his local computer which is executed using the internet to the cloud gaming server located globally. Game logic controller receives the command and sends it to graphical processing unit which performs the desired processing. The processed command is sent back to the user in the form of a rendered encoded video stream through a thin client (Figure 1). Table 1 shows different types of games and their delay tolerance which is acceptable for a good user experience. Cloud-based games may be divided into three categories based on their processing abilities (Table 1). First person shooter games require the fastest real-time response with a maximum delay of 100 ms–200 ms [4].

In role playing games the response can be typically much slower due to the different type of gameplay e.g. 350 ms–500 ms. Real time strategy games are multi-player and multi-environment. These games are strategy oriented, so they therefore have the highest level of delay tolerance i.e. 800 ms–1000 ms. The

Table 1. Types of games with delay tolerance.

Game Type	Perspective	Delay Tolerance	Examples
First Person Shooter	First Person	100 ms–200 ms	Call of Duty, IGI
Role Playing Games	Third Person	350 ms–500 ms	World of War Craft, Max Payne 3
Real Time Strategy	Omnipresent	800 ms–1000 ms	Star Craft, Freedom Fighters

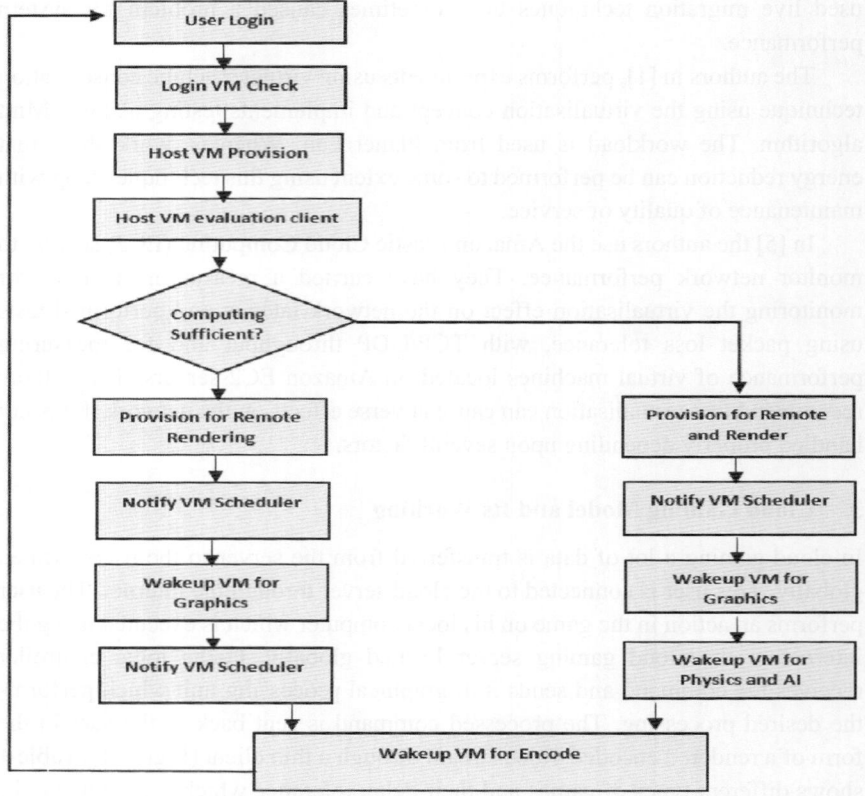

Figure 1. Flow chart for interactive cloud gaming operation.

tolerance level is calculated by game developers based on the architecture and user input speed.

Cloud based digital gaming is influenced by number of factors like tolerance interaction delay, bandwidth, geographical condition, gaming framework, video streaming and encoding [6]. All of these features have different impact on the quality of service [7]. They play a vital role in the implementation of cloud based digital games and have various pros and cons which are discussed below.

3.1. *Virtualization*

The introduction of virtual machines has made the cloud computing world more progressive as multiple tasks can now be performed on one server, addressing an issue which was present in previous models [6]. Initially, OnLive servers had limited scalability and were unable to run multiple user games. Therefore, virtualization technology (Figure 1) in a graphical processing unit can be improved to execute a game on each added VM separately [4]. In the gaming industry, game developers usually assume that all the processing is done in single GPU. Therefore, performance in multiple player gaming can be improved in general by introducing virtualization concept in GPU i.e. each game has its own GPU for graphics rendering reducing system overhead [8].

$$P(vm) = \frac{1}{VM(n)} \sum_{i=1}^{n} \frac{Pd(k)}{Cpu(k)}. \tag{1}$$

$P(vm)$ = performance because of virtual machine migration, $VM(n)$ = total number of virtual machines, $Pd(k)$ = level of degradation in the service of a particular virtual machine when it is migrated, $Cpu(k)$ = total utilization of CPU of particular virtual machine till it existence [1].

3.2. *Quality of experience*

Any service (video streaming, online search, gaming) that is provided to the user is required to be analysed for better customer satisfaction and quality maintenance [4]. In games, the user is very demanding in terms of speed, latency, and performance therefore, this factor cannot be ignored [9]. A number of different quality of experience models in the domain of cloud gaming should be designed to gauge the customer satisfaction level depending upon the system parameters and gaming players experience [2].

3.3. *Server selection*

In online multiple player games, users from all around the globe could be using the service by connecting to the servers [4]. Therefore, well-managed and robust server allocation models would be required to maintain the quality of service and quality of experience. There are many problems that could be addressed for improvement of the services e.g. closest server should be allocated to the user, load of the server should be calculated before allocation, availability of game on the server, location of each user in case of multi-player games for mapping, latency, performance and number of virtual machines required could be addressed in server design models [2].

3.4. *Parameter adaption*

Quality of service and quality of experience can be improved by designing an adaptable model depending upon latency, available bandwidth, server load etc.

[4]. This model can be configured in the runtime environment for efficient QoS and QoE. Different parameters e.g. packet loss, tolerance delay, graphics resolution could be exploited and trade-offs could be performed to deliver better services by using the available bandwidth and resources [8].

3.5. *Resource scheduling*

Allocation of resources is one the core issue in any game. If proper resource algorithms and techniques are implemented in the industry, cost for both user and companies could be reduced significantly [6]. Over-allocation can result in the consumption of resources thus making them scarce [10]. Therefore, efficient resource allocation models should be implemented for provisioning of resources (Figure 2). This is a challenging task and can help save over-provisioning or under-provisioning of resources in many cases, bringing profit to the gaming industry and users [2].

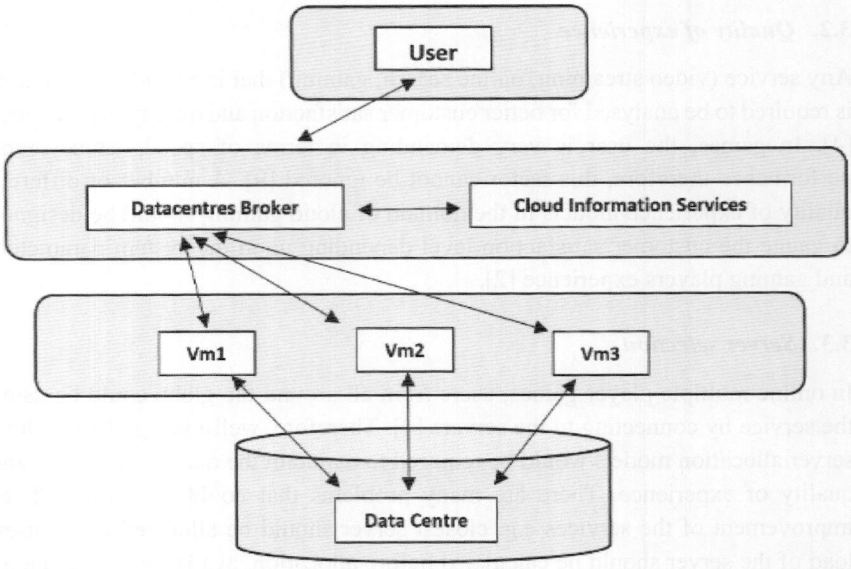

Figure 2. Cloud gaming scheduling operation.

4. Simulation Environment

The test for virtual machine consolidation has been performed in CloudSim using the gaming workload. CloudSim provides the user with an interactive platform where a user can design, test and implement the cloud environment. It provides experimental results relating different factor of quality of service, like energy

Table 2. System parameters for experimentation purpose.

System (HP ProLiant)	HP Xeon 3040 (ProLiant ML11OG4)	HP Xeon 3075 (ProLiant ML110G5)
CPU	1860 MHz	2660 MHz
Cores	Dual	Dual
RAM	4 GB	4 GB

consumption, service level agreements, creation of virtual machine with their performance measure. It uses the gaming workload from one of the multi-player game i.e. (World of War Craft) WoWAH. This test has been configured using Eclipse Luna and Java IDE. It consists of 800 physical hosts and 1000 virtual machines which are assigned with a workload. For experimental purposes, dual core, HP Xeon 3040 (ProLiant ML11OG4) and HP Xeon 3075 (ProLiant ML110G5) servers have been used as hosts. Both the systems have MIPS (instructions per second) rating of 1860 MHz and 2660 MHz, respectively. Detailed parameters are given below.

The defined model has 4 GB RAM and a bandwidth of 1 Gbit/sec. WoWAH a massively multi-player online game has been used for test purpose. It consists of data set of 3.5 GB, runtime of 1107 days, 91065 avatars, 667032 sessions etc. [11]. This will be used to analyse the quality of energy consumption, quality of service (QoS), service level agreement (SLAs), and virtual machine performance in the defined system.

The quality of service will be addressed using host virtualisation and how many service level agreements are achieved. Achievement of quality of service and service level agreement can be attained using virtualisation techniques. There may be hosts in the system which are being underutilised and will not be able to address another host while waiting for a response. This is wasteful of resources and consumes energy. Therefore, load can be shifted from one virtual machine to another to avoid underutilisation of resources. This will also help in improvement of quality of service and energy waste. Table 3 shows the detail of the host parameters.

Table 3. Host characteristics for the data center.

Host MIPS	Host RAM	Host Bw	Host PE(s)
1860	4096 MBs	1 Gbit/s	02
2660	4096 MBs	1 Gbit/s	02

4.1. Static threshold VM consolidation detection for under/over loaded host (STVMC)

In static threshold resource optimisation, a certain threshold for central processing unit utilisation and host scheduling is carried out within the defined limits. This works by differentiating the status of over-loaded and under-loaded hosts. When the procedure is called, it works by comparing the current CPU utilisation of the host with the defined threshold. The algorithm calculates the mean of the 'n' latest CPU utilisations. Therefore, if the threshold is over or under the defined limit it is considered as an over-loaded or under-loaded host, respectively. The hosts which are present on the system have minimum resource utilisation and therefore, to minimise resource wastage, virtual machines allocated to the host are moved, causing resource provisioning. So in this approach, the host node is not switched off or sent into sleep mode. Instead, it will remain active, increasing the probability of downtime and quality of service degradation. Thresholds are implemented with three VM selection policies i.e. minimum migration time policy (MMT), maximum correlation (MC) and minimum utilisation (MU).

Minimum Migration Time Policy: In this policy VMs are selected on the basis of the time it will take to migrate. The VM with the minimum migration time is selected for migration. This time is calculated by using RAM utilisation and space of network bandwidth.

$$\left(\frac{RAM_u(v)}{NETj}\right) \le \left(\frac{RAM_u(a)}{NETj}\right), v \in Vj \mid \forall a \in Vj. \tag{2}$$

Whereas, Vj shows the total number of VMs with host j, $RAM_u(a)$ is the amount of RAM that is used by the virtual machine (a), $NETj$ equals available bandwidth from host $'j'$.

Minimum Utilisation: In this policy, VMs are selected based on the utilisation criteria; the VM having minimum utilisation in the host will be selected for migration.

Maximum Correlation Policy: This policy decides the VM migration on the basis of the probability of maximum correlation i.e. the higher the value of the correlation between the values of resource utilisation the higher the probability will be of host overloading. The intra-VM correlation and CPU utilisation are estimated using the concept of multiple correlation coefficient (MCC). The dependent variables of real and predicted values have squared correlation between MCC coefficients [12].

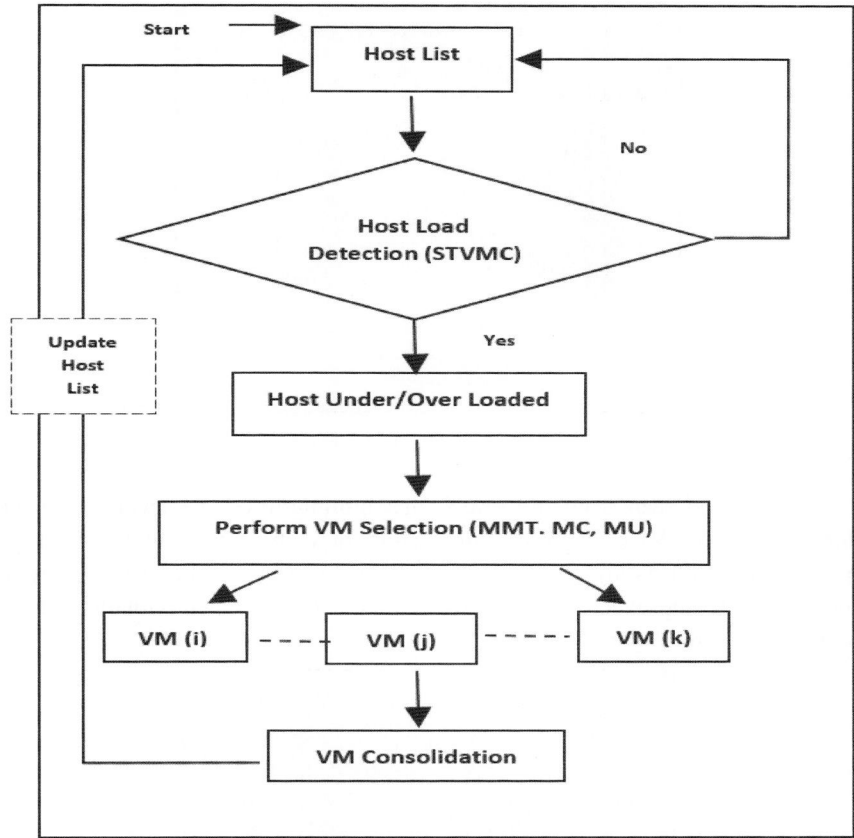

Figure 3. Flow chart for VM consolidation.

5. Experimental Results and Discussion

The results show comparisons using three approaches for virtual machine selection policies i.e. minimum time migration (MMT), minimum utilisation (MU) and maximum correlation (MC). The difference in the amount of energy consumption, service level agreement and quality of service degradation can be seen through the results which are estimated on the basis of CPU utilisation. From the results, it could be seen that if the STVMC algorithm is used with three different VM policies (MMT, MC, MU), the maximum correlation policy has better results for energy utilisation. It consumes 11% of less energy as compared to MMT and MU VM selection policies (Figure 4).

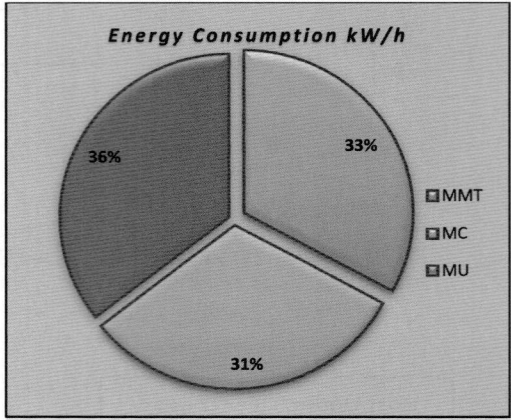

Figure 4. Energy utilisation comparison.

It could be seen from the results that minimum service level agreement degradation (SLAV) is achieved using the minimum utilisation policy. Therefore, by using the MU policy overall SLA violations can be reduced and quality of service can be improved by having less service level agreement violations in the proposed system (Figure 5).

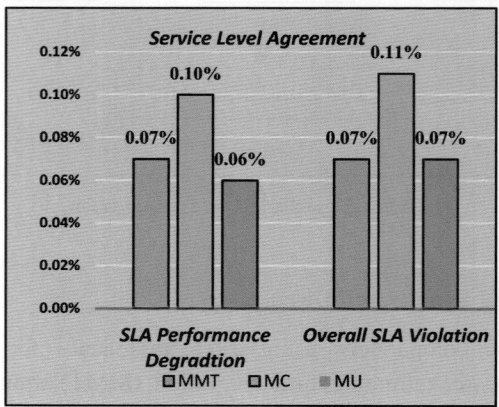

Figure 5. Service level agreement (SLAV).

Along with this, maximum number of VM are migrated from the system (Figure 6) using MU policy. Figure 7 shows different execution time using three different policies. It can be seen from the results that the selection of virtual machine takes minimum time whereas relocation of virtual machine requires more time than selection of host.

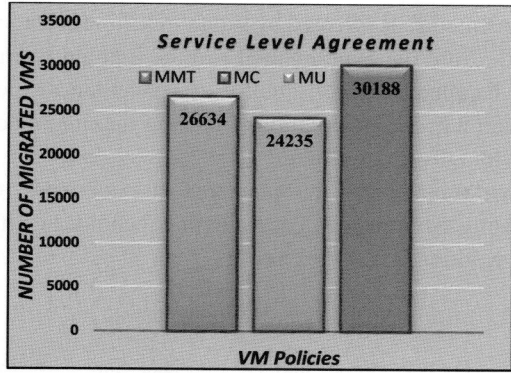

Figure 6. Number of VM migration.

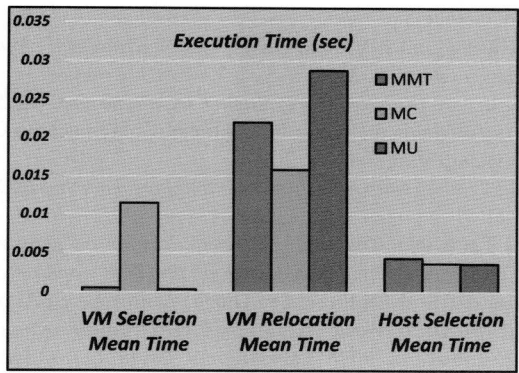

Figure 7: Execution time for VM migration.

Therefore, downtime can be reduced using proper virtual machine relocation technique. From the observed results of SLAV, VM migration, SLA violations and energy consumption effects could be seen, and we have predicted that minimum energy consumption in the system is achieved using maximum correlation.

6. Conclusion

In this paper, gaming workload has been used which have been implemented using a static threshold VM allocation policy along with three different VM selection policies i.e. maximum correlation, minimum utilisation, and minimum migration time. This helps to determine how different parameters of quality are affected by service level agreement utilisation, virtual machine migration, and downtime reduction. It was seen that resource optimisation and energy utilisation in a big scale infrastructure is affected by virtual machine selection and allocation policy being used. From the experimental results, it could be inferred that if all these factors are handled properly, quality of service and experience could be achieved with minimum SLAs violations by using maximum correlation virtual machine policy with STVMC technique.

As part of our future work, we plan to focus our work towards resource optimisation by using different VM consolidation techniques with both static and dynamic gaming workload.

References

1. Oikonomou, E., D. Panagiotou, and A. Rouskas, *Energy-aware Management of Virtual Machines in Cloud Data Centers, in Proceedings of the 16th International Conference on Engineering Applications of Neural Networks (INNS)*. 2015, ACM: Rhodes, Island, Greece. pp. 1–6.
2. Zhao, Z., K. Hwang, and J. Villeta, *GamePipe: A virtualized cloud platform design and performance evaluation.* 2012. 1–8.
3. Ahmad, B., et al., *Analysis of energy saving technique in CloudSim using gaming workload.* Proceedings of the Ninth International Conference on Cloud Computing, GRIDS, and Virtualization, IARIA, 2018.
4. Shea, R., et al., *Cloud gaming: architecture and performance.* IEEE Network, 2013. **27**(4): pp. 16–21.
5. Wang, G. and T.S.E. Ng, *The impact of virtualization on network performance of amazon EC2 data center, in Proceedings of the 29th conference on Information communications.* 2010, IEEE Press: San Diego, California, USA. pp. 1163–1171.
6. Song, J., et al. *FCM: Towards fine-grained GPU power management for closed source mobile games. in 2016 International Great Lakes Symposium on VLSI (GLSVLSI)*. 2016.
7. Huang, C.Y., et al., *Measuring the client performance and energy consumption in mobile cloud gaming, in 2014 13th Annual Workshop on Network and Systems Support for Games.* 2014. pp. 1–3.
8. Chen, K.T., C.Y. Huang, and C.H. Hsu. *Cloud gaming onward: research opportunities and outlook. in 2014 IEEE International Conference on Multimedia and Expo Workshops (ICMEW)*. 2014.
9. Ho Young, K. and K. Tag Gon. *Performance simulation modeling for fast evaluation of pipelined scalar processor by evaluation reuse. in Proceedings. 42nd Design Automation Conference, 2005*. 2005.
10. Janet, J., S. Balakrishnan, and E. Murali. *Improved data transfer scheduling and optimization as a service in cloud. in 2016 International Conference on Information Communication and Embedded Systems (ICICES)*. 2016.
11. Lee, Y.-T., et al., *World of warcraft avatar history dataset, in Proceedings of the second annual ACM conference on Multimedia systems.* 2011, ACM: San Jose, CA, USA. pp. 123–128.
12. Theja Perla, R. and S.K.K. Babu, *Evolutionary Computing Based on QoS Oriented Energy Efficient VM Consolidation Scheme for Large Scale Cloud Data Centers*, in *Cybernetics and Information Technologies*. 2016. p. 97.

The properties of soft product topological space

Fu Li

School of Mathematics and Statistics, Qinghai Nationalities University, PR China
fl0971@163.com

This paper defines the soft mappings over the soft product topological space, giving the representation of elements of soft product topology and discussing the relative properties of soft product topology, and also studies the soft connectedness, soft separation axioms and soft compactness in the soft product topological space.

Keywords: Soft topological space; soft product topology; soft connectedness; soft separation axioms; product space.

1. Introduction

Rough set[Pawlak1982], soft set[Molodsov1999, Maji2003] as new mathematical approaches, are widely applied in the uncertainty reasoning, and the related research results are flourishing, for example, Maji et al.[Maji 2001] presented the definition of fuzzy soft set and discussed the properties of fuzzy soft set. Topology[Euler1736] which is discussed in such fields by many authors who achieved lots of research results, is one of the greatest unifying ideas of mathematics. There are influent researches about soft topology (such as [Shbir2011, Li2015]), and in [Fu2017], researchers gave the definition of the soft product topology, but they did not discuss the soft connectedness, separation axiom over the soft product topological space.

This paper will define the product of soft topologies and further study its properties. The rest of this paper is organized as follows. In Section 2, we recall some basic concepts of soft sets and topology. In Section 3, the element character of soft product topology is given and the relative properties of soft product topology is also discussed. In Section 4, we discuss the soft separation axioms about the soft product topology, and discuss the soft connectedness and compactness of soft product topological spaces in Section 5. The conclusion is Section 6.

2. Basic Knowledge

Definition 2.1 (Molodsov1999). *Let U be an original universe set, E*

be parameters set. Let $A \subset E$, and $\mathcal{P}(U)$ be the power set of U. Then a pair (F, A) is called a **soft set** over U, where mapping F is from A to $\mathcal{P}(U)$.

We can consider a soft set (F, E) as the class of approximations, i.e. $(F, E) = \{F(e) \mid e \in E\} = \{(F(e), e) \mid e \in E\}$.

Definition 2.2 (Molodsov1999). Let (F_1, B_1) and (F_2, B_2) be the soft sets over U, E be an attributes set. $B_1, B_2 \subseteq E$, $F_i : B \to \mathcal{P}(U)(i = 1, 2)$.

(i) If $\forall x \in B_1 \subseteq B_2$, having $F_1(x) \subseteq F_2(x)$, then (F_1, B_1) is a **soft subset** of (F_2, B_2), denoted as $(F_1, B_1)\widetilde{\subset}(F_2, B_2)$. If $(F_1, B_1)\widetilde{\subset}(F_2, B_2)$, and $(F_2, B_2)\widetilde{\subset}(F_1, B_1)$, then (F_1, B_1) and (F_2, B_2) are **soft equal**, denote by $(F_1, B_1) = (F_2, B_2)$.

(ii) The **soft union** of (F_1, B_1) and (F_2, B_2) is the soft set $(H, C) = (H, B_1 \cup B_2)$, denoted as $(F_1, B_1)\widetilde{\cup}(F_2, B_2)$, where $\forall e \in C$

$$H(e) = \begin{cases} F_1(e), & if \ e \in B_1 - B_2 \\ F_2(e), & if \ e \in B_2 - B_1 \\ F_1(e) \cup F_2(e), & if \ e \in B_1 \cap B_2 \end{cases}$$

(iii) The **soft intersection** of (F_1, B_1) and (F_2, B_2) is denoted as $(F_1, B_1) \sqcap (F_2, B_2) = (H, C) = (H, B_1 \cap B_2)$, if $\forall e \in C, H(e) = F_1(e) \cap F_2(e)$.

(iv) The **relative soft complement** of (F, B) is denoted by $(F, B)^c = (F^c, B)$, and $F^c(x) = U - F(x), \forall x \in B$. Obviously, $((F, B)^c)^c = (F, B)$.

(v) (F, B) is a **relative null soft set**, denoted by $\mathcal{N}$, if $\forall x \in B, F(x) = \emptyset$. If $B = E$, then it is **absolute null soft set**, denoted as $\widetilde{\emptyset}$.

(vi) The **relative whole soft set** (denoted by $\widetilde{U}$), if $\forall x \in B, F(x) = U$.

Definition 2.3. Let (F_1, B_1) and (F_2, B_2) be the soft sets over U, E be an attributes set. $B_1, B_2 \subseteq E$, $F_i : B \to \mathcal{P}(U)(i = 1, 2)$, then the **soft product** of (F_1, B_1) and (F_2, B_2) is denoted by $(F_1, B_1) \times (F_2, B_2) = (H, C)$, where $H(e_1, e_2) = (F_1(e_1), F_2(e_2))$, $\forall (e_1, e_2) \in C \subseteq B_1 \times B_2$.

Definition 2.4 (Shbir2011). Let τ be the class of soft sets over U, E is a parameters set, then τ is a **soft topology** on U if
 (1) $\widetilde{\emptyset}, \widetilde{X}$ belong to τ.
 (2) $\tau_1 \subseteq \tau$, then $\bigcup_{(F,B) \in \tau_1}(F, B) \in \tau$.
 (3) $(F, B), (G, C) \in \tau$, then $(F, B) \sqcap (G, C)$ is also in τ.

Then (X, τ, E) is a **soft topological space** over U. The members of τ are **soft open sets**, $(F, B)^c = (F^c, B)$ is a **soft closed set**, if $(F, B) \in \tau$. If (F, B) is both soft open and soft closed, then (F, B) is a **soft open set**.

In the subsequence, (G, τ, M) is a soft topological space where G is objects set, M is parameters set. $B \subseteq M, F : B \to \mathcal{P}(G)$, (F, B) is soft set over G.

Definition 2.5 (Fu2014). $(F, B) \in \tau$, $B \subseteq M, x \in G$, then

(i) $(F, B)^\circ$ is **soft interior** of (F, B) which is the union of all soft open set contained in (F, B), i.e., $(F, B)^\circ = \bigcup \{(F_i, B_i) \mid (F_i, B_i) \widetilde{\subseteq} (F, B) \in \tau\}$.

(ii) $\overline{(F, B)}$ is **soft closure** of (F, B) which is the intersection of all soft closed super set of (F, B), i.e., $\overline{(F, B)} = \sqcap \{(F_i, B_i) \mid (F, B) \subseteq (F_i, B_i) \in \tau\}$.

(iii) If $\exists (F_1, B_1) \in \tau$, such that $x \in (F_1, B_1) \widetilde{\subseteq} (F, B)$, then x is a **soft interior point** of (F, B) and (F, B) is the **soft neighborhood** of x. If $\exists (F_1, B_1) \widetilde{\subseteq} (F, B)^\circ$, then (F, B) is a **soft neighborhood** of (F_1, B_1). If (F, B) is soft open (closed), then soft neighborhood is soft open (closed).

(iv) If $\exists (F, B) \in \tau, s.t. x \in F(e), \forall e \in B \subseteq M$, then $x \in (F, B))$, and x is a **soft point**. If $\exists e \in B$, s.t. $x \notin F(e)$, then $x \notin (F, B)$. $\mathcal{U}_x = \{(F, B') \mid (F, B') \in \tau, s.t. x \in (F, B')\}$ is the class of all the soft neighborhoods of x.

Remark 2.1. (i) $\mathcal{U}_x \neq \emptyset (\because x \in \widetilde{G})$;

(ii) If $(F_1, B_1), (F_2, B_2) \in \mathcal{U}_x$, then $(F_1, B_1) \sqcap (F_2, B_2)$ can be empty set.

For example: let $G = \{x_1, x_2\}, M = \{e_1, e_2\}, B_1 = \{e_1\}, B_2 = \{e_2\}$, $\tau = \{\widetilde{\emptyset}, (F_1, B_1), (F_2, B_2), \widetilde{G}\}$, and $(F_1, B_1) = \{(e_1, \{x_1\})\}$, $(F_2, B_2) = \{(e_2, \{x_1, x_2\})\}$, then $\mathcal{U}_{x_1} = \{(F_1, B_1), (F_2, B_2), \widetilde{G}\}$, $\mathcal{U}_{x_2} = \{(F_2, B_2), \widetilde{G}\}$, and $(F_1, B_1), (F_2, B_2) \in \mathcal{U}_{x_1}$, having, $(F_1, B_1) \sqcap (F_2, B_2) = \widetilde{\emptyset}$.

(iii) If $(F_1, B_1), (F_2, B_2) \in \mathcal{U}_x$, then $(F_1, B_1) \widetilde{\cup} (F_2, B_2)$ can not be in $\mathcal{U}_x$.

For the above example, $(F_1, B_1), (F_2, B_2) \in \mathcal{U}_{x_1}$, $(F_1, B_1) \widetilde{\cup} (F_2, B_2) = \{(e_1, \{x_1\}), (e_2, \{x_1, x_2\})\} \notin \mathcal{U}_{x_1}$.

Proposition 2.1. Let (G, τ, M) be a soft topological space, $\forall (F, B) \in \tau$, then $(F, B)^\circ = (\overline{(F, B)^c})^c$.

3. The Properties of Soft (Finite) Product Topological Space

In the sequence, when we are discussing product topological spaces, all the definitions and conclusions are used provided $n = 2$. Definitely, we can generate them to finite cases (using Mathematical Induction).

Definition 3.1 (Fu2017). Let (G_1, τ_1, M_1) and (G_2, τ_2, M_2) be soft topological spaces, then the soft topology τ over $G = G_1 \times G_2$ which has the base $\mathcal{B} = \{(F_1, B_1) \times (F_2, B_2) | (F_i, B_i) \in \tau_i, i = 1, 2\}$ is called **soft product topology**. And, (G, τ, M) is called **soft product topological space** of (G_1, τ_1, M_1) and (G_2, τ_2, M_2), denoted as $(G, \tau, M) = (G_1, \tau_1, M_1) \bigotimes (G_2, \tau_2, M_2)$.

Denotation $\tau = \{(F,B) | \forall (e_i, e_i') \in B_1 \times B_2, F(e_i, e_i') = F_1(e_i) \times F_2(e_i') = (F_1(e_i), F_2(e_i')) \subseteq G_1 \times G_2\} = \{((e_i, e_i'), (F_1(e_i), F_2(e_i'))) | (e_i, e_i') \in B_1 \times B_2, (F_1(e_i), F_2(e_i')) \subseteq G_1 \times G_2\}$. Obviously, $\widetilde{G}_1 \times \widetilde{G}_2 = \widetilde{G_1 \times G_2}$.

Example 3.1. Let $(G_1, \tau_1, M_1), (G_2, \tau_2, M_2)$ be two soft topological spaces in which $G_1 = \{h_1, h_2, h_3\}, M_1 = \{e_1, e_2\}, G_2 = \{h_1', h_2'\}, M_2 = \{e_1', e_2'\}$, and $\tau_1 = \{\emptyset, \widetilde{G}_1, (F_1, M_1), (F_2, M_1), (F_3, M_1)\}$ where $(F_1, M_1) = \{(e_1, \{h_2\}), (e_2, \{h_1\})\}, (F_2, M_1) = \{(e_1, \{h_2\}), (e_2, \{h_1, h_2\})\}, (F_3, M) = \{(e_1, \{h_1, h_2\}), (e_2, \{h_1, h_2, h_3\})\}; \tau_2 = \{\emptyset, \widetilde{G}_2, (F_1', M_2), (F_2', M_2)\}$ where $(F_1', M_2) = \{(e_1', \{h_2'\}), (e_2', \{h_1'\})\}, (F_2', M_2) = \{(e_1', \{h_1', h_2'\}), (e_2', \{h_1'\})\}$.

Clearly, τ_1, τ_2 are soft topologies over G_1, G_2 respectively.

Suppose that $((F,B) = (F_2, M_1) \times (F_1', M_2) \in \tau$, then $(F,B) = \{((e_1, e_1'), (\{h_2\}, \{h_2'\})), (e_1, e_2'), (\{h_2\}, \{h_1'\})), ((e_2, e_1'), (\{h_1, h_2\}, \{h_2'\})), ((e_2, e_2'), (\{h_1, h_2\}, \{h_1'\}))\}$.

If $(F_1, B_1) \times (F_2, B_2) = \widetilde{\emptyset}$, then (F_1, B_1) and (F_2, B_2) at least one is $\widetilde{\emptyset}$.

Lemma 3.1 (Fu2017). Let X, Y be sets, $A, B \subset X, C, D \subset Y$, then
(i) $(A \cap B) \times (C \cap D) = (A \times C) \cap (B \times D)$;
(ii) $(A \cup B) \times (C \cup D) = (A \times C) \cup (A \times D) \cup (B \times C) \cup (B \times D)$.

Definition 3.2. Let $f : (G_1, \tau_1, M_1) \longrightarrow (G_2, \tau_2, M_2)$ be a mapping.

(i) If $(F_1, B_1) \neq (F_2, B_2) \in \tau_1$, having, $f(F_1, B_1) \neq f(F_2, B_2) \in \tau_2$. Then f is a **soft injective mapping**. If $(F_2, B_2) \in \tau_2, \exists (F_1, B_1) \in \tau_1$, such that $f(F_1, B_1) = (F_2, B_2)$. Then f is a **soft surjective mapping**. If f is both soft surjective and soft injective, then f is **soft bijective**.

(ii) If $\forall (F_2, B_2) \in \tau_2, \exists (F_1, B_1) \in \tau_1$, such that $f^{-1}(F_2, B_2) = (F_1, B_1)$, then f is a **soft continuous mapping**. If $x \in G, \forall (F', B') \in \mathcal{U}_{f(x)}$, having $f^{-1}(F', B') \in \mathcal{U}_x$, then f is **soft continuous at** x.

(iii) If $\forall (F_1, B_1) \in \tau_1, \exists (F_2, B_2) \in \tau_2$, such that, $f(F_1, B_1) = (F_2, B_2)$, then f is a **soft open mapping**. If $\forall (F_1, B_1)^c \in \tau_1, \exists (F_2, B_2)^c \in \tau_2$, such that $f(F_1, B_1) = (F_2, B_2)$, then f is a **soft closed mapping**.

(iv) If f is a injective mapping, and f, f^{-1} are continuous mapping, then f is a **soft homeomorphism mapping**, and (G_1, τ_1, M_1) is homeomorphism to (G_2, τ_2, M_2) (denoted as $(G_1, \tau_1, M_1) \sim (G_2, \tau_2, M_2)$).

Remark 3.1. (1) For simplicity, we can omit "soft" in above definitions.

(2) If $f(F,B) = (F,B)$, then f is a soft identity mapping, and f is soft surjective, injective, bijective, continuous open (closed).

(3) f is continuous iff $\forall x \in G, f$ is continuous at x.

Lemma 3.2. Let $(G, \tau, M) = (G_1, \tau, M_1) \bigotimes (G_2, \tau_2, M_2)$, $p_i : (G, \tau, M) \to$

(G_i, τ_i, M_i) be a projection, $p_i : ((F_1, B_1) \times (F_2, B_2)) = (F_i, B_i), i = 1, 2$, $f : (G_i, \tau_i, M_i) \to (G, \tau, M)$, then

(1) p_i is a soft surjective, continuous open mapping;
(2) $p_i \circ f = p_i, i = 1, 2$ which is soft identity mapping.

Theorem 3.1. Let (G_i, τ_i, M_i) be soft topological spaces, $i = 1, 2, 3$, then
(1) $(G_1, \tau_1, M_1) \bigotimes (G_2, \tau_2, M_2) \sim (G_2, \tau_2, M_2) \bigotimes (G_1, \tau_1, M_1)$;
(2) $((G_1, \tau_1, M_1) \bigotimes (G_2, \tau_2, M_2)) \bigotimes (G_3, \tau_3, M_3) \sim (G_1, \tau_1, M_1) \bigotimes ((G_2, \tau_2, M_2) \bigotimes (G_3, \tau_3, M_3))$.

Proof. We only prove (1), the proof of (2) is similar to (1), we omit it.

(1) $f : ((G_1, \tau_1, M_1) \bigotimes (G_2, \tau_2, M_2)) \to ((G_2, \tau_2, M_2) \bigotimes (G_1, \tau_1, M_1))$, $\forall (F_1, B_1) \in \tau_1, (F_2, B_2) \in \tau_2, f((F_1, B_1) \times (F_2, B_2)) = (F_2, B_2) \times (F_1, B_1)$, then f is injective.

Define the projection $p_i : ((G_1, \tau_1, M_1) \bigotimes (G_2, \tau_2, M_2)) \to (G_i, \tau_i, M_i)$, in which $p_i((F_1, B_1) \times (F_2, B_2)) = (F_i, B_i), i = 1, 2$, and composition mapping $p_i \circ f$ as: $p_i \circ f(((F_1, B_1) \times (F_2, B_2)) = p_i(f((F_1, B_1) \times (F_2, B_2))$. Then $p_1 \circ f = p_1, p_2 \circ f = p_2$ which are all continuous, so f is continuous. Similarly, f^{-1} is also continuous. Hence, f is a homeomorphism mapping. $\square$

Theorem 3.2. Let $(G, \tau, M) = (G_1, \tau, M_1) \bigotimes (G_2, \tau_2, M_2)$, $\forall (F_1, B_1) \in \tau_1, (F_2, B_2) \in \tau_2$, then
(1) $\overline{(F_1, B_1) \times (F_2, B_2)} = \overline{(F_1, B_1)} \times \overline{(F_2, B_2)}$;
(2) $((F_1, B_1) \times (F_2, B_2))^\circ = (F_1, B_1)^\circ \times (F_2, B_2)^\circ$.

Proof. In here, we only prove (1).

$\forall x_1 \in G_1, x_2 \in G_2, (F_1, B_1) \in \tau_1, (F_2, B_2) \in \tau_2$ such that $x_1 \in (F_1, B_1), x_2 \in (F_2, B_2)$, denote $x = (x_1, x_2) \in G_1 \times G_2 = G$. $x \in \overline{(F_1, B_1) \times (F_2, B_2)} \Rightarrow \forall (F, B) \in \mathcal{U}_{x_1}, (F', B') \in \mathcal{U}_{x_2}, (F, B) \times (F', B') \in \mathcal{U}_x \Rightarrow x \in ((F, B) \times (F', B')) \sqcap ((F_1, B_1) \times (F_2, B_2)) \Rightarrow x \in ((F, B) \sqcap (F_1, B_1)) \times ((F', B') \sqcap (F_2, B_2)) \Rightarrow (F, B) \sqcap (F_1, B_1) \neq \widetilde{\emptyset}, ((F', B') \sqcap (F_2, B_2) \neq \widetilde{\emptyset} \Rightarrow x_1 \in \overline{(F_1, B_1)}, x_2 \in \overline{(F_2, B_2)} \Rightarrow x \in \overline{(F_1, B_1)} \times \overline{(F_2, B_2)}$.

Conversely, suppose that $x = (x_1, x_2) \in \overline{(F_1, B_1)} \times \overline{(F_2, B_2)}$, then $x_1 \in \overline{(F_1, B_1)}, x_2 \in \overline{(F_2, B_2)}, \forall (F, B) \in \mathcal{U}_x, \exists (F', B') \in \mathcal{U}_{x_1}, (F'', B'') \in \mathcal{U}_{x_2}$ such that $(F, B) = (F', B') \times (F'', B'')$. By $(F', B') \sqcap (F_1, B_1) \neq \widetilde{\emptyset}, ((F'', B'') \sqcap (F_2, B_2) \neq \widetilde{\emptyset}$, and $((F', B') \sqcap (F_1, B_1)) \times ((F'', B'') \sqcap (F_2, B_2)) = ((F', B') \times (F'', B'')) \sqcap ((F_1, B_1) \times (F_2, B_2)) = (F, B) \sqcap ((F_1, B_1) \times (F_2, B_2)) \neq \widetilde{\emptyset}$.

Hence, $\overline{(F_1, B_1) \times (F_2, B_2)} = \overline{(F_1, B_1)} \times \overline{(F_2, B_2)}$. $\square$

4. The Separation Axioms of the Soft Product Topology

In reference [7,8,10], authors defined and discussed soft T_i-spaces in detail which we do not review but demonstrate those referred to in the following.

Definition 4.1 (Fu2015). *(1) $x \neq y \in G$, if $\exists (F_1, M) \in \mathcal{U}_{x}, (F_2, M) \in \mathcal{U}_y$ s.t. $y \notin (F_1, M), x \notin (F_2, M)$, and $(F_1, M) \sqcap (F_2, M) = \tilde{\emptyset}$, then soft topological space (G, τ, M) over G is called a **soft T_2-space**, which is also called a soft **Hausdroff space**.*

*(2) $x \in G, (F, M)$ is a soft closed set, if $\exists (F_1, M), (F_2, M) \in \tau$, such that $x \in (F_1, M), (F, M) \subseteq (F_2, M)$, and $(F_1, M) \sqcap (F_2, M) = \tilde{\emptyset}$, then soft topological space (G, τ, M) is called a **soft regular** space.*

Definition 4.2. *(1) If $\exists (F, B) \in \tau$, such that $\overline{(F, B)} = \widetilde{G}$, then (G, τ, M) is a **soft separable topological space**, simply speaking, (G, τ, M) is a **soft separable**, and (F, B) is a **soft dense subset** of $\widetilde{G}$.*

*(2) Let $\mathcal{A} = \{(F, B) \mid (F, B) \text{ is an soft open set in } \tau\}$ be any soft open cover of G, if $\mathcal{A}$ has a soft countable subcover of G, then (G, τ, M) is a **soft Lindelöff topological space** over $\mathcal{T}$.*

Theorem 4.1. *If (G_1, τ, M_1) and (G_2, τ_2, M_2) are soft T_2-spaces (soft regular spaces, soft separable), $(G, \tau, M) = (G_1, \tau, M_1) \bigotimes (G_2, \tau_2, M_2)$, then (G, τ, M) is a soft T_2-space (soft regular spaces, soft separable).*

Proof. We only prove the case of soft soft T_2-spaces.

(1) $\forall (x_1, y_1) \neq (x_2, y_2) \in G_1 \times G_2$, then $x_1 \neq x_2$ or $y_1 \neq y_2$, without loss the generality, assume that $x_1 \neq x_2$, then $\exists (F_1, B_1) \in \mathcal{U}_{x_1}, (F_2, B_2) \in \mathcal{U}_{x_2}$, such that $x_1 \notin (F_2, B_2), x_2 \notin (F_1, B_1)$, and $(F_1, B_1) \sqcap (F_2, B_2) = \tilde{\emptyset}$, and $(x_1, y_1) \in (F_1, B_1) \times \widetilde{G_2}, (x_2, y_2) \in (F_2, B_2) \times \widetilde{G_2}, (x_2, y_2) \notin (F_1, B_1) \times \widetilde{G_2}, (x_1, y_1) \notin (F_2, B_2) \times \widetilde{G_2}$. And $((F_1, B_1) \times \widetilde{G_2}) \sqcap ((F_1, B_1) \times \widetilde{G_2}) = ((F_1, B_1) \times (F_2, B_2)) \sqcap \widetilde{G_2} = \tilde{\emptyset}$. So, (G, τ, M) is also a soft T_2-space. □

Remark 4.1. (1) We know, a soft T_2-space is also a soft T_1, T_0-space, so, the above conclusion also hods for soft T_1, T_0-space.

(2) If (G, τ, M) is a soft discrete topological space, G includes uncountable objects, then (G, τ, M) is not a soft separable topological space. In fact, by (G, τ, M) being soft discrete, $\forall (F, B) \in \tau, \overline{(F, B)} = (F, B) \neq \widetilde{G}$.

(3) Not all soft topological separation axioms can be hold in their soft product topological space. **For example**, $(\mathbb{R}_l, \tau, M)$ which soft base is $\mathcal{B} = \{[a, b) | a, b \in \mathbb{R}\}$ is a soft Lindelöff space, however, we can prove that $(\mathbb{R}_l, \tau, M) \bigotimes (\mathbb{R}_l, \tau, M)$ is not a soft Lindelöff space.

5. The Connectedness and Compactness of the Soft Product Topology

Definition 5.1. Let $(F_1, B_1), (F_2, B_2) \in \tau$.
(1) If $((F_1, B_1) \sqcap \overline{(F_2, B_2)}) \widetilde{\cup} ((F_2, B_2) \sqcap \overline{(F_1, B_1)}) = \widetilde{\emptyset}$, then (F_1, B_1) and (F_2, B_2) are **isolated soft subsets** of $\widetilde{G}$.
(2) $(F_1, B_1) \neq \widetilde{\emptyset}, (F_2, B_2) \neq \widetilde{\emptyset}$, and $(F_1, B_1), (F_2, B_2)$ be isolated soft subsets of $\widetilde{G}$, if $\widetilde{G} = (F_1, B_1) \widetilde{\cup} (F_2, B_2)$, then (G, τ, M) is **not connected**, otherwise, (G, τ, M) is connected.

Definition 5.2 (Fu2016). $\mathcal{A} = \{(F, B) \mid (F, B) \text{ is soft open set in } \tau\}$ is any soft open cover of G. If $\mathcal{A}$ has a soft finite sub-collection which covers G, then (G, τ, M) is a **soft compact topological space** (simply, **soft compact**). If $\mathcal{A}$ has a soft countable subcover of G, then (G, τ, M) is a **soft Lindelöff topological space** over $\mathcal{T}$.

Clearly, if (G, τ, M) is a soft compact topological space, then it must be soft Lindelöff, however, the converse does not hold. **For example**, let G include countable infinite points, then $\tau = \{(F_i, B_i) \mid F_i : B_i \to \wp(G)\}$ forms a discrete soft topology, and (G, τ, M) is a soft Lindelöff topological space, but it is not soft compact.

Theorem 5.1. Let $(G, \tau, M) = (G_1, \tau, M_1) \bigotimes (G_2, \tau_2, M_2)$, if (G_1, τ, M_1) and (G_2, τ_2, M_2) are soft connected (soft compact), then (G, τ, M) is soft connected (soft compact).

Proof. (1) Let $(F_1, B_1) \times (F_2, B_2), (F_1', B_1') \times (F_2', B_2') \in \tau$, that is, $(F_1, B_1)(F_1', B_1') \in \tau_1, (F_2, B_2)(F_2', B_2') \in \tau_2$. Suppose (G, τ, M) is not soft connected, then $\overline{(F_1, B_1) \times (F_2, B_2)} \sqcap ((F_1', B_1') \times (F_2', B_2')) = \widetilde{\emptyset} \Rightarrow (\overline{(F_1, B_1)} \times \overline{(F_2, B_2)}) \sqcap ((F_1', B_1') \times (F_2', B_2')) = \widetilde{\emptyset} \Rightarrow (\overline{(F_1, B_1)} \sqcap (F_1', B_1')) \times (\overline{(F_2, B_2)} \sqcap (F_2', B_2')) = \widetilde{\emptyset} \Rightarrow \overline{(F_1, B_1)} \sqcap (F_1', B_1') = \widetilde{\emptyset}$ or $\overline{(F_2, B_2)} \sqcap (F_2', B_2') = \widetilde{\emptyset}$. Which is contradict with (G_1, τ, M_1) and (G_2, τ_2, M_2) are soft connected topological spaces. Hence, (G, τ, M) is a soft connected topological space.
(2) Using the definition of soft compact, the proof omit it. $\square$

Remark 5.1. Not all soft topological properties can be hold in their soft product topological space.
For example, let $(\mathbb{R}_l, \tau, M)$ be soft real lower limit topological space which soft basis is $\mathcal{B} = \{[a, b) \mid a, b \in \mathbb{R}\}$, we know, $(\mathbb{R}_l, \tau, M)$ is a soft Lindelöff space, however, we can prove that $(\mathbb{R}_l, \tau, M) \bigotimes (\mathbb{R}_l, \tau, M)$ is not a soft Lindelöff space.

6. Conclusion

This paper discusses the product of soft topological space and the properties of soft product topology. Two points are mainly covered: Firstly, we give the element character of soft product, and some properties of soft product topology; Secondly, we discuss the soft separation axioms, the soft connectedness and soft compactness of soft product topological spaces. Our aim is to offer a new method for studying multi-objects formal context under the complex environment, hoping to offer a fresh idea for data process.

Acknowledgments

This work is supported by Qinghai Science Foundation (Grant No. 2018-ZJ-911), the National Natural Science Foundation of China (Grant No. 11641002, No. 61773019).

References

1. Z. Pawlak, *International Journal of computer and Information Sciences*, **11**, 341(1982).
2. D. Molodtsov, *Computers and mathematics with Applications*, **37**, 19(1999).
3. P.K. Maji, R. Biswas and A.R. Roy, *Computers and mathematics with Applications*, **45**, 555(2003).
4. P.K. Maji, R. Biswas, A.R. Roy, *Journal of Fuzzy Mathematics*, **9(3)**, 589(2001).
5. Euler, Leonhard, *Commentarii Academiae Scientiarum Imperialis Petropolitanae, reprint*, **8**, 128(1976).
6. Naim Cagman, Serkan Karatas, Serdar Enginoglu, *Computers and Mathematics with Applications*, **62**, 351(2011).
7. M. Shabir, M. Naz, *Computers and Mathematics with Applications*, **61**, 1786(2011).
8. Sabir Hussain, Bashir Ahmad, *Computers and Mathematics with Applications*, **62**, 4058(2011).
9. Bashir Ahmad, Sabir Hussain, *Mathematical science*, **64**, 1(2012).
10. Sabir Hussain, Bashir Ahmad, *Journal of Mathematics and Statistics*, **44**, 559(2015).
11. Li Fu, Hua Fu, *International Journal of Computers and Technologies*, **12**, 3536(2014).
12. Li Fu, Zhen Liu, *Int. J. of Computers System Science and Engineering*, **31**, 165(2015).
13. Li Fu, Yingchao Shao, *BMEI2015*, 810(2015).
14. Li Fu, Hua Fu, *British Journal of Mathematics and Computer Science*, **14**, 1(2016).
15. Li Fu, Hua Fu, Fei You, *The 2017 4th International Conference on Systems and Informatics*, 1513(2017).

Fuzzy TOPSIS: Violation of basic axioms

Boris Yatsalo* and Alexander Koborov

*Department of Information Systems,
Institute of Cybernetic Intelligent Systems of the National Research Nuclear University
MEPHI (IATE NRNU MEPHI), Obninsk - Moscow, Russian Federation*
*yatsalo@gmail.com

Luis Martínez

*Department of Computer Science, University of Jaén
Jaén, 23071, Spain*

The use of fuzzy numbers (FNs) for uncertainty treatment within Multi-Criteria Decision Analysis (MCDA) is a claiming approach for examination of multicriteria problems in fuzzy environment. For proper use of a fuzzy extension for the MCDA method (FMCDA), the fulfilment of corresponding axioms/requirements, associated with this method, should be analyzed. For empirical MCDA methods (e.g., TOPSIS, AHP, PROMETHEE) the basic axiom concerns the properties of generalized criterion for this method. The goal of this contribution is exploring violation of basic axioms for FMCDA methods for Fuzzy TOPSIS (FTOPSIS) as an example. The ways of adjustment of fuzzy MCDA methods to correct use is briefly discussed.

Keywords: MCDA; fuzzy number; ranking of fuzzy numbers; fuzzy TOPSIS; overestimation.

1. Introduction

Multi-Criteria Decision Analysis (MCDA)[1] aims at supporting decision makers to take effective or trade-off decisions more consistently. The most challenging difficulty in which decision analysis methods sometimes fail is the uncertainty of objective data and subjective judgments. In such cases, the use of fuzzy methods, that have been included in the multi-criteria decision analysis,[2] provides a useful approach for handling three main MCDA problems:[1] choosing, ranking and sorting.

The management and analysis of uncertainties by fuzzy modelling in real-world decision problems not only requires the use of fuzzy functions and models, but also inevitably leads to comparison of fuzzy quantities. Thus, in fuzzy decision analysis, ranking of fuzzy quantities plays a key role.[3,4]

For justified use of the value/utility based MCDA methods MAVT/MAUT, fulfilment of the specific axioms is required.[1,5] For empirical methods, such as, e.g., TOPSIS, PROMETHEE, and AHP, a key requirement/axiom concerns the *generalized* (or overall) criterion $V(.)$ (as a rule, benefit one, i.e., the more the better): if alternative A exceeds alternative B according to Pareto, then $V(A) \geq V(B)$.

To the best of our knowledge, there are no publications, where violation of basic axioms for FMCDA methods have been discussed. At the same time, there are axioms for ranking methods, and some of these intuitively understandable axioms are violated by all key popular ranking methods. These axioms and three ranking methods are considered in this paper.

A fuzzy extension of TOPSIS method, FTOPSIS, is considered in this contribution. This method is among of the most popular ones in applications.[6] Violation of the key axioms for generalized criterion (fuzzy coefficient of closeness) of FTOPSIS is explored for several ranking methods along with simplified and proper assessing functions of fuzzy variables based on transformation methods to overcome the overestimation problem.[7]

2. Preliminaries: Fuzzy Numbers and Their Ranking

This section reviews and fixes basic notions about fuzzy concepts, which are used in this contribution.

2.1. *Fuzzy numbers and fuzzy preference relations*

A fuzzy number (FN) is a convex, normal, and restricted fuzzy set in $\mathbb{R}$ with a continuous or piecewise upper-continuous membership function $\mu_Z(x)$. Therefore, we assume that there exist two real numbers $c_1, c_2 \in \mathbb{R}$, $c_1 \leq c_2$, such that:

$$Z = \{(x, \mu_Z(x)) : \mu_Z(x) > 0 \ \ x \in (c_1, c_2), \ \mu_Z(x) = 0 \ x \notin [c_1, c_2]\}. \quad (1)$$

$\mathbb{F}$ denotes the set of all FNs in accordance with Eq. (1).
If $c = c_1 = c_2$, then $Z = c$ is a singleton and $\mu_Z(c) = 1$. It should be added also, the condition $\mu_Z(c_1) = \mu_Z(c_2) = 0$ is, in general case, not necessary and is often used for convenience.

Let $Z \in \mathbb{F}$ be a FN and $\alpha \in (0, 1]$. An α-cut[8,9] of FN Z is defined as $Z_\alpha = \{x \in \mathbb{R} \mid \mu_Z(x) \geq \alpha\} = [A_\alpha, B_\alpha]$. If $\alpha = 0$, and we denote $[A_0, B_0] = [c_1, c_2]$, then a fuzzy number Z can be identified with the family of α-cuts: $Z = \{[A_\alpha, B_\alpha]\}$, $0 \leq \alpha \leq 1$.

Definition 2.1. Fuzzy preference relation R is a fuzzy relation on $\mathbb{F} \times \mathbb{F}$: $R = ((Z_i, Z_j), \mu_R(Z_i, Z_j))$, where membership function $\mu_R(Z_i, Z_j) \in [0, 1]$ indicates the degree of preference of Z_i over Z_j.

Definition 2.2. Let R be a fuzzy relation on $\mathbb{F} \times \mathbb{F}$. For any $Z_i, Z_j \in \mathbb{F}$, their fuzzy ranking is defined as:

$$Z_i \succeq Z_j \text{ if } \mu_R(Z_i, Z_j) \geq 0.5, Z_i \succ Z_j \text{ if } \mu_R(Z_i, Z_j) > 0.5, \text{ and } Z_i \sim Z_j \text{ if } \mu_R(Z_i, Z_j) = 0.5 \quad (2)$$

For the sake of clarity, let $\mathbf{Z} = \{Z_1, \ldots, Z_n\} \subset \mathbb{F}$ be a finite family of FNs, for a fuzzy relation $R(Z_i, Z_j)$ the following notations are used: $\mu_{ij} = \mu_R(Z_i, Z_j) = \mu_R(Z_i \geq Z_j) = \mu_R(Z_j \leq Z_i)$.

The ranking of FNs is a key stage in FMCDA. Below, a brief revision of different ranking approaches is provided.[9–11] The main classes of fuzzy ranking methods[10–12] (apart of linguistic approaches) are:

1. *Defuzzification based ranking methods.* Within these methods, FNs are represented by (defuzzified) real numbers with their subsequent ranking.[9,10] In this contribution, Centroid Index (CI) and Integral of Mean for α-cuts (IM) defuzzification based methods are used. According to the CI method, defuzzification of FN Z is based on the following formula:

$$CI(Z) = \int_S xZ(x)dx \Big/ \int_S Z(x)dx, \quad (3)$$

here, $Z(x) = \mu_Z(x)$, $S = supp(Z)$.

Within IM method, the following value for FN $Z = \{[A_\alpha, B_\alpha]\}$ is assessed:

$$IM(Z) = \int_0^1 (A_\alpha + B_\alpha)/2 \ d\alpha. \quad (4)$$

FN with higher Centroid Index/Integral of Mean value has a higher rank.

2. *Ranking fuzzy methods based on the distance to a reference set.* These methods define the reference (etalon) sets and evaluate each fuzzy number Z_i by computing and comparing its distance to the reference set (the less distance corresponds the higher rank).[9,10]

3. *Ranking fuzzy methods based on pairwise comparison.* The ordering of fuzzy quantities in this case consists of pairwise comparisons, and it is the most extensively studied approach.[11]

In this paper, the two defuzzification based ranking methods, Centroid Index (CI) and Integral of Mean values for α-cuts (IM), along with pairwise comparison method based on the Yuan's fuzzy preference relation are considered. The latest method is revised below.

Let $Z_i = \{[A_\alpha^i, B_\alpha^i]\}, Z_j = \{[A_\alpha^j, B_\alpha^j]\} \in \mathbb{F}$ be two FNs and $Z_{ij} = Z_i - Z_j = \{[A_\alpha, B_\alpha]\}$. Within the *Yuan's fuzzy preference relation*,[4] $R_Y = ((Z_i, Z_j), \mu_Y(Z_i, Z_j))$, the area, S_Y^+, is considered as a distance of the positive part of $Z_{ij} = \{[A_\alpha, B_\alpha]\}$ to the axis OY, that is computed as:

$$S_Y^+(Z_{ij}) = \int_0^1 (B_\alpha \theta(B_\alpha) + A_\alpha \theta(A_\alpha)) d\alpha, \tag{5}$$

here $\theta(x)$ is the Heaviside function: $\theta(x) = \{1, \ x \geq 0; \ 0, \ x < 0\}$. Consider $S_Y^-(Z_{ij}) = S_Y^+(Z_{ji})$. The *total adjusted* area under FN Z_{ij} is assessed as $S_Y(Z_{ij}) = S_Y^+(Z_{ij}) + S_Y^-(Z_{ij})$.

Definition 2.3. Let $Z_i, Z_j \in \mathbb{F}$ be two FNs and $Z_{ij} = Z_i - Z_j$. The Yuan's fuzzy preference relation R_Y, in which $\mu_Y(Z_i, Z_j)$ represents the degree of preference of Z_i over Z_j, is defined as:

$$\mu_Y(Z_i, Z_j) = S_Y^+(Z_{ij})/S_Y(Z_{ij}) \text{ if } S_Y(Z_{ij}) > 0. \tag{6}$$

For singleton FNs, $Z_i = c_i, Z_j = c_j, c_{ij} = c_i - c_j$,

$$\mu_Y(Z_i, Z_j) = \{1 \ if \ c_{ij} > 0; \ 0 \ if \ c_{ij} < 0; \ 0.5 \ if \ c_{ij} = 0\}. \tag{7}$$

The properties of Yuan's preference relation have been analyzed in.[4,11]

2.2. *Violation of axioms for raking methods*

Here, the basic axioms for ranking methods and the violation of some of them by the methods used in this contribution are considered. The following notations are commonly used in methods for ranking FNs of the set $\mathbb{S}$: $A \succ B, A \sim B$, and $A \succeq B$, that means, correspondingly, A has a higher ranking than B, the same ranking as B, and at least the same ranking as B. Seven axioms have been suggested[10] within ranking of FNs with the use of a ranking method $\mathbb{M}$. Among them, the axioms A_1 (reflexivity), A_2 (antisymmetry), A_3 (transitivity), A_4 (distinguishability), and A_5 (absence of ranks reversal) were analyzed. In this paper, the axioms A_6 and A_7 are considered:

- A_6: Let $A, B, A + C$ and $B + C$ be elements of $\mathbb{S}$. If $A \succeq B$ by $\mathbb{M}$ on $\mathbb{S}$, then $A + C \succeq B + C$ by $\mathbb{M}$ on $\mathbb{S}$;
- A_7: Let A, B, AC and BC be elements of $\mathbb{S}$ and $C \succeq 0$. If $A \succeq B$ by $\mathbb{M}$ on $\mathbb{S}$, then $AC \succeq BC$ by $\mathbb{M}$ on $\mathbb{S}$.

It was proved[10,11] that ranking of FNs by IM and R_Y satisfies the axioms $A_1 - A_6$ and both do not satisfy the axiom A_7. Ranking by CI satisfies the axioms $A_1 - A_5$ and does not satisfy the axioms A_6 and A_7.

Consider here three triangular FNs to demonstrate violation of the axiom A_7 for Yuan's preference relation: $A=(1, 4, 4.5)$, $B=(1, 1.2, 8.2)$, $C=(0.043, 0.1735, 1.3)$. According to the assessments, $\mu_Y(A \geq B) = 0.577$, and $\mu_Y(CB \geq CA) = 0.556$. Thus, $A \succeq_Y B \not\Rightarrow CA \succeq_Y CB$ (for FN C with positive support).

3. From TOPSIS to FTOPSIS

TOPSIS [13] is one of the most popular MCDA/MADM method in applications. [6,14] There are different variants of transition from classical TOPSIS to FTOPSIS, see. [15–18]

In this paper, a general approach to FTOPSIS is considered: fuzzy criterion values and fuzzy weight coefficients are FNs of the general type. Implementation of the suggested FTOPSIS model follows the below steps:
1. *Setting the list of alternatives*, a_i, $i = 1, ..., n$, and criteria, C_j, $j = 1, ..., m$, for the multicriteria problem under investigation. Evaluating criterion values $c_{ij} = \{[A_\alpha^{ij}, B_\alpha^{ij}]\}$ $i = 1, ..., n$, $j = 1, ...m$.
2. *Normalization of the criterion values.* For alternative $a_i = (c_{i1}, ..., c_{im})$, the following *linear* transformation of criterion values, $c_{ij} \to x_{ij}$, is applied: (i) for benefit criteria $\{j\}$: $x_{ij} = (c_{ij} - A_0^{0j})/(B_0^{0j} - A_0^{0j})$; (ii) for cost criteria: $x_{ij} = (B_0^{0j} - c_{ij})/(B_0^{0j} - A_0^{0j})$. Here $B_0^{0j} = max_i B_0^{ij}$, $A_0^{0j} = min_i A_0^{ij}$. Thus, for FNs x_{ij}, $supp(x_{ij}) \subseteq [0, 1]$, and in new dimensionless x-scale all the criteria are benefit ones. The suggested approach for normalization of criterion values is effective for computations within FTOPSIS.
3. *Choice of ideal and ant-ideal alternatives.* The approach to setting *global* ideal and anti-ideal alternatives is suggested in this FTOPSIS model: $I^+ = (1, ..., 1)$, $I^- = (0, ..., 0)$.
4. *Setting weight coefficients* $\{w_j, j = 1, ..., m\}$ for the criteria. There are several approaches to setting weight coefficients for TOPSIS and FTOPSIS based on subjective and objective approaches [19,20] that can be used here.
5. *Calculating the distances* of each alternative $x_i = (x_{i1}, ..., x_{im})$ to the ideal, D_i^+, and anti-ideal, D_i^-, alternatives:

$$D_i^+ = d(x_i, I^+) = (\sum_{k=1}^m w_k^2(1-x_{ik})^2)^{1/2}; \quad D_i^- = d(x_i, I^-) = (\sum_{k=1}^m w_k^2 x_{ik}^2)^{1/2} \quad (8)$$

6. Assessing generalized criterion D_i for alternative x_i (or the relative coefficient of closeness of alternative x_i to the ideal alternative):

$$D_i = \frac{D_i^-}{D_i^- + D_i^+} = \frac{(\sum_{k=1}^m w_k^2 x_{ik}^2)^{1/2}}{(\sum_{k=1}^m w_k^2(1-x_{ik})^2)^{1/2} + (\sum_{k=1}^m w_k^2 x_{ik}^2)^{1/2}}. \quad (9)$$

7. *Ranking alternatives* based on the chosen method(s) for ranking of FNs $\{D_i,\ i = 1,...,n\}$.

In fuzzy modeling with the use of standard fuzzy arithmetic, the overestimation problem[7] has often the place, and the level of overestimation depends on the form of fuzzy expression. For the FTOPSIS model (9), there is overestimation due to fuzzy values of weight coefficients and criterion values are both in nominator and denominator. To overcome the overestimation problem, the Transformation Method(s) (TM) may be used.[7] The overestimation for FTOPSIS model can be briefly described as follows: let $D_i(O)$ and $D_i(T)$ (correspondingly, FTOPSIS-O and FTOPSIS(T)) are the estimations of (9) based on standard fuzzy arithmetic and TM correspondingly, then $supp(D_i(T)) \subset supp(D_i(O))$.

4. Violation of the Basic Axioms for FTOPSIS Models

For TOPSIS/FTOPSIS models the basic axioms concern the generalized criterion $D = D(a_i)$ for evaluating alternatives $a_i, i = 1...,n$.

Consider Pareto domination notion: for a given ranking method, alternative a_i dominates alternative a_k according to Pareto, $a_i \succ_P a_k$, if a_i is not worse than a_k according to all criteria, i.e., $c_{ij} \succeq c_{kj}$ for all benefit criteria $\{j\}$ and $c_{ij} \preceq c_{kj}$ for all cost criteria $\{j\}$, and at least for one of the criteria, j_0, a_i exceeds a_k (i.e., at least one of the indicated inequalities is strong).

The basic and intuitively understandable axioms for the generalized criterion $D(.)$ of an FTOPSIS model, which is considered as a benefit one (i.e., the more the better), are as follows:

- Axiom A_{G1}: if alternative a dominates alternative b according to Pareto, $a \succ_P b$, and $D(.)$ is a generalized criterion for an FTOPSIS model, then $D(a) \succeq D(b)$;

- Axiom A_{G2}: if $a \sim b$ then $D(a) \sim D(b)$.

For classical MCDA methods, including TOPSIS, axioms A_{G1} and A_{G2}, evidently, have the place. Below, the fulfilment of these axioms for FTOPSIS model is explored.

FTOPSIS models are based on implementing the functions (8), and, for the model considered in this paper, on the expressions (9). These formulas include, among others, the following functions $f(x)$ and $f(x,y)$: xy, x^2, $\sqrt{x}$, x/y, $x+y$. The use of each of these functions can lead to "violation of original order of FNs" for ranking methods under consideration. E.g., for triangular FNs (TrFNs), $A = (1, 4, 4)$ and $B = (1, 1, 6.5)$,

$CI(A) = 3 > CI(B) = 2.83$, and $CI(B^2) = 13.51 > CI(A^2) = 10.44$.

Thus, taking into account violation of axioms A_6 and/or A_7 for three ranking methods under consideration along with the comment and example above), violation of the basic axioms A_{G1} and A_{G2} for FTOPSIS (both for Model-O and Model-T) is expected. To demonstrate this, consider three alternatives, a_1, a_2, and a_3 with the following criterion values: $a_1 = (c_{11}, 1, 1)$, $a_2 = (c_{21}, 1, 1)$, $a_3 = ((2, 2.7, 6), 0.9, 0.9)$, and weight coefficients: $w_1 = (0.183, 0.1915, 0.2)$ (w_1 is TrFN), $w_2 = 0.383$, and $w_3 = 0.426$; in the case $c_{11} = (1, 4, 4)$ and $c_{21} = (1, 1, 9)$, for ranking by Yuan's preference relation R_Y, $\mu_Y(c_{11}, c_{21}) = 0.54$, $a_1 \succ_{P(Y)} a_2$ and for generalized criterion $D(a)$, $D(a_2) \succ_Y D(a_1)$ as $\mu_Y(D(a_2), D(a_1)) = 0.505$ for FTOPSIS-O and $\mu_Y(D(a_2), D(a_1)) = 0.531$ for FTOPSIS-T model.

The similar examples of the axioms A_{G1} and A_{G2} violation for CI and IM ranking methods can be presented as for Yuan's preference relation.

In addition to the models FTOPSIS-O/T (9), a simplified triangular FTOPSIS model was also considered. Within this model, input data are TrFNs FNs, results of all functions within FTOPSIS are approximated by corresponding TrFNs; the resulting generalized criterion values D_i are TrFNs, and the defuzzification based method CI/IM is implemented for ranking D_i, $i = 1, ..., n$. Violation of the axioms A_{G1} and A_{G2} for such a simplified FTOPSIS/TrFNs model and CI-IM ranking methods is demonstrated similarly.

Thus, in the general case, when there is no restrictions on input data, the axioms A_{G1} and A_{G2} for FTOPSIS can be violated.

5. Conclusions

In this contribution, violation of the basic axioms A_{G1} and A_{G2} (Section 4) for generalized criterion of FTOPSIS models is demonstrated. In such a situation, the key question is: *are there any approaches to correct use of FTOPSIS model(s) under the additional requirements or constraints?*

There may be several approaches to overcoming this problem. One of them includes an extension of the notion of *equivalence* of two FNs (alternatives) to make the axiom A_{G2} more flexible. For this, Definition 2.2 for equivalent FNs can be modified. To hold the axiom A_{G1}, some restrictions on allowable input values (e.g., the use of distinguishable FNs[21] or FNs presented with linguistic variables) can be considered.

Thus, correctness of different Fuzzy MCDA methods regarding the fulfillment of basic axioms requires a careful exploring along with effective approaches to overcoming corresponding problems.

Acknowledgments

This work is partially supported by the Spanish National research project TIN2015-66524-P and ERDF, and Russian National research project RFBR-18-29-03166.

References

1. V. Belton and T. Stewart, *Multiple Criteria Decision Analysis: An Integrated Approach* (Kluwer Academic Publishers: Dordrecht, 2002).
2. W. Pedrycz, P. Ekel and R. Parreiras, *Fuzzy Multicriteria Decision-Making: Models, Methods and Applications* (John Wiley & Sons, Ltd. Chichester, UK, 2010).
3. R. Bellman and L. Zadeh, *Management Science* **17**, 141 (1970).
4. Y. Yuan, *Fuzzy Sets and Systems* **44**, 139 (1991).
5. R. Keeney and H. Raiffa, *Decisions with multiple objectives: preferences and value tradeoffs* (John Wiley & Sons, New York, 1976).
6. C. Kahraman, S. C. Onar and B. Oztaysi, *International Journal of Computational Intelligence Systems* **8**, 637 (2015).
7. M. Hanss, *Applied Fuzzy Arithmetic* (Springer-Verlag, 2005).
8. D. Dubois and H. Prade, *International Journal of Systems Science* **9**, 613 (1978).
9. X. Wang, D. Ruan and E. Kerre, *Mathematics of Fuzziness - Basic Issues*. (Springer-Verlag, 2009).
10. X. Wang and E. Kerre, *Fuzzy Sets and Systems* **118**, 375 (2001).
11. X. Wang and E. Kerre, *Fuzzy Sets and Systems* **118**, 387 (2001).
12. C. Kahraman and A. Tolga, *International Journal of Computational Intelligence Systems* **2**, 219 (2009).
13. H. C.-L. and Y. K., *Multiple Attribute Decision Making: Methods and Applications* (Lecture notes in economics and mathematical systems, 186. Springer-Verlag, Berlin, 1981).
14. M. Behzadian, S. Otaghsara, M. Yazdani and J. Ignatius, *Expert Systems with Applications* **39**, 13051 (2012).
15. S. Chen and C. Hwang, *Fuzzy Multiple Attribute Decision Making: Methods and Applications* (Springer-Verlag, Berlin, 1992).
16. C.-T. Chen, *Fuzzy Sets and Systems* **114**, 1 (2000).
17. T. Kayaand and C. Kahraman, *Expert Systems with Applications* **38**, 6577 (2011).
18. K. Rudnik and D. Kacprzak, *Applied Soft Computing* **52**, 1020 (2017).
19. H. Deng, C. Yeh and R. Willis, *Computers and Operations Research* **27**, 963 (2000).
20. D. Olson, *Mathematical and Computer Modelling* **40**, 721 (2004).
21. B. Yatsalo and L. Martínez, *IEEE Transactions on Fuzzy Systems* **0**, p. Submitted (2018).

s-grouping functions: A generalization of bivariate grouping functions by t-conorms

Hugo Zapata,[1,2] Graçaliz P. Dimuro,[3,4] Javier Fernández[4,5] and Humberto Bustince[4,5]

[1] *Facultad de Ciencias, Universidad Central de Venezuela*
Avenida Los Ilustres, Caracas 1020, Venezuela
[2] *Departamento de Matemáticas, Universidad de Sucre, Sincelejo, Colombia*
[3] *Centro Ciências Computacionais, Universidade Federal do Rio Grande*
Campus Carreiros, Rio Grande 96201-900, Brazil
[4] *Institute of Smart Cities, Universidad Publica de Navarra*
Campus Arrosadía, Pamplona 31006, Spain
[5] *Departamento de Automática y Computación, Universidad Publica de Navarra*
Campus Arrosadía, Pamplona 31006, Spain

This paper introduces a generalization of grouping functions by extending one of the boundary conditions of its definition. More specifically, instead of requiring that "the considered function is equal to one if and only if some of the inputs is equal to one", we allow the range in which some t-conorm is 1. We call such generalization by a s-grouping function with respect to such t-conorm. Then we analyze the main properties of s-grouping function and introduce some construction methods. Considering that the grouping functions depend only on the t-conorm, it is proposed in this work to extend this condition to any t-conorm.

Keywords: Aggregation function; grouping function; t-conorm.

1. Introduction

The notion of grouping function [1,4,7–10] has shown itself very useful to deal with situations in which it is necessary to determine up to what extent a given element belongs to one or several classes whose boundaries are not crisp. It has been used, e.g., in image processing, [12] classification problems [13,14] and decision making. [11]

Our goal here is to generalize the notion of grouping function by relaxing one of the boundary condition. In particular, instead of demanding that "the considered function is equal to 1 if and only if some of the inputs is equal to 1", we allow for some kind of threshold, defined in terms of a t-conorm S. We call such generalization by a s-grouping function with respect to S.

We notice that, this simple generalization allows us to state several interesting properties, which may allow for application in fuzzy rule-based system in order to discard bad rules when computing the compatibility degree. Section 2 presents some preliminary concepts. In Section 3, besides studying the main properties, we also propose some construction methods. Section 4 is the Conclusion.

2. Preliminaries

This section aims at introducing the background necessary to understand the paper.

Definition 1. A fuzzy negation is a function $N\colon [0,1] \to [0,1]$ satisfying: **(N1)** the boundary conditions: $N(0) = 1$ and $N(1) = 0$; **(N2)** N is decreasing: if $x \leq y$ then $N(y) \leq N(x)$.

A fuzzy negation N is said to be strong if: $\forall x \in [0,1] : N(N(x)) = x$ (the involutive property). The standard negation or the Zadeh's negation is given by $N_Z(x) = 1 - x$.

Definition 2. [3,15] A function $A : [0,1]^n \to [0,1]$ is said to be an n-ary aggregation operator if the following conditions hold:

(A1) A is increasing* in each argument: for each $i \in \{1,\ldots,n\}$, if $x_i \leq y$, then
$A(x_1,\ldots,x_n) \leq A(x_1,\ldots,x_{i-1},y,x_{i+1},\ldots,x_n)$;
(A2) A satisfies the Boundary conditions: $A(0,\ldots,0) = 0$ and $A(1,\ldots,1) = 1$.

Definition 3. Let $T : [0,1]^2 \to [0,1]$ a t-norm. A function $G_T : [0,1]^2 \to [0,1]$ is a t-overlap function with respect to T, if:

(G_T1) $G_T(x,y) = G_T(y,x)$,
(G_T2) $G_T(x,y) = 0 \Leftrightarrow T(x,y) = 0$,
(G_T3) $G_T(x,y) = 1 \Leftrightarrow x = y = 1$,
(G_T4) G_T is increasing,
(G_T5) G_T is continuous.

Definition 4. A t-conorm is a bivariate aggregation function $S : [0,1]^2 \to [0,1]$ satisfying the following properties, for all $x,y,z \in [0,1]$:

---

*In this paper, an increasing (decreasing) function does not need to be strictly increasing (decreasing).

(S1) Commutativity: $S(x,y) = S(y,x)$;
(S2) Associativity: $S(x, S(y,z)) = S(S(x,y), z)$;
(S3) Boundary condition: $S(x,0) = x$.
(S4) Increasing: $S(x,z) \leq S(y,z)$ si $x \leq y$.

Examples of t-conorms are $S(x,y) = x+y-xy$ and $S(x,y) = \min\{1, x+y\}$.

The main concern of this paper is the concept of grouping function.[1,4,7–9,12]

Definition 5. [4] An grouping function is a bivariate function $G\colon [0,1]^2 \to [0,1]$ satisfying the following properties, for all $x, y \in [0,1]$:

(G1) G is commutative: $G(x,y) = G(y,x)$;
(G2) $G(x,y) = 0$ if and only if $x = y = 0$;
(G3) $G(x,y) = 1$ if and only if $x = 1 \vee y = 1$;
(G4) G is increasing;
(G5) G is continuous.

3. Introducing s-Grouping Functions

This section generalizes the concept of grouping functions by changing the condition **(G3)** of Definition 5, namely, the property that requires that, for all $x, y \in [0,1]$ a grouping function $G : [0,1]^2 \to [0,1]$ it holds that $G(x,y) = 1 \Leftrightarrow x = 1 \vee y = 1$. In our generalization, we replace the max operation by a t-conorm $S : [0,1]^2 \to [0,1]$.

Definition 6. Let $G : [0,1]^2 \to [0,1]$ be a t-conorm. A function $G_S : [0,1]^2 \to [0,1]$ is said to be a s-grouping function with respect to S if the following conditions hold:

(G_S1) $G_S(x,y) = G_S(y,x)$,
(G_S2) $G_S(x,y) = 0 \Leftrightarrow x = y = 0$,
(G_S3) $G_S(x,y) = 1 \Leftrightarrow x = y \vee y = 1$,
(G_S4) G_S is increasing,
(G_S5) G_S is continuous.

Remark 1. Observe that, considering a fuzzy rule-based system, this generalization allows to discard bad rules when computing the compatibility degree. This is due to the fact that the membership degrees of the input with the antecedents would be low for bad rules and, consequently, s-grouping functions may return 1 instead of a low value, which can mislead the final prediction. Accordingly, we have maintained the second condition, since, intuitively, it is not interesting to give the same value to all the rules whose membership degrees are high, since it may imply a decrease in the predictive power.

The next theorem shows that by using strong negation, s-grouping functions can be constructed from t-overlap functions.

Theorem 1. Let G_S^T a t-overlap function and n a strong negation, then

$$G_G^S(x,y) = n(G_S^T(n(x), n(y)))$$

is a s-grouping function, where S is the dual t-conorm the t-norm T. Reciprocally, if G_G^S is a s-grouping function, then

$$G_S^T(x,y) = n(G_G^S(n(x), n(y)))$$

is a t-overlap function, where T is the dual t-norm of S.

Proof. $(G_G^S 1)$.

$$\begin{aligned} G_G^S(x,y) &= n(G_S^T(n(x), n(y))) \\ &= n(G_S^T(n(y), n(x))) \\ &= G_G^S(y, x) \end{aligned}$$

for all $x, y \in [0, 1]$, thus G_G^S is symmetric.

$(G_G^S 2)$.

$$\begin{aligned} G_G^S(x,y) = 0 &\Leftrightarrow n(G_S^T(n(x), n(y))) = 0 \\ &\Leftrightarrow G_S^T(n(x), n(y)) = 1 \\ &\Leftrightarrow T(n(x), n(y)) = 1 \\ &\Leftrightarrow n(S(x,y)) = 1 \\ &\Leftrightarrow S(x,y) = 0. \end{aligned}$$

$(G_G^S 3)$.

$$\begin{aligned} G_G^S(x,y) = 1 &\Leftrightarrow n(G_S^T(n(x), n(y))) = 1 \\ &\Leftrightarrow G_S^T(n(x), n(y)) = 0 \\ &\Leftrightarrow T(n(x), n(y)) = 0 \\ &\Leftrightarrow n(S(x,y)) = 0 \\ &\Leftrightarrow S(x,y) = 1. \end{aligned}$$

where S is the dual t-conorma of the t-norm T. The conditions $(G_G^S 4)$ and $(G_G^S 5)$ are directs, by this reasons G_G^S is a s-grouping function. Reciprocally, G_S^T is a t-overlap function: $(G_S^T 1)$.

$(G_S^T 2)$

$$\begin{aligned} G_S^T(x,y) = 0 &\Leftrightarrow n(G_G^S(n(x),n(y))) = 0 \\ &\Leftrightarrow G_G^S(n(x),n(y)) = 1 \\ &\Leftrightarrow S(n(x),n(y)) = 1 \\ &\Leftrightarrow nT(x,y) = 1 \\ &\Leftrightarrow T(x,y) = 0. \end{aligned}$$

$(G_S^T 3)$

$$\begin{aligned} G_S^T(x,y) = 1 &\Leftrightarrow n(G_G^S(n(x),n(y))) = 1 \\ &\Leftrightarrow G_G^S(n(x),n(y)) = 0 \\ &\Leftrightarrow S(n(x),n(y)) = 0 \\ &\Leftrightarrow nT(x,y) = 0 \\ &\Leftrightarrow T(x,y) = 1. \end{aligned}$$

$(G_S^T 4)$ and $(G_S^T 5)$ are directs. For all the above, it can be affirmed that G_S^T is a function t-overlap. □

Proposition 1. *Let $S : [0,1]^2 \to [0,1]$ be a continuous t-conorm. Then S is a s-grouping function with respect to itself.*

Proof. It is straightforward. □

Proposition 2. *Let $G : [0,1]^2 \to [0,1]$ be an grouping function and $S : [0,1]^2 \to [0,1]$ be a continuous t-conorm. Then the function $G_S : [0,1]^2 \to [0,1]$, defined, for all $x,y \in [0,1]$, by $G_S(x,y) = G(x,y)S(x,y)$ is a s-grouping function with respect to S.*

Proof. $(G_S 1)$ It is immediate.
$(G_S 2)$ For all $x,y \in [0,1]$, it follows that:

$$\begin{aligned} G_S(x,y) = 0 &\Leftrightarrow G(x,y)S(x,y) = 0 \\ &\Leftrightarrow G(x,y) = 0 \ \vee \ S(x,y) = 0 \\ &\Leftrightarrow x = y = 0 \text{by } \mathbf{(G2)}\ . \end{aligned}$$

(G_S3) For all $x, y \in [0, 1]$, it follows that:

$$\begin{aligned} G_S(x,y) = 1 &\Leftrightarrow G(x,y)S(x,y) = 1 \\ &\Leftrightarrow G(x,y) = 1 \wedge S(x,y) = 1 \\ &\Leftrightarrow x = 1 \vee y = 1 \wedge S(x,y) = 1 \text{ by } \mathbf{(G3)} \\ &\Leftrightarrow S(x,y) = 1. \end{aligned}$$

($G_S 4 - 5$) Since both G and S are continuous and increasing, then the results are immediate. □

The previous theorem may be generalized using a special s-conorm S' instead of the maximum function G and the t-conorm S with which the function G_S is a s-grouping with respect to S.

Theorem 2. Let $G_S^1, \ldots, G_S^n : [0,1]^2 \to [0,1]$ be s-grouping functions with respect to a t-conorm $S : [0,1]^2 \to [0,1]$ and $\omega_1, \ldots, \omega_n \in [0,1]$ be weights with $\sum_{i=1}^n \omega_i = 1$. Then the function $G_S : [0,1]^2 \to [0,1]$, defined, for all $x, y \in [0, 1]$, by $G_S(x, y) = \sum_{i=1}^n \omega_i G_S^i(x, y)$ is also a s-grouping function with respect to S.

Proof. ($G_S 1$) It is immediate.
($G_S 2$) For all $x, y \in [0, 1]$, it follows that:

$$G_S(x,y) = 0 \Leftrightarrow \sum_{i=1}^n \omega_i G_S^i(x,y) = 0$$
$$\Leftrightarrow \omega_i G_S^i(x,y) = 0, \forall i = 1, \ldots, n.$$

Since $\sum_{i=1}^n \omega_i = 1$, then there exists $k \in \{0, \ldots, n\}$ such that $\omega_k \neq 0$, and, thus $G_S^k(x,y) = 0$. By ($G_S 2$), it holds that $x = y = 0$. The reciprocal is analogous.
($G_S 3$) For all $x, y \in [0, 1]$, it follows that:

$$G_S(x,y) = 1 \Leftrightarrow \sum_{i=1}^n \omega_i G_S^i(x,y) = 1 = \sum_{i=1}^n \omega_i.$$

One has that $\sum_{i=1}^n \omega_i G_S^i(x,y) - \sum_{i=1}^n \omega_i = 0$, i.e., $\sum_{i=1}^n \omega_i(G_S^i(x,y) - 1) = 0$. This means that, for all $i = 1, \ldots, n$, it holds that $\omega_i G_S^i(x,y) - \sum_{i=1}^n \omega_i = 0$. However, since $\sum_{i=1}^n \omega_i \neq 0$, there exist $k \in \{1, \ldots, n\}$ such that $\omega_k \neq 0$. Thus, one has that $G_S^k(x,y) = 1$, and, by ($G_S 3$), it follows that $x = 1 \vee y = 1 \Leftrightarrow S(x,y) = 1$. The reciprocal is inmediate.

(O_T4-5) It is easy. □

Let $S : [0,1]^2 \to [0,1]$ be a t-conorm and denote $K_S = \{(x,y) \in [0,1]^2 \mid S(x,y) = 1\}$. Obviously, any s-grouping function with respect to a t-conorm S coincides with an grouping function if and only if $K_S = \{(x,y) \in [0,1]^2 \mid x = 1 \vee y = 1\}$.

Denote by Θ the set of all s-grouping functions with respect of any t-conorm S. The following result is immediate.

Theorem 3. *The ordered set $\mathfrak{S} = (\Theta, \leq_\Theta)$ is a lattice, where $\leq_\Theta$ is defined, for all $G_{S_1}, G_{S_2} \in \Theta$, by $G_{S_1} \leq_\Theta G_{S_2}$ if and only if $G_{S_1}(x,y) \leq G_{S_2}(x,y)$, for all $(x,y) \in [0,1]^2$.*

Theorem 4. *Let G be a s-grouping function with respect to the t-conorms $S_1, \ldots, S_n : [0,1]^2 \to [0,1]$ and let $\omega_1, \ldots, \omega_n \in [0,1]$ be weights such that $\sum_{i=1}^n \omega_i = 1$. If $S = \sum_{i=1}^n \omega_i S_i : [0,1]^2 \to [0,1]$ is a t-conorm, then G_{S_i} is a s-grouping function with respect to S.*

Proof. $(G_S 1)$ It is immediate.
$(G_S 2)$ ($\Rightarrow$) Since G is a s-grouping function with respect to the t-conorms $S_1, \ldots, S_n$, then, by $(G_S 2)$, for all $i = 1, \ldots, n$, it holds that whenever $G(x,y) = 0$ then $x = y = 0$.
$(G_S 3)$ $G(x,y) = 1 \Leftrightarrow S_i(x,y) = 1 \forall i = 1, \ldots, n$ then $S = \sum_{i=1}^n \omega_i S_i$ $(x,y) = \sum_{i=1}^n \omega_i = 1$. Now, if $S(x,y) = 1$ then $\sum \omega_i S_i(x,y) = \sum \omega_i = 1$ thus $\sum \omega_i - \sum \omega_i S_i(x,y) = \sum \omega_i(1 - S_i(x,y)) = 0$ $\Rightarrow \omega_i(1 - S_i(x,y)) = 0 \forall i = 1, \ldots, n$ as $\sum_{i=1}^n = 1$ then $\exists k \in \{1, \ldots, n\}$ such that $\omega \neq 0$ thus $1 - S_k(x,y) = 0$ which implies that $S_k(x,y) = 1$ so that $G(x,y) = 1$.
(O_T4-5) It is immediate. □

4. Conclusion

In this work, we generalized the concept of grouping functions, by relaxing the requirement that "one of its inputs must be 1 so that the grouping function is 1". For that, we considered grouping functions associated to t-conorms. Likewise, a method for constructing s-grouping functions based on certain simple conditions has been presented. Future work is concerned this generalization under an interval-valued approach, as in.[2,5,6]

Acknowledgments

Supported by Caixa and Fundación Caja Navarra of Spain, the Brazilian National Counsel of Technological and Scientific Development CNPq (Proc. 307781/2016-0), the Spanish Ministry of Science and Technology (TIN2016-77356-P).

References

1. Bedregal, B.C., Dimuro, G.P., Bustince, H., Barrenechea, E.: New results on overlap and grouping functions. Information Sciences 249, 148–170 (2013).
2. Bedregal, B.C., Dimuro, G.P., Santiago, R.H.N., Reiser, R.H.S.: On interval fuzzy S-implications. Information Sciences 180(8), 1373–1389 (2010).
3. Beliakov, G., Pradera, A., Calvo, T.: Aggregation Functions: A Guide for Practitioners. Springer, Berlin (2007).
4. Bustince, H., Fernandez, J., Mesiar, R., Montero, J., Orduna, R.: Overlap functions. Nonlinear Analysis: Theory, Methods & Applications 72(3-4), 1488–1499 (2010).
5. Dimuro, G.P.: On interval fuzzy numbers. In: 2011 Workshop-School on Theoretical Computer Science, WEIT 2011. pp. 3–8. IEEE, Los Alamitos (2011).
6. Dimuro, G.P., Bedregal, B.R.C., Reiser, R.H.S., Santiago, R.H.N.: Interval additive generators of interval t-norms. In: Hodges, W., de Queiroz, R. (eds.) Proceedings of the 15th International Workshop on Logic, Language, Information and Computation, WoLLIC 2008, Edinburgh, pp. 123–135. No. 5110 in LNAI, Springer, Berlin (2008).
7. Dimuro, G.P., Bedregal, B.: Archimedean overlap functions: The ordinal sum and the cancellation, idempotency and limiting properties. Fuzzy Sets and Systems 252, 39–54 (2014).
8. Dimuro, G.P., Bedregal, B.: On residual implications derived from overlap functions. Information Sciences 312, 78–88 (2015).
9. Dimuro, G.P., Bedregal, B., Bustince, H., Asiáin, M.J., Mesiar, R.: On additive generators of overlap functions. Fuzzy Sets and Systems 287, 76–96 (2016), theme: Aggregation Operations.
10. Dimuro, G.P., Bedregal, B., Bustince, H., Jurio, A., Baczyński, M., Miś, K.: QL-operations and QL-implication functions constructed from tuples (O,G,N) and the generation of fuzzy subsethood and entropy measures. International Journal of Approximate Reasoning 82, 170–192 (2017).
11. Garcia-Jimenez, S., Bustince, H., Hüllermeier, E., Mesiar, R., Pal, N.R., Pradera, A.: Overlap indices: Construction of and application to interpolative fuzzy systems. IEEE Transactions on Fuzzy Systems 23(4), 1259–1273 (2015).
12. Jurio, A., Bustince, H., Pagola, M., Pradera, A., Yager, R.: Some properties of overlap and grouping functions and their application to image thresholding. Fuzzy Sets and Systems 229, 69–90 (2013).
13. Lucca, G., Dimuro, G.P., Mattos, V., Bedregal, B., Bustince, H., Sanz, J.A.: A family of Choquet-based non-associative aggregation functions for application in fuzzy rule-based classification systems. In: 2015 IEEE International Conference on Fuzzy Systems (FUZZ-IEEE). pp. 1–8. IEEE, Los Alamitos (2015).
14. Lucca, G., Sanz, J.A., Dimuro, G.P., Bedregal, B., Asiain, M.J., Elkano, M., Bustince, H.: CC-integrals: Choquet-like copula-based aggregation functions and

its application in fuzzy rule-based classification systems. Knowledge-Based Systems 119, 32–43 (2017).
15. Mayor, G., Trillas, E.: On the representation of some aggregation functions. In: Proc. of IEEE Intl Sympl on Multiple-Valued Logic. pp. 111–114. IEEE, Los Alamitos (1986).

Upper and lower general aggregation operators based on a strong fuzzy metric

P. Orlovs[1] and S. Asmuss[1,2]

[1] *Department of Mathematics, University of Latvia*
Zellu str. 25, Riga, LV-1002, Latvia
[2] *Institute of Mathematics and Computer Science, University of Latvia*
Raina blvd. 29, Riga, LV-1459, Latvia
svetlana.asmuss@lu.lv, pavels.orlovs@gmail.com

The paper deals with general aggregation operators, which allow aggregate fuzzy sets taking into account the distance between elements of the universe given by a strong fuzzy metric. We introduce upper and lower aggregation operators based on an ordinary aggregation operator, t-norm and fuzzy metric. By using a strong fuzzy metric we include in the construction of such operators an additional parameter, which allows to develop the technique of aggregation. Finally, we consider upper and lower aggregation in the case when the applied fuzzy metric is induced by multiple usual metrics obtained accordingly to multiple criteria evaluation of elements of the universe.

Keywords: General aggregation operator; strong fuzzy metric; fuzzy equivalence relation.

1. Introduction

The paper is devoted to the special construction of a general aggregation operator, which is based on a fuzzy metric. The need for such operator may appear, for example, in decision making if fuzzy sets represent evaluation of some objects provided by several experts. In the case when objects under evaluation belong to a fuzzy metric space and the metric can be used to characterize the similarity of objects, it is important to include this metric in expert opinion aggregation process. The motivation is as follows: in order to obtain the evaluation of some object it is important to take into account how the experts evaluated similar objects.

Our approach here is based on the construction of upper and lower general aggregation operators introduced and developed in our previous papers[1,2] (i.e., upper and lower general aggregation operators based on a fuzzy equivalence relation). In this paper by including a strong fuzzy metric in the previously used construction instead of the corresponding fuzzy equivalence relation we introduce aggregation, which result depends on an additional parameter denoted by t. We show how this parameter can be effectively used by considering the case when the similarity of objects under consideration is defined on the basis of multiple criteria evaluation.

The structure of the paper is as follows. Sections 2 and 3 contain preliminaries on general aggregation operators and fuzzy metrics correspondingly. In Section 4 upper and lower aggregation operators defined in Subsection 2.3 are modified by including a strong fuzzy metric in their construction and the dependence of the result of aggregation on new parameter t is described. In Section 5 a special design of strong fuzzy metric is introduced in the case when the similarity of elements of the universe is defined on the basis of multiple criteria evaluation. Section 5 contains also a numerical illustration of the proposed technique of aggregation.

2. Preliminaries on Aggregation Operators

2.1. *Aggregation operator*

Aggregation is the process of combining several numerical values into a single representative value. Mathematically aggregation operator is a function that maps multiple inputs from a set into a single output from this set. In the classical case[3–5] aggregation operators are defined on interval $[0, 1]$.

Definition 2.1. A mapping $A : [0,1]^n \to [0,1]$ is called an aggregation operator if the following conditions hold:

(A1) $A(0, \ldots, 0) = 0$;
(A2) $A(1, \ldots, 1) = 1$;
(A3) for all $x_1, \ldots, x_n, y_1, \ldots, y_n \in [0,1]$ it holds

$$x_i \leq y_i, i = 1, \ldots, n \Longrightarrow A(x_1, \ldots, x_n) \leq A(y_1, \ldots, y_n).$$

2.2. *General aggregation operator*

The notion of general aggregation operator $\tilde{A}$ acting on $[0,1]^X$, where $[0,1]^X$ is the set of all fuzzy subsets of a set X, was introduced in 2003 by A. Takači.[6] We denote a partial order on $[0,1]^X$ by $\preceq$. In this paper we consider the case: $\mu \preceq \eta$ if and only if $\mu(x) \leq \eta(x)$ for all $x \in X$ (here $\mu, \eta \in [0,1]^X$). The least and the greatest elements of this order are denoted by $\tilde{0}$ and $\tilde{1}$, which are indicators of $\varnothing$ and X respectively, i.e., $\tilde{0}(x) = 0$ and $\tilde{1}(x) = 1$ for all $x \in X$.

Definition 2.2. A mapping $\tilde{A}\colon ([0,1]^X)^n \to [0,1]^X$ is called a general aggregation operator if and only if the following conditions hold:
($\tilde{A}1$) $\tilde{A}(\tilde{0},\ldots,\tilde{0}) = \tilde{0}$;
($\tilde{A}2$) $\tilde{A}(\tilde{1},\ldots,\tilde{1}) = \tilde{1}$;
($\tilde{A}3$) and for all $\mu_1,\ldots,\mu_n,\eta_1,\ldots,\eta_n \in [0,1]^X$ it holds

$$\mu_i \preceq \eta_i, i = 1,\ldots,n \Longrightarrow \tilde{A}(\mu_1,\ldots,\mu_n) \preceq \tilde{A}(\eta_1,\ldots,\eta_n).$$

The most simplest approach to construct a general aggregation operator $\tilde{A}$ based on an ordinary aggregation operator A is the pointwise extension:

$$\tilde{A}(\mu_1,\ldots,\mu_n)(x) = A(\mu_1(x),\ldots,\mu_n(x)).$$

2.3. Upper and lower general aggregation operators

Another approach is to construct a general aggregation operator by using a fuzzy equivalence relation. Let us recall the definition of two notions used in such construction.

Definition 2.3. Let T be a t-norm and E be a fuzzy relation on a set X, i.e., E is a fuzzy subset of $X \times X$. A fuzzy relation E is called a T-fuzzy equivalence relation if and only if for all $x,y,z \in X$ it holds
($E1$) $E(x,x) = 1$ (reflexivity);
($E2$) $E(x,y) = E(y,x)$ (symmetry);
($E3$) $T(E(x,y),E(y,z)) \leq E(x,z)$ (T-transitivity).

Definition 2.4. Let T be a left continuous t-norm. The residuum $\overrightarrow{T}$ of T is defined for all $x,y \in [0,1]$ by

$$\overrightarrow{T}(x|y) = \sup\{\alpha \in [0,1] \mid T(\alpha,x) \leq y\}.$$

Using the idea of upper and lower approximation operators we have introduced and investigated (see Ref. 1) upper and lower general aggregation operators based on a fuzzy equivalence relation.

Definition 2.5. Let $A\colon [0,1]^n \to [0,1]$ be an aggregation operator, T be a left continuous t-norm, $\overrightarrow{T}$ be the residuum of T and E be a T-fuzzy equivalence relation defined on a set X. The upper and lower general aggregation operators $\tilde{A}_{E,T}$ and $\tilde{A}_{E,\overrightarrow{T}}$ are defined respectively by

$$\tilde{A}_{E,T}(\mu_1,\ldots,\mu_n)(x) = \sup_{x' \in X} T(E(x,x'), A(\mu_1(x'),\ldots,\mu_n(x'))), \qquad (1)$$

$$\tilde{A}_{E,\overrightarrow{T}}(\mu_1,\ldots,\mu_n)(x) = \inf_{x' \in X} \overrightarrow{T}(E(x,x')|A(\mu_1(x'),\ldots,\mu_n(x'))), \qquad (2)$$

where $x \in X$ and $\mu_1,\ldots,\mu_n \in [0,1]^X$.

In Ref. 1 it was shown that operators $\tilde{A}_{E,T}$ and $\tilde{A}_{E,\vec{T}}$ actually are general aggregation operators.

3. Preliminaries on Fuzzy Metrics

3.1. *Fuzzy metric*

The concept of a fuzzy metric has been introduced by I. Kramosil and J. Michalek.[7] A. George and P. Veeramani[8] slightly modified the original concept. Their approach is accepted also in our paper.

Definition 3.1. Let T be a t-norm. A T-fuzzy metric on a set X is a fuzzy set $M\colon X \times X \times (0, +\infty) \to [0,1]$ such that:
$(FM1)$ $M(x,y,t) > 0$ for all $x,y \in X$ and all $t > 0$;
$(FM2)$ $M(x,y,t) = 1$ if and only if $x = y$;
$(FM3)$ $M(x,y,t) = M(y,x,t)$ for all $x,y \in X$ and all $t > 0$;
$(FM4)$ $M(x,z,t+s) \geq T(M(x,y,t), M(y,z,s))$ for all $x,y,z \in X, t,s > 0$;
$(FM5)$ $M(x,y,\cdot)\colon (0,+\infty) \to [0,1]$ is continuous for all $x,y \in X$.

3.2. *Strong fuzzy metric*

We use the notion of a strong fuzzy metric as it was introduced in Ref. 9. In the original definition of a strong fuzzy metric it was defined by using axioms $(FM1)$ - $(FM3)$, $(SFM4)$ and $(FM5)$. However, these axioms do not imply axiom $(SFM5)$ and hence a strong fuzzy metric does not need to be a fuzzy metric. It is a reason due to which axiom $(FM5)$ in the definition of a strong fuzzy metric has been replaced by axiom $(SFM5)$ (see Ref. 9).

Definition 3.2. A T-fuzzy metric $M\colon X \times X \times (0,+\infty) \to [0,1]$ is called strong if, in addition to the properties $(FM1)$ - $(FM5)$, the following stronger versions of axioms $(FM4)$ and $(FM5)$ are satisfied:
$(SFM4)$ $M(x,z,t) \geq T(M(x,y,t), M(y,z,t))$ for all $x,y,z \in X$ and $t > 0$;
$(SFM5)$ $M(x,y,\cdot)$ for all $x,y \in X$ is continuous and nondecreasing.

4. Upper and Lower Aggregation Operators Introduced via a Strong Fuzzy Metric

4.1. *Using of a strong fuzzy metric for data aggregation*

We propose to use $M(x,x',1/t)$ in the constructions (1)-(2) of upper and lower aggregation operators. Properties of the modified function $\tilde{M}$, where

$\tilde{M}(x,y,t) = M(x,y,1/t)$ for all $x,y \in X$ and for all $t > 0$, are very similar to the properties of M with the exception of property $(SFM5)$, instead of which the following modified property holds:

$(S\tilde{F}M5)$ $\tilde{M}(x,y,\cdot)$ for all $x,y \in X$ is continuous and nonincreasing.

Using $M(x,x',1/t)$ allows us to evaluate the closeness between x and x' in dependence on parameter $t > 0$. The role of this parameter could be described as follows: value t could be interpreted as amount of information used to evaluate the closeness between x and x' as $M(x,x',1/t)$. For example, in Section 5 we consider the case when value t is interpreted as number of criteria used to evaluate the closeness between x and x' (in this case it is enough to consider only integer values of t).

4.2. *Upper and lower aggregation introduced via a strong fuzzy metric*

Definition 4.1. Let $A\colon [0,1]^n \to [0,1]$ be an aggregation operator, T be a left continuous t-norm, $\overrightarrow{T}$ be the residuum of T and M be a strong T-fuzzy metric defined on a set X. The upper and lower operators $\tilde{A}_{M,T}$ and $\tilde{A}_{M,\overrightarrow{T}}$ are defined respectively by

$$\tilde{A}_{M,T}(\mu_1,\ldots,\mu_n)(x,t) = \sup_{x' \in X} T(M(x,x',1/t), A(\mu_1(x'),\ldots,\mu_n(x'))), \quad (3)$$

$$\tilde{A}_{M,\overrightarrow{T}}(\mu_1,\ldots,\mu_n)(x,t) = \inf_{x' \in X} \overrightarrow{T}(M(x,x',1/t)|A(\mu_1(x'),\ldots,\mu_n(x'))), \quad (4)$$

where $x \in X$, $t \in (0,+\infty)$ and $\mu_1,\ldots,\mu_n \in [0,1]^X$.

It is easy to see that for all $t \in (0,+\infty)$ operators $\tilde{A}_{M,T}(\cdot,\ldots,\cdot)(\cdot,t)$ and $\tilde{A}_{M,\overrightarrow{T}}(\cdot,\ldots,\cdot)(\cdot,t)$ actually are general aggregation operators such that

$$\tilde{A}_{M,\overrightarrow{T}}(\cdot,\ldots,\cdot)(\cdot,t) \le \tilde{A} \le \tilde{A}_{M,T}(\cdot,\ldots,\cdot)(\cdot,t),$$

where $\tilde{A}$ is the pointwise extension of aggregation operator A used in (3)–(4).

Let us also note that for all $\mu_1,\ldots,\mu_n \in [0,1]^X$ and all $x \in X$ function $\tilde{A}_{M,T}(\mu_1,\ldots,\mu_n)(x,\cdot)$ is nonincreasing and function $\tilde{A}_{M,\overrightarrow{T}}(\mu_1,\ldots,\mu_n)(x,\cdot)$ is nondecreasing on interval $(0,+\infty)$. For each $t \in (0,+\infty)$ operators $\tilde{A}_{M,\overrightarrow{T}}(\cdot,\ldots,\cdot)(\cdot,t)$ and $\tilde{A}_{M,T}(\cdot,\ldots,\cdot)(\cdot,t)$ can be considered as upper and lower approximations of the pointwise extension $\tilde{A}$, and the quality of such approximation depends on t: for all $\mu_1,\ldots,\mu_n \in [0,1]^X$ and all $x \in X$, considering bigger value t we will obtain smaller interval (subinterval)

$$[\tilde{A}_{M,\overrightarrow{T}}(\mu_1,\ldots,\mu_n)(x,t),\ \tilde{A}_{M,\overrightarrow{T}}(\mu_1,\ldots,\mu_n)(x,t)].$$

4.3. Upper and lower aggregation based on the standard fuzzy metric

Suppose we have a metric space (X, d). Let M_d be the function defined on $X \times X \times (0, +\infty)$ by the following formula

$$M_d(x, y, t) = \frac{t}{t + d(x, y)}. \tag{5}$$

By using formula (5) we could obtain a strong T-fuzzy metric for some t-norms T. In general, the condition $(SFM4)$ from the definition of a strong fuzzy metric could not fulfill for the minimum t-norm, but always holds for the product t-norm T_P and Lukasiewicz t-norm T_L, for example. If function M_d fulfills the condition $(SFM4)$ with respect to a t-norms T it is called the standard T-fuzzy metric induced by d.

Taking into account that

$$\lim_{t \to +0} M_d(x, y, t) = \begin{cases} 0, & \text{if } d(x, y) > 0, \\ 1, & \text{if } d(x, y) = 0, \end{cases}$$

one can easily notice that in the case of continuous t-norm T and finite set X the following equalities hold:

$$\lim_{t \to +\infty} \tilde{A}_{M_d, T}(\mu_1, \ldots, \mu_n)(x, t) = A(\mu_1(x), \ldots, \mu_n(x)),$$

$$\lim_{t \to +\infty} \tilde{A}_{M_d, \vec{T}}(\mu_1, \ldots, \mu_n)(x, t) = A(\mu_1(x), \ldots, \mu_n(x)).$$

5. Aggregation Operators Based on a Fuzzy Metric Introduced via Multiple Criteria Evaluation

5.1. Fuzzy metric introduced via multiple criteria

In this section we consider the case when the distances between elements of X are evaluated on the basis of k independent criteria. Let us denote by d_i the metric obtained by using the i-th criterion, $i = 1, 2, \ldots, k$. In this case we propose to use the strong fuzzy metric M such that

$$M(x, y, 1/t) =$$

$$= \begin{cases} (1 + t d_1(x, y))^{-1}, & \text{if } 0 < t \leq 1, \\ (1 + d_1(x, y) + (t - 1) d_2(x, y))^{-1}, & \text{if } 1 < t \leq 2, \\ (1 + d_1(x, y) + d_2(x, y) + (t - 2) d_3(x, y))^{-1}, & \text{if } 2 < t \leq 3, \\ \ldots \\ (1 + d_1(x, y) + \ldots + d_{k-1}(x, y) + (t - k + 1) d_k(x, y))^{-1}, & \text{if } k - 1 < t \leq k, \\ (1 + d_1(x, y) + \ldots + d_{k-1}(x, y) + d_k(x, y))^{-1}, & \text{if } k < t. \end{cases}$$

Value $M(x,y,1/t)$ corresponding to integer t characterizes the similarity between x and y obtained by using t first criteria. It is clear that M depends on the order of criteria. Modifications of the proposed fuzzy metric could be obtained by changing the order of criteria and including weights.

5.2. Example of upper and lower aggregation operators

Let us consider the discrete universe $X = \{x_1, x_2, x_3, x_4, x_5\}$ and let us take the following fuzzy subsets of X:

$$\mu_1 = \begin{pmatrix} 0.9 \\ 0.5 \\ 0.6 \\ 0.8 \\ 0.3 \end{pmatrix}, \mu_2 = \begin{pmatrix} 0.2 \\ 0.0 \\ 0.2 \\ 0.6 \\ 0.9 \end{pmatrix}, \mu_3 = \begin{pmatrix} 0.7 \\ 0.5 \\ 0.1 \\ 0.8 \\ 0.6 \end{pmatrix}, \mu_4 = \begin{pmatrix} 0.1 \\ 0.9 \\ 0.2 \\ 0.8 \\ 0.5 \end{pmatrix}.$$

We consider the case when the distances between elements of X are evaluated on the basis of 3 independent criteria. Metrics d_1, d_2 and d_3 obtained by using each criterion are given in a matrix form respectively by

$$\begin{pmatrix} 0 & 0.05 & 0.05 & 0.1 & 0.5 \\ 0.05 & 0 & 0.05 & 0.15 & 0.5 \\ 0.05 & 0.05 & 0 & 0.1 & 0.5 \\ 0.1 & 0.15 & 0.1 & 0 & 0.5 \\ 0.5 & 0.5 & 0.5 & 0.5 & 0 \end{pmatrix}, \begin{pmatrix} 0 & 0.01 & 0.1 & 0.2 & 1 \\ 0.01 & 0 & 0.1 & 0.2 & 0.1 \\ 0.1 & 0.1 & 0 & 0.3 & 1 \\ 0.2 & 0.2 & 0.3 & 0 & 1 \\ 1 & 1 & 1 & 1 & 0 \end{pmatrix}, \begin{pmatrix} 0 & 0.05 & 0.28 & 1.2 & 2.5 \\ 0.05 & 0 & 0.28 & 1.15 & 2.5 \\ 0.28 & 0.28 & 0 & 1.1 & 2.5 \\ 1.2 & 1.15 & 1.1 & 0 & 2.5 \\ 2.5 & 2.5 & 2.5 & 2.5 & 0 \end{pmatrix}.$$

Taking the arithmetic mean aggregation operator $A = AVG$ as an ordinary aggregation operator and fuzzy metric M as it was introduced in Subsection 5.1, we calculate the results of upper and lower general aggregation operators using the product t-norm T_P, Lukasiewicz t-norm T_L at points $t = 1$, $t = 2$ and $t = 3$ (the value of t is used as the upper index):

$$\tilde{A}^1_{M,T_P} = \begin{pmatrix} 0.68 \\ 0.65 \\ 0.68 \\ 0.75 \\ 0.58 \end{pmatrix}, \tilde{A}^2_{M,T_P} = \begin{pmatrix} 0.58 \\ 0.56 \\ 0.54 \\ 0.75 \\ 0.58 \end{pmatrix}, \tilde{A}^3_{M,T_P} = \begin{pmatrix} 0.48 \\ 0.48 \\ 0.33 \\ 0.75 \\ 0.58 \end{pmatrix},$$

$$\tilde{A}^1_{M,T_L} = \begin{pmatrix} 0.66 \\ 0.62 \\ 0.66 \\ 0.75 \\ 0.58 \end{pmatrix}, \tilde{A}^2_{M,T_L} = \begin{pmatrix} 0.52 \\ 0.49 \\ 0.46 \\ 0.75 \\ 0.58 \end{pmatrix}, \tilde{A}^3_{M,T_L} = \begin{pmatrix} 0.48 \\ 0.48 \\ 0.28 \\ 0.75 \\ 0.58 \end{pmatrix}.$$

$$\tilde{A}^1_{M,\vec{T}_P} = \begin{pmatrix} 0.29 \\ 0.29 \\ 0.28 \\ 0.3 \\ 0.41 \end{pmatrix}, \ \tilde{A}^2_{M,\vec{T}_P} = \begin{pmatrix} 0.32 \\ 0.32 \\ 0.28 \\ 0.39 \\ 0.58 \end{pmatrix}, \ \tilde{A}^3_{M,\vec{T}_P} = \begin{pmatrix} 0.39 \\ 0.39 \\ 0.28 \\ 0.69 \\ 0.58 \end{pmatrix},$$

$$\tilde{A}^1_{M,\vec{T}_L} = \begin{pmatrix} 0.32 \\ 0.32 \\ 0.28 \\ 0.37 \\ 0.58 \end{pmatrix}, \ \tilde{A}^2_{M,\vec{T}_L} = \begin{pmatrix} 0.41 \\ 0.41 \\ 0.28 \\ 0.56 \\ 0.58 \end{pmatrix}, \ \tilde{A}^3_{M,\vec{T}_L} = \begin{pmatrix} 0.48 \\ 0.48 \\ 0.28 \\ 0.75 \\ 0.58 \end{pmatrix}.$$

The result of upper aggregation decreases and the result of lower aggregation increases with respect to t for most of elements of X. For some elements the result does not change. It could be explained by the fact that there does not exist sufficiently close elements which could influence the result.

6. Conclusion

In the last section we described how the proposed construction of upper and lower aggregation operators could be applied in the case when the similarity of elements of the universe is defined on the basis of multiple criteria evaluation. A special design of strong fuzzy metric is introduced for this merit and illustrated with a simple example. Our future work will be devoted to development of this technique by modifying the formula for fuzzy metric accordingly to criteria structure and hierarchy. We hope that the proposed construction allows us to improve the method applied by the authors for aggregation of risk level assessments in Ref. 2

References

1. P. Orlovs and S. Asmuss, *Fuzzy Sets Syst.* **291**, 114 (2016).
2. P. Orlovs and S. Asmuss, *Advances in Intelligent Systems and Computing* **643**, 71 (2017).
3. M. Grabisch, R. M. J.-L. Marichal and E. Pap, *Aggregation Functions* (Cambridge University Press, 2009).
4. T. Calvo, G. Mayor and R. Mesiar, *Aggregation Operators: New Trends and Applications* (Physica-Verlag, Heidelberg, New York, 2002).
5. M. Detyniecki, *Fundamentals On Aggregation Operators* (Berkeley, 2001).
6. A. Takaci, *Novi Sad J. Math.* **33**, 67 (2003).
7. I. Kramosil and J. Michalek, *Kybernetika* **11**, 336 (1975).
8. A. George and P. Veeramani, *Fuzzy Sets Syst.* **64**, 395 (1994).
9. S. Grecova, A. Sostak and I. Uljane, *Appl. Gen. Topol.* **17(2)**, 105 (2016).

Part 2

(Fuzzy) Multi-Criteria Decision Analysis: Advanced Theory and Methods

Interval-valued Pythagorean fuzzy interaction aggregation operators and their application in multiple attribute decision making

Wei Yang* and Yongfeng Pang

*Department of Mathematics, Xi'an University of Architecture and Technology
Xi'an, Shaanxi 710055, China
yangweipyf@163.com

In this paper, the interaction operation laws have been defined between interval-valued Pythagorean fuzzy set. Based on the new operation laws, some aggregation operators have been defined including the interval-valued Pythagorean fuzzy weighted averaging operator and the generalized interval-valued Pythagorean fuzzy weighted averaging operator. A new multiple attribute decision making method has been proposed based on the new aggregation operators. Numerical example has been presented to illustrate the new algorithm.

Keywords: Pythagorean fuzzy set; aggregation; multiple attribute decision making.

1. Introduction

Pythagorean fuzzy set is developed by Yager,[1] which is the extension of intuitionistic fuzzy set and can model fuzzy and uncertain information more powerful and flexible since it has greater space than that of intuitionistic fuzzy set. Some Pythagorean fuzzy decision making methods have been developed. Peng and Yang[2] defined the Pythagorean fuzzy Choquet integral average operator. Zhang[3] developed Pythagorean fuzzy agglomerative hierarchical clustering approach. Wei[4] proposed some Pythagorean fuzzy interaction aggregation operators by considering interaction of membership and non-membership. Ren et al.[5] presented Pythagorean fuzzy TODIM method based on the prospect theory. Peng and Yang[6] gave interval-valued Pythagorean fuzzy aggregation operator. Rahman et al.[7] define the interval valued Pythagorean fuzzy weighted geometric (IPFWG) operator. All the existing methods haven't considered interaction between interval-valued membership and interval-valued non-membership of Pythagorean fuzzy set. Hence, we define the interaction operation laws for interval-valued Pythagorean fuzzy set in this paper.

2. Preliminaries

Definition 1.[1] Let X be a fixed set. A Pythagorean fuzzy set P on X can be represented as follows

$$P = \{< x, (\mu_P(x), \nu_P(x)) >| \ x \in X\}, \tag{1}$$

where $\mu_P(x) : X \to [0,1]$ is the membership function and $\nu_P(x) : X \to [0,1]$ is the non-membership function. For each $x \in X$, it satisfies the following condition $0 \leq (\mu_P(x))^2 + (\nu_P(x))^2 \leq 1$. $\pi_P(x) = \sqrt{1 - (\mu_P(x))^2 - (\nu_P(x))^2}$ is the indeterminacy degree of x to X. For simplicity, $(\mu_P(x), \nu_P(x))$ is called a Pythagorean fuzzy number (PFN), denoted by (μ_P, ν_P), where $\mu_P, \nu_P \in [0,1]$, $\pi_P = \sqrt{1 - (\mu_P)^2 - (\nu_P)^2}$ and $0 \leq (\mu_P)^2 + (\nu_P)^2 \leq 1$.

Definition 2.[6] Let X be a fixed set, then an interval-valued Pythagorean fuzzy set can be defined as

$$P = \{< x, (\tilde{\mu}_P(x), \tilde{\nu}_P(x)) >| \ x \in X\}, \tag{2}$$

where $\tilde{\mu}_P(x) = [\mu_P^L(x), \mu_P^U(x)]$, $\tilde{\nu}_P(x) = [\nu_P^L(x), \nu_P^U(x)]$ are interval membership and interval non-membership, respectively. For each $x \in X$, it satisfies the following condition $0 \leq (\mu_P^U(x))^2 + (\nu_P^U(x))^2 \leq 1$. $\tilde{\pi}_P(x) = [\pi_P^L(x), \pi_P^U(x)]$ is the interval indeterminacy degree x to X, where $\pi_P^L(x) = \sqrt{1 - (\mu_P^U(x))^2 - (\nu_P^U(x))^2}$, $\pi_P^U(x) = \sqrt{1 - (\mu_P^L(x))^2 - (\nu_P^L(x))^2}$. For simplicity, $([\mu^L, \mu^U], [\nu^L, \nu^U])$ is called a interval-valued Pythagorean fuzzy number (IVPFN), where $[\mu^L, \mu^U] \subseteq [0,1]$, $[\nu^L, \nu^U] \subseteq [0,1]$ and $0 \leq (\mu^U)^2 + (\nu^U)^2 \leq 1$.

Definition 3. Let $\tilde{\alpha} = ([\mu_\alpha^L, \mu_\alpha^U], [\nu_\alpha^L, \nu_\alpha^U])$, $\tilde{\alpha}_1 = ([\mu_{\alpha_1}^L, \mu_{\alpha_1}^U], [\nu_{\alpha_1}^L, \nu_{\alpha_1}^U])$ and $\tilde{\alpha}_2 = ([\mu_{\alpha_2}^L, \mu_{\alpha_2}^U], [\nu_{\alpha_2}^L, \nu_{\alpha_2}^U])$ be three IVPFNs, the interaction operation laws can be defined as
(1)

$$\tilde{\alpha}_1 \oplus \tilde{\alpha}_2 = ([\sqrt{(\mu_{\alpha_1}^L)^2 + (\mu_{\alpha_2}^L)^2 - (\mu_{\alpha_1}^L)^2 (\mu_{\alpha_2}^L)^2},$$
$$\sqrt{(\mu_{\alpha_1}^U)^2 + (\mu_{\alpha_2}^U)^2 - (\mu_{\alpha_1}^U)^2 (\mu_{\alpha_2}^U)^2}],$$
$$[\sqrt{(\nu_{\alpha_1}^L)^2 + (\nu_{\alpha_2}^L)^2 - (\nu_{\alpha_1}^L)^2 (\nu_{\alpha_2}^L)^2 - (\mu_{\alpha_1}^L)^2 (\nu_{\alpha_2}^L)^2 - (\nu_{\alpha_1}^L)^2 (\mu_{\alpha_2}^L)^2},$$
$$\sqrt{(\nu_{\alpha_1}^L)^2 + (\nu_{\alpha_2}^L)^2 - (\nu_{\alpha_1}^L)^2 (\nu_{\alpha_2}^L)^2 - (\mu_{\alpha_1}^L)^2 (\nu_{\alpha_2}^L)^2 - (\nu_{\alpha_1}^L)^2 (\mu_{\alpha_2}^L)^2}]),$$

(2)
$$\tilde{\alpha}_1 \otimes \tilde{\alpha}_2 = ([\sqrt{(\mu_{\alpha_1}^L)^2 + (\mu_{\alpha_2}^L)^2 - (\mu_{\alpha_1}^L)^2(\mu_{\alpha_2}^L)^2 - (\nu_{\alpha_1}^L)^2(\mu_{\alpha_2}^L)^2 - (\mu_{\alpha_1}^L)^2(\nu_{\alpha_2}^L)^2},$$
$$\sqrt{(\mu_{\alpha_1}^U)^2 + (\mu_{\alpha_2}^U)^2 - (\mu_{\alpha_1}^U)^2(\mu_{\alpha_2}^U)^2 - (\nu_{\alpha_1}^U)^2(\mu_{\alpha_2}^U)^2 - (\mu_{\alpha_1}^U)^2(\nu_{\alpha_2}^U)^2}]$$
$$[\sqrt{(\nu_{\alpha_1}^L)^2 + (\nu_{\alpha_2}^L)^2 - (\nu_{\alpha_1}^L)^2(\nu_{\alpha_2}^L)^2}, \sqrt{(\nu_{\alpha_1}^U)^2 + (\nu_{\alpha_2}^U)^2 - (\nu_{\alpha_1}^U)^2(\nu_{\alpha_2}^U)^2}])$$

(3)
$$\lambda\tilde{\alpha} = ([\sqrt{1 - (1 - (\mu_\alpha^L)^2)^\lambda}, \sqrt{1 - (1 - (\mu_\alpha^U)^2)^\lambda}], [((1 - (\mu_\alpha^L)^2)^\lambda - (1 - ((\mu_\alpha^L)^2 + (\nu_\alpha^L)^2))^\lambda)^{1/2}, ((1 - (\mu_\alpha^U)^2)^\lambda - (1 - ((\mu_\alpha^U)^2 + (\nu_\alpha^U)^2))^\lambda)^{1/2}]), \lambda > 0,$$

(4)
$$(\tilde{\alpha})^\lambda = ([\sqrt{(1 - (\nu_\alpha^L)^2)^\lambda - (1 - ((\mu_\alpha^L)^2 + (\nu_\alpha^L)^2))^\lambda},$$
$$((1 - (\nu_\alpha^U)^2)^\lambda - (1 - ((\mu_\alpha^U)^2 + (\nu_\alpha^U)^2))^\lambda)^{1/2}],$$
$$[\sqrt{1 - (1 - (\nu_\alpha^L)^2)^\lambda}, \sqrt{1 - (1 - (\nu_\alpha^U)^2)^\lambda}]), \lambda > 0.$$

Definition 4. Let $\tilde{\alpha} = ([\mu^L, \mu^U], [\nu^L, \nu^U])$, the score of $\tilde{\alpha}$ can be defined as

$$S(\tilde{\alpha}) = \frac{1}{2}((\mu^L)^2 + (\mu^U)^2 - (\nu^L)^2 - (\nu^U)^2), \ S(\tilde{\alpha}) \in [-1, 1],$$

the accuracy function can be defined as

$$A(\tilde{\alpha}) = \frac{1}{2}((\mu^L)^2 + (\mu^U)^2 + (\nu^L)^2 + (\nu^U)^2), \ A(\tilde{\alpha}) \in [0, 1].$$

For two IVPFNs $\tilde{\alpha}_1, \tilde{\alpha}_2$,
If $S(\tilde{\alpha}_1) > S(\tilde{\alpha}_2)$, then $\tilde{\alpha}_1 > \tilde{\alpha}_2$,
If $S(\tilde{\alpha}_1) = S(\tilde{\alpha}_2)$, then
    If $A(\tilde{\alpha}_1) > A(\tilde{\alpha}_2)$, then $\tilde{\alpha}_1 > \tilde{\alpha}_2$,
    If $A(\tilde{\alpha}_1) = A(\tilde{\alpha}_2)$, then $\tilde{\alpha}_1 \ \tilde{\alpha}_2$.

Definition 5. Let $\tilde{\alpha}_1 = ([\mu_{\alpha_1}^L, \mu_{\alpha_1}^U], [\nu_{\alpha_1}^L, \nu_{\alpha_1}^U])$ and $\tilde{\alpha}_2 = ([\mu_{\alpha_2}^L, \mu_{\alpha_2}^U], [\nu_{\alpha_2}^L, \nu_{\alpha_2}^U])$ be two interval-valued Pythagorean fuzzy numbers, the Hamming distance between $\tilde{\alpha}_1$ and $\tilde{\alpha}_2$ can be defined as follows

$$d(\tilde{\alpha}_1, \tilde{\alpha}_2) = \frac{1}{4}\Big(|(\mu_{\alpha_1}^L)^2 - (\mu_{\alpha_2}^L)^2| + |(\mu_{\alpha_1}^U)^2 - (\mu_{\alpha_2}^U)^2| + |(\nu_{\alpha_1}^L)^2 - (\nu_{\alpha_2}^L)^2|$$
$$+ |(\nu_{\alpha_1}^U)^2 - (\nu_{\alpha_2}^U)^2|\Big). \tag{3}$$

3. Interval-valued Pythagorean Fuzzy Interaction Weighted Averaging Operator

Definition 6. Let $T = (\tilde{\alpha}_1, \tilde{\alpha}_2, ..., \tilde{\alpha}_n)$ be a collection of IVPFNs, where $\tilde{\alpha}_i = ([\mu_i^L, \mu_i^U], [\nu_i^L, \nu_i^U])$ $(i = 1, 2, ..., n)$. The interval-valued Pythagorean fuzzy interaction weighted averaging (IVPFIWA) operator is defined as follows:

$$\text{IVPFIWA}(\tilde{\alpha}_1, \tilde{\alpha}_2, ..., \tilde{\alpha}_n) = w_1 \tilde{\alpha}_1 \oplus w_2 \tilde{\alpha}_2 \oplus ... \oplus w_n \tilde{\alpha}_n, \tag{4}$$

where $(w_1, w_2, ..., w_n)$ is the weight vector with $w_i \geq 0$ and $\sum_{i=1}^n w_i = 1$.

Theorem 1. Let $\tilde{\alpha}_i = ([\mu_i^L, \mu_i^U], [\nu_i^L, \nu_i^U])$ $(i = 1, 2, ..., n)$ be a collection of IVPFNs. The aggregated results of the IVPFIWA operator is still of a IVPFN, which has the following form:

$$\begin{aligned}
\text{IVPFIWA}(\tilde{\alpha}_1, \tilde{\alpha}_2, ..., \tilde{\alpha}_n) &= w_1 \tilde{\alpha}_1 \oplus w_2 \tilde{\alpha}_2 \oplus ... \oplus w_n \tilde{\alpha}_n \\
&= \Big([\sqrt{1 - \prod_{j=1}^n (1 - (\mu_j^L)^2)^{w_j}}, \sqrt{1 - \prod_{j=1}^n (1 - (\mu_j^U)^2)^{w_j}}], \\
&\quad [\sqrt{\prod_{j=1}^n (1 - (\mu_j^L)^2)^{w_j} - \prod_{j=1}^n (1 - ((\mu_j^L)^2 + (\nu_j^L)^2))^{w_j}}, \\
&\quad \sqrt{\prod_{j=1}^n (1 - (\mu_j^U)^2)^{w_j} - \prod_{j=1}^n (1 - ((\mu_j^U)^2 + (\nu_j^U)^2))^{w_j}}] \Big),
\end{aligned} \tag{5}$$

where $(w_1, w_2, ..., w_n)$ is the weight vector with $w_i \geq 0$ and $\sum_{i=1}^n w_i = 1$.

Proof.

$$w_j \tilde{\alpha}_j = \Big([\sqrt{1 - (1 - (\mu_j^L)^2)^{w_j}}, \sqrt{1 - (1 - (\mu_j^U)^2)^{w_j}}], [\sqrt{(1 - (\mu_j^L)^2)^{w_j} - (1 - ((\mu_j^L)^2 + (\nu_j^L)^2))^{w_j}}, \sqrt{(1 - (\mu_j^U)^2)^{w_j} - (1 - ((\mu_j^U)^2 + (\nu_j^U)^2))^{w_j}}] \Big),$$

$$\bigoplus_{j=1}^n w_j \tilde{\alpha}_j = \Big([\sqrt{1 - \prod_{j=1}^n (1 - (\mu_j^L)^2)^{w_j}}, \sqrt{1 - \prod_{j=1}^n (1 - (\mu_j^U)^2)^{w_j}}], [\sqrt{\prod_{j=1}^n (1 - (\mu_j^L)^2)^{w_j} - \prod_{j=1}^n (1 - ((\mu_j^L)^2 + (\nu_j^L)^2))^{w_j}}, \sqrt{\prod_{j=1}^n (1 - (\mu_j^U)^2)^{w_j} - \prod_{j=1}^n (1 - ((\mu_j^U)^2 + (\nu_j^U)^2))^{w_j}}] \Big).$$

Definition 7. Let $T = (\tilde{\alpha}_1, \tilde{\alpha}_2, ..., \tilde{\alpha}_n)$ be a collection of IVPFNs, where $\tilde{\alpha}_i = ([\mu_i^L, \mu_i^U], [\nu_i^L, \nu_i^U])$ $(i = 1, 2, ..., n)$. The generalized interval-valued Pythagorean fuzzy interaction weighted averaging (GIVPFIWA) operator is defined as follows:

$$\text{GIVPFIWA}(\tilde{\alpha}_1, \tilde{\alpha}_2, ..., \tilde{\alpha}_n) = \Big(w_1(\tilde{\alpha}_1)^\lambda \oplus w_2(\tilde{\alpha}_2)^\lambda \oplus ... \oplus (w_n \tilde{\alpha}_n)^\lambda \Big)^{1/\lambda}, \tag{6}$$

where $(w_1, w_2, ..., w_n)$ is the weight vector with $w_i \geq 0$ and $\sum_{i=1}^{n} w_i = 1$ and $\lambda > 0$.

Theorem 2. Let $\tilde{\alpha}_i = ([\mu_i^L, \mu_i^U], [\nu_i^L, \nu_i^U])$ $(i = 1, 2, ..., n)$ be a collection of IVPFNs. The aggregated results of the GIVPFIWA operator is still of an IVPFN, which has the following form:

$$\text{GIVPFIWA}(\tilde{\alpha}_1, \tilde{\alpha}_2, ..., \tilde{\alpha}_n) = \left(w_1(\tilde{\alpha}_1)^\lambda \oplus w_2(\tilde{\alpha}_2)^\lambda \oplus ... \oplus (w_n\tilde{\alpha}_n)^\lambda\right)^{1/\lambda}$$
$$= \Big(\big[\big((1 - \prod_{j=1}^n (1 - (1 - (\nu_j^L)^2)^\lambda + (1 - ((\mu_j^L)^2 + (\nu_j^L)^2))^\lambda)^{w_j} + \prod_{j=1}^n (1 - ((\mu_j^L)^2 + (\nu_j^L)^2))^{\lambda w_j}\big)^{1/\lambda} - \prod_{j=1}^n (1 - ((\mu_j^L)^2 + (\nu_j^L)^2))^{w_j}\big)^{1/2}, \big((1 - \prod_{j=1}^n (1 - (1 - (\nu_j^U)^2)^\lambda + (1 - ((\mu_j^U)^2 + (\nu_j^U)^2))^\lambda)^{w_j} + \prod_{j=1}^n (1 - ((\mu_j^U)^2 + (\nu_j^U)^2))^{\lambda w_j}\big)^{1/\lambda} - \prod_{j=1}^n (1 - ((\mu_j^U)^2 + (\nu_j^U)^2))^{w_j}\big)^{1/2} \big], \big[\big(1 - (1 - \prod_{j=1}^n (1 - (1 - (\nu_j^L)^2)^\lambda + (1 - ((\mu_j^L)^2 + (\nu_j^L)^2))^\lambda)^{w_j} + \prod_{j=1}^n (1 - ((\mu_j^L)^2 + (\nu_j^L)^2))^{\lambda w_j}\big)^{1/\lambda}\big)^{1/2}, \big(1 - (1 - \prod_{j=1}^n (1 - (1 - (\nu_j^U)^2)^\lambda + (1 - ((\mu_j^U)^2 + (\nu_j^U)^2))^\lambda)^{w_j} + \prod_{j=1}^n (1 - ((\mu_j^U)^2 + (\nu_j^U)^2))^{\lambda w_j}\big)^{1/\lambda}\big)^{1/2} \big] \Big),$$
(7)

where $(w_1, w_2, ..., w_n)$ is the weight vector with $w_i \geq 0$ and $\sum_{i=1}^{n} w_i = 1$.

Theorem 3 (Idempotency). Let $T = (\tilde{\alpha}_1, \tilde{\alpha}_2, ..., \tilde{\alpha}_n)$ be a collection of IVPFNs. If all $\tilde{\alpha}_k = ([\mu_k^L, \mu_k^U], [\nu_k^L, \mu_k^U])$ $(k = 1, 2, ..., n)$ are equal, i.e., $\tilde{\alpha}_k = \tilde{\alpha} = ([\mu^L, \mu^U], [\nu^L, \mu^U])$ $(k = 1, 2, ..., n)$, then

$$\text{GIVPFIWA}(\tilde{\alpha}_1, \tilde{\alpha}_2, ..., \tilde{\alpha}_n) = \tilde{\alpha}.$$

Theorem 4 (Monotonicity). Let $(\tilde{\alpha}_{p1}, \tilde{\alpha}_{p2}, ..., \tilde{\alpha}_{pn})$ and $(\tilde{\alpha}_{q1}, \tilde{\alpha}_{q2}, ..., \tilde{\alpha}_{qn})$ be two collection of IVPFNs, where $\tilde{\alpha}_{pk} = ([\mu_{pk}^L, \mu_{pk}^U], [\nu_{pk}^L, \mu_{pk}^U])$ $(k = 1, 2, ..., n)$ and $\tilde{\alpha}_{qk} = ([\mu_{qk}^L, \mu_{qk}^U], [\nu_{qk}^L, \mu_{qk}^U])$ $(k = 1, 2, ..., n)$. If $\mu_{qk}^L \leq \mu_{pk}^L$, $\mu_{qk}^U \leq \mu_{pk}^U$, $\nu_{qk}^L \geq \nu_{pk}^L$, $\nu_{qk}^U \geq \nu_{pk}^U$. Then

$$\text{GIVPFIWA}(\tilde{\alpha}_{q1}, \tilde{\alpha}_{q2}, ..., \tilde{\alpha}_{qn}) \leq \text{GIVPFIWA}(\tilde{\alpha}_{p1}, \tilde{\alpha}_{p2}, ..., \tilde{\alpha}_{pn}).$$

4. A New Method for Multiple Attribute Decision Making Based on the Proposed GIVPFIWA Operator

For a MAGDM problem with PFNs, let $\{A_1, A_2, ..., A_m\}$ be a collection of alternatives and $\{C_1, C_2, ..., C_n\}$ be a collection of attributes. $\tilde{D} = (\tilde{\alpha}_{ij})_{m \times n}$ is a decision matrix given by decision maker, where $\tilde{\alpha}_{ij} = ([\mu_{ij}^L, \mu_{ij}^L], [\nu_{ij}^L, \nu_{ij}^U])$ is the evaluation value of A_i with respect to attribute C_j. Let $(w_1, w_2, ..., w_n)$ be the weight vector of attribute with

$w_j \geq 0$ and $\sum_{j=1}^{n} w_j = 1$. The proposed method based on the GIVPFIWA operator is presented as follows:

Step 1. Decision maker give evaluation value of alternative A_i with respect to attribute C_j using Pythagorean fuzzy number $\tilde{\alpha}_{ij}$. Then the decision matrix is formed as $\tilde{D} = (\tilde{\alpha}_{ij})_{m \times n}$.

Step 2. Aggregate the alternatives evaluation into collective ones by using the GIVPFIWA operator.

$$\tilde{\alpha}_i = \text{GIVPFIWA} \left(\tilde{\alpha}_{i1}, \tilde{\alpha}_{i2}, ..., \tilde{\alpha}_{in} \right).$$

Step 3. Calculate the score values and accuracy values of the alternative collective evaluation values.

Step 4. Rank the collective evaluation values by using the method in Definition 4 and rank the alternatives accordingly.

5. An Illustrative Example

In this section, we present a numerical example to illustrate the feasibility and practical advantages of the new method.

An investment company wants to invest a large amount of money to the following five possible areas A_1−real estate, A_2−energy industry, A_3−gold, A_4−stock market, A_5−artificial intellectual company. The following attributes are considered: C_1−interest rate, C_2−growth potential, C_3−risk, C_4−inflation. The company invite expert to evaluate the alternatives and the proposed method is applied to select the area to invest.

Step 1. Decision maker evaluate alternatives with interval-valued Pythagorean fuzzy numbers and the decision matrix is formed as in Table 1.

Table 1. Decision matrix.

Alternative	C_1	C_2	C_3	C_4
A_1	([0.5,0.6],[0.4,0.5])	([0.8,0.9],[0.1,0.2])	([0.7,0.8],[0.2,0.3])	([0.3,0.4],[0.6,0.7])
A_2	([0.6,0.7],[0.3,0.4])	([0.5,0.6],[0.4,0.5])	([0.3,0.4],[0.6,0.7])	([0.7,0.8],[0.2,0.3])
A_3	([0.5,0.6],[0.4,0.5])	([0.5,0.6],[0.4,0.5])	([0.6,0.7],[0.3,0.4])	([0.8,0.9],[0.1,0.2])
A_4	([0.7,0.8],[0.2,0.3])	([0.2,0.3],[0.7,0.8])	([0.5,0.6],[0.4,0.5])	([0.6,0.7],[0.3,0.4])
A_5	([0.8,0.9],[0.1,0.2])	([0.6,0.7],[0.3,0.4])	([0.5,0.6],[0.4,0.5])	([0.3,0.4],[0.6,0.7])

Step 2. Assume the attribute weight vector is $(0.15, 0.35, 0.2, 0.3)$. Calculate the collective evaluation values of alternatives by using Eq. (7) to get $\tilde{\alpha}_1 = ([0.6807, 0.7894], [0.2772, 0.3547])$, $\tilde{\alpha}_2 = ([0.5818, 0.6851], [0.3518,$

$0.4408]$), $\tilde{\alpha}_3 = ([0.6568, 0.7643], [0.2669, 0.3595]), \tilde{\alpha}_4 = ([0.5468, 0.6537], [0.4305, 0.5085]), \tilde{\alpha}_5 = ([0.5935, 0.7009], [0.3557, 0.4402])$.

Step 3. The scores of $\tilde{\alpha}_i$ can be calculated as $S(\tilde{\alpha}_1) = 0.4419, S(\tilde{\alpha}_2) = 0.2449, S(\tilde{\alpha}_3) = 0.4075, S(\tilde{\alpha}_4) = 0.1412, S(\tilde{\alpha}_5) = 0.2616$.
Here $\lambda = 2$.

Step 4. The alternatives can be ranked according to the ranking of $S(\tilde{\alpha}_i)$ $(i = 1, 2, ..., 5)$ to get

$$A_1 > A_3 > A_5 > A_2 > A_4.$$

The optimal is A_1.

If other λ are considered, the results are shown in Table 2.

Table 2. Results of different λ.

Alternative	$S(\alpha_1)$	$S(\alpha_2)$	$S(\alpha_3)$	$S(\alpha_4)$	$S(\alpha_5)$	Ranking of the alternatives
$\lambda = 1$	0.3817	0.2093	0.3841	0.0587	0.2101	$A_3 > A_1 > A_2 > A_5 > A_4$
$\lambda = 2$	0.4419	0.2449	0.4075	0.1412	0.2616	$A_1 > A_3 > A_2 > A_5 > A_4$
$\lambda = 3$	0.4665	0.2630	0.4184	0.1879	0.2883	$A_1 > A_3 > A_5 > A_2 > A_4$
$\lambda = 5$	0.4882	0.2824	0.4306	0.2361	0.3178	$A_1 > A_3 > A_5 > A_2 > A_4$
$\lambda = 10$	0.5131	0.3117	0.4511	0.2879	0.3593	$A_1 > A_3 > A_5 > A_2 > A_4$

From the results we can see that different ranking can be got by using different λ. As λ increasing, A_1 becomes the optimal alternative, which have relative large evaluation values.

6. Conclusions

Interaction operation laws between interval-valued Pythagorean fuzzy numbers have been developed. Some interval-valued Pythagorean fuzzy aggregation operators have been defined including the interval-valued Pythagorean fuzzy weighted averaging operator and the generalized interval-valued Pythagorean fuzzy weighted averaging operator. A new multiple attribute decision making method based on the new operator has been proposed and numerical example has been presented to illustrate the new method.

Acknowledgments

This work is partly supported by National Natural Science Foundation of China (No. 11401457), Postdoctoral Science Foundation of China

(2015M582624) and Shaanxi Province Postdoctoral Science Foundation of China.

Author Contributions

Wei Yang conceived the research and wrote the paper. Yongfeng Pang Performed the experiments and revised the paper. All authors have read and approved the final version.

References

1. R.R. Yager and A.M. Abbasov, Pythagorean membership grades, complex numbers, and decision making. *Int J Intel Syst.* **28**, 436–452 (2013).
2. X.D. Peng and Y. Yang, Pythagorean Fuzzy Choquet Integral Based MABAC Method for Multiple Attribute Group Decision Making. *Int J Intel Syst.* **31**, 989–1020 (2016).
3. X.L. Zhang, Pythagorean Fuzzy Clustering Analysis: A Hierarchical Clustering Algorithm with the Ratio Index-Based Ranking Methods. *Int J Intel Syst.* in press.
4. G.W. Wei, Pythagorean fuzzy interaction aggregation operators and their application to multiple attribute decision making. *J Intell Fuzzy Syst.* **33**, 2119–2132 (2017).
5. P.J. Ren, Z.S. Xu, X.J. Gou, Pythagorean fuzzy TODIM approach to multi-criteria decision making. *Appl Soft Comput.* **42**, 246–259 (2016).
6. X.D. Peng, Y. Yang, Fundamental Properties of Interval-Valued Pythagorean Fuzzy Aggregation Operators. *Int J Intel Syst.* **31**(5), 444–487 (2016).
7. K. Rahman, S. Abdullah, M. Shakeel, M.S.A. Khan, M. Ullah, Interval-valued Pythagorean fuzzy geometric aggregation operators and their application to group decision making problem. *Cogent Mathematics* **4**, 1338638 (2017).

Approach for multi-attribute group decision making based on linguistic-valued aggregation operator[*]

Li Yao

School of Computer and Information Technology, Liaoning Normal University Dalian, 116081, China

Yiming Cao

School of Mathematics, Liaoning Normal University, Dalian, 116081, China

Pengsen Liu

University of Electric Science and Technology, Chengdu, 610054, China

Yingying Xu and Li Zou[†]

*School of Computer and Information Technology, Liaoning Normal University Dalian, 116081, China
[†]zoulicn@163.com*

In order to deal with decision making problem with fuzzy linguistic-valued information which is both comparable and incomparable, based on linguistic truth-valued intuitionistic fuzzy lattice, the model of multi-attribute group decision making which based on the linguistic-valued aggregation operator is established. Firstly, for the aggregation problem of fuzzy linguistic-valued information, linguistic truth-valued intuitionistic fuzzy weighted averaging operator (LTV-IFWA) and linguistic truth-valued intuitionistic fuzzy ordered weighted averaging operator (LTV-IFOWA) are proposed, and the properties of the proposed operators are discussed. Secondly, the group integrated attribute values are ranked, the best project is determined by the final ranking result. For the incomparable situation in the ranking process, reference nearness degree is presented. The algorithms of partial true reference nearness degree (β_T) and partial false reference nearness degree (β_F) are given. Decision makers can select algorithm by preference. Finally, an example is given to illustrate the reasonability and validity of the proposed group decision-making approach.

---

[*]This work is supported by National Natural Science Foundation of China (Nos. 61772250 and 61673320) and National Natural Science Foundation of Liaoning Province (No. 2015020059).

1. Introduction

Multi-attribute group decision making is a process where many experts participate in analysis together and formulate decision aiming at multiple attributes. Since people's judgement of things is imprecise in the real world, experts lead fuzzy set theory[1] to multi-attribute group decision making[2]. Atanassov proposed intuitionistic fuzzy set[3]. Xu put forward intuitionistic fuzzy weighted averaging operator, intuitionistic fuzzy ordered weighted averaging operator and intuitionistic fuzzy hybrid averaging operator[4].

Due to the complexity of objective things and the fuzziness of human thinking, people are difficult to use precise values to evaluate. Rodriguez et al. expressed linguistic values by hesitant fuzzy linguistic term sets, a group decision-making approach of dealing with comparable linguistic expression[5]. However, some linguistic value information in the decision-making problem is incomparable, but a simple chain structure cannot express this property of the language. Therefore, Xu proposed linguistic truth-valued lattice implication algebra[6]. Zou established linguistic truth-valued intuitionistic fuzzy lattice[7].

In order to deal with the multi-attribute decision-making problems, there are both comparable and incomparable linguistic value information, taking into account both positive and negative evidence, minimizing information loss, and realizing the intelligence of the decision-making process. Based on the intuitionistic fuzzy lattice of language truth, two language value aggregation operators are proposed for the aggregation of language values and the processing of incommensurable language values, and a multi-attribute based on language value aggregation operators is constructed. The group decision model explores the treatment of incomparable linguistic values so that decision makers can choose the most appropriate one based on their preferences when they encounter incomparable results.

The structure of this paper is organized as follows: Section 2 briefly reviews some basic concepts on intuitionistic fuzzy set, linguistic truth-valued intuitionistic fuzzy pairs and linguistic truth-valued intuitionistic fuzzy lattice. In Section 3, we propose linguistic-valued aggregation operator which are LTV-IFWA and LTV-IFOWA, and give reference nearness degree which are β_T and β_F. Section 4 establishes an approach of multi-attribute group decision making based on linguistic-valued aggregation operator. In Section 5, an illustrative example is provided to demonstrate the reasonability and the validity of the new group decision making method. The conclusions are drawn in Section 6.

2. 10-element Linguistic Truth-valued Intuitionistic Fuzzy Lattice LI_{10}

Definition 1[8]. Let U be a nonempty set, intuitionistic fuzzy set is defined in the following form:

$$A = \{(x, \mu_A(x), v_A(x)) \mid x \in U\},$$

where $\mu_A(x)$ and $v_A(x)$ are the membership degree and non-membership degree of the object $x \in U$ belonging to $A \subseteq U$, that is

$$\mu_A(x): U \to [0,1], x \in U \to \mu_A(x) \in [0,1],$$

$$v_A(x): U \to [0,1], x \in U \to v_A(x) \in [0,1],$$

and which satisfy the following condition $0 \leq \mu_A(x) + v_A(x) \leq 1$ for any $x \in U$. In addition, the expression $\pi_A(x) = 1 - \mu_A(x) - v_A(x) (\forall x \in U)$ is called the degree of indeterminacy of x in A.

Definition 2[7]. In the 2n-element linguistic truth-valued lattice implication algebra $L_{V(n \times 2)} = \{(h_i, t), (h_j, f) \mid h_i \in L_n, t, f \in L_2\}$, for any $(h_i, t), (h_j, f) \in L_{V(n \times 2)}$, $((h_i, t), (h_j, f))$ is a linguistic truth-valued intuitionistic fuzzy pair, then $S = \{((h_i, t), (h_j, f)) \mid i, j \in \{1, 2, ..., n\}\}$ is a 2n-element linguistic truth-valued intuitionistic fuzzy pair set.

Definition 3[7]. Based on 2n-element linguistic truth-valued lattice implication algebra, establish a linguistic truth-valued intuitionistic fuzzy lattice $LI_{2n} = (LI_{2n}, \cup, \cap)$. For any $((h_i, t), (h_j, f)), ((h_k, t), (h_l, f)) \in LI_{2n}$,
1) $((h_i, t), (h_j, f)) \cup ((h_k, t), (h_l, f)) = ((h_{max(i,k)}, t), (h_{max(j,l)}, f));$
2) $((h_i, t), (h_j, f)) \cap ((h_k, t), (h_l, f)) = ((h_{min(i,k)}, t), (h_{min(j,l)}, f)).$

Definition 4[7]. The relation of any element $((h_i, t), (h_j, f))$ and $((h_k, t), (h_l, f))$ as:
1) $((h_i, t), (h_j, f)) \geq ((h_k, t), (h_l, f))$, if and only if $i \geq k$ and $j \geq l$;
2) $((h_i, t), (h_j, f)) \,/\!/\, ((h_k, t), (h_l, f))$, if and only if $i > k$ and $j < l$, or $i < k$ and $j > l$.
Let mood word set be $L_5 = \{h_i \mid i = 1,2,3,4,5, h_1 = $ "slightly", $h_2 = $ "rather", $h_3 = $ "extremely", $h_4 = $ "very", $h_5 = $ "absolutely", $h_1 < h_2 < h_3 < h_4 < h_5\}$, element linguistic truth-valued set $L_2 = \{t, f \mid t = $ "true", $f = $ "false"$\}$, we get 10-element linguistic truth-valued intuitionistic fuzzy lattice LI_{10}. The Hasse diagram of LI_{10} is shown in Fig. 1.

3. Linguistic-valued Aggregation Operator and Reference Nearness Degree

Definition 5. Let S be 2n-element linguistic truth-valued intuitionistic fuzzy pair set, $a_1, a_2, ..., a_m$ be a collection of linguistic-valued data and $w = (w_1, w_2, ..., w_m)^T$ be a linguistic-valued weighted vector of a_j, where $a_j, w_j \in S$ ($j = 1,2,...,m$), then linguistic truth-valued intuitionistic fuzzy weighted averaging operator

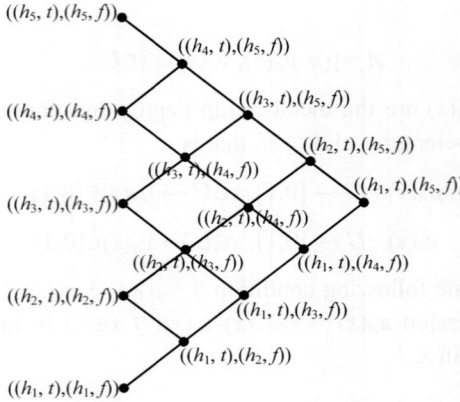

Fig. 1. Hasse diagram of LI_{10}.

LTV-IFWA: $S^n \to S$ is defined as:

$$\text{LTV-IFWA}(a_1, a_2, \ldots, a_m) = \bigcup_j \bigcap (w_j, a_j).$$

Definition 6. Let S be $2n$-element linguistic truth-valued intuitionistic fuzzy pair set, $a_1, a_2, \ldots, a_m$ be a collection of linguistic-valued data, b_j is the jth largest of $a_1, a_2, \ldots, a_m$, and $w = (w_1, w_2, \ldots, w_m)^T$ be a linguistic-valued weighted vector of b_j, where $a_i, w_j \in S$ ($i, j = 1,2,\ldots,m$), then linguistic truth-valued intuitionistic fuzzy ordered weighted averaging operator LTV-IFOWA: $S^n \to S$ is defined as:

$$\text{LTV-IFOWA}(a_1, a_2, \ldots, a_m) = \bigcup_j \bigcap (w_j, b_j).$$

Definition 7. Let S be $2n$-element linguistic truth-valued intuitionistic fuzzy pair set, for any $((h_i, t), (h_j, f)), ((h_k, t), (h_l, f)) \in S$, and $((h_i, t), (h_j, f)) // ((h_k, t), (h_l, f))$, $((h_i, t), (h_j, f)) \cup ((h_k, t), (h_l, f)) = ((h_p, t), (h_q, f))$, if β satisfies the following conditions:
1) $0 < \beta(((h_i, t), (h_j, f)), ((h_p, t), (h_q, f))) < 1$;
2) $\beta(((h_i, t), (h_j, f)), ((h_p, t), (h_q, f))) = \beta(((h_p, t), (h_q, f)), ((h_i, t), (h_j, f)))$;
3) If $\beta(((h_i, t), (h_j, f)), ((h_p, t), (h_q, f))) > \beta(((h_k, t), (h_l, f)), ((h_p, t), (h_q, f)))$, partializes $((h_i, t), (h_j, f))$, then the mapping $\beta: S \times S \to (0,1)$ is called reference nearness degree on the S.

Definition 8. Reference nearness degree β has the following two kinds of situation:
1) If give priority to true value, called partial true reference nearness degree, denoted as β_T, that is:

$$\beta_T(((h_i, t), (h_j, f)), ((h_p, t), (h_q, f))) = 1 - \frac{\max(i, p) + \min(j, q)}{2n}. \quad (1)$$

2) If give priority to false value, called partial false reference nearness degree, denoted as β_F, that is:

$$\beta_F(((h_i, t), (h_j, f)), ((h_p, t), (h_q, f))) = 1 - \frac{\min(i, p) + \max(j, q)}{2n}. \quad (2)$$

According to the definition of reference nearness degree β, we can get the following remark:

Remark 1. Let S be $2n$-element linguistic truth-valued intuitionistic fuzzy pair set, $((h_i, t), (h_j, f)), ((h_k, t), (h_l, f)) \in S$, and $((h_i, t), (h_j, f)) // ((h_k, t), (h_l, f))$, where $i > k, j < l$. Let $((h_i, t), (h_j, f)) \cup ((h_k, t), (h_l, f)) = ((h_p, t), (h_q, f))$, then
1) If reference nearness degree is β_T, partializes $((h_i, t), (h_j, f))$;
2) If reference nearness degree is β_F, partializes $((h_k, t), (h_l, f))$.

4. An Approach of Multi-attribute Group Decision Making Based on Linguistic-valued Aggregation Operator

On the $2n$-element linguistic truth-valued intuitionistic fuzzy lattice, based on linguistic-valued aggregation operators and reference nearness degree, this paper proposes an approach of multi-attribute group decision making, and the specific steps are as follows:

Step 1: For a multi-attribute group decision making, let $A = \{A_1, A_2, \ldots, A_n\}$ be a set of n possible alternatives, $G = \{G_1, G_2, \ldots, G_m\}$ be a set of m attributes and $E = \{e_1, e_2, \ldots, e_t\}$ be a set of t experts. Expert $e_k \in E$ ($k = 1, 2, \ldots, t$) expresses his (or her) linguistic assessment values of alternative $A_i \in A$ ($i = 1, 2, \ldots, n$) on an attribute $G_j \in G$ ($j = 1, 2, \ldots, m$) with a decision matrix $R_k = (r_{ij}^{(k)})_{n \times m}$, where $r_{ij}^{(k)} \in S$ and S is $2n$-element linguistic truth-valued intuitionistic fuzzy pair set.

Step 2: Utilizing LTV-IFWA operator to aggregate attribute values of the ith row in the decision matrix R_k, the integrated attribute value $z_i^{(k)}$ of alternative A_i provided by the expert e_k can be derived:

$$z_i^{(k)} = \text{LTV-IFWA}(r_{i1}^{(k)}, r_{i2}^{(k)}, \ldots, r_{im}^{(k)}) = \bigcup_j \bigcap \left(w_j, a_{ij}^{(k)}\right),$$

where $i = 1, 2, \ldots, n$, $j = 1, 2, \ldots, m$, $k = 1, 2, \ldots, t$, $w = (w_1, w_2, \ldots, w_m)^T$ is the linguistic-valued weighted vector of $a_{ij}^{(k)}$, and w_j is the jth element of w, $a_{ij}^{(k)}$ is the jth element of $\{r_{i1}^{(k)}, r_{i2}^{(k)}, \ldots, r_{im}^{(k)}\}$.

Step 3: Utilizing LTV-IFOWA to aggregate the integrated attribute value $z_i^{(k)}$ ($k = 1,2,\ldots,t$) of alternative A_i provided by t experts, the group integrated attribute value z_i of alternative A_i can be derived:

$$z_i = \text{LTV-IFOWA}(z_i^{(1)}, z_i^{(2)},\ldots, z_i^{(t)}) = \bigcup_k \bigcap \left(u_k, b_i^{(k)}\right),$$

where $i = 1,2,\ldots,n$, $\boldsymbol{u} = (v_1, v_2,\ldots, v_t)^T$ is the linguistic-valued weighted vector of $b_i^{(k)}$, u_k is the kth element of $\boldsymbol{u}$, and $b_i^{(k)}$ is the kth largest element of $(z_i^{(1)}, z_i^{(2)},\ldots, z_i^{(t)})$. In the ranking process of $z_i^{(1)}, z_i^{(2)},\ldots, z_i^{(t)}$, if appears incomparable situation, then according to corresponding situation, utilizing partial true reference nearness degree (β_T) or partial false reference nearness degree (β_F) to deal with incomparability.

Step 4: Use z_i ($i = 1,2,\ldots,n$) to rank all alternatives and choose the best.

In the ranking result, if appears incomparable situation, then according to corresponding situation, utilizing partial true reference nearness degree (β_T) or partial false reference nearness degree (β_F) to deal with it. Finally, according the ranking result, choose the best alternative.

5. Illustrative Example and Comparison Analysis

In the process of choose excellent teacher, firstly six assessment indexes of comprehensive quality: physical and psychological quality (G_1), cultural scientific quality (G_2), professional moral quality (G_3), instructional design and manage ability (G_4), scientific research ability (G_5), self-study and innovation ability (G_6). According to above indexes, the assessment results of four teachers A_1, A_2, A_3, A_4 provided by three experts e_k ($k = 1,2,3$) are listed in Table 1, Table 2 and Table 3 ("true" and "false" of linguistic truth-valued intuitionistic fuzzy lattice in this paper are equal to "strong" and "weak"), determined excellent teacher.

Table 1. The evaluation matrix R_1 of expert e_1.

Teachers	Attributes					
	G_1	G_2	G_3	G_4	G_5	G_6
A_1	$((h_3, t), (h_4, f))$	$((h_3, t), (h_3, f))$	$((h_3, t), (h_5, f))$	$((h_3, t), (h_4, f))$	$((h_4, t), (h_5, f))$	$((h_3, t), (h_4, f))$
A_2	$((h_3, t), (h_5, f))$	$((h_3, t), (h_3, f))$	$((h_4, t), (h_5, f))$	$((h_3, t), (h_4, f))$	$((h_2, t), (h_4, f))$	$((h_4, t), (h_5, f))$
A_3	$((h_4, t), (h_4, f))$	$((h_3, t), (h_4, f))$	$((h_3, t), (h_4, f))$	$((h_3, t), (h_5, f))$	$((h_3, t), (h_4, f))$	$((h_3, t), (h_3, f))$
A_4	$((h_4, t), (h_5, f))$	$((h_3, t), (h_5, f))$	$((h_2, t), (h_3, f))$	$((h_4, t), (h_4, f))$	$((h_4, t), (h_4, f))$	$((h_2, t), (h_3, f))$

Table 2. The evaluation matrix R_2 of expert e_2.

Teachers	Attributes					
	G_1	G_2	G_3	G_4	G_5	G_6
A_1	$((h_4, t), (h_4, f))$	$((h_3, t), (h_3, f))$	$((h_4, t), (h_4, f))$	$((h_4, t), (h_5, f))$	$((h_1, t), (h_5, f))$	$((h_2, t), (h_3, f))$
A_2	$((h_3, t), (h_4, f))$	$((h_2, t), (h_4, f))$	$((h_3, t), (h_4, f))$	$((h_4, t), (h_4, f))$	$((h_3, t), (h_5, f))$	$((h_3, t), (h_3, f))$
A_3	$((h_2, t), (h_3, f))$	$((h_2, t), (h_3, f))$	$((h_3, t), (h_3, f))$	$((h_4, t), (h_5, f))$	$((h_4, t), (h_4, f))$	$((h_4, t), (h_4, f))$
A_4	$((h_3, t), (h_3, f))$	$((h_3, t), (h_4, f))$	$((h_3, t), (h_5, f))$	$((h_3, t), (h_4, f))$	$((h_3, t), (h_5, f))$	$((h_3, t), (h_3, f))$

Table 3. The evaluation matrix R_3 of expert e_3.

Teachers	Attributes					
	G_1	G_2	G_3	G_4	G_5	G_6
A_1	$((h_3, t), (h_3, f))$	$((h_2, t), (h_3, f))$	$((h_5, t), (h_5, f))$	$((h_3, t), (h_4, f))$	$((h_4, t), (h_5, f))$	$((h_4, t), (h_4, f))$
A_2	$((h_3, t), (h_5, f))$	$((h_3, t), (h_4, f))$	$((h_4, t), (h_5, f))$	$((h_3, t), (h_4, f))$	$((h_3, t), (h_5, f))$	$((h_3, t), (h_5, f))$
A_3	$((h_3, t), (h_4, f))$	$((h_4, t), (h_4, f))$	$((h_2, t), (h_3, f))$	$((h_4, t), (h_4, f))$	$((h_2, t), (h_3, f))$	$((h_3, t), (h_3, f))$
A_4	$((h_3, t), (h_5, f))$	$((h_4, t), (h_5, f))$	$((h_3, t), (h_3, f))$	$((h_3, t), (h_5, f))$	$((h_3, t), (h_3, f))$	$((h_3, t), (h_3, f))$

Based on 10-element linguistic truth-valued intuitionistic fuzzy lattice, utilizing an approach of multi-attribute group decision making based on linguistic-valued aggregation operator to slove the problem:

Step 1: Transform Table 1, Table 2, Table 3 into assesment matrix R_1, R_2, R_3.

Step 2: Let $w = (((h_3, t), (h_5, f)), ((h_3, t), (h_3, f)), ((h_3, t), (h_4, f)), ((h_3, t), (h_5, f)), ((h_4, t), (h_4, f)), ((h_3, t), (h_3, f)))^T$, utilizing LTV-IFWA operator to aggregate attribute values of the ith row in the decision matrix R_k, the integrated attribute value $z_i^{(k)}(i = 1,2,3,4)$ of four teachers provided by the expert e_k ($k = 1,2,3$) can be derived:

$z_1^{(1)} = ((h_4, t), (h_4, f))$, $z_2^{(1)} = ((h_3, t), (h_5, f))$, $z_3^{(1)} = ((h_3, t), (h_5, f))$, $z_4^{(1)} = ((h_4, t), (h_4, f))$, $z_1^{(2)} = ((h_3, t), (h_5, f))$, $z_2^{(2)} = ((h_3, t), (h_4, f))$, $z_3^{(2)} = ((h_4, t), (h_5, f))$, $z_4^{(2)} = ((h_3, t), (h_4, f))$, $z_1^{(3)} = ((h_4, t), (h_4, f))$, $z_2^{(3)} = ((h_3, t), (h_5, f))$, $z_3^{(3)} = ((h_3, t), (h_4, f))$, $z_4^{(3)} = ((h_3, t), (h_5, f))$.

Step 3: Let $u = (((h_3, t), (h_4, f)), ((h_4, t), (h_4, f)), ((h_3, t), (h_5, f)))^T$, utilizing LTV-IFOWA to aggregate the integrated attribute value $z_i^{(k)}$ ($k = 1,2,3$) of four teachers provided by three experts. In this decision making, mainly consider membership degree of strong ability, then this paper chooses partial true reference nearness degree β_T to deal with incomparable situation. The group integrated attribute value z_i of four teachers can be derived:

$z_1 = ((h_4, t), (h_4, f))$, $z_2 = ((h_3, t), (h_5, f))$, $z_3 = ((h_3, t), (h_5, f))$, $z_4 = ((h_3, t), (h_4, f))$.

Step 4: Use z_i to rank four teachers, we can get:

$$A_1 \,//\, (A_2 = A_3) > A_4.$$

This example chooses the teacher of the strongest ability, mainly consider membership degree of strong ability, then this paper chooses partial true reference nearness degree β_T to deal with incomparable A_1 and A_2, A_3.

$$((h_3, t), (h_5, f)) \cup ((h_4, t), (h_4, f)) = ((h_4, t), (h_5, f)),$$

$$\beta_T(((h_3, t), (h_5, f)), ((h_4, t), (h_5, f))) = \frac{1}{10};$$

$$\beta_T(((h_4, t), (h_4, f)), ((h_4, t), (h_5, f))) = \frac{2}{10}.$$

Then partializes A_1. Hence, $A_1 > A_2 = A_3 > A_4$, excellent teacher is A_1. This example is given to illustrate the reasonability of the proposed group decision-making approach.

6. Conclusions

In order to solve aggregation problem with fuzzy linguistic values on the linguistic truth-valued intuitionistic fuzzy lattice, in the approach of multi-attribute group decision making which based on linguistic-valued aggregation operator, weights are linguistic values. Utilizing LTV-IFWA operator and LTV-IFOWA operator to aggregate fuzzy linguistic-valued information. For the incomparable situation in the aggregation process, experts can use reference nearness degree β to deal with it by preference, and according to the final ranking result to choose the best one. Moreover, we will deeply research determination method of linguistic-valued weights.

References

1. L.A. Zadeh, *Fuzzy sets[J]. Information and Control.* 8(3), 338-353(1965).
2. Y. Wu, H. Xu and C. Xu, *Uncertain multi-attributes decision making method based on interval number with probability distribution weighted operators and stochastic dominance degree[J]. Knowledge-Based Systems.* 113, 199-209(2016).
3. Atanassov K T, *Intuitionistic fuzzy sets[J]. Fuzzy Sets and Systems.* 20(1), 87-96(1986).
4. Z.S. Xu, *Intuitionistic fuzzy aggregation operators[J]. IEEE Transactions on Fuzzy Systems: A Publication of the IEEE Neural Networks Council.* 15(6), 1179-1187(2007).

5. R. M. Rodriguez, L. Martinez and F. Herrera, *A group decision making model dealing with comparative linguistic expressions based on hesitant fuzzy linguistic term sets[J]. Information Sciences.* 241, 28-42(2013).
6. Y. Xu and J. Ma, *Linguistic truth-valued lattice implication algebra and its properties[J]. The Proceedings of the IMACS Multi-Conference on Computational Engineering in Systems Applications.* 2, 1413-1418(2006).
7. L. Zou, X. Liu and Z. Pei, *Implication Operator on the Set of $\vee$-Irreducible Element in Linguistic Truth-Valued Intuitionistic Fuzzy Lattice. International Journal of Machine Learning and Cybernetics.* 4, 365-372(2013).
8. S. P. Wan, J. Xu and J. Y. Dong, *Aggregating decision information into interval-valued intuitionistic fuzzy numbers for heterogeneous multi-attribute group decision making[J]. Knowledge-Based Systems.* 113, 155-170(2016).
9. H. Zhu, J. Zhao and Y. Xu, *2-Dimension linguistic computational model with 2-tuples for multi-attribute group decision making[J]. Knowledge-Based Systems.* 103, 132-142(2016).
10. L. Zou, X. Wen and Y. Wang, *Linguistic truth-valued intuitionistic fuzzy reasoning with applications in human factors engineering[J]. Information Sciences.* 327, 201-216(2016).

Linguistic preference relation and its application in group decision making[*]

Xinzi Li

School of Computer Science and Technology, Dalian University of Technology
Dalian, Liaoning, China

Hongyue Diao

School of Mathematics, Liaoning Normal University
Dalian, Liaoning, China

Li Yao and Li Zou[†]

School of Computer and Information Technology, Liaoning Normal University
Dalian, Liaoning, China
[†]zoulicn@163.com

The group decision making problem with preference relation has shown definite advantages in many fields. To deal with the group decision making process with the linguistic-valued intuitionistic fuzzy information, this paper proposes the linguistic-valued intuitionistic fuzzy preference relation, linguistic-valued intuitionistic fuzzy 2-tuple model, and some linguistic-valued intuitionistic fuzzy aggregation operators, then develops a new approach of linguistic-valued intuitionistic fuzzy group decision making. Finally, a practical example is provided to illustrate the rationality.

1. Introduction

The group decision making procedure is widely used in the modern decision science. In the representation and processing of uncertain information [1], we often get across some information represented with linguistic values in natural language. The conventional linguistic-valued handling methodology is based on fuzzy set theory, which generally converts the linguistic information into a membership function, after which the generated fuzzy set will be transformed back into a word using linguistic approximation [2]. The process of converting

---

[*]This work is supported in part by the National Natural Science Foundation of P. R. China (Nos. 61772250, 61673320, 61672127), the Fundamental Research Funds for the Central Universities (No. 2682017ZT12), and the National Natural Science Foundation of Liaoning Province (No. 2015020059).

each other is mainly based on the extension principle, and it is usually time consuming, computationally complex, and involving the loss of information [3]. It would be more natural and reasonable to represent linguistic information in the symbolic way [4]. Because most symbolic approaches have the advantages that without the loss of information and computational simplicity by avoiding use of membership function [5]. Linguistic hedge algebra can be constructed as a partially ordered structure according to the natural meanings of the represented linguistic terms, and generally a lattice. [6]. Zou and other scholars established the linguistic truth-valued intuitionistic fuzzy lattice which is based on the linguistic truth-valued lattice implication algebra [7]. In order to solve the problem of information loss in the process of aggregation, Heerera [8] first proposed a linguistic 2-tuple model in 2000. The linguistic 2-tuple model and its aggregation operators have been applied to multi-attribute decision analysis [9] and many other fields.

Consider that, in some real-life situations, a decision maker may give his preference information by linguistic-valued intuitionistic fuzzy information with both positive and negative evidence. In this paper, we will pay attention to this issue, and investigate an approach to group decision making based on the linguistic-valued intuitionistic fuzzy preference relation. The remainder paper is organized as follows. Section 2 briefly gives the preliminaries. Section 3 introduces the concepts of linguistic-valued intuitionistic fuzzy preference relation. Section 4 introduces the concepts of linguistic-valued intuitionistic fuzzy 2-tuple model. Finally, Section 5 gives a practical example to illustrate the rationality.

2. Preliminaries

Definition 1 [1]: Let a crisp set X be fixed, and let $A \subset X$ be a fixed set. An intuitionistic fuzzy set is defined in the following form:

$$A = \{(x, \mu_A(x), v_A(x)) | x \in X\} \quad (1)$$

where $\mu_A(x)$ and $v_A(x)$ are the membership degree and non-membership degree of the element $x \in X$ to the set A, respectively, and for every $x \in X$, $0 \leq \mu_A(x) + v_A(x) \leq 1$ holds. In addition, the expression $\pi_A(x) = 1 - \mu_A(x) - v_A(x)$ is called the degree of indeter-minacy of x in A.

Definition 2 [7]: In the $2n$-element linguistic truth-valued lattice implication algebra, for any $(h_i, t), (h_j, f) \in L_{V(n \times 2)}$, if $((h_i, t), (h_j, f))$ satisfies $(h_i, t)' \geq (h_j, f)$, then $((h_i,t),(h_j,f))$ is defined as a linguistic truth-valued intuitionistic fuzzy pair, where the operation of "'" is order-reversing involution of the $L_{V(n\times 2)}$. $((h_i,t),(h_j,f))$ is a linguistic truth-valued intuitionistic fuzzy pair if and only if $i \leq j$.

Definition 3 [7]: Based on 2n-element linguistic truth-valued lattice implication algebra, establish a linguistic truth-valued intuitionistic fuzzy lattice $((h_i,t),(h_j,f)) = (LI_{2n}, \cup, \cap)$. For any $((h_i,t),(h_j,f)),((h_k,t),(h_l,f)) \in LI_{2n}$,

1. $((h_i,t),(h_j,f)) \cup ((h_k,t),(h_l,f)) = ((h_{\max(i,k)},t),(h_{\max(j,l)},f))$;
2. $((h_i,t),(h_j,f)) \cap ((h_k,t),(h_l,f)) = ((h_{\min(i,k)},t),(h_{\min(j,l)},f))$.

Definition 4 [9]: Let $S = \{s_0, s_1, \cdots, s_g\}$ be a linguistic term set and $\beta \in [0, g]$ be a value representing the result of a symbolic aggregation operation, then the 2-tuple that expresses the equivalent information to β is obtained with the following function:

$$\Delta : [0, g] \to S \times [-0.5, 0.5)$$

$$\Delta(\beta) = (s_i, \alpha), \text{with} \begin{cases} s_i, i = \text{round}(\beta) \\ \alpha = \beta - i, \alpha \in [-0.5, 0.5) \end{cases} \quad (2)$$

where round$(\cdot)$ is the usual round operation, s_i has the closest index label to "β", and "α" is the value of the symbolic translation.

There is always a Δ^{-1} function such that from a 2-tuple returns its equivalent numerical value $\beta \in [0, g] \subset R$.

$$\Delta^{-1} : S \times [-0.5, 0.5) \to [0, g]$$
$$\Delta^{-1}(s_i, \alpha) = i + \alpha = \beta \quad (3)$$

3. Linguistic-valued Intuitionistic Fuzzy Preference Relation

In the process of decision making, a decision maker generally needs to provide his preference for each pair of alternatives, and then constructs a preference relation. And in the real life, decision makers prefer to provide decision-making information by qualitative linguistic variables. Therefore when decision makers compare every two alternatives, they always express the decision information with the positive and negative linguistic information.

Definition 5: A linguistic-valued intuitionistic fuzzy preference relation B on the set X is represented by a linguistic-valued intuitionistic fuzzy preference relation matrix $B = (b_{ij})_{m \times m} \subset X \times X$, where $b_{ij} = \left(\left(h_{p_{ij}}, t \right), \left(h_{q_{ij}}, f \right) \right)$ is a linguistic truth-valued intuitionistic fuzzy pair, composed by the degree $\left(h_{p_{ij}}, t \right)$ to which x_i is preferred to x_j and the degree $\left(h_{q_{ij}}, f \right)$ to which x_i is non-preferred to x_j.

Furthermore, as for b_{ij} and b_{ji}, $p_{ij} = q_{ji}, p_{ji} = q_{ij}$ and $q_{ii} = p_{ii} = \frac{1}{2}(n+1)$.

4. Linguistic-valued Intuitionistic Fuzzy 2-tuple Model

Definition 6: Let $\left((h_i, c_1), (h_j, c_2) \right), \left((h_k, c_1), (h_l, c_2) \right) \in LI_{2n}$, $\alpha, \beta \in [1, n+1]$ be numerical values and $\alpha \leq \beta$, then the linguistic-valued intuitionistic fuzzy 2-tuple model can be defined as follows:

$$\Delta : [0, n] \times [0, n] \to LI_{2n} \times (-0.5, 0.5]$$

$$\Delta(\alpha, \beta) = \begin{cases} \left((h_i, c_1), (h_j, c_2) \right), & \text{if } \alpha, \beta \in N, \\ \left(\left((h_i, c_1), (h_j, c_2) \right), \gamma \right), & \text{otherwise}. \end{cases} \qquad (4)$$

where $i = round(\alpha), j = round(\beta)$, and $\gamma = \frac{1}{2}((\alpha - i) + (\beta - j))$ represents the deviation between $\Delta(\alpha, \beta)$ and $\left((h_i, c_1), (h_j, c_2) \right)$.

Let's take 10-element linguistic-valued intuitionistic fuzzy lattice for an example. If $\alpha = 3.3, \beta = 4.9$, then the linguistic-valued intuitionistic fuzzy 2-tuple model of α, β is that: $\Delta(\alpha, \beta) = \left(\left((h_3, c_1), (h_5, c_2) \right), 0.1 \right)$.

Definition 7: For any two linguistic-valued intuitionistic fuzzy 2-tuple models $\left(\left((h_i, c_1), (h_j, c_2) \right), \gamma_1 \right), \left(\left((h_k, c_1), (h_l, c_2) \right), \gamma_2 \right)$:

1. If $\left((h_i, c_1), (h_j, c_2) \right) < \left((h_k, c_1), (h_l, c_2) \right)$, then: $\left(\left((h_i, c_1), (h_j, c_2) \right), \gamma_1 \right) < \left(\left((h_k, c_1), (h_l, c_2) \right), \gamma_2 \right)$;

2. If $\big((h_i,c_1),(h_j,c_2)\big)=\big((h_k,c_1),(h_l,c_2)\big)$, then: $\begin{cases}\gamma_1<\gamma_2, \big(\big((h_i,c_1),(h_j,c_2)\big),\gamma_1\big)<\big(\big((h_k,c_1),(h_l,c_2)\big),\gamma_2\big); \\ \gamma_1=\gamma_2, \big(\big((h_i,c_1),(h_j,c_2)\big),\gamma_1\big)=\big(\big((h_k,c_1),(h_l,c_2)\big),\gamma_2\big).\end{cases}$

Property 1: For any numerical values $\alpha_1, \beta_1, \alpha_2, \beta_2 \in [1, n+1]$, if $\Delta(\alpha_1, \beta_1) = \big(\big((h_i,c_1),(h_j,c_2)\big),\gamma_1\big)$, $\Delta(\alpha_2,\beta_2) = \big(\big((h_k,c_1),(h_l,c_2)\big),\gamma_2\big)$, then we can get the following properties:

1. $\Delta(\alpha_1,\beta_1) = \Delta(\alpha_2,\beta_2)$, that is $\big(\big((h_i,c_1),(h_j,c_2)\big),\gamma_1\big) = \big(\big((h_k,c_1),(h_l,c_2)\big),\gamma_2\big)$ if and only if $\alpha_1 = \alpha_2, \beta_1 = \beta_2$;

2. $\Delta(\alpha_1,\beta_1) < \Delta(\alpha_2,\beta_2)$, that is $\big(\big((h_i,c_1),(h_j,c_2)\big),\gamma_1\big) < \big(\big((h_k,c_1),(h_l,c_2)\big),\gamma_2\big)$ if and only if $\alpha_1 < \alpha_2, \beta_1 < \beta_2$.

Definition 8: Let $r_k = \big((h_{a_k},t),(h_{b_k},f)\big), (k=1,2,\cdots,m)$ be m linguistic-valued intuitionistic fuzzy values, and let $\omega = (\omega_1, \omega_2, \cdots, \omega_m)^T$ be the weight vector of r_k, $\omega_k > 0$ and $\sum_{k=1}^{m} \omega_k = 1$, then some linguistic-valued intuitionistic fuzzy aggregation operators can be defined as follows:

1. Linguistic-valued intuitionistic fuzzy arithmetic averaging operator:

$$G_{LIAAO}(r_1, r_2, \cdots, r_m) = \Delta\left(\frac{1}{m}\sum_{k=1}^{m} a_k, \frac{1}{m}\sum_{k=1}^{m} b_k\right) = \left(\left(h_{\text{round}\left(\frac{1}{m}\sum_{k=1}^{m} a_k\right)}, t\right), \left(h_{\text{round}\left(\frac{1}{m}\sum_{k=1}^{m} b_k\right)}, f\right), \gamma\right)$$

$$\gamma = \frac{1}{2}\left(\left(\frac{1}{m}\sum_{k=1}^{m} a_k - \text{round}\left(\frac{1}{m}\sum_{k=1}^{m} a_k\right)\right) + \left(\frac{1}{m}\sum_{k=1}^{m} b_k - \text{round}\left(\frac{1}{m}\sum_{k=1}^{m} b_k\right)\right)\right) \quad (5)$$

2. Linguistic-valued intuitionistic fuzzy weighted arithmetic averaging operator:

$$G_{LIWAAO}(r_1, r_2, \cdots, r_m) = \Delta\left(\sum_{k=1}^{m} \omega_k a_k, \sum_{k=1}^{m} \omega_k b_k\right) = \left(\left(h_{\text{round}\left(\sum_{k=1}^{m} \omega_k a_k\right)}, t\right), \left(h_{\text{round}\left(\sum_{k=1}^{m} \omega_k b_k\right)}, f\right), \gamma\right)$$

$$\gamma = \frac{1}{2}\left(\left(\sum_{k=1}^{m} \omega_k a_k - \text{round}\left(\sum_{k=1}^{m} \omega_k a_k\right)\right) + \left(\sum_{k=1}^{m} \omega_k b_k - \text{round}\left(\sum_{k=1}^{m} \omega_k b_k\right)\right)\right) \quad (6)$$

3. Linguistic-valued intuitionistic fuzzy geometric averaging operator:

$$G_{LIGAO}(r_1,r_2,\cdots,r_m) = \Delta\left(\sqrt[m]{\prod_{k=1}^{m}a_k},\sqrt[m]{\prod_{k=1}^{m}b_k}\right) = \left(\left(h_{round\left(\sqrt[m]{\prod_{k=1}^{m}a_k}\right)},t,h_{round\left(\sqrt[m]{\prod_{k=1}^{m}b_k}\right)},f\right),\gamma\right)$$

$$\gamma = \frac{1}{2}\left(\left(\sqrt[m]{\prod_{k=1}^{m}a_k} - round\left(\sqrt[m]{\prod_{k=1}^{m}a_k}\right)\right) + \left(\sqrt[m]{\prod_{k=1}^{m}b_k} - round\left(\sqrt[m]{\prod_{k=1}^{m}b_k}\right)\right)\right) \quad (7)$$

4. Linguistic-valued intuitionistic fuzzy weighted geometric averaging operator:

$$G_{LIWGAO}(r_1,r_2,\cdots,r_m) = \Delta\left(\prod_{k=1}^{m}a_k^{\omega_k},\prod_{k=1}^{m}b_k^{\omega_k}\right) = \left(\left(h_{round\left(\prod_{k=1}^{m}a_k^{\omega_k}\right)},t,h_{round\left(\prod_{k=1}^{m}b_k^{\omega_k}\right)},f\right),\gamma\right)$$

$$\gamma = \frac{1}{2}\left(\left(\prod_{k=1}^{m}a_k^{\omega_k} - round\left(\prod_{k=1}^{m}a_k^{\omega_k}\right)\right) + \left(\prod_{k=1}^{m}b_k^{\omega_k} - round\left(\prod_{k=1}^{m}b_k^{\omega_k}\right)\right)\right) \quad (8)$$

Property 2: As for any linguistic-valued intuitionistic fuzzy aggregation operator $G_{LIO}(r_1,r_2,\cdots,r_m) \in \{G_{LIAAO}, G_{LIWAAO}, G_{LIGAO}, G_{LIWGAO}\}$, there are some properties:
1. If $r_k = r$, then $G_{LIO}(r_1,r_2,\cdots,r_m) = r$ $(k=1,2,\cdots,m)$.
2. If $p_k > q_k$, then $G_{LIO}(p_1,p_2,\cdots,p_m) > G_{LIO}(q_1,q_2,\cdots,q_m)$ $(k=1,2,\cdots,m)$.
3. $\min\{r_k\} \le G_{LIO}(r_1,r_2,\cdots,r_m) \le \max\{r_k\}$ $(k=1,2,\cdots,m)$.

5. Linguistic Preference Relation Group Decision Making Procedure

5.1. *The group decision making procedure*

Consider a group decision making problem, let $X = \{x_1,x_2,\cdots,x_m\}$ be a set of alternatives, and let $D = \{d_1,d_2,\cdots,d_s\}$ be the set of DMs. Let $\omega = (\omega_1,\omega_2,\cdots,\omega_s)^T$ be the weight vector of DMs, where $\omega_k > 0$ and $\sum_{k=1}^{s}\omega_k = 1$, $k=1,2,\cdots,s$.

Let mood word set be $L_5 = \{h_i \mid i = 1,2,3,4,5, h_1 =$ "slightly", $h_2 =$ "rather", $h_3 =$ "extremely", $h_4 =$ "very", $h_5 =$ "absolutely", $h_1 < h_2 < h_3 < h_4 < h_5\}$, element linguistic truth-valued set $L_2 = \{t, f \mid t =$ "true", $f =$ "false"$\}$, we get 10-element linguistic truth-valued intuitionistic fuzzy lattice LI_{10}.

Step 1: The DMs provides his linguistic intuitionistic fuzzy preference for each pair of alternatives, and constructs a linguistic intuitionistic fuzzy preference relation $B^{(k)} = \left(b_{ij}^{(k)}\right)_{m\times m}, i,j = 1,2,\cdots,m, k = 1,2,\cdots s$.

Step 2: Utilize the G_{LIAAO} or the G_{LIGAO} to aggregate all $b_{ij}^{(k)}$ corresponding to the alternative x_i and then get the averaged linguistic-valued intuitionistic fuzzy value $b_i^{(k)}$ of the alternative x_i over all the other alternatives.

Step 3: Utilize the G_{LIWAAO} or the G_{LIWAAO} to aggregate all $b_i^{(k)}$ corresponding to s DMs into a collective linguistic-valued intuitionistic fuzzy value b_i of the alternative x_i over all the other alternatives.

Step 4: Rank all b_i by the **Definition 7** and then rank all the alternatives x_i and select the best one according to the values of b_i.

5.2. *Illustrative example*

Now, consider a group decision making problem (adapted from [10]). Assume that a committee including three DMs $D = \{d_1, d_2, d_3\}$ has been built to interview the candidate exchange doctoral students from all over the world. The weights for the DMs are 0.2, 0.5, and 0.3, respectively. There are four students $X = \{x_1, x_2, x_3, x_4\}$ shortlisted as potential candidates.

Step 1: Each expert conducts pairwise comparisons with respect to these four students and determines the following linguistic-valued intuitionistic fuzzy preference relations:

$$B^{(1)} = \begin{pmatrix} ((h_3,t),(h_3,f)) & ((h_2,t),(h_4,f)) & ((h_2,t),(h_2,f)) & ((h_4,t),(h_2,f)) \\ ((h_4,t),(h_2,f)) & ((h_3,t),(h_3,f)) & ((h_5,t),(h_1,f)) & ((h_4,t),(h_2,f)) \\ ((h_2,t),(h_2,f)) & ((h_1,t),(h_5,f)) & ((h_3,t),(h_3,f)) & ((h_3,t),(h_2,f)) \\ ((h_2,t),(h_4,f)) & ((h_2,t),(h_4,f)) & ((h_2,t),(h_3,f)) & ((h_3,t),(h_3,f)) \end{pmatrix};$$

$$B^{(2)} = \begin{pmatrix} ((h_3,t),(h_3,f)) & ((h_3,t),(h_2,f)) & ((h_4,t),(h_2,f)) & ((h_2,t),(h_3,f)) \\ ((h_2,t),(h_3,f)) & ((h_3,t),(h_3,f)) & ((h_2,t),(h_1,f)) & ((h_3,t),(h_2,f)) \\ ((h_2,t),(h_4,f)) & ((h_1,t),(h_2,f)) & ((h_3,t),(h_3,f)) & ((h_4,t),(h_2,f)) \\ ((h_3,t),(h_2,f)) & ((h_2,t),(h_3,f)) & ((h_2,t),(h_4,f)) & ((h_3,t),(h_3,f)) \end{pmatrix};$$

$$B^{(3)} = \begin{pmatrix} ((h_3,t),(h_3,f)) & ((h_4,t),(h_2,f)) & ((h_5,t),(h_1,f)) & ((h_4,t),(h_2,f)) \\ ((h_2,t),(h_4,f)) & ((h_3,t),(h_3,f)) & ((h_2,t),(h_1,f)) & ((h_4,t),(h_1,f)) \\ ((h_1,t),(h_5,f)) & ((h_1,t),(h_2,f)) & ((h_3,t),(h_3,f)) & ((h_2,t),(h_2,f)) \\ ((h_2,t),(h_4,f)) & ((h_1,t),(h_4,f)) & ((h_2,t),(h_2,f)) & ((h_3,t),(h_3,f)) \end{pmatrix}.$$

Step 2: Utilize the G_{LIAAO} to aggregate all $b_{ij}^{(k)}$ and then get the averaged linguistic-valued intuitionistic fuzzy value $b_i^{(k)}$.

$$b_1^{(1)} = \left((h_3,t),(h_3,f),-\frac{1}{4}\right), \; b_2^{(1)} = ((h_4,t),(h_2,f),0), \; b_3^{(1)} = \left((h_2,t),(h_3,f),\frac{1}{4}\right), \; b_4^{(1)} = \left((h_2,t),(h_3,f),\frac{1}{4}\right);$$

$$b_1^{(2)} = \left((h_3,t),(h_2,f),\frac{1}{2}\right), \; b_2^{(2)} = ((h_3,t),(h_2,f),0), \; b_3^{(2)} = \left((h_2,t),(h_3,f),\frac{1}{4}\right), \; b_4^{(2)} = \left((h_2,t),(h_3,f),\frac{1}{2}\right);$$

$$b_1^{(3)} = ((h_4,t),(h_2,f),0), \; b_2^{(3)} = ((h_3,t),(h_2,f),0), \; b_3^{(3)} = \left((h_2,t),(h_4,f),-\frac{1}{4}\right), \; b_4^{(3)} = \left((h_2,t),(h_3,f),\frac{1}{4}\right).$$

Step 3: Utilize the G_{LIWAAO} to aggregate all $b_i^{(k)}$ into a collective linguistic-valued intuitionistic fuzzy value b_i.

$$b_1 = \left((h_3,t),(h_2,f),\frac{7}{10}\right), \; b_2 = \left((h_3,t),(h_2,f),\frac{1}{5}\right), \; b_3 = \left((h_2,t),(h_3,f),\frac{2}{5}\right), \; b_4 = \left((h_2,t),(h_3,f),\frac{3}{8}\right).$$

Step 4: According to the **Definition 7**, we can know that: $b_1 > b_2 > b_3 > b_4$.

Then the ranking of the alternatives is $x_1 > x_2 > x_3 > x_4$, which is as same as the result of the reference [10]. The result implies the first doctoral student is the best candidate for international exchanging.

6. Conclusions

The proposed linguistic-valued intuitionistic fuzzy preference relation and 2-tuple model of this paper can deal with the linguistic intuitionistic fuzzy information with both positive and negative evidence and reduce the information loss in the process of the decision making. In the future, we will further use the proposed concepts in many other fields, such as linguistic value uncertainty reasoning and linguistic evaluation.

References

1. L.A. Zadeh, *Fuzzy sets. Information and Control.* 8(3):338-353(1965).
2. N.C. Ho, *A topological completion of refined hedge algebras and a model of fuzziness of linguistic terms and hedges.* Elsevier North-Holland, Inc. (2007).

3. J.M. Mendel, *Computing with words and its relationships with fuzzistics*. Information Sciences, 177(4):988-1006(2007).
4. F. Herrera and L. Martinez, *The 2-tuple linguistic computational model. Advantages of its linguistic description, accuracy and consistency*. World Scientific Publishing Co. Inc. (2001).
5. Y. Xu, *Some views on information fusion and logic based approaches in decision making under uncertainty*. Journal of Universal Computerence, 16(1):3-19(2010).
6. B. Malakooti, *Ranking and screening multiple criteria alternatives with partial information and use of ordinal and cardinal strength of preferences*. IEEE Transactions on Systems, Man, and Cybernetics—Part A 30(3):355–368(2000).
7. L. Zou, P. Shi, Z. Pei and Y. Xu, *On an algebra of linguistic truth-valued intuitionistic lattice-valued logic*. Journal of Intelligent & Fuzzy Systems. 24: 447-456 (2013).
8. C. Kahraman, S.C. Onar and B. Oztaysi. *Fuzzy multicriteria decision-making: A literature review*. International Journal of Computational Intelligence Systems, 8(4):637-666 (2015).
9. F. Herrera and L. Martinez, *A 2-tuple fuzzy linguistic representation model for computing with words*. IEEE Transactions on Fuzzy Systems, 8(6):746-752(2000).
10. H. Liao and Z. Xu, *Some algorithms for group decision making with intuitionistic fuzzy preference information*. International Journal of Uncertainty, Fuzziness and Knowledge-Based Systems, 22(04):505-529(2014).

Managing consensus with an established goal in the DG model with an undirected graph: An approach based on minimizing adjusted opinions[*]

Xia Chen and Yucheng Dong

Business School, Sichuan University, Chengdu, 610065, China
chenxia@stu.scu.edu.cn, ycdong@scu.edu.cn

Zhaogang Ding

School of Public Management, Northwestern University, Xi'an, 710069, China
zgding@nwu.edu.cn

Nowadays, social networks facilitate individuals' expression of opinions about different political, economic and cultural issues, and firms or governments start to pay attentions on the management of public opinions, such as guiding the forming opinions to reach a specific consensus point. In this paper, we discuss that how to minimize the adjustments of individuals' initial opinions to reach a consensus with an established goal from the perspective based on opinion dynamics in social network.

1. Introduction

Opinion dynamics is widely used to model the fusion process of individual opinions based on the established fusion rules. A lot of opinion dynamics models have been proposed, such as voter model [11], the DeGroot model [3, 5], Friedkin and Johnsen model [13], bounded confidence model (e.g., [6, 12]).

Consensus process (e.g., [4, 10, 15-17]) has been received significant attentions in group decision making problems. A key issue in consensus process is to provide the adjustment suggestions. Recently, some consensus models with minimum adjustments have been proposed (e.g., [1, 2, 7, 19]).

The existing studies of opinion dynamics mainly focus on analyzing under what conditions individuals will reach a consensus or split. However, firms or governments may pay more attentions on guiding the forming opinions. In this paper, we plan to discuss that how to minimize the adjustments of individuals' initial opinions to reach a consensus with an established goal.

---

[*]This work is supported by the grant (No. 71571124) from NSF of China.

The rest of this paper is organized as follows. Section 2 introduces the basic model and investigates the consensus conditions. Then Section 3 proposes a procedure for reaching an established consensus goal. Section 4 provides a numerical example. Finally, Section 5 presents the concluding remarks.

2. Basic Model and Consensus Conditions

The social network can be depicted by a undirected graph $G(V,E)$, where $V = \{v_1, v_2, \ldots, v_n\}$ is the set of individuals and E is the set of edges. The edge $(v_i, v_j) \in E$ means the trust relation between individuals v_i and v_j. Let $A = (a_{ij})_{n \times n}$ be adjacency matrix of $G(V,E)$, where $a_{ij} = 1$ if $(v_i, v_j) \in E$ and $a_{ij} = 0$ if $(v_i, v_j) \notin E$.

Definition 1. In $G(V,E)$, a sequence of edges $(v_{o_1}, v_{o_2}), (v_{o_2}, v_{o_3}), \ldots, (v_{o_{n-1}}, v_{o_n})$ is called a path from v_i to v_j if $v_{o_1} = v_i$ and $v_{o_n} = v_j$.

If there exists at least one path for any two notes in $G(V,E)$, then $G(V,E)$ is connected. Let $R = (r_{ij})_{n \times n}$ be the accessibility matrix of $G(V,E)$, where $r_{ij} = 1$ if there exists at least one path between v_i and v_j; otherwise $r_{ij} = 0$. The accessibility matrix can be obtained by Warshall's Algorithm [18].

Let $x_i^t \in R$ be the opinion of the individual v_i $(i = 1, \ldots, n)$ at time t. The individual v_i gives the trust degree $p_i \in (0,1)$ to her own opinion. Meanwhile, the trust degree that individual v_i gives to the individual v_j $(j \in \{1, \ldots, n\}/\{i\})$ is

$$p_{ij} = \frac{(1-p_i)a_{ij}}{\sum_{j=1, j \neq i}^{n} a_{ij}}. \tag{1}$$

The opinion evolution of individual v_i $(i = 1, 2, \ldots, n)$ can be described by

$$x_i^{t+1} = p_i x_i^t + \sum_{j=1, j \neq i}^{n} p_{ij} \times x_j^t, \quad t = 0,1,2, \ldots, \tag{2}$$

based on Eq. (2), the opinion evolution of all individuals $V = \{v_1, \ldots, v_n\}$ is

$$X^{t+1} = P \times X^t, \quad t = 0,1,2, \ldots, \tag{3}$$

where $X^t = (x_1^t, x_2^t, \ldots, x_n^t)^T \in R^n$ and

$$P = \begin{bmatrix} p_1 & p_{12} & \cdots & p_{1n} \\ p_{21} & p_2 & \cdots & p_{2n} \\ \cdots & \cdots & \cdots & \cdots \\ p_{n1} & p_{n2} & \cdots & p_n \end{bmatrix}. \tag{4}$$

Theorem 1. Individuals will form a consensus if $G(V,E)$ is connected. Theorem 1 can be obtained from Lemmas 1, 3 and 4 in Dong et al. [8].

Theorem 2. When individuals in $G(V,E)$ form a consensus, there is a unique weights vector $\pi = (\pi_1, \ldots, \pi_n)$, where $\pi_i \geq 0$ and $\sum_{i=1}^{n} \pi_i = 1$, such that the consensus opinion

$$c = \sum_{i=1}^{n} \pi_i x_i^0, \tag{5}$$

where the weight of individual v_i's opinion

$$\pi_i = \frac{d_i}{1-p_i} / \sum_{i=1}^{n} \frac{d_i}{1-p_i}, \tag{6}$$

here d_i denotes the number of links (i.e., degree) of the individual v_i ($i = 1,2,\ldots,n$) in the social network $G(V,E)$.

Theorem 2 can be easily obtained from Theorem 3 in DeGroot [5].

3. A Procedure for Reaching an Established Consensus Goal

Let $G(V,E,X^0)$ denote a social network with individuals' initial opinions profile $X^0 = (x_1^0, x_2^0, \ldots, x_n^0)^T \in R^n$. Let $\overline{X^t} = (\overline{x_1^t}, \overline{x_2^t}, \ldots, \overline{x_n^t})^T \in R^n$ denote the adjusted initial opinions profile at time t. An optimization-based model for reaching an established consensus goal c with minimum adjustments is

$$\min_{\overline{X^0}} \sum_{i=1}^{n} \left| \overline{x_i^0} - x_i^0 \right| \tag{7}$$

$$\text{subject to } \lim_{t \to \infty} \overline{x_i^t} = c \ (i = 1,2,\ldots,n) \tag{8}$$

To solve the above optimization model, a two-step procedure is proposed:

(i) Network partition. A *Network Partition Algorithm* (see Table 1) is provided to divide $G(V,E,X^0)$ into connected sub-networks.

Table 1. Network Partition Algorithm.

Input: The social network $G(V,E,X^0)$ and its adjacency matrix A.
Output: The sub-networks $S = \{G_{(1)}(V_{(1)}, E_{(1)}, X_{(1)}^0), \ldots, G_{(m)}(V_{(m)}, E_{(m)}, X_{(m)}^0)\}$.
Step 1: Obtaining accessibility matrix $R = (r_{ij})_{n \times n}$ of $G(V,E,X^0)$ and let $k = 1$;
Step 2: If $V = \emptyset$ and $E = \emptyset$, go to Step 4; otherwise, go to next step;
Step 3: Without loss of generality, for any $v_i \in V$ we construct the sub-network $G_{(k)}(V_{(k)}, E_{(k)}, X_{(k)}^0)$ as follows:
$V_{(k)} = \{v_i\} \cup \{v_j | r_{ij} = 1 \text{ and } r_{ji} = 1; v_j \in V\}$,
$E_{(k)} = \{(v_\gamma, v_\tau) | (v_\gamma, v_\tau) \in E; v_\gamma, v_\tau \in V_{(k)}\}$,
$X_{(k)}^0 = \{x_\gamma^0 | x_\gamma^0 \in X^0; v_\gamma \in V_{(k)}\}$.
Let $V = V/V_{(k)}$, $E = E/E_{(k)}$, $k = k+1$ and go to Step 2;
Step 4: End

(ii) Minimizing adjustments of individuals' initial opinions. Let $c_{(k)}$ ($k = 1,2,...,m$) denote the consensus opinion reached by the kth connected sub-network $G_{(k)}(V_{(k)}, E_{(k)}, X_{(k)}^0)$ ($k = 1,2,...,m$), which can be explicitly calculated by Eq. (5). If there exists some $c_{(k)} \neq c$ for $k \in \{1,2,...,m\}$, then adjusting initial opinions of individuals in $G_{(k)}$.

The rules of adjusting initial opinions are described as follows:

Rule 3.1. For $c_{(k)} = c$, individuals' initial opinions in $G_{(k)}$ are not adjusted.

Rule 3.2. For $c_{(k)} > c$, individuals in $G_{(k)}$ are encouraged to decrease their opinions. According to the individuals' decreasing sequence that is determined by their weight values, we successively select an individual to decrease her/his initial opinions until the adjusted consensus opinion $\overline{c_{(k)}} = c$.

Rule 3.3. For $c_{(k)} < c$, individuals in $G_{(k)}$ are encouraged to increase their opinions. According to the individuals' decreasing sequence that is determined by their weight values, we successively select an individual to increase her/his initial opinions until the adjusted consensus opinion $\overline{c_{(k)}} = c$.

Following Theorem 2, it is natural that the adjustments of individuals' initial opinions based on Rules 3.1-3.3 are minimum.

4. Numerical Example

We considered 10 individuals in a social network $G(V, E, X^0)$ as Fig. 1.

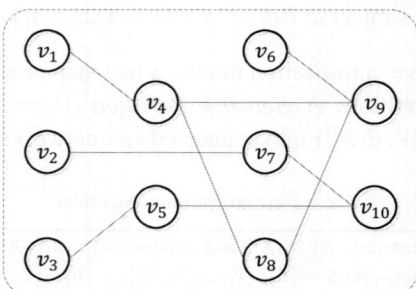

Fig. 1. The social network $G(V, E, X^0)$ in the example.

Initial opinions X^0 and self-trust degrees of individuals p are chosen randomly from the interval [0,1], and degrees of individuals d are obtained from Fig. 1.

$$X^0 = (x_1^0, x_2^0, ..., x_{10}^0)^T$$
$$= (0.13, 0.95, 0.66, 0.29, 0.48, 0.44, 0.73, 0.52, 0.67, 0.59)$$
$$p = (p_1, p_2, ..., p_{10})^T$$
$$= (0.28, 0.91, 0.85, 0.31, 0.57, 0.62, 0.08, 0.42, 0.72, 0.14)$$

$d = (d, d_2, \ldots, d_{10})^T = (1, 1, 1, 3, 2, 1, 2, 2, 3, 2)$

Note 1. The proposed consensus procedure is still feasible and effective when setting different values for the initial opinions, self-trust degrees and other social network structures.

(i) Network Partition

By *Network Partition Algorithm*, $G(V, E, X^0)$ is divided into two sub-networks (see Table 2). Based on this, according to Eq. (6), individuals' opinions weights

$$\pi = (\pi_1, \ldots, \pi_{10})^T$$
$$= (0.06, 0.41, 0.25, 0.19, 0.17, 0.12, 0.08, 0.15, 0.48, 0.09).$$

Furthermore, according to Eq. (5) the consensus opinion $c_{(k)}$ in the kth sub-network $G_{(k)}$ ($k = 1,2$) are obtained (see Table 2).

Table 2. The sub-networks obtained by *Network Partition Algorithm*.

Sub-networks	$V_{(k)}$	$c_{(k)}$
$G_{(1)}(V_{(1)}, E_{(1)}, X^0_{(1)})$	$\{v_1, v_4, v_6, v_8, v_9\}$	0.51
$G_{(2)}(V_{(2)}, E_{(2)}, X^0_{(2)})$	$\{v_2, v_3, v_5, v_7, v_{10}\}$	0.75

(ii) Adjusting initial opinions for reaching an established consensus goal

Here, we set the established goal $c = 0.7$.

(1) For sub-network $G_{(1)}$, because $c_{(1)} = 0.51 < c = 0.7$, then individuals in sub-network $G_{(1)}$ are encouraged to increase opinions. Specifically, weight values $\pi_9 > \pi_4 > \pi_8 > \pi_6 > \pi_1$. Accordingly,

1) When $\overline{x^0_9} = 1$ and $\overline{x^0_i} = x^0_i$ for $v_i \in V_{(1)}/v_9$, obtaining $\overline{c_{(1)}} = 0.67 < c = 0.7$, thus we must continue to increase the opinion of individual v_4;

2) When $\overline{x^0_9} = 1$, $\overline{x^0_4} = 1$ and $\overline{x^0_i} = x^0_i$ for $v_i \in V_{(1)}/(v_9, v_4)$, obtaining $\overline{c_{(1)}} = 0.81 > c = 0.7$, thus the finally adjusted opinions are $\overline{x^0_9} = 1$, $\overline{x^0_4} = 0.45$ and $\overline{x^0_i} = x^0_i$ for $v_i \in V_{(1)}/(v_9, v_4)$, obtaining $\overline{c_{(1)}} = 0.7$. In this case, the adjustments of $G_{(1)}$, i.e., $\sum_{v_i \in V_{(1)}} \left| \overline{x^0_i} - x^0_i \right| = 0.49$.

(2) For sub-network $G_{(2)}$, because $c_{(2)} = 0.75 > c = 0.7$, then individuals in sub-network $G_{(2)}$ are encouraged to decrease opinions. Specifically, weight values $\pi_2 > \pi_3 > \pi_5 > \pi_{10} > \pi_7$. Accordingly, when $\overline{x^0_2} = 0$ and $\overline{x^0_i} = x^0_i$ for $v_i \in V_{(2)}/v_2$, obtaining $\overline{c_{(2)}} = 0.36 < c = 0.7$. Thus the finally adjusted opinions are $\overline{x^0_2} = 0.84$, $\overline{x^0_i} = x^0_i$ for $v_i \in V_{(2)}/v_2$, obtaining $\overline{c_{(2)}} = 0.7$. In this case, the adjustments of $G_{(2)}$, i.e., $\sum_{v_i \in V_{(2)}} \left| \overline{x^0_i} - x^0_i \right| = 0.11$.

Based on this, it is easily to compute the minimum adjustments of all individuals' initial opinions $\sum_{i=1}^{n} \left| \overline{x_i^0} - x_i^0 \right| = 0.6$.

5. Conclusions

This paper discusses that how to minimize the adjustments of individuals' initial opinions to reach a consensus with an established goal.

We find that individuals will form a consensus if the social network is connected, and the consensus opinion is a weighted average value of individuals' initial opinions. Accordingly, we design a two-step procedure to reach an established consensus goal, which guarantees the minimum adjustments of individuals' initial opinions. Moreover, a numerical example is used to demonstrate the validity of our method.

In the future, we argue that it would be necessary to investigate the effect of different social network structures [9, 14, 20] on the minimum adjustments of individuals' initial opinions.

References

1. D. Ben-Arieh and T. Easton, *Decision Support Systems* **43**, 713 (2007).
2. D. Ben-Arieh, T. Easton and B. Evans, *IEEE Transactions on Systems, Man and Cybernetics, Part A: Systems and Humans* **39**, 210 (2009).
3. R.L. Berger, *Journal of the American Statistical Association* **76**, 415 (1981).
4. X. Chen, H.J. Zhang and Y.C. Dong, *Information Fusion* **24**, 72 (2015).
5. M.H. Degroot, *Journal of the American Statistical Association* **69**, 118 (1974).
6. J.C. Dittmer, *Nonlinear Analysis* **47**, 4615 (2001).
7. Y.C. Dong, X. Chen and F. Herrera, *Information Sciences* **297**, 95 (2015).
8. Y.C. Dong, Z.G. Ding, L. Martínez and F. Herrera, *Information Sciences* **397-398**, 187 (2017).
9. Y.C. Dong, M. Zhan, G. Kou, Z.G. Ding and H.M. Liang, *Information Fusion* **43**, 57 (2018).
10. Y.C. Dong, H.J. Zhang and E. Herrera-Viedma, *Knowledge-based Systems* **106**, 206 (2016).
11. R. Durrett, J.P. Gleeson, A.L. Lloyd, P.J. Mucha, F. Shi, D. Sivakoff, J.E.S. Socolarf and C. Varghese, *Proceedings of the National Academy of Sciences* **109**, 3682 (2012).
12. R. Hegselmann and U. Krause, *Journal of Artificial Societies and Social Simulation* **5**, 2 (2002).
13. N.E. Friedkin and E.C. Johnsen, *Journal of Mathematical Sociology* **15**, 193 (1990).
14. M. Jalili, *Physica A: Statistical mechanics and its applications* **392**, 959 (2013).
15. I. Palomares and L. Martínez, *IEEE Transactions on Fuzzy Systems* **22**, 762 (2014).
16. I. Palomares, L. Martínez and F. Herrera, *IEEE Transactions on Fuzzy Systems* **22**, 516 (2014).
17. J. Wu, F. Chiclana and E. Herrera-Viedma, *Applied Soft Computing* **35**, 827 (2015).

18. S. Warshall, *Journal of the Association for Computing Machinery* **9**, 11 (1962).
19. G.Q. Zhang, Y.C. Dong, Y.F. Xu and H.Y. Li, *IEEE Transaction on Systems, Man and Cybernetics, Part A: Systems and Humans* **41**, 1253 (2011).
20. K.J.S. Zollman, *Politics, Philosophy & Economics* **11**, 26 (2012).

A consensus model for large scale using hesitant information

R.M. Rodríguez

Department of Computer Science and A.I., University of Granada
Granada, 18071, Spain
rosam.rodriguez@decsai.ugr.es

L. Martínez

Department of Computer Science, University of Jaén
Jaén, 23071, Spain

G. De Tré

Department of Telecommunications and Information Processing, Ghent University
Ghent, B9000, Belgium

Nowadays due to the technological development, large-scale group decision making problems (LSGDM) are common and they often need to obtain accepted solutions for all experts involved in the problem. To do so, a consensus reaching process (CRP) is applied. A challenge in CRP for LSGDM is to overcome scalability problems. This paper presents a new consensus model to deal with LSGDM that is able to reduce the time cost of the CRP.

Keywords: Large-scale; consensus reaching process; scalability; hesitant fuzzy sets.

1. Introduction

Many real-world problems affect society and might require agreed decisions. In such cases, it is necessary to apply a CRP.[1] However, most of the results obtained in this area are focused on GDM dealing with a few number of experts, but due to the current demands in the society, it is necessary to propose CRPs to deal with LSGDM. Taking into account the main challenges of classical CRPs for LSGDM problems,[2] this contribution proposes a CRP model for LSGDM to overcome the scalability problem related to time cost. The proposal includes the following novelties:

- It detects subgroups of experts according to their preferences and computes the relevance for each subgroup considering its size and cohesion.

- Most of the CRPs aggregate experts' preferences in the early stages of the process, the aggregation might result in a loss of information. To avoid such situations, this proposal will model experts subgroup preferences by hesitant fuzzy sets (HFS).[3]
- It defines a new feedback process that guides the CRP according to the consensus degree achieved.

The contribution is structured as follows: Section 2 revises some concepts. Section 3 presents the new consensus model to deal with LSGDM problems using HFS. Finally, section 4 points out some conclusions.

2. Preliminaries

2.1. *Large scale group decision making*

The concept of LSGDM is quite similar to GDM, but differs because in the former the number of experts who express their assessments is much greater than in the latter.

LSGDM has used solution schemes similar to GDM, such as the selection process. However, this process does not always guarantee that the selected decision is accepted by all experts involved in the problem.[1] A solution to obtain an agreed decision is to apply a CRP which implies that experts modify their preferences making them closer to each other.[4]

In LSGDM, it is also necessary to apply CRPs. Thus, some proposals have been introduced.[4,5] Nevertheless, these proposals aggregate expert's preferences in early stages of the decision process, which implies a loss of information and do not consider different level of agreement across the CRP that can provoke high time cost.

2.2. *Hesitant information*

HFSs[6] model the uncertainty provoked by the doubt that might occur when an expert wants to assign the membership degree of an element to a set.

Definition 2.1.[3] Let X be a reference set, a HFS on X is a function $\mathfrak{h}$ that returns a subset of values in [0,1]: $\mathfrak{h} : X \to \wp([0,1])$.

This definition is completed with the mathematical representation, $A = \{\langle x, h_A(x)\rangle : x \in X\}$, where $h_A(x)$ is called Hesitant Fuzzy Element (HFE) and is a set of some values in [0,1].

Definition 2.2.[7] Let X be a reference set, a hesitant fuzzy preference relation (HFPR) on X is represented by a matrix $H = (h_{ij})_{n\times n} \subset X \times X$, where $h_{ij} = \{\gamma_{ij}^s | s = 1, 2, \ldots, \#h_{ij}\}(\#h_{ij}$ is the number of elements in h_{ij}) is a HFE that indicates the membership degrees that denote to which extent x_i is preferred to x_j. Additionally,

$$\gamma_{ij}^{\sigma(s)} + \gamma_{ji}^{\sigma'(s)} = 1, \quad h_{ii} = \{0.5\}, \quad \#h_{ij} = \#h_{ji}, \quad i, j = \{1, 2, \ldots, n\}$$
$$\gamma_{ij}^{\sigma(s)} < \gamma_{ij}^{\sigma(s+1)}, \quad \gamma_{ji}^{\sigma'(s+1)} < \gamma_{ji}^{\sigma'(s)},$$

where $\{\sigma(1), \ldots, \sigma(\#h_{ij})\}$ is a permutation of $\{1, \ldots, \#h_{ij}\}$, such that, $\gamma_{ij}^{\sigma(s)}$ is the sth smallest element in h_{ij}.

HFS computations sometimes require that the HFEs involved have the same number of elements, to solve this issue, a β-normalization is applied.[8] Let h_j be the shortest one, $h_j^- = min\{\gamma | \gamma \in h_j\}$ and $h_j^+ = max\{\gamma | \gamma \in h_j\}$, the value to be added a number of times such that its length becomes equal to the largest one, is obtained by $\gamma' = \eta h_j^+ + (1 - \eta)h_j^-$, $\eta(0 \leq \eta \leq 1)$.

3. A New Consensus Model for LSGDM

This section presents a novel consensus model for LSGDM that is able to deal with the scalability challenge of a CRP. It consists of 6 main phases.

3.1. *Gathering preferences*

Each expert $e_r \in E$, provides his/her opinions on X by means of a fuzzy preference relation (FPR), $P^r = (p_{ij}^r)_{n \times n}$, that is reciprocal $p_{ij}^r + p_{ji}^r = 1$, $i, j \in 1, \ldots, n$.

3.2. *Framework configuration*

In a LSGDM problem there are a set of possible alternatives $X = \{x_1, \ldots, x_n\}$ and a large number of experts $E = \{e_1, \ldots, e_m\}$ involved in the problem, being $m >> n$. Three parameters are established in the CRP: (i) a consensus threshold, $\vartheta \in [0, 1]$, (ii) a parameter used in the adaptive feedback process, $\delta \in [0, 1]$, $\delta < \vartheta$ and (iii) the maximum number of rounds, Max_round.

3.3. *Managing subgroups*

3.3.1. *Subgroups identification*

This phase detects subgroups of experts according to their similar preferences by using a k-means based algorithm.

(1) Initially, there is a cluster for each alternative, $C = \{C_1, \ldots, C_n\}$.

(2) A centroid c^l is computed for each cluster.
(3) The distance between each FPR, P^r, and the centroid c^l, $l \in \{1, \ldots, n\}$, is calculated by a distance measure, e.g. the Euclidean distance.
(4) The preference relation P^r is assigned to the cluster for which, the distance between P^r and the centroid, is minimum.
$C^{l(t)} = \{P^r : d(P^r, c^{l(t)}) \leq d(P^r, c^{z(t)}), \forall 1 \leq z \leq n\}$
(5) New centroids are computed, $c_{ij}^{l(t+1)} = \frac{1}{|C^{l(t)}|} \sum_{P^r \in C^{l(t)}} p_{ij}^r$, $i, j \in \{1, \ldots, n\}$, where $|C^{l(t)}|$ is the number of preference relations that belong to the cluster C^l during iteration t.
(6) Repeat steps (3)-(5) until the assignments to the clusters do not change.

3.3.2. Managing subgroups hesitation

To keep as much information as possible in the CRP, unlike of oversimplifying the preference modelling with aggregation procedures, our proposal considers that the different experts' preferences elicited in the subgroup, despite of being similar, show a kind of hesitation in the group. Let $G^l = \{e_1^l, \ldots, e_r^l\}$ be the subgroup of experts belonging to cluster, C^l, with FPRs, $P^{l1} = (p_{ij}^{l1})_{n \times n}, \ldots, P^{lr} = (p_{ij}^{lr})_{n \times n}$. A HFPR, $HP^l = (h_{ij}^l)_{n \times n}$, $l \in \{1, \ldots, n\}$, that fuses all experts' preferences in G^l, is built such that, $h_{ij}^l = \{p_{ij}^{lk} | k = 1, 2, \ldots, |G^l|\}$ where $|G^l|$ is the cardinality of G^l.

3.3.3. Weighting subgroups

The relevance of a subgroup is based on its size (i.e., number of experts in the subgroup) and its cohesion (i.e. the level of togetherness among experts).

A geometric approach to compute the *cohesion* is defined:

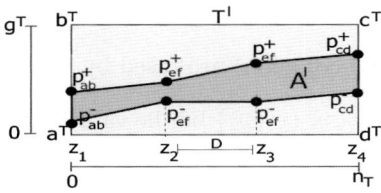

Fig. 1. Graphical representation of the cohesion of a subgroup.

(1) Let T^l be the total area of the rectangle consisting of the points a^T, b^T, c^T and d^T (see Fig. 1), i.e., $T^l = g^T \times n^T$.

(2) Let $I = \bigcup_{i,j \in n, i \neq j} \{(i,j)\}$ be a set with all the possible combinations over the set of alternatives $X = \{x_1, \ldots, x_n\}$. The minimum and maximum assessments for each pair of alternatives taking into account all the preferences of the subgroup G^l are obtained as follows.
$\gamma_{ij}^- = min\{\gamma_{ij}^1, \gamma_{ij}^2, \ldots, \gamma_{ij}^s\}, \forall (i,j) \in I$
$\gamma_{ij}^+ = max\{\gamma_{ij}^1, \gamma_{ij}^2, \ldots, \gamma_{ij}^s\}, \forall (i,j) \in I$
The first and last pair of alternatives considered on the X-axis are,
$\gamma_{ab}^- = min_{i,j \in I}\{\gamma_{ij}^-\}, (a,b) \in I$
$\gamma_{cd}^+ = max_{i,j \in I}\{\gamma_{ij}^-\}, (c,d) \in I$
A function f is defined to obtain the indexes of the pairs of alternatives, $f : \{z_1, z_2, \ldots, z_{n(n-1)}\} \to I$.
The area A^l, between the maximum and minimum assessments ordered on the X-axis by the minimum is computed by,

$$A^l = \left[\sum_{i,j \in I} (\gamma_{ij}^+ - \gamma_{ij}^-) - \frac{(\gamma_{ab}^+ - \gamma_{ab}^-) + (\gamma_{cd}^+ - \gamma_{cd}^-)}{2} \right] \cdot D \quad (1)$$

where D is the distance between z_i and z_{i+1}, which in our case is 1.

(3) Finally, the cohesion is given by, $cohesion(G^l) = 1 - \frac{A^l}{T^l}$.

The value of the *size* is obtained from the subgroup identification and is modelled through a fuzzy membership function μ_{size} as shown in Figure 2.

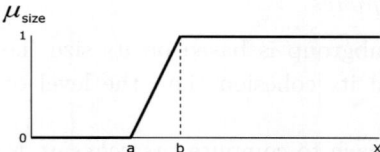

Fig. 2. Membership Function.

The values (size and cohesion) are fused by the convex combination.

Definition 3.1. Let $Y_{G^l} = \{y_1, y_2\}$ be the values obtained for the cohesion and size, $y_1, y_2 \in [0,1]$, of the subgroup G^l which are aggregated. The convex combination of Y_{G^l} is given by,

$$\varphi(Y_{G^l}) = y_1 \cdot \alpha_1 + y_2 \cdot \alpha_2 \quad (2)$$

being $\alpha = \{\alpha_1, \alpha_2\}$ a weighting vector, $\alpha_i \in [0,1], i = \{1,2\}$ and $\sum_i \alpha_i = 1$.

3.4. Computing the consensus degree

The consensus degree among experts is computed in 3 steps.

(1) For each pair of subgroups G^l and G^k, a similarity matrix $SM^{lk} = (sm^{lk}_{ij})_{n \times n}$ is computed, $sm^{lk}_{ij} = 1 - d(h^l_{ij}, h^k_{ij})$, where d is a distance measure for HFEs.[9]
(2) A consensus matrix, $CM = (cm_{ij})_{n \times n}$, is obtained by aggregating the similarity matrices by the arithmetic mean, $cm_{ij} = \frac{\sum_{l=1}^{n-1} \sum_{k=l+1}^{n} sm^{lk}_{ij}}{n(n-1)/2}$.
(3) The consensus degree is computed by, $cr = \frac{\sum_{i=1}^{n} \sum_{j=1, i \neq j}^{n} cm_{ij}}{n(n-1)}$.

3.5. Consensus control

The consensus degree cr is compared with the consensus threshold $\vartheta \in [0, 1]$. If $cr \geq \vartheta$, then the consensus process ends, otherwise more discussion rounds are necessary. The maximum number of discussion rounds is controlled by the parameter Max_round.

3.6. Adaptive feedback process

The rules for the advice generation depend on the consensus level achieved, cr, that determines whether the consensus level is "high" or "low". The feedback process consists of 3 steps.

- Obtain a collective matrix, HP^C, by aggregating the preferences represented by normalized HFPRs $\{\overline{HP}^1, \ldots, \overline{HP}^n\}$ (see Def. 2.2).
- Compute the proximity between each subgroup $\{\overline{HP}^1, \ldots, \overline{HP}^n\}$, and the collective matrix HP^C, by using a similarity measure for HFSs,[9] $pr^l = sim(HP^C, \overline{HP}^l) = 1 - d(HP^C, \overline{HP}^l)$.
- Adapt the feedback process according to reached consensus degree cr.

Group feedback process: If the consensus degree $cr < \delta$, this means that the consensus is "low". Then, all experts of the furthest subgroups are recommended to modify their preferences. This is done as follows.

(1) If $pr^l \leq \overline{pr}$, then the subgroup G^l is selected, $\overline{pr} = \frac{1}{n} \sum_{l=1}^{n} pr^l$.
(2) If $pr^l_{ij} \leq \overline{pr}_{ij}$, then the pair of alternatives (x_i, x_j) is selected, $\overline{pr}_{ij} = \frac{1}{n} \sum_{l=1}^{n} pr^l_{ij}$ and $pr^l_{ij} = 1 - d(h^C_{ij}, h^l_{ij})$

If $(v^l_{ij}) < (v^C_{ij})$, then all experts who belong to the subgroup G^l should increase their preferences over the pair of alternatives (x_i, x_j) and if $(v^l_{ij}) > (v^C_{ij})$, they should decrease them.

Let v_{ij}^l and v_{ij}^C be calculated as follows, $v_{ij}^l = \frac{1}{\#h} \sum_{s=1}^{\#h} \gamma_{ij}^{l,s}$ and $v_{ij}^C = \frac{1}{\#h} \sum_{s=1}^{\#h} \gamma_{ij}^{C,s}$.

Individual feedback process: If the consensus degree, $\delta \leq cr < \vartheta$, then the consensus level is "high".

(1) If $pr^l \leq \overline{pr}$, then the subgroup G^l is selected.
(2) If $pr_{ij}^l \leq \overline{pr}_{ij}$, then the pair of alternatives (x_i, x_j) is selected.
(3) If $pr_{ij}^{lr} \leq pr_{ij}^l = \{(r)|(1 - |v_{ij}^C - \gamma_{ij}^{lr}|) \leq pr_{ij}^l\}$, the expert e_r is selected.

- If $(\gamma_{ij}^{lr}) < (v_{ij}^C)$, then the expert e_r who belongs to the subgroup G^l should increase his/her preference over the pair of alternatives (x_i, x_j) and if $(\gamma_{ij}^{lr}) > (v_{ij}^C)$, then he should decrease it.
- If $(\gamma_{ij}^{lr}) = (v_{ij}^C)$, then it is not necessary to make changes.

4. Conclusions

LSGDM problems are common and agreed decisions are more appreciated, thus CRPs are necessary. Because current approaches are time consuming, scalability problems are a challenge. Therefore, a new CRP model dealing with LSGDM is proposed to overcome these problems.

Acknowledgments

The work was supported by the research project TIN2015-66524-P, Postdoctoral fellow (IJCI-2015-23715), Spanish mobility program Jose Castillejo (CAS15/00047) and ERDF.

References

1. S. Saint and J. R. Lawson, *Rules for Reaching Consensus. A Modern Approach to Decision Making* (Jossey-Bass, 1994).
2. Á. Labella, Y. Liu, R. M. Rodríguez and L. Martínez, *Applied Soft Computing*, p. DOI 10.1016/j.asoc.2017.05.045 (2017).
3. V. Torra, *International Journal of Intelligent Systems* **25**, 529 (2010).
4. I. Palomares, F. Estrella, L. Martínez and F. Herrera, *Information Fusion* **20**, 252 (2014).
5. I. Palomares, L. Martínez and F. Herrera, *IEEE Transactions on Fuzzy Systems* **22**, 516 (2014).
6. R. Rodríguez, B. Bedregal, H. Bustince, Y. Dong, B. Farhadinia, C. Kahraman, L. Martínez, V. Torra, Y. Xu, Z. Xu and F. Herrera, *Information Fusion* **29**, 89 (2016).
7. B. Zhu and Z. Xu, *Technol Econ Dev Eco* **19**, S214 (2013).
8. B. Zhu, Z. Xu and J. Xu, *IEEE Transactions on Cybernetics* **44**, 1328 (2014).
9. R. Rodríguez, L. Martínez, V. Torra, Z. Xu and F. Herrera, *International Journal of Intelligent Systems* **29**, 495 (2014).

An optimization-based approach to aggregating multi-granular hesitant fuzzy linguistic term sets

Zhen Zhang

Institute of Systems Engineering, Dalian University of Technology
Dalian, 116024, China
zhen.zhang@dlut.edu.cn

Wenyu Yu

Institute of Information Management and Information Systems,
Dalian University of Technology
Dalian, 116024, China
ywyeva@mail.dlut.edu.cn

> The hesitant fuzzy linguistic term set (HFLTS) is an effective tool for decision makers to elicit their linguistic assessments. In the literature, different aggregation operators have been proposed for HFLTSs. However, there are few studies that focus on the aggregation of multi-granular HFLTSs. In this paper, it is proposed an optimization model to aggregate multi-granular HFLTSs, which minimizes the deviation between the weighted collective assessment and the aggregation result. The aggregation result of the proposed approach is an HFLTS, which can provide interpretable results for decision makers. A numerical example is utilized to demonstrate the proposed approach.
>
> *Keywords*: Hesitant fuzzy linguistic term set; aggregation; multi-granular linguistic information; optimization.

1. Introduction

Computing with words (CW), which refers to a methodology for reasoning, computing and making decisions using information described in natural language,[1] has been widely studied in the literature. In the past few decades, different linguistic computational models have been proposed for CW, such as the membership function-based model,[2] the ordinal scale model,[3] the linguistic 2-tuple model,[4] the type-2 fuzzy set based model[5] and the numerical scale model.[6]

Classical linguistic computational models only allow decision makers to use a simple linguistic term to elicit their linguistic assessments, which cannot model decision makers' hesitancy for some complex decision making problems.[7] Rodríguez et al.[8] proposed the concept of hesitant fuzzy linguistic term sets (HFLTSs) which allowed the linguistic assessments elicited by decision makers to be a set of consecutive linguistic terms.

Recently, different aggregation models have been proposed to aggregate HFLTSs in the literature. For instance, Rodríguez et al.[8] defined the min_upper and max_lower operators to aggregate HFLTSs and proposed an approach to multi-criteria decision making based on HFLTSs. Liu and Rodríguez[9] defined the fuzzy envelope of an HFLTS which was a trapezoidal fuzzy number (TFN). Wei et al.[10] proposed some weighted averaging operators and ordered weighted averaging operators for HFLTSs based on the convex combination operation of linguistic labels. Zhang and Wu[11] defined some hesitant fuzzy linguistic aggregation operators based on the virtual linguistic model. However, Zhang and Guo[12] pointed out the drawbacks of the operators defined in Refs. 10–11, and developed some new operators which aggregated a collection of HFLTSs into a linguistic distribution assessment. Dong et al.[13] defined some novel operators based on the mixed 0-1 linear programming model to aggregate unbalanced HFLTSs into an extended HFLTS.

One can find that some progress has been made to aggregate HFLTSs. However, to the best of our knowledge, there are few studies focusing on how to aggregate HFLTSs when multi-granular linguistic term sets are used to elicit HFLTSs due to the difference of culture and knowledge background of decision makers.[14–16] Therefore, there is still a need to investigate the aggregation of multi-granular HFLTSs. In this paper, an optimization-based approach is proposed to carry out the aggregation process, which first transforms multi-granular HFLTSs into TFNs and obtains an HFLTS from a set of possible HFLTSs by minimizing the deviation between the aggregated TFN and the fuzzy envelope of each possible HFLTS. The main advantage of the proposed approach is that the aggregation result of multi-granular HFLTSs is also HFLTS, which provides interpretable aggregation results for decision makers and can be used to provide feedbacks in consensus reaching processes when multi-granular HFLTSs are elicited by decision makers for a group decision making problem.

The rest of this paper is organized as follows. Section 2 provides some preliminaries. In Section 3, an optimization-based approach is developed to aggregate multi-granular HFLTSs. In Section 4, a numerical example is utilized to demonstrate the proposed approach. Finally, this paper is concluded in Section 5.

2. Preliminaries

In this section, some preliminaries related to TFNs and HFLTSs are revised. First, let us recall the definition of a TFN.

Definition 2.1.[17] A fuzzy number $\tilde{A} = T(a, b, c, d)$ is called a TFN if its membership function is given by

$$\mu_{\tilde{A}}(x) = \begin{cases} (x-a)/(b-a) & a \leq x \leq b \\ 1 & b < x < c \\ (d-x)/(d-c) & c \leq x \leq d \\ 0 & \text{otherwise.} \end{cases} \quad (1)$$

Given two TFNs $\tilde{A}^1 = T(a^1, b^1, c^1, d^1)$, $\tilde{A}^2 = T(a^2, b^2, c^2, d^2)$ and a real number $\lambda > 0$, then two main operations of $\tilde{A}^1$ and $\tilde{A}^2$ can be given as follows:[17]

(1) $\tilde{A}^1 \oplus \tilde{A}^2 = T(a^1 + a^2, b^1 + b^2, c^1 + c^2, d^1 + d^2)$;
(2) $\lambda \otimes \tilde{A}^1 = T(\lambda a^1, \lambda b^1, \lambda c^1, \lambda d^1)$.

Moreover, the Hamming distance between two TFNs $\tilde{A}^1 = T(a^1, b^1, b^1, d^1)$ and $\tilde{A}^2 = T(a^2, b^2, c^2, d^2)$ can be defined as

$$d(\tilde{A}^1, \tilde{A}^2) = \frac{1}{4} \left(|a^1 - a^2| + |b^1 - b^2| + |c^1 - c^2| + |d^1 - d^2| \right). \quad (2)$$

Let $S = \{s_0, s_1, \ldots, s_g\}$ denote a linguistic term set with odd cardinality, the element s_i of which represents the i-th linguistic term in S, and $g+1$ is the cardinality of the linguistic term set S. The linguistic term set should satisfy the following characteristics:[18] (1) The set is ordered: $s_i > s_j$, if $i > j$; (2) There is a negation operator: $\text{Neg}(s_i) = s_j$, such that $j = g - i$.

Rodríguez et al.[8] defined the HFLTS as follows.

Definition 2.2. Let $S = \{s_0, s_1, \ldots, s_g\}$ denote a linguistic term set. An HFLTS H^S on S is an ordered finite subset of consecutive linguistic terms in S.

Liu and Rodríguez[9] proposed the fuzzy envelop of an HFLTS, which represents such linguistic expression using a TFN.

Definition 2.3. Let $H^S = \{s_i, s_{i+1}, \ldots, s_j\}$ be an HFLTS defined on a linguistic term set $S = \{s_0, s_1, \ldots, s_g\}$ such that $s_k \in S$, $k = i, i+1, \ldots, j$, then the fuzzy envelope of H^S is a TFN, denoted by

$$env_F(H^S) = T(a, b, c, d). \quad (3)$$

Details about how to calculate the fuzzy envelop of an HFLTS can be found in Ref. 9.

3. The Proposed Approach

In this section, an optimization-based approach is developed to aggregate multi-granular HFLTSs. Before proposing the model, some notations that will be used in the rest of this paper are defined as follows.

Let $\mathcal{S} = \{S^{g(1)+1}, S^{g(2)+1}, \ldots, S^{g(n)+1}\}$ be a set of linguistic term sets, where $S^{g(i)+1} = \{s_0^{g(i)+1}, s_1^{g(i)+1}, \ldots, s_{g(i)}^{g(i)+1}\}$ is a linguistic term set whose granularity is $g(i) + 1$, $i = 1, 2, \ldots, n$. Assume that there is a set of multi-granular HFLTSs $\{H_1^{S^{g(1)+1}}, H_2^{S^{g(2)+1}}, \ldots, H_n^{S^{g(n)+1}}\}$, where $H_i^{S^{g(i)+1}}$ is an HFLTS defined on the linguistic term set $S^{g(i)+1}$, $i = 1, 2, \ldots, n$, and there is also a weight vector $w = (w_1, w_2, \ldots, w_n)^{\mathrm{T}}$ such that $\sum_{i=1}^{n} w_i = 1$, $w_i \geq 0$, $i = 1, 2, \ldots, n$.

First, one can use the fuzzy envelop function to transform each HFLTS $H_i^{S^{g(i)+1}}$ into its fuzzy envelop $\tilde{A}_i$ as

$$\tilde{A}_i = T(a_i, b_i, c_i, d_i) = env_F(H_i^{S^{g(i)+1}}), i = 1, 2, \ldots, n. \tag{4}$$

Clearly, one can aggregate all the individual TFNs into a collective one $\tilde{A}$ by

$$\tilde{A} = T(a, b, c, d) = \oplus_{i=1}^{n} w_i \tilde{A}_i = T\left(\sum_{i=1}^{n} w_i a_i, \sum_{i=1}^{n} w_i b_i, \sum_{i=1}^{n} w_i c_i, \sum_{i=1}^{n} w_i d_i\right). \tag{5}$$

Given an integer N^l, which denotes the maximum number of the linguistic terms that the decision maker wants to use when eliciting HFLTSs based on the linguistic term set $S^{g(l)+1}$, one can generate the set of possible HFLTSs for $S^{g(l)+1}$. Let $\mathcal{H}_{N^l}$ be the set of possible HFLTSs for the l-th decision maker, then one has

$$\mathcal{H}_{N^l} = \{h \in H^{S^{g(l)+1}} | \#h \leq N^l\}. \tag{6}$$

Theorem 3.1. $\#\mathcal{H}_{N^l} = \dfrac{N^l(2g(l) + 3 - N^l)}{2}$.

Let us denote $\mathcal{H}_{N^l} = \{h_1, h_2, \ldots, h_{\#\mathcal{H}_{N^l}}\}$ and the set of the fuzzy envelops of the HFLTSs in $\mathcal{H}_{N^l}$ can be obtained by the fuzzy envelop function $env_F(\cdot)$ as $env(\mathcal{H}_{N^l}) = \{env_F(h_1), env_F(h_2), \ldots, env_F(h_{\#\mathcal{H}_{N^k}})\}$, where $env_F(h_j) = T(a^{h(j)}, b^{h(j)}, c^{h(j)}, d^{h(j)})$ is a TFN, $j = 1, 2, \ldots, \#\mathcal{H}_{N^k}$.

Theorem 3.2. *Let $\mathcal{H}_{N^l}$ and $env(\mathcal{H}_{N^l})$ be defined as before, then the fuzzy envelop of an HFLTS in $\mathcal{H}_{N^l}$ can be denoted by*

$$T(\bar{a},\bar{b},\bar{c},\bar{d}) = T\left(\sum_{j=1}^{\#\mathcal{H}_{N^l}} x_j a^{h(j)}, \sum_{j=1}^{\#\mathcal{H}_{N^l}} x_j b^{h(j)}, \sum_{j=1}^{\#\mathcal{H}_{N^l}} x_j c^{h(j)}, \sum_{j=1}^{\#\mathcal{H}_{N^l}} x_j d^{h(j)}\right), \tag{7}$$

where $x_j \in \{0,1\}$, $j = 1, 2, \ldots, \#\mathcal{H}_{N^l}$ and $\sum_{j=1}^{\#\mathcal{H}_{N^l}} x_j = 1$.

Proof. As $x_j \in \{0,1\}$, $j = 1, 2, \ldots, \#\mathcal{H}_{N^l}$, if $\sum_{j=1}^{\#\mathcal{H}_{N^l}} x_j = 1$, then there must exists an $x_k = 1$ and $x_j = 0$, $\forall j \neq k$. Therefore,

$$\bar{a} = \sum_{j=1}^{\#\mathcal{H}_{N^l}} x_j^{h(j)} = x_k a^{h(k)} + \sum_{j=1, j\neq k}^{\#\mathcal{H}_{N^l}} x_j^{h(j)} = a^{h(k)} + 0 = a^{h(k)}.$$

Similarly, we have

$$\bar{b} = b^{h(k)}, \bar{c} = c^{h(k)}, \bar{d} = d^{h(k)}.$$

As a result, $T(\bar{a},\bar{b},\bar{c},\bar{d}) = T(a^{h(k)}, b^{h(k)}, c^{h(k)}, d^{h(k)})$. □

As our aim is to obtain interpretable aggregation result, it is assumed that the output of the aggregation is an HFLTS $H^{S^{g(l)+1}}$, where l can be selected from $\{1, 2, \ldots, n\}$. For convenience, let $env_F(H^{S^{g(l)+1}}) = T(\bar{a},\bar{b},\bar{c},\bar{d})$. Clearly, it is hoped that the deviation between the collective assessment $\tilde{A} = T(a,b,c,d)$ and $env_F(H^{S^{g(l)+1}})$ should be kept as small as possible. Keeping this idea in mind, one can establish the following optimization model:

$$\min J = \frac{1}{4}(|a - \bar{a}| + |b - \bar{b}| + |c - \bar{c}| + |d - \bar{d}|)$$

$$\text{s.t. } a = \sum_{i=1}^{n} w_i a_i, b = \sum_{i=1}^{n} w_i b_i$$

$$c = \sum_{i=1}^{n} w_i c_i, d = \sum_{i=1}^{n} w_i d_i$$

$$\bar{a} = \sum_{j=1}^{\#\mathcal{H}_{N^l}} x_j a^{h(j)}, \bar{b} = \sum_{j=1}^{\#\mathcal{H}_{N^l}} x_j b^{h(j)} \tag{M-1}$$

$$\bar{c} = \sum_{j=1}^{\#\mathcal{H}_{N^l}} x_j c^{h(j)}, \bar{d} = \sum_{j=1}^{\#\mathcal{H}_{N^l}} x_j d^{h(j)}$$

$$\sum_{j=1}^{\#\mathcal{H}_{N^l}} x_j = 1, x_j \in \{0,1\}, j = 1, 2, \ldots, \#\mathcal{H}_{N^l}.$$

Theorem 3.3. *By introducing some auxiliary variables f_1, f_2, f_3 and f_4, the model* (M-1) *can be transformed into the following model:*

$$\min J = \frac{1}{4}(f_1 + f_2 + f_3 + f_4)$$

$$\text{s.t. } f_1 \geq a - \overline{a}, f_1 \geq \overline{a} - a$$

$$f_2 \geq b - \overline{b}, f_2 \geq \overline{b} - b$$

$$f_3 \geq c - \overline{c}, f_3 \geq \overline{c} - c$$

$$f_4 \geq d - \overline{d}, f_4 \geq \overline{d} - d$$

$$a = \sum_{i=1}^{n} w_i a_i, b = \sum_{i=1}^{n} w_i b_i \qquad \text{(M-2)}$$

$$c = \sum_{i=1}^{n} w_i c_i, d = \sum_{i=1}^{n} w_i d_i$$

$$\overline{a} = \sum_{j=1}^{\#\mathcal{H}_{N^l}} x_j a^{h(j)}, \overline{b} = \sum_{j=1}^{\#\mathcal{H}_{N^l}} x_j b^{h(j)}$$

$$\overline{c} = \sum_{j=1}^{\#\mathcal{H}_{N^l}} x_j c^{h(j)}, \overline{d} = \sum_{j=1}^{\#\mathcal{H}_{N^l}} x_j d^{h(j)}$$

$$\sum_{j=1}^{\#\mathcal{H}_{N^l}} x_j = 1, x_j \in \{0,1\}, j = 1, 2, \ldots, \#\mathcal{H}_{N^l}.$$

Proof. By the first four constraints in the model (M-2), we have $f_1 \geq |a - \overline{a}|$, $f_2 \geq |b - \overline{b}|$, $f_3 \geq |c - \overline{c}|$ and $f_4 \geq |d - \overline{d}|$. As the objective is to minimize $f_1 + f_2 + f_3 + f_4$, any solution that satisfies $f_1 > |a - \overline{a}|$, $f_2 > |b - \overline{b}|$, $f_3 > |c - \overline{c}|$ and $f_4 > |d - \overline{d}|$ is not the optimal solution to the model (M-2). Therefore, $f_1 = |a - \overline{a}|$, $f_2 = |b - \overline{b}|$, $f_3 = |c - \overline{c}|$ and $f_4 = |d - \overline{d}|$. By combining the rest constraints, the model (M-1) can be transformed into the model (M-2). □

The model (M-2) is a mixed 0-1 linear programming model, which can be easily solved by some optimization software packages. According to the solution, the corresponding HFLTS $H^{S^{g(l)+1}}$ can be derived.

4. A Numerical Example

Let $S^5 = \{s_0^5 : \text{poor}, s_1^5 : \text{slightly poor}, s_2^5 : \text{fair}, s_3^5 : \text{slightly good}, s_4^5 : \text{good}\}$, $S^7 = \{s_0^7 : \text{very poor}, s_1^7 : \text{poor}, s_2^7 : \text{slightly poor}, s_3^7 : \text{fair}, s_4^7 : \text{slightly good}, s_5^7 : \text{good}, s_6^7 : \text{very good}\}$, $S^9 = \{s_0^9 : \text{extremely poor}, s_1^9 : \text{very poor}, s_2^9 : \text{poor}, s_3^9 : \text{slightly poor}, s_4^9 : \text{fair}, s_5^9 : \text{slightly good}, s_6^9 : \text{good}, s_7^9 : \text{very good}, s_8^9 : \text{extremely good}\}$ be three linguistic term sets and $H_1^{S^{g(1)+1}} = H_1^5 = \{s_1^5, s_2^5\}$, $H_2^{S^{g(2)+1}} = H_2^{S^5} = \{s_2^5, s_3^5, s_4^5\}$, $H_3^{S^{g(3)+1}} = H_3^{S^7} = \{s_1^7, s_2^7, s_3^7\}$, $H_4^{S^{g(4)+1}} = H_4^{S^7} = \{s_5^7, s_6^7\}$,

$H_5^{S^{g(5)+1}} = H_5^{S^9} = \{s_6^9, s_7^9, s_8^9\}$. Moreover, let the weight vector be $w = (0.2, 0.3, 0.15, 0.1, 0.25)^T$.

First, one can transform multi-granular HFLTSs into their fuzzy envelops as

$$env_F(H_1^5) = T(0, 0.25, 0.5, 0.75), \ env_F(H_2^5) = T(0.25, 0.6875, 1, 1),$$

$$env_F(H_3^7) = T(0, 0.3, 0.3667, 0.6667), \ env_F(H_4^7) = T(0.6667, 0.9722, 1, 1),$$

$$env_F(H_5^9) = T(0.625, 0.9141, 1, 1).$$

Assume that decision makers who use the linguistic term set S^5 provide a value of the maximum number of the linguistic terms for an HFLTS as 3, then the set of possible HFLTSs for S^5 can be obtained as $\{\{s_0^5\}, \{s_1^5\}, \{s_2^5\}, \{s_3^5\}, \{s_4^5\}, \{s_0^5, s_1^5\}, \{s_1^5, s_2^5\}, \{s_2^5, s_3^5\}, \{s_3^5, s_4^5\}, \{s_0^5, s_1^5, s_2^5\}, \{s_1^5, s_2^5, s_3^5\}, \{s_2^5, s_3^5, s_4^5\}\}$.

By solving the model (M-2), one has that $x_8 = 1$ and $x_k = 0$, $\forall k \in \{1, 2, \ldots, 12\} \backslash 8$. As a result, the aggregation result is $\{s_2^5, s_3^5\}$.

Similarly, assume that the decision makers who use the linguistic term set S^7 provide a value of the maximum number of the linguistic terms for an HFLTS as 3, then one can obtain the aggregation result as $\{s_3^7, s_4^7, s_5^7\}$.

If the decision maker who use the linguistic term set S^9 provides a value of the maximum number of the linguistic terms for an HFLTS as 4, then one can obtain the aggregation result as $\{s_4^9, s_5^9, s_6^9, s_7^9\}$.

Obviously, the aggregation results obtained by the proposed method are HFLTSs, which provide interpretable results and can be returned to decision makers for reference in consensus reaching processes.

5. Conclusions

In this paper, a novel optimization model based approach is proposed to aggregate multi-granular HFLTSs. In the proposed approach, multi-granular HFLTSs are first transformed into TFNs using the fuzzy envelop function and then aggregated into a collective opinion. After that, an optimization model which minimizes the deviation between the collective opinion and the alternative HFLTS is developed. By solving the model, the aggregation result can be obtained as an HFLTS, which follows the paradigm of CW and can provide interpretable results for decision makers.

In terms of future research, we intend to extend the proposed method to aggregate HFLTSs defined on unbalanced linguistic term sets.[19,20] Also it is interesting to apply the proposed approach to consensus reaching in group decision making with multi-granular HFLTSs.[21,22]

Acknowledgments

This work was partly supported by the National Natural Science Foundation of China (Nos. 71501023, 71771034), the Funds for Creative Research Groups of China (No. 71421001), the China Postdoctoral Science Foundation (2015M570248) and the Fundamental Research Funds for the Central Universities (DUT17RC(4)11).

References

1. J. M. Mendel, L. A. Zadeh, E. Trillas, R. Yager, J. Lawry, H. Hagras and S. Guadarrama, *IEEE Comput. Intell. Mag.* **5**, 20 (2010).
2. Z. Zhang and C. Guo, *Knowledge-Based Syst.* **26**, 111 (2012).
3. R. R. Yager, *Int. J. Approx. Reasoning* **12**, 237 (1995).
4. F. Herrera and L. Martínez, *IEEE Trans. Fuzzy Syst.* **8**, 746 (2000).
5. J. M. Mendel and H. Wu, *IEEE Trans. Fuzzy Syst.* **15**, 301 (2007).
6. Y. Dong and E. Herrera-Viedma, *IEEE T. Cybern.* **45**, 780 (2015).
7. R. M. Rodríguez, A. Labella and L. Martínez, *Int. J. Comput. Intell. Syst.* **9**, 81 (2016).
8. R. M. Rodríguez, L. Martínez and F. Herrera, *IEEE Trans. Fuzzy Syst.* **20**, 109 (2012).
9. H. Liu and R. M. Rodríguez, *Inf. Sci.* **258**, 220 (2014).
10. C. Wei, N. Zhao and X. Tang, *IEEE Trans. Fuzzy Syst.* **22**, 575 (2014).
11. Z. Zhang and C. Wu, *J. Intell. Fuzzy Syst.* **26**, 2185 (2014).
12. Z. Zhang and C. Guo, New operations of hesitant fuzzy linguistic term sets with applications in multi-attribute group decision making, in *Proc. of the 2015 IEEE Int. Conf. on Fuzzy Systems*, (Istanbul, Turkey, 2015).
13. Y. Dong, C.-C. Li and F. Herrera, *Inf. Sci.* **367–368**, 259 (2016).
14. F. Herrera and L. Martínez, *IEEE Trans. Syst. Man Cybern. Part B-Cybern.* **31**, 227 (2001).
15. J. A. Morente-Molinera, I. J. Pérez, M. R. Ureña and E. Herrera-Viedma, *Knowledge-Based Syst.* **74**, 49 (2015).
16. Z. Zhang, C. Guo and L. Martínez, *IEEE Trans. Syst. Man Cybern. -Syst.* **47**, 3063 (2017).
17. A. Kaufmann and M. M. Gupta, *Introduction to Fuzzy Arithmetic: Theory and Application* (Van Nostrand Reinhold, New York, 1991).
18. R. M. Rodríguez and L. Martínez, *Int. J. Gen. Syst.* **42**, 121 (2013).
19. W. Yu, Q. Zhong and Z. Zhang, Fusing multi-granular unbalanced hesitant fuzzy linguistic information in group decision making, in *Proc. of the 2016 IEEE Int. Conf. on Fuzzy Systems*, (Vancouver, Canada, 2016).
20. W. Yu, Z. Zhang, Q. Zhong and L. Sun, *Comput. Ind. Eng.* **114**, 316 (2017).
21. I. Palomares, F. J. Estrella, L. Martínez and F. Herrera, *Inf. Fusion* **20**, 252 (2014).
22. Z. Zhang and C. Guo, *Int. J. Syst. Sci.* **47**, 2572 (2016).

The effect of opinion evolution in consensus reaching in group decision making

Quanbo Zha and Yucheng Dong

Business School, Sichuan University, Chengdu 610065, China

Consensus reaching process (CRP) is often a necessity in real-world group decision making (GDM) problems. Under the interactions among decision makers in CRP, the opinions of them would evolve. However, the extant CRPs have ignored the opinion evolution caused by interactions. In this paper a novel CRP is proposed, which considers the opinion evolution of decision makers with social relationships. Finally, a numerical analysis is used to illustrate the effect of opinion evolution to support our model.

1. Introduction

A group decision making (GDM) problem is the situation that a group of decision makers attempt to choose a best alternative from several alternatives [1,2], which generally includes two process [3-5]: consensus reaching process (CRP), and selection process. The CRP is often a necessity in GDM to achieve a consensus. The selection process is used to obtain the ranking of the alternatives.

The social network group decision making (SNGDM) is a hot topic now because of the growth of network technology. The CRP has been studied in the SNGDM problems [6-10]. However, the existing CRPs have not considered the interactions among the decision makers, which may cause opinion evolution [11]. In this paper, a novel CRP is proposed, which considers the opinion evolution at social networks.

The rest of this paper is arranged as follows. Section 2 presents social network analysis and opinion evolution. Then, Section 3 proposes a novel CRP, which considers opinion evolution at social networks. Next, a numerical analysis is provided in Section 4. Finally, the conclusions are shown in Section 5.

2. Preliminaries

This section presents the social network analysis in GDM and opinion evolution at social networks.

2.1. Social network analysis in GDM

The frequently used Definitions 1-2 regarding social network analysis [12-14] are presented as follows.

Definition 1. A directed graph $G(D,E)$ consisting of nodes and edges is defined as a social network, where the node $d_r \in D(r=1,2,\cdots,m)$ represents the decision makers, and the edge $(d_r, d_h) \in E$ represents the social relationships that decision maker d_r trusts d_h.

Definition 2. The adjacency matrix $A = (a_{ij})_{m \times m}$ defines the social relationships among decision makers, which is a 0-1 matrix. If d_r trusts d_h in $G(D,E)$, the (r,h)th entry is 1; otherwise, it is 0, i.e.,

$$a_{ij} = \begin{cases} 1, & (d_i, d_j) \in E \\ 0, & (d_i, d_j) \notin E \end{cases} \quad (1)$$

2.2. Opinion evolution at social networks

Opinion evolution is the process of a group of decision makers interact with each other in reforming opinions. The opinion evolution has been widely studied in a social network context, and we introduce the DeGroot model in the social network context in the following [15].

Let $X = \{x_1, x_2, \cdots, x_n\} (n \geq 2)$ be a finite set of alternatives and additive preference relation $P^r = (p_{lk}^r)_{n \times n}$ be the preference provided by d_r in $G(D,E)$, where $p_{lk}^r + p_{kl}^r = 1$ and $p_{lk}^r \in [0,1]$ means the preference degree of x_l over x_k, $\forall l, k \in \{1, 2, \cdots, n\}$. Let $P^{r,t} = (p_{lk}^{r,t})_{n \times n}$ be the preference relation of d_r at round t and $A = (a_{ij})_{m \times m}$ be the adjacency matrix associated to $G(D,E)$. Let $\theta_r \in (0,1)$ be the confidence of d_r in his preference. Thus, the opinion evolution of d_r can be presented as:

$$p_{lk}^{r,t+1} = \theta_r p_{lk}^{r,t} + \sum_{j=1, j \neq r}^{m} \frac{(1-\theta_r) a_{rj}}{\sum_{j=1, j \neq r}^{m} a_{rj}} \times p_{lk}^{j,t} \quad (2)$$

When $\theta_r (r=1,2,\cdots,m)$ and $a_{ij}(i, j=1,2,\cdots,m)$ are fixed, Eq. (2) is a DeGroot model [16,17].

3. Resolution of the SNGDM Problem

This section proposes a novel CRP to solve SNGDM problem.

3.1. The framework of proposed CRP

The SNGDM problem is the situation that a group of decision makers, who interact with each other based on their social relationships, try to obtain a collective solution from several alternatives. Let $X = \{x_1, x_2, \cdots, x_n\}$ be a finite set of alternatives, and $D = \{d_1, d_2, \cdots, d_m\}$ be the set of decision makers. $d_r (r = 1, 2, \cdots, m)$ provides the preference relation $P^r = (p_{lk}^r)_{n \times n}$.

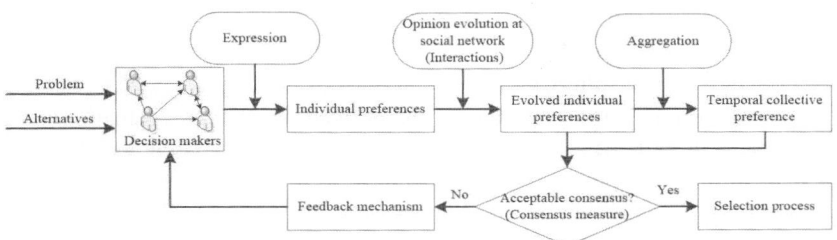

Fig. 1. The novel CRP framework.

The framework of the proposed CRP can be illustrated in Fig. 1. When the consensus is achieved, the selection process is activated.

3.2. *The proposed CRP*

The proposed CRP includes the following three key phases:

(1) Opinion evolution at social networks, described in Section 2.2, is used to simulate the interactions.

(2) Consensus measure is used to obtain the consensus level of the group. The collective preference relation $P^c = (p_{lk}^c)_{n \times n}$ can be computed by the weighted average operator f_π [18]:

$$p_{lk}^c = f_\pi(p_{lk}^1, p_{lk}^2, \cdots, p_{lk}^m) = \sum_{r=1}^{m} \pi_r \cdot p_{lk}^r, \text{ for } l, k = 1, \cdots, n \quad (3)$$

where $\{\pi_1, \pi_2, \cdots, \pi_m\}$ is the weights of the decision makers.

Then, the consensus index of d_r can be calculated as:

$$CI(d_r) = \frac{\sum_{l=1}^{n-1} \sum_{k=l+1}^{n} (1 - |p_{lk}^r - p_{lk}^c|)}{n \cdot (n-1)/2} \quad (4)$$

Thus the consensus index of the group can be obtained:

$$CI = \frac{\sum_{r=1}^{m} CI(d_r)}{m} \quad (5)$$

If $CI < \mu$, the decision makers are advised to modify their preferences, where $\mu \in [0,1]$ is a consensus threshold. Otherwise, the selection process is activated.

(3) Feedback mechanism is used to identify the decision maker who has the minimum consensus index, and ask him to modify his preference. The modification direction is shown as follows:

$$\begin{cases} \overline{p}_{lk}^r \in [\min(p_{lk}^r, p_{lk}^c), \max(p_{lk}^r, p_{lk}^c)], & 1 \le k \\ \overline{p}_{kl}^r = 1 - \overline{p}_{lk}^r, & 1 < k \end{cases} \quad (6)$$

The ranking of alternatives $(R_1, R_2, \cdots, R_n)$ obtained by the selection process when reaching a consensus is derived from the evaluation value $\{\sigma_1, \sigma_2, \cdots, \sigma_n\}$ [19,20]. Let $\{w_1, w_2, \cdots, w_m\}$ be the associated weights, then the evaluation value σ_l can be computed as:

$$\sigma_l = f_w(p_{l1}^c, p_{l2}^c, \cdots, p_{lm}^c) = \sum_{k=1}^{m} w_k \cdot p_{lk}^c \quad (7)$$

The algorithm of proposed CRP is shown in Table 1.

Table 1. The algorithm of the proposed CRP.

Input: $\{P^1, P^2, \cdots, P^m\}$, μ, and $G(D, E)$.
Output: $(R_1, R_2, \cdots, R_n)$.
Step 1: Let $t = 0$. Let $P^{r,t} = (p_{lk}^{r,t})_{n \times n} = (p_{lk}^r)_{n \times n} (r = 1, 2, \cdots, m)$.
Step 2: Obtain the evolved preferences $\{EP^{1,t}, EP^{2,t}, \cdots, EP^{m,t}\}$ from $\{P^{1,t}, P^{2,t}, \cdots, P^{m,t}\}$ using the Eq. (2) based on the adjacent matrix A corresponding to $G(D, E)$.
Step 3: Obtain $P^{c,t} = (p_{lk}^{c,t})_{n \times n}$ from $\{P^{1,t}, P^{2,t}, \cdots, P^{m,t}\}$ using Eq. (3).
Step 4: Calculate the consensus index CI^t based on $P^{c,t}$ and $\{EP^{1,t}, EP^{2,t}, \cdots, EP^{m,t}\}$ using Eqs. (4) and (5). If $CI^t \ge \mu$, go to Step 6; otherwise go to the next step.
Step 5: Obtain the adjusted individual preferences $\{P^{1,t+1}, P^{2,t+1}, \cdots, P^{m,t+1}\}$ from $\{EP^{1,t}, EP^{2,t}, \cdots, EP^{m,t}\}$ based on feedback mechanism. Let $t = t + 1$, then go to Step 2.
Step 6: Output $(R_1, R_2, \cdots, R_n)$ based on $\{\sigma_1, \sigma_2, \cdots, \sigma_n\}$ computed by Eq. (7).

4. Numerical Analysis

In this section, a numerical analysis is used to describe the resolution framework, where six decision makers express their preferences over four alternatives, and the initial preferences are shown as follows:

$$P^1 = \begin{pmatrix} 0.50 & 0.84 & 0.82 & 0.97 \\ 0.16 & 0.50 & 0.44 & 0.19 \\ 0.18 & 0.56 & 0.50 & 0.73 \\ 0.03 & 0.81 & 0.27 & 0.50 \end{pmatrix}, P^2 = \begin{pmatrix} 0.50 & 0.25 & 0.28 & 0.20 \\ 0.75 & 0.50 & 0.82 & 0.74 \\ 0.72 & 0.18 & 0.50 & 0.76 \\ 0.80 & 0.26 & 0.24 & 0.50 \end{pmatrix},$$

$$P^3 = \begin{pmatrix} 0.50 & 0.46 & 0.92 & 0.21 \\ 0.54 & 0.50 & 0.83 & 0.08 \\ 0.08 & 0.17 & 0.50 & 0.13 \\ 0.79 & 0.92 & 0.87 & 0.50 \end{pmatrix}, P^4 = \begin{pmatrix} 0.50 & 0.77 & 0.12 & 0.81 \\ 0.23 & 0.50 & 0.34 & 0.80 \\ 0.88 & 0.66 & 0.50 & 0.71 \\ 0.19 & 0.20 & 0.29 & 0.50 \end{pmatrix},$$

$$P^5 = \begin{pmatrix} 0.50 & 0.63 & 0.46 & 0.42 \\ 0.37 & 0.50 & 0.30 & 0.09 \\ 0.54 & 0.70 & 0.50 & 0.23 \\ 0.58 & 0.91 & 0.77 & 0.50 \end{pmatrix}, P^6 = \begin{pmatrix} 0.50 & 0.16 & 0.9 & 0.36 \\ 0.84 & 0.50 & 0.72 & 0.51 \\ 0.10 & 0.28 & 0.50 & 0.09 \\ 0.64 & 0.49 & 0.91 & 0.50 \end{pmatrix}.$$

Let $\{1/6,1/6,1/6,1/6,1/6,1/6\}$ be the weights in Eq. (3), and $\{1/4,1/4,1/4,1/4\}$ be the weights in Eq. (7). Three cases of social networks are adopted in this numerical analysis, and two of them are shown as Fig. 2. The Case 3 is the situation that the opinion evolution and social network are ignored. In this analysis, we suppose that $\theta_r = \theta(r=1,2,\cdots,6)$, and we set the Eq. (6) as an automatic process, which is formulated as follows:

$$\begin{cases} \overline{p}_{lk}^r = \theta_r \cdot p_{lk}^r + (1-\theta_r) \cdot p_{lk}^c, & l \leq k \\ \overline{p}_{kl}^r = 1 - \overline{p}_{lk}^r, & l < k \end{cases} \quad (8)$$

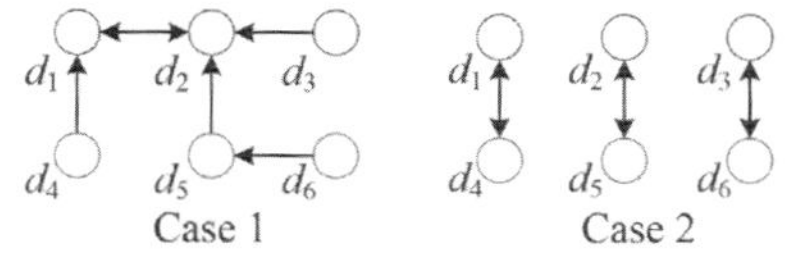

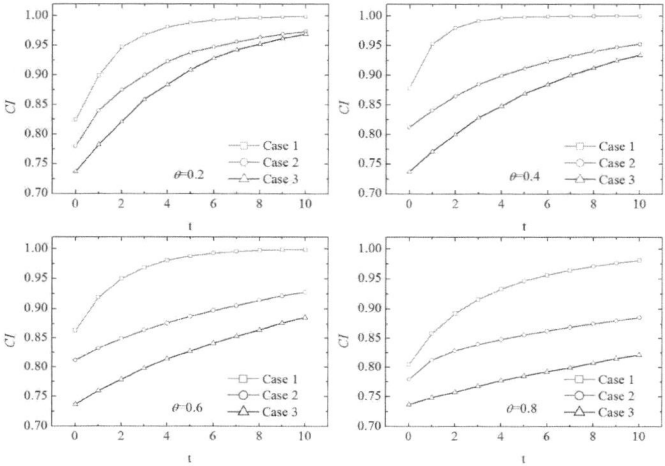

Fig. 2. The social networks in the numeral analysis.

Fig. 3. The consensus index CI of the three cases in t time under different θ values.

Fig. 3 illustrates the consensus index CI of the three cases under different θ values and μ close to 1. When setting different θ and μ values, Table 2 presents the rankings of the alternatives when $CI \geq \mu$.

From Fig. 3 and Table 2, we find that the opinion evolution have a great influence on the consensus reaching and decision result.

Table 2. The rankings of alternatives under under different θ and μ values.

Consensus threshold	Case	$\theta = 0.2$	$\theta = 0.4$	$\theta = 0.6$	$\theta = 0.8$
	Case 1	(1 3 2 4)	(2 1 3 4)	(1 2 4 3)	(1 2 4 3)
$\mu = 0.90$	Case 2	(2 3 4 1)	(3 2 4 1)	(2 3 4 1)	(2 3 4 1)
	Case 3	(2 3 4 1)	(2 3 4 1)	(2 3 4 1)	(2 3 4 1)
	Case 1	(1 3 2 4)	(2 1 3 4)	(1 2 3 4)	(1 2 3 4)
$\mu = 0.95$	Case 2	(2 3 4 1)	(3 2 4 1)	(3 2 4 1)	(3 2 4 1)
	Case 3	(2 3 4 1)	(2 3 4 1)	(2 3 4 1)	(2 3 4 1)

5. Conclusions

In this paper, a CRP which considers opinion evolution at social networks is proposed. We find that the opinion evolution at social networks have an effect on the consensus reaching and decision result in SNGDM problems.

In the future, we argue that it would be interesting to investigate the dynamical social networks in the CRP.

Acknowledgments

This work is supported by the grant (No. 71571124) from NSF of China.

References

1. W.D. Cook and M. Kress, *Management Science*. **31**, 26-32 (1985).
2. D.S. Hochbaum and A. Levin, *Management Science*. **52**, 1394-1408 (2006).
3. X. Chen, H.J. Zhang and Y.C. Dong, *Information Fusion*. **24**, 72-83 (2015).
4. E. Herrera-Viedma, S. Alonso, F. Chiclana and F. Herrera, *IEEE Transactions on Fuzzy Systems*. **15**, 863-877 (2007).
5. Y.C. Dong, H.J. Zhang and E. Herrera-Viedma, *Knowledge-Based Systems*. **106**, 206-219 (2016).
6. Q. Liang, X.W. Liao and J.P. Liu, *Knowledge-Based Systems*. **119**, 68-86 (2017).
7. Y. Liu, C.Y. Liang, F. Chiclana and J. Wu, *Knowledge-Based Systems*. **119**, 221-231 (2017).
8. Y.C. Dong, H.J. Zhang and E. Herrera-Viedma, *Decision Support Systems*. **84**, 1-15 (2016).
9. J. Wu, F. Chiclanab, H. Fujita and E. Herrera-Viedma, *Knowledge-Based Systems*. **122**, 39-50 (2017).

10. J. Wu, L.F. Dai, F. Chiclana, H. Fujita and E. Herrera-Viedma, *Information Fusion*. **41**, 232-242 (2017).
11. Y.C. Dong, M. Zhan, G. Kou, Z.G. Ding and H.M. Liang, *Information Fusion*. **43**, 57-65 (2018).
12. J.A. Bondy and U.S.R Murty, Graph Theory with Applications, the Macmillan press LTD. (1976).
13. R.A. Horn and C.R. Johnson, Matrix Analysis (2th ed.), Cambridge University Press (1994).
14. K.H. Rosen, Discrete Mathematics and its Application (6th ed.), McGraw-Hill (2007).
15. M.H. Degroot, *Journal of the American Statistical Association*. **69**(345), 118-121 (1974).
16. Y.C. Dong, Z.G. Ding, L. Martínez and F. Herrera, *Information Sciences*. **397-398**, 187-205 (2017).
17. R.L. Berger, *Journal of the American Statistical Association*. **76**(374), 415-418 (1981).
18. J. Aczél, *Aequationes Mathematicae*. **27**(1), 288-307 (1984).
19. Y.C. Dong, Y.T. Liu, H.M. Liang, F. Chiclana, and E. Herrera-Viedma, *Omega*. **75**, 154-164 (2018).
20. F. Herrera and E. Herrera-Viedma, *European Journal of Operational Research*. **120**(1), 144-161 (2000).

Improved score based decision making method by using fuzzy soft sets

Yaya Liu

School of Mathematics, Southwest Jiaotong University, Chengdu, Sichuan, PR China
Department of Computer Science, University of Jaen, Jaen, Spain
yayaliu@my.swjtu.edu.cn

Rosa M. Rodríguez

Department of Computer Science and Artificial Intelligence,
University of Granada, Granada, Spain
rosam.rodriguez@decsai.ugr.es

Keyun Qin

School of Mathematics, Southwest Jiaotong University, Chengdu, Sichuan, PR China
qinkeyun@home.swjtu.edu.cn

Luis Martínez

Department of Computer Science, University of Jaen, Jaen, Spain
martin@ujaen.es

Popular decision making methods by using fuzzy soft sets belong to two main categories, namely, score-based methods and fuzzy choice values based methods. It is necessary an application background to choose which one is better, since each of them makes sense in specific decision making situations and both of them could be improved. Therefore, in this contribution we focus on improving the former one. To improve the score based method, it is provided a novel adjustable algorithm by using fuzzy soft set that introduces thresholds when comparing two membership function values and afterwards coming up with new concepts of scores.

Keywords: Soft set; fuzzy soft set; decision making; scores.

1. Introduction

Many real word problems take place under uncertainties. Fuzzy set theory,[1] rough set theory,[2] vague set theory[3] and many others are some well known mathematical tools that could be used to handle these uncertainties. Lacking of parameterized tools is a common limitation of all previous models.

Molodstov[4] put forward the notion of soft set theory, which considers alternatives from different parameters aspects which successfully overcomes such a limitation. After the appearance of soft set theory, the researches concentrate on three main directions: 1) the operations and algebraic structure of soft sets;[5] 2) the fusion of soft sets with other existing models for dealing with uncertainty, various generalized soft set models have been proposed by researchers;[6] and 3) the applications of soft sets (and generalized soft sets) in practice situations,[7-10] especially in decision making (DM).[11-14]

Fuzzy soft sets were constructed from the combination of fuzzy sets and soft sets. One of the most popular DM method by using fuzzy soft sets was put forward by Roy and Maji[15] and built upon the concept of scores of alternatives, that is denoted by classical score based method. In such a method, by computing row and column sums of a comparison table the scores of alternatives could be obtained. In this contribution, a DM method by using fuzzy soft sets on the basis of several new concepts of scores for alternatives will be proposed. Since we follow Roy and Maji's idea[15] of using scores to achieve the optimal choice, it could be viewed as an improvement of classical score based method. Different comparison thresholds in forms of values or fuzzy sets could be adopted during the computation process of scores according to either the willing of decision makers or demands of the circumstances, in this way the decision results will be adjustable.

The remainder is organized as follows: Section 2 makes a brief review on soft sets and fuzzy soft sets. Section 3 introduces some new concepts for scores of alternatives based on fuzzy soft sets, and afterwards puts forward a decision making algorithm. Conclusions are given in Section 4.

2. Preliminaries

This section reviews basic definitions of soft sets and fuzzy soft sets.

A soft set is defined as a mapping from a parameter set to the power set of universe:

Definition 2.1.[4] Let U be the universe set and $P(U)$ be the power set of U. Let E be the parameter set and $A \subseteq E$. A pair (F, A) is called a soft set over U, where F is a mapping $F : A \longrightarrow P(U)$.

For any $e \in A$, $F(e)$ is $e-$approximate elements of the soft set (F, A), then a soft set is a parameterized family of subsets of U.

A fuzzy set[1] F in the universe U is defined as $F = \{(x, \mu_F(x))/x \in$

$U, \mu_F(x) \in [0,1]\}$, where $\mu_F(x)$ indicates the membership degree of alternative x determined by a membership function F.

Fuzzy soft sets are fuzzy generalizations of soft sets:

Definition 2.2.[16] Let U be the universe set and $F(U)$ be all fuzzy sets on U. A pair (F, A) is called a fuzzy soft set over U, where F is a mapping $F : A \longrightarrow F(U)$.

For any $e \in A$, $F(e)$ is a fuzzy subset of U. A fuzzy soft set will degenerate to a soft set when $F(e)$ degenerates to a subset of U for all $e \in A$. If $F(e)$ is a crisp subset of U for any $e \in A$, then the fuzzy soft set (F, A) degenerates to a crisp soft set. Denote the membership degree of $x \in U$ with respect to $e \in A$ as $\mu_{F(e)}(x)$, then $F(e)$ can be represented by $F(e) = \{< x, \mu_{F(e)}(x) > | x \in U\}$.

3. Improved Score Based Decision Making Method by Using Fuzzy Soft Sets

Maji and Roy introduced the notion of scores for alternatives and a way to compute the scores when solving DM problems by using fuzzy soft sets.[15] However, the way for determining scores should not be unique, because making use of different scores, different decisions may meet the demands of different applications. In this section, a novel method for DM will be given by introducing several novel concepts of scores for alternatives.

In the following, all concepts are constructed under the background of DM. Let $U = \{x_1, x_2, \ldots, x_n\}$ be a universe set that consists of n alternatives, E be a parameter set, $A \subseteq E$ and $A = \{e_1, e_2, \ldots, e_m\}$.

First, a measure called "d-level score" is introduced:

Definition 3.1. Suppose that (F, A) is a fuzzy soft set over U. For a value $d \in [0,1]$, d-level score of $x_i \in U$ with respect to $e_l \in A$ is defined as

$$S(x_i)(e_l)_d = R(x_i)(e_l)_d - T(x_i)(e_l)_d, \tag{1}$$

where $R(x_i)(e_l)_d = |x_j \in U : \mu_{F(e_l)}(x_i) - \mu_{F(e_l)}(x_j) \geq d|$ and $T(x_i)(e_l)_d = |x_j \in U : \mu_{F(e_l)}(x_j) - \mu_{F(e_l)}(x_i) \geq d|$.

Based on Eq. (1), the d-level score of x_i, denoted by S_i^d, is defined as

$$S_i^d = \sum_{l=1}^{m} S(x_i)(e_l)_d. \tag{2}$$

A tool called d-level score table could be constructed with rows labeled by parameters in A and columns corresponds to alternatives in U. For each entry position (i, j), there is a value $S(x_i)(e_l)_d$ computed by Eq. (1).

An example to illustrate d-level score table can be the following one:

Example 3.1. Let $U = \{x_1, x_2, \ldots, x_4\}$, $E = \{e_1, e_2, \ldots, e_6\}$ and (F, E) be a fuzzy soft set (see Table 1).

Table 1. Fuzzy soft set (F, E).

	e_1	e_2	e_3	e_4	e_5	e_6
x_1	0.6	0.8	0.1	0.4	0.2	0.9
x_2	0.9	0.2	0.4	0.3	0.7	0.6
x_3	0.6	0.7	0.4	0.8	0.1	0.2
x_4	0.2	0.3	0.1	0.3	0.4	0.1

Suppose that $d = 0.2$, then 0.2-level score table of (F, E) could be obtained (see Table 2).

Table 2. 0.2-level score table of (F, E).

	e_1	e_2	e_3	e_4	e_5	e_6
x_1	0	2	-2	-1	-2	3
x_2	3	-2	2	-1	3	1
x_3	0	2	2	3	-2	-2
x_4	-3	-2	-2	-1	1	-2

In detail, $S(x_2)(e_3)_{0.2} = R(x_2)(e_3)_{0.2} - T(x_2)(e_3)_{0.2} = 2 - 0 = 2$, then the value in entry position $(2,3)$ is 2. Values in other entries of Table 2 can be computed in a similar way.

Actually, $d \in [0, 1]$ is a comparison threshold used to compare two membership values for computing the d-level score. The methods for determining the threshold value (TV) d can be various, since it should be chosen depending on the requirement of DM situation.

The importance degrees for various parameters may be also different in decision makers' consideration, therefore the comparison TVs could be different for different parameters, which could be done by applying a function (a fuzzy set) instead of a value during the comparison process. Consider the specific structure of a fuzzy soft set, if we choose a TV within interval $[0, 1]$ for each parameter in A, all thresholds with respect to all parameters could form a fuzzy set.

In the following, we apply a fuzzy set θ as the threshold to introduce a concept called level score corresponding to a fuzzy set:

Definition 3.2. Let (F, A) be a fuzzy soft set over U and $\theta : A \longrightarrow [0,1]$ be a fuzzy set on A, called a comparison threshold fuzzy set (TFS). The level score of $x_i \in U$ on $e_l \in A$ corresponding to θ is denoted by $S(x_i)(e_l)_\theta$ and defined as

$$S(x_i)(e_l)_\theta = r(x_i)(e_l)_\theta - t(x_i)(e_l)_\theta \tag{3}$$

where $r(x_i)(e_l)_\theta = |x_j \in U : \mu_{F(e_l)}(x_i) - \mu_{F(e_l)}(x_j) \geq \theta(e_l)|$ and $t(x_i)(e_l)_\theta = |x_j \in U : \mu_{F(e_l)}(x_j) - \mu_{F(e_l)}(x_i) \geq \theta(e_l)|$. The level score of alternative x_i corresponding to θ is defined as

$$S_i^\theta = \sum_{l=1}^{m} S(x_i)(e_l)_\theta. \tag{4}$$

A tool called level score table corresponding to fuzzy set θ could be constructed with rows labeled by parameters in A and columns corresponding to alternatives in U. For each entry position (i, j), there is an input a value $S(x_i)(e_l)_\theta$ computed by Eq. (3). Obviously, the level score of alternative x_i corresponding to θ could be obtained by computing the row sums of the corresponding level score table.

For the comparison TFS θ, if $\theta(e_l) = t$ ($t \in [0,1]$ is a constant) for all $e_l \in A$, then the level score corresponding to θ degenerates to a d-level score. In other words, level score table corresponding to a fuzzy set degenerates to a d-level score table.

Based on (F, A), several commonly used TFSs are defined: (1) Mid TFS θ_F^{mid}: $\theta_F^{mid}(e_l) = \frac{1}{|U|}(max_{x_i \in U} \mu_{F(e_l)}(x_i) - min_{x_i \in U} \mu_{F(e_l)}(x_i))$; (2) Min TFS θ_F^{min}: $\theta_F^{min}(e_l) = min_{\{x_i, x_j \in U\}} |\mu_{F(e_l)}(x_i) - \mu_{F(e_l)}(x_j)|$; (3) Max TFS θ_F^{max}: $\theta_F^{max}(e_l) = max_{\{x_i, x_j \in U\}} (\mu_{F(e_l)}(x_i) - \mu_{F(e_l)}(x_j))$ for $e_l \in A$.

Level scores based DM method is carried out according to the below algorithm:

Algorithm 3.1.
Step 1. *Collect assessments on alternatives with respect to parameters and present the assessments as a fuzzy soft set (F, A).*
Step 2. *Select a comparison TV $d \in [0, 1]$ (or chose a comparison TFS θ) for (F, A).*
Step 3. *Construct the d-level score table (or level score table corresponding to fuzzy set θ) of (F, A).*
Step 4. *For each alternative x_i, calculate the d-level score of x_i, i.e. S_i^d (or level score corresponding to θ, i.e. S_i^θ).*

Step 5. Choose x_j as the optimal choice by $S_j^d = max_i S_i^d$ and denote the decision result by $D((F, A), t)$ (or select x_j as the optimal choice if $S_j^\theta = max_i S_i^\theta$ and denote the decision result by $D((F, A), \theta)$).

The optimal choice can be one or several alternatives. If there are too many selected alternatives as the decision result in step 5, decision makers could go back to step 2 to change the comparison threshold in order to limit the number of optimal choices. The application of TVs in DM benefits from idea of Feng et al.[17]

Example 3.2. In a DM problem, suppose that the assessments provided by decision makers form a fuzzy soft set (F, E) with its tabular representation (Table 1) in Example 3.1. If the problem is handled by using Algorithm 3.1, and the comparison TV is chosen as $d = 0.2$. From the 0.2-level score table of (F, E) (Table 2), the 0.2-level scores of alternatives are $S_1^{0.2} = 0$, $S_2^{0.2} = 6$, $S_3^{0.2} = 3$, and $S_4^{0.2} = -9$. Then, the optimal choice is x_2.

Theorem 3.1. *If the TV $d = 0$, then the d-level score of each alternative computed by Eq. (1) is the same as its score by using classical score based method.*

By Theorem 3.1, it is shown that the decision result obtained from Algorithm 3.1 will be the same as the result obtained by using classical score based method when the TV degenerates to zero. However, compared with classical score based method, the introduction of TVs could not only reflect the quantity, but also the quality to which degree alternatives satisfy parameters when two membership function values need to be compared, which makes the decision result adjustable and increases the flexibility of the classical method.

Theorem 3.2. *Suppose that a DM problem could be handled by using a fuzzy soft set (F, A). Let $D((F, A), \theta)$ be the decision result obtained from Algorithm 3.1, where θ is a comparison TFS. If $\theta(e_l) > \theta_F^{max}(e_l)$ for all $e_l \in A$, then $D((F, A), \theta) = U$.*

4. Conclusion

Explorations on DM methods by using fuzzy soft sets contribute to the development of soft set theory. A novel adjustable method has been introduced in this contribution. The proposed algorithm has the potential to be extended to deal with uncertain situations in which generalization models of fuzzy soft sets are applied.

Acknowledgments

This work is partially supported by the Spanish National Research Project TIN2015-66524-P, the Spanish Ministry of Economy and Finance Postdoctoral Fellow (IJCI-2015-23715), the National Natural Science Foundation of China (Grant Nos. 61473239, 61372187, and 61603307) and ERDF.

References

1. L. A. Zadeh, *Information and Control* **8**, 338 (1965).
2. Z. Pawlak, *International Journal of Computer and Information Sciences* **11**, 341 (1982).
3. W. L. Gau and D. J. Buehrer, *IEEE Transactions on Systems Man and Cybernetics* **23**, 610 (2002).
4. D. Molodtsov, *Computers and Mathematics with Applications* **37**, 19 (1999).
5. X. H. Zhang, C. Park and W. Supeng, *Journal of Intelligent and Fuzzy Systems* **34**, 559 (2018).
6. A. M. Abd El-Latif, *Journal of Intelligent and Fuzzy Systems* **34**, 517 (2018).
7. Y. Liu, K. Qin, C. Rao and M. A. Mahamadu, *International Journal of Applied Mathematics and Computer Science* **27**, 157 (2017).
8. J. Zhan and K. Zhu, *Soft Computing* **21**, 1923 (2017).
9. J. Zhan, M. I. Ali and N. Mehmood, *Applied Soft Computing* **56**, 446 (2017).
10. J. Zhan, Q. Liu and T. Herawan, *Applied Soft Computing* **54**, 393 (2017).
11. J. C. R. Alcantud and G. Santos-García, *International Journal of Computational Intelligence Systems* **10**, p. 394 (2017).
12. X. Ma, Q. Liu and J. Zhan, *Artificial Intelligence Review* **47**, 1 (2016).
13. F. Fatimah, D. Rosadi, R. F. Hakim and J. C. R. Alcantud, *Neural Computing and Applications*, 1 (2017).
14. F. Fatimah, D. Rosadi, R. B. F. Hakim and J. C. R. Alcantud, *Soft Computing*, 1 (2017).
15. A. R. Roy and P. Maji, *Journal of Computational and Applied Mathematics* **203**, 412 (2007).
16. P. K. Maji, R. Biswas and A. R. Roy, *Journal of Fuzzy Mathematics, Vol. 9, No. 3*, 589 (2001).
17. F. Feng, Y. B. Jun, X. Liu and L. Li, *Journal of Computational and Applied Mathematics* **234**, 10 (2010).

On group recommendation supported by a minimum cost consensus model

Raciel Yera

University of Ciego de Ávila
Ciego de Ávila, Cuba
ryera@unica.cu

Álvaro Labella

Computer Science Department, University of Jaén
Jaén, Spain
alabella@ujaen.es

Jorge Castro

Department of Computer Science and AI, University of Granada
Granada, Spain
School of Software, University of Technology Sydney, Sydney, NSW, Australia
jcastro@decsai.ugr.es

Luis Martínez

Computer Science Department, University of Jaén
Jaén, Spain
martin@ujaen.es

Group recommender systems (GRSs) have recently attracted the attention from researchers and industry. They focused on recommending items which satisfy the global preferences of a group, being TV programs and holidays packages typical examples of these scenarios. Although there have been established several basic approaches for GRSs, it has been also identified the limitation about dealing with conflicts about the recommendation within the groups and hence, the necessity of managing in a deeper way the consensus among the group members to improve the agreed satisfaction of the recommendations. The current contribution is focused on proposing the application of the minimum cost consensus model in the GRS scenario for achieving such objective. A case study will show that this consensus model positively influences the groups' satisfaction about the recommendations.

Keywords: Group recommender systems; minimum cost consensus reaching; recommendation performance.

1. Introduction

In the last few years, recommender systems (RSs) have become a necessary tool in several online scenarios to overcome the burden associated to overloaded search spaces, by providing users with the items that best fit their preferences and needs.[1] Therefore, RSs are frequently used in several domains such as e-commerce, e-learning, e-services, tourism, and so on.[2]

Furthermore, there is an important set of items, so called social items, which are usually consumed by a group of users and not by an individual. TV programs and holidays packages are clear examples of these kinds of items, where individual interests of the users in the same group can differ. In order to generate suitable item recommendations in these scenarios, group recommender systems (GRSs)[3,4] have recently attracted the attention from researchers and industry to recommend items which satisfy the global preferences of the group.

Specifically, the group recommendation task[3] has traditionally been performed using two main approaches as a extension of the individual recommendation task: (i) the rating aggregation approach, where individual's preferences are combined to obtain a unified profile that represents the group preference, and (ii) the recommendation aggregation approach, in which individual recommendations are generated at first and afterwards they are aggregated to obtain the final recommendation list.

Although these approaches have been extensively used in the last few years,[3,4] some recent works[5] have pointed out that it is necessary the development of researches beyond these basic aggregation approaches, because just aggregation could generate loss of information and biased recommendations, obtaining in turn low group satisfaction with recommendations. Therefore the study of managing the agreement among groups' members can improve group recommendations.

This contribution is focused on such an objective, by exploring the effect of applying a minimum cost consensus reaching model over the individual user preferences, that would obtain agreed group recommendations. Our aim is to process the individual user needs in a direct way, reducing the set of possible agreed recommendations by using the Borda voting system[6] and then limit our analysis to the recommendation aggregation approach.

The paper is structured as follows. Section 2 shortly reviews the necessary background on GRSs and consensus reaching process, regarding the current contribution. Section 3 presents an approach to integrate the mentioned consensus model in the GRS framework, which is evaluated in Section 4. Section 5 concludes the paper.

2. Background

This section briefly presents the necessary background for the development of the current research work. It is focused on GRSs, the Borda Count, and the Minimum Cost Consensus Model.

Group recommender systems, on its basic approaches, extend RSs for targeting recommendations to group of users ($G = \{g_1, ..., g_m\} \subseteq U$). Formally, GRSs focuses on finding the item (or set the items) that maximizes the preference predicted for the group of users:

$$Recommendation(I, G_a) = arg \max_{i_k \in I} Prediction(i_k, G_a) \quad (1)$$

There are two main approaches for group recommendation, supported by single user recommendation:[3]

- *Rating aggregation:* It is based on the creation of a pseudo-user profile that aggregates the preference of the group. This profile is used as the final target user for generating recommendations.
- *Recommendation aggregation:* It aggregates the individual recommendation list associated to each member, into a new recommendation list targeted on the group.

Borda count:[6] This well-known voting system in social choice computes a mapping from a set of individual ranking list associated to experts, to find a combined ranking of such lists. Specifically, each ranked item obtains 0 point for each last place vote received, 1 point for each next-to-last place vote, and so on, receiving M-1 points for each first place vote (being M the number of items). At last, items are downwardly ranked according to the sum of all associated ranks provided by each expert. Some of Borda count's advantages are its easy implementation, its intuitiveness, and its low computational cost.

Minimum cost consensus model:[7] This model is focused on minimizing costs associated to the modification of the independent experts' opinions to reach a consensus. Such minimum cost is obtained by solving a lineal programming model[8] (Equation 2).

$$\begin{cases} min \sum_{u=1}^{n} c_u |\bar{o}_u - o_u| \\ s.t. \bar{o} = \sum_{u=1}^{n} w_u \bar{o}_u \\ |\bar{o}_u - \bar{o}| \leq \epsilon, u = 1, 2, ..., n \end{cases} \quad (2)$$

The parameters involved in this model are:

- c_u: the cost of modifying the preferences of the expert u.
- o_u: the initial preferences of the expert u.
- $\bar{o}_u$: the final preferences of the expert u, after consensus reaching.
- $\bar{o}$: the collective preferences of the group of experts.
- ϵ: the maximum possible distance between collective and individual preferences.
- w_u: the weight of the expert u.

3. Minimum Cost Consensus in Group Recommendation

The use of the minimum cost consensus model in the context of group recommendation, to reach a higher consensus level in the final recommendations and improve the recommendation acceptance, consists of the following phases: (i) Individual recommendation generation, (ii) Borda count-based ranking, and (iii) Minimum cost consensus analysis.

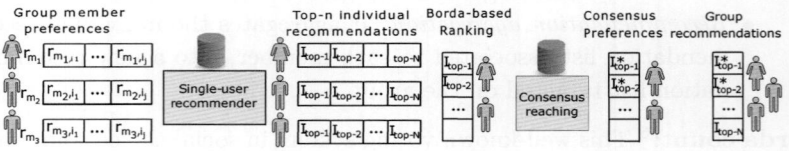

Fig. 1. General scheme of the proposal.

Individual recommendation generation: This phase computes the individual recommendations for each group's member. Here, it is applied a typical collaborative filtering recommendation approach to obtain such individual's recommendations. In our case study it will be considered the user-user and item-item neighborhood-based collaborative filtering approaches,[9] although any individual recommendation approach could be used in this scenario.

Borda count-based ranking: The Borda count-based ranking is applied here to shorten the possible set of recommended items that will be used in the consensus phase to reduce its computational cost. Therefore, only the top-k ranked common items according Borda will go to the consensus phase, to apply the minimum cost model over the individual's recommendations about these items.

The Borda count is applied to each group's member, and its individual ranking is obtained by downwardly sorting items according to the individual rating prediction performed in the previous step. Assuming that

$rank(i, R_u)$ returns the position of the predicted rating over the item i in the downwardly sorted list R_u of preference predicted for each user u member of the group G, then the average rank of i in the group G is formalized as follows:

$$Rank_i = \sum_{u \in G} rank(i, R_u). \qquad (3)$$

Considering the downwardly sort list I of items according to $Rank_i$, and j the item at position k in such list, then the top-k ranked common items would be formalized as:

$$Top_k = \{i | \forall_{i \in I} Rank_i \leq Rank_j\}. \qquad (4)$$

Minimum cost consensus model: This model is applied to the top-k common items obtained in the previous step. This phase receives individual's prediction values, and adjust the group recommendation to reach the consensus. As final output, the model recommends those items with highest agreed value to the group's members.

Specifically, the minimum cost consensus model is computed independently for each item i, considering the following assumptions for translating the consensus model notation into the GRS scenario:

- $\forall_{u \in G}(o_u = r_{ui})$ (each user preference on i is the expert's u opinion on i)
- $\forall_{u \in G}(c_u = 1)$ (the cost of modifying preferences of u is always 1)
- $\forall_{u \in G}(w_u = 1/n)$ (the expert's u weight is always $1/n$, being n the number of experts)
- $\epsilon = 0.2$.

The collective opinion $\bar{o}$ associated to each item i (see Equation 2) is considered as the group's preference prediction over such item.

4. Case Study

This section presents a case study developed over the well-known Movielens 100K recommender systems dataset[a] for evaluating the effect of applying the minimum cost consensus model in the final recommendation generation. Specifically, it is composed of 943 users, 1682 items, and 100000 ratings. The rating scale is in the range [1,5].

---

[a]https://grouplens.org/datasets/movielens/

In a similar way to previous works in GRS,[3,4] the group formation technique used is a random selection. Furthermore, a 20% item holdout is applied as validation technique, which is adjusted to be used in the GRS scenario by selecting the 20% of items evaluated by the current group as the test set, and the remaining ones as training set. Finally, Mean Absolute Error (MAE) and Area Under the receiver operating characteristic Curve (AUC)[10] are used as measures to evaluate the recommendation performance.

In the current case study, we focused our analysis in groups composed of five members, leaving to future works the evaluation with larger groups. In addition, the memory-based user-user (UKNN) and item-item (IKNN) collaborative filtering approaches are used as basic individual user recommenders,[9] needed in the phase 1 of the proposal.

We evaluate five group recommendation strategies to evaluate our proposal:

- *Cons Top-10:* It applies the minimum cost consensus model over the top-10 common items according to the Borda ranking, and obtains the final agreed rating value. For the rest of items, the aggregated value is obtained through the Mean aggregation strategy.
- *Cons Top-50*: The same strategy, but using the consensus model over the top-50 items.
- *Cons All*: It obtains the final agreed rating value through the minimum cost consensus model for all common items, disregarding the Borda ranking.
- *Mean*: It obtains the final aggregated rating value using the Mean aggregation strategy for all items.
- *Least Misery*: It obtains the final aggregated rating value using the Least Misery aggregation strategy for all items.[3]

For all cases, using the minimum cost model, consensus was reached in the optimization model.

Tables 1 and 2 present the performance associated to the five group recommendation strategies, according to AUC and MAE performance metrics. For both AUC and MAE measures, the best behavior was obtained for Cons Top-50 in the case of the UKNN individual recommender, and for Cons Top-10 in the case of the IKNN individual recommender. Globally, in the case of MAE the Least Misery strategy performs particularly worse. Another interesting finding was that for all cases the application of the consensus reaching only at some top-k ranked items according Borda, always leads to better performance in relation to the application of the consensus reaching to all items data.

Table 1. Evaluation results according to AUC. Larger values indicate better performance.

	Cons Top-10	Cons Top-50	Cons All	Mean	Least Misery
UKNN	0.6423	**0.6451**	0.6442	0.6428	0.6251
IKNN	**0.5576**	0.5498	0.5440	0.5527	0.5494

Table 2. Evaluation results according to MAE. Smaller values indicate better performance.

	Cons Top-10	Cons Top-50	Cons All	Mean	Least Misery
UKNN	0.7744	**0.7737**	0.7740	0.7742	0.8684
IKNN	**0.7969**	0.8032	0.8069	0.7995	0.9171

Overall, the performance values show that the consensus model could influence the GRS behavior, and that it could lead to the improvement of the recommendation performance. Further work will do a deeper exploration of the relation between the nature of the GRS data and the consensus reaching process to get a better justification of the obtained performance. Other consensus reaching models will be also explored.

5. Concluding Remarks

This paper introduced the use of the minimum cost consensus model in the GRS scenario. It is presented a case study that verifies that the proposal leads to a better recommendation accuracy in GRS. Our next future research will focus on the study of alternative consensus reaching models to be used in this scenario.

Acknowledgments

This research work was partially supported by the Research Project TIN2015-66524-P, a Scholarship for Young Doctors (2017) provided by the Advanced Center for Information and Communication Technologies at University of Jaén, Spain, and the Spanish FPU fellowship (FPU13/01151).

References

1. R. Yera and L. Martínez, *International Journal of Computational Intelligence Systems* **10**, 776 (2017).
2. J. Lu, D. Wu, M. Mao, W. Wang and G. Zhang, *Decision Support Systems* **74**, 12 (2015).
3. T. De Pessemier, S. Dooms and L. Martens, *Multimedia Tools and Applications* **72**, 2497 (2014).
4. J. Castro, R. Yera and L. Martínez, *Decision Support Systems* **94**, 1 (2017).
5. T. N. T. Tran, M. Atas, A. Felfernig and M. Stettinger, *Journal of Intelligent Information Systems*, 1 (2017).
6. J. C. Borda, *Histoire de l Academie Royale des Sciences* (1781).
7. D. Ben-Arieh, T. Easton and B. Evans, *IEEE Transactions on SMC-Part A: Systems and Humans* **39**, 210 (2009).
8. G. Zhang, Y. Dong, Y. Xu and H. Li, *IEEE Transactions on SMC-Part A: Systems and Humans* **41**, 1253 (2011).
9. G. Adomavicius and A. Tuzhilin, *IEEE Transactions on Knowledge and Data Engineering* **17**, 734 (2005).
10. A. Gunawardana and G. Shani, Evaluating recommender systems, in *Recommender Systems Handbook*, (Springer, 2015) pp. 265–308.

A new visualization for preferences evolution in group decision making

Á. Labella* and L. Martínez

Department of Computer Science, University of Jaén
Jaén, Spain
* alabella@ujaen.es

Consensus reaching processes (CRPs) in Group Decision Making (GDM) try to reach a mutual agreement among a group of decision makers before making a decision. To evaluate and understand the performance of a CRP is often complex due to, mainly, the presence of disagreement among decision makers. A clear, simple, correct and suitable visualization of the discussion consensus rounds is key for facilitating the analysis of such performance because, without a clear visualization, it is hard to understand the disagreements among experts. This paper proposes a new visualization related to experts' preferences and their evolution for CRPs based on the Principal Component Analysis (PCA).

Keywords: GDM; CRP; PCA.

1. Introduction

Decision making is a quotidian process in daily life. In Group Decision Making (GDM) problems, several decision makers or experts, with their own attitudes and opinions, need to reach a common solution, selecting the best alternative/s of a set of possible solutions.

Classically, the GDM resolution process, consisted in gathering the experts' assessments and choosing the best alternative/s according to the group's view. Nevertheless, many real-world problems might require consensualted decisions that are not ensured by the previous GDM process. For this reason, consensus reaching processes (CRPs), in which individuals/experts discuss and modify their preferences to reach a collective agreement before making decisions, have become an increasingly prominent research topic in GDM problems.[1-3] The resolution of GDM problems applying CRPs requires to take into account several aspects such as: conflicts between experts, detect non-collaborative experts, identify experts whose opinions are similar, etc. According to these aspects, to analyse and understand the evolution of a CRP is not a simple task. The visualization of the experts who participate in a CRP would facilitate the interpretability of the process, identifying easily the experts' behaviour, the advantages and drawbacks of a CRP model and bring to light either the correct or incorrect performance of the CRP.

The visualization of the experts' preferences to guide consensus in GDM problems presents an important challenge and, according to this, several proposals have been introduced.[4-7] Nevertheless, these proposals presents several aspects that should be improved/modified. On the one hand, they solely consider one type of preference relation to provide the experts' assessments; fuzzy preference relations (FPR)[8] in[5-7] and decision matrix in,[4] despite experts can provide their opinions using distinct preference relations such as linguistic preference relation (LPR),[9] hesitant preference relation[10] (HPR) or hesitant linguistic fuzzy preference relation[11,12] (HLPR). On the other hand, the visualization should be easy to compute and understand, allowing to obtain as much information as possible in the shortest possible time to facilitate the CRP analysis task. Furthermore, many of these proposals, develop/use a software focuses only on the experts' visualization. Nevertheless, it would be adequate to include such visualization in a framework that allows to consider other aspects related to the CRP for understanding the whole process in a proper way.

As it was aforementioned, the experts' preferences can be represented by means of multiple preferences relations, i.e. matrices compound by several rows and cols, that would require a multi-dimensional visualization that the human beings are not able to understand. The Principal Component Analysis (PCA)[13] technique has been successfully used for dealing with this problematic, since it allows to reduce the dimensionality of a set of data.

This paper presents a new visualization for experts' preferences and their evolution in CRPs using the PCA to reduce the dimensionality of the data and showing the preferences in a 2-D visualization space. Such visualization is integrated and validated in AFRYCA,[14,15] an analytic framework able to carry out analyses and studies in GDM problems resolution. The proposal is organized as follows: Section 2 makes a brief review on GDM, CRP and PCA. Section 3 introduces the new visualization for experts' preferences based on PCA, Section 4 shows an illustrative example using the AFRYCA 2.0 framework.[14] Finally, conclusions are given in Section 5.

2. Preliminaries

This section reviews several concepts about GDM problems CRPs, as well the main features of the PCA.

2.1. *Group decision making*

GDM is a process in which several experts participate in the selection of a common solution for a decision making problem, composed by a set of alternatives. Formally, a GDM problem is characterized[16] by n alternatives, denoted by $X = \{x_1, x_2, \ldots, x_n\}$ defined by a finite set of k criteria, $C = \{c_1, c_2, \ldots, c_k\}$, and a group of m experts, $E = \{e_1, e_2, \ldots, e_m\}$, who express their preferences over the alternatives, trying to reach a common solution. Each expert $e_i \in E$ express his/her opinion over distinct alternatives using distinct preference structures. The most common preference structures are introduced below:

- *Preference relations*: In a preference relation, experts express their opinions by means of pairwise comparisons between alternatives. These preferences are usually represented by symmetric matrices whose dimension is $n \times n$. Some examples are: FPR,[8] LPR,[9] HPR[10] and HLPR.[11,12]
- *Decision matrices*: In a decision matrix, experts express their opinions for each alternative over each criterion in function of its utility. These preferences are represented by matrices that might not symmetric whose dimension is $n \times k$.

Once the experts' preferences are gathered the classical process to solve a GDM problem is composed by two phases:[17] (i) Aggregation phase: experts' preferences are aggregated to obtain a collective assessments for the alternatives, (ii) Exploitation phase: an alternative or a subset of alternatives will be selected as solution for the problem.

2.2. *Consensus reaching processes*

Classically, the selection process in a GDM problem cannot guarantee the agreement between experts. Hence, several experts can feel that their opinion have not been sufficiently taken into account.[3] For this reason, CRPs were incorporated in the GDM problem resolution. CRP is an iterative and dynamic process in which experts change their opinions, trying to get closer each other, in order to reach a high agreement level after several rounds of discussion.[3] A CRP is composed by four phases:

(1) Gathering preferences: The experts provide their preferences.
(2) Consensus measurement: The group consensus degree is computed.
(3) Consensus control: The consensus degree obtained is compared with a predefined value which represents the minimum value of acceptable agreement. If the consensus degree is greater than the threshold value, the group starts the selection process, but, another discussion round would be carried out.
(4) Consensus progress: To increase the level of agreement throughout the discussion rounds of the CRP, experts have to modify their preferences.

2.3. *Principal component analysis*

Principal component analysis[13] is a multivariate technique that analyses a set of data whose observations are described by several inter-correlated quantitative dependent variables. Its main objective is to extract the most relevant information from the data, representing such information as a set of orthogonal variables called principal components. Finally, the similarity of the data and variables is computed and visualized by a reduced dimensional representation (e.g. 2D point). Starting from a matrix X composed by the initial data, the process consists of:

(1) Standardise the data subtracting the mean for each attribute.
(2) Compute the variance-covariance matrix of X, denote by C.

$$tC = X^T \cdot X \qquad (1)$$

where X^T denotes the transpose of X and t is a positive integer.
(3) Calculate the eigenvalues and eigenvector of C and select the r greater eigenvalues, being r the final data dimension.

$$C \cdot u = \lambda \cdot u \qquad (2)$$

(4) Represent the data in a U space with r dimension through a linear projection matrix denoted by M.

$$U = M \cdot X \qquad (3)$$

3. New Visualization for Experts' Preferences Evolution in GDM by Using PCA

It has been pointed out in Sec. 2.1, experts can provide their preferences using distinct preference structures in order to solve a GDM problem. These preferences structure are represented by a matrix P of changing dimension whose visualization might require a multi-dimensional representation hard to manage by human beings.

To visualize the experts' preferences in a interpretable way, we propose to utilize the PCA technique to reduce the dimensionality of the preferences into a 2 dimensional space. Taking into account that experts can use different preference structures to provide their preferences, a transformation of all of them into a decision matrix, denoted as X, by means of a dominance process [18] is carried out, being the preference structures processing the same for all the structures. The dominance values are noted as follows:

$$D_i = \{d^i_1, d^i_2, \ldots, d^i_n\} \quad i \in \{1, \ldots, m\} \quad (4)$$

Afterwards, the dominance vectors are standardised subtracting the arithmetic mean of such vector to its dominance values.

$$\overline{D_i} = \{\overline{d^i}_1, \overline{d^i}_2, \ldots, \overline{d^i}_n\}$$

$$\text{where} \quad \overline{d^i_j} = |d^i_j - \frac{1}{n}\sum_{k=1}^{n} d^i_k| \quad j,k \in \{1, \ldots, n\} \quad (5)$$

Starting from the standardised dominance vectors $\overline{D_i}$, the matrix X is composed as follows:

$$X = \begin{bmatrix} \overline{d^1_1} & \cdots & \overline{d^1_n} \\ \vdots & \ddots & \vdots \\ \overline{d^m_1} & \cdots & \overline{d^m_n} \end{bmatrix} \quad (6)$$

Then, the variance-covariance matrix of X is computed (see Eq. 2), obtaining the C matrix. Applying the decomposition of C, two eigenvectors denoted by u and v are calculated such that they have the greatest eigenvalues λ_1 and λ_2.

$$\begin{aligned} C \cdot u &= \lambda_1 \cdot u \\ C \cdot v &= \lambda_2 \cdot v \end{aligned} \quad (7)$$

The eigenvalues represent the amount of information contained in each principal component and their respective eigenvectors represent the direction of the principal component. Notice that we select two eigenvectors since the preferences will be represented in a 2D-map.

The coordinates of each expert in the (u, v) plane are given by:

$$(\overline{D^i}u, \overline{D^i}v) \qquad (8)$$

where $\overline{D^i}$ is the standardised dominance vector of the expert i in the original space.

This visualization process has been included in AFRYCA 2.0[14] to prove its good performance. This software provides different GDM problems and CRP model that facilitates the proposal incorporation. In addition, thanks to the architecture based on components in which the framework has been developed, such inclusion has been carried out easily.

4. Illustrative Example

In order to show the new proposal of visualization for experts' preferences in GDM, this section describes a GDM problem for selecting the best conference on Data Science in 2018, applying a CRP. The CRP simulation and the subsequent preferences visualization are carried out in AFRYCA 2.0,[14] an analytic framework able to carry out analyses and studies for GDM problems.

Let us suppose a group of renowned PhD $E = \{e_1, e_2, e_3, e_4, e_5, e_6, e_7\}$ that have to select the best conference in 2018 on Data Science and Knowledge Engineering among four possible alternatives $X = \{a_1, a_2, a_3, a_4\}$. The experts provide their preferences by using FPR, having been available the data set for public access in AFRYCA website.[a]

The evolution of the experts' preferences across the CRP are graphically represented in Fig. 1. Notice, the experts' preferences are represented with respect to the collective opinion thus, it is always represented in the center of the picture of each discussion round. The graphical information provided by the proposal facilitates the analysis of relevant aspects and difficulties found in GDM problems, such as conflicting opinions between experts, non-cooperative behaviours or disagreeing experts. In this example, it is easy to detect that the experts are receptive to the suggestions and no expert presents a non-cooperative behaviour, favouring the agreement between experts.

---

[a]http://sinbad2.ujaen.es/afryca/

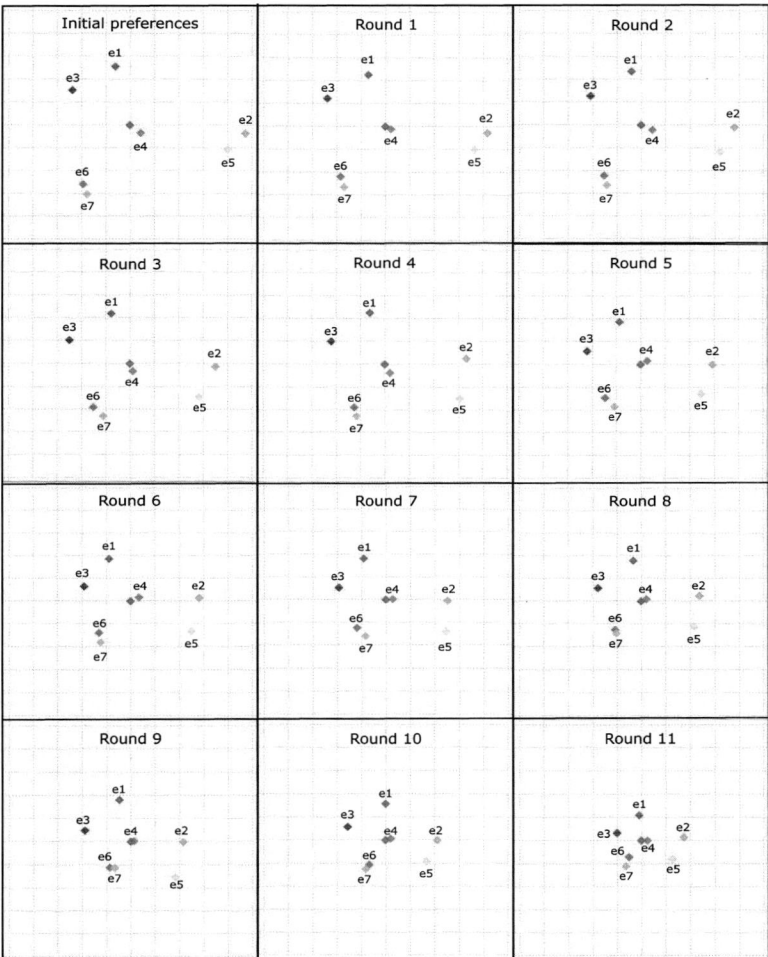

Fig. 1. CRP experts' preferences evolution.

5. Conclusions

Experts' preferences visualization in GDM is tremendously useful when a CRP is applied, since it allows to identify easily the experts' behaviour and evaluate the CRP performance. A novel visualization technique has been introduced in this contribution. The proposal is able to deal with multiple preference relations, reducing its dimensionality using the PCA technique and representing the experts' preferences in a bidimensional map.

Acknowledgments

This work is partially supported by the Spanish National Research Project TIN2015-66524-P, the Spanish Ministry of Economy.

References

1. R. Ureña, F. J. Cabrerizo, J. Morente-Molinera and E. Herrera-Viedma, *Information Sciences* **357**, 161 (2016).
2. E. Herrera-Viedma, F. Cabrerizo, J. Kacprzyk and W. Pedrycz, *Information Fusion* **17**, 4 (2014).
3. C. Butler and A. Rothstein, *On Conflict and Consensus: A Handbook on Formal Consensus Decision Making* (Food Not Bombs Publishing, 2006).
4. A. Ishizaka, S. Siraj and P. Nemery, *Energy* **95**, 602 (2016).
5. I. Palomares and L. Martínez, *Procedia Computer Science* **29**, 2090 (2014).
6. I. Palomares, L. Martínez and F. Herrera, *Knowledge-based Systems* **58**, 66 (2014).
7. S. Alonso, E. Herrera-Viedma, F. Cabrerizo, C. Porcel and A. G. López-Herrera, Using visualization tools to guide consensus in group decision making, in *Applications of Fuzzy Sets Theory*, (Springer Berlin Heidelberg, 2007).
8. S. Orlovsky, *Fuzzy Sets and Systems* **1**, 155 (July 1978).
9. R. M. Rodríguez, M. Espinilla, P. J. Sánchez and L. Martínez, *Internet Research* **20**, 296 (2010).
10. M. Xia and Z. Xu, *International Journal of Uncertainty, Fuzziness and Knowledge-Based Systems* **21**, 865 (2013).
11. R. Rodríguez, A. Labella and L. Martínez, *International Journal of Computational Intelligence Systems* **9**, 81 (2016).
12. R. M. Rodríguez, L. Martínez and F. Herrera, *Information Sciences* **241**, 28 (2013).
13. H. Hotelling, *Journal of educational psychology* **24**, p. 417 (1933).
14. Á. Labella, F. J. Estrella and L. Martínez, *Progress in Artificial Intelligence* **6**, 181 (2017).
15. I. Palomares, F. Estrella, L. Martínez and F. Herrera, *Information Fusion* **20**, 252 (2014).
16. J. Kacprzyk, *Fuzzy sets and systems* **18**, 105 (1986).
17. M. Roubens, *Fuzzy Sets and Systems* **90**, 199 (1997).
18. E. Czogała, A dominance of alternatives for decision making in probabilistic fuzzy environment, in *Cybernetics and Systems 86*, ed. R. Trappl (Springer, 1986) pp. 591–598.

An alternative core information of hesitant fuzzy linguistic term sets for linguistic decision making

Fangling Ren

College of Mathematics and Computer Science, Yan'an University
Yan'an, Shaanxi, China
renfangling203@163.com

Ying Qiao, Mingming Kong and Zheng Pei*

Center for Radio Administration & Technology Development, Xihua University
Chengdu, Sichuan, 610039, China
*pqyz@263.net

> The core information of hesitant fuzzy linguistic term sets is an important concept in hesitant fuzzy linguistic decision making, which is decided by "min_upper" and "max_lower" operators on hesitant fuzzy linguistic term sets. In the paper, we first provide t-norms and t-conorms on hesitant fuzzy linguistic term sets and discuss their properties. Then, we propose an alternative core information of hesitant fuzzy linguistic term sets based on t-norms and t-conorms. Finally, we utilize the new core information of hesitant fuzzy linguistic term sets for hesitant fuzzy linguistic decision making. We provide an example to illustrate the practicality of the new core information and comparison with the classical core information of hesitant fuzzy linguistic term sets, it seems that the new core information based on t-norms and t-conorms may be an alternative tool for hesitant fuzzy linguistic decision making.
>
> *Keywords*: Linguistic decision making; 2-tuple linguistic model; hesitant fuzzy linguistic term sets; t-norms; t-conorms.

1. Introduction

Nowadays, decision making methods with hesitant fuzzy linguistic term sets (HFLTSs) are hot in linguistic decision making (LDM). In many qualitative decision environments, experts think of several possible linguistic values or richer expressions than a single term for an indicator, alternative, variable, etc., accordingly, Rodríguez et al.[1] proposed the concept of HFLTSs to overcome the drawback of existed fuzzy linguistic approaches: the elicitation of single and very simple terms to encompass and express the qualitative information. HFLTSs allow us to use different and great flexible forms to represent decision makers' knowledge/preferences in LDM. To carry out HFLTSs decision making, Rodríguez et al. provided the "min_upper" and "max_lower" operators to obtain the core information of HFLTSs, which is the first method to combine HFLTSs and carry out hesitant fuzzy linguistic decision making. After then, many researchers have paid attention to HFLTSs decision making, such as Wei et al.[2] defined new negation, max-union and min-intersection closed operations for HFLTSs, then they proposed a hesitant fuzzy linguistic weighting averaging operator and a hesitant fuzzy linguistic ordered weighting averaging operator to deal with multi-criteria decision making problems with HFLTSs. Lee and Chen[3] proposed likelihood-based comparison relations of hesitant fuzzy linguistic term sets and several hesitant fuzzy linguistic aggregation operators to overcome the drawback of methods in.[1,2] Liu and Rodríguez[4] proposed fuzzy envelope of hesitant fuzzy linguistic term set for linguistic decision making with HFLTSs. Montserrat-Adell et al.[5] provided a lattice structure of the set of hesitant fuzzy linguistic term sets by means of the operations intersection and connected union and used the lattice in linguistic decision making with HFLTSs. Up to now, operations, extensions and measures of HFLTSs have been widely studied.[6-14]

In this paper, we focus on operators for HFLTSs, based on the new operators, we provide an alternative core information of HFLTSs and utilize it for hesitant fuzzy linguistic decision making. The rest of this paper is structured as follows: In Section 2, we briefly review the concept and the core information of HFLTSs. In Section 3, we propose t-norms and t-conorms of HFLTSs and analyze their properties, then we provide an alternative core information. In Section 4, we provide an example to show hesitant fuzzy linguistic decision making based on the new core information. We conclude the paper in Section 5.

2. Preliminaries

Definition 2.1. Let $S = \{s_0, \cdots, s_g\}$ be a linguistic term set, an HFLTS, H_S, is an ordered finite subset of the consecutive linguistic terms of S.

Basic operations on HFLTSs are: 1) Lower bound: $H_{S^-} = min(s_i) = s_j$, $s_i \in H_S$ and $s_i \geq s_j \forall i$; 2) Upper bound: $H_{S^+} = max(s_i) = s_j$, $s_i \in H_S$ and $s_i \leq s_j \forall i$; 3) Complement: $H_S^c = S - H_S = \{s_i | s_i \in S$ and $s_i \notin H_S\}$; 4) Union: $H_S^1 \cup H_S^2 = \{s_i | s_i \in H_S^1$ or $s_i \in H_S^2\}$; 5) Intersection: $H_S^1 \cap H_S^2 = \{s_i | s_i \in H_S^1$ and $s_i \in H_S^2\}$; 6) Envelope:

$env(H_S) = [H_{S^-}, H_{S^+}]$. Let $X = \{x_1, \ldots, x_n\}$ be a set of alternatives, $C = \{c_1, \ldots, c_m\}$ a set of criteria, $S = \{s_0, \cdots, s_g\}$ a linguistic term set and $\{H_S^j(x_i) | i \in \{1, \ldots, n\}, j \in \{1, \ldots, m\}\}$ a set of HFLTS. The min_upper operator consists of the following two steps.[1]

(1) Apply the upper bound H_{S^+} for each HFLTS that is associated with each alternative: $H_{S^+}(x_i) = \{H_{S^+}^1(x_i), \ldots, H_{S^+}^m(x_i)\}, i \in \{1, \ldots, n\}$.
(2) Obtain the minimum linguistic term for each alternative: $H_{S_{min}^+}(x_i) = min\{H_{S^+}^j(x_i) | j \in \{1, \ldots, m\}\}, i \in \{1, \ldots, n\}$.

The max_lower operator also consists of the following two steps.[1]

(1) Apply the low bound H_{S^-} for each HFLTS that is associated with each alternative: $H_{S^-}(x_i) = \{H_{S^-}^1(x_i), \ldots, H_{S^-}^m(x_i)\}, i \in \{1, \ldots, n\}$.
(2) Obtain the maximum linguistic term for each alternative: $H_{S_{max}^-}(x_i) = max\{H_{S^-}^j(x_i) | j \in \{1, \ldots, m\}\}, i \in \{1, \ldots, n\}$.

Let $H'_{max}(x_i) = max\{H_{S_{min}^+}(x_i), H_{S_{max}^-}(x_i)\}$ and $H'_{min}(x_i) = min\{H_{S_{min}^+}(x_i), H_{S_{max}^-}(x_i)\}$, the core information of the HFLTS aggregated is

$$H'(x_i) = [H'_{min}(x_i), H'_{max}(x_i)].$$

Rodríguez used preference degrees [15] between two core information to calculate the nondominance degree NDD_i of each alternative and obtained $X^{ND} = \{x_i | x_i \in X, NDD_i = max_{x_j \in X}\{NDD_j\}\}$.

From the algebraic operations point of view, t-norm and t-conorm serve as a natural generalization of min and max operations,[16] and various forms of t-norm or t-conorm have been used in fuzzy logics and many practical problems.[17-19] Formally, a t-norm is a binary operation T : $[0,1] \times [0,1] \rightarrow [0,1]$ such that for any $x, y, z \in [0,1]$, 1) $T(x,y) = T(y,x)$; 2) $T(x, T(y,z)) = T(T(x,y), z)$; 3) $T(x,y) \leq T(x,z)$, if $y \leq z$; 4) $T(x,1) = x$. A t-conorm is a binary operation $S : [0,1] \times [0,1] \rightarrow [0,1]$ such that for any $x, y, z \in [0,1]$, S satisfies 1)-3) and 4') $S(x,0) = x$. In 2-tuple linguistic representations,[21,22] Tao et al. used Archimedean t-norms and t-conorms to define new operations.[20] In the paper, we focus on the following t-norms and t-conorms shown in Table 1 to defined new operators of HFLTSs.

3. New Operators and Core Information of HFLTS

In this section, we provide t-norm and t-conorm on HFLTSs and obtain alternative core information.

Table 1. The most important t-norms and t-conorms ($\forall a, b \in [0, 1]$).

	t-norms			t-conorms
$T_\wedge$	$T_\wedge(a,b) = a \wedge b$	$S_\vee$		$S_\vee(a,b) = a \vee b$
$T.$	$T.(a,b) = a \cdot b$	$S.$		$S.(a,b) = a + b - a \cdot b$
T_E	$T_E(a,b) = \frac{a \cdot b}{1+(1-a)(1-b)}$	S_E		$S_E(a,b) = \frac{a+b}{1+ab}$
T_L	$T_L(a,b) = 0 \vee (a+b-1)$	S_L		$S_L(a,b) = 1 \wedge (a+b)$
T_n	$T_n(a,b) = \begin{cases} a \wedge b, & \text{if } a+b > 1, \\ 0, & \text{otherwise.} \end{cases}$	S_n		$S_n(a,b) = \begin{cases} a \vee b, & \text{if } a+b < 1, \\ 1, & \text{otherwise.} \end{cases}$
T_w	$T_w(a,b) = \begin{cases} a \wedge b, & \text{if } a \vee b = 1, \\ 0, & \text{otherwise.} \end{cases}$	S_w		$S_w(a,b) = \begin{cases} a \vee b, & \text{if } a \wedge b = 0, \\ 1, & \text{otherwise.} \end{cases}$

Definition 3.1. For t-norm T, t-conorm S and HFLTS $H = \{s_{i_1}, \cdots, s_{i_k}\}$ on $S = \{s_0, \cdots, s_g\}$, T and S operations on H are defined as follows.

$$T(H) = T(s_{i_1}, \cdots, s_{i_k}) = T(T(\cdots(T(s_{i_1}, s_{i_2}), \cdots), s_{i_{k-1}}), s_{i_k})$$
$$= \Delta(g \times T(T(\cdots(T(\frac{i_1}{g}, \frac{i_2}{g}), \cdots), \frac{i_{k-1}}{g}), \frac{i_k}{g})), \qquad (1)$$
$$S(H) = S(s_{i_1}, \cdots, s_{i_k}) = S(S(\cdots(S(s_{i_1}, s_{i_2}), \cdots), s_{i_{k-1}}), s_{i_k})$$
$$= \Delta(g \times S(S(\cdots(S(\frac{i_1}{g}, \frac{i_2}{g}), \cdots), \frac{i_{k-1}}{g}), \frac{i_k}{g})) \qquad (2)$$

where the function $\Delta : [0, g] \longrightarrow \{(s_i, \alpha_i) | s_i \in S, \alpha_i \in [-0.5, 0.5)\}$ is given by $\Delta(\beta) = (s_i, \alpha_i)$ with $i = round(\beta)$ and $\alpha_i = \beta - i$ for any $\beta \in [0, g]$.[21,22]

Because T and S are commutative and associative, $T(H)$ and $S(H)$ are well defined. Such as $T(\{s_3, s_4, s_5\}) = T(T(s_3, s_4), s_5) = \Delta(6 \times T(T(\frac{3}{6}, \frac{4}{6}), \frac{5}{6}))$ and $S(\{s_3, s_4, s_5\}) = S(S(s_3, s_4), s_5) = \Delta(6 \times S(S(\frac{3}{6}, \frac{4}{6}), \frac{5}{6}))$. If $T = \wedge$ and $S = \vee$, then $\wedge(\{s_3, s_4, s_5\}) = \Delta(6 \times \wedge(\wedge(\frac{3}{6}, \frac{4}{6}), \frac{5}{6})) = \Delta(6 \times \frac{3}{6}) = \Delta(3) = s_3$ and $\vee(\{s_3, s_4, s_5\}) = \Delta(6 \times \vee(\vee(\frac{3}{6}, \frac{4}{6}), \frac{5}{6})) = \Delta(5) = s_5$.

Proposition 3.1. *For any $S = \{s_0, \cdots, s_g\}$ and HFLTS $H = \{s_{i_1}, \cdots, s_{i_k}\}$ on S, 1) If $T = \wedge$, then $T(H)$ is lower bound of the HFLTS $H_S = \{s_j \in S | s_{i_1} \leq s_j \leq s_{i_k}\}$, i.e., $T(H) = H_{S-}$; 2) If $S = \vee$, then $S(H)$ is upper bound of the HFLTS $H_S = \{s_j \in S | s_{i_1} \leq s_j \leq s_{i_k}\}$, i.e., $S(H) = H_{S+}$.*

Proof. For $H = \{s_{i_1}, \cdots, s_{i_k}\}$, we have $\frac{i_1}{g} < \cdots < \frac{i_k}{g}$ due to $0 \leq i_1 < \cdots < i_k \leq g$, hence $T(H) = \wedge(s_{i_1}, \cdots, s_{i_k}) = \wedge(\wedge(\cdots(\wedge(s_{i_1}, s_{i_2}), \cdots), s_{i_{k-1}}), s_{i_k}) = \Delta(g \times \wedge(\wedge(\cdots(\wedge(\frac{i_1}{g}, \frac{i_2}{g}), \cdots), \frac{i_{k-1}}{g}), \frac{i_k}{g})) = \Delta(g \times \frac{i_1}{g}) = s_{i_1} = H_{S-}$ and $S(H) = \vee(s_{i_1}, \cdots, s_{i_k}) = \vee(\vee(\cdots(\vee(s_{i_1}, s_{i_2}), \cdots), s_{i_{k-1}}), s_{i_k}) = \Delta(g \times \vee(\vee(\cdots(\vee(\frac{i_1}{g}, \frac{i_2}{g}), \cdots), \frac{i_{k-1}}{g}), \frac{i_k}{g})) = \Delta(g \times \frac{i_k}{g}) = s_{i_k} = H_{S+}$. □

Proposition 3.1 means that $T(H)$ and $S(H)$ of HFLTS $H = \{s_{i_1}, \cdots, s_{i_k}\}$ on S are extensions of lower and upper bounds of HFLTS defined in.[1]

Definition 3.2. For a set of HFLTSs on S, $H = \{H_1, \cdots, H_r\}$, $H_{r'}(r' = 1, \cdots, r)$ is a HFLTS on S. The $T - S$ operator of H is defined as follows:

$$H_T^S = T(S(H_1), \cdots, S(H_r))$$
$$= T(T(\cdots(T(S(H_1), S(H_2)), \cdots), S(H_{r-1})), S(H_r)). \qquad (3)$$

The $S - T$ operator of H is defined as follows:

$$H_S^T = S(T(H_1), \cdots, T(H_r))$$
$$= S(S(\cdots(S(T(H_1), T(H_2)), \cdots), T(H_{r-1})), T(H_r)). \qquad (4)$$

Example 3.1. Let $S = \{s_0, s_1, s_2, s_3, s_4, s_5, s_6\}$. A set of HFLTSs on S is $H = \{H_1 = \{s_3, s_4, s_5\}, H_2 = \{s_4, s_5, s_6\}\}$, then $H_T^S = T(S(S(s_3, s_4), s_5), S(S(s_4, s_5), s_6))$ and $H_S^T = S(T(T(s_3, s_4), s_5), T(T(s_4, s_5), s_6))$. If $T = \wedge$ and $S = \vee$, then $H_T^S = s_5$ and $H_S^T = s_4$.

Proposition 3.2. *For any set* $H = \{H_1, \cdots, H_r\}$ *of HFLTSs on* S. *(1)* $H_\wedge^\vee$ *is the min_upper of the set of HFLTSs* $H_S = \{(H_1)_S, \cdots, (H_r)_S\}$, *i.e.,* $H_\wedge^\vee = H_{S_{min}^+}$ *; (2)* $H_\vee^\wedge$ *is the max_lower of the set of HFLTSs* $H_S = \{(H_1)_S, \cdots, (H_r)_S\}$, *i.e.,* $H_\vee^\wedge = H_{S_{max}^-}$.

Proof. According to Proposition 3.1, we have $H_\wedge^\vee = \wedge(\vee(H_1), \cdots, \vee(H_r)) = \wedge((H_1)_{S^+}, \cdots, (H_r)_{S^+}) = H_{S_{min}^+}$ and $H_\vee^\wedge = \vee(\wedge(H_1), \cdots, \wedge(H_r)) = \vee((H_1)_{S^-}, \cdots, (H_r)_{S^-}) = H_{S_{max}^-}$. □

Proposition 3.2 means that the $T-S$ and $S-T$ operators are extensions of the "min_upper" and "max_lower" operators. Hence, we can obtain alternative core information of the HFLTSs aggregated.

Definition 3.3. the $T - S$ core information of a set H of HFLTSs on S based on t-norms and t-conorms is defined as follows:

$$TS(H) = [T(H_T^S, H_S^T), S(H_T^S, H_S^T)]. \qquad (5)$$

According to Proposition 3.2, it is obvious that $TS(H) = [T(H_T^S, H_S^T), S(H_T^S, H_S^T)]$ is equal to $H'(x_i) = [H'_{min}(x_i), H'_{max}(x_i)]$[1] of HFLTSs on S if $T = \wedge$ and $S = \vee$, i.e., $[\wedge(H_\wedge^\vee, H_\vee^\wedge), \vee(H_\wedge^\vee, H_\vee^\wedge)] = [H'_{min}(x_i), H'_{max}(x_i)]$.

The ordering of t-norms and t-conorms [18] is $T_w < T < \wedge < \cdots < \vee < S < S_w$, i.e., the drastic product T_w (shown in Table 1) is the weakest t-norm and $\wedge$ (min) is the strongest t-norm, $\vee$ is the weakest t-conorm and drastic sum S_w is the strongest t-conorm.

Proposition 3.3. *For any* t-*norm* T, t-*conorm* S *and set* $H = \{H_1, \cdots, H_r\}$ *of HFLTSs on* S, $H_{r'}(r' = 1, \cdots, r)$ *is a HFLTS on* S. *If* $T(H_T^S, H_S^T)$

$\leq \wedge(H_\wedge^\vee, H_\vee^\wedge)$ and $\vee(H_\wedge^\vee, H_\vee^\wedge) \leq S(H_T^S, H_S^T)$, then the $T - S$ core information of a set H of HFLTSs on S satisfies $[H'_{min}(x_i), H'_{max}(x_i)] = [\wedge(H_\wedge^\vee, H_\vee^\wedge), \vee(H_\wedge^\vee, H_\vee^\wedge)] \subseteq [T(H_T^S, H_S^T), S(H_T^S, H_S^T)]$.

Proposition 3.3 means that if the core information in[1] represents the pessimistic and the optimistic, then $[T(H_T^S, H_S^T), S(H_T^S, H_S^T)]$ maybe the more pessimistic and the more optimistic, theoretically, $[T_w(H_{T_w}^{S_w}, H_{S_w}^{T_w}), S_w(H_{T_w}^{S_w}, H_{S_w}^{T_w})]$ maybe the most pessimistic and the most optimistic.

4. A Case Study

Let $A = \{a_1, a_2, a_3, a_4\}$ be a set of four alternatives, $C = \{c_1, c_2, c_3, c_4, c_5\}$ a set of criteria defined for each alternative and $S = \{$nothing (s_0), very low (s_1), low (s_2), medium (s_3), high (s_4), very high (s_5), perfect $(s_6)\}$. The assessments provided decision maker are shown in Table 2.

Table 2. Assessments of A with respect to criteria C.

	c_1	c_2	c_3	c_4	c_5
a_1	$\{s_1, s_2, s_3\}$	$\{s_4, s_5\}$	$\{s_4, s_5, s_6\}$	$\{s_3, s_4, s_5\}$	$\{s_2, s_3\}$
a_2	$\{s_2, s_3, s_4\}$	$\{s_3, s_4, s_5\}$	$\{s_3, s_4\}$	$\{s_2, s_3, s_4\}$	$\{s_3, s_4\}$
a_3	$\{s_4, s_5, s_6\}$	$\{s_1, s_2, s_3\}$	$\{s_2, s_3, s_4\}$	$\{s_3, s_4, s_5\}$	$\{s_4, s_5\}$
a_4	$\{s_4, s_5, s_6\}$	$\{s_2, s_3\}$	$\{s_4, s_5\}$	$\{s_4, s_5\}$	$\{s_1, s_2, s_3\}$

Based on Table 1 and Definition 3.3, all $T - S$ core information of alternatives can be calculated (shown in Table 3). Accordingly, preference degrees and nondominance degrees[15] between the four core information can be calculated (shown in Table 4). We obtain the set of nondominated alternatives as follows: 1) For $(T_\wedge, S_\vee)$, $a_1 = a_2 = a_3 = a_4$ by $NDD_1 = NDD_2 = NDD_3 = NDD_4 = 1$ (Cannot distinguish the preference order of alternatives[1]); For $(T_\cdot, S_\cdot)$ and (T_E, S_E), $a_3 \succ a_4 \succ a_1 \succ a_2$ due to $NDD_3 > NDD_4 > NDD_1 > NDD_2$; For (T_L, S_L), (T_n, S_n) and (T_w, S_w), $a_3 \succ a_1 = a_2 = a_4$, $a_2 \succ a_1 = a_3 = a_4$ and $a_1 = a_3 = a_4 \succ a_4$. It can be noticed that the $T - S$ core information maybe overcome drawback of the core information of HFLTSs defined by[1] in several cases.

5. Conclusions and Future Works

In this paper, we defined several new operations on HFLTS and a set of HFLTSs based on t-norms and t-conorms and analyze their properties, then we propose $T-S$ core information of a set of HFLTSs, which is an extension

Table 3. All $T-S$ core information of alternatives.

(t-norm, t-conorm)	$TS(H(a_1))$	$TS(H(a_2))$	$TS(H(a_3))$	$TS(H(a_4))$
$(T_\wedge, S_V)$	$[s_3, s_4]$	$[s_3, s_4]$	$[s_3, s_4]$	$[s_3, s_4]$
$(T, S.)$	$[(s_2, .35), (s_6, -.39)]$	$[(s_2, .39), (s_5, .29)]$	$[(s_3, .1), (s_6, -.3)]$	$[(s_2, .40), (s_6, -.25)]$
(T_E, S_E)	$[(s_3, -.19), (s_5, .45)]$	$[(s_4, -.24), (s_4, .39)]$	$[(s_4, -.09), (s_5, .39)]$	$[(s_3, -.29), (s_6, -.27)]$
(T_L, S_L)	$[s_5, s_6]$	$[s_2, s_6]$	$[s_6, s_6]$	$[s_5, s_6]$
(T_n, S_n)	$[s_0, s_6]$	$[s_6, s_6]$	$[s_3, s_6]$	$[s_0, s_6]$
(T_w, S_w)	$[s_4, s_6]$	$[s_0, s_6]$	$[s_4, s_6]$	$[s_4, s_6]$

Table 4. Nondominance degrees of alternatives.

(t-norm, t-conorm)	NDD$_1$	NDD$_2$	NDD$_3$	NDD$_4$
$(T_\wedge, S_V)$	1	1	1	1
$(T, S.)$	0.86	0.8	1	0.9
(T_E, S_E)	0.74	0.46	1	0.8
(T_L, S_L)	0	0	1	0
(T_n, S_n)	0	1	0	0
(T_w, S_w)	1	0.5	1	1

of the core information defined in,[1] from example, the $T-S$ core information can overcome the drawback of the core information[1] in several cases of linguistic decision making with HFLTSs.

Acknowledgments

This work has been partially supported by Research and Development of Yan'an (Grant No. 2017WZZ-03-02) and the National Natural Science Foundation of China (Grant No. 61372187).

References

1. R.M. Rodríguez, L. Martínez, F. Herrera, Hesitant fuzzy linguistic term sets for decision making, *IEEE Trans. Fuzzy Syst.* **20(1)**, 109–119 (2012).
2. C.P. Wei, N. Zhao, X.J. Tang, Operators and comparisons of hesitant fuzzy linguistic term sets, *IEEE Trans. Fuzzy Syst.* **22(3)**, 575–584 (2014).
3. L.-W. Lee, S.-M. Chen, Fuzzy decision making based on likelihood-based comparison relations of hesitant fuzzy linguistic term sets and hesitant fuzzy linguistic operators, *Inform. Sci.* **294**, 513–529 (2015).
4. H.B. Liu, R.M. Rodríguez, A fuzzy envelope of HFLTS and its application to multicriteria decision making, *Inform. Sci.* **258**, 220–238 (2014).
5. J. Montserrat-Adell, N. Agell et al., Modeling group assessments by means of hesitant fuzzy linguistic term sets, *J. Appl. Log.* (http://dx.doi.org/ 10.1016/j.jal.2016.11.005).
6. R.M. Rodríguez, L. Martínez, F. Herrera, A group decision making model dealing with comparative linguistic expressions based on hesitant fuzzy linguistic term sets, *Inform. Sci.* **241(1)**, 28–42 (2013).
7. R.M. Rodríguez, Á. Labella and L. Martínez, An overview on fuzzy modelling of complex linguistic preferences in decision making, *Int. J. Comput. Intell. Syst.* **9**, 81–94 (2016).
8. R.M. Rodríguez, L. Martínez, An Analysis of Symbolic Linguistic Computing Models in Decision Making, *Int. J. Gen. Syst.* **42 (1)**, 121–136 (2013).
9. K.-H. Chang, A more general reliability allocation method using the hesitant fuzzy linguistic term set and minimal variance OWGA weights, *Appli. Sof. Comp.* **56**, 589–596 (2017).
10. Z.S. Chen, K.S. Chin et al., Proportional hesitant fuzzy linguistic term set for multiple criteria group decision making, *Inform. Sci.* **357**, 61–87 (2016).
11. X.J. Gou, Z.S. Xu, H.C. Liao, Hesitant fuzzy linguistic entropy and cross-entropy measures and alternative queuing method for multiple criteria decision making, *Inform. Sci.* **388–389**, 225–246 (2017).
12. H.C. Liao, Z.S. Xu, Approaches to manage HFLTS information based on the cosine distance and similarity measures for HFLTSs and their application in qualitative decision making, *Exp. Syst. Appl.* **42**, 5328–5336 (2015).
13. F.L. Ren, M.M. Kong, Z. Pei, A new hesitant fuzzy linguistic topsis method for group multi-criteria linguistic decision making, *Symmetry* **2017**, *9*, 289.
14. Z. Li, C. Zhao, Z. Pei, Operations on HFLTS induced by archimedean triangular norms and conorms, *Int. J. Comput. Intell. Syst.* **11**, 514–524 (2018).
15. Y.M. Wang, J.B. Yang, D.L. Xu, A preference aggregation method through the estimation of utility intervals, *Comput. Oper. Rese.* **32**, 2027–2049 (2005).

16. V. Novák, I. Perfilieva, J. Močkoř, *Mathematical principles of fuzzy logic* (Kluwer Academic Publishers, 1999).
17. A. Mesiarová-Zemánková, Continuous additive generators of continuous, conditionally cancellative triangular subnorms, *Inform. Sci.* **339**, 53–63 (2016).
18. E.P. Klement, R. Mesiar, E. Pap, *Triangular Norms* (Kluwer Academic Publishers, Dordrecht, 2000).
19. E. Palmeira, B. Bedregal, R. Mesiar, J. Fernandez, A new way to extend t-norms, t-conorms and negations, *Fuzz. set. syst.* **240**, 1–21 (2014).
20. Z.F. Tao, H.Y. Chen, L.G. Zhou, J.P. Liu, On new operational laws of 2-tuple linguistic information using Archimedean t-norm and s-norm, *Knowl.-Based Syst.* **66**, 156–165 (2014).
21. F. Herrera, L. Martinez, A 2-tuple fuzzy linguistic representation model for computing with words, *IEEE Trans. Fuzzy Syst.* **8(6)**, 746–752 (2000).
22. L. Martínez, R.M. Rodriguez, F. Herrera (eds.), *The 2-tuple Linguistic Model-Computing with Words in Decision Making* (Springer International Publishing Switzerland, 2015).

An unbalanced linguistic terms transformation method for linguistic decision making

Jing Liu, Ying Qiao and Zheng Pei*

Center for Radio Administration & Technology Development, Xihua University Chengdu, Sichuan, 610039, China
*pqyz@263.net

In this paper, an unbalanced linguistic terms transformation method is proposed to transform multi-granularity unbalanced linguistic terms into 2-tuple linguistic values of any fixed linguistic term set, the method consists of transformations of linguistic terms and 2-tuple linguistic representations, transformation of linguistic terms can be used to obtain a unified information of multi-granularity unbalanced linguistic terms based on their triangular membership functions, transformation of 2-tuple linguistic representations can be used to transform the unified information as 2-tuple linguistic values of any fixed linguistic hierarchy. An example is utilized to illustrate the practicality of the new method and compare with the existed important transformation methods, it seems that the new method is an alternative transformation method for unbalanced linguistic decision making.

Keywords: Linguistic decision making; multi-granularity unbalanced linguistic terms; 2-tuple linguistic model; linguistic aggregation operator.

1. Introduction

Linguistic multi-criteria group decision making (LDM) based on 2-tuple fuzzy linguistic model[1] has been widely studied and applied in a wide range of fields and disciplines.[2–9] In most LDM problems, linguistic assessments are modeled by uniform and symmetrical linguistic term sets, such as comfortable degree of vehicles are often expressed by consumers using linguistic terms "poor, fair, good". In other cases, linguistic assessments are represented by linguistic term sets that are not uniform or symmetrically distribution, due to the nature of the linguistic variables that participate in the problem or preference needs a greater granularity on one side of the scale than on the other. In addition, due to different knowledge background and language habit, decision makers provide their linguistic information by means of multiple scales with different granularity of unbalanced linguistic term sets.[10–12] The modelling of unbalanced linguistic information was initially proposed in [13] by using the 2-tuple linguistic model and hierarchical linguistic terms, after then, many researchers studied unbalanced linguistic decision making problems. [14–18] To sum up, the main linguistic terms transformation methods can be described as follows: 1) Based on triangular membership functions: the method is based on triangular membership functions, in which, the max-min operator or linguistic symbolic are used to transform linguistic terms; 2) Based on linguistic hierarchy: The method is based on linguistic hierarchy, in which, a fixed linguistic hierarchical structure is provided to deal with multi-granularity linguistic terms and obtain precise results that can be expressed in any linguistic term set of the linguistic hierarchical structure.

In this paper, we provide an unbalanced linguistic terms transformation method for LDM. The rest of the paper is arranged as follows: In Section 2, we reviews several classical existing multi-granularity unbalanced linguistic information transformation methods. In Section 3, we propose the unbalanced linguistic terms transformation method. In Section 4, a practical example is used to show the proposed method and compare with existed transformation methods. We conclude the paper in Section 5.

2. Preliminaries

Let $S_p = \{s_0, \ldots, s_p\}$ be a linguistic term set, $\beta \in [0, p]$ a value representing the result of a symbolic aggregation operation. Then a 2-tuple linguistic value corresponding to β is defined as:[1] $\Delta : [0, p] \longrightarrow S_p \times [-0.5, 0.5)$ and $\Delta(\beta) \longmapsto (s_i, \alpha)$, where, $i = round(\beta)$ and $\alpha = \beta - i \in [-0.5, 0.5)$, $round(\cdot)$ is the usual round operation, s_i is the linguistic term that is mostly close to β and α represents the symbolic translation value. The main linguistic terms transformation methods can be formalized as follows: 1) Based on triangular membership functions: In the method, triangular membership functions of linguistic terms are defined in $[0, 1]$ and μ_{s_i} is expressed by a 3-tuple (a_i, b_i, c_i), where $a_i, b_i, c_i \in [0, 1]$, $\mu_{s_i}(b_i) = 1$, $\mu_{s_i}(a_i) = \mu_{s_i}(c_i) = 0$. Let $S_p = \{s_0, \ldots, s_p\}$ and $L_g = \{l_0, \ldots, l_g\}$, $g \geq p$, the transformation function from S_p to L_g be defined as follows:[2]

$$\tau_{S_p L_g} : S_p \to F(L_g), \tau_{S_p L_g}(s_i) \longmapsto \{(l_j, \alpha_j^i) | l_j \in L_g\}, \quad (1)$$

where, $\alpha_j^i = \max_y \min\{\mu_{s_i}(x), \mu_{l_j}(x)\}$ is the probability of s_i represented by l_j, $\mu_{s_i}(x)$ and $\mu_{l_j}(x)$ are membership degrees of $x \in [0, 1]$. Another transformation function[8] is $\chi : F(L_g) \to H(L_g)$ and $\chi(F(L_g)) = \Delta(\frac{\Sigma_{j=0}^{g} j\alpha_j}{\Sigma_{j=0}^{g} \alpha_j})$ = $\Delta(\beta) = (l, \alpha)$. 2) Based on linguistic hierarchy: Let $S_p = \{s_0, \ldots, s_p\}$

and $L_g = \{l_0, \ldots, l_g\}$ belong to different linguistic hierarchies, any 2-tuple linguistic term on S_p can be transformed to a 2-tuple linguistic term on L_g:[3] $TS_pL_g : H(S_p) \longrightarrow H(L_g)$ and $(s_i, \alpha_i) \longmapsto \Delta(\frac{\Delta^{-1}((s_i,\alpha_i))g}{p})$. It's advantages are that the computation is simple and there is no loss of information. However, the methodology is not suitable for the linguistic term sets with the same granular but different membership functions. For unbalanced linguistic terms transformation, we refer to[8,13] for more details.

3. An Unbalanced Linguistic Terms Transformation Method

In this section, we propose an unbalanced linguistic terms transformation method, the method consists of two major parts: 1) Transformation of linguistic terms; 2) Transformation of 2-tuple linguistic representations.

3.1. *Transformation of linguistic terms*

Suppose a linguistic term set S_p and their triangular membership functions (a_i, b_i, c_i) are defined on $[0, 1]$. Based on triangular membership functions of linguistic terms in S_p, we define transformation of $s_i \in S_p$ as follows.

Definition 3.1. For any $s_i \in S_p = \{s_0, \ldots, s_p\}$ and $\mu_{s_i} = (a_i, b_i, c_i)(i = 0, \ldots, p)$, the transformation of $s_i \in S_p$ is $\varphi(s_i) = (u_i, v_i)$, where

$$u_i = \frac{\sum_{j=0}^{i}(c_j - a_j)}{\sum_{j=0}^{p}(c_j - a_j)} \text{ and } v_i = b_i. \tag{2}$$

In transformation $\varphi(s_i)$ of $s_i \in S_p$, v_i is the kernel of μ_{s_i}. In addition, for each $\mu_{s_i} = (a_i, b_i, c_i)$, the area of μ_{s_i} is $\frac{1}{2} \times 1 \times (c_i - a_i)$, hence $u_i = \frac{\sum_{j=0}^{i}(c_j - a_j)}{\sum_{j=0}^{p}(c_j - a_j)} = \frac{\sum_{j=0}^{i}(\frac{1}{2} \times 1 \times (c_j - a_j))}{\sum_{j=0}^{p}(\frac{1}{2} \times 1 \times (c_j - a_j))}$. Membership function of (s_i, α_i) on S_p can be obtain by membership function of s_i, i.e., transformation of each $s_\beta = (s_i, \alpha_i) \in F(S_p)$ are $\varphi(s_\beta) = (u_{\beta_i}, v_{\beta_i})$, where $\beta_i = i + \alpha_i$, membership function of $s_\beta = (s_i, \alpha_i)$ is $\mu_{s_\beta} = (a_i + \frac{\alpha_i}{p}, b_i + \frac{\alpha_i}{p}, c_i + \frac{\alpha_i}{p})$, u_{β_i} and v_{β_i} are

$$u_\beta = \frac{(\sum_{j=0}^{i}(c_j - a_j)) + (c_i - a_i)}{(\sum_{j=0}^{p}(c_j - a_j)) + (c_i - a_i)} \text{ and } v_\beta = b_i + \frac{\alpha_i}{p}.$$

Such as $s_{4.3} = (s_4, 0.3)$ obtained from S_8 is shown in Fig. 1.

The advantage of transformation of 2-tuple linguistic value is to distinguish different linguistic term sets when the number of linguistic terms

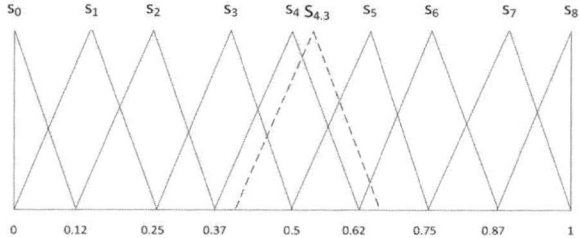

Fig. 1. Triangular membership functions of $S_8 = \{s_0, \ldots, s_8\}$.

are same. For example in Fig. 2. According to Eq. (2), transformations of $s_2^1 \in S_4^1$ is $\varphi(s_2^1) = (0.8125, 0.75)$, transformations of $s_2^2 \in S_4^2$ is $\varphi(s_2^2) = (0.803, 0.75)$, transformations $\varphi(s_i^1)$ and $\varphi(s_i^2)$ are different.

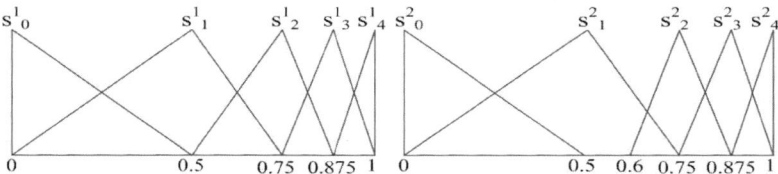

Fig. 2. Unbalanced linguistic term sets S_4^1 and S_4^2.

3.2. Transformation of 2-tuple linguistic representations

Here, we provide a method to represent transformation $\varphi(s_i) = (u_i, v_i)$ of $s_i \in S_p = \{s_0, \ldots, s_p\}$ as 2-tuple linguistic value on $L_g = \{l_0, \ldots, l_g\}$ and represent transformation $\varphi(s_\beta) = (u_{\beta_i}, v_{\beta_i})$ of any 2-tuple linguistic value $s_\beta = (s_i, \alpha_i) \in S_p$ as 2-tuple linguistic value on L_g. Denote transformation $\varphi(s_i) = (u_i, v_i)$ of $s_i \in S_p$ and $\varphi(s_j) = (u_j, v_j)$ of $s_j \in L_g$. The closeness degree of $\varphi(s_i) = (u_i, v_i)$ and $\varphi(s_j) = (u_j, v_j)$ is decided by

$$C_{ij} = max\{min\{\frac{u_j}{u_i}, \frac{u_i}{u_j}\}, min\{\frac{v_j}{v_i}, \frac{v_i}{v_j}\}\}, \quad (3)$$

in which, $C_{ij} \in [0, 1]$, and if $u_i = u_j$ or $v_i = v_j$, then $C_{ij} = 1$, especially, for $s_0 \in S_p$ or L_g, we limit $\frac{0}{0} = 1$ and $\frac{b_i}{0} = \infty$ if $b_i \neq 0$. For any fixed $s_i \in S_p = \{s_0, \ldots, s_p\}$, we denote

$$s_{j'} = min\{s_j \in L_g | C_{ij} = max\{C_{i0}, \ldots, C_{ig}\}\}. \quad (4)$$

Then 2-tuple linguistic value of $\varphi(s_i) = (u_i, v_i)$ on $L_g = \{l_0, \ldots, l_g\}$ is

$$s_\beta = \Delta^{-1}(j' + (max\{\frac{u_{j'}}{u_i}, \frac{v_{j'}}{v_i}\} \times v_i - v_{j'}) \times g). \tag{5}$$

Such as in Fig. 2, by Eq. (3), for $\varphi(s_1^1) = (0.625, 0.5)$, we have $C_{10} = max\{min\{\frac{0.263}{0.625}, \frac{0.625}{0.263}\}, min\{\frac{0}{1}, \frac{1}{0}\}\} \doteq 0.42$, $C_{11} = 1$, $C_{12} \doteq 0.78$, $C_{13} \doteq 0.67$ and $C_{14} \doteq 0.625$. According to Eq. (4), we have $s_1^2 = min\{s_j \in S_4^2 | C_{11} = max\{C_{10}, \ldots, C_{14}\}\}$. By Eq. (5), 2-tuple linguistic value of $\varphi(s_1^1) = (0.625, 0.5)$ on S_4^2 is $s_\beta = \Delta^{-1}(1 + (max\{\frac{0.658}{0.625}, \frac{0.5}{0.5}\} \times 0.5 - 0.5) \times 4) \doteq \Delta^{-1}(1 + (0.526 - 0.5) \times 4) = \Delta^{-1}(1.104) = (s_1^2, 0.104)$. For $\varphi(s_2^2) = (0.803, 0.75)$, we have $s_2^1 = min\{s_i \in S_4^1 | C_{22} = max\{C_{20}, \ldots, C_{24}\}\}$ and 2-tuple linguistic value of $\varphi(s_2^2) = (0.803, 0.75)$ on S_4^2 is $s_\beta = \Delta^{-1}(2 + (max\{\frac{0.8125}{0.803}, \frac{0.75}{0.75}\} \times 0.75 - 0.75) \times 4) \doteq \Delta^{-1}(2.035) = (s_2^1, 0.035)$.

4. Illustrative Example

A company intends to select an ERP system to implement from three candidates $A = \{a_1, a_2, a_3\}$. Three information officers $M = \{d_1, d_2, d_3\}$ from the three different departments of the company assesses the candidate ERP systems in terms of three criteria $C = \{c_1$ (potential cost), c_2 (function), and c_3 (operation complexity)$\}$. Three information officers give their assessments in three multi-granularity linguistic terms, their triangular membership functions are shown in Table 1. The weights of C and M are $\omega = (0.3, 0.5, 0.2)^T$ and $\lambda = (1/3, 1/3, 1/3)^T$. Linguistic assessments of alternatives provided by M with respect to C are shown in Table 2. We utilize the proposed method to transform linguistic terms into the same hierarchical linguistic terms. Suppose that S_8^1 is the basic linguistic term set. Based on Definition 3.1, we can obtain transformations of linguistic terms of S_8^1, S_6^2 and S_4^3 shown in Table 3. According to Eqs. (3)–(5) and

Table 1. Triangular membership functions of linguistic term sets.

d_1 and S_8^1	d_2 and S_6^2	d_3 and S_4^3
s_0^1 Worst/(0,0,0.125)	s_0^2 Worst/(0,0,0.2)	s_0^3 Worst/(0,0,0.5)
s_1^1 poor/(0,0.125,0.25)	s_1^2 poor/(0,0.2,0.6)	s_1^3 poor/(0,0.5,0.75)
s_2^1 Very bad/(0.125,0.25,0.375)	—	—
s_3^1 Little bad/(0.25,0.375,0.5)	s_2^2 Little bad/(0.2,0.6,0.7)	—
s_4^1 Middle/(0.375,0.5,0.625)	s_3^2 Middle/(0.6,0.7,0.8)	s_2^3 Middle/(0.5,0.75,0.875)
s_5^1 Little good/(0.5,0.625,0.75)	s_4^2 Little good/(0.7,0.8,0.9)	—
s_6^1 Very good/(0.625,0.75,0.875)	—	—
s_7^1 Preferably/(0.75,0.875,1)	s_5^2 Preferably/(0.8,0.9,1)	s_3^3 Preferably/(0.75,0.875,1)
s_8^1 Best/(0.875,1,1)	s_6^2 Best/(0.9,1,1)	s_4^3 Best/(0.875,1,1)

Table 2. Linguistic assessments of three alternatives.

	$d_1(S_8^1)$			$d_2(S_6^2)$			$d_3(S_4^3)$		
	a_1	a_2	a_3	a_1	a_2	a_3	a_1	a_2	a_3
c_1	s_3^1	s_3^1	s_7^1	s_2^2	s_2^2	s_1^2	s_3^3	s_4^3	s_0^3
c_2	s_7^1	s_3^1	s_7^1	s_2^2	s_0^2	s_6^2	s_0^3	s_3^3	s_4^3
c_3	s_8^1	s_1^1	s_0^1	s_2^2	s_4^2	s_6^2	s_1^3	s_4^3	s_1^3

Table 3. Transformations of linguistic terms.

	S_8^1	S_6^2	S_4^3
	$(0.0625, 0)$	$(0.1, 0)$	$(0.25, 0)$
	$(0.1875, 0.125)$	$(0.4, 0.2)$	$(0.625, 0.5)$
	$(0.3125, 0.25)$	—	—
	$(0.4375, 0.375)$	$(0.65, 0.6)$	—
$\varphi(s_i^k)$	$(0.5625, 0.5)$	$(0.75, 0.7)$	$(0.8125, 0.75)$
	$(0.6875, 0.625)$	$(0.85, 0.8)$	—
	$(0.8125, 0.75)$	—	—
	$(0.9375, 0.875)$	$(0.95, 0.9)$	$(0.9375, 0.875)$
	$(1, 1)$	$(1, 1)$	$(1, 1)$

Table 4. 2-tuple linguistic values of S_6^2 and S_4^3 in S_8^1.

S_8^1	S_6^2	S_4^3
$(s_0^1, 0)$	$(s_0^2, 0) \to (s_0^1, 0)$	$(s_0^3, 0) \to (s_0^1, 0)$
$(s_1^1, 0)$	$(s_1^2, 0) \to (s_3^1, 0)$	$(s_1^3, 0) \to (s_4^1, 0)$
$(s_2^1, 0)$	—	—
$(s_3^1, 0)$	$(s_2^2, 0) \to (s_5^1, 0.077)$	—
$(s_4^1, 0)$	$(s_3^2, 0) \to (s_6^1, 0.067)$	$(s_2^3, 0) \to (s_6^1, 0)$
$(s_5^1, 0)$	$(s_4^2, 0) \to (s_6^1, 0.118)$	—
$(s_6^1, 0)$	—	—
$(s_7^1, 0)$	$(s_5^2, 0) \to (s_7^1, 0.105)$	$(s_3^3, 0) \to (s_7^1, 0)$
$(s_8^1, 0)$	$(s_6^2, 0) \to (s_8^1, 0)$	$(s_4^3, 0) \to (s_8^1, 0)$

Table 3, linguistic terms of S_6^2 and S_4^3 are represented by 2-tuple linguistic values on S_8^1, which are shown in Table 4.

Based on Table 4, linguistic assessments of alternatives shown in Table 2 are transformed into 2-tuple linguistic values of the linguistic term set S_8^1. We utilize the weighted 2-tuple linguistic aggregation operator f_W [1] to obtain the final evaluation result of each alternative and candidate a_3 is the selected ERP system. Table 5 shows comparison of 2-tuple linguistic values in S_8^1 transformed from S_6^2 by using four methods. Utilizing the weighted 2-tuple linguistic aggregation operator f_W, we can obtain the decision making results of candidates based on Table 5, which are shown in Table 6. It seems that decision making results of three candidates based on our method are more distinguishable than methods.[2,13,17]

Table 5. Comparison of S_8^1 transformed from S_6^2.

S_6^2	2-tuple linguistic values in S_8^1			
	Method [17]	Method [13]	Method [2]	**Our method**
s_0^2	$(s_0^1, 0.38)$	$(s_1^1, -0.42)$	$(s_0^1, 0.00)$	$(s_0^1, 0.000)$
s_1^2	$(s_1^1, 0.45)$	$(s_2^1, 0.40)$	$(s_1^1, 0.30)$	$(s_3^1, 0.000)$
s_2^2	$(s_3^1, -0.33)$	$(s_4^1, 0.00)$	$(s_4^1, -0.30)$	$(s_5^1, 0.077)$
s_3^2	$(s_4^1, 0.00)$	$(s_6^1, -0.38)$	$(s_4^1, 0.00)$	$(s_6^1, 0.067)$
s_4^2	$(s_5^1, 0.33)$	$(s_6^1, 0.38)$	$(s_5^1, 0.30)$	$(s_6^1, 0.118)$
s_5^2	$(s_7^1, -0.50)$	$(s_7^1, 0.30)$	$(s_7^1, -0.30)$	$(s_7^1, 0.105)$
s_6^2	$(s_8^1, -0.22)$	$(s_8^1, -0.31)$	$(s_8^1, 0.00)$	$(s_8^1, 0.000)$

Table 6. Comparison of decision making results of three candidates.

A	2-tuple linguistic values in S_8^1			
	Method [17]	Method [13]	Method [2]	**Our method**
a_1	$(s_4^1, -0.055)$	$(s_5^1, -0.302)$	$(s_4^1, 0.166)$	$(s_5^1, -0.206)$
a_2	$(s_4^1, -0.224)$	$(s_4^1, 0.233)$	$(s_4^1, -0.077)$	$(s_4^1, 0.282)$
a_3	$(s_5^1, 0.279)$	$(s_6^1, -0.465)$	$(s_5^1, 0.33)$	$(s_6^1, -0.367)$

5. Conclusion

An unbalanced linguistic terms transformation method is proposed in this paper, the method consists of transformations of linguistic terms and 2-tuple linguistic representations, the triangular membership function of linguistic term is used to obtain the unified information (u_{s_i}, v_{s_i}). Then the unified information of linguistic term is transformed as 2-tuple linguistic values of any fixed linguistic hierarchy. Comparison analysis in a practical decision making problem shows that the method is an alternative method to transform multi-granularity unbalanced linguistic term sets.

Acknowledgments

This work has been partially supported by Sichuan project (2018GZ0256) and the National Natural Science Foundation of China (61372187).

References

1. F. Herrera, L. Martínez, A 2-tuple fuzzy linguistic representation model for computing with words, *IEEE Trans. Fuzzy Syst.* **8(6)**, 746–752 (2000).
2. F. Herrera, L. Martínez, A model based on linguistic 2-tuple for dealing with multi-granular hierarchical linguistic contexts in multi-expert decision-making, *IEEE Trans. Syst. Man. Cybe.* **31(2)**, 227–234 (2001).
3. F. Herrera, L. Martínez, P.J. Sánchez, Managing non-homogeneous information in group decision making, *Euro. J. Oper. Rese.* **166(1)**, 115–132 (2005).
4. Z. Pei, D. Ruan, J. Liu, Y. Xu, *Linguistic Values-based Intelligent Information Processing: Theory, Methods, and Applications* (Atlantis Press, 2010).

5. M.A. Abchir, I. Truck, An extension of the 2-tuple linguistic model to deal with unbalanced linguistic term sets, *Kybernetika* **49(1)**, 164–180 (2013).
6. F.J. Cabrerizo, I.J. Pérez, E. Herrera-Viedma, Managing the consensus in group decision making in an unbalanced fuzzy lingusitic context with incomplete information, *Knowl.-Based Syst.* **23(2)**, 169–181 (2010).
7. L. Martínez, F. Herrera, An overview on the 2-tuple linguistic model for computing with words in decision making: Extensions, applications and challenges, *Inform. Sci.* **207(1)**, 1–18 (2012).
8. L. Martínez, R.M. Rodriguez, F. Herrera, *The 2-tuple Linguistic Model-Computing with Words in Decision Making* (Springer International Publishing Switzerland, 2015).
9. L. Martínez, D. Ruan, F. Herrera, Computing with words in decision support systems: An overview on models and applications, *Inte. J. Comp. Inte. Syst.* **3(4)**, 382–395 (2010).
10. J.A. Morente-Molinera, I.J. Pérez, M.R. Ureña, E. Herrera-Viedma, On multi-granular fuzzy linguistic modeling in group decision making problems: A systematic review and future trends, *Knowl.-Based Syst.* **74**, 49–60 (2015).
11. L. Jiang, H.B. Liu, J.F. Cai, The power average operator for unbalanced linguistic term sets, *Info. Fusi.* **22**, 85–94 (2015).
12. C.C. Li, Y.C. Dong, Unbalanced linguistic approach for venture inverstment evaluation with risk attitudes, *Prog. Arti. Intel.* **3(1)**, 1–13 (2014).
13. F. Herrera et al., A fuzzy linguistic methodology to deal with unbalanced linguistic term sets, *IEEE Trans. Fuzzy Syst.* **16(2)**, 354–370 (2008).
14. F.J. Cabrerizo et al., Soft consensus measures in group decision making using unbalanced fuzzy linguistic information, *Sof. Comp.* **21:11**, 3037–3050 (2017).
15. J.A. Morente-Molinera, G. Kou et al., Solving multi-criteria group decision making problems under environments with a high number of alternatives using fuzzy ontologies and multi-granular linguistic modelling methods, *Knowl.-Based Syst.* **137**, 54–64 (2017).
16. Y.C. Dong et al., Connecting the linguistic hierarchy and the numerical scale for the 2-tuple linguistic model and its uses to deal with hesitant unbalanced linguistic information, *Inform. Sci.* **367–368**, 259–278 (2016).
17. Z.F. Chen, D. Ben-Arieh, On the fusion of multi-granularity linguistic label sets in group decision making, *Compu. Indu. Engi.* **51**, 526–541 (2006).
18. W. Yu, Z. Zhang, Q. Zhong, L. Sun, Extended TODIM for multi-criteria group decision making based on unbalanced hesitant fuzzy linguistic term sets, *Compu. Indu. Engi.* **114**, 316–328 (2017).

Absolute and relative preferences in AHP-like matrices

David Koloseni

Department of Mathematics, University of Dar es salaam, Tanzania
david.koloseni@his.se

Tove Helldin** and Vicenç Torra**
School of Informatics, University of Skövde, Sweden
**{*tove.helldin,vtorra*}*@his.se*

The Analytical Hierarchy Process (AHP) has been extensively used to interview experts in order to find the weights of the criteria. We call AHP-like matrices relative preferences of weights. In this paper we propose another type of matrix that we call a absolute preference matrix. They are also used to find weights, and we propose that they can be applied to find the weights of weighted means and also of the Choquet integral.

1. Introduction

Aggregation functions[1,2] have been used extensively in several contexts to combine and fuse information. There are functions for a large variety of data including numerical, categorical, partitions, sequences, and dendrograms.

Arithmetic and weighted means are the most well known functions. They assume (frequently implicitly) that the information we aggregate is independent. Other operators for numerical data include quasi-arithmetic mean, the ordered weighted operator (OWA),[3] the weighted OWA (WOWA),[4] and the fuzzy integrals.[5,6] Among the fuzzy integrals we find both the Choquet[7] and the Sugeno[8] integrals. One of the main characteristics of the Choquet and Sugeno integrals is that they permit to integrate numerical values from different information sources taking into account interactions between these sources. These interactions are expressed in their parameters.

In fact, most aggregation functions use parameters to express some background information either on the information sources or on the values to be aggregated (or both). For example, in the weighted mean we have the weights as parameters which are used to represent information about the importance of the information sources. Formally, the weights define a vector with one component for each information source. We have also a vector of weights in the OWA, and two vectors in the WOWA.

In fuzzy integrals, parameters are defined in terms of a fuzzy measure. A fuzzy measure is a set function over the set of information sources. That is, we have a kind of importance for each set of information sources, instead of considering the importance of single information sources.

In order to apply aggregation functions to real problems, a very much significant problem is the determination of the weights. Several approaches have been considered in the literature. Some are based on interviewing domain experts that help later to find the weights. Others are based on learning the weights from examples (or finding the weights that optimize a given error function).

The Analytical Hierarchy Process (AHP) developed by Saaty[9,10] is an example of the former. The idea is to question an expert about the relative importance of one information source with respect to the others. In this way, we obtain a matrix of comparisons: one value for each pair of information sources. From this matrix we extract the weights of each source.

In this paper we study and discuss the determination of fuzzy measures in a way inspired by AHP. In relation to the definition of the method, we also propose and discuss the difference between absolute preferences and relative preferences when building AHP-like matrices.

The structure of the paper is as follows. In Section 2 we give some preliminaries needed for the rest of the paper. In Section 3 we introduce and define relative and absolute preferences for AHP-like matrices. In Section 4 we discuss the problem of matrix elicitation. The paper finishes with some conclusions.

2. Preliminaries

This section gives a brief review of some concepts needed later. Due to page limitations, we refer the readers to[1,2] for details.

2.1. *Some aggregation functions*

Given data $a_1, \ldots, a_n$ the arithmetic mean is defined by $\sum a_i/n$. Let $w = (w_1, \ldots, w_n)$ be a weighting vector (i.e., $w_i \geq 0$ and $\sum w_i = 1$) with weights for each information source. The weighted mean is defined by $\sum_i w_i a_i$. When $w_i = 1/n$, the weighted mean reduces to the arithmetic mean.

The Choquet integral in discrete domains can be seen as a generalization of the weighted mean. The Choquet integral expresses the importance of sources by means of a fuzzy measure. A fuzzy measure is a monotonic set function. If we represent the set of information sources by X, a fuzzy measure μ is defined on subsets of X. As additivity is not required to μ, we can use fuzzy measures to represent interactions. Families of fuzzy measures have been defined in the literature. Sugeno λ-measures are some of them. They have the property that we can determine the measure from non-additive weights on the sources. See Section 5.3.1 in[1] for details.

The Choquet integral then takes into account these importances to aggregate the values a_i. When the measure is additive (i.e., criteria are independent), the Choquet integral reduces to a weighted mean.

2.2. The AHP method

The AHP method is a way to extract weights for a weighted mean. In a multicriteria decision making (MCDM) problem, this means to assign weights to criteria $c_1, \ldots, c_n$. The process starts with interviewing experts and asking them to evaluate each pair of criteria c_i, c_j and how many times c_i is more relevant than c_j. We define a matrix $\{a_{ij}\}$ with this information. Weights are then extracted from the $\{a_{ij}\}$. Note, if w_i and w_j are weights for criteria c_i and c_j, then $a_{ij} = w_i/w_j$. If experts are consistent $a_{ij} = 1/a_{ji}$ and then also $a_{ij} = a_{ik}a_{kj}$. In practice, these conditions do not hold. Alternative approaches exist to find weights that approximate the matrix.

3. On the Definition of the AHP Matrices

As we have explained in the previous section, AHP bases its definition on comparing pairs of criteria. For each pair of criteria c_i, c_j the expert is asked to what degree c_i is more important than c_j. Then, the value given by the expert, say a_{ij}, is put into the matrix.

In this process, it is implicitly assumed that there is an order on the importance of the criteria, and this order is the basis for the construction of the weights. Basically, if c_i is said to be more important than c_j, then the weight of c_i is presumed to be larger than the weight of c_j.

Looking at the problem from another perspective, we see that whichever criteria we take first, the matrix we obtain will be the same. At this point, of course, we can recall that the matrices are usually inconsistent and we would probably do not get exactly the same results, but the matrices would not differ substantially.

Let us call this type of matrix *absolute preference* to distinguish them in what follows. We think that this situation applies well when the criteria are independent.

Let us consider a different situation. We have a set of criteria, and we want to aggregate some data taking them into account. Nevertheless, some of these criteria are independent of each other, but some are not. Then, if we want to aggregate data so that alternatives rate well in independent criteria, we may result in matrices that do not represent absolute preferences but *relative preferences*.

In particular, let assume that c_1, c_2, c_3 are highly correlated variables, and c_4, c_5 are also highly correlated but that any variable of the first set of variables is not correlated with any of the variables of the second set. Then, depending on which criterion is selected first, the preference ratios will be different. If we start with c_2, the relevance to add c_2 will be zero and what is relevant is to take c_4 or c_5. If we consider c_1 the relevance of c_4 is high. Similarly, if we start with c_4 then the relevance of c_1 is also high. More particularly, we may expect that these two relevances are the same. In other words, we may expect that the matrix is symmetric. In addition, we may also expect that the diagonal is zero. Note that if we have already c_i, the importance or relevance of having c_i (again) is zero.

More specifically, this process defines the relative preference matrix $A = \{a_{ij}\}$ and the elements a_{ij} of the matrix correspond to answers to the question: *If we take attribute c_i, to which degree would you also include c_j?* The degree value is taken in the $[0, 1]$ interval.

Then, the relative preference matrix will be one in which the elements will be between 0 and 1 (i.e., $a_{ij} \in [0,1]$), where we expect symmetry ($a_{ij} = a_{ji}$) and in which the diagonal is zero ($a_{ii} = 0$). A matrix satisfying these properties will be called consistent.

In a way similar to absolute preferences in AHP, where matrices in practice are not fully consistent, we do not expect matrices in practice to satisfy all these properties. We may consider algorithms to find an approximate consistent matrix from an inconsistent one, and algorithms to find weights (or fuzzy measures) from the matrices.

Let us use for the sake of illustration the following matrix for the 5 criteria above: $C = \{c_1, c_2, c_3, c_4, c_5\}$.

$$A = \begin{pmatrix} 0 & 0.1 & 0.1 & 0.9 & 0.9 \\ 0.1 & 0 & 0.1 & 0.9 & 0.9 \\ 0.1 & 0.1 & 0 & 0.9 & 0.9 \\ 0.9 & 0.9 & 0.9 & 0 & 0.1 \\ 0.9 & 0.9 & 0.9 & 0.1 & 0 \end{pmatrix}$$

We now propose a method to find the weights of a matrix with relative preferences. We assume as above that the elements of the matrix are denoted by $\{a_{ij}\}$. The method defined below uses some randomness to select the alternative criteria.

The algorithm assumes that $a_{ij} \neq 0$ for $i \neq j$. When there are $a_{ij} = 0$, we may get into a situation where some criteria are selected but others cannot. If this is the case, all criteria that cannot be selected receive a weight equal to zero.

- **Step 1.** $k = 1$
- **Step 2.** $c(k) = i$; // Select criteria i at random
- **Step 3.** $w(i) = \alpha$; // Assign a high weight α to the ith criterion
- **Step 4.** while not all criteria are selected **loop**
  - **Step 4.1.** $k = k + 1$
  - **Step 4.2.** $c(k) =$ select a criteria connected to $c(k-1)$ not yet in c according to a probability distribution built from $a_{c(k-1),j}$.
  - **Step 4.3.** $w(c(k)) = \alpha \cdot \min_{j \in c((1),\ldots,c(k-1)} a_{j,c(k)}$;
- **Step 5 end while**

Let us look to Steps 4.2 and 4.3 in more detail.

Step 4.2 selects a new criteria among the ones that have not yet been selected (those not yet in the vector $c(\cdot)$). This selection is done randomly and the probability of a criterion being chosen is based on its relevance or eligibility. That is, if the last criterion considered is l and if $S^k = C \backslash c(\cdot)$ are the criteria pending to be selected at iteration k, we consider a probability distribution where for each criterion $c_j \in S^k$ its probability of being selected is a function of $a_{l,j}$. For example, we can just use $p(j) = a_{l,j} / \sum_{r \in S^k} a_{l,r}$.

Step 4.3 assigns a weight to the new selected criterion $c(k)$. If the new criterion is not relevant in relation to the ones already selected, it will have a low weight. For each criterion already included $j \in c(1), \ldots, c(k-1)$, we have that $a_{jc(k)}$ is the relevance of adding $c(k)$. The minimum of these relevances is the one that we select. E.g., if we have that for one criterion c_j it has null relevance to add $c_{c(k)}$, we do not add it.

Using this algorithm, and assuming that the iteration selects c_2, c_4, c_1, c_5, and c_3 in this order, we will get their weights equal to α, 0.9α, 0.1α, 0.1α, and 0.1α. That is, $w = (0.1\alpha, \alpha, 0.1, 0.9\alpha, 0.1\alpha)$.

Let us make some considerations on this approach.

- The value α has the largest importance for any criterion. When the values in the matrix are at most one, all the other weights will be at most α. This follows from Step 4.3 above. α is a parameter of the method.

- When all criteria are independent, the relative preference matrix is defined with $a_{ij} = 1$ for $i \neq j$. This implies that $w(i) = \alpha$ for all i.

When the aggregation operator to be used is the weighted mean, the weights obtained in the previous process can be normalized. Therefore, we will have $w'_i = w_i / \sum_j w_j$.

When the aggregation operator is the Choquet integral, and the measure is of the family of Sugeno λ-measures, we will use the approach proposed in,[11] see also,[12] and the above mentioned Section 5.3.1 in Reference.[1]

We now prove that if we build a Sugeno λ-measure, the definition is consistent with the interpretation of fuzzy measures.

Proposition 3.1. *Let $a_{ij} = 1$ for all $i \neq j$ and $a_{ii} = 0$; then, the following can be proven:*

- *When $\alpha = 1/n$ with n the number of criteria, we have that $\lambda = 0$ and the measure is additive. This produces the following measure: $\mu(\{c_i\}) = 1/n$ for all i.*
- *When $\alpha > 1/n$, then $\lambda < 0$, and the measure is a plausibility.*
- *When $\alpha < 1/n$, then $\lambda > 0$, and the measure is a belief.*

Proof. This follows from the conditions on λ and the results that show that when $\mu_1(\{x_i\}) \geq \mu_2(\{x_i\})$ for all i then $\lambda_1 < \lambda_2$. See e.g. Proposition 5.42 in.[1] □

Recall that an additive measure when combined with a fuzzy measure results into the arithmetic mean. This means that our construction is consistent with the meaning of fuzzy measures and integrals. That is, when we have that all criteria are equally relevant and eligible ($a_{ij} = 1$) and $\alpha = 1/n$, the whole process will produce a fuzzy measure that is just the arithmetic mean. In other words, equally relevant is understood here as independent, and independence leads to an additive measure that implies that the Choquet integral corresponds to the arithmetic mean.

4. Knowledge Elicitation from Experts

To extract knowledge in the form of variable weights and correlations from experts is not easy. This problem has been explored together with a variety of different methods, such as Bayesian theory,[13] AHP,[14] Conjoint Analysis,[15] Discrete Choice experiments[16] and Best-Worst Scaling.[17] Most methods suffer from the problem with how to best model the results from experts that vary in their subjective opinions, as well as that the context in which the questions are posed can highly influence the answers collected. Moreover, in many cases, such as decision-making in the medical domain, the results from the knowledge elicitation process need to be presented in a coherent and transparent way, making the presentation of model specific or complex mathematical expressions unsuitable for non-mathematicians.

As stated by Pecchia et al.,[14] the AHP approach towards decision-making enables a hierarchical process where a consistent framework can be built step-by-step, structuring a complex problem into smaller, less complicated ones that a decision maker can more easily solve and understand. The knowledge to be elicited is often extracted through interviews or questionnaires where the experts are asked to compare the relative importance of a set of variables. The questions posed are often phrased in the form of: "according to your experience, how important do you consider the element i compared to the element j?", where the participants are requested to answer using a 5-point Likert scale (see for instance[14]), thus hiding the more complex model assumptions from the experts, making it easier for them to express their knowledge and review their results.

5. Conclusions and Future Work

In this paper we have proposed the relative preference matrices as a way to elicit weights for weighted means and the Choquet integral and we have introduced an algorithm to set the weights from the matrices.

We have shown that we can infer from a relative preference matrix a fuzzy measure. More particularly, a Sugeno λ-measure. We have shown that this construction is consistent with the meaning of independence. Further properties need to be studied.

We initiated this problem considering how to build a fuzzy measure from a correlation matrix. We consider that it is possible to derive a relative preference matrix from preference matrices. We plan to work on this direction.

Some of the problems that have been considered in the literature for AHP matrices can be of relevance for relative preference matrices. See e.g., the problem of non complete matrices, matrices with other data than numeric, etc.

References

1. V. Torra, Y. Narukawa, Modeling decisions: information fusion and aggregation operators, (Springer, 2007).

2. G. Beliakov, A. Pradera, T. Calvo, Aggregation functions: A guide for practitioners, (Springer, 2008).
3. R. R. Yager, On ordered weighted averaging aggregation operators in multi-criteria decision making, IEEE Trans. on SMC 18 (1988) 183–190.
4. V. Torra, The weighted OWA operator, Int. J. Intel. Syst. 12 (1997) 153–166.
5. V. Torra, Y. Narukawa, M. Sugeno (eds.), Non-additive measures: theory and applications, (Springer, 2013).
6. E. Pap (ed.), Handbook of Measure Theory, (North-Holland, 2002).
7. G. Choquet, Theory of capacities, Ann. Inst. Fourier 5 (1953/54) 131–295.
8. M. Sugeno, Fuzzy measures and fuzzy integrals (in Japanese), Trans. of the Soc. of Instrument and Control Engineers 8:2 (1972).
9. T. L. Saaty, The Analytic Hierarchy Process, (McGraw-Hill, 1980).
10. R. W. Saaty, The analytic hierarchy process — what it is and how it is used, Mathematical Modelling 9:3–5 (1987) 161–176.
11. K. Leszczyski, P. Penczek, W. Grochulski, Sugeno's fuzzy measure and fuzzy clustering, Fuzzy Sets and Systems 15 (1985) 147–158.
12. H. Tahani, J. M. Keller, Information fusion in computer vision using the fuzzy integral, IEEE Trans. SMC 20:3 (1990) 733–741.
13. A. O'Hagan, C. E. Buck, A. Daneshkhah, J. R. Eiser, P. H. Garthwaite, D. J. Jenkinson, J. E. Oakley, T. Rakow, Uncertain judgements: eliciting experts' probabilities, (John Wiley & Sons, 2006).
14. L. Pecchia, J. L. Martin, A. Ragozzino, C. Vanzanella, A. Scognamiglio, L. Mirarchi, S. P. Morgan, User needs elicitation via analytic hierarchy process (AHP). A case study on a Computed Tomography (CT) scanner, BMC Medical Informatics and Decision Making 13:1 (2013) 2.
15. J. J. Louviere, T. N. Flynn, R. T. Carson, Discrete choice experiments are not conjoint analysis, Journal of Choice Modelling 3:3 (2010) 57–72.
16. E. W. de Bekker-Grob, M. Ryan, K. Gerard, Discrete choice experiments in health economics: a review of the literature, Health economics 21:2 (2012) 145–172.
17. A. C. Mühlbacher, A. Kaczynski, P. Zweifel, F. R. Johnson, Experimental measurement of preferences in health and healthcare using best-worst scaling: an overview, Health Economics Review 6:1 (2016) 2 (Jan 08).

Generalized hesitant fuzzy linguistic term sets for linguistic decision making

Li Yan

College of Science, Xihua University
Chengdu, Sichuan, 610039, China
15239181@qq.com

Mingming Kong and Zheng Pei

Center for Radio Administration & Technology Development, Xihua University
Chengdu, Sichuan, 610039, China
kongming000@126.com, pqyz@263.net

In this paper, the concept of generalized hesitant fuzzy linguistic term set based on 2-tuple linguistic model is provided, then several operators on generalized hesitant fuzzy linguistic term sets and their properties are proposed and analyzed for aggregating generalized hesitant fuzzy linguistic information in linguistic decision making. Especially, based on these new operators, the likelihood-based comparison relations of generalized hesitant fuzzy linguistic term sets are presented, and the generalized hesitant fuzzy linguistic weighted average operator is designed to fuse generalized hesitant fuzzy linguistic term sets in linguistic decision making. An example is used to illustrate the practicality of the aggregation operator based on generalized hesitant fuzzy linguistic term sets, result shows that the proposed method is an useful and alternative method for hesitant fuzzy linguistic decision making.

Keywords: 2-tuple linguistic model; hesitant fuzzy linguistic term sets; linguistic decision making.

1. Introduction

Because fuzzy linguistic terms are closest to human being's cognitive processes that occurs in real life and provide us a flexible and reliable form to represent qualitative information, many researchers pay attention to linguistic decision making (LDM), in which the 2-tuple linguistic model[1] owns the computational simplicity, no loss information, the accuracy, and understandability, many researchers have paid their attention to the 2-tuple linguistic model and widely employed it to solve linguistic decision making problems.[2-7] In real world practices, many qualitative decision making environments are associated with decision makers may be not sure about one single fuzzy linguistic term to assess the criteria or the alternatives, i.e., they hesitate about several fuzzy linguistic terms to assess the alternatives according to the criteria, Rodríguez et al.[8] proposed the concept of hesitant fuzzy linguistic term set (HFLTS) to represent the assessments when decision makers hesitate about several fuzzy linguistic terms in fuzzy linguistic decision making environments, up to now, linguistic decision making with HFLTS has attracted many researchers' attention.[9–20]

In this paper, we focus on generalized hesitant fuzzy linguistic term sets in the framework of 2-tuple linguistic model and their operations. In addition, inspired by the likelihood-based comparison relations of hesitant fuzzy linguistic term sets,[21] we propose the likelihood-based comparison relations of generalized hesitant fuzzy linguistic term sets for hesitant fuzzy linguistic decision making. The rest of this paper is structured as follows: In Section 2, we briefly review the 2-tuple linguistic model and HFLTS. In Section 3, we define generalized hesitant fuzzy linguistic term set in the framework of 2-tuple linguistic model, provide t-norms and t-conorms on the generalized hesitant fuzzy linguistic term sets and discuss their properties. In Section 4, we propose the likelihood-based comparison relations between generalized hesitant fuzzy linguistic term sets based on the concept of the 1-cut of generalized hesitant fuzzy linguistic term set. Then, we provide generalized hesitant fuzzy linguistic aggregation operator to fuse generalized hesitant fuzzy linguistic term sets in linguistic decision making. We conclude the paper in Section 5.

2. Preliminaries

Formally, let $S_g = \{s_0, \ldots, s_g\}$ be a linguistic term set, $\beta \in [0, g]$ be a value representing the result of a symbolic aggregation operation. Then a 2-tuple linguistic value corresponding to β is defined as:[1] $\Delta : [0, g] \longrightarrow S_g \times [-0.5, 0.5)$ and $\Delta(\beta) \longmapsto (s_i, \alpha)$, where, $i = round(\beta)$ and $\alpha = \beta - i \in [-0.5, 0.5)$, $round(\cdot)$ is the usual round operation, s_i is the linguistic term that is mostly close to β and α represents the symbolic translation value. Δ is an one-to-one mapping, it's inverse function transforms 2-tuple linguistic values to its equivalent numerical values, i.e., $\Delta^{-1} : S_g \times [-0.5, 0.5) \longrightarrow [0, g]$ and $\Delta^{-1}(s_i, \alpha) \longmapsto \beta = i + \alpha \in [0, g]$. Based on Δ and Δ^{-1}, the 2-tuple linguistic value (s_i, α) can be used to express the linguistic information on a universe of discourse, this means that using the 2-tuple linguistic model can effectively avoid the loss and distortion of information in linguistic information processing.

The concept of HFLTS[8] is proposed to serve as the flexible and useful tool for hesitant linguistic information by means of complex linguistic expressions. Formally, an HFLTS, H_S, is an ordered finite subset of the consecutive linguistic terms of S_g, where *consecutive* means that it does not make sense to hesitate among different linguistic terms and not hesitate in their middle linguistic terms. Basic operations on HFLTSs are:[8] 1) Lower bound: $H_{S-} = min(s_i) = s_j$, $s_i \in H_S$ and $s_i \geq s_j \forall i$; 2) Upper bound: $H_{S+} = max(s_i) = s_j$, $s_i \in H_S$ and $s_i \leq s_j \forall i$; 3) Complement: $H_S^c = S - H_S = \{s_i | s_i \in S \text{ and } s_i \notin H_S\}$; 4) Union: $H_S^1 \cup H_S^2 = \{s_i | s_i \in H_S^1 \text{ or } s_i \in H_S^2\}$; 5) Intersection: $H_S^1 \cap H_S^2 = \{s_i | s_i \in H_S^1 \text{ and } s_i \in H_S^2\}$; 6) Envelope: $env(H_S) = [H_{S-}, H_{S+}]$.

3. Generalized Hesitant Fuzzy Linguistic Term Sets

The character of Rodríguez's HFLTS is: 1) the consecutive linguistic terms of S_g; 2) the finite subset of a primary linguistic term set $S_g = \{s_0, \ldots, s_g\}$. Extended hesitant fuzzy linguistic term sets[20] has been proposed to deal with non-consecutive linguistic terms, which is an ordered subset of linguistic terms and obtained from the collective HFLTSs of decision makers.[16] Different with Wang's extended hesitant fuzzy linguistic term set, here we define the concept of the generalized hesitant fuzzy linguistic term set to deal with hesitant linguistic information provided by experts in 2-tuple linguistic decision making environments. In fact, the 2-tuple linguistic model allow experts to provide their linguistic assessments from $[s_0, s_g]$, naturally, the concept of HFLTS can be generalized from $L_g = \{s_0, \ldots, s_g\}$ to $[s_0, s_g]$. For convenience, we denote $s_\beta = (s_i, \alpha)$ and $L_{[0,g]} = [s_0, s_g]$ in the follows.

Definition 3.1. Let $L_g = \{s_0, \cdots, s_g\}$ be a primary linguistic term set. An generalized hesitant fuzzy linguistic term set (GHFLTS) $G_L = \{s_{\beta_1}, \cdots, s_{\beta_k}\}$ is an ordered finite subset of $L_{[0,g]}$ and $\{s_{round(\beta_1)}, \cdots, s_{round(\beta_k)}\}$ is an HFLTS on L_g.

Compared Definition 3.1 with the concept of HFLTS, an GHFLTS allows using $s_\beta = (s_i, \alpha)$ but not only s_i in an HFLTS on L_g, it is obvious that HFLTS is a special case of GHFLTS, *i.e.*, G_L is an HFLTS if $round(\beta_1) = i+1$, $round(\beta_2) = i+2$, $\cdots$ and $round(\beta_k) = i+k$. In addition, it seems that GHFLTS provide more flexible linguistic terms than HFLTS due to GHFLTS on $L_{[0,g]}$ and HFLTS on $L_g = \{s_0, \cdots, s_g\}$. Such as Let $L_6 = \{$nothing (s_0), very low (s_1), low (s_2), medium (s_3), high (s_4), very high (s_5), perfect $(s_6)\}$. $H_{L_6} = \{s_3, s_4\}$ is an HFLTS of L_6 and $G_{L_6} = \{(s_3, 0.3), (s_4, -0.4), (s_4, -0.2)\}$ is an GHFLTS on $L_{[0,6]}$.

Basic operations on GHFLTSs can be obtained from operations on HFLTSs, i.e., for any GHFLTSs G_L, G_L^1 and G_L^2 on $L_{[0,g]}$, $Ind(s_{\beta_{k'}})$ be the index $\beta_{k'}$ of a linguistic term $s_{\beta_{k'}}$ in $L_{[0,g]}$, and $Ind(G_L)$ the set of indexes of the linguistic terms in an GHFLTS G_L on $L_{[0,g]}$,

(1) Lower bound: $G_{L-} = min(s_{\beta_{k'}}) = s_{\beta_{k''}}, s_{\beta_{k'}} \in G_L$ and $s_{\beta_{k'}} \geq s_{\beta_{k''}}, \forall \beta_{k'}$;

(2) Upper bound: $G_{L+} = max(s_{\beta_{k'}}) = s_{\beta_{k''}}, s_{\beta_{k'}} \in G_L$ and $s_{\beta_{k'}} \leq s_{\beta_{k''}}, \forall \beta_{k'}$;

(3) Complement: $G_L^c = L_{[0,g]} - G_L = \{s_{\beta_{k'}} | s_{\beta_{k'}} \in L_{[0,g]}$ and $s_{\beta_{k'}} \notin G_L\}$;

(4) Union: $G_L^1 \cup G_L^2 = \{s_{\beta_{k'}} | s_{\beta_{k'}} \in G_L^1$ or $s_{\beta_{k'}} \in G_L^2\}$;

(5) Intersection: $G_L^1 \cap G_L^2 = \{s_{\beta_{k'}} | s_{\beta_{k'}} \in G_L^1$ and $s_{\beta_{k'}} \in G_L^2\}$;

(6) Envelope: $env(G_L) = [G_{L-}, G_{L+}]$;

(7) the negation of G_L: $\overline{G_L} = \{s_{g-\beta_{k'}} | \beta_{k'} \in Ind(G_L)\}$;

(8) the max-union: $G_L^1 \vee G_L^2 = \{max\{s_{\beta_{k'}}, s_{\beta_{k''}}\} | s_{\beta_{k'}} \in G_L^1, s_{\beta_{k''}} \in G_L^2\}$;

(9) the min-intersection: $G_L^1 \wedge G_L^2 = \{min\{s_{\beta_{k'}}, s_{\beta_{k''}}\} | s_{\beta_{k'}} \in G_L^1, s_{\beta_{k''}} \in G_L^2\}$.

Inspired by works in,[8,19] we can easily prove that operations on GHFLTSs satisfy the following propositions.

Proposition 3.1. *Let three GHFLTSs G_L^1, G_L^2 and G_L^3 on $L_{[0,g]}$, then*

(1) The complement of an GHFLTS is involutive, $((G_L^1)^c)^c = G_L^1$;

(2) Commutativity, $G_L^1 \cup G_L^2 = G_L^2 \cup G_L^1$, $G_L^1 \cap G_L^2 = G_L^2 \cap G_L^1$;

(3) Associative, $G_L^1 \cup (G_L^2 \cup G_L^3) = (G_L^1 \cup G_L^2) \cup G_L^3$, $G_L^1 \cap (G_L^2 \cap G_L^3) = (G_L^1 \cap G_L^2) \cap G_L^3$;

(4) Distributive, $G_L^1 \cap (G_L^2 \cup G_L^3) = (G_L^1 \cap G_L^2) \cup (G_L^1 \cap G_L^3)$, $G_L^1 \cup (G_L^2 \cap G_L^3) = (G_L^1 \cup G_L^2) \cap (G_L^1 \cup G_L^3)$.

Proposition 3.2. *Let three GHFLTSs G_L^1, G_L^2 and G_L^3 on $L_{[0,g]}$, then*

(1) The negation of an GHFLTS is involutive, $\overline{(\overline{G_L^1})} = G_L^1$;

(2) De Morgan law, $\overline{(G_L^1 \cup G_L^2)} = \overline{G_L^1} \cap \overline{G_L^2}$, $\overline{(G_L^1 \cap G_L^2)} = \overline{G_L^1} \cup \overline{G_L^2}$;

(3) Commutativity, $G_L^1 \vee G_L^2 = G_L^2 \vee G_L^1$, $G_L^1 \wedge G_L^2 = G_L^2 \wedge G_L^1$;

(4) Associativity, $G_L^1 \vee (G_L^2 \vee G_L^3) = (G_L^1 \vee G_L^2) \vee G_L^3$, $G_L^1 \wedge (G_L^2 \wedge G_L^3) = (G_L^1 \wedge G_L^2) \wedge G_L^3$;

(5) Distributivity, $G_L^1 \wedge (G_L^2 \vee G_L^3) = (G_L^1 \wedge G_L^2) \vee (G_L^1 \wedge G_L^3)$, $G_L^1 \vee (G_L^2 \wedge G_L^3) = (G_L^1 \vee G_L^2) \wedge (G_L^1 \vee G_L^3)$.

4. The Likelihood-based Comparison Relations on GHFLTS

Theoretically, comparison between two GHFLTSs is used to rank alternatives or select the best alternative in linguistic decision making with GHFLTSs. Let $L_g = \{s_0, \cdots, s_g\}$ be a primary linguistic term set, for any GHFLTSs $G_L^1 = \{s_{\beta_1}, \cdots, s_{\beta_i}\}(\beta_1 \leq \cdots \leq \beta_i)$ and $G_L^2 = \{s_{\gamma_1}, \cdots, s_{\gamma_j}\}(\gamma_1 \leq \cdots \leq \gamma_j)$ on $L_{[0,g]}$, making use of the 1-cut of the hesitant fuzzy linguistic term set defined in,[21] we can obtain the 1-cut of GHFLTS, such as $G_L^1(1) = [G_L^1(1)_l, G_L^1(1)_r] = [\beta_1, \beta_i]$, then we can obtain the likelihood-based comparison relation $p(G_L^1 \geq G_L^2)$ between G_L^1 and G_L^2 derived from:[21]

$$p(G_L^1 \geq G_L^2) = max\{1 - max\{\frac{G_L^2(1)_r - G_L^1(1)_l}{L(G_L^1(1)) + L(G_L^2(1))}, 0\}, 0\}, \quad (1)$$

where $G_L^1(1) = [G_L^1(1)_l, G_L^1(1)_r] = [\beta_1, \beta_i]$ and $G_L^2(1) = [G_L^2(1)_l, G_L^2(1)_r] = [\gamma_1, \gamma_j]$ are the 1-cut of G_L^1 and G_L^2, respectively, $L(G_L^1(1)) = G_L^1(1)_r - G_L^1(1)_l$ and $L(G_L^2(1)) = G_L^2(1)_r - G_L^2(1)_l$. For example, let $L = \{$nothing (s_0), very low (s_1), low (s_2), medium (s_3), high (s_4), very high (s_5), perfect $(s_6)\}$. For GHFLTSs $G_L^1 = \{(s_3, 0.3), (s_4, -0.2), (s_5, -0.2)\}$ and $G_L^2 = \{(s_2, 0.4), (s_3, 0.1), (s_4, 0.4)\}$ on $L_{[0,6]}$, then $G_L^1(1) = [G_L^1(1)_l, G_L^1(1)_r] = [3.3, 4.8]$ and $G_L^2(1) = [G_L^2(1)_l, G_L^2(1)_r] = [2.4, 4.4]$, $L(G_L^1(1)) = 4.8 - 3.3 = 1.5$, $L(G_L^2(1)) = 4.4 - 2.4 = 2$ and

$$p(G_L^1 \geq G_L^2) = max\{1 - max\{\frac{4.4 - 3.3}{1.5 + 2}, 0\}, 0\} = \frac{24}{35}.$$

As a special case, if $G_L^2 = L_g$, then $p(G_L^1 \geq L_g)$ has the more simple form,

$$p(G_L^1 \geq L_g) = \frac{G_L^1(1)_r}{L(G_L^1(1)) + g}. \quad (2)$$

Inspired by Lee' works,[21] here we provide the generalized hesitant fuzzy linguistic weighted average (GHFLWA) operator. Formally, if weights information of each GHFLTS or decision attribute is known, i.e., each GHFLTS or decision attribute with fixed weight, then we can define the following weighted operators to fuse GHFLTSs in linguistic decision making.

Definition 4.1. Let a set $G = \{G_L^1, \cdots, G_L^m\}$ of GHFLTSs on linguistic terms set L_g with fixed weights $W = \{w_1, \cdots, w_m\}$ such that $w_{m'} \geq 0$ ($m' = 1, \cdots, m$) and $\sum_{m'=1}^{m} w_{m'} = 1$. The generalized hesitant fuzzy linguistic weighted average (GHFLWA) operator is defined as

$$\text{GHFLWA}(G_L^1, \cdots, G_L^m) = 1 - \prod_{m'=1}^{m}(1 - p(G_L^{m'} \geq L_g))^{w_{m'}}. \quad (3)$$

In,[21] Lee and Chen used the 1-cut of the hesitant fuzzy linguistic term set to calculate the likelihood-based comparison relation $p(h_1 \geq S)$, according to Definition 3.1, the 1-cut of the hesitant fuzzy linguistic term set is a special case of the 1-cuts of GHFLTSs, and the hesitant fuzzy linguistic weighted average operator [21] is a special case of GHFLWA operator.

Example 4.1. Let $A = \{a_1, a_2, a_3, a_4\}$ be a set of four alternatives, $C = \{c_1, c_2, c_3, c_4, c_5\}$ be a set of criteria defined for each alternative and L_6 be the linguistic term set. The assessments provided in such multi-criteria linguistic decision making problem are shown in Table 1 and 1-cut $G_L^{ij}(1)$ are shown in Tables 2. According to Eq. (2), we can calculate each $p(G_L^{ij} \geq L_6)$ of G_L^{ij}, which are shown in Table 3. Let weights $W = \{w_1 = 0.2, w_2 = 0.1, w_3 = 0.3, w_4 = 0.1, w_5 = 0.3\}$ of criteria, based on Table 3 and GHFLWA operator (Eq. (3)), we can obtain final assessments of alternatives, which are GHFLWA(a_1) = GHFLWA(a_2) = 0.594 < GHFLWA(a_3) = GHFLWA(a_4) = 0.634, i.e., the preference order of alternatives is $a_4 = a_3 \succ a_2 = a_1$.

Table 1. Assessments of A with respect to criteria C.

	a_1	a_2	a_3	a_4
c_1	$\{s_1, s_2, s_3\}$	$\{s_1, (s_2, 0.4), s_3\}$	$\{s_4, s_5, s_6\}$	$\{s_4, (s_5, -0.2), s_6\}$
c_2	$\{s_4, s_5\}$	$\{s_4, (s_4, 0.2), s_5\}$	$\{s_1, s_2, s_3\}$	$\{s_1, (s_2, 0.4), s_3\}$
c_3	$\{s_4, s_5, s_6\}$	$\{s_4, (s_5, -0.2), s_6\}$	$\{s_2, s_3, s_4\}$	$\{s_2, s_3, s_4\}$
c_4	$\{s_3, s_4, s_5\}$	$\{s_3, (s_4, 0.2), s_5\}$	$\{s_3, s_4, s_5\}$	$\{s_3, (s_4, 0.2), s_5\}$
c_5	$\{s_2, s_3\}$	$\{s_2, s_3\}$	$\{s_4, s_5\}$	$\{s_4, s_5\}$

Table 2. $G_L^{ij}(1)$ of each assessment G_L^{ij}.

	c_1	c_2	c_3	c_4	c_5
a_1	[1, 3]	[4, 5]	[4, 6]	[3, 5]	[2, 3]
a_2	[1, 3]	[4, 5]	[4, 6]	[3, 5]	[2, 3]
a_3	[4, 6]	[1, 3]	[2, 4]	[3, 5]	[4, 5]
a_4	[4, 6]	[1, 3]	[2, 4]	[3, 5]	[4, 5]

5. Conclusions and Future Works

In this paper, we propose the concept of GHFLTSs in the framework of the 2-tuple linguistic model, define the 1-cut of GHFLTS, which is the generalization of the 1-cut of the hesitant fuzzy linguistic term set.[21] Then

Table 3. $p(G_L^{ij} \geq L_6)$ of $G_L^{ij} (i = 1,2,3,4, j = 1,2,3,4,5)$.

	$p(G_L^{i1} \geq L_6)$	$p(G_L^{i2} \geq L_6)$	$p(G_L^{i3} \geq L_6)$	$p(G_L^{i4} \geq L_6)$	$p(G_L^{i5} \geq L_6)$
a_1	$\frac{3}{8}$	$\frac{5}{7}$	$\frac{3}{4}$	$\frac{5}{8}$	$\frac{3}{7}$
a_2	$\frac{3}{8}$	$\frac{5}{7}$	$\frac{3}{4}$	$\frac{5}{8}$	$\frac{3}{7}$
a_3	$\frac{3}{4}$	$\frac{3}{8}$	$\frac{1}{2}$	$\frac{5}{8}$	$\frac{5}{7}$
a_4	$\frac{3}{4}$	$\frac{3}{8}$	$\frac{1}{2}$	$\frac{5}{8}$	$\frac{5}{7}$

we provide the likelihood-based comparison relations of GHFLTSs based on the 1-cuts of GHFLTSs and propose GHFLWA operator to fuse GHFLTSs in linguistic decision making. An example shows that GHFLWA operator ai alternative and useful method for hesitant fuzzy linguistic decision making.

Acknowledgments

This work has been partially supported by Xihua University Project (52jj2016-044) and the National Natural Science Foundation of China (61372187).

References

1. F. Herrera, L. Martinez, A 2-tuple fuzzy linguistic representation model for computing with words, *IEEE Trans. Fuzzy Syst.* **8(6)**, 746–752 (2000).
2. L. Martínez, R.M. Rodriguez, F. Herrera (eds.), *The 2-tuple Linguistic Model-Computing with Words in Decision Making* (Springer International Publishing Switzerland, 2015).
3. Z. Pei, D. Ruan, Y. Xu, J. Liu, *Linguistic values-based intelligent information processing: Theory, methods, and applications* (Atlantis Press, 2010).
4. R.M. Rodríguez, L. Martinez, An analysis of symbolic linguistic computing models in decision making, *Int. J. Gene. Syst.* **42(1)**, 121–136 (2013).
5. I. Beg, T. Rashid, Hesitant 2-tuple linguistic information in multiple attributes group decision making, *J. Intel. fuzz. syst.* **30(1)**, 109–116 (2016).
6. Z. Pei, L. Zou, L.Z. Yi, A linguistic aggregation operator including weights for linguistic values and experts in group decision making, *Int. J. Uncertainty, Fuzziness and Knowl.-Based Syst.* **21(6)**, 927–943 (2013).
7. D. Meng, Z. Pei, On weighted unbalanced linguistic aggregation operators in group decision making, *Inform. Sci.* **223**, 31–41 (2013).
8. R.M. Rodríguez, L. Martínez, F. Herrera, Hesitant fuzzy linguistic term sets for decision making, *IEEE Trans. Fuzzy Syst.* **20(1)**, 109–119 (2012).
9. R.M. Rodríguez, L. Martínez, F. Herrera, A group decision making model dealing with comparative linguistic expressions based on hesitant fuzzy linguistic term sets, *Inform. Sci.* **241(1)**, 28–42 (2013).
10. R.M. Rodríguez, Á. Labella and L. Martínez, An overview on fuzzy modelling of complex linguistic preferences in decision making, *Int. J. Comput. Intell. Syst.* **9**, 81–94 (2016).
11. K.H. Chang, A more general reliability allocation method using the hesitant fuzzy linguistic term set and minimal variance OWGA weights, *Appli. Sof. Comp.* **56**, 589–596 (2017).

12. Z.S. Chen et al., Proportional hesitant fuzzy linguistic term set for multiple criteria group decision making, *Inform. Sci.* **357**, 61–87 (2016).
13. Y.C. Dong et al., Connecting the linguistic hierarchy and the numerical scale for the 2-tuple linguistic model and its use to deal with hesitant unbalanced linguistic information, *Inform. Sci.* **367–368**, 259–278 (2016).
14. B. Farhadinia, Multiple criteria decision-making methods with completely unknown weights in hesitant fuzzy linguistic term setting, *Knowl.-Based Syst.* **93**, 135–144 (2016).
15. X.J. Gou, Z.S. Xu, H.C. Liao, Hesitant fuzzy linguistic entropy and cross-entropy measures and alternative queuing method for multiple criteria decision making, *Inform. Sci.* **388–389**, 225–246 (2017).
16. C. Wei, R.M. Rodríguez, L. Martinez, Uncertainty Measures of Extended Hesitant Fuzzy Linguistic Term Sets, *IEEE Trans. Fuzzy Syst.* (DOI 10.1109/TFUZZ.2017.2724023).
17. F.L. Ren, M.M. Kong, Z. Pei, A new hesitant fuzzy linguistic TOPSIS method for group multi-criteria linguistic decision making, *Symmetry* **9**, 289 (2017).
18. Z. Li, C. Zhao, Z. Pei, Operations on hesitant linguistic terms sets induced by archimedean triangular norms and conorms, *Int. J. Comput. Intell. Syst.* **11**, 514–524 (2018).
19. C.P. Wei, N. Zhao, X.J. Tang, Operators and comparisons of hesitant fuzzy linguistic term sets, *IEEE Trans. Fuzzy Syst.* **22(3)**, 575–584 (2014).
20. H. Wang, Extended hesitant fuzzy linguistic term sets and their aggregation in group decision making, *Int. J. Comput. Intell. Syst.* **8**, 14–33 (2015).
21. L.W. Lee, S.M. Chen, Fuzzy decision making based on likelihood-based comparison relations of hesitant fuzzy linguistic term sets and hesitant fuzzy linguistic operators, *Inform. Sci.* **294**, 513–529 (2015).

Type-2 intuitionistic fuzzy (IFS2) WASPAS

Cengiz Kahraman, Sezi Cevik Onar and Basar Oztaysi
*Istanbul Technical University, Industrial Engineering Department,
34367, Maçka, Beşiktaş, Istanbul, Turkey*

Esra Ilbahar
*Yildiz Technical University, Department of Industrial Engineering,
Besiktas 34349, Istanbul, Turkey*

Type 2 Intuitionistic sets (IFS2) have been introduced by Atanassov (1999). They let experts assign membership and nonmembersip degrees in a wider area than it is in ordinary intuitionistic fuzzy sets. In this paper, we employ IFS2 sets for the development of IFS2 WASPAS method. WASPAS is a multicriteria method integrating simple additive weighting and weighted product methods. Single valued IFS2 sets have been used in the IFS2 WASPAS method. We illustrate the application of the proposed IFS2 WASPAS through a selection problem among communication firms.

1. Introduction

Weighted Aggregated Sum Product Assessment (WASPAS) is a relatively new method, but it has been widely employed in the literature since its first introduction in 2012. WASPAS is a weighted combination of Weighted Sum Model (WSM) and Weighted Product Model (WPM) Extensions of WASPAS with fuzzy sets such as single-valued neutrosophic sets, interval valued intuitionistic fuzzy sets, and interval type-2 fuzzy sets have been also commonly studied.

In this paper we use single valued IFS2 WASPAS method for the selection among communication firms based on some criteria. The developed method is based on the arithmetic operations of IFS2 sets such as addition, multiplication, and division. After Atanassov's [1] introduction, IFS2 sets have been theoretically developed by many researchers (Peng and Yang [7]; Dick et al. [3]; Liu et al. [6]).

The rest of the paper is organized as follows. Section 2 presents the preliminaries of IFS2 sets. Section 3 gives the definition of WSM. Section 3 presents the definition of WPM. Section 5 involves the steps of the proposed IFS2 WASPAS method. Section 6 illustrates the application of the proposed method. Section 7 finalizes the paper with conclusions and suggestions for further research.

2. Type-2 Intuitionistic Fuzzy WASPAS: IFS2-WASPAS

2.1. *Preliminaries*

IFS2 sets were introduced by Atanassov [1]. Yager [9] called these sets Pythagorean fuzzy sets. They are defined by both membership and nonmembership degrees where their square sums should be equal and less than 1. IFS2 can be considered as a generalization of Intuitionistic Fuzzy Sets (IFS). In order to provide some basic knowledge of IFSs and IFS2s, Table 1 and Figure 1 are given.

Consider the data in Table 1. Figure 6 shows the relation between IFS and IFS2s based on Table 1. IFS2 is a dilation operation over IFS. Alternatively, we can say that IFS is a concentration operation over IFS2.

Table 1. Membership and nonmembership data.

IFS		IFS2		IFS		IFS2	
μ	v	μ	v	μ	v	μ	v
0	1	0	1	0.55	0.45	0.55	0.835
0.05	0.95	0.05	0.999	0.6	0.4	0.6	0.8
0.1	0.9	0.1	0.999	0.65	0.35	0.65	0.76
0.15	0.85	0.15	0.989	0.7	0.3	0.7	0.714
0.2	0.8	0.2	0.98	0.75	0.25	0.75	0.661
0.25	0.75	0.25	0.968	0.8	0.2	0.8	0.6
0.3	0.7	0.3	0.954	0.85	0.15	0.85	0.527
0.35	0.65	0.35	0.937	0.9	0.1	0.9	0.436
0.4	0.6	0.4	0.917	0.95	0.05	0.95	0.312
0.45	0.55	0.45	0.893	1	0	1	0
0.5	0.5	0.5	0.866				

An IFS2 is defined as follows:

$$\tilde{P} = \{x, P(\mu_P(x), v_P(x)) | x \in X\} \quad (1)$$

where $\mu_P: X \to [0,1]$ is the membership degree and $v_P: X \to [0,1]$ is the nonmembership degree. Then, Eq. (2) is valid:

$$0 \le (\mu_P(x))^2 + (v_P(x))^2 \le 1 \quad (2)$$

The degree of indeterminancy is defined as follows:

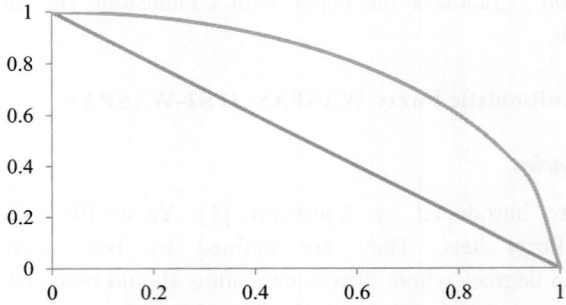

Figure 1. Limits of membership and nonmembership degrees in IFS and IFS2.

$$\pi_P(x) = \sqrt{1 - (\mu_P(x))^2 - (v_P(x))^2} \qquad (3)$$

For two IFS2, $\widetilde{P}_1 = \{x, P_1(\mu_{P_1}(x), v_{P_1}(x)) | x \in X\}$ and $\widetilde{P}_2 = \{x, P_2(\mu_{P_2}(x), v_{P_2}(x)) | x \in X\}$, the following arithmetic operations are valid:

$$\widetilde{P}_1 \oplus \widetilde{P}_2 = P\left(\sqrt{\mu_{P_1}^2 + \mu_{P_2}^2 - \mu_{P_1}^2 \mu_{P_2}^2}, \; v_{P_1} v_{P_2}\right) \qquad (4)$$

$$\widetilde{P}_1 \otimes \widetilde{P}_2 = P\left(\mu_{P_1} \mu_{P_2}, \; \sqrt{v_{P_1}^2 + v_{P_2}^2 - v_{P_1}^2 v_{P_2}^2}\right) \qquad (5)$$

$$\lambda \widetilde{P} = P\left(\sqrt{1 - (1 - \mu_P^2)^\lambda}, \; (v_P)^\lambda\right), \; \lambda \geq 0 \text{ and } \lambda \in R \qquad (6)$$

$$\widetilde{P}^\lambda = P\left((\mu_P)^\lambda, \; \sqrt{1 - (1 - v_P^2)^\lambda}, \right), \lambda > 0 \qquad (7)$$

Zhang and Xu (2014) defined the Euclidean distance between two IFS2 as in Eq. (8):

$$d(\widetilde{P}_1, \widetilde{P}_2) = \tfrac{1}{2}\left(|\mu_{P_1}^2 - \mu_{P_2}^2| + |v_{P_1}^2 - v_{P_2}^2| + |\pi_{P_1}^2 - \pi_{P_2}^2|\right) \qquad (8)$$

The Taxicab distance between two IFS2 is defined by Eq. (9):

$$T(\widetilde{P}_1, \widetilde{P}_2) = |\mu_{P_1} - \mu_{P_2}| + |v_{P_1} - v_{P_2}| + |\pi_{P_1} - \pi_{P_2}| \qquad (9)$$

Let $p_1 = (\mu_1, v_1)$ and $p_2 = (\mu_2, v_2)$ be two IFS2 and $\rho > 0$. The following operations are presented for IFS2 [7], [12].

$$\widetilde{P}_1 \ominus \widetilde{P}_2 = \left(\sqrt{\tfrac{\mu_1^2 - \mu_2^2}{1 - \mu_2^2}}, \tfrac{v_1}{v_2}\right), \text{ if } \mu_1 \geq \mu_2, v_1 \leq \min\left\{v_2, \tfrac{v_2 \cdot \pi_1}{\pi_2}\right\} \qquad (10)$$

$$\tfrac{\widetilde{P}_1}{\widetilde{P}_2} = \left(\tfrac{\mu_1}{\mu_2}, \sqrt{\tfrac{v_1^2 - v_2^2}{1 - v_2^2}}\right), \text{ if } \mu_1 \leq \min\left\{\mu_2, \tfrac{\mu_2 \cdot \pi_1}{\pi_2}\right\}, v_1 \geq v_2 \qquad (11)$$

Defuzzification of a simplified IFS2 set can be realized by Eq. (12):

$$Deff(\mu_P(x), v_P(x)) = \sqrt{(\mu_P(x))^2} - \sqrt{(v_P(x))^2} \qquad (12)$$

3. Simple Additive Weighting (WSM)

By analyzing the literature, we can see numerous MCDM-based approaches have been introduced in many areas. In this section, WSM and WPM methods are reviewed. WSM method is one of the most popular methods in MCDM [5].

The weighted product method is similar to the simple additive weighting method. However, multiplication operation is used instead of addition in this method. Eq. (13) presents the formula for the simple additive weighting method.

$$V_i = \sum_{j=1}^{n} w_j v_j(x_{ij}), \qquad i = 1, 2, \dots, m \qquad (13)$$

where $v_j(x_{ij})$ is the normalized value of the x_{ij}, which belongs to the jth criterion of ith alternative and w_j is the weight of the criterion j.

4. Weighted Product Method (WPM)

Similar to WSM method, WPM method is one of the most used methods in MCDM. This method is more efficient than other methods in problem solving because of shorter calculation times. WPM is simpler and easier to apply in the cases having high subjectivity. WPM has been used in many areas like optimal route selection, cyber operations evaluation, manufacturing, project manager selection, etc. [5].

Eq. (14) is for the weighted product method.

$$V_i = \prod_{j=1}^{n} x_{ij}^{w_j}, \qquad i = 1, 2, \dots, m \qquad (14)$$

In this method, no normalization process is needed. For cost criteria, the weight of the criterion is written with a negative sign. Besides, any x_{ij} must be larger than 1.0. Hence, all the criterion values involving any value less than 1.0 must be multiplied by 10^s in order to obtain all of the values larger than 1.0.

5. IFS2 WASPAS

Crisp WASPAS method is a combination of WSM and WPM as given in Eq. (15):

$$V = \lambda \sum_{j=1}^{n} w_j v_j(x_{ij}) + (1 - \lambda) \prod_{j=1}^{n} x_{ij}^{w_j} \qquad (15)$$

where w_j is the weight of the criterion j.

In the following, the steps of IFS2 WASPAS method are given:

Step 1. Expert judgments are obtained in linguistic form and converted to single valued IFS2 by using the scales given in Table 2.

Table 2. IFS2 linguistic scale.

Linguistic terms for Decision Matrix	Linguistic terms for Criteria	μ	v
Absolutely good (AG)	Absolutely important (AI)	1	0
Very good (VG)	Very important (VI)	0.9	0.436
Good (G)	Important (I)	0.75	0.661
Medium Good (MG)	Medium Importance (MI)	0.6	0.8
Fair (F)	Fairly important (FI)	0.45	0.893
Medium bad (MB)	Medium Low importance (MLI)	0.3	0.954
Very bad (VB)	Very low importance (VLI)	0.15	0.989
Absolutely bad (AB)	Absolutely low important (ALI)	0	1

Decision matrices of experts are aggregated by IFS2 Weighted Average operator in Eq. (16):

$$IFS2WA = (\sum_{i=1}^{n} w_i \mu_i, \sum_{i=1}^{n} w_i v_i) \quad (16)$$

or

$$IFS2WA = \left(\sum_{i=1}^{n} \sqrt{1-(1-\mu_i^2)^{w_i}}, \sum_{i=1}^{n} (v_i)^{w_i}\right) \quad (17)$$

Here, the summation sign means the addition operation for IFS2 sets given in Eq. (4). Then, for n terms

$$IFS2WA_n = \left(\sqrt{1-\prod_{i=1}^{n}(1-\mu_i^2)^{w_i}}, \prod_{i=1}^{n} v_i^{w_i}\right)$$

Step 2. The decision matrix is normalized using Eqs. (18)–(19).

$$\tilde{r}_{ij} = \frac{\tilde{x}_{ij}}{max_i p_{ij}} \text{ (for benefit criteria)} \quad (18)$$

$$\tilde{r}_{ij} = \frac{min_i p_{ij}}{\tilde{x}_{ij}} \text{ (for cost criteria)} \quad (19)$$

To determine the minimum and maximum values among IFS2 sets, the defuzzification equation given in Eq. (12)

Step 3. IFS2 weighted sum values are calculated by Eq. (20).

$$\tilde{Q}_i^{SAW} = \sum_{j=1}^{n} \tilde{r}_{ij} \tilde{w}_j \quad (20)$$

Step 4. IFS2 weighted product values are calculated by Eq. (21).

$$\tilde{Q}_i^{WP} = \prod_{j=1}^n \tilde{r}_{ij}^{\widetilde{w}_j} \quad (21)$$

Step 5. IFS2 weighted sum values and IFS2 weighted product values are combined by Eq. (22) or Eq. (23).

$$\tilde{Q}_i = \lambda \tilde{Q}_i^{SAW} + (1-\lambda)\tilde{Q}_i^{WP}, \lambda \in [0,1] \quad (22)$$

$$\tilde{Q}_i = \lambda \sum_{j=1}^n \tilde{r}_{ij}\widetilde{w}_j + (1-\lambda)\prod_{j=1}^n \tilde{r}_{ij}^{\widetilde{w}_j}, \lambda \in [0,1] \quad (23)$$

Step 6. The relative importance scores are defuzzified and the best alternative is determined.

6. Application

There are two well-known private communication firms in Turkey. A customer wants to select the most appropriate firm for himself / herself. Three experts fill in the decision matrices with linguistic terms as in Table 3. The determined alternatives are T and V. The evaluation criteria are service capabilities (SC), prices (P), and campaigns (C). The weights of the three experts are assigned with respect to their experiences in the communication sector. The weights of experts 1, 2, and 3 are 0.30, 0.25, and 0.45, respectively.

Table 3. Decision matrices with linguistic terms.

Alt.	SC	P	C	Alt.	SC	P	C	Alt.	SC	P	C
T	VG	G	AG	T	G	MG	VG	T	G	VG	VG
V	G	VG	MG	V	MG	G	G	V	G	AG	MG

The evaluations of the criteria with respect to three experts are given in Table 4.

Table 4. Evaluations of the criteria.

Experts	SC	P	C
E1	VI	MI	MI
E2	AI	I	AI
E3	VI	VI	VI

The result of aggregated decision matrices is given in Table 5.

Table 5. Aggregation of decision matrices.

	SC		P		C	
T	0.812003	0.583428	0.823288	0.574913	1	0
V	0.720314	0.693304	1	0	0.646574	0.762725

The result of aggregated weights is given in Table 6.

Table 6. Aggregated weights.

	SC		P		C	
Weights	0.745264	0.66682	1	0	0.9	0.583428

Because of the space constraints, we do not give the normalized and weighted normalized matrices. Based on the proposed IFS2 WASPAS method, the firm T is selected.

7. Conclusion

WASPAS method is a new MCDM method and it has been used for the solution of decision making problems in a short time period. The method has been also extended to its fuzzy versions such as hesitant fuzzy WASPAS, intuitionistic fuzzy WASPAS, and type-2 fuzzy WASPAS [8], [11], [4]. The classical WASPAS method has been first time extended to IFS2 WASPAS method in this paper. Single valued IFS2 sets have been preferred in the proposed method.

For further research, interval-valued IFS2 sets can be used in WASPAS method. Alternatively, triangular IFS2 or trapezoidal IFS2 sets may be used for the other possible extensions.

Acknowledgments

This work is supported by Scientific and Technological Research Council of Turkey (TÜBİTAK), TEYDEB 1505, Grant No: 5170012

References

1. K.T. Atanassov, (1999). Intuitionistic Fuzzy Sets: Theory and Applications, Springer.
2. S. Chakraborty, E. K. Zavadskas (2014). Applications of WASPAS method in manufacturing decision making, *Informatica* **25(1)**, 1-20.
3. S. Dick, R.R. Yager and O. Yazdanbakhsh. (2016). On Pythagorean and Complex Fuzzy Set Operations, *IEEE Transactions on Fuzzy Systems* **24(5)**, 1009-1021.
4. H. Garg (2016). A novel accuracy function under interval-valued Pythagorean fuzzy environment for solving multicriteria decision making problem, *Journal of Intelligent & Fuzzy Systems* **31(1)**, 529-540.
5. C. Kahraman, S. Birgün and V.Z. Yenen. (2008). Fuzzy Multi-Attribute Scoring Methods with Applications, *In Fuzzy Multi-Criteria Decision Making: Theory and Applications with Recent Developments* (Ed. C. Kahraman), Springer, 187-208.
6. W. L. Liu, J. Chang and X. He. (2016). Generalized Pythagorean fuzzy aggregation operators and applications in decision making, *Control and Decision* **31(12)**, 2280-2286.

7. X. Peng and Y. Yang. (2015). Some Results for Pythagorean Fuzzy Sets, *International Journal of Intelligent Systems* **30(11)**, 1133-1160.
8. P. Ren, Z. Xu and X. Gou. (2016). Pythagorean fuzzy TODIM approach to multi-criteria decision making, *Applied Soft Computing* **42**, 246-259.
9. R.R. Yager. (2016). Properties and Applications of Pythagorean Fuzzy Sets, In Imprecision and Uncertainty in Information Representation and Processing (Angelov, Plamen, Sotirov, Sotir, Eds.), Vol. 332, Springer International Publishing.
10. P. Angelov and S. Sotirov. (2016). Vol. 332, the Series Studies in Fuzziness and Soft Computing, Springer, 119-136.
11. S. Zeng, J. Chen and X. Li. (2016). A Hybrid Method for Pythagorean Fuzzy Multiple-Criteria Decision Making, *International Journal of Information Technology & Decision Making* **15(2)**, 403-422.
12. X. Zhang, Z. Xu. (2014). Extension of TOPSIS to Multiple Criteria Decision Making with Pythagorean Fuzzy Sets, International Journal of Intelligent Systems, 29, 1061–1078.

Part 3

Decision Analytics and Modeling under Uncertainty

Suppose two pieces of independent evidence e_1 and e_2 jointly support proposition θ. The combined degree of belief for θ is given by

$$P_{\theta,e(2)} = \begin{cases} 0 & \theta = \emptyset \\ \dfrac{\widehat{m}_{\theta,e(2)}}{\sum_{D\subseteq\Theta} \widehat{m}_{D,e(2)}} & \theta \subseteq \Theta, \theta \neq \emptyset \end{cases} \quad (2)$$

$$\widehat{m}_{\theta,e(2)} = \left[(1-r_2)m_{\theta,1} + (1-r_1)m_{\theta,2}\right] + \sum_{B\cap C=\theta} m_{B,1} m_{C,1} \quad (3)$$

where $\widehat{m}_{\theta,e(2)}$ consists of two parts: bounded sum of individual support and orthogonal sum of collective support.

4. Asthma Case Study

To identify asthma control level, four variables are considered: asthma causing daytime symptoms (D), asthma restricting exercise (E), asthma disturbing sleep (SL), and PEF(P), and correspondingly, four pieces of evidence is acquired [5]. Each piece of evidence has sub-evidence. For example, D has Asthma never causing daytime symptoms (D1), Asthma causing daytime symptoms 1 to 2 times per month (D2), Asthma causing daytime symptoms 1 to 2 times per week (D3), and Asthma causing daytime symptoms most (D4).

With four selected variables, only one proposed model is sufficient to identify asthma control step for a patient with any combination of variables. On the contrary, existing classification methods have to build multiple models. For example, if a new patient only has variable D, E and P, the model based on D, E, SL, and P cannot give a solution and a new model with and only with D, E and P should be built. Given four variables, it is necessary to build 11 ($= C_4^2 + C_4^3 + C_4^4$) models in any classification method. As shown in Table 2, besides the model with variable D, E, SL and P, some of the other models should be built with a very limited number of patients. For example, only 17 patients have and only have variable D and P. This small data size is not sufficient to build a model.

Table 2. The number of patients given a certain combination of variables.

Variables	# of patients	Variables	# of patients
D,E,SL	1953	D,SL	138
D,E,P	96	D,P	17
D,SL,P	82	E,SL	213
E,SL,P	151	E,P	18
D,E	76	SL,P	21

In the proposed model, firstly, D and E are aggregated as a new piece of evidence, which is aggregated with SL in the next step. The aggregation is a recursive process via ER as introduced in section 3.2. In the aggregation process, the reliability and weights are unknown parameters, which can be trained by minimising the difference between the final aggregated and observed probability distribution.

$$\min \quad f(R_{e_n}, W_{e_n}) = \frac{1}{m \times n} \sum_{j=1}^{m} \sum_{i=1}^{n} [P(e_{ij}) - P^*(e_{ij})]^2 \qquad (4)$$

s.t. $\quad 0 \leq R_{e_n} \leq 1 \ (n \geq 1)$
$\quad\quad 0 \leq W_{e_n} \leq 1 \ (n \geq 1)$

where

R_{e_n} is the reliability index of evidence e_n
W_{e_n} is the weight of evidence e_n
$P^*(e_{ij})$ is the joint likelihood of the ith aggregated evidence on asthma control step j
$P(e_{ij})$ is the aggregated likelihood on asthma control step j given the ith aggregated evidence
m is the total number of all pieces of aggregated evidence
n is the total number of asthma control steps

This model results in probability distributions on asthma control steps given combinations of evidence, e.g. given the combination of sub-evidence D1 and E1, the result is $\{(s_0, 0.385), (s_1, 0.194), (s_2, 0.157), (s_3, 0.144), (s_4, 0.120)\}$. Two parameters (reliability and weight) are trained [5] in this model with patients under 18 years old now, and model validation is conducted in patients who were under 18 years old when they visited GPs or hospitals. The mean absolute errors between the aggregated probability via the ER-based model and joint probability calculated from observations are 0.175, 0.061, and 0.091 for two, three and four pieces of evidence, respectively.

5. Conclusion

In this paper, authors proposed a new ER model that can work with missing data and does not depend on prior distribution. The developed model consists of a set of evidence, with each piece of evidence acquired from multiple sources and represented as a probability distribution on hypothesis space. Multiple pieces of evidence can be combined using the ER rule to generate more reliable results. The results from a case study show that the proposed ER-based prognostic model has desirable flexibility when dealing with multiple pieces of evidence from different data sources, especially when the prior distributions of asthma control steps are different given different pieces of evidence. The proposed model can be generalised as a probabilistic inference method which can be further applied in other classification or reasoning problems with uncertain prior.

References

1. British Thoracic Society. British Guideline on the Management of Asthma (2009).
2. H. K. Reddel, D. R. Taylor, E. D. Bateman, L. P. Boulet, H. A. Boushey, W. W. Busse, S. E. Wenzel. An official American Thoracic Society/European Respiratory Society statement: Asthma control and exacerbations — Standardizing endpoints for clinical asthma trials and clinical practice. *American Journal of Respiratory and Critical Care Medicine*, *180*(1), 59 (2009).
3. WHO. (2017). Asthma fact sheet. Retrieved from http://www.who.int/mediacentre/factsheets/fs307/en/
4. J. B. Yang, and D. L. Xu. Evidential reasoning rule for evidence combination. *Artificial Intelligence*, *205*, 1 (2013).
5. H. Zhu, J. B. Yang, D. L. Xu and C. Xu. Application of Evidential Reasoning rules to identification of asthma control steps in children. In *22nd International Conference on Automation and Computing (ICAC)* 444 (2016).

A further discussion on matching value for cooperative games and its application in profit allocation decision-making[*]

Fei Guan, Yuxin Cui and Hu Zhang

College of Mathematics & Statistics, Hebei University of Economics and Business, Shijiazhuang 050061, China

> Cooperative game is a mathematical model which describes a cooperative and conflict situation among n players. How to allocate the profits among each player after their cooperation is very important. In this paper, we mainly aim to give a further discussion on the matching value of cooperative games from the following aspects: individually rationality, collective rationality, uniqueness, symmetry, structure coalition monotonicity, computing complexity and so on. Furthermore, a concrete example is given to show the effectiveness of the matching value compared with the Core and the Shapley value. Thus this paper can promote the development of cooperative games and lay a sound base for the profit allocation decision-making problem.

1. Introduction

Since von Neumann and Morgenstern's pioneering work, game theory has been widely used to analyze some conflict and cooperative situations in economics, sociology and politics. This paper focuses on the profits allocation problem in classical cooperative game theory. In order to give an effective allocation method, several solution concepts were proposed, such as: von Neumann and Morgenstern [1] put forward the stable set from the dominance point of view; Gillies [2] gave the definition of the core set and discussed its properties; Aumann and Maschler [3] defined the negotiation set by the possible mutual negotiations between players; Shapley [4] proposed the Shapley value and proved its existence and uniqueness; Banzhaf [5] introduced the Banzhaf value in the context of voting games; Owen [6] presented the Owen value and proved its axiomatic method; Myerson [7] proposed the Myerson value to study some feasible coalitions; Tijs [8] defined the τ-value from the geometry point of view. In recent years, many scholars have conducted extensive researches on the

---

[*]This work is supported by the Natural Science Foundation of Hebei Province (F2017207010) and the Innovation ability enhancement project of Hebei province (174576455).

extended solutions, some success has been achieved. For example, Wang and Sun [9] defined a selfish coefficient function to describe the degrees of selfishness of players in different coalitions; Meng, Chen and Tan [10] proposed a generalized form of fuzzy games with interval characteristic functions; Li, Sun and Hou [11] discussed the position value for communication situations with fuzzy coalition.

In this paper, we mainly aim to have a further discussion on matching value of classical cooperative games which was defined in [12]. This paper is structured as follows: Preliminaries are introduced in Section 2. In Section 3, a series of properties of matching value are discussed. In Section 4, an example is given to verify the effectives of the matching value. The conclusions and further studies are discussed in Section 5.

2. Preliminaries

In this section, we will introduce some important Definitions and Theorems in [12, 13] which will be used in the following Sections.

2.1. Concepts in classical cooperative games

Definition 1. A classical cooperative game in coalition form is an ordered pair (N,v), where $N = \{1,2,\ldots,n\}$ is the set of players, and $v: 2^N \to R$ is a map, assigning to each coalition $S \in 2^N$ a real number $v(S)$, such that $v(\emptyset) = 0$. This function v is called the characteristic function of the game and $v(S)$ is called the worth (or value) of coalition S.

In the following paper, we denote by G^N the family of all classical cooperative games.

Definition 2. A game $v \in G^N$ is superadditive, if for $\forall R, T \subseteq N, R \cap T = \emptyset$,

$$v(R) + v(T) \leq v(R \cup T). \tag{1}$$

Definition 3. A game $v \in G^N$ is convex, if for any $R, T \subseteq N$, it holds

$$v(R) + v(T) \leq v(R \cup T) + v(R \cap T). \tag{2}$$

Definition 4. An imputation is a vector $x = (x_1, x_2, \ldots, x_n) \in R^n$, if

$$x_i \geq v(i), \quad i = 1, 2, \ldots, n. \tag{3}$$

$$x(N) = v(N). \tag{4}$$

Definition 5. The core of a game $v \in G^N$ is the set

$$C(v) = \{x \in R^n \mid v(S) - x(S) \leq 0 (\forall S \subseteq N), x(N) = v(N)\}. \tag{5}$$

Definition 6. Shapley value of a game $v \in G^N$ is defined as follows

$$\varphi_i(v) = \sum_{S \subseteq N\setminus\{i\}} \frac{|S|!(n-|S|-1)!}{n!} [v(S \cup i) - v(S)]. \tag{6}$$

Definition 7. The matching value of a game $v \in G^N$ is

$$\psi_i(v) = \frac{\psi_i^1 + \psi_i^2 + \cdots + \psi_i^u}{u}, \quad \forall i \in N. \tag{7}$$

u is the possible structure number of the grand coalition, $\psi_i^1, \ldots, \psi_i^u$ is the structure matching value of player i, and

$$\psi_i^u = v(i) + \sum_{j=1}^{k_i} \frac{v(\bigcup_{t=1}^{j} S_{it}^u \cup i) - v(S_{ij}^u) - v(\bigcup_{t=1}^{j-1} S_{it}^u \cup i)}{1 + \sum_{t=1}^{j} |S_{it}^u|}. \tag{8}$$

We call the set of all structure matching values as the matching value under total coalition structure, which is denoted as $M(v)$, that is, in a cooperative game (N,v)

$$M(v) = \bigcup_{u=1}^{C_n^2 * C_{(n-1)}^2 * \cdots * C_2^2} (\psi_1^u, \ldots, \psi_n^u). \tag{9}$$

$S_{ij} \in \mathcal{M}_i = \{S_{i1}, S_{i2}, \cdots, S_{ik_i}\}, \forall j \in \{1, 2, \cdots, k_i\}$ is called the structural units of player i, k_i is the number of the structural units.

Theorem 1. For a game $v \in G^N$, If $C(v) \neq \emptyset$, then $C(v)$ is the unique solution satisfying the individually rationality, the superadditivity and the converse reduced game property.

Theorem 2. For a game $v \in G^N$, Shapley value is the unique solution that satisfies the virtuality, anonymity, collective rationality and the additivity.

3. A Further Discussion on Matching Value

We defined a new solution concept called matching value in [12], as the length limits, some properties of it have been given without proof, so in the following paper, we will prove some properties in detail and discuss some other important properties of matching value.

Property 1. (Individual Rationality) Let $v \in G^N$, then $\psi_i(v) \geq v(i)$.
Property 2. (Efficiency) Let $v \in G^N$, then $\psi_1 + \psi_2 + \cdots + \psi_n = v(N)$.

Property 3. (Uniqueness) Matching value $\psi(v)$ is unique.

By the definition of matching value in Definition 7, the above three properties are clearly true.

Property 4. (Symmetry) If $v(S \cup i) = v(S \cup j), \forall i, j \notin S, S \subseteq N \setminus i, j$, then

$$\psi_i(v) = \psi_j(v). \tag{10}$$

Proof. For any $i, j \in N$, if $\forall S \subseteq N \setminus \{i, j\}$, then $v(S \cup i) = v(S \cup j)$, if there exists a cooperative unit $\{i, j\}$, for $v(i) = v(j)$, it is clear that $\psi_i(v) = \psi_j(v)$. If there does not exist a cooperative unit $\{i, j\}$, then $\{i\}$ and $\{j\}$ should match with other cooperative units, suppose there is a cooperative unit $R \subseteq N$ that $R \cup \{i\}$ is formed, we can get $i \notin R$. If $j \in R$, then $R \neq \{j\}$, thus $R \setminus \{j\} \neq \varnothing$, so $i, j \notin R \setminus \{j\}$. We have $v(R \setminus j \cup i) - v(R \setminus j) - v(i) = v(R \setminus j \cup i) - v(R \setminus j) - v(j)$ and $R \setminus j : i = j$. If $j \notin R$, we can obtain $v(R \cup i) = v(R \cup j)$ and $R : i = j$. So if we suppose the first structure unit of player i is $S_{i1}, S_{i2}, \ldots, S_{ik_i}$, then there exists a constant δ such that $j \in S_{i\delta}, 1 \leq \delta \leq k_i$. So there exists a structure of the grand coalition such that the structure unit of j is $S_{i1}, \ldots, S_{i(\delta-1)}, S_{i\delta} \cup i / j, S_{i(\delta+1)}, \ldots, S_{ik_i}$, suppose the number of the possible structures of the grand coalition is u, then $\psi_i^1 = \psi_j^u$. Thus $\psi_i^{1+t} = \psi_j^{u-t}, 0 \leq t \leq u-1$. And therefore

$$\psi_i(v) = \frac{\psi_i^1 + \psi_i^2 + \cdots + \psi_i^u}{u} = \frac{\psi_j^u + \psi_j^{u-1} + \cdots + \psi_j^1}{u} = \psi_j(v) \tag{11}$$

the proof of this theorem is complete.

Property 5. If player i is lagging behind, then matching value will be small.

Proof. Suppose player i is lagging behind one step, then S_{i1} and S_{i2} are firstly to constitute a unit link, $S_{i1} \cup S_{i2}$ and $\{i\}$ are going to constitute the unit link following, after that $S_{i1} \cup S_{i2} \cup \{i\}$ and S_{i3} will finish a perfect matching, the structure value of player i will be

$$\psi_i' = v(i) + \sum_{j=3}^{k_i} \frac{v(\bigcup_{t=1}^{j} S_{it} \cup i) - v(S_{ij}) - v(\bigcup_{t=1}^{j-1} S_{it} \cup i)}{1 + \sum_{t=1}^{j} |S_{it}|} + \frac{v(S_{i2} \cup S_{i1} \cup i) - v(S_{i2} \cup S_{i1}) - v(i)}{1 + |S_{i1}| + |S_{i2}|} \tag{12}$$

So we can compute

$$\psi_i - \psi_i' = \sum_{j=1}^{2} \frac{v(\bigcup_{t=1}^{j} S_{it} \cup i) - v(S_{ij}) - v(\bigcup_{t=1}^{j-1} S_{it} \cup i)}{1 + \sum_{t=1}^{j} |S_{it}|} - \frac{v(S_{i2} \cup S_{i1} \cup i) - v(S_{i2} \cup S_{i1}) - v(i)}{1 + |S_{i1}| + |S_{i2}|}$$

$$= (\frac{1}{1+|S_{i1}|} - \frac{1}{1+|S_{i1}|+|S_{i2}|}) \times [v(S_{i1} \cup i) - v(S_{i1}) - v(i)] + \frac{v(S_{i2} \cup S_{i1}) - v(S_{i2}) - v(S_{i1})}{1+|S_{i1}|+|S_{i2}|}$$

$$= \frac{|S_{i2}| \times [v(S_{i1} \cup i) - v(S_{i1}) - v(i)]}{(1+|S_{i1}|)(1+|S_{i1}|+|S_{i2}|)} + \frac{v(S_{i2} \cup S_{i1}) - v(S_{i2}) - v(S_{i1})}{1+|S_{i1}|+|S_{i2}|}$$

For the superadditivity, we can have

$$\psi_i - \psi_i' = \frac{|S_{i2}| \times [v(S_{i1} \cup i) - v(S_{i1}) - v(i)]}{(1+|S_{i1}|)(1+|S_{i1}|+|S_{i2}|)} + \frac{v(S_{i2} \cup S_{i1}) - v(S_{i2}) - v(S_{i1})}{1+|S_{i1}|+|S_{i2}|} \geq 0$$

That is $\psi_i' \leq \psi_i$, and this completes the proof of the theorem.

Property 6. Let (N,v) be a three-person convex cooperative game, then

$$(\psi_1(v), \psi_2(v), \psi_3(v)) \in C(v). \tag{13}$$

Proof. We can easily get that

$$\psi_1^1 = v(1) + \frac{v(1,2) - v(1) - v(2)}{2} + \frac{v(1,2,3) - v(1,2) - v(3)}{3}$$

$$\psi_2^1 = v(2) + \frac{v(1,2) - v(1) - v(2)}{2} + \frac{v(1,2,3) - v(1,2) - v(3)}{3}$$

$$\psi_3^1 = v(3) + \frac{v(1,2,3) - v(1,2) - v(3)}{3}$$

By the superadditivity of (N,v), we have

$$\psi_1^1 + \psi_2^1 = v(1,2) + \frac{2}{3}[v(1,2,3) - v(1,2) - v(3)] \geq v(1,2)$$

$$\psi_1^1 + \psi_3^1 = v(1) + v(3) + \frac{v(1,2) - v(1) - v(2)}{2} + \frac{2}{3}[v(1,2,3) - v(1,2) - v(3)]$$

Since (N,v) is convex and $v(1,2,3) - v(1,2) - v(3) \geq v(1,3) - v(1) - v(3)$, we can get: $\{1\}:\{2\} \succeq \{3\} \Leftrightarrow v(1,2) - v(1) - v(2) \geq v(1,3) - v(1) - v(3)$, thus

$$\psi_1^1 + \psi_3^1 \geq v(1) + v(3) + \frac{v(1,3) - v(1) - v(3)}{2} + \frac{2}{3}[v(1,3) - v(1) - v(3)]$$

$$= v(1,3) + \frac{v(1,3) - v(1) - v(3)}{6} \geq v(1,3)$$

Then we can obtain

$$\psi_2^1 + \psi_3^1 \geq v(2) + v(3) + \frac{v(2,3) - v(2) - v(3)}{2} + \frac{2}{3}[v(2,3) - v(2) - v(3)]$$

$$= v(2,3) + \frac{v(2,3) - v(2) - v(3)}{6} \geq v(2,3)$$

Therefore $(\psi_1^1,\psi_2^1,\psi_3^1) \in C(v)$ and $(\psi_1^i,\psi_2^i,\psi_3^i) \in C(v)$. For $C(v)$ is a convex set, then we can get $(\psi_1(v),\psi_2(v),\psi_3(v)) \in C(v)$.

Property 7. The computing complexity of matching value is less than $O(n^5)$.

Proof. To compute the matching value, we need to consider the grand coalition structure, in the first matching process, C_n^2 units will be formed, we know that sorting algorithm can go on as the speed of $O(M \cdot \log M)$, so the maximum computing complexity of matching value in the first matching process is

$$5 \cdot C_n^2 + C_n^2 \cdot \log C_n^2$$

thus the whole maximum computing complexity is

$$5C_n^2 + 5C_{n-1}^2 + \cdots + 5C_2^2 + C_n^2 \cdot \log C_n^2 + \cdots + C_2^2 \cdot \log C_2^2 \quad (14)$$

for $C_n^2 \leq n^2$ and $\log C_n^2 \leq 2\log n$, so

$$\lim_{x \to \infty} \frac{5C_n^2 + 5C_{n-1}^2 + \cdots + 5C_2^2 + C_n^2 \cdot \log C_n^2 + \cdots + C_2^2 \cdot \log C_2^2}{n^5} = 0$$

and this completes the proof.

4. A Case-Based Study

Suppose there are four small and medium-sized companies A, B, C and D in a city of china. Each of them has different scales and features. In order to break the regional restrictions and improve their market competitiveness, they decide to join together to produce a new product. By their prediction, the estimated profits after their cooperation are shown in Table 1. Please discuss that how to give a rational profit allocation scheme among these four companies.

Table 1. The estimated profit of four companies (unit: million).

A	0	AB	1	BD	0	ACD	0
B	0	AC	1	CD	0	ABC	1
C	0	BC	1	ABD	0	ABCD	1.5
D	0	AD	0	BCD	0		

Following we will choose some different solutions to get an allocation scheme. By Definition 2.5, we can obtain

$$C(v) = \{(0.5, 0.5, 0.5, 0)\}.$$

We can easily get that player 4 only get 0 million after their cooperation in the core solution, obviously it is not a reasonable profit allocation scheme. For v is not additive, thus shapely value cannot be used as a solution.

Next we will discuss the matching value, firstly, there are four cooperative units $\{A\}, \{B\}, \{C\}$ and $\{D\}$, we can get:

$$\{A\}:\{B\} = \{C\} \succ \{D\}$$
$$\{B\}:\{A\} = \{C\} \succ \{D\}$$
$$\{C\}:\{A\} = \{B\} \succ \{D\}$$
$$\{D\}:\{A\} = \{B\} = \{C\}$$

Suppose $\{A\}$ and $\{B\}$ constitute a coalition in the first phase of matching, then we get

$$\{A\}\cup\{B\}:\{C\} \succ \{D\}$$
$$\{C\}:\{A\}\cup\{B\} \succ \{D\}$$
$$\{D\}:\{A\}\cup\{B\} = \{C\}$$

then we can get the following three coalition structure of the grand coalition.

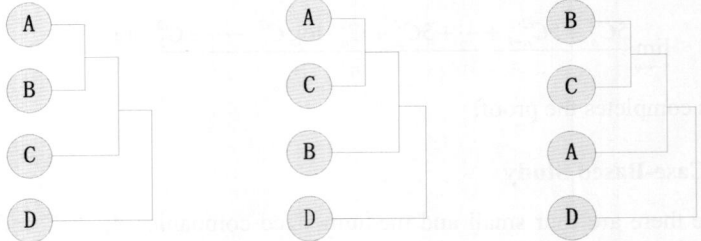

Figure 1. The possible coalition structure of the grand coalition.

We can compute the matching value

$$(\psi_1(v),\psi_2(v),\psi_3(v),\psi_4(v)) = (\frac{11}{24},\frac{11}{24},\frac{11}{24},\frac{1}{8})$$

We can see that the gained profits which allocated to A, B, C are the same, D will get a small part. This result not only reflects the full negotiation process but also supplies a stable and reasonable allocation result, so it will be a rational profit allocation scheme among these four companies.

5. Conclusions and Further Study

In this paper, we mainly have a further discussion on the matching value of cooperative games. We discuss the properties of matching value from the following aspects: individually rationality, collective rationality, uniqueness, symmetry, structure coalition monotonicity, computing complexity and so on. The discussion on the applications of the matching value will be our next step of work.

References

1. J. von Neumann, O. Morgenstern, Theory of Games and Economic Behavior [M]. *Princeton University Press*, (1944).
2. D. B. Gillies, Some theorems on n-person games [D]. *Department of Mathematics, Princeton University*, (1953).
3. R. Aumann, M. Maschler, The bargaining set for cooperative games [G]. *Advances in Game Theory*, (1964).
4. L. S. Shapley, A value for n-persons games [J]. *Annals of Mathematics Studies*, **307-318**, 28(7), (1953).
5. J. F. Banzhaf, Weighted voting does not work: a mathematical analysis [J]. *Rutgers Law Review*, **317-343**, 19(2), (1965).
6. G. Owen, Values of games with a priori unions. in: R. Hein, O. Moeschlin (Eds.), Essays in Mathematical Economics and Game Theory, *Springer Verlag, New York*. **76-88**, (1977).
7. R. B. Myerson, Graphs and cooperation in games [J]. *Math. Operations Research*, **225-229**, 2, (1977).
8. S. Tijs, G. Owen, Compromise values in cooperative game theory [C]. *Top*, **1-51**, 1(1), (1993).
9. W. Wang, H. Sun, *The family of ideal values for cooperative games [J]*. Social Science Electronic Publishing, **1-22**, (1), (2018).
10. F. Y. Meng, X. H. Chen, C. Q. Tan, *Cooperative fuzzy games with interval characteristic functions [J]*. Operational research, **1-24**, 16(1), 2016.
11. X. H. Li, H. Sun, D. S. Hou, On the position value for communication situations with fuzzy coalition [J]. *Journal of Intelligent and Fuzzy Systems*, **113-124**, 33(1), (2017).
12. F. Guan, D. Y. Xie, Q. Zhang, A matching value for cooperative games [J]. *Journal of Intelligent & Fuzzy Systems*. **201-212**, 31(1), (2016).
13. Xie Z., *The Game Theory [M]. Beijing: Science Press.* (2010).

Regret theory-based decision-making with evidential reasoning

Liuqian Jin

School of Economics and Management, Chongqing University of Posts and Telecommunications, Chongqing 400065, P. R. China
jinliuqian@163.com

Xin Fang

School of Business Planning, Chongqing Technology and Business University Chongqing 400067, P. R. China
qbo10086@163.com

In this paper, a method for uncertainty multi-attribute decision making based on regret theory is proposed, the uncertainty is represented by interval attributes. Firstly, we defined an interval utility function and regret-rejoice function, a computational formula for the perceived utility of alternative decisions. Then, we compute the decision makers optimal comprehensive perceived utility using evidential reasoning and obtain the ranking order of alternatives.

Keywords: Regret theory; evidential reasoning; decision making; interval number.

1. Introduction

In this paper we are concerned with decision making under uncertainty using regret theory and evidential reasoning. In particular, we are interested in the role that interval fuzzy sets might play in enhancing decision making.

Regret theory (RT) is an alternative behavioral decision theory which was presented by Bell [1] and Loomes and Sugden [2]. Recently, regret theory may be one of the most popular nonexpected utility models. The core idea is that the decision makers not only pay attention to the results obtained by the choice of the alternative, but also pay attention to the outcome of other alternatives, and avoid choosing the alternative that will make them regret [1,2]. Regret captures differences between the performance of a chosen product or service and the performance of a rejected product or service [3]. The psychological behavior of regret aversion of decision-makers in the decision-making process can be quantized [4]. Moreover, there have been some applications of regret theory such as investment choices [5,6], asset pricing [7], route choice [8], auctions [9], decision making [10,11].

Evidential Reasoning (ER) approach [12] is built by Yang. The ER approach involves the following steps [13]: (Step I) a set of well-organized grades are offered to evaluate attributes, (Step II) a distributed framework is established to describe uncertain assessment using a belief structure, (Step III) the Dempster-Shafer theory, developed by Dempster [14] and Shafer [15], is used to aggregate the assessment information, and (Step IV) the multi-attribute utility theory is applied to compare alternatives. There are three case of ER approach, the first one is ER recursive algorithm [16], the second one is ER analytical algorithm [17], and the other one is ER rule with weight [18]. So in reference, a new ER rule to combine evidence with weight is given, based on the weight handling method the ER recursive algorithm is simplified. The process of ER recursive algorithm is pellucid, ER analytical algorithm with low complexity, ER rule with weight is simpleness, they are appropriate for different problems.

The rest of this paper is organized as follows. Section 2 proposes a new method for uncertainty multi-attribute decision making using regret theory. Conclusions are drawn in Section 3.

2. Regret Theory-based Decision-making Method

2.1. *Problem statement*

The following elements and notations are used in the proposed regret theory-based decision-making method.

(1) $\mathbf{A} = \{A_i \,|\, i = 1, 2, \cdots, I\}$: the set of alternatives, where A_i ($i \in \{1, 2, \cdots, I\}$) represents the ith alternative.
(2) $\mathbf{X} = \{X_j \,|\, i = 1, 2, \cdots, J\}$: the set of attributes, where X_j ($j \in \{1, 2, \cdots, J\}$) represents the ith alternative.
(3) $\mathbf{W} = (W_1, \cdots, W_j, \cdots, W_J)$: the vector of attribute weights, where W_j ($j \in \{1, 2, \cdots, J\}$) represents the attribute weight of attribute X_j, satisfies $0 \leq W_j \leq 1$ and $\sum_{j=1}^{J} W_j = 1$.
(4) $x_{ij} = [x_{ij}^L, x_{ij}^U]$: the interval attribute value of alternative A_i under attribute X_j. The decision matrix is given as follows.

$$\mathbf{D} = \begin{bmatrix} x_{11} & x_{12} & \cdots & x_{1J} \\ x_{21} & x_{22} & \cdots & x_{2J} \\ \vdots & \vdots & & \vdots \\ x_{I1} & x_{I2} & \cdots & x_{IJ} \end{bmatrix}.$$

2.2. Decision-making method

In this subsection, the uncertainty multi-attribute decision making method with interval number using evidential reasoning and regret theory is proposed. First, the utility function and the regret-rejoice function should be given with regret theory.

Step 1. Construct utility function[19] of interval attribute value $x_{ij} = [x_{ij}^L, x_{ij}^U]$, calculate the value of alternative A_i $(i = 1, 2, \cdots, I)$ under attribute X_j $(j = 1, 2, \cdots, J)$.

$$V(x_{ij}) = \int_{x_{ij}^L}^{x_{ij}^U} v(x) f(x) \, dx.$$

Where for benefit attribute is defined as $v(x) = \frac{1-e^{-\alpha x}}{\alpha}, 0 < \alpha < 1$, for cost attribute is defined as $v(x) = \frac{1-e^{\beta x}}{\beta}, 0 < \beta < 1$[20]; $f(x)$ is the probability density function. In this paper, for $f(x)$, we take the form of normal distribution. When x obeys normal distribution $N(\mu, \sigma^2)$, according to the 3σ principle [21] in the knowledge of probability and statistics, x is covered by a probability of 99.73% in interval $[x^L, x^U]$, i.e., $\sigma = (x^U - x^L)/6$, $\mu = (x^L + x^U)/2$. The probability density of x is defined as follows.

$$f(x) = \frac{1}{\sqrt{2\pi}\sigma} e^{-\frac{(x-\mu)^2}{2\sigma^2}}.$$

Step 2. The regret-rejoice function $R(\Delta v)$ can be defined as follows.

$$R(\Delta v) = 1 - e^{-\gamma \Delta v}.$$

Where γ is risk aversion coefficient of decision maker. And the greater the γ, the greater the degree of risk aversion of decision maker. $\Delta v = v_1 - v_2$ denotes the difference in the utility value of two alternatives. When $R(\Delta v) > 0$, $R(\Delta v)$ denotes rejoice value. When $R(\Delta v) < 0$, $R(\Delta v)$ denotes regret value.

Step 3. Compute the utilize value $u(x_{ij})$ $(i = 1, 2, \cdots, I; j = 1, 2, \cdots, J)$ of alternative A_i under attribute X_j:

$$u(x_{ij}) = V(x_{ij}) + R(V(x_{ij}) - V(x_j^*)),$$

$$V(x_{ij}) = \begin{cases} \int_{x_{ij}^L}^{x_{ij}^U} \left(\frac{1-e^{-\alpha x}}{\alpha}\right) \left(\frac{1}{\sqrt{2\pi}\sigma_{ij}}e^{-\frac{(x-\mu_{ij})^2}{2\sigma_{ij}^2}}\right) dx, & X_j \text{ is benefit attribute} \\ \int_{x_{ij}^L}^{x_{ij}^U} \left(\frac{1-e^{\beta x}}{\beta}\right) \left(\frac{1}{\sqrt{2\pi}\sigma_{ij}}e^{-\frac{(x-\mu_{ij})^2}{2\sigma_{ij}^2}}\right) dx, & X_j \text{ is cost attribute} \end{cases}$$

$$R\left(V\left(x_{ij}\right) - V\left(x_j^*\right)\right) = 1 - e^{-\gamma(V(x_{ij}) - V(x_j^*))}$$

where $0 < \alpha < 1, 0 < \beta < 1, \sigma_{ij} = (x_{ij}^U - x_{ij}^L)/6$, $\mu_{ij} = (x_{ij}^L + x_{ij}^U)/2$,

$$x_j^* = \begin{cases} \max\{x_{1j}, \cdots, x_{ij}, \cdots, x_{Ij}\}, & X_j \text{ is benefit attribute} \\ \min\{x_{1j}, \cdots, x_{ij}, \cdots, x_{Ij}\}, & X_j \text{ is cost attribute} \end{cases} (j = 1, 2, \cdots, J),$$

γ is the risk aversion coefficient of decision maker.

Step 4. Calculate the combined utilize value $U(A_i)$ $(i = 1, 2, \cdots, I)$ of alternative A_i using evidential reasoning. With the evidential reasoning method, the basic assignment functions are given as follows.

$$m_j(\{A_i\}) = u(x_{ij}) W_j,$$

$$m_j(P(\{A_i\})) = 1 - u(x_{ij}) W_j,$$

$$\bar{m}_j(P(\{A_i\})) = 1 - W_j.$$

Let $m_{O(1)}(\{A_i\}) = m_1(\{A_i\})$, $m_{O(1)}(P(\{A_i\})) = m_1(P(\{A_i\}))$, $\bar{m}_{O(1)}(P(\{A_i\})) = \bar{m}_1(P(\{A_i\}))$. For the first $t \in \{2, 3, \cdots, J\}$ evidential, the combination formulas are as follows based on the interval certitude degree inference method.

$$m_{O(t)}(\{A_i\}) = m_{O(t-1)}(\{A_i\}) m_t(\{A_i\}) \\ + m_{O(t-1)}(\{A_i\}) m_t(P(\{A_i\})) + m_{O(t-1)}(P(\{A_i\})) m_t(\{A_i\})$$

$$m_{O(t)}(P(\{A_i\})) = m_{O(t-1)}(P(\{A_i\})) m_t(P(\{A_i\})),$$

$$\bar{m}_{O(t)}(P(\{A_i\})) = \bar{m}_{O(t-1)}(P(\{A_i\})) \bar{m}_t(P(\{A_i\})).$$

The combined interval prospect value is given as follows:

$$U(\{A_i\}) = \frac{m_{O(J)}(\{A_i\})}{1 - \bar{m}_{O(J)}(P(\{A_i\}))} = \frac{1 - \prod_{\tau=1}^{J}(1 - u(x_{i\tau}) W_\tau)}{1 - \prod_{\tau=1}^{J}(1 - W_\tau)}.$$

Step 5. Rank the order of alternatives. The larger the value of $U(\{A_i\})$ $(i = 1, 2, \cdots, I)$ is, the better the alternative is.

3. Conclusions

We described the regret theory-based decision-making problem in which the attribute values are interval number. We fused the uncertainty of the utilize value using evidential reasoning. Our objective was to select the alternative with the minimum effective regret. Firstly, the utility function and regret-rejoice function for the prospect theory with interval numbers are given. Then, the utilize values are fused using evidential reasoning approach. In future studies, we will try to compare with other methods, and give the illustrative example.

Acknowledgments

This work is supported by Doctor Project of Chongqing Social Sciences (Grant No. 2016BS032, 2016BS082), Project of Chongqing Municipal Education Commission (Grant No. KJ2016008391 44), Chongqing Commission of Science and Technology Research Projects (Grant No. KJ170616 5), Chongqing Engineering Research Center for Processing, Storage and Transportation of Char-acterized Agro-Products (Grant No. KFJJ2016026), Fundamental Science and Frontier Technol-ogy Research Project in Chongqing (cstc2017jcyjAX01301), National Natural Science Foundation of China (No. 71702015). This work is supported by National Science Foundation of China (Grant No. 61175055, 61305074), Sichuan Science and Technology Innovation Talent Project (Grant No. 2014-057).

References

1. D. E. Bell, Regret in decision making under uncertainty, Oper. Res. 30 (1982) 961-981.
2. G. Loomes, R. Sugden, Regret theory: An alternative theory of rational choice, T. Econ. J. 92 (1982) 805-824.
3. J. J. Inman, J. S. Dyer, J. Jia, A generalized utility model of disappointment and regret effects on post-choice valuation, Market. Sci. 16 (2) (1997) 97-111.
4. H. Bleichrodt, A. Cillo, E. Diecidue, A quantitative measurement of regret theory, Manage. Sci. 56 (1) (2010) 161-175.
5. A. Muermann, O. S. Mitchell, J. M. Volkman, Regret, portfolio choice, and guarantees in defined contribution schemes, Insur. Math. Econ. 39 (2006) 219-229.
6. S. Michenaud, B. Solnik, Applying regret theory to investment choices: Currency hedging decisions, J. Int. Money Finance 27 (2008) 677-694.

7. A. Dodonova, Y. Khoroshilov, Applications of Regret Theory to Asset Pricing, Technical report, University of Ottawa, 2005.
8. C. G. Chorus, Regret theory-based route choices and traffic equilibria, Transportmetrica 8 (2012) 291-305.
9. R. Engelbrecht-Wiggans, E. Katok, Regret and feedback information in first-price sealed-bid auctions, Manage. Sci. 54 (2008) 808-819.
10. X. Zhang, Z. P. Fan, F. D. Chen, Risky multiple attribute decision making regret aversion, J. Syst. Manage. 23 (2014) 111-117.
11. X. D. Peng, Y. Yang, Algorithms for interval-valued fuzzy soft sets in stochastic multi-criteria decision making based on regret theory and prospect theory with combined weight. Appl. Soft Comput. 54 (2017) 415-430.
12. J. B. Yang, M. G. Singh, An evidential reasoning approach for multiple-attribute decision making with uncertainty. IEEE Trans. on Syst., Man, and Cybern. 24(1) (1994) 1-18.
13. M. J. Zhang, Y. M. Wang, L. H. Li et al., A general evidential reasoning algorithm for multi-attribute decision analysis under interval uncertainty. Euro. J. of Oper. Res. 257 (2017) 1005-1015.
14. A. P. Dempster, Upper and lower probabilities induced by a multivalued mapping, Annals of Mathematical Statistics 38(2) (1967) 325-339.
15. G. Shafer, A mathematical theory of evidence. New Jersey/ Princeton: Princeton University Press, 1976.
16. J. B. Yang, D. L. Xu, On the evidential reasoning algorithm for multiple attribute decision analysis under uncertainty, IEEE Trans. on Syst., Man, and Cybern. 32(3) (2002) 289-304.
17. Y. M. Wang, J. B. Yang, D. L. Xu et al., The evidential reasoning approach for multiple attribute decision analysis using interval belief degrees. Euro. J. of Oper. Res. 175 (2006) 35-66.
18. J. B. Yang, D. L. Xu, Evidential reasoning rule for evidence combination, Artif. Intell. 205 (2013) 1-29.
19. X. Zhang, Z. P. Fan, F. D. Chen, Risky multiple attribute decision making regret aversion, J. Syst. Manage. 23 (2014) 111-117.
20. C. G. Chorus, Regret theory-based route choices and traffic equilibria, Transportmetrica 8 (2012) 291-305.
21. Z. Sheng, S. Q. Xie, C. Y. Pan, Probability Theory and Mathematical Statistics, Higher Education Press, Beijing, 2001.

Decision roughness measurement and application research based on data effect

Fachao Li[1], Xiao Zhang[2] and Chenxia Jin[3]

[1,3]*School of Economics and Management,*
[2]*School of Science,*
Hebei University of Science and Technology, Shijiazhuang, 050018, P. R. China
lifachao@tsinghua.org.cn; zhangxiao19931022@qq.com; jinchenxia2005@126.com

Roughness measurement, an important part of rough set theory, must be faced in building data-based decision-making method. Based on the deficiencies of the variable precision roughness measurement method, this paper introduces the concept of core data sets and establishes the decision roughness measurement based on data effect (DE-DRD); then, the properties of DE-DRD are discussed; Finally, as an application of DE-DRD, the attribute reduction method based on DE-DRD (DE-DRD-RM) is proposed, and combined with UCI data sets, the characteristics of DE-DRD-RM are analyzed. The theoretical analysis and practical application show that DE-DRD can integrate decision preference into measurement system, and has extensive application value in many fields.

1. Introduction

Rough set theory was proposed by Pawlak [1] in 1982. The basic idea is to consider the description of concepts by some divisions of the domain. Recently the theory and application are widely concerned in the academic field. In theoretical research, Chen and Xue et al. [2] proposed a neighborhood rough set model; Bonikowski et al. [3] provided a covering rough set model; Wei and Zhang [4] proposed a rough set model in probabilistic space. In application research, Pawlak [5] proposed the attribute reduction method based on the positive domain; Mi and Zhang [6] put forward the attribute reduction method based on the maximum distribution; Zhang and Mei et al. [7] proposed the attribute reduction method based on confidence degree.

The classification in the Pawlak rough set must be completely accurate, and it lacks of adaptability to noise data. To overcome this limitation, Ziarko [8] proposed the variable precision rough set model (VPRD). It is a method of dealing with uncertain information and effectively eliminating noise data. And some scholars have studied and generalized variable precision rough set, for example, Zhao and Tsang et al. [9] put forward the variable precision rough fuzzy set model; Huang and Li et al. [10] proposed a dynamic variable precision rough set model based on probability set valued information system; Park and Choi [11] proposed a variable precision rough set model based on information entropy. However, these discussions all have different forms of limitations and shortcomings, and one of the common problems is that they do not consider the characteristics of different sample data.

In order to solve this problem, this paper takes the knowledge hidden in the data system as the carrier, considers the function characteristics of different sample data, and discusses the reasonable roughness measurement mechanism. The structure of the paper is as follows: Section 2 proposes the basic concepts used in the follow-up discussion and provides the deficiencies of variable precision roughness measurement; Section 3 establishes the decision roughness measurement based on data effect (DE-DRD); Section 4 analyzes the basic properties of DE-DRD; Section 5 discusses the performance of DE-DRD with a practical example; Section 6 proposes an attribute reduction method based on DE-DRD; Section 7 summarizes the work of this paper.

2. Deficiencies of Variable Precision Roughness Measurement

Let (U, A, F, D) be a decision information system, $B \subseteq A$, $R_B = \{(x,y) | (x,y) \in U \times U$ and $f_a(x) = f_a(y)\}$ be the equivalence relation on U determined by B, $R_D = \{(x,y) | (x,y) \in U \times U$ and $f_D(x) = f_D(y)\}$ be the equivalence relation on U determined by D, $[x]_B$ be the R_B equivalence class of x. U/R_B is simplified as U/B, U/R_D is U/D; $|C|$ represents the number of elements in a finite set C; let $X, Y \subset U$, $I(X \subset Y) = |X \cap Y| / |X|$ is called the inclusion degree, and $|X| > 0$.

Definition 1. [1] Let $\Phi = (U, A, F, D)$ be a decision information system, $B \subseteq A, \beta \in (0.5, 1]$. For $X \subseteq U$,

$$\underline{R}_B^\beta(X, \Phi) = \{x \mid x \in U \text{ and } I([x]_B \subset X) \geq \beta\} \quad (1)$$

$$\overline{R}_B^\beta(X, \Phi) = \{x \mid x \in U \text{ and } I([x]_B \subset X) > 1 - \beta\} \quad (2)$$

$$\rho_B^\beta(X, \Phi) = 1 - |\underline{R}_B^\beta(X, \Phi)| / |\overline{R}_B^\beta(X, \Phi)| \quad (3)$$

1) $\underline{R}_B^\beta(X, \Phi)$ is called the β-lower approximation set of X, $\overline{R}_B^\beta(X, \Phi)$ is called the β-upper approximation set of X; 2) $\rho_B^\beta(X, \Phi)$ is called the β-roughness of X.

Definition 2. Let $\Phi = (U, A, F, D)$ be a decision information system, $B \subseteq A$, $U/D = \{D_1, D_2, \cdots, D_m\}$,

$$\rho_B(U/D, \Phi) = \sum_{j=1}^m \frac{|D_j|}{|U|} \cdot \rho_B(D_j, \Phi) \quad (4)$$

is called the decision roughness of D about B and Φ.

Variable precision rough set not only inherits the basic features of Pawlak rough set, but also has good noise tolerance. It is an effective tool for dealing with incomplete and imprecise information. But it is worth noting that: 1) Due to the VPRD reduces the classification standard, and the concept itself cannot be specifically described by the upper and lower approximation sets; 2) For $B \subset A$, U/B is coarser than U/A, but we may have $\underline{R}_A^\beta(X,\Phi) \subset \underline{R}_B^\beta(X,\Phi)$. Due to that the lower approximation set is the basic factor of the roughness measurement, the measurement results no longer satisfy the monotonicity; 3) Because there is no universally accepted criterion for precision, the VPRD has a locality and does not represent the role of the whole in the decision making process. How to systematically build roughness measurement method of rough set has important theoretical and applied value. For the deficiencies of VPRD, the reasonable roughness measurement mechanism will be built below.

3. Decision Roughness Measurement Based on Core Data

For the decision information system (U, A, F, D), $U/A = \{X_1, X_2, \cdots, X_n\}$, and $U/D = \{D_1, D_2, \cdots, D_m\}$. If the sample data $X_i \cap D_j$ that can not reach a certain threshold requirement r_{ij} is considered as noise or unimportant data, then removing data with $r_{ij} < \lambda$ from $X_i \cap D_j$ is a data deletion method based on knowledge reliability. $\lambda \in [0,1]$ is a reliable threshold, and $r_{ij} = I(X_i \subset D_j)$ is called the decision reliability.

Definition 3. Let $\Phi = (U, A, F, D)$ be a decision information system, $\lambda \in [0,1]$, $U/A = \{X_1, X_2, \cdots, X_n\}$, $U/D = \{D_1, D_2, \cdots, D_m\}$,

$$U_\lambda = \bigcup_{i=1}^n \bigcup_{j=1}^m \{X_i \cap D_j \mid I(X_i \subset D_j) \geq \lambda\} \tag{5}$$

is called λ core data set of U about Φ, $\Phi(\lambda) = (U_\lambda, A, F, D)$ is a λ subsystem of Φ.

Because $\Phi(\lambda)$ is only a local description of Φ in the background of decision knowledge, and λ is a quantitative index reflecting the reliability of the decision-making. The bigger (smaller) the λ, the stronger (weaker) the reliability of the decision made on the basis of the $\Phi(\lambda)$. Hence, if the impact of λ in the decision process is abstracted as a weight function $W(\lambda)$ (Effect Synthetic Function) from $[0,1]$ to $[0,+\infty)$, then it should satisfy the following basic conditions: 1) $W(\lambda)$ satisfies monotonicity; 2) $W(\lambda)$ satisfies continuity; 3) $\int_0^1 W(\lambda)d\lambda = 1$.

$$W(\lambda) = \begin{cases} 0, & \lambda \leq k, \\ \dfrac{n+1}{(1-k)^{n+1}}(\lambda-k)^n, & k < \lambda \leq 1. \end{cases} \quad (0 \leq k < 1, n \geq 0) \tag{6}$$

If effect synthetic function $W(\lambda)$ is used as the effect value of the threshold λ in the comprehensive decision process, then the roughness measurement of decision D about attribute B can be constructed:

$$\rho(U/D, B \oplus W(\lambda)) = \int_0^1 W(\lambda) \cdot \rho_B(U/D, \Phi(\lambda))d\lambda \qquad (7)$$

Called the decision roughness measurement based on data effect, DE-DRD.

It is not difficult to see that DE-DRD is a general roughness measurement model and can reflect the integrity of the rough measure. It can effectively eliminate noise data through decision reliability, and the concept itself can be specifically described by the upper and lower approximation sets. In practical problem, the decision maker chooses different $W(\lambda)$ under different research backgrounds, gets different measurement results and makes different decision processing consciousness.

Therefore, the DE-DRD and VPRD have differences in the measurement mechanism, and often have large differences in the measurement results. The DE-DRD lays a theoretical foundation for different decision-making.

4. The Basic Properties of DE-DRD

Theorem 1. Let $\Phi = (U, A, F, D)$ be a decision information system, $\Phi(\lambda)$ be a λ subsystem of Φ, $W(\lambda)$ be an effect synthetic function, $\varnothing \neq S \subseteq T \subseteq A$, then $\rho(U/D, S \oplus W(\lambda)) \geq \rho(U/D, T \oplus W(\lambda))$.

Proof: According to $S \subseteq T$, then $[x]_T^\lambda \subseteq [x]_S^\lambda$ is true for any $x \in U_\lambda$. And for any $\lambda \in [0,1]$, $\underline{R}_S(D_j^\lambda, \Phi(\lambda)) \subseteq \underline{R}_T(D_j^\lambda, \Phi(\lambda))$, $\overline{R}_S(D_j^\lambda, \Phi(\lambda)) \supseteq \overline{R}_T(D_j^\lambda, \Phi(\lambda))$ and $\rho_S(U/D, \Phi(\lambda)) \geq \rho_T(U/D, \Phi(\lambda))$ are true. So we can get $\rho(U/D, S \oplus W(\lambda)) = \int_0^1 W(\lambda) \cdot \rho_S(U/D, \Phi(\lambda))d\lambda \geq \rho(U/D, T \oplus W(\lambda)) \int_0^1 W(\lambda) \cdot \rho_T(U/D, \Phi(\lambda))d\lambda = \rho(U/D, T \oplus W(\lambda))$.

Theorem 1 shows that the finer the partition, the smaller the roughness. That is, known the more knowledge, the more reliable the information can get.

Theorem 2. Let $\Phi = (U, A, F, D)$ be a decision information system, $B \subseteq A$, $U/A = \{X_1, X_2, \cdots, X_n\}$, $U/D = \{D_1, D_2, \cdots, D_m\}$, $W(\lambda)$ be an effect synthetic function, $\Phi(\lambda)$ be a λ subsystem of Φ. If $0 = \lambda_0 < \lambda_1 < \cdots < \lambda_{t-1} < \lambda_t = 1$ is the sorting result of all the value in $\{I(X_i \subset D_j), 0, 1 | i = 1, 2, \cdots, n, j = 1, 2, \cdots, m\}$, then $\rho(U/D, B \oplus W(\lambda)) = \sum_{k=1}^t \varphi_k \cdot \rho_B(U/D, \Phi(\lambda))$, and $\varphi_k = \int_{\lambda_{k-1}}^{\lambda_k} W(\lambda)d\lambda$.

Proof: By the definition of U_λ, we know $\rho_B(U/D, \Phi(\lambda)) = \rho_B(U/D, \Phi(\lambda_k))$ for any $k \in \{1, 2, \cdots, t\}$ and $\lambda \in (\lambda_{k-1}, \lambda_k]$, then get $\rho(U/D, B \oplus W(\lambda)) = \int_0^1 W(\lambda) \cdot \rho_B(U/D, \Phi(\lambda))d\lambda = \sum_{k=1}^t \int_{\lambda_{k-1}}^{\lambda_k} W(\lambda) \cdot \rho_B(U/D, \Phi(\lambda))d\lambda = \sum_{k=1}^t \int_{\lambda_{k-1}}^{\lambda_k} W(\lambda) \cdot \rho_B(U/D, \Phi(\lambda_k))d\lambda = \sum_{k=1}^t \rho_B(U/D, \Phi(\lambda_k)) \cdot \int_{\lambda_{k-1}}^{\lambda_k} W(\lambda)d\lambda = \sum_{k=1}^t \varphi_k \cdot \rho_B(U/D, \Phi(\lambda_k))$.

Theorem 2 shows that the DE-DRD can be understood as the weighted average of the corresponding roughness measurement of multiple λ subsystems, so the calculation steps of $\rho(U/D, B \oplus W(\lambda))$ can be obtained:

Step 1 Input the decision information system (U, A, F, D);
Step 2 Determine $\{r_{ij}, 0, 1\}$ and $0 = \lambda_0 < \lambda_1 < \cdots < \lambda_{t-1} < \lambda_t = 1$;
Step 3 Determine $\rho_B(U/D, \Phi(\lambda_k))$, $k = 1, 2, \cdots, t$;
Step 4 Select the $W(\lambda)$, and calculate $\varphi_k = \int_{\lambda_{k-1}}^{\lambda_k} W(\lambda) d\lambda$;
Step 5 Output $\rho(U/D, B \oplus W(\lambda)) = \sum_{k=1}^{t} \varphi_k \cdot \rho_B(U/D, \Phi(\lambda))$.

5. Performance Analysis of DE-DRD

The following will combine a specific example to analyze the performance of DE-DRD from an intuitive perspective.

Example 5.1. Give a decision information system $\Phi = (U, A, F, D)$.

Table 1. Decision information system.

U	a_1	a_2	a_3	D
x_1	1	1	2	1
x_2	1	1	2	1
x_3	1	1	2	2
x_4	1	1	3	2
x_5	1	1	3	3
x_6	1	1	3	3
x_7	3	1	2	1
x_8	3	1	2	2
x_9	1	2	3	3
x_{10}	1	2	3	3

Let $U = \{x_1, x_2, \cdots, x_{10}\}$, $A = \{a_1, a_2, a_3\}$. It can be seen from Table 1: 1) $U/A = \{\{x_1, x_2, x_3\}, \{x_4, x_5, x_6\}, \{x_7, x_8\}, \{x_9, x_{10}\}\} \triangleq \{X_1, X_2, X_3, X_4\}$ and $U/D = \{\{x_1, x_2, x_7\}, \{x_3, x_4, x_8\}, \{x_5, x_6, x_9, x_{10}\}\} \triangleq \{D_1, D_2, D_3\}$, $r_{ij} = \{0, 1/3, 1/2, 2/3, 1\}$. 2) The U_λ can be obtained in Table 2.

Table 2. The subsystems of threshold λ.

λ	U_λ
[0, 1/3]	$\{x_1, x_2, x_3, x_4, x_5, x_6, x_7, x_8, x_9, x_{10}\}$
(1/3, 1/2]	$\{x_1, x_2, x_5, x_6, x_7, x_8, x_9, x_{10}\}$
(1/2, 2/3]	$\{x_1, x_2, x_5, x_6, x_9, x_{10}\}$
(2/3, 1]	$\{x_9, x_{10}\}$

Table 2 shows that with the increases of noise elimination intensity, the core data set is refined, and the reliability of decision knowledge increases. The decision roughness measurement results are given in Table 3.

Table 3. The decision roughness measurement results of D.

B	λ	$\rho_B(U/D,\Phi(\lambda))$	$\rho(U/D, B\oplus W(\lambda))$		
			$k=0, n=0$	$k=0.2, n=0.5$	$k=0.5, n=1$
$\{a_1, a_2\}$	$[0, 1/3]$	$9/10$	0.5685	0.3638	0.0864
	$(1/3, 1/2]$	$5/6$			
	$(1/2, 2/3]$	$7/9$			
	$(2/3, 1]$	0			
$\{a_1, a_2, a_3\}$	$[0, 1/3]$	$21/25$	0.3321	0.1076	0
	$(1/3, 1/2]$	$5/16$			
	$(1/2, 2/3]$	0			
	$(2/3, 1]$	0			

It can be seen from the above discussion, DE-DRD is a decision roughness measurement in the sense of entirety and is adapted to the rough processing consciousness of different backgrounds. The effect synthetic function $W(\lambda)$ is the core of DE-DRD, with the difference of $W(\lambda)$, the roughness results are different. Such as, if let $B_1 = \{a_1, a_2\}$, when $k = 0$, $n = 0$, the $\rho(U/D, B_1 \oplus W(\lambda)) = 0.5685$; when $k = 0.5$, $n = 1$, the $\rho(U/D, B_1 \oplus W(\lambda)) = 0.0864$. If let $B_2 = \{a_1, a_2, a_3\}$, and the U/B_1 is coarser than U/B_2, $\rho(U/D, B_1 \oplus W(\lambda)) > \rho(U/D, B_2 \oplus W(\lambda))$ is constant for any effect synthetic function $W(\lambda)$.

Therefore, there are essential differences between DE-DRD and VPRD in the measurement mechanism. DE-DRD can integrate decision consciousness into decision-making process from a structural perspective. If the decision idea is based on the higher reliability $\Phi(\lambda)$, the larger k and n can be used as the specific processing strategy; If the decision idea is to play down the role of different $\Phi(\lambda)$, the smaller k and n can be used as the specific processing strategy.

6. The Attribute Reduction Method Based on DE-DRD

If the decision information system (U_λ, B, F, D) is regarded as a decision knowledge base, then the $\rho(U/D, B \oplus W(\lambda))$ can be understood as a composite credibility measurement of the data effect. Therefore,

$$\Delta(A, B) \triangleq \rho(U/D, B \oplus W(\lambda)) - \rho(U/D, A \oplus W(\lambda)) \tag{8}$$

It reflects the difference between the knowledge contained in (U_λ, B, F, D) and (U_λ, A, F, D). An attribute reduction method can be constructed by $\Delta(A, B)$ not exceeding a given threshold δ ($0 \leq \delta \leq 0.05$), it is called decision roughness reduction method based on data effect, DE-DRD-RM.

In the following, the DE-DRD-RM and the Partition Reduction (PR-RM), the Distribution Reduction (DR-RM), the Maximum Distribution Reduction (MDR-RM) will be compared with four data sets from UCI database in Table 4.

Table 4. The description of the data sets.

Data sets	Samples	Condition set	Decision class
Iris	150	4	3
Ecoli	336	7	8
Hayes-Roth	160	4	3
Solar Flare	1389	10	3

All the experiments are run by MATLAB, and the reduction results are shown in Table 5.

Table 5. The reduction results with different reduction methods.

Data sets	PR-RM	DR-RM	MDR-RM	DE-DRD-RM		
				$W(\lambda)$	$\delta=0$	$\delta=0.05$
Iris	4	4	3	$n=0, k=0$	4	3
				$n=0.5, k=0.5$	3	3
Ecoli	7	7	7	$n=0.5, k=0.6$	7	5
				$n=1, k=0.9$	5	4
Hayes-Roth	4	4	4	$n=0, k=0$	4	4
				$n=1, k=0.8$	3	3
Solar Flare	10	6	5	$n=0, k=0$	6	2
				$n=1, k=0.8$	4	1

As can be seen from Table 5 that DE-DRD-RM is a kind of universal reduction method. When the δ is the same, the reduction results are different with different $W(\lambda)$. For example, the data set Ecoli, let $\delta=0$, when $n=0.5$ and $k=0.6$, the reduction result is seven; when $n=1$ and $k=0.9$, the reduction result is five. When the $W(\lambda)$ is same, the reduction results are different with different δ. For example, the data set Solar Flare, let $n=0$ and $k=0$, when $\delta=0$, the reduction result is six; when $\delta=0.05$, the reduction result is two. So, the $W(\lambda)$ and δ are directly affect the reduction result, and play a crucial role in the reduction process. DE-DRD-RM not only contains some other reduction methods, but also is essentially different from the existing methods. It can effectively reduce the amount of data storage, and integrate decision preference into the measurement system.

7. Conclusions

In this paper, we present a decision roughness measurement method based on data effect, named DE-DRD that considers the function characteristics of different sample data. DE-DRD not only makes up for the deficiencies of VPRD, but also can integrate the decision consciousness into the decision process. It has good structural characteristics and interpretability. Therefore, DE-DRD has wide application prospects in resource management, data mining, artificial intelligence and so on.

Acknowledgments

This work is supported by the National Natural Science Foundation of China (71771078, 71371064), and the Scientific Research Project Item of the Hebei Province Education Office (QN2017068).

References

1. Z. Pawlak, Rough sets, *IJICS*. **11**, 341-356 (1982).
2. Y. M. Chen, Y. Xue, Y. Ma and F. F. Xu, Measures of uncertainty for neighborhood rough sets, *Knowl-Based. Syst.* **120**, 226-235 (2017).
3. Z. Bonikowski, E. Bryniarski and U. Wybraniec, Extensions and intensions in the rough set theory[J], *Inf. Sci.* **107**, 149-167 (1998).
4. L. L. Wei and W. X. Zhang, Probabilistic rough sets characterized by fuzzy sets, *Int. J. Unc. Fuzz.* **12**, 47-60 (2004).
5. Z. Pawlak, Rough sets an intelligence data analysis, *Inf. Sci.* **147**, 1-12 (2002).
6. J. S. Mi, W. X. Zhang and W. Z. Wu, Knowledge reductions in inconsistent information systems, *Chin. J. Comp.* **26**, 12-18 (2003).
7. X. Zhang, C. L. Mei, D. G. Chen and J. H. Li, Multi-confidence rule acquisition oriented attribute reduction of covering decision systems via combinatorial optimization, *Knowl-Based. Syst.* **50**, 187-197 (2013).
8. W. Ziarko, Variable precision rough set model, *J. Comput. Syst. Sci.* **46**, 39-59 (1993).
9. S. Y. Zhao, E. C. C. Tsang and D. G. Chen, The model of fuzzy variable precision rough sets, *IEEE. Trans. Fuzzy. Syst.* **17**, 451-467 (2009).
10. Y. Y. Huang, T. R. Li and C. Luo, Dynamic variable precision rough set approach for probabilistic set-valued information systems, *Knowl-Based. Syst.* **122**, 131-147 (2017).
11. I. K. Park, G. S. Choi, A variable precision information entropy rough set approach for job searching, *Inf. Syst.* **48**, 279-288 (2015).

Semi-supervised transfer learning in Takagi-Sugeno fuzzy models

Hua Zuo, Guangquan Zhang and Jie Lu

Faculty of Engineering and Information Technology,
University of Technology Sydney, Sydney, NSW 2007, Australia

Transfer learning aims to leverage knowledge acquired from a related domain (called source domain) to improve the efficiency of completing a prediction task in the current domain (called target domain) which has different probability distribution from the source domain. Although transfer learning has been widely studied, most existing research on transfer learning has focused on classification tasks. This paper presents a fuzzy rule-based method that explores the information in the target domain to assist the regression tasks in transfer learning process.

1. Introduction

Traditional machine learning methods use learning models to extract knowledge from a large amount of labeled data. These methods work under an assumption that the training data and the testing data have the same feature space and the same probability distributions. However, if the feature space or the distribution of the target data changes, the models built from the source data become unsuitable for the target data and a new model needs to be rebuilt and trained.

Transfer learning [1] aims to solve the tasks in one domain (the target domain) much more quickly and effectively using previously acquired knowledge from a related domain (the source domain). Some successful applications in transfer learning include: using already-categorized English documents to help classify German documents [2]; detecting a user's current location based on the previously WiFi data [3]; and predicting the failure of banks in Australia using the data of banks in America [4].

Transfer learning belongs to one category of machine learning. Hence, its methods and models rely on many notable machine learning techniques, such as Bayesian models, SVM, neural networks, and deep learning [5-7]. These methods have had some success in handling transfer learning problems but ignore the inherent phenomenon of uncertainty — a crucial factor during the knowledge transfer process. There is a clear co-dependency between the level of certainty in learning a task and the amount of information that is available; problems with too little information have a high degree of uncertainty. If there are too few data with labels in the target domain, only a finite amount of information can be extracted, and this leads to a high degree of uncertainty. However, the introduction of fuzzy systems has shown promising results in overcoming this problem.

Most of these transfer learning methods using fuzzy systems are intended for classification problem [8, 9], yet a few concentrate on regression task. Some of our previous work [10, 11] concentrated on solving regression tasks, but ignore the exploration of the information in the target domain. To solve this issue, we develop a method that explores the information of the target data and exploits the interactive query strategy in active learning to correct imbalance in the knowledge to improve the generalization of the constructed models.

The reminder of this paper is structured as follows. Section 2 presents the preliminaries of this paper, including some important definitions in transfer learning, and the Takagi-Sugeno fuzzy model. Section 3 details the method of exploring the information in the target domain to improve the model's performance. The experimental results shown in Section 4 validate the proposed method. The final section concludes the paper and outlines the future work.

2. Preliminaries

This section begins with some basic definitions of transfer learning, followed by an introduction to the Takagi-Sugeno fuzzy model, which is the basic learning model used in our method.

2.1. Definitions

Definition 1 (Domain) [1]: A domain is denoted by $D = \{F, P(X)\}$, where F is a feature space, and $P(X)$, $X = \{x_1, \cdots, x_n\}$, are the probability distributions of the instances.

Definition 2 (Task) [1]: A task is denoted by $T = \{Y, f(\cdot)\}$, where $Y \in R$ is the response, and $f(\cdot)$ is an objective predictive function.

Definition 3 (Transfer Learning) [1]: Given a source domain D_s, a learning task T_s, a target domain D_t, and a learning task T_t, transfer learning aims to improve learning of the target predictive function $f_t(\cdot)$ in D_t using the knowledge in D_s and T_s where $D_s \neq D_t$ or $T_s \neq T_t$.

2.2. Takagi-Sugeno Fuzzy Models

The training model applied in this paper is the Takagi-Sugeno fuzzy model. It is composed of c fuzzy rules with the following representation:

$$\text{If } x \text{ is } A_i(x, v_i), \text{ then } y \text{ is } L_i(x, a_i) \quad i = 1, \dots, c \quad (1)$$

where v_i are the centers of the clusters that determine the layout of the fuzzy rules, and a_i defines the action of each rule on the input variables.

This fuzzy rule-based model is built using a set of instances $\{(x_1, y_1), \dots, (x_N, y_N)\}$ using a sequence of two procedures [12]: build condition parts of the rules using a fuzzy clustering algorithm, and construct the linear functions in the conclusions of the rules.

3. Fuzzy Transfer Learning Method

Consider two domains; a source domain with a large amount of labeled data, and a target domain with very little labeled data and massive amounts of unlabeled data. The dataset in the source domain is denoted as $D = \{(x_1^s, y_1^s), \dots, (x_{N_s}^s, y_{N_s}^s)\}$, where $x_k^s \in R^n$ ($k = 1, \dots, N_s$) is an n-dimensional input variable, the label $y_k^s \in R$ is a continuous variable, and N_s indicates the number of data pairs. The dataset in the target domain H consists of two subsets: one with labels and one without. $H = \{H_L, H_U\} = \{\{(x_1^t, y_1^t), \dots, (x_{N_{t1}}^t, y_{N_{t1}}^t)\}, \{x_{N_{t1}+1}^t, \dots, x_{N_t}^t\}\}$, where $x_k^t \in R^n$ ($k = 1, \dots, N_t$) is the n-dimensional input variable, $y_k^t \in R$ is the label only accessible for the first N_{t1} data. H_L includes the instances with labels, and H_U contains the data without labels. The number of instances in H_L and H_U are N_{t1} and $N_t - N_{t1}$ respectively, and satisfy $N_{t1} \ll N_t$, $N_{t1} \ll N_s$.

In this problem setting, because the labeled data are sufficient, a well-performing model can be built for the source domain. However, since the input distributions in two domains are quite different, the source model cannot be used directly to solve regression tasks in the target domain.

A fuzzy rule-based transfer learning method is presented to modify the source model for use with the target data. This method contains the following three steps.

Step 1: Construct a prediction model for the source domain

A source model M^s is built based on the source dataset D. The formulation for model M^s follows

$$\text{if } x_k^s \text{ is } A_i(x_k^s, v_i^s), \text{ then } y_k^s \text{ is } L_i(x_k^s, a_i^s) \qquad i = 1, \dots, c \qquad (2)$$

where v_i^s are the centers of the clusters that govern the conditions of the fuzzy rules, and a_i^s are the coefficients of the linear functions of the input variables that determine the conclusions of the fuzzy rules.

c is an important parameter in the model's construction. If the numbers of the fuzzy rules are obtained in two domains, suppose c_s and c_t respectively, we can make $c = \max\{c_s, c_t\}$. If there is no extra information about the number of fuzzy rules, we suppose the two domains have the same number of fuzzy rules. A certain range is given, the brute-force approach is adopted to choose the one that has the best performance.

Because the target domain contains different data distributions to the source data, the model M^s would perform poorly on the target data.

Step 2: Use active learning to augment the labeled target data

The purpose of this procedure is to increase the amount information in the target domain by actively selecting and labeling some of the data.

The procedure begins with an evaluation of the existing labeled target data. Suppose that we choose to construct the model with three fuzzy rules $c(=3)$. FCM clusters the input data for the target domain and gets the membership matrix U. The membership matrix for the labeled target data is denoted as U_L. The number of labeled data in each cluster is counted, with each instance counted in the cluster with the highest membership. The statistic result is supposed as follows

$$SU_L = [2 \quad 1 \quad 0] \qquad (3)$$

The first two clusters contain labeled data, but the third cluster does not, so active learning is used to augment the information in the target domain and populate this cluster.

"Informativeness" is the key criteria for which data to select for labeling in each cluster. Essentially, informativeness is a measure of information contained in the data, and the level of informativeness is highly dependent on the cluster it is grouped with, i.e., an instance will have a different level of informativeness in different clusters. A concrete instance x_k^t in the ith cluster is highly informative if x_k^t's membership to the ith cluster is high. Thus, the informativeness of x_k^t in cluster i is defined as the membership of x_k^t belonging to cluster i.

Suppose d determines the minimum number of labeled target data needed for each cluster. At the end of Step 2, the number of labeled target data has increased from N_{t1} to $3d$.

Step 3: Modify the existing fuzzy rules to fit the target data

The of changing the input variables is used to modify the existing fuzzy rules of source domain. Each input variable is assumed to be determined by some hidden features, so the different distributions of input variables in the two domains are due to the different hidden features or different weights of these features. The idea behind changing the input variables is to adjust the number and weights of the hidden features so that the changed input distribution is more compatible with the target data.

We apply the nonlinear mapping Φ proposed in our previous papers [10, 11] to modify the input space. The target model M^t is obtained under the mapping Φ.

if x_k^t is $A_i(\Phi(x_k^t), \Phi(v_i^s))$, then y_k^t is $L_i(\Phi(x_k^t), a_i^s)$ $\quad i = 1, \cdots, c$ (4)

These modifications are made through an optimization process, and the cost function is as follows.

$$S = \sqrt{\frac{1}{3d} \sum_{k=1}^{3d} \left(\sum_{i=1}^{c} \frac{A_i(\Phi(x_k^t), \Phi(v_i^s))}{\sum_{j=1}^{c} A_j(\Phi(x_k^t), \Phi(v_j^s))} L_i(\Phi(x_k^t), a_i^s) - y_k^t \right)^2} +$$
$$\lambda_1 \sqrt{\frac{1}{3d*h} \sum_{k=1}^{3d} \sum_{l=1}^{h} (y_k^t - y_k^t(l))^2 * \exp(-\|x_k^t - x_k^t(l)\|)} + \frac{\lambda_2}{2} w^T w$$
(5)

The cost function contains three terms. The first term intends to minimize the gap between the output of the constructed model and the target data's real labels using the target data with labels. The second term works on the assumption that data with less distance in the input space will have a similar label. Therefore, for each target data x_k^t in H_L, the h-nearest data $\{x_k^t(1) \cdots x_k^t(h)\}$ in H_U are found, and the outputs of $\{x_k^t(1) \cdots x_k^t(h)\}$ are expected to be close to the label of x_k^t. $\exp(-\|x_k^t - x_k^t(l)\|)$ determines that the data that are closer to the centre x_k^t, will have an output more approximate to the label of x_k^t. The third term is a structural risk of the cost function, and w is the vector of all the optimized parameters. Parameters λ_1 and λ_2 indicate the tradeoff among these three terms.

4. Experiments

The experiments include two parts: one part validates the effectiveness of the active learning technique in the proposed method; the other part shows the improvement of exploring the information in the unlabeled target data.

In the first part, three experiments are executed. In each experiment, the source and target datasets were generated with the same number of fuzzy rules (two, three, and four respectively), and all the labeled target data were selected from one cluster. The experimental results are shown in Table 1. Q1 is the performance of source model on the target data. 'Q2 (no active learning)' indicates the performance of the target model without using the active learning technique. 'Q2 (active learning)' represents the performance of the target model with active learning. All the models' construction follows five-fold cross validation, and the results are shown in the form of "mean±variance".

Table 1. Exploring the effect of the active learning technique.

Clusters	Q1	Q2 (no active learning)	Q2 (active learning)
2	1.37±0.00	1.46±0.02	1.07±0.01
3	0.35±0.00	0.93±0.03	0.88±0.02
4	2.03±0.00	2.19±0.07	1.75±0.13

Comparing values 'Q2 (no active learning)' and 'Q2 (active learning)' in the three experiments, we found that using the active learning technique significantly enhances the accuracy of the target model constructed using the proposed method.

Moreover, since target data without labels H_U are used to improve the performance of model M^t, Table 2 compares the RMSE of model M^t when built with and without H_U. In the first column, "5r to 4r" indicates that the source domain has five fuzzy rules, and the target domain has four fuzzy rules.

The results show that using target data H_U for training was better in five of six experiments — a clear performance improvement for M^t.

Table 2. Building using/not using H_U - model comparison.

Source to target dataset	model M^t (not using H_U)	model M^t (using H_U)
5r to 4r	1.0781±0.0004	1.0756±0.0004
5r to 3r	0.9352±0.0057	0.8962±0.0083
4r to 3r	2.1269±0.2059	2.0996±0.1718
3r to 4r	0.8876±0.0009	0.8457±0.0005
3r to 5r	2.5273±0.0007	2.5397±0.0014
4r to 5r	3.0755±0.0110	3.0614±0.0037

5. Conclusions and Future Work

This work presented a fuzzy rule-based method that explores the information of target domain to improve the knowledge transfer efficiency in transfer learning. The use of unlabeled target data helps enhance the accuracy of the target model, and the employment of active learning technique increase the information of target data.

In this work, the transfer learning problem is considered in a homogeneous space, i.e. the input feature space is the same. A more challengeable work, transfer learning in heterogeneous space, where the source domain and target domain have different feature spaces, will be considered in the future work.

Acknowledgments

This work was supported by the Australian Research Council under DP 170101623.

References

1. S. J. Pan and Q. Yang, "A survey on transfer learning," *IEEE Transactions on Knowledge and Data Engineering,* vol. 22, no. 10, pp. 1345-1359, 2010.
2. M. Xiao and Y. Guo, "Feature space independent semi-supervised domain adaptation via kernel matching," *IEEE Transactions on Pattern Analysis and Machine Intelligence,* vol. 37, no. 1, pp. 54-66, 2015.
3. S. J. Pan, V. W. Zheng, Q. Yang, and D. H. Hu, "Transfer learning for wifi-based indoor localization," in *Association for the Advancement of Artificial Intelligence (AAAI) Workshop*, 2008.
4. V. Behbood, J. Lu, and G. Zhang, "Long term bank failure prediction using fuzzy refinement-based transductive transfer learning," in *2011 IEEE International Conference on Fuzzy Systems (FUZZ)*, pp. 2676-2683, 2011.
5. M. Xiao and Y. Guo, "Feature space independent semi-supervised domain adaptation via kernel matching," *IEEE Transactions on Pattern Analysis and Machine Intelligence,* vol. 37, no. 1, pp. 54-66, 2015.
6. L. Bruzzone and M. Marconcini, "Domain adaptation problems: A DASVM classification technique and a circular validation strategy," *IEEE Transactions on Pattern Analysis and Machine Intelligence,* vol. 32, no. 5, pp. 770-787, 2010.
7. Y. Bengio, "Deep learning of representations for unsupervised and transfer learning," in *Proceedings of ICML Workshop on Unsupervised and Transfer Learning*, pp. 17-36, 2012.
8. V. Behbood, J. Lu, G. Zhang, and W. Pedrycz, "Multistep fuzzy bridged refinement domain adaptation algorithm and its application to bank failure prediction," *IEEE Transactions on Fuzzy Systems,* vol. 23, no. 6, pp. 1917-1935, 2015.
9. V. Behbood, J. Lu, and G. Zhang, "Fuzzy refinement domain adaptation for long term prediction in banking ecosystem," *IEEE Transactions on Industrial Informatics,* vol. 10, no. 2, pp. 1637-1646, 2014.
10. H. Zuo, G. Zhang, W. Pedrycz, V. Behbood, and J. Lu, "Fuzzy regression transfer learning in Takagi-Sugeno fuzzy models," *IEEE Transactions on Fuzzy Systems,* Vol. 25, No. 6, pp. 1795-1807, 2016.
11. H. Zuo, G. Zhang, W. Pedrycz, V. Behbood, and J. Lu, "Granular Fuzzy Regression Domain Adaptation in Takagi-Sugeno Fuzzy Models," *IEEE Transactions on Fuzzy Systems,* 2017.
12. M. L. Hadjili and V. Wertz, "Takagi-Sugeno fuzzy modeling incorporating input variables selection," *IEEE Transactions on fuzzy systems,* vol. 10, no. 6, pp. 728-742, 2002.

RNN-based traffic flow prediction for dynamic reversible lane control decision

Yang Song
School of Transportation, Tongji University, Shanghai, 201804, China

Jie Lu
*Faculty of Engineering and Information Technology,
University of Technology Sydney, NSW 2007, Australia*

Accurate and real-time traffic flow prediction is the basis of dynamic reversible lane control decision, which plays an important role in alleviating congestion. However, the existing methods could hardly memorize long-term dependencies to produce accurate result of sequence prediction. In the paper, a Long Short-Term Memory (LSTM) and a Gated Recurrent Units (GRU) Recurrent Neural Network (RNN) are adapted to accurately predict traffic flow. A dynamic reversible lane management model is then conducted based on the prediction outcome. The result shows that RNNs obviously outperform the existing feed forward neural network (FFNN) model in terms of accuracy, and therein the forecasting error of GRU is lower than LSTM. Moreover, the result presents that the dynamic reversible lane control plan based on GRU RNN could reduce the total travel time by 0.76% than LSTM and 9.30% than FFNN.

1. Introduction

To mitigate freeway traffic congestion, reversible traffic operations have been applied to practice since the 1920s [1]. Reversible lanes systems (RLS) can often significantly increase the capacity of a roadway, while often requiring little investment in roadway or control infrastructure. Their fundamental objective is to take advantage of underutilized lanes or shoulders in one direction of travel by reorienting the direction of traffic flow in the opposite direction, thereby increasing the overall capacity of the roadway [2]. Although the basic concept of RLS is simple, it can be complex to operate because the control and management of reversible lane require traffic flow information for the both directions. Otherwise, the traffic congestion may even be aggravated for unjustified allocation of lane resources.

For traffic flow prediction, the evolution process can be considered as a time series, which is the vehicle number passing the section of loop detector during a certain period of time. Due to the importance of traffic flow prediction in RLS management, numerous research on the prediction approaches has been conducted, which in general, can be classified into two categories, namely parametric approach and non-parametric approach.

Parameter models refer to the models with predetermined structure based on certain theoretical assumptions and the parameters that can be computed with empirical data, like Autoregressive Integrated Moving Average (ARIMA) model [3]. However, the existing parameter models cannot describe the non-linearity and uncertainty of traffic flow well so that they usually cause larger prediction error [4]. For the non-parameter models, they refer to the models with no fixed structure or parameters, like Support Vector Machine or FFNN [5, 9, 10]. Nevertheless, the above-mentioned non-parametric methods cannot memorize long-term dependencies, so it is not appropriate to deal with time series data like traffic flow.

Unlike other non-parametric models, RNNs work better on sequence modeling tasks such as time series prediction because it has cyclic connections over time so that the activations from each time step are stored in the internal state of the network to provide a temporal memory [6]. LSTM is a RNN architecture that could address the vanishing and exploding gradient problems of conventional RNNs for language models [7, 11]. Then to speed up training, GRU is presented to simplify LSTM. However, few research has been conducted to determine which RNN method is the most suited one for traffic flow prediction in terms of accuracy.

In this paper, traffic flow prediction based on RNNs is conducted to support dynamic reversible lane control decision. The performance of FFNN, LSTM RNN and GRU RNN is compared in terms of accuracy. Then, with prediction result, dynamic RLS management is conducted in a numerical example and the effect of different lane control plan is evaluated by the corresponding travel time.

The remainder of this paper is organized as follows: The architecture and equations of LSTM RNN and GRU RNN are detailed in Section 2. The design of traffic prediction experiment and the results analysis are given in Section 3. A numerical example of dynamic RLS management is presented in Section 4. Finally, concluding and the future envisions are proposed in Section 5.

2. LSTM and GRU Neural Network for Prediction

The structure of LSTM RNN and GRU RNN are described in the section.

2.1. *LSTM*

The structure of LSTM RNN is described as follows. It is composed of one input layer, one recurrent hidden layer whose basic unit is memory block instead of traditional neuron node, and one output layer. Memory blocks are a set of recurrently connected subnets. Each block contains one or more memory cells with self-connections storing the temporal state of the network, as well as four multiplicative units (input, input modulation, output and forget gates), to control the flow of information. With these multiplicative gates, LSTM memory cells could store and access information over long periods of time, thus mitigating gradient vanishing. Therein, the input gate takes a new input point from outside and processes newly coming data. While the input modulation gate gets input from the output of the LSTM RNN cell in the last iteration. Then, the optimal time lag for the input sequence is determined by forget gate, which decides when to forget the output results. Finally, the output gate calculates all the results and controls the output flow of the cell. Figure 1 illustrates the architecture of LSTM RNN prediction model with one memory block.

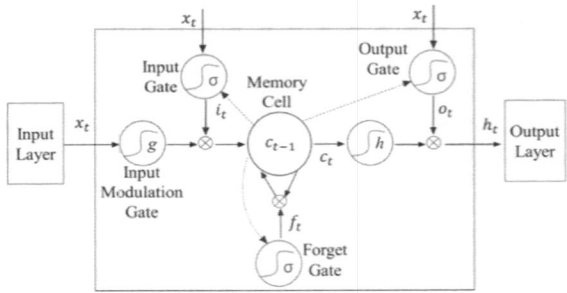

Figure 1. The architecture of LSTM RNN with one memory block.

The input time series are denoted as $X = (x_1, x_2, ..., x_n)$. The hidden vector sequence is $H = (h_1, h_2, ..., h_n)$. The output predicted sequence is $Y = (y_1, y_2, ..., y_n)$. Then, the computation is conducted as follows:

$$h_t = H(W_{xh}x_t + W_{hh}h_{t-1} + b_h) \qquad (1)$$

$$y_t = W_{hy}h_t + b_y \qquad (2)$$

where the W term represents weight matrices; b term denotes bias vectors; H is the hidden layer function, which is conducted by the following composite formulas.

$$i_t = \sigma(W_{xi}x_t + W_{hi}h_{t-1} + W_{ci}x_{t-1} + b_i) \qquad (3)$$

$$f_t = \sigma(W_{xf}x_t + W_{hf}h_{t-1} + W_{cf}c_{t-1} + b_f) \qquad (4)$$

$$c_t = f_t * c_{t-1} + i_t * g(W_{xc}x_t + W_{hc}h_{t-1} + b_c) \qquad (5)$$

$$o_t = \sigma(W_{xo}x_t + W_{ho}h_{t-1} + W_{co}c_t + b_o) \qquad (6)$$

$$h_t = o_t * h(c_t) \tag{7}$$

where σ stands for the standard logistic sigmoid function defined in Formula 8. The mark * refers to the scalar product of two matrices. The term g and h are the extends of function σ whose range are [-2, 2] and [-1, 1]. The i, f, o and c are the input gate, forget gate, output gate and cell activation vectors, all of which are the same size as the hidden vector. The squared error is the loss function, which is given by Formula 9.

$$\sigma(x) = \frac{1}{1+e^{-x}} \tag{8}$$

$$e = \sum_{t=1}^{n}(y_t - p_t)^2 \tag{9}$$

where y means the real output and p represents the predicted value. In this study, the algorism, Back Propagation Through Time (BPTT) is used to train the RNN model [7]. Meanwhile, Adam optimizer, which modifies stochastic gradient descent (SGD) optimizer with adaptive learning rates, is chosen to minimize training error and meanwhile avoid local optimal solution. Then, to prevent overfitting, dropout technology, an effective regularization method is applied to the training process [12].

2.2. GRU

GRU is a variant of LSTM, which is proposed by Cho et al. in 2014 [8]. It is simpler to compute because the forget gate and input gate in LSTM cell are combined into update gate in GRU cell. It also contains another gate, reset gate. The typical structure of GRU cells is illustrated in Figure 2

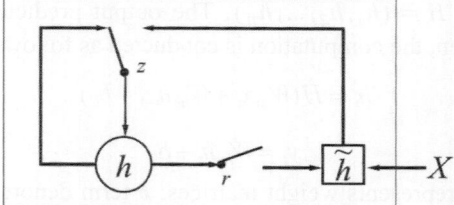

Figure 2. The structure of GRU cell.

where r denotes reset gate and z denotes update gate. GRU RNN is quite similar to LSTM RNN. For example, the hidden state output of GRU at time t is also calculated by the hidden state of time $t-1$ and the input time sequence value of time t. Also, reset gates in GRU play similar role as forget gates of LSTM. The detailed formulas of GRU can be referred to [8]. In the study, the regression method and optimizer for GRU RNN is the same as the abovementioned LSTM RNN.

3. Traffic Prediction Experiment

This section will provide the design of the traffic prediction experiment and the results analysis.

3.1. *Data Description and Experiments Design*

In this section, the data collected from one southbound lane of North-South Elevated Road in Shanghai is used to train and evaluate the traffic prediction models. The original data is collected per 5 minutes for one month by a loop detector. The dataset is divided into two parts, where the data of the first 28 days is the training set and the later 3 days is the testing set. Meanwhile, both traffic flow data and other influencing factors like workday, weather conditions and speed are used as the input, which is conjectured to produce more accurate training results. Since prediction accuracy may be affected by forecasting horizon and step, the prediction intervals of the experiments are set to 5 min, 10 min and 15 min. Then the original traffic flow data would be aggregated into the corresponding time interval. While a time sequence of the past 100 steps is used to predict the coming traffic flow in the next step.

In the experiment, the prediction of FFNN, LSTM and GRU RNN models are conducted and compared. FFNN model is the simplest one and serves as the baseline of the experiment. It is a classical neural network models with single hidden layer that consists of 20 units. Both LSTM RNN and GRU RNN models consist of two hidden layers with 100 units each in them. To evaluate prediction accuracy, two indices, mean square error (MSE) and mean absolute error (MAE) are used, which are defined as follows.

$$MSE = \frac{1}{n}\sum_{t=1}^{n}(y_t - p_t)^2 \qquad (10)$$

$$MAE = \frac{1}{n}\sum_{t=1}^{n}|y_t - p_t|^2 \qquad (11)$$

3.2. *Experiment Result*

The experiments are performed on an Intel(R) Core(R) i5-4200M @2.50 GHz CPU with 8 GB RAM and NVidia GeForce GTX 765M GPU. The networks were created in Python with Keras deep learning library using TensorFlow backend. The prediction error of the different prediction methods is presented in Table 1.

Table 1. Traffic flow prediction result.

Models	5min		10min		15min	
	MAE	MSE	MAE	MSE	MAE	MSE
FFNN	15.50	458.32	29.03	1603.33	44.71	3827.24
LSTM RNN	11.00	225.80	18.28	720.40	26.46	1555.61
GRU RNN	**10.75**	**194.08**	**17.69**	**574.72**	**26.13**	**1292.37**

From Table 1, the prediction error rises with the increase of time interval, which shows that the extension of prediction horizon would lead to low forecasting accuracy. Then, compared with FFNN, the MAE and MSE of both LSTM RNN and GRU RNN are much smaller. This demonstrates that RNNs could overcome the issue of the vanishing and exploding gradient problems through memory blocks, and thus shows superior capability for time series prediction with long temporal dependency. Therein, though the forecasting results are quite close, the prediction error of GRU RNN are all lower than that of LSTM RNN, which reveals that GRU cell could memorize long-term dependencies better and produce higher prediction accuracy.

4. Numerical Example of Dynamic RLS Management

Since the effectiveness of LSTM and GRU RNN in traffic flow prediction has been validated in the last section, the proposed dynamic RLS management method will be conducted based on these models by a numerical example in this section. A freeway with 4 ordinary lanes and 1 reversible lane is selected, as illustrated in Figure 3, where Lane 1 and Lane 2 are the northbound ordinary lanes; Lane 4 and Lane 5 are the southbound ordinary lanes; Lane 3 is the reversible lane whose direction may change according to the traffic flow of the two directions. The two-direction traffic flow data of the North-South Elevated Road is assigned to this theoretical freeway and is used to train the models. The prediction interval is set to 15 min and other configurations are the same as the experiment in Section 3.

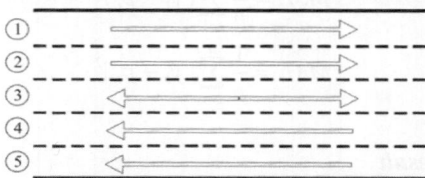

Figure 3. Numerical example illustration.

Based on the two-direction traffic flow predicted by the testing set, the RLS management can be implemented hourly. That is, if the northbound traffic volume is over the southbound, the direction of the reversible lane will be turned to the northward and vice versa. Then, since the objective of RLS is to mitigate traffic congestion, the index, travel time is used to evaluate the effect of different lane control results. The total travel time per hour of the freeway can be calculated by Bureau of public road (BPR) function, which is defined as follows.

$$t = \sum_{i=1}^{5} t_i^0 \times (1 + \alpha \times (v_i / c_i)^\beta) \qquad (12)$$

where t_i^0 is free-flow travel time on Lane i and equals 2 h. α and β are model constants, which generally equal 0.15 and 4 respectively. c_i represents the traffic capacity of Lane i and equals 1000 passenger car unit (pcu) per hour. v_i is the predicted traffic volume on Lane i per hour. For one day in the testing set, the comparison of observation and prediction traffic flow is shown in Figure 4, while based on different prediction results, the dynamic lane assignment plans are given in Figure 5, where lane management result based on observation value is the ideal case and serves as the baseline.

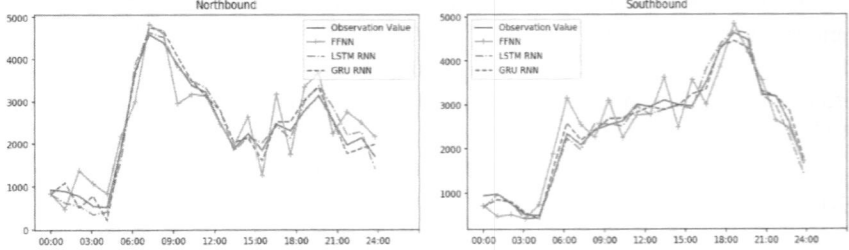

Figure 4. Comparison between observation and prediction value.

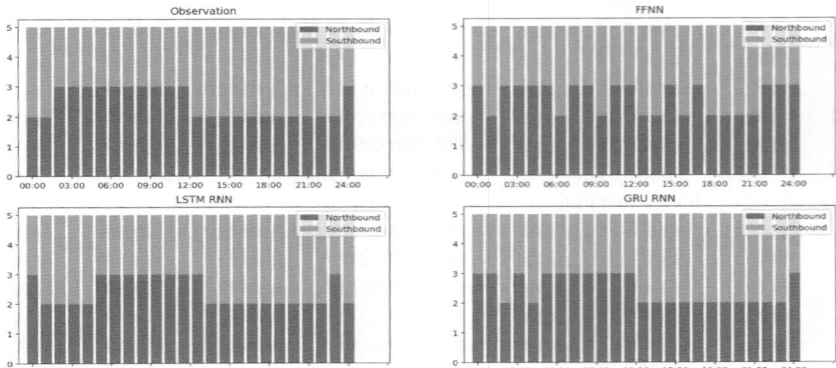

Figure 5. Dynamic lane assignment result.

From Figures 4 and 5, both the traffic flow prediction value and lane assignment results of FFNN are quite different from those of the observation values. While those of LSTM and GRU RNN are rather similar to the observation ones. Further, the total travel time of observation value, FFNN, LSTM and GRU RNN are 120.96 h, 120.97 h, 121.89 h and 132.22 h respectively. Hence, the control result of GRU is nearly that same as the ideal case and it reduces the travel time by 0.76% than LSTM and 9.30% than FFNN, which proves the effectiveness of dynamic RLS management based on the prediction result by RNNs.

5. Conclusion and Further Study

In the paper, traffic flow prediction based on RNNs is conducted to support dynamic reversible lane control decision. The prediction performance of FFNN, LSTM RNN and GRU RNN are compared under various time intervals and the result shows that RNNs obviously outperform FFNN model in terms of prediction accuracy, while therein the forecasting error of GRU is lower than LSTM. It proves RNNs could memorize long-term dependencies and have superior capacity in sequence prediction. Then, dynamic RLS management is conducted in a numerical example and it turns out that GRU RNN could reduce the total travel time by 0.76% than LSTM and 9.30% than FFNN, which validates the effectiveness of dynamic RLS management based on the prediction results by RNNs.

In our future work, the proposed methods will be tested by more real case studies to improve its generalization performance.

References

1. R. T. Dorsey, *Traffic*, **2**, 291-302 (1948).
2. B. Wolshon and L. Lambert, Journal of Transportation Engineering, 132, 933-944 (2006).
3. M. S. Ahmed and A. R. Cook, Transportation Research Record, 722, (1979).
4. R. Fu, Z. Zhang and L. Li, Youth Academic Annual Conference of Chinese Association of Automation, 2016, 324-328 (2016).
5. Y. Tian and L. Pan, IEEE International Conference on Smart City/SocialCom/SustainCom, 153-158 (2015).
6. A. Azzouni and G. Pujolle, Networking and Internet Architecture, (2017).
7. A. Graves, Springer Berlin Heidelberg, (2012).
8. K. Cho, B. Van Merrienboer, et al., Empirical Methods in Natural Language Processing, 1724-1734 (2014).
9. Castro-Neto, et al., Expert Systems with Applications An International Journal, 3, 6164-6173 (2009).
10. Vlahogianni, et al., *Transportation Research Part C*, **3**, 211-234 (2005).
11. Ma, X., et al., *Transportation Research Part C*, **54**, 187-197 (2015).
12. Hinton, et al., *Journal of Machine Learning Research*, **1**, 1929-1958. (2014).

Road traffic flow prediction using deep transfer learning

Bin Wang[1,2], Zheng Yan[2], Jie Lu[2], Guangquan Zhang[2] and Tianrui Li[1]

[1] *School of Information Science and Technology, Southwest Jiaotong University*
Chengdu 611756, China
[2] *Centre for Artificial Intelligence, FEIT, University of Technology Sydney*
NSW 2007, Australia
wangbin@my.swjtu.edu.cn, {yan.zheng, jie.lu, guangquan.zhang}@uts.edu.au,
trli@swjtu.edu.cn

Traffic flow prediction is a long-standing problem. Over the recent years, deep learning has gradually achieved a satisfying success on this task, but it depends on abundant historical traffic data. A realistic problem is that some new-established transportation networks only have few data which is not enough to train a robust deep learning model. To address this problem, we first explore and apply the transfer learning and fine-tuning to the field of transportation and propose a novel transferable traffic deep learning model, called TT-DL which can predict real-time traffic flow in data-strapped roads by transferring knowledge from data-rich roads. Our experimental results show that transfer learning is better than any other initialization methods. This indicates that traffic network has its special structure and there exists transferable knowledge between different traffic areas.

Keywords: Traffic flow prediction; deep learning; transfer learning.

1. Introduction

Forcasting traffic flow is of great importance to the public management and traffic planning.[1] Recently, researchers begin to apply deep learning to the traffic flow prediction problem, and prove that deep leaning can exceed traditional machine learning and statistic methods.[2,3] Nevertheless, nearly all existing deep traffic flow learning methods demand a massive data to train a robust model. In real world, sometimes this will not be satisfied. For instance, a newly established traffic area does not have enough historical data. In this case, deep learning will not achieve a great result. The similar problem also occurs in the computer vision and natural language processing tasks and that is when transfer learning comes into play. It refers to the situation where what has been learned in source domain is exploited to improve generalization in related target domain.[4,5] Recently, many researchers start exploring to combine transfer learning with other learning paradigms, e.g., fuzzy inference[6] and deep learning.[7,8] To transfer knowledge within deep learning, motivated by the phenomenon that many deep neural networks trained on images dataset learn features similar to Gabor filters, Yosinski et al. first verified the transferability of deep learning model for images of different domains.[7] They found that initializing a network with transferred features from almost any number of layers can produce a boost to generalization performance after fine-tuning to a new dataset. Hence this type of approach is parameter-based. On this basis, Long et al. firstly generalized deep convolutional neural networks to the domain adaptation scenario, which can transfer knowledge between the source and target domains in a common latent space and thus this type of method is features-based.[8]

Besides the great success of transfer learning in the field of computer vision, researchers began to apply this technique to solve other application problems, e.g., recommender system,[9] air quality classification[10] and crowd flow prediction.[11] In this paper, we explore to apply this technique to predict traffic flow and propose a transferable traffic deep learning model called TT-DL. The main difference between our study and [11] is that [11] utilized a feature-based approach but our method is parameter-based. The main contributions of this study are summarized as follows:

- We propose an innovative traffic flow forecasting method at the intersections of highways based on deep transfer learning. Existing methods can be well-behaved on a traffic network with rich data but our method can improve the precision of prediction when data are scarce on the target domain.
- We design a flow matrix to incorporate the spatio-temporal information of vehicle flows on highway intersections as the input. In this way, the complex representations can be automatically learned layer by layer.

The rest of the paper is organized as follows: we introduce our work by some backgrounds in Sec. 2. In Sec. 3, we explain our model configurations in details. Then, we experimentally evaluate our proposed prediction model and compare it with other baselines in Sec. 4. Finally, conclusions and future study directions are given in Sec. 5.

2. Background

2.1. Formulation of Traffic Flow Prediction

In this section, we give formal definitions of intersection traffic flows forecasting problem. To make things simplicity, all roads are one-way freeway.

Definition 1 (Outflow). For the target highway $\mathbb{R}$ of which we want to forecast traffic flow, we define its flow as the outflow, denoted by $O(\mathbb{R})$. And O^t indicates the outflow at time t.

Definition 2 (Inflow). For highways that can lead to the target highway such as $\mathbb{R}$, we define each of them as the inflow of $\mathbb{R}$, denoted by $I_i(\mathbb{R})$ where i means the ith inflow of $\mathbb{R}$ and I_i^t presents the ith inflow at time t.

To make this clearer, Fig. 1 illustrates that if we plan to predict the outflow of the highway with green arrow marked, three flows with red arrow marked will be regarded as inflows (in the driving right area).

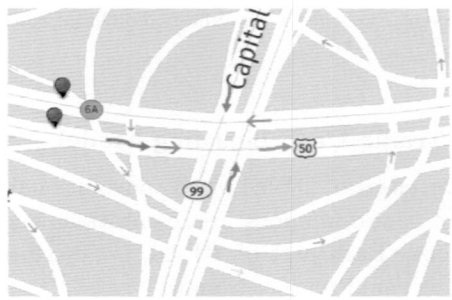

Fig. 1. Traffic network, inflow and outflow.

Definition 3 (Flow Matrix). With **Definition 1** and **Definition 2**, we can build the flow matrix $FM_{m+1,n+1}^t$ as follows:

$$FM_{m+1,n+1}^t = \begin{array}{c} \\ t-m \\ \vdots \\ t-1 \\ t \end{array} \begin{array}{c} I_n(\mathbb{R}) \quad I_{n-1}(\mathbb{R}) \quad \cdots \quad I_1(\mathbb{R}) \quad O(\mathbb{R}) \\ \left(\begin{array}{ccccc} I_n^{t-m} & I_{n-1}^{t-m} & \cdots & I_1^{t-m} & O^{t-m} \\ \vdots & \vdots & \ddots & \vdots & \vdots \\ I_n^{t-1} & I_{n-1}^{t-1} & \cdots & I_1^{t-1} & O^{t-1} \\ I_n^t & I_{n-1}^t & \cdots & I_1^t & O^t \end{array} \right) \end{array}$$

where m is the length of time sequence and n specifies the number of inflow highway.

Problem. Given the historical flow matrix $FM_{m+1,n+1}^t$, forecast O^{t+1}.

2.2. Traffic Transfer Learning

As shown in Fig. 2, we first train a source domain network, and then copy its layer weights as the initialized weights of the target domain network. Further, we backpropagate the errors from the target domain data to fine-tune the learned weights for the target task.

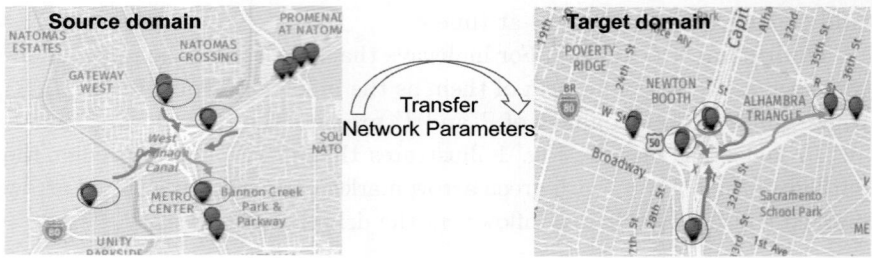

Fig. 2. Transfer knowledge between two different road networks.

3. Transferable Traffic Deep Learning Model

To verify the effectiveness of TT-DL, two convolutional operations are introduced to extract spatio-temporal representations and then followed by two fully connections. Each part of TT-DL is described below and more mathematical details could refer to Simard.[12]

(i) *Input layer.* The input matrix is constructed as **Definition 3**. In our case, we set $m = 3$ and $n = 3$ to construct a 4×4 flow matrix as input. This matrix naturally incorporates complicated spatio-temporal interactions information of traffic flow among the traffic network.

(ii) *Convolutional layers.* The primary purpose of convolution is to extract features from the input matrix. It can preserve the spatial-temporal relationship between roads by learning traffic flow features using kernel filters. Activation function here such as ReLU is to introduce nonlinearity for the model.

(iii) *Fully connected layers.* The function of the fully connected layers is to further implement deep features transformation and readily feed them into output layer.

(iv) *Output layer.* This layer finally predicts the scaled traffic flow value. For realistic meaning, the predicted value needs a re-scale operation.

(v) *Loss layer (a.k.a. objective function)*. It measures the difference between the prediction and target and performs error back-propagation for the integrated TT-DL. In our case, it is set as MSE (Mean Square Error).

The hyperparameters of TT-DL are presented in Table 1. The convolutions of Conv1 use 32 filters of size 3×3 with stride 1, and Conv2 uses a convolution with 64 filters of size 3×3 with stride 1. All activation functions are ReLU except the Sigmoid in the output layer. After convolutions, 2 fully connected layers follows to output the predicted value. Specially, the training epochs on source domain and target domain both are set as 10 epochs in our study.

Table 1. Configurations of TT-DL.

Layer type	Filter	Input	Output	Activation
Conv1	3x3, 1 strides	4x4x1	4x4x32	ReLU
Conv2	3x3, 1 strides	4x4x32	4x4x64	ReLU
Fully connection	-	1024	32	ReLU
Fully connection	-	32	1	Sigmoid

4. Experiments

4.1. Dataset

To evaluate our model, we use realistic traffic flow data provided by Caltrans Performance Measurement System (PeMS). In this paper, our source domain is from the urban intersection of Golden State Highway and Dwight D. Eisenhower Highway and totally includes 2004 sequences. The target domain data is from the urban intersection of Lincoln Highway and S Sacramento Freeway and includes only 852 sequences. For each dataset, the vehicle detectors ID indicates three input detector IDs and the output detector ID (* marked) and the time lag of sequence is 5 minutes. We split data into training set and test set as rate 9:1. To make TT-DL trained easily, we re-scale the inputs to [0,1] by max-min normalization. Table 2 show the details.

4.2. Results

In Table 3, we can conclude that: 1) a shallow model, i.e., *Support Vector Regression (SVR)*, has a poor performance due to the highly spatio-temporal property of the prediction task; 2) The proposed TT-DL performs

Table 2. PeMS datasets.

Dataset	Source domain	Target domain
Data type	Vehicle flow	Vehicle flow
Urban intersections	Golden State Hwy + DwightD. Eisenhower Hwy	Lincoln Hwy + S Sacramento Fwy
Vehicle detectors ID	318249, 316141, 319147, 31342*	312852, 312103, 312089, 312566*
Time span	01/Oct/2013-07/Oct/2013	05/Feb/2016-07/Feb/2016
Time interval	5 minutes	5 minutes
Available time interval	2,004	852

Table 3. Comparison among different baselines.

Baselines	Train MSE	Val MSE	Test MSE
SVR	9.44	-	11.96
random uniform	3.11	4.20	4.26
random normal	3.04	3.03	3.24
glorot uniform	2.86	2.66	2.59
glorot normal	3.09	3.14	2.74
no fine-tuning	-	-	3.57
TT-DL	**2.67**	**2.26**	**2.20**

better than other deep models with random weights initialization (*uniform, normal, glorot_normal and glorot_normal*); 3) *no fine-tuning* which is only trained on the source domain training data and directly used to predict the test data of target domain identifies the effect of fine-tuning. Therefore, it is shown that our method has achieved the best results.

In particular, we plot the training and validation loss of different initializations in Fig. 3 and Fig. 4, respectively. One interesting phenomenon is TT-DL has less loss in the first few epochs which befit from transfer learning. The prediction curve in Fig. 5 intuitively demonstrate, compared with *no fine-tuning*, *TT-DL* gets better performance during the real-time prediction.

5. Conclusion and Future Work

In this study, we first explored and applied the parameter-based deep transfer learning to the field of transportation and proposed a novel transferable traffic deep learning model which can predict real-time traffic flow in data-strapped roads by transferring knowledge from data-rich roads. Experimental results implied that our method is not only better than the shallow

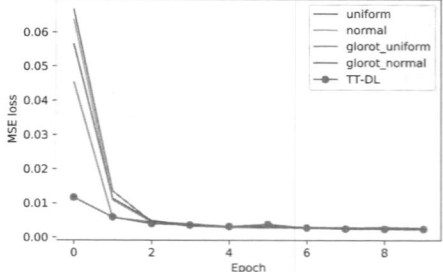

Fig. 3. The training loss of different initializers.

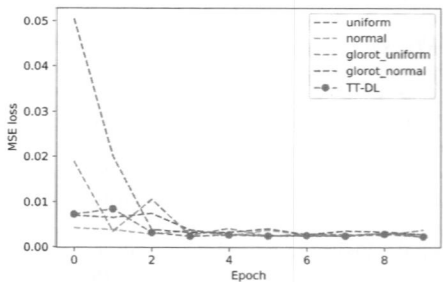

Fig. 4. The validation loss of different initializers during fine-tuning.

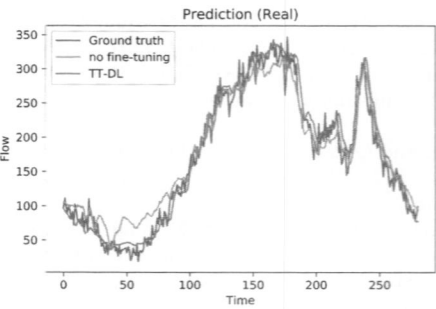

Fig. 5. Comparisons between *TT-DL* and *no fine-tuning* in real-time prediction.

model, but also than deep models with other initialization methods. In the future, we aim to validate transfer learning on the larger traffic networks convolutional kernels and fuse other transfer methods, e.g., domain adaptation, to further empower its capacity.

Acknowledgments

This work was supported by the National Natural Science Foundation of China (No. 61773324), the Fundamental Research Funds for the Central Universities (No. 2682015QM02), and the Australian Research Council (No. DP150101645).

References

1. Y. Zheng, L. Capra, O. Wolfson and H. Yang, Urban computing: concepts, methodologies, and applications, *ACM Transactions on Intelligent Systems and Technology (TIST)* **5** (ACM, 2014).
2. J. Zhang, Y. Zheng and D. Qi, Deep spatio-temporal residual networks for citywide crowd flows prediction, in *Proceedings of the AAAI Conference on Artificial Intelligence*, 2017.
3. Y. Lv, Y. Duan, W. Kang, Z. Li and F.-Y. Wang, Traffic flow prediction with big data: A deep learning approach, *IEEE Transactions on Intelligent Transportation Systems* **16** (IEEE, 2015).
4. S. J. Pan and Q. Yang, A survey on transfer learning, *IEEE Transactions on Knowledge and Data Engineering* **22** (IEEE, 2010).
5. J. Lu, V. Behbood, P. Hao, H. Zuo, S. Xue and G. Zhang, Transfer learning using computational intelligence: A survey, *Knowledge-Based Systems* **80** (Elsevier, 2015).
6. H. Zuo, G. Zhang, W. Pedrycz, V. Behbood and J. Lu, Fuzzy regression transfer learning in takagi–sugeno fuzzy models, *IEEE Transactions on Fuzzy Systems* **25** (IEEE, 2017).
7. J. Yosinski, J. Clune, Y. Bengio and H. Lipson, How transferable are features in deep neural networks?, in *Advances in Neural Information Processing Systems*, 2014.
8. M. Long, Y. Cao, J. Wang and M. Jordan, Learning transferable features with deep adaptation networks, in *International Conference on Machine Learning*, 2015.
9. Q. Zhang, D. Wu, J. Lu, F. Liu and G. Zhang, A cross-domain recommender system with consistent information transfer, *Decision Support Systems* **104** (Elsevier, 2017).
10. Y. Wei, Y. Zheng and Q. Yang, Transfer knowledge between cities, in *Proceedings of the 22nd ACM SIGKDD International Conference on Knowledge Discovery and Data Mining*, 2016.
11. L. Wang, X. Geng, X. Ma, F. Liu and Q. Yang, Crowd flow prediction by deep spatio-temporal transfer learning, *arXiv preprint arXiv:1802.00386*, 2018.
12. P. Y. Simard, D. Steinkraus, J. C. Platt *et al.*, Best practices for convolutional neural networks applied to visual document analysis, in *International Conference on Document Analysis and Recognition*, 2003.

Incremental possibilistic decision trees in non-specificity approach

Mohamed Sofien Boutaib* and Zied Elouedi

LARODEC, Institut Supérieur de Gestion de Tunis, Université de Tunis
41 Avenue de la Liberté, 2000 Le Bardo, Tunisie
*mohamedsofienboutaib@gmail.com
zied.elouedi@gmx.fr

Possibilistic decision trees are efficient tools that handle training instances in which the class values are stained by uncertainty. This paper proposes a new classification method that enables possibilistic decision trees to learn incrementally. Our proposed method called incremental possibilistic decision trees permits to adapt incrementally the new incoming instances by updating and restructuring an existing possibilistic decision tree instead of destructing it and rebuilding another one from scratch.

Keywords: Incremental induction; machine learning; decision tree; classification; uncertainty; possibility theory.

1. Introduction

Decision tree is considered as one of the most performant tools for learning from training instances, in which the classes are already known, in order to classify new incoming instances. Work in machine learning has demonstrated that, in several practical applications, uncertainty in data is often unavoidable. To cope with such data, various theories have been proposed. The possibility theory is one of them.[1,13] To deal with data which are characterized by uncertain classes, the possibility theory combined with the standard decision tree is used. This combination gives the so-called possibilistic decision tree in each of the three different approaches: a non-specificity approach[5] or a similarity approach[4] or a clustering approach.[7] The possibilistic decision tree is built in the non-incremental mode which requires the presence of all the training instances from the outset of the building process. Besides, the non-incremental possibilistic decision tree needs to be destructed and rebuilt from scratch whenever a new training instance occurs. Consequently, the concept of incremental learning[3,8,14] is the solution to overcome the drawbacks of the non-incremental method (called also the batch mode method) which has limits in an environment that requires a perpetual change. Some incremental decision trees under an uncertain environment are proposed: the incremental probabilistic decision tree,[12] the incremental fuzzy decision tree[11] and the incremental belief decision tree.[10] This paper proposes a new method called incremental possibilistic decision tree in the non-specificity[6] approach based on the adaptation of the incremental learning method to the possibilistic decision tree. After presenting the possibility theory basics in Section 2, we detail, in Section 3, the possibilistic decision tree in both the building and the classification procedures. Section 4 is devoted to describing our new method. Section 5 includes a comparison between the results of the experimentations conducted on incremental possibilistic decision tree and those on the non-incremental one. A conclusion is drawn in Section 6.

2. Possibility Theory

Let Ω represents the universe of discourse for the different states of knowledge $\{\omega_1, \omega_2, ..., \omega_n\}$. The possibility distribution denoted by π represents the fundamental tool of the possibility theory. It represents a function that maps $\pi : \Omega \to [0, 1]$ from Ω to the unit interval to obtain *possibility degree*. The extreme forms of knowledge are represented by: *Complete knowledge*: $\exists\, \omega_k \in \Omega$, $\pi(\omega_k) = 1$ and all remaining states $\omega : \pi(\omega) = 0$. *Total ignorance*: $\pi(\omega_k) = 1$, $\forall \omega_k \in \Omega$ (i.e. all states ω_k are completely possible). From the well-known source of uncertainty in possibility theory is *non-specificity*. The comparison between two possibility distributions is done after measuring the non-specificity for each one by using the so-called *U-uncertainty*.[2] This latter is defined by:

$$U(\pi) = \sum_{i=1}^{n} (\pi_{(i)} - \pi_{(i+1)}) \times \log_2(i) \qquad (1)$$

where $\pi_{(i)} - \pi_{(i+1)}$ represents the dispersion of the degrees of possibility where $\pi_{(i)}$ represents the degree of possibility of an element ω_i and n represents the number of possibility distributions. Note that $\pi_{(n+1)} = 0$.

3. Possibilistic Decision Tree

In this paper, for the building of the possibilistic decision tree, we deal with the non-specificity-PDT (NS-PDT)[5] approach This latter was chosen because it maintains the possibility distributions during all the building process.

3.1. *Building procedure of NS-PDT*

Let us start by introducing the notations of the NS-PDT approach: T is the training set with n observations (or instances), A_k represents an attribute, ω_q represents a set containing various possibility distributions and $D(A_k)$ represents the domain of an attribute A_k. The NS-PDT approach is composed of four fundamental components namely attribute selection measure, partitioning strategy, stopping criteria and structure of leaves.

Attribute selection measure: The *non-specificity Gain ratio* (denoted by $NSGr$)[5] is a selection measure of an attribute A in a set or subset of instances T in the NS-PDT approach. The next steps are followed to compute the $NSGr$ of an attribute A:

(1) Compute a representative possibility distribution π_{rep} for a training set or subset T after the calculation of an arithmetic average of nth of possibility distributions:

$$\pi_{AM}(\omega_q) = \frac{1}{n} \sum_{i=1}^{n} \pi_i(\omega_q) \qquad (2)$$

Then, π_{rep} is obtained by the normalization of π_{AM} as follows:

$$\pi_{rep}(\omega_q) = \frac{\pi_{AM}(\omega_q)}{\max_{q=1}^{|\Omega|}\{\pi_{AM}(\omega_q)\}} \qquad (3)$$

π_{rep} is a possibility distribution that represents the portion of different possibility degrees for different values.

(2) Compute the non-specificity of the π_{rep}^T (calculated above) using the U-uncertainty (see Equation 1) of an attribute A_k as follows:

$$U_{A_k}\left(\pi_{rep}^T\right) = \frac{1}{|D(A_k)|} \sum_{v \in D(A_k)} U\left(\pi_{rep}^{T_v^{A_k}}\right) \qquad (4)$$

(3) Compute the non-specificity Gain (NSG) for a training set or subset T of an attribute A_k in which possibility theory is the class values given as follows:

$$NSG(T, A_k) = U\left(\pi_{rep}^T\right) - U_{A_k}\left(\pi_{rep}^T\right) \qquad (5)$$

(4) Finally, the computation of *non-specificity Gain ratio* for an attribute A_k is given as follows:

$$NSGr(T, A_k) = \frac{NSG(T, A_k)}{SplitInfo(T, A_k)} \qquad (6)$$

where $SplitInfo(T, A) = - \sum_{v \in D(A_k)} \frac{|T_v^{A_k}|}{|T|} \log_2 \frac{|T_v^{A_k}|}{|T|}$ (7)

(5) Repeat for every attribute $A_k \in A$ and select the attribute having the highest *NSGr*.

Partitioning strategy: The partitioning strategy is similar to that of the standard decision tree. The different values of the attribute selected as a test attribute at a decision node will be the label at every different edge coming out from the decision node.

Stopping criteria: Ending the partitioning process. Generally, we stop the partitioning if one of the following conditions is reached:

(1) If no more attributes to test in the training partition.
(2) If the number of instances in a generated partition is equal to 0.
(3) If no information is gained, i.e., the *Non-Specificity Gain Ratio (NSGr)* ≤ 0.
(4) The partitioning process is stop is the case where the the training partition T_p is not pure enough, i.e., $(U(\pi_{rep}^{T_p}) \leq AvgU)$. *AvgU* (which represents the average of each instance) is computed as follows:

$$AvgU = \frac{1}{n} \sum_{i=1}^{n} U(\pi_i)$$ (8)

(5) If the partitioning includes only the possibility distributions which represent a total ignorance, i.e., if the following equation is satisfied: $\pi_{rep}^{T_p}$ is equal to $\log_2(c)$. c: represents the number of classes.

$$\pi_{AM}^{T_p} \equiv \pi_{rep}^{T_p} \left(U\left(\pi_{AM}^{T_p}\right) = U\left(\pi_{rep}^{T_p}\right) = \log_2(c) \right)$$ (9)

(6) The training partition is stopped in the case of complete knowledge:

$$\left(U\left(\pi_{rep}^{T_p}\right)\right) = 0$$ (10)

Structure of leaves: If one of the stopping criteria 1, 3, 4, 5 is met, a leaf will be created and labeled by $\pi_{rep}^{T_p}$. Otherwise, if the stopping criterion 2 is met, a leaf will be created and labeled by $\pi_{rep}^{T_p}$ of the preceding level. Then again, if the stopping criterion 6 is met, a leaf will be created and labeled by a fully specific possibility distribution $\pi_{rep}^{T_p}$.

3.2. Classification procedure of NS-PDT

This phase consists in making a decision about the object going from the root node until reaching a corresponding leaf by following the edges relative to the attribute values of the object. Once the leaves are labeled by the possibility distributions, the plausible class of the leaf reached by the object will be given to this latter.

4. Incremental Possibilistic Decision Trees in Non-specificity Approach IPDT

In this section, we propose a new method of learning incrementally called Incremental Possibilistic Decision Tree in non-specificity approach. It updates the existing tree whenever we have new uncertain training instances. So, we suggest following the next steps:

(1) Select the path that will be traversed by the new incoming instance till reaching the corresponding leaf.
(2) Add the new object to the set or subset of instances (with its corresponding possibility distribution), for each reached node depending on the path of the new incoming instances.
(3) Select the best attribute which has the highest *non-specificity Gain ratio* (see Equation 6) for each traversed decision node or leaf by calculating π_{rep} for a set containing different possibility degrees with different class values (see Equation 3). Then, we calculate the non-specificity (see Equation 4) of π_{rep} of each attribute.
(4) If each concerned test attribute remains the best one according to its position, update the node information (its possibility distribution),
(5) Otherwise,
   (a) Restructure the tree using Pull-up algorithm and the best attribute (which has the greatest *NSGr*) will be located at the root.
   (b) Recursively restore the best attribute at the root of each subtree.

Tree revision: The grouping of steps 1 and 2 represents the adding example task. This latter consists in choosing the path of a new instance depending on its attribute values and on the branch values of the tree. For every reached node, the new training instance is added to the set of the remaining instances that belong to this node. If the value of the new training instance does not exist in a reached node, our method adds a new branch to the node with the value of the attribute and puts the instance in the set of that node. The grouping of the remaining steps consists in updating and/or restructuring the tree task. In fact, a checking is required after the addition of the new instance. It is done by computing the *non-specificity Gain ratio* (denoted by *NSGr*) (see Equation 6) for each visited node. The calculation of the *NSGr* is based on the calculation of π_{rep} (see Equation 3) that represents the portion of different possibility degrees for different values. We can get one of the two following results:

- Each concerned test attribute is the best one according to its position. Thus, no change is required.
- One of the concerned test attributes is not the best, then the best of the existing attributes must be pulled at each subtree and transposed to the root of the tree and so the existing attribute at the root of the tree is moved to a lower level. This attribute must be compared with all the attributes that are below it and the best of them will be in the root of the subtree. This operation is repeated recursively until obtaining the best attribute at each node in the tree.

Once the updating and the restructuring are done, the tree is transformed to a compact form. To obtain the final incremental possibilistic decision tree, the stopping criteria as defined for NS-PDT must be checked at each leaf. If one of the stopping criteria is met, a π_{rep} is assigned to the leaf. In fact, the nodes on the tree must be checked: if all their existing values in the training set are represented as branches, π_{rep} is applied on all the leaves of every branch. If one of the values does not exist as a branch, a new branch containing the π_{rep} of the previous level is assigned to the node. Note that the partitioning strategy, the stopping criteria and the structure of the leaves are similar to those of the non-specificity possibilistic decision tree in the batch mode.

5. Experimental Results

Databases. Are taken from the U.C.I repository databases.[9] The used data sets are Balance Scale Weight, Car Evaluation, Congressional Voting, Tic-Tac-Toe Endgame, W. Breast Cancer. **Artificial Creation of Uncertain Databases.** The chosen databases are contaminated by uncertainty: the certain classes of the training instances take the value 1, the remaining classes take the level of the degrees of possibility: Low, Middle and High degrees of possibility are generated randomly respectively from]0, 0.3],]0.3, 0.6] and]0.6, 1]. **Evaluation criteria.** The percentage of the correctly classified instances (denoted by *PCC*) and the running time (RT) devoted to update a tree and calculated in terms of seconds. **Results of the IPDT.** Table 1 reports the *PCC* after applying the two modes in the NS-PDT approach. Note that in the following tables, low, middle, and high degrees of possibility are respectively denoted by PDT_{low}, PDT_{middle} and PDT_{high} in the batch NS-PDT approach. In the incremental NS-PDT approach, low, middle, and high degrees of possibility are respectively denoted by: $IPDT_{low}$, $IPDT_{middle}$ and $IPDT_{high}$. We can observe through Table 1 that the $IPDT$ generates the same PCC as the non-incremental PDT for all uncertainty levels. Note that this observation confirms what we said before about the similarity in the PCC. Table 2 shows the results obtained when we use the second evaluation criterion which is the running time. From the Table 2, we remark that the running time obtained in $IPDT$ in NS-PDT approach is shorter than the one obtained by the non-incremental PDT. Such difference is expected because the incremental mode modifies the existing PDT, while the non-incremental mode constructs a new PDT again from the outset, which requires a longer time. The experimentations were performed using an Intel Core i7 CPU 2.20 GHz PC with 8 MB of RAM.

Table 1. PCC results (%).

Databases	Non-incremental mode			Incremental mode		
	PDT_{low}	PDT_{middle}	PDT_{high}	$IPDT_{low}$	$IPDT_{middle}$	$IPDT_{high}$
Balance Scale Weight	68.71	75.32	90	68.71	75.32	90
Car evaluation	70.88	60.17	48.35	70.88	60.17	48.35
Tic-Tac-Toe Endgame	69.54	74.789	87.20	69.54	74.789	87.20
Congressional Voting	90.465	85.116	80.65	90.465	85.116	80.65
W. Breast Cancer	80.579	78.26	60	80.579	78.26	60

Table 2. Running time results (in seconds).

Databases	Non-incremental mode			Incremental mode		
	PDT_{low}	PDT_{middle}	PDT_{high}	$IPDT_{low}$	$IPDT_{middle}$	$IPDT_{high}$
Balance Scale Weight	3.31	3.12	2.84	0.81	0.64	0.36
Car evaluation	24.11	25.75	24.81	21.71	23.43	22.40
Tic-Tac-Toe Endgame	14.52	15.82	15.64	8.22	9.32	8.68
Congressional Voting	14.32	13.77	15.11	10.62	9.82	11.35
W. Breast Cancer	17.96	17.42	18.30	15.44	14.82	15.81

6. Conclusion

In this paper, we have proposed an Incremental Possibilistic Decision Tree in NS-PDT approach in order to reduce time and effort without rebuilding another tree from scratch. Besides, the aim of our approach is to obtain a PCC similar to the one of the batch mode and to obtain a better time (the lowest), when compared to the batch mode. We propose in future work, an incremental PDT based on either Similarity-PDT approach[4] or Clustering-PDT approach.[7]

References

1. D. Dubois and H. Prade. *Possibility Theory: An Approach to Computerized Processing of Uncertainty*. Plenum Press, New York, 1988.
2. M. Higashi and G. J. Klir. On the notion of distance representing information closeness: Possibility and probability distributions. *International Journal of General Systems*, pages 103–115, 1983.
3. Y. Huang, T. Li, C. Luo, H. Fujita, and S. Horng. Matrix-based dynamic updating rough fuzzy approximations for data mining. *Knowledge-Based Systems*, pages 273–283, 2017.
4. I. Jenhani, N. Ben Amor, S. Benferhat, and Z. Elouedi. Sim-pdt: A similarity based possibilistic decision tree approach. In *Proceedings of the 5th International Symposium on Foundation of Information and Knowledge Systems*, pages 348–364, 2008.
5. I. Jenhani, N. Ben Amor, and Z. Elouedi. Decision trees as possibilistic classifiers. *International Journal of Approximate Reasoning*, pages 784–807, 2008.
6. I. Jenhani, N. Ben Amor, Z. Elouedi, S. Benferhat, and K. Mellouli. Information affinity: a new similarity measure for possibilistic uncertain information. In *Proceedings of the 9th european conferences on symbolic and quantitative approaches to reasoning with uncertainty*, pages 840–852, 2007.
7. I. Jenhani, S. Benferhat, and Z. Elouedi. On the use of clustering in possibilistic decision tree induction. In *Proceedings of the 10th European Conference on Symbolic and Quantitative Approaches to Reasoning with Uncertainty*, pages 505–517, 2009.
8. T. Li, C. Luo, H. Chen, and J. Zhang. Pickt: a solution for big data analysis. In *International Conference on Rough Sets and Knowledge Technology*, pages 15–25, 2015.
9. P. M. Murphy and D. W. Aha. Uci repository of machine learning databases, 1996.
10. S. Trabelsi, Z. Elouedi, and M. El Aroui. Incremental induction of belief decision trees in averaging approach. In *International Conference on Database and Expert Systems Applications*, pages 454–461, 2014.
11. T. Wang, Z. Li, Y. Yan, and H. Chen. An incremental fuzzy decision tree classification method for mining data streams. In *Proceedings of MLDM 2007, LNCS (LNAI)*, pages 91–103, 2007.
12. C. Wanke and D. Greenbaum. Incremental, probabilistic decision making for en route traffic management. *Air Traffic Control Quarterly*, pages 299–319, 2007.
13. L. A. Zadeh. Fuzzy sets as a basic for a theory of possibility. *Fuzzy Sets and systems*, pages 3–28, 1978.
14. J. Zhang, J. Wong, Y. Pan, and T. Li. A parallel matrix-based method for computing approximations in incomplete information systems. *IEEE Transactions on Knowledge and Data Engineering*, pages 326–339, 2015.

Maintaining case knowledge vocabulary using a new evidential attribute clustering method

S. Ben Ayed and Z. Elouedi

LARODEC, Institut Supérieur de Gestion de Tunis, Université de Tunis
41 Avenue de la liberté, cité Bouchoucha, 2000 Le Bardo, Tunisia
safa.ben.ayed@hotmail.fr
zied.elouedi@gmx.fr

E. Lefevre

Univ. Artois, EA 3926, LGI2A,
62400 Béthune, France
eric.lefevre@univ-artois.fr

Maintaining the vocabulary of case knowledge within Case Based Reasoning (CBR) presents a crucial task to ensure a high-quality problem-solving and to improve retrieval performance for large-scale CBR systems. To do, we propose, in this paper, a method that manages uncertainty while selecting the best attributes characterizing case knowledge by using belief function theory. Actually, this method is based on a new evidential attribute clustering technique to eliminate redundant and noisy attributes describing cases.

Keywords: Case based reasoning; maintenance; case vocabulary; attribute clustering; belief function theory; feature selection.

1. Introduction

Case Based Reasoning is a methodology that aims to solve new problems through reusing the most similar past experiences.[1] Over the years, CBR has known a widespread interest in several domains thanks to its capability to learn incrementally. Actually, the arriving of a new problem triggers a cycle with four steps.[1] First, CBR *retrieves* from the case base the most similar one. Second, it *reuses* the corresponding solution to be adapted to the target problem. Third, the proposed solution is *revised*. Finally, the new case is *retained* in order to extend case base's capability in future problems resolution. To control this case knowledge growth along with preserving its competence, many works are interested in Case Base Maintenance (CBM) field.[2,3] However, knowledge containers[4] such as

Similarity measures, *Adaptation rules* and *Vocabulary* are also considerable maintenance targets. Accordingly, researches around Case-Based Reasoner Maintenance (CBRM)[5] are also directed. In this work, we are situated in the maintenance of the vocabulary knowledge container for structual CBR systems. Herein, a case is described using a number of attributes[a] which are mainly serving in matchmaking and case retrieval. Logically, the more a case is described, the best solution is offered. However, some application domains[12] describe cases with a very large number of features which leads to decrease problem-solving performance. Besides, the existence of irrelevant and noisy features can seriously reduce CBR systems' competence. To deal with these problems, we propose, in this paper, an approach that selects only the most 'informative' features using a new evidential attribute clustering method which is based on belief function theory[6,7] to manage all levels of uncertainty towards the membership of features to the different clusters.

The rest of this paper is organized as follows. Two among the used applicable concepts within vocabulary maintenance researches are reviewed in Section 2. The necessary background related to the used evidential clustering technique is briefly introduced in Section 3. Throughout Section 4, we describe the different steps of our proposed method for case knowledge vocabulary maintenance. Finally, Section 5 conducts experimental study on UCI data sets to evaluate our newly method.

2. Applicable Concepts for Vocabulary Maintenance

Many concepts within machine learning studies have been applied in the different methods for maintaining CBR systems. Among them, we review, in Subsections 2.1 and 2.2, *feature selection* and *attribute clustering* concepts. In Subsection 2.3, we explain our motivation behind this work.

2.1. *Applying feature selection for vocabulary maintenance*

The vocabulary of CBR systems defines the information towards the corresponding field and the way to express them. To maintain vocabulary with considering structured CBR systems, we should select only features that ensure accurate retrieval outcomes. Herein, the problem of Feature Selection (FS) arises. Actually, since FS is an NP-Hard problem aiming specially in irrelevant features elimination, we find numerous FS techniques where some

---

[a]In this paper, we use *attribute* and *feature* terms exchangeably.

of them were combined with CBR systems.[8,9] Besides, some techniques are leading to select features by assigning weights reflecting their relevance.[10]

2.2. Applying attribute clustering for feature selection

Attribute clustering is carried out in several researches[11,12] as a feature selection task. Like the standard objects' clustering, attributes belonging to the same cluster are similar and those belonging to different ones are dissimilar. However, the notion of similarity within attributes reflects the relation between them (*e.g in term of correlation, dependency, etc.*). Consequently, that leads us to eliminate dispensable features by selecting only representative one(s) for each cluster.

2.3. Motivation and discussion

Actually, using attribute clustering for maintaining case knowledge vocabulary has the advantage of preserving the relation between features which offers a better flexibility at the level of CBR framework. In fact, we can replace any selected representative feature by another one belonging to the same cluster. However, existing researches in this road even cannot manage uncertainty about attributes membership to clusters or they are not able to manage all levels of uncertainty, from the complete ignorance to the total certainty. For that reason, we propose to maintain cases vocabulary using a powerful tool for uncertainty management called belief function theory.[6,7]

3. Belief Function Theory

The belief function theory (or Evidence theory)[6,7] is a theoretical framework for reasoning with partial and unreliable information. Its basic concepts will first be recalled in Subsection 3.1, and the used evidential clustering algorithm will then summarized during Subsection 3.2.

3.1. Basic concepts

Let Ω be a finite set of events called the frame of discernment, and ω is a variable taking values in Ω. The basic belief assignment (*bba*) function m, from 2^Ω to $[0, 1]$, represents the partial knowledge towards the real value taken by ω verifying $\sum_{A \subseteq \Omega} m(A) = 1$. Complete ignorance corresponds to $m(\Omega) = 1$, and total certainty is achieved when $m(A) = 1$ and A is a singleton. The subset A is called focal element if $m(A) > 0$. Furthermore, a bba m can be represented by $bel(A)$ as the amount of support given only to the subset A. In addition, it can be represented by the plausibility $pl(A)$ which is the maximum amount of belief that can be assigned to A, and defined such that $pl(A) = \sum_{A \cap B \neq \emptyset} m(B)$ for all $A \subseteq \Omega$. Concerning the decision making process, choosing the highest pignistic probability $BetP$ presents one of the most powerful techniques, which is defined as follows:

$$BetP(A) = \sum_{B \subseteq \Omega} \frac{|A \cap B|}{|B|} \frac{m(B)}{1 - m(\emptyset)} \qquad \forall A \in \Omega \qquad (1)$$

3.2. Evidential dissimilarity data clustering

The aim of evidential dissimilarity data clustering is to construct a credal partition for dissimilarity data. Actually, the credal partition quantifies the uncertainty of n objects membership to clusters using bba functions where $\Omega = \{\omega_1, .., \omega_c\}$ denotes a set of c clusters. Among such techniques offering a credal partition, we enumerate RECM,[13] EVCLUS[14] and k-EVCLUS.[15] The two latter do not make assumption about the dissimilarity nature, although RECM assumes explicitly that the input dissimilarity is calculated as Squared Euclidean Distances.[13] For that reason, we centralize our work around k-EVCLUS which is an improvement of EVCLUS algorithm.

Let $D = (d_{ij})$ is $n \times n$ dissimilarity matrix where d_{ij} is the degree of dissimilarity between objects x_i and x_j. Besides, let $F_1, ..., F_f$ are f focal sets. Logically, the more two objects are similar, the more plausible that they belong to the same cluster. In fact, it is shown that $pl_{ij} = 1 - \kappa_{ij}$[14] with $\kappa_{ij} = \sum_{A \cap B = \emptyset} m_i(A) m_j(B)$ is the degree of conflict between m_i and m_j. Since similar objects should have mass functions with low degrees of conflict and conversely, the credal partition within k-EVCLUS, presented as a matrix M of size $n \times f$, is the result of the following stress function minimization which is solved using Iterative Row-rise Quadratic Programming (IRQP):

$$J(M) = \eta \sum_{i<j} (\kappa_{ij} - \delta_{ij})^2 \qquad (2)$$

where η is a normalizing constant and $\delta_{ij} = \varphi(d_{ij})$ are transformed dissimilarities. Using matrix notations, κ_{ij} is written herein as $m_i^T C m_j$, where C is a square $f \times f$ matrix with general term $C_{kl} = 1$ if $F_k \cap F_l = \emptyset$, and $C_{kl} = 0$ otherwise.

Moreover, k-EVCLUS eliminates redundancy of information within dissimilarity matrix (e.g. given x_1 and x_2 are two very similar objects. For any object x_3 dissimilar from x_1, it is then usually dissimilar from x_2) in order to reduce the complexity of stress criterion calculation such that:

$$J_k(M) = \eta \sum_{i=1}^{n} \sum_{r=1}^{k} (\kappa_{ij_r(i)} - \delta_{ij_r(i)})^2 \qquad (3)$$

with $j_1(i), ..., j_k(i)$ are k integers sampled randomly for $i = 1, .., n$.

Actually, $J_k(M)$ requires $O(nk)$ operations instead of $O(n^2)$ for EVCLUS.

4. Maintaining Case Knowledge Vocabulary in an Evidential Framework

The main purpose of our proposed method is to maintain case knowledge vocabulary by eliminating on the one hand redundant features which are so correlated, and on the other hand noisy features which lead to distort the problem-solving. Our method is thus summed up by the following steps.

4.1. Step 1: Creating cases' features relational matrix

Features' relationship that we take into account in our method reflects the amount of correlation between them. Given a case base CB with n objects and p features, we choose to use the *pearson's correlation coefficient*,[16] denoted by r, to measure the linear association between every two variables. Hence, $R = (r_{AB})$ is our relational matrix and r_{AB} is defined such that:

$$r_{AB} = \frac{\sum_{i=1}^{n}(a_i - \bar{a})(b_i - \bar{b})}{\sqrt{\sum_{i=1}^{n}(a_i - \bar{a})^2}\sqrt{\sum_{i=1}^{n}(b_i - \bar{b})^2}} \qquad (4)$$

where a_i and b_i are the values of every two attributes A and B respectively for object i, and $\bar{a}$ and $\bar{b}$ are their mean values.

4.2. Step 2: Generating cases' features dissimilarity matrix

Definition 1. Two features A and B are said to be similar if there is a high correlation between them, and conversely.

According to Definition 1, we can thus generate a matrix $D = (d_{AB})$ as a $p \times p$ dissimilarity matrix between features where $d_{AB} = f(r_{AB})$ with f is a function from $[-1, 1]$ to $[0, 1]$. Actually, we have $-1 < r_{AB} < 1$ where three *Situations* (S_i) are therefore arising:[16]

- S_1: If $r_{AB} \simeq -1 \Rightarrow$ High correlation (negative) $\Rightarrow$ High similarity.

- S_2: If $r_{AB} \simeq 1 \Rightarrow$ High correlation (positive) $\Rightarrow$ High similarity.
- S_3: If $r_{AB} \simeq 0 \Rightarrow$ No correlation $\Rightarrow$ High dissimilarity.

Within S_1 and S_2, A and B are offering the same information. Consequently, they are redundant, whereas it is not the case for S_3.

Now, it is straightforward to show that the dissimilarity between two features A and B is computed as follows:

$$d_{AB} = f(r_{AB}) = 1 - |r_{AB}| \qquad (5)$$

where r_{AB} represents the similarity between A and B.

4.3. Step 3: Evidential attribute clustering

After generating a square dissimilarity matrix for p features, we aim now to group them using a dissimilarity data clustering technique which is able to manage all levels of uncertainty within the input dissimilarity data. For that reason, we use an evidential technique called k-EVCLUS[15] as presented throughout Section 3 where we apply it on the already created dissimilarity matrix for p features during *Step 2*. The output of this attribute clustering procedure is the set of features detected as outliers as well as the credal partition of features' membership to the different clusters.

4.4. Step 4: Case knowledge vocabulary maintenance

Ultimately, we aim to define our strategy for case vocabulary maintenance. Actually, we eliminate all noisy features detected during the previous step since they distort the process of problem-solving. On the other hand, we also remove redundant features belonging to the same cluster and keeping only one as their representative. In fact, the membership of features to the different clusters is decided through the pignistic probability transformation from the credal partition as defined in Equation 1. Removing redundant features serves mainly in reducing the execution time of indexing and retrieving cases, which then conduct to improve CBR systems performance.

5. Experimental Study: Results and Analysis

To measure our method's efficiency, we developed it using R software, testing on UCI repository data sets, evaluating results via accuracy, which is calculated using 10-folds cross validation technique, and retrieval time criteria (Table 1). This is done after varying the number of clusters K from

3 to 7 and choosing then the most convenient one.[b] Finally, we compare results related to our method (AttEvClus-CBR) with those offered by the original non maintained case base (Original-CBR), as well as the updated case bases at the vocabulary level using ReliefF [10] (ReliefF-CBR) as one of the most known FS methods. Like we did with our method, we choose for RefiefF-CBR the most relevant attributes set offering the highest accuracy.[c]

Table 1. Evaluation of our proposed vocabulary maintaining method.

	Case bases	Original-CBR		ReliefF-CBR		AttEvClus-CBR	
		PCC[a]	Time[b]	PCC	Time	PCC	Time
1	Ionosphere	85.48	1.942	84.88	1.188	**88.33**	0.912
2	Glass	97.64	0.967	98.11	0.882	**98.59**	0.762
3	WDBC	60.16	1.710	96.33	1.112	**96.46**	1.013
4	German	64.6	1.812	**73.4**	1.213	73.25	1.211
5	Heart	57.5	2.103	62.45	1.091	**62.98**	1.028
6	Yeast	55.32	0.954	**99.05**	0.722	**99.05**	0.724

Note: [a]Percentage of correct classifications (%) offered by 5-NN algorithm (to be not sensible to noisy cases). [b]The retrieval time in seconds exerted in 5-NN.

Obviously, results offered by our proposed method, as shown in Table 1, ensure a high-quality case knowledge vocabulary maintenance. In term of accuracy (PCC), we note that our method has been able to increase the problem solving competence for all case bases (CB) comparing to the original ones. For instance, it increases the accuracy for "WDBC" data set from 60.16% to 96.46%. Comparing to ReliefF-CBR, our method has also competitive results by offering the best accuracies for almost all the CBs. These results can be explained by the high quality of the used evidential clustering technique in managing uncertainty and in noisy features detection. In term of retrieval time, our mainly objective is to provide lower values than those offered by the original CBR systems. Indeed, we note that there is a respectable time reduction for all the different CBs. For example, the time decreases from about 2.1s to about 1s for "Heart" data set. Besides, we note a slightly faster process for almost all the CBs comparing to ReliefF-CBR.

---

[b]The number of clusters (K) offering the highest accuracy for our method: Ionosphere ($K = 3$), Glass ($K = 5$), WDBC ($K = 4$), German ($K = 4$), Heart ($K = 4$), and Yeast ($K = 4$).
[c]The number of features (p) offering the highest accuracy for ReliefF: $p = 5$ for Ionosphere, Glass and WDBC. $p = 4$ for German and Heart. And $p = 3$ for Yeast.

6. Conclusion

In this paper, we have developed a method to maintain the vocabulary of CBR systems by eliminating irrelevant and redundant features. To do, we applied a new evidential attribute clustering technique that considers the correlation between features and manages uncertainty about their membership to clusters. Finally, it keeps only representative features for clusters.

References

1. A. Aamodt, E. Plaza, Case-based reasoning: Foundational issues, methodological variations, and system approaches. In *AI communications* (IOS press, 1994), pp. 39-59.
2. A. Smiti, Z. Elouedi, SCBM: soft case base maintenance method based on competence model. In *Journal of Computational Science* (Elsevier, 2017), DOI: 10.1016/j.jocs.2017.09.013.
3. S. Ben Ayed, Z. Elouedi, E. Lefevre, ECTD: Evidential Clustering and case Types Detection for case base maintenance. In *International Conference on Computer Systems and Applications* (IEEE, 2017), pp. 1462-1469.
4. M. M. Richter, M. Michael, Knowledge containers. In *Readings in Case-Based Reasoning* (Morgan Kaufmann, 2003).
5. D. C. Wilson, D. B. Leake, Maintaining Case-Based Reasoners: Dimensions and Directions. In *Computational Intelligence* (2001), pp. 196-213.
6. G. Shafer: A mathematical theory of evidence. In *Princeton university press* (Princeton university press, 1976).
7. A. P. Dempster, Upper and lower probabilities induced by a multivalued mapping. In *The annals of mathematical statistics* (1967), pp. 325-339.
8. N. Arshadi, I. Jurisica, Feature Selection for Improving Case-Based Classifiers on High-Dimensional Data Sets. In *FLAIRS Conference* (2005), pp. 99-104.
9. G. Zhu, J. Hu, J. Qi, J. Ma, Y. Peng, An integrated feature selection and cluster analysis techniques for case-based reasoning. In *Engineering Applications of Artificial Intelligence* (Elsevier, 2015), pp. 14-22.
10. I. Kononenko, Estimating attributes: analysis and extensions of RELIEF. In *European conference on machine learning* (Springer, 1994), pp. 171-182.
11. T. Hong, Y. Liou, Attribute clustering in high dimensional feature spaces. In *International Conference on Machine Learning and Cybernetics* (IEEE, 2007), pp. 2286-2289.
12. P. Maji, Fuzzy-rough supervised attribute clustering algorithm and classification of microarray data. In *Transactions on Systems, Man, and Cybernetics, Part B (Cybernetics)* (IEEE, 2011), pp. 222-233.
13. M. Masson, T. Denœux, RECM: Relational evidential c-means algorithm. In *Pattern Recognition Letters* (Elsevier, 2009), pp. 1015-1026.
14. T. Denœux, M. Masson, EVCLUS: evidential clustering of proximity data. In *Transactions on Systems, Man, and Cybernetics* (IEEE, 2004), pp. 95-109.
15. O. Kanjanatarakul, S. Sriboonchitta, T. Denœux, K-EVCLUS: Clustering large dissimilarity data in the belief function framework. In *International Conference on Belief Functions* (Springer, 2016), pp. 105-112.
16. K. Pearson, Mathematical contributions to the theory of evolution. III. Regression, heredity, and panmixia. In *Philosophical Transactions of the Royal Society of London.* (1896), pp. 253-318.

A workforce health insurance plan recommender system

Dianshuang Wu,[a] Jie Lu,[a] Farookh Hussain,[a] Craig Doumouras[b] and Guangquan Zhang[a]

[a]*Decision Systems and e-Service Intelligence Laboratory, Centre for Artificial Intelligence, Faculty of Engineering and Information Technology, University of Technology Sydney, Australia*
[b]*Workforce Health Assessors, Hobart, Tasmania, Australia*

Big data appearing in health domain bring great opportunities to develop a workforce health insurance plan recommender system, which can help workforce users select proper insurance plans efficiently. There are two challenges in the development of health insurance plan recommender system: (1) plan and user data present hierarchical tree structures; (2) user requirements are complex and hard to get accurately and completely. To handle both these challenges, this paper proposes a tree matching-based health insurance plan recommender system for workforce. It models the plans and user requirement as plan trees and user requirement trees, and develops user requirement tree construction method and a tree matching based recommendation method.

1. Introduction

The increasing development of information and communication technologies nowadays has brought us into the "big data" era [1]. In the health domain, big data, such as electronic health records, insurance records, pharmacy prescriptions, patient feedback, and responses, have also been accumulated recently [2, 3]. This provides great opportunities to develop various health information services [4, 5]. At the same time, the big volume and complexity of the health domain data brings the information overload problem [6]. Taking health insurance as an example, various insurance companies provide a large number of insurance plans with different prices, coverage benefits and quality, which are difficult to compare and select. Moreover, as the insurance plans are very complex, it is hard for users to specify their requirements clearly and accurately, which makes the plan selection harder for a particular user. This necessitates the personalized recommendation services in the health information service domain [7, 8]. A recommender system which aims to provide users with personalized product/service recommendations is helpful to handle the increasing information overload problem [9]. In this paper, we focus on developing a workforce health insurance plan recommender system, which is an important issue in health application domain.

There are two challenges to make proper health insurance recommendations to workforce of a business. 1) Hierarchical structured plan and user data. Various health insurance plans have different coverage aspects, and some coverage aspects have more detailed coverage items. It is difficult to use fixed vectors to represent the plan features, but to use more flexible hierarchical tree structures instead. It is not trivial to handle these complex hierarchical structured data. 2) Complex user requirement identification. Due to the complexity of health insurance plans, it is usually hard to get accurate and complete requirements from users.

To deal with the above challenges, a tree matching-based health insurance plan recommendation approach is proposed. In this approach, the health insurance plans, user profiles and user requirements are modelled as plan tree, user profile tree and user requirement tree respectively. A user requirement tree construction method is developed to infer user requirements from other similar users. A tree matching method is presented to evaluate the matching degree of plan tree to a user requirement tree, which is used to rank the plans and make recommendations. This paper has contributions on both the recommendation methods and practical applications.

The remainder of the paper is organized as follows. Section 2 presents the proposed tree matching-based health insurance plan recommendation approach in detail. Section 3 presents the design of the health insurance plan recommender system framework. Finally, the conclusion and further study are presented in Section 4.

2. Tree Matching-based Health Insurance Plan Recommendation Approach

This section presents the tree matching-based health insurance plan recommendation approach. It presents the user profile tree, plan tree and user requirement tree models, and their construction methods, and how to make plan recommendations by matching the plan trees to user requirement trees.

2.1. User profile tree

In the developed system, a user's (employee's) information consists of the following aspects: 1) the basic information, such as gender, age, position title and industry type; 2) health assessment records; 3) purchased health insurance plans; 4) explicitly expressed health insurance coverage requirements. For a specific user, not all these aspects of information are collected. Also, the health assessment questions for employees in different positions in different industries are probably different. Therefore, it is hard to represent the user data in a uniform vector. To model the user profile flexibly, a tree-structured data model, called user profile tree, is proposed. It is a rooted labelled tree. The root represents the user, and each sub tree represents one aspect of the profile features. Some features contain sub features, which are further represented by their sub trees. The labels of the internal nodes represent the semantic meaning of the feature and the label of the leaf node represents the value of relevant feature.

Taking the user profile tree T_p in Figure 1 as an example. It contains the user's basic information and health assessment records. The basic information includes "industry", "position", "gender" and "age", which are represented by nodes 2-5, and their values are represented by nodes 7-10. The health assessment records are represented by the sub tree under node 6.

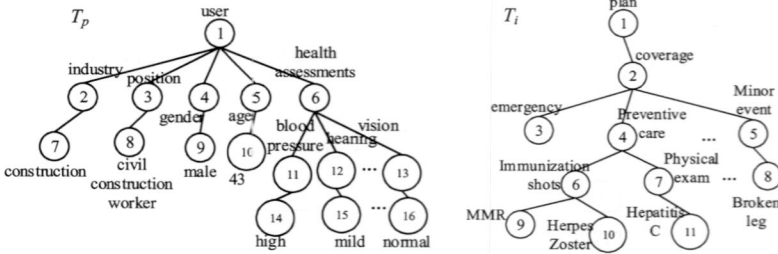

Figure 1. A user profile tree example T_p and a health insurance plan tree example T_i.

In our previous research, we have developed the similarity measure methods between tree structured data [10-12]. By use of the tree similarity measure, the similarity between two user profile trees T_{p1} and T_{p2}, denoted as $sim(T_{p1}, T_{p2})$, can be calculated. The similarity measure compares not only the feature structure of two trees but also their feature values, which evaluates the similarity between two users comprehensively. When computing the similarity, weights of sub trees can also be assigned by domain experts. Based on the similarity measure, the similar users of a target user can be selected.

2.2. Health insurance plan tree

The structures of various health insurance plans from different insurance companies are heterogeneous as these health insurance plans have different coverage or other benefit features. To flexibly represent the insurance plans, a tree structured model, called plan tree, is applied. It is a rooted labelled tree. The root represents a health insurance plan, and each sub tree represents one aspect of the plan features. The labels of the nodes represent the semantic meanings of the features.

Taking the plan tree T_i in Figure 1 as an example. This health insurance plan has three kinds of coverage: "preventive care", "emergence" and "minor event". Some have detailed descriptions, which are represented by sub trees.

2.3. User requirement tree

To make proper health insurance plan recommendations, it is important to identify the user requirement. Information about user requirements can essentially be obtained in two different ways: intentionally and extensionally. The intentionally expressed requirements refer to specifications by the user of what they desire in the health insurance plans under consideration. The extensionally expressed requirements refer to requirements that are inferred from other similar users. In this paper, the user requirement model covers both kinds of information.

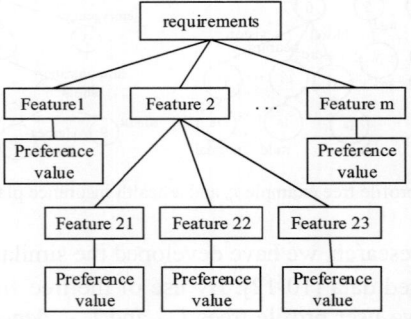

Figure 2. Intentionally expressed user preference.

The intentionally expressed preference is acquired directly from users. Because the features of the health insurance plans present tree, the requirements given by users should also be in tree structures, as shown in Figure 2. The requirement tree is a rooted labelled tree. The root represents a user's requirement, and each sub tree represents the user's requirement for an aspect of the features. The labels of the internal nodes represent the semantic meaning of the features. The labels of leaf nodes represent the desirability degree which is within [0, 1]. To express requirements, a user selects several features. For example, *Feature* 1, *Feature* 2, ..., *Feature m* are selected in Fig. 2. For each feature, there are two situations. First, the user can assign a preference value, such as *Feature* 1 in Fig. 2. Second, the user can drill down to detail and express preferences for finer features under the macro feature, as shown for *Feature* 2. Therefore, users' preference values can be expressed at different levels. In this study, the intentionally expressed preference focus on the health insurance coverage requirements.

In many situations, it is difficult to get the requirements from the user directly. It is more practical to infer the user requirements from other information like their health status and positions, which is easily obtained. It is assumed that similar users have similar requirements and will purchase similar health insurance plans. In the developed system, the users' health assessment data are collected from assessment centers, and some users' purchased insurance plan data are also accumulated. Suppose there are m users who have held a health insurance. Let their profile trees be $T_{p1}, T_{p2}, ..., T_{pm}$, and their health insurance plan tress be $T_{i1}, T_{i2}, ..., T_{im}$ respectively. Let the user profile tree of the target user be T_{pt}. We will use the similar users' health insurance plans to infer the target user's requirements. It takes the following three steps:

Step 1: calculate the user profile similarity between T_{pt} and T_{pj}, $1 \leq j \leq m$

By use of the similarity measure we have developed in [10], the similarity between user profile trees T_{pt} and T_{pj}, $1 \leq j \leq m$, are calculated, which are denoted as $sim(T_{pt}, T_{pj})$.

Step 2: select the K most similar users

Based on the similarity measures, the K most similar users to the target user will be selected, which are denoted as $T_{pj_1}, T_{pj_2}, ..., T_{pj_K}$. Their corresponding similarities are $sim_{pj_1}, sim_{pj_2}, ..., sim_{pj_K}$. Their health insurance plan trees are denoted as $T_{ij_1}, T_{ij_2}, ..., T_{ij_K}$ respectively.

Step 3: merge the features of $T_{ij_1}, T_{ij_2}, ..., T_{ij_K}$

The features of the insurance plans purchased by the users can reflect their requirements. Therefore, the user requirement tree can be constructed by merging these plan trees. In our previous research [10], we have developed a preference tree merging algorithm. By use of the tree merging algorithm, all the features included in the plan trees will be extracted and added into the requirement tree. The preference value to a specific feature is computed by aggregating the preferences to the plans that include the features. Here, the preference to a plan T_{ij_k} of the target user is reflected by the similarity value sim_{pj_k}.

Take the two plan trees T_{i1} and T_{i2} in Figure 3 as an example. Suppose the similarities between the target user and the users who purchased T_{i1} and T_{i2} are 0.8 and 0.6 respectively. By merging T_{i1} and T_{i2}, the requirement tree is obtained, which is shown as T_r in Figure 3. T_r contains all the features of T_{i1} and T_{i2}. The preference values for the features are also obtained from the user similarity. For some features which are only included in one tree, such as sub trees labelled as "emergence" and "Minor event", the preference values are assigned as the user similarity. For the features that are contained in both plan trees, the preference values are computed by aggregating the similarity values. In this example, the $max(\cdot)$ operation is applied. Therefore, the preference values for nodes labelled "MMR" and "Herpes Zoster" are both computed as $max(0.8, 0.6) = 0.8$.

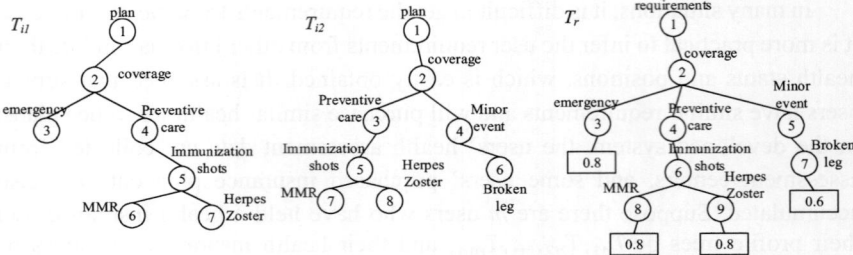

Figure 3. An example of merging two plan trees.

2.4. Tree matching-based recommendation

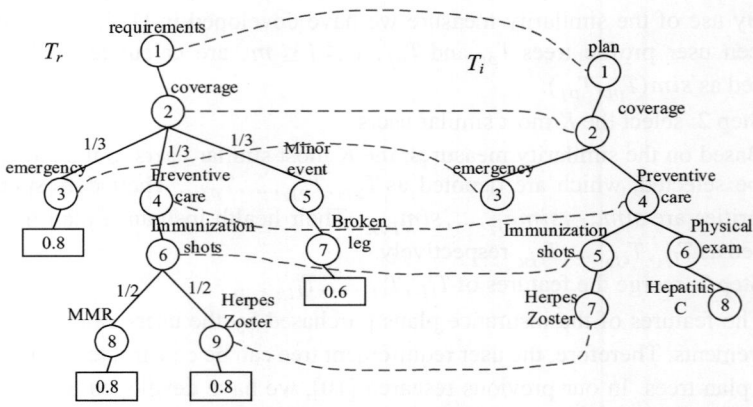

Figure 4. An example of matching a requirement tree and a plan tree.

Given a user requirement tree T_r and a health insurance plan tree T_i, the matching degree of the plan tree to the user requirement tree can be evaluated. We have developed a matching algorithm for tree structured data in [10]. By use of the tree matching algorithm, a maximum conceptual similarity tree mapping [10] can be constructed. The corresponding features of T_r and T_i can be identified. Let the matched part of T_r be T_{rm}. The matching degree of the plan to the user requirement is calculated by aggregating the preference values of the features in the matched sub tree. Let a node in the matched tree be denoted as tn, the child node set of tn be C_{tn}. The aggregated preference value of the sub tree rooted at tn is computed as:

$$pr(tn) = \begin{cases} pv(tn), & C_{tn} == null \\ \sum_{tn_i \in C_{tn}} w(tn_i) \cdot pr(tn_i), & C_{tn} \neq null \end{cases} \quad (1)$$

where $pv(tn)$ represents the preference value of node tn, $w(tn_i)$ represents the weight of node tn_i. The matching degree of T_i to T_r is then calculated by $pr(root(T_{rm}))$.

Take the user requirement tree T_r and health insurance plan tree T_i in Figure 4 as an example. Suppose that each sub tree in T_r is equally weighted. By use of the tree matching algorithm, a maximum conceptual similarity tree mapping is constructed, and the mapped nodes in the mapping are connected by the dashed lines. The matching degree of T_i to T_r is calculated by use of Formula (1) as $0.8 \times (1/3) + 0.8 \times (1/2) \times (1/3) = 0.4$.

To make recommendations, the matching degrees of all the potential plans to the user requirement tree are computed, and these potential plans are then ranked according to these matching degrees. The top N plans will be recommended.

3. A Health Insurance Plan Recommender System Framework

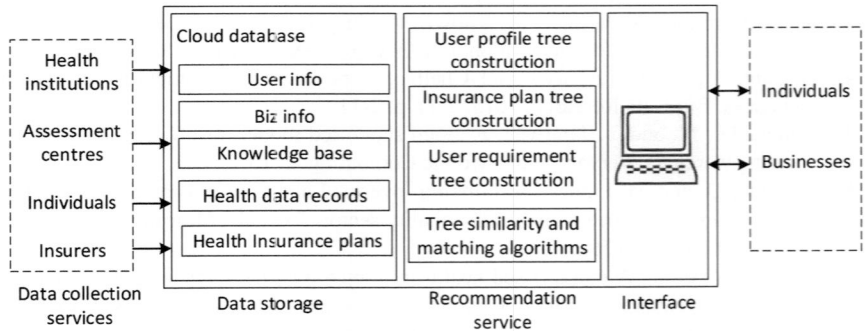

Figure 5. The framework of the health insurance plan recommender system.

This section presents the design of a health insurance plan recommender system based on the proposed recommendation approach in last section. The framework is shown in Figure 5. It consists of the following components: data collection services, data storage, recommendation services and user interface. The data collection services collect user information and health data records from individuals, health institutions and assessment centres, and collect health insurance plan data from insurance companies. The data storage component stores all the data in the system in a cloud database. Recommendation services implement the proposed tree matching based health insurance plan recommendation approach.

4. Conclusion and Further Study

This paper develops a tree matching-based workforce health insurance plan recommender system to help workforce users identify their health insurance requirements and select proper insurance plans effectively and efficiently. In the next step, the proposed method and framework will be applied and implemented in a real-world health information system. In the proposed system, the users' health data are from several different sources. How to guarantee the accuracy and consistency of the data is a big issue, which will be studied in the further study.

References

1. Li, Y. and Y. Guo, Wiki-Health: from quantified self to self-understanding. Future Generation Computer Systems, 2016. 56: p. 333-359.
2. Feldman, B., E.M. Martin, and T. Skotnes, Big Data in Healthcare Hype and Hope. October 2012. Dr. Bonnie, 2012. 360.
3. Wu, D., et al. A cloud-based comprehensive health inforamtion system framework. in Uncertainty Modelling in Knowledge Engineering and Decision Making: Proceedings of the 12th International FLINS Conference. 2016. World Scientific.
4. Huang, T., et al., Promises and Challenges of Big Data Computing in Health Sciences. Big Data Research, 2015. 2(1): p. 2-11.
5. Quwaider, M. and Y. Jararweh, A cloud supported model for efficient community health awareness. Pervasive and Mobile Computing, 2015.
6. Saxena, D. and M. Lamest, Information overload and coping strategies in the big data context: Evidence from the hospitality sector. Journal of Information Science, 2018: p. 0165551517693712.
7. Abbas, A., et al., A cloud based health insurance plan recommendation system: A user centered approach. Future Generation Computer Systems, 2015. 43: p. 99-109.
8. Elsweiler, D., et al., Second Workshop on Health Recommender Systems: (HealthRecSys 2017), in Proceedings of the Eleventh ACM Conference on Recommender Systems. 2017, ACM: Como, Italy. p. 374-375.
9. Lu, J., et al., Recommender system application developments: A survey. Decision Support Systems, 2015. 74: p. 12-32.
10. Wu, D., G. Zhang, and J. Lu, A fuzzy preference tree-based recommender system for personalized business-to-business e-services. IEEE Transactions on Fuzzy Systems, 2015. 23(1): p. 29-43.
11. Wu, D., J. Lu, and G. Zhang, A fuzzy tree matching-based personalized e-learning recommender system. IEEE Transactions on Fuzzy Systems, 2015. 23(6): p. 2412-2426.
12. Wu, D., J. Lu, and G. Zhang, Similarity measure models and algorithms for hierarchical cases. Expert Systems with Applications, 2011. **38**(12): p. 15049-15056.

New product development using disjunctive belief rule base[*]

Jianguo Xu, Mengjun Li, Leilei Chang[†] and JiangJiang

School of Systems Engineering, National University of Defense Technology
410073 Changsha, Hunan, China

Yuwang Chen

Alliance Manchester Business School, The University of Manchester
M139PL Manchester, England, UK

Longhao Yong

Decision Sciences Institute, Fuzhou University
350002 Fuzhou, Fujian, China

The new product development (NPD) problem is of great importance for company profits, even its survival. Generally, there are multiple factors to assess the popularity of the product. The main challenge of the problem lays in that the modeling and decision making process must be open for human involvement considering these factors simultaneously. In another word, experts and decision makers must be able to understand it so that (1) the knowledge and experience of experts and decision makers can be integrated, (2) the assessment results can be accepted, and (3) certain adjustments on the initial product development plan can be made as a feedback. In this study, the Belief Rule Base (BRB) is applied to solve the product development problem. The BRB under the disjunctive assumption can help further downsize BRB as well as maintain a high modeling accuracy. Moreover, a disjunctive BRB parameter learning model and an optimization algorithm are applied as well. A lemonade product case is studied in a comparative fashion to validate the efficiency of the proposed approach.

1. Introduction

The new product development (NPD) problem [1]-[4] is vital since it directly determines whether or not a product can attract customers and help the company gain profits or even its survival in the market.

Traditionally, multiple factors should be taken into consideration when developing a new product. It is almost impossible to construct an analytical model to directly connect the factors with the likeness of each product by the

---

[*]This work is supported by NSFC under Grants 71601180 and 71671186.
[†]Leilei Chang is the corresponding author.

customers. Normally, experts and decision makers are invited to grate each new product [1]. However, this is not 100% accurate since people's judgment may vary due to uncontrollable reasons [2]. Furthermore, complete data-driven approaches could not directly apply as well because it is impossible to understand and involve in its modeling and decision making process.

To summarize, the NPD problem has the following challenges [3, 4]:

1. Experts knowledge and experience must be taken into consideration as well as historic data. There could be multiple forms that the two types of information being taken into consideration.
2. The modeling and decision making process must be accessible. People must be involved in the process. Experts' knowledge must be integrated. Technicians and decision makers must be able to understand so that they can make sound decisions.
3. Although NPD is a complex problem, the model should not be too complex to understand. This is undoubtedly a dilemma.

BRB is an experts system which uses rule as a means to represent, integrate and further inference based on the knowledge and experience gathered from experts, technicians and decision makers. It has the ability of modeling multiple types of information under uncertainty. However, conventional BRB are constructed under the conjunctive assumption, which has to face the combinatorial explosion problem when there are over-numbered attributes and/or referenced values for the attributes. Therefore, a disjunctive BRB is used in this study to solve this problem. Moreover, the disjunctive BRB still inherits the advantages of the conjunctive BRB which can still be used to model the complex systems. Furthermore, an optimization model is introduced to improve the modeling accuracy of the disjunctive BRB.

2. New Product Development Using Disjunctive Belief Rule Base

2.1. *The new product development problem*

To determine whether a new product can occupy the market share, there are massive data of its predecessor products which was already in the market and their likings can be used as historic data for the NPD problem [3]. Furthermore, experts' knowledge, including technicians, workers and decision makers, can help fill up the gap when there is no quantitative data and/or no means to represent such information [4].

Therefore, to put a new product into the market, three types of prior information are existent: (1) multiple factors, (2) historic data of predecessor products of the same type, and (3) experts' knowledge on this type of products. The NPD problem is "a process of knowledge creation through the syndication of diverse streams of knowledge". So, the question is how to integrate multiple influential factors into the unified NPD problem assessment result as well as taking experts' knowledge/experience under consideration. Moreover, there would not be a lot of prior information which means that the limited information must be utilized.

2.2. Basics of disjunctive BRB

BRB is comprised of multiple disjunctive belief rules in the same belief structure [5]-[8]. The kth rule in a BRB system is described as:

$$R_k : \text{if} \quad \left(x_1 \text{ is } A_1^k\right) \vee \left(x_2 \text{ is } A_2^k\right) \vee \cdots \vee \left(x_M \text{ is } A_M^k\right),$$
$$\text{then} \quad \{(D_1, \beta_{1,k}), \cdots, (D_N, \beta_{N,k})\} \quad (1)$$
$$\text{with rule weight } \theta_k$$

where $x_m (m=1,\cdots M)$ denotes the mth attribute, $A_m^k (m=1,\cdots M; k=1,\cdots K)$ denotes the referenced values of the mth attribute, M denotes the number of the attributes, $\beta_{n,k} (n=1,\cdots N)$ denotes the belief for the nth degree, D_n, N denotes the number of the degrees. "$\vee$" denotes that a rule described in Eq. (1) follows the disjunctive assumption. The factors in the NPD problem are attributes of rules as in Eq. (1) and the assessment result is the conclusion of rules. The rules in BRB can be constructed complete using historic and/or present data gathering by records, sensors, questionnaires, etc. That is to say, the rules can be derived either from human knowledge or as an interpretation from historic/present information.

2.3. Disjunctive BRB parameter learning model for NPD problem

The principle of the parameters optimization model for conjunctive BRB is still applicable for the disjunctive BRB parameter optimization [6, 7].

$$\min MSE(A, \theta, \beta) \quad (2)$$

s.t.

$$lb_i \leq A_k^i \leq ub_i \quad (2a)$$
$$A_p^i = lb_i \quad (2b)$$
$$A_q^i = ub_i \quad (2c)$$
$$0 < \theta_k \leq 1 \quad (2d)$$
$$0 \leq \beta_{n,k} \leq 1 \quad (2e)$$

$$\sum_{n=1}^{N} \beta_{n,k} \leq 1 \quad (2f)$$

where, $k = 1, \cdots, K; i = 1, \cdots, M; n = 1, \cdots, N; p \neq q, p, q \in [1, \cdots, K]$.

Note that in 2(b/c), the two restraints are further relaxed than those in conjunctive BRB parameter optimization model: no need for the first and last rules to be comprised of the lower and upper bounds of the attributes. Although, in practical conditions, it is more rational and effective for people to derive two rules in the harshest and most relaxed conditions: two rules with the upper- and lower- bounds of the attributes are easier for people to understand and accept. Therefore, it is still recommended rather than compulsorily required to do so.

The optimization algorithm for the conjunctive BRB [7, 8] with Evolutionary Algorithms as the optimization engine as follows is applicable as well.

Step 3.1: Parameters initialization.

Step 3.2: Operations. Details of the optimization operations are determined by the specific evolutionary algorithms.

Step 3.3: Fitness calculation.

Step 3.3.1: BRB inference, including rule activation, matching degree activation and weight calculation [5]-[7].

Step 3.3.2: ER inference.

The ER algorithm is applied in Eqs (3)-(4) to integrate the activated L rules 8.

$$\beta_n = \frac{\mu[\prod_{k=1}^{L}(w_k \beta_{n,k} + 1 - w_k \sum_{n=1}^{N} \beta_{n,k}) - \prod_{k=1}^{L}(1 - w_k \sum_{n=1}^{N} \beta_{n,k})]}{1 - \mu[\prod_{k=1}^{L}(1 - w_k)]} \quad (3)$$

$$\mu = [\sum_{n=1}^{N} \prod_{k=1}^{L}(w_k \beta_{n,k} + 1 - w_k \sum_{s=1}^{N} \beta_{s,k}) - (N-1)\prod_{k=1}^{L}(1 - w_k \sum_{s=1}^{N} \beta_{s,k})]^{-1} \quad (4)$$

where β_n denotes the belief for the nth degree.

Step 3.3.3: Output by utility.

The integrated utility can be calculated by Eq. (5)

$$T = \sum_{n=1}^{N} U(D_n) \beta_n \quad (5)$$

where $U(D_n)$ denotes the utility of the nth degee D_n.

Step 3.4: Selection

Step 3.5: Stop check. If the stop criterion is not met, go to Step 3; if met, stop the optimization process.

3. Case Study

3.1. *Backgound*

A lemonade drink product development problem [5] is studied in a comparative fashion.

In this case, a total of 27 sets of data are gathered in which each set of data is with 12 influential factors and an average rating is given in [5]. The first 26 sets of data are he popular products in the market while the last one is the new product of the company which is about to be brought to the market. So the problem is to give the quality assessment result for the new product (the 27th product) by inferring from the 26 popular products in the market.

Upon investigating on 12 attributes, a hierarchy model is constructed in [5] and as shown in Fig. 1.

Firstly, the quality assessment of the lemonade products of divided into two sub-factors, namely taste and aroma. Then, three factors are categorized under taste and the rest nine factors are categorized under aroma. Furthermore, the nine aroma-related factors are transformed into three principles using Principle Component Analysis (PCA). Last, multiple sub-BRBs are constructed instead of a complete BRB with over-numbered attributes which would brought the combinatorial explosion problem.

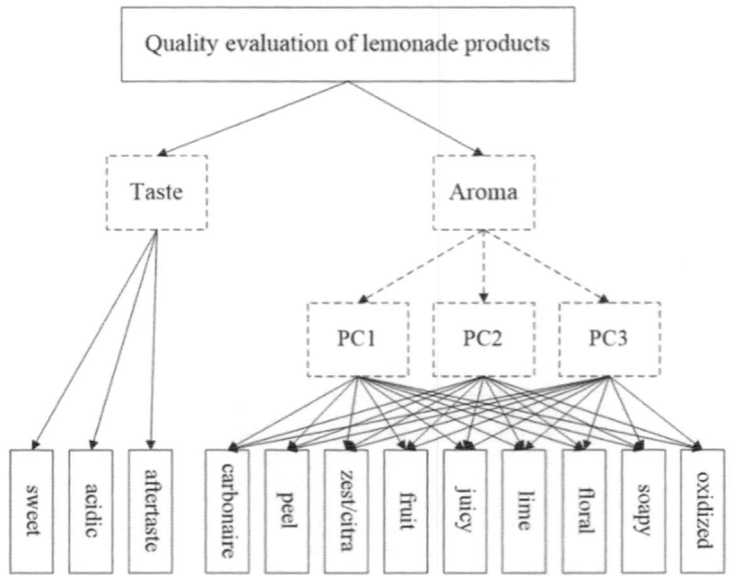

Figure 1. Conjunctive BRB for the lemonade drink product development.

However, it brings another question on the rationality of constructing such a model with multiple sub-BRBs instead of a complete one. Who should determine the hierarchy of the model for sub-BRB construction? How should the factors be categorized? What are the meanings of the components? This study attempts to directly construct a disjunctive BRB instead constructing multiple sub-BRBs in a hierarchy.

3.2. NPD assessment using disjunctive BRB

Using the proposed disjunctive BRB modeling approach as in Section II, ten experts are invited to produce the initial disjunctive BRB based on their knowledge as well as their read on the 26 sets of data. Furthermore, the 26 sets of data are used as the training data for the initial model and the 27th set of data is used as the testing data. 12 factors are disjunctively combined to inference its quality. Table 1 shows the optimized model with 5 rules, respectively.

Table 1. Disjunctive BRB with five rules.

No.	weights	attributes							
		sweet	acidic	aftertaste	carbonation	peely	zest	fruity	juicy
1	0.26	18.21	29.59	6.02	24.90	13.52	11.30	10.69	4.99
2	0.08	34.15	50.99	15.48	30.13	36.28	33.06	12.68	12.46
3	0.15	23.28	48.18	30.25	33.88	44.30	17.98	32.92	26.31
4	0.87	34.14	35.21	33.23	30.06	29.96	12.36	12.58	26.08
5	0.81	74.70	65.33	37.95	50.87	44.31	65.19	50.74	51.94

No.	attributes				conclusion				
	limey	floral	soapy	oxidized	1	2	3	4	5
1	7.23	6.33	9.29	10.67	0.99	0.00	0.00	0.00	0.00
2	13.26	37.06	34.16	25.81	0.06	0.78	0.01	0.13	0.00
3	13.25	19.28	15.32	45.86	0.00	0.00	0.00	0.00	0.99
4	44.38	37.08	44.53	17.33	0.99	0.00	0.00	0.00	0.00
5	45.63	43.57	49.80	47.29	0.07	0.00	0.00	0.00	0.92

Note: The utilities are 2/3.2/4.4/5.6/6.8 for 1/2/3/4/5 of the scales in the conclusion part, respectively.

3.3. Discussion

Table 2 shows the MSEs for the testing dataset by different approaches.

Table 2. MSEs by different approaches.

	method	accuracy
1	Multiple linear regression	87.09%
2	Neural network	97.46%
3	Conjunctive BRB (63 rules)	96.32%
4	Disjunctive BRB (3 rules)	87.32%
5	Disjunctive BRB (4 rules)	96.80%
6	Disjunctive BRB (5 rules)	98.13%

As in Table 2, it can be reduced to 3-5 rules by employing the disjunctive BRB rather than the conjunctive BRB. Moreover, the modeling accuracy can be improved as well.

Originally, the construction of a two-layer model and using PCA to transforming 9 factors into three principle components as "aroma" is to avoid the combinatorial explosion problem by adding a "hidden layer" which is comprised of two "artificial factors", namely "taste" (comprised of 3 direct factors) and "aroma" (comprised of 3 principle components and 9 direct factors). In the disjunctive BRB, it is not needs to add an additional "hidden layer". However, certain information may be lost due to the use of PCA. And by adding an additional layer and dividing the 12 direct factors into and "aroma", it may also affect the modeling ability of BRB.

Nevertheless, by applying the disjunctive BRB, the 12 factors can be directly applied, which may be the reason why the disjunctive BRB has shown superior modeling ability compared with that by the conjunctive BRB.

As for other approaches, as shown in Table 2 the Multiple Linear Regression (MLR) and the neural network (NN) [9] [10] have shown relative inferior performance. The reason is that, as pointed in [5], the nonlinearity modeling ability of MLP is rather unstable, and NN is essentially a black box and always suffers from the over-fitting problem.

As for the number of rules in BRB, it directly affects the modeling ability of the disjunctive BRB (so does the conjunctive BRB). From 3 to 4 and further 5 rules, the modeling accuracy can be improved by 87.32% to 96.80% and further 98.13%, which outperforms the neural network which is 97.46%. However, another factor also be taken into consideration, the modeling complexity. As the number of rules increased, the modeling complexity also increases. Therefore, it calls for a balance between the modeling complexity and the modeling accuracy. This is a very important topic for either conjunctive or disjunctive BRB modeling, which requires more work in the future.

4. Conclusion

The new product development problem has drawn attentions form multiple technicians, researchers, experts, decision makers in industries, academics, economics, etc. However, it has several characteristics that bring great challenges, e.g., experts' knowledge and experience must be taken into consideration as well as historic data, the modeling and decision making process should be open and accessible, and so on.

In this study, the disjunctive BRB is applied to solve this challenge. The BRB can help integrate multiple types of information including qualitative experts' knowledge and quantitative historic data. Moreover, the disjunctive assumption can help avoid the combinatorial explosion problem which is faced by the conventional conjunctive BRB.

Moreover, a parameter optimization model as well as an optimization algorithm is proposed to further improve the modeling accuracy.

Case study results show that, in comparison with other approaches and the conjunctive BRB, the disjunctive BRB not only can help reduce the size of BRB (from 63 rules of 3 sub-BRBs to 3/4/5 rules with one BRB) but also help improving the modeling accuracy (from 87.32%, 96.32% an 97.46% to 96.80/98.13%).

For future researches, more theoretical and practical cases using the disjunctive BRB should be studied to further validate the efficiency of the disjunctive BRB, as well as modifications to improve its modeling ability.

References

1. R. Madhavan and R. Grover, *Journal of Marketing* **62(4)**, 1-12 (1998).
2. E. Z. Yurtkulu, P. Hilletofth and G. Johansson, *EurOMA*, 2014.
3. B. Heyd and M. Danzat, *Lebensmittel-Wissenschaft und-Technologie* **31**, 607-611(1998).
4. S. Roper, P. Micheli, J. H. Love and P. Vahter, *Research Policy* **45(1)**, 319-329 (2016).
5. J. B. Yang, J. Liu, J. Wang, H. S. Sii, and H. W. Wang, *IEEE Transactions on Systems Man & Cybernetics Part A Systems & Humans* **36(2)**, 266-285(2006).
6. J. B. Yang and D. L. Xu, *IEEE Transactions on Systems Man & Cybernetics Part A Systems & Humans* **32(3)**, 376-393(2002).
7. J. B. Yang, Y. M. Wang, D. L. Xu, K. S. Chin and L. Chatton, *Expert Systems with Applications* **39(5)**, 4749-4759(2012).
8. Y. M. Wang, J. B. Yang, and D. L. Xu, *European Journal of Operational Research* **174(3)**, 1885-1913(2006).
9. G. S. El-taweel and A. K. Helmy, *Image Processing Iet* **7(5)**, 407-414(2013).
10. G. A. Carpenter, S. Martens and O. J. Ogas, *Neural Networks* **18(3)**, 287-295(2005).

Restricted multi-pruning of decision trees

Mohammad Azad,* Igor Chikalov and Mikhail Moshkov

*Computer, Electrical and Mathematical Sciences & Engineering Division
King Abdullah University of Science and Technology, Thuwal, Saudi Arabia
{*mohammad.azad, mikhail.moshkov}@kaust.edu.sa, igor.chikalov@gmail.com*

Shahid Hussain

*School of Science and Engineering, Habib University, Karachi, Pakistan
shahid.hussain@sse.habib.edu.pk*

The trade-off between the decision tree size and good classification accuracy is a research challenge. It can be achieved if we create multiple pruned trees from the set of Pareto optimal points using dynamic programming approach (multi-pruning process). However, this process can be extensively slow. We consider a modification of the multi-pruning process (restricted multi-pruning) that requires less memory and time but usually keeps the accuracy of the constructed classifiers.

Keywords: Decision trees; Pareto optimal points; dynamic programming.

1. Introduction

Decision trees serve as a basis for different kinds of classifier ensembles. The question how to construct decision trees with better predictive performance is of a great interest.[1,2] Previously, we have developed an approach (multi-pruning[3]) to analyze decision tables (datasets) and construct decision trees which can be used for classification and knowledge representation. Our approach applies to decision tables containing both categorical and numerical conditional attributes. It is based on an extension of dynamic programming and requires the construction of a directed acyclic graph (DAG) such that the nodes of this graph are subtables of the training subtable of the initial decision table given by restrictions on attribute values. This DAG describes a large set of CART(Classification and Regression Trees)-like decision trees.[3] Such trees use binary splits created on the base of conditional attributes. In contrast with standard CART,[1] which use the best splits among all attributes, CART-like trees use, additionally, the best splits for each attribute.

Extended dynamic programming approach allows us to describe, for this set of trees, the set of Pareto optimal points for bi-criteria optimization problem relative to the number of nodes and the number of misclassifications. We randomly derived some decision trees (five in our experiments) for each Pareto optimal point and found, based on the validation subtable, a decision tree with a minimum number of misclassifications among all derived trees. We evaluated the accuracy of prediction for this tree using test subtable. This process is similar to the usual pruning of a decision tree but here it is applied to many decision trees since the set of CART-like decision trees is closed under the operation of usual bottom-up pruning. We called this process multi-pruning. The classifiers constructed by the multi-pruning process often have better accuracy than the classifiers built by CART.

To make the dynamic programming approach more scalable we consider in this paper restricted multi-pruning procedure which deals with restricted CART-like decision trees: we use in each node of DAG only the best splits for a small number of attributes, instead of using the best splits for all attributes. The obtained DAGs with limited branching factor contain fewer nodes and edges and require less time for the construction and processing. However, the prediction accuracy of decision trees constructed by the restricted multi-pruning procedure is comparable to the case when we use the best splits for all attributes. For example, we did experiments with 15 decision tables from the UCI ML Repository.[4] For 11 tables, the trees constructed by the multi-pruning procedure which uses best splits for all attributes outperform the trees built by CART. The same situation is with decision trees constructed by the restricted multi-pruning procedure which uses only best splits for two attributes.

2. (m_1, m_2)-CART-like Decision Trees

We restrict the notion of CART-like decision trees[3] to the (m_1, m_2)-CART-like decision trees as follows. Let T be a decision table with n conditional attributes $f_1, \ldots, f_n$ and the decision attribute d. Let m_1 and m_2 be nonnegative integers such that $0 < m_1 + m_2 \leq n$. We now describe a way for construction of a set $S_{m_1,m_2}(\Theta)$ of admissible splits for a subtable Θ of T. Let $E(\Theta)$ be the set of all conditional attributes which are not constant on Θ, and $|E(\Theta)| = p$. For each attribute $f_i \in E(\Theta)$, we find a best split for Θ and the attribute f_i in the same way as in CART[1] (we consider binary splits that minimize the value of impurity function based on Gini index).

Let $s_1, \ldots, s_p$ be the obtained splits in order from best to worst. If $m_1 \geq p$ then $S_{m_1,m_2}(\Theta) = \{s_1, \ldots, s_p\}$. Let $m_1 < p$. Then $S_{m_1,m_2}(\Theta)$ contains splits $s_1, \ldots, s_{m_1}$ and $\min(p - m_1, m_2)$ splits randomly chosen from the set $\{s_{m_1+1}, \ldots, s_p\}$.

We consider (m_1, m_2)-CART-like decision trees for T in which each leaf is labeled with a decision (a value of the decision attribute d), each internal node is labeled with a binary split corresponding to one of the conditional attributes, and two outgoing edges from this node are labeled with 0 and 1, respectively. We correspond to each node v of a decision tree Γ a subtable $T(\Gamma, v)$ of T that contains all rows of T for which the computation of Γ passes through the node v. We assume that, for each internal node v, the subtable $T(\Gamma, v)$ contains rows labeled with different decisions, and the node v is labeled with a split from $S_{m_1,m_2}(T(\Gamma, v))$. We assume also that, for each leaf v, the node v is labeled with a most common decision for $T(\Gamma, v)$ (a decision which is attached to the maximum number of rows in $T(\Gamma, v)$).

3. Directed Acyclic Graph $G_{m_1,m_2}(T)$

We consider a directed acyclic graph (DAG) $G_{m_1,m_2}(T)$ which is used to describe the set of (m_1, m_2)-CART-like decision trees for T and to study this set. Nodes of the graph $G_{m_1,m_2}(T)$ are some subtables of the table T. We now describe an algorithm for the construction of the DAG $G_{m_1,m_2}(T)$.

Algorithm Construction of the DAG $G_{m_1,m_2}(T)$

Input: A decision table T with n conditional attributes, and non-negative integers m_1 and m_2 such that $0 < m_1 + m_2 \leq n$.
Output: The DAG $G_{m_1,m_2}(T)$

(1) Construct a graph which contains only one node T which is marked as not processed.
(2) If all nodes of the graph are processed then return it as $G_{m_1,m_2}(T)$ and finish. Otherwise, choose a node (subtable) Θ which is not processed yet.
(3) If all rows of Θ are labeled with the same decision, mark Θ as processed and proceed to step 2.
(4) Otherwise, construct set $S_{m_1,m_2}(\Theta)$ of admissible splits for Θ and, for each split s from $S_{m_1,m_2}(\Theta)$, draw two edges from Θ to subtables $\Theta_{s=0}$ and $\Theta_{s=1}$ (for $\delta = 0, 1$, the subtable $\Theta_{s=\delta}$ contains all rows from Θ for which the value of s is equal to δ), and label these edges with $s = 0$ and $s = 1$, respectively (this pair of edges is called an s-pair). If some

of the subtables $\Theta_{s=0}$ and $\Theta_{s=1}$ is not in the graph, add them to the graph. Mark Θ as processed and proceed to step 2. $\square$

4. Set of Pareto Optimal Points $POP_{m_1,m_2}(T)$

Let A be a finite set of points in two-dimensional Euclidean space. A point $(a,b) \in A$ is called a Pareto optimal point for A if there is no point $(c,d) \in A$ such that $(a,b) \neq (c,d)$, $c \leq a$, and $d \leq b$. We denote by $Par(A)$ the set of Pareto optimal points for A.

Let Γ be a decision tree from $DT_{m_1,m_2}(T)$ (the set of (m_1,m_2)-CART-like decision trees for T). We denote by $L(\Gamma)$ the number of nodes in Γ and by $mc(T,\Gamma)$ the number of misclassifications of Γ on rows of T. We correspond to each decision tree $\Gamma \in DT_{m_1,m_2}(T)$ the point $(mc(T,\Gamma), L(\Gamma))$. As a result, we obtain the set of points $\{(mc(T,\Gamma), L(\Gamma)) : \Gamma \in DT_{m_1,m_2}(T)\}$. Our aim is to construct, for this set, the set of all Pareto optimal points $POP_{m_1,m_2}(T) = Par(\{(mc(T,\Gamma), L(\Gamma)) : \Gamma \in DT_{m_1,m_2}(T)\})$.

We describe now an algorithm which attaches to each node Θ of the DAG $G_{m_1,m_2}(T)$ the set of Pareto optimal points.

Algorithm Construction of the set $POP_{m_1,m_2}(T)$

Input: The DAG $G_{m_1,m_2}(T)$ for a decision table T
Output: The set $POP_{m_1,m_2}(T)$

(1) If all nodes of $G_{m_1,m_2}(T)$ are processed then return the set $POP_{m_1,m_2}(T)$ attached to the node T and finish. Otherwise, choose a node Θ of $G_{m_1,m_2}(T)$ which is not processed yet and such that either all rows of Θ are labeled with the same decision or all children of Θ are already processed.

(2) If all nodes of Θ are labeled with the same decision then attach to Θ the set $POP_{m_1,m_2}(\Theta) = \{(0,1)\}$, mark Θ as processed, and proceed to step 1.

(3) If all children of Θ are already processed, and $S(\Theta)$ is the set of splits s such that an s-pair of edges starts in Θ then attach to Θ the set

$$POP_{m_1,m_2}(\Theta) = Par(\{(mc(\Theta, tree(\Theta)), 1)\}$$
$$\cup \bigcup_{s \in S(\Theta)} \{(a+c, b+d+1) : (a,b) \in POP_{m_1,m_2}(\Theta_{s=0}),$$
$$(c,d) \in POP_{m_1,m_2}(\Theta_{s=1})\}).$$

mark Θ as processed and proceed to Step 1. $\square$

For each subtable Θ and each Pareto optimal point from the set $POP_{m_1,m_2}(\Theta)$, we keep information about its construction which allows us to derive, for each point from $POP_{m_1,m_2}(T)$, decision trees with parameters corresponding to this point.

5. Restricted Multi-pruning

We divide the initial decision table T into three subtables: training subtable T_{train}, validation subtable T_{val}, and test subtable T_{test}. We construct the DAG $G_{m_1,m_2}(T_{\text{train}})$ and based on this DAG we construct the set of Pareto optimal points $POP_{m_1,m_2}(T_{\text{train}})$. For each point $(a, b) \in POP_{m_1,m_2}(T_{\text{train}})$, we derive randomly five decision trees $\Gamma_1, \ldots, \Gamma_5$ from $DT_{m_1,m_2}(T_{\text{train}})$ such that $(a, b) = (mc(T_{\text{train}}, \Gamma_i), L(\Gamma_i))$ for $i = 1, \ldots, 5$. Among these derived decision trees, we choose a decision tree Γ which has a minimum number of misclassifications on T_{val}. It will be used to get average misclassification error rate on T_{test}.

Table 1. Decision tables from the UCI ML Repository.

Data Set	Rows	Attributes	Type of Attributes
BALANCE-SCALE	625	5	Categorical
BANKNOTE	1372	5	Numerical
BREAST-CANCER	266	10	Categorical
CARS	1728	7	Categorical
GLASS	214	10	Numerical
HAYES-ROTH-DATA	69	5	Categorical
HOUSE-VOTES-84	279	17	Categorical
IRIS	150	5	Numerical
LYMPHOGRAPHY	148	19	Categorical
NURSERY	12960	9	Categorical
SOYBEAN-SMALL	47	36	Categorical
SPECT-TEST	169	23	Categorical
TIC-TAC-TOE	958	10	Categorical
WINE	178	13	Numerical
ZOO-DATA	59	17	Categorical

We did experiments with 15 decision tables from the UCI ML Repository[4] which were preprocessed since some of them had categorical conditional attributes with a unique value for each row. We removed such attributes. In some cases, there were identical rows, with possibly, different decisions. We replaced each group of such equal rows by a single row. We assign the most common decision for the group of rows as the decision of this single row. Similarly, missing values for an attribute were replaced with a most common value for that attribute. For each of 15 decision tables, Table 1 contains its name (Data Set), number of rows (Rows), number of conditional attributes (Attributes), and the type of attributes — categorical or numerical (Type of Attributes).

We compared quality of classifiers (decision trees) constructed by CART, quality of classifiers constructed by multi-pruning procedure ($m_1 = n$ and $m_2 = 0$ where n is the number of conditional attributes in the table), and quality of classifiers constructed by restricted multi-pruning procedure for all pairs (m_1, m_2) from the set $\{0, 1, 2, 3, 4, 5\}^2 \setminus \{(0, 0)\}$ (we will call such classifiers (m_1, m_2)-classifiers). We repeated 2-fold cross validation five times for each of the considered decision tables such that T_{train} contains 35% of rows, T_{val} contains 15% of rows, and T_{test} contains 50% of rows. After that we find average misclassification error rate of the constructed classifiers for each type of classifiers and for each of 15 decision tables. We also count average time to the construction and usage for each type of classifiers and each of 15 decision tables. The experiments were done on Intel Xeon CPU 2.4 GHz, 64 GB RAM, 64 bit Windows 7 OS.

Table 2. Average ranks of classifiers relative to the misclassification error rate.

$m_1 \backslash m_2$	0	1	2	3	4	5
0	N/A	31.87	24.80	23.80	17.47	17.33
1	21.63	21.57	22.23	17.40	15.30	16.27
2	**15.23**	15.90	16.83	18.23	19.43	15.93
3	**15.13**	15.40	17.43	18.10	15.57	16.43
4	17.00	17.07	19.00	17.70	16.87	17.03
5	15.90	15.83	**14.33**	16.77	16.13	17.07

Table 2 contains, the average ranks of (m_1, m_2)-classifiers relative to the average misclassification error rate for the considered 15 decision tables. For each decision table, we rank the classifiers based on their average misclassification error rate on this decision table, where we assign the best performing classifier the rank of 1, the second best rank 2, and so on. We break ties by computing the average of ranks. After that, we count, for each type of classifiers, the average rank among all 15 decision tables. The best three pairs are $(5, 2)$, $(3, 0)$, and $(2, 0)$. Table 3 contains, the average rank of (m_1, m_2)-classifiers relative to the average time for its construction and usage for the considered 15 decision tables. The best three pairs are $(3, 0)$, $(2, 0)$, and $(1, 0)$.

We will concentrate on the consideration of $(3, 0)$-classifiers and $(2, 0)$-classifiers only since $(5, 2)$-classifiers are too time-consuming, and

Table 3. Average ranks of classifiers relative to the time for construction and usage.

$m_1 \backslash m_2$	0	1	2	3	4	5
0	N/A	15.47	08.87	07.13	09.67	11.40
1	**04.53**	05.33	05.33	07.93	10.93	11.67
2	**04.33**	18.67	22.53	10.07	28.27	29.40
3	**03.40**	21.73	22.87	27.73	28.00	25.93
4	20.00	22.20	23.60	26.33	29.67	25.87
5	22.00	22.00	19.93	27.60	24.13	25.47

Table 4. Average misclassification error rates of classifiers.

Data Set	CART	MP	RMP	
			(3,0)	(2,0)
BALANCE-SCALE	23.81	23.68	24.70	25.83
BANKNOTE	3.20	1.91	1.96	2.14
BREAST-CANCER	29.17	29.32	29.77	28.12
CARS	5.16	5.08	5.03	4.86
GLASS	39.34	38.49	37.93	40.38
HAYES-ROTH-DATA	43.26	30.99	26.97	25.28
HOUSE-VOTES-84	6.81	6.67	6.60	7.03
IRIS	5.03	5.16	5.43	5.03
LYMPHOGRAPHY	27.97	24.86	28.92	24.86
NURSERY	1.40	1.44	1.39	1.28
SOYBEAN-SMALL	21.36	6.70	9.35	12.32
SPECT-TEST	5.21	4.74	4.74	4.74
TIC-TAC-TOE	10.21	7.81	5.97	7.52
WINE	12.47	11.80	11.35	11.57
ZOO-DATA	23.74	25.82	25.14	25.48
Average:	*17.21*	*14.97*	*15.02*	*15.10*

$(1,0)$-classifiers have bad misclassification error rate. Detailed results for $(3,0)$- and $(2,0)$-classifiers can be found in Tables 4 and 5 (note that we did not consider CART classifier when we compare the time because CART is a greedy algorithm which is usually essentially faster than others). One can see that multi-pruning classifiers outperform CART classifiers for 11 tables; restricted multi-pruning $(3,0)$-classifiers outperform CART classifiers for 10 tables; restricted multi-pruning $(2,0)$-classifiers outperform CART classifiers for 10 tables, and one table have the same average misclassification error rates. The average time of restricted multi-pruning for $(m_1, m_2) \in \{(3,0),(2,0)\}$ is almost five times less than the average time of multi-pruning.

Table 5. Average time in seconds for construction and usage of classifiers.

Data Set	MP	RMP	
		(3,0)	(2,0)
BALANCE-SCALE	149.19	60.94	76.90
BANKNOTE	66.30	13.31	17.39
BREAST-CANCER	114.20	24.46	34.35
CARS	194.69	71.83	100.09
GLASS	1387.40	24.41	29.93
HAYES-ROTH-DATA	23.54	9.49	13.71
HOUSE-VOTES-84	24.92	9.70	9.77
IRIS	12.48	7.41	4.41
LYMPHOGRAPHY	125.06	16.28	13.07
NURSERY	1399.62	419.75	362.96
SOYBEAN-SMALL	13.44	4.21	3.32
SPECT-TEST	30.04	7.34	6.92
TIC-TAC-TOE	324.67	68.67	54.73
WINE	204.31	6.47	6.45
ZOO-DATA	19.06	8.28	7.01
Average:	*272.59*	*50.17*	*49.40*

6. Conclusions

The obtained results show that our approach can be used in machine learning to improve predictive performance of individual trees. Ensembles based on such trees can have better accuracy of classification.

Acknowledgments

Research reported in this publication was supported by the King Abdullah University of Science and Technology (KAUST).

References

1. L. Breiman, J. H. Friedman, R. A. Olshen and C. J. Stone, *Classification and Regression Trees* (Wadsworth and Brooks, Monterey, CA, 1984).
2. T. Hastie, R. Tibshirani and J. H. Friedman, *The Elements of Statistical Learning: Data Mining, Inference, and Prediction* (Springer-Verlag, New York, 2001).
3. M. Azad, I. Chikalov, S. Hussain and M. Moshkov, Multi-pruning of decision trees for knowledge representation and classification, in *3rd IAPR Asian Conference on Pattern Recognition, ACPR 2015, Kuala Lumpur, Malaysia, November 3-6, 2015*, (IEEE, 2015).
4. M. Lichman, UCI Machine Learning Repository. University of California, Irvine, School of Information and Computer Sciences (2013).

Pipeline leak alarm decision via dynamic evidence fusion[*]

Xiaobin Xu,[1] Haiyang Xu,[1] Guo Li,[1] Pingzhi Hou,[3] Darong Huang[2] and Weifeng Liu[1]

[1]*School of Automation, Hangzhou Dianzi University, Hangzhou 100084, China*
[2]*School of Information Science and Engineering, Chongqing Jiaotong University Chongqing, 400074, China*
[3]*Hangzhou Yanshi Technologies Co. Ltd.
Hangzhou Shi, Zhejiang, China*

In order to solve pipeline leak detection, an alarm decision method is presented based on dynamic evidence fusion. Firstly, the flow difference between inlet and outlet is considered as an alarm variable. For measuring uncertainty of the alarm variable, it is transformed into an alarm evidence by using dual fuzzy thresholds. Secondly, the conditional evidence updating rule is used to dynamically fuse the current and historical evidence such that the fused evidence is more reliable for alarm decision. In fusion process, based on the similarity measure of evidence, a belief rule base (BRB) is designed to reason out the fusion weights of evidence. Finally, a liquefied petroleum gas (LPG) pipeline leak experiment demonstrates that the proposed method is flexible and superior for leak detection.

Keywords: Pipeline leak; alarm monitoring; belief rule base (BRB); evidence theory.

1. Introduction

Long distance pipelines are increasingly being adopted in various fields, such as petroleum transportation, complex chemical process, nuclear industry, agriculture. In the process of long-time running, the pipeline leaks are unavoidable and even frequent because of corrosion, geological disaster, equipment aging, and negligence in management and so on. Hence, it is important to study on the leak problem so as to decrease economic losses and environmental pollution. Leaks must be detected and located in time, at present, many methods and technologies have been proposed, such as mass/volume balance technique, static detection, negative pressure detection, transient model-based method, distributed optical fiber-based detection, acoustic waves-based detection [1–2]. Most of them are all

---

[*]This work was supported by the NSFC (No. 61433001, U1709215, 61573076, 61573275, 61771177), the University Students' Scientific and Technological Innovation Activity Plan of Zhejiang Province (Xin-Miao Talent Plan, No. 2017R407064), the Science and Technology Project of Zhejiang Province (No. 2018C04020, 2018C01031).

judge whether a leak happens or not by analyzing the extracted alarm variable in industrial processes. The most common practice for leak alarm decision is to compare the alarm variable with a given threshold namely a trip point, control limit [3–4].

For an operator to monitor an industrial plant, a satisfactory situation is to only give an alarm for any abnormal condition and take at least 10 minutes to deal with the alarm, namely, not receive more than 6 alarms per hour [5]. However, such requirement or standard is rarely met because of the noise, disturbance and uncertainty in industrial field. In practice, the received alarms usually include so many false or nuisance alarms which lead to low working efficiency. The American National Standards Institute has reported some classical alarm generation methods widely in industrial plants, and proposed three common indices "missed alarm rate (MAR)", "false alarm rate (FAR)" and "average alarm delay (AAD)" to measure the performance of alarm generation methods [6]. Complying with these indices, [3,7–9] studied on the optimal designs of the above classical methods under the constraint that the probabilistic properties of alarm variables are known. However, in practice, due to the uncertainties in the industrial equipment operation and state monitoring, it is difficult to ensure the excellent performance of alarm system.

Following our previous work on evidence updating fusion [4,10], we propose a data-driven alarm decision method based on dynamic evidence fusion without knowing the probabilistic statistical properties of alarm variables. Firstly, the flow difference between inlet and outlet is considered as an alarm variable. For measuring uncertainty of the alarm variable, it is transformed into an alarm evidence by using dual fuzzy thresholds. Secondly, the conditional evidence updating rule is used to dynamically fuse the current and historical evidence such that the fused evidence is more reliable for alarm decision. In fusing process, based on the similarity measure of evidence, a novel belief rule base (BRB) is designed to reason out the fusion weights of evidence. Finally, an actual industrial liquefied petroleum gas pipeline leak experiment demonstrates that the proposed method is flexible and superior for leak detection.

2. Case Study for LPG Pipeline Leakage

There is a typical industrial case about a LPG pipeline leak detection. The total distance of the LPG pipeline is more than 100 km, mass flow meters are installed at inlet and outlet for recording data with the sampling period $h = 10$ seconds. A leakage experiment is done and lasts 23.62 *hours*, $h = 10s$, $t = 1h, 2h, \ldots, 8505h$, such that the corresponding inlet flow $f_0(t)$ and outlet flow $f_1(t)$ can be collected as shown in Fig. 1. In the middle of the pipeline, a leak is simulated by opening a valve lasting 5.8 hours from $t = 122h$ to $t = 2209h$. In theory, the normal condition can be explained as $f_1(t) - f_0(t) = 0$, the leak (abnormal condition) as $f_1(t) - f_0(t) < 0$ hence we can let $x(t) = f_1(t) - f_0(t)$ as an alarm variable for representing running states. But unfortunately, Fig. 2 shows that $x_{tp} = 0$ is not optimal threshold at all since there are so many complex and uncertain factors leading to the irregular changes of fluid fluctuation. Obviously, these changes can hardly be modelled via the specific probability distributions about $x(t)$.

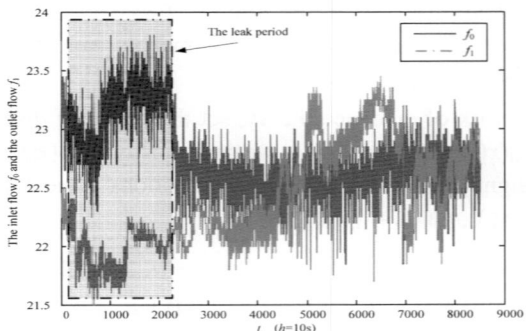

Fig. 1. The sampling data of $f_0(t)$ and $f_1(t)$.

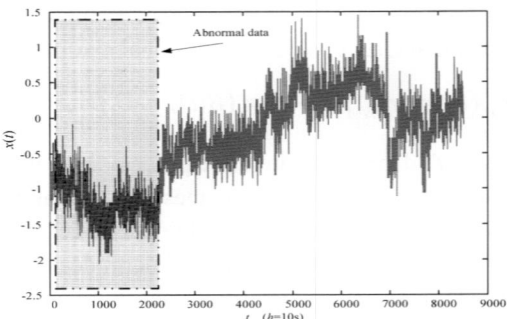

Fig. 2. The sampling data of $x(t)$.

3. Leak Alarm Decision via Dynamic Evidence Fusion and BRB Model

3.1. *Obtain alarm evidence from the alarm variable x*

Instead of the traditional absolute threshold x_{tp}, the corresponding relative dual fuzzy thresholds $\mu_A(x(t))$ and $\mu_{NA}(x(t))$ are designed to classifies $x(t)$ to two states "alarms(A)" and "no-alarms(NA)" respectively [4,11]

$$\mu_A(x(t)) = \begin{cases} 0, x(t) \leq x_{tp}^L \\ \dfrac{x(t)-x_{tp}^L}{x_{tp}^U-x_{tp}^L}, x_{tp}^L < x(t) < x_{tp}^U \\ 1, x(t) \geq x_{tp}^L \end{cases}, \quad \mu_{NA}(x(t)) = \begin{cases} 1, x(t) \leq x_{tp}^L \\ \dfrac{x(t)-x_{tp}^U}{x_{tp}^L-x_{tp}^U}, x_{tp}^L < x(t) < x_{tp}^U \\ 0, x(t) \geq x_{tp}^U \end{cases} \quad (1)$$

here $x_{tp}^L = x_{tp}(1-\zeta)$ and $x_{tp}^U = x_{tp}(1+\zeta)$ represent the lower and upper thresholds respectively with $\zeta \in \Re^+$. By plotting receiver operating characteristic (ROC) curve about FAR and MAR, the optimal x_{tp} can be determined when $G = \sqrt{FAR^2 + MAR^2}$ reaches to the minimum.

In the framework of evidence theory, set the frame of discernment as $\Theta = \{A, NA\}$. Since $\mu_A(x(t)) + \mu_{NA}(x(t)) = 1$, we can transform $\mu_A(x(t))$ and $\mu_{NA}(x(t))$ to a basic belief assignment (BBA) about the propositions A and NA, namely a piece of alarm evidence (X_t, m_t), $X_t = \{\{A\}, \{NA\}\}$

$$\begin{aligned} m_t(A) &= \mu_A(x(t)) / (\mu_A(x(t)) + \mu_{NA}(x(t))) \\ m_t(NA) &= \mu_{NA}(x(t)) / (\mu_A(x(t)) + \mu_{NA}(x(t))) \end{aligned} \quad (2)$$

3.2. Fuse alarm evidence using dynamic evidence updating rule

In this section, the linear updating rule of evidence is applied to fuse the current evidence and historical evidence to obtain the global alarm evidence $m_{0:t}$ [4,11]

$$m_{0:t}(B) = \tau_t m_{0:t-1}(B) + \upsilon_t m_t(B|D) \quad D, B = A, NA, \Theta \quad (3)$$

here, the subscript "$0:t$" denotes the iterative process from the initial step to the current step t, $t = 0h, 1h, 2h, \ldots$. $m_{0:t-1}(B)$ comes from the inertia of all historical alarm evidence. $m_t(B|D)$ is the current conditional belief mass, when $B = D$, $m_t(B|D) = 1$, otherwise $m_t(B|D) = 0$. The rule of alarm decision is if the $m_{0:t}(A) > m_{0:t}(NA)$, then the alarm system generates an alarm, otherwise, no alarm. The fusion weights υ_t and τ_t can be calculated by the following procedure. Firstly, Jousselme's distance between two pieces of evidence m_i and m_j is [12]

$$d_J(m_i, m_j) = \sqrt{\frac{1}{2}(\bar{m}_i - \bar{m}_j)^T \underline{\underline{D}} (\bar{m}_i - \bar{m}_j)} \quad (4)$$

d_J is a widely used measure for quantifying the similarity between BBAs [11]. Obviously, the similarity between m_i and m_j should decreases with the increase of $d_J(m_i, m_j)$, so the similarity can be directly defined as

$$Sim(m_i, m_j) = 1 - d_J(m_i, m_j) \quad (5)$$

Further, the support for the evidence m_i can be shown as

$$Sup_i = \sum_{j=1, j \neq i}^{N} Sim(m_i, m_j) \quad (6)$$

Based on Eqs. (4)–(6), Ref. [4] simply defines

$$\begin{cases} v_t = Sup_t \big/ (Sup_t + Sup_{0:t-1} + Sup_{0:t-2}) \\ \tau_t = 1 - v_t \end{cases} \quad (7)$$

Obviously, the linear model given in Eq. (7) is too simple to describing the complicated relationship between v_t and Sup_t, $Sup_{0:t-1}$, $Sup_{0:t-2}$ because the change of the alarm variable $x(t)$ or its m_t may be complex and even irregular as shown in Fig. 2. In order to solve the problem, in the next section, we will build a belief rule-based model based on expert knowledge and training data to describe the more complex nonlinear relationship between Sup_t, $Sup_{0:t-1}$, $Sup_{0:t-2}$ and v_t.

3.3. Online estimate fusion weights by the BRB model

The BRB methodology employs the extended *IF-THEN* belief rule-based systems to represent various types of information and knowledge with uncertainties. The physical meanings of the parameters in BRB (value and weight of attribute, belief distribution, rule weight, etc.) are never obscure, but clear and easy to be understood by experts or users. Hence the BRB model is fit for approximating the complicated nonlinear causal relationships across a wide variety of application areas, including fault diagnosis, system identification, risk and decision analysis [11]. In our context, we construct the BRB model to estimate the fusion weights.

3.3.1. Design the structure of BRB system

In the designed belief rule base, the kth rule R_k is constructed as follows [13]

R_k: If x_1 is $A_1^k \wedge x_2$ is $A_2^k \wedge \cdots \wedge x_M$ is A_M^k, then $\{(D_1, \beta_1^k), (D_2, \beta_2^k), \cdots, (D_N, \beta_N^k)\}$
with the rule weight θ_k and the attribute weight $\delta_1, \delta_1, ..., \delta_M$.

(8)

where A_i^k ($i = 1, 2, ..., M, k = 1, 2, ..., L$) is the referential value of the ith antecedent attribute, M and L are the total numbers of antecedent attribute and rule respectively. D_j ($j = 1, 2, ..., N$) is the jth referential value of consequent, $\beta_j^k \in [0,1]$ is the belief degree to which D_j is deemed as the consequent when $(x_1, x_2, ..., x_M) = (A_1^k, A_2^k, ..., A_M^k)$. The belief distribution $\{(D_1, \beta_1^k), (D_2, \beta_2^k), ..., (D_N, \beta_N^k)\}$ reflects uncertainties caused by the imprecise mapping relationship.

Table 1 shows the meanings of BRB parameters in the proposed evidence-based alarm system.

Table 1. The meanings of BRB parameters in evidence-based alarm system.

The BRB system	The correspondence in evidence-based alarm system
Antecedent attribute	Input variable $X=\{x_i\|i=1,\ldots,M, M=3\}$, $x_1=Sup_{0:t-2}$, $x_2=Sup_{0:t-1}$, $x_3=Sup_t$
Reference value set of antecedent attribute $A_i=\{A_{i,q}\| i=1,\ldots,M; q=1,\ldots,Q\}$	Reference value of input variable x_i
Antecedent of R_k, $k=1,2,\ldots,L$	Reference vector of X in R_k, $A^k=(A_1^k, A_2^k, A_3^k)$, $A_i^k \in A_i$
Consequent of R_k, $\{(D_1,\beta_1^k), (D_2,\beta_2^k),\ldots,(D_N,\beta_N^k)\}$ $\sum_{j=1}^{N}\beta_j^k = 1$	D_j is the reference value of output v_t when $X=A^k$; β_j^k is the belief degree of D_j, $j=1,2,\ldots,N$
Rule weight $\theta_k \in [0,1]$	Relative importance of the kth rule
Attribute weight $\delta_i \in [0,1]$	Relative importance of antecedent attributes

After understanding the structure of BRB system, we could estimate the output of the BRB system by the following steps: (1) Get the activation weight w_k of the rule R_k caused by the input $X=\{x_1, x_2, x_3\}$

$$w_k = \theta_k \prod_{i=1}^{M}(\alpha_i^k)^{\overline{\delta}_i} \Big/ \sum_{k=1}^{L}\theta_k \prod_{i=1}^{M}(\alpha_i^k)^{\overline{\delta}_i} \tag{9}$$

here $\overline{\delta}_i = \delta_i \big/ \max_{i=1,2,\cdots,M}\{\delta_i\}$, α_i^k is the matching degree that x_i matches $A_i^k \in A_i$

$$a_i^k = \begin{cases} \dfrac{(x_i - A_{i,q})}{(A_{i,q+1} - A_{i,q})}, & A_{i,q} < x_i \leq A_{i,q+1}, A_i^k = A_{i,q+1}, \\ \dfrac{(A_{i,q+1} - x_i)}{(A_{i,q+1} - A_{i,q})}, & A_{i,q} < x_i \leq A_{i,q+1}, A_i^k = A_{i,q}, \\ 1, & A_i^k = A_{i,Q} \leq x_i \text{ or } x_i < A_{i,1} = A_i^k, \\ 0, & \text{others} \end{cases} \tag{10}$$

(2) Use ER algorithm in [13] to fuse the belief distributions of all rules and then obtain the resulting distribution

$$O(X) = \{(D_j, \beta_j), j=1,2,\cdots,N\} \tag{11}$$

$$\beta_j = \left(\mu[\prod_{k=1}^{L}(\omega_k \beta_{j+1}^k - \omega_k \sum_{j=1}^{N}\beta_j^k) - \prod_{k=1}^{L}(1 - \omega_k \sum_{j=1}^{N}\beta_j^k)]\right) \Big/ \left(1 - \mu[\prod_{k=1}^{L}(1-\omega_k)]\right) \tag{12}$$

$$\mu = [\sum_{j=1}^{N}\prod_{k=1}^{L}(\omega_k \beta_j^k + 1 - \omega_k \sum_{j=1}^{N}\beta_j^k) - (N-1)\prod_{k=1}^{L}(1 - \omega_k \beta_j^k)]^{-1} \quad (13)$$

(3) Estimate outputs v_t and τ_t according to $O(X)$

$$v_t(X) = \sum_{j=1}^{N} D_j \beta_j, \tau_t(X) = 1 - v_t(X) \quad (14)$$

3.3.2. Optimization of the BRB parameters

It is possible to only use experts' alarm knowledge to construct an initial BRB, but when the historical data are available as $(x(t), m_t^I)$, $t = 1,\ldots, H$, we can optimize the parameters of the initial BRB by using these data such that the optimized BRB model has better performance. According to Table 1, we chose parameter set $P = \{\beta_j^k, \theta_k, \delta_i, D_j | i = 1,\ldots, M; j = 1,\ldots, N; k = 1,\ldots, L\}$ to be optimized.

The abnormal and normal states can be described as two absolute evidence $m_t^I = (m_t^I(A), m_t^I(NA), m_t^I(\Theta)) = (1,0,0)$ and $(0,1,0)$ respectively, after obtaining $m_{0:t}$ by (3), the optimal training objective can be designed to minimize the following Jousselme's distance defined in (4)

$$\min_P \xi(P) = \frac{1}{H}\sum_t^H d_J(m_{0:t}, m_t^I) \quad (15)$$

4. Experiments and Contrastive Analysis for the LPG Pipeline Leak Case

After introducing the evidence updating methods and build the BRB system to optimize the fusion weights in Section 3, here we do the LPG pipeline leak detection experiments using the historical data in Fig. 2 to illustrate the effectiveness of the proposed method.

Firstly, take the value of x_{tp} from the maximum to the minimum of the historical data and compare the historical $x(t)$ in Fig. 2 with each x_{tp}. By plotting ROC curve as explained in Section 3.1, we obtain the optimal $x_{tp} = -0.7$ and set $\zeta = 0.05$ to construct the dual fuzzy thresholds $\mu_A(x(t))$ and $\mu_{NA}(x(t))$. Thus, the historical $x(t)$ can be transformed to the correspond the alarm evidence (X_t, m_t) as shown in (2). The global evidence $m_{0:t}$ can be calculated by the linear updating Eq. (3) and then the alarm decisions can be made at each time by the decision rule given in Section 3.2. Here, we need to construct the BRB model to estimate v_t and τ_t. By analyzing the variation of alarm variable and the understanding of evidence updating mechanism, as experts, we can firstly give initial belief rules. In detail, we set 4 reference linguistic values respectively for inputs $Sup_{0:t-2}$, $Sup_{0:t-1}$, Sup_t, namely very small (VS), positive small (PS), positive medium (PM), large (L) and 4 reference linguistic values for output v_t, namely very small (VS), medium (M), high (H), very high (VH), their values are listed in following Table 2 and Table 3 respectively.

Table 2. Linguistic terms and reference values for $Sup_{0:t-2}$, $Sup_{0:t-1}$, Sup_t.

Linguistic term	$VS(A_{i,1})$	$PS(A_{i,2})$	$PM(A_{i,3})$	$L(A_{i,4})$
$x_1 = Sup_{0:t-2}$	0.7	1.0	1.4	2.1
$x_2 = Sup_{0:t-1}$	0.8	1.3	1.7	2.1
$x_3 = Sup_t$	0.001	0.8	1.5	2.1

Table 3. Linguistic terms and reference values for v_t.

Linguistic term	$VS(D_1)$	$M(D_2)$	$H(D_3)$	$VH(D_4)$
v_t	0.001	0.18	0.33	0.52

As a result, according to Tables 2–3, we can determine the parameters of BRB system in Table 4 as $M = 3$, $Q = 4$, $N = 4$, then there are totally $N^M = 64$ rules. Table 4 lists partial rules in which the initial values of β_j^k and θ_k are given respectively, and the initial values of attribute weight δ_i are all taken as 1.

Table 4. Partial initial belief rules given by experts.

k	θ_k	$Sup_{0:t-2} \wedge Sup_{0:t-1} \wedge Sup_t$	Belief distribution of v_t			
			β_1^k	β_2^k	β_3^k	β_4^k
1	1	VS ∧ VS ∧ VS	0.35	0.65	0	0
2	1	VS ∧ VS ∧ PS	0.62	0.38	0	0
⋮	⋮	⋮	⋮	⋮	⋮	⋮
6	1	VS ∧ PS ∧ PS	0.62	0.38	0	0
⋮	⋮	⋮	⋮	⋮	⋮	⋮
17	1	PS ∧ VS ∧ VS	0.64	0.36	0	0
60	1	L ∧ PM ∧ L	0	0	0.95	0.05
⋮	⋮	⋮	⋮	⋮	⋮	⋮
63	1	L ∧ L ∧ PM	0	0.10	0.90	0
64	1	L ∧ L ∧ L	0.64	0.36	0	0

It is necessary to fine tune rule base so that the performance of system can be improved. Based on the optimization model in Section 3.3.2, we use the historical data in Fig. 2 to train the parameter set $P = \{\beta_j^k, \theta_k, \delta_i \mid i = 1,..., 3; j = 1,..., 4; k = 1,..., 64\}$. As a result, Table 5 lists the partial trained belief rules.

Table 5. Partial trained belief rules belief rules.

k	θ_k	$Sup_{0:t-2} \wedge Sup_{0:t-1} \wedge Sup_t$	Belief distribution of v_t			
			β_1^k	β_2^k	β_3^k	β_4^k
1	0.424	VS ∧ VS ∧ VS	0.329	0.248	0.223	0.200
2	0.435	VS ∧ VS ∧ PS	0.321	0.248	0.226	0.205
⋮	⋮	⋮	⋮	⋮	⋮	⋮
6	0.427	VS ∧ PS ∧ PS	0.323	0.246	0.225	0.226
⋮	⋮	⋮	⋮	⋮	⋮	⋮
17	0.597	PS ∧ VS ∧ VS	0.655	0.156	0.109	0.080
60	0.528	L ∧ PM ∧ L	0.200	0.235	0.260	0.305
⋮	⋮	⋮	⋮	⋮	⋮	⋮
63	0.500	L ∧ L ∧ PM	0.192	0.232	0.264	0.312
64	0.503	L ∧ L ∧ L	0.196	0.233	0.261	0.310

After optimizing the relevant parameters, we could get better weights (v_t and τ_t) to replace the linear weights in (7), finally the alarm system get FAR = 3.41% and MAR = 0.62% for the training data.

In order to demonstrate the superiority of the proposed method, we decide to make 100 groups test. However, there is only one set of real data in Fig. 2, so we have to use these data as a reference set to generate 100 sets testing data by adding random disturbance obeying uniform distribution in interval [−0.5,0.5] to $x(t)$. Although the generated data do not completely replace the real data, to some extent they could reflect the dynamic variation of alarm variable. Table 6 shows the mean values of MAR and FAR of the 100 time tests respectively. By comparing with the method in [4] using the linear weight model in (7), the proposed BRB-based dynamic evidence fusion method has better performance.

Table 6. The average rate of "alarm", "no-alarm".

Methods	m(FAR)	m(MAR)
The proposed method	4.49%	0.285%
The method in [4]	5.28%	1.64%

5. Conclusion

In order to solve pipeline leak detection, this paper presents an effective alarm decision method based on the belief rule base and evidence updating. The main contribution lies in introducing the belief rule base and the corresponding inference technique to deal with the calculation of updating fusion weights. By the experiments of LPG pipeline leak detection, it is verified that the proposed

BRB-based evidence fusion method has better performance than the previous linear weight model-based fusion method.

References

1. Liu C., Li Y., Fang L., et al. Experimental study on a de-noising system for gas and oil pipelines based on an acoustic leak detection and location method[J]. International Journal of Pressure Vessels & Piping, 2017, 151:20–34.
2. Yang B. W., Recane M. Method and apparatus for pattern match filtering for real time acoustic pipeline leak detection and location: US, US 6389881 B1[P]. 2002.
3. J. Wang, F. Yang, T. Chen, & S.L. Shah. An overview of industrial alarm systems: main causes for alarm overloading, research status, and open problems. IEEE Trans. Automation Science and Engineering, vol. 13, pp. 1045–1061, 2016.
4. X. Xu, S. Li, X. Song, C. Wen, & D. Xu. The optimal design of industrial alarm systems based on evidence theory, Control Engineering Practice, vol. 46, pp. 142–156, 2016.
5. Engineering Equipment and Materials Users' Association (EEMUA). (2007). Alarm Systems-A Guide to Design, Management and Procurement- EEMUA Publication 191, version 2.
6. American National Standards Institute (ANSI), The International Society of Automation (ISA). (2009). Management of Alarm Systems for the Process Industries-ANSI/ISA-18.2.
7. Cheng, Y., Izadi, I., & Chen, T. (2013). Optimal alarm signal processing: Filter design and performance analysis. IEEE Transactions on Automation Science and Engineering, 10(2), 446–451.
8. Xu, J., & Wang, J. (2010, December). Averaged alarm delay and systematic design for alarm systems. In Decision and Control (CDC), 2010 49th IEEE Conference on (pp. 6821–6826).
9. Xu, J., Wang, J., Izadi, I., & Chen, T. (2012). Performance assessment and design for univariate alarm systems based on FAR, MAR, and AAD. Automation Science and Engineering, IEEE Transactions on, 9(2), 296–307.
10. Xu Xiaoxin, Zhang Zhen, Xu Dongling, Chen Yuwang, Interval-valued evidence updating with reliability and sensitivity analysis for fault diagnosis, International Journal of Computational Intelligence Systems, 2016, 9(3):396–415.
11. Xu Xiaoxin, Wen Chenglin, Jin Yindong, Sun Xinya, Evidence fusion and decision method in equipment fault diagnosis, Beijing: Chinese Science press, 2017.
12. Jousselme, A. L., & Maupin, P. (2012). Distances in evidence theory: Comprehensive survey and generalizations. International Journal of Approximate Reasoning, 53(2), 118–145.
13. J.-B. Yang, J. Liu, J. Wang, H.-S. Sii, H.-W. Wang. "Belief rule-base inference methodology using the evidential reasoning Approach-RIMER", IEEE Transactions on Systems Man & Cybernetics Part A Systems & Humans, Vol. 36, No. 2, pp. 266–285, 2006.

A belief rule-based inference model for prediction of R&D project success[*]

Ying Yang

*Department of Information Management and Information Systems,
Hefei University of Technology, 193 Tunxi Road Hefei 230009, China*

Yu-Xiao Li

*Department of Information Management and Information Systems,
Hefei University of Technology, 193 Tunxi Road Hefei 230009, China*

Yu-Wang Chen

*Alliance Manchester Business School, The University of Manchester
Manchester M15 6PB, UK*

Decision makers need to understand various risks in R&D projects so as to assess the likelihood of project success. In this paper, a recently-developed belief rule-based (BRB) inference methodology is adopted to further represent the non-linear relationships between risk factors and project success and improve the prediction accuracy of R&D project success. A hierarchical BRB inference model is firstly proposed to predict the likelihood of R&D project success. The model is validated using the data collected from 170 R&D projects in China. Comparative analysis results show the high performance of the proposed prediction model.

Keywords: R&D project management; performance prediction; risk analysis; belief rule-based inference.

1. Introduction

Projects are important for companies to achieve sustainable competitive advantages. As project delay or cost over-run is common in research and development (R&D), organizational project management is required to improve project performance. Failure to account for uncertainties and risks is a major cause for project over-run. Therefore, decision makers in organizational project management need to understand various risks and assess the likelihood of project success to seek continuous improvement.

---

[*]This work is supported by the National Nature Science Foundation of China (No. 71573071) and Key Projects of Nature Science Foundation in the universities of Anhui province (No. KJ2017A391).

Researchers have developed several reasoning models for risk analysis and performance prediction including multiple regression, artificial neural networks (ANNs) and Bayesian networks (BNs). ANNs focus on the achievement of high prediction accuracy with a black-box model. de Oliveira et al. developed a kind of ANNs, radial basis function (RBF) networks, to model the non-linear relationships between leadership and project performance [1]. BNs are powerful in reasoning about uncertainty and widely used in risk management. Causal relationships between risk events and project outcome can be represented in BNs through a white-box model or an interpretable model [2]. Yet et al. also proposed a Bayesian network framework for project cost, benefit and risk analysis [3].

However, the application of BNs in risk management is still limited owing to the complexity of the inter-relationships among risk factors and their causal structures. Moreover, the prediction accuracy of BNs is inferior to other artificial intelligent methods. In order to further represent the non-linear relationships between risk factors and project success and improve the prediction accuracy, we adopted the recently-developed belief rule-based (BRB) inference methodology for prediction of R&D project performance.

The remainder of the paper is organized as follows. Research variables and data collection are presented in Section 2. A BRB inference model for prediction is developed in Section 3. The model is evaluated by conducting a comparative analysis in Section 4. Conclusions and research perspectives are given in the last section.

2. Research Variables and Data Collection

2.1. *Research variables*

There are many risk events in R&D projects that may lead to project failure. Four critical risk factors including top management involvement (TMI), project manager competency (PMC), formalization of portfolio management (FPM) and project termination (PT) can be identified from the perspective of organizational project management.

TMI refers to the involvement of top managers who can devote sufficient time and efforts to R&D project management, including new project selection, resource allocation, process design and a series of planned activities [4]. MC refers to the competency of project managers who understand their responsibilities in project management. At the meanwhile, the managers are responsible for the implementation of projects [5]. FPM refers to that enterprises can follow the formal process and apply flexible methods in the decision making of project evaluation and selection. PT is used to measure the quality of project termination. The termination decision of projects can save resources and avoid continued losses arising from inappropriate projects and re-allocate resources to other important projects [5]. R&D project performance can be measured by time, cost, quality and customer satisfaction.

2.2. *Data collection*

These research variables are qualitative and cannot be observed directly. According to prior literature research, this study designs their measurement indicators and conducts a survey to collect sample data. For TMI, seven measurement indicators are designed. Five are designed for FPM. Project manager competency, project termination and project performance all have four measurement indicators respectively.

Each measurement indicator is measured by Likert 5 scales (1 = extremely inconformity, 5 = extremely conformity). 300 questionnaires were distributed to top managers, middle-level department managers and project managers engaging in project management or having rich experience in project management, 205 questionnaires were retrieved, 170 questionnaires were gained after deleting some invalid questionnaires, the efficiency is 82.9%.

3. A BRB Inference Model for Project Prediction

3.1. *Representation of causal relationships*

Compared with traditional rule, the belief rule extends knowledge representation using the belief structure to provide more flexibility and versatility for more precisely imitating human reasoning. The belief rule represents the causal relationships between antecedents and consequent using the if-then expressions with dedicated degrees [6]. It can be described as R_k ($k=1, ..., L$):

If $A_1^k \wedge A_2^k \wedge ... \wedge A_{T_k}^k$, then $\{(D_1, \beta_{1k}), (D_2, \beta_{2k}), ..., (D_N, \beta_{Nk})\}$ with the rule weight θ_k and attribute weights $\delta_1, \delta_2, ..., \delta_{T_k}, (\beta_{jk} \geq 0, \sum_{j=1}^{N} \beta_{jk} \leq 1)$ where A_i^k ($i =1, ..., T_k$) is the referential category of the ith antecedent attribute used in the kth rule. β_{jk} ($j = 1, ..., N, k = 1, ..., L$) is the belief degree assigned to consequent D_j and it can be obtained initially from domain experts. When the belief degree of consequents is 100% and only one consequent included in a rule, the belief rule base is transformed to a traditional IF-THEN rule. The parameters including rule weights, antecedent attribute weights and consequent belief degrees are embedded in the belief rules to represent expert knowledge.

Therefore, the causal relationships between risk factors and R&D project performance can be represented by a belief rule, whether linear or not. A hierarchical knowledge base is developed as shown in Fig. 1. It is composed of two sub-rule bases: sub-rule-base 1 models the causal relationship among project termination, top management involvement and formalization of portfolio management while sub-rule-base 2 represents relationships among project performance, project termination and project manager competency. Information is propagated from the bottom level up to the high level.

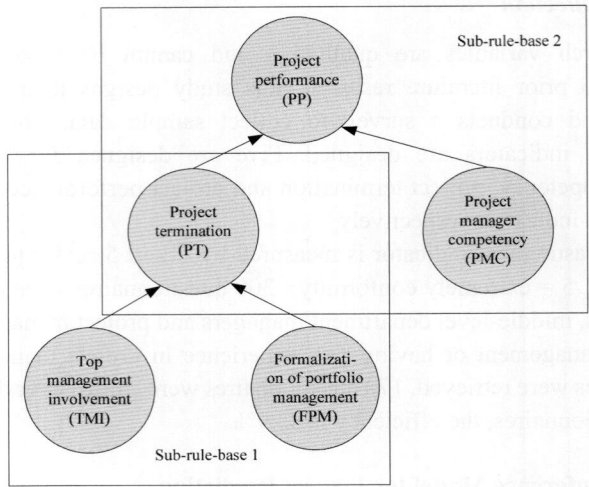

Fig. 1. A hierarchical rule base.

The rules are numbered and clustered in conjunction with the two sub-rule bases, where *H* stands for high, *M* for medium and *L* for low. The degrees of belief in the consequents were assigned by the domain experts initially. It could be trained using historical data and be updated once new evidence becomes available.

3.2. *Hierarchical rule-base inference*

The hierarchical belief rule-based inference model proposed in this paper is composed of several procedures such as discretization, transformation, activation and aggregation. The inputs of the model are the values of risk factors and the output is R&D project success.

3.2.1. *Discretization and transformation*

The final value of each risk factor is obtained by averaging the answers of each indicators. It is a continuous value and will be discretized into the counterparts represented in terms of the referential values: low, medium and high. Unsupervised equal frequency discretization method is used for data preparation here. All data is categorized into three classes with equal frequency of 33.3%.

Then the values of risk factors are transformed to belief structures in order to construct an initial belief rule base. The rule-based transformation technique is used for the quantitative data transformation [7].

3.2.2. Calculation of the activation weight of each belief rule

The rule activation weight is denoted by ω_k ($k=1, \ldots, L$) which represents the degree to which the packet antecedent A^k in the kth rule is activated by the inputs. It can be calculated as

$$\omega_k = \theta_k \alpha_k \Big/ \sum_{j=1}^{L} \theta_j \alpha_j, \quad k = 1, \ldots, L$$

where θ_k is the relative weight of the kth rule. The total degree of the input matching the packet antecedent A^k in the kth rule is denoted as α_k and can be obtained using the weighted multiplicative aggregation function.

$$\alpha_k = \prod_{i=1}^{T_k} (\alpha_i^k)^{\overline{\delta_{ki}}}$$

where

$$\overline{\delta_{ki}} = \frac{\delta_{ki}}{\max\{\delta_{ki}\}}, \quad 0 \le \overline{\delta_{ki}} \le 1$$
$$i = 1, \ldots, T_k$$

where δ_{ki} is the antecedent attribute weight representing the relative importance of the ith antecedent attribute in the kth rule. T_k is a total number of antecedent attributes in the kth rule and L is the total number of belief rules in the BRB.

3.2.3. Aggregation of the activated belief rules

Inference with the hierarchical BRB is implemented using the evidential reasoning (ER) algorithm. Pieces of evidence for the lowest level attributes (TMI, FPM) are aggregated as evidence for the second lowest level attributes.

Let β_j ($j = 1, \ldots, N$) represent the final combined belief degree associated with the corresponding consequent D_j ($j = 1, \ldots, N$). The analytic ER algorithm is given as follows:

$$\beta_j = \frac{\mu \times [\prod_{k=1}^{L}(\omega_k \beta_{jk} + 1 - \omega_k \sum_{j=1}^{N} \beta_{jk}) - \prod_{k=1}^{L}(1 - \omega_k \sum_{j=1}^{N} \beta_{jk})]}{1 - \mu \times [\prod_{k=1}^{L}(1 - \omega_k)]}$$

where

$$\mu = \left[\sum_{j=1}^{N}\prod_{k=1}^{L}(\omega_k \beta_{jk} + 1 - \omega_k \sum_{j=1}^{N} \beta_{jk}) - (N-1)\right.$$
$$\left. \times \prod_{k=1}^{L}(1 - \omega_k \sum_{j=1}^{N}\beta_{jk})\right]^{-1}$$

3.3. *BRB training process*

All the weights of antecedent attributes and rules are assumed to be equal to one initially in this study. They will be improved by data learning from training process.

The prediction model includes knowledge representation parameters: the attribute weight θ_k, the rule weight δ_k and the consequent belief degree β_{jk}. An optimal learning model is developed to adjust the parameters in order to minimize the difference between the observed project performance $\hat{y}_m$ and the predicted one y_m by the BRB system as follows:

$$\min \{\xi(P) = (1/M)\sum_{m=1}^{M}(y_m - \hat{y}_m)^2\}$$

s.t. $0 \le \beta_{jk} \le 1, j = 1,...,N, k = 1,...,L$

$0 \le \theta_k \le 1, k = 1,...,L$

$0 \le \delta_i \le 1, i = 1,...,T$

4. Model Evaluation

The proposed prediction model is trained and validated by all the sample data firstly. The parameters are all learned from data The attribute weights θ_k (k = 1,..., 4) of TMI, FPM, PT, PMC is updated to {1, 0.99, 0.55, .91} after training. The weights of rules and the consequent belief degrees after training are as shown in Table 1. The mean absolute percentage error (MAPE) between the predicted output and the observed output is 13.86% while it is 33.97% before training.

Then ten-fold cross-validation method is adopted to further evaluate the fit accuracy and reliability of the prediction model. 170 samples are randomly divided into 10 subsets with an equal size. A single fold out of the ten folds is retained as the validation set and the remaining nine folds are used as the training set. The average MAPE of ten times is 12.72% for training data and 15.07% for validation data.

Table 1. The knowledge representation parameters after training.

Rule No.	Weights	Consequent belief degrees		
1	0.98	1.0000	0.0000	0.0000
2	0.99	0.5574	0.4275	0.0151
3	1.00	0.5024	0.0017	0.4959
4	0.98	0.5422	0.4578	0.0000
5	1.00	0.3794	0.6206	0.0000
6	0.99	0.1373	0.6304	0.2323
7	0.99	0.2124	0.5128	0.2748
8	1.00	0.3750	0.3881	0.2369
9	0.99	0.2941	0.3859	0.3201
10	1.00	1.0000	0.0000	0.0000
11	0.73	0.6212	0.2120	0.1668
12	0.99	0.1673	0.5492	0.2835
13	0.87	0.4561	0.2201	0.3239
14	0.99	0.3219	0.3677	0.3104
15	0.00	0.8418	0.0054	0.1528
16	1.00	0.1362	0.4965	0.3674
17	1.00	0.3010	0.3216	0.3774
18	1.00	0.5256	0.0000	0.4744

Moreover, similar datasets are used to conduct comparative analysis between the proposed prediction model and other classic methods. Both tree augmented Naïve Bayes (TAN) and Bayesian networks (BN) are adopted to predict R&D project performance. The average prediction accuracy after ten-fold cross-validation for each method is listed in Table 2. It indicates that the proposed BRB inference model has a very high prediction performance for R&D projects.

Table 2. Comparative analysis of different models.

Prediction model	Training data	Validation data
TAN	75.30%	70.90%
BN	81.50%	72.20%
BRB	87.28%	84.93%

5. Conclusions

The paper presents a BRB inference model to predict R&D project performance. The model is compared to the most commonly used methods in structuring causal relationships in risk analysis, such as Bayesian networks and tree augmented Naïve Bayes. The ten-fold cross-validation methods is employed to validate the three prediction models. The average MAPE is used to compare their prediction accuracy. The results show that the proposed BRB inference model has robust and competitive performance. Moreover, the initial knowledge representation parameters can be further improved by learning from historical data.

Acknowledgments

This research was supported by the National Nature Science Foundation of China (No. 71573071) and Key Projects of Nature Science Foundation in the universities of Anhui province (No. KJ2017A391).

References

1. Oliveira, M. A. D., Possamai, O., Valentina, L. V. O. D., & Flesch, C. A. (2013). Modeling the leadership – project performance relation: radial basis function, gaussian and kriging methods as alternatives to linear regression. *Expert Systems with Applications, 40*(1), 272-280.
2. Hu, Y., Zhang, X., Ngai, E. W. T., Cai, R., & Liu, M. (2013). Software project risk analysis using bayesian networks with causality constraints. *Decision Support Systems, 56*(1), 439-449.
3. Yet, B., Constantinou, A., Fenton, N., Neil, M., Luedeling, E., & Shepherd, K. (2016). A bayesian network framework for project cost, benefit and risk analysis with an agricultural development case study. *Expert Systems with Applications, 60*(C), 141-155.
4. Patanakul, P. (2013). Key drivers of effectiveness in managing a group of multiple projects. *IEEE Transactions on Engineering Management, 60*(1), 4-17.
5. Jonas, D. (2010). Empowering project portfolio managers: how management involvement impacts project portfolio management performance. *International Journal of Project Management, 28*(8), 818-831.
6. Yang, J. B., Liu, J., Wang, J., Sii, H. S., & Wang, H. W. (2006). Belief rule-base inference methodology using the evidential reasoning approach-rimer. *IEEE Transactions on Systems Man & Cybernetics Part A Systems & Humans, 36*(2), 266-285.
7. Yang, J. B. (2001). Rule and utility based evidential reasoning approach for multi-attribute decision analysis under uncertainties. *European Journal of Operational Research*, 131(1), 31-61.

Causal inference with Gaussian processes for support of terminating or maintaining an existing program

Adi Lin, Junyu Xuan, Guangquan Zhang and Jie Lu

Centre for Artificial Intelligence, School of Software, Faculty of Engineering and Information Technology, University of Technology Sydney
Sydney, NSW 2007, Australia
Adi.Lin@student.uts.edu.au
Junyu.Xuan@uts.edu.au
Guangquan.Zhang@uts.edu.au
Jie.Lu@uts.edu.au

Decision makers often face a problem to decide to maintain or terminate an existing program with only observational data available. Within the potential outcomes framework, we show the problem could be modelled as a causal inference problem to estimate the conditional average treatment effect of the treatment on the treated (CATT). In this paper, we propose a model with separate Gaussian processes to estimate average treatment effect for the treated group and the control group. We conduct experiments on an empirical case and show that our method could contribute to decision making in the kind of problem described above.

Keywords: Causal inference; Gaussian process; data-driven; decision making.

1. Introduction

Data-driven decision making depends on the analysis of data. Some applications of data-driven decision making are credit-card fraud detection,[1] bank-telemarketing,[2] heart transplant survival,[3] global ranking that represents the social preference.[4] In this paper we focus on the kind of problem that a decision is needed to make to determine an existing program to be maintained or not. It is a problem to estimate CATT in causal inference.

Various work has been proposed to estimate conditional average treatment effect (CATE), which is a potential application in precision medicine. Tree-based methods,[5,6] Bayesian nonparametric models,[7–9] marked point process,[10] generative adversarial nets[11] and a lot of work based on neural networks[12–14] are developed to calculate CATE. For the methods using Gaussian processes, a nonparametric Bayesian method is proposed to estimate CATE using a multi-task Gaussian process[15] and a counterfactual Gaussian process to predict future trajectories in continuous-time.[16]

While in this paper, we focus on CATT, which is a key factor for decision makers to decide to maintain or terminate a program. Generally, it is not possible to estimate CATT with the treated group only without other data and extra assumptions. Our work estimates CATT with data combined the treated and control group under some assumptions. The contributions of this paper are: a) We propose two separate Gaussian processes, one for the treated group and another for the control group, to estimate CATT. We use Gaussian processes because the prediction is probabilistic, thus we can compute confidence intervals. And confidence intervals for prediction are often required for decision-making. b) We show it is possible for decision makers to make decisions from observational data without experimental data available under some assumptions. We use the Gaussian process trained for the control group to predict the outcomes of samples in the treated group if they are under the control group.

The remainder of the paper is organised as follows: Section 2 introduces related work briefly. Section 3 provides basic concepts in causal modelling and sets up the problem of CATT. Section 4 presents our algorithm to estimate CATT. Section 5 evaluates the algorithm with a case of health care and show how our research contributes to data-driven decision making. The last section concludes the paper and gives the future work.

2. Literature Review

The potential outcomes model[17,18] is a base to set up the problem of causal inference. The Neyman's doctoral dissertation(an English translated version[17]) used the idea of potential outcomes in randomised experiments. Without referring to Neyman's work, Rubin[18] argued the causal inference was possible from observational studies with the potential outcomes model. Randomised experiments and observational studies are defined according to the treatment assignment mechanism[19] in Rubin's work. Under the assumption of ignorability of the treatment assignment mechanism, many work estimates CATT for the assignment mechanism or for the response surface. And doubly robust methods combine a model for the assignment mechanism and a model for the response surface. In this paper, our work fit two models for the response surface.

The two kinds of methods for the treatment assignment mechanism are matching[20] and weighting.[21] Matching methods aim to balance the distribution of covariates in the treated and control groups. The matching methods could be nearest neighbor, Mahalanobis distance,[20] full matching[22] and optimal matching.[23] A review for matching methods could be found in Ref. 24. The goal of weighting is to create weighted empirical distributions of the treated and control that are as close as possible. The inverse probability of treatment weighting[25] is used mostly in causal inference. Recently, empirical calibration weighting is developed. Some are the work[26] estimating treatment effect with the negative Shannon entropy and the work[27] in which the semiparametric efficiency could be achieved in the empirical calibration estimators. Significantly, propensity score[28] is highly used in matching[29] and weighting.[21]

Another kind of methods is called response surface modeling, which treats estimation of CATT as a regression problem for the distribution of the counterfactual data. Response surface modeling is the most commonly used parametric method for estimating CATT. Some common methods are linear regression, parametric g-formula[30] and Bayes trees.[31]

Either a model for the treatment assignment mechanism or response surface modeling is sensitive for model misspecification. Doubly robust estimation combines a model for the assignment mechanism and a model for the response surface. An estimator is called doubly robust if it is consistent at least one of its two component models is correctly specified. Ref. 32 combines propensity score weighting and response surface modeling to estimate the average causal effect of a dichotomous treatment on an outcome. A sophisticated method is Targeted Maximum Likelihood.[33]

3. Problem Setup

We consider a group of subjects drawn from a distribution, where each subject i consists of a d-dimensional *feature* $X_i \in \mathcal{X}$, the treatment assignment indicator $W_i \in \{0,1\}$ (the value 1 indicates the subject i has received the treatment, otherwise the value is 0), the *potential outcomes* $Y_i(0), Y_i(1) \in \mathbb{R}$ ($Y_i(1)$ is the outcome if the subject i received the treatment, $Y_i(0)$ is the outcome if the subject i is in the control group). The setting is a classical one in potential outcomes framework.[17,18]

CATT is defined as

$$CATT = \sum_{\{i:W_i=1\}}^{n} \mathbb{E}[Y_i(1) - Y_i(0)|X_i]. \tag{1}$$

The estimand focuses on the average treatment effects on the treated of the sample and the treatment effect on the treated of each subgroup may be different. Often, decision makers also put interest on the subgroups. The treatment effect of a subgroup here is defined as the treatment effect at **x**, and it is

$$\tau(\mathbf{x}) = \mathbb{E}[Y_i(1) - Y_i(0)|X_i = \mathbf{x}]. \tag{2}$$

The estimand focuses on the average treatment effect on the treated of the subgroup which has value **x** on covariates X.

The main challenge is that we cannot observe both of two potential outcomes $Y_i(1)$ and $Y_i(0)$ for any subject i, but can only ever observe the outcome of the subject.[34] As shown in Table 1, it is impossible to observe the potential outcomes as treated for the subjects observed in the control group and potential outcomes as control for those observed in the treatment group. So we cannot calculate the unit-level treatment effect $Y_i(1) - Y_i(0)$. Fortunately, estimation of these treatment effects could be conceptualised as a regression problem, where $\mathbb{E}[Y(1)|X] = \mathbb{E}[Y|X, W = 1]$ and $\mathbb{E}[Y(0)|X] = \mathbb{E}[Y|X, W = 0]$, under the assumption *strong ignorability* of treatment assignment, consisting of both an unconfounded (or exchangeability) assumption and a overlap assumption. The assumption of unconfounded is also called strongly ignorable,[28]

$$P(W|X, Y(0), Y(1)) = P(W|X). \tag{3}$$

The overlap assumption (also called positivity) is a critical property in a randomised experiment — the unit-level probabilities of treatment assignment should be between 0 and 1,

$$0 < P(W_i = 1|X, Y) < 1. \tag{4}$$

Table 1. Potential outcomes.

Group	Y(1)	Y(0)
Treated group	Observed outcomes	Unobserved
Control group	Unobserved	Observed outcomes

4. Gaussian Processes for Estimating Causal Effects

We present a typical version of Gaussian process in Ref. 35. A Gaussian process is to used to describe a distribution over functions, we provide the definition of Gaussian processes in Def. 1.

Definition 1 (Gaussian Process). *A Gaussian process is a set of random variables, and any finite number of these random variables have a joint Gaussian distribution.*

Similar to the Gaussian distribution that is specified by its mean and covariance, a Gaussian process could be determined by its mean function and covariance function. The mean function $m(\mathbf{x})$ and the covariance function $k(\mathbf{x}, \mathbf{x}')$ of a real Gaussian process $g(\mathbf{x})$ are

$$m(\mathbf{x}) = \mathbb{E}[g(\mathbf{x})], \tag{5}$$

$$k(\mathbf{x}, \mathbf{x}') = \mathbb{E}[(g(\mathbf{x}) - m(\mathbf{x}))(g(\mathbf{x}') - m(\mathbf{x}'))]. \tag{6}$$

The $\mathbf{x}$ and $\mathbf{x}'$ are two vectors.

For regression, the function $\phi(\mathbf{x})$ maps a d-dimensional input vector $\mathbf{x}$ into an k dimensional feature space. And let the matrix $\Phi(\mathbf{x})$ be the aggregation of $\phi(\mathbf{x})$ for the whole training dataset. Now the model could be represented as

$$f(\mathbf{x}) = \Phi(\mathbf{x})^T \mathbf{w}, \tag{7}$$

where $\mathbf{w}$ is a prior,

$$\mathbf{w} \sim \mathcal{N}(0, \Sigma). \tag{8}$$

Σ is the covariance matrix. It is obvious that $\{f(\mathbf{x})\}$ is a Gaussian process. The covariance function used in this paper is the squared exponential kernel function

$$cov(\mathbf{x}, \mathbf{x}') = k(\mathbf{x}, \mathbf{x}') = \exp(-\tfrac{1}{2}\left|\mathbf{x} - \mathbf{x}'\right|^2). \tag{9}$$

Observations with noise could be represented as $\{(\mathbf{x_i}, y_i) | i = 1, \cdots, n\}$. Consider observations with additive noise, that is $y = f(\mathbf{x}) + \epsilon$. The test features are denoted as $\mathbf{x}_*$ and the test outputs are represented as f_*. Then the joint distribution of the observed outputs and the test outputs f_* are jointly Gaussian. It could be proved that the key function for regression — the conditional distribution $f_* | X_*, X, y$ is Gaussian. For more details about Gaussian processes for regression, see Ref. 35.

In this section, we focus on Gaussian processes to estimate the average treatment effect. We set two Gaussian processes, one for estimating $\mathbb{E}[Y(1)|X]$ and another one for estimating $\mathbb{E}[Y(0)|X]$. Then CATT could be calculated by

$$CATT = \sum_{\{i:W_i=1\}}^{n} \mathbb{E}[Y_i(1)|X_i] - \sum_{\{i:W_i=1\}}^{n} \mathbb{E}[Y_i(0)|X_i]. \tag{10}$$

5. Experiments

In observational datasets, the ground truth treatment effects are not available. To evaluate the algorithms, we use semi-synthetic experimental data. In semi-synthetic experimental setup, either the treatment assignment mechanism or the potential outcomes are simulated. The Infant Health and Development Program (IHDP)[7] is a semi-synthetic dataset for average treatment effect. The dataset consists 747 subjects (139 subjects for the treated group, 608 subjects for the control group) with 25 covariates (6 for continuous covariates, 19 for binary covariates). Potential outcomes are simulated with "Response Surface B" setting[7] (also used in Refs. 36 and 15).

Our method is compared with baseline methods — BART,[37] linear regression, propensity score matching,[29] and propensity score weighting.[21,38] For Gaussian processes, we use the typical version described in Section 4 and we did not try to tune the hyper-parameters to produce optimal results. Table 2 shows the results of root mean squared error (RMSE) and precision in estimation of heterogeneous effects (PEHE)[7] of each method from the 1000 realisations of the simulated potential outcomes. PEHE is defined as

$$\frac{1}{n_t} \sum_{\{i:w_i=1\}} (\hat{\mathbb{E}}[Y_i(1) - Y_i(0)|X = x_i]) - \mathbb{E}[Y_i(1) - Y_i(0)|X = x_i])^2, \quad (11)$$

the n_t is the number of the treated individuals. Our proposed method is competitive with other methods.

However, it is noted that our research does not aim to develop a method to outperform the state-of-the-art. The research aims to assist to make decisions from observational data. We take IHDP program as an example to show how our method could help to make decisions. First, the decision makers read estimate of cognitive and health status of children compared to cognitive and health status of these children had they not been taken care (decision makers read the estimate of CATT), then decision makers have ideas about the impact of the program for the children receiving the health care in the program. After that, the decision makers may put their interest on the impact of the health care program for the children with mothers who have high degrees (decision makers read the estimate of the average treatment effect on the treated of the subgroup who has mothers of high degrees, and the estimate has confidence intervals provided by Gaussian processes).

Table 2. Results on the IHDP dataset.

	RMSE	PEHE
Gaussian Process	0.3	2.6
BART	0.2	1.6
Propensity Score Matching	0.5	5.3
Propensity Weighted	0.4	5.2

6. Conclusion and Further Study

We proposed separate Gaussian processes for estimating average treatment effects in different treatment groups. Based on the experimental results, it has been shown that it is possible for decision makers to determine the future direction of a program from observational data. Our research has potential applications such as job training programs.

The future work is to develop algorithms for observational datasets with hidden confounding. A confounder may by hidden or unmeasured, but there may exist noisy proxies for confounders. The work will concentrate on building a latent-variable model to estimate the hidden confounders. Considering the real observational data of selection bias, the robust model is encouraged to develop in the future.

References

1. N. Carneiro, G. Figueira and M. Costa, *Decision Support Systems* **95**, 91 (2017).
2. S. Moro, P. Cortez and P. Rita, *Decision Support Systems* **62**, 22 (2014).
3. A. Dag, K. Topuz, A. Oztekin, S. Bulur and F. M. Megahed, *Decision Support Systems* **86**, 1 (2016).
4. A. Khetan and S. Oh, *Journal of Machine Learning Research* **17**, 1 (2016).
5. S. Wager and S. Athey, *Journal of the American Statistical Association* (2017).
6. S. Athey and G. Imbens, *Proceedings of the National Academy of Sciences* **113**, 7353 (2016).
7. J. L. Hill, *Journal of Computational and Graphical Statistics* **20**, 217 (2011).
8. M. Taddy, M. Gardner, L. Chen and D. Draper, *Journal of Business & Economic Statistics* **34**, 661 (2016).
9. Y. Xu, Y. Xu and S. Saria, A bayesian nonparametric approach for estimating individualized treatment-response curves, in *Machine Learning for Healthcare Conference*, (Los Angeles, USA, 2016).
10. E. Arjas and J. Parner, *Scandinavian Journal of Statistics* **31**, 171 (2004).
11. Anonymous, *International Conference on Learning Representations* (2018).
12. C. Louizos, U. Shalit, J. Mooij, D. Sontag, R. Zemel and M. Welling, Causal effect inference with deep latent-variable models (2017).
13. A. M. Alaa, M. Weisz and M. van der Schaar, *arXiv preprint arXiv:1706.05966* (2017).

14. F. Johansson, U. Shalit and D. Sontag, Learning representations for counterfactual inference, in *International Conference on Machine Learning*, (New York, USA, 2016).
15. A. M. Alaa and M. van der Schaar, *CoRR* **abs/1704.02801** (2017).
16. P. Schulam and S. Saria, *arXiv preprint arXiv:1703.10651* (2017).
17. J. Splawa-Neyman, D. M. Dabrowska and T. Speed, *Statistical Science*, 465 (1990).
18. D. B. Rubin, *Journal of educational Psychology* **66**, p. 688 (1974).
19. D. B. Rubin, *Journal of the American Statistical Association* **100**, 322 (2005).
20. A. Diamond and J. S. Sekhon, *Review of Economics and Statistics* **95**, 932 (2013).
21. P. C. Austin, *Multivariate behavioral research* **46**, 399 (2011).
22. B. B. Hansen, *Journal of the American Statistical Association* **99**, 609 (2004).
23. N. Kallus, A framework for optimal matching for causal inference, in *Artificial Intelligence and Statistics*, (Fort Lauderdale,Florida, USA, 2017).
24. E. A. Stuart, *Statistical science: a review journal of the Institute of Mathematical Statistics* **25**, p. 1 (2010).
25. J. M. Robins, *Synthese* **121**, 151 (1999).
26. J. Hainmueller, *Political Analysis* **20**, 25 (2012).
27. K. C. G. Chan, S. C. P. Yam and Z. Zhang, *Journal of the Royal Statistical Society: Series B (Statistical Methodology)* **78**, 673 (2016).
28. P. R. Rosenbaum and D. B. Rubin, *Biometrika* **70**, 41 (1983).
29. J. Hill and J. P. Reiter, *Statistics in Medicine* **25**, 2230 (2006).
30. E. J. Murray, J. M. Robins, G. R. Seage, K. A. Freedberg and M. A. Hernán, *American journal of epidemiology* **186**, 131 (2017).
31. D. P. Green and H. L. Kern, *Public opinion quarterly* **76**, 491 (2012).
32. H. Bang and J. M. Robins, *Biometrics* **61**, 962 (2005).
33. K. E. Porter, S. Gruber, M. J. Van Der Laan and J. S. Sekhon, *The International Journal of Biostatistics* **7**, 1 (2011).
34. P. W. Holland, *Journal of the American Statistical Association* **81**, 945 (1986).
35. C. E. Rasmussen and C. K. Williams, *Gaussian processes for machine learning* (MIT press Cambridge, 2006).
36. U. Shalit, F. Johansson and D. Sontag, *arXiv preprint arXiv:1606.03976* (2016).
37. H. A. Chipman, E. I. George, R. E. McCulloch et al., *The Annals of Applied Statistics* **4**, 266 (2010).
38. G. W. Imbens, *The review of Economics and Statistics* **86**, 4 (2004).

Social influenced decision making: A belief-based model analysis

Lei Ni, Yu-Wang Chen and Oscar de Bruijn

Alliance Manchester Business School
The University of Manchester, Manchester, M13 9SS, UK

Study on the effects of social influence on decision making is a hot research topic in the interdisciplinary field of decision making, complex networks and social impact. In this paper, a new perspective to characterise the effects of social influence is proposed grounded on belief theory, a promising tool in reasoning under uncertainty. It considers the situation of influence diffusion in networked structure by mapping belief values into diffusion process and attributing a parameter reflecting the openness of an individual to his/her social neighbour's influence to edge weight. By doing this, the evolving process of belief updates can be simulated as well as the trend in individuals' decision making.

1. Introduction

Social influence refers to the process where individuals shift their minds, update their beliefs, or alter their decisions as a consequence of interacting with others. It can be readily seen in collective decision-making problems. The study on social influence has a fairly-long history and tremendous achievements on theoretical works. The central theme of Social Influence Theory [1] emphasised on the distinction among three cognitive processes of influence (i.e., compliance, identification and internalisation). Social Impact Theory [2] further pointed out that individuals' acts on social influence were quantitatively associated with three factors, that is, strength, immediacy, and number of targets. Many scholars have demonstrated that exposure to information as social interactions occur constantly has great influence on a range of behavioural phenomena such as vaccination [3], stock purchases [4], and technology adoption [5]. Such information transmission from person to person in networks is also believed to cause social conformity through the process of awareness awakening, belief updating and hence, decision making [6]. Understanding the features of information propagation in networks is becoming a prevalent and subtle force in exploring social dynamics and seizing pre-emptive opportunities behind human interactions. It also provides a reasonable way of quantifying social influence which in turn determines the depth and width of information dissemination in social networks. Furthermore, research has been conducted extensively to better observe flows of information and understand social impact on individuals' behaviours [7-9]. However, the majority of them rely much on online networking data which has been criticised for its imprecision and imperfection [10]. Recent research also uses the theory of belief functions to evaluate social influence in networks [10-12]. It is believed that the use of belief functions can deal with the uncertainty existing in real world and the imperfection of data. Xia & Liu [11] designed a belief-based model to characterise the spread of awareness and its impact on individuals' vaccination decisions. Lu *et al.* [12] also used evidence theory to evaluate individual's influential power on psychological level in an opinion group.

In view of the benefits of evidence theory, we utilise a belief-based model [11] and enrich it with a coefficient measuring how open an individual is to the influence of another on his/her degrees of belief to characterise the effects of social influence on individuals' decision making. By mapping belief values into the process of influence spreading, we further simulate the trend in social influenced decision making. The rest of the paper is organised as follows. The description of the belief-based model is provided in Section 2. A case study is then exampled with preliminary simulation results in Section 3. Conclusion and discussion is made at the end.

2. Belief-based Model

To quantitatively analyse the effects of social influence on individuals' decision making, we consider that a group of individuals forms an interactive network in which they can communicate with each other through social links and they make decisions based on their degrees of belief which reflect the perceived opinions. Those individuals are assumed with limited prior knowledge and hence, may experience difficulty in making firm decisions. Instead of responding immediately, they tend to receive heterogeneous information across networks and revise their belief structures frequently based on the newly-arrived evidence until they can make decisions. As an exploratory research, self-awareness is not considered in this paper.

A belief-based model [11] is therefore, utilised to characterise this situation. To better reflect the evolving process of belief updates, we further introduce a coefficient describing the willingness of an individual to be influenced by another. As individuals and their interactive relationships are represented by a graph in terms of nodes and edges, this coefficient is also regarded as the edge weight in network analysis. For further understanding the pattern in social influenced decision making, we adopt the concept of belief boundary [13] as the turning point from belief updates to decision making. It assumes in the model that an individual will make a decision as soon as his/her belief value exceeds the corresponding boundary. From that point, individual's belief structure will no longer change. In general, the evolving process of belief under the circumstance of influence diffusion in social networks is formulated with four stages in this model. These are belief formulation, belief spreading and updating, weighted belief fusion, and belief-to-decision making.

2.1. Belief Formulation

According to evidence theory, for any binary decision problem such as 'to accept or to reject', it can be denoted by the power set $2^\Omega = \{\emptyset, \{accept\}, \{reject\}, \Omega\}$, where Ω is a finite set of possible conclusions. It satisfies the conditions that each element of the set Ω is mutually exclusive and collectively exhaustive.

Accordingly, the mass functions (also called belief structures) expressing the degree of belief that supports the claim on evidence for an individual's decision response will satisfy:

$$\begin{cases} m: 2^\Omega \to [0,1] \\ m(\emptyset) = 0 \\ m(\Omega) = 1 - m(accept) - m(reject) \end{cases} \quad (1)$$

All mass functions (i.e. m(accept), m(reject) & m(Ω)) range from 0 to 1. The former two represent an individual's certainty on his/her acceptance or rejection of the choice, respectively. m(Ω) indicates the uncertainty level a decision maker holds on whether to accept or reject. For this neutral state, it is named 'to wait' in the following section.

2.2. Belief Spreading and Updating

Beforehand, to describe that a bunch of individuals are socially connected and form a network where information can transmit via the links between each other, let the social network to be denoted by: $G = <N, E>$. It consists of a set of nodes (also referred to as individuals, vertices) $N = \{1, ..., n\}$ and edges (also referred to as links, ties) $E = \{e_{i,j} | 1 \le i, j \le n; i \ne j\}$.

Fig. 1 displays the mechanics of belief diffusion in a directed social network. When any node (in Fig. 1a) is fed (i.e. has firm attitude), he/she will turn to his/her adjacent neighbours and pass the information coupling with belief. Those being infected in Fig. 1b will revise their degrees of belief in accordance with the new evidence and then diffuse theirs in the next time step (shown in Fig. 1c). Fig. 1d shows the spreading process for time step 3.

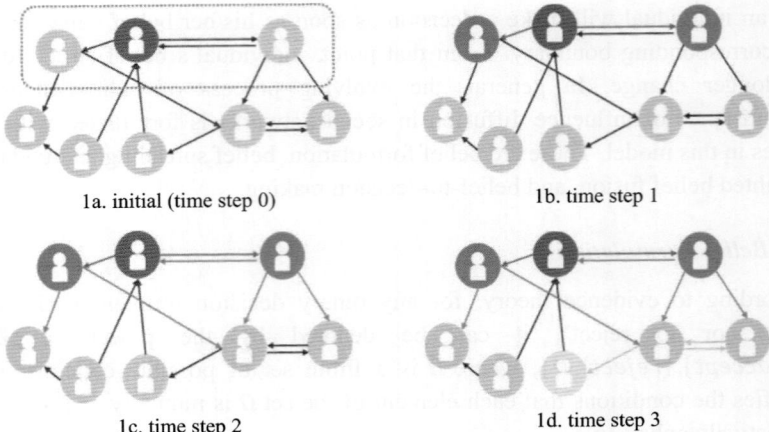

Fig. 1. Typical scenarios for belief diffusion in Social Network.

2.3. *Weighted Belief Fusion*

When information begins to spread along the connections, belief structures of those undecided nodes are presumed to update in accordance with three factors in this model, that is, what do they believe (i.e. individual's belief structures), what does the information source believe (i.e. social neighbour's belief structures), and how willing they are to accept his/her influence (i.e. interpersonal influence). The former two are believed to be affected by the initial belief status that each node holds. For the third one, to correspond to individual i's level of importance on j's certainties of belief, or individual j's openness to the influence of i, we consider a coefficient w_{ij} which is also regarded as the weight of edge $e_{i,j}$ in network analysis. Based on this coefficient, for any evidence transmitted from individual i to j, the mass functions can be described as follows:

$$\begin{cases} m_j^n(accept) = m_i(accept) \cdot w_{ij} \\ m_j^n(reject) = m_i(reject) \cdot w_{ij} \\ m_j^n(\text{wait}) = 1 - m_j^n(accept) - m_j^n(reject) \end{cases} \quad (2)$$

where $w_{ij} \in [0,1]$ represents how willing individual j is to rely on i's opinion or influence. For any directed network, $w_{ij} \neq w_{ji}$.

As assumed, individual j will continuously revise his/her beliefs by combining the present values m_j with the newly received evidence m_j^n until he/she makes up his/her mind. To realise this belief combination, denoted by $m'_j = m_j \oplus m^n_j$, according to Dempster's rule of combination, the updated belief structure is computed as follows:

$$m'_j(Z) = \frac{1}{1-K}\sum_{A \cap B = Z \neq \emptyset} m_j(A) \cdot m^n_j(B) \qquad (3)$$

where $A, B, Z \subseteq \Omega$. $\oplus$ is the combination operator. The denominator is also called normalisation factor with $K = \sum_{A \cap B = \emptyset} m_j(A) \cdot m^n_j(B)$. It measures the summation of the amount of conflicts between two evidences.

2.4. Belief-to-Decision Making

To further provide an insight into social influenced decision making, we consider the idea of belief boundary in this model. With respect to Ref. [13], boundary mechanism not only provides an ecological function to constrain the evidence needed for rendering a decision but also provides a mechanistic function to determine the termination of a decision process. Therefore, a threshold φ_n which represents individual's upper bound of uncertainty to certainty is assigned to each individual. The value is randomly taken from a uniform distribution $f(\varphi)$. It is assumed that an individual j will change his/her state to acceptance level if his/her mass function $m_j(accept) \geq \varphi_j$, vice versa. When an individual's state becomes stable (i.e. he/she has made a decision), his/her belief functions will no longer change.

3. A Case Study

To demonstrate the effectiveness and practicability of the model, we consider the following hypothetical scenario as a case study. A small company plans to launch a new pension strategy. They want to predict the attitudes across 46 employees and their acceptability. The dataset used in the analysis was originally collected by Cross & Parker [14]. It recorded the relationships among employees as well as their frequency of information or advice requests within the company. Here, we presume that an individual is more open to influence if he/she asks for advice more frequently. The weight coefficient is hence, determined by the frequency. Besides, this is a directed network and the arrow implies the direction to which information and influence may go.

3.1. Basic Statistics of Social Network

The characteristics of network are vital to information diffusion and knowledge transfer. Therefore, Table 1 summarises the basic statistics of network analysis. It implies a relatively dense network. This matches the background of the case that all nodes are the employees from one small company. A number of scholars have proved that network density will benefit the transmitter activity in social networks. In this regard, the spread of influence is expected to be very rapid in this network.

Table 1. Social Network Analysis Measures.

Network Structural Measures	Density	0.4236715
	Diameter	3
	Average path length	1.596591
Centrality measures	Degree centrality	NODE 2 & 20
	Betweenness centrality	NODE 12
	Eigenvector centrality	NODE 12 & 16

3.2. Example of Belief Diffusion Process in Social Networks

For the sake of illustration, an example of belief diffusion in networks is visualised in Fig. 2. We start from randomly assigning belief value and threshold value to each node and make sure no one will have firm attitude except for node 20 who is seeded in the beginning. As shown in Fig. 2a, individuals' degree of belief is attributed to the gradation of node's colour. Red colour or green for each node means that individual may have higher potential to accept or reject, respectively. For example, both node 12 and node 22 in Fig. 2a show red while the former is darker than the latter. This is because their belief of acceptance dominates the other mass variables in the beginning and node 12 has relative higher certainties on acceptance than node 22. In all, five gradations of colour are set for both red and green. Those who have firm attitudes of acceptance will show rouge colour like the single node in Fig. 2a and the majority in Fig. 2c.

In Fig. 2b, nodes who have direct connections with node 20 start to revise their beliefs because of the evidence obtained. This phenomenon continues and ends at Fig. 2d as almost everyone has made decisions during the diffusion process.

It is noticed that the majority nodes have turned to rouge colour in Fig. 2d except for those three on the right corner. The main reason is that node 15 and node 24 are only on the source list while node 30 will never be informed until node 24 is. In other words, unless node 15 and node 24 have their decisions by self-awareness, all three individuals will not make any decision during the process of diffusion.

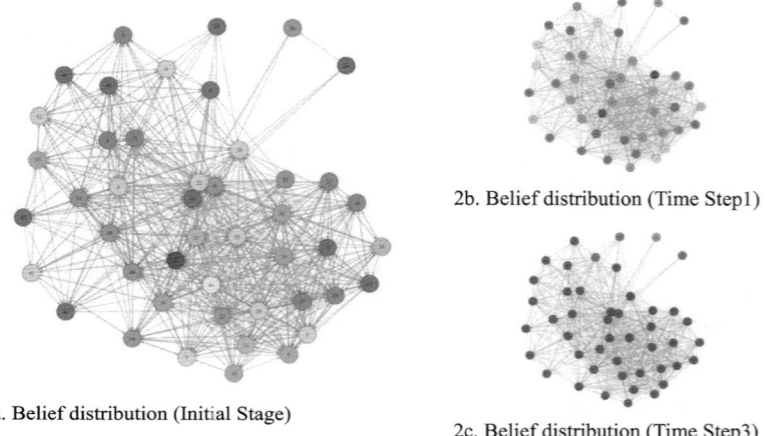

2a. Belief distribution (Initial Stage)

2b. Belief distribution (Time Step1)

2c. Belief distribution (Time Step3)

Fig. 2. An example of belief diffusion in networks.

4. Discussion

Social networks allow users to interact. Users express opinions in these interactive networks. Information propagation further brings about influence to spread. This interdependent relationship makes modelling the process of influence diffusion in networks popular in research domains. However, the complexity in mechanics leaves it a remaining challenge to conquer. In this paper, we provided a new perspective to visually and computationally characterise the spread of influence by using belief functions and its impact on individual's decision making. The weight coefficient can represent individual's openness to the influence obtained from others. To better illustrate the variation in social influenced decision making, the model is applied to a small case study.

References

1. Kelman, H.C., 1958. Compliance, identification, and internalization three processes of attitude change. *Journal of conflict resolution*, 2(1), pp. 51-60.
2. Latane, B., 1981. The psychology of social impact. *American psychologist*, 36(4), p. 343.
3. Bello-Orgaz, G., Hernandez-Castro, J. and Camacho, D., 2017. Detecting discussion communities on vaccination in twitter. *Future Generation Computer Systems, 66*, pp. 125-136.
4. Sprenger, T.O., Tumasjan, A., Sandner, P.G. and Welpe, I.M., 2014. Tweets and trades: The information content of stock microblogs. *European Financial Management*, 20(5), pp. 926-957.

5. Genius, M., Koundouri, P., Nauges, C. and Tzouvelekas, V., 2013. Information transmission in irrigation technology adoption and diffusion: social learning, extension services, and spatial effects. *American Journal of Agricultural Economics*, *96*(1), pp. 328-344.
6. Chamley, C., Scaglione, A. and Li, L., 2013. Models for the diffusion of beliefs in social networks: An overview. *IEEE Signal Processing Magazine*, *30*(3), pp. 16-29.
7. Garcia, D., Mavrodiev, P., Casati, D. and Schweitzer, F., 2017. Understanding popularity, reputation, and social influence in the twitter society. *Policy & Internet*, *9*(3), pp. 343-364.
8. Halberstam, Y. and Knight, B., 2016. Homophily, group size, and the diffusion of political information in social networks: Evidence from Twitter. *Journal of Public Economics, 143*, pp. 73-88.
9. Jones, J.J., Bond, R.M., Bakshy, E., Eckles, D. and Fowler, J.H., 2017. Social influence and political mobilization: Further evidence from a randomized experiment in the 2012 US presidential election. *PloS one, 12*(4), p. e0173851.
10. Jendoubi, S., Martin, A., Liétard, L., Hadji, H.B. and Yaghlane, B.B., 2017. Two evidential data based models for influence maximization in Twitter. *Knowledge-Based Systems, 121*, pp. 58-70.
11. Xia, S. and Liu, J., 2014. A belief-based model for characterizing the spread of awareness and its impacts on individuals' vaccination decisions. *Journal of The Royal Society Interface*, 11(94), p. 20140013.
12. Lu, X., Mo, H. and Deng, Y., 2015. An evidential opinion dynamics model based on heterogeneous social influential power. *Chaos, Solitons & Fractals*, 73, pp. 98-107.
13. Zhang, J., 2012. The effects of evidence bounds on decision-making: theoretical and empirical developments. *Frontiers in psychology*, *3*.
14. Cross, R., Parker, A., Christensen, C.M., Anthony, S.D. and Roth, E.A., 2004. *The hidden power of social networks*. Audio-Tech Business Book Summaries, Incorporated.

Pareto-smoothed inverse propensity weighing for causal inference

Fujin Zhu,[1,2] Adi Lin,[2] Guangquan Zhang,[2] Jie Lu[2] and Donghua Zhu[1]

[1] School of Management and Economics, Beijing Institute of Technology
Beijing 100081, China
[2] Center for Artificial Intelligence (CAI), School of Software, FEIT,
University of Technology Sydney, NSW 2007, Australia

Causal inference has received great attention across different fields ranging from economics, statistics, biology, medicine, to machine learning. Observational causal inference is challenging because confounding variables may influence both the treatment and outcome. Propensity score based methods are theoretically able to handle this confounding bias problem. However, in practice, propensity score estimation is subject to extreme values, leading to small effective sample size and making the estimators unstable or even misleading. Two strategies — truncation and normalization — are usually adopted to address this problem. In this paper, we propose a new Pareto-smoothing strategy to tackle this problem. Simulations and a real-world example validate the effectiveness.

1. Introduction

To minimize the confounding bias in observational causal inference, statistical "case-mix adjustment" techniques are frequently adopted. Among them, Rosenbaum and Rubin [1] introduced the propensity score to summarize the information required to control the confounders. The propensity score is the conditional probability of an individual to be assigned to the treatment group. Theoretically, one can account the difference between the treatment and control groups by directly modelling the assignment mechanism with propensity scores, and thus making the treated and control populations more comparable.

Though propensity score provides us a convenient solution to ease the issue of confounding, the true propensity scores are intrinsically unknown in pure observational studies. A practical concern is that the causal effect may be difficult to estimate precisely if the estimated propensity score is close to zero for a substantial fraction of the population [2]. This is a particular concern in setting with many covariates or the assignment mechanism is highly skewed.

When many of the estimated propensity scores are close the zero, the distribution of their reciprocals — the inverse propensity (IP) weights — can have a heavy right tail, which will lead to unstable inverse propensity weighting estimates, sometimes with infinite variance. To cope with this problem, methods including truncation and self-normalization have been proposed [3-5]. In this paper, we propose a new Pareto-smoothing strategy. Compared with truncation, our method is less biased. Compared with the normalization strategy, our experiment result shows that they both converge to the true value if we have enough data. One special merit of our method is that it is more stable in the small sample size cases, which are common in many real problems.

The reminder of the paper is organized as follows. In Section 2, we formalize the causal inference problem, introduce the concept of propensity score and two stabilization strategies for propensity score based estimators. Section 3 illustrates the proposed strategy and methods for parameter estimation. Experiments on simulated and real data are conducted in Section 4. Section 5 concludes the paper.

2. Causal Inference and Inverse Propensity Weighting

2.1. *Notation and Problem Formalization*

Suppose there are N units X_i ($i = 1, \ldots, N$), denote the treatment condition for unit i with A_i, where $A_i = 0$ indicating that unit i received the control treatment and $A_i = 1$ the active treatment. Let Y be the outcome variable of interest. $Y_i(A)$ is defined as the potential outcome of unit i had she received treatment A. We postulate the existence of a pair of *potential outcomes* for each unit, $(Y_i(0), Y_i(1))$, and the observed outcome $Y_i = Y_i(A_i) = A_i Y_i(1) + (1 - A_i) Y_i(0)$. With this notation, the individual treatment effect for unit i is $\tau_i = Y_i(1) - Y_i(0)$ and the average causal effect (aka, average treatment effect, ATE) is its expectation, i.e., $\tau = \mathbb{E}[\tau_i] = \mathbb{E}[Y_i(1)] - \mathbb{E}[Y_i(0)]$.

ATE measures the expected causal difference of a population if all of them were treated versus all were untreated, which is generally different from the conditional difference $\mathbb{E}[Y_i|A_i = 1] - \mathbb{E}[Y_i|A_i = 0]$. As a baseline, we also denote the empirical conditional difference as the naïve ATE estimator in Eq. (1)

$$\hat{\tau}_{naive} = \frac{1}{N_1} \sum_{i=1}^{N} A_i Y_i - \frac{1}{N_0} \sum_{i=1}^{N} (1 - A_i) Y_i \tag{1}$$

where $N_1 = \sum_{i=1}^{N} A_i$ is the number of treated and $N_0 = N - N_1$ the number of control.

Estimating ATE from observational data is generally impossible because of the fundamental problem of causal inference [4]. Under the conditional exchangeability (or unconfoundedness) condition, $Y_i(0), Y_i(1) \perp\!\!\!\perp A_i | X_i$, Pearl [6] proves that the ATE can consistently estimated by Eq. (2) as:

$$\tau = \int (\mathbb{E}[Y_i|A_i = 1, X_i = x] - \mathbb{E}[Y_i|A_i = 1, X_i = x]) dP(x) \tag{2}$$

This formula is also called the G-computation formula [7] and the back-door adjustment formula [6]. Although feasible for estimating ATE in principle, it is in practice infeasible to implement with many covariates. In the following section, we introduce the propensity score and its importance for solving this challenge.

2.2. Propensity Score and Inverse Propensity Weighting (IPW)

As discussed earlier, adjusting for all observed covariates to eliminate confounding bias may go out of the question. As the coarsest balancing score [4], the propensity score is a scalar proxy of them that suffices for removing the bias associated with imbalance in the pre-treatment covariates and is defined as:

Definition 1 (Propensity Score, PS). *The propensity score* $e(X_i)$ *is the conditional probability of an individual* X_i *to be assigned to the treatment group.*

Defining the inverse propensity weight (IP weight) for unit i as

$$w_i = \frac{1}{p(A_i|X_i)} = \frac{\mathbb{I}(A_i = 1)}{e(X_i)} + \frac{\mathbb{I}(A_i = 0)}{1 - e(X_i)} \quad (3)$$

where $\mathbb{I}(a_i = a)$ is the indicator function, we can build a balanced pseudo-population where the treatment assignments are randomized and all confounding is removed. The conditional difference in this super population consistently estimates τ by the inverse propensity weighted (IPW) estimator [8] [1]

$$\hat{\tau}_{IPW} = \frac{1}{N}\sum_{i=1}^{N}\frac{\mathbb{I}(A_i = 1)}{p(A_i = 1|X_i)}Y_i - \frac{1}{N}\sum_{i=1}^{N}\frac{\mathbb{I}(A_i = 0)}{p(A_i = 0|X_i)}Y_i \quad (4)$$

Note that the propensity scores $e_i = e(X_i)$ occur in the denominator of Eq. (3), we thus need to make the "positivity" or "overlapping" assumption, for all $i, 0 < e(X_i) < 1$, so that the IP weights are bounded, $w_i < \infty$. Theoretically, $\hat{\tau}_{IPW}$ is unbiased and consistent under this positivity assumption if we have infinite many observations. However, for finite data, the estimated propensities $\hat{e}(X_i)$ can be very close to zero for some $X_i = x$. An extreme case may occur that there are regions of covariate values observed in only one of the two treatment conditions. In this case, the IP weights w_i will be highly variable and even unbounded, thus estimation based on then will be unstable and misleading.

2.3. Stabilization by Truncation and Normalization

To remedy the issue of high variability, there are mainly two strategies for stabilization [5]: truncation (aka clipping) and normalization of the propensity score. The truncated IPW estimator for causal inference is given by

$$\hat{\tau}_{T-IPW} = \frac{1}{N}\sum_{i=1}^{N}\frac{\mathbb{I}(A_i = 1)}{g_i(A_i|X_i)}Y_i - \frac{1}{N}\sum_{i=1}^{N}\frac{\mathbb{I}(A_i = 0)}{g_i(A_i|X_i)}Y_i \tag{5}$$

with the estimated treatment probabilities truncated by a constant C:

$$g_i(A_i|X_i) = \begin{cases} C, & \text{constant} \quad \text{if } p(A_i|X_i) < \delta \\ p(A_i|X_i), & \text{else} \end{cases} \tag{6}$$

A consequence of PS truncation is the introduction of bias in the estimated PS, which in turn causes bias in PS-based causal estimators. Moreover, the cut-point δ is usually unknown and choosing it relies on experience or intuition. Recently, [9] propose a data-adaptive PS truncation algorithm which can select the optimal truncation threshold adaptively, but it is specially designed for target maximum likelihood estimators [10].

Alternatively, the normalized IPW estimator [5,11] divides the IP weights by the empirical mean of each treatment group and is given by

$$\hat{\tau}_{N-IPW} = \frac{\frac{1}{N}\sum_{i=1}^{N}\frac{\mathbb{I}(A_i = 1)}{p(A_i = 1|X_i)}Y_i}{\frac{1}{N}\sum_{i=1}^{N}\frac{\mathbb{I}(A_i = 1)}{p(A_i = 1|X_i)}} - \frac{\frac{1}{N}\sum_{i=1}^{N}\frac{\mathbb{I}(A_i = 0)}{p(A_i = 0|X_i)}Y_i}{\frac{1}{N}\sum_{i=1}^{N}\frac{\mathbb{I}(A_i = 0)}{p(A_i = 0|X_i)}} \tag{7}$$

For a set of IP weights $W = \{w_1, w_2, \ldots, w_N\}$, denote $\overline{W} = \frac{1}{N}\sum_{n=1}^{N} w_n$ and $\overline{W^2} = \frac{1}{N}\sum_{n=1}^{N} w_n^2$, we also define the effective sample size as $N_{eff} = \frac{N\overline{W}^2}{\overline{W^2}}$, which will be used as a measure of stability in the experiment sections. If the weights are highly imbalanced, they will have a high sampling variance, and the resulting estimate will be unreliable with a very small N_{eff}.

3. Pareto Smoothing for Causal Inference

Our method builds upon results in the extreme value theory [12]. The idea is simple, given the estimated IP weights $\{w_1, w_2, \ldots, w_N\}$, we fit a generalized Pareto distribution (GPD) on these extreme values, and replace them with order statistics of the fitted GPD. By this smoothing strategy, we try to stabilize the IP weights while keep the information of their relative order.

3.1. The Generalized Pareto Distribution

Among the series of extreme value distributions in the extreme value theory [12], the generalized Pareto distribution, named by Pickands [13], is a family of extreme value distributions that is often used to model the tails of another distribution. A GPD is specified by the location μ, scale $\sigma > 0$, and shape κ:

$$F(x) = \left[1 - \left(1 + \frac{\kappa(x-\mu)}{\sigma}\right)^{-\frac{1}{\kappa}}\right]\mathbb{I}(\kappa \neq 0) + \left(1 - e^{-\frac{x-\mu}{\sigma}}\right)\mathbb{I}(\kappa = 0) \qquad (8)$$

where the μ is a lower bound, i.e., $x \in (\mu, \infty)$. Pickands [13] proves that if an unknown distribution function $F(x)$ lies in the "domain of attraction" of some extremal distribution function, then $F(x)$ has a generalized Pareto upper tail.

3.2. Parameters Estimation

To fit the parameters $\theta = (\mu, \sigma, \kappa)$, we follow [14] and choose the location parameter μ so that the size of the *upper-tail* is

$$M = min(\lfloor 0.2S \rfloor, \lfloor 3\sqrt{S} \rfloor) \qquad (9)$$

Having decided the location μ, the other two parameters σ and κ can be estimated by maximum likelihood [12]. Given a random sample $X = \{x_1, x_2, \dots, x_M\}$, [15] reparametrize Eq. (8) by two parameters (α, κ), where $\alpha = \kappa/\sigma$, and the estimate $\hat{\alpha}$ is obtained by maximizing a *profile likelihood function* with a weakly informative prior, κ and σ are estimated by

$$\hat{\kappa} = \frac{1}{M}\sum_{i=1}^{M} log(1 - \hat{\alpha}x_i), \qquad \hat{\sigma} = \frac{\hat{\kappa}}{\hat{\alpha}} \qquad (10)$$

3.3. Summary of the Pareto-smoothed IPW Estimator

Given a set of N observations $\mathcal{D} = \{X_i, A_i, Y_i\}_{i=1}^{N}$, our proposed Pareto-smoothed IPW method can be easily implemented and proceeds as follows:

1. Estimate the propensity scores and get $\{e_i, i = 1, 2, \dots, N\}$;
2. Sort e_i descending, calculate M by Eq. (9) and choose the corresponding μ;
3. Let $\mu = 1/\mu$, and calculate the IP weights $\{w_i, i = 1, 2, \dots, N\}$;
4. Estimate σ and κ using the largest M IP weights by Eq. (10);
5. Replace the largest M weights with ordered statistics of the fitted GPD, and obtain the "Pareto-smoothed" weights $\{w_i^{PS}, i = 1, 2, \dots, N\}$;
6. Estimate the ATE using $\{w_i^{PS}, i = 1, 2, \dots, N\}$ by

$$\hat{\tau}_{PS-IPW} = \frac{1}{N}\left(\sum_{i:A_i=1} \frac{w_i^{PS}}{\overline{w^{PS}}_t} Y_i - \sum_{i:A_i=0} \frac{1 - w_i^{PS}}{\overline{w^{PS}}_c} Y_i\right) \qquad (11)$$

where $\overline{w^{PS}}_t = \frac{1}{N}\sum_{i:A_i=1} w_i^{PS}$ and $\overline{w^{PS}}_c = \frac{1}{N}\sum_{i:A_i=0}(1 - w_i^{PS})$.

4. Experimental Study

In this section, we validate our proposed method using simulated and semi-simulated data. In all the experiments, we use logistic regression to fit the propensity score model. The mean absolute error (MAE) $\epsilon_{ATE} = \frac{1}{n}|\sum_{i=1}^{n}(Y_i(1) - Y_i(0) - \hat{\tau}_i)| = \frac{1}{n}|\sum_{i=1}^{n}(\tau_i - \hat{\tau}_i)|$ will be reported. An application on a real world job training study is also conducted.

4.1. *Simulated and Semi-simulated Data*

The specific data-generating process of our simulation is: $X_{i,1} \sim Bernoulli(0.5)$, $X_{i,2} \sim Binomial(3, 0.5)$, $(A_i|X_i) \sim Bernoulli(\text{Sigmoid}(-1.3 - 3X_{i,1} + 3X_{i,2}))$, $(Y_i|X_i, A_i) \sim Bernoulli(\text{Sigmoid}(-2 - 2X_{i,1} + 3X_{i,2} + 3A_i + 2AX_i))$. We simulate data with sample size N ranging from 100 to 10^5, and run each simulation 10 times. Comparisons of the MAE and effective sample size are in Fig. 1. We know that on one hand, $\hat{\tau}_{PS-IPW}$ is less biased than $\hat{\tau}_{T-IPW}$. On the other hand, the estimate of $\hat{\tau}_{PS-IPW}$ converges together with $\hat{\tau}_{IPW}$ and $\hat{\tau}_{N-IPW}$ to the true estimate as the sample size gets large, say 10^4. Actually, both $\hat{\tau}_{IPW}$ and $\hat{\tau}_{N-IPW}$ are theoretically unbiased, but when the sample size is relatively small, their estimates are unstable compared with our Pareto-smoothed estimator. This indicates the advantage of our method in the small data cases. As to the effective sample size, since many of the IP weights are truncated to the same value, the effective sample size of $\hat{\tau}_{T-IPW}$ is supposed to be high. However, $\hat{\tau}_{PS-IPW}$ has higher effective sample size than $\hat{\tau}_{IPW}$ and $\hat{\tau}_{N-IPW}$ in general.

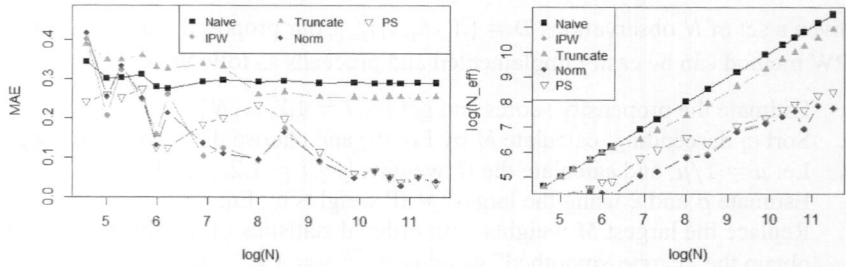

Figure 1. Comparison of the MAE (left) and log effective sample sizes (right) of different estimators.

Table 1. ATE estimates and effective sample size for the IHDP data.

	Naïve	IPW	T-IPW	N-IPW	PS-IPW
MAE	4.782	0.32	2.894	0.008	**0.0008**
N_{eff}	747	304.247	608.234	292.390	273.241

We also evaluated the performance of our algorithm through the semi-simulated IHDP dataset introduced in [16]. It is based on covariates from a real randomized experiment that evaluated the impact of the IHDP on the subjects' IQ test scores at the age of three while all outcomes are simulated. In total, the dataset consists of 747 subjects (139 treated, 608 control), and 25 covariates measuring properties of children and their mothers. The MAE and effective sample size results are listed in Table 1. Our proposed method outperforms other estimators regarding MAE. Actually, while the truncation strategy suffers a relatively high bias, the performances of $\hat{\tau}_{N-IPW}$ and $\hat{\tau}_{PS-IPW}$ are very close.

4.2. Real Data: NSW Job Training Study

As an application of the methods introduced in this paper, we use the randomized experiment data of [17], which is part of the "National support work" (NSW) demonstration programme implemented in the mid-1970s to study whether a systematic job-training programme would increase post-intervention income levels among workers [18]. In this paper, we simply use the nsw dataset in the R package ATE,[a] which provides LaLonde's original 722 observations (297 treated and 425 control). The kernel density fits of the estimated IP weights in Fig. 2 indicates the imbalance between the treatment and control group. The resulting estimates are $\hat{\tau}_{naive} = -537.803, \hat{\tau}_{IPW} = 3.696, \hat{\tau}_{T-IPW} = 736.033, \hat{\tau}_{N-IPW} = 798.488$, and $\hat{\tau}_{PS-IPW} = 805.881$. The result again validate the performance similarity between our Pareto-smoothing strategy and the normalization strategy.

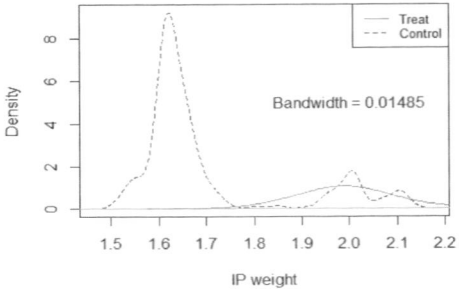

Figure 2. Comparison of density distribution of the estimated IP weights for the NSW dataset.

5. Conclusion

In this paper, we concluded two stabilization strategies for handling the problem of IP weights variability in PS-based causal inference, and proposed a new Pareto-smoothing strategy. Empirical results indicate that the proposed method has

---

[a]https://cran.r-project.org/package=ATE

appealing advantages, i.e., it is less biased than brute-force truncation and more stable than the normalization strategy in the small sample size setting. Though empirically appealing, our future work will be in its theoretical analysis as well as its applications in other causal effect estimators, for example, propensity score matching and balancing estimators.

References

1. P. R. Rosenbaum and D. B. Rubin, Biometrika **70** (1), 41 (1983).
2. S. Athey, G. Imbens, T. Pham, and S. Wager, arXiv preprint arXiv:1702.01250 (2017).
3. S. L. Morgan and C. Winship, *Counterfactuals and causal inference*. (Cambridge University Press, 2014).
4. G. W. Imbens and D. B. Rubin, *Causal inference in statistics, social, and biomedical sciences*. (Cambridge University Press, 2015).
5. M. A. Hernán and J. M. Robins, *Causal Inference*. (Boca Raton: Chapman & Hall/CRC, forthcoming, 2018).
6. J. Pearl, *Causality: Models, Reasoning and Inference*. (Cambridge University Press, 2000).
7. J. Robins, Mathematical modelling **7** (9-12), 1393 (1986); J. M. Robins, in *Latent variable modeling and applications to causality* (Springer, 1997), pp. 69.
8. D. G. Horvitz and D. J. Thompson, Journal of the American statistical Association **47** (260), 663 (1952).
9. C. Ju, J. Schwab, and M. J. van der Laan, arXiv preprint arXiv:1707.05861 (2017).
10. M. S. Schuler and S. Rose, American journal of epidemiology **185** (1), 65 (2017).
11. A. Swaminathan and T. Joachims, in *NIPS* (2015), pp. 3231.
12. S. Coles, J. Bawa, L. Trenner, and P. Dorazio, *An introduction to statistical modeling of extreme values*. (Springer, 2001).
13. J. Pickands Iii, the Annals of Statistics, 119 (1975).
14. A. Vehtari, A. Gelman, and J. Gabry, arXiv preprint arXiv:1507.02646 (2015).
15. J. Zhang and M. A. Stephens, Technometrics **51** (3), 316 (2009).
16. J. L. Hill, Journal of Computational and Graphical Statistics **20** (1), 217 (2011).
17. R. J. LaLonde, The American economic review, 604 (1986).
18. K. C. G. Chan, S. C. P. Yam, and Z. Zhang, Journal of the Royal Statistical Society: Series B (Statistical Methodology) **78** (3), 673 (2016).

Feature selection and weighing for case-based reasoning system using random forests

Booma Devi Sekar and Hui Wang

School of Computing, Ulster University
Northern Ireland, UK

Case-based reasoning has become a successful technique that uses the previous experience as a problem-solving paradigm. It adapts or reuses the solutions of a similar problem to solve a new one. In a case-based reasoning system, it is important to have a good similarity retrieval algorithm to retrieve the most similar cases to the query case. However, we also note that in a medical domain with increased use of electronic health records, the availability of patient cases and the related attributes have increased. Thus, as a pre-processing step or as part of the retrieval algorithm, it becomes critical to select the most informative features to improve the retrieval efficiency and accuracy in a case-based reasoning system. In this paper, we explore random forest, a popular method in machine learning, for feature selection and weighting in a case-based reasoning system and investigate the case retrieval accuracy.

1. Introduction

In recent years with the rapid increase in electronic health record (EHR) adoptation, there has been a growing expectation in the development of personalised medicine, which aims to customize treatment for an individual patient based on their likelihood of response to a therapy. The move towards personalized medicine is supported by various technological advancements, escpecially in the area of machine learning and artificial intelligence. One such pathway is the development of personalized diagnostic/predictive model based on patient similarity.[1] Such model aims to identify and derive insights from patients similar to the query (new) patient and then analyze the derived insights in the diagnostic/prediction model to provide personalized treatment/prediction to the query patient.

Case-based Reasoning (CBR), an artificial intelligent approach has become the most trustworthy methodology for developing personalized diagnostic/ predictive model that is very close to human reasoning.[2] CBR adapts a supervised learning algorithm, which trains on previous experience in form of resolved cases stored in the case base to provide a solution to the new problem. Thus, unlike various other AI approaches such as rule-based reasoning, or neural networks, that generate abstract representations from a set of training examples, CBR methodology adapts instance-based learning and uses previous similar cases as the basis for decision making. A generic CBR system is composed of four consecutive processes, known as the CBR cycle, including retrieval, reuse, revise and retain.[3] The first and most important step is 'retrieval' that applies similarity retrieval algorithm in search for the most similar cases from the case base. The subsequent step, 'adaptation' (reuse and revise) uses the information acquired from the retrieved case(s) to solve the query case. Finally, 'retain' learns from the problem-solving experience and stores the new knowledge in the case base for solving future new problems. Among these four steps, case retrieval efficiency and accuracy are topics of great interest among researchers. For which many researchers focused on improving the retrieval performance by developing different similarity measure algorithms.[4,5] However, an important step overlooked, that could improve the case retrieval performance is the selection of the informative features for CBR system.

Moreover, with the increased use of EHR and big data in healthcare, availability of a large number of patient cases and the relevant clinical attributes are becoming more common. Although such large case base and clinical attributes could increase the coverage of the application domain to provide a solution for the new query case, it does increase the possibility of having irrelevant and redundant features in the case base, which in turn affects the retrieval performance. Thus, as a pre-processing step or as part of the model, it becomes critical to select the most informative features for building any diagnostic/ predictive model. In CBR system, feature selection and weighting could determine the representative features required and remove the redundant ones.

In this paper, we investigate Random Forests (RF)[6] algorithm for feature selection and weighting for a CBR system. The contribution of the paper is the comparative assessment of the univariate, recursive feature elimination (RFE), RFE with cross validation (RFE-CV) and tree based feature selection methods with RF to compute the feature weighting for a CBR system. The main goal is to examine on whether the selected important feature variable using the above methods could improve the retrieval efficiency and accuracy of a CBR system.

In the following section, we present the technical background of the feature selection method and similarity retrieval measure applied. In Sec. 3, we evaluate the feature section method by analyzing the breast cancer database obtained from Breast Cancer Surveillance Consortium (BCSC). Finally, in Sec. 4, we present the conclusion drawn from the results obtained and propose future research work in such research area.

2. Background

2.1. *Feature Selection with Random Forest*

In general, feature selection can be categorized as a filter, wrapper, and embedded methods. The filter method, execute feature selection independent of the chosen predictor model. It treats feature selection as a pre-processing step, and are usually applied to remove some spurious features from the data set and are not much useful for measuring the feature importance. The wrapper, on the other hand, estimates the feature importance of variables by evaluating the model of interest. They are generally based on a black box evaluator and therefore are constrained by the given predictor model and search strategy. Finally, the embedded method performs feature selection as a part of the model building process and are generally found to be more beneficial, but face the challenge of over-fitting.

In order to combine the best properties of the above three methods, in this paper we explore hybrid feature selection methods. Study shows that with different combinations, hybrid methods could achieve higher efficiency and accuracy in feature selection. In this paper, we will investigate univariate chi-square,[7] recursive feature elimination (RFE)[8] and tree based feature selection[9] with RF for feature weighting in a CBR system.

RF is a non-parametric and highly flexible model and thus when applied for measuring variable importance, it could capture both linear and non-linear relations in the data. It has an embedded feature ranking technique: 'variable importance measure', which can be used as a tool to select the important features aiding the predictor model. A simple variable importance measure would count the number of times each variable is selected by all individual trees in the ensemble. In this paper, a Gini index, which measures the weighted mean of the individual trees improvement at the splitting point is used to measure the variable importance. Gini index based measure can be computed in RF using Eq. (1).

$$G(t) = 1 - \sum_{k=1}^{Q} p^2(k|t) \tag{1}$$

where, for a given node t and estimated class probabilities $p(k|t)$, $k = 1, \ldots, Q$ and Q is the number of classes.

2.2. *Similarity Retrieval Measure for Case Based Reasoning System*

In the proposed model, the similarity retrieval algorithm is defined using "local" and "global" similarity functions. Local similarity function measures the distance between the simple attributes, whereas the global similarity function applies the results from local similarity measures to compare the compound attributes. In the proposed CBR system, a patient case is represented as a compound attribute, composed of several simple attributes, including physiological and clinical variables. Thus, local similarity functions are first applied to compute the distance between simple attributes in the query case against the ones characterizing patient cases in the case base. The result of local similarity measures of all simple attributes is then aggregated using the global similarity function to select the patient case(s) that are most similar to the query case from the precedent ones present in the patient case base. K-Nearest Neighbour (k-NN) is computed as the global similarity function to retrieve the top k similar cases to the query case from the patient case base. Given the query patient case $X = x_i$ and the local similarity measure computed for the N number of patient cases in the case base $Y_j = y_{ji}$ the Euclidian distance between the query and case base is computed using Eq. (2).

$$d(X, Y_j) = \sum_{i=1}^{N} \sqrt{x_i^2 - y_{ji}^2} \qquad (2)$$

Based on the above result, the k nearest patient cases are first located in the case base. The k-NN similarity measure is then computed, which measures the arithmetic mean output across patient cases in the case base and returns a value between 0 ~ 1, with 0 and 1 indicating the retrieved case being less and most similar to the query case, respectively.

3. Evaluation

For evaluating the feature selection using RF, we analyse the breast cancer database obtained from BCSC. The database consists of 2,392,998 index-screening mammograms from women who were not diagnosed with breast cancer previously. In order to have a manageable size of the dataset, BCSC performed a cross-classification of risk factors and outcome and aggregated the patient cases based on the frequency of each combination. The reduced dataset, now with categorical variables consists of 280,660 records. Among which, 6274 were diagnosed with invasive or ductal carcinoma in situ in breast and 175629 with no cancer. Various research works have been conducted based on BCSC database, some of which include, pathology identification,[10] examining patterns in mammography for different ethnic group,[11] and genetic testing for breast cancer.[12] Table 1 shows the breast cancer patient variables and corresponding categorization (coding) defined in the dataset. Here the variables 1-12 were used as the input variables and variable 13 as the classification variable for the analysis. The dataset was split into 70% for training and 30% for testing.

Table 1. List of breast cancer patient variables in BCSC risk-estimate database.

Variable	Name	Coding
1	menopause	0 = premenopausal; 1=postmenopausal; 9 = unknown
2	agegrp	1 = 35-39; 2 = 40-44; 3 = 45-49; 4 = 50-54; 5 = 55-59; 6 = 60-64; 7 = 65-69; 8 = 70-74; 9 = 75-79; 10 = 80-84
3	density	BI-RADS codes: 1 = almost entirely fat; 2 = scattered fibroglandular densities; 3 = heterogeneously dense; 4 = extremely dense; 9 = unknown
4	race	1 = white; 2 = Asian/ Pacific islander; 3 = black; 4 = Native American; 5 = other/mixed; 9 = unknown
5	hispanic	0 = no; 1 = yes; 9 = unknown
6	bmi	Body mass index: 1 = 10-24.99; 2 = 25-29.99; 3 = 30-34.99; 4 = 35 or more; 9 = unknown
7	agefirst	Age at first birth: 0 = age < 30; 1 = age ≥ 30; 2 = Nulliparous; 9 = unknown
8	nrelbc	Number of first degree relatives with breast cancer: 0 = zero; 1 = one; 2 = 2 or more; 9 = unknown
9	brstproc	Previous breast procedure: 0 = no; 1 = yes; 9 = unknown
10	lastmamm	Result of last mammogram before the index mammogram: 0 = negative; 1 = false positive; 9 = unknown
11	surgmeno	Surgical menopause: 0 = natural; 1 = surgical; 9 = unknown or not menopausal
12	hrt	Current hormone therapy: 0 = no; 1 = yes; 9 = unknown
13	cancer	Diagnosis of invasive or ductal carcinoma in situ breast cancer within one year of the index screening mammogram: 0 = no; 1 = yes

Evaluating with RF, four tests were performed on the dataset. In the first test, no features were selected and thus all variables (1-12) in Table 1 were used to classify the patients with no cancer and invasive cancer or ductal carcinoma in situ breast cancer. In the second test, univariate feature selection using chi-square method with a number of features to be extracted 'k' = 7 was applied. This extracted the 7 best features (agegrp, race, brstproc, nrelbc, bmi, density, and surgmeno), which were then used to classify the patients using RF. In the third test – RFE is applied by pre-specifying the number of features to be extracted 'k' = 7. In this method, weights are initially assigned to each feature, it then recursively eliminate the features whose absolute weight is smallest and extracts the 'k' most important features. The 7 best features extracted using RFE were agegrp, density, race, bmi, agefirst, nrelbc and hrt. We note that, compared with the univariate feature selection method, RFE method extracted two different features, namely agefirst and hrt.

In the last test, RFE-cross validation (RFE-CV) was applied. With cross validation, the optimal number of features to obtain the best accuracy and the relevant important features were identified using RFE. Three optimal features, namely agegrp, density, and bmi were extracted using the RFE-CV method. Finally, using tree based feature selection with RF, the feature importance method was applied to eliminate the correlated feature in each iteration and list all the attributes according to its feature importance in solving the classification problem. Table 2 presents the classification results of the above four tests in terms of accuracy, sensitivity, specificity, true positive rate, true negative rate, false positive rate and false negative rate. Fig. 1 shows the sequence of feature importance of the input attributes using tree-based feature importance measure using RF.

Table 2. Statistical results of the four tests presented in Figure 1.

Method	Accuracy (%)	Sensitivity (%)	Specificity (%)	True Positive Rate	True Negative Rate	False Positive Rate	False Negative Rate
a	94.76	6.11	97.89	0.093	0.9671	0.021	0.9388
b	96.49	0	99.91	0	0.9657	0.0008	1
c	96.51	0	99.93	0.026	0.9658	0.0008	0.9994
d	96.50	0	99.92	0	0.9657	0.0008	1

Method: (a) RF with no feature selection (b) Univariate feature selection using Chi2, K = 7 and RF (c) RFE-RF with k=7 (d) REF-CV and RF: Optimal number of features = 3.

The results in Table 2 show that a better accuracy is achieved when feature selection method was applied. Having comparatively less number of cases (6274) with invasive cancer or ductal carcinoma in situ in the breast than the ones with no cancer (175629), the classification result show a poor sensitivity but a good specificity. The same can be observed for the true negative rate (closer to 1 is better) and false positive rate (closer to 0 is better) when compared to the true positive rate (closer to 1 is better) and false negative rate (closer to 0 is better).

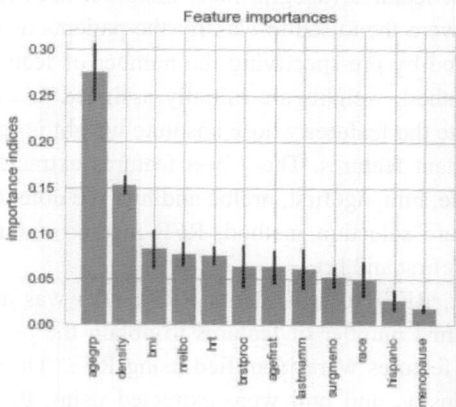

Figure 1. Variables listed according to the feature importance using RF.

As the main purpose of this paper is to evaluate on whether the feature importance using RF is useful for assigning weights to the description variables in the CBR system. As a case study, the features extracted using RF were applied to CBR system to evaluate on whether similar patient cases could be retrieved. Table 3 shows the results of similar cases retrieved for the same query case. Assigning $k = 3$ in the k-NN similarity measure, three similar cases were retrieved for each of the conditions presented above. Two tests were performed to evaluate on whether reducing the number of description variables in a CBR system could affect the performance of the case retrieval. In both the tests, wt = 1 was assigned to the important attributes, meanwhile, in the first test, the non-important attributes were eliminated by assigning wt = 0 and in the second test, a minimum wt = 0.1 was assigned to still include them in the computation of k-NN similarity measure.

Table 3 shows that in both the tests better accuracy was achieved when feature selection method was applied for feature weighting in the CBR system. However, when observing the attribute values which are different from the query case, we note that in the second test, in predominant of the cases, only one attribute was different from the query case. Whereas, in the first test, for the cases retrieved, many times two attributes were different from the query case, showing that it could not retrieve cases which were most similar to the query case.

4. Conclusion and Future Work

RF is a popular machine learning tool, frequently applied in solving various scientific problems, from feature selection, regression to classification. In this paper, we investigate RF for feature weighting the description variables in the CBR system and examine on whether feature selection could improve the case retrieval performance. Through evaluating the hybrid feature selection methods, including univariate, RFE, and tree based feature selection with RF on a breast cancer database obtained from BCSC, we conclude that feature selection with RF is a sensible approach for feature weighting in a CBR system. However, feature selection has to be done in two stages, first for a large dataset, all the spurious features have to be eliminated. This can be done by applying a filter method, such as the chi-square method, tested in this paper. Secondly, the feature importance method using RF or RFE-CV can be applied to select the important features. In terms of feature weighting for CBR system, to improve the case retrieval performance, it would be sensible to either use distributed weighting or assign minimum weight to the non-important attributes.

For future work, it would be worthwhile to test other feature selection methods such as Pearson correlation coefficient, mutual information, Gram-

Schmidt orthogonalization and hybrid feature selection methods. Also, evaluate on how they perform in comparison with the methods using RF presented for feature weighting in a CBR system. To assess the impact of feature selection methods, it would be useful to evaluate the performance of the feature selection methods using error rate and time scores performance matrices.

Table 3. Case retrieval results for a query case in a CBR system using the results of RF.

	Wt = 1 assigned to important attributes Wt = 0 is assigned to all other attributes			Wt = 1 is assigned to important attributes Wt = 0.1 is assigned to all other attributes		
	Case id	Retrieval accuracy (k-NN)	Attributes different from query case	Case id	Retrieval accuracy (k-NN)	Attributes different from query case
Query Patient	10000			10000		
Equal weight to all attributes (Wt = 1 assigned for all attributes)	10003	0.9166	agefirst	10003	0.9166	agefirst
	10495	0.9166	race	10495	0.9166	race
	10496	0.9166	race	10496	0.9166	race
Univariate - Chisquare, SelectKBest (K=7)	10003	1	agefirst	10003	0.9866	agefirst
	10002	1	agefirst, lastmamm	9999	0.9866	lastmamm
	10001	1	agefirst, nrelbc	9947	0.9866	hispanic
RFE with RF (K = 7))	9883	1	hispanic	9999	0.9866	lastmamm
	9947	1	hispanic	9947	0.9866	hispanic
	9946	1	hispanic, lastmamm	174378	0.973	hispanic, lastmamm
RFE with RF (with 3 Optimal features)	10001	1	agefirst, nrelbc	10003	0.9743	agefirst
	10002	1	agefirst, lastmamm	10495	0.9743	race
	10003	1	agefirst	10496	0.9743	race
RFE – CV (using distributed weight assigned according to variable importance measure with RF)	9947	0.981	hispanic	9947	0.981	hispanic
	10003	0.944	agefirst	10003	0.944	agefirst
	10495	0.944	race	10495	0.944	race

Acknowledgments

The DESIREE project has received funding from the European Union's Horizon 2020 research and innovation program under grant agreement No. 690238.

References

1. C.L. Parra-Calderón, Patient Similarity in Prediction Models Based on Health Data: A Scoping Review, *JMIR Med Inform.* 5(1), (2017).

2. X. Blanco, S. Rodríguez, J.M. Corchado, C. Zato, Case-based Reasoning applied to Medical Diagnosis and Treatment, *AISC,* 217 (2017).
3. J. Chen, Z. Teng, Z. Liu, A Review and Analysis of Case-based Reasoning Research, *ICITBS,* (2015).
4. D. Shasha, WALRUS: A Similarity Retrieval Algorithm for Image Databases, *IEEE TKDE,* (1999).
5. L. Liu, Z. Lin, L. Shao, Sequential Discrete Hashing for Scalable Cross-Modality Similarity Retrieval, *IEEE TIP,* 26(1), (2017).
6. L. Breiman, Random Forests, *Machine Learning,* 45, 5-32, (2001).
7. A.M. Bidgoli, M.N. Parsa, A Hybrid Feature Selection by Resampling, Chi-squred and Consistency evaluation Technique, IJCIE, 6(8), (2012).
8. X. Zeng et al., D.V. Alphen, Feature Selection using Recursive feature Elimination for Handwritten Digit Recognition, *IIH-MSP'09,* (2009).
9. G. Louppe et al., Understanding variable importance in forests of randomized trees, *Adv in Neural Information Processing Sys,* (2013).
10. D.L. Weaver et al., Pathologic findings from the Breast Cancer Surveillance Consortium, *Cancer,* 106(4), (2006).
11. F.D. Gilliland et al., Patterns of mammography using among Hispanic, American Indian and non-Hispanic White women in New Mexico, *Am J Epidemiol,* 152(5), (2000).
12. C.M. Velicer, Genetic testing for breast cancer: where are health care providers in the decision process, *Genet Med,* 3(2), (2001).

Comparative analysis on extended belief rule-based system for activity recognition

Long-Hao Yang

Decision Sciences Institute, Fuzhou University, Fuzhou, 350116, PR China
Department of Computer Science, University of Jaén, Jaén, Spain

Jun Liu

School of Computing and Mathematics, Ulster University, Northern Ireland, UK

Ying-Ming Wang

Decision Sciences Institute, Fuzhou University, Fuzhou, 350116, PR China
Key Laboratory of Spatial Data Mining & Information Sharing of Ministry of Education,
Fuzhou University, Fuzhou, 350116, PR China
msymwang@hotmail.com

Luis Martínez

Department of Computer Science, University of Jaén, Jaén, Spain

One of eminent activity recognition approaches is based on the extended belief rule-based system (EBRBS), which shows an excellent robustness comparing with many traditional approaches. For this reason, three versions of EBRBSs, namely original EBRBS (O-EBRBS), DRA-EBRBS, and Micro-EBRBS, are adapted in the comparative analysis to illustrate the performance of these EBRBSs. Results demonstrate that the Micro-EBRBS can produce a satisfied accuracy under less time of inference scheme and fewer numbers of activated rules comparing to the O-EBRBS and DRA-EBRBS.

Keywords: Extended belief rule-based system; activity recognition; comparative analysis.

1. Introduction

With the increasing number of elderly, researchers pay much more attention to the approaches of activity recognition, which can solve the key issue for the elderly when they stay as long as possible in their own homes with a healthy ageing and wellbeing. Hence, many approaches have been used for activity recognition in the past decade, such as data-driven approaches, which are based on machine learning techniques, and knowledge-driven approaches, which are based on the knowledge engineering and knowledge management techniques.

Recently, the extended belief rule-based system (EBRBS)[1], which can be data-driven, or knowledge-driven, or combining both, was used in the activity recognition to compare with the traditional approaches, including Naïve Bayes classifier, Nearest neighbor, Decision Table, and Support Vector Machines. The results suggest that the EBRBS can provide an encouraged performance against these approaches in terms of robustness.

However, the current application of the EBRBS on activity recognition is only based on the DRA-EBRBS[2], which is an extension of the original EBRBS (O-EBRBS for short). Apart from the O-EBRBS and DRA-EBRBS, a simplified version of the O-EBRBS in terms of time complexity and number of rules was proposed recently. Considering that the DRA-EBRBS has been successfully applied in activity recognition, it is valuable to investigate the applicability of O-EBRBS and Micro-EBRBS[3] for activity recognition.

Therefore, the previous three versions of EBRBSs, namely the O-EBRBS, DRA-EBRBS, and Micro-EBRBS are considered in this study to obtain the broadest possible overview. For this reason, a comparative analysis considering a popular activity recognition dataset is carried out to illustrate the potential performance of the three EBRBSs.

The remainder of this paper is organized as follows: Section 2 reviews the background of EBRBSs and their inference scheme for activity recognition. Section 3 provides the case study of comparative analysis for the three EBRBSs. Finally, Section 4 concludes this work.

2. EBRBSs for Activity Recognition

This section reviews briefly the components and differences of three versions of EBRBSs at first. Afterwards, the inference scheme of the O-EBRBS is introduced to illustrate the application on activity recognition.

2.1. *Background of EBRBSs*

As shown in the first version of EBRBS[1], there are two components as follows:

Component 1 (Extended belief rule base (EBRB)): This is a rule base composing of lots of extended belief rules, which include basic and generated parameters.

Component 2 (Inference scheme): This is the methodology applied to recognize activities of daily living based on the EBRB.

For the illustration of Component 1, suppose there are L extended belief rules, M antecedent attributes U_i ($i=1,…, M$) with J_i reference values $A_{i,j}$ ($j=1,…, J_i$), and one consequent attribute with N activities D_n ($n=1,…, N$). On the basis of these assumptions, experts are invited to determine the basic parameters, including utility value $u(A_{i,j})$ and attribute weight δ_i of each antecedent attribute U_i. Afterwards, the rule generation scheme[1] can be used to calculate the generated parameters, namely belief degrees $\alpha_{i,j}^k$ and $\beta_{n,k}$ and the rule weight θ_k of each extended belief rule R_k (k=1,..., L). Finally, the rule R_k is written as

$$IF\ U_1\ is\ \{(A_{1,j},\alpha_{1,j}^k); j=1,...,J_1\} \wedge \cdots \wedge U_M\ is\ \{(A_{M,j},\alpha_{M,j}^k); j=1,...,J_M\},$$
$$THEN\ D\ is\ \{(D_n,\beta_{n,k}); n=1,...,N\}, with\ \theta_k\ and\ \{\delta_1,...,\delta_M\} \quad (1)$$

For the illustration of Component 2, three versions of EBRBSs, namely the O-EBRBS, DRA-EBRBS, and Micro-EBRBS, are considered in this paper. Comparing to the O-EBRBS, the DRA-EBRBS has to determine the activated rules by using the dynamic method, and the Micro-EBRBS has to integrate the activated rules by using the classification ER algorithm, in which more details can be found in References 2 and 3, respectively. In the following subsection, the inference scheme of the O-EBRBS is introduced to illustrate the application of EBRBS on the activity recognition.

2.2. Inference scheme of O-EBRBS

While an EBRB is constructed based on expert knowledge and collected data, the O-EBRBS can be applied to recognize the activity of daily living based on Fig. 1 and the following steps.

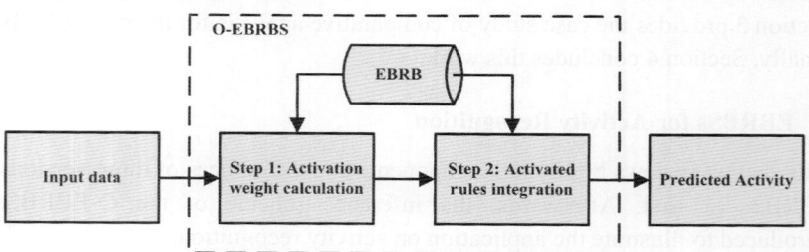

Fig. 1. Inference scheme of O-EBRBS for activity recognition.

Step 1 (Activation weight calculation): This step is to calculate activation weight for each rule in the EBRB. For each test input data x_i (i=1,..., M) in the input vector x, the belief distribution needs to be calculated by using the utility-based equivalence transformation technique[4] as follows:

$$S(x_i) = \{(A_{i,j},\alpha_{i,j}); j=1,...,J_i\} \quad (2)$$

where

$$\alpha_{i,j} = \frac{u(A_{i,j+1})-x_i}{u(A_{i,j+1})-u(A_{i,j})}\ and\ \alpha_{i,j+1}=1-\alpha_{i,j},\ if\ u(A_{i,j}) \leq x_i \leq u(A_{i,j+1}) \quad (3)$$

$$\alpha_{i,t} = 0, \text{ for } t = 1,\ldots, J_i \text{ and } t \neq j, j+1 \quad (4)$$

Afterwards, the similarity degree of the antecedent attribute set U in the kth ($k=1,\ldots, L$) rule is calculated by using the hamming distance:

$$S^k(\boldsymbol{x}, \boldsymbol{U}) = \begin{cases} 1, & \text{if } H^k(\boldsymbol{x}) = 0 \\ 0.2, & \text{if } H^k(\boldsymbol{x}) \leq 1 \\ 0.1, & \text{if } H^k(\boldsymbol{x}) \leq 2 \\ 0, & \text{if } H^k(\boldsymbol{x}) > 2 \end{cases}, \quad (5)$$

where

$$H^k(\boldsymbol{x}) = \sum_{i=1}^{M} \delta_i d_H^k(x_i) \quad (6)$$

$$d_H^k(x_i) = \begin{cases} 0, & \text{if } S(x_i) = S(x_{k,i}) \\ 1, & \text{otherwise} \end{cases} \quad (7)$$

where δ_i is the weight of the ith antecedent attribute, $S(x_i)$ is the belief distribution shown in Eq. (2), $S(x_{k,i})$ is the belief distribution of the sample input data used to generate the kth rule.

Finally, the activation weight of the kth extended belief rule is calculated by

$$w_k = \theta_k S^k(x_i, U_i) \quad (8)$$

where θ_k is the weight of the kth rule.

Step 2 (Activated rule integration): This step is to integrate the activated rules by using ER algorithm. After calculating activation weights based on Step 1, all activated rules (namely, $w_k > 0$ for $k = 1,\ldots, L$) should be integrated using the following analytical ER algorithm[5]:

$$\beta_n = \frac{\mu * \left[\prod_{k=1}^{L} \left(w_k \beta_{n,k} + 1 - w_k \sum_{i=1}^{N} \beta_{i,k} \right) - \prod_{k=1}^{L} \left(1 - w_k \sum_{i=1}^{N} \beta_{i,k} \right) \right]}{1 - \mu * \left[\prod_{k=1}^{L} (1 - w_k) \right]} \quad (9)$$

where $n = 1,\ldots,N$, w_k is given in Eq. (8), $\beta_{n,k}$ is given in Eq. (1) and

$$\mu = \left[\sum_{i=1}^{N} \prod_{k=1}^{L} \left(w_k \beta_{i,k} + 1 - w_k \sum_{j=1}^{N} \beta_{j,k} \right) - (N-1) \prod_{k=1}^{L} \left(1 - w_k \sum_{j=1}^{N} \beta_{j,k} \right) \right]^{-1} \quad (10)$$

Finally, the predicted activity of the EBRBS can be obtained by seeking the greatest belief degree from N activities D_n ($n = 1,\ldots, N$).

$$f(\boldsymbol{x}) = D_n, n = \arg\max_{i=1,\ldots,N} \{\beta_i\} \quad (11)$$

3. Comparative Analysis on Activity Recognition

In this section, the dataset of activity recognition and its simplified versions based on feature selection are illustrated firstly. Then, three versions of EBRBSs are investigated on the purpose of comparing their performances.

3.1. Activity recognition dataset

The dataset used in this paper is a popular activity recognition dataset[6], which was collected from the special smart environment that consists of a three-room apartment with 14 state-change sensors. By observing the daily living of a 26-year-old male in the smart environment, 245 data can be obtained within the 28 days in the apartment, in which these 245 data include 7 different activities, namely going to bed, using toilet, preparing breakfast, preparing dinner, getting a drink, taking a shower, and leaving the house.

Additionally, according to the two kinds of feature selection methods, two simplified versions of activity recognition datasets can be further obtained and the detailed information of these datasets is shown in Reference 7. Table 1 lists the main information of the above three datasets.

Table 1. Descriptions of activity recognition datasets.

Name	No. of sensors	No. of data	No. of activities
KasterenA-14	14	245	7
KasterenA-10	10	245	7
KasterenA-7	7	245	7

3.2. Comparative analysis for EBRBSs

In order to obtain the real performance of different EBRBSs, 5-fold cross-validation is applied to test each dataset, in which each dataset is divided into 5 blocks, with 4 blocks as a training dataset and remaining block as a testing dataset. Meanwhile, the result of each dataset and EBRBS is measured with: (1) accuracy, namely percentage of activities correctly recognized from the total; (2) Time of inference scheme (TIS), namely average time response of the EBRBS in milliseconds for each input data; (3) number of activated rules (NARs), namely average number of rules is activated for each input data. Table 2 shows the comparison results of EBRBSs for activity recognition.

As Table 2 illustrates, the Micro-EBRBS has the same accuracy as the O-EBRBS in three datasets, namely KasterenA-14, KasterenA-10, and KasterenA-7. However, the results of Micro-EBRBS are based on the less time of inference scheme and the fewer numbers of activated rules. For example, the time of O-EBRBS is 60 ms, 43 ms, and 59 ms for the three datasets.

Correspondingly, the time of Micro-EBRBS is 17 ms, 12 ms, and 7 ms. For the DRA-EBRBS, its best accuracy can be found in the dataset KasterenA-10 comparing to other EBRBSs. However, the time of the DRA-EBRBS is significantly more than the O-EBRBS and Micro-EBRBS. This is because the DRA-EBRBS has to perform the activation weight calculation multiple times owning to the dynamic method.

Table 2. Comparison of EBRBSs for activity recognition.

Data type	EBRBS type	Acc. (%)	TIS (ms)	NARs
KasterenA-14	O-EBRBS	**96.73**	60.0	93.29
	DRA-EBRBS	96.33	1355.0	40.81
	Micro-EBRBS	**96.73**	17.0	8.23
KasterenA-10	O-EBRBS	95.51	43.0	115.34
	DRA-EBRBS	**97.55**	1104.0	37.47
	Micro-EBRBS	95.51	12.0	7.47
KasterenA-7	O-EBRBS	**95.92**	59.0	113.99
	DRA-EBRBS	**95.92**	1176.0	40.27
	Micro-EBRBS	**95.92**	7.0	9.34

4. Conclusions

Activity recognition is the key issue to guarantee the independent lives of elderly in their homes for longer time. As an excellent rule-based system, the DRA-EBRBS has been successfully applied in activity recognition and shows an encouraged performance against the most popular approaches in terms of robustness.

In this study, apart from the DRA-EBRBS, the O-EBRBS and Micro-EBRBS are further investigated to compare with the DRA-EBRBS. The case study of comparative analysis shows that the Micro-EBRBS can produce a satisfied accuracy under the less time of inference scheme and fewer numbers of activated rules.

Acknowledgments

This research was supported by the National Natural Science Foundation of China (Nos. 61773123, 71371053, 71501047, and 71701050), the Humanities and Social Science Foundation of the Ministry of Education under Grant (No. 14YJC630056), the Natural Science Foundation of Fujian Province, China (No. 2015J01248), the Spanish National Research Project TIN2015-66524-P, the Spanish Ministry of Economy.

References

1. J. Liu, L. Martinez, A. Calzada, H. Wang, *Knowledge-Based Systems*, **53** (2013).
2. A. Calzada, J. Liu, H. Wang, A. Kashyap, *IEEE Transactions on Knowledge and Data Engineering*, **4**, 27, (2015).
3. L. H. Yang, J. Liu, Y. M. Wang, L. Martinez, *IEEE Transactions on Systems, Man, and Cybernetics: System*, (2018). (Submitted)
4. J. B. Yang, *European Journal of Operational Research*, **1**, 131, (2001).
5. Y. M. Wang, J. B. Yang, D. L. Xu, *European Journal of Operational Research*, **3**, 174, (2006).
6. T. Van Kasteren, A. Noulas, G. Englebienne, B. Krose, *in Proceedings of the 10th International Conference on Ubiquitous Computing, ACM*, pp. 1-9 (2008).
7. M. Espinilla, J. Medina, A. Calzada, J. Liu, L. Martinez, C. Nugent, *Microprocess and Microsystems*, **52**, (2017).

Part 4

Data Mining and Text Retrieval

On the reduction of biases in big data sets for the detection of irregular power usage

Patrick Glauner and Radu State

Interdisciplinary Centre for Security, Reliability and Trust, University of Luxembourg
1855 Luxembourg, Luxembourg
{patrick.glauner, radu.state}@uni.lu

Petko Valtchev

Department of Computer Science, University of Quebec in Montreal
H3C 3P8 Montreal, Canada
valtchev.petko@uqam.ca

Diogo Duarte

CHOICE Technologies Holding Sàrl
2453 Luxembourg, Luxembourg
diogo.duarte@choiceholding.com

In machine learning, a bias occurs whenever training sets are not representative for the test data, which results in unreliable models. The most common biases in data are arguably class imbalance and covariate shift. In this work, we aim to shed light on this topic in order to increase the overall attention to this issue in the field of machine learning. We propose a scalable novel framework for reducing multiple biases in high-dimensional data sets in order to train more reliable predictors. We apply our methodology to the detection of irregular power usage from real, noisy industrial data. In emerging markets, irregular power usage, and electricity theft in particular, may range up to 40% of the total electricity distributed. Biased data sets are of particular issue in this domain. We show that reducing these biases increases the accuracy of the trained predictors. Our models have the potential to generate significant economic value in a real world application, as they are being deployed in a commercial software for the detection of irregular power usage.

Keywords: Bias; class imbalance; covariate shift; non-technical losses.

1. Introduction

The contemporary Big Data paradigm can be summarized as follows: "It's not who has the best algorithm that wins. It's who has the most data."[1] However, in many cases, increasing the amounts of data is not a panacea since it can be biased: One frequently appearing bias results in training data and test data having different distributions, as depicted in Fig. 1. Learning from such training data leads to unreliable predictors that are not able to generalize to the test data. In the literature, this sort of bias is called covariate shift, sampling bias or sample selection bias. Covariate shift has been recognized as an issue in statistics since the mid-20th century.[2] In contrast, it has received only a limited attention in machine learning, mainly within the computational learning theory subfield, yet the situation is currently evolving.[3,4]

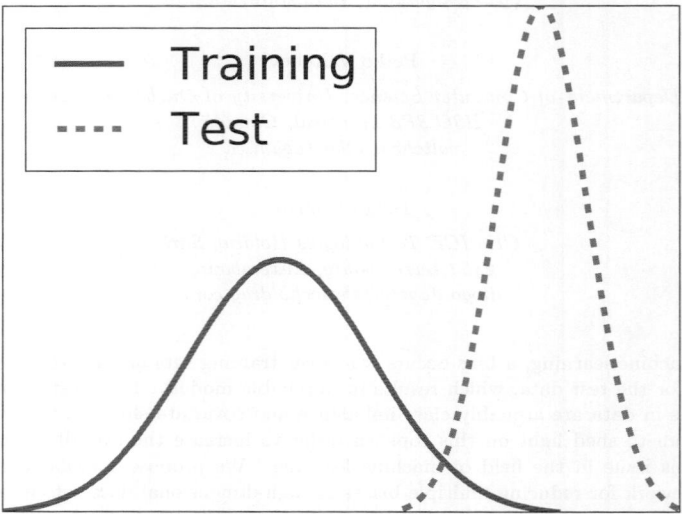

Fig. 1. Example of covariate shift: training and test data having different distributions.

Non-technical losses (NTL) appear in power grids during distribution and include, but are not limited to, the following causes: meter tampering in order to record lower consumptions, bypassing meters by rigging lines from the power source, arranged false meter readings by bribing meter readers or faulty or broken meters. NTL are more common in emerging countries, where electricity theft is the main contributor. NTL are reported to range up to 40% of the total electricity distributed in countries such as Brazil, India, Malaysia or Pakistan.[5,6] NTL are the source of major concerns for the electricity providers including financial losses and a decrease of stability and reliability in power grids. It is therefore crucial to detect customers that cause NTL. Recent research on NTL detection mainly uses machine learning models that learn anomalous behavior from customer data and known irregular behavior that was reported through on-site inspection results. In order to detect NTL more accurately, one may assume that having simply more customer and inspection data would help. We have previously shown that in many cases, the set of inspected customers is biased.[2] A reason for that is that past inspections have been largely focused on certain criteria and were not sufficiently spread across the population.

This paper builds on top of our previous contributions and aims at bias reduction in data, and further at more generalizable NTL predictors. Its main contributions are:

- We present a framework for reducing biases in data, such as class imbalance and covariate shift, in particular for spatial data.
- We propose a scalable novel methodology for reducing multiple biases in high-dimensional data sets at the same time.
- We report on how our method performs on the detection of NTL. Our method leads to a better detection of anomalous customers, subsequently reduces losses of electricity providers and thus increases stability and reliability of power distribution infrastructure.

2. Background and Related Work

In supervised learning, training examples $(x^{(i)}, y^{(i)})$ are drawn from a training distribution $P_{train}(X, Y)$, where X denotes the data and Y the label, respectively. The training set is biased if $P_{train}(X, Y) \neq P_{test}(X, Y)$. In order to reduce the bias, it has been shown that example $(x^{(i)}, y^{(i)})$ can be weighted during training as follows:[7]

$$w_i = \frac{P_{test}(x^{(i)}, y^{(i)})}{P_{train}(x^{(i)}, y^{(i)})}.$$

However, computing $P_{train}(x^{(i)}, y^{(i)})$ is impractical because of the limited amount of data in the training domain. It is for that reason that in the literature, predominantly two different types of biases are discussed: class imbalance and covariate shift. Class imbalance refers to the case where classes are unequally represented in the data. Therefore, we assume $P_{train}(X|Y) = P_{test}(X|Y)$, but $P_{train}(Y) \neq P_{test}(Y)$.[8] In contrast, for covariate shift, we assume $P_{train}(Y|X) = P_{test}(Y|X)$, but $P_{train}(X) \neq P_{test}(X)$.[9] Instance weighting using density estimation has been proposed for correcting covariate shift.[3] Furthermore, the Heckman method has been proposed to correct covariate shift.[10] However, the Heckman method only applies to logistic regression models. Other biases are reported in the literature, for example for change of functional relations, i.e. when $P_{train}(Y|X) \neq P_{test}(Y|X)$, or biases created by transforming the feature space.[7]

3. Reduction of Biases

We propose the following methodology: Given the assumptions made for class imbalance, we compute the corresponding weight for example i having a label of class k as follows:

$$w_{i,k} = \frac{P_{test}(x^{(i)}, y_k^{(i)})}{P_{train}(x^{(i)}, y_k^{(i)})} = \frac{P_{test}(x^{(i)}|y_k^{(i)})P_{test}(y_k^{(i)})}{P_{train}(x^{(i)}|y_k^{(i)})P_{train}(y_k^{(i)})} = \frac{P_{test}(y_k^{(i)})}{P_{train}(y_k^{(i)})}.$$

We use the empirical counts of classes for computing $P_{<dist>}(y_k)$. Given the assumptions made for covariate shift, we compute the corresponding weight for the bias in feature k of example i as follows:

$$w_{i,k} = \frac{P_{test}(x_k^{(i)}, y^{(i)})}{P_{train}(x_k^{(i)}, y^{(i)})} = \frac{P_{test}(y^{(i)}|x_k^{(i)})P_{test}(x_k^{(i)})}{P_{train}(y^{(i)}|x_k^{(i)})P_{train}(x_k^{(i)})} = \frac{P_{test}(x_k^{(i)})}{P_{train}(x_k^{(i)})}.$$

We use density estimation for computing $P_{<dist>}(x_k^{(i)})$.[11]

There may be a variety of biases in a learning problem that are far more than just class imbalance and covariate shift on a single dimension. We have shown previously that there may be multiple types of covariate shift, for example spatial covariate shifts on different hierarchical levels. There may be also covariate shifts for other master data, such as for the customer class or for the contract status.[2] We now aim to correct n different biases at a same time, e.g. for class imbalance as well as different types of covariate shift. As $x^{(i)}$ has potentially many dimensions with a considerable covariate shift, computing the joint $P_{<dist>}(x^{(i)})$ becomes impractical for an increasing number of dimensions. We propose a uniformed and scalable solution to combine weights for correcting the n different biases, comprising for example of class imbalance and different types of covariate shift. The corresponding weights per bias of an example are $w_{i,1}, w_{i,2}, ..., w_{i,n}$. The example weight w_i is the harmonic mean of the weights of the biases considered is computed as follows:

$$w_i = \frac{n}{\frac{1}{w_{i,1}} + \frac{1}{w_{i,2}} + \cdots + \frac{1}{w_{i,n}}} = \frac{n}{\sum_{k=1}^{n} \frac{1}{w_{i,k}}}. \tag{1}$$

As the different $w_{i,k}$ are computed from noisy, real-world data, special care needs to be paid to outliers. Outliers can potentially lead to very large values $w_{i,k}$ for the density estimation proposed above. It is for that reason that we choose the harmonic mean, as it allows to penalizes extreme values and give preference to smaller values.

4. Evaluation

The data used in this paper comes from an electricity provider in Brazil, from which we retain $M = 150,700$ customers. For these customers, we have a complete time series of 24 monthly meter readings before the most recent inspection. From each time series, we compute 304 features comprising generic time series features, daily average features and difference features, as detailed in Table 1. The computation of these features is explained in detail in our previous work.[12]

Table 1. Number of features before and after selection.

Name	#Features	#Retained features
Daily average	23	18
Fixed interval	36	34
Generic time series	222	162
Intra year difference	12	12
Intra year seasonal difference	11	11
Total	304	237

Next, we employ hypothesis tests to the features in order to retain the ones that are statistically relevant. These tests are based on the assumption that a feature x_k is meaningful for the prediction of the binary label vector y if x_k and y are not statistically independent.[13] For binary features, we use Fisher's exact test.[14] In contrast, for continuous features, we use the Kolmogorov-Smirnov test.[15] We retain 237 of the 304 features.

We previously found a random forest (RF) classifier to perform the best on this data compared to decision tree, gradient boosted tree and support vector machine classifiers.[12] It is for this reason that in the following experiments, we only train RF classifiers. When training a RF, we perform model selection by doing randomized grid search, for which the parameters are detailed in Table 2. We use 100 sampled models and perform 10-fold cross-validation for each model.

Table 2. Model parameters for random forest.

Parameter	Values
Max. number of leaves	$[2, 1000)$
Max. number of levels	$[1, 50)$
Measure of the purity of a split	{entropy, gini}
Min. number of samples required to be at a leaf	$[1, 1000)$
Min. number of samples required to split a node	$[2, 50)$
Number of estimators	20

We have previously shown that the location and class of customers have the strongest covariate shift.[2] When reducing these, we first compute the weights for the class imbalance, the spatial covariate shift and customer class covariate shift, respectively, as defined in Sec. 3. For covariate shift, we use randomized grid search for a model selection of the density estimator that is composed of the kernel type and kernel bandwidth. The complete list of parameters and considered values is depicted in Table 3.

Table 3. Density estimation parameters.

Parameter	Values
Kernel	{gaussian, tophat, epanechnikov, exponential, linear, cosine}
Bandwidth	$[0.001, 10]$ (log space)

Next, we use Eq. (1) to combine these weights step by step. For each step, we report the test performance of the NTL classifier in Table 4. It clearly shows that the larger the number of addressed biases, the higher the reliability of the learned predictor.

Table 4. Test performance of random forest.

Biases reduced	$\overline{AUC}$
None	0.59535
Class imbalance	0.64445
Class imbalance + spatial covariate shift	0.71431
Class imbalance + spatial covariate shift + customer class covariate shift	0.73980

Note: We use the area under the receiver-operating curve (AUC) metric. It is particularly useful for NTL detection, as it allows to handle imbalanced datasets and puts correct and incorrect inspection results in relation to each other.[5] $\overline{AUC}$ denotes the mean test AUC of the 10 folds of cross-validation for the best model.

5. Conclusions and Future Work

Biases appear in many real-world applications of machine learning and refer to the training data not being representative for the test data. The most common biases are class imbalance and covariate shift. In this work, we proposed a scalable model for reducing multiple biases in high-dimensional data at the same time. We applied our methodology to a real-world, noisy data set on irregular power usage. Our model leads to more reliable predictors, thus allowing to better detect customers that have an irregular power usage. Next, we aim to evaluate our methodology on other data sets, to derive models that reduce hierarchical spatial biases and to handpick a unbiased test set as ground truth for evaluation.

Acknowledgments

The present project is supported by the National Research Fund, Luxembourg under grant agreement number 11508593.

References

1. M. Banko and E. Brill, Scaling to very very large corpora for natural language disambiguation, in *Proceedings of the 39th annual meeting on association for computational linguistics*, 2001.
2. P. Glauner, A. Migliosi, J. A. Meira et al., Is big data sufficient for a reliable detection of non-technical losses?, in *2017 19th International Conference on Intelligent System Application to Power Systems (ISAP)*, Sept 2017.
3. H. Shimodaira, *Journal of statistical planning and inference* **90**, 227 (2000).
4. C. Cortes and M. Mohri, *Theoretical Computer Science* **519**, 103 (2014).
5. P. Glauner, J. A. Meira, P. Valtchev et al., *International Journal of Computational Intelligence Systems* **10**, 760 (2017).
6. J. L. Viegas, P. R. Esteves, R. Melício et al., *Renewable and Sustainable Energy Reviews* **80**, 1256 (2017).
7. J. Jiang, A literature survey on domain adaptation of statistical classifiers (2008).
8. N. Japkowicz and S. Stephen, *Intelligent data analysis* **6**, 429 (2002).
9. B. Zadrozny, Learning and evaluating classifiers under sample selection bias, in *Proceedings of the twenty-first international conference on Machine learning*, 2004.
10. J. J. Heckman, *Econometrica* **47**, 153 (1979).
11. F. Pedregosa, G. Varoquaux, A. Gramfort et al., *Journal of Machine Learning Research* **12**, 2825 (2011).
12. P. Glauner, N. Dahringer, O. Puhachov et al., Identifying irregular power usage by turning predictions into holographic spatial visualizations, in *Proceedings of the 17th IEEE International Conference on Data Mining Workshops (ICDMW 2017)*, November 2017.
13. P. Radivojac, Z. Obradovic, A. K. Dunker and S. Vucetic, Feature selection filters based on the permutation test, in *ECML*, 2004.
14. R. A. Fisher, *Journal of the Royal Statistical Society* **85**, 87 (1922).
15. F. J. Massey Jr, *Journal of the American statistical Association* **46**, 68 (1951).

Dynamic neighborhood-based decision-theoretic rough set

Xin Yang,[1,2] Tianrui Li,[1,*] Dun Liu,[3] and Qianqian Huang[1]

[1] *School of Information Science and Technology, Southwest Jiaotong University*
Chengdu 611756, China
*trli@swjtu.edu.cn
[2] *School of Computer Science, Sichuan Technology and Business University*
Chengdu 611745, China
[3] *School of Economics and Management, Southwest Jiaotong University*
Chengdu 610031, China

Dynamic neighborhood-based Decision-theoretic Rough Set is discussed in this paper. We propose the incremental methods to update approximations under the variation of attributes and the change of neighborhood threshold δ. To tackle six variational situations in such dynamic neighborhood decision table, the propositions are presented for efficiently updating the neighborhood relation matrix. Moreover, an illustration is conducted to interpret the procedure of incremental technology.

Keywords: Neighborhood; decision-theoretic rough set; incremental learning.

1. Introduction

Recently, neighborhood-based decision-theoretic rough set (DTRS)[1] was proposed to solve uncertain decision-making problems with numerical data in the context of rough set and three-way decisions.[2,3] How to efficiently compute lower and upper approximations through incremental leaning is a key issue for data mining in a dynamic neighborhood decision table. Zhang et al. presented a new dynamic method to maintain approximations under neighborhood rough sets.[4] Yang et al. introduced a unified model of dynamic DTRS to incrementally update three-way probabilistic regions,[5] and further proposed a unified framework of sequential three-way decisions and incremental process.[6] Wang et al. investigated the quantitative composite relation of DTRS in the dynamic composite information table which includes the various of data.[7,8] In this paper, our main motivation is to provide incremental technology in neighborhood-based DTRS to deal with dynamic numerical data under the change of threshold δ.

The paper is organized as follows: Section 2 briefly introduces the notion of Neighborhood-based DTRS. Section 3 presents the incremental updating propositions and Section 4 provides an illustration to interpret the incremental process. Finally, we conclude this paper and give the future plan in Section 5.

2. Neighborhood-based Decision-theoretic Rough Set

Suppose that $NT = (U, AT = C \bigcup D, V, f, \delta)$ is a neighborhood decision table. δ is the threshold of neighborhood.

Definition 2.1.[1] Given a NT. $\forall x_i \in U$ and $B \subseteq C$, the neighborhood $\delta_B(x_i)$ of x_i in B can be defined as $\delta_B(x_i) = \{x_j | x_j \in U, \Delta_B(x_i, x_j) \leq \delta\}$, where Δ is a distance function.

We adopt the Euclidean distance in this paper, namely, $\Delta_B(x_i, x_j) = (\Sigma_{k=1}^{|B|} |f(x_i, b_k) - f(x_j, b_k)|^2)^{1/2}$. Based on the concept of neighborhood granules, we give the definition of neighborhood relation matrix.

Definition 2.2.[1] Given a NT. $\forall x_i, x_j \in U$, N is a neighborhood relation and $D = (d_{ij} = \Delta(x_i, x_j))_{n \times n}$ is a distance matrix. The neighborhood relation matrix $M(N) = (r_{ij})_{n \times n}$ can be defined as $r_{ij} = \{\begin{smallmatrix}1, d_{ij} \leq \delta;\\ 0, otherwise.\end{smallmatrix}$

Definition 2.3.[1] Given a neighborhood approximation space $\langle U, N \rangle$, $B \subseteq U$. $\forall X \subseteq U$, the lower and upper approximations of X on B in the neighborhood-based DTRS model, are defined respectively as

$$\underline{N}_{(\alpha, \beta)}(X) = \{x_i \in U \mid P(X|\delta_B(x_i)) \geq \alpha\}, \quad (1)$$
$$\overline{N}_{(\alpha, \beta)}(X) = \{x_i \in U \mid P(X|\delta_B(x_i)) > \beta\},$$

where $P(X|\delta_B(x_i)) = \frac{|\delta(x_i) \cap X|}{|\delta(x_i)|}$ is the conditional probability, $0 \leq \beta < \alpha \leq 1$.

Based on the well-known Bayesian theory, the threshold α and β can be calculated with the minimum risk of classification. The detail derivation process can be referred in Yao's paper.[9]

3. Dynamic Neighborhood-based DTRS

In real-world applications, NT will evolve over time. Constantly, the useful evidence will arrive for obtaining more accurate decision-making. For instance, to better describe objects, more attributes are added into NT. In addition, a few attributes will be deleted from NT since they become redundancy or insignificance. Furthermore, the threshold δ also will be changed due to the variation of information evidence. We will select more reasonable thresholds to granulate the universe. In what follows, we will discuss the updating strategies to incrementally calculate the lower and upper approximations of neighborhood-based DTRS with the variation of condition attributes when δ is changed in NT.

3.1. The incremental approach with the addition of attributes when δ is changed

We assume that a set of condition attributes is added into NT from time t to time $t+1$. $\Delta C = C^{t+1} - C^t$ is a set of additional attributes, $D^t = (d_{ij}^t = \Delta(x_i, x_j)^t)_{n \times n}$ is a distance matrix at time t, $M^t = (r_{ij}^t)_{n \times n}$ is the neighborhood relation matrix at time t, $DD^t = (dd_{ij}^t = (\delta^t)^2 - (d_{ij}^t)^2)_{n \times n}$ is the difference matrix between threshold and distance at time t, $\Delta d_{ij}^{t+1} = \Sigma_{k=1}^{|\Delta C|} |f(x_i, \Delta C_k) - f(x_j, \Delta C_k)|^2$ is the square of additional distance at time $t+1$, and $\Delta \delta = \delta^{t+1} - \delta^t$ is the difference of thresholds from time t to time $t+1$.

Proposition 3.1. *Suppose a NT, and ΔC is added into NT. If $\Delta \delta = \delta^{t+1} - \delta^t = 0$, the neighborhood relation matrix $M(N) = (r_{ij})_{n \times n}$ will be changed as follows.*

$$r_{ij}^{t+1} = \begin{cases} 1, r_{ij}^t = 1 \wedge \Delta d_{ij}^{t+1} \leq dd_{ij}^t; \\ 0, r_{ij}^t = 0 \vee (r_{ij}^t = 1 \wedge \Delta d_{ij}^{t+1} > dd_{ij}^t). \end{cases} \quad (2)$$

Proposition 3.2. *Suppose a NT, and ΔC is added into NT. If $\Delta \delta = \delta^{t+1} - \delta^t > 0$, the neighborhood relation matrix $M(N) = (r_{ij})_{n \times n}$ will be changed as follows.*

$$r_{ij}^{t+1} = \begin{cases} 1, \Delta d_{ij}^{t+1} \leq \Delta \delta (2\delta^t + \Delta \delta) + dd_{ij}^t; \\ 0, \Delta d_{ij}^{t+1} > \Delta \delta (2\delta^t + \Delta \delta) + dd_{ij}^t. \end{cases} \quad (3)$$

Proposition 3.3. *Suppose a NT, and ΔC is added into NT. If $\Delta \delta = \delta^{t+1} - \delta^t < 0$, the neighborhood relation matrix $M(N) = (r_{ij})_{n \times n}$ will be changed as follows.*

$$r_{ij}^{t+1} = \begin{cases} 1, r_{ij} = 1 \wedge \Delta d_{ij}^{t+1} \leq \Delta \delta (2\delta^t + \Delta \delta) + dd_{ij}^t; \\ 0, r_{ij} = 0 \vee (r_{ij} = 1 \wedge \Delta d_{ij}^{t+1} > \Delta \delta (2\delta^t + \Delta \delta) + dd_{ij}^t). \end{cases} \quad (4)$$

When NT is varied, the granules divided by the neighborhood relation N will be changed. Furthermore, the approximations must be recomputed in traditional methods. Incremental strategies can avoid the repeated and redundant work. Under the change of attributes and threshold δ, updating neighborhood relation matrix M is a key step to calculate approximations in neighborhood-based DTRS model. According to Propositions 3.1, 3.2 and 3.3, we can update the neighborhood relation matrix M^{t+1} through incremental algorithms with respect to the addition of attributes and the change of δ from time t to time $t+1$. Finally, the approximations of neighborhood-based DTRS could be maintained in a less running time compared to the non-incremental algorithms.

3.2. The incremental approach with the deletion of attributes when δ is changed

We assume that a set of condition attributes is deleted in NT from time t to time $t+1$. $\Delta C' = C^t - C^{t+1}$ is a set of deleted attributes, $\Delta d_{ij}^{t+1} = \Sigma_{k=1}^{|\Delta C'|}|f(x_i, \Delta C_k') - f(x_j, \Delta C_k')|^2$ is the square of decreasing distance at time $t+1$, and other representations are the same with Subsection 3.1.

Proposition 3.4. *Suppose a NT, and $\Delta C'$ is deleted in NT. If $\Delta\delta = \delta^{t+1} - \delta^t = 0$, the neighborhood relation matrix $M(N) = (r_{ij})_{n\times n}$ will be changed as follows.*

$$r_{ij}^{t+1} = \begin{cases} 1, r_{ij}^t = 1 \vee (r_{ij}^t = 0 \wedge \Delta d_{ij}^{t+1} \geq |dd_{ij}^t|); \\ 0, r_{ij}^t = 0 \wedge \Delta d_{ij}^{t+1} < |dd_{ij}^t|. \end{cases} \quad (5)$$

Proposition 3.5. *Suppose a NT, and $\Delta C'$ is deleted in NT. If $\Delta\delta = \delta^{t+1} - \delta^t > 0$, the neighborhood relation matrix $M(N) = (r_{ij})_{n\times n}$ will be changed as follows.*

$$r_{ij}^{t+1} = \begin{cases} 1, r_{ij} = 1 \vee (r_{ij} = 0 \wedge dd_{ij}^t + \Delta d_{ij}^{t+1} + \Delta\delta(2\delta^t + \Delta\delta) \geq 0); \\ 0, r_{ij} = 0 \wedge dd_{ij}^t + \Delta d_{ij}^{t+1} + \Delta\delta(2\delta^t + \Delta\delta) < 0. \end{cases} \quad (6)$$

Proposition 3.6. *Suppose a NT, and $\Delta C'$ is deleted in NT. If $\Delta\delta = \delta^{t+1} - \delta^t < 0$, the neighborhood relation matrix $M(N) = (r_{ij})_{n\times n}$ will be changed as follows.*

$$r_{ij}^{t+1} = \begin{cases} 1, \Delta d_{ij}^{t+1} \leq \Delta\delta(2\delta^t + \Delta\delta) + dd_{ij}^t; \\ 0, \Delta d_{ij}^{t+1} > \Delta\delta(2\delta^t + \Delta\delta) + dd_{ij}^t. \end{cases} \quad (7)$$

Propositions 3.4, 3.5 and 3.6 provide an incremental method to update the neighborhood relation matrix M^{t+1} with the deletion of attributes and the change of δ. The distance matrix is not necessary to be recomputed. Based on such a dynamic updating technology, we can efficiently obtain the upper and lower approximations of neighborhood-based DTRS model.

4. An Illustration

Table 1: A neighborhood decision table NT.

U	a_1	a_2	a_3	d
x_1	0.2	0.3	0.2	1
x_2	0.3	0.2	0.1	1
x_3	0.1	0.1	0.3	2
x_4	0.2	0.2	0.2	2
x_5	0.1	0.3	0.1	1

In this section, we give an example to interpret the incremental process of our proposed methods. Table 1 is a neighborhood decision table NT. Suppose that $U = \{x_1, x_2, x_3, x_4, x_5\}$, $C = \{a_1, a_2\}$, $\delta = 0.15$. Firstly, we calculate the distance matrix D, the additional distance matrix ΔD, the difference matrix DD between threshold and distance, and the neighborhood relation matrix M. According to Definition 2.1, we can obtain the neighborhood granules, i.e., $\delta(x_1) = \{x_1, x_2, x_4, x_5\}$, $\delta(x_2) = \{x_1, x_2, x_4\}$, $\delta(x_3) = \{x_3, x_4\}$, $\delta(x_4) = \{x_1, x_2, x_3, x_4, x_5\}$, $\delta(x_5) = \{x_1, x_4, x_5\}$.

$$D = \begin{bmatrix} 0 & 0.141 & 0.224 & 0.100 & 0.100 \\ 0.141 & 0 & 0.224 & 0.100 & 0.224 \\ 0.224 & 0.224 & 0 & 0.141 & 0.200 \\ 0.100 & 0.100 & 0.141 & 0 & 0.141 \\ 0.100 & 0.224 & 0.200 & 0.141 & 0 \end{bmatrix}, \Delta D = \begin{bmatrix} 0 & 0.01 & 0.01 & 0 & 0.01 \\ 0.01 & 0 & 0.04 & 0.01 & 0 \\ 0.01 & 0.04 & 0 & 0.01 & 0.04 \\ 0 & 0.01 & 0.01 & 0 & 0.01 \\ 0.01 & 0 & 0.04 & 0.01 & 0 \end{bmatrix},$$

$$DD = \begin{bmatrix} 0.023 & 0.003 & -0.028 & 0.013 & 0.013 \\ 0.003 & 0.023 & -0.028 & 0.013 & -0.028 \\ -0.028 & -0.028 & 0.023 & 0.003 & -0.018 \\ 0.013 & 0.013 & 0.003 & 0.023 & 0.003 \\ 0.013 & -0.028 & -0.018 & 0.003 & 0.023 \end{bmatrix}, M = \begin{bmatrix} 1 & 1 & 0 & 1 & 1 \\ 1 & 1 & 0 & 1 & 0 \\ 0 & 0 & 1 & 1 & 0 \\ 1 & 1 & 1 & 1 & 1 \\ 1 & 0 & 0 & 1 & 1 \end{bmatrix}.$$

We assume the attribute a_3 is added into NT.

(1) $\Delta \delta = \delta^{t+1} - \delta^t = 0$. According to Proposition 3.1, if $r_{ij}^t = 0$, then $r_{ij}^{t+1} = 0$. If $r_{ij}^t = 1$, we only need to judge the size of Δd_{ij}^{t+1} and dd_{ij}^t. For instance, $r_{21}^t = 1$, then $r_{21}^{t+1} = 0$ since $\Delta d_{21}^{t+1} = 0.0100 > dd_{21}^t = 0.0025$.

(2) $\Delta\delta = \delta^{t+1} - \delta^t > 0$. According to Proposition 3.2, we must judge the size of Δd_{ij}^{t+1} and $\Delta\delta(2\delta^t + \Delta\delta) + dd_{ij}^t$. For instance, assume $\Delta\delta = 0.05$. We have $r_{21}^t = 1$. Because $\Delta d_{21}^{t+1} = 0.0100 < \Delta\delta(2\delta^t + \Delta\delta) + dd_{21}^t = 0.0175 + 0.0025 = 0.02$, we have $r_{21}^{t+1} = 1$.

(3) $\Delta\delta = \delta^{t+1} - \delta^t < 0$. According to Proposition 3.3, we have $r_{ij}^{t+1} = 0$ if $r_{ij}^t = 0$. when $r_{ij}^t = 1$, we should judge the size of Δd_{ij}^{t+1} and $\Delta\delta(2\delta^t + \Delta\delta) + dd_{ij}^t$. For instance, assume $\Delta\delta = -0.05$. We have $r_{21}^t = 1$. Because $\Delta d_{21}^{t+1} = 0.0100 > \Delta\delta(2\delta^t + \Delta\delta) + dd_{21}^t = -0.0175 + 0.0025 = -0.0015$, we have $r_{21}^{t+1} = 0$.

Similar to above incremental process, we also can efficiently update the neighborhood relation matrix with the deleted of attributes when threshold is changed according to Propositions 3.4, 3.5 and 3.6.

5. Conclusions

This paper introduced an incremental updating method when attributes and threshold δ were changed simultaneously in the neighborhood-based DTRS model. We will conduct comparative experiments to interpret the incremental mechanism and performance in our further work.

Acknowledgments

This work is supported by the National Science Foundation of China (Nos. 61573292, 61572406, 71571148).

References

1. W. W. Li, Z. Q. Huang, X. Y. Jia and X. Y. Cai, Neighborhood based decision-theoretic rough set models, *International Journal of Approximate Reasoning.* **69**, 1–17 (2016).
2. Y. Y. Yao, Three-Way Decisions and Cognitive Computing, *Cognitive Computation.* **8**, 543–554 (2016).
3. D. Liu, D. C. Liang, Three-way decisions in ordered decision system, Knowledge-Based Systems. **137**, 182–195 (2017).
4. J. B. Zhang, T. R. Li, D. Ruan and D. Liu, Neighborhood rough sets for dynamic data mining, *International Journal of Intelligent Systems.* **27**, 317–342 (2012).
5. X. Yang, T. R. Li, D. Liu, H. M. Chen and C. Luo, A unified framework of dynamic three-way probabilistic rough sets, *Information Sciences.* **420**, 126–147 (2017).
6. X. Yang, T. R. Li, H. Fujita, D. Liu and Y. Y. Yao, A unified model of sequential three-way decisions and multilevel incremental processing, *Knowledge-Based Systems.* **134**, 172–188 (2017).
7. L. N. Wang, X. Yang, L. Liu and P. Zhuo, Quantitative Composite Decision-theoretic Rough Set, in *Proc. Int. Conf. on ISKE*, (Nan Jing, China, 2017).

8. L. N. Wang, X. Yang, Y. Chen and L. Liu et al., Dynamic composite decision-theoretic rough set under the change of attributes, *International Journal of Computational Intelligence Systems.* **11**, 355–370 (2018).
9. Y. Y. Yao, S. K. M. Wong, A decision theoretic framework for approximating concepts, *International Journal of Man-Machine Studies.* **37**, 793–809 (1992).

Incremental updating of approximations in composite ordered decision systems under attribute generalization

Qianqian Huang, Tianrui Li,* Yanyong Huang and Xin Yang

School of Information Science and Technology, Southwest Jiaotong University
Chengdu, 611756, China
huangqnqn@126.com, trli@swjtu.edu.cn, yyhswjtu@163.com, yangxin2041@163.com

In this paper, we first propose a general composite ordered rough set model, which can deal with various types of attributes with a preference order. Then, we present a matrix method for computing approximations in composite ordered decision systems. Finally, we introduce an incremental method for updating approximations in composite ordered decision systems under the variation of attributes.

Keywords: Rough sets; decision systems; matrix; incremental learning.

1. Introduction

Dominance rough set approach (DRSA), proposed by Greco et al., is an important extension of rough sets which uses a dominance relation instead of the equivalence relation in classical rough set theory[1,2]. In DRSA, where all attributes are criteria and classes are preference ordered, the sets to be approximated are not the particular classes but upward unions and downward unions of the classes. Recently, DRSA has attracted much attention and been successfully extended to various types of decision systems, e.g., set-valued ordered decision systems[3], interval-valued ordered decision systems[4], incomplete ordered decision systems[5], and hybrid ordered decision systems[6,7]. However, when there are more than two different types of attributes simultaneously in an ordered decision system, the existing dominance rough set models are unsuitable for dealing with this issue. In this paper, we propose a novel composite dominance relation, which can be regarded as a generalization of the composite dominance relation proposed by Luo[8]. Then we redefine the approximations of upward unions and downward unions of decision classes based on the proposed composite dominance relation. Moreover, we present a matrix-based incremental approach for updating approximations when adding the attributes in composite ordered decision systems (CODS).

The remainder of this paper is structured as follows. The basic concepts of some extend dominance relations are briefly reviewed in Section 2. The novel composite ordered rough set model is introduced in Section 3. The matrix-based representation of approximations and the corresponding incremental method of updating approximations are presented in Sections 4 and 5, respectively. The results are concluded in Section 6.

2. Preliminaries

In this section, we briefly review some concepts of extended dominance relations in different types of ordered decision systems[2-6]. For the sake of convenience, let C_S, C_N, C_V, C_I, and C_M denote the categorical, numerical, set-valued, interval-valued and missing-valued attributes, respectively.

Definition 2.1.[2] A decision system $S = (U, A = C \cup \{d\}, V, f)$ is called the ordered decision systems if all attributes are criterions.

Next, we introduce the normalization method for numerical data[6].

Definition 2.2.[6] Let $x, y \in U$ and $P \subseteq C$. The normalized value of object x in terms of attribute $a_j \in C$ is denoted as:

$$\hat{f}(x, a_j) = \frac{f(x, a_j) - min(V_{a_j})}{max(V_{a_j}) - min(V_{a_j})},$$

where V_{a_j} denotes the domain of a_j, $min(V_{a_j})$ and $max(V_{a_j})$ are the minimum and maximum values of V_{a_j}, respectively. Moreover, the normalized distance between two objects x and y on P is defined as:

$$\hat{\Delta}_P(x, y) = (\sum_{j=1}^{|P|} |\hat{f}(x, a_j) - \hat{f}(y, a_j)|^q)^{\frac{1}{q}} |P|^{-\frac{1}{q}}.$$

In this paper, we let $q = 1$.

Definition 2.3.[2-6] Given a general ordered decision system $S = (U, C \cup \{d\}, V, f)$, where $C = C_k$, $k \in \{S, N, V, I, M\}$. Let $P_k \subseteq C_k$, then the dominance relation w.r.t. P_k is defined respectively as: (1) $D_{P_S} = \{(x, y) \in U^2 : f(x, a) \geq f(y, a), \forall a \in P_S\}$; (2) $D_{P_N} = \{(x, y) \in U^2 : \hat{\Delta}_{P_N}(x, y) \geq \delta \wedge \hat{f}(x, a) \geq \hat{f}(y, a), \forall a \in P_N\}$, where $0 \leq \delta \leq 1$; (3) $D_{P_V} = \{(x, y) \in U^2 : \max f(x, a) \geq \min f(y, a), \forall a \in P_V\}$; (4) $D_{P_I} = \{(x, y) \in U^2 : a^l(x) \geq a^l(y), a^u(x) \geq a^u(y), \forall a \in P_I\}$, where $f(x, a) = [a^l(x), a^u(x)]$; (5) $D_{P_M} = \{(x, y) \in U^2 : f(x, a) \geq f(y, a) \vee f(x, a) = * \vee f(y, a) = *$ for all $a \in P_M$ such that $f(x, a) \neq ?\}$.

Furthermore, the P_k-dominating sets and P_k-dominated sets w.r.t. $x \in U$ are denoted respectively as follows:

$$D_{P_k}^+(x) = \{y \in U : yD_{P_k}x\},$$
$$D_{P_k}^-(x) = \{y \in U : xD_{P_k}y\}.$$

The decision criterion d divides the universe U into a family of decision classes $Cl = \{Cl_t, t \in 1, \ldots, m\}$, and $Cl_m \succ \ldots \succ Cl_1$, where $\succ$ denotes a strong preference relation. The upward and downward union of decision classes are defined respectively as: $Cl_t^\geq = \bigcup_{s \geq t} Cl_s$ and $Cl_t^\leq = \bigcup_{s \leq t} Cl_s$, where $t, s \in \{1, \ldots, m\}$.

3. Composite Ordered Rough Set Model

In this section, we propose a new composite dominance relation to process the CODS and redefine the approximations of $Cl_t^\geq$ and $Cl_t^\leq$ under the proposed composite dominance relation.

Definition 3.1. Let $S = (U, A = C \cup \{d\}, V, f)$ be a CODS, where U is a non-empty finite set of objects. $C = \bigcup_k C_k (k \in \{S, N, V, I, M\})$ is the condition attribute set, and d is the decision criterion. $V = \bigcup_{C_k \subseteq C} V_{C_k}$, $V_{C_k} = \bigcup_{c \in C_k} V_c$, V_c is a domain of criterion c. $f : U \times A \to V$ is an information function, $f(x, a) \in V_a$ denotes the value of object x on criterion $a \in A$.

Table 1. A composite ordered decision system.

U	a_1	a_2	a_3	a_4	a_5	a_6	a_7	d
x_1	1	20	$\{0,1\}$	$[2.17, 2.86]$	2	15	2	1
x_2	3	85	$\{2\}$	$[3.55, 5.45]$	?	30	3	3
x_3	3	31	$\{1,2\}$	$[2.56, 4.10]$	2	10	2	2
x_4	1	60	$\{1\}$	$[2.49, 3.45]$	1	30	3	2
x_5	1	44	$\{0,2\}$	$[2.22, 3.07]$	*	15	2	1
x_6	2	74	$\{1,2\}$	$[2.58, 4.20]$	3	20	1	3

Example 3.1. Table 1 illustrates a CODS, where $U = \{x_i : 1 \leq i \leq 6\}$. $C = \{a_i : 1 \leq i \leq 7\}$, where $C_S = \{a_1, a_7\}$, $C_N = \{a_2, a_6\}$, $C_V = \{a_3\}$, $C_I = \{a_4\}$ and $C_M = \{a_5\}$, and d is the decision attribute.

Definition 3.2. Assume that $x, y \in U$ and $P = \bigcup P_k \subseteq C$, $P_k \subseteq C_k$. The composite dominance relation $CR_P^\succeq$ can be defined as

$$CR_P^\succeq = \{(x, y) \in U^2 : (x, y) \in \bigcap_{P_k \subseteq P} D_{P_k}\},$$

where $D_{P_k} \subseteq U^2$ is a dominance relation w.r.t. P_k on U. Furthermore, the dominating sets and dominated sets w.r.t. $x \in U$ can be denoted as:

$$CR_P^+(x) = \{y \in U : (y,x) \in CR_P^{\succeq}\} = \{y \in U : (y,x) \in D_{P_k}, \forall P_k \subseteq P\},$$
$$CR_P^-(x) = \{y \in U : (x,y) \in CR_P^{\succeq}\} = \{y \in U : (x,y) \in D_{P_k}, \forall P_k \subseteq P\}.$$

Proposition 3.1. *The following conclusions hold:*

(1) If $\delta = 0$, $CR_P^{\succeq}$ is reflexive, but not symmetric or transitive;
(2) If $0 < \delta \leq 1$, $CR_P^{\succeq}$ is not reflexive, symmetric or transitive.

Note that when $\delta = 0$, the composite dominance relation here degenerates to the composite dominance relation in Ref. 8. Therefore, it can be regarded as a generalization of the composite dominance relation in Ref. 8.

Definition 3.3. Let $S = (U, A = C \cup \{d\}, V, f)$ be a CODS, and $P \subseteq C$, $x \in U$, $Cl_t^{\geq}, Cl_t^{\leq} \subseteq U$ ($1 \leq t \leq m$). The lower approximations of $Cl_t^{\geq}$ and $Cl_t^{\leq}$ w.r.t. P are defined respectively as follows:

$$\underline{CR_P^{\succeq}}(Cl_t^{\geq}) = \left\{ x \in Cl_t^{\geq} : CR_P^+(x) \subseteq Cl_t^{\geq} \right\},$$
$$\underline{CR_P^{\succeq}}(Cl_t^{\leq}) = \left\{ x \in Cl_t^{\leq} : CR_P^-(x) \subseteq Cl_t^{\leq} \right\}.$$

By using the duality, the upper approximations of $Cl_t^{\geq}$ and $Cl_t^{\leq}$ w.r.t. P are defined respectively as follows:

$$\overline{CR_P^{\succeq}}(Cl_t^{\geq}) = Cl_t^{\geq} \cup \left\{ x \in Cl_t^{\leq} : CR_P^-(x) \cap Cl_t^{\geq} \neq \emptyset \right\},$$
$$\overline{CR_P^{\succeq}}(Cl_t^{\leq}) = Cl_t^{\leq} \cup \left\{ x \in Cl_t^{\geq} : CR_P^+(x) \cap Cl_t^{\leq} \neq \emptyset \right\}.$$

Obviously, when $\delta = 0$, Definition 3.3 can degrade into the original approximations of $Cl_t^{\geq}$ and $Cl_t^{\leq}$ in Ref. 2.

4. Matrix-based Representation of Approximations in CODS

In this section, we present the matrix representation of approximations of the upward and downward union of decision classes under $CR_P^{\succeq}$ in CODS. Due to the limitation of space, we only consider the approximations of $Cl_t^{\geq}$ as follows.

Definition 4.1. Let $S = (U, A = C \cup \{d\}, V, f)$ be a CODS, and $X \subseteq U$. $C(X) = [c_i]_{n \times 1}$ is the characteristic vector of X, where $c_i = 1$ if $x_i \in X$, otherwise $c_i = 0$.

Definition 4.2. Let $S = (U, A = C \cup \{d\}, V, f)$ be a CODS, and $P \subseteq C$. $M_P^+ = [m_{ij}]_{n \times n}$ represents the dominant matrix w.r.t. P, where $m_{ij} = 1$ if $x_j \in CR_P^+(x_i)$, otherwise $m_{ij} = 0$.

Proposition 4.1. Let $S = (U, A = C \cup \{d\}, V, f)$ be a CODS and $P \subseteq C$. The following items hold:

(1) If $\delta = 0$, then $m_{ii} = 1$; otherwise $m_{ii} = 0$.
(2) $M_P^- = (M_P^+)^T$.

Definition 4.3. The dominant diagonal matrix Λ_P^+ induced from $M_P^+ = [m_{ij}]_{n \times n}$ can be defined as follows:

$$\Lambda_P^+ = \text{diag}[\frac{1}{\lambda_1}, \cdots, \frac{1}{\lambda_n}], \text{ where } \frac{1}{\lambda_i} = \begin{cases} 0, & \sum_{j=1}^n m_{ij} = 0, \\ \frac{1}{\sum_{j=1}^n m_{ij}}, & \sum_{j=1}^n m_{ij} \neq 0. \end{cases}$$

Definition 4.4. The lower and upper vectors of $Cl_t^\geq$ under $CR_P^\geq$ can be defined respectively as follows:

$$L_P^+(Cl_t^\geq) = C(Cl_t^\geq) \odot \left(\Lambda_P^+ \cdot (M_P^+ \cdot C(Cl_t^\geq))\right),$$
$$U_P^+(Cl_t^\geq) = C(Cl_t^\leq) \odot \left(M_P^- \cdot C(Cl_t^\geq)\right),$$

where $C(Cl_t^\geq)$ and $C(Cl_t^\leq)$ denote the characteristic vectors of $Cl_t^\geq$ and $Cl_t^\leq$, "$\cdot$" and "$\odot$" denote the inner product and dot product of two matrices, respectively.

Corollary 4.1. Let $C(Cl_t^\geq) = [c_i]_{n \times 1}$, $C(Cl_t^\leq) = [\hat{c}_i]_{n \times 1}$, $L_P^+(Cl_t^\geq) = [l_i]_{n \times 1}$ and $U_P^+(Cl_t^\geq) = [u_i]_{n \times 1}$. Then

$$l_i = \frac{c_i \sum_{j=1}^n m_{ij} c_j}{\sum_{j=1}^n m_{ij}}, \quad u_i = \hat{c}_i \sum_{j=1}^n m_{ji} c_j, \quad 0 \leq l_i \leq 1, \ 0 \leq u_i \leq n, \ 1 \leq i \leq n.$$

Theorem 4.1. The lower and upper approximations of $Cl_t^\geq$ can be induced from $L_P^+(Cl_t^\geq)$ and $U_P^+(Cl_t^\geq)$, respectively.

$$\underline{CR_P^\geq}(Cl_t^\geq) = \left\{x_i \in U : l_i = 1, \forall l_i \in L_P^+(Cl_t^\geq)\right\},$$
$$\overline{CR_P^\geq}(Cl_t^\geq) = Cl_t^\geq \cup \left\{x_i \in U : u_i \geq 1, \forall u_i \in U_P^+(Cl_t^\geq)\right\}.$$

5. Incremental Maintenance of Approximations in CODS under the Attribute Generalization

In this section, we propose the approaches of incremental updating approximations in CODS with the variation of attributes. Let $S^r = (U, C^r \cup \{d\}, V^r, f^r)$ be a CODS at time r, where $C^r = \bigcup_k C_k^r$, $k \in \{S, N, V, I, M\}$. At time $r+1$, the original CODS is updated as $S^{r+1} = (U, C^{r+1} \cup \{d\}, V^{r+1}, f^{r+1})$ with the addition of an additional attribute set $\Delta C = \bigcup_k \Delta C_k$, where $C^{r+1} = C^r \cup \Delta C$, $C^r \cap \Delta C = \emptyset$.

Theorem 5.1. *Suppose $M_{C^r}^+ = [m_{ij}]_{n \times n}$ and $M_{C^{r+1}}^+ = [m'_{ij}]_{n \times n}$ denote the dominant matrices w.r.t. $CR_{C^r}^{\geq}$ and $CR_{C^{r+1}}^{\geq}$, respectively. When adding ΔC to C^r, the elements of the dominant matrix are updated as follows.*

(1) *If $m_{ij} = 0$,*
$$m'_{ij} = \begin{cases} 1, & x_j \in \bigcap_{k'} D_{C_{k'}^r}^+(x_i) \bigcap CR_{\Delta C}^+(x_i) \bigcap D_{C_N^{r+1}}^+(x_i), \\ m_{ij}, & otherwise, \end{cases}$$

(2) *If $m_{ij} = 1$,*
$$m'_{ij} = \begin{cases} m_{ij}, & x_j \in CR_{\Delta C}^+(x_i), \\ m_{ij}, & x_j \notin CR_{\Delta C}^+(x_i), \ x_j \in \bigcap_{k'} D_{\Delta C_{k'}}^+(x_i) \bigcap D_{C_N^{r+1}}^+(x_i), \\ 0, & otherwise, \end{cases}$$

where $k' \in \{S, V, I, M\}$.

Note that when $\delta = 0$, the updating rule of dominant matrix in Definition 5.1 can degrade into the updating rule of dominant matrix in Ref. 9.

Corollary 5.1. *Suppose $\Lambda_{C^r}^+ = \text{diag}[\frac{1}{\lambda_1}, \cdots, \frac{1}{\lambda_n}]$ and $\Lambda_{C^{r+1}}^+ = \text{diag}[\frac{1}{\lambda'_1}, \cdots, \frac{1}{\lambda'_n}]$ denote the dominant diagonal matrices induced from $M_{C^r}^+$ and $M_{C^{r+1}}^+$, respectively. Then we have*
$$\lambda'_i = \lambda_i + \sum_{j=1}^n (m_{ij} \oplus m'_{ij}) \cdot m'_{ij} - \sum_{j=1}^n (m_{ij} \oplus m'_{ij}) \cdot m_{ij},$$
where "$\oplus$" denotes XOR operation.

Corollary 5.2. *Suppose $M_{C^r}^+ \cdot C(Cl_t^{\geq}) = [\omega_i]_{n \times 1}$, $M_{C^r}^- \cdot C(Cl_t^{\geq}) = [\nu_i]_{n \times 1}$, $M_{C^{r+1}}^+ \cdot C(Cl_t^{\geq}) = [\omega'_i]_{n \times 1}$, and $M_{C^{r+1}}^- \cdot C(Cl_t^{\geq}) = [\nu'_i]_{n \times 1}$. Then*
$$\omega'_i = \omega_i + \sum_{j=1}^n (m_{ij} \oplus m'_{ij}) \cdot m'_{ij} c_j - \sum_{j=1}^n (m_{ij} \oplus m'_{ij}) \cdot m_{ij} c_j;$$
$$\nu'_i = \nu_i + \sum_{j=1}^n (m_{ji} \oplus m'_{ji}) \cdot m'_{ji} c_j - \sum_{j=1}^n (m_{ji} \oplus m'_{ji}) \cdot m_{ji} c_j.$$

Hence, according to Corollaries 5.1 and 5.2, the lower and upper vectors of $Cl_t^{\geq}$ can be updated directly when adding a new attribute set ΔC to C^r.

Corollary 5.3. *Suppose $L^+_{Cr+1}(Cl^{\geq}_t) = [l'_i]_{n\times 1}$ and $U^+_{Cr+1}(Cl^{\geq}_t) = [u'_i]_{n\times 1}$. Then, $l'_i = c_i\omega'_i/\lambda'_i$ and $u'_i = \hat{c}_i\nu'_i$.*

Based on the aforementioned computational process, we can effectively avoid unnecessary computations by utilizing previous data structures and results when the attributes change over time. Consequently, it's obviously that our proposed approaches greatly reduce the running time of calculating approximations in CODS with the variation of attributes.

Next, we illustrate the approaches of incremental updating approximations with a concrete example.

Example 5.1. (Continuous Example 3.1) Let $C = \{a_i : 1 \leq i \leq 5\}$ and $\Delta C = \{a_6, a_7\}$ be the original attribute set and additional attribute set, respectively. If $\delta = 0.04$, according to Definition 3.3, we have

$$M^+_C = \begin{bmatrix} 0 & 1 & 1 & 0 & 1 & 1 \\ 0 & 0 & 0 & 0 & 0 & 0 \\ 0 & 1 & 0 & 0 & 0 & 0 \\ 0 & 1 & 0 & 0 & 0 & 1 \\ 0 & 1 & 0 & 1 & 0 & 1 \\ 0 & 1 & 0 & 0 & 0 & 0 \end{bmatrix}, \quad M^+_{\Delta C} = \begin{bmatrix} 0 & 1 & 0 & 1 & 0 & 0 \\ 0 & 0 & 0 & 0 & 0 & 0 \\ 1 & 1 & 0 & 1 & 1 & 0 \\ 0 & 0 & 0 & 0 & 0 & 0 \\ 0 & 1 & 0 & 1 & 0 & 0 \\ 0 & 1 & 0 & 1 & 0 & 0 \end{bmatrix}.$$

When ΔC are added into C, by Theorem 5.1, we have

$$M^+_{C\cup\Delta C} = \begin{bmatrix} 0 & 1 & 0 & 0 & 1 & 0 \\ 0 & 0 & 0 & 0 & 0 & 0 \\ 0 & 1 & 0 & 0 & 0 & 0 \\ 0 & 1 & 0 & 0 & 0 & 0 \\ 0 & 1 & 0 & 1 & 0 & 0 \\ 0 & 1 & 0 & 0 & 0 & 0 \end{bmatrix}.$$

Because $Cl^{\geq}_2 = \{x_2, x_3, x_4, x_6\}$ and $Cl^{\leq}_2 = \{x_1, x_3, x_4, x_3\}$, by Definition 4.1, we have

$$C(Cl^{\geq}_2) = [0, 1, 1, 1, 0, 1]^T, \quad C(Cl^{\leq}_2) = [1, 0, 1, 1, 1, 0]^T.$$

According to Corollaries 5.1 and 5.2, it's easy to obtain that

$$\Lambda^+_{C\cup\Delta C} = \text{diag}[\frac{1}{2}, 0, 1, 1, \frac{1}{2}, 1];$$
$$M^+_{C\cup\Delta C} \cdot C(Cl^{\geq}_2) = [1, 0, 1, 1, 2, 1]^T;$$
$$M^-_{C\cup\Delta C} \cdot C(Cl^{\geq}_2) = [0, 3, 0, 0, 0, 0]^T.$$

From Corollary 5.3, we know that the lower and upper vectors can be updated as follows:

$$L^+_{C \cup \Delta C}(Cl_t^{\geq}) = [0,0,1,1,0,1]^T, \quad U^+_{C \cup \Delta C}(Cl_t^{\geq}) = [0,0,0,0,0,0]^T.$$

Hence, by Theorem 4.1, we have

$$\underline{CR^{\succeq}_{C \cup \Delta C}}(Cl_2^{\geq}) = \{x_3, x_4, x_6\}, \quad \overline{CR^{\succeq}_{C \cup \Delta C}}(Cl_2^{\geq}) = \{x_2, x_3, x_4, x_6\}.$$

6. Conclusions

In this paper, we proposed a general composite dominance relation and redefined the approximations of upward unions and downward unions of decision classes under the proposed composite dominance relation in CODS. Furthermore, we presented the incremental method for updating approximations in CODS with the variation of attributes. In our future work, we will verify the practicability of the proposed model in real-applications.

Acknowledgments

This work is supported by the National Science Foundation of China (Nos. 61573292, 61572406, 61603313).

References

1. S. Greco, B. Matarazzo, R. Slowinski, Rough sets theory for multicriteria decision analysis, *European Journal of Operational Research* **129**, 1–47 (2001).
2. S. Greco, B. Matarazzo, R. Slowinski, Rough approximation by dominance relations, *International Journal of Intelligent Systems* **17**, 153–171 (2002).
3. Y.H. Qian, C.Y. Dang, J.Y. Liang, D.W. Tang, Set-valued ordered information systems, *Information Sciences* **179**, 2809–2832 (2009).
4. Y.H. Qian, J.Y. Liang, C.Y. Dang, Interval ordered information systems, *Computers and Mathematics with Applications* **56**, 1994–2009 (2008).
5. W.S. Du, B. Q. Hu, Dominance-based rough set approach to incomplete ordered information systems, *Information Sciences* **346–347**, 106–129 (2016).
6. H.M. Chen, T.R. Li, Y. Cai, C. Luo, H. Fujita, Parallel attribute reduction in dominance-based neighborhood rough set, *Information Sciences* **373**, 351–368 (2016).
7. S.Y. Li, T.R. Li, J. Hu, Update of approximations in composite information systems, *Knowledge-Based Systems* **83**, 138–148 (2015).
8. C. Luo, T.R. Li, H.M. Chen, J.B. Zhang, Composite ordered information systems, *Journal of China Computer Systems* **35(11)**, 2523–2527 (2014). (in Chinese)
9. C. Luo, T.R. Li, H.M. Chen, Dynamical maintenance of approximations in set-valued ordered decision systems under the attribute generalization, *Information Sciences* **257**, 210–228 (2014).

Data-driven competitor identification using text mining and word network analysis[†]

Xueyan Zhong

*School of Transportation and Logistics, Southwest Jiaotong University
Chengdu 610031, Sichuan, China and
School of Computer Science, Southwest Petroleum University, Chengdu 610031, China*

Guoqing Chen

School of Economics and Management, Tsinghua University, Beijing 100084, China

Leilei Sun

School of Economics and Management, Tsinghua University, Beijing 100084, China

Lan Liu

*School of Transportation and Logistics, Southwest Jiaotong University
Chengdu 610031, Sichuan, China*

Competitive analysis plays an important role in product design and marketing strategies. However, in app markets, a large number of apps have been designed and released in a relatively short time, and the apps are being updated now and then. In this case, it is difficult for product managers to identify their competitors in real time, and it is also a big challenge for them to discover the functional overlaps of their products with other products. This paper provides a data-driven automatic competitor identification method, and also studies how to discover the key competitive attributes of apps. To identify competitors, we first represented the studied apps by vector space model according to their descriptive documents of their functionalities, then grouped the apps hierarchically, and last summarized the key attributes of apps at a group-level. For the purpose of discovering competitive domain, a word network mining method was proposed to analyze the associations of the app attributes. We conducted experiments on functional descriptions of 2451 traffic apps crawled from an app platform, experimental results validate the effectiveness of the proposed methods.

[†]Work partially supported by the scientific research starting project (No. 2014QHS010) and Humanities and Social Sciences Foundation (No. 2013RW007) of Southwest Petroleum University.

1. Introduction

Competitor identification plays an important role in product design and marketing strategies, which aims at helping product managers to find their competitors quickly and monitor their competitors closely. In the existing research, competitors were mainly identified manually by quantitative and qualitative approaches [1]. Competitors, competitive domains, competitive strength, and competitive evidences are key factors in competitive analysis [2,3]. Researches have used user generated contents for competitor mining from the consumer perspective [4]. Product instruction documents on the third party platforms (e.g., app store, google play, 360 mobile assistant) can provide even more accurate competitive information. These documents usually include abundant product functional attributes and features from the supplier perspective. However, such information have not been well explored for competitive analysis yet.

In this article, we propose Text Mining and Word Network (TMWN) analysis framework to automatically identify competitors and competitive domains according to product instructions on the third party platforms. Text mining [5] refers to the process of extracting useful, meaningful, and nontrivial information from unstructured text. Vector space model is used to convert the product instructions into structured vectors. Then, hierarchical clustering is used to divide the studied products into several clusters and to extract competitor groups according to clustering result. Additionally, word network is constructed and analyzed to discover the competitive domains and the relationship of product features [6]. Figure 1 presents the TMWN analysis framework, which consists of

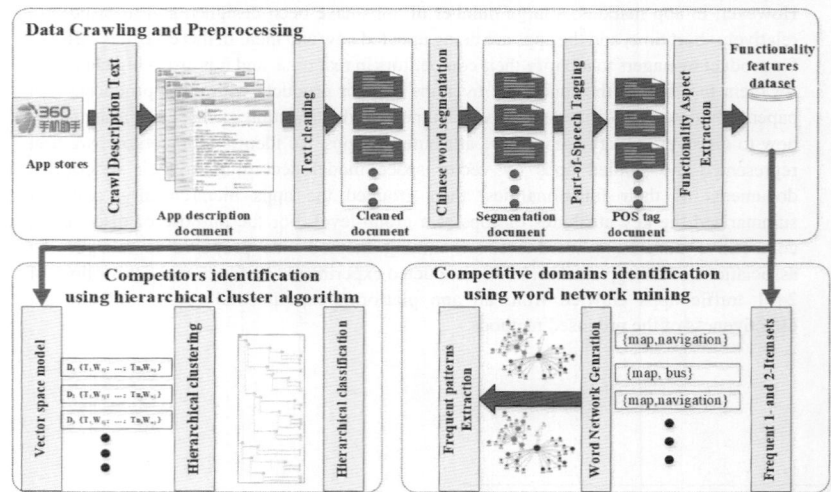

Figure 1. Text mining and word network (TMWN) analysis framework.

three parts: 1) data crawling and pre-processing, which crawls the raw data and represents the data as vectors. 2) Competitors identification, which uses hierarchical clustering algorithm to identify the competitive groups. 3) Competitive domain identification, which visualizes and analyses co-occurrence network relationships of the extracted product features.

The remainder of this paper is organized as follows. In Section 2 we present the detail of TMWN framework. Section 3 contains experiments and result. Section 4 outlines conclusions and future directions.

2. Methodology

2.1. *Data crawling and pre-processing*

The objective of this step is to crawl the original product texts and preprocess to extract product feature sets. It includes five main steps:

1. *Crawling product instruction* is to crawl information from web which contains products instruction, number of downloads, and other information.
2. *Text cleaning* is to remove useless information from the raw texts, such as punctuation, expression, symbols, space etc.
3. *Chinese word segmentation* is to divide text into meaningful units, such as words or phrases.
4. *Part-of-Speech tagging* is to produce the part-of-speech tag for each word (e.g. noun, verb, and adjective).
5. *Product feature extractor* is to extract nouns and verbs as feature candidates. For a single word, we preserve only nouns and verbs because most of the products feature information can be reflected by nouns [7] and verbs [8]. Some useless words are removed, for example, "ours, software, technology", these words do not reflect the functional characteristics of app products.
6. Then we get the product feature set for representing the products, which is defined as $D=\{d_j\}$, d_j is the representation of *j-th* product, $j \in \{1,2,...,n\}$, *n* is the number of products.

2.2. *Competitor identification using hierarchical clustering algorithm*

Competitor is discovered through clustering algorithm. Hierarchical clustering algorithm divides the product functional feature set *D* into several clusters. Every cluster has similar functional features. The competitor group is defined according the k-way cluster result. We can find the top-k competitors of a given item based on ranking in the cluster. It includes three main steps:

1. *Vector space model for representing each product*: we use *TF-IDF* term weighting model [9] to represent each product instruction as a vector. In order

to reduce the dimensionality of product instruction vectors, only nouns and verbs are preserved as candidate features, adverbs, prepositions are removed, we also set a minimum frequency threshold to reduce the number of features in a further step.
2. *Hierarchical clustering for finding competitor groups*: we first use cosine function to compute similarities of the feature vectors, then hierarchical clustering is chosen to divide the studied app products into several clusters as hierarchical clustering can build a hierarchy of the data objects [5].
3. *Extracting group features for summarizing competitor group*: the distinguished feature sets are extracted by selecting the attributes that dominates the similarities of the products.

2.3. Competitive domain identification using word network mining

In this paper, competition domains of Apps are identified by analyzing network structure. Firstly, we construct the product network, where nodes correspond to app features, while edges indicate the co-occurrences of functional features. The following are detailed steps of constructing product network and discover competitive domains of apps via structure analysis.

1. *Frequent 1-itemset and 2-itemset.* A k-itemset consists of k items. For example, the set {*map*} is a 1-itemset, while {*map*, *navigation*} is a 2-itemset. Given datasets D with extracted functional features of all products and a user predefined minimum support threshold $mini\_sup$. The *support* of an itemset X, denoted by $Dsupp(X)$, is the number of products in which X is contained. Mathematically, the *support* of an itemset X can be computed as $Dsupp(X) = \frac{\|X\|}{|D|}$, where $\|X\|$ is the number of products that contains itemset X, $|D|$ is the total number of products in D. Obviously, $0 \leq Dsupp(X) \leq 1$. All the *1-itemset and 2-itemset* that are frequent with respect to $mini\_supp$ are discovered. We choose FP-Growth algorithm [10] to find frequent itemsets.

2. *Word network generation.* According to the frequent 1- and 2-itemset patterns, we can generate a word network $G = (V, E)$, where the nodes $V = FP_T^{(1)}$ are frequent 1-itemsets, and the edges $E = FP_T^{(2)}$ are frequent 2-itemsets. The size of node $b \in V$ is proportional to $Dsupp(\{b\})$. For example, if $Dsupp(\{map\}) = 0.2$ and $Dsupp(\{navigation\}) = 0.3$, then the size of node *navigation* will be bigger than that of *map*. The width of edge $(u, v) \in E$ is defined as $w(u, v) = Dsupp(\{u, v\})$. It can be known that such a word network can well reflect the interconnections and popularity of the functional features.

3. *Network clustering for frequent patterns mining.* We employ network clustering analysis and visualization of word network by NodeXL (http://nodexl.codeplex.com/), which enables us to explore competitor domain relationships. Particularly, Girvan–Newman community clustering algorithm is used to divide the network [11].
4. *Max-confidence mining for dependent relationships.* We identify the dependent relation between a and b in the 2-itemset X = $\{a, b\}$ with a max-confidence measurement, which is defined [12] as formula:

$$MaxConf(a,b) = \max\{\frac{Dsupp(ab)}{Dsupp(a)}, \frac{Dsupp(ab)}{Dsupp(b)}\}$$

Max-confidence may be more efficient than cosine in evaluating the relation of dependency between any two products features [12]. The basic extraction principle is that: all the node b$j \in V, j = 1,...,$ n, $|V|$ should be extracted if the following relationships are satisfied: $MaxConf(b_k, b_j) \geq \theta$, $\theta \in [0,1]$, where θ is a predefined threshold. For example, map is one master node of navigation if $\{Dsupp(\{map\}) \geq Dsupp(\{navigation\})$. We can infer if one product have the function feature *"map"*, *"navigation"* is one possible extend function.

3. Experiments and Results

3.1. *Results of competitor group identification*

We crawled 2451 apps from "map tourism" category of 360 mobile assistant (zhushou.360.cn/list/index/cid/102231/) in May 17, 2017. 360 mobile assistant was the largest android application release platform in China. Python package of jieba was used for Chinese word segmentation. Stanford core natural language processing toolkit was used for part-of-speech tagging. Partition clustering algorithms *rbr* of CLUTO toolkit (glaros.dtc.umn.edu/gkhome/views/cluto) was used for clustering features. We labeled 20-way clustering result for competitor groups as Table 1. Rank of a given app in competitor group showed the position in the competitive domain. Ten experts in traffic field were invited to label and evaluate the result. Rank accuracy of top 5 in cluster competitor group reached 95%. Competitor group of *"navigation, map, taxi car, train, air ticket hotel"* were more popular than others. *Navigation* group was most popular according the download summary, which had the top download providers including *Gaode, Baidu, Google, Mapbar, Sogou, etc.*

Table 1. Competitive grouping and key features based hierarchical clustering.

Label of Cluster	Competition domain of Competition group	Apps number
Map navigation		
Navigation	Navigation,voice,map,car,phone,destination,electronic dog	136
Map offline	Map,offline,navigation,traffic,indoors,world,street,location	140
Parking spaces	Parking,car owners,navigation,maps	84
Traffic conditions	Traffic,road,travel,public,congestion,road,video	104
Positioning	Positioning,location,cell phone,friends,family,old man, map,GPS	110
Tourism attractions	Tourism,attractions,raiders,scenic spots, tour guides,food	183
Driving assistance		
GPS position	GPS,speed,satellite,tachometer,coordinates,position,elevation	103
Illegal owners	Violation,owner,vehicle,driving,fuel consumption, fault	111
Driving record	Record,traffic,video,compass,rack	64
Vehicle alarm	Vehicle,alarm,track,equipment,location,GPS,position,monitor	72
Rental sharing		
Taxi car	Drivers,taxis,passengers,car,orders,travel,platform	167
Carpool	Carpool,passenger,owners,downwind,platform,driver,shuttle	71
Car rental car	Car,rental,energy,charge,vehicle,platform	139
Freight truck	Freight,truck,logistics,supply,owner,business,goods	114
Bicycle rental car	Bicycles,bicycles,car rental,life,motorcycles,cars,cities,unlock	57
Bus and subway		
bus routes	Bus,line,site,transfer,inquiry,travel, local	276
Subway lines	Subway,line,bus,subway station,map,site,station,city	122
Ticket		
Train tickets	Train tickets,trains, timetables, trips,moments, inquiries	128
Air ticket hotel	Air tickets,flights,hotels,aviation,airports,specials	195
Car ticket	Bus ticket,bus,passenger, passenger station, ticket,car	69

3.2. *Results of competitive domain identification*

To analyze the competitive domains of traffic apps, we first construct the competitive domain graph as shown in Figure 2, where the nodes are frequent 1-itemsets as discussed in Subsection 2.3, width of edges between two nodes corresponds to frequency of frequent 2-itemset. We then utilizes Clauset-Newman-Moore cluster algorithm to divide the nodes into several clusters. Competitive domain relationships were showed with a Harel-Koren Fast Multiscale layout as Figure 2. We can find five competitive domain groups:

- *Navigation map* group with disk shape has important competitive features of *"discount information"*, *"customize information"*, *"attractions"*, *"hotel"*, *"delicious food"*.
- *Rental sharing* group with triangle shape has important competitive features of *"price"*, *"platform"*, *"order"*, *"easy"*, *"customer service"*.
- *Bus subway* with Solid Square shape has important competitive features of *"line"*, *"route"*, *"bus mode"*, *"position"*.

- *Ticket* group with solid diamond shape has important competitive features of "*train tickets*", "*air tickets*", "*timetables*", "*fight train*", "*inquire*".
- *Driving assistance* group with solid triangle shape has competitive features of "GPS", "recording", "track", "tool", "mode".

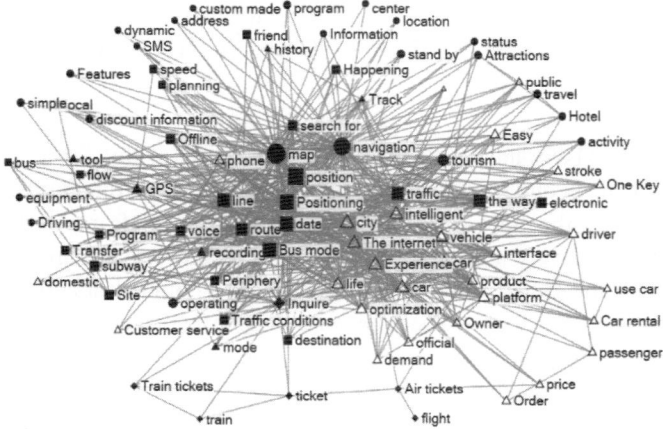

Figure 2. Harel-Koren fast multiscale network graph with $Dsupp(X) \geq 0.05$ and $Maxconf \geq 0$.

We used max-confidence measurement to evaluate the dependency relationship between any two function terms more efficiently. In Figure 3, given the conditions of $Dsupp(X) \geq 0.05$ and $Maxconf(a, b) \geq 0.6$, we could easily find the core functional features of "map", "navigation", "positioning", "address", "bus", "car" and so on. The dependent relationships around them were really clear. Given a feature of "*map*", we could find the dependent product features, such as "navigation", "voice", "search for", "address".

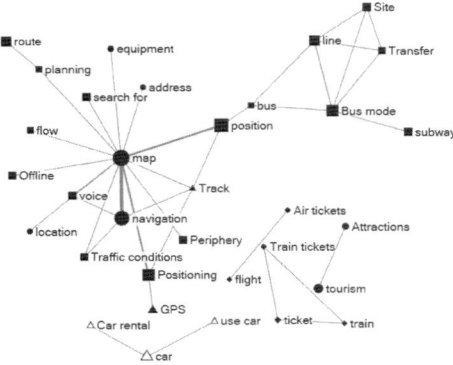

Figure 3. Harel-Koren fast multiscale layout with $Dsupp(X) \geq 0.05$ and $Maxconf(a, b) \geq 0.6$.

The result of network mining can help providers improving products. According to feed functional features, dependent product features can be found. The threshold of *Dsupp and Maxconf* can be personalized by users. Furthermore, given abundant product instruction texts in one filed, the dependency relationships between product features would be clearly presented. Suppliers can decide whether providing integrated service or focusing in a specific field.

4. Conclusion

In our effort, TMWN analysis framework is not only useful for a company to analyze its rivals in one filed, but also find the improvement directions. Our main contributions are the following: the proposal of a novel framework, TMNW, which could effectively mine competitors and competitive domains (fileds) from the third platform. Given a product, we can find the competitors group and rank competitors according to download. Given a competitive domain, product features relationships were dipicted and provide the impovement direction. The experimental results showed the proposed algorithm was highly effective.

Future research efforts will center on three directions. One is to try deep learning method to prune products features, including the process of generating word vectors through deep learning, expanding the seed dictionaries of specific domain functional features by the similarity between the seed dictionaries and candidate features. Two is to study more relationship of functional features by wordnet, try more clustering algorithm based the network structure, and predict the link to find the extend directions of product functionalities. The third is to extend the framework with competitive strength and competition dynamic analysis in a real-time analysis manner.

References

1. M. Bergen, M.A. Peteraf, Manag & Deci Econ 23(4), 157 (2002).
2. S.H. Bao, R. Li, Y. Yu, and Y.B Cao, Ieee T Knowl Data En 20(10), 1297 (2008).
3. D.D. Qiao, J. Zhang, Q. Wei, G.Q. Chen, Inform Manage-Amster (2016).
4. N. Oded, F. Ronen and G. Jacob, F. Moshe, Market Sci 31(3), 521 (2012)
5. M. Allahyari, S. Pouriyeh, M. Assefi et al., Proc Bigdas at KDD, 2017.
6. T.Y. Wu, Y. Chen and J. Han, Data Min Knowl Disc 21(3), 371 (2010).
7. M.Q. Hu, B. Liu. ACM SIGKDD, 168(2004).
8. G. Salton, A. Wong and C.S. Yang, Commu of the ACM 18(11), 613 (1975).
9. J.W. Han, J. Pei, Y. Yin and R. Mao, Data Min Knowl Disc 8(1), 53(2004).
10. M. Girvan, M.E.J. Newman, Proc Natl Acad Sci, 7821(2002).
11. T.Y. Wu, Y. Chen and J. Han, Data Min Knowl Disc 21(3), 371 (2010).

Parallel attribute reduction algorithms based on CUDA*

Yunmeng Hu, Tianrui Li, Jie Hu and Hongmei Chen

School of Information Science and Technology, Southwest Jiaotong University
Chengdu, 611756, China
huyunmeng@163.com, {trli, jiehu, hmchen}@swjtu.edu.cn

Attribute reduction is a hot topic in rough set theory. Many attribute reduction algorithms have been studied in previous literatures. However, existing algorithms are very time-consuming on massive data. To address this shortcoming, a parallel framework for attribute reduction is proposed. The divide-and-conquer methods for several typical attribute significances are presented. Then the corresponding parallel algorithms on CUDA programing model are developed. The experimental results on UCI datasets show that these proposed parallel algorithms have a good speedup performance.

Keywords: Attribute reduction; rough set theory; parallel algorithm; CUDA.

1. Introduction

Attribute reduction has become an important preprocessing step in areas such as pattern recognition, machine learning and data mining. Rough set theory as a powerful mathematical tool has been widely used in attribute reduction [1], [2], [3], which aims to retain the discernible ability of original features [4]. Attribute reduction for large data is still a challenging task and has attracted many interests. Wang *et al.* [5] developed a dimension incremental strategy for computing a reduct. Liang *et al.* [6] proposed an algorithm to find a new reduct in a shorter time when multiple objects are added to a decision table. Jing *et al.* [7] designed an incremental attribute reduction approach with a multi-granulation view. Since these methods are sequential which are unsuitable for massive data, it is vital to parallelize the traditional attribute reduction algorithms to improve their performance. Yang *et al.* [8] developed a method to parallelly compute the reduct for each sub-decision table and generated the reduct by deleting redundant attributes. Qian *et al.* [9] realized the parallel computation of equivalence classes and attribute significances using MapReduce. Zhang *et al.* [10] implemented a highly parallelizable method for large-scale reduction on Spark. Chen *et al.* [11]

---

*This work is supported by the National Science Foundation of China (Nos. 61573292, 61572406, 61603313) and the Soft Science Foundation of Sichuan Province (No. 2016ZR0034).

investigated approaches to attribute reduction in parallel using dominance-based neighborhood rough sets. In recent years, graphics processing units have quickly emerged as inexpensive parallel processors due to their high computation power and low price [12]. However, all of these existing parallel methods are based on MapReduce. This paper aims to develop a parallel framework to find a reduct based on CUDA.

The paper is organized as follows. Section 2 briefly reviews some basic concepts. Section 3 introduces a parallel framework for attribute reduction using CUDA. Section 4 demonstrates the experimental analysis. The paper ends with conclusions and future work in Section 5.

2. Preliminaries

Qian *et al.* [4] classified attribute significances into four categories: positive-region, Shannon's conditional entropy, Liang's conditional entropy and combination conditional entropy. We will use these four representative attribute significances to obtain a reduct in the following discussion.

Given a decision table $S = (U, C \cup D)$, $B \subseteq C$, the condition partition $U/B = \{E_1, E_2, \cdots E_e\}$ and the decision condition $U/D = \{D_1, D_2, \cdots D_d\}$ can be obtained. Through these notations, we review four types of attribute significances.

Definition 2.1 (PR)[4]**.** Let $S = (U, C \cup D)$ be a decision table, $B \subseteq C$. The attribute dependence degree of D with respect to B is defined as

$$\gamma_B(D) = \frac{|POS_B(D)|}{|U|} \tag{1}$$

where $POS_B(D) = \bigcup_{i=1}^{e}\bigcup_{j=1}^{d}\{E_i \mid E_i \subseteq D_j\}$, $|\cdot|$ denotes the cardinality of a set.

Definition 2.2 (SCE)[4]**.** Let $S = (U, C \cup D)$ be a decision table, $B \subseteq C$. The Shannon's conditional entropy of D with respect to B is defined as

$$H(D \mid B) = -\sum_{i=1}^{e} p(E_i) \sum_{j=1}^{d} p(D_j \mid E_i) \log(p(D_j \mid E_i)) \tag{2}$$

where $p(E_i) = \frac{|E_i|}{|U|}$, $p(D_j \mid E_i) = \frac{|E_i \cap D_j|}{|E_i|}$.

Definition 2.3 (LCE)[4]**.** Let $S = (U, C \cup D)$ be a decision table, $B \subseteq C$. The Liang's conditional entropy of D with respect to B is defined as

$$E(D \mid B) = \sum_{i=1}^{e}\sum_{j=1}^{d} \frac{|E_i \cap D_j|}{|U|} \frac{|D_j^c \cap E_i^c|}{|U|} \tag{3}$$

where D_j^c, E_i^c means the complement of the set D_j, E_i, respectively.

Definition 2.4 (CCE)[4]. Let $S = (U, C \cup D)$ be a decision table, $B \subseteq C$. The combination conditional entropy of D with respect to B is defined as

$$CE(D|B) = \sum_{i=1}^{e} \left(\frac{|E_i|}{|U|} \frac{C^2_{|E_i|}}{C^2_{|U|}} - \sum_{j=1}^{d} \frac{|E_i \cap D_j|}{|U|} \frac{C^2_{|E_i \cap D_j|}}{C^2_{|U|}} \right) \quad (4)$$

where $C^2_{|E_i|} = |E_i| \times (|E_i| - 1)/2$.

Definition 2.5 (Core)[4]. Let $S = (U, C \cup D)$ be a decision table. $Sig^{\Delta}_{inner}(a, C, D)$ is an inner significance of a. The core of attributes, denoted Core, is defined as

$$Core = \{a \mid Sig^{\Delta}_{inner}(a, C, D) > 0, a \in C\} \quad (5)$$

Definition 2.6 (Optimal Attribute)[4]. Let $S = (U, C \cup D)$ be a decision table, $B \subseteq C$. $Sig^{\Delta}_{outer}(a, B, D)$ is an outer significance of a. The a_{opt} is defined as

$$a_{opt} = \arg\max \{Sig^{\Delta}_{outer}(a, B, D), a \in C - B\} \quad (6)$$

All definitions above can be used to find a reduct in a forward greedy search strategy. A heuristic attribute reduction algorithm starts with a nonempty set and adds the attribute in core where satisfies inner significance. Then we keep adding optimal attribute with the highest outer significance at each iteration until the attribute subset meets the stopping criterion. Four representative inner significances and outer significances are shown in Table 1.

Table 1. The inner and outer significances of four representative methods.

Δ	$Sig^{\Delta}_{inner}(a, C, D), \forall a \in C$	$Sig^{\Delta}_{outer}(a, B, D), \forall a \in C - B$
PR	$\gamma_C(D) - \gamma_{C-\{a\}}(D)$	$\gamma_{B-\{a\}}(D) - \gamma_B(D)$
SCE	$H(D|C-\{a\}) - H(D|C)$	$H(D|B) - H(D|B \cup \{a\})$
LCE	$E(D|C-\{a\}) - E(D|C)$	$E(D|B) - E(D|B \cup \{a\})$
CCE	$CE(D|C-\{a\}) - CE(D|C)$	$CE(D|B) - CE(D|B \cup \{a\})$

3. Parallelization

From the analysis of attribute reduction algorithm, the most important and critical step is to calculate attribute significance. Hence, we show the decomposition of four attribute significances and propose a parallel attribute reduction framework using CUDA.

3.1. *Decomposition of Attribute Significance*

For the convenience, the multiset, denoted by K_{ij}, is defined as $K_{ij} = E_i \cap D_j$. Through this notation, we decompose four attribute significances as follows.

(1) Decomposition of PR-based attribute significance

$$\gamma_B(D) = \frac{|POS_B(D)|}{|U|} = \sum_{i=1}^{e}\sum_{j=1}^{d}\frac{|E_i|sng(K_{ij})}{|U|} \quad (7)$$

where $sng_{PR}(K_{ij}) = \begin{cases} 1, & K_{ij} = E_i \\ 0, & K_{ij} \neq E_i \end{cases}$.

(2) Decomposition of SCE-based attribute significance

$$\begin{aligned}
H(D|B) &= -\sum_{i=1}^{e} p(E_i)\sum_{j=1}^{d} p(D_j|E_i)\log(p(D_j|E_i)) \\
&= -\sum_{i=1}^{e}\frac{|E_i|}{|U|}\sum_{j=1}^{d}\frac{|E_i \cap D_j|}{|E_i|}\log(\frac{|E_i \cap D_j|}{|E_i|}) \\
&= -\sum_{i=1}^{e}\frac{|E_i|}{|U|}\sum_{j=1}^{d}\frac{|K_{ij}|}{|E_i|}\log(\frac{|K_{ij}|}{|E_i|}) = \sum_{i=1}^{e}\sum_{j=1}^{d}\left(-\frac{|E_i||K_{ij}|}{|U||E_i|}\log(\frac{|K_{ij}|}{|E_i|})\right)
\end{aligned} \quad (8)$$

(3) Decomposition of LCE-based attribute significance

$$\begin{aligned}
E(D|B) &= \sum_{i=1}^{e}\sum_{j=1}^{d}\frac{|E_i \cap D_j||D_j^c \cap E_i^c|}{|U|}\frac{}{|U|} = \sum_{i=1}^{e}\sum_{j=1}^{d}\frac{|E_i \cap D_j||E_i - D_j|}{|U|\,|U|} \\
&= \sum_{i=1}^{e}\sum_{j=1}^{d}\frac{|E_i \cap D_j||E_i|-|E_i \cap D_j|}{|U|\,|U|} = \sum_{i=1}^{e}\sum_{j=1}^{d}\left(\frac{|K_{ij}|}{|U|}\frac{|E_i|-|K_{ij}|}{|U|}\right)
\end{aligned} \quad (9)$$

(4) Decomposition of CCE-based attribute significance

$$\begin{aligned}
CE(D|B) &= \sum_{i=1}^{e}\left(\frac{|E_i|}{|U|}\frac{C^2_{|E_i|}}{C^2_{|U|}} - \sum_{j=1}^{d}\frac{|E_i \cap D_j|}{|U|}\frac{C^2_{|E_i \cap D_j|}}{C^2_{|U|}}\right) \\
&= \sum_{i=1}^{e}\sum_{j=1}^{d}\left(\frac{|E_i|}{d|U|}\frac{C^2_{|E_i|}}{C^2_{|U|}} - \frac{|E_i \cap D_j|}{|U|}\frac{C^2_{|E_i \cap D_j|}}{C^2_{|U|}}\right) \\
&= \sum_{i=1}^{e}\sum_{j=1}^{d}\left(\frac{|E_i|^2(|E_i|-1)}{|U|^2(|U|-1)d} - \frac{|E_i \cap D_j|^2 \times (|E_i \cap D_j|-1)}{|U|^2 \times (|U|-1)}\right) \\
&= \sum_{i=1}^{e}\sum_{j=1}^{d}\left(\frac{|E_i|^2(|E_i|-1)}{|U|^2(|U|-1)d} - \frac{|K_{ij}|^2 \times (|K_{ij}|-1)}{|U|^2 \times (|U|-1)}\right)
\end{aligned} \quad (10)$$

By above derivations, four attribute significances can be written as the form $\Theta^\Delta(D|B) = \sum_{i=1}^{e}\sum_{j=1}^{d}\theta^\Delta(S_{ij})$, where $S_{ij} \stackrel{def}{=} (E_i, D_j)$ and $\Delta = \{PR, SCE, LCE, CCE\}$.

Therefore, these attribute significances can be calculated in parallel because each attribute sub-significance $\theta^\Delta(S_{ij})$ can be calculated independently.

3.2. Parallel Attribute Reduction Based on CUDA

Definition 3.1 (Decision Matrix). Let $S = (U, C \cup D)$ be a decision table and $U/D = \{D_1, D_2, \cdots D_d\}$. Then the decision matrix $\mathbf{D}_{n \times d} = (d_{ij})_{n \times d} \in \{0,1\}$, where

$$d_{ij} = \begin{cases} 1, & x_i \in D_j \\ 0, & x_i \notin D_j \end{cases} \quad (11)$$

Definition 3.2 (Condition Matrix). Let $S = (U, C \cup D)$ be a decision table, $B \subseteq C$, $U/B = \{E_1, E_2, \cdots E_e\}$. Then the condition matrix $\mathbf{E}_{e \times n} = (e_{ij})_{e \times n} \in \{0,1\}$, where

$$e_{ij} = \begin{cases} 1, & x_j \in E_i \\ 0, & x_j \notin E_i \end{cases} \quad (12)$$

Definition 3.3 (Intersection Matrix). Let $S = (U, C \cup D)$ be a decision table. Given condition matrix $\mathbf{E}_{e \times n}$ and decision matrix $\mathbf{D}_{n \times d}$. The intersection matrix is defined as

$$\mathbf{K}_{e \times d} = \mathbf{E}_{e \times n} \times \mathbf{D}_{n \times d} = (k_{ij})_{e \times d} \quad (13)$$

From above definitions, the intersection matrix $\mathbf{K}$ can be calculated in parallel, because each k_{ij} can be computed independently.

Theorem 3.1. Let $S = (U, C \cup D)$ be a decision table, $B \subseteq C$, $U/B = \{E_1, E_2, \cdots E_e\}$ and $U/D = \{D_1, D_2, \cdots D_d\}$. Then the intersection matrix $\mathbf{K}_{e \times d} = \mathbf{E}_{e \times n} \times \mathbf{D}_{n \times d} = (k_{ij})_{e \times d}$, where $k_{ij} = |E_i \cap D_j|$.

Proof. According to E_i and D_j, $U = \{x_1, x_2, \cdots, x_n\}$ is partitioned into four exclusive crisp subsets $X_1 = \{x \mid x \in E_i \text{ and } x \notin D_j\}$, $X_2 = \{x \mid x \in E_i \text{ and } x \in D_j\}$, $X_3 = \{x \mid x \notin E_i \text{ and } x \in D_j\}$ and $X_4 = \{x \mid x \notin E_i \text{ and } x \notin D_j\}$. There are four different situations as follows.

(1) $\forall x_r \in X_1$, then $e_{ir} = 1, d_{rj} = 0$. Hence $e_{ir} \times d_{rj} = 0$.
(2) $\forall x_s \in X_2$, then $e_{is} = 1, d_{sj} = 1$. Hence $e_{ir} \times d_{rj} = 1$.
(3) $\forall x_t \in X_3$, then $e_{it} = 0, d_{tj} = 1$. Hence $e_{is} \times d_{sj} = 0$.
(4) $\forall x_l \in X_4$, then $e_{il} = 0, d_{lj} = 0$. Hence $e_{il} \times d_{lj} = 0$.

Therefore, we have

$$\begin{aligned} k_{ij} &= \sum_{p=1}^{n} e_{ip} \times d_{pj} \\ &= \sum_{x_r \in X_1} e_{ir} \times d_{rj} + \sum_{x_s \in X_2} e_{is} \times d_{sj} + \sum_{x_t \in X_3} e_{it} \times d_{tj} + \sum_{x_l \in X_4} e_{il} \times d_{lj} \\ &= \sum_{x_s \in X_2} e_{is} \times d_{sj} = |E_i \cap D_j| \end{aligned} \quad (14)$$

A GPU is a multi-core, multi-threaded highly parallel processor with enormous computing power. The NVIDIA Corporation released the CUDA framework to develop programs for GPUs [13]. By analyzing the parallelism of attribute significance, a parallel algorithm for computing attribute significance based on CUDA is designed as shown in Algorithm 1.

Algorithm 1. A parallel algorithm for calculating attribute significance.

Input: Decision table $S = (U, C \cup D)$ and attribute subset $B \subseteq C$;
Output: Attribute significance $\Theta^\Delta(D \mid B)$.
Step 1: Construct the decision matrix **D** in serial (CPU).
Step 2: Construct the condition matrix **E** in parallel and save **E** as CSR format (GPU).
Step 3: Construct the intersection matrix $\mathbf{K} = \mathbf{E} \times \mathbf{D}$ in parallel (GPU).
Step 4: Compute attribute sub-significance $\theta^\Delta(S_{ij})$ in parallel (GPU).
Step 5: Compute the sum of all attribute sub-significance in parallel (GPU).
Step 6: Output attribute significance (CPU).

After calculating decision matrix, the decision table, decision matrix and B are transformed to the global memory of GPU. Then we use CUDA Thrust library to parallelly sort the objects by values of B. Finally we obtain condition matrix and save it as a compressed sparse row (CSR) format to reduce the space. The intersection matrix is calculated by CUDA cuSPARSE library which provides the quickest and easiest route to high-performance linear algebra execution. A CUDA kernel is realized to compute attribute sub-significances in parallel and another CUDA kernel is used to parallelly compute the sum of all sub-significances.

Based on Algorithm 1, the inner and outer significances can be calculated in parallel and a parallel algorithm for attribute reduction is shown in Algorithm 2.

Algorithm 2. A parallel algorithm for attribute reduction.

Input: Decision table $S = (U, C \cup D)$;
Output: One reduct *red* .
Step 1: $red \leftarrow \varnothing$.
Step 2: For each $a_i \in C$, compute $Sig^\Delta_{inner}(a_i, C, D)$ by Algorithm 1.
Step 3: Put a_i into *red* , where $Sig^\Delta_{inner}(a_i, C, D) > 0$.
Step 4: Compute $\Theta^\Delta(D \mid red)$ and $\Theta^\Delta(D \mid C)$ by Algorithm 1.
Step 5: If $\Theta^\Delta(D \mid red) = \Theta^\Delta(D \mid C)$, then go to Step 8.
Step 6: $B \leftarrow red$, for each $a_i \in C - B$, compute $Sig^\Delta_{outer}(a_i, B, D)$ by Algorithm 1.
Step 7: Put a_{opt} into *red* where $a_{opt} = \arg\max\{Sig^\Delta_{outer}(a_i, B, D)\}$ and go to Step 5.
Step 8: Return *red* and end.

4. Experimental Analysis

In this section, we evaluate the proposed parallel algorithms based on four representative attribute significances. These algorithms are run on a computer with Windows10 and Intel(R)Xeon(R), 3.50GHz and 12GB memory. To test the performance of sequential and parallel algorithms, four datasets from UCI [14] and two NVIDIA GPUs are used. The CPU implementation is written in C++ and the GPU implementation is written in CUDA C with software Visual Studio 2013.

Table 2. Description of datasets.

No.	Dataset	Samples	Features	Class	Note
1	Adult	45222	14	2	small
2	Ticdata	5822	85	2	
3	Connect	67557	42	3	large
4	Covtype	581012	54	7	

The description of each data set is shown in Table2. The running time and speedup of four different datasets are presented in Table 3. For parallel algorithms, the running time includes all computation and memory transfer, etc. The speedup is defined as sequential execution time over parallel execution time.

Table 3. Running time and speedup on different data sets.

Method	Datasets	Running time(s)			Speedup		
		Serial	GT740M	TitanX	Serial	GT740M	TitanX
PR	Adult	69.52	3.59	1.60	1.00	19.36	43.45
	Ticdata	128.74	24.31	8.73	1.00	5.30	14.75
	Connect	2973.00	76.59	13.78	1.00	38.82	215.75
	Covtype	23011.00	161.84	30.57	1.00	142.18	752.73
SCE	Adult	88.56	3.77	1.75	1.00	23.49	50.61
	Ticdata	120.80	21.35	7.75	1.00	5.66	15.59
	Connect	3575.88	79.19	14.19	1.00	45.16	252.00
	Covtype	32186.10	162.87	37.42	1.00	197.62	860.13
LCE	Adult	97.17	3.72	1.69	1.00	26.13	57.55
	Ticdata	110.93	21.15	7.19	1.00	5.25	15.43
	Connect	3347.35	78.63	14.06	1.00	42.57	238.08
	Covtype	34003.10	161.45	39.25	1.00	210.61	866.32
CCE	Adult	99.269	3.73	1.63	1.00	26.59	60.58
	Ticdata	128.524	21.19	7.94	1.00	6.07	16.18
	Connect	3859.2	79.52	14.06	1.00	48.53	274.48
	Covtype	33238.6	162.4	36.06	1.00	204.67	921.76

As the results show, four parallel algorithms have a good speedup performance. GT740M has a lower speedup than TitanX, because GT740M only has 384 CUDA cores and TitanX has 3584 CUDA cores. As the size of the data increases, the speedup performs better. Furthermore, the CCE-based parallel algorithm implemented on TitanX achieves 921.76x over the CCE-based sequential algorithm on data set Covtype. Therefore, these parallel algorithms based on CUDA can deal with massive data efficiently.

5. Conclusions

In this paper, we combined CUDA programing model with traditional attribute reduction methods. A novel parallel framework based on four representative attribute significances was designed and the highest speedup of the parallel algorithm achieved 921.76x. Since the proposed parallel algorithm can only process symbolic data, we will present parallel algorithm combined with extended rough set models to deal with numerical, missing and set-valued data in the future.

References

1. Z. Pawlak, *Int. J. Comput. Inf. Sci.* **11**, 341 (1982).
2. G. Y. Wang, H. Yu and D. C. Yang, *Chin. J. Comput.* **25**, 759 (2002).
3. J. Qian, P. Lv, X. D. Yue, C. Liu, et al, *Knowl. Based. Syst.* **73**, 18 (2015).
4. Y. H. Qian, J. Y. Liang, W. Pedrycz and C. Dang, *Artif. Intell.* **174**, 597 (2010).
5. F. Wang, J. Y. Liang and Y. H. Qian, *Knowl. Based. Syst.* **39**, 95 (2013).
6. J. Y. Liang, F. Wang, C. Y. Dang and Y. H. Qian, *IEEE Trans. Knowl. Data. Eng.* **26**, 294 (2014).
7. Y. G. Jing, T. R. Li, H. Fujita, Z. Yu, et al, *Inf. Sci.* **411**, 23 (2017).
8. Y. Yang, Z. R. Chen, L. Zhang and G. Y. Wang, *RSK.* **6401**, 672 (2010).
9. J. Qian, D. Q. Miao, Z. H. Zhang and X. D. Yue, *Inf. Sci.* **279**, 671 (2014).
10. J. B. Zhang, T. R. Li and Y Pan, https://arxiv.org/abs/1610.01807 (2016).
11. H. M. Chen, T. R. Li, Y. Cai, C. Luo, et al, *Inf. Sci.* **373**, 351 (2016).
12. L. Shi, H. Chen and J. Sun, *IEEE Trans. Comput.* **61**, 804 (2012).
13. NVDIA CUDA, http://www.nvidia.cn/object/cudazone-cn.html (2018).
14. UCI Data Sets, http://archive.ics.uci.edu/ml/datasets.html (2018).

A multi-kernel spectral clustering algorithm based on incomplete views[*]

Wei Zhang, Yan Yang[†] and Jie Hu

School of Information Science and Technology, Southwest Jiaotong University
Chengdu, 611756, P. R. China
[†]yyang@swjtu.edu.cn

With the diversity of data sources, multi-view clustering algorithms are widely used. The traditional research routinely assumes that the multi-view data is complete, but the existing data may actually be missing. So the incomplete view clustering has become a hot research topic. In this paper, a multi-kernel spectral clustering algorithm based on incomplete views (IVMKSpec) is put forward. Firstly, the incomplete datasets are constructed with 10% to 90% of the loss rate, where they are clustered with the estimation of kernel and spectral clustering, and then the clustering results are evaluated by NMI and F-measure. Multi-kernel learning overcomes the defect that single kernel can effectively not handle data of heterogeneous and multiple data sources. Moreover, the multi-kernel spectral clustering is applied to incomplete datasets which improves the performance of incomplete clustering. Finally, the experimental results demonstrate that the proposed algorithm is robust and effective in most datasets.

Keywords: Multi-view clustering; spectral clustering; multi-kernel; incomplete view.

1. Introduction

Multi-view clustering has become a hot research for clustering multiple sources data. The traditional research assumes that these multi-view data is complete, but in many realistic scenarios, data in some views is not available. Therefore, some incomplete view clustering algorithms were introduced to solve these problems, such as incomplete view clustering (IVC) [1]. In addition, there was a study in subspace clustering with incomplete view [2]. In order to fill in the absent values, Trivedi et al. [3] put forward to the incomplete view clustering based on kernel correlation analysis (KCCA), which improved the stability of incomplete view clustering. With the incomplete dataset becomes bigger and bigger, an online multi-view clustering algorithm with incomplete views [4] was developed. To explore complementary information among multiple feature sets,

---

[*]This work is supported by the National Science Foundation of China (Nos. 61572407 and 61603313).

Zhao et al. [5] applied an incomplete clustering algorithm via deep sematic mapping.

More kernel methods are based on a single kernel method. Such as, Cai et al. [6] presented a weighted k-means multi-view clustering algorithm. And then, Tzortzis et al. [7] brought the spectral clustering in the above study. However, the multi-view contains heterogeneous information in the data sample characteristics and kernel function has different characteristics, thus a multiple kernel spectral clustering was presented in [8]. With the advent of incomplete view clustering, Liu et al. developed a multi-kernel k-means algorithm with incomplete kernels [9].

In order to exploit more potential information of incomplete views and enhance the performance of incomplete view clustering, a multi-kernel spectral clustering algorithm based on incomplete views (IVMKSpec) is put forward. It is the first time for the multi-kernel spectral clustering to apply to incomplete views in this paper. Compared to the single kernel function, multiple kernel learning integrates multiple kernel functions which accurately represent all the characteristics. Aiming at completing the incomplete kernel matrix, firstly mean filling is used, and then incomplete data is estimated by the overall data, moreover, IVMKSpec joints multi-kernel estimation and spectral clustering, in the end, experiment results verify the effectiveness of incomplete clustering.

The rest of the paper is shown below: In section 2, this paper introduces the multi-view spectral clustering algorithm. Then we present the details of proposed algorithm and the optimization process of IVMKSpec. Experimental results are presented in section 3. Section 4 summarizes the paper.

2. The Algorithm of IVMKSpec

2.1. *Multi-view spectral clustering (MVSpec)*

A study has been conducted to extend the spectral clustering to kernel clustering. Tzortzis et al. [7] put forward to the multi-view spectral clustering algorithm.

Given a multi-view data set $X = [x^1, x^2, \cdots, x^m] \in R^{d_v \times n}, 1 \leq v \leq m$, n is the number of samples, and m is the number of views, d_v is the dimension of each view. In order to measure the importance of each view, we lead in the vector of weight $(w^{(1)}, w^{(2)}, \cdots, w^{(m)})$, where $0 \leq w^{(v)} \leq 1$. The value of p is to control the sparse degree of view weight. The object function of multi-view spectral clustering is described as follows:

$$J = \min_{w_v,p,Y} \sum_{v=1}^{m} w_v^p (tr(K^{(v)}) - tr(Y^T K^{(v)} Y))$$

$$s.t. \ w_v \geq 0, \sum_{v=1}^{m} w_v = 1, p \geq 1, Y \in R^{n \times m}, Y_{ik} = \frac{\delta_{ik}}{\sqrt{\sum_{j=1}^{N} \delta_{jk}}} \quad (1)$$

where, $tr(K^{(v)}) - tr(Y^T K^{(v)} Y)$ is a trace difference of every view, it represents the intra-cluster variance. K is a single kernel matrix. It is expressed as $K_{ij} = K(x_i, x_j) = \phi(x_i)^T \phi(x_j) \ s.t. \ K \in R^{n \times n}$. Moreover, $\delta_{ik} = 1$ indicates that x_i belongs to the k cluster, otherwise $\delta_{ik} = 0$. Y is an indicator matrix that extracts the top M eigenvectors of kernel function K to optimize Y.

The value of Y is discrete, it is hard to optimize, we adopt the arbitrary orthogonal matrix U to replace, of which $U^T U = I$. The objective functions as below:

$$J = \min_{w_v,p,Y} \sum_{v=1}^{m} w_v^p (tr(K^{(v)}) - tr(U^T K^{(v)} U))$$

$$s.t. \ w_v \geq 0, \sum_{v=1}^{m} w_v = 1, p \geq 1, U \in R^{n \times m}, U^T U = I \quad (2)$$

2.2. *Optimization of IVMKSpec*

Firstly, it is important to construct multi-kernel matrix.

Kernel function. We assume $\vec{x}_k \in R^N (k = 1, 2, \ldots, K)$ is a set of samples in the input space, which use a nonlinear mapping ϕ to map this group of samples, and then a set of vectors in a high-dimensional space (also known as the feature space) is obtained: $\phi(\vec{x}_1), \phi(\vec{x}_2), \ldots, \phi(\vec{x}_K)$. The input space is represented by Mercer in the feature space:

$$K(\vec{x}_i, \vec{x}_j) = \phi(\vec{x}_i) \cdot \phi(\vec{x}_j) \quad (3)$$

At present, the common kernel functions are as follows:

Gaussian kernel:

$$K(\vec{x}_i, \vec{x}_j) = \exp(\frac{-\|\vec{x}_i - \vec{x}_j\|^2}{2\sigma^2}) \quad (4)$$

where $\sigma > 0$ is the width of the Gaussian kernel function.

Polynomial kernel:

$$K(\vec{x_i},\vec{x_j})=(\vec{x_i}\cdot\vec{x_j}+1)^d \qquad (5)$$

where d is the user-defined integer.

In general, the Gaussian kernel function is widely applied, because the characteristic space corresponding to the Gaussian kernel function is infinite dimension, and the finite sample must be linearly separable in this feature space, meantime, its local learning ability is very strong. Polynomial kernel function is also better in these kernel functions, which is very suitable for normalization of data sets. Thus we use the linear combination of Gaussian kernel and polynomial kernel function.

Generally, most of the data is filled with zero value for missing data. However it is not good to deal with the data clustering because there are enormous zero value lead to the singular value. Therefore, the initial data is processed by means of mean filling. Then, the multiple kernel learning is used to construct the kernel matrix of incomplete data.

Firstly, we need to normalize the multiple kernels.

$$L_v = D_v^{-1/2} K_v D_v^{-1/2}, v=1,2,...,m \qquad (6)$$

where D is a diagonal matrix of multi-kernel matrix, it satisfies $D_{ii} = \sum_j K_{ij}$. L is Laplacian matrix. Minimize the formula (2) is equal to maximize the formula (7).

$$\max_{U \in R^{n*k}} \sum_{v=1}^{m} w_v^p tr(U^T L_v U)$$

$$s.t.\ U^T U = I, v=1,2,...,m, p \geq 1 \qquad (7)$$

Now, the matrix is not a very complete matrix. As a result, we adopt the method by integrating the prediction kernel matrix and the clustering together in a unified optimization process. The kernel matrix is computed by clustering, and then the clustering is carried out through the updated kernel matrix, until the algorithm convergence.

The IVMKSpec algorithm is optimized by following three steps.

Step 1. Fix W, L, update U from formula (7).
Step 2. Fix U, W, update L. The detailed process is shown below.

From the formula (6) and formula (7), the formula (8) is calculated in the first.

$$\max_{U \in R^{n*k}} tr(U^T L_v U) \qquad (8)$$

According to [3], the U matrix is broken down into complete values and missing values. $U = \begin{bmatrix} U_c \\ U_m \end{bmatrix}$, U_c is the complete part, U_m is the missing part.

The kernel matrix is PSD (positive semi-definite), so we can decompose the kernel matrix to the multiplication of the two matrices. $L_v = A_v A_v^T, v = 1, 2, \ldots, m$. In the same way, the A matrix is broken down just like the U matrix. $A_p = \begin{bmatrix} A_v^c \\ A_v^m \end{bmatrix}$.

Therefore, the optimization of formula (8) is expressed as the optimization problem of formula (9).

$$\max_{U \in R^{n*k}} tr(U^T A_v A_v^T U) = \max_{U \in R^{n*k}} tr(UU^T A_v A_v^T) \tag{9}$$

After the formula (9), the formula (10) is obtained.

$$\max_{U \in R^{n*k}} tr\left(\begin{bmatrix} U_c \\ U_m \end{bmatrix} \begin{bmatrix} U_c \\ U_m \end{bmatrix}^T \begin{bmatrix} A_v^c \\ A_v^m \end{bmatrix} \begin{bmatrix} A_v^c \\ A_v^m \end{bmatrix}^T \right) \tag{10}$$

The value of A_v^m is calculated from formula (11).

$$A_v^m = -(U_c U_m^T + U_m U_c^T + 2 U_m U_m^T)^{-1} (U_m U_c^T A_v^c + U_c U_c^T A_v^c + U_m U_c^T A_v^c + U_m U_c^T A_v^c) \tag{11}$$

So, the value of L_v is shown as formula (12).

$$L_v = \begin{bmatrix} A_v^c \\ A_v^m \end{bmatrix} \begin{bmatrix} A_v^c \\ A_v^m \end{bmatrix}^T \tag{12}$$

Step 3. Fix U, L, update W from formula (7).

3. Experiment

3.1. *Experiment setting*

In our experiment, we select four classical clustering algorithms to compare the experiment, RMKMC[6], MVSpec[7], IMKKKM[9] and RMSC[10]. Moreover, we use the index NMI [11] (Normalized Mutual Information), F-measure [17] to evaluate the clustering performance of incomplete views. In the meantime, the six standard datasets are adopted. These datasets are complete, so we construct the incomplete datasets with 10% to 90% of the loss rate, the missing row or column is randomly selected according to a certain percentage. The information of the six multi-view clustering datasets used in the experiment is as Table 1 shows.

Table 1. Description of the multi-view datasets.

Datasets	Instances	Views	Clusters
3 Sources	169	3	6
Texas	187	2	5
Wisconsin	256	2	5
Washington	230	2	5
SenseIT	1000	3	2
Synth	1000	3	2

3.2. Experiment results

In the experiment, in order to avoid the influence of the initiation environment on the experimental results, all the final values of NMI and F-measure are obtained by calculating the mean value of the 30 times algorithm results.

The curves graph of the experimental results is shown in Figure 1 and Figure 2.

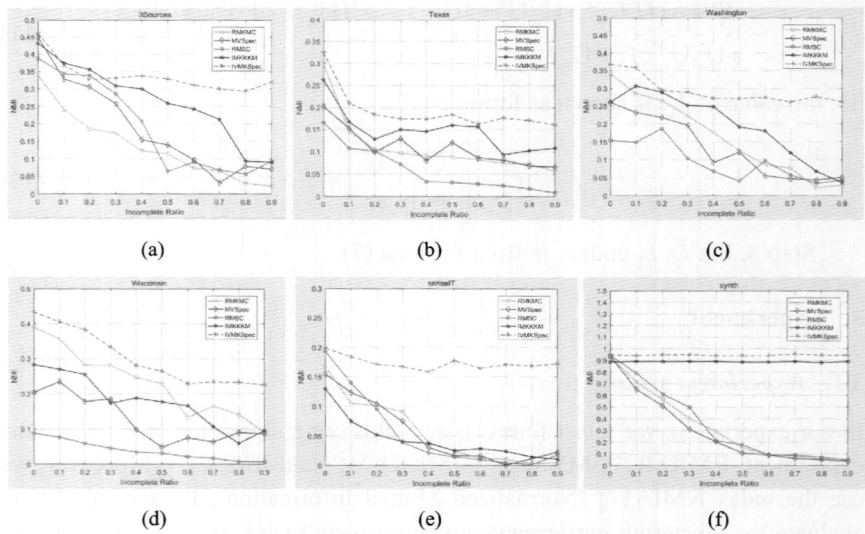

Figure 1. The NMI values of five algorithms on six datasets.

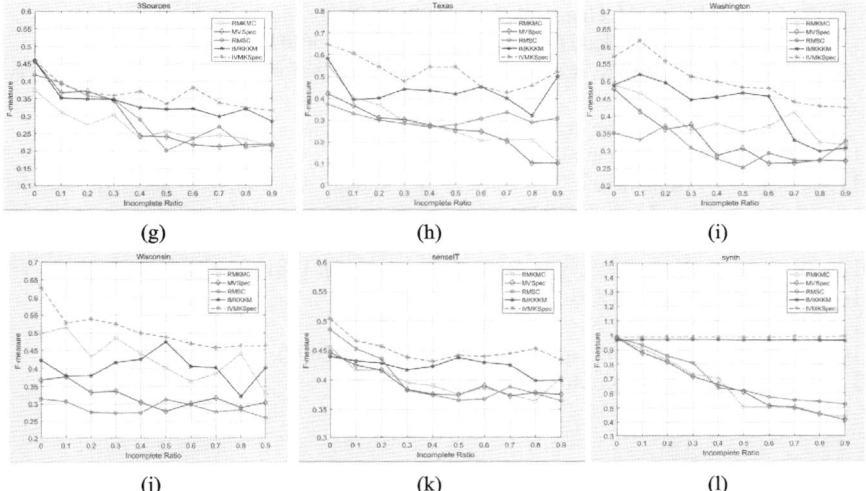

Figure 2. The F-measure values of five algorithms on six datasets.

Figure 1 and Figure 2 are comparing experimental results of the incomplete multi-view clustering algorithm. The red dotted curve represents the IVMKSpec algorithm proposed in this paper. It is noted that the clustering performance of the IVMKSpec algorithm is best in most datasets and appears more stable than other algorithms in the case of different miss rates.

4. Conclusion

This paper presents IVMKSpec, a multi-kernel spectral clustering based on incomplete views. It uses the valid combination of Gaussian kernel and polynomial kernel. Meanwhile, it computes the absent values and update by clustering. In addition, it uses multi-kernel spectral clustering to handle incomplete view datasets, which makes incomplete view clustering better. Experiment on six datasets shows that IVMKSpec outperforms the other algorithms on NMI and F-measure indexes. The proposed algorithm makes the operation particularly slow with too big datasets, thus we will extend the IVMKSpec algorithm to the cloud platform in the future work.

References

1. H. Gao, Y. Peng and S. Jian, Springer. 245(2016).
2. Q. Yin, S. Wu and L. Wang, ACM. 383(2015).
3. A. Trivedi, P. Rai, D. Hal, and S. L. DuVall, NIPS. (2010).
4. W. Shao, L. He, C. T. Lu et al., IEEE. 1012(2017).
5. L. Zhao, Z. Chen, Y. Yang et al., Neurocomputing. (2017).

6. X. Cai, F. Nie and H. Huang, AAAI. 2598(2013).
7. G. Tzortzis and A. Likas, Data Mining (ICDM). 675(2012).
8. D. Guo, J. Zhang, X. Liu et al., Pattern Recognition. IEEE Computer Society. 3774(2014).
9. X. Liu, M. Li, L. Wang et al., AAAI. (2017).
10. R. Xia, Y. Pan, L. Du et al., AAAI. 2149(2014).
11. J. Liu, J. Mohammed, J. Carter, S. Ranka, T. Kahveci and M. Baudis, Bioinformatics. **22**, 1971(2006).
12. S. Bhadra, S. Kaski and J. Rousu, Machine Learning. 1(2016).
13. W. Shao, S. Shi and P.S. Yu, Data Mining (ICDM). 1181 (2013).
14. S.Y. Li, Y. Jiang and Z.H. Zhou, AAAI. (2014).
15. W. Shao, L. He and P.S. Yu, Springer. 318 (2015).
16. M.R. Amini, N. Usunier and C. Goutte, NIPS. 28(2009).
17. Y. Yang, F. Jin and K. Mohamed. **25**, 1630(2008).

Attention-based bidirectional LSTM for Chinese punctuation prediction*

Jinliang Li, Chengfeng Yin, Zhen Jia, Tianrui Li and Min Tang

School of Information Science and Technology, Southwest Jiaotong University
Chengdu, 611756, China
lijinliang@my.swjtu.edu.cn, meg_yin@163.com, {zjia, trli}@swjtu.edu.cn,
tangmin007@foxmail.com

Punctuation prediction is an important task in Chinese automatic proofreading system which aims to tell whether the words and punctuations we use are right or not. The proofreading problem can be transformed into a prediction task. In this study, we propose an attention-based bidirectional Long Short-Term Memory (LSTM) model to predict punctuations. We use not only the sentence before the punctuation as the model's input, but also the sentence after the punctuation and the properties of the words. Experimental results show that the proposed LSTM model can achieve a good performance in punctuation prediction.

Keywords: Punctuation prediction; bidirectional LSTM; attention mechanism; proofreading

1. Introduction

Automatic text proofreading is one of the main applications of natural language processing. Punctuation plays an important role in text. The correct use of punctuation can improve the readability of text. A viable approach to proofread the punctuations is to predict which punctuation should be used and to compare it with the original punctuation. This paper will focus on the prediction of four different kinds of punctuations: comma, period, question mark and exclamation mark. They are the most frequently used punctuations in text.

Much research on punctuation prediction has been done. The earliest work used language model to deal with it. Beeferman et al. [1] used trigram language model and Viterbi algorithm for common punctuation symbols prediction. Gravano et al. [2] presented an approach to punctuation and capitalization restoration for English using purely text-based *n*-gram language models. Punctuation prediction could also be considered as an instance of sequence

---

*This paper is partially supported by the Demonstration Project of Science and Technology Service Industry, Sichuan Province (No. 2016GFW0167).

labeling tasks. Conditional Random Fields (CRF) and maximum entropy (ME) have been proved to be effective for automatic punctuation annotation. Lu and Ng [3] proposed a simple dynamic CRF method jointly perform both sentence boundary and punctuation prediction on speech utterances. And on the basis of [3], Wang et al. [4] combined prosodic, lexical and n-gram score features into a dynamic CRF framework which performs better in punctuation prediction. Huang and Zweig [5] used maximum entropy model (ME) for punctuation insertion in English text. In the past few years, due to their good performance, recurrent neural network (RNN) and its variants have been widely used in a variety of sequential labelling tasks, e.g., speech recognition [6], language modeling [7] and machine translation [8]. Recently, LSTM, a variety of RNN, was used in [9] for punctuation prediction. Xu et al. [10] proposed a bidirectional LSTM based on [9] and added a CRF layer to model the output. The results showed that the CRF model decreased the model's performance.

In this paper, we propose an attention-based bidirectional LSTM model to deal with punctuation prediction. We compare different LSTM models' performance on the People's Daily corpus. We also show the contribution of the sentence after the punctuation.

The rest of this paper is organized as follows. Section 2 outlines our proposed method. The details of experiments, results and analysis are discussed in Section 3. Section 4 presents some conclusions and possible directions of future work.

2. Method

In this section, we propose a Bi-LSTM model based on attention mechanism in detail. The general architecture of our model is shown in Figure 1. The whole model consists of four layers: input layer, LSTM layer, attention layer and output layer.

2.1. Input layer

The input of the network is the sentence before each punctuation. Every sentence contains several words $S = \{w_1, w_2, \ldots, w_T\}$. Each word w_t in S can be represented as a fixed size vector v_t from pre-trained word embedding dictionary. Property is an important feature of word. We also add properties of the words to the input. There are 14 kinds of properties in total. So we use one-hot encoding to deal with the properties. Each property can be represented as a 14-dimensional vector p_t which only has one single high (1) bit and all the other low (0). The network input x_t is concatenated by w_t and p_t.

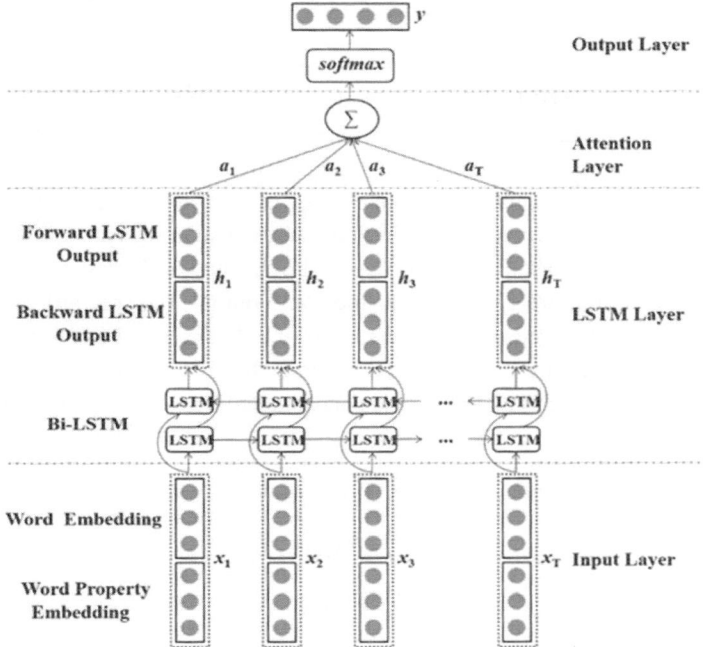

Figure 1. The architecture of attention-based bidirectional LSTM.

2.2. *LSTM layer*

LSTM units are firstly proposed by Hochreiter and Schmidhuber in 1997 [11]. The LSTM units can learn when to forget historical information and update stored information, thus avoiding the problem of disappearance of gradients encountered in traditional RNNs. A common LSTM unit is composed of a cell, an input gate, an output gate and a forget gate. The cell which is also called memory cell is used to store values over arbitrary time intervals. These gates control the behaviors of memory cells. At each time step, current cell state c_t is determined by three parts, which are inputs x_t, previous hidden state h_{t-1} and previous state c_{t-1}. For an input sequence $x = \{x_1, x_2, \cdots x_T\}$, each cell in LSTM can be computed as follows:

$$f_t = \sigma_g(W_f x_t + U_f h_{t-1} + b_f) \tag{1}$$

$$i_t = \sigma_g(W_i x_t + U_i h_{t-1} + b_i) \tag{2}$$

$$o_t = \sigma_g(W_o x_t + U_o h_{t-1} + b_o) \tag{3}$$

$$c_t = f_t \circ c_{t-1} + i_t \circ \sigma_c(W_c x_t + U_c h_{t-1} + b_c) \tag{4}$$

$$h_t = o_t \circ \sigma_h(c_t) \tag{5}$$

where σ is the sigmoid function, f_t, i_t, o_t, c_t and h_t are the vectors of forget gate, input gate, output gate, cell state and hidden state at time t, the W, U and b are weight matrices and bias vector parameters which need to be learned during training, and the operator ∘ denotes element wise multiplication.

Regular LSTM can only make use of previous input information. Bidirectional LSTM overcomes the limitation by considering both past and future inputs [12]. Its realization principle is as follows: The two LSTM networks with the opposite timing are connected to the same output, the forward LSTM can obtain information from past inputs, and the backward LSTM can obtain information from future inputs. Model accuracy has been greatly enhanced. The hidden state h_t of Bi-LSTM at time t is concatenated by forward h_{ft} and backward h_{bt}:

$$h_{ft} = \overrightarrow{LSTM}(x_t, h_{t-1}, c_{t-1}) \tag{6}$$

$$h_{bt} = \overleftarrow{LSTM}(x_t, h_{t+1}, c_{t+1}) \tag{7}$$

$$h_t = [h_{ft}, h_{bt}] \tag{8}$$

2.3. Attention layer

The standard LSTM cannot detect which input is more important. It only uses the hidden state at final time step. In order to address this issue, we propose an attention-based LSTM that can capture the key part of inputs. The attention layer output o can be computed as

$$o = \sum_{t=1}^{T} a_t h_t \tag{9}$$

where

$$a_t = \frac{\exp(e_t)}{\sum_{t=1}^{T} \exp(e_t)} \tag{10}$$

$$e_t = \tanh(W_A h_t + b_A) \tag{11}$$

where a_t is the attention weights. W_A and b_A are the weight matrix and bias in the attention mechanism to be learned in training step.

2.4. Output layer

We feed the attention layer output vector o to a softmax classifier to predict which punctuation should be used after the sentence:

$$\hat{y} = \text{softmax}(V^T o + b) \qquad (12)$$

where V and b are parameters to be learned. We train the model by minimizing the cross-entropy between the prediction punctuation y and the truth punctuation y.

3. Experiments

3.1. Corpus and experimental setup

We perform experiments on People's Daily corpus. The corpus has total 761000 sentences and about 6 million words in total. We use SWJTU segmentation tools to do the word segmentation and POS tagging. We only focus on four common punctuations. The normalization of punctuations is shown in Table 1.

Table 1. The normalization of punctuations.

Original Punctuation	Mapped Punctuation
，、	，
。．；：：	。
！！	！
？？	？
all the other punctuations	remove

In our task, we consider four kinds of punctuations to predict. They are comma (,), period (。), question mark (?) and exclamation mark (!). And we use precision (*prec.*), recall (*rec.*) and F1-measure (F_1) to assess the performance of the punctuation prediction tasks.

To find out whether the future inputs and attention mechanism are effective, we carry out four groups of experiments: standard LSTM (LSTM), attention-based LSTM (ALSTM), bidirectional LSTM (BLSTM) and attention-based bidirectional LSTM (ABLSTM). We can easily find that punctuation we use in the text not only depends on the sentence before the punctuation, but also the sentence after the punctuation. So we also do another experiment (ABLSTM-2S) to evaluate the performance of the two sentences around the punctuation. The input of ABLSTM-2S is the splicing of two sentences around the punctuation, while ABLSTM only use the sentence before the punctuation as the input. We use back-propagation algorithm to train all the LSTM networks. Network parameters are updated for every batch by using Adam optimizer [13].

3.2. Result and analysis

Table 2 presents the experimental results of different LSTM models mentioned above. We notice that compared to LSTM, BLSTM improves the F_1 score of three kinds of punctuations. This result supports our view that both past and future inputs are useful for punctuation prediction. It can be seen that ALSTM outperforms LSTM, and ABLSTM also outperforms BLSTM, which proves that the effectiveness of the attention mechanism. The ABLSTM model has the best performance. Surprisingly, compared to ABLSTM, the F_1 score of comma, period and exclamation mark are improved 15.8%, 10% and 5.8% by ABLSTM-2S, respectively, and ABLSTM-2S performance as good as ABLSTM on question mark. It shows that the sentence after punctuation plays an important role in punctuation prediction. This is in line with people's common sense. In addition, all the models' performances of question mark are the best among the four punctuation marks. The reason is that most question sentences have distinctive feature words which benefit the prediction.

Table 2. Experimental results of different LSTM models (%).

		LSTM	ALSTM	BLSTM	ABLSTM	ABLSTM-2S
comma	prec.	62.5	61.8	65.7	63.3	**81.7**
	rec.	61.9	63.0	60.2	63.5	**76.8**
	F_1	62.2	62.4	62.8	63.4	**79.2**
period	prec.	56.5	55.6	57.0	61.1	**68.6**
	rec.	62.6	57.8	58.9	58.1	**70.4**
	F_1	59.4	56.7	57.9	59.5	**69.5**
question mark	prec.	74.4	78.3	**82.1**	81.3	79.6
	rec.	**83.4**	79.9	77.5	80.8	80.5
	F_1	78.6	79.1	79.7	**81.0**	80.1
exclamation mark	prec.	**72.4**	69.0	65.5	64.5	71.1
	rec.	59.2	63.4	73.3	68.4	**73.3**
	F_1	65.1	66.1	69.2	66.4	**72.2**

4. Conclusions

In this study, we proposed different LSTM models for punctuation prediction. The best performing model is ABLSTM-2S, its key idea is to let every time step's hidden state participate in computing attention weights and join the two sentences around the punctuation as the inputs. The proposed model can concentrate on different parts of the sentences, so that it achieves a good performance in punctuation prediction. As the future work, adding the syntax analysis information of the sentences to the LSTM model would be a valuable research direction.

References

1. D. Beeferman, A. Berger and J. Lafferty, *ICASSP.* **2**, 689 (1998).
2. A. Gravano, M. Jansche and M. Bacchiani, *ICASSP.* 4741 (2009).
3. W. Lu and H. T. Ng, *EMNLP.* 177 (2010).
4. X. Wang, H. T. Ng and K. C. Sim, *INTERSPEECH.* 1384 (2012).
5. J. Huang and G. Zweig, *INTERSPEECH.* 917 (2012).
6. Y. Miao, M. Gowayyed and F. Metze, https://arxiv.org/abs/1507.08240.
7. K. Tran, A. Bisazza and C. Monz, *NAACL-HLT.* 321 (2016).
8. K. Cho, B. Merrienboer, C. Gulcehre, et al, https://arxiv.org/abs/1406.1078.
9. O. Tilk and T. Alumäe, *INTERSPEECH.* 683 (2015).
10. K. Xu, L. Xie and K. Yao, *ISCSL.* 1 (2016).
11. S. Hochreiter and J. Schmidhuber, *NEURAL COMPUT.* **9**, 1735 (1997).
12. M. Schuster and K. K. Paliwal, *IEEE Trans. Signal Processing.* **45**, 2673 (1997).
13. D. P. Kingma and J. Ba, https://arxiv.org/abs/1412.6980.

Clutter reduction of parallel coordinates based on an approximate measure of line crossing[†]

Yunlong Li, Tianrui Li, Shengdong Du and Xun Gong

School of Information Science and Technology, Southwest Jiaotong University
Chengdu 611756, China
liyunlong_swjtu@163.com, trli@swjtu.cn, dsd2000@126.com, xgong@swjtu.cn

Parallel coordinates are a significant way of visualizing multidimensional data. With their assistance, people can easily and intuitively observe and analyze the data. However, parallel coordinates may cause unnecessary visual clutter due to the improper arrangement of their attribute dimensions, which affecting people's perception of the data. In this paper, we propose a method by using representative lines to estimate the number of crossings between lines in parallel coordinates and take it as the basis to optimize the arrangement of attribute dimensions. The experimental results show that the proposed method can arrange the attribute dimensions fast and achieve better performance on the analysis of data.

Keywords: Parallel coordinates; line crossing; visual clutter; representative line.

1. Introduction

Parallel coordinates [1] are widely used for the visualization of multidimensional data due to their ease of use and other features. However, the original parallel coordinates do not specify the arrangement of their attribute dimensions, which may decrease people's ability to observe and understand the data.

A good arrangement of attribute dimensions can help people find out the overall trend of the data and the law of change between attributes, which has been extensively studied in recent years. Ankerst *et al.* [2] proposed a method based on a similarity measure, which arranges the attribute dimensions with similar properties in a close position. Yang *et al.* [3] presented an approach based on principal component analysis to order dimensions according to the degree of importance. Lu *et al.* [4] developed a similar approach based on singular value decomposition. Based on the distance measure, Peng *et al.* [5] optimized the arrangement of attribute dimensions by minimizing the number of outliers, but the computational pressure reduces the practicability of this method. From a

---

[†]This paper is partially supported by the Soft Science Foundation of Sichuan Province (No. 2016ZR0034).

visual intuition, Ellis and Dix [6] used the percentage of plotted pixels as a measure. Tatu et al. [7] proposed a method based on Hough transform to find the line clusters that are similar in position and orientation. Other measures, e.g., the number of line crossings, cross angles, and mutual information, can also be used to optimize the arrangement of attribute dimensions [8].

Although there are many measures for the visualization effect of parallel coordinates, most of them have high computational complexity. In view of the complexity of the existing methods for calculating the number of line crossings, this paper aims to provide a quick estimation method to help users quickly achieve the optimal arrangement of attribute dimensions.

The structure of this paper is organized as follows. Section 2 introduces the measure of visual clutter of parallel coordinates. Section 3 achieves the optimal dimension arrangement by estimating the number of line crossings. Experimental results are presented in Section 4 and followed by the discussions in Section 5. This paper ends with conclusions and future work in Section 6.

2. Measure of Visual Clutter

The purpose of this paper is to provide the users with a clear and intuitive view of data in parallel coordinates, and the line crossings are directly related to the visual intuition. Therefore, we choose the number of line crossings as the measure of visual clutter of parallel coordinates. By changing the arrangement of dimensions and reducing the number of line crossings in parallel coordinates, the goal of reducing visual clutter and making the relations between dimensions clearer is finally achieved [8].

A simple observation shows that the reason for producing line crossings is that the relative sizes of the data on adjacent axes have changed. Thus, judging whether the lines of two samples $d_i, d_j (1 \leq i, j \leq m)$ cross between axes A_p, $A_q (1 \leq p, q \leq n)$ is to judge whether their size relation has changed:

$$Cross(d_{i,p}, d_{i,q}, d_{j,p}, d_{j,q}) = \begin{cases} 1, & (d_{i,p} - d_{j,p})(d_{i,q} - d_{j,q}) < 0 \\ 0, & (d_{i,p} - d_{j,p})(d_{i,q} - d_{j,q}) \geq 0 \end{cases} \quad (1)$$

Based on formula (1), it can be easy to calculate the total number of all line crossings in parallel coordinates, as shown in formula (2).

$$Clutter(A_p, A_q)_{total} = \sum_{i=1}^{m-1} \sum_{j=i+1}^{m} Cross(d_{i,p}, d_{i,q}, d_{j,p}, d_{j,q}) \quad (2)$$

In addition, for data with class labels, users can pay more attention to the differences between different classes of the data by color coding, ignoring the number of internal line crossings within the same class. Relatively, we call it the external line crossings. The method of calculating the number of external line crossings is:

$$Clutter(A_p, A_q)_{ext} = \sum_{i=1}^{m-1} \sum_{j=i+1}^{m} Cross(d_{i,p}, d_{i,q}, d_{j,p}, d_{j,q}) Class(d_i, d_j) \quad (3)$$

where $Class(*)$ is the function of judging whether the two samples are in the same class:

$$Class(d_i, d_j) = \begin{cases} 0, & d_i, d_j \text{ are in the same class} \\ 1, & \text{otherwise} \end{cases} \quad (4)$$

It should be pointed out that the effect of multiple overlapping crossed points on visual clutter is essentially equivalent to one single crossed point, but in order to facilitate the calculation, the number of the crossed points overlapped in positions are repeatedly calculated in formulae (2) and (3). Nevertheless, it is still a slow process to accurately calculate the number of line crossings between two adjacent axes, with an $O(m^2)$ time complexity. We will show a faster estimation method in Section 3.

3. Methodology

This section is to achieve the dimension arrangement with the minimum approximate number of line crossings by the following steps: 1) The samples are divided according to the class attribute of the data; 2) The representative lines are selected for each class based on the distribution characteristics of the samples; 3) The number of line crossings of the original lines is estimated according to the crossings of the representative lines and its corresponding proportions; 4) The dimensions are arranged in the order of the minimum number of line crossings. Specific steps are as follows.

3.1. Data Clustering

As a basis for selecting representative lines, class labels are the prerequisite work for this paper. For datasets that do not contain the class attribute, existing clustering methods can be employed to get the class labels. In order to improve the efficiency, k-means [9] is used to cluster data without class labels.

3.2. Representative Lines Selection

Polylines in parallel coordinates are connected by the points on adjacent axes. Therefore, selecting representative lines requires selecting representative points in each dimension first.

In order to reflect the distribution characteristics of different classes $C_s (1 \leq s \leq k)$ on dimension $A_p (1 \leq p \leq n)$, we select the mean point $\bar{d}_{s,p}$ and its two 2-times standard deviation points $d_{s,p}^u$, $d_{s,p}^l$ as the representative points. $\bar{d}_{s,p}$, $d_{s,p}^u$ and $d_{s,p}^l$ are calculated as:

$$\bar{d}_{s,p} = \frac{1}{|C_s|} \Sigma_{d_i \in C_s} d_{i,p}. \quad d_{s,p}^u, d_{s,p}^l = \bar{d}_{s,p} \pm 2\sigma_{s,p} \tag{5}$$

where $|C_s|$ is the number of samples of class C_s, and $\sigma_{s,p}$ is the standard deviation of the samples on A_p, calculated as formula (6).

$$\sigma_{s,p} = \sqrt[2]{\frac{1}{|C_s|} \Sigma_{d_i \in C_s} (d_{i,p} - \bar{d}_{s,p})^2} \tag{6}$$

where $\alpha_{s,p}$, $\beta_{s,p}$ and $\gamma_{s,p}$ are the representative proportions of points $\bar{d}_{s,p}$, $d_{s,p}^u$ and $d_{s,p}^l$, respectively. This paper takes a fixed value for the proportions: $\alpha_{s,p} = 0.5$ and $\beta_{s,p} = \gamma_{s,p} = 0.25$ for any data class C_s and any dimension A_p.

After getting the representative points, the lines connected by a Cartesian combination of the representative points of the same class on any two dimensions, namely, the representative lines. Their representative proportions are the product of the proportions of the connected representative points, as shown in Figure 1.

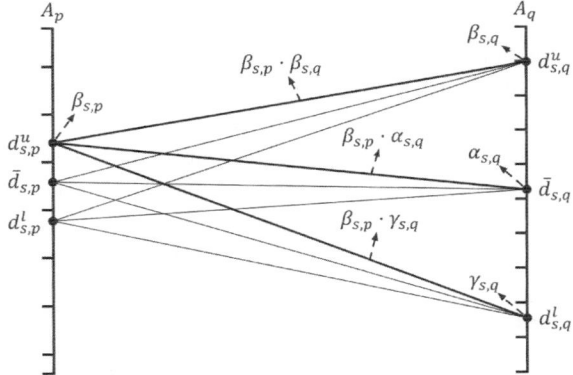

Figure 1. Representative lines connected by the representative points. Arrows refer to their proportions.

3.3. *Estimation of the Number of Line Crossings*

For different classes of the data, the paper calculates the number of line crossings based on the crossings of their representative lines with the same Cartesian combination. Therefore, the number of line crossings between the dimensions A_p, A_q can be estimated according to the sum of the number of line crossings of all different class representative lines:

$$\begin{aligned}
Clutter&(A_p, A_q)_{appr} = \\
\sum_{i=1}^{k-1} &|C_i| \sum_{j=i+1}^{k} |C_j| (\alpha_{i,p}\alpha_{i,q}\alpha_{j,p}\alpha_{j,q} Cross(\bar{d}_{i,p}, \bar{d}_{i,q}, \bar{d}_{j,p}, \bar{d}_{j,q}) \\
&+ \alpha_{i,p}\beta_{i,q}\alpha_{j,p}\beta_{j,q} Cross(\bar{d}_{i,p}, d^u_{i,q}, \bar{d}_{j,p}, d^u_{j,q}) \\
&+ \alpha_{i,p}\beta_{i,q}\alpha_{j,p}\beta_{j,q} Cross(\bar{d}_{i,p}, d^u_{i,q}, \bar{d}_{j,p}, d^u_{j,q}) \\
&+ \beta_{i,p}\alpha_{i,q}\beta_{j,p}\alpha_{j,q} Cross(d^u_{i,p}, \bar{d}_{i,q}, d^u_{j,p}, \bar{d}_{j,q}) \\
&+ \beta_{i,p}\beta_{i,q}\beta_{j,p}\beta_{j,q} Cross(d^u_{i,p}, d^u_{i,q}, d^u_{j,p}, d^u_{j,q}) \\
&+ \beta_{i,p}\gamma_{i,q}\beta_{j,p}\gamma_{j,q} Cross(d^u_{i,p}, d^l_{i,q}, d^u_{j,p}, d^l_{j,q}) \\
&+ \gamma_{i,p}\alpha_{i,q}\gamma_{j,p}\alpha_{j,q} Cross(d^l_{i,p}, \bar{d}_{i,q}, d^l_{j,p}, \bar{d}_{j,q}) \\
&+ \gamma_{i,p}\beta_{i,q}\gamma_{j,p}\beta_{j,q} Cross(d^l_{i,p}, d^u_{i,q}, d^l_{j,p}, d^u_{j,q}) \\
&+ \gamma_{i,p}\gamma_{i,q}\gamma_{j,p}\gamma_{j,q} Cross(d^l_{i,p}, d^l_{i,q}, d^l_{j,p}, d^l_{j,q}))
\end{aligned} \quad (7)$$

In the above formula, when all the representative lines do not cross, the calculation result is 0, indicating that there is no line crossing in the original plot. The sum of the proportions of all the representative lines is 1, which means that when all the representative lines are crossed, it is equivalent in the original plot that any line of one class crosses the lines of another class, and the number of line crossings is the product of the number of samples of the two classes.

3.4. *Dimension Arrangement*

According to Section 3.3, it can get an $n \times n$ clutter matrix similar to the similarity matrix in [2], where $Clutter_{p,q} = Clutter_{q,p}$ for any two dimensions A_p, A_q, and $Clutter_{p,q} = 0$ when $p = q$.

Under the clutter matrix, the dimension arrangement with the minimum clutter can be obtained. However, to get this optimal solution, we have to try every permutation because this is an NP-complete problem [2]. Thus, we adopt an ant colony optimization strategy [2] [10] when dealing with high-dimensional data.

4. Experimental Results

To demonstrate the validity of the proposed method of dimension arrangement, we conduct experiments on several real-world datasets. All experiments are implemented by D3.js [11], and processed on a MacBook Air laptop with 1.6 GHz Intel Core i5 CPU and 4GB of memory.

4.1. *Iris Dataset*[12]

The *Iris* dataset contains 4 attribute dimensions and one class attribute for a total of 150 samples. We color the samples according to the class attribute and get the original parallel coordinates plot as shown in Figure 2(a). The effect of our method is shown in Figure 2(b). Figure 2(c) is the same result of the optimal dimension arrangement obtained by accurately calculating the number of line crossings according to formulae (2) and (3) in Section 2. By comparison, it can be found that although the approximate method in this paper cannot guarantee the optimal dimension arrangement, it is still a good result, which can fully reduce the visual clutter and make the relationship between attribute dimensions more obvious. At the same time, it can also reduce the calculation and then improve the efficiency. The specific calculation results are shown in Tables 1 and 2.

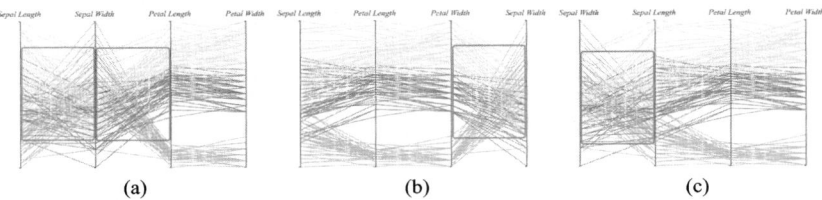

Figure 2. Optimal dimension arrangement on dataset *Iris* by different measures of line crossings: (a) The original arrangement. (b) Approximate measure of the paper. (c) Accurate measure of total (or external) line crossings.

4.2. Cars Dataset[13]

In order to further verify the effectiveness of the proposed method, we test it in a slightly larger dataset *Cars*. The dataset has 392 records of complete data and contains 7 attribute dimensions. The original plot and our plot with the optimal arrangement of minimum number of approximate line crossings are shown in Figure 3(a) and (b), respectively. From Figure 3, we can get a conclusion similar to the one in Section 4.1: The number of line crossings in the plot is reduced (as shown in the rectangular frames), and the dimensions with similar characteristics are arranged together to make the data clearer.

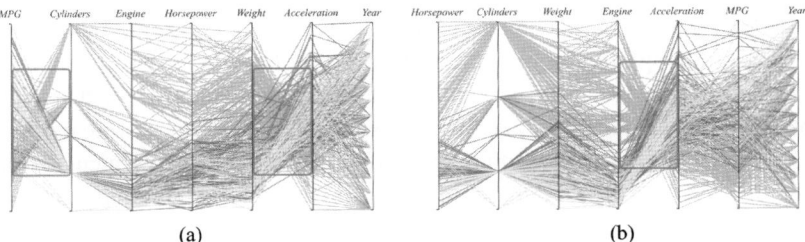

Figure 3. The results of dimension arrangement on dataset *Cars*: (a) The original. (b) Our method.

4.3. Wine-quality-white Dataset[12]

To demonstrate the high efficiency of the proposed method when processing large amounts of data, this experiment is conducted on a *Wine-quality-white* dataset with 4898 samples and 11 attribute dimensions. The results are shown in Tables 1 and 2. It is clear that although the data amount is more than 20 times of *Iris* and the dimension is nearly three times of *Iris*, the calculation of the clutter matrix can still be completed in a very short period of time, and its calculation speed is more than 4,000 times the original accurate calculation, which shows the high efficiency of the proposed method.

Table 1. Calculation results of the number of line crossings on the datasets for different measures.

Method	Dimension Arrangement		Iris	Cars	Wine-quality-white
Total		Original	12216	140686	51687364
		Optimal	7560	93645	44280691
External		Original	9605	78481	34900087
		Optimal	5245	51202	29803263
Approximate		Original	10000	120399	32397348
	Optimal	Total	7770	107625	49853268
		External	5604	58215	33417819
		Approximate	5000	68595	21944239

Table 2. Running time of calculating the clutter matrix in milliseconds.

Method	Iris	Cars	Wine-quality-white
Total	126	1530	567873
External	105	890	399764
Approximate	3	7	93

5. Discussions

The experimental results show that the proposed method can reduce the visual clutter in the original parallel coordinates and reduce the phenomenon of changing the size relations between clusters. The similar dimensions are arranged in the adjacent positions and the relationship among the dimensions is to be clearer.

Compared with the methods of accurately calculating the number of line crossings, although the proposed method may not be able to get the dimension arrangement with the actual minimum number of line crossings of the original plot, it should be pointed out that the purpose of this paper is to provide users with a clear and intuitive visualization of parallel coordinates. Experimental results show that for most cases, dimension arrangement with the minimum number of line crossings according to the approximate measure in this paper can help users reduce the visual clutter in parallel coordinates effectively and provide users with a good way to analyze. Besides, the method can bring about a great efficiency improvement. In fact, the time complexity of the method for calculating the visual clutter between any two dimensions is $O(m + k^2)$. When the number of samples m is much larger than that of classes k, it is close to linear efficiency and can therefore be used as a prerequisite for many analytical tasks.

6. Conclusions and Future Work

The goal of this paper is to minimize the number of line crossings by rearranging the attribute dimensions to provide the users with a clear parallel coordinates plot with the minimal visual clutter. Considering the complexity of the existing methods of calculating the number of line crossings, this paper presented a method to estimate the number of line crossings in the original plot by using representative lines. The results showed that the proposed method can accomplish the computational task fast, while ensuring a certain degree of visual clutter reduction.

Though the method in this paper can effectively calculate the number of line crossings, some work remains to be done, *e.g.*, the selection of representative points and representative lines. If the actual distribution of data is known, sufficiently using the probability characteristics to select the representative points and lines should be able to make the estimation of the number of line crossings closer to the accurate value. The same applies to their representative proportions.

On the other hand, we need to find a balance between the number of representative lines and the computational efficiency. The efficiency of this method derives from using a small number of representative points and lines to estimate the overall situation. A fixed number of representative lines may produce large deviations in the face of large amounts of data. Therefore, it is also worth studying to improve the accuracy of the calculation results without any significant performance reduction.

References

1. A. Inselberg, *Vis Comput.* **1**, 69 (1985).
2. M. Ankerst, S. Berchtold and D. A. Keim, *IEEE InfoVis.* 52 (1998).
3. J. Yang, W. Peng, M. O. Ward and E. A. Rundensteiner, *IEEE InfoVis.* 105 (2003).
4. L. F. Lu, M. L. Huang, J. Zhang, *J. Vis. Lang. Comput.* **33**, 3 (2016).
5. W. Peng, M. O. Ward and E. A. Rundensteiner, *IEEE InfoVis.* 89 (2004).
6. G. Ellis and A. Dix, *IEEE Trans. Vis. Comput. Graph.* **12**, 717 (2006).
7. A. Dasgupta and R. Kosara, *IEEE Trans. Vis. Comput. Graph.* **16**, 1017 (2010).
8. A. Tatu, G. Albuquerque, M. Eisemann, P. Bak, *et al.*, *IEEE Trans. Vis. Comput. Graph.* **17**, 584 (2011).
9. S. Lloyd, *IEEE Trans. Inf. Theory.* **28**, 129 (1982).
10. M. Dorigo and L. M. Gambardella, *IEEE Trans. Evol. Comput.* **1**, 53 (1997).
11. UCI Datasets, http://archive.ics.uci.edu/ml/datasets.html (2018).
12. StatLib Datasets, http://lib.stat.cmu.edu/datasets/ (2018).
13. D3.js, https://d3js.org/ (2018).

A novel preprocessing approach for imbalanced learning in software defect prediction

Kamal Bashir,[1,3] Tianrui Li,[1] Chubato Wondaferaw Yohannese,[1] Mahama Yahaya,[2] and Tayseer Ali

[1] *School of Information Science and Technology, Southwest Jiaotong University Chengdu 611756, China*
[2] *School of Transport and Logistics Engineering, Southwest Jiaotong University Chengdu 611756, China*
[3] *Department of Information Technology, College of Computer Science and Information Technology, Karary University, Omdurman 12304, Sudan kamalbashir1@yahoo.com, trli@swjtu.edu.cn, freewwin@yahoo.com, yahayamahama448@yahoo.com, tays_all@yahoo.com*

Software Defect Prediction (SDP) models are developed to predict modules that are prone to defect. It is achieved by mining datasets from historical software depositories. When one acquires data through this approach, it often includes class imbalance. Several methods including the Synthetic Minority Over-Sampling Technique (SMOTE) have been designed to address this problem. However, recent studies claim that class imbalance is not a problem in itself and that the deterioration of model performance is also linked with other factors related to the distribution of the data. The presence of noisy and borderline examples is one of these factors. To balance data only using the SMOTE is insufficient to handle these problems. Hence, this study proposes a novel preprocessing method that applies Fuzzy-Rough Instance Selection (FRIS) and Iterative Noise Filter based on the Fusion of Classifiers (INFFC) after SMOTE to overcome these problems. The experimental results show that the new method significantly outperforms the existing methods compared in this study.

Keywords: Software defect prediction; data sampling; fuzzy rough set; noise filtering.

1. Introduction

A software defect refers to incorrect step, process, or data definition in a computer program[1] which causes it to output wrong results and behave in ways unexpected. An effective approach to prediction of Fault-Prone (FP) software modules is necessary for appropriate testing to reduce costs. However, real-world data sets are often predominately composed of ordinary instances with only a negligible percentage of odd or interesting examples,[2] thus resulting in imbalance class representation. Standard machine learning algorithms consider that the training samples are uniformly distributed among different classes. Therefore, the classifier performs better when the classification approach is applied to data that has approximately equal class representation.[3]

The imbalanced data problem occurs when the examples in one class significantly outnumber the instances of the other class.[4] The minority class is usually the one that represents the concept to be learned. The problem of learning from data with imbalanced class distribution has gained more attention in recent times and several methods have been proposed to address it. Re-sampling techniques are independent of classifiers. Generally, the object is to modify the data distribution by changing the balance between classes, taking the local characteristics of examples into account. Several studies have been conducted in which their advantages have been discussed.[3,5-8] Among these techniques, Synthetic Minority Over-Sampling Technique (SMOTE) is one of the most well-known. It generates new artificial minority class instances by interpolating among several minority class examples that lie together.

With the SMOTE approach,[2] synthetic data are generated along the line segment that links minority class examples. SMOTE has the disadvantage of clustering the separation between minority and majority with noisy and borderline examples.[7] To overcome these problems, two different approaches are followed in the literature.

1. **Modifications of SMOTE** (hereafter called change-direction methods). These guide the creation of positive examples performed by SMOTE towards specific parts of the input space, taking into account specific characteristics of the data. Within this group, Safe level SMOTE,[9] Borderline-SMOTE,[10] ADASYN[11] and ADMOS[12] methods are found, which try to create positive examples close to areas with a high concentration of positive examples or only inside the boundaries of the positive class.
2. **Extensions of SMOTE**. It means that SMOTE is integrated with filtering techniques, e.g., SMOTE Tomek Links (TL),[13] SMOTE-ENN [Edited Nearest Neighbor Rule (ENN)],[13] SMOTE-RSB,[14] SMOTE-FRST[8] and SMOTE-IPF[7] as a post-processing step after using SMOTE.

In this study, we propose a new preprocessing method in which Fuzzy-Rough Instance Selection (FRIS) and Iterative Noise Filter based on the Fusion of Classifiers (INFFC) are implemented after SMOTE. This approach intends to overcome the problem produced by borderline and noisy examples in imbalance datasets. The FRIS in this study is to select useful instances from the training set, whereas the INFFC eliminates the noisy instances. This process referred to as SMOTEFRIS-INFFC will take care of two major challenges: (1) The influence of imbalanced class distribution in predicting the minority class examples: (2) Noisy and borderline examples.

2. A New Hybrid Preprocessing Method

In this section, we describe SMOTEFRIS-INFFC, our new hybrid preprocessing method for imbalanced datasets. The algorithm consists of three stages, which are implemented progressively to ensure that the noisy and borderline examples are eliminated from the imbalanced training set. First, to balance the training set, we apply SMOTE to introduce new synthetic minority class instances. Secondly, we remove synthetic instances, or majority class instances, for which the membership to the fuzzy-rough positive region of the training set falls below a given threshold $\tau(\tau = 1)$. Finally, the INFFC is implemented to remove the noisy examples originally presented in the dataset and those created through the implementation of SMOTE.

Algorithm 2.1 SMOTEFRIS-INFFC

Input: Set of D with $\{min, maj\}$, where min corresponds to a minority class, maj denotes majority class; α is the granularity parameter; τ is a selection threshold; $R_{min} : R_{maj}$ denotes desired balanced ration.
Output: enhanced dataset $D_{enhanced}$.
begin
1: Create S $\ni |D| = |S|$.
2: Create synthetic examples from minority class using SMOTE algorithm.
3: $\gamma = (maj * R_{\min})/R_{maj}$ /* the ratio between classes*/
4: $\varepsilon_0 = \gamma - \min$
5: $D = D + \varepsilon_0$
6: **foreach** $x \in D$.
7: **if** $(POS_A^{\alpha,D}(x) < \tau$ **then**
8: $\quad Y \leftarrow D - \{x\}$
9: Extract (min) from S and replace it in D
10: $D_{enhanced} \leftarrow INFFC(D)$
11: **return**
end

This study assumes that some of the synthetic minority class instances that are artificially introduced may not positively impact learning, and hence should be removed. Also, majority class instances in the original training data that do not satisfactorily belong to the fuzzy-rough positive region are removed. Since the minority class instances are rare, this procedure does not apply them.

3. Experimental Framework

In this work, we partition data by implementing 10-fold cross-validation. Then we build models on three learners viz. C4.5, Naive Bayes (NB) and Random Forest (RF). To create and examine empirical results, WEKA version 3.6.13, MATLAB R2016a, KEEL software tool and IBM SPSS Statistic23 are adopted. This study assesses the performance of classifiers using the AUC metric. Finally, several tests are carried out to evaluate the statistical significance of performance of each preprocessing method against the proposed method.

Datasets: A summary of the data sets is presented in Table 1, where JDT, LC, ML, PDE are from [15] and the rest are publicly accessible from a repository of software projects database. [16]

Table 1. Characteristics of datasets.

Datasets	#Modules	#Attribute	FP	NFP	Defect Ratio	Datasets	#Modules	#Attribute	FP	NFP	Defect Ratio
JDT	997	62	206	791	3.84	ant-1.7	745	21	166	579	3.48
LC	691	62	64	627	9.79	arc	234	21	27	207	7.67
ML	1862	62	245	1617	6.6	berek	43	21	16	27	1.68
PDE	1497	62	209	1288	6.16	camel-1.6	965	21	188	777	4.13
cm1	327	38	42	285	6.78	prop-5	8516	21	1299	7217	5.55
jm1	7782	22	1672	6110	3.65	prop-6	660	21	66	594	7.71
mc1	1988	39	46	1942	42.21	tomcat	858	21	77	781	10.14
pc1	705	38	61	644	10.55						

4. Results and Discussions

To examine the results obtained, some statistical tests are executed to analyze the difference between our approach and the method without preprocessing, as well as its comparison with ten existing preprocessing methods selected from the literature. Fig. 1 and Table 2 show the accuracy results of three classifiers (C4.5, NB and RF) when each of the ten preprocessing methods considered in this study is adopted for data preprocessing.

Table 2. The classification performance over all datasets for the three classifiers.

Dataset	Normal			SMOTE			Borderline_SMOTE			Safe_Level_SMOTE			SMOTE_ENN			SMOTE_IPF			SMOTE_RSB		
	C4.5	RF	NB	C4.5	RF	NB	C4.5	RF	NB	C4.5	RF	NB	C4.5	RF	NB	C4.5	RF	NB	C4.5	RF	NB
JDT	0.666	0.885	0.809	0.833	0.942	0.791	0.872	0.972	0.828	0.788	0.826	0.781	0.857	0.972	0.845	0.841	0.947	0.793	0.820	0.953	0.807
LC	0.631	0.814	0.762	0.865	0.973	0.730	0.929	0.988	0.812	0.742	0.794	0.659	0.782	0.958	0.938	0.873	0.974	0.661	0.885	0.981	0.729
ML	0.640	0.829	0.703	0.843	0.945	0.706	0.905	0.979	0.716	0.724	0.760	0.679	0.906	0.984	0.715	0.861	0.956	0.871	0.854	0.965	0.717
PDE	0.593	0.793	0.729	0.802	0.933	0.728	0.868	0.975	0.773	0.706	0.747	0.697	0.863	0.977	0.761	0.805	0.936	0.732	0.816	0.955	0.742
ant-1.7	0.696	0.837	0.811	0.780	0.901	0.805	0.828	0.948	0.833	0.739	0.771	0.791	0.816	0.937	0.867	0.793	0.925	0.829	0.780	0.912	0.818
arc	0.550	0.706	0.683	0.826	0.915	0.729	0.916	0.966	0.886	0.756	0.770	0.727	0.838	0.954	0.775	0.850	0.946	0.741	0.792	0.903	0.718
berek	0.860	0.976	0.928	0.915	0.970	0.969	0.943	0.985	0.977	0.807	0.942	0.919	0.965	1.000	0.969	0.915	0.970	0.969	0.913	0.978	0.963
camel-1.6	0.614	0.740	0.679	0.738	0.856	0.677	0.834	0.951	0.711	0.671	0.688	0.664	0.832	0.947	0.751	0.770	0.878	0.682	0.751	0.890	0.682
cm1	0.565	0.737	0.689	0.783	0.917	0.722	0.872	0.976	0.814	0.696	0.710	0.727	0.800	0.945	0.751	0.801	0.929	0.726	0.713	0.909	0.723
jm1	0.617	0.696	0.637	0.703	0.834	0.630	0.796	0.934	0.678	0.654	0.626	0.632	0.761	0.901	0.732	0.759	0.884	0.661	0.723	0.859	0.627
mc1	0.582	0.898	0.727	0.865	0.973	0.749	0.985	0.999	0.823	0.943	0.963	0.668	0.952	0.998	0.746	0.958	0.996	0.750	0.940	0.996	0.718
pc1	0.675	0.874	0.785	0.850	0.957	0.780	0.940	0.992	0.858	0.718	0.769	0.714	0.823	0.964	0.764	0.862	0.962	0.782	0.826	0.955	0.763
prop-5	0.630	0.731	0.694	0.821	0.875	0.689	0.911	0.938	0.741	0.749	0.742	0.679	0.946	0.991	0.755	0.936	0.980	0.745	0.871	0.905	0.692
prop-6	0.576	0.744	0.686	0.878	0.935	0.693	0.928	0.971	0.830	0.788	0.802	0.682	0.876	0.948	0.709	0.927	0.975	0.714	0.857	0.925	0.693
tomcat	0.638	0.826	0.812	0.834	0.936	0.797	0.930	0.987	0.891	0.800	0.816	0.794	0.780	0.939	0.950	0.857	0.952	0.803	0.844	0.952	0.810
Average	0.636	0.806	0.742	0.822	0.924	0.746	0.897	0.971	0.811	0.752	0.782	0.721	0.853	0.961	0.802	0.854	0.947	0.764	0.826	0.936	0.747

Dataset	SMOTE_FRS			SMOTE_TL			ADASYN			ADOMS			INFFC			Proposed method		
	C4.5	RF	NB	C4.5	RF	NB	C4.5	RF	NB	C4.5	RF	NB	C4.5	RF	NB	C4.5	RF	NB
JDT	0.826	0.948	0.803	0.854	0.968	0.826	0.828	0.939	0.749	0.859	0.967	0.807	0.799	0.971	0.911	0.890	0.982	0.873
LC	0.900	0.983	0.826	0.863	0.985	0.750	0.857	0.968	0.745	0.921	0.985	0.751	0.918	0.989	0.983	0.928	0.993	0.883
ML	0.851	0.964	0.780	0.865	0.975	0.724	0.841	0.943	0.686	0.899	0.974	0.720	0.692	0.987	0.862	0.906	0.988	0.826
PDE	0.822	0.957	0.794	0.826	0.968	0.744	0.793	0.933	0.703	0.876	0.965	0.728	0.732	0.977	0.957	0.865	0.979	0.851
ant-1.7	0.828	0.938	0.868	0.804	0.935	0.846	0.747	0.874	0.766	0.820	0.949	0.809	0.942	0.992	0.973	0.929	0.991	0.952
arc	0.851	0.943	0.851	0.871	0.965	0.775	0.801	0.891	0.703	0.861	0.976	0.729	0.871	0.977	0.874	0.914	0.982	0.903
berek	0.958	0.998	0.987	0.876	0.974	0.977	0.869	0.941	0.916	0.943	0.980	0.973	0.982	0.979	0.966	0.958	0.998	0.987
camel-1.6	0.788	0.923	0.753	0.801	0.937	0.730	0.737	0.849	0.663	0.816	0.945	0.693	0.791	0.984	0.985	0.839	0.970	0.828
cm1	0.830	0.960	0.808	0.784	0.945	0.772	0.760	0.900	0.713	0.862	0.976	0.712	0.727	0.945	0.870	0.878	0.976	0.838
jm1	0.707	0.945	0.775	0.740	0.904	0.682	0.681	0.818	0.601	0.813	0.923	0.647	0.841	0.984	0.959	0.879	0.999	0.997
mc1	0.956	0.995	0.885	0.951	0.999	0.757	0.953	0.994	0.739	0.979	0.994	0.769	0.000	0.000	0.000	0.979	0.999	0.900
pc1	0.924	0.989	0.872	0.856	0.981	0.801	0.846	0.957	0.763	0.902	0.957	0.779	0.808	0.987	0.833	0.952	0.996	0.890
prop-5	0.892	0.985	0.797	0.902	0.957	0.739	0.816	0.871	0.676	0.920	0.945	0.707	0.948	1.000	0.995	0.933	0.995	0.824
prop-6	0.820	0.948	0.787	0.922	0.976	0.722	0.866	0.929	0.664	0.910	0.929	0.691	0.500	1.000	0.833	0.920	0.989	0.905
tomcat	0.892	0.977	0.888	0.861	0.961	0.822	0.832	0.934	0.775	0.899	0.934	0.812	0.892	0.996	0.991	0.952	0.994	0.918
Average	0.857	0.964	0.832	0.852	0.962	0.778	0.815	0.916	0.724	0.885	0.960	0.755	0.763	0.918	0.866	0.915	0.989	0.892

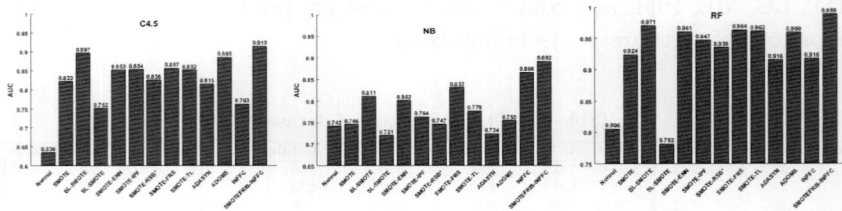

Fig. 1. Average AUC over all datasets for the three classifiers.

4.1. Test Accuracy Results

For all the learners (C4.5, SVM and RF), SMOTEFRIS-INFFC is the best method for the majority of the datasets used. For the 15 datasets used, the results of C4.5 demonstrate that the proposed method performs better in 8 datasets. It is also observed that, in situations where extreme imbalance occurs such as mc1, the BL-SMOTE demonstrates relative superiority over all the preprocessing techniques. The INFFC also shows outstanding performance over all the preprocessing approaches in three datasets. The average performance of ADMOS and SMOTE-FRS over all the datasets, considering AUC measure for the C4.5 classifier, are also significant even though lower than the proposed method. Also, the performance of

(Normal) ought to be reported. The average learning performance for the C4.5 is shown to be least when no preprocessing technique is applied (Normal). The specific case of the berek dataset is exceptional. In this instance, learning with Normal shows a better performance relative to SL-SMOTE. However, its behavior worsens with the increasing imbalance ratio. With NB, the AUC measure demonstrates improved performance in favor of the proposed method considering the number of datasets in which it outperforms the other selected methods. For the 15 datasets adopted, it performs better in 12. It is interesting to note that, with NB, the proposed approach performs better when applied on the datasets with the highest class imbalance ratio (mc1) as well as datasets with fairly balanced class ratio (berek), unlike the C4.5. The results of the RF classifiers built show better performance of SMOTEFRIS-INFFC in 9 out of the 15 datasets deployed. Besides our approach, the INFFC also demonstrates better performance in 5 of the datasets. For the berek dataset in which the class distribution is naturally approximately balanced (IR=1.68), the SMOTEENN performs better than all the preprocessing techniques. On the other hand, where the imbalance ratio is high as in mc1, the proposed method together with SMOTE-TL and BL-SMOTE demonstrate the best results, whereas the INFFC demonstrate relatively poor performance. The analysis presented in this study show that INFFC is not suitable for application in highly imbalance data. In such situations, the approach proposed in this study is proven to be more appropriate as demonstrated in the experimental results. Apart from the capability of our strategy to maintaining stability at different imbalance levels, its suitability for various learning algorithms (NB and RF) has also been noted. Thus, for imbalance learning, besides the influence of data balance techniques, choosing appropriate learning algorithm is crucial for improving SDP performance.

4.2. *Statistical Test*

Table 3 presents Wilcoxons signed ranks statistical test results for the comparison between SMOTEFRIS-INFFC versus Normal (no preprocessing) and INFFC considering the AUC measure for the C4.5, NB and RF classifiers. As the p-values (PWilcoxon) and the sums of ranks (R+ and R-) indicates, the application of SMOTEFRIS-INFFC improves the results acquired concerning no preprocessing (Normal) or preprocessing only with SMOTE which integrates the synthetic imbalanced datasets with borderline examples. In the case of INFFC, it is found that even though our approach consistently outperforms in terms of rank, the difference is not statistically significant as reflected in the p-value.

Concerning the assessment with other preprocessing techniques considering the 15 selected software metrics datasets, Table 4 presents the ranks and PHochberg values of the aligned Friedmans method (Rank column) for each technique (change-direction and the filtering-based methods) given the AUC for the three classifiers (C4.5, NB and RF). Considering the Friedmans aligned rank test, the preprocessing approach with the highest average ranking amongst all the datasets is regarded as the best. For the fact that SMOTEFRIS-INFFC has demonstrated the best aligned Friedmans rank, it indicates the superiority of the approach. The significance of the differences found by the aligned Friedmans test (PAlignedFriedman row) reflected in the p-value reaffirms the effectiveness of the proposed method. Similarly, the adjusted p-values for the Hochbergs tests are very low for all comparison and indicate that SMOTEFRIS-INFFC significantly outperforms all the methods compared in this study.

Table 3. Wilcoxons test results for the comparison of SMOTEFRIS-INFFC versus Normal, SMOTE and INFFC for the three learners on all datasets.

ALGORITHM	C4.5			NB			RF		
	R+	R-	PWilcoxon	R+	R-	PWilcoxon	R+	R-	PWilcoxon
SMOTEFRIS-INFFC vs Normal	8	0	0.001	8	0	0.001	8	0	0.001
SMOTEFRIS-INFFC vs INFFC	9.25	3	0.004	8.8	6.4	0.111	8.8	6.4	0.111
SMOTEFRIS-INFFC vs SMOTE	8	0	0.001	8	0	0.001	8	0	0.001

Table 4. Multiple statistical comparison with SD datasets.

	ALGORITHM	C4.5		NB		RF	
		RANK	PHochberg	RANK	PHochberg	RANK	PHochberg
	SMOTEFRIS-INFFC	4.6		4.9		4.9	
	BL-SMOTE	4.1	0.035	3.8	0.001	3.8	0.001
Change Direction	SL-SMOTE	1	0.001	1.1	0.001	1.1	0.001
	ADASYN	2	0.001	2.1	0.001	2.1	0.001
	ADOMS	3.3	0.002	3.1	0.001	3.1	0.001
	PAlignedFriedman	0		0		2.89E-11	
	SMOTEFRIS-INFFC	5.6		5.87		5.87	
	SMOTE-ENN	3.7	0.000805	3.8	0.001	3.8	0.001
Filtering	SMOTE-IPF	3.4	0.000979	2	0.001	2	0.001
	SMOTE-RSB	1.67	0.000655	1.87	0.001	1.87	0.001
	SMOTE-FRS	3.37	0.001206	3.53	0.001	3.53	0.001
	SMOTE-TL	3.27	0.003872	3.93	0.001	3.93	0.001
	PAlignedFriedman	0.000002		0		6.21E-09	

5. Conclusion and Future Work

The presence of noisy and borderline examples is a crucial and contemporary research issue for learning from imbalanced data. In this paper, we proposed a hybrid preprocessing technique for imbalanced datasets, SMOTEFRIS-INFFC, in which SMOTE is implemented to generate synthetic examples, FRIS is used to remove synthetic minority instances as well as original majority instances that have a small membership degree to the fuzzy positive region, and finally INFFC is applied to clean the whole data. We employ the proposed technique to real-world software datasets and build classifiers for the prediction of defective modules. The performance of our proposed method was compared with SMOTE and INFFC. The Wilcoxon rank test results demonstrated statistically significant improvement in favor of the proposed method. The experimental results also showed that our proposed technique outperforms all the methods compared in this study. For the future work, we plan to apply boosting to improve performance of the proposed method for SDP further.

Acknowledgments

This paper is partially supported by the Soft Science Foundation of Sichuan Province (No. 2016ZR0034).

References

1. J. Radatz, A. Geraci and F. Katki, *IEEE Std* **610121990**, p. 3 (1990).
2. N. V. Chawla, K. W. Bowyer, L. O. Hall and W. P. Kegelmeyer, *Journal of Artificial Intelligence Research* **16**, 321 (2002).
3. K. Bashir, T. Li, C. W. Yohannese and Y. Mahama, Enhancing software defect prediction using supervised-learning based framework, in *12th International Conference on Intelligent Systems and Knowledge Engineering (ISKE)*, (Nanjing, china, 2017).
4. P. Branco, L. Torgo and R. P. Ribeiro, *ACM Computing Surveys* **49**, 1 (2016).
5. C. W. Yohannese and T. Li, *International Journal Of Computational Intelligence Systems* **10** (2017).
6. C. W. Yohannese, T. Li, M. Simfukwe and F. Khurshid, Ensembles based combined learning for improved software fault prediction: A comparative study, in *International Conference on Intelligent Systems and Knowledge Engineering (ISKE)*, (Nanjing, china, 2017).
7. J. A. Saez, J. Luengo, J. Stefanowski and F. Herrera, *Information Sciences* **291**, 184 (2015).
8. E. Ramentol, N. Verbiest, R. Bello, Y. Caballero, C. Cornelis and F. Herrera, in *Uncertainty Modeling in Knowledge Engineering and Decision Making*, 2012, pp. 800–805.
9. C. Bunkhumpornpat, K. Sinapiromsaran and C. Lursinsap, Safe-level-smote: Safe-level-synthetic minority over-sampling technique for handling the class imbalanced

problem, in *Pacific-Asia Conference on Knowledge Discovery and Data Mining*, (Bangkok, Thailand, 2009).
10. H. Han, W.-Y. Wang and B.-H. Mao, Borderline-smote: a new over-sampling method in imbalanced data sets learning, in *International Conference on Intelligent Computing*, (Hefei, China, 2005).
11. H. He, Y. Bai, E. A. Garcia and S. Li, Adasyn: Adaptive synthetic sampling approach for imbalanced learning, in *IEEE International Joint Conference on Neural Networks*, (Hong Kong, China, 2008).
12. S. Tang and S.-P. Chen, The generation mechanism of synthetic minority class examples, in *International Conference on Information Technology and Applications in Biomedicine*, (Corfu, Greece, 2008).
13. G. E. Batista, R. C. Prati and M. C. Monard, *ACM SIGKDD Explorations Newsletter* **6**, 20 (2004).
14. E. Ramentol, Y. Caballero, R. Bello and F. Herrera, *Knowledge and Information Systems* **33**, 245 (2011).
15. M. DAmbros, M. Lanza and R. Robbes, *Empirical Software Engineering* **17**, 531 (2011).
16. K. R. P. D. Menzies, T., *http://openscience.us/repo/*. North Carolina State University, Department of Computer Science (2015).

A new tool for static and dynamic Android malware analysis

A. Martín,* R. Lara-Cabrera* and D. Camacho*

Computer Science Department,
Universidad Autónoma de Madrid
Madrid, Spain
{alejandro.martin,raul.lara,david.camacho}@uam.es
www.uam.es

AndroPyTool is a tool for the extraction of both, static and dynamic features from Android applications. It aims to provide Android malware analysts with an integrated environment to extract multi-source features able of modelling the behaviour of a sample and that can be used to discern its nature, whether malware or goodware. AndroPyTool integrates well known tools in this field, such as AndroGuard, DroidBox, FlowDroid, AVClass, VirusTotal or Strace, which allow to obtain a wide set of features including Application Programming Interface (API) calls, permissions, labels obtained from the different antivirus engines included in VirusTotal, Source-Sinks data connections, API calls invoked in real time, accessed files, files operations and many others. AndroPyTool is an open source tool that can be used via both the source code and a Docker container, in just three stages (pre-static, static and dynamic analysis).

Keywords: Cybersecurity; Android malware analisys; static analysis; dynamic analysis.

1. Introduction

In recent years, malware analysis has acquired great importance positioning itself as a strategic and promising research field. The continuous apparition of new shapes of malware always goes hand in hand with a deep study in order to prevent the effects of these new signatures and to mitigate the damage. Android has focused many researches in this respect, mostly due to the amount of malware designed specifically for this platform. Many of these researches have make an effort on designing malware detection tools, which typically follow the same procedural steps: firstly, a feature extraction step is carried out, to procure information able to describe the behaviour and intentions from each particular sample; secondly, to build a features space all samples can be represented, lastly, to train a classification method able to create a separation between malware and goodware (also known as benignware) samples in this space.

Focusing on the first step, many tools have been proposed. These are aimed at extracting several features, each one focused on modelling the behaviour of a sample from a different perspective. For instance, the well-known Androguard tool is designed for extracting static features from files contained in the package file. DroidBox, another widely used tool, is employed to run an application in an emulator while all the actions undertaken by the sample under analysis are captured. However, if a complete and multilevel analysis is required (*i.e.* combining the output of different static and dynamic features), it is necessary to deploy and execute separately all these tools.

In this paper, we present AndroPyTool, a framework for integrating the process of extracting varied static and dynamic features. AndroPyTool comprehends the most important Android analysis tools, performs code inspection in order to retrieve a wide set of characteristics and processes all the information gathered. The main objective of this tool is to provide those researchers interested in Android malware with an all-in-one tool that allows to perform the whole malware analysis cycle of a suspicious sample, thus making this process faster and easier. As a result, a detailed report for each sample is generated containing all the collected features. Furthermore, it provides a scalable solution and enables to easily integrate new tools. AndroPyTool is publicly available.[a]

2. Background

The 2017 *State of Malware Report* from Malwarebytes LABS illustrates and evidences the current problem stated by mobile malware. Furthermore, it evidences new concerns due to the novel characteristics found in the latest Android malware samples found in the wild, which are becoming smarter, and due to the significant presence of ransomware for this platform. Recently, the increase in the number of malware detected specifically designed for the Android platform[1] has been accompanied by the corresponding efforts for tackling this problem. Thus, multiple mechanisms, tools and frameworks appeared to model the behaviour of malicious and benign samples, building classification tools for detecting malware or defining new kinds of features among many others. Specifically, all the tools

---

[a]https://github.com/alexMyG/AndroPyTool

proposed so far are typically grouped into two different categories:[2] static and dynamic analysis, although it is also possible to establish a previous stage named pre-static analysis. The *pre-static analysis* is focused on extracting meta-information, features from the compressed *apk* file which do not require code inspection. The *static analysis* step is employed to extract more detailed features, such as declared Application Programming Interface (API) calls or permissions required. Finally, the *dynamic analysis* step monitors the application when it is executed in order to capture the real behaviour and actions performed.

There are different tools widely employed for analysing Android applications. In this sense, Apktool[3] is commonly used to decompress and decode the application executable. It allows to read the different files contained, such as the Android Manifest and also the code extracted from Dalvik byte-code to *smali*. The *Androguard* tool[b] is a Python library to obtain dozens of features from an *apk* file. Dex2jar[4] is used to decompile Dalvik bytecode to Java compiled code (a *.jar file*) which can be later read as Java using *jd-gui*. Other tools allow to obtain deeper characteristics. For instance, FlowDroid[5] runs taint analysis to detect information flows. In the dynamic analysis research area, DroidBox[6] connects to an Android emulator to obtain dynamically extracted information.

All these instruments try to model the behaviour of Android samples in order to, for instance, develop malware detection frameworks such as RevealDroid,[7] which combines static features like API calls and taint analysis traces obtained with FlowDroid, to later train a machine learning classifier (including the C4.5 decision-tree classifier and the 1-nearest-neighbor algorithm) to segregate samples between malware and goodware. Drebin[8] makes malware classification based on permissions requested or API calls invoked. MOCDroid[9] employs a genetic algorithm to build an Android malware classifier. Other example is ADROIT,[10] which detects Android malware analysing meta-information and where a set of machine learning classifiers are trained. Other examples such as Droid-Sec[11] include the use of deep learning to detect Android malware.

3. AndroPyTool

AndroPyTool integrates different analysis tools and Android applications processing tools, in order to deliver fine-grained reports drawing their

---

[b]https://github.com/androguard/androguard

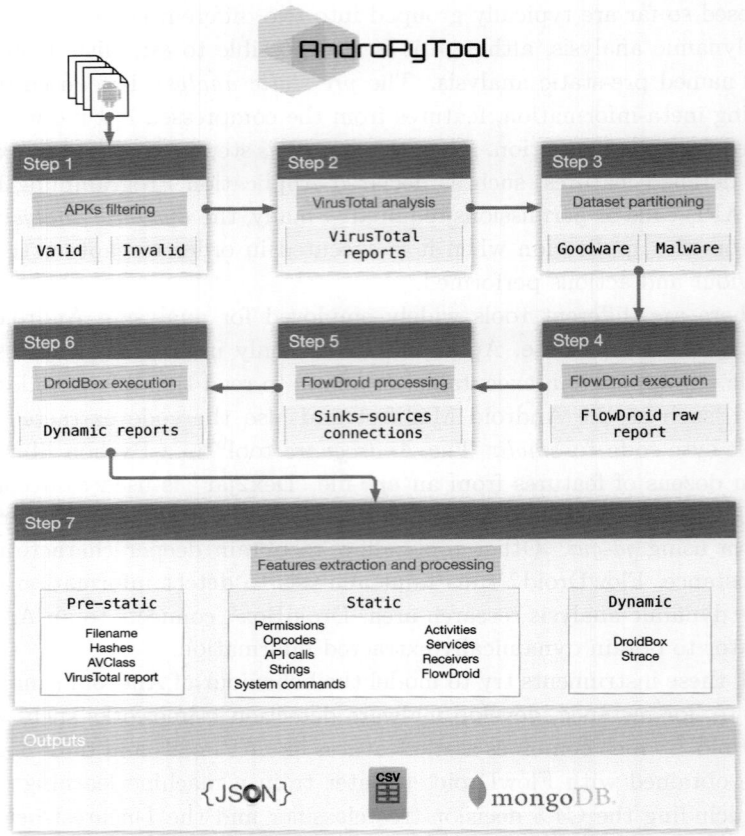

Fig. 1. AndroPytool architecture.

individual behaviour and characteristics. This framework, which has been implemented as a Python project, takes the *apk* file to be analysed as input and follows up a seven-step process (see Fig. 1) in which different static and dynamic analysis tools are executed, the code is automatically inspected to obtain different features, and finally concludes generating both, an application independent report including all the features extracted and processed, and a secondary report including all samples analysed.

- **Step 1: APKs filtering.** This first step is in charge of inspecting each sample with Androguard in order to check if the sample is a true and a valid Android application.

- **Step 2: VirusTotal analysis.** A report for each sample is retrieved from the VirusTotal portal. It contains the scan result and analysis date from more than 60 different antivirus engines.
- **Step 3: Dataset partitioning.** A categorisation based on the VirusTotal report is used to split the set of samples into two sets: malware samples (at least one antivirus tests for positive) and goodware samples.
- **Step 4: FlowDroid execution.** The taint analysis based FlowDroid framework is executed for each sample.
- **Step 5: FlowDroid outputs processing.** The output delivered by FlowDroid for each sample is processed in order to extract the connections between sources and sinks found.
- **Step 6: DroidBox execution.** In this step, the Android dynamic analysis tool DroidBox is executed for each sample. The version executed of DroidBox includes some modifications as integrating the *Strace*.
- **Step 7: Features extraction and processing.** Finally, the rest of the features are extracted and all the information is processed and handled in order to put it together. A report for each sample and global output including all the samples analysed is generated in three different formats: *Comma-separated Values (CSV)*, *JavaScript Object Notation (JSON)* and as a MongoDB database.

3.1. *Features description*

Through the seven-step process, AndroPyTool executes different tools in order to extract a wide set of features from an input set of Android samples. All these features, and the information that they represented, are organised in three different categories (pre-static, static and dynamic), both the features and how they are extracted are described below:

3.1.1. *Pre-static analysis*

It involves extracting information without code inspection and allows to identify and to track the sample. This category includes: the file name, MD5, SHA1 and SHA256 hashes, a report obtained from the VirusTotal[c] portal with the scan result from more than 60 antivirus engines, and

---
[c]https://www.virustotal.com/

consensual label determined by the AVClass.[12] It also includes the package name and the main activity name, which are obtained with *Androguard*.

3.1.2. *Static analysis*

This category includes those features that are retrieved by analysing the application at the code level. They can be used to build a representation about the expected behaviour of the sample. In this category, features such as API calls, activities, opcodes or permissions can be found. More specifically the features extracted are:

- **API calls.** They allow to model the expected behaviour of the sample. Certain API calls are commonly found among malware samples. These API calls are extracted by parsing the *smali* files and searching for *invoke* opcodes.
- **Opcodes.** Collecting a counter for each Dalvik opcode that appears in the *smali* files can help to understand the complexity and behaviour of the sample. Again, the *smali* files are parsed in order to retrieve all the opcodes declared.
- **Permissions.** Android permissions shape barriers between the application and the use of certain device functionalities and must be declared in Manifest. They are obtained using *Androguard*.
- **Intents.** An intent expresses an action to be performed. In many cases, malware often listens to some specified intents. The *AndroidManifest.xml* file is manually analysed to obtain this feature.
- **Receivers.** This application component allows applications to receive intents broadcast by the system or other applications. We focus on Receivers declared in the Android Manifest. A list of Intents declared for each Receiver is included.
- **Services.** This component allows performing operations without the user interaction or exposes functionalities to other applications to use. This feature is obtained by parsing the Manifest XML file. A list of Intents declared for each Service is included.
- **Activities.** Activities implement interfaces that allow the user to interact with the application. All activities must be declared in the *AndroidManifest.xml*, which can be parsed using *Androguard*. A list of Intents declared for each Activity is included.
- **Strings.** An analysis over the strings, which appear in the code as constants, becomes really interesting, as they can contain code to be loaded in runtime, network addresses or threatening messages.

String variables are declared in the *smali* files as Dalvik opcodes.
- **System commands.** System commands are indicators of certain activities which must be controlled, like privilege escalation or the process of rooting the device. These system commands can be stored as strings.

FlowDroid has also been integrated in AndroPyTool. It is a taint analysis based framework that detects potential data leaks within the life cycle of an application, which can be later used to differentiate malicious behaviours. Basically, it detects information flow that starts in a source (*i.e.* invoking an API call that returns certain data) and that end in a sink (*i.e.* another API call that receives this data). According to the authors, FlowDroid[5] is *"fully context, flow, field and object-sensitive"*. It considers the whole Android application life-cycle as well as the UI widgets of the application. As result of executing FlowDroid, a list counting the number of connections between each source and sink is obtained.

3.1.3. *Dynamic analysis*

Finally, the last set of features are extracted by monitoring the application execution in a controlled environment (*i.e.* an emulator). We have chosen the DroidBox tool for this purpose, which allows to dynamically obtain diverse information in real time. This enables to make a fine-grained analysis of the application behaviour actually exhibited when it is executed. Furthermore, we have built a modification of the DroidBox framework to integrate Strace. The information gathered by the DroidBox tool includes: the use of *cryptographic* functions, loaded *DEX classes* in run time, *files accessed* and the kind of *operation*, *network connections*, *SMS*, *phone calls*, *started services*, *enforced permissions* and information leaks detected.

4. Conclusions

With AndroPyTool, our main goal is to provide researchers and malware analysts with a powerful and integrated tool for extracting multi source features from Android applications. Thus, this tool puts together the most important and employed tools in this domain, in order to build a framework that automates the process of extracting meta-information, executing static and dynamic analysis, processing all the data collected, to finally generating formatted outputs for later use. In future work, we aim to integrate more analysis tools into AndroPyTool and to improve the data processing stages, in order to give more functionalities to the users.

Acknowledgments

This work has been co-funded by the following research projects: Comunidad Autonoma de Madrid (CAM) grant S2013/ICE-3095 (CIBERDINE: Cybersecurity, Data and Risks), EphemeCH (TIN2014-56494-C4-4-P) and DeepBio (TIN2017-85727-C4-3-P) by Spanish Ministry of Economy and Competitivity, under the European Regional Development Fund FEDER, and Justice Programme of the European Union (2014-2020) 723180 – RiskTrack – JUST-2015-JCOO-AG/JUST-2015-JCOO-AG-1. The contents of this publication are the sole responsibility of their authors and can in no way be taken to reflect the views of the European Commission.

References

1. M. Lindorfer, M. Neugschwandtner, L. Weichselbaum, Y. Fratantonio, V. Van Der Veen and C. Platzer, Andrubis–1,000,000 apps later: A view on current android malware behaviors, in *Building Analysis Datasets and Gathering Experience Returns for Security (BADGERS)*, (Wroclaw, Poland, 2014).
2. K. Tam, A. Feizollah, N. B. Anuar, R. Salleh and L. Cavallaro, *ACM Computing Surveys (CSUR)* **49**, p. 76 (2017).
3. R. Winsniewski, Android–apktool: A tool for reverse engineering android apk files (2012).
4. B. Alll and C. Tumbleson, *Octeau D, Jha S, McDaniel R Retargeting*.
5. S. Arzt, S. Rasthofer, C. Fritz, E. Bodden, A. Bartel, J. Klein, Y. Le Traon, D. Octeau and P. McDaniel, *Acm Sigplan Notices* **49**, 259 (2014).
6. P. Lantz, A. Desnos and K. Yang, Droidbox: Android application sandbox (2012).
7. J. Garcia, M. Hammad, B. Pedrood, A. Bagheri-Khaligh and S. Malek, *Obfuscationresilient, efficient, and accurate detection and family identification of android malware*, tech. rep., Department of Computer Science, George Mason University (2015).
8. D. Arp, M. Spreitzenbarth, M. Hubner, H. Gascon and K. Rieck, Drebin: Effective and explainable detection of android malware in your pocket., in *NDSS*, (San Diego, USA, 2014).
9. A. Martín, H. D. Menéndez and D. Camacho, *Soft Computing* **21**, 7405 (2017).
10. A. Martín, A. Calleja, H. D. Menéndez, J. Tapiador and D. Camacho, Adroit: Android malware detection using meta-information, in *Computational Intelligence (SSCI), 2016 IEEE Symposium Series on*, December 2016.
11. Z. Yuan, Y. Lu, Z. Wang and Y. Xue, Droid-sec: Deep learning in android malware detection, in *Proceedings of the 2014 ACM Conference on SIGCOMM*, SIGCOMM '14 (ACM, New York, NY, USA, August 2014).
12. M. Sebastián, R. Rivera, P. Kotzias and J. Caballero, Avclass: A tool for massive malware labeling, in *International Symposium on Research in Attacks, Intrusions, and Defenses*, (Evry, France, 2016).

A super-hyper network model based on matrix operation*

Shengjiu Liu

School of Information Science and Technology, Southwest Jiaotong University
Chengdu 611756, China, liushengjiu2008@163.com

Tianrui Li

School of Information Science and Technology, Southwest Jiaotong University
Chengdu 611756, China, trli@swjtu.edu.cn

Yunlong Li

School of Information Science and Technology, Southwest Jiaotong University
Chengdu 611756, China, liyunlong_swjtu@163.com

In this paper, a super-hyper network model based on Khatri-Rao Product on the correlation matrix of a series of hypergraphs is proposed. Both marginal and joint node degree, node hyperdegree, hyperedge degree and their corresponding polynomials are introduced to describe this super-hyper network model. It is shown that it is fractal since its correlation matrix is a fractal matrix. The fractal parameter is then provided. It is also validated that it is small-world for its diameter won't exceed twice the summation of the diameter of primitive hypergraphs. By a novel product of either marginal or joint node degree polynomial, node hyperdegree polynomial and hyperedge degree polynomial, the corresponding marginal and joint node degree, node hyperdegree and hyperedge degree are obtained.

Keywords: Super-hyper network; matrix; fractal; small-world; node degree; node hyperdegree; hyperedge degree.

1. Introduction

Network theory is an important tool for describing and analysing complex systems throughout the social, biological, physical, information and engineering sciences. ER/BA/WS/NW network models have been proposed one by one. Based on generating functions in [1], Newman et al. developed the theory of random graphs with arbitrary degree distributions and derived exact expressions for the position of the phase transition [2].

---

*This work is supported by grants 61573292 and 61262058 of the National Science Foundation of China and grant 2016ZR0034 of the Soft Science Foundation of Sichuan Province.

In a graph, a link only relates to a pair of nodes. However, the edges of the hypergraph, namely, hyperedges, can relate to groups of more than two nodes [3]. Hu et al. built a new evolving hypernetwork model and presented some basic topological properties [4]. Many hypernetworks in the real world have been presented and studied. For example, Liu et al. proposed a new hypernetwork model based on Tracy-Singh Product on the correlation matrix of hypergraph [5].

Nowadays, research on complex systems has become increasingly essential to move beyond simple graphs [6]. Multilayer network is able to encapsulate a much more detailed description of a system than monoplex network. There has been considerable interest in generalizing concepts from monoplex networks to multilayer networks, and it is very important to do so. Mikko et al. pointed out that there are many networks that can be classified into multilayer network [7]. Liu et al. proposed a supernetwork model based on Khatri-Rao Product and Khatri-Rao Sum on adjacency matrix of graph [8]. It is also a kind of multilayer network since it contains multiple layers.

In this paper, we construct a super-hyper network model inherited from hypernetwork and super network based on Khatri-Rao Product on correlation matrix of a series of primitive hypergraphs by an iterative approach. It is shown that it is fractal since its correlation matrix is a fractal matrix. Its fractional dimension is also provided. In addition, it is small-world because its diameter won't exceed twice the summation of the diameter of primitive hypergraphs. Finally, we obtain its both marginal and joint node degree, node hyperdegree, and hyperedge degree by a novel product of corresponding marginal and joint node degree polynomial, node hyperdegree polynomial and hyperedge degree polynomial, respectively.

2. Preliminaries

2.1. *Hypergraph*

A hypergraph is a generalization of a graph in which an edge can connect any number of vertices. Formally, a hypergraph H is a pair $H = (X, E)$ where X is a set of elements called nodes or vertices, and E is a set of non-empty subsets of X called hyperedges or edges.

Definition 2.1[5]. The correlation matrix of a hypernetwork $H = (X, E)$ is a matrix $C(H) = (c_{ij})_{|X| \times |E|}$ with $|X|$ rows and $|E|$ columns, where $c_{ij} = 1$ if $x_i \in e_j$, and $c_{ij} = 0$ otherwise.

Since every row is in correspondence to one node and every column represents one hyperedge, the correlation matrix is usually considered as a partitioned matrix.

Definition 2.2[4]. The node degree of a hyperedge e_i, denoted as $d_{Hd}(e_i)$, is the number of nodes that e_i contains.

Definition 2.3[4]. The node hyperdegree of a node x_j, denoted as $d_{Hhd}(x_j)$, is the number of hyperedges which contain x_j.

Definition 2.4[5]. The hyperedge degree of a hyperedge e_k, denoted as $d_{Hed}(e_k)$, is the number of hyperedges adjacent to e_k.

Note that Definition 2.4 is different from that in [4] since $d_{Hed}(e_k)$ in this paper refers to the number of hyperedges adjacent to e_k including e_k itself while $d_{Hed}(e_k)$ in [4] is the number of hyperedges adjacent to e_k without including e_k.

By using the generating function in [1], node degree polynomial, node hyperdegree polynomial and hyperedge degree polynomial can be obtained.

2.2. Supernetwork

A supernetwork is constructed by Khatri-Rao Product and Khatri-Rao Sum on adjacency matrices of graphs [8]. Super network is a kind of multilayer network with multiple layers. We proposed marginal degrees and marginal degree distribution polynomial to describe local properties of supernetwork, as well as joint degrees and joint degree distribution polynomial to describe global properties of supernetwork.

Definition 2.5[8]. The marginal degree of a supernetwork S is the degree of node in every single layer of the supernetwork.

Definition 2.6[8]. The joint degree of a supernetwork S is the degree of node in all layers of the supernetwork.

Definition 2.7[8]. The marginal degree polynomial of a supernetwork S, denoted as $Poly_{M(i)}(S)$, is the summation of monomial with the degree of node as degree and cardinality of node as coefficient in i-th layer of the supernetwork.

Definition 2.8[8]. The joint degree polynomial of a supernetwork S, denoted as $Poly_j(S)$, is the summation of monomial with degree of node as degree and cardinality of node as coefficient in all layers of the supernetwork.

2.3. Matrix Operation

Definition 2.9[9]. If A is an $m \times n$ matrix and B is a $p \times q$ matrix, then the Kronecker Product on A and B, denoted as $A \otimes B$, is an $mp \times nq$ block matrix.

Definition 2.10[9]. The Khatri-Rao Product is defined as

$$A * B = \left(A_{ij} * B \right)_{ij} = \left(\left(A_{ij} \otimes B_{kl} \right)_{kl} \right)_{ij} \tag{1}$$

which means that the (ij)-th subblock of the $mp \times nq$ product $A * B$ is the $m_i p \times n_j q$ matrix $A_{ij} * B$, of which the (kl)-th subblock equals to the $m_i p_k \times n_j q_l$

matrix $A_{ij} \otimes B_{kl}$. The Khatri-Rao Product essentially is the pairwise Kronecker Product for a pair of partitions corresponded in the two matrices.

2.4. Fractal Theory

An important result in statistical physics was the generation of fractal geometries, the structures of which look the same on all length scales. Recently, self-similarity has received much more attention since it has been considered as the third property of complex network following small-world and scale-free. Song et al. [9] used the concept of renormalization as a mechanism for the growth of fractal and non-fractal modular networks and showed that the key principle that gives rise to the fractal architecture of networks is a strong effective "repulsion" between the most connected nodes on all length scales. Zhang [10] presented a model of growing trees and showed that the evolving process and the emergence of self similar structure can be completely expounded by a simple exponent law, called as a self-organized critical state which governs the ordered evolving process and leads to the emergence of the layer-by-layer similar structure. Fractal dimension is the most important description of fractal parameter. Hausdorff dimension is a useful method to calculate fractal dimension of fractal pattern, e.g., Cantor Set, Koch Snow, and Menger Sponge.

Definition 2.11[11]. Fractal Matrix is the name given to the class of unbounded matrices obtained by the continued application of the recursion

$$E_{m+1} = E_1 \oplus E_m, m=2, 3, \ldots \qquad (2)$$

E_1 is a $r \times s$ matrix of integers $e_1(i, j)$, $1 \leqslant i \leqslant r$ and $1 \leqslant j \leqslant s$. It is called the generator matrix. $A \oplus B$ denotes the so-called bigsum operation between two matrices A and B: it is the matrix obtained by replacing each element $a(i, j)$ in A by the matrix B to which the first $a(i, j)$ has been added to all its elements. These matrices also form an extension of a special subset in the class of iterated Kronecker Product[12].

3. A Super-hyper Network Model

For the correlation matrix $C(SH) = \{C(H_{(1)})_{M \times N}, \ldots, C(H_{(i)})_{M \times N}, \ldots, C(H_{(n)})_{M \times N}\}$ of a series of hypergraphs $SH = \{H_{(1)}, \ldots, H_{(i)}, \ldots, H_{(n)}\}$, the result of Khatri-Rao Product between $C(SH)$ and $C(SH)$ is a new matrix given by:

$$C^{(1)}\left(SH^{(1)}\right) = \left\{C^{(1)}(H^{(1)}_{(1)})_{M^2 \times N^2}, \cdots, C^{(1)}(H^{(1)}_{(i)})_{M^2 \times N^2}, \cdots, C^{(1)}(H^{(1)}_{(n)})_{M^2 \times N^2}\right\} \qquad (3)$$

Eq. (3) can be regarded as the correlation matrix of a new network called super-hyper network combined with hypernetwork and supernetwork. By $C^{(k+1)}(SH^{(k+1)}) = C^{(k)}(SH^{(k)})*C(SH)$, a series of matrices and super-hyper networks can be obtained, where the number of nodes of $S^{(k)}$ in every signal layer as hypernetwork is M^{k+1} and the number of hyperedges is N^{k+1}.

Definition 3.1. The correlation matrix of a super-hyper network $SH = \{H_{(1)}, ..., H_{(i)}, ..., H_{(n)}\}$ is a partitioned matrix $C(SH) = \{C(H_{(1)})_{M\times N}, ..., C(H_{(i)})_{M\times N}, ..., C(H_{(n)})_{M\times N}\}$ with n partitions, and every partition with M rows and N columns. M is the cardinality of nodes and N is the cardinality of hyperedges.

Definition 3.2. The marginal node degree of a hyperedge in a super-hyper network is the number of nodes this hyperedge contains in every single layer of the super-hyper network.

Following Definition 3.2, we can obtain the marginal node hyperdegree of a node and the marginal hyperedge degree of a hyperedge in the super-hyper network.

Definition 3.3. The joint node degree of a hyperedge in the super-hyper network is the number of nodes this hyperedge contains in all layers of the super-hyper network.

Following Definition 3.3, we can get the joint node hyperdegree of a node and the joint hyperedge degree of a hyperedgethe in the super-hyper network.

Definition 3.4. The marginal node degree polynomial of a super-hyper network SH, denoted as $Poly_{HdM(i)}(SH)$, is the summation of monomial with marginal node degree of a hyperedge as degree and cardinality of hyperedges as coefficient in i-th layer of the super-hyper network.

The marginal node hyperdegree polynomial and the marginal hyperedge degree polynomial of a super-hyper network SH, denoted as $Poly_{HhdM(i)}(SH)$ and $Poly_{HedM(i)}(SH)$, respectively, can be obtained.

Definition 3.5. The joint node degree polynomial of a super-hyper network SH, denoted as $Poly_{HdJ}(SH)$, is the summation of monomial with joint node degree of a hyperedge as degree and cardinality of hyperedges as coefficient in all layers of the super-hyper network.

The joint node hyperdegree polynomial and the joint hyperedge degree polynomial of a super-hyper network SH, denoted as $Poly_{HhdJ}(SH)$ and $Poly_{HhdJ}(SH)$ and $Poly_{HedJ}(SH)$, respectively, can be obtained.

3.1. Fractional Dimension

The super-hyper network is exhibited as a multilayer network and combined with supernetwork and hypernetwork since every layer of super-hyper network is hypernetwork and sharing the properties of multiple layers with supernetwork.

The fractional dimension of super-hyper network can be obtained from fractional dimension of hypernetwork and supernetwork.

For hypernetwork H, the fractional dimension is [5]:

$$FD(H) = \frac{2\log \sum_{i=1, j=1}^{i=|X|, j=|E|} |c_{ij}|}{\log |X||E|} \tag{4}$$

For supernetwork S, the fractional dimension is [8]:

$$FD(S) = 1 + \left(\prod_{i=1}^{n} FD(G_i) \right)^{\frac{1}{n}} \tag{5}$$

Compared with super-hyper network and hypernetwork, by overlaying hypernetwork on hypernetwork, hypernetwork will advance to super-hyper network. Moreover, compared with super-hyper network and supernetwork, by replacing every layer of the supernetwork as graph with hypergraph, supernetwork will evolve to super-hyper network. As super-hyper network is combined with hypernetwork and supernetwork, the fractional dimension of super-hyper network can also be obtained by a combination of fractional dimension of hypernetwork in Eq. (4) with the fractional dimension of supernetwork in Eq. (5). By replacing fractional dimension of graph $FD(G_i)$ in Eq. (5) with the fractional dimension of hypergraph $FD(H)$ in Eq. (4), the fractional dimension of super-hyper network SH is obtained as follows:

$$FD(SH) = 1 + \left(\prod_{i=1}^{n} FD(H_i) \right)^{\frac{1}{n}} \tag{6}$$

Theorem 1. The fractional dimension of self-similarity super-hyper network is no less than 1 and no more than 3.

Proof. Since the fractional dimension of hypernetwork $FD(H_i)$ is no less than 0 and no more than 2 [5], Theorem 1 holds by the following two inequalities.

$$\begin{cases} FD(SH) = 1 + \left(\prod_{i=1}^{n} FD(H_i) \right)^{\frac{1}{n}} > 1 + \left(\prod_{i=1}^{n} 0 \right)^{\frac{1}{n}} = 1 \\ FD(SH) = 1 + \left(\prod_{i=1}^{n} FD(H_i) \right)^{\frac{1}{n}} < 1 + \left(\prod_{i=1}^{n} 2 \right)^{\frac{1}{n}} = 3 \end{cases} \tag{7}$$

3.2. Diameter

Theorem 2. The diameter of self-similarity super-hyper network won't exceed twice the summation of diameter of primitive hypergraphs.

Proof. Since the fractional dimension of self-similarity supernetwork $FD(S)$ won't exceed twice the summation of diameter of primitive graphs [5], and the super-hyper network can be obtained by replacing every layer of supernetwork as a graph with hypergraph, the diameter of self-similarity super-hyper network won't exceed twice the summation of diameter of primitive hypergraphs.

3.3. Polynomial

Theorem 3. The marginal node degree of self-similarity super-hyper network can be obtained by a novel production of marginal node degree polynomial with a product on both degrees and coefficients, namely,

$$Poly_{HdM(i)}(SH) = NP(\sum_{i=1}^{|E_a|} x^{d_{Hda(i)}(e_i)}, \sum_{j=1}^{|E_b|} x^{d_{Hdb(i)}(e_j)}) = \sum_{i=1}^{|E_a|}\sum_{j=1}^{|E_b|} x^{d_{Hda(i)}(e_i) \times d_{Hdb(i)}(e_j)}$$

(8)

Proof. For correlation matrices $C^{(k+1)}(SH^{(k+1)})$, $C^{(k)}(SH^{(k)})$ and $C(SH)$, $C^{(k+1)}(SH^{(k+1)})$ can be obtained from Khatri-Rao Product on $C^{(k)}(SH^{(k)})$ and $C(SH)$. As $C^{(k+1)}(SH^{(k+1)})$ is generated from the product of every element of both $C^{(k)}(SH^{(k)})$ and $C(SH)$, the marginal node degree polynomial of $SH^{(k+1)}$ can be obtained from the marginal node degree polynomial of $SH^{(k)}$ and SH with the product on both degrees and coefficients of both $Poly(SH^{(k)})$ and $Poly(SH)$.

Properties of the marginal node hyperdegree and marginal hyperedge degree are the same as Theorem 3.

Lemma 1. The marginal node hyperdegree of self-similarity super-hyper network can be obtained by a novel product of marginal node hyperdegree polynomial with a product on both degrees and coefficients, namely,

$$Poly_{HhdM(i)}(SH) = NP(\sum_{i=1}^{|X_a|} x^{d_{Hhda(i)}(e_i)}, \sum_{j=1}^{|X_b|} x^{d_{Hhdb(i)}(e_j)}) = \sum_{i=1}^{|X_a|}\sum_{j=1}^{|X_b|} x^{d_{Hhda(i)}(e_i) d_{Hhdb(i)}(e_j)}$$

(9)

Proof. It directly follows the definition of marginal node hyperdegree and Theorem 3.

Lemma 2. The marginal hyperedge degree of self-similarity super-hyper network can be obtained by a novel product of marginal hyperedge degree polynomial with a product on both degrees and coefficients, namely,

$$Poly_{HedM(i)}(SH) = NP(\sum_{i=1}^{|X_a|} x^{d_{Heda(i)}(e_i)}, \sum_{j=1}^{|X_b|} x^{d_{Hedb(i)}(e_j)}) = \sum_{i=1}^{|X_a|}\sum_{j=1}^{|X_b|} x^{d_{Heda(i)}(e_i)d_{Hedb(i)}(e_j)}$$

(10)

Proof. It directly follows the definition of marginal hyperedge degree and Theorem 3.

Theorem 4. The joint node degree of self-similarity super-hyper network can be obtained by a novel product of joint node degree polynomial with a product on both degrees and coefficients, that is to say,

$$Poly_{HdJ}(SH) = NP(\sum_{i=1}^{|E_a|} x^{d_{Hda}(e_i)}, \sum_{j=1}^{|E_b|} x^{d_{Hdb}(e_j)}) = \sum_{i=1}^{|E_a|}\sum_{j=1}^{|E_b|} x^{d_{Hda}(e_i) \times d_{Hdb}(e_j)} \quad (11)$$

Proof. For correlation matrices $C^{(k+1)}(SH^{(k+1)})$, $C^{(k)}(SH^{(k)})$ and $C(SH)$, $C^{(k+1)}(SH^{(k+1)})$ can be obtained from Khatri–Rao Product on $C^{(k)}(SH^{(k)})$ and $C(SH)$. As $C^{(k+1)}(SH^{(k+1)})$ is generated from the product of every element of both $C^{(k)}(SH^{(k)})$ and $C(SH)$, the joint node degree polynomial of $SH^{(k+1)}$ can be obtained from the joint node degree polynomial of $SH^{(k)}$ and SH with the product on both degrees and coefficients of both $Poly(SH^{(k)})$ and $Poly(SH)$.

Properties of the joint node hyperdegree and joint hyperedge degree of self-similarity super-hyper network are similar to Theorem 4.

Lemma 3. The joint node hyperdegree of self-similarity super-hyper network can be obtained by a novel product of joint node hyperdegree polynomial with a product on both degrees and coefficients, namely,

$$Poly_{HhdJ}(SH) = NP(\sum_{i=1}^{|X_a|} x^{d_{Hhda}(e_i)}, \sum_{j=1}^{|X_b|} x^{d_{Hhdb}(e_j)}) = \sum_{i=1}^{|X_a|}\sum_{j=1}^{|X_b|} x^{d_{Hhda}(e_i)d_{Hhdb}(e_j)} \quad (12)$$

Proof. It directly follows the definition of joint node hyperdegree and Theorem 4.

Lemma 4. The joint hyperedge degree of self-similarity super-hyper network can be obtained by a novel product of joint hyperedge degree polynomial with a product on both degrees and coefficients, namely,

$$Poly_{HedJ}(SH) = NP(\sum_{i=1}^{|X_a|} x^{d_{Heda}(e_i)}, \sum_{j=1}^{|X_b|} x^{d_{Hedb}(e_j)}) = \sum_{i=1}^{|X_a|}\sum_{j=1}^{|X_b|} x^{d_{Heda}(e_i)d_{Hedb}(e_j)} \quad (13)$$

Proof. It directly follows the definition of joint hyperedge degree and Theorem 4.

4. Conclusions

This paper presented a super-hyper network model with properties of both self-similarity and small-world by Khatri-Rao Product on the correlation matrix of primitive super-hyper network. The property of self-similarity is originated from the fractal matrix based on fractal theory. The property of small-world is originated from the fact that the diameter won't exceed twice the summation of the diameter of primitive hypergraphs. The fractional dimension of this kind of self-similarity networks is also given by Hausdoff Dimension combined with the fractional dimension of hypernetwork and supernetwork. Finally, a novel product of polynomials was presented to calculate marginal node degree, marginal node hyperdegree, marginal hyperedge degree and joint node degree, joint node hyperdegree, joint hyperedge degree of super-hyper networks theoretically through their polynomial expressions. Our future research work will continue to study the properties of this self-similarity and small-world super-hyper network model and try to discover more inherent properties to better describe the real world.

References

1. H. S. Wilf, London: Academic. 1994.
2. M. E. J. Newman, S. H. Strogatz, et al, *Phys. Rev.* **E64**, 261181(2001).
3. E. Estrada and V. Rodrigues, *Phys. Rev.* **E71**, 1(2005).
4. F. Hu, H. X. Zhao and X. J. Ma, *Sci Sin-Phys Mech Astron*, **43**, 16(2013).
5. S. J. Liu and T. R. Li, *ISKE*, **11**, 176(2015).
6. A. Cardillo, M. Zanin, et al, *Eur. Phys. J. Special Topics*, **215**, 23(2013).
7. K. Mikko, A. Alex, et al, *J. Complex Networks*, **2**, 203(2014).
8. S. J. Liu, T. R. Li, et al, *CAAI Trans on Intel Syst.*(in press).
9. C. Song, S. Havlin and H. Makse, *Nature Phys*, **2**, 275(2006).
10. S. Y. Zhang, *Complex Syst Complex Sci.* **3**, 41(2006).
11. M. B. Andre, *Visual Comput.* **9**, 233(1993).
12. T. Emmerich, A. Bunde and S. Havlin, *Phys. Rev.* **E89**, 62806(2014).

Visualization methods for tracking the dynamics of coronal dimmings

Yuhang Yang, Bo Peng, Tianrui Li and Yunmeng Hu

*School of Information Science and Technology, Southwest Jiaotong University
Chengdu, 611756, China
yuhang_yang@foxmail.com; bpeng@swjtu.edu.cn; trli@swjtu.edu.cn;
huyunmeng@163.com*

Coronal dimming, as an important phenomenon on the Sun and one of the main drivers of the space weather, has been widely studied for decades. An essential research topic for the coronal dimming phenomena is tracking their dynamics. In this paper, we propose a method to visualize the dynamics of coronal dimming phenomena directly from the original images of the Sun. Specifically, our method consists of two components. First, we extract coronal dimming-related features, e.g. variance and entropy, from the original image data. Second, based on the extracted features, we propose some visualization methods to track coronal dimmings. Experimental results, based on the collected datasets from the real world, demonstrate the effectiveness of our visualization methods.

Keywords: Coronal dimming; visualization; feature extraction.

1. Introduction

Coronal dimming is an eruptive phenomenon on the Sun which is regarded as one of the main drivers of the space weather.[1] Coronal dimmings appear relatively suddenly on timescales of minutes, and will lead to the decreases in intensity in soft X-rays[2,3] and EUV data[4] of some areas on the Sun.

Due to the importance of the coronal dimming, for the last few decades, several coronal dimming detection algorithms have been proposed. One of the basic detection approaches is thresholding pixel intensity values in base difference images.[5] Podladchikova et al.[6] analyzed statistical properties of the distribution of the pixels in running difference images, where variance and kurtosis were employed as indicators to detect the occurrences of dimmings and the beginning of an eruption was characterized by a sudden increase of variance. Attrill et al.[7] explored the augmented statistical signatures of dimmings in sub-images to make it efficient to detect those small dimming events. Besides, feature-based classifiers have also been used to the automated detection of small-scale dimmings.[8] Instead of using difference images, Krista et al.[9] developed a method using the original, non-difference EUV images and the corresponding magnetograms to have a comprehensive study for dimming regions. Krista et al.[10] improved the precision of dimming detection by limiting the identification to footpoint-dimmings and by automating the effort as much as possible. Yang et al.[11] proposed some multi-label learning algorithms to simultaneously detect eight phenomena associated with the coronal dimming eruption phenomenon.

In this paper, different with the previous detection algorithms, we propose visualization methods to track the dynamics of coronal dimming phenomena on the Sun as visualization technologies have been applied to many fields to enhance data presentation effect. For example, Nguyen et al.[12] visualized the multifaceted features learned by each neuron in deep neural networks to have a better understanding of deep neural networks. Cui et al.[13] proposed a new framework to visualize hyperspectral images. Panta et al.[14] proposed a web-based approach for quick visualizations of big data from brain magnetic resonance imaging scans, allowing researchers to rapidly identify and extract meaningful information from big datasets.

Our visualization methods can help astronomy physicists monitor the dynamic status of coronal dimmings, e.g. the occurences and evolutions of coronal dimmings. Specifically, as shown in Fig. 1, our work is comprised of two components. First, we extract coronal dimming-related features, e.g. variance and entropy, from the images of the Sun. Second, we propose methods, based on an open access visualization tool d3.js,[a] to visualize the extracted features, which can present the dynamics of the coronal dimmings on the Sun.

Fig. 1. The framework of our visualization process.

The rest of this paper is organized as follows. In Sec. 2, we introduce the structure of the image data of the Sun in real world. In Sec. 3, we detail how to extract features (i.e. variance and entropy) from the image data. Sec. 4 studies the visualization methods and obtains the visualization results. In the end, this paper is concluded in Sec. 5.

---

[a]https://d3js.org

2. Dataset

We use the datasets collected from the real world. The original images observed by SOHO/EIT is obtained from the website[b] with a cadence of about twelve minutes. As demonstrated in Fig. 2, the original EIT images on 12 May 1997 from 04:34 UT to 07:34 UT make up a image sequence. There are n continuous time slots, and at each time slot t, there is an image of the Sun, which can be denoted by a matrix $P(\cdot,\cdot,t)$. For the matrix $P(\cdot,\cdot,t)$, each entry $P(i,j,t)$ refers to the pixel value of the location (i,j) in the image (i.e. on the Sun) at time slot $t(t \geq 1)$. The area inside the red circle is the Sun-disk (see the left figure in Fig. 2). After extracting related features from these images, we can visualize the dynamics of the coronal dimming at each time slot, using our proposed visualization methods.

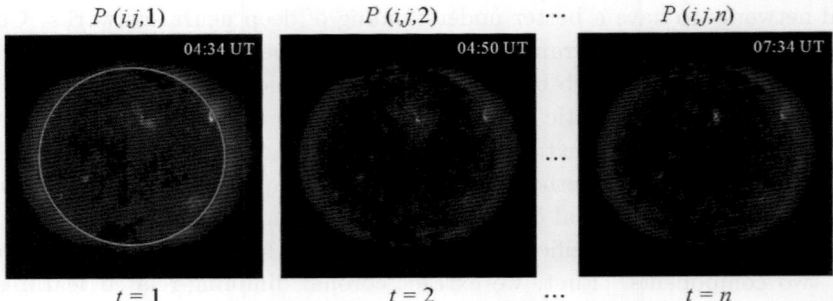

Fig. 2. Original EIT images of the Sun on 12 May 1997.

3. Feature Extraction

In this section, we extract features related with the coronal dimming, from the original EIT images of the Sun. Specifically, from the image data, we extract two types of features, i.e. variance and entropy. Since variance has been used to detect coronal dimmings by Podladchikova et al.[6] and Attrill et al.,[7] it can effectively reflect the dynamics of the coronal dimming. And entropy can also make sense. Below, we detail these two features.

---

[b]https://umbra.nascom.nasa.gov/eit/eit-catalog.html

3.1. *Variance*

Variance can reflect the fluctuation of pixel values in images. Since coronal dimming phenomena will lead to the fluctuation of the pixel values, for each image, areas inside a coronal dimming are likely to be with higher variances than other areas. Specifically, based on the image data, we can formulate two kinds of variances: 1) Spatio-Temporal (ST) variance and 2) Running Difference-based (RD-based) variance.

3.1.1. *ST Variance*

In image series, the pixel value at a particular location could changes from time to time due to the eruption, evolution and disappearance of a coronal dimming. In Ref. 15, the change of the pixel values from one time slot to another is considered as the state transition of the pixel. For example, in a 256-level gray image, each pixel has 256 states. Along with time, the pixel's state would change from one to another. Thus, the fluctuation of the pixels at each location can be used to characterize the intensity of transformation at the position, i.e. the dynamics of the coronal dimming at the location.

With the above observation, the ST variance of pixel values at location (i,j), denoted by $V^{ST}(i,j)$, can be defined as:

$$V^{ST}(i,j) = \frac{1}{n-1} \sum_{t=1}^{n} [P(i,j,t) - \overline{P}(i,j)]^2, \qquad (1)$$

where n denotes the total number of images in image series, $P(i,j,t)$ refers to the pixel value at location (i,j) at time slot t, and $\overline{P}(i,j)$ is the average pixel value of $P(i,j,t)$ over all time slots, i.e. $\overline{P}(i,j) = \frac{1}{n}\sum_{t=1}^{n} P(i,j,t)$.

Apparently, the larger the ST variance $V^{ST}(i,j)$, the more likely there is a coronal dimming phenomenon nearby location (i,j) on the Sun during the n time slots. In other words, the ST variance can effectively reflect the coronal dimmings on the Sun.

3.1.2. *RD-based Variance*

Besides the ST variance, we can also define the running difference-based variance. The running difference-based difference is a three-dimension matrix, denoted by V^{RD}. For each location (i,j) in images at time slot t, in math, $V^{RD}(i,j,t)$ is formulated as:

$$V^{RD}(i,j,t) = \sum_{t_0=t}^{t+1} [P(i,j,t_0) - \overline{P}(i,j,t)]^2, \qquad (2)$$

where $\overline{P}(i,j,t)$ is the average of $P(i,j,t)$ and $P(i,j,t+1)$, i.e. $\overline{P}(i,j,t) = \frac{P(i,j,t)+P(i,j,t+1)}{2}$. Thus, $V^{RD}(i,j,t)$ reflects the fluctuation of the pixel values at location (i,j) between time slot t and $t+1$. Similarly, if there is a coronal dimming phenomenon nearby location (i,j) at time slot t, the $V^{RD}(i,j,t)$ could be very large.

3.2. *Entropy*

For images, the entropy is also a matrix and is denoted by E. Specifically, for each location (i,j) and each time slot t, $E(i,j,t)$ is formulated as

$$E(i,j,t) = -\sum_{m=0}^{255} p_m \log p_m, \qquad (3)$$

where p_m denotes the ratio of the number of the pixels being m to the total number of pixels in the sector area containing the location (i,j), as shown in Fig. 3. The center line (highlighted by the red dash line) of the sector area covers the location (i,j) and the sector area has an angle of θ (e.g. $\theta = 5°$ in this work).

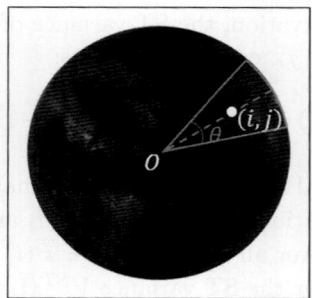

Fig. 3. Sector area when calculating the entropy of pixel values nearby (i,j).

The smaller the entropy $E(i,j,t)$, the more probable there exists a coronal dimming phenomenon nearby location (i,j) at time slot t. The reason is that pixel values for areas outside a coronal dimming are usually randomly distributed, i.e. having bigger entropies, while pixel values inside a coronal dimming are more concentrated, i.e. having smaller entropies.

4. Visualization and Results

In this section, with the extracted futures above, we show how to apply visualization methods to visualize these features and visualize the dynamics of the coronal dimming phenomena on the Sun, based on the d3.js visualization tool.

In this work, without loss of generality, we employ our visualization method to track a sample dimming event occurred on 12 May 1997, to evaluate our visualization method. The image data contains the original EIT images of this dimming from 04:34 UT to 07:34 UT. Specifically, there are 10 images with around 17 minutes as one time slot, i.e. $n = 10$. Note that it is easy to apply our visualization methods to other dimmings.

Using the data, we extract features and based on the d3.js visualization tool, we can obtain the visualization results as below.

4.1. Variance

4.1.1. ST Variance

The ST variance is a two-dimension matrix V^{ST}, as calculated by Eq. (1). For each location (i,j) on the Sun, we have the ST variance $V^{ST}(i,j)$. The visualization result is demonstrated in Fig. 4(b), in which different colors denote different values of $V^{ST}(i,j)$. For example, the yellow color corresponds to high ST variances. In contrast, the purple color denotes low ST variances. Clearly, based on the Fig. 4(a), we easily find the sector area (highlighted by the yellow color) where a coronal dimming phenomenon happens. This result is highly consistent with the location where the coronal dimming actually happens. As shown in Fig. 4(a), the darker area in the difference image, computed by using image on 05:07 UT minus that on 04:34 UT, is the dimming area in this dimming event happend on 12 May 1997.

Note that in Fig. 4(b), we apply a slide window (a sector area) to smooth the ST variance matrix, such that we can easily find the sector area containing the coronal dimming. Specifically, as shown in Fig. 4(b), for each location (i,j), we use the average ST variance of the pixels in the sector area (highlighted by the red line) as its new ST variance.

4.1.2. RD-based Variance

Different with the ST variance which is a two-dimension matrix, the RD-based variance is a three-dimension matrix V^{RD}, which is able to reflect the dynamics of the coronal dimming. The visualization result is demonstrated in Fig. 5(a). The Fig. 5(a) consists of $n-1$ rings, each of which denotes the RD-based variance at a time slot. In this way, we can visualize the three-dimension result in the two-dimension space.

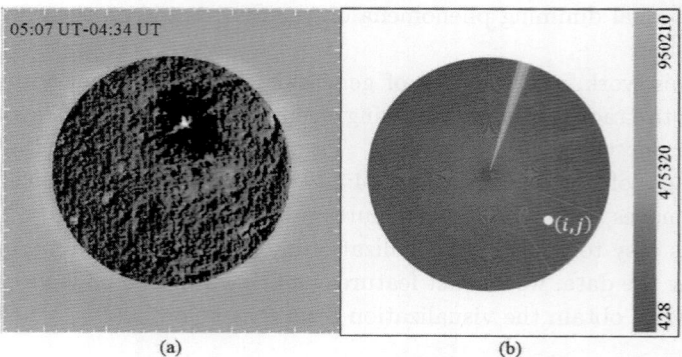

Fig. 4. Visualization based on the ST variance.

From Fig. 5(a), we can easily track the evolution of the coronal dimming phenomenon over the entire time slots. For example, for the last three time slots, the coronal dimming phenomenon vanishes gradually.

4.2. *Entropy*

The visualization result based on the entropy is similar to that based on the RD-based variance, as they are both three-dimensional. Fig. 5(b) presents the visualization result. The Fig. 5(a) consists of n rings, each of which denotes the Entropy at a time slot. As mentioned before, areas with smaller entropies are more likely to be in a coronal dimming. Clearly, the entropy is also an effective feature to track the dynamics of coronal dimmings on the Sun.

5. Conclusion

In this paper, we proposed visualization methods to track the dynamics of the coronal dimming phenomena on the Sun. First, we extracted dimming-related features from the original EIT image data, including variance(ST variance, RD-based variance) and entropy. Second, based on the extracted features, we applied visualization methods to visualize the dynamics of coronal dimmings on the Sun. Experimental results using the real-world image data of the Sun demnonstrated the effectiveness of our visualization methods.

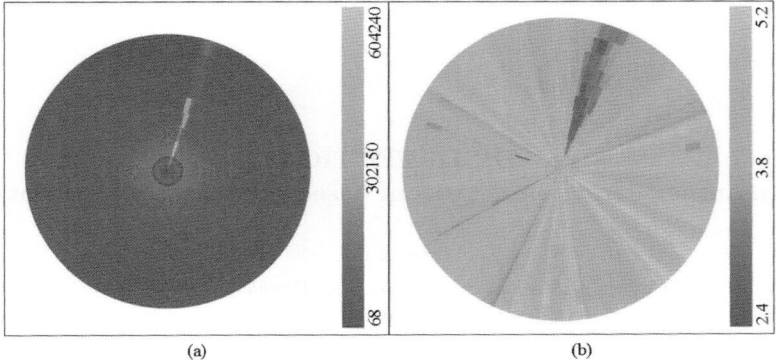

Fig. 5. Visualizations based on the RD variance (a) and entropy (b).

Acknowledgments

This work is supported by the National Science Foundation of China (Nos. 61573292, 61773324, 61772453) and the Fundamental Research Funds for the Central Universities (No. 2682015QM02).

References

1. E. Kraaikamp and C. Verbeeck, *SWSC* **5** (2015).
2. H. S. Hudson, L. W. Acton and S. L. Freeland, *ApJ* **470** (1996).
3. A. C. Sterling and H. S. Hudson, *ApJ* **491** (1997).
4. B. J. Thompson, S. P. Plunkett, J. B. Gurman, J. S. Newmark, O. C. S. Cyr and D. J. Miche, *GRL* **25** (1998).
5. D. Bewsher, R. A. Harrison and D. S. Brown, *A&A* **478** (2008).
6. O. Podladchikova and D. Berghmans, *Solar Physics* **228** (2005).
7. G. D. R. Attrill and M. J. Wills-Davey, *Solar Physics* **262** (2010).
8. N. Alipour, H. Safari and D. E. Innes, *ApJ* **746** (2012).
9. L. D. Krista and A. A. Reinard, *ApJ* **762** (2013).
10. L. D. Krista and A. A. Reinard, *ApJ* **839** (2017).
11. Y. H. Yang, H. M. Tian, B. Peng, T. R. Li and Z. X. Xie, *Solar Physics* **292** (2017).
12. A. Nguyen, J. Yosinski and J. Clune, *Arxiv* (2016).
13. M. Cui, A. Razdan, J. Hu and P. Wonka, *TGRS* **47** (2009).
14. S. R. Panta, R. Wang, J. Fries, R. Kalyanam, N. Speer, M. Banich, K. Kiehl, M. King, M. Milham, T. D. Wager, J. A. Turner, S. M. Plis and V. D. Calhoun, *TGRS* **10** (2016).
15. Y. F. Ma and H. J. Zhang, Detecting motion object by spatio-temporal entropy, in *ICME'01*, (Tokyo, Japan, 2001).

DP_DETECTION:
An outlier detection algorithm based on density of big data

Xiaodi Li[a]

College of Mechanical Engineering, Donghua University
Shanghai, Shanghai 201620, China
xiaodi_li327@mail.dhu.edu.cn

Ping Deng[a]

The School of Information Science and Technology, Southwest Jiaotong University
Chengdu, Sichuan 611756, China
dengping609@gmail.com

Ming Huang

Health Sciences Research, Mayo Clinic, Rochester, MN 55905, USA
huang.ming@mayo.edu

Dingcheng Li

AI and Big Data Lab, Baidu Research, Bellevue, WA 98004, USA
lidingcheng@baidu.com

Hongjun Wang*

The School of Information Science and Technology, Southwest Jiaotong University
Chengdu, Sichuan 611756, China
wanghongjun@swjtu.edu.cn

The purpose of outlier detection is to isolate regular data points from polluting ones named outliers. So far, there are three main approaches in the outlier detection: statistical-based, depth-based and distance-based. We chose distance-based one because it is most appropriate to handle large datasets with dimensions more than 4. In our work, we propose a new algorithm called DP_DETECTION by adapting the well-known DP clustering algorithm to solve high-dimensional large-scale outlier detection. Unlike other algorithms, we design a Map Reduce framework to wrap up DP algorithm based on density of big data, we focus DP algorithm on optimizing outlier detection rather than clustering accuracy and add a regulation parameter based on the proportion

---

[a]Equal contribution authors.
*Corresponding author.

between the density and the distance to the original DP algorithm to increase the robustness of our method. To test the effectiveness of DP_DETECTION, we ran our algorithm on both virtual and real-world datasets. The experimental results show that our proposed DP_DETECTION makes great performance in the time comparison with traditional algorithms and achieves exceedingly high F1-score.

Keywords: Outlier detection; DP algorithm; map reduce framework; big data.

1. Introduction

Outlier detection aims at separating a core of regular observations from polluting ones, called "outliers". There are three approaches in this detection: statistical-based, depth-based and distance-based. Statistical-based one focuses on discordancy/outlier tests, depending on: (i) the data distribution, (ii) whether the distribution parameters are known, (iii) the number of expected outliers, and (iv) the type of expected outliers. Nevertheless, most of these tests are univariate and all of them are distribution-based. Meanwhile, depth-based methods only offer acceptable performance for $k \leq 2$ (k is the number of dimensions). In contrast, distance-based methods, win more popularity due to their flexibility in defining distance matrix[1] and our proposal is based on this. But it will be more difficult in a big-data situation. Usually, outliers are implicitly regarded as background noises, however, it becomes increasingly hard to estimate the multidimensional distributions of the data points.

In this work, we propose a new algorithm called DP_DETECTION to handle high-dimensional large scale outlier detection. There are three innovations in our approach: (i) We design a map reduce algorithm to wrap up the efficient clustering by fast search-and-find of density peaks (DP) so that those datasets can be processed in a real time fashion. (ii) Unlike other clustering algorithm focusing on optimizing clustering accuracy, we focus on optimizing outlier detection. (iii) A regulation parameter based on the proportion between density and distance is added to the original DP algorithm to enhance the robustness at borderlines or in extreme cases.

2. Related Work

2.1. *Clustering*

Clustering is a technique to organize data points in a way that data points from the same class are more related to one another. Data can be clustered in various algorithms. Among them, $k - means$, $k - medoids$ and

$DBSCAN$ are the most representative ones. For example, $k-means$[2] partitions the n points into k classes where each point goes to the class with the minimum mean value. $DBSCAN$ is a basic density-based clustering method[3] featuring the density-reachability cluster model. Nevertheless, they all have shortnesses. For instance, neither $k-means$ nor $k-medoids$ is sensible to noises. $DBSCAN$ may drop some density points at border regions.

2.2. DP Algorithm

DP Algorithm 1 Density estimation
Inputs: d_c, the cutoff distance
D, $n*n$ distance matrix.
Outputs: ρ, n length density vector
for $i \leftarrow 1:n$ do
    $\rho(i) \leftarrow Count(D(i, i^-)) < d_c$
end for

Clustering by fast search-and-find of density peaks (DP) was proposed by Alex et al.[4] DP is based on the assumption that a cluster center is a high dense point compared with its neighbors and located farther from other cluster centers. For every given data point i, DP estimates its density ρ_i and distance δ_i. The effectiveness of DP algorithm highly depends on the evaluation of ρ and the cutoff distance d_c. The essential parameter d_c is utilized to estimate the densities, and define border points and noises. A decision graph utilized by DP can be used to identify cluster centers.

DP algorithm is composed of two basic algorithms. The first one is density estimation as shown in Algorithm.2.2 In this algorithm, the local density ρ_i of data point i is defined as $\rho_i = \sum_j \chi(d_{ij} - d_c)$, where $\chi(d_{ij} - d_c) = 1$ if $d_{ij} - d_c < 0$ and $\chi(d_{ij} - d_c) = 0$ otherwise, and d_c is a cutoff distance. Basically, ρ_i is equal to

DP Algorithm 2 Distance from higher local density points (δ_i)
Inputs: D, $n*n$ Distance matrix; ρ, n size density vector
D, $n*n$ distance matrix.
Outputs: δ, N, N distance vector of n objects from nearest higher density; N, N neighbor, index vector of nearest neighbor of each element i
for $i \leftarrow 2:n$ do
    $\delta(sorted_\rho(i)) \leftarrow max(D)$
    for $j \leftarrow 1:n-1$ do
        if $D(sorted_\rho(i), sorted_\rho(j)) < \delta(sorted_\rho(i))$ then
            $\delta(sorted_\rho(i)) \leftarrow D(sorted_\rho(i), sorted_\rho(j))$
            N $Nneightbor(sorted_\rho(i)) \leftarrow \delta(sorted_\rho(i))$
        end if
    end for
end for

the number of points that are closer than d_c to point i. The second one is the distance calculation from higher local density points. δ_i is measured by computing the minimum distance between the point i and any other point with higher density: $\delta_i = min_{j:\rho_j>\rho_i}(d_i j)$, for the point with highest density; $\delta_i = max_{j:\rho_j>\rho_i}(d_i j)$ is employed rather than min. In this way, cluster centers are guaranteed to have both high values in both δ_i and ρ_i.

2.3. Map Reduce Algorithm

In order to handle extreme large data set, we plan to wrap DP algorithm with map reduce framework. MapReduce is a programming model and an associated implementation for processing and generating large data sets.[5] Users specify a map function that processes a key/value pair to generate a set of intermediate key/value pairs, and a reduce function that merges all intermediate values associated with the same intermediate key. In addition, the MapReduce framework allows for the definition of *combiners* and *partitioners*. Combiners perform local aggregation on the key value pairs after map function helping reduce the size of intermediate data transferred. Partitioners control how messages are routed to reducers.

3. Proposed Method

We design a map-reduce algorithm in order to discover outliers from large scale dataset illustrated as Figure 1. In this workflow, three stages can be seen. The first stage is computing local outliers and local centers for each mapper, which results from three steps of computations: i) compute $M_{d_{ij}}$, ii) select radius d_c and iii) update ρ_i and δ_i. The second one aims at computing global-partial outliers based on local outliers and local centers passed from mappers. Each reducer is composed of four steps, i) compute $d_{O_{loc},C_{loc}}$, ii) rank $d_{O_{loc},C_{loc}}$, iii) select top $O_{global-partial(gp)}$ and iv) compute partial global center $C_{global-partial(gp)}$. The third stage is a driver which comprises of three steps, i) compute $d_{O_{gp},C_{gp}}$, ii) rank $d_{O_{gp},C_{gp}}$ and iii) select top $Y_{O_{global-final(gf)}}$.

3.0.1. Mapper: Update $\mathbf{O}_{loc}$ and $\mathbf{C}_{loc}$

For the map function 3.0.1, the inputs include key-value pairs where the key is the block ID, ranging from 1 to M, the number of blocks and the value is a subset of the whole data. Before the mapper starts, a few parameters are entered into a distributed cache as predefined ones. Besides

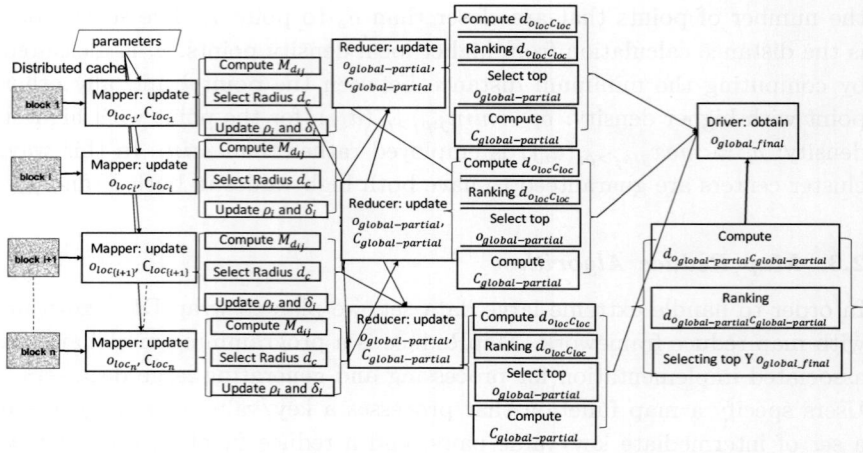

Fig. 1. Map Reduce Framework for Large Scale Dataset.

M, others include Y, the number of outliers and K, the cluster number. The outputs from mapper include an outlier set, a cluster center set and D, a $n \times n$ distance matrix. We firstly make a rough estimation on the data distributions and determine the block number as M. Then, for each block, we construct a distance matrix between each data point. Each matrix is a square symmetric matrix with rows and columns as data points. The mapper computes the distance between data points of each block.

Algorithm 3.0.1 illustrates the detailed procedure of the Map function. Four steps are involved there. The first step is to calculate the distance matrix between data points. In our work, *Mahalanobis distance* is employed and the reason is explained in 3.0.5. The second step is to select a radius distance d_c and the third step is to run DP program. As the usual DP program, *density estimation for ρ_i* and *distance from higher density points δ_i* are calculated. Different from DP program, we focus more on selection of outliers guided by both ρ and δ.

3.0.2. *Partitioner: Evenly Distribute Workloads*

The Map function in Algorithm 3.0.1 emits local results of outliers and centers for updating partial global outliers. These outliers and centers are keyed by a composite key set $< p_{left}, p_{right} >$. These keys take two forms: tuples of outliers and data identifiers or tuples of centers and data identifiers. A partitioner is required to ensure that messages from the

Distributed DP Algorithm 1 Mapper
Inputs:
KEY-block ID $b \in [1, M]$, where M = number of blocks.
VALUE-a subset/block of the whole data
Configure:
Load in parameters from *distributed cache*, including,
M: the number of blocks; Y: the number of outliers; K: cluster number;
Outputs: outlier set and cluster center set; D, $n \times n$ distance matrix.
Map:
(1) Calculate the distance matrix between data points for each mapper
(2) Select a radius distance d_c for each mapper as well
(3) Run DP program at each mapper then
(a) Density estimation
for $i \leftarrow 1 : n$ do
$\rho(i) \leftarrow Count(D(i, i^-)) < d_c$
end for
(b) Distance from higher local density points (δ_i)
δ, N, N distance vector of n objects
from nearest higher density; N, N neighbor,
index vector of nearest neighbor of each element i
for $i \leftarrow 2 : n$ do
$\delta(sorted_\rho(i)) \leftarrow max(D)$
for $j \leftarrow 1 : n-1$ do
if $D(sorted_\rho(i), sorted_\rho(j)) < \delta(sorted_\rho(i))$ then
$\delta(sorted_\rho(i)) \leftarrow D(sorted_\rho(i), sorted_\rho(j))$
$N\ Nneightbor(sorted_\rho(i)) \leftarrow \delta(sorted_\rho(i))$
end if
end for
end for
(4) In light of the number of outliers, select those with low densities but
far from the center as outliers

mappers are sent to the appropriate reducers. Each reducer is responsible for updating the partial global outlier. This is accomplished by ensuring the partitioner sorts on $d_{O_{loc}, C_{loc}}$ values.

3.0.3. Reducer: Update O_{gp} and C_{gp}

The reducer function updates O_{gp} associated with a bunch of O_{loc} and C_{loc} and meanwhile computes corresponding C_{gp} as well. Our final goal is to find the global outliers, however, local outliers discovered by mappers may not be real ones since the splitting of smaller blocks may separate some closely related data points into bunches of isolated ones being considered as outliers. Then recalculations need to be done.

Distributed DP algorithm 2 Reducer
Inputs: KEY-key pair $< p_{left}, p_{right} >$ VALUE-an iterator I over sequence of values. Outputs: top Y outliers Reduce:
(1) Calculate the distance from each local center point to each local outlier
(2) Rank all distance values
(3) Based on the number of input outliers, select top Y outliers as global partial results
(4) Compute global partial centers for each reducer

Essentially, the equation of computing global-partial outliers is $dist_{O_{loc}, \mathbf{C}_{loc}} = \sum_{j=0}^{count(C_{loc})} dist(O_{loc}, C_{loc_j})$.

Namely, $dist_{O_{loc}, \mathbf{C}_{loc}}$, the distance between each local outlier and all local centers can be obtained by the summation of distances between each local outlier and each local center. Once all $dist_{O_{loc}, \mathbf{C}_{loc}}$ are obtained, they are sorted as a set of O_{gp} and top Y is selected for each reducer.

Since top $Y_{O_{gp}}$ is computed and selected from each reducer where not full set of local centers are employed, the third stage of computation is needed as illustrated in Figure 1. The first thing we need to do is to merge local centers. There are two approaches: (i) assign each local center of each reducer an integer number as a reference label. (ii) match local centers based on their similarity. We employ the second one because it is a more complicated, but more robust way as illustrated in Figure 2.

Each mapper outputs K centers and L outliers and each reducer receives a few sets of $\mathbf{C}_{loc}$, with the size of K. Then, at each reducer, computation of C_{gp} is done with the following steps. (i) Compute distance matrix, $dist_{C_{loc_i}, C_{loc_j}}$, where i and $j \in M$, the number of blocks assigned to each reducer. (ii) Sort all $dist_{C_{loc_i}, C_{loc_j}}$ from closer to farther. (iii) Use sorted distance matrix as the similarity matrix to cluster close local centers as global-partial centers. (iv) Disambiguate those global-partial centers by comparing less closer centers.

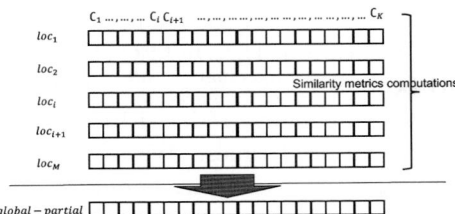

Fig. 2. Illustration of Local Center Merging in Reducer.

3.0.4. Driver: Update O_{gf}

The last step is to go through the first three steps of the reducer stage globally as shown in Figure 1. What we need to be aware about in the driver step is that unlike mappers and reducers, driver step is finished all at once. Mappers and reducers can be done asynchronously. The driver will not start until all mappers and reducers are finished.

3.0.5. Distance Calculations

We can employ *Manhattan distance* (L_1) and *Euclidean Metric* (L_2) to calculate distances. We assume that the data follows normal distributions and utilize *Mahalanobis distance* for our purpose. Specifically, for our D-dimensional vector **x**, they follow multivariate Gaussian distribution taking the form $N(x|\mu, \Sigma) = \frac{1}{(2\pi)^{D/2}} \frac{1}{|\Sigma|^{1/2}} exp\left\{-\frac{1}{2}(x-\mu)^T \Sigma^{-1}(x-\mu)\right\}$, where μ is a D-dimensional mean vector, Σ is a $D \times D$ covariance matrix, and $|\Sigma|$ denotes the determinant of Σ. The functional dependence of the Gaussian on their data points **x** is through the quadratic form $\Delta^2 = (x-\mu)^T \Sigma^{-1}(x-\mu)$, which appears in the exponent of the distribution. The quantify Δ is called the *Mahalanobis distance* from μ to **x** and reduces to the Euclidean distance when Σ is the identity matrix.

4. Experiments

4.1. Experiment Setting

We compared our algorithm (DP_DETECTION) with iForest algorithm. iForest algorithm is a state-of-the-art algorithm meeting the requirement of Big Data processing. We employ AUC (Area Under Roc Curve) and TIME (Elapsed Time) as evaluation index. AUC is in direct proportion to diagnosis accuracy. TIME of DP_DETECTION is the sum of mean time in mapper stage and time in reducer stage, namely, $\frac{time(mapper)}{sum(mapper)+time(reducer)}$. Finally, we recorded F1-Score of DP_DETECTION algorithm separately. The definition of F1-Score is $\frac{2 \times precision \times recall}{precision + recall}$.

In experiment one, we randomly produced 108000 bars of two-dimension data in Gaussian distribution 9 times. We regarded those data owning less than 1/1000 of confidence coefficient as outliers, employed AUC and mean TIME as items of the experiment result and ran our experiment in computers configuring Windows 7 and RAM 8GB. In experiment two, we applied our experiment in a real-world datasets displayed in Table 1, and ran our experiment in computers configuring Windows Server 2008 and RAM 256GB.

Table 1. Real-world datasets and results in experiment two.

Data	Real-World Datasets				AUC Comparison	
					DP_DETECTION	iForest
Card	UCI	30000	23	2	0.7972	0.9978
Epileptic	UCI	11500	178	5	0.5123	0.9991
HTRU2	UCI	17898	8	2	0.5046	0.9716
Post-Operative	UCI	780000	36	5	0.5351	0.9998
USPS	UCI	9298	256	10	0.7488	0.9998
Covert	UCI	581012	10	7	0.8461	0.9965
Human	UCI	10299	561	6	0.6086	0.9995
Orig	UCI	70000	784	10	0.5736	0.9996
Sensorless	UCI	13239	48	3	0.5016	0.9984
Shut	UCI	43500	9	7	0.6201	0.9994

4.2. *Experimental Results*

The experimental result of experiment one is shown in Table 3.

In the result of experiment one, DP_DETECTION spent about 3.70% of iForest's time, while its accuracy is poorer. In experiment two, we applied DP_DETECTION on 10 datasets from UCI to test its performance in real world. The result is shown in Tables 1 and 2. From Tables 1 and 2, we can make the same conclusion that our algorithm performed better in TIME comparison while worse in AUC comparison. Satisfyingly, our algorithm made a great performance in F1-Score. It means that DP_DETECTION has both high precision and recall value.

Table 2. Real-world datasets and results in experiment two.

Data	TIME Comparison DP_DETECTION	iForest	F1-Score DP_DETECTION
Card	1.7084	18.6784	0.9762
Epileptic	0.4623	8.5140	0.9191
HTRU2	0.7229	13.6978	0.9390
Post-Operative	2.3759	37.6617	0.9995
USPS	0.2978	10.8343	0.9998
Covert	352.0684	241.1078	0.9970
Human	0.7766	10.0950	0.9966
Orig	2.1913	39.3425	0.9970
Sensorless	0.4775	10.9019	0.9784
Shut	1.4031	21.5159	0.9976

Table 3. Experiment result of experiment one.

Evaluation Index	DP_DETECTION	iForest
AUC	0.8735	0.9991
TIME	1.5063	40.6566

5. Conclusion

In this work, we propose a new algorithm called DP_DETECTION by adapting the well-known DP clustering algorithm for detecting outliers from large-scale datasets. Specifically, we design a Map Reduce framework to wrap up DP algorithm for efficient outlier discovery based on density of big data. Experiments show that our framework yields promising results in TIME comparisons while dose not have dominance in AUC comparisons. In future, extra efforts will be put onto deriving more efficient distance functions and finding more efficient distributive architectures for expanding current framework.

Acknowledgments

This work is partially supported by Blockchain Technology Fund of Hi-Tech Information Technology Research Institute of Chengdu (No. 2018H01207), the Online Education Research Center of the Ministry of Education Online Education Research Fund (No. 2016YB158) and National College Students Innovation and Entrepreneurship Training Program (No. 201510638047).

References

1. E. M. Knox and R. T. Ng, Algorithms for mining distancebased outliers in large datasets, in *Proceedings of the International Conference on Very Large Data Bases*, 1998.
2. J. MacQueen *et al.*, Some methods for classification and analysis of multivariate observations, in *Proceedings of the fifth Berkeley symposium on mathematical statistics and probability*, (14) 1967.
3. M. Ester, H.-P. Kriegel, J. Sander, X. Xu *et al.*, A density-based algorithm for discovering clusters in large spatial databases with noise., in *Kdd*, (34) 1996.
4. A. Rodriguez and A. Laio, *Science* **344**, 1492 (2014).
5. J. Dean and S. Ghemawat, *Communications of the ACM* **51**, 107 (2008).

An iterative multi-criteria optimization of product snippets enhanced by feature extraction from online reviews

Yao Mu, Qiang Wei, Guoqing Chen and Xunhua Guo

School of Economics and Management, Tsinghua University
Beijing 100084, China
{muy.12, weiq, chengq, guoxh}@sem.tsinghua.edu.cn

Product snippets on e-commerce platforms are expected to attract consumers with appealing featured information about products. Related researches focus mainly on search traffic optimization, which, however, is usually less of attraction, i.e., showing weakness on catching consumers' eyebrows. Meanwhile, online reviews contain versatile features that consumers highlight, which are found potentially useful to enhance the attraction of snippets. This paper proposes an iterative multi-criteria optimization method, by taking both snippet keywords and review features weighted by search intensity and review intensity, respectively, into consideration. The method can output enhanced snippets with optimized potential attraction and search traffic.

Keywords: Product snippet; online reviews; multi-criteria optimization; feature extraction.

1. Introduction

Product snippet, as a special type of online text ads, is known as a short text briefly describing product features.[1] As an important information for preliminary understanding, snippets greatly affect consumers' decision on whether the product suits their interest or not.[2] It has been shown that informative snippets covering product features attractive to consumers tend to get more clicks, and further to increase the probability of purchases.[3]

Product snippets are generally designed from the perspective of sellers, and their qualities are thus limited by the unilateral understanding from sellers. On one hand, generated from discrete lines of knowledge, snippets of a same product may be variable. On Amazon.com, one seller's snippet for Canon Powershot SX530 HS digital camera indicates "50x optimal image stabilized zoom", while another emphasizes it's "Wi-Fi & NFC enabled". Obviously, both of them try to attract consumers, which, however, may just ignore consumers' perspective. That is, this stream of snippets cannot guarantee the existence of appealing product features for consumers. On the other hand, due to the competition on fetching search traffic, many sellers usually output similar snippets for a product, i.e., including generic hot keywords, which leads to lack of discrimination and attraction to consumers. For example, on Taobao.com (i.e., the largest e-commerce platform in China), the top ranked sellers of Cuckoo CCRP-G1052FR e-cooker frequently highlight the generic keywords like "Korea", "authentic", "high-pressure", "voice control", etc., crowding out the charactered features like "easy-to-clean", "free gift", etc., which may be more possible to target consumers' personalized intents. In a sentence, state of art product snippet generation does not involve consumers' perspective into consideration, which enlightens our study on enhancing product snippets.

Recent years witness the rapidly accumulated wealth of online reviews of products. The abundant user-generated reviews can to a large extent mirror consumers' intents and preferences, which become a rich source of understanding and capturing key features of products from consumers' perspective. Clearly, consumers comments mainly on product features they care most about,[4] i.e., implicating collective intelligence.[5] This offers a more economical and effective manner to approach consumers. Concretely, review intensity, i.e., number of reviews, on a certain feature could be deemed as a measure on attraction of the feature. For instance, reviews on Canon Powershot SX530 on Amazon.com frequently indicate the features like "compact and portable", "perfect for travel", etc., which are quite attractive to consumers whereas missed by sellers. Therefore, integrating the useful information from online reviews could benefit product snippet generation.

Related works to our study focus on two topics, i.e., snippet optimization and online review analysis. As stated, snippets evidently influence the PV (Page View), CTR (Click Trough Rate) and CVR (Click Value Rate) of ads,[3] so snippet optimization has aroused wide attention from the academia and the industry. Existing methods stress more on query-dependent snippet optimization for search engines, based on positional index,[6] entity path selection,[7] etc., which benefit probabilities of ads' appearance on search result pages, i.e., improve search traffic, but pay little attention on snippets' attraction to consumers. Although some e-commerce platforms (e.g., Amazon.com) provide guides for constructing appealing snippets, the guidances depend highly on sellers' observations and experiences, while lacking scientific quantization and systematical improvement of snippets' attraction. In order to fill the gap, data mining techniques can be used to analyze online reviews and then to extract valuable product features as fragments for improving snippets. Aiming to get a step closer to consumers' psychology and further to assist business strategies, researchers have conducted studies on online review analysis, such as feature extraction,[8] summarization[9] and ranking.[10] Despite of the maturity of the methods, they were rarely utilized for the product snippet optimization task, and the attraction of review messages has not yet been synthetically evaluated or optimized with search traffic.

To approach sellers' actual purpose for improving both search traffic and consumer attraction of snippets, this paper designs an iterative multi-criteria optimization method of enhancing product snippet with feature extraction from online reviews to maximize the promotion of total utility weighted by both search traffic and consumer attraction. Section 2 describes the details of the proposed method. Experimental results are discussed in Section 3. Section 4 presents the final conclusion.

2. The Iterative Multi-Criteria Optimization Method

For a product, its text snippet is commonly composed of keywords representing features which sellers consider important, while as stated, online reviews contain features which consumers care about. Without loss of generality, the set of keywords/features in a product's snippet can be denoted as T and the set of features mentioned in its online reviews can be denoted as W. Naturally, the two sets overlap each other and comprise a set L consisting of informative features, i.e., $L = T \cup W$. In accordance with existing work, a feature j in set L has a search intensity s_j reflecting its capability in drawing search traffic, i.e., search volume index, and also a review intensity r_j reflecting the number of reviews commenting the feature, which is positively correlated with consumers' concern or its attraction.[11] Given the feature's search frequency s'_j in search history and commenting frequency r'_j in online reviews, the two degrees of intensities can be calculated as:

$$s_j = \frac{s'_j - \min_{i \in L}\{s'_i\}}{\max_{i \in L}\{s'_i\} - \min_{i \in L}\{s'_i\}} \quad (1-a) \qquad r_j = \frac{r'_j - \min_{i \in L}\{r'_i\}}{\max_{i \in L}\{r'_i\} - \min_{i \in L}\{r'_i\}} \quad (1-b)$$

where s_j and r_j both take values in range $[0, 1]$.

Since the length of a snippet is usually limited within m words, to exclude the influence of snippet's length variation,[3] a keyword substitution strategy is adopted to enhance snippet. For increasing attraction of currently traffic-oriented snippet, a straightforward process is to replace the feature having lowest search intensity in T with that having highest review intensity in L. That is to seek the largest gain in consumer attraction with the smallest loss in search traffic, but fails to achieve comprehensive optimization on both of the two criteria, i.e., search traffic and consumer attraction.

Consequently, it is more reasonable to construct a utility function to jointly evaluate both search traffic and consumer attraction of a feature. Moreover, to better support sellers' flexible manipulation on these two criteria, a tuning parameter α in range $[0, 1]$, could be used to represent the degree of preference from traffic to attraction. Thus, the utility function u_j for a feature j could be defined as follow.

$$u_j = (1-\alpha)s_j + \alpha r_j, \quad \alpha \in [0,1] \qquad (2)$$

Suppose that the original snippet is composed of features p_i ($i = 1, \cdots, m$) and the optimized one is composed of features q_i ($i = 1, \cdots, m$), for simplicity, an one-step optimization model could be constructed as $(3-a)$, i.e., maximizing the utility by substituting only one feature. Clearly, our purpose is to continue multiple substitutions until reaching global optimum, i.e., no further substitutions can improve the total utility. Thus, our method, in form of an iterative optimization model, is as shown in $(3-b)$.

$$\max \sum_{i=1}^{m} u_{q_i}$$
s.t. $u_{p_k} \leq u_{p_i}, \forall i = 1, \cdots, m$
$q_i = p_i, \forall i \neq k$
$s_{q_k} \geq s_{p_k}$
$r_{q_k} \geq r_{p_k}$
$q_k \neq p_i, \forall i \neq k$
$q_k \in L$

$$\max \sum_{i=1}^{m} u_{q_i}$$
s.t. $s_{q_k} \geq s_{p_k}, \forall k = 1, \cdots, m$
$r_{q_k} \geq r_{p_k}, \forall k = 1, \cdots, m$
$q_k \neq p_i, \forall k = 1, \cdots, m, \forall i \neq k$
$q_k \in L, \forall k = 1, \cdots, m$

$$(3-a) \qquad\qquad (3-b)$$

Based on the utility measure, a direct manner to optimize a snippet could be to eliminate the features in current snippet with low utility and to fill in others with higher utility chosen from L, which, nevertheless, is not suitable for e-commerce environment. Since it is notable that the actual online shopping process consists of two successively separate phases (i.e., searching out a category of products and clicking a specific product snippet). To align with the process, the optimization is supposed not to sacrifice snippets' performance in either phase, but to improve above the baseline with some propensity. To this end, the filled-in features' intensities are constrained to be no less than those of the eliminated ones.

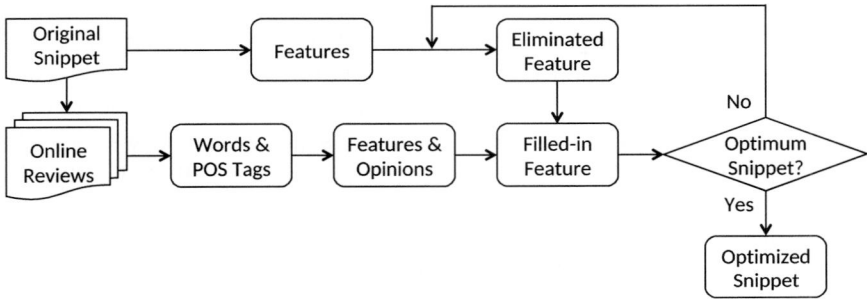

Fig. 1. The proposed method framework.

Figure 1 depicts the framework of the iterative optimization. For an original product snippet, product features in it can be obtained by word parsing. Function (2) can be used to pick the feature with lowest utility as the eliminated one. Inspired by,[12,13] online reviews can be processed in three steps. First, NLP tools are used to do text parsing and POS tagging of the crawled reviews. Second, nouns and adjectives are extracted as candidate features along with recorded commenting frequencies. Third, the most adjacent adjectives to a noun are also picked out as corresponding opinions on the feature. Based on relevant vocabularies, further operation is to remove the stop words and the features with negative opinions in majority, as well as merging the synonyms. Eligible feature with highest utility in the remaining part is chosen as the filled-in, which ulteriorly replaces the eliminated one to finish substitution of one feature. If aiming to the global optimum snippet, feature with the second lowest utility in the snippet is then chosen as the eliminated and the filled-in feature should be selected from the updated set of candidates meeting the constraints. The process would iteratively continue until no qualified filled-in features could be further detected to improve total utility of the whole snippet.

For illustrative purpose, taking the snippet "Summer New-style Female Student T-shirt Fashion Casual Bottoming-shirt" with 8 features as an example. From the crawled 9,607 reviews, 1,054 features were extracted. After removing stop words like "clothes" and negatively-commented features like "color", we further merged synonyms like "vogue", "fashion" and "in trend". The remaining feature with highest frequency of 2,013 was "comfortable", which stated the aspect consumers care most. Various attempts in weights α generated diverse results of one-step optimization. When $\alpha = 0.3$, the feature "new-style" was substituted by "light-and-thin", which increased both search intensity and review intensity about 150%. As α increased to 0.7, the snippet took the filled-in feature "super-discount" instead of "fashion", since the former is obviously more attractive according to the reviews. Search intensity 0.035 of "super-discount" was slightly higher than 0.032 of "fashion", while review intensity 0.410 of the filled-in was far larger than 0.024 of the eliminated. That is, despite simultaneous improvements of the two criteria, a larger α lays more stress on attraction, and thus usually leads to a more significant increase in review intensity of the snippet. Setting $\alpha = 0.7$ and iteratively optimizing the snippet, we got the final "Summer Floral-printed Female Light-and-thin T-shirt Super-discount Casual Bottoming-shirt", where 3 features were substituted. Total utility of the snippet increased by 44.7% in this iterative process, which is more remarkable than the 31.9% brought by one-step optimization.

3. Experiments

Since online reviews of experience goods usually show more informative and influential than those of search goods,[14] three typical categories of experience goods (i.e. dresses, shoes and bags) on Taobao.com were involved in the real-world data experiments. For each category, we chose five products with more than 5,000 reviews and randomly picked one top-ranked text snippet for optimization. The total 120,384 online reviews were preprocessed by ICTCLAS (v2016), a commonly used tool for Chinese lexical analyzing, into words and tags. Java programs were also developed to automatically extract features from the fragments and further screen the candidates. Real-world data used were crawled in April 2016, while data processing and user experiments were conducted within the following two months. All our data processing was run on a PC with two dual-core Intel Core i5 processors (1.8GHz each), 4GB of main memory, running Windows 7 (32-bit).

The first experiment was to evaluate the performance of the one-step optimization, where 120 subjects (i.e., senior students randomly recruited from a university) were averagely divided into 4 groups (30 subjects each) with different α values, i.e., 0, 0.3, 0.7 and 1.0, respectively. The 4 chosen values were considered representative enough for user experiments. From 0 to 1, when increasing α by 0.1 for each time, we found that the keyword substitution did not vary continuously, and substitutions on almost all the α values were the same as one of the 4 chosen. In respect of increase in review intensity, α from 0.1 to 0.4 deviated less than 6% from $\alpha = 0.3$, α from 0.5 to 0.8 deviated less than 7% from $\alpha = 0.7$, and $\alpha = 0.9$ deviated less than 2% from $\alpha = 1.0$, which implied that the subtle distinctions could hardly be perceived by users in nature. Each subject was asked to vote which snippet of the two he/she was more willing to click without knowing whether they were original or one-step optimized, i.e., a blind-test. Results of average votes and corresponding ANOVA tests are shown in Table 1. Clearly, on all α values, the one-step optimized result outperforms the original one.

Table 1. Results of the first experiment with one-step optimization.

α value	% Increase in review intensity[a]	Average votes for original	Average votes for optimized	P-value of ANOVA test
0.0	4.33	14.67	15.33	0.409
0.3	26.94	14.07	15.93	0.040
0.7	77.00	12.93	17.07	0.009
1.0	7.88	14.60	15.40	0.305

Note: [a]% Increase in review intensity = (Optimized−Original)/Original.

In order to test the superiority of the iterative method, the second blind-test experiment guided another two groups of 30 subjects to compare the iteratively optimized snippet with the original as well as the one-step optimized separately on each of the 15 products. This experiment was conducted with $\alpha = 0.7$. Results are listed in Table 2, revealing the superiority of the proposed iterative optimization method.

Table 2. Results of the second experiment with iterative optimization.

Baseline snippet	% Increase in review intensity	Average votes for baseline	Average votes for optimized	P-value of ANOVA test
Original	128.41	12.67	17.33	0.000
One-step Optimized	27.68	14.13	15.87	0.029

4. Conclusion

By introducing extracted features consumers care about from online reviews, the paper proposes a novel iterative optimization method to enhance product snippets to improve consumers' willings to click. The proposed method reasonably quantizes both search traffic and consumer attraction into a multi-criteria optimization model, and designs an iterative optimization strategy to significantly generate more appealing snippets. Experimental results also demonstrate the benefit of the proposed method.

References

1. Q. Wei, M. Ren, J. W. Lei and J. Zhang, How can product text snippets benefit from online customer reviews, in *Pacific Asia Conf. on Info. Sys. (PACIS 2014)*, (Chengdu, China, 2014).
2. N. Zotos, P. Tzekou, G. Tsatsaronis, L. Kozanidis, S. Stamou and I. Varlamis, To click or not to click? the role of contextualized and user-centric web snippets, in *SIGIR Workshop on Focused Retrieval*, (Amsterdam, the Netherlands, 2007).
3. D. Maxwell, L. Azzopardi and Y. Moshfeghi, A study of snippet length and informativeness: Behaviour, performance and user experience, in *Int. ACM SIGIR Conf. on Res. and Dvpt. in Info. Retrieval (SIGIR '17)*, (Tokyo, Japan, 2017).
4. V. Dhar and A. Ghose, *Info. Sys. Res.* **21**, 760 (2010).
5. X. Ni, Y. H. Sai, A. N. Wang and Q. Zhang, Intelligent discovery of notable product features by mining large scale online reviews, in *IEEE Int. Conf. on Data Sci. in Cyberspace (DSC 2017)*, (Shenzhen, China, 2017).
6. H. Bast and M. Celikik, *ACM Trans. on Info. Sys.* **32**, 6:1 (2014).
7. Z. Liu, Y. Huang and Y. Chen, *ACM Trans. on Database Sys.* **35**, 19:1 (2010).
8. Y. Zhang and W. Zhu, Extracting implicit features in online customer reviews for opinion mining, in *Int. World Wide Web Conf. (WWW 2013)*, (Rio de Janeiro, Brazil, 2013).
9. A. S. Muhammad, P. Damaschke and O. Mogren, Summarizing online user reviews using bicliques, in *Int. Conf. on Current Trends in Theory and Practice of Computer Sci. (SOFSEM 2016)*, (Harrachov, Czech Republic, 2016).
10. H. Y. Hsieh and S. H. Wu, Ranking online customer reviews with the svr model, in *IEEE Int. Conf. on Info. Reuse and Integration (IRI 2015)*, (Redwood City, CA, USA, 2015).
11. R. Decker and M. Trusov, *Int. J. of Res. in Marketing* **27**, 293 (2010).
12. M. Hu and B. Liu, Mining and summarizing customer reviews, in *ACM SIGKDD Conf. on Knowledge Discovery and Data Mining (KDD '04)*, (Seattle, WA, USA, 2004).
13. M. Hu and B. Liu, Mining opinion features in customer reviews, in *Natl. Conf. on Artificial Intelligence (AAAI-04)*, (San Jose, California, 2004).
14. P. Huang, N. H. Lurie and S. Mitra, *J. of Marketing* **73**, 55 (2009).

Using fuzzy representation in educational data mining and learning analytics

Jun Ma,* Jie Yang* and Sarah K. Howard**

*SMART Infrastructure Facility, University of Wollongong
Wollongong, NSW 2522, Australia
jma,jiey@uow.edu.au
**School of Education, University of Wollongong
Wollongong, NSW 2522, Australia
sahoward@uow.edu.au

Carlos Gonzalez and Dany Lopez

Facultad de Educación, Pontificia Universidad Católica de Chile
Campus San Joaquín,
Avda. V. Mackenna 4860, Macul, 7820436 Santiago, Chile
cgonzalu@uc.cl, danylopfiqui@gmail.com

Technique progress is driving the adoption of educational data mining and learning analytics in education research and education environment. A big challenge when implementing the methods of them is lack of effective knowledge representation and processing. In this paper, we discuss the features of education data used in education research and demonstrates the possibility of using fuzzy set and its extension as a tool to get richer information and insight from the data.

Keywords: Learning analytic; educational data mining; uncertainty; fuzzy set.

1. Background

The last two decades have witnessed the dramatic changes in education area with widely application of digital techniques. Data generation and collection becomes more easier. The huge volume of data in teaching and learning provides opportunities to researcher to extracting valuable information and get insight into education practices. Recently, educational data mining and learning analytics techniques and methods have been used in education research to discover previously unseen patterns in education activities, probe the underlying reasons of them and design better education systems.[1–4]

Educational data mining and learning analytics aim to extract useful and valuable information and measurements from the collected data, analyse learning and teaching processes, and provide appropriate suggestions and recommendations to educators and learners. The collected data covers a wide range aspects related to education such as classroom video observations, library loan and devices usage, subject quizzes and test logs, etc. Representing and processing uncertainty in this data is a big challenge before the data can be effectively used in the following analysis processes.

Many methods for representing and processing uncertainty have been presented and successfully used in other fields such as management and decision science, computer sciences and engineering. These techniques and methods includes such as probability, belief theory, and fuzzy sets.[5,6] Considering the features of the data in education background, we choose fuzzy set and its extension as they can provide a more natural and understandable way to describe information in data, particularly, the semantics in it. This paper presents a few discussions and study cases of why and how we use fuzzy sets for representing and processing uncertainty in our work in educational data mining and learning analytics.

The remainder of the paper is outlined as follows. Section 2 discusses the features of collected data in education applications and the reason we choose fuzzy set and its extension. Section 3 briefly overviews definition of fuzzy set. Section 4 lists several study cases where we are using in our work. Section 5 presents some conclusions and outlines future works.

2. Education Data

The educational data shares some common features of data used in other research fields. Some of these features can be summarised as: 1) Education data is multi-sourced. Typically, the data is generated and collected from different systems such as library management, student information management system, online interactive learning platform, etc. With the usage of new digital techniques and devices, more and more data is collected from classrooms, laboratories, and networks. 2) Education data is information-enriched. The collected data covers a broad range of aspects related to teaching and learning. It can contain a student's basic demographic information, history of using learning resources in library, activity in subject quizzes and exams, and learning performance measurement. 3) Education data is diversified. The collected data is commonly in a various forms, such as numbers, texts, images, audio and video. The data diversity is driven by the rapid progress of information and communication techniques as well as the requirement of in-depth research in education. 4) Education data is big data. It is of massive volume and generated in high velocity. It is valuable for a variety of users, from educators, learners to public communities.

Besides aforementioned features, education data has some distinctions. The value of insight derived from the raw data is more interested than that of the raw data itself. In other words, the value of the education data lies in explaining and exploring education problems rather than building accurate data models. This is determined by the essence of education research as a social science. Given a student's data of accessing online learning resources and their learning performance measurement, education researcher may focus on problems such as "if the student is of academic integrity" or "is the online resources useful" rather than the problem "if a predictive model is accurate enough to predict learning performance from resource accessing frequency". Education data is often composed of a large number of short-period, high-randomness, one-off individual data. For examples, a student's learning activity in a subject spans only a few month; and a student's learning performance measurement only exists in 3-4 years. Individual data often displays bigger randomness. Two different individuals seldom have same or "very similar" behaviours. Individual's data is personalised. Education data exhibits a certain periodic characteristic. For example, library study room's usage reaches its peak at the time of end semester and falls during school holiday; while students have more online activity 1-2 days before and after a subject quiz. The periodicity exists in both individual and whole population.

Considering the features of education data, we tried to use fuzzy set and its extension to represent and analyse the data used. Firstly, fuzzy set was presented to handle human being perception of a concept in a given context. In education research, it is more reasonable to describe information, knowledge or data using human perceptible forms such as natural language rather than formalised languages. Secondly, context changes may affect the expression forms of a formalised language but the natural language. Hence, semantic conveyed in the data in one context can be relatively accurately transferred to another context with less loss. Thirdly, research in many other areas has shown the power of fuzzy set and its extensions in handling the uncertainty. Based on these considerations, we used fuzzy set and its extensions in analysis education data in several projects. Section 4 gives some study cases.

3. Fuzzy Set and Its Extensions

Fuzzy set was firstly presented as a tool to address issues of representation and processing uncertainty in engineering applications. The uncertainty relies in the unclear boundary of concepts and descriptions. A commonly used illustrative example is the concept of "middle aged". Given a person's age, people can understand if the person is middle aged or not; however, it is hard for people to clear set a clear range of ages for "middle aged". Examples of this kind are easily to find in real life.

Formally, a fuzzy set $\tilde{A}$ is defined on a set X of objects, $X = \{x_1, \ldots, x_n\}$; and for any $x_i \in X$, A associates a real number $\mu_i \in [0, 1]$ with x_i. Because μ_i is used to indicate to what extend the object x_i belongs to $\tilde{A}$, μ_i is called the membership degree of x_i to $\tilde{A}$.

Since it was presented, fuzzy set has gain considerable attentions and development. On the one hand, techniques and methods based on them are widely used in many applications to solve problems with uncertainty (fuzziness), particularly, in control systems and decision management; on the other hand, theoretical research on fuzzy set and its extensions to enrich their processing and representing power for uncertainty has been reported.

4. Study Cases

This section describes two cases of combining fuzzy sets and data mining techniques for analysing education data.

4.1. Survey questionnaire analysis

The first case is analysing the student survey questionnaire data. The data is taken from a state-level secondary student survey as part of Australian federal Digital Education Revolution initiative. It contains questions about students' school engagement, PC/laptop usages, learning experiences, beliefs about information and communication technologies (ICTs) in school teaching, and future plans for study. In the survey data, some questions can be better understood if converting them to use fuzzy set representation. For example, a question is about the frequency of PC/laptop usage; and the given options include "once a day", "1-3 times a month", "once a term", etc. By using fuzzy set, we defined concepts "frequent user", "occasion user" and "infrequent user". By this way, the inherent semantic of those options is easily summarised and aggregated. For other questions with similar characters we defined other fuzzy representations accordingly.

Using this method, we are able to combine the obtained insight with data mining techniques to find useful information. Figure 1 shows an outcome of using fuzzy representation and association analysis[7] of students who reported had positive ICT engagement. However, we noted two centred nodes "EPC=L" and "EC=L" in the outcomes which indicates many situations lead to students have "Low" (L) efficacy in "creation" (EC=L) and "processing" (EPC=L). This is a very interesting finding which contradict to previous commonsense that they should have higher efficacy.

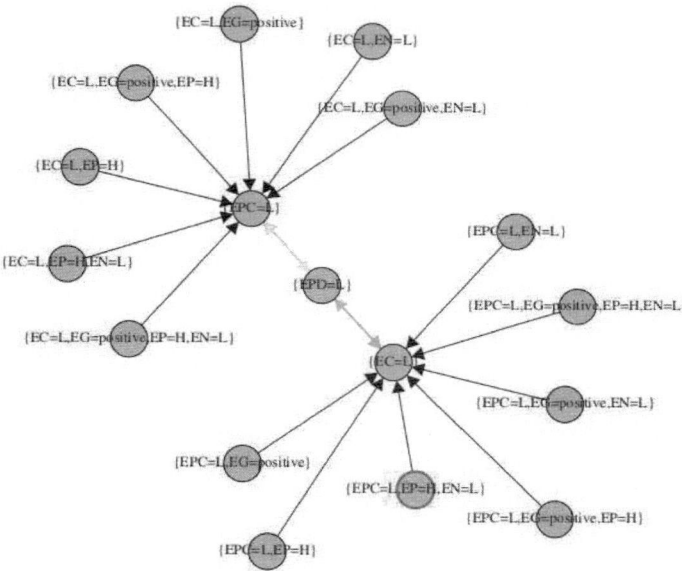

Fig. 1. An output of combining fuzzy representation and association analysis.

4.2. *Online learning activity analysis*

The second study case provides an example of how other analyses of students online educational data can be extended using data mining approaches. The project included three datasets from a Chilean university: library loan data, electronic journal access and student academic information, which cover an 8-year time frame. The initial analysis of this data was presented in Ref. 8. Using stepwise linear regression, the authors found that the factor with the largest effect on student performance was Standardized Average Course Score. They found little effect of student borrowing on overall performance and, overall, the models explained only a small amount of variance. However, the identified that there are two groups of students, based on library resource activity: those that do and those that do not access electronic resources. In both groups, accessing library materials changes over the time at university, but they show different trajectories of change. The authors have identified that it is difficult to identify which factors have an effect on student performance, given the large number of factors in the datasets.[8] This analysis will be extended using fuzzy set, by defining a broader range of library usage concepts. The same will be done for other key factors identified. Combining this will data mining approaches, more nuanced patterns within the high and low usage groups can be identified. It can be hypothesized that, given it is known higher performing students have different study habits to lower performing students, that different patterns will be extracted from the two datasets.

5. Conclusions

Uncertainty in education data has enriched semantics. Using fuzzy set and its extensions can represent and handle this semantics for better education decision making. In this paper, we discussed the main features of education data and pointed out the necessity of representing uncertainty in education data using fuzzy set and its extensions. Through two/three study cases, we demonstrated the benefit of this method. Recently using data mining techniques in education research becomes popular. Through the combination of fuzzy representation and data mining, more and more previously unseen patterns have been discovered. As we have mentioned in previous sections, the education data has some specific features and how to represent the data including the uncertainty in it using a natural and understandable way is and will still be a critical issue in education data mining and learning analytics.

References

1. R. S. Baker and P. S. Inventado, *Educational Data Mining and Learning Analytics*, in *Learning Analytics: From Research to Practice*, eds. J. A. Larusson and B. White (Springer New York, New York, NY, 2014), New York, NY, pp. 61–75.
2. L. Calvet Liñán and Á. A. Juan Pérez, *International Journal of Educational Technology in Higher Education* **12**, 98 (Jul 2015).

3. A. Mishra, R. Bansal and S. N. Singh, Educational data mining and learning analysis, in *2017 7th International Conference on Cloud Computing, Data Science Engineering - Confluence*, Jan 2017.
4. P. Ihantola, A. Vihavainen, A. Ahadi, M. Butler, J. Borstler, S. Edwards, E. Isohanni, A. Korhonen, A. Petersen, K. Rivers, M. Rubio, J. Sheard, B. Skupas, J. Spacco, C. Szabo and D. Toll, *Educational data mining and learning analytics in programming: Literature review and case studies*, in *Proceedings of the 20th Annual Conference on Innovation and Technology in Computer Science Education (ITiCSE 2015)*, eds. N. Ragonis and P. Kinnunen (Association for Computing Machinery (ACM), 2015), pp. 41–63.
5. D. Dubois and H. Prade, *Formal Representations of Uncertainty*, in *Decision-making Process*, (ISTE, 2010), pp. 85–156.
6. R. R. Yager, *IEEE Transactions on Systems, Man, and Cybernetics, Part B (Cybernetics)* **32**, 13(Feb 2002).
7. S. K. Howard, J. Ma and J. Yang, *Computers & Education* **101**, 29 (2016).
8. M. Montenegro, P. Clasing, N. Kelly, C. Gonzalez, M. Jara, R. Alarcn, A. Sandoval and E. Saurina, *The Journal of Academic Librarianship* **42**, 551 (2016).

Compressive sensing based feature selection: A case study for commuter behaviour modelling

Jie Yang* and Jun Ma

SMART Infrastructure Facility, Faculty of Engineering and Information Sciences University of Wollongong, Northfields Ave, Wollongong, NSW 2522, Australia
*jiey@uow.edu.au
jma@uow.edu.au

Xiangqian Wang

School of Economics and Management,
Anhui University of Science and Technology, Huainan 232001, China
xqwaust@163.com

We present a novel algorithm to address the problem of feature selection using the compressive sensing (CS) model. In the proposed algorithm, the candidate feature set is first regarded as a basis dictionary in CS, and then features that minimize the output error are selected. As a result, the selected features have a direct correspondence to the target samples, thereby achieving a better prediction performance. Compared to traditional dimensionality reduction algorithms, the proposed algorithm neither uses the problem-dependent parameters nor requires additional computation for the eigenvalue decomposition. Experimentally, the proposed algorithm is evaluated using a real-life problem for commuter behaviour modelling. The experimental results show that the proposed method is better or competitive when compared to existing feature selection methods.

Keywords: Dimensionality reduction; feature selection; compressive sensing; commuter behaviour modelling.

1. Introduction

The problem of the curse of dimensionality has attracted much research effort over the past decades. One simple reason is that the target datasets are generally characterised by a large number of features, in which only a small subset of them may contribute significantly to the problem. Additionally, many machine learning algorithms become unwieldy and computationally expensive if the number of features for training is too large.

There is not a theoretical formula giving clear insight for how to reduce the dimensionality. Searching for the appropriate features is typically problem-dependent, and usually specialised prior knowledge is required. The difficulty is to balance the number of required features and the training performance. For instance, a large number of features will not only result in over-fitting and poor generalisation, but also increase the computational overhead. On the other hand, a small set of features may not be sufficient for the modelling purpose due to lack of information.

Dimensionality reduction approaches, therefore, are usually employed to reduce the input features without affecting the modelling performance. As a result, fewer but discriminative features could lead to less computational complexity and better generalization ability. Dimensionality reduction methods can be broadly categorized as *feature extraction* (such as principal component analysis (PCA) and linear discriminant analysis (LDA)) and *feature selection*.[1-3] Traditionally, feature extraction methods transform the original data from the high-dimensional space to a space with fewer dimensions. By contrast, feature selection algorithms aim to determine the significance of available features, thereby selecting a small set of informative features from the original data. Consequently, the computational requirement for feature selection algorithms is always lower than that of extraction algorithms since they usually depend on a binary transformation.

In this paper, we propose a novel algorithm for feature selection based on compressive sensing, termed herein *compressive feature selection* (CFS) algorithm. The compressive sensing model is used to reconstruct original signals using a small number of basis functions from a given dictionary. Similarly, we consider the discriminative features as a subset of the full features. The aim then is to find the sparse features that explain the observations. To this end, the problem of the feature selection is first formulated as the compressive sensing model. Finding discriminative features is equivalent to searching for the sparse representation of full features. We further apply the proposed algorithm to solve a dimensionality-reduction problem for commuter behaviour modelling.

The remainder of the paper is organized as follows. Section 2 and Section 3 present a brief review of existing dimensionality reduction algorithms and the compressive sensing model, respectively. Section 4 establishes the link between feature selection and compressive sensing, and presents the CFS algorithm. Section 5 presents experimental results based on a real-life problem, and compares the performance of the CFS algorithm with those of existing algorithms. Section 6 presents concluding remarks.

2. Feature Selection

Dimensionality reduction is a fundamental technique for large-scale data processing. By reducing the redundant dimensions, the salient information from the original data is maintained while the number of required features is reduced. Generally, dimensionality reduction approaches are cast into two categories: feature extraction and feature selection.

- Feature extraction aims to extract features by projecting the original high-dimensional data to a lower-dimensional space through an algebraic transformation. Examples of feature extraction include *principal component analysis* (PCA) and *linear discriminant analysis* (LDA).
- Feature selection finds the most representative features from the original data using a binary transformation matrix. The most popular feature selection methods are *Laplacian Score* (LS),[1] *Mutual Information*-based methods,[2] and Genetic Programming (GP) based method.[3]

In this paper, we focus on the feature selection based technique to develop the dimensionality reduction approach. The general procedure of feature selection is to measure all available features and rank them according to some significance indexes; features are later selected based on their rankings.

Suppose we have Q pairs of input vectors $\boldsymbol{p}_i$ and the corresponding output z_i, $i = 1, \cdots, Q$. Assuming the input vector $\boldsymbol{p}_i$ is multidimensional, let $\boldsymbol{f}_j$ denote a vector containing the j-th feature from all input patterns:

$$\boldsymbol{f}_j = [p_{1,j}, p_{2,j}, \ldots, p_{Q,j}]^\mathrm{T}, \qquad (1)$$

where $p_{i,j}$ is the j-th component of $\boldsymbol{p}_i$. Feature selection is performed by assigning to each feature $\boldsymbol{f}_j$ its significance of relevance $R(\boldsymbol{f}_j)$. Several feature selection methods have been proposed to evaluate the significance of features, some of which are summarized below.

(1) Laplacian Score (LS) method evaluates a feature based on its locality preserving power.[1] The LS method generates a nearest neighbour graph with Q nodes, with each node representing an input pattern. The significance of $\boldsymbol{f}_j$ is computed as follows:

$$R^{LS}(\boldsymbol{f}_j) = \frac{\widehat{\boldsymbol{f}}^\mathrm{T} M_l \widehat{\boldsymbol{f}}}{\widehat{\boldsymbol{f}}^\mathrm{T} (M_w + M_l) \widehat{\boldsymbol{f}}}, \qquad (2)$$

where

$$\widehat{\boldsymbol{f}} = \boldsymbol{f}_j - \frac{\boldsymbol{f}_j^\mathrm{T}(M_w + M_l)\boldsymbol{1}}{\boldsymbol{1}^\mathrm{T}(M_w + M_l)\boldsymbol{1}}, \qquad (3)$$

M_w is the weight matrix of the graph, M_l is the graph Laplacian matrix,[4] and $\boldsymbol{1}$ is a column vector with all elements equal to 1.

(2) Mutual information (MI)-based feature selection approach is introduced in.[2] Two algorithms have been proposed using the Max-Relevance (MAX_R) and Min-Redundancy (MIN_R) measures. First, the mutual information is introduced to measure the level of similarity between two discrete random vectors f and z:

$$I(f,z) = \sum_{f \in f} \sum_{z \in z} p(f,z) \log \left(\frac{p(f,z)}{p(f)p(z)} \right), \quad (4)$$

where $p(f,z)$ is the joint probability function of f and z, and $p(f)$ and $p(z)$ are the marginal probability functions of f and z, respectively. Then the MAX_R measures the mutual information between the selected features and all the desired outputs:

$$\text{MAX}_R = \frac{1}{|S|} \sum_{j \in S} I(f_j, z), \quad (5)$$

where S is the set of indices of the selected features. Moreover, the MIN_R measure considers the mutual information between the j-th and k-th feature:

$$\text{MIN}_R = \frac{1}{|S|} \sum_{j,k \in S} I(f_j, f_k). \quad (6)$$

Two methods have been proposed to optimize Max-Relevance and Min-Redundancy simultaneously. One of them is the *mutual information difference* (MID) method that searches features that maximize the difference between MAX_R and MIN_R,

$$R^{\text{MID}}(f_j) = \text{MAX}_R - \text{MIN}_R.$$

Another method is called *mutual information quotient* (MIQ), which selects important features by maximizing the ration between MAX_R and MIN_R,

$$R^{\text{MIQ}}(f_j) = \text{MAX}_R / \text{MIN}_R.$$

(3) Genetic Programming (GP) based method selects discriminative features by combining distinct selection metrics, such as Information Gain, Odds Ratio.[3] Intuitively, one feature might be with different importance according to various selection criteria. Therefore, the GP based method is employed to search for a possible combination of a set of "basic" selection metrics to yield one "compound" metric to evaluate feature importance. Let the terminals (leaf nodes in GP) an non-terminals be T, NT, respectively. Accordingly, for the GP-based method, the feature significance then is computed as follows:

$$R^{\text{GP}}(f_j) = (T, NT) = \left((\cap, \cup, \backslash), (R_1(f_j), R_2(f_j), R_3(f_j), ...) \right),$$

where $\cap$, $\cup$, $\backslash$ represents set operations intersection, union, difference, and $R_i(f_j)$ can be any one "basic" selection metrics, such as Information Gain, Odds Ratio.

3. Compressive Sensing Model

Recently, the compressive sensing (CS) model has drawn increasing attention in the signal processing research community.[5,6] The CS model aims to reconstruct one signal using less available measurements than the dimension of the original signal. One main advantage of the CS model is that it offers a rigorous mathematical framework to analyse high-dimensional data: few coefficients are capable of representing the majority information from the target signals.

Mathematically, the aim of the CS model is to recover a sparse signal $x \in \mathbb{R}^N$ from a few linear measurements $y \in \mathbb{R}^M$ by solving below programming problem:

$$P: \quad \min S(x) \quad \text{subject to} \quad y = \mathcal{D}x, \tag{7}$$

where $S(x)$ denotes a sparsity measure, and $\mathcal{D} \in \mathbb{R}^{M \times N}$ is known as the *dictionary*. An *atom* of a dictionary is one column vector from the dictionary.

One simple strategy for solving Eq. (7) is to minimize the l_0-norm of x which is the cardinality or number of non-zero elements in x, i.e., $S(x) = \|x\|_0$. Thus Eq. (7) can be rewritten as

$$Q: \quad \min \|x\|_0 \quad \text{subject to} \quad y = \mathcal{D}x. \tag{8}$$

4. Compressive Sensing-based Feature Selection Algorithm

In this section, we describe the proposed compressive sensing feature selection (CFS) algorithm. Herein the feature selection process is formulated as a compressive sensing model, in which discriminative features are selected by finding the sparse representation for the entire feature set.

Suppose that we have a full data set consisting of Q samples (p_i, z_i), $i = 1, \ldots, Q$, in which the input vector p_i represents the i-th input and z_i is the i-th actual output. We again use f_j to stand for the j-th feature (see Eq. (1)). To build up a classification model, we aim to extract a decision rule subject to the following constraint:

$$z_i = f(p_i) + e_i, \tag{9}$$

where $f(\cdot)$ is an unknown decision function to be estimated and e_i is the corresponding error. In general $f(p_i)$ can be expressed as

$$f(p_i) = x^T \Psi(p_i), \tag{10}$$

where x is an $N \times 1$ weight vector and $\Psi(\boldsymbol{p}_i)$ is an N-dimensional feature vector derived from the input $\boldsymbol{p}_i$. For instance, in support vector machines (SVMs), $\Psi(\cdot)$ represents the user-defined kernel function, whereas in feed-forward neural networks, $\Psi(\cdot)$ is generated by the hidden layers of the network.

Let $P = [\boldsymbol{p}_1, \boldsymbol{p}_2, \cdots, \boldsymbol{p}_Q]^\mathrm{T}$ denote the matrix containing all inputs arranged into rows and $z = [z_1, z_2, \cdots, z_Q]^\mathrm{T}$ be the corresponding vector of desired outputs ($z \in \mathbb{R}^Q$) for a given TMC dataset. Combining Eq. (9) and (10) for all training samples, we obtain the constraint equation in the following form:

$$z = \Psi(P)x + e, \qquad (11)$$

where e is the error vector with e_i as the i-th element. The main problem with Eq. (11) is that all N features, i.e., columns of $\Psi(P)$, contribute to the final output even though some of them may be redundant or irrelevant. To select significant features, we can manipulate elements within x as follows: when a feature f_j is selected, its corresponding weight x_j is assigned a non-zero value; when a feature is eliminated, its corresponding weight is assigned to zero. Consequently, the feature selection process is formulated as searching for a sparse representation for the weight vector x, i.e. solving the following model:

$$\min\ S(x) \quad \text{subject to} \quad \|z - \Psi(P)x\|_2 \leq \varepsilon, \qquad (12)$$

where $S(x)$ is the vector sparsity measurement and ε is the bound of the error. Compared to the compressive sensing model in Eq. (7), the matrix $\Psi(P)$ plays the role of a dictionary, and the weight vector x contains the sparse decomposition coefficients. Accordingly, the *compressive sensing feature selection* (CFS) algorithm is proposed by solving the CS model in Eq. (12), in which only the features that contribute most to the minimization of the prediction error are selected.

In addition, we should note that the proposed algorithm can be employed to the input data directly if a simple linear mapping between input samples and feature vectors is adopted, *i.e.*, $\Psi(P) = P$. In this case, a linear input-output relation, $z = Px$, is assumed during feature selection.

5. Commuter Behaviour Modelling Case Study

In this section, a real-world commuter behaviour modelling data set is employed for the evaluation purpose. This data set comes from a household travel survey for the Sydney Greater Metropolitan area,[7] that is collected through face-to-face interviews with approximately 3000-3500 households between July 2006 to July 2011. During this period, a total of 67299 records are collected. Meanwhile, there are five travel modes in this dataset: Car (61.6%), Walk (29.2%), Bus (4.4%),

Train (3.3%), and Others (1.5%). In addition, Table 1 shows details about available features from the data set, including departure time, travel distance, origin and destination, and other socio-demographic features such as household income, and size.

Table 1. Description of the considered TMC features.

TMC features	Details	TMC features	Details
TripID	Index of one trip	SumWorkingHome	Person working from home
Day	Travel day of week	SumResident	Household size
Hf	Household type	SumTrips	Number of total trips
Occupancy	Household occupancy	Purpose	Travel purpose
Veh_parking	Number of parked vehicles	TripDist	Trip distance
Income	Household income	PreMode	Previous travel mode
SumLicence	Number of driving licenses	DepartTime	Trip departure time
SumStudent	Number of students	TripTime	Trip spending time

To conduct the TMC prediction, we randomly partitioned the entire data set into two independent sets: a training and testing set. The size of the training and testing set is 70% and 30%, respectively. Then the proposed algorithm is employed to compare with existing feature selection algorithms, *i.e.*, Laplacian score (LS),[1] mutual information difference (MID), mutual information quotient (MIQ),[2] and Genetic Programming (GP).[3]

Table 2 list the average classification accuracy on the test set over 20 runs, along with the number of selected features when the best prediction result is achieved. As observed, the proposed algorithm outperforms other feature selection algorithms by achieving the best classification accuracy on average. In addition, the proposed approach also selects fewest number of features to predict the travel modes (only 8 out of 16).

Table 2. Number of selected features from different algorithms.

Algorithms	Testing	#. features	Top 8 features
LS	66.22%	12	Purpose, Hf, SumTrips, Income, TripID, Day, SumResident,PreMode
MID	86.88%	11	PreMode, TripDist, Hf, Purpose, occupancy, DepartTime, SumTrips, Veh_parking
MIQ	86.88%	11	PreMode, TripDist, Hf, Purpose, occupancy, Veh_parking, DepartTime, SumTrips
GP	81.99%	13	DepartTime, TripTime, SumResident, TripDist, SumWorkingHome, PreMode, SumTrips, Hf
CFS	87.16%	8	TripDist, PreMode, SumWorkingHome, TripTime, Veh_parking, SumLicence, TripID, SumResident

In conclusion, it is empirically confirmed that the proposed CFS method obtains better improvement compared to existing feature selection methods. Note that since the proposed algorithm selects features that minimize the output error, the feature selection process has a direct correspondence with the classification accuracy. By contrast, the criteria used by other features selection methods are not directly related to the classification purpose.

6. Conclusions

In this paper, we have presented a novel algorithm for feature selection, termed compressive sensing based feature selection (CFS). In the proposed algorithm, original features are used as the dictionary in the compressive sensing model. Then the CFS algorithm calculates the sparse solution to select features that minimize the residual output error. The proposed algorithm was evaluated on a real-life problem for travel mode choice prediction. Statistical results show that the proposed method is superior to existing feature selection methods; it achieves better classification accuracy with fewer number of selected features.

References

1. X. He, D. Cai and P. Niyogi, Laplacian score for feature selection, in *Proceedings of the 18th International Conference on Neural Information Processing Systems*, (Cambridge, MA, USA, 2005).
2. H. Peng, L. Fulmi and C. Ding, *IEEE Transactions on Pattern Analysis and Machine Intelligence* **27**, 1226 (2005).
3. F. Viegas, L. Rocha, M. Gonalves, F. Mouro, G. S. T. Salles, G. Andrade and I. Sandin, *Neurocomputing* **273**, 554 (2018).
4. F. R. K. Chung, *Spectral Graph Theory* (Regional Conference Series in Mathematics. Providence, RI: American Mathematical Society (AMS)., 1997).
5. D. L. Donoho, *IEEE Transactions on Information Theory* **52**, 1289 (2006).
6. J. Yang and J. Ma, *Knowledge-Based Systems* **109**, 61 (October 2016).
7. Australian Bureau of Statistics, Household Travel Survey.

ALES: An Arabic Legal query Expansion System

Imen Bouaziz Mezghanni* and Faiez Gargouri

*University of Sfax, ISIMS, MIRACL Laboratory,
PO. Box 242. 3021. Sakiet Ezzeit, Sfax, Tunisia*
*imen_bouaziz_miracl@yahoo.com

To overcome short queries and word mismatching problems, several query expansion techniques were proposed to help users formulate a better query to obtain the appropriate information satisfying their needs. This paper introduces a query expansion system called Arabic Legal query Expansion System (ALES) for information retrieval in Arabic legal databases. The developed approach exploits information coming from Arabic domain ontology.

Keywords: Legal information retrieval; Arabic query expansion; domain ontology; semantic search.

1. Introduction

In the legal domain, while solving legal problems, lawyers have to deal with an enormous mass of information based primarily on huge amounts of electronically available legal texts stored in legal databases. The latter often pose problems by making the user unable to find the pertinent and useful information. This inability is generally due to two key factors: ***human factor*** and ***language factor***.

It is natural that the accuracy of any information retrieval system depends highly on the users' queries which need to be more definite in order to find relevant documents. This task is still complicated as the user's needs are often vague since the majority of users are non-professionals and cannot express their needs with appropriate words in the query. Even the search itself is usually varied. In fact, it can be a search for a known element, for general information and for specific information or simply an exploration. Furthermore, users generally do not use the same terms in the documents as search terms, which raise a major problem of term mismatch in information retrieval. It is also observed that users typically tend to submit very short queries. Usually, a short query cannot cover many useful search terms due to the lack of sufficient words. In fact, the short query problem can adversely affect the performance of the search performance.

On the other hand, the queries are formulated in natural languages which are complex, ambiguous, redundant, often fuzzy, vague or inexact in meaning and not well formalized. These features make the information retrieval a non-trivial task.

To solve these problems, it is very important to retrieve as much information as possible from these short queries in order to understand both the researcher's intention and the contextual meaning of the terms of his/her query.

Moreover, conventional web search engines, like Google, are widely used nowadays to retrieve relevant information on the Internet. But, the problem is that, the documents and contents can be retrieved only based on keywords. These engines may not provide the most relevant and useful content related to the user query (a great number of irrelevant documents is generated) because the semantics of the query is not considered. These problems can be handled by performing the query expansion (QE) process whose main goal is to add new meaningful terms or phrases to the initial query in order to increase the chances of capturing more relevant documents.[1] Although QE was addressed by several studies in different languages, Arabic QE was given less attention. In this paper, we implement a semantic QE system called ALES.

The remainder of this paper is organized as follows: The next section discusses the related works dealing with Arabic semantic Web search engines. In Section 3, we highlight the real problems of semantic search. Section 4 details our query expansion system for Arabic language. Finally, we end up this paper with a conclusion and some proposals to improve the Arabic semantic search in the future.

2. Related Work

Undoubtedly, the current search engines, such as Google, Gmail, Yahoo, etc., are powerful and popular. However, most of these engines do not know the meaning of the terms and expression used in the web pages and the relationship between them. On the other hand, semantic search integrates the technologies of Semantic Web and search engine to improve the search results gained by current search engines and evolves to next-generation search engines built on Semantic Web. Some of these engines, like Bing,[a] Google,[b] Swoogle[c] and Evi,[d] support Arabic language while others, such as Hakia,[e] SenseBot[f] and Cognition,[g] do not.

---

[a] https://www.bing.com/
[b] https://www.google.com/
[c] https://swoogle.umbc.edu/
[d] https://www.evi.com/
[e] https://www.hakia.com/
[f] https://www.sensebot.net/
[g] https://www.cognition.com/

Nevertheless, a web-based multilingual tool for information retrieval, including the Arabic language, was proposed.[2] The author built domain ontology in the legal Arabic domain. By applying the Arabic ontology, the recall and precision were improved from 115 to 1230 and from 2 to 7, respectively. Moreover, ISWSE is an "Islamic Semantic Web Search Engine" system based on Islamic Ontology.[3] It uses Azhary[4] as a lexical ontology for the Arabic language. The experiments were carried out in the Quran Prophets stories as the most detailed part of the Islamic Ontology. The average precision and recall of the proposed system were 98.5% and 97% respectively for 30 executed queries. Indeed, the system retrieval relies on classifying the concepts in the ontology. Moreover, a tool "Quran Search for a Concept" was developed to search, in the Quran topics, index from an academic source: "Tafsir Ibn-Kathir" and book of "Mushaf Altajweed topics". The system used 1,217 Quranic concepts.[5] We can additionally cite CASONTO,[6] IBRI-CASONTO[7] and CASENG.[8]

3. Semantic Search and Arabic Language

Information retrieval systems may improve their effectiveness by applying QE process which adds new terms to the original query posed by a user. QE consists in enriching the initial user query through adding new words deemed to be somehow (usually semantically) connected to those included in the initial query.

3.1. Semantic QE

Semantic QE is based on expanding the input query relying on the semantic analysis of their contents using natural language processing to retrieve the exact information.[9] Indeed, the semantic search ensures more smart and relevant results according to the ability of understanding terms context and their synonyms rather than the keyword matching. Adding further terms into query can either be manual, automatic or user-assisted. Manual QE relies on user expertise to make decisions about which terms should be included in the new query (interactive system). In the case of automatic query expansion, weights are calculated for all terms and those having the highest weight are added to the initial query. Various weight functions produce different results. Therefore, retrieval performance depends on how the weighs are calculated. With the user-assisted QE, the system generates the possible terms and the user selects the appropriate one. The new terms resulting from the chosen term selection method should provide contextual information for the initial query to improve the retrieval results. The contextual information can be acquired from relevance feedback and term co-occurrence. More recently, it has been derived from knowledge models such as ontologies.

3.2. *Arabic Language Peculiarities*

In fact, the Arabic language is integral to the majority of the population of the Middle East and the rituals of Muslims because it is their mother tongue and the religious language of all Muslims of various ethnicities around the world. It is also a Semitic language containing 28 alphabets. Moreover, Arabic is considered as one of the six official languages of the United Nations and the mother language of more than 400 million people.

Despite the importance of this language, Arabic semantic web and Arabic search engines were not significantly improved. Due to the specificities of Arabic,[10] handling this language is not an easy task, compared to other languages, especially Indo-European ones. Arabic is also an agglutinative language in which the clitics are agglutinated to words. Besides, unlike Indo-European languages, Arabic does not have capital letters, which makes the task of segmenting Arabic texts into sentences harder than segmenting texts written in other languages, such as English, where the capital letters are used as cues for text splitting. Moreover, Arabic texts can be diacritized, partially diacritized, or totally non diacritized, with the majority of current Arabic documents being not diacritized.

Certainly, all the above-cited peculiarities affect considerably the availability of compatible and harmonious semantic Web applications within this language. Regardless of the afore-mentioned difficulties, our objective consists in developing a semantic-based Arabic information retrieval system which semantically expands the input query using domain ontology.

4. The Proposed System: ALES

It is important to point out that the development of the ALES system is part of an application-based evaluation of the learning of a domain ontology called ALO describing the Arabic legal field.[11]

Our purpose is to develop a system that provides the user with a list of terms related to his/her query to improve his/her search. Indeed, we think, on the one hand, that the user is the best one who is able to determine which terms are useful to reformulate his/her query. On the other hand, we believe that automatic query changes, when done without users' knowledge, can be disruptive to them, especially with the "frustrating" returned results.

Indeed, reformulation is a method applied to solve the problems of differences existing between the search engine user's query language and the language of the documents being queried. Thus, the problem consists in matching the terms of the query with the terms of the documents. Matching these two languages can be operated at different levels: at the level of the database documents as a function of the query language; at the query level by knowing the links between the query language and the document language; and finally both at the query and document level by producing a standardized representation covering both languages. In fact, our approach, having a knowledge base about the language of the documents database and thus on the domain language (the ontology), deals with the second case in order to act at the query level to make it resembles to this language. Therefore, we used our ALO ontology as the only source of knowledge for query reformulation (considered, in this paper, as synonym to expansion) as illustrated in Figure 1.

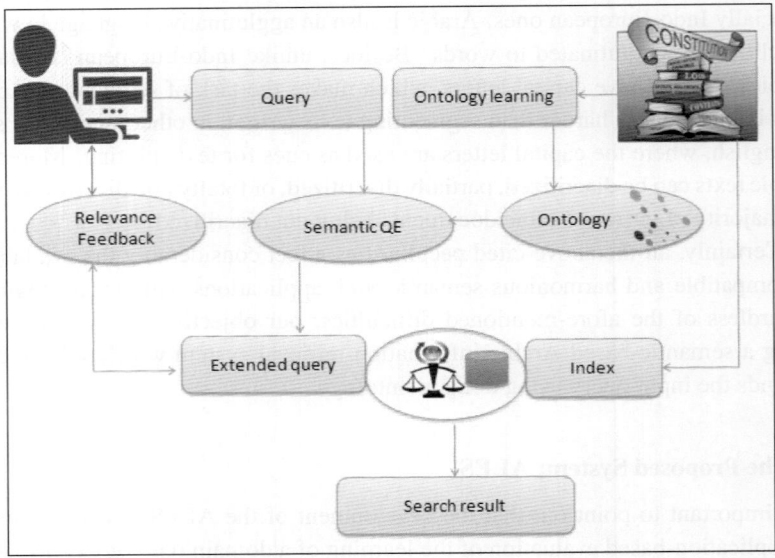

Fig. 1. The general process of ALES.

Actually, the architecture of our system consists of two phases: inside phase and outside phase representing respectively the right part and the left part presented in Figure 1. In the former, the index of the Arabic information retrieval system was created and maintained for the Arabic corpus based on vector space model. Besides, the Arabic domain ontology ALO was integrated in the system. In the outside phase, the user query was expanded using ALO ontology. Then, the search results were retrieved and ranked. The present paper focuses on the outside phase. The different steps of this phase by which our system proceeds are the following:

- The query breakdown: In this step, the query is analyzed word by word. If it is composed of a sequence of words, new terms are produced from these words. These terms are constructed as single words or combinations of them while keeping the order of the words chosen by the user;
- The identification of the known terms: Starting from the list of words formed on the basis of the query entered by the user in the previous step, our system identifies those that are known in the ontology. In the list of words formed on the basis of the query, only known words in the ontology are retained;
- The reference of the known terms and their related terms in the ontology: All terms related to known terms are collected in the ontology. In fact, each term and the list of terms related to it in the ontology are specified to the user. It should be noted that the terms of the list presented to the user are not weighted;
- The query reformulation: From these lists of terms, the user chooses the terms that he/she considers useful to formulate a new query. The query, sent to the database and corresponding to the reformulated query, uses the Boolean operators "AND" and "OR";
- The information retrieval: It is based on VSM and ranking which matches the expanded query vector versus the document vectors to compute the similarity between them.

The proposed semantic-based Arabic information retrieval system relies on a JEE web platform using MySQL as popular Open Source relational database management system (RDBMS) commonly used in web applications thanks to its quick processing, proven reliability, ease and flexibility of use and NetBeans as IDE. The system was experimented on the corpus of the Tunisian criminal case law. The number of the tested documents was 100. Two experiments were conducted: without QE (word-based search) and with QE based on the ALO ontology. The first experience did not modify the entered query, while the second extended the query using the ontology concepts. For example: If we have a query ضابطة (known in the ontology), the system seeks, in the ontology, ضابطة. Then, it references all the terms related to ضابطة and those related to it in the ontology, as shown in Figure 2. The word-based search method retrieved only 09 documents. However, the ontology-based technique retrieved a total of 42 documents that are semantically related to the input query. Preliminary results were very promising and showed that there was a significant improvement in precision and recall.

In addition, the proposed system was tested using ten queries presented in Figure 3. The obtained results were compared to those based on the words of the

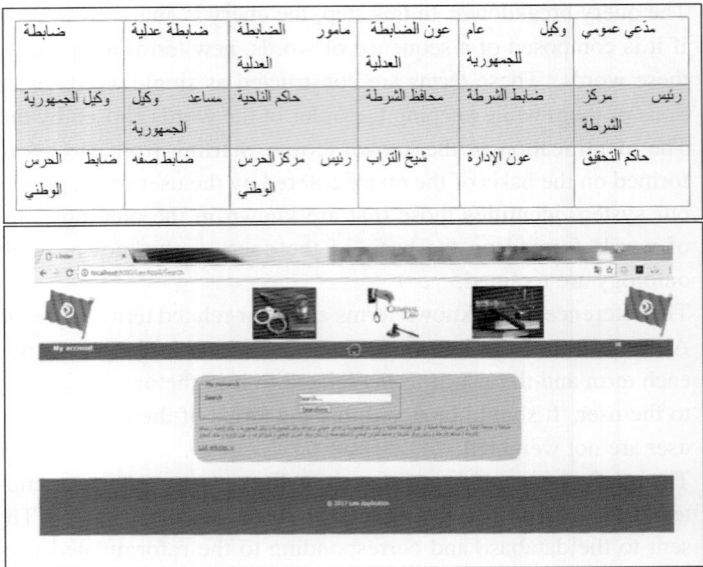

ضابطة	ضابطة عدلية	مأمور الضابطة العدلية	عون الضابطة العدلية	وكيل عام للجمهورية	مدعي عمومي
رئيس مركز الشرطة	وكيل الجمهورية	وكيل الجمهورية مساعد	حاكم الناحية	محافظ الشرطة	ضابط الشرطة
حاكم التحقيق	ضابط الحرس الوطني	ضابط صفه	رئيس مركز الحرس الوطني	شيخ التراب	عون الإدارة

Fig. 2. Expansion of the query ضابطة by all the returned terms through ALES.

initial query. Therefore, the results of the proposed system were evaluated against the results of word-based search method and revealed that the developed system improved the precision as well as the recall as shown in Figure 3.

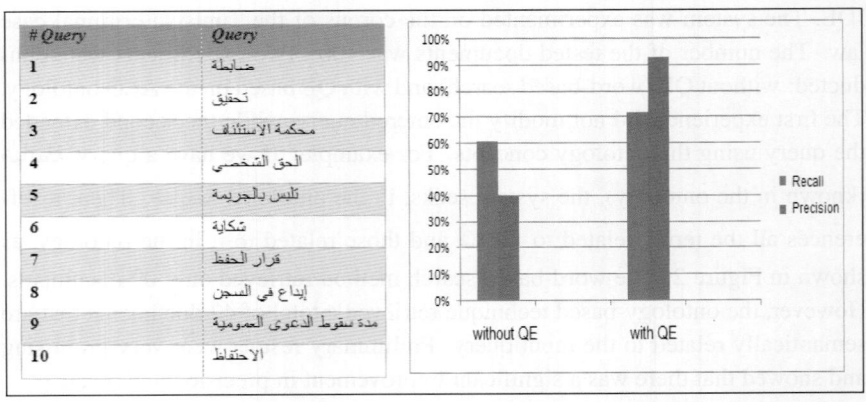

# Query	Query
1	ضابطة
2	تحقيق
3	محكمة الاستئناف
4	الحق الشخصي
5	تلبس بالجريمة
6	شكاية
7	قرار الحفظ
8	إيداع في السجن
9	مدة سقوط الدعوى العمومية
10	الاحتفاظ

Fig. 3. Evaluation results.

The average precision and recall rate obtained by applying our system were 93% and 72% respectively for 10 executed queries, which improved the rates provided by the word-based search method by around 53% and 11% respectively. Precision is the percentage of relevant documents in the search results, while recall is the percentage of retrieved relevant documents over the total number of relevant documents.

5. Conclusion

In this paper, we presented a semantic-based Arabic information retrieval system which semantically expanded the input query using a domain ontology, which allowed providing pertinent results. The findings obtained by the proposed approach outperformed those provided by using word-based method in terms of precision and recall. In our future work, we intend to more develop our system and apply it in other domains using corresponding domain ontology.

References

1. X. Jinxi and C. W. Bruce, Query expansion using local and global document analysis, in *Proceedings of the 19th Annual International ACM SIGIR Conference on Research and Development in Information Retrieval (SIGIR '96)*, 1996.
2. S. Zaidi, A cross-language information retrieval based on an arabic ontology in the legal domain, in *Proceedings of the 1st International Conference on Signal-Image Technology and Internet-Based System (SITIS '05)*, 2005.
3. H. Ishkewy and H. Harb, *International Journal of Computer Applications* **112**, 37 (2015).
4. H. Ishkewy, H. Harb and H. Farahat, *International Journal of Web & Semantic Technology* **5** (2014).
5. N. Abbas, Quran 'search for a concept' tool and website, PhD thesis, School of Computing, University of Leeds, (Leeds, United Kingdom, 2009).
6. A. Sayed and A. Almaqrashi, *International Journal of Web Information Systems* **12**, 242 (2016).
7. A. Sayed and A. A. Muqrishi, *Egyptian Informatics Journal* **18**, 181 (2017).
8. A. A. Muqrishi, A. Sayed and M. A. Muqrishi, *Journal of Theoretical and Applied Information Technology* **75**, 148 (2015).
9. E. M. Voorhees, Query expansion using lexical-semantic relations, in *Proceedings of the 17th Annual International ACM SIGIR Conference on Research and Development in Information Retrieval (SIGIR '94)*, 1994.
10. I. B. Mezghanni and F. Gargouri, Information retrieval from unstructured arabic legal data, in *Proceedings of the 14th Pacific Rim International Conference on Artificial Intelligence (PRICAI '16): Trends in Artificial Intelligence*, 2016.
11. I. B. Mezghanni and F. Gargouri, Towards an arabic legal ontology based on documents properties extraction, in *12th IEEE/ACS International Conference of Computer Systems and Applications (AICCSA '15)*, 2015.

Extracting a diverse information subset by considering information coverage and redundancy simultaneously

Baojun Ma

School of Economics and Management, Beijing University of Posts and Telecommunications, Beijing 100876, P. R. China
mabaojun@bupt.edu.cn.

Qiang Wei, Guoqing Chen and Qiongwei Ye

China Retail Research Center, School of Economics and Management, Tsinghua University, Beijing 100084, P. R. China

> Information overload has been a big challenge for web users to find the information they want or are interested in. To extract or provide a small set of diverse result subset is valuable and important to both information providers and users. This paper proposes a heuristic algorithm named *CovRedSA-Select* by considering information coverage and redundancy simultaneously based on the strategy of simulated annealing. Furthermore, the comparative experiments reveal performances advantageous over other related methods.

1. Introduction

Information overload has been a big challenge for web users to find the information they want or are interested in, especially in the big data era. Though the tools of search engines or recommendation systems have been developed to help web users to find their potential interested information, the size of the result set could still be very huge. Therefore, to extract or provide a small set of result subset with high quality will be valuable and important to both information search service providers and users[1].

Under the circumstance of information overload, the information quality of extracted small subsets with respect to the original dataset could be understood from two perspectives of information diversity[2]: on one hand, subsets with higher quality should cover as many useful information in original set as possible (i.e., information coverage); on the other hand, subsets with higher quality should possess as little redundant information as possible (i.e., information redundancy). Although in recent years various research efforts have focus on diverse subsets extraction from perspectives of search results diversification[3-8] and clustering-based strategies[9], there is little work conducting good and satisfactory tradeoff between the two objectives of high information coverage and low redundancy. Moreover, even if in our previous work, we have proposed a novel diversity-based extraction algorithm, the perspective of "information redundancy" had not been considered in the extracting process, which motivates the effort of this study.

2. Related Work

This section will review the related work on diversity-oriented extraction methods which will be used as benchmark methods in our comparative experiments section.

For diversity-oriented extraction, mainly there are two streams, i.e., search result diversification methods (SRD) and clustering-based methods. Search result diversification methods can be categorized to be either implicit or explicit, depending on how they account for the different aspects underlying a query[8].

Implicit search result diversification methods assume that similar documents cover similar aspects of a query and should be denoted in the final results. Among the implicit SRD methods, the maximal marginal relevance (MMR) strategy proposed by [3] is the most typical one. The general idea of MMR is to trade off document similarity with respect to the query and document dissimilarity with respect to the already selected documents. Subsequent implementations of this idea include the method of [4] to use the portfolio theory in finance to diversify documents ranking, in which two documents were compared based on the correlation of their relevance scores. In addition, [5] proposed MCDC algorithm by solving a graph partition problem on a weighted dissimilarity graph.

In contrast, explicit search result diversification methods explicitly model the aspects underlying a query. For instance, [6] employed a classification taxonomy over queries and documents to represent query aspects (called IA-Select). The method iteratively promotes documents that share a high number of classes with a query, while demoting those with classes already well represented in the ranking. Similarly, [7] proposed a probabilistic method (i.e., FM-LDA) to maximize the coverage of the retrieved documents with respect to the aspects of a query by modeling these aspects as topics identified from the top ranked documents using LDA[10]. In addition, [11] introduced the xQuAD probabilistic framework for search result diversification, which explicitly represents different query aspects as "sub-queries". They defined a diversification objective based on the estimated relevance of documents to multiple sub-queries, as well as on the relative importance of each sub-query considering the initial query.

Moreover, clustering-based methods possess the relatively consistent objectives with diversity. The popular and common-used clustering methods, such as K-means[12] and agglomerative hierarchical clustering (i.e., AHC)[13], with cluster centroid extraction, could provide diverse subsets to some extent. Moreover, differently from the above clustering methods, [9] proposed a framework to extract diverse subsets (called RR) based on query-specific clustering as well as cluster ranking.

It is worth noting that, though many efforts have been made in diversity extraction from different angles, existing methods take little aspect of structure coverage into consideration, while structure coverage is regarded meaningful and important for web users.

Furthermore, in our previous work, we proposed a fast approximation heuristic extraction method called *FastCov$_{C+S}$-Select* to obtain a diverse result set when considering information coverage metric from a combined perspective of content and structure[1]. Although this method could extract relative "diverse" results in an effective, efficient and robust manner, which has been demonstrated by evaluation experiments, the perspective of "information redundancy" had not been considered in the extracting process directly.

3. Algorithm Design Considering Coverage and Redundancy

In this section, a heuristic extraction algorithm is introduced based on the idea of optimizing the information coverage and information redundancy of the extracted subsets with regards to the original data set simultaneously.

Given an original set D of n documents and an extracted small set D' with size $= k$, in our previous work, two metrics called $Cov(D', D)$ and $Red(D')$ have been proposed to evaluate the information coverage of the extracted subset D' with respect to the original set D as well as the information redundancy of the extracted subset D' itself [2]. The detailed definitions of the two metrics are as follows:

$$Cov(D',D) = \begin{cases} Cov_C(D',D) = \frac{1}{n}\sum_{d \in D} sim(d_1',d) & \text{if } k=1 \\ Cov_C(D',D) \times Cov_S(D',D) = \\ \frac{1}{n}\sum_{d \in D} \max_{d' \in D'}\{sim(d',d)\} \times \left\{-\frac{1}{\log_2 k}\sum_{j=1}^{k} \frac{n_j^v}{n^v} \cdot \log_2\left(\frac{n_j^v}{n^v}\right)\right\} & \text{if } k>1 \end{cases} \quad (1)$$

$$Red(D') = \frac{1}{k} \times \sum_{d_i \in D'}\left(1 - \frac{1}{\sum_{d \in D'} sim(d_1,d)}\right) \quad (2)$$

In Eq. (1) and Eq. (2), $sim(*)$ is the similarity function between two documents. For more information about the two metrics, please refer to the reference[2].

As we discussed in the Introduction, optimizing the information coverage and information redundancy of the extracted subsets with regards to the original data set simultaneously is a multi-objective optimization problem (i.e., diversity problem). To deal this problem, we combine the information coverage and information redundancy as one optimization objective by utilizing the users' preference on the two perspectives. Specifically, the diversity maximization problem could be formulated as follows:

Definition 1 (*maxDiv(k, λ)*). *Given an original set* $D = \{d_1, d_2, \ldots, d_n\}$, *the similarity between any two documents in D (i.e., $sim(d_i, d_j)$), an integer k (i.e., $1 < k < n$) and a positive number λ representing the users' preference, the diversity maximization problem (i.e., maxDiv(k,λ)) is to find a subset of documents $D'_* \subseteq D$ with $|D'_*| = k$ such that*

$$Div(D'_*, \lambda) = Cov(D'_*, D) - \lambda \cdot Red(D'_*)$$

$$= \max_{D' \subseteq D, |D'|=k} \left\{ \frac{\sum_{d \in D} \max_{d' \in D'} \{sim(d', d)\}}{n} \times \left\{ -\frac{1}{\log_2 k} \sum_{j=1}^{k} \frac{n_j^v}{n^v} \cdot \log_2(\frac{n_j^v}{n^v}) \right\} - \lambda \left\{ \frac{1}{k} \times \sum_{d_i \in D'} (1 - \frac{1}{\sum_{d \in D'} sim(d_i, d)}) \right\} \right\} \quad (3)$$

Notably, but not too surprisingly, by mapping it into a classical NP-hard problem named Max Coverage[14], it could be observed that the desired objective of *maxDiv(k,λ)* is also a NP-hard problem. However, it could be easily proved that the combined optimization objective function *Div(D',λ)* is not a submodular function, thus it can hardly apply the simple greedy strategy to guarantee the *Div(D',λ)* value (i.e., *Div*) of the resultant subset to be or be very close to global optimum. Instead, the idea of simulated annealing is adopted, which is complex but effective.

Simulated annealing (SA) is a compact and robust technique, which provides excellent solutions to single and multiple objective optimization problems with a substantial reduction in computation time, which is a kind of stochastic search algorithm based on Monte-Carlo iterations and has been proven to possess the characteristic of asymptotic optimality[15].

In our study, by applying the stochastic search strategy used in simulated annealing, we propose a heuristic algorithm named *CovRedSA-Select* to solve the *maxDiv(k,λ)* problem, whose pseudo-code is illustrated in Figure 1. In *CovRedSA-Select*, differently from traditional simulated annealing method, a memory state variable (i.e., D_{max}) is introduced to avoid the problem of missing the best solution at the time of certain iteration due to the implementation step of acceptance probabilities. Given the original set D of size n, the extraction size k, users' preference coefficient λ, the initial solution (state) D_0 from the output of *Cov$_C$-Select* algorithm (i.e., the basic extraction algorithm in [1] to obtain a subset with relative high coverage based on the idea of submodularity and greedy), the similarity value between any two documents $sim(d_i, d_j)$ as well as the initial cooling temperature T_0 and final temperature T_{min} [15], *CovRedSA-Select* tries to extract a set containing k documents with maximum value of *Div*. In the initialization stage, the memorial variable D_{max} is introduced to record the best solution with the highest value of *Div* at the time after each iteration (line 1). The iteration does not terminate until the current temperature T drops to the minimum temperature T_{min}. In each iteration, the extracted document with the highest redundancy value would be selected as the candidate to be replaced (line 7-10) and the document in the remaining set (i.e., D_p) making the replaced extracted subset (i.e., D'_d) possessing highest value of *Div* would be selected as the candidate to replace (line 11-16). If the new document replacement could improve the value of *Div*, we accept this replacement; otherwise, we would utilize the idea of stochastic search strategy in simulated annealing and the acceptance probability to decide whether to accept the current replacement (line 17-27). It can be seen that *CovRedSA-Select* is an algorithm of typical simulated annealing nature and the probability that it terminates with a global optimal solution approaches 100%[15]. Furthermore, by known that the computational complexity of Cov_C-*Select* is $O(k^2n^2)$[1], it is easy to infer that the total computation complexity for *CovRedSA-Select* is $O(k^2n^2) + O(T_0kn^2)$, in which T_0kn^2 is the main influence factor since $k \ll n$ and $k \ll T_0$ generally.

4. Data Experiments

Duo to space limit and most of the effectiveness, efficiency and parameter robustness experiments are very similar between this study and our previous work in[1], we only report the following two distinct experiments.

As we know, the parameter λ in $maxDiv(k,\lambda)$ problem represents the users' preference on information coverage and information redundancy and is actually a penalty factor for information redundancy in the optimization objective. The first experiment aims to exam the influence of λ values on the information coverage and information redundancy of the final extracted subsets. The experimental results are shown in Fig. 2.

In Fig. 2(a), the value of λ changes from 0 to 10 and it is shown that when the value of λ changes from 0 to 1, both information coverage and redundancy of the extracted subset decreases rapidly. While with λ increases continuously, information coverage of the extracted subset remains at a lower level of stability with a higher level of information redundancy, which is not the results we want and implying that higher value of λ is detrimental to the trade-off between coverage and redundancy. Fig. 2(b) shows the detailed results when $0 < \lambda < 1$ and we can infer that even if we only use a relative lower value of λ (e.g., 0.5), information redundancy of the extracted subset could decrease to about 10% value of that obtained by *FastCov$_{C+S}$-Select* [1] by only sacrificing around 10% value of information coverage.

CovRedSA-Select Algorithm
Input: D, D_0, k, λ, $sim(d_i, d_j)$, T_0, T_{min}
Output: set of k documents D'

1. $D' = D_0$, $D_{max} = D_0$, $T = T_0$, $N = 1$;
2. while $T > T_{min}$ do
3. $\quad Div_0(\lambda) \leftarrow Cov(D', D) - \lambda Red(D')$;
4. $\quad$ for $d' \in D'$ do
5. $\quad\quad Red_{d'} \leftarrow Red(d', D')$
6. $\quad$ end for
7. $\quad Red_{max_R} \leftarrow max\{Red_{d'}\}$, $max_R = argmax_{d'}\{Red_{d'}\}$;
8. $\quad$ if $Red_{max_R} = avg\{Red_{d'}\}$ do
9. $\quad\quad max_R = Random(1, k)$;
10. $\quad$ end if
11. $\quad D_p = D \setminus D'$;
12. $\quad$ for $d \in D_p$ do
13. $\quad\quad D'_d = D' \setminus \{d'_{max_R}\} + \{d\}$;
14. $\quad\quad Div_d(\lambda) \leftarrow Cov(D'_d, D) - \lambda Red(D'_d)$;
15. $\quad$ end for
16. $\quad Div_{max_D}(\lambda) \leftarrow max\{Div_d(\lambda)\}$, $max_D = argmax_d\{Div_d(\lambda)\}$;
17. $\quad \Delta Div(\lambda) = Div_{max_D}(\lambda) - Div_0(\lambda)$;
18. $\quad$ if $\Delta Div(\lambda) \geq 0$ do
19. $\quad\quad D' \leftarrow D' \setminus \{d'_{max_R}\} + \{d_{max_D}\}$;
20. $\quad$ else do
21. $\quad\quad acceptProb = e^{\Delta Div(\lambda)/T}$;
22. $\quad\quad RandomProb = Random(0,1)$;
23. $\quad\quad$ if $acceptProb \geq RandomProb$ do
24. $\quad\quad\quad D' \leftarrow D' \setminus \{d'_{max_R}\} + \{d_{max_D}\}$;
25. $\quad\quad$ end if
26. $\quad$ end else
27. $\quad$ end if
28. $\quad T \leftarrow T/log(1+N)$;
29. $\quad N \leftarrow N+1$;
30. $\quad$ if $Div_{D'}(\lambda) > Div_{D_{max}}(\lambda)$ do
31. $\quad\quad D_{max} \leftarrow D'$;
32. $\quad$ end if
33. end while
34. $D' \leftarrow D_{max}$;
35. return D';

Figure 1. Pseudo-code of *CovRedSA-Select* algorithm.

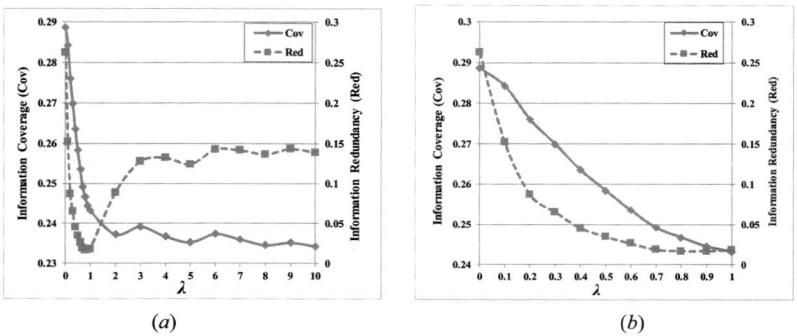

Figure 2. Experimental results on the value of λ of *CovRedSA-Select* algorithm.

To comprehensively evaluate the performance of the proposed algorithm, we used Google search snippet results of 111 queries raised in KDD Cup 2005 as the dataset, which is widely used in related studies[1]. Specifically, the 111 queries were chosen as the search keywords in Google. Then all the snippets of each query were crawled from Google using Apache Lucene, HTML parser and HTTP client packages and APIs, in which the number of snippets for each query is roughly 1,000. In total, the evaluation experiments were conducted on the dataset of 111,000 snippets. In the experiments, the $Cov(D', D)$ in Eq. (1) and $Red(D')$ in Eq. (2) were used to evaluate the information coverage and redundancy of extracted subsets respectively. Hence the average value of Cov and Red of all the 111 Google queries, denoted as $avgCov$ and $avgRed$, were used as the evaluation metrics in the second experiment. We compared our newly proposed *CovRedSA-Select* algorithm ($\lambda = 0.5$) with other 10 diversity-oriented extraction methods introduced in Related Work Section (see Table 1), in which $FastCov_{C+S}$-*Select* is our previous proposed algorithm only considering information coverage[1].

Table 1 shows that from the perspective of $avgCov$, our proposed extractioon algorithms, i.e., *CovRedSA-Select* and $FastCov_{C+S}$-*Select* performed significantly better than all the other 9 methods in different extraction sizes. Meanwhile, *CovRedSA-Select* algorithm could extract subsets with rather lower information redundancy while only sacrificing little information coverage compared to $FastCov_{C+S}$-*Select*, implying that *CovRedSA-Select* performs satisfactorily on balancing information coverage and information redundancy.

Table 1. $avgCov$ and $avgRed$ gap on Google search results dataset (111 queries).

Category	ID	Method	k=10		k=20		k=30	
			avgCov	avgRed	avgCov	avgRed	avgCov	avgRed
Proposed	1	CovRedSA-Select	0.2584	0.0349	0.3330	0.1519	0.3892	0.2866
	2	$FastCov_{C+S}$-Select	0.2881	0.3272	0.3662	0.4320	0.4165	0.5098
Implicit SRD	3	MMR	0.1607	0.0778	0.1919	0.1130	0.2426	0.1667
	4	Portfolio	0.1792	0.2614	0.2235	0.5259	0.2693	0.6051
	5	MCDC	0.1773	0.3535	0.2003	0.4598	0.2208	0.5333
Explicit SRD	6	IA-Select	0.1775	0.4440	0.2283	0.5860	0.2625	0.6352
	7	FM-LDA	0.1629	0.6430	0.2093	0.6738	0.2533	0.7077
	8	xQuAD	0.1835	0.5346	0.2482	0.5735	0.2862	0.6535
Clustering-based	9	Kmeans-based	0.2045	0.1993	0.2770	0.3259	0.3241	0.4326
	10	AHC-based	0.1895	0.1023	0.2512	0.1834	0.2998	0.2698
	11	RR	0.1799	0.3397	0.2345	0.5616	0.2796	0.6966

5. Conclusions and Future Work

This paper proposes a heuristic algorithm named *CovRedSA-Select* by considering information coverage and redundancy simultaneously based on the strategy of simulated annealing. We also show the performances advantageous over other related methods by real data experiments. Future work will focus on utilizing more datasets, more benchmark methods and human evaluation experiments to further exam the performance of our proposed algorithm.

Acknowledgments

This work is partly supported by National Natural Science Foundation of China (71402007/71772017/71372044) and the Beijing Municipal Social Science Foundation (17GLB009). Thanks are also due to the support from Yunnan Science and Technology Fund (2017FA034), Yunnan Province Young Academic and Technical Leader Candidate Program (2018HB), Yunnan Provincial E-Business Entrepreneur Innovation Interactive Space (2017DS012) & Kunming Key Laboratory of E-Business and Alternative Finance (KGF[2018]18).

References

1. B. Ma, Q. Wei, G. Chen, J. Zhang and X Guo, Content and Structure Coverage: Extracting a Diverse Information Subset, *Informs J. Comput.* **29**, 660-675 (2017).
2. B. Ma and Q. Wei, Measuring the coverage and redundancy of information search services on e-commerce platforms, *Electron. Commer. Res. Appl.* **11**, 560-569 (2012).
3. J. Carbonell and J. Goldstein, The use of MMR, diversity-based reranking for reordering documents and producing summaries, *Proceedings of the 21st annual international ACM SIGIR conference on Research and development in information retrieval*. Melbourne, Australia: ACM, 335-336 (1998).
4. J. Wang and J. Zhu, Portfolio theory of information retrieval, *Proceedings of the 32nd international ACM SIGIR conference on Research and development in information retrieval*. Boston, MA, USA: ACM, 115-122 (2009).
5. S. Krishnan and K. Goldberg, The Minimum Conductance Dissimilarity Cut (MCDC) Algorithm to Increase Novelty and Diversity of Recommendations. (2015). http://goldberg.berkeley.edu/pubs/krishnan-recsys-final2.pdf.
6. R. Agrawal, S. Gollapudi, A. Halverson and S. Ieong, Diversifying search results, *Proceedings of the Second ACM International Conference on Web Search and Data Mining*. Barcelona, Spain: ACM, 5-14 (2009).
7. B. Carterette and P. Chandar, Probabilistic models of ranking novel documents for faceted topic retrieval, *Proceeding of the 18th ACM conference on Information and knowledge management*. Hong Kong, China: ACM, 1287-1296 (2009).
8. R. Santos, C. Macdonald and I. Ounis, Selectively diversifying web search results, *Proceedings of the 19th ACM international conference on Information and knowledge management*. Toronto, ON, Canada: ACM, 1179-1188 (2010).

9. J. He, E. Meij and Md. Rijke, Result diversification based on query-specific cluster ranking. *J. Am. Soc. Inf. Sci. Tec.* **62**, 550-571 (2011).
10. D. Blei, A. Ng and M. Jordan, Latent dirichlet allocation, *J. Mach. Learn. Res.* **3**, 993-1022 (2003).
11. R. Santos, C. Macdonald and I. Ounis, Intent-aware search result diversification, *Proceedings of the 34th international ACM SIGIR conference on Research and development in Information Retrieval.* Beijing, China: ACM, 595-604 (2011).
12. J. MacQueen, Some methods for classification and analysis of multivariate observations, *Proceedings of the fifth berkeley symposium on mathematical statistics and probability.* 281-297 (1967).
13. H. Malik, J. Kender, D. Fradkin and F. Moerchen, Hierarchical document clustering using local patterns, *Data. Min. Knowl. Disc.* **21**, 153-85 (2010).
14. D. Hochba, Approximation Algorithms for NP-Hard Problems, *ACM SIGACT News.* **28**, 40-52 (1997).
15. T. Cormen, C. Leiserson, R. Rivest and C. Stein, *Introduction to Algorithms. 3nd ed.* Cambridge, Massachusetts London, England: The MIT Press (2009).

Term frequency occurrences on web pages for textual information retrieval

Karthika Sivapathasundaram,[*] Xiaochun Cheng[†] and Miltos Petridis[‡]

Department of Computer Science, Middlesex University, London, UK
[*]ks1135@live.mdx.ac.uk, [†]x.cheng@mdx.ac.uk, [‡]m.petridis@mdx.ac.uk

The number of pages on web is gradually increasing. It needs to classify web pages so that user can easily locate the relevant information. There exists method used to classify web page with visual information which includes headings, links, author, and navigation bar. When web page is loaded based on the user request, some junk web pages could display which consume server load and the execution time. Segmentation method is used to overcome those problems. Term frequency inverse document frequency (TF-IDF) is used for calculating term weight based on number of occurrences in the document. Term weight mostly used in information retrieval, text mining and to filtering words from different fields such as text summarization and classification. The web page is analysed using the weighting term to improve the results in the proposed system.

Keywords: Information retrieval; text mining; segmentation; term frequency; inverse document frequency.

1. Introduction

A collection of textual and visual information are present on web. A variety of search engines attempt to develop image search and web content search to improve the performance and user satisfaction [1]. For example, Google has more than 11.8 billion indexed images (Google, 2004). Normally, the textual information, images, and videos are retrieved based on the user keywords and some irrelevant links and information are also retrieved from web page [4]. It is important to analyze how the queries, images, and videos are retrieved from the web page and understand the difficulties of web search engine in dealing with a variety of textual information and images. The visual features are efficiently measured and has two major advantages such as fast query response- Retrieval process can complete quickly, and low storage consumption. The storage of high dimensional features can easily reduce [15]. The purpose of this study was to analyze the term frequency of web pages and further investigate the information retrieval based on semantic analysis.

When a user searches any information, the result must be meaningful and relevant to their queries. In the web page, mostly keyword-based and content-based image retrieval are present but many user handling keyboard-based search techniques. The content-based concept has some visual attributes such as colour, texture, and shape. Many search engines are developed for text, image and video search. To improve the performance of searching, system handling many techniques such as use of weighting mechanism and relevance feedback from user. The query analysis is used to retrieve information on the web page and most of the user's input short queries and view few search results [3]. The crawler is a process to download all the web pages. Through crawler, an index will be created, and all the data and user Ids can store. When a user searches some query, the response to the user query is going to the index and based on query it will be ranked. So this process is called query processing. Thousands of results are present in the search engine based on the user query. But the user always focuses on few results which are placed on the top. Web search engine able to provide millions of results based on a collection of other web pages.

2. Related Work

Data is classified in a structured format for representing web page content. Web page has mainly text information and visual information. An HTML code used for getting the visual information, and it includes information about visual blocks [7]. In this visual block, text information also presents. It is used to represent the document by using the bag-of-words. The document is represented by Term Frequency (TF) can easily identify individual users for rating. The representation of document is applied only for textual information and not for visual information. The text plays a major role in web page classification. By using HTML tags can develop term weighting. HTML tags permit some properties such as font size, font colour and by using these properties can alter the weights of text terms. It would not affect the visual properties [10].

The web page is classified based on link context. The author [12], has checked whether the link is related to the topics or not. Web crawler consists of some web pages. It is implemented to fetches the web page to process. Neural network is used to represent the topics as well as for filtering topics. It is mainly used on web page for getting information. It can be used in many ways such as web page classification, search engine and topical crawler [13]. Topical crawler is used to specify which links are closely related to the topics. It follows the links and then retrieves web page based on the related topic. The link context is analyzing whole web page as the context. The link which is present in the web page should be more close to the user query.

3. Methodology

3.1. *Page segmentation method*

Segmentation methods are used only for visual blocks, not for textual elements. It is used to separate each visual features which includes main content, advertisement, navigation bar, and other content. The advertisement is mostly present on the web page; it is related to the topic. Web pages give priority to the important page, and then it will display top of the result. The content present on the web is difficult to understand sometimes false, and error information occurs to affect the accuracy of results. The author [8], introduces Link Information Categorization to solve this type of problem. The links are present on the web page, it goes to other websites, and can retrieve more information relevant to the topics. In this case, some of the irrelevant links are present depends on the user query.

3.2. *Classification*

It is used to classify dataset into a small part of categorizing. Every classification has easily analyzed based on the user query. The two outcomes for the results of classification are classified data- it provide the information about the text and Performance- the accuracy of data model, error rate and amount of time consumed and the memory space is computed during the classification [6]. Text mining is used to process millions of information and resources are provided on the web. The user can easily access the information because most of the content present on the web is text. User is access all the queries and information on the web is highly improved. It has many techniques such as data collection, data selection, and data cleaning.

An information retrieval used to store, organizes the data for retrieving the relevant information based on the user query. It has three main components such as collection of document, user information and retrieving relevant content. It stores all the documents and information by extracting a set of keywords in the document. User query may transform into its information content by extracting keywords that related to exact document features [11].

3.3. *Calculation of term weight*

The words are represented with Term frequency (TF). The number of times the word can appear in the document is called term frequency. In the equation below, t represent the term and d represent the document [3] [11].

$$TF(t,d) = \frac{0.5 + 0.5 * f(t,d)}{\text{MaxFreq}(d)} \quad (1)$$

where MaxFreq(d) is the maximum Frequency of document. Inverse Document Frequency (IDF) calculation and output of TF/IDF are given as:

$$TF/IDF(t,d) = TF(t,d) * \left(1 + \log\left(\frac{n}{k}\right)\right) \quad (2)$$

where n – count of all the documents, k – number of times the document containing the term t.

3.4. Results

Term frequency is used to count the words in a document. Single word counter have analysed for both the web pages such as www.bbc.co.uk and www.independent.co.uk to calculate TF-IDF.

Table 1. Analysis of single word counter on www.bbc.co.uk.

www.bbc.co.uk (Static Dataset)			
Term	Documents	Term count	TF*IDF
Brexit	100	663	0.99
Trump	78	32	0.07
Parliament	56	50	0.13
European	89	298	0.49
Union	100	193	0.29

The above table shows an analysis of single word counter on www.bbc.co.uk for counting the term frequency. TF is calculated by number of times that particular term is appears in the document divided by total number of terms in the document. Here the word "Brexit" term count is 663 and the total number of terms is 1236. So TF = (663/1236 = 0.54) and inverse document frequency is calculated by log of total number of documents divided by number of documents has the particular term. So IDF = log e(423/100 = 1.84). Term Frequency-Inverse document frequency is calculated by multiply both TF*IDF = (0.54*1.84 = 0.99).

Table 2. Analysis of single word counter on www.independent.co.uk.

www.independent.co.uk (Static Dataset)			
Term	Document	Term count	TF*IDF
Trump	78	32	0.23
2018	89	98	0.64
Passport	44	17	0.22
Control	78	32	0.23
Government	89	100	0.66

The above table shows an analysis of single word counter on www.independent.co.uk for counting the term frequency. TF is calculated by number of times that particular term is appears in the document divided by total number of terms in the document. Here is the word "Trump" term count is 32 and the total number of terms is 279. So TF = (32/279 = 0.11) and inverse document frequency is calculated by log of total number of documents divided by number of documents has the particular term. So IDF = log e(378/78 = 2.10). Term Frequency-Inverse document frequency is calculated by multiply both TF*IDF = (0.11*2.10 = 0.23).

4. Conclusion

Link context is used to classify the web page and information retrieval. The topical crawler used to check whether the links are related to the topic or not. Some of the links are irrelevant to the actual content. The filtering method is used to get higher precision and high speed of the process. In this research, we use segmentation method to analyze the visual blocks and get the results accurately. Term weight is used to filter words from different fields such as text summarization and classification. A single word counter have analysed for both web pages such as www.bbc.co.uk and www.independent.co.uk to count number of times the particular word occur in the document. From the term weight, the web page will provide the relevant information based on the user query.

References

1. Ahmadi, A., Fotouhi, M., Khaleghi, M. (2011), "Intelligent Classification of Web Pages Using Contextual and Visual Features", *Applied soft computing*, vol. 11, Issue. 2, pp. 1638-1647.
2. Aliakbary, S., Abolhassani, H., Rahmani, H. and Nobakht, B. (2009). "Web Page Classification Using Social Tags", *International conference on computational science and Engineering,* pp. 588-593.
3. Bartik, V. (2010), "Text-Based Web Page Classification with Use of Visual Information", *International Conference on Advances in Social Networks Analysis and Mining*, pp. 416-420.
4. Burget, R., and Rudolfova, I. (2009). "Web Page Element Classification Based on Visual Features", *First Asian conference on intelligent information and database systems*, pp. 67-72.
5. Columbus, C., Jayapriya, K. and Santhanakumar, M. (2018), "Multi term based co-term frequency method for term weighting in information retrieval", Int. J. of business information system, Vol. 28, No. 1.
6. Dasondi, V., Pathak, M. and Rathore, N. (2016). "An Implementation of Graph Based Text Classification Technique for Social Media," *Symposium on Colossal Data Analysis and Networking* (CDAN), pp. 1-7.

7. Golub, K. (2006), "Automated Subject Classification of Textual Web Documents", *Journal of Documentation,* Vol. 62, No. 3, pp. 350-371.
8. Gong, J., and Song, J. (2016). "Research on the Performance of Segmentation of Text Classification Based on Chinese News Information Classification and Code (CNICC)", *IEEE ICIS 2016,* pp. 1-3.
9. Hammami, M., Chen, L. and Chahir, Y. (2005), "Using Visual Content-Based Analysis with Textual and Structural Analysis for Improving Web Filtering", *International journal of web information systems*, Vol. 1, issue: 4 pp. 241-254.
10. Khanchana, R., and Punithavalli, M. (2011). "An Efficient Web Page Prediction Based on Access Time-Length and Frequency," *IEEE Electronics Computer Technology*, pp. 273-277.
11. Ibrahim, O. and Landa-Silva, Dario (2016) *Term frequency with average term occurrences for textual information retrieval.* Soft Computing, 20 (8). pp. 3045-3061. ISSN 1433-7479.
12. Pandey, S., Khanna, P., Yokota, H. (2016), "A Semantics and Image Retrieval System for Hierarchical Image Databases," pp. 571-591, *Information process, and management.*
13. Qi, X., and Davison, B. (2009). "Web Page Classification: Features and Algorithms", *ACM computing surveys*, Vol. 41, No. 2, Article 12, pp. 12-31.
14. Rajkumar, K., and Kalaivani, V. (2012). "Dynamic Web Page Segmentation Based on Detecting Reappearance and Layout of Tag Patterns for Small Screen Devices," pp. 508-513.
15. Zhu, L., Shen, J., Xie, L. and Cheng, Z. (2017). "Unsupervised Visual Hashing with a Semantic Assistant for Content-Based Image Retrieval", *IEEE Transaction Knowledge and Data Engineering,* Vol. 29, No. 2, pp. 472-486.
16. Xiang, C., Zhao, X., Xu, G., and Yang, G. (2012). "Tibetan Web page classification Based on Column Navigator," *International Conference on Intelligent Systems Design and Engineering Application*, pp. 610-612.
17. Xu, Z., Yan, F., Qin, J. and Zhu, H. (2011). "Web Page Classification Algorithm Based on the Link Information", *10th International symposium on distributed computing and applications to business, engineering and science*, pp. 82-86.
18. Yildirim, Y., and Yazici, A. (2013). "Automatic Semantic Content Extraction in Videos Using a Fuzzy Ontology and Rule-Based Model", *IEEE transactions on knowledge and data engineering*, Vol. 25, Issue: 1, pp. 47-61.
19. Zhang, X., and Wu, B. (2015). "Short Text Classification Based on Feature Extension Using the N-Gram Model," *12th International Conference on Fuzzy Systems and Knowledge Discovery* (FSKD), pp. 710.

Community finding in dynamic networks using a genetic algorithm improved via a hybrid immigrants scheme

A. Panizo,* G. Bello-Orgaz,* A. Ortega* and D. Camacho*

*Computer Science Department,
Universidad Autónoma de Madrid
Madrid, Spain*
*{angel.panizo, gema.bello, alfonso.ortega, david.camacho,}@uam.es
http://aida.ii.uam.es*

Due to the temporal nature of real-world networks, the interest in community detection problems on dynamic networks have experienced an increasing attention over the last years. Genetic Algorithms, and other bio-inspired methods, have been successfully applied to tackle the community finding problem in static networks. However, few research works have been done related to the improvement of these algorithms for temporal or dynamic domains. This paper is focused on the design, implementation, and empirical analysis of a new Genetic Algorithm based on a *hybrid-immigrants scheme*, whose main goal is to improve the algorithm convergence when it is applied to identify communities on dynamic networks.

Keywords: Dynamic community finding; genetic algorithms; graph computing; network analysis.

1. Introduction

Community detection is a highly relevant problem when it is applied to disciplines such as sociology, biology, marketing, or computer science, among others. A community can be defined as a set of nodes, whose interactions between the nodes, belonging to this set, are stronger than the interactions to the rest of nodes that belong to other communities. There are two basic approaches to this problem, the first one (named non-overlapping) considers that one node only can belong to a single community.[1] The second one, (named overlapping) allows each node to belong to several communities at time.[2]

In recent years, motivated by the temporal nature of real-world networks such as social networks or mobile communications, new methods and algorithms have been proposed which can handle networks that evolve in time. The communities found by these algorithms are called *dynamic communities*. The evolution of a graph is usually modeled using a sequence of snapshots, where each snapshot represents the graph at a given point in time. A dynamic community can be modeled in two different ways.[3] The first one is a sequence of static communities, where there is a set of subgraphs representing the communities detected for each snapshot. On the other hand, in the second model, each dynamic community is represented as an initial static community and a sequence of modifications over time.

Four main approaches have been used in the literature to detect dynamic communities:[3] The first approach detects static communities for each snapshot independently and then match communities between consecutive snapshots.[4] The second approach processes all the sequence of snapshots simultaneously, a single community detection algorithm is applied once creating a set of communities which contains nodes from different snapshots.[5] The third approach detects the static communities on the first snapshot, then the remaining snapshots are processed following a chronological order, and using the communities detected at the previous snapshot to identify the communities at current snapshot.[6] Finally, the fourth approach finds the static communities in the first snapshot, after that, in an iterative process and following a chronological order, the communities detected are updated using only the modifications of the network between consecutive snapshots.[7]

Genetic Algorithms (GAs) are a stochastic meta-heuristic inspired in the biological principles of Darwinian theory of evolution and Mendel's genetics. GAs have been used to detect both static[8,9] and dynamic communities.[4] Many techniques have been developed to adapt GAs to dynamic environments so when a change occurs, if the new problem is similar to the previous one, the knowledge acquired for the previous solutions can be used to solve the new ones in order to decrease the computational effort. Some of these techniques are *hypermutation*[10] (increase the mutation rate of the population when a change occurs in the environment), *memory schemes*[11] (maintain a memory of individuals during all the GA execution, and when a change occurs, to use this memory to influence the actual population) or *immigrants*[5] (insert new individuals into the population when a change occurs in order to improve the population's diversity). These techniques seem to work well together with the dynamic community detection approaches previously mentioned. For example, Yang and Tinós[12] introduced a new algorithm based on the *immigrants scheme* that aims to reduce the convergence time of GAs when applied to solve problems in dynamic environments.

However, few research works are focused on the application of these approaches to find communities into a temporal or dynamic network. In particular, for detecting dynamic communities, Keehyung et al.[5] presented a Multi-objective GA that uses different immigrants schemes. This algorithm follows the second approach, processing all the sequence of snapshots simultaneously, generating the same set of communities for all the snapshots in the network. Immigrants are inserted into the population for each generation of the algorithm, and the algorithm has 20 generations to adapt after each change.

Our paper presents a new algorithm for community finding on dynamic networks based on the third approach previously mentioned, instead of following the second approach proposed in Keehyung et al.[5] Therefore, the new GA proposed generates a different set of communities for each snapshot of the network. Instead of inserting immigrants in the population at each generation, the proposed algorithm inserts immigrants only into the initial population of each GA trying to reduce the convergence time. In addition, the main contribution of these algorithms is that they keep the information of the changes in communities overtime. Finally, an empirical analysis of the new GA based on a immigrants scheme is carried out to evaluate the achieved improvements.

The rest of the paper is structured as follows: Section 2 describes the dynamic community detection problem. Section 3 presents the procedure followed to test the algorithm and discusses the experimental results. Finally, the last section draws some conclusions and the future research lines of work.

2. Proposed Algorithm

2.1. *Problem formulation*

Let $DN = \{G^t | \forall t \in 0..n\}$ be a dynamic network with n snapshots, each one modeled as a graph G^t. The goal of the algorithm is to find a dynamic community identification $C_{dynamic}$. Which is a sequence of partitions P^t, one for each $G^t \in DN$, where the components in each P^t have more interactions inside them than between them. Components in the same P^t do not overlap, in other words, the intersection between them is empty.

2.2. *Genetic algorithm and immigrants scheme*

The proposed algorithm uses an immigrants scheme that extends a standard GA to detect dynamic communities for each snapshot (G^t) on the dynamic network (DN). As shown in 2.1 (lines 2 to 12), in the first snapshot (G^0), the algorithm works in the same way as the standard GA. However, for the rest of the snapshots ($\{G^t | t > 0\}$), instead of evolving a new random population, the algorithm evolves a population composed of a mixture of random individuals, and individuals from the previous snapshot's population.

Algorithm 2.1 GA for detecting dynamic communities

```
 1: function DYNAMICCOMMUNITIESGA(DN)
 2:     C_dynamic ← ∅
 3:     C_last ← null
 4:     for all G^t ∈ DN do
 5:         if t = 0 ∨ standardGA then
 6:             P_initial ← makeRandomPopulation(G^t, popSize)
 7:         else
 8:             P_initial ← makeImmigrantPopulation(C_last, G^t, popSize, randomRate)
 9:         C_actual ← elitistGA(P_initial, G^t, elitism, crxRate, mutateRate, nGenerations)
10:         C_last ← C_actual
11:         C_dynamic ← C_dynamic ∪ C_actual
12:     return C_dynamic
13: function MAKEIMMIGRANTPOPULATION(C_last, G^t, popSize, randomRate)
14:     P_random ← makeRandomPopulation(G^t, popSize * randomRate)
15:     P_elite ← selectBest(C_last, popSize * (1 − randomRate))
16:     for all individual ∈ P_elite do
17:         individual ← repair(individual, G^t)
18:     return P_elite ∪ P_random
19: function REPAIR(individual, G^t)
20:     for all i ∈ [1, size(individual)] do
21:         if individual[i] ∉ neighbors(i, G^t) then
22:             individual[i] ← selectRandom(neighbors(i, G^t))
23:     return individual
```

Each individual in the population represents one possible partition(P^t) of the current snapshot(G^t) using the Locus-based adjacency representation.[13] According to this encoding, each individual of the initial random population (line 6) is generated by filling each position in the chromosome with a random neighbor, in G^t, of the node that the gen codifies. On the other hand, for generating the individuals of the initial immigrant population related to the snapshots after G^0, the *Immigrants Selection* and *Repair* functions are applied to pass individuals between populations (line 8). The first function (line 15) selects the N-best individuals from the old population that pass to the new one, if the new population is not full, as many individuals are randomly chosen and added as needed (line 14). The second function (line 16 to 17) repairs the individuals, from the old population, that represent invalid solutions in the new population. This function (lines 20 to 22) checks if a gen inside an individual's chromosome points to a node that no longer is connected to the node with the same id as the gen position. When this occurs the gen value is changed to the id of a random neighbor of the node that the gen codifies.

Once the initial population (random or immigrant) has been created, it evolves using an elitist genetic algorithm (line 9) with the next functions: *Uniform Crossover, Roulette wheel selection, modularity*[1] as *Fitness function* and for *Mutation* a function that randomly selects a series of gens, and change them with a random neighbor of the corresponding node.

3. Experimental Study

To study the effectiveness of our approach we have compared it against a standard GA without immigrants measuring the number of generations that both need to reach the optimum modularity value. We have used a synthetic benchmarks generator [14] that allows to create dynamic networks that evolve in a periodic manner, more specifically, a dynamic network with 100 snapshots, 512 nodes, 17205 edges and a periodicity of 100 time steps have been created.

The degree of change between snapshots in the generated network is not homogeneous. For measuring this degree we have calculated the percentage of neighbors of each node that change between consecutive snapshots. In order to have datasets with different degree of change we have selected three sets of 10 consecutive snapshots from the 100 snapshots of the dynamic network. Fig. 1 shows the median percentage of neighbors that change for each snapshot in the datasets. All the experimentation will be done over this three sets.

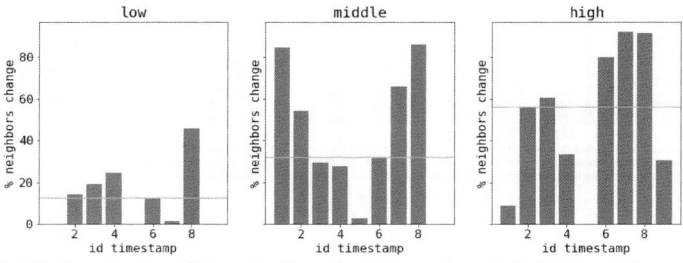

In blue bars, median of the percentage of neighbors of a node that change between consecutive time steps. Orange horizontal line shows the median for all values.

Fig. 1. Percentage of neighbors that change for each snapshot in the datasets.

The set-up of the GA which has been used during the experimentation phase is: crossover rate 0.8, mutation rate 0.2 and elite reproduction 10% of the population size. The population size is 300 and the *stop condition* is fulfilled when the number of generations reach 120. The different parameters of the algorithm have been tuned up experimentally.

In order to select the optimal mixture of random and previous individuals for the initial population (parameter randomRate in the line 9 of 2.1), we have tested the next random-elite ratios: 0%-100%, 20%-80%, 40%-60%, 60%-40%, 80%-20% and 100%-0%, being the last one the standard GA without immigrants. Each experiment has been performed 10 times. To measure the efficiency of the technique, we compare the median number of generations that the algorithm takes to reach the highest fitness value using the different immigrants rates. Results can be seen in the Fig. 2.

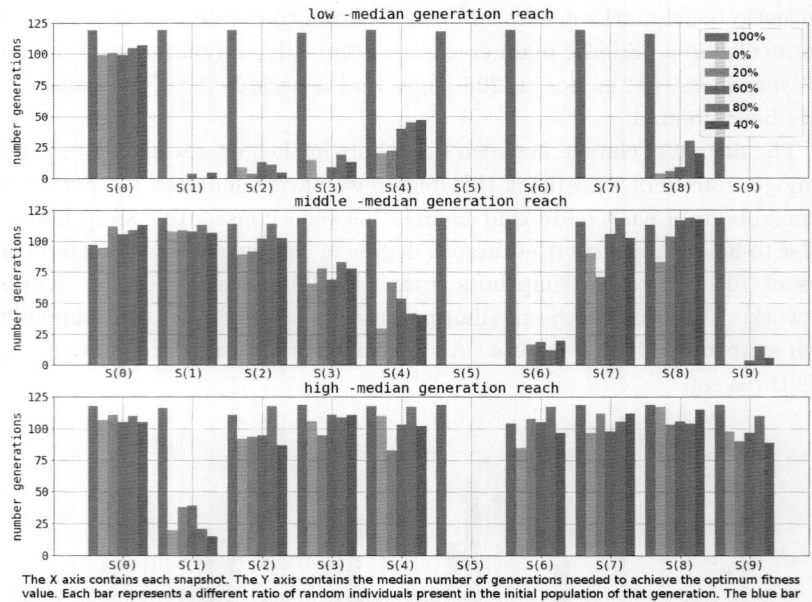

Fig. 2. Result of applying different immigrants rates for each dataset.

Analyzing these results, we can conclude that convergence speed improves if a percentage of individuals from the previous execution of the GA passes to the next snapshot and the variation between snapshots is low. In addition, comparing these results (Fig. 2), and the degree of change shown in Fig. 1, we can draw a parallelism between the results obtained for each snapshot of the different datasets. For example looking at the *middle* dataset, in snapshots 3, 4 and 5, as the degree of change decreases the efficiency of the immigrants scheme improves. We can see the opposite effect if we check snapshots 6, 7 and 8 of the same dataset. An extreme example can be seen between snapshots 8 and 9. At the beginning of the snapshot 8 we have a degree of change over 80%, which makes the immigrants scheme nearly as good as the standard GA. On the other side, at the beginning of snapshot 9, we have a degree of change of 0% making the immigrants scheme very effective. This behavior can be explained examining the *repair* operator. Our operator randomizes the parts of an individual that change between snapshots. When a big part of an individual changes between snapshots the operator is unable to reuse the knowledge generated in the previous snapshot and it transforms the initial population into a random population even thought individuals from the previous snapshot are present. Therefore, we have not found a clear random-elite ratio to improve the overall performance. However, our results suggest that a reasonable choice could be in the range 0-20.

4. Conclusions and Future Work

This paper presents a new evolutionary approach based on a GA using an immigrant scheme to detect communities on dynamic networks. This approach executes a GA to identify the communities belonging to each snapshot of the network. In order to improve the convergence speed, a combination of the best individuals from the previous execution of the GA, and random individuals, is used as the initial population of the next execution of GA. Experiments on synthetic datasets show that this approach is effective in networks where the degree of change is low. Taking into account the experimental results, it can be concluded that the repair operator used to keep individuals along generations is too basic and it fails when the change between consecutive snapshots is high. Future research will be focused on the design of new repair operators that could better transfer information between snapshots despite the degree of change between them.

Acknowledgments

This work has been co-funded by the following research projects: EphemeCH (TIN2014-56494-C4-4-P) and DeepBio (TIN2017-85727-C4-3-P) projects (Spanish Ministry of Economy and Competitivity, under the European Regional Development Fund FEDER).

References

1. M. E. Newman and M. Girvan, *Physical review E* **69**, p. 026113 (2004).
2. G. Bello-Orgaz, H. D. Menéndez and D. Camacho, *International journal of neural systems* **22**, p. 1250018 (2012).
3. R. Alhajj and J. Rokne, *Encyclopedia of social network analysis and mining* (Springer Publishing Company, Incorporated, 2014).
4. F. Folino and C. Pizzuti, *IEEE Transactions on Knowledge and Data Engineering* **26**, 1838 (2014).
5. K. Kim, R. I. McKay and B.-R. Moon, Multiobjective evolutionary algorithms for dynamic social network clustering, in *Proceedings of the 12th annual conference on Genetic and evolutionary computation*, 2010.
6. D. Chakrabarti, R. Kumar and A. Tomkins, Evolutionary clustering, in *Proceedings of the 12th ACM SIGKDD international conference on Knowledge discovery and data mining*, 2006.
7. N. Aston and W. Hu, *Communications and Network* **6**, p. 124 (2014).
8. C. Pizzuti, Ga-net: A genetic algorithm for community detection in social networks, in *International Conference on Parallel Problem Solving from Nature*, 2008.
9. G. Bello-Orgaz and D. Camacho, Evolutionary clustering algorithm for community detection using graph-based information, in *Evolutionary Computation (CEC), 2014 IEEE congress on*, 2014.
10. H. G. Cobb and J. J. Grefenstette, *Genetic algorithms for tracking changing environments.*, tech. rep., Naval Research Lab Washington DC (1993).
11. A. Simões and E. Costa, An immune system-based genetic algorithm to deal with dynamic environments: diversity and memory, in *Artificial Neural Nets and Genetic Algorithms*, 2003.
12. S. Yang and R. Tinós, *International Journal of Automation and Computing* **4**, 243 (2007).
13. Y. Park and M. Song, A genetic algorithm for clustering problems, in *Proceedings of the third annual conference on genetic programming*, 1998.
14. C. Granell, R. K. Darst, A. Arenas, S. Fortunato and S. Gómez, *Physical Review E* **92**, p. 012805 (2015).

Using graph-theoretic methods for text classification

Niloofer Shanavas, Hui Wang, Zhiwen Lin and Glenn Hawe

*School of Computing, Ulster University, Jordanstown,
Northern Ireland, BT37 0QB, UK*

Recent research has utilized graph-based algorithms for several natural language processing tasks to improve their effectiveness. The structural information in text can be modeled using graphs. Graphs can take into account the order of words, co-occurring words, and association between terms that contribute to the meaning of text. In this paper, we introduce the application of core decomposition of graphs for text classification to rank the terms in each class in order to calculate the supervised term weight. In addition to the number of direct connections, it considers sub-graphs of high connectedness to identify the important nodes in the class graphs built from the training documents. The decomposition method also helps in reducing the graph size by eliminating the outer cores without affecting the performance of text classification. The experiments on the benchmark text classification datasets show the superior performance of the proposed approach based on graph-theoretic techniques for text classification.

1. Introduction

Automatic text classification has gained importance with the ever-increasing quantity of text documents making it tedious to organize the data manually. Machine learning is used for automatic text classification mainly due to its effectiveness and reduction in expert labour [1]. Text classification has a wide variety of applications including document indexing, document organization, spam filtering, sentiment analysis and language identification.

The performance of text classification is significantly influenced by the text representation model. The bag-of-words model is generally used for text representation. It considers a text document as a set of independent terms, and represents it as a vector of numerical values corresponding to the weights of the terms in the document. The weights are usually based on the frequencies of the terms in text. It is simple and fast, and does not take into account the dependencies between the different terms in text. However, the meaning of a text document depends on the order of terms and the association between the different terms. The structural information in a text document can be captured accurately using a graph. In the graph-based representation of text, the nodes represent the text units and the edges correspond to the relationships between the text units. The graph-based representation of text has been applied in several areas including text classification, text clustering, information retrieval, document summarization and keyword extraction.

Recent research has shown the connection between graph theory and natural language processing [2]. The effectiveness and efficiency of many natural language processing tasks can be improved using graph-theoretic methods. The increased growth in research on graph-based representations of text and graph algorithms for text processing applications is due to its potential to advance the state-of-the-art in natural language processing. Graph-based representation of text is much more expressive than the bag-of-words model. The performance of graph-based text processing application depends on how it makes use of the rich information in the graph-based text representation. In this paper, we explore the use of graph decomposition in our graph framework for text classification based on graph-theoretic methods to improve the performance of text classification. In the graph framework, undirected co-occurrence graphs are used to represent a document to be classified and the documents in a class; these are called the *document graph* and the *class graph* respectively [3]. The k-core decomposition [4, 5] is applied to the class graphs in order to assign score to node based on the region of the class graph the node is located instead of only relying on the number of its local connections. It is based on the idea that the important nodes in class graphs have not only many connections but also belong to the highly cohesive sub-graphs (or the densest parts of the graph). The number of direct connections and the cohesiveness information obtained by the k-core decomposition of a graph are used to rank the nodes in the class graphs. The relevance of a term to the text classification task is then calculated based on the ranks of the node denoting the term in the class graphs. Also, further processing required need not be done on the entire graph and can be focused on the important cores, which saves time.

The rest of the paper is organized as follows. Section 2 gives an overview of the related works. Section 3 explains core decomposition of graphs and how we utilize it for text classification. Section 4 describes the experiments and results. Finally, Section 5 concludes the paper.

2. Related Work

This section focuses on the works published in the area of core decomposition of graphs for natural language processing applications. The k-core decomposition has been applied recently in different text processing applications including summarization, sub-event detection, keyword extraction, clustering algorithms, and text classification.

k-core decomposition has been employed for extractive summarization of text documents [6]. The extract contains the sentences representing the nodes of the main core and also includes sentences of the lower cores until the compression rate limit is reached. The decomposition-based technique has also been utilized to generate extractive summaries of text documents represented as word co-occurrence graphs [7]. In this approach, the terms are scored based on the positional information in the graph obtained by k-core decomposition before scoring and selecting sentences. Graph decomposition has been applied to detect real time sub-events in twitter messages and generate a summary of the sub-events [8]. A set of tweets is represented as a graph where a tweet is a completely connected sub-graph. k-core decomposition is applied to the weighted graph representing the tweets. Real time sub-events are detected based on the scores assigned to the terms using the core numbers.

The concept of k-core decomposition has been used to extract keywords from documents represented as co-occurrence graphs [9, 10]. The nodes of the main core correspond to the keywords in their approach as it is the most cohesive part of the graph [9]. k-truss decomposition was also explored to extract keywords [10]. Since all the keywords need not be in the main core and can be found in the lower cores, they developed three algorithms to select the best cores to extract keywords. The first two algorithms to retain the best cores were based on the density of the nodes in the cores and the variation in the shell sizes respectively. The third algorithm to retain the best cores decreases the granularity from sub-graph level to node level by using the neighborhood coreness measure proposed in [11] that assigns to each node the sum of the core numbers of its neighbors.

A clustering framework has been developed based on the k-core decomposition method [12]. It can be used with any graph clustering algorithm in order to improve the quality of the results and reduce the time complexity. The k-core structure helps in selecting the densest cores as the starting point for the clustering algorithms. The k-core decomposition has been utilized to speed up the graph-based feature extraction process for text classification where documents are represented by the frequent sub-graphs mined from the graph-based representation of text document [13]. Using k-core decomposition, the main core is extracted before mining the frequent sub-graphs which also reduces the number of sub-graphs extracted.

As k-core decomposition has not been explored much for text classification, we focused our study on using graph decomposition for text classification. We utilize the information obtained by k-core decomposition to locate the densest parts of the class graph and rank the nodes in the class graph based on their location in the graph. This information obtained is then used to calculate supervised term weights. Supervised term weighting was introduced to weight terms by learning from training data the strength of the terms to classify the documents accurately [14]. The division of the class graphs into cores also helps in reducing the size of the class graphs, thereby eliminating the unimportant words in each class.

3. Decomposition of Graph for Text Classification

The k-core decomposition technique was introduced to study network cohesion [4, 5]. If G is a graph and H is a maximal connected sub-graph within G with a minimum degree of H greater than or equal to k, the H is called a core of order k or k-core [5]. It produces a hierarchy of nested sub-graphs and the set of all cores of the graph is called the k-core decomposition of the graph. As the order increases, the connectedness of the sub-graph increases. The main core is the core of highest order in the graph. The core number of a node is the order of the highest order core that the node belongs to. An example of k-core decomposition is illustrated in Figure 1.

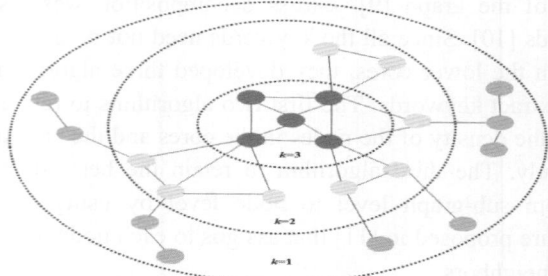

Figure 1. Example of k-core decomposition. k denotes the order of the core. The nodes with the same core number have the same colour.

In the proposed approach, we have used the graph framework for text classification developed in our previous work to explore k-core decomposition to obtain cohesiveness information and utilize it for text classification [3]. The first step in the text classification pipeline is the preprocessing of documents to remove stop words and stem the remaining words. The documents are then represented as co-occurrence graphs where each node in the graph represents a unique term and the edges link terms that co-occur within a fixed sliding window. Class graphs are co-occurrence graphs built for every class from the pre-classified training documents. We have experimented with k-core decomposition to calculate the relevance of terms to text classification using the core number of nodes and also to reduce the graph to discard the unimportant terms. It is explained in detail in the below two sections.

3.1. *Supervised term weight based on k-core decomposition*

k-core decomposition is applied to the class graphs in order to identify the highly cohesive components of the graph. It divides the graph into a set of cores based on cohesiveness. The higher the core number of the node, the higher is the connectedness of the node with the other nodes in the graph. With this approach, we can consider the cohesiveness of sub-graph the node belongs to in addition to the number of direct connections while calculating the ranks of nodes. Each node's rank is the sum of the core number of the node and the core number of its neighbors. Since we take into account the core number of its neighbors, it also gives importance to the number of local connections. Hence, it considers the node's local and global importance in the class graphs. The calculation of the rank of node denoting the term *t* in the *i*-th class graph is given below in Eq. (1). $CN_i(t)$ is the core number of node denoting the term *t* in the *i*-th class graph (C_i) and $N_i(t)$ is the set of neighbors of the node.

$$R_i(t) = CN_i(t) + \sum_{n \in N_i(t)} CN_i(n) \tag{1}$$

The node rank is normalized by the maximum value of the rank in the class graph. These ranks are then used to calculate the supervised term weight factor, class relevance centrality (crc), proposed in our previous work that determines the terms relevance to the text classification task [3]. This supervised term weighting scheme takes into account the information obtained from the class graph to compute the term's relevance to the classification task. We use crc in this work to calculate the variation in the rank of the nodes in order to increase the weights of those terms that help in discriminating the documents in different classes. The calculation of crc using the rank of the nodes, denoted as crc(cor), is shown below.

$$M(t) = \max_{1 \le i \le |S|} R_i(t) \tag{2}$$

$$\lambda(t) = \sum_{i=1}^{|S|} R_i(t) \tag{3}$$

$$L(t) = |\{C_i : R_i(t) > 0, \forall i \in [1,|S|]\}| \tag{4}$$

$$A(t) = \frac{\lambda(t) - M(t)}{L(t) - 1} \tag{5}$$

$$\text{crc(cor)} = \log_2(2 + \frac{M(t)}{\max(minc, A(t))} * \frac{|S|}{L(t)}) \tag{6}$$

where $|S|$ is the total number of classes and *minc* is the minimum of the node ranks in the class graphs.

3.2. *Reducing the size of the graph using k-core decomposition*

The *k*-core decomposition builds a hierarchy of nested sub-graphs with cohesiveness reducing as core number decreases. The lower levels of the hierarchy can be removed to reduce the size of the graph and eliminate the unimportant words. Our method of core elimination reduces the size of the graph without affecting the performance. We set maximum thresholds p and u for the percentage of nodes and the rank of nodes that can be removed respectively. Let n be the number of nodes in a core, e be the number of eliminated nodes, r be the rank of a node and t be the total number of nodes in the graph. The core elimination starts from the lowest *k*-core and moves towards the higher k-cores. The core elimination stops if $n+e$ is greater than or equal to $(p*t)/100$ or if any node in the core has a rank r that is greater than or equal to u. An optimum value of p and u ensures that the number of terms is not reduced considerably and it does not remove important terms by eliminating a core. The advantage of removing the nodes is that if further processing is needed, it need not be done on the entire collection of nodes, saving time.

4. Experiments and Results

We have explored *k*-core decomposition and its effect on text classification by experimenting on four standard text classification datasets — webKB, R8, R52 and 20 Newsgroup (20NG). We obtained a pre-processed version of the dataset that contains a train and test split.[a] The class graphs are built from the documents in the training set. The documents to be classified are represented as graphs. The class graphs and document graphs are undirected co-occurrence graphs with edges that link terms that co-occur within a sliding window of size 2. The *k*-core decomposition is applied to the class graphs. It is fast and computed in linear time. The core number of each node in the graph is utilized to determine the ranks of the nodes. This rank score is then used to calculate the supervised term weight factor crc. Hence, we utilize the rich information in the class graphs to determine the term's discriminatory power. It is based on the idea that important terms are located within the densest parts of the graph. The final weight of each term is the product of tw and crc. tw is based on the importance of the node in the document graph. We have used degree centrality measure to calculate tw.

---

[a]http://ana.cachopo.org/datasets-for-single-label-text-categorization

The precision, recall and f1 scores have been calculated to evaluate the performance of text classification with SVM and are given in Tables 1, 2 and 3. We have denoted the crc calculated using the ranks based on core numbers as crc(cor) which is explained in Section 3.1. The crc calculated using degree and closeness centrality measures are denoted as crc(deg) and crc(cl) respectively [3]. tw(deg)-crc(cor) significantly outperforms the baselines tf, tf-idf and the graph-based unsupervised term weighting schemes. The advantage of applying k-core decomposition is that we can eliminate the nodes having less connectedness. We used the proposed graph reduction method explained in Section 3.2. The thresholds p and u are set to 50 and upper quartile of the set of ranks (sorted in ascending order) respectively. Table 4 shows the number of nodes after the reduction. The crc(cor-red) is the supervised term weight calculated using the reduced graph. Tables 1, 2 and 3 show that the performance of text classification is maintained even after the removal of the nodes. This shows that there are many irrelevant terms in text that do not have a role in classifying the documents and can be discarded.

Table 1. Precision scores.

	tf	tf-idf	tw	tw-idf	tw(deg)-crc(deg)	tw(deg)-crc(cl)	tw(deg)-crc(cor)	tw(deg)-crc(cor-red)
webKB	0.8762	0.8459	0.8961	0.8757	**0.9149**	0.9118	**0.9136**	0.9128
R8	0.9620	0.9622	0.9689	0.9758	**0.9762**	**0.9768**	0.9759	0.9753
R52	0.9235	0.9200	0.9031	0.9389	0.9551	**0.9564**	**0.9560**	0.9558
20NG	0.7636	0.7813	0.7845	0.8353	0.8375	**0.8414**	**0.8405**	0.8399

Table 2. Recall scores.

	tf	tf-idf	tw	tw-idf	tw(deg)-crc(deg)	tw(deg)-crc(cl)	tw(deg)-crc(cor)	tw(deg)-crc(cor-red)
webKB	0.8768	0.8467	0.8933	0.8746	**0.9148**	0.9119	**0.9133**	0.9126
R8	0.9621	0.9621	0.9685	0.9758	**0.9762**	**0.9767**	0.9758	0.9753
R52	0.9283	0.9190	0.9186	0.9435	0.9544	**0.9556**	**0.9548**	**0.9548**
20NG	0.7599	0.7763	0.7814	0.8335	0.8352	**0.8390**	0.8379	**0.8382**

Table 3. F1 scores.

	tf	tf-idf	tw	tw-idf	tw(deg)-crc(deg)	tw(deg)-crc(cl)	tw(deg)-crc(cor)	tw(deg)-crc(cor-red)
webKB	0.8764	0.8462	0.8902	0.8693	**0.9146**	0.9112	**0.9132**	0.9124
R8	0.9616	0.9618	0.9681	0.9756	**0.9761**	**0.9766**	0.9757	0.9752
R52	0.9240	0.9146	0.9037	0.9370	0.9515	**0.9522**	**0.9519**	0.9517
20NG	0.7591	0.7760	0.7740	0.8301	0.8341	**0.8372**	**0.8371**	0.8370

Table 4. Number of nodes before and after k-core decomposition.

	Before decomposition	After decomposition
webKB	7288	4982
R8	14575	10863
R52	16145	12695
20NG	54573	37833

5. Conclusion

In this paper, we conducted experiments with k-core decomposition for text classification. It helped in locating the densest and highly connected parts of the class graphs. This information is used to calculate the supervised term weight and also to eliminate the terms that do not have an important role in the text classification task. The experimental results show that the graph-theoretic methods like node centrality measures and k-core decomposition can be utilized to improve the performance of text classification.

References

1. F. Sebastiani, 'Machine learning in automated text categorization', ACM Computing Surveys, vol. 34, no. 1, pp. 1–55, 2002.
2. R. Mihalcea and D. R. Radev, 'Graph-Based Natural Language Processing and Information Retrieval', Cambridge University Press, New York, NY, USA.
3. N. Shanavas, H. Wang, Z. Lin, and G. Hawe, 'Centrality-Based Approach for Supervised Term Weighting', IEEE International Conference on Data Mining Workshops (ICDMW), pp. 1261–1268, 2016.
4. B. Bollobás, 'Extremal graph theory', Academic Press, 1978.
5. S. B. Seidman, 'Network structure and minimum degree', Social Networks, vol. 5, no. 3, pp. 269–287, 1983.
6. L. Antiqueira, O. N. O. Jr., L. da Fontoura Costa, and M. das Graças Volpe Nunes, 'A complex network approach to text summarization', Inf. Sci., vol. 179, no. 5, pp. 584–599, 2009.
7. A. J.-P. Tixier, P. Meladianos, and M. Vazirgiannis, 'Combining Graph Degeneracy and Submodularity for Unsupervised Extractive Summarization', in NFiS@EMNLP, pp. 48–58, 2017.
8. P. Meladianos, G. Nikolentzos, F. Rousseau, Y. Stavrakas, and M. Vazirgiannis, 'Degeneracy-Based Real-Time Sub-Event Detection in Twitter Stream', in ICWSM, pp. 248–257, 2015.
9. F. Rousseau and M. Vazirgiannis, 'Main Core Retention on Graph-of-Words for Single-Document Keyword Extraction', in ECIR, vol. 9022, pp. 382–393, 2015.
10. A. J.-P. Tixier, F. D. Malliaros, and M. Vazirgiannis, 'A Graph Degeneracy-based Approach to Keyword Extraction', in EMNLP, pp. 1860–1870, 2016.
11. J. Bae and S. Kim, 'Identifying and ranking influential spreaders in complex networks by neighborhood coreness', Physica A: Statistical Mechanics and its Applications, Volume 395, pp. 549–559, 2014.

12. Giatsidis, F. D. Malliaros, D. M. Thilikos, and M. Vazirgiannis, 'CoreCluster: A Degeneracy Based Graph Clustering Framework', in AAAI, pp. 44–50, 2014.
13. F. Rousseau, E. Kiagias, and M. Vazirgiannis, 'Text Categorization as a Graph Classification Problem', in ACL (1), pp. 1702–1712, 2015.
14. F. Debole and F. Sebastiani, 'Supervised Term Weighting for Automated Text Categorization', Text Mining and its Applications: Results of the NEMIS Launch Conference, Springer Berlin Heidelberg, pp. 81–97, 2004.

Argumentation system for intelligent assistants using fuzzy-based reasoning

T. Koivuaho, M. Ibrahim, F. Ummul and M. Oussalah

Centre for Ubiquitous Computing, Faculty of Information Technology and Electrical Engineering, University of Oulu
Oulu, 90014- Finland

> This paper addresses the issue of building intelligent assistant that is able to maintain sustainable conversation with the user while taking into account his emotional, personality aspects, and, at the same time, maintaining high level focus. The approach makes use four meta-features; namely, topic, emotion, personality and dialogue-act. A sequence-to-sequence recurrent neural network approach was used to learn answer prototypes from Reddit.com sport corpus, while an ANFIS based approach was developed to extrapolate from the limited configurations used by the neural network to various dialogue utterances.

1. Introduction

The growth of e-commerce as well as the multiplicity of the smart city projects has led to the development of online systems that allow corpora, authorities and SMEs to interact with their users and potential customers, efficiently with minimal operator intervention [9]. Watson et al. [15] investigated mobile marketing platforms that enhance consumers' relationship with a brand through text messaging, mobile advertisements, m-commerce and permission based marketing. Such marketing is found to be useful for companies to easily reach consumers at relatively low cost [10].

The emergence of (intelligent) conversational agents, also known as chatbots utilizing chat, messaging, or other natural language interfaces (i.e. voice) to interact with people, brands, or services in a bidirectional asynchronous context [8], is at the heart of the success story in e-commerce and e-democracy like applications. In this respect, as pointed out by Shopify [14], consumers can chat with company representatives, get customer support, ask questions, get personalized recommendations, read reviews, and click to purchase all from within messaging apps.

Despite the potential of chatbots, their developments are challenged by inherent barriers, which are often rooted back to the limitation of the software agent to comprehend human language. The inherent barrier to understand the motives and the feelings of the user / customer sometimes creates psychological to ensure full user satisfaction and continuation of services. Long conversations are hard to automate and if in an open domain, the conversation can go in any direction [1]. In most of current systems, human intervention is still needed when things get complex [4]. Besides, since pioneer work of Turing in early fifties, the interaction between computers and humans via natural language is a topic that is extensively researched in information processing community, artificial intelligence and humanities and is a complex task. Besides, issues of cross-domain portability, evaluation and validation have been reported as immature in the field (e.g. Kuligowska [7]). Yet, as chatbots gradually start to expand to the messenger interface, a different research approach is required to quantify its acceptability. On the other hand, the identification of user's reasoning and rationality using standard natural language processing is at its infancy, and more research is still needed in order to open new horizons for chatbots.

This paper aims to contribute to the ongoing research in this direction where a fuzzy based approach is developed in order to leverage the capacity of chatbot to identify user's argumentation while accounting for his emotional state and feelings.

Strictly speaking, investigating recent works on the issue from both academia and commercial like applications reveals that the quasi-majority of such promising prototypes advocate the use of machine-learning like approach where a dedicated configuration of neural network learns the best sentences from a large dataset of question-answers corpus. For instance, Sutskever et al. [13] and Zhou et al. [17] promoted a sequence-to-sequence neural network, where a multi-layer model has been successfully used in translation task. Nevertheless, despite their limited success, such models often ignore the input and produce highly generic responses such as "I don't know what you are talking about" answer to be shortsighted and ignore their influence on future outcomes, which, trivially, negatively impact the ability of the underlying chatbot to calm users down, and built trust relationship as pointed out in [11]. This calls further research on the issue.

The key reasoning advocated in our methodology of handling is to distinguish five key patterns that are extracted from the user's textual statements, namely, personality, emotion, topical discussion, sentiment and dialogue act, using a combination of natural language processing, machine learning and commonsense-based reasoning. These patterns are next integrated into a fuzzy rule based system in order to yield the prototype of potential chatbot answer. The exact answer is randomly selected among those sentences belonging to the prototype class. Without loss of generality and for evaluation purpose, we deliberately restricted to the sport topic. Three different approaches using full rule-based system, deep learning and hybrid (rule-based and deep learning) are compared.

2. Methodology

2.1. *General approach and motivation grounds*

As pointed out in the introduction section, the general approach relies first on the identification of key discussion patterns, which are used as control variables to derive the answer prototype through a combination of fuzzy rule based and learning like approaches, and then randomly select a candidate answer from the class. The rationale behind the proposal is the following.

First, the fact that the system outputs a prototype answers instead of a single unique "statement" answer agrees with the socio-linguistic findings and commonsense reasoning that any statement can be equivalently conveyed through a set of fully equivalent and distinct other statements. For instance, the statement of agreement can be translated using statements like "OK", "Sure", "I agree", "Fully agree", etc. Besides, any attempt to force a unique single answer may create a state of boringness that would impact negatively the subsequent dialogue.

Second, the multiplicity of control variables employed in the sequel is motivated by the complexity of comprehending human language as well as the complexity of human emotion, which can change drastically with respect to tiny change of words. For instance, any miss-understanding of user language can trigger strong emotion. More specifically indicators on discussion topics (gathered using Drichlet latent modelling), dialogue act (using corpus and support vector machine), five-trait personality score using the cumulative discussion of the user up to the current time, emotion as quantified using Linguistic Inquiry Word Count features, were employed as control variables.

Third, the use of fuzzy based reasoning [5] in the process of integration of these control variables is motivated by the inherent subjectivity and absence of accurate uncertainty model pervading these control variables. Therefore, a fuzzy inference based strategy was employed. On the other hand, interestingly, unlike commonly employed fuzzy inference system, the universe of discourses ascribed to the aforementioned mentioned variables are not all defined on interval scale, but some rather use only a nominal scale. For instance, both topical and dialogue act variables use only a set of words to define their universe of discourse. This brings our reasoning more close to Zadeh's concept of "computing with words" [16]. More explicitly, the detail description of the features and control variables are given below.

2.2. Dialogue act classifier

For the sake of simplicity and intuitive interpretation, we imitate the 42-dialogue act of the switchboard corpus [12]. Examples of dialogue act include Statement, Backchannel / Acknowledge, Opinion, Appreciation, Yes-No question, Yes answer, No answer). Non-verbal, uninterpretable and third-party talk labels were left out, as they were not seen to be relevant for our case, which leaves 39-dialogue act to be effectively considered. The classifier was trained using the Switchboard Telephone Speech Corpus by Linguistic Data Consortium [3] and then implemented a logistic regression for the purpose of classifying the 39 various dialogue acts. The training and testing sets contained roughly 129 000 and 43 000 utterances, respectively. Following good accuracy rate obtained elsewhere using the same corpus, the features, extracted for each utterance, and employed in the above classification are the following: i) First 10 tokens, lemmatized, including punctuation, and padded with blank if fewer; ii) First 10 POS-tags, including punctuation, padded with blank if fewer; iii) Presence of a question mark; iv) Previous speaker; v) Previous dialogue act; vi) Predicate verb and respective POS-tag; vii) Subject and respective POS-tag; iix) Object and respective POS-tag.

Lemmatization and POS-tagging were done using SpaCy such that the predicate, the subject and the object were determined using SpaCy's dependency parser. The macro accuracy of our classifier using the specified testing material is 0.77.

2.3. Emotion

We used WordNet-Affect, an extension of WordNet domains that concerns a subset of synsets suitable to represent affective concepts correlated with affective words. Especially, the affective concepts representing emotional state are individuated by synsets marked with the a-label *emotion*. Distinct labels were also attributed to those concepts representing moods, situations eliciting emotions, or emotional responses. It is organized as an xml file providing a tree in the set of words. This tree first describes the root of human behavior which can be broken into physical state, behavior, trait, sensation, situation and mental state, while emotions are considered part of the mental states. Divided into positive and negative emotions, the tree holds most of the words that can have an emotional meaning. This approach matches the words in the input text with these words in the tree, when matched with a high similarity, the module returns the emotion behind these words and then returns the root feeling whether it is positive or negative one. This occurs for every word in the input text, as the module tries to match every word out of the input to the tree in the domains files. The result will be for a set of different feelings, then by taking average, we can get a relative estimation about the actual feeling from the text.

The open source WNAffect (implementation of WordlNet-Affect) tries to match the adjectives and nouns entered by the user to the wordnet-domains. JJ, an adjective and NN, nouns are called.

2.4. Personality

Taking advantage of the accumulated input text from same user, our purpose was to identify the user personality in terms of the five-personality model. The latter encompasses the following five personality traits:

- Extroversion, which indicates how the person can engage with the world. E.g., he may enjoy being with people and tend to be more enthusiastic, action-oriented and like to assert himself. A high level of emotional stability means that the person is more confident and sure of himself with less self-doubt.
- Agreeableness, means how much the person is liked, respected and sensitive to others needs. It shows how the person is generous, helpful and friendly. It correlates in a weak way with extroversion.
- Conscientiousness, where people with high values in conscientiousness are likely to value order, duty, achievement, and self-discipline, and consciously practice deliberation and work towards increased competence.
- Openness to experience, which is the tendency of people to try new experiences and knowledges, usually linked to creativity and imaginative thinking.
- Neuroticism, which represents the tendency to experience unpleasant emotions such as anger, anxiety, depression and vulnerability. It is sometimes referred as the degree of emotion stability and impulse control.

We used an application developed by University of Sheffield in cooperation with University of Arizona that uses MRC Psycholinguistic Database, where the Linguistic Inquiry and Word Count (LIWC) is used to get emotions, thinking styles, social concerns, and even parts of speech. At each new utterance of the user, the application outputs the percentage of each of the five personality trait up to the current state of the human-agent interaction.

2.5. Topic

Latent Dirichlet Allocation (LDA) [1] is employed to detect the topics that are present at individual utterances or corpus. The LDA generative model assumes that documents contain a combination of topics, and that topics are a distribution of words; since the words in a document are known, the latent variable of topics can be estimated through Gibbs sampling. We used an implementation of the LDA algorithm provided by the Mallet package [6] adjusting one parameter (alpha~0:30) to favor fewer topics per document, since individual utterance updates tend to contain fewer topics than the typical documents (newspaper or encyclopedia articles) to which LDA is applied. Besides, in order to avoid subsequent unnecessary complexity burden while ensuring high interpretability of the results, we also reduced the size of the vocabulary of the words used by the LDA to the most frequent 500 words, excluding the stopword elements in the Reddit corpus. The algorithm is fully unsupervised, which means no human input is necessary only need a corpus of plain text documents. The purpose of using this model in our application is to identify underlying "topics" that user talked about with our chat agent. Here, we have used separate documents for each user, so that we can find about which topic they are discussing with our chat agent.

2.6. Prototype answers

Without loss of generality and avoiding the complexity of open debate, we restricted our analysis to sport field in Reddit corpus (www.reddit.com/r/sport).

Due to its proven performance in other studies, we used neural machine translation model, seq2seq recurrent neural networks [17] which works with encoder decoder principle in which encoder generates "thought vector" which decoder then decodes into vector representation of words. Our neural network also uses attention mechanism in the decoder phase as well as beam search. Our neural machine translation model had 800 000 of sentence-translation pairs as its training material with the vocabulary size of 20 000. All the training material was extracted from reddit.com.

On the other hand, the five-meta features (word topic, personality vector, emotion and dialogue act) are extracted from the original (Reddit) sentence-inputs. This will enable our neural network system to learn the answer prototype according to various input configuration of the meta-features.

Beam search is used for getting best 10 outputs of the neural net instead of one, provided the outputs are sufficiently reliable, up to a threshold level on average classification accuracy obtained on training dataset, otherwise, the prototype will include a small number of instances (less than 10 answers).

Therefore, for each combination of the meta-feature vectors, a prototype answer, constituted of a class of up to 10 instances, is generated. A tensor implementation of the seq2seg recurrent neural network has been used.

3. Fuzzy Based Reasoning

Given that the meta-features vectors obtained during the prototype answer generation in previous step are far to be complete as many other vector configuration can be generated when using our utterance corpus as quite distinct from the Reddit corpus, therefore the question of which prototype answer to use is well open. For this purpose, and given the already proven results in terms of universal approximation results. This motivates our approach of using a fuzzy inference system as a universal approximator in order to extrapolate new unseen meta-feature configurations from known configurations. More specifically, an ANFIS based methodology [4] was employed. This allows us to learn and optimize the various parameters of the inference system.

In order to ascribe a fuzzy like quantification to each attribute, we considered the inherent classification score generated by the corresponding algorithm for each of the meta-features. For instance in case of topic meta-feature, the LDA algorithm generates a score value for each word representing the topic, e.g.,

Topic A → {Word 1 (0.6), Word 2 (0.4), Word 3 (0.5)}

Therefore, attributes of Topic A will be described, for instance, as

Topic A → {Word 1 is High AND Word 2 is Medium AND Word 3 is High}.
Similarly, a configuration of meta-feature — personality such that:
Personality A → {Openness (0.7), Conscientiousness (0.4), Extraversion
(0.44), Agreeableness (0.2), Neuroticism (0.6)}
is translated into
Personality A → {Openness is High AND Conscientiousness is Medium AND Extraversion is Medium AND Agreeableness is Low AND Neuroticism is High}

Similar reasoning applies to dialogue act as well, although we only restrict to the two classes (first and second ranked dialogue act class).

In the case of emotion meta-feature, the WNAffect only outputs the set of emotion words as traversed through WordNet hierarchical taxonomy. Therefore, all the generated words are considered equally high.

The exact boundaries as well as the number of partition is left open as part of the optimization-process of ANFIS system.

Therefore, the accumulated (meta) fuzzy rule looks like
If Topic is T AND Emotion is E AND Personality is P AND DialogueAct is D THEN Answer is A.
where the T, E, P and D are rather defined in a multidimensional space. While, the consequent part A is chosen to be crisp (non-fuzzy), which bring the reasoning close to Zero-degree Takagi-Sugeno fuzzy inference system. The detailed description of the system will be reported at other publications forums.

4. Implementation

ChatInstance-class is used for integrating the functional components of the system, and for maintaining the necessary data structures of a single conversation. ChatInstance implements getter-methods for the different features of the argument vector as well as connection to the NMT-model. ChatInstance is an access point for the user to use lower level functions. A top method called converse calls other methods responsible for getting features, which are: get_emotion, get_personality, get_topic, get_dialogueAct. A user interface is provided in Fig. 1.

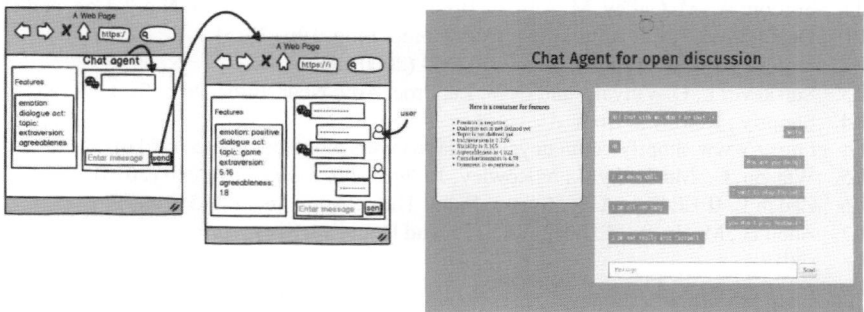

Fig. 1. User interface.

5. Conclusion

This paper describes a fuzzy based approach for intelligent assistant or chatboot design. Especially, four types of meta-features were employed and integrated as part of the methodology. This consists of topic (using LDA reasoning), emotion, personality and dialogue-act. A sequence-to-sequence recurrent neural network approach was used to learn answer prototypes from Reddit.com sport corpus, while an ANFIS based approach was developed to extrapolate from the limited configurations used by the neural network to various dialogue utterances.

Acknowledgments

This work is (partially) funded by the Marie Skłodowska-Curie Actions (645706-GRAGE).

References

1. Blei D.M., Ng A.Y., Jordan M.I., J Mach Learn Res 3: 993 (2003).
2. Britz, D. Retrieved at http://www.wildml.com/2016/04/deep-learning-forchatbots-part-1- introduction/ (2016).
3. Calhoun, S., Carletta, J., Brenier, J. M., Mayo, N., Jurafsky, D., Steedman, M., & Beaver, D. Lang. Res. Eval. 44(4), 387 (2010).

4. Jang JSR, IEEE Trans Sys Man Cybern 23: 665 (1993).
5. Klir G., Fuzzy Sets and Fuzzy Logic. Theory and Applications, Pren. Hall Pub. (1995).
6. McCallum AK, Available at: http://mallet.cs.umass.edu (2002).
7. Kuligowska, K., *PCBR*, *2*(02), 1 (2015).
8. Messina, C., Retrieved from https://medium.com/chris-messina/2016-will-be-the-yearof-conversational- commerce-1586e85e3991#.bsdskkyj (2016).
9. Pavlou, P. A., Lie, T., & Dimoka, A., SSRN, http://dx.doi.org/10.2139/ ssrn.2380712 (2007).
10. Persaud, A., & Azhar, I., *Mark. Int. & Plan. 30*(4), 418 (2012).
11. Sordoni A., M. Galley, M. Auli, C. Brockett, Y. Ji, M. Mitchell, J. Nie, J. Gao, and B. Dolan, Proc. North Amer. Chap. Ass. Comp. Ling. 196 (2015).
12. Stolcke A. et al., Comp. Ling. 26(3), 339 (2000).
13. Sutskever I., O. Vinyals, and Q. V. Le, Proc. Adv. Neur. Infor. Proc. Syst. 27 (2014).
14. Shopify., Retrieved from https://www.shopify.com/encyclopedia/ conversationalcommerce (2016).
15. Watson, C., McCarthy, J., & Rowley, J., *Int. J. Inf. Man. 33*(5), 840 (2013).
16. Zadeh L. IEEE Tran. Circ. Syst. I: Fund. Theor. Appl. 46(1), 105 (1999).
17. Zhou H., M. Huang, T. Zhang, X. Zhu, and B. Liu., arXiv preprint arXiv:1704.01074, (2017).

A fuzzy based approach for wordsense disambiguation using morphological transformation and domain link knowledge

F. Farooghian

iHR, Aston University, B4 7ET, Birmingham, UK

M. Oussalah and E. Gilian

Centre for Ubiquitous Computing, Faculty of Information Technology and Electrical Engineering, University of Oulu, Oulu, 90014- Finland

This paper describes a fuzzy-based methodology in order to aggregate outcomes of distinct wordsense disambiguation algorithms. The latter are derived from standard Lesk algorithm, its WorldNet extension and new interpretations of the set-intersection that accounts for various WordNet domain knowledge and part-of-speech conversion. The fuzzy preference model imitates the fuzzy Borda voting scheme. The developed algorithms are evaluated according to SenseEval 2 competition dataset, where a clear improvement to the baseline algorithm has been testified.

Keywords: Word sense disambiguation; fuzzy preference; semantic similarity; WordNet.

1. Introduction

Word sense ambiguity is inherent to human language and prevalent in all natural languages where a single can convey multiple meanings. For instance, "bank" may stand for a financial institution, objects (materials) grouped together in rows, high mass/ mound of a particular substance, or a land near river / lake. The correct sense of an ambiguous word can be selected based on the context where it occurs [1].

The appropriate handling of word sense disambiguation (WSD) task can potentially provide a major breakthrough in the information retrieval systems where identification of correct sense of query terms yields a milestone breakthrough in document retrieval systems, question-answering systems, among others [2].

Simultaneous interest in the linguistic community to research the structure of corpus resulted in various types of manually annotated corpora that populated senseval/Semeval evaluation[a] where several research competitions are held towards designing new algorithms for disambiguation task. Lesk algorithm [3],

---

[a]http://www.senseval.org

based on the amount of overlapping between the gloss of the target word (for each sense) and the glosses of the context words, and its various extensions, e.g., Banerjee and Pederson [4] where WordNet was used as a source of glosses, have often setup a standard in the field.

This paper contributes to the wordsense disambiguation effort in the following way. First, new enhanced Lesk-like algorithms are put forward using different interpretations of set-intersection. The former assumes a metric viewpoint calculated using the path-length measure of WordNet hierarchical synset structure. While the latter utilizes the WordNet domains links and extends both the set intersection and available domain hierarchical distance metric accordingly. Especially, in order to benefit from the dense hierarchical structure of noun-category in WordNet lexical database [5], a word morphology transformation is employed, which then serves as basis for subsequent semantic similarity. Second, a fuzzy preference based strategy is employed in order to aggregate the outcomes of various disambiguation algorithms. The performances of the suggested algorithms are evaluated using both Senseval-2 and SemCor datasets where a systematic improvement over the baseline has been noticed. Section 2 of this paper provides background and related work. Our methodology is detailed in Section 3, while testing results are reported in Section 4.

2. Background and Related Work

A pioneer work in word sense disambiguation is the Lesk algorithm [3] where the particular sense of a target word corresponds to the sense whose gloss (definition of the sense) shares the largest number of words with glosses of the words in the phrase to be disambiguated.

More formally, let W_1, W_2 be two words whose meanings are defined in the dictionary $N_{W_1} = \{W_1^1, W_1^2, .. W_1^{n_1}\}$ and $N_{W_2} = \{W_2^1, W_2^2, .. W_2^{n_2}\}$, repectively. The W_1, W_2 are therefore assigned senses W_1^i and W_2^j, $i \in \{1,2,...,n_1\}$, $j \in \{1,2,...,n_2\}$ (in the context of a phrase containing W_1 and W_2) such that

$$\left|W_1^i \cap W_2^j\right| = \max_{1 \leq k \leq n_1, 1 \leq l \leq n_2} \left|W_1^k \cap W_2^l\right| \tag{1}$$

Typically, N_{W_1} and N_{W_2} rely on glosses found in traditional dictionaries, e.g., Oxford English dictionary.

The complexity search for (1) increases exponentially with the number of words in the phrase. A known approximated solution to this problem, referred to as a simplified Lesk algorithm consists of restricting the overlapping operation, for each sense of the target word, to words surrounding the target word. More formally, using the above notations and given a sentence /phrase where two words W_1 and W_2 co-occurs, say

$$S = <W_0, W_1, W_2, W_3, \ldots, W_m>$$

W_1, W_2 are assigned senses W_1^i and W_2^j, respectively, such that

$$\left|W_1^i \cap W_2 \cap W_3 \cap \ldots \cap W_m\right| = \max_k \left|W_1^k \cap W_2 \cap W_3 \cap \ldots \cap W_m\right| \quad (2)$$

$$\left|W_2^j \cap W_1 \cap W_3 \cap \ldots \cap W_m\right| = \max_k \left|W_2^k \cap W_1 \cap W_3 \cap \ldots \cap W_m\right| \quad (3)$$

With the emergence of lexical databases, especially, WordNet where word senses are grouped into synsets organized in a hierarchical organization that creates semantic relations, Lesk's methodology has been extended in various directions. Banerjee and Pedersen [4] suggested to use the phrases that appear at each synset (sense) pertaining to individual word as a counterpart of glosses in expression (1); namely, W_1^i, W_2^j would stand for all wording involved in describing i^{th} and j^{th} synset of word W_1 and W_2, respectively. Agirre and Martinez [6] proposed to use a (WordNet-based) semantic similarity in order to identify the correct sense. The latter corresponds to the senses that maximize the semantic similarity of the two words as in (4).

$$(W_1, W_2) \rightarrow (W_1^i, W_1^j): Sim(W_1^i, W_1^j) = \max_{k,l} Sim(W_1^k, W_1^l) \quad (4)$$

This approach works only if the two words belong to the same part-of-speech. Mihalcea and Moldovan [7] extended this concept to pairs of different part-of-speech, especially for noun-verb connected via syntactic relations such as verb-object, noun-adverb. Agirre and Rigau [8] introduced the concept of "conceptual density" defined as the overlap between the semantic concept hierarchy C (root of the hierarchy) and words in the same context.

3. Method

3.1. *Part-of-speech category conversion*

In order to deal with the discrepancy of semantic information available for distinct part-of-speech where noun category has much richer hierarchy structure than other categories in WordNet lexical database, our approach consists of using a morphological transformation in order to transform all non-noun entities (identified through an initial part-of-speech tagging) into their corresponding noun entities. For this purpose, we used the Categorial Variation Database (CatVar) [9-10].

The PoS conversion augmented with CatVar is accomplished by finding the database cluster containing the word to be converted and replacing it with the target word. In case of multiple nouns that can be associated to the given word, the algorithm picks up the first noun that induces the smallest Edit distance with the original word, which favours transformations that preserve as much of the original wording as possible. Others words whose entry cannot be found in WordNet are left unchanged, e.g., named-entities.

In the same spirit as Banerjee and Pedersen [4], the process of word sense disambiguation of an individual word, say W with senses $W^1,...,W^n$, in the context constituted of a sentence $S = \{W_1, W_2, ..., W_m\}$, where each component W_i can be assigned sense in $\{W_i^1, W_i^2,...,W_i^{n_i}\}$ (n_i stands for the number of senses (synsets) of word W_i), involves the following. First, translating the non-noun senses into noun-sense using the aforementioned CatVar transformation, yielding for each W_i, $\{N_i^1, N_i^2,...,N_i^{n_i}\}$. Second, calculating, for each sense W^k ($k = 1, n$) of target word W, its associated score:

$$Score(W^k) = \sum_i \max_j Sim(N^k, N_i^j) \qquad (5)$$

where N^k stands for the noun-counterpart, if required, of the sense W^k of the target word, and $Sim(.,.)$ stands for Wu and Palmer semantic similarity measure [11]. The sense W^{k*} of the target word is then selected such that

$$W^{k*} = \arg\max_{W^k} Score(W^k) \qquad (6)$$

Inspired by SSI (structural semantic interconnection) algorithm [12], the implementation of (5-6) can be rendered simple using an iterative process by first selecting words S' in S that are monosemous, say:

$$S' = \{W_i: \text{senses}(W_i) = \{W_i^1, W_i^2,...,W_i^{n_i}\}, n_i = 1\}.$$

So that the counterpart of (5) becomes

$$Score(W^k) = \sum_{j: W_j \in S'} Sim(N^k, N_j) \qquad (7)$$

Expression (5-6) or their SSI implementations, if any, allows us to select the appropriate sense of the target word W that maximizes the overall semantic similarity in the sense of Wu and Palmer WorldNet similarity measure with all words of the context sentence S.

3.2. Use of WordNet domain category

Motivated by the existence of domain categorization in WordNet domains project,[b] the key idea is to utilize such information in the disambiguation task. Strictly speaking, WordNet domain project contains more than 100,000 domain links, where individual synset of noun, verb or adjective is assigned one or more Subject Field Codes (e.g., $doctor_n^1$ is tagged with the Medicine domain), or domain labels similar to the field labels used in dictionaries (e.g., Medicine, Engineering or Architecture). The domain labels are based on the Dewey Decimal Classification system and are arranged into a topic hierarchy [13]. We hypothesize that synsets that share the largest number of domain links are likely to have matching senses. Otherwise, if no common domain exists, the synsets that share the closest common subsumer in domains hierarchy are assumed to have coherent senses. Using a more formal representation, for a given synset W_i^j (j = 1 to n_i) of word W_i, let $D_i^j = \{d_{ij}^1, d_{ij}^2, ..., d_{ij}^{l_{ij}}\}$, j = 1 to n_i, i = 1 to m, be the set of domain links associated to synset W_i^j. Similarly, let $D_0^k = \{d_k^1, d_k^2, ..., d_k^{p_k}\}$, k = 1 to n, be the domain links associated to synset W_k of the target word W, then an alternative to semantic similarity based disambiguation (5-6) is

$$Score\_d(W^k) = \left| \bigcap_{i=0,m} \left(\bigcap_j D_i^j \right) \right| \quad (8)$$

Therefore, the sense k* is chosen so that

$$W^{k*} = \arg\max_{W^k} Score\_d(W^k) \quad (9)$$

In case where all cardinalities |.| in (8) vanish because there is no common domain link, an alternative to cardinality would be to explore the hierarchical structure of the domain links and compute the path-length $dist(.,.)$ of the underlying nodes, which draws some analogy with WordNet Wu and Palmer semantic similarity such that:

$$Score\_d'(W^k) = \min_{j,k} \sum_{i=0,m-1} dist(D_i^j, D_{i+1}^k) \quad (10)$$

Therefore, the associated sense is determined as:

$$W^{k*} = \arg\min_{W^k} Score\_d'(W^k) \quad (11)$$

---

[b]http://wndomains.fbk.eu/

Especially, (10-11) expressions are triggered only if expression (8) yields zero-value for all senses W^k.

Interestingly, domain links-based reasoning does not require the word-part of speech transformation because the domain links exist for various part-of-speech category, and provide a sound alternative approach to wordsense disambiguation. On the other hand, as far as our testing is concerning, one should notice that most of synsets are rather assigned one single domain link, therefore, the hierarchical distance based scoring function (10-11) is the most applied one in the subsequent reasoning.

3.3. *Fuzzy Borda voting scheme*

In the classical Borda count each expert gives a mark to each alternative, according to the number of alternatives worse than it. The fuzzy variant [16] is a natural extension that allows the experts to show numerically how much some alternatives are preferred to the others, evaluating their preference intensities from 0 to 1. More specifically, let $R^1, R^2,.., R^m$ be the fuzzy preference relations of m experts over n alternatives, say, $x_1,..,x_n$, yielding a preference matrix intensity for each expert k: $\left[r_{ij}^k \right]_{i,j=1,n}$, where $r_{ij}^k = \mu_{R^k}(x_i, x_j)$ being the membership function of R^k, quantifying the degree of confidence in which the k-expert prefers alternative x_i to alternative x_j. The score assigned for k-th expert to alternative x_i is aggregated as:

$$r_k(x_i) = \sum_{j=1,n \ \& \ r_{i,j}^k > 0.5} r_{i,j}^k \qquad (12)$$

Taking into account the score of each individual expert, the overall score of a given alternative x_i will be:

$$r(x_i) = \sum_{k=1}^{m} r_k(x_i) \qquad (13)$$

On the other hand, a practical eliciting of the individual (fuzzy) preference from individual expert estimation w_i of the quantity of interest as suggested in [16]:

$$r_{i,j}^k = \frac{w_i}{w_i + w_j} \qquad (14)$$

Application to disambiguation

The key in applying fuzzy Borda voting scheme to the aforementioned problem of wordsense disambiguation is first to assume the aforementioned methodologies for wordsense disambiguation as an expert in the sense of Borda voting scheme. More specifically, we shall consider four distinct experts corresponding to following:

R^1: Lesk-WordNet as in expression (4); R^2: Lesk-WordNet-CatVar as in (5)

R^3: Lesk-WordNet-Monosemous as in (7); R^4:Lesk- domain category as in (8,10)

Second, the various alternatives x_i, correspond to the various senses of the target word to be disambiguated. Third, the estimation score yielded by each of the above disambiguation method R^i with respect to specific sense will be used through (14) to elicit the membership grade $r_{i,j}^k$. Fourth, the outcome of the voting scheme corresponds to the sense x_j that yields the highest score in the sense of (13).

4. Evaluation

We used the test data from English lexical sample task used in Senseval-2 [14] comparative evaluation of word sense disambiguation systems. It contains a total of 4,328 test instances divided among 29 nouns, 29 verbs and 15 adjectives. Each test instance contains a sentence with a single target word to be disambiguated, and one or two surrounding sentences that provide additional context. The results in terms of precision, recall and runtime are reported in Table 1, together with comparison with some of the state of art approaches. The Lesk's algorithm is taken as a baseline for this analysis.

The results in Table 1 demonstrate the feasibility and high performance of our developed wordsense disambiguation algorithms. The performance achieved by CatVar semantic similarity based approach as well as Catvar –semantic similarity with syntactic features outperform the baseline by more than 19% in both precision and recall evaluations. Among the four algorithms introduced in this paper, the CatVar-semantic similarity shows a marginal improvement over the use of domain category, monosemous and WordNet based Lesk's extension. On the other hand, the use of fuzzy Borda voting scheme is also shown to improve, although, sometimes marginally the precision and recall performances with respect to the individual disambiguation algorithms (R^i, i = 1,4).

Table 1. Classification results of the developed disambiguation algorithms on SenseEval 2 competition dataset.

Algorithm	Noun(%)		Verbs (%)		Runtime (s)
	P	R	P	R	
R^1	71	68	58	55	0.071
R^2	73	69	60	54	0.082
R^3	68	59	54	53	0.051
R^4	73	66	62	53	0.042
Lesk (baseline)	61	60	21	18	0.0049
Simple Lesk	32	28	29	28	0.0143
Adapted Lesk	34	29	23	27	0.0182
Fuzzy Borda Voting	74	69	69	58	0.0983

5. Conclusion

This paper contributes to the hot topic of wordsense disambiguation and four new extensions of standard Lesk's algorithm have been provided based on the use of CatVar morphological transformation, domain link and monosemous. Next, a fuzzy-Borda voting scheme has been adapted in order to combine the outcomes of the various individual disambiguation algorithms to provide a global result. Although, the results are promising, the study prompted several interesting issues that will be further enhanced. For instance, the individual disambiguation algorithms are not fully independent from each other, which triggers interesting scenarios of accounting for the dependency level in the fuzzy Borda voting scheme. On the other hand, further theoretical results and convergence properties are still required in order to guarantee the superiority of the voting outcome over individual expert assessment.

Acknowledgments

This work is (partially) funded by the Marie Skłodowska-Curie Actions (645706-GRAGE).

References

1. R. Navigli, ACM Comp. Surv.41(2), 10 (2009).
2. T. Berners-Lee, J. Hendler, and O. Lassila. Sci. Amer. 284(5) 28 (2001).
3. M. Lesk, Proc. 5th ann. Int. Conf. Syst. Docu. ACM Press, 24 (1986).
4. S. Banerjee, T. Pedersen, Proc. 3rd Int. Conf. Intel. Text Proc. Comp. Ling. 136 (2002).
5. C. Fellbaum. WordNet – An Electronic Lexical Database, MIT Press, 1988.
6. E. Agirre and D. Martine, Proc. SENSEVAL-2 Work. ACL'2001/EACL'2001.
7. R. Mihalcea and D. I. Moldovan, Proc. NAACL WordNet and Other Lex. Res. 95 (2001).
8. E. Agirre, G. Rigau, Proc. 16th Int. Conf. Comp. Ling. 16 (1996).

9. N. Habash and B. Dorr, *Proc. North Amer. Chap. Assoc. Comp. Ling. Hum. Lang. Tech. 1*, 17, (2003).
10. M. Muhidin, and M. Oussalah. Proc. *COLING'14 37* (2014).
11. Z. Wu and M. Palmer, Proc. 32nd, Ann. Meet. Assoc. Comp. Ling. 133 (1994).
12. R. Navigli and P. Velardi, Proc. 3rd Int. Work. Eval. of Syst. Sem. Anal. Text, 179 (2004).
13. B. Magnini and G. Cavagli, Proc. 2nd Int. Conf. Lang. Res. Eval. (LREC 2000), 1413 (2000).
14. P. Edmonds and S. Cotton. Proc. 2nd Int. Work. Eval. Word Sense Disam. Syst. 1 (2001).
15. S. Atkins. Tools for computer-aided lexicography: The Hector project. Acta Ling. Hung. 41, 5 (1993).
16. H. Nurmi, *Group, Dec. Negot.* 10(2), 177 (2001).

A novel path planning approach for unmanned ships based on deep reinforcement learning[*]

Chen Chen

School of Computer Science and Technology, Wuhan University of Technology
1040, Heping Avenue, Wuhan, Hubei, China

Feng Ma, Jia-Lun Liu, Xin-Ping Yan

Intelligent Transportation System Center, Wuhan University of Technology
1040, Heping Avenue, Wuhan, Hubei, China

Xian-Qiao Chen

School of Computer Science and Technology, Wuhan University of Technology
1040, Heping Avenue, Wuhan, Hubei, China

Intelligent ships which are the representatives of the under-actuated agents need to navigate safely with finite power and energy, therefore the requirements about path planning rises. This paper proposed an approach based on reinforcement learning method algorithm to set up an online artificial intelligent system to plan path for the under-actuated water surface agents. A parallel simulation meshing platform is constructed to predict and feedback the situation around ships and the ships' own motion. Facing with the giant feature space, the value function is connected with determination of path planning, and its searching speed is raised for improvement of energy, safety and efficiency, by using many theories including navigation rules and shipping experience. Eventually, this online artificial intelligent system is expected to be built by exercising parameters and structures of network and value function. This proposed approach will be capable of autonomous shipping effective, which will have a comparable result to manual shipping, and been capable of keeping the path planning, motion control, motion rules and motion experience synchronized.

1. Introduction and Background

Since the cost has risen rapidly in recent years, shipping companies have been under much pressure. Human cost in shipping enterprises accounted for a large proportion, reached between 30 and 40 percent of the total operating cost. As a result, ships need to reduce crew to reduce costs. At the same time, the crew members are easily fatigued during the long-term work at sea. According to

---

[*]The first author Chen Chen is financed by the China Scholarship Council under Grant 201706950028. This work is supported by the National Natural Science Foundation of China (Grant no. 61503289).

statistics, 85 % of ship safety accidents are caused by human factors. Therefore, autonomous ship driving provides a new method for enhancing the safety of ships and reducing navigation costs. On the other hand, studies have pointed out that the unmanned transport of ships can not only reduce carbon emissions by 20% but also increase operational efficiency by 20% [1][2]. Nearly three years, the United Kingdom, Norway, and France have all launched different shipping research projects and are planning to operate a driverless cargo ship around 2030. For the above multiple requirements of economic benefits, navigation safety, energy conservation and emission reduction of shipping enterprises, the "autonomous driving" of ships is of great research significance. Compared with the autonomous control of water surface intelligent robot, it has a good research foundation in the fields of military investigation, search and rescue, hydrological survey and so on.

Under the promotion of the 'e-Navigation' strategy suggested by the IMO (International Maritime Organization), technologies such as perception, interaction and automatic control of ship navigation environment have been greatly improved. Considering the under-driving characteristics of ships, a reasonable path must be planned under the constraints of safety, economy and reliability on the premise of communication or understanding the intention of encounter ships. Path planning has been well studied in the field of robotics. Many algorithms have been studied in the field of robotics in order to improve the speed and reliability of path planning, such as a PRM (Probabilistic Road Maps) algorithm and a RRT (Rapidly-exploring Random Tree) algorithm [3][4][5]. PRM is a multiple-query method, which uses historical data samples to create a road map in the configuration space, and then uses traditional a A* algorithm to query path on the road map. Therefore, the core of this method is not to find the path, but how to sample and build the road map. This method can find the best path in very large space, not only in two-dimensional space, but also in three-dimensional space. Due to the only drawback is the large amount of computation, some researchers have transformed it into a single-query method. But because of the non-uniform sampling leads to phenomenon often occurs that some paths exist but cannot be planned. This kind of problem is called "narrow passage". To address the shortcomings of the PRM, the RRT was proposed in 2000. This algorithm takes the starting point and the target point in the configuration space as the root respectively, generates two trees at the same time, and gives them a random increment every time, so that they grow towards each other until they are merged together. As mentioned above, these algorithms have their own advantages, but the lack of comprehensive evaluation index of path rationality, cannot fully consider the ship maneuverability, safety, economy, and are not suitable for under actuated ships.

This paper introduces reinforcement learning to solve the problem of path planning, which regards this problem as a continuous path selection problem under the trade-off of profit and loss. In 1956, Bellman proposed the dynamic programming method [6], which laid the foundation for reinforcement learning. Watkins [7] suggested a Q-learning algorithm in 1989, which is also the most common reinforcement learning algorithm. A standard reinforcement learning setup consists of an agent, which can interact with an environment [8]. The agent in different states has different actions, and the agent will receive different rewards given by the environment. When this process is repeated enough, the agent generates memory that can handle choosing the most reasonable action under different states. However, this algorithm was neglected with the limited computer performance until 2015, when the Deep Mind Technologies of Google proposed the Deep Q-Network algorithm [9]. In the same year, an artificial intelligent team in America which is major in driverless car applied this algorithm to the intelligent driving field. Facing smart cars in real-time, they make use of the Convolution Neural Network to learn features, reduce the information dimension, and then take advantage of the Monte Carlo Tree method to improve the search efficiency of the value function [10]. It is helpful to provide the best results more quickly without manual assistance. The typical applications are Google driverless car and Tesla car [11], which use LIDAR with SLAM technology to locate and reconstruct maps in any unknown environment. In addition, laser radars, cameras and other sensors cooperate with each other to solve the problem of perception in road environment.

However, there are many uncertain factors that need to be considered about ships, such as waterway environment, encounter ships and their own maneuverability and controllability. These factors constitute the understanding of navigation conditions, called as "navigation conditions". It can be said that using deep reinforcement learning to solve the problem of ship route planning is a worthwhile exploration direction.

2. A Proposed Method

Markov decision processes formally describe an environment for reinforcement learning, and the environment is fully observable [12]. When an agent performs a task, it interacts with surrounding environment by an action. With the action and the environment, the agent generates a new state and the environment gives a return. After several iterations of learning, intelligent physical finally learned to select actions to maximize total future reward. The process of reinforcement learning is shown as Figure 1.

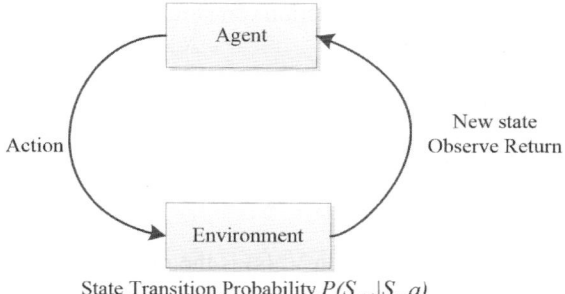

Figure 1. The process of reinforcement learning.

Suppose a computational agent moving around in some discrete finite world, selecting one from a finite set of actions at each state. This world constitutes a Markov decision process, which all states are Markov [13]. At each state S_t, the agent executes action A_t and receives a scalar reward R_t, whose reward value R_t, just depends on the state S_t an d action A_t. The state transition probability matrix is represented as in Eq. (1).

$$P^a_{ss'} = P[S_{t+1} = s' | S_t = s, A_t = a] \tag{1}$$

The goal of the agent is to find the optimal strategy for a Markov decision process. The so-called policy refers to maps from states to actions typically represented by π, which refers to a distribution over the set of actions given the state, as in Eq. (2).

$$\pi(a|s) = P[A_t = a | S_t = s] \tag{2}$$

Value function is a prediction of future reward, which is used to evaluate the goodness or badness of states, presented by Eq. (3).

$$\begin{aligned} V_\pi(s) &= E[G_t | S_t = S] \\ &= E_\pi[R_{t+1} + \gamma R_{t+2} + \gamma^2 R_{t+3} + \ldots | S_t = s] \\ &= E[R_{t+1} + \gamma(R_{t+2} + \gamma R_{t+3} + \ldots) | S_t = s] \\ &= E[R_{t+1} + \gamma G_{t+1} | S_t = s] \\ &= E[R_{t+1} + \gamma V(S_{t+1}) | S_t = s] \end{aligned} \tag{3}$$

The discount $\gamma \in [0,1]$ is the present value of future rewards. The state-value function can be decomposed into immediate reward plus discounted value of successor state. The action-value function can similarly be decomposed as in Eq. (4).

$$q_\pi(s,a) = E_\pi\left[R_{t+1} + \gamma q_\pi(S_{t+1}, A_{t+1}) \mid S_t = s, A_t = a\right] \quad (4)$$

The theory of Dynamic Programming (Bellman & Dreyfus, 1962; Ross, 1983) assures that there exists an optimal policy π^*, which is better than or equal to all other policies.

The optimal state-value function V*(s) is the maximum value function over all policies, as in Eq. (5)

$$V_*(s) = \max_\pi V_\pi(s) \quad (5)$$

The optimal action-value function $q*(s,a)$ is the maximum action-value function over all policies, as in Eq. (6)

$$q_*(s,a) = \max_\pi q_\pi(s,a) \quad (6)$$

An optimal policy can be found by maximizing over $q*(s,a)$, as in Eq. (7).

$$\pi_*(a \mid s) = \begin{cases} 1 & \text{if } a = \arg\max_{a \in A} q_*(s,a) \\ 0 & \text{otherwise} \end{cases} \quad (7)$$

In order to solve the path planning problem with reinforcement learning, it is necessary to make the real world similar to the electronic game, which is mainly reflected as follows:

1. Abstract ships and waterways into symbols to meet input requirements of reinforcement learning. In this respect, the ship has a unique advantage over the car.

2. Reinforcement learning is currently discussing more model-free prediction, which is quite different from the regular traffic world. How to guide and enhance learning and improve training effect based on rules is the focus of this study.

In the field of ships, the abstraction of the real world becomes simple. Because there are few obstacles in waterway except for ships and icebergs. Moreover, most ships are equipped with AIS, radar and other sensors, and also have electronic charts to describe the environment. Finally, the navigation of ships is relatively free, which avoidance space is larger than the vehicle.

In this paper, the ship navigation environment is abstracted as a virtual scene similar to "electronic game". On the basis of reinforcement learning, the evaluation and calculation of path "return value" is formed on the basis of safety constraints. Finally, using the artificial operation data and "simulation environment", training step by step, and establish the online intelligence of autonomous path planning in the end.

3. A Field Testing

This research uses Visual Studio 2014 to simulate with the experiment, which is called "Ship Manipulation v0.11", as shown in Figure 2. For simplicity, the target ship model and the other ships models that have the same shape and size, which are set in the platform.

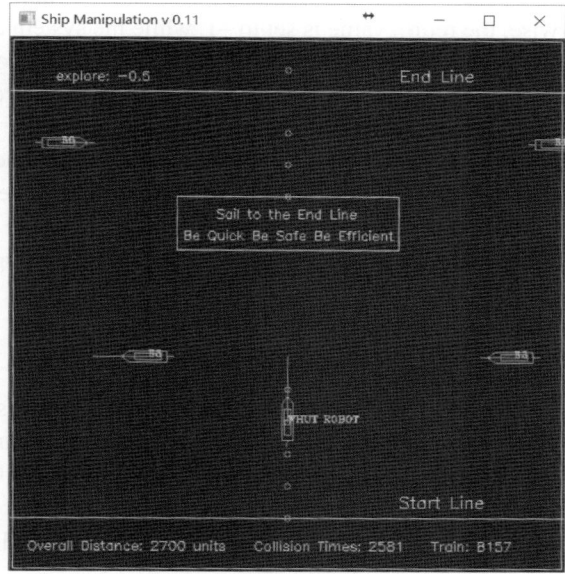

Figure 2. The interface of Ship Manipulation v0.11.

In this platform, ships with different orientations and speeds are random passing through two simulated channels with 1.0 to 5.0 units per seconds. In this condition, the maneuvered ships need an assistance to cross the channels which can help it to plan path correctly. Q-Learning is introduced to build artificial intelligent to decide whether to cross the channel or to wait.

An "explore" parameter is signed in the top left corner and has two states, including random "explore" and Q-Table as described in the previous section. The "explore" state includes exploring the position of maneuvered ship and other ships, and the speed of other ships. The status information may be recorded or the q table updated when the maneuvered vessel reaches the boundary. If the maneuvered ship collides with another ship, it will be recorded as experience to fill in the Q-Table. At the same time, this experience is quantified as a value to evaluate the collision trend. It is recommended that the target ships to change their direction next time. If the "explore" value is +0.5, it means that the target ship is navigating based on learning experience. There is a "return" value which indicates the trend in deciding to how to take actions going forward or stopping. At the bottom right of the interface, record "train times" and "collision times" represent times of training path planning algorithm and the number of collision respectively.

At each step, the maneuvered ship observes the position, speed and direction of the other ships, and then makes acceleration or deceleration action. If it does not hit another ship, the return value is set to the current cumulative distance forward. Otherwise, the return value is set to -1, while the system ends the game and begins learning again. After many times of learning, the agent can choose the best action according to the state of other ships in the current environment.

The experiment result shows that that the obstacle avoidance rate is 19.2% when the maneuvered is trained 8,157 times. A following experiment shows that the manipulated ship might avoided with obstacles at 8.2% percent when train times is 21,087, which was run on a computer for two days. These state data and return value data are recorded in Q-Table as t as a basis for selecting actions from the operation space.

4. Conclusions

In summary, Q-Learning might be a practical principle for establishing artificial intelligence in ship route planning. However, this research is preliminary, which does not take into account the characteristics of the ship. The following issues may need to be discussed in the future.

1. More complex routes such as turning and stopping should be taken into consideration;

2. Large space and scenarios are also needed in the following research, and too much coordinates and ship status may lead to "dim disaster". The training of Q-Table will continue forever. Deep Q-Network proposed by Deep Mind might be practical in this case.

Acknowledgments

This work is supported by the National Natural Science Foundation of China (Grant no. 61503289).

References

1. J. Luo, et al., Anti-disturbance control for an underwater vehicle in shallow wavy water, *Procedia Engineering*, 15(1) (2011).

2. J. Luo, H. Liu, C. Huang et al., Denoising and tracking of sonar video imagery for underwater security monitoring systems, *IEEE International Conference on Robotics and Biomimetics*, (2013).
3. C. Watkins, J.C. Hellaby, Learning from delayed rewards, Diss. King's College, Cambridge, (1989).
4. L. Kavraki, P. Svestka, J. Latombe et al., Probabilistic roadmaps for path planning in high-dimensional configuration spaces, *IEEE Trans on Robotics & Automation*, (1994).
5. S. Lavalle. Rapidly-exploring random trees: a new tool for path planning, *Algorithmic & Computational Robotics New Directions*, (1998).
6. R.E. Bellman, S.E. Dreyfus, Applied dynamic programming, *Journal of the American Statistical Association*, (1962).
7. C.J. Watkins, P. Dayan, Technical note: q-learning, *Machine Learning*, (1992).
8. A.G. Barto, S.P. Singh, On the computational economics of reinforcement learning, *Connectionist Models*, (1991).
9. V. Mnih, K. Kavukcuoglu, D. Silver et al., Human-level control through deep reinforcement learning, *Nature*, (2015).
10. F.P. Such, V. Madhavan et al., Deep neuroevolution: genetic algorithms are a competitive alternative for training deep neural networks for reinforcement learning, *Neural and Evolutionary Computing* (2018).
11. T.H. Pearl, Fast & furious: the misregulation of driverless cars, *Social Science Electronic Publishing*, (2017).
12. M. Sato, K. Abe, H. Takeda, Learning control of finite Markov chains with an explicit trade-off between estimation and control, *Systems Man & Cybernetics IEEE Transactions on*, (1988).
13. M.L. Puterman, *Markov Decision Processes: Discrete Stochastic Dynamic Programming*, John Wiley & Sons, Inc., (1994).
14. R.J. Williams, Simple statistical gradient-following algorithms for connectionist reinforcement learning, *Reinforcement Learning*, Springer US, (1992).

Part 5

Data Analytics for Classification

Naive Bayes clusterer

Mujiexin Liu, Hongjun Wang,* Tian Rui Li and Ping Deng

School of Statistics, University of Minnesota
Minneapolis, MN 55414, USA
* liux1780@umn.edu
www.umn.edu

School of Information Science and Technology
Chengdu, Sichuan 610031, China
wanghongjun@swjtu.edu.cn

It is common practice to process labeled data with classifiers and unlabeled ones with clusterers. The idea of combining supervised and unsupervised learning methods to process data in a more efficient way leads to semi-supervised learning methods, which effectively utilizes both labeled and unlabeled data. However, the thought of processing unlabeled data with a combination of supervised and unsupervised methods to achieve better efficiency and accuracy calls for the proposal of a brand new clustering model that we call Naive Bayes Clusterer Model (NBCM). In this paper, we built NBCM based on an optimization problem of conditional probability and solved it by partitioning the hyperedges consists of learnt label groups from unsupervised learning. Extensive experiments were conducted and results on real data show that NBCM outperforms all other base clustering algorithms compared with the highest level of accuracy.

Keywords: Naive Bayes; unsupervised learning; similarity matrix.

1. Introduction

Differentiating collected data into distinctive groups or clusters has been an objective for almost all businesses, also a direction of research for statisticians and computer scientists. Multiple classification methods have been developed for data sets with marked labels, such as Naive Bayes Classifier,[1] K-nearest-neighbors method,[2] Random Forests,[3] etc. Unlabeled data have not left untouched, since clustering algorithms are developed to solve this problem. A classic example is K-means, first proposed by Stuart Lloyd in 1957 and modified by E.W. Forgy in 1965, was eventually published in public in 1982.[4,5] K-medoids, alongside K-means, an optimized method replacing cluster centroids with exemplars, fixes the problem of highly inaccurate centroid when dealing with large noise.[6] Both methods try to minimize distance between points to be labeled to a cluster to its centroid/exemplar and are widely used for its simplicity and low computational complexity.

Despite the obvious advantages, whenever a classifier is applied, the operational cost, randomness and objectiveness for manual labeling process are high. The approach of combining supervised learning and unsupervised learning to build a more accurate classifier that can automatically obtain labels from base clusterings is a potential applicable solution. Raw data are first processed by multiple cluster algorithms (K-means,[7] K-Medoids[8] and FCM[9]) from which a collection of labels are returned, put in an indicator matrix and used as hyper-edges[10] by METIS[11,12] to output an agreed set of labels. This label is then used to train Naive Bayes Classifier[13] model to increase quality and efficiency of classification results. The resulting Naive Bayes Clusterer is ground-breaking in several contexts: 1) Clustering with a classifier: In the entire process, all labels used in either training or testing stages are extracted from base-clusterings and combined by a consensus function, all of which consists of unsupervised learning methods. No true label is involved unless for calculating accuracy for parallel comparison. 2) Using data with the highest level of confidence: only data with matching labels of value 1 are selected for our similarity matrix S. We will show in the experiment section that this selection process greatly improves the accuracy of clustering results. 3) Ability to process diversified data in real-time: Raw data with or without true labels can be directly thrown in the NB cluster and obtain a consensus label. Consider the ensembling of base clusters based on shared information, the more diversified its features are, the more accurate and robust the result cluster is.

The rest of the paper is organized as follows. In Section 2, we present the Naive Bayes Clusterer Model and its algorithm. In Section 3, we show the basic information about our data sets and present an empirical study for NBCM along with experiment results. Finally, the paper ends with conclusions and further research topics in Section 4.

2. Naive Bayes Clusterer Model

In this section we propose Naive Bayes Clusterer Model (NBCM) and provide a detailed inference. Also its algorithm is described.

Given a data set $X = \{x_i^f, i = (1, 2, ..., N), f = (1, 2, ..., M)\}$, and

$C = \{c_j, j = (1, 2, ..., K)\}$ is the center set. NBCM can be represented as

$$C^* = arg\max_C \sum_{i=1}^{N}\sum_{j=1}^{K} p(c_j|x_i^f),$$

where $p(c_j|x_i^f)$ is the probability that the point x_i^f belongs to c_j. According to Bayes theorem,

$$p(c_j|x_i^f) = \frac{P(x_i^f|c_j)P(c_j)}{P(x_i^f)},$$

where $P(x_i^f)$ is constant for all centers, so

$$p(c_j|x_i^f) = P(x_i^f|c_j)P(c_j).$$

Assume that attributes are conditionally independent.

$$P(x_i^f|c_j) = \prod_{f=1}^{M} P(x_i^f|c_j).$$

Then,

$$p(c_j|x_i^f) = P(c_j)\prod_{f=1}^{M} P(x_i^f|c_j).$$

But the problem is that $P(c_j)$ and $P(x_i^f|c_j)$ is unknown.

If the data points are with corresponding labels, the above problem is solved. Unsupervised learning is used to obtain part of labels among data points, and the data points with learnt labels are as the seeds to boot NBCM. It is obvious that the possibilities of data points belong to a certain cluster are negatively related to their distances between.

To maximize $p(c_j|x_i^f)$, we want to minimize the distance from point x_i^f to cluster center/exemplar of clusters C_j. To reach this goal, we developed the Naive Bayes Clusterer model based on clustering methods and Naive Bayes Classifier. The base-clustering methods with varying degrees of efficiency is integrated unsupervised learning. The learnt labels are then put in the form of an indicator matrix, with which we can check the cross-sections of the learnt labels in binary form. Similarity matrix S of indicator matrix is then derived: $S = \frac{1}{r}\mathbf{HH}^\mathsf{T}$ where $\mathbf{H}$ is the indicator matrix consists of labels from base clustering models, and r is the number of attributes in a data set.

With the S matrix, we can now pick out those labels that are 'globally accepted', which are those labels with value=1 in S:

$$S_{ij} = 1 \Rightarrow S_{ij} \in C$$

where C, as the final consensus label, is a vector containing all satisfactory S_{ij}s. This combined label is highly informational and is best used as training data, along with its corresponding cluster. The easiest and most efficient model to train is the Naive Bayes Classifier. Looking back to the objective function of our proposed model, the conditional probability on the denominator, which is also the likelihood of data point x_i^f belonging to cluster C_j, is obtained from the Naive Bayes Model:

$$p(C_j|x_i^f) = \frac{p(x_i^f|C_j)p(C_j)}{p(x_i^f)}.$$

Since C_j is a part of C, we can plug in the consensus label vector C obtained from the previous step and get the resembling likelihood $p(x_i^f|C_j)$:

$$p(x_i^f|C_j) = \prod_{k=1}^{i} p(\lambda_i|C_j)$$

where λ_i are the features of x_i^f, $p(\lambda_i|C_j)$ and $p(C_j)$ can be calculated by counting the number of labels in each grouping. And the input priori probability of x_i^f is:

$$p(x_i^f) = \prod_{k=1}^{i} p(\lambda_i).$$

With both parameters and priori probability ready, raw data can now be used as an input to test our model. The algorithm of NBCM is shown in the subsection below. We discovered that when training Naive Bayes with C, the accuracy of the trained model is noteworthy comparing with K-means, Density Peak (DP), Affinity Propagation (AP). The comparison results will be shown in-detail in the next section.

2.1. Algorithm In-depth View

Naive Bayes Clusterer Algorithm:
Input: Raw data points with or without true labels.
Output: A consensus label of a combined cluster.

(1) Input raw data;
(2) Data discretization;
(3) Apply base-clustering;
(4) Input learnt labels to indicator matrix H and calculate matrix S;
(5) If $S_{ij}==1$, map S_{ij} to new matrix S^*;
(6) Input S^* to METIS.[11]
(7) Return consensus label C.
(8) Train Naive Bayes with consensus label C.

As we have mentioned before, the entire learning process is unsupervised. We choose the highest valued attributes in S and map them to a new matrix S^* so that the resulting consensus label has the maximum level of shared information among the learnt labels, so that we can maximize the accuracy and reliability of results from NBCM with this consensus label as input.

3. Empirical Study for Naive Bayes Clusterer Model

In this section we show the experiments done with discrete data sets under base clustering methods K-means, K-medoids and FCM. In this experiment, we compare the performance of NBCM with that of K-means, Density Peak,[14] and Affinity Propagation.[15] K-means is very representative as a classic base clusterer while DP and AP methods are optimized methods with less noise. The diversified data sets we used in are summarized in Table 1.

3.1. Data Source

Table 1 shows the source of the data sets used in our experiment along with their instances, sources and categories.

All data used in this experiment are real data. Most data sets are discrete data sets, since Naive Bayes Classifier can only work with discrete data.

Table 1. The sources and the number of the instances, features and categories in each dataset.

Dataset Name	Rename	Data Source	Instances	Features	Categories
Cmap	D1	UCI	300	27	6
Phishing	D2	UCI	1353	9	3
syncon	D3	UCI	600	60	6
wbdc	D4	UCI	569	30	2
ionosphere	D5	UCI	351	34	2
hepa	D6	UCI	155	19	2
wine	D7	UCI	178	13	3
biodeg	D8	UCI	1055	41	2
dermatology	D9	UCI	366	34	6
weapon	D10	Microsoft	858	899	3
aerosol	D11	Microsoft	905	892	3

3.2. Results

Table 2 shows the comparison of performance among NBCM and three other models. Since this is a simple experiment, we only use accuracy of models as our evaluation indicator. We use Micro-precision[16] as indicator of accuracy. It is calculated as shown:

$$MP = \sum_{k=1}^{K} \frac{N_k}{N}$$

Table 2. Accuracy of different models compared with 11 diverse data sets. Results are ranked from most accurate to least accurate.

Datasets \ Models	NBCM	K-Means	DP	AP
D4	**0.9332**	0.8541	0.7908	0.8541
D8	**0.8033**	0.7022	0.7078	0.6685
D3	**0.7883**	0.7050	0.5466	0.5633
D9	**0.7270**	0.5886	0.4407	0.4407
D5	**0.7236**	0.7122	0.4871	0.7037
D7	**0.6387**	0.5612	0.4516	0.5548
D1	**0.6133**	0.6033	0.4933	0.4266
D6	**0.5218**	0.2540	0.3415	0.2486
D2	0.5166	0.6104	**0.6415**	0.0753
D11	**0.4347**	0.3123	0.4347	0.2202
D10	0.4000	0.3458	**0.4828**	0.2662

As we can see, NBCM achieved the highest accuracy for all datasets except for D2 and D10. Also, there are no strict relation between the dimension/number of feature of a dataset and the corresponding accuracy of results.

4. Conclusion

In this paper we proposed a model that can output clustering labels with the Naive Bayes classifier and have proven that it has the highest accuracy when applied on all but two data sets under the same environment. To the best of our knowledge, it is the first study to solve the cluster ensemble problem with NB model, which is a supervised learning model. The input to the trained model is the combined label chosen from multiple base-clusterers that covered the most information among clusters, and the output is a classifier-turned cluster model. As of cluster criterion, the distance between a data point and its possible cluster centers is converted into its level of association to the cluster. Empirical studies on data sets from UCI and Microsoft are conducted to evaluate the effectiveness of our model along with Affinity Propagation (AP), Density Peak (DP) and K-means. The results demonstrate that our model out-performs all other methods in the experiment and it can output a consensus cluster with high accuracy and low cost.

Even though our current model is proven significant, its accuracy and robustness can still be greatly improved. In future studies, we wish to test and train our model on larger data sets and those with higher dimensions to further optimise NBCM.

Acknowledgments

This paper and corresponding research is partially supported by Blockchain Technology Fund of Hi-Tech Information Technology Research Institute of Chengdu (No. 2018H01207), the Online Education Research Center of Ministry of Education Online Research Fund (No. 2016YB158) and National College Students Innovation and Entrepreneurship Training Program (No. 201510638047).

References

1. S. M. Stigler, *The American Statistician* **37**, 290 (1983).
2. B. W. Silverman and M. C. Jones, *International Statistical Review / Revue Internationale de Statistique* **57**, 233 (1989).
3. T. G. Dietterich, *Machine learning* **40**, 139 (2000).
4. S. Lloyd, *IEEE transactions on information theory* **28**, 129 (1982).
5. E. W. Forgy, *Biometrics* **21**, 768 (1965).
6. D. E. Goldberg, J. Richardson *et al.*, Genetic algorithms with sharing for multimodal function optimization, in *Genetic algorithms and their applications: Proceedings of the Second International Conference on Genetic Algorithms*, 1987.

7. J. MacQueen *et al.*, Some methods for classification and analysis of multivariate observations, in *Proceedings of the fifth Berkeley symposium on mathematical statistics and probability*, (14) 1967.
8. L. Kaufman, P. K. Hopke and P. J. Rousseeuw, *Using a parallel computer system for statistical resampling methods* (University of Technology, 1987).
9. J. C. Bezdek, R. Ehrlich and W. Full, *Computers & Geosciences* **10**, 191 (1984).
10. F. Makedon and S. Tragoudas, *Algorithmic Aspects of VLSI*, 133 (1993).
11. G. Karypis and V. Kumar, *Journal of Parallel and Distributed computing* **48**, 96 (1998).
12. G. Karypis and V. Kumar, hmetis: A hypergraph partitioning package, november 1998.
13. M. Bayes and M. Price, *Philosophical Transactions (1683-1775)*, 370 (1763).
14. A. Rodriguez and A. Laio, *Science* **344**, 1492 (2014).
15. B. J. Frey and D. Dueck, *science* **315**, 972 (2007).
16. Z.-H. Zhou and W. Tang, *Knowledge-Based Systems* **19**, 77 (2006).

Semi-supervised cluster ensemble based on density peaks

Kadhim Mustafa, Hongjun Wang,* Yuan Zhou and Jian Song

School of Information Science and Technology,
Southwest Jiaotong University
Chengdu, Sichuan 60031, China
*wanghongjun@swjtu.edu.cn

In this paper, we propose a clustering ensemble method based on density peaks and proved Semi-supervised affinity propagation (SSAP) to get the base clustering gives better results. Further, we investigate different similarity matrices (Rapid Maximal Information Coefficient (RapidMic), Cosine, and Earth Mover's (EMD), Euclidean) to find out the ones which can best conjugate the proposed method. Our method enhances the density peak clustering by using semi-supervised learning qualities and extends it to clustering ensemble. Henceforth, we call this technique Semi-supervised Cluster Ensemble based on Density Peaks (SSEDP). We prove Density Peaks can be an effective semi-supervised clustering ensemble that gives high accuracy and consistent performance when used with SSAP, cosine similarity, and pairwise constraints to redefine the similarity matrices of the base clustering. We used real-world datasets to evaluate the method. The experimental results demonstrate its functionality and it entirely outperforms some of the previous ensemble methods.

Keywords: Clustering ensemble; density peaks; semi-supervised clustering; affinity propagation.

1. Introduction

Cluster ensembles conjunct various clusterings of a variety of objects into a unified clustering, which is called consensus function. Further, it is vastly used in the scope of artificial intelligence, utilized in many domains of applications, and studied in various areas of pattern recognition, applied statistics, machine learning, and information theory.[1] The cluster ensemble proposed by Strehl and Ghosh[2] that introduced three consensus functions (CSPA), (HGPA), and (MCLA). However, many consensus functions existed such as Quadr mutual information consensus function (QMI) and Mixture Model (EM),[3] graph partitioning with multi-granularity link analysis (GP-MGLA),[4] probability trajectory based graph partitioning (PTGP),[5] Bayesian Cluster Ensembles (BCE).[6] These functions have got different shortcomings.

Recently many clustering algorithms have been proposed, the most popularly used is Density peaks (DP).[7] The knowledge of DP is based on "cluster centers are distinguishing through a higher density than other neighbors and a comparatively large distance of higher density points".

Several Clustering algorithms and their ensembles have gained good accuracy while the semi-supervised framework has been incorporated[8-10] which effectively enhances the results than the original unsupervised versions. The most commonly used technique in semi-supervised framework is the instance-level constraints that dictates whether a couple of data objects can be in the same group or not, this is known as *must-link (ML)*, and *cannot-link, (CL)* respectively.

In this paper, we utilized Semi-supervised Affinity Propagation clustering[11] to get base clustering results, and amended the unsupervised DP algorithm using the instance-level constraints[8] to achieve our the proposed model, henceforth termed as Semi-supervised cluster ensemble based on Density peaks (SSEDP). Standard datasets are chosen to evaluate the proposed model and the other consensus functions and perform the Micro-precision to get the accuracies.

The rest of the paper is organized as follows. Section 2 explains the proposed model Semi-supervised cluster ensemble based on Density Peaks. Section 3 reports the experimental steps and results. Section 4 discusses the conclusion and future work.

2. Semi-Supervised Cluster Ensemble Based on Density Peak

2.1. *Unsupervised density peaks*

It is the most popular evolution of unsupervised clustering that integrates both density and distance. In another word, density peaks algorithm (DP) is based on the fundamental hypothesis that cluster centers hold higher density and encompassed them with lower density data objects. The similarity between cluster centers is comparatively longer. The algorithm calculates the density ρ_i and the minimum distance δ_i then finds the cluster centers by the plotting ρ_i against δ_i and allocates each remaining object to the same cluster as its closest neighbor with higher density.

Some experiments on DP have got different unsuitable outcomes specially with varying large densities, this leads to critically inaccurate and incorrect clustering outcomes.[10,12] Thus, to solve this issue, some researchers have improved the clustering performance by combining the semi-supervised information that we discuss in Section 2.2.

2.2. Proposed method

As we mentioned earlier this paper integrates the advantages of semi-supervised clustering algorithms and cluster ensemble to acquire best clustering accuracy. Figure 1 describes the steps of the proposed method.

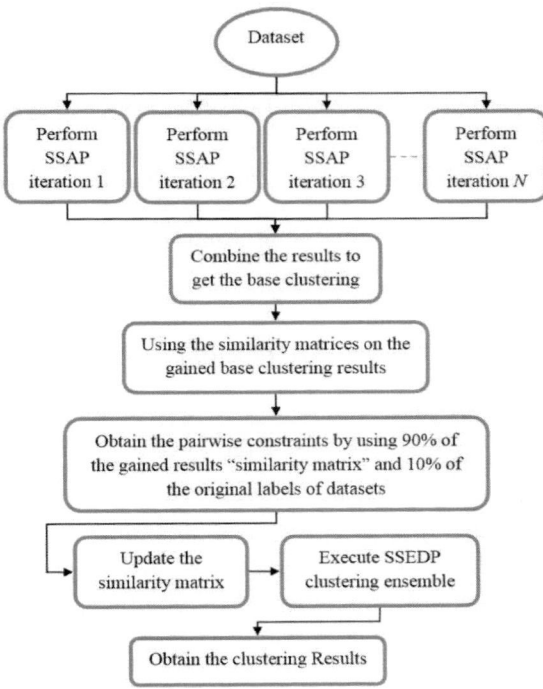

Fig. 1. Semi-supervised Cluster Ensemble Based On Density Peaks.

We choose Semi-supervised Affinity propagation algorithm (SSAP)[11] to get the base clustering result. The SSAP does not need to specify the number of clusters since all the data points are used as potential clustering centers.

We select four distance measurements like Rapid Maximal Information Coefficient (RapidMic),[13] Cosine,[14] Earth Mover's (EMD),[15] and Euclidean[16] to survey which similarity matrix is the best compatible with our proposed method.

Our proposed method includes several steps to meet the major modification. First, perform the SSAP to gain the base clustering results $BaseCstr$

then utilize the similarity matrices (Rapid, Cosine, Euclidian, EMD) on the base clustering results *BaseCstr* and set the results in *disM*. Secondly, get 10% of the original data labels randomly *rl* to Preform the pairwise constraints on *rl* then update the value of the data points of the similarity matrix *disM*. Finally, our proposed method the Semi-supervised Cluster Ensemble based on Density Peaks (SSEDP) concludes by utilizing the Density Peaks to give the ensemble outcomes. Algorithm 2.1 illustrates the procedure of our method. The experimental results prove the modification caused good performance and increased the accuracy of the cluster ensemble outcomes.

Algorithm 2.1. *Semi-Supervised Cluster Ensemble Based On DP*

Require: *Distance matrix (disM), cutoff distance (cd), Dataset original labels (labels), number of clusters (K)*
Ensure: *the consensus clustering C_k.*
1: *calculate ls, ls = size(labels);*
2: *calculate rv, rv = genRandomInt[ls*10%, ls];*
3: *calculate rl, rl = labels(rv);*
4: **for** *(i = 1 : size(rl))* **do**
5:    **for** *(j = i + 1 : size(rl))* **do**
6:       **if** *($rl_i = rl_j$)* **then**
7:          *Constraintsstatus = 1;*
8:       **else**
9:          *Constraintsstatus = 0;*
10:      **end if**
11:      *disM(i, j) = Constraintsstatus*
12:      *disM(j, i) = Constraintsstatus*
13:    **end for**
14: **end for**
15: *Density calculation $p_i = \sum_{j \in D} x(dist(i,j) - Eps)$*
16: *Minimum similarity calculation $\delta_i = \min_{j: j \in D, p_j > p_i}(dist(i,j))$*
17: *Find γ_i to choose the cluster centers $\gamma_i = \delta_i \times p_i$*
18: *Allocated each remaining object to the same cluster as its closest neighbor of higher density*

In the algorithm above from point 4 to 14 content the procedure of applying pairwise constraints, point 6 means if rl_i equals rl_j so they must belong to same cluster (*ML*), otherwise point 9 means rl_i and rl_j cannot belong to same cluster (*CL*), then update the similarity matrix to perform the rest steps of the algorithm. The clustering results can be evaluated by micro-precision [17] (*MP*), the largest value of micro-precision indicates the best clustering performance. The formula of MP is expressed as $\frac{1}{m} \sum_{t=1}^{c} a_t$.

3. Experimental Steps and Results

In this section, we discuss our experimental results and clarify on the compatible distance measure with our method to gain better effectiveness. We used thirteen different real-world datasets from UCI repository. The main characteristics of the datasets are summarized as dataset "name(number of objects, Features, classes)" the datasets are indianliver(583, 10, 2), plrx(182, 12, 2), heartdisseaseh(294, 13, 5), hvwntr(606, 100, 2), indi(583, 8, 2), monster(793, 899, 3), gun(822, 899, 3), drawing(827, 899, 3), pointer(828, 899, 3), diamond(831, 899, 3), egg(833, 899, 3), bow(834, 899, 3), biodeg(1055, 41, 2).

We first used the datasets as an input to the SSAP to get the base clustering results then performed the consensus functions to compared them with SSEDP, furthermore, we ran the similarity matrices to find out the best compatible one with SSEDP. We performed the experiment in ten iterations for each dataset and evaluated the outcomes by Micro-Precision to get the cluster ensemble results.

Table 1 shows our experimental results. We can notice the proposed method has got the best results in all cases. The SSEDP obtained eight best results with cosine (SSEDP-Cosine), two with Earth mover's (SSEDP-Emd), two with Euclidean (SSEDP-Euclidean), and one with RapidMic (SSEDP-Rapid). Other consensus functions have got best results as the SSEDP such as the QMI and SSEDP-Cosine obtained 0.7080 in *gun* dataset, on the other hand, QMI and SSEDP-Euclidean gained 0.7253 in *plrx* dataset as best accuracy. The SSEDP attained the best results with the remaining datasets.

From the experiment we indicate our proposed method has the best results among all the consensus functions and increased the outcomes accuracy. The SSAP is getting low accuracy results because the number of clustering always gained automatically and it gets ten clusters or more.

4. Conclusion and Future Work

In the experiment we compared the performance of SSEDP with other consensus functions using real-world datasets. We deduced from the performance that SSEDP is consistently better than other consensus functions, this indicates the advantages of semi-supervised learning. We also proved that the cosine similarity matrix is more suitable for SSEDP than other similarity matrices. The future work, it ought to extend clustering ensembles to another clustering algorithms like semi-supervised kernel learning using relative constraints algorithm.

Table 1. Average performance on Database Evaluated by Micro-Precision with 10 iterations.

Datasets	S	SAP-A	SSAP-M	CSPA	HGPA	MCLA	SSEDP	EM	QMI	GPMGLA	PTGP
indianliver	0.0364	0.0789	0.6449	0.6501	0.6398	**0.7153**-C	0.6364	0.7067	0.7015	0.7050
plrx	0.0214	0.0879	0.5549	0.5824	0.5769	**0.7253**-Eu	0.5495	**0.7253**	0.6648	0.7253
heartdisseaseh	0.0755	0.1633	0.2857	0.2891	0.2789	**0.6259**-C	0.2687	0.6054	0.5782	0.5952
hvwntr	0.5087	0.5116	0.4934	0.4934	0.4934	**0.5330**-C	0.4934	0.5066	0.4934	0.5083
indi	0.0449	0.0978	0.6415	0.6432	0.6415	**0.7170**-E	0.5489	0.7084	0.7067	0.7101
monster	0.0082	0.0567	0.3594	0.3796	0.3859	**0.7238**-C	0.3707	0.7201	0.7024	0.7175
gun	0.0101	0.0718	0.3905	0.3990	0.3881	**0.7080**-C	0.3796	**0.7080**	0.6740	0.7056
drawing	0.0209	0.1076	0.3700	0.3869	0.3785	**0.9178**-R	0.3785	0.9166	0.7969	0.9166
pointer	0.0128	0.0954	0.4867	0.4650	0.4771	**0.5229**-C	0.3659	0.5024	0.4807	0.4940
diamond	0.0071	0.0481	0.4404	0.4501	0.4549	**0.6438**-C	0.3670	0.6378	0.6173	0.6354
egg	0.0091	0.0504	0.3601	0.3902	0.3733	**0.4886**-Eu	0.3878	0.4742	0.4790	0.4778
bow	0.0086	0.0600	0.3789	0.3897	0.3717	**0.4796**-C	0.3669	0.4676	0.4664	0.4664
biodeg	0.0540	0.1118	0.5981	0.6218	0.6142	**0.6815**-E	0.5545	0.6654	0.6616	0.6616
AVG	0.0629	0.1186	0.4619	0.4723	0.4672	**0.6525**	0.4360	0.6419	0.6172	0.6399

*Notice: R = RapidMic, C = Cosine, Eu = Euclidean, E = (EMD), A = Average, M = Maximum

Acknowledgments

This work is partially supported by the Online Education Research Center of the Ministry of Education Online Education Research Fund (No. 2016YB158), and by Blockchain Technology Fund of Hi-Tech Information Technology Research Insitute of Chengdu (No. 2018H01207).

References

1. S. Sarumathi, N. Shanthi and G. Santhiya, *Int. Journal of Computer Applications* **65** (2013).
2. A. Strehl and J. Ghosh, *Journal of machine learning research* **3**, 583 (2002).
3. A. Topchy, A. K. Jain and W. Punch, *IEEE transactions on pattern analysis and machine intelligence* **27**, 1866 (2005).
4. D. Huang, J.-H. Lai and C.-D. Wang, *Neurocomputing* **170**, 240 (2015).
5. D. Huang, J.-H. Lai and C.-D. Wang, *IEEE transactions on knowledge and data engineering* **28**, 1312 (2016).
6. H. Wang, H. Shan and A. Banerjee, *Statistical Analysis and Data Mining: The ASA Data Science Journal* **4**, 54 (2011).
7. A. Rodriguez and A. Laio, *Science* **344**, 1492 (2014).
8. H. Wang, R. Nie, X. Liu and T. Li, *Knowledge-Based Systems* **36**, 315 (2012).

9. S. Basu, A. Banerjee and R. J. Mooney, Active semi-supervision for pairwise constrained clustering, in *Proc. of the 2004 SIAM int. Conf on Data Mining*, (PA, USA, 2004).
10. W. Q. Fan, C. D. Wang and J. H. Lai, Sdenpeak: Semi-supervised nonlinear clustering based on density and distance, in *IEEE Second Int. Conf. on Big Data Computing Service and Applications*, (Oxford, UK, 2016).
11. K.-J. Wang, J. Li, J.-Y. Zhang and C.-Y. Tu, *Jisuanji Gongcheng/ Computer Engineering* **33**, 197 (2007).
12. W. F. Li, X. T. Li, Y. M. Ye, Y. Li and E. K. Wang, A novel density peak based semi-supervised clustering algorithm, in *Int. Conf. on Machine Learning and Cybernetics*, (Jeju, South Korea, 2016).
13. D. Tang, M. Wang, W. Zheng and H. Wang, *Evolutionary bioinformatics online* **10**, p. 11 (2014).
14. A. F. R. Hernandez and N. Y. G. Garcia, *IEEE Latin America Transactions* **14**, 2857 (2016).
15. J. Xu, B. Lei, Y. Gu, M. Winslett, G. Yu and Z. Zhang, *IEEE Transactions on Knowledge and Data Engineering* **27**, 2148 (2015).
16. L. Liberti, C. Lavor, N. Maculan and A. Mucherino, *Siam Review* **56**, 3 (2014).
17. X. Xie and S. Sun, Multi-view clustering ensembles, in *2013 Int. Conf on Machine Learning and Cybernetics (ICMLC)*, (Lanzhou, China, 2013).

Ensemble Evidential Editing k-NNs through rough set reducts

Asma Trabelsi

Université de Tunis, Institut Supérieur de Gestion de Tunis, LARODEC, Tunisia
Univ. Artois, EA 3926, Laboratoire de Génie Informatique et d'Automatique de
l'Artois (LGI2A), Béthune, F-62400, France
*trabelsyasma@gmail.com

Zied Elouedi

Université de Tunis, Institut Supérieur de Gestion de Tunis, LARODEC, Tunisia
*zied.elouedi@gmx.fr

Eric Lefèvre

Univ. Artois, EA 3926, Laboratoire de Génie Informatique et d'Automatique de
l'Artois (LGI2A), Béthune, F-62400, France
*eric.lefevre@univ-artois.fr

Ensemble classifier is one among the machine learning hot topics and it has been successfully applied in many practical applications. Since the construction of an optimal ensemble remains an open and complex problem, several heuristics for constructing good ensembles have been introduced for several years now. One alternative consists of integrating rough set reducts into ensemble systems. To the best of our knowledge, almost existing methods neglect knowledge imperfection, knowing that several real world databases suffer from some kinds of uncertainty and incompleteness. In this paper, we develop an ensemble Evidential Editing k-Nearest Neighbors classfier (EEk-NN) through rough set reducts for addressing data with evidential attributes. Experimentations in some real databases have been carried out with the aim of comparing our proposal to another existing approach.

Keywords: Ensemble classifiers; rough set reducts; Evidential Editing k-Nearest Neighbors classfier; evidence theory.

1. Introduction

Ensemble system has attracted a great attention since 1990s thanks to its prediction performance ability.[3] Diversity between classifiers represents a key element for designing good successful ensembles.[10] Manipulating the input feature space has been theoretically and experimentally defined as a sufficient way for establishing high diversity between base classifiers.[1,5,8,23]

The choice of the most suitable feature subsets for constructing ensemble systems is still an open question. Recently, feature subsets yielded through rough set reducts [12] have been successfully introduced into ensemble systems. [14,16,17,24] It must be emphasized that almost all real world data are vulnerable to incompleteness, inconsistency and imprecision. This imperfection may pervade either the attribute values, the class labels or both of them. Despite its importance, little attention has been drawn to extract reducts from a such kind of data. In this paper, we are only interested to data with uncertain attribute values represented within the evidence theory [20] and we aim to construct an ensemble of the Evidential Editing k-Nearest Neighbor classifier (EEk-NN) [22] through rough set reducts to process uncertainty. The remaining of this paper is organized as follows: Section 2 is dedicated to recall some basic concepts of the evidence theory. We describe, in Section 3, our novel ensemble system framework. We present, in Section 4, our experimentations on several synthetic databases. Finally, the conclusion and our main future work directions are reported in Section 5.

2. Basic Concepts of the Evidence Theory

The frame of discernment Θ constitutes a finite non empty set of elementary hypotheses. [15] An expert's belief over a given subset of Θ has to be represented by the so-called basic belief assignment m (bba) fulfilling:

$$\sum_{A \subseteq \Theta} m(A) = 1 \tag{1}$$

The simple support function (ssf) is a special case of the basic belief assignments. It has two focal elements: the frame of discernment Θ and a strict subset of Θ which named the focus of the ssf. [18]

The evidence theory provides a set of combination rules for merging distinct information sources. Dempster's rule is one of the best known rules. Given two information sources S_1 and S_2 with respectively m_1 and m_2 as bbas, Dempster's rule, denoted by $\oplus$, will be set as:

$$m_1 \oplus m_2(A) = \frac{1}{1 - \sum_{B \cap C = \emptyset} m_1(B) m_2(C)} \sum_{B \cap C = A} m_1(B) m_2(C), \quad \forall A \subseteq \Theta \tag{2}$$

3. Classifier Ensemble Through Rough Set Reducts

In this paper, we present a new classifier ensemble framework for processing imperfect knowledge. More concretely, we propose an ensemble of our EEk-NN classifier [22] through rough set reducts for handling data described by evidential attributes. The proposed framework is detailed in Algorithm 3.1. It consists of two main levels. The first one concerns the generation of reducts from a given uncertain data, while the second one selects reducts enabling the construction of a successful EEk-NN ensemble. We present in what follows each of these steps.

3.1. *A novel framework for generating reducts from uncertain data*

A number of solutions has been proposed for dealing with multiple reduct generation problems. The Rosetta software is well known to be among the most effective alternative.[9] It provides a set of algorithms for multiple reduct extraction. An example includes the SAVGenetic Reducer that implements a genetic algorithm for searching approximate hitting sets, meaning approximate reducts.[4] One limitation of this latter is its inability to process uncertainty. In this paper, we propose an extension of the SAVGenetic algorithm for addressing data with uncertain attribute values that are expressed in terms of evidence. In accordance with the standard SAVGenetic reducer, our proposal starts by computing a discernability matrix from a given data. We have developed in a previous work,[20] a novel algorithm allowing the computation of a belief discernability matrix Λ' from data with evidential attributes. Let $O=\{O_1,\ldots,O_N\}$ be a given data described by a finite non empty set of N objects. Each object i ($i \in \{1,\ldots,N\}$) is defined by a set of n uncertain attributes $uA = \{A_1,\ldots,A_n\}$ with values $uV^i = \{uv_1^i,\ldots,uv_n^i\}$ and a certain class label $Y_i \in C = \{c_1,\ldots,c_Q\}$. Suppose that Θ_k denotes the frame of discernment of the attribute A_k ($k \in \{1,\ldots,n\}$). Every uncertain attribute value uv_k^i of an instance O_i is represented by a basic belief assignment $m_i^{\Theta_k}$. Assume that S refers to a tolerance threshold (i.e. S is set to 0.1 with the aim of maximizing the search space) and *dist* reflects the Jousselme distance.[7] The entries of the belief discernibility matrix Λ' are computed as follows:

$$\Lambda'(O_i, O_j) = \{A_k \in uA | Jousselme\_Dist(m_i^{\Theta_k}, m_j^{\Theta_k}) > S \text{ and } Y_i \neq Y_j\}$$
(3)

The non empty set of Λ' will then be stored in a multiset ζ'. The approximate hitting sets of ζ' correspond to the approximate reducts. For picking out the approximate hitting sets, we relied on the genetic algorithm with the following fitness function for each subset $B \in 2^n$:

$$f(B) = (1-\alpha) \times \frac{|uA| - |B|}{|B|} + \alpha \times min\{\varepsilon, \frac{[F \in \zeta'|F \cap B = \emptyset|]}{|\zeta'|}\} \quad (4)$$

The fitness function $f(B)$ consists mainly on two terms. The former one rewards subsets with shortest size and the latter one rewards subsets that are hitting sets (i.e. meaning subsets having a non empty intersection with all elements of the discernability matrix). Herein, $\alpha \in [0,1]$ refers to the adaptive weighting between the two parts and ε reflects the minimal hitting set fraction.

3.2. Reduct selection for ensemble learning

An ensemble system with rough Set feature reducts has been viewed for some years as a valid alternative for getting optimal performance.[6] Since several reducts may be generated for a given data set, the choice of the appropriate ones remains a field of research to further develop. Herein, we draw our inspiration from a study conducted in [11] for finding out the suitable reducts for an ensemble of EEk-NN classifiers when relied on both the accuracy and the diversity of base classifers. That is an appropriate trade-off between the diversity of classifiers and the accuracy of each individual classifier is really sufficient for yielding good performance. The assessment function that balances the accuracy and the diversity of base classifiers is as follows:

$$Fitness(f, L) = Accuracy(f, L) + \omega \times Diversity(f, L) \quad (5)$$

where L is the number of classifiers, $Accuracy(f, L)$ reflects the average accuracy of the base classifiers, $Diversity(f, L)$ represents the diversity between base classifiers and ω corresponds to the parameter that balances Accuracy and Diversity. It is worth noting that there are several classifier diversity measures. Authors in [10] have distinguished pairwise and non-pairwise diversity measures. The choice of the most convenient one remains unanswered question. In this paper, we relied on the disagreement measure, which is a pairwise one, for computing classifier diversity. Concerning the parameter ω, it has to be adjusted automatically for maximizing the fitness function value.[11] In addition to the accuracy and diversity of the base classifiers, ensuring diversity between reducts has also been regarded as a substantial key element when designing ensemble systems. In fact, we aim to reduce the searching space of reducts by taking into consideration the diversity measure proposed in.[2] It is set to:

$$Div_{R_k} = 1 - \frac{\frac{R_k \cap Selected\_Red}{R_k \cup Selected\_Red}}{NB\_Selected\_Reduct} \qquad (6)$$

where R_k is the candidate reduct, $Selected\_Reduct$ reflects the selected reducts and $Nb\_Selected\_Reduct$ states the number of selected reducts. The candidate reducts with a diversity measure smaller than a threshold T will then be removed from the search space.

Algorithm 3.1 Successful rough set ensemble framework

1: **Input:** An uncertain data, M is the maximum chosen reducts.
2: **Output:** ensemble system.
3: *% Subsection 3.1*
4: Find multiple reducts *Reducts*
5: *% Subsection 3.2*
6: $Selected\_Red \leftarrow \emptyset$, $Ens\_Classifier \leftarrow \emptyset$
7: Choose the reduct R_1 with the lowest weight from the reduct pool *Reducts*, $Selected\_Red \leftarrow \{Selected\_Red, R_1\}$, $NB\_Selected\_Reduct \leftarrow 1$
8: $Reducts \leftarrow Reducts$-$R_1$
9: $Ens\_Classifier \leftarrow \{Ensemble\_Class, f_1\}$
10: **do**
11: Compute the diversity between $R_k \in Reducts$ and $Selected\_Red$
12: $Reduct\_To\_Remove \leftarrow$ all $R_k \in Reducts$ fulfilling $Div_{R_k} < T$
13: $Reducts \leftarrow Reducts - Reduct\_To\_Remove$
14: Choose a new reduct R_j from *Reducts* satisfying:
15: $Fitness(f_j, Ens\_Classifier) = \max_{R_k \in Reducts}(Fitness(f_k, Ens\_Classifier))$
16: $Ens\_Classifier \leftarrow \{Ensemble\_Class, f_j\}$, $Selected\_Red \leftarrow \{Selected\_Red, R_j\}$, $NB\_Selected\_Reduct \leftarrow NB\_Selected\_Reduct+1$
17: **until** $NB\_Selected\_Reduct = M$ **or** isempty(*Reducts*)=true
18: Ensemble system merged through the Dempster operator

---

4. Experimentations Settings and Results

Throughout this paper, we propose to construct an ensemble of EEk-NN classifiers from data with evidential attributes. Since real world applications suffer from incompleteness and uncertainty, there is a lack of datasets that take imperfection into consideration. With the aim of evaluation of proposed approach, we propose to generate synthetic databases. The underling idea consists of injecting an uncertainty level P to some real categorical databases delivered by the the UCI machine learning repository.[13] Table 1 describes the used databases for experimentations. Getting inspiration from the method proposed in,[19] four uncertainty levels P have been considered: an certain case when P=0, a Low uncertainty case when $(0 < P < 0.4)$, a Middle uncertainty case when $(0.4 \leq P < 0.7)$ and a High uncertainty case $(0.7 \leq P \leq 1)$. So that, each attribute value has to be expressed by a simple support mass function, meaning P has to be assigned be the focus reflecting the true attribute value and $1-P$ has to be allocated to the frame of discernment of that attribute.

Table 1. Description of databases.

Databases	#Instances	#Attributes	#Classes
Voting Records	435	16	2
Monks	432	7	2
Lymphography	148	18	4
Tic-Tac-Toa	958	9	2

Our ensemble EEk-NN classifiers through rough set reducts is evaluated and compared to an ensemble of 25 EEk-NN classifiers through Random Subspaces presented in [21] and we have following a 10-fold cross validation approach. Taking k=3 as nearest neighbors, the obtained Percentage of Correct Classifications (PCCs) are presented in Table 2 where ERR and ERS reflect respectively the ensemble EEk-NN through rough set reducts and the ensemble EEk-NN through random subspaces and $size$ represents the size of an ensemble constructed using the ERR approach.

Table 2. PCCs results.

	No		Low		Middle		High	
	ERS	ERR	ERS	ERR	ERS	ERR	ERS	ERR
Voting Records	91.62	**93.02**	91.92	**95.35**	91.39	**95.12**	89.53	**90.11**
Monks	60.26	**100**	59.49	**100**	60.26	**100**	53.68	**68.18**
Lymphography	82.85	**87.90**	75.14	**79.29**	82.85	**84.12**	62.85	**75.66**
Tic-Tac-Toa	61.15	**72.53**	55.78	**59.05**	56	**74.95**	57.68	**57.76**

The obtained results have proven the efficiency of the rough set reduct method over the random subspace approach. In fact, the yielded PCCs through the rough set techniques are strictly higher than those obtained using the random subspace method. Let us take the Monks database with uncertainty equals $High$ as an example. The PCC derived by ensemble rough sets is equal to 68.18 %, while that achieved by the random subspace method equals 53.68 %.

5. Conclusion

We have proposed a novel framework for classifier ensemble for addressing data with evidential attribute values. Precisely, we have developed an ensemble of the EEk-NN classifier by relying on some rough set techniques for generating suitable feature subsets. For the purpose of assessing our novel approach, we have made a comparative study with an ensemble EEk-NN constructed via random subspaces. The achieved PCC results have proven the efficiency of our novel framework over that generated with random subspaces. Although, there are several combination operators within the evidence theory, in this paper we have merged classifier using the Dempster rule as it is very well known, in future work, we look forward to paying more attention to the combination procedure. Notably, we intend to pick out the best combination rule within the context of ensemble evidential classifiers.

References

1. R. Bryll, R. Gutierrez-Osuna, and F. Quek. Attribute bagging: improving accuracy of classifier ensembles by using random feature subsets. *Pattern recognition*, 36(6):1291–1302, 2003.
2. E. Debie, K. Shafi, C. Lokan, and K. Merrick. Reduct based ensemble of learning classifier system for real valued classification problems. In *IEEE Symposium on Computational Intelligence and Ensemble Learning (CIEL)*, pages 66–73. IEEE, 2013.
3. T. G. Dietterich. Machine-learning research. *AI magazine*, 18(4):97, 1997.
4. M. A. El-Monsef, M. Seddeek, and T. Medhat. Classification of sand samples according to radioactivity content by the use of euclidean and rough sets techniques. *EG EG0600123*, page 266, 2003.
5. S. Günter and H. Bunke. Feature selection algorithms for the generation of multiple classifier systems and their application to handwritten word recognition. *Pattern recognition letters*, 25(11):1323–1336, 2004.
6. Y. Guo, L. Jiao, S. Wang, S. Wang, F. Liu, K. Rong, and T. Xiong. A novel dynamic rough subspace based selective ensemble. *Pattern Recognition*, 48(5):1638–1652, 2015.
7. A. Jousselme, D. Grenier, and E. Bossé. A new distance between two bodies of evidence. *Information fusion*, 2(2):91–101, 2001.
8. Y. Kim. Toward a successful crm: variable selection, sampling, and ensemble. *Decision Support Systems*, 41(2):542–553, 2006.
9. J. Komorowski, A. Øhrn, and A. Skowron. The rosetta rough set software system. *Handbook of data mining and knowledge discovery*, pages 2–3, 2002.
10. L. I. Kuncheva and C. J. Whitaker. Measures of diversity in classifier ensembles and their relationship with the ensemble accuracy. *Machine learning*, 51(2):181–207, 2003.
11. D. W. Opitz. Feature selection for ensembles. *AAAI/IAAI*, 379:384, 1999.
12. Z. Pawlak. Rough sets. *International Journal of Computer & Information Sciences*, 11(5):341–356, 1982.
13. P. Murphy and D. Aha.

14. S. Saha, C. Murthy, and S. K. Pal. Classification of web services using tensor space model and rough ensemble classifier. In *International Symposium on Methodologies for Intelligent Systems*, pages 508–513. Springer, 2008.
15. G. Shafer. *A mathematical theory of evidence*, volume 42. Princeton university press, 1976.
16. L. Shi, X. Ma, L. Xi, Q. Duan, and J. Zhao. Rough set and ensemble learning based semi-supervised algorithm for text classification. *Expert Systems with Applications*, 38(5):6300–6306, 2011.
17. L. Shi, L. Xi, X. Ma, M. Weng, and X. Hu. A novel ensemble algorithm for biomedical classification based on ant colony optimization. *Applied Soft Computing*, 11(8):5674–5683, 2011.
18. P. Smets. The canonical decomposition of a weighted belief. In *IJCAI*, volume 95, pages 1896–1901, 1995.
19. N. Sutton-Charani, S. Destercke, and T. Denoeux. Learning decision trees from uncertain data with an evidential EM approach. In *Proceedibgs of the 12th International Conference on Machine Learning and Applications, ICMLA 2013, 2013, Volume 1*, pages 111–116, 2013.
20. A. Trabelsi, Z. Elouedi, and E. Lefevre. Feature selection from partially uncertain data within the belief function framework. In *proceesings of the 16th International Conferenc Information Processing and Management of Uncertainty in Knowledge-Based IPMU, Part II*, pages 643–655, 2016.
21. A. Trabelsi, Z. Elouedi, and E. Lefevre. Ensemble enhanced evidential k-nn classifier through random subspaces. In *European Conference on Symbolic and Quantitative Approaches to Reasoning and Uncertainty*, pages 212–221. Springer, 2017.
22. A. Trabelsi, Z. Elouedi, and E. Lefevre. A novel k-nn approach for data with uncertain attribute values. In *Proceedings of the 30th International Conference on Industrial Engineering and Other Applications of Applied Intelligent Systems, IEA/AIE 2017, Part I*, pages 160–170, 2017.
23. K. Tumer and N. C. Oza. Input decimated ensembles. *Pattern Analysis & Applications*, 6(1):65–77, 2003.
24. S.-L. Wang, X. Li, S. Zhang, J. Gui, and D.-S. Huang. Tumor classification by combining PNN classifier ensemble with neighborhood rough set based gene reduction. *Computers in Biology and Medicine*, 40(2):179–189, 2010.

A new evidential collaborative filtering: A hybrid memory- and model-based approach

R. Abdelkhalek,* I. Boukhris and Z. Elouedi

*LARODEC, Institut Supérieur de Gestion de Tunis,
Université de Tunis, Tunis, Tunisia
41 rue de la Liberté, 2000 Le Bardo, Tunisie
* abdelkhalek_raoua@live.fr*

One of the most promising approaches in the field of Recommender Systems (RSs) is Collaborative Filtering (CF). CF techniques are commonly divided in the two general classes of memory-based and model-based. A wise strategy would be to combine these two methods to increase their performance while leveling out the weakness of each one. Otherwise, the uncertainty pervaded throughout the different steps of the recommendation process should not be ignored. Handling uncertainty is very challenging and important for more reliable and intelligible predictions. That is why, we propose in this paper a new CF approach which combines these two categories under the belief function theory while dealing with the uncertainty pervaded in the prediction process. The effectiveness of our proposal is validated on a real-word data set and compared to state-of-the-art CF approaches under certain and uncertain frameworks.

Keywords: Recommender systems; collaborative filtering; memory-based; model-based; uncertainty; belief function theory.

1. Introduction

Collaborative Filtering (CF) approaches[1] have achieved a great success and popularity in today's Recommender Systems (RSs).[2] Their main goal is to predict the user's preferences based on his past ratings. CF approaches can be either model-based, which learn a model to make predictions, or memory-based which compute the similarities between users (user-based) or items (item-based) and then select the most similar ones for recommendations. However, performances of CF are usually limited by data imperfection issues. Dealing with the uncertainty arising throughout the prediction process is considered as a crucial challenge in the RSs area. Different kinds of uncertainties can be represented under the belief function framework[3–5] which is able to deal with partial or even total ignorance in a flexible way. In fact, memory-based CF has been extended under this framework where the K-similar items have been considered as different sources of evidence leading to the final predictions.[6,7] Model-based CF has been also exploited under this theory where the Evidential c-Means technique has been adopted to cluster items based on their ratings. The prediction was an average of the ratings corresponding to all the members of the same cluster.[8]

Unlike these approaches, we propose in this paper an evidential hybrid CF that takes into account the intuition of both memory- and model-based methods under an uncertain context using the belief function tools.

This paper is organized as follows: Section 2 provides an overview of the belief function framework. In Section 3, we recall the Collaborative Filtering recommender. Section 4 describes our proposed approach. Section 5 gives the experimental results. Finally, Section 6 concludes the paper.

2. Belief Function Framework

Assume Θ be the frame of discernment representing the set of n elementary events such that: $\Theta = \{\theta_1, \theta_2, \cdots, \theta_n\}$. It contains hypotheses concerning the given problem. The power set of Θ, denoted by 2^Θ, represents all the possible values that can be taken by each subset of Θ. The mapping $m : 2^\Theta \to [0, 1]$ is the basic belief assignment (bba) such that $\sum_{E \subseteq \Theta} m(E) = 1$. $m(E)$ is the basic belief mass (bbm) which states the part of belief exactly committed to the event E. The subsets E of Θ having $m(E) > 0$ are called the focal elements. A simple support function (ssf) corresponds to the bba having at most two focal elements namely Θ and a strict subset of Θ, refereed to as the focus of the ssf. It is defined as follows:
$$m(X) = \begin{cases} w & \text{if } X = \Theta \\ 1 - w & \text{if } X = E \text{ for some } E \subseteq \Theta \\ 0 & \text{otherwise} \end{cases}$$
where E is the focus and $w \in [0,1]$.

The reliability of the piece of evidence can be evaluated through a discounting mechanism as following: $m^\alpha(E) = (1 - \alpha) \cdot m(E), \forall E \subset \Theta$; $m^\alpha(\Theta) = \alpha + (1 - \alpha) \cdot m(\Theta)$ where $\alpha \in [0,1]$ is the discounting factor.

The fusion of two bba's m_1 and m_2 derived from two reliable and distinct sources of evidence can be ensured using Dempster's rule of combination defined as: $(m_1 \oplus m_2)(E) = k \cdot \sum_{F,G \subseteq \Theta: F \cap G = E} m_1(F) \cdot m_2(G)$ where $(m_1 \oplus m_2)(\varnothing) = 0$ and $k^{-1} = 1 - \sum_{F,G \subseteq \Theta: F \cap G = \varnothing} m_1(F) \cdot m_2(G)$. To make decisions, beliefs can be transformed into a pignistic probability $BetP(E)$ as follows:

$BetP(E) = \sum_{F \subseteq \Theta} \frac{|E \cap F|}{|F|} \frac{m(F)}{(1-m(\emptyset))}$ for all $E \in \Theta$. Several machine learning techniques have been proposed under this theory to process uncertain data, such as the Evidential K-Nearest Neighbors[9] allowing a credal classification of the objects and the Evidential c-Means (ECM)[10] where an object may belong to more than only one cluster.

3. Collaborative Filtering Recommender

CF approaches are categorized into memory-based and model-based. The two categories basically differ in how they process the user-item matrix. Model-based CF relies on a model learned from the past ratings to make predictions. Different methods can be adopted in the model building process, such as clustering techniques, which can first assign items to clusters based on their ratings and then perform recommendations accordingly. On the other hand, memory-based, also referred to as neighborhood-based, does not require to learn and maintain a model to make predictions. It exploits the entire rating matrix and computes the similarities between users or items according to the past ratings. Pearson and Cosine are the widely used similarity measures[2] in the neighborhood-based approaches. Once the neighbors are selected, their ratings are aggregated to generate the final predictions. Memory-based approaches are simple, reasonable and highly effective which explains their great applicability. However, these CF methods can become computationally expensive, in terms of both time and space complexity, as they need to search the whole user-item space to compute similarities. Yet, some CF approaches[11–13] have been proposed in this context aiming to unify the two CF categories to improve the recommendation performance. Nevertheless, these approaches do not take into consideration the uncertainty pervaded throughout the recommendation process. That is why, our new CF approach will be centered around memory- and model-based CF while handling uncertainty under the belief function framework.

4. EHMM: Evidential Hybrid Memory- and Model-based

Our proposed evidential hybrid memory- and model-based approach, that we denote by EHMM, is based on the intuition of both memory- and model-based methods commonly adopted by the CF recommender. We embrace the belief function theory in order to quantify and represent the uncertainty involved during the clusters assignments as well as during the prediction process. Based on the rating matrix, we first construct a clustering model using ECM. Then, to perform predictions from the learned model, we select the K-similar items and generate the final predictions represented by basic belief assignments (*bba's*) as shown in Fig. 1.

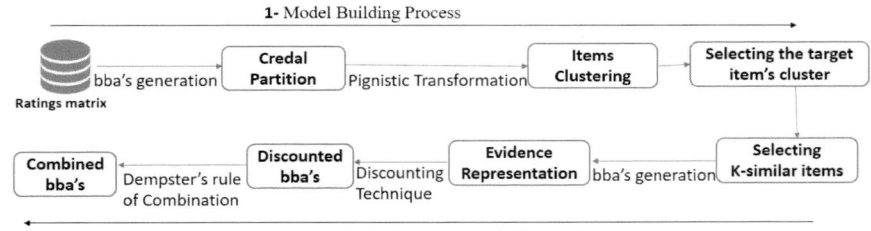

Fig. 1. Combining memory- and model-based CF under the belief function theory.

Step 1: Model Building

This phase corresponds to the evidential items clustering. We define $\Omega_1 = \{c_1, c_2, \ldots, c_m\}$ where m is the number of clusters c. Following,[10] we randomly initialize the cluster centers commonly referred to as prototypes. Then, we compute the Euclidean distance between the items and the non empty subsets of Θ. By exploiting the entire user-item matrix, we aim to generate a credal partition of the items. A credal partition is defined as a general extension of the crisp, fuzzy and possibilistic ones where an item can belong to any subsets of Ω_1. To this end, we involve ECM, an efficient soft clustering technique, which allows us to allocate, for each item in the rating matrix, a mass of belief not only to single clusters, but also to any subsets of Ω_1 as illustrated in Table 1 where $c=2$.

Table 1. Example of credal partition.

	∅	$\{c_1\}$	$\{c_2\}$	Ω_1
$Item_1$	0	0.7	0.3	0
$Item_2$	0	0.2	0	0.8
$Item_3$	0	0	1	0
$Item_4$	0	0	0	1
$Item_5$	1	0	0	0

Finally, each *bba* is transformed into a pignistic probability and each item is assigned to the cluster having the highest pignistic value. Once the clusters are assigned, only the items belonging to the same cluster as the target item are considered in the next step.

Step 2: Evidential Predictions

We define $\Omega_2 = \{r_1, r_2, \cdots, r_n\}$ where n is the number of the possible ratings r. To make predictions, we first compute the distances between the target item and the items belonging to the same cluster as follows: $D(a,b) = \frac{\sqrt{\sum_{u \in (u_a \cap u_b)} (r_{u,a} - r_{u,b})^2}}{|u_a \cap u_b|}$. $r_{u,a}$ and $r_{u,b}$ are the ratings of the user u for the target item a and the item b. u_a and u_b are the users who rated both items a and b. According to the computed distances, the K-similar items are picked up and their ratings are transformed into a bba represented through a simple support function as follows:[9] $m_{a,b}(\{r_i\}) = \alpha_0 \exp^{-(\gamma_{r_i}^2 \times (d(a,b))^2)}$; $m_{a,b}(\Omega_2) = 1 - \alpha_0 \exp^{-(\gamma_{r_i}^2 \times (d(a,b))^2)}$. α_0 is initialized to the value 0.95[9] while γ_{r_i} is the inverse of the mean distance between each couple of items sharing the same ratings. Once the bba's are generated, the discounting technique[3] is applied to quantify the reliability of each similar item. The key idea is to weight most heavily the evidence of the items having the lowest distances and conversely for the less reliable ones. We define the discounting factor β as: $\beta = d(a,b)/max(d)$ where $max(d)$ corresponds to the maximum value of the computed distances. The discounted bba's are: $m_{a,b}^\beta(\{r_i\}) = (1-\beta) \cdot m_{a,b}(\{r_i\})$; $m_{a,b}^\beta(\Omega_2) = \beta + (1-\beta) \cdot m_{a,b}(\Omega_2)$. After discounting the evidence of each similar item, the obtained bba's are aggregated into a unique final one as follows:[9] $\forall r_i \in \{r_1, \cdots, r_N\}$ $m^\beta(\{r_i\}) = \frac{1}{Z}(1 - \prod_{i \in S_K}(1 - \alpha_{r_i})) \cdot \prod_{r_j \neq r_i} \prod_{i \in S_K}(1 - \alpha_{r_j})$; $m^\beta(\Omega_2) = \frac{1}{Z} \prod_{i=1}^{N}(1 - \prod_{i \in S_K}(1 - \alpha_{r_i}))$ where S_K is the set containing the K-nearest neighbors of the target item over the user-item matrix. N is the number of the ratings provided by the similar items, α_{r_i} is the belief committed to the rating r_i, α_{r_j} is the belief committed to the rating $r_j \neq r_i$, Z is a normalized factor defined by: $Z = \sum_{i=1}^{N}(1 - \prod_{i \in S_K}(1-\alpha_{r_i})) \prod_{r_j \neq r_i} \prod_{i \in S_K}(1-\alpha_{r_j}) + \prod_{i=1}^{N}(\prod_{i \in S_K}(1-\alpha_{r_j})))$. The final predictions are represented through basic belief assignments as illustrated in Table 2.

5. Experimental Analysis

To evaluate our proposal, we rely on the well-known MovieLens[a] data set. It contains 1682 movies rated by 943 users. We follow the methodology[14]

---

[a]http://movielens.org

Table 2. Example of evidential predictions.

	$\{r_1\}$	$\{r_2\}$	$\{r_3\}$	$\{r_4\}$	$\{r_5\}$	Ω_2
$Item_1$	-	0.2	-	-	0.7	0.1
$Item_2$	0.1	-	0.5	-	0.3	0.1
$Item_3$	0.6	-	0.1	-	0.1	0.2
$Item_4$	0.1	-	0.3	0.4	0.1	0.1
$Item_5$	-	0.6	-	0.2	0.1	0.1

which consists in ranking the movies rated by the 943 users according to the number of the total ratings such as: $Nb_{user}(movie_1) \geq Nb_{user}(movie_2) \geq \cdots \geq Nb_{user}(movie_{1682})$ where $Nb_{user}(movie_i)$ is the number of users who rated the $movie_i$. Then, 10 different subsets are extracted by increasing progressively the number of the missing rates from 53.8 % to 95.9%.

Evaluation Measures

We used two evaluation metrics: The *Mean Absolute Error* (MAE) defined as: $MAE = \frac{1}{\|\widehat{R}_{u,i}\|} \sum_{u,i} |\widehat{R}_{u,i} - R_{u,i}|$ and the precision defined by: $Precision = \frac{IR}{IR+UR}$. $R_{u,i}$ is the real rating for the user u on the item i and $\widehat{R}_{u,i}$ is the predicted value. $\|\widehat{R}_{u,i}\|$ is the total number of the predicted ratings. IR indicates that an interesting item has been correctly recommended while UR indicates that an uninteresting item has been incorrectly recommended. The lower the MAE is, the more accurate the predictions are while the highest precision indicates a better recommendation quality.

Experimental Results

We perform experiments over the 10 subsets while switching each time the number of clusters c. For each experiment, we use the values $c=2$, $c=3$, $c=4$ and $c=5$. Then, we compute the MAE and the precision for each value and we note the overall results. We fix $\alpha_0=0.95$, as in,[9] for the bba's generation. For all the obtained clusters, the number of the K-nearest neighbors K is set to the value $\|c\| - 1$ since most of the best results were achieved with this value. $\|c\|$ corresponds to the number of items in the cluster c. The performance of the proposed approach is compared against five traditional item-based CF systems: The evidential model-based CF (ECL),[8] the two evidential memory-based CF namely, the evidential item-based CF (EV)[6] and the discounting-based item-based CF (DE).[7] Finally, we run the standard Pearson item-based CF (P) and Cosine item-based CF (C). The obtained results are depicted in Table 3.

Table 3. Overall MAE and precision.

Measures	Subsets	Sparsity	EV	C	P	DE	ECL	EHMM
MAE	S_1	53%	0.751	0.824	0.839	0.711	0.749	0.740
Precision			0.79	0.778	0.774	0.774	0.792	0.830
MAE	S_2	56.83%	0.84	0.87	0.936	0.802	0.8	0.851
Precision			0.76	0.739	0.737	0.748	0.74	0.755
MAE	S_3	59.8%	0.761	0.825	0.863	0.836	0.747	0.779
Precision			0.77	0.749	0.752	0.711	0.785	0.732
MAE	S_4	62.7%	0.763	0.876	0.905	0.743	0.793	0.75
Precision			0.763	0.745	0.746	0.775	0.782	0.764
MAE	S_5	68.72%	0.831	1	0.990	0.802	0.845	0.793
Precision			0.741	0.69	0.707	0.787	0.752	0.757
MAE	S_6	72.5%	0.851	0.917	0.976	0.843	0.8	0.845
Precision			0.735	0.733	0.732	0.74	0.813	0.743
MAE	S_7	75%	0.744	0.877	0.943	0.736	0.733	0.703
Precision			0.78	0.745	0.752	0.783	0.805	0.792
MAE	S_8	80.8%	0.718	0.848	0.927	0.723	0.762	0.711
Precision			0.778	0.718	0.729	0.821	0.755	0.779
MAE	S_9	87.4%	0.840	0.978	0.958	0.839	0.873	0.798
Precision			0.707	0.654	0.665	0.74	0.73	0.737
MAE	S_{10}	95.9%	0.991	1.13	0.913	0.978	0.83	0.87
Precision			0.513	0.509	0.463	0.431	0.55	0.66
Overall MAE			0.809	0.914	0.925	0.789	0.793	**0.784**
Overall Precision			0.733	0.706	0.706	0.743	0.75	**0.755**

The hybrid evidential method acquires the lowest average MAE and the highest precision compared to the single evidential memory- and model-based approaches. The EHMM achieves better results in term of MAE with a value of 0.784 compared to 0.925 and 0.914 for both Pearson and Cosine CF, 0.809 for EV, 0.789 for DE and 0.793 for ECL. Similarly, the average precision of the new approach (0.755) outperforms EV (0.733), DE (0.743), ECL (0.75) as well as Pearson and Cosine approaches (0.706).

6. Conclusion

In this paper, we have proposed a new evidential CF approach combining both model-based and memory-based CF methods. The idea is to build a model from the rating matrix and then performing predictions based on the learned model while taking into account the uncertainty that occurs in the different steps of the recommendation process. The fusion framework is effective in improving the prediction accuracy under an uncertain context. As a future work, we intend to perform more comparisons with other hybrid CF approaches and to use other combination rules in the prediction phase such as the evidential reasoning rule. Besides, we tend to rely on the different baa's corresponding to the different clusters rather than the most significant one in the model building phase.

References

1. X. Su, M. TM. Khoshgoftaar, A survey of collaborative filtering techniques. In *Advances in artificial intelligence* (Hindawi Publishing, 2009), pp. 1-19.
2. F. Ricci, L. Rokach, B. Shapira, Recommender systems: introduction and challenges. In *Recommender Systems Handbook* (Springer, 2015), pp. 1-34.
3. A. P. Dempster, A generalization of Bayesian inference. In *Classic works of the dempster-shafer theory of belief functions* (Springer, 2008), pp. 73-104.
4. G. Shafer, A mathematical theory of evidence. In *Princeton university press* (Princeton university press, 1976).
5. P. Smets, The transferable belief model for quantified belief representation. In *Quantified Representation of Uncertainty and Imprecision* (Springer, 1998), pp. 267-301.
6. R. Abdelkhalek, I. Boukhris and Z. Elouedi, Evidential item-based collaborative filtering. In *International Conference on Knowledge Science, Engineering and Management* (Springer, 2016), pp. 628-639.
7. R. Abdelkhalek, I. Boukhris and Z. Elouedi, Assessing Items Reliability for Collaborative Filtering Within the Belief Function Framework. In *International Conference on Digital Economy* (Springer, 2017), pp. 208-217.
8. R. Abdelkhalek, I. Boukhris and Z. Elouedi, A Clustering Approach for Collaborative Filtering Under the Belief Function Framework. In *European Conference on Symbolic and Quantitative Approaches to Reasoning and Uncertainty* (Springer, 2017), pp. 169-178.
9. T. Denoeux, A K-nearest neighbor classification rule based on Dempster-Shafer theory. In *Transactions on Systems, Man and Cybernetics* (IEEE, 1995), pp. 804-813.
10. M. H. Masson, T. Denoeux, ECM: An evidential version of the fuzzy c-means algorithm. In *Pattern Recognition* (Elsevier, 2008), pp. 1384-1397.
11. R. Al Mamunur, S. K. Lam, G. Karypis, J. Riedl, ClustKNN: a highly scalable hybrid model- and memory-based CF algorithm. In *Proceeding of WebKDD* (2006).
12. S. Gong, H. Ye, H. Tan, Combining memory-based and model-based collaborative filtering in recommender system. In *Pacific-Asia Conference on Circuits, Communications and Systems* (IEEE, 2009), pp. 690-693.
13. D. M. Pennock, E. Horvitz, S. Lawrence, C. L. Giles, Collaborative filtering by personality diagnosis: A hybrid memory- and model-based approach. In *International conference on Uncertainty in Artificial Intelligence* (Morgan Kaufmann Publishers, 2000), pp. 473-480.
14. X. Su, T. M. Khoshgoftaar, Collaborative filtering for multi-class data using bayesian networks. In *International Journal on Artificial Intelligence Tools* (World Scientific, 2008), pp. 71-85.

Improving the activities of a robotic guide by taking into account human reactions*

J. Javier Rainer, Fernando López Hernández

Universidad Internacional de La Rioja (UNIR)
Av. De la Paz, 137
26006 Logroño, La Rioja, Spain
javier.rainer@gmail.com / javier.rainer@unir.net, fernando.lopez@unir.net

Ramón Galán

Centre for Automation and Robotics UPM - CSIC. Universidad Politécnica de Madrid
C/ José Gutiérrez de Abascal 2. 28006, Madrid, Spain

Traditionally robots are autonomous systems that plan and execute actions without taking into account any external feedback. This paper studies to what extent a robot is able to alter its behavior based on observing a series of positive or negative human reactions. We describe the result of introducing these human reactions in a plan of actions by means of fuzzy logic analysis of the reactions. In this way, the election of the actions to perform is based on a combination of the initial plan of actions and of a fuzzy logic weighting of the utility of previously executed actions.

1. Introduction

The issue of how machines integrate the perceived human sentiments into their behaviour represents an open challenge for evolutionary/developmental robotics [1-4]. One of the challenges is the observation and integration of human reactions, and subsequently changing the robot behaviour based on this feedback. The main objective of this paper is to describe a suitable way to integrate fuzzy logic analysis into the planned actions.

Some of the most recent works about planning with human reaction are described in [5-9]. These works propose different architectures and methodologies than those presented here. The integration of fuzzy logic in robotic planning has been studied for objectives such as robot navigation [10]. The growing popularity of integrating fuzzy logic in planning is due to the fact

---

*This work is partially funded by Universidad Internacional de La Rioja UNIR, under the Research Support Strategy, Research Group: Artificial Intelligence and Robotics Group http://gruposinvestigacion.unir.net/ia/.

that they allow approximating reasoning in uncertain environments in a simple and inexpensive way, reducing the mathematical complexity required [10]. When designing fuzzy systems, the knowledge from domain experts takes precedence over mathematical models. Therefore, fuzzy modelling can be applied when knowledge of expert is significant enough to define the objective function and decision variables [11].

2. Urbano Robot Overview

We have developed our own interactive mobile robot called Urbano specially designed to be a tour guide in exhibitions [2]. This Section introduces a high level view if the Urbano robot, its hardware, software and the experience. Urbano has been the result of a series of development and refinements, until its actual state.

Urbano is a B21r platform from iRobot, equipped with a four wheeled synchro drive locomotion system. It is a SICK LMS200 laser scanner horizontally mounted on top, used for navigation and SLAM, with a mechatronic face and a robotic arm used to express emotions such as happiness, sadness, surprise or anger. The platform has also two on-board PCs and one touch screen.

The software is structured in several executable modules (connected via TCP/IP), which allow for a decoupled and independent development. Most of these executable modules are conceived as servers and service providers as face control, arm control, navigation systems voice synthesis and recognition, and web server. A client-server paradigm is used, the central module being the only client called the Urbano Kernel. It is responsible of managing the whole system [2].

URBANO has a technology based on distributed application software. The most recent version is an agent-based architecture, which uses a specific CORBA (*Common Object Request Broker Architecture*) approach as an integration tool.

3. Urbano DDM Agent

The architecture of Urbano is based on the agents. The DMM (Decision Making Mechanism) is the most significant agent since it is the one that decides and performs the sequence of actions to accomplish the activities taking into account the external information given by the environment. Therefore, this agent constitutes the core of the action planner of Urbano.

When the agenda is decoded, the knowledge server provides all the information about the available actions, including the series of restrictions must be considered to decide the order or the actions. The DMM optimizes the actions to perform within the multiple choices generated when establishing the daily agenda.

4. The Planner

Currently our Urbano robot is able to plan and execute the tourist guide *activities* saved in the *agenda* or the robot. Each activity is composed of one or more *actions* [1] (see Figure 1(a)). The robot is able to select the sequence of actions (a *plan*) that achieve the activities of the day. Figure 1(b) shows an example of an activity and corresponding actions.

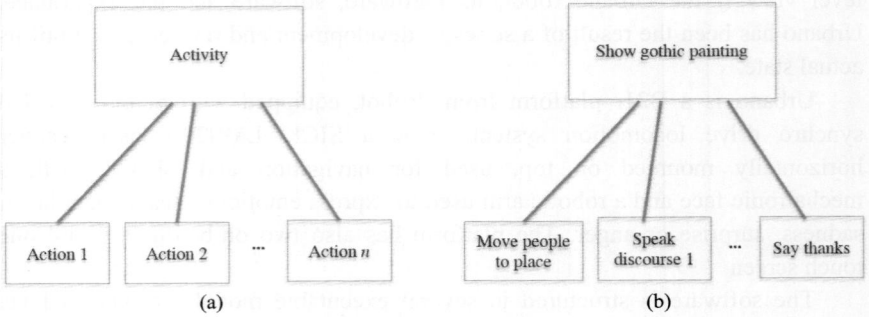

Figure 1: Activities and actions.

The activities area created or updated on a daily basic, therefore the system must regularly check if any new activity has come along. Each action to perform might also be composed by several actions on its own. The output plan defines the information required in order to know when, where and in which order to execute the actions.

In the first phase we search the potential set of plans in the agenda. The agenda highlights the items that belong to each action to perform. For each item, these elements are established: its identification, its priority, and its numerical order.

The feasible actions are stored in the *knowledge database* (KB). The simplest actions correspond to basic actions the robot can perform, with their own parameters; i.e.: action: "spin", with a certain rotating "degrees" as a parameter. More elaborate actions include movement: go on to a point, walk to the left, walk to the right, etc.

5. Fuzzy Decision Making

Decision-making is a part of the paradigm proposed by Zadeh [12] that has been currently examined in [13] to select a course of action from multiple alternatives. In this work we use fuzzy logic to weight the reaction of the user to a specific action. This positive and negative reactions to calculate the quality index of each executed action. In this way, if we detect a negative reaction of the user to an action (e.g. a joke), the robot will look for ways to undertake the activity avoiding repeating this action again, or vice versa.

All information available at the moment about the quality criteria and in the knowledge database (KD): its influence on the quality index and corresponding ontology.

The fuzzyfication phase uses the function of membership. The exit variable quality_index is also modelled with five terms and triangular functions. The technique of centroid method is used in the defuzzyfication phase.

This enables to obtain one quality index for each action, making it preferable to accomplish activities using actions with the one which scores a highest quality index.

We have tested these quality indexes with three classical alternative state-space searches. The first one uses "brute force" to generate all the possible plans. Then we assign each plan a numeric value indicating the "quality criteria" of the paragraphs that form the presentation, and then, using a set of fuzzy rules, it estimates the quality index. It selects the agenda with the highest index.

The second alternative uses "best-first search" state-space search so that as it goes along, it takes the option that partially presents the best index. This alternative is, without a doubt, the fastest one, but it cannot guarantee the selection of the best option.

The third alternative is here described and it consists in calculating a global quality index alone for each one of the alternative possible agendas to accomplish each day, which is generated from all the combinations within every action. The agenda chosen will be the one with a higher quality index, according the fuzzy logic.

6. Reinforcement Learning Phase

The most important contribution of the proposal is reinforcement learning, i.e., the ability of the robot to learn by maximizing rewards and minimize penalties. Initially the robot has a small number of quality criteria available to evaluate some actions as rewards and penalties, corresponding to the minimum level of knowledge. As the robot learns for the environment, it best organizes the time and agenda, maximizing the performance of the quality index.

To ensure the making decisions mechanism works properly, test have been conducted with an Urbano at a Museum, where it should guide a visit. Once the visit is over, a simple questionnaire has been designed that the public can fill out. It asks for an evaluation of each quality criteria known at the time, indicating whether the robot should spend more or less time on each item, and a percentage evaluation of what the visits consider valuable in the presentation.

Since the robot beliefs, on how to execute the actions, might not meet the "external reality", it is very important to obtain this information of the visitors and feed it back to the robot, so that, in time, its beliefs will match with the opinion of the visitors on the correct actions performance.

7. Conclusions

This paper shows how to integrate fuzzy logic and reinforcement learning in a classical planner to implement a flexible decision-making mechanism. The agenda of the robot is dynamic because the museum operators tend to vary the actions and activities throughout the day, as well as to prioritize its important.

The learning phase enables the continuous improvement and the adaptation to a dynamic environment, i.e., one in which the information changes. Also the environmental knowledge that the robot has must meet the "external world".

This optimization has to be based on the continuous contrast of beliefs states and external events (rewards and penalties). Measuring these external events enables to learn and improve the quality index of the actions to accomplish.

The proposed mechanism is exportable to other mechanisms of autonomous robot.

References

1. G. Bekey, Autonomous Robots: from biological inspiration to implementation and control, *MIT Press books*. (2005).
2. D. Rodríguez-Losada, F. Matia, R. Galán, M. Hernando, J. M. Montero, and J. M. Lucas, Urbano, an Interactive Mobile Tour-Guide Robot. *Advances in Service Robotics*. Ed. H. Seok. In-The. 229-252, (2008).
3. Schaal S, The new robotics: towards human-centered machines, *HFSP Journal*. **1** (2): 115-126, (2007).
4. Weng J.J., McClelland J., Pentland A., Sporns O., Stockman I., Sur M. and Thelen E, Autonomous mental development by robots and animals, *Science*. **291**:599–600, (2001).
5. Pomerol, Jean-Charles and F. Adam. Studies in Computational Intelligence. Cap: Understanding Human Decision Making-A Fundamental Step Towards Effective Intelligent Decision Support. **97**: 3-40. Springer. (2008).

673

6. Velásquez, J. When Robots Weep: Emotional Memories and Decision-Making. In: American Association for Artificial Intelligence (AAAI-98 Proceedings), 70-75. AAAI Press and The MIT Press USA. (1998).
7. Yuchul Jung and Yoonjung Choi. Integrating Robot Action Scripts with a Cognitive Architecture for Cognitive Human-Robot Interactions. In: IEEE International Conference on Information Reuse and Integration (IRI 2007), 152-157. IEEE, USA. (2007).
8. Zhang, Yuli, J. Weng and W-S. H. Auditory Lerning: A Development Method. In: IEEE Transactions on Neural Networks, **16**(3): 601-616. IEEE press, USA. (2005).
9. Zachary, W. W., J. M. Ryder, and J. H. Hicinbothom. Cognitive action analysis and modelling of decision making in complex environments In: Decision making under stress: Implications for training and simulation (J. Cannon-Bowers & E. Salas), 315-344. American Psychological Association, Washington, USA. (1998).
10. Jaya, A. S. M., Hashim, S. Z. M., & Rahman, M. N. A. (2010). Fuzzy logic-based for predicting roughness performance of TiAlN coating. In *2010 10th International Conference on Intelligent Systems Design and Applications* (pp. 91–96). https://doi.org/10.1109/ISDA.2010.5687284
11. Seraji, H., & Howard, A. (2002). Behavior-based robot navigation on challenging terrain: A fuzzy logic approach. IEEE Transactions on Robotics and Automation, 18(3), 308–321. https://doi.org/10.1109/TRA.2002.1019461
12. L. A. Zadeh, "Fuzzy sets," Information and Control, **8**: 338–353, (1965).
13. L. A. Zadeh, "Is there a need for fuzzy logic?" Information Sciences, **178**(13): 2751–2779, (2008).

Hierarchical classification learning: A novel two-layer framework for multiclass classification[*]

Yuanyuan Liu,[1] Xiaoshuang Qiao,[2] Gongde Guo[2]

[1]*Department of Mathematics and Information, Fujian Normal University*
Fuzhou, P.R. China
[2]*Department of Mathematics and Information, Fujian Normal University*
Key Lab of Network Security and Cryptology, Fuzhou, P.R. China

Hui Wang[3]

[3]*Department of Computing and Mathematics, University of Ulster at Jordanstown*
Northern Ireland, UK

This paper reports on how to transform a multiclass classification problem into a set of simpler classification problems and then combine the solutions to the simpler problems into a solution to the original multiclass classification. A novel two-layer framework is presented, called *Hierarchical Classification Learning*. Different machine learning algorithms can be employed as the base classifier in this classification learning framework. First of all, the multiclass data set is reformed for every pair of classes, resulting in multiple 3-class *sensor data sets* — two classes for the pair of classes and the third class for any data instance that not belong to any of the two classes. A classification model, called *sensor model*, is constructed for each of the sensor data sets using a machine learning algorithm, or the *base classifier*. Then every data instance of the original data set is put into all the sensor models, generating a set of sensor outputs or *secondary features* which compose a *sensor vector*. Every sensor vector is viewed as a reformed version of the original data that has the same class label as the original, so put all sensor vectors together and get a new *reformed data set*. A classification model, *decision model*, is constructed from the new reformed data set. At last, extensive experiments have been conducted to evaluate the framework, and neural network, decision tree, random tree, and support vector machine are used as the base classifiers. Experiment results on UCI datasets and some popular face classification datasets show that the classification learning framework has achieved superior performance than their corresponding base classifiers.

Keywords: Hierarchical classification learning; sensor data; sensor model; reformed data; secondary features.

---

[*]This work is supported by the National Natural Science Foundation of China (61672157), the Project of Network and Information Security Key Theory and Technological Innovation Team in Fujian Normal University (IRTL1207).

1. Introduction

Hierarchical classification learns to classify data, taki[ng] hierarchical structure in data. Fan et al. (2017) proposed a classifier for video face recognition to solve the problem that scatter might even be higher than the between-class one [1]. Banos et al. (2015) developed a sensor hierarchical weighting classifier for activity recognition. Recently [2]. Sun et al. (2016) proposed a novel hierarchical classification framework that is used for emotion recognition, which combines the feature-level and decision-level fusion strategy for all of the extracted multimodal features [3]. However, the hierarchical structures exploited by existing hierarchical classification methods tend to be task specific.

In this paper, we consider the problem of task-independent hierarchical classification and present a simple but effective two-layer hierarchical classification learning (HCL) framework which constructs one layer of 3-class classifiers (sensor models) to generate discriminative and informative higher-level features (secondary features) from low-level (original) features, and employs existing classification learning methods to build classifier (decision model) on the new features. Thus the proposed classification framework is feature-based and applicable to a broad range of tasks, in particular, face recognition.

Our work is related to some recent work (Samara et al., 2016; Kan et al., 2013; Kumar et al., 2010). In Samara et al. (2016), the authors proposed a two-stage classification method where a set of binary classifiers (resulting in higher-level features) are built from original dataset, one for each class, and a classification model is built by using the outputs of the binary classifiers [4]. Our work is different to Samara et al. (2016) in that we construct one 3-class classifier for every pair of classes, not a binary classifier for every class. In Kan et al. (2013), the higher-level features are extracted as the outputs of a set of pre-learnt SVM models [5]. In Kumar et al. (2010), their framework represents each sample in the weakly labeled data set as a (higher-level) feature vector with each entry being the corresponding decision value from the classification hyperplane of one SVM model [6]. Experimental evaluations show that the proposed framework yields much better result for multi-class classification than some of the existing machine learning algorithms.

The rest of the paper is organized as follows. Section 2 describes the proposed hierarchical classification learning framework in details followed by experimental results and conclusions in Section 3 and Section 4, respectively.

2. Hierarchical Classification Learning Framework

2.1. *The Notation*

We begin by introducing the notation used in this paper. We denote a training dataset by $<X,Y>$ where $X = \{x_1, x_2, ..., x_N\} \in R^{D \times N}$ and $Y = \{y_1, y_2, ..., y_N\} \in R^{1 \times N}$, D is the number of features and N is the number of training samples. A labelled data sample is $<x_i, y_i>$ where $x_i = \langle x_{i1}, x_{i2}, ..., x_{iD} \rangle$ is a D-dimensional vector and $y_i \in C$ is its class label. We use n for the number of sensor models and $c = |C|$ for the number of classes. Similarly we denote a test dataset by $<X^*, Y^*>$ where $X^* = \{x_1^*, x_2^*, ..., x_M^*\} \in R^{D \times M}$, $Y^* = \{y_1^*, y_2^*, ..., y_M^*\} \in R^{1 \times M}$ and M is the total number of test samples. The reformed training dataset is $<Z,Y>$ where $Z = \{z_1, z_2, ..., z_N\} \in R^{f \times N}$ where each data sample has f new secondary features. The reformed test dataset is $<Z^*, Y^*>$ where $Z^* = \{z_1^*, z_2^*, ..., z_M^*\} \in R^{f \times M}$. Note that the reformed datasets have the same class labels as their original counterparts and the number of features in the reformed datasets equals to the number of sensor models. How the reformed datasets are generated will be explained in detail.

2.2. *The Framework*

The HCL framework works in two stages. In the first stage, the multiclass dataset is transformed into a set of 3-class *sensor datasets*, one for each pair of classes resulting in $c*(c-1)/2$ sensor datasets — two classes for the pair of classes and the third class being created to consist of any data instances not belonging to any of the two classes. A *sensor model* is constructed from each of the sensor datasets, using a *base classifier*, with one model employed for distinguishing between each pair of classes. All data instances are put through the sensor models, resulting in *sensor vectors* or the *reformed dataset*. In the second stage, a classification model, called *decision model* (DM), is constructed from the reformed dataset to predict the class label of a data instance based on the decisions from all of the sensor models. The HCL framework is illustrated in Figure 1.

2.2.1. *The Training Process*

The first step is to build sensor models from the class-relabelled training datasets for all pairs of classes by applying a machine learning algorithm. Every data instance in the original training dataset is put through the aforementioned sensor models, converting it into a sensor vector with output of each sensor model

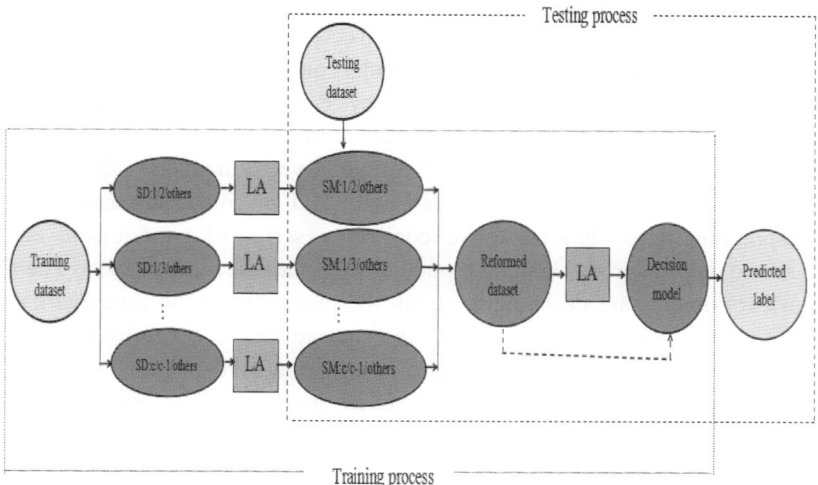

Figure 1. The HCL framework. LA: learning algorithm, SD: sensor dataset, SM: sensor model.

being a secondary feature in the sensor vector. Putting all sensor vectors together results in a new training dataset, the reformed training dataset. Note that the number of sensor models n and the number of classes c in dataset satisfy the following relationship: $n=c*(c-1)/2$ and the number of secondary features f equals to the number of sensor models n, i.e. $f=n$. In our work, for simplicity, we assume that one sensor model produces one feature for the new reformed training dataset. The second step is to build a decision model from the reformed training dataset by applying a machine learning algorithm.

2.2.2. *The Testing Process*

Firstly, we put the original test dataset through the sensor models, which are built in the training process, to construct the reformed test dataset. Every instance in the original test dataset goes through the sensor models resulting in a set of sensor outputs. The sensor outputs are arranged as a vector, sensor vector, serving as a reformed version of the original test data instance. Collecting all sensor vectors together results in a reformed test dataset. Secondly, we put the reformed test dataset through the decision model, which is also built in the training process, to predict the class labels for all test data instances.

3. **Experiments and Discussion**

In this section we present our experimental evaluation of the HCL framework. In our experiment we consider four algorithms — neural networks (NN), random tree (RT), decision tree (DT), and support vector machine (SVM). We use the base classifier to build both the sensor models and the decision model, and we denote this instance of the framework by e.g., experiment sample neural networks (ESNN), experiment sample random tree (ESRT), experiment sample decision tree (ESDT), experiment sample support vector machine (ESSVM).

We evaluate ESSVM, ESDT, ESNN and ESRT by comparing them with the corresponding base classifiers. We conduct a series of experiments on 14 real-world datasets from UCI and 3 popular face classification datasets. The first set of experiments are designed to verify the feasibility of the framework on a wide range of real-world datasets. The second set of experiments are to demonstrate the effectiveness of the framework on facial datasets.

We choose 10-fold cross-validation as the evaluation framework and consider estimated mean accuracy (EMA) and Macro-Averaged F1-Measure (Macro-F1) as evaluation metrics (Yang, 1999) [7].

Our experimental platform is Matlab R2015b on intel core i7 @3.6GHZ with 8.00GB RMB.

3.1. Datasets

3.1.1. The UCI Datasets

14 datasets from UCI Data Repository (Blake *et al.*, 1998) are selected [8]. All datasets are normalized by min-max normalization before training and testing (Han *et al.*, 2012) [9]. Some general information about these datasets is shown in Table 1 lists the details of the UCI datasets.

Table 1. Description of the UCI real-world data sets.

Datasets	#Instances	#Attributes	#Categories	Class proportion
Ecoli	336	8	8	143/77/52/35/20/5/2/2
Zoo	101	17	7	41/20/5/13/4/8/10
Yeast	1484	8	10	463/429/244/163/51/44/37/30/20/5
Wine	178	13	3	59/71/48
Car	1728	6	4	384/69/1210/65
Iris	150	4	3	150/150/150
Glass	214	9	6	70/76/17/13/8/29
Vehicle	846	18	4	212/217/218/199
Balance	625	4	3	49/288/288
Breast	277	9	2	81/196
DNA	2000	180	3	464/485/1051
Msplice	3175	240	3	1648/765/762
Thyroid	7200	21	3	6666/368/166
Segment	2310	18	7	330/330/330/330/330/330/330

3.1.2. Face Recognition

We use three face databases in the experiment. For face verification we use Labeled Faces in the Wild (LFW) (Huang et al., 2007) [10]. We randomly select 200 matched face pairs and 200 mismatched face pairs and crop each image of 80 by 150 pixels as in Kan et al. (2013) [5]. We thus have 24000 features for each pair of images. Some general information about this subset is shown in Table 2.

For face identification we use the ORL face database (Samaria et al., 1994) [11] and AR face database (Martinez et al., 2001) [12]. They are commonly used to evaluate face recognition algorithms. In our experiment, we use a subset of the AR database which are face images of 9 persons and 7 non-occluded face images of each person from the first session. Besides, we crop the face part of the image and resize all images to a standard image size of 80 by 100 pixels. This will yield a 8000-dimentional feature space.

Table 2. Description of the face image datasets used in the experiments.

Datasets	#instances	#attributes	#categories	class proportion
LFW	400	24000	2	200/200
ORL	90	10304	9	10/10/10/10/10/10/10/10/10
AR	63	8000	9	7/7/7/7/7/7/7/7/7

3.2. Experiment Results and Discussion

Ten runs of cross-validation by each algorithm are carried out for each dataset, and the average cross-validation results (EMA and Macro-F1) are reported for each dataset by each algorithm, as shown in Table 3 and Table 4. The best results for each dataset are marked in bold typeface. We can see from Table 3 that ESSVM, ESDT ESNN and ESRT outperform their counterparts, SVM, DT, NN and RT respectively. ESRT gets best performance on 10 out of 14 datasets in terms of EMA and Macro-F1. We list the state of the art performance on some of the datasets from recent literature (Gregor et al., 2012)[13] in Table 3 (column 'recent'). Clearly, ESRT is still competitive when compared with the state when art classification algorithms are used as the base classifier. In particular, the outperformance of ESRT over the benchmark classifiers on Zoo, Wine, Iris, Vehicle, Glass is significant. For example, ESRT/EMA on Zoo is 99.00% whereas RT is only 96.09%. This suggests that the proposed hierarchical classification learning framework has advanced the state of the art in classification.

Table 3. Experimental results on UCI datasets. Note 'nan' in the table indicates that the value is undefined.

Methods / Data	SVM	NN	DT	RT	ESS	ESNN	ESDT	ESRT	R
	EMA(%)								
	Macro-F1								
Ecoli	85.42	76.24	81.52	**85.73**	81.5	81.16	81.89	85.67	8
	nan	nan	nan	nan	nan	nan	0.547	0593	—
Zoo	73.18	88.64	95.64	96.09	73.2	90.48	97.89	**99.00**	9
	nan	0.719	0.665	0.891	nan	0.741	0.844	**0.981**	—
Yeast	59.84	53.96	52.88	60.83	59.7	58.89	53.72	**61.51**	—
	0.554	nan	0.476	0.542	0.53	nan	0.446	**0.545**	—
Wine	66.02	97.25	89.79	95.88	67.9	98.02	90.38	**98.24**	9
	0.713	0.972	0.901	0.962	0.72	0.977	0.905	**0.984**	—
Car	76.57	72.75	82.05	84.44	83.6	81.30	83.52	**84.67**	—
	0.740	0.586	0.642	0.731	0.69	0.557	0.698	**0.732**	—
Iris	94.67	94.67	94.67	95.33	95.3	**97.33**	96.00	97.33	9
	0.947	0.947	0.947	0.953	0.95	**0.974**	0.941	0.967	—
Glass	67.28	51.71	65.08	75.72	69.1	61.40	68.79	**79.39**	6
	0.621	nan	0.612	0.700	0.66	nan	0.638	**0.742**	—
vehicle	71.14	72.37	70.85	74.00	71.8	**77.11**	71.40	76.32	7
	0.712	0.721	0.708	0.737	0.71	**0.771**	0.715	0.761	—
Balance	89.94	89.87	77.23	83.51	90.1	**91.19**	78.39	83.45	—
	nan	0.752	0.554	0.596	nan	**0.804**	0.558	0.666	—
Breast	68.62	72.92	67.28	**73.65**	69.4	71.29	66.84	72.32	7
	0.487	0.629	0.599	**0.651**	0.50	0.583	0.593	0.640	—
DNA	59.55	90.25	91.20	93.00	59.9	91.35	91.05	**93.65**	—
	0.574	0.892	0.899	0.921	0.57	0.905	0.897	**0.926**	—
Msplice	60.75	94.17	93.10	95.40	60.1	94.45	93.35	**95.68**	—
	0.595	0.936	0.922	0.948	0.58	0.938	0.925	**0.952**	—
Thyroid	94.76	93.93	99.51	99.61	94.7	94.36	99.60	**99.68**	—
	0.678	0.632	0.972	0.978	0.67	0.697	0.977	**0.980**	—
Segment	95.11	88.35	95.80	98.05	94.8	94.94	96.67	**99.14**	9
	0.952	0.883	0.958	0.981	0.95	0.950	0.967	**0.985**	—

Table 4. Experimental results on face datasets. Note 'nan' in the table indicates that the value is undefined.

Methods Data	SVM	NN	DT	RT	ESS	ESNN	ESD	ESR
	EMA(%)							
	Macro-F1							
LFW	45.00	55.00	58.00	**66.75**	44.50	58.25	58.25	63.25
	0.449	0.550	0.580	**0.668**	0.445	0.588	0.588	0.633
ORL	0	60.02	70.04	96.67	0	75.38	75.38	**98.89**
	nan	0.470	0.653	0.968	nan	0.691	0.691	**0.989**
AR	0	70.74	68.67	94.44	0	74.00	74.00	**96.11**
	nan	0.589	0.586	0.915	nan	0.565	0.565	**0.926**

The experiments on face recognition also get good results. For instance, ESRT gets best performance on ORL, an increase to 98.89% from 96.67% by RT on EMA, and an increase to 0.989 from 0.968 on Macro-F1. However, ESRT performs less well on LFW which is a verification task. The results on face recognition experiments provide further evidence to suggest that ESRT has superior performance.

The outperformance of the framework over the base classifier is not significant on Breast and LFW, which are binary datasets. This suggests that the framework does not apply to binary classification.

4. Conclusion

In this paper, we present a novel hierarchical classification learning framework for multiclass classification in order to improve the performance of existing classifiers. A given dataset is reformed via multiple feature-sensing classification models, one for each pair of classes, before a decision classification model is constructed from the reformed dataset. Any classification algorithm can be used as the base classifier in the framework. Extensive experiments using 14 UCI datasets and 3 face datasets have shown that the framework leads to improved performance in almost all cases when four state of the art classification algorithms are used as the base classifier — SVM, NN, RT, DT. Since the base classifier is used multiple times in the framework, there are time costs associated with the framework,the computational time is more than when it is used only once. There are also limitations. First of all, the outperformance is not always significant and underperformance does happen, albeit very rarely. For example, ESSVM underperforms SVM on Ecoli. Secondly, when the number of classes is very large, the time cost of building sensor models will be too high. When the number of classes is moderate, the performance gain will outweigh the cost of additional time.

As future work we will test the HCL framework on more datasets and use more learning algorithm as the base classifier. We will further study the framework used in novelty class detection and apply it to some real-world applications.

References

1. Z. Fan, S. Weng, Y. Zeng, et al. Misclassified Samples based Hierarchical Cascaded Classifier for Video Face Recognition[J]. Ksii Transactions on Internet & Information Systems, 11(2017).
2. O. Banos, M. Damas, H. Pomares, et al. Human activity recognition based on a sensor weighting hierarchical classifier[J]. Soft Computing, 17(2):333-343(2013).
3. B. Sun, L. Li, X. Wu, et al. Combining feature-level and decision-level fusion in a hierarchical classifier for emotion recognition in the wild[J]. Journal on Multimodal User Interfaces, 10(2):125-137(2016).
4. A. Samara, L. Galway, R. Bond, et al. Sensing Affective States Using Facial Expression Analysis[C]// International Conference on Ubiquitous Computing and Ambient Intelligence. Springer International Publishing, 341-352(2016).
5. M. Kan, D. Xu, S. Shan, et al. Learning prototype hyperplanes for face verification in the wild. [J]. IEEE Transactions on Image Processing, 22(8):3310-3316(2013).
6. N. Kumar, A. C. Berg, P. N. Belhumeur, et al. Attribute and simile classifiers for face verification[C]// IEEE, International Conference on Computer Vision. IEEE, 365-372(2010).
7. Y. M. Yang, An evaluation of statistical approaches to text categorization, information retrieval, 1(1), 69-90(1999).
8. C. Blake, E. Keogh, C. J. Merz, UCI Repository of Machine Learning Database (1998).
9. J. Han, M. Kamber, J. Pei, Data Mining: Concepts and Techniques,Third Edition, Beijing: China Machine Press, 74(2012).
10. G. B. Huang, M. Mattar, T. Berg, et al. Labeled Faces in the Wild: A Database forStudying Face Recognition in Unconstrained Environments[J]. Month (2007).
11. F. S. Samaria, A. C. Harter, Parameterisation of a stochastic model for human face identification[C]// Applications of Computer Vision, Proceedings of the Second IEEE Workshop on. IEEE, 138-142(1994).
12. A. M. Martinez, A. C. Kak, PCA verus LDA, IEEE Transaction on Pattern Analysis and Machine Intelligence, 23(2), 228-233(2001).
13. S. Gregor, K. Simon, P. Igor, et al. Comprehensive Decision Tree Models in Bioinformatics[J]. Plos One, 7(3):e33812(2012).

A rough set based hybrid approach for classification

Ahmed Saad Hussein, Tianrui Li, Noora Sabah Jaber and
Chubato Wondaferaw Yohannese

School of Information Science and Technology, Southwest Jiaotong University
Chengdu, 611756, China
husseinsaad187@yahoo.com, trli@swjtu.edu.cn, noorasja2016@gmail.com,
freewwin@yahoo.com

Uncertainty defined as a situation with inadequate information can be of three types: inexactness, unreliability, and ignorance, and not merely the absence of knowledge. However, uncertainty can prevail in cases where a considerable amount of information is available. In this regard, Rough Set Theory (RST) is a new mathematical model that deals with uncertain information. Thus, in this paper, we propose a new hybrid approach that combines RST together with Machine Learning (ML) algorithm to improve the classification performance efficiently. We use three ML algorithms, e.g., K-Nearest Neighbors (KNN), Naive Bayes (NB), and Support Vector Machine (SVM), for experimental validation. The results confirm that the proposed method can achieve a remarkable classification performance.

Keywords: Rough sets; uncertainty; machine learning; classification.

1. Introduction

Machine learning techniques are widely used in various applications in business, government, and education. Examples include banking, bioinformatics, environmental modeling, epidemiology, finance, marketing, medical diagnosis, and meteorological data analysis.[1,2] Data is often associated with uncertainty because of measurement inaccuracy, sampling discrepancy, legacy data conversion, or other errors.[1-3] In addition, uncertainty can be caused by our limited conception or comprehension of reality (e.g., limitations of the surveillance equipment; limited resources to collect, store, or understand data); furthermore, it can also be inherent in nature (e.g., due to preconception). Therefore, uncertainty is linked with randomness, vagueness, roughness, and imperfect knowledge. The theories of probability, fuzzy set, rough set, and so on have been used for uncertainty analysis. Rough Set Theory (RST) is a new method that has gained an increasing attention from both the theoretical and the applied domains of data mining. Machine Learning (ML) has made great progress in data analysis research.[3-6] In recent years, many studies[3,7-9] have been conducted on the management of uncertain data in databases, such as the description of uncertainty in databases and querying data with uncertainty. Many factors affect uncertainty in real-world situations. Therefore, examining and analyzing the characteristics of uncertainty from various viewpoints is vital in intelligent information processing. Thus, the objective of this study is to improve the classification performance by tackling uncertainty with the combined implementation of RST and ML. Its idea is to use ML techniques to deal with uncertain rules obtained from RST which is different with previous work.[10,11] Hence, a hybrid approach is proposed and the experimental validation is carried out using datasets from UCI and open ML data repository. The result shows the robustness of the combined technique.

The rest of the paper is organized as follows. In Section 2, the basic concepts are described. Section 3 demonstrates the proposed approach by a combination of RST and ML. Section 4 presents the experimental results. Finally, concluding remarks and future work are stated in Section 5.

2. Preliminaries

In this section, we will review some basic concepts of classical RST.

Definition 1.[12,13] Let $S = \{U, AT = C \cup D, V, f\}$ be an information system, where U is a non-empty finite set of objects, called the universe; AT is a non-empty finite set of attributes including condition attributes C and decision attributes D; $V = \cup_{a \in AT} V_a$ and V_a is a domain of attribute a; $f: U \times AT \to V$ is an information function such that $f(x, a) \in V_a$ for every $a \in AT, x \in U$. $\forall X \subseteq U$ and $B \subseteq A$, the lower and upper approximations of X with respect to the equivalence relation R_B are respectively defined as:

$$\underline{R_B}(X) = \{X \mid [x]_{R_B} \subseteq X\}, \tag{1}$$

$$\overline{R_B}(X) = \{X \mid [x]_{R_B} \bigcap X \neq \phi\}, \tag{2}$$

where $[x]_{R_B} = \{y|(x,y) \in R_B\}$ is the equivalence class determined by the equivalence relation $R_B = \{(x,y) \in U \times U | f(x,b) = f(y,b), \forall b \in B\}$. According to the lower and upper approximations, the positive, negative and boundary regions of X are easy to get as follows.

$$\begin{cases} POS_B(X) = \underline{R_B}(X), \\ NEG_B(X) = U - \overline{R_B}(X), \\ BND_B(X) = \overline{R_B}(X) - \underline{R_B}(X). \end{cases}$$

Its boundary region causes the uncertainty of a concept. The larger the boundary region, the greater the degree of uncertainty of the concept.

3. A Rough Set Based Hybrid Approach for Classification

Several scholars have applied different theories and algorithms to obtain good classification performance.[4,7,10,14] This study aims to improve the classification performance by tackling uncertainty with the implementation a hybrid of RST with ML-based methods. The proposed approach is a supervised learning model. As shown in Fig. 1, it consists of the following steps, namely preprocessing, rough set analysis, generating certain and uncertain rules, dealing with uncertain rules by ML algorithms and finally output rules for classification.

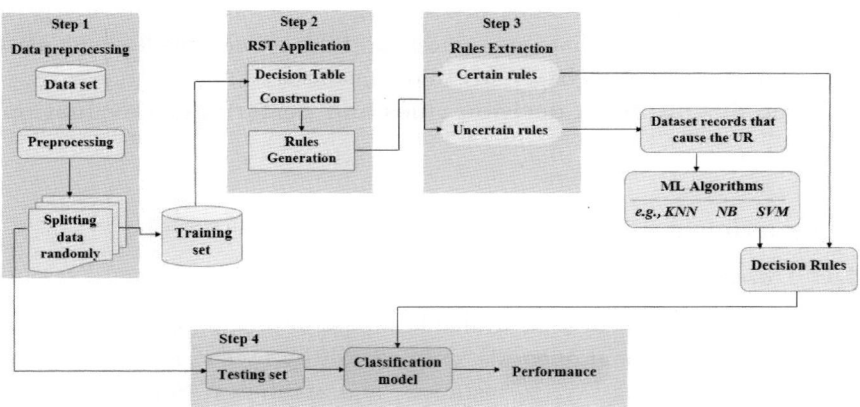

Fig. 1. The proposed framework of combined RST with ML algorithms.

The preprocessing part reads the dataset records, specific condition attributes, and decision attributes, and initializes the attributes value to a particular range. RST analysis performs the main task of the proposed approach, i.e., that generates certain and uncertain rules which extracts from the lower approximation and boundary region. The accuracy of a deterministic rule is 1, which implies that it is certain, whereas that of a non-deterministic rule lies exclusively between 0 and 1, which implies that it is uncertain. Uncertain rules are processed again by applying ML algorithms (e.g., KNN, NB, and SVM, denoted as RST-KNN, RST-NB, RST-SVM, respectively) on the data record that causes the Uncertain Rule (UR) to achieve better classification performance. The proposed algorithm, namely, a hybrid of RST with ML-based methods, is illustrated in Algorithm 3.1.

Algorithm 3.1 The proposed hybrid algorithm
Input: A training data set.
Output: Rules for classification.

1: **Step 1: Data preprocessing**
2: **Step 2: RST analysis**
3: Construct elementary sets
4: Calculate discernibility matrix
5: Calculate Core and reducts of attributes (absorption law)
6: Calculate Core and reducts of attribute values
7: Output certain and uncertain rules
8: **Step 3: Uncertain rules processing**
9: Insert data record that cause uncertain rules to machine learning algorithms (e.g., KNN, NB, SVM) to obtain rules
10: **Step 4: Output of decision rules and performance evaluation**

---

4. Experimental Evaluation

4.1. *Datasets*

Among the different datasets available, we perform experiments using six datasets from UCI and open ML data repository by considering only the essential features, e.g., types, instances, and attributes. Some basic facts about the datasets are listed in Table 1.

4.2. *Results*

The experiments are carried out using Visual Studio 2015 and MATLAB R2014a. We randomly split each data set into a training set and a testing set. The training set accounts for 80% of the entire dataset and the remaining 20% is used for the testing set. Using the above-mentioned scheme for

Table 1. Description of datasets.

Datasets	Data Types	Instances	Attributes
BN authentication	Multivariate	1372	5
CMC	Multivariate	1473	9
Mammographic Mass	Multivariate	961	6
seeds	Multivariate	210	7
jEdit	Binary	274	9
RMFTSA	Binary	508	11

experiments, the average of obtained results is taken. The test is repeated ten times independently. The result is the average of the ten repetitions. The experimental results of accuracy, precision, and f-measure are listed in Tables 2–4. The average performance of the proposed method and others on six datasets (e.g., accuracy, precision, f-measure, and sensitivity detection rate) are depicted in Figs. 2–3, and Table 5, respectively. It is clear that RST-SVM achieves the highest accuracy of 96.52%, and RST-NB and RST-KNN reach accuracy of 86.08% and 77.07%, respectively. On the other hand, the accuracy of the conventional method only using RST is 71.14%, which is lower than those of a hybrid of RST with all three ML-based methods.

Table 2. Accuracy (%).

Datasets	RST	RST-KNN	RST-NB	RST-SVM
BN authentication	88.11	92.20	96.30	99.42
CMC	44.75	46.74	57.61	98.37
Mammographic Mass	77.67	81.55	92.56	91.91
seeds	79.49	89.74	93.59	97.44
jEdit	55.87	68.62	88.23	94.11
RMFTSA	80.95	83.60	88.24	97.88

Table 3. Precision.

Datasets	RST	RST-KNN	RST-NB	RST-SVM
BN authentication	0.73	0.85	0.93	0.99
CMC	0.44	0.68	1	0.97
Mammographic Mass	0.74	0.88	0.95	1
seeds	0.96	1	1	1
jEdit	0.68	0.70	0.80	0.94
RMFTSA	0.83	0.88	0.89	0.96

Table 4. F-measure.

Datasets	RST	RST-KNN	RST-NB	RST-SVM
BN authentication	0.83	0.9	0.95	0.99
CMC	0.5	0.68	0.78	0.97
Mammographic Mass	0.77	0.83	0.92	0.92
seeds	0.87	0.91	0.92	0.95
jEdit	0.60	0.68	0.87	0.94
RMFTSA	0.82	0.84	0.89	0.97

Table 5. Average performance for six datasets.

	Accuracy(%)	Precision	F-measure	Sensitivity rate
RST	71.14	0.73	0.73	0.75
RST-KNN	77.07	0.83	0.81	0.80
RST-NB	86.08	0.93	0.89	0.87
RST-SVM	96.52	0.97	0.96	0.94

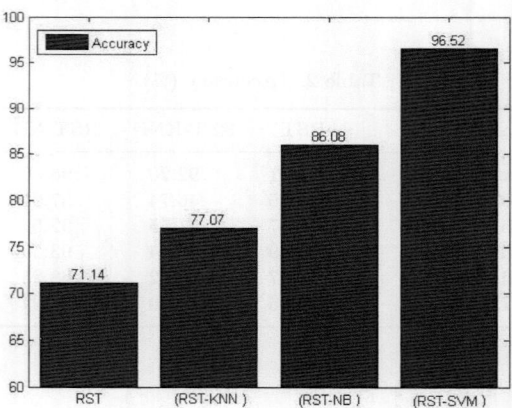

Fig. 2. Average accuracy for six datasets.

Clearly, our proposed hybrid approach performs better on most aspects compared with RST. The following lines explain these results in detail:

(1) The accuracy of our proposed method in Fig. 2 shows an excellent performance of the classifier.

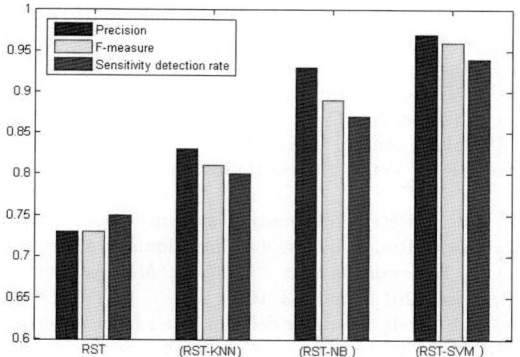

Fig. 3. Average performance for six datasets.

(2) Our proposed method increases the precision of classification (see Table 5).
(3) It can be seen from Fig. 3 that the precision of RST-KNN, RST-NB and RST-SVM are better than RST. Further, F-measure of our proposed method is also higher which indicates a better performance.
(4) Our proposed method increases sensitivity detection rate (see Table 5).

5. Conclusion

This paper presented a rough set based hybrid approach that combines ML algorithms to improve the classification performance. It was validated by using six different datasets from the UCI ML Repository and Open ML. The experimental results showed that it achieved a good performance in classification. For the future work, we will focus on how to hybrid other methods, e.g., neural network, with RST to further improve the classification performance.

References

1. C. K.-S. Leung, *Wiley Interdisciplinary Reviews: Data Mining and Knowledge Discovery* **1**, 316 (2011).
2. G. Suresh, S. Shaik, E. Reddy and U. A. Shaik, *International Journal of Computer Science and Information Security* **8**, 111 (2010).
3. M. Chau, R. Cheng, B. Kao et al., Uncertain data mining: A new research direction, in *Proc. of the Workshop on the Sciences of the Artificial*, (Hualien, Taiwan, 2005).
4. A. P. D. Silva, *European Journal of Operational Research* **261**, 772 (2017).
5. C. W. Yohannese and T. Li, *International Journal of Computational Intelligence Systems* **10**, 647 (2017).

6. Y. Wang and N. Zhang, *The Scientific World Journal* **2014** (2014).
7. J. Liang, Uncertainty and feature selection in rough set theory, in *Int. Conf. on Rough Sets and Knowledge Technology*, (Banff, AB, Canada, 2011).
8. G. V. Suresh, E. V. Reddy and E. S. Reddy, Uncertain data classification using rough set theory, in *Proc. Int. Conf. on Information Systems Design and Intelligent Applications*, (Visakhapatnam, India, 2012).
9. T. Beaubouef and F. E. Petry, *International Journal of Intelligent Systems* **15**, 389 (2000).
10. T. Beaubouef and F. Petry, Information systems uncertainty design and implementation combining: Rough, fuzzy, and intuitionistic approaches, in *Flexible Approaches in Data, Information and Knowledge Management*, eds. P. Olivier and Z. Sawomir (Springer, 2014) pp. 143–164.
11. S. Trabelsi and Z. Elouedi, Learning decision rules from uncertain data using rough sets, in *Int. FLINS Conf. on Computational Intelligence In Decision And Control*, (Madrid, Spain, 2008).
12. Q. Zhang, Q. Xie and G. Wang, *CAAI Transactions on Intelligence Technology* **1**, 323 (2016).
13. Z. Pawlak, *European Journal of Operational Research* **136**, 181 (2002).
14. Y. Huang, T. Li, C. Luo, H. Fujita and S.-J. Horng, *Knowledge-Based Systems* **122**, 131 (2017).

A Fuzzy Gamma Naive Bayes classifier

Ronei Marcos de Moraes

Departament of Statistics, Federal University of Paraiba
Joao Pessoa, Paraiba, Brazil
ronei@de.ufpb.br

Elaine Anita de Melo Gomes Soares

Graduate Program in Decision Models and Health, Federal University of Paraiba
Joao Pessoa, Paraiba, Brazil
elaineanita1@gmail.com

Liliane dos Santos Machado

Departament of Computer Science, Federal University of Paraiba
Joao Pessoa, Paraiba, Brazil
liliane@di.ufpb.br

Classifiers based on Gamma statistical distribution can be found in the scientific literature, but they assume data collected without errors. However, in some cases precision of information can not be guaranteed. This paper presents a proposal of a new classifier named Fuzzy Gamma Naive Bayes network (FGamNB). The theoretical development is presented, as well as results of its application on simulated multidimensional data. A brief comparison among FGamNB, a classical Gamma Naive Bayes classifier and a Naive Bayes classifier was performed. The results obtained showed that the FGamNB produced the best performance, according to the Overall Accuracy Index, Kappa and Tau Coefficients.

Keywords: Gamma statistical distribution; fuzzy classification; fuzzy statistics; Naive Bayes.

1. Introduction

Classification methods have been applied for several purposes in different human knowledge areas, such as: pattern recognition,[1] data mining,[2] assessment of training based on virtual reality,[3] and others. Many methods are designed to work with a specific statistical distribution and they obtain their best results when applied to data that follow it,[4] as for instance the Bayesian networks.[5]

Gamma is a statistical distribution defined by two parameters given by positive real numbers. The Gamma distribution is a general distribution and other statistical distributions are particular cases from it (Exponential, Poisson, Erlang and Chi-squared distributions, among others).[6] For instance, the necessary time until the occurrence of the first event of interest in a Poisson process follows the Exponential distribution. In this same case, the time necessary until the occurrence of the second event of interest, i.e., the sum of two time intervals follows the Gamma distribution.

The Gamma distribution is used for survival and reliability studies.[7] Moreover, it is used for many applications as pattern recognition on SAR images,[8] identify peak-callers, and classify mutation types of cancer in genomics,[9,10] studies on inter-spike intervals distribution in neuroscience,[11] accumulated rainfall,[12] stock markets calibration,[13] among others.

Several types of applications can be modeled using Gamma distribution. However, there are cases in which variables can not be measured with complete accuracy, thus it can be interesting to model them as fuzzy events. For this reason, this paper presents a new fuzzy supervised classifier based on Gamma statistical distribution, structured as a Bayesian network. Furthermore, Zadeh's probability formulation for fuzzy events[14] was used in order to provide the new Fuzzy Gamma Naive Bayes network (FGamNB).

2. Selected Fundamentals

In order to make the study self-contained, some pertinent prerequisites are provided in this section. Gamma distribution is explained in details, as well as the classical Naive Bayes Classifier and its version using Gamma distribution.

2.1. *Gamma Statistical Distribution*

In general, the Gamma statistical distribution for a random variable X can be written as:

$$P(X|\alpha, \beta) = \frac{X^{\alpha-1}}{\beta^\alpha \Gamma(\alpha)} e^{\frac{-X}{\beta}}, \quad \forall X > 0, \tag{1}$$

where $\alpha > 0$ is the shape parameter, $\beta > 0$ is the scale parameter, and Γ is the Gamma function, which is given by:

$$\Gamma(\alpha) = \int_0^\infty t^{(\alpha-1)} e^{-t} \, dt. \tag{2}$$

It is worth noticing that if α is a positive integer, the equation (2) become $\Gamma(\alpha) = (\alpha - 1)!$, i.e. the factorial function of $(\alpha - 1)$.

When $\alpha = 1$, Gamma distribution is named Exponential distribution, which is able to model how long time is necessary until some event occurs. If α is a positive integer, the distribution is known as Erlang distribution, which is used to model how long apart are α events occurring. When $\alpha = k/2$, where k is a positive integer, and $\beta = 2$, it is named χ^2-distribution with k degree of freedom, which it is widely used in inferential statistics.[6]

Different parametrizations for Gamma distribution can be found in the literature, with $\alpha > 0$ and $\beta > 0$, such as: (i) with shape parameter α, and scale parameter β, as that one provide by equation (1); (ii) with shape parameter α, and rate parameter $\theta = 1/\beta$; and (iii) with shape parameter α, and mean parameter $\theta = \alpha/\beta$.

A Gamma distribution using the parametrization presented in the equation (1) is denoted by $Ga(\alpha, \beta)$. The mean or expected value for X, which follows that distribution, is $E(X) = \alpha\beta$ and its variance is $V(X) = \alpha\beta^2$.

2.1.1. *Parameters Estimation*

Several approaches for the estimation of Gamma distribution parameters can be found in the scientific literature.[6,12,15] Consequently, it is also valid for the parameter estimation for classification methods based on that statistical distribution. In this paper, the approach provide by Minka[16] is used.

Using the generalized Newton, a fast interactive approximation can be obtained to estimate α, denoted by $\hat{\alpha}$, from the log-likelihood function. Using sample data, *generally* it can converge in about four iterations:[16]

$$\frac{1}{\hat{\alpha}^{new}} = \frac{1}{\hat{\alpha}} + \frac{\overline{log(x)} - log\bar{x} + log\hat{\alpha} - \Psi(\hat{\alpha})}{\hat{\alpha}^2[1/\hat{\alpha} - \Psi'(\hat{\alpha})]} \quad (3)$$

where $\Psi(\hat{\alpha}) = log(\hat{\alpha}) - \frac{1}{\hat{\alpha}}$. A good starting point for $\hat{\alpha}$ is:

$$\hat{\alpha} \approx \frac{0.5}{log\bar{x} - \overline{logx}} \quad (4)$$

The maximum likelihood estimator for β parameter is:

$$\hat{\beta} = \frac{\bar{x}}{\hat{\alpha}} \quad (5)$$

2.2. Naive Bayes Classifier

The Naive Bayes network (NB) is a classical method used for data classification, decision making, data mining, among others. This method is based on the Bayes' theorem. It assumes that there is no dependency between variables of a system. Although assuming this hypothesis is not realistic in some situations, the NB classifier is able to provide satisfactory results.[17] An advantage of that assumption is that NB classifier is able to classify data for which it was not trained for.[18]

Formally, let there be a space of decisions $\Omega = \{1, ..., M\}$, where M is the total number of classes. Let there be w_i, $i \in \Omega$, the most probable decision class to be chosen. It is possible to determine this decision class given a data vector $X = \{X_1, X_2, ..., X_n\}$. The method is then expressed as:

$$P(w_i|X) = \frac{P(w_i)}{S} * \prod_{k=1}^{n} P(X_k|w_i) \qquad (6)$$

where $P(w_i)$ is the probability of this vector to belong to w_i, S is a scale factor, and $P(X_k|w_i)$ is the probability of X given w_i. This classifier assumes multinomial distribution, but it can be used for continuous variables, after discretizing the information. However, this procedure can lead to loss of information.[17]

The classification is a maximization process. So, as the scale factor S is constant, it can be suppressed. Thus, the probability $P(X_k|w_i)$ can be rewritten as:

$$P(w_i|X) = P(w_i) \times \prod_{k=1}^{n} P(X_k|w_i) \qquad (7)$$

The classification rule for the Naive Bayes Classifier is given by:
select class w_i for the vector X if

$$P(w_i|X) \geq P(w_j|X) \quad \text{for all i} \neq j \quad i, j \in \Omega \qquad (8)$$

where P is expressed by the equation (7). The parameters for NB classifier can be estimed by maximum likelihood, Laplace's estimator or m-estimate.[19]

2.3. Gamma Naive Bayes Classifier

The Gamma Naive Bayes Classifier (GamMNB) is an instance of the Naive Bayes Classifier, in which it is assumed the Gamma distribution density (equation 1) for the probability $P(w_i|X)$ in equation (7). Thus, that equation can be rewritten as following:

$$P(w_i|X) = P(w_i) \times \prod_{k=1}^{n} \left(\frac{X_k^{\alpha-1}}{\beta^\alpha \Gamma(\alpha)} e^{\frac{-X_k}{\beta}} \right) \quad (9)$$

where α and β are estimated from the given data through equations (4) and (5). Furthermore, the decision rule for this classifier is the same as the one presented in equation (8).

3. The New Fuzzy Gamma Naive Bayes Classifier

In 1968, Zadeh introduced the concept of probability for fuzzy events, from the classical probability formulation. In a formal way, let B be a σ-field of Borel subsets in R^n and P be a probability measure over Ω. Let F be a fuzzy event in B with membership function $\mu_F : R^n \to [0,1]$. The probability of F is defined by the integral of Lebesque-Stieljes:[14]

$$P(F) = \int_{F \subseteq R^n} \mu_F(x) dP = E(\mu_F) \quad (10)$$

i.e., the probability of a fuzzy event F is the mathematical expectation of its membership function. It can be rewritten as:

$$P(F) = \int_{F \subseteq R^n} \mu_F(x) P(x) dP \quad (11)$$

Applying the Zadeh's Probability concept on equation (9), we create the Fuzzy Gamma Naive Bayes Classifier (FGamNB) and define it as:

$$P(w_i|X) = P(w_i) \times \prod_{k=1}^{n} \left(\frac{X_k^{\alpha-1}}{\beta^\alpha \Gamma(\alpha)} e^{\frac{-X_k}{\beta}} \times \mu_i(X_k) \right) \quad (12)$$

Using logarithm function is possible to simplify the equation above in order to reduce its computational complexity by replacing multiplications for additions.[4]

$$g_f(w_i, X) = \log P(w_i|X) = \log[P(w_i)] + (\alpha - 1) \sum_{k=1}^{n} \log(X_k)$$
$$- n\alpha \log(\beta) - n\log[\Gamma(\alpha)] - \frac{1}{\beta} \sum_{k=1}^{n} X_k + \sum_{k=1}^{n} \log[\mu_i(X_k)] \quad (13)$$

where g_f is the new classification function. The necessary parameters should be learned from sample data. The Gamma parameters α and β

are estimated through equations (4) and (5) and the membership functions $\mu_i(X_k)$ can be estimated using histograms from sample data.

The better estimation for class of the vector X is obtained from the highest values of the classification function g_f. So, the classification rule for FGamNB is:

$$X \in w_i \text{ if } g_f(w_i, X) > g_f(w_j, X) \qquad (14)$$

for all $i \neq j \in \Omega$ and the functions g_f are given by equation (13).

4. Simulation

In order to assess the new FGamNB classifier, a Monte Carlo simulation was performed. Three classes composed by 3-dimensional Gamma distributed random variables each one were randomly generated. The parameters used for this simulation are presented in the following: (i) $class1 = [Ga(2.0, 4.0); Ga(8.0, 3.0); Ga(3.0, 5.0)]$; (ii) $class2 = [Ga(16.0, 2.0); Ga(12.0, 4.0); Ga(25.0, 2.0)]$; and (iii) $class3 = [Ga(20.0, 3.0); Ga(16.0, 3.0); Ga(35.0, 2.0)]$.

In total, 40 double databases were created, where the first one is for training and the second one is for testing. The same Gamma parameters were used to create both of them. In these simulations, it was assumed 200 observations for all three cases. As result, we gathered the Overall Accuracy Index,[20] Kappa Coefficient,[21] Tau Coefficient,[22] and confusion matrices in order to compare and assess the methods.

5. Results and Discussions

From those double databases created, the files with training samples were used to estimate the parameters for the FGamNB. The files with testing samples were used to evaluate the performance of that classifier. The results were then compared with known classification, which was generated from the simulation. The CPU time in the classifications tasks were also measured using a Core 2 Duo PC compatible with 2GB of RAM.

The best result obtained with FGamNB, according to Kappa Coefficient, was $K = 96.25\%$ with variance 9.1387×10^{-4}. Using Tau Coefficient, the aggrement was $T = 96.25\%$ with variance 9.1387×10^{-5}. Furthermore, the Overall Accuracy resulted in $OA = 97.50\%$ with variance 4.0625×10^{-5}. The FGammaNB made mistakes in 15 cases. That performance is excellent and it shows the adequacy of FGamNB in the solution of this kind of problem. The computational performance of the FGammaNB classifier was 1.2080 seconds for the 200 cases.

The same procedure was used in order to evaluate NB and GamNB classifiers, i.e., the same samples of training were used to obtain the parameters for all classifiers, and the same testing samples were used for a controlled and impartial comparison among them.

The NB classifier using the m-estimate produced the following coefficients: $K = 69.50\%$ with variance 5.925×10^{-4}, $T = 69.50\%$ with variance 6.0746×10^{-4}, and $OA = 79.67\%$ with variance 2.6998×10^{-4}. It demanded 1.9040 seconds of CPU. In this case, there were 122 misclassifications. The GamNB classifier produced Kappa and Tau coefficients $K = 94.75\%$ with variances 1.2665×10^{-4} and 1.2667×10^{-4}, respectively. The Overall Accuracy obtained was 96.50% with variance 5.6292×10^{-5}. Additionally, it demanded 0.0400 seconds of CPU. In this case, there were only 21 misclassifications. Table 1 summarizes all results for the three classifiers in this comparison.

Table 1. Comparison results obtained from each classifier in the simulation.

Classifier	Overall Acc.	Kappa Coeff.	Tau Coeff.	CPU Time	Misclassif.
FGamNB	97.50%	96.25%	96.25%	1.208 sec	15
NB	79.67%	69.50%	69.50%	1.904 sec	122
GamNB	96.50%	94.75%	94.75%	0.040 sec	21

It is worth noticing, from the Kappa coefficients, that the performance of the FGamNB classifier is better than both other classifiers. However, the computational time taken by the new classifier is not as good as the GamNB. The main reason for that is related to the fact that the FGamNB also needs to compute the membership functions. The NB classifier provide worst results and worst computational time when compared to the previous ones. The reasons are that NB must discretize all variables before computing their probabilities and perform the classification task. As mentioned before, the discretization procedure of variables leads to loss of information.

6. Conclusion

In this paper was proposed a new classifier named Fuzzy Gamma Naive Bayes network. It is based on Gamma statistical distribution under Naive Bayes hypothesis and is able to classify data obtained from fuzzy events.

The classifier proposed was able to provide satisfactory results when compared to the other two classifiers. In this case, even expending more processing time, the number of misclassifications is lower than the others and show the potential of this classifier in terms of classification accuracy.

Acknowledgments

This project is partially supported by grants 132170/2017-5, 308250/2015-0 and 310561/2012-4 of the National Council for Scientific and Technological Development (CNPq) and is related to the National Institute of Science and Technology Medicine Assisted by Scientific Computing (465586/2014-4) also supported by CNPq.

References

1. A. R. Webb and K. D. Copsey, *StatisticalPattern Recognition*, 3rd edn. (Wiley, 2011).
2. I. H. Witten, Eibe Frank and M. A. Hall, *Data Mining Practical Machine Learning Tools and Techniques*, 3rd edn. (Morgan Kaufmann Pub., 2011).
3. R. M. Moraes and L. Machado, *Knowledge Based Systems* **70**, 97 (2014).
4. R. M. Moraes and L. S. Machado, A fuzzy exponential naive bayes classifier, in *Uncertainty Modelling in Knowledge Engineering and Decision Making: Proceedings of the 12th International FLINS Conference (FLINS 2016)*, 2016.
5. D. Koller and N. Friedman, *Probabilistic Graphical Models* (MIT Press, 2009).
6. C. Forbes, M. Evans, N. Hastings and B. Peacock, *Statistical Distributions*, 4th edn. (John Wiley New Jersey, 2011).
7. A. Papoulis and S. U. Pillai, *Probabilistic Random Variables and Stochastic Processes*, 4th edn. (McGraw Hill Boston, 2002).
8. G. Gao, K. Ouyang, Y. Luo, S. Liang and S. Zhou, *IEEE Transactions on Geoscience and Remote Sensing* **55**, 1812 (2017).
9. M. Mendoza-Parra, M. Nowicka, W. V. Gool and H. Gronemeyer, *BMC Genomics* **14** (2013).
10. W. Zhang, A. Edwards, W. Fan, P. Deininger and K. Zhang, *BMC Genomics* **12** (2011).
11. M. Wright, I. Winter, J. Forster and S. Bleeck, *Hearing Research* **317**, 23 (2014).
12. G. J. Husak, *International Journal of Climatology* **27**, 935 (2007).
13. D. T. Nguyen, S. P. Nguyen, U. H. Pham and T. D. Nguyen, A calibration-based method in computing bayesian posterior distributions with applications in stock market, in *International Conference of the Thailand Econometrics Society*, 2018.
14. L. A. Zadeh, *Journal of mathematical analysis and applications* **23**, 421 (1968).
15. P. Ricci, *Fitting Distributions with R*, cran, CRAN (Vienna, Austria, 2005).
16. T. P. Minka, *Estimating a Gamma distribution*, microsoft research technical report, Microsoft (Cambridge, UK, 2002).
17. R. M. Moraes and L. Machado, *Mathware & Soft Computing* **16**, 123 (2009).
18. M. Ramoni and P. Sebastiani, *Artificial Intelligence* **125**, 209 (2001).
19. L. Jiang and C. Li, *Journal of Software* **6**, 1368 (2011).
20. R. G. Congalton and K. Green, *Assessing the accuracy of remotely sensed data: Principles and practices* (Lewis Publishers New York, 1999).

21. J. Cohen, *Educational and Psychological Measurement* **20**, 37 (1960).
22. Z. Ma and R. L. Redmond, *Photogrammetric Engineering and Remote Sensing* **61**, 435 (1995).

Spectral data classification using locally weighted partial least squares classifier

Weiran Song,[1] Hui Wang,[1] Paul Maguire[2] and Omar Nibouche[1]

[1]School of Computing, [2]School of Engineering, Ulster University
BT37 0QB, Newtownabbey, Co. Antrim, UK

Partial least squares discriminant analysis (PLS-DA) is an effective chemometric method for handling ill-conditioned problems in data matrices, such as small-sample-size, high dimensionality and high collinearity. Although PLS-DA has been widely used in the classification of spectral data, it is often confronted with performance degradation when physical and chemical properties of a testing object have complex effects on spectra, such as detector-based and chemical-based nonlinearity. Locally weighted partial least squares (LW-PLS) is a variant of PLS for regression to address nonlinearity in data. It utilizes the Euclidean distance based similarity to weight training samples and then constructs local PLS models for prediction. However, using LW-PLS for classification is still blank and its classification performance has yet to be reported. In this paper, we extend LW-PLS for the classification of spectral data, resulting LW-PLSC. Experimental results on ten UCI benchmark and two spectral datasets show that LW-PLSC can outperform five baseline methods, achieving the highest classification accuracies most of the time.

Keywords: Partial least squares; locally weighted; classification; spectral data.

1. Introduction

The analysis of high dimensional spectral data provides a powerful means for non-invasively exploring the chemical constituents of a material. This type of data is usually obtained from spectroscopy and represented as a spectrum, a plot of the intensity of energy detected versus the wavelength or frequency of the energy. Recently, there has been an upward trend in research effort towards fast and non-destructive detection of object identities by using low-cost spectroscopy coupled with chemometric methods. A major challenge in this task is to efficiently handle data complexity caused by reasons such as varied sampling conditions and inadequate measures.

Partial least squares (PLS) regression is one of the most commonly used techniques in spectral data analysis, which searches for linear combinations of independent variables, namely latent variables (LVs), that maximize the covariance between the latent variable and the response. It has been adapted for classification, namely PLS discriminant analysis (PLS-DA), by transforming categorical vector into numerical responses via dummy matrix coding [1]. PLS is practically suitable for handling ill-conditioned problems in data matrices, i.e., small-sample-size, high dimensionality and high collinearity [2, 3]. However, it yields unsatisfactory results under nonlinear conditions, which are common in spectral data for reasons such as stray light, detector-based and chemical-based nonlinearity. Kernel PLS (KPLS) and locally weighted PLS (LW-PLS) are two typical variants of PLS for nonlinearity [4, 5]. KPLS maps the original data into Hilbert feature space, where a linear PLS model is constructed. The nonlinear relationship among variables in the original space becomes linear after mapping. Thus, KPLS can efficiently capture the nonlinearity and improve prediction performance. However, it is not directly possibly to see the contribution of each variable with respect to the prediction model [6]. Also, kernel approaches are more prone to overfitting than their non-kernel counterparts if the dataset has a small number of samples [2].

The other approach, LW-PLS, constructs a local regression model based on the similarity between a given query and training samples. The contribution of neighboring samples for the query is enlarged, while the influence of remote samples is lessened. As a result, the global nonlinearity can be lessened. To our knowledge, local PLS classifiers are seldom investigated in chemometrics and machine learning. A recent study on local PLS uses k-nearest neighbors (k-NN) to select local samples for a query, and then builds a weighted PLS-DA model for classification, namely LW-PLS-DA [7]. This method utilizes local classification strategy; however, it is not based on the original LW-PLS and requires extensive work for searching the optimal number of nearest neighbors. Therefore, using LW-PLS for classification is still blank and its classification performance on machine learning and spectral data has yet to be reported.

This paper presents work on a modification of LW-PLS for classification, namely LW-PLSC, and reports an evaluation on twelve benchmark datasets, which cover highly complex data structures such as high dimensionality, nonlinearity and imbalance. We demonstrate LW-PLSC outperforms five baseline methods, achieving the highest classification accuracies most of the time.

The remainder of this paper is organized as follows. Section 2 gives the algorithm of LW-PLSC. Experiments on UCI and spectral datasets are presented in Section 3, including datasets description, parameter optimization and results. Conclusions are drawn in Section 4.

2. The Algorithm

The nth sample ($n = 1, 2, ..., N$) of input and output variables is expressed as

$$x_n = [x_{n1}, x_{n2}, ..., x_{nM}]^T \quad (1)$$

$$y_n = [y_{n1}, y_{n2}, ..., y_{nL}]^T \quad (2)$$

where M and L denote the number of input and output variables, respectively. Let $X \in \Re^{N \times M}$ and $Y \in \Re^{N \times L}$ be the input and output variable matrices whose nth row are x_n^T and y_n^T. To predict the output of a given query x_q, the similarity ω_n between x_q and x_n is calculated, and then a local PLS model is built by weighting samples with a similarity matrix $\Omega \in \Re^{N \times N}$ defined by

$$\Omega = \text{diag}(\omega_1, \omega_2, ..., \omega_N) \quad (3)$$

where diag (·) represents a diagonal matrix and ω_n is defined on the basis of the Euclidean distance as follows:

$$\omega_n = \exp\left(-\frac{\varphi d_n}{\sigma_d}\right) \quad (4)$$

$$d_n = \sqrt{(x_n - x_q)^T(x_n - x_q)} \quad (5)$$

where φ is a localization parameter and σ_d is a standard deviation of $\{d_n\}$. The predicted output $\hat{y}_q$ is calculated through the following procedure.

1. Set K to the desired number of latent variables, and initialize $k = 1$.
2. Calculate the similarity matrix Ω according to Eqs. (3)–(5).
3. Calculate X_k, Y_k, and $x_{q,k}$

$$X_k = X - \mathbf{1}_N[\bar{x}_1, \bar{x}_2, ..., \bar{x}_M] \quad (6)$$

$$Y_k = Y - \mathbf{1}_N[\bar{y}_1, \bar{y}_2, ..., \bar{y}_L] \quad (7)$$

$$x_{q,k} = x_q - [\bar{x}_1, \bar{x}_2, ..., \bar{x}_M]^T \quad (8)$$

$$\bar{x}_m = \frac{\sum_{n=1}^N \omega_n x_{nm}}{\sum_{n=1}^N \omega_n} \quad (9)$$

$$\bar{y}_l = \frac{\sum_{n=1}^N \omega_n y_{nl}}{\sum_{n=1}^N \omega_n} \quad (10)$$

where $\mathbf{1}_N \in \Re^N$ is a vector of ones.

4. Derive the kth latent variable of X.

$$t_k = X_k w_k \quad (11)$$

where w_k is the eigenvector of $X_k^T \Omega Y_k Y_k^T \Omega X_k$, which corresponds to the maximum eigenvalue.

5. Derive the kth loading vector of X and the kth regression coefficient vector.

$$p_k = \frac{X_k^T \Omega t_k}{t_k^T \Omega t_k} \qquad (12)$$

$$q_k = \frac{Y_k^T \Omega t_k}{t_k^T \Omega t_k} \qquad (13)$$

6. Derive the kth latent variable of x_q.

$$t_{q,k} = x_{q,k}^T w_k \qquad (14)$$

7. If $k = K$, finish the output estimate

$$\hat{y}_q = [\bar{y}_1, \bar{y}_2, ..., \bar{y}_L]^T + \sum_{k=1}^{K} t_{q,k} q_k \qquad (15)$$

Otherwise, set

$$X_{k+1} = X_k - t_k p_k^T \qquad (16)$$

$$Y_{k+1} = X_k - t_k q_k^T \qquad (17)$$

$$x_{q,k+1} = x_{q,k} - t_{q,k} p_k \qquad (18)$$

8. Set $k = k + 1$ and go to step 4.

In order for the LW-PLS regression to be used for classification, we transform the categorical vector into output variable matrix Y via dummy coding. The Y matrix is initialized as a zero matrix which contains as many rows and columns as the number of samples N and classes L, respectively. If a sample x_i belongs to the lth class, the lth element in corresponding dummy vector y_i is equals to 1. LW-PLSC is a variant of PLS which projects variables to low-dimensional latent space. Thus, it can effectively handle small-sample-sized and high dimensional classification problems. Moreover, LW-PLSC sets distance-based weights for queries and constructs local models to reduce the influence of global nonlinearity.

3. Experiments

3.1. *Datasets*

We check the performance of LW-PLSC on twelve datasets, including ten UCI datasets and two spectral datasets [8, 9, 10]. The UCI datasets are selected to demonstrate the diversity and efficiency of LW-PLSC, while the spectral datasets are mainly used to present the outperformance of LW-PLSC in the prime application of PLS-DA. The information about two spectral datasets are provided as follows.

- NIR-apple: This dataset was obtained by using a portable near infrared (NIR) spectroscopy. A total of 182 apples were used to distinguish non-organic and organic apples, respectively 96 and 86 samples of each class. Each spectrum has 512 variables in the wavelength of 901.06-1721.24 nm.
- FTIR-oil: A total of 120 authenticated extra virgin olive oils were used to distinguish the country of their origins: Greece, Italy, Portugal and Spain (respectively 20, 34, 16 and 50 samples of each). Each spectrum contains 570 variables in 798.89 to 1896.81 nm.

To improve the classification performance on spectral datasets, we directly apply the same pre-processing steps as in [11]: the raw data matrix was centred by subtracting the mean spectrum, scaled by standard deviation, and processed by the Savitzky-Golay first-order derivative (5-point moving window and second-order polynomial). The raw and pre-processed NIR-apple spectra are shown in Figure 1.

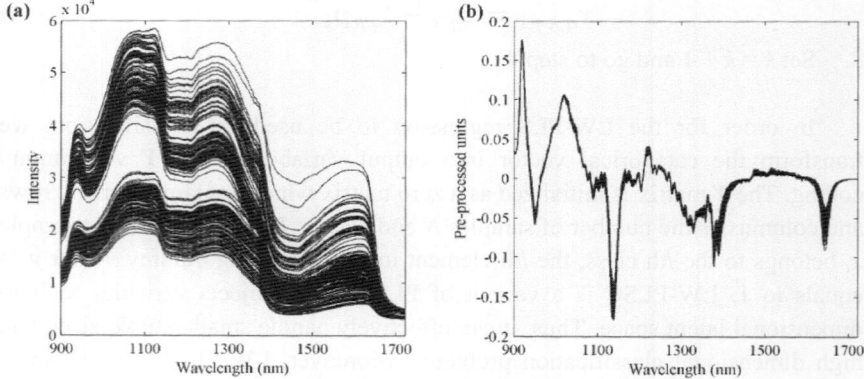

Figure 1. Raw spectra (a) and pre-processed spectra (b) of NIR-apple dataset.

3.2. *Experimental settings*

The proposed LW-PLSC is compared to five baseline methods, including k-NN, PLS-DA, KPLS-DA, collaborative representation based classification (CRC) and nearest regularized subspace classification (NRS). We set proper ranges to tune the hyper-parameters of these algorithms via 10-fold and leave-one-out cross validation, respectively on UCI and spectral datasets. The nearest neighbors in k-NN is set from 1 to 49 with an interval of 2. To prevent overfitting, the range of LVs is varied from 1 to 10 if the minimum number between N and M is above 10, otherwise, from 1 to min (N, M). Gaussian kernel function is used in KPLS-DA with its width σ adjusting from 10^{-3} to 10^5 on a logarithmic scale. CRC and NRS search a linear combination of within-class training samples to approximate a query, and then attributes the query to the class which yields the least approximation error. Their regularization parameter λ is mostly set from 10^{-7} to 10^2 on a logarithmic scale. The optimal value of the localization parameter φ in LW-PLSC is usually found in the range of 0 to 10 [12], we only adjust φ to these values: 0.1, 0.5, 1, 5 and 10.

For a fair comparison between algorithms, we use the same training and test sets for all algorithms. Using the optimal parameter(s) for the corresponding algorithm, the classification performance is evaluated by 10-fold cross validation. We repeat this procedure for 10 times to obtain the average classification accuracy.

3.3. Results

We demonstrate a grid search of the optimal number of LVs and φ for NIR-apple dataset via leave-one-out cross validation, which is shown in Figure 2a, as a mesh plot. LW-PLSC achieves the highest accuracy of 95% when LVs and φ equals to 9 and 1, respectively. It also presents high performance (> 89%) when LVs is over 5. The performance of three PLS-based methods over the parameter LVs are shown in Figure 2b, by setting other parameters to their optimal values. LW-PLSC not only obtains comparably high accuracies using small LVs but also outperforms PLS-DA and KPLS-DA for each LV.

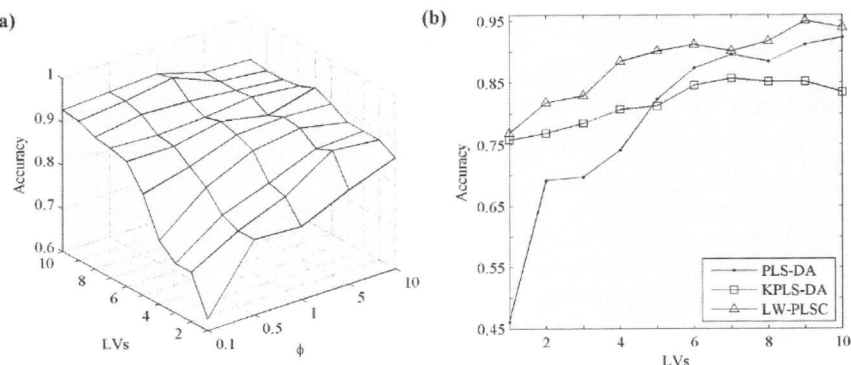

Figure 2. The accuracy of leave-one-out cross validation of LW-PLSC with varying parameters (a) and the performance of three PLS-based algorithms (b) on NIR-apple dataset.

Table 1. Average classification accuracy (%) of different algorithms for 10 UCI and 2 spectral (NIR-apple and FTIR-oil) datasets.

Datasets	k-NN	PLS-DA	KPLS-DA	CRC	NRS	LW-PLSC
Breast tissue	81.6	83.5	84.8	81.6	81.6	**87.3**
Ecoli	87.4	84.6	87.5	83.7	86.7	**87.8**
Frost types	88.8	86.0	89.0	89.4	**90.5**	89.8
Glass	73.1	59.8	68.0	61.5	**74.8**	74.4
Ionosphere	86.7	87.0	93.3	80.9	92.9	**93.8**
Leaf	58.6	49.4	48.5	55.6	74.3	**74.5**
Parkinsons	85.1	85.6	85.7	83.7	86.8	**87.9**
Sonar	82.6	76.0	84.1	80.9	85.7	**87.4**
Spectf	79.7	78.4	81.9	79.4	79.0	**82.0**
Wine	87.3	96.6	95.7	92.5	92.5	**97.1**
NIR-apple	84.9	91.6	84.8	90.9	90.5	**92.7**
FTIR-oil	91.9	93.2	92.8	**94.8**	90.3	93.3
Average	83.1	80.1	82.2	81.6	86.6	**88.1**

4. Conclusion

This paper adapts LW-PLS for nonlinear classification tasks. Termed LW-PLSC, the LW-PLS extension enlarges the contribution of neighboring samples for a given query, and reduces the impact of remote samples. Thus, the global nonlinearity can be handled by a combination of local PLS models. Our experiments on 12 benchmark datasets show that LW-PLSC outperforms five baseline methods in classification. In our future work, we will improve the performance of LW-PLSC by optimizing weighting scheme and using kernel extension.

References

1. M. Barker, W. Rayens, Partial least squares for discrimination, *J. Chemom.* 17 (2003) 166–173.
2. T.B. Blank, S.D. Brown, Nonlinear Multivariate Mapping of Chemical-Data Using Feedforward Neural Networks, *Anal. Chem.* 65 (1993) 3081–3089.
3. F. Despagne, D. Luc Massart, P. Chabot, Development of a robust calibration model for nonlinear in-line process data, *Anal. Chem.* 72 (2000) 1657–1665.
4. R. Rosipal, L.J. Trejo, Kernel partial least squares regression in reproducing kernel Hilbert space, *J. Mach. Learn. Res.* 2 (2002) 97–123.
5. S. Kim, M. Kano, H. Nakagawa, S. Hasebe, Estimation of active pharmaceutical ingredients content using locally weighted partial least squares and statistical wavelength selection, *Int. J. Pharm.* 421 (2011) 269–274.
6. G.J. Postma, P.W.T. Krooshof, L.M.C. Buydens, Opening the kernel of kernel partial least squares and support vector machines, *Anal. Chim. Acta.* 705 (2011) 123–134.
7. M. Bevilacqua, F. Marini, Local classification: Locally weighted-partial least squares-discriminant analysis (LW-PLS-DA), *Anal. Chim. Acta.* 838 (2014) 20–30.

8. K. Bache, M. Lichman, UCI Machine Learning Repository, Univ. Calif. Irvine Sch. Inf. 2008 (2013) 0.
9. W. Song, H. Wang, P. Maguire, O. Nibouche, Nearest clusters based partial least squares discriminant analysis for the classification of spectral data, *Anal. Chim. Acta.* 1009 (2018) 27–38.
10. H.S. Tapp, M. Defernez, E.K. Kemsley, FTIR Spectroscopy and Multivariate Analysis Can Distinguish the Geographic Origin of Extra Virgin Olive Oils, *J. Agric. Food Chem.* 51 (2003) 6110–6115.
11. W. Zheng, X. Fu, Y. Ying, Spectroscopy-based food classification with extreme learning machine, *Chemom. Intell. Lab. Syst.* 139 (2014) 42–47.
12. T. Uchimaru, M. Kano, Sparse Sample Regression Based Just-In-Time Modeling (SSR-JIT): Beyond Locally Weighted Approach, *IFAC-PapersOnLine.* 49 (2016) 502–507.

Attention-based bidirectional LSTM for Chinese punctuation prediction*

Jinliang Li, Chengfeng Yin, Zhen Jia, Tianrui Li and Min Tang

*School of Information Science and Technology, Southwest Jiaotong University
Chengdu, 611756, China
lijinliang@my.swjtu.edu.cn, meg_yin@163.com, {zjia, trli}@swjtu.edu.cn,
tangmin007@foxmail.com*

Punctuation prediction is an important task in Chinese automatic proofreading system which aims to tell whether the words and punctuations we use are right or not. The proofreading problem can be transformed into a prediction task. In this study, we propose an attention-based bidirectional Long Short-Term Memory (LSTM) model to predict punctuations. We use not only the sentence before the punctuation as the model's input, but also the sentence after the punctuation and the properties of the words. Experimental results show that the proposed LSTM model can achieve a good performance in punctuation prediction.

Keywords: Punctuation prediction; bidirectional LSTM; attention mechanism; proofreading.

1. Introduction

Automatic text proofreading is one of the main applications of natural language processing. Punctuation plays an important role in text. The correct use of punctuation can improve the readability of text. A viable approach to proofread the punctuations is to predict which punctuation should be used and to compare it with the original punctuation. This paper will focus on the prediction of four different kinds of punctuations: comma, period, question mark and exclamation mark. They are the most frequently used punctuations in text.

Much research on punctuation prediction has been done. The earliest work used language model to deal with it. Beeferman et al. [1] used trigram language model and Viterbi algorithm for common punctuation symbols prediction. Gravano et al. [2] presented an approach to punctuation and capitalization restoration for English using purely text-based *n*-gram language models. Punctuation prediction could also be considered as an instance of sequence

---

*This paper is partially supported by the Demonstration Project of Science and Technology Service Industry, Sichuan Province (No. 2016GFW0167).

labeling tasks. Conditional Random Fields (CRF) and maximum entropy (ME) have been proved to be effective for automatic punctuation annotation. Lu and Ng [3] proposed a simple dynamic CRF method jointly perform both sentence boundary and punctuation prediction on speech utterances. And on the basis of [3], Wang et al. [4] combined prosodic, lexical and n-gram score features into a dynamic CRF framework which performs better in punctuation prediction. Huang and Zweig [5] used maximum entropy model (ME) for punctuation insertion in English text. In the past few years, due to their good performance, recurrent neural network (RNN) and its variants have been widely used in a variety of sequential labelling tasks, e.g., speech recognition [6], language modeling [7] and machine translation [8]. Recently, LSTM, a variety of RNN, was used in [9] for punctuation prediction. Xu et al. [10] proposed a bidirectional LSTM based on [9] and added a CRF layer to model the output. The results showed that the CRF model decreased the model's performance.

In this paper, we propose an attention-based bidirectional LSTM model to deal with punctuation prediction. We compare different LSTM models' performance on the People's Daily corpus. We also show the contribution of the sentence after the punctuation.

The rest of this paper is organized as follows. Section 2 outlines our proposed method. The details of experiments, results and analysis are discussed in Section 3. Section 4 presents some conclusions and possible directions of future work.

2. Method

In this section, we propose a Bi-LSTM model based on attention mechanism in detail. The general architecture of our model is shown in Figure 1. The whole model consists of four layers: input layer, LSTM layer, attention layer and output layer.

2.1. *Input layer*

The input of the network is the sentence before each punctuation. Every sentence contains several words $S = \{w_1, w_2, \ldots, w_T\}$. Each word w_t in S can be represented as a fixed size vector v_t from pre-trained word embedding dictionary. Property is an important feature of word. We also add properties of the words to the input. There are 14 kinds of properties in total. So we use one-hot encoding to deal with the properties. Each property can be represented as a 14-dimensional vector p_t which only has one single high (1) bit and all the other low (0). The network input x_t is concatenated by w_t and p_t.

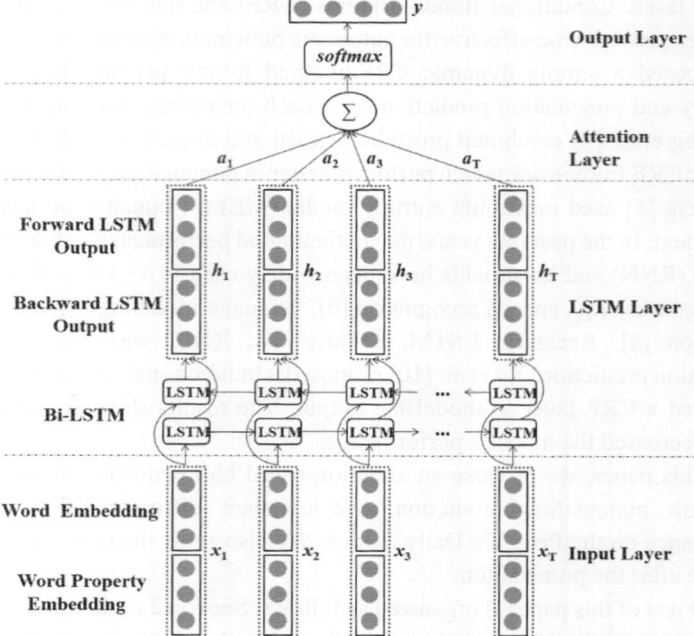

Figure 1. The architecture of attention-based bidirectional LSTM.

2.2. LSTM layer

LSTM units are firstly proposed by Hochreiter and Schmidhuber in 1997 [11]. The LSTM units can learn when to forget historical information and update stored information, thus avoiding the problem of disappearance of gradients encountered in traditional RNNs. A common LSTM unit is composed of a cell, an input gate, an output gate and a forget gate. The cell which is also called memory cell is used to store values over arbitrary time intervals. These gates control the behaviors of memory cells. At each time step, current cell state c_t is determined by three parts, which are inputs x_t, previous hidden state h_{t-1} and previous state c_{t-1}. For an input sequence $x = \{x_1, x_2, \cdots x_T\}$, each cell in LSTM can be computed as follows:

$$f_t = \sigma_g(W_f x_t + U_f h_{t-1} + b_f) \tag{1}$$

$$i_t = \sigma_g(W_i x_t + U_i h_{t-1} + b_i) \tag{2}$$

$$o_t = \sigma_g(W_o x_t + U_o h_{t-1} + b_o) \tag{3}$$

$$c_t = f_t \circ c_{t-1} + i_t \circ \sigma_c(W_c x_t + U_c h_{t-1} + b_c) \tag{4}$$

$$h_t = o_t \circ \sigma_h(c_t) \tag{5}$$

where σ is the sigmoid function, f_t, i_t, o_t, c_t and h_t are the vectors of forget gate, input gate, output gate, cell state and hidden state at time t, the W, U and b are weight matrices and bias vector parameters which need to be learned during training, and the operator ∘ denotes element wise multiplication.

Regular LSTM can only make use of previous input information. Bidirectional LSTM overcomes the limitation by considering both past and future inputs [12]. Its realization principle is as follows: The two LSTM networks with the opposite timing are connected to the same output, the forward LSTM can obtain information from past inputs, and the backward LSTM can obtain information from future inputs. Model accuracy has been greatly enhanced. The hidden state h_t of Bi-LSTM at time t is concatenated by forward h_{ft} and backward h_{bt}:

$$h_{ft} = \overrightarrow{LSTM}(x_t, h_{t-1}, c_{t-1}) \tag{6}$$

$$h_{bt} = \overleftarrow{LSTM}(x_t, h_{t+1}, c_{t+1}) \tag{7}$$

$$h_t = [h_{ft}, h_{bt}] \tag{8}$$

2.3. Attention layer

The standard LSTM cannot detect which input is more important. It only uses the hidden state at final time step. In order to address this issue, we propose an attention-based LSTM that can capture the key part of inputs. The attention layer output o can be computed as

$$o = \sum_{t=1}^{T} a_t h_t \tag{9}$$

where

$$a_t = \frac{\exp(e_t)}{\sum_{t=1}^{T} \exp(e_t)} \tag{10}$$

$$e_t = \tanh(W_A h_t + b_A) \tag{11}$$

where a_t is the attention weights. W_A and b_A are the weight matrix and bias in the attention mechanism to be learned in training step.

2.4. Output layer

We feed the attention layer output vector o to a softmax classifier to predict which punctuation should be used after the sentence:

$$\hat{y} = \text{softmax}(V^T o + b) \tag{12}$$

where V and b are parameters to be learned. We train the model by minimizing the cross-entropy between the prediction punctuation $\hat{y}$ and the truth punctuation y.

3. Experiments

3.1. Corpus and experimental setup

We perform experiments on People's Daily corpus. The corpus has total 761000 sentences and about 6 million words in total. We use SWJTU segmentation tools to do the word segmentation and POS tagging. We only focus on four common punctuations. The normalization of punctuations is shown in Table 1.

Table 1. The normalization of punctuations.

Original Punctuation	Mapped Punctuation
，、 ——	，
。．；；：：	。
！！	！
？？	？
all the other punctuations	remove

In our task, we consider four kinds of punctuations to predict. They are comma (，), period (。), question mark (？) and exclamation mark (！). And we use precision (*prec.*), recall (*rec.*) and F1-measure (F_1) to assess the performance of the punctuation prediction tasks.

To find out whether the future inputs and attention mechanism are effective, we carry out four groups of experiments: standard LSTM (LSTM), attention-based LSTM (ALSTM), bidirectional LSTM (BLSTM) and attention-based bidirectional LSTM (ABLSTM). We can easily find that punctuation we use in the text not only depends on the sentence before the punctuation, but also the sentence after the punctuation. So we also do another experiment (ABLSTM-2S) to evaluate the performance of the two sentences around the punctuation. The input of ABLSTM-2S is the splicing of two sentences around the punctuation, while ABLSTM only use the sentence before the punctuation as the input. We use back-propagation algorithm to train all the LSTM networks. Network parameters are updated for every batch by using Adam optimizer [13].

3.2. Result and analysis

Table 2 presents the experimental results of different LSTM models mentioned above. We notice that compared to LSTM, BLSTM improves the F_1 score of three kinds of punctuations. This result supports our view that both past and future inputs are useful for punctuation prediction. It can be seen that ALSTM outperforms LSTM, and ABLSTM also outperforms BLSTM, which proves that the effectiveness of the attention mechanism. The ABLSTM model has the best performance. Surprisingly, compared to ABLSTM, the F_1 score of comma, period and exclamation mark are improved 15.8%, 10% and 5.8% by ABLSTM-2S, respectively, and ABLSTM-2S performance as good as ABLSTM on question mark. It shows that the sentence after punctuation plays an important role in punctuation prediction. This is in line with people's common sense. In addition, all the models' performances of question mark are the best among the four punctuation marks. The reason is that most question sentences have distinctive feature words which benefit the prediction.

Table 2. Experimental results of different LSTM models (%).

		LSTM	ALSTM	BLSTM	ABLSTM	ABLSTM-2S
comma	prec.	62.5	61.8	65.7	63.3	**81.7**
	rec.	61.9	63.0	60.2	63.5	**76.8**
	F_1	62.2	62.4	62.8	63.4	**79.2**
period	prec.	56.5	55.6	57.0	61.1	**68.6**
	rec.	62.6	57.8	58.9	58.1	**70.4**
	F_1	59.4	56.7	57.9	59.5	**69.5**
question mark	prec.	74.4	78.3	**82.1**	81.3	79.6
	rec.	**83.4**	79.9	77.5	80.8	80.5
	F_1	78.6	79.1	79.7	**81.0**	80.1
exclamation mark	prec.	**72.4**	69.0	65.5	64.5	71.1
	rec.	59.2	63.4	73.3	68.4	**73.3**
	F_1	65.1	66.1	69.2	66.4	**72.2**

4. Conclusions

In this study, we proposed different LSTM models for punctuation prediction. The best performing model is ABLSTM-2S, its key idea is to let every time step's hidden state participate in computing attention weights and join the two sentences around the punctuation as the inputs. The proposed model can concentrate on different parts of the sentences, so that it achieves a good performance in punctuation prediction. As the future work, adding the syntax analysis information of the sentences to the LSTM model would be a valuable research direction.

References

1. D. Beeferman, A. Berger and J. Lafferty, *ICASSP.* **2**, 689 (1998).
2. A. Gravano, M. Jansche and M. Bacchiani, *ICASSP.* 4741 (2009).
3. W. Lu and H. T. Ng, *EMNLP.* 177 (2010).
4. X. Wang, H. T. Ng and K. C. Sim, *INTERSPEECH.* 1384 (2012).

5. J. Huang and G. Zweig, *INTERSPEECH*. 917 (2012).
6. Y. Miao, M. Gowayyed and F. Metze, https://arxiv.org/abs/1507.08240.
7. K. Tran, A. Bisazza and C. Monz, *NAACL-HLT*. 321 (2016).
8. K. Cho, B. Merrienboer, C. Gulcehre, et al, https://arxiv.org/abs/1406.1078.
9. O. Tilk and T. Alumäe, *INTERSPEECH*. 683 (2015).
10. K. Xu, L. Xie and K. Yao, *ISCSL*. 1 (2016).
11. S. Hochreiter and J. Schmidhuber, *NEURAL COMPUT*. **9**, 1735 (1997).
12. M. Schuster and K. K. Paliwal, *IEEE Trans. Signal Processing*. **45**, 2673 (1997).
13. D. P. Kingma and J. Ba, https://arxiv.org/abs/1412.6980.

A new map for combination of spatial clustering methods

Danielly Cristina De Souza Costa Holmes

Laboratory of Applied Statistics to Image Processing and Geoprocessing
Department of Statistics, Federal University of Paraíba, João Pessoa/PB, Brazil
daniellycristina9@gmail.com

Rodrigo De Pinheiro De Toledo Vianna

Department of Nutrition, Federal University of Paraíba,
João Pessoa/PB, Brazil
vianna@ccs.ufpb.br

Ronei Marcos De Moraes

Laboratory of Applied Statistics to Image Processing and Geoprocessing
Department of Statistics, Federal University of Paraíba, João Pessoa/PB, Brazil
ronei@de.ufpb.br

Recently, combination of spatial clustering methods was proposed, based on a similar idea of combining classifiers. However, this combination is able produce a binary map on which is not possible to know how many methods contributed for the final result. The goal of this paper is provide a new soft map for combination of spatial clustering methods when voting is used as combiner. A case study based on 2011 dengue fever epidemiological data from municipalities in the state of Paraíba, Brazil and combining five spatial clustering methods, using voting rule is presented. The soft map was provided considering the number of votes received for each municipality. This map can help a health manager about decision making with respect to each sub-region and its epidemiological priority.

Keywords: Decision map; spatial clustering methods; dengue fever; soft map.

1. Introduction

Classifier Combination is multistage classification process, where individual classifiers provide class measures in a fisrt stage and those measures are features for a second stage classification scheme[1]. In the scientific literature, there are two justifications for combining classifiers: efficiency and accuracy[1]. It can improve the final results with respect to each individual classifiers[2,4]. Combining classifiers are divided in three types of structures: sequential or linear, parallel and hierarchical. It can provide good solutions for pattern recognition problems[3,4] as such as remote sensing, character recognition, geophysical prospecting, speech recognition, and medical applications among others[14]. There are different types of combining rules as for instance, Maximum, Median, Mean, Minimum, Product, Bagging, Boosting, among others[6,7].

In the area of spatial analysis, spatial clustering methods have as objective the identification of significant and non-significant (in the statistical point of view) spatial clusters of a phenomenon in the study of a specific geographical region[7]. There are several applications for these methods. For example, disease spatial clusters are used in Epidemiology[14], and determining high homicide risk areas in Public Security[15]. Some spatial clustering methods are: Scan Statistic[7], Getis-Ord statistic[9] and the Besag and Newell method[10,11], M statistic[11] and Tango's statistic[12].

According to Kulldorff and Nagarwalla[7] "each method works with different methodologies and provides different results with respect to the others. In addition to these issues, as there is no reference information about the real clusters, it is not possible to check the similarity between the results produced by one method and the true result".

In the face of the presented problems, a possible solution in order to obtain better results would be the combination of spatial clustering methods[6]. Nevertheless, such a combination would produce a binary ("hard") map on which is not possible to know how many methods contributed for the final result. This information could be relevant for a manager in order to choose priorities in the decision-making process. The goal of this paper is to provide a new soft map for combination of spatial clustering methods when voting is used as combiner.

2. Methodology

Figure 1 presents the scheme for combining spatial clustering methods. A georeferenced database contains information about the spatial coordinates of the distinct sub-regions (named geo-objects), which compose a geographical region of interest, as for example number of cases registered of a disease for each geo-object and number of inhabitants for the same geo-objects. This dataset is used as input for n spatial clustering methods, which will produce n maps in which all geo-objects are associated to their specific p-value, whose values are in the interval $[0,1]$ that denote the significance of clusters for each geo-object. Subsequently, all the p-values are binarized where zero means "not significant" and one means "significant". Those geo-objects with their correspondent binary values, from each spatial clustering method, are inputs for the combination rule. In this process, the values $\{0,1\}$ of each geo-object are combined in order to generate a decision map, which is the final product of the proposed scheme[6]. This scheme is general and allows using different kinds of spatial clustering methods, as well as, combinations rules, as for instance: majority voting e plurality voting[16].

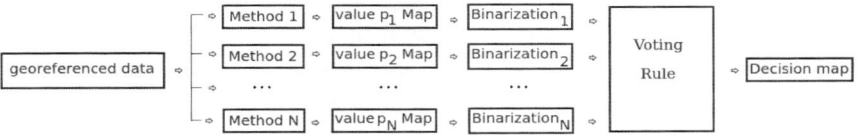

Figure 1. Scheme for combining spatial clustering methods.

From a formal point of view, this scheme can be seen as a set of functions. Let $G \in Z^2$ be a set of spatial coordinates, over which the geo-objects are presented. Let D be a data set with at least the following variables: population and number of occurrences (both subsets of Z). Thus, $f: G \to D$ is a mapping from a set of data on geographical coordinates. Each element of f can be denoted as an ordered pair $c = (g, f(g))$, where $g \in G$ and $f(g) \in D$.

Each k, $k = 2, ..., N$, spatial clustering methods (for instance: Getis Ord, Scan Statistic, Besag and Newell Method, M Statistic, Tango method, etc.) uses the structure f as input in order to provide k outputs, which are mappings from p-values over geographical coordinates $p_k: G \to [0,1]$, in order to create a geographical map of p-values. Each element of p_k can be denoted by an ordered pair $c' = (g, p_k(c))$, where $p_k(c) \in [0,1]$.

A binarization method transforms all p-values in one of the two Boolean labels, where 1 is significant and 0 is not significant is applied to each element of structure p_k where the result is a new structure given by the function $t_k: G \to \{0, 1\}$. Each element of t_k is ordered pair, denoted by $c' = (g, t_k(c'))$, where $t_k(c') \{0,1\}$.

In the following, the k structures given by t_k are aggregated by a voting rule in order to produce soft labels for geo-objects. This function is denoted by $h: G \to \{d_1, ..., d_m\}$, where $m \geq 2$ and d_m denotes the decision by the number of votes that the geo-object had. It can be interpreted in different ways according to the application area. From the epidemiological point of view, for instance, the number of votes received by a geo-object in a geographical region can provide information about its priority or not. Each element of h is denoted by the ordered pair $(g, h(c''))$, and the function h has its codomain given set of possible decisions $\{d_1, ..., d_m\}$, generating a decision map. Thus, taking into account the definition of c'', the function h can be rewritten as a decision-making function, which is given by:

$$h(c') = \Diamond\ [\ g,\ t_1\ (c'),\ t_2\ (c'),\ ...,\ t_k\ (c')\ ] \tag{1}$$

where $\Diamond$ denotes the combination rule used and $k=1, 2, ..., N$.

In a different way of Holmes et al. [6], the final decision map is soft and considers the number of votes received. From these votes is possible know the relevance degree of each geo-object, according to the epidemiological priority. The manager can utilize soft map for final decision, as it was produced. However, the manager can also define a threshold according to the problem context convenience in order to justify a decision. Depending on the threshold chosen, it is possible to represent the same map presented by Holmes et al.[6] or to produce other binary map in order to choose epidemiological priority for each geo-object contained in it.

3. Case Study

The study area in the state of Paraíba, Brazil. Paraíba is located in the Northeast region, a total surface area of about 56,469,744 km², a population of 3,972,202 people, a tropical climate and is formed by 223 municipalities.

In the state of Paraíba, 26,646 cases were recorded of dengue fever from 2009 to 2011. The number of registered cases were: 1,597 for 2009, 8,678 for 2010 and 16,371 for 2011.

In Figure 2 (left) were identified 45 spatial clusters. They are more concentrated in the west region of the State. In Figure 2 (right) were identified 20 municipalities with clusters distributed in all region. Analyzing the maps can be observed differences in the identification of spatial clusters. Tango's statistic map presents a concentration of clusters in the West region of the State, but the clusters are distributed in all regions of state in the Besag Newell's map.

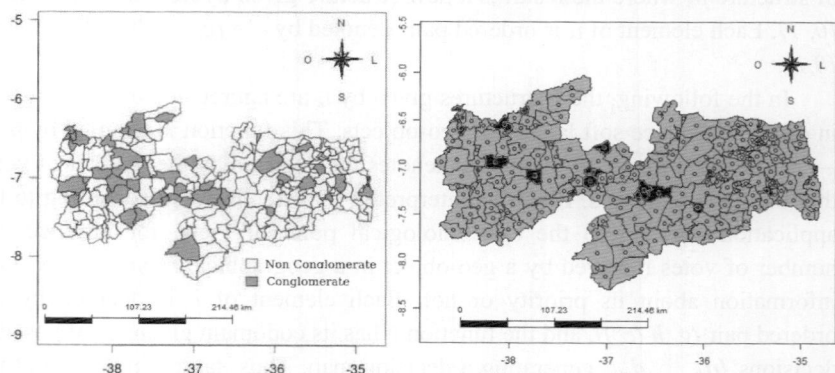

Figure 2. Tango's statistic map (left) and (right) Besag Newell statistic of dengue fever for the state of Paraíba in 2011.

In the map provided by the M Statistic for the year of 2011, the clusters are heterogeneously distributed over all the state of Paraíba. In Figure 3 (left), were identified 126 spatial clusters. In the map provided by the Scan statistic were identified 48 spatial clusters concentrated in only one region of the state.

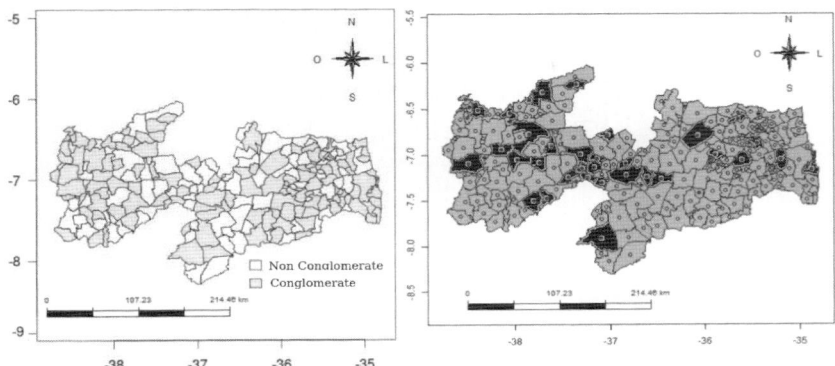

Figure 3. Statistic M map (left) and (right) Scan statistic map of dengue fever for the state of Paraíba in 2011.

Figure 4 presents the Decision map with the combination of spatial clustering methods (Scan statistic, Getis and Ord, Besag and Newell method, Tango's statistic and M Statistic), with six voting categories, i.e., from zero to five votes. In this map, can be observed the number of votes that each municipality (geo-object) has received. The Decision map of the combination of spatial clustering methods identified that great part of the municipalities received one vote, in other words, only one method among five spatial clustering methods detected those municipalities. From of point of view of state regions, the West and East regions presented concentration of municipalities with two and four votes. In Figure 4 only one municipality located in the North region achieved five votes.

The health manager's final decision can use directly the map of the Figure 4. However, the manager can define a threshold according to with health and/or financial policies of the state. This allows to justify decisions about priorities to the manager's superiors and eventually, to the others municipalities too. In a practical situation, considering the financial availability, others cities can be added to priority municipalities list, taking into account that number of votes which they have received. On the other hand, if there are financial restrictions, that threshold could be higher and produce a final decision more restrictive in terms of epidemiological priority.

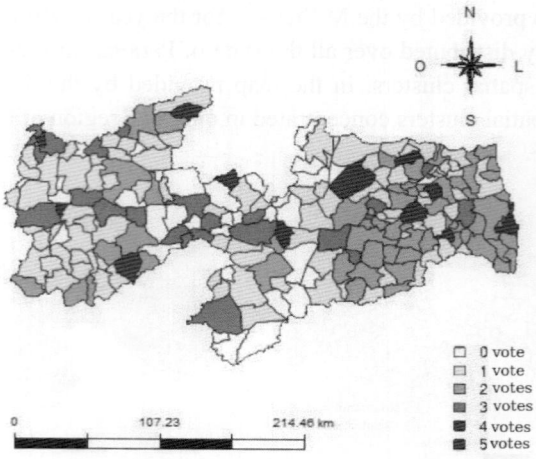

Figure 4. Decision map of the combination of spatial clustering methods (Scan statistic, Besag Newell method, Tango's statistic and M Statistic) (right) presenting the total number of votes dengue fever for each municipality in the state of Paraíba in 2011.

4. Conclusion

The contribution of this work is a new soft map for the visualization of the combination of spatial clustering methods, identifying the number of votes that each geo-object has received. This methodology is general and can be used for any epidemiological indicator or in the analysis of any spatial random variable.

In the decision map of the combination of spatial clustering methods, it is possible noting the several decision levels presented in a same map, which can be selected using a threshold.

An epidemiological case study is provided, in which it was analyzed dengue fever in the state of Paraíba, Brazil. Five spatial clustering methods were used and a voting rule was used as combiner. The soft map was provided considering the number of votes received for each municipality. This map can help a health manager about decision making with respect to each sub-region and its epidemiological priority.

Acknowledgments

This project is partially supported by CAPES. It is also partially supported by grants 132170/2017-5, 308250/2015-0 and 310561/2012-4 of the National Council for Scientific and Technological Development (CNPq) and is related to the National Institute of Science and Technology Medicine Assisted by Scientific Computing (465586/2014-4) also supported by CNPq.

References

1. J. Kittler, R. P. W. Duin and J. Matas, *IEEE Transactions on Pattern Analysis and Machine Intelligence* **20**, 226 (1998).
2. R. M. Moraes and L. S. Machado, in *Safety, Health and Environment World Congress (SHEWC'2012)*, 2012.
3. J. J. Hull, A. Commike and T. K. Ho. *IEEE Transactions on Pattern Analysis and Machine Intelligence* **16**, 550 (1994).
4. L. I. Kuncheva, J. C. Bezdek and R. P. W. Duin, *Pattern Recognition*. **34**, 299 (2001).
5. L. I. Kuncheva, *Combining Pattern Classifiers: Methods and Algorithms*. (Wiley, 2004).
6. D. C. S. C. Holmes, R. M. Moraes and R. P. T. Vianna, in *Seventh International Conferences on Pervasive Patterns and Applications* (PATTERNS'2015).
7. M. Kulldorff and N. Nagarwalla. *Statistics in Medicine* **14** 799 (1995).
8. J. Besag and J. Newell, *Journal of the Royal Statistical Society*. **154**, 143 (1991).
9. M. A, Costa and R. M. Assunção, *Environmental and Ecological Statistics*. **12** 301 (2005).
10. P. A. Rogerson, *Geographical Analysis*. **33**, 3 (2001).
11. T. Tango. *In Disease Mapping and Risk Assessment for Public Health (Lawson A. et al. Eds)* (John Wiley & Sons, 1999).
12. L. Xu, A. Krzyzak and C. Y. Suen, *IEEE Transactions on Systems, Man, and Cybernetics*. **22** 418 (1992).
13. L. H Duczmal, G. J.P Moreira, D. Burgarelli, R H. C Takahashi, F. C. O Magalhães and E. C Bodevan, *International Journal of Health Geographics*. **10**, 29 (2011).
14. R. Minamisava, S. S. Nouer, O. L. M. Neto, L. K. Melo and A. L. S, *International Journal of Health Geographics*. **8**, 66 (2009).
15. Z-H. Zhou, *Ensemble Methods: Foundations and Algorithms*. (CRC Press, 2012).

Part 6

Logic and Automated Reasoning

Part 6

Logic and Automated Reasoning

Distinctive features of the contradiction separation based dynamic automated deduction

Yang Xu[1,3], Shuwei Chen[†,1,3], Jun Liu[2,3], Xiaomei Zhong[1,3] and Xingxing He[1,3]

[1]*School of Mathematics, Southwest Jiaotong University, Chengdu 610031, China*
[2]*School of Computing and Mathematics, Ulster University, Northern Ireland, UK*
[3]*National-Local Joint Engineering Laboratory of System Credibility Automatic Verification, Southwest Jiaotong University, Chengdu 610031, China*
[†]*swchen@swjtu.edu.cn*

Contradiction separation based dynamic automated deduction is a novel development of the standard static (i.e., fixed) binary resolution into a dynamic multi-clause synergized contradiction separation based inference rule. In this paper, we consider some distinctive features/advantages of this novel automated deduction mechanism, including multi-clause involvement, dynamic deduction, synergized deduction, robustness, exchangeability, controllability, scaling, repeatability, integrity and flexibility, along with some illustrative examples.

Keywords: Resolution; automated deduction; contradiction; contradiction separation.

1. Introduction

Resolution [1] and its refinements has been very successful in the field of automated reasoning [2], and so has served as the main inference mechanism of most of today's leading automated reasoning systems, e.g., Vampire [3], E [4], Spass [5]. However, there are still a good number of problems unsolved as illustrated in TPTP (Thousands of Problems for Theorem Provers) [6]. The efficiency and versatility of contemporary automated deduction depend on inference rule and techniques that may go beyond the pure resolution calculus, especially go beyond binary resolution [7, 8].

Aiming at the above objective, we have proposed a novel, multi-ary, dynamic, contradiction separation based inference rule for automated deduction [9, 10]. This short paper summarizes the featured advantages of this novel contradiction separation based dynamic automated deduction mechanism, in order to provide certain guidance for the development and implementation of the corresponding automated deduction systems.

2. Contradiction Separation Based Dynamic Automated Deduction

Some necessary concepts and conclusions of contradiction separation based dynamic automated deduction are provided in this section. The readers are referred to [9, 10] for more details about contradiction separation deduction.

Definition 2.1 Let $S = \{C_1, C_2, ..., C_m\}$ be a clause set in propositional or first-order logic. If $\forall (x_1, \cdots, x_m) \in \prod_{i=1}^{m} C_i$, x_i ($i=1,...,m$) is a literal, there exist at least one complementary pair of literals among $\{x_1,..., x_m\}$, then $S = \wedge_{i=1}^{m} C_i$ is called a *standard contradiction*.

Definition 2.2 Let $S = \{C_1, C_2, ..., C_m\}$ be a clause set in first-order logic. Without loss of generality, assume that there does not exist the same variables among $C_1, C_2, ..., C_m$. The following inference rule that produces a new clause from S is called *a standard contradiction separation rule*:

For each C_i ($i=1, 2, ..., m$), firstly applying a substitution σ_i to C_i (σ_i could be an empty substitution but not necessary the most general unifier), denoted as $C_i^{\sigma_i}$; then separate $C_i^{\sigma_i}$ into two sub-clauses $C_i^{\sigma_i -}$ and $C_i^{\sigma_i +}$ such that

(1) $C_i^{\sigma_i} = C_i^{\sigma_i -} \vee C_i^{\sigma_i +}$, where $C_i^{\sigma_i -}$ and $C_i^{\sigma_i +}$ have no common literals;

(2) $C_i^{\sigma_i +}$ can be an empty clause itself, but $C_i^{\sigma_i -}$ cannot be an empty clause;

(3) $\wedge_{i=1}^{m} C_i^{\sigma_i -}$ is a standard contradiction, that is $\forall (x_1,..., x_m) \in \prod_{i=1}^{m} C_i^{\sigma_i -}$, there exists at least one complementary pair among $\{x_1,..., x_m\}$.

The resulting clause $\vee_{i=1}^{m} C_i^{\sigma_i +}$, denoted as $\mathcal{C}_m^{s\sigma}(C_1,..., C_m)$, $\sigma = \cup_{i=1}^{m} \sigma_i$, σ_i is a substitution to C_i, $i=1,..., m$), is called a standard contradiction separation clause of $C_1,..., C_m$, and $\wedge_{i=1}^{m} C_i^{\sigma_i -}$ is called a separated standard contradiction.

Definition 2.3 Let $S = \{C_1, C_2, ..., C_m\}$ be a clause set in first-order logic. $\Phi_1, \Phi_2, ..., \Phi_t$ is called *a standard contradiction separation based dynamic deduction sequence* from S to a clause Φ_t, denoted as $\mathcal{D}^s$, if

(1) $\Phi_i \in S$, $i=1, 2, ..., t$; or

(2) there exist $r_1, r_2,..., r_{k_i} < i$, $\Phi_i = C_{r_{k_i}}^{s\theta_i}(\Phi_{r_1}, \Phi_{r_2}, ..., \Phi_{r_{k_i}})$, where $\theta_i = \cup_{j=1}^{k_i} \sigma_j$, σ_j is a substitution to Φ_{r_j}, $j=1,..., k_i$.

Remark 2.1 The above concepts apply to propositional logic as well without considering substitution.

Theorem 2.1 (Soundness Theorem) Let $S = \{C_1, C_2, ..., C_m\}$ be a clause set in propositional or first-order logic. $\Phi_1, \Phi_2, ..., \Phi_t$ is a standard contradiction separation based dynamic deduction from S to a clause Φ_t. If Φ_t is an empty clause, then S is unsatisfiable.

Theorem 2.2 (Completeness Theorem) Let $S = \{C_1, C_2, ..., C_m\}$ be a clause set in propositional or first-order logic. If S is unsatisfiable, then there exists a standard contradiction separation based dynamic deduction from S to an empty clause.

3. Distinctive Features of Contradiction Separation Based Dynamic Automated Deduction

This section, compared with resolution principle based automated deduction, discusses some featured advantages of the novel contradiction separation based dynamic automated deduction.

3.1. Multi-clause deduction

Different from the binary resolution which allows only two clauses containing a complementary pair of literals from the clause set to participate the resolution deduction each time, more than two, even all of the clauses in the clause set, can be used in one deduction step in the contradiction separation based dynamic automated deduction. Furthermore, it does not necessarily require the existence of complementary literals in every two involved clauses.

Example 3.1 Let $C_1 = \sim l_3 \vee \sim l_7$, $C_2 = l_2 \vee l_3 \vee l_5 \vee \sim l_6$, $C_3 = l_1 \vee \sim l_2 \vee l_5 \vee \sim l_7$, $C_4 = l_1 \vee l_3 \vee \sim l_5$, $C_5 = l_3 \vee l_4 \vee l_6$, $C_6 = \sim l_1 \vee l_3$, and $C_7 = l_7$. As shown in Table 1, it follows from the contradiction separation rule (Definition 2.2) that the contradiction separation clause $\mathcal{C}_7(C_1, C_2, C_3, C_4, C_5, C_6, C_7) = l_4$ can be obtained in one deduction step.

Table 1. The sub-clauses C_i^- and C_i^+ for $C_1, C_2, C_3, C_4, C_5, C_6, C_7$.

	C_1	C_2	C_3	C_4	C_5	C_6	C_7
C_i^+					l_4		
C_i^-	$\sim l_3 \vee \sim l_7$	$l_2 \vee l_3 \vee l_5 \vee \sim l_6$	$l_1 \vee \sim l_2 \vee l_5 \vee \sim l_7$	$l_1 \vee l_3 \vee \sim l_5$	$l_3 \vee l_6$	$\sim l_1 \vee l_3$	l_7

It can be seen that all the 7 clauses are involved in one deduction step, which reflects the feature of multi-clause involvement. However it will need a number of steps of binary resolutions on $(C_1, C_2, C_3, C_4, C_5, C_6, C_7)$ to obtain l_4.

3.2. Dynamic deduction

In binary resolution, only fixed number of two clauses are allowed to take part in the resolution deduction each time. On the contrary, the number of clauses involved in each deduction step varies dynamically during the deduction process for contradiction separation based automated deduction.

Example 3.2 Let $S = \{C_1, C_2, \ldots, C_{13}\}$ be a clause set in propositional logic, where

$C_1: \sim l_4 \vee l_6$, $C_2: l_6 \vee \sim l_7$, $C_3: \sim l_6 \vee l_7$, $C_4: \sim l_6 \vee \sim l_7$, $C_5: l_1 \vee l_2 \vee l_3$, $C_6: l_1 \vee l_2 \vee \sim l_3$
$C_7: \sim l_1 \vee l_2 \vee l_3$, $C_8: \sim l_1 \vee \sim l_2 \vee l_3$, $C_9: \sim l_1 \vee \sim l_2 \vee \sim l_3$, $C_{10}: l_4 \vee \sim l_5 \vee l_7$
$C_{11}: l_1 \vee \sim l_2 \vee l_3 \vee l_4$, $C_{12}: l_1 \vee \sim l_2 \vee \sim l_3 \vee l_5$, $C_{13}: \sim l_1 \vee l_2 \vee \sim l_3 \vee l_6$.

Using the contradiction separation rule on the clauses C_5, C_6, C_7, C_8, C_9, C_{11}, C_{12}, C_{13}, we obtain a contradiction separation clause involving 8 clauses:
$C_{14} = \mathcal{C}_8 \, (C_5, C_6, C_7, C_8, C_9, C_{11}, C_{12}, C_{13}) = l_4 \vee l_5 \vee l_6$.

Furthermore, using the contradiction separation rule for 3 clauses C_1, C_{10}, and C_{14}, we obtain another contradiction separation clause involving 3 clauses:
$C_{15} = \mathcal{C}_3 \, (C_1, C_{10}, C_{14}) = l_6 \vee l_7$.

Finally, we can deduce the empty clause using 4 clauses:
$C_{16} = \mathcal{C}_4 \, (C_2, C_3, C_4, C_{15}) = \varnothing$.

During the deduction process in Example 3.2, the number of involved clauses is not fixed, i.e., 8, 3 and 4 respectively, which reflects the contradiction separation based deduction is a dynamic process. This dynamic feature provides the contradiction separation deduction, compared with binary resolution, more chances to find possible paths for proof search, and therefore to improve the efficiency of the corresponding algorithms and implementations.

3.3. *Synergized deduction*

Generally speaking, the logical relations among the clauses, literals, constants and functions in a clause set are very complex. It would be much more complex to judge its overall satisfiability/unsatisfiability through a series of smaller local deductions (resolution between two clauses) than bigger local deductions (more clauses involved contradiction separation deduction). The contradiction consisting of more clauses can better reflect the overall logical relation among these clauses, while this kind of logical relation cannot usually be reflected by the logical relations between a series of two clauses.

Take still Example 3.2 to illustrate the synergized feature of contradiction separation deduction from the following three points:

(1) The number of literals deleted through the standard contradiction separation process is usually much more than one binary resolution process, where only two literals can be deleted in each binary resolution.

(2) The number of literals in a standard contradiction separation clause is normally much less than that in the binary resolvent. For example, the third standard contradiction separation clause is an empty clause, while it is impossible to obtain an empty binary resolvent no matter selecting whichever two clauses from C_2, C_3, C_4, C_{15} for resolution.

(3) Each deduction reflects the synergized effects of all the clauses involved and the multiple steps binary resolution is reduced significantly, so it could go beyond the binary resolution in terms of efficiency. For example, it is not easy to obtain the contradiction separation clause $C_{15} = l_6 \vee l_7$ from C_1, C_{10},

C_{14} using multiple steps of binary resolution, while, as illustrated above, C_{15} play an important role in obtaining the empty clause $C_{16}= \emptyset$.

3.4. Robustness

The unsatisfiability of the constructed contradiction, as well as the contradiction separation clause, will not change by adding some literals to or deleting some literals from the contradiction following certain rules.

Example 3.3 Adding or deleting some literals in the contradiction constructed in Example 3.1 will not change the deduction result as illustrated in the following tables.

Table 2. Adding some literals to the contradiction.

	C_1	C_2	C_3	C_4	C_5	C_6	C_7
c_i^+					l_4		
c_i^-	$\sim l_3 \vee \sim l_7$	$l_2 \vee l_3 \vee l_5 \vee \sim l_6$	$l_1 \vee \sim l_2 \vee l_5 \vee \sim l_7$	$l_1 \vee l_3 \vee \sim l_5$	$l_3 \vee l_6$ $\vee l_1 \vee l_5$	$\sim l_1 \vee l_3$ $\vee \sim l_7$	l_7

Table 3. Deleting some literals from the contradiction.

	C_1	C_2	C_3	C_4	C_5	C_6	C_7
c_i^+					l_4		
c_i^-	$\sim l_3 \vee \sim l_7$	$l_2 \vee \cancel{l_3} \vee \cancel{l_5} \vee \sim l_6$	$\cancel{l_1} \vee \sim l_2 \vee l_5 \vee \sim l_7$	$l_1 \vee \cancel{l_3} \vee \sim l_5$	$l_3 \vee l_6$	$\sim l_1 \vee l_3$	l_7

3.5. Exchangeability

According to the definition, the unsatisfiability of contradiction and the contradiction separation clause are regardless of the ordering of involved clauses. It means that the order of the clauses participated in the contradiction separation deduction is exchangeable, and the roles of the clauses used to construct the contradiction are equal. This coincides with the exchangeability of the two clauses in binary resolution.

For example, the first contradiction separation clause in Example 3.2,
$C_{14}= \mathcal{C}_8 (C_5, C_6, C_7, C_8, C_9, C_{11}, C_{12}, C_{13}) = l_4 \vee l_5 \vee l_6$.

The corresponding separation clause is:
$(l_1 \vee l_2 \vee l_3) \wedge (l_1 \vee l_2 \vee \sim l_3) \wedge (\sim l_1 \vee l_2 \vee l_3) \wedge (\sim l_1 \vee \sim l_2 \vee l_3) \wedge (\sim l_1 \vee \sim l_2 \vee \sim l_3) \wedge (l_1 \vee \sim l_2 \vee l_3) \wedge (l_1 \vee \sim l_2 \vee \sim l_3) \wedge (\sim l_1 \vee l_2 \vee \sim l_3)$.

We can change the order of the 8 clauses arbitrarily, while the unsatisfiability of the contradiction and the contradiction separation clause remains the same.

3.6. Controllability

The number of the involved clauses during the contradiction separation clause is totally controllable, so is the size of standard contradiction and the corresponding literals. As a consequence, the number of literals and even which literals in contradiction separation clause are controllable. Normally, the more clauses involved in the standard contradiction, the less literals in the corresponding contradiction separation clause.

Example 3.4 Let $S = \{C_1, C_2, \ldots, C_7\}$ be a clause set in propositional logic, where $C_1 = \sim l_3 \vee \sim l_7$, $C_2 = l_2 \vee l_3 \vee l_5 \vee \sim l_6$, $C_3 = l_1 \vee \sim l_2 \vee l_5 \vee \sim l_7$, $C_4 = l_1 \vee l_3 \vee \sim l_5$, $C_5 = l_3 \vee l_4 \vee l_6$, $C_6 = \sim l_1 \vee l_3 \vee l_2$, $C_7 = l_7 \vee l_6$.

If the contradiction is constructed as $(\sim l_3 \vee \sim l_7) \wedge (l_2 \vee l_3 \vee l_5 \vee \sim l_6) \wedge l_7 \wedge (l_1 \vee \sim l_2 \vee l_5 \vee \sim l_7) \wedge (l_1 \vee l_3 \vee \sim l_5) \wedge (l_3 \vee l_6) \wedge (\sim l_1 \vee l_3)$, then the contradiction separation clause is obtained as $\mathcal{C}_7(C_1, C_2, C_3, C_4, C_5, C_6, C_7) = l_2 \vee l_4 \vee l_6$. On the other hand, if the contradiction is constructed as $(\sim l_1 \vee l_3) \wedge (l_1 \vee l_3) \wedge l_7 \wedge (\sim l_3 \vee \sim l_7)$, then the contradiction separation clause is $\mathcal{C}_4(C_1, C_4, C_6, C_7) = l_2 \vee \sim l_5 \vee l_6$.

3.7. Scaling

In general, the bigger the constructed standard contradiction (more clauses and literals are involved), the less literals in the corresponding contradiction separation clause that is usually more useful for the proof search. Therefore, we expect the constructed contradiction to be bigger. For a clause set $S = \{C_1, C_2, \ldots, C_m\}$, if we have already obtained the standard contradiction as $\bigwedge_{i=1}^{m} C_i$, then we can extend it to a bigger standard contradiction $\bigwedge_{i=1}^{m}(C_i \vee l) \bigwedge_{i=1}^{m}(C_i \vee \sim l)$, where l is an arbitrary literal in the clause set S. Furthermore, for any clause set $S_2 = \{D_1, \ldots, D_n\}$, the constructed standard contradiction $\bigwedge_{i=1}^{m} C_i$ can be extended to a bigger standard contradiction $\bigwedge_{i=1}^{m} C_i \wedge \bigwedge_{j=1}^{n} D_j$.

On the other hand, the obtained standard contradiction can also be made smaller containing less clauses according the requirements of the deduction process. For example, given the standard contradiction $\bigwedge_{i=1}^{m} C_i$ constructed from a clause set $S = \{C_1, C_2, \ldots, C_m\}$, it can be made to a smaller standard contradiction $\bigwedge_{i=1}^{m}(C_i - \{l, \sim l\})$, where l or $\sim l$ is an arbitrary literal in the clause set S.

3.8. Repeatability

In binary resolution, the two clauses used for each resolution usually cannot be repeated, especially in propositional logic. However, in contradiction separation deduction, the same clause can be used repeatedly in the same contradiction construction process, which allows more opportunities to choose suitable clauses and therefore can speed up the proof search process.

3.9. Integrity

For a non-redundant clause set S, its property (satisfiability or unsatisfiability) is determined totally by all the clauses in it. Therefore, it helps to improve the efficiency for judging its property if we can use up all the clauses in S during the deduction process. For binary resolution deduction, this must be realized via deduction sequence, while for contradiction separation deduction, this can be realized in one deduction step.

Example 3.5 For the clause set S shown in Example 3.2, we can construct a standard contradiction as in Table 4. From this standard contradiction, the overall logical property, unsatisfiablility, can be concluded in one contradiction separation deduction step.

Table 4. A standard contradiction with empty contradiction separation clause.

	C_1	C_2	C_3	C_4	C_5	C_7	C_8	C_9	C_{10}	C_{11}	C_{12}	C_{13}
C_i^+												
C_i^-	$\sim l_4 \vee l_6$	$l_6 \vee \sim l_7$	$\sim l_6 \vee l_7$	$\sim l_6 \vee \sim l_7$	$l_1 \vee l_2 \vee l_3$	$\sim l_1 \vee l_2 \vee l_3$	$\sim l_1 \vee \sim l_2 \vee l_3$	$\sim l_1 \vee \sim l_2 \vee \sim l_3$	$l_4 \vee \sim l_5 \vee l_7$	$l_1 \vee \sim l_2 \vee l_3 \vee l_4$	$l_1 \vee \sim l_2 \vee \sim l_3 \vee l_5$	$\sim l_1 \vee l_2 \vee \sim l_3 \vee l_6$

3.10. Flexibility

Dynamic selection of different numbers of clauses during the deduction process provides much flexibility and enhances the adaptive behavior of the automated deduction. We can then able to select different number of clauses according to real-time requirements and the deduction strategies during the dynamic deduction process. The flexibility in selecting the number of clauses in the proposed contradiction separation-based dynamic deduction actually provides an effective way to overcome the two clause restriction and continue the proof search using multiple paths as shown in Example 3.2.

4. Conclusions

This paper has discussed ten distinctive features/advantages of the novel contradiction separation based dynamic automated deduction, which is essentially a good extension of the classical resolution rule and its refinements. For the concrete contradiction separation based automated deduction methods and/or algorithms, there are still some features, e.g., parallelism, goal-guidance and high efficiency etc., which will be further studied in the future.

Acknowledgments

This work is partially supported by the National Natural Science Foundation of China (Grant No. 61673320) and the project of Department of Education of Sichuan Province (Grant No. 18ZB0589).

References

1. J.A. Robinson, A machine oriented logic based on the resolution principle, J. ACM, 12(1), pp. 23-41, 1965.
2. M. P. Bonacina, Automated Reasoning for Explainable Artificial Intelligence, The First International ARCADE Workshop (in association with CADE-26), Gothenburg, Sweden, August 6, 2017.
3. L. Kovács and A. Voronkov, Frist-order theorem proving and vampire, http://vprover.org/cav2013.pdf, 2013.
4. S. Schulz, System Description: E1.8, Lecture Notes in Computer Science, 51(11), pp. 1927-1938, 2013.
5. C. Weidenbach, D. Dimova, A. Fietzke, et al. SPASS Version 3.5, Proceedings of 22nd International Conference on Automated Deduction, pp. 140-145, 2009.
6. TPTP: http://www.cs.miami.edu/~tptp/
7. D. Plaisted, History and prospects for first-order automated deduction, Automated Deduction – CADE-25, August 1-7, 2015, pp. 3-28.
8. J. Gozny and B.W. Paleo, Towards the compression of first-order resolution proofs by lowering unit clauses, Automated Deduction – CADE-25, August 1-7, 2015, pp. 356-366.
9. Y. Xu, J. Liu, S.W. Chen and X.M. Zhong, A novel generalization of resolution principle for automated deduction. Proc. of 12th International FLINS Conference, August 24-26, 2016, ENSAIT, Roubaix, France, World Scientific, pp. 483-488, 2016.
10. Y. Xu, J. Liu, S.W. Chen and X.M. Zhong and X.X. He, Contradiction separation based dynamic multi-clause synergized automated deduction, to appear, 2018.

The empirical study of imported genetic algorithm combined with ant colony algorithm based on 3-SAT problems

Huimin Fu[†1#], Yang Xu[2], Xinran Ning[1] and Wuyang Zhang[1#]

[1]School of Information Science and Technology, Southwest Jiao Tong University
Chengdu, 610031, China
[#]fuhm6688@qq.com, 793859003@qq.com
[2]System Credibility Automatic Verification Engineering Lab of Sichuan Province,
School of Mathematics, Southwest Jiao Tong University
Chengdu, 610031, China

> The genetic algorithm and ant colony algorithm have the ability of random global searching, and many research and application based on them have been reported. This paper combines improved genetic algorithm with ant colony algorithm making them complementary advantages. Through mathematical analysis of improved algorithm, some genetic and ant colony factors have great influence for solving 3-SAT problems. So, this article studied the experimental of improved algorithm based on 3-SAT problem to analyze the influence of the initial population and heuristic factor on improved algorithm.

1. Introduction

1.1. The Description of 3-SAT Problems

Boolean satisfiability problem (SAT) refers to the satisfiability problem of the Conjunctive Normal Form, which is a basic problem of logic. SAT problem was the first non-deterministic polynomial complete (NPC) problem [1] whose difficulty in the course of the development of computer science and intelligent science has become one of the major scientific and technological problems in the cross century and the core issue of computer science. SAT problem have a wide range of applications [2].

3-SAT is the problem of determining whether all of the collection of 3-Literal disjunctions of Boolean variables is true for some truth assignment to the variables [3]. 3-SAT is a special case of SAT which is the problem of determining whether all of a collection of clauses are true for some truth assignment to the variables

---

This work is supported by the National Natural Science Foundation of China (Grant No. 61673320) and the Fundamental Research Funds for the Central Universities (Grant No. 2682017ZT12, 268201 6CX119).

contained in those clauses. So, algorithms for solving SAT Problem can also be used to solve 3-SAT Problem. 3-SAT problem is also a NPC problem [4].

There are many optimization algorithms to solve the SAT problems, which are divided into two categories: one is complete, the other is incomplete. The complete method adapts to the application class problem and the combination problem [5]. The disadvantage is that with the increasing of the scale of the problem, the computational efficiency is very low. The incomplete algorithm is effectively used in many categories of satisfied instances, and in some instances the performance is far higher than the complete algorithm. The incomplete method adapts to the stochastic generative problem, and it has been favored by many scholars, and developed and applied rapidly [6].

1.2. Genetic Algorithm

Genetic Algorithm (GA), first developed by Holland [7], is a global optimization probability search algorithm, which simulates the process of genetic evolution in the biological world. Compared with other optimization algorithms, it has many advantages, such as multi point search, parallel computing, scalability, robustness, and so on, so it has been extensively developed in theory and application.

GA gives some possible solutions to the problem by coding, and the encoded solutions are regarded as individuals (also known as chromosomes) in a group. The evaluation function of the individual's adaptability to the environment is the objective function of the problem. In addition, GA simulates the evolution operator of crossover, mutation and replication in genetics and determines the search direction based on the selection method of survival of the fittest.

1.3. Ant Colony Algorithm

Ant colony algorithm is a stochastic optimization method just born in recent years. It is a new bionic algorithm from nature. It is the Italian scholar M. Dorigo et al. [8] first proposed this algorithm.

Ant colony algorithm is mainly achieved through the transmission of information between ant groups to achieve the purpose of optimization, also known as ant colony optimization (ACO). At present, many researchers use ant colony algorithm to study the traveling salesman problem and scheduling problem, and a series of better experimental results have been obtained.

Because the genetic algorithm is not enough to use the feedback information in the system, a large number of redundant iterations are often done when a certain range is solved, and the efficiency is low, and the disadvantage of ant colony algorithm is the lack of initial pheromone and slow solution speed. So this paper aims to combine improved genetic algorithm [6] with ant colony algorithm to make up for each other's shortcomings, so that it can solve random 3-SAT instances faster.

2. Preliminaries

2.1. Basic Knowledge of 3-SAT Problem

The 3-SAT problems are a satisfiability problem, which is that a given Boolean value formula can be found to satisfy this problem [9]. Firstly, we introduce some related notions and then the 3-Satisfiability Problem in detial.

The symbol x_i, $i \in \{1,2,\cdots,n\}$ in the text represents a boolean variable. Let $X_n = \{x_1, x_2, \cdots, x_n\}$ symbols a collection of boolean variables. The symbol $l_1, l_2, \cdots$ stands for literals. The symbol $c_1, c_2, ..., c_m$ represents clauses. Let $C_m = \{c_1, c_2, ..., c_m\}$ be a collection of clauses. Let F be CNF formula.

1. Boolean variable: Its value is either true or false. In the algorithm, 1 are true, and 0 is false in general.
2. Value Assignment: Value assignment defined on the variable set X_n is a function $\mu : X_n \to \{true, false\}$.
3. Literal: Boolean variable x_i or the negation of boolean variable $\neg x_i$ represents literal, $i \in \{1,2,\cdots,n\}$.
4. Clause: A Clause is made up of a disjunction of some literals. It can be expressed by the form of $l_1 \vee l_2 \vee \cdots \vee l_n$, and these n literals are different.
5. Conjunctive normal form (CNF): A CNF means a conjunction of some clauses. Here, we suppose the number of clauses is m, so, it can be expressed by the form of $F = c_1 \wedge c_2 \wedge \cdots \wedge c_m$.
6. 3-SAT problems: If each $c_j \in F$, $j \in \{1,2,\cdots,m\}$, is a disjunction of the three literals, the function F is 3-SAT formula. The 3-SAT problems are whether there is a truth assignment that makes it true. If there are such truth assignments, the 3-SAT problems can be solved.

2.2. Genetic Algorithm for 3-SAT

Using genetic algorithms to solve 3-SAT problems mainly include three aspects: problem transformation, chromosome encoding and genetic manipulation design.

The core of the problem transformation is defining the fitness function f the 3-SAT problems are transformed into the optimization problem of the extremum of the corresponding fitness function. Using the binary string to represent the true value assignment is the most intuitive chromosome cod method, which takes full advantage of the characteristics of SAT itself, easy to calculate the fitness function and design a variety of genetic operations. Because chromosomes are encoded by binary code, cross operation can be accomplished by truncating and stitching binary strings. Mutation manipulation simulates the mutation of a chromosome gene in a biological evolution that flips each chromosome at a certain probability, i.e. $0 \to 1$, $1 \to 0$.

2.3. Ant Colony Algorithm for 3-SAT

For an example, the number of variables is n and the number of clauses is m. Suppose the variable set is $X_n = \{x_1, x_2, \cdots, x_n\}$ and the clause set is $C_m = \{c_1, c_2, \cdots, c_m\}$ Construct a structure diagram as shown in Figure 1. There are two types of values for each variable x_i, $i \in \{1, 2, \cdots, n\}$, i.e. 1 or 0, and the structure Figure 1 has $2*n$ edges $\{(x_i, x_{i+1})^1, (x_i, x_{i+1})^0 \mid i \in \{1, 2, \cdots, n\}\}$ [10]. The $(x_i, x_{i+1})^1$ indicates that true value assignment of the variable x_i takes 1, and the $(x_i, x_{i+1})^0$ means that true value assignment of the variable x_i takes 0. An ant has to traverse the n edge to get a set of true value assignment.

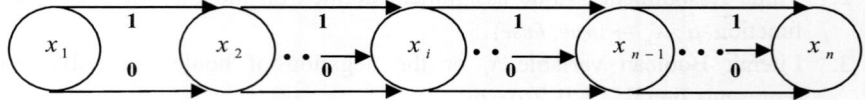

Fig. 1. n variables structure diagram.

In the search process, the ant calculates the transition probability based on the amount of information in each path and the information of the path. The ant goes from vertex $x_i, i \in \{1, 2, \cdots, n\}$ to vertex x_{i+1}, and there are two paths: 1 and 0. The probability of x_i taking 1 and 0 is respectively.

3. Improved Genetic Algorithm Combined with Ant Colony Algorithm

According to the advantages and disadvantages of genetic algorithm and ant colony algorithm, we make full use of their advantages in this work so that they can complement each other. That is, the improved genetic algorithm is used to generate the pheromone distribution for the following improving ant algorithm. A description of the improved genetic algorithm combined with ant colony algorithm (IGA) is as follows. maxGenerations represents the maximum limit of evolutionary generation, and α is a heuristic factor of pheromone, and β is the expected heuristic factor, and p is the pheromone residue factor, and Q is a constant related to the pheromone quantity released by ants.

Algorithm 1: IGA
Input: CNF-formula F
1: Parameters maxAnts=$n/2$, maxGenerations, α, β, p, Q
Output: A satisfying assignment σ of F, or "Ultimately the best assignment σ"
2: begin
3:   Perform an improved genetic algorithm;
4:   σ := a better truth assignment generated by genetic algorithm;
5:   if σ satisfies F then return σ;
6:   Initialize heuristic information;
7:   Initialize pheromone by σ;
8:   for step := 1 to maxGenerations do
9:       calculates the transition probability;
10:      for step := 1 to maxAnts do
11:          P := Get a new truth assignment σ by roulette rules according to transition probability; // α *and* β *are used in transition probability*
12:          if σ satisfies F then return σ;
13:      end for
14:      Update pheromone; // p *and* Q *are used in update pheromone*
15: end for
16: Return "no solution found";
17: end

4. Experimental Result

The experimental environment as: Processor (Intel (R) Core(TM)i5-3337 CPU@ 1.8GHz 2.7GHz), RAM(2.00GB). A series of benchmark instances have used in this paper all come from the network database called SATLIB [11]. Numerical experiments of IGA algorithm based on 3-SAT problem with 20 variables and 91 clauses are performed.

By trying different values of maxGenerations, α, β, p, Q respectively, we found that the a, p, Q has a very small impact on the algorithm, so we only analyze the three factors of maxGenerations, β and population size N in the following experiments, and T represents maxGenerations, and success rate is the number of successful solving instances divided by the number of total instances.

Table 1 shows the different success rates with N=100, and we can find for different β, when T=2500, the success rate of IGA algorithm is to reach the maximum, and when β=5, T=2500, IGA algorithm has the highest success rate. So in order to further improve the performance of the algorithm, we test the effect of different N on the performance of the algorithm on the basis of β=5.

Table 1. Comparison of success rates on different heuristic factor β and T.

	$\beta = 0$	$\beta = 1$	$\beta = 5$	$\beta = 10$	$\beta = 20$
T=0	0.865	0.865	0.865	0.865	0.865
T=500	0.866	0.867	0.876	0.869	0.865
T=1000	0.867	0.869	0.878	0.867	0.865
T=1500	0.870	0.874	0.885	0.873	0.868
T=2000	0.866	0.874	0.883	0.869	0.863
T=2500	0.874	0.885	**0.893**	0.881	0.873
T=3000	0.873	0.883	0.893	0.877	0.868

Table 2. Comparison of success rates on different heuristic factor N and T.

	N = 60	N = 100	N = 150	N = 200	N = 250	N = 300
T=0	0.853	0.865	0.865	0.890	0.900	0.898
T=500	0.866	0.876	0.881	0.892	0.901	0.900
T=1000	0.880	0.878	0.878	0.904	0.903	0.900
T=1500	0.883	0.885	0.883	0.896	0.911	0.899
T=2000	0.883	0.883	0.880	0.900	0.906	0.908
T=2500	0.888	0.893	0.883	0.904	0.909	0.907
T=3000	0.885	0.893	0.886	0.900	**0.916**	0.913

Table 2 shows the different success rates with $\beta=5$, and we can find for different N, the performance of the IGA algorithm is almost always better with the increase of T, so T plays an important role in the performance of IGA. When we do not consider T parameter, that is, T=0, the success rate of IGA is obviously different for different N, indicating that N has an important role in the IGA algorithm. When $\beta=5$, N=250 and T=3000, IGA algorithm has the highest success rate which is 0.916.

5. Conclusion

The IGA algorithm parameters, such as the population size, heuristic factor β, Maximum iteration number T of ant colony algorithm. From two tables, we can conclude that the population size has a great influence. The success rate of IGA algorithm changes with the change of population size. Considering of heuristic factor only has little impact on IGA algorithm, but with the combination of T, heuristic factor has a great impact on IGA algorithm, which indirectly, T plays an important role in IGA algorithm for solving 3-SAT problems.

Acknowledgments

This work is supported by the National Natural Science Foundation of China (Grant No. 61673320) and the Fundamental Research Funds for the Central Universities (Grant No. 2682017ZT12, 2682016CX119).

References

1. S. A. Cook, *The complexity of theorem-proving procedures*, Annual ACM Symposium on Theory of Computing, 151(1991).
2. Y. Diao, X. Wei, T. K. Lan and Y. L. Wu. *Coupling reverse engineering and SAT to tackle NP-complete arithmetic circuitry verification in ~ O (# of gates)*, Asia and South Pacific. IEEE on Design Aut. Conf. (ASP-DAC), 139(2016).
3. R.T. Faizullin, V. I. Dulkeyt and Y. Y. Ogorodnikov, *Hybrid method for the approximate solution of the 3-satisfiability problem associated with the factorization problem*, Trudy Instituta Matematiki i Mekhaniki UrO RAN, **19**(2) 285(2013).
4. D.M. Zhu and S.H. Ma, *Design and Analysis of Algorithm*, Higher Education Press (2009).
5. Audemard G., Lagniez J. M., Simon L. *Improving Glucose for Incremental SAT Solving with Assumptions: Application to MUS Extraction*, Theory and Applications of Satisfiability Testing – SAT 2013. Springer Berlin Heidelberg, 309(2013).
6. Huimin Fu, Yang Xu, Guangfeng Wu and Xinran ning, *An Improved Genetic Algorithm for Solving 3-SAT Problems Based on Effective Restart and Greedy Strategy*, Intelligent Systems and Knowledge Engineering (ISKE), 472(2017).
7. J.H. Holland, *Adaptation in natural and artificial system*, (The University of Michigan Press, Ann Arbor, Michigan), (1975).
8. Marco Dorigo, GambardeUa, Luca Maria. *Ant colonies for the traveling salesman problem*, Biosystems, **43**(2), 73(1997).
9. D. Gamarnik and M. Sudan, *Performance of the Survey Propagation-guided decimation algorithm for the random NAE-K-SAT problems*, arXiv preprint arXiv:1402.0052, (2014).
10. Fu Wang, Yuren Zhou and Li Ye, *Ant colony algorithm combined with survey propagation for satisfiability problem*, Computer Science, **39**(4), 227 (2012).
11. www.cs.ubc.ca/ ~ hoos/SATLIB/benchm.html.

Verifying Deutsch-Schorr-Waite algorithm in first-order logic with arithmetic

Bo Yang* and Sheng Liang

School of Mathematics and Information Science, Guiyang University
Guiyang, 550005, China
*99byang@163.com

Ying Zhang and Mingyi Zhang[†]

Guizhou Academy of Science
Guiyang, 550001, China
†zhangmingyi045@aliyun.com

This paper shows how to verify programs manipulating recursive data structures in first-order theories with arithmetic. By extending an existed formalism, mutable data structures can be described and reasoned about more uniformly. We take Deutsch-Shorr-Waite as an example, and show that the resulting first-order theory enable us to reason about its properties in a natural way, without resorting to the loop invariants.

Keywords: Program verification; first-order axioms; recursive structure.

1. Introduction

When a computer program is represented in a more abstract system, e.g. first-order logic, its behaviors can be captured more succinctly, and properties concerned can be verified in a more rigorous way. Lin presented a formalism to translate an imperative program into a set of first-order axioms.[1] With these axioms, correctness of a program can be verified in first-order logic, just relying on simple mathematical induction and recurrence, without resorting to the loop invariants.[2] To reason about mutable data structures, i.e. recursive structures such as linked list or tree whose shapes are apt to be changed by inserting or deleting, program is extended with list and tree.[3] Then programs with simple pointer operations were translated and discussed, but mutable data structures were not considered.[4] Deutsch-Schorr-Waite (DSW) algorithm[5] performs in-order traversal on a binary tree, and because of the complicated pointer operations it is called "the first mountain that any formalism for pointer analysis should climb".[6]

This paper tries to describe programs with pointer operations and mutable data structures in a unified way. DSW algorithm is a proper example to justify this task. Section 2 introduces program syntax and DSW algorithm, then reviews important concepts and notations presented in Lin's formalism, and shows how main program instructions are axiomatized. In Section 3, we translate the DSW algorithm into a set of axioms and establish the partial correctness and termination of it. Some related works are introduced briefly in Section 4. Finally we conclude this work.

2. Preliminaries

2.1. *Programs*

This paper focuses on programs manipulating mutable structures. We assume that programs are constructed according to the following syntax:

```
E::=ID | PID | IE | PE | B    ID::=id | #PE
PID::=ptr | ptr.ptr   IE::=ID | PID.id | operator(IE, ..., IE)
PE::=PID | &id | ptr-op(PE)    B::=E=E | bool-op(E,...,E)
P::=ID:=IE | PID:=PE | PID:=NULL | P;P |if B then P else P |
    while B do P
```

Symbols defined in these BNF, such as E, ID, PID, IE, denote program components — expression, identifier, pointer identifier, integer expression, and so on. For simplicity we just consider programs manipulating integers and nodes of a mutable structure. So here id means Integer program variable and ptr means pointer variable pointing to another identifier of Integer or other structured type. operator and bool-op represent arithmetical and boolean operations in programs respectively. ptr-op is the arithmetical operations on a pointer variable. #PE is the value of a variable whose address stored in PE, and &id the address of identifier id. ptr.ptr is used to denote any sub-structure of an recursive structure.

Example 1. According to the above syntax, DSW algorithm can be represented as follows:

```
  Deutsch-Schorr-Waite(Node *root)
1 { if root=Null                9      cur.r:=prev;
2      return;                  10     prev:=cur;
3   else                        11     cur:=next;
4    { prev:=-1;                12     if cur=Null
5      cur:=root;               13      { cur:=prev;
```

```
6          while cur!=-1            14          prev:=Null;
7          {next:=cur.l;            15          }
8          cur.l:=cur.r;            16          }}}
```

DSW algorithm visits each node of the tree *root* 3 times. Pointer *cur* is used to refer to the current note that is to be visited, and *next* always points to the next node to be visited. Every time *cur* moves to the next node, *prev* records *cur*'s previous location. The original value of *prev*(-1) works as a guard to promise the termination of the algorithm. For node x with xP as its parent and xL, xR as left child and right childs, *cur* points to x for the first time after xP is visited (1 or 2 times). Now x's left and right pointers are "rotated" counterclockwise once, i.e. $x.l = \&xR$ and $x.r = \&xP$. The second visit to x is happened after all nodes on x's left sub-tree are visited 3 times. At this time *cur* moves to x from xL, then x's left and right pointers rotate again, i.e. $x.l = \&xP$ and $x.r = \&xL$. The third time x is visited after all nodes on its right sub-tree are visited 3 times, *cur* moves to x from xR, then x's left and right pointers rotate again, i.e. $x.l = \&xL$ and $x.r = \&xR$, $x.l$ and $x.r$ get their original values.

2.2. Representation of program components

This paper follows Lin's formalism[1,4] and makes some extension if necessary. A multi-sorted first-order language L with discrete linear order is used. There are three sorts: *nat*, *int* and *loc*. Sort *nat* represents natural numbers including constant 0, linear ordering relation < (and ≤), successor function $s(n) = n + 1$. Sorts *int* and *loc* represent integers and memory locations respectively. A special value Null in *loc* means no value.

We view a variable name as a location in sort *loc*. Function $val : loc \to int \cup loc$ is defined to model values of variables during program execution. For integer variable X, $val(X)$ is a integer. For pointer variable X, $val(X)$ is a memory location that stores location address of another program variable. Variable in form of $X.field$ is modelled by function $field(X)$ and $val(X.field) = val(field(X))$. We use $val(X)$ and $val'(X)$ to denote values of X at the beginning and the end of the program respectively. In loops, values of X in the nth iteration is represented as $val(X, n)$.

Two shorthands are also used:[1,4] (1) $e_1 =$ *if* φ *then* e_2 *else* e_3 means $\forall \vec{x}.\varphi \to e_1 = e_2$ and $\forall \vec{x}.\neg\varphi \to e_1 = e_3$, where $\vec{x}$ denotes all free variables in φ and $e_i (i=1,2,3)$. (2) $smallest(e, n, \varphi)$ is a shorthand for $\varphi(n/e) \wedge \forall m.m < e \to \neg\varphi(n/m)$. It says that e is the smallest natural number satisfying $\varphi(n)$. $\varphi(n/e)$ is the result of replacing n in $\varphi(n)$ by e, similarly for $\varphi(n/m)$.

2.3. Axiomatizing a program

Given a program P and a set $\vec{X}$ of program variables occurring in P, the set of first-order axioms for P and $\vec{X}$, written as $\Pi_P^{\vec{X}}$, is generated according to some rules.[4] Here we just list those relating to our task in this paper.
(1) If P is: V:=E, where V is an integer or pointer variable, E an integer or pointer expression, then $\Pi_P^{\vec{X}}$ includes the following axioms:

$val'(x) = $ if $x = V$ then $val(E)$ else $val(x)$, where x ranges over loc.

(2) If P is: P1;P2, then $\Pi_P^{\vec{X}}$ includes:

$\varphi(val'/tmp)$, for each $\varphi \in \Pi_{P1}^{\vec{X}}$, $\varphi(val/tmp)$, for each $\varphi \in \Pi_{P2}^{\vec{X}}$.

$\Pi_{P1}^{\vec{X}}$ and $\Pi_{P2}^{\vec{X}}$ should have no common temporary function names. And tmp is a new function of the same arity as val to replace val and val'.
(3) If P is: if B then P1 else P2, then $\Pi_P^{\vec{X}}$ is:

$B \to \varphi$, for each $\varphi \in \Pi_{P1}^{\vec{X}}$, $\neg B \to \varphi$, for each $\varphi \in \Pi_{P2}^{\vec{X}}$.

(4) If P is: while B do P1, $\Pi_P^{\vec{X}}$ is generated by introducing an index parameter n to record their values after the nth execution of $P1$.

$\varphi[n]$, for each $\varphi \in \Pi_{P1}^{\vec{X}}$, $smallest(N, n, \neg B[n])$,

$val(X) = val(X, 0), val'(X) = val(X, N)$, for each $X \in \vec{X}$

where n is a new natural number variable not occurring in φ, N a new constant not already used in $\Pi_{P1}^{\vec{X}}$.

3. Verification of Deutsch-Schorr-Waite Algorithm

3.1. Translating

In DSW algorithm, $\vec{X} = \{root, prev, cur, next, x.l, x.r\}$, where x ranges over sort loc. The whole body of the algorithm is a conditional, line 2 is the only statement in if-branch(denoted as P_{if}) and line 4 – line 16 form the else-branch P_{else}. Line 2 does nothing but ending the execution of the algorithm. We have $\Pi_{P_{if}}^{\vec{X}}$ as: $val'(X) = val(X)$, for each $X \in \vec{X}$.

P_{else} is in form of $P_1; P_2$, while P_2 is a loop containing line 6 – line 15, P_1 is a sequence $line4; line5$. We first translate P_1:
For line 4: prev:=-1; $\Pi_{line4}^{\vec{X}}$ is:

$val'(root) = val(root),$ $val'(prev) = -1,$ $val'(cur) = val(cur),$
$val'(next) = val(next),$ $val'(l(x)) = val(l(x)),$ $val'(r(x)) = val(r(x)).$

For line 5: `cur:=root`; $\Pi_{line5}^{\vec{X}}$ is:

$val'(root) = val(root),\ val'(prev) = val(prev),\ val'(cur) = val(root),$
$val'(next) = val(next),\ val'(l(x)) = val(l(x)),\ val'(r(x)) = val(r(x)).$

To generate $\Pi_{P_1}^{\vec{X}}$, we introduce 6 temporary functions $tmp_1 - tmp_6$ to substitute $val(root), val(prev), val(cur), val(next), val(l(x)), val(r(x))$ in $\Pi_{line5}^{\vec{X}}$ and their primed versions in $\Pi_{line4}^{\vec{X}}$ according to the substituting rules for sequence statement listed in section 2.3. After getting rid of temporary functions, we get:

$val'(root) = val(root),\quad val'(prev) = -1,\quad val'(cur) = val(root),$
$val'(next) = val(next),\ val'(l(x)) = val(l(x)),\ val'(r(x)) = val(r(x)).$

In the same way, we get the following axioms for line 7 – line 15:

$val'(root) = val(root),\quad val'(next) = val(l(cur)),$
$val'(prev) = \text{if } val(l(cur)) = Null \text{ then } Null \text{ else } val(cur),$
$val'(cur) = \text{if } val(l(cur)) = Null \text{ then } val(cur) \text{ else } val(l(cur)),$
$val'(l(x)) = \text{if } x = cur \text{ then } val(r(cur)) \text{ else } val(l(x)),$
$val'(r(x)) = \text{if } x = cur \text{ then } val(prev) \text{ else } val(r(x)).$

By introducing index parameter n, $\Pi_{P_2}^{\vec{X}}$ is generated as follows:

$val(X, 0) = val(X)$, for each $X \in \vec{X}$,
$val(root, n+1) = val(root, n),$
$val(prev, n+1) = \text{if } val(l(val(cur, n)), n) = Null \text{ then } Null \text{ else } val(cur, n),$
$val(cur, n+1) = \text{if } val(l(val(cur, n)), n) = Null \text{ then } val(cur, n)$
$\hspace{5cm} \text{else } val(l(val(cur, n)), n),$
$val(next, n+1) = val(l(val(cur, n)), n),$
$val(l(x), n+1) = \text{if } x = cur \text{ then } val(r(val(cur, n)), n) \text{ else } val(l(x), n),$
$val(r(x), n+1) = \text{if } x = cur \text{ then } val(prev, n) \text{ else } val(r(x), n),$
$smallest(N, n, \neg(val(cur, n) \neq -1)),$
$val'(X) = val(X, N)$, for each $X \in \vec{X}$.

Since $root$ does not change, we can omit some irrelative axioms from $\Pi_{P_2}^{\vec{X}}$, and then combine with $\Pi_{P_1}^{\vec{X}}$ and $\Pi_{if}^{\vec{X}}$, we finally get $\Pi_{DSW}^{\vec{X}}$ as follows:

$val(prev, 0) = -1 \wedge val(cur, 0) = val(root) \wedge val(next, 0) = val(next)$

$$\wedge\, val(l(x), 0) = val(l(x)) \wedge val(r(x), 0) = val(r(x)), \quad (1)$$
$$val(prev, n+1) = \text{if } val(l(val(cur, n)), n) = Null \text{ then}$$
$$Null \text{ else } val(cur, n), \quad (2)$$
$$val(cur, n+1) = \text{if } val(l(val(cur, n)), n) = Null \text{ then}$$
$$val(cur, n) \text{ else } val(l(val(cur, n)), n), \quad (3)$$
$$val(next, n+1) = val(l(val(cur, n)), n), \quad (4)$$
$$val(l(x), n+1) = \text{if } x = cur \text{ then } val(r(val(cur, n)), n)$$
$$\text{else } val(l(x), n), \quad (5)$$
$$val(r(x), n+1) = \text{if } x = cur \text{ then } val(prev, n) \text{ else } val(r(x), n), \quad (6)$$
$$val(cur, N) = -1 \wedge \forall m.m < N \rightarrow val(cur, m) \neq -1, \quad (7)$$
$$val'(X) = \text{if } val(root) = Null \text{ then } val(X) \text{ else } val(X, N),$$
$$\text{for each } X \in \vec{X}. \quad (8)$$

3.2. Proving correctness of DSW algorithm

According to analysis on DSW algorithm in Section 2.1, we define predicate $checked(t)$ to characterize the effect of traversal on binary tree t:

$$checked(t) \equiv val'(cur) = val(prev) \wedge val'(next) = val(prev) \wedge$$
$$val'(prev) = t \wedge \forall x.val'(l(x)) = val(l(x)) \wedge val'(r(x)) = val(r(x))$$

Proof. We prove DSW algorithm by induction on the structure of a tree.
Base. Let t be a binary tree with just one node, namely the one referred to by t, then we have $val(l(t)) = val(r(t)) = Null$. Starting from conjunction (1) in $\Pi_{DSW}^{\vec{X}}$, we solve the recurrence and get:

$$val(cur, 1) = val(cur, 2) = t \neq -1, \quad val(cur, 3) = -1$$
$$val'(prev) = val(prev, 3) = t, \quad val'(next) = val(next, 3) = -1,$$
$$val'(l(t)) = val(l(t), 3) = Null, \quad val'(r(t)) = val(r(t), 3) = Null.$$

Thus $check(t)$ holds.
Induction. For any two trees t_1 and t_2, assume that $checked(t_1)$ and $checked(t_2)$ hold, thus there are two natural numbers N_{t_1}, N_{t_2} such that:

$$val'(prev_{t_i}) = t_i \wedge val'(cur_{t_i}) = val(prev_{t_i}) \wedge val'(next_{t_i}) = val(prev_{t_i})$$
$$\wedge \forall x.val'(l_{t_i}(x)) = val(l_{t_i}(x)) \wedge val'(r_{t_i}(x)) = val(r_{t_i}(x)), \quad (9)$$
$$val'(cur_{t_i}, N_{t_i}) = val(prev_{t_i}) \wedge$$
$$\forall m.m < N_{t_i} \rightarrow val(cur_{t_i}, m) \neq val(prev_{t_i}).(\text{where } i = 1, 2) \quad (10)$$

If there is a tree s with $l_s(s) = t_1$ and $r_s(s) = t_2$, initially we have:

$val(prev_s, 0) = -1, \quad val(cur_s, 0) = s, \quad val(next_s, 0) = val(next_s),$
$val(l_s(x), 0) = val(l_s(x)), \quad val(r_s(x), 0) = val(r_s(x))$

By using axioms (2)–(6), we get:

$val(prev_s, 1) = $ if $val(l_s(val(cur_s, 0)), 0) = Null$ then $Null$ else $val(cur_s, 0)$
$\qquad\qquad\quad = $ if $val(l_s(s), 0) = Null$ then $Null$ else $s = s,$
$val(cur_s, 1) = $ if $val(l_s(cur_s), 0) = Null$ then $val(cur_s, 0)$ else $val(l_s(cur_s), 0)$
$\qquad\qquad\quad = $ if $val(l_s(s), 0) = Null$ then s else $val(l_s(s)) = t_1,$
$val(next_s, 1) = val(l_s(val(cur_s, 0)), 0) = val(l_s(s), 0) = t_1,$
$val(l_s(x), 1) = $ if $x = cur_s$ then $val(r_s(val(cur_s, 0)), 0)$ else $val(l_s(x), 0)$
$\qquad\qquad\quad = $ if $x = cur_s$ then t_2 else $val(l_s(x)),$
$val(r_s(x), 1) = $ if $x = cur_s$ then $val(prev_s, 0)$ else $val(r_s(x), 0)$
$\qquad\qquad\quad = $ if $x = cur_s$ then -1 else $val(r_s(x))$

Now $val(cur_s, 1) = t_1$ means cur refers to t_1 for the first time. By induction assumption, $checked(t_1)$ holds, that is we have axiom (9) for $i = 1$. From $val(cur_s, 0) = s$ and $val(cur_s, 1) = t_1$ we know that cur_s moves to t_1 from s, namely s is the previous position of cur_s, then we have $val(prev_{t_1}) = val(cur_s, 0) = s$.

Since t_1 is the left subtree of s, then all nodes of t_1 is also nodes of s, and the $(n+1)^{th}$ recurrence for s is actually the nth one for t_1. So for all nodes except s, we have $val(l_{t_1}(x)) = val(l_s(x))$, $val(r_{t_1}(x)) = val(r_s(x))$ and:

$val(prev_s, n+1) = val(prev_{t_1}, n), \quad val(cur_s, n+1) = val(cur_{t_1}, n),$
$val(next_s, n+1) = val(next_{t_1}, n), \quad val(l_s(x), n+1) = val(l_{t_1}(x), n),$
$val(r_s(x), n+1) = val(r_{t_1}(x), n).$

Thus we have:

$val(prev_s, N_{t_1} + 1) = val(prev_{t_1}, N_{t_1}) = t_1,$ \hfill (11)
$val(cur_s, N_{t_1} + 1) = val(cur_{t_1}, N_{t_1}) = val(prev_{t_1}) = s,$ \hfill (12)
$val(next_s, N_{t_1} + 1) = val(next_{t_1}, N_{t_1}) = val(prev_{t_1}) = s,$ \hfill (13)
$val(l_s(x), N_{t_1} + 1) = val(l_{t_1}(x), N_{t_1}) = val(l_s(x)), x \neq s$ \hfill (14)
$val(r_s(x), N_{t_1} + 1) = val(r_{t_1}(x), N_{t_1}) = val(r_s(x)), x \neq s.$ \hfill (15)

As for s, we have gotten $val(l_s(s), 1) = t_2$ and $val(r_s(s), 1) = -1$. For t_1, predicate *smallest* holds, that is:

$$val(cur_{t_1}, N_{t_1}) = val(prev_{t_1}), \forall m.m < N_{t_1} \to val(cur_{t_1}, m) \neq val(prev_{t_1})$$

Thus $val(cur_s, n)(1 \leq n \leq N_{t1})$ can not be s, so when $val(l_s(x))$ and $val(r_s(x))$ are computed iteratively using (5) and (6), $val(l_s(x), n+1)$ and $val(r_s(x), n+1)$ always take the else-branch, so $val(l_s(s), N_{t_1} + 1) = val(l_s(s), 1) = t_1$, $val(r_s(s), N_{t_1} + 1) = val(r_s(s), 1) = -1$. Together with (14) and (15), we can get:

$$val(prev_s, N_{t_1} + 1) = t_1, val(cur_s, N_{t_1} + 1) = s, val(next_s, N_{t_1} + 1) = s,$$
$$val(l_s(x), N_{t_1} + 1) = \text{if } x = s \text{ then } t_2 \text{ else } val(l_s(x)),$$
$$val(r_s(x), N_{t_1} + 1) = \text{if } x = s \text{ then } -1 \text{ else } val(r_s(x))$$

Keep iterating one more time by using (2)–(6):

$$val(prev_s, N_{t_1} + 2) = s, val(cur_s, N_{t_1} + 2) = t_2, val(next_s, N_{t_1} + 2) = t_2,$$
$$val(l_s(x), N_{t_1} + 2) = \text{if } x = s \text{ then } -1 \text{ else } val(l_s(x)),$$
$$val(r_s(x), N_{t_1} + 2) = \text{if } x = s \text{ then } t_1 \text{ else } val(r_s(x)).$$

Now $val(cur_s, N_{t_1} + 2) = t_2$, cur refers to t_2 for the first time. Its previous position is $val(cur_s, N_{t_1} + 1) = s$. Since t_2 is right subtree of s, and t_1 and t_2 are disjoint, we can conclude the following axioms through similar steps for t_1:

$$val(prev_s, N_{t_2} + N_{t_1} + 2) = val(prev_{t_2}, N_{t_2}) = val(cur_{t_2}) = t_2,$$
$$val(cur_s, N_{t_2} + N_{t_1} + 2) = val(cur_s, N_{t_2}) = val(prev_{t_2}) = s,$$
$$val(next_s, N_{t_2} + N_{t_1} + 2) = val(next_{t_2}, N_{t_2}) = val(prev_{t_2}) = s,$$
$$val(l_s(x), N_{t_2} + N_{t_1} + 2) = \text{if } x = s \text{ then } -1 \text{ else } val(l_s(x)),$$
$$val(r_s(x), N_{t_2} + N_{t_1} + 2) = \text{if } x = s \text{ then } t_1 \text{ else } val(r_s(x))$$

Keep iterating one more time with axioms (2)–(6), we can get:

$$val(prev_s, N_{t_2} + N_{t_1} + 3) = s, \quad val(cur_s, N_{t_2} + N_{t_1} + 3) = -1,$$
$$val(next_s, N_{t_2} + N_{t_1} + 3) = -1,$$
$$val(l_s(x), N_{t_2} + N_{t_1} + 3) = \text{if } x = s \text{ then } t_1 \text{ else } val(l_s(x)),$$
$$val(r_s(x), N_{t_2} + N_{t_1} + 3) = \text{if } x = s \text{ then } t_2 \text{ else } val(r_s(x))$$

We have known that $val(l_s(s)) = t_1$ and $val(r_s(s)) = t_2$, so finally we get:

$$val'(prev_s) = val(prev_s, N_{t_2} + N_{t_1} + 3) = s,$$

$$val'(cur_s) = val(cur_s, N_{t_2} + N_{t_1} + 3) = -1,$$
$$val'(next_s) = val(next_s, N_{t_2} + N_{t_1} + 3) = -1,$$
$$val'(l_s(x)) = val(l_s(x), N_{t_2} + N_{t_1} + 3) = val(l_s(x)),$$
$$val'(r_s(x)) = val(r_s(x), N_{t_2} + N_{t_1} + 3) = val(r_s(x))$$

This means $checked(s)$ holds, thus we showed that DSW algorithm traversed each node of tree s, and there was a natural number $N_{t_2} + N_{t_1} + 3$ promising termination of the algorithm. □

4. Related Work

Separation Logic[7] is a remarkable extension of Hoare Logic for reasoning about pointer-based in C-like imperative language.[8-10] The key of Separation Logic is the introduction of a novel logic operator *separating conjunction*, donated as "$*$". Operation $P * Q$ asserts that P and Q hold for disjoint regions of the heap, then each of them can be reasoned about independently. Reynolds showed how the list-reverse algorithm was proven easily in Hoare Logic style.[7] Yang proved the Schorr-Waite algorithm with an older version of Separation Logic as well.[11] The two proofs still depended on loop invariants, although the invariants are quite concise.

Loginov, Reps and Sagiv introduced an automated verification of the partial correctness and termination of DSW algorithm.[5] Based on three-valued logic, an abstract is defined to capture the invariants, which can be automatically synthesized during analysis.

5. Conclusion

In this paper we extended an existed formalism to reason about programs involving recursive structures in a more unified style. In this way, program manipulating any mutable data structures can be translated into first-order theory. With the theory, concerned properties of program can be reasoned about effectively just relying on simple mathematical skills, without the need of loop invariants and complicated concepts or operations. As an example, DSW algorithm is translated and verified in first-order logic. This shows that our method is valid for describing and reasoning about recursive structures.

Acknowledgments

This paper is supported by the Natural Science Foundation of China (No. 11761016) and Natural Science Foundation of Guizhou Province

(No. LH[2014]7214). And the authors would like to thank Fangzhen Lin for useful advices related to this subject.

References

1. F. Lin, A formalization of programs in first-order logic with a discrete linear order, in *Principles of Knowledge Representation and Reasoning: Proceedings of the Fourteenth International Conference, KR 2014, Vienna, Austria, July 20-24, 2014*, 2014.
2. C. Hoare, An axiomatic basis for computer programming, *Communications of the ACM* **12**, 576 (1969).
3. F. Lin and B. Yang, *Reasoning about Mutable Data Structures in First-Order Logic with Arithmetic: Lists and Binary Trees*, technical report, Hong Kong University of Science and Technology (2015).
4. F. Lin, A formalization of programs in first-order logic with a discrete linear order, *Artificial Intelligence* **235**, 1 (2016).
5. A. Loginov, T. Reps and M. Sagiv, Automated verification of the deutsch-schorrwaite tree-traversal algorithm, in *Static Analysis*, (Springer, 2006) pp. 261–279.
6. R. Bornat, Proving pointer programs in hoare logic, in *Mathematics of program construction*, 2000.
7. J. Reynolds, Separation logic: A logic for shared mutable data structures, in *Proceedings of 17th Annual IEEE Symposium on Logic in Computer Science*, 2002.
8. J. Berdine, C. Calcagno and P. W. Ohearn, Symbolic execution with separation logic, in *Programming Languages and Systems*, (Springer, 2005) pp. 52–68.
9. B. Cook, C. Haase, J. Ouaknine, M. Parkinson and J. Worrell, Tractable reasoning in a fragment of separation logic, in *CONCUR 2011–Concurrency Theory*, (Springer, 2011) pp. 235–249.
10. M. Parkinson and G. Bierman, Separation logic for object-oriented programming, in *Aliasing in Object-Oriented Programming. Types, Analysis and Verification*, (Springer, 2013) pp. 366–406.
11. H. Yang, An example of local reasoning in bi pointer logic: the schorr-waite graph marking algorithm, in *Proceedings of the 1st Workshop on Semantics, Program Analysis, and Computing Environments for Memory Management*, (Citeseer, 2000).

Look-ahead clause selection strategy for contradiction separation based automated deduction

Shuwei Chen,[†,1,3] Yang Xu,[1,3] Jun Liu[2,3] and Feng Cao[3]

[1] *School of Mathematics, Southwest Jiaotong University, Chengdu, 610031, China*
[2] *National-Local Joint Engineering Lab of System Credibility Automatic Verification, Southwest Jiaotong University, Chengdu, 610031, China*
[3] *School of Computing and Mathematics, Ulster University, Northern Ireland, UK*
[†] *swchen@swjtu.edu.cn*

Contradiction separation based dynamic automated deduction is a novel logic based automated deduction framework, which extends the static binary resolution inference rule to a dynamic multiple contradiction separation based automated deduction mechanism. The efficient implementation of this contradiction separation deduction lays, to a good extent, on how to select appropriate clauses and/or literals to construct the contradiction. This paper proposes a clause or literal selection strategy, so-called look-ahead strategy, which consider mainly the synergized effect of multi-clauses during the deduction process. Technical analysis along with some examples are provided to illustrate the feasibility of the proposed strategy.

Keywords: Automated reasoning; logic; contradiction separation deduction; clause selection; literal selection.

1. Introduction

Automated reasoning, as the foundation of Artificial Intelligence, has been an active research area for a long time.[2,3] Resolution based inference scheme, as an important automated reasoning mechanism, has been very successful for over five decades.[5,6] However, there are still a lot of real problems unsolved or not solved efficiently as illustrated in TPTP (Thousands of Problems for Theorem Provers).[8] The research on logic based automated reasoning, as indicated by A. Voronkov,[9] has encountered a bottleneck where a breakthrough in theory, especially go beyond binary resolution,[3,10,11] is necessary.

Motivated by the idea to provide a solution to break this bottleneck, Xu et al.[10,11] have proposed a novel contradiction separation (CS) based inference rule for automated deduction. It extends the binary resolution to dynamic and multiple (two or more) clauses handling in a synergized way and takes binary resolution as its special case. The coordination of multi-clauses during the deduction process is the most important feature, because it reflects the synergized logical relationship among these involved clauses, while this kind of logical relationship usually cannot be reflected by a series of logical relations between two clauses.[1] Although loads of strategies or heuristics for selecting clauses/literals have been implemented successfully in state-of-the-art provers, e.g., total selection, maximal selection, quality selection, symbol counting (weight), goal-directed heuristics and so on.[4,7] They usually cannot reflect the synergized effect among multiple clauses. We[1] have proposed some strategies in order to reflect the synergized effect among the multiple clauses. This paper further proposes a so-called look-ahead strategy for clause/literal selection during the automated deduction process, which considers mainly the coordinated effect of the clause to be selected and the clauses/literals already selected for constructing the contradiction. Some examples are put forward to illustrate the feasibility of this proposed strategy.

The remainder of this paper is structured as follows. In Section 2, we briefly review some preliminaries about the key concepts and results of contradiction separation based deduction. The look-ahead clause selection strategy is then proposed in Section 3, along with some illustrative examples. Concluding remarks and future work are drawn in Section 4.

2. Preliminaries

Some necessary concepts and conclusions of contradiction separation based dynamic automated deduction are provided in this section. The readers are referred to Ref. 10 and Ref. 11 for more details about contradiction separation deduction.

2.1. *Contradiction separation deduction in propositional logic*

Definition 2.1 Assume a clause set $S = \{C_1, C_2, \cdots, C_m\}$ in propositional logic. If $\forall (x_1, \cdots, x_m) \in \prod_{i=1}^{m} C_i$, x_i $(i = 1, \cdots, m)$ is a literal, there exist at least one complementary pair of literals among $\{x_1, \cdots, x_m\}$, then $S = \bigwedge_{i=1}^{m} C_i$ is called a *standard contradiction*, where C_i is regarded as a set of literals $(i = 1, \cdots, m)$.

Definition 2.2 Let $S = \{C_1, C_2, \cdots, C_m\}$ be a clause set in propositional logic. The following inference rule that produces a new clause from S is called a *contradiction separation rule*, in short, a CS rule: For each C_i

($i = 1, \cdots, m$), C_i can be separated into two sub-clauses C_i^- and C_i^+ such that

(1) $C_i = C_i^- \vee C_i^+$, where C_i^- and C_i^+ share no common literal;
(2) C_i^+ can be an empty clause itself, but C_i^- cannot be an empty clause;
(3) $\bigwedge_{i=1}^m C_i^-$ is a standard contradiction.

The resulting clause $\bigvee_{i=1}^m C_i^+$, denoted as $\mathcal{C}_m(C_1, C_2, \cdots, C_m)$, is called a *contradiction separation clause* (CSC) of $C_1, C_2, \cdots, C_m$, and $\bigwedge_{i=1}^m C_i^-$ is called a *separated standard contradiction*.

Definition 2.3 Suppose a clause set $S = \{C_1, C_2, \cdots, C_m\}$ in propositional logic. $\Phi_1, \Phi_2, \cdots, \Phi_t$ is called a *contradiction separation based dynamic deduction sequence* (or CS based dynamic deduction sequence) from S to a clause Φ_t, denoted as $\mathcal{D}$, if

(1) $\Phi_i \in S$, $i \in \{1, 2, \cdots, t\}$;
(2) there exist $r_1, r_2, \cdots, r_{k_i} < i$, $\Phi_i = \mathcal{C}_{k_i}(\Phi_{r_1}, \Phi_{r_2}, \cdots, \Phi_{r_{k_i}})$.

Theorem 2.1 (Soundness Theorem of the CS-Based Deduction in Propositional Logic) Suppose a clause set $S = \{C_1, C_2, \cdots, C_m\}$ in propositional logic. $\Phi_1, \Phi_2, \cdots, \Phi_t$ is a CS based deduction sequence from S to a clause Φ_t. If Φ_t is an empty clause, then S is unsatisfiable.

Theorem 2.2 (Completeness Theorem of the CS-Based Deduction in Propositional Logic) Suppose a clause set $S = \{C_1, C_2, \cdots, C_m\}$ in propositional logic. If S is unsatisfiable, then there exists a CS based deduction sequence from S to an empty clause.

2.2. *Contradiction separation deduction in first-order logic*

Definition 2.4 Suppose that $S = \{C_1, C_2, \cdots, C_m\}$ is a clause set in first-order logic. Without loss of generality, assume that there does not exist the same variables among $C_1, C_2, \cdots, C_m$. The following inference rule that produces a new clause from S is called a *standard contradiction separation rule*, in short, an S-CS rule, in first-order logic:

For each C_i ($i = 1, 2, \cdots, m$), firstly applying a substitution σ_i to C_i (σ_i could be an empty substitution but not necessary the most general unifier), denoted as $C_i^{\sigma_i}$; then separate $C_i^{\sigma_i}$ into two sub-clauses $C_i^{\sigma_i-}$ and $C_i^{\sigma_i+}$ such that

(1) $C_i^{\sigma_i} = C_i^{\sigma_i-} \vee C_i^{\sigma_i+}$, where $C_i^{\sigma_i-}$ and $C_i^{\sigma_i+}$ do not share the same literal;
(2) $C_i^{\sigma_i+}$ can be an empty clause, $C_i^{\sigma_i-}$ cannot be an empty clause;

(3) $\bigwedge_{i=1}^{m} C_i^{\sigma_i -}$ is a *standard contradiction*, that is, $\forall (x_1, \cdots, x_m) \in \prod_{i=1}^{m} C_i^{\sigma_i -}$, there exists at least one complementary pair of literals among $x_1, \cdots, x_m$.

The resulting clause $\bigvee_{i=1}^{m} C_i^{\sigma_i +}$, denoted as $\mathcal{C}_m^s(C_1^{\sigma_1}, C_2^{\sigma_2}, \cdots, C_m^{\sigma_m})$ is called as a *standard contradiction separation clause* (S-CSC) of $C_1, C_2, \cdots, C_m$, and $\bigwedge_{i=1}^{m} C_i^{\sigma_i -}$ is called a *separated standard contradiction* (S-SC).

Definition 2.5 Suppose a clause set $S = \{C_1, C_2, \cdots, C_m\}$ in first-order logic. $\Phi_1, \Phi_2, \cdots, \Phi_t$ is called a *standard contradiction separation based deduction sequence* (or a S-CS based deduction sequence from S to a clause Φ_t, denoted as $\mathcal{D}_s$, if

(1) $\Phi_i \in S$, $i \in \{1, 2, \cdots, t\}$;
(2) there exist $r_1, r_2, \cdots, r_{k_i} < i$, $\Phi_i = \mathcal{C}_{r_{k_i}}^{s\theta_i}(\Phi_{r_1}, \Phi_{r_2}, \cdots, \Phi_{r_{k_i}})$, where $\theta_i = \bigcup_{j=1}^{k_i} \sigma_j$, σ_j is a substitution to Φ_{r_j}, $j = 1, \cdots, k_i$.

Theorem 2.3 (Soundness Theorem of the S-CS Based Deduction in First-Order Logic) Suppose a clause set $S = \{C_1, C_2, \cdots, C_m\}$ in first-order logic. $\Phi_1, \Phi_2, \cdots, \Phi_t$ is a S-CS based deduction from S to a clause Φ_t. If Φ_t is an empty clause, then S is unsatisfiable.

Theorem 2.4 (Completeness Theorem of the S-CS Based Deduction in First-Order Logic) Suppose a clause set $S = \{C_1, C_2, \cdots, C_m\}$ in first-order logic. If S is unsatisfiable, then there exists a S-CS based deduction from S to an empty clause.

3. Look-ahead Clause Selection

The construction of the separated standard contradiction is the crucial part of the CS-based automated deduction, and the contradiction is determined by the selected clauses. Furthermore, there is one literal from each selected clause in the standard contradiction that plays an important role on determining the separated contradiction, and therefore is called as *decision literal*.[1] Denote the set of decision literals as D_l. It can seen that the size of D_l, i.e., the number of decision literals in it, changes dynamically. There will be more decision literals in D_l if more clauses are used for constructing the standard contradiction, and at the beginning, $D_l = \emptyset$, denoted as D_{l_0}. The literals in the subsequent clauses after certain substitutions that are complementary to the previous decision literals will be "pull" into the standard contradiction, and the remaining literals in the clauses, if there is any, will contribute to the S-CSC.

The look-ahead strategy in the CS-based automated deduction proposed here is mainly to select the clause that not only has less literals left after eliminating the literals complementary to the decision literals, but also can work together with the current decision literals to find better subsequent clauses. It is essentially different to the look-ahead selection that tries to select the literals that result in the smallest number of children.[4]

Suppose a clause set $S = \{C_1, C_2, \cdots, C_m\}$ ($m \geq 2$) and we have already selected i ($1 \leq i < m-1$) clauses to construct the standard contradiction, and got a set of decision literals D_{l_i}, and we proceed to select the subsequent clauses from the clause set S and the corresponding decision literals. There are different ways to select the starting decision literal(s).[1] For example, to choose the literal whose complementary literal after certain substitution appears most in the clause set. We focus here mainly on the determination of the subsequent clauses/literals based on the current decision literals.

Denote $N(C)$ as the number of literals in a clause C, and $C_D(\sigma) = \{\sim l \in C | \exists$ substitution θ_j, s.t. $, l^\sigma = l_j^{\theta_j}$, where $l_j \in D_{l_i}, 1 \leq j \leq i\}$, for a given substitution σ, then $N(C) - N(C_D)$ is the number of literals in C that don't have complementary literal in D_l after substitution.

Step 1. Provisional selection of clause C_{i+1}. Select the clause C_{i+1} where the number of literals in C_{i+1} that don't have complementary literal in D_l after substitution is least in the clauses, i.e., $N(C_{i+1}) - N(C_{(i+1)D}) = \min\{N(C_j) - N(C_{jD}) | C_j \in S\}$.

At the same time, select another clause C'_{i+1}, where $N(C'_{i+1}) - N(C'_{(i+1)D}) = \min\{N(C_j) - N(C_{jD}) | C_j \in S, C_j \neq C_{i+1}\}$, i.e., the number of literals in C'_{i+1} that don't have complementary literal in D_l after substitution is least in the clauses excluding C_{i+1}.

Step 2. Provisional selection of decision literal l_{i+1}.

Case 1. If $N(C_{i+1}) - N(C_{(i+1)D}) = 0$, it means that we obtain an empty S-CSC, then it can be concluded that the clause set is unsatisfiable, and the deduction process terminates.

Case 2. If $N(C_{i+1}) - N(C_{(i+1)D}) = 1$, it means that there is only one literal left in C_{i+1} after eliminating the literals that are complementary to the current decision literals, then this literal, denoted as l_{i+1}, can be added into D_{l_i} as a new decision literal, and go to Step 3 with the updated decision literal set $D_{l_{i+1}}$.

Case 3. If $N(C_{i+1}) - N(C_{(i+1)D}) > 1$, it means that there are more than one literals left in C_{i+1} after eliminating the literals that are complementary to the current decision literals, then choose the literal from the remaining literals as l_{i+1} whose complementary literal after certain substitution appears most in the clause set, goto Step 3 with the updated decision literal set $D_{l_{i+1}}$.

Step 3. Determination of clause C_{i+1}. Choose the literal from the remaining literals in C'_{i+1} after eliminating the literals that are complementary to the current decision literals, as l'_{i+1} whose complementary literal after certain substitution appears most in the clause set. We can have another provisional decision literal set $D'_{l_{i+1}}$.

Compute $N(C_{i+2}) - N(C_{(i+2)D})$ and $N(C'_{i+2}) - N(C'_{(i+2)D})$ based on the updated $D_{l_{i+1}}$ and $D'_{l_{i+1}}$. Compare $N(C_{i+2}) - N(C_{(i+2)D}) + N(C_{i+1}) - N(C_{(i+1)D})$ with $N(C'_{i+2}) - N(C'_{(i+2)D}) + N(C'_{i+1}) - N(C'_{(i+1)D})$, and choose the clause C_{i+1} or C'_{i+1} corresponding to the smaller value as the determined C_{i+1} and the corresponding literal as l_{i+1}. Goto Step 1 with the determined decision literal set $D_{l_{i+1}}$.

Remark 3.1 Of course, the above process can be applied to look-ahead for more than two clauses, i.e., to select the clause C_{i+3} with the provisional selected C_{i+1} and C_{i+2}, and then determine C_{i+1} and/or C_{i+2} based on C_{i+3}. However, it will usually be too complex and source consuming by keeping more than two clauses at hand. Therefore, we discuss here just the above process to look-ahead for one more step to determine the clause C_{i+1}.

The CS-based automated deduction method by applying the *look-ahead strategy* is sound, i.e., we have the following theorem.

Theorem 3.1 Let $S = \{C_1, C_2, \cdots, C_m\}$ be a clause set in first-order logic. $\Phi_1, \cdots, \Phi_t$ is a S-CS based deduction from S to a clause Φ_t by applying the *look-ahead strategy*. If Φ_t is an empty clause, then S is unsatisfiable.

Proof. Actually, because that the *look-ahead strategy* just restricts the generation of certain clauses, the new clauses generated during the deduction process are still generated by the S-CS based automated deduction, and the S-CS based deduction sequence $\Phi_1, \Phi_2, \cdots, \Phi_t$ by applying the *look-ahead strategy* is of course still a S-CS based deduction sequence. Therefore, if we can generate a S-CS based deduction sequence from S to an empty clause by applying the *look-ahead strategy*, it means we have a S-CS based deduction sequence from S to an empty clause, and as a consequence, S is certainly unsatisfiable according to Theorem 2.3.

The following example illustrates how the proposed *look-ahead strategy* works.

Example 3.1 Let $S = \{C_1, C_2, C_3, C_4, C_5, C_6, C_7\}$ be a clause set in first-order logic, where
$C_1 = P_1(a)$, $C_2 = \sim P_2(a, b)$, $C_3 = P_3(a, f(c), f(b))$, $C_4 = P_3(x_1, x_1, f(x_1))$,

$C_5 = \sim P_3(x_2, x_3, x_4) \vee P_3(x_3, x_2, x_4)$, $C_6 = \sim P_3(x_5, x_6, x_7) \vee P_2(x_5, x_7)$,
$C_7 = \sim P_1(x_8) \vee \sim P_3(x_9, x_{10}, x_{11}) \vee \sim P_2(x_8, x_{11}) \vee P_2(x_8, x_9) \vee P_2(x_8, x_{10})$.

Step 1. In this example, by applying the *unit clauses in a whole* strategy proposed in,[1] C_1, C_2, C_3 and C_4 are chosen firstly in a whole to form the starting part of the separated contradiction, and the corresponding decision literal set is $D_{l_4} = \{P_1(a), \sim P_2(a,b), P_3(a, f(c), f(b)), P_3(x_1, x_1, f(x_1))\}$.

Step 2. It can be computed that $N(C_5) - N(C_{5D}) = 1$, $N(C_6) - N(C_{6D}) = 1$, and $N(C_7) - N(C_{7D}) = 1$ based on the decision literal set D_{l_4}, so we can keep all the three clauses as the provisional clause, and proceed to the next step to choose which one should be determined and what the corresponding decision literal is.

For C_5, the corresponding provisional decision literal should be $P_3(x_1, x_1, f(x_1)) = C_4$, which is redundant, or $P_3(f(c), a, f(b))$. For C_6, the corresponding provisional decision literal should be $P_2(x_1, f(x_1))$ or $P_2(a, f(b))$. For C_7, the corresponding provisional decision literal should be $\sim P_2(a, f(b))$.

Step 3. If we choose C_5 with its decision literal $P_3(f(c), a, f(b))$, the next step clause can be selected from C_6 and C_7. It can then be computed that $N(C_6) - N(C_{6D}) = 1$, and $N(C_7) - N(C_{7D}) = 1$ based on the updated decision literal set $D_{l_5^1} = \{P_1(a), \sim P_2(a,b), P_3(a, f(c), f(b)), P_3(x_1, x_1, f(x_1)), P_3(f(c), a, f(b))\}$.

If C_6 is chosen, the updated decision literal set will be $D_{l_5^2} = \{P_1(a), \sim P_2(a,b), P_3(a, f(c), f(b)), P_3(x_1, x_1, f(x_1)), P_2(a, f(b))\}$, and the next step clause can be selected from C_5 and C_7. It can be computed easily that $N(C_5) - N(C_{5D}) = 1$, and $N(C_7) - N(C_{7D}) = 0$. Therefore, it can be concluded that the clause set S is unsatisfiable with C_1, C_2, C_3, C_4, C_6 and C_7 consisting the standard contradiction.

Of course, if we choose C_7 with the updated decision literal set $D_{l_5^3} = \{P_1(a), \sim P_2(a,b), P_3(a, f(c), f(b)), P_3(x_1, x_1, f(x_1)), \sim P_2(a, f(b))\}$, we can also have $N(C_6) - N(C_{6D}) = 0$ and obtain the unsatisfiability conclusion.

4. Conclusion

This paper has proposed a so-called look-ahead strategy for clause/literal selection in the contradiction separation based automated deduction mechanism, which considers the coordinated effect of the clauses to be selected with the clauses already selected and the subsequent determined clause as well, for constructing the contradiction. The feasibility of this strategy has been analyzed technically and based on example illustration. Further study will be done to improve the efficiency of the proposed strategy, and then implement it along with other strategies in the CS-based prover.

Acknowledgments

This work has been partially supported by the National Natural Science Foundation of China (Grant No. 61673320) and the project of Department of Education of Sichuan Province (Grant No. 18ZB0589).

References

1. S. W. Chen, Y. Xu, J. Liu et al., Some synergized clause selection strategies for contradiction separation based automated deduction, the 12th International Conference on Intelligent Systems and Knowledge Engineering (ISKE2017), Nanjing, China, November 24-26, 2017, pp. 143-148.
2. J. Harrison, *Handbook of Practical Logic and Automated Reasoning*, Cambridge University Press, 2009.
3. D. Plaisted, History and prospects for first-order automated deduction, Automated Deduction - CADE-25, August 1-7, 2015, pp. 3-28.
4. G. Reger, M. Suda, A. Voronkov, and K. Hoder, Selecting the selection. The 8th International Joint Conference on Automated Reasoning - IJCAR 2016, LNAI 9706, June 27-July 2, 2016, pp. 313-329.
5. J. A. Robinson, A machine oriented logic based on the resolution principle, *J. ACM*, 12(1): 23-41, 1965.
6. J. A. Robinson and A. Voronkov, *Handbook of Automated Reasoning*, Vol. 1 and 2, the MIT Press and North Holland, 2001.
7. S. Schulz and M. Mohrmann, Performance of clause selection heuristics for saturation-based theorem proving, The 8th International Joint Conference on Automated Reasoning - IJCAR 2016, LNAI 9706, June 27-July 2, 2016, pp. 330-345.
8. G. Sutcliffe. The TPTP problem library and associated infrastructure: the FOF and CNF parts, v3.5.0. *Journal of Automated Reasoning*, 43(4):337-362, 2009.
9. A. Voronkov, Automated reasoning: Past story and new trends, In: Gottlob, G., Walsh, T. (eds.) IJCAI 2003, pp. 1607-1612.
10. Y. Xu, J. Liu, S. W. Chen, and X. M. Zhong, A novel generalization of resolution principle for automated deduction, The 12th International FLINS Conference on Uncertainty Modelling in Knowledge Engineering and Decision Making (FLINS2016), ENSAIT, Roubaix, France, August 24-26, 2016, pp. 483-488.
11. Y. Xu, J. Liu, S. W. Chen, X. M. Zhong, and X. X. He, Contradiction separation based dynamic multi-clause synergized automated deduction, *Information Sciences*, to appear, 2018.

α-Generalized resolution method based on linguistic truth-valued first-order logic system

Weitao Xu*

College of Information Science and Engineering, Henan University of Technology
Zhengzhou, 450001, P.R. China
*hnxmxwt@haut.edu.cn

Yang Xu

National-Local Joint Engineering Laboratory of System
Credibility Automatic Verification
Southwest Jiaotong University
Chengdu, 610031, P.R. China
xuyang@home.swjtu.edu.cn

This paper proposes an α-generalized resolution method in linguistic truth-valued lattice-valued first-order logic system $\mathcal{L}_{V(n\times 2)}F(X)$ based on linguistic truth-valued lattice implication algebra $\mathcal{L}_{V(n\times 2)}$. By establishing a lift lemma, both soundness and weak completeness theorems for α-generalized resolution method are given in $\mathcal{L}_{V(n\times 2)}F(X)$. This work will provide a fundament for extending α-generalized resolution method under linguistic truth-valued level in a set of general generalized clauses.

Keywords: Automated reasoning; linguistic truth-valued lattice-valued first-order logic; general generalized clause; α-generalized resolution.

1. Introduction

Since Robinson presented resolution principle in 1965 [1], many scholars have extensively studied resolution-based automated reasoning. A lot of important applications of such systems have been found in many areas. Many resolution methods based on classical resolution principle have been presented from the different ways. Some typical resolution methods are studied extensively, such as linear resolution, semantic resolution and lock resolution [2, 3].

With the development of non-classical logics, which can deal with uncertain or incomparable information in the real world, resolution-based automated reasoning have attracted a lot of interests in the framework of non-classical logic.

As the use of non-classical logics becomes increasingly important in information science and AI, the development of automated theorem proving based on non-classical logic is currently an active area of research. Lattice-valued logic is an important non-classical logic, and plays an important role for dealing with comparability and incomparability in the real word. In the course of information processing, in order to deal with fuzziness and incomparability of processed object itself and uncertainty, Xu presented lattice implication algebra by combining lattice with implication algebra, and established the corresponding theories and methods [4]. Later, Xu et al. established lattice-valued propositional logic $LP(X)$ based on lattice implication algebra [4]. Further, Xu et al. extended lattice-valued propositional logic $LP(X)$ into lattice-valued first-order logic $LF(X)$ based on lattice implication algebra [4]. $LP(X)$ and $LF(X)$ not only have strict syntax proof but also sound semantic interpretation, and provide a scientific and reasonable logical foundation for automated theorem proving in AI. Xu et al. presented α-resolution principle in lattice-valued propositional logic system $LP(X)$ based on lattice implication algebra [4]. Under the framework of lattice-valued first-order logic system, α-resolution principle in lattice-valued first-order logic system $LF(X)$ based on lattice implication algebra had also established [4].

Xu constructed a class of linguistic truth-valued lattice implication algebras in order to treat with linguistic-truth value information [6]. In fact, it provides an efficient approach by a bijection for obtaining a linguistic truth-valued lattice implication algebra. In linguistic truth-valued lattice-valued propositional logic system and linguistic truth-valued lattice-valued first-order logic system, which truth-value fields are a linguistic truth-valued lattice implication algebras, linguistic truth-valued resolution automated reasoning was also focused [8]. Xu et al. gave the determination of generalized literals in lattice-valued logic system [10]. These work will provide a important foundation for $\alpha-$resolution of generalized literals.

This paper is organized as follows: Section 2 as a preliminary gives an overview of some basic concepts of linguistic truth-valued lattice implication algebra,and α-resolution principle in lattice-valued first-order logic. Section 3 as a major work proposes α-generalized resolution method for linguistic truth-valued lattice-valued first-order logic system $\mathcal{L}_{V(n\times 2)}F(X)$. Conclusions and future researches are presented in Section 4.

2. Preliminaries

Definition 2.1 [4] (Lattice implication algebra) Let $(L, \vee, \wedge, O, I)$ be a bounded lattice with an order-reversing involution $'$, I and O the greatest and the smallest element of L respectively, and $\to: L \times L \to L$ be a mapping. $\mathcal{L} = (L, \vee, \wedge, ', \to, O, I)$ is called a lattice implication algebra if the following conditions hold for any $x, y, z \in L$,

(1) $x \to (y \to z) = y \to (x \to z);$
(2) $x \to x = I;$
(3) $x \to y = y' \to x';$
(4) $x \to y = y \to x = I$ implies $x = y;$
(5) $(x \to y) \to y = (y \to x) \to x;$
(6) $(x \vee y) \to z = (x \to z) \wedge (y \to z);$
(7) $(x \wedge y) \to z = (x \to z) \vee (y \to z).$

Definition 2.2 [6] (Linguistic truth-valued lattice implication algebra) Let $AD_n = \{a_1, a_2, \cdots, a_n\}$ be a set with n modifiers and $a_1 < a_2 < \cdots < a_n$, $MT = \{f, t\}$ be a set of meta truth values, $f < t$. Denote

$$L_{V(n \times 2)} = AD_n \times MT.$$

Define a mapping g as

$$g: L_{V(n \times 2)} \to \mathcal{L}_n \times \mathcal{L}_2,$$

and

$$g((a_i, mt)) = \begin{cases} (d'_i, b_1) & \text{when } mt = f, \\ (d_i, b_2) & \text{when } mt = t. \end{cases}$$

then g is bijection, denote its inverse mapping as g^{-1}. For any $x, y \in L_{V(n \times 2)}$, define

$$x \vee y = g^{-1}(g(x) \vee g(y)),$$

$$x \wedge y = g^{-1}(g(x) \wedge g(y)),$$

$$x' = g^{-1}((g(x)),$$

$$x \to y = g^{-1}((g(x) \to g(y)).$$

We call $\mathcal{L}_{V(n \times 2)} = (L_{V(n \times 2)}, \vee, \wedge, ', \to, (a_n, f), (a_n, t))$ a linguistic truth-valued lattice implication algebra generated by AD_n and MT. Its elements are called linguistic truth-values, and g is an isomorphic mapping from $(L_{V(n \times 2)}, \vee, \wedge, ', \to, (a_n, f), (a_n, t))$ to $\mathcal{L}_n \times \mathcal{L}_2$ (see Fig. 1).

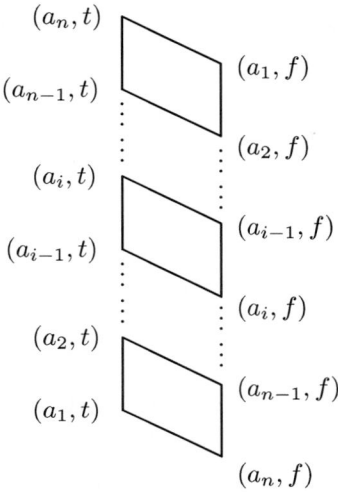

Fig. 1. Hasse diagram of linguistic truth-valued lattice implication algebra.

In the above definition, both $\mathcal{L}_n$ and $\mathcal{L}_2$ are Lukasiewicz implication algebras, i.e.,

$$\mathcal{L}_n = (L_n, \vee_{(L_n)}, \wedge_{(L_n)}, '^{(L_n)}, \rightarrow_{(L_n)}, d_1, d_n).$$
$$\mathcal{L}_2 = (L_2, \vee_{(L_2)}, \wedge_{(L_2)}, '^{(L_2)}, \rightarrow_{(L_2)}, b_1, b_2).$$

$\mathcal{L}_n \times \mathcal{L}_2$ is a lattice implication algebra generated by $\mathcal{L}_n$ and $\mathcal{L}_2$, i.e.,

$$\mathcal{L}_n \times \mathcal{L}_2 = (L_n \times L_2, \vee, \wedge, ', \rightarrow, (d_1, b_1), (d_n, b_2)).$$

Definition 2.3 [4] In lattice-valued first-order logic system $\mathcal{L}F(X)$, let $\mathcal{L}$ be a lattice implication algebra, $\alpha \in L$, G_1 and G_2 two generalized clauses of the form

$$G_1 = g_1 \vee \cdots \vee g_i \vee \cdots \vee g_m,$$

$$G_2 = h_1 \vee \cdots \vee h_j \vee \cdots \vee h_n,$$

where $g_i (i = 1, 2, \cdots, m)$ and $h_j (j = 1, 2, \cdots, n)$ are generalized literals in G_1 and G_2 respectively. If there exists a substitution ρ such that

$$g_i^\rho \wedge h_j^\rho \leq \alpha,$$

then

$$g_1^\rho \vee \cdots \vee g_{i-1}^\rho \vee g_{i+1}^\rho \vee \cdots \vee g_m^\rho \vee h_1^\rho \vee \cdots \vee h_{j-1}^\rho \vee h_{j+1}^\rho \vee \cdots \vee h_n^\rho$$

is called an α-resolvent of G_1 and G_2, denoted by $R_\alpha(G_1, G_2)$, and (g_i, h_j) is called an α-resolution pair, denoted by $(g_i, h_j) - \alpha$.

Theorem 2.1 [4] Suppose a generalized conjunctive normal form $S = C_1 \wedge C_2 \wedge \cdots \wedge C_n$ in $LF(X)$, $\alpha \in L$, $D_1, D_2, \cdots, D_m$ is an α-resolution deduction from S to a generalized clause D_m. If D_m is $\alpha - \diamond$, then $S \leq \alpha$, i.e., if $D_m \leq \alpha$, then $S \leq \alpha$.

Theorem 2.2 [4] Let S be a generalized conjunctive normal form in $LF(X)$, $\alpha \in L$, α a daul numerator and $\bigvee_{\alpha \in L}(a \wedge a') \leq \alpha < I$. Suppose that there exists $\beta \in L$ such that $\beta \wedge (\beta \to \beta')$ is neither less than nor equal to α. If $S \leq \alpha$, then there exists an α-resolution deduction from S to $\alpha - \diamond$.

3. α-Generalized Resolution Method Based on Linguistic Truth-Valued Lattice-Valued First-order Logic System

In the following, generalized literal and generalized clause are respectively denoted by g-literal and g-clause.

Definition 3.1 In linguistic truth-valued lattice-valued first-order logic system $\mathcal{L}_{V(n \times 2)} F(X)$, let $g_1, g_2, \cdots, g_n$ be g-literals, $\Psi(g_1, g_2, \cdots, g_n)$ is a lattice-valued logical formula, which is composed of $g_1, g_2, \cdots, g_n$ by using "$\vee, \wedge, ', \to, \leftrightarrow$", we called $\Psi(g_1, g_2, \cdots, g_n)$ is a general g-clause, denoted by Ψ.

Definition 3.2 In linguistic truth-valued lattice-valued first-order logic system $\mathcal{L}_{V(n \times 2)} F(X)$, let Ψ be a general g-clause, $\alpha \in \mathcal{L}_{V(n \times 2)}$, $\alpha < (a_n, t)$.

(1) If there exists an interpretation $\mathcal{I}_D = <D, \mu_D, \nu_D>$, and $\nu_D(\Psi) > \alpha$, then Ψ is called α-satisfiable.

(2) For any interpretation $\mathcal{I}_D = <D, \mu_D, \nu_D>$, and $\nu_D(\Psi) > \alpha$, then Ψ is called α-true.

(3) For any interpretation $\mathcal{I}_D = <D, \mu_D, \nu_D>$, and $\nu_D(\Psi) \leq \alpha$, then Ψ is called α-false.

Example 3.1 In linguistic truth-valued lattice-valued first-order logic system $\mathcal{L}_{V(9 \times 2)} F(X)$, $\alpha \in \mathcal{L}_{V(9 \times 2)}$, $\alpha = (Ex, T)$, and

$$\Psi = q(y) \to (Ve, T),$$

Then Ψ is α-false.

Definition 3.3 In linguistic truth-valued lattice-valued first-order logic system $\mathcal{L}_{V(n \times 2)} F(X)$, let S be a set of general g-clause, $x_1, x_2, \cdots, x_n$ are free variables occurring in S, then logical formula

$$\forall_{x_1} \forall_{x_2} \cdots \forall_{x_n} S^*$$

is called universality respect to S, i.e., $\forall_{x_1}\forall_{x_2}\cdots\forall_{x_n}S^*$ is a universal close formula.

Definition 3.4 Let Ψ be a general g-clause in linguistic truth-valued lattice-valued first-order logic system $\mathcal{L}_{V(n\times 2)}F(X)$, if there exists a most general unifier σ for some local extremely g-literals $g_i(i=1,2,\cdots,r)$, then Ψ^σ is a factor of Ψ.

Definition 3.5 In linguistic truth-valued lattice-valued first-order logic system $\mathcal{L}_{V(n\times 2)}F(X)$, let

$$\Phi(g_1,g_2,\cdots,g_i,\cdots,g_n)$$

and

$$\Psi(h_1,h_2,\cdots,h_i,\cdots,h_n)$$

two general g-clauses. There have no common free variables in Φ and Ψ, $\alpha \in \mathcal{L}_{V(n\times 2)}$. If there exist the most general unifier σ,τ and a substitution ρ, satisfy:

(1) Φ^σ is a factor by unifying local extremely complex g-literals $g_{i_1},g_{i_i},\cdots,g_{i_s}$ of Φ,

(2) Ψ^τ is a factor by unifying local extremely complex g-literals $h_{j_1},h_{j_i},\cdots,h_{j_s}$ of Ψ,

(3) ρ is a substitution of $g_{i_1}^\sigma$ and $h_{j_1}^\tau$, and $g_{i_1}^{\sigma\rho} \wedge h_{j_1}^{\tau\rho} \leq \alpha$, then

$$\Phi^{\sigma\rho}(g_{i_1}^{\sigma\rho}=\alpha) \vee \Psi^{\tau\rho}(h_{j_1}^{\tau\rho}=\alpha)$$

is an α-generalized resolvent of Φ and Ψ, denoted by

$$R^f_{\alpha\in\mathcal{L}_{V(n\times 2)}}(\Phi,\Psi).$$

Theorem 3.1 In linguistic truth-valued lattice-valued first-order logic system $\mathcal{L}_{V(n\times 2)}F(X)$, let Φ and Ψ two general g-clauses, $\alpha \in \mathcal{L}_{V(n\times 2)}$, $\alpha < (a_n,t)$. If there exist two local extremely complex g-literals $g \in \Phi$ and $h \in \Psi$, and a substitution ρ such that $g^\rho \wedge h^\rho \leq \alpha$, $R^f_{\alpha\in\mathcal{L}_{V(n\times 2)}}(\Phi,\Psi)$ is an α-generalized resolvent of Φ and Ψ, then

$$\Phi \wedge \Psi \leq R^f_{\alpha\in\mathcal{L}_{V(n\times 2)}}(\Phi,\Psi).$$

Definition 3.6 Let S be a set of general g-clauses in linguistic truth-valued lattice-valued first-order logic system $\mathcal{L}_{V(n\times 2)}F(X)$, $\alpha \in \mathcal{L}_{V(n\times 2)}$, $\alpha < (a_n,t)$. $\omega = \{D_1,D_2,\cdots,D_m\}$ is called an α-generalized resolution deduction from S to D_m, if the following conditions hold,

(1) $D_i \in S$ or

(2) $D_i = R^f_{\alpha\in\mathcal{L}_{V(n\times 2)}}(D_j{}^0,D_k{}^0)$, $j<i$, $k<i$,

where,
$D_j{}^0 = D_j$ or $D_j{}^0$ is an instance of D_j,
$D_k{}^0 = D_k$ or $D_k{}^0$ is an instance of D_k.

Theorem 3.2 Let S be a set of general g-clauses in linguistic truth-valued lattice-valued first-order logic system $\mathcal{L}_{V(n\times 2)}F(X)$, $\alpha \in \mathcal{L}_{V(n\times 2)}$, $\alpha < (a_n, t)$.

$$\{D_1, D_2, \cdots, D_i, \cdots, D_m\}$$

is an α-generalized resolution deduction from S to D_m. If D_m is an $\alpha - \diamond$, then $S < \alpha$, i.e., if $D_m \leq \alpha$, then $S \leq \alpha$.

Theorem 3.3 In linguistic truth-valued lattice-valued first-order logic system $\mathcal{L}_{V(n\times 2)}F(X)$, $\alpha \in \mathcal{L}_{V(n\times 2)}$, $\alpha < (a_n, t)$. Let Φ_1^0 and Φ_2^0 be the instances of Φ_1 and Φ_2, and Φ^0 is an α-generalized resolvent of Φ_1^0 and Φ_2^0, then there exists an α-generalized resolvent Φ of Φ_1 and Φ_2 satisfied that Φ^0 is an instance of Φ.

Theorem 3.4 Let S be a general g-clause in linguistic truth-valued lattice-valued first-order logic system $\mathcal{L}_{V(n\times 2)}F(X)$, let S be a set of general g-clauses, $\alpha \in \mathcal{L}_{V(n\times 2)}$, $\alpha < (a_n, t)$, $(n+1)/2 \leq k < (2n+1)/3 (n \geq 7, n \neq 8)$, H_S is a set of local extremely complex g-literals from the set S of general g-clauses. If the following conditions hold,

(1) $S \leq \alpha$,

(2) h_1 and h_2 are local extremely complex g-literals in Φ and Ψ, $h_1^\rho \wedge h_2^\rho \leq \alpha$ by using a substitution, h_1^ρ and h_2^ρ are independent with $S^{\rho*}$ respectively, where $H_{S^{\rho*}} = H_{S^\rho} - \{h_1^\rho, h_2^\rho\}$, then there exists an α-generalized resolution deduction from S to $\alpha - \diamond$.

4. Conclusions

In this paper α-generalized resolution method is established in linguistic truth-valued lattice-valued first-order logic system $\mathcal{L}_{V(n\times 2)}F(X)$ based on linguistic truth-valued lattice implication algebra $\mathcal{L}_{V(n\times 2)}$. In a resolution process, the most general unify and substitution are implemented in two general g-clauses, and α-generalized resolution literals are replaced by resolution level α. By using the lift lemma, both soundness and weak completeness theorems are also obtained.

Acknowledgments

This work is partially supported by the National Natural Science Foundation of P. R. China (Grant No. 61673320), the Fundamental Research Funds for the Central Universities (Grant No. 2682017ZT12), the High-level Talent Foundation of Henan University of Technology (No. 2012BS012).

References

1. J.A. Robinson, A machine-oriented logic based on the resolution principle, J. ACM, 12(1), 23-41(1965).
2. C.L. Chang, R.C.T. Lee, Symbolic Logic and Mechanical Theorem Proving, Academic Press, New York(1973).
3. X.H. Liu, Resolution-based Automated Reasoning, Academic Press of China, Beijing, China(1994).
4. Y. Xu, D. Ruan, K. Qin and J. Liu, Lattice-Valued Logic: An alternative approach to treat fuzziness and incomparability. Berlin: Springer-Verlag(2003).
5. J. Liu, D. Ruan, Y. Xu, Z.M. Song, A resolution-like strategy based on lattice-valued logic, IEEE Transactions on Fuzzy Systems, 11(4), 560-567(2003).
6. Y. Xu, S.W. Chen and J. Ma, Linguistic Truth-valued Lattice Implication Algebra and Its Properties, in: IMACS Multiconference on Computational Engineering in Systems Applications, 1413-1418(2006).
7. V. Sofronie-Stokkermans, C. Ihlemann, Automated reasoning in some local extensions of ordered structures, Journal of Multiple-Valued Logic and Soft Computing, 13(4-6), 397-414(2007).
8. Y. Xu, S.W. Chen, J. Liu, D. Ruan, Weak Completeness of Resolution in a Linguistic Truth-Valued Propositional Logic, in: Proc. IFSA2007 on Theoretical Advances and Applications of Fuzzy Logic and Soft Computing, 358-366(2007).
9. Y. Xu, W.T. Xu, X.M. Zhong, X.X. He, α-Generalized Resolution Principle Based on Lattice-Valued Porpositional Logic LP(X), in: Proc. FLINS2010 on Foundations and Applications of Computational Intelligence, Chengdu, China, 66-71(2010).
10. Y. Xu, J. Liu, D. Ruan, X.B. Li, Determination of α-Resolution in Lattice-Valued First-order Logic LF(X), Information Sciences, 181(10), 1836-1862(2011).

Deductive control strategies based on contradiction separation rule

Feng Cao

School of Information Science and Technology, Southwest Jiao Tong University
Chengdu 610031, China
caofeng19840301@163.com

Yang Xu

School of Mathematics, Southwest Jiao Tong University, Chengdu 610031, China

Xinran Ning

School of Information Science and Technology, Southwest Jiao Tong University
Chengdu 610031, China

Xuecheng Wang

Department of Computer Science and Technology, Tang Shan University
Tang Shan 063000, China

The contradiction separation rule is a kind of multi-dynamic, flexible reasoning method, it can be used to perform "elimination literals" well, thus generates clauses with few literals. This paper proposes deductive control strategies based on contradiction separation rule, it includes numbers of deductive clauses control strategy, more generated clauses strategy in a contradiction separating deduction, reuse deductive clauses strategy, generated clause control strategy, etc. With the CADE 2017 international competition problems as test object, the first-order logic theorem prover based on contradiction separation rule and this deductive control strategies solves numbers of problems exceeding some famous prover, such as Princess, Zipperpin, Prover9, Beagle etc., and time efficiency is similar to Prover9.

Keywords: Contradiction separation; deductive control; time efficiency.

1. Introduction

Since the resolution principle [1] was proposed in 1965, it has become the core method of first-order logic theorem proving. The state-of-the-art first-order logic theorem provers almost adopt this resolution method or its refinements as the inference method such as binary resolution, UR resolution [2], equality resolution and hyper-resolution [3], but the essence of these resolution methods are binary resolution. Binary resolution is easy to implement, but the resolvent of a clause with n literals and another clause with m literals has n+m-2 literals, making the generated clause is too long and limiting the efficiency of deduction [4], so first-order logic theorem prover improves the deductive efficiency through a large number of strategies [5,9].

Contradiction separating deduction [10] which called S-CS rule for first-order logic is a multi-ary, dynamic method. It has important characteristics of relevance with more clauses, so as to eliminate more literals, and easily generates clauses with less numbers of literals or unit clauses, effectively improve the shortcomings of too long generated clause by binary resolution. Contradiction separating deduction is a general method, it is very flexible. How to control numbers of clauses involved in the deduction, how to make good use of the process of separating contradiction, how to reuse deductive clauses, and how to control the length of generated clause, these are very important for designing a first-order logical theorem prover.

This paper focuses on the deductive control strategies based on contradiction separation rule and proposes the definition of middle-generated clause and related strategies. Section 2 introduces the deductive control strategies; Section 3 outlines the performance of first-order logic theorem prover based on this deductive control strategies. The paper is concluded in Section 4.

2. Deductive Control Strategies

The deduction based on contradiction separation rule is a multi-dynamic deduction method. The number of clauses participating in deduction is controllable. In order to effectively describe clauses, literals and clause set, we describe the main properties as follows:

(1) Clause complexity

Clause complexity is used to describe the statistics of terms in each clause. We describe each clause complexity in three ways. When the weight of variable term is 1, the weight of non-variable item (function item and constant item) is 1, and the complexity of a clause (is called clause complexity 1) is obtained by counting the weight of each term. When the weight of variable term is 1, the weight of non-variable item is 2, and the complexity of a clause (is called clause complexity 2) is obtained by counting the weight of each term. When the weight of variable term is 2, the weight of non-variable item is 1, and the complexity of a clause (is called clause complexity 3) is obtained by counting the weight of each term.

Clause complexity in the original clause set can be used to guide the deduction based on the contradiction separation rule, and as a basis for setting the complexity threshold for generated clauses.

(2) The largest numbers of variable terms in the clause

Count largest numbers of variable terms in each clause, when different literals in the same clause contains the same variable term, the variable term is marked as shared variable term. When a clause doesn't contain variable term, mark the clause as a ground clause.

(3) Literal complexity

Because different clauses contain different numbers of literals, in order to describe complexity of clauses more accurately, we also use three ways to describe complexity of literals like clause complexity.

2.1. Numbers of Deductive Clauses Control Strategy

Numbers of deductive clauses is called deductive depth based on the contradiction separating deduction, and the largest numbers of deductive clauses is controlled by setting threshold. This threshold reflects the ability of playing synergistic deduction of clauses. The larger the threshold, the more able to play the synergy of clauses. When the largest numbers of clauses participating in the deduction exceeds the threshold, the contradiction separating deduction will be stopped, and the final generated clause is obtained.

The threshold of this strategy can be adopted by default value, that is, the largest numbers of clauses participating in deduction is not limited. However, this strategy is influenced by the generated clause control strategy. When related threshold setting by generated clause control strategy is exceeded, the deduction will also be stopped.

2.2. More Generated Clauses Strategy

Because the process of contradiction separating deduction is choosing clause continually to separate contradiction, each process of deduction is also contradiction separation rule. Therefore, new clauses can be generated by each process of deduction.

Definition 2.1 New clause generated by intermediate process of deduction based on the contradiction separating deduction is called middle-generated clause.

Example 2.1 Let $C_1 = l_1(a) \vee l_2(b)$, $C_2 = \sim l_2(x) \vee l_3(y)$, $C_3 = \sim l_2(b) \vee \sim l_3(c) \vee l_4(z)$, $C_4 = \sim l_2(b) \vee \sim l_3(c) \vee \sim l_4(d) \vee \sim l_5(a)$. Following Tables 1–3 below, then it follows from S-CS rule that the generated clause $C_4^s(C_1, C_2, C_3, C_4) = l_1(a) \vee \sim l_5(a)$. We can generate the middle-generated clauses $C_2^s(C_1, C_2) = l_1(a) \vee l_3(y)$, $C_3^s(C_1, C_2, C_3) = l_1(a) \vee l_4(z)$.

Table 1. S-CS rule for C1, C2, C3, C4.

	C_4	C_3	C_2	C_1
C_y^+		$\sim l_5(a)$		$l_1(a)$
C_y^-	$\sim l_2(b) \vee \sim l_3(c) \vee \sim l_4(d)$	$\sim l_2(b) \vee \sim l_3(c) \vee l_4(d/z)$	$\sim l_2(b/x) \vee l_3(c/y)$	$l_2(b)$

Table 2. Generate middle-generated clauses $C_2^s(C_1, C_2)$.

	C_2	C_1
C_y^+	$l_3(y)$	$l_1(a)$
C_y^-	$\sim l_2(b/x)$	$l_2(b)$

Table 3. Generate middle-generated clauses $C_3^s(C_1, C_2, C_3)$.

	C_3	C_2	C_1
C_y^+	$l_4(z)$		$l_1(a)$
C_y^-	$\sim l_2(b) \vee \sim l_3(c)$	$\sim l_2(b/x) \vee l_3(c/y)$	$l_2(b)$

Lemma 2.1 Contradiction separating deduction involved more than two clauses can generate middle-generated clause.

Whether to allow the generation of middle-generated clauses in process of deduction is decided by more generated clause strategy. If the deduction is allowed to generate middle-generated clauses, in order to avoid generating too many new clauses, the conditions for generating middle-generated clauses are set by this strategy. For example, we can limit numbers of literals, only generate middle-generated clauses which are unit clauses or numbers of literals doesn't exceed 2; we can also decide whether to generate middle-generated clauses according to clause complexity.

2.3. *Reuse Deductive Clauses Strategy in a Contradiction Separation Deduction*

Contradiction separation rule is very flexible, it allows to reuse the deductive clauses, make full use of the unified substitution of variable term in a clause, search for more deductive path.

Example 2.2 Let $C_1 = l_1(x)$, $C_2 = l_2(b)$, $C_3 = \sim l_1(a) \vee \sim l_1(b) \vee \sim l_2(b)$. Following Table 4 below, then it follows from S-CS rule that the generated clause $C_3^s(C_1, C_2, C_3) = \sim l_1(b)$.

Table 4. S-CS rule for C1, C2, C3.

	C_3	C_2	C_1
C_y^+	$\sim l_1(b)$		
C_y^-	$\sim l_1(a) \vee \sim l_2(b)$	$l_2(b)$	$l_1(a/x)$

We reuse clause C1, the empty clause is generated by a standard contradiction separating deduction. Following Table 5 below, then it follows from S-CS rule that the generated clause $C_4^S(C_1, C_1, C_2, C_3) = \varnothing$ ($\varnothing$ is empty clause).

Table 5 S-CS rule for C1, C1, C2, C3.

	C_4	C_3	C_1	C_1
C_y^+				
C_y^-	$\sim l_1(a) \vee \sim l_1(b) \vee \sim l_2(b)$	$l_2(b)$	$l_1(b/x)$	$l_1(a/x)$

Reuse deductive clauses strategy is as follows:

Set threshold for reusing the deductive clauses. If deductive clause is repeated over the threshold, the clause is no longer reused in this deduction, but reusing this clause still counts from 0 until the threshold in next deduction.

In general, we reuse deductive clauses with variable term. The purpose of reusing clauses is to repeatedly use the variable term of the clauses to substitute different terms during a contradiction separating deduction. In process of deduction, if a clause has variable term substitution, the clause is allowed to be reused.

2.4. *Generated Clause Control Strategy*

Generated clause control strategy mainly include numbers of literals control, size of function term layer control and generated clause complexity control.

Numbers of literals control strategy: set numbers of literals threshold by strategy, when numbers of literals in generated clause is beyond the threshold, the deduction will be stopped, get the final generated clause. Due to different clause sets, the original clause contains a large difference maximal numbers of literals. In order to avoid the original clause with large numbers of literals can't effectively participate in deduction because of too small threshold, the threshold is automatically adjusted to ensure that the original clause containing a large number of literals can be effectively deduced by comparing largest numbers of literals in clause set.

Size of function term layer control strategy: set the size of function term layer threshold by strategy. When the size of function term layer in generated clause exceeds the threshold, the deduction will be stopped and get the final generated clause. In order to avoid that the original clauses with larger size of function term layer can't effectively participate in the deduction because of too small threshold, the threshold is also automatically adjusted by the same way as above.

Generated clause complexity control strategy: In order to avoid generating too much complexity of generated clause due to unified substitution of variable term in the first-order logic formula, set complexity threshold of generated clause complexity by strategy. Once exceeds threshold, the deduction will be stopped and get the final generated clause.

3. Performance of the Prover Base on Deductive Control Strategies

The 2017 CADE competition theorems was tested (total 500) by the first-order logic theorem prover base on this deductive control strategies on a computer by time limit of 300s (standard time), and the test hardware environment is 3.4 GHz Inter (R) Core i3-3240 processing, 4GB memory, operating system using Ubuntu15.04 64-bit.

3.1. Solved Numbers

Following Table 6 below, MCE-SCS 0.1 (our prover, based on deductive control strategies) solves numbers of problems is 187, outperformed the well-known prover such as Princess, Zipperpin, Prover9 and Beagle.

Table 6. Solved numbers by our prover compared with other famous provers.

MCE-SCS 0.1	Princess	Infinox	Fampire	Darwin	Zipperpin
187	184	182	180	176	157
Prover9	Metis	Equinox	Beagle	Scavenger	Otter
140	136	130	128	71	10

3.2. Solved Problems Compared with Prover9

Following Table 7 below, the 187 problems that our prover solves, Prover9 solves 72. In the solved problems according to the difficulty rating, it shows that our prover has the ability to solve more difficult problems. Our prover solves a problem with rating coefficient more than 0.9, and all competitive provers can't solve it.

Table 7. Solved problems compared with Prover9.

Prover\Rating	0.2≤R <0.3	0.3≤R <0.4	0.4≤R <0.5	0.5≤R <0.6	0.6≤R <0.7	0.7≤R <0.8	0.8≤R <0.9	0.9≤R <1
MCE-SCS 0.1	120	21	6	27	4	7	1	1
Prover9	44	11	3	13	0	1	0	0

3.3. *Time Efficiency Compared with Prover9*

Figure 1 shows the time spent on solving the 2017 CADE competition theorems between MCE-SCS 0.1 and Prover9. As we can see from the figure, numbers of solved problems by MCE-SCS 0.1 is equivalent to that solved by Prover9 in 60 seconds; after 60 seconds, numbers of solved problems by MCE-SCS 0.1 is gradually more than Prover9. To solve the same numbers of problems (140), the average time of MCE-SCS 0.1 is 26.38 seconds, and the average time of Prover9 is 26 seconds.

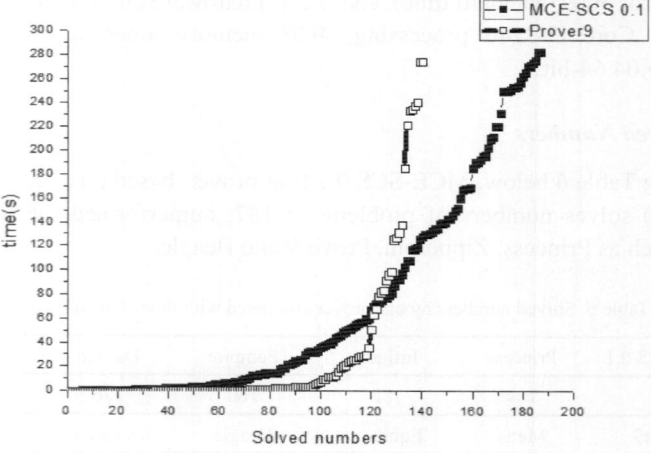

Figure 1. Time efficiency of MCE-SCS 0.1 compared with Prover9.

4. Conclusions

Contradiction separating deduction is a multi-ary, dynamic method, this paper mainly introduces deductive control strategies based on contradiction separation rule. The first-order logic theorem prover based on this strategies solves the 2017 CADE competition problems more than some well-known provers and has a certain time efficiency. Primary version of MCE-SCS has a certain capacity, and further refinement of relevant deductive control strategies is an important way to optimize the next version.

Acknowledgments

This paper is supported by the National Natural Science Foundation of China (Grant No. 61673320), the Fundamental Research Funds for the Central Universities (Grant No. 2682017ZT12); National College Students Innovation and Entrepreneurship Training Program (Grant No. 201510638047).

References

1. J.A. Robinson, A machine oriented logic based on the resolution principle, J. ACM, 12, No. 1, 1965, pp. 23-41.
2. J. McCharen, R. Overbeek, and L. Wos. Complexity and related enhancements for automated theorem-proving programs. Computers and Mathematics with Applications, 2:1–16, 1976.
3. J.A. Robinson, Automatic deduction with hyper-resolution. International Journal of Computer Mathematics, 1(1965): 227-234.
4. Slaney J., Paleo B. W. Conflict Resolution: A First-Order Resolution Calculus with Decision Literals and Conflict-Driven Clause Learning. Journal of Automated Reasoning, 2016:1-24.
5. Reger G., Tishkovsky D., Voronkov A. Cooperating Proof Attempts. In: Felty A., Middeldorp A. (eds.) CADE 2015.LNCS, vol. 9195, pp. 339-355. Springer, Heidelberg (2015).
6. Schulz S., Möhrmann M. Performance of Clause Selection Heuristics for Saturation-Based Theorem Proving. In: Olivetti N., Tiwari A. (eds.) IJCAR 2016. LNCS, vol. 9706, pp. 330-345. Springer, Heidelberg (2016).
7. Hoder K., Reger G., Suda M., Voronkov A. Selecting the Selection. In: Olivetti N., Tiwari A. (eds.) IJCAR 2016. LNCS, vol. 9706, pp. 313-329. Springer, Heidelberg (2016).
8. Schulz, S.: Simple and efficient clause subsumption with feature vector indexing. In: Bonacina, M.P., Stickel, M.E. (eds.) Automated Reasoning and Mathematics. LNCS, vol. 7788, pp. 45–67. Springer, Heidelberg (2013).
9. Khasidashvili Z., Korovin K. Predicate Elimination for Preprocessing in First-Order Theorem Proving. In: Creignou N., Le Berre D. (eds.) SAT 2016. LNCS, vol. 9710, pp. 361-372. Springer, Heidelberg (2016).
10. Y. Xu, J. Liu, S.W. Chen, and X.M. Zhong, A novel generalization of resolution principle for automated deduction, The 12th International FLINS Conference on Uncertainty Modelling in Knowledge Engineering and Decision Making (FLINS2016), August 24-26, 2016, ENSAIT, Roubaix, France.

Preliminary framework to combine contradiction separation based automated deduction with superposition and given-clause algorithm[*]

Xinran Ning,[1] Yang Xu[1] and Feng Cao[1]

National-Local Joint Engineering Laboratory of System Credibility Automatic Verification, Southwest Jiaotong University
Chengdu, Sichuan, China

Jun Liu[2]

School of Computing, Ulster University
Northern Ireland, UK

Given-clause algorithm is a saturation-based algorithm applied by many well-known first-order theorem provers, where superposition calculus based on binary resolution is also applied mostly. A new first-order automated deduction method, called *a contradiction separation based dynamic multi-clause synergized automated deduction*, in short CS-DMS deduction, has been developed recently. In the paper, binary resolution in superposition is proposed to be replaced by the contradiction separation and the rest parts of superposition are remained the same, and the new calculus is proposed to be achieved under the framework of given-clause algorithm, this forms a preliminary step of combination.

1. Introduction

A contradiction separation based dynamic multi-clause synergized automated deduction (CS-DMS) has been developed recently [2, 7], which is an extension and a variant of binary resolution. The method breakthroughs the constraint in binary resolution that there should be and only be two clauses involved at each deduction. The contradiction concept does not imply only one complementary pair among the clause set. Here "synergized" means cooperative interaction among multiple clauses that creates a combined effect; "dynamic" means that the number of clauses involved in the contradiction separation in each deduction can be varied in the whole deduction process, so it is regarded as a dynamic deduction process. Now that the important roles of superposition and given-clause algorithm in many state of art first-order theorem provers, this paper aims at proposing some

---

[*]This work is supported by the National Natural Science Foundation of P. R. China (Grant No. 61673320) and the Fundamental Research Funds for the Central Universities of China (Grant No. 2682017ZT12, 2682016CX119).

preliminary ideas and framework in order to combine the contradiction separation method with superposition and then integrate it into the given-clause algorithm.

2. Preliminaries

2.1. *Brief of contradiction separation based dynamic multi-clause synergized (CS-DMS) automated deduction*

This section gives a brief overview of the CS-DMS automated deduction [2, 7].

Definition 2.1 Suppose a clause set $S=\{C_1, C_2,\ldots, C_m\}$ in first-order logic. A new clause is called a *standard contradiction separation clause* (S-CSC) of $C_1, C_2,\ldots, C_m$, denoted as $\Re_m^s(C_1, C_2, \ldots, C_m)$ (here "s" means "standard"), if the following conditions hold:

1) There does not exist the same variables among $C_1, C_2,\ldots, C_m$ (if there exists the same variables, there will be a rename substitution which makes them different);

2) For any C_i, $i=1,2,\ldots,m$, a substitution σ_i can be applied to C_i (σ_i could be an empty substitution) and the same literals merged after substitution, denoted as $C_i^{\sigma_i}$; in addition, $C_i^{\sigma_i}$ can be partitioned into two sub-clauses $(C_i^{\sigma_i})^+$ and $(C_i^{\sigma_i})^-$ such that

- $C_i^{\sigma_i} = (C_i^{\sigma_i})^+ \vee (C_i^{\sigma_i})^-$, where $(C_i^{\sigma_i})^+$ and $(C_i^{\sigma_i})^-$ do not share the same literal, $C_i^{\sigma_i^+}$ can be an empty clause, $C_i^{\sigma_i^-}$ cannot be an empty clause; moreover,

- For any $(x_1, \ldots, x_m) \in \prod_{i=1}^m C_i^{\sigma_i^-}$, there exists some complementary pair of literals among $x_1,\ldots,x_m, \wedge_{i=1}^m C_i^{\sigma_i^-}$ or S is called a *standard contradiction*;

- $\vee_{i=1}^m (C_i^{\sigma_i})^+ = \Re_m^s(C_1, C_2, \ldots, C_m)$.

The inference rule that produces a new clause $\Re_m^s(C_1, C_2, \ldots, C_m)$ is called a *standard contradiction separation rule* in first-order logic, in short, a S-CS rule.

Definition 2.2 Suppose a clause set $S=\{C_1, C_2,\ldots, C_m\}$ in first-order logic. $\Phi_1, \Phi_2, \ldots, \Phi_t$ is called a *standard contradiction separation based dynamic deduction sequence* (or a *S-CS based dynamic deduction sequence*) from S to a clause Φ_t, denoted as $\mathcal{D}_s$, if

1) $\Phi_t \in S, i = 1,2,\ldots,t$;
2) there exist $r_1, r_2, \ldots, r_{k_i} < i$, $\Phi_i = \Re_m^s(\Phi_{r_1}, \Phi_{r_2}, \ldots, \Phi_{r_{k_i}})$.

When Φ_t is an empty clause, the deduction is a refutation or a proof of S.

2.2. Brief of superposition calculus

The superposition calculus as implemented in modern theorem provers usually derives from the work of Bachmair and Ganzinger [1]. The superposition calculus is a calculus for reasoning in equational first-order logic. It was developed in the early 1990s and combines concepts from first-order resolution with ordering-based equality handling as developed in the context of (unfailing) Knuth–Bendix completion. It can be seen as a generalization of binary resolution. As most first-order calculi, superposition tries to show the unsatisfiability of a set of first-order clauses, i.e. it performs proofs by refutation.

Resolution
$$\frac{A \vee C_1 \quad \neg A' \vee C_2}{(C_1 \vee C_2)\theta}$$

Factoring
$$\frac{A \vee A' \vee C}{(A \vee C)\theta}$$

where, for both inferences, $\theta = mgu(A, A')$ and A is not an equality literal

Superposition
$$\frac{l=r \vee C_1 \quad L[s]_p \vee C_2}{(L[r]_p \vee C_1 \vee C_2)\theta} \quad \text{or} \quad \frac{l=r \vee C_1 \quad t[s]_p \otimes t' \vee C_2}{(t[r]_p \otimes t' \vee C_1 \vee C_2)\theta}$$

where $\theta = mgu(l, s)$ and $r\theta < l\theta$ and, for the left rule $L(s)$ is not an equality literal, and for the right rule $\otimes$ stands either for $=$ or $\neq$ and $t'\theta = t[s]\theta$ (the notation $<$ means ordering)

EqualityResolution
$$\frac{s \neq t \vee C}{C\theta}$$

EqualityFactoring
$$\frac{s=t \vee s'=t' \vee C}{(t \neq t' \vee s'=t' \vee C)\theta}$$

where $\theta = mgu(s, s')$, $t\theta < s\theta$ and $t'\theta < s'\theta$

Fig. 1. The rules of the superposition and resolution calculus.

2.3. Brief of given-clause algorithm

Given-clause algorithm is a saturation-based algorithm applied by many well-known theorem provers, including the most successful theorem provers Vampire, E and Spass [4]. In the given-clause algorithm, there are two clause sets, one called *passive* and the other one called *active*. Initially, all the input clauses will be put into the set *passive* while the set *active* is empty. Then it will conduct a loop repeatedly: (1) A clause called *current* will be selected form the set *passive* according to a certain heuristic strategy and put into the set *active*; (2) all the inference rules will be implemented between the clause *current* and clauses in the set *active*; (3) putting all the new produced clauses into the set passive (the simplification is ignored) [5]. Fig. 2 below provides a step-by-step data flow of the given clause algorithm.

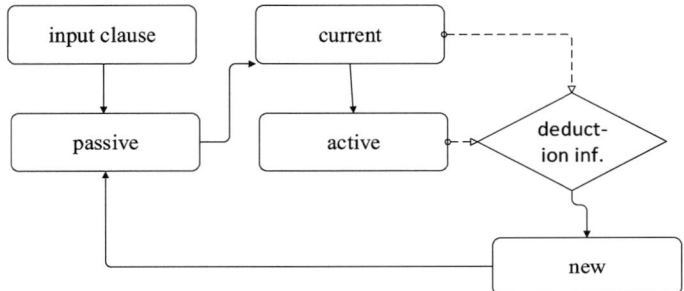

Fig. 2. Dataflow in the given-clause algorithm.

3. Combining the Contradiction Separation with Superposition and Incorporating Them into the Given-Clause Algorithm

The combination scheme of CS-MDS, superposition and the given clause algorithm is summarized as follow:

Step 1: we substitute binary resolution in superposition calculus by the standard contradiction separation method. Compared with the superposition calculus, the combined scheme can overcome the constraint every time there can be and only be two clauses in a deduction.

Standard contradiction separation rule (S-CS)
$$\frac{C_1, C_2, \ldots, C_m}{\Re_m^s(C_1, C_2, \ldots, C_m)}$$

Factoring
$$\frac{A \vee A' \vee C}{(A \vee C)\theta}$$

Where, for both inferences, $\theta = mgu(A, A')$ and A is not an equality literal

Superposition
$$\frac{l=r \vee C_1 \quad L[s]_p \vee C_2}{(L[r]_p \vee C_1 \vee C_2)\theta} \quad \text{or} \quad \frac{l=r \vee C_1 \quad t[s]_p \otimes t' \vee C_2}{(t[r]_p \otimes t' \vee C_1 \vee C_2)\theta}$$

Where $\theta = mgu(l, s)$ and $r\theta < l\theta$ and, for the left rule $L(s)$ is not an equality literal, and for the right rule $\otimes$ stands either for $=$ or $\neq$ and $t'\theta = t[s]\theta$ (the notation $<$ means ordering)

EqualityResolution
$$\frac{s \neq t \vee C}{C\theta}$$

EqualityFactoring
$$\frac{s=t \vee s'=t' \vee C}{(t \neq t' \vee s' = t' \vee C)\theta}$$

where $\theta = mgu(s, s')$, $t\theta < s\theta$ and $t'\theta < s'\theta$

Fig. 3. Combination scheme of the contradiction separation method and the superposition calculus.

Step 2: if we achieve the substitution in the given-clause algorithm, the operation of contradiction separation method in the innovative given-clause algorithm is similar with the operation of binary resolution, except that the standard contradiction separation needs multiple clauses to be involved in while binary resolution only needs two. Fig. 4 illustrates the all the necessary steps.

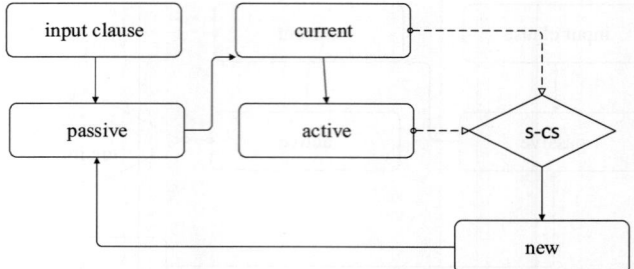

Fig. 4. Dataflow of the standard contradiction separation rule applied in the given-clause algorithm.

4. Specific Implementation of the Standard Contradiction Separation Method in the Given-Clause Algorithm

This section details how the S-CS rule can be implemented inside the given clause algorithm. Suppose each time a clause *current* is selected from the set *passive*, put into the set *active* and involved in making a contradiction separation with other clauses in the set *active*, the implementation process is detailed as follows.

Assuming that the clause *current* is $C' = L \vee L_1 \vee ... \vee L_m$, and $L_1 \vee ... \vee L_m$ can be an empty set. And there are n clauses currently in the set *active* S_a, that is, $C_1, C_2, ..., C_n$. When we make a standard contradiction separation, C' is always the beginning clause of the standard contradiction separation.

Step 1: Initially, let set A be an empty set and then select one literal L of C' to add it into the set A, $A = \{L\}$.

Step 2: Search the clauses which has literals enable to resolve with L in S_a. Let the set of those clauses be $S_{a1} = \{C_{11}, C_{12}, ..., C_{1k_1}\}$, without loss of generality, let C_{1i} be selected as the clause to participate in the contradiction body according to a certain strategy. Assuming that $C_{1i} = \neg L' \vee L_{11} \vee ... \vee L_{1m_{1i}}$, $L_{11} \vee ... \vee L_{1m_{1i}}$ can be empty. Let the most general unifier (mgu) between L and L' be θ_1:

- If $(L_1 \vee ... \vee L_m)\theta_1 \vee (L_{11} \vee ... \vee L_{1m_{1i}})\theta_1 = \varnothing$ ($\varnothing$ means the empty set), the procedure is terminated, the clause set is unsatifiable.
- If $(L_{11} \vee ... \vee L_{1m_{1i}})\theta_1 \neq \varnothing$:
  ➤ If the contradiction body is constructed continually according to a certain strategy, without loss of generality, add $(L_{11})\theta_1$ into the set A, then $(C_{1i}^{\theta_1})^- = (\neg L' \vee L_{11})\theta_1$, $A = \{L, (L_{11})\theta_1\}$. After that, go to Step 3.
  ➤ If the contradiction body is ended according to a certain strategy, then $\Re_2^s(C', C_{1i}) = (L_1 \vee ... \vee L_m)\theta_1 \vee (L_{11} \vee ... \vee L_{1m_{1i}})\theta_1$, $\Re_2^s(C', C_{1i})$ is the new produced clause.

- If $(L_1 \vee ... \vee L_m)\theta_1 \neq \varnothing$ and $(L_{11} \vee ... \vee L_{1m_{1i}})\theta_1 = \varnothing$, the contradiction body is ended, and the new produced clause is $\Re_2^s(C', C_{1i}) = (L_1 \vee ... \vee L_m)\theta_1$.

Step 3: search the clauses which has literals enable to resolve with L_{11} in S_a, let the set of those clauses be $S_{a2} = \{C_{21}, C_{22}, ..., C_{2k_2}\}$, without loss of generality, let C_{2i} be selected as the clause to participate in contradiction body according to a certain strategy. Assuming that $C_{2i} = \neg L'_{11} \vee L_{21} \vee ... \vee L_{2m_{2i}}$, $L_{21} \vee ... \vee L_{2m_{2i}}$ can be empty. Let the most general unifier (mgu) between $(L_{11})\theta_1$ and L'_{11} be θ_2. Then check whether there exist literals in $C_{2i}\theta_2$ which are complementary with literals in the set $A - \{(L_{11})\theta_1\}$. Let S'_{2i} be the set of complementary literals (S'_{2i} can be empty), at the mean time, $S'_{2i} \subseteq S_{2i}^{\theta_2^-}$.

- If $\left(\bigvee_{x \in (C_{2i} - S'_{2i} - (\neg L'_{11}))\theta_2} x\right) \vee ((L_1 \vee ... \vee L_m)\theta_1 \vee (L_{11} \vee ... \vee L_{1m_{1i}})\theta_1)\theta_2 = \Phi$, the process is terminated and the clause set is unsatisfiable.
- If $\bigvee_{x \in (C_{2i} - S'_{2i} - (\neg L'_{11}))\theta_2} x \neq \varnothing$:
  ➢ If the contradiction body is constructed according to a certain strategy, without loss of generality, add $(L_{21})\theta_2$ into the set A, then $(C_{2i}^{\theta_2})^- = (\neg L'_{11})\theta_2 \vee ((L_{21})\theta_2 \vee (\bigvee_{x \in S'_{2i}} x)$, and $(C_{2i}^{\theta_2})^+ = \left(\bigvee_{x \in ((C_{2i})\theta_2 - S'_{2i} - (\neg L'_{11})\theta_2 - (L_{21})\theta_2)} x\right)$, $A = \{L, (L_{11})\theta_1, (L_{21})\theta_2\}$. After that, transfer to Step 4.
  ➢ If the contradiction body is ended according to a certain strategy, then $\Re_3^s(C', C_{1i}, C_{2i}) = \left(\bigvee_{x \in ((C_{2i})\theta_2 - S'_{2i} - (\neg L'_{11})\theta_2} x\right) \vee ((L_1 \vee ... \vee L_m)\theta_1 \vee (L_{11} \vee ... \vee L_{1m_{1i}})\theta_1)\theta_2$ is the new produced clause.
- If $\bigvee_{x \in (C_{2i} - S'_{2i} - (\neg L'_{11}))\theta_2} x = \Phi$ and $((L_1 \vee ... \vee L_m)\theta_1 \vee (L_{11} \vee ... \vee L_{1m_{1i}})\theta_1)\theta_2 \neq \Phi$, the contradiction body is ended. And the new produced clause is $\Re_3^s(C', C_{1i}, C_{2i}) = ((L_1 \vee ... \vee L_m)\theta_1 \vee (L_{11} \vee ... \vee L_{1m_{1i}})\theta_1)\theta_2$.

……

Step h+1. When the step is $h + 1$, search the clauses which has literals enable to resolve with $L_{(h-1)1}$ in S_a, let the set of those clauses be $S_{ah} = \{C_{h1}, C_{h2}, ..., C_{hk_h}\}$, without loss of generality, let C_{hi} be selected as the clause to participate in contradiction body according to a certain strategy. Assuming that $C_{hi} = \neg L'_{(h-1)1} \vee L_{h1} \vee ... \vee L_{hm_{hi}}$, $L_{h1} \vee ... \vee L_{hm_{hi}}$ can be empty. Let the most general unifier (mgu) between $(L_{(h-1)1})\theta_{h-1}$ and $L'_{(h-1)1}$ be θ_h. Then check whether there exist literals in $C_{hi}\theta_h$ which are complementary with literals in the set $A - \{L_{(h-1)1})\theta_{h-1}\}$. Let S'_{hi} be the set of complementary literals (S'_{hi} can be empty), at the meantime, $S'_{hi} \subseteq (C_{hi}^{\theta_h})^-$.

- If the disjunction of literals $\left(\vee_{x\in((C_{hi})\theta_h - S'_{hi} - (\neg L'_{(h-1)1})\theta_h} x\right) \vee$ $\left(\left(\vee_{x\in((C_{(h-1)i})\theta_{h-1} - S'_{(h-1)i} - (\neg L'_{(h-2)1})\theta_{h-1} - (L_{(h-1)1})\theta_{h-1}} x\right) \vee \ldots \vee (L_{11} \vee \ldots \vee L_{1m_{1i}})\theta_1)\theta_2)\cdots)\theta_h\right) = \varnothing$, the procedure is terminated. The clause set is unsatisfiable.

- If $\vee_{x\in((C_{hi})\theta_h - S'_{hi} - (\neg L'_{(h-1)1})\theta_h} x \neq \varnothing$,

  ➢ If the contradiction body is constructed continually according to a certain strategy, without loss of generality, add $L_{h1})\theta_h$ into the set A, then $S_{hi}^{\theta_h^-} = (\neg L'_{(h-1)1})\theta_h \vee ((L_{h1})\theta_h \vee (\vee_{x\in S'_{hi}} x)$, $(C_{hi}^{\theta_h})^+ = \left(\vee_{x\in((C_{hi})\theta_h - S'_{hi} - (\neg L'_{(h-1)1})\theta_h - (L_{h1})\theta_h} x\right)$, $A = \{L, (L_{11})\theta_1, (L_{21})\theta_2, \ldots, (L_{h1})\theta_h,\}$. After that, transfer to Step (h+2).

  ➢ If the contradiction body is ended according to a certain strategy, then

  $\Re_h^s(C', C_{1i}, \ldots, C_{hi}) = \left(\vee_{x\in((C_{hi})\theta_h - S'_{hi} - (\neg L'_{(h-1)1})\theta_h} x\right) \vee$ $\left(\left(\vee_{x\in((C_{(h-1)i})\theta_{h-1} - S'_{(h-1)i} - (\neg L'_{(h-2)1})\theta_{h-1} - (L_{(h-1)1})\theta_{h-1}} x\right) \vee \ldots \vee (L_{11} \vee \ldots \vee L_{1m_{1i}})\theta_1)\theta_2)\cdots)\theta_h\right)$ is the produced clause.

- If $\vee_{x\in((C_{hi})\theta_h - S'_{hi} - (\neg L'_{(h-1)1})\theta_h} x = \varnothing$ and $\left(\left(\vee_{x\in((C_{(h-1)i})\theta_{h-1} - S'_{(h-1)i} - (\neg L'_{(h-2)1})\theta_{h-1} - (L_{(h-1)1})\theta_{h-1}} x\right) \vee \ldots \vee (L_{11} \vee \ldots \vee L_{1m_{1i}})\theta_1)\theta_2)\cdots)\theta_h\right) \neq \varnothing$. The contradiction body is ended. And the new produced clause is $\Re_h^s(C', C_{1i}, \ldots, C_{hi}) = \left(\left(\vee_{x\in((C_{(h-1)i})\theta_{h-1} - S'_{(h-1)i} - (\neg L'_{(h-2)1})\theta_{h-1} - (L_{(h-1)1})\theta_{h-1}} x\right) \vee \ldots \vee (L_{11} \vee \ldots \vee L_{1m_{1i}})\theta_1)\theta_2)\cdots)\theta_h\right)$.

Above is the specific implementation process of the contradiction separation method under the framework of the given-clause algorithm, which is quite different from binary resolution achieved under the framework of the given-clause algorithm. For binary resolution, new clauses are obtained by resolving *current* clause with clauses in *active* set for one time in each loop, then they will be sent to set *passive* may contain thousands of clauses and wait for being chosen next time. However, for the contradiction separation method, new clauses are obtained by making contradiction body between clause *current* and clauses in set *active* in each loop of given-clause algorithm, as a result, deduction reasoning can go further in each loop and if the clause *current* is the right one, we can get empty clause within one loop which is the advantage over the original superposition calculus in the given-clause algorithm.

5. Conclusions

The paper present some preliminary ideas and framework about combining the recently developed contradiction separation based dynamic first-order automated deduction method with the superposition calculus and then being applied in the well-known given-clause algorithm. Since the contradiction separation method did not cover the equality dealing, it is necessary to increase the proving efficiency, capability and applicability by combining it with superposition. What's more, because the contradiction separation method breakthroughs the constraint in the binary resolution that there needs to be and only be two clauses in every deduction, the proposed combination scheme provided a good basis to enhance the binary resolution and flexibility compared with original superposition. There are more research spaces left for the future work about the combination schemes, including further detailed algorithms, heuristic strategies involved, implementation and performance evaluation.

References

1. L. Bachmair, H. Ganzinger, Rewrite-based equational theorem proving with selection and simplification. J. Log. Comput. 4(3)(1994): 217-247.
2. Y. Xu, J. Liu, S.W. Chen, X.M. Zhong, A novel generalization of resolution principle for automated deduction. Proc. Int. Conf. on Uncertainty Modelling in Knowledge Engineering and Decision Making, pp. 483-488 (FLINS 2016)
3. K. Hoder, G. Reger, M. Suda, A. Voronkov, Selecting the Selection. IJCAR 2016, Coimbra, Portugal, 27 June-2 July, pp. 313-329.
4. A. Riazanov, New Implementation Framework for Saturation-Based Reasoning. Artificial Intelligence (cs.AI); Logic in Computer Science (cs.LO), arXiv preprint arXiv:0802.2127 (2008).
5. X.R. Ning, Y. Xu, X.X. He, A novel goal-directed strategy for ATP based on the similarity between clauses. J. of fuzzy mathematics, 25(1) (2017), pp. 247-254.
6. S. Schulz, M. Möhrmann, Performance of clause selection heuristics for saturation-based theorem proving. IJCAR 2016, pp 330-345.
7. Y. Xu, J. Liu, S.W. Chen, X.M. Zhong, and X.X. He, Contradiction Separation Based Dynamic Multi-Clause Synergized Automated Deduction, submittted to Information Science under review.

Chaotic mappings in symbol space

I. Bula*

*Department of Mathematics, University of Latvia,
Institute of Mathematics and Computer Science of the University of Latvia,
Riga, Latvia
*ibula@lanet.lv
www.lu.lv/eng/*

Models with chaotic mappings are not predictable in a long-term, therefore it is important to recognize such mappings. We study mappings in the symbol space (set of one-sided infinite sequences (or infinite words) of symbols). A well known chaotic mapping in the symbol space is a shift mapping. However, exist other chaotic mappings in the symbol space. We define a blinking mapping and show that it is chaotic. The decision maker has to decide on which mapping to make predictions.

Keywords: Symbol space; infinite sequence; shift mapping; chaotic mapping; blinking mapping.

1. Introduction

Information is often represented as a sequence of discrete symbols drawn from a fixed finite set. The technique of characterizing the orbit structure of a dynamical system via infinite sequences of symbols is known as symbolic dynamics.

Symbolic dynamics study dynamical systems on the basis of symbol sequences obtained for a suitable partition of a state space. The basic idea behind symbolic dynamics is to divide the phase space into a finite number of regions and to label each region by an alphabetic symbol. This approach exploits the property that system dynamics reduce to a shift operation in symbol space. The technique requires the knowledge the current symbolic state of the system and selection it future symbols. It is known that a shift mapping in the symbol space is a chaotic mapping (Ref. 10–12, 16). Our purpose is to show that exist other chaotic mappings in the symbol space too.

2. Preliminaries

The terminology comes from the combinatorics on words (for example, Ref. 3, 14, 15).

First of all we give some notations:

$$Z \text{ — set of integers,}$$
$$Z_+ = \{x | x \in Z \,\&\, x > 0\},$$
$$N = Z_+ \cup \{0\}.$$

With A we denote a finite *alphabet*, i.e., a finite nonempty set $\{a_0, a_1, a_2, ..., a_n\}$ whose elements are called *letters*. We assume that A contains at least two symbols. By A^* we denote the set of all finite sequences of letters, or finite *words*, this set contains empty word (or sequence) λ too. Let $A^+ = A^* \setminus \{\lambda\}$. A word $\omega \in A^+$ can be written uniquely as a sequence of letters as $\omega = \omega_1\omega_2\omega_3...\omega_l$, with $\omega_i \in A$, $1 \le i \le l$. The integer l is called the *length* of ω and denoted $|\omega|$. The length of λ is 0. An extension of the concept of a finite word is obtained by considering infinite sequences of symbols over a finite set. *One-sided* (from left to right) infinite sequence or word, or simply *infinite word*, over A is any total map $\omega : N \to A$. The set A^ω contains all infinite words. Let $A^\infty = A^* \cup A^\omega$. If the word $u = u_0u_1u_2... \in A^\omega$, where $u_0, u_1, u_2, ... \in A$, then a finite word $u_0u_1u_2...u_n$ is called the *prefix* of u of a length $n + 1$. The empty word λ is assumed to be a prefix of u of the length 0.

$$Pref(u) = \{\lambda, u_0, u_0u_1, u_0u_1u_2, ..., u_0u_1u_2...u_n, ...\}$$

is a set of all prefixes of the word u.

We introduce a metric d in A^∞ as follows.

Definition 2.1 (Ref. 15). *Let $u, v \in A^\infty$. The mapping $d : A^\infty \times A^\infty \to R$ is called a metric (prefix metric) in the set A^∞ if*

$$d(u,v) = \begin{cases} 2^{-m}, & u \ne v, \\ 0, & u = v, \end{cases}$$

where $m = \max\{|\omega| \,|\, \omega \in Pref(u) \cap Pref(v)\}$.

The term "chaos" in reference to functions was first used in Li and Yorke's paper "Period three implies chaos" (Ref. 13, 1975). We use the following definition of R. Devaney Ref. 8.

Definition 2.2. *Let (X, ρ) be metric space and $A \subset X$. The function $f : A \to A$ is chaotic if*

a) the periodic points of f are dense in X,
b) f is topologically transitive and
c) f exhibits sensitive dependence on initial conditions.

We explain the notions in Definition 2.2.

A point $x \in A$ is *periodic* for f if $f^n(x) = x$ for some $n \geq 1$ and we say that x has period n under f. If x is periodic, the smallest integer n for which $f^n(x) = x$ is a prime period of x. If $f(x) = x$, then x is called a *fixed point* for f.

Let $F \subset A$. We say that the set F is *dense* in A (Ref. 8, 10, 16) if for each point $x \in A$ and each $\varepsilon > 0$, there exists $y \in F$ such that $d(x,y) < \varepsilon$.

We say that the function f is *topologically transitive* on A (Ref. 8, 10, 16) if for any two points x and y in A and any $\varepsilon > 0$, there is $z \in A$ such that $d(z,x) < \varepsilon$ and $d(f^n(z), y) < \varepsilon$ for some n.

We say that the function $f : A \to A$ exhibits *sensitive dependence on initial conditions* (Ref. 8, 10, 16) if there exists $\delta > 0$ such that for any $x \in A$ and any $\varepsilon > 0$, there is $y \in A$ and a natural number n such that $d(x,y) < \varepsilon$ and $d(f^n(x), f^n(y)) > \delta$.

Devaney's definition is not the only classification of a chaotic map. For example, another definition can be found in Ref. 16. Also mappings with only one property — sensitive dependence on initial conditions — frequently are considered as chaotic (see, Ref. 9). Banks, Brooks, Cairns, Davis and Stacey Ref. 2 have demonstrated that for continuous functions, the defining characteristics of chaos are topological transitivity and the density of periodic points.

Theorem 2.1 (Ref. 2). *Let A be an infinite subset of metric space X and $f : A \to A$ to be continuous. If f is topologically transitive on A and the periodic points of f are dense in A, then f is chaotic on A.*

The example of a chaotic mapping in the symbol space A^ω is shift mapping $\sigma : A^\omega \to A^\omega$

$$\forall x = x_0 x_1 x_2 x_3 ... \in A^\omega \quad \sigma(x_0 x_1 x_2 x_3 ...) = x_1 x_2 x_3 x_4 ....$$

A notion of increasing mapping is introduced in Ref. 4. The shift mapping is a special case of increasing mapping in the symbol space A^ω. In Ref. 5 we considered a k-swith mapping but in Ref. 6 we considered a jump mapping. All three mappings change in word $x \in A^\omega$ finite number of symbols. All three mappings are chaotic.

3. Blinking Mapping

We consider other class of mappings that are not increasing mappings, not k-swith and not jump mappings. Our new mapping — a blinking mapping — can be changed infinitely many symbols.

Definition 3.1. The mapping $\beta : A^\omega \to A^\omega$ is called a blinking mapping if $\forall s = s_0 s_1 s_2 s_3 ... s_k s_{k+1} ... \in A^\omega : \beta(s) = \overline{s_1 s_2 s_3} ... \overline{s_k s_{k+1} s_{k+2}} ...$, where $\overline{s_i}$, $i = 1, 2, ...$, is the same symbol (letter) as s_i or by bijection rule symbol s_i change to other symbol to alphabet A.

In other words, at first, blinking mapping is a shift and secondly, this mapping changes all symbols by a defined rule. If this rule is such that $\overline{s_i} = s_i$ for all $i = 1, 2, ...$, then it is a classical shift mapping. But if there is at least one symbol $a_i \in \{a_0, ..., a_n\}$ that changes to other $a_j \in \{a_0, ..., a_n\}$, $i \neq j$, and then a_j changes to a_i, then we have different mapping as a shift.

Generally, if we consider alphabet A that contains at least three symbols $A = \{a_0, a_1, a_2, ..., a_n\}$, then a rule which a_i changes to a_j and for every $a_j \in A$ exists only one $a_i \in A$ with this rule.

For example, let $A = \{0, 1\}$, $0 \to 1$ and $1 \to 0$. If $s = 11111....$, then $\beta(s) = 0000....$, or if $s = 11001011....$, then $\beta(s) = 0110100....$.

For example, if $A = \{a, b, c\}$ and $a \to b$, $b \to c$ and $c \to a$, then $\beta(aaaaaaaaaa...) = bbbbbbbbb....$ or $\beta(ababababab...) = cbcbcbcbc....$. Or another rule: $a \to a$, $b \to c$ and $c \to b$, then $\beta(aaaaaaaaaa...) = aaaaaaaaa....$ or $\beta(ababacabab...) = cacabacac....$.

We show three important properties of the blinking mapping.

Theorem 3.1. *The blinking mapping $\beta : A^\omega \to A^\omega$ is continuous in the set A^ω.*

Proof. We fix word $u \in A^\omega$ and $\varepsilon > 0$. We need to prove that there is $\delta > 0$ such that whenever $d(u, v) < \delta$, $v \in A^\omega$, then $d(\beta(u), \beta(v)) < \varepsilon$.

We choose m such that $2^{-m} < \varepsilon$. Let $0 < \delta < 2^{-(m+1)}$. If $d(u, v) < \delta$, then by definition of a prefix metric follows that $u_i = v_i$ for all $i = 0, 1, ..., m$. From definition of blinking mapping $\overline{u_i} = \overline{v_i}$, $i = 1, 2, ..., m$, therefore $d(\beta(u), \beta(v)) \leq 2^{-m} < \varepsilon$. □

A^ω is a infinite metric space and blinking mapping is continuous therefore by Theorem 2.1 for the chaotic mapping it is sufficient to prove that this mapping is topologically transitive on A^ω and the periodic points of mapping are dense in A^ω.

Theorem 3.2. *The blinking mapping* $\beta : A^\omega \to A^\omega$ *is topologically transitive in the set* A^ω.

Proof. We fix words $u, v \in A^\omega$ and $\varepsilon > 0$. We need to prove that there is word $z \in A^\omega$ such that $d(z, u) < \varepsilon$ and that there exists iteration k such that $d(\beta^k(z), v) < \varepsilon$.

We choose m such that $2^{-m} < \varepsilon$. Then there exist a word $z \in A^\omega$ such that $z_i = u_i$, $i = 0, 1, ..., m-1$, and therefore $d(z, u) \leq 2^{-m} < \varepsilon$.

We choose $k = m$. By m iterations

$$\beta^m(z_0 z_1 ... z_{m-1} z_m z_{m+1} ...) = \overline{z_m z_{m+1}} ... \overline{z_{2m-1} z_{2m}} ...,$$

We need to find equivalent word such that $\overline{z_m z_{m+1} ... z_{2m-1}} = v_0 v_1 v_2 ... v_{m-1}$. Since changing rule is bijective we can m times backward iterate symbols $v_0, v_1, v_2, ..., v_{m-1}$ and find the corresponding symbols $z_m, z_{m+1}, ..., z_{2m-1}$.

We demonstrate this assertion with example. Let $A = \{a, b, c\}$ and $a \to b$, $b \to c$ and $c \to a$. Let $m = 4$, $u = abbcaacbabbb...$ and $v = bbccaaabac...$. This means that the first 4 symbols in the word z are the same as in word u: $z = abbcz_4 z_5 z_6 z_7...$. Symbols z_4, z_5, z_6, z_7 we have to find from inequality $d(\beta^4(z), v) < 2^{-4}$. This means that word $v_0 v_1 v_2 v_3$ is equal with first 4 symbols in $\beta^4(z)$. We iterate 4 times backward $v_0 v_1 v_2 v_3 = bbcc$: $b \leftarrow a \leftarrow c \leftarrow b \leftarrow a$ and $c \leftarrow b \leftarrow a \leftarrow c \leftarrow b$. Hence $z = abbcaabbz_8 z_9...$, the symbols z_8, z_9 and others are any from A.

We proved that there exists z such that $d(z, u) < \varepsilon$ and $d(\beta^k(z), v) < \varepsilon$. $\square$

Theorem 3.3. *The set of periodic points of blinking mapping* $\beta : A^\omega \to A^\omega$ *is dense set in the set* A^ω.

Proof. Let $\varepsilon > 0$. Then there exists m such that $2^{-m} < \varepsilon$. We assume that $u \in A^\omega$ and $v = u_0 u_1 ... u_{m-1}$ is a prefix of word u of length m. Then there exists point (infinite word) $x \in A^\omega$ with the same prefix as v and this x is periodic point for a blinking mapping with period m. In this case $d(u, x) \leq 2^{-m} < \varepsilon$.

We construct the searched word x in the following way. Let $x = u_0 u_1 u_2 ... u_{m-1} x_m x_{m+1} x_{m+2} ...$. The basic idea comes from iterations and fact that changing rule is bijective. Since x is a periodic point with period m then

$$\beta^m(x) = \overline{x_m x_{m+1} x_{m+2} ...} = x = u_0 u_1 u_2 ... u_{m-1} x_m x_{m+1} x_{m+2} ....$$

We need to find symbols $x_m, x_{m+1}, x_{m+2}, ..., x_{2m-1}$ such that in m-th iteration by mapping β are equal with $u_0, u_1, ..., u_{m-1}$.

The changing rule is bijective therefore we can m times backward iterate symbols $u_0, u_1, \ldots, u_{m-1}$ and find corresponding symbols $x_m, x_{m+1}, x_{m+2}, \ldots, x_{2m-1}$. Then in a similar way we find next symbols $x_{2m}, x_{2m+1}, \ldots, x_{3m-1}$. Since the alphabet is finite such searching procedure is finite — in one moment we find the same symbols as the first $u_0, u_1, \ldots, u_{m-1}$. This means that word x consists for infinity many repeated finite sequences of symbols that were found previously considered backward iterations. □

We demonstrate a periodic point construction algorithm of Theorem 3.3 with an example. Let $A = \{a, b, c\}$ and $a \to b$, $b \to c$ and $c \to a$. We assume that first 4 symbols in a word x is $abbc$. We find next symbols such that word x is periodic point of β with period 4. We iterate backward 4 times symbols a, b, c: $a \leftarrow c \leftarrow b \leftarrow a \leftarrow c$, $b \leftarrow a \leftarrow c \leftarrow b \leftarrow a$ and $c \leftarrow b \leftarrow a \leftarrow c \leftarrow b$, therefore $x_4 x_5 x_6 x_7 = caab$. From backward iteration of c, a, a, b we obtain $x_8 x_9 x_{10} x_{11} = bcca$ but from backward iteration of b, c, c, a we obtain $x_{12} x_{13} x_{14} x_{15} = abbc$. So word x is in the form

$$x = abbc\, caab\, bcca\, abbc\, caab\, bcca\, abbc\, caab\, bcca\, \ldots .$$

Fixed points and periodic points are an interesting topic itself. We remark that for the blinking mapping $\beta : A^\omega \to A^\omega$ with an alphabet $A = \{0, 1\}$ there are two fixed points $01010101\ldots$ and $10101010\ldots$ but periodic points with prime period two are $00000000\ldots$ and $11111111\ldots$. With prime period three exist six points: $000111\,000111\ldots$, $001110\,001110\ldots$, $100011\,100011\ldots$, $011100\,011100\ldots$, $110001\,110001\ldots$ and $111000\,111000\ldots$.

Generally, the periodic point for the blinking mapping with even period k is in form $x_0 x_1 \ldots x_{k-1}\, x_0 x_1 \ldots x_{k-1} \ldots$ but with odd period k is in form $x_0 x_1 \ldots x_{k-1}\, \overline{x_0 x_1 \ldots x_{k-1}}\, x_0 x_1 \ldots x_{k-1}\, \overline{x_0 x_1 \ldots x_{k-1}} \ldots$ where $x_i \in A = \{0, 1\}$ and $\overline{x_i} = 1$, if $x_i = 0$ and $\overline{x_i} = 0$, if $x_i = 1$.

Finally we conclude

Theorem 3.4. *The blinking mapping $\beta : A^\omega \to A^\omega$ is chaotic in the set A^ω.*

4. Conclusion

The notion of periodicity for symbolic sequences is also important (Ref. 1). Biologists are interested in determining structural characteristics of various genetic texts (e.g. DNA, protein, amino-acid, nucleotide sequences), expressed as words written in an alphabet. One such characteristic is periodicity.

Chaotic mapping implies that if we are using an iterated function to model long-term behavior (in economics or biology) and the function exhibits sensitive dependence, then any error in measurement of the initial conditions may result in large differences between the predicted behavior and the actual behavior of the system we are modelling. Since all physical measurements include error, this condition may severely limit the utility of our model. Chaotic function gives uncontrolled long term prediction. But there are other areas of science in which chaotic mappings might be useful, for example, cryptography (Ref. 7, 17, 18). We will offer chaotic mappings for designing the keystream generator.

References

1. N. Atreas, *Detecting Hidden Periodicities on Symbolic Sequences*, Journal of Interdisciplinary Mathematics **12**, 639-646 (2009).
2. J. Banks, J. Brooks, G. Cairns, G. Davis, P. Stacey, *On Devaney's definition of chaos*, Amer. Math. Monthly **99**, 29-39 (1992).
3. V. Berthé, M. Rigo (Eds.), *Combinatorics, Words and Symbolic dynamics*, Encyclopedia Math. Appl. **159** (Cambridge University Press, 2016).
4. I. Bula, J. Buls, I. Rumbeniece, *Why can we detect the chaos?* J. of Vibroengineering **10**, 468-474 (2008).
5. I. Bula, J. Buls, I. Rumbeniece, *On new chaotic mappings in symbol space*, Acta Mech Sin **27**, 114-118 (2011).
6. I. Bula, *New class of chaotic mappings in symbol space*, World Academy of Science, Engineering and Technology **67**, 305-309 (2012).
7. J. Buls, *Construction of Pseudo-random Sequences from Chaos*, Proc. of 2nd Int. Conf. Control of Oscillations and Chaos **3**, 558-560 (2000).
8. R. Devaney, *An introduction to chaotic dynamical systems* (Benjamin Cummings: Menlo Park, CA, 1986).
9. D. Gulick, *Encounters with chaos* (McGraw-Hill, Inc., 1992).
10. R. A. Holmgren, *A first course in discrete dynamical systems*, 2nd edition (Universitext, Springer-Verlag, 1996).
11. B. P. Kitchens, *Symbolic Dynamics. One-sided, two-sided and countable state Markov shifts* (Springer-Verlag, 1998).
12. D. Lind, B. Marcus, *An introduction to symbolic dynamics and coding* (Cambridge University Press, 1995).
13. T. Y. Li, J. A. Yorke, *Period three implies chaos*, Amer. Math. Monthly **82**, 985-992 (1975).
14. M. Lothaire, *Combinatorics on Words. Encyclopedia of Mathematics and its Applications*, **17** (Addison-Wesley, Reading, MA, 1983).
15. A. de Luca, S. Varricchio, *Finiteness and regularity in semigroups and formal languages* (Monographs in Theoretical Computer Science, Springer-Verlag, 1999).
16. C. Robinson, *Dynamical systems. Stability, symbolic dynamics, and chaos*, 2nd edition (CRS Press, 1999).
17. P.K. Shukla, A. Khare, M.A. Rizvi, S. Stalin, S. Kumar, *Applied cryptography using chaos function for fast digital logic-based systems in ubiquitous computing*, Entropy **17**, 1387-1410 (2015).

18. A.A. Zaher, A. Abu-Rezq, *On the design of chaos-based secure communication systems*, Communications in Nonlinear Science and Numerical Simulation **16**, 3721-3737 (2011).

A new strategy for preventing repeated and redundant clauses in contradiction separation based automated deduction

Xingxing He[1], Y. Xu[1], J. Liu[2], Xiaoping Qiu[3]*, Yingfang Li[4]

[1] *School of Mathematics, Southwest Jiaotong University,*
Chengdu 610031, Sichuan, PR China
[2] *School of Computing and Mathematics, University of Ulster, Northern Ireland, UK*
[3] *School of Transportation and Logistics, Southwest Jiaotong University,*
Chengdu 610031, Sichuan, PR China
[4] *School of Economics Information Engineering, Southwestern University of Finance*
and Economics, Chengdu 611130, Sichuan, PR China
x.he@swjtu.edu.cn

Contradiction separation based deduction is a key generalized inference rule of binary resolution, and has unique ability for automated theorem proving. Different inference rules need different proof searching strategies. This paper proposes two methods to avoid the repeated deductions and redundant clauses generated in the contradiction separation based deduction, that is, an improved clauses and literals weighted based method, and a redundant clauses prejudging and backtracking strategy. This work provides fundamental methods for contradiction separation based deduction on practical implementation.

Keywords: Contradiction separation based automated deduction; proof searching; repeated deduction; redundancy.

1. Introduction

Generally, inference rules and proof searching strategies are two main factors which play a fundamental role in automated theorem proving. Contradiction separation (CS-) based deduction,[13] as a key extension of resolution,[9] is a dynamic multi-clause synergized CS-based inference method, which can delete not only a complementary pair, but also an unsatisfiable clauses set from the involved clauses. The remaining literals form a new generated clause called as the CS-clause. Compared to binary resolution, it has some unique features such as dynamics, synergistic, multiplicity, etc.

---

*Corresponding author

Proof searching strategies guide the deduction direction, and hence a good strategy can quickly lead to a contradiction from a given unsatisfiable clauses set. Many scholars proposed many useful methods from different ways to solve existed problems. For example, refined resolution [2,8,10,12,14] like lock resolution, semantic resolution, hyper-resolution, and resolution with set of support, model-based reasoning with backtracking, [7] goal-sensitivity, [1] resolvents evaluation, [8] machine learning based for literals and clauses selection especially for deep network guided proof searching [3,5] and so on. These strategies have been also implemented on theorem provers successfully such as E, [3,11] Vampire, [4] Isabella, [6] etc.

However, the searching methods should be adapted to inference rules for an efficient prover. For the CS-based deduction, many clauses are involved in a contradiction, and hence this rule brings some difficulties to control the proof searching compared to binary resolution. Meanwhile, it indeed deletes more literals simultaneously, but more clauses are involved, hence how to balance the contradiction and involved clauses is an important issue for its efficient implementation. Furthermore, Redundancy including tautology clauses and subsumed clauses are useless for deriving an empty clause, and should avoid them in the deduction. Unlike binary resolution, if we just delete them after they generated in the CS-based deduction, then it wastes too much time and space. In this sense, this paper proposes a weighted method based CS-deduction to prevent the redundant clauses and repeated deduction generated, and also gives a redundant clause prejudging and backtracking strategy for efficient contradictions constructing.

2. Preliminaries

In this section, we only recall some elementary definitions and properties needed in the following discussions, more detailed notations and results about contradiction separation based deduction can be seen in the related references. [2,8,13]

Definition 2.1. [13] Let $S = \{C_1, \cdots, C_m\}$ be a clause set. If for any $(x_1, \cdots, x_m) \in \prod_{i=1}^{m} C_i$, there exists at least one complementary pair among $\{x_1, \cdots, x_m\}$, then $S = \bigwedge_{i=1}^{m} C_i$ is called a standard contradiction. If $\bigwedge_{i=1}^{m} C_i$ is unsatisfiable, then $S = \bigwedge_{i=1}^{m} C_i$ is called a quasi-contradiction.

Definition 2.2. [13] Suppose a clause set $S = \{C_1, C_2, \cdots, C_m\}$ in first-order logic. Without loss of generality, assume that there does not exist the same variables among $C_1, C_2, \cdots, C_m$ (if there exists the same variables, there exists a rename substitution which makes them different).

The following inference rule that produces a new clause from S is called a standard contradiction separation rule, in short, an S-CS rule: For each $C_i (i=1,2,\cdots,m)$, firstly applying factoring to C_i, i.e., using a substitution σ_i to C_i (σ_i could be an empty substitution), denoted as $C_i^{\sigma_i}$. Then separate $C_i^{\sigma_i}$ into two sub-clauses $C_i^{\sigma_i^-}$ and $C_i^{\sigma_i^+}$ such that

(1) $C_i^{\sigma_i} = C_i^{\sigma_i^-} \vee C_i^{\sigma_i^+}$, where $C_i^{\sigma_i^-}$ and $C_i^{\sigma_i^+}$ have no common literals;
(2) $C_i^{\sigma_i^+}$ can be an empty clause itself, but $C_i^{\sigma_i^-}$ cannot be an empty clause;
(3) $\bigwedge_{i=1}^m C_i^{\sigma_i^-}$ is a standard contradiction, that is $(x_1,\cdots,x_m) \in \prod_{i=1}^m C_i^{\sigma_i^-}$, there exists at least one complementary pair among $\{x_1,\cdots,x_m\}$.

The resulting clause $\bigvee_{i=1}^m C_i^{\sigma_i^+}$, denoted as $\mathscr{C}_m^{s\sigma}(C_1,\cdots,C_m)$ (here "s" means "standard"), is called a standard contradiction separation clause (S-CSC) of $C_1,\cdots,C_m$, and $\bigwedge_{i=1}^m C_i^{\sigma_i^-}$ is called a separated standard contradiction (S-SC).

Definition 2.3.[13] Suppose a clause set $S = \{C_1, C_2,\cdots,C_m\}$ in first-order logic. $\Phi_1, \Phi_2,\cdots,\Phi_t$ is called a standard contradiction separation based deduction sequence (or a S-CS based deduction sequence from S to a clause Φ_t, denoted as $D^{s\sigma}$ (where σ is a substitution), if

(1) $\Phi_i \in S, i=1,2,\cdots,t$. Or
(2) there exist $r_1, r_2,\cdots,r_{k_i} < i$, $\Phi_i = \mathscr{C}_{r_{k_i}}^{s\sigma}(\Phi_{r_1},\Phi_{r_2},\cdots,\Phi_{r_{k_i}})$.

3. Literals and Clauses Weighted Strategies

In the S-CS based deduction, many clauses are involved for one separated standard contradiction constructing, and then generate a new S-CSC. An S-CS based refutation is an S-CS based deduction sequence where the last S-CSC is an empty clause. The structure of linear resolution deduction is equivalent to its of S-CS based deduction, hence we use the linear resolution as an example to clarify our proposed searching control strategies.

First we introduce some notations of our methods. Let S be a set of clauses $S = C_1 \wedge C_2 \wedge \cdots \wedge C_n$, P_i a literal of C_i. Suppose $NL(C_i)$ is the length of C_i, that is, denotes the number of literals in C_i, $NA(C_i)$ and $NP(C_i)$ denotes the number of the clause C_i actively and passively participates in the deduction respectively, similarly, $NA(P_i)$ and $NP(P_i)$ denotes the number of the literal P_i actively and passively participates in the deduction respectively.

Initially, $NA(C_i) = 0$ where $C_i \in S$. During the deduction process, $NA(C_i)$ is computed as follows.

(1) If $C_i \in S$, and actively participates in an S-CS based deduction, then $NA(C_i) = NA(C_i) + 1$.
(2) If C_i is an S-CSC of $C_{i_1}, C_{i_2}, \cdots, C_{i_m}$, then $NA(C_i) = (NA(C_{i_1}) + NA(C_{i_2}) + \cdots + NA(C_{i_m}))/m$.

Remark 3.1. Similarly, $NA(P_i) = 0$ where $P_i \in S$ initially. During the deduction process, $NA(P_i)$ is computed as follows.

(1) If $P_i \in S$, and actively participates in an S-CS based deduction, then $NA(P_i) = NA(P_i) + 1$.
(2) If P_i is a literal of an S-CSC of $C_{i_1}, C_{i_2}, \cdots, C_{i_m}$, then $NA(P_i)$ equals to the value of itself in C_{ij}.

Furthermore, initially, $NP(C_i) = 0$ where $C_i \in S$. During the deduction process, $NP(C_i)$ is computed as follows.

(1) If $C_i \in S$, and passively participates in an S-CS based deduction, then $NP(C_i) = NP(C_i) + 1$.
(2) If C_i is an S-CSC of $C_{i_1}, C_{i_2}, \cdots, C_{i_m}$, then $NP(C_i) = (NP(C_{i_1}) + NP(C_{i_2}) + \cdots + NP(C_{i_m}))/m$.

Remark 3.2. The number $NP(P_i)$ can be defined similar to $NA(P_i)$.

Definition 3.1. Let S be set of clauses $S = \{C_1, C_2, \cdots, C_n\}$, $\mathscr{C}_m^{s\sigma}(C_1, \cdots, C_m)$ an S-CSC of $C_1, \cdots, C_m (m \leq n)$, $P_{i_1}, P_{i_2}, \cdots, P_{i_m}$ the involved literals, σ a substitution. If $P_{ij_1}, P_{ij_2}, \cdots, P_{ij_{n_0}} \in \mathscr{C}_m^{s\sigma}(C_1, \cdots, C_m)$, such that $P_{ij_1} \in C_{ij_1}, P_{ij_2} \in C_{ij_2}, \cdots, P_{ij_{n_0}} \in C_{ij_{n_0}}$ are complementary literals, or subsumed literals, then $P_{ij_2}, \cdots, P_{ij_{n_0}}$ are called as the real redundant literals, and the clauses $C_{ij_1}, C_{ij_2}, \cdots, C_{ij_{n_0}}$ are called as the real redundant clauses.

Definition 3.2. Let S be set of clauses $S = \{C_1, C_2, \cdots, C_n\}$, $\mathscr{C}_m^{s\sigma}(C_1, \cdots, C_m)$ an S-CSC of $C_1, \cdots, C_m (m \leq n)$. If $P_{i_1}, P_{i_2}, \cdots, P_{i_m}$ are the real redundant literals, and the $C_{ij_1}, C_{ij_2}, \cdots, C_{ij_{n_0}}$ are the real redundant clauses, then

(1) $NA(C_{i_k}) = NA(C_{i_k}) + 2$, $NA(P_{i_k}) = NA(P_{i_k}) + 2$,
(2) $NP(C_{i_k}) = NP(C_{i_k}) + 2$, $NP(P_{i_k}) = NP(P_{i_k}) + 2$,

where $k = j_1, j_2, \cdots, j_{n_0}$.

Remark 3.3. Increasing the values of the NA and NP of the real redundant literals and clauses is to decrease the possibility of choosing these literals and clauses in an S-CS based deduction.

From the values of definitions above, there are three attributes for each literals and clauses, that is, $P_i(NL, NA, NP)$, and $C_i(NL, NA, NP)$. In the S-CS based deduction, we can choose different literals and clauses according to the different order of three metrics.

In order to record the information of literals and clauses more accurately, an improved weighted method is proposed here.

Definition 3.3. Let S be a set of clauses, P_i, P_j the literals in S. $NA_2(P_i, P_j)$ and $NP_2(P_i, P_j)$ are denoted as the numbers of P_i actively participating with P_j, and P_i passively participating with P_j in the deduction, respectively.

Algorithm 3.1. A weighted method for a standard contradiction constructing.

Step 0. Let S be a set of clauses $S = \{C_1, C_2, \cdots, C_n\}$. Set $i = 1$. Set $NA_2(P_i, P_j) = 0$ and $NP_2(P_i, P_j) = 0$ if P_i and P_j in S.

Step 1. If $P_i \in C_i$ actively participates with $P_j \in C_j$ in the deduction, and the remaining literals in C_i are real redundant literals, then $NA_2(P_i, P_j) = NA_2(P_i, P_j) + 2$. Furthermore, the remaining literals in C_i lead to the new generated clause as a redundant one several times, then $NA_2(P_i, P_j) = NA_2(P_i, P_j) + 8$.

Step 2. If the remaining literals in C_j are real redundant literals, then $NP_2(P_i, P_j) = NP_2(P_i, P_j) - 2$. Furthermore, the remaining literals in C_j lead to the new generated clause as a redundant one several times, then $NP_2(P_i, P_j) = NP_2(P_i, P_j) - 8$.

Step 3. Otherwise, $NA_2(P_i, P_j) = NA_2(P_i, P_j) + 1$, $NP_2(P_i, P_j) = NP_2(P_i, P_j) + 1$.

Step 4. If $i = n$, or no extended clause exists in S, then stop, get an S-CSC $R = \bigvee_{j=1}^{n} D_j^{\sigma_j^+}$, the standard contradiction is $\bigwedge_{j=1}^{n} D_j^{\sigma_j^-}$. Otherwise, $i = i + 1$, go to Step 1.

4. A Redundant Clauses Backtracking and Prejudging Strategy

Proposition 4.1. Let C_i, R_j be two clauses, then the following conclusions hold.

(1) If all the terms in C_i are the same ones in the same positions in R_j, then there exists a substitution σ, such that $C_i^\sigma \subseteq R_j$.

(2) If the terms in C_i correspond to the different ground terms in the same positions in R_j, then there doesn't exist a substitution σ, such that $C_i^\sigma \subseteq R_j$.

Proposition 4.2. *Let C_i, R_j be two clauses. There exist two substitutions σ, and σ_1 such that $C_i^\sigma \subseteq R_j^{\sigma_1}$ if and only if one of the following conditions exists.*

(1) all the terms are variables in both C_i and R_j, and all the terms in R_j are not the same.

(2) all the terms are in ground ones in C_i, and all the terms are variables in R_j.

(3) all the terms only include variables and ground ones in C_i, and all the terms are variables in R_j.

(4) all the terms are function symbols in C_i, and all the terms are variables in R_j.

In the Propositions 4.1 and 4.2, these conditions can be seen as the sufficient ones for judging the redundancy, hence we can get the following backtracking rules during S-CS constructing without loss of completeness as follows.

(1) If all the literals in $\mathscr{C}_m^{s\sigma}(C_1,\cdots,C_m)$ are ground literals, and there exists C_i in S and a substitution σ_1, such that $C_i^{\sigma_1} \subseteq \mathscr{C}_m^{s\sigma}(C_1,\cdots,C_m)$, then stop, and backtrack to the real leading literals, and choose other clauses and literals for S-CS constructing.

(2) If C_i is a unit ground clause in S, and there exists a substitution σ_1, such that $C_i^{\sigma_1} \subseteq \mathscr{C}_m^{s\sigma}(C_1,\cdots,C_m)$, then stop, and backtrack to the real leading literals, and choose other clauses and literals for S-CS constructing.

Definition 4.1. *Let $\bigwedge_{i=1}^m D_i^{\sigma_i^-}$ be an standard contradiction, $\bigvee_{j=1}^m D_j^{\sigma_j^+}$ an S-CSC, where D_i is the involved clause in standard contradiction, σ_i is a substitution of D_i. σ_i is called an inverse substitution if σ_i changes the literals in $D_i, D_{i-1}, \cdots, D_1$.*

Proposition 4.3. *σ_i is an inverse substitution in $D_i, D_{i-1}, \cdots, D_1$ if it satisfies one of the following conditions.*

(1) The same variables exist in $D_i^{\sigma_i^-}$ and $D_{i+1}^{\sigma_{i+1}^-}$.

(2) Not all the terms in $D_j^{\sigma_i^-}$ are ground terms.

Inverse substitution is an important feature of the S-CS based deduction, which also leads to the complexity of the searching control. According to the backtracking rules above, we can get a method to avoid redundant clauses generated with same literals or clauses in standard contradiction constructing.

Algorithm 4.1. Standard contradiction constructing without redundancy.

Step 0. Let S be a set of clauses $S = \{C_1, C_2, \cdots, C_n\}$. Set $i = 1$.

Step 1. Reorder the clauses and literals in S according to their attributes computed by **Algorithm 3.1**, and choose the first clause as extended clause D_i in S, and the first literal as extended literal P_i in D_i. $R_i = \bigvee_{j=1}^{i} D_j^{\sigma_j^+}$, where σ_j is a substitution.

Step 2. If there exists a substitution σ and a clause C in S, such that $C^\sigma \subseteq R_i$, or R_i is a tautology, then go to Step 3. Otherwise, go to Step 5.

Step 3. If all the real redundant literals in R_i are ground literals or no same variable symbol between the remaining literals and extended literal P_i of the same position in R_i and D_i, respectively, then go to Step 4. Else go to Step 5.

Step 4. Backtrack to the literal P_i, and choose other literals from remaining literals in D_i for standard contradiction constructing. Go to Step 1.

Step 5. If $i = n$, or no extended clause exists in S, then stop, get an S-CSC $R = \bigvee_{j=1}^{n} D_j^{\sigma_j^+}$, the standard contradiction is $\bigwedge_{j=1}^{n} D_j^{\sigma_j^-}$. Otherwise, $i = i + 1$, go to Step 1.

Remark 4.1. Algorithm 4.1 can prejudge the redundant clause during the S-CS based deduction, and also be handle back subsumption in the standard contradiction constructing.

Proposition 4.4. *No redundant clause generates by Algorithm 4.1.*

5. Conclusion

This paper proposed two methods to reduce the repeated deductions and redundant clauses in the contradiction separation based deduction, that is, the backtracking and prejudging strategy, and the weighed method. For the strategy, we can get the accurately backtracking position, which means that it can avoid too many redundant pathes, and also preserve its completeness. The further research will be concentrated on controlling the proof search more accurately such as resolvents evaluation, clauses and literals metric.

Acknowledgments

This work is partially supported by the National Natural Science Foundation of China (Grant No. 61100046 and 61673320), the Fundamental Research Funds for the Central Universities (Grant No. 2682018ZT10, 2682016CX053), the science and technology project of Sichuan province(Grant no. 2017GZ0371).

References

1. M.P. Bonacina, D.A. Plaisted, Semantically-Guided Goal-Sensitive Reasoning: Inference System and Completeness, Journal of Automated Reasoning, 59(2), pp. 165-218, 2017.
2. C.L. Chang, R.C.T. Lee, Symbolic logic and mechanical theorem proving, Academic Press, USA, 1997.
3. C. Kaliszyk, S. Schulz, J. Urban, System description: ET 0.1, Proc. CADE-25, Berlin, pp. 389-398, 2015.
4. L. Kovács, A. Voronkov, First-order theorem proving and Vampire. In International Conference on Computer Aided Verification, pp. 1-35, 2013.
5. D. Kühlwein, J.C. Blanchette, C. Kaliszyk, MaSh: machine learning for sledgehammer International Conference on Interactive Theorem Proving. Springer Berlin Heidelberg, pp. 35-50, 2013.
6. T. Nipkow, L.C. Paulson, M. Wenzel, Isabelle/HOL: a proof assistant for higher-order logic. Springer Science and Business Media, 2002.
7. D.A. Plaisted, History and Prospects for First-Order Automated Deduction, A.P. Felty and A. Middeldorp (Eds.): CADE-25, LNAI 9195, pp. 3-28, 2015.
8. A. Robinson and A. Voronkov, Handbook of Automated Reasoning, Vol. 1 and 2, the MIT Press and North Holland, 2001.
9. J.A. Robinson, A machine oriented logic based on the resolution principle, Journal of the ACM, 12(1), pp. 23-41, 1965.
10. J.A. Robinson, Automatic deduction with hyper-resolution. Int. J. Comput. Math. 1, pp. 227-234, 1965.
11. S. Schulz, System Description: E 1.8, Proceedings of the 19th LPAR, Stellenbosch, pp. 477-483, 2013.
12. J.R. Slagle, Automatic theorem proving with renamable and semantic resolution. J. ACM 14(4), pp. 687-697, 1967.
13. Y. Xu, J. Liu, S.W. Chen, X.M. Zhong, A novel generalization of resolution principle for automated deduction, In Uncertainty Modeling in Knowledge Engineering and Decision Making: Proceedings of the 12th International FLINS Conference, pp. 483-488, 2016.
14. L. Wos, D. Carson, G. Robinson, Efficiency and completeness of the set of support strategy in theorem proving. J. ACM, 12, pp. 536-541, 1965.

Logical difference of propositional theories

Yisong Wang and Hong Liu

Key Laboratory of Intelligent Medical Image Analysis and Precise Diagnosis of Guizhou Province,
Department of Computer Science and Technology, Guizhou University
Guiyang, Guizhou, China
yswang@gzu.edu.cn

Ying Zhang, Mingyi Zhang and Danning Li

Guizhou Academy of Science, No 40 East Yanan,
Guiyang, Guizhou, 550001, China
zhangmingyi045@aliyun.com, lidn121@foxmail.com

> Logical difference is an important notion to distinct different versions of knowledge-base systems. To capture such difference in terms of logic consequence, clause consequence and prime clause consequence respectively, this paper proposes three notions of difference over relevant signatures — logical difference, clausal difference and prime difference. They are closely related to forgetting. It is generally intractable to compute such differences, even for Horn theories. Preliminary experimental results on clausal difference and prime difference illustrate an interesting phase transition phenomenon over random 3-CNF theories.
>
> *Keywords*: Clause difference; prime difference; forgetting; complexity.

1. Introduction

The notion of logical difference plays an important role in evolving logical systems, and knowledge bases in particular, which generally suffers from dynamic changes continuously. For example, to capture version difference of an ontology system, Konev et al. proposed three notions of difference for description logic $\mathcal{EL}$ — syntactic difference, structural difference and logical difference.[1,2] Their notion of logical difference is closely related to that of forgetting in description logic[3,4] and inseparability.[5,6] Thus, it is useful for ontology engineering as well. The notion of logical difference is also extended to fuzzy $\mathcal{EL}^+$ ontologies.[7]

The notion of difference is also useful in many other domains, such as decision making and negotiation. Let us consider the following scenario:

Example 1.1. Bob is negotiating with his daughter Annie on her breakfast — milk, oat meal and cookie. Bob prefers to good healthy breakfast, while Annie prefers to good tasty one. Both agree to have only two breakfast items. Bob has the following knowledge: if the breakfast includes milk and oat meal then it is good healthy, if the breakfast includes milk and cookie then it is good tasty, and if the breakfast includes oat meal and cookie then it is either good healthy or good tasty. Now Annie wants to change her ordinary breakfast choice of milk and (oat meal or cookie) to oat meal and (milk or cookie). Suppose what Bob really concerns on breakfast is its quality, *i.e.*, any consequence on "good healthy" and "good tasty". Should Bob agree with her request? A positive answer will be given, cf. Example 3.1.

Although Eiter *et al.* proposed a difference operator for propositional logic, it is in the view of model difference.[8] Formally, they define the difference of two formulas φ and ψ as $\varphi \wedge \neg \psi$. Evidently, this notion of difference cannot capture the difference of logical consequences since $\neg \psi$ may be not a logical consequence of φ.

To capture logical difference of propositional theories (knowledge bases), we propose three notions of difference in the paper. Informally, given two formulas φ and ψ, and a relevant signature A,

- the *logical difference* of φ from ψ over A consists of the formulas over A that are logical consequence of φ but not of ψ;
- the *clausal difference* of φ from ψ over A consists of the clauses over A that are logical consequence of φ but not of ψ;
- the *prime difference* of φ from ψ over A consists of the prime clauses in the clausal difference of φ from ψ over A.

Interestingly, these notions of difference are also closely related with the notion of *forgetting* of propositional logic.[9,10] The computational complexities on deciding if there are difference over a signature between two formulas are shown to be intractable in general, even if formulas are Horn ones. Preliminary experiments on clausal difference and prime difference demonstrate an interesting phase transition phenomenon on random 3-CNF theories, which is very similar to the satisfiability of random 3-SAT.

2. Preliminaries

We assume a propositional language $\mathcal{L}$ with a underlying finite nonempty signature $\mathcal{A} = \{x_1, \ldots, x_n\}$, whose elements are called *propositional atoms* or *variables*. The notions of *formula, clause, Horn clause, interpretation, model, entailment, equivalence* etc. of $\mathcal{L}$ are defined in the standard manner. By $Var(\xi)$ we denotes the set of atoms/variables occurring in expression ξ. Let $A \subseteq \mathcal{A}$, α be a clause and φ be a formula.

- α is an *implicate* of φ if $\varphi \models \alpha$.
- α is an *A-implicate* of φ if it is an implicate of φ and $Var(\alpha) \subseteq A$.
- α is a *prime implicate* of φ whenever it is an implicate of φ and no other implicate α' of φ satisfying $\alpha' \subset \alpha$.
- α is a *A-prime implicate* of φ whenever it is an A-implicate of φ and no other A-implicate α' of φ satisfying $\alpha' \subset \alpha$. The set of all of the A-prime implicates of φ is denoted by $PI_A(\varphi)$.

Now we recall the notion of forgetting and relevance in propositional logic.[9,10] Forgetting a set V of atoms from a formula φ, written $\mathsf{Forget}(\varphi, V)$, is defined as follows:

- $\mathsf{Forget}(\varphi, \emptyset) = \varphi$.
- $\mathsf{Forget}(\varphi, \{x\}) = \varphi[\top/x] \vee \varphi[\bot/x]$ where $\varphi[t/x]$ is the formula obtained from φ by replacing every occurrence of x with t;
- $\mathsf{Forget}(\varphi, V \cup \{x\}) = \mathsf{Forget}(\mathsf{Forget}(\varphi, \{x\}), V)$.

Let $A \subseteq \mathcal{A}$. A formula φ is *A-irrelevant* if there is a formula φ' such that $Var(\varphi') \cap A = \emptyset$ and $\varphi \equiv \varphi'$. It is proved that φ A-irrelevant if and only if $\varphi \equiv \mathsf{Forget}(\varphi, A)$.[10]

By the definitions of forgetting and A-prime implicate, the next lemma follows.

Lemma 2.1. *Let φ be a formula, $A \subseteq \mathcal{A}$ and α a clause. Then* $\mathsf{Forget}(\varphi, \overline{A}) \models \alpha$ *iff* $\exists \beta \in PI_A(\varphi)$ *such that* $\beta \models \alpha$.

3. Differences

In this section we present the formal definitions of three notions of difference, and explore their properties.

3.1. *Logical difference and clausal difference*

Firstly, let us starting with logical difference.

Definition 3.1 (Logical difference). *Let φ, ψ be two formulas, and $A \subseteq \mathcal{A}$. The logical difference of φ from ψ over A, written $\mathit{Diff}^*_A(\varphi, \psi)$, is the* set of formulas ξs such that $Var(\xi) \subseteq A$, $\varphi \models \xi$ and $\psi \not\models \xi$, i.e.,

$$Diff_A^*(\varphi, \psi) = \{\xi | Var(\xi) \subseteq A, \varphi \models \xi, \psi \not\models \xi\}. \tag{1}$$

Example 3.1. [Continued from Example 1.1] For simplicity let us denote "milk", "oat meal", "cookie", "good healthy" and "good tasty" by m, ot, c, gh and gt respectively. Now Bob's knowledge/theory Σ about the breakfast can be encoded as follows:

$$m \wedge ot \supset gh \wedge \neg c, \quad m \wedge c \supset gt \wedge \neg ot,$$
$$ot \wedge c \supset (gt \vee gh) \wedge \neg m, \quad m \wedge (ot \vee c).$$

The above last line encodes the knowledge about Annie's ordinary breakfast. If Bob agrees Annie's request then the knowledge Σ is changed to Σ' by replacing the above last line with $ot \wedge (m \vee c)$. It is not difficult to check that $\Sigma \models (gh \vee gt) \wedge (\neg c \vee \neg ot)$ and $\Sigma' \models (gh \vee gt) \wedge (\neg c \vee \neg m)$. Furthermore, over the signature $A = \{gh, gt\}$, $Diff_A(\Sigma, \Sigma') = Diff_A(\Sigma', \Sigma) = \emptyset$. It means that the quality of breakfast does not change even if Bob agrees with Annie's request. So that Bob should agree with Annie to change her breakfast choice.

From practice point of view, propositional knowledge bases are generally represented in some regular forms, such as CNF, Horn and so on. In this sense, one may interested in only clauses in logical difference. This motivates the following notion of clausal differences.

Definition 3.2 (Clausal differences). Let φ, ψ be two formulas, and $A \subseteq \mathcal{A}$. The clausal difference of φ from ψ over A, written $Diff_A(\varphi, \psi)$, is the set of clauses ξs such that $Var(\xi) \subseteq A$, $\varphi \models \xi$ and $\psi \not\models \xi$, i.e.

$$Diff_A(\varphi, \psi) = \{\xi \text{ is a clause} | Var(\xi) \subseteq A, \varphi \models \xi, \psi \not\models \xi\}. \tag{2}$$

The following proposition illustrates the relationship between logical difference and clausal difference.

Proposition 3.1. Let φ, ψ be two theories and $A \subseteq \mathcal{A}$.

(1) $Diff_A(\varphi, \psi) \subseteq Diff_A^*(\varphi, \psi)$.
(2) $Diff_A(\varphi, \psi) = \{\xi \text{ is a clause} | \xi \in Diff_A^*(\varphi, \psi)\}$.
(3) $Diff_A^*(\varphi, \psi) = \emptyset$ iff $Diff_A(\varphi, \psi) = \emptyset$.

Proof. (1) It follows directly from definitions of 3.1 and 3.2.
 (2) Let α be a clause such that $Var(\alpha) \subseteq A$. We have $\alpha \in Diff_A(\varphi, \psi)$ iff $\varphi \models \alpha$ and $\psi \not\models \alpha$
iff $\alpha \in Diff_A^*(\varphi, \psi)$ since α is a clause.

(3) By (1) of the proposition, it is sufficient to show that $\text{Diff}_A(\varphi, \psi) = \emptyset$ implies $\text{Diff}_A^*(\varphi, \psi) = \emptyset$. Suppose $\text{Diff}_A(\varphi, \psi) = \emptyset$ but $\text{Diff}_A^*(\varphi, \psi) \neq \emptyset$, i.e., there is a non-tautology formula α with $\text{Var}(\alpha) \subseteq A$ in $\text{Diff}_A^*(\varphi, \psi)$. Let the conjunctive normal form α' of α is $\alpha_1 \wedge \cdots \wedge \alpha_k$. It follows that $\alpha' \in \text{Diff}_A^*(\varphi, \psi)$, i.e., $\varphi \models \alpha_i$ for each i ($1 \leq i \leq k$) and $\psi \not\models \alpha_j$ for some j ($1 \leq j \leq k$). It implies $\alpha_j \in \text{Diff}_A(\varphi, \psi)$, i.e., $\text{Diff}_A(\varphi, \psi) \neq \emptyset$, a contradiction. □

The next proposition illustrates a close connection between clausal difference and forgetting.

Proposition 3.2. *Let φ, ψ be formulas and $A \subseteq \mathcal{A}$. Then*

(1) $\text{Diff}_A(\varphi, \psi)$ is $\overline{A}$-irrelevant.
(2) $\text{Forget}(\varphi, \overline{A}) \models \text{Diff}_A(\varphi, \psi)$.
(3) $\text{Diff}_A(\varphi, \psi) = \emptyset$ iff $\psi \models \text{Forget}(\varphi, \overline{A})$.

Proof. (1) It follows from the definition of clausal difference since every clause in $\text{Diff}_A(\varphi, \psi)$ does not mention any atom from $\overline{A}$.

(2) It follows from Definition 3.2 and the fact that $\text{Forget}(\varphi, V) \models \alpha$ if and only if $\varphi \models \alpha$ if α is V-irrelevant.

(3) It follows from (2) of the proposition and Definition 3.2. □

Though clausal differences are not logically closed, removing subsumed ones from a clausal difference still preserves logical equivalence. This motivates the notion of prime difference.

Definition 3.3. *Let φ, ψ be two formulas, and $A \subseteq \mathcal{A}$. The prime difference of φ from ψ over A, written $\text{Diff}_A^{pd}(\varphi, \psi)$, consists of*

$$\xi \in \text{Diff}_A(\varphi, \psi) \text{ and } \not\exists \xi' \in \text{Diff}_A(\varphi, \psi) \text{ s.t. } \xi' \subset \xi. \tag{3}$$

The next proposition demonstrates a close connection between prime difference and clausal difference.

Proposition 3.3. *Let φ, ψ be two formulas and $A \subseteq \mathcal{A}$.*

(1) $\text{Diff}_A^{pd}(\varphi, \psi) \equiv \text{Diff}_A(\varphi, \psi)$.
(2) $\text{Diff}_A^{pd}(\varphi, \top) = PI_A(\varphi)$.

Proof. (1) It is evident. (2) It follows from (1) of the proposition, (3) of Proposition 3.1 and Lemma 2.1. □

4. Computational Complexities

In this section we consider computational complexities for clausal difference and prime difference. Readers are referred to the textbook[11] for various notations on computational complexity. We concentrate on the following problems. Given two formulas φ and ψ, a clause ξ, a literal l and $A \subseteq \mathcal{A}$,

VACUITY: $\mathit{Diff}_A(\varphi, \psi) = \emptyset$?
RELEVANCE: $l \in \alpha$ for some $\alpha \in \mathit{Diff}_A(\varphi, \psi)$?
NECESSITY: $l \in \alpha$ for every $\alpha \in \mathit{Diff}_A(\varphi, \psi)$?
MEMBERSHIP: $\xi \in \mathit{Diff}_A(\varphi, \psi)$?

The next theorem shows that these problems are all intractable.

Theorem 4.1. *Let φ and ψ be formulas, ξ a clause, l a literal and $A \subseteq \mathcal{A}$.*

(1) MEMBERSHIP *is BH_2-complete.*
(2) VACUITY *is Π_2^p-complete.*
(3) RELEVANCE *is Σ_2^p-complete.*
(4) NECESSITY *is Π_2^p-complete.*

Proof sketch: We consider the proof for the hardness of (2) only. Let $\mathit{Var}(\varphi) = X \cup Y$ and $X \cap Y = \emptyset$. Then
$\mathit{Diff}_X(\varphi, \top) = \emptyset$
iff $\top \models \mathsf{Forget}(\varphi, Y)$
iff $\mathsf{Forget}(\varphi, Y) \equiv \top$
iff $\forall X \exists Y \varphi$ is valid, which is Π_2^p-complete. Thus, this problem is Π_2^p-hard.

The next theorem shows that, for Horn theories, the complexity of VACUITY, RELEVANCE, NECESSITY and MEMBERSHIP is one level below that of general case.

Theorem 4.2. *Let φ and ψ be Horn theories, ξ a clause, l a literal and $A \subseteq \mathcal{A}$.*

(1) MEMBERSHIP *is in P for Horn theories.*
(2) VACUITY *is co-NP-complete for Horn theories.*
(3) RELEVANCE *is NP-complete for Horn theories.*
(4) NECESSITY *is co-NP-complete for Horn theories.*

Proof sketch: We consider the proof for the hardness of (3). Let φ, ψ be two Horn formulas over $X \cup Y$ with $X \cap Y = \emptyset$, x a fresh atom, $X' = X \cup \{x\}$ and φ' be the Horn formula obtained from φ by replacing each Horn clause α of φ by $\alpha \vee \neg x$. It is clear that φ' is a Horn formula. It can be shown

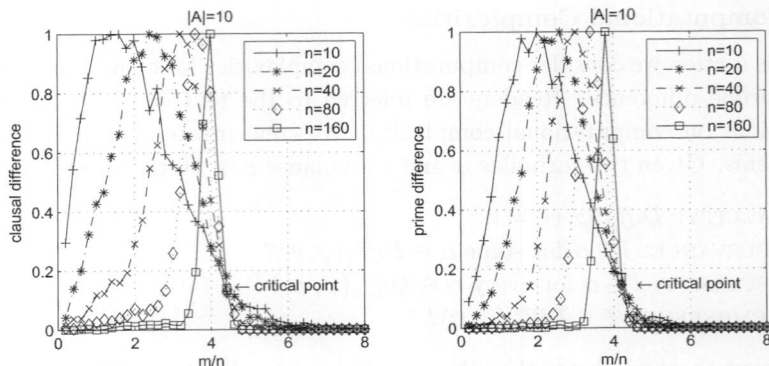

Fig. 1. Normalized aggregate difference.

that $\alpha \vee \neg x \in \mathit{Diff}_{X'}(\varphi', \psi)$ iff $\mathit{Diff}_X(\varphi, \psi) \neq \emptyset$ where α is a non-tautology reduced clause. Thus, this problem is NP-hard.

5. Experiments on Random 3-CNFs

The notions of clausal/prime difference allow us to explore the phase transition of random 3-SAT from the perspective of logical difference. Experiments are conducted on a workstation running Linux 3.19.5 with GCC 4.9.2, 32GB memory and 8 Intel(R) Core(TM) i7-4770K CPUs @ 3.50GHz.

We report experimental results on clausal and prime differences of random 3-CNF theories with $|\mathcal{A}| = n$, which ranges from 10, 20, 40, 80 to 160 over a random relevant signature A with $|A| = 10$. The sizes of tested theories are chosen according to the ration m/n from .2 to 8 with interval .2, where m is the number of clauses. In each case, we do 100 trails and report experimental results in average.

The normalized aggregate number of clausal difference and prime difference is reported in Figure 1. As can be seen, with n increasing, the aggregate clausal difference and prime difference approach to and then deviate from peaks sharply. Secondly, the peaks of aggregated clausal difference and prime difference shift from lower ration m/n to higher ratio m/n. So, it illustrates a phase transition for the overall clausal and prime difference.

6. Conclusion and Future Work

In this paper three notions of difference for propositional logic were proposed — logical difference, clausal difference and prime difference. Their properties, computational complexities and preliminary experiments were explored. Further work includes theoretical extension on first order logic, more detailed experiments and deep insights of its phase transition phenomena.

Acknowledgments

We thank reviewers for their helpful comments. This work is partially supported by NSFC under grants 63170161, 61562009 and 61262029, Stadholder Fund of Guizhou Province under grant (2012)62, Outstanding Young Talent Training Fund of Guizhou Province under grant (2015)01 and Science and Technology Fund of Guizhou Province under grant [2014]7640.

References

1. Boris Konev, Dirk Walther, and Frank Wolter. The logical difference problem for description logic terminologies. In Alessandro Armando, Peter Baumgartner, and Gilles Dowek, editors, *IJCAR*, volume 5195 of *Lecture Notes in Computer Science*, 259–274. Springer, 2008.
2. Boris Konev, Michel Ludwig, Dirk Walther, and Frank Wolter. The logical difference for the lightweight description logic $\mathcal{EL}$. *Journal of Artificial Intelligence Research*, 44:633–708, 2012.
3. Zhe Wang, Kewen Wang, Rodney W. Topor, and Jeff Z. Pan. Forgetting for knowledge bases in DL-Lite. *Annuals of Mathematics and Artificial Intelligence*, 58(1-2):117–151, 2010.
4. Roman Kontchakov, Frank Wolter, and Michael Zakharyaschev. Logic-based ontology comparison and module extraction, with an application to DL-Lite. *Artificial Intelligence*, 174(15):1093–1141, 2010.
5. Boris Konev, Carsten Lutz, Dirk Walther, and Frank Wolter. Model-theoretic inseparability and modularity of description logic ontologies. *Artificial Intelligence*, 203:66–103, 2013.
6. Elena Botoeva, Roman Kontchakov, Vladislav Ryzhikov, Frank Wolter, and Michael Zakharyaschev. When are description logic knowledge bases indistinguishable? In *Proceedings of IJCAI 2015*, 4240–4246. AAAI Press, 2015.
7. Shasha Feng, Yonggang Zhang, Dantong Ouyang, Haiyan Che, and Jie Liu. The logical difference for fuzzy EL+ ontologies. In *Proceedings of DL 2010*, volume 573 of *CEUR Workshop Proceedings*, 457-463. CEUR-WS.org, 2010.
8. Thomas Eiter, Toshihide Ibaraki, and Kazuhisa Makino. On the difference of horn theories. *Journal of Computer Systtem Science*, 61(3):478–507, 2000.
9. Fangzhen Lin and Ray Reiter. Forget it! In *In Proceedings of the AAAI Fall Symposium on Relevance*, 154–159, New Orleans, US, 1994.
10. Jérôme Lang, Paolo Liberatore, and Pierre Marquis. Propositional independence: Formula-variable independence and forgetting. *Journal of Artificial Intelligence Research*, 18:391–443, 2003.
11. Christos H. Papadimitriou. *Computational complexity*. Addison-Wesley, 1994.

Reducing Answer Set Programs under partial assignments

Jianmin Ji[1]

[1] School of Computer Science and Technology,
University of Science and Technology of China, Hefei, China
jianmin@ustc.edu.cn

Hai Wan[2,3,]*

[2] School of Data and Computer Science, Sun Yat-sen University, Guangzhou, China
[3] Guangdong Key Laboratory of Big Data Analysis and Processing, Guangzhou, China
*wanhai@mail.sysu.edu.cn

Given a propositional language, a partial assignment is of a pair of sets of atoms, (P, N) such that P and N are the sets of atoms that have been assigned true and false, respectively. Such partial assignments are of interest for many reasons. One is that they are the states explored by DPLL-based solvers for SAT and ASP (Answer Set Programming). In this paper, we consider ASP and study the problem of how to decide if a partial assignment can be extended to an answer set of a given logic program. For this, we propose a notion of reduction of a logic program under a partial assignment.

Keywords: Answer Set; logic programming; partial assignment.

1. Introduction

This paper studies Answer Set Programming (ASP), a constraint-based problem solving paradigm that has found applications in many areas.[1] Over the years, there has been much work on developing efficient ASP solvers (e.g.[2–4]). Most of these solvers use depth-first search like DPLL algorithm[5] for SAT that make incremental assignments on atoms (variables). In other words, these solvers search through states of *partial assignments* which are pairs of sets of atoms, (P, N) such that P and N are the sets of atoms that have been assigned true and false, respectively. Thus it is of interest in detecting early on whether a partial assignment can or cannot reduce the program. In this paper we consider this problem.

2. Preliminaries

2.1. *Answer Set Programming*

In this paper, we only consider finite logic programs based on a propositional language $\mathcal{L}$. A *(disjunctive) logic program* (DLP) is a finite set of (disjunctive) rules of the form

$$a_1 \vee \cdots \vee a_k \leftarrow a_{k+1}, \ldots, a_m, \textit{not } a_{m+1}, \ldots, \textit{not } a_n \qquad (1)$$

where $n \geq m \geq k \geq 1$ and $a_1, \ldots, a_n$ are atoms. If $k = 1$, it is a *normal rule*. A *normal logic program* (NLP) is a finite set of normal rules. A *constraint* is of the form

$$\leftarrow a_1, \ldots, a_m, \textit{not } a_{m+1}, \ldots, \textit{not } a_n \qquad (2)$$

where $n \geq m \geq 1$ and $a_1, \ldots, a_n$ are atoms. We consider a constraint c of form (2) as a shorthand of the normal rule $f_c \leftarrow a_1, \ldots, a_m, \textit{not } a_{m+1}, \ldots, \textit{not } a_n, \textit{not } f_c$ where f_c is an atom in $\mathcal{L}$ that does not appear in other rules of the program.

We will also write a rule r of form (1) as $head(r) \leftarrow body(r)$ where $head(r)$ is $a_1 \vee \cdots \vee a_k$, $body(r) = body^+(r) \wedge body^-(r)$, $body^+(r)$ is $a_{k+1} \wedge \cdots \wedge a_m$, and $body^-(r)$ is $\neg a_{m+1} \wedge \cdots \wedge \neg a_n$, and we identify $head(r)$, $body^+(r)$, $body^-(r)$ with their corresponding sets of atoms, and $body(r)$ the set $\{a_{k+1}, \ldots, a_m, \neg a_{m+1}, \ldots, \neg a_n\}$. We use $tr(r)$ to denote the propositional formula $body(r) \supset head(r)$ and $tr(\Pi) = \bigwedge_{r \in \Pi} tr(r)$.

Given a DLP Π, we denote by $Atoms(\Pi)$ the set of atoms in it. A *literal* is either an atom p or its negation $\neg p$. Given a literal l, the *complement* of l, written $\bar{l}$ below, is $\neg a$ if l is a and a if l is $\neg a$, where a is an atom. For a set L of literals, we let $\bar{L} = \{\bar{l} \mid l \in L\}$. A *partial assignment* of a DLP Π is a pair (P, N), where P and N are sets of atoms in Π. (P, N) is *consistent* if $P \cap N = \emptyset$. (P, N) is *complete* if $P \cup N = Atoms(\Pi)$. The answer sets of a DLP are defined in.[6] Given a DLP Π and a set S of atoms, the Gelfond-Lifschitz reduct of Π on S, written Π^S, is obtained from Π by deleting: (1) each rule that has a formula *not p* in its body with $p \in S$, (2) all formulas of the form *not p* in the bodies of remaining rules. A set S of atoms is an *answer set* of Π if S is a minimal set (in the sense of subset relationship) satisfying Π^S. We use $AS(\Pi)$ to denote the set of answer sets of Π.

2.2. Loops, Loop Formulas, and Completion

Ref. 7 extended the notions of loops and loop formulas[3] to DLPs. For a DLP Π, the *positive dependency graph* of Π, written G_Π^p, is a directed graph whose vertices are atoms in Π, and there is an arc from p to q if there is a rule $r \in \Pi$ such that $p \in head(r)$ and $q \in body^+(r)$. A set L of atoms is a *loop* of Π, L-induced subgraph of G_Π^p is strongly connected. Note that, a singleton is also a loop of the program. We use $Loop(\Pi)$ to denote the set of loops of Π. Π is *tight*, if every loop of Π is a singleton.

Given a DLP Π and a nonempty set $E \subseteq Atoms(\Pi)$, a rule $r \in \Pi$ is an *external support* of E under Π if $head(r) \cap E \neq \emptyset$ and $E \cap body^+(r) = \emptyset$. Let $R^-(E, \Pi)$ be the set of external support rules of E under Π. The *loop formula* of E under Π, written $LF(E, \Pi)$, is the following implication:

$$\bigwedge_{p \in E} p \supset \bigvee_{r \in R^-(E,\Pi)} \left(body(r) \wedge \bigwedge_{q \in head(r) \setminus E} \neg q \right).$$

The *completion* of a DLP Π,[7] written $Comp(\Pi)$, is defined to be the formula $tr(\Pi) \wedge \bigwedge_{p \in Atoms(\Pi)} LF(\{p\}, \Pi)$. When Π is an NLP, $Comp(\Pi)$ is equivalent to the Clark's completion of Π.[8]

Theorem 2.1 (Theorem 1 in[7]). *For a DLP Π and a set S of atoms, $S \in AS(\Pi)$ iff $S \cup \overline{Atoms(\Pi) \setminus S}$ satisfies $tr(\Pi)$ and $LF(E, \Pi)$ for every nonempty set $E \subseteq Atoms(\Pi)$ iff $S \cup \overline{Atoms(\Pi) \setminus S}$ satisfies $tr(\Pi)$ and $LF(L, \Pi)$ for each loop L of Π.*

3. Reduce Programs under Partial Assignments

We provide a reduction Π' of a logic program Π under a partial assignment (P, N), so that Π' does not contain atoms in $P \cup N$. We provide a condition for Π and (P, N), under which $Comp(\Pi)$ has a model satisfying $P \cup \overline{N}$ iff $Comp(\Pi')$ is consistent. We also provide a condition for Π and (P, N), under which Π has an answer set under (P, N) iff Π' has an answer set.

We first introduce a reduction of a propositional formula ϕ under a partial assignment (P, N). We denote $\phi_{(P,N)}$ to be the formula obtained from ϕ by replacing each occurrence of an atom of P with $\top$ and each occurrence of an atom of N with $\bot$. Given $\psi = (a \vee d) \wedge (b \vee c)$, then $\psi_{(\{a\},\{c\})} = (\top \vee d) \wedge (b \vee \bot)$, which is equivalent to the formula b.

Proposition 3.1. *Let ϕ be a propositional formula and (P, N) a partial assignment. If (P, N) is consistent, then ϕ has a model I s.t. $P \cup \overline{N} \subseteq I$ iff $I \setminus (P \cup \overline{N})$ is a model of $\phi_{(P,N)}$.*

Let (P, N) be a partial assignment of a DLP Π. We denote $\Pi_{(P,N)}$ to be the DLP obtained from Π by deleting:

(1) each rule r s.t. $body^+(r) \cap N \neq \emptyset$, $body^-(r) \cap P \neq \emptyset$, or $head(r) \cap P \neq \emptyset$,
(2) all formulas of the form p in the heads of the remaining rules with $p \in N$,
(3) all formulas of the form p in the bodies of the remaining rules with $p \in P$,
(4) all formulas of the form $not\ p$ in the bodies of the remaining rules with $p \in N$.

Example 3.1. Consider the DLP Π_1:

$$a \vee c \leftarrow not\ b. \quad a \leftarrow b, not\ c. \quad b \leftarrow a, not\ c. \quad a \leftarrow a. \quad b \leftarrow b.$$

$\Pi_{1(\{a\},\{c\})}$ is $\{b \leftarrow .\ b \leftarrow b.\}$, $\Pi_{1(\{c\},\emptyset)} = \{a \leftarrow a.\ b \leftarrow b.\}$, and $\Pi_{1(\{a,c\},\emptyset)} = \{b \leftarrow b.\}$.

Proposition 3.2. *Let (P, N) be a partial assignment of a DLP Π.*

- *If I is a model of $Comp(\Pi)$ and $P \cup \overline{N} \subseteq I$, then $I \setminus (P \cup \overline{N})$ is a model of $Comp(\Pi_{(P,N)})$.*
- *If $S \in AS(\Pi)$, $P \subseteq S$, and $N \cap S = \emptyset$, then $S \setminus P \in AS(\Pi_{(P,N)})$.*

$Comp(\Pi_{(P,N)})$ is consistent does not imply that $Comp(\Pi)$ has a model satisfying $P \cup \overline{N}$ and $\Pi_{(P,N)}$ has an answer set does not imply that Π has an answer set under (P, N).

Example 3.1 (Continued). $Comp(\Pi_{1(\{a,c\},\emptyset)})$ *is consistent, however, $Comp(\Pi_1)$ does not have a model satisfying $\{a, c\}$. Moreover, $AS(\Pi_{1(\{a\},\{c\})}) = \{\{b\}\}$, however, $AS(\Pi_1) = \{\{c\}\}$, Π_1 does not have an answer set under $(\{a\}, \{c\})$.*

3.1. Extend partial assignments to models of completion

Here we introduce a condition for a DLP Π and a partial assignment (P, N), under which $Comp(\Pi)$ has a model satisfying $P \cup \overline{N}$ iff $Comp(\Pi_{(P,N)})$ is consistent. Then we provide a procedure to compute a model of $Comp(\Pi)$ based on such a partial assignment (P, N). We first introduce a binary relation $\sqsubseteq$ between partial assignments. Given two partial assignments, $(P_1, N_1) \sqsubseteq (P_2, N_2)$ iff $P_1 \subseteq P_2$ and $N_1 \subseteq N_2$. We introduce two mappings of a DLP Π, T_Π^+ and T_Π^-, for a partial assignment (P, N):

$T_\Pi^+(P, N) = \{p \mid \text{there is a rule } r \in \Pi \text{ such that } p \in head(r),$
$body^+(r) \subseteq P, \text{ and } body^-(r) \cup (head(r) \setminus \{p\}) \subseteq N\},$
$T_\Pi^-(P, N) = \{p \mid \text{for every rule } r \in \Pi, \text{ if } p \in head(r) \text{ then}$
$body^+(r) \cap N \neq \emptyset \text{ or } (body^-(r) \cup (head(r) \setminus \{p\})) \cap P \neq \emptyset\}.$

Then we define the mapping T_Π for a DLP Π and a partial assignment (P, N): $T_\Pi(P, N) = (T_\Pi^+(P, N), T_\Pi^-(P, N))$.

$T_\Pi \uparrow^0 (P, N) = (P, N);$
$T_\Pi \uparrow^k (P, N) = T_\Pi(T_\Pi \uparrow^{k-1} (P, N)). \qquad k > 0$

Similar to the discussion of the corresponding three-valued operator in,[9] we provide some properties of T_Π.

Proposition 3.3. *Let (P, N) be a partial assignment of a DLP Π. The following statements hold:*

- *If (P, N) is consistent, then $T_\Pi(P, N)$ is consistent.*
- *T_Π is monotonic, i.e., for any partial assignments (P_1, N_1) and (P_2, N_2), $(P_1, N_1) \sqsubseteq (P_2, N_2)$ implies $T_\Pi(P_1, N_2) \sqsubseteq T_\Pi(P_2, N_2)$.*
- *If (P, N) is consistent and $(P, N) \sqsubseteq T_\Pi(P, N)$, then there exists a consistent partial assignment $(P', N') = T_\Pi \uparrow^\infty (P, N)$ such that $(P', N') = T_\Pi(P', N')$ and $(P, N) \sqsubseteq (P', N')$.*
- *If (P, N) is a complete and consistent partial assignment of Π and $(P, N) = T_\Pi(P, N)$, then $P \cup \overline{N}$ is a model of $Comp(\Pi)$.*
- *If I is a model of $Comp(\Pi)$, then $(I \cap Atoms(\Pi), \overline{I} \cap Atoms(\Pi))$ is a complete and consistent partial assignment of Π and it is a fixed point of T_Π.*

Theorem 3.1. *Let (P, N) be a partial assignment of a DLP Π. If (P, N) is consistent and $(P, N) = T_\Pi(P, N)$, then $Comp(\Pi)$ has a model I such that $P \cup \overline{N} \subseteq I$ iff $I \setminus (P \cup \overline{N})$ is a model of $Comp(\Pi_{(P,N)})$.*

Given a consistent partial assignment (P, N) for a DLP Π with $(P, N) \sqsubseteq T_\Pi(P, N)$, from Proposition 3.3, we can compute a consistent fixed point (P^*, N^*) of T_Π via $T_\Pi \uparrow^\infty (P, N)$ in polynomial time. Then we can compute a model I^* of $Comp(\Pi_{(P^*, N^*)})$. From Theorem 3.1, $I^* \cup P^* \cup \overline{N^*}$ is a model of $Comp(\Pi)$.

Example 3.1 (Continued). $T_{\Pi_1}(\{a, c\}, \emptyset) = (\{a\}, \{c\})$ and $T_{\Pi_1}(\{c\}, \emptyset) = (\emptyset, \emptyset)$. $T_{\Pi_1}(\{a\}, \{c\}) = (\{a, b\}, \{c\})$ and $T_\Pi \uparrow^\infty (\{a\}, \{c\}) =$

($\{a,b\},\{c\}$), then $\{a, b, \neg c\}$ is a models of $Comp(\Pi_1)$. On the other hand, $T_{\Pi_1}(\{c\},\{a,b\}) = (\{c\},\{a,b\})$ and $\{\neg a, \neg b, c\}$ is also a model of $Comp(\Pi_1)$.

3.2. Extend partial assignments to answer sets

Here we introduce a condition for a DLP Π and a partial assignment (P, N), under which Π has an answer set under (P, N) iff $\Pi_{(P,N)}$ has an answer set. Then we provide a procedure to compute an answer set of Π based on a partial assignment (P, N).

First, we review the notions of unfounded sets and the well-founded operator.[10,11] Let (P, N) be a partial assignment of a DLP Π. A set X of atoms is an *unfounded set* of Π w.r.t. (P, N) if for each atom $p \in X$ and each rule $r \in \Pi_{(P,N)}$ with $p \in head(r)$, $X \cap body^+(r) \neq \emptyset$. If Π is an NLP, the union of two unfounded sets is also an unfounded set. (P, N) is *unfounded-free* for a DLP Π if $P \cap X = \emptyset$ for every unfounded set X of Π w.r.t. (P, N). If (P, N) is unfounded-free, then the union of two unfounded sets of Π w.r.t. (P, N) is also an unfounded set, thus there exists the greatest unfounded set of Π w.r.t. (P, N). We use $U_\Pi(P, N)$ to denote such greatest unfounded set, if exists. Note that, $T_\Pi^-(P, N) \subseteq U_\Pi(P, N)$. Now we define the well-founded operator: $W_\Pi(P, N) = \big(T_\Pi^+(P, N), U_\Pi(P, N)\big)$.

$$W_\Pi \uparrow^0 (P, N) = (P, N);$$
$$W_\Pi \uparrow^k (P, N) = W_\Pi(W_\Pi \uparrow^{k-1}(P, N)). \qquad k > 0$$

Notice that, for an NLP Π, $W_\Pi \uparrow^\infty (\emptyset, \emptyset)$ is the *well-founded model* of Π.[10]

Similar to the discussion of the corresponding three-valued operator in,[9] we provide some properties of W_Π.

Proposition 3.4. *Let (P, N) be a partial assignment of an NLP Π. The following statements hold:*

- W_Π *is monotonic, i.e., for any partial assignments (P_1, N_1) and (P_2, N_2), $(P_1, N_1) \sqsubseteq (P_2, N_2)$ implies $W_\Pi(P_1, N_2) \sqsubseteq W_\Pi(P_2, N_2)$.*
- *If $(P, N) \sqsubseteq W_\Pi(P, N)$, then there exists $(P', N') = W_\Pi \uparrow^\infty (P, N)$ such that $(P', N') = W_\Pi(P', N')$.*
- *If (P, N) is a complete and consistent partial assignment of Π and $(P, N) = W_\Pi(P, N)$, then P is an answer set of Π.*
- *If S is an answer set of Π, then $(S, Atoms(\Pi) \setminus S)$ is a complete and consistent partial assignment of Π and it is a fixed point of W_Π.*

Note that, given a partial assignment (P, N) of a DLP Π, (P, N) is consistent and $(P, N) \sqsubseteq W_\Pi$ do not imply that $W_\Pi \uparrow^\infty (P, N)$ is consistent.

Example 3.2. Consider the NLP Π_2:

$$a \leftarrow b. \qquad\qquad b \leftarrow a. \qquad\qquad a \leftarrow c.$$

$(\{a,b\}, \emptyset)$ is consistent and $W_{\Pi_2}(\{a,b\}, \emptyset) = (\{a,b\}, \{c\})$. However, $W_{\Pi_2}\uparrow^\infty(\{a,b\}, \emptyset) = (\{a,b\}, \{a,b,c\})$.

Theorem 3.2. *Let (P, N) be a partial assignment of a DLP Π. If (P, N) is consistent and $(P, N) = W_\Pi(P, N)$, then Π has an answer set S such that $P \subseteq S$ and $S \cap N = \emptyset$ iff $S \setminus P$ is an answer set of $\Pi_{(P,N)}$.*

Given a consistent and unfounded-free partial assignment (P, N) for a DLP Π, we can first expand (P, N) by W_Π. In specific, we define:

$$W_\Pi^*(P,N) = (P \cup T_\Pi^+(P,N), N \cup U_\Pi(P,N)); \quad W_\Pi^*\uparrow^0(P,N) = (P,N);$$
$$W_\Pi^*\uparrow^k(P,N) = W_\Pi^*(W_\Pi^*\uparrow^{k-1}(P,N)). \qquad\qquad k > 0$$

W_Π^* is monotonic and $(P, N) \sqsubseteq W_\Pi^*(P, N)$. Then we can compute a fixed point (P^*, N^*) of W_Π^* via $W_\Pi^*\uparrow^\infty(P, N)$ in polynomial time. Note that, $W_\Pi(P^*, N^*) \sqsubseteq (P^*, N^*)$, then we can compute a fixed point (P', N') of W_Π via $W_\Pi\uparrow^\infty(P^*, N^*)$ in polynomial time. If (P', N') is consistent, then we can compute an answer set S' of $\Pi_{(P,N)}$. From Theorem 3.2, $S' \cup P'$ is an answer set of Π.

Example 3.1 (Continued). $W_{\Pi_1}(\{a\}, \{c\}) = (\{a,b\}, \{c\})$ and $W_\Pi\uparrow^\infty(\{a\}, \{c\}) = (\{a,b,c\}, \{a,b,c\})$ which is not consistent. $W_{\Pi_1}(\{c\}, \emptyset) = (\emptyset, \{a,b\})$, $W_\Pi^*\uparrow^\infty(\{c\}, \emptyset) = (\{c\}, \{a,b\})$, and $(\{c\}, \{a,b\})$ is a consistent fixed point of W_{Π_1}, then $\{c\}$ is an answer set of Π_1.

4. Conclusion

This paper introduces the notion of the reduction of a logic program under a partial assignment. Given that partial assignments are states that most of the ASP solvers explore, this work is related to the expansion operators of DPLL-based ASP solvers for NLPs[2] and DLPs,[12] forgetting in ASP,[13,14] splitting,[15,16] and simplifying a logic program by its consequences.[17] Furthermore, a partial assignment (P, N) can be considered a three valued model that assigns atoms in P true, those in N false and all the others unknown. So this work is related to three-valued semantics for logic programs.

Acknowledgments

Jianmin Ji's research was partially supported by the Guangdong Province Science and Technology Plan projects (No. 2017B010110011). Hai Wan's research was in part supported by the National Natural Science Foundation of China (No. 61573386), Natural Science Foundation of Guangdong Province (No. 2016A030313292), Guangdong Province Science and Technology Plan projects (No. 2016B030305007 and 2017B010110011), Guangzhou Science and Technology Project (No. 201804010435), and Sun Yat-sen University Young Teachers Cultivation Project under grant (No. 16lgpy40) and GF Cultivation Project.

References

1. M. Gelfond and V. Lifschitz, The stable model semantics for logic programming, in *Proceedings of the 5th International Conference on Logic Programming (ICLP-88)*, 1988.
2. P. Simons, I. Niemelä and T. Soininen, *Artificial Intelligence* **138**, 181 (2002).
3. F. Lin and Y. Zhao, *Artificial Intelligence* **157**, 115 (2004).
4. M. Gebser, B. Kaufmann, A. Neumann and T. Schaub, Conflict-driven answer set solving, in *Proceedings of the 20th International Joint Conference on Artificial Intelligence (IJCAI-07)*, 2007.
5. M. Davis and H. Putnam, *Journal of the ACM (JACM)* **7**, 201 (1960).
6. M. Gelfond and V. Lifschitz, *New generation computing* **9**, 365 (1991).
7. J. Lee and V. Lifschitz, Loop formulas for disjunctive logic programs, in *Proceedings of the 19th International Conference on Logic Programming (ICLP-03)*, 2003.
8. K. L. Clark, Negation as failure, in *Logic and Databases*, eds. H. Gallaire and J. Minker (Plenum Press, New York, 1978) pp. 293–322.
9. M. Denecker, V. W. Marek and M. Truszczyński, *Information and Computation* **192**, 84 (2004).
10. A. Van Gelder, K. Ross and J. S. Schlipf, *Journal of the ACM* **38**, 620 (1991).
11. N. Leone, P. Rullo and F. Scarcello, *Information and computation* **135**, 69 (1997).
12. N. Leone, G. Pfeifer, W. Faber, T. Eiter, G. Gottlob, S. Perri and F. Scarcello, *ACM Transactions on Computational Logic* **7**, 499 (2006).
13. J. P. Delgrande and K. Wang, A syntax-independent approach to forgetting in disjunctive logic programs, in *Proceedings of the 29th AAAI Conference on Artificial Intelligence (AAAI-15)*, 2015.
14. J. Ji, J.-H. You and Y. Wang, On forgetting postulates in answer set programming, in *Proceedings of the 24th International Joint Conference on Artificial Intelligence (IJCAI-15)*, 2015.
15. V. Lifschitz and H. Turner, Splitting a logic program, in *Proceedings of the 11th International Conference on Logic Programming (ICLP-94)*, 1994.
16. J. Ji, H. Wan, Z. Huo and Z. Yuan, Splitting a logic program revisited, in *Proceedings of the 29th AAAI Conference on Artificial Intelligence (AAAI-15)*, 2015.
17. J. Ji, H. Wan, Z. Huo and Z. Yuan, Simplifying a logic program using its consequences, in *Proceedings of the 24th International Joint Conference on Artificial Intelligence (IJCAI-15)*, 2015.

Improvements of categorical propositions on quantification and systemization

Yinsheng Zhang

Institute of Scientific & Technical Information of China
Beijing, 100038, China

Categorical propositions, as widely used forms in natural languages since Aristotle's definition and usage, remain two limitations that (a) the particular quantifier is ambiguous for its restricted reading by Euler as "non-empty but not universal", and the unrestricted reading as "non-empty and possibly universal" by Gergonne; (b) the predicate (right-hand) term formally lacks a modifier, rendering unclear and insufficient expressing its quantity. So, expanded categorical propositions (ECPs) with dyadic and generalized quantifiers are proposed for overcoming the two limitations. ECPs are proved to be calculated logically and operated mathematically; moreover, these two kinds of computations are correspondingly, i.e., ECPs and their mathematical models are isomorphic, providing ECPs' logic calculi with computing essentials.

1. Introduction

Natural languages are used to express and infer knowledge, sometimes with troubles due to ambiguities. Categorical propositions, making up syllogisms hence widely having been used by Aristotle, have been encountering some ambiguities.

Definition 1 An Aristotelian categorical proposition (ACP) is formalized as the pattern "QA be/be not B"; or as the follows structure

$$Q \text{ Term copula Term,} \qquad (1)$$

where Q refers to a universal or particular quantifier.

ACPs' classification is listed in Table 1.

Table 1. Classification of Aristotelian categorical propositions (ACPs) [1].

Names of ACPs	Expression in English	Quantifier	Quantifier's type
A	Every (For all) S is P	Every (For all)	Universal
E	Every (For all) S is not P	Every (For all)	Universal
I	Some S is P	Some	Particular
O	Some S is not P	Some	Particular

Against ACPs, Euler and Gergonne respectively made different illustrations as in Table 2 [2][3][4][5].

Table 2. Different illustrations on ACPs by Euler and Gergonne.

Relations	Being Included	Coincidence	Outer Separation	Intersection	Inclusion
Relation No.	$\pi = 1$	$\pi = 2$	$\pi = 3$	$\pi = 4$	$\pi = 5$
Symbols	$\subset$	●	☒	$\supset$	‖
Topologies of X and Y	Y(X)	(XY)	(X)(Y)	X(Y)	(X)(Y)
Euler Diagrams	All x are y		Some x are y. Some x are not y		All x are not y.
Gergonne Diagrams	All x are y			Some x are not y.	All x are not y.
		Some x are y.			

Obviously, there is an inconsistency between Euler diagrams and Gergonne diagrams on ACPs. Besides Gergonne, some other researchers like Venn [6], Peirce [7], William Kneale and Martha Kneale, Sun-Joo Shin [8] have criticized Euler diagrams for partly depicting the universal and particular quantifications. This ambiguity problem posed, e.g., by Keith Stenning and Michiel Van Lambalgen: "How could a brilliant mathematician like Euler make such a fundamental mistake? [9]"

The present paper aims to disambiguate and harmonize ACPs, creating improved categorical propositions expressed and inferred both in formal and natural languages for computations.

2. Factor Analysis of the Ambiguities

Before analysis, we make a definition of a Gergonne space.

Definition 2 Let one Euler circle denote a non-empty set of discourse, and all the topological samples of two Euler circles are called a Gergonne space G.

It is easy to cognize that a Gergonne space constitutes all topological relations of two sets, which are symbolized by $\subset, \bullet, \varnothing, \supset, \|$ and numbered by a variable $\pi = 1, 2, 3, 4, 5$, as shown in Table 2.

Comparing the interpretations between Euler and Gergonne on A, E, I, O, we find that there exists the consent on E, which we do not need to discuss furthermore; there is a little bit discrepancy on A — that is, a coincidence relation in G is lost in Euler, and should be added so as to remove the discrepancy; while, the difference matters deeply in a contradictory interpretation of the particular quantifier. To identify it, Definition 3 is given as follows.

Let X be a set which consists of all the elements of x_i, $i=1, 2, ..., n$, $n \geq 2$.
UN(X) is the universal set of x_i, i.e., UN(X) = $\{x_i | i=1,2, ..., n\}$.
PT(X) is a proper subset of UN(X), i.e., PT(X) = $\{x_i | i=1, ..., m, m<n\}$.
ID(X) is a subset of UN(X), i.e., ID(X) = $\{x_i | i=1, 2, ..., m, m \leq n\}$.
NU(X) is null set, i.e., NU(X) = ø.

Definition 3 Of the expanded quantifiers: $\forall x := $ UN(X), $\exists x := $ ID(X), $\top\!\!\!\!\bot\, x := $ PT(X). "$\top\!\!\!\!\bot$" reads "the partial".

Proposition 1 The particular quantifier $\bar{\exists}$ interpreted by Euler and Gergonne is inconsistent between a restricted reading (non-empty but not universal) and an unrestricted reading (non-empty and possibly universal), which give rise to contradiction or inconsistency in quantification. Say,

$$\text{Intrpr}_1 (\bar{\exists}) = \text{PT} \tag{2}$$

$$\text{Intrpr}_2 (\bar{\exists}) = \text{ID} \tag{3}$$

Here, *Intrpr* denotes an interpretation with an independent logic value PT or ID. In short, formula (2) is the restricted interpretation by Euler excluding the scheme of the universal discourse, i.e., only referring to a (proper) part, a closed set, meaning "non-empty but not universal". While formula (3) is the unrestricted interpretation by Gergonne including the scheme of the universal discourse, i.e., referring to an open set "non-empty and possibly universal". Therefore, formula (2) and formula (3) contradict each other; so, the particular quantifier is ambiguous.

Besides the default, stated in Proposition 1, of ACPs, there is another default — both Euler and Gergonne diagrams put X and Y to same semantic roles, but Y is not modified in forms of ACPs. In such a case, the right-hand term Y, unlike X is ambiguous for missing a quantifier to indicate the topological illustration.

Therefore, some measures should be taken to get rid of contradiction of interpretation on the particular quantifier, and to vanish the ambiguity of monadic quantifier forms in ACPs.

3. Formal Improvements on ACPs

In Definition 3, $\forall$, $\exists$ and $\top\!\!\!\!\bot$ cover Euler and Gergonne's interpretations of quantifiers of ACPs, which can be a correction of contradiction in quantification. The partial quantifier $\top\!\!\!\!\bot$ can be regarded as one of generalized quantifiers, which proceeds from the advances of studies about generalized quantifiers. For recent years, many researchers such as Mostovski [10], Montague [11] [12], Barwise and Cooper [13] [14] created some forms of categorical propositions with generalized quantifiers, and established corresponding mathematical models to compute these generally quantified propositions.

As an ACP is of a monadic proposition form, that is, there is only one quantifier binding the left term in an ACP, causing quantitative vagueness of another term, so logicians have been trying to make up this shortcoming. G. Bentham and W. Hamilton [15] [16] first came up with a "two-place" quantitative proposition form (as "$Q_1 A$ is/is not $Q_2 B$", hereafter "dyadic proposition"), but lacking the existential quantifier among their proposed quantifiers. J. Bentham [17] established models of dyadic categorical propositions.

Therefore, to overcome the shortage of the quantifiers for Euler and Gergonne quantifier systems, and the lack of right-hand quantifier in an ACP, a new system of expanded categorical proposition is proposed here.

Definition 4 An expanded categorical (atomic) proposition (ECP) is a dyadic categorical proposition expressed by the following structure:

$$Q_1 \ Term_1 \ Copula \ Q_2 \ Term_2 \qquad (4)$$

and satisfy that (a.) $Q_1, Q_2 \in \{\top, \forall, \exists\}$; (b.) $copula \in \{+, -\}$ (to denote "be", "be not"; and + can be omitted.); (c.) $Q_2 \notin \top, \exists$, if the copula is "–". For example, "All birds are not stones" is an ECP, which omits a right-hand quantifier "all"; so is "All birds are partial animals"; but "All birds are not partial animals", "All birds are not existential animals" are not allowed.

ECPs are given as in Table 3, where, some propositions were proposed by G. Bentham and W. Hamilton [15] [16]. However, the existential quantifier was lost in their forms, that is, only universal and partial quantifiers were designated for dyadic categorical propositions.

Expression (4) covers all the samples of $Q_1, Q_2 \in \{\top, \forall, \exists\}$ and $copula \in \{+, -\}$ with exceptions of "$-\top$" and "$-\exists$" permutations — i.e., denying a partial or existential predicate term is forbidden according to natural language traditions. (4) has merged Euler's and Gergonne's schemes on interpretations of the particular quantifier, and removed the ambiguity resulting from losing the right-hand quantifier. In quantification, (4) describes all the states from uncertainty ($\exists X$ or $\neg \exists X$) to the certainty ($\exists$): from the partial ($\top$) to all ($\forall$) in a first quantification partition in case the existential state is ascertained. So (4) should be formal and natural propositions without ambiguity for knowledge expression and inference.

Evidently, according to Definition 1 and semantics of natural languages we have Axiom 1 and Axiom 2.

Table 3. Expanded categorical propositions (ECPs).

Names of propositions	Forms of propositions (Σ)	Combinations of relations of sets of X and Y ($C[H_{(x,\eta,\delta)}] \cdot n$)	Expressions in a natural language
a	$\exists X \exists Y$	$\{\bullet,\square,\supset,\subset\},\{\square,\supset,\subset\},$ $\{\bullet,\square\},\{\supset,\subset\}$	Existential X are at least one Y
b	$\forall X \exists Y$	$\{\bullet,\subset\}$	All X are at least one Y
c	$\top X \exists Y$	$\{\square,\supset\}$	Partial X are at least one Y
d	$\exists X \forall Y$	$\{\bullet,\supset\}$	Existential X are all Y
e	$\exists X \top Y$	$\{\square,\subset\}$	Existential X are partial Y
f	$\forall X \forall Y$	$\{\bullet\}$	All X are all Y
g	$\forall X \top Y$	$\{\subset\}$	All X are/is partial Y
h	$\top X \forall Y$	$\{\supset\}$	Partial X are all Y
i	$\top X \top Y$	$\{\square\}$	Partial X are Partial Y
x	$\exists X \neg \forall Y$	$\{\|,\square,\supset\},\{\|,\square\},\{\|,\supset\}$	Existential X are not Y
y	$\forall X \neg \forall Y$	$\{\|\}$	All X are not Y
z	$\top X \neg \forall Y$	$\{\square,\supset\}$	Partial X are not Y

Axiom 1

$$\text{ID}(X) \subseteq \text{UN}(X); \text{ i.e., } \{x\}_{i=1}^{m \leq n} \subseteq \{x\}_{i=n} \tag{5-1}$$

$$\text{PT}(X) \subset \text{UN}(X); \text{ i.e., } \{x\}_{i=1}^{m < n} \subset \{x\}_{i=n} \tag{5-2}$$

Axiom 2

$$\exists \equiv \top | \forall \tag{6-1}$$

$$\forall \equiv \exists \wedge \neg \top \tag{6-2}$$

$$\top \equiv \exists \wedge \neg \forall \tag{6-3}$$

where, $\exists$ is the existential quantifier, reads "at least one"; | reads "not both true or false", namely, "Xor". So $\exists$ states "either partial or all"; $\forall$ indicates "existential but not partial"; $\top$ conveys "existential and not all"; $\equiv$ was first used by Russell, referring to equivalent substituition or logic equivalence "$\leftrightarrow$".

Theorem 1 The calculus rules of ECPs' quantifiers satisfy the basic logic calculus $\rightarrow$, |, $\wedge$, $\neg$ according to (6-1), (6-2) and (6-3), and the calculus results are listed in Tables 4, 5, 6 and 7.

Tables 4, 5, 6, 7. The calculi ($\rightarrow$, |, $\wedge$, $\neg$) rules of the quantifiers in ECPs.

$\rightarrow$	$\top$	$\forall$	$\exists$	|	$\top$	$\forall$	$\exists$	$\wedge$	$\top$	$\forall$	$\exists$		$\equiv$	
$\top$	T	$\varnothing$	T	$\top$	$\top$	$\varnothing$	$\exists$	$\top$	$\top$	$\varnothing$	$\top$		$\top$	$\forall \vdash \exists$
$\forall$	$\varnothing$	T	T	$\forall$	$\varnothing$	$\forall$	$\exists$	$\forall$	$\varnothing$	$\forall$	$\forall$	$\neg$	$\forall$	$\top \vdash \exists$
$\exists$	$\varnothing$	$\varnothing$	T	$\exists$	$\exists$	$\exists$	$\exists$	$\exists$	$\top$	$\forall$	$\exists$		$\exists$	$\neg \top \wedge \neg \forall \equiv \forall \neg$

Proof Tables 4, 5, 6 and 7 are directly obtained from Definition 3, Axiom 1, Axiom 2 and basic logic calculi. Here are some simple proofs for simplicity of presentation. Given, take note that, a certain expression of ECP, the relations of double terms, i.e., the elements in G, are random-event samples, so the calculus "$\vee$" for a disjunct relation becomes "Xor", i.e., "$|$".

For $\neg\exists$: $\neg(\exists\equiv\overline{\top}|\forall) \Rightarrow \neg\exists\equiv\neg(\overline{\top}|\forall)\equiv\neg\overline{\top}\wedge\neg\forall$

For $(\forall\wedge\overline{\top})$: $(\forall\equiv\exists\wedge\neg\overline{\top})\wedge(\overline{\top}\equiv\exists\wedge\neg\forall)\Rightarrow$
$(\forall\wedge\overline{\top})\equiv\exists\wedge\neg\overline{\top}\wedge\exists\wedge\neg\forall\equiv\exists\wedge\neg\exists\equiv\emptyset$

For $(\overline{\top}\to\forall)\to\emptyset$: $\overline{\top}\to\forall\equiv\neg\overline{\top}\vee\forall\equiv(\forall|\neg\exists)\vee\forall\equiv(\forall|\neg\exists)$
$\equiv((\exists\wedge\neg\overline{\top})|\neg\exists)\equiv(\exists|\neg\exists)\wedge(\neg\overline{\top}|\neg\exists)$
$\equiv(\exists|\neg\exists)\wedge\neg(\overline{\top}\wedge\exists)\equiv T\wedge\neg\overline{\top}\equiv\neg\overline{\top}$
$\Rightarrow \overline{\top}\to\forall\to\neg\overline{\top}\Rightarrow\overline{\top}\to\neg\overline{\top}\Rightarrow(\overline{\top}\to\forall)\to\emptyset.$

4. Mathematic Models of Logic Calculus of ECPs

At the moment, it follows an important problem that whether ECPs can be mathematically modeled, so that ECPs and their inferences are based on mathematical operations — solving which is a goal for logicians since Boole made logic based on algebra, like said logic historians William Kneale and Martha Kneale: "Boole's success in constructing an algebra which included all the theorems of traditional logic led some logicians to assume that all logic must be capable of presentation in algebraic form, and attempts were made in the next generation to work out a logic of relations in the same fashion as the logic of classes.[18]" In fact, ACPs and their derivatives from the generalized quantifier theories as well as the dyadic proposition ideas above mentioned, are difficult to be modeled by algebraic forms for the discrepancy between Euler and Gergonne, and for the complex combines of relations in G. If ECPs could overcome these difficulties, ECPs would continue to achieve "Boole's success". To do this, we will use isomorphism, by which Godel once confirmed that Heyting Arithmetic (HA) functions act as the "agent" of Peano Arithmetic (PA), manifesting that isomorphism is a proper method to express a primary logic, e.g., ECPs' calculi, by their mathematic functions' operations [19] [20], like operations of sets of G-relations.

Let $<X_\pi, Y_\pi>$ denote a sample of relations of X and Y in G, π be an ordinal number of a sample, i.e., $\pi \in \{1, 2, 3, 4, 5\}$.

Let $H_{(X, Y)}$ be the universal set of the samples of G, namely

$$H_{(X, Y)} = \{<X_\pi, Y_\pi>|\pi=1, 2, 3, 4, 5\}.$$

Proposition 2 According to the definition of Gergonne space, there is an injection h:

$$h: \mathrm{H}_{(X,\,Y)} \to G, \text{ or}$$

$$h: \{<X_\pi, Y_\pi>\} \to \{\subset, \bullet, \varnothing, \supset, \|\}.$$

Proposition 2 posits that every ECP has, by a f function, a unique combination (or a unique Xor of the combinations) of G-elements, which means that ECPs logic calculi correspond to operations of the combinations of G-elements. Let $A, B \in \Sigma$, and $C[\mathrm{H}_{(X,\,Y)},\delta]$. n denote a δ-combination of $\mathrm{H}_{(X,\,Y)}$, $\delta \leq 4$; n is a code of δ-combinations, we can, according to Axiom 2, set up an algebraic structure $<\Sigma, \to, |, \wedge, \neg>$, and consequently, to build an algebraic structure $<\{C[\mathrm{H}_{(X,\,Y)},\delta].\,n\}, \subset, \cup, \cap, ->$.

Theorem 2 There is an isomorphism between logic calculi of ECPs and relation operations of G-element combinations. That is,

$$<\Sigma, \to, |, \wedge, \neg> \cong <\{C[\mathrm{H}_{(X,\,Y)}, \delta].\,n\}, \subset, \cup, \cap, -> \qquad (7)$$

where, $\cong$ denotes isomorphism; $\subset$, $\cup$, $\cap$, $-$ are the set operations of G-elements, and the last operator "$-$" represents the calculus "difference set".

Theorem 2.1 (ECPs Implication Theorem): Given $A, B, M \in \Sigma$, iff $A \to B$, $C[\mathrm{H}_{(X,\,Y)},\delta_1].\,n_1 \subset C[\mathrm{H}_{(X,\,Y)},\delta_2].\,n_2$.

Proof $A \to B$ shows that an ECP A implies B, and equivalently means that if A is TRUE then B must TRUE, and if B is FALSE then A must be FALSE. All the ECPs satisfying this conditions are listed in Table 8 and Table 9 by Table 4. For example, **f→b**, i.e., $\forall X \forall Y \to \forall X \exists Y$ says, "all X are all Y" implies "all X are at least one Y"; **e→a**, i.e., $\exists X \top Y \to \exists X \exists Y$ says, "the existential X are partial Y" implies "the existential X are at least one Y".

Furthermore, the h functions of ECPs, those combinations, or "|" operations of the combinations, of G-elements, have the operations of "$\subset$". For example,

$\mathbf{b} \to \mathbf{a}$;
$f(\mathbf{b}) = \{\bullet, \subset\}$; $f(\mathbf{a}) = \{\bullet, \varnothing\} | \{\supset, \subset\} | \{\bullet, \varnothing, \supset, \subset\} | \{\varnothing, \supset, \subset\}$;
$\{\bullet, \subset\} \subset \{\bullet, \varnothing, \supset, \subset\} \subset \{\{\bullet, \varnothing\} | \{\supset, \subset\} | \{\bullet, \varnothing, \supset, \subset\} | \{\varnothing, \supset, \subset\}\}$;
$f(\mathbf{b}) \subset f(\mathbf{a})$;

So, $<\{\mathbf{b}, \mathbf{a}\}, \to> \cong <\{\{\bullet, \subset\}, \{\bullet, \varnothing\} | \{\supset, \subset\} | \{\bullet, \varnothing, \supset, \subset\} | \{\varnothing, \supset, \subset\}\}, \subset>$.

Tables 10, 11 just bear out the isomorphic relations with Tables 8, 9 (for ease of making tables, $\to$ and $\subset$ are denoted as their reverse orientations $\leftarrow$ and $\supset$). In the same way, we have Theorem 2.2 ~Theorem 2.4.

Theorem 2.2 (ECPs Xor Theorem): Let

$A, B, M \in \Sigma$, $f(A) = C[\mathrm{H}_{(X,\,Y)},\delta_1].\,n_1$, $f(B) = C[\mathrm{H}_{(X,\,Y)},\delta_2].\,n_2$, $f(M) = C[\mathrm{H}_{(X,\,Y)},\delta_3].\,n_3$. Iff $A|B \equiv M$, $C[\mathrm{H}_{(X,\,Y)},\delta_1].\,n_1 \cup C[\mathrm{H}_{(X,\,Y)},\delta_2].\,n_2 = C[\mathrm{H}_{(X,\,Y)},\delta_3].\,n_3$.

Theorem 2.3 (ECPs Conjunction Theorem): Let

$A, B, N \in \Sigma$, $f(A) = C[\mathrm{H}_{(X,\,Y)},\delta_1].\,n_1$, $f(B) = C[\mathrm{H}_{(X,\,Y)},\delta_2].\,n_2$, $f(N) = C[\mathrm{H}_{(X,\,Y)},\delta_3].\,n_3$. Iff $A \wedge B \equiv N$, $C[\mathrm{H}_{(X,\,Y)},\delta_1].\,n_1 \cap C[\mathrm{H}_{(X,\,Y)},\delta_2].\,n_2 = C[\mathrm{H}_{(X,\,Y)},\delta_3].\,n_3$.

Tables 8, 9. The implication calculi of ECPs.

ECP with two "∃"	ECP with one "∃"	ECP without any "∃"		ECP with one "∃"	ECP without any "∃"	
∃X∃Y a	∀X∃Y b	∀X∀Y	f	∃X–∀Y x	∀X–∀Y	y
		∀X⊤̄Y	g		⊤̄X–∀Y	z
	⊤̄X∃Y c	⊤̄X∀Y	h	←		
		⊤̄X⊤̄Y	i			
	∃X∀Y d	∀X∀Y	f			
		⊤̄X∀Y	h			
	∃X⊤̄Y e	∀X⊤̄Y	g			
		⊤̄X⊤̄Y	i			
	←					
←						

Tables 10, 11. "⊃" operations of functions of ECPs.

ECP with two "∃"	ECP with one "∃"		ECP without any "∃"		ECP with one "∃"		ECP without any "∃"	
{●,□,⊃,⊂}, {□,⊃,⊂}, {●,□}, {⊃,⊂}			{●}	f(f)	{∥,□,⊃}, {∥,□}, {∥,⊃}	f(x)	∀X–∀Y	f(y)
	{●,⊂}	f(b)	{⊂}	f(g)			⊤̄X–∀Y	f(z)
	{□,⊃}	f(c)	{⊃}	f(h)	⊃			
			{□}	f(i)				
	{●,⊃}	f(d)	{●}	f(f)				
			{⊃}	f(h)				
	{□,⊂}	f(e)	{⊂}	h(g)				
			{□}	h(i)				
			⊃					
⊃								

Theorem 2.4 (ECPs Negation Theorem): Let $A, B \in \Sigma$. Iff $A \equiv \neg B$, or $\neg A \equiv B$, $C[H_{(X, Y)}, \delta] \cdot n_1 = G - C[H_{(X, Y)}, \varepsilon] \cdot n_2$. Or, $C[H_{(X, Y)}, \varepsilon] \cdot n_2 = G - C[H_{(X, Y)}, \delta] \cdot n_1$.

All the proofs of Theorem 2.2~Theorem 2.4 are roughly the same with the proof of Theorem 2.1, which we omit for simplicity of presentation.

5. Conclusions

ACPs, as the forms of knowledge expressions and atomic propositions for inferences in natural languages, remain two drawbacks. (a.) the particular quantifier has two interpretations: the restricted reading by Euler as non-empty but not universal ($\exists \wedge \neg \forall$); in parallel, the unrestricted reading as non-empty and possibly universal (either partial or universal, $\top|\forall$) by Gergonne, which is inconsistent, i.e., violates the law of identity. (b.) There is only one quantifier manifestly modifying one (left-hand) term, losing exact and expressive indication of quantification of the right-hand term, as well as of the dyadic-terms relations. These two limitations block ACPs from algebraic expressions. Euler narrowed interpretations of ACPs, neglecting some samples of the possible relations of the dyadic sets of the double terms. Although Gergonne proposed interpretations by systematic relations of the two sets, but did not reform the categorical propositions, which are still flawed by the two kinds of interpretations. Whereas generalized quantifiers theories have added quantifiers which are short for ACPs, but have not harmonized conflictions between Euler and Gergonne, and basically no quantifier has been set for right-hand term of an ACP. Therefore, a categorical proposition system, removing the two flaws and unifying logic rules and mathematical operations correspondingly, are needed to be remedied.

ECPs are designed to be expressions in both formal and natural languages, and to overcome the two drawbacks. ECPs are systemized, firstly, by the systemized quantifiers — the existential, the universal and the partial, they cover all the steps to roughly partition a part of a set: the primary partition is of certitude (by $\exists$ existence is apart from the non-existent), and the further partition is of exactitude (discriminating $\forall$ or $\top$). Secondly, ECPs are quantification-improved by having dyadic places covered by all the systemized quantifiers $\exists, \forall$ and $\top$.

As a results, every ECP maps some certain relations of its double-terms articulating sets, and these relations' operations are corresponded by those logic calculations of ECPs. So ECPs system can wholly and mathematically modeled to meet the algebraic conventions of logic. Besides, the mathematical expressions of ECPs with their calculi make a way to be computed.

References

1. W.E. Preece and P.W. Goetz, *Syllogistic*, The new Encyclopaedia Britannica (in 32 volumes, 15 th Edition), pp. 455, 129(1974-2012).
2. L. Euler, *Lettres à une Princesse d'Allemagne*, St. Petersberg; l' Academie Imperiale des Science (1768).

3. W. Kneale and M. Kneale. *The Development of Logic.* Oxford at the Clarendon Press, 353(1963).
4. J.D. Gergonne, Essai de dialectique rationelle, *Annales des mathématiques qures et appliqués*, 7 189-228(1816-1817).
5. W. Kneale and M. Kneale, *The Development of Logic*, Clarendon Press, Oxford, 350-352(1962).
6. J. Venn, M.A., *Symbolic Logic*, Macmillan and Co., London, 15-16(1881).
7. Peirce, The Simpest Mathematics, The Collected Papers of Charles Sanders Peirce, Vol. 4, Virginia, USA, Paragraph 353 (1994).
8. S.-J. Shin, The Logical Status of Diagrams, Cambridge University Press, Cambridge, 13(2006).
9. K. Stenning, M. Van Lambalgen, Human Reasoning and Cognitive Science, The MIT Press, Cambridge MA, 302~303(2008).
10. A. Mostowski. On a generalization of quantifiers, Fundementa Mathematicae 44 17-36(1957).
11. Richard Montague, The Proper Treatment of Quantification in Ordinary English. Formal Semantics,The essential Readings. Edited by Paul Portner and Babara H. Partee. Blackwell Publishing. 17-34(1974).
12. Barbara Hall Partee. The structure of meaning. Lecture 13: Noun Phrases and Quantification
http://people.umass.edu/partee/NZ_2006/NZ13%20Noun%20Phrases%20and%20Quantification.pdf
13. J. Barwise and Robin Cooper. Generalized quantifiers and natural language. Linguistics and Philosophy, Volume 4, Number 2 / 159-219 (1981).
14. Jon Barwise and Robin Cooper General Quantifiers and Natural Language. Formal Semantics, The essential Readings. Edited by Paul Portner and Babara H. Partee. Blackwell Publishing. 75—126(1974).
15. William Kneale and Martha Kneale. The Development of Logic. Oxford At the Clarendon Press, 353(1963).
16. D.J. Bennett. Logic Made Easy — How to Know When Language Deceives You, New York., 71(2003).
17. J. Benthem. Polyadic quantifiers. Linguistics and Philosophy, Springer Volume 12 No. 437-464(1989).
18. W. Kneale and M. Kneale. The Development of Logic. Oxford At the Clarendon Press, 427(1963).
19. G. Kurt. Über eine bisher noch nicht benützte Erweiterung des finiten Standpunktes. dialecticaVolume 25 Issue 4:280-287(1958).
20. J. Avigag and S. Feferman. Godel's Functional ("Dialectica") Interpretation, Samual Buss Editor, Handbook of Proof Theory, Elsevier Press, 342-397(1998).

Non-prioritised belief revision for DL-Lite TBoxes

Quan Yu*

School of Mathematics and Statistics,
Qiannan Normal University for Nationalities, China
yuquanlogic@126.com

Zhiqiang Zhuang, Zhe Wang and Kewen Wang

School of Information and Communication Technology, Griffith University, Australia
{z.zhuang, zhe.wang, k.wang}@griffith.edu.au

Previous studies of belief revision for description logics mostly assume that the new information are to be fully accepted. This assumption is not suitable for applications where new information is not always fully trustworthy. In this paper, we propose revision operators for description logics that do not give priority to the new information and provide representation theorems.

Keywords: Belief revision; description logic.

1. Introduction

The dominant approach in belief change is the so called AGM framework,[1,2] where the main strategies are to articulate principles called *rationality postulates* capturing the intuitions behind rational belief changes and to specify change mechanisms called *construction methods* for the operations. It is commonly accepted that the AGM framework provides the best set of postulates and well motivated construction methods.

While belief change is traditionally studied under propositional logic, over the years, many have attempted defining belief change operators for description logics (DLs).[3-9] Existing approaches of belief revision for DLs often assume that the new information is fully accepted. This assumption is unsuitable for scenarios where new information is not always fully trustworthy, thus can only be partially accepted or even rejected.

Defining change operations that do not give priority to new information was investigated under the name of *non-prioritised belief revision*.[10-14] The focus is a mechanism to evaluate the new information such that it could be rejected, partially accepted, or fully accepted and the integration of the mechanism with standard revision operation. In this paper, we adapt non-prioritised belief revision to DL-Lite$_{core}$ which is the core language of the DL-Lite family[15] that underlies the OWL 2 QL profile of OWL 2. We start with an existing revision operator[16] for incorporating new information into a logically closed DL-Lite$_{core}$ TBoxes and integrate it with features of non-prioritised belief revision. We provide representation theorems for the proposed revision operators.

2. DL-Lite and Type Semantics

DL-Lite$_{core}$ is the core of the family of DL-Lite languages. Complex concept and role expressions can be built on atomic concept A and atomic role P as follows:

$$B \to A \mid \exists R \qquad C \to B \mid \neg B \qquad R \to P \mid P^-$$

where A denotes an *atomic concept*, P an *atomic role*, P^- the *inverse* of the atomic role P. B denotes a *basic concept* which can be either an atomic concept or an unqualified existential quantification. C denotes a *general concept* which can be either an basic concept or its negation. We also include $\bot$ denoting the empty set and $\top$ denoting the whole domain. Let $\mathcal{B}$ and $\mathcal{R}$ be the sets of all concepts B and all roles R. For an inverse role $R = P^-$, we write R^- to represent P for the convenience of presentation. In this paper, we assume $\mathcal{B}$ and $\mathcal{R}$ are finite. A DL-Lite$_{core}$ TBox is a finite set of *concept inclusion axioms* of the form $B \sqsubseteq C$, $B \sqsubseteq \bot$, or $\top \sqsubseteq C$. That is only basic concept or $\top$ can appear on the left-hand side of a concept inclusion.

The semantics of DL-Lite$_{core}$, the entailment relation which is denoted $\models$, logical equivalence which is denoted $\equiv$, consistency, and coherence are defined in the standard way.[15] We use $\models \phi$ to denote that ϕ is a tautology such as $A \sqsubseteq A$. We use $\{\top \sqsubseteq \bot\}$ to denote the (unique) inconsistent TBox. The closure of a TBox $\mathcal{T}$, denoted as $cl(\mathcal{T})$, is the set of all TBox axioms ϕ such that $\mathcal{T} \models \phi$. In the upcoming sections all TBoxes are assumed to be closed DL-Lite$_{core}$ TBoxes and by DL-Lite we mean DL-Lite$_{core}$.

In this paper, we will work with an alternative semantics for DL-Lite called *type semantics*.[16] Central to the semantics is the notion of *types*. A *type* is a possibly empty subset of $\mathcal{B}$ and we denote the universal set of types as Ω. A type τ is a type model of a TBox $\mathcal{T}$ if τ satisfies $\mathcal{T}$ propositionally and $\exists R^- \notin \tau$ whenever $\mathcal{T} \models \exists R \sqsubseteq \bot$. We denote the type models of a TBox $\mathcal{T}$ and an axiom ϕ as $|\mathcal{T}|$ and $|\phi|$ respectively. Type models of the negation of ϕ, denoted by $\neg \phi$, is defined as $\Omega \setminus |\phi|$.

Type semantics can be used to capture the major reasoning tasks in DL-Lite.

Proposition 1. [16] Let $\mathcal{T}$ be a TBox and ϕ a TBox axiom. Then
1. $\mathcal{T} \models \phi$ iff $|\mathcal{T}| \subseteq |\phi|$.
2. $\mathcal{T}$ is consistent iff $|\mathcal{T}| \neq \emptyset$.

In comparison with DL semantics, type semantics has the clear advantage of being more succinct. More importantly, given any DL-Lite TBox, the number of type models is always finite.

For convenience, we extend the notion of coherence to single TBox axioms and sets of types. An axiom ϕ (a set of types M) is coherent if and only if $\{\phi\} \not\models B \sqsubseteq \bot$ (resp. $M \not\subseteq |B \sqsubseteq \bot|$) for all $B \in \mathcal{B}$. Given a set of types M there may not be a TBox $\mathcal{T}$ whose set of type models is M. A corresponding TBox $\mathcal{T}$ for a set of types M is a TBox such that $M \subseteq |\mathcal{T}|$ and there is no TBox $\mathcal{T}'$ such that $M \subseteq |\mathcal{T}'| \subset |\mathcal{T}|$. It is shown that any coherent set of types satisfies the condition for uniqueness.[16] In fact, the sets of types we will encounter in defining revision operators for DL-Lite$_{core}$ are coherent and thus there is always a unique corresponding TBox. In the upcoming sections, for any coherent set of types M, we use $\mathcal{T}_{core}(M)$ to denote the unique TBox that corresponds to M.

3. Non-prioritised Revision Operators

In this section, we define non-prioritised revision operators for DL-Lite TBoxes.

3.1. Rationality Postulates

Zhiqiang et al.[16] defined revision operators for DL-Lite$_{core}$ TBoxes in a model-theoretic way that is based on type semantics. They first clarified a fundamental difference between AGM revision and DL revision. AGM revision aims to incorporate a new formula into a belief set while resolving any inconsistency caused. DL revision goes beyond inconsistency resolving. In addition to consistency, meaningful DL TBoxes have to be coherent, thus DL revision has to resolve both the inconsistency and incoherence caused in incorporating new axioms. By replacing conditions on consistency with coherence, AGM revision postulates have been reformulated as follows for revision over DL-Lite TBoxes.

$(T*1)$ $T*\phi = cl(T*\phi)$
$(T*2)$ $\phi \in T*\phi$
$(T*3)$ $T*\phi \subseteq cl(T \cup \{\phi\})$
$(T*4)$ If $T \cup \{\phi\}$ is coherent, then $cl(T \cup \{\phi\}) \subseteq T*\phi$
$(T*5)$ If ϕ is coherent, then $T*\phi$ is coherent
$(T*6)$ If $\phi \equiv \psi$ then $T*\phi = T*\psi$
$(T*f)$ If ϕ is incoherent then $T*\phi = \{\top \sqsubseteq \bot\}$

$(T*1)$–$(T*6)$ correspond to the six AGM revision postulates. The failure postulate $(T*f)$ is dedicated to the limiting case when the revising axiom is incoherent. Based on the above postulates, we present our postulates for credibility limited TBox revision on coherent and closed TBox.

$(T \circ 1)$ $T \circ \phi = cl(T \circ \phi)$
$(T \circ 2)$ Either $T \circ \phi = T$ or $\phi \in T \circ \phi$
$(T \circ 3)$ $T \circ \phi \subseteq cl(T \cup \{\phi\})$
$(T \circ 4)$ If $T \cup \{\phi\}$ is coherent and $\phi \in T \circ \phi$, then $cl(T \cup \{\phi\}) \subseteq T \circ \phi$
$(T \circ 5)$ $T \circ \phi$ is coherent
$(T \circ 6)$ If $\phi \equiv \psi$ then $T \circ \phi = T \circ \psi$

The main change is the weakening of $(T*2)$. $(T \circ 2)$, the counterpart of $(T*2)$, is also termed as *relative success*.[11] The the counterpart of $(T*5)$ is $(T \circ 5)$, which we call *strong coherence* and is adapted from the so called *strong consistency* postulate.[11]

Definition 1. An operator $\circ$ is a *non-prioritised TBox revision operator* if it satisfies $(T \circ 1)$–$(T \circ 6)$.

3.2. Construction Methods

In this subsection, we first recall the construction method proposed by Zhiqiang et al.,[16] and then present ours.

A function γ is a selection function if $\gamma(M)$ is a non-empty subset of M unless M is incoherent. A selection function γ is *faithful* with respect to a TBox $\mathcal{T}$ if it satisfies:

(1) if M is coherent then $|\mathcal{T}| \cap M \subseteq \gamma(M)$, and
(2) if $|\mathcal{T}| \cap M$ is coherent then $\gamma(M) = |\mathcal{T}| \cap M$.

In revising $\mathcal{T}$ by ϕ, condition 1 deals with the case when models of $\mathcal{T}$ overlaps with those of ϕ which means $\mathcal{T} \cup \{\phi\}$ is consistent. In line with the principle of minimal change, in this case, the selection function has to pick all the overlapping models to preserve as much as possible the original TBox axioms. Condition 2 deals with the case that not only the overlapping exists but also it is coherent. Since there is no incoherence to resolve, the revision boils down to a set union operation (i.e., $cl(\mathcal{T} \cup \{\phi\})$). The selection function therefore picks all the overlapping models and no others. Central to the revision, the selection function has to guarantee the type models picked are coherent. A selection function γ is *coherent preserving* if for all $B \in \mathcal{B}$ there is $\tau \in \gamma(M)$ such that $B \in \tau$. Then a TBox revision operator $*$ is defined as $\mathcal{T} * \phi = \mathcal{T}_{core}(\gamma(|\phi|))$ where γ is a selection function that is coherent preserving and faithful with respect to $\mathcal{T}$.[16] The operator can be characterised by $(T*1)$–$(T*6)$ and $(T*f)$.

Our construction is a combination of the above revision operator and that of Hansson et al.[11]

Definition 2. A non-prioritised TBox revision operator $\circ$ is *induced* by a (prioritised) TBox revision operator $*$ and a set Θ of TBox axioms if for each axiom ϕ, $\mathcal{T} \circ \phi = \mathcal{T} * \phi$ whenever $\phi \in \Theta$ and $\mathcal{T} \circ \phi = \mathcal{T}$ otherwise.

The intuition is that an agent decides whether the input axiom ϕ is trustworthy and it revises its beliefs set only if ϕ is trustworthy, that is $\phi \in \Theta$.

Definition 3. Let $\circ$ be a non-prioritised revision operator induced by a revision operator $*$ and a set of axioms Θ. Then $\circ$ is an *endorsed credible* operator if the following hold: $\mathcal{T} \subseteq \Theta$ and $\phi \in \Theta$ iff there is $\tau \in |\Theta|$ such that τ satisfies ϕ. $\circ$ satisfies *outcome credibility* if $|\mathcal{T} \circ \phi| \cap |\Theta| \neq \emptyset$.

The following are some reasonable conditions on Θ:

(Θ_1) If $\phi \equiv \psi$ and $\phi \in \Theta$, then $\psi \in \Theta$.
(Θ_2) If $\phi \in \Theta$, then $cl(\{\phi\}) \subseteq \Theta$.
(Θ_3) If $\phi \in \Theta$, then ϕ is coherent.
(Θ_4) For each coherent and closed TBox $\mathcal{T}$, each prioritised revision function $*$ and each axiom ϕ, if $\phi \in \Theta$ then $\mathcal{T} * \phi \subseteq \Theta$.

3.3. Representation Theorems

In this subsection, we provide representation theorems for the operators defined in the previous subsection.

Theorem 1. *An operator $\circ$ satisfies $(T \circ 1)$–$(T \circ 6)$ iff $\circ$ is induced by a prioritised revision function $*$ and a set of axioms Θ, where Θ satisfies $\mathcal{T} \subseteq \Theta$, (Θ_1) and (Θ_3).*

On the one hand, Theorem 1 presents a way to construct a non-prioritised TBox revision operator. That is, suppose $\mathcal{T}$ is a coherent and closed TBox, and Θ satisfies $\mathcal{T} \subseteq \Theta$, ($\Theta_1$) and ($\Theta_3$), then we can use a prioritised revision function $*$ and Θ to induce a non-prioritised TBox revision operator. On the other hand, given any coherent and closed TBox $\mathcal{T}$ and a set of axioms Θ satisfying $\mathcal{T} \subseteq \Theta$, ($\Theta_1$) and ($\Theta_3$), if $\circ$ is induced by Θ and a prioritised revision function $*$, then $\circ$ is a non-prioritised TBox revision operator, i.e., $\circ$ satisfies $(T \circ 1)$–$(T \circ 6)$.

We present two additional postulates, which are adapted from the *strict improvement* and *regularity* postulates proposed by Hansson et al.[11]

$(T \circ st)$ If $\phi \in T \circ \phi$ and $|\phi| \subseteq |\psi|$, then $\psi \in T \circ \psi$
$(T \circ r)$ If $\psi \in T \circ \phi$, then $\psi \in T \circ \psi$

Theorem 2. *An operator $\circ$ satisfies $(T \circ 1)$–$(T \circ 6)$ and $(T \circ st)$ iff $\circ$ is induced by a prioritised revision function $*$ and a set of axioms Θ, where Θ satisfies $\mathcal{T} \subseteq \Theta$, ($\Theta_1$), ($\Theta_2$) and ($\Theta_3$).*

Theorem 3. *Let $\circ$ be an endorsed credible revision function induced by a prioritised revision function $*$ and a set of axioms Θ where Θ satisfies (Θ_4). Then $\circ$ satisfies $(T \circ r)$ iff $\circ$ satisfies outcome credibility.*

4. Related Work

Zhiqiang et al.[16] defined model-based contraction and revision operators for DL-Lite$_{core}$ TBoxes and provided representation theorems for the operators. Noticeably, instead of DL semantics, their approach is based on type semantics. This work[16] was later extended[9] to the more expressive DL-Lite$_\mathcal{R}$ as well as to knowledge bases.

Fermé and Hansson[10] proposed selective belief revision in which it is possible to partially accept new information. Hansson et al.[11] studied different ways of constructing non-prioritised belief revision of which the most general one is called credibility-limited revision.

Booth et al.[12] also studied credibility-limited revision in which iterated belief revision is considered. Aaron and Richard[13] developed an approach to trust-sensitive belief revision. They argued that trust-sensitive revision is a specialization of selective revision. Dmitriy et al.[14] studied trust-sensitive evolution of DL-lite knowledge bases in which trust-sensitive model-based evolution and algorithms are proposed.

5. Conclusion

With the aim of handling belief revision scenarios for DLs in which new information is not always trustworthy, we proposed non-prioritised belief revision for DL-Lite$_{core}$ TBoxes. We provide postulates for such non-prioritised revision and proved representation theorems. For future work, we plan to study non-prioritised revision for more expressive DLs. Also we plan to study non-prioritised contraction for DLs.

Acknowledgments

This work is supported by NSF of China under Grant No. 61463044, industrial technology program of Qiannan under Grant No. 201710, and Grant No. qnsyrc201715 from Qiannan Normal University for Nationalities.

References

1. Carlos E. Alchourrón, Peter Gärdenfors, and David Makinson. On the logic of theory change: Partial meet contraction and revision functions. *Journal of Symbolic Logic*, 50(2):510–530, 1985.
2. Peter Gärdenfors. *Knowledge in Flux: Modelling the Dynamics of Epistemic States*. MIT Press, 1988.
3. F. Baader, D. Calvanese, D. McGuinness, D. Nardi, and P. Patel-Schneider, editors. *The Description Logic Handbook*. CUP, Cambridge, UK, 2003.
4. Guilin Qi and Jianfeng Du. Model-based revision operators for terminologies in description logics. In *Proc. IJCAI-2009*, pages 891–897, 2009.
5. Márcio M. Ribeiro and Renata Wassermann. Base revision for ontology debugging. *Journal of Logic and Computation*, 19(5):721–743, 2009.
6. Zhe Wang, Kewen Wang, and Rodney W. Topor. A new approach to knowledge base revision in dl-lite. In *Proc. AAAI-2010*, 2010.
7. Bernardo Cuenca Grau, Ernesto Jimenez Ruiz, Evgeny Kharlamov, and Dimitry Zhelenyakov. Ontology evolution under semantic constraints. In *Proceedings of the 13th International Conference on Principles of Knowledge Representation and Reasoning (KR-2012)*, pages 137–147, 2012.
8. Zhe Wang, Kewen Wang, Zhiqiang Zhuang, and Guilin Qi. Instance-driven ontology evolution in DL-Lite. In *Proceedings of the Twenty-Ninth AAAI Conference on Artificial Intelligence (AAAI-2015)*, pages 1656–1662, 2015.
9. Zhiqiang Zhuang, Zhe Wang, Kewen Wang, and Guilin Qi. DL-Lite contraction and revision. *Journal of Artificial Intelligence Research*, 56:329–378, 2016.
10. Eduardo L. Fermé and Sven Ove Hansson. Selective revision. *Studia Logica*, 63(3):331–342, 1999.
11. Sven Ove Hansson, Eduardo L. Fermé, John Cantwell, and Marcelo A. Falappa. Credibility limited revision. *J. Symb. Log.*, 66(4):1581–1596, 2001.
12. Richard Booth, Eduardo Fermé, Sébastien Konieczny, and Ramón Pino Pérez. Credibility-limited revision operators in propositional logic. In *Principles of Knowledge Representation and Reasoning: Proceedings of the Thirteenth International Conference*, 2012.

13. Aaron Hunter and Richard Booth. Trust-sensitive belief revision. In *Proceedings of the 24th International Joint Conference on Artificial Intelligence (IJCAI-2015)*, pages 3062–3068, 2015.
14. Dmitriy Zheleznyakov, Evgeny Kharlamov, and Ian Horrocks. Trust-sensitive evolution of dl-lite knowledge bases. In *Proceedings of the Thirty-First AAAI Conference on Artificial Intelligence (AAAI-2017)*, pages 1266–1273, 2017.
15. D. Calvanese, G. De Giacomo, D. Lembo, M. Lenzerini, and R. Rosati. Tractable reasoning and efficient query answering in description logics: The DL-Lite family. *Journal of Automatic Reasoning*, 39(3):385–429, 2007.
16. Zhiqiang Zhuang, Zhe Wang, Kewen Wang, and Guilin Qi. Contraction and revision over DL-Lite TBoxes. In *Proceedings of the Twenty-Eighth AAAI Conference on Artificial Intelligence (AAAI-2014)*, pages 1149–1156, 2014.

Strongest necessary and weakest sufficient conditions in S5

Renyan Feng, Yisong Wang* and Panfeng Chen

Key Laboratory of Intelligent Medical Image Analysis and Precise Diagnosis of Guizhou Province, Department of Computer Science and Technology, Guizhou University, Guiyang, China
*yswang@gzu.edu.cn

Jincheng Zhou

School of Mathematics and Statistics, Qiannan Normal University for Nationalities, Duyun, China

Modal logic S5 is an important formalism for epistemic reasoning in agent domain. This paper extends the notions of (strongest) necessary condition and (weakest) sufficient condition from classical logic into S5. These two notions are closely related to *abduction*, *definability* and *forgetting*, that play an important role in artificial intelligence. It shows that the two notions are dual and can be extended to arbitrary formulas in S5. In order to compute strongest necessary and weakest sufficient conditions in terms of forgetting, an algorithm based on resolution is proposed to compute knowledge forgetting in S5.

Keywords: Strongest necessary conditions; weakest sufficient conditions; forgetting.

1. Introduction

Epistemic reasoning concerns the problem of how to reason about agents' epistemic states (knowledge) in a dynamic environment. Modal logic S5 is an import formalism for epistemic reasoning in multi-agent domain.[1,2] The strongest necessary condition (SNC in short) and the weakest sufficient condition (WSC in short) are two important notions to capture agents' epistemic states about propositions.

Informally speaking, the SNC of a proposition under a theory is the most general consequence deduced from the proposition under the given theory, and the WSC of a proposition is the most general abduction that we can make from the proposition under the given theory.[3,4] It is well-known these notions provide a good way to compute successor state axioms in basic action theory[5] and Dijkstra's weakest precondition is a special case of WSC.[6,7] These notions play an important role in artificial intelligence, database and so forth.

These notions of SNC and WSC have extensively explored in classical logic. However, to our best knowledge, there is little literature about these notions in modal logic. In this paper, we extend these two notions from classical logic to propositional modal logic **S5** and show that they share many interesting properties that hold in classical logic, including the duality of the two notions. We also show that these two notions are closely related to the knowledge forgetting in **S5**. An approach for computing SNC and WSC is also proposed in terms of computing knowledge forgetting,[8] thanks to the modal resolution.[9,10]

2. Preliminaries

In this section, we recall some basic notations of propositional modal logic **S5** and the notion of forgetting,[8,11] which plays a pivotal role in the paper.

2.1. *The syntax and semantic of S5*

Let $\mathcal{A}$ be a finite set of propositional atoms (called *variables* or *propositions*). A formula ϕ of the underlying language $\mathcal{L}$ of propositional modal logic **S5** over $\mathcal{A}$ is defined recursively as:

$$\phi ::= \bot \mid \top \mid p \mid \neg\phi \mid \phi \supset \phi \mid \mathrm{K}\phi \mid \phi \vee \phi \mid \phi \wedge \phi \tag{1}$$

where $p \in \mathcal{A}$, $\bot$ means "falsity" and $\top$ means "tautology". The modal operator K is for "necessity" (in a possible world), and the modal operator B for "possibility" is defined as $\mathrm{B}\phi = \neg\mathrm{K}\neg\phi$, and the formula $\phi \leftrightarrow \psi$ stands for $(\phi \supset \psi) \wedge (\psi \supset \phi)$. A formula (of $\mathcal{L}$) is *objective* if it contain no modal operator. A *knowledge set* is a finite set of formulas. A *literal* is an atom p or its negation $\neg p$ where $p \in \mathcal{A}$.

Let $S \subseteq \mathcal{A}$. We denote $\overline{S} = \mathcal{A} - S$. Given an expression ξ, we denote by $Var(\xi)$ the set of atoms occurring in ξ, and denote by $\xi(p/q)$ the expression obtained from ξ by replacing every occurrence of p in ξ by q, where ξ may be a formula or a set of formulas and $p, q \in \mathcal{L}$.

An *interpretation* is an assignment $v : \mathcal{A} \to \{0, 1\}$. It is usually written as a set of atoms, which assigns 1 to every atom in the set, and 0 to others. A *(possible) world* is an interpretation or a set of atoms. A K-*interpretation* $\mathcal{M}$ is a tuple $\langle W, w \rangle$ where W is a set of possible worlds, and $w \in W$ is called the *actual world* of $\mathcal{M}$.

The *satisfiability relationship* between a K-interpretation $\mathcal{M} = \langle W, w \rangle$ and a formula ϕ of $\mathcal{L}$, written $\mathcal{M} \models \phi$, is recursively defined as:

$\mathcal{M} \models \top$; $\mathcal{M} \not\models \bot$; $\langle W, w \rangle \models p$ if $p \in w$;
$\mathcal{M} \models \neg\phi$ if $\mathcal{M} \not\models \phi$; $\mathcal{M} \models \phi \wedge \psi$ if $\mathcal{M} \models \phi$ and $\mathcal{M} \models \psi$;
$\mathcal{M} \models \phi \vee \psi$ if $\mathcal{M} \models \phi$ or $\mathcal{M} \models \psi$; $\mathcal{M} \models \phi \supset \psi$ if $\mathcal{M} \not\models \phi$ or $\mathcal{M} \models \psi$;
$\langle W, w \rangle \models \mathrm{K}\phi$ if $\langle W, w' \rangle \models \phi$ for each $w' \in W$.

Given a knowledge set Γ, $\mathcal{M} \models \Gamma$ if $\mathcal{M} \models \phi$ for each $\phi \in \Gamma$. Let ξ be a knowledge set or a formula. A K-interpretation $\mathcal{M}$ is a K-*model* of ξ if $\mathcal{M} \models \xi$. We say that ξ is K-*satisfiable* (resp. K-*valid*) if it has a K-model (resp. every K-interpretation is a K-model of ξ). By $Mod(\xi)$ we denote the set of K-models of ξ. By $\phi \models \psi$ we mean $Mod(\phi) \subseteq Mod(\psi)$, and by $\phi \equiv \psi$ we mean $Mod(\phi) = Mod(\psi)$, where ϕ and ψ are formulas or knowledge sets. A formula or knowledge set ξ is *irrelevant* to a set $V \subseteq \mathcal{A}$ if there is a formula or knowledge set ξ' such that $\xi \equiv \xi'$ and $Var(\xi') \cap V = \emptyset$.

2.2. Knowledge forgetting in S5

Let w and w' be two worlds, and $V \subseteq \mathcal{A}$. We denote $w \simeq_V w'$ if $w \cap \overline{V} = w' \cap \overline{V}$.

Definition 2.1. Let $\mathcal{M} = \langle W, w \rangle$ and $\mathcal{M}' = \langle W', w' \rangle$ be two K-interpretations, and $V \subseteq \mathcal{A}$. We say that $\mathcal{M}$ and $\mathcal{M}'$ are V-*bisimilar*, denoted by $\mathcal{M} \leftrightarrow_V \mathcal{M}'$, if

(1) $w \simeq_V w'$;
(2) $\forall w_1 \in W, \exists w_2 \in W'$ such that $w_1 \simeq_V w_2$ (the forth condition); and
(3) $\forall w_2 \in W', \exists w_1 \in W$ such that $w_2 \simeq_V w_1$ (the back condition).

Note that $\mathcal{M}$ and $\mathcal{M}'$ may have different number of worlds even if $\mathcal{M} \leftrightarrow_V \mathcal{M}'$. By the above definition, $\leftrightarrow_V$ is obviously an equivalence relation.

Lemma 2.1 (Lemma 2 of[11]). *Let ϕ be a formula and $V \subseteq \mathcal{A}$ s.t. $Var(\phi) \cap V = \emptyset$. If two K-interpretations $\mathcal{M}_1$ and $\mathcal{M}_2$ are V-bisimilar then $\mathcal{M}_1 \models \phi$ if and only if $\mathcal{M}_2 \models \phi$.*

Definition 2.2 (Knowledge forgetting). *Let Γ be a knowledge set and $V \subseteq \mathcal{A}$. A knowledge set Γ' is a result of knowledge forgetting V from Γ if*

$$Mod(\Gamma') = \{\mathcal{M}' \mid \exists \mathcal{M} \in Mod(\Gamma) \text{ s.t. } \mathcal{M} \leftrightarrow_V \mathcal{M}'\}. \tag{2}$$

It is proven that such knowledge forgetting results always exist, and they are equivalent in **S5**. In this sense, we denote it by $\text{KF}(\Gamma, V)$.

Lemma 2.2. *Let Γ be a knowledge set, α a formula and $q \in \overline{Var(\Gamma \cup \{\alpha\})}$. Then $\text{KF}(\Gamma \cup \{q \leftrightarrow \alpha\}, q) \equiv \Gamma$.*

Proof. Let $\Gamma' = \Gamma \cup \{q \leftrightarrow \alpha\}$. We show $\text{KF}(\Gamma \cup \{q \leftrightarrow \alpha\}, q) \models \Gamma$ only.
($\Rightarrow$) $\mathcal{M} \models \text{KF}(\Gamma \cup \{q \leftrightarrow \alpha\}, q)$
$\Rightarrow \exists \mathcal{M}' \models \Gamma \cup \{q \leftrightarrow \alpha\}$ s.t. $\mathcal{M} \leftrightarrow_{\{q\}} \mathcal{M}'$
$\Rightarrow \mathcal{M}' \models \Gamma$
$\Rightarrow \mathcal{M} \models \Gamma$ since $q \notin Var(\Gamma)$ and $\mathcal{M} \leftrightarrow_{\{q\}} \mathcal{M}'$ by Lemma 2.1. □

3. Strongest Necessary and Weakest Sufficient Conditions

In this section, we present the notions of necessary and sufficient conditions in **S5**. Let $V \subseteq \mathcal{A}$. A formula ϕ is called a *formula of V* if $Var(\phi) \subseteq V$.

Definition 3.1 (SNC). *Let Γ be a knowledge set, $V \subseteq Var(\Gamma)$, and $q \in Var(\Gamma) - V$. A S5 formula φ of V is a necessary condition of q on V under Γ if $\Gamma \models q \supset \varphi$. It is a* strongest necessary condition (SNC) *of q on V under Γ if $\Gamma \models q \supset \varphi$ and $\Gamma \models \varphi \supset \varphi'$ for any φ' of V with $\Gamma \models q \supset \varphi'$.*

The (weakest) sufficient condition is similar defined in the following.

Definition 3.2 (WSC). *Let Γ be a knowledge set, $V \subseteq Var(\Gamma)$, and $q \in Var(\Gamma) - V$. A S5 formula φ of V is a sufficient condition of q on V under Γ if $\Gamma \models \varphi \supset q$. It is a* weakest sufficient condition (WSC) *of q on V under Γ if $\Gamma \models \varphi \supset q$ and $\Gamma \models \varphi' \supset \varphi$ for any φ' of V with $\Gamma \models \varphi' \supset q$.*

A proposition p is *definable* on V under a knowledge set Γ if there is a formula ϕ of V such that $\Gamma \models q \leftrightarrow \phi$. The following proposition shows that the problem of definability can be solved by computing SNC and WSC.

Proposition 3.1. *A knowledge set Γ defines a proposition q on $V \subseteq \mathcal{A}$ iff $\Gamma \models \varphi \supset \phi$, where φ and ϕ is respectively any SNC and WSC of q on V under Γ.*

Proof. ($\Rightarrow$) Let ψ be a formula such that $\Gamma \models q \leftrightarrow \psi$. We have the following
$\Gamma \models q \leftrightarrow \psi$
$\Rightarrow \Gamma \models q \supset \psi$ and $\Gamma \models \psi \supset q$
$\Rightarrow \psi$ is a sufficient and necessary condition of q on V under Γ
$\Rightarrow \Gamma \models \varphi \supset \psi$ since φ is a SNC of q on V under Γ and, $\Gamma \models \psi \supset \phi$ since ϕ is a WSC of q on V under Γ
$\Rightarrow \Gamma \models \varphi \supset \phi$.

($\Leftarrow$) Let φ and ϕ be a SNC and WSC of q on V under Γ respectively. We have
$\Rightarrow \Gamma \models q \supset \varphi$ and $\Gamma \models \phi \supset q$
$\Rightarrow \Gamma \models \phi \supset \varphi$
$\Rightarrow \Gamma \models \phi \leftrightarrow \varphi$ by $\Gamma \models \varphi \supset \phi$
$\Rightarrow \Gamma \models q \leftrightarrow \phi$ since $\Gamma \models q \supset \varphi$ and $\Gamma \models \phi \supset q$
$\Rightarrow q$ is definable on V under Γ. □

The above (strongest) necessary condition and (weakest) sufficient conditions can be extended to formulas.

Definition 3.3. *Let Γ be a knowledge set, α a formula, and $V \subseteq Var(\Gamma \cup \{\alpha\})$. A formula φ of V is a necessary condition of α on V under Γ if $\Gamma \models \alpha \supset \varphi$. It* is a *strongest necessary condition (SNC)* of α on V under Γ if $\Gamma \models \alpha \supset \varphi$ and, $\Gamma \models \varphi \supset \varphi'$ for any φ' with $\Gamma \models \alpha \supset \varphi'$.

The (weakest) sufficient conditions for formulas can be similarly given. The next proposition shows that computing the SNC and the WSC of a formula can be reduced to that of a proposition.

Proposition 3.2. *Let Γ, V, and α be as in Definition 3.3. A formula φ of V is an SNC (WSC) of α on V under Γ iff it is an SNC (WSC) of q on V under $\Gamma' = \Gamma \cup \{q \leftrightarrow \alpha\}$, where q is a new proposition not in $Var(\Gamma \cup \{\alpha\})$.*

Proof. Let φ be a formula such that $Var(\varphi) \subseteq V$. Note that
$\Gamma \cup \{q \leftrightarrow \alpha\} \models q \supset \varphi$
iff $\Gamma \cup \{q \leftrightarrow \alpha\} \models \alpha \supset \varphi$
iff $\text{KF}(\Gamma \cup \{q \leftrightarrow \alpha\}, q) \models \alpha \supset \varphi$ since $q \notin Var(\Gamma \cup \{\alpha\})$
iff $\Gamma \models \alpha \supset \varphi$ by Lemma 2.2.

Now we have that: φ is the SNC of α on V under Γ
iff $\Gamma \models \alpha \supset \varphi$ and $\Gamma \models \varphi \supset \psi$ for any $\Gamma \models \alpha \supset \psi$ and $Var(\psi) \subseteq V$
iff $\Gamma' \models q \supset \varphi$ and $\Gamma' \models \varphi \supset \psi$ for any $\Gamma' \models q \supset \psi$ and $Var(\psi) \subseteq V$
iff φ is the SNC of q on V under Γ'. □

The next proposition shows that the notions of SNC and WSC are dual in **S5**.

Proposition 3.3. *A formula ψ is the SNC (WSC) of q on V under the knowledge set Γ iff $\neg\psi$ is the WSC (SNC) of $\neg q$ on V under the knowledge set Γ.*

Proof. A formula ψ is the SNC of q on V under Γ
iff $\Gamma \models q \supset \psi$ and $\Gamma \models \psi \supset \psi'$ for any $\Gamma \models q \supset \psi'$ with $Var(\psi') \subseteq V$
iff $\Gamma \models \neg\psi \supset \neg q$ and $\Gamma \models \neg\psi' \supset \neg\psi$ for any $\Gamma \models \neg\psi' \supset \neg q$ with $Var(\psi') \subseteq V$
iff $\neg\psi$ is the WSC of $\neg q$ on V under Γ. □

4. Computing Strongest Necessary and Weakest Sufficient Conditions

In this section we present an algorithm for computing SNC and WSC in terms of computing knowledge forgetting result.

Theorem 4.1. *Let Γ be a knowledge set, $V \subseteq \mathcal{A}$ and $q \in Var(\Gamma) \setminus V$.*

(i) *The SNC of q on V under Γ is $\text{KF}(\Gamma \cup \{q\}, Var(\Gamma) \setminus V)$.*
(ii) *The WSC of q on V under Γ is $\neg\text{KF}(\Gamma \cup \{\neg q\}, Var(\Gamma) \setminus V)$.*

Proof. (i) Let $S = Var(\Gamma) \setminus V$ and $\phi = \text{KF}(\Gamma \cup \{q\}, S)$. It is clear $Var(\phi) \subseteq V$.

Firstly, $\Gamma \cup \{q\} \models \text{KF}(\Gamma \cup \{q\}, S)$
$\Rightarrow \Gamma \cup \{q\} \models \phi$
$\Rightarrow \Gamma \models q \supset \phi$
$\Rightarrow \phi$ is a necessary condition of q on V under Γ.

Secondly, let ψ be a formula of V such that $\Gamma \models q \supset \psi$. We have
$\Gamma \cup \{q\} \models \psi$
$\Rightarrow \text{KF}(\Gamma \cup \{q\}, S) \models \psi$ by $\text{Var}(\psi) \subseteq V$ and $V \cap S = \emptyset$
$\Rightarrow \phi \models \psi$
$\Rightarrow \Gamma \models \phi \supset \psi$
$\Rightarrow \phi$ is the SNC of q on V under Γ.

(ii) It follows from (i) and Proposition 3.3. □

It shows that one can compute SNC and WSC by computing knowledge forgetting result. For this purpose, we propose an algorithm to compute knowledge forgetting in terms of resolution in **S5**.

A *clause* is a disjunction of literals, which is usually written as a set of the literals occurring in the clause. An *S5 literal* is a literal, $\bot$, $\text{K}\phi$ or $\text{B}\psi$, where ϕ is a clause and ψ is a conjunct of clauses. An *S5 clause* is a disjunction of **S5** literals. Recall that every **S5** formula can be equivalently transformed into *modal conjunctive normal form*, which is a conjunction of formulas of the form

$$\alpha_0 \vee \text{K}\alpha_1 \vee \cdots \vee \text{K}\alpha_{n-1} \vee \text{B}\alpha_n \qquad (3)$$

where α_i $(0 \leq i \leq n)$ are objective formulas.[12] The following lemma is evident.

Lemma 4.1. *Any S5 formula can be equivalently transformed into a conjunction of S5 clauses.*

In terms of the above lemma, we can assume that each **S5** clause has the form (3) where $\alpha_i s$ $(0 \leq i \leq n-1)$ are clauses and α_n is a conjunction of clauses.

Definition 4.1. Let $\varphi_1 = C \vee \text{K}D_1 \vee \text{K}D_2 \vee \ldots \vee \text{K}D_n \vee \text{B}E$ and $\varphi_2 = C' \vee \text{K}D'_1 \vee \text{K}D'_2 \vee \ldots \vee \text{K}D'_m \vee \text{B}E'$ be two **S5** clauses. We say that φ_1 *subsumes* φ_2 if

- C subsumes C', i.e. $C \subseteq C'$;
- $\forall D_i (1 \leq i \leq n), \exists D'_j (1 \leq j \leq m)$ s.t. D_i subsumes D'_j;
- for each conjunct e' of E', E has a conjunct e such that e subsumes e'.

Let $V \subseteq \mathcal{A}$ and ϕ a clause. The *suppressing* ϕ *on* V, written $Supp(V, \phi)$, is defined as

$$Supp(V, \phi) = \begin{cases} \top, & \text{if } p \text{ or } \neg p \text{ occurs in } \phi \text{ for some } p \in V; \\ \phi, & \text{otherwise.} \end{cases}$$

Let ϕ a **S5** clause of the form (3). The *suppressing* ϕ *on* V, written $Supp(V,\phi)$, is the **S5** clause

$$Supp(V,\alpha_0) \vee \left(\bigvee_{1 \leq i \leq n-1} \text{K} Supp(V,\alpha_i) \right) \vee \text{B} \left(\bigwedge_{\beta \text{ is a conjunct of } \alpha_n} Supp(V,\beta) \right).$$

Patrice et al.[9] proposed a sound and complete resolution system RS5 for **S5**, which consists of the following resolution rules for **S5** clauses:

(KB) $\dfrac{C \vee \text{K}(l \vee D) \quad C' \vee \text{B}(\neg l \vee D', E)}{C \vee (C' \vee \text{B}(D \vee D', \neg l \vee D', E)}$; (K$\bot$) $\dfrac{C \vee \text{K}\bot}{C}$;

(KK) $\dfrac{C \vee \text{K}(p \vee D) \quad C' \vee \text{K}(\neg p \vee D')}{C \vee C' \vee \text{K}(D \vee D')}$; (B$\bot$) $\dfrac{C \vee \text{B}(\bot, E)}{C}$;

(K) $\dfrac{C \vee \text{K}(l \vee D) \quad C' \vee \neg l}{C \vee C' \vee D}$; (Clas) $\dfrac{C \vee p \quad C' \vee \neg p}{C \vee C'}$;

(B) $\dfrac{C \vee \text{B}(p \vee D, \neg p \vee D', E)}{C \vee \text{B}(D \vee D', p \vee D, \neg p \vee D', E)}$; (Fact) $\dfrac{E[D \vee D \vee C]}{E[D \vee C]}$.

Here, $\text{B}(S)$ stands for the formula $\text{B}(\bigwedge S)$ for a set S of clauses, and $E[\psi]$ means that ψ is a subformula of the formula E. According to the resolution system RS5, an approach for computing knowledge forgetting is proposed in Algorithm 4.1. In Algorithm 4.1, step 4 eliminates those **S5** clauses that are in the form $p \vee C'$ or $\neg p \vee C'$ where $p \in V$. In this way one can avoid to generate "no useful" consequents of resolutions, because those consequents will be eliminated at step 7. And step 3 the soundness of the algorithm follows from that of RS5. This algorithm has been implemented in SWI-Prolog.

5. Conclusion and Future Work

In this paper, we extended the concepts of SNC and WSC from classical first-order logic into **S5**. It was showed that they share many interesting properties as that of classical logic. An resolution based algorithm for computing knowledge forgetting was proposed and implemented in Prolog, which can be used to computing SNC and WSC. We shall consider experiments on computing the two conditions.

Acknowledgments

We thank reviewers for their helpful comments. This work is partially supported by NSFC under grant 63170161 and 61562009, Stadholder Fund of Guizhou Province under grant (2012)62, Outstanding Young Talent Training Fund of Guizhou Province under grant (2015)01 and Science and Technology Fund of

Algorithm 4.1 Computing knowledge forgetting based on resolution

Input: A knowledge set Γ and $V \subseteq \mathcal{A}$
Output: $\text{KF}(\Gamma, V)$

Step 1: Transform Γ into a set Γ' of **S5** clauses; Let Γ_2 be the set of **S5** clauses in Γ' which contains no atom from V and $\Gamma_1 = \Gamma' - \Gamma_2$.
Step 2: If V is empty, then go to Step 6, otherwise select a proposition p from V and delete it from V.
Step 3: Do exhaustive resolution on p for the **S5** clauses in Γ_1 using RS5.[9]
Step 4: Simplify Γ_1 by removing the **S5** clause C if it is subsumed by some other C' in Γ_1, or it is in the form $p \vee D$ or $\neg p \vee D$ where D is a **S5** clause.
Step 5: Go back to Step 2.
Step 6: Suppress V in Γ_1: replacing each **S5** clause $\phi \in \Gamma_1$ by $Supp(V, \phi)$.
Step 7: Return $\Gamma_2 \cup \Gamma_1$.

Guizhou Province under grant [2014]7640. Jincheng's work is partially supported by the industrial technology program of Qiannan under Grant No. 201710.

References

1. Liangda Fang, Yongmei Liu, and Hans van Ditmarsch. Forgetting in multi-agent modal logics. In *Proceedings of IJCAI 2016, USA*, 1066–1073. IJCAI/AAAI Press, 2016.
2. Thomas Caridroit, Jean-Marie Lagniez, Daniel Le Berre, Tiago de Lima, and Valentin Montmirail. A sat-based approach for solving the modal logic s5-satisfiability problem. In *Proceedings of AAAI 2017, California, USA.*, 3864–3870. AAAI Press, 2017.
3. Fangzhen Lin. On strongest necessary and weakest sufficient conditions. *Artificial Intelligence*, 128(1):143–159, 2001.
4. Pierre Marquis and Pierre Marquis. *Propositional independence: formula-variable independence and forgetting.* AI Access Foundation, 2003.
5. Fangzhen Lin. Compiling causal theories to successor state axioms and strips-like systems. *Journal of Artificial Intelligence Research*, 19:279–314, 2003.
6. Edsger W Dijkstra and Carel S Scholten. *Predicate calculus and program semantics.* Springer Science & Business Media, 2012.
7. Marcus Bjäreland and Lars Karlsson. Reasoning by regression: Pre- and postdiction procedures for logics of action and change. In *Proceedings of IJCAI 97, Nagoya, Japan, 1997, 2 Volumes*, 1420–1425. Morgan Kaufmann, 1997.
8. Yan Zhang and Yi Zhou. Knowledge forgetting: Properties and applications. *Artificial Intelligence*, 173(16):1525–1537, 2009.
9. Patrice Enjalbert and Luis Fariñas Del Cerro. Modal resolution in clausal form. *Theoretical Computer Science*, 65(1):1–33, 1989.
10. Andreas Herzig and Jérôme Mengin. Uniform interpolation by resolution in modal logic. In Proceedings of *JELIA 2008, Dresden, Germany*, 219–231. Springer, 2008.
11. Yan Zhang and Yi Zhou. Properties of knowledge forgetting. In *Proceedings of NMR-2008*, 68–75, 2008.
12. M.J. Cresswell and G.E. Hughes. *A new introduction to modal logic* Routledge, 1996.

The number of minimal preference contraction operation

Maonian Wu

Huzhou University, Huzhou, 313000, China
Guizhou University, Guiyang, 550025, China

Shaojun Zhu,* Bo Zheng and Weihua Yang

Huzhou University, Huzhou, 313000, China
**zhushaojun@zjhu.edu.cn*

Mingyi Zhang

Guizhou Academy of Sciences, Guiyang, 550001, China

> In recent years, preferences change is central both to individual decision making and to strategic interactions between rational agents. It is widely used in artificial intelligence and social choice theory and so on. Hansson defined the minimal preference contraction in the spirit of AGM model of belief revision. Alechina, Liu and Logan gave a linear time algorithm which implements minimal contraction by a single preference. Is there any other minimal preference contraction operation? In addition, whether there exists a linear time algorithm to compute a minimal preference contraction operation? How many minimal preference contraction operations are there? These problems are hot topics in this field. We give another linear time algorithm to compute a minimal preference contraction operation. The maximal and minimal number of minimal preference contraction operations contracting a preferences set by a single preference is showed in the paper. It will promote the study of the number of minimal belief change operations on some knowledge base.
>
> *Keywords*: Preference change; minimal contraction; the number of operation; algorithm.

1. Introduction

Preferences change is central both to individual decision making, and to strategic interactions between rational agents in recent years.[7,9] They have been studied in artificial intelligence, algebraic structure, social choice theory, and philosophy and so on (a comprehensive collection of research papers in preference change from different fields can be found in.[7]) Since the 1990s, the dynamics of preference change has become a hot topic in this field. For example, Hansson proposed postulates for several preference change operations in the spirit of the Alchourrn, Gardenfors and Makinson (AGM) model of belief revision;[1,9] Van Benthem & Liu formulated a dynamic epistemic logic (DEL)-based dynamic preference logic to model the changes of some informational events;[16] Alechina, Liu and Logan focus on the dynamics of preference and take a more computationally oriented approach.[2]

Since a revision can be defined by contraction following Levi identity,[12] we focus on the operation of an agent's set of preferences contracting by a single preference. The agent wishes to remove one preference from its preferences set, together with any preference from which the target preference can be derived by transitivity. A minimal preference contraction is a contraction that removes the smallest number of preferences necessary to make the target preference underivable.

Alechina, Liu and Logan proposed a representation theorem for minimal preference contraction and gave a linear time algorithm which implements minimal contraction by a single preference.[2,13,14]

Naturally, we hope to know whether there is any other minimal preference contraction operation? In addition, whether there exists a linear time algorithm to compute a minimal preference contraction operation? Finally, it is important to know that how many minimal preference contraction operations exist. These problems are hot topics for researchers in this field recently. We first proposed a novel linear time algorithm which implements another minimal contraction by a single preference. We showed maximal and minimal number of minimal preference operators contracting a preference set by a single preference.

The paper is organized as follows. In section 2, we introduce preliminaries to this paper, such as preference relations, preference sets and preference inference system. In section 3, we consider the problem of minimal preference contraction, and give a set of postulates and a representation theorem. In section 4, we show linear time algorithms for minimal preference contraction. In section 5, the maximal/minimal number of operations contracting a set of preferences by a single preference is formulated. We conclude this paper in section 6.

2. Preliminaries

We assume that an agent's preferences are given by a set of binary relations over some $finite$ set of alternatives $\mathcal{A}$. An agent's preference state is represented by a **preference set** consisting of preference sentences (or simply preferences) which are atomic statements involving preference relations. Here we assume that we have a set of preference relations (and corresponding connectives in atomic sentences). The relations are $<$ (where $A < B$ means that B is strictly preferred to A), $\equiv$ (where $A \equiv B$ means that A and B are equally preferred), and $\#$ (where $A \# B$ means that A and B are incomparable).

In addition to atomic sentences built using preference relations, a preference set may contain a special sentence $\bot$, which is used to indicate a contradiction (derivability of an inconsistency). We assume that a preference set may be incomplete (it is possible that no relation holds between some alternatives A and B), but that the agents are rational, i.e., they don't accept $A < B$ and $B < A$ or $A \# B$ at the same time, and that they can complete their preference sets using transitivity of $<$ and $\equiv$ and symmetry of $\#$, etc. (so the sets are deductively closed with respect to the corresponding rules). We postulate the following natural set of rational reasoning rules or integrity constraints in the sense of preference relations.[7]

1. $A \# B \Rightarrow B \# A$
2. $A \equiv A$
3. $A \equiv B \Rightarrow B \equiv A$
4. $A \equiv B, B \equiv C \Rightarrow A \equiv C$
5. $A < B, B < C \Rightarrow A < C$
6. $A < B, B < A \Rightarrow \bot$
7. $A \equiv B, A < B \Rightarrow \bot$
8. $A \equiv B, A \# B \Rightarrow \bot$
9. $A \# B, A < B \Rightarrow \bot$

Rule 1 states that $\#$ is symmetric, rules 2–4 state that $\equiv$ is an equivalence relation, rule 5 states that $<$ is transitive, and the remaining rules state that at most one of $\#, \equiv, <$ can hold between two alternatives.

We denote by $C_n(S)$ the closure of a set S under the rules above. Formally, $C_n(S)$ is the set of preferences which contains S, $A \equiv A$ for every $A \in \mathcal{A}$, and in addition for every rule $p_1, ..., p_n \Rightarrow p$ above, if $p_1, ..., p_n \in C_n(S)$, then $p \in C_n(S)$. A set of preferences S is deductively closed iff $S = C_n(S)$.

Occasionally, we will use the notation $S \vdash p$ to say that p can be derived from S and the reasoning rules above by application of the following inference rule (where $n \geq 2$):

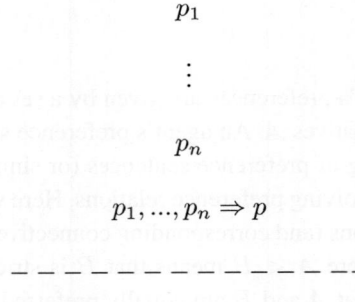

Clearly for any p, $\vdash p$ (p is derivable from an empty set) if, and only if, p is of the form $A \equiv A$. We do not assume any logical connectives or any other inference rules. In what follows, we assume that the agent's set of preferences S is deductively closed. The set of preferences is consistent if and only if it does not contain $\bot$.

3. Minimal Contraction Definition and Postulates

We recall the minimal preference contraction operation and its postulates.

Definition 1. (Minimal contraction[2]) Given a preference set S and a preference p, such that $\nvdash p$, a **minimal contraction** of S by p is any operation — that returns a set $S - p$ such that:
(1) $S - p \subseteq S$,
(2) $S - p \nvdash p$ and
(3) for any other subset S' of S such that $S' \nvdash p$, it holds that $|S'| \leq |S - p|$.

Before stating the postulates characterising minimal contraction, we introduce the following abbreviations:

We denote $\{C : A < C \in S\}$ by $A_S^<$, $\{C : C < A \in S\}$ by $A_S^>$ and $\{C : A \equiv C \in S\} \setminus \{A\}$ by $A_S^{\equiv}$.

The **cost** $c_S(p)$ of $p \in S$ (intuitively, the number of preferences a contraction by p has to remove from S) is defined as follows:
(1) $c_S(A < B) = |A_S^< \cap B_S^>| + 1$
(2) $c_S(A \equiv B) = 2 * |A_S^{\equiv}|$
(3) $c_S(A \# B) = 2$

The following six classic contraction postulates characterize minimal preference contraction in.[9]

(C-Closure) $S - p = C_n(S - p)$;
(C-Inclusion) $S - p \subseteq S$;
(C-Vacuity) If $S \nvdash p$, $S - p = S$;
(C-Success) If p is not of the form $A \equiv A$, then $p \notin S - p$;
(C-Equivalence) If $C_n(p_1) = C_n(p_2)$, then $S - p_1 = S - p_2$;
(C-Minimality) If $p \in S$, then $|S - p| = |S| - c_S(p)$.

The postulates of C-Closure, C-Inclusion, C-Vacuity, C-Success and C-Equivalence are standard postulates for contraction of beliefs.[1] The C-Minimality postulate characterized specifically minimal contraction of preferences, because for preference it is possible to predict the cardinality of the resulting set.[2] The following representation theorem described minimal preference contraction.[2]

Theorem 1. *Any minimal preference contraction satisfies contraction postulates above, and every contraction satisfying these postulates is a minimal preference contraction.*

It is hot topic that whether there exists a linear time algorithm to compute one or all minimal preference contraction operation or not?

4. Minimal Contraction Algorithms

Alechina, Liu and Logan has given a concrete polynomial time algorithm for a contraction operation by a preference.[2] We give another polynomial time algorithm for a contraction operation by a preference.

The main idea of their algorithm is deleting each preference which adjoins **start alternative** A of the preference $A < B$ from candidate set of preferences. Based on the idea, we give a similar concrete polynomial time algorithm for contraction operation by a preference $A < B$ if we deleting each preference which adjoins **terminal alternative** B from candidate set of preferences.

It is not difficult to prove the following theorem.

Theorem 2. *Algorithm 1 computes one minimal preference contraction operation.*

Evidently there is only one minimal contraction operation for contracting a set S of preferences by preference $A \equiv B$ and $A \# B$. But it is difficult to answer how many minimal contraction operations for contracting a set S of preferences by a preference $A < B$?

5. Number of Minimal Contraction Operation

In the section, we first answer partially the problem for special set S of preference or preference $A < B$. Secondly, we give the maximal and minimal number of such minimal preference contraction operations.

Firstly we define linear set of preference. It is a common conception in algebra, specially in lattice theory.[3]

Definition 2. Preference set S is linear if $A_1 < A_2 \in S$ or $A_2 < A_1 \in S$ for every two different $A_1, A_2 \in \{C \in \mathcal{A} | C < B \in S\} \cup \{C \in \mathcal{A} | B < C \in S\}$.

If preference set S is linear then we compute the number of minimal contraction operation which contracting S by any preference $p := A < B$. If $S \nvdash p$ then there is only one minimal preference contraction $S - p = S$ based C-Vacuity. The

Algorithm 1: Compute another minimal preference contraction

Input: S, p
Output: S

1 **case** $p \notin S$ **do**
2   **return**;
3 **end**
4 **case** $p := A < B$ **do**
5     $A^< := \{C | A < C \in S\}$
6     $B^> := \{C | C < B \in S\}$
7   **for** each $C \in A^< \cap B^>$ **do**
8        $\mathbf{S := S \setminus \{C < B\}}$
9   **end**
10    $S := S \setminus \{A < B\}$
11 **end**
12 **case** $p := A \equiv B$ **do**
13     $A^\equiv := \{C | A \equiv C \in S, A \neq C\}$
14   **for** each $C \in A^\equiv$ **do**
15        $S := S \setminus \{A \equiv C, C \equiv A\}$
16   **end**
17 **end**
18 **case** $p := A \# B$ **do**
19     $S := S \setminus \{A \# B, B \# A\}$
20 **end**

following theorem discusses the number of minimal contraction operations when S is linear and $p \in S$.

Theorem 3. *There are $n + 1$ operations of minimal preference contraction for contracting a linear preference set S by a preference $A < B \in S$ where $|A_S^< \cap B_S^>| = n$.*

We will discuss the operation $S - p$ contracting S by p in 3-layer preference.

Definition 3. Preference $A < B \in S$ is 3-layer preference of S if there is no $D \in \mathcal{A}$ such that $A < D \in S \,\&\, D < C \in S$ or $C < D \in S \,\&\, D < B \in S$ for each $C \in A^< \cap B^>$.

For a 3-layer preference p of preferences set S, we compute the number of minimal contraction operations contracting S by p.

Theorem 4. *There are 2^n operations of minimal preference contraction for contracting a preference set S by a 3-layer preference $A < B$ of S where $|A_S^\leq \cap B_S^\geq| = n$.*

Based on the method of cut edge in discrete mathematics,[8] we can divide $\{C_0, C_1, ..., C_{n+1}\}$ into two parties $\{C_0, C_1, ..., C_i\}$ and $\{C_{i+1}, ..., C_n, C_{n+1}\}$ by deleting edge $<C_0, C_{n+1}>$ and $\{<C_i, C_j> \mid j \geq i+1\} \cup \{<C_j, C_i+1> \mid j \leq i-1\}$. We use $MPC(S,p)$ to denote the number of minimal preference contraction operations which contracts preference set S by preference p. Evidently $MPC(S,p) = 2^n$ for a 3-layer preference p of preferences set S.

Lemma 1. *Let S_1, S_2 be preference set and $p := A < B$ preference with $A_{S_1}^\leq \cap B_{S_1}^\geq = A_{S_2}^\leq \cap B_{S_2}^\geq$. If $S_1 \subseteq S_2$ then $MPC(S_1, p) \geq MPC(S_2, p)$.*

Based Theorem 3, Theorem 4 and Lemma 1, the maximal and minimal number of such minimal preference contraction operations was formulated.

Theorem 5. *Let S be preference set and preference $A < B \in S$. $n + 1 \leq MPC(S, A < B) \leq 2^n$ where $|A_S^\leq \cap B_S^\geq| = n$.*

6. Conclusion and Future Works

In this paper, we propose a novel linear time algorithm which implements minimal contraction by a single preference. The maximal and minimal number of such minimal preference contraction operations is formulated.

In future, we will study the maximal and minimal number of minimal belief change operations on some knowledge base. In recent years, local belief revision and Horn belief revision are two hot topics in the field of belief revision.[4–6,10,11,15,17–19] Maybe the maximal and minimal number of minimal belief change operations in local belief revision, and Horn belief revision is some future study topics.

Acknowledgments

We are grateful to two anonymous reviewers for their insightful suggestions. Supported partially by the National Science Foundation of China under Grant No. 61262029, 61370161 and Zhejiang Provincial Natural Science Foundation of China under Grant No. LY16F020015, Science and Technology of Zhejiang Province under grant No. LQ18F020002, LGF18H120003, and National Science Foundation of Huzhou under grant No. 2016YZ02.

References

1. Carlos E. Alchourrón, Peter Gärdenfors, and David Makinson. On the logic of theory change: Partial meet contraction and revision functions. *J. Symb. Log.*, 50(2):510–530, 1985.
2. Natasha Alechina, Fenrong Liu, and Brian Logan. Minimal preference change. *Journal of Logic and Computation*, 2015.
3. R.B.J.T. Allenby. *Rings, fields and groups: An introduction to abstract algebra*. McGraw-Hill, Hodder Arnold, 1983.
4. Richard Booth, Thomas Meyer, and Ivan José Varzinczak. Next steps in propositional Horn contraction. In *IJCAI-09*, pages 702–707, 2009.
5. James P. Delgrande. Horn clause belief change: Contraction functions. In *KR-08*, pages 156–165, 2008.
6. James P. Delgrande and Renata Wassermann. Horn clause contraction functions: Belief set and belief base approaches. In *KR-10*, 2010.
7. T. Grne-Yanoff and S. O. Hansson eds. *Preference Change: Approaches from Philosophy, Economics and Psychology. Theory and Decision Library*. Springer, 2009.
8. Rosen K. H. *Discrete mathematics and its applications(5th ed.)*. McGraw-Hill, Oxford, 2003.
9. S. O. Hansson. Changes in preference. *Theory and Decision*, 38:1–28, 1995.
10. George Kourousias and David Makinson. Parallel interpolation, splitting, and relevance in belief change. *J. Symb. Log.*, 72(3):994–1002, 2007.
11. Marina Langlois, Robert H. Sloan, Balázs Szörényi, and György Turán. Horn complements: Towards Horn-to-Horn belief revision. In *AAAI-08*, pages 466–471, 2008.
12. Isaac Levi. Subjunctives, dispositions and chances. *Synthese*, 34:423–455, 1997.
13. F. Liu N. Alechina and B. Logan. Minimal preference change. In *Proceedings Fourth International Workshop on Logic, Rationality and Interaction (LORI-IV)*, pages 15–26, 2013.
14. F. Liu N. Alechina and B. Logan. Postulates and a linear-time algorithm for minimal preference contraction. In *In Workshop on Logical Aspects of Multi-Agent Systems (LAMAS 2014)*, Informal Proceedings, 2014.
15. Rohit Parikh. Beliefs, belief revision, and splitting languages. *Logic, language, and Computation*, 2:266–278, 1999.
16. J. van Benthem and F. Liu. Dynamic logic of preference upgrade. *Journal of Applied Non-Classical Logic*, 17:157–18, 2007.
17. Maonian Wu, Dongmo Zhang, and Mingyi Zhang. Language splitting and relevance-based belief change in horn logic. In *Proceedings of the Twenty-Fifth AAAI Conference on Artificial Intelligence, AAAI 2011, San Francisco, California, USA, August 7-11, 2011*, 2011.
18. Maonian Wu and Mingyi Zhang. A constructive method of the finest splitting of belief set. In *The 8th International FLINS Conference on Computational Intelligence in Decision and Control (FLINS2008)*, pages 319–324, 2008.
19. Maonian Wu and Mingyi Zhang. Algorithms and application in decision-making for the finest splitting of a set of formulae. *Knowl.-Based Syst.*, 23(1):70–76, 2010.

Protection zone designation of railway radio environment based on TD-LTE system[*]

Peng Xu[1,4], Meirong Chen[2], Lifang Feng[3], Fangli Ma[4,5], Guanfeng Wu[4,5] and Yang Xu[1,4]

[1]School of Mathematics, Southwest Jiaotong University, Chengdu, 610031, China
[2]Electronic Information and Electric Engineering Department, Chengdu Textile College, Chengdu 611731, China
[3]School of Computer and Communication Engineering, University of Science and Technology Beijing, Beijing 100083, China
[4]National-Local Joint Engineering Laboratory of System Credibility Automatic Verification, Southwest Jiaotong University, Chengdu, 610031, China
[5]School of Information Science and Technology, Southwest Jiaotong University, Chengdu, 610031, China

With the gradual development of radio services and the wide application of wireless communication technologies, the radio environment along railway lines is becoming more and more complicated. Different wireless communication systems inevitably experience mutual signal interference with each other, including intra-frequency, adjacent channel, intermodulation and other types, which can cause wireless system communication interruption, blockage and other system anomalies, and further affect the safe operation of the railway. In this paper, from the perspective of radio environmental factors, the mutual interference between TD-LTE railway private network and FDD-LTE system in the 1785-1805MHz frequency band had been researched, the calculation method of protection zone along the railway line was also given.

1. Introduction

Electromagnetic environment refers to the sum of all the electromagnetic phenomena existing in given places, including the entire time and the entire spectrum. Regional electromagnetic environment is a certain administrative area or geographical electromagnetic environment. The radio environment is the electromagnetic environment in the radio frequency range and is the sum of the electromagnetic phenomena generated by all radio transmitters in operation in a given location [1].

---

[*]This work is supported by National Science Foundation of China (Grant No. 61673320), the project of Department of Education of Sichuan Province (Grant No. 18ZB0589), the Radio Association of China project (Grant No. T/RAC 015-2016), the Fundamental Research Funds for the Central Universities (Grant No. 2682017ZT12, 2682016CX119).

Railways usually cross a number of cities, the geo-environment along the route is complex and there are many frequency-consuming services, especially in the sections that cross the cities and transport hubs. The coexistence of radio systems such as mobile communication networks, wireless LANs, cluster systems, leads a spectrum shared phenomenon [2][3]. With reference to the Radio Division Regulation [4], a variety of wireless systems have resulted in increasingly crowded and overlapping spectrum, with significant increases in background noise and interference, which in turn constitute a complex radio environment. That directly affect the quality of radio equipment, possibly causing serious impact on the railway mobile communications platform, in extreme circumstances resulting in wireless network interruption, launcher equipment failure. At present, the railway system is gradually adopting the 1785-1805MHz TD-LTE private network system as the service platform for railway communications. Combining the frequency division rules, it is known that the McWill system (1785MHz-1805MHz) exists in the same frequency band. In adjacent channel, there are the GSM1800 system, the TD-SCDMA system, WCDMA / HSPA system, CDMA2000 system, IMT system, which inevitably produce the same frequency, adjacent frequency, intermodulation and other types of interference. The establishment of interference protection zone in TD-LTE private network system and non-LTE system has drawn much attention [5]. However, researches between LTE systems are still few.

Therefore, how to set the railway radio environment protection zone in TD-LTE system, is the necessary step to ensure the safe operation of railway and the user experience of other wireless systems, and also provide technical support for the construction of railway wireless system.

2. Influencing Factor of Protection Zone

There are co-channel, adjacent channel and intermodulation interference among wireless communication systems. Co-channel interference refers to interference signals and useful signals working in the same channel, and the interference signal bandwidth is less than the useful signal bandwidth to form co-channel interference. Adjacent interference refers to the interference signal and the useful signal center frequency is different but similar, and interference signal part of the spectrum fall into the receiver bandwidth caused interference. Intermodulation Interference refers to two or more radio signals into non-linear devices to generate new frequency components, causing interference to useful signals. Blocking interference is that when the transmitter is operating near the receiver, the useful and unwanted signals transmitted by reducing the dynamic range of the receiver, resulting in decreased sensitivity and the formation of the receiver desensitization.

When the interference signal is strong and the useful signal is weak, the receiver is saturated and can not receive information normally. Spurious emission interference means that the spurious emission of the transmitter falls within the receiving band of other receivers and causes interference to the receiver.

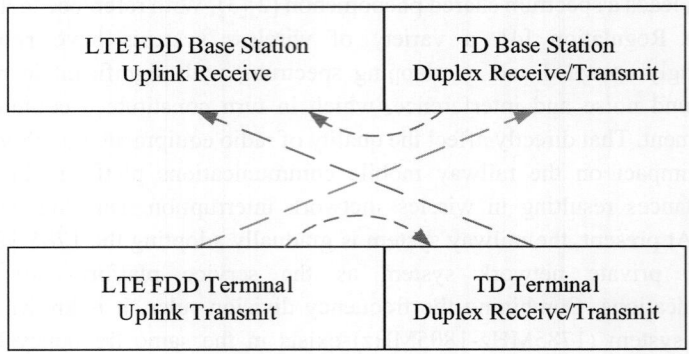

Fig. 1. Interference scenario.

Specific interference scenarios include four types, including base station interference base stations, base station interference terminals, terminal interference base stations, and terminal interference terminals. An interference scenario of TD-LTE and FDD-LTE is shown in Fig. 1. The interference between the base station and the terminal is divided into the uplink interference and the downlink interference. Such interference distance is far, the obstacles are large, the loss is large and the interference is small. For the interference between the terminal and the terminal, due to the low transmitting power of the terminal, also with uncertain positions and the quantities, referring to the low probability of close to each other, make the inter-terminal interference is low. Among base stations, base stations have high transmission power, strong transmission time continuity, relatively good propagation environment, usually introducing significant interference. In addition, base stations have relatively fixed positions and low sensitivity, and are more susceptible to interference. Therefore, radio environment analysis mainly consider the interference scene between base stations.

From the analysis of the interference scenario, both the jamming and the spurious jamming are mainly considered in the inter-system interference analysis. TD-LTE system is used for railway-specific system, and the main parameters of interference analysis are shown in Table 1. FDD system parameters are shown in Table 2.

Table 1. TD-LTE system parameters.

TD-LTE Base Station	Parameters
Frequency band	1785-1805MHz
System bandwidth	5MHz 10MHz 15MHz 20MHz
Max transmitting power	43dBm(5MHz) 46dBm(10/15/20MHz)
Antenna gain	15dBi
Antenna height	15m
Noise coefficient	5dB
Base Station background noise	-109dBm/MHz
Interference protection ratio	9dB

Table 2. FDD LTE system parameters.

FDD Base Station	Parameters
Frequency band	1805-1880MHz, 1710-1780MHz
Transmitting power	46dBm
Carrier wave bandwidth	200MHz
Noise coefficient	5dB
Background noise	-109dBm/MHz

3. Protection Zone Designation

The existence of a variety of interference will seriously affect the efficient and safe operation of the railway, set up radio environment protection zone along the railway in order to achieve the maximum isolation of interference signals, as shown in Fig. 2.

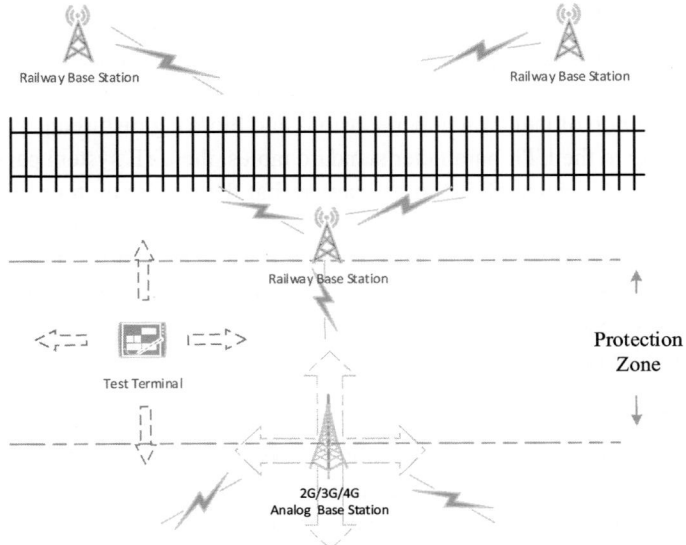

Fig. 2. Protection zone.

According to methods and procedures for calculating distance and frequency separation in the context of an acceptable interference level, as proposed in Recommendation ITU-R SM.337-6, protection zone can be obtained.

1) Calculate the jammer input interference level, including the spectrum factor and space factor

2) Interference criteria

$$P_d - P_i \geq \alpha \qquad (1)$$

where P_d indicates the desired signal level (dBW), P_i indicates the interference signal level (dBW), and α indicates the protection ratio (dB).

$$P_i = P_t + G_r - L_p - OCR(\Delta f) \qquad (2)$$

where P_t represents the interfering transmitter's equivalent e.i.r.p. (dBW), G_r represents the gain (dBi) of the receiver antenna relative to the omnidirectional antenna, L_p represents the propagation path loss, OCR (Δf) represents the frequency spacing Frequency offset rejection factor at Δf.

3) Calculate the relationship between d and Δf

As recommended by ITUR SM337, the protection zone calculation model uses the Enhanced Minimum Coupling Loss method to introduce the link availability (coverage quality) in the interfering system as an input to the calculation in terms of the maximum transmit power of the interferer.

4. Numerical Calculations

The interference types of LTE system to wireless signals of railway system are mainly spurious emission interference and blocking interference.

1) Calculation of isolation under spurious and blocking disturbances

According to the LTE parameter standard tables, the LTE system parameters and the GSM05.05 protocol, it can be learned that the maximum transmit power of the LTE system is 43dBm, the spurious emission power of the LTE system in the LTE-R system is -36dBm / 100khz, Sensitivity -104dBm, LTE-R system interference protection ratio of 9dB, N is 0.8dB. Therefore, in order to avoid spurious interference, the required space isolation is calculated by the formula of enhanced minimum coupling loss [6].

$$L_d = P_{INT} + dB_{BW} + G_{VICT} + G_{INT} - (S_{VICT} - C/I_{VICT}) + f(dBc_{INT}, P_{INT}) - 10\log(10^{N/10} - 1) \qquad (3)$$

where P_{INT} indicates interference source maximum transmit power, dB_{BW} indicates bandwidth conversion factor between interferer and victim, G_{VICT} indicates disturbed antenna gain, G_{INT} indicates interference sources antenna gain, S_{VICT} indicates receiver sensitivity of the victim, C/I_{VICT} indicates disrupted victim protection ratio. $f(dBc_{INT}, P_{INT})$ represents relative maximum interference power source transmit power, broadband noise power and frequency offset function. N represents Interfered system link availability factor.

$$L_d = -36 - 10\lg(100/200) - (-104-9) - 10\log(100.8/10-1) = 86.95 \text{dB}$$

To avoid blocking interference, the required spatial isolation is calculated as follows.

$$L_d = P_{INT} + G_{VICT} + G_{INT} - f(B_{VICT}, S_{VICT}) - 10\log(10^{N/10}-1) \qquad (4)$$

where $f(B_{VICT}, S_{VICT})$ represents the jammer's performance as a function of frequency offset.

$$L_d = 43 - 5 - 10\log(100.8/10-1) = 44.9 \text{dB}$$

2) Protection distance

According to the analysis, the spatial isolation needed to avoid spurious interference and blocking interference is $f = 86.95$ dB. f is the working frequency band of the interfered system LTE, and the working frequency band of the LTE system is 1785-1805MHz, taking $f = 2000$MHz. To avoid spurious emission interference and blocking interference caused by the LTE system, the radio environment protection distance required by the railway system wireless signal is calculated as follows.

$$L_d = 22 + 20\lg(\frac{S}{\lambda}) - (G_{VICT} + G_{INT}) \qquad (5)$$

where S indicates space isolation distance, λ indicates the wavelength of the transmitted signal.

$$S = \lambda * 10^{\frac{L_d + G_{VICT} + G_{INT} - 22}{20}} \qquad (6)$$

$S = 3*10^8 / (1790*10^6) * 10^{(44.9+15+15-22)/20} = 74 \text{ (m)}$

5. Conclusions and Discussions

Based on the complex radio environment along the railway, the interference between different wireless systems were analyzed, and the calculation method of setting the protection distance along the railway line is given, which provides technical support for the railway construction and security.

References

1. National Radio Interference Standardization Technical Committee, National Electromagnetic Compatibility Standardization Technical Committee. Electromagnetic Compatibility Standard Implementation Guide. Beijing: China Standard Press, 2010.

2. X. Ge, S. Tu, G. Mao, C. X. Wang and T. Han. 5G Ultra-Dense Cellular Networks. *IEEE Wireless Communications*, vol. 23, no. 1, pp. 72-79, February 2016.
3. M. Ayyash *et al*. Coexistence of WiFi and LiFi toward 5G: concepts, opportunities, and challenges. *IEEE Communications Magazine*, vol. 54, no. 2, pp. 64-71, February 2016.
4. Administration of Radio Industry of Ministry of Industry and Information Technology. Regulation on Radio Frequency Allocation of the People's Republic of China. Beijing: People's Posts and Telecommunications Press, 2010.
5. Wang Ping. The Designation of Electromagnetic environment protection zone of Railway Wireless Communication Systems. Tianjin University, 2013.
6. Recommendation ITU-R SM.1269. Classification of direction finding bearings. 2007.

Study on transition*

Long Hong

Computer School, Nanjing University of Posts and Telecommunications
Nanjing 210003, China

Jianyang Zhao

Faculty of Computer and Software Engineering, Huaiyin Institute of Technology
Huaian, 223003, China

Transition is a universal phenomenon in nature and our everyday life, and should be one of basic studying objects of artificial intelligence. There exist many papers studied the transition in specific topics, but most of them are not easy to address the general law of transition phenomenon. This paper attempt describing the transition itself and studying its basic characteristic. We first briefly introduce the concept of interval adjacency that is important for discussing transition, and then define transition in the field of real number; discuss transition area, transition point, direction of transition and so on. Since transition correlating of AI, the results got in this paper could be beneficial to the research and development of AI.

1. Introduction

Transition means a gradually changing process from one state to another [1], and it is a universal phenomenon in nature and our everyday life. For example, dawn is the transition period from night to day; middle age is a transition segment from youth to old age. These examples are unidirectional transition. Yet another type of transition is bidirectional such as gray is the transition range from black to white or from white to black. In the development of artificial intelligence, uncertainty is always a key issue [2] and randomicity and fuzziness are emblematical expressed form of uncertainty. From the view point of transition, probability for processing random phenomenon grasps the transition from cause to effect; membership grade dealing with fuzzy phenomenon is the model for calculating the scale of the neither here nor there. Therefore, transition is one of basic studying objects of AI.

---

*This work was supported by the National Natural Science Foundation of China under Grant No. 61170322 and the Open Fund of the State Key Laboratory of Smart Grid Protection and Control under Grant No. SGPC 201610.

Researchers around the world have dedicated tremendous amount of effort to study transition phenomenon, and have made important contributions in information science [3, 4], pattern recognition [5, 6], computer science [7, 8], physics [9], psychology [10], chemistry [11], etc. However, in spite of a number of contributions in dealing with transitions, most of these papers studied the transition in specific topics so that it was not easy to address the general law of transition phenomenon.

In this paper, we shall describe the transition itself and study its basic characteristic. The remainder of this paper is organized as follows: Section 2 briefly introduce interval adjacency that is important for discussing transition. Section 3 defines transition in the field R and study its general laws, and conclude our paper in Section 4.

2. Interval Adjacency

Two intervals R_i and R_{i+1} are called ordered if for any $a \in R_i$ and $b \in R_{i+1}$, we have $a \leq b$, where $i \in N$. If every two blocks are ordered in the partition, then the partition is called ordered partition. Only finite ordered interval partition is in this paper.

The left-hand symbol and right-hand symbol of an interval are respectively denoted by 〚 and 〛, namely 〚∈{ [, (} and 〛 ∈{),] }.

Definition 1. If $R_i =$ 〚a, b〛 and $R_{i+1} =$ 〚b, c〛, then R_i and R_{i+1} are called neighboring interval each other; the group of three symbols 〛, 〚 in the middle of 〚a, b〛, 〚b, c〛 is called interval adjacency, and b is called adjacent element.

In interval adjacency 〛, 〚, the symbols 〛 and 〚 are general matching symbols of interval, and the symbol of comma , is a connector between them.

The sequence of neighboring interval constructed by every two neighboring intervals is

〚a_1, a_2〛, 〚a_2, a_3〛, ..., 〚a_{n-1}, a_n〛, 〚a_n, a_{n+1}〛.

Moreover, there are $n-1$ interval adjacencies in the sequence consisting of n intervals.

The adjacency is called a **TypeI** adjacency if the both 〛 and 〚 in the adjacency are closed or both are open; otherwise, it is called a **TypeII** adjacency.

Theorem 1. Assume $R_1, R_2 \subseteq (r_i, r_j)$; R_1 and R_2 are neighboring interval.

(1) The interval adjacency of the neighboring interval is **TypeI** adjacency if and only if R_1 and R_2 are not partition blocks of (r_i, r_j);

(2) The interval adjacency of the neighboring interval is **TypeII**adjacency if and only if R_1 and R_2 are blocks of (r_i, r_j).

3. Representation of Transition in Real Number Field

Definition 2. Given neighboring interval sequence A consisting of R_i, where $R_i \subset \boldsymbol{R}$ and $i \in I$, and if $a \in R_{i-1}$, $b \in R_i$, $c \in R_{i+1}$, then $a < b < c$. If variable v going through $[\![a, c]\!]$ is monotonic and continuous, the variational process of v in R_i is called the transition of v being between R_{i-1} nd R_{i+1}. The v is called transition variable; R_i is called the transition area between R_{i-1} and R_{i+1}, and it is denoted as α; the first point of a transition area reached by v is called a beginning point of transition and denoted as β.

We may use a ordered triple $T(v, \alpha, \beta)$ to express transition, and discuss transition according to it.

Theorem 2. There exists a transition variable *iff* there exists a transition area.

Proof. Let R_{i-1}, R_i, $R_{i+1} \subset \boldsymbol{R}$ and they are in neighboring interval sequence, and given transition variable v between R_{i-1} and R_{i+1}. Suppose v has no transition area R_i, then v is not transition variable between R_{i-1} and R_{i+1} by definition 2, so it is a contradiction.

Given transition area R_i. Assume to contradict that there is no v between R_{i-1} and R_{i+1}, which shows R_i is not transition area. #□

Theorem 3. A transition variable changes in unique transition area.

Proof. There are two steps. (1) One transition has only one corresponding transition area. Supposed there are two transition areas, R_i and R_j, in one transition, where $R_i \neq R_j$. Then, there are two change processes of a transition variable v in R_i and R_j respectively, which implies there are two transitions that are v being between R_{i-1} and R_{i+1} and between R_{j-1} and R_{j+1}. This is a contradiction. Similarly, we can prove that one transition has no above two transition areas.

(2) One transition has merely one transition variable. By (1) and definition 2, it is obtained immediately.

Synthesize (1) and (2), the proof is completed. #□

Corollary 4. There is unique beginning point in a transition area.

By theorem 3 and definition 2, corollary 4 is obtained directly.

Definition 3. $T(v, R_i, P_f)$ is called the increasing transition if transition variable v changes from any point of R_{i-1} to R_{i+1}, and the R_i is called an increasing transition area denoted as ^+R_i. Otherwise, T is called a decreasing transition and the decreasing transition area denotes ^-R_i.

In general, we use $^+(R_i \cup R_{i+1} \cup ... \cup R_{i+n})$ to denote the increasing transition area from any point of R_{i-1} to R_{i+n+1}, and $^-(R_i \cup R_{i+1} \cup ... \cup R_{i+n})$ the decreasing transition area from one of R_{i+n+1} to R_{i-1}.

The transitions discussed in the following are increasing transitions unless otherwise stated.

Theorem 5. Given $R_{i-1}= [\![r_a, r_1]\!]$, $R_i= [\![r_2, r_3]\!]$, $R_{i+1}= [\![r_4, r_b]\!]$. If R_i is the transition area between R_{i-1} and R_{i+1}, then $R_{i-1} \cup R_i \cup R_{i+1} = [\![r_a, r_b]\!]$.

Proof. For the sake of contradiction, suppose $R_{i-1} \cup R_i \cup R_{i+1} \neq [\![r_a, r_b]\!]$. We distinguish two cases:

(1) $R_{i-1} \cup R_i \neq [\![r_a, r_3]\!]$ but $R_i \cup R_{i+1} = [\![r_2, r_b]\!]$. Then, there exists the gap between R_{i-1} and R_i;

(2) $R_i \cup R_{i+1} \neq [\![r_2, r_b]\!]$ but $R_{i-1} \cup R_i = [\![r_a, r_3]\!]$. Namely there exists the interstice between R_i and R_{i+1}.

Above (1) and (2) show v can not continuously change between R_{i-1} and R_i or between R_i and R_{i+1}, which contradicts that R_i is the transition area between R_{i-1} and R_{i+1}. Hence, $R_{i-1} \cup R_i \cup R_{i+1} = [\![r_i, r_j]\!]$. #□

Theorem 6. Given $R_{i-1}= [\![r_a, r_1]\!]$, $R_i= [\![r_2, r_3]\!]$, $R_{i+1}= [\![r_4, r_b]\!]$. If R_i is the transition area between R_{i-1} and R_{i+1}, then $(R_{i-1} \cup R_{i+1}) \cap R_i = \emptyset$.

Proof. $(R_{i-1} \cup R_{i+1}) \cap R_i = R_{i-1} \cap R_i \cup R_{i+1} \cap R_i$. Assume $R_{i-1} \cap R_i \neq \emptyset$, then $r_2 \leq r_1$ which contradicts the condition '$\alpha < \beta < \gamma$' in definition 2, so $R_{i-1} \cap R_i = \emptyset$. The argument of $R_i \cap R_{i+1} = \emptyset$ is similar. Hence, $(R_{i-1} \cup R_{i+1}) \cap R_i = \emptyset$. #□

Corollary 7. Adjacencies of intervals forming any transition are all **TypeII**adjacencies.

Proof. Let $R_{i-1}= [\![r_i, r_1]\!]$, $R_i= [\![r_2, r_3]\!]$ and $R_{i+1}= [\![r_4, r_j]\!]$. We have $R_{i-1} \cup R_i = [\![r_i, r_3]\!]$ according to theorem 5 and $R_{i-1} \cap R_i = \emptyset$ by theorem 6, so it is none but $r_1 = r_2$. By the same argument, $r_3 = r_4$. Hence, ']' , '[' must be '),[' or '],(.'. Therefore, interval adjacencies among R_{i-1}, R_i and R_{i+1} are **TypeII**adjacencies. #

Corollary 8. The group of three intervals R_{i-1}, R_i and R_{i+1} is a partition of $R_{i-1} \cup R_i \cup R_{i+1}$ if there exists a $T(v, R_i, P_f)$.

By Corollary 7 and Theorem 1, Corollary 8 follows.

Theorem 9. Given $R_{i-1}= [\![r_{i-1}, r_i)$, $B=[r_i+\varepsilon, r_{i+1})$ and $R_{i+1}=[r_{i+1}, r_{i+2}]\!]$. The B is not a transition area between R_{i-1} and R_{i+1} if $\varepsilon \neq 0$.

Proof. We distinguish two cases:
(1) If $\varepsilon > 0$, then $R_{i-1} \cup B \neq [\![r_{i-1}, r_{i+1})]\!]$; further, $R_{i-1} \cup B \cup R_{i+1} \neq [\![r_{i-1}, r_{i+2}]\!]$. By Theorem 5, B is not a transition area between R_{i-1} and R_{i+1};
(2) If $\varepsilon < 0$, then $R_{i-1} \cap B \neq \emptyset$ leading to $(R_{i-1} \cup R_{i+1}) \cap B \neq \emptyset$, so B is not a transition area between R_{i-1} and R_{i+1} by Theorem 6. #□

This theorem shows that there are no catastrophe and overlap in the transition discussed in this paper.

Remark 1. Let $R_{i-1}= [\![r_{i-1}, r_i)$, $R_{i+1}=[r_{i+1}, r_{i+2}]\!]$. If R_i is none but r_i, then $R_i=[r_i, r_i]$, which is called single point transition, and r_i is called transition point. As particular single point transition is, it is common in science research and in real world. For example, zero is the transition point between positive number and negative number.

Now, we may specify that R_i is the interval from r_i to r_{i+1}.

Having defined the beginning point of transition, we have to answer a question that is where is the ending point.

Let $R_{i-1} = [\![r_{i-1}, r_i]\!]$, $R_i = [\![r_i, r_{i+1})]\!]$, $R_{i+1} = [\![r_{i+1}, r_{i+2}]\!]$. By definition of beginning point and Corollary 7, we have (1) $R_i=[r_i, r_{i+1})$, $R_{i+1}=[r_{i+1}, r_{i+2}]\!]$ or (2) $R_i = [r_i, r_{i+1}]$, $R_{i+1} = (r_{i+1}, r_{i+2}]\!]$. In (1), there is only ending point without beginning point in transition area; but in (2), there are both beginning point and ending point. However, the (2) doesn't meet Definition 2.

Theorem 10. There is no ending point in transition area.

Proof. Let R_i be the transition area from R_{i-1} to R_{i+1} and R_{i+1} is the transition area from R_i to R_{i+2}. Now suppose there is a ending point in R_i, then $R_i=[r_i, r_{i+1}]$, and further, $R_{i+1}=(r_{i+1}, r_{i+2}]\!]$ by Corollary 7. So there is no beginning point in R_{i+1} as the transition area, which is a contradiction to Corollary 4. #□

Remark 2. From this paper's view, one after another is one of main properties of transition. In general, therefore, the ending point of this transition is the beginning point of next transition. In an increasing transition, two minimal elements in ^+R_i and R_{i+1} are beginning point and ending point, respectively; likewise, maximal elements in ^-R_i and R_{i-1} are beginning point and ending point in a decreasing transition, respectively.

Corollary 11. (1) There is only an adjacency constructed by a group of three symbols), [in an increasing transition. (2) In a decreasing transition, the only possible adjacency is],(.

Denote $R_[$ as $\{R_i \mid [r_i, r_{i+1}) \ \& \ 1 \leq i \leq n\}$ and $R_]$ as $\{R_i \mid (r_i, r_{i+1}] \ \& \ 1 \leq i \leq n\}$.

Corollary 12. (1) $R_{i+1} \in R_[$ if ^+R_i; (2) $R_{i-1} \in R_]$ if ^-R_i.

Proof. (1) Assume to the contrary that $R_{i+1} \notin R_[$ for ^+R_i, then $R_{i+1} = (r_{i+1}, r_{i+2}]$, which shows there is no minimal element in R_{i+1}; namely there is the beginning point in increasing transition. This is a contradiction to Corollary 4.

(2) Can be proved similarly. #□

Theorem 13. (1) $^+(R_i \cup R_{i+1} \cup \cdots \cup R_{i+n-1})$ if ^+R_j; (2) $^-(R_i \cup R_{i+1} \cup \cdots \cup R_{i+n-1})$ if ^-R_j. Where $i \leq j \leq i+n-1$.

Proof. As $^+R_i, ^+R_{i+1}, \ldots, ^+R_{i+n-1}$ and by Corollary 12, $R_i, R_{i+1}, \cdots, R_{i+n} \in R_[$. Hence,

$R_{i-1} \cup (R_i \cup \cdots \cup R_{i+n-1}) \cup R_{i+n}$
$= [r_{i-1}, r_i) \cup ([r_i, r_{i+1}) \cup \cdots \cup [r_{i+n-1}, r_{i+n})) \cup [r_{i+n}, r_{i+n+1})$
$= [r_{i-1}, r_{i+n+1})$.

Obvious, $(R_{i-1} \cup R_{i+n}) \cap (R_i \cup \cdots \cup R_{i+n-1}) = \emptyset$.
So, $^+(R_i \cup R_{i+1} \cup \cdots \cup R_{i+n-1})$.
(2) Omitted here. #□

Example. In the visible spectrum, the wavelengths of various color ranges are as follows: blue is 450-495nm; green is 495-570nm; yellow is 570-590nm; orange is 590-620nm and so on [12]. From the view of wavelength, green is the transition area between blue and yellow; while yellow is the transition area between green and orange. Let s be the shifting of the spectral line, then transition triple SR(s, [495, 570] ,ℓ) expresses the transition with green as transition area. If $450 \leq \ell < 495$, then SR is the transition from blue into yellow, namely $^+$ [495, 570] ; if $^-$ [490, 570] , then SR is the transition from yellow into blue, that is $570 < \ell \leq 590$.

Of course, It can also be regarded that the color range of green and yellow is the transition area between blue and orange, and both $^+(\![495, 570)\cup[570, 590]\!])$ and $^-(\![495, 570)\cup[570, 590]\!])$ are two types of this transition.

4. Conclusions

We have studied the first notion of transition in field R, and its essentials are summarized as follows:

Transition means the process of gradually changing from one state to another, and the process may be a state. A transition is the process of motion that possesses beginning point and ending point, and has directions that are distinguished positive or negative.

The sudden change, jumping and mutation don't belong to transition described in this paper.

To clearly describe transition, we have briefly introduced the concept of interval adjacency.

References

1. www.oxforddictionaries.com
2. J. Pearl, *Cambridge University Press*, (2009).
3. S. Roy and R. Dhal, *IEEE J. STSP.* **9**, 304(2015).
4. C. Cartis and A. Thompson, *IEEE Trans. Inform. Theory*, **61**, 2019 (2015).
5. H. Othman and T. Aboulnasr, *IEEE Pattern and Mach. Intell.* **25**, 1229 (2003).
6. B. Chen et al., *IEEE Trans. Syst. Man Cybern: Syst.* **43**, 1279(2013).
7. H. Zheng et al., *IEEE Trans. Comput.*, **64**, 1607(2015).
8. N.K. Ure et al., *Proc. ECML PKDD.* **II**, 99(2012).
9. W.S. Bakr et al., *Science*, **329**, 547(2010).
10. C. Dyer, *Oxford University Press*, (2005).
11. A. Westgren, *Metallwirtschaft*, **9**, 919(1930).
12. J. Bruno and D. Svoronos, *CRC Press*, (2005).

A scale-free model for random ASP programs

Lian Wen, Kaile Su and Zhe Wang

School of ICT, Griffith University, Brisbane, Australia
l.wen, k.su, zhe.wang@griffith.edu.au

Scale-free property, which means a small number of nodes dominate a network with far more number of connections than other nodes, has been observed in many large-scale real-world networks and many large software systems. However, this is not investigated for answer set programming (ASP), a major paradigm of declarative problem solving. This paper first presents a generator (model) for randomly generating ASP programs that demonstrate scale-free property. Then reports that all the 28 real ASP programs from different domain showing clear scale-free property. Finally significant experiments are conducted to demonstrate that random generated ASP programs could be solved in polynomial time if they are scale-free, while general random ASP programs are NP hard problems. The results may help to understand the nature of large ASP programs.

Keywords: Answer set program; logic systems; scale-free networks; non-monotonic logic.

1. Introduction

Most complex system problems can be presented in complex networks traditionally modeled in random networks.[1] In random networks, edges are generated randomly to connect a set of given vertices and the number of edges attached to each vertex (the degree) is shown to follow a normal distribution as shown in the left cell of Fig. 1.

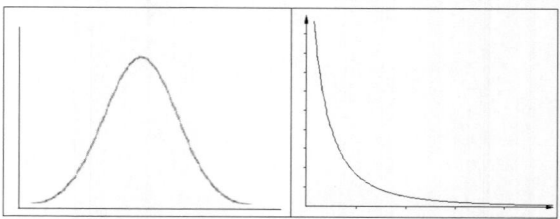

Fig. 1. The left cell shows a normal degree distribution which is typical for random networks, while the right cell shows a power-law degree distribution in typical scale-free networks.

However, many real large networks such as human social networks, WWW, research collaboration networks,[1] cellular metabolisms,[2] chemical reaction networks[3] are discovered not fit in the random network model. One conspicuous feature of those real networks is that their degree distributions are usually close to a power-law function as in the right cell of Fig. 1. Networks with this feature are called scale-free networks.

Scale-free networks are also discovered in software engineering; previous research indicates that for functional programs, the class dependency networks are scale-free.[4] The fact that so many different kinds of systems show strikingly similar topological properties has spawned interest across many disciplines and has caused researchers to postulate the existence of some basic organisational principle. Interesting properties are discovered in scale-free networks. In sorting algorithm, the scale-free property emerges when the algorithm is optimized while low efficient algorithm results more like random networks.[5] Experiments also show minimum dominating set (MDS) is in linear scaling respect to the network size when the network is scale-free.[6] However the true nature of scale-free property and the fundamental reason behind the pervasive occurrence of scale-free networks is still unknown.

Compared to functional programming, answer set programming (ASP) is a form of non-monotonic logic that is devised to capture and represent defeasible inferences. It supports negation as failure (NAF). This feature makes it closer to human cognitive process or common sense reasoning. For example, consider a simple logic system described by two statements *if there is fire, the student needs to evacuate, if the student is not evacuated, he is studying*. Let f denote "to have a fire", e denote "to evacuate", and s denote "to study". In the proposition logic, we can use two propositional formulae $f \rightarrow e$, $\neg e \rightarrow s$ to represent the system. From this system, we can not infer if the student is studying or not, because the system has five valid models: $\{f,e,s\}$, $\{f,e\}$, $\{e,s\}$, $\{e\}$,$\{s\}$ with three of them show the student is studying and two of them not. The above example can be encoded as an ASP program: $e \leftarrow f$. $s \leftarrow$ not e. In this program, because there is no evidence to show there is a fire, so by default, there is no fire. As there is no fire, there is no evidence to support to evacuate, so the student is not evacuated. Then based on the second rule, because the student is not evacuated, the student is studying.

For an ASP program, if we treat each distinct atom as a node, each rule (formula) as a hyper-link among all the atoms appearing in this rule, the program is mapped to an atom-rule hyper-graph. In this paper, we make the first attempt to investigate scale-free properties of the atom-rule hyper-graph associated with an ASP program. To simplify the notion, when we say an ASP program shows scale-free property, we mean the program's atom-rule hyper-graph shows scale-free property. Our contributions in the paper can be summarized as follows:

(1) We tested 28 real ASP programs from different domains and found that all these real ASP programs tend to be scale-free.
(2) We proposed a new way to generate random ASP programs with scale-free properties. The new algorithm is more general than previous algorithms used to generate scale-free networks. The detail of the algorithm is explained in Subsection 3.1.
(3) From our experiments, we are able to corroborate an interesting propositions of random ASP programs generated by our generator: If ASP programs generated by our model with strong enough scale-free property, then they can be solved in polynomial time. This is an interesting result as it is well known that ASP is NP-complete in general.

The paper is organized in the following way. Section 2 provides more background knowledge about ASP and scale-free properties. Section 3 presents our model to generate random ASP programs with scale-free properties and also introduces the proposition of the relationship between scale-free properties and the hardness of random ASP programs. Section 4 contains our experimental results. Finally, a short conclusion is given in the last section.

2. ASP Programs

This section review some basics of ASP programs. A rule r of ASP is of the form

$$a \leftarrow b_1, b_2, ..., b_{t_p}, \text{not } c_1, \text{not } c_2, ..., \text{not } c_{t_n}. \tag{1}$$

Here $t_p, t_n \in \mathbb{Z}_0$; $a, b_i, c_j \in A$ ($i \leq t_p, j \leq t_n$) are atoms. a is called the head of r, denoted as $\text{head}(r)$. If a is empty, it is called a constraint, otherwise it is called a normal rule. Define $\text{body}^+(r) = \{b_1, b_2, ..., b_{t_p}\}$, $\text{body}^-(r) = \{c_1, c_2, ..., c_{t_n}\}$. If both $t_p = 0$ and $t_n = 0$, it is called a fact. If $t_n = 0, t_p > 0$, it is called a positive rule.

The semantics of an ASP program P is defined in terms of its *answer sets* (*stable models*)[7] as follows. Given an interpretation S, the *reduct* of P on S is defined as $P^S = \{head(R) \leftarrow body^+(R) \mid R \in P, body^-(R) \cap S = \emptyset\}$. Then we say S is an *answer set* of P, if S is the least model of P^S. By AS(P) we denote the collection of all answer sets of P. P is said to be *consistent*, if it has at least one answer set.

This paper considers ASP programs with only normal rules and constraints. Each program must contain negative literals in some rules. Each rule (constraint), based on the number of positive literals t_p and negative literals t_n in the body, is classified into a unique body pattern. And the set of all the rules (constraints) are denoted as $\mathcal{R}(t_p, t_n)$ (and $\mathcal{C}(t_p, t_n)$).

Definition 1. Let $t_n \in \mathbb{Z}_0$ and $t_p \in \mathbb{Z}_0$, define $\mathcal{R}(t_p, t_n)$ (and $\mathcal{C}(t_p, t_n)$) as the set of all normal rules (constraints) with t_p positive literals and t_n negative literals in their body. We call such a rule (constraint) with **body pattern** (t_p, t_n).

For example, if r is a rule like $a \leftarrow b_1, b_2, \text{not } c$, then r is with a body pattern of $(2,1)$ and $r \in \mathcal{R}(2,1)$.

For a random grounded ASP P with only normal rules and constraints, the key statistic properties of P include the total number of distinct atoms, the number of rules and also distribution of its rule and constraint patterns.

Definition 2. Let P be a grounded ASP with only normal rules and constraints, n be the total number of distinct atoms, l be the total number of rules. Then the **rule-atom rate** of P is defined as $\zeta_p = l/n$.

Definition 3. Let P be a grounded ASP with only normal rules and constraints, $t_p \in \mathbb{Z}_0$ and $t_n \in \mathbb{Z}_0$. Define the number of rules with the body pattern of (t_p, t_n) as $l_r(P, t_p, t_n) = |\{r | r \in P, r \in \mathcal{R}(t_p, t_n)\}|$; and the number of constraints with the body pattern of (t_p, t_n) as $l_c(P, t_p, t_n) = |\{r | r \in P, r \in \mathcal{C}(t_p, t_n)\}|$.

Definition 4. Let P be a grounded ASP with only normal rules and constraints, n is the number of distinct atoms in P, $\mu \in \mathbb{R}$, which is called the **order** of the program, and $\mu \geq 1$, the **pattern distribution** of P is defined as a function τ_P^μ from $\mathsf{C} \times \mathbb{Z}_0 \times \mathbb{Z}_0$ to $\mathbb{R}$, while $\mathsf{C} = \{\mathbf{r}, \mathbf{c}\}$. So

$$\tau_P^\mu(e, t_p, t_n) = \begin{cases} l_r(P, t_p, t_n) n^{-\mu} & \text{if } e = \mathbf{r} \\ l_c(P, t_p, t_n) n^{-(\mu-1)} & \text{if } e = \mathbf{c} \end{cases} \quad (2)$$

If $\mu = 1$, we may omit μ and say the program is linear under τ. If $\mu = 2$, we call the program a quadratic program under τ^2.

A pattern distribution function can be presented as a set of 4-tuples such as
$$\tau_P^\mu = \{(e_1, t_{p_1}, t_{n_1}, c_1), ..., (e_k, t_{p_k}, t_{n_k}, c_k)\},$$
which means $\tau_P^\mu(e_i, t_{p_i}, t_{n_i}) = c_i$.

For example, if a random ASP P contains 100 atoms, 400 rules in the form as $a \leftarrow \text{not } b$, and 5 constraints in the form as $\leftarrow c_1, \text{not } c_2$. If we select the order of P as $\mu = 1$, then the pattern distribution of P is $\tau_P^1 = \{(\mathbf{r}, 0, 1, 4), (\mathbf{c}, 1, 1, 5)\}$ (because $\mu = 1$, τ_P^1 can also be written as τ_P). Obviously, a program P may have different pattern distribution under different orders. If two programs P and P', under the same order μ, have the same pattern distribution function, we call the two programs have the same pattern distribution under μ.

3. Properties of Scale-free Programs

3.1. *Algorithm to Generate Random ASP with Scale-Free Property*

This paper proposes a new algorithm to generate random ASP with Scale-Free Property. The algorithm is different from existing scale-free network generation algorithms in literature.[1,6] There are two major differences. The first is that the old algorithms were designed to generate scale-free networks with typical graphs, where each edge connecting exact two vertices while the new algorithm can generate scale-free property for hyper-graphs. The second difference is that the new algorithm introduces a scale-free coefficient ρ, which controls the algorithm to generate networks with different proportion of scale-free feature. When $\rho = 0$, it is the same as the traditional random model; when $\rho = 1$, the degree distribution will be a typical power-law function; when $0 < \rho < 1$, the degree distribution will be in the middle of a normal distribution and a power-law distribution.

To generate an ASP program P, let A_n ($|A_n| = n$) be the set of all atoms for the program. For $\forall x \in A_n$, let $n(x)$ be the number of occurrences of x in the new program P. Because initially P is empty, $n(x) = 0$ for all x. Then we define a weight function $w(x) = 1 + \rho n(x)$. Then we start to generate rules one by one. Each rule, based on the class of P, contains a number of atoms randomly selected from A_n. The probability for any $a \in A_n$ to be selected is $p(a) = w(a)/\Sigma w(x)$. Once a particular atom a is selected, the software needs to check if a is a valid selection. If a is not a valid selection, we will drop a and make another selection until a valid selection if found. If a is a valid selection, we will update $n(a) = n(a) + 1$, also update the weight function $w(a)$, that might affect the probability distribution for the entire A_n. One by one, until all rules in P are generated, the program P is then saved.

3.2. *Scale-Free*

To solve ASP programs in general is a NP-complete program. When we map an ASP program as a hypergraph, with each atom as a vertex and each rule as a hyperedge, our experiments show that once the corresponding hypergraph is a scale-free network, the program is much easier (in polynomial time) to be solved by existing solvers.

Proposition 1. *There exists a fixed preference coefficient $\exists \rho_0$, for all distribution functions $\forall \tau$, any program order $\forall \mu \geq 1$, any preference coefficient $\rho \geq \rho_0$, let P be randomly generated ASP with pattern distribution as τ^μ, and selection preference coefficient as ρ. Let n be the total number of atoms in P, then there exists an algorithm that can solve P in a polynomial time regarding n.*

4. Experiments

Our random program generator can generate various random logic programs based on user-input parameters. Once a random program or a set of random programs are generated, the tool evokes an ASP solver to compute the answer sets of those programs and then the results returned by the ASP solver are recorded and processed. The ASP solver used in our experiments is CLASP.[8] Based on our experiments, CLASP is faster than DLV[9] and SMODELS.[10]

We have conducted a huge amount of experiments and all of them corroborate the properties stated in the last section. We only present some representative results. More complete results and experiment software can be downloaded from the paper website.[11]

4.1. Real ASP Programs

We first present an experimental result based on 28 real ASP programs from different domains. 26 of them are from 2013 ASP competition benchmark suite, and the other two are from colleagues with one for robot control and another for a board game. All those programs are publicly available.[11,12]

From our experiments, we found that all those real ASP programs tend to be scale-free. We present the degree distribution of two programs in Fig. 2. The x-axes represent the degree (number of occurrences) of an atom in a program; the y-axes represent the number of atoms. The left cell shows the degree distribution of N19 (Abstract-Dialectical-Frameworks-Well-founded-Model) problem in 2013 ASP competition, and the right cell for N18 (Chemical-Classification). N19 has 72 rules with 65 atoms while N18 has 199340 rules with 11829 atoms. Even though the two programs are significantly different in size, the degree distributions are remarkably similar and scale-free.

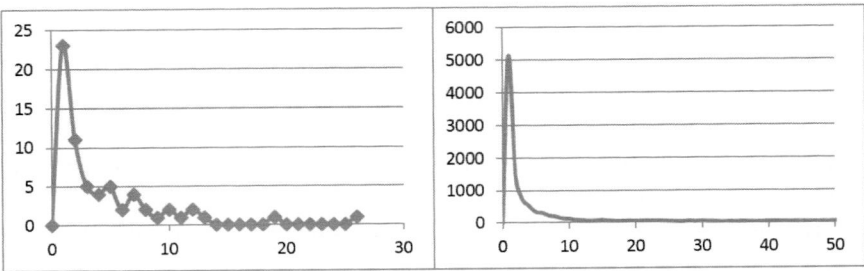

Fig. 2. The degree distribution of two ASP of real applications.

4.2. *Experiments for Scale-Free — Impacts of ρ on ASP Hardness*

For all random programs generated under different pattern distribution functions τ, different orders μ, we observed that once we adjust the preference coefficient ρ, it has noticeable impact on the probability of stable and the expected number of answer sets, but not dramatic. However, its impact on the hardness of the program is remarkable. It will significantly reduce the hardness of the program. Particularly, once the preference coefficient ρ reaches 0.5, all the tested classes of ASP problem can be solved in a polynomial time regarding the number of atom n by CLASP.[8]

Due to the page limit, we only present a typical result in this paper that shows the impact of ρ on the hardness of two literal negative programs with the pattern distribution function $\tau = \{(r, 0, 1, 20)\}$.

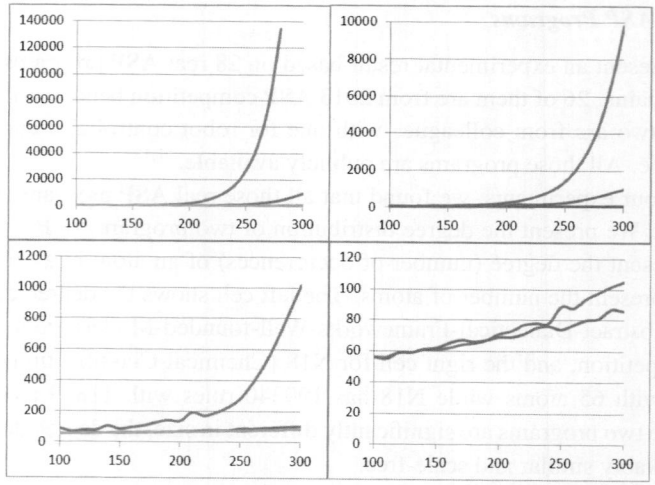

Fig. 3. The impact of scale-free on the hardness of normal two literal negative programs.

For normal two literal negative programs, we select ρ=0, 0.1, 0.2, 0.5, 1, for each ρ, let n=100, 110, 120,...,300. For each combination of ρ and n, 5,000 independent programs were generated and tested (when $\rho = 0$ and n=290 and 300, it took too long for the solver to solve the programs, so we do not have the results for those two combinations).

The results are presented in Fig. 3. The x-axes represent the number of atoms n and y-axes represent the average time to solve a program in millisecond. The top left cell shows five curves with ρ=0, 0.1, 0.2, 0.5, and 1 respectively. Because the curve of $\rho = 0$ increases so quickly, it dominates the figure and compresses the other curves almost to the x-axis with only a slight warp for $\rho = 0.1$ around $n = 300$. The top left cell removes the curve of $\rho = 0$, the bottom left cell further more removes $\rho = 0.1$; while the last cell only shows the curves of ρ=0.5 and 1. Obviously, when ρ increases the time to solve a program decreases remarkably. Also we note that when ρ reaches 0.5, the time complexity is reduced from an exponential order to a polynomial order. Interestingly, after $\rho > 0.5$, further increasing of ρ only has relatively marginal impact on the average time to solve a program.

5. Conclusions

This paper has made the first attempt to investigate scale-free properties of random ASP programs. We first propose a new model to generate random ASP programs with scale-free property. We then observed that all the real ASP programs are scale-free. Finally we discovered that random ASP programs with stronger scale-free property are easier to be solved; especially when the scale-free coefficient ρ is larger enough, the random ASP programs can be solved in polynomial time. In the future, we plan to provide a theoretical proof for the above stated property.

References

1. A.-L. Barabási, *Linked: the new science of networks* (Perseus Books Group, United States, 2002).
2. H. Jeong, B. Tombor, R. Albert, Z. Oltvai and A. Barabási, *Nature* **407**, 651 (2000).
3. H. Jeong, S. Mason, A. Barabási and Z. Oltvai, *Nature* **411**, 41 (2001).
4. L. Wen, D. Kirk and G. Dromey, Software systems as complex networks, in *Proceedings of The 6th IEEE International Conference on Cognitive Informatics*, (IEEE CS Press, August 2007).
5. L. Wen, G. Dromey and D. Kirk, *IEEE Transactions on Systems, Man, and Cybernetics, Part B: Cybernetics* **39(4)**, 845 (2009).
6. F. Molnár, S. Sreenivasan, B. Szymanski and G. Korniss, *Scientific Reports* **3(1736)** (2013).
7. M. Gelfond and V. Lifschitz, *Proceedings of the Fifth International Conference and Symposium of Logic Programming*, 1070 (1988).
8. M. Gebser, B. Kaufmann and T. Schaub, The conflict-driven answer set solver clasp: Progress report, in *Proceedings of the 10th International Conference on Logic Programming and Nonmonotonic Reasoning (LPNMR-09)*, (Sep, 2009).
9. N. Leone, G. Pfeifer, W. Faber, T. Eiter, G. Gottlob, S. Perri and F. Scarcello, *ACM Transactions on Computational Logic* **7**, 499 (2006).
10. T. Syrjänen and I. Niemelä, *Proceedings of the 6th International ConferenceLogic Logic Programming and Nonmonotonic Reasoning (LPNMR-01)*, 434 (2001).
11. ASP, Experimental results http://www.beworld.org/asp/' (January 2017).
12. ASP, 2013 answer set programming competition official problem suite, https://www.mat.unical.it/aspcomp2013/officialproblemsuite (2013).

Procedural extensions for executable ontologies in conventional software development

Selena S. Baset

Information Management Institute, University of Neuchatel, A.L. Breguet 2
Neuchatel, 2000, Switzerland

Kilian Stoffel

Information Management Institute, University of Neuchatel, A.L. Breguet 2
Neuchatel, 2000, Switzerland

Ontologies have gone lengths in various areas of knowledge engineering, yet they are falling short of reaching an equal position as formal domain models in the landscape of enterprise software development. In this paper, we present an approach for integrating ontologies into the code space of conventional software. We argue that the limited adoption of ontologies in software development is partially due to the lack of imperative programming capabilities. We propose extending ontologies with procedural extensions by expressing them in an executable form. Finally, we discuss the advantages of this representation and the possibilities for further improvements.

1. Introduction

Despite their proven utility in various areas of knowledge engineering, ontologies are falling short of reaching an equal position as formal domain models in the landscape of enterprise software development. This shy adoption of ontologies in software engineering communities is partially due to the opposing semantics of logic-based formalisms when compared to those of object-oriented constructs. The technical stacks behind ontological applications and conventional enterprise software are also different; from languages and editors to infrastructure support. Furthermore, in comparison to other modeling languages such as UML, ontologies are more suitable for capturing the static aspects of modeling such as taxonomies and relations between concepts, but they lag behind when it comes to capturing the dynamic behavior of the interacting components in the domain under consideration. For that purpose, UML still offers better options with a variety of diagrams: state, collaboration and sequence diagrams.

All these reasons make it difficult for developers to overcome the difficulties in integrating ontologies into the development life cycle. In this paper, we address these issues and we propose an approach to integrate ontologies directly into the code space of conventional software by expressing ontologies in a general-purpose programming language instead of the custom-build ontological languages such as OWL. Expressing ontologies in a general-purpose programming language reinforces the synergies between software engineering and knowledge engineering. It accentuates a more declarative style of coding by adhering to the semantics of ontological modeling yet at the same time, it allows the developer the possibility to switch to an imperative style, when necessary, to add a procedural extension that can capture the dynamic behavior of the system.

As this article is mostly about enforcing the synergies between ontological modeling and object-oriented programming, we will start by providing a brief background on frame-based systems as a common "ancestor" of both paradigms and how it gradually led to the immersion of ontological languages. We then look at the current diverging directions of development in each side before we present our approach on how to bring them back closer together. We discuss some key points and relate to the research work done in this area before finally concluding with reflections and future work.

2. Background

2.1. *Frame-based systems*

Frames stand for a knowledge representation technique that evolved from the cognitive intuition of semantic networks as a more structured mechanism to represent real-world entities [1]. Knowledge in a frame-based system is represented as a collection of frames in which each frame has a number of slots that can be filled with links to other frames. We can differentiate between generic frames denoting a class of objects and instance frames denoting individuals. Inheritance mechanism is supported between generic frames, as well as the possibility of an instance frame to override a default slot value provided by its generic frame. One substantial difference that frames have in comparison with semantic networks is their inferential capabilities powered by a procedural extension to the declarative nature of the slots. This extension takes the form of triggers (Demons) attached to the slot as a means of controlling slot-level manipulations; they serve as a kind of slot self-checking mechanism to ensure frame correctness and consistency. Scripts are another procedural extension that frames have. A script is a type of frame that is used to indicate the steps needed to complete a task in a defined order.

2.2. Logic-based formalisms

Confronted with the lack of precise semantics in semantic networks, the pursuit of a means of providing semantics to representational structures grew larger. That was the moment when the role of logic in knowledge representation got into the spot light. Logic offered a rich palette of declarative languages that are well suited for knowledge representation tasks. Owing to its simplicity, propositional logic was one of the first candidates for the job but at the same time, this simplicity imposed many constraints on the kind of knowledge propositional logic can express. This limitation was overcome by First Order Logic (FOL) with its existential and universal quantifiers. FOL could express all kinds of axioms that prepositional logic could not express, but, as it turns out, this enhanced expressiveness comes at a high computational price. Consequently, the typical trade off that is usually present when using logic-based formalisms is between expressiveness and tractability of reasoning procedures and this is the reason behind the emersion of Description Logics DL; a particular family of logics derived from FOL that offers the right balance between expressiveness and reasoning complexity [2]. Today, the most dominant ontology language OWL is a manifestation of using logic-based formalisms to express ontologies. OWL has a profile dedicated for DL, and it is the language used in most knowledgebase systems nowadays [3] [4].

3. Approach

Even though not formally descending from frames, object-oriented programming follows the same line of thinking of frame-based systems [5] and most concepts in OOP have equivalents in frames as we can see in the summary of Table 1.

Table 1. Terms used in frame-based systems and their counterparts in object-oriented programming and the web ontology language OWL.

Frames (late 1970s)	OOP (1980s)	OWL (2000s)
Generic frames	Classes	owl:class axiom
Generic frames inheritance	Class inheritance	rdfs:subclass axiom
Individual frames	Instances of a class	Individual axioms
Slots	Class properties	Data type property axioms
Instance frame override	Method override	--
Triggers (demons)	Access modifiers	--
Scripts	Methods	--

Building on some of the ideas behind frame-based systems and adhering to the logic-based formalisms of OWL, we have a common ground to integrate ontologies and conventional object-oriented programs. The idea is to rely on ontologies to model the static aspects of a system (classes and properties) and then extend the model with object-oriented methods to cover up for the missing procedural control. To allow this kind of extensions to ontologies, they need to be executable, i.e. expressed in a programming language as we shall see in the following sections.

3.1. *Modeling the static aspects*

By executable ontologies, we denote ontologies that are expressed in a general-purpose programming language and can be compiled to produce the corresponding machine code. Ideally, this executable representation should not impact the semantics of the concepts and relations of an ontology. This requirement imposes many challenges on the process of producing a semantically-equivalent executable form of an ontology. Formal programming languages are more restrictive when it comes to expressing ontological axioms that deviate from classical programming paradigms. When expressing ontologies in a programming language, one thus needs to consider the possible alternatives for expressing certain OWL axioms like `owl:disjointWith` or `owl:equivalentClass` that have no native counterparts in languages like Java or C#.

When applying our approach, we used C# as the only language for expressing as well as querying ontologies. To overcome the semantic gap difficulty, we relied on a meta-property layer of code to map between OWL axioms and the missing C# counterparts. In the case of a disjoint class axiom, for example, a static property of the target class will keep track of all disjoint classes; and so is the case for other description or constraint axioms. This meta-property code layer serves as an initial scheme and is accessible from all subsequent concepts (i.e. classes) of an ontology. In other words, it forms the basis on which all executable C# ontologies in our approach are built. Figure 1. depicts the structure and properties of the meta-property layer.

For the sake of reusability, in most cases, we did not start writing executable ontologies from scratch. Instead, we tried to make use of existing ontologies and automatically translate them into executables. For that, we developed a translation tool that makes use of metaprogramming techniques to generate and compile C# code. More on the OWL to C# mapping details and the translation routine can be found in our earlier work [6].

3.2. *Adding procedural extensions*

Once in their executable form, it is now possible to extend ontologies by defining and implementing relevant static or instance methods that will cover up for the missing procedural control. In an ontology for a travel agency, for example, tasks like making a reservation or managing a banking transaction require a more

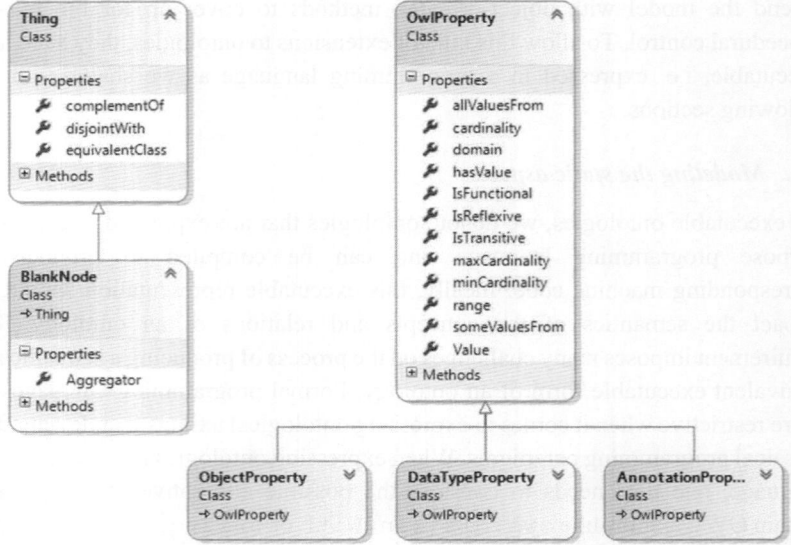

Figure 1. Meta-property layer of executable ontologies.

imperative style to capture and to manage the dynamic exchange of the interacting components. These aspects are not supported by the declarative nature of ontologies but in the executable form, the developer can add the necessary procedural extensions by implementing the proper object-oriented methods.

4. Discussion

Except for Horn-like rules such as the ones provided by SWRL extension [7], which amount to some imperative programming, ontologies are mostly declarative. While this adherence to the declarative paradigm is commonly perceived as a positive practice, we argue that it is one of the reasons behind the limited utility of ontologies in conventional software development. From software development perspective, when modeling a target domain, it is essential to capture all aspects of the entities of that domain including their behavior and their possible interactions taking into account potential parallelism or sequential order. This can only be translated into code artifacts by giving the developer more control on the flow of program execution even if in some cases, this means indulging into an imperative style of programming which may not fit into the philosophical perspective of ontologies. On the other hand, the proposed approach is not a naïve encouragement for a more imperative style. It is still important to adhere to the declarative essence of ontologies as they provide more reusability and interoperability. Imperative procedures should be spared to situations where they are indispensable like in transaction management or scenarios involving end-user interaction.

Programming paradigm aside, while working with executable ontologies, we could also perceive some other advantages and possibilities for further improvements.

- Enforced modularity and extended portability: packed as dynamic linking libraries (.dll) or executables (.exe), ontologies now enjoy extended portability as these formats can be loaded and used directly in more languages (C, C++, Visual Basic, C# or F#).
- Discrepancy between the intended model and the code implementation, which is usually a recurrent problem in software development, is much less evident with executable ontologies. I.e. code is the model.
- Using the programming language to express ontologies eliminates the need for external editors or persistence modules; ontologies are now code namespaces residing in a code repository either locally or in the cloud. This has permitted us to substantially reduce the cost of maintaining a separated technical stack for ontology management.

5. Related Work

The integration of ontologies into software engineering realm has been of interest for researchers and practitioners alike. The perspectives and the techniques applied in this area can vary considerably but they almost all agree on the potentials of the integration and the unneglectable semantic gap between two distinct schools of modeling; the school of ontological modeling and that of object-oriented design. In this section we briefly cover some of the efforts that are related to our proposed approach.

In [8], the primary intention is to provide guidance on how to build real-world semantic web applications. The authors draw analogy between deploying ontologies as high-level models in software development and the approach used in Model Driven Architecture MDA. They suggest a software architecture for web services and agents for the semantic web driven by domain ontologies. According to the authors of [9], the combined use of ontologies with standard programming practices would enable the development of semantic-rich enterprise applications and they suggest a framework for translating some ontology constructs into Enterprise Java Beans.

The authors of [10] proposed a hybrid modeling software framework that combines the object-oriented representation of a domain with its ontological representation after analyzing the advantages and disadvantages of such hybrid modeling approach. OWL to UML mapping has also a got a good share in the literature: [11] presents a UML-based visualization of OWL DL ontologies while the work done in [12] provides a rigorous comparison between UML and OWL as two flagship languages for artificial intelligence and software engineering communities; the authors argue that based on the core definitions of ontologies and models, none of the informal distinctions made between the two terms is actually justifiable. Instead, ontologies themselves are to be regarded as models.

6. Conclusion and Future Work

Adopting ontologies in software engineering practices might be gaining momentum as a standalone research topic. However, shifting a bit from research towards the circles of conventional software developers reveals a rather conservative mindset on how ontologies can be actively integrated into the software development life cycle. In this paper, we proposed integrating ontologies into conventional code repositories by expressing them directly using a general-purpose programming language. An approach that permits the developers to add procedural extensions to the merely declarative model expressed by the ontology.

For future work, we are interested in studying the possibilities to exploit the language compiler support for performing entailment-reasoning tasks over the ontologies in their executable form.

References

1. M. Minsky, "A Framework for Representing Knowledge," Massachusetts Institute of Technology, Cambridge, MA, USA, 1974.
2. F. Baader, The description logic handbook: Theory, implementation and applications, Cambridge university press, 2003.
3. B. Motik, P. F. Patel-Schneider and B. C. Grau, "Owl 2 web ontology language direct semantics," *W3C recommendation,* vol. 27, 2009.
4. B. Motik, B. C. Grau, I. Horrocks, Z. Wu, A. Fokoue, C. Lutz and others, "Owl 2 web ontology language: Profiles," *W3C recommendation,* vol. 27, p. 61, 2009.
5. O. Lassila, "Frames or Objects, or Both?," in *Workshop Notes from the 8th National Conference on AI (AAAI-90): Object-Oriented Programming in AI, Boston (MA),* Boston, 1990.
6. S. Baset and K. Stoffel, "OntoJIT: Parsing Native OWL DL into Executable Ontologies in an Object Oriented Paradigm," in *International Experiences and Directions Workshop on OWL,* 2016.
7. Horrocks, Ian and Patel-Schneider, Peter F and Boley, Harold and Tabet, Said and Grosof, Benjamin and Dean and Mike and others, "WRL: A semantic web rule language combining OWL and RuleML," *W3C Member submission,* vol. 21, no. 21, p. 79, 2004.

8. H. Knublauch, "Ontology-driven software development in the context of the semantic web: An example scenario with Protege/OWL," in *1st International Workshop on the Model-Driven Semantic Web (MDSW2004)*, 2004.
9. I. N. Athanasiadis, F. Villa and A.-E. Rizzoli, "Ontologies, JavaBeans and Relational Databases for enabling semantic programming," in *Computer Software and Applications Conference, 2007. COMPSAC 2007. 31st Annual International*, 2007.
10. C. Puleston, B. Parsia, J. Cunningham and A. Rector, "Integrating object-oriented and ontological representations: a case study in Java and OWL," in *International Semantic Web Conference*, 2008.
11. S. Brockmans, R. Volz, A. Eberhart and P. Loffler, "Visual modeling of OWL DL ontologies using UML," in *International Semantic Web Conference*, 2004.
12. C. Atkinson, M. Gutheil and K. Kiko, "On the Relationship of Ontologies and Models," *WoMM,* vol. 96, pp. 47-60, 2006.

Graphic deduction based on set (I):
Venn graphic representation of judgment J^* *

Xia He

Department of Foreign Languages, Huaiyin Institute of Technology
Huaian, 223003, China
Philosophy Department, Chinese Academy of Science
Beijing, 100732, China

Guoping Du

Philosophy Department, Chinese Academy of Science
Beijing, 100732, China

Long Hong

Computer School, Nanjing University of Posts and Telecommunications
Nanjing, 210003, China

Based on basic concept of symbolic logic and set theory, this paper focuses on judgments and attempts to provide a new method for the study of logic. The paper contains two parts. Part One introduces knowledge of predicate and set theory, establishes the formal language of the extension of judgment J^*, and formally describes **a, e, i, o** judgments. It then develops set theory representation and graphic representation that enables to distinguish universal judgments and particular judgments. In Part Two, graphic statement form of connectives based on set theory will be presented and the graphic deduction of J^* will be carried out.

1. Introduction

Logic is the general science of reasoning[1] and the basic tool for the study of artificial intelligence. *Formal logic* mainly involves *judgment* and *deduction*. *Affirmative, negative, subject, predicate, universal, particular, singular* are the key words of formal logic that can all be accurately described with *predicate calculus*. As the interpretation of predicate calculus depends on *domain of individuals*, that judgment and deduction can be shifted to predicate can broaden

---

*This work was supported by the Fund for Major Bidding Project of National Social Science of China under Grant No. 14ZDB014 and the National Natural Science Foundation of China under Grant No. 61170322.

the application range of formal logic on the one hand and can transform the content of a *non-modal judgment* into the equivalent concept of a *set* on the other. Therefore, *non-modal deduction* can be deduced by the *Venn Diagram* based on set.

Based on the basic concepts of *symbolic logic* and *set theory*, this paper focuses on *judgment* and attempts to provide a new method for the study of logic. The rest of the paper runs in outline as follows: Section 2 briefly introduces predicate and set theory employed in this article; Predicate description of the main concept of judgment and the establishment of the judgment expansion J^* are presented in Section 3; Section 4 describes the set theory description of J^*, and Section 5 goes into a conclusion.

2. A Brief Introduction to Predicates and Set Theory

2.1. *Predicates and Quantifiers*

Predicates are used to characterize *nature* and *relation* of individuals. For example, $P(x)$ means 'x has a property P', where P is the symbol which stands for a predicate and x an *individual variable*. For example, let P stand for 'animal' and a stand for 'cow' (a is an *individual constant*), then $P(a)$ means 'cow is an animal'.

Two kinds of *quantifiers* have been introduced into symbolic logic: The symbol $\forall$ translates 'all', and is called a *universal quantifier*; $\exists$ translates 'exists' and is called an *existential quantifier*. Then $\forall x$ means 'all x' and $\exists x$ means 'some x'.

Intuitively, the semantics of 'some people are not students' and 'not everyone is a student' in natural language are the same. This makes us feel that there must be some relation between *universal quantifier* and *existing quantifier*. Let predicate H represent 'human' and A represent 'animal', then the above two sentences may be translated into symbols as:

$$(1)\ \exists x(H(x) \wedge \neg S(x))$$
$$(2)\ \neg \forall x(H(x) \rightarrow S(x))$$

(2) can be transformed into:

$$(2')\ \neg \forall x \neg (H(x) \wedge \neg S(x))$$

Obviously, $\exists x$ in (1) is replaced by $\neg \forall x \neg$ in (2). Therefore, "$\exists x$" and "$\neg \forall x \neg$" are equivalent.

2.2. *A Brief Introduction to Naive Set Theory*

Set refers to collection of things with the same nature, which is represented by curly braces. Contents enclosed in curly braces separated by commas are called *elements* of the set. For example, if the set of *positive integers* is represented as *I*, then *I*= {1, 2, 3, ... }.

There are two kinds of basic symbols in set theory. One is *relator* between sets, and the other is *operator* of sets. There are two basic relators. One is '∈', which means that an individual is an element of a set. The other is '⊇' or '⊆', meaning that all the elements of a set are elements of another set.

There are two basic operators. One is '∪', which means to combine the elements of the two sets together. The other is '∩', meaning that same elements in both sets are taken out and put together.

There is also a symbol ∅, called empty set, which means that there is no element in the set.

Figure 1 shows the *Venn Diagram* representation of basic relation and operation between sets. 'E' represents *universal set*, which means all the sets in the domain. ~A of Figure 1e is read as the *complementary set* of A, the part that does not contain A. '~'is read as *complementary operation*: ~A=E–A. A-B of Figure1f contains only the part of A, and does not contain the part of B: A–B=A∩~B; '–' is read as *difference operation*, and '+' of Figure 1g is read as *symmetric difference*.

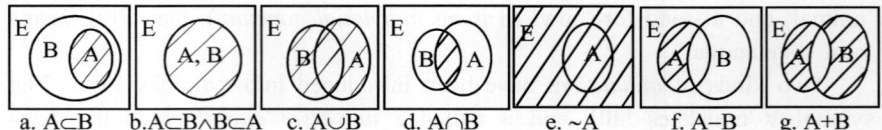

a. A⊂B   b. A⊂B∧B⊆A   c. A∪B   d. A∩B   e. ~A   f. A–B   g. A+B

Figure 1. The Venn graphic representation of sets.

3. Judgment *J* and *J* *

3.1. *Nature Judgment J*

'Judgment' here refers to non-modal judgments in formal logic, which include universal affirmative judgment (**A**), universal negative judgment (**E**), particular affirmative judgment (**I**), and particular negative judgment (**O**). In formal logic, the description of **A**, **E**, **I** and **O** are as follows[2]:

The form of *universal affirmative judgment* is 'All S are P' (represented by A). The form of *universal negative judgment* is 'All S is not P' (represented by E). The form of *particular affirmative judgment* is 'S is P' (represented by I). The form of *particular negative judgment* is 'S is not P' (represented by O).

Translating **A**, **E**, **I** and **O** into symbols, we get:

$$\textbf{A}: \forall x(S(x) \to P(x))$$
$$\textbf{E}: \forall x(S(x) \to \neg P(x))$$
$$\textbf{I}: \exists x(S(x) \wedge P(x))$$

O: $\exists x(S(x) \wedge \neg P(x))$

Additionally, there are two other nature judgments concerning *individuals*, which are described as follows:

The judgment of a *singular affirmative judgment* is the judgment of a certain individual thing having a certain character. The judgment of a *singular negation judgment* is the judgment of a certain individual thing not having a certain character[2].

Let *a* be a thing, where *a* is an *individual constant*, then S(a) means '*a* has a property of S". This is the symbolic description of a singular positive judgment. Similarly, the symbolic description of a singular negative judgment is represented as $\neg S(a)$.

We put together the above symbolic description of judgments, and make it *J*.

3.2. *J* and Its Argument*

A, E, I, O in *J* have a clear definition, which limits the scope of its research and application. To meet the extension of *J*, we need to design a new symbol. We let the four new symbols having the same semantics with **A, E, I** and **O** be **a, e, i, o**, and call them judgment words and make *J** an *extension* of *J*.

The formal language of *J** consists of two parts.

(1) *symbols*

    a. *constants:* $a, b; a_1, a_2, \ldots, a_n$
    b. *variables:* $x, y, z; x_1, x_2, \ldots, x_n$
    c. *predicates:* $P, Q, S; P_1, P_2, \ldots, P_m$
    d. *logical connectives:* $\neg, \vee, \wedge, \rightarrow, \leftrightarrow$
    e. *judgment words:* **a, e, i, o**
    g. *technical symbols:*), (.

(2) Generate rules of *well-formed formula*

    a. If $p \in \{P, Q, S, P_1, P_1, \ldots, P_m\}$, $c \in \{a, b; a_1, a_2, \ldots, a_n\}$, $v \in \{x, y, z; x_1, x_2, \ldots, x_n\}$, $\alpha \in \{p(c), p(v)\}$, then α is a well-formed formula.

    b. if α is a well-formed formula, then $\neg \alpha$ is a well-formed formula.

    c. if α, β are well-formed formulas, then $\alpha \wedge \beta, \alpha \vee \beta, \alpha \rightarrow \beta, \alpha \leftrightarrow \beta$ are well-formed formulas.

    d. if α, β are well-formed formulas, then $\alpha \textbf{ a } \beta, \alpha \textbf{ e } \beta, \alpha \textbf{ i } \beta, \alpha \textbf{ o } \beta$ are well-formed formulas.

    e. All well-formed formulas are merely *a, b, c* and *d*.

Built on this, the number of statement forms with judgment words can be expanded from 4 kinds of *J* to 32 kinds of *J**, which are then constrained to 8 kinds. We call these statement forms judgment patterns. To further extend the modes of the eight kinds of statement forms, we get:

α **a** β: $\forall x(\alpha(x) \to \beta(x))$ (1)
α **e** β: $\forall x(\alpha(x) \to \neg \beta(x))$ (2)
α **i** β: $\exists x(\alpha(x) \wedge \beta(x))$ (3)
α **o** β: $\exists x(\alpha(x) \wedge \neg \beta(x))$ (4)
$\neg\alpha$ **a** β: $\forall x(\neg\alpha(x) \to \beta(x))$ (5)
$\neg\alpha$ **e** β: $\forall x(\neg\alpha(x) \to \neg \beta(x))$ (6)
$\neg\alpha$ **i** β: $\exists x(\neg\alpha(x) \wedge \beta(x))$ (7)
$\neg\alpha$ **o** β: $\exists x(\neg\alpha(x) \wedge \neg\beta(x))$ (8)

4. Set representation of J^*

As predicates can construct sets, set theory can be adopted to study the above eight kinds of statement forms.

If set A={$x|\alpha(x)$}, B={$x|\beta(x)$}, then (1), (2), (5), and (6) can be easily rewritten to be equivalent *set expressions*. However, it is not that easy to rewrite the rest forms. For example, if we rewrite (3) as $\exists x\ (x \in A \wedge x \in B)$, then it can be rephrased either as 'Some x belong to both A and B' or as: 'There are some x that belong neither to nor belong to B.' Let us take A for analysis. Set A consists of all x with property α. In other words, an element either belongs to A or does not belong to A, and there is no saying as part of x belonging (not belonging) to A. Having this contradiction, we have the following resolution:

Subset A'(composed by part of elements of set A) is contained in A. For example, let A={1, 2, 3, 4, 5}, A'={2, 3}. Obviously, A'⊆A. Also, let $y \in$ A', then $y \in$ A. Likewise, we can rewrite '$\exists x(\alpha(x) \wedge \beta(x))$' into '$\exists y(y \in A \wedge y \in B)$'. Here we get a reasonable and clear expression to avoid contradiction.

Now, let us rewrite logic statement forms (1) to (8) into equivalent set expressions:

$\forall x(x \in A \to x \in B)$ (1')
$\forall x(x \in A \to x \notin B)$ (2')
$\exists y(y \in A \wedge y \in B)$ (3')
$\exists y(y \in A \wedge y \notin B)$ (4')
$\forall x(x \notin A \to x \in B)$ (5')
$\forall x(x \notin A \to x \notin B)$ (6')
$\exists y(y \notin A \wedge y \in B)$ (7')
$\exists y(y \notin A \wedge y \in B)$ (8')

Here A={$x|\alpha(x)$}, B={$x|\beta(x)$} ; $y \in$A'⊆A, or $y \in$B'⊆B.

Figure 2 shows Venn diagrams of Formulas (1') to (8'). In Venn diagrams with existential quantifier statements, circles representing two sets are crossed to show the feature of 'part'. Slashes represent the state of element x in A and B, and the symbol '×' represents the state of element y in A and B.

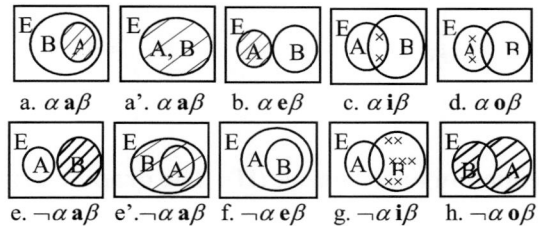

a. $\alpha a \beta$ a'. $\alpha a \beta$ b. $\alpha e \beta$ c. $\alpha i \beta$ d. $\alpha o \beta$

e. $\neg \alpha a \beta$ e'. $\neg \alpha a \beta$ f. $\neg \alpha e \beta$ g. $\neg \alpha i \beta$ h. $\neg \alpha o \beta$

Figure 2. Venn graphic representation of **a, e, i** and **o** modes.

Figure 2a' shows A=B, which is a special case of Figure 2a; Although the two sets in Figure 2f are empty sets, they clearly indicate a inclusion relation.

5. Conclusion

Set theory is recognized as a basic theory, which is widely used in many disciplines. Applying set theory to the study of logic will receive succinct, natural, and reliable results. It also provides a new method for the application of logic in artificial intelligence. This paper establishes a set theory representation and a graphic representation for the expansion of judgment J^*. In particular, it presents a reasonable method for describing existential quantifiers using sets. The work will lay foundation for the deduction of graphics.

Acknowledgments

The first author would like to thank Department of Foreign Languages of Huaiyin Institute of Technology where she serves as a teacher and Graduate school of Chinese Academy of Social Science where she studies for a doctoral degree.

References

1. S Blackburn, *Oxford University Press* (1996).
2. Yue-lin Jin, *People's Publishing House* (2006).

Graphic deduction based on set (II): The graphic argument of J^**

Xia He

*Department of Foreign Languages, Huaiyin Institute of Technology
Huaian, 223003, China
Philosophy Department, Chinese Academy of Science
Beijing, 100732, China*

Guoping Du

*Philosophy Department, Chinese Academy of Science
Beijing, 100732, China*

Long Hong

*Computer School, Nanjing University of Posts and Telecommunications
Nanjing, 210003, China*

Based on basic concept of symbolic logic and set theory, this paper focuses on judgment and attempts to provide a new method for the study of logic. The paper contains two parts. On the basis of Part One, Part Two develops graphic statement form of connectives based on set theory. According to the content of non-modal deductive reasoning in formal logic, it gives weakening theorem, strengthening theorem and a number of typical graphical representation theorem (graphic theorem), where graphic deduction is carried out. Graphic deduction will be beneficial to the research of artificial intelligence, which is closely related to judgment and deduction in logic.

1. Introduction

Logic is the general science of reasoning [1] and the basic tool for the study of artificial intelligence. Graphic deduction will be beneficial to the research of artificial intelligence, which is closely related to judgment and deduction in logic.

In Part One, knowledge of predicate and set theory, the formal language of the extension of judgment J^*, and formal description of a, e, i, o judgments are

---

*This work was supported by the Fund for Major Bidding Project of National Social Science of China under Grant No. 14ZDB014 and the National Natural Science Foundation of China under Grant No. 61170322.

presented. It develops set theory representation and graphic representation that enables to distinguish universal judgments and particular judgments.

On the basis of Part One, Part Two presents graphic statement form of connectives based on set theory and carries out the graphic deduction of J^*. According to the content of non-modal deductive reasoning in formal logic, it gives weakening theorem, strengthening theorem and a number of typical graphic theorem, where graphic deduction is carried out.

The rest of the paper runs in outline as follows: Section 2 introduces the basic form of graphic statement as well as graphic deduction of judgment patterns; the graphic deduction of J^* is taken up in Section 3 and Section 4 goes into a conclusion.

2. Graphic Statement and Its Argument

To effectively demonstrate its graphic argument, we need to create graphic statement forms that fit the well-formed formula of J^*.

Definition 1. (DF1)

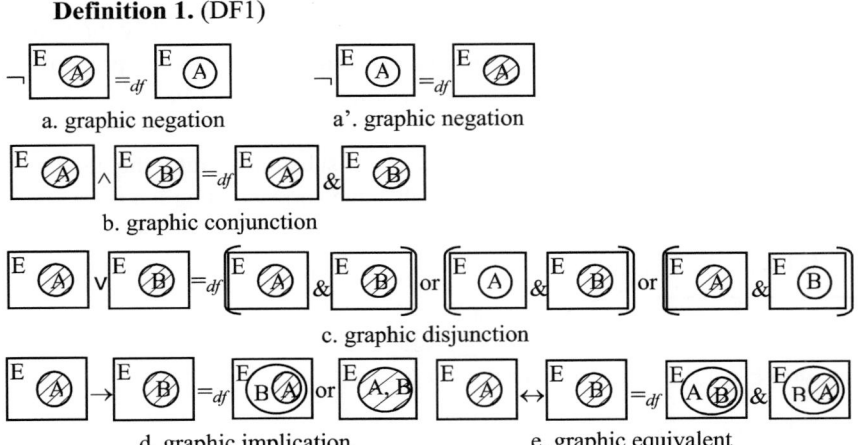

a. graphic negation   a'. graphic negation

b. graphic conjunction

c. graphic disjunction

d. graphic implication   e. graphic equivalent

Here symbol "&" is the separator between premises.

Let us first observe Figures in [2], and Fig. 2a and Fig. 2c are both affirmative judgments. The slashes in Figure 2a occupy the entire A, while the slashes in Figure 2c occupy only part of A. This shows that: **i** is less certain than **a**. See Figure 2b and Figure 2d, which are both negative judgments. We can get: **o** is less negative than **e**. A similar analysis of Figure 2e and Figure 2g, Figure 2f and Figure 2h also leads to a corresponding conclusion. Due to the fact that the affirmative (negative) degree of universal judgments is weaker than particular judgments, we have the following conclusions:

Weakening theorem. If the premise and the conclusion of **a** judgment have the same antecedent and consequent, then **a** judgment pattern directly introduces **i** judgment pattern and **e** judgment pattern directly introduce **o** judgment pattern. According to weakening theorem, GT1 (Graphic Theorem) deduced from the graph can be obtained, where the symbol ' ⊢' means 'introduce'.

GT1.

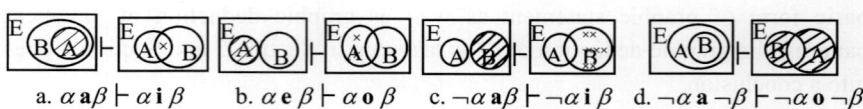

a. $\alpha\,a\beta \vdash \alpha\,i\,\beta$ b. $\alpha\,e\,\beta \vdash \alpha\,o\,\beta$ c. $\neg\alpha\,a\beta \vdash \neg\alpha\,i\,\beta$ d. $\neg\alpha\,a\,\neg\beta \vdash \neg\alpha\,o\,\neg\beta$

GT1d has a special premise, as it consists of two empty sets. Therefore, as for the degree of 'empty', the degree of conclusion is weaker than that of the premise.

For clarity, let us use ST to represent a theorem corresponding to the symbolic description of GT. For example, ST1 corresponds to GT1.

As it is an individual deduction, existential quantifiers are easily thought of to be involved. For example, $\alpha(b)$ means that individual b has the property α. Since individual b is the instantiation of variable x, $\exists x\alpha(x)$ is true.

Strengthening theorem. If the individual is substituted into the judgment antecedent (consequent), so that it is true, then the corresponding **i**, **o** is true. According to strengthening theorem, there are:

GT2.

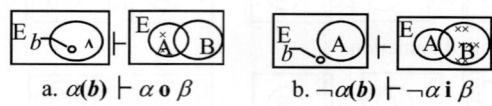

a. $\alpha(b) \vdash \alpha\,o\,\beta$ b. $\neg\alpha(b) \vdash \neg\alpha\,i\,\beta$

GT3. (ST3. $\delta\,a\,\beta,\ \alpha\,a\delta \vdash \alpha\,a\,\beta$)

Since there is one more predicate, we need to construct the set: $C=\{x|\delta(x)\}$. According to *irrelevance theorem*, there must be a connection between the premises, so it can be argued. First, draw the Venn diagram according to Figure 2 in [2].

Analysis: There is $A\subseteq C \wedge C \subseteq B$, so $A\subseteq B$.

GT4. (ST4. $\delta\,e\,\beta,\ \alpha\,a\delta \vdash \alpha\,e\,\beta$

We first give a graphic theorem and then an analysis.

Analysis: There is $A \subseteq C \wedge C \cap B = \varnothing$, so $A \cap B = \varnothing \wedge A \cup B = A$.

GT3 and GT4 are basic forms of argument of *Aristotle syllogism*, so the use of graphical deduction can deduce others out.

We can use theorems we have obtained to deduce and arrive at new theorems.

GT5. (ST5. $\delta a\beta, \alpha a\delta \vdash \alpha i \beta$)

Proof:

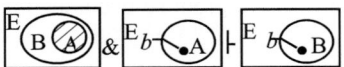

The text on the symbol '$\vdash$' in the proof provides the basis for the next graph, where 'g' means 'the *premise graphic* in the front of the symbol '$\vdash$''. The numbers that follow are serial numbers of graphs from left to right.

GT6. (ST6. $\alpha a\beta, \alpha(b) \vdash \beta(b)$))

This is the famous *syllogism*, except that GT6's argumentation is simpler than the *normative syllogism*.

GT7. (ST7. $\alpha a\beta \vdash \neg \alpha o\beta$)

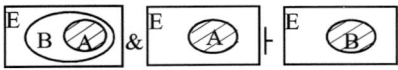

3. Graphic Deduction of *J** Statement Form

Based on the well-formed formula of the *J** form language, let us extend the range of graphic deduction.

GT8.

where $1 \leq i \leq n$.

The correctness of GT8 is obvious, given that the premise in the argument form must be true and *deduction fidelity* must be guaranteed.

GT9. (ST9. $\alpha a\beta, \alpha \vdash \beta$)

Analysis: There is $A \subseteq B \wedge A \neq \varnothing$, so $B \neq \varnothing$.

GT10. (ST10. $\alpha \vdash \beta \Rightarrow \alpha \mathbf{a} \beta$)

GT11. (ST11. $\neg \alpha \mathbf{a} \neg \beta \vdash \beta \mathbf{a} \alpha$)

GT12. (ST12. $\neg \alpha \vdash \beta, \neg \alpha \vdash \neg \beta \Rightarrow \alpha$)

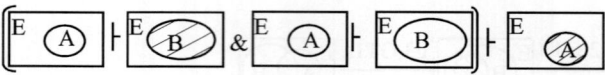

Proof:

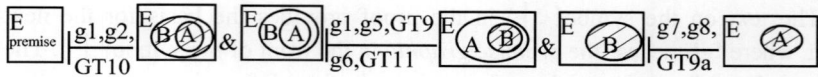

GT12 is called *proof by contradiction*. When the negative premise introduces two contradictory sub-conclusions, the positive premise becomes the conclusion.

GT13. (ST13. $\alpha \vdash \beta, \alpha \vdash \neg \beta \Rightarrow \neg \alpha$)

Proof:

GT13 is called *reductio ad absurdum*; When an affirmative premise introduces two mutual-contradictory substatements, the negative premise becomes a conclusion. GT12 and GT13 are often used in proofs.

GT14. (ST14. $\alpha \mathbf{a} \beta, \alpha \mathbf{e} \beta \vdash \neg \alpha$)

Proof:

A hypothesis (assuming A is true) is added to GT14's proof. The conclusion is drawn using the method of reductio ad absurdum.

GT15. (ST15. $\alpha \leftrightarrow \beta, \neg \beta \vdash \neg \alpha$)

Proof:

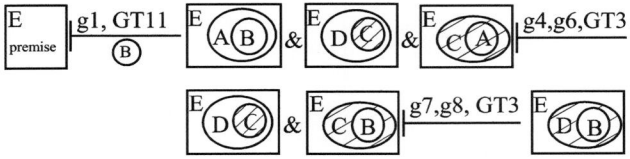

GT16. (ST16. α a β, δ a γ, $\neg \alpha$ a $\delta \vdash \neg \beta$ a γ)

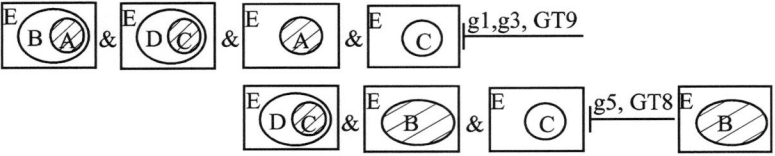

where $D=\{x|\gamma(x)\}$.

Proof:

The equivalent form to ST16 is: α a β, δ a γ, $\alpha \vee \delta \vdash \beta \vee \gamma$. This is a *dilemma*. If α, δ are true, the theorem is true. However, α, δ may not be both true. The use of graphic argument contains two situations: $\alpha=T \wedge \delta=F$, $\alpha=F \wedge \delta=T$. The following is a proof with $\alpha=T \wedge \delta=F$ as the premise.

Another case is similar, omitted.

4. Conclusion

Set theory is recognized as a basic theory, which is widely used in many disciplines. Applying set theory to the study of logic will receive succinct, natural, and reliable results. In artificial intelligence, the understanding of natural language and the representation of knowledge are its basic research contents, which are closely related to judgment and deduction in form logic.

Graphic theorem proposed in this paper involves non-modal deduction of form logic. Therefore, graphic deduction will be beneficial to the research of AI.

Acknowledgments

The first author would like to thank Department of Foreign Languages of Huaiyin Institute of Technology where she serves as a teacher and Graduate school of Chinese Academy of Social Science where she studies for a doctoral degree.

References
1. S. Blackburn, *Oxford University Press* (1996).
2. Xia He, Guoping Du, Long Hong, *Proceeding of The 13th Flins* (2018).

Parallel Monte Carlo for calculating π value[*]

Hui Zong and Jianyang Zhao
Faculty of Computer and Software Engineering, Huaiyin Institute of Technology
Huaian, 223003, China

Long Hong
Computer School, Nanjing University of Posts and Telecommunications
Nanjing, 210003, China

Monte Carlo method helps to get the approximate result of a given problem by adopting random numbers and statistics. It has been applied to science, engineering and other fields. This paper designs a parallel algorithm PAlgoMonteCarlo-π for computing π value based on non-virtual machine pattern in UCS, and discusses the selection principle of random number scale as well as the relationship between random number scale and calculation accuracy. The programming experiment results show that PAlgoMonteCarlo-π is effective and has a high speedup ratio.

1. Introduction

To solve certain problem by Monte Carlo method, people must first build a model that relates to the problem with probability theory as its theoretical basis, then the model is randomly sampled or observed through statistics to get the approximate solution of the given problem. Because its velocity of convergence has nothing to do with the dimension of the problem, and can get the approximate value to meet the requirements, Monte Carlo method has been applied in science, technology, engineering and social management. For example, Monte Carlo algorithm is a general algorithm for solving six-dimensional integral equations in nuclear science and nuclear engineering [1]. It is used for large prime number detection and game design [2], distributed generation optimization distribution [3], biomedicine [4], etc. However, few reports about the Monte Carlo parallel algorithm are found.

This paper introduces a Monte Carlo parallel algorithm for computing π and discusses the details of the algorithm. The remaining parts are arranged in this

---

[*]This work was supported by the Jiangsu Natural Fund of China under Grant No. BK20161302 and the National Natural Science Foundation of China under Grant No. 61170322.

way: Section 2 establishes a model for calculating π, and designs serial and parallel algorithms for this model; Section 3 discusses the performance and calculation accuracy of the algorithm, and Section 4 gives a conclusion.

2. Algorithm for Calculating π Value

2.1. *Calculation Model*

Let the length of the side of the square be d, and there is an inscribed circle within the square. The A_S denotes the area of the square and the A_C denotes the area of the circle, then the ratio of them is: $A_C/A_S = (\pi d^2/4)/d^2$, so $\pi = 4A_C/A_S$.

Making a finite number of random numbers in the square, the ratio of the number of the points within a circle and within a square can represent the ratio of the area of the circle and the square; the error between the calculated value of π and the true value of π also decreases as the random number increases. Denote the number of points in the square as x_s, and in the inscribed circle as x_c. Obviously, x_c is a function of x_s, and x_c increases as x_s increases. Suppose the random number generator quality meet the general criterion, then

$$\pi = \lim_{x_s \to \infty} 4\frac{x_c}{x_s} \tag{1}$$

In order to calculate the π value on the computer using Monte Carlo method, we use a square with a side length of 2 and calculate the quarter of its area. The coordinate of the image range taken in the algorithm is shown in Fig. 1.

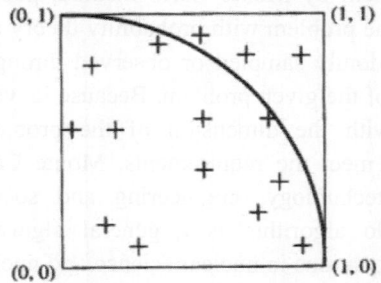

Figure 1. The coordinate of the image range taken in the algorithm.

2.2. *Algorithm Design*

In order to get the coordinate points in the graph, two random numbers must be generated at [0, 1] at the same time to make them the coordinate points (r_x, r_y). The following AlgoMonteCarlo-π is a serial algorithm that randomly generates

$n1*n2$ coordinate points and counts the number of coordinate points of the inscribed circle.

AlgoMonteCarlo-π $(k[l]=0)$ // Randomly generate $n1*n2$ coordinate points to calculate π

begin

  for $(i=0; i<n1; i++)$

  {

    for $(j=0; j<n2; j++)$

    {

      call random number generator to get r_x, r_y.

      if (r_x, r_y) is in the inscribed circle, then $k[l]++$

    }

  }

end

The serial algorithm SAlgoMonteCarlo-π for calculating π using the Monte Carlo method is as follows:

SAlgoMonteCarlo-π

begin

  $l=1$;

  AlgoMonteCarlo-π $(k[l]=0)$

  output $\pi=4k[l]/(n1*n2)$

end

According to (1) and its related conditions, the accuracy of π is proportional to the number of points generated x_s. However, it takes a longer time to calculate π value with serial algorithm. Therefore, we introduce a parallel algorithm for calculating π values based on AlgoMonteCarlo-π.

PAlgoMonteCarlo-π

begin

  Parfor (l=1; l<m; l++)

    {

      AlgoMonte-π ($k[l]$=0)

    }

  output π=4*(k[1]+k[2]+…+k[m])/(m*$n1$*$n2$)

end

The parallel statement '**Parfor** (l=1; l<m; l++)' in PAlgoMonteCarlo-π enables the parallel operation of m homomorphic subprocesses AlgoMonteCarlo-π.

3. Discussion

3.1. *Calculation Accuracy*

According to (1), the accuracy of the calculation is related to the scale of random points. As the premise of (1) is that the number generated by the random number generator conforms to a uniform distribution, and the probability of the occurrence of a same sequence is very low, the ideal π value can be obtained when a suitable random number generator and scale of random points is chosen.

If the number generated by the random number generator be represented by an n-bit binary number, then the generator can generate 2^{2n} different coordinate points. Let the scale of the coordinate point be S. According to *the pigeonhole principle*, when $S > 2^{2n}$, duplicate coordinate points must appear. Therefore, the scale of the coordinate points satisfying the basic requirements of Monte Carlo method is

$$S \leq 2^{2n} \tag{2}$$

So, AlgoMonteCarlo-π should satisfy

$$n1*n2 \leq 2^{2n} \tag{3}$$

And PAlgoMonteCarlo-π should satisfy

$$m*n1*n2 \leq 2^{2n}. \tag{4}$$

Under normal circumstances, if the scale of the coordinate points does not satisfy (2), the calculation accuracy will not increase. When $S-2^{2n}$ points fall within the inscribed circle, the calculated value increases, whereas the calculated value is decrease. Since it is assumed that the number generated by the random number generator conforms to a uniform distribution, the possibility that calculated value is increase is $\pi/4$. The analysis is basically supported by the experimental data in Table 1, and the data are gotten by using the mt19937 function.

Table 1. The experimental data when scale of the coordinate points increase.

Number of coordinate points	Calculation results	Standard value of π	Absolute error
2^{32}	3.1416209	3.1415926	$2.83*10^{-4}$
$2^{32}+2*10^7$	3.1416531	3.1415926	$6.05*10^{-4}$
$2^{32}+4*10^7$	3.1415571	3.1415926	$3.55*10^{-4}$
$2^{32}+8*10^7$	3.1416349	3.1415926	$4.15*10^{-4}$
$2^{32}+16*10^7$	3.1416404	3.1415926	$4.78*10^{-4}$
$2^{32}+32*10^7$	3.1416469	3.1415926	$5.70*10^{-4}$

3.2. *Calculation Performance of PAlgoMonteCarlo-π*

The accuracy of Monte Carlo method is directly proportional to the size of the random number. The more random numbers, the greater the amount of calculation and the more time it takes. By replacing the serial algorithm with a parallel algorithm, we can get a calculation value with the same precision as the serial algorithm in a short time, or get a higher precision value in the same amount of time. In order to ensure the even distribution of the coordinate points generated by the *m* threads, *m* different random number generator seeds need to be generated when programming the parallel algorithm to generate *m* different random number sequences.

Based on non-virtual machine pattern in the UCS environment, we implemented SAlgoMonteCarlo-π and PAlgoMonteCarlo-π using a single thread programming and multithread programming, respectively. The experimental results are shown in Table 2.

Table 2. The time of computation from difference threads under 2^{32} coordinate points based on on-virtual machine pattern in the UCS environment, and CPU is intel E5-2630 V3, 2.39GHz.

Number of threads	Calculation results	Time of computation (ms)	Radio of speed-up
1	3.1416209	4165192	1
4	3.1415570	1159175	3.59
8	3.1416657	575344	7.24
16	3.1416387	273734	15.22

The data in Table 2 was obtained from the average number of three laboratory results from the same programme. It is can be seen that speed-up ratio is as high as close to the number of logical cores.

4. Concluding Remarks

Taking constant π value as an example, this paper studies the parallel Monte Carlo method from the perspective of computational speed, which intends to promote the wide application of Monte Carlo method to help adapt to the natural world full of uncertainty.

Due to limited space, the relationship between the Monte Carlo parallel method and the way to obtain random numbers, in this paper, is not discussed. However, this issue should be very interesting.

References

1. A. Dubi, *John Wiley & Sons* (2000).
2. B. Bouzy, B. Helmstetter, *Springer* (2004).
3. A. Martinez, G. Guerra, *IEEE Trans. Power Syst.*, **29**, 2926(2014).
4. Vahid Moslemi; M. Ashoor, *IEEE Trans. Nuclear Sci.*, **64**, 2578(2017).

Learnt clause deletion based on clause deepness

Zhonghe Du and Zhenming Song[*]

National-Local Joint Engineering Laboratory of System Credibility Automatic Verification, Southwest Jiaotong University, Chengdu, Sichuan, 610031, China
[*]*zhmsong@swjtu.edu.cn*

> Learnt clause deletion is a key component of SAT (Satisfiability Problem) solvers. First, this paper introduces clause activity based deletion and LBD (Literal Blocks Distance) based deletion respectively. Then, the deepness of learnt clauses is defined, and a new clause deletion method based on deepness of learnt clauses is proposed to accelerate the BCP (Boolean Constraint Propagation) procedure. Last, experimental results show that deepness based clause deletion method can improve the efficiency of solvers Minisat and Glucose.

1. Introduction

Satisfiability problem (SAT) is the first NP-complete problem that has been proved[1]. It is the crucial problem in the fields of artificial intelligence and computer science. Furthermore, SAT problem is widely used in practice, such as software development and verification, large scale integrated circuit design, and so on. There exist no algorithms of polynomial time complexity for SAT due to its NP completeness. Thus, it is a challenge to solve SAT problem with high efficiency. At present, algorithms to solve SAT problem mainly can be divided into complete algorithms and incomplete algorithms. The former one is based on branching and backtracking, while the latter one is based on random local searching. Algorithms like DP[2], DPLL (Davis-Putnam-Logemann-Loveland)[3], GRASP[4], rel_sat[5] are representatives of complete algorithms. Besides using DPLL, the widely used solver CDCL (Conflict Driven Clause Learning)[8] adopts some pivotal techniques, such as conflict based clause learning[4], lazy data structure[6], learnt clause deletion[7], periodically restart[9] and others.

Recently, many efficient CDCL SAT solvers come into being, such as zChaff[12], BerkMin[7], Minisat[10], Glucose[11] and so on. The most prominent feature of these solvers is the learning scheme. Once a conflict occurs, a learnt clause will be derived from the conflict and added to the clause database to aim at guiding the subsequent search. It is obvious that the number of learnt clauses grows monotonously, which can result in the sharp increase of memory consumption. Therefore, learnt clause deletion technique is an important part of CDCL SAT solvers.

In this paper, two widely used techniques are introduced. One is the clause activity based method, the other is the clause LBD based method. Then, the different variable decision levels of a learnt clause is explored, that is, the clause deepness is defined to evaluate the quality of a learnt clause. Meanwhile, a learnt clause deletion technique based on the clause deepness is proposed. Lastly, experimental results show the comparison among the clause activity based technique, the LBD based technique and the clause deepness based technique.

2. Learnt Clause Deletion

Technically, it is unnecessary to store learnt clauses in the memory since they are resolved from original clause database. However, learnt clauses can prevent the solver from searching subspace where there is no solution and contribute to the proof when the problem is unsatisfiable. Although learnt clauses can guide the search process after the conflict, too many learnt clauses will occupy a large proportion of computer memory, and then they can slow down BCP procedure. Therefore, they affect the whole efficiency. Hence, part of learnt clauses should be deleted to improve the performance of the solver.

2.1. *Clause activity based deletion technique*

The developers of solver BerkMin proposed a learnt clause deletion technique base on clause activity[7]. Learnt clauses are stored in a stack, so the latest learnt clause will be in the top position, called the top clause. Learnt clauses can be partitioned into young clauses and old clauses. A clause is considered to be young if the distance of the clause from the top clause is less than 15/16 of the stack size. In addition, each clause C has a counter to store the number of conflicts that has been responsible, called as the clause activity. Clauses deleted by BerkMin can be divided into the following three parts:

1. Clauses that are satisfied by the retained assignments.
2. Young clauses whose length $(C) > 43$ and clause activity $(C) < 7$.
3. Old clauses whose length $(C) > 9$ and clause activity $(C) <$ threshold.

The threshold is initially equal to 60 and then it is gradually increased. BerkMin deletes learnt clauses according to their activities and lengths. The activity of a clause reflects its contribution to previous conflicts. Instead of deletion, learnt clauses of small length are kept in the database regardless of their activity. Let us consider the following special case. If all the learnt clauses have a length more than 42 and their activities are less than 7, then all of them will be deleted from the database. So it is possible that the solver removes and then deduces the same clause set. To figure out this problem, a learnt clause is marked and forbidden to be deleted. It is partially implemented in BerkMin by forbidding to delete the top clause of the stack. Besides, the solver guarantees the number of marked clauses in the database increases monotonically.

2.2. LBD based deletion technique

Keeping too many learnt clauses will slow down the BCP procedure, but deleting too many of them will weaken the benefit from learning. Therefore, it is important to identify which part of learnt clauses are with high quality. That is to say, these clauses are highly relevant with subsequent search. Authors of the solver Glucose proposed a method to evaluate the quality of a learnt clause, called literal blocks distance (LBD)[13]. After that, a high-efficiency clause deletion policy based on LBD is proposed.

Definition 1[13] Given a clause C, and a partition of its literals into n subsets according to the current assignment, $s.t.$ literals are partitioned with regard to their decision levels. The LBD of C is exactly n.

Example 1 Given a clause $C = x_1 \lor x_2 \lor x_3 \lor x_4$, and the decision levels of x_1, x_2, x_3 and x_4 are 2, 2, 3 and 4, respectively. The LBD of C is 3.

The LBD value of a learnt clause shows the decision numbers to derive it. The larger the LBD, the more decisions. It is not difficult to understand the importance of learnt clause of LBD 2. Extensive experimental results show that clauses with smaller LBD are more useful than those with larger LBD in the process of conflict analysis and BCP procedure. The clause deletion technique implemented in Glucose is described as follows:

No matter the size of the initial formula, half of the learnt clauses are removed (asserting clauses are kept) for every 20000+500*x conflicts (x is the number of times this action was previously performed). Obviously, clauses with smaller LBD are kept while those with higher LBD are deleted.

3. Learnt Clause Deletion Based on Clause Deepness

Learnt clause deletion is an important component of CDCL SAT solvers. Good clause deletion strategies can guarantee the efficiency of conflict analysis and BCP procedure while the usage of memory is reasonable. One of the challenge of nowadays deletion techniques is that they may delete some clauses that are related to the following research after conflicts. There are two major problems we concern. One is which part of clauses should be deleted, the other is when the deletion should happen. This section gives the definition of clause deepness, which is based on the different variable decision levels in a learnt clause. After that, a clause deletion technique based on clause deepness is proposed.

The decision level of a variable is the depth of the branching variable according to the branching heuristic in the binary tree. The decision level of the first branching variable is 1. Variables whose assignment can be implied from a given variable through BCP procedure have the same decision level with that variable. Decision levels reflect the assign order of variables throughout the search. A variable with smaller decision level is assigned before another variable with bigger decision level. Every time the solver reaches a conflict, a learnt clause is derived and the variables in the learnt clause are all assigned. On the basis of different decision levels of learnt clause, we give the definition of clause deepness as follows:

Definition 2 Given a clause $C = x_1 \vee x_2 \vee ... \vee x_n$, and the decision level of x_i is d_i, then the clause deepness of C is defined as $deepness(C) = (d_1+d_2+...+d_n)/n$.

Example 2 Let $C = x_2 \vee \neg x_3 \vee \neg x_5 \vee x_6$ be a learnt clause, and the decision levels of x_2, x_3, x_5 and x_6 are 3, 3, 4 and 5, respectively, then the clause deepness of clause C is $deepness(C) = (3+3+4+5)/4 = 3.75$.

In some sense, clause deepness is a reflection of the number of decisions to derive the learnt clause. On the other hand, clause deepness can be used to estimate the time quantum to derive the learnt clause. It takes more time to obtain learnt clauses whose deepness are relatively bigger, so this kind of clauses should be stored for a long time from their appearance. In this way, the time consumption in the unsatisfiable subspace could be shorten, and the search proceeds towards the direction where the solution probably exists. A clause deletion technique based on clause deepness is described as follows:

Learnt clauses are stored in a stack (stackOfLearntClause). Once a learnt clause is generated, compute its LBD. In case of LBD < 6, the learnt clause is pushed into stack and its deepness is computed. In case of LBD ≥ 6, the learnt clause is deleted immediately. For every 5000+1000*x conflicts (x is the number of times this action was previously performed), we check the deepness of clauses belongs to 10% of the stack size from the bottom. If the deepness of a learnt clause is less than the average deepness of all learnt clauses in the stack, then we delete it.

```
1    if (Unit Propagation() == conflict)        // A conflict occurred
2    {
3      derive a learnt clause c;
4      if ( c.LBD < 6 )
5      {
6        StackOfLearntClause.puch(c);
7        sumDeepness += deepness(c);
8      }
9      conflict number +=1;
10     if ( conflict number == 5000+ 1000 * times of clause deletion occured so far )
11     {
12       for ( clause belongs to stackOfLearntClauseFromBottom * 0.10 * sizeof(stack) )
13         if ( deepness(clause) < sumDeepness / conflict number )
14           clause.delete;
15     }
16   }
```

The proposed deletion technique is different from the clause activity based one and LBD based one. It can be regarded as twice optimizations of learnt clause set. Firstly, as clauses with smaller LBD is more important than those with bigger LBD, and there is a judgement from the generation of the learnt clause to its store. Only clauses with LBD < 6 are stored into the stack, and other clauses with bigger LBD are deleted in advance. As a result, the time spent on deleting these clauses is saved. Secondly, whenever the conflict number reaches a threshold value, clauses which belong to 10% of the stack from the bottom are checked. Learnt clauses whose deepness are less than the average deepness of all learnt clauses are deleted. This skill is significant to clauses with same LBD because they may have different clause deepness, and whether they are kept or deleted is decided by their clause deepness.

4. Experimental Results

Experiments are performed to evaluate the impact of clause deepness based deletion technique on CDCL SAT solver, and the results are listed in this section.

We replace the clause activity based deletion technique in Minisat solver with our clause deepness based deletion technique. Similarly, the LBD based deletion technique in Glucose solver is replaced with clause deepness based deletion technique as well. We do not change other components of the solvers, such as branching heuristics, restart policies and so on. Note that Minisat is a classical solver as it has been a template for many widely used CDCL SAT solvers from its presentation, and Glucose is the winner of numerous SAT competitions in recent years. All experiments are carried out on a machine with Intel(R) Pentium(R) 64-bit CPU N3700 @1.60GHz CPU 4.00GB RAM Windows10. Timeout for each instance is 3600 seconds. We use 'minisat_org', 'minisat_dbcd', 'glucose_3.0' and 'glucose_dbcd' to represent Minisat 2.2.0, Minisat 2.2.0 with deepness based clause deletion, Glucose 3.0 and Glucose 3.0 with deepness based clause deletion, respectively.

Table 1 lists the number of solved instances in SAT competition 2015 (300 instances in total) of four solvers. Figure 1 shows the relationship between CPU time and the successfully solved instances of Minisat and Minisat 2.2.0 with deepness based clause deletion. Figure 2 shows the same relationship of Glucose 3.0 and Glucose 3.0 with clause deepness based deletion.

Table 1. Numbers of solved instances in SAT competition 2015.

	minisat_org	minisat_dbcd	glucose_3.0	glucose_dbcd
SAT	110	114	98	101
UNSAT	64	66	78	80
TOTAL	174	180	176	181

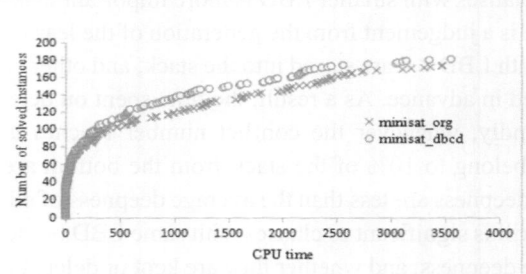

Figure 1. Test results on Minisat.

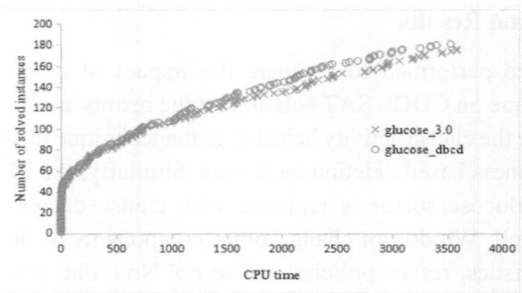

Figure 2. Test results on Glucose.

As shown in Table 1, both Minisat and Glucose can solve more instances when they are embedded with deepness based clause deletion technique. From Figure 1 and Figure 2, we can see that the solving time of the improved solvers minisat_dbcd and glucose_dbcd is shorten, which means their efficiency is improved with comparison with the unmodified solvers. These experimental results show that deepness based clause deletion technique can evaluate learnt clauses well, and guide the learnt clause deletion stage effectively, thus improve the performance of the whole solver.

5. Conclusion and Future Work

This paper firstly introduced clause activity based deletion technique and clause LBD based deletion technique in detail. Then, clause deepness was defined according to the different variable decision levels of a learnt clause. After that, the deepness based clause deletion technique was proposed and corresponding algorithm was generated. Finally, experimental results indicated that Minisat with deepness based clause deletion and Glucose with deepness based clause deletion outperform the unmodified ones in solving time and the number of successfully solved instances. In the future study, we will do a deeper research about the features of clause deepness, and how the clause deepness and LBD work together and affect each other when evaluate the clause quality.

Acknowledgments

This work is partially supported by the National Natural Science Foundation of China (Grant Nos. 61305074, 61603307, and 61673320), the Fundamental Research Funds for the Central Universities of China (Grant No. 2682015CX052).

References

1. Cook S. A. The complexity of theorem-proving procedures[C]. Proc3rd ACM Symposium on Theory of Computing, 1971: 151-158.
2. Davis M., Putnam H. A computing procedure for quantification theory[J]. ACM J, 1960, 7(3): 201-215.
3. Davis M., Logemann G., Loveland D. A machine program for the theorem proving[J]. Communications of the ACM, 1962, 5 (7): 394-397.
4. J. P. Marques-Silva and K. A. Sakallah. GRASP: A new search algorithm for satisfiability. In International Conference on Computer Aided Design, pages 220-227, November 1996.
5. Bayardo, R. and Schrag, R.: Using CSP look-back techniques to solve real-world SAT instances, in Proc. of the 14th Nat.(US) Conf. on Artificial Intelligence (AAAI-97), AAAI Press/The MIT Press, 1997, pp. 203-208.

6. Moskewicz, M.W., Madigan, C.F., Zhao, Y., Zhang, L., Malik, S.: Chaff: engineering an efficient SAT solver. In: Proceedings of the 38th Design Automation Conference, DAC 2001, pp. 530-535. ACM, Las Vegas, June 18-22, 2001.
7. Goldberg, E. and Novikov, Y.: BerkMin: A fast and robust SAT-solver[C]. Discrete Applied Mathematics, 2007, 155(12): 1549-1561.
8. Marques-Silva, J.P., Lynce, I., Malik, S.: Conflict-driven clause learning SAT solvers. 131-153.
9. J. Huang. The effect of restarts on clause learning. In International Joint Conference on Artificial Intelligence, pages 2318-2323, 2007.
10. Eén, N., Sörensson, N.: An extensible SAT-solver. In: Giunchiglia, E., Tacchella, A. (eds.) SAT 2003. LNCS, vol. 2919, pp. 502-518. Springer, Heidelberg, 2004.
11. Audemard G, Simon L. GLUCOSE: a solver that predicts learnt clauses quality[J]. SAT Competition, 2009: 7-8.
12. Mahajan Y S, Fu Z, Malik S. Zchaff2004: An Efficient SAT Solver[C]// Theory and Applications of Satisfiability Testing, International Conference, SAT 2004, Vancouver, Bc, Canada, May 10-13, 2004, Revised Selected Papers. DBLP, 2004: 360-375.
13. Audemard G, Simon L. Predicting Learnt Clauses Quality in Modern SAT Solvers.[C]// IJCAI 2009, Proceedings of the International Joint Conference on Artificial Intelligence, Pasadena, California, Usa, July. DBLP, 2009: 399-404.

Part 7

Applied Computing Intelligence

Part 7

Applied Computing Intelligence

Compound trajectory optimization methodology for parafoil delivery system based on quantum genetic algorithm[*]

Hao Sun, Qinglin Sun, Shunzhen Luo, Zengqiang Chen and Wannan Wu

College of Computer and Control Engineering, Nankai University
No. 38, Tongyan Road, Jinnan District, Tianjin 300350, China

Jin Tao

Department of Electrical Engineering and Automation, Aalto University
Espoo 02150, Finland

In this paper, a novel trajectory optimization methodology based on the quantum genetic algorithm for parafoil delivery system is presented. The optimized trajectory is composed with the standard straight lines and circles. It not only facilitates the implement in the experiment, but also can achieves an optimal trajectory under multiple constraints. Meanwhile, the quantum genetic algorithm is applied considering its computation speed. At last the simulation is carried out. Comparing with the chaos particle swarm optimization method, the proposed methodology shows huge improvement and advantage.

1. Introduction

Parafoil delivery system is a kind of unique unmanned air vehicle (UAV) with specific aerodynamic characteristic. It consists of a payload and parafoil. Comparing with other kinds of UAVs, it has huge advantage on its load capacity and flight stability. So it is always applied to execute airdropping task [1]. The system must transport its payload to the target site securely and accurately. Considering the wind influence, the terrain constraints and other unknown disturbance, the landing precision is tremendously challenged. So, the trajectory optimization of the homing control is an important research of the parafoil delivery system.

In recent years, there are two kinds of the trajectory optimization theory of homing control: optimal homing and multiphase homing theory. Researches of the optimal homing theory like a terminal guidance method from Slegers [2], a

---

[*]This work is partially supported by grant 61273138, 61573197 of the National Natural Science Foundation of China, grant 2015BAK06B04 of the National Key Technology R&D Program and grant 14ZCZDSF00022 of the Key Technologies R&D Program of Tianjin.

Bezier curve path planning from Rogers [3]. Meanwhile, the multiphase homing theory is more focused on the application. Like the X-38 aircraft recovery project [4]. The theories of the multiphase homing are similar [5]. But from the previous researches, we can observe that both theories have merits and demerits. The optimal homing theory is hard to realize in the practical experiment. And the traditional multiphase homing is hard to consider the multiple constraints.

For solving the above problems of the traditional homing theory, in this paper, a novel trajectory planning methodology for parafoil delivery system based on quantum genetic algorithm (QGA) is designed. Firstly, a 3-degree of freedom (DOF) model is set from the airdrop experiment data. Then, in the trajectory planning, all the important influences are considered, like wind disturbance, terrain avoidance, the smooth degree of the trajectory and so on. At last, the proposed methodology is compared with the chaos particle swarm optimization method in the simulation experiment. It proves the practicability and the performance of the proposed methodology.

2. Data Analysis and Dynamic Model

2.1. *Data analysis of the airdrop experiment*

In the airdrop experiment, the experiment of the model identification is explored.

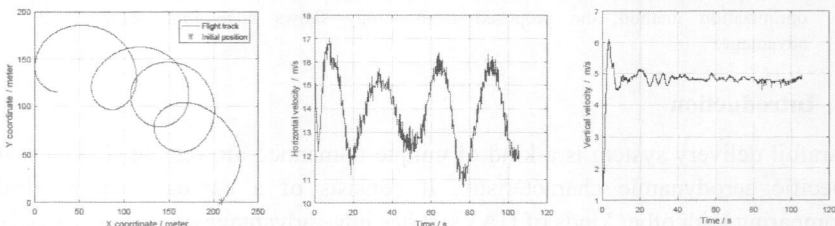

Fig. 1. Model identification: (a) Flight trajectory (b) Horizontal velocity (c) Vertical velocity.

In the experiment, in Fig. 1(c), it is obvious that the vertical velocity can nearly maintain constant in flight. Further, by analyzing Fig. 1(b), it also can be seen that the horizontal speed will be dramatically changed by the constant wind. So, considering these results, the wind disturbance and the realizability of the planned trajectory are the factors that must be considered in the homing control.

2.2. 3-DOF dynamic modeling

For the sake of the computation speed, the 3-DOF model is applied in this paper. Based on that data of the experiment, the 3-DOF model can be expressed as:

$$\begin{cases} \dot{x} = v_s \cos\psi \\ \dot{y} = v_s \sin\psi \\ \dot{\psi} = u \\ \dot{z} = v_z \end{cases} \quad (1)$$

where $\dot{x}$ and $\dot{y}$ denotes the horizontal velocity in x and y coordinate, v_s denotes the horizontal velocity, ψ and u denote the control quantity, v_z denote the vertical velocity. And the velocity is deduced from the experiment data.

3. Trajectory Planning Methodology

3.1. Quantum genetic algorithm

The problem of the trajectory planning is to find an optimal trajectory under the proposed homing strategy. In this paper, the QGA is applied. It is a combination of the genetic algorithm and the quantum computation [6]. It highlights the advantage of the quantum computation, like quantum parallelism and quantum entanglement. It applied the quantum bit for the coding scheme and quantum rotation gate for the update operation. Further, considering its computation speed, QGA also satisfy the requirement of the computation time.

3.2. Homing strategy

Based on QGA, the objective function is designed in this section. The constraints are all consider, especially the landing precision, wind disturbance and the especially.

1. Virtual target landing site

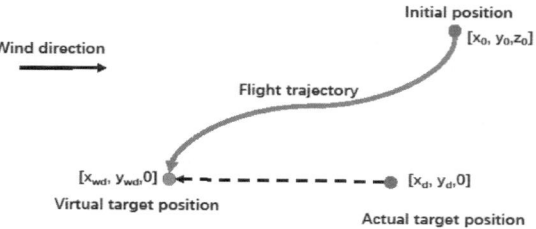

Fig. 2. Virtual target landing site in windy environment.

In this strategy, firstly, since it is known that the wind speed from the GPS device by the wind identification method [7], the target landing site can be changed by analyzing the wind condition, as shown in Fig. 2. If the landing position of the parafoil is designed as as: $[x_d, y_d, 0]$. In windy environment, the revised target landing site can be obtained by the dropping altitude:

$$\begin{cases} x_{wd} = x_d + v_{wx}\dfrac{z_0}{v_z} \\ y_{wd} = y_d + v_{wy}\dfrac{z_0}{v_z} \end{cases} \quad (2)$$

where wd denotes the velocity of the wind.

2. Optimized parameters

The methodology is applied to arrive the target position at the time t_f. In this method, the planned trajectory is separated into some straight lines. So, design l_i as the flight length and θ_i as the flight direction. The optimized parameters are designed as follows: $[l_1, \theta_2, l_2, \theta_2, l_3, \theta_4, l_4, \ldots, \theta_n, l_n]$. Since it is known that the initial flight direction, only the flight distance is considered in the first vector. So, the planned trajectory is set as some straight lines. Then, for the control and implement, the straight line is smoothed in the switch position. In this way, the planned trajectory consists of several arcs and lines.

3. Objective function

The practicability and correctness of the planned trajectory include the landing precision, terrain avoidance and the landing direction. All the constraints will be considered synthetically and achieve an optimal trajectory. The objective function can be expressed as:

$$J = \begin{cases} 3.5 - \dfrac{1}{1+J_p}, J_p \neq 0 \\ 2.5 - \dfrac{1}{1+J_u} - \dfrac{1}{1+J_L+J_\theta} - \dfrac{0.5}{1+J_a}, J_p = 0 \end{cases} \quad (3)$$

where $J_u = \int_{t_0}^{t_f} u^2 dt$ denotes the consumption of the control quantity; And $J_\theta = |\theta_n - \theta_w|$ denote the direction of the flared landing; if we define L_p as the distance of the impossible trajectory, the objective function of the terrain avoidance is designed as: $J_p = L_p$.

J_L denotes the objective function of the landing precision:

$$J_L = \left| R_{\min} \times (\sum_{i=1}^{n-1} \omega_i) + \sum_{i=1}^{n} L_{li} - f \times z_0 \right| \quad (4)$$

where $r_{\min}$ denotes the minimum turning radius, by analying the experiment data and its aerodynamics characteristics, it is set as 100 meters; l_{li} denotes the length of the corrected lines; f denotes the glide ratio;

It is clear that this objective function can consider all the important constraints in the homing of the parafoil delivery system. And it has huge improvements from the traditional one. After considering all the constraints, the robustness of the trajectory planning will be improved largely.

4. Simulation and Results Analysis

In the simulation, from the flight data, the parameters of the parafoil delivery system are shown in Table 1. The initial state is changed in each experiment. The wind disturbance of the first case is $[3m/s, 3.14rad]$ and $[3m/s, 0rad]$ for the second.

Table 1. Parameters of the parafoil delivery system.

Parameters	Value
Span	10.5 m
Chord	3.1 m
Mass of canopy	10 kg
Mass of payload	80 kg
Vertical velocity	4.6 m/s
Horizontal velocity	13.6m/s

In Case 1, as shown in Fig. 3, it is obvious that the optimized trajectories of the proposed methodology achieve great performance in different initial position. The constraints, like the terrain avoidance and flared landing, are all satisfied. The error of the landing position of all the conditions is less than 2 meters. With the proposed methodology, the trajectory is obviously realizable and smooth.

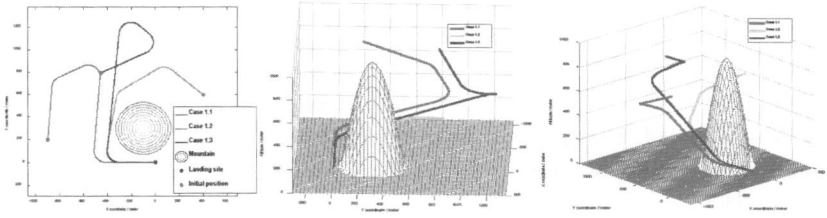

Fig. 3. Flight trajectory for Case 1: (a) Horizontal trajectory, (b) and (c) 3D trajectory.

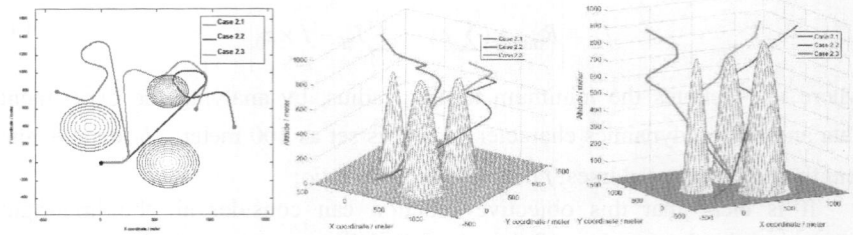

Fig. 4. Flight trajectory for Case 2: (a) Horizontal trajectory, (b) and (c) 3D trajectory.

In Case 2, since the terrain is more complicated, the planned trajectory also has more segments. In Fig. 4, it is clear that all the trajectories achieve successful based on the terrain avoidance. All the constraints are satisfied as well. It proves that the proposed methodology is suit for the complex terrain environment.

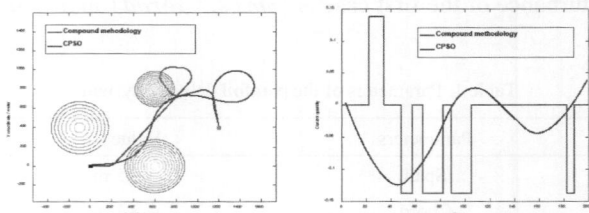

Fig. 5. Comparison with CPSO: (a) Horizontal trajectory, (b) Control quantity.

As shown in Fig. 5, the comparison is also carried out. In this experiment, the flight paths of these two methods are not similar, after consuming the redundant altitude, the flight paths of these two methods are analogous. The consumption of the proposed methodology is 1.294 and the CPSO is 1.198. They have little difference. Considering the proposed compound methodology is designed with standard lines and arcs, it has better realizability than the traditional optimal homing theory in practice. Meanwhile, it also proves that the compound homing methodology can achieve an optimal trajectory with simple flight path.

5. Conclusion

In this paper, a novel trajectory planning methodology from parafoil delivery system is designed. Firstly, the 3-DOF model of the parafoil delivery system is built. By analyzing the flight data in the airdrop experiment, the model is reliable. Then, the compound trajectory planning method based on QGA is introduced. By considering the multiple constraints, it not only obtain high landing precision in windy environment, but also achieve all the important control objective, like terrain avoidance, flared landing and so on. Further, the planned trajectory consists of the straight line and standard arc. It shows great realizable and wide application prospects.

References

1. D. Carter, S. George, P. Hattis, M. W. McConley, S. Rasmussen, L. Singh, S. Tavan, Autonomous large parafoil guidance, navigation, and control system design status, *in 19th AIAA Aerodynamic Decelerator Systems Technology Conference and Seminar*, (2007).
2. N. Slegers, O. A. Yakimenko, Optimal control for terminal guidance of autonomous parafoils, *in 20th AIAA Aerodynamic Decelerator Systems Technology Conference and Seminar* (2009).
3. J. Rogers, N. Slegers, Terminal guidance for complex drop zones using massively parallel processing, *in Aerodynamic Decelerator Systems Technology Conferences, Institute of Aeronautics and Astronautics*, (2013).
4. J. Valasek, D. Ito, D. Ward, Robust dynamic inversion controller design and analysis for the X-38, *in AIAA Guidance, Navigation and Control Conference and Exhibit*, (2001).
5. J. Tao, Q.L Sun, Z.Q. Chen, Y. P. He, *Control Theory. App.* **33(12)**, 1630-1638, (2016, in Chinese).
6. J.C Lee, W.M Lin, G.C Liao, T.P Tsao, *Int J. of elec power.* **33(2)**, 189-197, (2011).
7. P.L Tan, Q.L Sun, H.T Gao, *Acta Aeronautica et Astronautica*, **37(7)**, 2286-2294, (2016, in Chinese)

Multiple μ-stability of complex-valued Cohen-Grossberg neural networks with unbounded time-varying delays

Yunfeng Liu, Manchun Tan* and Desheng Xu

College of Information Science and Technology, Jinan University
Guangzhou 510632, P.R. China
*tanmc@jnu.edu.cn

The multistability problem is studied for the complex-valued Cohen-Grossberg neural networks (CVCGNNs) with unbounded time-varying delays. By using the fixed point theorem and other analytical tools, some novel algebraic criteria are established to guarantee that the existence of equilibrium points and μ-stability of the CVCGNNs. The obtained results generalize some previous works in the literature. Finally, one numerical example is presented to show the validity of the theoretical results.

Keywords: Complex-valued Cohen-Grossberg neural networks; multistability; μ-stability; unbounded time delays.

1. Introduction

Recently, the multistability of neural networks has been studied for its applications in pattern recognition, associative memory and so on (see Refs. 1–4 and the references therein). The μ-stability has received much attention of researchers (see, e.g., Refs. 5–8). As studied in most previous literature, the time delays are bounded. However, it is difficult to measure delays precisely because delays are variable and in fact are unbounded in some practical applications.[9] In this paper, we study the multiple μ-stability of complex-valued Cohen-Grossberg neural networks with unbounded time-varying delays.

2. Model of Neural Network

Consider the following complex-valued Cohen-Grossberg neural networks with time-varying delays:

$$\dot{z}_i(t) = a_i(z_i(t))\left[-b_i z_i(t) + \sum_{j=1}^n h_{ij} f_j(z_j(t)) + \sum_{j=1}^n w_{ij} f_j(z_j(t-\tau_{ij}(t))) + J_i\right], \quad (1)$$

for $t \geq 0$, $i = 1, 2, ..., n$, where $z(t) = (z_1(t), z_2(t), \cdots, z_n(t))^T \in \mathbb{C}^n$, $z_i(t)$ is the state of the ith neuron at time t; $a_i(z_i(t))$ represents an amplification function at time t; $b_i \in \mathbb{R}$ with $b_i > 0$; $h_{ij}, w_{ij} \in \mathbb{C}$ denote the connection strengths; $J_i \in \mathbb{C}$ corresponds to the external bias. $\tau_{ij}(t) \leq \tau(t)$ represents the time-varying transmission delay; $f_j(\cdot) : \mathbb{C} \to \mathbb{C}$ is the activation function.

Throughout this paper, we make the following assumptions to simplify the system (1):

Assumption 2.1. *Each function $a_i(z_i)$ is continuous: $\mathbb{C} \to \mathbb{R}^+$ and $0 < \check{a}_i < a_i(z_i)$ holds for all $z_i \in \mathbb{C}$, where $\check{a}_i$ is a positive constant, $i = 1, 2, ..., n$.*

Assumption 2.2. *Let $z = x + iy$ with $x, y \in \mathbb{R}$, then the real-imaginary-type activation functions $f_j(z)$ can be expressed by its real and imaginary parts with $f_j(z) = f_j^R(x) + i f_j^I(y)$, where $f_j^R(x) = 0.5(|x+1| - |x-1|)$ and $f_j^R(y) = 0.5(|y+1| - |y-1|)$ $(j = 1, 2, ..., n)$.*

According to the Assumptions 2.1 and 2.2, the system (1) can be rewritten as follows for $i = 1, 2, ..., n$,

$$\begin{cases} \dot{x}_i(t) = a_i(z_i(t)) \Big\{ -b_i x_i(t) + \sum_{j=1}^n \left[h_{ij}^R f_j^R(x_j(t)) - h_{ij}^I f_j^I(y_j(t)) \right] \\ \qquad + \sum_{j=1}^n \left[w_{ij}^R f_j^R(x_j(t - \tau_{ij}(t))) - w_{ij}^I f_j^I(y_j(t - \tau_{ij}(t))) \right] + J_i^R \Big\}, \\ \dot{y}_i(t) = a_i(z_i(t)) \Big\{ -b_i y_i(t) + \sum_{j=1}^n \left[h_{ij}^R f_j^I(y_j(t)) + h_{ij}^I f_j^R(x_j(t)) \right] \\ \qquad + \sum_{j=1}^n \left[w_{ij}^R f_j^I(y_j(t - \tau_{ij}(t))) + w_{ij}^I f_j^R(x_j(t - \tau_{ij}(t))) \right] + J_i^I \Big\}. \end{cases} \quad (2)$$

The initial conditions of neural networks (2) are given as follows:

$$x_i(\theta) = \phi_i(\theta), \quad y_i(\theta) = \varphi_i(\theta), \quad \theta \in (-\infty, 0], \quad (3)$$

where $\phi_i(\theta) \in C((-\infty, 0], \mathbb{R})$, $\varphi_i(\theta) \in C((-\infty, 0], \mathbb{R})$, $i = 1, ..., n$.

Notably, from (2) and Assumption 2.1, the stationary equation for system (1) can be written as

$$\begin{cases} G_i(Z) = -b_i x_i + \sum_{j=1}^n \left[h_{ij}^R f_j^R(x_j) - h_{ij}^I f_j^I(y_j) \right] + \sum_{j=1}^n \left[w_{ij}^R f_j^R(x_j) - w_{ij}^I f_j^I(y_j) \right] + J_i^R = 0, \\ H_i(Z) = -b_i y_i + \sum_{j=1}^n \left[h_{ij}^R f_j^I(y_j) + h_{ij}^I f_j^R(x_j) \right] + \sum_{j=1}^n \left[w_{ij}^R f_j^I(y_j) + w_{ij}^I f_j^R(x_j) \right] + J_i^I = 0, \end{cases} \quad (4)$$

where $Z = (z_1, z_2, ..., z_n)^T$, $z_i = x_i + i y_i$.

For the convenience of description, we denote that
$$\Omega^\alpha = \{(x_1, x_2, ..., x_n, y_1, y_2, ..., y_n) \in \mathbb{R}^{2n} \,|\, x_i \in \Omega_i^{R(\alpha_i)}, y_i \in \Omega_i^{I(\alpha_{n+i})},$$
$$\text{for } i = 1, 2, ..., n\}$$
where $\alpha = (\alpha_1, \alpha_2, ..., \alpha_{2n})$, $\alpha_i = 1$ or 2 or 3 for $i = 1, 2, ..., 2n$ and
$$\Omega_i^{R(1)} = \{x \in \mathbb{R} \,|\, -\infty < x < -1\}, \quad \Omega_i^{R(2)} = \{x \in \mathbb{R} \,|\, -1 \leq x \leq 1\},$$
$$\Omega_i^{R(3)} = \{x \in \mathbb{R} \,|\, 1 < x < +\infty\}, \quad \Omega_i^{I(1)} = \{y \in \mathbb{R} \,|\, -\infty < y < -1\},$$
$$\Omega_i^{I(2)} = \{y \in \mathbb{R} \,|\, -1 \leq y \leq -1\}, \quad \Omega_i^{I(3)} = \{y \in \mathbb{R} \,|\, 1 < y < +\infty\}.$$
$$\widehat{\Omega}^\alpha = \{(x_1, x_2, ..., x_n, y_1, y_2, ..., y_n) \in \mathbb{R}^{2n} \,|\, x_i \in \Omega_i^{R(1)} \text{ or } \Omega_i^{R(3)},$$
$$y_i \in \Omega_i^{I(1)} \text{ or } \Omega_i^{I(3)} \quad \text{for } i = 1, 2, ..., n\}.$$
It is obvious that in $\mathbb{C}^n$, there are 9^n such kind of regions as Ω^α.

Definition 2.1. Let (x^*, y^*) be the equilibrium point of (2) and $(x(t), y(t))$ be an arbitrary solution of (2). Suppose that $\mu(t)$ is a positive continuous function and satisfies $\mu(t) \to +\infty$ as $t \to +\infty$. If there exist scalars $M > 0$ and $T \geq 0$ such that $|(x_i(t), y_i(t)) - (x_i^*, y_i^*)| \leq \frac{M}{\mu(t)}$ holds for $t \geq T$ and $i = 1, 2, ..., n$, then the equilibrium point (x^*, y^*) is said to be μ-stable.

3. Main Result

Lemma 3.1. *For any subset region Ω^α, suppose that $w_{ii}^R > 0$ and*

$$\begin{cases} b_i - h_{ii}^R - w_{ii}^R + \sum_{j=1, j\neq i}^{n} |h_{ij}^R + w_{ij}^R| + \sum_{j=1}^{n} |(h_{ij}^I + w_{ij}^I| + J_i^R < 0, \\ -b_i + h_{ii}^R + w_{ii}^R - \sum_{j=1, j\neq i}^{n} |h_{ij}^R + w_{ij}^R| - \sum_{j=1}^{n} |(h_{ij}^I + w_{ij}^I| + J_i^R > 0, \\ b_i - h_{ii}^R - w_{ii}^R + \sum_{j=1, j\neq i}^{n} |h_{ij}^R + w_{ij}^R| + \sum_{j=1}^{n} |h_{ij}^I + w_{ij}^I| + J_i^I < 0, \\ -b_i + h_{ii}^R + w_{ii}^R - \sum_{j=1, j\neq i}^{n} |h_{ij}^R + w_{ij}^R| - \sum_{j=1}^{n} |h_{ij}^I + w_{ij}^I| + J_i^I > 0 \end{cases} \quad (5)$$

hold. There exists one equilibrium point of system (1) in each Ω^α and region $\widehat{\Omega}^\alpha$ is positively invariant.

Proof. The equilibria of system (1) are the roots of (4) according to the Assumption 2.1. Let Ω^α be one of these regions. For any given $z = x + iy \in$

Ω^α, we solve for ξ_i and η_i in

$$\begin{cases} \overline{G}_i(\xi_i) := -b_i\xi_i + (h_{ii}^R + w_{ii}^R)f_i^R(\xi_i) + \sum_{j=1,j\neq i}^{n}(h_{ii}^R + w_{ii}^R)f_j^R(x_j) - \sum_{j=1}^{n}(h_{ii}^I + w_{ii}^I)f_j^I(y_j) \\ \qquad + J_i^R = 0, \\ \overline{H}_i(\eta_i) := -b_i\eta_i + (h_{ii}^R + w_{ii}^R)f_i^R(\eta_i) + \sum_{j=1,j\neq i}^{n}(h_{ii}^R + w_{ii}^R)f_j^I(y_j) + \sum_{j=1}^{n}(h_{ii}^I + w_{ii}^I)f_j^R(x_j) \\ \qquad + J_i^I = 0, \end{cases} \qquad (6)$$

where $i = 1, 2, ..., n$. From Condition (4), we can have that

$$\overline{G}_i(-\infty) > 0, \ \overline{G}_i(-1) < 0, \ \overline{G}_i(1) > 0, \ \overline{G}_i(+\infty) < 0;$$

$$\overline{H}_i(-\infty) > 0, \ \overline{H}_i(-1) < 0, \ \overline{H}_i(1) > 0, \ \overline{H}_i(+\infty) < 0.$$

One can always find nine solutions to (6) in Ω^α for each i. Denote the solution of Eq. (6) as $z_i = x_i^* + iy_i^*$. Obviously, $z^* = x^* + iy^* = (x_1^* + iy_1^*, x_2^* + iy_2^*, \cdots, x_n^* + iy_n^*) \in \Omega^\alpha$. Then, for any given $z \in \Omega^\alpha$, we define a mapping $T_{\Omega^\alpha} : \Omega^\alpha \to \Omega^\alpha$ by $T_{\overline{\Sigma}}(z) = z^* = (x_1^* + iy_1^*, x_2^* + iy_2^*, \cdots, x_n^* + iy_n^*)^T$, where x_i^* and y_i^* are the solutions of (6). The mapping T_{Ω^α} as defined is continuous, since f_i^R and f_i^R are continuous. By virtue of the Brouwer's fixed point theorem, it is obvious that the mapping T_{Ω^α} has one fixed point $z^* \in \Omega^\alpha$, which is just an equilibrium of system (1) in Ω^α.

For any given region $\widehat{\Omega}^\alpha$. Consider any initial condition $(\phi, \varphi) \in \widehat{\Omega}^\alpha$, we claim that the solution $z(t; (\phi, \varphi))$ of system (1) remains in $\widehat{\Omega}^\alpha$ for all $t \geq 0$. If this is not true, there exists an index $i_1 \in \{1, 2, ..., n\}$ and a time point t_1 when $z(t)$ firstly escapes from $\widehat{\Omega}^\alpha$. Without loss of generality, suppose the real part of the i_1th component z_{i_1} of $z(t)$ is the first one that escapes from $(-\infty, -1)$ or $(1, +\infty)$ at the time t_1. That is, there exists a sufficiently small constant $\varepsilon_0 > 0$ such that either $x_{i_1}(t_1) = -1 - \varepsilon_0$, $\dot{x}_{i_1}(t_1) \geq 0$ and $x_{i_1}(t) \leq -1 - \varepsilon_0$ for all $t \in (-\infty, t_1]$ or $x_{i_1}(t_1) = 1 + \varepsilon_0$, $\dot{x}_{i_1}(t_1) \leq 0$ and $x_{i_1}(t) \geq 1 + \varepsilon_0$ for all $t \in (-\infty, t_1]$. For the first case, due to $w_{i_1 i_1}^R \geq 0$, the monotonicity of $f_{i_1}^R$ and Condition (5), we derive from (2) that

$$\dot{x}_{i_1}(t_1) = a_{i_1}(z_{i_1}(t_1))\Big\{-b_{i_1}x_{i_1}(t_1) - \sum_{j=1}^{n}(h_{i_1 j}^R - h_{i_1 j}^I) - \sum_{j=1}^{n}(w_{i_1 j}^R - w_{i_1 j}^I) + J_{i_1}^R\Big\}$$

$$\leq a_{i_1}(z_{i_1}(t_1))\Big\{b_{i_1} - h_{i_1 i_1}^R - w_{i_1 i_1}^R + \sum_{j=1,j\neq i_1}^{n}|h_{i_1 j}^R + w_{i_1 j}^R| + \sum_{j=1}^{n}|h_{i_1 j}^I + w_{i_1 j}^I| + J_{i_1}^R\Big\}$$

$$< 0.$$

This yields a contradiction to $\dot{x}_{i_1}(t_1) \geq 0$. With the same argument, the second case is not true either. Therefore, $\widehat{\Omega}^\alpha$ is a positively invariant under the flow generated by system (1). This completes the proof.

Theorem 3.1. *Assume the condition* (5) *and* $w_{ii}^R \geq 0$ *hold for all* $i = 1, 2, ..., n$. *Furthermore, assume that there exists a nondecreasing function* $\mu(t) > 0$ *with*

$$\lim_{t \to +\infty} \mu(t) = +\infty, \qquad \lim_{t \to +\infty} \frac{\dot{\mu}(t)}{\mu(t)} = \alpha, \qquad -b_i + \check{a}_i \alpha < 0, \qquad (7)$$

where α is nonnegative constants. There exist 4^n local μ-stable equilibria located in $\widehat{\Omega}^\alpha$ for system (1) *with unbounded time-varying delays.*

Proof. By Lemma 3.1, any given region $\widehat{\Omega}^\alpha$ is a positive invariant set for system (1) and system (1) has a unique equilibrium point z^* lying in $\widehat{\Omega}^\alpha$. With translation $\tilde{z}(t) = z(t) - z^*$, system (1) can be changed to

$$\dot{\tilde{z}}_i(t) = a_i(\tilde{z}_i(t) + z_i^*)\Big[-b_i \tilde{z}_i(t) + \sum_{j=1}^n h_{ij} \tilde{f}_j(\tilde{z}_j(t)) + \sum_{j=1}^n w_{ij} \tilde{f}_j(\tilde{z}_j(t - \tau_{ij}(t)))\Big], \tag{8}$$

where $\tilde{f}_j(\tilde{z}_j(t)) = f_j(z_j(t)) - f_j(z_j^*) = f_j^R(x_j(t)) + \mathrm{i} f_j^I(y_j(t)) - f_j^R(x_j^*) - \mathrm{i} f_j^I(y_j^*)$ ($j = 1, 2, ..., n$). Denote by $\tilde{x}_i(t)$ and $\tilde{y}_i(t)$ the real part and the imaginary part of $\tilde{z}_i(t)$, respectively. Denote by $\tilde{f}_i^R(\tilde{x}_i(t))$ and $\tilde{f}_i^I(\tilde{y}_i(t))$ the real part and the imaginary part of $\tilde{f}_i(\tilde{z}_i(t))$, respectively. Then $\tilde{x}_i(t) = x_i(t) - x_i^*$, $\tilde{y}_i(t) = y_i(t) - y_i^*$, $\tilde{f}_i^R(\tilde{x}_i(t)) = f_i^R(x_i(t)) - f_i^R(x_i^*)$, $\tilde{f}_i^I(\tilde{y}_i(t)) = f_i^I(y_i(t)) - f_i^I(y_i^*)$, and $\tilde{f}_j^R(\tilde{x}_j(t)) = \tilde{f}_j^R(\tilde{x}_j(t - \tau_{ij}(t))) = 0$, $\tilde{f}_j^I(\tilde{y}_j(t)) = \tilde{f}_j^I(\tilde{y}_j(t - \tau_{ij}(t))) = 0$ for any given region $\widehat{\Omega}^\alpha$, therefore system (8) can be written as

$$\begin{cases} \dot{\tilde{x}}_i(t) = -a_i(z_i(t))b_i \tilde{x}_i(t), \\ \dot{\tilde{y}}_i(t) = -a_i(z_i(t))b_i \tilde{y}_i(t). \end{cases} \tag{9}$$

Let $X_i(t) = \mu(t)\tilde{x}_i(t)$, $Y_i(t) = \mu(t)\tilde{y}_i(t)$ and $Z(t) = (X(t)^T, Y(t)^T)^T$ with $X(t)^T = (X_1(t), X_2(t), ..., X_n(t))^T$ and $Y(t)^T = (Y_1(t), Y_2(t), ..., Y_n(t))^T$.

Define

$$M(t) = \sup_{s \leq t} \|Z(s)\|_\xi, \quad t \geq T, \tag{10}$$

where $\|Z(s)\|_\xi = \max\{\|X(s)\|_{\xi^R}, \|Y(s)\|_{\xi^I}\}$ with $\|X(s)\|_{\xi^R} = \max_{1 \leq i \leq n}\{\xi_i^R |X_i(s)|\}$ and $\|Y(s)\|_{\xi^R} = \max_{1 \leq i \leq n}\{\xi_i^I |Y_i(s)|\}$ and ξ_i^R, ξ_i^I are positive numbers.

We claim that $M(t)$ is bounded. According to the definition of $M(t)$, it is obvious that $\|Z(t)\|_\xi \leq M(t)$ for all $t \geq T$. If at some special time point t such that $\|Z(t)\|_\xi = M(t)$, then there are two possible cases:

Case 1: There exists an index $i^R = i^R(t)$ depending on t such that
$$||Z(t)||_\xi = ||X(t)||_{\xi^R} = \xi_{i^R}^R |X_{i^R}(t)|. \tag{11}$$

Case 2: There exists an index $i^I = i^I(t)$ depending on t such that
$$||Z(t)||_\xi = ||Y(t)||_{\xi^I} = \xi_{i^I}^I |Y_{i^I}(t)|. \tag{12}$$

For the case 1, by virtue of (7), (9), (11) and Assumption 2.1, it follows that

$$\begin{aligned}(\xi_{i^R}^R)^{-1} D^- ||Z(t)||_\xi &= D^- |X_{i^R}(t)| \\ &= \text{sign}(\tilde{x}_{i^R})\, \dot{\mu}(t)\, \tilde{x}_{i^R}(t) - \text{sign}(\tilde{x}_{i^R}) \mu(t) a_{i^R}(z_{i^R}(t)) b_{i^R} \tilde{x}_{i^R}(t) \\ &\leq \frac{\dot{\mu}(t)}{\mu(t)} |X_{i^R}| - a_{i^R}(z_{i^R}(t)) b_{i^R} |X_{i^R}(t)| \\ &\leq a_{i^R}(z_{i^R}(t))\Big(- b_{i^R} + \frac{\dot{\mu}(t)}{\check{a}_{i^R}\mu(t)}\Big)(\xi_{i^R}^R)^{-1} ||X(t)||_\xi \\ &\leq a_{i^R}(z_{i^R}(t))\big(- b_{i^R} + \check{a}_{i^R}\alpha\big)(\xi_{i^R}^R)^{-1} M(t) \\ &\leq 0. \end{aligned} \tag{13}$$

For case 2, from (7), (9), (12) and Assumption 2.1, we can similarly derive that

$$\begin{aligned}(\xi_{i^I}^I)^{-1} D^- ||Z(t)||_\xi &= D^- |Y_{i^I}(t)| \\ &= \text{sign}(\tilde{x}_{i^I})\, \dot{\mu}(t)\, \tilde{y}_{i^I}(t) - \text{sign}(\tilde{y}_{i^I}) \mu(t) a_{i^I}(z_{i^I}(t)) b_{i^I} \tilde{x}_{i^I}(t) \\ &\leq \frac{\dot{\mu}(t)}{\mu(t)} |Y_{i^I}| - a_{i^I}(z_{i^I}(t)) b_{i^I} |Y_{i^I}(t)| \\ &\leq a_{i^I}(z_{i^I}(t))\Big(- b_{i^I} + \frac{\dot{\mu}(t)}{\check{a}_{i^I}\mu(t)}\Big)(\xi_{i^I}^I)^{-1} ||Y(t)||_\xi \\ &\leq a_{i^I}(z_{i^I}(t))\big(- b_{i^I} + \check{a}_{i^I}\alpha\big)(\xi_{i^I}^I)^{-1} M(t) \leq 0. \end{aligned} \tag{14}$$

From (13) and (14), we can conclude that $D^- ||Z(t)||_\xi \leq 0$ which implies that $||Z(t)||_\xi$ will be nonincreasing when time passes the point t. Through the above analyses, we can conclude that $M(t)$ is bounded. Thus, there exists a scalar $\lambda > 0$ such that $M(t) < \lambda$ for all $t \geq T$, namely

$$|\tilde{x}_i(t)| < \frac{\lambda^R}{\mu(t)}, \qquad |\tilde{y}_i(t)| < \frac{\lambda^I}{\mu(t)}, \tag{15}$$

where $t \geq T$, $\lambda^R = \lambda \max_{1\leq i \leq n} \{(\xi_i^R)^{-1})\}$ and $\lambda^I = \lambda \max_{1 \leq i \leq n} \{(\xi_i^I)^{-1})\}$. Therefore the equilibrium point (x^*, y^*) is locally μ-stable located in $\widehat{\Omega}^\alpha$. For $\widehat{\Omega}^\alpha$ chosen arbitrarily, system (1) has 4^n locally μ-stable equilibrium point with unbounded time-varying delays. This completes the proof.

Corollary 3.1. *The μ-stability can be specified with respect to some particular time delays,[6] which can lead to multiple exponential stability, multiple power stability, multiple log-stability, multiple log-log-stability.*

4. Illustrate Example

Consider the following system consisting of two neurons:

$$\begin{cases} \dot{z}_1(t) = (0.4 + |z_1(t)|)\Big\{ -3z_1(t) + (4+\mathrm{i})f_1(z_1(t)) + (1+0.2\mathrm{i})f_2(z_2(t)) \\ \qquad\qquad + (5+0.5\mathrm{i})f_1(z_1(0.6t)) + (0.8-0.6\mathrm{i})f_2(z_2(0.6t)) \Big\}, \\ \dot{z}_2(t) = (0.5 + |z_2(t)|)\Big\{ -4z_2(t) + (1+0.5\mathrm{i})f_1(z_1(t)) + (5+0.5\mathrm{i})f_2(z_2(t)) \\ \qquad\qquad + (0.5+0.3\mathrm{i})f_1(z_1(0.6t)) + (3+0.4\mathrm{i})f_2(z_2(0.6t)) \Big\}, \end{cases} \tag{16}$$

where $|z_j| = \sqrt{x_j^2 + y_j^2}$ and $f_j^R(\xi) = \frac{1}{2}(|\xi+1|-|\xi-1|)$, $f_j^I(\xi) = \frac{1}{2}(|\xi+1|-|\xi-1|)$ $(j=1,2)$.

It is easy to see that conditions (5), (7) and $w_{ii}^R \geq 0$ hold for $i = 1,2$. Thus, by Theorem 3.1, the system (16) can have 81 equilibrium points, 16 of them are locally μ-stable. When $\tau(t) = 0.4t$, the 16 equilibrium points are locally power stable.

The dynamics of system (16) are depicted in Fig. 1 to show the effectiveness of our results, where evolutions of 160 random initial conditions have tracked.

5. Conclusions

In this paper, we consider the multiple μ-stability of CVCGNNs with unbounded time-varying delays. Sufficient conditions have been established to guarantee the existence of 9^n equilibria and local μ-stability of 4^n equilibria for the CVCGNNs. The obtained results generalize some previous results. One example with simulations has been provided to clarify the validity of the obtained results.

Acknowledgments

The research is supported by grants from the National Natural Science Foundation of China (No. 61572233 and No. 11471083), and the Science and Technology Program of Guangzhou, China (No. 201707010404).

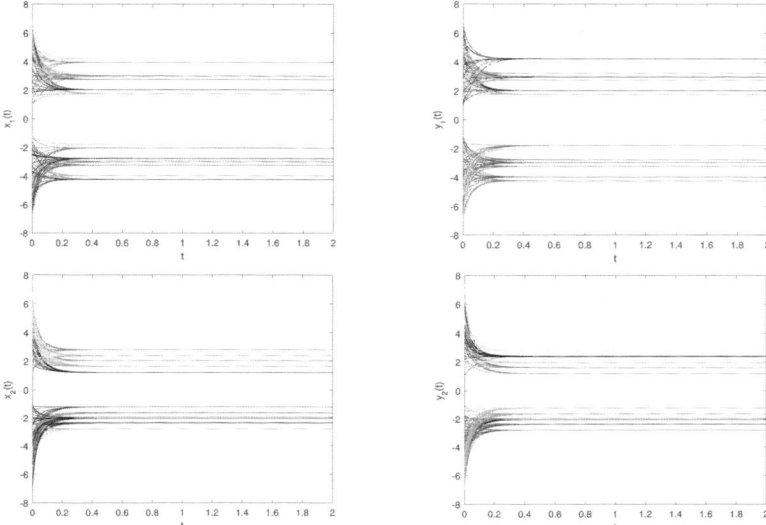

Fig. 1. Transient behavior of the state variable $x_i(t)$ and $y_i(t)$, $i = 1, 2$.

References

1. C.Y. Cheng, K.H. Lin and C. Shih, Multistability in recurrent neural networks, *SIAM J. Appl. Math.* **66**(4):1301–1320 (2006).
2. Y.J. Huang, H.G. Zhang and Z.S. Wang, Multistability of complex-valued recurrent neural networks with real-imaginary-type activation functions, *Appl. Math. Comput.* **229**:187–200 (2014).
3. X.F. Chen, Z.J. Zhao, Q.K. Song and J. Hu, Multistability of complex-valued neural networks with time-varying delays, *Appl. Math. Comput.* **294**:18–35 (2017).
4. X.B. Nie and W.X. Zheng, Multistability of neural networks with discontinuous non-monotonic piecewise linear activation functions and time-varying delays, *Neural Networks* **65**:65–79 (2015).
5. T.P. Chen and L.L. Wang, Global μ-stability of delayed neural networks with unbounded time-varying delays, *IEEE Trans. Neural Networks* **18**(6):1836–1840 (2007).
6. L.L. Wang and T.P. Chen, Multiple μ-stability of neural networks with unbounded time-varying delays, *Neural Networks* **53**:109–118 (2014).
7. R. Rakkiyappan, G. Velmurugan and J.D. Cao, Multiple μ-stability analysis of complex-valued neural networks with unbounded time-varying delays, *Neurocomputing* **149**:594–607 (2015).
8. M.C. Tan and D.S. Xu, Multiple μ-stability analysis for memristor-based complex-valued neural networks with nonmonotonic piecewise nonlinear activation functions and unbounded time-varying delays, *Neurocomputing* **275**:2681–2701 (2018).
9. Y. Zhang, P.A. Heng and K.S. Leung, Convergence analysis of cellular neural networks with unbounded delay, *IEEE Trans, Circuits Syst I* **48**(6):680–687 (2001).

SIFT-based textile defects detection by using adaptive neural-fuzzy inference system

Xueqing Zhao,*,[1,2] Xinjuan Zhu[1,2] and Tao Xue[1,2]

[1] *Shaanxi Key Laboratory of Clothing Intelligence, Xi'an Polytechnic University*
Xi'an, Shaanxi, 710048, China
[2] *School of Computer Science, Xi'an Polytechnic University*
Xi'an, Shaanxi, 710048, China
* cherzhao@hotmail.com
www.xpu.edu.cn

Kaixuan Liu[3,4] and Yongmei Deng[3,4]

[3] *Shaanxi Key Laboratory of Clothing Intelligence, Xi'an Polytechnic University*
Xi'an, Shaanxi, 710048, China
[4] *School of Apparel and Art Design, Xi'an Polytechnic University*
Xi'an, Shaanxi, 710048, China

The quality of produced textile plays very important role in the textile industry, and the detection scheme is the key problem. In this paper, we propose a scale-invariant feature transform-based (SIFT for short) adaptive neural-fuzzy inference system (ANFIS for short) textile defects detecting method. Image features of textile images are extracted by using SIFT, and the ANFIS system is trained to detect the defects of textiles. Moreover, the proposed SIFT-based ANFIS textile defects detector can be simulated on the TILDA textile defect database. Experimental results show that the proposed method can get good performance on detecting the quality of textile.

Keywords: Textile defects; SIFT; ANFIS.

1. Introduction

With the rapidly development of textile science, the textile industry is moving forward to the industrial 4.0 era, the intelligence and greening has become an inevitable trend in the development of the textile industry.[1] Usually, in the actual production of textiles, the quality of produced textile is very important for a company in competitive markets, at the same time, it also adds extra cost to enterprise products. Therefore, in order to better serve the textile industry, automatic defect detection system applied for industrial product quality inspection has been developed, which is exactly detecting non-quantifiable items, like crack, wear, bump and stain, and distinguishing the qualified and defective products, it has a high scientific and applied value in the textile industry.[2]

In the textile weaving stage, how to automatically detect defects on the surface of textiles and then do further processing is a big problem plaguing the textile industry. The purpose of textiles defect detection is to locate anomalous local textures and visualize the area,[3] recently, many textile defect detection techniques based on machine vision have been proposed. The most representative ones are divided into two categories,[4] frequency domain-based and space domain-based defect detection assessment methods. The former is the image is converted to the frequency domain to describe the defect area, the most representative of these methods include Fourier transform, wavelet transform and Gabor filter transform. Hu et al.[5] proposed an unsupervised textiles fiber defect detection and assessment method based on Fourier transform and wavelet threshold shrinkage, Malek et al.[6] proposed the automatic detection and assessment of textile defects by fast Fourier combined with cross-correlation method, these Fourier transform-based textiles defect detection methods are ineffective in detecting the defect of the image in the space domain, and can not detect the defect of the random textiles fiber texture images. Li et al.[7] proposed a textiles fiber detection method based on multiscale wavelet transform and Gaussian mixture model, although this method can detect the defect area, but the calculation of wavelet coefficients is very strict, and the computational complexity is higher; Lucia et al.[8] proposed a textiles defect detection based on Gabor filter combined with principal component analysis, Hu et al.[9] proposed the optimization of Gabor filter based on simulated annealing algorithm (SA) for textiles fiber defect detection, which makes the filter transform to a specific frequency and direction to match the feature of non-defect image. However, Gabor filter parameters of the Gabor filter based on the method of calculation requires very precise and meticulous, and computationally intensive. The latter textiles defect detection method is concentrate on in the spatial domain, where the textile image pixel directly operated to detect textiles fiber defect. The most representative methods include morphological operation method, gray level co-occurrence matrix method, local binary pattern method and dictionary learning method. Celik et al.[10] proposed a textiles fiber defect detection method combined with linear filtering and morphological operations, which could detect the normal area, but its very strict threshold setting requirements and poor generalization performance; Raheja et al.[11] proposed a textiles defect detection method based on gray level co-occurrence matrix, which requires intensive calculation, the computational cost is very large, and the large-scale textiles fiber defects can not be very good detection; Jing et al.[11] proposed a defect detection and classification method based on Gabor filters and local binary patterns, however, such mode depends strongly on the threshold setting, if the threshold is not set properly, some normal textiles texture regions are also detected as defect regions; Jing et al.[12] proposed a fiber texture defect detection based on single-scale dictionary learning, however, due to the high texture complexity of the textile surface, the detection efficiency is lower.

In this paper, we introduced a SIFT-based textile defects detection by using adaptive neural-fuzzy inference system, we introduce the proposed method detailed in section 2, include SIFT, ANFIS and SIFT-based ANFIS Textile defects detector. In section 3, we give the experiment, include some important functions and parameters used in our proposed SIFT-based ANFIS textile defects detector. Finally, we present some conclusions in section 4.

2. The Proposed SIFT-based ANFIS Textile Defects Detector

The general scheme of the mentioned proposed detection method mainly based on textile SIFT feature,[13] and its structures are described in Figure 1. The implementation process is given below.

Firstly, the textile image pixels are obtained, see in the Figure 1, one textile image block with the size of 3×3.

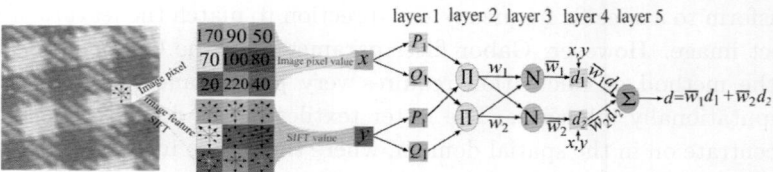

Fig. 1. The SIFT-based ANFIS detector structure, which has been used in this paper.

Secondly, the textile image feature is extracted. Texture of textile is an important visual cue, which can be widely found in the surface of various textiles. So, we use the SIFT to get the texture feature, which can be described into the following 4 steps:

Step 1: Detecting the extreme point in the scale space, which can be obtained by applying different Gaussian linear transform to the original image $I(x,y)$ with the scale σ, where computed by the following formula (1):

$$L(x,y,\sigma) = G(x,y,\sigma) * I(x,y) \qquad (1)$$

where $*$ is convolution operation, (x,y) is the location of the image pixel. $G(x,y,\sigma)$ is Gaussian function, and its obtained by:

Step 2: Accurately locate feature points. By fitting the three-dimensional quadratic function to accurately determine the location and scale of the feature points, at the same time, low-contrast feature points and unstable edge response points are filtered, to enhance the matching stability and improve the anti-noise ability.

Step 3: Feature point distribution. In order to make the descriptor rotation invariant, each feature point is assigned a direction by using the local features of the image, the value of gradient $v(x,y)$ and direction $\theta(x,y)$ are obtained by using the characteristics of the gradient and the direction distribution pixels in their neighborhood:

$$v(x,y) = \sqrt{(L(x+1,y) - L(x-1,y))^2 + (L(x,y+1) - L(x,y-1))^2} \qquad (2)$$

$$\theta(x,y) = tan^{-1}\frac{L(x,y+1) - L(x,y-1)}{L(x+1,y) - L(x-1,y)} \qquad (3)$$

Step 4: Generate feature descriptors. In order to ensure the rotation invariance of the eigenvectors, firstly, the coordinate axes are rotated to the main direction of the feature points, Gaussian circular windows are used to carry out Gaussian weighting on the gradient, the size of window is 16×16, and then the window is divided into sub-regions with the size of 4×4, each region is calculated by gradient histogram in eight directions, at last, a 128-dimensional SIFT feature vector could be obtained which is invariant to translations, rotations and scaling transformations.

Thirdly, SIFT-based ANFIS Textile defects detector is designed. ANFIS is a fuzzy inference system implemented in the framework of adaptive networks,[14] which serves as a basis for constructing a set of fuzzy if-then-else rules with appropriate membership functions to generate the stipulated input-output pairs. The ANFIS structure used in this method is illustrated in Figure 1, which possesses two inputs (x,y) and one output. Here, x and y denote the textile image pixel value and the SIFT value, respectively. Each input of the ANFIS structure has two different triangular membership functions and the rule base contains a total of 4 rules for 2 inputs, which are as follows:

$$\begin{cases} Rule1: if\ x\ is\ A_1\ and\ y\ is\ B_1, & then\quad f_{11} = p_{11}x + q_{11}y + r_{11} \\ Rule2: if\ x\ is\ A_1\ and\ y\ is\ B_2, & then\quad f_{12} = p_{12}x + q_{12}y + r_{12} \\ Rule3: if\ x\ is\ A_2\ and\ y\ is\ B_1, & then\quad f_{21} = p_{21}x + q_{21}y + r_{21} \\ Rule4: if\ x\ is\ A_2\ and\ y\ is\ B_2, & then\quad f_{22} = p_{22}x + q_{22}y + r_{22} \end{cases} \quad (4)$$

3. Experimental Results

In this section, we will present some experimental results from our proposed SIFT-based ANFIS textile defects detector on the TILDA textile defect image database created at the University of Freiburg, Germany. In TILDA, textile samples are grouped as defect-free or having a certain type of defect, some of samples list in Figure 2. All the experiments are run with Matlab code on a work station with 3.10 GHz CPU, RAM 8.00 GB.

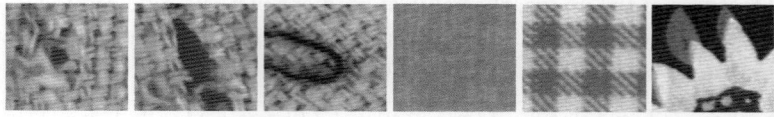

Fig. 2. Textile defect image samples.

Some important functions and parameters used in our proposed SIFT-based ANFIS textile defects detector are given in the following Table 1.

Table 1. Some important functions used in our proposed SIFT-based ANFIS textile defects detector.

Layers of SIFT-based	Important functions ANFIS detector	Instructions and parameters
Layer 1	$\mu_i(x) = max(min(\frac{x-a_i}{b_i-a_i}, \frac{c_i-x}{c_i-b_i}), 0)$	triangular membership functions
Layer 2	$w_{ij} = \mu_{A_i}(x)\mu_{B_j}(y)$	for all the rules
Layer 3	$\bar{w}_{ij} = \frac{w_{ij}}{w_{11}+w_{12}+w_{21}+w_{22}}$	$\bar{w}_{ij}$ normalized
Laye 4	$\eta_{ij} = \bar{w}_{ij}f_{ij} = \bar{w}_{ij}(p_{ij}x + q_{ij}y + r_{ij})$	linear
Laye 5	$d = \sum_{i=1}^{2}\sum_{j=1}^{2}\eta_{ij}$	summation

In the proposed SIFT-based ANFIS textile defects detector, the size of training and testing window is 25×25. We normalized the images of textile is 128×128, see the Figure 3(a); when the error is 0, the mean epoch of the training is 451, at this point, the network reaches convergence, see the Figure 3(b); the testing data listed in Figure 3(c); in our simulation experiments, we use 3 trigonometric membership functions, for 2 input, the number of fuzzy rules is 3^2, see Figure 3(d) and (e).

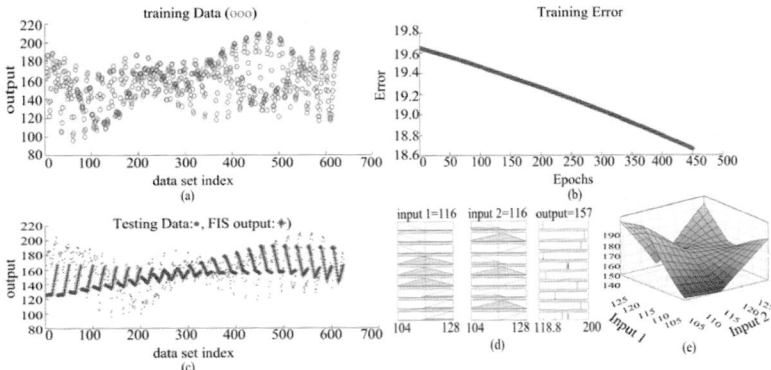

Fig. 3. SIFT-based ANFIS textile defects detector training and testing results.

The training and testing results are listed in the Table 2, the Regression Mean Squared Error (RMSE) is used to monitor the training errors during the ANFIS training process. The training data adjusts the input parameters through the neural network to find the relationships between input/output, and to minimize errors. RMSE is used and defined as:

$$RMES = \sqrt{\frac{1}{N} \sum_{j=1}^{N} (y_j - \hat{y}_j)^2} \qquad (5)$$

where N denote the number of prediction, $\hat{y}_j$ is the predicted data sets, and y_j is the real data sets.

The Regression Mean Squared Error (RMSE) of the proposed are shown in Table 2.

Table 2. RMSE value of the proposed SIFT-based ANFIS textile defects detector.

	Samples	RMSE
Training	50	18.65
Testing	5	18.46

4. Conclusion

In this paper, we present a SIFT-based ANFIS method to detect the textile defects, a new SIFT-based ANFIS model is designed and simulated on the TILDA textile defect image database. The result shows that the proposed method can get good performance on detecting the quality of textile. Based on the proposed method, we can realize textile image SIFT-based ANFIS detector. The detector will improve the production efficiency of textile defects significantly.

Acknowledgments

This paper was financially supported by the Special Scientific Research Project of Education Department of Shaanxi Provincial Government (No. 16JK1328), the Natural Science Research Plan in Shaanxi Province of China (Youth Programs, No. 2017JQ6071), and the Scientific and technological innovation team of Xi'an Polytechnic University (No. TD-12).

References

1. D. Ruan and X. Zeng, *Intelligent sensory evaluation: methodologies and applications*, Springer Science and Business Media, 2013.
2. S. Susan and M. Sharma, *Neurocomputing*, **239**, 232 (2017).
3. J. Jing, X. Fan, P. Li, *Opt. Eng.*, **5**, 445 (2016).
4. Y. Henry, K. Grantham, H. Nelson, *Image and Vision Comput.*, **29**, 442 (2011).
5. G. Hu, Q. Wang, G. Zhang, *Appl. Optics*, **54**, 2963 (2015).
6. A. Malek, J. Drean, L. Bigue, *Text. Res. J.*, **83**, 256 (2013)
7. P. Li, H. Zhang, J. Jing, R. Li, J. Zhao, *J. Text. I.*, **106**, 1381 (2015).
8. B. Lucia, B. Giuseppe, P. Pisana, R. Elisa, S. Andrea, V. Paolo, *J. Vis. Commun. Image R.*, **24**, 838 (2013).
9. G. Hu, *Proceedings of Conference on Information Science, Electronic and Electrical Engineering (ISEEE)*, Vol. 2, p. 860, 2014.
10. H. Celik, L. Canan, T. Mehmet, *Indian J. Fiber Text*, **39**, 254 (2014).
11. J. Raheja, S. Kumar, A. Chaudhary, *Optik*, **23**, 6469 (2013).
12. J. Jing, H. Zhang, J. Wang, P. Li, J. Jia, *J. Text. I.*, **104**, 18 (2013).
13. T. Lindeberg, "Scale invariant feature transform." *Scholarpedia*, **7**, 10491 (2012).
14. X. Wang, X. Zhao, F. Guo et al. *Aeu-Int. J. Electron. C.*, **65**, 429 (2011).

Fuzzy explicit simplified MPC with adjustment parameter

J.M. Escaño,* K. Witheephanich,[†] S. Roshany-Yamchi[†] and C. Bordons[‡]

*Universidad Loyola Andalucía, [†]Cork Institute of Technology, [‡]Universidad de Sevilla
*jmescano@uloyola.es

In this work, a novel methodology is presented to reduce the computational complexity of applying explicit solution of model based predictive control. The methodology is based on applying the *functional principal component analysis*, providing a mathematically elegant approach to reduce the complexity of rule-based systems, like piecewise affine systems, allowing the reduction of the number of consequents and combining and merging the antecedents. The proposed design has been validated using an industrial system model.

Keywords: Piece wise affine; functional principal component analysis; model predictive control; fuzzy control.

1. Introduction

In traditional constrained *model predictive control* (MPC), the control problem is solved in an on-line optimisation, i.e. MPC explicitly uses a dynamic model of the process in the optimisation problem to predict the future evolution of the output over a finite horizon to define the optimal control action at each time step by minimising an objective function while satisfying the system constraints.

To solve the MPC problem, there are many available and reliable *quadratic programming* (QP) algorithms, e.g. *active set, feasible direction, pivoting methods*, etc.[1] They all use an iterative algorithm, which means that due to the computational burden they are not suitable for every hardware platform. As such, it is well recognised that the main drawback of MPC is that it is restricted to relatively slow response process applications with sampling times in the order of minutes and requires powerful computing platforms. In,[2] the implementation aspects are presented and the restriction of the horizon on limited-resource hardware such as fixed-point arithmetics is addressed. Fast model predictive control can be done on-line and off-line. On-line fast MPC is based on the formulation of MPC as a QP problem and an appropriate variable reordering to apply the interior point optimisation algorithms, the active set strategies or the first-order fast gradient methods. Faster on-line optimisation techniques are described in.[3–5] Off-line MPC or explicit MPC is based on solving the optimisation problem that is parameterised by state, i.e. parametric optimisation, to obtain a pre-computed control law as function of state. Several explicit MPC algorithms have been proposed.[6,7] In,[6] an explicit form of the MPC controller has been proposed by solving a multi-parametric quadratic program (mpQP) transformed from the constrained MPC QP problem, hence on-line computation is dramatically reduced, obtaining a *piecewise affine* (PWA) controller. Such a function is then composed of numerous distinct affine feedback laws defined over a set of polytopic regions. For further results on explicit MPC, the reader is referred to[8] and the references therein. However, the use of explicit MPC can lead to an explosion of regions when it comes to high-order models (e.g. more than 10 states). In this work, *functional principal component analysis* (FPCA) has been applied to reduce the number of consequents in the explicit solution. In[9] this technique has been applied to fuzzy systems, reducing the number of rules in an analytical way. This article is organised as follows: An introduction to PCA and FPCA is given in section 2. Section 3 presents the application of FPCA to PWA systems. In section 4, an application is used to illustrate the proposed method. Conclusions are given in section 5.

2. Functional Principal Component Analysis

FPCA is the functional extension of the PCA, which works in a space of functions. Let $f_1(x), f_2(x), ..., f_n(x)$ be functions in separable Hilbert space endowed with inner product:

$$\langle f_i | f_j \rangle = \int_0^X f_i(x) f_j(x) dx \qquad \forall f_{i,j} \in L^2[0, X]. \tag{1}$$

If each function $f_i(x)$ may be decomposed in:

$$f_i(x) = \sum_{l=1}^{L} c_{il} \theta_l(x) = \mathbf{c_i}^T \Theta(x), \tag{2}$$

the mean and covariance functions of f_i, will be:

$$\bar{f}(x) = E(f(x)) = \bar{\mathbf{c}}^T \Theta(x), \tag{3}$$

$$Cov[f(x), f(s)] = \Theta(x)^T cov(\mathbf{C}) \Theta(s), \tag{4}$$

where $\mathbf{C} = \{c_{il}, i = 1, ..., n, l = 1, ..., L\}$. Defining the covariance operator as:

$$C(f(x)) = \int_0^X Cov[f(x), f(s)]f(s)ds,$$
$$\forall f \in L^2[0, X], \forall x, s \in [0, X], \quad (5)$$

where the kernel $Cov[f(x), f(s)]$ is the covariance function. The covariance operator is positive, selfadjoint and compact,[11] thus, using Mercer's Theorem,[12] we may write:

$$Cov[f(x), f(s)] = \sum_{i=1}^{\infty} \lambda_i \xi_i(x) \xi_i(s), \quad \forall x, s \in [0, X], \quad (6)$$

where $\lambda_1 > \lambda_2 > ... > 0$ is an enumeration of the eigenvalues of C, and the corresponding orthonormal eigenfunctions are $\xi_1, \xi_2, ...$ Thus, they form a complete orthonormal set of solutions of the Fredholm equation:

$$\int_0^X Cov[f(x), f(s)]\xi_i(s)ds = \lambda_i \xi_i(x). \quad (7)$$

3. FPCA Applied to PWA Systems

In general, a PWA system can be formulated as:

$$\text{IF } \mathbf{x} \in \Theta_1 \text{ THEN } u = f_1(\mathbf{x})$$
$$\text{ELSIF } \mathbf{x} \in \Theta_2 \text{ THEN } u = f_2(\mathbf{x})$$
$$\vdots$$
$$\text{ELSIF } \mathbf{x} \in \Theta_N \text{ THEN } u = f_N(\mathbf{x})$$
$$\text{ENDIF} \quad (8)$$

Each region Θ_i is defined by a polytope. Let δ_i be functions defined as:

$$\delta_i(\mathbf{x}) = \begin{cases} 1 & if \ \mathbf{x} \in \Theta_i \\ 0 & else \end{cases} \quad (9)$$

If N is the number of regions, the PWA system can be expressed as:

$$u = \sum_{i=1}^{N} \delta_i(\mathbf{x}) f_i(\mathbf{x}) \quad (10)$$

with $\sum_{i=1}^{N} \delta_i = 1$.

Considering each $f_i(\mathbf{x})$ as a linear combination of the inputs,

$$u = \sum_{i=1}^{N} \delta_i(\mathbf{x})(\rho_{0i} + \rho_{1i}x_1 + ... + \rho_{ni}x_n), \qquad (11)$$

where n is the number of inputs. Thus, the expression (11) takes a form as:

$$u(\mathbf{x}) = \tilde{g}_0(\mathbf{x}) + \tilde{g}_1(\mathbf{x})x_1 + ... + \tilde{g}_n(\mathbf{x})x_n, \qquad (12)$$

where:

$$\tilde{g}_i(\mathbf{x}) = \sum_{j=1}^{N} \delta_j(\mathbf{x}) \cdot \rho_{ji}. \qquad (13)$$

And for all the regions, $\tilde{\mathbf{g}}$ is:

$$\tilde{\mathbf{g}}(\mathbf{x}) = \begin{bmatrix} \tilde{g}_0(\mathbf{x}) \\ \tilde{g}_1(\mathbf{x}) \\ \vdots \\ \tilde{g}_n(\mathbf{x}) \end{bmatrix} = \begin{bmatrix} \rho_{10} & \rho_{20} & \cdots & \rho_{N0} \\ \rho_{11} & \rho_{21} & \cdots & \rho_{N1} \\ \vdots & & & \\ \rho_{1n} & \rho_{2n} & \cdots & \rho_{Nn} \end{bmatrix} \cdot \begin{bmatrix} \delta_0(\mathbf{x}) \\ \delta_1(\mathbf{x}) \\ \vdots \\ \delta_N(\mathbf{x}) \end{bmatrix}$$

$$\tilde{\mathbf{g}}(x) = \mathbf{R} \cdot \Delta(x). \qquad (14)$$

The covariance functions of $\tilde{\mathbf{g}}(x)$ are:

$$Cov[\tilde{\mathbf{g}}(\mathbf{x}), \tilde{\mathbf{g}}(\mathbf{s})] = \Delta(\mathbf{x})^T cov(\mathbf{R}) \Delta(\mathbf{s}). \qquad (15)$$

Supposing that the eigen functions are:

$$\Gamma(x) = \begin{bmatrix} \gamma_1(x) \\ \gamma_2(x) \\ \vdots \\ \gamma_h(x) \end{bmatrix} = \Delta(x)^T \cdot \mathbf{b}. \qquad (16)$$

Thus, taking into account (15):

$$\int_0^X Cov[\tilde{\mathbf{g}}(x), \tilde{\mathbf{g}}(s)] \cdot \gamma(s) ds =$$

$$\int_0^X \Delta(x)^T cov(\mathbf{R}) \Delta(s) \cdot \Delta(s)^T \cdot \mathbf{b} ds = \Delta(x)^T cov(\mathbf{R}) \cdot \mathbf{W} \cdot \mathbf{b}$$

$$cov(\mathbf{R}) \cdot \mathbf{W} \cdot \mathbf{b} = \lambda \cdot \mathbf{b}, \qquad (17)$$

where:

$$\mathbf{W} = \int_0^X \Delta(s) \cdot \Delta(s)^T ds \qquad (18)$$

being $\gamma(x)$ orthogonal, then $\langle \gamma_i(x), \gamma_j(x) \rangle = b_i^T \cdot \mathbf{W} \cdot b_j = 0$.

Matrix **W** is symmetric by definition, thus, defining $\mathbf{p} = \mathbf{W}^{\frac{1}{2}}\mathbf{b}$,

$$\mathbf{W}^{\frac{1}{2}} \cdot cov(\mathbf{R}) \cdot \mathbf{W}^{\frac{1}{2}} \cdot \mathbf{p} = \lambda \cdot \mathbf{p}. \tag{19}$$

A symmetric eigenvalue problem now remains to be solved. Since only one region will be active in each state vector, equation (16) will give only one of the values within **b** parameter,

$$\gamma_i(x) \in \{b_{i1}, b_{i,2}, ..., b_{iN}\}. \tag{20}$$

It may be that there are repeated b_{ij} values, in such case, the regions associated to those values can be merged, hence reducing the number of regions, i.e. if $b_{ij} = b_{ik}$ with $j \neq k$, then $\gamma_i(x) = b_{i1}\delta_1(\mathbf{x}) + ... + b_{ij}(\delta_i(\mathbf{x}) + \delta_j(\mathbf{x})) + ... + b_{iN}\delta_N(\mathbf{x})$, being:

$$\delta_i(\mathbf{x}) + \delta_j(\mathbf{x}) = \begin{cases} 1 & if \ \mathbf{x} \in \Theta_i \bigcup \Theta_j \\ 0 & else \end{cases} \tag{21}$$

4. Example of Explicit MPC: Distillation Column

To illustrate the performance of a reduced PWA system, a high purity distillation column is used as an example. The distillation process typically works around an operating point, being identifiable by a linear model. The example is be carried out using the model shown in,[13] where the linear model in an operational point is defined as:

$$\begin{bmatrix} \dot{x}_1 \\ \dot{x}_2 \end{bmatrix} = \begin{bmatrix} -0.0133 & 0 \\ 0 & -0.0133 \end{bmatrix} \begin{bmatrix} x_1 \\ x_2 \end{bmatrix} + \begin{bmatrix} 0.0117 & 0.0115 \\ 0.0144 & 0.0146 \end{bmatrix} \begin{bmatrix} u_1 \\ u_2 \end{bmatrix},$$

$$\begin{bmatrix} y_1 \\ y_2 \end{bmatrix} = \begin{bmatrix} x_1 \\ x_2 \end{bmatrix}, \tag{22}$$

where y_1 and y_2 are the top and bottom product compositions, respectively, and the inputs, u_1 and u_2, are the reflux flow rate and the boil-up, respectively, as shown in Fig. 1.[14] Considering the reference tracking problem, i.e., the problem of driving the output (product composition) y to track a given reference signal $r \in \mathcal{R}^p$ by adjusting the control inputs (reflux and boiler flow rates) u under the control input and control increment constraints. The tuning parameters used for deriving the explicit MPC controller are as follows: $N_y = 20$, $N_u = 3$, $\mathbf{Q} = \mathbf{R} = \mathbf{I}$, with a sampling time of 10 min. The control input constraints are given as $-2 \leq u_1 \leq 2$ and $-1 \leq u_2 \leq 2$, respectively. The number of regions obtained for the control law is 140. Applying the developed FPCA above to this application, with the same

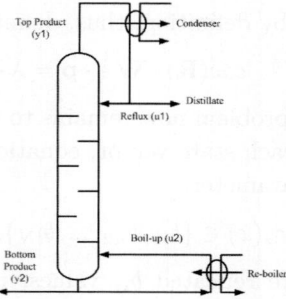

Fig. 1. Distillation column.

tuning, just 5 consequents are obtained for the first manipulated variable and 6 for the second. Associating regions with the same parameter as it has been seen in (21), the arithmetic operations of antecedents are reduced from 140 to 68. An important drawback of the explicit solution of the MPC is the lack of on-line tuning parameters to the designed controller. It is a weakness in terms of implementation. One possible solution is to design different PWA controllers for different settings parameters and linearly interpolating the control action of the different controllers, e.g. by the inference of a fuzzy controller having the control actions of the various PWA designed for different parameter values as inputs. The problem with this is an increase in complexity, due to the increasing the number of rules. Figure 2 shows the scheme of this structure. A Takagi-Sugeno fuzzy system[15] has been chosen to interpolate the different control actions for the distillation column. Only the input λ is taken for fuzzyfication, choosing three membership functions (mf1, mf2 and mf3 according to three precalculated values of λ, 1, 5 and 10). Once the FPCA is applied, a huge reduction in the number of consequents can be observed, and following (21), an

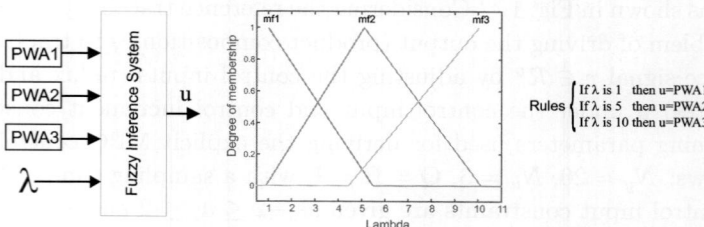

Fig. 2. FMPC with explicit solution subject to constraints.

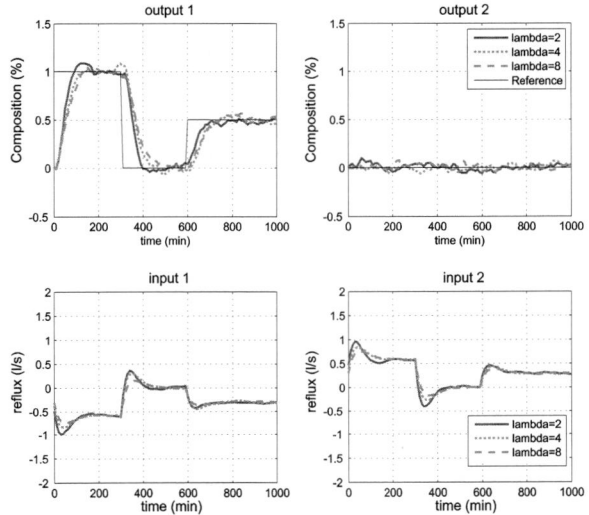

Fig. 3. FMPC for distillation column depending on lambda.

association of regions can be done, reducing the antecedents between 45 and 51%. Following the structure of the system proposed above (see Fig. 2), the parameter λ (move suppression) can be adjusted in order to give more or less aggressiveness to the system. Figure 3 shows the performance of the controller when λ takes different values.

5. Conclusions and Future Works

A novel technique has been applied for PWA systems to reduce the number of consequents and to associated regions in the antecedents of the rules. The technique is based on the application of FPCA to the structure of the PWA. Providing a tuning parameter for explicit MPC could be done by using a PWA controller per each adjusted parameter, increasing more the system complexity. The application of this technique, together with the use of a fuzzy system to combine the different PWA controllers, has succeeded in designing a MPC scheme with constraints and an on-line adjustment parameter, with drastically reducing the number of regions of consequents, improving programming time (down 96% the number of consequents to add into the code). The technique applied to the PWA also allows a new merge of regions (not necessarily adjacent), opening the door to future research on reducing their complexity.

References

1. E. F. Camacho and C. Bordons, "Model Predictive Control". Advanced Textbooks in Control and Signal Processing. Advanced Textbooks in Control and Signal Processing (Springer, 2007).
2. P. Zometa, M. Kogel, T. Faulwasser and R. Findeisen, "Implementation aspects of model predictive control for embedded systems". *American Control Conference (ACC), 2012*, June 2012.
3. Y. Wang and S. Boyd, "Fast Model Predictive Control Using Online Optimization". *IEEE Transactions on Control Systems Technology* **18**, 267(March 2010).
4. J. L. Jerez, P. J. Goulart, S. Richter, G. Constantinides, E. C. Kerrigan and M. Morari, "Embedded Online Optimization for Model Predictive Control at Megahertz Rates". *IEEE Transactions on Automatic Control* **59**, 3238 (2014).
5. M. Herceg, C. N. Jones and M. Morari, "Dominant speed factors of active set methods for fast MPC". *Optimal Control Applications and Methods* **36**, 608 (2015).
6. A. Bemporad, M. Morari, V. Dua and E. N. Pistikopoulos, "The explicit linear quadratic regulator for constrained systems". *Automatica* **38**, 3 (2002).
7. E. N. Pistikopoulos, "Perspectives in multiparametric programming and explicit model predictive control". *AIChE Journal* **55**, 1918 (2009).
8. A. Alessio and A. Bemporad, "Nonlinear Model Predictive Control: Towards New Challenging Applications". Springer Berlin Heidelberg, Berlin, Heidelberg, 2009, Berlin, Heidelberg, ch. A Survey on Explicit Model Predictive Control, pp. 345–369.
9. J. M. Escaño and C. Bordons, "Complexity Reduction in Fuzzy Systems Using Functional Principal Component Analysis". *Fuzzy Modeling and Control: Theory and Applications* (Atlantis Press, Paris, 2014), Paris, pp. 49–65.
10. H. L. Shang, "A survey of functional principal component analysis". *AStA Advances in Statistical Analysis* **98**, 121 (2013).
11. J. Deville, "Méthodes statistiques et numériques de l'analyse harmonique". *Annales de l'inséé*, 3, 5(Jan.-Apr. 1974).
12. J. Mercer, "Functions of Positive and Negative Type, and their Connection with the Theory of Integral Equations". *Philosophical Transactions of the Royal Society of London A: Mathematical, Physical and Engineering Sciences* **209**, 415 (1909).
13. S. Skogestad, M. Morari and J. Doyle, "Robust control of ill-conditioned plants: high-purity distillation". *IEEE Transactions on Automatic Control*, **33**, 1092(Dec 1988).
14. S. Roshany-Yamchi, R. R. Negenborn, M. Cychowski, B. De Schutter, J. Connell and K. Delaney, "Distributed model predictive control and estimation of large-scale multi-rate systems", in *Proceedings of the 18th IFAC World Congress*, 2011.
15. T. Takagi and M. Sugeno, "Fuzzy Identification of Systems and Its Applications to Modeling and Control". *IEEE Transactions on Systems, Man, and Cybernetics* **15**, 116(February 1985).

A control to soybean aphid via fuzzy linear programming

M. S. Peixoto,* S. M. S. Carvalho

DFQM, CCTS, Universidade Federal de São Carlos
Sorocaba, São Paulo, Brasil
magda@ufscar.br, silviamsc@ufscar.br

L. C. Barros, R. C. Bassanezi, E. E. Laureano

DMA, IMECC, Universidade Estadual de Campinas
Campinas, São Paulo, Brasil
laeciocb@ime.unicamp.br, rodney@ime.unicamp.br, eelaureano@ime.unicamp.br

W. A. Lodwick

DM, University of Colorado
Denver, Colorado, USA
wlodwick@math.cudenver.edu

This work proposes a chemical control to soybean aphid by fuzzy linear programming. The soybean aphid, *Aphis glycines* (Hemiptera: Aphididae), is an invasive herbivore to North America. In this paper we have propose a chemical control in the plantation when the prey population exceeds the economic damage threshold. On the other hand, the soybean aphid has become the most devastating insect pest of soybean in the United States. Brazil is the second largest exporter of soybean at present, after the USA and before Argentina. According to the Bureau of Agriculture of the USA, it has been estimated that Brazil will be the largest soybean exporter in 2023.

Keywords: Fuzzy set; fuzzy linear programming; control.

1. Introduction

The soybean aphid, *Aphis glycines* (Hemiptera: Aphididae), is an invasive herbivore to North America [1]. It was first discovered in North America in Wisconsin in late July 2000 infesting soybean crop. Natural enemies have been observed to attack this pest. An economic threshold was developed for chemical control, i.e., when an pesticide treatment is warranted. Economic

---

*Financial support by São Paulo Research Foundation (FAPESP): project number 2016/04299–9.

thresholds for the soybean aphid have been developed and vary from 250 to 273 aphids per plant and aphids population increases [2]. The control measures are not introduced before reaching the economic threshold.

In this work we propose a chemical control to soybean aphid in the plantation by fuzzy linear programming [3], that is the optimization of use of pesticide in chemical control to soybean aphid when the prey population exceeds the economic damage threshold and aphids population increases.

The model includes a fuzzy predator-prey system in order to describe the interaction between the prey, *Aphis glycines* (Hemiptera: Aphididae) — the soybean aphid, and its predator, *Orius insidiosus* (Hemiptera: Anthocoridae) considering biotic (predator) and abiotic (temperature) factors, which affect the soybean aphid population dynamics, and a comparison between the fuzzy model and real data reported in the literature [4, 5]. This mathematical model is very useful to evaluate the evolution of the population of the soybean aphid over time. Although this pest is not currently present in South America, the model is important to simulate possible scenarios in a soybean plantation.

We proposed in [5] a fuzzy biological control to soybean aphid, that is, the model provides how often and how much to add the predators in the plantation by fuzzy rule-based system, instead using pesticides.

We proposed in [6] a fuzzy chemical control to soybean aphid, that is, the model provides how often and how much to apply the pesticide on the plants in a simple, intuitive and a direct way.

On the one hand, the soybean aphid has still not found in Brazil. Therefore, before any eventual invasion, a predictive model to enhance control program is desirable. On the other hand, the soybean aphid has become the most devastating insect pest of soybean in the United States. Brazil is the second largest exporter of soybean at present, after the USA and before Argentina. According to the Bureau of Agriculture of the USA, it has been estimated that Brazil will be the largest soybean exporter in 2023.

2. Fuzzy Linear Programming

The classical linear programming problem is to find the minimum (or maximum) values of the linear function under constraints represented by linear inequalities or equations, that is,

maximize (or minimize) $c^T x$
subject to $\quad Ax \leq b$
$\quad x \geq 0,$

where x is a vector of variables, A is called a constraint matrix, and the vector b is called a right-hand-side vector. As is well known, many practical problems can be formulated as linear programming problems [3].

Fuzzy linear programming is a family of optimization problems in which the optimization model parameters are not well defined, that is, the objective function and/or constraint coefficients are not exactly known and that some of the inequalities involved may also be subject to unsharp boundaries [7].

The optimization model associated with a linear programming problem in which only the right-hand-side numbers B are fuzzy numbers [8, 9] is formulated as follows:

$$\max \text{ (or min)} \sum_{j=1}^{n} c_j x_j$$

$$\text{subject to } \sum_{j=1}^{n} a_{ij} x_j \leq B_i \quad (i \in \mathbb{N}_m)$$

$$x_j \geq 0 \quad (j \in \mathbb{N}_n)$$

where $x_j \in \mathbb{R}$.

In this case, fuzzy numbers B_i typically have the form

$$B_i(x) = \begin{cases} 1, & x \leq b_i \\ \dfrac{b_i + p_i - x}{p_i}, & b_i < x \leq b_i + p_i \\ 0, & x > b_i + p_i, \end{cases} \quad (1)$$

In general, fuzzy linear programming problems are first converted into equivalent crisp linear or nonlinear problems, which are then solved by standard methods. The final results of a fuzzy linear programming problem are thus real numbers, which represent a compromise in terms of the fuzzy numbers involved. Next, we determine the fuzzy set of optimal values. This is done by calculating the lower and upper bounds of the optimal values first.

The lower bound of the optimal values, z_l, is obtained by solving the standard linear programming problem:

$$\max \text{ (or min)} \ z = cx$$

$$\text{subject to } \sum_{j=1}^{n} a_{ij} x_j \leq b_i \quad (i \in \mathbb{N}_m)$$

$$x_j \geq 0 \quad (j \in \mathbb{N}_n).$$

The upper bound of the optimal values, z_u, is obtained by a similar programming problem in which each b_i is replaced with $b_i + p_i$ [3].

Now, we have the following classical optimization problem:

max (or min) λ

subject to $\lambda(z_u - z_l) - cx \leq -z_l$ $\quad (\lambda \in \mathbb{R})$

$\lambda p_i + \sum_{j=1}^{n} a_{ij} x_j \leq b_i + p_i$ $\quad (i \in \mathbb{N}_m)$

$\lambda, x_j \geq 0$ $\quad (j \in \mathbb{N}_n)$.

3. The Mathematical Model

In this work, let x denote the number of aphids population in the plant and z the quantity of pesticide (%) of the quantity recommended by manufacturer. By the fuzzy sets of the number of aphid defined in [6] and the economic thresholds for the soybean aphid, the problem can be formulated as the following fuzzy linear programming problem:

min $z = 0.2x$

subject to $x \leq B$

$x \geq 0$

and B is defined by

$$B(x) = \begin{cases} 1, & x \leq 250 \\ \dfrac{700 - x}{450}, & 250 < x \leq 700. \\ 0, & x > 700 \end{cases} \qquad (2)$$

First the lower and upper bounds of the objective function $z_l = 0$ and $z_u = 100$, respectively.

Then, the fuzzy set of optimal, G, which is a fuzzy subset of $\mathbb{R}$, is defined by

$$G(x) = \begin{cases} 0, & 0.2x \leq 0 \\ \dfrac{0.2x}{100}, & 0 < 0.2x \leq 100. \\ 1, & 0.2x > 100 \end{cases} \qquad (3)$$

Now, by (2) and (3), the fuzzy linear programming problem becomes the following classical optimization problem:

min λ

subject to $100\lambda - 0.2x \leq 0$

$450\lambda + x \leq 700$

$x, \lambda \geq 0$.

4. Results and Conclusions

The problem has been modelled via Simplex Method in MATLAB®. Solving this classical optimization problem, we find that the minimum $\lambda = 0.7368$, is obtained for $x = 368.4211$. The quantity of pesticide z is then calculated by

$$z = 0.2x = 73.6842$$

This model suggests that the quantity of pesticide was 73.7% of the quantity recommended by manufacturer in the plantation with about 370 aphids per plant. In this way, the model suggests that the quantity of pesticide may be lower than the quantity recommended by manufacturer in plantations of soybean damaged by soybean aphid.

The concern about the environment has been increasingly important. Currently, actions aimed at sustainable management of natural resources are goals. In general, pesticides are toxic, harmful to human health and the environment. One of the most common problems is the contamination of soil, groundwater, rivers and lakes. When the pesticide is used, it intoxicates all life present. Studies show the decrease in the number of pollinating bees and the destruction of bird habitat in environments where pesticides are used. The abusive use of pesticides may lead to an increase number of pests because pests become more resistant, requiring stronger pesticides that will damage the environment even more and will kill the pests' natural predators [10].

In this way, the model suggests that the quantity of pesticide may be lower than the quantity recommended by manufacturer. Besides, low-quantities of pesticides from those recommended by the manufacturer may be effective. On one hand, there are costs for each application. On the other hand, there should be a concern about the environmental damage caused by the abusive use of pesticides in the plantations.

We will develop further studies on simple and specific method using a fuzzy rule-based system to help the implementation of an integrated pest management system.

Acknowledgments

The authors acknowledge São Paulo Research Foundation (FAPESP), projects numbers 2016/04299–9, 2010/06822–4 and 2013/24148–7, and the National Council for Scientific and Technological Development (CNPq), project numbers 306546/2017–5, for the financial support.

References

1. B. P. McCornack, D. W. Ragsdale and R. C. Venette, *J. Ec. Entomology* **97**, 854 (2004).
2. D. W. Ragsdale, B. P. McCornac, R. C. Venette, B. D. Potter, I. V. MacRae, E. W. Hodgson and M. E. O'Neal, *Journal of Economic Entomology* **100**, 1258 (2007).
3. G. J. Klir and B. Yuan, *Fuzzy Sets And Fuzzy Logic: Theory and Applications* (Prentice Hall, N. Jersey, 1995).
4. T. Hunt, *Soybean aphid management in Nebraska*, tech. rep., NebFacts. Nebraska Cooperativa Extension IARN-UNL (Lincoln, NE, 2005).
5. M. S. Peixoto, L. C. Barros, R. C. Bassanezi and O. A. Fernandes, *Applied Mathematics* **7**, 2149 (2016).
6. M. S. Peixoto, L. C. Barros, R. C. Bassanezi and O. A. Fernandes, An approach via fuzzy systems for dynamics and control of the soybean aphid, in *Proceedings of 9th IFSA World Congress and 20th NAFIPS International Conference, Jul 25-28*, (Paris: Atlantis Press, Gijón, Asturias, Spain, 2015).
7. W. Pedrycs and F. Gomide, *An Introduction to Fuzzy Sets: Analysis and Design* (Massachusets Institute of Technology, 1998).
8. L. C. Barros, R. C. Bassanezi and W. A. Lodwick, *A First Course in Fuzzy Logic, Fuzzy Dynamical Systems, and Biomathematics* (Springer, 2017).
9. L. A. Zadeh, *Information and Control* **8**, 338 (1965).
10. D. Pimentel and H. Lchman, *The Pesticide question: environment, economics, and ethics* (Chapman and Hal, New York, 1993).

An approach to recognize coral species on the coast of Brazil using image analysis and fuzzy associative memories based on equivalent measures

E. Esmi* and J. B. Florindo

Department of Applied Mathematics, University of Campinas
Campinas, São Paulo, Brazil, 13083-859
* eelaureano@ime.unicamp.br
jbflorindo@ime.unicamp.br

F. Pérez and M. Barbeitos

Federal University of Paraná
Curitiba, Paraná, Brazil, 3360-5000
flavioperez72@hotmail.com
msbarbeitos@gmail.com

This article presents an approach to classify images of corals obtained on the coast of Brazil into three distinct species, namely *Siderastrea siderea*, *Siderastrea stellata*, and *Siderastrea radians*. To this end we first employ the well-known Completed Local Binary Patterns method to extract a vector of image descriptors from each image. Subsequently, we use a fuzzy associative memory based on equivalent measures to associate each feature vector to one of the species.

Keywords: Classification of coral species; image analysis; fuzzy associative memory; equivalent measures; completed local binary patterns.

1. Introduction

The scleractinian (hard) corals Siderastrea stellata, Siderastrea radians and Siderastrea siderea form the so-called "Atlantic Siderastrea Complex". These species are fundamental for reef ecosystems because they are important reef builders along the Brazilian coast. In this species complex, morphological variation poses challenges to classification due to the overlap of quantitative diagnostic traits and differences in opinions among specialists with respect to how characterize each species. Thus, the interspecific morphological limits in this group remain controversial.[1] For instance, Veron does not recognize the presence of S. radians in Brazil and considers S. stellata to be an endemic species,[2] whereas Cairns considers the latter to be an invalid species. Forsman et al. found great genetic proximity between S. radians from Panama and S. stellata from Pernambuco,[3] while Neves et al. report reproductive isolation between S. radians and S. stellata in Northeastern Brazil[4] and Nunes et al. suggest that S. stellata is actually a hybrid of the other two species.[5] This controversy has direct implications in conservation. S. radians and S. siderea are not reported for Brazil in IUCN's Red Book (http://www.iucnredlist.org) whereas S. stellata is reported as an endemic species (in opposition to Veron) and listed as "data deficient". It is impossible to design conservation policies for poorly described species. Besides, the complex may hide geographically restricted cryptic species that may be highly threatened. Hence, it is of primary importance to find objective ways of characterizing species from morphological data. The use of software for image processing is a valuable tool that makes possible the recognition and classification of biological structures with minimal human input, minimizing error due to subjectivity in variable interpretation.

In this article we present an approach to associate coral images into one of three species of corals, namely *Siderastrea siderea*, *Siderastrea stellata*, and *Siderastrea radians*. The proposed method comprises two steps. The first one consists of extracting image descriptors from each obtained image using a well-known method of texture image analysis called Completed Local Binary Patterns (CLBP).[6] Second, we design a fuzzy associative memory based on equivalent measures, which associates each feature vector with a particular coral group. We tested our approach in a data set of 370 labeled coral images collected along the coast of Brazil.

2. Completed Local Binary Patterns

CLBP descriptors essentially combines different strategies to compute local binary features. The first and most well-known of such strategies are the classical Local Binary Patterns (LBP).[7] Let g be a binary image of $W \times H$ dimensions, the LBP code of the reference pixel $g_c := g(i_c, j_c)$ is defined in terms of its P neighbors in the radius R:

$$LBP_{P,R}(i_c, j_c) = \sum_{p=0}^{P-1} s(g_p - g_c)2^p, \qquad (1)$$

where $g_p = g(i_c + R\cos(2\pi p/P), j_c + R\sin(2\pi p/P))$, $p = 1, \ldots, P$, and $s(x) = 0$ if $x \geq 0$, otherwise, $s(x) = 1$ if $x < 0$. The values of those points that fall outside the grid of pixels in the discrete domain of the image are obtained by interpolation. The LBP descriptors are given by the histogram of LBP codes:

$$H(k) = \sum_{i=1}^{W}\sum_{j=1}^{H} \delta(LBP_{P,R}(i,j), k), \qquad k \in [0, K],$$

where $\delta(x,y) \in \{0,1\}$ with $\delta(x,y) = 1 \Leftrightarrow x = y$ and K is the maximum value assigned to an LBP code.

Another important definition in LBP theory is the U value, corresponding to transitions between bits 0 and 1 in the LBP code:

$$U(LBP_{P,R}) = |s(g_{P-1} - g_c) - s(g_0 - g_c)| + \sum_{p=1}^{P-1} |s(g_p - g_c) - s(g_{p-1} - g_c)|.$$

Given this, the locally rotation invariant binary pattern is defined for each point (i_c, j_c) by

$$LBP_{P,R}^{riu2}(i_c, j_c) = \begin{cases} \sum_{p=0}^{P-1} s(g_p - g_c), & \text{if } U(LBP_{P,R}) \leq 2 \\ P+1, & \text{otherwise.} \end{cases} \qquad (2)$$

The same circular neighborhood can also provide other interesting features. An important example is the local difference vector $[d_0, \cdots, d_{P-1}]$, where $d_p = g_p - g_c$, for $0 \leq p \leq P-1$.. A useful property of d_p is its robustness to illumination changes. Besides, it can be decomposed into a sign (s_p) and a magnitude (m_p) component $d_p = s_p * m_p$, where $s_p = \text{sign}(d_p)$ and $m_p = |d_p|$. This operation provides us with the sign vector $[s_0, \cdots, s_{P-1}]$ and the magnitude vector $[m_0, \cdots, m_{P-1}]$. These vectors can also give rise to local codes in a similar manner to that employed in the classical LBP method. Those local codes are named completed LBP (CLBP) in.[6] The magnitude vector gives rise to the $CLBP\_M$ code:

$$CLBP\_M_{P,R} = \sum_{p=0}^{P-1} t(m_p, c) 2^p,$$

where $t(x,c)$ is a threshold function: $t(x,c) = 1$ is $x \geq c$ and $t(x,c) = 0$ otherwise, where c is the mean value of m_p over the whole image. Similarly, one can define $CLBP\_S$ over the s_p vector, but this coincide with the classical LBP code defined in (1). Finally, we have the $CLBP\_C$ code, generated by the gray value of the reference pixel (g_c):

$$CLBP\_C_{P,R} = t(g_c, c_I),$$

where c_I is the average gray level of the entire image.

$CLBP\_M$, $CLBP\_S$ and $CLBP\_C$ can be summarized by histograms like in the classical LBP descriptors and those histograms can be combined in two ways: by concatenation or in a three-dimensional joint histogram. Here we adopt the second strategy, which yields the best results in most scenarios.

3. Fuzzy Associative Memory Based on Equivalent Measures

An associative memory (AM) is a mapping $\Phi : X \to Y$ designed to store a set of pairs of data $\mathcal{M} = \{(\mathbf{x}^\xi, \mathbf{y}^\xi) \in X \times Y \mid \xi = 1, \ldots, p\}$ called *fundamental memory set* or the set of *fundamental memories*. Ideally, an AM Φ satisfies $\Phi(\mathbf{x}^\xi) = \mathbf{y}^\xi$ for $i = 1, \ldots, p$ and additionally is endowed with a certain type of tolerance: $\Phi(\tilde{\mathbf{x}}^\xi) = \mathbf{y}^\xi$ if $\tilde{\mathbf{x}}^\xi$ stands for a corrupted or noisy version of $\mathbf{x}^\xi$. In practice, many AM models are not able to store all fundamental memories and present limited correction capacity, retrieving approximately $\mathbf{y}^\xi$ for corrupted or noisy versions of $\mathbf{x}^\xi$. A fuzzy associative memory (FAM) is a type of AM that is also a fuzzy neural network,[8] i.e., an artificial neural network whose inputs or weights are fuzzy.

Θ-fuzzy associative memories (Θ-FAMs) consist of a subclass of fuzzy associative memories having a competitive hidden layer whose calculation of the ξth hidden neuron is given by a function $\Theta^\xi : \mathbb{L} \to [0, 1]$, where the symbol $\mathbb{L}$ denotes a bounded lattice, that is, a partial ordered set with maximal and minimal elements such that the infimum and supremum of any two elements of $\mathbb{L}$ exist and belong to $\mathbb{L}$.[9] Particular cases of Θ^ξ functions are given by fuzzy subsethood or equivalent measures, leading to (weighted) subsethood, dual subsethood, and equivalent measure FAMs.[10–12] Given a finite set $\{(\mathbf{x}^\xi, B^\xi) \in \mathbb{L} \times \mathcal{F}(Y) : \xi = 1, \ldots, p\}$ where $\mathbb{L}$ is a bounded lattice and $\mathcal{F}(Y)$ denotes the class of fuzzy sets of an arbitrary universe Y. Let $\Theta^\xi \mathbb{L} \to [0, 1]$ be functions such that $\Theta^\xi(\mathbf{x}^\xi) = 1$ for $\xi = 1, \ldots, p$ and $\mathbf{v} \in \mathbb{R}^p$, the Θ-*FAM based on* Θ^ξ *and* $\mathbf{v}$, for short Θ-FAM, is a function $\mathcal{O} : \mathbb{L} \to \mathcal{F}(Y)$ defined for each $\mathbf{x} \in \mathbb{L}$ by:[10,11]

$$\mathcal{O}(\mathbf{x}) = \bigcup_{j \in I_\mathbf{v}(\mathbf{x})} B^j, \tag{3}$$

where $I_\mathbf{v}(\mathbf{x}) = \{j \in \{1, \ldots, p\} : v_j \Theta^j(\mathbf{x}) = \max_{\xi=1,\ldots,p} v_\xi \Theta^\xi(\mathbf{x})\}$. Sufficient conditions for $\mathcal{O}(\mathbf{x}^\xi) = B^\xi, \xi = 1, \ldots, p$, and a characterization of the basins of attraction around each $\mathbf{x}^\xi$ can be found in.[11]

In this paper, we focus on equivalent measure FAMs (E-FAMs). We should recall that an equivalent measure on a bounded lattice $\mathbb{L}$ is a function $E : \mathbb{L}^2 \to [0, 1]$ that satisfies the following conditions:[10,13,14]

E1) $E(x,y) = E(y,x)$ for all $x, y \in \mathbb{L}$;
E2) $E(0_\mathbb{L}, 1_\mathbb{L}) = 0$;
E3) $E(x,x) = 1$ for all $x \in \mathbb{L}$;
E4) if $x \leq y \leq z$, then $E(x,z) \leq E(x,y)$ and $E(x,z) \leq E(y,z)$.

Let E^ξ be equivalent functions on $\mathbb{L}$, the corresponding E-FAM is obtained by taking $\Theta^\xi(\mathbf{x}) = E^\xi(\mathbf{x}^\xi, \mathbf{x})$ for each $\xi = 1, \ldots, p$.

Let $\mathbb{L} = [a_1, b_1] \times \ldots \times [a_n, b_n]$, $a_i, b_i \in \mathbb{R}$ with $a_i \leq b_i$ for $i = 1, \ldots, n$, and let $\lambda \in (0,1]^n$ and $\mathbf{w} \in [0,1]^n$ such that $\sum_{i=1}^n w_i = 1$. The function $E_{\lambda, \mathbf{w}} : \mathbb{L}^2 \to [0,1]$ given by

$$E_{\lambda, \mathbf{w}}(\mathbf{x}, \mathbf{y}) = \sum_{i=1}^n w_i \max\left(0, 1 - \frac{|x_i - y_i|}{\lambda_i |b_i - a_i|}\right), \forall \mathbf{x}, \mathbf{y} \in \mathbb{L}, \quad (4)$$

is an equivalent measure on $\mathbb{L}$.[10,15] Other examples of equivalent measures can be found in[10,11,13,16,17] and references therein.

4. Identification of Coral Species from Images

In order to obtain a method to recognize the species to which a given coral image pertains, we assume that each input image has a fixed size and is a photo of one of the species Siderastrea siderea (SD), Siderastrea stellata (SS), and Siderastrea radians (SR). Moreover, we suppose that we have at hand a set of p images of coral that were labeled in one of these three species by an expert. Under this hypotheses, our strategy to identify the coral species for a given coral image comprises twos steps. The first one consists of applying a method of image analysis that is scale and translation-invariant, to know, CLBP, described in Section 2, since the photos may be taken from different positions and distances. Thus, the first step consists of associating each image with a feature vector (image descriptors), that is, with a point of $\mathbb{L} = [a_1, b_1] \times \ldots \times [a_n, b_n]$, $a_i, b_i \in \mathbb{R}$ with $a_i < b_i$ for $i = 1, \ldots, n$, where n is the number of descriptors or features extracted using CLBP method. The second step consists of designing an E-FAM geared to associate each feature vector to one of these three species. To this end, we consider $Y = \{SD, SS, SR\}$ such that each class label $y \in Y$ is associated with the fuzzy number of $\mathcal{F}(Y)$ given by the characteristic function of $\{y\}$. Thus, the given labeled coral images leads us to a set of p fundamental memories that we use to obtain an E-FAM.

We test our approach using 370 coral images with size 1280×960 collected along the coast of Brazil, with 92 images belonging to the species Radians, 72 to Siderea, and 206 to Stellata. One sample from each species

Radians   *Siderea*   *Stellata*.

Fig. 1. One image sample from each coral species.

is illustrated in Figure 1. In the first step, we use $R = 3$ and $P = 24$ since this values produced the best experimental results in.[6] Thus, the ξth image was converted to a vector in $\mathbf{z}^{\xi} \in \mathbb{Z}^{1352}$, for $\xi = 1, \ldots, 370$. Subsequently, in order to reduce the computational effort, we apply the well-known Principal Component Analysis[18] (PCA) method to associate each $\mathbf{z}^{\xi}$ with a vector $\mathbf{x}^{\xi} \in \mathbb{Z}^{35}$. In the second stage, we use an E-FAM based on equivalent measures given in Equation (4) with $\lambda = (\frac{1}{35}, \ldots, \frac{1}{35})$, $\mathbf{w} = (1, \ldots, 1) \in [0,1]^{35}$, where $a_i = \min_{\xi=1,\ldots,35} x_i^{\xi}$ and $b_i = \min_{\xi=1,\ldots,35} x_i^{\xi}$, $\xi = 1, \ldots, 370$. In 100 experiments using 5-fold cross-validation, the corresponding E-FAM produced an average percentage of images correctly classified (accuracy) of $90,4\%(\pm 3.5)$. Table 1 lists the accuracy compared with other well-established classifiers in the literature, to know, Multi-layer Perceptron,[19] Naive-Bayes,[18] K-Nearest-Neighbor[18] and Support Vector Machine.[20]

Table 1. Classification accuracy.

Method	Accuracy
Multi-layer Perceptron	83.2±1.8%
Naive-Bayes	66.5±1.5%
K-Nearest-Neighbors	82.6±1.6%
Support Vector Machine	78.1±1.1%
Proposed method	90.4±3.5%

4.1. Discussion

This result represents an important step forward in the identification of coral species. We have here an example of an automatic approach confirming the prediction of a specialist with high accuracy. We intend to investigate those samples that are assigned to the incorrect species in the future, but for now it is remarkable how a well-planned scheme of machine learning as the reported here can be of great interest in helping specialist to attenuate the effect of subjective evaluations in this process.

It is also rather interesting to observe the role of fuzzy classifiers in this problem. We have here a typical situation where ambiguities arise even among experts in the area. The use of fuzzy structures in the classifier demonstrated to be more precise and robust in the adequate modelling of the expert judgment. Such result was expected from the nature of fuzzy logic, in which the parameters are more complete and can capture nuances that a conventional classifier could not represent. Another outcome that is worth to mention is the association of texture descriptors with the fuzzy classifier. Those descriptors are known to have high efficiency in describing images with patterns statistically well-defined. Nevertheless, the intrinsic randomness present in natural structures makes the direct (crisp) mapping between the descriptors and the classes in the training set employed by classical classifiers inefficient in various situations. In this context, this work opens the opportunity for deeper studies on the application of fuzzy neural networks for image classification in general, especially associated to texture descriptors.

5. Final Remarks

In this paper we introduced an approach to identify coral species based on a photographed image employing CLBP descriptors and an equivalent measure fuzzy associative memory. In preliminary experiments we obtained a satisfactory performance in the classification accuracy using E-FAM. In future works, we intend to test our proposal with other descriptors and classifiers.

Acknowledgments

J. B. Florindo was supported by CNPq (National Council for Scientific and Technological Development, Brazil) (Grant # 301480/2016-8), and FAPESP (The State of São Paulo Research Foundation) (Process 2016/16060-0). E. Esmi was supported by FAPESP (Process 2016/26040-7). M. Barbeitos was supported by Boticário Group Foundation for Nature Protection under grant no. 1040_20151.

References

1. N. M. d. Menezes, E. G. Neves, F. Barros, R. K. P. d. Kikuchi and R. Johnsson, **13**, 108 (2013).

2. J. E. N. Veron, *Corals of the World* (Australian Institute of Marine Sciences, 2000).
3. Z. H. Forsman, H. M. Guzman, C. A. Chen, G. E. Fox and G. M. Wellington, **24**, 343.
4. E. G. Neves, S. C. S. Andrade, F. L. d. Silveira and V. N. Solferini, **132**, 243 (2007).
5. F. L. D. Nunes, R. D. Norris and N. Knowlton, **6**, p. e22298 (2011).
6. Z. Guo, L. Zhang and D. Zhang, *IEEE Transactions on Image Processing* **19**, 1657 (2010).
7. T. Ojala, M. Pietikäinen and T. Mäenpää, *IEEE Transactions on Pattern Analysis and Machine Intelligence* **24**, 971 (2002).
8. J. J. Buckley and Y. Hayashi, *Fuzzy Sets and Systems* **66**, 1 (1994).
9. G. Birkhoff, *Lattice Theory*, 3rd edn. (American Mathematical Society, Providence, 1993).
10. E. Esmi, P. Sussner and S. Sandri, *Fuzzy Sets and Systems* **292**, 242 (2016).
11. E. Esmi, P. Sussner, H. Bustince and J. Fernández, *IEEE Transactions on Fuzzy Systems* **23**, 313 (2015).
12. P. Sussner, E. L. Esmi, I. Villaverde and M. Graña, *Journal of Mathematical Imaging and Vision* **42**, 134 (2012).
13. H. Bustince, M. Pagola and E. Barrenechea, *Information Sciences* **177**, 906 (2007).
14. J. Fodor and R. Roubens, *Fuzzy Preference Modelling and Multicriteria Decision Support* Fundamental Theories of Physics, Fundamental Theories of Physics (Springer, 1994).
15. F. T. Martins-Bedé, L. Godo, S. Sandri, L. V. Dutra, C. C. Freitas, O. S. Carvalho, R. J. Guimarães and R. S. Amaral, Classification of Schistosomiasis prevalence using fuzzy case-based reasoning, in *Proceedings of the 10th International Work-Conference on Artificial Neural Networks: Part I: Bio-Inspired Systems: Computational and Ambient Intelligence*, IWANN '09 (Springer-Verlag, Berlin, Heidelberg, 2009).
16. H. Bustince, E. Barrenechea and M. Pagola, *Fuzzy Sets and Systems* **157**, 2333 (2006).
17. J. Fan, W. Xie and J. Pei, *Fuzzy Sets and Systems* **106**, 201 (1999).
18. R. O. Duda and P. E. Hart, *Pattern Classification and Scene Analysis* (Wiley, New York, 1973).
19. S. Haykin, *Neural Networks: A Comprehensive Foundation*, 2nd edn. (Prentice Hall PTR, Upper Saddle River, NJ, USA, 1998).
20. C. Cortes and V. Vapnik, *Machine Learning* **20**, 273 (1995).

Tasks scheduling in computational grids: A proposal considering an uncertainty regime

Bruno M. P. Moura, Guilherme B. Schneider,[*]
Adenauer C. Yamin, Mauricio L. Pilla and Renata H. S. Reiser

*Federal University of Pelotas (UFPEL)
Centre for Technological Development (CDTEC),
Laboratory of Ubiquitous and Parallel Systems (LUPS),
Pelotas, RS, Rua Gomes Carneiro, 1 - 96010-610, Brazil*
{*bmpdmoura, gbschneider, adenauer, pilla, reiser*}*@inf.ufpel.edu.br*

Among the research challenges in computational grids, we need to make the system robust both the uncertainties of the different measures extracted from the computational infrastructure and the imprecision of the calculations related to the decision making. Scheduling tasks is a known NP-Hard problem. This paper provides an approach with Fuzzy Type-2 logics to treat uncertainties and dynamic behavior for scheduling tasks in grid environments, named Int-fGrid. The scheduler was validated through simulations in the SimGrid framework with a model of the GridRS architecture. Our results show that the Fuzzy Type-2 approach provides makespans up to 17.60% better than the best alternative tested scheduler XSufferage.

Keywords: Fuzzy logic; scheduling tasks; grid computing; bag-of-tasks.

1. Scope, Motivation and Objective

Nowadays, with the spread of high-speed networks, it is advisable to employ distributed computing environments Grid Computing (GC), consisting of a set of computers or interconnected heterogeneous processors via the Internet, motivating users and administrators clusters and supercomputers to project applications these environments [1].

In this context, it is necessary to implement robust scheduling systems to model the uncertainties extracted from the infrastructures used in GC [1]. Computational Power (CP), Communication Cost (CC) and Energy Consumption (EC) are among the factors that imply uncertainties in the scheduling of tasks in the GC.

The vague and imprecise association of the PC, CC, and EC factors

---

[*]Scholarship student AT/CNPq.

produce new interval membership functions, so the resulting Type-2 Fuzzy Sets (T2FSs), in this work, Interval-valued Fuzzy Sets (IvFSs), promote a description combining the expert-modeled uncertainty obtained through simulations.

This work aims at presenting *Int-fGrid*, a module for decision-making on the Scheduling of Tasks in Computational Grids (STCGs) using Interval-valued Fuzzy Logic (IvFL). This initiative extends previous works [2, 3] by exploring a T2FSs approach in the analysis of Machine Priority (P) to allocate the tasks in computational resources while dealing with the uncertainties and inaccuracies associated with the CP and CC variables.

The section 1 deals with the contextual foundations of work. Section 2 introduces basic concepts of type-2 fuzzy logics. In section 3, details of *Int-fGrid* component and its conception are discussed, including database, fuzzification, rule base, inference and defuzzification. Section 4 describes the experimental evaluation. Finally, section 5 presents the conclusions and further work.

2. Type-2 Fuzzy Logic Related Aspects to the Int-fGrid Framework

Type-2 Fuzzy Logic (T2FL) was introduced by Lotf Zadeh in 1975 as an extension of the traditional Fuzzy Logic [4] modeling the inherent uncertainties related to the antecedent and consequent membership functions, enabling the manipulation of imprecise terms throughout its fuzzy inference system [5].

T2FSs can be used in situations where there exists uncertainty about the degrees, forms or parameters of the membership functions [6], providing potential strategy on the uncertainty treatment in information models obtained from distinct specialists and/or extracted from simulators.

Moreover, IvFL is based on IvFSs, with an purpose of treating the problem by allowing to specify only an interval $X = [\underline{\mu_A}(x), \overline{\mu_A(x)}]$ as the membership degree of such element x in a IvFS A [7]. Thus, by complementing FSs theory, IvFSs theory can model vagueness with an additional ability to consider imprecision (non-specificity) as two important aspects of uncertainty reflecting the length of the interval membership degree.

Let $\mathbb{U}$ be the set of all real intervals in the unitary interval $U = [0, 1]$ and the partial order: *Product order*: $X \leq Y$ iff $\underline{X} \leq \underline{Y}$ and $\overline{X} \leq \overline{Y}$ [8].

By [7], a function $\mathbb{T}(\mathbb{S}) : \mathbb{U}^2 \to \mathbb{U}$ qualifying fuzzy union, is an **interval-valued t-norm (t-conorm)** if it is commutative, associative, monotonic w.r.t. the product order and has $\mathbf{1} = [1,1]$ ($\mathbf{0} = [0,0]$) as the neutral element.

An interval function $\mathbb{N} : \mathbb{U} \to \mathbb{U}$ is an **interval-valued fuzzy negation** if, for all $X, Y \in \mathbb{U}$, it holds that: (i) N1: $\mathbb{N}(\mathbf{0}) = \mathbf{1}$ and $\mathbb{N}(\mathbf{1}) = \mathbf{0}$; (ii) N2: If $X \geq Y$ then $\mathbb{N}(X) \leq \mathbb{N}(Y)$; and (iii) N3: If $X \subseteq Y$ then $\mathbb{N}(X) \subseteq \mathbb{N}(Y)$.

A system based on IvFL can estimate input and output functions by using heuristic and interval techniques. Its main blocks are described below:

1. **Fuzzification Interface**: The fuzzification process based on IvFL is performed according to the nature and definition of such type-2 set, associating an input value with an interval function and not simply with a single value of U. In other words, it is inserted to the mechanism of inference the uncertainty regarding the input membership function. Thus, for each IvFS A and $n \in \mathbb{N}$, the transformation of an input vector $\mathbf{x} = (x_1, x_2, ..., x_n)$ to a pair of vectors in $\mathbb{U}^n$ is expressed as $\left(\overline{\mu_A(x_1)}, \ldots, \overline{\mu_A(x_n)}\right), \left(\underline{\mu_A(x_1)}, \ldots, \underline{\mu_A(x_n)}\right)$.
2. **Rule Base (RB)**: Composed by rules that classify the Linguistic Variables (LVs) according to the IvFSs;
3. **Logic Decision Unity**: Executing inference operations between the input data and the rules defined in the RB to obtain a performed by the system action;
4. **Defuzzification**: Two main stages of IvFSs are considered:

   (i) **Type Reducer** has the function of transforming a IvFSs into fuzzy sets, that is, it tries the best fuzzy set that represents the type-2 fuzzy set, and that must satisfy the following premise: When all uncertainties disappear, the result of System Based on Fuzzy Rules 2 (SBFR2) is reduced to a System Based on Fuzzy Rules 1 (SBFR1) [9, 10];

   (ii) **Defuzzification**: An defuzzified output of SBFR2 is given by the average of limits points y_L and y_R:

$$y(x) = \frac{y_L + y_R}{2}, \forall x \in \chi, \tag{1}$$

when values y_L and y_R can be calculated using the iterative method of Karnik and Mendel (KM algorithm) [11]. Therefore, defuzzification step can still be obtained through the use of a conventional method such as the centroid to get the final value of the inference.

3. Int-fGrid: Modeling Type-2 Fuzzy System

The *Int-fGrid* is responsible for the scheduling of Bag-of-Task (BoT) homogeneous tasks. *Int-fGrid* system considers a Rule Base is acting on three steps: Fuzzification, Inference, and Defuzzification, returning as output the priority of each machine. The modeling of the type-2 fuzzy system was performed using the Interval Type-2 Fuzzy Logic System Toolbox (IT2FLT) module [12, 13].

In BoT applications [14], each task is independent of the others, and communication only occurs when the task is deployed and when it ends, returning results. Therefore, the order in which tasks are executed does not affect the final output of the system.

Int-fGrid Data Base - Membership Functions:

Through the study of variables with a specialist, each one of LVs was associated with three distinct FSs, using the trapezoidal graphical representation to corresponding membership functions. The setting reading of the simulated grid computing environment is performed to measure the CP and CC. These values are then applied to a standard scale adopted, considering the interval $[0; 10]$, as shown in Figure 1(a) for CP and Figure 1(b) CC to obtain their degrees of membership.

The Linguistic Terms (LTs) defining the FSs of this variable CP are stated as follows: "Limited" (CPL), "Reasonable" (CPR) and "High" (CPH - best case). Being $CP = a$ and $a \in [0; 10]$.

The communication is measured between the machine containing the expected application that will be sent to the grid and a machine of each Grid cluster. The LTs to the FSs defined for this variable are: "Small" (SCC - best case), "Average" (ACC) and "Big" (BCC). Being $CC = b$ and $b \in [0; 10]$. It is considered that the communication costs among many processors of the same cluster are equal.

The machines output (Priority) is also adapted to a standard scale, as shown in Figure 1(d), and the LTs for FSs used are: "Low" (LP), "Medium" (MP) and "High" (HP - best case). Being $P = c$ and $c \in [0; 10]$.

(i) **Fuzzification:** mapping the input values (already set for an observed scale in the section 3) to the fuzzy domain.

(ii) **Rule Base:** considering that T2FL in system performance is subjected to the rules that describe the consistent control strategy [15]. The RB, takes three factors for its construction: (1) the LVs that appoint FSs, making modeling closer to the real world; (2) it is also considered that logical connections of the "AND" type are used to create the relationship between the input variables; (3) the resulted implications are *modus ponens*

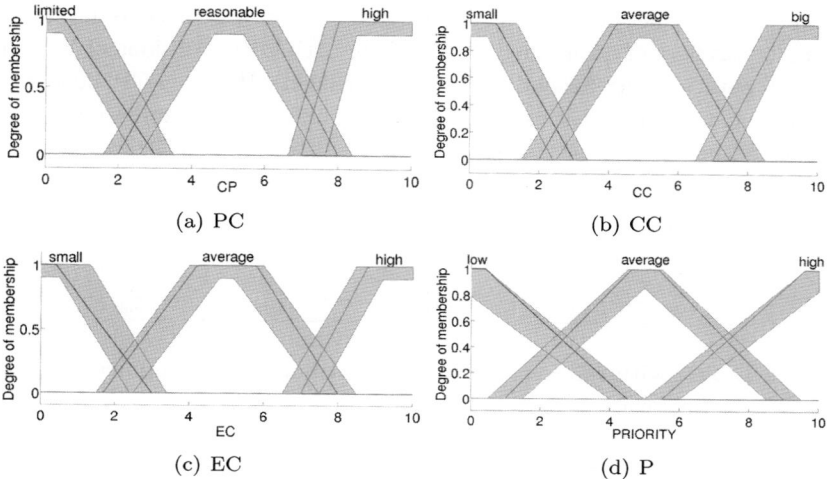

Fig. 1. Membership Function for LV's: (a) PC, (b) CC and (c) Priority to default scale.

type (affirmative): "if X is A, then Y is B".

(iii) **Inference:** disposing transactions between the FSs, combining rules background with their implications using the *generalized modus ponens* operator. The process occurs in three steps: (i) Operation Fuzzy Application; (ii) Implication Fuzzy Method Application; (iii) Aggregation of Fuzzy Method Application.

(iv) **Defuzzification:** acting in the research progress, the region transformation happens to be the result of the inference in a discrete value (which is the priority). The defuzzification technique in the Int-fGrid model was the center of the area. This method calculates the centroid (x) of the area consisted of the output of fuzzy inference system (connection of all contributions rules show in the section). The centroid is calculated by the following formula:

$$u = \frac{\sum_{i=1}^{N} u_i \mu_{OUT}(u_i)}{\sum_{i=1}^{N} \mu_{OUT}(u_i)} \qquad (2)$$

4. Proposal Evaluation

Providing test cases which were developed considering the GridRS infrastructure in SimGrid [16]. The structure of GridRS is composed of clusters of four Brazilian universities: UFPEL, PUCRS, UFRGS, and UFSM. Each of these clusters contains a set of homogeneous machines, but the characteristics vary among sites, thus making the structure heterogeneous. In

Table 1 the values used in the grid configuration are presented: *(i)* machine quantifiers of each cluster, *(ii)* CP in MFLOPS (Millions of Floating Point Operations per Second), *(iii)* Standardized CP = $(10*CP/3000)$, *(iv)* communication channels used in the interconnection of simulated GridRS infrastructure.

Table 1. GridRS features.

Institution	Machines	CP (MFLOPS)	Standardized CP	Channels
UFPEL	12	2713	9.04	Link 1
UFRGS	17	648	2.28	Link 2
PUCRS	15	1763	5.87	Link 3
UFSM	12	1942	6.47	Link 4

See Table 2 with values related to the attributes of communication channel: *(i)* Bandwidth, *(ii)* Latency.

Table 2. Communication channel characteristics.

Link	From	To	Bandwidth	Latency (us)
Link 1	UFPEL	UFPEL	1GBps	2.241
Link 2	UFPEL	UFRGS	60MBps	14.2016
Link 3	UFPEL	PUCRS	40MBps	12.169
Link 4	UFPEL	UFSM	30MBps	16.361

In the *Int-fGrid* evaluation, variations were considered quantifying: *(i)* tasks that constituted the applications together with corresponding, *(ii)* computational cost, and *(iii)* communication cost, as set out in Table 3. By reproducing perturbations in the fuzzy system input, fluctuations in simulated GridRS communication channels were applied, varying in the range of 10% to 40% of their capacities. The perturbation generated for CP was of 10% representing the load of the middleware for grid management. The execution of the evaluations are simulated in the SimGrid for *Round-Robin*, *Small-Latency* and *XSufferage* as three well-known algorithms and additionally, for *Int-fGrid*. The results obtained with the use of *Int-fGrid* to assist the scheduler in the decision making are in agreement with the results generated through the *fGrid* [2, 3], yielding as output more approximate values about the configurations applied in the framework SimGrid.

However, in the evaluations, the execution times for the *fGrid* are not highlighted about the *Int-fGrid*, since they were the same in all executions, generating as output a priority list in the same order as provided through the application of *Int-fGrid*, but with different values. Based on the IvFL approach, *Int-fGrid* produces the list of nodes with respective priorities through the application of Fuzzification, Inference, and Defuzzification.

Additionally, time executions obtained in the simulations are shown in the Table 3 for *Int-fGrid*, Round-Robin, Small-Latency, and XSufferage.

Table 3. Task configuration and evaluation numeric results.

Execution	Tasks	CP	CC	Int-fGrid	Round-Robin	Small-Latency	XSufferage
1	100	5	2	3.79713	5.05237	3.79713	4.02844
2	200	10	5	21.6356	31.3835	29.1177	26.2032
3	300	50	10	65.6118	94.9732	89.7107	79.7996
4	400	5	2	17.3394	23.9912	23.1075	20.4506
5	500	10	5	56.3208	78.8722	77.994	67.7267
6	1000	50	10	225.419	317.5	314.673	272.982
7	2000	5	2	90.3008	126.81	125.889	108.914
8	3000	10	5	337.995	481.639	480.702	411.749
9	4000	50	10	905.775	1277.49	1277.34	1098.42
10	5000	5	2	227.827	319.146	318.371	274.171
11	6000	10	5	677.287	963.451	962.963	824.151
12	7000	50	10	1586.84	2239.14	2243.83	1925.85

CP = Computational Power of Task (MFLOPS)
CC = Communication Cost of Task (MBps)

Figure 2 depicts the makespan barplot in seconds for the executions considering the configuration of the cases described in the Table 3, applying *Int-fGrid*, Round-Robin, Small-Latency, and XSufferage. Each column shows the makespan of all executions for a given test case using one of the schedulers. The barplot is grouped by category, where each represents the executions of each scheduler. For all results, the IvFL approach of *Int-fGrid* produces smaller makespan than the three other schedulers. *Int-fGrid* provides a makespan 29.3% smaller than average for Round-Robin, 29.11% for Small-Latency and 17.6% smaller than average for XSufferage for the 12th test case.

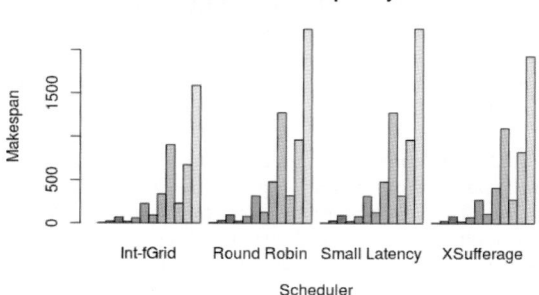

Fig. 2. Makespan for Int-fGrid, Round-Robin, Small-Latency and XSufferage.

5. Conclusion

In this work, we presented *Int-fGrid*, a scheduler for Computational Grids using Type-2 Fuzzy Logics. We simulated it and compared results against three well-known scheduling algorithms, Round-Robin, Small-Latency and XSufferage using SimGrid. The main advantage of *Int-fGrid* is to handle the uncertainties directly in the schedulers avoiding the rescheduling and migration of tasks. In most cases, *Int-fGrid* achieved better makespans compared to the XSufferage schedules, achieving up to 17.60% better results, compared to Small-Latency made gains of up to 29.28% and finally to Round-Robin achieved up to 29.10% better makespans.

Future work we intend to carry out simulations considering the new EC variable in the SimGrid, as well as implement the scaling module in a real GC environment.

References

1. T. S. Somasundaram, K. Govindarajan, U. Kiruthika and R. Buyya, *The Journal of Supercomputing* **68**, 509 (2014).
2. B. Moura, Y. Soares, L. Sampaio, R. Reiser, A. Yamin and M. Pilla, *Fuzzy System Modeling for Task Scheduling in Computational Grids* (World Scientific, Roubaix, France, 2016), Roubaix, France, pp. 806–811.
3. B. Moura, Y. Soares, L. Sampaio, R. Reiser, A. Yamin and M. Pilla, fGrid: Uncertainty variables modeling for computational grids using fuzzy logic, in *2016 IEEE International Conference on Fuzzy Systems (FUZZ-IEEE)*, (IEEE, July 2016).
4. L. Zadeh, *Information Sciences* **8**, 199 (1975).
5. J. M. Mendel, Fuzzy sets for words: a new beginning, in *Fuzzy Systems, 2003. FUZZ '03. The 12th IEEE International Conference on*, (IEEE, 2003).
6. N. N. Karnik and J. M. Mendel, Introduction to type-2 fuzzy logic systems, in *1998 IEEE International Conference on Fuzzy Systems Proceedings. IEEE World Congress on Computational Intelligence*, (IEEE, 1998).
7. M. Gehrke, C. Walker and E. Walker, *International Journal of Intelligent Systems* **11**, 751 (1996).
8. E. Klement, R. Mesiar and E. Pap, *Fuzzy Sets and Systems* **143**, 5 (2004).
9. W. W. Tan and T. W. Chua, *IEEE Computational Intelligence Magazine* **2**, 72(Feb 2007).
10. D. Wu and M. Nie, Comparison and practical implementation of type-reduction algorithms for type-2 fuzzy sets and systems, in *FUZZ-IEEE*, (IEEE, 2011).
11. P. Rizol, L. Mesquita and O. Saotome, *Revista Sodebras* **6**, 27 (2011).
12. Institute of Technology and Baja California Autonomous University, *"Users Guide, Interval Type-2 Fuzzy Logic Toolbox For Use with MATLAB"*, (2005-2008).
13. J. R. Castro, O. Castillo and L. G. Martínez, *Engineering Letters* **15**, 89 (2007).
14. G. Terzopoulos and H. D. Karatza, Bag-of-tasks load balancing on power-aware clusters, in *2016 24th Euromicro International Conference on Parallel, Distributed, and Network-Based Processing (PDP)*, Feb 2016.
15. G. J. Klir, *Uncertainty and Information: Foundations of Generalized Information Theory* (Wiley-Interscience, 2005).
16. H. Casanova, A. Giersch, A. Legrand, M. Quinson and F. Suter, *Journal of Parallel and Distributed Computing* **74**, 2899(June 2014).

Safety helmet recognition based on deep convolution neural networks

Ningning Zhou

School of Computer, Nanjing University of Posts and Telecommunications
Nanjing, Jiangsu, 210023, China

Guofang Huang

State Key Laboratory of Smart Grid Protection and Control Laboratory
Nanjing, Jiangsu, 210023, China

Shaodong Shi

School of Computer, Nanjing University of Posts and Telecommunications
Nanjing, Jiangsu, 210023, China

This paper introduces deep learning for safety helmet recognition. First, a pedestrian detection algorithm based on HOG+SVM is discussed. Then, a safety helmet recognition method based on deep convolution neural networks is presented. Simulation results show that compared with a traditional neural network method, the proposed method improves the automatic recognition accuracy of helmets and shows better adaptability to the environment.

1. Introduction

In industrial production, a safety helmet is one of the most common and practical personal protective measures that can effectively prevent and mitigate the risk of external harm to the human head. However, some operators lack safety awareness, especially in terms of wearing basic protective equipment (such as helmets), which greatly increases operational risk. But security is not only guaranteed by the system, but also requires some corresponding technical approaches. With the widespread application of video surveillance systems in many production fields, it's meaningful and feasible to detect the wearing of safety helmets and other protective equipment through video. Many researchers [1], [2] have shown an interest in this area in recent years.

Deep learning [3], is one of the most important breakthroughs in the field of artificial intelligence in the last ten years. The difference from traditional shallow machine learning is that deep learning pays more attention to the layers of model structure and the importance of feature learning. The primary deep learning models include the Auto-encoder [4], RBM [5] (Restricted Boltzmann Machine), DBNs [6] (Deep Belief Networks), CNNs [7] (Convolution Neural Networks), and Biological Heuristic Model [8]. Among them, the convolution operation in CNNs preserves the spatial information of the image, which makes it more suitable for expressing the image.

This paper introduces deep learning for safety helmet recognition. Considering the relationship between the human and the safety helmet. First, a pedestrian detection algorithm based on HOG+SVM is discussed. Then, a safety helmet recognition method based on DCNNs (Deep Convolution Neural Networks) is presented. Experimental results show that compared with a traditional neural network method, the proposed method improves the automatic recognition accuracy of helmets and shows better adaptability to the environment.

2. Pedestrian Detection Algorithm Based on HOG+SVM

Identifying specific objects in an image is difficult because the object may vary in appearance. Since the local object appearance and shape can be described by the distribution of intensity gradients or edge directions, we use an algorithm based on HOG+SVM to detect the pedestrian. First, we divide the image into small connected regions called cells; for pixels within each cell, a histogram of gradient directions is generated. Concatenation of these histograms forms the HOG descriptor. For improved accuracy, local histograms can be contrast-normalized by calculating a measure of intensity across a larger region of the image, called a block, and using this value to normalize all cells within the block. This normalization results in a higher invariance to changes in illumination and shadowing. This descriptor is then applied to the SVM classifier. A flowchart of the pedestrian detection algorithm based on HOG+SVM is shown in Figure 1.

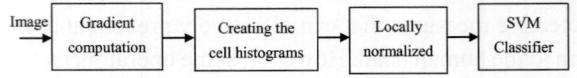

Figure 1. Flowchart of pedestrian detection algorithm.

3. Safety Helmet Recognition Based on DCNNs

3.1. *Structural design and parameter optimization of DCNNs*

(1) Analysis of network structure and parameter settings

Based on studies of the recognition and training process for DCNNs, we know that network structure selection and the determination of related parameters are key issues of the recognition system. Among them, the structure of the hidden layer is the main problem in network structure. Parameters include size of convolution kernel, number of feature maps, initialization of weights and bias, number of samples, and number of iterations [9] [10].

① Convolution kernel size and network layer number

The convolution kernel, weight matrix W, is used in the convolution operation (W is denoted by size N×N). The size of matrix W is the same as the image region dot-multiplied. N is odd, usually N = 3, 5, 7, 9, 11, etc.

Network depth is the sum of the convolution layers, pooling layers, and full connection layers. Generally, in a certain range, increasing the neural network depth can effectively improve recognition accuracy. But beyond this range, as the number of layers increases, not only will the network structure become complex, but the recognition accuracy will also gradually decrease. Compared with selecting more network layers, setting the appropriate hidden layer nodes is a more suitable approach for reducing the error rate. Feature extraction is the core problem in image recognition. A simple description is a suitable set of data that describe the image. It is determined by the image size. Simpler is more representative, and therefore better. Complex means too many invalid features that may hide information. As a result, the network structure is determined by the size of the training sample and the size of the convolution kernel.

② Number of feature maps in hidden layer

The number of feature maps in the hidden layer is the number of feature maps in the convolution and pooling layers. Selecting the appropriate number of feature maps in hidden layers is a more suitable approach for reducing the error rate than selecting the network structure. It is easier to adjust and improve the training process. By considering the accuracy requirement and complexity of the network, we use two convolution layers and two pooling layers. The pooling operation is performed by convoluting the convolution layer with a 2×2 convolution kernel. The pooling layer is determined by the upper convolution layer feature map. In general, the number of feature maps of the convolution layer is set based on experience.

③ Initialization of weights

Weight initialization is the initialization of weights for the convolution kernels and the weight matrix in the fully connected layer. The initial weight has an impact on training speed and recognition accuracy. A range of initialization that is too large will make training impossible. The initial weight value is a random number in [-1,+1]. The weight is adjusted continuously during the reverse optimization process until an optimal solution is achieved.

④ Number of batch samples

In order to reduce the number of iterations and improve the recognition rate, training is done by grouping. The number of samples is the number of samples in each group. In general, under the same total experimental data conditions, the fewer the number of batch samples, the more training that is needed, the higher the recognition accuracy, and the more iterations. The number of batch samples is selected based on the recognition accuracy and efficiency.

⑤ Number of iterations

The number of iterations is the number of training runs for all training data. The lower the iteration number, the lower the recognition rate, and the shorter the time. The higher the iteration number, the stronger the generalization ability, the higher the recognition accuracy, and the longer the time. However, too many iterations will lead to overfitting, which will cause robustness deterioration. Selecting the correct recognition accuracy is also a key point to training.

It is important to determine the DCNN structure by the complexity of the classification model which aims to obtain the highest recognition accuracy. In order to determine the proper convolution neural network structure, this paper applies a large number of comparative experiments to determine the size of the convolution kernel, number of feature graphs, and depth of the convolution neural network. Based on experimental results, we select the DCNNs shown in Figure 2.

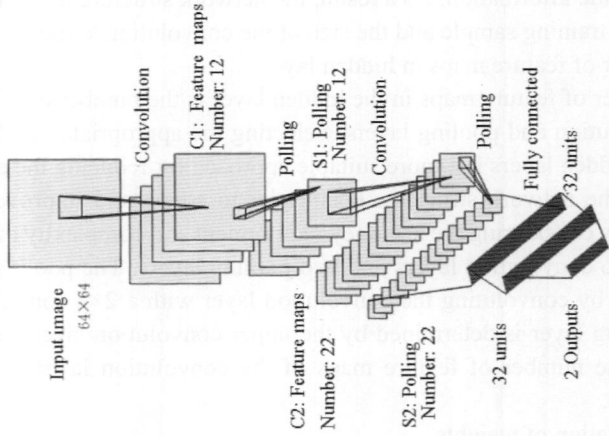

Figure 2. Structure of DCNNs in this study.

3.2. Cost function

Variance is typically used as a cost function. In order to overcome shortcomings of the traditional cost function whose learning speed is slow, this paper introduces cross-entropy [11], [12] as a cost function:

$$C = -\frac{1}{n}\sum_{x}[y\ln a + (1-y)\ln(1-a)] \qquad (1)$$

This cost function satisfies ① function value is greater than 0 and ② when $a = y$, cost $= 0$.

The derivative of equation (1) gives:

$$C=-\frac{1}{n}\sum_{x}[y\ln a + (1-y)\ln(1-a)] \Longrightarrow \frac{\partial C}{\partial w_j} = \frac{1}{n}\sum_{x}(\sigma(z) - y) \qquad (2)$$

Equation (2) shows that the learning speed of the network depends on C, i.e., the output error. The larger the error, the more updates and the faster the study. When the training error is relatively small, learning is slow.

3.3. Regularization constraint

We use L2 [13] regularization constraints, as shown in equation (3). L2 regularization is achieved by directly penalizing the square of all parameters in the target:

$$C = C_0 + \frac{\lambda}{2n}\sum_{w} w^2 \qquad (3)$$

where C_0 denotes the original cost function and n nis the number of training samples. λ is the regularization strength used to regulate the proportion of the regular term.

The derivative of equation (3) gives:

$$\frac{\partial C}{\partial w} = \frac{\partial C_0}{\partial w} + \frac{\lambda}{n}w$$

$$\frac{\partial C}{\partial b} = \frac{\partial C_0}{\partial b} \qquad (4)$$

According to equation (4), adding regular entries does not affect bias b updates. However, for weight w, there exist subtle changes. For the stochastic gradient descent method based on mini-batch, the update of weight w and bias b is different.

L2 regularization has the intuitive explanation that it severely penalizes the peak weight vector and prefers the discrete weight vector. Since weight multiplies the input, this encourages the use of all input rather than some input.

During the process of the gradient descent parameter update, using L2 regularization means that each weight linearly attenuates towards 0. The weight will become smaller by regularization constraint. In a sense, this can reduce network complexity and fit data better. Experimental results show that fitting by L2 regularization is better than with non-regularized neural networks.

3.4. *Activation function*

Before the ReLu [13] function is considered, the sigmoid and tanH functions are usually used as activation functions. When there are more layers in the neural network, the learning rate becomes smaller during the reverse updating. The problem is that the gradient is not stable. In order to overcome these shortcomings, ReLu which is shown in Figure 3 has become popular. Compared with activation functions similar to sigmoid, ReLu is faster and more efficient for complex neural network deep learning on large and complex data sets. But ReLu can cause neuronal necrosis, i.e., neurons may go to a relatively bad state and become inactive to all input. In this state, no gradient flows back through the neuron, so the neuron becomes permanently stuck. This greatly reduces model capacity.

Leaky ReLus is used to solve the problem of ReLu necrosis [14]. In Leaky ReLus, when x < 0, the output is not 0, but a very small value [15] as shown in Figure 3 (right), where a is a small constant.

$$Leaky\ ReLus(x) = \begin{cases} x & x > 0 \\ ax & x \leq 0 \end{cases} \quad (5)$$

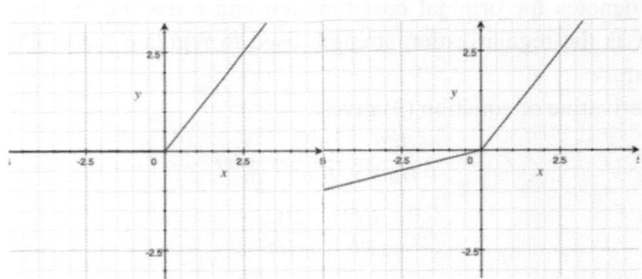

Figure 3. ReLu (left) and Leaky ReLus (right).

In this approach, the data distribution is optimized. The value of a partial negative axis can be reserved so that the information of a negative axis is not completely lost. Therefore, we select Leaky ReLu as the activation function, where a is 0.02.

4. Experimental Results and Analysis

4.1. *Data set*

The data set used in this paper is a self-built data set. The data set comes from two sources. One is from web crawling using "workers", "construction workers", and "safety helmet workers" as key words. The other is taken by us. All images are normalized to 64×64 and labeled first. Images in the data set have complex backgrounds and helmets in the images have different sizes, lighting, poses, and angles. The data set is divided into two parts: training set and testing set. The training set contains 9000 samples, 4500 of which are positive samples (workers wearing helmets) and 4500 negative samples. The testing set contains 1000 images.

4.2. *Training process for DCNNs*

First, use HOG+SVM to extract the image region that may contain the target (64×64) from the input image. Then, input the image region to the DCNNs as shown in Figure 3. The training process is as follows:

(1) Set classification labels on data set;

(2) Based on size of image region (64×64) in input layer, set convolution kernel size and network layer number as follows: convolution kernel size is 7×7, two convolution layers, and two polling layers;

(3) Compress weight w and bias b to [0,1] by normal distribution rule;

(4) Remove first batch of data and input to training network;

(5) Compare output vectors with vectors in labels and calculate errors;

(6) Compute weight adjustment and offset adjustment according to error and learning rate;

(7) Update weight and offset based on errors;

(8) If it does not satisfy the iteration stop law, continue to learn;

(9) Training is complete.

The neural network is trained by the above steps and we obtain a basic classifier. The basic training classifier is used for intensive scanning on a large number of test sets. Erroneous samples are added to the training set. We use the training set to train the neural network again to obtain the final classifier.

4.3. *Experimental results*

4.3.1. *Training times*

With increasing training times, the error in the training data is less and the error in the testing data is more, as shown in Figure 4. This shows that the neural network is overfitting. We use a minimum error of testing samples to end the iterations. In addition, to prevent falling into a local minimum, when 50 consecutive errors are less than 0.01, we stop the iterations.

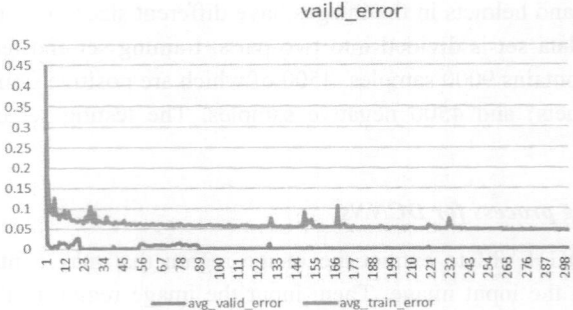

Figure 4. Training error and valid error for different iteration times.

4.3.2. *Recognition results*

The Linear Regression and BP neural networks were used in comparison to evaluate the performance of our method. Recognition accuracy rates (average of 10 times) for 1000 testing samples are shown in Table 1. Results show that the proposed method outperforms the Linear Regression and BP neural network.

Table 1. Recognition accuracy rates.

	Recognition accuracy rates
Linear Regression	96.34%
BP	98.75%
DCNNs	99.37%

5. Conclusion

In order to improve the automatic detection accuracy of safety helmets in videos, a method of pedestrian detection based on HOG+SVM was studied. A safety helmet recognition method based on DCNNs was presented. The structure of DCNNs, parameter settings, cost function, regularization constraint, and activation function were discussed in detail. Experimental results show that compared with a traditional neural network method, the proposed method improves the automatic recognition accuracy of helmets and has better adaptability to the environment.

Acknowledgments

This work is supported by the State Key Laboratory of Smart Grid Protection and Control of China (No. 9) and the National Natural Science Foundation of China No. 61170322, No. 61373065, and No. 61302157.

References

1. X. Liu and Xining Y. Journal of East China University of Science and Technology (natural Science Edition), V40, 365(2014).
2. G. C. Feng, Y. Y. Chen. Machine Design and Manufacturing Engineering, V44, 39(2015).
3. G. E. Hinton, R. R. Salakhutdinov. Science, V313, 504(2006).
4. Bourlard H., Kamp Y. Biological Cybernetics, V59, 291(1988).
5. D. Rumelhart, J. Mcclelland. J. MIT Press, 194(1986).
6. B. G. E. Hinton, S. Osindero et al. Neural Computation, V18, 1527(2014).
7. Y. Lecun, L. Bottou et al. Proc. IEEE, V86, 2278(1998).
8. Y. Huang, K. Huang et al. IEEE Transactions on Systems Man & Cybernetics V41, 1668(2011).
9. L. J. Ba, R. Caruana. Advances in Neural Information Processing Systems, 2654(2013).
10. Y. Y. Li. Shenyang University of Thechnology, 2016.
11. M. Kline, L. Berardi. Neural Computing & Applications, V14, 310(2005).
12. F. ltiparmak, B. Dengiz. European Journal of Operational Research, V199, 542(2009).
13. M. C. De, E.V. De et al. Journal of Complexity, V25, 201(2009).
14. W. Shang, K. Sohn et al. Proc. International Conference on Machine Learning. 2217(2016).
15. K. He, X. Zhang et al. Proc. IEEE International Conference on Computer Vision. IEEE, 1026(2015).

Research on optimal trajectory curve of intelligent vehicles based on neighborhood system

Xing Wang, Hailiang Zhao and Zhigang Wang

Department of Information & Computation Science, School of Mathematics
Southwest Jiaotong University, Chengdu, 610031, China

In order to select the optimal trajectory curve of intelligent vehicles, the theory of the neighborhood system is adopted. Firstly, we transform the complex dynamic control process into series ones of simple static control. And the trajectory planning of the intelligent vehicle is simplified to a superposition of the trajectory planning within a series of standard feasible neighborhoods. Secondly, the bending resistance of a curve is proposed by using integral of the curve curvature for the search of the optimal trajectory curve in a standard feasible neighborhood. Then an evaluation model is presented for the optimal trajectory curve in the neighborhood based on the bending resistance index. Finally, a satisfactory trajectory curve is obtained by using interpolation method. Simulation result shows that the neighborhood system theory is a feasible method and can effectively select the satisfactory trajectory of intelligent vehicles.

Keywords: Intelligent vehicle; automatic driving; neighborhood system; optimal trajectory; obstacles avoidance.

1. Introduction

The research on intelligent vehicle is various, and the trajectory planning theory[1-3] of intelligent vehicles is an important part. For the trajectory planning of intelligent vehicles, how to find the optimal trajectory curve is the core of this problem.

Based on neighborhood system control theory, a decision-making and select process for an optimal trajectory of an intelligent vehicle, which is dynamic and complex from a macroscopic point of view, are decomposed into a series of static and simple ones from a microscopic point of view. Its main idea is to reduce the large circumstance of a vehicle in motion to a local small standard feasible neighborhood of the vehicle, which can be regarded as a stable neighborhood in a short time. Therefore, the movement of the vehicle in a complex road is transformed into a superposition of ones in a series of standard feasible neighborhoods.

---

*This work is supported by the National Natural Science Foundation of China (61473239) and (61402382).

To plan an optimal trajectory in a standard feasible neighborhood, we present some concepts that can characterize the features of a curve, such as the bending resistance and the length index for a segment of a curve, etc. The integral of the curve curvature is used to express the curve bending resistance index, and the length index of the curve is expressed by the arc length. These indexes can be used to indicate the feelings of passengers on vehicles. According to the indexes above, an optimal trajectory curve model is proposed. And finally, a trajectory curve based on Hermite interpolation method is obtained. Comparing result with other commonly used trajectory curves shows that the curve established in this paper is a satisfactory trajectory curve.

2. Neighborhood System for Intelligent Vehicle

Intelligent vehicles will be equipped with sensors[4] of all kinds because of the various needs[5] of the people. When moving on a road, these sensors give various information of the current road to determine a safe range of the vehicle, which is a feasible neighborhood for the vehicle, defined as follows:

Definition 2.1 (Feasible Neighborhood[6]**)** Assume $Q(x_0, T)$ is a bounded open region, if $\forall z \in Q(x_0, T)$, $\exists \{x(t); x_0, T\} \in \Gamma(x_0)$, such that $z \in \{x(t); x_0, T\}$ and $\{x(t); x_0, T\} \subseteq Q(x_0, T)$, then call $Q(x_0, T)$ a feasible neighborhood of state x_0, while T is called holding time of the feasible neighborhood, $\Gamma(x_0)$ is a set of trajectory whose initial state is x_0.

It is difficult to describe the shapes and math expressions of feasible neighborhoods. Therefore, it is necessary to establish a standard feasible neighborhood of vehicle. There is no unified definition of the standard feasible neighborhood. The standard feasible neighborhood of this paper is as follows:

Standard feasible neighborhood: a limited number of specific neighborhoods in feasible neighborhoods.

The standard feasible neighborhood is proposed in order to facilitate the model building and mathematical description in the process of computer simulation. According to different purposes, standard feasible neighborhood of different shapes and sizes can be established.

The concept of neighborhood system control of intelligent vehicles [6-9] mainly consists of two stages. First, the control scope is reduced to the local feasible neighborhood, and then the control in the locally feasible neighborhood is simplified to a limited number of standard feasible controls in the neighborhood. Specific steps are as follows:

1) Select the appropriate neighborhood system according to the characteristics of observed variables in the control process;

2) To determine a feasible neighborhood in the current state through some optimization methods;

3) In a feasible neighborhood, according to some planning methods to give a satisfactory decision and the implementation of new decision-making behavior;

4) Keep the same behavior for the duration of the feasible neighborhood until the next decision is made;

5) The above process is a complete decision cycle. When a cycle is completed, a new cycle begins with steps (2) through (4) in turn;

It can be seen that the control process in each neighborhood is repeatedly superimposed, thus constituting the entire control process. The model decomposes the dynamic decision-making process in complex macroscopic environment into a series of simple static decision-making processes in the neighborhood and simplifies the whole decision-making process so that it faces a limited and local simple environment instead of infinite and complex world.

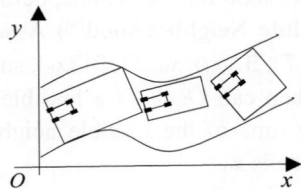

Figure 1. Intelligent vehicle dynamic trajectory planning.

As shown in Figure 1, the driving status of the intelligent vehicle at three different times is given. Each rectangle is a rectangular standard feasible neighborhood established by the intelligent vehicle at this moment. The trajectory planning of the intelligent vehicle on the road is the cumulative process of trajectory planning within every standard feasible neighborhood.

3. Theoretical Analysis of Trajectory Curve

3.1. *Evaluation of optimal trajectory curve*

As shown in Figure 2, three curves are given, where the large rectangle is the rectangular standard feasible neighborhood, the small rectangle is the outline of the car body.

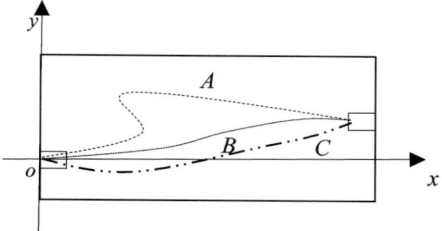

Figure 2. Example.

For the trajectory equation $y = y(x)$ of the trajectory curve L, the following two indicators are given in this paper:

Definition 3.1 (Length Index) Define

$$J = J[y(x)] = \int_{x_0}^{x_1} \sqrt{1+(y'(x))^2}\, dx \tag{1}$$

J is called the length index of the curve L.

Definition 3.2 (Bending Resistance Index) Define

$$K = K[y(x)] = \int_{x_0}^{x_1} \frac{|y''(x)|}{\sqrt{(1+y'(x))^3}}\, dx \tag{2}$$

K is called the bending resistance index of the curve L.

Specific description is as follows:

The length index measures the driving length of different trajectories that are same at the beginning and end positions. Generally speaking, for a curve with a given start and end point, the shorter the trajectory length is, the better the driving efficiency along the curve is.

Bending resistance index mainly depends on the smooth running trajectory, easy to know the smaller the curvature, the smoother the trajectory. Therefore, the bending resistance index can measure whether the curves of different trajectory drive smoothly.

From the definition, we can see that the bending resistance index depends only on the inherent characteristics of the curve. Therefore, the smaller the K value of the trajectory curve, the smoother the travel trajectory, which is not related to driving habits.

In order to compare the merits of the indicators more reasonably, the indicators need to be in the same order of magnitude, so that it is convenient for the comprehensive comparison and evaluation and subsequent data processing. Therefore, the above indicators need to be standardized.

Definition 3.3 (Distance Level) Define

$$J^*_{min} = \frac{(x_1 - x_0)}{1 + (x_1 - x_0)} \quad (3)$$

$$J^* = \frac{1}{(1+J)(1-J^*_{min})} \quad (4)$$

J^* is called the distance level of the curve L.

J^* is the standardization of the length index. The interval $[x_0, x_1]$ represents the interval where the first and last position of the intelligent car trajectory curve is located, and the shortest distance $d = (x_1-x_0)$ is the straight line distance. J^*_{min} is the standardized form of the shortest distance d. The standardization process is to map the length index from $[d, +\infty]$ to $[0,1]$. When the length index J is the shortest distance d, the value of (4) is taken as 1; when the length index is close to $+\infty$, the closer the value of (4) to 1.

Definition 3.4 (Smoothness Level) Define

$$K^* = \frac{1}{1+K} \quad (5)$$

K^* is called the smoothness level of the curve L.

The smoothness level of the curve L is the standardization process of the bending resistance index. Considering that the curvature of the straight line is 0, The standardization process is to map the bending resistance index from $[0, +\infty]$ to $[0,1]$. When the bending resistance index K is 0, the value of (5) is taken as 1, and when the bending resistance index is close 0, Therefore, the closer the value of (5) to 1.

For drivers with different driving styles, in order to judge the requirements of different drivers on the optimal trajectory curve comprehensively, a comprehensive evaluation index for the optimal trajectory curve is introduced.

Definition 3.5 (Comprehensive Evaluation Index) Define

$$C^* = w_1 \cdot J^* + w_2 \cdot K^* \quad (6)$$

C^* is called the comprehensive evaluation index of the curve L, where w_1 and w_2 are the weights of J^* and K^* indicators for drivers of different driving styles, which can be given by experts.

To sum up, we can see that selecting an optimal curve among many trajectory curves requires only comparing the values of the comprehensive evaluation indexes of different trajectory curves, and the curve with the largest numerical value is the desired one.

3.2. Satisfied trajectory curve

This section first establishes the multi-objective optimization model of the optimal trajectory curve, which needs to meet the three needs of short time, short walking trajectory and passengers feel comfortable, so the model is as follows:

(Optimal Trajectory Curve Model) (OTC Model) For the set of trajectories $\Gamma(x_0)$ in the standard feasible neighborhood Q, find the optimal trajectory curve so that the following objective function can obtain the extreme value

$$\begin{cases} min\,T \\ min\,J[y(x)] \\ min\,K[y(x)] \end{cases} \quad (7)$$

$$\text{subject to } \begin{cases} y(x_0) \subset \Gamma(x_0) \subset Q \\ T \leq T_{max} \end{cases} \quad (8)$$

where $\Gamma(x_0)$ is the set of trajectories whose initial state is x_0 in the standard feasible neighborhood Q; J, K is the length index and bending resistance index of the curve trajectory. T is the total time taken by the intelligent vehicle to complete the planned trajectory in a standard feasible neighborhood; T_{max} is the longest time that can be accepted to complete the planned trajectory. The optimal solution of the objective function is the optimal trajectory curve. This model can be abbreviated as *OTC model*.

As shown in Figure 3, where the midpoint of the bottom of the vehicle is at a distance s from the midpoint of the bottom of the feasible neighborhood, the neighborhood length is H and the vehicle body length is l.

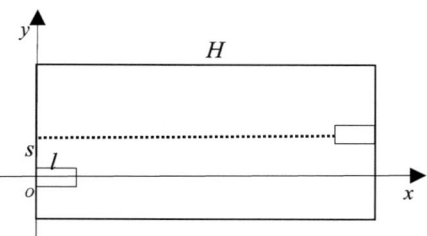

Figure 3. Trajectory planning diagram.

Solve *OTC Model*, the application of Hermite interpolation method is used as follows:

a) Select the initial data to determine the three interpolation nodes, i.e. the first position of the curve, the end position and the midpoint position, respectively;

b) Determine the initial value of the derivative at the midpoint of the curve;

c) Calculate the trajectory curve;

d) Calculate the length index and the bending resistance index;

e) Determine the satisfaction index, calculate comprehensive evaluation index, adjust the derivative of the midpoint of the curve, go to step c.

According to the above algorithm, the result is not unique because only a satisfactory curve is required. In this paper, the comprehensive index of satisfaction index of 0.8, weight $w_1 = 0.4$, $w_2 = 0.6$, the calculated curve midpoint coordinates and derivatives are as follows:

a) The node coordinates are $\left(\dfrac{H-l}{2}, \dfrac{s}{2}\right)$

b) The first derivative of the node is $\dfrac{2s}{H-l}$.

4. Simulations

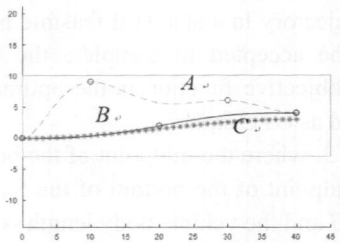

Figure 4. Example comparison diagram.

As shown in Figure 4, it is a few trajectory curve of the intelligent vehicle in a standard feasible neighborhood. The curve A is a polynomial curve represented by the graph "-". The curve B is the cubic Hermite interpolation curve constructed in this paper, which is very close to the curve C. The curve C is triangular function curve, represented by the graphic "*". The equation of the three trajectories is as follows:

$$y_A(x) = -1.21 \cdot 10^{-8} \cdot x^7 + 2.08 \cdot 10^{-6} \cdot x^6 - 1.40 \cdot 10^{-4} \cdot x^5$$
$$+ 4.65 \cdot 10^{-3} \cdot x^4 - 7.63 \cdot 10^{-2} \cdot x^3 + 5.09 \cdot 10^{-1} \cdot x^2$$
$$y_B(x) = 3.13 \cdot 10^{-7} \cdot x^5 - 3.13 \cdot 10^{-5} \cdot x^4 + 8.75 \cdot 10^{-4} \cdot x^3 - 2.50 \cdot 10^{-3} \cdot x^2$$

$$y_C(x) = \frac{3}{2} \cdot \sin\left(\frac{\pi \cdot x}{40} - \frac{\pi}{2}\right) + \frac{3}{2}$$

Conduct a comprehensive evaluation, set the J^*, K^* indicators take the weight $w_1 = 0.3$, $w_2 = 0.7$, the result is as follows:

$$C^*[y_A(x)] = 0.433 \quad C^*[y_B(x)] = 0.797 \quad C^*[y_C(x)] = 0.649$$

Inequality $C^*[y_A(x)] < C^*[y_C(x)] < C^*[y_B(x)]$ holds. So the interpolation curve B in this paper is the optimal trajectory curve of the three curves.

As shown in Figure 5, the trajectory planning of an intelligent vehicle on the road is the process of superposition of the trajectory planning in every local standard feasible neighborhood. The trajectory curve of this paper has good stability and smoothness on the road.

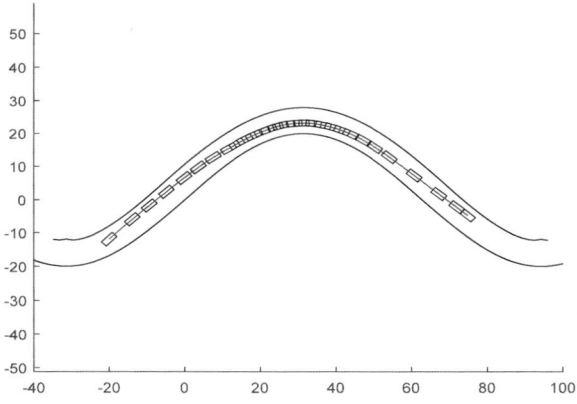

Figure 5. Road trajectory simulation.

5. Conclusions

In this paper, based on the theory of neighborhood system, the process of complex macro decision-making is decomposed into a series of simple static decision-making processes. The bending resistance index and the length index of curve are proposed. The model of the optimal trajectory in the neighborhood is established, and a satisfactory trajectory curve is constructed. The simulation result shows the effectiveness of the proposed method.

Acknowledgments

This work is supported by the National Natural Science Foundation of China, No. 61473239 and No. 61402382.

References

1. K. L. Zhang, D. F. Zhang et al. State-driven priority scheduling mechanisms for driverless vehicles approaching intersections. *IEEE Transactions on Intelligent Transportation Systems*. **16**, 5(2015).
2. K. Zhang, A. Yang et al. Unified modeling and design of reservation-based cooperation mechanisms for intelligent vehicles. *IEEE International Conference on Intelligent Transportation Systems*. 1192(2016).
3. G. P. Gwon, W. S. Hur et al. Generation of a precise and efficient lane-level road map for Intelligent Vehicle Systems. *IEEE Transactions on Vehicular Technology*. **66**, 6(2017).
4. C. M. Martinez, M. Heucke et al. Driving Style Recognition for Intelligent Vehicle Control and Advanced Driver Assistance: A Survey. *IEEE Transactions on Intelligent Transportation Systems*. **99**, 1(2017).
5. W. Chen, M. Wu et al. Research on auto-calibration technology of intelligent vehicle camera based on machine vision. *IEEE International Conference on Image, Vision and Computing*. 684(2017).
6. H. L. Zhao. A dynamic optimization decision and control model based on neighborhood systems. *IEEE International Congress on Image and Signal Processing*. **3**, 1319(2014).
7. H. M. Fu, H. L. Zhao. A control method for intelligent car based on neighborhood system. *Fuzzy Systems and Mathematics*. **30**, 2(2016). (*In Chinese*).
8. S. H. Xiong, H. L. Zhao. Simulations on movement of intelligent vehicle based on rectangular safe neighborhood. *Application Research of Computers*. **30**, 12(2013). (*In Chinese*).
9. W. G. Yang, H. L. Zhao. An easy method on fuzzy interpolative control for automatic truck backer-upper. *Acta Scientiarum Naturalium Universitatis Sunyatseni*. **51**, 1(2012). (*In Chinese*).

977

Goodwin model via p-fuzzy system

D. Sanchez

Patagonia Campus, University Austral of Chile
Coyhaique, Aysén 5950-000, Chile
daniel@ime.unicamp.br

L. C. Barros,* E. Esmi

Department of Applied Mathematics, University of Campinas
Campinas, São Paulo 13083-859, Brazil
*laeciocb@ime.unicamp.br; eelaureano@ime.unicamp.br

A. D. Miebach

Business School, Pontifical Catholic University of Rio Grande do Sul
Porto Alegre, Rio Grande do Sul 90619-900, Brazil
alessandro.miebach@pucrs.br

In this work we propose a Mamdani fuzzy controller based on a fuzzy rule base whose fuzzy rules represent economic rules inspired by the Goodwin model. This model describes the dynamical interaction between the distribution of income and the employment level in the economy. Computational simulations reveal that the numerical solution of the proposed (p-fuzzy) system produces increasing cycles similar to those of the classical the Goodwin model. Finally, we discuss the capability of the p-fuzzy approach as approximator and we apply this methodology using historical economic data from Germany.

Keywords: Goodwin model; fuzzy numbers; fuzzy rule; p-fuzzy systems.

1. Introduction

The dynamical model of Goodwin is considered the major contribution of the economist Richard Goodwin.[1] Roughly speaking, this model expresses a mathematical formulation of the process of capitalist accumulation described by Karl Marx in his book *Capital: Critique of Political Economy*, representing the dynamical interaction between the employment rate v and the distribution of income in the economy given by the labour share u.[2]

Conceptually, this economical model is designed by Goodwin over the following premises. If the wages increasing, then the profits decreasing. The reduction of the profits causes the reduction of savings and investments, arresting the emergence of new jobs. However, the workforce grows up from the incorporation of new contingents of workers, as well as from the liberation of workers as a result of technical progress. Thus, wages lag behind productivity growth and profits grow. Accumulation accelerates again. This causes the gradual reduction of unemployment that leads to higher wages, restarting the process and characterizing the cyclical nature of the model.[2]

The literature involving the Goodwin model can be divided in two branches. The first one consists of proposals of modifications of this model in order to extend its scope. For example, in,[3] Desai discusses the presence of structural stability and Veneziani and Mohun[4] establish alternative interpretations for this model. The second branch consists of empirical studies for specific countries.[5,6] These studies have identified that the income distribution of several countries exhibit evidence of behavior consistent with the model.

Since the original formulation of the Goodwin model coincide with the Lotka-Volterra model, we design a fuzzy rule-based system of prey-predator type[7] in order to describe the temporal dynamic of the employment rate v (interpreted as prey) and the labour share u (interpreted as predator). Based on this fuzzy rule-based system, we determine a solution of the Goodwin economical model via a p-fuzzy system. Therefore, this work investigates the application of fuzzy theory and describes a simple methodology to study an economic model which can be used for any expert.

2. Mathematical Background

2.1. Basic Concepts of Fuzzy Sets

A fuzzy subset of A of an universal set X is characterized by a function $\varphi_A : X \to [0, 1]$ called membership function of A such that $\varphi_A(x)$ represents the membership degree of x in X.[8] For notation convenience, we also use the symbol $A(x)$ instead of $\varphi_A(x)$ and we will use a particular class of fuzzy sets, called fuzzy numbers. In particular, a trapezoidal fuzzy number A, denoted by a quadruple $(a; m; n; b)$, with $a, m, n, b \in \mathbb{R}$ and $a \leq m \leq n \leq b$, consists of a fuzzy number whose membership function is given by[8]

$$A(x) = \begin{cases} \frac{x-a}{m-a}, & \text{if } x \in [a, m), \\ 1, & \text{if } x \in [m, n], \\ \frac{b-x}{b-n}, & \text{if } x \in (n, b], \\ 0, & \text{otherwise.} \end{cases}$$

In the case where $m = n$, we speak of triangular fuzzy number and it is denoted by the symbol $(a; m; b)$ instead of $(a; m; m; b)$.[8]

2.2. *Fuzzy Rule-Based Systems*

Fuzzy Rule-Based Systems (FRBS) have four components: an *fuzzification* module, a fuzzy rule base, a fuzzy inference method, and an *defuzzification* module.[8,9]

In the fuzzification module, real-valued inputs are translated into fuzzy sets of their respective universes. In the general case, expert knowledge plays an important role to build the membership functions for each fuzzy set associated with the inputs.[9] The most basic fuzzifier method is the canonical inclusion that consists to associate each real number with its its characteristic function.

Here, we consider a fuzzy rule base given by a collection of fuzzy conditional rules of the form "if x_1 is A_{i1} and x_2 is A_{i2} then y is B_i", for $i = 1, \ldots, r$, where A_{ij} and B_i, $i = 1, \ldots, r$ and $j = 1, 2$, are fuzzy sets that represent linguistic terms and are called respectively antecedents and consequent of each fuzzy rule.[8]

In this work we use the Mamdani inference with canonical inclusion fuzzifier method. In this case, for a given input (x_1, x_2), the Mamdani inference produces the following as output a fuzzy set B given by:[8]

$$B(y) = \max_{i=1,\ldots,r} \min\{\, A_{i1}(x_1)\,,\, A_{i2}(x_2)\,,\, B_i(y)\,\}, \ \forall y \in \mathbb{R}. \tag{1}$$

Defuzzification module consist of a process that allows us to represent a fuzzy set by a real value. In this paper, we adopt a typical defuzzification scheme, namely center of gravity method.[9]

2.3. *P-Fuzzy Systems*

A *partially* fuzzy system or, for short, a *p*-fuzzy system, is a dynamical system where the direction field is given by FRBS based on a partially a priori known of the direction field. Furthermore, the state variables and their variations are considered linguistic. Thus, the state variables are correlated to their variations by means of fuzzy rules where the state variables are the input and the variations are outputs. Since in such methodologies, processes of defuzzification are expected, the final solution of a *p*-fuzzy system is deterministic.[8] Here, we use *p*-fuzzy systems to deal with autonomous initial value problems (IVPs) of the form

$$\begin{cases} \dfrac{dx}{dt} = f(x,y)\,, & x(0) = x_0\,, \\ \dfrac{dy}{dt} = g(x,y)\,, & y(0) = y_0\,. \end{cases} \tag{2}$$

where the functions f and g are partially known.

To obtain the solution of the IVP (2) via a p-fuzzy system or at least an approximation of it, without knowing the field f and g explicitly, we take advantage of the qualitative information available to design a fuzzy rule base which represent the properties that characterize the phenomenon.[10] Thus, the solution $(x(t), y(t))$ of (2) can be estimated by a sequence (x_n, y_n) of a p-fuzzy system obtained by means of numerical methods for the ordinary differential equations (ODE) such as Euler and Runge Kutta methods, or by means of numerical methods for integration.[8] In this work we use the Euler method (adapted) for functions (f and g) representing specific variations. More precisely, we use the formulas

$$x_{n+1} = x_n + h\, x_n\, f(x_n, y_n) \quad \text{and} \quad y_{n+1} = y_n + h\, y_n\, g(x_n, y_n), \qquad (3)$$

where h is the step (in time) and $f(x_n, y_n)$ and $g(x_n, y_n)$ are specific variations obtained by FRBSs. Note that, in general, a Mamdani fuzzy controller yields a function f_r^* (and g_r^*) where r denotes the number of rules in the fuzzy rule base. Thus, it seems reasonable to assume that the adjusted function f_r^* (and g_r^*) approximates f (and g) when the number of data r increases. In other words, the more the information of f (and g) the better its approximation f_r^* (and g_r^*).[10]

3. Description of the Goodwin Model

The Goodwin model is structured from seven assumptions:[1]

1) *steady technical progress;*
2) *steady growth in the labour force;*
3) *only two factors of production, labour and capital, both homogeneous and non specific;*
4) *all quantities real and net;*
5) *all wages consumed, all profits saved and invested;*
6) *a constant capital output ratio;*
7) *a real wage rate which rises in the neighbourhood of full employment.*

Each one of theses assumptions characterizes the parameters of two dynamic equations, one for the employment rate, (v), and one for the labour share, (u).[3] In order to obtain a dynamic equation for the labour share u, Goodwin used a linear approximation of the specific rate in the real wage dynamics $\frac{1}{\omega}\frac{d\omega}{dt} \approx (-\gamma + \rho v)$, and this does quite satisfactorily for moderate movements of v near the point $+1$.[1] In this way, defining the share of labour in national income u as the ratio of the real wage to average product per worker, that is $u = \frac{w}{a}$, and taking the rate of growth of labour productivity $\frac{1}{a}\frac{da}{dt} = \alpha$ as a constant. In the dynamic equation for the employment rate we use β as the exogenous growth rate in the labour force and σ as the capital output ratio (inverse of capital productivity). Thus, the Goodwin model is given by

$$\begin{cases} \dfrac{dv}{dt} = \left(\dfrac{1-u}{\sigma}\right)v - (\alpha + \beta)v, \\ \dfrac{du}{dt} = -(\gamma + \alpha)u + \rho v u. \end{cases} \qquad (4)$$

The system (4) is a Lotka-Volterra model with u and v playing the role of predator and prey, respectively.[11] The non-trivial steady states of the Goodwin model is given by the values

$$v^* = \frac{(\gamma + \alpha)}{\rho} \quad \text{and} \quad u^* = 1 - (\alpha + \beta)\sigma. \qquad (5)$$

Note that the stability analysis of the Goodwin model predicts oscillations about the this steady state (u^*, v^*) wich as in the predator-prey model.[11]

4. Solution of the Goodwin Model via P-Fuzzy System

The model of Goodwin given in (4) is based on the seven premises listed at the beginning of Section 3. These equations present the following fundamental property:

"given a value of a variable, the specific growth of another is constant." (6)

For instance, for each fixed v in the second equation of (4), the specific rate $\frac{1}{u}\frac{du}{dt}$ is constant. Thus, the fuzzy system is based on the property (6) since the seven premises described in Section 3 are also based on this property.

Both input variables V and U can be classified as *low* (A_1 and B_1), *average low* (A_2 and B_2), *average high* (A_3 and B_3), and *high* (A_4 and B_4). The output variables, the specific growths of the input variables $\frac{1}{V}\frac{dV}{dt}$ and $\frac{1}{U}\frac{dU}{dt}$, can assume the fuzzy terms *high negative* (N_2), *low negative* (N_1), *low positive* (P_1), and *high positive* (P_2). The seven premises of Section 3 together with the property (6) can be translated into a set of fuzzy rules that play the role of a direction field. For example, we obtain a fuzzy rule of the form: *if U is low (B_1) then the specific growth of V ($\frac{1}{V}\frac{dV}{dt}$) is high positive (P_2)*. Figure 1 exhibits a graphical representation of the obtained fuzzy rule base where the arrows represent the direction and magnitude of the specific growths.

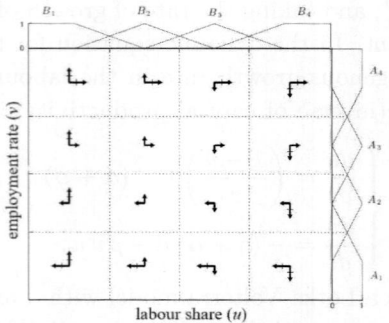

Fig. 1. Graphic representation of the fuzzy rules as direction vectors for Goodwin model. The arrow on the right/up (left/down) side indicates positive (negative) specific growth rate. The length of the arrow indicates the magnitude of the specific growth rate.

4.1. *Results*

We use the proposed p-fuzzy system to simulate the dynamic behaviour of the variables u and v for the German economy. We compare the obtained solution with the real economic data series extracted from.[5] In our simulation, we model all linguistic terms by trapezoidal and/or triangular fuzzy numbers. Based on the historical data, the antecedents are defined using the *maximum, minimum* and *mean* values whereas the consequents are defined in terms of *standard deviations*. In order to fit the consequents fuzzy numbers, we test a set of parameters and we select that one whose the difference of the average of employment and labour share from the corresponding p-fuzzy system and the historical data are less than 3%.

Figure 2 presents the phase diagrams of the oscillatory behaviour for the p-fuzzy solution (for u and v), and Germany's historical economic data (from 1956 to 1994), showing that the qualitative behaviour of the solution is consistent with the data and with the previous results obtained by Harvie.[5] We observe that the proposed p-fuzzy is able to represent the observed behavior of the variables at the same time that it identifies the behavior predicted by same model. Finally, Figure 3 exhibits the dynamic trajectories of the employment rate v and the labour share u from p-fuzzy system and from the historical economic data of Germany.

5. Concluding Remarks

The main contribution of this work is to present a methodology based on Fuzzy Rule-Based Systems (FRBS) to produce a solution to Goodwin's

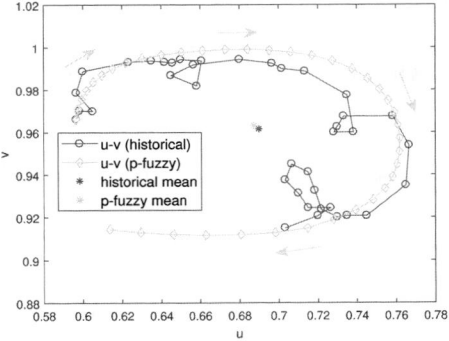

Fig. 2. The Goodwin growth cycle: historical data and solution via p-fuzzy system to Germany economy between 1956 and 1994.

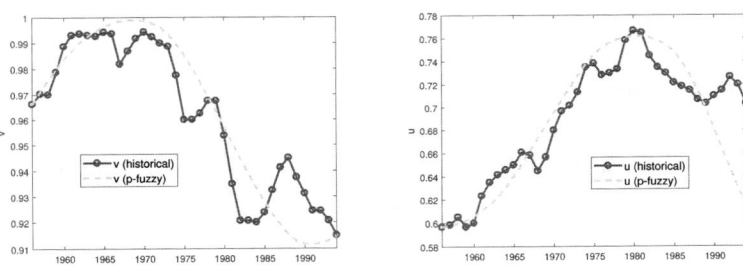

Fig. 3. Dynamic trajectories of a) employment rate v and b) labour share u from the historical data and from p-fuzzy system with respect to period from 1956 to 1994 of the Germany economy.

economic model. The solution obtained by our approach is (quantitative and qualitative) similar to the ones produces using differential equations theory. In addition, note that our proposal can be used by any economist and does not require previous experience with differential equations.

This research adopts a FRBS, with the Mamdani inference method, allowing us to represent the dynamics of the Goodwin economic model. Such a methodology, called p-fuzzy, made it possible to represent the growth cycles of the Goodwin economic model for historical data from Germany between 1956 and 1994.

Initially, it was observed that the (original) Goodwin model, considering the employment rate v as prey and the labour share u as predator, is analogous to the classic prey-predator model of Lotka-Volterra that establishes a cycle (closed) in the phase diagram between the species. We have seen the same qualitative behaviour for the phase diagram in the solution of the

p-fuzzy system from the historical data of Germany (see Figure 2). Thus, it is possible to infer that the relative stability in the dynamic trajectory of the variables reflects the institutional stability of the country for the period that includes the so-called "German miracle" of the 1950s and 1960s until German reunification in 1990.

From a numerical point of view, since the structural instability of the original model (prey-predator type) is known, we perceive a great deal of sensitivity in the linguistic terms (fuzzy numbers) of the variables used in the FRBS. Future research will be needed to establish general indicative notions to discover the linguistic terms that interact in the rule-base.

Finally, it should be noted that the use of the p-fuzzy system is based on universal approximation capability, which means that it is a good estimator of theoretical problems. It is also easily interpreted and implemented by experts from various fields of science such as economics.

Acknowledgments

This research was partially supported by CONICYT of Chile, FAPESP under grant no. 2016/26040-7 and CNPq under grant no. 306546/2017-5.

References

1. R. M. Goodwin, *A growth cycle model*, in *Socialism, Capitalism & Economic Growth*, ed. C. H. Feinstein (Cambridge University Press, 1967), pp. 54–58.
2. A. D. Miebach, Goodwin's growth cycle: A model of non-linear economic dynamics (in Portuguese), Master's thesis, Pontifícia Universidade Católica do Rio Grande do Sul (2011).
3. M. Desai, B. Henry, A. Mosley and M. Pemberton, *Journal of Economic Dynamics and Control* **30**, 2661 (2006).
4. R. Veneziani and S. Mohun, *Structural Change and Economic Dynamics* **17**, 437 (2006).
5. D. Harvie, *Cambridge Journal of Economics* **24**, 349 (2000).
6. S. Mohun and R. Veneziani, *Goodwin cycles and the US economy, 1948-2004*, in *Mathematical Economics and the Dynamics of Capitalism: Goodwin's Legacy Continued*, eds. P. Flaschel and M. Landesmann (Routledge, 2006), pp. 107–130.
7. M. S. Peixoto, L. C. Barros and R. C. Bassanezi, *Ecological Modelling* **214**, 39 (2008).
8. L. C. Barros, R. C. Bassanezi and W. A. Lodwick, *A First Course in Fuzzy Logic, Fuzzy Dynamical Systems, and Biomathematics* (Springer, 2016).
9. R. M. Jafelice, L. C. Barros, R. C. Bassanezi and F. Gomide, *Bulletin of Mathematical Biology* **66**, 1597 (2004).
10. M. R. B. Dias and L. C. Barros, Differential equations based on fuzzy rules, in *Proc. 2009 IFSA/EUSFLAT Conf.*, (Lisbon, Portugal, 2009).
11. N. F. Britton, *Essential Mathematical Biology* (Springer Science & Business Media, 2012).

Part 8

(Fuzzy) Multi-Criteria Decision Analysis and Intelligent Support System for Marketing Decisions

The new product design problem using novel preference approaches

Leyva-Lopez, Juan Carlos and Leon-Santiesteban Martín
Economic and Management Science Department, Universidad Autónoma de Occidente
Culiacán, México

Ahumada-Valenzuela, Omar
CONACYT Research Fellow, Management Science Doctoral Program,
Universidad Autónoma de Occidente, Culiacán, México

Solano-Noriega, Jesus Jamie
Economic and Management Science Department, Universidad Autónoma de Occidente
Culiacán, México

The present research develops a new product design method to find the best combination of attributes and levels to maximize frequency-based consumer choice shares based on consumer preference and multicriteria decision analysis. The product is designed using a genetic algorithm that maximizes frequency based choice shares for the new product, combining attributes and levels that better match the stated preferences of consumers.

1. The New Product Design Problem

The importance of new product design cannot be overstated, since it is critical for the long-term profitability and competitiveness of the companies and of the value chain. Despite the fact that successful new products are crucial to profitability and growth, the actual success rate of new products introductions are fairly disappointing, which make new product development a very important and risky process for companies [1].

In the New Product Design (NPD) Problem, a company tries to find the best combination of attributes and levels that can better satisfy consumers desires and or requirements, to increase their profits through the sale of that product in the marketplace [2]. A new product design is based on consumer's preferences and needs (considering the market for that product), but also on the feasibility of making the product at a profit. The new product must bring enough new revenue to justify designing, producing and distributing to the selected market segment.

This problem is related to product positioning, which consists in choosing the characteristics of a new product in a given attribute space so as to maximize the number of customers patronizing the product [3]. Also, we can think of it as how to position the product in a way that maximizes the utility or the preference from a consumer or a group of consumers (market segment). Product positioning and product design are generally viewed as closely related important problems in marketing research, which deal with the generation of "promising" options for a company that plans to extend or modify its existing product lines [4].

Usually the objective in this type of problem is to gain market share, or optimizing market share, which is used as a proxy to profit optimization, since it is associated to higher revenue levels. Other related objectives could be customer satisfaction or product acceptance.

Traditional approaches assume that consumers behave like utility maximizers, with little uncertainty with respect to their preferences [5]. But our research considers alternative methods of designing new products, such as Multicriteria Decision Analysis (MCDA), by using directly the stated consumer preferences (e.g. outranking methods), since these methods are usually less demanding in terms of assumptions and are more flexible than traditional methods, which we explain in the rest of this section.

2. General Methodology

The envisioned planning framework for New Product Design NPD in its first step, involves obtaining direct consumer preferences for products, by performing surveys from consumers. The second step involves the segmentation and preference disaggregation of consumers. In the segmentation part, we cluster consumers according with their demographic and or preference criteria. In the preference disaggregation part, we determine those factors that are more relevant for each consumer, based on their stated preferences (we estimate the parameters for the consumer preference model). For this research, we explore the use of outranking relations, based on the ELECTRE III Methodology [6]. The output from the second step is a preference disaggregation model for single consumers, and for groups of consumers.

The third step involves using the outranking relations and selecting a choice model (also known as brand selection model) that best fits the consumer choice, per their preferences in the outranking relation, and determine their estimated market share for the current market (estimation the purchasing decisions of consumers). Those market shares need to be validated with current market shares, so that the potential new products evaluated are consistent with current preferences of consumers, and later, we can feel confident with their estimated market shares.

The fourth step involves preference simulation, which can be used to determine the market share of new products or changes to existing products. The output of the simulation is an estimated market share, for products that are not yet on the market or for novel combinations like new products for an existing brand.

The fifth step is new product design, here the decision maker (DM) can input a new product design manually or could use an optimization algorithm that uses the outranking relations, to find those products that provide the best market share within the boundaries selected by the DM. Those market shares are converted to sales and based on costs for the development and introduction of the product, we can simulate the design and introduction of different potential products. This procedure could be highly iterative, given the nature of the problem, particularly in from the third step to the fifth.

3. Problem Formulation

Let $A = \{a_1, a_2, ..., a_n\}$ be a set of products available to the consumers, which can be analyzed by family of criteria $G = \{g_1, g_2, ..., g_n\}$, where $g_j(a_i)$ represents the evaluation of product a_i by criterion g_j. We can further extend this formulation to the levels within each criterion and for each customer: $g_{jm}^k(a_i)$: Evaluation of product a_i on criterion j with level m for consumer k. These evaluations are based on the customers' judgements and should be directly collected from them.

We assume that the aggregation model of preferences for each one of the consumers k_i, $i = 1,2,...,r$ can be represented by a fuzzy outranking relations $S_A^{\sigma_{ki}}$, and that each consumer k_i can rank the products in decreasing order of preference R_{ki}, producing a total order, total preorder, or a partial pre-order depending on the characteristics of the products and preferences of consumers. We also assume that consumers can be segmented by their preferences $S = \{s_1, s_2, ..., s_t\}$, or could also be grouped based on the socioeconomic characteristics of consumers. Finally, let us assume that we know in advance the market share of each one of the products in A. Our research can be described by the following problems:

Assume that the market consists of N competitive products with known configurations, including the candidate item for the company (note that $N=m+1$). Then $\Xi = (1, 2, ..., N)$ is the set of products that comprise the market (or at least the market under study).

$\Omega = (1, 2, ..., n)$ is the set of n attributes that comprise the product.

$\Phi_r = (1, 2, ..., J_r)$ is the set of J_r levels of attribute r.

$\Theta = (1, 2, ..., K)$ is the set of K consumers.

We also have that w_{ktr} is the assessment that consumer $k \in \Theta$ assigns to level $t \in \Phi_r$ of attribute $r \in \Omega$, where

$$x_{tr} = \begin{cases} 1 : \text{if the level of attribute } r \text{ of the new product is } t \\ 0 \text{ otherwise} \end{cases}$$

$T^{(k)}(i,j), (1 \leq i,j \leq N)$ is the number of times that a product a_j was found at a certain place in the ranking of the individual associated to the members of the final restricted Pareto for a consumer k.

$w_i^{(k)}$ = weights according to the importance of the products to be ranked for a consumer k. $(1 \leq i \leq N)$

$B_{kj} = \sum_{i=1}^{m} w_i^{(k)} T^{(k)}(i,j), \quad j = 1,2,...,m$ is the total importance (performance value) that consumer k assign to product a_j.

We assume that each of the N products has a certain "relative frequency" to be selected, which is calculated with the use of a frequency choice model, and it is based on the position in a ranking of alternatives.

The weighted preference relative frequency that consumer k chooses product j is estimated as follows:

$$F_{kj} = \frac{B_{kj}}{\sum_{i \in \Xi} B_{ki}} \quad \text{Where} \quad B_{kj} = \sum_{i=1}^{m} w_i^{(k)} T^{(k)}(i,j) \text{ is the total performance value the}$$

consumer assigns to product a_j.

With the use of the frequency choice model, we calculate the weighted preference relative frequency that consumer k chooses the new product, denoted as F_k.

In this context, the problem of new product design is formulated as the following non-linear program [7]:

$$Max\ f = \sum_{k \in \Theta} F_k$$

subject to:

$$\sum_{t \in \Phi_r} x_{tr} = 1, \quad r \in \Omega \quad (1)$$

$$x_{tr} = 0,1 \text{ integer number}$$

which implies maximizing the frequency-based choice share for the new product. The expected outcome of this model is a recommendation to the decision maker, of a product that maximizes the objectives of the company.

The output consists of a set of values of the binary variables, for the design of a new product:

$$x_{tr} = \begin{cases} 1 : \text{if the level of attribute } r \text{ of the new product is } t \\ 0 \text{ otherwise} \end{cases}$$

For this case, the choice of an optimization technique, given the non-linearity of the objective function, is a metaheuristic based on genetic algorithms to deal with the structure of the problem.

4. Evolutionary Methodology

This section describes the details of the proposed evolutionary process for determining the binary variables for the design of a new product.

4.1. *Representation and fitness criterion*

Let's for instance assume that a product consists of the attributes g_1 with levels (g_{11}, g_{12}, g_{13}), g_2 with levels $(g_{21}, g_{22}, g_{22}, g_{24})$, g_3 with levels (g_{31}, g_{32}, g_{33}). Then a product with characteristics (g_{12}, g_{23}, g_{33}) will be represented by the chromosome $C = \{2,3,3\}$.

The fitness of an individual is calculated according to a given fitness function. Fitness measures "how good" an individual is regarding solving the new product design problem and is used during the selection and replacement phases to determine which individuals are selected for reproduction and replacement respectively. The value returned by the fitness function will determine the individual's fitness and hence its chance of survival and reproducing. The general rules for defining a fitness function is that it should provide an efficient mapping of an individual's chromosome representation to a single value, and that the fitness should reflect the "quality" of a solution in the given solution space.

An initial population of P chromosomes, $W(0) = \{W_1, W_2, ..., W_P\}$ is then generated in a totally random manner.

The metric $f(W)$ is used as an objective function to estimate the performance of all chromosomes W_i $i = 1, 2, ..., P$ in each population. However, a fitness function is used to map objective values to fitness values, following a rank-based normalization scheme. Chromosomes W_i are ranked in descending order of $f(W)$, since the objective function is to be maximized. Let $rank(W_i) \in \{1, 2, ..., P\}$ be the rank of chromosome W_i $i = 1, 2, ..., P$, (rank = 1 corresponds to the best chromosome and rank = P to the worst). Defining an arbitrary fitness value F_B for the best chromosome, the fitness $F(W_i)$ of the *i-th* chromosome is given by the linear function:

$$F(W_i) = F_B - [rank(W_i) - 1]D, \quad i = 1, 2, ..., P \tag{2}$$

where D is a decrement rate. The major advantage of the rank-based normalization is that, since fitness values are uniformly distributed, it prevents the generation of super chromosomes, avoiding premature convergence to local minima. Furthermore, by simply adjusting the two parameters F_B and D, it is very easy to control the selective pressure of the algorithm, effectively influencing its convergence speed to a global minimum.

4.2. Parent selection method

To produce a child, two k-ary tournaments are held, each of which produces one parent string. These two parent strings are then combined to produce a child.

4.3. Crossover and mutation operators

Many crossover techniques exist in the literature (e.g. [8]), but, when working with real numbers in the encoding with different genetic material, it is necessary to create both crossover and mutation operators that are specifics to this form of encoding. The main difficulty encountered when using non-standard representations is the design of a suitable crossover operator, which must combine relevant characteristics of the parent solutions into a valid offspring solution. In this paper, we make use of the uniform crossover operator for real number encoding [9] with the probability p_c taking the value 0.45. In this operator, each parent gene is a potential crossover point.

The uniform crossover [10] is used because it is unbiased with respect to the positioning of genes and can generate any combination of alleles from the two parents in a single crossover [8]. For the mutation operator, the Uniform Mutation is employed. The Uniform Mutation operator requires a single parent and produces a single offspring. This operator randomly selects a gen g_r from the individual W_i and randomly alters its allele value s to produce an offspring, where $r \in \{1, 2, ..., n\}$ is a random value with a uniform probability distribution. Therefore, a mutation probability pm (usually 1%) is used to randomly determine if the gene will be muted. That is, a random number $t \in [0,1]$ is generated for the gene and replacement takes place if $t < p_m$; otherwise the gene remains intact.

4.4. Population replacement scheme and stopping criterion

This GA part defines how new chromosomes will be put into the existing population. Once new chromosomes have been generated for a given population $W(n)$ $n \geq 0$, the next generation population, $W(n + 1)$, is formed by inserting these new chromosomes into $W(n)$ and deleting an appropriate number of older chromosomes, so that each population consists of P members. The exact number of old chromosomes to be replaced by new ones defines the replacement strategy of the GA and greatly affects its convergence rate. An elitist strategy has been selected for replacement, where a small percentage of the best chromosomes is copied into the succeeding generation together with their offspring, improving the convergence speed of the algorithm. From the reproduced chromosomes, plus the offspring plus the mutated chromosomes, only the N fittest are maintained to the next generation, and the algorithm iterates until a stopping criterion is met. In this work, we employ a moving average rule, where the algorithm terminates when the percentage change in the average fitness of the best three chromosomes over the five previous generations is less than 0.2% (convergence rate).

4.5. *Summary of our GA for optimizing the model's parameters*
1) Generate an initial population of P random solutions.
2) Evaluate fitness of individuals in the population;
3) Repeat
4) Select two individuals W_i and W_j from the population using the k-ary tournament selection method.
5) Combine W_i and W_j to form two new offspring W_i' and W_j' using the Uniform crossover operator.
6) Apply mutation to the newly created chromosomes using the Uniform mutation operator.
7) Evaluate fitness of the children.
8) Replace all the population by the children except a small percentage of the best chromosomes using an elitist replacement strategy.
9) Until the best chromosome fitness remains constant for many generations.

5. Conclusions and Future Research

An outranking approach of the problem is used to develop a non-linear programming formulation of the problem, and to develop a genetic algorithm-based metaheuristic. The focus in this paper is on the problem of identifying a single multicriteria new product design.

Our research provides an initial methodology to develop new procedures for developing new conceptual products that take consumer preferences, without imposing demanding conditions from the responses presented by the consumer, thus providing a flexible framework for new product evaluation and development.

Future research will include a test with real data and the comparison of the proposed methodology with the current state of the art.

Acknowledgments

This work is part of the "Development of a decision support system based on intelligent agents for the design of new agro-industry products", research project #2015-01-162, financed by the Mexican National Council for Science and Technology (Conacyt).

References

1. K. G. Grunert and H. C. M. van Trijp, "Consumer-Oriented New Product Development," in *Encyclopedia of Agriculture and Food Systems*, 2014, pp. 375–386.
2. N. F. Matsatsinis and A. P. Samaras, "Brand choice model selection based on consumers' multicriteria preferences and experts' knowledge," *Comput. Oper. Res.*, vol. 27, no. 7–8, pp. 689–707, 2000.
3. P. Hansen, B. Jaumard, C. Meyer, and J.-F. Thisse, "New algorithms for product positioning," *Eur. J. Oper. Res.*, vol. 104, no.1, pp. 154–174, 1998.
4. D. Baier and W. Gaul, "Optimal product positioning based on paired comparison data," *J. Econom.*, vol. 89, pp. 365–392, 1998.
5. S. Tsafarakis, E. Grigoroudis, and N. Matsatsinis, "Consumer choice behaviour and new product development: an integrated market simulation approach," *J. Oper. Res. Soc.*, vol. 62, no. 7, pp. 1253–1267, 2011.
6. B. Roy, "The outranking approach and the foundations of ELECTRE methods," in *Readings in multiple criteria decision aid*, Springer, 1990, pp. 155–183.
7. S. Tsafarakis, K. Lakiotaki, A. Doulamis, and N. Matsatsinis, "A probabilistic choice model for the product line design problem," *2008 IEEE Int. Conf. Syst. Man Cybern.*, pp. 1361–1366, 2008.
8. D. Whitley, "A Genetic Algorithm Tutorial by Darrell Whitley," *Stat. Comput.*, vol. 4, pp. 65–85, 1994.
9. M. Srinivas and L. M. Patnaik, "Genetic Algorithms: A Survey," *Computer (Long. Beach. Calif).*, vol. 27, no. 6, pp. 17–26, 1994.
10. G. Syswerda, "A Study of Reproduction in Generational and Steady State Genetic Algorithms," *Found. Genet. Algorithms*, vol. 1, pp. 94–101, 1991.

An intelligent decision support system for the design of new products*

J. F. Figueroa Pérez

Facultad de Ingeniería Mochis - UAS, Fuente de Poseidón y Ángel Flores S/N,
Los mochis, Sinaloa, México
juanfco.figueroa@uas.edu.mx
http://fim.uas.edu.mx/

J. C. Leyva López

Universidad de Occidente, Blvd Lola Beltran S/N,
Culiacán, Sinaloa, México
juan.leyva@udo.mx
www.udo.mx

L. C. Santillán Hernández

Universidad de Occidente, Blvd Lola Beltran S/N,
Culiacán, Sinaloa, México
luis.santillan@udo.mx
www.udo.mx

E. O. Pérez Contreras

Universidad de Occidente, Blvd Lola Beltran S/N,
Culiacán, Sinaloa, México
pece78@gmail.com
www.udo.mx

In modern manufacturing companies, new product design is a matter of great importance that can directly affect their profitabilities. Nowadays, customers demand higher quality products, lower prices, and better performance in delivery time. The intense competition of companies in global markets stimulates a significant change in the way products are designed, manufactured and delivered. These situations are forcing to designers and manufacturing engineers to consider the use of tools for the process of new product design. In this paper, we propose an intelligent decision support system (DSS) based on a proposed distributed multi-agent architecture. This DSS implements a New Product

---

*This work is part of the "Development of a decision support system based on intelligent agents for the design of new agro-industry products", research project #2015-01-162, financed by the Mexican National Council for Science and Technology (Conacyt).

Design Multicriteria Methodology based on consumer preferences. The system is composed by elements of Marketing Decision Support Systems, agent technologies, multi-objective evolutionary algorithms and multicriteria methods. Finally, we show a Marketing Intelligent Decision Support System prototype to support new product design decisions which is a combination of MDSS, agent technologies, multi-objective evolutionary algorithms and multicriteria ELECTRE III method.

Keywords: Product design; decision support system; multi-agents, DSS.

1. Introduction

Globalization of the world economy has led to a large number of companies facing a level of global competition to export, sell or even keep their products on the market. In this way, the ability of a company to respond or adapt their products to market changes can be determinant for the survival of the business.

The product design problem is a significant part of the new product development problem and one of the most crucial decisions for an company [2]. In modern manufacturing companies, new product development process is becoming a decentralized task. In this new environment, activities are distributed among several working groups, which are organized in distributed and hierarchical structures, where specialists are divided by disciplines. Thus, industries are facing new challenges: 1) designing products in collaboration with other partners in a decentralized environment; 2) designing products that meet customer needs and expectations, which sometimes conflict; 3) reducing product development time and costs through simultaneous optimization [3].

In this paper, we propose a Marketing Intelligent Decision Support System (MIDSS) for new product design based on consumer preferences. The system is intelligent because it employs intensively artificial intelligence methods such as multi-objective evolutionary algorithms and intelligent agents. It means that the MIDSS was developed using intelligent software agents and their models were implemented through multi-objective evolutionary algorithms which are mainly based-on the multicriteria outranking methods.

The paper is structured as follows. Section 2, presents a brief literature review about Decision Support Systems and New Product Design. Section 3, has a short description of a New Product Design Multicriteria Methodology. Section 4, describes the proposed Marketing Intelligent Decision Support System. Finally, section 5 shows the conclusions and future work.

2. Decision Support Systems and New Product Design

The literature, have been reported some applications to automate product design activities using agent technologies. In Matsatsinis et al. [4] is presented an intelligent software agents based system implementing an original consumer-based methodology for product penetration strategy selection in real world situations. Likewise, a method to address uncertain in the decisions involved in the new product and technology development and positioning with the implication of hybrid intelligent systems is proposed in Banerjee et al. [5]. A hybrid decision model and a multiagent framework for collaborative decision support in the design process is presented by Zha et al. [6]. A three-layer system structure composed of decision customer layer, decision core layer and decision resource layer of a multi-intelligent-agent technology to develop a distributed marketing decision support system is described in Ai et al. [1]. Finally, an architecture for a multi-agent Decision Support System (DSS) for e-commerce and a prototype system for making on-line investment decisions are presented in Vahidov and Fazlollahi [7].

While these developments in agent-based decision support for product design applications are very encouraging, there are few systems that jointly consider the different elements involved in the decision process of product design, such as the participating actors distribution, the multiple criteria involved and uncertain information. Moreover, in this group of systems, few of them contemplate the consumer satisfaction to issue recommendations and those who do assume that consumer preferences are complete and transitive, which is not always true. Additionally, these systems do not consider the importance assigned by customers to each criterion when evaluate a product, ignoring with it valuable information. This situation can directly affect the accuracy of the recommendations issued by the systems. This indicates that important elements of the new product design decision process are being left out in the existing MDSS and new developments must consider them. Next section presents a New Product Design Multicriteria Methodology implemented computationally by MIDSS.

3. New Product Design Multicriteria Methodology

A New Product Design Multicriteria Methodology (NPDMM) based on consumer preferences is implemented computationally by the proposed MIDSS.

Here, consumer satisfaction is modeled using flexible multicriteria outranking approaches through a new aggregation/disaggregation preferences model, a market segmentation model and a brand choice model. This way, unlike others proposals in literature, when a product is evaluated the principle of transitivity does not lie and the importance of each criterion in the evaluation is considered. It is believed that this will allow to represent more appropriately consumer preferences in the software.

On the other hand, NPDMM is directly related to the Simon's Decision-Making Process (SDMP) [8] and the generic Product Design Process (PDP) [9]. A summary with the relation between SDMP, PDP and the NPDMM is presented in Table 1.

Table 1. Relation between Simon's problem-solving model, Generic Product Design Process and the New Product Design Multicriteria Methodology.

SDMP Stages	Generic PDP	NPDMM	Comments
Intelligence	Planning and clarifying the task	Project definition	Finding occasions calling for decision. Product idea. Opportunity discovery. Problem description. Design specifications: criteria, scales. Identify participants
Design	Concept Design Embodiment Design	Market Studies, Aggregation. Disaggregation Method, Market Segmentation Method, Brand Choice Model	Identifying, Developing and Analyzing all possible alternatives. Generate and evaluate rough design layouts. Preliminary specifications of layout. Get consumer preferences and market segments
Choice	Detail Design	Optimal product recommendation, product simulations with different scenarios	Selecting from the available choices. Get specification of production from optimal recommendation or simulations

4. Marketing Intelligent Decision Support System

4.1. *System Architecture*

MIDSS architecture (Fig. 1) is based in a combination of the well-known proposal of agent types presented in Sycara et al. [10] and the component integration based on Simons decision-making process presented in Vahidov

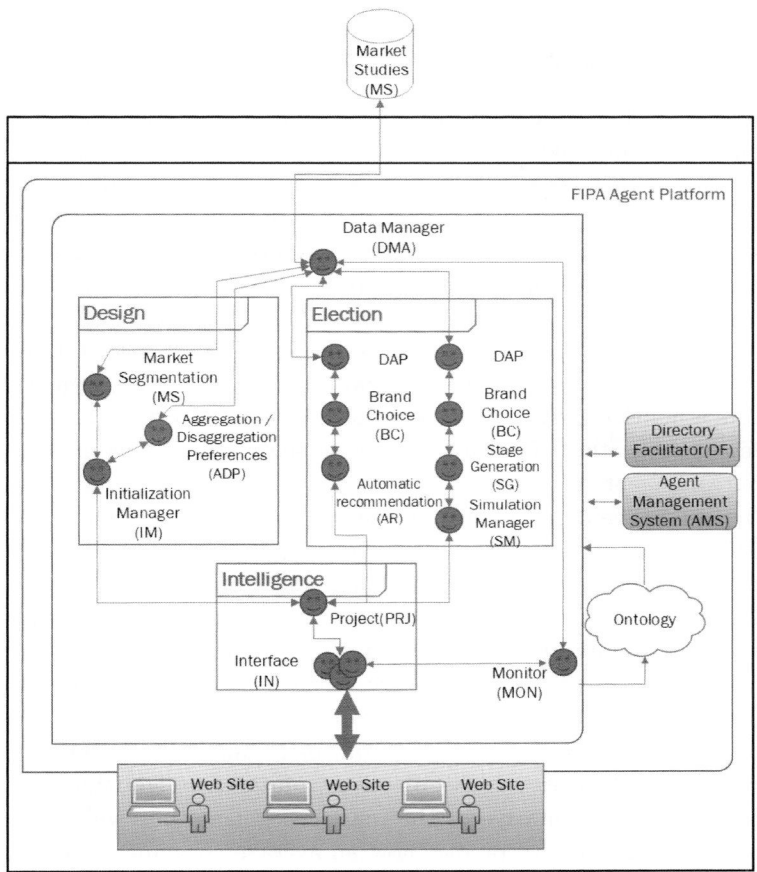

Fig. 1. MIDSS multi-agent architecture.

and Fazlollahi [7]. In the proposed architecture, agents are classified in three different types: interface agents, task agents and information agents [10]. The main components of the architecture include Intelligence Team, Design Team and Choice Team. The suggested naming reflects the phases of Simons decision-making process that the corresponding agents aim to support [7].

The intelligence team have two information agents: the interface agent and the project agent. Its main responsibilities include to coordinate the registration of initial necessary elements for new product design project definition. Here are defined the new product problem description, criteria, participants, alternatives and the survey that will be applied to carry out the market study.

The design team incorporates information and task agents: initialization manager (information), preferences aggregation/disaggregation agent (task) and market segmentation agent (task). Its main responsibilities include the initialization of the environment for new product design, optimal recommendation or simulation. It obtains consumer behaviour using multiobjective evolutionary algorithms based on NSGA-II and ELECTRE III method, which are incorporated in the preferences aggregation/disaggregation agent. On the other hand, market segmentation agent implements a multi-objective evolutionary algorithm that segment the market from a multicriteria analysis point of view forming classes of consumers indifferent each other. Output of both elements are used as input for the choice team agents.

The choice team consist of five agents: simulation manager agent (information), scenario generator agent (task), brand choice agent (task), automatic recommendation agent (task) and aggregation/dissagregation preferences agent (task). The simulation manager agent stores information about the state of the simulation of a new product design. Scenario generator agent creates the necessary scenarios according to the type of simulation to be carried out, which can be with the initial criteria, initial criteria and price, or initial criteria, price and other criteria. Brand choice agent implements a multiobjective evolutionary algorithm that gives an individual representative of the market segment that will be used in the simulation process to predict how an average consumer will value the design of a new product. Automatic product recommendation agent implements a multiobjective evolutionary algorithm that obtain an optimal product configuration recommendation according to the market study used, the alternatives and criteria. The DM can accept it or go to the simulations process to evaluate other configurations. The simulation agent coordinates an iterative process to simulate different configurations for a new product. It is an iterative process that obtain the position of a new product under design with respect to its competitors in a market segment.

4.2. *System Prototype*

In order to show our approach, we have developed a MIDSS prototype to support new product design decisions. Fig. 2 shows sample screenshots of

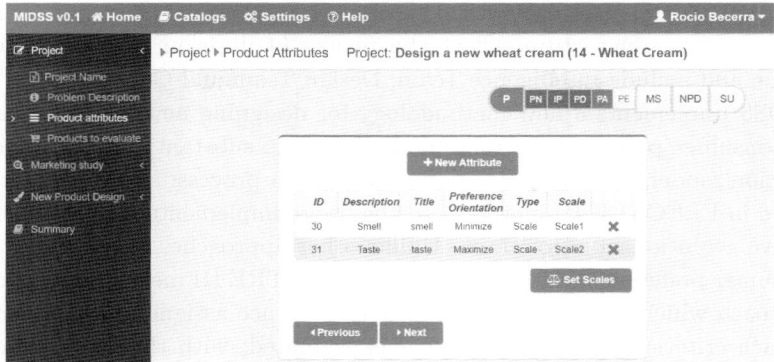

Fig. 2. MIDSS market study description graphic interface.

user interface. The interface is a PHP Web site and provides interaction between the user and MIDSS agent platform to view information, perform useful calculations, what-if analysis and carry out other related tasks. The models, data and user interface are traditional DSS components. Data is collected from applied surveys of market studies and is stored in a relational MS Sql Server database. At this moment, the prototype incorporates the agents of intelligence, design and choice teams. Intelligence allows to define the new product problem description, criteria, participants, alternatives and the survey that will be applied to carry out the market study; Design team agents prepare initial information stored in the market study to be used by simulation team agents. Finally, in the choice team can be obtained an optimal product configuration recommendation or can be simulate different product configurations according to the market study, alternatives and criteria used. It is an iterative process that obtain the position of a new product under design with respect to its competitors in a market segment. The DM can accept it or go to evaluate different simulated until find a satisfactory one.

5. Conclusions and Future Work

This paper presents a distributed agent-based architecture for a new decision support system. It is argued that a combination of MDSS, agent technologies, multi-objective evolutionary algorithms, ELECTRE III method to support decision making in marketing applications will prove to be a powerful tool. The MIDSS architecture has three different agent types:

interface agents, task agents and information agents. The major components of the architecture reflect the phases of decision making stage of Simon and include Intelligence Team, Design Team and Choice Team. The MIDSS implements a new methodology for designing new products based on consumer preferences. Such methodology is substantiated on Simon's decision model and generic new product design process. Their models are based in ELECTRE III method and they were implemented using multiobjective evolutionary algorithms. Unlike other approaches in the literature, consumer preferences are modeled using ELECTRE III method as a flexible approach which allows to consider the importance assigned by a customer to each criterion to when a product is evaluated; with this, we intend to model more appropriately consumer preferences with respect to the current proposals. Future work includes using the MIDSS in a distributed real-world context.

References

1. W. Ai, J. Sun and H. Li, A distributed marketing decision support system based on multi-intelligent-agent, in *Proceedings of 2004 International Conference on Machine Learning and Cybernetics (IEEE Cat. No.04EX826)*, Aug 2004.
2. G. Alexouda, *Decis. Support Syst.* **38**, 495(January 2005).
3. C. Baril, S. Yacout and B. Clément, *Optimization and Engineering* **13**, 121(Mar 2012).
4. N. F. Matsatsinis, P. Moraitis, V. Psomatakis and N. I. Spanoudakis, *Applied Artificial Intelligence* **17**, 901 (2003).
5. C. G. Soumya Banerjee, Ajith Abraham, *International Journal of System Management* **3**, 51 (2005).
6. X. F. Zha, R. D. Sriram and W. F. Lu, Knowledge intensive collaborative decision support for design process, in *International Design Engineering Technical Conferences and Computers and Information in Engineering Conference*, 2003.
7. R. Vahidov and R. Fazlollahi, *Journal of Computer Information Systems* **44**, 87 (2004).
8. H. Simon, *Administrative Behavior, 4th Edition* (Free Press, 2013).
9. T. Hasenkamp, T. Adler, A. Carlsson and M. Arvidsson, *Total Quality Management & Business Excellence* **18**, 351 (2007).
10. K. Sycara, A. Pannu, M. Williamson, D. Zeng and K. Decker, *IEEE Expert: Intelligent Systems and Their Applications* **11**, 36(December 1996).

A multicriteria and multiobjective approach for the market segmentation problem[*]

Gastelum-Chavira, Diego Alonso
Management Science Doctoral Program, Universidad Autonoma de Occidente
Culiacan, Mexico

Leyva-Lopez, Juan Carlos
Economic and Management Science Department, Universidad Autonoma de Occidente
Culiacan, Mexico

Larreta-Ramirez, Elsa Veronica
Management Science Doctoral Program, Universidad Autonoma de Occidente
Culiacan, Mexico

Within for-profit organizations, market segmentation plays an important role in marketing. Theoretically, individuals within market segments share similarity in some variables considered in a market study. In this paper, a market segmentation model is presented under a multicriteria outranking approach and modeled as a multiobjective optimization problem. This problem is addressed by using a multi-objective evolutionary algorithm and two empirical cases are presented as results.

1. Introduction

As a general concept, segmentation consists in dividing something into smaller parts. Inside organizations, especially in for-profit organizations, segmentation plays an important role in marketing because regardless if the market is physical or not, typically there are heterogeneous individuals which have diverse points of view and preferences regarding products and services.

Segments in a good segmentation must be homogeneous and at the same time, be heterogeneous to each other. However, there are some elements to be considered during the segmentation process; according with Kotler and Keller,[1] the marketer must identify the number and nature of segments and then, he/she needs to decide the target segments. Additionally, it is necessary to process information related with expectations and preferences of a set of customers; where

---

[*]This work was supported by the Mexican National Council of Science and Technology.

such information is often subjective. Moreover, as note by Maričić and Đorđević,[2] possibly information processing is based on different statistical methodologies that often do not provide same results. In papers about market segmentation problem, it is classical to find methods such as k-means, hierarchical clustering, conjoin analysis as well as other approaches: Fuzzy Clustering methods in Chan et al.,[3] Simulated Annealing Heuristic in Brusco et al.,[4] multicriteria decision making and clustering in Cho et al.,[5] etc.

On the other hand, and due to its nature, the marketing segmentation problem, can be addressed as multiobjective optimization problem, since individuals within a segment are similar to each other and, at the same time, are dissimilar to individuals that belong to other segments.

The aim of this work is to present a method based on the multicriteria outranking approach ELECTRE[7] and a multiobjective evolutionary algorithm for the market segmentation problem. The paper is organized as follows: in Section 2, the market segmentation problem is defined from the multicriteria and multiobjective approaches. Section 3 includes an empirical evaluation with two instances. Finally, Section 4 is intended for conclusions and future work.

2. The Market Segmentation Problem from Multicriteria and Multiobjective Perspectives

From a multicriteria outranking point of view, the market segmentation problem can be defined as follows:

Let $A = \{a_1, a_2, ..., a_m\}$ be a set of alternatives that represents consumers of a market, which is valued by a set of criteria $G = \{q_1, g_2, ..., g_n\}$. In addition, let S_A^σ be a fuzzy outranking relation defined on $A \times A$, which integrates the preferences of consumers and a decision maker on the multiple criteria that describe A. This S_A^σ is generated using ELECTRE III method, but it could be created with another outranking method such as PROMETHEE.

The problem is to exploit S_A^σ in order to obtain a market segmentation P_A; where consumers belonging to a specific segment $C_r \in P_A$ are indifferent each other; at the same time, they are not indifferent to consumers of another segment $C_q \in P_A$.

Given two alternatives $a_i, a_j \in A$, $a_i S_A^\sigma a_j$ means that "a_i is at least as good as a_j". By using a credibility or cut level $\lambda-cut$ on S_A^σ, a crisp outranking relation S_A^λ is obtained. This relation is used to accept or reject the predicate "a_i is at least as good as a_j" with credibility level $\lambda-cut$. For each $a_i, a_j \in A$, S_A^λ is deduced by the following relations:

- Indifference:
$$I_A : a_i I_A a_j \leftrightarrow a_i S_A^\sigma a_j \wedge a_j S_A^\sigma a_i \qquad (1)$$

- Not Indifference $\neg I_A$:

$$\neg I_A : a_i \neg I_A a_j \leftrightarrow a_i S_A^\sigma a_j \wedge a_j \neg S_A^\sigma a_i \text{ or}$$
$$a_i \neg I_A a_j \leftrightarrow a_i \neg S_A^\sigma a_j \wedge a_j S_A^\sigma a_i \text{ or} \qquad (2)$$
$$a_i \neg I_A a_j \leftrightarrow a_i \neg S_A^\sigma a_j \wedge a_j \neg S_A^\sigma a_i$$

where $\neg$ is the logical NEGATION operator and $\wedge$ is the logical AND operator.

In addition to the multicriteria approach, the marketing segmentation problem was modeled as a multiobjective optimization problem. From this approach, S_A^σ is exploited to obtain that segmentation using a Multiobjective Evolutionary Algorithm (MOEA), which was created for this purpose. The MOEA compares consumers in terms of preferences. Among all possible crisp outranking relations S_A^λ obtained with the MOEA, it tries to find the best segmentations which are the most compatible with the preference information contained in S_A^σ; i.e., it tries to identify the S_A^λ that minimize the following inconsistencies:

i. $a_i, a_j \in C_r$ while $a_i \neg I a_j$ in S_A^σ; $C_r \in P_A$,

ii. $a_i \in C_r$ and $a_j \in C_q$ with $C_r \neg I_A C_q$ while $a_i I_A a_j$ in S_A^σ; $C_r, C_q \in P_A$,

iii. Given two crisp outranking relations $S_{A_i}^\lambda$ and $S_{A_j}^\lambda$ with different $\lambda - cut$, the segmentation with the lower $\lambda - cut$ is chosen.

where P_A is a consumer market segmentation and $\neg I_A$ is the Not-Indifference relation defined previously.

In order to describe the violation of these three situations for a given relation S_A^λ, a set of three indicators were defined: heterogeneity *HG*, incoherence *IC* and trustworthiness *TW*. The *HG* indicator is related with the inconsistency number *I*, meanwhile the *IC* indicator is regarding with the *ii* inconsistency and the *TW* indicator is with inconsistency number *iii*. Thus, the multiobjective optimization problem is summarized as follows:

$$Min(HG(S_A^\lambda)), Min(IC(S_A^\lambda)), Max(TW(S_A^\lambda)) \qquad (3)$$

subject to:

$$S_A^\lambda \in \Omega; \ \lambda \in [0,1], \lambda \geq 0.5.$$

where Ω is the set of reflexive crisp outranking relations of segments of customers of A and λ is the level of credibility.

3. An Empirical Validation

As a first approximation to know the performance of the MOEA, two ad hoc instances for the multicriteria segmentation problem were created. These instances were done by hand and each one corresponds to a fuzzy outranking relation. The first relation was created without inconsistencies, i.e. no loops, no missing information and it has a perfect representation in a segmentation solution for a $\lambda = 0.8$. On the other hand, the second relation was created without loops and without missing information but it has at least two representations in a segmentation solution for a $\lambda = 0.8$. Every fuzzy outranking relation represents the preferences of a decision maker on a set of criteria which evaluate a set consumers. The first instance is shown in Figure 1.

a) Fuzzy Outranking Relation

	a_0	a_1	a_2	a_3	a_4	a_5
a_0	1.00	0.80	0.80	0.80	0.80	0.80
a_1	0.80	1.00	0.80	0.80	0.80	0.80
a_2	0.80	0.80	1.00	0.80	0.80	0.80
a_3	0.00	0.00	0.00	1.00	0.80	0.80
a_4	0.00	0.00	0.00	0.80	1.00	0.80
a_5	0.00	0.00	0.00	0.00	0.00	1.00

b) Crisp Outranking Relation obtained from a) with a cut level of 0.80.

	a_0	a_1	a_2	a_3	a_4	a_5
a_0	1	1	1	1	1	1
a_1	1	1	1	1	1	1
a_2	1	1	1	1	1	1
a_3	0	0	0	1	1	1
a_4	0	0	0	1	1	1
a_5	0	0	0	0	0	1

Figure 1. Instance without inconsistencies for the segmentation problem.

Considering the crisp outranking relation shown in Figure 1b, the expected segmentation which should be obtained by the MOEA is presented in Figure 2:

Alternative	a_0	a_1	a_2	a_3	a_4	a_5
Segment of the alternative	0	0	0	1	1	2

Figure 2. Expected segmentation to be obtained by the MOEA.

Once the MOEA was executed, it obtained the results shown in Figure 3:

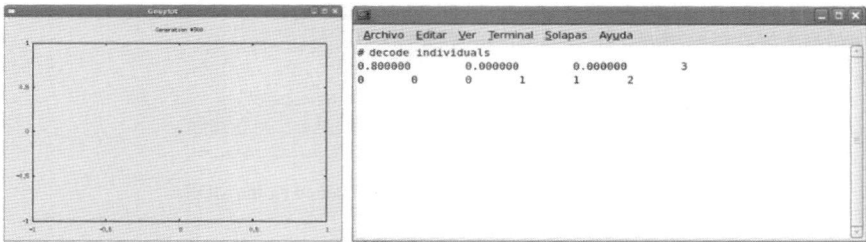

Figure 3. Results obtained by the MOEA.

The interpretation of the obtained result (Figure 3) is a solution without inconsistencies. In the plot, a (0,0) point is presented. It means that there is not any inconsistency within and outside the segments. On the other hand, in the right side of Figure 3, 0.80000 value means a $\lambda - cut = 0.8$, 0.00000 means non-inconsistencies within the segments, next 0.00000 means non-inconsistencies outside the segments, the number 3 means that 3 segments were created. Next values 0, 0, 0, 1, 1, 2 means that alternatives a_0, a_1 and a_2 belong to segment 0; meanwhile alternatives a_3, a_4 belong to segment 1; and finally, alternative a_5 belongs to segment 2. This result is equal to the expected segmentation shown in Figure 2.

On the other hand, the second instance was provided to the MOEA to obtain a segmentation. The fuzzy outranking relation and the crisp outranking relation are shown in Figure 4.

a) Fuzzy Outranking Relation						b) Crisp Outranking Relation obtained from a) with a cut level of 0.75.					
	a_0	a_1	a_2	a_3	a_4	a_5		a_0	a_1	a_2	a_3	a_4	a_5
a_0	1.00	0.80	0.80	0.80	0.80	0.80	a_0	1	1	1	1	1	1
a_1	0.80	1.00	0.80	0.80	0.80	0.80	a_1	1	1	1	1	1	1
a_2	0.80	0.80	1.00	0.80	0.80	0.80	a_2	1	1	1	1	1	1
a_3	0.80	0.80	0.50	1.00	0.80	0.80	a_3	1	1	0	1	1	1
a_4	0.50	0.50	0.80	0.80	1.00	0.80	a_4	0	0	1	1	1	1
a_5	0.50	0.50	0.50	0.50	0.50	1.00	a_5	0	0	0	0	0	1

Figure 4. Instance without inconsistencies and two representations for the segmentation problem.

Again, considering the crisp outranking relation shown in Figure 4b, it is expected that the MOEA finds at least, one of the solutions shown in Figure 5:

Solution 1:

Alternative	a_0	a_1	a_2	a_3	a_4	a_5
Segment of the alternative	0	0	0	0	1	2

Solution 2:

Alternative	a_0	a_1	a_2	a_3	a_4	a_5
Segment of the alternative	0	0	0	1	1	2

Figure 5. Expected segmentation to be obtained by the MOEA.

In this case in point, Solution 1 was found by the MOEA, which is shown in Figure 6.

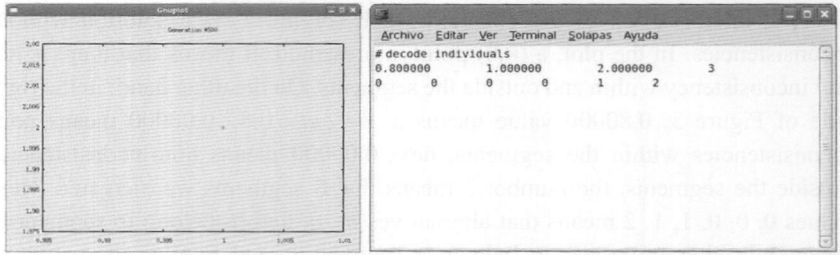

Figure 6. Solution with inconsistencies obtained by the MOEA.

The MOEA found a segmentation composed by three segments and three inconsistencies for $\lambda = 0.8$. One inconsistency with-in the segments (1.00000) and two inconsistencies outside the segments (2.000000). This result is equal to one of the expected segmentation shown in Figure 5, where: a_0, a_1, a_2 and a_3 belong to segment 0; a_4 and a_5 belong to segment 1 and segment 2 respectively.

4. Conclusions and Future Work

The main objective of this contribution was to present a method for the marketing segmentation problem from a multiobjective and multicriteria points of view. From our perspective, a method for this problem must to consider both: the preferences of consumers and decision makers. These preferences can be integrated in a fuzzy outranking relation, following an outranking relational approach such as ELECTRE or PROMETHEE. Later, it can be exploited to obtain a segmentation. The quality of the segmentation must be valued considering the number of inconsistencies with-in and outside the segments.

In this work, two empirical cases were performed following the presented approach. The obtained results by our MOEA showed to be consistent with the information aggregated in the fuzzy outranking relations. Nevertheless, it is strongly required to expand the number of test in order to know the proposed method's performance, while carrying out simulations with different number of consumers and segments; and later, a real case must be performed.

Acknowledgments

The work described on this contribution was partially supported by the National Council of Science and Technology (CONACyT), Mexico.

References

1. Kotler, P. & Keller, K. L. *Marketing Management*. (Pearson, 2016).
2. Maričić, B. R. & Đorđević, A. Članci / Papers Strategic Market Segmentation. *Marketing* 243–252 (2011).
3. Chan, K. Y., Kwong, C. K. & Hu, B. Q. Market segmentation and ideal point identification for new product design using fuzzy data compression and fuzzy clustering methods. *Appl. Soft Comput.* **12**, 1371–1378 (2012).
4. Brusco, M. J., Cradit, J. D. & Stahl, S. A Simulated Annealing Heuristic for a Bicriterion Partitioning Problem in Market Segmentation. *J. Mark. Res.* **39**, 99–109 (2002).
5. Cho, I., Kim, J. K., Park, H. & Lee, S. M. Industrial management & data systems article information. *Ind. Manag. Data Syst.* **114**, 1360–1377 (2014).
6. Liu, Y., Ram, S., Lusch, R. F. & Brusco, M. Multicriterion Market Segmentation: A New Model, Implementation, and Evaluation. *Mark. Sci.* **29**, 880–894 (2010).
7. Roy, B. The outranking approach and the foundations of electre methods. *Theory Decis.* **31**, 49–73 (1991).

A new disaggregation preference method for new products design[*]

Pavel Anselmo Alvarez[†] and Juan Carlos Leyva Lopez
Management and Economic Science, Universidad Autónoma de Occidente
80120 Culiacan, Sinaloa, Mexico

Pavel Lopez Parra
Doctoral Program of Management Science, Universidad Autónoma de Occidente
80120 Culiacan, Sinaloa, Mexico

> The consumer satisfaction is an interesting research field for new products design. Indirect procedure is an important approach to extract consumer preference and construct preferential model. The contribution of the paper is a method to elicit parameters indirectly by using holistic information provided by the consumer through a reference ranking set and additional preference information. In this paper, we propose an aggregation-disaggregation approach for ELECTRE III based on a genetic algorithm to deal with inherent issues of the problem. The disaggregation method finds a set of parameters obtained by the genetic algorithm for the construction of a valued outranking relation model that best restore the reference ranking of the consumer.

1. Introduction

Analysis of consumer behavior and market seems to be of interest for new products design methodologies. It seems that adapt swiftly to changes in consumer satisfaction will obtain competitive advantages [1]. In this paper, a New Product Design Multicriteria Methodology (NPDMM) based on consumer preferences is considered to propose a new disaggregation preference method for new products design. The disaggregation analysis regards overall preferences that for real problems could be infeasible, not just for the complexity of mathematical properties of the problem modeling, also because the inconsistency of data when consumer express their preferences to attributes of the products, thereby methods are required to deal with inherent characteristic of the multicriteria decision aids.

---

[*]This work is supported by Mexican National Council for Science and Technology (CONACYT) and to the Strengthening of Educational Quality Program (PFCE-2016-2017).
[†]Corresponding author: Pavel Anselmo Alvarez, pavel.alvarez@udo.mx, (+52) 6677591300, ext. 2309.

The presented preference disaggregated analysis takes the consumer preference in the form of ranking of some stimuli as reference set that is called preference actions A^*. The presented method corresponds to a genetic algorithm of preference disaggregation as an indirect approach to determine the preference parameters, which are required by ELECTRE III [2] to construct a valued outranking relation that represent the consumer's preference. Thereby, the consumer's preferential model is as consistent as possible with their holistic preference. The main objective of the present paper is to contribute in the development of new product design with a new disaggregation method for consumer's preferences, concerning consumer satisfaction and market trends, applying a study case of local cold tea market. The method disaggregates overall preferences of the consumer from a preorder (complete or partial) in a set of inter-criteria parameters.

The paper is organized as follow. Section 2 contains the preference disaggregation procedure for consumer choice. The general model for disaggregates consumer preference for new product design is presented in Section 3. Section 4 shows a case study of disaggregating consumer preference for a local tea market. Finally, Section 5 draws the main conclusions.

2. Preference Disaggregation of Cold Tea Consumer Choice

Consumer behavior analysis is accomplished utilizing multicriteria preference analysis for each consumer separately in combination with data analysis methods and techniques concerning consumer profiles in general. A consumer selection policy can be externalized by means of a set of reference products which he/she, the consumer, either has or can rank through simple questionnaires, familiar decision making situations, and so on [3].

The disaggregation analysis problem here addressed is about how to estimate for a given consumer-outranking model, which should be as consistent as possible with the consumer ranking of these reference products. During the survey, each consumer is asked to express his/her estimation for a set of products $A = \{a_1, a_2, ..., a_m\}$, and evaluate the products for each attribute based in her/his satisfaction level. A Preference Disaggregation Analysis (PDA) with ELECTRE III is used to support consumer to construct an aggregation preference model. Due the original data can be imprecise and uncertain; the proposed technique seems appropriate as a part of the multicriteria decision analysis approach for the consumer preference to products market. The presented method corresponds to a genetic algorithm developed as disaggregation approach, which may produce a representative compatible set of preference parameters for ELECTRE III. The fuzzy outranking relation containing the preferential model is derived from the inferred parameters, and should be as consistent as possible with the ranking reference of consumers.

2.1. The ELECTRE III method

The outranking method used for the preference disaggregation analysis is ELECTRE III which constructs a fuzzy outranking relation expressed as σ (a_i, a_l) [2]. First ELECTRE III constructs the comprehensive concordance index from the partial concordance index. C_j (a_i, a_l) is the partial concordance index, a fuzzy measure that evaluates if "action a_i is at least as good as action a_l" on the criterion g_i. C (a_i, a_l) is the comprehensive concordance index that measures the performances on all criteria of the pair of alternatives (a_i, a_l) ∈ $A \times A$ to evaluate in what grade the criteria support the assertion "a_i outranks a_l". Then, the method estimates the credibility index $\sigma(a_i,a_l); (0 \leq \sigma(a_i,a_l) \leq 1)$ which assesses the strength of the assertion that "a_i is at least as good as a_l", $a_i S a_l$.

$$\sigma(a_i,a_l) = \begin{cases} C(a_i,a_l) & \text{if } \overline{F}(a_i,a_l) = 0 \\ C(a_i,a_l) \times \prod_{j \in \overline{F}(a_i,a_l)} \dfrac{1-d_j(a_i,a_l)}{1-C(a_i,a_l)} & \text{if } \overline{F}(a_i,a_l) \neq 0 \end{cases} \quad (1)$$

where d_j (a_i, a_l) is the discordant index and $\overline{F}(a_i,a_l)$ is the set of pairs where d_j (a_i, a_l) > C (a_i, a_l).

The distillation-ranking algorithm is used in the exploitation procedure of the fuzzy outranking relation. It is based on the degree of credibility of each pair of actions in order to get a final partial preorder, resulting from the intersection of two complete preorders. The ascending and descending distillations construct each a complete preorder and the combination (intersection) of the two preorders gives the final ranking.

2.2. Model to minimize inconsistencies of the consumer preference

The proposed model is designed to take into account the consumer preference to minimize the inconsistencies between those preferences and the preferential model, constructed from some inter-criteria parameters inferred in the disaggregation process.

In order to use an ELECTRE III method, one must specify the following parameters:

i) The weight vector $W = (w_1, w_2, ..., w_n)$, $w_i \geq 0$ y $\sum_{i=1}^{n} w_i = 1$.

ii) The vector of indifference thresholds $q=(q_1,q_2,...,q_n)$, $q_i \geq 0$ $i=1,2,...,n$.
iii) The vector of preference thresholds $p=(p_1,p_2,...,p_n)$, $p_i \geq 0$, $i=1,2,...,n$.
iv) The vector of veto thresholds $v=(v_1,v_2,...,v_n)$, $v_1 > p_i$, $i=1,2,...,n$.

Solution representation and problem formulation

The direct representation of the parameters regarding the preference, indifference and veto thresholds, does not allow an efficient search of the solution space. Therefore, an indirect representation scheme is employed, adopted from [4]. The above parameters are defined for each criterion j as follows:

$$g_j(b) - q_j = g_j(b) - p_j + y_j \quad \forall j = 1,2,...,n \tag{2}$$

$$g_j(b) + p_j = g_j(b) - q_j + z_j \quad \forall j = 1,2,...,n \tag{3}$$

$$g_j(b) - v_j = g_j(b) - p_j - x_j \quad \forall j = 1,2,...,n \tag{4}$$

The parameters of the ELECTRE III method can be easily expressed in terms of the decision variables x_j, y_j, z_j. In particular from (3), it follows that $p_j = z_j - q_j$ substituting this to Eq. (2) leads to

$$q_j = \frac{z_j - y_j}{2} \tag{5}$$

To ensure the non-negativity of the indifference threshold, the variables y_j and z_j should satisfy the constraint $z_j \geq y_j$. With Eq. (5), the preference threshold can now be expressed as

$$p_j = \frac{z_j + y_j}{2} \tag{6}$$

From Eqs. (5) and (6), it follows that the preference and indifference thresholds always satisfy $p_j \geq q_j$. To obtain the veto threshold, one can substitute (6) on (4) to get the following expression:

$$g_j(b) - v_j = g_j(b) - (\frac{z_j + y_j}{2}) - x_j \tag{7}$$

Finally, (6) can be used to obtain the veto threshold from Eq. (4):

$$v_j = x_j + \tfrac{1}{2} y_j + \tfrac{1}{2} z_j \tag{8}$$

From this equation, together with (6), it can be easily seen that $v_j \geq p_j$.

Thus, given that $x_j, y_j, z_j \geq 0$ and $z_j > y_j$, the selected representation ensures that $v_j \geq p_j \geq q_j \geq 0 \; \forall j =1,2,\ldots n$.

To reduce the search space, we also impose the condition $p_j \leq x_j^* - x_{j*}$, where $x_j^* = \max_{i \in A^*}\{x_{ij}\}$ and $x_{j*} = \min_{i \in A^*}\{x_{ij}\}$. This condition is ensured with the constraint:

$$y_j + z_j \leq 2(x_j^* - x_{j*}) \tag{9}$$

The solution vector has n x, y, z variables, n variables corresponding to the criteria weights, one variable corresponding to the cutting point λ and n binary variables $t_j \in \{0,1\}$ introduced to indicate whether a criterion j can pose veto ($t_j = 1$) or not ($t_j = 0$). Thus, overall the solution vector has $5n+1$ decision variables. We use a value-encoding scheme to represent a solution vector (a potential solution) $\tilde{p}$. Let $\tilde{p} = p_1 p_2 \cdots p_{5n+1}$ be the schematic representation of an individual's chromosome. $\tilde{p} \in \prod_{i=1}^{5n+1} C_i$, where C_i is the set of values that p_i can takes. This set of values is dependent of the problem.

The general form of the optimization problem can be expressed as follows:

$$\text{Min} \quad K(\vec{w},\vec{x},\vec{y},\vec{z},\vec{t},\lambda)$$
s.t.
$$y_j + z_j \leq 2(x_j^* - x_{j*}) \quad j=1,2,\ldots,n$$
$$w_1 + w_2 + \ldots + w_n = 1 \tag{10}$$
$$\vec{z} \geq \vec{y}$$
$$\vec{w},\vec{x},\vec{y},\vec{z},\vec{t} \geq \vec{0}$$
$$t \in \{0,1\}, \lambda \in (0,1), \lambda > 0.5$$

where $\vec{w},\vec{x},\vec{y},\vec{z},\vec{t}$ are the vectors of the decision variables w_j, x_j, y_j, z_j, t_j $j = 1,2,\ldots n$, and $K(\vec{w},\vec{x},\vec{y},\vec{z},\vec{t},\lambda)$ denotes the inversion of the ranking, minimizing this error increase the ranking accuracy of the model corresponding to a specific solution $(\vec{w},\vec{x},\vec{y},\vec{z},\vec{t},\lambda)$.

3. Case Study for Disaggregating Consumer Preference for a Local Tea Market

We tested the new method with data collected by a pilot survey questionnaire in a small city in the northwest of Mexico. The pilot survey was done for improve the questionnaire. However, as aim of this work is to test the method of disaggregation of preference, so it was done with the answers of a single consumer.

3.1. Criteria definition for the tea consumer preference

In [5], it is studied attributes which contribute to the role on perceived quality of tea and coffee, and consumers' attitudes regarding the perceived quality of tea and coffee drinks. It seems that direct measurements of perceived intensities of target attributes are appearance, color, aroma, taste and texture [6]. In this sense, some attribute for consumer satisfaction evaluation were defined and used as decision criteria as listed below.

Attributes of cold tea products defined as decision criteria: Flavor (C1), Net content (C2), Price (C3), Design of the product (C4), Color (C5), Odor (C6), Healthy (C7). Five different brands were considered in the problem. Table 1 shows the evaluation of a single consumer to the five products regarding different brands each (*P1, P2, P3, P4, P5*). The consumer was required by a questionnaire to respond her/his satisfaction level for each criterion for every product in a scale of [1,5], where 1 is the best level of satisfaction and 5 is not satisfaction at all.

Table 1. Evaluation of products by consumer.

	C1	C2	C3	C4	C5	C6	C7
P1	2	2	3	1	3	4	1
P2	1	2	3	1	3	3	3
P3	3	3	3	3	3	3	3
P4	3	2	3	3	3	3	3
P5	4	4	3	3	4	4	4

The consumer is required to express a ranking reference of products and express the importance order of criteria as secondary preference information. Primary preference information is shown in Table 2. The consumer defined strict preference in first three positions in the ranking, in the third position P3 and P4 are defined indifferent and P5 is incomparable with P3 and P4 (see Table 2a)). Table 2b) represents the preference matrix from consumer, where four binary relation are considered for pair (a_i, a_l): *P* is strict preference, *P-* is inverse preference, *I* is indifference and *R* is incomparability. Both a) and b) represent the same preference information, but the genetic algorithm uses the format from b). The secondary preference information is shown below as complete order of criteria: *C1>C4>C7>C5>C6>C2>C3*.

A set of inter-criteria parameters was inferred from the preference disaggregation genetic algorithm and model in Eq. (10). The result of the disaggregation process for the consumer preference is shown in Table 3.

Table 2. Consumer's rank reference of beverage for cold tea.

a)

Position	Preference	Incomparable
1	P1	
2	P2	
3	P3, P4	P5
4		
5		

b)

	P1	P2	P3	P4	P5
P1	I	P	P	P	P
P2	P-	I	P	P	P
P3	P-	P-	I	I	R
P4	P-	P-	I	I	R
P5	P-	P-	R	R	I

Table 3. Solution from disaggregated process of consumer preference.

	C1	C2	C3	C4	C5	C6	C7
w	0.525	0.029	0.028	0.278	0.048	0.033	0.059
q	1	0	0	0	0	0	1
p	1	0	0	1	0	0	1
v	2			2			2

Table 4a) shows the corresponding fuzzy outranking relations representing the consumer preference model and the derived ranking is in Table 3b) column "PDA". We can observe some difference presented in the consumer choice from the ranking reference (column "Consumer choice") and the derived ranking (column "PDA"). The ranking generated with the exploitation of the generated consumer preferential model presents one inversion in the ranking in comparison with the consumer choice in the ranking format. An inversion is considered when one ranking shows the preference pair as a_iPa_l and the other ranking shows a different preference as $a_iP\text{-}a_l$, a_iIa_l, a_iRa_l. The distance in terms of [7] between consumer ranking and ranking representing the consumer model is 6 as total divergence (see Table 4b)). The consumer choice shows $P3RP5$ and $P4RP5$, while PDA shows $P3PP5$ and $P4PP5$. This divergence regards the penalization for the inversion of $(P3, P5)$ and $(P4, P5)$ between partial rankings.

Table 4. Valued outranking relation for tea consumer preference and corresponding ranking.

a)

	P1	P2	P3	P4	P5
P1	1	0.75	0.93	0.93	1
P2	0	1	1	1	1
P3	0	0.4	1	0.97	1
P4	0	0.43	1	1	1
P5	0	0	0	0	1

b)

Position	Consumer choice		PDA	
	Preference	Incomparable	Preference	Incomparable
1	P1		P1	
2	P2		P2	
3	P3, P4	P5	P3, P4	
4			P5	

4. Conclusion

The paper addresses consumer preference Disaggregation Analysis Approach to construct a preference model from ranking products. Particularly, the disaggregation procedure constructs a consumer's preferential model as consistent as possible with their holistic preference. The interesting of this approach is avoiding some problems can be presented in direct methods, because the aggregation phase requires elicitation of parameters that are far different from natural terms in which consumer express his/her preferences and expertise. The process of disaggregation of genetic algorithm deals with some problems presented when consumer express their preferences about products, as imprecision and uncertainty. Because the reference ranking of consumer is disaggregated in inter-criteria parameters and those are continually evaluated in terms of matching the holistic preference. In that case if the data do not present consistency with the holistic preference, the disaggregation process constructs the fuzzy outranking relation as consistent as possible with her/his preferences concerning data and holistic preference of the consumer.

References

1. N. F. Matsatsinis, and Y. Siskos, "New product development methodology," *Intelligent support systems for marketing decisions*, N. F. Matsatsinis and Y. Siskos, eds., Norwell, MA.: Kluwer Academic Publisher, 2003.
2. B. Roy, "The Outranking Approach and the Foundations of ELECTRE Methods," *Reading in Multiple Criteria Decision Aid*, C. A. Bana e Costa, ed., pp. 155-183, Berlin: Springer-Verlag, 1990.
3. Y. Siskos, E. Grigoroudis, N. F. Matsatsinis, and G. Baourakis, "Disaggregation analysis in agricultural product consumer behavior," *Advances in Multicriteria Analysis*, P. M. Pardalos, Y. Siskos and C. Zopounidis, eds., pp. 185-202: Kluwer Academic Publishers, 1995.
4. M. Doumpos, Y. Marinakis, M. Marinaki, and C. Zopounidis, "An evolutionary approach to construction of outranking models for multicriteria classification: The case of the ELECTRE TRI method," *European Journal of Operational Research*, vol. 199, no. 2, pp. 496-505, 2009.
5. I. M. Monirul, and J. H. Han, "Perceived Quality and Attitude Toward Tea & Coffee by Consumers," *International Journal of Business Research and Management*, vol. 3, no. 3, pp. 100-112, 2012.
6. R. N. Bleibaum, H. Stone, T. Tan, S. Labreche, E. Saint-Martin, and S. Isz, "Comparison of sensory and consumer results with electronic nose and tongue sensors for apple juices," *Food Quality and Preference*, vol. 13, no. 6, pp. 409-422, 2002.
7. B. Roy, and R. Slowinski, "Criterion of distance between technical programming and socio-economic priority," *RAIRO - Operations Research*, vol. 27, no. 1, pp. 45-60, 1993.

A choice model for the product design problem based on the outranking approach

Leyva-Lopez, Juan Carlos and Leon-Santiesteban Martín

Economic and Management Sciences Department, Universidad Autónoma de Occidente
Culiacán, México

Ahumada-Valenzuela, Omar and Romero-Serrano, Alma Montserrat

Economic and Management Sciences Department, Universidad Autónoma de Occidente
Culiacán, México

> New product design is critical for the long-term profitability and competitiveness of companies. Even though many optimization algorithms have been applied for solving the problem, most of them adopt choice models based on utilities to simulate the consumer's choice behavior. However, there could also be other choice models that are not dependent on utilities. In this paper, we propose a choice model based on the outranking approach that can be used with algorithms that solve the optimal product design problem, using the share of choices criterion. Our model deals with preference heterogeneity among consumers, while the model's predictive accuracy is optimized using Genetic Algorithms.

1. Introduction

In today's competitive environment the development of appropriate new products is necessary for the survival of a company. The optimal product design problem is a significant part of the new product development problem and one of the most crucial decisions for a company [1].

Several intelligent models and algorithms have been applied to the product design area, contributing greatly to its formalization and automation [2]. Normally, these models and algorithms have been incorporated into Marketing Decision Support Systems (MkDSS) in [3]; [1]; [4] and [5]. While the optimization part of the problem has been abundantly studied, little attention has been paid to the models simulating the consumer's choice process.

In this paper, we propose a choice model based on the outranking approach for handling in algorithms that solve the optimal product design problem, using the share of choices criterion. The use of Genetic Algorithms allows for the representation of preference heterogeneity among consumers with the highest possible accuracy, while exceptionally improves the model's predictive performance.

2. The Product Design Problem

In the Product Design Problem, a company tries to find the best combination of attributes and levels that can better satisfy consumers desires and or requirements, to increase their profits through the sale of that product in the marketplace. A new product design is based on consumer's preferences and needs (considering the market for that product), but also on the feasibility of making the product at a profit. The new product must bring enough new revenue to justify designing, producing and distributing to the selected market segment.

Usually the objective in this type of problem is to gain market share, or optimizing market share, which is used as a proxy to profit optimization, since it is associated to higher revenue levels. But our research considers alternative methods of designing new products, such as Multicriteria Decision Aid (MCDA), by using directly the stated consumer preferences (e.g. Outranking methods).

The aggregation model of preferences must represent at least three sources of consumer heterogeneity [6]: response heterogeneity, perceptual heterogeneity, structural heterogeneity, and form heterogeneity.

3. Proposed Model

3.1. *Preference Disaggregation*

To model preference heterogeneity at the consumer level, we apply the preference disaggregation method developed by [7] to each consumer. The method estimates a fuzzy outranking relation defined on $A \times A$, where A is the set of products under consideration in a specific product design problem. This means that we associate with each ordered pair $(a, b) \in A \times A$ a real number $\sigma(a, b)$, $(0 \leq \sigma(a, b) \leq 1)$ reflecting the degree of strength of the arguments favoring the crisp outranking relation aSb. $\sigma(a, b)$ can be interpreted as the credibility degree of the predicate "a is at least as good product as b."

3.2. *Frequency Based Choice Model*

The fuzzy outranking relation S_A^σ inferred from the application of ALG-GEN-ELECTRE represents the first three sources of consumer heterogeneity. We exploit the fuzzy outranking relation for proposes of recommendation and analysis. The exploitation phase transforms the global information included in S_A^σ into a global ranking of the elements of A. the exploitation phase could be treated as a multi-objective optimization problem [8], in this way a finite amount of solutions (rankings) can be found which provide the decision-maker with insight into the characteristics of the problem.

A fuzzy outranking relation S_A^q can be exploited with the multi-objective evolutionary algorithm proposed by [9]. As an output, we obtain a restricted Pareto front $PF_{known}^{restricted}$ and the associated final set of solutions (ranking) returned by the MOEA at the termination $P_{known}^{restricted}$.

The number $T(i,j)$, ($1 \leq i, j \leq m$), of times that a product a_i was found at a certain place in the ranking of the individual associated to the members of the final restricted Pareto front is given in Table 1. This table summarizes the results of all the individual preference over a set of products in an evolutionary algorithm.

Table 1. Number of times that a product was found at a certain place in the ranking.

	Weight w_i	Rank	a_1	a_2	...	a_m
	w_1	1	$T(1,1)$	$T(1,2)$	...	$T(1,m)$
	w_2	2	$T(2,1)$	$T(2,2)$	...	$T(2,m)$
$B_j = \sum_{i=1}^{m} w_i T(i,j)$	.	.	.	.	.	.
	.	.	.	.	.	.
	.	.	.	.	.	.
	w_{m-1}	$m-1$	$T(m-1,1)$	$T(m-1,2)$	...	$T(m-1,m)$
	w_m	m	$T(m,1)$	$T(m,2)$	...	$T(m,m)$
Borda Count	$B_j = \sum_{i=1}^{m} w_i T(i,j)$		B_1	B_2	...	B_m

where
$$\sum_{i=1}^{m} T(i,j) = \left|P_{known}^{restricted}\right| \quad \forall \ j=1,2,...,m \quad (1)$$

Given that the ranking of the products provides the information regarding the individual preferences of consumers, for a given product, the number of times that a product is found at a certain place in the ranking is relevant to find their relative preference for that consumer.

To capture the importance of products, we implement a Borda Count (based on Table 1), using the next procedure: the number of times that a product is found at a certain place in the ranking is weighted according to the importance of the relative position of the products to be ranked. It is reasonable to conclude that in certain cases; the rank of the products would not be of equal importance.

The total importance (performance value) of product a_j denoted as B_j, is defined as the following weighted sum:

$$B_j = \sum_{i=1}^{m} w_i T(i,j), \quad j=1,2,...,m \quad (2)$$

where the Borda Count B_j is the relative preference of the product, given the individual consumer obtained from the non-dominated rankings from the outranking relations. We could consider this as the market preferences of individual consumers.

But there could be a differentiated consumer response, based on the consistency of a brand, the quality or other non-tractable features. To account for this, and to each customer, we estimate the products' preference relative weighted frequencies as proxy to market sales:

$$F_{kj} = \frac{B_{kj}}{\sum_{i \in C} B_{ki}} \quad (3)$$

where F_{kj} is the preference relative frequency that customer k chooses product j:

$$B_{kj} = \sum_{i=1}^{m} w_i T(i,j) \quad (4)$$

is the total performance value the consumer assigns to product a_j, C is the set of the considered products.

3.3. Frequency Based Choice Model Calibration

We fit the value of the parameters w_i, $i = 1, 2, ..., m$, individually, to model to better simulate his choice process. Where the values of w_i, $i = 1, 2, ..., m$ are optimized using a method described in this section, so that our model shows the highest possible predicting performance.

3.3.1. Frequency based choice shares estimation

The application of the calibrated preference model to each individual consumer, results into a vector of weighted preference relative frequencies $[F_{i1}, F_{i2}, ..., F_{im}]$, where $i = 1, 2, ..., n$, the number of customers, and $j = 1, 2, ..., m$ the number of products. The total weighted preference relative frequency for a product j results from the integration of the weighted preference relative frequency for the product across all consumers:

$$CF_j = \sum_{i=1}^{n} F_{ij} .$$

From the total weighted preference relative frequency, we estimate the frequency-based choice share for the product j:

$$MS_j = 100 \times \frac{CF_j}{\sum_{k=1}^{m} CF_k} \% \quad (5)$$

3.3.2. *Optimization of the model's parameters*

The primary goal of every choice model is to simulate with the highest possible accuracy the costumers' choice behavior. For this reason, we optimize the values of the parameters w_i, $i = 1, 2, ..., m$, so that the model's predicted choice shares most closely resemble the "external shares". This is a global optimization problem formulated as follows:
Find w_i, $i = 1, 2, ..., m$ that minimize $f = \sum |MS_j - ES_j|$, $j = 1, 2, ..., m$ subject to $w_1 > w_2 > ... w_m$ \hfill (6)

where MS_j represents the shares estimated by the model and ES_j the external shares.

Unfortunately, the complexity for an exhaustive search of the m variables to obtain the minimum value of f is practically unfeasible due to the tremendous computational cost. For this purpose, we adopt the evolutionary paradigm to resolve this problem by means of a genetic algorithm.

3.4. *The Genetic Algorithm*

<u>Representation and Fitness criterion</u>
It seems natural to code the parameter adjustment problems with a string of real numbers. These are simply positive real numbers, which describe the value of a set of parameters. In this case, a potential solution to a parameter adjustment problem may be represented by chromosomes whose genetic material consists of frame numbers (indices). Chromosomes are thus represented by index vectors $W = (w_1, w_2, ..., w_m)$ assigning a real valued weight as the encoding scheme, that is, using real numbers for the representation of chromosome elements (genes) w_i, $i = 1, 2, ..., m$ with $w_1 > w_2 > ... w_m$ and $1 \leq w_i \leq 100$, $i = 1, 2, ..., m$. Using these numbers (instead of binary) representation is that all genetic operators, such as crossover and mutation, should only be applied to genes w_i, and not to arbitrary bits of their binary representation.

An initial population of P chromosomes, $W(0) = \{W_1, W_2, .., W_p\}$ is then generated by selecting P sets of frames whose feature vectors reside in extreme locations of the feature vector trajectory, as described in the temporal variation approach. Since we do have some knowledge about the distribution of local optima, the above approach exploits the temporal relation of feature vectors and increases the possibility of locating sets of feature vectors with small correlation within the first few GA cycles.

The metric f(W) is used as an objective function to estimate the performance of all chromosomes W_i $i = 1, 2, ..., P$ in each population. However, a fitness function is used to map objective values to fitness values, following a rank-based normalization scheme. Chromosomes W_i are ranked in ascending order of f(W), since the objective function is to be minimized. Let $rank(W_i) \in \{1, 2, ..., P\}$ be the rank of chromosome W_i $i = 1, 2, ..., P$, (rank = 1 corresponds to the best chromosome and rank = P to the worst). Defining an arbitrary fitness value F_B for the best chromosome, the fitness $F(W_i)$ of the i-th chromosome is given by the linear function:

$$F(W_i) = F_B - [rank(W_i) - 1]D, \quad i = 1, 2, ..., P \qquad (7)$$

where D is a decrement rate. The major advantage of the rank-based normalization is that, since fitness values are uniformly distributed, it prevents the generation of *super chromosomes*, avoiding premature convergence to local minima. Furthermore, by simply adjusting the two parameters F_B and D, it is very easy to control the *selective pressure* of the algorithm, effectively influencing its convergence speed to a global minimum.

3.4.1. *The genetic algorithm's parameters*

Parent selection method
To produce a child, two k-ary tournaments are held, each of which produces one parent string. These two parent strings are then combined to produce a child.

Crossover and Mutation Operators
In this paper, we make use of a uniform crossover and mutation operators for ordinal encoding.

4. Example and Results

To show numerically the proposed model, we illustrate it with an example of 5 consumers and 6 different products ($a_1, a_2, a_3, a_4, a_5, a_6$). We assume that after running the genetic algorithm we obtained the fitted weights for the first consumers, which are:

$w_1 = 30.23$, $w_2 = 21.70$, $w_3 = 13.66$, $w_4 = 2.04$, $w_5 = 1.73$, $w_6 = 1.31$

with the following Borda Counts for Consumer 1:

$$B_{11} = 260.25, B_{12} = 350.66, B_{13} = 647.20$$

$$B_{14} = 567.28, B_{15} = 478.65, B_{16} = 384.67$$

Then for that Consumer ($k = 1$) and the first Product ($j = 1$) we calculate from (3) the following:

$$F_{11} = \frac{260.25}{260.25 + 350.66 + 647.28 + 567.28 + 478.65 + 647.28} = \frac{260.28}{2688.79} = 0.096$$

Now suppose that for consumers $k = 2,...,5$ and the same product a_1 we have:

$$F_{21} = 0.261, F_{31} = 0.312, F_{41} = 0.054, F_{51} = 0.075$$

Then the total weighted relative frequency for product 1 is:

$$CF_1 = \sum_{k=1}^{5} F_{k1} = 0.096 + 0.261 + 0.312 + 0.054 + 0.075 = 0.798$$

We also can suppose that the total weighted relative frequency for the other products are as follows:

$$CF_2 = 0.986, CF_3 = 1.101, CF_4 = 0.701, CF_5 = 0.768, CF_6 = 0.304$$

Then from (5) we can calculate the frequency-based choice share for product 1 as:

$$MS_1 = 100\left(\frac{0.798}{0.798 + 0.986 + 1.101 + 0.701 + 0.768 + 0.304}\right)\% = 100\left(\frac{0.798}{4.658}\right)\% = 17.13\%$$

5. Conclusions and Future Research

In this paper, we presented a frequency-based choice model that meets suitable conditions, to using it as a tool in the optimal product design problem. Choice models were calibrated for each consumer, through the calculation of best fitting individual w_i *weights*, allowing for the model to adequately represent the consumer heterogeneity. This process is implemented through a *genetic algorithm*, which permitted the (near) optimization of the weights in tractable time. In a future work, we will carry out experiments concerning the predictive performance.

Acknowledgments

This work is part of the "Development of a decision support system based on intelligent agents for the design of new agro-industry products", research project #2015-01-162, financed by the Mexican National Council for Science and Technology (Conacyt).

References

1. G. Alexouda, "A user-friendly marketing decision support system for the product line design using evolutionary algorithms," *Decis. Support Syst.*, vol. 38, pp. 495–509, 2005.

2. A. Kusiak and F. A. Salustri, "Computational intelligence in product design engineering: Review and trends," *IEEE Transactions on Systems, Man and Cybernetics Part C: Applications and Reviews*, vol. 37, no. 5. pp. 766–778, 2007.
3. B. Besharati, S. Azarm, and P. K. Kannan, "A decision support system for product design selection: A generalized purchase modeling approach," *Decis. Support Syst.*, vol. 42, no. 1, pp. 333–350, 2006.
4. P. V. (Sundar) Balakrishnan and V. S. Jacob, "Triangulation in decision support systems: Algorithms for product design," *Decis. Support Syst.*, vol. 14, no. 4, pp. 313–327, 1995.
5. N. F. Matsatsinis and Y. Siskos, "MARKEX: An intelligent decision support system for product development decisions," *Eur. J. Oper. Res.*, vol. 113, no. 2, pp. 336–354, 1999.
6. W. S. DeSarbo *et al.*, "Representing Heterogeneity in Consumer Response Models 1996 Choice Conference Participants," *Mark. Lett.*, vol. 8, no. 3, pp. 335–348, 1997.
7. P. Alvarez Carrillo, J. Leyva Lopez, and O. Ahumada Valenzuela, *Deriving Parameters and Preferential Model for a Total Order in ELECTRE III*. 2017.
8. E. Fernandez and J. C. Leyva, "A method based on multiobjective optimization for deriving a ranking from a fuzzy preference relation," *Eur. J. Oper. Res.*, vol. 154, no. 1, pp. 110–124, 2004.
9. J. C. Leyva López, M. A. Aguilera Contreras, J. C. Leyva-Lopez, and M. A. Aguilera-Contreras, "A multiobjective evolutionary algorithm for deriving final ranking from a fuzzy outranking relation," *Evol. Multi-Criterion Optim.*, vol. 3410, no. Third International Conference, EMO 2005, Lecture Notes in Computer Science, pp. 235–249, 2005.

Interval valued neutrosophic CODAS method for renewable energy selection

Eda Bolturk

Department of Industrial Engineering, Istanbul Technical University
Macka, 34367/Istanbul, Turkey

Ali Karasan

Institute of Natural and Applied Sciences, Yildiz Technical University
Davutpasa, 34347/Istanbul, Turkey

Renewable energy selection is a critical issue for manufacturing companies. Neutrosophic sets are characterized by three components: truth membership, indeterminacy membership, and falsity membership. COmbinative Distance-based ASsessment (CODAS) method which is a new multi criteria decision making (MCDM) technique introduced by Ghorabaee et al. [1]. The aim of this paper is to develop the neutrosophic CODAS method with an application to select the renewable energy selection.

1. Introduction

Nowadays, the effects of the global warming are becoming progressively felt more than ever before due to water, air, and soil pollutions which are caused by the unsystematic increase of the human population. A number of precautions are being taken against the global warming and these precautions aim to reduce the effects of global warming. The most important of these effects is changing usage habits of consumable energy sources which are considered one of the main causes of these pollution to renewable sources of energies. To determine the most effective renewable energy source alternatives for a certain region, many criteria are considered as appraisal factors in the literature. For this reason, in order to deal with the multi-criteria and alternatives, MCDM methods are proposed as the most appropriate techniques. MCDM is a tool that supports decision-makers to subjectively appraise the scores of alternatives with respect to criteria based on existing data [2]. CODAS is one of the newly proposed MCDM method and based on distance [3]. Appraisal score of an alternative is calculated by using its Euclidean distance to the negative-ideal point. If the Euclidean distance is lower than the threshold parameter, Hamming distance is added to the calculations as a secondary measure to obtain appraisal score. Ranks of the alternatives are determined by utilizing the descending order of overall scores. Because real life conditions are not often precise, they cannot be described by crisp or deterministic models. In order to avoid incapableness of ordinary fuzzy and its extensions, Smarandache [4] introduced neutrosophic logic and neutrosophic sets (NSs). The neutrosophic sets are defined as the set where each element of the universe has a degree of truthiness, indeterminacy and falsity which are between]-0,1+[the non-standard unit interval [5]. In neutrosophic sets, uncertainty is represented as truth (degrees of ownership), and falsity (non-ownership) values; inconsistency is represented as indeterminacy (degree of hesitancy) value. Since all types of fuzzy sets agreed on the values of any element in a set has absolute value, this cannot reflect the indeterminacy of decision makers. But in neutrosophic sets, since indeterminacy value is assigned for the distinguishing relativity and absoluteness of decision makers' preferences it is the superiority of neutrosophic sets over the ordinary fuzzy sets and its extensions. Inteval-valued neutrosophic sets are special type of neutrosophic sets.

In this study, an interval-valued neutrosophic CODAS method is developed and applied to determine the most appropriate renewable energy alternatives. The rest of this paper is organized as follows: the preliminaries of single valued neutrosophic sets and interval-valued neutrosophic sets are given in Section 2. In Section 3, our proposed methodology is presented with all its details. In Section 4, an application to determine the best renewable energy source is given by using the proposed interval-valued neutrosophic CODAS method and a sensitivity analysis is also given to check the robustness of the decisions. The paper ends with the conclusions and suggestions for further research with Section 5.

2. Preliminaries

2.1. *Neutrosophic Sets*

Bolturk and Kahraman [9] proposed the symbol $\tilde{\tilde{A}}$ for the neutrosophic set A, that the three dots represent the elements of a neutrosophic set; T, I, F and tilde substitute for fuzzy set.

Definition 1. [6-8] Let E be a universe. A neutrosophic sets A in E are characterized by a truth-membership function T_A, a indeterminacy-membership function I_A and a falsity-membership function F_A. $T_A(x)$; $I_A(x)$ and $F_A(x)$ are real standart elements of [0,1]. It can be written as;

$$\tilde{\tilde{A}} = \{< x, (T_A(x), I_A(x), F_A(x)) >: x \in E, (T_A(x), I_A(x), F_A(x) \in]^-0,1[^+\}. \quad (1)$$

The sum of $T_A(x)$; $I_A(x)$ and $F_A(x)$ is; $0^- \leq T_A(x) + I_A(x) + F_A(x) \leq 3^+$.

2.2. Interval-Valued Neutrosophic Sets

Definition 2. [10] X be a universe of discourse. An interval-valued neutrosophic set N in X is independently defined by a truth-membership function $T_N(x)$, an indeterminacy-membership function $I_N(x)$, and a falsity-membership function $F_N(x)$ for each $x \in X$, where $T_N(x) = [T_{N(x)}^L, T_{N(x)}^U] \subseteq [0,1]$ $I_N(x) = [I_{N(x)}^L, I_{N(x)}^U] \subseteq [0,1]$ and $F_N(x) = [F_{N(x)}^L, F_{N(x)}^U] \subseteq [0,1]$. Also they meet the condition $0 \leq T_N^L(x) + I_N^L(x) + F_N^L(x) \leq 3$. So, the interval-valued neutrosophic set $\widetilde{\widetilde{N}}$ can be shown as:

$$\widetilde{\widetilde{N}} = \{\langle x, [T_N^L(x), T_N^U(x)], [I_N^L(x), I_N^U(x)], [F_N^L(x), F_N^U(x)]\rangle | x \in X\}. \quad (2)$$

Definition 3. Deneutrosophication formula is given in Eq. (3) [9];

$$\mathfrak{D}(x) = \left(\frac{T_x^L + T_x^U}{2} + \left(1 - \frac{I_x^L + I_x^U}{2}\right) * (I_x^U) - \left(\frac{F_x^L + F_x^U}{2}\right) * (1 - F_x^U)\right) \quad (3)$$

where $\widetilde{\widetilde{x}}_j = \langle [T_x^L, T_x^U], [I_x^L, I_x^U], [F_x^L, F_x^U]\rangle$.

Definition 4. [11,12] Let $\widetilde{\widetilde{a}} = \langle [T_a^L, T_a^U], [I_a^L, I_a^U], [F_a^L, F_a^U]\rangle$ and $\widetilde{\widetilde{b}} = \langle [T_b^L, T_b^U], [I_b^L, I_b^U], [F_b^L, F_b^U]\rangle$ be two interval-valued neutrosophic numbers and the relations of them are given below:

1. $$\widetilde{\widetilde{a}}^c = \langle [T_a^L, T_a^U], [1 - I_a^U, 1 - I_a^L], [F_a^L, F_a^L]\rangle \quad (4)$$

2. $\widetilde{\widetilde{a}} \oplus \widetilde{\widetilde{b}} = \langle [T_a^L + T_b^L - T_a^L T_b^L, T_a^U + T_b^U - T_a^U T_b^U], [I_a^L I_b^L, I_a^U I_b^U], [F_a^L F_b^L, F_a^U F_b^U]\rangle$ (5)

3. $\widetilde{\widetilde{a}} \otimes \widetilde{\widetilde{b}} = \langle [T_a^L T_b^L, T_a^U T_b^U][I_a^L + I_b^L - I_a^L I_b^L, I_a^U + I_b^U - I_a^U I_b^U], [F_a^L + F_b^L - F_a^L F_b^L, F_a^U + F_b^U - F_a^U F_b^U]\rangle$ (6)

Definition 5. Neutrosophic Euclidean and Hamming distances of two neutrosophic numbers are defined as in Eqs. (7) and (8):

$$E_i = \sqrt{(T_x^L - T_y^L)^2 + (T_x^U - T_y^U)^2 + (I_x^L - I_y^L)^2 + (I_x^U - I_y^U)^2 + (F_x^L - F_y^L)^2 + (F_x^U - F_y^U)^2} \quad (7)$$

$$H_i = (|T_x^L - T_y^L| + |T_x^U - T_y^U| + |I_x^L - I_y^L| + |I_x^U - I_y^U| + |F_x^L - F_y^L| + |F_x^U - F_y^U|) \quad (8)$$

Definition 6. The weighted aggregation operation (INNWA) for interval-valued neutrosophic numbers is given in Eq. (9) [11]:

$$INNWA_w(A_1, A_2, \ldots, A_n) = \langle [1 - \prod_{i=1}^{n}(1 - \inf T_{A_i})^{w_i}, 1 - \prod_{i=1}^{n}(1 - \sup T_{A_i})^{w_i}], [\prod_{i=1}^{n}(\inf I_{A_i})^{w_i}, \prod_{i=1}^{n}(\sup I_{A_i})^{w_i}], [\prod_{i=1}^{n}(\inf F_{A_i})^{w_i}, \prod_{i=1}^{n}(\sup F_{A_i})^{w_i}]\rangle \quad (9)$$

where $W = (w_1, w_2, \ldots, w_n)$ is the weight vector of $A_j (j = 1, 2, \ldots, n)$, with $w_j \in [0,1]$ and $\sum_{j=1}^{n} w_j = 1$.

3. Proposed Interval-Valued Neutrosophic CODAS Method

Step 1. Construct the neutrosophic decision-making matrix ($\tilde{\tilde{X}}_l$) of each decision maker as in Eq. (10):

$$\tilde{\tilde{X}}_l[\tilde{\tilde{x}}_{ijl}]_{n \times m} = \begin{bmatrix} \tilde{\tilde{x}}_{11l} & \cdots & \tilde{\tilde{x}}_{1ml} \\ \vdots & \ddots & \vdots \\ \tilde{\tilde{x}}_{n1l} & \cdots & \tilde{\tilde{x}}_{nml} \end{bmatrix} \quad (10)$$

where $\tilde{\tilde{x}}_{ijl} = \langle [T^L_{ijl}, T^U_{ijl}], [I^L_{ijl}, I^U_{ijl}], [F^L_{ijl}, F^U_{ijl}] \rangle$ denotes the neutrosophic evaluation score of ith ($i \in \{1,2, \ldots, n\}$) alternative with respect to j^{th} criterion ($j \in \{1,2, \ldots, m\}$) and l^{th} ($l \in \{1,2, \ldots, q\}$) decision maker,

Step 2. Compute the aggregated neutrosophic decision matrix ($\tilde{\tilde{X}}$) by using Definition 6 as in Eq. (11)

$$\tilde{X}[\tilde{x}_{ij}]_{n \times m} = \begin{bmatrix} \tilde{\tilde{x}}_{11} & \cdots & \tilde{\tilde{x}}_{1m} \\ \vdots & \ddots & \vdots \\ \tilde{\tilde{x}}_{n1} & \cdots & \tilde{\tilde{x}}_{nm} \end{bmatrix} \quad (11)$$

where $\tilde{\tilde{x}}_{ij} = \langle [T^L_{ij}, T^U_{ij}], [I^L_{ij}, I^U_{ij}], [F^L_{ij}, F^U_{ij}] \rangle$ shows the aggregated neutrosophic score of ith alternative with respect to j^{th} criterion.

Step 3. Obtain the neutrosophic weight of each criterion ($\tilde{\tilde{w}}_j$) from each decision maker:

$$\tilde{\tilde{W}}_l = [\tilde{\tilde{w}}_{jl}]_{1 \times m} \quad (12)$$

where $\tilde{\tilde{w}}_{jl}$ denotes the neutrosophic weight of jth criterion ($j \in \{1,2, \ldots, m\}$) with respect to l^{th} decision maker ($l \in \{1,2, \ldots, q\}$),

Step 4. Compute the aggregated neutrosophic decision matrix ($\tilde{\tilde{W}}$) by using Definition 6 as in Eq. (13):

$$\tilde{\tilde{W}} = [\tilde{\tilde{w}}_j]_{1 \times m} \quad (13)$$

where $\tilde{\tilde{w}}_j$ shows the average neutrosophic weight of jth criterion.

Step 5. Calculate the neutrosophic weighted aggregated decision matrix ($\tilde{\tilde{R}}$):

$$\tilde{\tilde{R}} = [\tilde{\tilde{r}}_{ij}]_{n \times m} \quad (14)$$

$$\tilde{\tilde{r}}_{ij} = \tilde{\tilde{w}}_j \otimes \tilde{\tilde{n}}_{ij} \quad (15)$$

where $\tilde{\tilde{w}}_j$ denotes the neutrosophic weight of jth criterion, and $0 < \mathfrak{H}(\tilde{w}_j) < 1$. $\mathfrak{H}(\tilde{w}_j)$ [9] shows the deneutrosophicated value of $\tilde{w}_j$.

Step 6. Determine the neutrosophic negative ideal solution ($\widetilde{\widetilde{NS}}$):

$$\widetilde{\widetilde{NS}} = [\widetilde{\widetilde{ns}}_j]_{1 \times m} \quad (16)$$

$$\widetilde{\widetilde{ns}}_j = \min_i \tilde{\tilde{r}}_{ij} \quad (17)$$

where $\min_i \tilde{\tilde{r}}_{ij} = \{\tilde{\tilde{r}}_{ij} | \mathfrak{H}(\tilde{\tilde{r}}_{ij}) = \min_i \left(\mathfrak{H}(\tilde{\tilde{r}}_{ij})\right), k \in \{1,2,\ldots,n\}\}$ for Benefit criteria.

$$\widetilde{\widetilde{ns}}_j = \max_i \tilde{\tilde{r}}_{ij} \tag{18}$$

where $\max_i \tilde{\tilde{r}}_{ij} = \{\tilde{\tilde{r}}_{ij} | \mathfrak{H}(\tilde{\tilde{r}}_{ij}) = \max_i \left(\mathfrak{H}(\tilde{\tilde{r}}_{ij})\right), k \in \{1,2,\ldots,n\}\}$ for Cost criteria.

Step 7. Calculate the weighted Euclidean Distance (ED_i) and weighted Hamming Distance (HD_i) of alternatives from the neutrosophic negative ideal solution as given by Eqs. (7) and (8):

$$ED_i = \sum_{j=1}^m d_E(\tilde{\tilde{r}}_{ij}, \widetilde{\widetilde{ns}}_j) \tag{19}$$

$$HD_i = \sum_{j=1}^m d_D(\tilde{\tilde{r}}_{ij}, \widetilde{\widetilde{ns}}_j) \tag{20}$$

Step 8. Determine the relative assessment matrix (RA):

$$RA = [p_{ik}]_{n \times n} \tag{21}$$

$$p_{ik} = (ED_i - ED_k) + \left(t(ED_i - ED_k) \times (HD_i - HD_k)\right) \tag{22}$$

where $k \in \{1,2,\ldots,n\}$ and t is a threshold function that is defined as follows:

$$t(x) = \begin{cases} 1 & if \ |x| \geq \theta \\ 0 & if \ |x| < \theta \end{cases} \tag{23}$$

The threshold parameter (θ) of this function can be set by decision maker. In this study, we use $\theta = 0.02$ in our calculations.

Step 9. Calculate the assessment score (AS_i) of each alternative:

$$AS_i = \sum_{k=1}^n p_{ik} \tag{24}$$

Step 10. Rank the alternatives according to the decreasing values of assessment scores and select the alternative with the maximum assessment score.

4. Application

In our problem, we have 5 alternatives with 4 criteria and 3 decision maker (DM). The constructed decision matrix based on linguistic terms with respect to decision makers' expertise is given in Table 1.

Table 1. Decision matrix based on linguistic terms with respect to decision makers' expertise.

Alter-natives	DM1 (0.42)				DM2 (0.3)				DM3 (0.28)			
	C1	C2	C3	C4	C1	C2	C3	C4	C1	C2	C3	C4
	Be	Be	Co	Co	Be	Be	Co	Co	Be	Be	Co	Co
AL1	AA	H	AA	L	A	BA	L	H	BA	A	CL	BA
AL2	H	H	L	L	BA	A	H	H	BA	A	H	BA
AL3	VH	H	CL	VL	AA	A	L	L	CL	CL	BA	H
AL4	AA	H	L	L	H	H	BA	BA	CH	CH	L	H
AL5	CH	CH	L	L	H	AA	L	VL	VH	H	VL	L

Be: Benefit, Co: Cost, CL: Certainly Low, VL: Very Low, L: Low, AA: Above Average, A: Average, BA: Below Average, H: High, VH: Very High, CH: Certainly High, DM: Decision Maker

Table 2. Aggregated decision matrix.

	AL1	AL2	AL3	AL4	AL5
C1	<[0.47,0.62],[0,0.27],[0.43,0.58]>	<[0.5,0.66],[0.13,0.34],[0.39,0.55]>	<[0.57,0.75],[0.23,0.45],[0.31,0.49]>	<[0.69,1],[0.18,0.4],[0.18,0.36]>	<[0.78,1],[0.3,0.5],[0.11,0.28]>
C2	<[0.52,0.68],[0,0.3],[0.37,0.53]>	<[0.55,0.7],[0,0.27],[0.35,0.51]>	<[0.47,0.64],[0,0.36],[0.42,0.58]>	<[0.72,1],[0.24,0.45],[0.16,0.33]>	<[0.74,1],[0.22,0.44],[0.14,0.32]>
C3	<[0.35,0.51],[0.18,0.4],[0.54,0.7]>	<[0.52,0.68],[0.2,0.4],[0.37,0.54]>	<[0.2,0.36],[0.22,0.44],[0.69,0.85]>	<[0.28,0.43],[0.16,0.37],[0.62,0.77]>	<[0.22,0.37],[0.22,0.43],[0.68,0.83]>
C4	<[0.43,0.59],[0.16,0.37],[0.47,0.63]>	<[0.43,0.59],[0.16,0.37],[0.47,0.63]>	<[0.36,0.53],[0.24,0.44],[0.53,0.69]>	<[0.42,0.58],[0.16,0.37],[0.47,0.63]>	<[0.22,0.37],[0.23,0.43],[0.68,0.83]>

Table 3. Weighted aggregated decision matrix.

	AL1	AL2	AL3	AL4	AL5
C1	<[0.25,0.45],[0.41,0.65],[0.59,0.78]>	<[0.26,0.48],[0.49,0.68],[0.56,0.77]>	<[0.3,0.54],[0.55,0.74],[0.5,0.73]>	<[0.36,0.73],[0.52,0.71],[0.41,0.67]>	<[0.41,0.73],[0.59,0.76],[0.35,0.62]>
C2	<[0.3,0.57],[0.24,0.56],[0.48,0.73]>	<[0.32,0.58],[0.24,0.54],[0.46,0.71]>	<[0.27,0.53],[0.24,0.6],[0.52,0.76]>	<[0.42,0.83],[0.42,0.66],[0.3,0.61]>	<[0.43,0.83],[0.4,0.65],[0.28,0.6]>
C3	<[0.09,0.24],[0.52,0.72],[0.78,0.92]>	<[0.14,0.32],[0.54,0.72],[0.71,0.88]>	<[0.05,0.17],[0.55,0.74],[0.86,0.96]>	<[0.08,0.2],[0.51,0.7],[0.82,0.94]>	<[0.06,0.18],[0.55,0.73],[0.85,0.95]>
C4	<[0.18,0.38],[0.35,0.59],[0.66,0.84]>	<[0.18,0.38],[0.35,0.59],[0.66,0.84]>	<[0.16,0.34],[0.41,0.64],[0.7,0.87]>	<[0.18,0.37],[0.35,0.59],[0.66,0.84]>	<[0.1,0.24],[0.4,0.63],[0.8,0.93]>

Table 4. Distances of the alternatives with final scores, and ranks.

	Criterion	AL1	AL2	AL3	AL4	AL5
Euclidean Distance	C1	0.000	0.096	0.217	0.385	0.473
	C2	0.079	0.117	0.000	0.469	0.472
	C3	0.128	0.000	0.247	0.189	0.233
	C4	0.000	0.000	0.099	0.008	0.239
	Total	0.207	0.213	0.563	1.052	1.417
Hamming Distance	C1	0.000	0.194	0.500	0.852	1.116
	C2	0.175	0.260	0.000	1.055	1.055
	C3	0.258	0.000	0.501	0.397	0.471
	C4	0.000	0.000	0.237	0.019	0.536
	Total	0.433	0.454	1.239	2.323	3.179
	Score	-7.9	-7.7	-2.1	5.79	11.9
	Rank	5	4	3	2	1

4.1. Sensitivity Analysis

One-at-a-time sensitivity analysis based on each criterion is performed to demonstrate effects of criteria on the results. To do this, we develop a pattern which is given in Table 5. We used CLI, AI, and CHI linguistic terms as reference points. All sets columns present the ranks of alternatives based on each criterion test variable, respectively.

Table 5. Pattern for the sensitivity analysis.

Pattern		Cases with respect to criteria			
		Case -1	Case -2	Case -3	Case -4
Test Variables	CLI	Ranks	...	...	Ranks
	AI	⋮	...	⋱	⋮
	CHI	⋮	...	⋱	⋮

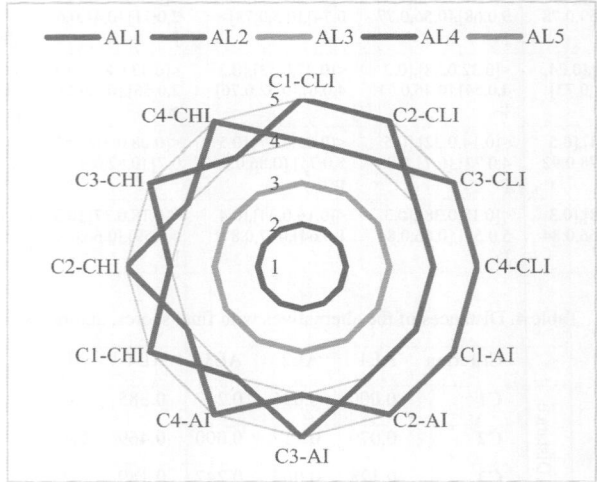

Figure 1. Results of the sensitivity analysis.

5. Conclusion

It is a hard decision which renewable energy selection should be selected since several uncertainties exist in this problem. Neutrosophic Logic presents an excellent tool to capture the vagueness and indefiniteness existing in problems. We have extended CODAS method to Neutrosophic CODAS in order to select the best renewable energy selection under fuzziness. Neutrosophic CODAS method produces meaningful results and can be used as an alternative MCDM method under fuzziness. For further research, we suggest other extensions of fuzzy sets to be applied.

References

1. M. Keshavarz Ghorabaee, E.K. Zavadskas, L. Olfat and Z. Turskis, *Multi-criteria inventory classification using a new method of Evaluation Based on Distance from Average Solution (EDAS) Informatica*, **26**, 3, 435–451 (2015).
2. E.K. Zavadskas, Z. Turskis and S. Kildienė, *State of art surveys of overviews on MCDM/MADM methods, Technological and Economic Development of Economy*. **20**, 165-179 (2014).
3. M. Keshavarz Ghorabaee, E. K. Zavadskas, Z. Turskis ve J. Antucheviciene, *A New Combinative Distance-Based Assessment (CODAS) Method for Multi-Criteria Decision-Making Economic Computation & Economic Cybernetics Studies & Research*. **50**, 3, 25-44, (2016).
4. F. Smarandache, Neutrosophic logic and set, mss. (1995).
5. U. Rivieccio, Neutrosophic logics: prospects and problems. Fuzzy Sets Syst 159, 14, 1860–1868 (2008).
6. F Smarandache, Neutrosophy Neutrosophic Probability. Set, and Logic. Amer Res Press Rehoboth. 12-20 (1998).
7. P. Biswas, S. Pramanik, Giri, *A New Methodology for Neutrosophic Multi-attribute Decisionmaking with Unknown Weight Information. Neutrosophic Sets and Systems*. **3**, 42-50 (2014).
8. H. Wang, F. Smarandache, Y. Zhang, R. Sunderraman, *Single valued neutrosophic sets. Multispace Multistruct*. **4**, 410–413 (2010).
9. E. Bolturk, and C. Kahraman, *A Novel Interval-Valued Neutrosophic AHP with Cosine Similarity Measure*. https://doi.org/10.1007/s00500-018-3140-y (2018).
10. Y. Li, Y. Wang, P. Liu, *Multiple attribute group decision-making methods based on trapezoidal fuzzy two-dimension linguistic power generalized aggregation operators. Soft Computing*. **20**, 7, 2689-2704 (2016).
11. H.-y. Zhang, J-q Wang, X-h. Chen, *Interval neutrosophic sets and their application in multicriteria decision-making problems. The Scientific World Journal*. **2014**, doi:10.1155/2014/645953 (2014).
12. A. Karasan and C. Kahraman, *Interval-valued neutrosophic extension of EDAS method. Advances in Intelligent Systems and Computing*. **642**, 343-357 (2018).

Natural gas technology selection using Pythagorean fuzzy CODAS

Eda Bolturk[1,2]

[1]Department of Industrial Engineering, Istanbul Technical University
Macka, 34367/Istanbul, Turkey
[2]Information Technologies Department, Istanbul Takas ve Saklama Bankası A.S.
Borsa Istanbul Street, 34467, Sarıyer, Istanbul, Turkey

Cengiz Kahraman

Department of Industrial Engineering, Istanbul Technical University
Macka, 34367/Istanbul, Turkey

Natural gas technology selection is a critical issue for manufacturing companies. Pythagorean fuzzy sets (PFSs) are characterized by a membership degree and a nonmembership degree satisfying the condition that their square sum is equal to or less than 1. COmbinative Distance-based ASsessment (CODAS) method which is a new multi criteria decision making (MCDM) technique introduced by Ghorabaee et al. [6]. The aim of this paper is to apply the Pythagorean fuzzy extension of CODAS method with an application to select the natural gas technology. The proposed method provides a larger definition area for membership and non-membership degrees.

1. Introduction

The natural gas industry has been able to keep pace with growing demand and produce greater amounts of natural gas through technological innovation. Technological innovation in the exploration and production sector has equipped the industry with the equipment and practices necessary to continually increase the production of natural gas to meet rising demand. These technologies serve to make the exploration and production of natural gas more efficient, safe, and environmentally friendly.

In real life problems, decision making processes require vague and imprecise evaluations rather than exact and numerical evaluations. Fuzzy logic and extensions have been used in MCDM in order to overcome this issue such as type-2 fuzzy sets, Intuitionistic Fuzzy Sets (IFSs) and Hesitant Fuzzy Sets (HFSs). CODAS method combines two different scoring methods, namely simple additive weighting method (SAW) and weighted product method (WPM).

The rest of the paper is organized as follows. In Section 2, natural gas technology techniques are summarized and their performance criteria are given. In Section 3, the preliminaries of PFSs are given. In Section 4, Pythagorean Fuzzy CODAS method, developed by Bolturk [14], is presented. In Section 5, an application is presented for a natural gas technology problem. In Section 6, the conclusion is given and suggestions for further research are presented.

2. Natural Gas Technologies & Performance Criteria

There are several natural gas technologies in the industrial market. We briefly summarize them as follows. **3-D and 4-D Seismic Imaging (SI)** technology uses traditional seismic imaging techniques, combined with powerful computers and processors, to create a three-dimensional model of the subsurface layers. **CO_2-Sand fracturing (SF)** involves using a mixture of sand proppants and liquid CO_2 to fracture formations, creating and enlarging cracks through which oil and natural gas may flow more freely. **Coiled Tubing (CT)** technologies replace the traditional rigid, jointed drill pipe with a long, flexible coiled pipe string. In **Hydraulic Fracturing (HF)**, a liquid mix that is 99 percent water and sand is injected into the rock at very high pressure, creating fractures within the rock that provide the natural gas a path to flow to the wellhead [10]. The criteria for selecting the best technology for a certain area are determined as follows. **Drilling Costs (DC):** The technologies above require different drilling costs. **Exploration Time (ET):** Shorter exploration times reduce the other costs such employees' salaries and rent costs. **Environmental Damage (ED):** The natural gas technologies cause different levels of environmental damage. **Number of Dry Holes Drilled (DHD):** One of the performance indicators of a natural gas technology is the percentage of successful holes over unsuccessful holes. **Geological Structure Suitability (GSS):** Natural gas technologies are applied based on appropriate geological structures.

3. Pythagorean Fuzzy Sets

PFSs are the extension of IFSs and they let decision makers use a wider range of membership and nonmembership values but still with a constraint that the square sum of these values must be at most equal to 1. Many papers have been published on Pythagorean fuzzy MCDM since it appeared first time in 2014. Kahraman et al. [2] developed Pythagorean fuzzy scoring methods based on classical SAW and WPM methods. Mohagheghi et al. [8] evaluated last aggregation group decision-making process by ordered weighted based aggregation method. Garg [3] used confidence Pythagorean fuzzy weighted and ordered weighted operators by ordered weighted based aggregation method. Mohd & Abdullah [9] proposed Pythagorean fuzzy AHP method. PFSs introduced by Yager [7] can be defined by both membership and nonmembership degrees where the square sum of membership degree and nonmembership degrees should be equal or less than 1. PFSs, which were originally introduced by Atanassov as Type-2 IFS, can be considered as a generalization of Intuitionistic Fuzzy Sets (IFS) [4]. Alternatively, we can say that IFSs are a kind of concentration operation over PFSs [1]. A PFS is defined as follows:

$$\tilde{P} = \{x, P(\mu_P(x), v_P(x)) | x \in X\} \quad (1)$$

where $\mu_P: X \to [0,1]$ is the membership degree and $v_P: X \to [0,1]$ is the nonmembership degree and Eq. (2) is valid and the degree of indeterminancy is defined in Eq. (3):

$$0 \leq (\mu_P(x))^2 + (v_P(x))^2 \leq 1 \quad (2)$$

$$\pi_P(x) = \sqrt{1 - (\mu_P(x))^2 - (v_P(x))^2} \quad (3)$$

For two PFSs, $\tilde{P}_1 = \{x, P_1(\mu_{P1}(x), v_{P1}(x)) | x \in X\}$ and $\tilde{P}_2 = \{x, P_2(\mu_{P2}(x), v_{P2}(x)) | x \in X\}$, the following operations are valid:

$$\tilde{P}_1 \oplus \tilde{P}_2 = P\left(\sqrt{\mu_{P_1}^2 + \mu_{P_2}^2 - \mu_{P_1}^2 \mu_{P_2}^2}, v_{P_1} v_{P_2}\right) \quad (4)$$

$$\tilde{P}_1 \otimes \tilde{P}_2 = P\left(\mu_{P_1} \mu_{P_2}, \sqrt{v_{P_1}^2 + v_{P_2}^2 - v_{P_1}^2 v_{P_2}^2}\right) \quad (5)$$

Zhang and Xu [11] defined the Euclidean distance between two PFSs as in Eq. (6):

$$d(\tilde{P}_1, \tilde{P}_2) = \frac{1}{2}\left(|\mu_{P_1}^2 - \mu_{P_2}^2| + |v_{P_1}^2 - v_{P_2}^2| + |\pi_{P_1}^2 - \pi_{P_2}^2|\right) \quad (6)$$

The Taxican distance between two PFSs is defined by Eq. (7):

$$T(\tilde{P}_1, \tilde{P}_2) = |\mu_{P_1} - \mu_{P_2}| + |v_{P_1} - v_{P_2}| + |\pi_{P_1} - \pi_{P_2}| \quad (7)$$

Let $p_1 = (\mu_1, v_1)$ and $p_2 = (\mu_2, v_2)$ be two PFNs and $\rho > 0$. The following operations are presented for PFNs [11, 12].

$$\tilde{P}_1 \ominus \tilde{P}_2 = \left(\sqrt{\frac{\mu_1^2 - \mu_2^2}{1 - \mu_2^2}}, \frac{v_1}{v_2}\right), \text{if } \mu_1 \geq \mu_2, v_1 \leq \min\left\{v_2, \frac{v_2 \pi_1}{\pi_2}\right\} \quad (8)$$

$$\frac{\tilde{P}_1}{\tilde{P}_2} = \left(\frac{\mu_1}{\mu_2}, \sqrt{\frac{v_1^2 - v_2^2}{1 - v_2^2}}\right), \text{if } \mu_1 \leq \min\left\{\mu_2, \frac{\mu_2 \pi_1}{\pi_2}\right\}, v_1 \geq v_2 \quad (9)$$

4. Pythagorean Fuzzy CODAS

Ghorabaee et al. [6] developed CODAS for complex MCDM problems and the other paper extended the CODAS with fuzzy logic [5]. The steps of Pythagorean

Fuzzy CODAS are presented in the following:
Step 1. Construct the Pythagorean Fuzzy decision matrix as given in Eq. (10).

$$\tilde{X} = [\tilde{x}_{ij}]_{n \times m} = \begin{bmatrix} \tilde{x}_{11} & \cdots & \tilde{x}_{1m} \\ \vdots & \ddots & \vdots \\ \tilde{x}_{n1} & \cdots & \tilde{x}_{nm} \end{bmatrix} \quad (10)$$

where $\tilde{x}_{ij} \geq 0$ and $\tilde{x}_{ij} = (\mu_P, v_P)$ and $0 \leq (\mu_P(x))^2 + (v_P(x))^2 \leq 1$.

Step 2. Calculate the Pythagorean fuzzy normalized matrix using linear normalization as in Eq. (11).

$$\tilde{r}_{ij} = \begin{cases} \dfrac{\tilde{x}_{ij}}{\max\limits_i \tilde{x}_{ij}} & if\ j \in N_b \\ \dfrac{\min\limits_i \tilde{x}_{ij}}{\tilde{x}_{ij}} & if\ j \in N_c \end{cases} \quad (11)$$

where N_b and N_c represent the sets of benefit and cost criteria, respectively.

Definition 1. Find the maximum and minimum values with respect to a certain criterion, the closeness index given in Eq. (12) [13].

$$D_P = \frac{1 - v_P^2}{2 - \mu_P^2 - v_P^2} \quad (12)$$

where μ_P is the membership degree and v_P is the nonmembership degree.

Step 3. Calculate the Pythagorean fuzzy weighted normalized matrix. The Pythagorean fuzzy weighted normalized performance values are calculated as in Eq. (13):

$$\tilde{u}_{ij} = \tilde{w}_j \tilde{r}_{ij} \quad (13)$$

where $w_j(0 < w_j < 1)$ denotes the weight of jth criterion, and $\sum_{j=1}^m w_j = 1$.

Step 4. Determine the Pythagorean fuzzy negative ideal solution as given in Eqs. (14) and (15):

$$\widetilde{ns} = [\widetilde{ns}_j]_{1 \times m} \quad (14)$$

$$\widetilde{ns}_j = \min_i \tilde{u}_{ij} \quad (15)$$

Step 5. Calculate the Pythagorean fuzzy Euclidean and Taxicab distances of alternatives from the negative ideal solution as Eqs. (16) and (17):

$$E_i = \sqrt{\sum_{j=1}^m (\tilde{u}_{ij} - \widetilde{ns}_j)^2} \quad (16)$$

$$T_i = \sum_{j=1}^m |\tilde{u}_{ij} - \widetilde{ns}_j| \quad (17)$$

Step 6. Construct the relative assessment matrix based on the Pythagorean fuzzy Euclidean and Taxicab distances as given in Eqs. (18) and (19):

$$Ra = [h_{ik}]_{n \times n} \tag{18}$$

$$h_{ik} = (E_i - E_k) + (\psi(E_i - E_k) \times (T_i - T_k)) \tag{19}$$

where k∈ {1,2, ..., n} and ψ denotes a threshold function to recognize the equality of the Euclidean distances of two alternatives as given in Eq. (20):

$$\psi(x) = \begin{cases} 1 \; if \; |x| \geq \tau \\ 0 \; if \; |x| < \tau \end{cases} \tag{20}$$

In our application, $\tau = 0.02$ is selected.

Step 7. Calculate the assessment score of each alternative as given in Eq. (21):

$$H_i = \sum_{k=1}^{n} h_{ik} \tag{21}$$

Step 8. Rank the alternatives according to the decreasing values of assessment score (H_i). The alternative with the highest H_i is the best alternative among the alternatives.

5. Application

The natural gas is generally sold by governments in the world because of their high investment costs. In the countries that open up natural gas markets to private sector representatives, there has been a rapid decline in trade volume based on long-term contracts. In particular, energy exchanges and clearinghouses that provide central counterpart and/or central clearing services are effective in this decline. The private firms in the natural gas sector, market participants began to take place in various value chains in Turkey since 2001 with the liberalization in the gas sector.

In this respect, Takasbank is the main player that manages central settlement organizations, exchanges and collateral management services. With the central clearing, central risk and collateral management mechanisms to be provided by Takasbank, it is envisaged that the liquidity in the market will increase after the participation of the private sector. Takasbank signed an agreement with the Scientific and Technological Research Council of Turkey (TUBITAK) on Natural Gas Market Cash Settlement and Collateral Management Project which belongs to Technology and Innovation Support Programs Presidency of TUBITAK. Our problem is to select the best natural gas production technology among the HF, SF, CT, and SI methods. With the new technology, it is desired for the involvement of private sector participants in the natural gas sector to increase.

The application steps of the proposed Pythagorean Fuzzy CODAS method are given in the following:

Step 1. The scale for Pythagorean Fuzzy CODAS method is given in Table 1 and the Pythagorean fuzzy compromised decision matrix method is presented in Table 2.

Table 1. Scale of Pythagorean fuzzy CODAS.

Linguistic Term	μ	v	Linguistic Term	μ	v
Absolutely Weakly Satisfactory (AWS)	0.1	0.95	Strongly Satisfactory (SS)	0.7	0.4
Very Weakly Satisfactory (VWS)	0.25	0.85	Very Strongly Satisfactory (VSS)	0.85	0.25
Weakly Satisfactory (WS)	0.4	0.7	Absolutely Satisfactory (AS)	0.95	0.1
Fairly Satisfactory (FS)	0.55	0.55			

Step 2. The normalized matrix using linear normalization is obtained as in Table 3 by using Eq. (11).

Step 3. The weighted normalized Pythagorean fuzzy matrix is obtained by using equal weights (Pythagorean fuzzy number (1,0)). It is the same as Table 3.

Step 4. The negative ideal solution on the basis of criteria is given in Table 4.

Table 2. Pythagorean fuzzy compromised decision matrix.

Alternatives	Cost DC		Cost ET		Cost ED		Cost DHD		Benefit GSS	
	μ_1	v_1	μ_2	v_2	μ_3	v_3	μ_4	v_4	μ_5	v_5
HF	0.27	0.70	0.29	0.61	0.1	0.8	0.3	0.71	0.44	0.61
SF	0.30	0.67	0.3	0.6	0.19	0.78	0.39	0.63	0.45	0.6
CT	0.35	0.65	0.33	0.57	0.3	0.75	0.5	0.6	0.49	0.58
SI	0.39	0.63	0.34	0.56	0.45	0.70	0.67	0.59	0.65	0.57

Table 3. Normalized decision matrix.

Alternatives	DC		ET		ED		DHD		GSS	
	μ_1	v_1	μ_2	v_2	μ_3	v_3	μ_4	v_4	μ_5	v_5
HF	1.000	0.000	1.000	0.000	1.000	0.000	1.000	0.000	0.677	0.066
SF	0.900	0.045	0.967	0.017	0.526	0.026	0.769	0.127	0.692	0.050
CT	0.771	0.077	0.879	0.070	0.333	0.067	0.600	0.183	0.754	0.017
SI	0.692	0.111	0.853	0.089	0.222	0.143	0.448	0.203	1.000	0.000

Table 4. Negative ideal solution for alternatives.

Alternatives	DC	ET	ED	DHD	GSS
HF	1.000	1.000	1.000	1.000	**0.648**
SF	0.840	0.938	0.580	0.707	0.657
CT	0.711	0.814	0.528	0.602	0.698
SI	**0.655**	**0.785**	**0.508**	**0.545**	1.000

Step 5. Euclidean and Taxicab distances of alternatives are given in Table 5 and Table 6, respectively.

Table 5. Euclidean distances for alternative pairs.

Alternative Pairs and Euclidean Distances										Total Distance	Euclidean Distance
SI-HF	0.521	SI-HF	0.272	SI-HF	0.951	SI-HF	0.800	HF-HF	0.000	2.543	1.595
SI-SF	0.331	SI-SF	0.207	SI-SF	0.228	SI-SF	0.391	HF-SF	0.021	1.178	1.085
SI-CT	0.116	SI-CT	0.045	SI-CT	0.062	SI-CT	0.160	HF-CT	0.110	0.492	0.701
SI-SI	0.000	SI-SI	0.000	SI-SI	0.000	SI-SI	0.000	HF-SI	0.542	0.542	0.736

Table 6. Taxicab Distances for alternative pairs.

Alternatives Pairs and Taxicab Distances										Taxicab Distance
SI-HF	1.132	SI -HF	0.751	SI-HF	1.885	SI-HF	1.626	HF-HF	0.000	5.394
SI-SF	0.553	SI -SF	0.445	SI-SF	0.536	SI-SF	0.642	HF-SF	0.044	2.221
SI-CT	0.195	SI -CT	0.087	SI-CT	0.211	SI-CT	0.264	HF-CT	0.202	0.959
SI-SI	0.000	SI- SI	0.000	SI-SI	0.000	SI-SI	0.000	HF-SI	1.122	1.122

Step 6. The relative assessment matrix is given in Table 7.

Table 7. Relative assessment matrix.

h_{ik}	HF	h_{ik}	SF	h_{ik}	CT	h_{ik}	SI
HF-SF	2.127	SF-HF	1.107	CT-HF	3.069	SI-HF	2.810
HF-CT	4.856	SF-CT	0.868	CT-SF	0.101	SI-SF	0.035
HF-SI	4.527	SF-SI	0.733	CT-SI	-0.029	SI-CT	0.040

Step 7. The assessment score of each alternative is calculated. The alternative score for HF is 11.510, the alternative score for SF is 2.708, the alternative score for CT is 3.140 and the alternative score for SI is 2.885. The ranking of the alternatives is HF >CT> SI>SF. Hydraulic Fracturing is the best alternative among the alternatives with the highest H_i.

6. Conclusion

It is a hard decision which natural gas technology should be selected since several uncertainties exist in this problem. PFSs present an excellent tool to capture the vagueness and uncertainty existing in this problem. The literature review shows that PFSs provide a new way to extend MCDM methods under fuzzy environment. We have extended CODAS method to Pythagorean fuzzy CODAS in order to select the best natural gas technology under fuzziness. Pythagorean fuzzy CODAS method produces meaningful results and can be used as an alternative MCDM method under fuzziness. For further research, we suggest other extensions of ordinary fuzzy sets to be used.

Acknowledgments

As a specialist in Information Technologies Department, Eda Bolturk thanks Istanbul Takas ve Saklama Bankası A.S. (Takasbank) for getting support for this study.

References

1. C. Kahraman., S.C. Onar, B Oztaysi. *Present Worth Analysis Using Pythagorean Fuzzy Sets. In: J. Kacprzyk, E. Szmidt, S. Zadrożny, Atanassov K., M. Krawczak (eds.) Advances in Fuzzy Logic and Technology 2017. IWIFSGN 2017, EUSFLAT 2017.* Advances in Intelligent Systems and Computing, **642** Springer (2018).
2. C. Kahraman, B. Oztaysi, and S. C. Onar. *Multicriteria Scoring Methods using Pythagorean Fuzzy Sets.* Advances in Intelligent Systems and Computing. **642**. (2018).
3. H. Garg, *Confidence levels based Pythagorean fuzzy aggregation operators and its application to decision-making process.* Computational and Mathematical Organization Theory. **23(4)**, 546-571 (2017).
4. K. T. Atanassov, *Intuitionistic Fuzzy Sets Theory and Applications.* Springer (1999).
5. M. K. Ghorabaee, M. Amiri, E. K. Zavadskas, R. Hooshmand and J. Antuchevičienė, *Fuzzy extension of the CODAS method for multi-criteria market segment evaluation.* Journal of Business Economics and Management. **18(1)**, 1-19 (2017).
6. M. K. Ghorabaee., E. K. Zavadskas, Z. Turskis, and J. Antucheviciene, *A new combinative distance-based assessment (CODAS) method for multi-criteria decision-making,* Economic Computation and Economic Cybernetics Studies and Research. **50(3)**, 25-44 (2016).
7. R. R., Yager, *Pythagorean membership grades in multicriteria decision making.* IEEE Transactions on Fuzzy Systems. **22(4)**, 958-965 (2014).
8. V. Mohagheghi, S. M. Mousavi and B. Vahdani, *Enhancing decision-making flexibility by introducing a new last aggregation evaluating approach based on multi-criteria group decision making and Pythagorean fuzzy sets,* Applied Soft Computing Journal. **(61)** 527-535 (2017).
9. W. R. W. Mohd and L. Abdullah, *Pythagorean fuzzy analytic hierarchy process to multi-criteria decision making,* AIP Conference Proceedings 2017. (2017).
10. http://naturalgas.org/environment/technology/
11. X. Zhang and, Z. Xu, *Extension of TOPSIS to multiple criteria decision making with pythagorean fuzzy sets.* International Journal of Intelligent Systems. **29(12)**, 1061-1078 (2014).
12. X. Peng and Y. Yang, *Some results for Pythagorean fuzzy sets,* Int. J. Intell. Syst. **30(11)** 1133–1160 (2015).
13. X. Zhang, *Multicriteria Pythagorean fuzzy decision analysis: A hierarchical QUALIFLEX approach with the closeness index-based ranking methods.* Information Sciences. **330**, 104-124 (2016).
14. E. Bolturk, *Pythagorean Fuzzy CODAS and its application to supplier selection in A Manufacturing Firm.* Journal of Enterprise Information Management. In press. (2018).

A multicriteria model for launching a new product for a group of decision-makers

Pedro Flores Leal
Management and Economic Science, Universidad Autónoma de Occidente
80120 Culiacan, Sinaloa, Mexico

Pavel Anselmo Alvarez
Management and Economic Science, Universidad Autónoma de Occidente
80120 Culiacan, Sinaloa, Mexico

Diego Alonso Gastelum Chavira
Management and Economic Science, Universidad Autónoma de Occidente
80120 Culiacan, Sinaloa, Mexico

Juan Carlos Leyva Lopez
Management and Economic Science, Universidad Autónoma de Occidente
80120 Culiacan, Sinaloa, Mexico

Group decision-making processes have a wide application in different activities in organizations. This process requires the coordination of individual activities as well a series of procedures and aggregation methods that incorporate individual member's activities to reach a high-quality group decision. This paper presents a model for launching a new product by the implementation of a multicriteria group decision-making based on an individual preferences aggregation scheme. The model is tested in a real case study with data collected in a regional food sector company.

1. Introduction

Decisions on launching a new product have a major impact on the financial success of the new product [2]. The strategy for launching a new product is a crucial decision made by Decision-Makers (DM). Effective market testing and research are critical in eliciting information regarding customers at the time of launching a new product [3]. Launching a new product have different stages, the first one is to design a new product based on parameters of target markets, taking the basis of production restrictions of the company. This process requires analyzing information to make decisions in an environment of uncertainty. DM must take the challenges of analyzing information, participating and reaching consensus with other team members [2].

The case presented in this article refers to the problem of a company for making decisions to launch new products. The aim of this work is to propose a multicriteria model for launching a new product for a group of decision-makers. In section 2 the model for launching a new product is briefly described. In section 3 the real case study is carried out with data from a regional food sector company. Finally, conclusion is described on section 4.

2. Multicriteria Model

Given a set of solution alternatives A, in Multicriteria Decision Analysis (MCDA), four types of problems are identified: description, choice, ranking and sorting. For these problems, several approaches and methods have been developed. One of these approaches is the known as outranking approach [5]. In this approach, given two alternatives $a, b \in A$; $aS_A^\sigma b$, in a global sense, means that "a is at least as good as b".

In outranking approach, two general steps are identified: Aggregation and Exploitation. The first one is related with the DM preference modeling, where alternatives, criteria, weights, among others are defined. Then, by following the outranking approach, an integral preference model of the DM is obtained. This model is a pairwise comparison among alternatives, where given two alternatives $a, b \in A$, we can say how a is respect to b: indifferent (aIb), preferred (aPb), non-preferred ($a\neg Pb$) or incomparable (aRb) for a give credibility level λ. Then, it is possible to say if "a is at least as good as b" or not.

The multicriteria model for launching a new product for group decision-making it was divided in four phases based on the aggregation schemes [1], that is shown in Figure 1. The Phase 1 (P1) corresponds to the problem definition for launching a new product regarding company's goals, and the product to be evaluated as an apart of the decision process. The Phase 2 (P2) correspond to the definition of individual preference; criteria, pseudo-criteria, evaluation of new products, construction of individual preference model and ranking of products as individual DM. In Phase 3 (P3), individual results are aggregated in a group preference model. In the Phase 4 (P4) the temporal collective solution is generated and if a certain consensus level is reached, the final group solution is presented.

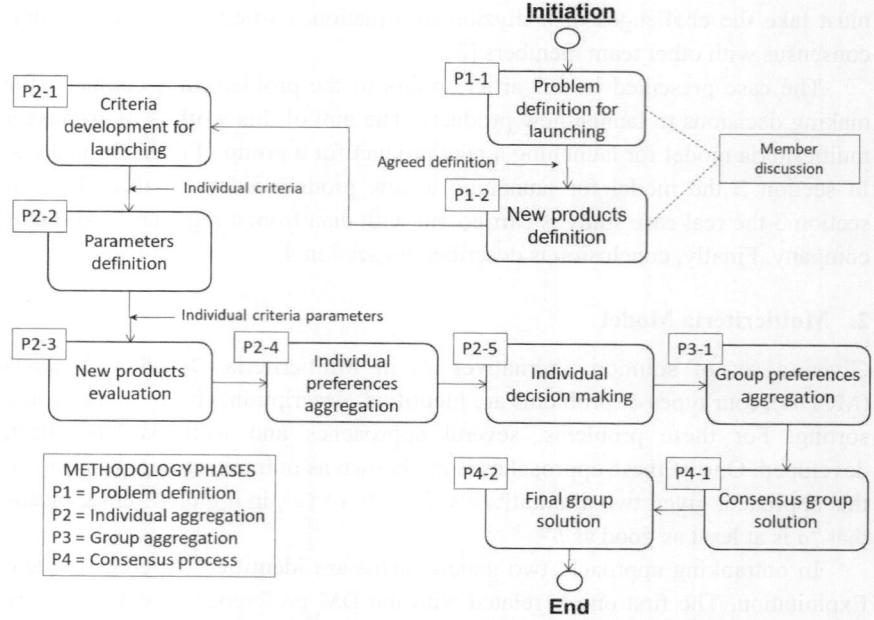

Figure 1. Multicriteria model for launching a new product for group decision-making.

A brief description of the model is presented below: in the first phase for a set of products (P1-2) $A = \{a_1, a_2, ..., a_m\}$, each DM $D = \{d_1, d_2, ..., d_k\}$ defines a set of criteria (P2-1) $G = \{g_1, g_2, ..., g_n\}$, and a set of pseudo-criteria parameters in P2-2. These criteria are used to evaluate the set of alternatives A. Then, alternatives are valued by the DM in (P2-3). Using ELECTRE-III method, a fuzzy outranking relation S_A^σ is obtained for each DM (P2-4). Then, by using a Multiobjective Evolutionary Algorithm, a complete ranking r_i; $i = 1, ..., |D|$ is obtained for each S_A^σ in P2-5. Each r_i is presented to the correspondent DM d_i to be analyzed. If the a d_i is not agree with its r_i, the analyst suggest to the DM to change values in weighs and/or thresholds to again generate a new fuzzy outranking relation S_A^σ and a new ranking r_i. If all the DM agree with their respective rankings, the process is continued to (P3-1).

In Phase P3-1, each S_A^σ and r_i are provided to a method called ELECTRE GD to generate a global fuzzy outranking relation GS_A^σ. The new relation represents the global preferences of the group [4].

In Phase P4-1, the GS_A^σ is exploited to obtain a global ranking R_{Global}. Each individual ranking r_i is compared with R_{Global} to get k distances d_{ri}; $i = 1, ..., k$; $k = |D|$. These distances are computed by using the Kendall's Weighted distance [4]. Then an average D_{avg} of these distances is computed. This average represents the group consensus level (P3-1). Given a predefined consensus level χ where $0.70 \leq \chi \leq 0.85$; if $D_{avg} \geq \chi$, the consensus level has been reached and the process is finished (P4-2); otherwise, it is necessary to return to a previous steps to modify weights and thresholds of DM whose rankings are discordant with the global ranking R_{Global} to generate news individual rankings r_i and continue with the process again.

3. Case Study

3.1. *Problem description*

The company needs to launch new products based on the application of a model that use techniques to order and grant preferences in a series of alternatives. The company is in the sector of nutritional supplements based on nutraceutical compounds. For this analysis, five products were proposed to be launched based on their commercial relevance. Technical profiles of the five products were integrated as alternatives, which were shared with the DMs. The company haven't the capacity for launching the five products to the market, so it requires applying a model with techniques and methods of support to the decision making to launch of new products that respond better to the criteria of preference of the market segments of greater relevance for the growth of the company.

The challenge associated with this case study is the prioritization of company's five new products for launching stage, dealing with a multicriteria group decision making problem. The resolution of this problem allows the company to plan its investment and response capacity with greater opportunities for success marketing in target market segments.

Six people were involved as DM, each with different functions in the company value chain: sales manager (DM1-SaM), marketing manager (DM2-MaM), marketing manager (DM3-MaM), store manager (DM4-StM), customer service manager (DM5-CuSM), research and development manager (DM6-RDM). All DMs have at least one year working in the company and have experience in launching products in the market with successful results. DMs are women from 29 to 45 years old, with a university degree, with at least five years of experience. The criteria were proposed by the DMs based on their practice in the industry. Table 1 presents the criteria description. Table 2 presents the alternatives proposed by the DMs.

Table 1. Criteria description.

Criteria	Description	Direction
C1	Address the current needs of the market (procured in stores)	Maximize
C2	Offer high value products (cover new ailments)	Maximize
C3	Create new markets with innovative products (new segments)	Maximize
C4	Growth of the company (have products that are in the industry)	Maximize
C5	Vanguard in functional products (use of new active ingredients)	Maximize
C6	Improve existing products (with new attributes)	Maximize
C7	Increase the product catalog (extend lines)	Maximize

Table 2. Alternative description.

Alternative	Description
A	Protein supplement of high biological value and intermediate absorption
B	100% natural soluble prebiotic fiber with a soft green apple flavor
C	Magnesium supplement used for multiple biochemical functions in the body
D	Potassium supplement to regulate water balance and blood pressure
E	Zinc supplement, essential for growth and multiple metabolic functions

3.2. Development and results

At Table 3 the performance matrix presents five products and the above defined criteria for each DM. As consequence of the iteration of experts with analyst, the elicitation of the inter-criteria parameters results is show in Table 4 and Table 5. Table 4 shows direction and relative importance of criteria. Table 5a) shows the indifference and preference thresholds for DM1, DM2, DM3 and DM4. Table 5b) shows the indifference and preference thresholds for DM5 and DM6.

The ELECTRE III method is used to generate the preferential model in the format of a fuzzy outranking relation (O_k) for every DM_K (see Table 6). Once we obtained the individual preferential model a MOEA is used for the exploitation phase, and generate an individual ranking. 5 ranking were generated for each DM to propose individual solutions. A weighted sum (WS) was applying to propose a final individual ranking for each DM. Every time an alternative appears in the position one, a score of value 5 is accumulated, the WS method is applied. Last position of the ranking assigns a score of value 1. The WS is used to propose an individual ranking based on the previous analysis. Table 7 suggests the following ranking for DM1 $C \succ D \succ B \succ E \succ A$.

Final ranking for each DM is showed at Table 8. Once we obtained the individual ranking we carry out the GDM process. The ELCTRE-GD use as input the preferential model (O_k) and ranking (R_k) of every DM. In the GDM process, ELECTRE-GD constructs a collective fuzzy outranking relation (O_G). Once we obtained her/his valued matrix the exploitation phase is again performed and the collective ranking generated is $C \succ A \succ B \succ D \succ E$.

Table 3. Performance matrix for each DM.

	C1	C2	C3	C4	C5	C6	C7
DM1							
A	0.4	0.2	0.6	0.6	0.4	1	0.8
B	0.4	0.4	0.4	0.8	0.8	1	1
C	1	0.8	0.8	1	1	1	1
D	0.6	0.2	0.6	1	1	1	1
E	0.6	0.2	0.4	1	0.8	0.8	1
DM2							
A	0.4	0.6	0.8	0.8	1	1	0.6
B	0.2	1	1	0.8	0.8	1	1
C	0.6	0.8	0.8	0.8	0.8	1	0.6
D	0.6	0.8	0.8	0.8	0.8	1	0.6
E	0.6	0.8	0.8	0.8	0.8	1	0.6
DM3							
A	0.6	0.4	1	1	0.8	0.2	0.2
B	0.8	1	1	1	0.4	1	0.6
C	1	0.6	0.8	0.8	0.4	0.8	0.2
D	1	0.6	0.8	0.8	0.4	0.6	0.2
E	1	0.6	0.8	0.8	0.4	0.4	0.2
DM4							
A	1	1	0.8	1	1	1	0.6
B	1	0.8	0.8	0.6	1	0.8	0.6
C	1	1	1	0.8	0.8	0.8	0.6
D	1	1	1	0.8	0.8	0.8	0.6
E	1	1	1	0.8	0.8	0.8	0.6
DM5							
A	1	1	1	1	1	0.8	0.6
B	1	1	1	1	1	0.8	0.6
C	1	1	0.8	0.8	1	0.8	0.6
D	1	0.8	0.8	0.8	1	0.8	0.6
E	1	0.8	0.8	0.8	1	0.8	0.6
DM6							
A	1	0.4	0.6	0.2	1	0.8	1
B	0.4	0.2	0.4	0.2	0.4	0.8	1
C	1	1	0.6	0.2	1	0.8	1
D	1	1	0.6	0.2	1	0.8	1
E	1	1	0.6	0.2	1	0.8	1

Table 4. Weights and direction of criteria.

	C1	C2	C3	C4	C5	C6	C7
	Max	Max	Max	Max	Max	Max	Max
DM1	21.4	25	10.8	7.2	17.8	14.2	3.6
DM2	15.1	26.4	9.5	15.1	9.5	20.7	3.7
DM3	16.1	16.1	24.5	20.3	11.9	7.7	3.4
DM4	17.8	14.2	21.4	3.6	7.2	25	10.8
DM5	17.7	17.7	17.7	22.5	13	8.1	3.3
DM6	24	24	24	12.6	12.6	1.4	1.4

Table 5. Indifference and preference thresholds.

		C1	C2	C3	C4	C5	C6	C7
a)								
DM1, DM2. DM3.DM4	q	0	0	0	0	0	0	0
	p	0.2	0.2	0.2	0.2	0.2	0.2	0.2
b)								
DM5	q	0	0	0	0	0	0.2	0.4
	p	0.2	0.2	0.2	0.2	0.2	0.4	0.6
DM6	q	0	0	0	0.2	0	0	0
	p	0.2	0.2	0.2	0.4	0.2	0.2	0.2

Table 6. Fuzzy outranking relation for DM1.

	A	B	C	D	E
A	1	0.46	0.14	0.5	0.5
B	0.89	1	0.18	0.43	0.71
C	1	1	1	1	1
D	1	0.75	0.43	1	1
E	0.75	0.61	0.11	0.57	1

Table 7. Rankings for DM1.

Position	R1	R2	R3	R4	R5	Final
1	C	C	C	C	C	C
2	D	D	D	D	D	D
3	B	B	B	E	B	B
4	E	E	E	B	E	E
5	A	A	A	A	A	A

Table 8. Rankings for group.

Position	DM1	DM2	DM3	DM4	DM5	DM6	G
1	C	B	B	A	A	E	C
2	D	C	A	C	B	D	A
3	B	E	C	D	C	C	B
4	E	D	D	E	E	A	D
5	A	A	E	B	D	B	E
Disagrees	4	3	3	5	3	7	
Weighted Kendall	0.562	0.7	0.714	0.6	0.7	0.238	0.586

Table 8 Column G shows the collective ranking generated from the individual preference information obtained by DMs. We can find some smooth inversions between individual and collective ranking shown in the disagreement row. A Weighted Kendall metric is used to rate the individual agreement between DM and collective preference. DM2, DM3 and DM5 show the higher agreement 0.7, 0.714 and 0.7, respectively. DM1, DM4 and DM6 show the lower agreement 0.562, 0.6 and 0.238, respectively.

Based on the individual agreement we computed a collective agreement (0.586). The preferences of DM and their differences in opinion are shown in the comparison between their individual ranking against the collective ranking. A low collective agreement is reached because important difference between individual ranking, even the collective ranking express the individual preference, those preferences are in conflict and the collective ranking reach the best compromise between them, showing some similarities and difference between individual and collective ranking.

4. Conclusion

The paper presents a multicriteria model for launching a new product for group decision making. The process carried out for the integration of information, as well as the steps within the four phases of development, provide the company an objective guide to reach a suitable consensus level and a group final decision. The calculation of individual agreement helps to identify those DMs that are farthest from the group's preference, as well as presenting that if the DM improves their individual result the level of consensus is raised. DMs of the company recognize the usefulness of integrating a multicriteria analysis model for the launch of new products, in the group decision making process based in individual preference aggregation scheme.

References

1. Álvarez Carrillo, P. A., Leyva López, J. C., & Sánchez Castañeda, M. D. L. D. (2015). An Empirical Study of the Consequences of Coordination Modes on Supporting Multicriteria Group Decision Aid Methodologies. Journal of Decision Systems, 24(4), 383-405.
2. Gross, U. (2014). Fighting the fire: Improvisational behavior during the production launch of new products. International Journal of Operations & Production Management, 34(6), 722-749.
3. Ledwith, A., & O'Dwyer, M. (2008). Product launch, product advantage and market orientation in SMEs. Journal of Small Business and Enterprise Development, 15(1), 96-110.
4. Leyva López, J. C., & Alvarez Carrillo, P. A. (2015). Accentuating the rank positions in an agreement index with reference to a consensus order. International Transactions in Operational Research, 22(6), 969-995.
5. Roy, B. (1990). The outranking approach and the foundations of ELECTRE methods. In Readings in multiple criteria decision aid (pp. 155-183). Springer, Berlin, Heidelberg.

Ground handling services firm evaluation based on neutrosophic MULTIMOORA method

Serhat Aydin
Industrial Engineering Department, National Defense University Air Force Academy
Istanbul, 34149, Turkey

Mehmet Yörükoğlu
Industrial Engineering Department, National Defense University Air Force Academy
Istanbul, 34149, Turkey

Ground Handling Services (GHS) include all the services an aircraft needs during the period it remains on the ground. In this study we focus on passenger services for disabled passengers and it is aimed to evaluate the GHS firms for passenger services based on neutrosophic MULTIMOORA method which is a very newly developed method. We evaluate three firms according to eight different criteria. Criteria have different weights and determined by Analytic Hierarchy Process. Then, we apply Neutrosophic MULTIMOORA method's algorithm to the problem. Finally, the conclusion is given. The originality of this paper is evaluating GHS firms with the Neutrosophic MULTIMOORA method for the first time.

1. Introduction

United Nations' population report points out that the World's population reached 7.6 billion in mid-2017 and the population is expected to be 8.6 billion in 2030, and 11.2 billion in 2100 [1]. By 2030, the urbanization rates are projected to increase 60 per cent over the World [2]. Annual traffic of passenger will increase to 80 trillion passenger-kilometers (%50 increase) by 2030 [3]. Air transport has an important effect on globalization and economy by accelerating the goods, information, capital and people flows [4]. The International Air Transport Association (IATA) announced that passenger traffic data for the year 2016 showed a 6.3% increase in demand compared to 2015. IATA's 20-year estimation reports that the number of air passengers will double by 2034 [5].

For air passengers, the journey itself is served as main service, but a series of complementing services are given to the passengers before and after the flight [6]. Services provided before and after the flight are called "Ground Handling Services" (GHS) [7], despite GHS has many similar definitions [8, 9, 10, 11] it has no formal or official definition. Some airlines receive GHS from independent firms, while others prefer to handle GHS themselves [12, 13]. GHS are located at the intersection of all activities at the airport where air transportation services are performed. GHS play a vital role in ensuring that air transportation is effective, safe and cost-effective [12]. The scope of GHS is based on the principles determined by international organizations and aviation authorities around the World [6]. The "Standard Ground Handling Agreement" document of IATA draws the framework of the services to be provided by GHS firms. In Turkey, GHS are determined by Turkish Directorate General of Civil Aviation in this framework [14].

Customer Satisfaction (CS) is feeling of customer on received services or products [15], CS depends on the quality of the product or the customer's personal expectation and perception [16]. GHS given to passengers (General Services, Departure Services, Arrival Services, Inter-modal Transportation Services by rail, road or sea) [8, 17] plays a direct role in determining CS [18]. Although all GHS providers take care of all air passengers, some passengers differ from the services they need. These are minors travelling alone, infants and children, groups, disabled passengers and passenger requiring medical clearance [19]. Disabled passengers are handled separately by all aviation authorities as a special and separate group [20, 21, 22]. The number of disabled people increases with population growth and ageing, with 15% of the population forming the largest minority in the World [23]. "Accessibility" in air transport services for disabled passengers is the basic criterion and it means "the use of buildings, open spaces, transportation information services and information and communication technologies can be accessed and used safely and independently for the disabled" [21, 22]. Therefore, detailed services for disabled passengers (meeting with assistance, accompanying, check-in, security control, social needs, boarding gate, getting on/off the plane) are specifically defined by the aviation authorities [21, 22, 24] and mainly supplied by GHS companies [24].

We aim to evaluate the GHS firms according to the determined criteria. GHS firms' evaluation problem can be named multi-objective decision making problem. Multi-objective decision making methods optimize both beneficial and non-beneficial criteria. MULTIMOORA method is one of the effective multi-objective optimization solution methods developed by Brauers and Zavadskas [25]. The method uses simple calculation operations, gives effective results, and is not affected by the introduction of any extra parameters [26].

Crisp data are generally unavailable and difficult to be exposed in real-life problems, decision makers usually need to take into account the vagueness of the information. Because of this need, the fuzzy set theory was developed by Zadeh [27] and widely used in order to solve multi-objective decision making problems. In recent years, fuzzy set has been extended to new types. [28]. Neutrosophic set is one of the extensions of fuzzy sets and it uses indeterminacy-membership for the first time. A neutrosophic set is expressed by three parameters, which are called *"truthiness"*, *"indeterminacy"* and *"falsity"*.

Neutrosophic set and MULTIMOORA method were combined in Neutrosophic MULTIMOORA by Stanujkic *et al.* [29]. Combining Neutrosophic set and MULTIMOORA method provides flexibility and easy computational operations to overcome multi-objective decision making problems in vagueness environment.

The rest of the paper is organized as follows: in Section 2, a literature review is given. In Section 3, the application steps of the Neutrosophic MULTIMOORA are presented. In Section 4, an application is given and the conclusions are presented in the final section.

2. Literature Review

In the literature section, we focus on passenger services in aviation transportation, disabled passenger services in aviation transportation and services given to disabled passengers by GHS firms.

Gourdin [30] presented the price, time schedule, food beverage, baggage carriage, security and seating comfort in terms of CS. Chen *et al.* [16] showed for air passengers fast, reliable, comfortable and appropriate services are basic needs expected from service providers. Alodhaibia *et al.* [31] indicated that passengers' satisfaction increases as time shorten in the system and they demand personalized self-service options with latest technologies. Correia *et al.* [32] identified "total time of service" and "total distance of walking" as two important indicators showing the level of air passenger service by using a psychometric scaling technique. IATA's 2017 global passenger survey shows [33] that on board service, boarding, bag collection and border control/immigration issues have more impact on overall passengers' satisfaction. Kandampully [34] pointed out some innovations like social media communication are new service opportunities for air passengers. Shaw and Coles [35] met with 24 disabled people and highlighted their problems with air travel. Murray and Sproats also [36] pointed out behavioral, physical and economic barriers are the main obstacles [35, 36]. Cavinato and Cuckovich [37] showed accurate information interaction is a facilitator for disabled passengers at air transportation services.

Tretheway and Markhvida [38] indicated GHS firms are one of the main value creator actors in the air transportation such as airport operators, airlines, aircraft manufacturers, aviation infrastructure providers and leasing firms. Oostveen and Lehtonen [39] studied whether disabled passengers prefer to use automatic systems in place of assistance agent services by interviews and a survey of disabled passengers. Chang and Chen [40] determined the disabled air passengers' preferential service needs as barrier-free ramp and lift, slip resistant floors, wheelchair services, kindly attitudes of staffs, check-in and boarding priority by using a two-stage survey. Yörükoğlu and Kayakutlu [41] analyzed the services given to disabled passengers by GHS companies with cognitive mapping technique.

3. Neutrosophic MULTIMOORA Method

Because of the space limit, we will explain only Neutrosophic MULTIMOORA method. If any researcher wants to examine classical MULTIMOORA and Single Valued Neutrosophic sets, can look the paper by Stanujkic et al. [29]. The Neutrosophic MULTIMOORA Method consists of Neutrosophic MOORA-Ratio Method, Neutrosophic Moora-Reference Point Method and Neutrosophic MOORA-Full Multiplicative Form. Because of the space limit only Neutrosophic MOORA- Ratio Method will be explained, the other methods can be found in the paper by Stanujkic et al. [29].

3.1. *Neutrosophic MOORA-Ratio Method*

Step 1. Calculate Y_i^+ and Y_i^- by using the Single Valued Neutrosophic Weighted Average Operator, as follows:

$$Y_i^+ = \left(1 - \prod_{j \in \Omega_{max}} (1-t_j)^{w_j}, \prod_{j \in \Omega_{max}} (i_j)^{w_j}, \prod_{j \in \Omega_{max}} (f_j)^{w_j}\right). \tag{1}$$

$$Y_i^- = \left(1 - \prod_{j \in \Omega_{min}} (1-t_j)^{w_j}, \prod_{j \in \Omega_{min}} (i_j)^{w_j}, \prod_{j \in \Omega_{min}} (f_j)^{w_j}\right) \tag{2}$$

where Y_i^+ and Y_i^- denote the importance of the alternative i obtained based on the benefit and cost criteria, respectively; Y_i^+ and Y_i^- are Single Valued Neutrosophic Numbers.

Step 2. Find Score Function of Y_i^+ and Y_i^- as follows:

$$y_i^+ = s(Y_i^+) \qquad y_i^- = s(Y_i^-) \tag{3}$$

$$s(Y_i) = (1 + t_{y_i} - 2i_{y_i} - f_{y_i})/2 \tag{4}$$

$$s(Y_i) \in [-1,1]$$

Step 3. The overall importance of each alternative can be calculated as follows:

$$y_i = y_i^+ - y_i^- \tag{5}$$

Step 4. Alternatives can be ranked according to value of Y_i in descending order and the alternative with the highest value is the best alternative.

4. Application

In this section, three Turkish GHS companies were evaluated according to eight determined criteria from Directorate General of Civil Aviation document as follows: C_1-**Meeting with assistance** (benefit), C_2-**Accompanying** (benefit), C_3-**Check-in and baggage Registration** (benefit), C_4-**Security** (benefit), C_5-**Social needs** (benefit), C_6-**Boarding gate** (non-benefit), C_7-**Getting on/off the plane** (benefit), C_8-**Total service time** (cost). First, we determined the criteria weights by Classical Analytical Hierarchy Process [42]. The weights are; *0.14, 0.16, 0.14, 0.15, 0.13, 0.12, 0.11, 0.05,* respectively. Alternatives were evaluated by the disabled passengers with spinal cord injury using the airline for transportation. The obtained ratings are shown in Table 1. When he was not sure to assign a neutrosophic number to evaluate alternatives according to criterion, we got help from Yörükoğlu and Kayakutlu [43] study for assigning a neutrosophic number into the decision matrix.

Table 1. Decision matrix for the GHS firm.

	C_1 0.14 *max*	C_2 0.16 *max*	C_3 0.14 *max*	C_4 0.15 *max*	C_5 0.13 *max*	C_6 0.12 *max*	C_7 0.11 *max*	C_8 0.05 *min*
A_1	(0.8,0.2,0.2)	(0.8,0.1,0.2)	(0.7,0.2,0.2)	(0.6,0.1,0.2)	(0.6,0.3,0.2)	(0.7,0.2,0.3)	(0.8,0.2,0.1)	(0.8,0.2,0.2)
A_2	(0.8,0.3,0.3)	(0.5,0.3,0.2)	(0.9,0.1,0.2)	(0.8,0.1,0.2)	(0.7,0.3,0.3)	(0.7,0.3,0.3)	(0.7,0.3,0.3)	(0.5,0.3,0.3)
A_3	(0.7,0.1,0.2)	(0.7,0.1,0.2)	(0.8,0.2,0.1)	(0.8,0.1,0.2)	(0.6,0.2,0.1)	(0.6,0.2,0.1)	(0.6,0.2,0.1)	(0.8,0.1,0.1)

Then Neutrosophic MOORA-Ratio method was applied in order to get ranking results. The overall performances of alternatives are shown in Table 2.

Table 2. Overall performances of the alternatives.

	Y_i^+	Y_i^-	y_i^+	y_i^-	y_i	Rank
A_1	(0.60,0.26,0.31)	(0.08,0.92,0.92)	0.383	-0.845	1.228	3
A_2	(0.65,0.30,0.34)	(0.03,0.94,0.94)	0.355	-0.895	1.251	2
A_3	(0.61,0.22,0.26)	(0.08,0.89,0.89)	0.453	-0.798	1.252	1

Then we applied Neutrosophic MOORA-Reference Point method. Table 3 shows the reference point and Table 4 shows the deviation from the reference points.

Table 3. Reference points.

	C_1	C_2	C_3	C_4	C_5	C_6	C_7	C_8
r_j^*	⟨0.8,0.1,0.2⟩	⟨0.8,0.1,0.2⟩	⟨0.9,0.1,0.1⟩	⟨0.8,0.1,0.2⟩	⟨0.7,0.2,0.1⟩	⟨0.7,0.2,0.1⟩	⟨0.8,0.2,0.1⟩	⟨0.5,0.1,0.3⟩

Table 4. Weighted deviations from the reference point.

	r_1^*	r_2^*	r_3^*	r_4^*	r_5^*	r_6^*	r_7^*	r_8^*	d_i^{max}	Rank
A_1	0.00	0.00	0.03	0.03	0.01	0.00	0.00	0.01	0.030	2
A_2	0.00	0.05	0.00	0.02	0.00	0.00	0.01	0.00	0.048	3
A_3	0.01	0.02	0.01	0.00	0.01	0.01	0.02	0.01	0.022	1

Then we applied Neutrosophic MOORA-Full Multiplicative form method. Table 5 shows utility value for each alternative.

Table 5. Utility values of alternatives.

	A_i	B_i	a_i	b_i	u_i	Rank
A_1	⟨0.77,0.13,0.15⟩	⟨0.99,0.01,0.01⟩	0.678	0.978	0.693	2
A_2	⟨0.79,0.17,0.18⟩	⟨0.97,0.02,0.02⟩	0.636	0.956	0.664	3
A_3	⟨0.79,0.10,0.12⟩	⟨0.99,0.10,0.12⟩	0.730	0.987	0.740	1

Finally, we applied dominance theory in order to get the final ranking of three different methods. The final ranking of alternative is shown in Table 6.

Table 6. The final ranking order.

	NMRM	NMRPM	NMFMF	Rank
A_1	3	2	2	2
A_2	2	3	3	3
A_3	1	1	1	1

As seen in Table 6, the ranking of GHS firms on the dominance theory is as follows: A_3, A_1, A_2.

5. Conclusion

The existence of sustainable developments in the business world can be ensured by the improvement of both social approaches and scientific methods. This article aims to better identify the services provided to the disabled air travelers in air transport, one of the locomotive sectors of the globalized world, and thus the service providers.

Evaluation of GHM firms is a decision-making problem based on humans' thoughts and judgments. Fuzzy methodology can handle such problems consisting of human's thoughts and judgments. Neutrosophic sets are one of the extensions of fuzzy sets and it uses indeterminacy-membership and carries more information than other fuzzy sets. MULTIMOORA method uses simple equations and gives effective results in multi-objective decision making problems. Neutrosophic MULTIMOORA method can handle multi-objective decision making problem easily and effectively under uncertainty.

In this paper, GHM firm evaluation multi-objective problem is handled and solved by Neutrosophic MULTIMOORA method. Three Turkish GHM firms were evaluated according to eight different criteria contains both benefit and non-benefit. Finally, firms are ranked according to dominance theory. The originality of the paper is using Neutrosophic MULTIMOORA method to evaluate GHS firms for the first time in literature. And, this study will provide a new perspective on determining and improving aviation ground services and similar processes.

In the future studies, the same problem can be solved by different multi-objective methods and the results can be compared.

References

1. United Nation, *World Population Prospects*. **The 2017 Revision**, 12 (2017).
2. United Nation, *Data Booklet*. **The World's Cities in 2016**, 1 (2017).
3. The World Bank, *Tracking Sector Performance*. **Global Mobility Report 2017**, 6 (2017).
4. P. Niewiadomski, *Geoforum. 87*, 4 (2017).
5. IATA, *Report*. **Annual Review 2016**, 31 (2016).
6. M. Yörükoğlu and G. Kayakutlu, *ICCIIS (INDE) 2011*. **Vol II WCE**, 1083-1088 (2011).
7. Conference Secretariat, *Conference on the Economics of Airports and Air Navigation Services*. **ANSConf-WP/10**, 1 (2000).
8. IATA, *Standard Ground Handling Agreement SGHA*. **AHM 810**, 5 (2013).
9. The Council of the European Union, *Council Directive 96/67/EC of 15 October 1996*. **Official Journal L 272**, 36 (1996).
10. ICAO, *Airport Economic Manual*. **Doc 9562**, 2-12 (2013).
11. **UNECE**, *Europa Glassory, F. Air Transport*. **2018/01/24 EC-DG**, 6 (2016).
12. M. Studic, A. Majumdar, W. Schuster and W.Y. Ochieng, *Trans. Res.* **C 74**, 245-260 (2017).
13. HAVAS, *Ground Handling Services*. http://www.havas.net/en/OurServices/GroundHandlingServices/Pages/GroundHandlingServices.aspx (available at 07.02.2018).
14. IATA, *Standard Ground Handling Agree. (SGHA)*. **AHM 810**, 5 (2004).
15. HAVAS**,** *Standard Ground Handling Agreement (SGHA)*. **IATA SGHA 2013-2008–2004**, 244 (2013).
16. J.K.C. Chen, and J. Batnasan, *Tech. in Soci.* **43**, 219-230 (2015).
17. **HAVAS**, *Standard Ground Handling Agreement (SGHA)*. **IATA SGHA 2013-2008–2004**, (2013).
18. S.B. Schmidberger, L.E. Hartman and C. Jahns, *Int. Jour. of Prod. Eco.* **117**, 104–116 (2009).
19. IATA, *Ground Operations Manual*. **IGOM**, 136 (2014).

20. ICAO, *Facilitation.* **Annex 9**, 90 (2011).
21. The European Parliament and the Council of the European Union, *Regulation (EC).* **No 1107/2006**, 8 (1996).
22. SHGM (DGCA), *Talimat-Engelsiz.* **SHT**, 10 (2015).
23. A.M. Oostveen and P. Lehtonen, *Tech. in Soci.* **Art. In Press**, 1-10 (2017).
24. Dalaman Airport, *Prosedür.* **PR.OP.17**, 27 (2016).
25. A.W.K.M. Brauers and E. K. Zavadskas, *Technological and Economic Development of Economy*, **17:1**, 174-188 (2011).
26. T.K. Jana, B.S. Paul, B. Sarkar, and J. Saha, *Journal of Manufacturing Systems.* **32**, 801-819, (2013).
27. T.K.L. Zadeh, *Information Control.* **8 (3)**, 338–353 (1965).
28. C. Kahraman, B. Oztayşi, S. Cevik, *International Journal of Computational Intelligence Systems*, **3**, 24 (2016).
29. D. Stanujkic, E.K. Zavadskas, F. Samarandache, W. Brauers, *Informatica.* **28 1**, 181-192 (2017).
30. K. Gourdin, *Trns. Jour.* **27(3)**, 23–29 (1988).
31. S. Alodhaibia, R.L. Burdettb and P. Yarlagaddaa, *Proce. Eng.* **174**. (2017).
32. A.R. Correia, S.C. Wirasinghe and A.G. Borros, Trans. Rese. Part A **42**, 330–346 (2008).
33. IATA, *Highlights.* Global Passenger Survey, 129 (2017).
34. J. Kandampully, *Eur. J. Innov. Manag.* **5 (1)**, 18-26 (2002).
35. G. Shaw and T. Coles, Tour. Manag. **25**, 397–403 (2004).
36. M. Murray and J. Sproats, *Jour. of Tour. Stud.* **1(1)**, 6-15 (1990).
37. J. L. Cavinato and M. L. Cuckovich, *Trans. Jour.* **31(3)**, 46-53 (1992).
38. M.W. Tretheway and K. Markhvida, *Air Trans. Manag.* **41**, 3-16 (2014).
39. A.M. Oostveen and P. Lehtonen, *Tech. in Soci.* Art. In Press, 1-10 (2017).
40. Y.C. Chang and C.F. Chen, *Tour. Manag.* **32**, 1214-1217 (2011).
41. M. Yörükoğlu and G. Kayakutlu, *ICOVACS 2010.* Spain, 15-17 (2010).
42. T.L. Saaty, McGraw-Hill, New York (1980).
43. M. Yörükoğlu, G. Kayakutlu and S. Ercan, *Jour. of Aero. and Space Tec.* **7/1**, 1-23 (2014).

Creating alternatives for the eggplant waste problem in a horticultural company

Larreta-Ramirez, Elsa Veronica
Management Science Doctoral Program, Universidad Autonoma de Occidente
Culiacan, Mexico

Gastelum-Chavira, Diego Alonso
Management Science Doctoral Program, Universidad Autonoma de Occidente
Culiacan, Mexico

Leyva-Lopez, Juan Carlos
Economic and Management Science Department, Universidad Autonoma de Occidente
Culiacan, Mexico

Valdez-Lafarga, Octavio
Morrison School of Agribusiness, W.P. Carey School of Business,
Arizona State University, Arizona, USA

Worldwide, a large amount of food that is produced for human consumption is wasted. This situation is due by several factors such as short shelf life, overproduction, consumer preferences, among others. Such is the case of eggplant, whose production generates high volume income due to export activities. Despite the previous fact, demand a quality levels in the international target market eggplant waste. In this contribution, the methodology of Value-Focused Thinking was applied to obtain solution alternatives to avoid eggplant waste in a Mexican horticultural company. Also, and market research was conducted in Sinaloa, Mexico to find out consumer preferences regarding a set of possible eggplant products.

1. Introduction

Nowadays, new product design is an important component for the growth and survival of profit organizations, which face complex and highly competitive scenarios. The new product design is one of the tasks of greatest risk and uncertainty for organizations. It has been shown by the 24%-55% commercial fail rate range when a product is launched in the market [1-3].

Thus, the decision-making process for new product design is complex, because the prediction about future markets and the demand for operation has a high degree of uncertainty.

Given this situation, organizations try to introduce ideas in their innovation processes to design new products; mainly considering consumer preferences, which are increasingly diverse and changing. Such is the case of companies in the food sector, which operate in dynamic and competitive market; where demand and production trends of agricultural food are increasing worldwide. However, the food sector is entering a new era, shown by the scarcity of resources, the increase in demand and greater volatility risks [4].

Moreover, some regions or countries produce different types of products, which are not consumed by its population due to cultural roots. Such is the case of Sinaloa, Mexico, which produces 94.1% of eggplant nationwide. In 2016, farmers obtained $24.18 million USD by eggplant exportation. The main target of this production is the United States of American, with 99%. Although, when eggplant does not fulfill the quality standards of the market, it is wasted despite being in optimal conditions for human consumption, because it is not preferred by Mexican consumers.

In that sense, the horticultural sector in Sinaloa faces challenges related to the design of new products, because even though it is an important player in the USA market, its exportations are mainly in fresh. Also, this sector faces problems of economic uncertainty due to its dependence on intermediaries, increased competition, lack of financing and training. Within the food sector, horticulture is an emblematic economic activity in Sinaloa, not only for being the main source of exports in the state, but for the quality and variety of products it spans.

In this contribution, the multicriteria methodology of Value-Focused Thinking (VFT) [5], was applied in a horticultural company located in Sinaloa, Mexico to generate possible solution alternatives to avoid eggplant waste. Also, a set of possible eggplant products were submitted to potential consumer through a market research. These results will be used in a new multicriteria methodology for the design of new products that is under development. The rest of the document is organized as follows: Section 2 presents a description of the elements to take in to account when a set of solution alternatives needs to be created. On the other hand, the case study of how to structure a problem, how to generate a set of solution alternatives by using VFT is presented in Section 3. Finally, section 4 is meant for conclusions.

2. Creating Alternatives for the New Product Design Problem

In a competitive market, companies in order to design and develop successful new products to get competitive advantage to survive and grow. These new products must fulfill preferences and needs of customers at competitive prices.

To improve the potential success of a new product, decision makers need to consider the product's, attributes, customer's requirements and market's competing factors. These considerations lead decision maker to analyze a large amount of data, which is typically highly unstructured and includes some degree of uncertainty. These data are usually obtained from surveys; which must be designed to gather consumer's preferences, while paying attention to questions writing and sequences as well as chosen measurement scales to be used. In addition to consumer's preferences, it is necessary to know decision makers preferences and company's capabilities is to develop the new product.

In literature, there exist different methodologies and methods for design and development of new products, such as the Stages-Gates methodology, Quality Function Deployment (QFD), Conjoint Analysis, a consumer-based methodology for New Product Design proposed by [6]. However, in this work, the Value Focused Thinking methodology was used to structure the eggplant waste problem and to generate possible products to avoid such waste.

The Value Focused Thinking (VFT) is a multicriteria methodology used to know how values can be used to improve the decision-making process. It provides a decision frame, work encompasses the decision context and the fundamental objectives that define a set of alternatives appropriate for a decision situation. VFT methodology has been applied in a wide range of contexts to identify the objectives of decision makers (DMs), such as: is to understand the values of education available by mobile technology in [7]; in the plaster waste disposing problem from building sites in [8]; getting a set of values mobile support to transform virtual teams in organizations [9].

In general, a VFT application can divided in six stages: 1) Semi-structured Interviews, 2) Problem Identification, 3) Identifying and Structuring Objectives, 4) Generating a Hierarchy of Fundamental Objectives, 5) Defining attributes to measure fundamental objectives and 6) Creating Solution Alternatives.

The obtained alternatives (potential products) could be included in surveys to conduct out a market research. These surveys should be designed considering stakeholders' opinions such as decision makers, specialty stores employees regarding these potential products. Likewise, surveys should be designed according to the type of analysis that will be carried out with collected data once the surveys have been applied. Then, surveys should be applied to a set of possible consumers to get their preferences about these types of products and these consumers profiles. The results of this could be used in methodologies for the design of new products to help companies determine consumer preferences about product price and presentation.

In this contribution, a case study is presented in which a set of potential products were generated. This case study was carried out considering input data, which are required by a new multicriteria methodology for new products design, which is currently under development. This methodology consists of four main modules: Multicriteria market segmentation, Multicriteria preferences disaggregation, Multicriteria market consumer's choice and New product simulation. The decision maker preferences are included in all modules. A schema of this methodology is presented in Figure 1.

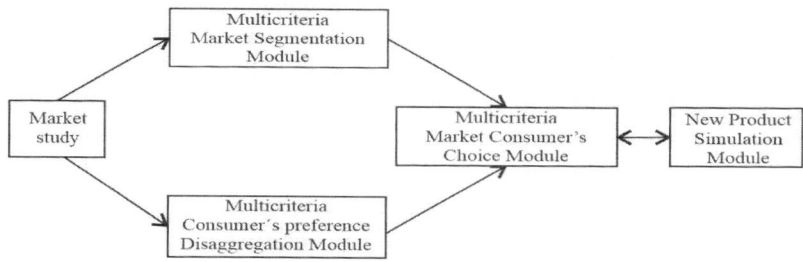

Figure 1. Schema of the methodology for new product design under development.

In Figure 1, it is noted that the process for a new product design, starts with a market research. In which typically, it includes a consumer's preferences survey regarding a set of products in the market and other elements of interest. This stage corresponds precisely to the work presented here, where a set of potential consumers was questioned about their consumption of eggplant and some potential products derived from this vegetable. The set of these potential products were obtained following the VFT methodology and traditional interviews with stakeholders of these kind of products.

3. Case Study

This case is about how the multicriteria VFT methodology was applied in a horticultural company to obtain a set of alternatives for eggplant waste. This company is in Sinaloa, Mexico, it produces and exports bell pepper and eggplant. Most of its production is exported to the United States of America (USA). Export products are marketed through distributors, which are its main customers. Nowadays, the company's market share of eggplant is between 15% and 19% in the Nogales, Arizona port of entry. Company's owners want to increase this market share and diversify their eggplant-based products, either fresh or processed. In addition, they want to open new domestic markets for processed eggplant products, because there exists a variety of export barriers for them in the USA. Specially for eggplant that does not fulfill the export quality standards because it is considered waste due to it is low consumption in Mexico.

By using the VFT methodology, the eggplant waste problem was structured and some alternatives for this problem were obtained. The following is a summary of the VFT application in the afore mentioned:

- First and second stages: a set of meetings and semi-structured interviews where applied to the decision maker of the company. Here, information related with the company's context and problematics was identified.
- Third stage: a list of fundamental and means-end objectives was identified. Among the first ones are overhead costs, eggplant yield, economic loss due to eggplant depletion. Likewise, some of the second ones are production costs, seedling greenhouse yield and use of depletion.
- Fourth stage: the list of fundamental and means-end objectives was sorted to generate a hierarchy of them. Also, for each objective, a maximization or minimization direction was defined, e.g. *Minimize overhead costs*.
- Fifth stage: for each fundamental objective, a set of attributes were defined to measure them e.g. for the overhead costs objective; *Production cost* (per box packed in pesos) and *Costs in rates of interests* (suppliers, banks) were set.
- Sixth stage: here, a variety of ideas as potential solution alternatives were identified. Some of them are: *design and develop a new product derived from eggplant, production planning adapted to the markets,* seedling *in adequate conditions* and *increase eggplant shelf life.*

3.1. Obtaining the Potential Products

The set of potential solution alternatives was showed to the decision maker of the company. Then, the SMARTER [10] method was used to choose the most prominent alternative solution. However, the decision maker considered to work in parallel with some of the potential solution alternatives, emphasizing eggplant new product design. In part, because the company has been able to produce some types of products in its own pilot plant. Hence, the question is: what product should be produced?

To answer this question, a survey was designed and applied to a set of 87 potential consumers. This amount is due to another complementary research, not provided here, which carried out with them at the same time; where the total time for both studies was approximately 45 minutes.

The survey contained some typical sociodemographic questions, which are going to be used for the segmentation purposes. Moreover, it has questions about eggplant consumption and possible eggplant products. The list of eggplant products included in the survey was obtained from different interviews with nutriologists, employees of naturist and specialty stores, a production manager of a nutraceutical products company and the decision maker of the previously stated horticultural company. In all the interviews, VFT was partially used to obtain the list of possible eggplant products. The set of these possible products obtained in this case study, is presented below:
- Dehydrated eggplant as a snack
- Marmalade
- Eggplant water as a diuretic
- Flour for bakery
- Food Supplement
- Medallions for hamburgers
- Eggplant tea

In the survey, among other questions of interest, the following question was included to get at least the three most prominent products of eggplant: *If you are interested in consuming eggplant products, please indicate, which of the following ones would you like to consume?* In this question, respondent ordered the most preferred products afore mentioned list.

3.2. Results

Once the survey was applied, 87 preference rankings of the potential products were obtained. Then, by using Borda's count, a global ranking of products was derived. The global ranking was obtained considering the number $T(i,j)$, ($1 \leq 3$, $1 \leq 7$) of times that a product was found at a certain place in the ranking of each respondent. That is because each place in the ranking has its own importance. Thus, each place in the ranking is considered as a weight w. Hence, it is possible to compute a weighted sum $\sum w_i T(i,j)$; $i = 1, ..., 3$; $j = 1, ..., 7$. By using this weighed sum, a global ranking in decreasing order of preference is obtained. This global ranking represents a recommendation for the decision maker. Table 1 shows the number of times that each product had the first, second and third preference place in the 87 rankings.

Table 1. Number of times that a product was found in a certain place in the 87 rankings and the Borda's count computation.

Weight	Rank	Dehydrated	Supplement	Water	Tea	Medallion	Flour	Marmalade
3	1	44	12	5	2	5	4	10
2	2	7	11	3	12	6	9	21
1	3	3	13	14	10	4	5	9
$\sum_{i=1}^{3} w_i T(i,j); j = 1,..,7$		149	71	35	40	31	35	81

From Table 1, the global ranking of products in a decrease order of preference is shown below:
1. Dehydrated eggplant as a snack
2. Marmalade
3. Food Supplement
4. Eggplant tea
5. Eggplant water as a diuretic and Flour for bakery
6. Medallions for hamburgers

4. Conclusions

In literature, when a problem is solved by a method, typically a set of alternatives are given and the way of how such alternatives are obtained is not provided. This could lead to the idea that creating alternatives is an easy task. However, it is a complex problem where several elements are interrelated and it is necessary to structure the problem to understand what the problem is. For that, VFT methodology was used because it is suitable to help decision makers understand and structure the problem. VFT can be used to carry out the intelligence stage of the decision process of Herbert A. Simon.

Contributions of the paper were, in first instance, how to apply VFT methodology to generate potential solution alternatives to the problem of eggplant waste, and how to generate potential eggplant products considering not only consumer preferences but different stakeholders, including the decision maker. From the case study results, the first and third potential products are of interest for the decision maker, because a process of dehydration of another product is currently being implemented thus, these new products can be produced using the current equipment of the pilot plant with minor investment.

Acknowledgments

This work is part of the "Development of a decision support system based on intelligent agents for the design of new agro-industry products", research project #2015-01-162, supported by the Mexican National Council of Science and Technology (CONACyT).

References

1. R. G. Cooper, "Identifying industrial new product success: Project NewProd," *Industrial Marketing Management*, vol. 8, pp. 124-135, 1979.
2. A. Griffin and J. R. Hauser, "The voice of the customer," *Marketing science*, vol. 12, pp. 1-27, 1993.
3. M. Adams, "Findings from the PDMA research foundation CPAS benchmarking," *PDMA Foundation*, 2004.
4. W. E. FORUM. (2011). *Desarrollar una nueva visión para la agricultura:* Available: http://www3.weforum.org/docs/WEF_FB_NewVisionAgriculture_RoadMap_2011_SP.pdf
5. R. L. Keeney, *Value-focused thinking: A path to creative decisionmaking*: Harvard University Press, 2009.
6. Y. Siskos and N. Matsatsinis, "A DSS for market analysis and new product design," *Journal of Decision Systems*, vol. 2, pp. 35-60, 1993.
7. H. Sheng, K. Siau, and F. F.-H. Nah, "Understanding the values of mobile technology in education: a value-focused thinking approach," *ACM SIGMIS Database*, vol. 41, pp. 25-44, 2010.
8. L. H. Alencar, C. M. de Miranda Mota, and M. H. Alencar, "The problem of disposing of plaster waste from building sites: problem structuring based on value focus thinking methodology," *Waste management*, vol. 31, pp. 2512-2521, 2011.
9. K. Siau and M. Ling, "Mobile Collaboration Support for Virtual Teams: The Case of Virtual Information Systems Development Teams," *Journal of Database Management (JDM)*, vol. 28, pp. 48-69, 2017.
10. W. Edwards and F. H. Barron, "SMARTS and SMARTER: Improved simple methods for multiattribute utility measurement," *Organizational behavior and human decision processes*, vol. 60, pp. 306-325, 1994.

Interval-valued intuitionistic fuzzy MULTIMOORA approach for new product development[*]

Orhan Feyzioğlu
Department of Industrial Engineering, Galatasaray University
Çırağan Cad. No. 36 Ortaköy, 34349, Istanbul, Turkey

Fethullah Gocer
Department of Industrial Engineering, Galatasaray University
Çırağan Cad. No. 36 Ortaköy, 34349, Istanbul, Turkey

Gulcin Buyukozkan[†]
Department of Industrial Engineering, Galatasaray University
Çırağan Cad. No. 36 Ortaköy, 34349, Istanbul, Turkey

This study aims at refining the quality of decision-making processes and improving the effectiveness of new product development. The objective of a decision-making process is basically to weight several alternatives, each aiming to attain some of the required objectives and to pick the best one in the complete set. New product introduction has long been categorized as the core function of corporations if they want to stay competitive in an increasingly competitive global market. Nowadays rapidly changing and highly uncertain environment makes the new product development extremely complex and vague in nature. In order to bring a solution to this complex decision, this study aims to identify decision points in the process of new product development and vagueness factors affecting the process. Therefore, the MULTIMOORA (Multi-objective Optimization by Ratio Analysis plus the Full Multiplicative Form) approach based on Interval-Valued Intuitionistic Fuzzy (IVIF) is the proposed approach to shape the decisions. An application is presented to demonstrate its potential. A comparative analysis of the results validates the proposed method.

1. Introduction

New Product Development (NPD) is quite simple in concept - it is basically the conversion of global market demands into a reality-based solution, the creation of the tangible from the intangible and new ways of wealth. The success in NPD

---

[*]This work is financially supported by Galatasaray University Research Fund (Project No:18.402.008).
[†]Corresponding author; e-mail: gulcin.buyukozkan@gmail.com Tel: +902122274480, Fax: +902122595557.

efforts basically needs a considerable amount of resources, where failures or lackluster results are common. Due to high risk of failures, it is significant to comprehend why and how NPD is important for corporations. NPD is the lifeblood of corporations. Developed products can be in a physical type, from the hair drier to the automobile, airplane to smartphone, etc. They can also be services such as software, a novel marketing concept, or a manufacturing process. If corporations want to move from survival mode to thriving, they have little choice but grow and improve. However, similar to all decision-making processes, an NPD process involves a significant amount of vagueness affecting factors to reach the aimed performance. Vagueness arises from internal or external sources like commercial, technical and management issues. Therefore, the key is to use a structured methodology which could minimize the risk in NPD. In order to help corporate managers and decision makers (DMs) in expressing their judgments, linguistic terms can be used to assess the needs and ratings of NPD alternatives. The fuzzy set theory is a useful tool to accomplish this goal.

Decision-making is all about selecting the best among the set of available alternatives. Multiple-criteria decision-making (MCDM) methods are being developed to consider the impact of many criteria at the same time since the early 1970s. MCDM approaches are divided into Multi-Objective approaches and Multi-Attribute approaches. The fundamental difference among these is based on the determination of the alternatives. In Multi-Objective, the alternatives are not a fixed set. Instead, predetermined objectives are optimized for a fixed set of constraints. In Multi-Attribute, the set of alternatives is fixed and these predetermined alternatives are evaluated against a set of criteria. Many different types of MCDM approaches have been proposed but most of them run with crisp or fuzzy numbers, which can be insufficient in certain environments. To address the inexactness and lack of precision of the conventional crisp or fuzzy sets, the interval-valued intuitionistic fuzzy (IVIF) concept is developed. The major advantage of IVIF sets over the crisp or classical fuzzy sets is that IVIF sets differentiate the positive and the negative indication for an element's interval membership and non-membership in the set. To the best of our knowledge, there is no study that uses IVIF values under group decision making (GDM) for assessing and rating the NPD alternatives. Therefore, in this study, the MULTIMOORA (Multi-objective Optimization by Ratio Analysis plus the Full Multiplicative Form) MCDM methodology under GDM IVIF environment was proposed to make more rational selection decisions. The MULTIMOORA methodology is chosen for its robustness over many other competing methodologies.

Organization of this paper is as follows. Section 2 presents an overview of IVIF MULTIMOORA. This is followed with section 3, a detailed description of the proposed model for NPD analysis. Experimental results involving NPD alternatives are given in the following section 4. The last section concludes the study with some future research suggestions.

2. Overview of IVIF MULTIMOORA

This section briefly introduces IVIF sets, and the MULTIMOORA approach as the main concepts of the proposed MCDM methodology. The Intuitionistic Fuzzy sets developed by Atanassov [1] are an extension of the fuzzy set theory developed by Zadeh [2]. IVIF sets are the extended versions of IF sets [3]. It basically designates a degree of membership on each element. A ⊂ E be a crisp set. An IVIF set, $\tilde{A}$, in E is an object of the following form; an ordered pair, characterized by an interval valued membership value $\tilde{\mu}_{\tilde{A}}(x)$ and an interval valued non-membership value $\tilde{v}_{\tilde{A}}(x)$. $\tilde{A}$ is called IVIF set, where $\tilde{\mu}_{\tilde{A}}(x) \subset [0,1]$ and $\tilde{v}_{\tilde{A}}(x) \subset [0,1]$, $\forall x \in E$ with the condition of $0 \leq \sup(\tilde{\mu}_{\tilde{A}}(x)) + \sup(\tilde{v}_{\tilde{A}}(x)) \leq 1$. For convenience, the IVIF set's lower and upper end points are denoted by $\tilde{A} = [\mu_{\tilde{A}}^L, \mu_{\tilde{A}}^U], [v_{\tilde{A}}^L, v_{\tilde{A}}^U]$. Arithmetic operations in IVIF sets [3] are defined as follow:

$$\tilde{A} = \{\langle x, \tilde{\mu}_{\tilde{A}}(x), \tilde{v}_{\tilde{A}}(x) \rangle | x \in E\} \quad (1)$$

$$\tilde{A} \oplus \tilde{B} = \begin{pmatrix} [\mu_{\tilde{A}}^L(x) + \mu_{\tilde{B}}^L(x) - \mu_{\tilde{A}}^L(x)\mu_{\tilde{B}}^L(x), \mu_{\tilde{A}}^U(x) + \mu_{\tilde{B}}^U(x) - \mu_{\tilde{A}}^U(x)\mu_{\tilde{B}}^U(x)], \\ [v_{\tilde{A}}^L(x)v_{\tilde{B}}^L(x), v_{\tilde{A}}^U(x)v_{\tilde{B}}^U(x)] \end{pmatrix} \quad (2)$$

$$\tilde{A} \otimes \tilde{B} = \begin{pmatrix} [\mu_{\tilde{A}}^L(x)\mu_{\tilde{B}}^L(x), \mu_{\tilde{A}}^U(x)\mu_{\tilde{B}}^U(x)], \\ [v_{\tilde{A}}^L(x) + v_{\tilde{B}}^L(x) - v_{\tilde{A}}^L(x)v_{\tilde{B}}^L(x), v_{\tilde{A}}^U(x) + v_{\tilde{B}}^U(x) - v_{\tilde{A}}^U(x)v_{\tilde{B}}^U(x)] \end{pmatrix} \quad (3)$$

$$\tilde{A} \otimes \tilde{B} = \begin{pmatrix} [\mu_{\tilde{A}}^L(x)\mu_{\tilde{B}}^L(x), \mu_{\tilde{A}}^U(x)\mu_{\tilde{B}}^U(x)], \\ [v_{\tilde{A}}^L(x) + v_{\tilde{B}}^L(x) - v_{\tilde{A}}^L(x)v_{\tilde{B}}^L(x), v_{\tilde{A}}^U(x) + v_{\tilde{B}}^U(x) - v_{\tilde{A}}^U(x)v_{\tilde{B}}^U(x)] \end{pmatrix} \quad (4)$$

$$\tilde{A} \oslash \tilde{B} = \left(\left[\frac{\mu_{\tilde{A}}^L(x)}{\mu_{\tilde{B}}^L(x)}, \frac{\mu_{\tilde{A}}^U(x)}{\mu_{\tilde{B}}^U(x)}\right], \left[\frac{v_{\tilde{A}}^L(x) - v_{\tilde{B}}^L(x)}{1 - v_{\tilde{B}}^L(x)}, \frac{v_{\tilde{A}}^U(x) - v_{\tilde{B}}^U(x)}{1 - v_{\tilde{B}}^U(x)}\right] \right) \quad (5)$$

Table 1. Scale for IVIF linguistic terms [4].

Terms	EP	VP	P	F	G	VG	EG
$\mu_{\tilde{A}}^L(x)$	0.00	0.10	0.30	0.50	0.60	0.80	0.95
$\mu_{\tilde{A}}^U(x)$	0.00	0.15	0.35	0.55	0.75	0.90	1.00
$v_{\tilde{A}}^L(x)$	0.95	0.75	0.55	0.40	0.25	0.05	0.00
$v_{\tilde{A}}^U(x)$	1.00	0.85	0.65	0.45	0.25	0.10	0.00

EP: Extremely Poor, VP: Very Poor, P: Poor, F: Fair, G: Good, VG: Very Good, EG: Extremely Good

The MOORA (Multi-Objective Optimization on the basis of Ratio Analysis) approach was first introduced by Brauers and Zavadskas [5] involving two components, the ratio system, and the reference point. It was further developed and the full multiplicative form is added, which constitutes today's MULTIMOORA [6], which is a robust and effective Multi-Attribute approach. Compared to other techniques, MULTIMOORA uses a vector normalization procedure and is quite efficient and easy to use. IVIF numbers provide an opportunity for a much more adequate modeling and solving complex problems. The originality of this study comes from its strength in exhibiting the extension of MULTIMOORA, a novel GDM-based IVIF MCDM technique, where the performance ratings of alternatives are expressed in IVIF numbers, as well as by adopting a similar procedure studied in [7]. In this method, 'm' alternatives $A_1, A_2, \ldots, A_m$, based on 'n' criteria $C_1, C_2, \ldots, C_n$ and the weight vectors $w_1, w_2, \ldots, w_n$, are evaluated with $w_j \geq 0$, $j = 1, 2, \ldots, n$, and $\sum_{j=1}^{n} w_j = 1$. The steps of the IVIF MULTIMOORA methodology is as follow:

Step 1: Taking the DMs' judgments on each factor as a linguistic term.

Step 2: Transforming the linguistic terms into IVIF values that are listed in Table 1.

Step 3: Determining each DM's weights $\lambda^k (1 < k < K)$ by the Eq. (6) [8].

$$\lambda^k = \frac{\sqrt{\frac{1}{2}\left[\left(1-\pi_{\tilde{A}}^L(x)^k\right)^2 + \left(1-\pi_{\tilde{A}}^U(x)^k\right)^2\right]}}{\sum_{l=1}^{K}\sqrt{\frac{1}{2}\left[\left(1-\pi_{\tilde{A}}^L(x)^l\right)^2 + \left(1-\pi_{\tilde{A}}^U(x)^l\right)^2\right]}} \quad (6)$$

Step 4: Calculating the criteria weights by using the Eq. (7).

$$\breve{w}_i = \frac{\sum_{j=1}^{n} \frac{w_j\left(\mu_{\tilde{A}ij}^L + \mu_{\tilde{A}ij}^U\right)}{2}}{\sqrt{\sum_{j=1}^{n} \frac{w_j\left(\mu_{\tilde{A}ij}^{L^2} + \mu_{\tilde{A}ij}^{U^2} + v_{\tilde{A}ij}^{L^2} + v_{\tilde{A}ij}^{U^2}\right)}{2}}}, W_i = \frac{1-\breve{w}_i}{n - \sum_{i=1}^{n} \breve{w}_i} \quad (7)$$

Step 5: Constructing the aggregated matrix. IVIF weighted averaging (IIFWA) operator [9] is used for aggregation, as shown in Eq. (8).

$$IIFWA = \left(\begin{bmatrix} 1 - \prod_{j=1}^{n}(1-\mu_{\tilde{A}}^L)^{\lambda^k}, \\ 1 - \prod_{j=1}^{n}(1-\mu_{\tilde{A}}^U)^{\lambda^k} \end{bmatrix}, \begin{bmatrix} \prod_{j=1}^{n}(v_{\tilde{A}}^L)^{\lambda^k}, \\ \prod_{j=1}^{n}(v_{\tilde{A}}^U)^{\lambda^k} \end{bmatrix}\right) \quad (8)$$

Step 6: Establishing normalized decision matrix [7] $\left(\tilde{r}_{ij} = [a_{ij}^L, a_{ij}^U], [b_{ij}^L, b_{ij}^U]\right)$.

Step 7: Establishing weighted-normalized decision matrix $\left(\dot{\tilde{r}}_{ij} = \tilde{r}_{ij} * w_i\right)$ using the criteria weights calculated in Step 4.

Step 8: Calculating the score for ratio system ($\tilde{y}_i$), reference point approach ($\tilde{p}_i$), and overall utility function ($\tilde{U}_i$) IVIF values and r_j is the best criteria values of each alternative by the Eq. (9).

$$\tilde{y}_i = \sum_{j=1}^{g} \dot{\tilde{r}}_{ij} - \sum_{j=g+1}^{n} \dot{\tilde{r}}_{ij}, \tilde{P}_i = \min_{(i)}\left\{\max_{j}|r_j - \dot{\tilde{r}}_{ij}|\right\}, \tilde{U}_i = \frac{\tilde{A}_i}{\tilde{B}_i} \quad (9)$$

Step 9. Rank alternatives by the dominance theory [10].

3. Application of Proposed Methodology

This section presents the NPD selection factors, implements the methodology that is used to measure these factors and analyzes alternatives to make use of this methodology. This is done for the evaluation of the NPD process. NPD alternatives and criteria are adapted from the studies in [11], [12].

Table 2. NPD selection criteria and their description.

Criteria	Description
Profitability (C_1)	Profitability in NPD or market share for new products under the current competitive environment.
Efficiency (C_2)	Efficiency in NPD reduces costs and time to reach to market
Business Impact (C_3)	Business Impact considers the nature of the business and the overall motivation in NPD.
Strategic Value (C_4)	Strategic Value in NPD brings better products with faster without too much effort to create value stream
Financial Risk (C_5)	Financial Risk in NPD makes responses to the strategies, levels, and alteration in price
Technical Risk (C_6)	Technical Risk in NPD required to get better targeted and more innovative products
Managerial Risk (C_7)	Managerial Risk is the competitive reactions to NPD strategies
Personnel Risk (C_8)	Personnel Risk is the impact of NPD on profits or market share

Three DMs, an academic, an engineer, and a top-level manager, are selected to evaluate 5 NPDs, a new software development. These new products are submitted by different customers and employees. Software development case is an exceptional setting for NPD, considering that there exists a notable difference between traditional and software production. Consequently, to evaluate NPD alternatives in a feasible manner, benefit and cost factors should be taken into account. Therefore, the authors have identified 8 criteria for the assessment process of the study. Table 2 presents the identified criteria and their simple description. As it could be inferred, criteria 1 through 4, and 5 through 8 are regarded as benefit and cost criteria, respectively.

Step 1: DMs opinions on the criteria as a linguistic term are displayed in Table 3. Alternatives linguistic evaluations are scaled from the studies in [11], [12] as displayed in Table 4;

Step 2: Using the linguistic scale in Table 1, IVIF value transformation of DMs' opinions is done;

Step 3: Utilizing Eq. (4) to determine the DMs' weights, the values in Table 3 are calculated with the highest weight of the first DM;

Table 3. Linguistic evaluations and respective importance weights for each DM and each criterion.

	DM_1	DM_2	DM_3		DM_1	DM_2	DM_3		DM_1	DM_2	DM_3
DMs	EG	VG	G	C_1	VG	VG	G	C_2	G	VG	F
λ^k	0.37	0.34	0.29	w_1		0.154		w_2		0.144	
C_3	F	VG	VG	C_4	G	F	F	C_5	VG	G	VG
w_3		0.128		w_4		0.098		w_5		0.148	
C_6	G	F	VG	C_7	G	G	F	C_8	F	F	G
w_6		0.116		w_7		0.120		w_8		0.092	

Step 4: Using the Eq. (5), criteria weights are computed as displayed in Table 3 with the highest importance C_4 criterion;

Step 5: Using the IIFWA operator in Eq. (6), the aggregated decision matrix is obtained. Due to limited space, all data could not be displayed here;

Step 6: Aggregated results are normalized by a similar procedure as in [7];

Step 7: By utilizing the criteria weights obtained in step 4, the weighted normalized decision matrix is constructed;

Step 8: Using Eq. (7), the score of alternative for the ratio system, the reference point, and the full multiplicative form are found, respectively. Table 5 presents the final IVIF values and their defuzzifications;

Step 9: The alternatives are ranked according to the dominance theory. The ranking result indicates that A_3 is the best one, A_1 is the second, A_5 is the third, A_4 is the third, and A_2 is the last in rank; $A_3 > A_1 > A_5 > A_4 > A_2$. Table 5 presents the detailed numerical values. Table 6 compares the rankings for each alternative with Fuzzy MULTIMOORA and IF MULTIMOORA.

Table 4. Linguistic evaluations of alternatives for each DM.

	A_1			A_2			A_3			A_4			A_5		
DM	1	2	3	1	2	3	1	2	3	1	2	3	1	2	3
C_1	EG	P	VG	VG	F	G	P	F	VP	VG	F	G	G	VG	VG
C_2	VG	G	F	G	G	F	P	F	P	VG	G	F	F	P	F
C_3	VG	G	VG	F	VG	G	F	G	VP	F	G	VG	F	P	G
C_4	F	G	VG	F	P	EG	F	F	P	VG	VP	G	VP	P	VG
C_5	F	P	VG	F	G	VG	G	VP	F	VP	G	P	F	G
C_6	G	VP	F	G	F	G	EG	VG	G	VG	G	F	VP	F	F
C_7	F	P	VG	P	F	G	F	P	G	VG	VG	F	P	G	P
C_8	EG	VG	VG	F	VP	G	F	VG	F	G	F	VP	P	G	F

Table 5. Final values and rankings for each alternative.

		[μ^L,	μ^U],	[v^L,	v^U]		Deff	Rank
$\tilde{y}_i$	A_1	0.139	0.163	0.465	0.490		0.788	5
	A_2	0.018	0.051	0.571	0.588		0.877	1
	A_3	0.027	0.055	0.573	0.587	Y_i	0.874	2
	A_4	0.130	0.142	0.489	0.510		0.802	4
	A_5	0.117	0.125	0.580	0.604		0.833	3
$\tilde{P}_i$	A_1	0.035	0.036	0.926	0.933		0.963	2
	A_2	0.052	0.057	0.918	0.927		0.952	5
	A_3	0.035	0.039	0.930	0.936	P_i	0.964	1
	A_4	0.045	0.041	0.921	0.930		0.959	4
	A_5	0.039	0.043	0.925	0.944		0.962	3
$\tilde{U}_i$	A_1	0.000	0.001	0.996	0.997		0.9989	2
	A_2	0.000	0.001	0.994	0.996		0.9985	3
	A_3	0.000	0.001	0.997	0.998	U_i	0.9991	1
	A_4	0.000	0.001	0.991	0.993		0.9977	5
	A_5	0.000	0.001	0.993	0.995		0.9982	4

Table 6. Comparison of rankings for each alternative with fuzzy and IF sets.

	A_1	A_2	A_3	A_4	A_5
Fuzzy MULTIMOORA	2	4	3	5	1
IF MULTIMOORA	1	3	2	5	4
IVIF MULTIMOORA	2	5	1	4	3

4. Conclusion

In this paper, the IVIF MULTIMOORA methodology is used to assess NPD alternatives. The aim of this study is to use an MCDM approach which integrates MULTIMOORA and IVIF to assess a set of available NPD alternatives for the sake of deciding on the best qualified one. Linguistic terms in the evaluation process are represented as IVIF numbers for each alternative. In the case of the NPD selection process, the proposed method is useful to rank the most suitable NPD. With the help of a literature survey and a team of DMs, the decision criteria are identified, which need to be taken into account while selecting the most suitable alternative. In the light of these attributes, all NPD alternatives are ranked with the proposed approach. An application is illustrated to endorse the proposed approach and a comparison is carried out to validate the results. This method is capable of dealing with similar types of uncertain situations in MCDM problems. For future research, other types of MCDM methods can be integrated with IVIF values or other environments can be combined with MULTIMOORA to solve the problem and compare to these outcomes.

Acknowledgments

The authors acknowledge the contribution of the experts without which this study could not be accomplished.

References

1. K. Atanassov, *Fuzzy Sets and Systems*, 20th ed. (1986).
2. L. A. Zadeh, *Inf. Control*, vol. 8, no. 3, pp. 338–353, Jun. (1965).
3. K. Atanassov and G. Gargov, *Fuzzy Sets Syst.*, no. 3, pp. 343–349, (1989).
4. G. Büyüközkan, F. Göçer, and O. Feyzioğlu, *Adv. in Int. Syst.*, vol. 641, pp. 318–329, (2018).
5. W. K. M. Brauers and E. K. Zavadskas, *Control Cybern.*, vol. 35, no. 2, pp. 445–469, (2006).
6. W. K. M. Brauers and E. K. Zavadskas, *Technol. Econ. Dev. Econ.*, vol. 16, no. 1, pp. 5–24, (2010).
7. G. Buyukozkan, F. Gocer, and O. Feyzioglu, in *2017 IEEE International Conference on Fuzzy Systems*, pp. 1–6, (2017).
8. F. Zhang and S. Xu, *Gr. Decis. Negot.*, vol. 25, no. 6, pp. 1261–1275, (2016).
9. Z. Xu and X. Cai, *Intuitionistic Fuzzy Information Aggregation*. (2012).
10. W. K. M. Brauers and E. K. Zavadskas, *Technol. Econ. Dev. Econ.*, vol. 17, no. 1, pp. 174–188, Mar. (2011).
11. G. Büyüközkan and O. Feyzıoğlu, *Int. J. Prod. Econ.*, pp. 27–45, (2004).
12. C. Kahraman, G. Büyüközkan, and N. Y. Ateş, *Inf. Sci. (Ny).*, vol. 177, no. 7, pp. 1567–1582, Apr. (2007).

Analysis of companies' digital maturity with hesitant fuzzy linguistic MCMD methods[*]

Merve Güler

Industrial Engineering Department, Galatasaray University
Ciragan Cad., 36 Ortaköy-Istanbul, 34349, Turkey

Esin Mukul

Industrial Engineering Department, Galatasaray University
Ciragan Cad., 36 Ortaköy-Istanbul, 34349, Turkey

Gülçin Büyüközkan[†]

Industrial Engineering Department, Galatasaray University
Ciragan Cad., 36 Ortaköy-Istanbul, 34349, Turkey

Hesitant fuzzy linguistic term sets (HFLTS) is a technique that can be used to facilitate Decision Makers' (DMs) judgment processes in complex and uncertain situations. In real-world problems, DMs might hesitate while giving their opinions. HFLTS technique gives DMs the possibility to use linguistic expressions with comparative judgments and to overcome the hesitancy. This paper presents a decision framework based on HFLTS, Hesitant Fuzzy Linguistic (HFL) Analytic Hierarchy Process (AHP) and Additive Ratio ASsessment (ARAS) method for the first time. There is a need for an analytical tool to determine the criteria that affect a company's level of digitalization, to analyze the significance of these criteria to construct the Digital Maturity Models (DMMs) and to rank the companies according to their digital maturity. This paper provides a scientific method that helps to determine the most important criteria for companies' DMM and to rank enterprises. A case study in the banking sector is illustrated to verify the applicability of this methodology.

1. Introduction

Investing in technology does not exactly mean Digital Transformation (DT). Technology is guided by DT, its purpose being to reshape the business. DT is the reorganization of technology, business models, and processes to more effectively compete in a continuously changing digital economy [1].

---

[*]This work is financially supported by Galatasaray University Research Fund (Project No: 18.402.001).
[†]Corresponding author; e-mail: gulcin.buyukozkan@gmail.com

Digital Maturity Model (DMM) offers a practical approach to transformation. In 2005, a methodology was proposed for the main stages of capability model development [2]. In 2009, a maturity model was constructed for hospital information systems and digital government concept [3-4]. Since 2017, the popularity of the maturity model has increased. The digital maturity levels of companies are proposed in different ways for various sectors [5-9].

There is a variety of factors that affect the maturity level of the companies. However, the importance of these factors on firms' digitalization is unknown. Thus, the purpose of this study is twofold. First one is to propose a DMM and to propose an analytical tool to help companies to understand the importance of the factors affecting their digital maturity level. The second aim is to rank the companies.

The importance of factors will be determined by calculating their weights with the Hesitant Fuzzy Linguistic (HFL) Analytic Hierarchy Process (AHP) method with fuzzy envelope technique. The alternatives will be ranked with the HFL Additive Ratio ASsessment (ARAS) method. The uncertainty and vagueness of information are reflected in this study by using fuzzy logic [10]. Moreover, experts can hesitate while expressing their opinions. This is where the hesitant fuzzy linguistic term sets (HFLTS) technique becomes helpful in solving this problem [11]. In 2014, Hesitant Fuzzy (HF) AHP method was introduced [12]. In the past, HFL AHP methodology was integrated with Group Decision Making (GDM) approach, Quality Function Deployment and TOPSIS method [13-19]. In 2013, ARAS method was proposed [20]. Fuzzy ARAS methodology was integrated with Fuzzy TOPSIS method and applied in different fields like sustainable building assessment [21-24]. There is a lack of research on implementing HFLTS with ARAS method. Thus, in this study, HFL AHP, combined with HFL ARAS methodology is implemented for the first time.

The structure of the paper consists of 4 sections. The next section explains the research methodology. In Section 3, application of the methodology is provided. The last section summarizes the concluding remarks and the perspective for future studies.

2. Research Methodology

The first phase of the problem is composed of problem definition and determination of the DMM factors (criteria) and alternatives. The second phase is about calculating the criteria weights with HFL AHP method. The last phase consists of ranking the alternatives with HFL ARAS method.

2.1. The Hesitant Fuzzy Linguistic Term Set

In 2009, Hesitant Fuzzy Sets (HFS) were first presented [25]. In 2012, a model that presents linguistic expressions by a set of HFLTS was introduced [11]. Please refer to [11, 25] for further information.

Definition 1: E_{GH} is a function that transforms linguistic phrases into HFLTS. This function is applied to convert comparative linguistic phrases into HFLTS [11].

Definition 2: A fuzzy envelope for HFLTS is built to transform HFLTS into trapezoidal fuzzy numbers [26]. The factors of this function are denoted by $A = (α, β, γ, δ)$. This model is implemented in this study for the linguistic expressions *"at least s_i"*, *"at most s_i"*, and *"between s_i and s_j"*.

2.2. The Hesitant Fuzzy Linguistic AHP Method

Step 1. Pairwise comparison matrices are constructed, and linguistic expressions are converted into HFLTS as described in *Definition 1*. Please refer to [15].
Step 2. The fuzzy envelope for HFLTS is aggregated as described in *Definition 2*.
Step 3. The pairwise comparison matrix $(\tilde{C})$ is obtained. The reciprocal values are obtained as:

$$\tilde{c}_{ij} = (\frac{1}{c_{iju}}, \frac{1}{c_{ijm2}}, \frac{1}{c_{ijm1}}, \frac{1}{c_{ijl}}) \tag{1}$$

Step 4. The fuzzy geometric mean $(\tilde{r}_i)$ of the matrix $\tilde{C}$ is calculated as:

$$\tilde{r}_i = (\tilde{c}_{i1} \otimes \tilde{c}_{i2} ... \otimes \tilde{c}_{in})^{1/n} \tag{2}$$

Step 5. The fuzzy weight $(\tilde{w}_i^{CR})$ of every main criterion is computed as:

$$\tilde{w}_i^{CR} = \tilde{r}_i \otimes (\tilde{r}_1 \otimes \tilde{r}_2 ... \otimes \tilde{r}_n)^{-1} \tag{3}$$

Step 6. The fuzzy global weights of sub-criteria are computed.

$$\tilde{w}_{ij}^G = \tilde{w}_i^{CR} \times \tilde{w}_j^{CR} \tag{4}$$

where $\tilde{w}_{ij}^G$ is the global weight of sub-criteria.
Step 7. Trapezoidal fuzzy numbers $\tilde{w}_{ij}^G$ are defuzzified and these values are normalized as:

$$w_{ij}^G = \frac{α+2β+2γ+δ}{6} \tag{5}$$

$$w_{ij}^N = \frac{w_{ij}^G}{\Sigma_i \Sigma_j w_{ij}^G} \tag{6}$$

2.3. The Hesitant Fuzzy Linguistic ARAS Method

Step 1: The decision matrix with linguistic statements is constructed and these expressions are converted into HFLTS. Please refer to [27] for details.
Step 2: The matrix is normalized as:
For the criteria with maxima preferable values:

$$\tilde{\tilde{x}} = \frac{\tilde{x}_{ij}}{\sum_{i=0}^{m} \tilde{x}_{ij}} \quad (7)$$

For the criteria with minima preferable values:

$$\tilde{x}_{ij} = \frac{1}{x_{ij}^*}, \quad \tilde{\tilde{x}}_{ij} = \frac{\tilde{x}_{ij}}{\sum_{i=0}^{m} \tilde{x}_{ij}} \quad (8)$$

Step 3: The weighted normalized matrix is constructed as:

$$\hat{\tilde{x}}_{ij} = \tilde{\tilde{x}}_{ij} \tilde{w}_j, \quad i=0,1,\ldots,m \quad (9)$$

w_j is the jth criterion's weight and:

$$\sum_{j=1}^{n} w_j = 1 \quad (10)$$

Step 4: The value of optimality function of ith alternative is determined as:

$$\tilde{S}_i = \sum_{j=1}^{n} \hat{\tilde{x}}_{ij}, \quad i=0,1,\ldots,m \quad (11)$$

Step 5: In order to find the result, the center of area method is applied as:

$$S_i = 1/3(S_{i\alpha} + S_{i\beta} + S_{i\gamma}) \quad (12)$$

Step 6: The utility degree of alternative is determined as:

$$K_i = \frac{S_i}{S_0}, \quad i=0,1,\ldots,m \quad (13)$$

Where S_0 is the most ideal criterion value.

3. Implementation of the Proposed Methodology

In recent years, the banking sector in Turkey has taken significant steps in terms of digitization. For this reason, the applicability of the proposed methodology will be tested on the banking sector. In order to select the most digital bank, different banking companies are identified and evaluated by using the proposed HFL AHP-HFL ARAS methodology. Evaluation factors are determined based on research, reports, and advice of industry experts. These factors are summarized in Table 1 and are mainly based on Forrester's DMM [28]. Banking alternatives are identified based on press, academic papers, and white papers. For privacy concerns, the names of these banks are named as A, B, C, and D. The most digital company will be declared at the end of the study. Therefore, four possible alternatives are: A1 is A, A2 is B, A3 is C and A4 is D.

Table 1. The DMM factors.

Culture (C1)	Organization (C2)	Technology (C3)	Insights (C4)
Competitive strategy's dependency on digital (C11)	Prioritization of customer journeys (C21)	Having fluid technology budget to allow shifting priorities (C31)	Having clear and quantifiable goals (C41)
Board and C-level executives' support on digital strategy (C12)	Assignment of right resources to digital strategy, governance, execution (C22)	Co-working of marketing and technology resources (C32)	Recognition of the importance of every employee (C42)
Leaders' attitude towards digital strategy (C13)	Best qualified staff in digital functions (C23)	Flexible and collaborative approach in technology development (C33)	Using customer-centric metrics (C43)
Targeted digital education and training (C14)	Having digital skills embedded throughout the organization (C24)	Leveraging the modern architectures (C34)	Measurement of channels' together work (C44)
Communication of digital vision internally and externally (C15)	Encouraging cross-functional teams (C25)	Measuring technology teams by business outcomes time (C35)	Driving digital strategy by customer insights (C45)
Taking measured risks to enable innovation (C16)	Having defined and repeatable processes for digital programs (C26)	Using customer experience assets to steer the technology design (C36)	Digital design and development powered by customer insights (C46)
Prioritization of the customer experience with the performance of any channel (C17)	Having vendor partners that deliver value to digitalization (C27)	Implementation of digital tools to promote employee (C37)	Feeding lessons learned from digital programs back into our strategy (C47)

DMs evaluated DMM factors with comparative linguistic terms according to their insights and experience in this domain. The linguistic terms, abbreviations and the triangular fuzzy numbers used in this paper for HFL AHP technique are displayed in Table 2.

The evaluations about main criteria are displayed in Table 3 and Table 4.

Considering the evaluation data given in Table 3, Table 4 and the evaluation data for Culture (C1), Organization (C2), Technology (C3), Insights (C4) factors, (1)-(6) are employed. The result of the study is given in Table 5. As a result, the driving digital strategy by customer insights (C45) has the most importance in terms of DMM.

Table 2. The linguistic scale for HFL AHP method [15].

Linguistic term	s_i	Abb.	Triangular fuzzy number
Absolutely High Importance	s_{10}	(AHI)	(7,9,9)
Very High Importance	s_9	(VHI)	(5,7,9)
Essentially High Importance	s_8	(ESHI)	(3,5,7)
Weakly High Importance	s_7	(WHI)	(1,3,5)
Equally High Importance	s_6	(EHI)	(1,1,3)
Exactly Low Importance	s_5	(EE)	(1,1,1)
Equally Low Importance	s_4	(ELI)	(0.33,1,1)
Weakly Low Importance	s_3	(WLI)	(0.2,0.33,1)
Essentially Low Importance	s_2	(ESLI)	(0.14,0.2,0.33)
Very Low Importance	s_1	(VLI)	(0.11,0.14,0.2)
Absolutely Low Importance	s_0	(ALI)	(0.11,0.11,0.14)

Table 3. The evaluations about main criteria with linguistic expressions.

	C1	C2	C3	C4
C1	EE	Between VLI and WLI	At least ESHI	At least VHI
C2		EE	Between ESLI and ELI	Between VLI and WLI
C3			EE	At least VHI
C4				EE

Table 4. The evaluations about main criteria with s_i.

	C1	C2	C3	C4
C1	$\{s_5\}$	$\{s_1, s_2, s_3\}$	$\{s_8, s_9, s_{10}\}$	$\{s_9, s_{10}\}$
C2		$\{s_5\}$	$\{s_2, s_3, s_4\}$	$\{s_1, s_2, s_3\}$
C3			$\{s_5\}$	$\{s_9, s_{10}\}$
C4				$\{s_5\}$

Table 5. The weights of criteria resulted by HFL-AHP method.

Criteria	C11	C12	C13	C14	C15	C16	C17	C21	C22	C23
Weight	0.069	0.034	0.017	0.020	0.020	0.013	0.012	0.056	0.016	0.074
Ranking	5	10	19	17	18	21	24	7	20	4
Criteria	C24	C25	C26	C27	C31	C32	C33	C34	C35	
Weight	0.013	0.005	0.024	0.009	0.104	0.044	0.058	0.031	0.026	
Ranking	22	28	15	27	1	9	6	12	14	
Criteria	C36	C37	C41	C42	C43	C44	C45	C46	C47	
Weight	0.012	0.012	0.089	0.094	0.055	0.026	0.104	0.044	0.058	
Ranking	23	25	3	2	8	13	1	9	6	

The linguistic terms, abbreviations and the triangular fuzzy numbers used for HFL ARAS technique are given in Table 6.

DMs evaluated banking companies by using comparative linguistic terms based on their knowledge of these companies. The evaluation data alternatives are taken into consideration and (7)-(13) are employed. Finally, the ranking result of the study is given in Table 7. The results show that the most digitized bank is A1 (K_1:0.786) among these alternatives and A3 (K_3:0.702) is ranked as the second.

Table 6. The linguistic scale for HFL ARAS method [20].

Linguistic term	s_i	Abb.	Triangular fuzzy number
Perfect	s_6	(P)	(0.83,1,1)
Very Good	s_5	(VG)	(0.67,0.83,1)
Good	s_4	(G)	(0.5,0.67,0.83)
Medium	s_3	(M)	(0.33,0.5,0.67)
Bad	s_2	(B)	(0.17,0.33,0.5)
Very Bad	s_1	(VB)	(0,0.17,0.33)
Nothing	s_0	(N)	(0,0,0.17)

Table 7. The ranking of alternatives resulted by HFL ARAS method.

A_i	$S_{i\alpha}$	$S_{i\gamma}$	$S_{i\beta}$	S_i	K_i	Ranking
A0	0.245	0.280	0.336	**0.287**	1.000	-
A1	0.155	0.222	0.299	0.225	0.786	1
A2	0.114	0.167	0.242	0.174	0.607	3
A3	0.138	0.196	0.270	0.201	0.702	2
A4	0.085	0.137	0.219	0.147	0.512	4

4. Conclusion

DMM provides a systematic approach by introducing different levels, dimensions, and factors on companies' DT journey. Therefore, in this study, the first aim was to construct a DMM framework and to propose an analytical tool to obtain the importance of criteria for guiding managers. The second aim was to rank companies according to their digital maturity level. Therefore, the integrated HFL AHP-HFL ARAS methodology is employed. This paper contributes to the ARAS-related literature by extending the method to the hesitant fuzzy environment. At the end of the study, the most digital bank is found as A1, Yapi Kredi Bankasi. The actual result agrees with the real result. The perspective for future work can be to examine DMs' different expression styles to extend our analysis by hesitant GDM approach with personalized individual semantics [29].

Acknowledgments

The authors kindly express their appreciation for the support of industrial experts.

References

1. B. Solis, Cognizant Report: The Six Stages of Digital Transformation Maturity, (2017).

2. T. De Bruin, T. Freeze, R. Kaulkarni and M. Rosemann, Understanding the Main Phases of Developing a Maturity Assessment Model, in: ACIS, United States, (2005).
3. R. Vandewetering and R. Batenburg, A PACS maturity model: A systematic meta-analytic review on maturation and evolvability of PACS in the hospital enterprise, *Int. J. Med. Inform.* **78**, 127-140 (2009).
4. P. Gottschalk, Maturity levels for interoperability in digital government, *Gov Inf Q.* **26**, 75-81 (2009).
5. S. Berghaus and A. Back, Stages in Digital Business Transformation: Results of an Empirical Maturity Study, in: MCIS, Cyprus, 22 (2016).
6. C. Danjou, L. Rivest and R. Pellerin, Douze positionnements stratégiques pour l'Industrie 4.0: entre processus, produit et service, de la surveillance à l'autonomie. in: CIGI, France, (2017).
7. C. Grange, S. Ricoul, Organisations : quel est votre degré de maturité numérique?, *Gestion,* **42**, 86-89 (2017).
8. J. Hägg and S. Sandhu, Do or Die: How Large Organizations Can Reach a Higher Level of Digital Maturity, (2018).
9. B. Tavakoli and I. Mohammadi, Digital Maturity within Distribution, (2017).
10. L. Zadeh, Fuzzy Sets, *Inform. Control,* **8**, 338-353 (1965).
11. R. Rodriguez, L. Martinez and F. Herrera, Hesitant Fuzzy Linguistic Term Sets for Decision Making, *IEEE Trans. Fuzzy Syst.* **20**, 109-119 (2012).
12. S.M. Mousavi, H. Gitinavard, and A. Siadat, A new hesitant fuzzy analytical hierarchy process method for decision-making problems under uncertainty, in: IEEM, Malaysia, 622-626 (2014).
13. B. Zhu and Z. Xu, Analytic hierarchy process-hesitant group decision making, *Eur. J. Oper. Res.* **239**, 794-801 (2014).
14. A. Başar, Hesitant fuzzy pairwise comparison for software cost estimation: a case study in Turkey, *Turk. J. Elec. Eng. & Comp. Sci.* **25**, 2897-2909 (2017).
15. S. Ç. Onar, G. Büyüközkan, B. Öztayşi and C. Kahraman, A new hesitant fuzzy QFD approach: an application to computer workstation selection, *Appl. Soft Comput.* **46**, 1-16 (2016).
16. W. Zhou and Z. Xu, Asymmetric hesitant fuzzy sigmoid preference relations in the analytic hierarchy process, *J. Inf. Sci.* **358-359**, 191-207 (2016).
17. B. Zhu, Z. Xu and M. Hong, Hesitant analytic hierarchy process, *Eur. J. Oper. Res.* **358**, 191-207 (2016).
18. M. Çolak and İ. Kaya, Prioritization of renewable energy alternatives by using an integrated fuzzy MCDM model: A real case application for Turkey, *Renew. Sust. Energ. Rev.* **80**, 840-853 (2017).
19. F. Tüysüz and B. Şimşek, A hesitant fuzzy linguistic term sets-based AHP approach for analyzing the performance evaluation factors: an application to cargo sector, *Complex and Intell. Syst.* **3**, 1-9 (2017).
20. Beg and R. Rashid, TOPSIS for Hesitant Fuzzy Linguistic Term Sets, *Int. J. Intell. Syst.,* **28**, 1162-1171 (2013).
21. V. Keršulienė and Z. Turskis, An integrated multi-criteria group decision making process: selection of the chief accountant, *Procedia Soc. Behav. Sci.* **110**, 897-904 (2014).

22. V. Kutut, E.K. Zavadskas and M. Lazauskas, Assessment of priority options for preservation of historic city centre buildings using MCDM (ARAS), *Procedia Eng.* **57**, 657-661 (2013).
23. M. Medineckiene, E.K. Zavadskas, F. Björk and Z. Turskis, Multi-criteria decision-making system for sustainable building assessment/certification, *Arch. Civ. Mech. Eng.* **15**, 11-18 (2015).
24. E.K. Zavadskas, Z. Turskis, T. Vilutienė, T. and N. Lepkova, Integrated group fuzzy multi-criteria model: Case of facilities management strategy selection, *Expert Syst. Appl.* **82**, 317-331 (2017).
25. V. Torra and Y. Narukawa, On hesitant fuzzy sets and decision, in: Fuzzy Systems, Fuzz-IEEE, (2009).
26. R. Rodriguez and H. Liu, A fuzzy envelope for hesitant fuzzy linguistic term set and its application to multi-criteria decision making, *J. Inf. Sci.* **258**, 220-238 (2014).
27. E.K. Zavadskas, Z. Turskis and V. Bagočius, Multi-criteria selection of a deep-water port in the Eastern Baltic Sea, *Appl. Soft Comput.* **26**, 180-192 (2015).
28. M. Gill and S. VanBoskirk, Forrester Report: The Digital Maturity Model 4.0, (2016).
29. C.C. Li, R. Rodriguez, L. Martinez, Y.C. Dong, F. Herrera, Personalized individual semantics based on consistency in hesitant linguistic group decision making with comparative linguistic expressions. *Know.-Based Syst.* **145**, 156-165 (2018).

A multi-criteria decision support system for multi-UAV mission planning

C. Ramirez-Atencia, V. Rodríguez-Fernández and D. Camacho

Department of Computer Engineering, Universidad Autónoma de Madrid
Madrid, 28049, Spain
cristian.ramirez@inv.uam.es, {victor.rodriguezf, david.camacho}@uam.es
www.aida.ii.uam.es

The Multi-UAV Mission Planning problem is focused on the search of a set of solutions that satisfy several constraints on the mission scenario and has some variables to be optimized, such as the makespan, the cost of the mission or the risk. Thus, there could exist a large number of solutions to the problem. It turns a big issue for the operator to select the final solution to execute among the many obtained. In order to reduce the operator workload, this work proposes a Multi-Criteria Decision Support System, which consists of a ranking function that sorts the solutions obtained. Several ranking functions have been tested in real mission scenarios with different operator profiles. Expert operators have evaluated the solutions returned in order to compare the different ranking systems and demonstrate the usefulness of the proposed approach.

Keywords: Multi-criteria decision making; multi-objective optimization; unmanned air vehicles; mission planning.

1. Introduction

The advent of Unmanned Air Vehicles (UAVs) and rapid development of their capabilities have paved the way towards new military and commercial applications.[1] Mission Planing for a swarm of UAVs has become a highlight goal in this context, since it is a complex problem that hardens UAV operators workload. Currently, UAVs are controlled remotely by human operators from Ground Control Stations (GCSs), using rudimentary planning systems, such as pre-configured plans or classical planners.

The Multi-UAV Cooperative Mission Planning Problem (MCMPP) usually deals with several criteria to quantify the quality of a solution, such as the fuel consumption, the makespan, the cost of the mission, the number of UAVs or GCSs to employ, and different risk factors that could compromise the mission. Thus, the problem is considered a Multi-Objective Optimization Problem (MOP), for which an estimation of the

Pareto Optimal Frontier (POF) must be inferred so as to get a portfolio of solutions (mission plans) differently albeit optimally balancing the considered conflicting objectives. In a previous work,[2] this problem was solved using a Multi-Objective Evolutionary Algorithm (MOEA) combined with a Constraint Satisfaction Problem (CSP) inside the fitness of the algorithm.

One critical point in this problem is that sometimes the entire POF comprises a large number of solutions. In this situation, the process of decision making to select one solution among them becomes a difficult task for the operator. In some cases, the operator can provide a priori information about his/her preferences, which can be used in the optimization process. However, most of the times the operator does not provide this information, so a posteriori approaches must be considered. The contribution of this work lies in developing a Decision Support System (DSS) to guide the operator in the process of selecting the best solutions, which becomes essential in this context. This DSS provides a ranking system to sort the solutions obtained by the optimization algorithm, and sometimes a filtering system that reduces the number of solutions provided.

In a previous work,[3] a Knee-Point MOEA approach was used to reduce the number of solutions obtained by the algorithm. In this work, we depart from these solutions and provide a Multi-Criteria DSS consisting of a ranking system. To test this approach, several Multi-criteria Decision Making (MCDM) algorithms have been assessed over a number of realistic mission scenarios with different operator profiles. Expert operators have evaluated the solutions provided for each mission through an evaluation application. With this evaluation, the MCDM algorithms have been scored. The framework used in this work is represented in Figure 1.

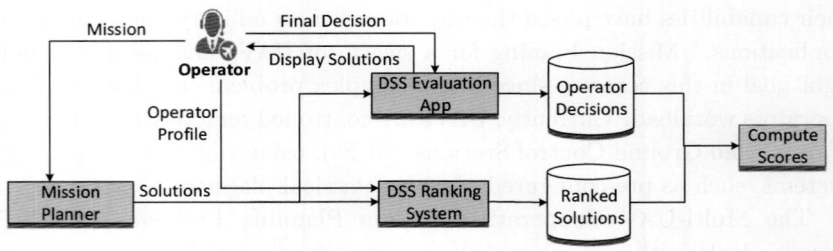

Fig. 1. Architecture of the framework designed to score the DSS Ranking systems.

The rest of the paper is structured as follows: Section 2 describes the UAV mission planning problem and the multiple criteria considered for optimization. Section 3 presents the proposed DSS, the ranking algorithms considered and the evaluation criteria used to score the algorithms. Section 4 discusses the experimental results. Finally, the last section draws some conclusions and the future research lines of work.

2. Multi-Objective Optimization for Mission Planning

The MCMPP can be defined as a set of *tasks* to be performed by a swarm of UAVs within a specific time interval. Each mission is performed in a specific mission scenario. In addition, several GCSs control the swarm of UAVs. A basic mission planner comprises the assignment of each task to one or more specific UAVs, as well as each UAV to a specific GCS, ensuring that the mission can be successfully performed within a time frame. More information about this problem can be found in.[2]

When solving a MCMPP, there exist several constraints (e.g. fuel constraints, time dependencies, path constraints, etc.) that must be evaluated for each solution in order to ensure whether they are valid or not. This is performed by a CSP, which in our approach[3] is located inside the fitness function of the evolutionary algorithm. In addition, when valid solutions are found, this algorithm considers the multiple criteria of the problem (see Table 1) in order to select the fittest ones that are inside the POF.

Table 1. Description of the criteria considered in the MCMPP problem.

#	Criterion
1	Total cost of the mission.
2	End time of the mission or makespan (hours), i.e. the time at which the last vehicle returns to the base.
3	Total fuel consumption (kg), i.e. the sum of the fuel consumed by every UAV.
4	Total flight time (hours), i.e. the sum of the flight time of every UAV.
5	Total distance traversed (km), i.e. the sum of the distance traversed by every UAV.
6	Risk of fuel usage (%), which indicates how risky is the mission according to the UAVs that finish it with low fuel.
7	Risk of distance to ground (%), which indicates how risky is the mission according to the UAVs that fly near to the ground somewhere in their path.
8	Risk of distance between UAVs (%), which indicates how risky is the mission according to the UAVs that fly close between them.
9	Risk of coverage loss (%), which indicates how risky is the mission according to the UAVs that fly out of the coverage of the GCSs.
10	Number of UAVs used in the mission.
11	Number of GCSs used in the mission.

3. Decision Support System

The MCDM problem is found across many domains. Decision makers have to select, assess or rank alternative solutions of a problem, according to the weights of multiple criteria, which usually are in conflict with each other.[4]

There are several MCDM algorithms[5] that have been developed for this purpose. In this work, we have considered the following: Weighted Sum Model (WSM), VIKOR,[6] Reference Ideal Method (RIM),[7] Multiplicative Multi-Objetive Optimization by Ration Analysis (MOORA)[8] and Technique for the Order of Prioritisation by Similarity to Ideal Solution (TOPSIS)[9] with the vectorial normalization procedure. On the other hand, the weights of the criteria needed in the algorithms are expressed according to their importance for the operator. This leads to the definition of *operator profiles*, where each criterion is ranked in five degrees of importance: *Very low (1)*, *Low (2)*, *Medium (3)*, *High (4)* and *Very high (5)*.

3.1. Evaluation Criteria

In order to perform an external evaluation of the quality of a ranking, and to compare different ranking algorithms objectively, we have created a "ground truth" dataset based on collective human judgement. Human judgement as a way to create ground truth data is a common approach used in many domains such as image recognition or performance analysis.[10]

In this work, the evaluation (or ground truth) data is created by asking expert UAV operators to decide, given a set of unranked mission plans of the same scenario, which of them is the best, i.e., which of them they would choose to be executed in a real environment. For each decision submitted by the operator, the following data is saved: the operator profile, an identifier of the mission being evaluated and an identifier of the chosen plan.

The process of gathering the evaluation data has been automated by the use of a web-app, shown in Figure 2. For each mission scenario to evaluate, a table with several possible solutions (plans) is displayed, one per row. Each column shows the value of a specific risk factor. To allow an easier comparison between the solutions, each cell is shaded with a two-coloured progress bar. The green part of the bar represents the relative quality of the risk factor with respect to the rest of solutions in the table. The red part is only used to fill the rest of the cell.

With the evaluation dataset created, we can give a score to the ranks created by a MCDM algorithm in the context of this work. Let a be a ranking algorithm, m a mission scenario and op an operator profile, then

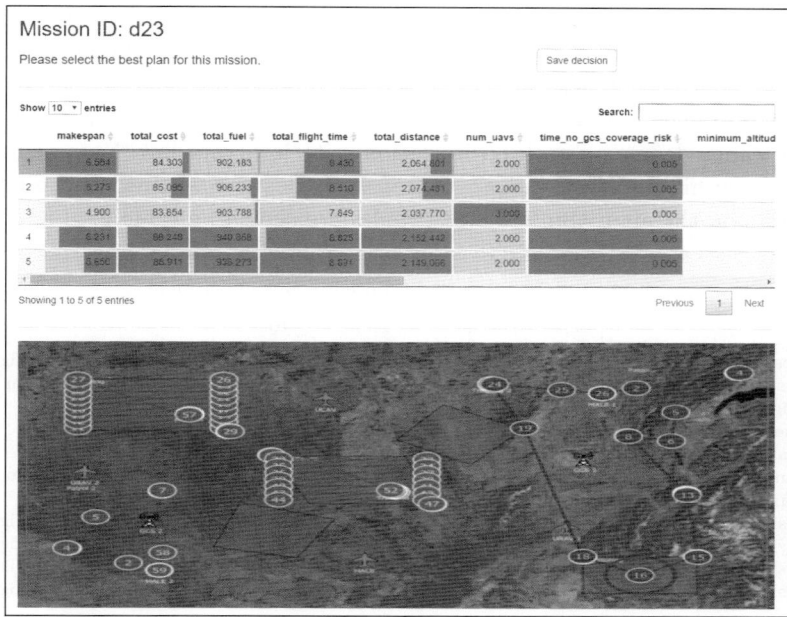

Fig. 2. Screenshot of the app developed to retrieve the evaluation data.

$L(a, m, \mathrm{op})$ denotes the ordered list of solutions (plans) returned by the algorithm in that context. Having an ordered list L, $r(L,p)$ denotes the rank position of plan p in the list L, where 1 means "best". On the other hand, let op be an operator profile, we refer to $S(\mathrm{op}, m)$ as the selection of best plan made by the profile op in the context of mission m. This selection is retrieved from the evaluation dataset. With this, we can define the *score* of an algorithm for a given operator profile and a given mission as follows:

$$Score(a, \mathrm{op}, m) = \frac{\mathrm{num\_solutions}(m) - r(L(a, m, \mathrm{op}), S(\mathrm{op}, m))}{\mathrm{num\_solutions}(m) - 1}, \quad (1)$$

As it can be seen, the value of the score is bounded on $[0, 1]$, where 1 represents the best possible matching (the selected plan is the first listed by the algorithm) and 0 the worst.

4. Experiments

As mentioned above, five MCDM algorithms are used in this work: WSM, VIKOR, RIM, Multiplicative MOORA and TOPSIS. In order to compare their performance, four real mission planning problems have been provided, each one of them in a different scenario with a different number of tasks, UAVs and GCSs. These problems were previously solved using the Knee-Point approach.[3] Finally, six UAV operators were asked to assign their importance degree on each optimization criterion (very low, low, medium, high, very high). With this, six operator profiles were defined and used as weights on the algorithms. These profiles are shown in Table 2.

Table 2. Operator profiles used in this experiment.

	Balanced	Cost	Time	Risk	Resources	RiskCost
Cost	Medium	Very High	Medium	Low	High	Very High
Distance	Medium	Medium	Medium	Low	Medium	Medium
Flight Time	Medium	Low	High	Medium	Low	Low
Fuel	Medium	High	Medium	Medium	High	High
Makespan	Medium	Low	Very High	Medium	Low	Low
Num GCSs	Medium	Medium	Medium	High	Very High	Medium
Num UAVs	Medium	High	Medium	High	Very High	High
Risk Distance Ground	Medium	Medium	Low	Very High	Medium	Very High
Risk Distance UAVs	Medium	Medium	Low	Very High	Medium	Very High
Risk Fuel Usage	Medium	Medium	Low	Very High	Medium	Very High
Risk Out of Coverage	Medium	Low	Low	Very High	Medium	Very High

The experiments have been performed in three steps:

(1) Each algorithm has been executed for every mission with the different operator profiles as weights of the optimization criteria.
(2) The operators have selected the best plan using the evaluation app presented in previous section.
(3) The score metric (see Eq. (1)) is applied to every triplet ⟨mission, profile, algorithm⟩ obtained after finishing the previous steps.

In order to give a global score for an algorithm, we take the average value of Eq. (1) over every mission scenario available in the evaluation dataset. The results are aggregated into groups of operator profiles and algorithms, in order to compare the performance of the algorithms over the different profiles. The comparative can be seen in Figure 3. Note that the last set of columns, "Global" does not refer to an specific operator profile, but to the average score over every profile tested.

It is clearly appreciable that *Risk* is the most difficult profile to evaluate, followed by the *RiskCost* and the *Balanced* profiles. As these two consider the risk factors as important criteria, it can be concluded that the risk factors are the most difficult features to evaluate.

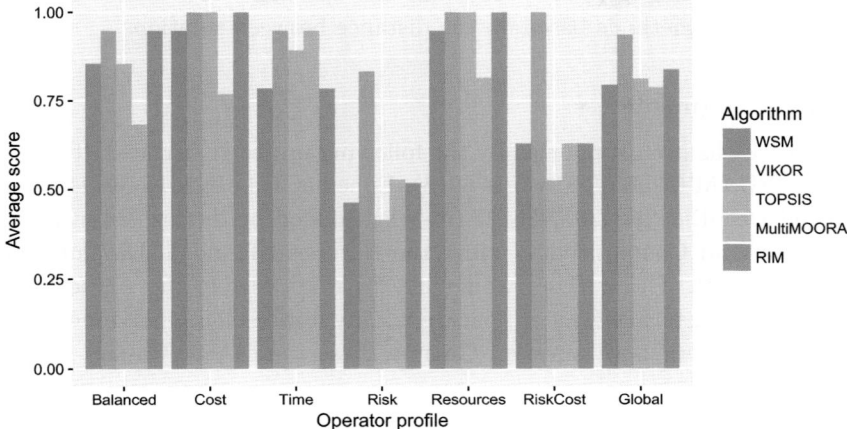

Fig. 3. Average scores of the different MCDM algorithms with the different operator profiles, and global scores.

On the other hand, it can be seen that VIKOR obtained the best results for all the profiles, being the only one able to get good results in the *Risk* profile. In the global picture, it is also the one that got the best results, followed far behind by RIM. VIKOR method provides a compromise solution that maximizes the best values for every criteria at the same time that minimizes the worst values. This approach proves to be the most suitable to apply in the context of this work.

5. Conclusions and Future Work

In this work, we have studied the performance of different Multi-Criteria Decision Making algorithms for ranking the solutions of a Mission Planning Problem, according to a set of 11 criteria. In the experimental phase, we have compared 5 MCDM algorithms, executed over 4 different mission scenarios, with 6 different operator profiles. The algorithms have been scored using the evaluation of expert operators, who selected the best solution according to their profile for every mission. Results show that VIKOR obtained outperforms the rest of the algorithms in all cases.

In future works, we will test more algorithms with a bigger benchmark of Mission Planning Problems of different complexity, using more operator profiles. In addition, it would be interesting to express the contents of Table 2 as fuzzy sets, making this problem a fuzzy MCDM. Finally, to improve the Decision Making (DM) process for the operator, we will also provide some filtering methods based on the distance between solutions.

Acknowledgments

This work has been supported by the following projects: Airbus Defence & Space (FUAM-076914/FUAM-076915); EphemeCH (TIN2014-56494-C4-4-P) and DeepBio (TIN2017-85727-C4-3-P), funded by Spanish Ministry of Economy and Competitivity, under the European Regional Development Fund FEDER. The authors would like to acknowledge the support obtained from Airbus Defence & Space, specially from Savier Open Innovation project members: José Insenser, Gemma Blasco and César Castro.

References

1. C. Ramirez-Atencia, V. Rodriguez-Fernandez, A. Gonzalez-Pardo and D. Camacho, New Artificial Intelligence Approaches for Future UAV Ground Control Stations, in *2017 IEEE Congress on Evolutionary Computation (CEC 2017)*, (IEEE, 2017).
2. C. Ramirez-Atencia, G. Bello-Orgaz, M. D. R-Moreno and D. Camacho, Solving complex multi-UAV mission planning problems using multi-objective genetic algorithms, *Soft Computing* **21**, 4883 (2017).
3. C. Ramirez-Atencia, S. Mostaghim and D. Camacho, A Knee Point based Evolutionary Multi-objective Optimization for Mission Planning Problems, in *Genetic and Evolutionary Computation Conference (GECCO 2017)*, (ACM, 2017).
4. E. K. Zavadskas, A. Zakarevicius and J. Antucheviciene, Evaluation of ranking accuracy in multi-criteria decisions, *Informatica* **17**, 601 (2006).
5. E. Triantaphyllou, *Multi-criteria Decision Making Methods: A Comparative Study* (Springer US, 200).
6. S. Opricovic and G.-H. Tzeng, Compromise solution by MCDM methods: A comparative analysis of VIKOR and TOPSIS, *European Journal of Operational Research* **156**, 445 (2004).
7. E. Cables, M. Lamata and J. Verdegay, RIM-reference ideal method in multicriteria decision making, *Information Sciences* **337-338**, 1 (2016).
8. W. Brauers and E. Zavadskas, Project management by MULTIMOORA as an instrument for transition economies, **16**, 5 (2010).
9. M. S. Garca-Cascales and M. T. Lamata, On rank reversal and TOPSIS method, *Mathematical and Computer Modelling* **56**, 123 (2012).
10. V. Rodríguez-Fernández, H. D. Menéndez and D. Camacho, Analysing temporal performance profiles of UAV operators using time series clustering, *Expert Systems with Applications* **70**, 103 (2017).

The use of decision support systems for new product design: A review

J. F. Figueroa Pérez

Facultad de Ingeniería Mochis - UAS, Fuente de Poseidón y Ángel Flores S/N,
Los mochis, Sinaloa, México
juanfco.figueroa@uas.edu.mx
http://fim.uas.edu.mx/

J. C. Leyva López

Universidad de Occidente, Blvd Lola Beltran S/N,
Culiacán, Sinaloa, México
juan.leyva@udo.mx
www.udo.mx

L. C. Santillán Hernández

Universidad de Occidente, Blvd Lola Beltran S/N,
Culiacán, Sinaloa, México
luis.santillan@udo.mx
www.udo.mx

E. O. Pérez Contreras

Universidad de Occidente, Blvd Lola Beltran S/N,
Culiacán, Sinaloa, México
pece78@gmail.com
www.udo.mx

In modern organizations, the decision making for new product design is a very important issue that can directly affect the profitability of the company or allow to achieve a competitive advantage in the current markets. Nowadays, has been reported in literature a significant number of Decision Support Systems (DSSs) to automate product design activities that have contributed to the evolution of the knowledge in this area. Thereby, in order to advance in the field, it is useful to determine its current state of the art. So, the aim of this paper is to conduct a systematic review of the literature of DSSs that focus on the design of new products. The review is focused particularly on those contributions published in the period from 1998 to 2017 that take into account different elements involved in this decision process, such as collaboration with other partners in a distributed environment and consideration of the customer needs to issue their recommendations. The paper performs an analysis of DSS's development characteristics, such as the models they implemented and the main technological resources that were used for constructing them.

Keywords: Literature review; product design; decision support systems.

1. Introduction

The development of the globally integrated economy represents new challenges for manufacturing companies related to how to stay in the market. In this way, the ability they have to adapt the marketing strategy will have a strong impact on the success or failure of the company's business [1].

Design of new products play an important role in the process of development new products. Experts in Marketing have pointed out the importance of the design of new products for the viability of enterprises [2]. In this sense Baril et al. [3] point out that the competitiveness of industries depends, among other things, on their ability to identify the customer's needs and to create products that meet these needs. At the same time, Lei and Moon [4] indicate that to maintain and enhance the level of profitability in an increasingly competitive and transparent market place, a company must continuously reposition and redesign its existing products or introduce new products to specific market segments.

A search for literature reviews yielded a number of reviews that have been performed in the design of new products area. As far as we are aware of, no systematic review on the topic of Decision Support Systems (DSSs) for the design of new products has been performed yet. Given the absence of a literature overview on DSSs for new product design, the goal of the research presented here is to perform a systematic literature review (SLR) of papers dealing with DSSs for new product design. In particular, the goal of this SLR is to provide an inventory of what has been done in previous years in the context of DSS guidelines for design new products.

The paper is structured as follows. Section 2 presents the systematic review methodology. Section 3 presents conclusions.

2. Systematic Review Methodology

According to Cooper [5] a systematic and objective review contains a five-staged main structure. The first stage is the formulation of the problem. In the second stage, the data collection strategy is determined. The third stage revolves around evaluating the retrieved data. In the fourth stage, an analysis and interpretation of the literature is reviewed. Finally, the results are presented in the fifth stage.

2.1. Formulation of the problem

In order to learn more in depth about the topics that interest us about the DSSs built in recent years to support New Product Design Process, we sought to address the following questions:

(a) How is decision support implemented in the revised DSSs? Do they take into account consumer satisfaction? and if so, what methods do they use to model them?
(b) Which of the revised DSSs for designing new products take into account the participation of multiple stakeholders in a distributed work environment? and if so, how they were implemented the support for them?
(c) Which DSSs take into account consumer satisfaction and a distributed work environment in an integrated way?

2.2. Data collection strategy

Scientific papers related to the research questions were recovered through a computerized search from libraries and web robots. The review covers the period 1998-2017. The sources included publishing, databases and search tools. Among them we noted at CONRICYT[a] (including ACM Digital Library, Cambridge University Press, EBSCO, Elsevier, Emerald, IEEE, Springer and Thomson-Reuters), Google Scholar, Google and Microsoft Academic. The search included the terms "DSS" or "decision support system" and "product design". From the relevant results of these searches it was conducted the analysis presented in the following sections.

2.3. Evaluating the retrieved data

From the set of publications ([7]-[64]) obtained using the strategy described above, 58 were selected since they had the more relevant titles (a detailed reference of selected papers collected can be found in goo.gl/Kynf3G) from a total of 370 reviewed. The selected items are those that meet the specified terms and conform to our research questions. The rest of the results were not considered because they do not have relation to the subject matter.

Selected articles show mainly DSS activity for supporting new product design, although some also include other features. The revised systems may

---

[a]The National Consortium of Scientific and Technological Information Resources (CONRICyT) allows us access to digital databases of Bibliographic Resources, in order to improve the capacities of Higher Education Institutions and Research Centers.

or may not consider consumer satisfaction and / or may not be prepared to operate in a distributed environment, make use of artificial intelligence techniques and / or multi-criteria analysis methods used in their models.

To answer the questions presented above, the following criteria were used:

(a) Decision support: Consideration of consumer satisfaction, how consumer satisfaction is modeled?
(b) Distribution support: Support for operational processes working in distributed environments, how is the support for decision-making implemented in a distributed environment?
(c) Consideration of both consumer satisfaction and distributed environment.

2.4. *Analysis and interpretation of the literature*

To further analyze the distribution of papers across various attributes, we use the technique of Formal Concept Analysis [6]. This technique allows to group the papers along the different dimensions that are addressed into a lattice and to visualize the commonality of certain attributes, that is to say, the level to which papers address the same or different attributes. Each node in the lattice identifies an attribute and the number of papers addressing specifically this attribute. In addition, upward lines denote a subset relationship in which each paper addresses the attributes of its node and all the attributes of the upward nodes. This allows to easily seeing how often a topic is addressed: the higher the node of a topic, the more often it is addressed. Most interestingly, the graph also visualizes which topics are often addressed together. Nodes with explicitly attached quality attributes have a full color fill, while half-filled nodes collect papers that (only) combine the quality attributes attached to higher nodes. The size of the nodes is proportional to the number of papers attached to the node.

Figure 1 shows the notation of the attributes used in the formal context of the study. They were derived from the defined criteria to answer the questions in the formulation of the problem:

Where US: UTASTAR, TS: TOPSIS; AH: AHP, PR: PROMETHEE, EA: Evolutionary Algorithms, NN: Neural Networks, HA: Heuristic Search Algorithms, MA: Multiobjective Algorithm, RN: Relation Network, AN: Artificial Networks; OH: Other Heuristics, KB: Knowledge Based System, HS: Hybrid System, ES: Expert System, NE: No specified, Y: Yes, N: No, MN: Monolitic, LY: Layered, SA: Software Agents, NE: No Specified.

Papers	Attributes																				
	Decision support															Distribution support					
	Consumer satisfaction											Others				Implemented		Environment architecture			
	Multicriteria method		Artificial intelligence																		
	U S	T S	A H	P R	E A	N N	H A	M A	R N	A N	O H	K B	H S	E S	N E	Y	N	M N	L Y	S A	N E

Fig. 1. Attributes used in the formal context.

2.5. Results of the Review

In this section we show the results of the review. These results are structured in decision support, distribution support and the consideration of consumer satisfaction and distributed work environment.

2.5.1. Decision support

Regarding to implementation of decision support, it was found that its implementation is diverse and, among others, it is carried out through models based on artificial intelligence techniques or to a lesser extent models based on multicriteria methods. Other techniques include knowledge based systems, expert systems, etc. Consumer satisfaction was modeled in 23(39%) of the 58 reviewed DSSs. Among these, 9 modeled them using multicriteria analysis methods and 14 using different artificial intelligence techniques, Figure 2 shows the results. All of them assume that consumer preferences are always complete and transitive in their consumer satisfaction modeling, which is not always the case in consumer's minds.

2.5.2. Distribution support

Support for operational processes in distributed environments was considered in 22(37%) of the 58 revised DSS and it was found that it is implemented mainly through multilayer architectures (2-tier, 3-tier, n-tier) or distributed intelligent agents, Figure 3 shows the results.

Among the 22 DSS with distribution support, the following lattice shows us that 9(41%) have a software agent-based architecture, 8(36%) have a layered architecture and in 5(23%) of them it is not specified. Figure 4 shows resulting lattice of distribution support and its environment architecture.

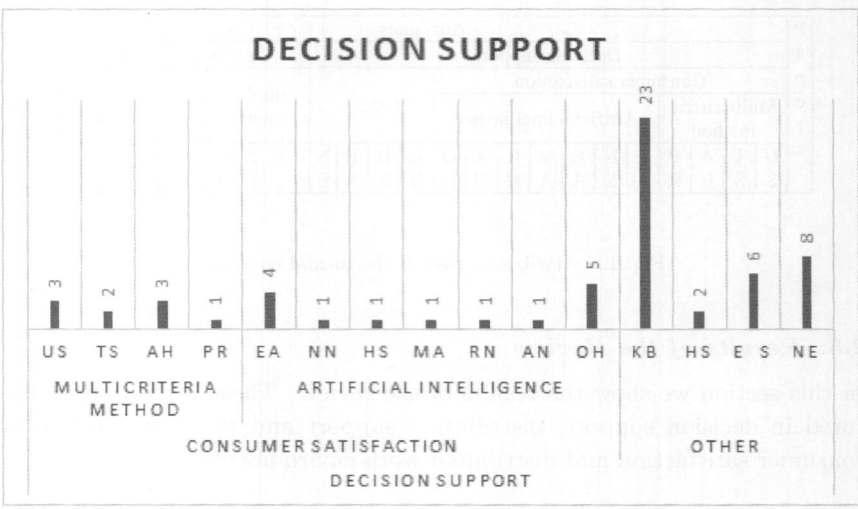

Fig. 2. Decision support.

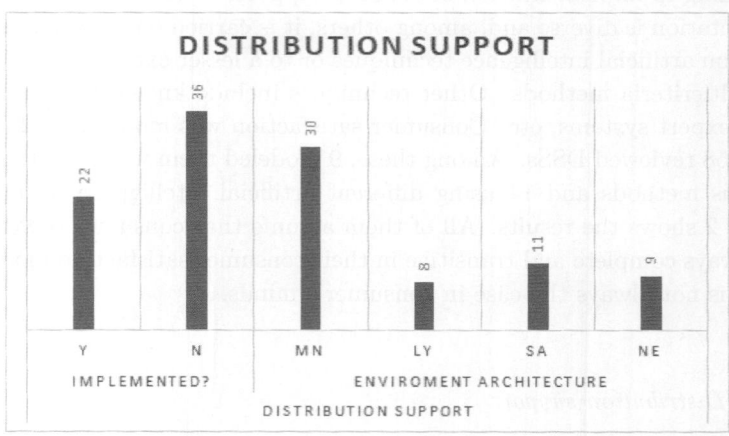

Fig. 3. Distribution support.

2.5.3. *Consideration of both consumer satisfaction and distributed environment*

Figure 5 shows resulting analysis lattice of consideration of both consumer satisfaction and distributed environment. Here we can see that only 4(7%)

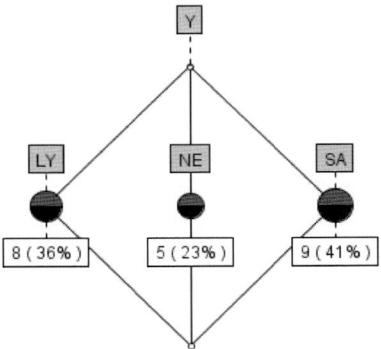

Fig. 4. Lattice diagram of distribution support and environment architecture.

of the 58 revised DSS considered together consumer satisfaction and a distributed work environment.

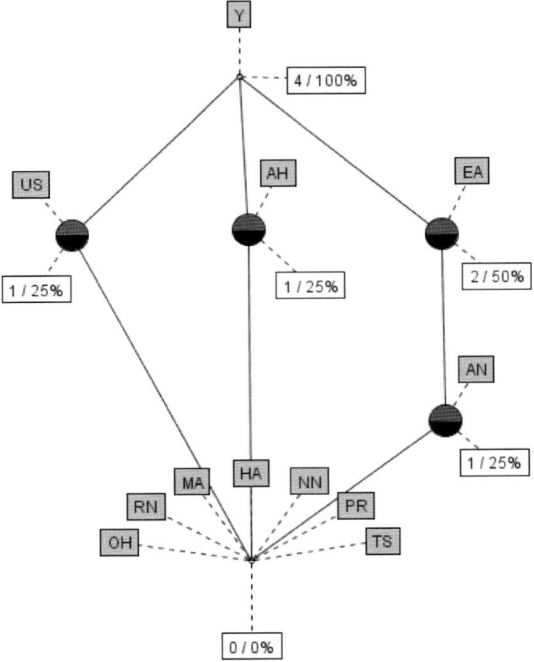

Fig. 5. Lattice diagram of distribution support and environment architecture.

3. Conclusions

In modern organizations, the process of designing new products has become a complex distributed decision-making task which is influenced by multiple factors among which are consumer satisfaction. However, in the literature, only 7% of developed DSS to support this process contemplating both elements together were found. This indicates that important elements of this decision process are being left out in the systems developed so far. In this way, we believe that future developments must considered both them so that they are more consistent with the current nature of this process and can support it more effectively.

References

1. W.-G. Ai, J. Sun and H. Li, A distributed marketing decision support system based on multi-intelligent-agent, in *Proceedings of 2004 International Conference on Machine Learning and Cybernetics (IEEE Cat. No.04EX826)*, Aug 2004.
2. N. F. Matsatsinis, P. Moraitis, V. Psomatakis and N. I. Spanoudakis, An agent-based system for products penetration strategy selection *Applied Artificial Intelligence* **17**, 2003.
3. C. Baril, S. Yacout and B. Clément, An interactive multi-objective algorithm for decentralized decision making in product design *Optimization and Engineering* **13**, Mar 2012.
4. N. Lei and S. K. Moon, A decision support system for market-driven product positioning and design *Decision Support Systems* **69**, 2015.
5. H. M. Cooper, *Integrating research: A guide for literature reviews*, 2nd edn. (Sage Publications, Thousand Oaks, CA, 1989).
6. R. Wille, Formal concept analysis (Springer-Verlag, Berlin, Heidelberg, 2005) pp. 1–33.

Prioritization of the requirements for collaborative feedback platform for course contents using Pythagorean fuzzy sets

Basar Oztaysi
Department of Industrial Engineering, Istanbul Technical University
Besiktas, Istanbul, 34367, Turkiye

Sezi Cevik Onar
Department of Industrial Engineering, Istanbul Technical University
Besiktas, Istanbul, 34367, Turkiye

Cengiz Kahraman
Department of Industrial Engineering, Istanbul Technical University
Besiktas, Istanbul, 34367, Turkiye

The education system is very important for the society. In this study, we try to define and prioritize the requirements of an n online platform which can be used as a source for collecting feedback from stakeholders of an educating system. The involved stakeholders are the course content generators, teachers, students and families of the students. In the application three experts used linguistic terms to evaluate the performance indicators and the weights are calculated using Pythagorean fuzzy Analytic Hierarch Process (AHP) Method.

1. Introduction

There are various types of feedbacks to analyze the effects of a lecture. But a formal system needs to continuously get feedback from the stakeholders and provide input to course content generators. The main aim of the study is to find the expectations of different stakeholders and prioritize their needs form this system.

For this scope we follow the six steps combining qualitative and quantitative methods for determining and prioritizing the features of the system through a collaborative process:

1. Identifying target groups: In this study we collaborated with TEGV (Turkey Educational Volunteers), a Non-profit organization (NGO) working on education all over Turkey. After analyzing the educational system of TEGV, we

identify the related stakeholders as; the content generators, teachers, students and families of the students.

2. Determining the Good Practices: We reviewed the literature about existing systems and examples of good practices to determine the base features of the architecture design.

3. Determining the potential system requirements through interviews, focus group studies and brainstorming method: We made several sessions with the members of TEGV, to formulate the problems encountered during gathering feedbacks after a course. The agenda of the brainstorming session was set by the themes obtained from interviews and literature review. We organized the ideas into an affinity diagram to list all possible requirements of the system design.

4. Filtering the requirements: System designers and end users negotiated together to filter out the requirement list to exclude the basic (i.e. basic elements required in all systems) and trivial features (too sophisticated to be effective). We ended up with a simplified feature list.

5. Prioritization of the system needs: The remaining system features are prioritized using Pythagorean Fuzzy Analytic Hierarchy Process. The degree of importance of each system requirement is determined.

Prioritization of the system needs is very critical for developing a system. The problem is a multicriteria decision making (MCDM) problem since there can be various numbers of criteria and subcriteria. In the field of MCDM, Analytical Hierarchy Process (AHP) is commonly used for complex decision processes. Analytic Hierarchy Process (AHP) is a method developed firstly by Saaty (1980) which disintegrates the complex decision problems according to different perspectives and deals with them in a hierarchical manner. Hence, complex problems are simplified into easier sub-problems. AHP method has an appropriate structure to make a decision by considering both the quantitative and qualitative factors.

Recently, fuzzy extensions of fuzzy sets are also proposed (Kahraman et al., 2016) Fuzzy sets are used in the literature to provide more realistic and accurate results through the decision making process. While regular fuzzy sets provide a better representation of linguistic terms, since they have some limitations extensions of fuzzy sets are proposed in the literature. Type2 fuzzy sets, hesitant fuzzy sets, intuitionistic fuzzy sets fuzzy multisets are some of the extensions which have been applied to MCDM problem (Oztaysi, 2015; Oztaysi et al., 2015; Kahraman et al., 2017; Kahraman et al., 2016; Oztaysi et al., 2017). Pythagorean fuzzy sets (PFSs) proposed by Yager (2013) are an extension of intuitionistic fuzzy sets (IFSs) satisfying the condition $\mu^2 + \nu^2 \leq 1$. Thus, PFSs enable the decision makers to use a larger range for both membership and nonmembership degrees. PFSs have been used to extend several multicriteria decision making methods to their fuzzy cases such as Cevik Onar (2018), Ilbahar et al. (2018), Chen (2018), Zhang (2016) Ren et al. (2016), Zhang and Xu (2014).

In this study, to better represent expert opinions and reach more accurate results, we propose using interval-valued intuitionistic set based AHP method for the prioritization problem.

2. Pythagorean Fuzzy AHP

In this paper, we propose using Pythagorean Fuzzy Analytic Hierarch Process (PFAHP) based on Buckley's (1985) fuzzy AHP. The methodology is composed eight steps and ends with obtaining the criteria using expert judgements.

A Pythagorean fuzzy set is defined as follows:

$$\tilde{P} = \{x, P(\mu_P(x), v_P(x)) | x \in X\} \quad (1)$$

where $\mu_P: X \to [0,1]$ is the membership degree and $v_P: X \to [0,1]$ is the nonmembership degree. Then, Eq. (2) is valid:

$$(\mu_P(x))^2 + (v_P(x))^2 \le 1 \quad (2)$$

The degree of indeterminancy is defined as follows:

$$\pi_P(x) = \sqrt{1 - (\mu_P(x))^2 - (v_P(x))^2} \quad (3)$$

The steps of PFAHP is as follows:

Step 1. Linguistic pairwise comparison matrices are formed according to the decision model, and decision makers fill the matrices using linguistic scale given in Table 1.

Table 1. Linguistic scale and its corresponding Pythagorean fuzzy number.

Linguistic Scale	Abbreviation	Pyth. Fuzzy Num.
Absolute	A	[0.91;0.11]
Very Strong	VS	[0.81;0.21]
Fairly Strong	FS	[0.71;0.31]
Weak	W	[0.61;0.41]
Equal	E	[0.5;0.5]
Recoprical Weak	RW	[0.41;0.61]
Recoprical Strong	RFS	[0.31;0.71]
Recoprical Strong	RVS	[0.21;0.81]
Recoprical Absolute	RA	[0.11;0.91]
Equally Important (EI)	EI	[0.55;0.55]

Step 2. The linguistic pairwise matrices are converted to their corresponding Pythagorean fuzzy sets using the scale given in Table 1 to obtain Pythagorean pairwise comparison matrix ($\tilde{R}$).

$$\tilde{R} = \left(C_j(x_i)\right)_{m \times n} = \begin{bmatrix} P(u_{11}, v_{11}) & \cdots & P(u_{1n}, v_{1n}) \\ \vdots & \ddots & \vdots \\ P(u_{m1}, v_{m1}) & \cdots & P(u_{mn}, v_{mn}) \end{bmatrix} \quad (4)$$

Step 3. For each row of $\tilde{R}$, geometric means are obtained. To this end first, the values in the same row are multiplied using Eq. (5) and then geometric mean is alculated using Eq. (6).

$$\tilde{P}_1 \otimes \tilde{P}_2 = P\left(\mu_{P_1}\mu_{P_2},\ \sqrt{v_{P_1}^2 + v_{P_2}^2 - v_{P_1}^2 v_{P_2}^2}\right) \quad (5)$$

$$\tilde{P}_1 = P\left(\mu_{P_1}{}^n,\ \left((1 - v_{P_1}^2)^n\right)^{1/2}\right) \quad (6)$$

Step 4. Geometric mean values are normalized. To this end, sum of the geometric means are calculated using Eq. (7), then each geometric mean is divided by this sum using Eq. (8).

$$\tilde{P}_1 \oplus \tilde{P}_2 = P\left(\sqrt{\mu_{P_1}^2 + \mu_{P_2}^2 - \mu_{P_1}^2 \mu_{P_2}^2},\ v_{P_1} v_{P_2}\right) \quad (7)$$

$$\frac{\tilde{P}_1}{\tilde{P}_2} = \left(\frac{\mu_{P_1}}{\mu_{P_2}}, \sqrt{\frac{v_{P_1}^2 - v_{P_2}^2}{1 - v_{P_2}^2}}\right), if\ \mu_{P_1} \leq \min\left\{\mu_{P_2}, \frac{\mu_{P_2} \cdot \pi_{P_2}}{\pi_{P_2}}\right\},\ v_{P_1} \geq v_{P_2} \quad (8)$$

Step 5. Steps 1-4 are repeated for all pairwise matrices according to the decision model. The weights of the subcriteria are obtained by multiplying the local weight of a subcriteria with the associated criterion.

Step 6. The Pythagorean fuzzy weights are defuzzified using Eq. (9), and the subcriteria weights are ranked.

$$P = \left(\frac{1 - v^2}{2 - v^2 - \mu^2}\right) \quad (9)$$

The normalized weights obtained in Step 6 are the final results that represent the priority of the system requirements of the users.

3. Application

In this study, we collaborated with TEGV which is one of the most active NGO on education area operation all over Turkey. Within the scope of the study, we conducted a focus group and interviews with different stakeholders.

Elicited system criteria may be grouped into five groups (i) Information flow: this group underlines the type of data collected from the system users (ii) Interaction among the users: This group underlines the main functionalities to encourage system usage by maintaining interaction (iii) Access: This group is about the different stake holders that can use the system, and quality of information gathered from the users. (iv) Security and Privacy: One of the main concerns of end users is the privacy issue. This criterion underlines the criteria related to privacy and security issues (v) Management: From the managerial perspective, resources needed for running a system is very critical. This criterion underlines the managerial perspectives to alternative systems.

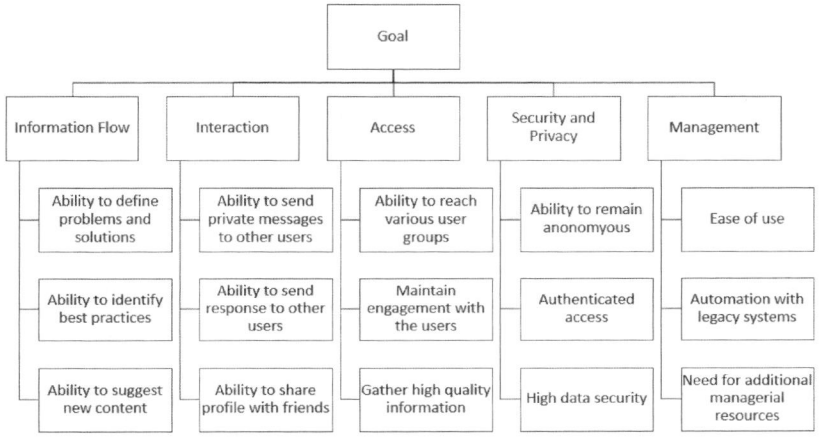

Figure 1. Hierarchy of the system requirements.

The users pairwise compare the five main criteria and the compromised evaluations are given in Table 2.

Table 2. Linguistic evaluations of the three experts.

	C1	C2	C3	C4	C5
C1	E	FS	W	VS	VS
C2	RFS	E	RW	FS	FS
C3	RW	W	E	W	FS
C4	RVS	RFS	RW	E	W
C5	RVS	RFS	RFS	RW	E

Later, the linguistic variables are transformed into pythagorean fuzzy sets as given in Table 3.

Table 3. Pythagorean fuzzy evaluations.

	C1	C2	C3	C4	C5
C1	(0.5;0.5)	(0.71;0.31)	(0.61;0.41)	(0.81;0.21)	(0.81;0.21)
C2	(0.31;0.71)	(0.5;0.5)	(0.41;0.61)	(0.71;0.31)	(0.71;0.31)
C3	(0.41;0.61)	(0.61;0.41)	(0.5;0.5)	(0.61;0.41)	(0.71;0.31)
C4	(0.21;0.81)	(0.31;0.71)	(0.41;0.61)	(0.5;0.5)	(0.61;0.41)
C5	(0.21;0.81)	(0.31;0.71)	(0.31;0.71)	(0.41;0.61)	(0.5;0.5)

In the next step, the geometric mean of each raw is calculated. To this end, first Pythagorean multiplication is applied to all elements of a row, and the root of the result is obtained to find the geometric mean. The geometric mean of each row is divided into sum of geometric means to find Pythagorean weights. Later the fuzzy weights are defuzzified and normalized. Table 4 represents the geometric means, Pythagorean fuzzy weights and crisp weights.

Table 4. Fuzzy and crisp weights.

	Row Multip.	Geometric mean	Pythagorean weights	Defuzzified Weights	Normalized Weights
C1	(0.142;0.696)	(0.522;0.445)	(0.756;0.435)	0.65	0.3
C	(0.032;0.9)	(0.318;0.651)	(0.46;0.646)	0.42	0.2
C3	(0.054;0.84)	(0.378;0.578)	(0.548;0.572)	0.49	0.23
C4	(0.008;0.966)	(0.201;0.771)	(0.292;0.768)	0.31	0.14
C5	(0.004;0.98)	(0.161;0.811)	(0.233;0.809)	0.27	0.13

By repeating the steps for each sub criteria, the local weights of the subcriteria are obtained. Then the local subcrieria weight is multiplied by the weight of the related criteria to find local weights of each subcriterion. Table 5 shows the local and global weights of the system requirements.

Table 5. Local and general weights of the system requirements.

	Local Significance	General Significance		Local Significance	General Significance
Inf. Flow	0.305		**Sec. & Priv.**	0.144	
R1.1		0.384	0.117	R1.1	0.253	0.037
R1.2		0.333	0.102	R1.2	0.227	0.033
R1.3		0.282	0.086	R1.3	0.519	0.075
Interaction	0.198		**Management**	0.125	
R2.1		0.254	0.050	R1.1	0.358	0.045
R2.2		0.321	0.064	R1.2	0.321	0.040
R2.3		0.425	0.084	R1.3	0.321	0.040
Access	0.228				
R2.1		0.333	0.076		
R2.2		0.333	0.076		
R2.3		0.333	0.076		

R1.1 (ability to define problems), R1.2 (ability to identify best practices), and R1.3 (ability to suggest new content) are the most important requirements among the system features.

4. Conclusions

The education system is very important for the society. In this study, we try to define and prioritize the requirements of an online platform which can be used as a source for collecting feedback from stakeholders. The involved stakeholders are the course content generators, teachers, students and families of the students. In the application linguistic terms are used to evaluate various system requirements. Pythagorean fuzzy AHP method is used to prioritize the requirements. The most important requirements are found to be: the ability to define problems, ability to identify best practices and ability to suggest new content.

References

1. Buckley J. (1985), Fuzzy Hierarchical Analysis, Fuzzy Sets and Systems, 17 (3), 233-247.
2. Cevik Onar S., Oztaysi B., Kahraman C. (2018) Multicriteria Evaluation of Cloud Service Providers Using Pythagorean Fuzzy TOPSIS, Journal of Multiple-Valued Logic & Soft Computing 30.
3. Chen T.Y. (2018) Remoteness index-based Pythagorean fuzzy VIKOR methods with a generalized distance measure for multiple criteria decision analysis, Information Fusion, Volume 41, Pages 129-150.
4. Ilbahar E., Karaşan A., Cebi S., Kahraman C. (2018) A novel approach to risk assessment for occupational health and safety using Pythagorean fuzzy AHP & fuzzy inference system, Safety Science, Volume 103, 2018, Pages 124-136.
5. Kahraman C., Ghorabaee M.K., Zavadskas E.K, Cevik Onar S., Yazdani M., Oztaysi B. (2017) Intuitionistic fuzzy EDAS method: an application to solid waste disposal site selection, Journal of Environmental Engineering and Landscape Management 25 (1), 1-12.
6. Kahraman C., Oztaysi B., Çevik Onar S. (2016) A comprehensive literature review of 50 years of fuzzy set theory, International Journal of Computational Intelligence Systems 9 (sup1), 3-24.
7. Oztaysi B. (2015) A Group Decision Making Approach Using Interval Type-2 Fuzzy AHP for Enterprise Information Systems Project Selection, Journal of Multiple-Valued Logic & Soft Computing 24 (5), 2015.
8. Öztaysi B., Cevik Onar S., Boltürk E., Kahraman C. (2015) Hesitant fuzzy analytic hierarchy process, Fuzzy Systems (FUZZ-IEEE), 2015 IEEE International Conference on, 1-7.
9. Oztaysi B., Cevik Onar S., Goztepe K., Kahraman C. (2017) Evaluation of research proposals for grant funding using interval-valued intuitionistic fuzzy sets, Soft Computing 21 (5), 1203-1218.
10. Ren P., Xu Z., Gou X. (2016) Pythagorean fuzzy TODIM approach to multi-criteria decision making, Applied Soft Computing, Volume 42, Pages 246-259.
11. Saaty T.L. (1980) The analytic Hierarchy Process, McGraw-Hill International Book Company, New York.
12. Yager, R. R. (2013). Pythagorean fuzzy subsets, in: IFSA World Congress and NAFIPS Annual Meeting (IFSA/NAFIPS), 57–61.

13. Zadeh, L. A. (1965). Fuzzy sets, Information and control, 8 (3), 338–353.
14. Zhang (2016), Multicriteria Pythagorean fuzzy decision analysis: A hierarchical QUALIFLEX approach with the closeness index-based ranking methods, Information Sciences, Volume 330, Pages 104-124.
15. Zhang, X., Xu, Z. (2014). Extension of TOPSIS to Multiple Criteria Decision Making with Pythagorean Fuzzy Sets. Int. J. Intell. Syst., 29: 1061–1078.

Solar energy project selection by using hesitant Pythagorean fuzzy TOPSIS

Veysel Çoban, Sezi Çevik Onar, Basar Oztaysi and Cengiz Kahraman

Istanbul Technical University, Istanbul, Turkey

The use of fossil-based energy resources as primary energy sources to meet rising energy demands reveals economic, environmental and social problems at a global scale. Solar energy is the most important renewable energy source as an alternative to fossil energy sources. As with all renewable energy systems, solar energy facilities also have a high initial cost and low operating costs. Therefore, solar energy investments with a high initial cost necessitate a sensitive and accurate decision. Environmental and social factors, as well as technical and economic factors, have an essential influence in choosing the right site for solar energy plant projects. The hesitant Pythagorean fuzzy TOPSIS method is applied to select the most suitable plant among the alternative solar power plant projects with different environmental and social characteristics in the Pythagorean fuzzy ambient. The TOPSIS method, based on the Pythagorean fuzzy set theory, provides a more realistic solution to decision-making problems in the Pythagorean fuzzy environment.

1. Introduction

Governments tend towards the alternative energy sources along with fossil-based energy sources to meet increasing energy demand [1]. Renewable energy sources emerge as the most important alternative energy sources because of their environmental, social and economic impacts [2]. Solar energy is the most important renewable energy source with its high efficiency and economic advantages. Solar energy systems require high installation costs, even though they have low operating costs [3]. It is necessary to determine the most suitable locations and conditions for solar energy systems with a high initial cost in order to obtain high energy yield and to provide price advantage against competitors.

Determination of the most suitable solar power plant project is a valuable decision-making process for investors and managers. The technical, economic, environmental and social factors that influence the decision-making process may result in complex situations where the implication is very hard [4]. Direct identification and measurement of environmental and social factors affecting the selection of a solar power plant site are important for decision-making models. Hesitant Pythagorean fuzzy set [5] is preferred since it helps to deal with the uncertainties encountered in describing model factors and in examining the effects of factors in the model. In this study, the environmental and social factors which are useful in the selection of the solar power plant are defined, and these factors are applied in the Pythagorean fuzzy based TOPSIS decision-making model.

The topics followed in the rest of the study are as follows: The second section refers to the environmental and social factors that are effective in site selection for solar energy plant. The third section explains Hesitant Pythagorean fuzzy sets (HPFSs) and the HPFS based TOPSIS multi-criteria decision-making model and the process steps of this model. The fourth section refers to a practical application of solar energy project selection using the extended TOPSIS model. The study is concluded with the conclusion section and future studies.

2. Site Selection for Solar Power Plants

Environmental and social factors play an essential role in determining the energy power site together with technical and economic requirements. Therefore, it is necessary to include the environmental and social factor in the evaluating and decision-making process of solar power plant projects.

2.1. *Environmental interaction*

The physical, climatic and plant cover characteristics of the project site can create positive or negative interactions between the environment and the power plant. The environmental factors evaluated in the decision-making process are as follows [6, 7]:

- **Radiation/heating effect:** System equipment with collective and reflective characteristics causes an increase in temperature in the plant environment. The change of temperature can interfere with the natural habitat of plants and animals around the plant.
- **Protected environment:** An area considered suitable for technical and economic criteria may be protected due to its historical, cultural or natural life characteristics. These characteristics may create limitations or obstacles to the use of the appropriate area.
- **Used land:** The selection of areas used for agricultural and livestock activities for solar energy production causes these activities to be damaged. This situation, which affects the daily life activities of the people in the environment, also causes social and ecological changes.
- **Water resources:** Although solar power plants do not produce industrial waste, changing ambient temperature and ecological impact indirectly affect water resources. In addition, thermal solar power systems which use water to generate energy should be installed close to water sources.

2.2. Social interaction

Social factors examine the interaction of the installation of the solar energy plant on the social structure of the facility environment. Social reactions and prejudices should be included in the decision-making process, taking into account the legal regulations [6, 8].

- **Economic effects:** The economic returns to the settlements and businesses in the environment and the impact on energy prices affect the social acceptance positively. Projects that increase social, economic level are preferred by investors and are supported by local and national governments.
- **Legal regulations:** Local and national governments publish incentive and support programs for regions that have difficulties in meeting energy demands. It is easier for investors to accept solar power plant projects supported by incentives.
- **Social acceptance:** Social prejudice and perception that occur outside of economic and regulatory influences affect the acceptance of solar energy projects. The social acceptance and reaction in the project environment arise as an essential factor for the establishment and development of the plant.

3. Preliminaries

3.1. Hesitant Pythagorean fuzzy sets (HPFSs)

The evaluation of the membership degree and nonmembership degree values defined in the Pythagorean Fuzzy Network (PFN) with different values makes it necessary to expand the PFN with HFS, and this developed structure is called HPFS [9]. Because HPFS returns two sets of membership and nonmembership degrees, membership and nonmembership degrees in PFN are described as Hesitant Fuzzy Elements (HFEs). HPFS, P on a fixed set X is defined as:

$$P = \{< x, P(h_P(x), g_P(x))|x \in X >\} \quad (1)$$

$h_P(x)$ represents the possible Pythagorean membership degree and $g_P(x)$ represents the possible Pythagorean nonmembership degree of the element x to the set P. They are two sets of some values in [0,1] with requirements as follow:

$$0 \leq \varphi, \phi \leq 1, 0 \leq (\varphi^+)^2 + (\phi^+)^2 \leq 1 \quad (2)$$

where $\varphi^+ = \max_{\varphi \in h_P(x)}\{\varphi\}$ and $\phi^+ = \max_{\phi \in g_P(x)}\{\phi\}$ for $\varphi \in h_P(x)$, $\phi \in g_P(x)$, and all $x \in X$. In HPFS consisting of two elements, Pythagorean membership hesitancy function and Pythagorean nonmembership hesitancy function, $h_P(x)$ and $g_P(x)$ are HFEs and their numbers are expressed as $\#h_P(x)$

and $\#g_P(x)$. Some basic operations for HPFNs $p = P(h,g)$, $p_1 = P(h_1,g_1)$, and $p_2 = P(h_2,g_2)$.

$$\varphi^- = min_{\varphi \in h}\{\varphi\}, \varphi^+ = max_{\varphi \in h}\{\varphi\}, \phi^- = min_{\phi \in g}\{\phi\}, \phi^+ = max_{\phi \in h}\{\phi\} \quad (3)$$

Equations represent the minimum and the maximum of the element h and g respectively [9].

$$p_1 \oplus p_2 = \cup_{\varphi_1 \in h_1, \varphi_2 \in h_2, \phi_1 \in g_1, \phi_2 \in g_2} \left\{\left\{\sqrt{(\varphi_1)^2 + (\varphi_2)^2 - (\varphi_1)^2(\varphi_2)^2}\right\}, \{\phi_1\phi_2\}\right\} \quad (4)$$

$$p_1 \otimes p_2 = \cup_{\varphi_1 \in h_1, \varphi_2 \in h_2, \phi_1 \in g_1, \phi_2 \in g_2} \left\{\{\varphi_1\varphi_2\}, \left\{\sqrt{(\phi_1)^2 + (\phi_2)^2 - (\phi_1)^2(\phi_2)^2}\right\}\right\} \quad (5)$$

$$np = \cup_{\varphi \in h, \phi \in g} \left\{\left\{\sqrt{1-(1-\varphi^2)^n}\right\}, \{\phi^n\}\right\}, where\ n \in R\ and\ n \geq 0 \quad (6)$$

$$\pi = \sqrt{1 - \left(\frac{1}{\#h}\Sigma_{\varphi \in h}\varphi^2 + \frac{1}{\#g}\Sigma_{\phi \in g}\phi^2\right)} \quad (7)$$

where π represents the degree of indeterminancy of p. The normalization process is applied on each unequal corresponding element of Hesitant Pythagorean Fuzzy Elements (HPFEs) [10]. The added membership degree and the nonmembership degree are defined as $\bar{\varphi} = \zeta\varphi^+ + (1-\zeta)\varphi^-$ and $\bar{\phi} = \zeta\phi^+ + (1-\zeta)\phi^-$ for a HPFE $p = P(h,g)$. Parameter ζ defined in the [0,1] interval is determined by decision maker according to risk level. Bigger ζ values refer the optimistic conditions for decision makers [10]. Distance measure, which is very important in the MCDM applications, generated for the normalized HPFEs is defined as follows [10]:

- The distance between two HPFEs $p_1 = P(h_1,g_1)$ and $p'_2 = P(\{1\},\{0\})$ that is the normalized set of p_2 as $d(p_1,p'_2)$;

$$d(p_1,p'_2) = \left(\frac{1}{2}\left(\frac{\frac{1}{\#h_1}\Sigma_{\varphi \in h_1}(1-\varphi^2)^\alpha + \frac{1}{\#g_1}\Sigma_{\phi \in g_1}\phi^{2\alpha} +}{\left(1-\left(\frac{1}{\#h_1}\Sigma_{\varphi \in h_1}\varphi^2 + \frac{1}{\#g_1}\Sigma_{\phi \in g_1}\phi^2\right)\right)^\alpha}\right)\right)^{1/\alpha} \quad (8)$$

- The distance between two HPFEs $p_1 = P(h_1,g_1)$ and $p'_2 = P(\{0\},\{1\})$ that is the normalized set of p_2 as $d(p_1,p'_2)$;

$$d(p_1,p'_2) = \left(\frac{1}{2}\left(\frac{\frac{1}{\#h_1}\Sigma_{\varphi \in h_1}\varphi^{2\alpha} + \frac{1}{\#g_1}\Sigma_{\phi \in g_1}(1-\phi^2)^\alpha +}{\left(1-\left(\frac{1}{\#h_1}\Sigma_{\varphi \in h_1}\varphi^2 + \frac{1}{\#g_1}\Sigma_{\phi \in g_1}\phi^2\right)\right)^\alpha}\right)\right)^{1/\alpha} \quad (9)$$

where α constant is greater than zero and defined by decision maker.

HPFSs which is a class of non-standard fuzzy sets, is preferred because it allows for some lack of commitment with uncertainty in the process of defining expressions. Thus, HPFS has the ability to comprehensively model uncertainties of decision makers in the models.

3.2. HPFSs based TOPSIS model

The TOPSIS method is an important decision-making method that produces solutions to MCDM problems. This method determines the optimal alternative which is the closest to the positive ideal solution and the farthest alternative to the negative ideal solution [11]. A new MCDM method is defined by expanding the TOPSIS method with HPFSs to produce more realistic solutions to the decision-making problems encountered in the hesitant Pythagorean fuzzy environment [9]. Multicriteria decision making the procedure of Pythagorean fuzzy TOPSIS method for HPFEs is stepped as follow [9]:

Step 1: The discrete set of alternatives $A = \{a_1, a_2, \ldots, a_k\}$, the attributes of alternatives $C = \{c_1, c_2, \ldots, c_l\}$, the weight vector of alternatives $W = (w_1, w_2, \ldots, w_l)^T$, ζ and α values are determined. The evaluation value, $P(h_{ij}, g_{ij})$ is defined for each alternative, a_i according to their criterion, c_j and the hesitant Pythagorean fuzzy decision matrix, $P = \left(P(h_{ij}, g_{ij})\right)_{k \times l}$ is generated ($i = 1,2,\ldots k; j = 1,2,\ldots,l$).

Step 2: Decision matrix P is normalized according to defined ζ value. Obtained new decision matrix described as, $P' = \left(P(h'_{ij}, g'_{ij})\right)_{k \times l}$.

Step 3: The positive ideal solution, x^+ and the negative ideal solution, x^- are defined as:

$$x^+ = (v_1^+, v_2^+, \ldots, v_l^+) = \left(P(\{1\},\{0\}), P(\{1\},\{0\}), \ldots, P(\{1\},\{0\})\right) \quad (10)$$

$$x^- = (v_1^-, v_2^-, \ldots, v_l^-) = \left(P(\{0\},\{1\}), P(\{0\},\{1\}), \ldots, P(\{0\},\{1\})\right) \quad (11)$$

Step 4: The distance between the alternative a_i and positive and negative ideal solution are calculated for each alternative.

$$d(a_i, x^+) = \sum_{j=1}^{l} w_j d\left(P(h'_{ij}, g'_{ij}), v_j^+\right) = \sum_{j=1}^{l} w_j d\left(P(h'_{ij}, g'_{ij}), P(\{1\},\{0\})\right)$$

$$= \sum_{j=1}^{l} w_j \left(\frac{1}{2} \left(\left(\frac{1}{\#h'_{ij}} \sum_{\varphi \in h'_{ij}} (1-\varphi^2)^\alpha + \frac{1}{\#g'_{ij}} \sum_{\phi \in g'_{ij}} \phi^{2\alpha} + \right. \right.\right.$$
$$\left.\left.\left. \left(1 - \left(\frac{1}{\#h'_{ij}} \sum_{\varphi \in h'_{ij}} \varphi^2 + \frac{1}{\#g'_{ij}} \sum_{\phi \in g'_{ij}} \phi^2\right)\right)^\alpha \right)\right)^{1/\alpha}\right) \quad (12)$$

$$d(a_i, x^-) = \sum_{j=1}^{l} w_j d\left(P(h'_{ij}, g'_{ij}), v_j^-\right) = \sum_{j=1}^{l} w_j d\left(P(h'_{ij}, g'_{ij}), P(\{0\},\{1\})\right)$$

$$= \sum_{j=1}^{l} w_j \left(\frac{1}{2} \left(\left(\frac{1}{\#h'_{ij}} \sum_{\varphi \in h'_{ij}} \varphi^{2\alpha} + \frac{1}{\#g'_{ij}} \sum_{\phi \in g'_{ij}} (1-\phi^2)^\alpha + \right. \right.\right.$$
$$\left.\left.\left. \left(1 - \left(\frac{1}{\#h'_{ij}} \sum_{\varphi \in h'_{ij}} \varphi^2 + \frac{1}{\#g'_{ij}} \sum_{\phi \in g'_{ij}} \phi^2\right)\right)^\alpha \right)\right)^{1/\alpha}\right) \quad (13)$$

Step 5: The relative closeness of each alternative, a_i is calculated according to the positive ideal solution, x^+ and calculated as:

$$RC(a_i) = \frac{d(a_i,x^-)}{d(a_i,x^-)+d(a_i,x^+)}, i = 1,2,\ldots,k \quad (14)$$

Step 6: All alternatives are ranked according to their relative closeness, $RC(a_i)$ (i=1,2,…,k).

4. Application for Solar Energy Project Selection

Determination of the most suitable solar energy projects in uncertain and risky environments is necessary to provide the energy strategy of a country. Environmental and social factors, as well as technical and economic factors, play an active role in determining the most appropriate project.

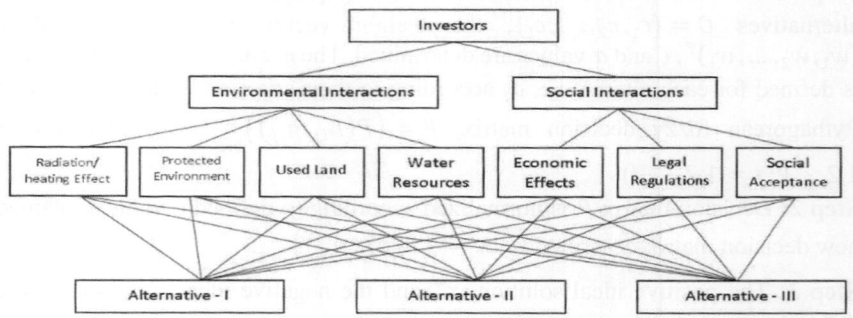

Fig. 1. The hierarchical structure of solar energy plant selection.

In this application, alternative sites recommended for solar energy plant installation are evaluated under environmental and social criteria using hesitant Pythagorean fuzzy TOPSIS method. While radiation/heating effect, protected environment and used land are defined as cost criteria, water resources, economic effects, legal regulations and social acceptance defined as benefit criteria in PF-TOPSIS method. Attributes used to evaluate the alternative projects are defined as: c_1: radiation/ heating effect, c_2: protected environment, c_3: used land, c_4: water resources, c_5: economic effects, c_6: legal regulations, c_7: social acceptance ($C = (c_1, c_2, c_3, c_4, c_5, c_6, c_7)$) (Fig. 1).

The criteria's weight vector is defined as W= (0.15, 0.16, 0.13, 0.12, 0.17, 0.13, 0.14)^T by experts' consensus. Three alternative project areas ($A = \{a_1, a_2, a_3\}$) are evaluated in this example application. Alternative project sites are evaluated by experts with HPFEs in hesitant Pythagorean fuzzy conditions and hesitant Pythagorean fuzzy matrix, P is obtained (Table 1).

Table 1. HPF decision matrix P.

	c1	c2	c3	c4
A1	P({0.3,0.4,0.5},{0.6,0.7})	P({0.3,0.45,0.8},{0.3,0.4})	P({0.3,0.6,0.8},{0.2,0.3,0.7})	P({0.4,0.6},{0.1,0.3,0.7})
A2	P({0.5,0.7,0.8},{0.4,0.5})	P({0.3,0.6,0.7},{0.4,0.7,0.85})	P({0.2,0.4,0.5},{0.3,0.5})	P({0.4,0.5,0.7},{0.3,0.5})
A3	P({0.6,0.7},{0.5,0.7,0.8})	P({0.4,0.45},{0.5,0.6,0.55})	P({0.3,0.4},{0.5,0.6})	P({0.3,0.8},{0.5,0.6})
	c5	c6	c7	
A1	P({0.4,0.6,0.7,0.9},{0.2,0.5})	P({0.4,0.5,0.6},{0.2,0.3,0.4})	P({0.3,0.4,0.6,0.7},{0.2,0.45,0.6})	
A2	P({0.5,0.6,0.7},{0.4,0.7})	P({0.3,0.5,0.7},{0.4,0.5})	P({0.3,0.6,0.7},{0.4,0.6})	
A3	P({0.4,0.8},{0.3,0.6,0.7})	P({0.4,0.7},{0.3,0.4,0.7})	P({0.4,0.5},{0.3,0.5,0.6})	

ζ is defined as 0.5 to normalize the decision matrix P and matrix P' is obtained (Table 2). α is defined as 2 to calculate the geometric distances of each alternative to positive (x^+) and negative (x^-) ideal solutions. Ideal solutions are defined with normalized form as follows:

$$x^+ = (P(\{1,1,1,1\},\{0,0,0\}), P(\{1,1,1,1\},\{0,0,0\}), P(\{1,1,1,1\},\{0,0,0\}), P(\{1,1,1,1\},\{0,0,0\}),$$
$$P(\{1,1,1,1\},\{0,0,0\}), P(\{1,1,1,1\},\{0,0,0\}), P(\{1,1,1,1\},\{0,0,0\}))$$

$$x^- = (P(\{0,0,0,0\},\{1,1,1\}), P(\{0,0,0,0\},\{1,1,1\}), P(\{0,0,0,0\},\{1,1,1\}), P(\{0,0,0,0\},\{1,1,1\}),$$
$$P(\{0,0,0,0\},\{1,1,1\}), P(\{0,0,0,0\},\{1,1,1\}), P(\{0,0,0,0\},\{1,1,1\}))$$

Table 2. Normalized and ordered decision matrix P'.

	c1	c2	c3	c4
A1	P({0.3,0.4,0.4,0.5},{0.6,0.65,0.7})	P({0.3,0.45,0.55,0.8},{0.3,0.35,0.4})	P({0.3,0.55,0.6,0.8},{0.2,0.3,0.7})	P({0.4,0.5,0.5,0.6},{0.1,0.3,0.7})
A2	P({0.5,0.65,0.7,0.8},{0.4,0.45,0.5})	P({0.3,0.5,0.6,0.7},{0.4,0.7,0.85})	P({0.2,0.35,0.4,0.5},{0.3,0.4,0.5})	P({0.4,0.5,0.55,0.7},{0.3,0.4,0.5})
A3	P({0.6,0.65,0.65,0.7},{0.5,0.7,0.8})	P({0.4,0.425,0.425,0.45},{0.5,0.55,0.6})	P({0.3,0.35,0.35,0.4},{0.5,0.55,0.6})	P({0.3,0.55,0.55,0.8},{0.5,0.55,0.6})
	c5	c6	c7	
A1	P({0.4,0.6,0.7,0.9},{0.2,0.35,0.5})	P({0.4,0.5,0.5,0.6},{0.2,0.3,0.4})	P({0.3,0.4,0.6,0.7},{0.2,0.45,0.6})	
A2	P({0.5,0.6,0.6,0.7},{0.4,0.55,0.7})	P({0.3,0.5,0.5,0.7},{0.4,0.45,0.5})	P({0.3,0.5,0.6,0.7},{0.4,0.5,0.6})	
A3	P({0.4,0.6,0.6,0.8},{0.3,0.6,0.7})	P({0.4,0.55,0.55,0.7},{0.3,0.4,0.7})	P({0.4,0.45,0.45,0.5},{0.3,0.5,0.6})	

Calculated geometric distance values are as follows: $d(a_1, x^+) = 0.649$, $d(a_2, x^+) = 0.614$, $d(a_3, x^+) = 0.644$, $d(a_1, x^-) = 0.720$, $d(a_2, x^-) = 0.662$ and $d(a_3, x^-) = 0.624$. Relative closeness of each alternative is calculated according to geometric distances and the results obtained as follows: $RC(a_1) = 0.526$, $RC(a_2) = 0.519$ and $RC(a_3) = 0.492$. The $a_1 > a_2 > a_3$ result is obtained when the alternatives are sorted in such a way that the positive ideal result is closest to the closest negative result. The order of $a_1 > a_2 > a_3$ is obtained when the relative distances of the alternatives are calculated according to the negative ideal result. This ranking reveals that the first project among solar energy project alternatives is the most appropriate project based on environmental and social factors.

5. Conclusion

Although solar energy plants have low operating costs, the high initial cost is required for plant installation. Therefore, the solar power plant installation projects must be evaluated at the economic, technical, social and environmental aspects and the correct site must be determined for plant installation.

Environmental and social factors, which are difficult to define directly with crisp values, have an important influence on the installation of the solar energy plant. The hesitant Pythagorean fuzzy TOPSIS model is proposed to select the most suitable alternative under the defined attributes in the Pythagorean fuzzy environment with ambiguity and uncertainty. In this study, alternative projects proposed for the installation of the solar energy plant are modeled under environmental and social criteria, and the model is solved by the TOPSIS multi-criteria decision-making method extended with HPFSs. In application, three alternative solar energy projects are evaluated under seven environmental and social criteria. There is no significant difference between alternative projects according to the TOPSIS relative closeness values, and the first alternative is selected as the most suitable project. HPF TOPSIS decision making model solves real-life problems by incorporating hesitancy and uncertainties into the calculations and helps the investor in the decision making process. In future studies, other decision making methods (e.g. VIKOR, AHP, ELECTRE) can be expanded with Pythagorean or hesitant Pythagorean fuzzy sets to develop new decision making methods for uncertain real-life problems.

References

1. BP, *Statistical Review of World Energy* June 2017, (2017).
2. International Energy Agency (IEA), *Renewable energy*, [cited 2018 January 10]; Available from: www.iea.org/about/faqs/renewableenergy/, (2018).
3. P. Mir-Artigues and P. Del Río, The Economics and Policy of Solar Photovoltaic Generation, *Springer*, (2016).
4. G.M. Crawley, *Solar Energy*, World Scientific Publishing Co., (2016).
5. R.R. Yager, Pythagorean fuzzy subsets. in IFSA World Congress and NAFIPS Annual Meeting (IFSA/NAFIPS), *2013 Joint, IEEE*, (2013).
6. M. Vafaeipour et al., Assessment of regions priority for implementation of solar projects in Iran: New application of a hybrid multi-criteria decision making approach, *Energy Conversion & Management*, **86**:p.653-663, (2014).
7. M. Zoghi et al., Optimization solar site selection by fuzzy logic model and weighted linear combination method in arid and semi-arid region: A case study Isfahan-IRAN, *Renewable and Sustainable Energy Reviews*, (2015).
8. Y. Wu et al., Decision framework of solar thermal power plant site selection based on linguistic Choquet operator, *Applied Energy*, **136**: p. 303-311, (2014).
9. D. Liang and Z. Xu, The new extension of TOPSIS method for multiple criteria decision making with hesitant Pythagorean fuzzy sets, *Applied Soft Computing*, **60**: p. 167-179, (2017).
10. B. Zhu and Z. Xu, Some results for dual hesitant fuzzy sets, *Journal of Intelligent & Fuzzy Systems*, **26**(4): p. 1657-1668, (2014).
11. C.-T. Chen, Extensions of the TOPSIS for group decision-making under fuzzy environment, *Fuzzy sets and systems*, **114**(1): p. 1-9, (2000).

Interval-valued intuitionistic fuzzy based QFD application for smart hospital design[*]

Deniz Uztürk

Department of Business Administration, Galatasaray University, 34349 İstanbul, Turkey

Gülçin Büyüközkan[†]

Department of Industrial Engineering, Galatasaray University, 34349 İstanbul, Turkey

Ahmet Fahri Negüs

Department of Business Administration, Galatasaray University, 34349 İstanbul, Turkey

M. Yaman Öztek

Department of Business Administration, Galatasaray University, 34349 İstanbul, Turkey

> Quality Function Deployment (QFD) method is a process that is widely used in manufacturing or service sectors to project the customers' will into the production phase. Its easy computational steps give the power of natural transformation of needs into engineering requirements. In this study, a QFD method is proposed in an Interval-Valued Intuitionistic Fuzzy (IVIF) environment to make use of its ability to deal with uncertain data. To test the applicability of the proposed technique, IVIF based QFD is applied to a case study where a hospital building design is introduced based on smart and ecological requirements.

1. Introduction

Decision-making is an every-day phenomenon in our lives. It relies on either ranking the alternatives or selecting the suitable alternative(s) after a general assessment. Since real-life environments can bring high complexity, decision-making processes can become extensively complicated. The evaluation of different aspects, and consideration of various constraints together leads us to a multi-criteria decision-making (MCDM) problem setting. In MCDM, a decision

---

[*]This work is financially supported by Galatasaray University Research Fund (Project No: 18.102.001).
[†]Corresponding author; e-mail: gulcin.buyukozkan@gmail.com Tel: +902122274480, Fax: +902122595557.

should be made under the consideration of a set of criteria that affects the process. Also, in real-life applications, due to extensive and complex information caused by unstable socio-economic environments, one decision maker may not be enough to reflect the optimal decision for the process. At this point, group decision-making (GDM) is a reliable solution as a balanced way to reach objective decisions. Even though GDM is an appropriate way to overcome subjectivity of the decision process, it is not always sufficient to deal with uncertain data from hazy socio-economic environments. Therefore, fuzzy logic is also employed to handle the vagueness [1] of data. In the literature, there are various decision-making tools that are based on fuzzy-based models [2].

Intuitionistic fuzzy (IF) sets introduced by Atanassov in 1986 can successfully represent the vagueness of human judgments over the objects [3]. Thanks to its ability to reflect the hesitancy of human opinions and to deal with uncertainty, IF sets are commonly used together with decision-making problems [4], [5], [6]. Moreover, Interval-Valued Intuitionistic Fuzzy (IVIF) sets, which are an extension of IF sets [7], are commonly used to handle complex problems [8]. The benefits of IVIF sets over IF and classical sets is their ability to show negative and positive notions of interval values for membership and non-membership function values to model complex decision-making problems [9], [10], [11].

In this study, IVIF sets are chosen for a decision-making problem, to be used together with the Quality Function Deployment (QFD) method. QFD is well-known for its ability to balance customer needs in a service or a production environment [12], which can be used to design service or production processes by detecting priorities according to customer expectations. In the literature, this technique is used with IF sets to take advantage of its power to operate with impreciseness and vagueness. In 2014 it used for a selection problem for a knowledge management system [13] and in 2016, it is applied to construct an evaluation mode of innovative design for green products [14].

In this paper, the aim is to integrate the QFD method with IVIF sets to create an adequate environment for handling imprecise data for a decision-making problem. The method is then applied to a hospital design case to demonstrate its practical usefulness and technical abilities.

2. Methodology

This study proposes an IVIF set-integrated QFD method for decision-making problems. For this purpose, both techniques are explained in this section.

2.1. Interval-Valued Intuitionistic Fuzzy Sets

Atanassov [3] developed IF sets, where each element contains a membership and non-membership value to reflect the level of hesitancy. Amid of these sets, their extension IVIF sets are represented as follows [7]:

$$\tilde{A} = \{\langle x, \tilde{\mu}_{\tilde{A}}(x), \tilde{v}_{\tilde{A}}(x) \rangle | x \in E\} \qquad (1)$$

where $\tilde{\mu}_{\tilde{A}}(x)$ is an interval-valued membership value and $\tilde{v}_{\tilde{A}}(x)$ an interval-valued non-membership value. Also, $\tilde{A}$ is called IVIF set, where $\tilde{\mu}_{\tilde{A}}(x) = [\mu_{\tilde{A}}^L, \mu_{\tilde{A}}^U] \subset [0,1]$ and $\tilde{v}_{\tilde{A}}(x) = [v_{\tilde{A}}^L, v_{\tilde{A}}^U] \subset [0,1]$, $\forall x \in E$, $\tilde{\mu}_{\tilde{A}}(x)$ and $\tilde{v}_{\tilde{A}}(x)$ are intervals with the condition of $\mu_{\tilde{A}}^L = \inf \tilde{\mu}_{\tilde{A}}(x)$, $\mu_{\tilde{A}}^U = \sup \tilde{\mu}_{\tilde{A}}(x)$, $v_{\tilde{A}}^L = \inf \tilde{v}_{\tilde{A}}(x)$, $v_{\tilde{A}}^U = \sup \tilde{v}_{\tilde{A}}(x)$, $0 \leq \sup(\tilde{\mu}_{\tilde{A}}(x)) + \sup(\tilde{v}_{\tilde{A}}(x)) \leq 1$.

Additionally, $\tilde{\pi}_{\tilde{A}}(x) := [\pi_{\tilde{A}}^L, \pi_{\tilde{A}}^U]$, $x \in X$, where

$$\pi_{\tilde{A}}^L = 1 - \mu_{\tilde{A}}^U - v_{\tilde{A}}^U, \pi_{\tilde{A}}^U = 1 - \mu_{\tilde{A}}^L - v_{\tilde{A}}^L. \qquad (2)$$

Moreover, the upper and lower points of IVIF sets are shown in the following equation:

$$\tilde{A} = \langle [\mu_{\tilde{A}}^L, \mu_{\tilde{A}}^U], [v_{\tilde{A}}^L, v_{\tilde{A}}^U], [\pi_{\tilde{A}}^L, \pi_{\tilde{A}}^U] \rangle \qquad (3)$$

The comparison of different IVIF sets are made according to the following relation:

$$\tilde{A} \leq \tilde{B} \Leftrightarrow \mu_{\tilde{A}}^L \leq \mu_{\tilde{B}}^L, \mu_{\tilde{A}}^U \leq \mu_{\tilde{B}}^U, v_{\tilde{B}}^L \leq v_{\tilde{A}}^L, v_{\tilde{B}}^U \leq v_{\tilde{A}}^U \qquad (4)$$

where $\tilde{B}$ is another IVIF set defined as: $\tilde{B} = \langle [\mu_{\tilde{B}}^L, \mu_{\tilde{B}}^U], [v_{\tilde{B}}^L, v_{\tilde{B}}^U] \rangle$.

In addition to all these relations, arithmetic operations also are available between IVIF sets, as mentioned in [7].

2.2. IVIF-Integrated QFD

In this study, the primary House of Quality (HoQ) is proposed as a solution to a design problem (Figure 1). HoQ is the commonly used matrix of QFD, which aims to identify the weight of each design requirement based on customer requirements [15].

Steps for IVIF integrated QFD are explained next:
1. Determine the purpose of the study and identify the customer requirements (CRs) related to it. Then decide on the number of decision-makers (DMs) and a linguistic scale for the evaluation of the problem.
2. Let the DMs provide the importance of each CR and aggregate them with the IIFWA operator [16] as in Eq. (5):

$$IIFWA_{C_j} = \left(\left[1 - \prod_{k=1}^{n}(1-\mu_{\tilde{A}}^{L})^{\lambda^k}, \atop 1 - \prod_{k=1}^{n}(1-\mu_{\tilde{A}}^{U})^{\lambda^k} \right], \left[\prod_{k=1}^{n}(v_{\tilde{A}}^{L})^{\lambda^k}, \atop \prod_{k=1}^{n}(v_{\tilde{A}}^{U})^{\lambda^k} \right] \right) \quad (5)$$

where λ^k is the weight of the kth DM; n is the number of DM and $\tilde{\mu}_{\tilde{A}}(x)$ interval-valued membership value and $\tilde{v}_{\tilde{A}}(x)$ an interval-valued non-membership value for the evaluations of each DM. Then defuzzify them to obtain their crisp weights. Subsequently, normalize them.

3. Identify the Design Requirements (DRs) related to the subject and CRs.
4. Let DMs establish the relations between each CR and DR in their linguistic scale, aggregate them with Eq. (5), and defuzzify them to obtain crisp relation values for each CR-DR pair.
5. Calculate DRs' priorities with Eq. (6):

$$I_j = \frac{\sum_{j=1}^{l} \sum_{i=1}^{m} nw_i \times r_{ij}}{m} \quad (6)$$

where I_j is the importance of the jth DR; nw_i is the aggregated crisp normalized weight of the ith CR; r_{ij} is the aggregated crisp value of the relations between i^{th} CR and j^{th} DR and m is the number of CRs; l is the number of DRs.

Then normalize them with the following relation:

$$NI_j = \frac{I_j}{\sum_{j=1}^{l} I_j} \quad (7)$$

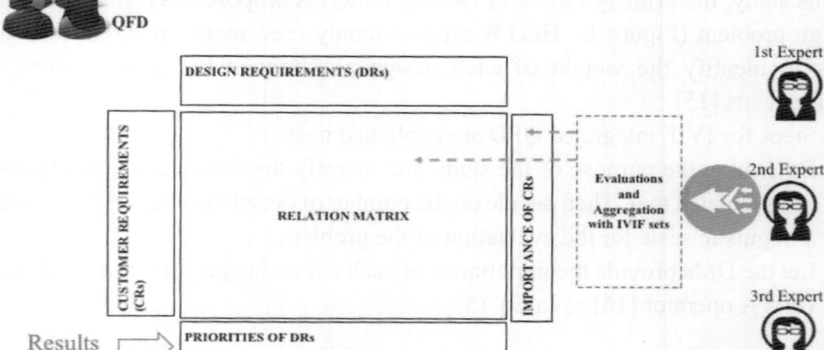

Figure 1. IVIF integrated QFD.

3. Application: Smart Hospital Design

This paper proposes a method that is based on IVIF integrated QFD for an MCDM problem with GDM. To test its plausibility, it is applied to a smart building design problem. The steps are as follows:

1. The purpose of the decision problem is defined as hospital design according to smart requirements. Then, CRs are determined accordingly as: natural lighting (C1), natural ventilation (C2), renewable energy (C3), natural resources (C4), healing environment (C5), strategic landscaping (C6), use of environment-friendly materials (C7), use of non-toxic materials (C8), sustainable and innovative building design (C9), safety/security (C10), flexibility (C11), connectivity (C12), energy optimization (C13), comfort (C14) and responsiveness (C15). Three DMs make evaluations, and their linguistic scales are represented in Table 1.

Table 1. Linguistic scale for DMs [17].

s_5 linguistic scale	$[\mu^L$	$\mu^U]$	$[v^L$	$v^U]$	$[\pi^L$	$\pi^U]$
None (N)	[0.01	0.05]	[0.85	0.9]	[0.05	0.14]
Low (L)	[0.25	0.3]	[0.6	0.65]	[0.05	0.15]
Medium (M)	[0.45	0.5]	[0.4	0.45]	[0.05	0.15]
High (H)	[0.75	0.8]	[0.1	0.15]	[0.05	0.15]
Very High (VH)	[0.85	0.9]	[0.01	0.05]	[0.05	0.14]

2. Three different DMs assess the CRs by their importance; assessments are aggregated with Eq. (5). The normalized values are represented in Table 2.
3. DRs for a hospital design are defined according to the CRs as: camera systems (DR1), emergency escape systems (DR2), maintenance systems (DR3), sensors (DR4), IoT technologies (DR5), connection of buildings systems with each other (DR6), light control systems (DR7), management information systems (DR8), sustainable and innovative architecture (DR9), cabling plans and records (DR10), smart waste systems (D11) energy management systems (DR13), meters (DR14), cyclic command systems (DR15), building forms and dimensions (DR16), types, sizes, and shapes of openings (DR17), fixed light windows for skylight (DR18), local infrastructure (DR19), low flow plumbing fixing (DR20).
4. Three different DMs provide their judgments on the CR-DR relations with their linguistic scales. Their evaluations are aggregated with Eq. (5) with equal importance of DMs. Then, they are defuzzified to obtain the crisp values.
5. Eq. (6) is used to calculate their priorities. Then, they are normalized with Eq. (7) to obtain their ranking (Table 2).

Table 2. HoQ for hospital design.

	C1	C2	C3	C4	C5	C6	C7	C8	C9	C10	C11	C12	C13	C14	C15	Imp.
Imp.	0.07	0.08	0.05	0.05	0.08	0.02	0.05	0.08	0.05	0.08	0.07	0.08	0.08	0.07	0.05	
3rd Ex.	M	VH	L	H	VH	M	L	VH	H	VH	VH	VH	VH	M	L	
2nd Ex.	VH	VH	H	L	VH	N	H	VH	L	VH	M	VH	VH	VH	H	
1st Ex.	M	VH	M	M	VH	L	M	VH	M	VH	H	VH	VH	H	M	
DR24	N	N	N	H	L	N	N	N	N	N	N	N	H	N	N	0.0563
DR23	N	N	N	N	H	N	N	N	N	N	N	N	H	N	N	0.0507
DR22	N	N	N	N	N	N	VH	VH	N	N	N	N	N	N	N	0.0228
DR21	N	N	N	N	N	N	H	N	N	N	N	N	N	N	N	0.0210
DR20	N	N	N	H	N	N	N	N	N	N	N	N	N	N	N	0.0210
DR19	N	N	N	N	N	L	N	N	N	N	N	N	N	N	N	0.0192
DR18	VH	N	N	N	N	N	N	N	N	N	N	N	N	N	N	0.0222
DR17	VH	VH	VH	N	N	N	N	N	N	N	N	N	N	N	N	0.0228
DR16	H	M	M	N	M	L	N	N	L	N	N	N	N	M	N	0.0441
DR15	L	L	L	H	M	N	N	N	N	N	H	N	VH	N	N	0.0620
DR14	N	L	L	N	H	N	N	N	N	N	H	N	VH	N	H	0.0694
DR13	N	L	L	N	N	N	N	N	N	N	N	N	VH	N	N	0.0618
DR12	N	N	N	N	L	VH	N	N	N	N	N	N	N	L	N	0.0299
DR11	N	N	N	L	N	N	N	M	N	H	N	N	N	N	N	0.0209
DR10	N	N	N	N	N	N	N	N	N	N	H	N	N	N	H	0.0443
DR9	M	M	M	N	M	N	N	N	VH	N	N	N	N	N	N	0.0228
DR8	L	N	N	N	N	N	N	N	L	M	N	N	N	N	N	0.0199
DR7	N	N	N	VH	M	N	N	N	N	VH	N	N	N	N	N	0.0227
DR6	N	N	N	N	M	N	N	N	N	M	VH	VH	H	M	VH	0.0726
DR5	N	N	N	N	M	N	N	N	N	H	VH	VH	H	M	VH	0.0726
DR4	L	L	N	H	M	N	N	N	N	L	H	N	VH	M	H	0.0739
DR3	N	N	N	L	L	N	N	N	N	H	N	N	N	H	H	0.0623
DR2	N	N	N	N	N	N	N	N	N	N	VH	N	N	N	N	0.0226
DR1	N	N	N	N	L	N	N	N	N	H	N	N	N	H	H	0.0623

4. Results and Discussions

IVIF-integrated QFD is applied to a hospital design problem. Initially, CRs are defined. Then, DMs are given their relative importance.

Later, DRs are identified. Once CR-DR are assessed, their ranking is obtained. The ranking and importance of DRs are shown in Table 2. The most important three requirements are highlighted in red in Table 2. The most essential five requirements which should be considered at the first place to achieve smart hospitals are "Sensors," "IoT technologies," "Connection of building systems with each other," "Meters" and "Camera systems".

Table 2 provides three different experts' evaluations for each CR and their aggregated importance obtained from IVIF. Moreover, it provides the importance of DRs. The first expert's evaluation of the CR-DR relations are provided as an example in the same table.

When CRs' importance are investigated, the most important ones highlighted by DMs are energy and health-related requirements, as in Table 2. Their concordance becomes more visible when they are compared to the related importance of DRs. The most essential DRs are those requirements that have crucial roles in energy management systems and health-related systems. The model is successful to represent CRs on DRs for hospital design. Thanks to IVIF sets, non-linguistic data are used to better reflect DMs' knowledge on the decision-making process.

5. Conclusions

In this paper, the aim is to create a stable decision-making environment under vagueness and imprecision. For that propose, IVIF sets, which are an extended form of IF sets, are chosen to be integrated with an MCDM method.

As an MCDM technique, QFD is preferred due to its power to translate customer needs into engineering requirements. The same technique is applied to a hospital design problem, where QFD is chosen because smart requirements act like customer expectations that need to be heard in the buildings' design phase. This is a complicated process, where various criteria should be considered. While assessing these criteria, experts' opinions cannot always be precise and crisp. Therefore, the QFD method, integrated with IVIF sets, can create a flexible and reliable decision-making approach. To illustrate the plausibility of the proposed model, it is applied to a hospital design problem. Integration of QFD with IVIF empowered the ability of the method to handle uncertain linguistic data. In addition, IVIF-integrated QFD is employed as a novelty in the literature.

As a result, the model shows that linguistic variables can provide flexibility for DMs for their assessments and IVIF calculation for evaluations create consistent results.

For future studies, larger scales can be used for evaluations for each DM. Each DM can have different linguistic scales according to their experience on the subject. Different aggregation techniques can be applied to better reflect the group decision.

References

1. L. Zadeh, "Fuzzy Sets," *Information and Control*, vol. 8, no. 3, pp. 338-353, 1965.
2. Z. Xu and N. Zhao, "Information fusion for intuitionistic fuzzy decision making: An overview," *Information Fusion*, vol. 28, pp. 10-23, 2016.
3. K. Atanassov, "Intuitionistic fuzzy sets," *Fuzzy Sets and Systems*, vol. 20, no. 1, pp. 87–96, 1986.
4. J. Zhao, X. You, H. Liu and S. Wu, "An Extended VIKOR Method Using Intuitionistic Fuzzy Sets and Combination Weights for Supplier Selection," *Symmetry-Basel*, vol. 9, no. 9, 2017.
5. J. Ye, "Aggregation Operators of Trapezoidal Intuitionistic Fuzzy Sets to Multicriteria Decision Making," *International Journal of Intelligent Information Technologies*, vol. 13, no. 4, pp. 1-22, 2017.
6. W. Zhou and Z. Xu, "Extended Intuitionistic Fuzzy Sets Based on the Hesitant Fuzzy Membership and their Application in Decision Making with Risk Preference," *International Journal of Intelligent Systems*, vol. 33, no. 2, pp. 417-443, 2018.
7. K. Atanassov and G. Gargov, "Interval valued intuitionistic fuzzy sets," *Fuzzy Sets and Systems*, vol. 31, pp. 343-349, 1989.
8. G. Büyüközkan and F. F. O. Göçer, "Cloud Computing Technology Selection Based on Interval Valued Intuitionistic Fuzzy COPRAS," in *Advances in İntelligent Systems and Computing - EUSFLAT 2017*, Warsaw, 2018.
9. V. Nayagam, S. Jeevaraj and P. Dhanasekaran, "An intuitionistic fuzzy multi-criteria decision-making method based on non-hesitance score for interval-valued intuitionistic fuzzy sets," *Soft Computing*, vol. 21, no. 23, pp. 7077-7082, 2017.
10. S. Chen and W. Han, "A new multiattribute decision making method based on multiplication operations of interval-valued intuitionistic fuzzy values and linear programming methodology," *Information Sciences*, vol. 429, pp. 421-432, 2018.
11. S. Cheng, "Autocratic multiattribute group decision making for hotel location selection based on interval-valued intuitionistic fuzzy sets," *Information Sciences*, vol. 427, pp. 77-87, 2018.
12. Y. Akao, "Quality Function Deployment: Integrating Customer Requirements into Product Design (G.H. Mazur)," *Productivity Press*, 1990.
13. M. Li, L. Jin and J. Wang, "A new MCDM method combining QFD with TOPSIS for knowledge management system selection from the user's perspective in intuitionistic fuzzy environment," *Applied Soft Computing*, vol. 21, pp. 28-37, 2014.
14. C.-H. Wang, "An intuitionistic fuzzy set–based hybrid approach to the innovative design evaluation mode for green products," *Advances in Mechanical Engineering*, vol. 8, no. 4, pp. 1-16, 2016.

15. J. R. Hauser and D. Clausing, "The House of Quality," *Harvard Business Review*, May 1988.
16. Z. XU and X. Cai, Intuitionistic Fuzzy Information Aggregation: Theory and Applications, 2012.
17. L. Yu, L. Wang and Y. Bao1, "Technical attributes ratings in fuzzy QFD by integrating interval-valued intuitionistic fuzzy sets and Choquet integral," *Soft Comput*, vol. 22, pp. 2015–2024, 2016.

Part 9

Qualitative and Quantitative Decision Making in Management and Social Science

Causality and inference identification method of individual poverty*

Xiaohong Liu
College of Management, Southwest Minzu University, Chengdu, 610041, P.R. China

Xianyi Zeng
The ENSAIT Textile Institute, 9 rue de l'Ermitage, Roubaix, F-59100, France

Poverty and anti-poverty are difficult problems in the world. As we know, identifying the causes of poverty is a basic premise in the process of anti-poverty. Individual poverty is numerous and scattered, so it is the focus and difficulty of a country or region in anti-poverty. In order to improve the method of identifying individual poverty, the causality of individual poverty is put forward in this paper. According to the rough set principle, a method to identify individual poverty is established, that is that the causes of individual poverty are included in the decision information table, and the key causes leading to individual poverty are summed up by reducing the decision information table. This will provide useful information for accurate poverty reduction.

1. Introduction

Poverty and anti-poverty with the main history of human social development is a common global problem. China is the largest developing country in the world and once it is one of the most impoverished countries in the world. After forty years of hard work since 1978, China has made remarkable achievements in anti-poverty. In November 2013, Chinese national chairman Xi Jinping put forward the concept and requirements of *precision poverty alleviation*. In January 2015, Xi Jinping re-emphasized the concept of *precision poverty alleviation* and *precision poverty reduction*. On the issue of anti-poverty, China's national practice and successful experience based on the concept of precision poverty alleviation and precision detachment have in fact provided an alternative Chinese wisdom and Chinese plan for the world anti-poverty problem. The causes of poverty all over the world are highly complicated, and generally include many uncertain factors, such as history and reality, humanities

---

*This work is partially supported by National Social Science Fund Project of China (grant No. 15BGL209), State Bureau of Foreign Expert's Affairs Project of China (grant No. 2017-031), and Special fund project of basic scientific research expenses in central universities of Southwest Minzu University (grant No. 2018SZD04).

and nature, development and mutation, individuals and families. In a certain region and period, correctly identifying the causes of poverty is an important prerequisite for the country to effectively carry out anti-poverty. Therefore, the study of the causes of poverty has important theoretical and practical significance.

Scholars have been studying poverty for a long time. The British scholar Benjamin (1901) began to define poverty in Britain by income. American economist Mollie Orshanskych (1963) began to define poverty in the United States by income. In 1981, the World Bank began to measure the consumption and income poverty of all developing countries. Amartya Senn (1982) proposed the theory of capability approach to poverty and won the Nobel Prize in economics in 1998. In 1990, the United Nations Development Programme's Human Development report published its first HDI index. In 2010, the Human Development report of the United Nations Development Program published the Multidimensional Poverty Index (MPI) for the first time, which expanded the theory of human development to measure poverty. Harrell Rodgers (2012) analyzes the changes of the poverty line in the United States, the causes of poverty and the related theories of poverty, and the welfare legislation of successive American governments to solve the problem of poverty. For China's poverty and anti-poverty issues, scholars have carried out a more systematic study, such as Xi Jinping's thoughts on poverty alleviation (Lei Ming Li, 2018), precision poverty alleviation (Zhuang Tianhui, etc., 2016), path Choice of precise Poverty Eradication based on shared Development Vision (Li Peng, 2017) and so on. The existing research results not only prove the importance and necessity of carrying out accurate and accurate poverty alleviation in the process of anti-poverty in China, but also explain the complexity and difficulty of the anti-poverty problem.

This paper mainly consists of four parts. According to the two-dimensional relationship between poverty itself and poverty identification, the possibility of error recognition and its adverse effects are analyzed. Secondly, the causality of individual poverty is put forward from the perspective of social system, which provides a logical framework for analyzing the causes of individual poverty. Thirdly, with the help of rough set theory, a reasoning method of individual poverty formation is established, the main purpose of which is to improve the efficiency of individual poverty identification. Finally, an example is given to illustrate the feasibility of the reasoning method for individual poverty identification.

2. Background of the Issue

At the Millennium Summit on September 2000, world leaders made a commitment to eradicate poverty. The United Nations believes that *internal and external troubles* are the main causes of extreme poverty in these countries.

In short, the world faces not only an extremely severe anti-poverty situation, but also a long way to go. Among them, poverty causes identification is the premise of effective anti-poverty. This paper is mainly based on the following main problems: there are truth and falsehood from poverty itself, and the recognition results are true and false. Therefore, the relationship between poverty itself and poverty identification results is as shown in Figure 1.

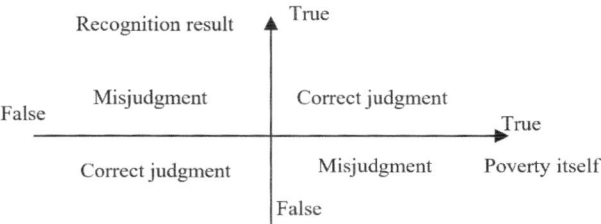

Fig. 1. The relationship between poverty itself and poverty identification results.

According to Figure 1, poverty recognition includes correct judgment and wrong judgment, and there are four kinds of results:

Correct judgment: (1) correct acceptance, i.e. true-true poverty; (2) correct rejection, i.e. false-false poverty.

Misjudgment: (1) error acceptance, i.e. true-false poverty; (2) error rejection, i.e. false- true poverty.

The practical problems that need to be pointed out are: (1) error acceptance will lead to waste of resources for accurate poverty alleviation and accurate poverty eradication; (2) error rejection will result in omission of the object of poverty alleviation.

In particular, misjudgment is more likely to occur in the case of family or individual poverty, which involves a large number of scattered families. Therefore, it is necessary to study how to improve the accuracy and effectiveness of poverty identification from the perspective of social system and with the aid of reasoning, so as to reduce the incidence of poverty recognition errors.

3. Causality of Individual Poverty

Individual poverty is a result of many factors, such as families or individuals, regions and emergencies in a certain period of time. Therefore, the main factors affecting family or personal poverty include internal factors, external factors and cross factors. The internal factors mainly include personal factors and family factors, the external factors are regional humanistic factors and regional natural factors, and the cross factors mainly refer to emergencies. The factors contributing to the formation of individual poverty are shown in Figure 2:

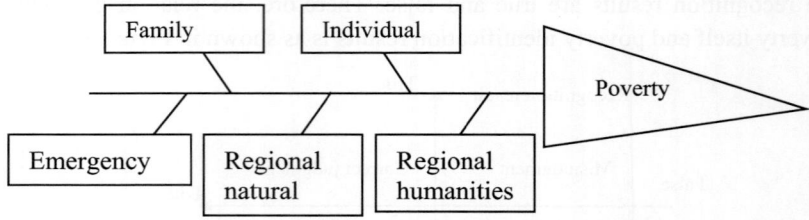

Fig. 2. Factors affecting the formation of individual poverty.

According to Figure 2, the main factors affecting the formation of individual poverty are:

Personal factors include: age, sex, health, education, skills, industry, means of production and so on.

Family factors include: parental age, parental relationship, parental education, parental occupation, family education, number of members, means of production, family events and so on.

Regional humanistic factors are political, policy, economic, cultural, social, technical and legal factors that affect the development of a region.

Regional natural factors refer to the natural environment, such as altitude, temperature polarization, soil quality, average precipitation and per capita cultivated land (grassland), which are formed naturally in a region.

Emergency: refers to the family or the personal external influence causes the poverty emergency, mainly includes the natural disaster, the war, the traffic accident, the sudden illness and the force majeure accident and so on.

4. A Reasoning Model for the Formation of Individual Poverty

4.1. *Rational representation of individual poverty formation*

Let the evaluation set of individual poverty be P, a partial order expressed by Likert's five-level quantization, that is,

$$P = \{p_1, p_2, p_3, p_4, p_5\}, p_1 \succ p_2 \succ p_3 \succ p_4 \succ p_5$$

The five factors that affect individual poverty, such as individual factors, family factors, regional human factors, regional natural factors and cross-factors, constitute the matrix of primary evaluation index, using V_1, V_2, V_3, V_4 and V_5 to denote respectively.

$$V_i = \begin{pmatrix} v_{i1} \\ v_{i2} \\ \dots \\ v_{ik_i} \end{pmatrix}, i = 1,2,3,4,5$$

k_i is the number of secondary indexes corresponding to the ith first grade evaluation index.

Then the inferential concept of the degree of poverty in individual j can be expressed as:

$$F_j : V_1 \times V_2 \times V_3 \times V_4 \times V_5 \to P$$

In general, the following reasoning relationships illustrate the effects of individual poverty factors:

(1) The factors influencing the formation of individual poverty are significant:

$$V_1 \times V_2 \times V_3 \times V_4 \times V_5 \to p_1$$

(2) The factors influencing the formation of individual poverty are obviously related to:

$$V_1 \times V_2 \times V_3 \times V_4 \times V_5 \to p_2$$

(3) The factors influencing the formation of individual poverty are uncertain:

$$V_1 \times V_2 \times V_3 \times V_4 \times V_5 \to p_3$$

(4) There is no significant relationship between the factors of individual poverty formation and the formation of individual poverty:

$$V_1 \times V_2 \times V_3 \times V_4 \times V_5 \to p_4$$

(5) The factors that affect the formation of individual poverty have no effect on it and can be completely deleted:

$$V_1 \times V_2 \times V_3 \times V_4 \times V_5 \to p_5$$

4.2. *Rules of reasoning*

Based on rough set theory, the decision information table and inference rules of individual poverty forming influence relation are established.

The 1st step is to construct the original decision information table. The conditional attributes of individual poverty are V_1, V_2, V_3, V_4 and V_5 respectively, and the evaluation set of individual poverty is P. The decision information about the relationship between individual poverty and poverty is shown in Table 1.

Table 1. Decision-making information table for the impact of individual poverty.

U \ A	V_1		V_2		V_3		V_4		V_5		P
	v_{11}	v_{1k1}	v_{21}	v_{2k2}	v_{31}	v_{3k3}	v_{41}	v_{4k4}	v_{51}	v_{5k5}	
u_1	5	5	5	5	5	5	5	5	5	5	p_1
...											
u_n	1	1	1	1	1	1	1	1	1	1	P_5

The 2nd step is to determine the conditional attribute values of the decision information table. The conditional attributes were discredited according to the Lekert five scales and the preference relationship of the decrease of satisfaction degree (or degree of achievement) from left to right and according to the principle of maximum membership degree. A decision information table with the numbers 1,2,3,4 and 5 to indicate satisfaction (or degree of achievement) from low to high is established.

For example, the discretization of the age evaluation index of individual factors, according to the principle of 30-40 years as the best age, the discretization information of the age is established. The age is divided into 20-30, 30-40, 40-50, 50-60, over 60, the corresponding evaluation result of v_{11} is 4, 5, 3, 2, 1.

A similar approach can be taken to deal with discrete information of other indicators.

The 3rd step is to reduce the decision information table. According to rough set method, the redundant lines of repeated conditions and decision information are first deleted. Secondly, the kernel set of the decision table is obtained by deleting the columns which cannot support the decision table. Thirdly, the attribute value reduction, the decision rules are analyzed one by one, the decisive attribute value is selected, and the redundant attribute value is removed.

The 4th step is to establish the inference rules. According to the decision attributes, the inference rules obtained in the third step are further sorted out, and the inference rules which accord with the characteristics of the object set are established.

5. A Numerical Example

Generally speaking, for a region where political economy and society are relatively stable, its humanistic and natural factors have a relatively stable impact on the formation of individual poverty in this region. It may not be considered when assessing the impact of individual poverty. This example only discusses the influence of individual factors on the formation of individual poverty.

Table 2. The original decision table of individual factors of individual poverty.

U \ A	V_1							P
	V_{11}	V_{12}	V_{13}	V_{14}	V_{15}	V_{16}	V_{17}	
1	1	2	3	1	4	5	1	2
2	1	2	4	2	3	4	2	3
3	2	3	5	3	4	3	3	4
4	3	4	1	4	1	3	4	3
5	4	5	2	5	5	2	5	4
6	5	2	3	5	4	1	5	5
7	1	2	4	4	5	1	4	3
8	1	1	5	3	4	3	3	4
9	2	1	1	2	4	4	2	3
10	3	5	2	1	2	5	1	4

By reducing Table 2, the decision table after the reduction of individual poverty caused by individual factors in Table 3, and the relative advance rules are established.

Table 3. Decision-making table of individual poverty reduction.

U \ A	V_1							P
	V_{11}	V_{12}	V_{13}	V_{14}	V_{15}	V_{16}	V_{17}	
3	-	-	5	-	-	-	-	4
5	-	-	-	5	5	-	-	4
6	5	-	-	5	-	-	-	5
10	-	-	-	-	-	5	-	4

The reasoning rules are:

$$v_{11}v_{14} \rightarrow p_1$$
$$v_{13} \vee v_{14}v_{15} \vee v_{16} \rightarrow p_2$$

In this example, first of all, it shows that age and education is the significant cause of individual poverty in the sample; Secondly, health, education and skills, industriousness, are the main causes of individual poverty.

6. Conclusion

Up to now, the quantitative basis of poverty determination is mainly disposable income, but it is relatively difficult to identify the qualitative basis of ability or opportunity. Therefore, the identification method of individual poverty formation needs to be improved. We proposed a method to identify the causes of individual poverty, which aimed to attract attention to the study and analysis of the internal and external causes of poverty. The research in this paper will help to provide useful premise information for accurate poverty alleviation and precision poverty reduction. Because of the complexity and arduousness of anti-poverty, it is necessary to study the coordination of anti-poverty policy, the main body of participation, the effective path and the actual effect.

References

1. D. Tandia and M. Havard, The evolution of thinking about poverty: exploring the interactions, Working Papers 55.6 (1999):957-63.
2. Rodgers, Harrell R. American poverty in a new era of reform. M.E. Sharpe, 2000.
3. M.E. Santos and K. Ura, Multidimensional Poverty in Bhutan: Estimates and Policy Implications, Ophi Working Papers (2008).
4. F. Bourguignon and S. R. Chakravarty, the Measurement of Multidimensional Poverty, DELTA (Ecole normale supérieure), 2008:25-49.
5. A. Sabina and X.L. Wang, Measurement of Multidimensional Poverty in China: Estimation and Policy Implications, *China Agricultural University Journal of Social Sciences Edition* (2013):4-10.
6. Y.Q. Tan, Z. Wang, and Y. Chen, Measurement, Decomposition, and Policy Implications of Multidimensional Poverty in Wuling Mountainous Areas, *Journal of Jishou University* (2015).
7. S. Alkire and M. E. Santos, Acute Multidimensional Poverty: A New Index for Developing Countries, Queen Elizabeth House, University of Oxford, 2011.
8. W.L. Hanandita and G. Tampubolon. Multidimensional Poverty in Indonesia: Trend Over the Last Decade (2003–2013), Social *Indicators Research* 128.2(2016):559-587.
9. S.K. Mohanty et al., Multidimensional Poverty in Mountainous Regions: Shan and Chin in Myanmar, *Social Indicators Research* 3(2017):1-22.

The influence of ecological protection in poverty-stricken areas of western China*

Haozhen Liu

School of Computer and Communication, Lanzhou University of Technology Lanzhou, 730050, P.R. China

Bin Luo

College of Chemistry and Environment, Southwest Minzu University Chengdu, 610041, P.R. China

Accelerating the coordinated development in poverty-stricken areas is a difficult problem which we face in the today's world. In the course of building a moderately prosperous society in an all-round way in 2020, on the one hand the poverty-stricken areas in western China are faced with the ecological pressure of getting rid of poverty accurately, and on the other hand, they should be focused on the long-term ecological goal of the construction of beautiful ecological civilization. On the basis of the analysis of the requirements for coordinated development of regional socio-economic ecosystems, this paper presents the problem of ecological protection in the poverty-stricken areas in western China. According to the requirements for coordinated development, this paper discusses the influence relationship of ecological protection in the poverty-stricken areas of western China, which provides a new perspective for ecological protection.

1. Introduction

In October 2017, the 19th National Congress of the Communist Party of China proposed that socialism with Chinese characteristics has entered a new era. The main contradiction in Chinese society in the new era is the contradiction between the people's growing need for a better life and imbalanced and inadequate development. Because of the influence of Chinese modern history, regional natural environment, as well as the influence of the national development strategy of "let some people get rich first, lead the rich before the rich", there is some poverty-stricken areas in western China. In 2020, the poverty-stricken areas in western China became the key and difficult areas to win the well-to-do society in an all-round way. These areas have many demands

---

*This work is partially supported by National Social Science Fund Project of China (grant No. 15BGL209).

for improving the people's livelihood, a single mode of economic development, and great pressure of ecological protection. The socioeconomic ecosystem is relatively fragile, and the task of coordinated development of the socioeconomic ecosystem is arduous.

From different perspectives, scholars have carried out systematic researches on the concept, connotation and requirements of China's precision poverty alleviation, as well as the policies, contents and applications of ecological protection in western China's ethnic minority areas. there are the following documents: connotation of precision poverty alleviation (Zhuang Tianhui et al., 2016), the evaluation and countermeasures of the process of precision poverty alleviation in minority areas of China (Yang Hao et al., 2016), precision poverty alleviation and poverty relief in Tibet and four provinces of China (Li Hong, 2016), application of modern ecological protection concept in ecological protection (Wang Lixia, 2017), the protection of the source of Sanjiang in China (Shao Quanqin et al., 2017; Li Fen et al., 2017) and so on. The existing research results not only prove the importance and necessity of ecological protection in the process of speeding up the precision poverty alleviation in the poverty-stricken areas of western China, but also explain the complexity and arduousness of the ecological protection problems in the poverty-stricken areas of western China. Ecological protection in poverty-stricken areas in western China should be placed within the framework of socio-economic ecosystems, but we should be faced with two basic problems. The first problem is that the existing socio-economic ecosystems in the areas are not only weak in self-adaptive ability, but also have poor internal coordination and stability. The second problem is that in the process of achieving the goal of poverty alleviation and in the process of realizing the goal of synchronizing the development of a well-off society with the whole country, it will have a new impact on the socio-economic ecosystem. In particular, it will have new adverse effects on ecological subsystems, which will lead to new problems and requirements in ecological protection in the region.

On the basis of the analysis of the concept of coordinated development of the regional socioeconomic ecosystem, this paper puts forward the problem of ecological protection in the poverty-stricken areas of western China. Based on the coordination of socio-economic ecosystems, this paper discusses the dynamic model of ecological protection in the poverty-stricken areas of western China, which provides a new analytical perspective for strengthening ecological protection in this region.

2. The Requirement of Harmonious Development of Regional Socio-Economic Ecosystem

According to the view of social system theory, any region has its own specific socio-economic ecosystem, this system in a certain human environment (including national policies, economic development, social concepts, technology development and law) and natural conditions, has its specific ability to adapt to the environment and coordination and stability within the system.

The theory of coordination and coordinated development in the theory of systems science is mainly focused on the methodology of coordination. General System Theory holds that the system is an organic whole with certain functions, which is composed of several elements connected in a certain form. With the basic characteristic of "the whole is greater than the sum of the parts", it emphasizes the whole, open and dynamic principle of the system, which is the internal structure of the system in order to understand and coordinate the development from the whole and the whole. It provides new ideas, methods and analytical framework for coordination and development from the perspective of the relationship between function and system and external.

The regional socio-economic ecosystem consists of three subsystems: social, economic and ecological. The core functions of the three subsystems are livelihood improvement, endogenous development and environmental protection. The coordinated development between them is shown in Figure 1:

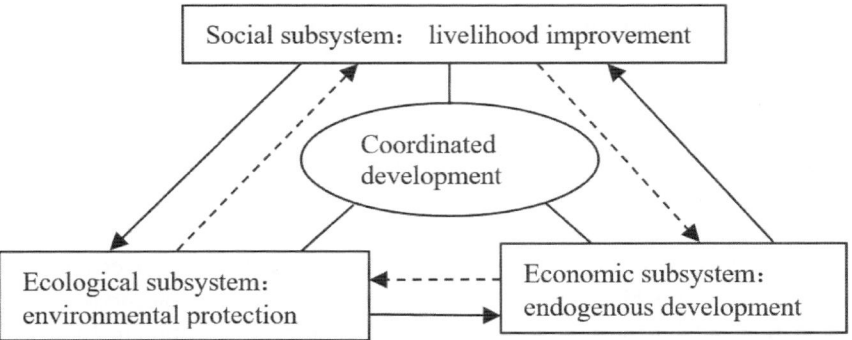

Figure 1. Conceptual model of regional socio-economic ecosystem.

In short, the requirements for coordinated development of the socio-economic ecosystems in the poverty-stricken areas of western China are: The general goal of regional development is to realize the overall development of human being. On the basis of following the objective law of development, two kinds of coordination relations are established: The first is the coordination relationship among the social, economic and ecological subsystems and the overall regional development system. The two is the mutual coordination of each subsystem and coordination relationship of internal elements of the subsystem.

3. The Pressure of Ecological Protection in the Poverty-stricken Areas of Western China

At present and in the next period of time, China's poverty-stricken areas are mainly concentrated in the western region. The major objective causes of poverty in these areas are the leapfrogging development of historical stage, the bad natural conditions and the destruction of ecological environment. The western part of China consists of 12 provinces, municipalities and autonomous regions, namely five provinces and regions in Southwest China (Sichuan, Yunnan, Guizhou, Tibet, Chongqing) and five provinces and regions in Northwest China (Shanxi, Gansu, Qinghai, Xinjiang, Ningxia) and Inner Mongolia, Guangxi. The total area of the western part of China is about 6.86 million square kilometers, accounting for about 72% of the total area of China. The desert, Gobi and desertification in western China cover a total area of 153.3 million square kilometers. An analysis of the causes of desertification in China shows that the rapid spread of decertified land is mainly due to unreasonable activities in human history, such as excessive agriculture, overgrazing, over cutting and improper utilization of water resource. If special protective measures are not taken, desertification in these areas will continue to develop.

As a result, precision poverty alleviation in western China is faced with unfavorable natural conditions and ecological environment. If the work of precision poverty alleviation exceeds the carrying capacity of the ecological subsystem, it will not only greatly reduce the actual effect of the work of precision poverty alleviation, but also may cause new adverse effects on the ecological environment.

According to China's national development strategy, the poverty-stricken areas in western China will be faced with the dual tasks of accurate poverty alleviation and the construction of beautiful ecological civilization in 2020. This is not only a severe and realistic challenge to the ecological protection of the region, but also a rare historical opportunity to promote the regional ecological protection.

4. A Method of Expressing the Output of the Socio-economic Ecosystem in the Poverty-stricken Areas of Western China

In order to establish a comparable relationship among social, economic and ecological subsystems in a socio-economic ecosystem, we propose a representation method of the output results of the three systems based on fuzzy sets and use the triangular principle to represent the coordination relationship model among social, economic and ecological subsystems.

Set the output domain of social subsystem is S, $S \in [0, +\infty)$. The fuzzy membership function of the total output of the social subsystem is $s(x)$, $s(x) \in [0,1]$. According to the law of diminishing marginal utility in economics, the marginal utility of social subsystem output is $ms(i)$ ($ms(i) \in [0,1]$), as shown in Figure 2.

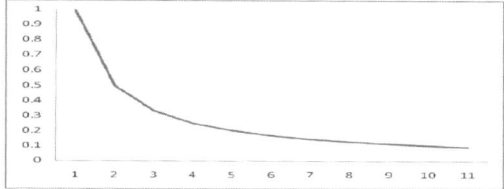

Figure 2. Marginal utility function of social subsystem output.

And has:

$$ms(i) = \frac{1}{i}, i \in N. \tag{1}$$

$$s(x) = \frac{\sum_{i=1}^{x} ms(i)}{\sum_{i=1}^{n} ms(i)}, x = 1,2,\ldots, n \tag{2}$$

Set the domain of economic subsystem is E, $E \in [0,+\infty)$. The fuzzy membership function of the total output of the economic subsystem is $e(y)$, $e(y) \in [0,1]$. The income sources of the economic subsystem include the region itself and the external transfer, set up E_i and set up E_o respectively. Set the degree of support of economic subsystem to social development and ecological protection is $es(y)$ and $ec(y)$ respectively, $es(y) \in [0,1]$, $ec(y) \in [0,1]$

$$E = E_i + E_o \tag{3}$$

$$e(y) = es(y) \circ ec(y) \tag{4}$$

In type (4), $\circ$ is a fuzzy set operator.

Set the domain of the ecologic subsystem is C, $C \in [0,+\infty)$. The fuzzy membership function of the total output of ecological subsystem is $c(z)$, $c(z) \in [0,1]$. Ecological subsystem income sources include existing and post development output in the region, set is C_i and C_o respectively. The ecological subsystem includes the existing ecological quality and the changing ecological quality, set is $cp(z)$ and $cf(z)$ respectively, $cp(z) \in [0,1]$, $cf(z) \in [0,1]$.

$$C = C_i \times C_o \tag{5}$$

$$c(z) = cp(z) \circ cf(z) \tag{6}$$

5. The Influence model of Ecological Protection in the Poverty-stricken Areas of Western China

5.1. Basic model

After a long period of natural evolution and normal development, the socio-economic ecosystems in the poverty-stricken areas of western China have the attribute of coordinated development. The main performance is that three subsystems of the social, economic and ecological can take into account their respective development and mutually support development. In this regard, the triangular principle is used to express this relationship. The fuzzy membership function values produced by the social, economic and ecological subsystems can form a triangle, as shown in Figure 3:

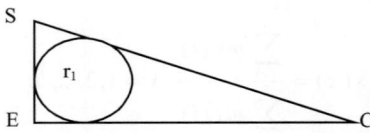

Figure 3. Coordination of social and economic ecosystems in poor areas of western China.

In Figure 3, the ΔSEC indicates that the socio-economic ecosystems of the poverty-stricken areas in western China have coordination, and the three sides of the triangle SC, EC and SE is the output values of the three subsystems, namely, social, economic and ecological, respectively.

5.2. Main strategies

In accordance with the new requirements for building a beautiful China, the poverty-stricken areas in western China bears the important new task of ecosystem protection. There are not only the sources of the Yangtze River, the

Yellow River and Lancang River, but also the centralized Nature reserves in China. Therefore, the task of ecological construction here is quiet arduous.

In general, the root of poverty in the poor areas of western China is the lag of development, which mainly includes the poor accumulation of material in history, the weak economic base, and the inadequacy of endogenous development ability and the man-made destruction of ecology and so on. Therefore, anti-poverty and regional coordinated development is intrinsically linked.

Poverty-stricken areas in western China are faced with the dual tasks of precision poverty alleviation and the construction of beautiful ecological civilization. In the short term, there is a conflict between precision poverty alleviation and the construction of beautiful ecological civilization. Precision poverty alleviation may increase the stress of regional ecological environment. However, construction of beautiful ecological civilization needs to maintain or increase the carrying capacity of the regional ecological environment. In order to alleviate the short-term contradiction between precision poverty alleviation and the construction of beautiful ecological civilization, the state needs to further increase preferential policies and financial support, and help to improve the people's livelihood and to make new breakthroughs in the ecological protection in the poverty-stricken areas of western China.

The realization of regional coordinated development is not only the key to solve the poverty problem caused by lagging development, but also an effective way to solve the precision poverty alleviation and the construction of beautiful ecological civilization. From the historical experience, it is difficult to achieve coordinated development by mainly relying on the regional economic capacity and policy coordination. Therefore, it is suggested that the state further perfect the ecological compensation system, especially increase the financial support and policy coordination, which will be an important system guarantee for the task of realizing the beautiful ecological construction in the poverty-stricken areas of western China.

6. Conclusion

It is an important political task to get rid of poverty accurately in the short-term in the poverty-stricken areas of western China. Therefore, the social subsystems of these regions have the strong function of guidance and goal. Because of the strong support of the central government and local governments at all levels, it is easy to ignore the role of ecological subsystem. Based on this, this paper presents: according to China's goal of winning an all-around well-off society in 2020, combining with the new requirements for dual tasks of precision poverty alleviation and beautiful ecological construction in the poverty-stricken areas of western China, promoting the coordinated development of the social system and the ecosystem is the strategy of laying the foundation and managing the long-term. Because of the complexity of social system and ecosystem and the gradual change of function and output of ecological subsystem, it is necessary to study the policy coordination, effective path and effect evaluation of ecological protection systematically.

References

1. Matthew J. Perkins et al., Conserving intertidal habitats: What is the potential of ecological engineering to mitigate impacts of coastal structures? Estuarine, Coastal and Shelf Science 167(2015):504-515.
2. M.H. Gong et al., Delineating the ecological conservation redlines based on the persistence of key species: Giant pandas (Ailuropoda melanoleuca)) inhabiting the Qinling Mountains, *Ecological Modelling* 345 (2017):56-62.
3. Brèchignac, F., and M. Doi, Challenging the current strategy of radiological protection of the environment: arguments for an ecosystem approach, *Journal of Environmental Radioactivity* 100.12(2009):1125.
4. X.L. Wang, Y.Q. Yangin and C.Y. Zeng, Definition and Division of Ecological Environmental Stakeholders: Based on the Perspective of Environmental Externality, *Yunnan Environmental Science*, 32, 3(2013).
5. D.C. Cui and X.D. Hao, Research of Public Participation Method of Environmental Impact Assessment Based on Stakeholder Theory, *Environmental Protection Science*, 41, 3(2015).
6. J. P. Arrigoni1, G. Paladino and F. Laos, Feasibility and Performance Evaluation of Different Low-Tech Composter Prototypes, *International Journal of Environmental Protection*, 5, 1(2015).
7. Y.M. Li, and H. Qin, Ecological Conservation and Construction in China: Progress and Situation, *Modern Economy*, 100.10(2014):5938-43.

An interval valued intuitionistic fuzzy location based recommendation system utilizing social platforms

S. Ceren Öner
Istanbul Technical University, Industrial Engineering Department, Turkey

Başar Oztaysi
Istanbul Technical University, Industrial Engineering Department, Turkey

Mahir Öner
Istanbul Technical University, Industrial Engineering Department, Turkey

The revealing of advancements in mobile technologies and wide application of location based services have derived the emergence of location based services considering user similarities. The demand for location based systems integration with recommendation systems is extracted as a need of understanding the process of customer preferences considering location, time and actual needs. The needs can be derived from consumer life style, demographical information, consumption behavior and reaction to previously sent messages. On the other hand, these factors don't solely reflect the final purchasing decision which can cause imprecise environment for recommendation systems. Thus, researchers try to search other indicators that can reflect customer characteristics such as geographical data, digital participation in social media and search history of products for better understanding of the changes in customers purchasing tendency. Thus, in this paper, an intuitionistic fuzzy set theory based recommendation system is constructed by integrating three widely used social platforms: Trip Advisor, Zomato and Foursquare to implement restaurant offerings to proper social platform users. First, a sentiment analysis is adapted to selected restaurants and number of negative, positive and neutral comments are gathered. After that, restaurant and location information are emerged according to several criteria and user and location clustering are adapted separately via fuzzy clustering. Finally, multi criteria interval valued intuitionistic fuzzy recommendation system is adapted for restaurant recommendations to similar customer groups.

1. Introduction

Because of the increasing penetration of mobile technologies such as GPS and Bluetooth 4.0, location based systems can reflect users' private concerns in a specific location for a specific time [7]. Thus, location based systems aim to provide content providers to send proper services to customers when they visit a specific location at a specific time [6]. The main concepts of these services are tagging, tracking, navigating and mobile commerce. All these services rely on consumer interests for the developing of targeted location based systems.

In recent years, location determination and prediction of future visits have become important and integrated with recommender systems in targeted advertisement management, especially making private suggestions which can improve companies' communication channels to customers with constricted advertisement budgets [5]. Location based systems imply on time and preference based services especially following up the current position of mobile user and defining proper suggestions by utilizing possible visiting places gathered from previously visited places [8]. On the other hand, customers are not generally interested in reading or looking at the instant messages because of previously sent substantial number of irrelevant messages or redundancy of the message context [9]. The other reasons of not following these instant messages are location privacy concerns as a significant barrier to the penetration of location based services and erroneous keystrokes [10]. Additionally, user interests and needs continuously change and differentiation of these needs and interests can be problematic to perform efficient recommender systems. Thus, companies are prone to search and apply more efficient approaches that provide both personalized contexts and broad alternatives to their recommender systems. To overcome irrelevancy problem, it is suggested that location visiting information for recommender systems is necessary for the determination of customer tendency to buy goods. For this reason, possible location determination based recommendation systems exist as the most substantial topic for evaluating user visiting tendency as a characteristic of customer preferences or needs [7].

In this respect, the main research question of this study is constructing an intuitionistic fuzzy set theory adaptation to recommendation systems by integrating three widely used social platforms, Zomato, Trip Advisor and Foursquare to implement restaurant offerings to proper social platform users. First, a sentiment analysis is adapted to selected restaurants and number of negative, positive and neutral comments are gathered. After that, restaurant and location information are emerged according to several criteria and user and location clustering are adapted separately. Finally, multi criteria interval valued intuitionistic fuzzy recommendation system is adapted for restaurant recommendations to similar customer groups. This approach will enable social media effect to location based systems using imprecise information.

2. Interval Valued Intuitionistic Fuzzy Sets and Interval Valued Intuitionistic Fuzzy Location Based System Methodology

2.1. *Preliminaries*

The idea of intuitionistic fuzzy sets (IFSs) is introduced by Atanassov, when a non-membership degree is assigned to each element of the set. The interval-valued intuitionistic fuzzy set (IVIFS) is an extended form of IFSs. Let $D[0,1]$ be a subinterval of interval $[0,1]$ and let X be a given set which is not an empty set. An interval valued intuitionistic fuzzy set in X can be defined as $\tilde{S} = \{\langle (x, \mu_{\tilde{s}}(x), v_{\tilde{s}}(x) | x \in X) \rangle\}$ that $\mu_{\tilde{s}}: X \to D[0,1]$ is the membership degree and $v_{\tilde{s}}: X \to D[0,1]$ is the non-membership degree of the element of x considering $0 < \mu_{\tilde{s}}(x) + v_{\tilde{s}}(x) \leq 1$. Here, $\mu_{\tilde{s}}(x)$ and $v_{\tilde{s}}(x)$ are closed intervals with lower and upper bounds as $\mu_{\tilde{s}}(x)^-$, $\mu_{\tilde{s}}(x)^+$ for $\mu_{\tilde{s}}(x)$ and also for $v_{\tilde{s}}(x)$ as $v_{\tilde{s}}(x)^-$ and $v_{\tilde{s}}(x)^+$ where $0 < \mu_{\tilde{s}}(x)^+ + v_{\tilde{s}}(x)^+ \leq 1$ and $\mu_{\tilde{s}}(x)^-, v_{\tilde{s}}(x)^- \geq 0$. Thus, the distance measure between $\tilde{S}_1 = ([a_1, b_1], [c_1, d_1])$ and $\tilde{S}_2 = ([a_2, b_2], [c_2, d_2])$ is given in the following:

$$d(\tilde{S}_1, \tilde{S}_2) = \frac{1}{4}(|a_1 - a_2| + |b_1 - b_2| + |c_1 - c_2| + |d_1 - d_2|) \quad (1)$$

such that $0 \leq d(\tilde{S}_1, \tilde{S}_2) \leq 1$.

Let $X = (\tilde{x}_{ij})_{mxn}$ be a matrix. If all $\tilde{x}_{ij}$s are IVIFNs for $i=1,2,\ldots,m$ and $j=1,2,\ldots,n$ as $b_{ij} + d_{ij} \leq 1$, then this matrix will be called interval valued intuitionistic fuzzy matrix (IVIFM). The distance measure between X_1 and X_2 can be defined as follows:

$$d(X_1, X_2) = \frac{1}{4mn} \sum_{i=1}^{m} \sum_{j=1}^{n} (|a_{ij}^{(1)} - a_{ij}^{(2)}| + |b_{ij}^{(1)} - b_{ij}^{(2)}| + |c_{ij}^{(1)} - c_{ij}^{(2)}| + |d_{ij}^{(1)} - d_{ij}^{(2)}|) \quad (2)$$

such that $0 \leq d(X_1, X_2) \leq 1$.

The formula of similarity measure for IVIFNs $\tilde{S}_1, \tilde{S}_2$ can be defined as below:

$$sim(\tilde{S}_1, \tilde{S}_2) = \frac{d(\tilde{S}_1, \tilde{S}_2^c)}{d(\tilde{S}_1, \tilde{S}_2) + d(\tilde{S}_1, \tilde{S}_2^c)} \quad (3)$$

where $\tilde{S}_2^c = ([c_2, d_2], [a_2, b_2])$ and $0 \leq sim(\tilde{S}_1, \tilde{S}_2) \leq 1$.

In similar manner, the similarity measure can be modified for IVIFN $\tilde{x}_{ij}^{(1)}$ and $\tilde{x}_{ij}^{(2)}$ as below:

$$sim(X_1, X_2) = \frac{\sum_{i=1}^{m} \sum_{j=1}^{n} d\left(\tilde{x}_{ij}^{(1)}, \left(\tilde{x}_{ij}^{(2)^c}\right)\right)}{\sum_{i=1}^{m} \sum_{j=1}^{n} d\left(\tilde{x}_{ij}^{(1)}, \tilde{x}_{ij}^{(2)}\right) + \sum_{i=1}^{m} \sum_{j=1}^{n} d\left(\tilde{x}_{ij}^{(1)}, \left(\tilde{x}_{ij}^{(2)^c}\right)\right)}$$

(4)

where, $\tilde{x}_{ij}^{(2)^c} = \left(\left[c_{ij}^2, d_{ij}^2\right], \left[a_{ij}^2, b_{ij}^2\right]\right)$ and $0 \leq sim(X_1, X_2) \leq 1$ [3].

2.2. Presented methodology

In this work, data is needed to be transformed to interval valued intuitionistic fuzzy number in order to apply the recommended algorithm. Additionally, text reviews and comments should be separated as negative, positive and neutral comments and these comments should be recorded as "number of positive comments", "number of negative comments" and "number of neutral comments" according to each restaurant.

2.2.1. Sentiment analysis

Sentiment Analysis, in other name opinion mining, is the process of specifying whether a written text contains positive, negative or neutral comments in order to derive the opinion or attitude of a user about a particular topic. First, the corpus which is a large and structured set of texts that are prepared for further analysis, is obtained with text cleaning process. Secondly, associating a polarity (positive or negative) to each word and query both the synonyms and antonym should be processed by seed words which could be described by user or automatically. After that, converting the unstructured data into term-document matrix (TDM) which is a structured for the representation of the corpus is adapted. The TDM is a matrix form in which the documents represented by the rows and terms represented by the columns. After the creation of the TDM and dealing with the dimensionality problem, the next task is to extract novel and beneficial patterns from this matrix. In order to determine the themes of comments, classification and clustering methods could be used as used in this study, k-means clustering. Finally, number of positive, negative and neutral comments could be gathered [5].

2.2.2. Multi criteria interval valued intuitionistic fuzzy recommendation system

Definition 1. *Multi criteria Interval valued Intuitionistic Fuzzy recommendation system (IVIFRS)*: Let P, S, D be users, locations and restaurants respectively. Each user P_i ($\forall\, i \in \{1, ..., n\}$) and location S_j ($\forall\, j \in \{1, ..., m\}$) are assumed to have the same features. Recommendation system matching for each user to a restaurant will be based on X and Y which consist of s interval valued intuitionistic linguistic labels. Thus, restaurants, D_i ($\forall i \in \{1, ..., k\}$) will have s interval valued intuitionistic linguistic labels. The utility function, T is a mapping specified on (X, Y) as follows:

$$R: X \times Y \rightarrow D_1 \times D_2 \times ... \times D_k$$

$$\left\langle \begin{matrix} [\mu_{1X}(x)^-, \mu_{1X}(x)^+], [v_{1X}(x)^-, v_{1X}(x)^+] \\ [\mu_{2X}(x)^-, \mu_{2X}(x)^+], [v_{2X}(x)^-, v_{2X}(x)^+] \\ ... \\ [\mu_{sX}(x)^-, \mu_{sX}(x)^+], [v_{sX}(x)^-, v_{sX}(x)^+] \end{matrix} \right\rangle \times \left\langle \begin{matrix} [\mu_{1Y}(y)^-, \mu_{1Y}(y)^+], [v_{1Y}(y)^-, v_{1Y}(y)^+] \\ [\mu_{2Y}(y)^-, \mu_{2Y}(y)^+], [v_{2Y}(y)^-, v_{2Y}(y)^+] \\ ... \\ [\mu_{sY}(y)^-, \mu_{sY}(y)^+], [v_{sY}(y)^-, v_{sY}(y)^+] \end{matrix} \right\rangle$$

$$\rightarrow \left\langle \begin{matrix} [\mu_{1D}(D_1)^-, \mu_{1D}(D_1)^+], [v_{1D}(D_1)^-, v_{1D}(D_1)^+] \\ [\mu_{2D}(D_1)^-, \mu_{2D}(D_1)^+], [v_{2D}(D_1)^-, v_{2D}(D_1)^+] \\ ... \\ [\mu_{sD}(D_1)^-, \mu_{sD}(D_1)^+], [v_{sD}(D_1)^-, v_{sD}(D_1)^+] \end{matrix} \right\rangle \times ... \times \left\langle \begin{matrix} [\mu_{1D}(D_k)^-, \mu_{1D}(D_k)^+], [v_{1D}(D_k)^-, v_{1D}(D_k)^+] \\ [\mu_{2D}(D_k)^-, \mu_{2D}(D_k)^+], [v_{2D}(D_k)^-, v_{2D}(D_k)^+] \\ ... \\ [\mu_{sD}(D_k)^-, \mu_{sD}(D_k)^+], [v_{sD}(D_k)^-, v_{sD}(D_k)^+] \end{matrix} \right\rangle$$

(5)

Multi criteria IVIFRS contains the fundamental function given below:

a) *Prediction*: Define the values of $[\mu_{lD}(D_i)^-, \mu_{lD}(D_i)^+], [v_{lD}(D_i)^-, v_{lD}(D_i)^+], \forall l \in \{1, ..., s\}, \forall i \in \{1, ..., k\}$.

b) *Recommendation*: Select $i^* \in [1, s]$ to provide $i^* = \arg max_{i=\overline{1,s}} \left\{ \sum_{j=1}^{k} w_j \left([\mu_{iD}(D_j)^-, \mu_{iD}(D_j)^+], [v_{iD}(D_j)^-, v_{iD}(D_j)^+] \right) \right\}$ where $w_j \in [0,1]$ is the weight of D_j that $\sum_{j=1}^{k} w_j = 1$ [1].

Now, considering the Eq. 5, suppose that X_1 and X_2 are two IVIFMs in multi criteria based IVIFRSs. The interval valued intuitionistic fuzzy similarity degree (IVIFSD) is defined as below:

$$SIM(X_1, X_2) = \alpha \sum_{i=1}^{s} w_{1i} \tilde{S}_{1i} + \beta \sum_{i=1}^{s} w_{2i} \tilde{S}_{2i} + \delta \sum_{h=3}^{t} \sum_{i=1}^{s} w_{hi} \tilde{S}_{hi} \quad (6)$$

where $\tilde{S}$ is the IVIFSM between X_1 and X_2. $W = (w_{ij})(\forall i \in \{1, ..., t\}, \forall j \in \{1, ..., s\})$ is the weight matrix of IVIFSM between X_1 and X_2 that provides $\sum_{i=1}^{s} w_{1i} = 1$, $\sum_{i=1}^{s} w_{2i} = 1$, $\sum_{i=1}^{s} w_{hi} = 1$ $\forall h \in \{3, ..., t\}$ and $\alpha + \beta + \delta = 1$. These formulas can be generalized for user based, item based and rating based similarity degrees in recommendation systems when $\beta = \delta = 0$, $\delta = \alpha = 0$ and $\alpha = \beta = 0$, respectively.

The formulas to predict the values of linguistic labels of user P_u ($\forall u \in \{1,...,n\}$) to location S_j ($\forall j \in \{1,...,m\}$) according to the restaurants $D_1, D_2, ... D_k$ in multi criteria based IVIFRS are defined as follows:

$$\left[\mu_{iD}^{P_u}(D_j)^-, \mu_{iD}^{P_u}(D_j)^+\right] = \frac{\sum_{v=1}^{n} SIM(P_u, P_v) \times \left[\mu_{iD}^{P_v}(D_j)^-, \mu_{iD}^{P_v}(D_j)^+\right]}{\sum_{v=1}^{n} SIM(P_u, P_v)} \quad (7)$$

where $\forall i \in \{1,...,s\}, \forall j \in \{1,...,k\}, \forall u \in \{1,...,n\}$.

$$\left[v_{iD}^{P_u}(D_j)^-, v_{iD}^{P_u}(D_j)^+\right] = \frac{\sum_{v=1}^{n} SIM(P_u, P_v) \times \left[v_{iD}^{P_v}(D_j)^-, v_{iD}^{P_v}(D_j)^+\right]}{\sum_{v=1}^{n} SIM(P_u, P_v)} \quad (8)$$

where $\forall i \in \{1,...,s\}, \forall j \in \{1,...,k\}, \forall u \in \{1,...,n\}$ ([1]; [2]).

3. Application

Since the increasing potential of online marketing and sales operations, social networks have become an essential tool for the extension of communication with the customers. Thus, a diversified types of online congregations, firms, blogs and forums focused on the development of their social network attraction. Therefore, data gathered from social networks should be evaluated for constructing location based systems. In addition to that, the need of competitive analysis of the shopping malls should be extracted from the differentiation as positive and negative and the contribution of the feelings of the people to recommendation system will be beneficial especially in offering specialized alternatives (here restaurants) [4].

Our dataset is manually collected from three well-known websites, Foursquare, Zomato and Trip Advisor for extracting location information (Shopping mall), restaurant information and user information. Dataset contains 251 restaurants, 98 users and 36 places. Location information includes shopping mall names in Istanbul, recommended time to be spent in the relevant shopping mall, ranking of the location, number of comments, type of the comments (Excellent, Very good, Moderate, Bad and Terrible) and language of the comments (foreign and native) from Trip Advisor website. In addition to location information, restaurant information obtains data from appeared in Trip Advisor web site such as ranking of the restaurant, total score, scores for service, value and food, number of comments in Zomato, menu prices for two people, average point on Zomato, number of votes, number of negative, positive and neutral comments, places nearby the related restaurant, average point of places nearby the restaurants, Foursquare price score, Foursquare average point, number of point scorings in Foursquare. User information is obtained from Trip Advisor, Foursquare and

Zomato comprising years of membership, level of participation, number of contributions, number of visited places, number of useful votes, number of added photos, comment distribution (Excellent, Very Good,...etc.), Foursquare saved places, average scores of visited places, number of followers, number of followed people, number of comments in Zomato, number of followers in Zomato. Note that some of the users do not have Trip Advisor, Zomato or Foursquare account. We assumed that if user data appears in Zomato but does not appear in Foursquare and Trip Advisor, the columns are signed as "Null" for better data processing.

In the first stage, sentiment analysis is adapted for extracting number of negative, positive and neutral comments for locations from Trip Advisor. After text cleaning process in which misspelled words are eliminated, seed words are selected for noticing positive comments as ('Perfect', 'excellent', 'nice', 'good', 'beautiful', 'very beautiful', 'great'), negative comments as ('puff', 'none', 'bad', 'terrible', 'awful') and neutral ('do not know', 'no comment'). The number of extracted comments for some of the locations (shopping malls) are given in the following Table 1.

Table 1. Number of positive, negative and neutral comments from Trip Advisor.

	Akasya	Kanyon	City's	Istinye	Buyaka	Zorlu	Aqua Florya	Astoria	Istanbul Marina	Ozdilek	Nautilus	Palladium	Canpark	Capitol
Positive	2398	2540	4426	4880	4839	2109	2220	1841	2366	2298	5151	4123	4574	1497
Negative	111	1234	3456	53	6738	10566	5757	1799	7477	3753	9099	420	2905	5573
Neutral	456	976	881	672	106	232	101	267	130	433	749	690	558	593

As mentioned before, location similarity and grouping is necessary as the initial step for location prediction of moving customers. Thus, in the second stage, users and locations are grouped via fuzzy c means clustering. In order to determine the optimum number of groups, Xie-Beni index values for different values of c parameter are determined and four clusters are gathered to perform whole dataset of users. Second, cluster centers are randomly assigned and distance vector to cluster centers are calculated. Finally, locations are grouped in similar manner using R package.

In the final step, multi criteria interval valued intuitionistic fuzzy recommendation system (IVIFRS) procedure is adapted to user and location groups considering transformed IVIFN based data of restaurants. Recommendation system matching for each user group to a specific restaurant is applied by the determination of similarity measures and predicting the values of linguistic labels of user P_u to location S_j. Results are listed as seen from Table 2.

Table 2. Restaurant recommendation results to each user group with each location group.

User-location group	Recommended restaurant name	Location name	User-location group	Recommended restaurant name	Location name
P1-S1	Burger King	Capitol	P1-S3	HD Iskender	Nautilus
P2-S1	Sait Efendi	Ozdilek	P2-S3	Bursa Kebapevi	Canpark
P3-S1	Gunaydın Steakhouse	Istinye Park	P3-S3	Plus Kitchen	Zorlu
P4-S1	Bodrum Mantı	Buyaka	P4-S3	Günaydın Kofte Doner	Palladium
P1-S2	Big Chefs	Astoria	P1-S4	Tike	Istanbul Marina
P2-S2	Midpoint	Aqua Florya	P2-S4	Kirpi Café	Buyaka
P3-S2	Happy Moon's	Akasya	P3-S4	Balık Ev	City's
P4-S2	Tazele	Kanyon	P4-S4	Pilezza	Ozdilek

According to Table 2, restaurants are recommended to user and location groups. As seen from the results, users who visit Location 1 generally prefer most crowded shopping malls. In the similar manner, users who generally prefer to go Location 3 are prone to eat "middle income" based restaurants. Users in cluster 3 are exactly opposite of the people who visit Location 1 and Location 3 type of locations because of the favored restaurants which are "esoteric" that involve "high income" group.

4. Conclusions

Since location based technologies have been improved to facilitate marketing operations, mobile-wireless technologies are increasingly applied to send proper messages to customers. Before sending messages or giving an advertisement, customer purchasing behavior and probability should be appropriately investigated considering diversified characteristics such as previous visits, location data etc. In this respect, information gathered from social media is very useful for location clustering and user grouping before the implementation of offering alternative locations. In this study, a novel use of social application information is presented and initial results from a real world case study is conducted. To this end, data is collected from Trip advisor, Zomato and Foursquare, preprocessed for sentiment analysis in order to determine positive, negative and neutral comments before clustering procedure. On the other side, users and location clustering are implemented using fuzzy clustering. Finally, IVIFRS methodology is adapted for recommendation of alternative restaurants with respect to specific locations. Results indicate that using such this social data considering various locations has the potential to show customers' life style and interests. As a result, user group-restaurant matching based on user location data enables a high potential for getting insight about each individual user before implementing personalized recommender systems.

References

1. L.H. Son and N.T. Thong, Intuitionistic fuzzy recommender systems: An effective tool for medical diagnosis, *Knowledge-Based Systems, Volume 74*, pp.133-150. (2015).
2. L.C. Cheng and H.A. Wang, A fuzzy recommender system based on the integration of subjective preferences and objective information, *Applied Soft Computing, Volume 18*, pp. 290-301, (2014).
3. Z. Yue, Deriving decision maker's weights based on distance measure for interval-valued intuitionistic fuzzy group decision making, *Expert Systems with Applications 38*, pp. 11665–11670, (2011).
4. S.C. Öner and B. Öztayşi. An Interval Valued Hesitant Fuzzy Clustering Approach for Location Clustering and Customer Segmentation, *Advances in Fuzzy Logic and Technology*, pp. 56-70, (2017).
5. B. Oztaysi, C. Öner, D.H. Beyhan, Market Analysis Using Computational Intelligence: An Application for GSM Operators Based on Twitter Comments, *Intelligent Techniques in Engineering Management*, pp. 483-501, (2015).
6. I.P. Tussyadiah, A concept of location-based social network marketing, *Journal of Travel & Tourism Marketing, 29 (3)*, pp. 205–220, (2012).
7. S. Fan, R. Y.K. Lau, J. L. Zhao, Demystifying Big Data Analytics for Business Intelligence Through the Lens of Marketing Mix, *Big Data Research, Volume 2, Issue 1*, pp. 28-32, (2015).
8. C.H. Wu, S.C. Kao, C.C. Wu, S. Huang, Location-aware service applied to mobile short message advertising: Design, development, and evaluation, *Information Processing & Management, Volume 51, Issue 5*, pp. 625–642, (2015).
9. W. Shin, and T.C. Lin, Who avoids location-based advertising and why? Investigating the relationship between user perceptions and advertising avoidance, Computers in Human Behavior 63, pp. 444-452, (2016).
10. R. Abbas, K. Michael, M.G. Michael, The regulatory considerations and ethical dilemmas of location-based services (LBS): A literature review, *Information Technology & People, 27 (1)*, pp. 2–20 (2014).

The effects of CEO's facial trustworthiness on the investment decision[†]

Jinping Gao[1], Si Long[1], Chang Yuan[1], Jing Fan[2] and Yan Wan[1]

[1]*School of Economics and Management, Beijing University of Posts and Telecommunications, Beijing, China*
[2]*International Business School, Beijing Foreign Study University, Beijing, China*

With incomplete and uncertain knowledge, CEO's photo as an important visual information in the company's annual report, can influence cognitive process and bring unexpected effect. Extracting the facial characteristic of CEO's photo disclosed in the chairman's report of company's annual report as object, dual process theory as foundation, this paper explores the effects of CEO's facial trustworthiness on the investment decision through behavioral experiments. We find that investors are more willing to invest in the company with high-facial-trustworthiness CEO, and the influence process is mediated by the positive emotions of investors. Therefore, it provides direct evidence for the market value of CEO's innate characteristic and theoretical basis for corporate information disclosure strategy.

1. Introduction

Since the 1990s, the financial scandals such as Enron have been exposed, a large number of accountants and auditors have been questioned and examined, not only shaken the trust of financial statements and the entire financial system. Thence stakeholders put forward higher requirements for transparency in information disclosure in the business field, and encouraged the release of higher quality and more comprehensive information through various media [1]. Among them, the company's annual report is considered as the main information disclosure channel and important information exchange tool.

Annual report contains mandatory disclosure (financial information) and non-mandatory disclosure information. The usefulness of financial information has been widely recognized [2,3]. Among the non-mandatory disclosure information, the chairman's report, as the most prominent and widely read part, can reduce information asymmetry between managers and shareholders [4], thus affecting individual investor decisions. Visual information, such as photos, helps

---

[†]Work partially supported by National Science Foundation of China (71372193, 71573022, 71471019).

to draw the reader's attention, paint a blueprint for the company's future prospects, and stimulate readers' emotional responses [5], thus conveying corporate priorities, values, and culture. The disclosed photos of executives who take responsibility of authenticity and reliability of financial information are linked to organizational credibility [6]. Some even believe that the quality of photos represents the accuracy of accounting data, executives need to look tidy and reliable so readers can trust the reliability of their annual report. In the ups and downs of the stock market, the exposure of CEO's photo with high-facial-trustworthiness undoubtedly in a strong agent, stimulate emotional experience. While the rational system operates, the intuition system, which relies on emotion and experience, makes a quick sensing decision and affects its investment decision.

Based on the above, extracting the facial characteristic of CEO's photo disclosed in the chairman's report as the research object, dual process theory as the research foundation, this paper explores the influence of CEO's facial trustworthiness on the investment decision through behavioral experiments and discusses the mediating effect of emotional experience.

2. Literature Review and Hypotheses

2.1. *Facial trustworthiness and investment decision*

Behavioral strategy merges cognitive and social psychology with strategic management theory and practice [7]. Many researchers apply the cognitive research to business strategy decisions, to begin a discussion of the emerging field of Behavioral Strategy [8-9]. Signaling Theory believe that the higher quality of information, the more willing to conduct voluntary corporate information disclosure and transmit good corporate governance signals to the public, which can enhance the image and value of the company. As the most important executive of business financial activities, CEO's photos often represent the accuracy of accounting figures and reliability of the annual report, sending important signals to the capital market about its credibility and quality of company's financial statements, thus affecting the response to capital markets [10]. Halford & Hsu (2014) found that better-looking CEOs create higher value for shareholders, are easier to be accepted in negotiations, and are more likely to be favored by investors. Shen Yifeng (2017) also show that there is a "beauty premium" among CEOs in the IPO market. Investors prefer companies with good-looking CEO, leading to low IPO subscription acceptance rate rate, low first day turnover rate and low discount rate. According to the control theory, the relationship between investor and executive belongs to the cooperative principal-agent relationship. In collaborative behavior, people tend to make their own trust judgment based on their face and make decisions about whether to trust and cooperate with them [11]. In the trust game, participants expressed a clear sense about the economic value of face in economic decisions-making [12]. Therefore, we believe that in the rapidly changing stock market, the CEO's facial trustworthiness conveys signals to investors, enhances the image of company, improve investors' confidence and cooperative intention, thereby promoting the investment decision. The first hypothesis is stated as:

Hypothesis 1: Investors are more willing to invest in the company with high-facial-trustworthiness CEO.

2.2. Mediating effect of emotional experience

Dual-process theory holds that when people make decisions, the brain has an intuitive system characterized by heuristic processing and a rational system characterized by analytical processing to complete human reasoning and decision-making, which emphasizing the role of emotional and intuitive irrational factors in decision-making. Rational factors are important, but in uncertain and incomplete decision-making conditions, people tend to use the emotional experience jointly. In the judgments of risks and benefits, individual perceived risks and benefits are always associated with individual affect evaluation, if the individual feels an activity is "like", then tends to believe the activity is low risk and high benefits; high risk and low benefits if an activity "dislikes" [13]. After investors read the company's annual report, the operating performance and development potential is an important reference for investors, and ultimately affect the investment decision and behavior, which is the rational system processing with careful and critical analysis. At the same time, the positive emotional valuation, stimulated by the CEO's high- trustworthiness photo, makes consumers to bear the risk of higher degree, overestimate return, underestimate risk and improve investment [14]; conversely for the low-trustworthiness photo, negative sentiment increases the degree of investors "loss avoidance". Thus, we propose that CEO's face will trigger positive and/or negative emotional experience, making intuition system operate by emotional judgments, affecting their investment decision and behavior. The following hypothesis is tested:

Hypothesis 2: The effect of CEO's facial trustworthiness on the investment decision is mediated by positive emotions of investors.

Hypothesis 3: The effect of CEO's facial trustworthiness on the investment decision is mediated by negative emotions of investors.

3. Methods

3.1. *Participants*

Participants were 90 graduate students of Beijing University of Posts and Telecommunications (40 men, 50 women). The ages of participants ranged from 22 to 25. All participants had stock investment experience. Finally, 84 effective data were collected (36 men, 48 women).

3.2. *Task materials*

CEO's photos were chosen from the real disclosures of HKEX and got standardized. 58 undergraduates (13 men, 45 women) were asked to rate how trustworthy they thought each photo and selected a high trustworthiness photo (M = 4.534, SD = 0.9499) and a low trustworthiness photo (M = 2.966, SD = 0.9861). Genders had no significant influence on their assessments.

The text parts of Chairman's report were also chosen from the real one. These reports classified into 3 groups according to their profit status: profit, loss and break-even. The undergraduates were asked to rate their investment decision. From each group, we selected 2 reports between which there were no significant difference of investment decision. We combined 6 text parts with high/low facial trustworthiness CEO's photos to form 6 companies' Chairman's reports. All of the reports' font, size, format and layout were standardized.

3.3. *Task procedures*

We simulated the real investment process and divided the experiment into 3 groups: profit (A&B), loss (C&D) and break-even (E&F). Each participant was required to participate in all 3 groups randomly and assumed to have 100,000 yuan to invest and investigated their investment decision through experiments. We used E-prime 2.0 to conduct our experiment. Before each group, participants were required to read 2 Chairman's reports, filled in the feedback and start their tasks after experimental instruction. In each trail, company's CEO's photo and financial indicators were shown on the left of the screen randomly. On the right of screen, there was a random investment proposal between 10,000 and 100,000 yuan (Common difference is 10000). Participants could choose to "accept" or "reject" this investment proposal with time unlimited. After choosing, this trail end and they were required to stare at a fixation point to pay attention again. Then, next trail started. 10 investment proposal and 2 companies formed 20 combinations. Each combination was repeated five times. Therefore, each participant completed 100 trials in one group. Participants could have a rest and then participated in the next group. When all experiments were completed, they were required to fill in assessment materials and scales.

4. Results

4.1. Facial trustworthiness and investment decision

The data was analyzed using Weibull program ($y = 1-e^{-(x/a)^k}$; x: the investment amount, including 0.1, 0.2, ..., 0.9, 1; y: the 'accept' ratio of participants for each amount; a, k: distributed parameters). Through the fitting analysis, we calculated the just-receivable amount, 'a' (the amount that a participant has 50% accept or reject probability). The higher the just-receivable amount, the more likely participants were to accept the investment proposal. Finally, we used SPSS22 to analyze the effective experiment data.

According to the results of paired T test, CEO's Facial Trustworthiness had significant influence on the investment amount. The investment amount of CEO's photo with high-facial-trustworthiness was significantly higher than the low facial trustworthiness photo in 3 groups (profit: $t = 3.389$, $p = 0.001$; break-even: $t = 3.426$, $p = 0.001$; loss: $t = 2.838$, $p = 0.006$), as shown in Figure 1. Therefore, H1 was supported. Regardless of profit status, investors are more willing to invest in the company with high-facial-trustworthiness CEO.

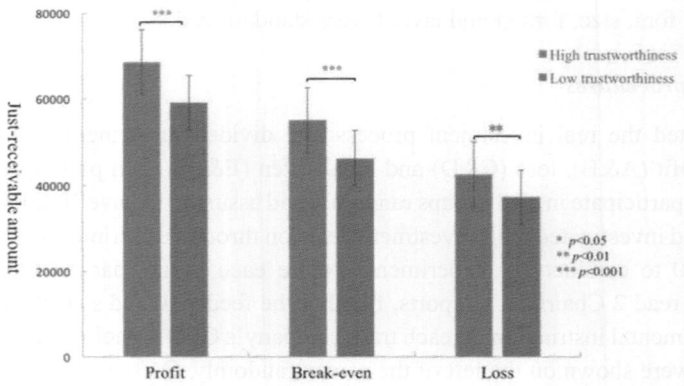

Figure 1. Investment amount under different surpluses.

4.2. Mediating effect of emotional experience

Both the positive emotion (M-high trustworthiness = 4.843, SD = 0.7438; M-low trustworthiness = 3.557, SD = 1.0746) and negative emotion (M-high trustworthiness = 1.883, SD = 0.6565; M-low trustworthiness = 2.933, SD = 1.2869) evoked by 2 different trustworthiness photos significantly differed (positive: t = 9.218, p = 0.000; negative: t = −7.410, p = 0.000). There were significant relations between the different trustworthiness photos and the positive and negative emotion they evoked (r = 0.610, p = 0.000; r = −0.513, p = 0.000).

Bootstrapping analysis indicated that 0 is not include in the indirect path regardless of profit status (profit: LLCI = 0.1086, ULCI = 0.5003; break-even: LLCI = 0.0535, ULCI = 0.4728; loss: LLCI = 0.0417, ULCI = 0.3670). These results meant that, positive and negative emotion mediated the relationship between CEO's facial trustworthiness and investment amount. An examination of the specific indirect effects indicated that only positive emotion was a mediator (positive: 0.0779, 0.3813; break-even: 0.0604, 0.3992; loss: 0.1008, 0.3708) and the mediating effect of negative emotion was not significant (positive: −0.0243, 0.2065; break-even: −0.0883, 0.1624; loss: −0.1471, 0.0643). The results are shown in Table 1. H2 is supported, H3 isn't supported. The results indicated that the relationship between CEO's facial trustworthiness and investment amount was mediated by investors' positive emotion and was not mediated by negative emotion regardless of profit status.

Table 1. Mediating effect of positive and negative emotion.

Status	Path	Effect	Boot SE	Boot LLCL	Boot ULCI
profit	direct	0.0901	0.1402	-0.1868	0.3670
	indirect	0.2964	0.0981	0.1086	0.5003
	positive emotion	0.2121	0.0777	0.0779	0.3813
	negative emotion	0.0843	0.0584	-0.0243	0.2065
break-even	direct	0.0782	0.1549	-0.2276	0.3840
	indirect	0.2477	0.1046	0.0535	0.4728
	positive emotion	0.2164	0.0868	0.0604	0.3992
	negative emotion	0.0313	0.0631	-0.0883	0.1624
loss	direct	0.0944	0.1323	-0.1668	0.3556
	indirect	0.1844	0.0803	0.0417	0.3670
	positive emotion	0.2232	0.0676	0.1008	0.3708
	negative emotion	-0.0388	0.0527	-0.1471	0.0643

5. Conclusion

Few empirical studies have been conducted on the CEO's photo which is an important visual information in the company's annual report. Without complete and certain information, CEO's photo would trigger intuitive emotional experience, influence cognitive process, which may bring unexpected effect. In this paper, we explore the value of CEO's face in individual investment behavior under different profit status through behavioral experiment, which provides direct evidence for the impact of human innate characteristics on the company's financial behavior. First of all, behavioral experiments, which designed to simulate the scenarios, found that investors are more willing to invest in the company with high-facial-trustworthiness CEO regardless of profit status, that is, the face exerts potential "premium". Secondly, according to the dual process and heuristic theory of emotion, we explore that the effect of CEO's facial trustworthiness on the investment decision is mediated by positive emotions of investors.

Therefore, the company can effectively use this tool to send positive signals to investors and stakeholders, improving the company's credibility and value recognition, enhancing the effectiveness of financial information, and promoting market and enterprise development healthier and more effective.

References

1. Uyar, A., M. Kilic, and N. Bayyurt, Association between firm characteristics and corporate voluntary disclosure: Evidence from Turkish listed companies. Intangible Capital. **9(9)**: p. 1080-1112(2013).
2. Cheng, M., D. Dan, and Y. Zhang, Does investment efficiency improve after the disclosure of material weaknesses in internal control over financial reporting? Journal of Accounting & Economics. **56(1)**: p. 1-18(2013).
3. Jung, B., W.J. Lee, and D.P. Weber, Financial Reporting Quality and Labor Investment Efficiency. Contemporary Accounting Research. **31(4)**: p. 1047-1076(2014).
4. G. Boesso and K. Kumar, Drivers of corporate voluntary disclosure. Accounting Auditing & Accountability Journal. **20(2)**: p. 269-296(2007).
5. J. Davison, Photographs and accountability: cracking the codes of an NGO. Accounting Auditing & Accountability Journal. **20(1)**: p. 133-158(2007).
6. Y. Benschop and H.E. Meihuizen, Keeping up gendered appearances: representations of gender in financial annual reports. Accounting Organizations & Society. **27(7)**: p. 611-636(2002).
7. T C. Powell, L. Dan and C R. Fox. Behavioral strategy. Strategic Management Journal. **32(13)**:1369-1386(2011).
8. G. Gavetti. Toward a Behavioral Theory of Strategy. Organization Science. **23(1)**: 267-285(2012).
9. J. E. Schrager and A. Behavioral Madansky, strategy: a foundational view. Journal of Strategy and Management. **6(1)**: 81-95(2013).
10. Y. Zhang and M.F. Wiersema, Stock market reaction to CEO certification: the signaling role of CEO background. Strategic Management Journal. **30(7)**: p. 693–710(2009).
11. M. Van't Wout, and A.G. Sanfey, Friend or foe: The effect of implicit trustworthiness judgments in social decision-making. Cognition. **108(3)**: p. 796-803(2008).
12. C.C. Eckel, and R. Petrie, Face Value. American Economic Review. **101(4)**: p. 1497-1513(2011).
13. P. Slovic, and E. Peters, Risk Perception and Affect. Current Directions in Psychological Science. **15(6)**: p. 322-325(2010).
14. U. Malmendier, and G. Tate, CEO Overconfidence and Corporate Investment. Strategic Direction. **60(5)**: p. 2661-2700(2006).

The application of evaluation method based on prospect theory in stock investment[†]

Chang Yuan[1], Yujia Sui[1], Jinping Gao[1], Yu Pan[2] and Li Gao[2]

[1]*School of Economics and Management,*
Beijing University of Posts and Telecommunications, Beijing, China
[2]*College of International Business, Shanghai International Studies University*
Shanghai, China

Individual investors in the decision whether to buy and long-term hold a stock, assess the value of the stock according to their own preferences. This paper fully considers investors' finite rationality in the process of stock investment, applying the prospect theory to the analytic hierarchy process. It introduces the positive and negative ideal points, better describes the actual perceived value of the decision makers, and has stronger reality. Different decision makers mean different assumed reference points and risk avoidance factors when the prospect value is calculated. Obviously, the order of the stock investment value based on the prospect value is also different.

1. Introduction

Since the value of stock investment has caught the attention of investors, many scholars have studied the review of stock investment value [1]. There are traditional discount models, Graham's "intrinsic value" model, and value models based on book value and "residual income" [2]. There are many factors involved in the evaluation of stock investment value, such as profitability index, growth index and risk index [3]. However, when individual investors face with the choice of more and complex structure of the stock investment problems, they are often difficult to make judgments. In the 1870s, the American operational research experts, Saaty proposed the Analytic Hierarchy Process (AHP) of systematic decision-making method, which was applied to the determination of multi-attribute weighting factors, so that it was widely used in various fields [4]. In the past, the study of stock investment decision-making methods rarely took into account the influence of investors' subjective emotions and finite rationality. Based on finite reality, Kahneman put forward the prospect theory which was more in line with the real behavior of people, and effectively explains the Allais paradox and the Ellsberg paradox [5]. Due to different venture investors have

---

[†]Work partially supported by National Science Foundation of China (71372193, 71573022).

different risk appetite, prospect theory argues that decision makers pay more attention to the amount of change rather than the final amount. Besides, the emotional experience is different when facing to the same gains and losses, because the pain of losing is greater than the joy of getting. With the recognition of the importance of psychological factors, the decision-making method based on the prospect theory has gradually entered people's life.

This paper considered in real life the individual investors' selection of the stock investment, long-term holding, and no consideration of the occasional factors. By combining the prospect theory with the analytic hierarchy process, it evaluated the investment value of the stock, and thought about the decision model of the investors with different risk preferences. It used the analytic hierarchy process to determine the weight of each criterion and defined the virtual scheme. According to the characteristics of the decision maker, it assumed the reference point, introduced prospect theory to calculate the prospect value, and set the size of the prospect value as the basis for sorting.

2. Decision-making Method Based on AHP and Prospect Theory

2.1. *Step 1: Using the Analytic Hierarchy Process to determine the combined weight vector for each criterion*

a. Establish the hierarchical model of decision-making problems.

b. Constructing judgment matrix. The relative importance of each factor in each level is judged by a numerical form, and the judgment matrix is constructed. Any judgment matrixes should be satisfied $b_{ij} = 1$

$$b_{ij} = \frac{1}{b_{ij}} \ (i,j = 1,2,\ldots,n)$$

The matrix b_{ij} represented the relative importance of B_i and B_j for A_k. Saaty showed that the 1-9 scaling method could better quantify the thinking judgment. When constructing the judgment matrix, the 1-9 scale value was used. The judgment of index value in the matrix is based on the measured data, the document data and the expert opinion generated from overall weighting.

c. Single hierarchical arrangement and consistency checking. Using the root method to find the largest eigenvalue of the judgment matrix and its corresponding eigenvector, the single hierarchical order and its consistency checking are carried out. The consistency checking means firstly calculating the maximum eigenvalue λ_{max} of the judgment matrix and CI the consistency index, and then judging CR the consistency ratio. When $CR < 0.1$, the judgment matrix passes the consistency checking. For the n-order judgment matrix, CI and CR are calculated as in Equation (1).

$$CI = \frac{\lambda_{max}-n}{n-1} \quad CR = \frac{CI}{RI} \tag{1}$$

Value of random consistency index RI is shown in Table 1.

Table 1. Value of random consistency index RI.

n	1	2	3	4	5	6	7	8	9
RI	0.00	0.00	0.58	0.90	1.12	1.24	1.32	1.41	1.45

d. Total taxis of hierarchy and consistency checking. After weight vector for factors in each layer, the total hierarchy and consistency checking are carried out. Then the combined weight vector is calculated to obtain the target combined weight vector for all factors above the case layer. Considering the cumulative effect of inconsistency, the consistency checking of the hierarchical total order is also based on the single order consistency checking, so as to determine whether the combined weight vector can be used as the final sorting basis. There are n factors in the p-th layer. Setting the weight vector of the $p-1$ layer is w^{p-1}. The consistency index of the j-th factor is CI_j^p and the corresponding random consistency index is RI_j^p. Then we get the p-th hierarchical total order consistency index CI_j^p and random consistency index RI_j^p.

$$CI^p = [CI_1^p, CI_2^p, \ldots, CI_n^p] * w^{p-1} \tag{2}$$

$$RI^p = [RI_1^p, RI_2^p, \ldots, RI_n^p] * w^{p-1} \tag{3}$$

The hierarchical consistency ratio of the p-th layer is

$$CR^p = \frac{CI^p}{RI^p} \tag{4}$$

If $CI_j^p < 0.1$, it is considered to pass through the hierarchical consistency test. The combined weight vector of the p-th layer is

$$w^p = D^p * w^{p-1} \tag{5}$$

Among them, J-th row of D^p is weight vector of all factors in p-th layer under j-th factor in $p-1$ layer.

2.2. Step 2: Obtaining the positive ideal point P_U and the negative ideal point P_L by comparing the selection

Under each criterion, a positive ideal is selected by one of the best or nearly best case from each scenario. At the same time, a negative ideal is selected by a worst or nearly worst case. The positive ideal point is the best under each criterion, and the negative ideal point is the worst under each criterion. The role of positive and negative ideal point is a standard for judgment and comparison.

2.3. Step 3: Building the judgment matrix from each scheme and the positive and negative ideal points, and then getting the normalized order vector

With the positive ideal point as the first program, negative ideal point is the last one. With the number of programs as n and B as the judgment matrix when the ideal point is not considered, the new judgment matrix is obtained after considering the ideal point.

$$A = \begin{bmatrix} 1 & U_1 & a_{1(n+2)} \\ U_2 & B & L_2 \\ a_{(n+2)1} & L_1 & 1 \end{bmatrix} \quad (6)$$

U_1 is the row vector obtained by comparing the positive ideal point with other programs, and L_1 is the row vector obtained from the negative ideal point and the other programs. Besides, U_2 is the comparison between the other programs and the positive ideal point, and L_2 is the comparison between the other programs and the negative ideal point. The new judgment matrix has such characteristics: $a_{1j} \geq 1$, $a_{j1} \leq 1$. The consistency vector test is carried out by using Equation (1), and then the matrix w of the positive eigenvector corresponding to the largest eigenvalue is solved. The eigenvector is normalized by maximum difference dormalization method as follows:

$$w_{ij}^* = \frac{w_{ij} - min_j(w_{ij})}{max_j(w_{ij}) - min_j(w_{ij})} \quad (7)$$

The normalized order vector matrix w^* under each criterion is obtained according to Equation (7). Thus, each program P_i corresponds to a row vector, and the positive rational point P_U corresponds to the unit row vector, as well as the negative ideal point P_L corresponds to the zero row vector.

2.4. Step 4: Selecting the reference point vector to calculate the foreground value for the scenario

In selection of a reasonable reference point vector, the prospect value of the program is calculated. The size of prospect value is used for program order basis, so as to get a more reliable and satisfactory program.

Prospect theory suggests that personal utility is consists of decision weights and value function values:

$$V = v(x)\pi(p) \qquad (8)$$

Among them, $v(x)$ is the value function of decision, and $\pi(p)$ is the corresponding decision weight.

According to the prospect theory, the subjective feelings of loss curve is significantly steeper than the subjective feelings of income curve. In risk investment decisions, decision makers pay more attention to losses. In the selection of the value function, the following quantitative analysis is used. Assuming that the value function is a two-stage function, the value function $v(x)$ is in the form of:

$$v(x) = \begin{cases} (\Delta x)^\alpha & \Delta x \geq 0 \\ -\lambda(-\Delta x)^\beta & \Delta x < 0 \end{cases} \qquad (9)$$

In addition, Δx is the gain and loss of the reference point; α and β are the risk attitude coefficients. The larger the α and the β are, the greater the risk that the venture investors are willing to take. When $\alpha, \beta = 1$, it is neutral; when $1 < \alpha$ and $\beta < 1$ represent the degree of sensitivity of the profit, the loss to the value respectively; as λ is the risk aversion coefficient, $\lambda > 1$ indicates that the venture investor is more sensitive to the loss. The selection of the reference point depends on the characteristics of the decision maker. According to Kahneman's standard, when $\lambda = 2.25$ and it is very closely to the empirical data, so this paper also uses this data.

3. Cases Analysis

An investor is more optimistic about the development of the Internet, longing to invest and long-term hold a company's stock in the Internet industry. After the initial screening, he decided to assess and choose in the five companies which are Letv P_1 (300104), East Money P_2 (300059), Sanqi Interactive Entertainment P_3 (002555), King network P_4 (002517) and Kunlun Worldwide P_5 (300418). Although there are a lot of indicators to evaluate the value of stock investment, in fact, the more factors you consider are not so better. There are three main reasons: firstly, the negative aspect of accumulated errors may be greater than the positive aspect of considering; secondly, the individual in the actual stock investment, often only concerned about a few important indicators, so that selecting the key indicators to assess is more realistic; thirdly, it is suitable for the presumption of finite rationality in this paper. Therefore, according to relevant literature and expert assessment, the paper has selected most concerned indicators in four investors: B_1 as earnings per share (yuan), B_2 as net assets per share (yuan), B_3 as net profit growth rate (%), B4 as tradable shares (billion) [6]. B_1, B_2, B_3, B_4 are all positive indicators. The hierarchical model of the problem is shown in Figure 1.

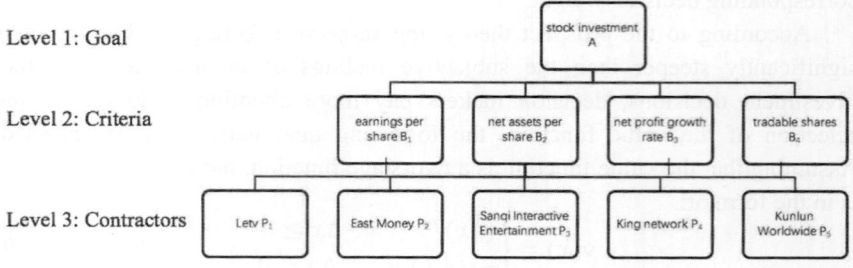

Figure 1. Hierarchy of the project example.

Doing the step 1: The construction of judgment matrix. In AHP model of the stock selection shown in Figure 1, different types of stocks have different weights of the criteria. For Internet stocks, it uses 1 to 9 scale. The judgment matrix of B_1, B_2, B_3, and B_4 under the total target A is:

$$A = \begin{bmatrix} 1 & 2 & 2 & 3 \\ 1/2 & 1 & 1/2 & 1 \\ 1/2 & 2 & 1 & 2 \\ 1/3 & 1 & 1/2 & 1 \end{bmatrix}$$

Single hierarchical order and consistency checking. According to the constructed judgment matrix, the maximum feature root $\lambda_{max} = 4.046$ of the judgment matrix A is obtained by the root method. According to the formula (1), the consistency index CI = 0.015 and the random consistency ratio CR = 0.017 are obtained. Because of CR < 0.1, it can be considered that the judgment matrix has satisfactory consistency. The normal feature vector corresponding to the largest eigenvalue is calculated and normalized with result in formula (10).

$$w = (0.423, \ 0.161, \ 0.270, \ 0.145) \tag{10}$$

Through comparison of two pairs, it got 5 stocks in the judgment matrix under four criteria:

$$B_1 = \begin{bmatrix} 1 & 2 & 1/2 & 1/4 & 1/3 \\ 1/2 & 1 & 1/3 & 1/5 & 1/4 \\ 2 & 3 & 1 & 1/3 & 1/2 \\ 4 & 5 & 3 & 1 & 2 \\ 3 & 4 & 2 & 1/2 & 1 \end{bmatrix} \quad B_2 = \begin{bmatrix} 1 & 3 & 5 & 2 & 4 \\ 1/3 & 1 & 3 & 1/2 & 2 \\ 1/5 & 1/3 & 1 & 1/4 & 1/2 \\ 1/2 & 2 & 4 & 1 & 3 \\ 1/4 & 1/2 & 2 & 1/3 & 1 \end{bmatrix}$$

$$B_3 = \begin{bmatrix} 1 & 4 & 1/4 & 3 & 1/3 \\ 1/4 & 1 & 1/7 & 1/3 & 1/6 \\ 4 & 7 & 1 & 5 & 2 \\ 1/3 & 3 & 1/5 & 1 & 1/4 \\ 3 & 6 & 1/2 & 4 & 1 \end{bmatrix} \quad B_4 = \begin{bmatrix} 1 & 1/3 & 5 & 6 & 5 \\ 3 & 1 & 6 & 7 & 6 \\ 1/5 & 1/6 & 1 & 2 & 1 \\ 1/6 & 1/7 & 1/2 & 1 & 1/2 \\ 1/5 & 1/6 & 1 & 2 & 1 \end{bmatrix}$$

Under the criteria, it selected the best and worst program, and got the positive ideal point and negative ideal point:

$$P_U = (P_4, P_1, P_3, P_2), \quad P_L = (P_2, P_3, P_2, P_4)$$

After considering the ideal point, a new judgment matrix A_1, A_2, A_3, A_4 is obtained according to Equation (6).

$$A_1 = \begin{bmatrix} 1 & 4 & 5 & 3 & 1 & 2 & 5 \\ 1/4 & 1 & 2 & 1/2 & 1/4 & 1/3 & 2 \\ 1/5 & 1/2 & 1 & 1/3 & 1/5 & 1/4 & 1 \\ 1/3 & 2 & 3 & 1 & 1/3 & 1/2 & 3 \\ 1 & 4 & 5 & 3 & 1 & 2 & 5 \\ 1/2 & 3 & 4 & 2 & 1/2 & 1 & 4 \\ 1/5 & 1/2 & 1 & 1/3 & 1/5 & 1/4 & 1 \end{bmatrix} \quad A_2 = \begin{bmatrix} 1 & 1 & 3 & 5 & 2 & 4 & 5 \\ 1 & 1 & 3 & 5 & 2 & 4 & 5 \\ 1/3 & 1/3 & 1 & 3 & 1/2 & 2 & 3 \\ 1/5 & 1/5 & 1/3 & 1 & 1/4 & 1/2 & 1 \\ 1/2 & 1/2 & 2 & 4 & 1 & 3 & 4 \\ 1/4 & 1/4 & 1/2 & 2 & 1/3 & 1 & 2 \\ 1/5 & 1/5 & 1/3 & 1 & 1/4 & 1/2 & 1 \end{bmatrix}$$

$$A_3 = \begin{bmatrix} 1 & 4 & 7 & 1 & 5 & 2 & 7 \\ 1/4 & 1 & 4 & 1/4 & 3 & 1/3 & 4 \\ 1/7 & 1/4 & 1 & 1/7 & 1/3 & 1/6 & 1 \\ 1 & 4 & 7 & 1 & 5 & 2 & 7 \\ 1/5 & 1/3 & 3 & 1/5 & 1 & 1/4 & 3 \\ 1/2 & 3 & 6 & 1/2 & 4 & 1 & 6 \\ 1/7 & 1/4 & 1 & 1/7 & 1/3 & 1/6 & 1 \end{bmatrix} \quad A_4 = \begin{bmatrix} 1 & 3 & 1 & 6 & 7 & 6 & 7 \\ 1/3 & 1 & 1/3 & 5 & 6 & 5 & 6 \\ 1 & 3 & 1 & 6 & 7 & 6 & 7 \\ 1/6 & 1/5 & 1/6 & 1 & 2 & 1 & 2 \\ 1/7 & 1/6 & 1/7 & 1/2 & 1 & 1/2 & 1 \\ 1/6 & 1/5 & 1/6 & 1 & 2 & 1 & 2 \\ 1/7 & 1/6 & 1/7 & 1/2 & 1 & 1/2 & 1 \end{bmatrix}$$

Similarly, the maximum characteristic root λ_{max} of each judgment matrix is obtained by using the root method, and the consistency index CI and the random consistency ratio CR are obtained according to Equation (1). The results are shown in Table 2.

Table 2. Value vector and consistency checking.

	A_1	A_2	A_3	A_4
λ_{max}	7.1000	7.1000	7.2570	7.2131
CI	0.0167	0.0167	0.0428	0.0355
CR	0.0126	0.0126	0.0325	0.0269

When CR values are less than 0.1, it could be considered that the judgment matrix have a satisfactory consistency. At the same time, we got the positive eigenvector corresponding to the largest eigenvalue: W_{A1} = (0.2771, 0.0706, 0.0446, 0.1111, 0.2771, 0.1750, 0.0446), W_{A2} = (0.2771, 0.2771, 0.1111, 0.0446, 0.1750, 0.0705, 0.0446), W_{A3} = (0.2927, 0.0991, 0.0309, 0.2928, 0.0601, 0.1935, 0.0309), W_{A4} = (0.3203, 0.1817, 0.3203, 0.0546, 0.0343, 0.0546, 0.3642). And then obtained the numerical representation vector of the five stocks by the formula (7):

$$\begin{bmatrix} P_1 \\ P_2 \\ P_3 \\ P_4 \\ P_5 \end{bmatrix} = \begin{bmatrix} 0.1119 & 1 & 0.2607 & 0.5154 \\ 0 & 0.2862 & 0 & 1 \\ 0.2862 & 0 & 1 & 0.0712 \\ 1 & 0.5608 & 0.1115 & 0 \\ 0.5608 & 0.1119 & 0.6210 & 0.0712 \end{bmatrix}$$

Doing the step 4, assume that the reference point P_r (0.5, 0.5, 0.5, 0.5) is selected, the prospect value of each value could be calculated by formula (8), (9), (10), see Table 3.

Table 3. Prospect value.

Companies	P1	P2	P3	P4	P5
Prospect Value	-0.495	-0.862	-0.45	-0.198	-0.234

According to the order by the size of the prospect value: P4 > P5 > P3 > P1 > P2, the most satisfied stock is P4 as the reference point 0.5. The choice of reference points varies from person to person, may be could selected accord to personal risk preferences in a realistic environment and the prospect value of the stock depends on the choice of the reference point.

4. Conclusion

According to the results of this study, the following conclusions and policy opinions can be drawn:

(1) It is more systematic and more realistic to evaluate the stock value by the combination of prospect theory and analytic hierarchy process, which provides a quantitative investment evaluation mind and scientific decision-making means for stock investment decision-making. It helps investors more accurately evaluate the value of stock investment in accordance with their own preferences, increases the efficiency of stock investment decision-making, and further improves the stock value evaluation method system.

(2) Investors with different degrees of risk aversion have different opinions to assess stock value. Venture investors are bounded rationally and are influenced by factors such as their own risk appetite, emotional state, and so on. Therefore, venture investors should take full account of their own characteristics when choosing stocks, especially considering the affordability of the results. In real life, people are finite rational, almost no rational investors. The evaluation of prospect theory promotes the reality and effective venture control of intrinsic value of the stock, and is conducive to the guidance of investment decision-makers assessment work to improve decision-making utility and experience utility.

This paper applies AHP and prospect theory to the evaluation of long-term holding value of stock. This method takes into account the finite rationality of the investment decision-makers in real life. It is ordered by prospect value for each decision-making programs, but not by the weight ratio of the traditional analytic hierarchy process. Thus, it is more suitable to the reality, because different decision makers have different risk aversions and different selected reference points as well as various sorting of the value of the stock investment.

References

1. L. Hongyi, C. Zhang and D. Zhao, Stock Investment Value Analysis Model Based on AHP and Gray Relational Degree. Management Science & Engineering, **4(4)**:1-6(2010).
2. W.H. Beaver, Perspectives on recent capital market research. Accounting Review, **77(2)**, 453-474(2002).
3. C.A. Guo, Study of the Methods for Evaluating the Entropy Weight Coefficient of the Investment Value of Stocks. Nankai Economic Studies, **(5)**, 65-67(2001).
4. T.L. Saaty, A scaling method for priorities in hierarchical structures. Journal of Mathematical Psychology, **15(3)**: 234-281(1977).
5. D. Kahneman and A. Tversky, Prospect theory: An analysis of decision under risk. Econometrica, **47(2)**: 140-170(1979).
6. Y. Hao and Q. Zhang, A New Method for Stock Investment Value Analysis Based on the Attibute Synthetic Evaluation System and Vague Sets. Journal of Management Science (2005).

Sustainable supplier selection using two-phase QFD and TOPSIS within PD-HFLTS context

Zhen-Song Chen[†]

School of Civil Engineering, Wuhan University, Wuhan 430072, China

To incorporate sustainable customer requirements (CRs) into sustainable supplier selection (SSS) with uncertainty, in this paper we propose a novel model for SSS which integrates two-phase quality function deployment (QFD) approach and TOPSIS in the context of possibility distribution-based hesitant fuzzy linguistic term sets (PD-HFLTS). The two-phase QFD approach is utilized to translate CRs into sustainable supplier assessment criteria (ACs) through sustainable product development strategies (DSs), and it is further used to determine the importance ratings (IRs) of ACs. Afterwards, the Technique for Order Preference by Similarity to Ideal Solution (TOPSIS) can be adapted to the selection of the optimal alternative. To validate the feasibility and efficiency of the proposed model, we present an example of car seat supplier selection for an automobile manufacturing enterprise.

1. Introduction

The increasing pressure of environmental and social issues from the stakeholders including customers, non-government organization, government, and market competitors, forces more and more enterprises in the market to introduce sustainable development into enterprise production and management process. Sustainable supplier selection (SSS) plays an important role in sustainable development [1,2]. Bearing in mind the triple bottom line (TBL) of sustainable development [1], purchasers should take into account the economic, environmental and social performances of suppliers simultaneously during SSS, such as cost, pollutant emission, resource saving, employment practices [3]. It is noteworthy that its downstream customer requirements drive all the production and operation activities of an enterprise along the sustainable supply chain. Incorporating sustainable customer requirements (CRs) into SSS can help to exactly select the most appropriate supplier with which the enterprise wishes to develop the stable business relationship to maximize the customer satisfaction. To

---

[†]Work supported by the Theme-based Research Projects of the Research Grants Council (T32-101/15-R) and the Fundamental Research Funds for the Central Universities (2042018kf0006).

achieve it, the purpose of this paper is to develop a novel model for incorporating CRs into SSS.

For any supplier selection problem, the determination of supplier assessment criteria and their importance ratings (IRs) known as criteria weights plays a basic role. During the production and management process of an enterprise, sustainable supplier assessment criteria (ACs) are determined by its sustainable product development strategies (DSs), while CRs further drive DSs. The relationships between SRs, DSs, and AC (Fig. 1) fit very well with the quality function deployment(QFD) approach which is a requirement-driven product development toll which can translate CRs into the design requirements [4]. The typical QFD consists of four linked phases, in which each phase's outputs generated from the phase's inputs are converted into the new inputs of the next phase. And in each phase, the link between inputs and outputs is House of Quality (HOQ), which consists of inputs, IRs of inputs, outputs, relationship matrix between inputs and outputs and IRs of outputs [4]. Within HOQ, IRs of outputs can be obtained by other elements. Inspired by the QFD analysis method, we will propose a two-phased approach to determine the ACs and their IRs.

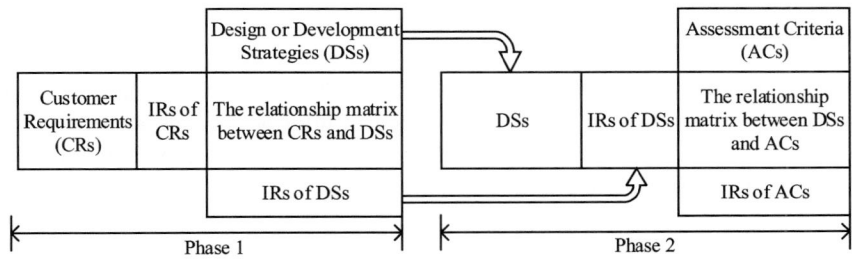

Fig. 1. The proposed two-phase QFD method.

After the determination of ACs and their IRs, the alternative sustainable suppliers should be identified, and the ranking or decision method need to be used to determine the best one. Up till now, several multi-criteria decision making (MCDM) methods have been proposed to help to select the optimal supplier, such as VIKOR, TOPSIS, AHP, ELECTRE, PROMETHEE, etc. Among the aforementioned decision-making techniques, TOPSIS (Technique for Order Preference by Similarity to Ideal Solution) approach is a widely accepted method because of its advantages of the untrammeled range of criteria and alternatives involved and the lower computation complexity. In the model to be proposed, TOPSIS will be introduced to determine the optimal supplier. Also, during the SSS, there are a significant number of subjective judgements need to be made. Because of the incompleteness of information needed and the ambiguity of people cognition on qualitative objectives, it is very hard for decision makers (DMs) responsible for SSS to provide accurate assessment information. Actually, in practical application, DMs prefer to utilize linguistic information to model and manage the uncertainty. Starting with the pioneering work of Zadeh in 1975, the fuzzy linguistic approaches have garnered considerable attention. To date, different extensions and improvements have been successively proposed [5]. Possibility distribution based HFLTS (i.e. possibility distribution-based HFLTS, PD-HFLTS) [6], which is developed on the basis of context-free grammar judgements (CFGJs) and hesitant fuzzy linguistic term set (HFLTS) [5], can provide a new more flexible and accurate way to characterize and manage uncertainty. In the proposed model, PD-HFLTS are utilized to deal with the uncertainty in SSS.

The paper is organized as follows. In Section 2, we develop a two-phase QFD methodology to determine the weights of AC and propose a novel TOPSIS integrated PD-HFLTS for SSS. In Section 3, an example is proposed to illustrate the feasibility of the proposed method. Finally, a discussion of the proposed approach and the conclusions are presented in Section 4.

2. Proposed Model

In this section, we introduce a novel model integrating two-phase QFD method with TOPSIS for SSS, in which the assessment information can be elicited and handled in the use of PD-HFLTS. For the relevant definitions and operation rules of PD-HFLTS, please refer to the literature [5,6].

2.1. Two-phase QFD methodology

2.1.1. Phase 1: CRs and DSs

The customers involved are asked to identify their requirements $\{CR_1,\ldots,CR_i,\ldots,CR_I\}$ for sustainable product and to provide their judgements for the IRs of CRs using CFGJs with the aid of the linguistic term set (LTS) $S^1 = \{s_0^1, \cdots, s_l^1, \cdots, s_g^1\}$ expressed as

$$S^1 = \{s_0^1 = \text{very unimportant}, s_1^1 = \text{unimportant}, s_2^1 = \text{weakly unimportant},$$
$$s_3^1 = \text{fair}, \ s_4^1 = \text{weakly important}, s_5^1 = \text{important}, s_6^1 = \text{very unimportant}\} \quad (1)$$

Then the judgements can be transferred into the HFLTSs embedded with possibility distribution and represented as $A = (a_1, \cdots, a_i, \cdots, a_I)$ where a_i

represents the IR of CR_i, and its possibility distribution can be denoted by $P^{CR-i} = \{p_0^{CR-i}, \cdots, p_l^{CR-i}, \cdots, p_g^{CR-i}\}$.

Secondly, the DMs responsible for SSS specify $\{DS_1, \ldots, DS_j, \ldots, DS_J\}$ to capture CRs and further construct the relationship matrix $\mathbf{R}^{CR-DS} = \left(r_{ij}^{CR-DS}\right)_{I \times J}$ utilizing CFGJs with the aid of the LTS $S^2 = \{s_0^2, \cdots, s_l^2, \cdots, s_g^2\}$ depicted by formula (2), in which r_{ij}^{CR-DS} represents the relationship level between CR_i and DS_j and can be transferred into HFLTS embedded with possibility distribution which is denoted by $P_{ij}^{CR-DS} = \{p_0^{(CR-DS)ij}, \cdots, p_l^{(CR-DS)ij}, \cdots, p_g^{(CR-DS)ij}\}$.

$$S^2 = \{s_0^2 = \text{very low}, s_1^2 = \text{low}, s_2^2 = \text{weakly low}, s_3^2 = \text{medium},$$
$$s_4^2 = \text{weakly high}, s_5^2 = \text{high}, s_6^2 = \text{very high}\} \quad (2)$$

Thirdly, to calculate the IRs of DSs, the matrix $\mathbf{R}^{CR-DS}$ should be transformed into the normalized version $\bar{\mathbf{R}}^{CR-DS} = \left(\bar{r}_{ij}^{CR-DS}\right)_{I \times J}$ based on the definition of the expected value of PD-HFLTS [6], where $\bar{r}_{ij}^{CR-DS}$ is a real number satisfying $0 \leq \bar{r}_{ij}^{CR-DS} \leq 1$ and $\sum_{j=1}^{J} \bar{r}_{ij}^{CR-DS} = 1$. Then the IRs of DSs $\mathbf{B} = (b_1, \cdots, b_j, \cdots, b_J)$ can be obtained using the hesitant fuzzy linguistic weighted average (HFLWA) [6], in which b_j is an HFLTS embedded with possibility distribution $P^{DS-j} = \{p_0^{DS-j}, \cdots, p_l^{DS-j}, \cdots, p_g^{DS-j}\}$ determined by

$$b_j = \text{HFLWA}\{a_1, \cdots, a_I\} = \text{HFLWA}\{P^0, \cdots, P^I\} = \{p_0^{DS-j}, \cdots, p_l^{DS-j}, \cdots, p_g^{DS-j}\} \quad (3)$$

where $p_l^{DS-j} = \sum_{i=1}^{I} \bar{r}_{ij}^{CR-RS} p_l^{CR-i}$.

2.1.2. Phase 2: DSs and ACs

After the determination of DSs and their IRs, DMs are asked to identify ACs to capture DSs and to construct the relationship matrix between DSs and ACs to determine the IRs of ACs in the same way. Denote the ACs by $\{AC_1, \ldots, AC_n, \ldots, AC_N\}$ and the relationship matrix by $\mathbf{R}^{DS-AC} = \left(r_{jn}^{DS-AC}\right)_{J \times N}$ which is constructed utilizing CFGJs with the assistance of the LTS S^2. Analogously, $\mathbf{R}^{DS-AC} = \left(r_{jn}^{DS-AC}\right)_{J \times N}$ is transformed into the normalized matrix $\bar{\mathbf{R}}^{DS-AC} = \left(\bar{r}_{jn}^{DS-AC}\right)_{J \times N}$ in which $\bar{r}_{jn}^{DS-AC}$ is a real number satisfying

$0 \leq \bar{r}_{jn}^{DS-AC} \leq 1$ and $\sum_{n=1}^{N} \bar{r}_{jn}^{DS-AC} = 1$. Then, integrating the IRs of DSs **B** and the normalized relationship matrix $\bar{R}^{DS-AC}$ using HFLWA, the IRs of ACs $C = (c_1, \cdots, c_n, \cdots, c_N)$ can be obtained, in which c_n is an HFLTS embedded with possibility distribution $P^{AC-n} = \{p_0^{AC-n}, \cdots, p_l^{AC-n}, \cdots, p_g^{AC-n}\}$ and can be determined by

$$c_n = \text{HFLWA}\{b_1, \cdots, b_J\} = \text{HFLWA}\{P^0, \cdots, P^J\} = \{p_0^{AC-n}, \cdots, p_l^{AC-n}, \cdots, p_g^{AC-n}\} \quad (4)$$

where $p_l^{AC-n} = \sum_{j=1}^{J} \bar{r}_{ij}^{DS-AC} p_l^{DS-j}$.

For the convenience of calculation, the IRs of ACs are normalized as $\{\bar{c}_1, \cdots, \bar{c}_n, \cdots, \bar{c}_N\}$ where $\bar{c}_n$ is a real number satisfying $0 \leq \bar{c}_n \leq 1$ and $\sum_{n=1}^{N} \bar{c}_n = 1$.

2.2. *TOPSIS based on PD-HFLETS*

Following the determination of IRs of ACs, DMs identify the potential sustainable suppliers $\{SP_1, \ldots, SP_m, \ldots, SP_M\}$ and give their judgements on each supplier's performances on each AC utilizing CFGJs with the assistance of the LTS S^3 depicted by the formula (5). The results are denoted by the decision matrix $E = (e_{mn})_{M \times N}$, where e_{mn} denotes the performance of the alternative SP_m on the criteria AC_n and is an HFLTS embedded with possibility distribution transformed from CFGJ, in which the possibility distribution embedded in e_{mn} can be denoted by $P_{mn} = \{p_0^{mn}, \cdots, p_l^{mn}, \cdots, p_g^{mn}\}$.

$$S^3 = \{s_0^3 = \text{very poor}, s_1^3 = \text{poor}, s_2^3 = \text{slightly poor}, s_3^3 = \text{fair},$$
$$s_4^3 = \text{slightly good}, s_5^3 = \text{good}, s_6^3 = \text{very good}\} \quad (5)$$

Then, the positive ideal solution (PIS) and the negative ideal solution (NIS) can be determined using the following formula:

$$PIS = \{e_n^+ | \max_m \{e_{mn}\}, n = 1, 2, \cdots, N\}, NIS = \{e_n^- | \min_m \{e_{mn}\}, n = 1, 2, \cdots, N\} \quad (6)$$

where e_n^+ and e_n^- are HFLTSs embedded with their possibility distributions denoted by $P_n^+ = \{p_0^{n+}, \cdots, p_l^{n+}, \cdots, p_g^{n+}\}$ and $P_n^- = \{p_0^{n-}, \cdots, p_l^{n-}, \cdots, p_g^{n-}\}$, respectively.

And then, the separation measures of each alternative from the PIS and the NIS, E_m^+ and E_m^-, can be calculated based on the weighted distance measure by the following equations

$$E_m^+ = \sum_{n=1}^{N} \overline{c}_n d\left(e_{mn}, e_n^+\right), \quad E_m^- = \sum_{n=1}^{N} \overline{c}_n d\left(e_{mn}, e_n^-\right) \quad (7)$$

where $d\left(e_{mn}, e_n^+\right)$ and $d\left(e_{mn}, e_n^-\right)$ are the distance measures of PD-HFLTS whose definition was given by the literature [6].

Finally, the closeness coefficient of each alternative to the PIS can be obtained by the equation

$$CC_m = \frac{E_m^-}{E_m^+ + E_m^-}. \quad (8)$$

Based on the descending order of CC_m, we can rank the alternatives and determine the optimal supplier.

3. Illustrations

In this section, an illustrative example of the car seat supplier selection for an automobile manufacturing enterprise is presented.

Firstly, customers provided their requirements $\{CR_1, CR_2, CR_3, CR_4, CR_5\}$ presented in Table 1 and gave their judgements on the IRs of CRs expressed as {at least s_5^1, between s_2^1 and s_4^1, between s_4^1 and s_5^1, between s_5^1 and s_6^1, at most s_4^1}. Then DMs specified six development strategies $\{DS_1, DS_2, DS_3, DS_4, DS_5, DS_6\}$ also presented in Table 1 and constructed the relationship matrix $\boldsymbol{R}^{CR-DS} = \left(r_{ij}^{CR-DS}\right)_{5\times 6}$ presented in Table 2. Then, the IRs of DSs were determined and also presented in Table 2. After that, seven supplier assessment criteria $\{AC_1, AC_2, AC_3, AC_4, AC_5, AC_6, AC_7\}$ were identified and presented in Table 1. And then the relationship matrix $\boldsymbol{R}^{DS-AC} = \left(r_{jn}^{DS-AC}\right)_{6\times 7}$ was constructed, and the IRs of ACs were obtained, which are presented in Table 3. Finally, the DMs identified four potential suppliers $\{SP_1, SP_2, SP_3, SP_4\}$ and further gave the assessment information $E = \left(e_{mn}\right)_{4\times 7}$ showed in Table 4. After the implementation of TOPSIS, the closeness coefficients of all the alternatives were determined, and the final ranking order of alternatives was obtained with the result being $sp_2 \succ sp_1 \succ sp_4 \succ sp_3$.

Table 1. The five CRs, the six DSs and the seven ACs for the illustrative example.

CRs	content	DSs	content	ACs	content
CR_1	Safely and comfortably	DS_1	Ergonomics	AC_1	Costs
CR_2	High performance-price ratio	DS_2	Safety for drivers and passengers	AC_2	Quality
CR_3	Durability	DS_3	Coordination with the overall design of the car	AC_3	Time and flexibility
CR_4	Heathy and environment friendly	DS_4	Economic	AC_4	Technology capability
CR_5	Responsible for society	DS_5	Environment	AC_5	Eco-design
		DS_6	Manufacturing process responsible for society	AC_6	Environmental management system
				AC_7	Social measures

Table 2. The relationship matrix between CRs and DSs.

	DS_1	DS_2	DS_3	DS_4	DS_5	DS_6
CR_1	at least s_5^2	at least s_5^2	s_0^2	at most s_2^2	s_0^2	s_2^2
CR_2	at most s_3^2	at most s_2^2	at most s_1^2	s_6^2	s_3^2	s_0^2
CR_3	at most s_3^2	between s_2^2 and s_4^2	s_0^2	at least s_4^2	at most s_1^2	s_0^2
CR_4	s_0^2	between s_4^2 and s_5^2	at most s_1^2	between s_2^2 and s_4^2	s_6^2	s_0^2
CR_5	s_0^2	between s_4^2 and s_5^2	at most s_1^2	at most s_3^2	s_0^2	s_6^2
IRs of DSs	(0,0,0.042,0. 042,0.117,0. 271,0.196)	(0.072,0.072, 0.1,0.1,0.25, 0.507,0.357)	(0.008,0.008,0. 022,0.022,0.02 2,0.018,0.018)	(0.024,0.024,0. 191,0.191,0.44 1,0.393,0.143)	(0,0,0.083,0. 083,0.108,0. 239,0.214)	(0.096,0.096,0. 096,0.096,0.09 6,0.071,0.071)

Table 3. The relationship matrix between DSs and ACs.

	AC_1	AC_2	AC_3	AC_4	AC_5	AC_6	AC_7
DS_1	s_3^2	between s_4^2 and s_5^2	s_0^2	at least s_5^2	s_0^2	s_0^2	s_0^2
DS_2	s_0^2	s_5^2	s_0^2	s_2^2	s_2^2	s_0^2	s_3^2
DS_3	s_3^2	between s_4^2 and s_5^2	s_6^2	s_5^2	s_3^2	s_0^2	s_0^2
DS_4	s_6^2	between s_4^2 and s_5^2	between s_3^2 and s_5^2	s_3^2	s_4^2	s_0^2	s_0^2
DS_5	s_0^2	s_0^2	s_0^2	s_3^2	s_6^2	s_6^2	s_4^2
DS_6	s_0^2	s_0^2	s_0^2	s_0^2	at most s_3^2	at most s_3^2	s_6^2
B	0.115	0.217	0.056	0.236	0.159	0.066	0.152

Table 4. The relevant decision information during the process of TOPSIS.

	AC_1	AC_2	AC_3	AC_4	AC_5	AC_6	AC_7	E_m^+	E_m^-	CC_m
sp_1	s_4^3	s_4^3	s_5^3	s_4^3	s_5^3	s_4^3	between s_4^3 and s_5^3	0.89	1.875	0.678
sp_2	at most s_2^3	s_5^3	s_3^3	s_5^3	at least s_5^3	at least s_4^3	at least s_4^3	0.656	2.109	0.763
sp_3	s_3^3	s_3^3	s_4^3	s_3^3	s_3^3	between s_2^3 and s_3^3	between s_2^3 and s_3^3	2.253	0.511	0.185
sp_4	at least s_5^3	s_4^3	s_6^3	s_5^3	s_4^3	at most s_3^3	at most s_3^3	1.283	1.482	0.536
PIS	at least s_5^3	s_5^3	s_6^3	s_5^3	at least s_5^3	at least s_4^3	at least s_4^3			
NIS	at most s_2^3	s_3^3	s_3^3	s_3^3	s_3^3	at most s_3^3	at most s_3^3			

4. Conclusion

In this paper, we propose a novel model for SSS integrating the two-phase QFD and TOPSIS within PD-HFLTS context, and provide an illustrative example to validate the feasibility and efficiency of the model. The main contributions of the paper are the following two aspects: the two-phase QFD approach is introduced to incorporate CRs into the SSS; PD-HFLTS is utilized to handle the uncertainty.

References

1. J. Elkington, *Environmental Quality Management*, 8(1), 37-51 (1998).
2. E. Hassini, C. Surti, C. Searcy, *International Journal of Production Economics*, 140(1), 69-82 (2012).
3. J. Dai, J. Blackhurst, *International Journal of Production Research*, 50(19), 1-17 (2012).
4. L. K. Chan, M. L. Wu, *European Journal of Operational Research*, 143(3), 463-497 (2002).
5. R. M. Rodríguez, L. Martínez, F. Herrera, *Information Sciences*, 241(12), 28-42 (2013).
6. Z. B. Wu, J. P. Xu, *IEEE transactions on cybernetics*, 46(3), 694-705 (2016).

Neutrosophic AHP and prioritization of legal service outsourcing firms/law offices

Cengiz Kahraman, Başar Öztayşi, Sezi Çevik Onar and Eda Boltürk
Department of Industrial Engineering, Istanbul Technical University
34367, Macka, Besiktas, Istanbul, Turkey

Neutrosophic sets are an extension of intuitionistic fuzzy sets. Neutrosophic sets provided a new approach to uncertainty with their three components: Truth, Indeterminacy, and Falsity. The performance of law offices can be comparatively measured by multicriteria decision making methods. Linguistic assessments can be used in this process rather than exact numerical evaluations. In this paper, we used a neutrosophic analytic hierarchy process (AHP) for comparing the performances of law offices. The illustrative problem hierarchy includes four criteria and four alternatives are given.

1. Introduction

In this study, our aim is to develop a novel Analytic Hierarchy Process (AHP) under neutrosophic environment and apply it to the performance assessment of law firms. Law firms come in a variety of shapes and sizes, ranging from single-attorney law practices to multi- state, multi-staffed legal organizations. When a company needs legal service outsourcing, it can apply to these law firms.

Neutrosophic sets handle the uncertainty through three concepts: Truth, Indeterminacy and Falsity. Different from intuitionistic fuzzy sets, neutrosophic AHP enables decision makers to take into account their falsity in defining a membership function. Theory of Neutrosophic sets have been developed by many researchers and employed in multicriteria decision making methods in a short time ([1], [2], [3], [5], [6], [7], [9], [10]).

There are many extensions of ordinary fuzzy AHP: Fuzzy AHP with type I fuzzy sets, fuzzy AHP with type II fuzzy sets, fuzzy AHP with intuitionistic fuzzy sets, fuzzy AHP with hesitant fuzzy sets. Neutrosophic AHP publications are relatively few and should be further researched.

The organization of the remaining paper is as follows. Section 2 gives some information about the types of law offices. Section 3 presents the preliminaries for neutrosophic sets. Section 4 includes the proposed neutrosophic AHP. Section 5 gives the application of the proposed method. The last section presents the conclusions and suggestions for further research.

2. Law Offices and Performance Criteria

There are a variety of law firms to choose from, generally broken down by size, type of practice, (for example, litigation, criminal defense, or transactional), location, or legal topic (like personal injury law, family law or tax law.) While there is no one-size-fits all solution to solving legal problems, choosing the right law firm can make the difference between a successful outcome and missed opportunity. Knowing which law firm to hire will depend on a number of factors — including your finances, geographical location, personal work preferences, and your specific legal challenge or need. Below is a summary of the various types of law firms available in most areas.

Solo Law Firms: Solo law firms are run by a single lawyer. These "solo practitioners" typically handle general legal matters on a variety of topics — ranging from personal injury law to family law, but may also specialize in one particular area of law, like patent law. Small Law Firms: Small law firms generally employ from two to ten attorneys. They often allow the lawyers an opportunity to collaborate with other lawyers on complicated or related legal matters. Large Law Firms: Large law firms can range in size from several dozens of lawyers and employees, to several thousands of employees and can exist in multiple cities, states, and even countries. Large law firms specialize in all areas of the law and typically have big legal departments, such as corporate, employment, and real estate groups. Litigation vs. Transactional Law Firms: Law firms are sometimes broken down by the type of legal services they offer. For example, a law firm might only focus on litigation, representing clients in court cases or it can focus on transactional matters involving heavy paperwork relating to disputes over money, property, and insurance. Criminal Law Firms: Law firms specializing in criminal defense against crimes such as securities fraud, DUI and other crimes often focus on representing private clients who can afford their own criminal defense attorney.

The performance of a law firm can be measured by the following criteria: loan collection success in the assigned files ($), early loan collection success (%), difficulty coefficient of the assigned files, and conformance to their goals.

3. Neutrosophic Sets

Since the introduction of fuzzy logic, many systems have been developed in order to deal with approximate and uncertain reasoning; among the latest and most general proposal is neutrosophic logic, introduced by Smarandache [8] as a generalization of fuzzy logic and several related systems [4]. Neutrosophic logic is based on neutrosophy. Fuzzy logic extends classical logic by assigning a membership between 0 and 1 to variables. As a generalization of fuzzy logic, neutrosophic logic introduces a new component called "indeterminacy", and carries more information than fuzzy logic.

Definition 1. Let X be a space of points (objects), with a generic element in X denoted by x. A neutrosophic set A in X is characterized by a truth-membership function $T_A(x)$, an indeterminacy-membership $I_A(x)$ and a falsity-membership function $F_A(x)$. The functions $T_A(x), I_A(x)$ and $F_A(x)$ are real standard or nonstandard subsets of $]0^-,1^+[$, that is $T_A(x): X \to]0^-,1^+[, I_A(x): X \to]0^-, 1^+[, F_A(x): X \to]0^-, 1^+[$. A simplification of A is denoted by

$$A = \{x, T_A(x), I_A(x), F_A(x) | x \in X\} \quad (1)$$

There is no restriction on the sum of $T_A(x), I_A(x)$ and $F_A(x)$, so $0^- \leq \sup T_A(x) + \sup I_A(x) + F_A(x) \leq 3^+$.

Definition 2. Let A and B be two simplified neutrosophic sets (SNSs), the following operations can be true

$.A+B = \{<x, T_A(x) + T_B(x) - T_A(x).T_B(x), I_A(x) + I_B(x) - I_A(x).I_B(x), F_A(x) + F_B(x) - F_A(x).F_B(x) > [x \in X\}; \quad (2)$

$A.B = \{<x, T_A(x).T_B(x), I_A(x).I_B(x), F_A(x).F_B(x) > [x \in X\}; \quad (3)$

$\lambda.A = \{<x, 1-(1 - T_A^{(x)})^\lambda, 1-(1 - I_A^{(x)})^\lambda, 1-(1 - F_A^{(x)})^\lambda > [x \in X\}, \lambda > 0; \quad (4)$

$A^\lambda = \{< x, T_A^\lambda(x), I_A^\lambda(x), F_A^\lambda(x) > [x \in X\}, \lambda > 0 \quad (5)$

Definition 3. Aggregation of SNSs can be made by using Eq. (2):
For a SNS $A_j (j = 1,2, \ldots\ldots, n)$,

$$F_w(A_1, A_2, \ldots, A_n) = \langle 1 - \prod_{j=1}^n \left(1 - T_{A_j}(x)\right)^{w_j}, 1 - \prod_{j=1}^n (1 - I_{A_j}(x))^{w_j}, 1 - \prod_{j=1}^n (1 - F_{A_j}(x))^{w_j} \rangle \quad (6)$$

where $W = (w_1, w_2, \ldots, w_n)$ is the weight vector of $A_j (j = 1,2, \ldots, n), w_j \in [0,1]$ and $\sum_{j=1}^n w_j = 1$.

4. Neutrosophic AHP

In this study, we extended Buckley's fuzzy AHP using interval neutrosophic sets. Below, the steps of the extended Neutrosophic AHP are presented.

Step 1: Define the multicriteria decision making problem with its criteria and alternatives.

Step 2: Collect evaluations from experts using a questionnaire form including pairwise comparisons of criteria and alternatives.

Step 3: Transform the experts' linguistic judgments into neutrosophic sets using the following proposed interval neutrosophic scale in Table 1.

Table 1. Proposed simplified neutrosophic scale.

Linguistic Terms	Simplified NSs
Exactly equal (EE) importance	(1,1,1)
Medium High (MH) importance	(0.7, 0.4, 0.3)
High (H) importance	(0.8, 0.3, 0.2)
Very High (VH) importance	(0.9, 0.2, 0.1)
Absolutely High (AH) importance	(1, 0, 0)

To the best knowledge of the authors, this scale is first time proposed in the literature presenting linguistic values with its corresponding simplified neutrosophic sets.

Step 4: Produce the weights of criteria and alternatives from pairwise comparisons matrices.

Step 4.1. Check the consistency of pairwise comparison matrices by employing Eq. (7)

$$dv = 0.6 + 0.4T - 0.2I - 0.4F \tag{7}$$

where dv is the deneutrosophicated value.

Step 4.2. Aggregate the evaluations of experts using Eq. (6).

Step 4.3. Calculate geometric mean for criteria and sub-criteria. The geometric means of the parameters of SNS are calculated by applying Eq. (8):

$$T_1 = [1 \times T_{12} \times \ldots \times T_{1n}]^{1/n}$$
$$\ldots$$
$$T_n = [T_{n1} \times T_{n2} \times \ldots \times 1]^{1/n} \tag{8}$$
$$I_{1m} = [1 \times I_{12m} \times \ldots \times I_{1nm}]^{1/n}$$
$$\ldots$$
$$I_{im} = [I_{n1m} \times I_{n2m} \times \ldots \times 1]^{1/n}$$
$$F_{1m} = [1 \times F_{12u} \times \ldots \times F_{1nu}]^{1/n}$$
$$\ldots$$
$$F_{im} = [F_{n1m} \times F_{n2m} \times \ldots \times 1]^{1/n}$$

Step 4.4. Obtain neutrosophic weights of criteria and sub-criteria by dividing T, I and F values with the sum of the geometric means in the row for lower, medium and upper parameters as presented in Eq. (9).

$$c_{1u} = [1 \times c_{12u} \times \ldots \times c_{1nu}]^{1/n}$$
$$\ldots \tag{9}$$
$$c_{iu} = [c_{n1u} \times c_{n2u} \times \ldots \times 1]^{1/n}$$

Assume that the sums of the geometric mean values in the row are a_{1s} for lower parameters; a_{2s} for medium parameters; and a_{3s} for upper parameters. Finally $\tilde{r}_{ij}$ matrix is obtained by using a_{ij} values obtained above by Eq. (10):

$$\tilde{r}_{ij} = \begin{cases} \left(\dfrac{a_{1l}}{a_{3s}}, \dfrac{b_{1m}}{a_{2s}}, \dfrac{c_{1u}}{a_{1s}}\right) \\ \left(\dfrac{a_{2l}}{a_{3s}}, \dfrac{b_{2m}}{a_{2s}}, \dfrac{c_{2u}}{a_{1s}}\right) \\ \vdots \\ \left(\dfrac{a_{il}}{a_{3s}}, \dfrac{b_{im}}{a_{2s}}, \dfrac{c_{iu}}{a_{1s}}\right) \end{cases} \quad (10)$$

Step 4.5. Repeat the steps summarized above for the pairwise comparisons of alternatives.

Step 5: Combine the neutrosophic weights of criteria with the neutrosophic performance scores of the alternatives using Eq. (6).

The neutrosophic weights and neutrosophic performance scores are aggregated as follows:

$$\tilde{U}_i = \sum_{j=1}^{n} \widetilde{w}_j \, \tilde{r}_{ij}, \forall i. \quad (11)$$

where $\tilde{U}_i$ is the fuzzy utility of alternative i; $\widetilde{w}_j$ is the weight of the criterion j and $\tilde{r}_{ij}$ is the performance score of alternative i with respect to criterion j.

Step 6. Rank the alternatives with respect to their weighted scores.

5. Application

The problem involves four criteria and four law firm alternatives. There are three experts E1, E2, and E3 whose weights are 0.3, 0.2, and 0.5, respectively. The pairwise comparison matrices for the criteria and alternatives are presented in Table 2 and Table 3.

Table 2. Pairwise comparison matrices for criteria by three DMs.

DM1	C1	C2	C3	C4	C5	DM2	C1	C2	C3	C4	C5	DM3	C1	C2	C3	C4	C5
C1	EE	MH	H	VH	MH	C1	EE	MH	VH	AH	H	C1	EE	MH	VH	AH	MH
C2	ML	EE	H	VH	MH	C2	ML	EE	MH	H	MH	C2	ML	EE	MH	VH	MH
C3	L	L	EE	MH	ML	C3	VL	ML	EE	MH	ML	C3	VL	ML	EE	MH	L
C4	VL	VL	ML	EE	VL	C4	AL	L	ML	EE	L	C4	AL	VL	ML	EE	VL
C5	ML	ML	MH	VH	EE	C5	L	ML	MH	H	EE	C5	ML	ML	H	VH	EE

Table 3. Pairwise comparison matrices for the criteria and alternatives.

	WRT C1					WRT C2			
DM1	LF-1	LF-2	LF-3	LF-4	DM1	LF-1	LF-2	LF-3	LF-4
LF-1	EE	H	MH	VH	LF-1	EE	ML	MH	L
LF-2	L	EE	ML	H	LF-2	MH	EE	H	L
LF-3	ML	MH	EE	H	LF-3	ML	L	EE	AL
LF-4	VL	L	L	EE	LF-4	H	H	AH	EE
DM2	LF-1	LF-2	LF-3	LF-4	DM2	LF-1	LF-2	LF-3	LF-4
LF-1	EE	VH	EE	AH	LF-1	EE	L	MH	ML

Table 3. (Cont'd)

LF-2	VL	EE	ML	VH	LF-2	H	EE	VH	MH
LF-3	EE	MH	EE	H	LF-3	ML	VL	EE	L
LF-4	AL	VL	L	EE	LF-4	MH	ML	H	EE
DM3	LF-1	LF-2	LF-3	LF-4	DM3	LF-1	LF-2	LF-3	LF-4
LF-1	EE	AH	MH	H	LF-1	EE	VL	ML	L
LF-2	AL	EE	ML	ML	LF-2	VH	EE	H	MH
LF-3	ML	MH	EE	H	LF-3	MH	L	EE	ML
LF-4	L	MH	L	EE	LF-4	H	ML	MH	EE
		WRT C3					WRT C4		
DM1	LF-1	LF-2	LF-3	LF-4	DM1	LF-1	LF-2	LF-3	LF-4
LF-1	EE	MH	VH	ML	LF-1	EE	L	MH	ML
LF-2	ML	EE	H	L	LF-2	H	EE	H	MH
LF-3	VL	L	EE	AL	LF-3	ML	L	EE	MH
LF-4	MH	H	AH	EE	LF-4	MH	ML	ML	EE
DM2	LF-1	LF-2	LF-3	LF-4	DM2	LF-1	LF-2	LF-3	LF-4
LF-1	EE	VH	MH	ML	LF-1	EE	VL	ML	L
LF-2	VL	EE	ML	VL	LF-2	VH	EE	H	MH
LF-3	ML	MH	EE	L	LF-3	MH	L	EE	ML
LF-4	MH	VH	H	EE	LF-4	H	ML	MH	EE
DM3	LF-1	LF-2	LF-3	LF-4	DM3	LF-1	LF-2	LF-3	LF-4
LF-1	EE	VH	MH	ML	LF-1	EE	AL	ML	L
LF-2	VL	EE	L	AL	LF-2	AH	EE	VH	H
LF-3	ML	H	EE	L	LF-3	MH	VL	EE	ML
LF-4	MH	AH	H	EE	LF-4	H	L	MH	EE

Table 4 presents the aggregated evaluations of decision makers for the criteria based on Eq. (6).

Table 4. Aggregated evaluations for the criteria.

0.500	0.500	0.500	0.600	0.400	0.400	0.774	0.231	0.231	0.877	0.131	0.131	0.622	0.381	0.381
0.400	0.400	0.600	0.500	0.500	0.500	0.633	0.372	0.372	0.783	0.221	0.221	0.600	0.400	0.400
0.231	0.231	0.774	0.372	0.372	0.633	0.500	0.500	0.500	0.600	0.400	0.400	0.352	0.352	0.654
0.131	0.131	0.877	0.221	0.221	0.783	0.400	0.400	0.600	0.500	0.500	0.500	0.221	0.221	0.783
0.381	0.381	0.622	0.400	0.400	0.600	0.654	0.352	0.352	0.783	0.221	0.221	0.500	0.500	0.500

Table 5 gives the aggregated pairwise comparison matrix of alternatives with respect to C1. Because of the space constraints, we donot give the rest of the aggregated matrices of alternatives with respect to the other criteria.

Table 5. Aggregated matrix of alternatives with respect to C1.

0.500	0.500	0.500	0.840	0.185	0.185	0.582	0.421	0.421	0.787	0.234	0.234
0.185	0.185	0.840	0.500	0.500	0.500	0.400	0.400	0.600	0.609	0.334	0.457
0.421	0.421	0.582	0.600	0.400	0.400	0.500	0.500	0.500	0.700	0.300	0.300
0.234	0.234	0.787	0.457	0.334	0.609	0.300	0.300	0.700	0.500	0.500	0.500

The result of Eq. (8) for the criteria is given in Table 6. The result of Eq. (8) for the alternatives with respect to C1 is given in Table 7. Applying Eqs. (9-10), the ranking is obtained as LF4 > LF2 > LF1 > LF3.

Table 6. Geometric means of the parameters in Table 4.

T	I	F
0.662	0.297	0.297
0.569	0.366	0.397
0.390	0.360	0.577
0.264	0.264	0.694
0.523	0.359	0.429

Table 7. Geometric means of the parameters in Table 5.

T	I	F
0.719	0.391	0.391
0.468	0.415	0.649
0.616	0.479	0.511
0.437	0.411	0.700

6. Conclusion

Neutrosophic sets present a new point of view to uncertainty introducing a new parameter, namely falsity. They can be viewed as the extension of intuitionistic fuzzy sets. Neutrosophic decision making has been quickly penetrated into the literature. Neutrosophic AHP has been proposed in this paper and applied to the performance comparison of law firms successfully. Our proposal is based on Buckley's ordinary fuzzy AHP method since it had the least criticism in the past. Simplified neutrosophic sets can be replaced by interval-valued neutrosophic sets in the proposed method and suitable operations can be developed for further research.

Acknowledgments

This work is supported by Scientific and Technological Research Council of Turkey (TÜBİTAK), TEYDEB 1505, Grant No: 5170012.

References

1. Bausys, R., Zavadskas, E.K. (2015). Multicriteria decision making approach by VIKOR under interval neutrosophic set environment, Economic Computation & Economic Cybernetics Studies & Research, 49(4), 33-48.
2. Bausys, R., Zavadskas, E.K., Kaklauskas, A. (2015). Application of neutrosophic set to multicriteria decision making by COPRAS, Economic computation and economic cybernetics studies and research / Academy of Economic Studies 49(2), 1-15.
3. Broumi S., Ye. J., Smarandache. F. (2015). An extended TOPSIS method for multiple attribute decision making based on interval neutrosophic uncertain linguistic variables, Neutrosophic Sets and Systems 8, 22-31, 2015.
4. Kharal, A. (2013) A Neutrosophic Multicrriteria Decision Making Method, New Mathematics and Natural Computation, Creighton University, USA, 2013.

5. Liu, P., Zhang, L. (2018). The Extended VIKOR Method for Multiple Criteria Decision Making Problem Based on Neutrosophic Hesitant Fuzzy Set, General Mathematics, 1-13, http://fs.gallup.unm.edu/TheExtendedVIKORMethod.pdf.
6. Peng, J.-J., Wang, J.-Q., Wu, X.-H., Wang, J., Chen, X.-H. (2015). Multi-valued Neutrosophic Sets and Power Aggregation Operators with Their Applications in Multi-criteria Group Decision-making Problems, International Journal Of Computational Intelligence Systems, 8(2), 345-363.
7. Peng, J.-J., Wang, J.-Q., Zhang, H. Y., Chen, X.-H. (2014). An outranking approach for multi-criteria decision making problems with simplified neutrosophic sets, Applied Soft Computing, 25, 336-346.
8. Smarandache, F. (1998) A unifying field in logics neutrosophy: neutrosophic probability, set and logic. American Research Press, Rehoboth.
9. Tian, Z.-P., Zhang, H.-Y., Wang, J., Wang, J.Q., Chen, X.-H. (2015). Multi-criteria decision- making method based on a cross-entropy with interval neutrosophic sets, International Journal of Systems Science 47(15), 3598-3608.
10. Wang, Z., Liu, L. (2016). Optimized PROMETHEE Based on Interval Neutrosophic Sets for New Energy Storage Alternative Selectio, Rev. Téc. Ing. Univ. Zulia. 39(9), 69-77.

Evaluation of smart cities with integrated hesitant fuzzy linguistic AHP–COPRAS method[*]

Esin Mukul

Industrial Engineering Department, Galatasaray University
Çırağan Cad. 36, Ortaköy/Istanbul, 34349, Turkey

Merve Güler

Industrial Engineering Department, Galatasaray University
Çırağan Cad. 36, Ortaköy/Istanbul, 34349, Turkey

Gülçin Büyüközkan[†]

Industrial Engineering Department, Galatasaray University
Çırağan Cad. 36, Ortaköy/Istanbul, 34349, Turkey

Smart cities are based on the principle of self-management of transportation, infrastructure, and networks, which shall be managed coherently for rational solutions of city problems. The mixed structure of these solutions involves various contradictory criteria. When information is of uncertain nature, it is difficult to decide on and rank smart cities. The hesitant fuzzy linguistic term set technique overcomes the uncertainty- and hesitation-related difficulties of this problem. The aim of the study is to introduce an integrated hesitant fuzzy linguistic (HFL) multi-criteria decision-making approach to evaluate smart cities for the first time. The criteria are weighted with HFL Analytic Hierarchy Process. Then, smart cities are evaluated with the HFL COmplex PRoportional ASsessment method. Finally, the potential of this approach is presented through a case study.

1. Introduction

Rapidly growing populations and migration from rural areas to urban space have considerable impact on the problems of cities. These difficulties not only negatively affect the economic and social life in cities. It also deteriorates the life quality of their inhabitants. The new smart city concept presents significant potential as a rational solution to urban problems. It is a development vision for urban planning that integrates urban assets and resources by utilizing information

---

[*]This work is financially supported by Galatasaray University Research Fund (Project No: 18.402.001).
[†]Corresponding author; e-mail: gulcin.buyukozkan@gmail.com.

technologies. Works have been accelerated in recent years to develop smart cities that will raise the level of social prosperity in a complex network of living spaces [1-3].

Smart cities, which are based on the principle of self-management of transportation, infrastructure, and networks, are composed of many components. In this study, the evaluation of smart cities is approached as a multi-criteria decision making (MCDM) problem [4]. The mixed structure of smart cities evaluations involve various and contradictory criteria. It is difficult to decide on, and rank smart cities when information is of uncertain nature. Sometimes decision makers (DMs) may have difficulties in expressing their thoughts by numbers because these quantitative values are not similar to how they actually think in daily life. Furthermore, DMs can express their opinions more comfortably with words, instead of crisp numbers. The hesitant fuzzy linguistic term set (HFLTS) [5] overcomes the uncertainty of such MCDM problems.

Analytic Hierarchy Process (AHP) has a simple hierarchical structure and the ability to deal with complex decision problems [6]. Many possible values are used in HFL AHP to describe the hesitancy of DMs' assessments [7, 8]. COmplex PRoportional ASsessment (COPRAS) is another useful method that requires restricted subjective information from the DMs and allows the evaluation of both qualitative and quantitative criteria [9]. In the literature, the hesitant fuzzy hierarchical COPRAS method is applied for renewable energy selection [10]. For the first time in the literature, this study proposes a COPRAS method that is based on HFLTS with fuzzy envelope and HFL AHP.

In this study, a hesitant fuzzy linguistic (HFL) MCDM approach is presented to evaluate smart cities. In the first step, the HFL AHP method is implemented for computing the weights of criteria. In the second step, the HFL COPRAS method is applied to rank and evaluate smart cities in terms of their importance and benefit ratings. This paper contributes to smart cities literature by proposing a novel integrated HFL MCDM method.

The study has been organized as follows. The second section presents the proposed methodology. In Section 3, an application is illustrated to demonstrate the effectiveness of the methodology. Finally, the last section concludes the study.

2. The Proposed Methodology

The proposed methodology in this study consists of three basic steps:

Step 1. Determination of the criteria and alternatives for evaluating smart cities.

Step 2. Computation of the evaluation criteria weights with the HFL AHP method.

Step 3. Evaluation of smart cities by HFL COPRAS method according to the criteria.

2.1. The Hesitant Fuzzy Linguistic Term Set

Decision problems encountered in actual life are often intertwined with uncertainty and complexity. Linguistic information can help to represent uncertainty. Torra and Narukawa [5] introduced the hesitant fuzzy set (HFS) to address this challenge. Alcantud and Torra [16] proved a decomposition theorem for hesitant fuzzy sets. The degree of membership of an element in these sets may have many possible values between zero and one. Please refer to [11, 12] for the preliminaries of HFLTS.

Liu and Rodriguez [13] present an MCDM model where DMs express their evaluations with linguistic expressions. This model presents these expressions by representing a set of HFLTS.

S is defined as a set of linguistic terms, $S = \{s_0, \ldots, s_g\}$. An HFLTS, H_s, is an ordered finite subset of S.

The upper bound H_{s+} and lower bound H_{s-} of the HFLTS are described as

$$H_{s+} = \max(s_i) = s_j,\ s_i \in H_S \text{ et } s_i \leq s_j \forall_i \tag{1}$$

$$H_{s-} = \min(s_i) = s_j,\ s_i \in H_S \text{ et } s_i \leq s_j \forall_i \tag{2}$$

The envelope of the HFLTS, $env(H_S)$, is a linguistic interval with the upper bound (max) and the lower bound (min) as shown below:

$$env(H_S) = [H_s^-, H_s^+],\ H_s^- \leq H_s^+ \tag{3}$$

2.2. The HFL AHP Method for Determining the Relative Importance of Criteria

The following steps of HFL AHP are used to calculate the weights of criteria (please refer to [14] for the methodological steps of the HFL AHP method):

Step 1. First, pairwise comparison matrices are created and the compromise evaluations from the DMs are obtained with HFLTS, which are found with linguistic terms in Table 1.

Step 2. Using the OWA operator, the fuzzy envelope for HFLTS is aggregated and established [13]. This aggregation gives a trapezoidal fuzzy number as a result.

Step 3. The pairwise comparison matrix $(\widetilde{C})$, which consists of the aggregated fuzzy numbers generated in Step 2 with $\widetilde{c_{ij}} = (c_{ijl}, c_{ijm1}, c_{ijm2}, c_{iju})$, is obtained. The reciprocal values are obtained as:

Table 1. Linguistic scale for HFL AHP and HFL COPRAS.

Linguistic terms	si	Abb.	Triangular fuzzy number
Absolutely high importance	s10	(AHI)	(7,9,9)
Very high importance	s9	(VHI)	(5,7,9)
Essentially high importance	s8	(ESHI)	(3,5,7)
Weakly high importance	s7	(WHI)	(1,3,5)
Equally high importance	s6	(EHI)	(1,1,3)
Exactly low importance	s5	(EE)	(1,1,1)
Equally low importance	s4	(ELI)	(0.33,1,1)
Weakly low importance	s3	(WLI)	(0.2,0.33,1)
Essentially low importance	s2	(ESLI)	(0.14,0.2,0.33)
Very low importance	s1	(VLI)	(0.11,0.14,0.2)
Absolutely low importance	s0	(ALI)	(0.11,0.11,0.14)

$$\widetilde{c_{ij}} = (\frac{1}{c_{iju}}, \frac{1}{c_{ijm2}}, \frac{1}{c_{ijm1}}, \frac{1}{c_{ijl}}) \qquad (4)$$

Step 4. For each row ($\tilde{r}_i$) of the matrix $\tilde{C}$, the fuzzy geometric mean is calculated using Eq. (5).

$$\tilde{r}_i = (\tilde{c}_{i1} \otimes \tilde{c}_{i2} \ldots \otimes \tilde{c}_{in})^{1/n} \qquad (5)$$

Step 5. The fuzzy weight ($\widetilde{w}_i^{CR}$) of each main criterion is computed with ($\tilde{r}_i$) values as shown below:

$$\widetilde{w}_i^{CR} = \tilde{r}_i \otimes (\tilde{r}_1 \otimes \tilde{r}_2 \ldots \otimes \tilde{r}_n)^{-1} \qquad (6)$$

Step 6. The fuzzy global weights of sub-criteria are computed where $\widetilde{w}_{ij}^G$ is the global weight of sub-criteria.

$$\widetilde{w}_{ij}^G = \widetilde{w}_i^{CR} \times \widetilde{w}_j^{CR} \qquad (7)$$

Step 7. The trapezoidal fuzzy numbers $\widetilde{w}_{ij}^G$ are defuzzified using Eq. (8) and the defuzzified values are normalized using Eq. (9).

$$w_{ij}^G = \frac{\alpha + 2\beta + 2\gamma + \delta}{6} \qquad (8)$$

$$w_{ij}^N = \frac{w_{ij}^G}{\Sigma_i \Sigma_j w_{ij}^G} \qquad (9)$$

2.3. *The HFL COPRAS Method for Evaluating Smart Cities*

With this method, the ratings of alternatives are evaluated by using HFLTS. The HFL COPRAS method steps are as follows:

Step 1. Initially, the DMs evaluate the criteria and alternatives with regards to each other by using the linguistic scale given in Table 1. These linguistic expressions are converted to fuzzy numbers with fuzzy envelope. The decision matrices composed of the HFS formed by the DMs are defuzzified into crisp numbers with the center of area method. The crisp value x_{ij} can be found using Eq. (10).

$$x_{ij} = \frac{[(U_{x_{ij}} - L_{x_{ij}}) + (M_{x_{ij}} - L_{x_{ij}})]}{3} + L_{x_{ij}} \tag{10}$$

Step 2. Normalize the decision matrix with Eq. (11).

$$x_{ij}^* = \frac{x_{ij}}{\sum_{i=1}^{m} x_{ij}} \tag{11}$$

Step 3. Determine the weighted normalized decision matrix by using Eq. (12).

$$d_{ij} = x_{ij}^* \cdot w_j \tag{12}$$

Step 4. The sums S_i^- and S_i^+ of weighted standardized values are computed using the following equations for both beneficial and non-beneficial criteria separately:

$$S_{i+} = \sum_{j=1}^{k} d_{ij} \tag{13}$$

$$S_{i-} = \sum_{j=k+1}^{n} d_{ij} \tag{14}$$

Step 5. The Q_i values are relative importance values for each alternative. These values are calculated using Eq. (15). The best alternative is identified with the highest relative importance value.

$$Q_i = S_{i+} + \frac{\sum_{i=1}^{m} S_{i-}}{S_{i-} * \sum_{i=1}^{m} \frac{1}{S_{i-}}} \tag{15}$$

Step 6. The highest relative priority (Q_{max}) value is found.

Step 7. The performance index (P_i) of each alternative is computed with Eq. (16).

$$P_i = \left[\frac{Q_i}{Q_{max}}\right] \times 100\% \tag{16}$$

3. Application

Smart cities are one of the popular topics on the public agenda. There are many academic and industrial studies on this subject. According to the literature and industrial reports, 6 'smart' criteria are determined [15], which are likely to be highly relevant: economy, people, governance, mobility, environment and living. These criteria can successfully characterize a smart city. They can be broken down into 24 relevant sub-criteria (see in Table 2), which reflect the most important aspects of every smart characteristic. In order to support the ranking in the reports, smart cities are evaluated with analytical methods. There are six possible alternatives: A1 is Aalborg, A2 is Tampere, A3 is Luxembourg, A4 is Aarhus, A5 is Odense and A6 is Turku.

Table 2. Criteria for evaluation of smart cities.

Main criteria	Sub-criteria
Smart economy (C1)	• Ability to transform (C11) • Economic image& trademarks (C12) • Productivity (C13) • Entrepreneurship (C14)
Smart people (C2)	• Qualification level (C21) • Affinity to lifelong learning (C22) • Flexibility/ Creativity (C23) • Participation in public life (C24)
Smart governance (C3)	• Participation in decision-making (C31) • Public and social services (C32) • Transparent governance (C33) • Political strategies & perspectives (C34)
Smart mobility (C4)	• Local accessibility (C41) • International accessibility (C42) • Availability of ICT-infrastructure (C43) • Sustainable, innovative and safe transport systems (C44)
Smart environment (C5)	• Attractivity of natural conditions (C51) • Pollution (C52) • Environmental protection (C53) • Sustainable resource management (C54)
Smart living (C6)	• Cultural facilities (C61) • Health conditions (C62) • Education facilities (C63) • Touristic attractivity (C64)

Table 3. Pairwise comparisons of the main criteria of smart cities.

	C1	C2	C3	C4	C5	C6
C1	EE	Between ESHI and AHI	Between EHI and WHI	Between ESLI and ELI	Between ESLI and ELI	Between ELI and EHI
C2		EE	Between WHI and ESHI	Between ELI and EHI	Between ELI and EHI	Between ESLI and ELI
C3			EE	Between ALI and VLI	Between ESLI and ELI	Between ELI and EHI
C4				EE	Between ELI and EHI	Between EHI and WHI
C5					EE	Between EHI and WHI
C6						EE

Table 3 shows the pairwise comparisons of the main criteria, filled by DMs' evaluations by using the linguistic scale given in Table 1.

Considering the evaluation of DMs with the help of the linguistic scale given in Table 1, steps of HFL AHP are employed and the weights of criteria are computed. These weights are given in Table 4.

Table 4. Weights of criteria.

Criteria	Weights	Criteria	Weights	Criteria	Weights	Criteria	Weights
C11	0.048	C23	0.028	C41	0.019	C53	0.084
C12	0.024	C24	0.022	C42	0.066	C54	0.068
C13	0.052	C31	0.002	C43	0.066	C61	0.069
C14	0.041	C32	0.009	C44	0.102	C62	0.012
C21	0.017	C33	0.025	C51	0.029	C63	0.037
C22	0.067	C34	0.017	C52	0.082	C64	0.015

Considering the evaluation of DMs with the help of the linguistic scale given in Table 1, steps of HFL COPRAS are applied and smart cities alternatives are evaluated. Evaluation and ranking of smart cities are given in Table 5.

Table 5. Evaluation of smart cities with respect to criteria.

Alternatives	Si+	Si-	Qi	Pi	Ranking
A1	3.767	0.175	3.922	93.212	4
A2	3.557	0.119	3.785	89.955	6
A3	4.055	0.179	4.207	100	1
A4	3.974	0.172	4.132	98.220	2
A5	3.716	0.175	3.871	92.001	5
A6	3.932	0.179	4.084	97.074	3

As a result, Luxembourg (A3) is found to be the "smartest" city among the six alternatives with the final performance value.

4. Conclusion

Smart cities are based on the idea of restructuring cities that maximize the efficiency for people and nature. Smart cities aim for human-focused, strategic, environment-friendly management approaches, service areas, and increased living standards. In this study, the evaluation of smart cities is considered as an integrated HFL MCDM problem. First, criteria weights are calculated by using the HFL AHP method. Then, the selected smart cities are evaluated with the HFL COPRAS method. An application is illustrated to demonstrate the effectiveness of the methodology and its results are given. The smartest city is found as Luxembourg (A3).

For future research, the same problem can be solved using aggregation operators for group decision making to aggregate DMs' evaluations. As a second perspective, both HFL AHP and COPRAS methodologies can be applied with intuitionistic fuzzy sets.

Acknowledgments

The authors kindly express their appreciation for the support of industrial experts.

References

1. R. G. Hollands, "Will the real smart city please stand up? Intelligent, progressive or entrepreneurial ?", City, **12(3)**, 303-320 (2008).
2. A. Cocchia, "Smart and digital city: A systematic literature review", In Smart city, Springer International Publishing, 13-43 (2014).
3. T. Nam and T. A. Pardo, "Conceptualizing smart city with dimensions of technology, people, and institutions", In Proceedings of the 12th annual international digital government research conference: digital government innovation in challenging times, 282-29 (2011, June).
4. C. L. Hwang and K. Yoon, "Multiple Attribute Decision Making — Methods and Applications", Springer-Verlag, Berlin (1981).
5. V. Torra and Y. Narukawa, "On hesitant fuzzy sets and decision. In Fuzzy Systems", FUZZ-IEEE 2009, 1378-1382 (2009, August).
6. T. L. Saaty, "The Analytic Hierarchy Process", McGraw-Hill, New York (1980).
7. S. Cevik Onar, B. Oztaysi and C. Kahraman, "Strategic decision selection using hesitant fuzzy TOPSIS and interval type-2 fuzzy AHP: a case study", Int. J. Comput. Int Sys., **7(5)**, 1002-1021 (2014).
8. B. Zhu and Z. Xu, "Analytic hierarchy process-hesitant group decision making", Eur. J. Oper. Res., **239(3)**, 794-801 (2014).
9. E. K. Zavadskas, A. Kaklauskas and V. Sarka, "The new method of multicriteria complex proportional assessment of projects", Technological and economic development of economy, **1(3)**, 131-139 (1994).
10. M. Mousavi and R. Tavakkoli-Moghaddam, "Group decision making based on a new evaluation method and hesitant fuzzy setting with an application to an energy planning problem", International Journal of Engineering-Transactions C: Aspects, **28(9)**, 1303 (2015).
11. R. M. Rodriguez, L. Martinez and F. Herrera, "Hesitant fuzzy linguistic term sets for decision making", IEEE Trans. on Fuzzy Syst., **20(1)**, 109-119 (2012).
12. V. Torra, "Hesitant fuzzy sets", International Journal of Intelligent Systems, **25(6)**, 529-539 (2010).
13. H. Liu and R. M. Rodríguez, "A fuzzy envelope for hesitant fuzzy linguistic term set and its application to multicriteria decision making", Information Sciences, **258**, 220-238 (2014).
14. S. Çevik Onar, G. Büyüközkan, B. Öztayşi and C. Kahraman, "A new hesitant fuzzy QFD approach: An application to computer workstation selection", Appl. Soft Comput., **46**, 1-16 (2016).
15. R. Giffinger and H. Gudrun, "Smart cities ranking: an effective instrument for the positioning of the cities?", ACE: Architecture, City and Environment, **4(12)**, 7-26 (2010).
16. J. C. R. Alcantud and V. Torra, "Decomposition theorems and extension principles for hesitant fuzzy sets", Information Fusion, **41**, 48-56 (2018).

Weighting performance indicators of law offices by using interval valued intuitionistic fuzzy AHP[*]

Basar Oztaysi

Department of Industrial Engineering, Istanbul Technical University
Besiktas, Istanbul, 34367, Turkey

Sezi Cevik Onar

Department of Industrial Engineering, Istanbul Technical University
Besiktas, Istanbul, 34367, Turkey

Cengiz Kahraman

Department of Industrial Engineering, Istanbul Technical University
Besiktas, Istanbul, 34367, Turkey

Performance measurement (PM) is defined as the process of collecting, analyzing and/or reporting information regarding the performance of individuals, teams or the whole organization. The first and one of the most important steps in performance measurement is determining the performance indicators and their weights. In this study, we focus on performance measurement of law offices which deal with follow-up of unpaid bills. We propose a decision model for weighting the performance indicators using interval valued intuitionistic fuzzy AHP.

1. Introduction

Performance measurement (PM) is used in different levels of management. Top management level use PM to evaluate the overall results of the past activities and identify the future position of the company. At the individual level, it is used for identifying the deficiencies and encouraging for the future activities (Meyer, 2002). PM is also critical for outsourced business processes. Outsourcing is the action of contracting a specific task, function or process to an external company instead of doing it with organization's own resources. Starting from 1980s, outsourcing has become a key strategic factor which allows companies to only

---

[*]This work is supported by Scientific and Technological Research Council of Turkey (TÜBİTAK), TEYDEB 1505, Grant No.: 5170012.

focus on their core competencies and benefit from other companies' expertise (Efendigil et al., 2008).

The literature provides several studies from various industries such as: tourism (Sainaghi et al., 2017; Huang and Coelho, 2017), carbon emission (Hu et al., 2017), energy (Ke at al., 2017; Oztaysi et al., 2018), healthcare (Otay et al., 2017) customer relationship (Kim and Kim, 2009; Oztaysi et al., 2011), supply chain (Maestrini et al., 2017; Oztaysi and Surer, 2014), manufacturing companies (Huang and Badurdeen, 2017; Oztaysi and Sari, 2012), insurance companies (Felicio and Rodrigues, 2015; Ak and Oztaysi, 2009), aftersales service (Lau et al., 2017; Cevik Onar et al., 2016),

Obtaining the weights of performance indicators is very critical for calculating overall performance score. The problem is a multicriteria decision making (MCDM) problem since there can be various numbers of criteria and sub-criteria involved into the problem. In the field of MCDM, Analytical Hierarchy Process (AHP) is commonly used for complex decision processes. AHP is a structured technique for organizing and analyzing complex decisions using pairwise comparisons and hierarchical structure. Recently, fuzzy extensions of fuzzy sets are also proposed (Kahraman et al., 2016). Fuzzy sets are used in the literature to provide more realistic and accurate results through the decision making process. While regular fuzzy sets provide a good representation of linguistic terms, since they have some limitations, new extensions of fuzzy sets are proposed in the literature. Type2 fuzzy sets, hesitant fuzzy sets, intuitionistic fuzzy sets fuzzy multisets are some of the extensions. In this study, in order to better represent expert opinions and reach more accurate results we propose using interval-valued intuitionistic set based AHP method for performance measurement model.

2. Methodology

In this paper, we use an interval-valued intuitionistic AHP method for performance measurement. The methodology is composed of nine steps and ends by ranking the alternatives based on criteria weights obtained in the first phase and expert judgements. In the following, we present the steps of our proposed method.

Step 1. Linguistic pairwise comparison matrices are formed according to the decision model and decision makers fill the matrices using linguistic scale given in Table 1.

Table 1. Linguistic scale and its corresponding IVIFS.

Linguistic Terms	Membership & Non-membership values
Absolutely Low (AL)	([0.10, 0.25], [0.65, 0.75])
Very Low (VL)	([0.15, 0.30], [0.60, 0.70])
Low (L)	([0.20, 0.35], [0.55, 0.65])
Medium Low (ML)	([0.25, 0.4]), [0.50, 0.60])
Approximately Equal (AE)	([0.45, 0.55], [0.30, 0.45])
Medium High (MH)	([0.50, 0.60], [0.25, 0.40])
High (H)	([0.55,0.65], [0.20, 0.35])
Very High (VH)	([0.60,0.70], [0.15,0.30])
Absolutely High (AH)	([0.65,0.75], [0.10,0.25])
Exactly Equal (EE)	([0.5, 0.5], [0.5, 0.5]).

Step 2. The linguistic pairwise matrices are converted to their corresponding interval-valued intuitionistic fuzzy sets using the scale given in Table 1 in order to obtain intuitionistic pairwise comparison matrices and aggregated pairwise comparison matrix ($\widecheck{R}_g$).

$$\widecheck{R}_g = \begin{bmatrix} \left(\left[\mu^-_{g_{11}}, \mu^+_{g_{11}}\right], \left[v^-_{g_{11}}, v^+_{g_{11}}\right]\right) & \cdots & \left(\left[\mu^-_{g_{1n}}, \mu^+_{g_{1n}}\right], \left[v^-_{g_{1n}}, v^+_{g_{1n}}\right]\right) \\ \vdots & \ddots & \vdots \\ \left(\left[\mu^-_{g_{n1}}, \mu^+_{g_{n1}}\right], \left[v^-_{g_{n1}}, v^+_{g_{n1}}\right]\right) & \cdots & \left(\left[\mu^-_{g_{nn}}, \mu^+_{g_{nn}}\right], \left[v^-_{g_{nn}}, v^+_{g_{nn}}\right]\right) \end{bmatrix} \quad (1)$$

Step 3. Score judgement matrices ($\tilde{S}$) are formed using the scoring function given in Eq. (8).

$$\tilde{S} = \begin{bmatrix} \left[\mu^-_{g_{11}} - v^+_{g_{11}}, \mu^+_{g_{11}} - v^-_{g_{11}}\right] & \cdots & \left[\mu^-_{g_{1n}} - v^+_{g_{1n}}, \mu^+_{g_{1n}} - v^-_{g_{1n}}\right] \\ \vdots & \ddots & \vdots \\ \left[\mu^-_{g_{n1}} - v^+_{g_{n1}}, \mu^+_{g_{n1}} - v^-_{g_{n1}}\right] & \cdots & \left[\mu^-_{g_{nn}} - v^+_{g_{nn}}, \mu^+_{g_{nn}} - v^-_{g_{nn}}\right] \end{bmatrix} \quad (2)$$

Step 4. Interval exponential matrices ($\tilde{A}$) are calculated as given in Eq. (3).

$$\tilde{A} = \begin{bmatrix} \left[e^{(\mu^-_{g_{11}} - v^+_{g_{11}})}, e^{((\mu^+_{g_{1n}} - v^-_{1n}))}\right] & \cdots & \left[e^{(\mu^-_{g_{1j}} - v^+_{g_{1j}})}, e^{(\mu^+_{g_{1j}} - v^-_{g_{1j}})}\right] \\ \vdots & \ddots & \vdots \\ \left[e^{(\mu^-_{g_{n1}} - v^+_{g_{n1}})}, e^{((\mu^+_{g_{n1}} - v^-_{n1}))}\right] & \cdots & \left[e^{(\mu^-_{g_{nn}} - v^+_{g_{nn}})}, e^{(\mu^+_{g_{nn}} - v^-_{g_{nn}})}\right] \end{bmatrix}$$

$$= \begin{bmatrix} [\tilde{a}^-_{11}, & \tilde{a}^+_{11}] & \cdots & [\tilde{a}^-_{1n}, & \tilde{a}^+_{1n}] \\ \vdots & & \ddots & \vdots \\ [\tilde{a}^-_{n1}, & \tilde{a}^+_{n1}] & \cdots & [\tilde{a}^-_{nn}, & \tilde{a}^+_{nn}] \end{bmatrix} \quad (3)$$

Step 5. Priority vectors of the interval exponential matrices are calculated using Eq. (4).

$$\widetilde{w}_i = \left[\frac{\sum_{j=1}^n \tilde{a}^-_{ij}}{\sum_{i=1}^n \sum_{j=1}^n \tilde{a}^+_{ij}}, \frac{\sum_{j=1}^n \tilde{a}^+_{ij}}{\sum_{i=1}^n \sum_{j=1}^n \tilde{a}^-_{ij}}\right] = [w_i^-, w_i^+], i = 1, \ldots, n \quad (4)$$

Step 6. Possibility degree matrices are obtained using Eq. (5) and Eq. (6).

$$P(\widetilde{w}_i > \widetilde{w}_j) = p_{ij} = \frac{max\left(0, w_i^+ - w_j^-\right) - max\left(0, w_i^- - w_j^+\right)}{(w_i^+ - w_i^-) + (w_j^+ - w_j^-)} \quad (5)$$

$$P(\widetilde{w}_j > \widetilde{w}_i) = p_{ji} = \frac{max\,(0,w_j{}^+ - w_i{}^-) - max\,(0,w_j{}^- - w_i{}^+)}{(w_i{}^+ - w_i{}^-) + (w_j{}^+ - w_j{}^-)} \quad (6)$$

Step 7. Possibility degrees are prioritized using Eq. (7).

$$w_i = \frac{\sum_{j=1}^{n} p_{ij} - 1}{n} + 0.5 \quad (7)$$

Step 8. The weights are normalized as given in Eq. (8).

$$w_i^T = \frac{w_i}{\sum_{i=1}^{n} w_i} \quad (8)$$

Step 9. The steps are repeated for each criteria with respect to the goal. Finally the global weight of each criterion is calculated.

3. Application

A company wants to build a performance measurement system for the Law Offices which make the follow-up of the unpaid bills. The first step of the performance measurement system is to determine the criteria and their weights. To this end, first a literature review is maintained and the list of the criteria is created with the decision makers as given in Figure 1.

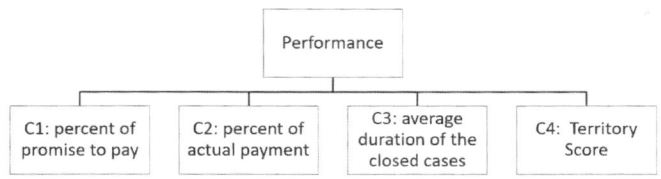

Figure 1. Performance indicators.

In this application the four performance criteria are prioritized by using Intuitionistic Fuzzy AHP. First three decision makers first evaluate the four alternatives using the linguistic scale.

Table 2. Linguistic Evaluations of the three experts.

DM1	C1	C2	C3	C4	DM2	C1	C2	C3	C4	DM3	C1	C2	C3	C4
C1	EE	ML	H	MH	C1	EE	L	MH	MH	C1	EE	L	MH	H
C2		EE	VH	VH	C2		EE	H	VH	C2		EE	H	H
C3			EE	ML	C3			EE	E	C3			EE	ML
C4				EE	C4				EE	C4				EE

After the linguistic variables are transformed into interval intuitionistic fuzzy sets, they are aggregated to reach aggregated pairwise comparison matrix ($\widetilde{R}_g$) as mentioned in Step 2. The aggregated pairwise comparison matrix is given in Table 3.

Table 3. Aggregated pairwise comparison matrix.

	C1	C2	C3	C4
C1	([0,5,0,5],[0,5,0,5])	([0,22,0,37],[0,53,0,63])	([0,52,0,62],[0,23,0,38])	([0,52,0,62],[0,23,0,38])
C2	([0,53,0,63],[0,22,0,37])	([0,5,0,5],[0,5,0,5])	([0,57,0,67],[0,18,0,33])	([0,58,0,68],[0,17,0,32])
C3	([0,23,0,38],[0,52,0,62])	([0,18,0,33],[0,57,0,67])	([0,5,0,5],[0,5,0,5])	([0,25,0,45],[0,42,0,55])
C4	([0.23.0.38],[0,52,0,62])	([0,17,0,32],[0,58,0,68])	([0,48,0,58],[0,27,0,42])	([0,5,0,5],[0,5,0,5])

Next, the score judgement index is obtained using Eq. (8). The score judgement index is given in Table 4.

Table 4. Score judgement index.

	C1	C2	C3	C4
C1	[0,0]	[0,0]	[0.17,0.42]	[0.27,0.52]
C2	[0,0]	[0,0]	[0.17,0.42]	[0.17,0.42]
C3	[-0.42,-0.17]	[-0.42,-0.17]	[0,0]	[0.07,0.25]
C4	[-0.52,-0.27]	[-0.42,-0.17]	[-0.22,-0.06]	[0,0]

The interval exponential matrix is calculated as given in Eq. (3) and the weights are calculated using Eq. (4). The results are given in Table 5.

Table 5. Interval exponential matrix and the weights.

	C1	C2	C3	C4	weights
C1	[1,1]	[0.66,0.85]	[1.14,1.47]	[1.14,1.47]	[0.215,0.317]
C2	[1.18,1.52]	[1,1]	[1.26,1.63]	[1.31,1.68]	[0.259,0.386]
C3	[0.68,0.88]	[0.62,0.79]	[1,1]	[0.74,1.03]	[0.166,0.245]
C4	[0.68,0.88]	[0.6,0.77]	[1.07,1.37]	[1,1]	[0.183,0.266]

Possibility degree matrix is formed using Eqs. (5) and (6) and then possibility degrees are prioritized and normalized to find the criteria weights.

Table 6. Possibility degree matrix, weights and normalized weights.

	C1	C2	C3	C4	Weights	Normalized weights
C1	0.50	0.25	0.83	0.73	1.363	0.262
C2	0.75	0.50	1.00	0.97	1.543	0.297
C3	0.17	0.00	0.50	0.38	1.110	0.213
C4	0.27	0.03	0.62	0.50	1.184	0.228

The results reveal that C2 (percent of actual payment) is the most important criteria and it is followed by C1 (percent of promise to pay). *Territory Score* takes the third place and the least important factor is found as *average duration of the closed cases*.

4. Conclusions

One of the most important step in performance measurement is obtaining the performance indicators and their priorities. In this study we focus on weighting the indicators in a real world performance measurement system. In the application three experts used linguistic terms to evaluate the performance indicators and the weights are calculated using interval intuitionistic fuzzy AHP Method.

Acknowledgments

This work is supported by Scientific and Technological Research Council of Turkey (TÜBİTAK), TEYDEB 1505, Grant No.: 5170012.

References

1. Ak. R., Oztaysi B. (2009) Performance measurement of insurance companies by using balanced scorecard and ANP. In proceedings of 10th Annual International Symposium on the Analytical Hierachy Process.
2. Cevik Onar S., Oztaysi B., Kahraman C. (2017) Dynamic intuitionistic fuzzy multi-attribute aftersales performance evaluation, Complex & Intelligent Systems, Volume 3, Issue 3, pp 197–204.
3. Felicio J.A., Rodrigues R. (2015) Organizational factors and customers' motivation effect on insurance companies' performance. J. Bus. Res. 68(7), 1622-1629.
4. Hu X., Si T. Liu C. (2016) Total factor carbon emission performance measurement and development, Journal of Cleaner Production 142.
5. Huang, Y., Coelho, V.R. (2017) Sustainability performance assessment focusing on coral reef protection by the tourism industry in the coral triangle region. Tour. Management, 59, 510–527.
6. Kahraman C., Oztaysi B., Cevik Onar S. (2016) A comprehensive literature review of 50 years of fuzzy set theory, International Journal of Computational Intelligence Systems, 9 (sup1), 3-24.
7. Ke, M.T., Yeh C.H., Su, C.J. (2017) Cloud computing platform for real-time measurement and verification of energy performance. Applied Energy 188, 497-507.
8. Kim H.S., Kim Y.G. (2009) A CRM performance measurement framework: Its development process and application, Industrial Marketing Management, 38, (4), pp. 477-489.
9. Lau P.Y.Y., Tong J.L.Y.T., Lien B.Y., Hsu Y., Chong C. L. (2017) Ethical work climate, employee commitment and proactive customer service performance: Test of the mediating effects of organizational politics, Journal of Retailing and Consumer Services, Volume 35, pp 20-26.
10. Maestrini, V., Luzzini, D., Maccarrone, P. and Caniato, F. (2017), "Supply chain performance measurement systems: a systematic review and research agenda", International Journal of Production Economics, Vol. 183, pp. 299-315.

11. Meyer, M.W.: Rethinking Performance Measurement: Beyond the Balanced Scorecard. Cambridge University Press, New York (2002).
12. Otay I., Oztaysi B., Cevik Onar S., Kahraman C. (2017) Multi-expert performance evaluation of healthcare institutions using an integrated intuitionistic fuzzy AHP&DEA methodology, Knowledge-Based Systems 133, 90-106.
13. Oztaysi B., Çevik Onar S., Kahraman C., Karaşan A. (2018) Fuzzy Sets Based Performance Evaluation of Alternative Wind Energy Systems. In: Kahraman C., Kayakutlu G. (eds.) Energy Management — Collective and Computational Intelligence with Theory and Applications. Studies in Systems, Decision and Control, vol 149. Springer.
14. Oztaysi B., Kaya T. Kahraman C. (2011) Performance comparison based on customer relationship management using analytic network process, Expert Systems with applications, 38, 9788-9798.
15. Oztaysi B., Sari İ.U. (2012) Performance measurement of a manufacturing company using fuzzy analyric network process, International Journal of Applied Management Sciences, 4(4).
16. Oztaysi B., Surer O (2014) Supply Chain Performance Measurement Using SCOR Based Fuzzy VIKOR Approach, in (Eds. Cengiz Kahraman, Başar Öztayşi) Supply Chain Management Under Fuzziness: Recent Developments and Techniques, pp. 199-223.
17. Sainaghi R., Phillips P., Zavarrone E. (2017) Progress in Tourism Management Performance measurement in tourism firms: A content analytical meta-approach, Tourism Management 59:36-56.
18. Efendigil T., Önüt S. and Kongar E. (2008), A holistic approach for selecting a third-part y reverse logistics provider in the presence of vagueness, Comput. Indust. Eng. 54., 269–287.

Multi-criteria evaluation of law firms by using dynamic intuitionistic fuzzy sets

Sezi Çevik Onar
Industrial Engineering, Istanbul Technical University
Macka, Istanbul, 34367, Turkey
cevikse@itu.edu.tr

Basar Oztaysi
Industrial Engineering, Istanbul Technical University
Macka, Istanbul, 34367, Turkey

Cengiz Kahraman
Industrial Engineering, Istanbul Technical University
Macka, Istanbul, 34367, Turkey

Performance measurement in the service industry is a critical issue for identifying how well is the company working. In manufacturing companies, performance measurement primarily relies on capacity and the number of products produced without defects. How close is the final product to the targeted product is a tool for observing the effectiveness of production process. In the banking industry, the performance depends on revenue - cost analysis. Companies in service industry need to understand how successful their business goes; several techniques developed for measuring their performance. Most of those techniques need too many parameters to observe and prioritize for reflecting relative effects of individual criteria into the analysis. It is essential to specify each of them because more criteria mean more dimension and high accuracy. The criteria involved in law firms' performance measurement are uncertain, vague and collected at different periods. The purpose of this study is to observe longitudinal performance evaluations in law firms by using dynamic intuitionistic fuzzy sets.

1. Introduction

The increase in legal service businesses supported developments in an economy based on the service industry. Larger law firms fed by enlarged lawyer community and increased their market share. Measuring the performance of law firms and the lawyers in these firms become very important. Several studies focus on measuring the performance of law firms. Wang (2000) adapted stochastic production frontier function on law firms. In this study, legal service performance is considered as an output of various criteria that can be formulated as follows:

$$\ln Y_i = \beta_0 + \beta_1 \ln X_{1i} + \beta_2 (\ln X_{1i})^2 + \beta_3 \ln X_{2i}$$
$$+ \beta_4 (\ln X_{2i})^2 + \beta_5 \ln X_{3i} + \beta_6 (\ln X_{3i})^2$$
$$+ \beta_7 \ln X_{4i} + \beta_8 (\ln X_{4i})^2 + \beta_9 \ln X_{5i}$$
$$+ \beta_{10} (\ln X_{5i})^2 + \beta_{11} (\ln X_{1i})(\ln X_{5i}) + \beta_{12} D_i + \varepsilon_i \qquad (1)$$

where revenue of law company i (Yi) depends on the number of lawyers (X1i), associate-to-partner ratio (X2İ), number legal assistants (X3i), partners' legal experience (x4İ), success in an assessment(X5i), scale of the company (Di; national:1 or local:0) and the error term (εi).

According to the results of the study while the number of lawyers, paralegals, associates per partner ratio have a supportive effect on revenue, their quadratic terms have an adverse effect. On the other hand, legal experience has an unpredictable negative coefficient while it is quadratic equation has positive. When the effect of lawyers and experience observed together, their effect is adverse on revenue at 10% level. Company scale does not have a statistically significant effect on revenue (Wang, 2000).

It is hard to measure the output of a law firm because of the task variety. Total revenue can be considered as one of the performance indicators since the lack of access to detailed legal service reports of the firms. Other factors, such as the size of the firm may also have a substantial impact on performance evaluation.

The performance evaluation of law firms involves multiple criteria that should be measured at different periods. The nature of the performance evaluation often necessitates uncertain and imprecise evaluations. Intuitionistic fuzzy sets are excellent tools for dealing with uncertainty and imprecision. Therefore, in this study, we used dynamic intuitionistic fuzzy sets that enable dealing with both longitudinal and uncertain characteristics of law firms' performance evaluation.

The rest of the study is organized as follows: In Section 2, multi-criteria decision-making methods for performance evaluation are analyzed. Section 3 gives the details of dynamic intuitionistic fuzzy multi-attribute decision making. The considered law firms' performances are evaluated in Section 4 and the last section we conclude and give further suggestions.

This paper is organized as follows: Section 2 summarizes aftersales performance evaluation systems in literature. Section 3 presents basic concepts of the dynamic intuitionistic fuzzy sets. In Section 4 aftersales service performance of an electronics company is measured. The last section concludes the paper and gives some perspectives.

2. Multi-Criteria Decision Making Methods for Performance Evaluation

In literature, multi-criteria decision making techniques based performance evaluation systems applied on different areas. Multi-criteria decision making (MDCM) methods mostly provide statistical data support for decision makers to select the best among a set of criteria. Referring linguistic variables while dealing with problems is a conventional approach (Li et al., 2015). For evaluating more complex environments, fuzzy approaches for MCDM are preferred to reflect uncertainty into the decision making process.

Evaluating performance in the higher education sector is a real issue because it depends on intangible factors as, teachers' knowledge, experience, and qualifications. So that, there is not such a universal framework to apply all education sector. Different MCDM applications are used for estimating performance. Das et al. (2015) preferred to use fuzzy analytic hierarchy process (AHP) because it is beneficial to define complex systems' criteria and also determine their relative importance. Then apply MOOSRA (Multi-objective optimization on the basis of simple ratio analysis) which is simple to implement, gives robust results and requires less time for calculations (Das et al., 2015). Johnes (1996) selected to use data envelopment analysis (DEA) due to its' convenience of overcoming multiple inputs and multiple outcomes. As a result, he showed that personal characteristics and efforts of both institutions and individuals have an impact on aggregate level DEAs (Johnes, 1996).

Performance of air transportation industry depends on offers and service quality of the company. Indeed, service quality is a distinctive advantage for airlines companies. In order to increase service quality, they need to find out the critical point to improve that can add more value among others (Mardani et al., 2015).

Birgün et al. (2014) identify the facility location for the call center by using AHP, GRA and VIKOR methods. Selecting location among a set of different alternatives is effecting the whole performance of a company. Since it is very costly and difficult to change a decision when it has already implemented, Birgün et al. suggested to give a convenient decision with MCDM methods (Birgün et al., 2014).

SERVQUAL is a common approach for generating criteria's importance in healthcare industry; criteria determinations are done based on this approach. In the article of Shieh et al. (2010), criteria are compared according to their influences on each other with the DEMATEL method (Shieh et al., 2010).

Tourism and hospitality sector is another sector that MCDM has been popularly used (Mardani et al., 2016). Lin et al. (2009) preferred to use critical incident technique (CIT) and fuzzy AHP to find most important criteria of destination tour operators' (DTO) in order to increase service quality. After criteria are determined, relative weight of each of them are calculated with fuzzy AHP method for finding the most effective criteria overall.

3. Dynamic Intuitionistic Fuzzy MCDM

Intuitionistic fuzzy sets developed by Atanassov (1986) are one of the most used fuzzy extensions where the uncertainty can be better represented with both membership and non membership degrees. An intuistionistic fuzzy set $\tilde{S}$ can be defined as follows:

$$\tilde{S} = \{\langle x, \mu_{\tilde{S}}(x), v_{\tilde{S}}(x), \pi_{\tilde{S}}(x)\rangle; x \epsilon X\}, \qquad (2)$$

where $0 \leq \mu_{\tilde{S}}(x) + v_{\tilde{S}}(x) \leq 1$, $\pi_{\tilde{S}}(x) = 1 - \mu_{\tilde{S}}(x) - v_{\tilde{S}}(x)$ and $\mu_{\tilde{S}}(x): X \to [0,1]$, $v_{\tilde{S}}(x): X \to [0,1]$, $\pi_{\tilde{S}}(x): X \to [0,1]$ for every $x \epsilon X$.

Xu and Cai (2012) define the dynamic intuitionistic fuzzy sets where the evaluation time is added as an additional dimension. A dynamic Intuitionistic fuzzy set $\widetilde{S(t)}$ can be defined as follows:

$$\widetilde{S(t)} = \{\langle x, (\mu_{\widetilde{S(t)}}, v_{\widetilde{S(t)}}, \pi_{\widetilde{S(t)}})\rangle; x \epsilon X\}, \qquad (3)$$

where t is a time variable, $\mu_{\widetilde{S(t)}} \in [0,1]$, $v_{\widetilde{S(t)}} \in [0,1]$, $\mu_{\widetilde{S(t)}} + v_{\widetilde{S(t)}} \leq 1$ and $\pi_{\widetilde{S(t)}} = 1 - \mu_{\widetilde{S(t)}} - v_{\widetilde{S(t)}}$.

In this study, we utilize the dynamic intuitionistic MCDM method developed by Xu and Yager (2008). This approach enables evaluating longitudinal fuzzy performance therefore have been used in various studies Cevik Onar et al. (2016, 2017a, b). The steps of this method can be defined as follows (Xu and Yager, 2008):

Step 1. Define the criteria and the law firm alternatives.
Step 2. Evaluate alternatives longitudinally.
Step 3. In order to define the weights of the periods use Eq. (4) (Xu, 2008).

$$\omega(t_{k+1}) - \omega(t_k) = c, \omega(t_k) = \eta + (k-1)c \qquad (4)$$

where $\omega(t) = \big(\omega(t_1), \omega(t_2), \dots, \omega(t_p)\big)$ is the weights of the periods and c is a constant.

Step 4. Use dynamic weighted fuzzy averaging (DIFWA) operator for aggregating the performances collected at different periods (Xu and Yager, 2008).

$$\overline{DIFWA}_{\omega(t)}^{(\tilde{I}(t_1),\tilde{I}(t_2),...,\tilde{I}(t_n))} = \left(1 - \prod_{k=1}^{n}(1-\mu_{\tilde{I}(t_k)})^{\omega(t_k)}, \prod_{k=1}^{n} v_{\tilde{I}(t_k)}^{\omega(t_k)}, \prod_{k=1}^{n}(1-\mu_{\tilde{I}(t_k)})^{\omega(t_k)} - \prod_{k=1}^{n} v_{\tilde{I}(t_k)}^{\omega(t_k)}\right)$$

(5)

Step 5. Define the dynamic intuitionistic fuzzy positive and negative ideal solutions respectively, $Y^+ = (I_1^+, I_2^+, ..., I_m^+)^T$, $Y^+ = (I_1^-, I_2^-, ..., I_m^-)^T$ where $I_i^+ = (1,0,0)$ is the largest and $I_i^- = (0,1,0)$ is the smallest Intuitionistic fuzzy numbers.

Step 6. Calculate the closeness coefficient of each alternative using the following equation:

$$C_i = \frac{\sum_{j=1}^{m} w_j(1-v_{ij})}{\sum_{j=1}^{m} w_j(1+\pi_{ij})} \quad (6)$$

where $i = 1,2,...,n$ and w_j is the weight of the jth attribute.

4. Longitudinal Performance Evaluation of Law Firms

A GSM operator wants to evaluate the performance of the law firms that provides service to their company. There are five law firms that needs to be evaluated the evaluation criteria are the overall collection performance (OC), regional collection performance (RC), legal service quality (LS), attitude (AT), adaptation to the company procedures (CP) and the customer relations (CR). The performance of the law firms are evaluated by the managers of GSM operator technical team responsible for legal services. The evaluations are done by a quarterly basis. At each quarter, after multiple discussions, the compromise evaluations are achieved. The importance of the criteria are set as follows: The weight of collection performance (OC) is 35%, regional collection performance (RC) is 20%, legal service quality (LS) is 10%, attitude (AT) is 10%, adaptation to the company procedures (CP) is 10% and the customer relations (CR) is 15%. The results of the compromise evaluations are given in Table 1.

The evaluations collected at different periods are aggregated. The final aggregated evaluations are given in Table 2.

The closeness coefficients are achieved as in Table 3.

The results indicate that the law firm 4 shows the highest performance. In order to check the robustness of the evaluation a sensitivity analysis is performed. In this sensitivity analysis, the weights of the selected criterion have been increased incrementally from 0 to 1. In order to keep total sum of the criteria weights to 1, the weights of other criteria have been changed proportional to the initial weights. The results of the sensitivity analysis is given in Figure 1.

Table 1. Longitudinal evaluations of law firm performances.

	OC	RC	LS	t1 AT	CP	CR
Law firm 1	(0.4,0.3,0.3)	(0.4,0.5,0.1)	(0.7,0.1,0.2)	(0.1,0.6,0.3)	(0.9,0.1,0)	(0.5,0.1,0.4)
Law firm 2	(0.7,0.1,0.2)	(0.8,0.1,0.1)	(0.5,0.1,0.4)	(0.7,0.3,0)	(0.9,0,0.1)	(0.9,0,0.1)
Law firm 3	(0.5,0.4,0.1)	(0.8,0.1,0.1)	(0.8,0.1,0.1)	(0.9,0,0.1)	(0.4,0.1,0.5)	(0.7,0.3,0)
Law firm 4	(0.9,0.1,0)	(0.3,0.3,0.4)	(0.7,0.2,0.1)	(0.4,0.3,0.3)	(0.2,0.7,0.1)	(0.5,0.1,0.4)
Law firm 5	(0.6,0.3,0.1)	(0.1,0.1,0.8)	(0.4,0.4,0.2)	(0.8,0.1,0.1)	(0.9,0.1,0)	(0,0.3,0.7)

	OC	RC	LS	t2 AT	CP	CR
Law firm 1	(0.6,0.3,0.1)	(0.6,0.1,0.3)	(0.7,0.2,0.1)	(0.9,0.1,0)	(0.6,0.2,0.2)	(0.6,0.3,0.1)
Law firm 2	(0.4,0.1,0.5)	(0.7,0.2,0.1)	(0.2,0.2,0.6)	(0.4,0.2,0.4)	(0.1,0.8,0.1)	(0.2,0.1,0.7)
Law firm 3	(0,0.3,0.7)	(0.8,0.1,0.1)	(0.6,0.4,0)	(0.5,0.3,0.2)	(0.4,0.4,0.2)	(0.3,0.2,0.5)
Law firm 4	(0.3,0.5,0.2)	(0.7,0.1,0.2)	(0.8,0.1,0.1)	(0.2,0.2,0.6)	(0.2,0.7,0.1)	(0.5,0.5,0)
Law firm 5	(0.3,0.4,0.3)	(0.7,0.3,0)	(0.7,0.2,0.1)	(0.8,0.1,0.1)	(0.4,0.5,0.1)	(0.4,0.1,0.5)

	OC	RC	LS	t3 AT	CP	CR
Law firm 1	(0.7,0.1,0.2)	(0.6,0.1,0.3)	(0.7,0.1,0.2)	(0.3,0.4,0.3)	(0.1,0.2,0.7)	(0.4,0.1,0.5)
Law firm 2	(0.6,0.1,0.3)	(0.9,0.1,0)	(0.5,0.3,0.2)	(0.9,0.1,0)	(0.2,0.6,0.2)	(0.6,0.1,0.3)
Law firm 3	(0.2,0.6,0.2)	(0.4,0.1,0.5)	(0.5,0.3,0.2)	(0.7,0.3,0)	(0.8,0.1,0.1)	(0.4,0.6,0)
Law firm 4	(0.7,0.3,0)	(0.7,0.1,0.2)	(0.2,0.8,0)	(0.2,0.6,0.2)	(0.2,0.4,0.4)	(0.9,0,0.1)
Law firm 5	(0,0.3,0.7)	(0.7,0,0.3)	(0.6,0.3,0.1)	(0.1,0.1,0.8)	(0.2,0.6,0.2)	(0.6,0.4,0)

	OC	RC	LS	t4 AT	CP	CR
Law firm 1	(0.7,0.2,0.1)	(0.6,0.1,0.3)	(0.5,0.4,0.1)	(0.7,0.2,0.1)	(0.9,0.1,0)	(0.9,0.1,0)
Law firm 2	(0.8,0.1,0.1)	(0.5,0.5,0)	(0.7,0.2,0.1)	(0.5,0.2,0.3)	(0.6,0.2,0.2)	(0.8,0.1,0.1)
Law firm 3	(0.7,0.1,0.2)	(0.8,0.1,0.1)	(0.2,0.6,0.2)	(0.8,0.2,0)	(0.1,0.1,0.8)	(0.1,0.7,0.2)
Law firm 4	(0.9,0.1,0)	(0.7,0.2,0.1)	(0.4,0.4,0.2)	(0.9,0.1,0)	(0.9,0.1,0)	(0.9,0.1,0)
Law firm 5	(0.8,0,0.2)	(0.5,0.4,0.1)	(0.7,0.2,0.1)	(0.8,0.2,0)	(0.5,0.3,0.2)	(0.1,0.4,0.5)

Table 2. Aggregated evaluations.

	Aggregated		
	OC	RC	LS
Law firm 1	(0.659,0.183,0.158)	(0.583,0.117,0.3)	(0.632,0.2,0.168)
Law firm 2	(0.681,0.1,0.219)	(0.746,0.219,0.035)	(0.552,0.211,0.237)
Law firm 3	(0.461,0.245,0.294)	(0.722,0.1,0.178)	(0.473,0.376,0.151)
Law firm 4	(0.795,0.192,0.013)	(0.673,0.147,0.18)	(0.51,0.348,0.142)
Law firm 5	(0.554,0,0.446)	(0.589,0,0.411)	(0.649,0.242,0.109)
	AT	CP	CR
Law firm 1	(0.653,0.239,0.108)	(0.745,0.141,0.114)	(0.735,0.125,0.14)
Law firm 2	(0.696,0.169,0.135)	(0.496,0,0.504)	(0.697,0,0.303)
Law firm 3	(0.747,0,0.253)	(0.492,0.132,0.376)	(0.321,0.478,0.201)
Law firm 4	(0.662,0.219,0.119)	(0.652,0.272,0.076)	(0.838,0,0.162)
Law firm 5	(0.686,0.132,0.182)	(0.492,0.367,0.141)	(0.342,0.295,0.363)

Table 3. Closeness coefficients.

	Closeness coefficient
Law firm 1	0.711
Law firm 2	0.726
Law firm 3	0.619
Law firm 4	0.746
Law firm 5	0.660

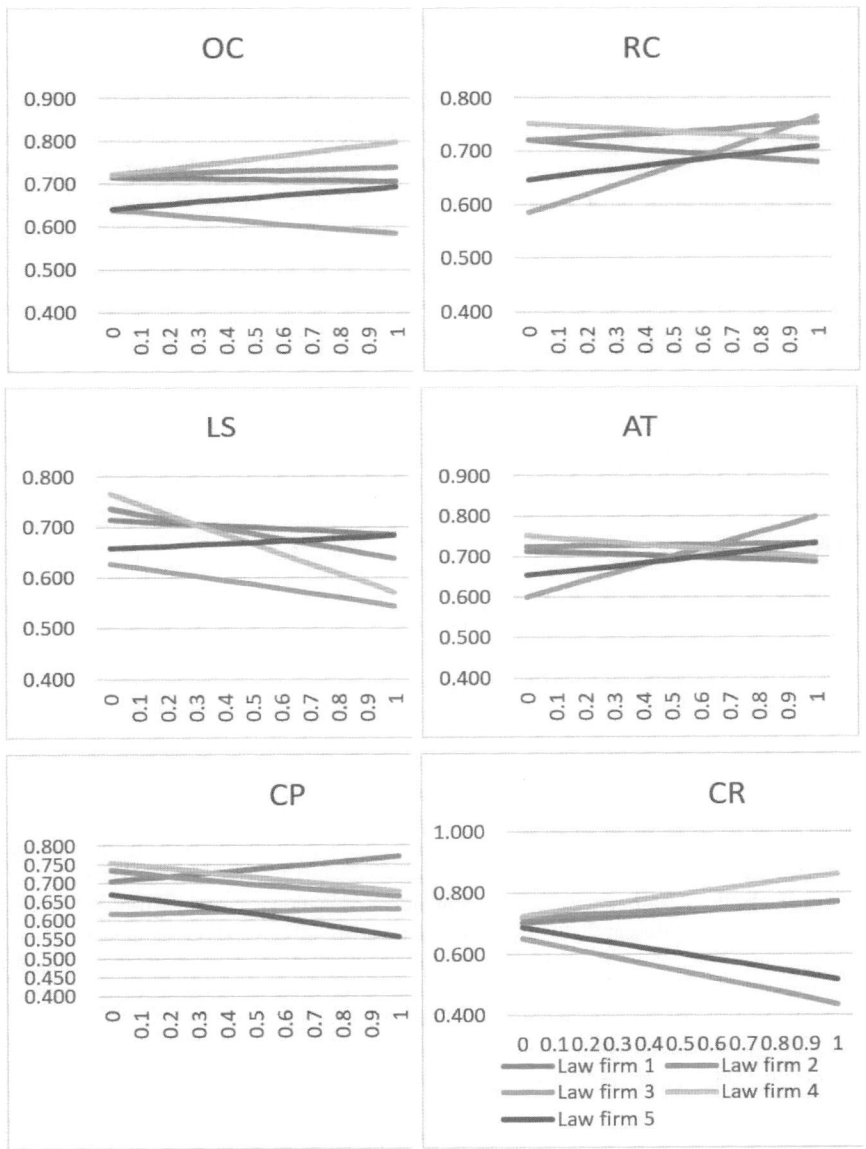

Figure 1. Sensitivity analysis.

The performance rankings of the law firms change based on the weights of the criteria. This shows that the results are sensitive and the weights of the criteria should be selected carefully.

5. Conclusion and Further Suggestions

The results indicate that the law firm performance evaluations change based on the weights of the performance evaluation criteria. Thus, defining the weight of the criteria has a crucial role in performance evaluation. Although utilized dynamic intuitionistic fuzzy MCDM enable dealing both with longitudinal and uncertain information, it is vulnerable to the changes in the weights. The evaluation of the performances is a detailed and time-consuming process whereas the weights are assigned with a simplistic approach. Further studies can improve the methodology by enhancing the criteria weight assignment. The same dynamic approach can be used in the other extensions of fuzzy sets.

Acknowledgments

This work is supported by Scientific and Technological Research Council of Turkey (TÜBİTAK), TEYDEB 1505, Grant No.: 5170012.

References

Atanassov, K. (1986), Intuitionistic fuzzy sets, Fuzzy Sets Syst. 20, 87–96.

Birgün, S., & Güngör, C. (2014). A multi-criteria call center site selection by hierarchy grey relational analysis. Journal of Aeronautics and Space Technologies (Havacilik ve Uzay Teknolojileri Dergisi), 7(1).

Cevik Onar S., Oztaysi B., Kahraman C. (2016) Aftersales Service Performance Measurement Using Dynamic Intuitionistic Fuzzy Multi-Attribute Decision Making , Proceedings of the 12th International FLINS Conference, 24-26 August 2016, Roubaix, France, pp 845-850.

Das, M. C., Sarkar, B., & Ray, S. (2015). A performance evaluation framework for technical institutions in one of the states of India. Benchmarking: An International Journal, 22(5), 773–790. doi:10.1108/bij-02-2013-0019.

Johnes, J. (1996). Performance assessment in higher education in Britain. European Journal of Operational Research, 89(1), 18-33. doi:10.1016/s0377-2217(96)90048-x.

Li, G., Kou, G., & Peng, Y. (2015). Dynamic fuzzy multiple criteria decision making for performance evaluation. Technological and Economic Development of Economy, 21(5), 705–719. doi:10.3846/20294913.2015.1056280.

Lin, C.-T., Lee, C., & Chen, W.-Y. (2009). Using fuzzy analytic hierarchy process to evaluate service performance of a travel intermediary. The Service Industries Journal, 29(3), 281–296.

Mardani, A., Jusoh, A., Zavadskas, E. K., Khalifah, Z., & Nor, K. M. (2015). Application of multiple-criteria decision-making techniques and approaches to evaluating of service quality: A systematic review of the literature. Journal of Business Economics and Management, 16(5), 1034–1068.

S. Cevik Onar, B. Oztaysi, C. Kahraman (2017) Dynamic intuitionistic fuzzy multi-attribute aftersales performance evaluation, Complex & Intelligent Systems, 1-8.

S.C. Onar, B. Oztaysi, C. Kahraman, Dynamic Intuitionistic Fuzzy Evaluation of Entrepreneurial Support in Countries, Advances in Fuzzy Logic and Technology 2017, Proceedings of: EUSFLAT- 2017 – The 10th Conference of the European Society for Fuzzy Logic and Technology, September 11-15, 2017, Warsaw, Poland.

Shieh, J.-I., Wu, H.-H., & Huang, K.-K. (2010). A DEMATEL method in identifying key success factors of hospital service quality. Knowledge-Based Systems, 23(3), 277–282.

Wang, W. (2000). Evaluating the technical efficiency of large US law firms. Applied Economics, 32(6), 689–695.

Xu Z. and Cai, X. Interval-Valued Intuitionistic Fuzzy Information Aggregation, Springer, Berlin (2012).

Xu Z. S. and Yager, R. Dynamic intuitionistic fuzzy multiple attribute decision making. International Journal of Approximate Reasoning, 48, 246–262 (2008).

Analysis on the effectiveness of education in institutions of higher learning based on cultural confidence

Yanyan Ding[†]

*School of Civil and Commercial Law, Northwest University of Political Science and Law
Xi'an, Shaanxi, China*

Xiaohong Liu

College of Management, Southwest Minzu University, Chengdu, 610041, China

The completion of the Communist Youth League's (CYL's) core mission of ideological and political work cannot be done with the absence of culture in Institutions of Higher Learning. However, if we want to strengthen cultural education, we must first adhere to cultural confidence. With the help of the related scales, this paper analyzes the present situation of cultural education advocated by the CYL, and finds out that the value perception and satisfaction of college students to the overall service of the CYL are relatively low. Meanwhile, the problems of lack of value soil, imbalance between supply and demand, and strong administrative color exist in the process of achieving cultural education advocated by the CYL.

1. Introduction

"There is an invisible cultural spirit behind any business that supports and sustains the success or failure of this undertaking."[1] This indicates that the completion of the CYL's core mission of ideological and political work cannot be done with the absence of culture. In June 6, 2017, the CYL Central Committee and the Ministry of Education issued The Suggestions on Strengthening and Improving the CYL's Ideological and Political Work in Institutions of Higher Learning under the New Circumstances. It clearly pointed out that we should strengthen cultural education and strive to carry forward the main theme and spread positive energy on campus. This confirms the importance of culture in the work of CYL. However, to strengthen cultural education, adhere to the cultural confidence has become an inevitable premise. As a more basic, broader and deeper self-confidence, cultural confidence is the

---

[†]Work supported by the fund project of the research for the work of college counselors in Shaanxi in 2017 "The research on the realization path of the identity of the socialist core values of college students under the network culture ecology" (project number: 2017FKT24).

cornerstone of nurturing and educating people in the real sense. Only by internalizing the essence of excellent Chinese traditional culture and revolutionary culture, advanced socialist culture as well as healthy and upward campus culture can college students enhance strong faith in Communist Party of China and likely to become qualified socialist builders and reliable successors with Chinese characteristics. At present, the post-95s has become the main group of college students in China. Their value orientation and development options are becoming more and more diversified. So how do college students view the CYL and its activities at present? Are there any cognitive deviations between the starting points of the CYL and its image in the eyes of college students? The exploration of these issues plays a key role in the effective implementation of CYL's promotion of cultural education.

2. An Analysis of the Present Situation of Cultural Education Advocated by CYL

2.1. *Study design*

Based on "The evaluation index system of service satisfaction of the CYL" compiled by Li Chenglong et al. (2009), this paper revises and compiles "The value perception scales of the service of the CYL" and "The satisfaction evaluation scale of the service of CYL". Among them, "The value perception scale of the service of the CYL" contains 17 questions, which involves the evaluation of the overall service of the CYL, as well as the evaluation of the concrete activities carried out by the CYL. "The satisfaction evaluation scale of the service of CYL" contains 38 items, involving the overall service satisfaction rating of the CYL, and the satisfaction evaluation of the service of ideological and political education, the service of extracurricular activities of science and technology, the service of social practice, the service for the maintenance of the rights and interests of the league members, the guidance service of employment and entrepreneurship and the service of quality development provided by the CYL. Likert5 scale is used in all scales, "very disagree" "disagree" "general" "agree" "strongly agree" and "very dissatisfied" "dissatisfied" "general" "satisfied" "very satisfied" are recorded as 1-5 points. In this paper, a total of 400 questionnaires are issued, and 378 valid questionnaires are collected, and the effective recovery rate of the questionnaire is 94.5%.

2.2. The current situation of college students' perception of the service value of the CYL

The average value recognition of the overall service of CYL is 3.05, and the average value recognition of specific activity is 3.25. At the same time, the proportions that agree with and strongly agree with "The overall service work of the CYL in our school is of great value to the students" "The CYL of our school has a great influence on the students" "The service work of the CYL in our school is of great help to students' work, study and ability improvement" are 49.8%, 48.5%, 49.3% respectively. They are the three items with the smallest proportion in "The value perception scale of the service of the CYL". And the proportions that agree with and strongly agree with "It is very meaningful to carry out 'three kinds of going to the countryside' social practice activities of college students in the winter and summer vocations" "It is very meaningful to carry out the program of college students volunteering to serve the West" "It is meaningful to organize the activities of college students' innovation and entrepreneurship competition" "The organization of the debate is very meaningful" are respectively 60.2%, 61.5%, 61.2%, 62.0%, which are obviously higher than the value recognition of the overall service of CYL.

2.3. The present situation of college students' satisfaction evaluation on the service of the CYL

The overall satisfaction mean of the service of the CYL is 3.17, the ideological and political education service is 3.23, the service of the extracurricular activities of science and technology is 3.26, the social practice service is 3.31, the maintenance service for the rights and interests of the members of the group is 3.29, the employment and entrepreneurship guidance service is 3.34, the quality development service is 3.35. It can be seen that there is also the same phenomenon that the overall satisfaction of the CYL service is lower than the satisfaction of their specific services. At the same time, it can also be seen in the evaluation of the specific service work satisfaction of the CYL, the highest degree of satisfaction is the quality development service, the second is the employment and entrepreneurship guidance service, the third is social practice service, the fourth is the maintenance service for the rights and interests of the members of the group, the fifth is the service of the extracurricular activities of science and technology, and the sixth is the ideological and political education service.

The correlation analysis on the overall satisfaction of the service of the CYL and its influencing factors in ideological and political education service, service of the extracurricular activities of science and technology, social practice service, maintenance service for the rights and interests of the members of the group, employment and entrepreneurship guidance service, quality development service. It can be concluded that the overall satisfaction of the service of the CYL is highly correlated with its influencing factors (see Table 1).

Table 1. The relationship between overall satisfaction of the service of the CYL and its influencing factors.

	ideological and political education	extracurricular activities of science and technology	social practice	maintenance for the rights and interests of the members of the group	employment and entrepreneurship guidance	quality development
CYL	.884**	.794**	.810**	.802**	.808**	.774**

Taking the overall satisfaction of all factors affecting the service of CYL as the abscissa and the correlation coefficient as the ordinate (see Figure 1), it can be concluded that the quality development service is the advantage of the service of the CYL of Northwest University of Political Science and Law, and the correlation coefficient is low but the student satisfaction is high, so it should be maintained. The maintenance service for the rights and interests of the members of the group, social practice service, employment and entrepreneurship guidance service are its core work, students have higher satisfaction and higher correlation coefficient, so they should be further improved. For the service of the extracurricular activities of science and technology, student satisfaction is low, and the correlation coefficient is also low. Therefore, it can be postponed strengthen in the case of limited energy, but it can not be ignored. For the ideological and political education service, the correlation coefficient is high but the students' satisfaction is low. So it is a weak link, and measures should be taken to make great improvement.

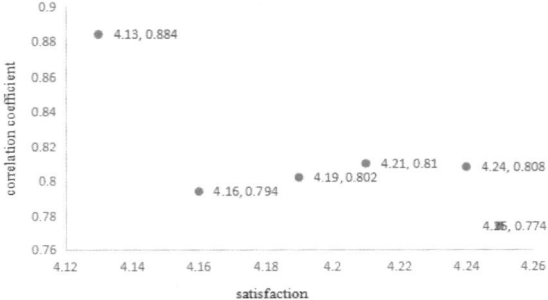

Figure 1. Satisfaction of factors affecting the overall satisfaction of CYL service - correlation coefficient.

3. The Problems Existing in the Process of Realizing Cultural Education of the CYL and Its Causes Analysis

3.1. *Lack of value soil, culture anaemia phenomenon appears*

The survey of "the overall service of the CYL in our school is of great value to the students" depicts that the students who very disagree with and disagree with it account for 33.2%. And data collected in "the CYL of our school has a great influence on the students" shows that the students who very disagree with and disagree with it account for 32.4%. At the same time, the students' value recognition for the overall service of the CYL is lower than its specific activities, indicating that the current university students' value recognition of the CYL is weakened, and that the CYL is lack of value soil in the process of promoting cultural confidence to realize cultural education. The CYL tried its best to advocate "elite culture". However, such "elite culture" usually emphasizes the connotative experience. Therefore, it cannot cause resonance. At the same time, the "mass culture", represented by the popular culture of youth, attracts the attention of the students with its entertaining and contemporary features. However, this kind of "mass culture", often lacks value connotations. Therefore, on the one hand, the "mass culture" keen on by students is lack of "nutritional value". On the other hand, the "elite culture" advocated by the CYL can not trigger a student's resonance. As a result, the CYL lacks the value soil and the phenomenon of cultural anaemia appeared.

3.2. *Imbalance between supply and demand, the cultural generation gap is caused*

The data in "your evaluation of the CYL service in our school" says that very dissatisfied and dissatisfied students account for 28.7%. With respect to the figure in "from your actual perception compared with your expectation, your satisfaction with the service of the CYL of our school", students who are very dissatisfied with and dissatisfied with the work account for 29.3%. At the same time, the overall satisfaction of students with the CYL's service is lower than the satisfaction of its specific services, indicating that the current in the process of promoting the cultural confidence to achieve the goal of educating people with culture, there is a generation gap between the cultural supply of the CYL and the cultural demand of college students. Contemporary college students are mostly post-95s, the thinking mode, behavior, value orientation are obviously different from that of the post-70s and 80s. However, most of the CYL cadres are the post-70s and 80s, and they use "first in authority" to put their own value orientation throughout the activities carried out. Such a top-level design lacks the basis of students' participation. Therefore, there is an imbalance and asymmetry between the culture provided by the CYL and the cultural needs of students.

3.3. *Strong administrative color and prominent cultural distance*

18.5% of students say that "they do not participate in any student body or community during college". 4% of students say "they do not want to get more information and help from the CYL". At the same time, 37.8% of students say "they often participate in activities organized by the CYL". It shows that the current college students are highly involved in the related organizations of CYL organizations and have strong willingness to participate in the activities they carry out. However, their actual participation in specific activities is not high, and there is a cultural gap between the CYL and college students. The CYL is influenced by its own environment and has a certain administrative color. In the process of promoting cultural confidence and realizing cultural education, most of its information transmission belongs to the top - down communication path. The working methods tend to give orders to students based on the administrative system. However, this information communication and working methods obviously do not apply to the post-95s. Even more worrying is that this concept of "official rank standard" is inherited by the relevant student organizations of the CYL, resulting in student organizations "flashy" and "lost students' support", which severely weaken the credibility and voice of the CYL. This phenomenon resulted in the emotional distance between students and the CYL.

4. The Countermeasures and Suggestions to Improve Cultural Education Advocated by the CYL

4.1. *Realizing the popularization of elite culture*

The construction of a rational and style campus culture is inseparable from the participation of the elite culture[2]. Therefore, in the process of promoting cultural confidence to achieve cultural education, we must continue to vigorously carry forward the excellent traditional culture of China. However, at the same time, it is necessary to make the elite culture popular by means of concrete activities, that is, to achieve the goal of "finding a meeting point between guiding young people, caring for the characteristics of youth and the reasonable needs of young people, and making meaningful things interesting"[3]. According to the results of Figure 1, due to the students' satisfaction with the service of social practice and the guidance of employment and entrepreneurship, their correlation coefficient with the overall satisfaction of service of the CYL is also high, therefore, when the CYL carrying out these two types of services, based on the existing work experience, can combine ideological and political education services with the characteristics of contemporary college students, to achieve elite culture popularization. In this way, on the one hand, it ensures the university students' satisfaction with the CYL service and on the other hand, it can expand the audience of the elite culture.

4.2. Realize complementary needs and achieve a balance between supply and demand

At present, the CYL cadres have rich experience and inherit the revolutionary traditional culture, but their thinking is relatively closed. At the same time, college students are creative, but their thinking is still immature. Therefore, the CYL and students should learn to respect each other and complement each other's advantages. But it is known that the achieving of these complementary needs cannot be separated from the carrier of specific activities, combined with the analysis of Figure 1, due to the students' low satisfaction with the service of the extracurricular activities of science and technology, indicating that students have higher expectation and greater demand for this service. Therefore, when carrying out this service, it not only needs to integrate ideological and political education services into it, but also needs more empathy. This can not only improve the students' satisfaction with the service of the CYL, but also realize complementary needs between the CYL and students, and achieve the balance between supply and demand.

4.3. Getting rid of the administration, eliminating cultural distance

The administrative color has produced a cultural distance between the CYL and students. This kind of distance can easily increase the gap between them. Therefore, the CYL must go to get rid of the administration. The cadres of the CYL should respect students, in-depth student "circle of friends" to narrow the distance between students. Combined with the results of the analysis in Figure 1, due to the high level of satisfaction of the students with the service of quality development, this service belongs to the superiority of the service provided by the CYL of Northwest University of Political Science and Law. At the same time, the students are also more satisfied with the maintenance service for the rights and interests of the members of the group. However, the evaluation of the satisfaction of the maintenance service for the rights and interests of the members of the group is closely related to the "sense of distance" between the CYL and students. Therefore, the CYL can start from these two services to narrow the distance with the students. When carrying out these two services, the

CYL not only need to integrate ideological and political education services into them, but also make use of the previous high-level evaluation of satisfaction and cosying up to the communities so as to let students appreciate the feeling of being respected, valued in the services.

In short, if we want to improve the effectiveness of cultural confidence and achieve cultural education, we must eliminate the cultural conflicts between the CYL and students. However, in order to eliminate this cultural conflict, the CYL must learn to foster strengths and circumvent weaknesses. For the advantages of services, it should be made to play a more positive role as much as possible, and strengthen all-round and various fields of the weak links.

References

1. Wang Shuguo. The social responsibility and scientific development of the university [J]. Journal of the National Institute of Education Administration, 2009 (12): 12-14.
2. Zou Qiao. On the cultural conflict and elimination of the Communist Youth League organizations in colleges and universities [J]. Youth Exploration, 2014(04):89-92.
3. Liu Junyan. A brief discussion on the activities of the Communist Youth League [J]. Chinese Youth Studies, 2013 (08): 41-47.
4. Li Chenglong, Ren Xiaojie, Gao Jiajia. The construction of evaluation system for the satisfaction degree of the Communist Youth League service in colleges and universities [J]. Ideological and Theoretical Education, 2009(11):67-70.
5. Wang Bin. An investigation report on the current situation of the Communist Youth League work in colleges and universities — Taking Southeast University as an example [J]. Chinese Youth Studies, 2011(02):50-53.

2-tuple combined group decision making methodology for climate change strategy selection[*]

Gülçin Büyüközkan[†]

Department of Industrial Engineering, Galatasaray University, 34349 İstanbul, Turkey

Deniz Uztürk

Department of Business Administration, Galatasaray University, 34349 İstanbul, Turkey

Decision-making is all about making choices. Using a number of experts can present an excellent solution to better analyze and reflect the constraints of decision-making processes. Nonetheless, the aggregation of different points of view remains a challenge. In this study, a 2-Tuple-based multi-criteria decision-making (MCDM) framework is proposed. First, the 2-Tuple based Simple Additive Weighting Approach (SAW) is applied to detect the importance of the decision criteria. Then, the 2-Tuple based Axiomatic Design (AD) is employed for ranking the decision alternatives. To test the applicability of the proposed methodology, it is utilized to select the best nature-based climate adaptation strategy for Istanbul.

1. Introduction

Expert decisions with more than one decision maker play an essential role in many fields. In such settings, different experts contribute their points of view in the decision process. As a result, a balanced decision environment can be established. In multi-criteria decision-making (MCDM), the decision is determined based on a set of criteria. Having more than one decision maker can provide an objective environment to overcome the limitations of many decision-making processes. Having different perspectives about a subject can also give an in-depth evaluation of the problem at hand. However, at this point, the key is the aggregation of these various opinions under a common, meaningful form so that the results can be valuable. In this study, the 2-Tuple linguistic method, developed by Herrera and Martinez [1], is chosen to use with Simple Additive Weighting Approach (SAW) and Axiomatic Design (AD) to overcome the different granularity results from more than one decision maker.

---

[*]This work is financially supported by Galatasaray University Research Fund (Project No: 18.102.001).
[†]Corresponding author; e-mail: gulcin.buyukozkan@gmail.com

SAW is a simple-to-use and popular MCDM method based on the concept of average weighting. An assessment score is calculated for each alternative by multiplying the weights assigned to the alternatives by decision makers. Then, products of all alternatives are summed up [2]. The benefit of this technique is its proportional linear conversion of raw data, implying that the relative order of magnitude of the standardized scores remains the same [2]. In this study, SAW is proposed to weigh the selected criteria. It is a highly preferable method with easy computational steps for ranking criteria. Due to its benefits and convenience, SAW is chosen as a weighting tool in this paper.

Then, AD is proposed for the selection problem, based on the 2-Tuple linguistic information to overcome the multi-granularity arising from multiple experts. AD is a technique that is first introduced by Suh [3]. It uses two axioms: the first one is the independence axiom, which sets out that Function Requirements (FRs) must be independent, and the second one is the information axiom, which sets out that the design with the minimum information content is better than all the other designs that satisfy the FRs [3]. The AD method is generated first for design problems. Today, it has many applications for MCDM problems. In this paper, the best alternative is to be selected by providing each criterion with minimum information. AD is preferred due to that requirement.

Many different MCDM methods are applied for selection problems in the literature. In this study, AD and SAW are chosen thanks to their power of integrating with linguistic data. Furthermore, the 2-Tuple method can be combined with these MCDM techniques to augment their capability to integrate with multi-granularity linguistic data.

This study is organized as follows: Next part is a literature review about MCDM techniques with their 2-Tuple extensions and nature-based solutions for Istanbul. The part after that provides the application of the proposed method for a climate change strategy selection for Istanbul. Finally, the results, discussions, and conclusions are presented.

2. Literature Review

2.1. *2-Tuple Linguistic Model and MCDM*

As mentioned in the previous section, the 2-Tuple method is a technique to overcome the imprecision and multi-granularity stemming from group decision-making (GDM) [4]. Due to its properties, it is integrated with different MCDM techniques. In recent years, many studies have integrated the 2-Tuple with different MCDM methods. In the literature, VIKOR is combined with 2-Tuple for different purposes, such as HR evaluation [5], material selection [6], site selection in waste management [7], and personal selection [8] subjects. Another example is QFD, which is extended with a 2-Tuple model in supplier evaluation and selection [9] and sustainable warehouse design [10] subjects. TOPSIS, another MCDM method, is combined with the 2-Tuple method in robot evaluation and selection [11], supplier evaluation and selection [12] and healthcare waste treatment technology selection subjects [13]. 2-Tuple based AHP is used for supplier segmentation [14] and ANP is employed with ELECTRE II for supplier selection problem [14]. Also, DEA is integrated with the 2-Tuple for multi-attribute group decision problems where the weights of decision makers are known [15]. Amid these several of 2-Tuple, this study focuses on AD, which is a robust method for selecting the most appropriate alternative with minimum information.

2.2. *Nature-based Solutions for Istanbul*

Nature-based (NB) solutions can be defined as measures for protecting, sustainably managing, and restoring natural or modified ecosystems that effectively and adaptively address social challenges, also providing well-being and biodiversity benefits to people [16]. They improve resilience to climate change, support sustainable development and protect biodiversity [17].

In this study, the essential aim is to prioritize possible NB solutions for Istanbul according to their ability to adapt to climate change impacts. First, expected climate change impacts are gathered from Istanbul Municipality's report about climate scenarios, which have been prepared for the climate action plan of Istanbul. They are identified as FRs for 2-Tuple based AD: general temperature increase due to global warming (FR1), temperature increase due to urbanization (FR2), higher increase in summer temperature compared to winter (FR3), increase in hot temperature extreme weather events (FR4), augmentation of vaporization (FR5), augmentation of dry seasons (FR6), increase in precipitation during heavy rainy days (FR7), decrease in summer rains (FR8), almost no temperature drops below 0 °C (FR9) and evolution of climate towards the typical Mediterranean climate (FR10) [18].

To overcome these effects, NB solutions are identified as: green structures (S1), green-blue corridors (S2), "heat attention maps" for urban developers or planners (S3), sustainable and climate-proof districts in the city (S4), sustainable and climate-proof buildings (S5), innovative design principals combined with technical measures to create climate-proof green infrastructure (S6), revitalization of peatlands in and around the city (S7), embrace eco-urbanism for new developments and renovations for city (S8), permeable surfaces to reduce the adverse effects of severe rainfall (S9), converting abandoned industrial sites into urban parks (S10), rain gardens (S11), increasing the number of inner-city lanes to make space for greenways (S12) and urban agriculture (S13) [19], [20], [21].

3. Application of Proposed Methodology: NB Climate Change Adaptation Strategy Selection

To test the applicability of the proposed framework, the 2-Tuple-based combined methodology is applied for the NB strategy selection for climate change adaptation for Istanbul.

The steps of the proposed methodology are described next:
Step 1: Detecting FRs for the AD. The chosen subject is climate change adaptation strategy. The FRs are generated for the selection problem, as mentioned in the previous section.
Step 2: Identification of alternatives. NB solutions for Istanbul aim to diminish the impacts of climate change, as presented in the previous section.
Step 3: Deciding on the number of experts and detecting their linguistic scales for evaluations during the decision-making process. Three experts are chosen as a decision makers in this application. All of them have different backgrounds, so different scales are given them for assessments, as represented in Table 1.

Table 1. Linguistic scales for each expert.

1st Expert	S_i^5	2nd Expert	S_i^7	3rd Expert	S_i^9
Very Low (VL)	S_0^5	Very Low (VL)	S_0^7	Very Low (VL)	S_0^9
Low (L)	S_1^5	Low (L)	S_1^7	Low (L)	S_1^9
Medium (M)	S_2^5	Medium Low (ML)	S_2^7	Medium Low (ML)	S_2^9
High (H)	S_3^5	Medium (M)	S_3^7	Almost Medium (AM)	S_3^9
Very High (VH)	S_4^5	Medium-High (MH)	S_4^7	Medium (M)	S_4^9
		High (H)	S_5^7	Medium-High (MH)	S_5^9
		Very High (VH)	S_6^7	High (H)	S_6^9
				Very High (VH)	S_7^9
				Perfect (P)	S_8^9

Step 4: Applying 2-Tuple-based SAW to weight the FRs for the AD.

Step 4.1: Differently-scaled evaluations from experts, as presented in Table 2, are normalized under the same linguistic term set for each criterion. First, linguistic term sets are converted into the 2-Tuple form by adding a zero value, as shown in the following relation:

$$S_i \in S \Rightarrow (S_i, 0) \qquad (1)$$

Here, S is the linguistic term set.

Table 2. Different granulated assessments of each expert.

	1st Expert	2nd Expert	3rd Expert		1st Expert	2nd Expert	3rd Expert
FR1	M	M	ML	FR6	M	M	AM
FR2	VH	VH	H	FR7	VH	H	P
FR3	L	ML	MH	FR8	VL	VL	ML
FR4	H	H	P	FR9	L	L	ML
FR5	VH	H	M	FR10	L	L	L

Then, by taking the intersection of the fuzzy membership function of each linguistic term sets (Figure 1) with the one with the highest granularity [22] and by applying the following relation, they are normalized under the same granularity:

$$X: F(S_{g+1}) \to [0, g]$$
$$X: (\tau(I)) = X\left(F(S_{g+1})\right) = X\left[(s_i, \alpha_i)\, i \in [0,1,\ldots g]\right] \quad (2)$$
$$= \sum_{i=0}^{g} i\alpha_i \Big/ \sum_{i=0}^{g} \alpha_i = \beta = \Delta(\beta) = (s, \alpha)$$

Here, α_i is the membership function of the intersected points of different levels, i is the level number, and g is the level of the linguistic set with the highest granularity.

Figure 1 illustrates the intersection of S^5 (purple) and S^9 (red) as an example. The same intersection is applied for seven levels and nine levels as well.

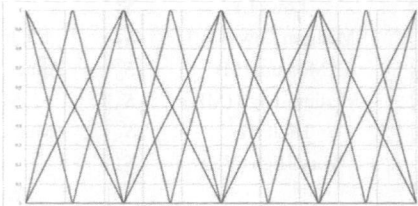

Figure 1. The intersection of five level linguistic term set and nine level linguistic term set.

Step 4.2: Aggregating the normalized linguistic variables with the *Weighted Aggregation Operator* (WAO) using the following relation [23]:

$$\bar{x} = \left(\frac{\sum_{i=1}^{n} \Delta^{-1}(r_i, \alpha_i) \times \Delta^{-1}(w_i, \alpha_i)}{\sum_{i=1}^{n} \Delta^{-1}(w_i, \alpha_i)}\right) = \Delta\left(\frac{\sum_{i=1}^{n} \beta_i \times w_i}{\sum_{i=0}^{n} w_i}\right) \quad (3)$$

Here, (r_i, α_i) is the relative aggregated importance assigned to each FR by each expert; (w_i, α_i) stands for the weights of experts, n represents the number of experts, β_i is the β values for the ith FR importance, and w_i stands for the β values obtained for the ith expert with Eq. (4).

To apply the weighted AD, each criterion must possess a numerical weight, instead of a linguistic one. At this point, the 2-Tuple's reverse equation is used to represent linguistic variable with numerical form (Eq. (4)) as β values. Then, they are normalized with Eq. (5).

$$\Delta^{-1}(S_i, \alpha) = i + \alpha = \beta \qquad (4)$$

$$Normalized\ Weights = \beta_i \Big/ \sum_1^m \beta_i \qquad (5)$$

Here, β_i is the β value for the ith criterion's importance, m is the number of criteria. Obtained normalized weights are shown in Table 3.

Table 3. Weighted information content for each alternative.

	S1	S2	S3	S4	S5	S6	S7	S8	S9	S10	S11	S12	S13
FR1	0.19	0.30	0.23	0.23	0.14	0.14	0.19	0.15	0.18	0.19	0.23	0.19	0.33
FR2	0.37	0.59	0.54	0.37	0.59	0.59	0.59	0.37	0.34	0.54	0.94	0.37	0.44
FR3	∞	0.19	0.19	0.23	0.19	∞	0.19	0.19	0.23	0.21	0.30	0.23	0.49
FR4	0.48	0.53	∞	0.77	0.34	0.34	0.95	0.95	0.34	0.55	0.34	∞	0.48
FR5	1.17	0.47	0.45	0.29	0.86	0.50	0.38	0.42	0.45	0.38	0.39	0.45	0.38
FR6	0.25	0.65	0.25	0.25	0.65	0.33	0.65	0.33	0.43	0.30	0.19	0.40	0.30
FR7	0.98	0.35	∞	0.79	0.35	0.35	∞	0.79	0.35	0.35	0.35	∞	0.54
FR8	0.06	0.08	0.07	0.09	0.23	0.09	0.09	0.08	0.08	0.08	0.11	0.17	0.17
FR9	0.12	0.34	0.12	0.11	0.25	0.13	0.11	0.13	0.11	0.11	0.12	0.11	0.13
FR10	0.10	0.09	0.09	0.09	0.20	0.09	0.09	0.09	0.09	0.12	0.09	0.09	0.10
Sum		3.58		3.23	3.79			3.50	2.60	2.83	3.05		3.36
Rank		7		4	8			6	1	2	3		5

Step 5: Detecting the System Ranges (SRs) for each FR and Design Ranges (DRs) for each alternative. Then, form an aggregated evaluation matrix with SRs and DRs.

Step 5.1: Each expert evaluated with their scales. First, their assessments are converted into the 2-tuple representation by adding a 0 value, as in Eq. (1). Then, they are normalized as described previously in Step 4.1 by using Eq. (2).

Step 6: Calculating the Information Content (I) with (Eq. (6)) by taking the intersection of the aggregated SRs and aggregated DRs for each strategy. Then, I is weighted by multiplying the weights generated from the 2-Tuple-based SAW.

$$I = \log_2(intersected\ area\ of\ DR\ and\ SR) \qquad (6)$$

Step 7: Making the selection of alternatives according to their weighted I values.

4. Results and Discussions

The final ranking is given in Table 3. According to the AD, each alternative must meet the minimum requirement for each criterion. I values are calculated for each alternative according to the FRs. Alternative (strategy) with the I value converges to zero cannot fulfill the minimum requirements for the criteria. They are shown with ∞ in Table 3. For that reason, the strategies S1, S3, S6, S7 and S12 are eliminated. When the rest of the strategies are ranked in decreasing order, S9, which is the "Permeable surfaces to reduce the adverse effects of severe rainfall", is the most suitable adaptation strategy for Istanbul in the first place.

5. Conclusions

This paper proposes a 2-Tuple-based SAW and AD methodology for a selection problem. SAW and AD techniques are integrated with a 2-Tuple method to overcome the multi-granular information gathered from different decision makers. Then, as a test of the proposed process, it is applied to a climate change adaptation strategy selection problem for Istanbul city. In this application, the 2-Tuple-based SAW is used to weigh the criteria for the AD. Then, the 2-Tuple-based AD is used to choose with a group of experts. As a result, "permeable surfaces to reduce the adverse effects of severe rainfall" seems to be a suitable strategy for enhancing the climate change resilience of Istanbul.

2-Tuple extensions of both methods have been used for the first time in the literature for this application. In addition, an application of climate change strategy ranking has been done for Istanbul city for the first time in literature.

Different granulated data should be collected under the same form and granularity to achieve successful group decision-making. The properties and advantages of the 2-Tuple provide the homogeneity for this decision-making problem. For further studies, different granulated data from experts can be used with different 2-tuple-integrated MCDM techniques, and their comparison can be made.

References

1. F. Herrera and L. Martinez, "A 2-tuple fuzzy linguistic representation model for computing with words," *IEEE Transactions on Fuzzy Systems*, vol. 8, no. 6, pp. 746–752, 2000.
2. A. Afshari, M. Mojahed and R. M. Yusuff, "Simple Additive Weighting approach to Personnel Selection problem," *International Journal of Innovation*, vol. 1, no. 5, pp. 511–515, 2010.
3. N. Suh, Axiomatic Design: Advances and Applications vol. 4, New York: Oxford University Press, 2001.

4. L. Martínez, R. M. Rodríguez and F. Herrera, The 2-tuple Linguistic Model. Computing with Words in Decision Making, Springer International Publishing, 2015, p. 54.
5. P. Liu and X. Wu, "A competency evaluation method of human resources managers based on multi-granularity linguistic variables and VIKOR method," *Technological and Economic Development of Economy*, vol. 18, no. 4, pp. 696–710, 2012.
6. H.-C. Liu, L. Liu and J. Wu, "Material selection using an interval 2-tuple linguistic VIKOR method considering subjective and objective weights," *Materials and Design*, no. 52, pp. 158–167, May 2013.
7. H.-C. Liu, J.-X. You, X.-J. Fan and Y.-Z. Chen, "Site selection in waste management by the VIKOR method using linguistic assessment," *Applied Soft Computing*, no. 21, pp. 453–461, April 2014.
8. H.-C. Liu, J.-T. Qin, L.-X. Mao and Z.-Y. Zhang, "Personnel Selection Using Interval 2-Tuple Linguistic VIKOR Method," *Human Factors and Ergonomics in Manufacturing & Service Industries*, vol. 25, no. 3, pp. 370–384, 2015.
9. E. E. Karsak and M. Dursun, "An integrated fuzzy MCDM approach for supplier evaluation and selection," *Computers & Industrial Engineering*, vol. 82, pp. 82–93, 2015.
10. G. Büyüközkan and D. Uztürk, "Combined QFD TOPSIS Approach with 2-Tuple Linguistic Information for Warehouse Selection," in *International Conference on Fuzzy Systems (FUZZ-IEEE)*, Naples, 2017.
11. H.-C. Liu, M.-L. Ren, J. Wu and Q.-L. Lin, "An interval 2-tuple linguistic MCDM method for robot evaluation and selection," *International Journal of Production Research*, vol. 52, no. 10, pp. 2867–2880, 2014.
12. E. Karsak and M. Dursun, "An integrated fuzzy MCDM approach for supplier evaluation and selection," *Computers & Industrial Engineering*, vol. 82, pp. 82–93, 2015.
13. C. Lu, J.-X. You, H.-C. Liu and P. Li, "Health-Care Waste Treatment Technology Selection Using the Interval 2-Tuple Induced TOPSIS Method," *INTERNATIONAL JOURNAL OF ENVIRONMENTAL RESEARCH AND PUBLIC HEALTH*, vol. 13, no. 6, 2016.
14. L. Santos, L. Osiro and R. Lima, "A model based on 2-tuple fuzzy linguistic representation and Analytic Hierarchy Process for supplier segmentation using qualitative and quantitative criteria," *Expert Systems with Applications*, vol. 79, pp. 53–64, 2017.
15. X. Geng, H. Qiu and X. Gong, "An extended 2-tuple linguistic DEA for solving MAGDM problems considering the influence relationships among attributes," *Computers & Industrial Engineering*, vol. 112, pp. 135–146, 2017.
16. IUCN, "International Union for Conservation of Nature annual report 2016," Hawai, 2016.
17. ICLEI, "Local Governments for Sustainability," 2017.
18. Istanbul Municipality, "Istanbul için İklim Değişikliği Senaryolarının Değerlendirilmesi," İBB, İstanbul, 2017.
19. European Comission, "Nature-Based Solutions & Re-Naturing Cities," Eurepean Union, 2015.
20. M. Scott, M. Lennon, D. Haase, A. Kazmierczak, G. Clabby and T. Beatley, "Nature-based solutions for the contemporary city/Re-naturing the city/Reflections on urban

landscapes, ecosystems services and nature based solutions in cities/Multifunctional green infrastructure and climate change adaptation: brownfield greening as an ad," *Planning Theory & Practice*, vol. 17, no. 2, pp. 267–300, 2016.
21. ICLE, "Nature-based solutions for sustainable urban development," 2008.
22. M. Li, "The Extension of Quality Function Deployment Based on 2-Tuple Linguistic Representation Model for Product Design under Multigranularity Linguistic Environment," Mathematical Problems in Engineering, vol. 2012, 2012.
23. F. Herrera and L. Martínez, "A 2-Tuple Fuzzy Linguistic Representation Model for Computing with Words," *IEEE Transactions on Fuzzy Systems*, vol. 8, no. 5, pp. 746–752, 2000.

Analysis of balance between reform of judicial system and supervision to power in China

Cheng-Gao Liu and Yun Yuan
College of Management, Southwest Minzu University
Chengdu, Sichuan Province, P.R. China

Zhao-Yi Zhang
College of Foreign Language, Southwest Minzu University
Chengdu, Sichuan Province, P.R. China

The core of the rule of law is to build a judicial system of equity, probity and high efficiency. However, the key point of the construction of judicial structure is based on every judicial authority that carries out dividing the boundaries of duty and power in various fields and links. According to practical proofs, this paper provides a detailed analysis of balance between reform of judicial system and supervision to power in China. We claim that the way of dealing with dissimilation in power is to build a thorough system, meanwhile, using the law to regulate these systems, and with the force of law to guarantee the implementation effectively. Owing to all kinds of shortcomings existing in the reform of judicial system, it causes many mistrials and irregular perversions now and then which corrodes the prestige and equity of our national judicial system, also increasing the sense of urgency in the reform of judicial system.

1. Theory of the top-level design and the balance of rights in the general plan of the Rule of Law

In October, 2014, the CPC in the fourth plenary session of the 18th CPC Central Committee puts the Rule of Law into a high level of social economic development, the governing capacity of the CPC, the Chinese nation's revitalization. The meeting aims at the judicial equity that will have an important leading role in social equity, on the contrary, the judicial injustice has the deadly devastating effect on social justice. Therefore, we must improve the judicial management system and the operating mechanism of judicial power, regulate the judicial behavior and strengthen the supervision of judicial activity, which make all efforts for the masses who can feel the justice and equity in every judicial legal cases.

Proved in practice, it's hard to depend on personal moral cultivation and self-discipline to achieve good management, when losing the power of constraint and checks and balances. The essential way of coping with the power of dissimilation is to establish a complete perfect system, furthermore, using the law to regulate these systems, with the force of law to guarantee the implementation effectively. Hence, the key point of the construction of judicial structure is based on every judicial authority that carries out dividing the boundaries of duty and power in various fields and links.

The earliest power balance theory can be gone back to French ideologist, Montesquieu (1689-1755) in the early 18th century, that is to say, Montesquieu can be regarded as the genuine inaugurator in western power balance theory. He thinks, the private property, as the natural right of people, is inviolability. Besides, he thinks, from the nature of thins, if we want to prevent against abusive power, we must live up to restrain power by power. Based on that, Montesquieu also expounds systematically the idea of separation and the three powers, clearly coming up with the power balance theory, namely, the theory on legislative authority, executive authority, judicial authority which can not only tie down each other, but also coordinate in advance.*

2. Analysis of Chinese national judicial system's condition at present

2.1. *Basic framework of the present judicial system in China*

The judicial system is an important part of the national law system. To put law into effective practice, adequate judicial system is requisite. Now, the judicial organs in China are made up of People's Court, People's procuratorate, judicial administrative organ which is supervised by People's government and is composed by public security, jail, justice bureau, supervision and so on. According to the function distribution, people's court is the judgmental organ of China that is in charge of the judgmental work of China.

As the organizational system of judicial department, judicial system in inevitably reflects the settings leadership or inspective system in judicial department, the division of powers between judicial departments and the management system of judicial departments. In our country, to be specific, the settings of judicial departments mainly include the setting between judicial, procuratoral and public security organ and the settings of the interior of the organ. The leadership of judicial organ or supervision system mainly includes the correlation between judicial organ and other national organ, the relation between each judicial organs and the relation between superior and subordinate. These relationships can be mainly shown in the relation between leading and being led, supervising and being supervised. The division of powers between judicial organs mainly includes the jurisdiction and specific powers of judicial

---

*Montesquieu, The Spirit of Laws, the commercial press (Xu Minglong), in May, 2015.

organs. The management system of judicial system includes the business, personnel, funds, equipment and other management systems of judicial organ.[†]

2.2. Major disadvantages of Chinese national judicial system at present

Due to history, culture, the stage of economic development and all kinds of reasons, nowadays, Chinese national judicial system mainly has these disadvantages as follows:

- *Overlapping of judicial power and administrative power is serious, and the independence of judicial power is not strong*

In 1982, provided by the 135th provision of the Constitution, in handling criminal cases, the people's courts, the people's procuratorates and the public security organs shall work in accordance with their respective responsibilities, and shall be in accordance with each other so as to ensure the accurate and effective enforcement of the law. But in reality, in addition to the public security organs, the people's courts and procuratorates, there are political committees and the judicial organs which need to be coordinated and leaded, and that, the position in charge of the public security organ is usually occupied by the secretary of the political science and law commission or the same level leader.

- *Judicial organs lack the mechanism of restriction and supervision*

Judicial independence does not mean that justice can be free from restraint, on the contrary, because of the independence and importance of its implement of rights, it is urgent to supervise and balance effectively.

- *The professional level of judicial personnel is low, and the internal management model of the judiciary is rigid*

In recent years, with the improvement of professional quality of judicial personnel is increasing, the education level of judges and crown counsels has been improved, but owing that the judicial organs have no actual independent personnel power, leading a number of people from the personnel department of the government or the army of professional recruitment, in contrast, with bachelor degree master, even PhD graduates are often difficult to enter the court's procuratorate through the distribution directly, what's more, many lawyers who have professionally legal ability, it's more rare to transfer the position to become judges.

Due to the uneven quality of individual judges, as well as the administrative intervention, caused the same case, the same origin of case but the consequence is quite different, it will not only damage the judicial justice, lead to the higher

---

[†]Research on several issues of China's judicial system, Tan Shigui, Legal Research, session 3, 2011.

proportion of case on appeal, but also cause the waste of judicial resources and the litigation cost increase for the parties.

2.3. *Necessity and urgency of judicial system reform*

Although, in reality, because the judicial predicament also forces us to make some adjustments in the judicial field, most of the existing judicial reform is on their own behalf, mostly the key points are the judicial organs to improve the internal business judicial system, various judicial organs for their own benefits and achievements cause all kinds of cases that are unjust and false from time to time, which have already eroded the existing judicial system of our country's prestige and justice, as well as strengthened the sense of urgency for the reform of judicial system that we must do.

3. Reform of judicial system centered on checks and balance of power

The reform of judicial system is a long way to go, it involves a wide range and influence greatly, thus, the reform could begin on the following sides:

3.1. *Supervising by power-to-power, and establishing the judicial supervisory and restraining system by CPC and NPC*

Supervising by power-to-power mainly includes legislative power, administrative power, judicial power, and the national watch dog, and the checks and balance to the powers form the CPC. Supervising by power-to-power is the most powerful and strongest supervision, and it also the core step to check and counterbalance public rights.

- *Strengthening and improving the party's leading position in the administration of justice*

The reform of the judicial system shall be implemented so as to more effectively carry out the Party's road, guiding principle and policy, which can promote the party and the government's ability of governance, thus better serving the nation and social economy. Therefore, strengthening and improving the party's leadership is not contradictory to strengthening judicial independence and supervision.

- *Perfect the people's congress to supervise the terms of reference*

As a state authority, the National People's Congress and its Standing Committee are legally supervised on behalf of the state and the people. The purpose is to ensure that the Constitution and the law are properly implemented and that the executive and judicial powers are properly exercised and that citizens, the legitimate rights and interests of other organizations are respected and maintained.

But now, due to the constraints of institutional and institutional factors, the National People's Congress's supervisory function has been long-term imagination, the NPC on the "two high" is basically limited to work reports and major personnel elections and appointments, in a large number of daily judicial process, the NPC's supervisory power cannot be effectively fulfilled.

Hence, it is necessary to establish and perfect the daily supervision mechanism of the National People's Congress and set up a special judicial supervision institution to normalize and institutionalize the judicial supervision of the judicial organs.

3.2. *Establishing the independent operation and administration system of justice centered on judge*

- *Judicial system to de-administration and de-localization, to achieve the classification management among the judges, prosecutors and civil servants*

To solve the problems of de-administration and de-localization of the judicial administration system from the constitutional framework design and legal system perfect, and to establish the independent operation and administration system of justice centered on judge, and to adjust the system of people's court and people's procuratorate, especially on the fiscal system and supervision system.

Moreover, to clear the core status of legislative power of the NPC in the judicial system, so as to secure the statues of state adjudicatory organization all judicial courts at all levels, and to break the 2 in 1 administrative system: the complete overlap of judicial districts and administrative districts. To establish the independent judicial districts according to the practical need, and to insure the judges have independent judicial personality, and could make a judgment in an objective and just way.

- *Establishing the provincial direct governing mechanism of the judicial system and the independent financial budget and final accounts*

Due to the wide geographical distribution, and the imbalance of economic development in different areas and other factors, the national judicial system of central co-ordination is not realistic. Thus, the mechanism of provincial governing could be take first, which could maintain the functions of the Supreme People's court: to unify the standards of application of law through policy guidance, rulemaking, and judicial proceedings.

Specifically, administration authority of personal effects administrated by central government and NPC and Supreme People Procuratorate should be established, whereas the provincial judicial system, and the sub-provincial-level could be administrated by provincial institutions; as for the unified management mode, the local Party committees, people's Congress, the provincial court, procuratorate should be responsible for it. To explore the establishment of cross regional judicial court and judicial administration system handling of the circuit court and judicial management system, and to prevent the restriction of administrative division based on saving the cost of justice, so as to make full use of the impact of judicial resources and judicial independence to solve the off-site jurisdiction, remote execution and other difficult issues.

3.3. *Improving domestic supervision and counterbalance mechanism of the judicial organs, and enhance judicial openness and transparency*

- *Domestic supervision and counterbalance among public security organs*

The investigation power, procurator power and judicial power should be more reasonable and optimized, so as to prevent the abuse of judicial power. The correlation between legal examination and criminal procedure, such as criminal case review, prosecution and trial, cannot affect the independence of the judges and procurators, which is also the core requirement of the balance of powers and the protection of procedural justice. For example, the executive power from the people's Court can be implemented by administrative organs, so as to gradually reduce the decision-making power and administrative power of the public security organs. As a result, to separate the power of custody from the public security organs, in order to prevent the abuse of investigation power caused by seeking department or individual interests, and also to prevent the abuse of compulsory disposal power. For the procuratorial organs, the point is to strengthen the supervision on their action and the public prosecution instituted by them, and seriously control the approval power and power to decision. To give a full play to the role of lawyers, to supervise and counterbalance, in the judicial operation, and exclude all kinds of savage intervention from administrative organs, judicial organs on the normal practice and defense of lawyers. And to allow lawyers express independent legal opinion. To specify the channels for layers' actions, such as meeting, taking evidence, and inquiry and taking evidence to the judiciary officials. Thus to prevent the randomness in the justice, and elaborate a dynamic role of layers in the cases of securing legitimate interests of parties, and the judicial fair.

- *To promote judicial democracy and judicial openness*

Judicial openness is a project that is crucial to the judicial reform, and it is also an important means to supervise and counterbalance public rights. Efforts to build an open and transparent judicial mechanisms, to promote judicial publicity, and to public judicial enforcement, judicial basis, procedures, process, results and effective legal documents. To strengthen the interpretation of legal facts and legal evidence, to establish effective legal instruments, and to unify Internet access and public inquiry system. In particular, the court trial should be the most important breach of judicial openness, and the trial should be gradually recorded in voice and video, and then be kept in archives. The whole video recording could constraint on the trial activities of judges, and promote litigation participants to exercise their rights according to law. As long as it does not involve personal privacy and state secrets cases, the trial should be free to attend, so as to show the confidence and transparency of the administration of justice. The hearing system for major cases should be implemented, and the press release system should be adopted regularly to establish an interactive platform between the judiciary and the public.

3.4. Accelerating the reform of the judicial personnel system

- *Professional admittance and withdrawal system of judicial personnel*

As it mentioned earlier, the serious problems: low degrees of education background, professional cultivation and skills, and professional mismatch, are among China's current judges, prosecutors and judicial personnel, which has become a bottleneck of improving judicial ability of China's judicial organs, and also one of the important reasons to the failure of power supervision and counterbalance system.

The judge cultivation mechanism of China basically takes the "clerk – Assistant Judge – judge" training mode, which means there is no one from layers. Once they become judges and prosecutors, they have "secure jobs", basically with no withdrawal mechanism. Therefore, the entrance and withdrawal system of judicial officers should be built seriously, especially the judges occupation training system, and occupation evaluation mechanism of judges and prosecutors and judicial personnel database, so as to ensure the quality of judicial personnel. To build the professional entrance and withdrawal mechanism, which means the judicial personnel have serious negligence and dereliction of duty should give punishment even repaying occupation. Attention should be paid to the construction of professional transformation channels from lawyers to judges and prosecutors, so as to enhance the overall professionalism of judicial personnel.

- *Employment security mechanism for judicial officials*

From its occupation characteristics, judges and prosecutors should not only master the professional knowledge of law, but also have some working experience and social experience, and most importantly, they must have occupation ethics guided by the legal faith. All of these must be built on the employment security mechanism, therefore, the employment security mechanism for judicial officers, especially for judges, is of great importance.

Employment security mainly includes several elements: identity protection, namely the judges have independent judgment rights, and could not be removed or transferred without any legal fact or process, so as to secure the comparatively high social position of judges. Economic security, to secure the judges and procurators have corresponding economic treatment to their occupation, and link the paybacks of judicial officials to their professional skills, case-handling capacity and professional integrity, so as to abolish administrative ranks of judges and procurators.

Safety and security, to ensure that judges and prosecutors and their family would not be threatening by doing their jobs; the government must deal seriously cases that the judges and procurators are retaliated, which means the government should build up the public awe to the judges and procurators.

Acknowledgments

The work is partially supported by the fund on the particular item of colleges and universities of the CPC (the Project Number: 2015SYB35).

References

1. Montesquieu, *The Spirit of Laws*, the Commercial Press, in May, 2015.
2. H.L. Jiang, Specific problems faced by the reform of the judicial system in the future, *Finance and Economics*, 2013, 12-04.
3. S.G. Tan, Research on several issues of China's judicial system, *Legal Research*, Session 3, 2011.
4. W.D. Chen, Judicial "de-localization", Logic, challenges and Countermeasures of judicial constitutional reform, *Global Law Review*, Session 1, 2014.
5. Y.P. Xie, W. Shi, Configuration and operation, On the power relation in criminal procedure, *Social Science*, Session 1, 2007, page 81.

Influence factors of enterprises' social responsibility performance in China[*]

Yun Yuan

College of Management, Southwest Minzu University
The 1st Ring Road, Chengdu, 610041, P.R. China

Cheng-Gao Liu

College of Management, Southwest Minzu University
The 1st Ring Road, Chengdu, 610041, P.R. China

Yin-Ye Liao

College of Foreign Languages, Chengdu Normal University
People's South Road, Chengdu, 610041, P.R. China

In recent years, the corporate social responsibility (CSR) has got a rapid expansion in China, for which there are numerous reasons encompassing the active promotion from government and social organizations, the demonstration of some enterprises at home and abroad, the facilitation of public opinion, and finally and most importantly, the practical needs of sustainable development problems such as resource consumption and environmental protection at the present stage. However, the current situation is not optimistic: enterprises pay less attention to social responsibility. Setting out to explore the factors that affect enterprises' CSR performance, this paper chooses A share (RMB common stocks) listed companies in China for 2012-2014 as the sample, takes Rankins CSR Ratings (RKS) as the standards, and investigates the impact of company size, industry characteristics, regional attributes and financial performance on enterprises' CSR performance. Thus, this article provides a basis for the further effective commitment of Chinese enterprises to social responsibility.

Keywords: Listed company; corporate social responsibility (CSR); influence factor.

1. A Literature Review of Prior Studies

In China and other countries, there are many research achievements on corporate social responsibility (CSR), dealing with the content of CSR, the form of CSR report, the evaluation of companies' CSR performance, and so on. As for the study of influence factors of companies' CSR performance, it is mainly focused on the

---

[*]This work is supported by Scientific Research Fund of Resource-dependent Cities Research Center (ZYZX-YB-1404).

following aspects: enterprise management performance, enterprise scale, industry attribute and other influencing factors.

1.1. *Corporate financial performance and CSR performance*

Early studies conducted by Bragdon and Marlin (1972), Bowman and Haire (1975), and Chen and Mectalf (1980) have proved that there is a positive correlation between CSR performance and the financial performance of the enterprise. Chen and Metcalf (1980) used the price-earnings ratio index to measure business performance and they all found that there is a positive correlation between CSR and financial performance. Based on social responsibility disclosure report and corporate financial report issued by 134 companies in the United States, Cowen, Ferreri and Parker (1987) found that there is a positive relationship between the profitability of the enterprise and the disclosure of CSR information, and the positive relationship is significant. Considering the effects of risk, time, and environment, Mc Guire *et al.* (1988) used accounting and market indicators to analyze the sample. And the results of the study also showed that the performance of the company has a positive impact on the CSR information. What's more, after comparing the social performance and financial performance of the 67 major American companies in ten years, Preston and O. Bannon (1997) found that there is a positive correlation between business performance and CSR.

In China, Wan Shouyi (2011) carried out an empirical study of A-share (RMB common stocks) listed companies and took the social responsibility report issued by those companies as a breakthrough point. It is found that the profitability and operating ability of the corporate financial characteristics have a positive impact on the disclosure of CSR information. And it is also found that some characteristic variables of the company passed the saliency test. Those variables include two-job-in-one (with the chairman and the general manager being one person), state-owned holding and enterprise size.

1.2. *Enterprise size and CSR performance*

Foo and Tan (1988) studied nearly 300 of the 305 companies in Malaysia and Singapore that were listed on the Singapore Stock Exchange in late 1985. They finally found that whether scale is measured by market equity, asset value, pretax profit or business income, large scale enterprises are more likely to disclose the impact of corporate behavior on society than small businesses. Similarly, Patten (1991), taking the annual report of 128 companies in the United States in 1985 as the research object, proved that the scale of the enterprise does have a significant impact on the level of information disclosure of social responsibility. Through

Brammer and Pavelin's analysis (2004), it is concluded that large scale enterprises are more vulnerable to adverse factors because of their wide audience, so their public visibility is higher and the social responsibility information disclosed is more.

1.3. Industry attributes and CSR performance

Patten's findings (1991) suggest that enterprises in a highly visible and politically sensitive industry will disclose more information on social responsibility. Deegan and Gordon (1996) investigated environmental reports from 197 industries sampled randomly from 50 industries, and found that the environment sensitive industry enterprises had more positive environmental disclosures. Clarke and Gibson-Sweet (1999) analyzed the environmental information disclosed in the UK's 100 largest corporate annual reports. They found that there are only a few companies that have not disclosed environmental information in the oil, gas and nuclear industries, but all enterprises in the mine and chemical industry have disclosed environmental information. This proves that there is a close relationship between the disclosure of social responsibility information and the industry that the enterprise belongs to. Jenkins and Yakovleva (2006) found that large mining companies have revealed more information on social and environmental performance, health, safety and morality.

Through the investigation and analysis of factors affecting agricultural CSR of 225 agricultural enterprises, Zhang and Yan (2012) found that the agricultural enterprise social responsibility is mainly affected by internal factors and enterprise profitability, management ability, managers' educational level, product export ability and innovation ability are five key factors. Wang Xuemei and Bu Hua (2012) chose the A-share listed companies in Shenzhen and Shanghai in the food and beverage industry as a research sample. His research concluded that: in order to promote the fulfillment of social responsibility, the beverage industry must also attach importance to the executive incentive, the two roles of chairman and general manager, and the positive role of CSR.

1.4. Geological factors and CSR performance

Taking the Shanghai and Shenzhen 300 index enterprises as an example and then setting up a multiple linear regression model, Xu Yingjie and Shi Ying (2014) studied the influencing factors of strategic social responsibility of Chinese listed companies, and concluded that the strategic social responsibility of employees, communities and environment in the developed areas is better.

1.5. Corporate governance and CSR performance

Abdul and Ibrahim (2002) issued a questionnaire to senior executives in Malaysia and empirically studied the impact of managerial attitude on social responsibility. Their study found that managers' attitudes had a significant impact on the performance of social responsibility. To further explain the management attitude and the potential impact of the government on CSR, Haniffa and Cooke (2005) took Malaysia enterprises as samples, and surveyed the managers of enterprises. The results showed that CSR information disclosure is closely related to management attitude. Chen Zhi (2011) conducted an empirical analysis from the perspective of corporate governance. The following conclusions can be drawn: corporate governance is the main factor influencing the fulfillment of social responsibility; effective internal and external corporate governance is conducive to CSR; CSR can also promote effective improvement and healthy development of corporate governance.

1.6. Other aspects

Gelb and Strawser (2001) proved that the quality of information disclosure of social responsibility is significantly related to the sense of CSR. That is to say, if the enterprise bears more social responsibility, then the information disclosure will provide more relevant information. Therefore, the increase of the information disclosure of social responsibility is the expression of CSR. Many enterprises are aware of the importance of maintaining relationships with stakeholders. Accordingly, they will provide more social responsibility information to meet the information needs of stakeholders. Yang Chunfang's research (2009) found that external factors (such as the intensification of market competition, the strengthening of the level of government intervention and the improvement of the legal environment) had no significant influence on CSR behavior of Chinese enterprises, and their internal factors (such as export behavior, innovation ability, management level and its own financial situation) all affected CSR performance.

Through reviewing the relevant literature at home and abroad, it can be proved that the research on the influencing factors of CSR in China's outside areas started earlier and the research results are very rich; the research in China started late. Although China's research has gradually begun to deepen, but the analysis of the influencing factors is still lack of systematic and in-depth investigation. Therefore, the conclusion of the study remains to be further deepened. Especially at the present stage, more and more Chinese enterprises are beginning to realize the importance of social responsibility, so the research of large sample is particularly necessary. At the same time, as different scholars usually adopt their own social responsibility evaluation criteria, the conclusions are not unified. In view of this, this paper adopts the widely accepted social responsibility evaluation index as the standard of CSR, so as to further explore the influencing factors of CSR, and provide more abundant reference data for the research of CSR in China.

2. Research Design

2.1. *Hypotheses*

2.1.1. *Enterprise size*

According to the principal-agent theory, the larger the scale of the enterprise, the higher the agency cost. Therefore, in order to reduce the agency costs, large-scale enterprises will disclose more information, especially the good news of the development of the enterprise. In this way, stakeholders will know the contribution of enterprises to the society, that is, the social responsibility to fulfill, so as to further establish a good image of the company in the public mind, and enhance corporate reputation and invest in enterprises. From this, the paper suggests Hypothesis 1.

Hypothesis 1: The greater the scale of the enterprise, the better the performance of its social responsibility.

2.1.2. *Profitability*

The stronger the profitability and the better financial performance of an enterprise is, the more energy and willingness of the enterprise to fulfill its social responsibilities, and the more active information disclosure of its social responsibility, because companies are willing to improve their reputation. From this, the paper suggests Hypothesis 2.

Hypothesis 2: The stronger the profitability of an enterprise, the better the performance of its social responsibility.

2.1.3. *Industry attributes*

The social responsibilities and concerns of different industries are different. The stakeholders' demand for social responsibility in different industries is also not the same. For instance, some heavy polluting industries and environmental sensitive enterprises will tend to disclose social responsibility information in order to maintain their own image and to show their efforts in social responsibility. Therefore, the credibility of those industries' social responsibility report will be highly concerned.

According to the "Guide to Environmental Information Disclosure of Listed Companies" issued by the Ministry of Environmental Protection in 2010, the following 16 categories of industries are classified as heavy polluting industries. They are thermal power, building materials, iron and steel, cement, electrolytic aluminum, paper making, brewing, petrochemical industry, pharmaceutical industry, coal, metallurgy, chemical industry, fermentation, textile, leather and mining. These industries have a high degree of attention. Once social responsibility appears to be a problem, their reputation will be seriously affected. Therefore, in order to prevent the occurrence of this phenomenon, the enterprises will be better to fulfill the social responsibility. From this, the paper suggests Hypothesis 3.

Hypothesis 3: Heavy polluting industries are likely to fulfill their social responsibility better than other kinds of enterprises.

2.1.4. *Geological factors*

It is generally believed that the enterprises in the developed areas will develop better, receive more attention and are more willing to fulfill their social responsibilities. From this, the paper suggests Hypothesis 4.

Hypothesis 4: Enterprises in developed areas perform their social responsibilities better than those in other regions.

2.2. Research design

2.2.1. *The resource of data and definition of variables*

This paper chooses 2001 CSR reports, which were released in 2012-2014 by A-share listed companies on Shanghai and Shenzhen stock exchange as the samples. All the assurance information of those CSR reports come from Rankins CSR Ratings (RKS) database, and the finance information comes from the database of CSMAR. After getting rid of ST companies and reports with incomplete data, there remain 1900 reports with valid assurance.

2.2.2. *Variables and model constructing*

- Dependent Variables

CSR performance is taken as the dependent variable. Its data comes from corporate comprehensive marks scored by Rankins CSR Ratings and the full mark is 100.

- Independent Variables

Independent variables consist of enterprise size, profitability, industry attributes and geological factors. First, enterprise size is measured in terms of the natural logarithm of enterprise's total assets. Second, profitability is measured in terms of enterprise's return on equity (Roe). High Roe means strong profitability. According to the "Guide to Environmental Information Disclosure of Listed Companies" issued by the Ministry of Environmental Protection in 2010, 16 categories of industries are classified as heavy polluting industries. For the category of industry, if the enterprise under investigation belongs to the heavy polluting industry, it is assigned as 1. Otherwise, it is assigned as 0. And for the category of region, if the enterprise is in the developed region, it is assigned as 1. Otherwise, it is assigned as 0. The definition of economically developed areas is based on the GDP ranking in 2014. The eight provinces and cities such as Beijing, Tianjin, Shanghai, Guangdong, Fujian, Zhejiang, Jiangsu and Shandong are economically developed areas.

- Control Variables

Of diverse determinants for CSR performance, this paper selects the following three as control variables: debt outstanding, Ownership structure and Growth capability. First, debt outstanding is measured in terms of asset-liability ratio. Second, Ownership structure is measured in terms of Ownership nature of controlling shareholders. If the actual controlling shareholder is state-owned, it is assigned as 1. Otherwise, it is assigned as 0. Last, growth capability is judged by revenue growth rate. The higher revenue growth rate is, the higher reputation the company enjoys and the higher rank the company holds, which impels the company to better CSR performance.

Table 1. Variables.

Variable	Meaning of variable	Variable symbol	Method
Dependent variable	Corporate social responsibility performance	CSR	Rankins CSR Ratings, RKS
Independent variable	Enterprise size	Size	The logarithm of total assets
	Profitability	Roe	Net margin/profit
	Industry attributes	Industry	Dummy variables, industry polluting heavily = 1, or, = 0
	Geological factors	Province	Location in developed area = 1, or, = 0
Control variable	Debt outstanding	Rda	Asset-liability ratio
	Ownership structure	State	State share = 1, or, = 0
	Growth capability	Growth	Revenue increase rate

- Model Constructing

Based on the above hypotheses, the following regression model can be constructed.

$$\text{CSR} = \beta_0 + \beta_1 \text{ Size} + \beta_2 \text{ Roe} + \beta_3 \text{ Industry} + \beta_4 \text{ Province} + \beta_5 \text{ Rda} + \beta_6 \text{ State} + \beta_7 \text{ Growth} + \varepsilon \quad (1)$$

Among them, the β_i (i = 1,2,...7) is the regression coefficient of each research variable, and ε is the error term.

2.3. Descriptive statistics

Table 2. Descriptive statistics.

	Number	Minimum	Maximum	Mean	Std. deviation
CSR	1900	15.12	88.45	38.96	12.74
Size	1900	19.54	30.66	23.21	1.77
Roe	1900	-1.86	3.83	0.08	0.14
Industry	1900	0	1	0.53	0.50
Province	1900	0	1	0.66	0.48
Rda	1900	0.007	1.11	0.52	0.22
State	1900	0	1	0.24	0.43
Growth	1900	-0.88	4.65	0.13	0.36

Table 2 provides a descriptive statistical analysis of all the variables.

In Table 2, the mean value of CSR reporting level (CSR) is 38.96, the minimum value is 15.12, and the maximum value is 88.45, all of which suggest that the social responsibility gap issued by A-Share listed companies in China in 2012-2014 is still larger.

According to the table, the maximum value of enterprise size is 30.66, and minimum value is 19.54; the minimum value of profitability (Roe) is −1.86, and the maximum value is 3.83, the mean value of Roe is 0.08. Together, they show that there is a great difference in the profitability of the selected sample enterprises.

The average industry attribute (Industry) is 0.53, indicating that 53% of the enterprises that publish the social responsibility report belong to the heavy pollution industry. The mean value of regional factor (Province) is 0.66, suggesting that 66% of the sample enterprises are in the economically developed areas. The average value of the debt scale (Rda) is 0.52, showing that the asset liability ratio of the sample enterprises is at a reasonable level. The mean value of equity (State) is 0.24, presenting that 24% of the enterprises studied are state-owned shares or state-owned corporate shares. The minimum value of development ability is -0.88, and the maximum value is 4.65, which indicates that the development ability of the sample enterprises is quite different.

3. An Empirical Study on the Influence Factors of Enterprises' Social Responsibility Performance in China

3.1. *Correlation*

Table 3. Correlation.

	CSR	Size	Roe	Industry	Province	Rda	State	Growth
CSR	1							
Size	0.504**	1						
Roe	0.066**	0.117**	1					
Industry	-0.064**	-0.142**	-0.061**	1				
Province	0.085**	0.063**	0.066**	-0.106**	1			
Rda	0.162**	0.393**	-0.041	-0.063**	0.007	1		
State	0.06**	0.108**	0.033	-0.027	-0.024	0.085**	1	
Growth	0.017	0.016	0.055*	-0.031	0.042	-0.015	0.085**	1

Notes: ***Correlation is significant at the 0.01 level (2-tailed). **Correlation is significant at the 0.05 level (2-tailed). *Correlation is significant at the 0.1 level (2-tailed).

The data of Table 3 shows the correlation between the variables.

From the perspective of firm size, the scale of corporate social responsibility and the disclosure of social responsibility report were significantly positively correlated with the level of 5%, indicating that larger enterprises are more willing to issue social responsibility information. Hence, hypothesis 1 is confirmed.

From the point of view of profitability, the profitability of the enterprise and the level of social responsibility report are positively related to the level of 5%. This shows that the stronger the profitability of the enterprise, the more willing to publish the social responsibility report, the good news of the enterprise profit to the stakeholders. Therefore, hypothesis 2 is confirmed.

From the perspective of industry attributes, the enterprises that choose to disclose the information of social responsibility and belonging to the heavy pollution industry have a negative correlation with the level of social responsibility reporting disclosure. This shows that the social responsibility consciousness of the heavy pollution industry at the present stage of our country is still to be strengthened, and the level of social responsibility report issued by it is still low. This is just the opposite of hypothesis 3.

From the perspective of regional factors, the two levels is positively correlated at 5% level, which indicates that enterprises in economically developed areas are more willing to disclose social responsibility information and prefer to fulfill their responsibilities to society.

3.2. Analysis of regressions

Table 4 is the regression line value obtained by SPSS, in which the P value of the enterprise scale is 0 and the P value of the regional factors is 0.09, indicating that the significance is strong. R^2 is 0.260, Adjusted R^2 is 0.257, indicating that the fitting degree of regression equation is general. It shows that CSR report will also receive other factors. This is also the direction for further research in the future.

Table 4. Independent variable's regression.

	Unstandardized coefficients		Typical coefficient	t	Sig.
	B	Standard error			
β_0	-47.550	3.494		-13.610	.000
Size	3.726	.158	.519	23.602	.000
Roe	.069	1.868	.001	.037	.971
Industry	.372	.512	.015	.725	.468
Province	1.399	.535	.052	2.617	.009
Rda	-2.345	1.268	-.040	-1.849	.065
State	.287	.598	.010	.481	.631
Growth	.200	.713	.006	.280	.780
	$R^2 = 0.260$		Adjusted $R^2 = 0.257$		

3.3. Research conclusions

The following conclusions are drawn from the above empirical tests:

First, there is a significant positive relationship between the level of CSR and the size, region of the enterprise. That is to say, the greater the scale of the enterprise is, the more it pays attention to its social responsibility. And the enterprises in the economically developed areas pay more attention to social responsibility than those in undeveloped regions. Generally, large enterprises pay more attention to the reputation and their social image, and they can improve their reputation by taking more social responsibilities. In economically developed areas, the public has a higher demand for social benefits for enterprises in addition to their economic performance expectations. Meanwhile, various NGOs in the developed areas are more perfect, all of which push enterprises to fulfill more social responsibilities to a certain extent.

Second, the more profitable businesses are, the more they are willing to publish social responsibility reports. They expect to convey the good news of CSR to stakeholders, so as to get more investment and promote a greater degree of profitability.

Third, The CSR level of different industry attributes is different. Especially, the level of CSR released by heavy polluting industries is still low. The awareness of social responsibility of these industries needs to be strengthened. However, due to the continuous enhancement of the whole society's ecological consciousness and the promotion of relevant policies and regulations, such as the beginning to levy environmental tax, heavy polluting enterprises will also enhance their social responsibility performance under the dual role of internal and external factors.

Acknowledgments

This work is supported by Scientific Research Fund of Resource-dependent Cities Research Center (ZYZX-YB-1404).

References

1. Bragdon J., Marlin J. Is Pollution Profitable? Risk Management. 1972(19):9-18.
2. Bowman E. H., Haire M., A Strategic Posture toward Corporate Social Responsibility. California Management Review, 1975(18):49-58.
3. Chen K. H., Metcalf R. W., The Relationship between Pollution Control Record and Financial Indicators Revisited. Accounting Review, 1980(55):168-177.
4. Cowen S.S., Ferreri L.B, Parker L.D. The Impact of Corporate Characteristics on Social Responsibility Disclosure: A Typology and Frequency-based Analysis. Accounting, Organizations and Society, 1987, 12(2):111-122.
5. Foo S.L, Tan M.S. A Comparative Study of Social Responsibility Reporting in Malaysia and Singapore. Singapore Accountant, 1988.
6. Deegan C., Gordon B. A Study of the Environmental Disclosure Practices of Australian Corporations. Accounting and business research, 1996, 26(3): 187-199.
7. Clarke J., Gibson-Sweet M. The Use of Corporate Social Disclosures in the Management of Reputation and Legitimacy: A Cross Sectorial Analysis of UK Top 100 Companies. Business Ethics: A European Review, 1999, 8(1): 5-13.
8. Jenkins, H. and Yakovleva, N. (2006). Corporate social responsibility in mining industry: Exploring trends in social and environmental disclosure. Journal of Cleaner Production, 14, 271-284.
9. S.R. Zhang and N. Yan (2012), Empirical Studies on the Economic Support of Emigrants to the Aged in Developing Rural Areas: Take Zhongzhai Brigade of Yidu Village in Defang County as an Example." Journal of Tsinghua University (Philosophy and Social Sciences) 27: 46–54.
10. Abdul M.Z., Ibrahim S. Executive and Management Attitudes towards Corporate Social Responsibility in Malaysia. Corporate Governance, 2002, 2(4): 10-16.
11. Haniffa R.M., Cooke T E. The Impact of Culture and Governance on Corporate Social Reporting. Journal of accounting and public policy, 2005, 24(5): 391-430.
12. Mc Guire J.B., Sundgren A., Schneeweis T. Corporate Social Responsibility and Firm Financial Performance. Accounting of Management Journal, 1988, (31):854-872.
13. Lee E. Preston, Douglas P. O'Bannon. The Corporate Social- Financial Performance Relationship- A Typology and Analysis. Business &Society, 1997(36).
14. Brammer S., Pavelin S. Building a Good Reputation. European Management Journal, 2004, (22):704-713.
15. Patten D.M. Exposure, Legitimacy, and Social Disclosure. Journal of Accounting and Public Policy, 1991, (10):297-308.

16. Gelb, D., Strawer, J.A., Corporate Social Responsibility and financial disclosures: an alternative explanation for increased disclosure. Journal of Business Ethics, 2001(33).
17. Wang Xuemei, BU Hua. Research on the Relationship between Corporate Governance and Corporate Social Responsibility, Friends of Accounting, 2012(5).
18. Xu Yingjie, Shi Ying. A Research on the Factors Influencing the Strategic Social Responsibility of Chinese Listed Companies. Reform of Economic System, 2014(4).
19. Wan Shouyi, Liu Wei. A Research on the Influencing Factors of Corporate Social Responsibility Information Disclosure. Friends of Accounting, 2011(6).
20. Yang Chunfang. An Empirical Study on the Factors Influencing the Social Responsibility of Chinese Enterprises. Economist, 2009(1).
21. Chen Zhi, Xu Guangcheng. A Research on the Factors Influencing the Social Responsibility of Chinese Enterprises. Enterprise Management, 2011(4).

Study on management decision of China's dry cleaning enterprises based on the current situation of industry development

Ming Jiang

College of Management, Southwest Minzu University, Chengdu, 610041, P.R. China

Caijuan Zhang

Law school, Southwest Minzu University, Chengdu, 610041, P.R. China

In recent years, China's dry cleaning industry is in a phase of rapid development. However, in the process, it also faces many challenges and bottlenecks. For example, washing equipment does not meet the requirements of environmental protection; practitioners have a lower degree of specialization; washing quality is lower; the frequency of dry-cleaning accidents is higher and so on. Based on this, the author intends to sort out the relevant research literature, through the investigation and study, to analyze the problems and causes of China's dry cleaning industry. Then, based on the chain management concept and the "cost-benefit" model, this paper proposes the business decision-making and development approaches for dry-cleaning enterprises in China.

1. Introduction

The dry-cleaning is a way to remove stains by using solvents other than water as a medium. The dry-cleaning technology mainly originated in the mid-19th century in France, commonly known as "French dry cleaning". With a modern way to wash and ironing clothing, it has a hundred years of history. China's dry cleaning industry started in the late 19th century early 20th century. Around the 1950s, China's dry-cleaning industry had some development in large and medium-sized cities in the eastern coastal areas. After 1978, with the continuous improvement of people's living standards and the development of the textile industry, China's apparel fabrics started to grow from cotton wool and silk to various chemical fibers, which promoted the development of the market for washes. In recent years, China's dry-cleaning chain has made great development. Whether it is the entry of laundry brands in other countries or the rise of local washing companies, China's dry-cleaning industry has all along been able to achieve the goal of rapid expansion by way of franchising. According to the statistics of the National Bureau of Statistics of China, the number of chain stores nationwide has increased from 2,600 in 1998 to more than 8,000 in 2016. The percentage increases for both single laundries and franchise stores have been significantly changed. The former dropped from 85.6% in 1998 to 40.4% in 2016, while the franchise chain increased from 14.4% to 59.6%.

Faced with the development of the domestic economy and the changes in the international market, the Chinese dry-cleaning industry is about to undergo a new round of reshuffle. All dry cleaning companies should make new strategic choices and business decisions. At present, most academics focus on research from the perspectives of dry cleaning technology and environmental protection, but rarely do systematic and systematic studies on the dry-cleaning industry from the perspectives of business management decisions and strategic choices. Based on this, the author intends to sort out the relevant research literature, through the investigation and study, to analyze the problems and causes of China's dry cleaning industry. Then, based on the chain management concept and the "cost-benefit" model, this paper proposes the business decision-making and development approaches for dry-cleaning enterprises in China.

2. The development status of dry cleaning industry in China

2.1. The development characteristics

According to the National Bureau of Statistics of China (NBS), there are a total of 1.43 million laundries (including guesthouses and laundries), employing as many as 7 million people. During the ten years from 2007 to 2016, the number of domestic laundries increased by 300,000. At present, all the dry cleaners in China have an annual washing volume of more than 4 billion pieces, with an annual operating income of about 30 billion yuan (Table 1).

Table 1. The statistical table of development of dry-cleaning industry in China in 2016.

Number of laundry	Number of employees	Number of the dry cleaning machine	Annual washing capacity	Annual revenue
1.43 million laundries	7 million people	41 million units	More than 40 billion pieces	About 300 billion yuan

According to the statistics, since the mid-to-late 1990s, the laundry chain has grown by about 1,000 or so annually. There is no doubt that franchising has become a bright spot in the development of China's dry cleaning industry.

2.2. The bottlenecks and causes

First, washing equipment can not meet the environmental requirements. In the end of 2016, 95% of the 410,000 dry-cleaning machines in China were PCE solvent dry-cleaning machines and the rest were petroleum solvent dry-cleaning machines. In addition, there are a very small number of machines that use fluorocarbon or hydrocarbon solvent dryers. In the domestic dry cleaning machines of PCE solvent, the open dry-cleaning machines's market share reached 90%. In 2010, the national laundry added 20,000 dry cleaning machines, of which 17,000 are open dry-cleaning machines. In the 200,000 dry-cleaning machines, about 180,000 are working properly and achieving a washing effect.[1]

PCE dry cleaning machine running, it has a very serious air and human hazards. If people often touch the PCE, there will be dizziness, headache, vertigo, nausea and vomiting and other symptoms. If acute inhalation PCE, people will show eyes, nose, throat, pharyngeal irritation symptoms.[2] A studies have shown that children are particularly sensitive to high levels of chloride in dry cleaners; the effect of dry cleaners on male sexual function has been demonstrated.[3] The International Agency for Research on Cancer (IARC) has included PCE as a potential carcinogen, and the Cleaner Alternative Technology Evaluation (CTSA) developed by the U.S. Environmental Protection Agency (EPA) also links the use of PCE to cancer and neurotoxicity. According to open dry-cleaning machines on the PCE recovery rate of 92% and closed dry-cleaning machine recovery rate of 97% to calculate the average daily dry-cleaning machine emissions of about 0.56 kg, then the national daily emissions of tetrachlor in the air Ethylene about 18,000 kilograms, which seriously affected the air quality and personal health.

Second, the washing quality is low. At present, the overall level of China's laundry is still very low, the competition between the washing market is very fierce, there is a difference between the size of the laundry equipment and investment. The standard laundry costs high due to the need to invest in high-end equipment and high-quality washing materials and standard washing operations. They can not compete with the simple equipment, cutting corners and washing small shops in water for price competition, which leads to chaos in the operating environment of the laundry market.

Third, the industry's awareness of self-protection is poor. As the dry cleaning industry is still in the primary stage of development in our country, the development of dry-cleaning management in all parts of the country is unbalanced and even the dry-cleaning shop layout in some area of the same city is unbalanced. In this case, easily lead to vicious competition. According to the survey, at present, Chinese consumers wash an entire set of clothes every 50 days. Therefore, it is speculated that there can be a living space for laundry every 3000 people. However, at present, the number of laundries in some areas is too large, meanwhile, the distance between these laundries is too close, resulting in duplication of investment in the same industry.

Fourth, the frequency of dry cleaning accidents is higher. The dry cleaning is one of the most accidents in China's service industry. In recent years, the frequency of accidents caused by the improper treatment of steam boiler and dry cleaning machine is higher than that of the steam boiler and dry cleaning machine. Some laundry shops use 50% of pressure vessels without boiler safety production license. These washing enterprises generally do not have annual inspection means, resulting in equipment aging. Therefore, accidents often happen. At present, most of the domestic washing machinery and equipment are not marked with the use of time and years. We can often see that many of the expired equipment appear in production.

Fifth, the degree of professionalization of the employees is low. According to the statistics of the international washing up association, among the employees of the dry cleaning industry in China, colleges and universities are abound. Although a large number of graduate students are engaged in the industry, less than one percent of those who have been trained by vocational qualification training. According to the China government to simplify the approval procedures for business registration without preconditions, as long as there are sufficient funds, the appropriate location, you can register, so that anyone can be employed, and legitimate, that no one can control the laundry market regulations or powerful organization.

3. The strategic choice and management decision of dry cleaning enterprises in China

3.1. Strategic choice — chain model

The chain operation is actually a consortium. It refers to several stores that run similar commodities, which form a whole through certain ties and according to certain rules. Chain stores carry out specialized division of labor under the overall planning of managers, and implement standardized operation on this basis, so as to simplify complex business activities and improve business income and scale efficiency.[4]

1) Chain management strategy is an unstoppable trend

Laundry service is a promising job. But with the specialization, intensification and integration of market economy, the survival space of single laundries is getting smaller and smaller because of less funds and lack of management technology. In order to cope with the fierce market competition, chain management has become the fastest and most effective way. ILSA (Italy laundry brand) and FORNET (France laundry brand) opened the first stage of China's laundry industry franchise. The development of China's laundry chain operation and even the whole industry has undergone tremendous changes. Almost all the business activities in the laundry industry are affected by the chain operation. Nationwide, ILSA from Italy, FORNET of France, GEP of the US, SANYO of Japan, Beijing's old PRIDE and new BONNYCH have developed well.

2) Scale efficiency analysis of the chain management

Scale advantage is the most prominent feature of dry-cleaning chain. Its fundamental meaning is that when a sales volume of a dry-cleaning chain (store) reaches a certain scale, it has the lowest cost or the highest profit. Philip Kotler (American economist) has a more systematic explanation of the number of organizations that have economies of scale in the chain industry. He believes that the number of chain stores should reach about 11, and on this basis, the formation of the lowest cost or maximum operating profit. (Figure 1)

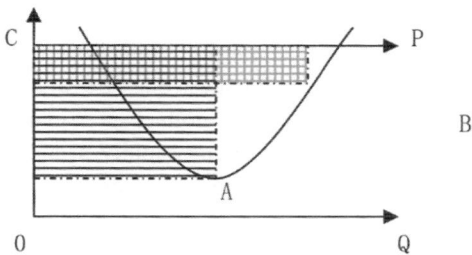

Figure 1. The economies of scale of the dry-cleaning chain.

In the Figure 1, some abbreviations are as following:
C: Cost;
P: Price;
A: The lowest cost point;
B: The biggest profit point;
Q: scale

The scale advantage of dry-cleaning chain operations is fundamentally the scale and logistics brought by the highly organized and chain following many single stores, resulting in the low-cost operation of the entire chain. In economic terms, the scale is essentially to speed up the turnover of goods and improve the utilization of resources. After the expansion of the scale, all dry cleaners can achieve resource sharing, reduce the unit washing service costs (including advertising costs, information resources development costs, operating expenses, etc.). The chain operation scale, the degree of specialization of dry cleaners will be higher, not only can improve labor productivity, reduce the cost of enterprise long-term average (Figure 2), but also can play the benefits manager to enlarge occupation, so as to further promote the enterprise profit growth.

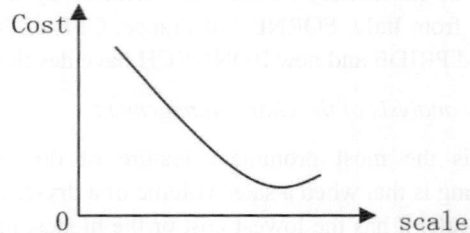

Figure 2. The relationship between long-term average cost and the size of the business.

3) Efforts to cultivate brand-name dry cleaning business

Franchise chain is a formal form of formal development to a certain stage of the advanced form, it is through the overall transfer of intellectual property to achieve network-based chain operations. At present, franchising is more common in developed countries. According to statistics, out of every U.S. dollar spent on daily spending, one out of every three U.S. dollars spent in a variety of franchised chains. Franchising in Germany is growing rapidly at an annual rate of 25%. According to a survey conducted by the U.S. Commerce Department, in the service and retail sectors, the failure rate of a single shop is 85%, while the franchise's success rate is 90%. China's dry-cleaning chain has greatly developed, there have been some well-known local brands, but so far not a large-scale franchise chain dry cleaning company with foreign ILSA and FORNET to contend.

The experience of chain operation in Europe and the United States, as well as the practice of chain management in our country, show that in the initial stage of the development of chain management, direct store development is a reasonable choice. However, after completing the functions of direct demonstration, it must inevitably be transferred to the franchise chain. May really form a large-scale operation, to create a brand-name business. Needless to say, Franchise chain stores will become the mainstream of the laundry market in the future.

3.2. *The management decision — green dry cleaning*

1) The green dry cleaning concept and presenting background

The so-called green dry cleaning means that dry cleaners must provide services that are environmentally friendly in their operation. By adopting environmentally friendly technologies and pollution prevention measures in dry cleaning equipment, washing solvents, washing processes and services. It is possible to eliminate or reduce the waste generated from various (production and use) processes and to maximize the recovery and reuse of wastes that can not be eliminated in the above process (e.g., dry cleaners). For non-reusable waste pollutants, enterprises should properly dispose of such waste (such as landfill and discharge) to reduce their impact on human health and the environment while fully ensuring environmental safety [5]. In recent years, the emergence of various new detergents has brought environmental pollution. Some bad washing methods not only affect the ecological environment, but also bring adverse effects on the residents' health. The concept of green dry cleaning is proposed in this context.

2) Green dry cleaning is a necessary requirement for the sustainable development of dry cleaning enterprises

Traditional dry cleaning mainly open PCE dry cleaning machines as the main equipment. Although PCE has in the past provided good service to the dry cleaning industry with its excellent cleaning efficiency and the industry has been working hard to reduce the amount of PCE used with more efficient equipment and better laundry training, the dry cleaning industry is still facing the public Questioning the health and environmental hazards of PCE. As a volatile ingredient known to be detrimental to the Earth's ozone layer, PCE can adversely affect the green consumption of textiles through supply chains, product flows and service delivery.

More than 80% of the existing dry-cleaning machines in our country are open-type or semi-open-type equipment without any recovery device. If the organic solvent can not be fully recovered, there will be emission of washing liquid waste, waste gas and waste residue, and the environmental protection and Safety related issues.(Table 2) "Chinese Occupational Medicine" 2003 the third period "Lanzhou dry cleaners in the air pollution situation of tetrachlorethylene" article pointed out that in Lanzhou City randomly selected 24 dry cleaners a total of 33 workers, of whom 10 were due to PCE Often have headache, dizziness, respiratory and eye irritation symptoms, most of the dry cleaners hygiene is not standard ventilation poor.

Table 2. Wastes from the dry cleaning process.

Process	Washing process	Dry cleaning process
Laundry program	drain	Waste organic solvent / organic solvent waste gas
Drying procedures	exhaust	Organic solvent exhaust
Ironing procedure	Exhaust system for work environment	Operating environment of the steam system

At present, Germany, Japan, the United States and Canada have included the inclusion of halogenated olefin cleaners, including PCE, in the control of ecotextiles. With a view to resolving the problem of textile clean production, it can eliminate the potential environmental pollution caused by the use of textiles influences.

3) The green dry cleaning path to achieve

According to the meaning of green dry cleaning, dry cleaning companies must implement a comprehensive prevention and control strategy for the service process. Dry cleaning solvents and washing process should be in line with environmental standards, businesses to the extent possible, reducing the amount of hazardous waste generated during the dry cleaning process. It includes reducing emissions and recycling of pollutants, and requires that raw materials and energy be saved in the production process, toxic materials should be removed, and emissions should be reduced. For example, an enterprise should reduce the impact of the entire product life cycle on human health and the environment in terms of product services (Figure 3). In other words, the dry cleaning industry's green service must provide more environmentally friendly services in its marketing process. Businesses eliminate or reduce as much as possible of wastes generated from a wide range of (production, use) processes by taking environmentally friendly technologies and pollution prevention measures in dry cleaning equipment, washing solvents, washing processes and services to maximize recovery Use waste that can not be eliminated in the above process (e.g., dry cleaners). For non-reusable waste pollutants, dry-cleaning enterprises should properly dispose of them (such as landfill and discharge) to ensure their environmental health and reduce their impact on human health and the environment.

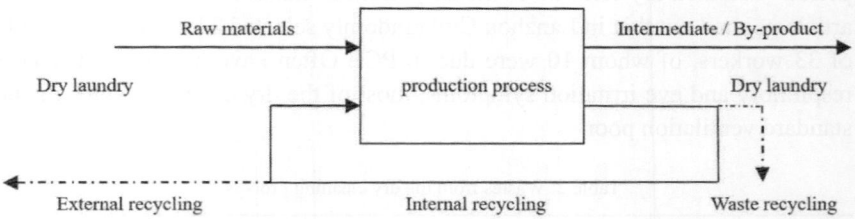

Figure 3. Green dry cleaning process.

4. Conclusion

In short, chain management and green dry cleaning because of its own economies of scale and environmental protection advantages, will be the future development of dry-cleaning business trends. As international dry cleaning giants such as TREVIL, ILSA, FORNET and RRADIANT make major inroads into China, traditional dry cleaning methods characterized by high consumption, high emission and pollution are gradually being replaced by green dry cleaning. The author believes that in the future more and more dry-cleaning enterprises will choose green strategies, promote green technological innovation and promote the sound development of dry-cleaning industry.

References

1. Yuan Xunxia. Toxicity of chemicals. Shanghai Science and Technology Literature Press. pp. 120-124. 1990.
2. D.A. Keys, D.G. Wallace, R.B. Conolly. Source reduction of chlorinated solvents: dry cleaning of fabrics. California Department of Toxic Substances Control, Alternative Technology Division. Sacramento, California. pp. 23~5. January 2010.
3. T.B. Gray, S.D. Gangolli. Aspects of the testicular toxicity and environmental impact of OCs, Environ. Health Perspect. pp. 229-235. 2006.
4. Qinghua Zhang. Chain management principles. Lixin press. pp. 55-56. 2006.
5. Jun Chen. Green cleaning — the direction of sustainable development of dry cleaning industry in China. Cleaning Technology. July 2014.

A study on the use and development strategy of official Weibo of Minzu universities in China

Siyuan Song
College of Management, Southwest Minzu University
Chengdu, Sichuan Province, P.R. China

Ming Jiang
Sichuan Research Center for National Education Development,
Southwest Minzu University, Chengdu, Sichuan Province, P.R. China

> With the development of Internet technology, Weibo has become the major means of social and information dissemination for college students due to its excellent convenience, timeliness, and interactivity. It brings huge opportunities to Minzu universities meanwhile some challenges followed. So it has become a realistic and urgent topic on how to run the official Weibo in Minzu universities. This article explains the significance of opening Weibo accounts for Minzu universities and takes 10 Minzu universities in China as examples to analyze the situations and relative problems of their official Weibo, raise some advices for the construction and management of Weibo in Minzu universities in three aspects: information post, brand establishing and public relations maintenance. The main purpose of this dissertation is to provide certain constructive suggestions for the future development of the official Weibo of Minzu universities in China.

1. Introduction

Since the 19th National Congress of the CPC, the new central leadership with Jinping Xi as the core has attached great importance to the construction of the Internet and new media. In October 2017, President Jinping Xi noted that we must make use of new media and new technologies to make our work more active and enhance the sense of age. The 19th National Congress of the Communist Party emphasized that we need to strengthen the construction of the contents of the Internet, build a comprehensive network management system and create a clean and bright cyberspace. It gave the direction for colleges to strengthen the construction and management of new media.

With the rapid development of economy in the internet era and continuous advancement of digitalization and informationization in the campus of colleges, Weibo with its large capacity, new carrier, fast speed and strong interaction has become one of the most convenient means of communication in the information age. Until June 2017, according to the 40th Statistical Report on Internet Development in China, the number of Chinese Internet users has reached 751 million, which covered one fifth of the total number of Internet users in the world. The statistics of financial report on Q3 Weibo in 2017 showed that Weibo reached 376 million monthly active users until September 2017, which increased by 27% over the same period of 2016. However, users with higher education were always the main users of Weibo. Under the new trend of the prosperous development of Weibo application, the Chinese national universities have taken the opportunities to establish official Weibo though they are in the initial stage of development. So there are still some problems such as lack of innovation, single contents and insufficient communication. National universities in the background of Internet era must make scientific use of official microblogs, integrate campus culture, national culture and network culture to create unique ways of propaganda and discourse system for national colleges, only in these ways can they adapt to the general trend of mobility and enhance their influence and public image.

2. Significance of Opening Weibo Accounts for Minzu Universities

2.1. *Information propagation window*

Compared with the traditional universities, the nationalities universities in the new media era are no longer closed. Due to the popularity and development of Weibo, the way of college students to obtain information is gradually changed from the single way with unvaried content and low speed to multiple channels with rich content and fast speed.

For Minzu universities, Weibo is the most economical and convenient window for information publicity. Weibo publishes information by visual means of text, pictures and videos with a wide range of dissemination, fast speed, timeliness and strong interaction. At the same time, because of its fission style communication, it is not a one way transmission, which means Weibo can transmit campus information to every audience who needs to be transmitted. Therefore, more and more Minzu universities use Weibo as the window for information propagation to spread campus information, admission information and social hot spots, making the information dissemination multi-angle and diversified.

2.2. *Channels for crisis public relations*

The crisis public relation is a type of public relation when universities as origination bodies encounter sudden accidents which related to admission, student status management, daily administration and etc. So that the normal operation system of universities is disturbed, causing the damage of the university images. The basic composition of the university crisis has two parts, one is the sudden event itself, and the other is the impact of this incident on the university.

Weibo, known for its timely and convenient features, can publish the latest progress of unexpected events without delay, which can satisfy the audience's desire to learn and help constitute an orderly mechanism of public participation. Second, the reporting and handling of emergencies of Weibo is conducted in a way that is popularized and equal. All teachers and students as well as the public have the right to learn about the events and to participate in the discussions which is of great significance to weaken the external resistance and hostile consciousness to correctly guide public opinions. Therefore, when Minzu universities are conducting the crisis public relation, they can not only gain the trust of students and social media with the fastest speed and the lowest cost, but also respond to the crisis effectively and reshape their images by using Weibo.

2.3. *Carrier of cultural inheritance and innovation*

Every Minzu university has its unique ethnic culture and academic atmosphere as well as its own cultural expression ways. It is an advantage for Minzu universities to present campus culture and inherit innovations in ethnic styles. Under the new condition, it's hard for single and homogeneous carriers to meet the requirements of campus cultures inheritance and innovations. Weibo, as a new carrier, it has the functions of following, reposting and comment, and it shows a diversified and integrated trend. Weibo can effectively make up for the shortcomings of traditional carriers, break the constraints in traditional cultural construction, and lead the direction of campus cultural construction. With the help of Weibo, Minzu universities can post campus cultural dynamics in time, collect suggestions for campus culture construction, realize cultural exchange and interaction, and ensure the timely and effective campus culture inheritance and innovation.

3. Investigations on the Use of Weibo of Minzu Universities in China

3.1. *Research objects and methods*

In this paper, the investigation time is from November 1st 2017 to January 1st 2018. Ten representative Minzu universities in China are selected as the objects of investigation. Quantitative and systematic statistics of all kinds of information data of these articles are carried out by using content analysis method.

This paper analyzes the current situation of Weibo in Minzu universities from three categories of "information post form", "content composition" and "communication effect". In the construction of specific categories, the analysis of "information post form" can be divided into four variables: "text", "text + picture", "text + video" and "amount of likes"; "Content composition" is divided into "originating", "forwarding", "integration and reorganization", "main content" and "Communication effect" is investigated from four aspects of the number of followers, the number of fans, the number of tweets and interaction rate.

3.2. Reliability analysis

The reliability of content analysis refers to the degree of consistency of the same materials judged by two or above researchers according to the same analysis dimension. The higher the consistency, the higher the reliability, and vice versa. It is an important index to ensure the reliability, rationality and objectivity of the content analysis results.

The reliability formula based on content analysis: $R = n \times K/1 + (N - 1) \times K$, $K = 2M/N1 + N2$, of which R is reliability, K is average agreement degree, N1 and N2 is the number of categories analyzed by the two judges other than the principal judge. M is the number of categories approved by both judges. The average agreement degree calculated by the formula: $K = KAB + KAC + KBC/n = 0.83$, thus $R = 3 \times 0.83/1 + (3-1) \times 0.83 = 0.94$, which is greater than the standard reliability 0.8, indicating that the category construction of this paper matches reliability standard, and the reliability degree is high. Thus the next data analysis can be conducted based on this category.

4. Text Analysis

4.1. Analysis of information post form in official Weibo of Minzu universities

The form of information post is an important part of Weibo content, and it is also the key factor that affects the first impression of the audiences. This part investigates and analyzes the information post form of official Weibos in 10 Minzu universities, finds out that the form of "text + picture" usage rate is 88% which is the most frequent used among the three forms of information post of Weibo accounts in Minzu universities. However, the lowest is in "plain text" form because the average frequency is only 2%. There are 6 universities use "text + picture" frequency of more than 90% among the 10 Minzu universities, while "plain text" usage rate is generally between 1% and 4%, and "text + video" usage rate is between 5% and 30%. For example, the all information in Weibo of Inner Mongolia University within two months is all in the form of "text + picture" with a usage rate of 100%! It can be seen from the above that the official Weibo information post model in Minzu universities and colleges is relatively single and imbalanced (see Table 1).

Table 1. Information post form in official Weibo of Minzu universities.

National Colleges and Universities	Pure text	Text + Video	Text + pictures
Minzu University of China	1%	9%	90%
South-Center University for Nationalities	4%	5%	91%
Southwest Minzu University	1%	11%	88%
Northwest Minzu University	1%	18%	81%
North Minzu University	3%	6%	92%
Dalian Minzu University	1%	19%	80%
Yunnan Minzu University	1%	7%	92%
Guangxi University for Nationalities	1%	5%	94%
Inner Mongol University for Nationalities	0%	0%	100%
Qinghai Nationalities University	4%	22%	74%
Total	17%	102%	882%
Average	2%	10%	88%

In addition, 10 tweets of each information post form are selected from each school randomly, consisting 100 tweets in each form, a total of 300 posts as variables of the survey and analysis for counting the amount of likes. By analyzing the frequency of each information post form and the number of likes, it can be concluded that the form of information post has a certain effect on the popularity of Weibo and the post forms of "text + Picture" and "text + Video" can get a larger number of likes (see Fig. 1).

Fig. 1. The amount of likes in official Weibo of Minzu universities.

4.2. Analysis of the content composition of official Weibo of Minzu universities

Through statistics on the content sources of official Weibos of 10 Minzu Universities, the number of original blogs is found to be the largest, followed by the blogs of "integration and reorganization" and "forwarding".

In comparison of the "main content", although Minzu universities have more original blogs than other universities, in general, articles have attracted less attention with a low quality. Many blogs about greeting such as "Good morning or Good evening MingDa (short for Minzu universities)" and all sorts of inspirational articles account for more than 55% of all blogs, while news, examinations, information consultation, and others account for only 18%, as well as the amount of likes is commonly less than 20. Indicating articles were overall lower in quality and paid less attention (see Fig. 2).

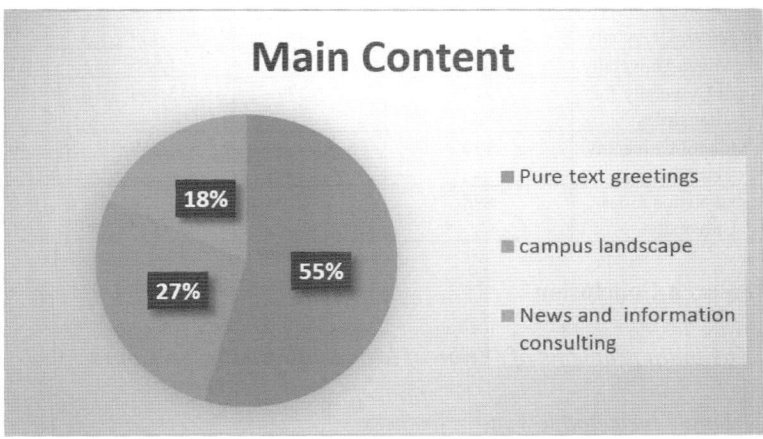

Fig. 2. Main content of official Weibo of Minzu universities.

4.3. Analysis of the communication effect of official Weibo of Minzu universities

Based on the statistics from the follower number, fan number, tweet number and the interaction rate about comments, reply and repost of Weibo. It's obvious that the fan number of Weibo of different Minzu university varies hugely.

40% of universities have more than 10,000 fans with 60,000 at its maximum and 900 at its minimum; There are 6 universities possessing 3,000 plus tweets, of which Southwest Minzu University posted 7,780 tweets and lists No. 1, while at the bottom only 249 tweets were posted by Inner Mongolia University for Nationalities. From the number of followers, Qinghai Nationalities University is the only one to have more than 500 followers while other universities is below 500. When it comes to the interaction frequency, the average rate of the interaction between the official Weibo and the fans is 13% and 60% of Minzu universities had no interaction with fans at all in two months (see Table 2).

Table 2. Communication effect of official Weibo of Minzu universities.

National Colleges and Universities	The number of followers	The number of fans	The number of tweets	Interaction rate
Minzu University of China	322	64,010	6,145	0%
South-Center University for Nationalities	134	34,585	3,735	0%
Southwest Minzu University	252	23,385	7,780	8%
Northwest Minzu University	145	9,800	1,840	0%
North Minzu University	406	16,060	6,032	27%
Dalian Minzu University	112	5,414	729	0%
Yunnan Minzu University	321	5,287	1,555	57%
Guangxi University for Nationalities	219	8,279	3,844	39%
Inner Mongol University for Nationalities	130	956	249	0%
Qinghai Nationalities University	898	9,612	7,114	0%

5. Research Conclusion

5.1. *The main problems of Weibo of Minzu universities in China*

5.1.1. *Single form and low attractiveness*

At present, the information post of the nationalities universities is no longer in traditional pure text form, instead the information posting combined pictures and text is more popular. Yet with the progress of Internet technology, and the upgrade of the social needs, this simple combination of images and words can no longer meet the needs of the audience.

When some Minzu universities post messages, the information presentation form is single with too many combinations of pictures and text. When it comes to practice, the problem occurs that the pictures selected are monotonous and similar and the pictures don't match the text. As a consequence, viewers find it hard to be interested in these tweets and easily become tired, which can explain why some articles have a large viewing number but with a small amount of likes. In addition, the combination of "text + video" is not often used. Most videos are reposted and only a few are original works, the new factors and new technology about school information are seldom added into, thus they are not able to provide fresh feelings and excitement to audience. Therefore, to utilize specific information post advantages to obtain the attention of the public in the Internet era becomes one core problem, so that nationalities universities Weibo accounts should attach great importance to the future development.

5.1.2. *Lack of characteristics and innovation*

Compared with other non-nationalities universities, Minzu universities have unique campus characteristics and ethnic advantages.

However, most Weibo of Minzu universities have failed to take advantage of their own ethnic characteristics to release tweets, so that these tweets are lack of innovation and personalities. In recent years, although more original tweets have appeared, problems such as serious content, heavy bureaucratic taste, few sections, characteristics deficiency and low quality still commonly exist. Take Southwest Minzu University as an example, only 3 tweets show ethnic characteristics among the 50 tweets that released recently, and the rest are mostly serious news reports or school affairs information. For the integrated and reorganized tweets, school features are rarely injected into the cited articles, so these tweets are lack of innovation and difficult to meet the diverse needs of the audiences. For the reposted articles, most are commonly copied or excerpted from other websites without any processing which are lack of characteristics and creativeness. As time goes by, for the extreme low viewing quantity of these tweets, fans will stop following.

5.1.3. *Insufficient communication and low rate of interaction*

One prominent feature of Weibo is strong interaction. But because Weibos of Minzu universities are non-profitable, they mainly focus on the unilateral information release but yet rarely repost, reply or comment on tweets of their fans. So it hard to embody the equal communication of two parties.

As time passes, the followers will lose their interest to comment or repost the tweets of the school, which will affect its secondary transmission. For example, there are six universities for nationalities, such as Northwest Minzu University and Dalian Minzu University, did not reply or repost fans' tweets within two months, and did not give timely feedback on the comments they received. The rate of interaction with fans was zero. The average number of retransmissions of each Weibo at these universities was less than 15 and the average number of comments was only four. It can be seen that although official Weibo of many

Minzu universities are in a certain number of fans, they are not welcomed by students because of lack of interaction with fans and followers.

5.2. The development strategy of Weibo for China's Minzu universities

Currently, the Internet world has entered We-media age of where "everyone is a communicator". As the most important tool of communication and essential type of social media in this Internet era, Weibo has become a new carrier of Internet public opinions with fastest rising speed and broadest spreading range after Tieba, forum and blog. In face of these opportunities and challenges, Minzu universities should positively meet opportunities, face challenges, enhance their influence and public relations image in the light of their own ethnic characteristics.

5.2.1. Information post strategy

Minzu universities use the diversified information post form to spread information. It already becomes an effective measure to attract the attention of the audience.

In the process of information post, the official operation team of Weibo accounts should not only be able to simplify the language and express it beautifully, but also respond actively to the work requirements of "transform for things, advance for times, and update for conditions" put forward by President Xi. The team should attach great importance to the construction and innovation of communication means, make full use of the active time of the fans, and create topics that students are interested in, such as study consultation, entertainment activities, current affairs, hot issues, job hunting and recruitment to meet the psychological needs of the audience. Besides, fashionable Internet language is more favorable than stereotyped tones. It's a good way to approach students as well as enhance the communication power and credit of the Weibo. On top of that, the official team should collect more original materials of text, pictures and videos and use them more accurately so that the content of Weibo can be fully expressed and more people will be attracted.

5.2.2. Brand building strategy

Each Minzu university has its own unique ethnic characteristics and cultural brand. The content of their Weibo accounts should not only distinguish themselves from the non-nationalities universities, but also show their prominent individuality among other Minzu universities. This not only helps the construction of ethnic brand and image, but also benefits the wide dissemination of information.

So, the Weibos of Minzu universities should broaden their orientation and create a brand on campus network culture with ethnic characteristics. Combining the characteristics of most minority students to set up more sub-columns on ethnic style presentation, campus ethnic culture, ethnic festival activities, etc. Enriching the content of related sections on school affair consultation, community activities, admission and recruitment. Second, when releasing the content, it is necessary to take into account the particularities of certain ethnic cultures, such as taboos of Hui, Uygur and Tibetan nationalities, they should publish Weibo content in the correct language to avoid pushing sensitive topics. Moreover, they should improve the quality and creativity of tweets and create a school network brand. All of these methods are crucial to improve the impression and credibility in the heart of students and public audience.

5.2.3. *Public relations maintenance strategy*

To Minzu universities, the true value of Weibo is reflected in its function of raising influence and building up public relations. In addition to the teachers and students of the university, the audience of Weibo include past schoolfellows, other public audience and cooperative organizations.

If Minzu universities want to maintain a good public relation image on Weibo platform, they need to carefully maintain the relationship with fans and various audience, build a delicate relationship network bridge. To do so, first of all, Weibo of Minzu university should go into the lives of fans, construct regular interactive mechanism through surveys, interviews and voting, so as to improve the function of consulting service. Selectively answer fans' questions, and pay special attention to the interactive communication about the ethnic topics, avoid forming the cold impression that "I comment you but you ignore me" in order to build the friendly and gentle service image. Second, in the event of public emergencies, Minzu universities should make an effective voice through Weibo platform, timely announce the truth of the incident, and do a good job in emergency management. In addition, they should continue to track the development of the incident, guide public opinions, and answer questions after releasing the message. By doing so, we can create a good public opinion environment for the school and create a good image of public relations, so as to enhance the visibility and influence of the school.

6. Conclusion

In the Internet era, Minzu universities should not limit themselves in the traditional form of thinking and operation, but instead of conforming to the trend of the times, seizing the opportunities brought by Weibo in the Internet age and daring to take the challenges to make full use of the positive effect of Weibo. It means that Minzu universities should utilize Weibo as a platform to spread information, promote unique ethnic brand construction and improve its core competitiveness. On this platform, the information posting form should be improved to highlight the features. What's more, it's necessary to create official Weibo brands and complete users' interactive mechanism in order to promote the experiences and satisfaction of Weibo users and make the Weibo of nationalities universities more useful in presenting the school image, serving teachers and students and attracting social attention. Only in this way, Minzu universities can seize the competitive opportunities and win long-term development among the competition with non-nationalities universities and even similar universities for nationalities.

References

1. Xinhuanet. The 39th Statistical Report on the Development of China's Internet. (2017).
2. Tie Zheng. University News Propaganda in the era of New Media [M]. Beijing: China Literature and History Press. (2014).
3. Feng Xiaotian. Social Research Methods [M]. Beijing:Renmin University Press, China. (2013).
4. Wan Dailin, Nian Yongqi. Investigation and Research on the Current Situation and Influence of WeChat Public Platform in Universities [J]. New Media Studies, 2017, 3(17):26–27+40.
5. Bai Yi. Operational strategies of the government WeChat public number [J]. Audio-visual, 2017(12): 129–130.

Part 10

Decision Making Under Uncertainty in Health Care Systems

Frequency domain analysis of telephone helpline call data

Alexander Grigorash, Raymond R. Bond, Maurice D. Mulvenna
School of Computing, Ulster University, Shore Road, Newtownabbey, BT37 0QB, UK

Siobhan O'Neill, Cherie Armour, Colette Ramsey
School of Psychology, Ulster University, Cromore Road, Coleraine, BT52 1SA, UK

The paper presents a frequency domain analysis of call data records of a telephone helpline for those seeking mental health and wellbeing support and for those who are in a suicidal crisis. A call data record dataset provided by Samaritans Ireland helpline is used. Fourier series is used to ascertain periodicity in the call volume. The main findings from the paper indicate that strong repetitive intra-day and intra-week patterns are found, while intra-month repetitions are conspicuously absent.

1. Introduction

Telephone helplines provide a significant service that continues to underpin mental wellbeing and suicide prevention efforts [1]. This work is part of a research project that involves analysis of digital telephony data provided by Samaritans Ireland, a charity with a helpline to provide emotional support to anyone in distress or at risk of suicide.

Data were provided for all calls made to Samaritans in Ireland from April 2013 to December 2016, a total of 3.449 million inbound calls. Each data record carries the date-time stamp of the call arrival. The arrival time stamps are resolved to the last second of time. There are no simultaneous arrivals.

Informal observations by Samaritans personnel suggested that certain callers dial in at regular intervals. Inspecting the data revealed apparent peaks and troughs in the call volume, *i.e.* the number of calls, arriving to the call centre (Figure 1).

However, seeing cyclic fluctuations in data does not necessarily mean seeing seasonal, *i.e.*, periodic, regularities, see *e.g.*, [2]. Thus, it became interesting to explore what period lengths (hourly, daily, weekly, monthly, etc.), if any at all, find support in the dataset?

The following referenced, explanatory quote summarizes the intentions and sets out relevant terminology, with our own emphases: "Put simply, a *time-*

domain graph shows how a signal changes over time, whereas a *frequency-domain* graph shows how much of the signal lies within each given frequency band over a range of frequencies. ... A given function or signal can be converted between the time and frequency domains with a pair of mathematical operators called a *transform*. An example is the *Fourier transform*, which converts the time function into a sum of sine waves of different frequencies, each of which represents a frequency component. The 'spectrum' of frequency components is the frequency domain representation of the signal. The *inverse Fourier transform* converts the frequency domain function back to a time function" [3].

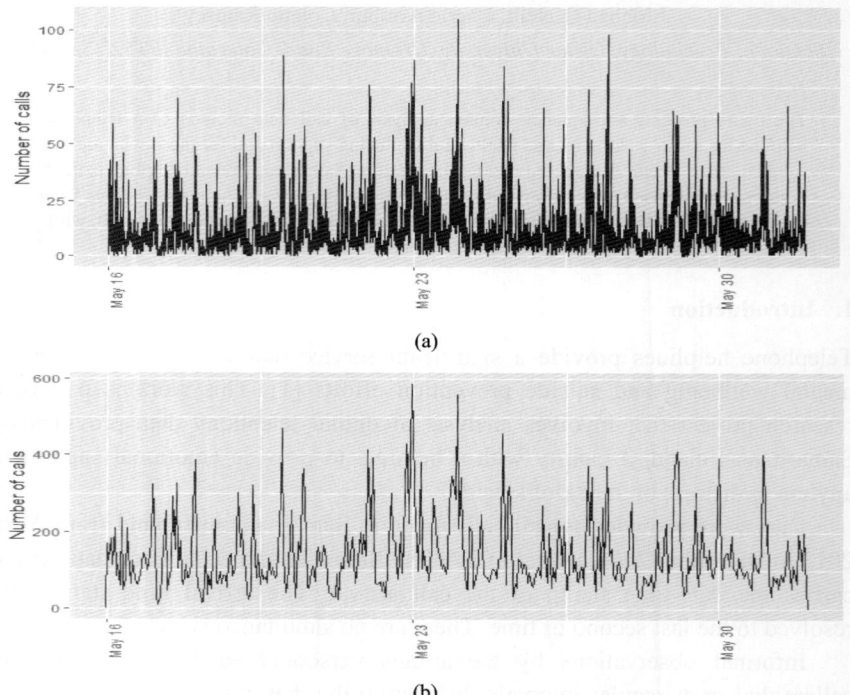

Figure 1. Samaritans call arrival data (two weeks in May 2016) viewed through two different aggregation intervals: (a) 5 minutes, (b) 1-hour.

Recent literature surveys [4, 5, 6] show a common approach to converting the point process of incoming calls into a time series. The timeline is split into time buckets, typically 5 minutes or 15 minutes or 30 minutes long. Call arrival times (points) are counted for each bucket. This procedure generates a sequence of call volume values (call counts) versus times, *i.e.*, a time series

Relevant time domain techniques include seasonal moving average, multiplicative seasonal ARIMA and Holt-Winters exponential smoothing [7]. Irregular spikes in call volume are filtered out with machine learning approaches [8] thereby enabling the use of call volume models that expect the data to be reasonably smooth. An older approach relies on statistical analysis alone, that involves judiciously censoring the data set: call arrivals deemed atypical by the researchers are removed from consideration [9].

Surprisingly, sources on frequency domain analysis in the context of call center data are rare. A single Australian conference paper was identified where the authors analyzed 52 weeks of call data records of a police assistance line [10]. They sampled at a resolution of 48 intervals per day, and discovered 11 dominant frequencies, from 1 cycle per week to 4 cycles per day that explained daily and weekly fluctuations of the call volume.

2. Methods

In order to extract frequency components, 2-year-long subset of the data, covering years 2015 and 2016 were used. This time span was chosen because no generic trend that could mask periodic activity was obvious.

In order to convert the call arrival point process into a call count time series, the chosen time span was split into a large number of sampling intervals. The number of calls arriving within each of these intervals (buckets) was counted, and these became the values of the time series. The sampling interval (the bucket size for aggregating calls) was 30 minutes long. This particular size offered the smallest number of sampling intervals at a resolution capable of capturing oscillations as frequent as once per hour.

The fundamental period, i.e., the range of data supporting the analysis, was from 01 January 2015 00:00 to 31 December 2016 23:59, lasting 30,588 sampling intervals exactly. The maximum meaningful frequency detectable at this resolution (Nyquist frequency), equals 17,544 cycles over 2 years, or 731 cycles per month, or 24 cycles per day, or about 168 cycles per week.

During 2015 and 2016, the overall trend in the number of calls per unit of time appeared to be practically static. In order to confirm this, a linear trend line was fitted to the data. The fitted value at the start of the time span was 63.9 calls per sampling interval, the fitted value at the end of the time span was 70.8 calls. This amounts to a very gradual rate of increase, giving about 3.5 calls yearly difference, in line with the perception of a static trend.

R programming language and R Studio were used for data wrangling and to implement the analysis. An open online tutorial by Joao Neto [12]. provided

3. Results

Figures 2-4 depict spectral plots of the Fourier transform. The frequencies of oscillations are plotted against the amplitudes that measure the peak number of calls for each harmonic. The spectral plots are shown after de-trending which amounted to removing the non-oscillatory component that would have shown as frequency of 0 cycles. The strongest set of dominant frequencies corresponds to intra-day oscillations, at 1, 2, 3, 4, 5, 6, 7, 8, and 24 cycles per day (Figure 2).

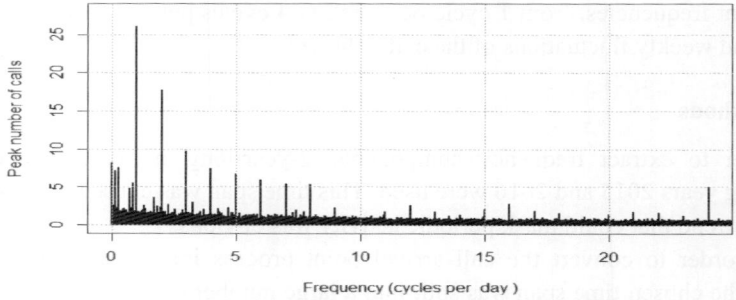

Figure 2. Spectral plot for complete frequency range.

Figure 3 depicts a closer view of the range 0 to 3 cycles per day, expressed using cycles per week. On this plot, the secondary seasonality group of frequencies, associated with intra-week oscillations, is clearly seen. These secondary dominant frequencies are situated at 1 cycle per week, 2, 5 and 6 cycles per week. These frequencies correspond to a natural human inclination to repeat tasks weekly, or twice weekly, or every working day, or every day except *e.g.*, Sunday. The amplitudes of these harmonics range from 5 to 8 calls. The large amplitude at 7 cycles per week shows exactly the same oscillation as the once-a-day amplitude in Figure 2.

Remarkably, few callers seem to regularly call using an 'every other day' pattern. The frequency of 0.5 cycles per day, equivalent to 3.5 cycles per week, is indistinguishable from noise.

Further to the right on the plot on Figure 3, the 12 cycles a week frequency can be visually set apart from the noise. However, its strength amounts to about 3.5 calls only, which is between 2 and 2.5 times inferior to the weakest of the previously identified dominant frequencies. Therefore, weaker frequencies in any of the two identified seasonalities are not included.

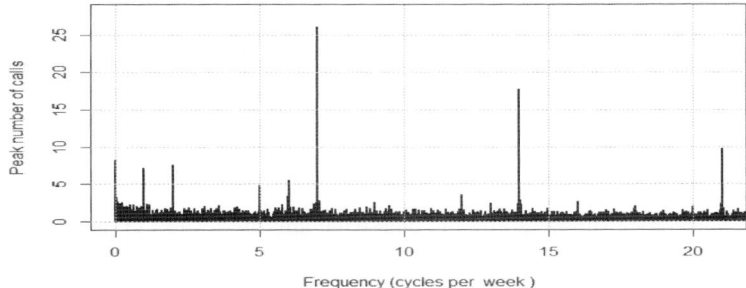

Figure 3. Spectral plot showing the lowest 1/8 of the frequency range, here frequencies are expressed in cycles per week.

A closer view at the range 0 to 1 calls per week, expressed as 0 to 4.5 cycles per month, is shown in Figure 4.

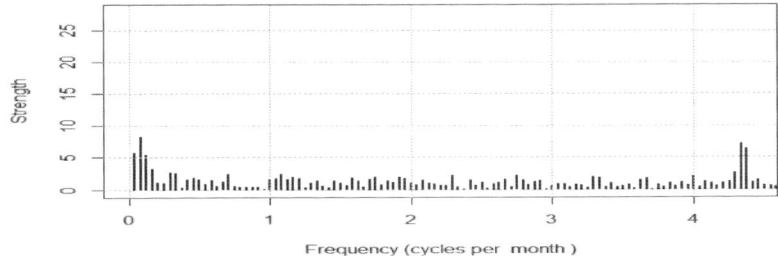

Figure 4. Spectral plot showing the lowest 1/24 of the frequency range, here frequencies are expressed in cycles per month.

On the right of this plot there is an amplitude spike around 4.3 cycles per month, which corresponds to 1 cycle per week. On the left of this plot there is a prominent amplitude of 8 calls situated at 0.2 cycles per month, or about twice a year. The same amplitude is visually shown at near 0 cycles per week on the plot Figure 3. Other frequencies from monthly and quarterly ranges are lacking prominence.

4. Discussion

The volume of calls to Samaritans exhibits strong intra-day and intra-week repetitive patterns, while intra-month repetitions are absent. This double seasonality effect was described in the literature [7, 5]. It appears that the double seasonality is part of the nature of a telephone helpline *per se*, regardless the applied area: mental health and wellbeing helplines, banking [7], sales [8], law enforcement [10].

The once-an-hour, or 24 cycles per day, frequency was part of our intra-day seasonality pattern. As this frequency sits right at the limit of our current resolution, it would be worthwhile re-sampling the time series of calls at a higher rate and re-running the frequency analysis to see if any higher frequencies contribute to the pattern.

The harmonics supported by the data have amplitudes in the range between 4.9 and 27 calls per sampling interval. This makes the seasonal components comparable in magnitude to the linear trend which ranged between 64 and 71 calls per sampling interval.

Future work may involve transforming data in the frequency domain back into the time domain, allowing for call volume modelling and forecasting. However, in this first instance, the primary interest was to test whether periodic activity is present. The answer to that turned out to be a definitive yes.

Acknowledgments

Financial support for this research was provided by Samaritans Ireland with support from Ireland's National Office for Suicide Prevention.

References

1. S. Howe, B. Meakin, and F. Islam-Barrett. (2014). Helplines at the Frontline of Health and Social Care. *Helplines*. Available: https://www.helplines.org/wp-content/uploads/2014/12/Helplines-at-the-Frontline-of-Health-and-Social-Care.pdf
2. R. J. Hyndman. (14 December 2011), Cyclic and seasonal time series, *Hyndsight blog*. Available: https://robjhyndman.com/hyndsight/cyclicts/
3. Frequency domain, *Wikipedia*. (Accessed 1 Dec 2017) Available: https://en.wikipedia.org/wiki/Frequency_domain
4. R. Ibrahim, H. Ye, P. L'Ecuyer and H. Shen, *Int. J. Forecast.* **32(3)**: 865 – 874 (2016).
5. N. Meade and T. Islam, *Int. J. Forecast.* **31(4)**: 1105–1126 (2015).
6. J. Taylor, *Manage. Sci.* **54(2)**: 253–265 (2008).
7. D. Barrow, *J. Bus. Res.* **69(12)**: 6088–6096 (2016).
8. D. Barrow and N. Kourentzes, *Eur. J. Operat. Res.* **264(3)**: 967-977 (2016).
9. L. Brown et al., *J. Am. Statist. Assoc.* **100(469)**: 36–50 (2005).
10. B. Lewis, R. Herbert, and R. D. Bell, *Proc. Int. Cong. Model. Simul. (MODSIM03)*, Townsville, Australia, 1281–1286 (2003). Available: https://www.mssanz.org.au/MODSIM03/Volume_03/B10/06_Lewis.pdf
11. J. Neto. (March 2013). Fourier Transform: A R Tutorial. Available: http://www.di.fc.ul.pt/~jpn/r/fourier/fourier.html

Predicting assistive technology adoption for people with Parkinson's disease using mobile data from a smartphone

Jonathan Greer, Ian Cleland, Sally McClean

School of Computing, Ulster University, Newtownabbey, Co. Antrim, BT37 0QB, UK

This paper presents results from an investigation into the ability of various classification methods to model adoption of assistive technologies. Using data from the mPower study on Parkinson's disease, three classification algorithms were evaluated: Lazy IBk (k-Nearest-Neighbours), Naïve Bayes, and J48 Decision Trees. J48 Decision Trees and Lazy IBk were found to give the best performance, with an accuracy/standard deviation of 76.98% (5.40) and 73.36% (5.23) respectively. The suitability of the classifiers in the context of future application to assistive technologies is also discussed. Furthermore, this paper investigates the use of survival analysis to gain insights into technology adoption. The Kaplan-Meier survival curve showed a general trend similar to other apps within the same timeframe. Cox Regression analysis showed that the most significant factors contributing to adoption were whether users had used a smartphone to take part in a video call, previously smoked or not, employment status, gender, and the date when they were diagnosed with Parkinson's disease. Future work will investigate how technology adoption may change with a more user-centred approach to app design.

1. Introduction and Related Work

Assistive technologies (AT's) are an increasingly common topic in modern research, promising numerous benefits including a greater level of independence and autonomy for individuals suffering from chronic conditions such as Dementia and Parkinson's [1, 2].

The effectiveness of any AT, however, will ultimately be limited by whether a user is able to integrate it fully into their daily life. Recently, a number of studies have attempted to understand the factors which affect technology adoption [3, 4, 12]. If an individual is unlikely to adopt a technology in the long term, it may be preferable to pursue other options rather than investing time and resources into an intervention which will not be suitable. Previous research by the authors, has shown promise in the ability to model adoption of AT's in people with dementia (PwD) using socio-demographic and health information [5, 6, 7, 8].

Possibly the most relevant example is the ongoing TAUT [6] project (Technology Adoption and Usage Tool), which has focused on modelling technology adoption for people with dementia using an assistive reminding app.

This report aims to build upon the previous work of the TAUT study in modelling technology adoption by applying the techniques to other domains — namely people diagnosed with Parkinson's disease (PD) in addition to a control group with no prior diagnosis. Additionally, this study seeks to incorporate survival analysis to gain insights into the prediction of technology adoption.

2. Methods and Results

This project uses data from the mPower mobile Parkinson's Disease study [9], namely data from 4 app activities and a survey on demographics. The mPower study itself was focused on improving the current state of knowledge concerning the variations of PD and to investigate whether mobile devices can prove useful when measuring its progression. Data was accessed by permission of the original authors and in line with the outlined ethical procedures.

Both the survey and activity data contain Pseudonymized User ID's to protect the privacy of the study participants. The demographics survey contains environmental, social, and physical data concerning the participants, with a full list of the variables given in Table 1. User engagement with the app was tracked, with a unique instance created every time the individual completed any of the four activities. More detail on the mPower app can be found in [9]. Each instance contains timestamps used to identify individual usage instances, in addition to data specific to the activity itself, such as accelerometery or audio data.

Table 1. Demographic and health information collected from participants through the surveys. This data is used as input features for the technology adoption model.

ROW_ID	ROW_VERSION	recordID
healthCode	createdOn	appVersion
phoneInfo	age	are.caretaker
deep.brain.stimulation	Diagnosis.year	Education
employment	gender	health.history
healthcare.provider	home.usage	last.smoked
marital.status	medical.usage	medical.usage.yesterday
medication.start.year	onset.year	packs.per.day
past.participation	phone.usage	professional.diagnosis
race	smartphone	smoked
surgery	video.usage	years.smoking

2.1. Classification

Following on from previous work [5, 6, 7, 8], this paper uses machine learning models to classify app users into two groups: Adopters and Non-Adopters, with each group containing 74 and 307 individuals respectively. Figure 1 provides a

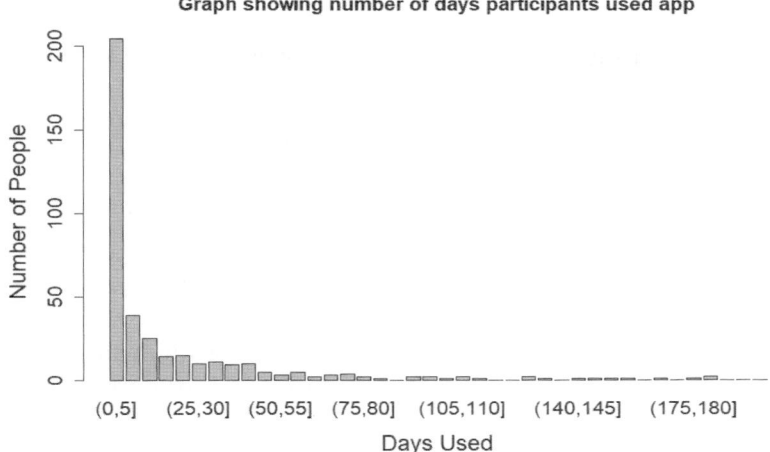

Figure 1. Graph showing the total number of days the participants used the 'Tapping' activity on the app, to the nearest five. Significantly fewer people used the app for greater than 60 days, owing to a combination of late-joiners to the study, and attrition of previous users.

visualization of how long, in days, the participants used the 'tapping' activity of the app.

The 'Adopters' group was selected from the users who used the app for 30 days or more, with 'Non-Adopters' being those who dropped out before this point. The '30-day' point was chosen as this was considered long enough for 'Non-Adopters' to drop out, whilst also being short enough to minimize the misrepresentation of those who joined later into the study. A metric for distinguishing between adopters vs non-adopters is difficult to find within existing scientific literature. Researchers within the mobile intelligence community have, however, taken to including the 30-day point when describing long term retention curves for apps, albeit not within a machine learning context [13].

Feature selection was performed to reduce the dimensionality of the data and hence improve the accuracy of the machine learning models. Firstly, any attributes deemed irrelevant were removed manually. These are features which contained data of no relevance to the study or its participants, such as 'ROW_ID' and 'appVersion'. An attempt was then made to condense potential duplicate attributes into one. A number of feature ranking techniques were then applied using 'Feature Elimination' and 'Extra Trees Classifier' within the Python scikit-learn library alongside 'InfoGainAttributeEval' and 'CorrelationAttributeEval' within Weka. The five highest scoring features were selected for the classification, as they were found to consistently perform better than the other attributes across all the methods used: 'age', 'education', 'employment', 'gender', and 'marital status'.

SMOTE resampling was used in order to avoid a biased dataset affecting the classification. The Adopters group was boosted from 74 to 296.

Three types of classification were performed: J48 pruned decision tree (DT), Naive Bayes, and Lazy IBK (k-nearest neighbour). These were chosen due to their generally good performance with data mining [10], and to provide a comparison to past research in the field of modelling technology adoption [5, 6, 7, 8]. These classifiers were then assessed based on both the accuracy to which they classified the data, and their calculated f-measures. These assessments were performed using 10-fold cross-validation and 10 repetitions, with the tests themselves taking the form of an 'Accuracy' and a 'Rank' test; the results of which are shown in Table 2. The F-measures for the three algorithms were 0.78 for J48 DT, 0.73 for Lazy IBk and 0.72 for Naïve Bayes.

Table 2. Classification Results. A ZeroR classifier was also included as a baseline to which the other classifiers could be compared. The ZeroR classifier is the least complex algorithm which can be run; it ignores the predictor variables and simply predicts the majority class.

Classifier	ACCURACY TEST		RANK TEST		
	% Accuracy (Std. Dev.)	F-measure	Score	Wins	Losses
J48 DT	76.98 (5.40)	0.78	2	2	0
Lazy IBk	73.36 (5.23)	0.73	1	1	0
Naïve Bayes	69.05 (6.37)	0.72	0	1	1
ZeroR	50.91 (0.70)	-	-3	0	3

2.2. Survival analysis

The survival analysis consists of two main components. Firstly, a Kaplan-Meier survival curve was created using the mPower study data, with the aim of obtaining a descriptive overview of technology adoption among the study participants. Secondly, a Cox proportional hazards regression model [11] was used to assess the effect of the various features on survival time.

Figure 2 shows the Kaplan-Meier survival curve created from the mPower data. The most obvious feature is the very high drop in survival probability early into the study, with the curve predicting a 50% survival probability within only 3-10 days. As mentioned previously, this kind of dropout has been observed to be common among mobile apps [13], with the dropout becoming much less pronounced as time progresses.

The Cox Regression consisted of both a multivariate and univariate analysis. The hazard ratios and confidence intervals are shown in Table 3. From the results of the multivariate analysis, 4 of the 15 covariates used were observed to have a statistically significant effect on survival: 'video.usage', 'smoked', 'employment'

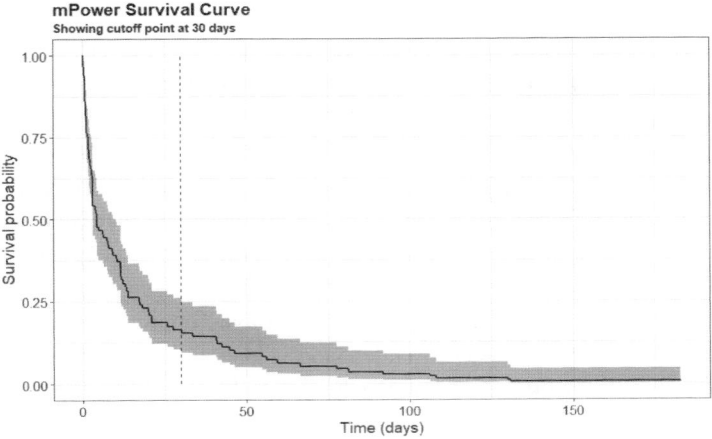

Figure 2. Kaplan-Meier survival curve, showing the 30-day adopter/non-adopter cut-off point.

and 'gender'. The univariate analysis resulted in two covariates which were statistically significant: 'video.usage' and 'onset.year'. The tests performed on the model itself showed statistical significance: the Likelihood ratio, Wald, and Score (logrank) tests gave p values of 0.01709, 0.02043 and 0.01307, respectively.

Table 3. Results of Cox Regression analysis.

Covariates	Univariate Analysis			Multivariate Analysis		
	Hazard Ratio	Lower .95	Upper .95	Hazard Ratio	Lower .95	Upper .95
Age	0.996178	0.9899	1.003	0.986905	0.966210	1.0080
Are.caretaker	1.3128	0.8169	2.11	1.168195	0.486462	2.8053
Deep.brain.stimulation	1.2304	0.704	2.15	2.357134	0.961846	5.7765
Home.usage	1.3120	0.3265	5.273	3.071779	0.167899	56.1993
Medical.usage	1.002154	0.4739	2.119	0.856620	0.189613	3.8700
Video.usage	0.7042	0.5208	0.9522	0.324279	0.182705	0.5756
Onset.year	0.989498	0.9802	0.9989	0.987832	0.972456	1.0035
Professional.diagnosis	0.8454	0.6471	1.105	1.509186	0.887545	2.5662
Smoked	1.4069	0.1968	10.06	0.053834	0.003205	0.9042
Smartphone	1.01662	0.9248	1.118	0.972844	0.821400	1.1522
Education	1.002836	0.952	1.056	0.982082	0.870740	1.1077
Employment	0.95565	0.8925	1.023	0.826639	0.715118	0.9556
Gender	1.1344	0.897	1.435	2.438021	1.360024	4.3705
maritalStatus	1.03924	0.9742	1.109	0.990637	0.788618	1.2444
Race	1.0007965	0.9963	1.005	0.992531	0.981105	1.0041

3. Conclusions

The classification results lead to the conclusion that the J48 DT was the most effective classifier, but also that the difference compared to the Lazy IBk classifier was not statistically significant.

Previous research has also shown kNN and DTs to be effective algorithms with regards to modelling technology adoption for AT's [7, 8]. The results of this study and insights from prior papers lead to the conclusion that DTs seem to be the most suitable algorithm for modelling technology adoption of ATs. DTs have been shown to have comparable accuracy to both kNN and logistic regression algorithms, while being superior to a Naïve Bayes approach. The additional advantage of DTs being easier to understand is also an important benefit, as its end users will likely be medical staff who naturally have little experience with these types of classification algorithms.

Compared to the previous work of the TAUT study [6], the f-measure for the J48 DT was slightly lower but not significantly, at 0.78 compared to the C4.5 DT score of 0.79 in TAUT. The features selected for the classification in this study differed from the TAUT study: 'age', 'education', 'employment', 'gender', and 'marital status' were used, while TAUT found those with the most predictive power to be genome information (APE0/APE04) and the type of dementia.

This genome information included indicators linked to the presence of Dementia, and as such would not be applicable to people with PD. It is possible that similar genetic indicators could be used to predict technology adoption for PD, but as this data was taken from a mobile study, this information is not available. The features used for this current study have the advantage that they can potentially be applied across multiple domains instead of just one target group, although further research is needed to confirm this hypothesis.

The Kaplan-Meier survival curve offered some insights into the dropout rates of the participants. Most apparent was the high risk of individuals leaving after only a few days, but with those who survived past the 30-day point largely continuing to use the app in the long term. However, the survival probability at this point was only shown to be 25% at best.

From a practical standpoint, these findings mean that efforts will need to be taken to improve retention of users if AT's are to be utilized effectively. With regards to the use of apps, previous papers have attested to the importance of user-centred design, with an emphasis on involving future users as part of the design process [2, 12]. Another potential approach would be the use of notifications within apps to improve technology adoption. On the other hand, over-reliance on prompting methods, such as notifications, has been decried by some sources as short-sighted with regards to the larger picture surrounding technology adoption

— that notifications will only boost user retention if the user wants to use the app in the first place [13]. It can be argued that a greater focus should be placed upon improving the user's initial experience with the app. BJ Fogg discusses some behaviours and responses of humans towards machines and how these can be used to elicit a desired response [14].

The mPower study also lists some common user responses from their feedback survey. When asked what they liked most about participating, common responses were: "I feel like I am making a difference", "I can participate whenever I want" and "I like hearing about the research". When asked for what additions they would like to see added, common responses were: "More and different activities", "Positive reinforcement" and "More information about Parkinson's disease." These improvements may have an impact on engagement and sustained usage of the app. One interesting avenue for future study could be to observe how the survival curve may change in cases where these methods have been adopted as part of the design process.

From the results of both the multivariate and univariate analysis, it was found that individuals were less likely to drop out of the study if they had used a smartphone to take part in a video call, smoked (either past or present), were not in full-time employment, were female, or developed PD more recently.

It is interesting to note that video.usage was a much more effective predictor than the other smartphone related variables: 'home.usage' (whether they use email and internet at home), 'medical.usage' (whether they use their smartphone to look up medical related information online) and 'smartphone' (how easy they thought it was to use their smartphone). This can possibly be attributed to video calls being a more advanced smartphone feature than simple e-mail or internet searches, and hence being indicative of greater familiarity with mobile devices. This may potentially prove to be an effective method of determining the future assistive technology adoption of an individual, as any assistive technology is likely to be more complex when compared to normal mobile device usage.

Acknowledgments

The data used in this report were contributed by users of the Parkinson mPower mobile application as part of the mPower study developed by Sage Bionetworks and described in Synapse [doi: 10.7303/syn4993293].

References

1. L. Robinson *et al.*, "Keeping in Touch Everyday (KITE) project: developing assistive technologies with people with dementia and their carers to promote independence", International Psychogeriatrics, **21:494-502**, (2009).

2. A. Bharucha et al., "Intelligent Assistive Technology Applications to Dementia Care: Current Capabilities, Limitations and Future Challenges", American journal of geriatric psychiatry, **17:88-104**, (2009).
3. S. Kowalewski et al., "Accounting for User Diversity in the Acceptance of Medical Assistive Technologies", Electronic Healthcare, **69:175-183**, (2010).
4. D. Yen et al. "Determinants of users' intention to adopt wireless technology: an empirical study by integrating TTF with TAM", Computers in Human Behaviour, **26:906-915**, (2010).
5. P. Chaurasia et al., "Modelling assistive technology adoption for people with Dementia", Journal of Biomedical Informatics, **63:235-248**, (2016).
6. I. Cleland et al., "Predicting Technology Adoption in People with Dementia; initial results from the TAUT project", Ambient Assisted Living and Daily Activities, **8868:266-274**, (2014).
7. S. Zhang et al., "A Predictive Model for Assistive Technology Adoption for People with Dementia", *IEEE* JBHI, **18:375-383**, (2014).
8. S. Zhang et al., "Prediction of Assistive Technology Adoption for People with Dementia", Health Information Science, **7798:160-171**, (2013).
9. B. Bot et al., "The mPower Study, Parkinson Disease Mobile Data Collected Using ResearchKit", Scientific Data, **3:160011**, (2016).
10. X. Wu et al., "Top 10 algorithms in data mining", KAIS, **14:1-37**, (2008).
11. D. R. Cox, "Regression models and life tables," Journal of the Royal Statistical Society, **34:187**, (1972).
12. J. R. Thorpe et al., "Pervasive Assistive Technology Adoption for People with Dementia", Healthcare Technology Letters, **3:297–302**, (2016)
13. A Chen. "New data shows losing 80% of mobile users is normal and why the best apps do better", andrewchen.co, (2015).
14. B. J. Fogg. "Persuasive technology: using computers to change what we think and do", Magazine Ubiquity, **2002:5**, (2002).

Machine learning using synthetic and real data: Similarity of evaluation metrics for different healthcare datasets and for different algorithms

Rachel Heyburn[†], Raymond R. Bond, Michaela Black, Maurice Mulvenna, Jonathan Wallace, Deborah Rankin, Brian Cleland

Faculty of Computing, Engineering and the Built Environment, Ulster University, Shore Road, Newtownabbey, Co. Antrim, BT37 0QB, UK

Sharing data is often a risk in terms of security and privacy especially if the data is sensitive. Algorithms can be used to generate synthetic data from an original raw dataset in order to share data that are considered more 'privacy preserving', and that increase the level of anonymity. In this paper, we carry out an experiment to study the validity of conducting machine learning on synthetic data. We compare the evaluation metrics produced from machine learning models that were trained using synthetic data with metrics yielded from machine learning models that were trained using the corresponding real data.

1. Introduction

The volume of data being generated every year is growing exponentially. A report from IBM[1] in 2013 said that 90% of the world's data was produced over the last two years and a more recent report from IBM[2] titled "10 Key Marketing Trends for 2017" said that we create over 2.5 quintillion bytes of data every day. Data scientists are availing of this huge mass of data to solve real world problems for the greater good of society and data science has already proven its worth in areas such as policing, target marketing and in new technologies like self-driving cars. We know data science also has the potential to hugely improve areas such as healthcare and cyber security — but why have these improvements not been observed already? The answer lies in an issue that faces many data scientists: the availability of data. Privacy concerns over health care data, for example, mean that although the data exists, it is deemed too sensitive to be available for sharing outside of specialised servers for public use. Also, in light of the forthcoming GDPR, data sharing and data use will demand careful governance. In fraud detection, instances of fraud may be so rare that

---
[†]Work partially supported by the European Union as part of the Meaningful Integration of Data, Analytics and Services (MIDAS) project from the European Union's Horizon 2020 programme.

there is simply not enough data to which data science techniques can be applied. Machine Learning models, for example, rely on examples of fraud from which to learn, so that when they are faced with a previously unseen set of data they can accurately predict whether something should be classed as fraudulent or not fraudulent. One way to overcome the issue of data availability is to use synthetic data rather than real data[3-5]. Synthetic data is generated from real data by using the underlying statistical properties of the real data to produce synthetic datasets which exhibit these same statistical properties. Some work has been done to ascertain whether synthetic data can preserve hidden complex patterns that data mining can uncover in the same way it would when mining the original dataset[6]. A good synthetic dataset should replace sensitive values and provide stronger guarantees of privacy and anonymity. Synthetic data can be used in two ways:
1. To increase the size of a dataset, for times when a dataset is unbalanced due to the limited occurrence of an event.
2. To generate a full synthetic dataset that is representative of the original dataset, for times when data is not available due to its sensitive nature.

The aim of this work is to explore whether synthetic data can be a reliable replacement for real-world data used by machine learning algorithms. This paper looks at ways to generate synthetic datasets and evaluates their performance when they are used to train machine learning models.

2. Methodology

2.1. *Dataset selection*

For this work, synthetic datasets were generated for two datasets from the UCI Repository. The first was the Breast Cancer Wisconsin dataset which has numeric variables, 699 rows and ten attributes, plus the class attribute. Each instance belongs to one of two classes: benign, represented by a 2 in the dataset, or malignant, represented by a 4 in the dataset. The second was the Nursery dataset which has categorical variables, 12,960 rows and eight attributes, plus the class attribute. Each instance belongs to one of five classes — 'not_recom', 'recommend', 'very_recom', 'priority' or 'spec_prior'. It was not difficult to find data to work with for this project as the synthesis of data can be demonstrated on most datasets. However, the reason for choosing these two datasets was to determine if the variable type or the size of the dataset had any bearing on the synthesis of data. The original Breast Cancer dataset, along with the synthetic datasets subsequently generated from it, are the datasets used to train the machine learning models for which we will compare the evaluation metrics.

2.2. Generating synthetic data

Generating synthetic data for the purposes of balancing datasets requires the SMOTE (Synthetic Minority Over-Sampling Technique)[7] function, which uses a K-nearest neighbour algorithm, for example, to generate synthetic observations of the rare event. The SMOTE function is available in Weka and R. Python provides a module called 'Imbalance Learn' which has a similar function. R offers a convenient approach to generating a full synthetic dataset using a library called 'Synthpop' which is "a tool for producing synthetic versions of microdata containing confidential information so that they are safe to be released to users for exploratory analysis"[8]. This tool takes the variables in the dataset and, in turn, generates synthetic values using classification/regression trees or parametric models, depending on the type of variable. Synthetic data is produced using a syn() function which provides the user with control over which method should be used; either the default method or a parametric method. Two full synthetic datasets were generated for the Breast Cancer dataset using the Synthpop library. The first synthetic dataset was generated using the default method in the syn() function and the second using the 'parametric' method in the syn() function. The way in which synthetic data was generated for each column in the synthetic Breast Cancer datasets is shown in Table 1.

Table 1. Table showing which model was used to generate synthetic data in each column of the Breast Cancer dataset, using the 'default' and 'parametric' methods in the syn() function.

	Default	Parametric
Sample Code #	Sample	Sample
Clump Thickness	Cart	Norm Rank
Uniformity of Cell Size	Cart	Norm Rank
Uniformity of Cell Shape	Cart	Norm Rank
Marginal Adhesion	Cart	Norm Rank
Single Epithelial Cell Size	Cart	Norm Rank
Bare Nuclei	Cart	Polyreg
Bland Chromatin	Cart	Norm Rank
Normal Nucleoli	Cart	Norm Rank
Mitoses	Cart	Norm Rank
Class	Cart	Norm Rank

In both methods, the synthetic unique identifiers 'sample code number' are generated using a random sample from the observed data. In the default method the rest of the synthetic variables are generated by drawing from conditional distributions fitted to the original data using classification/regression tree models.[6] In the 'parametric' method, the synthetic values are found using 'normrank': normal linear regression preserving the marginal distribution and 'polyreg': unordered polytomous regression. Figure 1 shows how the distributions of the two synthetic Breast Cancer datasets look compared to the original dataset. Observing the distributions of both synthetic Breast Cancer datasets, it is clear that they exhibit similar underlying statistical properties as that of the original dataset. In addition, this project involved the development of a new method to generate synthetic data for the Breast Cancer dataset. It used the underlying distributions of each variable and a machine learning algorithm to generate the new synthetic values. This involved randomly sampling the unique identifier 'sample code number'. Then, for each column in the dataset, apart from the last, we determined the weight of each value within the column and used the random.randint() function in Python to generate a new synthetic column of values that was representative of the original column. This provided a synthetic dataset containing all but the class variable. To determine this class variable, a decision tree classifier was trained using the original dataset and used it to predict the class of each instance in my synthetic dataset. The distribution of this synthetic dataset compared to the original Breast Cancer dataset is shown in Figure 2.

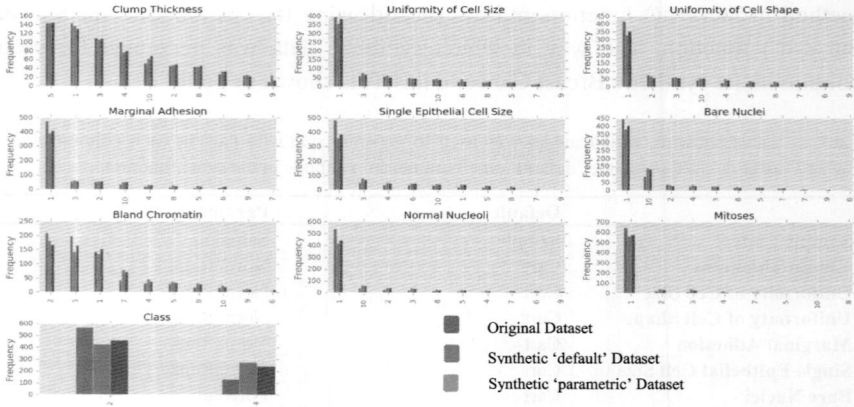

Figure 1. Figure showing how the distributions of each variable in the two synthetic Breast Cancer datasets compare to those in the original Breast Cancer dataset.

To understand whether the type of variable impacts the synthesis of data in the Synthpop library in R, the two methods in the syn() function was used to generate synthetic data for the categorical Nursery dataset. Table 2 shows how the synthetic data was generated for each column in both of the synthetic datasets.

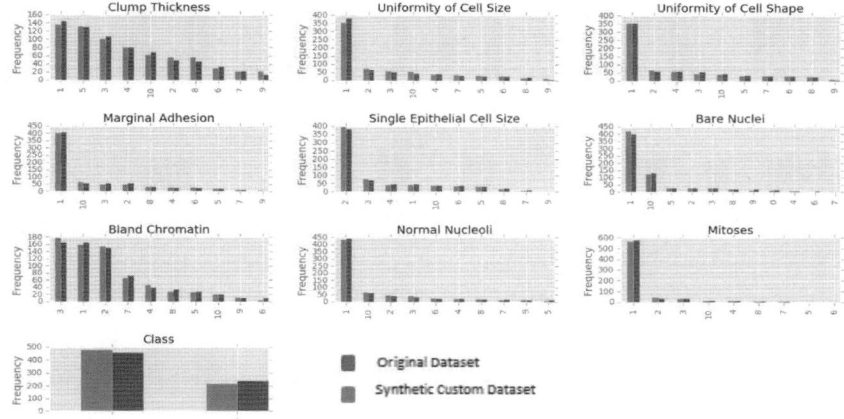

Fig. 2. Figure showing how the distributions of each variable in the third synthetic Breast Cancer dataset compared to those in the original Breast Cancer dataset.

Table 2. Table showing which model was used to generate synthetic data in each column of the Nursery dataset, using the 'default' and 'parametric' methods in the syn() function.

	Default	Parametric
Parents	Sample	Sample
Has_Nurs	Cart	Polyreg
Form	Cart	Polyreg
Children	Cart	Polyreg
Housing	Cart	Polyreg
Finance	Cart	Logreg
Social	Cart	Polyreg
Health	Cart	Polyreg
Class	Cart	Polyreg

Synthetic data is generated identically for both the Breast Cancer dataset and the Nursery dataset using the 'default' method in the syn() function. Using this method, we see that categorical or numerical variables have no bearing on how the synthetic data is generated. However, we see a difference when we use the 'parametric' method. In the numerical Breast Cancer dataset, most of the synthetic variables are generated using the normal linear regression, with one generated using polytomous (multinomial) regression. In the categorical Nursery dataset, most synthetic variables are generated using polytomous regression and one generated using logistic regression. Figure 3 shows how the distributions of the two synthetic Nursery datasets look compared to the original dataset.

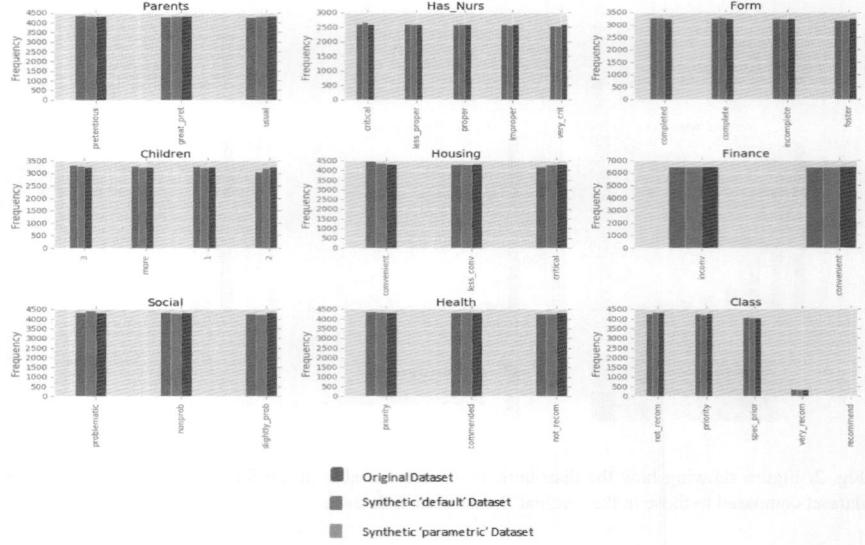

Figure 3. Figure showing how the distributions of each variable in the two synthetic Nursery datasets compare to those in the original Nursery dataset.

We observe that the distributions of both synthetic Nursery datasets are almost identical to the distribution of the original dataset and can observe that the statistical properties of the original dataset have been preserved. We also observe that the type of variable or the size of the dataset has no bearing on the distributions of synthetic data which has been generated using the syn() function in R.

2.3. Machine learning using real and synthetic data

To evaluate whether synthetic datasets can be used in place of real datasets in machine learning models, different classification models were trained with the original and synthetic Breast Cancer datasets. For this part of the work Python's machine learning library, Scikit – Learn was used, as it has a wide selection of algorithms and a consistent API. For the Breast Cancer dataset with binary classification a Linear Classification model, a Decision Tree Classifier, a K-Nearest Neighbour Classifier, a Support Vector Machine Classifier and a Random Forest Classifier were used. A combination of simple and more complex algorithms was purposively chosen to see how well each model performed when trained with the synthetic and real datasets. For training and testing, 10-fold cross validation (CV) was used, as this is a more sophisticated holdout training and testing procedure than simply splitting the data into one training and test set, and makes better use of limited data. To implement the Linear Classification, the SGDClassifier with default parameters and loss = *'hinge'* and *random_state = 0 were used.* To implement the Decision Tree Classification the DecisionTreeClassifier with default parameters and *criterion = 'gini', max_depth = 10, random_state = 0 were used.* To implement the k-Nearest Neighbour Classifier the KNeighborsClassifier with default parameters with leaf_size = 30, metric = 'minkowski', n_jobs = 2, n_neighbors = 10, p = 2 and weights = 'uniform' were used. To implement the Random Forest Classifier, the RandomForestClassifier with default parameters and *n_estimators = 10, criterion = 'gini', max_depth = 10, min_samples_split = 2* and *random_state = 1 was used.* Finally, to implement the Support Vector Machine Classifier, SVC with default parameters and *C = 1.0, kernel = 'rbf', degree = 3, probability = True* and *random_state = None were used.* The parameters in the Machine Learning models were the same when training the original Breast Cancer dataset and the three synthetic Breast Cancer datasets, to enable precise comparison of the evaluation metrics.

3. Results

To compare the performances of each model after being trained with the original and synthetic datasets, a variety of evaluation metrics were used. The first, most obvious evaluation metric to compare was the accuracy of each model. Table 3 compares the accuracy each model achieved after being trained by the three datasets.

Table 3. Table comparing the accuracy scores achieved by each model as trained by each dataset.

Dataset	Linear Model	Decision Tree	KNN	Random Forest	SVM
Original	0.971428	**1.0**	0.969999	0.998571	0.995714
Default Synthetic	0.922753	**0.997142**	0.942795	0.989999	0.99
Parametric Synthetic	0.894161	**0.998571**	0.952836	0.989999	0.985714
Custom Synthetic	0.6882194	0.998550	0.864099	**0.998571**	**0.998571**

We see that the most accurate model for both the original and default and parametric synthetic datasets is the Decision Tree. It achieves a perfect accuracy score when trained with the original dataset and very high accuracy for each of the synthetic datasets. However, the synthetic dataset which was trained using the 'parametric' method performs slightly better. The most accurate models for the custom synthetic dataset are the Random Forest and SVM, followed closely by the Decision Tree. The least accurate model for the original dataset was the k-Nearest Neighbour classifier, while for the three synthetic datasets the least accurate model was the Linear Model. The Linear Model also provides the largest variation in accuracy score between the four datasets. We observe that the accuracy score in all other models does not vary significantly across the four datasets. Accuracy can often be too simplistic, so it is vital that we use other evaluation metrics to fully understand how the models are performing. Precision scores, recall scores and the F1 measure evaluation metrics should provide more insight into model performance. These evaluation metrics are shown in Table 4, Table 5 and Table 6.

Table 4. Table comparing the precision scores of each model after being trained by each dataset.

Dataset	Linear Model	Decision Tree	KNN	Random Forest	SVM
Original	0.972	1.000	0.970	0.999	0.996
Default Synthetic	0.930	0.997	0.943	0.990	0.990
Parametric Synthetic	0.899	0.999	0.954	0.990	0.986
Custom Synthetic	0.786	0.999	0.873	0.999	0.984

Table 5. Table comparing the recall scores of each model after being trained by each dataset.

Dataset	Linear Model	Decision Tree	KNN	Random Forest	SVM
Original	0.971	1.000	0.970	0.999	0.996
Default Synthetic	0.923	0.997	0.943	0.990	0.990
Parametric Synthetic	0.894	0.999	0.953	0.990	0.986
Custom Synthetic	0.688	0.999	0.864	0.999	0.984

Table 6. Table comparing the F1 scores of each model after being trained by each dataset.

Dataset	Linear Model	Decision Tree	KNN	Random Forest	SVM
Original	0.971	1.000	0.970	0.999	0.996
Default Synthetic	0.923	0.997	0.943	0.990	0.990
Parametric Synthetic	0.879	0.999	0.951	0.990	0.985
Custom Synthetic	0.562	0.999	0.855	0.999	0.984

We see that precision, recall and F1 scores for each model for each dataset offer the same insight into model performance as the accuracy score. In terms of these evaluation metrics, the Decision Tree is still the best classifier for all datasets, with the Random Forest also performing well for the custom synthetic dataset. We observe that the Linear Model provides the largest variation in precision, recall and F1 scores between the four datasets. Although precision, accuracy and F1 measures are summaries of the confusion matrix in some form; it is still beneficial to separate out the decisions made by the model, to show where one class is being misclassified for another. Figure 4 shows the confusion matrices for the Decision Tree, trained by the original dataset and the three synthetic datasets.

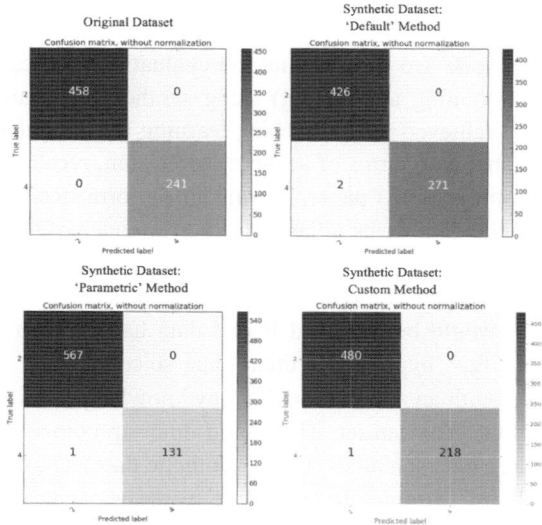

Figure 4. Confusion matrices for the Decision Tree Classifier after being trained by each dataset.

Figure 4 shows that the Decision Tree Classifier correctly predicted every instance in the original dataset. The Decision Tree Classifier for each synthetic dataset both incorrectly predicted the class as being a '2' when in fact its true class was a '4'; however the 'default' synthetic dataset did this twice as often as the 'parametric' and custom datasets. This means that twice in the 'default' synthetic dataset a tumour was predicted to be non-cancerous when in fact it was cancerous. This highlights the importance of using more evaluation metrics than just an accuracy score. The accuracy score makes no distinction between false positive and false negative errors and makes an assumption that they are equally important. This kind of misclassification could be very dangerous in the real world; classifying a tumour as non-cancerous when the tumour is cancerous is more serious than classifying the tumour as cancerous when it is non-cancerous. Therefore, while the synthetic datasets can be a very close match to the original data in terms of their distributions and is a feasible solution to producing a dataset that enhances privacy, when the class variable has such high importance, we need to err on the side of caution if we wish to use them to train machine learning models. We also need to ensure the use of multiple evaluation metrics, not just the simple accuracy score which can be misleading in terms of how the model is actually performing. In this instance, the use of synthetic data to predict whether a tumor is cancerous or non-cancerous may not be recommended as the consequences to classifying an instance incorrectly are too serious.

4. Conclusion

In this very early work, we can see that the evaluation metrics achieved when machine learning (training and testing) using synthetic data are similar to the evaluation metrics achieved when machine learning (training and testing) using the real datasets. That is in terms of accuracy, precision, recall and F1 scores. In this limited case study in this paper, the model performance when trained and tested using synthetic data was similar to the performance of the model that was trained and tested with the real data. This is only one case study, but this may suggest that the evaluation of models built using synthetic data maybe reflective of the results that would be achieved if real data had of been used. If further research supports this hypothesis, then data scientists can mine synthetic healthcare datasets with an assumption that any knowledge elicited is very likely to be reflected in the real dataset. This could open up competitions and health data mining to more data scientists. Using synthetic datasets to facilitate privacy preserving machine learning to discover patterns and viable predictive modelling without giving away raw sensitive data maybe a useful process to minimise risk. We recognise that this work is primitive and limited since there was no cross-testing of the models. Future work would include testing a machine learning model that was built using synthetic data on real data to ascertain if a model trained using synthetic data would perform just as well with real-world cases.

Acknowledgments

We acknowledge support from the 'Meaningful Integration of Data, Analytics and Services (MIDAS)' project as funded by the European Union's Horizon2020 programme-H2020-SC1-2016-CNECT SC1-PM-18-2016-Big Data Supporting Public Health Policies, Grant Agreement No. 727721.

References

1. www-01.ibm.com. (2018). *IBM What is big data? Bringing big data to the enterprise - India*. [online] Available at: https://www-01.ibm.com/software/in/data/bigdata/ [Accessed 29 Jan. 2018].
2. www-01.ibm.com. (2018). *10 Key Marketing Trends for 2017*. [online] Available at: https://www-01.ibm.com/common/ssi/cgi-bin/ssialias?htmlfid=WRL12345USEN [Accessed 29 Jan. 2018].
3. C.C. Aggarwal, and S.Y. Philip, A general survey of privacy-preserving data mining models and algorithms. In *Privacy-preserving Data Mining*, Springer, Boston, MA, pp. 11–52 (2008).
4. X. Qi, and M. Zong, An overview of privacy preserving data mining. *Procedia Environmental Sciences*, *12*, pp. 1341–1347 (2012).
5. Y.A.A.S Aldeen, M. Salleh, and M.A. Razzaque, A comprehensive review on privacy preserving data mining. *SpringerPlus*, *4*(1), p. 694 (2015).
6. J. Eno, CW. Thompson, Generating synthetic data to match data mining patterns. *IEEE Internet Computing*, 12(3), pp. 78–82 (2008).
7. NV. Chawla, KW. Bowyer, LO. Hall, WP. Kegelmeyer. SMOTE: synthetic minority over-sampling technique. *Journal of Artificial Intelligence Research*, 16, 321–57 (2002).
8. Cran.r-project.org. (2018). *CRAN - Package synthpop*. [online] Available at: https://cran.r-project.org/web/packages/synthpop/index.html [Accessed 29 Jan. 2018].

Fuzzy framework for activity recognition in a multi-occupant smart environment based on wearable devices and proximity beacons[*]

Macarena Espinilla[†], Javier Medina Quero

Computer Science, University of Jaén, Campus Las Lagunillas, Jaén, 23071, Spain

Naomi Irvine, Ian Cleland, Chris Nugent

School of Computing, Ulster University, Jordanstown Co. Antrim, BT37 0QB, UK

A fuzzy framework is proposed for activity recognition in a multi occupancy smart environment based on wearable devices and Bluetooth Low Energy (BLE) beacons. In this context, there is a set of Received Signal Strength Indicators (RSSIs), which is received by wearable devices worn by the inhabitants, from a set of BLE beacons deployed in objects within a smart environment. The RSSIs generated by the BLE beacons are influenced by external factors in radio waves, which are presented in the RSSI as important fluctuations. The main contribution in this work has been to propose the use of fuzzy logic and the fuzzy linguistic approach to model indicators of fluctuation. A new framework is introduced that includes a Mamdani fuzzy inference system to recognize the activity that is being carried out in real time for each inhabitant based on data gleaned from their wearable device. The fuzzy rules in addition to the fuzzy linguistic terms used to model the RSSIs can be customized for each inhabitant according to their habits, customs or routines. A case study of the proposed fuzzy framework in the smart lab of the University of Jaén is presented to demonstrate the effectiveness of the approach.

1. Introduction

An emerging trend in the context of ubiquitous environments is the process of activity recognition. This process considers that it is possible to recognize the current activity carried out by an inhabitant by using data generated from a set of sensors deployed in the environment [1].

Usually, a set of binary sensors are deployed within the environment to monitor an inhabitant's behavior. In a multi-occupancy smart environment using binary sensors it can be very difficult to identify the individual who has interacted with a specific objects and who has carried out a specific activity. To avoid these challenges with binary sensors video cameras coupled with computer vision

---

[*]This work is supported by European Unions Horizon grant agreement No. 734355 together the Spanish government by research project TIN2015-66524-P.
[†]mestevez@ujaen.es.

techniques have been used as a method of distinguishing between different inhabitants within the smart environments [2]. This approach, however, presents important concerns around violation of privacy [3].

With the aim of safeguarding inhabitants' privacy, the use of Bluetooth Low Energy (BLE) beacons presents themselves as an excellent option to replace binary sensors within a smart environment. By combining these beacons with wearable devices worn by inhabitants, it is possible to overcome the challenges of multiple occupancy and to distinguish different inhabitants. When the wearable device comes within close proximity of the BLE beacon the information recorded can be sued to infer that the inhabitant is interacting with the object or is present in the location near the beacon, and thus it is possible to recognize the activity that is being performed.

The operation the beacons can be explained as follows. Each beacon has a broadcasting power with which it broadcasts its signal. Wearable devices have the capability to read the Received Signal Strength Indicators (RSSIs) from the beacons when they are in range of them. The proximity between a wearable device and a BLE beacon impacts upon the RSSI. So, for example, the greater the RSSI received by the wearable device, the smaller the distance between the wearable device and the BLE beacon [4].

Due to external factors influencing radio waves such as absorption, interference, or diffraction, the RSSI tends to fluctuate. The RSSI can therefore be considered unstable [4]. For this reason, proposed activity recognition systems in the literature combine BLE beacons with data provided by other kinds of sensors for example accelerometer or PIR modules [5, 6]. In these systems, the data collected by the wearable device needs to be sent to other devices or a server, thereby generating concerns about security and privacy issues. Usually, an activity is recognized in the server by using a general trained classifier that does not consider the adaptation of each user.

These issues motivated us to investigate the use of fuzzy logic and a fuzzy linguistic approach to model RSSI for the purposes of activity recognition in a multi-occupant smart environment in an effort to overcome the limitation presented when using BLE beacons. The fuzzy logic and fuzzy linguistic approach have previously demonstrated their ability for modeling imperfect information in numerous applications and is therefore considered as being an appropriate choice of technique in activity recognition [7]. In this work, a novel fuzzy framework for activity recognition is proposed. The framework can recognize activities in a multi-occupant smart environment with BLE beacons and inhabitants using wearable devices. The framework is composed of two phases. The first phase is the definition of a smart environment for activity recognition and all necessary elements, including the linguistic processing of RSSIs. The second phase is the definition of a Mamdani fuzzy inference system (MFIS) in the wearable device of each inhabitant to recognize the activity that is being carried out [8]. We have chosen this type of inference system due to its advantages: to be semantically clear and interpretable [1]. Furthermore, the set of rules can be adapted to each inhabitant in their own wearable device.

Therefore, the main contributions of our proposal are the following: i) interactions with objects and their locations in a smart environment with BLE beacons can be uniquely and exclusively identified by wearable devices worn by the inhabitants according to the RSSIs. ii) The recognition is performed in the same wearable device where signals are collected, reducing problems related to security and privacy. iii) The set of rules of each MFIS in the wearable device can be customized for each inhabitant.

This paper is organized as follows: Section 2 provides a review of fuzzy concepts necessary to understand the proposed activity recognition framework. Sections 3 presents the proposed fuzzy framework for activity recognition. Section 4 presents an instance of the proposed fuzzy framework in a real smart environment. Finally, conclusions are drawn in Section 5.

2. Fuzzy background

This section provides a concise summary of the main concepts related to the proposed solution.

Fuzzy logic is a logic of imprecision and approximate reasoning [7]. Fuzzy logic can be explained as a system that models non-linear functions and a fuzzy set is a class of objects with a continuous degree of membership [7]. This set is characterized by a membership function, which assigns each object a value between 0 and 1. So, a fuzzy set is prescribed by vague or ambiguous properties; hence, its boundaries are ambiguously specified. On the contrary, a classical set is defined by crisp boundaries; it is characterized by its membership function and may take only two values 0 and 1. The main idea of the Mamdani method is a way of mapping an input space to an output space by means of linguistic variables and to use these variables as inputs to control rules [8].

A MFIS involves developing membership functions and defining the subsequent rules. The rules match the input variables with the output variables and are based on the fuzzy state description that is obtained by the definition of the linguistic variables [7]. A MFIS has three main component that are described in detail as follows: *Fuzzification phase* that involves a domain transformation where crisp values (inputs) are transformed into linguistic terms by means of membership functions of the system linguistic variables. *The knowledge base* that is composed by fuzzy linguistic rules that can be defined by expert knowledge or by data. A fuzzy linguistic rule is a simple IF-THEN rule where conditions and the conclusion are linguistic terms. The inference function takes the membership degrees of the fuzzification supported in the rule base to generate the output of the fuzzy system. *Defuzzification phase* is used to obtain an output from the previous fuzzy set.

Medina [9], previously presented how to model data sensor streams, considering fuzzy temporal windows. To do so, a sensor data stream of a sensor s^j is denoted by st^j that has a set of measures $st^j = \{m_i^j\}$. Each measure is represented by $m_i^j = \{d_i^j, t_i^j\}$, where d_i^j represents the data provided by the sensor s^j that depends on the nature of each sensor, for example, temperature, humidity, and t_i^j represents the timestamp when the data was provided by the sensor s^j.

First, a fuzzy linguistic value variable for each sensor s^j is defined V^j with linguistic terms $V^j = \{V_0^j, \ldots, V_k^j, \ldots, V_{gv}^j\}$ in which each linguistic term has its fuzzy membership function $\mu V_k^j(d_i^j)$ associated with it. The fuzzy membership function $\mu V_k^j(d_i^j)$ is interpreted as the degree of membership of data d_i^j in the linguistic value term $v_k^j \in V^j$. Second, it is necessary to define a set of linguistic temporal measures $T_{t_0}^j = \{T_0^j, \ldots, T_h^j, \ldots, T_{gt}^j\}$ to consider a sub-set of measures in the data sensor stream $st^j = \{m_i^j\}$ regarding to the current time t_0. So, a linguistic temporal term T_h^j has an associated fuzzy membership function $\mu T_h^j(t_i^j)$ that is interpreted as the degree of membership of a timestamp t_i^j in the linguistic temporal term T_h^j regarding to t_0. Finally, the relevance of a measure m_i^j in a linguistic value term V_k^j in a linguistic temporal term T_h^j is defined by an intersection operation to fuse both degrees of membership [9]. The relevance of a sub-set of measures in the sensor data stream that are associated with T_h^j, i.e., $\mu T_h^j(t_i^j) > 0$ are aggregated using the union operator in order to obtain a single degree of the degrees implied in a linguistic value term in the linguistic temporal term [9].

3. A fuzzy framework for activity recognition in a multi occupancy smart environment based on wearable devices and BLE beacons

This section introduces a novel fuzzy framework for activity recognition within multi-occupant smart environments with wearable devices. The aim of this new fuzzy framework is to establish an approach to define the necessary phases for the activity recognition in a multi occupancy smart environment in which BLE beacons have been deployed and inhabitants are using a wearable device. The framework is composed of two phases: i) Smart environment definition and ii) MFIS definition. These phases are further detailed in the following subsections.

Smart environment definition

The following notions and terminology are presented in the proposed fuzzy framework for multi occupant activity recognition with wearable devices.

1. Let a smart environment in which a set of **activity classes** exist be defined as $A = \{A^1, \ldots, A^i, \ldots, A^{Al}\}$.
2. Let a **set of objects or areas** $O = \{O^1, \ldots, O^j, \ldots, O^J\}$ with which an inhabitant interacts or stays in the realization of activities be defined.
3. Let a **set of BLE beacons** $S = \{s^1, \ldots, s^j, \ldots, s^J\}$ be defined that is associated with a set of objects or areas, respectively. For each BLE beacon the following elements are defined: A **transmit power** denoted by tp_s^j that is the power with which the beacon broadcasts its signal. So, each object or location has associated with it a BLE beacon with a specific transmit power. The transmit power is defined according to its nature. For example, the proximity beacon associated with a *toothbrush* should have a weak transmit power so that it should emit only in a small range. The proximity beacon associated with the *bed* or *sofa* should have a higher transmission power in order to cover a larger range. The **broadcast frequency** denoted by f_s^j is defined as the interval between each transmission. The shorter the interval, the more stable the signal. A **linguistic value variable** with a set of linguistic value terms and its membership functions $V^j = \{V_0^j, \ldots, V_k^j, \ldots, V_{K-1}^j\}$. A **linguistic temporal variable** with a set of fuzzy linguistic temporal terms and its membership functions $T_{t_0}^j = \{T_0^j, \ldots, T_h^j, \ldots, T_H^j\}$.
4. Let a set of **wearable devices** $D = \{D^1, \ldots, D^k, \ldots, D^K\}$ that are associated with a **set of inhabitants** $I = \{I^1, \ldots, I^k, \ldots, I^K\}$ be defined. Each inhabitant has an associated wearable device. So, inhabitant I^k has an associated wearable device D^k. The set of wearable devices can be a smart phone or a smart watch, the only requirement is that it can receive the RSSI provided by the BLE beacons deployed in the smart environment. Each device D^k defines the frequency of reading of RSSIs, these readings generate the **stream of RSSI signals** St_{RSSI}^k. In this work, this stream is defined by a set of measures $St_{RSSI}^k = \{m_i^k\}$ where each measure is defined by a 3-tuple $m_i^k = \{s_i^{jk}, d_i^k, t_i^k\}$, with s_i^{jk} being the BLE beacon reading by the wearable device D^k with an RSSI value of d_i^k in the time stamp t_i^k.

Mamdani fuzzy inference definition

In order to carry out the process of activity recognition, we propose a Fuzzy Rule-Based Inference Engine. The rule-based inference engine is carried out in the wearable device of each inhabitant. The fuzzy rules in addition to the fuzzy linguistic terms can be customized for each inhabitant to their habits, customs or routines. The proposed Rule-Based Inference is based on a collection of fuzzy logic rules in the form of IF (Antecedents) THEN (Consequent). Each antecedent includes, at least, a linguistic variable according to a BLE beacon s^j, with a linguistic value term in a linguistic temporal term in the form: V^j IS V_k^j T_h^j that it is computed by $V_k^j \cap T_h^j$ (St_{RSSI}^k). The consequence, represented as I^K IS A^i, is interpreted like the inhabitant I^K is carrying out the activity A^i in the smart environment. Finally, the inference engine determines the degree of matching of antecedents, based on the stream of RSSI of each inhabitant, assigning the activity consequence in real time.

4. Case study in the smart lab of the University of Jaén

In this Section, a description of the proposed fuzzy framework is shown in the smart lab of the University of Jaen called UJAmI (University of Jaén and Ambient Intelligence) smart lab.

The UJAmI smart lab is a fully furnished apartment that is organized into the following spaces: a lobby, a workplace, a living room, a kitchen and a bedroom with an integrated bathroom. There are more than 130 smart devices deployed in this apartment that allow the analysis of the behavior of its inhabitants. In this case study, eight BLE beacons were deployed with the features and locations that are shown in Table 1.

Table 1. BLE beacons deployed in the smart lab and their locations.

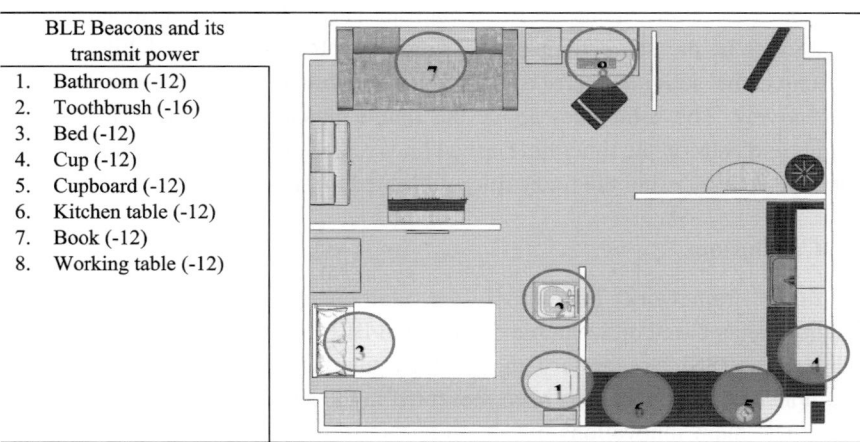

BLE Beacons and its transmit power
1. Bathroom (-12)
2. Toothbrush (-16)
3. Bed (-12)
4. Cup (-12)
5. Cupboard (-12)
6. Kitchen table (-12)
7. Book (-12)
8. Working table (-12)

The description of the fuzzy rules of the MFIS that have been defined by a knowledge expert are shown in Table 2. The linguistic value variable and the linguistic temporal variable that model the RSSI streams are shown in Figure 1.

Table 2. Knowledge base with fuzzy rules.

	Antecedents		Activity
IF	Bathroom IS near AND Time is Now	THEN	toileting
IF	Toothbrush IS touch AND Time is Now	THEN	tooth brushing
IF	Toothbrush IS touch AND Time is Now	THEN	sleeping
IF	Cup IS immediate AND Time is Now AND Cupboard IS near and AND Time is Now	THEN	cooking
IF	Cup IS immediate AND Time is A while AND Cupboard IS immediate AND Time is A while AND Kitchen table IS touch AND Time is Now	THEN	eating
IF	Book IS touch AND Time is Recently	THEN	reading
IF	Working table IS inmediate and AND Time is Recently	THEN	working

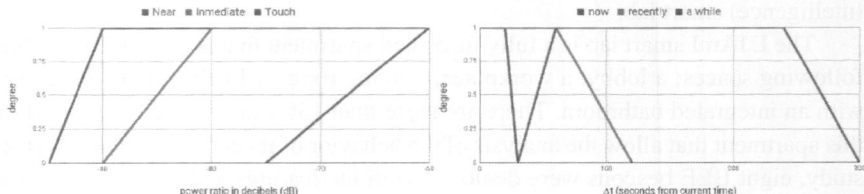

Figure 1. Linguistic variables to model the RSSI streams.

The case study includes 14 human activities that are developed by two inhabitants at the same time in the smart lab. In Figure 2 and Figure 3 are presented the activities carried out by the inhabitant A and the inhabitant B, respectively. The solid lines above the figure indicate when each real activity begins and ends on the timeline. Dashed lines represent the degree of membership for each activity using the fuzzy rules presented in Table 2 on the timeline.

5. Conclusions

A fuzzy framework for activity recognition in multi occupancy smart environments based on wearable devices and BLE beacons has been proposed in which the RSSI streams are modeled by the fuzzy linguistic approach. An example of the proposal has been presented in the smart lab of the University of Jaén. The main innovations of our proposal are: i) the use exclusively of BLE beacons in order to identify the interactions. ii) The recognition is done in the same wearable device where signals are collected. iii) The set of rules in the wearable device can be customized for each inhabitant. Our future works are focused on using a data driven approach to train the fuzzy classifier to define the membership functions of the involved fuzzy linguistic variables.

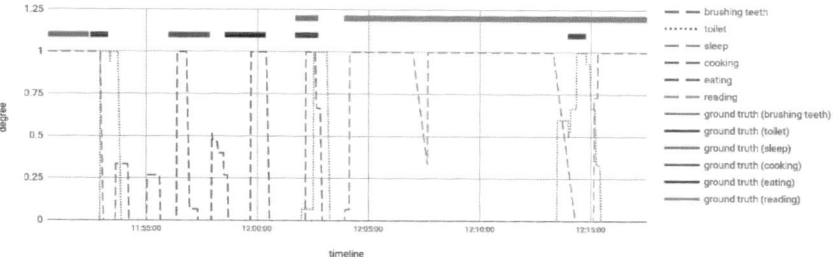

Figure 2. Timeline of membership degrees of activities developed by person A.

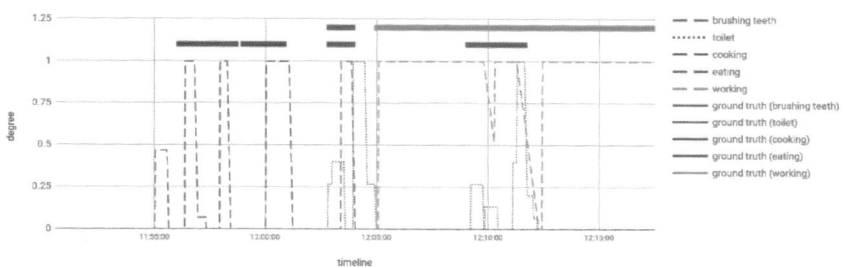

Figure 3. Timeline of membership degrees of activities developed by person B.

References

1. L. Chen, J. Hoey, C. Nugent, D. Cook, Z. Yu, *Syst. Man Cybern.* **C42** (6) (2012).
2. Shewell, C., Medina-Quero, J., Espinilla, M., Nugent, C., Donnelly, M., Wang, H. Int J Commun Syst, **30**, 5, (2017).
3. M. Amiribesheli, A. Benmansour, and A. Bouchachia, *J. Ambient Intell. Humaniz. Comput.* **6**, 4 (2015).
4. G. de Blasio, A. Quesada-Arencibia, C. R. García, J. Molina-Gil, and C. Caballero-Gil, *Sensors,* **6**, 1299 (2017).
5. A. Filippoupolitis, W. Oliff, B. Takand and George Loukas. *Sensors*, **17**, 6, (2017).
6. G. Mokhtari, Q. Zhang, G. Nourbakhsh, S. Ball and M. Karunanithi. IEEE Sensors Journal, **17**, 5 (2017).
7. L. Zadeh, *Information Sciences*, **8** (1975).
8. E. Mamdani and S. Assilian, *International Journal of Man-Machine Studies*, **7** (1975).
9. J. Medina, M. Rosa Fernández-Olmo, M.D. Peláez and M. Espinilla, Sensors, **17**, 2892 (2017).

Impact of dataset quality on the performance of data-driven approaches for human activity recognition

Naomi Irvine[1], Chris Nugent[1], Shuai Zhang[1], Hui Wang[1], Wing W. Y. Ng[2], Ian Cleland[1] and Macarena Espinilla[3]

[1]*School of Computing, Ulster University, Co. Antrim, Northern Ireland, UK*
[2]*School of Computer Science and Engineering, South China University of Technology Guangzhou, China*
[3]*Department of Computer Science, University of Jaén, Spain*

This paper discusses the impact of data quality on activity classification using data-driven approaches. Data was collected by 141 undergraduate students at Ulster University using a triaxial accelerometer. A clearly defined data collection protocol was provided for the participants as data collection occurred in an unsupervised setting. Results produced by four common classifiers highlight the effects of noisy data by comparing the classification performances of raw and subsequently cleaned data. Results also highlight the importance of following a data collection protocol attentively to improve overall classification performance. The Naïve Bayes classifier improved most significantly with a 12.967% increase in performance.

1. Introduction

Human Activity Recognition (HAR) serves as a key component in many Ambient Assisted Living and/or connected health applications, ranging from promoting physical activity to monitoring of long term chronic conditions. It is concerned with the ability to automatically recognise and interpret human activity using computational methods, which remains a highly complex task [1]. The recognition of activities or actions can be determined through the interpretation of human body motion or gestures via sensors such as body-worn accelerometers, though despite the comprehensive effort in the community HAR still requires significant improvement [22].

In this work, a range of data-driven approaches have been applied to generate models for activity classification. Data-driven approaches rely on good training data. This has been one of the motivating factors for the current work. A secondary factor is the importance of a clear data collection protocol to ensure the prevalence of noise and outliers are minimalised.

Section 2 provides an overview on data cleansing and the effects data quality can have on classification. Following this, Section 3 describes the dataset collected and utilised in addition to the methods used within the activity recognition process. Results are then presented in Section 4 highlighting comparisons between the classification accuracies of raw and cleaned data.

2. Background

Human activity recognition is a fundamental component to a broad range of application areas including ambient assistive living, connected health and pervasive computing [2]. It is commonly used in monitoring the activities of elderly residents to support the management, and also the prevention, of chronic disease. Another common application area is HAR within smart homes. It is used in this scenario to monitor the health and wellbeing of inhabitants by tracking their daily activities [1].

HAR can be generally classified as belonging to two categories: either sensor-based or vision-based. Most notably, sensor-based activity recognition has attracted considerable research interest in pervasive computing due to advancements with sensor technologies and wireless sensor networks [3]. A frequently utilised wearable sensor for monitoring human activities is the accelerometer, which is particularly effective in observing movements such as walking, running, standing, sitting, and ascending stairs [2].

Within the realms of data-driven approaches to HAR, data quality is a significant consideration. Data cleansing can be defined as the process of removing errors or inconsistencies such as noise and/or outliers from a collection of data [4]. The presence of noise and outliers in data is an important issue to address as these can have a substantial influence on the results produced by data driven techniques, according to [5]. Nevertheless, [6] states the border between normal and abnormal (noise/outlier) data is often unclear, where a large "gray area" may exist.

Noise can be presented at a class or attribute level in supervised learning, with previous effort made by [7] to evaluate their impacts on classification accuracy on 17 datasets, collected from various domains. Various levels of noise were manually applied to each dataset, with conclusions stating that as noise levels increase, classifier performance decreases further, and that class noise is usually more harmful than attribute noise. Further to this, [8] compared numerous classifiers to evaluate how well they performed with noisy data, with conclusions stating that performance and robustness to noisy data varied substantially between various algorithms. The Random Forest classifier proved to be most resilient to noise, followed by kNN. Other studies into data quality include investigations made by [9-11]. In [9] a Fuzzy Relevance Vector Machine (FRVM) is used to learn from unbalanced data and noise, with experiments demonstrating reasonable and more robust performance than a regular RVM. In [10] and [11] respectively, noise detection and handling mechanisms were explored in relation to class noise.

Some outlier detection methodologies include the use of probabilistic and statistical models [21, 23], linear regression models [24], or distance-based models [25].

3. Methodology

The dataset utilised in the current study was collected by 141 students enrolled in the Pervasive Computing in Healthcare module at Ulster University using a triaxial accelerometer [21]. The students were each assigned an AR scenario containing 3 activities, and were subsequently tasked to collect, process, and classify data as part of the module assessment. In total, there were 6 scenarios and 18 activities recorded amongst the cohort. To investigate the impact of noise on HAR performance a subset of the data is considered which consists of the Self-care scenario involving hair grooming, hand washing and teeth brushing activities. This scenario was chosen for initial consideration as [21] states hair grooming and hand washing activities are most difficult to discriminate between of the 18 activities recorded. The selected scenario contained recordings produced by 24 participants, thus there were 72 activity files as each participant individually recorded the 3 activities specified in the self-care scenario.

3.1. *Data Acquisition*

Data used was previously collected as reported in [21]. The students were instructed to follow a data collection protocol and were given video examples clearly demonstrating how the activities should be recorded and how the sensor should be configured. The protocol specified that each activity was to be 2 minutes in duration.

Data was collected using the Shimmer wireless sensor platform, with the device placed on the dominant wrist to record data for each activity. Prior to recording, the device was calibrated by the participants and subsequently configured with a sampling rate of 51.2Hz and a sensitivity range of ±1.5g. With a sampling rate of 51.2Hz, this indicates the accelerometer will record 51 samples per second, and during each activity recording the data for the x, y and z axis is streamed and stored in a .csv file format along with a timestamp. Due to this, approximately 6120 samples were expected to be recorded per activity, per person as each of the participants were tasked to collect 2 minutes of data per activity.

for training and testing the models, and accuracy was measured to assess their performance.

Naïve Bayes is a simple, probabilistic approach to classification that is generated through applying Bayes theorem [14]. Even in its simplest form, this algorithm is known to be surprisingly effective and rather robust [19]. Decision trees resemble a hierarchical approach that generate a set of rules for ultimately making classification decisions [20]. They are popular for inductive inference [14] and have been widely used for a range of classification problems. kNN determines activity classification by evaluating the majority of the k-nearest neighbours that relate to the given activity [20]. This approach is easy to comprehend and implement, though despite its simplicity it can perform substantially well in many scenarios [19]. MLP neural networks are essentially feed-forward NNs with one or more layers between the input and output layers. They are capable of learning complex decision regions and non-linear relationships; however, they can require high computational power.

Model performance was first evaluated on the raw data with the four algorithms using 10-fold cross validation, where both the training and test sets included noise (N-N). Following this, the classification models were retrained with cleaned data (through applying the cleaning method described in Section 3.2), again using 10-fold cross validation with the four algorithms identified. In this case, both the training and test sets consisted of data with improved quality (C-C). Nevertheless, to simulate a real-life application another case was introduced to ascertain if the models were able to retain their capability of generalizing; consequently, each model was trained on a cleaned set and tested on a noisy set (C-N). Another case was initially considered where the models would be trained on noisy data and tested on cleaned data (N-C), however this was not employed as the N-C combination isn't a valid scenario for our evaluation.

Significance testing was applied to determine whether the comparisons made were of statistical importance using t-testing with a 95% confidence. A p-value of less than 0.05 was believed to be statistically significant.

4. Results and Discussion

Table 1 provides the classification accuracies produced by four algorithms for the activity recognition problem where N-N indicates a noisy training set paired with a noisy test set, C-C indicates a cleaned training set with a cleaned test set, and C-N indicates a cleaned training set paired with a noisy test set.

Table 1. Accuracies of four algorithms for the classification of activities included in the **Self-Care** scenario. N-N indicates a noisy training set with a noisy test set, C-C indicates a cleaned training set with a cleaned test set, and C-N indicates a cleaned training set with a noisy test set.

	N-N (%)	C-C (%)	C-N (%)
MLP	89.688	90.939*	89.272
Decision Tree	84.516	85.935	85.502
kNN	90.862	92.202*	92.522**
Naïve Bayes	67.240	80.207***	75.887***

Note: The table displays the comparisons between whether there was a significant difference between N-N & C-C cases, and N-N & C-N cases through T-Testing. *p<0.05. **p<0.01. ***p<0.001.

Results in Table 1 show that the kNN classifier performed the best in terms of the accuracies presented, with the Naïve Bayes classifier performing the least effectively across all cases (N-N, C-C, and C-N).

Cleaning the data improved the performance of all four classifiers when generating comparisons between the N-N & C-C cases, with significant improvements made by the MLP, kNN and the Naïve Bayes model. The Naïve Bayes model improved classification accuracy from 67.240% to 80.207% (an increase of 12.967%) in comparing the N-N to C-C cases. This proves the benefits of training and testing models on cleaned data, and suggests that noise has an apparent negative impact on classification accuracy.

A more realistic case perhaps to consider is the C-N circumstance as a model can be trained on cleaned data, though the prevalence of a clean test set is highly unlikely in a more naturalistic, real-world setting. This case evaluates whether the models are capable of generalising if tested with noisy data. When generating comparisons between the N-N & C-N situations, the Naïve Bayes and kNN classifiers improved significantly and retained their ability to generalise. The MLP and decision tree didn't significantly improve, yet they didn't perform worse or lose their generalisation capability even after being trained on a smaller dataset (as 9.44% of instances were removed during cleaning). As previously stated, there may be an unclear border between normal and abnormal data, which indicates some instances may have been removed that were seemingly outliers though may have actually been useful to remain within the dataset; therefore, the MLP and Decision Tree may have been affected as a result.

5. Conclusion

This paper has presented an attempt at generating comparisons between noisy and cleaned data for activity recognition, using a dataset collected by multiple participants in an unsupervised setting. Since noise is introduced during the data acquisition stage, the results highlight the importance of following a data collection protocol attentively and ensuring the recordings contain high quality data for classification purposes.

Automating the data cleaning process is considered for further work, for example outliers may be detected automatically through utilising statistical models. Outliers could be deemed as values that exist beyond a specified number of absolute deviations away from the median (Median Absolute Deviation) [21,23]. Additionally, thresholds could be set with upper and lower boundaries to detect issues such as signals with a range outside the measurable capability of the sensor, and to identify portions of recordings that include brief time delays between starting/ending the recordings, and performing the target activity.

Future work will also consist of identifying an optimal subset of features to enhance the performance of the models generated during the classification stage. In addition, further scenarios of the full dataset acquired will be evaluated to investigate the impact of noise on the additional AR scenarios. It is anticipated that the findings from this work can be used to refine the data collection protocol for further studies, as the presence of noise is largely due to participants failing to follow protocol attentively.

Acknowledgments

This work was partially supported from the REMIND Project from the European Union's Horizon 2020 research and innovation programme under the Marie Skłodowska-Curie grant agreement No. 734355.

References

1. J. K. Aggarwal, L. Xia, O.C. Ann and L. B. Theng, Human Activity Recognition: A Review, *Pattern Recognition Letters*, 28-30 (2014).
2. L. Chen, J. Hoey, C. D. Nugent, D. J. Cook, Z. Yu and S member, Sensor-based Activity Recognition, 42(6),790–808. (2012).
3. T. Gu, L. Wang, Z. Wu, X. Tao and J. Lu. A Pattern Mining Approach to Sensor-based Human Activity Recognition. *IEEE Transactions on Knowledge and Data Engineering*, 23(9), 1359–1372. (2011).
4. O. Maimon and L. Rokach. Data Mining and Knowledge Discovery Handbook. *Springer*, 1, 23-25. (2005).
5. F. Gorunescu. Data Mining: Concepts, Models and Techniques. *Springer*, 1, 50-53. (2011).
6. J. Han, M. Kamber and J. Pei, Data Mining: Concepts and Techniques. *Elsevier Inc*, 3, 544-558. (2012).
7. X. Zhu and X. Wu. Class Noise vs. Attribute Noise: A Quantitative Study. *Artificial Intelligence Review*, 22(3) 177-210 (2004).
8. A. Folleco, T. M. Khoshgoftaar *et al.*, Identifying Learners Robust to Low Quality Data. *IEEE Information Reuse and Integration*, 191-195 (2008).

9. D. F. Li, W. C. Hu, W. Xiong and J. Yang. Fuzzy Relevance Vector Machine for Learning from Unbalanced Data & Noise. *Pattern Recognition Letters*, 29(9), 1175-1180. (2008).
10. J. Van Hulse and T. M. Khoshgoftaar, Class Noise Detection Using Frequent Itemsets. *Intelligent Data Analysis*, 10(6), 487-507 (2006).
11. X. Zhu and X. Wu. Cost-Guided Class Noise Handling for Effective Cost-Sensitive Learning. *IEEE Data Mining*, 299-304 (2004).
12. N. C. Krishnan and D. J. Cook. Activity Recognition on Streaming Sensor Data. *Pervasive and Mobile Computing.* 10(B), 138-154 (2014).
13. O. Banos, J. Galvez, M. Darnas, H. Pomares and I. Rojas Window Size Impact on Human Activity Recognition, *Sensors* 14(4) 6474-6499 (2014).
14. D. J. Cook and N. C. Krishnan. Activity Learning: Discovering, Recognizing and Predicting Human Behavior from Sensor Data. *Wiley*, 1, 21-55 (2015).
15. M. Hoogendoorn and B.Funk. Machine Learning for the Quantified Self: On the Art of Learning from Sensory Data. *Springer*, 1, 64-67 (2018).
16. A. Bulling, U. Blanke and B. Schiele. A Tutorial on Human Activity Recognition Using Body-Worn Inertial Sensors. *ACM Computing Surveys* 26(3) (2014).
17. A.M. Khan, Y. K. Lee and S. Y. Lee. Accelerometer's Position Free Human Activity Recogntion Using A Hierarchical Recognition Model. *IEEE HealthCom* (2010).
18. A. Mannini, M. Rosenberger, W. Haskell, A. Sabatini and S. Intille. Activity Recognition in Youth Using a Single Accelerometer Placed at the Waist or Ankle. *Med Sci Sports Exerc*, 49(4) 801-812 (2017).
19. X. Wu, V. Kumar *et al.*, Top 10 Algorithms in Data Mining. *Knowledge and Information Systems*, 14(1) 1-37 (2008).
20. S. J. Preece, J. Y. Goulermas *et al.*, Activity Identification using Body-Mounted Sensors – A Review of Classification Techniques. *Physiological Meas*, 30 (2009).
21. I. Cleland, M. P. Donnelly, C. D. Nugent, J. Hallberg and M. Espinilla. Collection of a Diverse, Naturalistic and Annotated Dataset for Wearable Activity Recognition. PerCom *in press* (2018).
22. S. C. Mukhopadhyay. Wearable Sensors for Human Activity Monitoring: A Review. *IEEE Sensors*, 15(3) 1321-1330 (2015).
23. C. Leys, C. Ley, O. Klein, P. Bernard and L. Licata. Detecting Outliers: Do not use standard deviation around the mean, use absolute deviation around the median. *Journal of Experimental Social Psychology*, 49(4) 764-766 (2013).
24. Aggarwal, C. Outlier Analysis. *Springe Int. Publishing*. 2 ed. 12-15 (2017).
25. Niu. Z, Shi, S, Sun, J, He, X. A Survey of Outlier Detection Methodologies and Their Applications. *Artificial Intelligence and Comp. Intelligence.* 380-387 (2011).

Features selection and improving for trauma outcomes prediction models

Fatima Almaghrabi
Alliance Manchester Business School, University of Manchester
Booth St E, Manchester, M13 9SS, UK

Professor Dong-Ling Xu
Alliance Manchester Business School, University of Manchester
Booth St E, Manchester, M13 9SS, UK

Professor Jian-Bo Yang
Alliance Manchester Business School, University of Manchester
Booth St E, Manchester, M13 9SS, UK

Various demographic and medical factors have been linked with mortality after suffering from traumatic injuries such as age and post-injury disability. A considerable amount of literature has been published on the building of trauma prediction models. However, few analyse the features selection criteria. Patient records comprise a large amount of data and numerous variables, and some are more important than others. Highlighting the most influential variables and their correlations would assist in the better use of it. The intention of this study is to clarify several aspects of demographic and medical factors that could affect the outcome of trauma in order to exhibit the interaction between these factors and to represent their relationships. In addition, the aim is to use ranking and feature weights to select the features that increase accuracy and lead to better results.

Introduction

Trauma is a major public health issue and a major cause of mortality and disability worldwide. In 2010, there were 17,201 injury-related deaths, in England and Wales. For every death following an injury, there are approximately ten people who survive, potentially with serious permanent disabilities. In addition, trauma is the leading cause of mortality in the under-40 age group [2]. Several demographic and medical factors, such as age and post-injury health conditions, have been associated with the death of patients with traumatic injuries. These factors are extremely interconnected, and finding the most accurate model that represents the correlations between these factors and patient outcomes is not easy. Outcome prediction techniques and different models have received considerable attention in trauma literature from the early 1970s.

The ability to predict the outcome of a trauma patient with a high degree of accuracy would have a huge impact when it comes to clinicians transferring trauma patients with severe injuries to trauma centres and offering them urgent care. Moreover, the variables in a model that would predict a negative effect on patient outcomes would guide the medical trauma team in their rapid attempts to serve trauma victims [2].

Many different factors have been linked with a bad outcome for patients after traumatic injuries. However, identifying the most relevant factors would provide a better model as more emphasis will be placed on the most known key features and computing time will be reduced. The model could also be used to assist other medical decision-making tasks.

Firstly, the research will review features selection methods previously used to identify the key important variables. Secondly, feature ranking will be proposed, using an out of bag error after training random forest algorithm and ReliefF Algorithm. The resultant features will be tested on new data.

1. Features Selection Methods

There are four features selection techniques wrappers, filters, embedded, and hybrid. Each has its advantages and disadvantages. The wrappers techniques compare different combinations of the proposed features and return the most accurate one. Therefore, it's computerised consumption. The filter methods rank the features based on their relationship with the outcome. However, they are not model oriented. Finally, the embedded methods select the features by learning the model.

1.1. *ReliefF*

Relief algorithm was introduced by Kira & Rendell [6] to calculate variables' importance based on how well their values distinguish between instances that are near to each other. After that, ReliefF was introduced as an extension from ReliefF that can handle multi class problems and help reduce the noise in the data [5]. The features' weight provided by the algorithm ranges from -1 to 1 with large positive weights assigned to important variables.

Machine learning methods

Machine learning (ML) offers a range of methods to deal with different modelling issues such as nonlinear relations that most medical data encounter.

ML algorithms have been proven to be effective in dealing with these problems. Moreover, overfitting could be avoided, by using techniques such as cross-validation and pruning. Therefore, ML methods have been used in many prediction models in different fields such as remote sensing-based estimations [4], stock price direction [1], and health and medical diagnosis [9].

1.2. Random forest algorithm

The random forest (RF) algorithm uses a subset of the observations through bootstrapping techniques in each random binary tree. The algorithm monitors the error rate for observations left out of the bootstrap for each tree grown which is called an out-of-bag (OOB) error rate [3]. The mean decrease in accuracy is determined during the calculation of the OOB error, which indicates the importance of each predictor in the resulting RF [7]. This would help in the ranking of independent variables and variable selection for improving the model. The importance range starts from zero, which means that that variable is not important and assign positive values for each variable.

2. Data

The data was collected from the TARN database, which is the largest trauma database in Europe and includes the data of trauma patients from all of the hospitals in England and Wales [2]. All trauma patients, irrespective of age, that arrived at any hospital in England or Wales alive from 1 January 2012 to 31 December 2015 suffering from a traumatic injury were included for the analysis which conclude to 186969 patient records.

Participating TARN patients were included in the study dataset if their final outcomes had been recorded. The number of patients in the category of dead patients for the outcome data is much lower compared to the number in the category for patients who were still alive, which would affect the prediction accuracy results, 1. 92.85% of trauma patients' outcome was alive; while, 7.15% was dead.

2.1. Model variables

Choosing the most appropriate variable is important to minimize or prevent complications in building the model. This research involved an analysis of the prediction variables based on variables importance produced from out-of-bag-error and ReliefF algorithm. 55 variables were included such as demographic variables and medical tests. Patients' outcome at 30 days or discharge, whichever comes first, was considered as the outcome.

2.2. Cross validation

The cross-validation procedure is a sample partitioning strategy used with machine learning models. It prevents the over-fitting problem (Hsu, Chang, Lin, 2003) and it improves the prediction according to bias and/or variance [8].

2.3. Data analysis

Using MATLAB 2017b, data was divided into training and testing sets, with the training set containing 60% of patient data and being used to calculate the variables importance. The rest of the data was used to test accuracy of the prediction based on the most important 10,20,30,55 predictors using RF. Ten-folds cross validation was used to split the data before applying RF and the mean importance of each variable was considered. AUC and accuracy were used to compare the results. (ten-folds cross validation have been performed twice)

3. Results

The three most important variables resulted from the out-of-bag-error are total length of stay at the hospital, total LOS in ICU and Pulse rate measured at the ED. Figure 1 shows the importance of each variable. After testing different no. of variables based on the outcome of the training data. The result shows that the highest accuracy can be reached after using the most 20 important variables which is equal to 96.55%. When 30 variables included, the prediction accuracy decreased slightly and continue to decrease when more variables were added. The area under the curve on the other hand is 0.88 when 10 variables were included then it increased to 0.94 after adding 10 variables more. The AUC didn't change significantly after adding more than 20 variables.

Figure 2 shows that the weight of each variables after applying ReliefF algorithm. Glasgow coma scale(GCS), total LOS in ICU and Time to CT (hours). The accuracy of using all the 55 variables is 96.43% which is slightly higher than applying less chosen variables based on the ReliefF algorithm. The accuracy of applying the most important 20 variables and 30 variables is almost the same with a small difference toward the 20 variables which is 96.27%. The results of the AUC are almost the same as the accuracy with the highest AUC achieved by including all the variables.

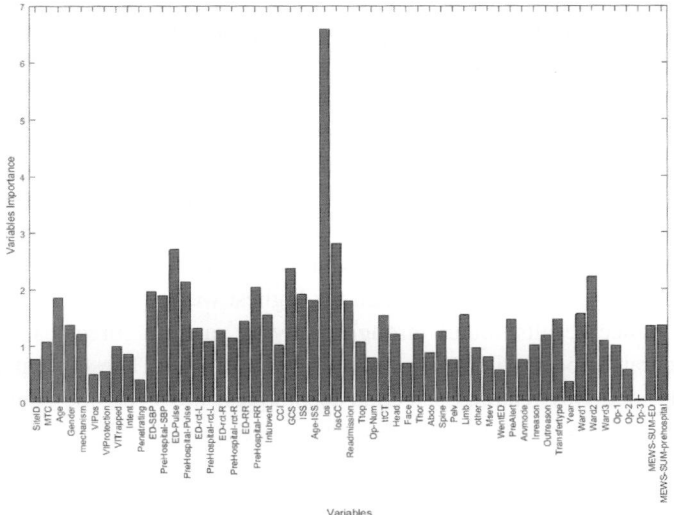

Figure 1. The variable importance (VI) measurements provided by the RF classifier.

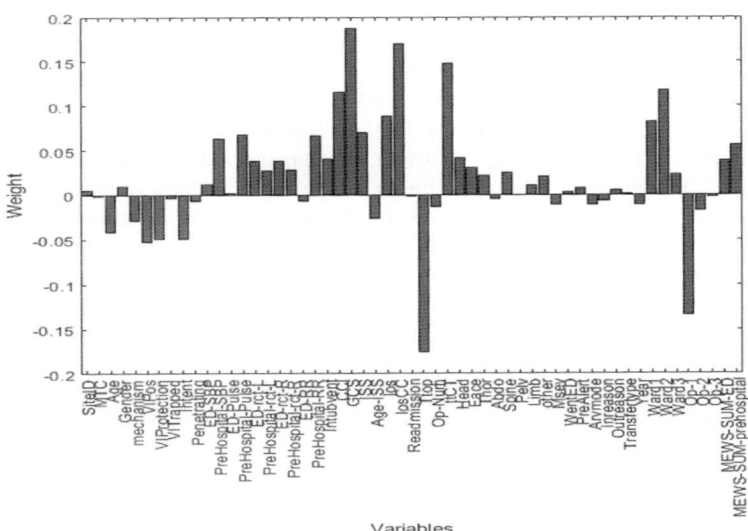

Figure 2. The variable weight provided by the ReliefF algorithm.

4. Conclusion

The present research was designed to investigate different trauma outcome predictors, estimate the importance for each variable and propose the most accurate combination of features. Random forest was implemented to calculate variables importance and the most important first 10, 20 and 30 were compared to the whole set of data on new dataset. The results showed that choosing 20 variables achieves the highest accuracy and a good AUC. Another features selection approach has been implemented which is the ReliefF algorithm. The findings were suggesting that there is no huge difference in accuracy after choosing 10, 20 and 30 variables with the highest weight. We can conclude that the importance results found through applying random forest can determine more accurate model with less yet important variables. The current study has only examined two features selection techniques; however, the results can be compared to other features selection techniques. Moreover, experts' subjective opinion can be added to create quantitative and qualitative judgement.

References

1. Ballings, M., Van Den Poel, D., Hespeels, N., & Gryp, R. (2015). Evaluating multiple classifiers for stock price direction prediction. Expert Systems with Applications, 42(20), 7046-7056.
2. Bouamra, O., Jacques, R., Edwards, A., Yates, D. W., Lawrence, T., Jenks, T., & Lecky, F. (2015). Prediction modelling for trauma using comorbidity and 'true'30-day outcome. Emerg Med J, emermed-2015.
3. Breiman, L. (2001). Random forests. *Machine Learning*, 45(1), 5–32.
4. Fassnacht, F. E., F. H., Latifi, H., Berger, C., Hernández, J., Corvalán, P., & Koch, B. (2014). Importance of sample size, data type and prediction method for remote sensing-based estimations of aboveground forest biomass. Remote Sensing of Environment, 154, 102-114.
5. Hall, M. (1999). Correlation-based Feature Selection for Machine Learning. *Methodology*, 21i195-i20(April), 1–5.
6. Kira, K., & Rendell, L. A. (1992). *A practical approach to feature selection*. Proceedings of the ninth international workshop on Machine learning.
7. Pourghasemi, H. R., & Kerle, N. (2016). Random forests and evidential belief function-based landslide susceptibility assessment in Western Mazandaran Province, Iran. Environmental Earth Sciences, 75(3), 1–17.
8. Santafe, G., Inza, I., & Lozano, J. A. (2015). Dealing with the evaluation of supervised classification algorithms. Artificial Intelligence Review, 44(4), 467–508. http://doi.org/10.1007/s10462-015-9433-y
9. Stylianou, N., Akbarov, A., Kontopantelis, E., Buchan, I., & Dunn, K. W. (2015). Mortality risk prediction in burn injury: Comparison of logistic regression with machine learning approaches. Burns, 41(5), 925-934.

Computer aided diagnostic tool for prostate cancer with rule extraction from Support Vector Machines

Guanjin Wang[*1], Jie Lu[2], Jeremy Yuen-Chun Teoh[3] and Kup-Sze Choi[1]

[1] *Centre for Smart Health, School of Nursing,*
The Hong Kong Polytechnic University
Hong Kong SAR, China
* guanjin.br.wang@connect.polyu.hk, thomasks.choi@polyu.edu.hk

[2] *Centre for Artificial Intelligence, School of Software,*
University of Technology Sydney
Broadway, NSW, 2007, Australia
Jie.Lu@uts.edu.au

[3] *Division of Urology, Department of Surgery,*
The Chinese University of Hong Kong
Hong Kong SAR, China
jeremyteoh@surgery.cuhk.edu.hk

Prostate cancer is a common malignancy among men, necessitating accurate and timely diagnosis at an early stage. With the advent of Artificial Intelligence (AI) technologies in the health field, support vector machines (SVMs) as one of the most well-known machine learning methods have been widely applied for prostate cancer detection. They have good generalization performances but no interpretability on the learned patterns, which bring difficulties for health professionals to understand the inner working of the predictive model. In this paper, we aim to build a computer aided diagnostic tool for prostate cancer using the SVMs where rule extraction is enabled. Experimental results on a real-world prostate cancer dataset collected in a Hong Kong hospital show that the proposed model not only had the ability for rule generation but also achieved better prediction results compared with decision tree, exhibiting a potential to assist physicians with clinical decision support in future.

Keywords: Diagnosis of prostate cancer; clinical decision support; rule extraction; support vector machines; decision tree.

1. Introduction

Prostate cancer is one of the most common malignancies in males. In 2017 in the United States, there were an estimated 161,360 new prostate cancer cases (nearly 10% of all the new cancer cases), and 26,730 deaths.[1]

Definitive diagnosis of prostate cancer is biopsy operation which needs to remove a small piece of prostate gland followed by examination under the microscope. Such invasive procedure may bring side effects and risks to the patients. Moreover, many diagnosed prostate cancer cases are insignificant which in fact do not affect patients' long-term survival. Therefore, instead of biopsy, a safer and faster diagnostic tool using the pre-biopsy information is in demand for the effective prostate cancer detection. In the past years, several risk calculators have been built by exploiting the available preliminary investigations like prostate-specific antigen (PSA) test and digital rectal examination (DRE) on the Caucasian population to serve the purpose, including European Randomized Study of Screening for Prostate Cancer (ERSPC) risk calculator,[2] the Prostate Cancer Prevention Trial (PCPT) risk calculator[3] and the Sunnybrook risk calculator.[4] Asians are genetically and physiologically different from the Caucasians, and thus require a new model to predict the outcome of prostate biopsy.

Benefiting from advances in Artificial Intelligence (AI), the current new trend in the health field is to apply machine learning techniques to construct predictive models where the patient data are intelligently analyzed to discover the hidden patterns associated with the outcome of the interest. Artificial Neural Networks (ANNs) and Support Vector Machines (SVMs) are the two most extensively used techniques for prostate cancer prediction.[5] However, they are typical black-box models which lack transparency and interpretability of learned rules or models. This may become a big barrier for health professionals to understand the emerging technology and apply it in the clinical practice. Therefore, the computer-aided diagnostic tool should be able to explain how it reaches a predicted outcome for end-users' references.

In this paper, we present a new diagnostic tool with rule extraction for prostate cancer detection. In our proposed method, well-trained SVMs are first constructed to obtain support vectors (SVs). Then the original labels of these SVs are replaced by the new labels predicted by the trained SVMs model. After that, the re-constructed SVs are put into decision tree for rule extraction. The experimental results show that the proposed tool can provide satisfactory prediction results, and moreover help health professionals to assess individual patient based on interpretable rules.

This paper is structured as follows. Section 2 reviews three main types of methods for rule extraction from SVMs. Section 3 presents the proposed method. Section 4 describes the real-world prostate cancer dataset and gives experimental analysis. Section 5 concludes the work.

2. Rule Extraction from SVMs

SVMs have been successfully applied in various health care applications to improve diagnostic decisions.[6-10] However, there is a significant downside of SVMs is that it cannot give transparent rules of the trained model. That is, they are typical black-box models. In cancer diagnosis, it is especially useful to supply explanatory information to health professionals to understand constructed models. Therefore, different techniques have been created to enable rule extraction from SVMs. The overall idea of them is to reveal the learned knowledge either from the black-box model or embedded in the structure of the model, such as SVs.[6] In general, techniques from literature can fall into three categories:[11] (i) the pedagogical approach, where the SVMs are remained as a black box for model construction. The resulting patterns from SVMs are used to fed into the decision tree for rules generation. (ii) the decomositional approach, where rules are extracted from SVs and the decision hyperplane of the trained SVMs. (iii) the eclectic approach integrates the traits of the first two approaches, where rules are extracted only from SVs or artificial data based on SVs. In this work, the presented model belongs to the last category. Chaves et al.[12] proposed a fuzzy rule extraction method from SVMs where the rule's antecedents are associated with fuzzy sets. Barakat et al.[13] proposed a novel algorithm called SQRex-SVMs to extract rules directly from the SVs using a modified sequential covering algorithm. Rules are generated according to a sequenced search of the most discriminative features. Wang et al.[14] proposed a hybrid rule extraction method for the detection of epilepsy using EEG signals. SVs are produced to generate rules using tree based learning methods.

3. Proposed Diagnostic Tool with Rule Extraction from SVMs

Fig. 1 demonstrates the learning process of the proposed diagnostic tool for prostate cancer detection. The whole process can be divided into three steps. First, the training data are used to build SVMs model to achieve an acceptable performance by tuning the parameters. The trained model is regarded to be ideal to predict the output. Second, the corresponding SVs of the trained SVMs are determined. The original labels of SVs are replaced by the new labels predicted by inputing the SVs in the constructed model. The motivation of generating artificial SVs is to ensure the future extracted rules will mimic the predictions of SVMs as much as possible. This is based on an assumption that the constructed SVMs are perfect

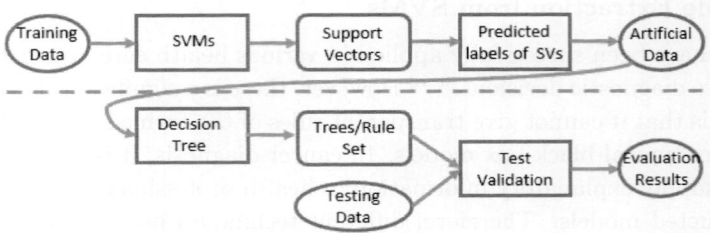

Fig. 1. Learning process of the rule extraction approach from SVMs.

and have the capability to better represent the underlying patterns in the dataset than the artificial data set. After the substitution of SV labels, some noises caused by the class overlap in the dataset can be removed.[15] In the last step, the artificial data are fed into the decision tree to generate rules which are evaluated using the testing data.

3.1. SVMs

Support Vector Machines (SVMs) were proposed by Cortes and Vapnik.[16] It aims to project the original data to a higher dimensional feature space where an optimal hyperplane can be found between different classes with the maximum margin. Kernel trick is used to perform in the high dimensional feature spaces without explicit feature mapping and computation. Instead, we only need to compute the inner products over pairs of data points in the original feature space. As demonstrated in Fig. 2, the SVM classifier builds a maximum margin optimal hyperplane to separate the data points from two classes. The data points identifying the hyperplanes are called SVs. The quadratic programming optimization problem of SVMs is represented as

$$\begin{aligned}
\text{minimize} \quad & \frac{1}{2}\vec{w}^2 + C\sum_{i=1}^{N}\xi_i \\
\text{s.t.} \quad & y_i(\vec{w}^T\vec{x}_i + b) \geq 1 - \xi_i \\
& \xi_i \geq 0, i = 1, 2, \cdots, N
\end{aligned} \quad (1)$$

where ξ_i ($i = 1, 2, \cdots, N$) is the slack variable to allow some errors. C is the regularization parameter to control the trade-off between a low training error and a low testing error. To solve the problem in Eq. (1), the primal problem is transformed into the dual problem in Eq. (2) by introducing the dual Lagrange multiplier $\vec{\alpha}$ as below:

$$\text{maximize} \quad \sum_{i=1}^{N} \alpha_i - \frac{1}{2}\sum_{i=1}^{N}\sum_{j=1}^{N} \alpha_i\alpha_j y_i y_j \varphi(\vec{x}_i)^T \varphi(\vec{x}_j) \qquad (2)$$

$$\text{s.t.} \quad \sum_{i=1}^{N} \alpha_i y_i = 0, 0 \leq \alpha_i \leq C, i = 1, 2, \cdots, N$$

For solving $\vec{\alpha}$ using gradient decent algorithm, $\vec{w}$ and b can be correspondingly calculated. The SVs are determined by the training samples with nonzero Lagrange multiplier. The decision function is completely defined by the SVs and thus can be represented as:

$$f(x) = \sum_{i=1}^{N} \alpha_i y_i \varphi(\vec{x}_i)^T \varphi(\vec{x}_j) + b \qquad (3)$$

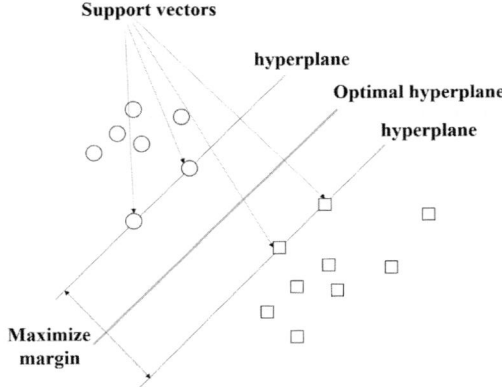

Fig. 2. An example of the SVMs optimal hyperplane to separate two classes (circles and squares) with maximum margin.

3.2. Decision tree

Decision tree [17] is a traditional non-parametric supervised learning method for classification and pregression. it is the tree-like graph or model to learn the hidden rules by testing on the input variables. Each branch of the tree denotes a test on a variable and each leaf node denotes an output label. The links from the root to leafs represent the decision rules. Therefore, decision tree is regarded as the white-box model which can be easily interpret and understood. In contract, kernel-based methods such as SVMs are more difficult to explain. This motivates us to inherit the interpretability from decision tree to SVMs, such that the end-users without technical background can have comprehensible and presentable solutions to support good decision making on the cancer diagnosis.

4. Experiments

4.1. *Dataset*

In the experiment, a real-world prostate cancer dataset retrieved from a transrectal ultrasound (TRUS) guided prostate biopsy database in a hospital in Hong Kong was adopted. During data preprocessing, the patient records with missing values were removed. Totally, there are 1306 Chinese patient records remained in the dataset. The prostate biopsy outcome indicates whether the participant is healthy or with prostate cancer. In general, 1110 (84.99%) subjects are healthy and 196 (15.01%) have prostate cancer. In addition to PSA level, patient information such as age, prostate volumes, DRE and TRUS findings are recorded. Our goal is to build the proposed diagnostic tool to detect prostate cancer using the pre-biopsy information such that unnecessary invasive biopsies can be avoided.

4.2. *Results and analysis*

In the experiment, the proposed diagnostic model is compared with SVMs, decision tree and back-propagation neural network (BPNNs) on the adopted prostate cancer dataset in terms of precision, recall and F-score. 10-fold cross validation is used to produce averaged estimations of different performance metrics. The classification results on the testing dataset were shown in Table 1. It can be seen that SVMs achieved better performances than decision tree and BPNNs regarding three evaluation metrics, proving that our motivation of rule extraction from SVMs is reasonable. The proposed model achieved a precision, a recall and a F-score of 0.9190, 0.9050 and 0.9022, respectively, which outperformed decision tree, but is comparatively less powerful than the SVMs and BPNNs in terms of precision and F-score. However, the proposed model has the superior advantage on the rule extraction from SVMs such that physicians can make better decision making guided by the interpretable patterns. The rule extracted such as: (1) If PSA $>=$ 37.85 and TRUS volume $>$ 109.5, then cancer.

(2) If PSA > 11.05 and PSA < 37.85 and TRUS volume < 36.25 and age >= 71.5, then cancer. (3) If PSA >= 37.85 and DRE findings = 'normal', then non-cancer. Conversely, the SVMs and BPNNs cannot reveal the inner workings which may reduce the end-users' confidence over the model and become very difficult to promote in the clinical practice. As the result, we conclude that the proposed model can obtain an acceptable learning capability and at the same time inherit the interpretability from decision tree after inputing artificial SVs. On the other hand, the decision rules generated from the proposed model are produced by SVs, therefore they are less complicated and time-consuming compared with those using the whole dataset.

Table 1. Performance results for the testing set for 10 runs.

	Precision	Recall	F-score
Proposed model	0.9190±0.0019	0.9050±0.0062	0.9022±0.0032
Decision Tree	0.8930±0.0094	0.8790±0.0087	0.8730±0.0043
SVMs	0.9917±0.0085	0.8764±0.0172	0.9304±0.0089
BPNNs	0.9802±0.0162	0.8721±0.0146	0.9229±0.0112

5. Conclusions

In this work, we proposed a computer-aided diagnostic tool for prostate cancer. The predictive model is built by SVMs, where rule extraction from the constructed SVMs is enabled to provide second opinion for diagnosis. More specific, first, SVMs are used for constructing the predictive model, where SVs are determined. Then the labels of SVs are replaced by the new labels predicted after inputing SVs into the constructed model. The new artificial SVs are fed into decision tree to generate rules from SVMs. At last, the extracted rules are evaluated on the testing dataset. The rule generation is particularly useful in cancer research which can be regarded as the supplementary information for non-technical end-users when screening undiagnosed individuals. A prostate cancer dataset collected from a local hospital in Hong Hong is employed in the experiment. The results show that the proposed model achieved good diagnostic performances with interpretability, exhibiting a potential to assist clinical decisions in the health field.

In future, we plan to add more patient features for model construction and investigate on the feature importance from the extracted rules to identify how sensitive these risk factors are in screening prostate cancer for individuals. Moreover, the diagnosed prostate cancer cases can be further subdivided into 'insignificant' and 'significant'. It is a big challenge to improve the generalization performance of the proposed model for multi-class classification in cancer diagnosis.

Acknowledgments

The work was supported by the Research Grants Council of the Hong Kong SAR (PolyU 152040/16E) and G. Wang is supported by the YC Yu Scholarship for the Centre for Smart Health.

References

1. Cancer stat facts: Prostate cancer https://seer.cancer.gov/statfacts/html/prost.html, note = Accessed: 2018-02-01.
2. R. Kranse, M. Roobol and F. H. Schröder, *The Prostate* **68**, 1674 (2008).
3. D. P. Ankerst, J. Hoefler, S. Bock, P. J. Goodman, A. Vickers, J. Hernandez, L. J. Sokoll, M. G. Sanda, J. T. Wei, R. J. Leach et al., *Urology* **83**, 1362 (2014).
4. R. K. Nam, A. Toi, L. H. Klotz, J. Trachtenberg, M. A. Jewett, S. Appu, D. A. Loblaw, L. Sugar, S. A. Narod and M. W. Kattan, *Journal of Clinical Oncology* **25**, 3582 (2007).
5. J. A. Cruz and D. S. Wishart, *Cancer Informatics* **2**, p. 117693510600200030 (2006).
6. N. Barakat, A. P. Bradley and M. N. H. Barakat, *IEEE transactions on Information Technology in Biomedicine* **14**, 1114 (2010).
7. W. Yu, T. Liu, R. Valdez, M. Gwinn and M. J. Khoury, *BMC Medical Informatics and Decision Making* **10**, p. 16 (2010).
8. A. T. Azar and S. A. El-Said, *Neural Computing and Applications* **24**, 1163 (2014).
9. R. J. Klement, M. Allgäuer, S. Appold, K. Dieckmann, I. Ernst, U. Ganswindt, R. Holy, U. Nestle, M. Nevinny-Stickel, S. Semrau et al., *International Journal of Radiation Oncology, Biology and Physics* **88**, 732 (2014).
10. Y. Sakumura, Y. Koyama, H. Tokutake, T. Hida, K. Sato, T. Itoh, T. Akamatsu and W. Shin, *Sensors* **17**, p. 287 (2017).
11. X. Fu, C. Ong, S. Keerthi, G. G. Hung and L. Goh, Extracting the knowledge embedded in support vector machines, in *Proceedings of 2004 IEEE International Joint Conference on Neural Networks*, 2004.
12. A. C. Chaves, M. M. Vellasco and R. Tanscheit, Fuzzy rule extraction from support vector machines, in *Fifth International Conference on Hybrid Intelligent Systems (HIS)*, 2005.
13. N. H. Barakat and A. P. Bradley, *IEEE Transactions on Knowledge and Data Engineering* **19**, 729 (2007).
14. G. Wang, Z. Deng and K.-S. Choi, *Neurocomputing* **228**, 283 (2017).
15. D. Martens, B. Baesens and T. Van Gestel, *IEEE Transactions on Knowledge and Data Engineering* **21**, 178 (2009).
16. C. Cortes and V. Vapnik, *Machine Learning* **20**, 273 (1995).
17. J. R. Quinlan, *Machine Learning* **1**, 81 (1986).

Map-based medical practice behavior analysis: Methodology and a case study on Australia's medical practices[*]

Yi Zhang, Wei Wang, Junyu Xuan, Jie Lu, Guangquan Zhang

Centre for Artificial Intelligence, Faculty of Engineering and Information Technology, University of Technology Sydney, Australia

Hua Lin

MedicalDirector Inc., Sydney, Australia

> This paper constructs an integrated system for analyzing the behaviors of Australia's medical practices, in which 1) weblog mining techniques are developed to retrieve and identify the communities of medical practices and their behavior patterns; 2) concentrating on specific diseases, text analytics are used to profile the research hotspots of specific diseases from external medical databases; and 3) visualization techniques are then involved in presenting identified intelligence in an interpretable way and detecting the potential trends of specific diseases in Australian regions. This study provides significance to support decision-making in healthcare-related government and industry sectors.

1. Introduction

Indicated by Mckinsey, the vastly increased information is leading a big data revolution in healthcare sectors [1]. The rapid development of information technologies (ITs) further enhances the ability of collecting and analyzing healthcare data for decision support. However, the way to incorporate information systems and data analytics to analyze the behaviors of medical practices to effectively identify medical intelligence and interpretably cohere with real-world business value is still elusive.

This paper, with the aim of addressing the above concern, concentrates on weblog information retrieved from an Australian medical information system X and constructs a system for Australian medical practice-oriented behavior analysis. A series of data analytic techniques are involved: 1) we exploit weblog mining techniques to retrieve meaningful features and statistical models are used to identify the communities of medical practices and their behavior patterns;

---

[*]This work is partially supported by the Australian Research Council under Discovery Grant DP150101645.

2) oriented to some specific diseases, we collect supplementary information from external medical databases and engage text analytics to profile the research hotspots of given diseases. In parallel, focusing on behavior patterns and geographical distribution, the potential trend of certain diseases (usually acute diseases) can be detected; 3) visualization techniques (e.g., science mapping) are involved to present and help interpret identified intelligence to stakeholders.

The innovation and significance of this paper include: 1) a system effectively integrating weblog mining, text analytics, and visualization techniques is designed and developed. 2) The system adapts to diverse healthcare data sources and holds the ability in investigating a wide range of chronic and acute diseases, which provides solutions to real-world healthcare issues.

The structure of this paper is organized as follows: Section 2 reviews previous studies. Section 3 presents the research framework of the map-based medical practice behavior analysis. Section 4 follows, in which a case study on analyzing and visualizing the behaviors of Australia's medical practices is given. Finally, we conclude limitations and future study in Section 5.

2. Related Work

Weblog mining can be a specific task of web mining, which is to discover the patterns of user behaviors by analyzing weblog entries [2]. A large number of new techniques (e.g., association rules and fuzzy sets) have been involved [3], and its applications have been widely extended from website evaluation to behavior analysis and pattern recognition [4].

The increasing amounts of text data have created a need for advanced analytic algorithms, and text analytics are considered to be such a tool. Related techniques such as topic models have been well-developed and discussed [5]. As a part of text analytics, in bibliometrics term-based analysis has become a crucial sub-area and introduces new angles for topic analysis [6, 7].

3. Methodology

Oriented to the weblog information retrieved from an Australian medical information system X, this paper constructs an integrated system, including a weblog analytic function, a bibliometrics-based topic analytic function, and a visualization function. Its research framework is given in Fig. 1.

3.1. Weblog Analytics

The medical information system X provides mechanisms to record user activities in its servers and the records contain sufficient information for filtering and

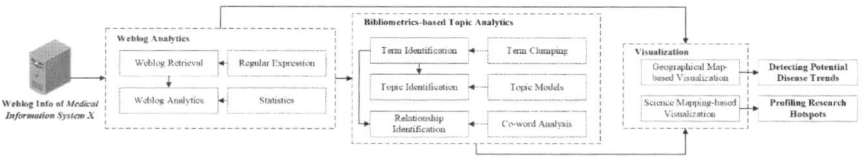

Fig. 1. Research framework of map-based medical practice behavior analysis.

analyzing the behavior of medical practices in Australia. By doing that, it is feasible to use regular expression to obtain information on certain important features for behavior analysis, such as user profiles, target domains (e.g., databases, journals, and articles), and time stamps.

Aiming to further reorganize and explore underlying information behind the data, statistics, especially classification techniques, are involved. Based on the geographical information of user features, we organize all features by the geographic distribution. For example, the use of search words is classified according to the Australia's states and cities, and a user-term matrix is then constructed, which can be decomposed to sub-state matrices and sub-city matrices. Such classification enables us to focus on the behavior of medical practices in specific regions (e.g., one or more states, and one or more cities). Further analysis can be conducted from two aspects:

- Topic identification – we use the same topic model that will be introduced in the bibliometrics-based topic analytic function to identify topics;
- Serial analysis – the serial evolution of topics is then tracked to determine whether emergent topics appear or old topics are fading away. Under this circumstance, the changes of diseases (especially acute diseases) and their possible treatments can be detected.

3.2. Bibliometrics-based Topic Analytics

Concentrated on the search words of specific chronic diseases, bibliometrics-based topic analytics are applied to explore supplementary intelligence from external academic databases. The analysis can be conducted by the following steps:

- Term identification: a search strategy is proposed for collecting relevant scientific articles from external databases. The title and abstract of each article will be combined. The entire corpus will be used for topic models-based topic identification, and there is no need for term identification at this stage. We apply natural language processing techniques to retrieve terms, including both single-word terms and multiple-word terms. A term clumping model [8] is used to remove noise and consolidate synonyms. A list of core technological terms will be generated for co-word-based relationship identification.

- Topic identification: we introduce a latent Dirichlet allocation (LDA) model [9] to identify topics, in which a topic is represented by a set of terms and a number of topics can be used to profile the hotspots of a given research area.
- Relationship identification: a latent semantic analysis (LSA)-based co-word analysis [10] is used for identifying relationships between core terms and also between the topics represented by these terms. Such relationships can shed light on understanding the interactions among symptoms, causes, and possible treatments, and can hold great interest to related medical practices.

3.3. *Visualization*

A series of maps are involved in visualization. Specifically, geographical maps will be used to present the communities of medical practices and their behavior patterns, e.g., indicating the potential trends of diseases in a given region or area. Science mapping techniques will illustrate the research hotspots of selected diseases and their interactions.

4. Case Study

4.1. *Data*

The cast study concentrates on the weblog information of the medical information system X, and we collected a sample with more than 15 million records (all private information has been removed) covering the period from August 1, 2016 to August 31, 2016.

As indicated in Australia's Health 2016[†], an annual report published by the Australian Institute of Health and Welfare, diabetes is one of the top chronic diseases in Australia and more than 1.2 million Australian are suffering from diabetes. Under this circumstance, we selected diabetes as our target disease in the topic analytics, and 30,628 diabetes-related scientific articles published in 2016 was collected from the Medline database integrated in the Web of Science[‡].

4.2. *Detecting Potential Diseases Trends in Australia*

The weblog analytic function was first applied. A number of behavior features are collected, e.g., search words, user profile (e.g., the location of users), and time stamps, and are classified based on their locations. After the topic identification, we selected a topic concentrated on "flu" and generated a geographical map to illustrate the dynamics of this topic in the entire Australia, given in Fig. 2.

[†]http://www.aihw.gov.au/WorkArea/DownloadAsset.aspx?id=60129555788
[‡]https://www.webofknowledge.com/

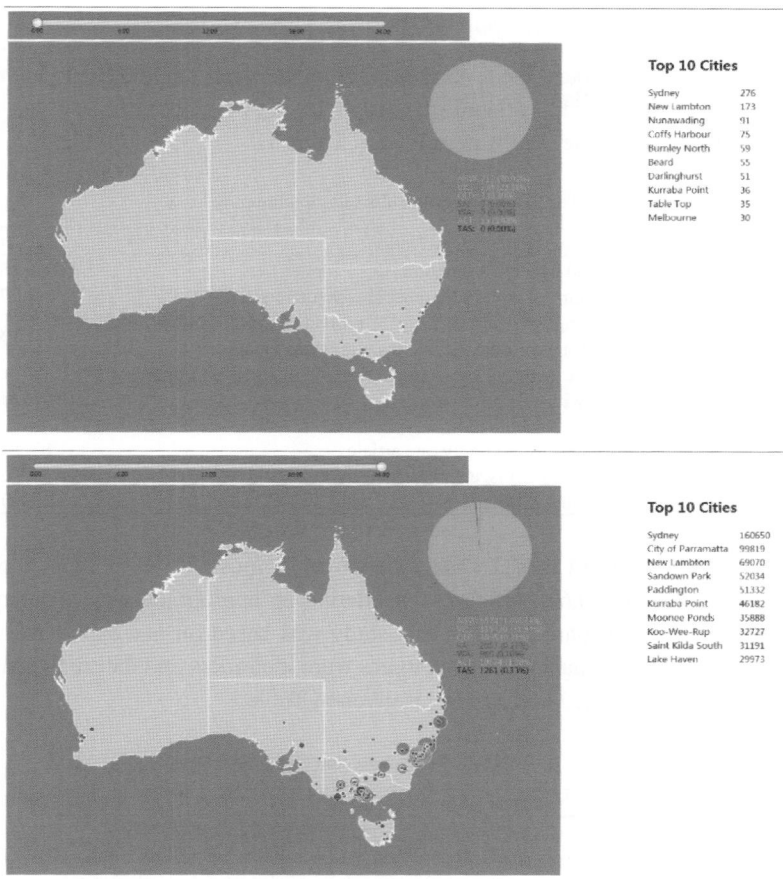

Fig. 2. (i) The geographical map on flu-related search words in Australia (on August 1, 2016) and (ii) the geographical map on flu-related search words in Australia (on August 31, 2016).

When a brief summary is given to reveal the general information of different Australian states, the search frequency of the topic is identified on a city level. The size of the circle represents the frequency, and red color indicates a rapid increase of the search frequency while the default color is blue. A threshold is used to trigger the color, which can be considered as a warning mechanism. As shown in the two figures, in August 2016 people suffering flu concentrated in the cities of the New South Wales (NSW) State, in particular Sydney and Parramatta. Such warnings also can be noticed in Melbourne (the Victoria State) and some other cities in NSW. Definitely, medical practices, local healthcare bodies, and domestic governments in related states and even the federal government can then manage their resources to handle the issue.

4.3. Profiling Research Hotspots of Diabetes

Concentrating on diabetes, the term clumping model was first applied to identify core technological terms related to diabetes, e.g., symptoms, causes, and treatments. The stepwise results are given in Table 1.

Table 1. Stepwise results of term clumping.

Step	Description	No. Term
0	Raw terms retrieved by a NLP function.	650,161
1	Removing terms starting with non-alphabetic characters, e.g., "1.5%."	541,772
2	Removing meaningless terms, e.g., pronouns, prepositions, and conjunctions and common scientific terms, e.g., "introduction."	525,185
3	Consolidating terms with the same stem (e.g., the singular and plural of a noun) and terms with the same words but different orders, e.g., "type 2 diabete milltus" and "diabete milltus type 2."	475208
4	Removing terms appearing in only one article	83,964
5	Removing single-word terms, e.g., "patient."	67,223
6	Top 1000 high-frequency terms	1000

In parallel, the LDA model was applied to the corpus with the combined titles and abstracts of the 30,628 articles. Considering an appropriate way to interpret results, the setting of parameters in this study was based on our experience and expert knowledge. 13 diabetes-related topics were identified, and their information including the word-based descriptions is given in Table 2.

Table 2. Diabetes-related topics.

No.	Topic	Description
1	gene	variants, polymorphism, DNA, genotype, SNPs
2	healthcare	management, data, intervention, medical, education,
3	insulin	treatment, therapy, glucose, glycemic, hypoglycemia
4	depression	exercise, physical, sleep, cognitive, activity
5	liver	disease, infection, NAFLD, transplantation, syndrome
6	kidney	renal, disease, CKD, pressure, hypertension
7	women	GDM, pregancy, gestational, maternal, birth
8	retinpathy	foot, retinal, neuropathy, wound, nerve
9	heart	coronary, stroke, artery, cardiovascular, myocardial
10	risk	factors, age, prevalence, population, mortality
11	dietary	intake, acid, diet, consumption, food
12	cells	panceatic, islet, beta, pancreas, immune
13	expression	protein, endothelial, activation, stress, vascular

Some impressive findings are observed from Table 2: 1) the relationships between diabetes and factors have been investigated in previous studies; 2) the symptoms of diabetes can be detected from other organs. Diseases in these organs might be the causes of diabetes; and 3) demographical statistics on diabetes are also one interesting topic for related research.

Considering the appealing of a term correlation map, we selected the top 1000 high-frequency terms in Step 6 of Table 1 for further co-word analysis. With the help of a science mapping platform VOSViewer, the term correlation map on the research frontier of diabetes was generated, shown in Fig. 3.

In this paper, detailed clusters and descriptive terms are given to delve into diabetes-related research frontiers, and some interesting observations include: 1) the details of some experiments can be addressed (blue nodes); 2) Indicators, symptoms, and related diseases are given (yellow and green nodes); 3) Potential causes (in particular daily habits) that relate to diabetes are grouped (red nodes),; and 4) certain high-relevant chronic diseases can be found surrounding terms "diabetes mellitus", "diabetes type 2" and "diabetes type 1". Science maps visualization provides a good solution to profile the research hotspots of specific diseases with diverse granularities (e.g., clusters and descriptive labels).

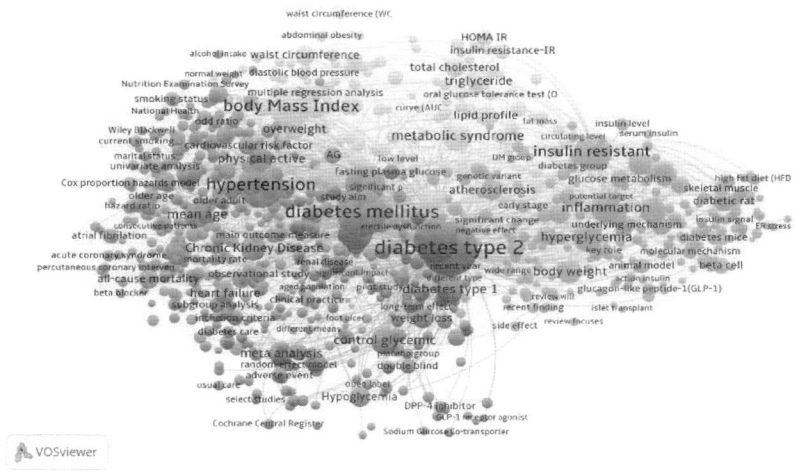

Fig. 3. Term correlation map on the research frontiers of diabetes.

5. Discussion and Conclusions

This paper constructs an integrated system that provides a solution to analyze and visualize the behaviors of Australia's medical practices. The weblog information of the medical information system X was utilized and analyzed, by which we developed approaches to retrieve and identify the behavior patterns of Australia's medical practices, and geographical map-based visualization would detect potential disease trends in given regions or areas in an interpretable way.

Empirical insights on the detection of the potential trend of flu in Australia in August 2016 and the research hotspots of diabetes are given via the case study, and the results also demonstrate the feasibility of our methodology.

Several future directions of our research can be pursued. 1) Integrating multiple data sources (e.g., patient information, and medical prescription) will greatly benefit the progress of precision medicine. 2) Engaging multiple indicators to delve into the hidden mechanism behind the behavior of medical practices can create new angles to support decision-making in related government and industry.

References

1. B. Kayyali, D. Knott, and S. Van Kuiken, "The big-data revolution in US health care: Accelerating value and innovation," Mckinsey & Company 2013.
2. L. Zhou, Y. Liu, J. Wang, and Y. Shi, "Utility-based web path traversal pattern mining," in Data Mining Workshops, 2007. Seventh IEEE International Conference on, 2007, pp. 373-380.
3. S. G. Matthews, M. A. Gongora, A. A. Hopgood, and S. Ahmadi, "Web usage mining with evolutionary extraction of temporal fuzzy association rules," Knowledge-Based Systems, vol. 54, pp. 66-72, 2013.
4. M. Spiliopoulou, "Web usage mining for web site evaluation," Communications of the ACM, vol. 43, pp. 127-134, 2000.
5. J. Xuan, J. Lu, G. Zhang, R. Y. Da Xu, and X. Luo, "Doubly nonparametric sparse nonnegative matrix factorization based on dependent indian buffet processes," IEEE Transactions on Neural Networks and Learning Systems, 2017.
6. Y. Zhang, G. Zhang, D. Zhu, and J. Lu, "Science evolutionary pathways: Identifying and visualizing relationships for scientific topics," Journal of the Association for Information Science and Technology, vol. 68, pp. 1925-1939, 2017.
7. Y. Zhang, G. Zhang, H. Chen, A. L. Porter, D. Zhu, and J. Lu, "Topic analysis and forecasting for science, technology and innovation: Methodology and a case study focusing on big data research," Technological Forecasting and Social Change, vol. 105, pp. 179-191, 2016.
8. Y. Zhang, A. L. Porter, Z. Hu, Y. Guo, and N. C. Newman, "Term clumping for technical intelligence: A case study on dye-sensitized solar cells," Technological Forecasting and Social Change, vol. 85, pp. 26-39, 2014.
9. J. Xuan, J. Lu, G. Zhang, and X. Luo, "Topic model for graph mining," IEEE Transactions on Cybernetics, vol. 45, pp. 2792-2803, 2015.
10. N. van Eck, L. Waltman, E. Noyons, and R. Buter, "Automatic term identification for bibliometric mapping," Scientometrics, vol. 82, pp. 581-596, 2010.

Part 11

Decision Model and Intelligent System for Risk and Security Analysis

Part II

Analysis of sentencing reasoning on the traffic accident crime in China[*]

Caijuan Zhang

College of Law, Southwest Minzu University, Chengdu, 610041, P.R. China

Ming Jiang

College of Management, Southwest Minzu University, Chengdu, 610041, P.R. China

> Due to the complexity of the legal sentencing circumstances of the traffic accident crime, the legal provisions of various countries are not the same. China is a country with relatively more traffic accidents in the world. No clear regulation on sentencing in multi-circumstance cassias led to the uncertainty of sentencing in these case. It leads to the lack of equilibrium in the results of specific cases. The if-then rules can take full consideration of the circumstances of the case and adapt to the sentencing of the crime with circumstances. Accordingly, the sentencing model establishment in this paper can be used for the analysis of sentencing reasoning. The application of the reasoning rules of the penalty for traffic accident crime can improve the scientificity of the sentencing and promote the justice of the judiciary.

1. Introduction

It is generally agreed that the basic characteristics of the law are certainty and predictability, which is also the basic requirement of the rule of law society. Traffic accident crime is a common crime, and the circumstances of sentencing are complicated. The criminal law and relevant judicial interpretations show that, the traffic accident crime sentencing plot consisted of the number of deaths, the number of serious injuries, the proportion of responsibility, and loss of property, whether or not to escape, and so on. Due to traffic accident crime sentencing provisions of criminal law for the rough, and specific sentencing rules are not clear, so the judge's discretion is excessive, and the scientific and impartiality of the traffic accident crime is not enough.

In order to solve the problem of sentencing effectively, scholars gradually began to attach importance to and strengthen the quantitative research of

---

[*] This work is partially supported by special fund project of basic scientific research expenses in central universities of Southwest Minzu University (grant no. 2016SZYQN12).

sentencing. By developing a quantifiable sentencing model, we found the inherent relationship between punishment and sentencing circumstances. Bai Jianjun (2016) believed that the result of sentencing is the addition to each sentencing plot. The influential factors put forward by Bao Guowei (2017) include slight injury, light injury, chief criminal, accessory criminal, recidivism, attempted to surrender, surrender oneself, render meritorious service, and so on. Max Weber (1988) firmly opposed this legal formalism, and the judge was not only the operator of the machine. The pure sentencing model provides a certain convenience for the practice of sentencing. But in fact, any case is the result of the discretion of the judge in accordance by law. Influenced by various objective and subjective factors, the sentencing result of a particular case is not absolute and unique, but a certain degree of fuzziness. In other words, the process of the judge's sentencing is actually a process of fuzzy decision. Therefore, the pure mathematical model cannot be the main basis for the sentencing. The scientific sentencing should depend on the new rules of reasoning. The if-then rule is a common reasoning rule, and an integrated system [4]. It is widely used in the fields of artificial intelligence, management, linguistics and other related fields. In this paper, the fuzzy control tool of if-then is used to amend the sentencing of the crime of traffic accident, so as to adapt to the uncertainty of the law.

2. Analysis of the Circumstances of Sentencing of the Traffic Accident Crime

2.1. *Legal sentencing elements for the traffic accident crime*

Traffic accident behavior is a global problem. The criminal laws of major countries such as the United States, Germany and Japan all stipulates the conviction and sentencing of traffic accident crime. Due to different national conditions, the standard of sentencing of traffic accident crime is not completely unified in various countries. China, as a world power, is also the country with the largest car ownership. According to the data released by the traffic administration bureau of the Ministry of public security, by the end of 2017, the number of motor vehicles in China has reached 310 million, and the number of motor vehicle drivers has reached 385 million. Therefore, China attaches great importance to the conviction and sentencing of the crime of traffic accident.

The criminal law in China provide three kinds starting point penalty of traffic accident crime: First, three years of fixed-term imprisonment or criminal detention. The sentencing factors include criminal facts, the deaths, the serious injuries, and property loss; Second, three to seven years of imprisonment. The sentencing factors include hit-and-run or other especially serious circumstances; thirdly, seven to fifteen years of imprisonments. The sentencing factor is death caused by escape.

According to China's criminal law and relevant judicial interpretations, Sentencing factors of traffic accident crime can be divided into two types: the first is constitutive elements of a crime including the number of deaths, number of serious injuries, the degree of liability, property damage size, whether the behavior of escape, whether overload etc. The facts above mainly affect the starting point punishment; The second is the legal sentencing plot, such as surrender, meritorious service, confession, criminal form, and attitude. The plot above mainly determines the determination of the benchmark penalty.

In addition, there are many possible plot combinations in practice. They could be as follows: full responsibility + one person death + surrender, main responsibility + two person injury + escape, main responsibility + two seriously injured + unlicensed driving + serious overload. It is obvious that the legal plot of traffic accident crime is very complicated, and the combination of multiple plots is very common. Even the legal circumstances identical cases, May also be some sentence because of the discretion of the judge.

2.2. *Sentencing steps of traffic accident crime*

After the conviction, the judge should make a sentence according to the plot of the crime. Usually this sentencing involves three steps: the first step is to determine the starting point of the sentencing. The second step is to determine the benchmark penalty, and the third step is to determine the announced penalty.

(1) The starting point of sentencing is determined according to the basic crime fact in the corresponding legal penalty range. The starting point is the basis of the sentencing.

(2) The benchmark penalty is determined by increasing the amount of penalty, according to the accident liability, the number of people seriously injured, the number of death or the amount of property loss and other plots that affect the constitution of a crime.

(3) The announced penalty is proposed after adjusting the benchmark penalty according to the circumstances of the sentencing. Comprehensive consideration of the whole case, the announced penalty shall be declared in accordance with the law.

2.3. Standardized sentencing implementation of the traffic accident crime

The Supreme People's Court issued "the common crime sentencing guidance". Subsequently the high court every province enacted "the common crime sentencing guidance on Implementation Rules", which refined the fifteen kinds of common crime sentencing. It provides a standardized sentencing basis for local court. Although the result of sentencing is not exactly the same, it has a certain similarity in the technical level of rule making. Based on the "detailed rules for the implementation of sentencing guidelines on common crimes" formulated by the Sichuan Provincial Higher People's court, the sentencing situation of traffic accident crime in China is sorted out in Table 1.

Table 1. A list of the sentencing of the crime of traffic accident in China.

Term of sentence	Starting point of sentencing			Benchmark penalty	
	Facts	Responsibility	Result years	Increased plot	Increased sentence
less than 3 years; detention	1 died or 3 seriously injured	primary responsibility	0.5-1.5	1 seriously injured	0.5-1 years
		full responsibility	1-2		
	3 died	coequal responsibility	1-2	1 died	0.5-1 years
	loss 300 thousand	primary responsibility	0.5-1.5	50 thousand	1-3 months
		full responsibility	1-2		
	3 seriously injured and one of the 6 cases	primary responsibility	0.5-1.5	1 seriously injured	0.5-1 years
		full responsibility	1-2	1 seriously injured	0.5-1 years
3 to 7 years	escape	\	3-5	\	\
	2 died or 5 seriously injured	primary responsibility	3-4	1 died	9 months-1 year
				1 seriously injured	3-6 months
		full responsibility	4-5	1 died	1-1.5 years
				1 seriously injured	0.5-1 years
	6 died	coequal responsibility	4-5	1 died	0.5-1 years
	loss 600 thousand	primary responsibility	3-4	50 thousand	1-4 months
		full responsibility	4-5		
7 to 15 years	1 died of escape	\	7-10	1 died	3-5 years

Remarks: If someone escapes the scene of the accident in order to escape legal investigation, the sentence term should be increased 0.5-1 years.

Through the above table, we find that the higher people's Court has refined the specific rules of sentencing for traffic accident crime, and standardized starting point of sentencing and the benchmark penalty respectively. First of all, the starting point is made by the basic crime fact within the range of statutory penalty; Secondly, according to the number of deaths and injuries and property loss of the amount of the plot, the judge can decide the benchmark sentence with the amount of increasing the penalty; Finally, if there are drugs, drunk driving, driving without a license, unlicensed or scrapped cars, serious overloading, bad social impact and other circumstances in the case, the benchmark sentence should be increased of 10%. If a case embraced two spots above, the benchmark sentence should be increased of not more than 50%. If someone has the behavior of protecting the scene and rescuing the injured, his benchmark sentence should be reduced less than 20%.

It can be seen that the rules are more detailed and standardized. The rules have a good guiding role in standardizing the sentencing of the crime of traffic accident. However, as we know, the law has certain uncertainty. The sentencing process of judges is not a single process of thinking, and the result of sentencing is not the only one. Therefore, this paper will use if-then inference rule to explain the sentencing of traffic accident crime, so as to improve the scientific nature of sentencing.

3. Reasoning Method for Sentencing of Traffic Accident Crime

3.1. *The establishment of if-then rules*

Rule-based reasoning is a main form of scientific cognition, and the if-then rule is a main form of representation in rule reasoning. All relevant knowledge about fuzzy inference methods based on fuzzy if-then rule bases will be represented [5]. Generally speaking, the if-then reasoning rules of traffic accident crime sentencing include two steps: the first step is to collect relevant fuzzy information, which is sentencing legal circumstances can impacting the traffic accident crime. The task has been completed in the last part; the second step are the establishment and application of if-then rules. Because of the complexity of sentencing plots, sentencing is a process of dealing with complex information. This process is not an one-way thinking process, but a parallel process of dealing with complex circumstances. The if-then rule is the simplest control statement, which is often used to make decisions and to change the control processes. The if-then rule can be expressed as follows: if the conditions are satisfied, then the results will be executed. Assuming that X is a sentencing plot, A is specific information, Y is the result of sentencing, and B is a term of sentence. The basic expression of this rule is:

if $X(A) = A1$ and $X(A) = A2$ and $X(A) = A3$ ... and $X(N) = An$, then $Y = B$

In addition, if-then-else is the development of the if-then rules, which is more suitable for the sentencing of the multiple plots crime. The sentencing process of traffic accident crime is multilevel and multi-angle. No matter how complex, it can be described in if-then-else rule.

3.2. Application of if-then rules

The If-then rule is easy to understand because it is expressed by a deterministic factor [6], which is more consistent with the determinism of the sentencing factor. According to the requirements of the law on sentencing, the if-then rule will be applied to the starting point and the benchmark penalty of traffic accident crime in turn.

3.2.1. *The starting penalty rule of traffic accident crime*

(1) A term of imprisonment of less than 3 years and criminal detention

IF 1 people died or 3 seriously injured, and the main responsibility, THEN: 0.5-1.5 years of imprisonment

IF 1 people died or 3 seriously injured, and all responsibility, THEN: 1-2 years of imprisonment

IF 3 people died, and equal liability, THEN: 1-2 years of imprisonment

IF loss 300 thousand, and main responsibility, THEN: 0.5-1.5 years of imprisonment

IF loss 300 thousand, full responsibility of and, THEN:1-2 years of imprisonment

IF1 people seriously injured and one of the 6 cases, and the primary responsibility, THEN: 0.5-1.5 of imprisonment

IF 1 people seriously injured, and has of the 6 cases, and all responsibility, THEN: 1-2 years of imprisonment

(2) a term of imprisonment for a term of imprisonment of 3-7 years

IF escape, THEN: 3-5 years imprisonment

IF 2 people died, or seriously injured 5 people, and main responsibility, THEN: 3-4 years imprisonment

IF 2 people died, or seriously injured 5 people, and all responsibility, THEN: 4-5 years imprisonment

IF 6 people died, and equal liability, THEN: 4-5 years imprisonment

IF property loss 600 thousand, and main responsibility, THEN: 3-4 years imprisonment

IF property loss of 600 thousand, and full responsibility, THEN: 4-5 years imprisonment

(3) A term of imprisonment of more than 7 years
    IF escaped to 1 people died, THEN: 7-10 years imprisonment

3.2.2. Benchmark penalty rules for the crime of traffic accident

On the basis of sentencing starting point, we should increase the amount of penalty according to the accident liability, the number of people seriously injured, the number of death or the amount of property loss and other factors that affect the constitution of a crime, and determine the benchmark penalty.
(1) A term of imprisonment of less than 3 years and criminal detention
    IF serious injury to 1 people, THEN: $Y = B + (0.5-1.5)$
    IF death increased by 1, THEN: $Y = B + (0.5-1.5)$
    IF property loss increased by 50 thousand, THEN: $Y = B + (1-3 \text{ months})$ a term of imprisonment of 3-7 years
    IF died 2 people, or seriously injured 5 people, and main responsibility, and death 1 people, THEN: $Y = B + (9 \text{ months-1 year})$
    IF died 2 people, or seriously injured 5 people, and major responsibility, and seriously injured 1 people, THEN: $Y = B + (3-6 \text{ months})$
    IF died 6 people, and equal responsibility, and death increased 1 person, THEN: $Y = B + (0.5-1)$
    IF property loss increased by 50 thousand, THEN: $Y = B + (1-4 \text{ months})$
(2) A term of imprisonment of more than 7 years
    IF death increased by 1, THEN: $Y = B + (3-5 \text{ years})$

3.2.3. Determination of the rules of declaration of punishment

On the basis of determining the benchmark penalty, we should examine whether there is a statutory penalty for mitigating, lightening or avoiding punishment. After that, with the discretion of the judge we will decide the declared penalty according to the whole case. The rules are as follows: If the adjustment result of benchmark punishment is in the range of statutory penalty, and suiting punishment, we can directly determine that as declared punishment; If the case has a mitigating circumstances, we shall determine declared punishment within the minimum statutory penalty punishment; If the case has only extenuating circumstances, we shall determine the minimum statutory penalty for punishment to be declared criminal by law; If sentence of the regulation be declare goes beyond the scope statutory penalty, we shall determine the last declaration of punishment in accordance with the statutory maximum penalty; Considering the whole case, if there is a guilt of the crime of torture within the sentence of the regulation be declared, we can adjust it within range of 20%. Of course, we can also submit the case to the judicial committee, which declared criminal finally; If the crime plot is minor and criminal suspect should not be punished, he can be exempted from criminal punishment; If the criminal suspect should be sentenced to three years imprisonment, detention and comply with the applicable conditions of probation, he can be suspended sentence according to law; If the criminal suspect is under 18 years of age, pregnant women and over 75 years of age, it can be suspended sentence according to law.

4. Conclusion

We found that the pure quantitative study is obviously not completely suitable for the field of social science. It is not consistent with the uncertainty of some degree of law. The if-then rule is a common fuzzy decision making method, which can satisfy both the requirement of quantitative research and the qualitative research. The rule meets the empirical common sense of the judge's sentencing, and is adapted to the demand of the criminal sentencing with complex criminal plots. At the same time, the rule is more harmonious and unified with the current standardized sentencing system. In this paper, the if-then rule is introduced into the sentencing process of traffic accident crime based on reasoning method. The rule can offer an intuitive and descriptive reasoning mechanism, which provides a new idea for expanding the scientific sentencing method.

References

1. Bai J. J., Sentencing Prediction Research Based on Judges' Collective Experience. Law Research, pp. 140-154, 2016.
2. Bao G. W., Academy P. O. An Empirical Study on the Sentencing of Crime of Affray. Journal of Political Science & Law, pp. 128-138, 2017.
3. Max Weber. On the law of the economy and society. Zhang Naigen translation, China Encyclopedia press, pp. 355, 1998.
4. Nguyen T. T., Liew W. C., To C. et al., Fuzzy If-Then Rules Classifier on Ensemble Data. International Conference on Machine Learning and Cybernetics. pp. 362-370, 2014.
5. Vilém Novák, Stephan Lehmke. Logical structure of fuzzy IF-THEN rules. Fuzzy Sets and Systems, pp. 2003-2009, 2006.
6. Zhang Hua, Li Xiaoxia. The sentencing characteristics and sentencing model construction of drunken driving crime: An Empirical Study Based on 4782 random sampling judgments in China. Chinese Journal of criminal law, pp. 99-108, 2014.

Ranking road safety risk factors using preference structures and fuzzy preference structures

Yongjun Shen
School of Transportation, Southeast University
Sipailou 2, Nanjing 210096, China

Elke Hermans
Transportation Research Institute, Hasselt University
Wetenschapspark 5 bus 6, 3590 Diepenbeek, Belgium

Qiong Bao
School of Transportation, Southeast University
Sipailou 2, Nanjing 210096, China

Traditionally, road safety research focused on describing, explaining and predicting the number of crashes and casualties. In addition to the investigation of road safety outcomes, essential underlying risk factors are worthwhile studying. Insight into the importance of each risk factor provides policymakers with valuable information about the kind of measures most urgently needed to improve road safety. Road safety risk factors may be ranked in several ways. Classical preference structures and fuzzy preference structures, both well-studied mathematical structures in the theory of preference modeling, are introduced in this paper and applied to an European data set. These techniques prove to be promising for the road safety risk context.

1. Introduction

Traffic volume steadily increased during the past decades, thereby causing environmental as well as safety problems. On a worldwide scale this results in 50 million injured persons per year of which 1.2 million fatalities [1]. These outcome figures indicate the enormous safety problem. Road safety policymaking aims at reducing the number of crashes and casualties. Therefore, risk factors leading to crashes and casualties need to be identified. In case we get insight into the most important risk factors, necessary measures can be taken.

This study handles about the ordering of risk factors in terms of importance. In this respect, a risk factor is considered to be more important in case it has a stronger link with road safety outcomes. Assessing an ordering in the risk factors of road safety can be done in several ways. A common manner is to use the opinion of experts. However, the subjectivity of the results is a major drawback of this kind of methods.

In this study, risk factors will be ordered by means of classical preference structures and fuzzy preference structures, which have been widely investigated in multi-criteria optimization and decision analyses [2]. So rather than using a qualitative approach (e.g., expert opinions), the ordering of the risk factors results from the data. This requires the quantification of each risk factor. Therefore, indicators will be used as proxies for the risk factors and data will be gathered.

In the subsequent section, the indicator and fatality data will be discussed. In Section 3, a description of the preference structure and fuzzy preference structure methodology and application are given, followed by a discussion. This paper closes with concluding remarks and topics for further research.

2. Data

From literature (e.g., [3]) the following domains are agreed upon to be the main causes of crashes and casualties: alcohol/drugs, speed, protective systems, vehicle, infrastructure and trauma management. For each risk domain, several possible indicators exist, but due to the unavailability of reliable and comparable indicator data we are obliged to use best available indicators instead of ideal ones.

In this research, five road safety risk factors are considered within the above risk domains. They are: speed, seat belt, vehicle, road and emergency medical service (EMS). For illustrative purposes, one indicator is used for each of these risk factors. For speed, the percentage of drivers exceeding the speed limit on urban roads is the indicator (I_1); The seat belt factor is represented by the daytime seat belt wearing rate in front seats of light vehicles (< 3.5 tons) (I_2); The annual renewal rate of passenger cars is the selected vehicle indicator (I_3); The share of motorways and national roads in total road length describes the road factor (I_4) and for EMS the number of EMS stations per 1000 km^2 is the selected indicator (I_5).

From international data sources (see [4]) average values related to 2006-2008 were obtained (imputed) for these five safety performance indicators for 28 European countries. To eliminate the effect of the measurement unit and the scale of each indicator, the values presented in Table 1 have been normalized by the distance to a reference approach [5], with a higher value representing a better performance (thus a lower risk). The last column in Table 1 gives an indication about the average number of road fatalities per million inhabitants in the 28 countries between 2006 and 2008, which is also rescaled. For this variable a low score also implies a bad performance, therefore a high value should be aimed at.

Table 1. Data on five road safety risk indicators and fatalities.

	Speed I_1	Seat belt I_2	Vehicle I_3	Road I_4	EMS I_5	Fatality per inhabitants
AT	0.254	0.904	0.664	0.329	0.727	0.510
BE	0.222	0.799	0.983	0.275	0.711	0.445
BG	0.239	0.870	0.177	0.098	0.285	0.318
CY	0.205	0.819	0.532	0.893	0.280	0.393
CZ	0.563	0.908	0.377	0.155	0.348	0.397
DK	0.228	0.932	0.704	0.125	0.467	0.631
EE	0.308	0.880	0.468	0.161	0.169	0.326
FI	0.323	0.911	0.497	0.499	0.106	0.644
FR	0.318	1.000	0.614	0.060	0.753	0.588
DE	0.246	0.986	0.738	0.241	0.738	0.731
EL	0.255	0.701	0.533	0.281	0.013	0.300
HU	0.230	0.727	0.534	0.118	0.334	0.368
IE	0.223	0.901	0.858	0.165	0.602	0.572
IT	0.383	0.696	0.612	0.158	0.647	0.494
LV	0.176	0.809	0.277	0.069	0.094	0.257
LT	0.318	0.609	0.112	0.187	0.134	0.218
LU	0.226	0.819	1.000	1.000	0.699	0.544
NL	0.234	0.959	0.631	0.110	0.177	1.000
NO	0.259	0.937	0.509	0.842	0.089	0.833
PL	0.165	0.799	0.181	0.207	0.097	0.303
PT	0.360	0.881	0.438	0.324	0.750	0.484
RO	0.237	0.666	0.713	0.588	0.504	0.329
SK	0.269	0.696	0.411	0.251	1.000	0.395
SI	0.163	0.874	0.617	0.119	0.931	0.338
ES	0.279	0.884	0.641	0.113	0.883	0.525
SE	0.259	0.973	0.622	0.140	0.088	0.900
CH	1.000	0.887	0.668	0.072	0.979	0.879
UK	0.277	0.942	0.743	0.366	0.577	0.875

From Table 1 it can be seen that in terms of speed Switzerland performs best while Slovenia worst. In general, low scores on the risk indicators correspond to a low score in the final column and vice versa. In the next section, the method used in this study for ordering the five risk factors is described.

3. Preference Structures: Theory and Application

Preference structures are well-studied mathematical structures in the theory of preference modeling. The technique is often applied in multi-criteria decision making to describe the links between two alternatives. However, only values zero and one are allowed in classical preference structures to express the strict preference, indifference and incomparability relation, which led to the advent of fuzzy preference structures. In this paper, both classical preference structures and fuzzy preference structures are introduced and used to evaluate 28 European countries on their performance of five road safety risk factors.

3.1. Preference Structures and Fuzzy Preference Structures

A preference structure on a set of alternatives A is a triplet (P, I, J) of binary relations in A — which denote a strict preference, indifference and incomparability relation, respectively — describing a decision maker's preferences. At the same time, the relations P, I and J should satisfy some conditions [6].

As preference structures are based on classical set theory and are therefore restricted to classical relations, they do not allow to express degrees of strict preference, indifference or incomparability. This is seen as an important drawback to the practical use of these structures. With the introduction of fuzzy set theory, and its immense impact, fuzzy relations have become heavily involved in preference models as the generalization of the concept of classical preference structures. In literature [6], the historical development of the concept of fuzzy preference structures is summarized and the existence, construction and reconstruction of fuzzy preference structures are elaborated [7]. In addition, a substantial number of studies using fuzzy preference modeling are available (see e.g., [8-10]).

3.2. Application Based on Classical Preference Structures

To get the priority ranking of five road safety risk factors, we construct the preference relations of five indicators and one road safety outcome (i.e., the number of fatalities per million inhabitants). Taking the speed indicator as an example, whose preference relations are partially shown in Table 2, it can be seen that for each criterion, the one-dimensional set of values from Table 1 is transformed into a two-dimensional matrix (28x28) specifying the preference structures. Moreover, only values zero and one are possible in the preference relations. A value equal to one in cell (i,j) indicates strict preference of country j over country i. For example, in the matrix presented in Table 2, the column of Cyprus contains only zeros indicating its relatively bad or even worst performance in terms of speed. Analogously, we obtain P_{SB}, P_V, P_R, P_{EMS} and P_F, which present the preference relations of the other four indicators (seat belt, vehicle, road and EMS) and the number of fatalities per million inhabitants.

Subsequently, we pairwise compare the relations P_S, P_{SB}, P_V, P_R and P_{EMS} with P_F, thereby using the following comparison rule for the preference relation:

$$X \succ Y \Leftrightarrow \frac{|P_X \cap P_F|}{|P_X \Delta P_F|} \succ \frac{|P_Y \cap P_F|}{|P_Y \Delta P_F|} \tag{1}$$

where X and Y are two of the five criteria, P_X and P_Y are related preference relations, $\cap$ denoting intersection and Δ symmetric difference, both ordinary. The basic idea of formula (1) is that a greater similarity between P_X and P_F compared to the similarity between P_Y and P_F implies a greater importance of criterion X than of criterion Y. Therefore, it allows us to get the following priority ranking of the five risk factors: Seat belt (4.058) > Vehicle (2.439) > Speed (1.513) = EMS (1.513) > Road (1.119).

Table 2. Preference relations of the speed indicator.

P_S	AT	BE	BG	CY	CZ	DK	EE	FI	...
AT	0	0	0	0	1	0	1	1	...
BE	1	0	1	0	1	1	1	1	...
BG	1	0	0	0	1	0	1	1	...
CY	1	1	1	0	1	1	1	1	...
CZ	0	0	0	0	0	0	0	0	...
DK	1	0	1	0	1	0	1	1	...
EE	0	0	0	0	1	0	0	1	...
FI	0	0	0	0	1	0	0	0	...
...	...	...	...	...	...	...	...	...	...

3.3. Application Based on Fuzzy Preference Structures

In order to better express the non-binary reality, fuzzy preference structures are applied for the priority ranking of the five road safety risk factors. Again, several steps need to be taken. Firstly, the fuzzy preference relations of the five indicators and the number of fatalities per million inhabitants are constructed. Still focusing on the speed indicator, whose fuzzy preference relations are partially shown in Table 3, the value of fuzzy preference in speed (FP_S) between Austria and Belgium is computed as follows:

$$FP_S(AT, BE) = \max\{x_S^{AT} - x_S^{BE}, 0\} \tag{2}$$

where x_S^{AT} and x_S^{BE} are the values of the speed indicator for Austria and Belgium presented in Table 1. Formula (2) presents one possible way of fuzzification to get the fuzzy preference relations [7].

Comparing Table 3 to Table 2, we see that the calculation based on formula (2) also results in zeros on the diagonal of the matrix, but instead of value 1, the real difference in the speed values is given in cell (i,j) in case country j has a better performance than country i. For example, cell (2,1) in Table 3 has a value of 0.032, meaning that the score for the speed indicator of Austria is somewhat higher than the one of Belgium.

Table 3. Fuzzy preference relations of the speed indicator.

FP_S	AT	BE	BG	CY	CZ	DK	EE	FI	...
AT	0.000	0.000	0.000	0.000	0.309	0.000	0.054	0.070	...
BE	0.032	0.000	0.017	0.000	0.340	0.006	0.086	0.101	...
BG	0.015	0.000	0.000	0.000	0.324	0.000	0.069	0.084	...
CY	0.049	0.017	0.034	0.000	0.357	0.023	0.103	0.118	...
CZ	0.000	0.000	0.000	0.000	0.000	0.000	0.000	0.000	...
DK	0.026	0.000	0.011	0.000	0.335	0.000	0.080	0.096	...
EE	0.000	0.000	0.000	0.000	0.254	0.000	0.000	0.015	...
FI	0.000	0.000	0.000	0.000	0.239	0.000	0.000	0.000	...
...	...	...	...	...	...	...	...	...	...

Analogously, we obtain FP_{SB}, FP_V, FP_R, FP_{EMS} and FP_F, which represent the fuzzy preferences of the other four indicators and the number of fatalities per million inhabitants. Next, a comparison of the fuzzy preference relations of two criteria — in accordance with the one of classical preference relations in formula (1) — is defined in the following way:

$$X \succ Y \Leftrightarrow \frac{\sum_{i,j}^{m,m}|FP_X(x_{ij}) \cap FP_F(x_{ij})|}{\sum_{i,j}^{m,m}|FP_X(x_{ij}) \Delta FP_F(x_{ij})|} \succ \frac{\sum_{i,j}^{m,m}|FP_Y(x_{ij}) \cap FP_F(x_{ij})|}{\sum_{i,j}^{m,m}|FP_Y(x_{ij}) \Delta FP_F(x_{ij})|} \quad (3)$$

where X and Y are criteria, $FP_X(x_{ij})$, $FP_Y(x_{ij})$ and $FP_F(x_{ij})$ are the values of the cell in the i-th row and j-th column of the matrix describing the fuzzy preference relations related to criterion X, Y and the number of fatalities per million inhabitants, respectively, m is the number of alternatives (here we focus on 28 European countries) and $i, j \in \{1, 2, ..., m\}$.

Moreover, to simplify the above formula, a basic triangular norm (t-norm for short), i.e., the minimum t-norm is used to replace the intersection $\cap$ and Zadeh's absolute difference is used for the symmetric difference Δ. Therefore, we get the following new formula:

$$X \succ_{T_M} Y \Leftrightarrow \frac{\sum_{i,j}^{m,m}\min(FP_X(x_{ij}), FP_F(x_{ij}))}{\sum_{i,j}^{m,m}|FP_X(x_{ij}) - FP_F(x_{ij})|} \succ \frac{\sum_{i,j}^{m,m}\min(FP_Y(x_{ij}), FP_F(x_{ij}))}{\sum_{i,j}^{m,m}|FP_Y(x_{ij}) - FP_F(x_{ij})|} \quad (4)$$

Based on formula (4), the ordering of the five road safety risk factors can be fully deduced as follows: Seat belt (0.474) > Vehicle (0.447) > EMS (0.256) > Speed (0.221) > Road (0.196).

4. Discussion

The methodology used in this study, i.e., classical and fuzzy preference structures, results in valuable outcomes. They both present the one-dimensional values of a risk factor or the number of fatalities per million inhabitants in a two-dimensional matrix, thereby better indicating the relationships between countries. Since the indicators are expected to have some predictive power in terms of crashes and casualties, it is valuable to compare the relative indicator scores with the relative fatality numbers for a large group of countries. In case there is a high degree of overlap between the indicator matrix and the fatality matrix, a high evaluation result is obtained for this indicator because it is able to represent the differences in fatalities between countries to a considerable extent.

Moreover, in both techniques, the different steps are easy to follow and to compute. Based on the evaluation result of each risk factor an ordering can be obtained. Comparing the priority ordering of five road safety risk indicators based on both approaches, we find limited dissimilarity, especially for EMS and speed, which resulted in the same priority score in the classical preference structures (therefore no ranking), but different in the fuzzy preference structures. This occurs mainly because different degrees of preference, instead of only 0 and 1 are considered in the fuzzy preference structures, thereby better discriminating power.

5. Concluding Remarks

Within the context of road safety, the ordering of essential risk factors is a relevant topic. Having knowledge about the relative importance of these risk components, appropriate action can be taken thereby reducing the number of crashes and casualties in an effective way. In order to overcome the major drawback of subjectivity of ordering results based on expert methods, the approaches of classical preference structures and fuzzy preference structures were introduced in this study for the ordering of road safety risk factors. Classical preference structures are easy to compute whereas fuzzy preference structures better express the reality as they take the size of the differences between the values into account. Both techniques have proven to be useful with encouraging first results. Using a basic t-norm, the ordering of the five road safety risk factors — represented by one best available indicator here — results in Seat belt > Vehicle > EMS > Speed > Road.

In the future, a further exploitation of this and related research is worthwhile. For example, the impact of the use of other t-norms on the final ordering can be assessed; the ranking of risk factors based on expert knowledge on the one hand and a ranking based on fuzzy preference structures on the other hand could be aggregated by transforming the input information in multiplicative preference relations; and the use of fuzzy rather than crisp indicator data are all relevant research challenges.

Acknowledgments

This research was supported by the National Natural Science Foundation of China (Grant No. 71701045), and the Fundamental Research Funds for the Central Universities (Grant No. 2242018K40002).

References

1. World Health Organization (WHO). Global Status Report on Road Safety 2015. WHO, Geneva (2015).
2. Franco, C.A., On the analytic hierarchy process and decision support based on fuzzy-linguistic preference structures, *Knowledge-Based Systems*, 70, 203 (2014).
3. Hakkert, A.S., Gitelman, V. and Vis, M.A. (Eds.), Road Safety Performance Indicators: Theory. Deliverable D3.6 of the EU FP6 project SafetyNet (2007).
4. Shen Y. Inter-national benchmarking of road safety performance and development using indicators and indexes: Data envelopment analysis based approaches, PhD thesis, Hasselt University (2012).
5. Shen, Y., Hermans, E., Brijs, T. and Wets, G. Data envelopment analysis for composite indicators: A multiple layer model. *Social Indicators Research*, 114(2), 739 (2013).
6. De Baets, B. and Fodor, J. Twenty years of fuzzy preference structures (1978-1997). *Decisions in Economics and Finance*, 20(1), 45 (1997).
7. Hüllermeier, E., Brinker, K. Learning valued preference structures for solving classification problems. *Fuzzy sets & systems*, 159 (18), 2337 (2008).
8. Llamazares, B. Characterization of fuzzy preference structures through Łukasiewicz triplets, *Fuzzy sets & systems*, 136 (2), 217 (2003).
9. Gheorghe, R., Bufardi, A. and Xirouchakis, P., Construction of global fuzzy preference structures from two-parameter single-criterion fuzzy outranking relations, *Fuzzy sets & systems*, 153 (3), 303 (2005).
10. Khalid, A. and Beg, I. Incomplete interval valued fuzzy preference relations, *Information Sciences*, 348, 15 (2016).

Chaos in hydrology: A case study in Konya Basin, Turkey

Didem Odabasi Cingi
Department of Construction Technologies, Istanbul Aydin University
Istanbul, Besyol/Kucukcekmece 34295, Turkey

Ergun Eray Akkaya
Mechatronics Engineering Department, Istanbul Gelisim University
Istanbul, Avcilar 34315, Turkey

Dilek Eren Akyuz
Civil Engineering Department, Istanbul University, Istanbul, Avcilar 34320, Turkey

Every natural behaviour is non-linear, but not always is chaotic. This paper aims to investigate low dimensional chaotic behaviour of study area: Konya Basin by using non-linear time series techniques with three stages: i) Mutual Information, ii) False Nearest Neighbour (FNN) algorithm, iii) Stretching Exponential. These techniques calculate the delay time, the embedding dimensions and the maximal positive Lyapunov exponent respectively. The data set consists of daily average flow rates of three stations in Konya Basin through the study period between 1968 and 2014. Analysed data implied that these time series have shown chaos. This information helps catchment manager to forecast future and the extreme flow rates such as droughts and floods.

1. Introduction

Konya Basin is one of the 25 river basins in Turkey. It has approximately 7% of Turkey with an area of 5.5 million hectares. 2% of the available surface water resources in Turkey is Konya Basin. On the other hand, it has about 17% of the groundwater potential of our country due to its large closed basin [1]. In this study, we analysed the observed data from three stations located in the east part of Konya Basin (Figure 1). The 1968-2015 Konya Basin daily average flow rates time series reveals significant inter-annual and inter-decadal fluctuations. Such wet or dry periods are associated with regional climatic variability and are vital for understanding or predicting drought and the long-term accessibility of water. In this case, Konya Basin is known as semi-arid. Therefore, water management concepts such as water treatment and pollution control are important for water quality and sustainability of the basin.

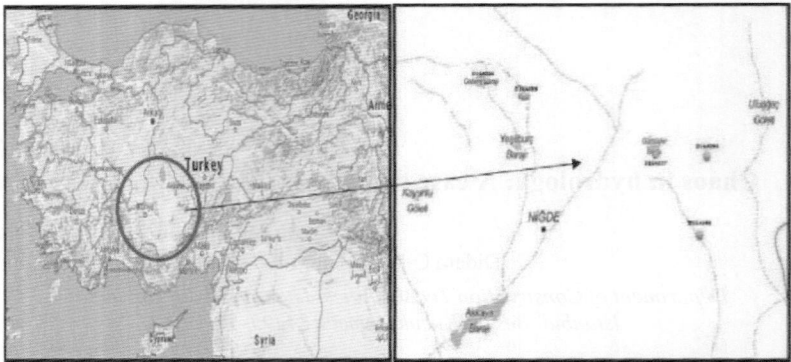

Figure 1. Location of Konya Basin and stations.

A multitude of factors indicate the climatic state and hence the Konya Basin flow rate. The integrating effect of the basin may lead to its fluctuations being determined largely by a few, unknown dynamical variables that may be complex, nonlinear functions of the physical variables. In this paper we test the hypothesis that the dynamics of Konya Basin can be described by a small set of variables, that is, it is low-dimensional. Such evidence may be useful for developing low-order models that explain at least a part of Konya Basin variability and perhaps can be ultimately related to low-frequency climatic variability.

Some current studies have demonstrated that low-dimensional deterministic strategies can be connected as an elective technique for modelling and the results are promising [2]. A sophisticated behaviour in nature can be distinguished as deterministic and disorderly or non-deterministic and randomized, subject to its base dynamics. Therefore, it is quite difficult to distinguish a chaotic deterministic system from a purely random system. Since both of them can produce irregular and apparently unpredictable temporal and/or spatial variability. The theory of chaos in hydrological sciences was introduced in the late 1980s [3, 4]. Since then, applications of chaos theory in hydrological sciences have been significantly advanced [5, 6, 7]. As things stand, chaotic and low dimensional deterministic time series techniques verifies the obtained results in the field of research, modelling and predicting possible river flow dynamics.

All the time an adequate measure of care is practiced in applying low-dimensional deterministic methods and in evaluating the results, such methods can be valuable in studying dynamics of river flow [5]. Furthermore, late articles which make prediction on river flow by using chaos theory, can reveal the number of factors that impact the river flow dynamics [2, 8]. This situation shows that; non-linear analysis is picking up significance and convenience for river flow analysis.

2. Theory and Data

In this study, three gauging stations were selected for investigation of chaotic behaviour within the closed basin. The data set contains the daily average flow rates of the stations between years 1968 and 2014. The methods shown schematically in Figure 2 were used to analyse time series data.

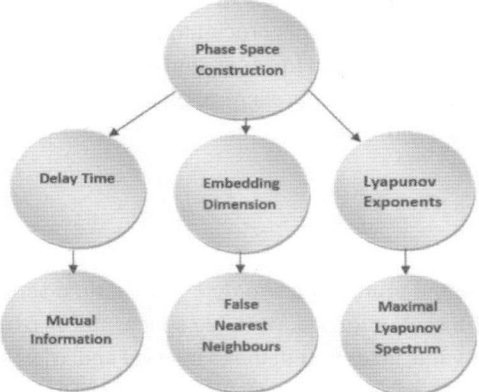

Figure 2. Data analysis steps in time series.

2.1. *Phase Space Reconstruction*

The reconstruction of a d-dimensional phase space is a vital step to analyse the underlying dynamics of a scalar time series. From the phase-space configuration of a system, we can obtain some information about the asymptotic properties (*e.g.* positive Lyapunov exponents) of the studied system, indicating how chaotic a system is, and the topological dimension of an attractor. Since attractors contain geometrical and dynamical properties of the original phase space, study properties of attractors signify study properties of the system. So, phase space reconstruction enables studying unobserved variables. In reconstructed phase space, we can determine the relation between current and future situation. For time series of scalar flow rate $s_{nn}(1)$, $s_{nn}(2)$, $s_{nn}(3)$, ..., $s_{nn}(n)$ the delay of the reconstructed phase space $\vec{y}_{nn}(k)$ is defined by,

$$\vec{y}_{nn}(k) = [\, s_{nn}(k), s_{nn}(k+\tau), s_{nn}(k+2\tau), ..., s_{nn}(k+(d-1)\tau)\,] \quad (1)$$

Where, $k = 1, 2, ..., n$; τ and d are the time delay and the embedding dimension respectively. Thus, the two parameters, time delay τ and embedding dimension d, need to be selected before reconstructing the phase space. Time delay can be measured from the first zero of correlation function (linear criterion) or first minimum of average mutual information [9].

2.2. Mutual Information

Some of the past studies recommend the use of the autocorrelation function to choose the time delay τ [4, 10]. But in fact, autocorrelation function is a linear statistic and does not agree the nonlinear correlations. However, in 1986, Fraser and Swinney suggested using the mutual information method for the determination of the optimum choice of the time delay for the phase space reconstruction [9]. At this point, it is crucial to state that, for some attractors, the method choice does not matter. However, for others, the prediction of τ might depend exceedingly on the approach applied. If nonlinear correlation is considered, the mutual information method is more accurate regarding the proper choice of τ [11].

$$S = -\sum_{ij} P_{ij}(\Delta t) \ln \frac{P_{ij}(\Delta t)}{P_i P_j} \tag{2}$$

where; P_i and P_j are the probabilities to determine a time series value into the ith and j^{th} intervals of a section of the available space, and $P_{ij}(\Delta t)$ is the probability that an observation value takes part in the ith interval and, after some observational delay time Δt, later takes part in the jth interval [12]. The first minimum of the mutual information is widely used for its quality to capture the nonlinear correlation in time series data [9]. And, it can be calculated by hand.

2.3. False Nearest Neighbours

The False Nearest Neighbours (FNN) method is the most popular tool for choosing the minimum embedding dimension. In the phase space, the attractor trajectory is directly influenced by embedding dimension. Thus, it is important for the neighbourhood of the points. FNN is based on the hypothesis that nearly located two points in a sufficient embedding dimension should remain close when dimension increases. If the embedding dimension is too small, selected points can be seen as neighbours but in fact they are distant to each other. If the embedding dimension is too large, this situation causes undesirable consequences such as redundancy on chaotic data and decreasing performance of many algorithms. When these cases are taken into consideration, the false neighbours should be corrected. In this case, FNN percentage should drop to zero when the optimum embedding dimension value is reached. This process is calculated as follows:

$$R_i = \frac{|R_{i+1} - R_{j+1}|}{\|\vec{R_i} - \vec{R_j}\|} \tag{3}$$

where; $\vec{R_i}$ is a constructed vector by using delay time, $\vec{R_j}$ is the nearest neighbour of $\vec{R_i}$ vector in a random dimension. This process should be iterated for all consecutive vectors to obtain R_i value. If the point of data is selected as a false neighbour in distance, R_i crosses above the certain threshold [13].

2.4. Maximal Lyapunov Exponent

Lyapunov exponent measures the exponential divergence and convergence of initially close state-space trajectories and quantities the amount of chaos in the system. So, Lyapunov exponent shows the chaotic nature of attractor. If exponential growth rate of nearby trajectories is positive, this state signs a chaos. This exponential growth rate is called as maximal Lyapunov exponent. In that case, maximal Lyapunov exponent physically can be considered as the rate of divergence of close trajectories. The stretching of the trajectories is calculated as follows:

$$S(\epsilon, m, t) = < \ln \frac{1}{u_n} \sigma s_n \epsilon u_n |s_{n+t} - s_{n'+t}| > \qquad (4)$$

Where; $s_{n'}$ is a neighbouring point, ϵ is the box size, after t time, the distance between the selacted points will be $s_{n+t} - s_{n'+t}$. This equation calculates the growth of this distance in time. If S (ϵ,m,t) is linear, the maximal Lyapunov exponent can be calculated from the slope.

2.5. Evaluation of Data

This study includes the dynamics of the daily average flow rates of 3 gauging stations located in Konya Basin. Data were obtained from Directorate General for State Hydraulic Works (DSI). We tried to select convenient combinations of time delay and embedding dimension of reconstructed phase space. In this case, we calculated statistical properties of the time series such as minimum, average, maximum, skewness and standard deviation values (Table 1).

Table 1. Statistical properties of time series.

Station Number	Length (Day)	Observation Years	Average (m³/s)	Max (m³/s)	Min (m³/s)	Skewness	Std. Dev.
D16A066	13128	1969-2014	0.061	0.640	0.000	1.182	0.100
D16A077	13877	1968-2014	0.107	0.620	0.000	1.120	0.161
D16A089	13150	1971-2014	0.116	3.700	0.000	3.676	0.179

By the reason of identifying and interpreting the relationships in data, we generated time series plots for each station (Figure 3). And Table 2 demonstrates the calculated values such as mutual information, embedding dimension and Lyapunov k (maximal Lyapunov exponent of a given scalar data set) [14].

Table 2. Nonlinear time series analysis results of Konya Basin.

Station Number	Time Delay (Days)	Embedding Dimension (m)	Lyap_k (values)
D16A066	20	3	0.0106
D16A077	17	3	0.0360
D16A089	18	3	0.0286

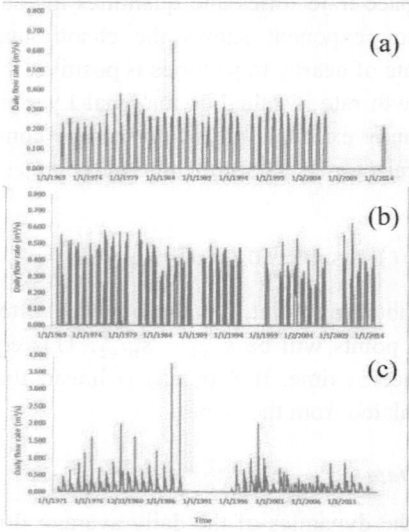

Figure 3. Flow rates time series for a) D16A066 station, b) D16A077 station, c) D16A089 station.

3. Application

First, we analysed the raw data as depicted in Figure 3. And we saw the variation of autocorrelation coefficient of the original data series of daily average flow rates. It also indicates seasonal variation with some quick decay in some lags. Then we marked zero-crossing point. This value gives information about downward tendency of the linear correlation. Thereafter the result obtained from the first minimum of the mutual information output which is shown in Figure 4(a) was compared with the zero-crossing point. It is sign of the nonlinear information of descent value of the data at lag 18. This state shows that the nonlinear correlation is more acceptable than the linear correlation in the daily average flow rates data series. Then, we decided to apply the chaotic time series prediction based on non-linear time series techniques to the daily average flow rates of three stations in Konya Basin.

This study consists of three steps. In the first step, mutual information is calculated by using TISEAN package [12] to determine time delay. In the second step, we computed false nearest neighbours to find the embedding dimension. In the last step, we used these values to obtain maximal positive Lyapunov exponent for the stations (Table 2).

As seen in Figure 4, slope of the maximal positive Lyapunov exponent of D16A066 station is increasing. This situation indicates the chaotic behaviour. Figure 4 also shows tendency is chaotic. In Table 2; time delay, embedding dimension and maximal Lyapunov exponents of three stations are presented as computed values. In this case, other stations sign nearly same results. They also have chaotic behaviour.

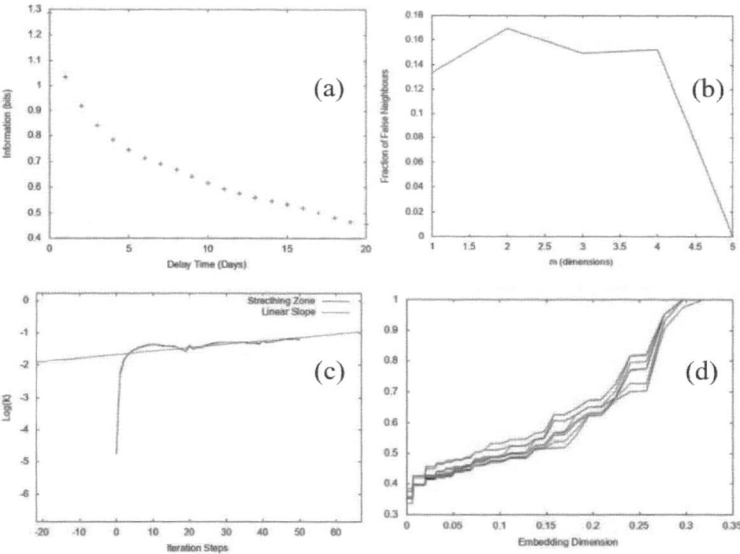

Figure 4. Chaotic time series analysis for D16A066 station a) Mutual information, b) False nearest neighbour, c) Lyapunov exponent by Kantz algorithm, d) Dimensional correlation.

4. Conclusion

Low-dimensional chaos is defined as a dynamical system indicating one positive Lyapunov Exponential when the embedding dimension of the system is less than or equal to 3. Maximal Lyapunov Exponent is defined as the rate of divergence of close trajectories and it is counterbalanced by the other effects so the system is considered as stable. In this case, the attractor is limited and for small perturbation of the trajectories we remain in the attractor.

There is another concept called as hyper-chaos. Since it would be impossible to counterbalance more than one positive Lyapunov exponent, hyper-chaos cannot be seen in a low-dimensional dynamical system. If embedding dimension of a dynamical system is greater than 3 and there is one positive Lyapunov exponent in this dynamical system, this is called as Higher Dimensional Chaos. When a Higher Dimensional Dynamical System (with embedding dimension is greater than 3) contents more than one positive Lyapunov Exponent, it is called as hyper-chaos [15].

In the light of this information, we analysed chaotic behaviour for three stations. For each station, there is only one positive Lyapunov exponent was calculated. The embedding dimension of each station was obtained as 3. But embedding dimension of the station D16A077 was calculated at the limit value (3). This means that our study area could be Higher Dimensional Dynamical System. In this case, low-dimensional non-linear techniques would be useless for this station. We would use other methods that appropriate to HD chaos such as HD Lorenz, Duffing, Rössler and Van der Pol oscillators, modified canonical Chua's circuits.

This study is the first step of chaotic time series estimation. By defining the chaotic behaviour, making forward and backward estimates will be easier and the missing data of each station can be defined. In this way, extreme values can be calculated such as flood and drought estimation. These results will help us to understand and evaluate the real world much better in mathematical terms.

References

1. wwf.org, (2018).
2. R. Khatibi, B. Sivakumar, M.A. Ghorbani, O. Kisi, K. Koçak, D.F. Zadeh, *Journal of Hydrology* **414-415**, 108–117 (2012).
3. A. Hense *Beitr Phys Atmos* **60**(1):34–47 (1987).
4. I. Rodriguez-Iturbe, F.B. De Power, M.B. Sharifi, K.P. Georgakakos. *Water Resour Res* **25**(7):1667–1675 (1989).
5. B. Sivakumar, (2003).
6. H. Tongal, M.C. Demirel, M.J. Booij, *Stoch Environ Res Risk Assess* **27**:489–503 (2013).
7. M.S. Kyoung, H.S. Kim, B. Sivakumar, V.P. Singh, K.S. Ahn, *Stoch Environ Res Risk Assess* **25**(4):613–625 (2011).
8. N.H. Adenan and M.S. Noorani, *Journal of Korean Society of Civil Engineering*, **18**, 2268-2274 (2014).
9. A.M. Fraser and H.L. Swinney, *Physical Review,* **A33**, 1134 (1986).
10. Q. Wang and T.Y. Gan, *Water Resour. Res.* **34**(9), 2329–2339 (1998).
11. A.A Tsonis and J.B. Elsner, *Nature* **333**, 545–547 (1988).
12. R. Hegger, H. Kantz and T. Schreiber, *Chaos*, **9**, 413 (1999).
13. H. Delafrouz, A. Ghaheri and M.A. Ghorbani, *Soft Computing*, **22**(7):2205-2215 (2018).
14. H. Kantz, *Physics Letters,* **A185**, 77 (1994).
15. F. Takens, *Lecture Notes Math*, **898**, 366-381 (1981).

Intuitionistic fuzzy decision making:
Multi-drug resistant tuberculosis risk assessment[*]

Elif Dogu

Industrial Engineering Department, Galatasaray University
34349, Besiktas, Istanbul, Turkey

Y. Esra Albayrak

Industrial Engineering Department, Galatasaray University
34349, Besiktas, Istanbul, Turkey

The experience of physicians is the key factor in the success of multi-drug resistant tuberculosis treatment. Even if the disease is the same, the experience degrees of physicians are not equal everywhere around the world. Especially for the treatment planning, a decision support system is needed in order to cope with this infectious disease. The decisions must be taken rapidly in order to gain time and advantage. At this point, the purpose of this study is to build a framework and establish a mathematical model that will help decision makers (physicians) while estimating the risk of multi-drug resistance when a new tuberculosis patient arrives, using intuitionistic fuzzy cognitive maps.

1. Introduction

Decision analysis is one of the operation research fields, however critical decision moments occur anytime and anywhere in life. Even if senior managers, top executives, financial experts who control large amounts, people working with major risk factors etc. make more stressful decisions, the most difficult decisions to be made are about someone's health. A manager's decision is a profit or loss issue for a company but a physician's decision is a matter of life or death for a person. In the decision-making process, physicians need previous in-depth researches, analyzed statistics, current information on the subject, patient's test results and medical history which mean a well-organized healthcare information system is required. In Turkey, this information system is being installed at the moment, and it is not completed. Therefore, physicians make these decisions alone or just get help from their colleagues who are working at the same hospital. With important decisions, comes great responsibility. The

---

[*]This work is supported by Galatasaray University Research Fund, 17.402.010.

physician makes these decisions according to his/her own experience and research and takes full responsibility for possible results, which bring stress and anxiety to the working environment. The decision, which may occur once in a patient's lifetime, is confronted by a physician every day. In Turkey, considering their working conditions, physicians need all kinds of support while making decisions about the patient's condition.

Tuberculosis (TB) is a global burden and one of the leading causes of morbidity and mortality [1]. TB is known for centuries, has a vaccine, also a standard treatment that takes six months. The medicine solved the problem of TB decades ago by finding the cure; it is not a problem of medicine anymore, it is a problem of management.

Tuberculosis would disappear from the world long ago if all the people in the world had the same health-care conditions and education level. If all people could afford the proper amount of drugs for the treatment and if infected people had not dropped their treatment prematurely, there would not be a disease called Resistant TB. Unfortunately, Turkey is one of the countries with lower education level. As a natural result, Resistant TB has the potential to become a major problem of Turkey.

Multi-Drug Resistant Tuberculosis (MDR-TB) is a type of TB that does not respond to the two most powerful first-line anti-TB drugs: rifampicin and isoniazid. MDR-TB can be treated with second-line TB drugs with an extensive treatment up to two years. About 480.000 people worldwide developed MDR-TB in 2015 [1]. When a new TB case is diagnosed, it is vital to capture the drug resistance risk. The focus of this study is to construct a risk assessment model for MDR-TB using intuitionistic fuzzy cognitive mapping (IFCM) which is an effective decision-making tool in medical problems considering the hesitation degrees of decision makers.

2. Cognitive Maps in Medical Decision Making

In the last two decades, cognitive mapping (CM) has been widely used in medical decision-making especially accompanied by fuzzy logic. Since medical decisions involve vagueness, ambiguity, and fuzziness, FCM and its extensions as IFCM are used in detection, diagnosis and treatment planning.

In 2005, CM is used for important communication skills that physicians need over the course of caring for a person with cancer [2]. In 2008, CM is applied for reshaping the diagnostic process and improving the management of digital imaging [3]. Lastly, in 2010, CM is used again for medical diagnosis support [4]. Apart from these, conventional CM has not been used in the last five years in medical problems.

FCM is widely used for detection, diagnosis and treatment objectives in medical decision-making as shown in Table 1.

Table 1. FCM in medical decision making.

FCM Publications	Objective
An Integrated Two-Level Hierarchical System For Decision Making In Radiation Therapy Based On Fuzzy Cognitive Maps [5]	Treatment
A Fuzzy Cognitive Map Approach To Differential Diagnosis Of Specific Language Impairment [6]	Diagnosis
Advanced Soft Computing Diagnosis Method For Tumor Grading [7]	Diagnosis
Brain Tumor Characterization Using The Soft Computing Technique Of Fuzzy Cognitive Maps [8]	Diagnosis
Fuzzy Cognitive Maps Structure For Medical Decision Support Systems [9]	Detection
A Novel Approach On Constructed Dynamic Fuzzy Cognitive Maps Using Fuzzified Decision Trees And Knowledge-Extraction Techniques [10]	Detection
Intuitionistic Fuzzy Cognitive Maps For Medical Decision Making [11]	Detection
A New Methodology For Decisions In Medical Informatics Using Fuzzy Cognitive Maps Based On Fuzzy Rule-Extraction Techniques [12]	Treatment
Application Of Evolutionary Fuzzy Cognitive Maps To The Long-Term Prediction Of Prostate Cancer [13]	Detection
A Fuzzy Cognitive Map Of The Psychosocial Determinants Of Obesity [14]	Detection
Design Of Activation Functions For Inference Of Fuzzy Cognitive Maps: Application To Clinical Decision Making In Diagnosis Of Pulmonary Infection [15]	Diagnosis
Development Of A Decision Making System For Selection Of Dental Implant Abutments Based On The Fuzzy Cognitive Map [15]	Diagnosis
Fuzzy Cognitive Map Software Tool For Treatment Management Of Uncomplicated Urinary Tract Infection	Treatment
Application Of Evolutionary Fuzzy Cognitive Maps For Prediction Of Pulmonary Infections [16]	Detection
Formalization Of Treatment Guidelines Using Fuzzy Cognitive Maps And Semantic Web Tools [17]	Treatment
A Fuzzy Grey Cognitive Maps-Based Decision Support System For Radiotherapy Treatment Planning [18]	Treatment
Support System For Decision Making In The Identification Of Risk For Body Dysmorphic Disorder: A Fuzzy Model [19]	Detection
Application Of Probabilistic And Fuzzy Cognitive Approaches In Semantic Web Framework For Medical Decision Support [20]	Treatment
Intuitionistic Fuzzy Cognitive Maps [21]	Detection
Modeling Of Parkinson's Disease Using Fuzzy Cognitive Maps And Non-Linear Hebbian Learning [22]	Detection
Time Dependent Fuzzy Cognitive Maps For Medical Diagnosis [23]	Diagnosis
A Fuzzy Information-Based Approach For Breast Cancer Risk Factors Assessment [24]	Detection

Intuitionistic Fuzzy Set (IFS) is an extended approach to define a fuzzy set which can present hesitation degrees of the decision makers in the mathematical model.

Let $X = \{x_1, x_2, ..., x_n\}$ be a finite universal set. An IF set A in X is defined as: $A = \{\langle x_l, \mu_A(x_l), \upsilon_A(x_l)\rangle | x_l \in X\}$ with the functions;

$$\mu_A : X \rightarrow [0,1], x_l \in X \rightarrow \mu_A(x_l) \in [0,1] \text{ and}$$

$$\upsilon_A : X \rightarrow [0,1], x_l \in X \rightarrow \upsilon_A(x_l) \in [0,1]$$

defining the degree of membership ($\mu_A(x_l)$) and the degree of non-membership ($\upsilon_A(x_l)$) of the element $x_l \in X$ to the set $A \subseteq X$ and for every $x_l \in X$, $0 \leq \mu_A(x_l) + \upsilon_A(x_l) \leq 1$.

$\pi_A(x_l) = 1 - \mu_A(x_l) - \upsilon_A(x_l)$ is Atanassov's intuitionistic fuzzy index, the degree of indeterminacy membership, of the element x_l in the set A and for every $x_l \in X$, $0 \leq \pi_A(x_l) \leq 1$.

IFCM is an extension of conventional CM by implementing IFS. It has the same concept nodes and relation edges structure as FCM however its concept value updating function is different. In this study, the IFCM equation proposed in [11] is adopted:

$$s_i^{k+1} = f\left(s_i^k + \sum_{\substack{j=1 \\ j \neq i}}^{N} s_j^k \cdot w_{ji}^\mu \cdot (1 - w_{ji}^\pi)\right) \quad (1)$$

where s_i^k is the value of concept i at iteration k, w_{ji}^μ is the influence weight, w_{ji}^π is the hesitancy weight and f is the sigmoid function.

3. MDR-TB Risk Factors

In order to evaluate the risk of multi-drug resistance, the factors that are effective in the resistance development process are determined. These factors are not the risk factors of being infected with TB, but the risk factors of developing resistance. First, an in-depth literature research is conducted, and then three chest diseases experts are interviewed. Nine factors are determined:

- Age [25-27]: The patient's age in years.
- Substandard housing conditions [28, 29]: Substandard means homelessness, excessive household crowding etc.
- BMI [26, 28]: Low Body-Mass Index represents potential risks for the disease.
- History of MDR-TB Exposure [30, 31]: MDR-TB is reported as an infectious disease.
- Presence of comorbidities [32, 33]: Especially HIV is associated with TB resistance risks.
- Previous use of antibiotics [32]: There exist the risk of resistance development for each antibiotic previously used.

- Being an immigrant [34]: Migration represent many risks like low income, poor health-care, etc.
- History of imprisonment [29, 31]: Prisons are dangerous in terms of infectious diseases.
- History of travel outside the country [35]: Especially travel to countries with high MDR-TB risk increases the risk of exposure.

4. Concluding Remarks

This is an ongoing study and the numerical application has not done yet. Future research direction will be as follows: Required information will be gathered from a full-fledged chest diseases hospital in Turkey, physicians will be interviewed, diagnosis and treatment processes of current patients will be monitored and according to this information a mathematical model will be established in order to facilitate the decision making process for the physicians. Chest diseases experts will evaluate the factors and determine the direction and strength of causal relationships and they will express their hesitation degrees using linguistic variables. Linguistic variables will be represented by intuitionistic fuzzy sets through a pre-defined scale. Intuitionistic fuzzy weight matrix will be constructed and IFCM of the system will be obtained. With the iterative process, the final values of the factors will be calculated. Thus, the strength of all factors on the MDR-TB risk will be revealed which will aid the physicians while making treatment decisions.

Acknowledgments

This study is supported by Galatasaray University Research Fund, 17.402.010.

References

1. WHO, "Global Tuberculosis Report," World Health Organization Press, Geneva, Switzerland 2016. Available: http://www.who.int.
2. A. L. Back, R. M. Arnold, W. F. Baile, J. A. Tulsky, and K. Fryer-Edwards, *Ca-A Cancer Journal for Clinicians*, vol. 55, no. 3, pp. 164-177, 2005.
3. E. Lettieri, C. Masella, and P. Zanaboni, *International Journal of Healthcare Technology and Management*, vol. 9, no. 1, pp. 45-59, 2008.
4. W. Froelich and A. Wakulicz-Deja, *Control and Cybernetics*, vol. 39, no. 2, pp. 439-456, 2010.
5. E. I. Papageorgiou, C. D. Stylios, and P. P. Groumpos, *IEEE Transactions on Biomedical Engineering*, vol. 50, no. 12, pp. 1326-1339, 2003.
6. V. C. Georgopoulos, G. A. Malandraki, and C. D. Stylios, *Artificial Intelligence in Medicine*, vol. 29, no. 3, pp. 261-278, 2003.
7. E. Papageorgiou, P. Spyridonos, C. D. Stylios, P. Ravazoula, P. P. Groumpos, and G. Nikiforidis, *Artificial Intelligence in Medicine*, vol. 36, no. 1, pp. 59-70, 2006.

8. E. I. Papageorgiou et al., *Applied Soft Computing*, vol. 8, no. 1, pp. 820-828, Jan 2008.
9. C. D. Stylios and V. C. Georgopoulos, in *Studies in Fuzziness and Soft Computing* vol. 218, 2008, pp. 151-174.
10. E. I. Papageorgiou, in *Studies in Fuzziness and Soft Computing* vol. 247, 2010, pp. 43-70.
11. D. K. Iakovidis and E. Papageorgiou, *IEEE Transactions on Information Technology in Biomedicine*, vol. 15, no. 1, pp. 100-107, 2011, Art. no. 5640672.
12. E. I. Papageorgiou, *Applied Soft Computing Journal*, vol. 11, no. 1, pp. 500-513, 2011.
13. W. Froelich, E. I. Papageorgiou, M. Samarinas, and K. Skriapas, *Applied Soft Computing Journal*, vol. 12, no. 12, pp. 3810-3817, 2012.
14. P. J. Giabbanelli, T. Torsney-Weir, and V. K. Mago, *Applied Soft Computing Journal*, vol. 12, no. 12, pp. 3711-3724, 2012.
15. I. K. Lee, H. S. Kim, and H. Cho, *Healthcare Informatics Research*, vol. 18, no. 2, pp. 105-114, 2012.
16. E. I. Papageorgiou and W. Froelich, *IEEE Transactions on Information Technology in Biomedicine*, vol. 16, no. 1, pp. 143-149, 2012, Art. no. 06080733.
17. E. I. Papageorgiou, J. D. Roo, C. Huszka, and D. Colaert, *Journal of Biomedical Informatics*, vol. 45, no. 1, pp. 45-60, 2012.
18. J. L. Salmeron and E. I. Papageorgiou, *Knowledge-Based Systems*, vol. 30, pp. 151-160, 2012.
19. M. J. A. De Brito et al., *International Journal of Medical Informatics*, vol. 82, no. 9, pp. 844-853, 2013.
20. E. I. Papageorgiou, C. Huszka, J. De Roo, N. Douali, M. C. Jaulent, and D. Colaert, *Computer Methods and Programs in Biomedicine*, vol. 112, no. 3, pp. 580-598, 2013.
21. E. I. Papageorgiou and D. K. Iakovidis, *IEEE Transactions on Fuzzy Systems*, vol. 21, no. 2, pp. 342-354, 2013.
22. A. P. Anninou and P. P. Groumpos, *International Journal on Artificial Intelligence Tools*, vol. 23, no. 5, 2014, Art. no. 14500109.
23. E. Bourgani, C. D. Stylios, G. Manis, and V. C. Georgopoulos, Time Dependent Fuzzy Cognitive Maps for Medical Diagnosis, in *Artificial Intelligence: Methods and Applications*, eds. A. Likas, K. Blekas, and D. Kalles, Lecture Notes in Artificial Intelligence, vol. 8445, 2014, pp. 544-554.
24. A. Buyukavcu, Y. E. Albayrak, and N. Goker, (in English), *Applied Soft Computing*, vol. 38, pp. 437-452, Jan. 2016.
25. F. Talay, S. Kumbetli, and S. Altin, *Japanese Journal of Infectious Diseases*, vol. 61, no. 1, p. 25, 2008.
26. K. Chung-Delgado et al., *PLoS One*, vol. 6, no. 11, p. e27610, 2011.
27. M. da Silva Garrido et al., *PLoS One*, vol. 7, no. 6, p. e39134, 2012.
28. M. F. Franke et al., *Clinical Infectious Diseases*, vol. 46, no. 12, pp. 1844-1851, 2008.
29. O. Aibana et al., *BMC Infectious Diseases*, vol. 17, no. 1, p. 129, 2017.
30. M. Gler, L. Podewils, N. Munez, M. Galipot, M. Quelapio, and T. Tupasi, *International Journal of Tuberculosis and Lung Disease*, vol. 16, no. 7, pp. 955-960, 2012.

31. N. M. Shariff, S. A. Shah, and F. Kamaludin, *International Journal of Mycobacteriology*, vol. 5, no. 1, pp. 51-58, 2016.
32. G. M. Bastos, M. C. Cezar, F. C. d. Q. Mello, and M. B. Conde, *Jornal Brasileiro de Pneumologia*, vol. 38, no. 6, pp. 733-739, 2012.
33. D. B. Tierney *et al.*, *PLoS One*, vol. 9, no. 9, p. e108035, 2014.
34. O. S. Elmi, H. Hasan, S. Abdullah, M. Z. M. Jeab, Z. B. Alwi, and N. N. Naing, *Journal of Infection in Developing Countries*, vol. 9, no. 10, pp. 1076-1085, 2015.
35. M. Dessalegn, E. Daniel, S. Behailu, M. Wagnew, and J. Nyagero, *Pan African Medical Journal*, vol. 25, no. Suppl. 2, 2016.

Software fault prediction using data reduction approaches

Chubato Wondaferaw Yohannese, Tianrui Li, Kamal Bashir, Macmillan Simfukwe and Ahmed Saad Hussein

School of Information Science and Technology, Southwest Jiaotong University Chengdu, 611756, China
freewwin@yahoo.com, trli@swjtu.edu.cn, kamalbashir1@yahoo.com, macsims85@gmail.com, husseinsaad187@yahoo.com

Building accurate Software Fault Prediction (SFP) model has been a challenge in software engineering research. Though, Machine Learning (ML) algorithms have been used for SFP. However, those algorithms require good data quality. Most research on ML assumes that training instances are free from outliers; and all features and instances are equally important. Yet, real world defect datasets suffer from irrelevant instances, irrelevant and redundant features and existence of outliers. Blindly applying such ML techniques on defect datasets may fail to make accurate predictions, resulting in poor inference and inefficient resource management. Therefore, this paper proposes eleven single and combined data reduction approaches for better SFP. Accordingly, the experimental results show that, single data reduction approaches improves prediction performance independently. However, the excellent performance achievements are attributed to the combined approach that removes outlier prior, handles irrelevant and redundant features. And then selects useful instances. Therefore, as shown in this study, data quality challenges must be carefully investigated and the combined data reduction approach should be considered in order to obtain robust performance for SFP.

Keywords: Software fault prediction; data reduction approach; outlier analysis; feature selection; instance selection.

1. Introduction

The quality of data in the fault proneness prediction of software module is important to obtain quality knowledge and accurate Software Fault Prediction (SFP). As observed in the literature,[1-11] the performance of predictors is directly influenced by the quality of the datasets used. In most datasets, the data values often contain outliers that have unusually large or small values when compared with others in the dataset.[5] Thus outliers are analyzed to find data points that are unusual and that may cause a negative effect in SFP. This paper uses a filter for detecting outliers based on Inter

Quartile Ranges (IQR) rule, a widely used outlier detection rule to find data points that are unusual.[10] Moreover, in defect datasets, several instances are stored but some of them are not useful for fault proneness prediction of software module. Therefore, it is possible to get acceptable or better prediction accuracy ignoring non useful instances. Thus, this paper uses main Instance Selection (IS) algorithms[10] reported in the literature that are based on estimating probabilities from the k-nearest neighbor patterns of an instance, in order to obtain more compact edited training sets while maintaining the classification rate. In addition to that, Correlation-Based Feature Selection (CFS) technique[7,12] is used to remove less important features from the datasets that suffer from a large number of irrelevant and redundant features. CFS is used in conjunction with Best First (BF) and Evolutionary Search (ES) methods. Thus, we design a robust SFP framework based on three data reduction methods, namely, Outlier Analysis (OA),[3–6] IS[8–10] and Feature Selection (FS)[1,7,13] methods. Eleven approaches are developed based on three strategies. Five ML algorithms are used as evaluator, namely Bagging (BG), Random Forest (RF), J48 Decision Tree, Naive Bayes (NB) and Decision Table (DT). Different software metrics, namely, McCabe and Halstead Static Code Metrics,[11,14] Chidamber and Kemerer's Object Oriented Metrics[15] and AEEEM datasets (which combine six different groups of software metrics)[16] are used (see Table 1). The purpose of using five evaluators, multiple software metrics and two search methods are to avoid the bias to make generalized decision based on only a single evaluator, software metrics and search method. It also aims to show the robustness of combined approach when talking the three challenges mentioned above.

2. Related Works

In this section, we focus on the studies that have attempted to address classification problems using data reduction techniques to build SFP models. In this regard, Shivaji et al.[1] investigated multiple FS. Liu et al.[17] made a comprehensive survey of FS algorithms. Chubato et al.[12] investigated the contribution of FS with combined SFP. In general, the experimental result shows that, feature selection is beneficial for improving the performance of learning algorithms. On the other hand, Rathore and Kumar[6] investigated several challenges including OA as defect data quality problem that has to be resolved for accurate SFP. Goel et al.[3] and Singh et al.[4] performed outlier analysis on object-oriented software metric. Cao et al.[5] performed

OA for source and target data. These studies confirmed that analyzing outlier helps to find data points that are over influential and avoiding them found to be essential. Furthermore, Olvera-Lpez et al.[8] reviewed several IS methods. Ryu et al.[9] tackled the major challenges of cross-project defect prediction with hybrid IS method. Vázquez et al.[10] extended the original Wilson's editing method and proposed Edited Nearest Neighbor (ENN). These studies confirmed that in ML, IS is an important task and experimental result shows that IS helps to achieve high overall performance.

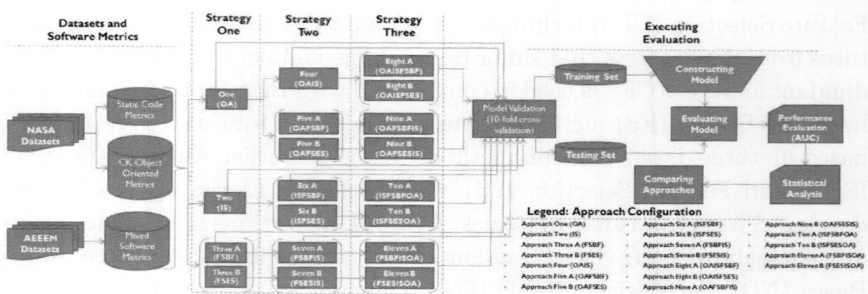

Fig. 1. A data reduction based framework for improving SFP.

3. A Data Reduction Based Framework for Improving SFP

A data reduction based framework for improved SFP is shown in Fig. 1. Our framework hybridizes three data reduction preprocessing methods. OA is performed before and after combined implementation of IS and FS to see the effect of the presence and absence of outliers on selecting useful instances and features as well as on prediction performance. IS and FS are performed one after the other to find the approaches which contribute better for accurate SFP. In strategy one, we perform individual data reduction methods. OA, IS and FS are performed using IQR, ENN and CFS, respectively. Therefore, the input of the framework for this strategy is outlier free data, useful instances and best features selected. In addition to that, we perform performance evaluation before any data cleaning process (Normal) (see Table 2) to benchmark the best approach. Then we construct and evaluate SFP models on the cleaned and normal datasets. This strategy serves as to compare the performance of single data reduction approaches. In strategy two, we develop a pair approach by combining single data reduction methods to see the contribution of resolving those challenges in a combined form.

This strategy serves as to realize more efficient approaches. In strategy three, we combine all three data reduction methods together to maximize the benefit of combining all individual methods. This strategy serves as to generate better data reduction approaches for more accurate prediction. Moreover, we make comparison between eleven approaches within strategy one, two and three based on statistical results. The purpose of this comparison is manifold: to prove the performance improvements; realize more efficiently performing approach; and see the significant contribution of resolving those challenges in single and combined form.

Table 1. Description of datasets.

No.	Dataset	#Attr	#Ins	#NFP	#FP	%NFP	%FP	No.	Dataset	#Attr	#Ins	#NFP	#FP	%NFP	%FP
1	EQ	61	324	195	129	60.19%	39.81%	5	ant-1.6	21	351	259	92	73.79%	26.21%
2	JDT	61	997	791	206	79.34%	20.66%	6	PC4	37	1458	1280	178	87.79%	12.21%
3	ML	61	1862	1617	245	86.84%	13.16%	7	prop-4	23	8718	7878	840	90.36%	9.64%
4	KC3	39	194	158	36	81.44%	18.56%								

Experimental design: During the experiments, the performance evaluation of each approach is carried out by running a 10-fold cross-validation using 7 datasets (see Table 1). Each cross-validation run is repeated 10 times to avoid bias. Then the results are captured using AUC. Experiments are performed using Mathlab 2014a, KEEL version 3.0 and WEKA version 3.8.

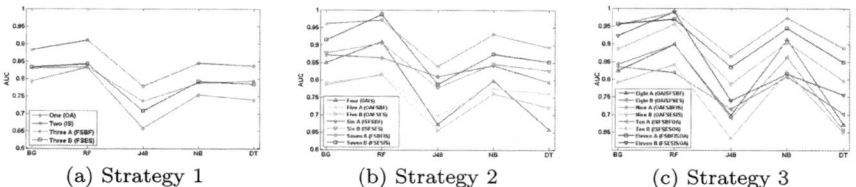

(a) Strategy 1   (b) Strategy 2   (c) Strategy 3

Fig. 2. Prediction performance of data reduction approaches.

4. Analysis and Discussions

In this section, the performance comparison among eleven approaches are presented in figures and tables. Using performance results, approaches in each strategy are compared by performing a one-way ANOVA F-test to examine the significance level of the differences in each approach. The factor of interest considered in our ANOVA experiment are the eleven approaches (see Table 2). The null hypothesis for the ANOVA test is that all the group population means (in our case approaches in each strategies) are the same, while the alternate hypothesis is that at least one pair of means is different. Multiple pairwise comparison is conducted by using Tukey's Honestly Significant Difference (HSD) criterion to find out which pairs of means (in each strategy) are significantly different, and which are not. The significance level for ANOVA experiment and Tukey's HSD test is $\alpha = 0.05$.

4.1. Results: Single Data Reduction Approaches

Average performance results of single data reduction in approach one (OA), two (IS) and three (FS) using five evaluators (BG, RF, J48, NB and DT) and seven datasets (Table 1) are presented in Figs. 2(a), 3(a) and Table 2, respectively. In terms of AUC, approach two shows great performance with all evaluators. Following that, approach three demonstrated better result than approach one with all evaluators. However, the performance of approach three with BF and ES search methods varies. In this strategy, approach three with ES method outperforms that with three evaluators (BG, RF, and NB) whereas BF outperforms that with two evaluators (J48 and DT) (see Fig. 2(a) and Table 2). Therefore, the result reflects the presence of unnecessary instances in defect datasets, is one of the major challenges and resolving those instances greatly contributes for better fault proneness prediction of software modules.

Table 2. Classification results of data reduction approaches.

Single Data Reduction Approaches (Strategy 1)						Two Data Reduction Approaches (Strategy 2)						Three Data Reduction Approaches (Strategy 3)					
Approaches	BG	RF	J48	NB	DT	Approaches	BG	RF	J48	NB	DT	Approaches	BG	RF	J48	NB	DT
Normal	0.830	0.848	0.687	0.761	0.786	Four (OAIS)	0.850	0.910	0.673	0.798	0.658	Eight A (OAISFSBF)	0.824	0.899	0.698	0.913	0.672
One (OA)	0.794	0.833	0.658	0.754	0.740	Five A (OAFSBF)	0.792	0.814	0.698	0.776	0.761	Eight B (OAISFSES)	0.843	0.899	0.691	0.862	0.655
Two (IS)	0.884	0.911	0.779	0.845	0.837	Five B (OAFSES)	0.787	0.816	0.655	0.761	0.721	Nine A (OAFSBFIS)	0.953	0.989	0.865	0.973	0.888
Three A (FSBF)	0.830	0.834	0.736	0.788	0.794	Six A (ISFSBF)	0.872	0.864	0.810	0.842	0.794	Nine B (OAFSESIS)	0.886	0.955	0.787	0.905	0.796
Three B (FSES)	0.833	0.843	0.708	0.792	0.785	Six B (ISFSES)	0.878	0.906	0.778	0.846	0.827	Ten A (ISFSBFOA)	0.836	0.819	0.716	0.808	0.702
						Seven A (FSBFIS)	0.961	0.971	0.839	0.931	0.893	Ten B (ISFSESOA)	0.791	0.842	0.633	0.823	0.648
						Seven B (FSESIS)	0.915	0.987	0.788	0.874	0.852	Eleven A (FSBFISOA)	0.956	0.969	0.835	0.944	0.848
												Eleven B (FSESISOA)	0.923	0.990	0.740	0.816	0.756

Strategy one of Table 3, shows the ANOVA results for single data reduction methods. The p-value is greater than the value of α, indicating that the alternate hypothesis is rejected, meaning that, even though, approach two shows better performance, however, the performance differences are not statistically significant (see also Fig. 3(a)).

Table 3. One-way ANOVA results of three strategies and 11 approaches.

Steratrgy One					Steratrgy Two					Steratrgy Three					Eleven Approaches				
Source	SS	df	MS	F	Sig.	Source	SS	df	MS	F	Sig.	Source	SS	df	MS	F	Sig.	Source	SS	df	MS	F	Sig.
Columns	0.023	3	0.008	2.75	0.0767	Columns	0.121	6	0.020	4.73	0.0019	Columns	0.154	7	0.022	2.82	0.0207	Columns	0.314	18	0.017	3.21	0.0002
Error	0.045	16	0.003			Error	0.120	28	0.004			Error	0.249	32	0.008			Error	0.413	76	0.005		
Total	0.068	19				Total	0.241	34				Total	0.403	39				Total	0.728	94			

4.2. Results: Combined Two Data Reduction Approaches

In Figs. 2(b), 3(b) and Table 2, the performance comparison of combined approaches of strategy two are presented. The result of our second experiment suggests that approach seven yields better result than others. Here again, the performance of approach seven with BF and ES search methods varies. However, in this strategy, approach seven with BF search outperforms that with four evaluators (BG, J48, NB and DT) whereas ES search outperforms that with only one evaluator (RF). Following that, approach six demonstrates better result than approaches four and five except that approach four yields better result using RF. When observing the performance of approach six with BF and ES search, ES search method helps to yield better result with four evaluators (BG, RF, NB and DT) whereas BF outperforms that with only one evaluator (J48). In addition to that, approach four outperforms approach five with three evaluators (BG, RF and NB), however, approach five outperforms with J48 and DT except that approach five of OAFSES yields low result than approach four when using J48 evaluator. Furthermore, in approach five, the performance with BF search (BG, J48, NB and DT) yields better result than that with ES search (RF). This result reflects that the presence of unnecessary instances and irrelevant and redundant features in defect datasets. Therefore, resolving these challenges following approach seven, by first selecting useful features and then removing unnecessary instances increase data quality and improve SFP.

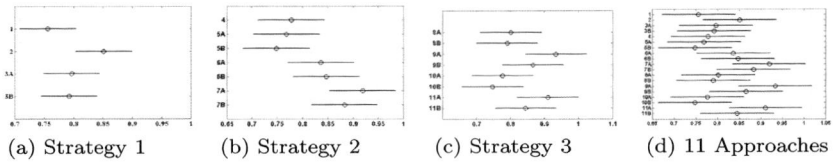

(a) Strategy 1   (b) Strategy 2   (c) Strategy 3   (d) 11 Approaches

Fig. 3. Multiple pairwise comparisons of 3 strategies and 11 approaches.

Strategy two of Table 3, shows the ANOVA results for combined two data reduction approaches. The p-value is less than the value of α indicating that the alternate hypothesis is accepted, meaning that. Fig. 3(b) show the multiple pairwise comparison results of strategy two. Accordingly, approach seven of (FSBFIS) significantly outperforms approach four and both approach five. Following that, approach seven of (FSESIS) significantly outperforms approach five of (OAFSES).

4.3. Results: Combined Three Data Reduction Approaches

In Figs. 2(c), 3(c) and Table 2, the performance comparison of combined approach of strategy three are presented. The result of our third experiment suggests that the combined approach nine of (OAFSBFIS) yields better result than others with three evaluator (J48, NB and DT). Here again, the performance of approach nine with BF (with all evaluators) outperforms that with ES search method. Following that, approach eleven of (FSBFISOA) and (FSESISOA) outperform approach nine with BG and RF evaluators, respectively. Furthermore, approach eleven yields better result with BF (BG, J48, NB and DT) than that with ES search method (RF). In addition to that, in this strategy, the competitive performance of approach eight and ten are observed. Approach eight of (OLEISFSBF) and approach ten of (ISFSBFOA) outperforms in two and three evaluators (RF, NB) and (BG, J48, DT), respectively. However, approach eight of (OAISFSES) outperforms approach ten of (ISFSBFOA) with all five evaluators. The results reflects that analyzing outliers prior to implementing other data reduction approaches benefits more. Furthermore, the outperformance of approach nine confirms that combining three data reduction techniques contributes more for data quality while greatly reducing the runtime and removing outliers helps to select more precise instances and features. Thus, it significantly contributes for software quality estimation.

Strategy three of Table 3 shows the ANOVA results for combined three data reduction approaches. The p-value is less than the value of α indicating that the alternate hypothesis is accepted. Thus, Fig. 3(c) show the multiple pairwise comparison results of strategy three. Even though, there is performance difference among strategy three approaches, except approach nine and ten, there are no approaches that are statistically significantly different from each other. Approach nine of (OAFSBFIS) significantly outperforms approach ten of (ISFSESOA).

4.4. Comparison of Eleven Data Reduction Approaches

As expected, removing outliers, selecting useful instances and features have seen proved to be useful and contribute for more accurate SFP. With proposed framework, experimental result shows the achieved performance improvements and statistical significance (see Fig. 3(d) and Table 3 of eleven approaches). Accordingly, the more efficiently performed approach is found to be combined approach nine of (OAFSBFIS) (it also perform much better than normal (see Table 2)), which is statistically significantly different from three approaches (approaches one (OA), five of (OAFSES) and Ten of (ISFSESOA)). This can be explained that removing outliers prior helps to select useful instances and features more precisely. Following that, approach seven of (FSBFIS) outperforms other approaches and statistically significantly different from two approaches (approach five of (OAFSES) and ten of (ISFSESOA)). One interesting finding observed from this experiment is that the combination of FS with BF search method followed by IS shows better performance, even when combining three data reduction methods.

5. Conclusion and Future Works

This study proposed data reduction approaches for SFP. Thus, 11 approaches are generated and empirically evaluated by following three strategies. On the tested datasets, we observed that, the single data reduction method improved performance of the predictors independently. However, the excellent performance achievements attributed to the combined approach that removes outlier prior, handle irrelevant and redundant features, and then select useful instances. Thus, dealing with the challenges of SFP mentioned in this study, our proposed combined approach confirms remarkable prediction performance which lays the pathway to produce quality software products. For the future work, we plan to employ more preprocessing methods and realize how the proposed framework helps to identify more efficient approaches for better fault proneness prediction.

Acknowledgments

This paper is partially supported by the Soft Science Foundation of Sichuan Province (No. 2016ZR0034).

References

1. S. Shivaji, E. J. Whitehead, R. Akella and S. Kim, *IEEE Transactions on Software Engineering* **39**, 552 (2013).
2. C. W. Yohannese, T. Li, M. Simfukwe and F. Khurshid, Ensembles based combined learning for improved software fault prediction: A comparative study, in *Int. Conf. on Intelligent Systems and Knowledge Engineering*, 2017.
3. B. Goel and Y. Singh (Springer Berlin Heidelberg, 2008).
4. Y. Singh, A. Kaur and R. Malhotra, *Software Quality Journal* **18**, 3 (2010).
5. Q. Cao, Q. Sun, Q. Cao and H. Tan, Software defect prediction via transfer learning based neural network, in *Int. Conf. on Reliability Systems Engineering*, 2016.
6. S. S. Rathore and S. Kumar, *Artificial Intelligence Review*, 1 (2017).
7. M. A. Hall, *Correlation-based Feature Selection for Machine Learning*, tech. rep. (1999).

8. J. A. Olvera-Lpez, J. A. Carrasco-Ochoa, J. F. Martnez-Trinidad and J. Kittler, *Artificial Intelligence Review* **34**, 133 (2010).
9. D. Ryu, J. I. Jang, J. Baik, Member and IEEE, *Journal of Computer Science and Technology* **30**, 969 (2015).
10. F. Zquez, J. S. Nchez and F. Pla, *Lecture Notes in Computer Science* **3523**, 35 (2005).
11. R. Malhotra, *Applied Soft Computing Journal* **27**, 504 (2015).
12. C. W. Yohannese and T. Li, *International Journal of Computational Intelligence Systems* **10**, 647 (2017).
13. K. Bashir, T. Li, C. W. Yohannese and Y. Mahama, Enhancing software defect prediction using supervised-learning based framework, in *Int. Conf. on Intelligent Systems and Knowledge Engineering*, 2017.
14. T. J. McCabe, *IEEE Transactions on Software Engineering* **2**, 308 (1976).
15. D. Radjenovi, M. Heriko, R. Torkar and A. ivkovi, *Information & Software Technology* **55**, 1397 (2013).
16. M. D'Ambros, M. Lanza and R. Robbes, *Empirical Software Engineering* **17**, 531 (2012).
17. H. Liu and L. Yu, *IEEE Transactions on Knowledge and Data Engineering* **17**, 491 (2005).

The sensitive data leakage detection model based on Bayesian convolution neural network

Chunliang Zhou, Zhengqiu Lu, Yangguang Liu

College of Information Engineering, Ningbo Dahongying University
Xue Yuan Road, Ningbo, Zhejiang Province, China

In order to improve the leaked problem for the user sensitive information, a sensitive data leakage detection model based on Bayes Convolution neural networks (CNN) is proposed. At first, the data is divided in this method, and the leakage degree is detected with the hierarchical clustering algorithm, which the packets are clustered to generate signatures. Then, the frequent items in data acquisition are collected combined with Bayes Convolution neural networks, and access correlation is calculated between the sensitive data and normal data. Finally, through the simulation experiment, the performance of the method and other methods is deeply studied. The results show that this method has good adaptability.

Keywords: Sensitive data; information leakage; Bayes; neural networks; hierarchical clustering.

1. Introduction

With the rapid development of Internet and e-commerce, a growing number of user data information is input in the database, and the illegal users who obtain the sensitive information results in more and more serious damages to the users, so the information security issue has become an issue that should not be overlooked [3, 7, 5, 8, 1]. The emerging information security objective is to maintain three features of information management system: confidentiality, integrity and availability. Therefore, it is required to adopt a certain permissions to protect sensitive information.

The traditional way for sensitive information is to use public key to encrypt data so as to protect the sensitive information documents [2, 6], but such method will produce new sensitive information in plain text, but it still cannot protect the sensitive information in real sense. The security issues of sensitive information IY will be attached great importance from all aspects. For example, the issue that the user-sensitive information may be revealed is a barrier to popularize the cloud computing services. Literature [9] proposes an Android malware detection method based on permission sequential pattern mining algorithm, and designs the PApriori mining algorithm to conduct the permission sequential detection for 49 malware applications, and makes early warning for malicious use of sensitive information, but with relatively high false alarm rate. Literature [10] proposes the sliding window model based on time stamp and the mining algorithm of frequent item set as the traditional sliding window model based on affair is hard to describe the dynamic changes in data distribution in data flow with the time by introducing the concept of type change boundary, and makes delay process for the item set in accordance with the changes in size of the sliding window. Literature [11] guarantees the security of the sensitive information from both processing and transmission respectively according to the sensitive logic and dynamic identification technology of data flow trace and the sensitive operation isolation execution technology based on virtualization. The experimental results indicate that sensitive logics in the applications account for less than 2%, and the protection overhead for executing relevant pages when executing sensitive logics isolatedly. Literature [4] establishes an integrated access control mechanism based on use of control technology and attribute encryption mechanisms according to the data security and distribution issue in sensitive information sharing application, addresses the problem in permission abuse of shared users, and conducts the simulation experiment for the security and performance of the mechanism, and results reveals that it could reduce the workload of server significantly. The sensitive information are of same importance in other information areas, and the database uses the encryption algorithms to protect the characteristics of sensitive information, such as the transparent data encryption is used to encrypt the sensitive information in the database, but the access control is still dependent on the authorization of external function modules.

On the basis of the above-mentioned study, this paper studies the sensitive information base on the Bayes convolutional neural network model, determines and identifies effectively the disclosure of the sensitive information, file encryption, etc., protects the sensitive information and non-sensitive information, simplifies the manual setting and the control over access rules to a maximum extent, and verifies the effectiveness of such method by simulation experiment.

2. Sensitive Information Disclosure Identification Model

The types of sensitive information include mainly UDIDs (IMEI, serial ID of IMSI SIM card, etc.), hash value of UDIDs (ID MD5, ID SHA1) and name of operator (CARRIER) etc., and the identification model established by this paper clusters and generates characteristics signature according to HTTP data package.

Assume that on a given network, $dstip_n$ indicates an IPV4 address, $dsport_n$ indicates the port number, $host_n$ indicates the host domain name, then a certain data package in the network can be expressed as $p_n = \{dstip_n, dsport_n, host_n\}$. Here, X and Y of two HTTP data package are defined as P_x and P_y, respectively, the distance between the two packages is as follows through calculation:

$$D_{dst}(P_x, P_y) = \frac{lmatch(ip_x, ip_y)}{32} + match(port_x, port_y) + \frac{ed(host_x, host_y)}{\max(len(host_x), len(host_y))} \quad (1)$$

$$\begin{cases} \dfrac{lmatch(ip_x, ip_y)}{32} \in [0,1] \\ match(port_x, port_y) \in [0,1] \\ \dfrac{ed(host_x, host_y)}{\max(len(host_x), len(host_y))} \in [0,1] \end{cases} \quad (2)$$

Whereas, *lmatch* is the bit function shared jointly by two IP addresses, *match* indicates the function that when the port number is matched as 1, it does not equal to 0 vice versa; *ed* is the function that computes and edits the distance, *len* is the function to get the string length of the host domain name, and max refers to the maximum function to select the string length of host domain name.

Meanwhile, two sensitive clusters C_x and C_y are determined pursuant to the above-mentioned variables, the evaluation distance of the clusters is:

$$D_{group}(C_x, C_y) = \frac{1}{|C_x||C_y|} \sum_{P_x \in C_x} \sum_{P_y \in C_y} D_{dst}(P_x, P_y) \quad (3)$$

Cluster the sensitive data based on the *Kolmogorov*-based complexity standards and generate characteristic signatures *S*:

$$S(x, y) = \frac{C(xy) - \min(C(x), C(y))}{\max(C(x), C(y))} \quad (4)$$

The frequent item set refers to the item set that the data come up less than the user-defined thresholds, $\eta = \{i_1, i_2, \ldots, i_n\}$ is used to represent the different data sets, whereas $|\eta| = n$ represents the length, the given data $X = \{x_1, x_2, \ldots, x_m\}$ and the sensitive data set $D = \{T_1, T_2, \ldots, T_m\}$, and each T_i represents the affair of one η, defining the objective function of the frequent item set. $\eta(X)$

$$\eta(X) = \sum_m D_T^{-\alpha \eta^T (T_m, \tilde{T}_m)} \eta^0(x_n, \tilde{x}_n, s_n) + \beta v_n \eta^T (T_m, \tilde{T}_m) \quad (5)$$

Whereas, s_n is the confidence coefficient of sample, if the sample is of source data, and then $s_n = 1$, otherwise $s_n \in [0,1)$. v_n is the data parameter of the sample, if the sample is the target data, then $v_n = 1$; otherwise $v_n = 0$. Also, set the error function between the computing characteristic data and the reconstructing characteristic data:

$$\eta^T(T_m, \tilde{T}_m) = [-y_m \lg \tilde{y}_m - (1 - y_m) \lg(1 - \tilde{y}_m)] \| T_m - \tilde{T}_m \|^2 \qquad (6)$$

Use similar cosine formula to calculate the similarity of the two eigenvectors V_1 and V_2, and obtain the simple threshold *rate* with the lowest false positive rate according to the cosine of the sensitive data:

$$\cos\theta = \frac{\sum_{i=1}^{t} V_{1i} V_{2i}}{\sqrt{V_1 V_1} \| V_2 \|} \qquad (7)$$

$$\text{rate} = \frac{(B+C)}{(A+B+C+D)} \qquad (8)$$

Whereas, A refers to the quantity that is identified as security data, B refers to the quantity that is identified as the sensitive data wrongly, C refers to the quantity that is identified as security data incorrectly, and D refers to the quantity that is properly recognized as sensitive data. Set $\theta_{ij}(r)$ as the model for the difference between the indicator data vector $x_i(r)$ and $x_j(r)$, reflect the distance by included angle, and establish the data correlation model of the similarity angle and the approximate data:

$$|x_i(r) - x_j(r)| = ((x_i(r,1) - x_j(r,1)^2) + ... + (x_i(r,n) - x_j(r,n))^2)^{1/2} \qquad (9)$$

3. Solution Model Based on Bayes CNN

Bayes algorithm is a value-added algorithm supporting neural network, and this paper collects the characteristics of the permissions relevancy of sensitive data, frequent item sets and others at first together with neural network in order to address the problem of sensitive information disclosure. Meanwhile, it creates characteristics layer for the identification of the sensitive data disclosure by Bayes algorithm and the depth convolutional neural network (DCNN), and each layer will obtain the tower calculation of the characteristics layer calculated and output by the previous layer. The Bayes CNN proposed by this paper mainly consists of five layers like the convolutional layer c, the sample layer s, etc., and designs that the c layer uses the convolution kernel to filter and strengthen the data and operate the Bayes convolution, and each layer is activated and then outputs a characteristics layer α_j^t, which defined as:

$$\alpha_j^t = f(\sum_{i \in P_j} k_{i,j}^t * \alpha_j^{t-1} + b_j^t) \tag{10}$$

Whereas, f is the activation function of the basic *Sigmoid* function, t indicates the number of layer, $k_{i,j}$ is the convolution kernel, * indicates the operation of *2D* convolution, b_j is the offset, and P_j indicates the set of the input characteristic patter selected. The s layer is to reduce the feature dimension of *c*layer, and in order to facilitate the calculation of probability ratio that the sensitive data and other special samples occur, each $n \times n$ sample in the layer is averaged and maximized to calculate the quantity of the special samples and mark the sensitive data:

$$\alpha_j^{\prime t} = f(down(\alpha_i^{t-1}) \cdot \alpha_i^{t-1} + b_j^t) \tag{11}$$

Whereas, w is weight, *down(·)* is the lower sampling function. In order that the identification model can simulate the neural network and search targets rapidly in data flow, the characteristics are integrated in layer C *and* S according to Formulas (12) and (13) to realize the multiple characteristics expression of sensitive data, which are obtained from candidate target area and integrated by multiple characteristics.

$$\begin{cases} BY(c,s) = |(B(c)-Y(c)) \oplus (Y(s)-B(s))| \\ RG(c,s) = |(R(c)-G(c)) \oplus (G(s)-R(s))| \end{cases} \tag{12}$$

$$O(c,s,\theta) = |O(c,\theta) \oplus O(s,\theta)| \tag{13}$$

Whereas, $R = r - (g + b)/2$, $G = g - (r + g)/2$, $Y = (r + g)/2 - |r - g|/2 - b$, c and s represent the central scale and the periphery scale of the database formed by data, and θ is the direction of the filter *Gabor*. As to the disclosed sensitive data, considering the contribution scale of each characteristic to target location, and the outstanding disclosure area W is determined in combination of significant feature weight β^j:

$$\beta^j = \frac{P(O \mid Fsali^j)}{P(O \mid Bsali^j)} = \frac{P(Fsali^j \mid O)P(O)}{P(Fsali^j)} \cdot \frac{P(Bsali^j)}{P(Bsali^j \mid O)P(O)} = \frac{P(Fsali^j \mid O)P(Bsali^j)}{P(Fsali^j)P(Bsali^j \mid O)} \tag{14}$$

$$W = \sum_{j \in \{Co, In, Or\}} \beta^j \sum_{k=1}^{N} sali_j^k \tag{15}$$

Whereas, *Sali* represents the significance value, *Co*, *In* and *Or* represents the color significance value, the luminance significance value and the orientation significance value, k represents different dimension, and $P(O)$ represents the prior probability that object O occurs. If $n \times m$ groups of unknown sample data sets $A = (c_1, c_2, ..., c_n)$ are collected, and $C = (a_1, a_2, ..., a_m)$, the Bayes CNN algorithm is used to calculate the probability value $P(a_m)$ containing the sensitive characteristic a_m in the sample:

$$P(a_m) = \sum_{m=1,n=1}^{AUC} P(a_m|c_n)P(c_n) = \sum_{m=1,n=1}^{AUC} \frac{\sum_{m=1,n=1}^{\infty} S(\exists a_m \forall c_n)}{\sum_{n=1}^{C} N(c_n)} P(c_n) \quad (16)$$

Finally, the sample to be identified is obtained y_i, and the obtained probability is normalized and calculated to get the posterior probability Mp:

$$Mp = \arg\max_{\exists m \in M} P(c_n|y_i) = \arg\max_{\exists m \in M} \frac{\prod_{m=1}^{n} P(y_i|c_n)P(c_n)}{P(y_i)} = \arg\max_{\exists m \in M} \prod_{i=1}^{n} P(y_i|c_n)P(c_n) \quad (17)$$

$$|c_n(n) - a_m(n)| = ((c_n(n,1) - a_m(n,1)^2) + ... + (c_n(i,m) - a_m(i,m))^2)^{1/2} \quad (18)$$

The model based on prior knowledge of Bayesian and convolutional neural networks can not only exert the advantages of neural networks, but also use the results of systematically analyzing the results in advance. The Bayesian CNN introduces a feedback path, corrects prior knowledge by post-knowledge, and reduces systematic errors due to inaccuracy and incompleteness of prior knowledge. Under the Bayesian analysis framework, the model parameters are regarded as uncertain quantities, and the explicit probability distribution assumptions are used to enter the model and analyzed and inferred. The prior knowledge of unknown variables is quantified by prior distribution. Error data will be interpreted as a likelihood function is defined, and the regularization may correspond to a prior probability distribution rights on the network, Bayesian convolutional neural network is formed by the prior distribution is assumed to integrate by a given observation data Adjusting the search for the posterior probability distribution of the weighting variable, the network prediction is based on the Bayesian inference of the posterior distribution.

This paper uses the correlation coefficient model to calculate the permission correlation according to the division of the neural network, the nerve cell in input layer receives the input signal, and each nerve cell in the hidden layer and the input layer is connected with all the adjacent nerve cell, while the nerve cell of the same layer does not connect with each other so as to identify the data through such model. The specific algorithms are described as follows:

(1) Initialization:Input the sample value, set relevant parameters, determine the weight β^j, and identify and calculate the sensitive data disclosure based on the five-layer structure of Bayes CNN.

(2) The first layer simulation samples and divides the data types, including the sensitive data group, the generic data group and the unknown data group,

and conducts the random sampling for 28 channels × 60, with the number of channel set to be N, sampling point of T, expressed as $I_{N,T}$, and then calculate the convolution of the data obtained from each layer.

(3) The second layer is the convolutional layer, which convolves and activates the characteristic pattern of the previous layer, if not activated, skip to Step (2); otherwise, start to calculate an output the characteristic pattern according to Formula (18) (the convolution kernel of [28*1] is k_m^2, and $b_m^2(j)$ is the offset):

$$y_m^2 = f\left(\sum_{i=1}^{i\leq 28} I_{i,j} * k_m^2 + b_m^2(j)\right) \tag{19}$$

(4) Determine if the output value of the convolutional characteristic pattern on the second layer is less than or equals to T, if not, skip to Step (3); otherwise, calculate the convolution of the third layer, and determine the output characteristic pattern (k_m^3 is the convolutional kernel of [1*10], and $b_m^3(j)$ is the offset), integrate the data that occur frequently and calculate its frequent item sets:

$$y_m^3(j) = f\left(\sum_{i=1}^{i\leq 10} y_m^2((j-1)\times 10+i) * k_m^3 + b_m^3(j)\right) \tag{20}$$

(5) Connect the nerve cell on the fourth layer with that on the third layer (the connection weight of the nerve cells on third layer and the fourth layer is $w_i^4(p)$ and $b^4(j)$ is the offset), an calculate and output the characteristic pattern of such layer:

$$y^4(j) = f\left(\sum_{i=1}^{i\leq 40}\sum_{p=1}^{p\leq 6} y_i^3(p)w_i^4(p) + b^4(j)\right) \tag{21}$$

Cluster the data package of frequent item sets that the fourth layer obtains from the calculation of the third layer, and generate the characteristics signature according to Formula (4).

Meanwhile, in order to calculate the probability of the sensitive item sets within the nerve cell, this paper introduces the Bayes algorithm elements, and calculates the probability with sensitive item sets in combination of nerve cell distribution, and the Bayes CNN can solve the problem that the preprocessing capacity of DCNN is not that satisfactory, and calculate the probability of sensitive characteristics in this layer:

$$P(y^4(j)) = \sum_{i=1, p=1}^{NUT} \frac{\sum_{i=1}^{i\leq 40} w_i^4(p)}{\sum_{p=1}^{p\leq 6} p_i^3(p)} + b^4(j) \qquad (22)$$

(6) Connect all nerve cells in the fifth layer with those in the fourth layer ($w^5(i)$ indicates the connection weight of the fourth layer and the fifth layer, $b^5(j)$ is the offset of the fifth layer), and calculate the permission correlation between the normal data and the sensitive data through the correlation coefficient model obtained base on Formula (18):

$$y^5(j) = f\left(\sum_{i=1}^{i\leq 100} y^4(i)w^5(i) + b^5(j)\right) \qquad (23)$$

(7) Train and converge the data in the fifth layer, and the fifth layer simulates the collecting box outside the network, and calculate the output detection rate according to the posterior probability obtained from Formula (17), and such data serve as the sensitive data disclosure probability.

(8) The algorithm finishes.

4. Mathematical Simulation

In order to verify the effectiveness of the sensitive data disclosure detection based on DCNN, this paper conducts the simulation experiment in MATLAB. At first, 20,000 data package samples, which are captured on the website, include the sensitive data package, the generic data package and unknown data package, among which the unknown data package consists of sensitive data package and the generic data package. This experiment selects N data packages randomly from the samples, and uses Bayes CNN to calculate and generate the characteristics signature and obtain the probability of the frequent item set. 200 of the sensitive data set, generic data set and unknown data set each are selected for the calculation of frequent item sets, and the frequent item sets of the generic data package account for the lowest proportion, and those of the sensitive data are the highest, and there are a certain sensitive data in the unknown sets.

As shown in Figure 1, there are a certain frequent item sets in the sensitive data, and if there are sensitive data in the generic data, the probability that the frequent item sets occur will be increased. If the permission correlation is known, the characteristics weight is integrated to describe the permission correlation, for example, it will start up the software automatically from the background to obtain information, and increase the utilization space for operating the contents while stealing the sensitive data, etc.

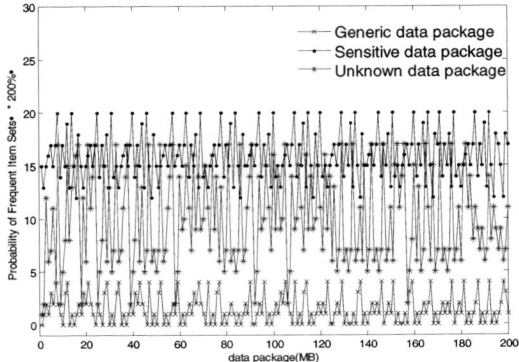

Figure 1. Results on probability of frequent item sets.

Meanwhile, Figure 2 illustrates the comparison results about the permission correlation between the generic data and the sensitive data. The permission serves as a distinguishing feature, the software permission involved by sensitive data is higher than normal value in general, and the permission correlation in the generic data package is commonly low; therefore, the security performance of the software involved by generic data package is relatively large, it can obtain the strength of the sensitive information involved by software through the permission correlation. It can be seen from Figure 2, it can distinguish the generic data from the sensitive data effectively.

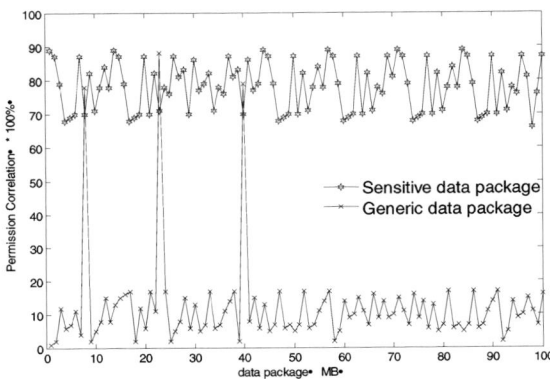

Figure 2. Comparison of permission correlation.

While the misdetection rate and the erroneous judgment rate can embody the accuracy of the identification model laterally, including if the identified sensitive data are disclosed or not under different permissions, and the erroneous judgment rate for the disclosure of the sensitive data in different frequent item sets. As it can be seen from Figure 3 and Figure 4, with the increase in permission correlation or the increase in the quantity of the frequent item sets, the disclosure of the sensitive data also worsens accordingly, the erroneous judgment rate and the omission rate from Bayes CNN are low and stable, and the Bayes CNN algorithm integrates the two features of the Naive Bayer and DCNN, with powerful adaptability and strong stability.

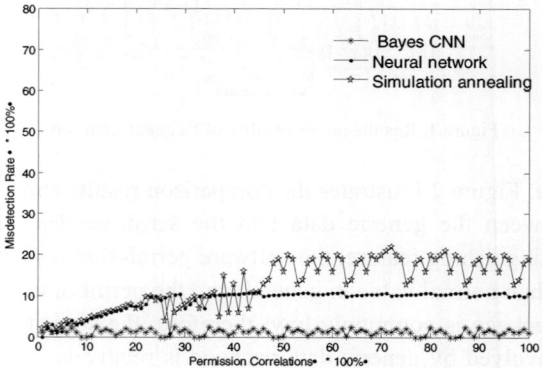

Figure 3. Comparison on misdetection rate of different permission correlations.

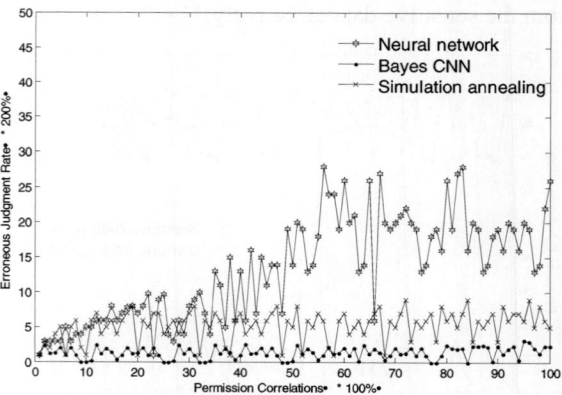

Figure 4. Comparison on erroneous judgment rate.

5. Conclusions

In order to address the problem that the identification model does not have satisfactory detection performance for network sensitive data disclosure, this paper proposes a sensitive data disclosure detection model based on Bayes CNN. It simulates the sampling layer of the neural network for filtering and reinforcement by the operations combining the Bayes algorithm and the DCNN, and maps the sensitive data, calculates the sensitive data in the convolutional layer, and compares the characteristics of sthe normal data packages and the sensitive data packages. This paper integrates the Naive Bayes algorithm and the DCNN, and reduces the complexity of the sensitive data effectively.The possibility experiment that can reduce the disclosure of the sensitive information significantly proved that the detection method proposed by this paper is featured with better adaptation compared with the neural network and the simulated annealing algorithm based on the feature that the detection model of DCNN has invariance property, powerful robustness and fault tolerant capacity, easy to train and optimize the network structure. In the following studies, it can be considered to improve the sensitive data disclosure identification model by integrating multiple evaluation indicators.

References

1. Dong Liang, Zhuang Yi, Gao Yang, Bu Yang. Research on Real Time Trigger System for Sensitive Data Safe Destruction [J]. Journal of Chinese Computer System, 2010, 31(7): 1323-1327.
2. Li Haifeng, Zhuang Ning, Zhu Jianming, Cao Huaihu. Frequent Itemset Mining over Time Sensitive Streams [J]. Chinese Journal of Computers, 2012, 35(11): 2283-2293.
3. Liu Jing, Zhou Mingtian. Key Management and Access Control for Large Dynamic Multicast Groups [J]. Journal of Software, 2002, 13(2): 291-297.
4. Sui Tingting, Wang Xiaofeng. Convolutional Neural Networks with Candidate Location and Multi feature Fusion [J]. ACTA Automatica Sinica, 2016, 42(6): 875-882.
5. Tian Chunzi, Shao Xiaokang. Location Sensitive Data in Database Mining Technology Research [J]. Computer Simulation, 2016, 33(9): 454-457.
6. Wang Dongqi, Chen Zhaofeng, Zhang Huilin. Study of a Protection Method of Sensitive Data Web Servers in the Cloud [J]. Journal of Tsinghua University (Science and Technology Edition), 2016, 56(1): 51-57.
7. Wang Zhiqiang, Zhang Yuqing, Liu Qixu. A Method to Detect Android Malicious Behaviors. Journal of Xidian University [J] 2015, 42(3): 8-14.
8. Yang Huan, Zhang Yuqing, Hu Yupu, Liu Qixun Android Malware Detection Method Based on Permission Sequential Pattern Mining Algorithm [J]. Journal on Communications, 2013, 34(1): 106-115.

9. Yang Lei, Cao Cuiling, Sun Jianguo, Zhang Liguo. Study on Improved Naive Bayes Algorithm in Spam Filtering [J]. Journal on Communications, 2017, 38(4): 140-148.
10. Yan Xixi, Geng Tao. Fused Access Control Scheme for Sensitive Data Sharing [J]. Journal on Communications, 2014, 35(8): 71-77.
11. Zhou Feiyuan, Jin Linpeng, Dong Jun. Review of Convolutional Neural Network [J]. Chinese Journal of Computers, 2017, 40(7): 1-23.

Digital supply chain risk analysis with intuitionistic fuzzy cognitive map[*]

Gulcin Buyukozkan

*Department of Industrial Engineering, Galatasaray University
Çırağan Cad. No. 36, Ortaköy, 34349, Istanbul, Turkey
gulcin.buyukozkan@gmail.com*

Fethullah Gocer

*Department of Industrial Engineering, Galatasaray University
Çırağan Cad. No. 36, Ortaköy, 34349, Istanbul, Turkey*

The digital era is evolving extremely fast in almost all areas of life. Any organization that fails to adapt to this change and does not adopt digitalization will not likely have high chances of survival in the future. For supply chains, digital evolution goes beyond just using digital enablers. Digital Supply Chain (DSC) necessitates the evolution of digitalization, technology implementation and supply chain management in a harmony. In order to sustain competency and achieve continuous customer satisfaction, DSC must distinguish, manage, rank, and evaluate its DSC risks. These risk components and DSC factors are connected with each other via indirect or direct relationship in a highly complicated manner. Thus, this research offers a logical way for analyzing DSC risks by utilizing the proposed Interval-Valued Intuitionistic Fuzzy Cognitive Map (IVIFCM) approach. This study proves that IVIFCM is particularly advantageous for decision-making in which several numbers of controllable and uncontrollable decision factors are causally interrelated.

1. Introduction

Digital Supply Chain (DSC) is a popular concept that is implemented in supply chains with the aim to lower costs in of the supply chain, drive visibility or quality across distribution channels. DSC enablers provide capabilities to quickly deliver high-quality products and make it easier to maintain regulatory compliance. DSC helps organizations to become more customer-centric. This enables organizations to manage multiple-product-lines, across multiple-distribution-channels, in multiple-regions and with multiple-collaborative-partners all from a centralized location [1]. DSC is described as a customer-centric platform that is modeled to

---

[*]This work is financially supported by Galatasaray University Research Fund (Project No.: 18.402.003).

capture and maximize the utilizations of real-time information emerging from a variety of sources, enabling demand stimulation, sensing, matching, and management in order to have an optimized performance and minimized risk [2].

Decision-making is a daily life activity which is mostly associated with selecting the best among many. During the past quarter-century, real-life decision-making has gained in popularity and complexity, which can be overcome by taking the impact of multiple criteria into account. A set of objects having common features are stated as a crisp set, in which the membership values are presented in binary terms. Crisp sets can be insufficient to manage real problems, given the fuzziness in decision-making. Zadeh presented the concept of the fuzzy set theory [3], by assigning a degree of membership to each element with the help of a membership function generating a value between 0-1. Atanassov extended it into the Intuitionistic Fuzzy set theory by assigning degrees of membership, non-membership, and hesitancy to each element in the set [4]. Interval-Valued Intuitionistic Fuzzy (IVIF) sets are the extended versions of Intuitionistic Fuzzy sets [5].

Interval-Valued Intuitionistic Fuzzy Cognitive Map (IVIFCM) has more advanced capabilities than traditional fuzzy cognitive maps in modeling states described in terms of major concepts and their cause and effect relationships. One of the best practical features of the IVIFCM is its potential for handling decision-making processes as a prediction tool [6]. IVIFCM utilizes interval values to express ambiguity and vagueness associated with the context or to the lack of model precision. IVIFCM simulates its development over time to forecast the future state by providing an initial behavior for a system, represented by a set of values of its constituent concepts. These aspects make IVIFCM as a fairly appealing tool for DSC risk analysis. The scalability and resource requirements of this methodology allow using it in practice. To the best of authors' knowledge, there is no study that applies IVIFCM for DSC risk assessment.

The flow of this study is as follows: Section 2 presents an overview of IVIFCM. This section is followed by the detailed description of the proposed model for DSC risk analysis. Its application involving IVIFCM scenarios are presented in section 4. The last section concludes the paper with some future suggestions.

2. Overview of IVIF Sets and CM Approach

This section briefly introduces IVIF sets, and Cognitive Maps (CM), thus describing the main concepts of IVIFCM. IVIF sets basically designate a degree of membership on each element. $A \subset E$ is a crisp set. An IVIF set $\tilde{A}$ is characterized by an interval valued membership value $\tilde{\mu}_{\tilde{A}}(x)$ and an interval valued non-membership value $\tilde{v}_{\tilde{A}}(x)$, as an ordered pair in the object of the following form;

$$\tilde{A} = \{\langle x, \tilde{\mu}_{\tilde{A}}(x), \tilde{v}_{\tilde{A}}(x)\rangle | x \in E\} \quad (1)$$

$\tilde{A}$ is called an IVIF set, where $\tilde{\mu}_{\tilde{A}}(x) \subset [0,1]$ and $\tilde{v}_{\tilde{A}}(x) \subset [0,1]$, $\forall x \in E$ with the condition of $0 \le \sup(\tilde{\mu}_{\tilde{A}}(x)) + \sup(\tilde{v}_{\tilde{A}}(x)) \le 1$. For convenience, an IVIF set's lower end points and upper-end points are denoted by $\tilde{A} = [\mu_{\tilde{A}}^L, \mu_{\tilde{A}}^U], [v_{\tilde{A}}^L, v_{\tilde{A}}^U]$. Arithmetic operations in IVIF sets [5] are defined as follow:

$$\tilde{A} \oplus \tilde{B} = \begin{pmatrix} [\mu_{\tilde{A}}^L(x) + \mu_{\tilde{B}}^L(x) - \mu_{\tilde{A}}^L(x)\mu_{\tilde{B}}^L(x), \mu_{\tilde{A}}^U(x) + \mu_{\tilde{B}}^U(x) - \mu_{\tilde{A}}^U(x)\mu_{\tilde{B}}^U(x)], \\ [v_{\tilde{A}}^L(x)v_{\tilde{B}}^L(x), v_{\tilde{A}}^U(x)v_{\tilde{B}}^U(x)] \end{pmatrix} \quad (2)$$

$$\tilde{A} \otimes \tilde{B} = \begin{pmatrix} [\mu_{\tilde{A}}^L(x)\mu_{\tilde{B}}^L(x), \mu_{\tilde{A}}^U(x)\mu_{\tilde{B}}^U(x)], \\ [v_{\tilde{A}}^L(x) + v_{\tilde{B}}^L(x) - v_{\tilde{A}}^L(x)v_{\tilde{B}}^L(x), v_{\tilde{A}}^U(x) + v_{\tilde{B}}^U(x) - v_{\tilde{A}}^U(x)v_{\tilde{B}}^U(x)] \end{pmatrix} \quad (3)$$

$$negation(\tilde{A}) = \{\langle x, [v_{\tilde{A}}^L(x), v_{\tilde{A}}^U(x)], [\mu_{\tilde{A}}^L(x), \mu_{\tilde{A}}^U(x)]\rangle | x \in E\} \quad (4)$$

CM was introduced in the 70s as a type of directed graphs used to capture and understand relationships of cause and effect in complex causal systems and facilitate to understand the inter-connections within the elements of the concepts [7]. Fuzzy CM is an extension of the traditional CM, introduced in the 80s that includes the concepts to be represented in linguistic terms with a related fuzzy value rather than demanding them to be crisp [8]. The value of each concept is computed by applying the following Equation (5):

$$A_i^{(k+1)} = f\left(A_i^k + \sum_{j=1}^{N} A_j^{(k)} w_{ji}\right) \quad (5)$$

Here, $A_i^{(k+1)}$ is the value of concept i at iteration $k + 1$, $A_i^{(k)}$ is the value of the interconnected concept j at iteration k, w_{ji} is the weighted arc and f is the threshold function. Intuitionistic Fuzzy CM is introduced in 2009 as an extension of CM and Fuzzy CM [9]. Since then, Intuitionistic Fuzzy CM has been used in several studies [6], [9]–[13]. The powerful feature of Intuitionistic Fuzzy CM is its ability to cope with hesitations. Intuitionistic Fuzzy CM also has the iteration-based system same as Fuzzy CM. A similar version of IVIFCM is first mentioned in [14] as an extension of Intuitionistic Fuzzy CM, in which the generalization of Fuzzy CM is proposed in IVIF environment to cope with complex links among concepts for supplier selection processes.

3. IVIFCM Model for DSC Risk Analysis

The proposed IVIFCM model improves the existing CM approach by applying IVIF arithmetic. The results prove that the proposed methodology outperforms other approaches for the given application requirements. The following Equation (6) proposed by the authors presents the IVIFCM calculation process:

$$A_i^{(k+1)} = f\left(Deff\left(\oplus, \left(\bigoplus_{j=1, j \neq i}^{N} \left(\begin{Bmatrix}[\mu_A^L(c), \mu_A^U(c)], \\ [v_A^L(c), v_A^U(c)]\end{Bmatrix}_i^k \otimes \begin{Bmatrix}[\mu_A^L(c), \mu_A^U(c)], \\ [v_A^L(c), v_A^U(c)]\end{Bmatrix}_j^k \otimes \begin{Bmatrix}[\mu_A^L(w), \mu_A^U(w)], \\ [v_A^L(w), v_A^U(w)]\end{Bmatrix}_{ji}\right)\right)\right)\right) \quad (6)$$

$$Deff = 1 - \frac{\mu_A^L(x) + \mu_A^U(x) + \left(1 - v_A^L(x)\right) + \left(1 - v_A^U(x)\right) + \mu_A^L(x) * \mu_A^U(x) - \sqrt{\left(1 - v_A^L(x)\right) + \left(1 - v_A^U(x)\right)}}{4} \quad (7)$$

Here, for negative influence, negation operator is applied. $\{[\mu_A^L(w), \mu_A^U(w)], [v_A^L(w), v_A^U(w)]\}$ represents the membership and non-membership interval influence weights and $\{[\mu_A^L(c), \mu_A^U(c)], [v_A^L(c), v_A^U(c)]\}$ represents the membership and non-membership interval values of concepts. Here, f is the sigmoid function of Deff, which defuzzifies IVIF values, in other words resulting value is defuzzified before it is functioned by sigmoid. The proposed IVIFCM method is as follows:

Step 1: The concepts are defined by the decision makers (DMs).

Step 2: The interactions between concepts are specified and their strengths are determined by the DMs in linguistic terms.

Step 3: By the IVIFCM Equation (6), each concept is calculated to get the final result.

4. IVIFCM Application for DSC Risk Analysis

A group of DMs is gathered in order to outline the concepts which influence DSC risks and develop IVIFCM on these factors. These factors are identified with a literature review and DMs' opinions; relationships, strength, and direction between them are refined based on a consensus of the DMs. The panel was made up of 4 DMs including academics and industry experts. There are 19 factors in DSC risk analysis [15], [16]. Volatility (C_1), Collaboration (C_2), Visibility (C_3), Integration (C_4), Flexibility (C_5), Efficiency (C_6), Innovation (C_7), Automated Execution (C_8), Silver Bullet Chase (C_9), Demand Forecast (C_{10}), Agility (C_{11}), Responsiveness (C_{12}), Alignment (C_{13}), Over Confidence (C_{14}), Analytic Adaptability (C_{15}), Information Sharing (C_{16}), Lack of Knowledge (C_{17}), and Personalized Experience (C_{18}) concepts are defined to analyze the risk factors in order to effectively implement the DSC structure (C_{19}).

MATLAB software is used to code the algorithm. The code runs freely until 100 iterations but stops if the values converge before than that. The tolerance of the convergence is taken as the ten to the power of minus six. Considering the hesitancy degrees' negative effect on the interrelations, the linguistic scale must be chosen carefully, as given in Table 1. Utilizing the initial values of the concepts as the initial state IVIF values are taken as VH; $A_i^0 = VH, \forall i = 1, \dots, 19$ by the DMs. The strengths of influence and directions are displayed in Figure 1, red lines in the graph represent negative relations, the IVIFCM are simulated in order to reach a steady state. The sigmoid function is used as activation functions.

Table 1. Scale for IVIFCM linguistic terms.

Terms	EL	VL	L	M	H	VH	EH
$\mu_A^L(x)$	0.00	0.10	0.30	0.50	0.60	0.80	0.95
$\mu_A^U(x)$	0.00	0.15	0.35	0.55	0.75	0.90	1.00
$v_A^L(x)$	0.95	0.75	0.55	0.40	0.20	0.05	0.00
$v_A^U(x)$	1.00	0.85	0.65	0.45	0.25	0.10	0.00

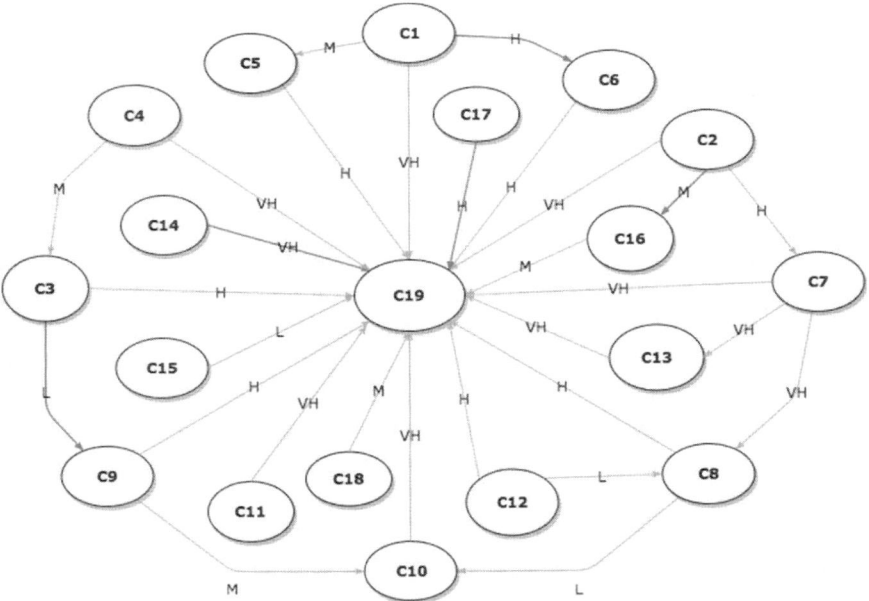

Figure 1. Casual relationships, directions, and strengths of DSC risk analysis.

The IVIFCM experiment converged in 5 iterations in MATLAB coding and the resulting value of the output concept is computed as C_{19}: 0.7760. The slight escalation in the C_{14} concept highly negatively affects the most "effectively implement the DSC structure". The rest of the values of each concept are displayed in Table 2. Compared to the initial state values, the average hesitation of IVIF values increases over time, corresponding to the emergence of uncertainty about the effective implementation of the DSC structure in the future. Although the resulting convergence is intensely related to the topology of IVIFCM, it is observed that the resulting outcome of concepts for DSC structure obtained by IVIFCM differs primarily in DSC risk analysis. This suggests that DSC risks are contemplated the most significant category of concepts in the IVIFCM model. Instead of traditional CM approaches, the IVIFCM model is based on a more general concept of IVIF sets, which provides a successful tool for dealing with robust ambiguity in the values of concepts and their causal relationships.

Table 2. Final values for each concept.

C_1:	0.649	C_2:	0.639	C_3:	0.662	C_4:	0.653	C_5:	0.701	C_6:	0.701
C_7:	0.629	C_8:	0.680	C_9:	0.667	C_{10}:	0.678	C_{11}:	0.677	C_{12}:	0.680
C_{13}:	0.678	C_{14}:	0.769	C_{15}:	0.742	C_{16}:	0.718	C_{17}:	0.753	C_{18}:	0.718

5. Conclusion

In this paper, the IVIFCM approach is presented to analyze DSC risks. IVIFCM is based on IVIF centered-graph-structure to represent the causal reasoning. IVIF allows the representation of interval values among causalities between several uncontrollable or controllable factors. IVIF centered-graph-structure makes the systematic causal propagation happen, in other words, backward or forward linkage. IVIFCM is established to discern the components associated to the vulnerabilities of DSC. Since the DMs' causal reasoning is unavoidably subject to the limitations of human cognitive processes, it is most likely to be biased. Additionally, DMs' remembrance is variable and finite. IVIFCM can assist with easing the effect of cognitive boundaries through providing data on the suitable direction leading to the success of DSC. The major contribution and the originality of the proposed study is the deepening the knowledge of the cognitive image of the DSC by introducing IVIFCM technique for the first time, which provides insight regarding the digitalization challenges.

The bottom line of IVIFCM is that it is highly promising. The proposed approach, therefore, bestows a feasible methodology to establish consequence propagation. The authors are committed to further developing IVIFCM to meet DSC's necessities. In the future, DSC risk factors can be elaborated and the research area can be widened to include other fuzzy environments in the comparison.

Acknowledgments

The authors are grateful to the experts who have been very helpful and supportive during the evaluation process of this research.

References

1. G. Buyukozkan and F. Gocer, in *2017 IEEE International Conference on Fuzzy Systems*, pp. 1–6, (2017).
2. The Digital Supply Chain Initiative, Sanya, China, (2015).
3. L. A. Zadeh, *Inf. Control*, vol. 8, no. 3, pp. 338–353, Jun. (1965).
4. K. Atanassov, *Fuzzy Sets and Systems*, 20th ed. (1986).
5. K. Atanassov and G. Gargov, *Fuzzy Sets Syst.*, vol. 31, no. 3, pp. 343–349, Jul. (1989).
6. D. K. Iakovidis and E. Papageorgiou, *IEEE Trans. Inf. Technol. Biomed.*, vol. 15, no. 1, pp. 100–107, Jan. (2011).
7. R. Axelrod, *Structure of Decision*. NewJersey: Princeton University Press, (1976).
8. B. Kosko, *Int. J. Man. Mach. Stud.*, vol. 24, no. 1, pp. 65–75, Jan. (1986).
9. E. I. Papageorgiou and D. K. Iakovidis, in *2009 9th International Conference on Information Technology and Applications in Biomedicine*, pp. 1–4, (2009).
10. Guo-Fang Zhang, Li-Hui He, Yu-Ting Jiang, and Yue-Wang, in *2013 International Conference on Machine Learning and Cybernetics*, pp. 188–193, (2013).
11. E. I. Papageorgiou and D. K. Iakovidis, *IEEE Trans. Fuzzy Syst.*, vol. 21, no. 2, pp. 342–354, Apr. (2013).
12. D. K. Iakovidis and E. I. Papageorgiou, in *2011 IEEE International Conference on Fuzzy Systems*, no. 4, pp. 821–827, (2011).
13. E. Dogu, T. Gurbuz, and Y. E. Albayrak, vol. 641, no. 36. *Cham: Springer International Publishing*, (2018).
14. P. Hajek and O. Prochazka, in *Intelligent Decision Technologies*, vol. 73, Cham: Springer International Publishing, pp. 207–217, (2018).
15. A. Ganeriwalla, G. Walter, L. Kotlik, R. Roesgen, and S. Gstettner, *bcg perspective*, (2016).
16. M. Dougados, B. Felgendreher, CapGemini, *Infor, and GTNexus*, (2016).

Development of the approach to check the correctness of workflows

Alexander Afanasyev, Nikolay Voit, Maria Ukhanova, Irina Ionova

Ulyanovsk State Technical University
32, Severny Venetz str., Ulyanovsk, 432027, Russia

The paper deals with an interesting approach for checking business process workflows. Business processes workflows are presented as a diagram based on graphical languages such as eEPC, UML, BPMN, IDEF0, and etc. We offer the approach based on a temporal grammar, a timed automaton and an ontology. It allows narrowing the semantic gap between business process analysis and business process execution. We propose to check the structural and semantic errors. Semantic errors are checked by the ontological model. The proposed approach can detect 23 errors, and the results are provided in visual form. The approach is illustrated by an example.

1. Introduction

Workflow is a trace for executing a set of business process tasks, taking into account time constraints and data flows. It is necessary to identify and correct errors in the processes in order to avoid failures. Although errors can occur in cause-effect relationships among tasks, we focus on the workflow execution's semantic errors, especially on denotative and significative semantics. Denotative semantics determines the errors of antonymy, words' synonymy in the workflow's business events. Significative semantics reveals workflow structural errors on the basis of trace isomorphism and homomorphism. Ad-hoc is an add-on in the workflow and makes the process not so strict, thereby it violates the canonical rules of the process. Such a workflow execution can lead to a customer's satisfaction decrease, an employee overload increase, a brand image decrease, a profits decrease, and a significant management time expenditure. Thus, it is important for business to identify and correct semantic errors in workflows.

The workflow should be conceptually presented in the formal language for analysis and expertise before deployment into a real business environment. This view is also useful when transferring workflow tasks between designers, users, process engineers, managers and technical personnel. In addition, process models in the presentation can be tested by approaches that have a corresponding formal language to determine a workflow. Conceptual representations can be performed using Workflow Nets (WF-nets), Workflow Graphs, Object Coordination Nets

(OCoNs), Adjacency Matrix, Unified Modeling Language (UML) diagrams, Evolution Workflow Approach and Propositional Logic. Today test algorithms exist for WF-nets, Workflow Graphs, UML diagrams, Propositional Logic and Adjacency Matrix representations. And popular algorithms are those that are based on WF-nets and Workflow Graphs. WF-nets are based on Petri nets, and many formal methods for analyzing Petri nets are used to obtain theoretical solutions for problems encountered in the design of WF networks. Although many complicated structures of process language that are useful in a business environment can be implemented via WF-nets, the Workflow Management Council (WfMC) uses only six basic structures of process language. WfMC has adopted this approach to keep the simulation very simple and clear.

For a business event, a subset of workflow tasks is performed in accordance with the object data (customer data, environment data, business process data, and business domain data), for example, such as ordering. This subset of tasks, together with the workflow used to execute the business process, is called an instance. Until now, most workflow management systems (WfMSs) provide only modeling tools for testing workflow models via a trial and error method [1]. These modeling tools can be used to perform a subset of workflow instances to check for structural conflicts that may occur in the respective scenarios. However, workflows can have many instances, and the verification task becomes difficult for all instances.

Check for structural and semantic errors in workflows is a computational task, so different formal approaches and languages can be used for this. However, the approach taken for verification should support the language of the workflow description. Because of the computational complexity of a task (polynomial, exponential), only a few approaches successfully cope with the verification of workflows, taking into account constraints, including time constraints, for all types of workflow graphs.

The paper has the following structure. The list of standard problems with workflows is given in Introduction. The Related works paragraph has an overview of works on this topic. In Temporal grammar, Timed automaton, Ontology and List of errors, we describe the approach. In Elaborate example, the Implementation presents our proposed approach. Results and the further directions of researches are in Conclusion.

2. Related Works

We have studied many research works considered with the workflows' specification, verification and translation. Some of them focus on formal semantics and workflows verification methods using Petri nets, process algebra, and abstract state machine [5], [6]. Decker and Weske offer a formalism based on Petri Nets to define such properties as reliability and promptness, and a method for testing these two properties. However, they only describe the synchronous relationship and do not have any research comparisons for high-level interaction modeling languages as BPMN. Lohmann and Wolff offer the analysis using existing templates and monitoring them using compatible templates. In [3], the authors draw attention to the translation of BPMN into the process algebra for analyzing choreographies using the help model and checking equivalence. The Woflan tool was developed by H.W.M. Verbeek and W.M.P. Van der Aalst for checking structural conflict errors in WF-nets. The Woflan tool can also be used to test inheritance.

The main limit of the methods considered is that they do not work in different types of diagrams at the same time; it means that the input diagrams cannot be analyzed in some cases.

3. Temporal Grammar

Temporal grammar (RVT-grammar) is defined as the tuple

$$G = (V, \Sigma, \tilde{\Sigma}, C, E, R, \tau, r_0). \tag{1}$$

where $V = \{v_e, e = \overline{1.L}\}$ is an additional alphabet for the operation onto a memory; $\Sigma = \{(a_l, t_l), l = \overline{1.T}\}$ is an alphabet (words) of events; $\tilde{\Sigma} = \{(\tilde{a}_n, \tilde{t}_n), n = \overline{1.\tilde{T}}\}$ is a quasi-term alphabet, extending Σ; $C = \{c_i, c_i = c_i + t_{l-1}, i \in N\}$ is a set of a time identifier, and a beginning $c_i = 0$; E is a set of the temporal relations as $\{c_i \sim t_l\}$, where c is a variable (a time identifier), $\sim \in \{=, <, \leq, >, \geq\}$; $R = \{r_i, i = \overline{0.I}\}$ is a rule of this grammar G (a set of production rule's complexes), where this complex r_i has a subset P_{ij} of the production rule $r_i = \{P_{ij}, j = \overline{1.J}\}$; $\tau = \{t_l \in [0; +\infty], l = \overline{1.T}\}$ is a set of timestamps, where $c_i \in \tau \times \sim \times \tau$; $r_0 \in R$ is an axiom of this grammar (a name of the first production rule), $r_k \in R$ is the last production rule. The production rule $P_{ij} \in r_i$ has a view as

$$(a_l, t_l) \xrightarrow{\{W_\gamma(v_1, \ldots, v_n)|E\}} r_m. \tag{2}$$

where $W_\gamma(v_1, \ldots, v_n)$ is n-relation, that defines a type of an operation over memory, depending on $\gamma = \{0,1,2,3\}$ (0 – operation is not performed, 1 – write, 2 – read, 3 – compare); (a_l, t_l) is a word as a pair of an event and a timestamp; $r_m \in R$ is a name of a target production rule. The language $L(G)$ of this grammar has words as (a_l, t_l) and presents a trace $\sigma = \{a_0, 0\} \to \{a_l, t_l\} \to \{a_k, t_T\}$.

4. Timed Automaton

The timed automaton *TimedAutomaton* is represented by following components:

$$TimedAutomaton = (V, \Sigma, C, E, \delta, S_0, S, S_k) \qquad (3)$$

where $V = \{v_e, e = \overline{1.H}\}$ is an auxiliary alphabet (the alphabet of operations over internal memory); $\Sigma = \{(a_l, t_l), l = \overline{1.L}\}$ is a terminal alphabet of a language; $C = \{c_i, c_i = c_i + t_{l-1}, i \in N\}$ is a finite set of clock identifiers, and a beginning $c_i = 0$; E – is a set of time expressions C (clock limitation and clock reset), is limited by the following expressions: onwards $\{c_i \sim t_l\}$, and c_i is a variable, and t_l is a constant, $\sim \in \{=, <, \leq, >, \geq\}$; $S = \{S_i, i = \overline{0.I}\}$ is a set of states; $S_0 \in S$ is a beginning state; $S_k \in S$ is a ending state; the state transition function of automaton $\delta: S_i \times (a_l, t_l) \xrightarrow{\{W_\gamma(v_1, ..., v_n) | E\}} S_m$ is the ratio of transitions, where $W_\gamma(v_1, ..., v_n)$ is a *n*-th relation, which determines the type of operation over the internal memory depending on $\gamma \in \{0,1,2,3\}$ (respectively, 0 – operation is not performed, 1 – record, 2 – read, 3 – compare); $v_1, ..., v_n \in V$; $r_i \in R$ is the name of the complex of a production rule's source; $S_m \in S$ is the name of the state of a production rule's successor.

5. Ontology

The ontology is presented as follows:

$$O = (Class, Property, Relation, Axiom). \qquad (4)$$

where *Class* is a set of concepts (classes) defined for a particular subject domain; *Property* is a set of concept properties; *Relation* is a set of semantic links defined among concepts in *Class*. A set of relation types is the following: one to one, one to many and many to many. A set of basic relations is presented by: synonymy, a kind of something, part of something (*f*), instance of something, property of something (*property of*); *Axiom* is a set of axioms. An axiom is a real fact or a rule that determines the cause-effect relationship.

6. List of Errors

We can detect the following errors: 1. The cyclic link; 2. Mutually exclusive links; 3. Multiple links; 4. Remote context error; 5. Control transfer failure; 6. Input multiplicity error; 7. Output multiplicity error; 8. Invalid link; 9. Link error; 10. Access level error; 11. Message transmission error; 12. Control Transfer Error; 13. A quantitative error of diagram's elements; 14. Excluding links of a wrong type; 15. A call directed to the life line; 16. Dead link; 17. Dependency multiplicity violation; 18. Mutually exclusive links; 19. A synchronous call until a response; 20. Great synonymy; 21. Objects' antonymy; 22. Conversion of relationships; 23. Inconsistency of objects.

7. Elaborate Example

Business processes as workflows are presented as a diagram based on graphical languages such as eEPC, UML, BPMN, IDEF0 using software of IBM, Whitestien Technologies, ARIS, OMG, etc.

The temporal property of this design process (workflow) is very important for designing and manufacturing, especially if we can manage time limits of production. The ARIS eEPC methodology gives all temporal properties of a design process as against UML, IDEF0, BPMN. Events and Functions are the main object in this methodology. Therefore, Events are represented by moments, and Functions are represented by decision-making processes. Events have a Frequency folder with attributes as the frequency of an event. Functions have a Simulation folder with attributes as a time period for making decision. The logical operations as AND, OR, XOR are control functions of workflows in this methodology.

Let's take a look at Figure 1 where it is described a sample of an approval process of design documentations. It usually takes three days to approve a design documentation for an assembly. This process consists of the Begin, Processing, Checking, Improve, Approval, End phases which are described on the basis of eEPC methodology.

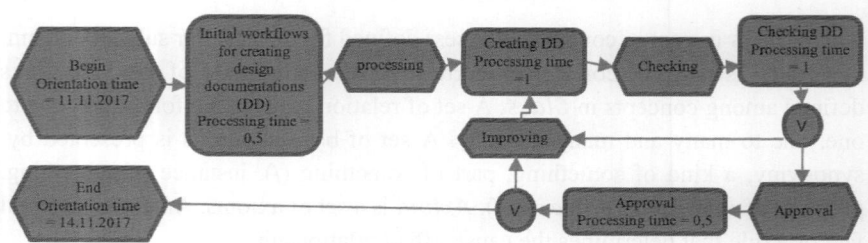

Figure 1. The approval of a design documentation based on eEPC methodology.

The Processing time attribute is presented in the above-mentioned folders (the sum of them is 3). The Orientation time attribute is presented in the Begin and End folders. We have developed a mathematical tool in order to analyze the temporal property of workflows by the temporal grammar. This tool helps us to dynamically reorganize parallel workflows for automated systems' lifecycle in a large enterprise. The goal is to reduce the idle time in manufacturing and time spent on an assembly.

Let's write grammar for Figure 1. The one automated store m and the one tape 1^t are used by this grammar as the internal memory. The timestamp tm is recorded on a tape. In Table 1, we can see the written RVT-grammar for Figure 1.

Table 1. RVT-grammar for Figure 1.

Source production rule's complex	Quasi-term	Target production rule's complex	Relation
r_0	B	r_1	$W_1(1^{1m})$, $W_1(tm^{1t})$
r_1	W	r_2	$W_1(2^{1m})/W_3(c<=tm^{1t})$
r_2	C	r_3	$W_1(3^{1m})/W_3(c>tm^{1t})$
r_2	C	r_4	$W_3(c<=tm_{1t})$
r_3	I	r_4	$c=0$, $W_2(3^{1m})/W_3(c<=tm^{1t})$
r_4	A	r_3	$W_2(3^{1m})/W_3(c>tm^{1t})$
r_4	A	r_k	$W_2(2^{1m})$, $W_2(1^{1m})$
r_k	E	–	–

Let's write a timed automaton for RVT-grammar. The alphabet of an event process is a set of Σ= {Begin, Processing, Checking, Improve, Approval, End}. S = {B, W, C, I, A, E}. Let's define the state transition function of automaton δ that has been formulated in a section Timed automaton (Table 2).

Table 2. A matrix of the state transition function of automaton δ.

Constraint	B	W	C	I	A	E
B	$W_1(1^{1m})$, $W_1(tm^{1t})$					
W		$W_1(2^{1m})/W_3(c<=tm^{1t})$				
C			$W_1(3^{1m})/W_3(c>tm^{1t})$	$W_3(c<=tm^{1t})$		
I				$c=0$, $W_2(3^{1m})/W_3(c<=tm^{1t})$		
A			$W_1(3^{1m})/W_3(c>tm^{1t})$			$W_2(2^{1m})$, $W_2(1^{1m})$
E						

Let's depict an ontology for this timed automaton. We transform the automaton into an ontology, replacing the *States* with *Classes*, adding properties to the notions (*Property*). We get a graphical representation of the ontology with class properties (Figure 2).

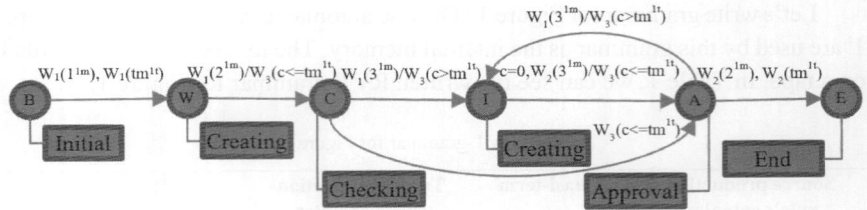

Figure 2. The ontology for Figure 1.

8. Discussion

In Figure 2, we see that *Class* W and I have the *Creating* property, so you can say that the properties of these classes are synonymous and you need to establish a synonymy relationship among these classes. Also, classes B and E are antonyms, so an antonymy relationship can be established between them. Thus, it is possible to structure an ontology and identify semantic errors, including isomorphism and homomorphism, according to List of errors.

9. Conclusion and Future Works

We have developed an approach in order to analyze errors according to the list of errors in business processes workflows. This research work is different from existing ones in the fact that it has checked not only structural errors but also semantic errors. The proposed temporal finite-state grammar has a linear characteristic of time analysis of a workflow, takes into account the process description language and can be applied to any diagram. A timed automaton allows modeling a process in visual form. Analysis for denotative and significative errors in workflows is based on the ontological model. We developed a list of structural and semantic errors encountered in workflows.

Our future works will present examples of the approach application in industry, training, cyber-physical systems, in the development of automated systems.

Acknowledgments

The reported study was funded by RFBR according to the research project № 17-07-01417.

References

1. H. H. Bi, J. L. Zhao, *Information Technology and Management*, **5**, 293 (2004).
2. P. Poizat, G. Salaün, Checking the Realizability of BPMN 2.0 Choreographies, in *Proc. of* SAC'12 (2012).

3. A. N. Afanasyev, N. N. Voit and S. Y. Kirillov, Development of RYT-grammar for analysis and control dynamic workflows, in *Proc. of International Conference on Computing Networking and Informatics (ICCNI)*, (Lagos, Nigeria, 2017). doi: 10.1109/ICCNI.2017.8123797, URL: http://ieeexplore.ieee.org/stamp/stamp.jsp?tp=&arnumber=8123797&isnumber=8123766.
4. A. Afanasyev, N. Voit, R. Gainullin, The analysis of diagrammatic models of workflows in design of the complex automated systems, in *Proc. of International conference on Fuzzy Logic and Intelligent Technologies in Nuclear Science (FLINS2016)*, (France, Roubaix, 2016).
5. Y. Wang, Y. Fan, Using Temporal Logics for Modeling and Analysis of Workflows, in *Proc. of E-Commerce Technology for Dynamic E-Business, IEEE International Conference on*. doi: 10.1109/CEC-EAST.2004.72 (2004).
6. N. Saeedloei and G. Gupta, Timed definite clause ω-grammars, In *Proc. of/ Technical Communications of the International Conference on Logic Programming* (2010). URL: http://www.floc-conference.org/ICLP-home.html

Estimation of fuzzy reliability:
A case study for flash vessel in ammonia storage tank

Satish Salunkhe

*Computer Engineering Department, Terna Engineering College, Mumbai University
Navi Mumbai, Maharashtra, India*

Ashok Deshpande

*Berkeley Initiative Soft Computing–Special Interest Group (SIG)-Environment
Management Systems (EMS), UC Berkeley, CA 94720-5800, USA*

Fault tree analysis is one of the most effective techniques for estimating the frequency of occurrence of hazardous events in probabilistic risk assessment study. The uncertainty present in fault tree evaluation can also be modeled using fuzzy set-theoretic operations. In this paper, we have attempted to estimate reliability of hazardous event via Top Event Probability (FTEP) in fuzzy fault tree analysis. The case study relates to flash vessel in an ammonia tank in a large Fertilizer complex in Mumbai, India.

1. Introduction

Major industrial disaster during the last four decades has brought to fore the adverse consequences of industrial development with the scientific assessment. The Flixborough disaster (1974), the Bhopal Gas Tragedy (1985), Mexico LPG disaster (1984), and many such events have underlined the need for a priori Probabilistic Risk assessment (PRA). It is a systematic approach to evaluating likelihood, consequences, and risk of adverse events and includes Fire and Explosive Index & Toxicity Index (FETI), Consequence analysis, event tree development and fault tree construction and evaluation. Fault tree analysis (FTA) is a top-down, deductive failure analysis in which an undesired state of a system is analyzed using Boolean logic to combine a series of lower-level events. This analysis method is mainly used in the fields of safety engineering and reliability engineering to understand how systems can fail, to identify the best ways to reduce risk or to determine (or get a feeling for) event rates of a safety accident or a particular system level (functional) failure.

Ferdous et al., In a recent paper [2], present some of the methods of uncertainty handling in FTA for process system risk analysis with a focus on fuzzy set theory and Dempster-Shafer Theory of evidence. In this paper, apart from the application of some of the formulations based on fuzzy set theory [5][6], an attempt has been made to present a relatively simple and straightforward formalism in the estimation of the imprecise probability of a fuzzy event p. Purba et al., [3] suggested a different method for the computation of top event probability in FTA. The subtask of the method includes failure function granulation and designing, basic event failure evaluation, expert justified aggregation, basic failure event calculation (defuzzification), and fuzzy top event probability calculation.

The contents of the paper are formalized as follows: Section 1 is an introduction and covers the various facets of PRA while Section 2 is a write up on the conventional procedure of fault tree construction and evaluation. Section 3 describes fuzzy fault tree analysis (FPFTA) approach, which incorporates the concept of failure possibilities, fuzzy numbers with a case study. The computation of fuzzy top event probability and reliability of the hazardous event flash vessel empty in ammonia plant are included the concluding remarks and further scope for research for an integral part of Section 4.

2. Fuzzy Fault Tree Analysis (FFTA) Approach

The evaluation of risk, arising out of an undesired event, comprises an estimation of the *expected frequency of undesirable events per unit time* (f) and *expected damage* (d)[4]. Then a customary definition of *risk*, R, regarding the *expected frequency of occurrence* and the *expected damage D*, is:

$$R = f * d \qquad (1)$$

All those techniques that help in determining the frequency or probability of events with adverse or potentially adverse effects can be considered as means of quantifying hazards. Hazard quantification may be performed after or in conjunction with hazard identification. Ideally, it should also assist identification of those parts/components of a system which contribute most to the enhancement of risk levels. The credibility of hazard quantification largely depends on the element of uncertainty introduced by those hazards which have not been identified. In our view, it is improper to evaluate fault tree using conventional probability approach.

2.1. *Preliminaries and notations*

This section covers some of the important elements used in the computation of fuzzy top event probability (or possibility using FST). It may be a practical and logical approach of characterizing uncertainty involved in failure data analysis using the concept of the interval of confidence- termed as *man-machine precision* in *computing with words* methodology and level of presumption expressed as membership function in fuzzy set theory. The concept of the fuzzy number of uncertainty is, therefore, suggested. Fuzzy number as shown in Figure 1 could be triangular (TFN) or trapezoidal (TrFN) which is possibility distribution, meaning thereby "whatever is improbable may not be impossible."

2.2. *Tanaka fuzzy arithmetic operators*

2.2.1. *Multiplication*

According to the extension principle the computational procedure for multiplication of two fuzzy sets E_i and E_j is quite involved and therefore, Tanaka suggested an approximation of the multiplication procedure by defining as follows

$$E_i \bullet E_j = (q_i^l q_j^l, p_i^l p_j^l, p_i^r p_j^r, q_i^r q_j^r) \qquad (2)$$

Multiplying the corresponding elements of fuzzy probabilities of event X_i and X_j respectively.

2.2.2. *Complementation*

The complementation of any fuzzy set E_i is given as:

$$\overline{E_j} = (1 - q_j^r, 1 - p_j^r, 1 - p_j^l, 1 - q_j^l) \qquad (3)$$

Multiplying the corresponding elements of fuzzy probabilities of event X_i and X_j respectively.

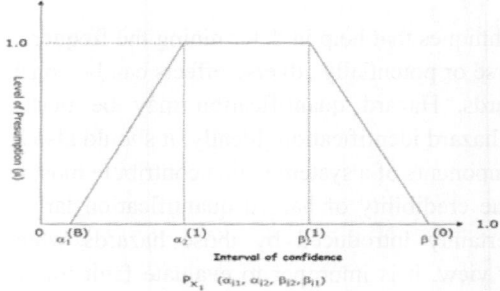

Figure 1. Fuzzy Probability-Trapezoidal representation.

For each basic event, the lower bound (q_l) and upper bound (q_r) of failure probability is given. We need to calculate the lower bound, most likely bound and upper bound of all intermediate events.

The steps to follow to compute top event probability is shown below.

<u>Step 1</u> – Compute the lower bound (α_{i1}) and upper bound (β_{i1}) of failure probability for each intermediate node (N_k) using Breadth First Traversal of given Fault Tree (FT) depending on the type of operators.

(a) For AND operator

$$\text{Lower Bound, } \alpha_{i1} = q_i^l q_j^l \qquad (4)$$

$$\text{Upper Bound, } \beta_{i1} = q_i^r q_j^r \qquad (5)$$

(b) For OR operator

$$\text{Lower Bound, } \alpha_{i1} = (1 - q_i^r) * (1 - q_j^r) \qquad (6)$$

$$\text{Upper Bound, } \beta_{i1} = (1 - q_i^l) * (1 - q_j^l) \qquad (7)$$

<u>Step 2</u> – Compute the interval of confidence [α_{i2}, β_{i2}] of middle bound of failure probability for each intermediate node (N_k) using Breadth First Traversal of given Fault Tree (FT) depending on the type of operators.

For AND operator

$$\text{Left Middle Bound, } \alpha_{i2} = p_i^l p_j^l \qquad (8)$$

$$\text{Right Middle Bound, } \beta_{i2} = p_i^r p_j^r \qquad (9)$$

For OR operator

$$\text{Left Middle Bound, } \alpha_{i1} = (1 - p_i^r) * (1 - p_j^r) \qquad (10)$$

$$\text{Right Middle Bound, } \beta_{i1} = (1 - p_i^l) * (1 - p_j^l) \qquad (11)$$

<u>Step 3</u>- Compute the most likely Middle Bound of failure probability for each intermediate node (N_k), divide the entire range of failure probability [α_{i1}, β_{i1}] of each corresponding child nodes (E_i) into some m number of strips $S_{E_k}^{(p)}$ of fixed window width size w.

$$ES_i = \{S_{E_i}^{(1)}, S_{E_i}^{(2)} \cdots, S_{E_i}^{(m)}\} \qquad (12)$$

$$ES_j = \{S_{Ej}^{(1)}, S_{Ej}^{(2)} \cdots, S_{Ej}^{(m)}\} \qquad (13)$$

Perform Cartesian Product of strip vector of each child nodes to get all possible middle bound values of intermediate node (N_k).

3. Case Study

The case study relates to flash vessel of ammonia storage tank [1][7]. Ammonia from the plant consists of inert which are to be removed before it is taken to main storage tank. Failure of flash vessel may lead to over pressurization of the tank and could result in excessive load on the refrigeration system. Pressure control valve at the upstream of flash vessel would regulate the pressure so that proper flashing can take place in flash vessel leading to separation of liquid and gas phase. The inert are removed from the top of the vessel and liquid from the bottom which is further taken to ammonia tank. Level control valve at the downstream in flash vessel controls the level of liquid ammonia in flash vessel. Figure 2 presents fault tree of Flash vessel in ammonia plant with top event as "Flash vessel empty."

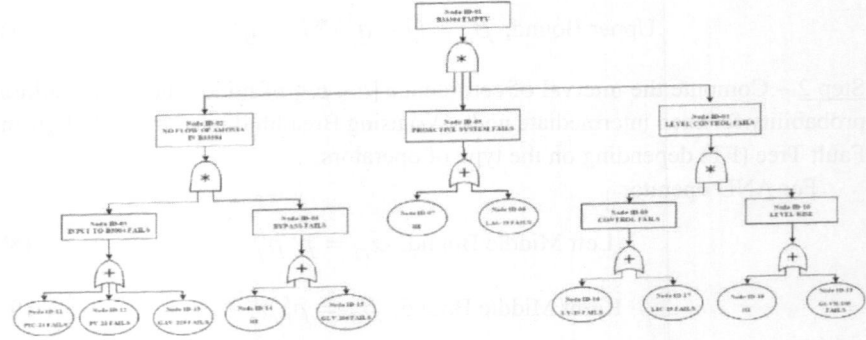

Figure 2. Fault tree for B33304 empty.

Input in the form failure probability of process control instruments including human failure is presented in table. In this study, the concept of triangular fuzzy number is used.

Table 1. Failure probability of basic events.

Node_ID	Basic Event	Failure Probability	
		Lower Bound	Upper Bound
11	Pressure Indicator and Controller (PIC21)	0.1	0.21
12	Pressure Value (PV21)	0.45	0.8
13	Global Valve (GAV219)	0.0007	0.0015
14, 18, 7	Human Error Operator (HE)	0.4	0.7
15	Manual Valve (Globe) GLVR106	0.00085	0.0012
8	Level Alarm Low (LAL19)	0.1	0.3
16	LV19	0.55	0.8
17	LIC19	0.09	0.19
19	GLVR106 Fails	0.1	0.22

3.1. Results and discussion

Figure 3 presents the outcome the fuzzy top event probability of flash vessel empty using fuzzy simulation approach wherein Tanaka operators are used. For intermediate Node ID-5 i.e. "Input to B3004 Fails" there are three basic child events namely Node ID-11, 12, and 13 with operator OR. Applying the procedure of FFTA the lower and upper bound of failure probability is 5.05×10^{-5}, 8.42×10^{-5} and the most likely value ranges between 5.59×10^{-5} to 8.3×10^{-5}.

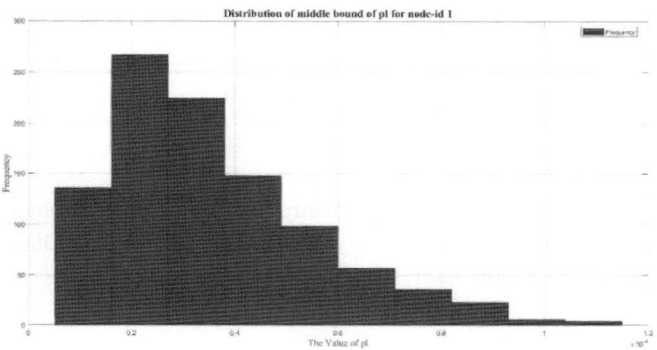

Figure 3. Distribution of most likely value of fuzzy top event probability.

The Top Event Probability (TEP) of hazardous event (accidental release of ammonia) using conventional FTA approach has been estimated as 1×10^{-5} /yr. Random combination of the fuzzy probability of primary events was used for $\mu = 1$, and the range of Fuzzy Top Event Probability (FTEP) was estimated. It could be inferred that the FTEP ranges between 1.14×10^{-6} and 5.1×10^{-4} /yr. with most likely value between 3.5×10^{-5} for $\mu = 1$. Employing conventional approach, TEP is 1.00×10^{-5} /yr., i.e., a value contained in the interval $(5 \times 10^{-6} - 5 \times 10^{-5})$ obtained through FFTA.

Table 2. Probability and reliability of fuzzy top event.

	Optimistic	Most Likely	Pessimistic
Probability of top event	0.0000051	0.000035881128918	0.00011489670000000
System Reliability	0.9999949	0.99996412	0.99988511

We believe that the results obtained using fuzzy simulation approach would be better as it signifies not probability but fuzzy probability for a hazardous event. We consider the top event is fuzzy, and therefore computed reliability of flash vessel as a fuzzy event is 99.996412%.

4. Concluding Remarks and Future Research

The major success of PRA study is based on the computation of the probability or more importantly fuzzy probability/possibility of undesired event and the possibility of risk of a fuzzy event. The occurrence of such an event might lead to undesirable consequences resulting into human or property loss or contribute towards overall environmental degradation. To further enhance the credibility of PRA, need for the estimation of error possibility (which is an aggregation of instrument error and human error possibility). It is strongly believed that the quantification of risk could be expressed as fuzzy fatality probability instead of fatality probability. However, much remains to be done as the biggest room in this world is the room for improvement.

Acknowledgments

The wholehearted assistance for the construction and computations of fault tree and the development from R.S. Olaniya and Mrs. B. Dabir and Dhananjay Raje for the collection of failure data is gratefully acknowledged.

References

1. NEERI Report on Hazard Study and Quantitative Risk Assessment of RCF Complex of Chembur Bombay, Vol. 1, New Ammonia Plant (Trombay) Nov. (1990).
2. Ferdous, R., Khan, F., Sadiq, R., Amyotte, P., and Veitch1, B., Fault, and Event Tree Analyses for Process Systems Risk Analysis: Uncertainty Handling Formulations, Risk Analysis DOI: 10.1111/j.1539-6924.2010.01475.x.
3. Purba, J. H., Lu, J., and Zhang, G., Failure possibilities for nuclear safety assessment by fault tree analysis, Special issue on "Nuclear Applications of AI and Computational Intelligence at FLINS2010" in Int. J. of Nuclear Knowledge Management.
4. Norman, J. McCormick, Reliability and Risk Analysis – Methods and Nuclear Power Applications, Academic Press, New York 1981 (Book).
5. Mishra, K. B., and Weber G. G. Use of fuzzy set theory for Level 1 study in probabilistic risk assessment, Fuzzy sets, and systems, 27, pp 139-160, 1990.
6. Kaufman, A, and Gupta M. M., Introduction to Fuzzy Arithmetic Theory and Applications, Van Nostrand, New York 1984 First Edition.
7. Deshpande, A. W., Deshpande, U. A. and Khanna P., Fuzzy Fault tree Analysis –A Case Study, paper presented at the 2nd International Conference on Fuzzy logic and Neural network. (IIZUKA, 92) July 17-22 Japan.

A fuzzy telematics data-driven approach for vehicle insurance policyholder risk assesment

Mohammad Siami, Anahita Namvar, Mohsen Naderpour, Jie Lu

Decision Systems and e-Service Intelligence Laboratory
Centre for Artificial Intelligence (CAI), Faculty of Engineering and IT
University of Technology Sydney (UTS)
Sydney, NSW, Australia
Mohammad.SiamiNamini@student.uts.edu.au;
{Anahita.Namvar, Mohsen.Naderpour, Jie.Lu}@uts.edu.au

> Recent technological advances in telematics devices are providing companies with opportunities to offer new products and services to their customers. Many insurance companies are exploiting the value of telematics devices to develop new insurance models based on use. To help these companies assess the risk associated with policyholders, we have developed a new fuzzy approach that analyzes complex driving patterns using big telematics data. A fuzzy risk score is calculated for each driver according to the frequency of their risky driving behaviors and the average speed of individual trips. The aggregate of these scores reflects the risk level of each policyholder and, thus, the potential exposure for insurers. The approach has multiple benefits for insurance companies, particularly in setting appropriate premiums for individual policyholders to counter some of that risk. We tested the proposed approach on a dataset containing the driving habits of more than 2500 drivers. The proposed approach can estimate the risk level of each policyholder based on their driving features.
>
> *Keywords*: Risk assessment; fuzzy logic; usage-based insurance; big data.

1. Introduction

Competition plays a vital role in the insurance market, and gaining a competitive edge through new products and services is an important part of this landscape. Insurance companies underwrite risk; hence, gathering sufficient knowledge about the risk profiles of their customers is a crucial element in managing their organizational exposure and capital adequacy rates. In vehicle insurance, the most beneficial source of information in developing a driver's risk profile is the historical data on their previous insurance claims. Traditional risk assessment methodologies tend to rely heavily on such information to estimate the real levels of risk in different timeframes, locations, and driving conditions, but these methods break down when information about a customer's driving behavior is scarce. However, recent technological improvements in telematics are providing new opportunities for insurance companies to use in-vehicle devices to track a policyholder's driving habits [1].

Telematics devices record vehicle movements – location, speed, acceleration – and the changes in these parameters under various conditions. The big data captured by these devices forms an accurate picture of a driver's habits [2]. However, the data captured is big data and comes with all the benefits and impediments that big data brings; it is very valuable but also very challenging due to its volume, velocity, variety, and veracity. The emerging issues insurance companies are facing, as a result, give rise to a number of research questions, which form the focus of this study. 1) What is the characteristics of an expert system that can be assessed the risk of policyholders based on driving habits? 2) How can the trip's risk be evaluated in an uncertain situation and which parameters are the most important in calculating risk? 3) How can the dangerous driving patterns of vehicle insurance policyholders be detected by considering the value of unstructured telematics data?

To answer these questions, we developed a new fuzzy risk assessment approach that generates a profile for each customer that reflects the risk associated with their policy based on their driving patterns. The outcomes of this research make several contributions to the literature on assessing vehicle insurance risk with telematics. First, we proposed an expert system to investigate the drivers' exposure, the proposed risk assessment process provides insurance companies with the ability to evaluate the risks associated with each policyholder based on their driving habits. Second, we proposed fuzzy risk assessment methodology, which calculates the risk score of driver in one trip based on frequency of dangerous behaviors and average speed. Third, our solutions offer a new way of detecting dangerous driving patterns in insurance by using trajectory data. As risk assessment includes many areas of uncertainty, the approach relies on fuzzy logic to evaluate the unknown or ambiguous factors in a situation. To the best of our knowledge, no study has considered the risk of vehicle policyholders using a fuzzy risk assessment scheme.

The rest of this paper is organized as follows. Section 2 provides a review of the literature on assessing driving risk. Section 3 details the proposed fuzzy expert system for risk assessment with vehicle insurance policyholders. Sections 4 and 5 present the experimental results, conclusion and future work.

2. Literature Review

Vaia et al. [3] proposed the first example of usage-based-insurance with telematics as a result of collaborations with Unipol, one of Italy's largest insurers, and Octo Telematics, a technology provider. They introduced telematics devices as a technology for data gathering and proposed a methodology for risk assessment based on driving behaviors, i.e., mileage and travel times. According to Husnjak et al. [4], with such data, insurers can calculate premiums for usage-based insurance with pay-as-you-drive and pay-as-how-you-drive measures and ask customers to pay based on their driving styles.

A considerable amount of literature has been published on investigations into driving conditions, but there are relatively few studies that have investigated driving conditions for risk assessment. Handel et al. [5] proposed a measurement methodology for driving characteristics. They used geospatial data to analyze vehicle movements and presented a value for usage-based-insurance. They introduced acceleration, velocity, trip distance and duration, the time of day, and the number of maneuvers and their harshness, as measures for a customer's risk assessment. Eboli et al. [6] explored the relationship between velocity and acceleration to distinguish dangerous driving conditions, finding a correlation between instantaneous velocity and acceleration in dangerous drivers. Based on their findings; a driver's behavior is risky when the value of acceleration is bigger than the set threshold in Equation 1.

$$|\bar{a}| = g \cdot \left[0.198 \cdot \left(\frac{V}{100}\right)^2 - 0.592 \cdot \left(\frac{V}{100}\right)^2 + 0.569 \right] \quad (1)$$

where $|\bar{a}|$ is the instantaneous acceleration norm, and V is the value of velocity(km/h) in each second. g denotes gravity, which is equal to 9.18(m/s2). According to this equation, when the value of acceleration is more than $|\bar{a}|$(m/s2) the driver is engaging in risky behavior.

Telematics devices stream data are unstructured and very big, and without analytics, the value of these data cannot be leveraged. Big data analytics with telematics data is an interesting research topic, but so far very few studies have investigated its pertinence to the risks associated with vehicle policyholders for insurers. Our previous research presented a big data-driven conceptual model for estimating the risk of each vehicle policyholder [7], but the lack of a quantitative method for risk assessment in this area motivated us to propose a fuzzy risk assessment approach for use by insurance companies.

3. The Fuzzy Telematics Data-driven Risk Assessment Approach

The research aims to propose an expert system for assessing the risk of drivers based on their driving characteristics. In many real-world situations, experts assess the risk of events based on their opinions and previous experiences, and their opinions have a linguistic structure. However, linguistic structures are often imprecise and involve uncertainty. Hence, the risk assessment system presented in this study employs fuzzy logic to eliminate the uncertainties and imprecision of real-world situations [8].

The proposed expert system has three levels. At Level 1, dangerous driving patterns by considering the velocity and acceleration according to Equation 1 have been detected, and the frequency of these behaviors for each trip is calculated. At Level 2, the fuzzy risk score of each trip is calculated with a fuzzy risk assessment. At Level 3, the final risk score of each policyholder will be calculated by aggregating the individual risk scores for each trip into one risk score. This score depicts the risk level of a vehicle policyholder based on driving habits.

3.1. Dangerous behaviors

Eboli et al. [6] found the relationship between instantaneous velocity and acceleration as an indicator of dangerous driving. They show that a driving pattern is unsafe when acceleration exceeds $|\bar{a}|$, as reflected in Equation 1. To the best of our knowledge, no study has considered this relationship as a metric for vehicle insurance risk assessment.

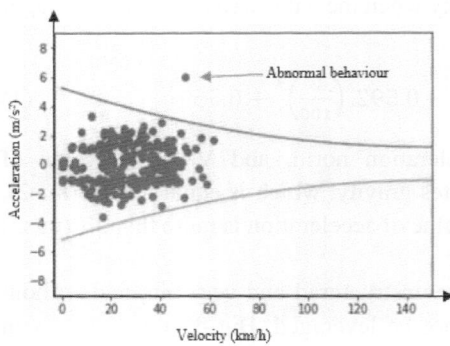

Figure 1 illustrates an example relationship between acceleration and velocity for one trip as a two-dimensional plot. The plot only contains one abnormal (unsafe) behavior, which is a low frequency of risky positions. Hence, the driver has behaved safely during this trip.

Figure 1. Dangerous behaviors.
The x-axis measures velocity in km/h. The y-axis measures acceleration in m/s2. The two curved lines indicate the safe range of acceleration based on velocity.

3.2. Trip risk score

To obtain the risk level of each trip, we proposed a fuzzy risk assessment process, and we used the scikit fuzzy package in python to develop the fuzzy model.

1) Linguistic variables

The model includes three fuzzy linguistic variables. The input variables are speed and frequency. The output variable is risk. Each variable is explained below. The fuzzy sets appear in Table 1.

Fuzzy input variable 1: according to transportation research [9] and expert opinions in the area, a safe value for velocity in urban areas ranges from 0 to 180 (km/h). The four fuzzy linguistic values defined for this range are low, medium, high, and very high.

Fuzzy input variable 2: the frequency of a driver's risky behavior, i.e., the frequency fuzzy variable, can be categorized as non-frequent, low-frequent, and high-frequent [10].

Fuzzy output variable: the fuzzy values for the output risk variables are defined as Acceptable (A), Tolerable-Acceptable (TA), Tolerable-Not-Acceptable (TNA), Not-Acceptable (NA), following [11].

2) Membership functions

A number of membership functions can be used to determine fuzzy linguistic variables, such as triangular, trapezoidal, and Gaussian. Selecting a suitable membership function fundamentally depends on: the characteristics of the variable; the information available; and expert knowledge. We chose a combination of trapezoidal and triangular functions defined according to the shapes of the membership functions. The triangular functions simplify calculations, and the trapezoidal functions increase the sensitivity of the bounds [12]. Table 1 shows the membership functions for the speed, frequency, and risk variables.

Table 1. Fuzzy linguistic variables and parameters.

Speed		Frequency		Risk	
L.V	P.	L.V.	P.	L.V	P.
Very high	(0,0,40,60)	Non- Frequent	(0,0,2,5)	Acceptable (A)	(0,0,2)
High	(40,60,80,100)	Low- Frequent	(2,5,8)	Tolerable-Acceptable (TA)	(2,5,8)
Medium	(80,100,120,140)	High- Frequent	(5,8,100,100)	Tolerable-Not-Acceptable (TNA)	(2,4,6)
Low	(120,140,180,180)			Not-Acceptable (NA)	(4,6,6)

3) Fuzzy rules

Defining a risk matrix is a major step in developing a fuzzy expert system. Table 2 shows the risk matrix proposed in our approach, including the 12 fuzzy rules defined for this application. An example rule follows.

IF the velocity is very high AND frequency is high-frequent, then the risk is NA.

The fuzzy rules defined in the risk matrix should fill the gap between the input and output variables [11]. The 12 rules we defined are based on expert opinions and relevant studies on transportation and accident risk analysis. The key insights derived from these studies follow.

Table 2. Fuzzy risk matrix.

dangerous behaviours		Low	Medium	High	Very High
	High Frequent	NA	NA	NA	NA
	Low Frequent	TNA	TA	TNA	NA
	Non Frequent	A	A	A	NA
		_____ Velocity _____			

- Velocities in excess of 120 km/h reflect unsafe driving patterns.
- Trips with a frequency of dangerous behaviors exceeding 8% are considered unsafe [16].
- The probability of an accident for drivers with mid-range velocity (60-90 km/h) is less than low (0-30 km/h) and high (+90 km/h) intervals [9].
- The probability of an accident increases as variance in speeds increases on all types of roads [13].

The centroid method is used for defuzzification.

3.3. Customer risk score

The overall risk score of a driver is calculated with Equation 2.

$$Customer\ Risk\ Score = \frac{\sum_{i \in tr} trScore_i}{N} \quad (2)$$

where trScore is the risk score of each trip, and N is the total number of trips during the period studied.

4. Experiments

To evaluate the performance of the model, we used a dataset collected by an insurance company containing trip data for over 500,000 journeys for more than 2500 drivers. We have 200 trips for each driver in this dataset [14].

4.1. Calculating the trip risk score

The risk score of each driver for each trip was calculated according to the defined fuzzy variables and risk matrix, and then the results compared with the pattern of velocity and acceleration proposed by [6] to differentiate between safe and unsafe driving habits.

Figure 2 illustrates the resulting driving patterns of four sample trips (A, B, C, and D). Table 3 lists the estimated risk scores for each trip.

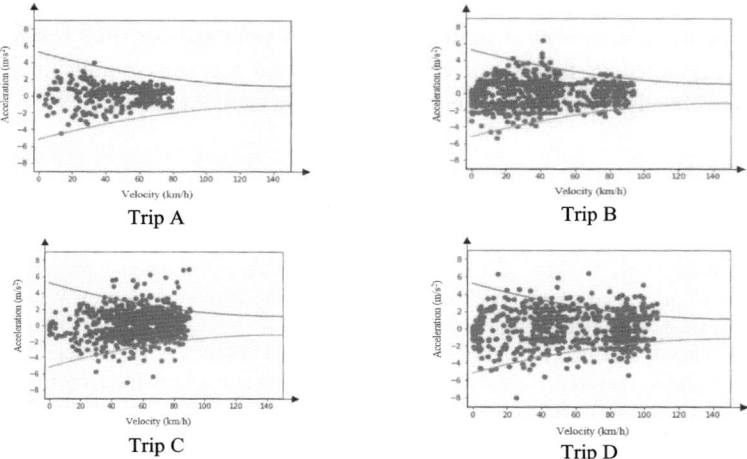

Trip A Trip B

Trip C Trip D

Figure 2. Driver behaviors.

These figures depict the driving habits of four sample trips according to proposed method by Eboli et al. [6]. The results indicate that trip A shows very few risky behaviors and trip D shows a large number of risky behaviors.

From Table 3, we can see the proposed risk score of the four sample trips from the selected drivers are in line with the abnormal behaviors defined in [6].

Table 3. Sample trips risk scores.

Trip	Risk Level	Risk score	Abnormal behaviors
A	A	0.66	0.34 %
B	TA	2	2.59%
C	TNA	4	7.41%
D	NA	5.33	15.07 %

5. Conclusion and Future Works

Technological improvements in recent years have introduced telematics devices that can be used for many purposes. These devices generate useful data for describing a driver's characteristics. However, to date, a quantitative way to calculate a driver's risk score based on the actions reflected in telemetric data has not been studied. This gap in the literature motivated us to develop a risk assessment methodology based on telemetric data. Our model has three levels. The first step is to count the occurrence of dangerous driving habits in each individual vehicle trip. What constitutes a dangerous driving habit is defined according to previous studies on transportation research. Then we proposed a fuzzy expert system, which leads to a 'per trip risk score'. Finally, these individual scores are then aggregated into a final risk score that reflects the estimated risk exposure associated with a particular driver or policyholder. For insurers, the model presented in this paper could be a useful tool for establishing policy premiums levels, decision making, and managing capital adequacy.

References

1. P. Baecke and L. Bocca, "The Value of Vehicle Telematics Data in Insurance Risk Selection Processes," Decision Support Systems, 2017.
2. J. Wahlström, I. Skog, and P. Händel, "Driving behavior analysis for smartphone-based insurance telematics," in Proceedings of the 2nd workshop on Workshop on Physical Analytics, 2015, pp. 19-24.
3. G. Vaia, E. Carmel, W. DeLone, H. Trautsch, and F. Menichetti, "Vehicle Telematics at an Italian Insurer: New Auto Insurance Products and a New Industry Ecosystem," MIS Quarterly Executive, vol. 11, 2012.
4. S. Husnjak, D. Peraković, I. Forenbacher, and M. Mumdziev, "Telematics system in usage based motor insurance," Procedia Engineering, vol. 100, pp. 816-825, 2015.
5. P. Handel, I. Skog, J. Wahlstrom, F. Bonawiede, R. Welch, J. Ohlsson, et al., "Insurance telematics: Opportunities and challenges with the smartphone solution," IEEE Intelligent Transportation Systems Magazine, vol. 6, pp. 57-70, 2014.
6. L. Eboli, G. Mazzulla, and G. Pungillo, "Combining speed and acceleration to define car users' safe or unsafe driving behaviour," Transportation research part C: emerging technologies, vol. 68, pp. 113-125, 2016.
7. M. Siami, M. Naderpour, and J. Lu, "Generating a Risk Profile for Car Insurance Policyholders: A Deep Learning Conceptual Model," presented at the Australasian Conference on Information Systems, 2017.
8. L. A. Zadeh, "Fuzzy sets," Information and control, vol. 8, pp. 338-353, 1965.
9. J. Paefgen, T. Staake, and E. Fleisch, "Multivariate exposure modeling of accident risk: Insights from Pay-as-you-drive insurance data," Transportation Research Part A: Policy and Practice, vol. 61, pp. 27-40, 2014.
10. L. Eboli, G. Mazzulla, and G. Pungillo, "How to define the accident risk level of car drivers by combining objective and subjective measures of driving style," Transportation research part F: traffic psychology and behaviour, vol. 49, pp. 29-38, 2017.
11. A. S. Markowski and M. S. Mannan, "Fuzzy risk matrix," Journal of hazardous materials, vol. 159, pp. 152-157, 2008.
12. M. Naderpour, J. Lu, and G. Zhang, "A fuzzy dynamic bayesian network-based situation assessment approach," in Fuzzy Systems (FUZZ), 2013 IEEE International Conference on, 2013, pp. 1-8.
13. N. J. Garber and R. Gadirau, "Speed Variance and Its Influence on Accidents," 1988.
14. W. Dong, J. Li, R. Yao, C. Li, T. Yuan, and L. Wang, "Characterizing Driving Styles with Deep Learning," arXiv preprint arXiv:1607.03611, 2016.

Part 12

Soft Computing Methods in Image and Dynamic Data Processing

Detection of structural breaks and perceptionally important points in time series

Vilém Novák

Institute for Research and Applications of Fuzzy Modeling
University of Ostrava
NSC IT4Innovations, 30. dubna 22, 701 03 Ostrava, Czech Republic
vilem.novak@osu.cz

In this paper we suggest to use special fuzzy modeling techniques for detection of structural breaks and perceptionally important points in time series, namely the *fuzzy (F-)transform* and one method of *Fuzzy Natural Logic* (FNL). The idea is based on application of the F^1-transform which makes it possible to estimate effectively slope of time series over an imprecisely specified area (ignoring its possible volatility) and its evaluation by a suitable evaluative linguistic expression. The method is computationally very effective.

Keywords: Fuzzy transform; F-transform; evaluative linguistic expressions; fuzzy natural logic; time series.

1. Introduction

This paper addresses interesting problem related to the area of mining information from time series (cf.[1]), namely detection of structural breaks and perceptionally important points (PIP). These are specific local changes of the course of time series, its unexpected shifts or sudden change of its volatility. They can be caused, e.g., by changes in the economic development, global shifts in capital and labor, changes in resource availability due to war or natural disaster, discovery or depletion of natural resources, a change in the political system, etc.

There are various statistical methods for detection of structural breaks in time series, for example.[2–4] In this paper, however, we suggest to use special "fuzzy" techniques for detection of structural breaks in time series, namely the *fuzzy transform* (F-transform) and one method belonging to *Fuzzy Natural Logic* (FNL). Application of these techniques to the analysis and forecasting of time series was described in several papers and also in the book.[5]

Application of the F-transform has a few outcomes: it provides analysis of the time series, makes it possible to extract its trend and/or trend-cycle with high fidelity, and enables to estimate average derivatives of the time series (taken as a function) over an imprecisely specified area, while volatility of the time series in this area is ignored. These features enable us to apply methods of the fuzzy natural logic to forecasting of the future behavior of the time series, to generate automatically linguistic description of specific parts of it (cf.[6,7]), and to mine various other kinds of information from it.

In this paper, we suggest method how the above mentioned techniques can be used for detection of structural breaks and perceptionally important points in time series. The idea is to estimate the average tangent using the F-transform and then apply the theory of evaluative linguistic expressions (part of FNL). The main outcomes of our method are its *relative simplicity, transparency* and *computational effectiveness* because the complexity of the F-transform is linear.

2. F-transform and fuzzy natural logic in time series analysis

A time series is usually modeled as a stochastic process (see[8,9]) $X : \mathbb{T} \times \Omega \longrightarrow \mathbb{R}$ where Ω is a set of elementary random events and $\mathbb{T} = \{0, \ldots, p\} \subset \mathbb{N}$ is a finite set of numbers interpreted as time moments. Application of our methods is based on the assumption that the time series can be decomposed as follows:

$$X(t, \omega) = Tr(t) + C(t) + S(t) + R(t, \omega), \qquad t \in \mathbb{T}, \tag{1}$$

where Tr is a *trend*, C is a *cyclic* component, S is a *seasonal* component that is a linear combination of periodic functions and R is a random *noise*, i.e., a sequence of (possibly independent) random variables $R(t)$ such that for each $t \in \mathbb{T}$, the $R(t)$ has zero mean and finite variance. Note that Tr, C, S are ordinary functions not having stochastic character. The model (1) is often simplified by joining trend and cycle into one component $TC(t) = Tr(t) + C(t)$ called *trend-cycle*.

The F-transform is a procedure applied, in general, to a bounded real continuous function $f : [a, b] \longrightarrow [c, d]$. It is based on the concept of a *fuzzy partition* that is a set $\mathcal{A} = \{A_0, \ldots, A_n\}$, $n \geq 2$, of fuzzy sets fulfilling special axioms. The fuzzy sets are defined over nodes $a = c_0, \ldots, c_n = b$. The *direct* F-transform assigns to each A_k a component $F_k[f]$. We distinguish zero degree F-transform whose components $F_k[f]$ are numbers and first degree F-transform whose components have the form $F_k^1[f](x) = \beta_k^0[f] + \beta_k^1[f](x - c_k)$. The coefficient $\beta_k^1[f]$ provides estimation of an average value of the tangent (slope) of f over the area characterized by the fuzzy set $A_k \in \mathcal{A}$.

From the given components we can form a function $Inv_\mathcal{A}(f)$ that is called *inverse* F-transform of f. The function $Inv_\mathcal{A}(f)$ approximates the original function f.

Because of the lack of space, we omit more detailed description of the F-transform and refer the reader, e.g., to[5,10,11] where it has been proved that the trend-cycle TC in (1) can be estimated with high fidelity using the inverse F-transform as $TC \approx Inv_A(X)$. Moreover, the F-transform also provides a good estimation of the trend of the time series X. For a given fuzzy set $A_k \in \mathcal{A}$, direction of the trend is determined by a value of the first derivative $\beta_k^1[X]$.

The structural breaks and PIP are determined by means of evaluation of the trend using special expressions of natural language — the, so called, *evaluative linguistic expressions*. The trend can be evaluated by sentences of the form

$$Trend \text{ is } \langle direction \rangle \tag{2}$$

where

$$\langle direction \rangle := \text{stagnating}|\langle special\ hedge \rangle\langle sign \rangle, \tag{3}$$
$$\langle sign \rangle := \text{increasing}|\text{decreasing} \tag{4}$$

and

$$\langle special\ hedge \rangle := \emptyset|\text{negligibly}|\text{slightly}|\text{somewhat}|$$
$$\text{clearly}|\text{roughly}|\text{sharply}|\text{significantly}.$$

To generate (2), we must first specify a *context* in which the (direction of) the trend is evaluated. This is done by means of three distinguished values v_L, v_S, v_R of the tangent where v_R is the extreme increase (decrease), v_S is typical medium and v_L is the smallest one (usually set to $v_L = 0$). The result is the linguistic context $w_{tg} = \langle v_L, v_S, v_R \rangle$ for the trend. Moreover, because we distinguish between increase and decrease of time series, we must also distinguish positive context $w_{tg}^+ = \langle v_L^+, v_S^+, v_R^+ \rangle$ from the negative one $w_{tg}^- = \langle v_R^-, v_S^-, v_L^- \rangle$. We usually put $v_L^- = v_L^+ = 0$.

Let $X|\bar{\mathbb{T}}$ be a time series (1) considered over a time interval $\bar{\mathbb{T}} \subset \mathbb{T}$ and let $\beta^1[X|\bar{\mathbb{T}}]$ be value of the tangent (slope) of X in the area determined by a fuzzy set A with the support $\bar{\mathbb{T}}$. Then the evaluative predication (2) can be construed by means of the following standard form:

$$Trend\ of\ X|\bar{\mathbb{T}}\ \text{is}\ \pm Ev[X|\bar{\mathbb{T}}] \tag{5}$$

where $Ev[X|\bar{\mathbb{T}}]$ is a canonical evaluative expression, i.e., having the form "⟨hedge⟩small, ⟨hedge⟩medium, ⟨hedge⟩big" or "zero"). This is obtained using the special function of local perception

$$\pm Ev[X|\bar{\mathbb{T}}] := LPerc(\beta^1[X|\bar{\mathbb{T}}], w_{tg}^- \sqcup w_{tg}^+) \tag{6}$$

where $\sqcup$ denotes join of the negative and positive contexts. Expression (6) must be translated into the form (2), (3) and (4). A possible translation is suggested in Table 1.

Table 1. Translation table between special and canonical evaluative expressions.

special hedge, direction of trend	$Ev[X\|\bar{\mathbb{T}}]$ from (5)
stagnating	extremely small or zero (Ex Sm or Ze)
negligibly	significantly small (Si Sm)
slightly	very small (Ve Sm)
somewhat	rather small (Ra Sm)
clearly	very roughly small or medium (VR Sm or Me)
roughly	very roughly big (VR Bi)
fairly large	roughly big (Ro Bi)
quite large	rather big (Ra Bi)
large	big (Bi)
sharply	very big (Ve Bi)
significantly	significantly big (Si Bi)
huge	extremely big (Ex Bi)

3. Detection of structural breaks and PIP in time series

Definition 3.1. Let a realization $\{X(t) \mid t \in \mathbb{T}\}$ of the time series X be given and w_{tg} be a context for the trend of X. Let $\bar{\mathbb{T}}_1, \bar{\mathbb{T}}_2 \subset \mathbb{T}$ be two subsequent intervals and $\pm Ev[X|\bar{\mathbb{T}}_1], \pm Ev[X|\bar{\mathbb{T}}_2]$ be evaluations (6). A *structural break* in an interval $\bar{\mathbb{T}}_2$ is a sudden change of the values $X(t)$ in such a way that $Ev[X|\bar{\mathbb{T}}_2] \in \{VR\ Bi, Ro\ Bi, Ra\ Bi, Bi, Ve\ Bi, Si\ Bi, Ex\ Bi\}$ and $\pm Ev[X|\bar{\mathbb{T}}_1]$ takes either one of the latter values with sign opposite to $\pm Ev[X|\bar{\mathbb{T}}_2]$ or $Ev[X|\bar{\mathbb{T}}_1] \in \{Ze, Ex\ Sm, Si\ Sm, Ve\ Sm, Ra\ Sm, VR\ Sm, Me\}$.[a]

The interval $\bar{\mathbb{T}}_2$ should be appropriately "short". According to the experiments, its proper length should be $|\bar{\mathbb{T}}| \in \{4, \ldots, 10\}$.

The method for detection of structural breaks or PIP using fuzzy modeling methods is the following:

(i) Specify the fuzzy partition $\mathcal{A}$ over $\mathbb{T}$ and wet the context w_{tg} for evaluation of the trend.
(ii) Compute the F^1-transform $\mathbf{F}^1[X] = (F_1^1[X], \ldots, F_{n-1}^1[X])$ over the fuzzy partition determined in step (i).
(iii) Locate all the components $F_k[X]$ such that the coefficients $\beta_{k-1}^1[X], \beta_k^1[X]$ form a structural break according to Definition 3.1.
(iv) The located components point to the areas of structural breaks or PIP in the time series. The areas are determined by the corresponding fuzzy sets $A_k \in \mathcal{A}$ from the given fuzzy partition.

---

[a]We apply her the special ordering $\lll$ defined in.[5]

The above described method enables us to find abrupt changes in the course of the time series X which can indicate either structural break of PIP. However, we can meet also other kind of structural break that manifests itself in abrupt change of the volatility of the time series (cf. Figure 2). The volatility is usually characterized using standard deviation computed over a shifting time horizon. In this paper, we will estimate volatility using the F-transform (cf.[12]).

Definition 3.2. Let X be a time series and $\mathcal{A}$ a fuzzy partition. Let $V(t) = |X(t) - Inv_{\mathcal{A}}(X)(t)|$ be a function, $t \in \mathbb{T}$. Then the *volatility* of X is a function $\hat{V}$ computed using the formula

$$\hat{V}(t) = Inv_{\mathcal{A}}^1(V)(t), \qquad t \in \mathbb{T}. \tag{7}$$

Note that in (7), the first degree F-transform is considered. Let $F_k^1[V](x) = \beta_k^0[V] + \beta_k^1[V](t - c_k)$ be the k-th component. Then $\beta_k^1[V]$ is estimation of the average tangent over an area determined by the fuzzy set $A_k \in \mathcal{A}$.

Let w_V^+, w_V^- be positive and negative contexts for the tangent of V. Then, for a given $A_k \in \mathcal{A}$, we can find an evaluative expression

$$\pm Ev[V]_k := LPerc(\beta_k^1[V], w_V^- \sqcup w_V^+) \tag{8}$$

that linguistically characterizes the slope of V in the area determined by A_k. If $Ev[V]_{k+r} \in \{Ra\ Sm, ML\ Sm, Sm, Ve\ Sm, Si\ Sm, Ex\ Sm, Ze\}$ for $r \in \{-1, +1\}$ and $Ev[V]_k \in \{VR\ Bi, Ro\ Bi, Ra\ Bi, Bi, Ve\ Bi, Si\ Bi, Ex\ Bi\}$ then the area A_k is a place of a structural break due to big change of the volatility.

4. Demonstration

In this section we will demonstrate the above described method. We prepared 4 artificial time series with typical structural breaks. The time series were obtained by combination of few real time series taken from the INDUSTRY subset of time series on a monthly basis from the M3-Competition published on the Internet.[b]

Each time series in Figure 1 is depicted together with the fuzzy partition.[c] All the fuzzy partitions are equidistant and formed by triangular basic functions with the support equal to 8. The linguistic contexts for the time series were obtained by setting $v_R = \sigma/8$ where the standard deviation σ is for the given time series $\sigma_{(A)} = 978$, $\sigma_{(B)} = 1711$, $\sigma_{(C)} = 6307$, respectively. Further parameters of the

---

[b] https://forecasters.org/resources/time-series-data/m3-competition/
[c] Figure 1 was prepared using the experimental software FT-studio which makes it possible to apply fuzzy transform to a function given either by a precise formula, or by the data. The program was developed in the Institute for Research of Applications of Fuzzy Modeling of the University of Ostrava, Czech Republic. Its author is Radek Valášek.

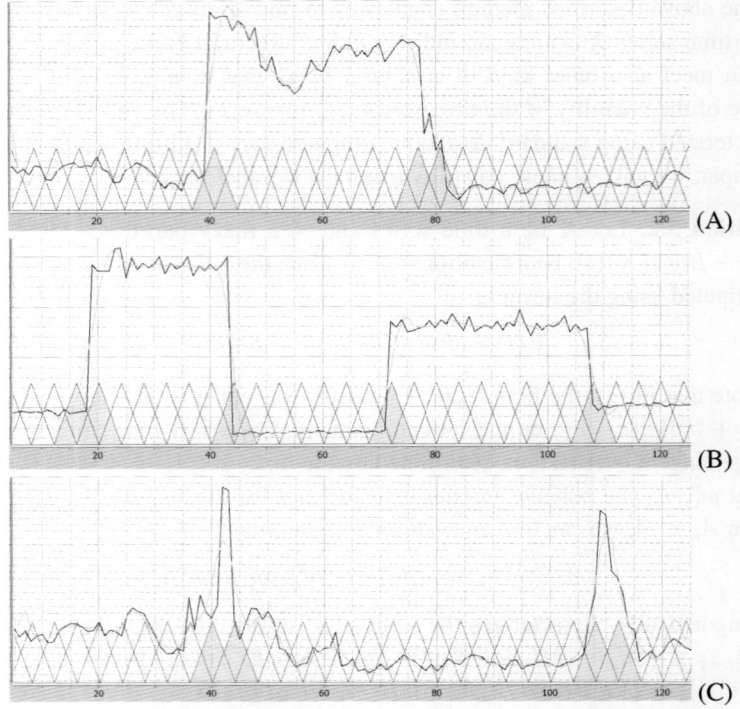

Fig. 1. Artificial time series with a fuzzy partition and marked basic functions detecting structural breaks. Note that the fuzzy partition is constructed over the interval $\mathbb{T}$. The evaluative expression over the interval determined by the marked basic functions is always *huge increase (decrease)*. The green line is approximation of the time series using the inverse F-transform.

context are set to $v_L = 0$ and $v_S = 0.4\, v_R$. Alternatively (and, perhaps, more naturally) we can set $v_R = |\max_{\mathbb{T}}(X) - \min_{\mathbb{T}}(X)|/8$.

One can see in Figure 1 that the places of structural breaks are well detected. Moreover, we also immediately know whether it is jump up or down (in correspondence with the sign of β^1):

(A) There are 2 structural breaks determined by 3 basic functions. The values of the corresponding coefficients are $\beta^1_{10} = 813$, $\beta^1_{19} = -592$ and $\beta^1_{20} = -465$ (from left-to-right) with peaks at the time moments $t = 41, 77, 81$, respectively. It is clear from the signs of β^1 that the structural break is shift up in the area around $t = 41$ and shift down at $t \in \{77, 81\}$.

Linguistic evaluation of the course of the time series between the structural breaks (the interval [41, 77]) is "stagnating".[d]

(B) There are 4 structural breaks determined by 5 basic functions. The values of the corresponding coefficients are $\beta_4^1 = 1430$, $\beta_5^1 = 2299$, $\beta_{11}^1 = -3898$, $\beta_{18}^1 = 2408$ and $\beta_{27}^1 = -1842$ with peaks at the time moments $t = 16, 20, 44, 52, 108$, respectively.

Linguistic evaluations of the course of time series between the structural breaks are in both intervals [20, 44] and [52, 108] "stagnating".

(C) There are 2 narrow structural breaks, each determined by 2 basic functions, one for shift up and the second one for shift down. The values of the corresponding coefficients are $\beta_{10}^1 = 616$, $\beta_{11}^1 = -695$, $\beta_{27}^1 = 762$ and $\beta_{28}^1 = -628$ with peaks at the time moments $t = 40, 44, 108, 112$, respectively.

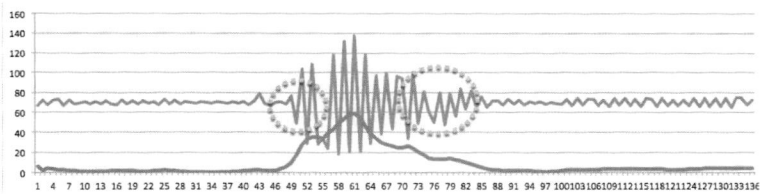

Fig. 2. Artificial time series with changing volatility. The lower graph shows estimation of the volatility using F-transform. By circles are denoted areas over which values of the tangent of volatility are very big, i.e., the volatility is abruptly changing.

The fourth kind of structural break is in Figure 2. It consists of a sudden large change of the volatility. The goal is thus to locate the position of this change. Similarly as above, we will define the context w_V on the basis of the difference between maximal and minimal values of V. Thus, we set $v_R = \pm 7.3$, $v_L = 0$ and $\pm v_S = 2.9$. Values of $\beta^1[V]$ in the areas marked by circles in Figure 2 are 7.2 and 6.2, respectively. W.r.t. the given context, they are evaluated as "extremely big" and "rather big". The other values before and after these areas vary in $[0.1, 0.3]$ and so, they are evaluated as "small" (or around) which means that the circled areas mark the places of structural breaks due to abrupt change of the volatility.

---

[d]Evaluation of the course of time series was obtained using the software LFL Forecaster developed in the Institute for Research of Applications of Fuzzy Modeling of the University of Ostrava, Czech Republic. Its author is Viktor Pavliska.

5. Conclusions

In this paper we suggested the method how structural breaks in time series can be detected using methods of fuzzy modeling. The method is based on application of the first degree fuzzy transform that provides estimation of the direction of trend of time series in an imprecisely determined area and its evaluation using evaluative linguistic expressions. The method is robust w.r.t. the starting position of the fuzzy partition. It is also important to emphasize that the detection is extremely fast because *time complexity of the fuzzy transform is linear*. Because of the lack of space we did not demonstrate recognition of PIP. Let us remark, however, that this works also with remarkable robustness and, at the same time, sensitivity.

Acknowledgments

The paper has been supported by the grant 18-13951S of GAČR, Czech Republic.

References

1. T.-C. Fu, *Engineering Applications of Artificial Intelligence* **24**, 164 (2011).
2. S. De Wachter and D. Tzavalis, *Computational Statistic Data Analysis* **56**, 3020 (2012).
3. P. Preuss, R. Puchstein and H. Detter, *Journal of American Statistical Association* **110**, 654 (2015).
4. P. Fischer and A. Hilbert, Fast detection of structural breaks, in *Proc. of 21th International Conference on Computational Statistics*, (Lisbon, Portugal, 2014).
5. V. Novák, I. Perfilieva and A. Dvořák, *Insight into Fuzzy Modeling* (Wiley & Sons, Hoboken, New Jersey, 2016).
6. V. Novák, *International Journal of Approximate Reasoning* **78**, 192 (2016).
7. V. Novák, *Fuzzy Sets and Systems* **285**, 52 (2016).
8. J. Anděl, *Statistical Analysis of Time Series* (SNTL, Praha, 1976 (in Czech)).
9. J. Hamilton, *Time Series Analysis* (Princeton, Princeton University Press, 1994).
10. V. Novák, I. Perfilieva, M. Holčapek and V. Kreinovich, *Information Sciences* **274**, 192 (2014).
11. L. Nguyen and V. Novák, Trend-cycle forecasting based on new fuzzy techniques, in *Proc. Int. Conference FUZZ-IEEE 2017*, (Naples, Italy, 2017).
12. L. Troiano, E. Mejuto and P. Kriplani, An alternative estimation of market volatility based on fuzzy transform, in *Proc. IFSA-SCIS*, 2017).

Post-processing in edge detection based on segments

P. A. Flores-Vidal* and N. Martínez

Statistics and Operational Research,
Faculty of Mathematics of Complutense University, Madrid, Spain
*pflores@ucm.es
www.ucm.es

D. Gómez

Statistics and Operational Research II,
Faculty of Statistics of Complutense University, Madrid, Spain
dagomez@estad.ucm.es

In this paper we propose a new post-processing technique for edge detection problems. We use a novel Fuzzy Clustering algorithm to performance a global evaluation over binary images. These binary images were obtained from outputs of edge detection algorithms as the one of Sobel or Canny. In a first step the edges identified by the detection algorithm are modeled as candidates to be edge. After that, the segments are built connecting those candidates that are connected. The next step requires a Fuzzy Clustering in order to select the good segments that will be considered as final edges. We show the effectiveness of this technique over classical edge detectors in many scenarios, even without applying smoothing.

Keywords: Image processing; edge detection; post-processing; global evaluation; segments.

1. Introduction

Edge detection [1] is considered one of the main techniques in Image Processing. [2] This technique traditionally consists in localizing significant changes in the intensity function of the image.

Edge detection is quite useful in many fields. For instance, the recognition of different pathologies for medical diagnosis. [3] As well it is being used in agriculture, military industry, law enforcement, among others's. [4–7]

Edge detectors are image processing algorithms which analyze the spectral information of an image, which commonly means analyzing the intensity luminosity of each pixel. When a luminosity variation between two neighboring pixels in the image is located, an edge of the region (or boundary) containing one of these pixels is detected.[2] Among the best known edge detection algorithms are Sobel's[8] and Canny's.[9]

Some of them try to detect if a certain pixel could be an edge using only information provided by adjacent pixels -*neighbours*-. This is called Local Edge Evaluation. Due to limitations of this Local Evaluation, Global Evaluation approach emerged as a more advanced strategy. In,[10] the concept of *edge segment* was formalised (see Section 2).

This work is based on the use of *edge segments*, and the methodology for creating them was introduced in previous works by Flores-Vidal et al.[11,12] The algorithm proposed in this paper presents a modification that allows it to be applied as a post-processing technique over others edge detection algorithms. In this sense, our algorithm has no intend of competing with other algorithms but improving them.

The remaining of this paper is organized as follows: The next section is dedicated to the preliminaries. Section 3 is focused on our proposal. The last two sections are dedicated to the comparatives and results, and conclusions respectively.

2. Preliminaries

Let us denote by I a digital image, and by (i,j) the pixel coordinates of the *spatial domain*. For notational simplification the coordinates are integers, where each point (i,j) represents a pixel with $i = 0,\ldots,n$ and $j = 0,\ldots,m$. Therefore, the size of an image, $n \times m$, is the number of its horizontal pixels multiplied by its number of verticals.

Let us denote by $I_{i,j}$ the spectral information associated with each pixel (i,j). For a given pixel (i,j) we will denote by $X_{ij}^1, \ldots X_{ij}^k$ the k extracted characteristics. Aggregating the information of these k features together into a single value is denoted as *edginess*. Let us denote by

$$I^{bf} = \phi(X^1, \ldots, X^k)$$

the aggregation result of these characteristics, where ϕ is the aggregation function. Then, for a given pixel (i,j), the value $I_{i,j}^{bf}$ represents the total variation of this pixel. It is common to represent this matrix as a grayscale image, where each pixel has its degree of *edginess* associated.

In previous works[11,12] were defined these four key concepts:

- *Edge candidates pixels*
- *Set of edge candidates pixels*

- *Edge segments*
- *Segment features*

Let c be a candidate to edge pixel. Then c has to meet two conditions:

(1) If (i_c, j_c) is the position of c in the spatial domain then $I^{bf}_{i_c,j_c} > 0$. In other words, it has to be a non-zero intensity pixel.
(2) If there are three adjacent pixels to c that meet (1), then it is not possible that they set up a square shape. Therefore this is a *thinned image*. This condition can be seen as well as an statement about edge thickness.

From previous definition we are able to define the set that contains these pixels. Let $C = \{c_1, ..., c_m\}$ be the set of all the *edge candidate pixels* in an image.

The third definition in which is based our methodology is the one of edge segment. Let $S = (c_1, ..., c_n) \subset C$ be a subset of *edge candidate pixels set*, then we will call it an *edge segment* if and only if:

(1) S is connected, i.e., $\forall c_a, c_b \in S$ there is a path through adjacent pixels $(c_i)_{i \in \{1,...,n\}} \subset S$ from c_a to c_b.
(2) S is maximal, i.e., if $S' \subset C$ is another connected set of *edge candidate pixels*, then $S \subset S' \Rightarrow S = S'$.

Once the edge segments are built we can proceed to the feature extraction. For each segment were used the following features:

- Length. For each segment S_l, $Length_l = |S_l|$.
- Median of the edginess. For each segment S_l, $Median(Bf_{ij})$.
- Maximum edginess. For each segment S_l, we obtained $Max(Bf_{ij})$.
- Standard deviation of the intensity. For each segment S_l,

$$\sigma_l = \frac{\sum_{(i,j) \in S_l} (I^{bf}_{i,j} - IM_l)^2}{Length_l}$$

where IM_l is the Intensity Mean. For each segment S_l,

$$IM_l = \frac{\sum_{(i,j) \in S_l} I^{bf}_{i,j}}{Length_l},$$

- Ordinal variation index [13] of the intensity. For each segment S_l,

$$IOV_l = \sum_{(i,j),(i',j') \in S_l} f_{i,j} f_{i',j'} dist((i,j),(i'j')),$$

where $f_{i,j}$ is the relative intensity of the pixel (i,j) (i.e. $\frac{Bf_{ij}}{\sum_{(r,s)\in S_l} Bf_{r,s}}$), and $dist((i,j),(i'j'))$ is the Manhattan distance between the pixels.
- Gravity center. For each segment S_l, we obtained the coordinates of the pixel that occupies the central position in the segment: $(i_c, j_c) = Central_l$; where i_c is the average vertical position and j_c is the average horizontal position of the pixels in S_l, i.e., $i_c = \frac{\sum_{(i,j)\in S_l} i}{Length_l}$ and $j_c = \frac{\sum_{(i,j)\in S_l} j}{Length_l}$.
Once the gravity center was computed we used its euclidean distance to the four intersection points following the rule of thirds, which is an standard in photography composition (see [14]). This rule is used to place the important objects of an image in the most relevant positions for being detected for humans. The intuition behind using this feature it is because any picture taken by a human tends to place the relevant visual information close to these points. As well, the importance of this rule was experimentally confirmed.

3. Fuzzy Cluster of Segments as Post-processing Algorithm

Our proposal of algorithm [11] can be resumed in:

(1) Given an already blended and thinned grayscale image I^{bf}, we have to obtain the set C and the segments set $\boldsymbol{S} = \{S_l : l = 1, ..., s\}$ of the image I^{bf}.
(2) For each segment S_l, obtain the segment's features. Such features can be normalized and thus be measured as values in $[0,1]$. Let us denote by x_i^l the i-th associated characteristic of segment S_l, for $l = 1, ..., s$; $i = 1, ..., f$ where f is the number of features extracted for each segment. Thus the space of segment features can be defined as $F = [0,1]^f$ and $\mathbf{x}^l$ the vector of characteristics of segment S_l.
(3) On the space F we then apply a fuzzy clustering algorithm based on relevance, redundancy and covering concepts [15] obtaining two clusters. From the defuzzification of this fuzzy clustering solution we obtain the classification between *bad* and *good* segments that will give the final solution. The bad segments are considered those that are mainly noise.

Once the algorithm has been defined, it is time to use our proposal as a post-processing technique. For doing so we need to obtain as an input for the FCS algorithm the set of candidates C that was created with a edge detection algorithm. In this paper we have used three of them: Sobel, Canny and Automated-Sobel (see next section).

4. Comparatives and Results

In order to confirm if our FCS post-processing algorithm was able to improve the classic algorithms proposed by Sobel and Canny, we run the four algorithms with six different threshold values (see Table 1). One reason for doing this way it is because each image requires a different threshold value in order to obtain a satisfactory binarized image with their significant edges. For example, images with too much detail tend to need a high threshold in order to extract the significant edges -or at least part of them-. By contrast, less detailed images tend to require a lower threshold (for example α between 0.05 and 0.10). We worked with the first 25 images of the training set images from the Berkeley Segmentation Data Set of 500 images (BSDS500).[16] We are aware of BSDS is non specific for edge detection but for segmentation and contours, but it has been used for edge detection evaluation with interesting results.[17] As the BSDS ground truth images are references from four to seven different humans these F comparatives where aggregated using the mean -other typical aggregations are the maximum and the minimum-.

As we can see in Table 1, our FCS post-processing algorithm reached higher values of F measures at low and medium thresholds ($\alpha = 0.05$ to $\alpha = 0.20$). It is inside this threshold interval where our post-processing proposal proved to be a powerful tool for improving edge detection and noise removal. To be sure if this interval was relevant for most of the images, we decided to compare our FCS post-processing algorithm with an automated edge detection algorithm, in this case a version of Sobel's using the automatic thresholding computed by the Rosin method.[18] The Rosin method can be considered as semi-supervised, as it uses the information of each image selecting the optimum threshold by means of its luminosity histogram. We checked that our FCS algorithm improved as well this automated version of Sobel that was using Rosin method. Using this automated Sobel for the non-smoothing case ($\sigma_{smooth} = 0$) it resulted a $F_{Mean} = 0.31$, but it reached a higher value ($F_{Mean} = 0.35$) when working with our Fuzzy Clustering algorithm. As well we found a better performance for the Gaussian smoothing case ($\sigma_{smooth} = 1$) that is commonly applied for removing noise in the original images.

5. Conclusions

We are aware of that our algorithm's performance didn't improve the results for all the possible thresholds (only in the range from $\alpha = 0.05$ to

$\alpha = 0.20$). We found that the higher was the threshold applied, FCS worked more poorly, as it happened for $\alpha = 0.25$ and $\alpha = 0.30$. This bad performance at higher thresholds is due to that an important amount of segments are needed for FCS to work properly, and normally the higher the threshold applied the less candidates to be edge are left and then less possible segments. We can appreciate this in the scatterplots of the Figure 1. As the threshold increases, there are less good segments left and more bad segments -noise- remain.

In resume, our algorithm needs a big amount of segments to perform a high quality clustering, specially in a such unbalanced conditions (the proportion of good segments against the bad one's goes from 1/8 to 1/10). This limitation could be overcome if some segments were broken, generating more segments in the process, which eventually would led to improvements in the clustering. Future research could follow this "breaking segments" approach.

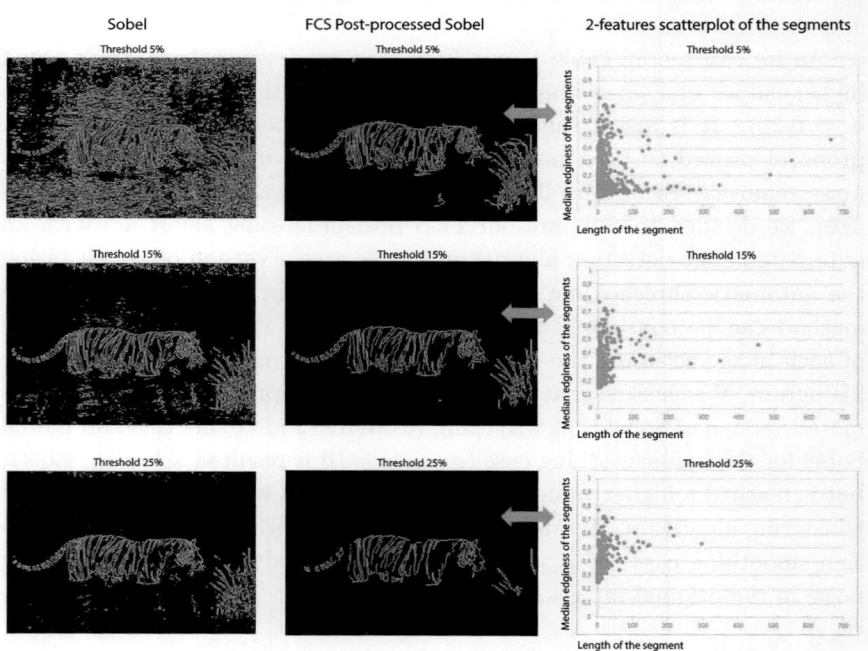

Fig. 1. Binarized image with Sobel and FCS post-processing Sobel.

Table 1. F mean comparatives: Sobel and Canny without and with FCS post-processing.

	Threshold (Thr) (for Canny Thr_{sup})	Sobel	Canny $\sigma=2$	Canny $\sigma=4$	FCS-Sobel	FCS-Canny $\sigma=2$	FCS-Canny $\sigma=4$
$\sigma_{smooth}=0$	0.05	0.21	0.24	0.32	**0.32**	**0.44**	**0.47**
$\sigma_{smooth}=0$	0.10	0.27	0.28	0.35	**0.36**	**0.44**	**0.46**
$\sigma_{smooth}=0$	0.15	0.32	0.33	0.39	**0.36**	**0.40**	**0.44**
$\sigma_{smooth}=0$	0.20	0.34	0.37	0.42	**0.34**	**0.37**	**0.43**
$\sigma_{smooth}=0$	0.25	0.35	0.40	0.44	0.34	0.32	0.39
$\sigma_{smooth}=0$	0.30	0.35	0.42	0.46	0.30	0.31	0.37
$\sigma_{smooth}=1$	0.05	0.22	0.25	0.32	**0.34**	**0.45**	**0.47**
$\sigma_{smooth}=1$	0.10	0.30	0.30	0.36	**0.39**	**0.44**	**0.46**
$\sigma_{smooth}=1$	0.15	0.35	0.34	0.39	**0.40**	**0.40**	**0.45**
$\sigma_{smooth}=1$	0.20	0.38	0.38	0.42	**0.39**	**0.38**	**0.46**
$\sigma_{smooth}=1$	0.25	0.38	0.41	0.44	0.37	0.32	0.39
$\sigma_{smooth}=1$	0.30	0.38	0.43	0.46	0.34	0.33	0.43

Note: Numbers in bold letters mean higher or equal values for FCS than for the classic algorithms.

Acknowledgments

This research has been partially supported by the Government of Spain, grant TIN2015-66471-P.

It has been strongly helpful for the conducting of our research the code of KERMIT Research Unit [19] (Ghent University), The Kermit Image Toolkit (KITT), B. De Baets, C. Lopez-Molina (Eds.), Available on-line at www.kermitimagetoolkit.net.

References

1. D. Marr and E. Hildreth, *Proceedings of the Royal Society of London - Biological Sciences* **207**, 187 (1980).
2. C. Guada, D. Gómez, J. T. Rodríguez, J. Yá'nez and J. Montero, *International Journal of Computational Intelligence Systems* **9**, 43 (2016).
3. M. Sonka, *IEEE Transactions on Medical Imaging* **33** (2014).
4. O. Monga, R. Deriche, G. Malandain and J. P. Cocquerez, *Image and Vision Computing* **9**, 203 (1991).
5. M. Fathy and M. Y. Siyal, *Pattern Recognition Letters* **16**, 1321 (1995).
6. T. Zielke, M. Brauckmann and W. Vonseelen, *CVGIP: Image Understanding* **58**, 177 (1993).
7. S. K. Pal and R. A. King, *IEEE Transactions on Pattern Analysis and Machine Intelligence* **PAMI-5**, 69 (1983).
8. I. Sobel, *Artificial Intelligence* **5**, 185 (1974).
9. J. Canny, *IEEE Transactions on Pattern Analysis and Machine Intelligence* **PAMI-8**, 679 (1986).
10. S. Venkatesh and P. L. Rosin, *Graphical Models and Image Processing* **57**, 146 (1995).

11. P. Flores-Vidal, D. Gómez, P. Olaso and C. Guada, *A new edge detection approach based on fuzzy segments clustering*, Advances in Intelligent Systems and Computing, Vol. 642 2018.
12. P. A. Flores-Vidal, J. Montero, D. Gómez and G. Villarino, Classifying segments in edge detection problems, in *Proceedings 12th Int. Conf. Intelligent Systems and Knowledge Engineering, ISKE 2017, At Nanjing (China)*, 2018.
13. J. Blair and M. G. Lacy, *Perceptual and motor skills* **84**, 411 (1996).
14. E. Goldstein, *Sensación y percepción Sexta ed.,Thomson Editores Spain* 2009.
15. A. del Amo, D. Gómez, J. Montero and G. S. Biging, *Improving fuzzy classification by means of a segmentation algorithm*, Studies in Fuzziness and Soft Computing, Vol. 220 2008.
16. D. Martin, C. Fowlkes, D. Tal and J. Malik, A database of human segmented natural images and its application to evaluating segmentation algorithms and measuring ecological statistics, in *Proceedings of the IEEE International Conference on Computer Vision*, 2001.
17. I. Perfilieva, P. Hodáková and P. Hurtík, *Fuzzy Sets and Systems* **288**, 96 (2016).
18. P. L. Rosin, *Pattern Recognition* **34**, 2083 (2001).
19. B. de Baets and C. López-Molina, The Kermit Image Toolkit (KITT), Ghent university http://www.kermitimagetoolkit.net, (2016), [Online; accessed 19-January-2018].

Noise influence in FzT+JPEG image compression

P. Hurtik and I. Perfilieva

*Institute for Research and Applications of Fuzzy Modeling, University of Ostrava
30. dubna 22 701 03 Ostrava, Czech Republic
petr.hurtik@osu.cz, irina.perfilieva@osu.cz*

Based on our recent research, where we the designed hybrid image compression algorithm combining F-transform and JPEG called FzT+JPEG, we now analyze the influence of noise on image compression. Our analysis is focused on the impact of a noise to a size of a compressed image and on the ability to decompress an image without noise by both FzT+JPEG and JPEG algorithms. On some benchmarks, we show that FzT+JPEG algorithm rapidly outperforms the pure JPEG in both aspects.

Keywords: Fuzzy transform; image compression; JPEG; noise filtering.

1. Introduction

Image compression in the sense of reduction of stored image size is a crucial step in image handling because it can save space, transmission time, and therefore even electricity. In this work, we are focusing on a lossy compression, namely, JPEG. This paper is a continuation of our original research[1] where hybridization of JPEG with the F-transform[2] was designed and where we showed that the hybrid algorithm reaches higher decompressed-image quality for a strong compression. In this paper, we use the hybrid algorithm and investigate the side-effect of noise filtering, i.e., the influence of a noise on the compressed image size and decompressed image quality. The importance of this research is explained by the fact that a noise is caused by an electronic manner in sensors and circuits in cameras, so all images taken by cameras include a variable amount of noise. Therefore, the noise filtering by a compression algorithm as its side effect is valuable.

2. JPEG Compression

The general idea of the JPEG[3] compression is to transform an image from a spatial domain to a frequency domain and suppress high frequencies. In the first step, an image is transformed from the RGB color model into $YCbCr$[4] one. The fact that the differences in Cb and Cr channels are negligible for a human eye is used in JPEG in the step chroma sub-sampling. If a strong compression is required, Cb and Cr are down-sampled, i.e., only one value is used to describe 2×2 block per channel instead of four ones while Y channel remains fully described by four values. The further processing is realized for each of the three channels independently in the same way. An input image f is decomposed into non-overlapping 8×8 sub-images (blocks) f'_i such that $f'_1 \cup f'_2 \cup \cdots \cup f'_c = f$. Using DCT II transformation,[5] the sub-images are transformed from spatial domain into frequency domain D so that:

$$D_{k,\ell} = \sum_{n=0}^{7} \sum_{m=0}^{7} f'(n,m) \cos[0.125\pi k(n+0.5)] \cos[0.125\pi \ell(m+0.5)]. \quad (1)$$

From this equation, it can be observed that the higher values k, ℓ are used, the higher is the sampling rate and therefore the higher frequencies are handled. In the frequency domain, $D_{k,\ell} = D_{k,\ell}/Q_{k,\ell}$ is applied, where

$$Q = \begin{bmatrix} 16 & 11 & 10 & 16 & 24 & 40 & 51 & 61 \\ 12 & 12 & 14 & 19 & 26 & 58 & 60 & 55 \\ 14 & 13 & 16 & 24 & 40 & 57 & 69 & 56 \\ 14 & 17 & 22 & 29 & 51 & 87 & 80 & 62 \\ 18 & 22 & 37 & 56 & 68 & 109 & 103 & 77 \\ 24 & 35 & 55 & 64 & 81 & 104 & 113 & 92 \\ 49 & 64 & 78 & 87 & 103 & 121 & 120 & 101 \\ 72 & 92 & 95 & 98 & 112 & 100 & 103 & 99 \end{bmatrix}$$

is so-called general quantization matrix[6] whose particular values are dependent on compression strength $q \in \{1, \ldots, 100\}$ given by an user. The final used quantization matrix is derived from the general one as $Q_{k,\ell} = (S * Q_{k,\ell} + 50)/100$, where $S = 5000/q$ if $q < 50$; otherwise $S = 200 - 2q$. The smaller q, the bigger values are in the quantization matrix and the smaller values are in the corresponding output block after the division. In the output quantized block, all sufficiently small values are rounded to zero. So the output block includes several non-zero values and a lot of zeros. Values in the output block are usually encoded by Huffman coding[7] with the so-called zig-zag scheme[8] in order to encode the zeros effectively. Finally, values of each block, for each channel, are stored into a file using JFIF standard describing the file structure.

3. Fuzzy Transform and Image Compression

The F-transform (originally, *fuzzy transform*,[2]) or FzT in short, is a particular integral transform whose peculiarity consists in using a *fuzzy partition* of a universe of discourse (usually, the set of reals $\mathbb{R}$). In the F-transform method, each fuzzy set in a fuzzy partition is considered as a weight function that determines a weighted orthogonal projection of an object from $L_2(\mathbb{R})$ onto a space of object features. The F-transform has two phases: direct and inverse. The direct F-transform (dFzT) is applied to functions from $L_2(\mathbb{R})$ and maps them linearly onto sequences (originally finite) of numeric/functional components. Each component is a weighted orthogonal projection of a given function on a certain linear subspace of $L_2(\mathbb{R})$. The inverse F-transform (iFzT) is applied to a sequence of components and transforms it linearly into a function from $L_2(\mathbb{R})$. In general, it is different from an original one. In,[2] it has been shown that the iFT can approximate a continuous function with an arbitrary precision.

The problem of a full reconstruction from the dFzT components was discussed in,[9] where it has been proved that a smooth and band-limited function can be reconstructed by a specially designed iFzT based on the so-called *adjoint partition* which differs from that used in the direct phase. The proposed compression/decompression method[1] is based on the direct/inverse FzT computed with respect to different adjoint partitions. In order to increase a quality of a decompressed image, we propose to combine FzT based compression with JPEG. In particular, we propose to apply JPEG to the sequence of dFzT-components and by this, make both compressions stronger. This observation follows from the fact that the resulting compression ratio is equal to the product of the two particular ones. The proposed compression and decompression hybrid algorithms depend on the F-transform and the JPEG settings, namely: on chosen fuzzy partition connected with the choice of basic functions and the JPEG compression quality q. In our previous work,[1] we came to the conclusion that the best results were obtained by the basic functions in the uniform fuzzy partition generated by a raised cosine with such setting which guarantees that an image will be downscaled into 0.25 of its original size before JPEG will be executed.

4. Benchmarks

To compare JPEG and FzT+JPEG robustness against noise, we will discuss two matters: a) impact of noise in an input image to the size of a compressed image; b) ability to decompress image without noise. We assume that the original image f and the noisy one are connected by $f'(x,y) = f(x,y) + n(x,y)$, where $n(x,y) \sim N(0, \sigma^2)$, see Figure 1 for illustration.

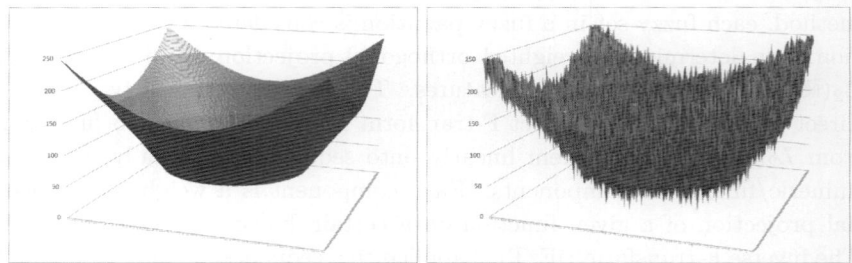

Fig. 1. Charted values of original image (left) and noisy one (right).

In the case of matter a), we analyze Formula (1). A noise creates discontinuities in signal/images, i.e., it gives additional high frequencies. It can be observed from (1) that on the step of DCT, JPEG does not suppress or enhance any frequencies. The suppressing of higher frequencies occurs when the operation of division by the quantization matrix is applied. Moreover, it is easy to see that high frequencies are not suppressed, if their values are big enough. Further, if their values are big enough and not omitted, i.e., not set to zero, the entropy coding is less effective and therefore the compressed file has a large size. Contrary to this, in the combination of FzT and JPEG, FzT filters out high-frequencies before JPEG is applied in this hybrid combination and therefore, the compressed output should not be affected by a noise.

Figure 2 supports the above-given reasoning by the demonstration of three 8×8 blocks after DCT II transform with quantization. On the left side of Figure 2, values of the original image without noise are illustrated - it is clear that there is a lot of zeros. Then, the output of FzT+JPEG for the noisy image is shown in the middle and finally, the right side of Figure 2 shows the JPEG output for the same noisy image with the same q setting. The influence to the image size is also shown in the graph in Figure 3 where the size of a compressed file according to a quantity of noise is analyzed. In the case of FzT+JPEG, dependency is linear and much lower, therefore more suitable for compression of noisy images than in the case of the pure JPEG where the compressed file size grows much faster.

In the second case of matter b) — noise filtering ability — we compress

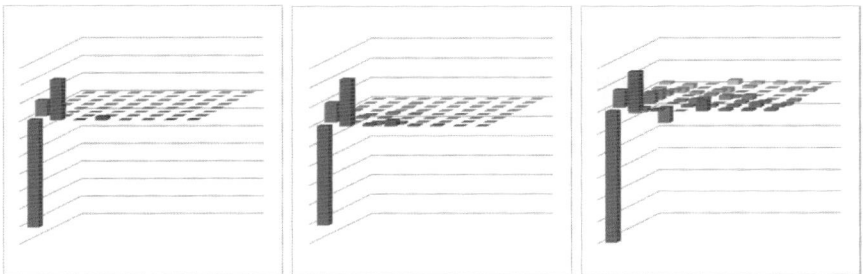

Fig. 2. Values of 8 × 8 output after DCT II with quantization. From left: original without noise, FzT+JPEG applied to a noisy image, JPEG applied to a noisy image. FzT filters out high-frequencies before JPEG, therefore the FzT+JPEG output includes more zeros and can be more efficiently encoded.

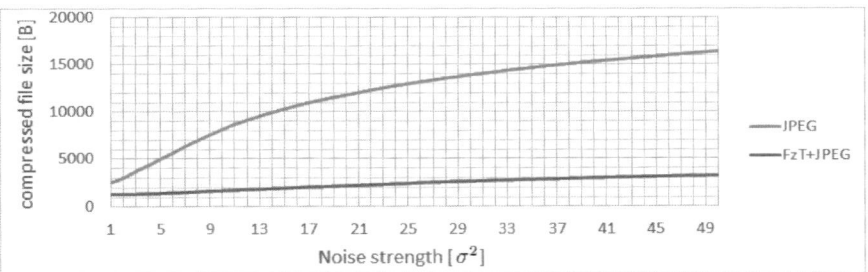

Fig. 3. Dependency of the amount of noise to the size of a compressed file. FzT+JPEG is less sensitive to noise than JPEG and therefore is more suitable to compress noisy images.

noisy images under various q settings and measure two standard variables bpp and SSIM in image compression field. As the first option, the number of bits per pixel (bpp) is used as the ratio between the compressed file size and image resolution. As the second option, the value of structure similarity (SSIM, implemented via $imagecompression.info/lossy/mssim.zip$) between the decompressed noisy image and the original image without noise is used. The result for three noise strength ($\sigma^2 = \{5, 15, 30\}$) is shown in the form of graph in Figure 4. The graph shows that FzT+JPEG can achieve stronger compression (smaller bpp) and that the proposed combination achieves significantly higher quality than pure JPEG for an arbitrary bpp, i.e., it can filter out the noise effectively.

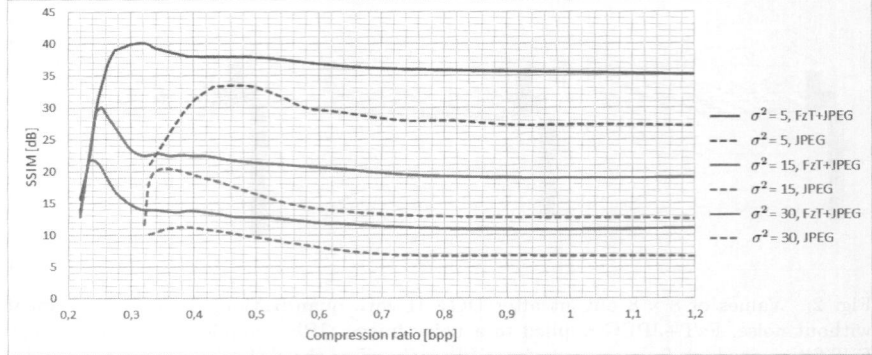

Fig. 4. The SSIM versus bpp correspondence for various noise strength. The proposed algorithm FzT+JPEG has better SSIM quality than pure JPEG.

5. Conclusion

To investigate noise influence, the JPEG compression was analyzed with the accent to DCT II that transforms an image from a spatial to a frequency domain and matrix quantization that suppresses high frequencies in the frequency domain. Further, we showed that when high frequencies (i.e., noise in our case) are big enough, they are not filtered-out and they decrease JPEG compression efficiency. We presented the recently proposed FzT+JPEG hybrid algorithm which at first applies the F-transform and by this, filters-out high frequencies before JPEG is executed. Finally, we showed on some benchmarks that the proposed combination achieves significantly higher-quality of a decompressed image than the pure JPEG.

Acknowledgments

This research was supported by the project "LQ1602 IT4Innovations excellence in science".

References

1. P. Hurtik and I. Perfilieva, A hybrid image compression algorithm based on jpeg and fuzzy transform, in *2017 IEEE International Conference on Fuzzy Systems (FUZZ-IEEE)*, 2017.
2. I. Perfilieva, Fuzzy transforms: Theory and applications *Fuzzy Sets and Systems* **157** (Elsevier, 2006).
3. W. B. Pennebaker and J. L. Mitchell, *JPEG: Still image data compression standard* (Springer Science & Business Media, 1992).
4. K. Saarinen, Comparison of decimation and interpolation methods in case of multiple

repeated rgb-ycbcr colour image format conversions, in *Circuits and Systems, 1994. ISCAS'94., 1994 IEEE International Symposium on*, 1994.
5. K. R. Rao and P. Yip, *Discrete cosine transform: algorithms, advantages, applications* (Academic press, 2014).
6. J. D. Kornblum, Using jpeg quantization tables to identify imagery processed by software *Digital Investigation* **5** (Elsevier, 2008).
7. D. A. Huffman, A method for the construction of minimum-redundancy codes *Proceedings of the IRE* **40** (IEEE, 1952).
8. G. K. Wallace, The jpeg still picture compression standard *IEEE transactions on consumer electronics* **38** (IEEE, 1992).
9. I. Perfilieva, M. Holčapek and V. Kreinovich, A new reconstruction from the f-transform components *Fuzzy Sets and Systems* **288** (Elsevier, 2016).

Synthetic dataset for compositional learning

Vojtech Molek and Jan Hula

Institute for Research and Applications of Fuzzy Modelling, University of Ostrava
Ostrava, 701 03, Czech Republic
vojtech.molek@osu.cz
jan.hula@osu.cz
irafm.osu.cz

This contribution presents a framework for a generation of synthetic images. The framework is built on top of the Unreal Engine 4, a software kit capable of rendering realistic images. Besides image data, additional label information, such as depth, normal maps and object components masks, are generated. Hierarchical nature of generated labels corresponds to hierarchical representations which we want to be captured by the neural network. Such labels enable training of deep models in a compositional manner. This leads to the better understanding of the internal representations of the models and acceleration of the learning procedure. The framework allows users to render arbitrary scenes and objects according to their specific domain.

Keywords: Dataset; synthetic data; unreal engine; compositional learning.

1. Introduction

In the last decade, deep neural networks have overcome traditional approaches in tasks such as regression or classification.[1,2] Such progress was enabled by new processing devices, new algorithms and last but not least availability of large labeled datasets.[3-6] Large datasets are key to proper supervised learning, however, the creation of such datasets is very tedious and time-consuming. Moreover, a lot of information about data is not explicitly contained in the labels (such as edges, contours, depth maps, segmentation masks, bounding boxes, occlusion, orientation etc. in case of image data). In some instances, it is even practically impossible to create all such labels manually. Having more control over data gathering process and more sensor types would lead to additional label information, although it could be impractical or even impossible in real world.

One way to simplify the process of data gathering is to use data from simulation instead of real world. Instead of gathering, we *generate* synthetic data. During generation of synthetic images, we generate different labels, including labels for object components — i.e., paws, body, tail, ears, eyes, and whiskers which together form a concept of a cat. Obviously, this hierarchy has variable complexity. We discuss the relation of hierarchical labels to training neural networks in section 2. In section 3 we describe our synthetic dataset and issues it potentially solves. In the last section 4, we introduce our framework for generating synthetic datasets.

2. Idea of Compositional Learning

During the learning process, Deep Learning[7] models learn a hierarchy of representations (in form of *features*) where some of the learned features have recognizable meaning and correspond to components of objects. For example, to recognize the aforementioned cat, a deep model will learn to detect components of a cat, even without explicitly assigned labels to them. The hierarchy of features was previously demonstrated by various features visualization methods.[8–13]

The main idea of *Compositional Learning* is to help a model in searching the parameter space by explicitly forcing hierarchical parts of the model to learn the features we believe to be important for recognition. In the context of image recognition, this is achieved by providing additional labels for the components of the object (paws, body, tail, etc.) and extending the objective function to take them into account. Concretely, certain convolutional kernels in the layers of a convolutional network would be forced to detect given components of the object. Thus, the objective function measures the accuracy of the whole network as well as how the said kernels detect the assigned components.

3. Synthetic Dataset

For the purposes of testing the idea of Compositional Learning, we have created synthetic dataset of images of plants. Because of the complete control over the 3D environment and complete knowledge about the objects it contains, we have generated the images with labels for every pixel in a trivial manner.

The 3D models used for the rendering may be obtained in multiple ways. The 3D models can be created by artists or by scanning real objects with a 3D scanner. However, such 3D models, which are not procedural, have fixed geometry. Therefore the rendered images will not contain any variation in the geometry. Another issue with scanned 3D models is the need to manually separate components of a 3D object if one wants to render different variations of textures for each component separately.

A much better approach is to create the geometry using a generative modeling. A generative modeling consists of programs for generating the geometry according to some parameters. By changing the parameters, the program generates different variations of the same object. For our purposes, we have decided to use generative models of plants as plant generation is a well-developed area of computer graphics and therefore many generative models of various plants are available. By using these 3D models, we generate variations of the same plant species with separate components of the plant that can also vary.

4. Synthetic Data Engine

For the purposes of rendering images from 3D models, we have chosen Unreal Engine 4 (UE4). UE4 is a game engine with open source code, capable of realistic rendering.[a] Realistic rendering is crucial to bridge the gap between synthetic data and real-world data. Moreover, it contains design tools for creating 3D sceneries. This enables to create realistic background context for rendered images.

Because our goal was to generate data as fast as possible, ideally in real time, we have used UE4 with NVIDIA *Voxel Global Illumination*[14] technology (VXGI) publicly available at Github.[b] The VXGI is global illumination approximation technique able to illuminate scenes in roughly real time and without pre-computations. This enables us to dynamically modify scenes during a game play.

To control the environment in UE4, we have extended a project called *UnrealCV*.[15] UnrealCV is UE4 plugin that let users manipulate certain aspects of UE4 through pre-defined commands such as camera manipulation, objects masking and rendering the images. These commands are invokable from Python which makes integration with popular machine learning frameworks easy and convenient.

4.1. *Extending UnrealCV*

The UnrealCV uses server-client architecture. The server is embedded into the running game and communicates with the client. The user uses the client to send commands and receive responses. The set of featured commands, however, does not allow to load objects into a scene during a game

---

[a]https:
//docs.unrealengine.com/latest/INT/Resources/Showcases/RealisticRendering/
[b]https://github.com/NvPhysX/UnrealEngine/

play. Therefore a user is not able to add and/or remove objects from a scene.

In our work, we have added a possibility to load objects during the game play with the command: `vget /load [example]` where `example` is a name of a configuration file. The configuration file is in the JSON format and includes the following information about the object: labels, meshes, materials, lights, and properties such as scale, location, rotation and material characteristics. All data has to be in a proprietary format (*uasset*), and we, therefore, convert the models from *fbx* to uasset file format with our file converter.

4.2. *Generating data*

Let us consider an example, where we generate image data of a sunflower. We start with picking scene. The unreal engine has a wide range of assets (paid and free) available at its online shop. We pick adequate scene[c] (Fig. 1) where the flower fits in naturally. In the next step, we load our model(s) through JSON file (Fig. 2). The loaded sunflower consists of multiple separated parts (Fig. 3). These parts can be visualized with view mode `object_mask` and such representation serves as ground truth labels for these parts. The generated data are shown in Fig. 3.

The source code used to generate example at Fig. 3 is hosted at Github.[d] The repository will include an example of Compositional learning as well as the generated dataset.

5. Conclusion

In our contribution, we have described how 3D graphics and its sophisticated tools can help to create and expand training datasets. We have extended UnrealCV by ability of dynamically load object into scene during game play. The object loading is done by adding new command to UnrealCV. Properties of the loaded object is specified in JSON file and has to be in the proprietary format called uasset. For this purpose we have created a format converter. We have also provided helper method to render images of an object with labels from different angles. This way, we have made the whole process of generating synthetic data easily manageable. The combination of user-friendly Python programming language, rich

---

[c]Publicly available at
https://www.unrealengine.com/marketplace/open-world-demo-collection.
[d]https://github.com/Jan21/unrealcv

Fig. 1. 3D scene of the forest.

Fig. 2. 3D scene of the forest with loaded flower.

assets of UE4, and easy access to control rendering is a powerful tool for continuous synthetic data generation during/for training. Further, we have discussed how Compositional Learning can take advantage of synthetic data generation and its rich labels. The future work will include experimental usage of Compositional Learning and expansion of the dataset generating framework.

References

1. K. He, X. Zhang, S. Ren and J. Sun, *CoRR* **abs/1512.03385** (2015).
2. K. He, G. Gkioxari, P. Dollár and R. B. Girshick, *CoRR* **abs/1703.06870** (2017).
3. O. Russakovsky, J. Deng, H. Su, J. Krause, S. Satheesh, S. Ma, Z. Huang, A. Karpathy, A. Khosla, M. S. Bernstein, A. C. Berg and F. Li, *CoRR* **abs/1409.0575** (2014).

Fig. 3. Columns: Sunflower from 0°, 120° and 240° with 0° and 30° rotation in upward direction (along y axis); Rows: First row are semantic labels (masks), second row are images of fully lit sunflower, third row are images of normal maps of a model and last row is depth of scene.

4. J. Johnson, B. Hariharan, L. van der Maaten, L. Fei-Fei, C. L. Zitnick and R. B. Girshick, *CoRR* **abs/1612.06890** (2016).
5. R. Krishna, Y. Zhu, O. Groth, J. Johnson, K. Hata, J. Kravitz, S. Chen, Y. Kalantidis, L. Li, D. A. Shamma, M. S. Bernstein and F. Li, *CoRR* **abs/1602.07332** (2016).
6. T. Lin, M. Maire, S. J. Belongie, L. D. Bourdev, R. B. Girshick, J. Hays, P. Perona, D. Ramanan, P. Dollár and C. L. Zitnick, *CoRR* **abs/1405.0312** (2014).
7. I. Goodfellow, Y. Bengio, A. Courville and Y. Bengio, *Deep learning* (MIT press Cambridge, 2016).
8. D. Erhan, Y. Bengio, A. Courville and P. Vincent, *University of Montreal* **1341**, p. 1 (2009).
9. K. Simonyan, A. Vedaldi and A. Zisserman, *arXiv preprint arXiv:1312.6034* (2013).
10. A. Mahendran and A. Vedaldi (2015).
11. A. Nguyen, J. Yosinski and J. Clune, *arXiv preprint arXiv:1602.03616* (2016).
12. D. Bau, B. Zhou, A. Khosla, A. Oliva and A. Torralba, Network dissection: Quantifying interpretability of deep visual representations, in *Computer Vision and Pattern Recognition (CVPR), 2017 IEEE Conference on*, 2017.
13. A. Nguyen, A. Dosovitskiy, J. Yosinski, T. Brox and J. Clune, Synthesizing the preferred inputs for neurons in neural networks via deep generator networks, in *Advances in Neural Information Processing Systems*, 2016.
14. A. Panteleev, *ACM SIGGRAPH 2014 presentations* (2014).
15. W. Qiu, F. Zhong, Y. Zhang, S. Qiao, Z. Xiao, T. S. Kim and Y. Wang, Unrealcv: Virtual worlds for computer vision, in *Proceedings of the 2017 ACM on Multimedia Conference*, 2017.

Differentiation of organic and non-organic apples using image processing — A cost-effective approach

Jiang Nan Feng

School of Mathematics and Informatics, Fujian Normal University, Fuzhou, China

Wang Hui

School of Computing, Ulster University, UK

Guo Gong De

School of Mathematics and Informatics, Fujian Normal University, Fuzhou, China

With the increase of expectation for higher quality of life, consumers have higher demands for quality food. Food authentication is the technical means of ensuring food quality, which is intended to confirm "food is what it says on the tin". A popular approach to food authentication is based on spectroscopy analysis which has been widely used for identifying and quantifying the chemical compositions of an object. Such approach is non-destructive and effective but expensive. This paper presents an image-based approach to food authentication using image processing and pattern recognition techniques. In this approach, flashlight is used to illuminate apples and images of the illuminated apples are captured by a smartphone. These images are represented by LBP (local binary pattern) image descriptors. Data pre-processing algorithms are used to prepare the image representations, and pattern recognition algorithms including k-nearest neighbors and support vector machine are used for classification. This approach is evaluated in a food differentiation (to separate organic apples from non-organic ones) experiment using a reasonable collection of apple samples, resulting in the highest classification accuracy of 86.7%. It is shown that this low-cost approach has potential to lead to a viable solution to empower consumers in food authentication.

1. Introduction

The demand for organic food products has increased rapidly worldwide in recent years, and organic food products can now be found in more and more stores. Consumer studies show that expectations concerning health benefits of organic food are the strongest motives for consumers to buy organic foods [1]. However, due to the higher value of organic foods, food frauds have arisen in terms of the integrity of food items when fraudsters use non-organic foods as organic foods

[2]. It is therefore necessary to authenticate foods with respect to their organic status as well as their origin and ingredient.

There has been a growing trend towards fast and non-destructive approaches for food authentication. A popular approach is to differentiate one food type from another by using spectroscopic techniques such as near-infrared (NIR), Fourier-transform infrared (FTIR). This approach has been investigated in many food quality studies, including identifying varieties, pesticide residuals [3]. However, to the best of our knowledge, such an approach is seldomly used to detect organic products which have higher value than non-organic ones but are less distinctive in surface morphology and easier for adulteration. Recent studies attempt to use portable NIR spectroscopy for differentiating organic food from non-organic ones, such as apples [4]. These studies have demonstrated that portable spectroscopy together with chemometrics is a potential alternative to laboratory-based spectroscopy due to its classification efficiency. However the cost of using a portable spectroscopic sensor for differentiating organic food is expensive and exceeds the expectation of consumers.

In this paper, we present a low-cost sensor system based on computer vision techniques which can be used for differentiating organic apples from non-organic ones. The hardware of this system is designed to acquire data sample of food rapidly and non-destructively, and consists of flashlight and a smartphone. Using flashlight to illuminate an apple and reflect off the apple sample, reaching the camera of the smartphone to be recorded as camera images. The software of the sensor system is designed to analyze these images and build classification models. More specifically, the images are represented as feature vectors according to the local binary pattern (LBP) image descriptor [5], pre-processed in a way to allow building effective classification models. The pre-processed data is then used for model construction and analysis. This system is consumer-friendly in that it does not require expert knowledge and is cheap. Importantly, it has classification capability that is comparable to portable NIR spectrometers in differentiating organic apples from non-organic ones.

2. Measurements

2.1. *Sensor System*

The proposed system aims to acquire image data from certain objects, i.e., organic and non-organic apples by coupling low-cost measurements with computer vision techniques. Using flashlight to illuminate an apple, the apple image is captured by camera. Then we apply computer vision techniques and image pre-processing, convert the apple image into a sample vector for analysis. The overall system is shown in Figure 1.

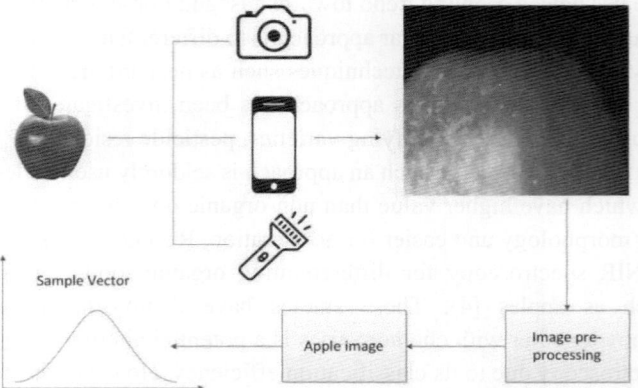

Figure 1. The overall architecture of the proposed sensor system.

2.2. *Image Acquisition*

In this paper, we used flashlight to illuminate apple samples and use smartphone to capture each apple image.

To reduce the influence of ambient light and produce images of high quality, the experiment was conducted in a dark environment. There is no surface contamination or damage in each apple and no surface preparation was carried out prior to image acquisition. We place light source 10 centimeters away from the apple, which can ensure that the light source is effectively focused on the apple surface. After generating images, smartphone is used to photograph the whole experimental environment.

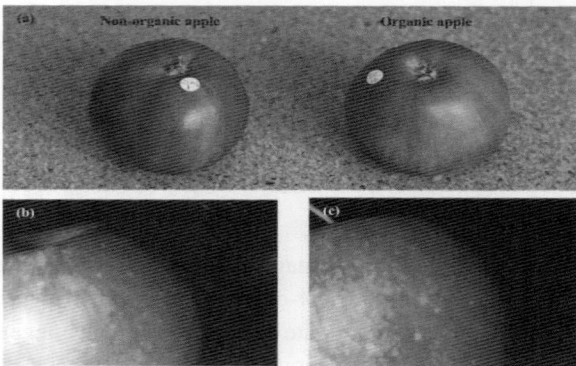

Figure 2. (a) Non-organic apple (left) and organic apple (right); (b) The image of a non-organic apple; (c) The image of an organic apple.

Figure 2 shows two apples from different classes (organic vs. non-organic) and their images. Basically, the two apples cannot be visually identified by their physical appearances. If we compare the images, it is still difficult to tell the difference between the organic and non-organic apples. In order to differentiate organic apples from non-organic ones precisely, we convert each rainbow image into sample vector for further analysis.

2.3. *Image Denoising*

Digital images are generally affected by noise from the imaging instrument and the external environment during digitization and transmission. In this paper, we use the median filter for image denoising. The main idea of median filtering is to replace the value of a point in a digital image or digital sequence with the median of neighboring points. As a result, the surrounding pixel values are close to their real value and the isolated noise points are eliminated. The median filtering is especially useful for denoising which can efficiently preserve edges.

2.4. *Feature Vector Representation*

Local Binary Pattern (LBP) texture operator was first introduced as a complementary measure to the local contrast of the image [4]. The original LBP operator labels the image pixels with 8-bit binary codes, which are obtained by the each pixel. For each given pixel, an 8-bit binary codes is obtained by concatenating all these binary values in a clockwise direction, which starts from the one of its top-left neighbor. The LBP images obtained by LBP operator is divided into subspace to calculate histograms, then the feature vector of image is made by concatenating histograms. In this paper, after obtaining the apple image, We convert the image into the LBP feature vector, then analysis them by pattern recognition approach.

3. Data Analysis

3.1. *The Nonlinear Problem*

To obtain an overview of the distinctions of species and types, raw apple data were subjected to principal component analysis (PCA) [6], as shown in Figure 3. Different species of apple are nonlinearly distributed. The next step is differentiating organic apples from non-organic ones by using a pattern recognition framework.

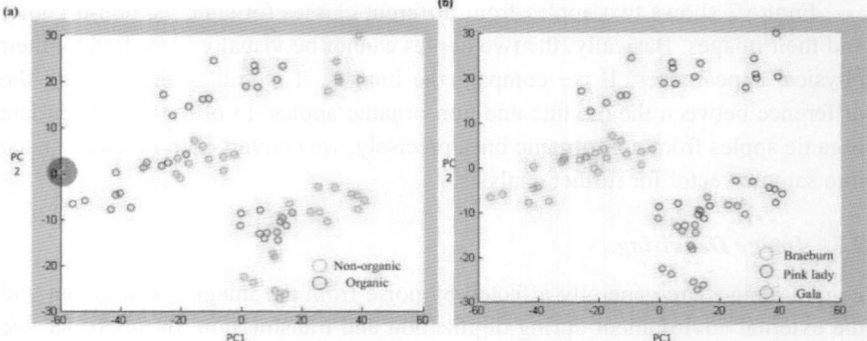

Figure 3. Score plots of the first two dimensions of PCA of apple data: (a) Non-organic and organic types. (b) Braeburn, Gala and Pink lady species.

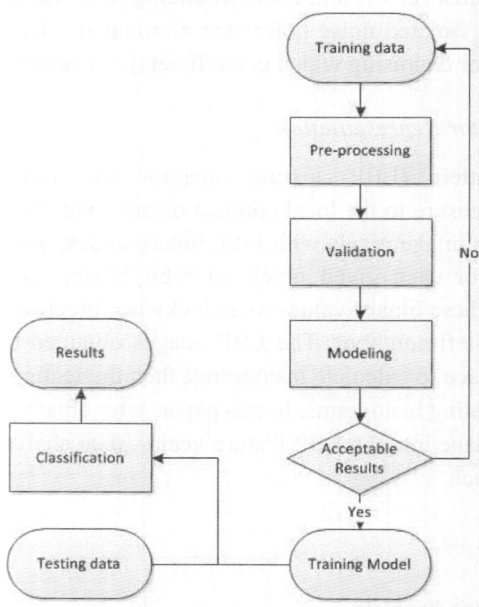

Figure 4. A pattern recognition framework for classifying organic and non-organic apple data.

3.2. *Pattern Recognition Framework*

The pattern recognition framework for the classification of organic and non-organic apple data consists mainly of pre-processing, modelling, validation and classification procedures, as shown in Figure 4. It firstly applies pre-process techniques to reduce noise effects and unwanted variations existed in raw data which are caused by instrumental and experimental artifacts. In this paper we used

Savitzky-Golay smoothing for pre-processing [7]. Modelling procedure, then uses classification algorithms to reveal the relationship between training data and their corresponding classes. To achieve the highest classification accuracy possible, the parameter(s) of a classifier requires to be optimized in validation procedure. Finally, the optimal model is selected to classify testing data.

3.3. Classifiers

Two classification algorithms, namely, k-NN and SVM, were used to classify raw and pre-processed apple data. k-NN is an effective and efficient classification algorithm that is commonly used in pattern recognition as a baseline. SVM is comparably complex in modelling and generally provide better prediction performance under nonlinear conditions. These algorithms are summarized in following sections.

3.3.1. k-Nearest Neighbors (k-NN)

The k-NN classification algorithm is a popular method which classifies a query depending on the classes of its neighboring samples. If most of the k closest samples belongs to a certain class, the query will be assigned to this class. Specifically, k-NN directly assigns the query to the class of its nearest neighbor when k equals to 1 [8].

3.3.2. Support Vector Machine (SVM)

The SVM classifier aims to find an optimal hyperplane which correctly separates the samples of the different classes while maximizing the shortest distances from the hyperplane to the nearest samples for each class [9]. It can be extended to nonlinear classification by mapping the input data into feature space via kernel functions.

3.4. Performance Evaluation

We firstly partition the apple data into training and testing sets by using DUPLEX splitting with the ratio of 2:1. DUPLEX splitting maintains the same diversity in both sets, so that the data in each set follows the statistical distribution of the overall data. Then leave-one-out cross validation is implemented to obtain the optimal parameter(s) of each algorithm and the validation accuracy. With the selected optimal parameter(s), each algorithm constructs a classification model on training set and use it to predict samples from testing set. The classification results are finally evaluated by overall accuracy and per-class accuracy.

4. Results and Discussion

4.1. *Parameter Optimization and Classification Performance*

The optimal parameters of five algorithms are set by leave-one-out cross validation on training set. The number of nearest neighbors in k-NN is chosen from 1 to 49 with an interval of 1 and We set the value of penalty parameter in SVM from 1 to 8 and select kernel function (poly). The validation results of two algorithms and their corresponding optimal parameters are provided in Table 1. SVM obtain around 86% validation accuracy.

Table 1. The overall and per class classification accuracy (%) of the different algorithms on apple data.

	Training		Testing			Parameters	
	Raw	Pre-processed	Overall	Non-organic	Organic		
k-NN	72	83	84	75	85	NN: 1	
SVM	82	85	86	83	86.7	C: 4	Kernel: poly

5. Conclusion

This paper presents a new sensor system, which consists of low-cost hardware and pattern recognition software, for differentiating organic apples from non-organic ones. The sensor system can effectively acquire apple images from food samples. Experiments show that the proposed sensor system has achieved the highest classification accuracy of 86.7% for identifying organic and non-organic apples, demonstrating the potential of the new sensor system as a rapid, non-destructive and low-cost solution for food authentication.

References

1. Vindigni, G., Janssen, M., Jager, W., Organic food consumption: a multi-theoretical framework of consumer decision making. *British Food Journal, 104*(8) (2002) 624-642.
2. Nandwani, D., *Organic Farming for Sustainable Agriculture.* Springer International Publishing. (2016).
3. Wu, X., Zhu, J., Sun, J., Dai, C., Discrimination of tea varieties using ftir spectroscopy and allied gustafson-kessel clustering. *Computers & Electronics in Agriculture.* 147 (2018) 64-69.
4. Song, W., Wang, H., Maguire, P., Nibouche, O., Differentiation of organic and non-organic apples using near infrared reflectance spectroscopy — A pattern recognition approach. *IEEE Sensors.* (2016) 1-3.
5. Guo, L., Li, J., Zhu, Y. H., Tang, Z. Q., A novel Features from Accelerated Segment Test algorithm based on LBP on image matching. *IEEE, International Conference on Communication Software and Networks.* (2011) 355-358.

6. Moore, B., Principal component analysis in linear systems: controllability, observability, and model reduction. *IEEE Transactions on Automatic Control.* 26(1) (2003) 17-32.
7. Savitzky, A., Golay, M. J. E., Smoothing and differentiation of data by simplified least squares procedures. *Analytical Chemistry.* 36(8) (1964) 1627-1639.
8. Kevin, B., Jonathan, G., Raghu, R., When Is "Nearest Neighbor" Meaningful? *International Conference on Database Theory.* Springer. (1999) 217-235.
9. Cherkassky, V., The nature of statistical learning theory. *IEEE Transactions on Neural Networks.* 38(4) (2002) 409-409.

Analysis of senile dementia from the brain magnetic resonance imaging data with clustering

Xiaobo Zhang, Yan Yang*, Hongjun Wang, Ping Deng

*School of Information Science and Technology, Southwest Jiaotong University
Chengdu 611756, China*

The senile dementia is a common disease on clinical diagnosis. In this study, the brain magnetic resonance image (MRI) data with Alzheimer's disease is extracted from a publication data sets. Clustering algorithms of K-Means, K-Medoids, Gaussian Mixture Model (GMM), Affinity Propagation (AP), Density Peaks (DP) and clustering ensemble algorithm of Cluster-based Similarity Partitioning Algorithm (CSPA) are applied to the MRI data sets for analyzing the senile dementia diagnose. The experimental results show that the GMM clustering algorithm yields better results in terms of Acc (Accuracy) evaluation and the AP gives the better values of Pur (Purity) quota, while the CSPA is the best for MRI data in medical science with highest Fm(F-measure) information.

Keywords: Senile dementia; brain magnetic resonance image (MRI); clustering.

1. Introduction

With the rapid artificial intelligence development, disease diagnosis using medical images has been studied in many disciplines of computer engineering. The magnetic resonance image is an advanced medical imaging technique, which can be used to produce high quality images of the parts contained in the human body.[8]

The Alzheimer's disease[2] was studied from the MRI data with computer neural network technology.[12,14] Brain tumour was detected and extracted from patient's MRI scan images of the brain using K-Means clustering.[9] Meanwhile, the MRI segmentation was realized with some other clustering algorithms.[7] More and more magnetic resonance image data sets were also investigated by neuroimaging and deep learning methods.[5,6] However, the research of senile dementia from the MRI data based on clustering technology have not been found yet.

In this study, a series of MRI data sets we used is from the Open Access Series of Imaging Studies (OASIS), which is publicly available for study and analysis. The initial data set consists of a ross-sectional collection of 416 subjects that are

---

*Corresponding author: yyang@swjtu.edu.cn

all right-handed and include both men and women aged from 18 to 96.[4] We applied the 235 complete marking of 416 subjects to study the Alzheimer's disease by MRI. The first step is processing the initial data set for capturing useful information. And then the clustering methods of K-Means,[13] K-Medoids,[3] GMM,[15] AP,[1] DP[10] and CSPA[16] were employed to detect the effectiveness for the senile dementia of each subject. In the experiment, Acc (Accuracy), Pur (Purity) and Fm (F-measure) are used for the performance comparison.[11,17,19]

2. Methodology

2.1. *The K-Means Algorithm*

The K-Means Algorithm is one of the basic ways used in the clustering problems. It's a method that is based on the idea that the gravity centers of the cluster elements represent the cluster.[13] The K-Means algorithm divides the data cluster entered into the system by the user and consisting of n number of the data into k number of clusters that are entered by the user again.

2.2. *The K-Medoids Algorithm*

The K-Medoids Algorithm calculates the distance matrix once and uses it for finding new medoids at every iterative step. It makes the sum of the distance between the center and the rest of the cluster be minimized.[3]

2.3. *The GMM Algorithm*

The GMM is to estimate the probability density distribution of the sample. The estimated model is the weighted sum of several Gaussian models. Each Gaussian model represents a cluster. We could get the probability on each class separately from the projection of sample data on the model of Gaussian, and select the class with the highest probability as the decision result.[15] The GMM can be defined as

$$p(x) = \sum_{k=1}^{K} \pi_k P(x \mid k),$$

where the parameter K is the number of models, π_k is the weight of Gaussian, and $p(x/k)$ is the probability density of the Gaussian sorted to k.

2.4. *The AP Algorithm*

The AP method takes as input measures of similarity between pairs of data points. Real-valued messages are exchanged between data points until a high-quality set of exemplars and corresponding clusters gradually emerges.[1] The responsibilities are computed using the following rule

$$r(i,k) = s(i,k) - max_{k' \neq k}(a(i,k) + s(i,k'))$$

where $r(i,k)$ is set to the input similarity between point i and point k as its exemplar, minus the largest of the similarities between point i and other candidate exemplars.

The availability are computed using the following rule

$$a(i,k) = \begin{cases} min\{0, r(k,k) + \sum_{i' \notin (i,k)} max(0, r(r',k))\}, & i \neq k, \\ \sum_{i' \neq k} max(0, r(i',k)), & i = k. \end{cases}$$

where the availability $a(i,k)$ is set to the self-responsibility $r(k,k)$ plus the sum of the positive responsibilities candidate exemplar k receives from other points.

2.5. *The DP Algorithm*

The DP Algorithm has its basis in the assumptions that cluster centers are surrounded by neighbors with lower local density, and cluster centers are at a relatively large distance from any points with a higher local density.[10]

2.6. *The Clustering Ensemble Algorithm of CSPA*

Clustering ensemble consists of generating a set of clusterings from the same dataset and combining them into a final clustering that is able to improve the quality of individual data clusterings by combining different clustering results, and the CSPA constructs a $n \times n$ similarity matrix, which can be regarded as the adjacency matrix of a fully connected graph, where the nodes are elements of the set X, an edge between two objects has an associated weight with the same times if these objects are in the same cluster. Clustering ensemble has proved to be a good alternative when facing cluster analysis problems.[16]

2.7. *Acc (Accuracy), Pur (Purity) and Fm (F-measure)*

Acc is the proximity value of the clustering results. Micro-precision (MP) could evaluate the accuracy of the cluster. The MP is defined as[19]

$$MP = \sum_{k=1}^{K} \frac{N_k}{N} \qquad (0 \leq MP \leq 1),$$

where N_k is the number of data items that are correctly assigned to all the classes. The bigger the value of MP, the better the clustering performance.

Pur is a common and transparent evaluation measure, which is defined as [17]

$$Purity = \frac{1}{n} \sum_{k=1}^{r} max_{1 \leq l \leq q} n_k^l,$$

where n_k^l is the number of samples in the cluster that belong to original class l. A larger purity value indicates better clustering performance.

F_m penalizes false negatives more strongly than false positives. It's the rule given as follows. [11]

$$F_\beta = \frac{(\beta^2 + 1)PR}{\beta^2 P + R} \qquad (0 \leq \beta \leq +\infty),$$

where β is a parameter that controls a balance between P and R. When $\beta = 1$, F_1 is equivalent to the harmonic mean of P and R. If $\beta > 1$, F becomes more recall-oriented, and if $\beta < 1$, it becomes more precision-oriented.

3. Experiments

In this section, we use the MRI data derived from OASIS, which is a project aimed at making MRI data sets of the brain freely available to the scientific community.[4] We implement the K-Means, K-Medoids, GMM, AP, DP and CSPA clustering algorithms on MRI data sets, and demonstrate the clustering performance by the evaluation of Acc, Pur and Fm.

3.1. Datasets

The MRI data set is used for the experiments. We process the initial data sets with Python scripts. There are 416 samples in the MRI data set, but only 235 samples of them have complete feature information and labels. One hundred of the included subjects older than 60 years have been clinically diagnosed with very mild to moderate Alzheimer's disease. The characteristics of each subject are AGE, M/F, EDUC, MMSE, eTIV, ASF and nWBV, and the data label is CDR,[4] all of which are displayed in the Table 1. More visually, eTIV and nWBV are drawn with ages in Figure 1 and Figure 2 for the data included in the samples.

3.2. Experimental Setup

All experiments are performed by a PC Server (Intel(R) Core(TM) i5-3337U CPU @ 1.80GHz, memory 4GB). Firstly, we implement ActivePython-2.7.13.2716

Table 1. Imaging measures in the datasets.

Age	Age at time of image acquisition (years)
Sex	male or female
Edu	Years of education
MMSE	Ranges from 0 (worst) to 30 (best)
ASF	Atlas scaling factor
eTIV	Estimated total intracranial volume (cm3)
nWBV	Expressed as the percent of all voxels in the atlas-masked image
CDR	0 = no dementia, 0.5 = very mild, 1 = mild, 2 = moderate

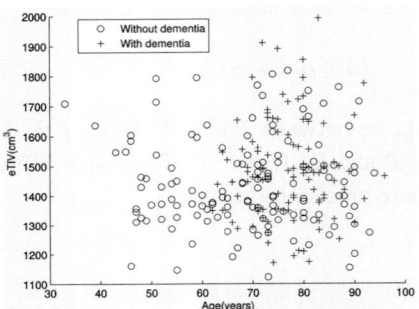

Fig. 1. Plots of automated anatomical measures with eTIV and age.

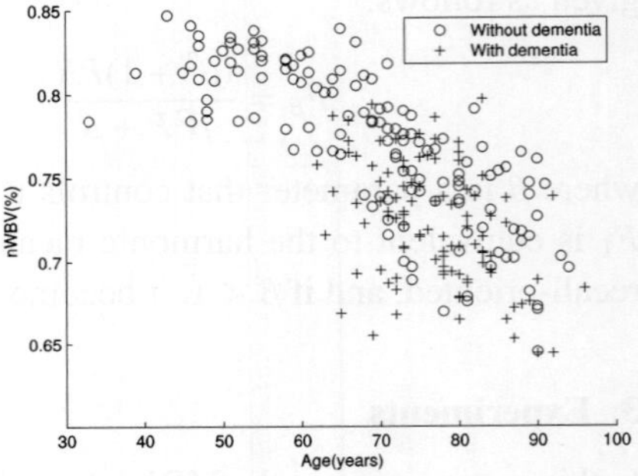

Fig. 2. Plots of automated anatomical measures with nWBV and age.

software and Python script to process the initial data sets. Secondly, we select the 235 samples with data labels. Thirdly, we choose the K-Means, K-Medoids, GMM, AP and DP clustering algorithms and apply Acc, Pur and Fm as the evaluation of the performance in service of the Matlab R2014a software. At last, we compare the results of each evaluation from the six unsupervised learning algorithms, i.e., K-Means, K-Medoids, GMM, AP, DP and CSPA.

3.3. Results and Discussion

In this subsection, we give the results of the comparisons after the experiments mainly with the five basic algorithms and the clustering ensemble algorithm of CSPA combining with the results of the AP and DP algorithms on the MRI data. The values of Acc, Pur and Fm of each algorithm with 100 running times for MRI dataset are shown in Table 2, respectively, which contain mean and standard deviation.

In Table 2, the highest Acc, Pur and Fm have been highlighted. It is noted that the GMM has the highest Acc value and the AP has the most excellent Pur value.

Table 2. The evaluation results of Acc, Pur, Fm and Tt.

	K-Means	K-Medoids	GMM	AP	DP	CSPA
Acc	35.21±0.56	34.74±5.17	**52.56±3.50**	27.23±0.00	42.13±0.00	31.91±0.00
Pur	68.53±0.09	68.54±0.25	62.87±4.54	**68.77±0.00**	68.60±0.00	68.36±0.00
Fm	28.89±7.03	31.96±6.18	25.87±16.13	30.91±0.00	22.58±0.00	**34.27±0.00**

Also, the CSPA has the best performance on Fm. More intuitively, Acc, Pur and Fm comparison results are shown in Fig. 3, Fig. 4, and Fig. 5, respectively.

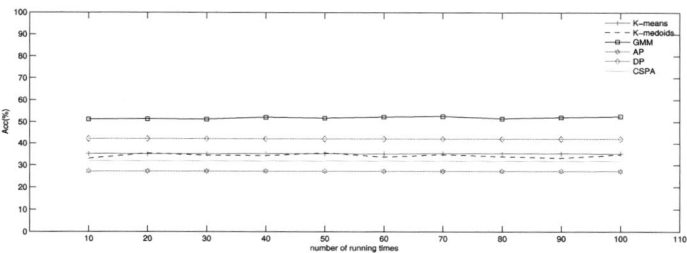

Fig. 3. The Acc comparison from different clustering algorithms.

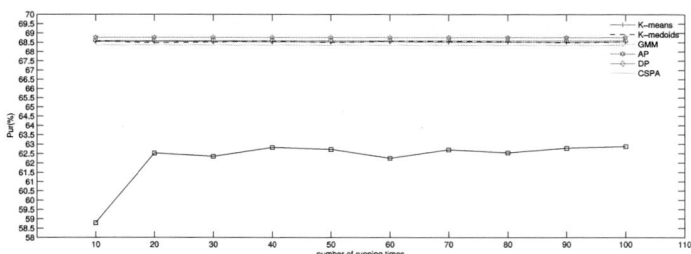

Fig. 4. The Pur comparison from different clustering algorithms.

A comparative analysis of Fig. 3 shows that the Acc of GMM algorithm is much better than other algorithms, Fig. 4 shows that the Pur of AP is the highest one, and the Pur of others approaches to the highest value except GMM. However, Fig. 5 shows that the Fm of CSPA is apparently higher than others.

It is more significant for the value of Fm for MRI data, because Acc is mainly used to evaluate the correctness and error of clustering, but the Fm focuses on punishing misjudgment. In medical treatment, the cost of miscarriage of justice is often higher, and the impact of diagnostic errors is very serious. Therefore, the Fm evaluation is more persuasive in this paper.

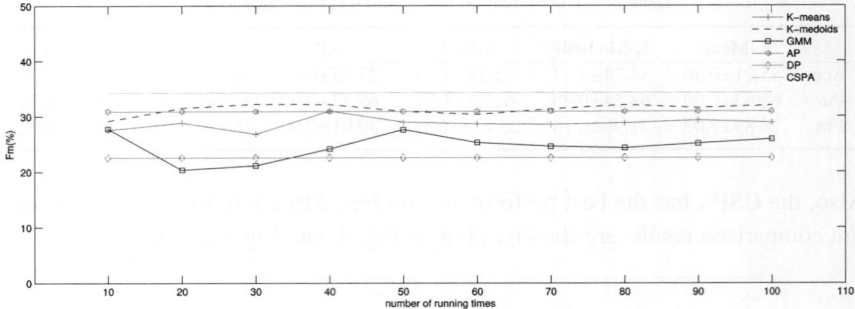

Fig. 5. The Fm comparison from different clustering algorithms.

4. Conclusion

Generally data label is needed for the classification methods, but the medical image data in reality does not usually have a specific label, hence we choose clustering methods to handle them. This paper first adapts clustering methods to analyze MRI data, and realizes the function of unlabeled medical image data clustering to diagnose disease. For the analysis of senile dementia from MRI data, six different unsupervised learning clustering algorithms are applied to analysis and contrast the results in the experiments. The Pur evaluation of the AP clustering algorithm gets the best result. In the respect of Acc evaluation, the GMM clustering algorithm has the highest value comparing with the real MRI data corresponding Alzheimer's disease. More important, the ensemble CSPA we designed has a best performance in terms of the FM evaluation for the clustering of Alzheimer's disease in MRI data. In the future this study will be done better with optimized GMM and CSPA algorithms. These MRI data will also be analyzed by multi-view clustering, or multi-task multi-view clustering methods.[18]

Acknowledgments

The data set in this paper was provided by OASIS, that is made available by the Washington University Alzheimer's Disease Research Center, Dr. Randy Buckner at the Howard Hughes Medical Institute (HHMI) at Harvard University, the Neuroinformatics Research Group (NRG) at Washington University School of Medicine, and the Biomedical Informatics Research Network (BIRN). This work is supported by the National Natural Science Foundation of China (Grant No. 61572407).

References

1. B.J. Frey, Delbert Dueck. Clustering by Passing Messages Between Data Points. Science. 2007, 135: 972-976.
2. D. Harman. Alzheimer's Disease Pathogenesis. Annals of the new york academy of sciences. 2007, 1067: 454-560.
3. C.H. Jun. A simple and fast algorithm for K-medoids clustering. Expert Systems with Applications. 2009, 36(2): 3336-3341.
4. D.S. Marcus, T.H. Wang, J. Parker, J.G. Csernansky, J.C. Morris, R.L. Buckner. Open Access Series of Imaging Studies (OASIS): Cross-sectional MRI Data in Young, Middle Aged, Nondemented, and Demented Older Adult. Journal of Cognitive Neuroscience. 2007, 19(9): 1498-1507.
5. F. Milletari et al., Hough-CNN: deep learning for segmentation of deep brain regions in MRI and ultrasound. Computer Vision and Image Understanding. 2017, 164: 92-102.
6. T.E. Nichols et al., Best practices in data analysis and sharing in neuroimaging using MRI. Nature neuroscience. 2017, 20(3): 299.
7. E. küçükkülahlı, P. Erdogmus, K. Polat. MRI Segmentation based on Different Clustering Algorithms. International Journal of Computer Applications. 2016, 155(3): 37-40.
8. Y.T. Liu. Magnetic Resonance Imaging. Current Laboratory Methods in Neuroscience Research. 2013: 249-270.
9. A.K. Panda, M. Kumar, M.K. Chaudhary, A.A.K. Gupta. Brain Tumour Extraction from MRI Images Using K-Means clustering. InternationalL Journal of Innovative Research in electrical, electronics, instrumentation and control engineering. 2016, 4(4): 356-359.
10. A. Rodriguez, A. Laio. Clustering by fast search and find of density peaks. Science. 2014: 1492-1496.
11. Y. Sasaki. The truth of the F-measure. Teach Tutor mater, 2007. 1(5).
12. C. Sorg, V. Riedl, M. Muhlau, V.D. Calhoun et al., Selective changes of resting-state networks in individuals at risk for Alzheimer's disease. Proceedings of the National Academy of Sciences, 2007, 104(47): 18760-18765.
13. D. Steinley. K-Means clustering: A half-century synthesis. British Journal of Mathematical and Statistical psychology. 2006, 59(1): 1-34.
14. M. Tahmasian, J.M. Shao, C. Meng, T. Grimmer et al., Based on the Network Degeneration Hypothesis: Separating Individual Patients with Different Neurodegenerative Syndromes in a Preliminary Hybrid PET/MR Study. Nucl Med. 2016, 57: 410-415.
15. N. Vlassis. Aristidis Likas. A Greedy EM Algorithm for Gaussian Mixture Learning. 2002, 15(1): 77-87.
16. S. Vega-Pons, J. Ruiz-Shulcloper. A survey of clustering ensemble algorithms. International Journal of Pattern Recognition and Artificial Intelligence, 2011, 25(03): 337-372.
17. Z.R. Yang, E. Oja. Linear and nonlinear projective nonnegative matrix factorization. IEEE Transactions on Neural Networks. 2010, 21(5): 734-749.
18. Y. Yang, H. Wang, Multi-view Clustering: A Survey, Big data mining and analytics, 1(2): 83-107, 2018.
19. Z.H. Zhou, W. Tang. Clusterer ensemble. Knowledge-Based Systems. 2006, 19(1): 77-83.

Cross-domain image description generation using transfer learning[*]

Philip Kinghorn
School of Engineering and Computing, Durham University
Durham, DH1 3LE, UK

Li Zhang
Department of Computer and Information Sciences,
Faculty of Engineering and Environment, Northumbria University
Newcastle, NE1 8ST, UK

In this research, we present image description generation using deep and transfer learning techniques for cross-domain images. The proposed system is able to conduct image captioning for out-of-scope images via fine-tuning and transfer learning. Tested with several cross-domain datasets, the work shows impressive capabilities for describing out-of-domain CCTV surveillance, artwork and healthcare images.

1. Introduction

Image captioning/description has been an active research area for computer vision, recently gaining attraction from the deep learning community [1-3]. Image captioning is the direct result of a computer system that can receive an input image, then correctly and automatically annotate and describe its contents. Image captioning can therefore cover many existing computer vision domains in order to fulfil this criterion. For example, for these systems to recognize and locate objects, object classification must be explored, and for natural sounding output descriptions, utilising techniques and methods within the field of Natural Language Processing (NLP) becomes necessary.

Therefore, in our recent research [1], we studied the above aspects and proposed a regional deep learning based image description generation system. The proposed system is composed of object detection and recognition, scene classification, attribute prediction and language generation. The details of the proposed system are presented in [1]. Specifically in our previous system, region-based object detection, attribute classification and relationship

---

[*]This work is supported by Higher Education Innovation Fund & Northumbria PhD Studentship.

identification allowed for efficient descriptive capabilities and incorporated regional details into image description generation.

In this research, we aim to extend the capabilities of our previous system by exploring aspects of transfer learning in order to describe out-of-scope images efficiently. Transfer learning is utilised when there is not sufficient data in an intended domain, as well as when the typical training procedure is adhered to over fitting and the model performs unsatisfactorily. To this end, related data can be utilised from either the target or another related domain to include information learned from both domains. This is similar to the belief in which children learn. For example a child can only store and recognise a certain number of objects within a day. Anything new they see or learn is heavily weighted on knowledge and understanding from those existing learned classes.

Transfer learning can take many forms and has many different approaches in order to implement the knowledge transfer, for example, cross-domain and cross-view knowledge transfer. Cross domain identifies the gap between source and target data that can be entirely different to another, for example cars. This domain could utilise data from wheels or car radiators to the unrelated such as laptops, whereas cross view aims to account for the near infinite available viewpoints of any one object [4].

In this research, we are dedicated to the cross-domain transfer learning for image description generation in order to enable our previous system to deal with image captioning for out-of-domain (e.g. artwork and CCTV) images. In comparison to our previous work, the updated system possesses new domain knowledge and shows great capabilities in dealing with diverse out-of-scope challenging images. It is also able to produce more refined descriptions over other existing holistic methods [5] by recovering human and object attributes.

2. Related Work

For image classification [1-6], Convolutional Neural Network (CNN) has been used as a feature extractor for image representation. By removing the output layers, this would therefore indicate that the network would produce a high dimensional vector rather than label confidence scores. For example, on the AlexNet CNN, removing the output (or fully connected) layers of the network, can produce a 4096-dimensional feature vector. It has been common cases for CNNs pre-trained on the incredibly large ImageNet dataset to be used as a generic feature extractor in many other areas, such as scene and object recognition.

Features of images can be collected from something as simple as pixel values. Collecting each value in an image would lead to the feature vectors being incredibly large due to the increase in mainstream camera quality etc. CNNs can be used as a feature extractor, by learning the features from a small grid and using this trained detector at different locations across the whole of the image, dramatically scaling down the size of the collected features making them much more manageable, combining this with pooling layers and then alternating creates a very discriminative feature extractor that can be utilised on high resolution images. Also, pooling would be done in a very similar fashion however simply collecting the highest value within its sliding window or adding them all together or averaging among others.

In related research [1-3], sentences and descriptions for image description have been constructed with methods such as templates, substituting entities into the variables within a fixed sentence and Conditional Random Fields (CRFs) which can produce longer sentences but can sound robotic and rigid. There have also been works in which the closest relating sentence/description has been retrieved from the corpus based on the image features. In a bid to overcome this issue in recent research the use of Recurrent Neural Networks (RNN) has been the main methodology for generating sentences from images.

RNNs are based on the proposition of time steps. As most neural networks accept the current input example, RNNs take the current example and the input at the previous time step into account. This can be simply defined as the decision a RNN makes. The timestep t-1 affects the decision of the network at timestep t. These are the two inputs of RNNs, i.e. the present moment (t), and the immediate past moment (t-1). This can act as a form of context and allow the correct reactions to new data, much in the same way humans also respond to similar circumstances.

This is different from typical neural networks due to this feedback loop, revising their own outputs as inputs immediately after producing them. This adds memory to neural networks, collecting and storing the hidden information and patterns within sequences of data.

RNNs can suffer from a number of issues which can dramatically affect the training and in the end how successful the model is at its given task. The largest issue is the vanishing and exploding gradients problem. These mathematical problems are on the basis that the scale that the matrices within these networks go through the multiplication process is larger than 1, thus the values can quickly become immense, the reverse when the multiplication is less than 1. This makes the weights either immeasurably large, which can be truncated or squashed. However incredibly small (vanishing) gradients propose a much larger challenge [5]. These RNNs can generate sentences and descriptions word by word, or on an individual character based level, and are used to overcome the limitation of the fixed input and output vector length of neural networks.

Long Short-Term Memory Units (LSTM) was proposed after RNNs as a potential and usable solution to the problems of vanishing gradients as previously discussed. This is achieved by their memory that can help to store and maintain a more constant error. This allows the RNN it is linked with to iterate and learn over many more time steps. LSTMs differ slightly to those nodes typically found within a RNN, as the gated cells within LSTMs allow data to be stored, written and read from. The decision to perform each action is performed from within the cell. The gates within the LSTM are activated or blocked based on the input and its confidence which is determined and if necessary filtered based on its own set of internal weights. These weights are also adjusted during the RNN training procedure. The above-mentioned CNN and LSTM models have been used in our previous and this research.

3. The Proposed System Using Transfer Learning

In our previous deep image captioning system [1], firstly R-CNN is used to perform object detection, which is a traditional CNN but with extra outputs that predict bounding box coordinates. RNN-based attribute classification is subsequently conducted. Holistic CNN features are also extracted for scene classification. Two RNNs with an encoder-decoder structure are employed for language generation. One of the initial aims of the proposed study was to allow the created system to be utilized on any given image. This can be interpreted in many ways. Therefore, we aim to conduct further experimentation in order to address some of these possibilities.

In order to extend our system's capabilities in cross-domain image captioning, transfer learning is conducted and embedded in the above deep network. Specifically, in this research, we perform fine-tuning of the last three layers of R-CNN to equip it with new knowledge from cross-domain images. The overall system shows great computational efficiency and requires an average of 0.5 seconds for the description generation for each test image.

We could test the ability of our system with transfer learning for coping with images in the wild, meaning any image captured on a camera, or that it is within the domain of images that the system is already trained on. Another possibility is that the system can cope with any image, e.g. photographs, artwork, or even domain specific images. We conduct some additional experiments on dramatically differing datasets, to test the upgraded system with fine-tuning and transfer learning on the above given domains.

Four transfer learning experiments were conducted. The first experiment conducted is on the Pandora dataset [7]. This database is available in either 7K or 18K formats, each equipped with approximately the number of images in its name. In this experiment, a test set of ~1000 images of the 7740 in the first database are taken and passed through the entirety of the architecture. We conduct multiple experiments on this dataset, investigating the effects of transfer learning.

The results from the Pandora dataset in Table 1 show that the system is detecting and reporting many entities within the image correctly. For example, describing the scene and scope of the valley, as well as the colour of a person's clothes and gender within a full and rounded descriptive sentence. However, it fails to annotate other aspects at all. This shows that there is promise within this technique and its subsequent methodology to allow such a system the ability to annotate and accurately describe these kinds of paintings.

Table 1. Initial transfer learning experiments on the Pandora dataset in which paintings were passed through our fine-tuned framework and their associated results.

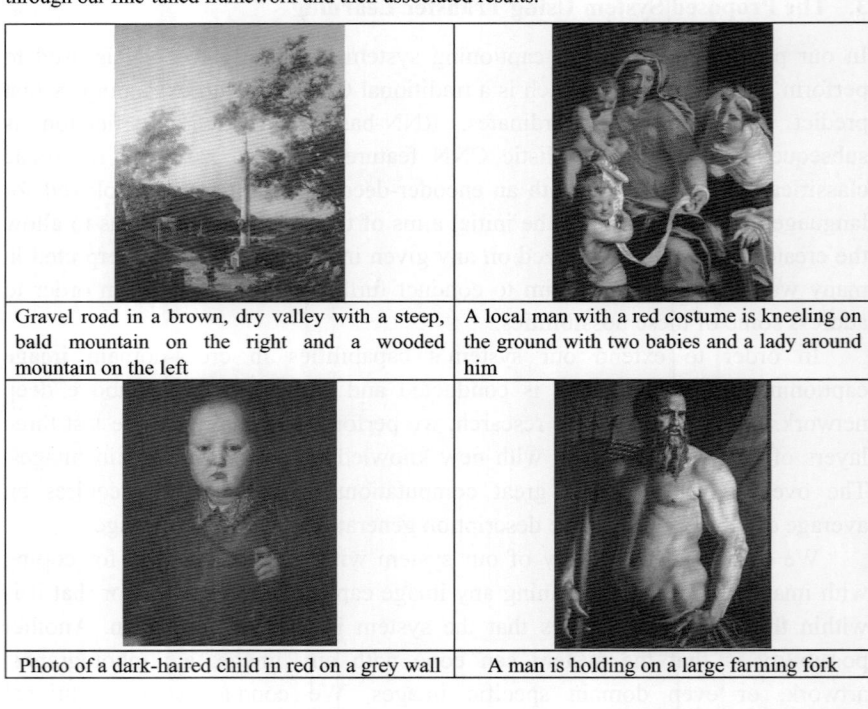

Gravel road in a brown, dry valley with a steep, bald mountain on the right and a wooded mountain on the left	A local man with a red costume is kneeling on the ground with two babies and a lady around him
Photo of a dark-haired child in red on a grey wall	A man is holding on a large farming fork

For a different, larger more diverse painting dataset, the application is tested in similar fashion as the previous experiment, however, this time on the WikiArt dataset [8]. This dataset contains a dramatically diverse range of art styles, ranging from realism to incredibly abstract works of art. Examples of the generations provided by this system are provided in Table 2.

Table 2. Initial transfer learning experiments on the WikiArt dataset in which a painting was passed through our fine-tuned framework and its associated results.

A green coast with the leaves of a palm tree

This research was also extended with the regional descriptive capabilities on the person re-identification dataset (iLids) [9]. This dataset contains readily cropped images of people on relatively low resolution CCTV images, from 2 cameras at 2 different angles. This dataset also consists of 300 different pedestrians observed from 2 distinct camera positions. The dataset is also split into static and image sequences. The image sequences are a series of images ranging from 23 to 192 individual captured camera frames, with an average sequence length of 73. This dataset is claimed to be extremely challenging due to the clothing similarities among the subjects, as well as lighting and viewpoint changes due to the multiple angles, cluttered and busy backgrounds and random unavoidable image occlusions.

For evaluation of this dataset we collect a random set of 400 images from both cam1 and cam2. This creates a relatively small test subset of 800 images. These images are then passed into the architecture described earlier, however the final translation stage is removed. This is simply as there is no need for a 'sentence' to be inferred, as the regional person descriptions are just as if not more accurate in this given domain. As the previous experiment on the Pandora dataset, the effects of fine-tuning and transfer learning are explored. The initial experiment consists of the model and architecture as described. These results can be seen in Table 3.

The iLids person re-identification dataset was tested with this description capability, in a bit to address how such as system could process and annotated relatively low resolution images of people. To this end, the whole system framework was not utilized. This is due to the nature of the sentences and descriptions produced. These would more than likely be irrelevant or include information that is distracting as well as potentially incorrect. To this end, the whole system with the absence of the sentence/descriptive layers was utilized for this particular experiment.

Table 3. Initial transfer learning experiments in which person re-identification datasets were tested upon.

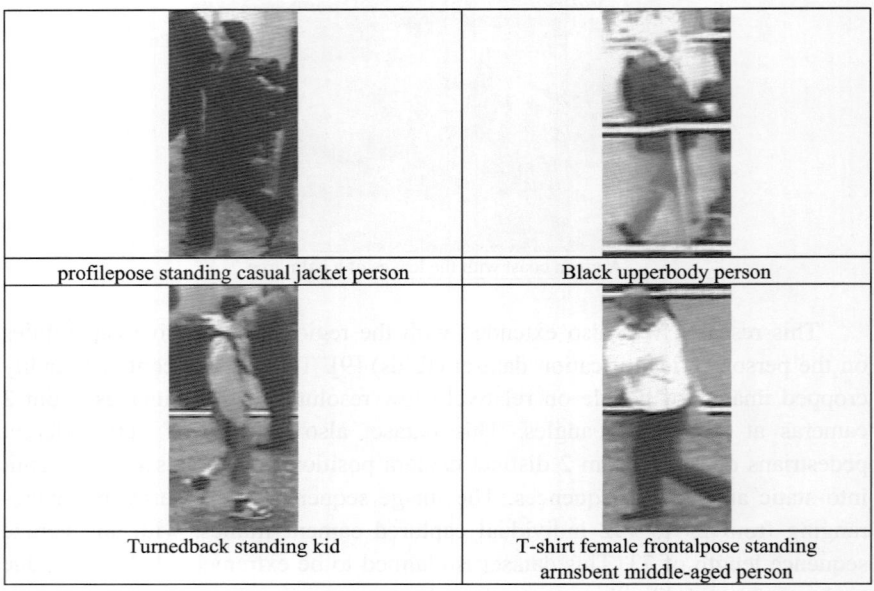

profilepose standing casual jacket person	Black upperbody person
Turnedback standing kid	T-shirt female frontalpose standing armsbent middle-aged person

As shown in Table 3, the results from the iLids person re-identification experiment without the description framework show the success of the framework with minimal alterations.

Finally, we have also evaluated the system with 50 cross-domain out-of-scope healthcare images (e.g. fall actions and hazards on the floor). Our system with transfer learning shows great robustness and achieves a BLEU score of 38 in comparison to 28 obtained by an existing system, i.e. NeuralTalk [5].

4. Conclusion

In this research, we have extended our previous system and tested it within a transfer learning experiment, to determine its ability to annotate and describe images from an irrelevant domain, as well as a re-identification task. The out of domain images consist of paintings from different artists, periods and styles making it a difficult task. The re-identification task was conducted with only part of the overall system. With fine-tuning and transfer learning, the system is able to show the power of attribute prediction and image captioning for cross-domain images. The system can accurately label scenes, genders, clothes and estimate age bands for the artwork and the low-resolution iLids datasets. It also shows great efficiency in dealing with healthcare images. For future work, we also aim to evaluate the proposed system for medical microscopic image annotation and museum exhibition description generation.

References

1. P. Kinghorn, L. Zhang and L. Shao. A region-based image caption generator with refined descriptions. *Neurocomputing*. **272** (2018) 416-424.
2. P. Kinghorn, L. Zhang and L. Shao. A Hierarchical and Regional Deep Learning Architecture for Image Description Generation. *Pattern Recognition Letters*. (2018).
3. P. Kinghorn, L. Zhang and L. Shao. Deep learning based image description generation. *In: 2017 International Joint Conference on Neural Networks (IJCNN)*. IEEE, Piscataway, 919-926. (2017).
4. L. Shao, F. Zhu and X. Li. Transfer Learning for Visual Categorization: A Survey, *IEEE Transactions on Neural Networks and Learning systems*, **26** (5) 1019-1034. (2015).
5. A. Karpathy and F.F. Li, Deep Visual-Semantic Alignments for Generating Image Descriptions. *In: IEEE Conference on Computer Vision and Pattern Recognition*. 3128-3137. (2015).
6. Y. Shen, L. Zhang and L. Shao. Semi-supervised vision-language mapping via variational learning. *In: IEEE International Conference on Robotics and Automation (ICRA)*. IEEE, 1349-1354. (2017).
7. C. Florea et al., Pandora: Description of a Painting Database for Art Movement Recognition with Baselines and Perspectives. Available at: http://arxiv.org/abs/1602.08855 (Accessed: 7 September 2017). (2016).
8. W.R. Tan, C.S. Chan, H.E. Aguirre and K. Tanaka. Ceci n'est pas une pipe: A deep convolutional network for fine-art paintings classification. *In: International Conference on Image Processing*. 3703-3707. (2016).
9. T. Wang, S. Gong, X. Zhu and S. Wang. Person Re-Identification by Discriminative Selection in Video Ranking, *IEEE Transactions on Pattern Analysis and Machine Intelligence*, **38** (12), 2501-2514. (2016).

Heart sound de-noising using wavelet and empirical mode decomposition based thresholding methods

Shaocan Fan
Electrical and Computer Engineering, University of Macau
Macau, China

Booma Devi Sekar
School of Computing, Ulster University
Northern Ireland, UK

Peng Un Mak
Electrical and Computer Engineering, University of Macau
Macau, China

Sio Hang Pun, Mang I Vai
State Key Lab of Analog and Mixed-Signal VLSI, University of Macau
Macau, China

Heart sound de-noising is considered as an important signal pre-processing step in developing computer assisted heart auscultation model. In this paper, we investigate three white noise reduction methods, namely wavelet transform, wavelet packet transform, and empirical mode decomposition for heart sound de-noising. The de-noised signals are evaluated using signal-to-noise ratio and root mean square error. The results show wavelet transform and empirical mode decomposition methods outperform the wavelet packet transform in heart sound de-noising. The wavelet transform method with 'dmey' wavelet provides a better result for most of the heart sound records. These three de-noising methods are useful to attenuate the white Gaussian noise. It can provide a high quality signal for further signal processing and classifying the heart sound signal.

1. Introduction

Heart sound (HS) by auscultation is one of the most traditional biomedical signal primarily monitored to diagnose cardiovascular diseases (CVDs). In recent decades, intelligent computerized cardiac auscultation, recorded as phonocardiogram (PCG) has provided a convenient approach to monitor CVDs. Literature show that recent evolution of various key technologies, including signal processing, traditional machine learning and deep learning has now made computerized HS analysis and

CVD diagnosis possible[1]. However, applying such methods becomes challenging when the HS signal under analysis is polluted by external and internal noise. It becomes critical to attenuate the noise before further signal processing and classification methods could be applied.

Many sources of noise, including ambient noise, lung and breath sound, fetus heart sound (if the subject is pregnant), muscle tremors, noise from the recording device etc., can pollute the HS signal. As the frequencies of these signals overlap with the HS signal frequency, HS de-noising becomes an absolute necessary pre-processing step in the development of computer aided HS analysis model.

Literature show that various methods of HS de-noising have been developed using soft and hard computing. Some of these include, adaptive noise canceller for real-time HS enhancement[2], interference suppression via spectral comparison technique based on STFT to attenuate ambient noise[3], and adaptive line enhancer (ALE) filter to filter out HS signal from lung signal[4]. Study show that the soft de-noising methods, namely wavelet based de-noising (wavelet de-noising (WTD) and wavelet packet de-noising (WPTD))[5] and empirical mode decomposition de-noising (EMD)[6] are relatively more efficient to white noise reduction.

In this paper, the WTD, WPTD, and EMD methods are investigated and compared for HS de-noising.

1.1. *Wavelet Transform*

Wavelet transform (WT) can obtain a good time and frequency resolution simultaneously by using a variable size window region instead of constant window size as in FT or STFT. The continuous wavelet transform (CWT)[7] is described as Eq. (1), where $\psi\left(\frac{t-b}{a}\right)$ is obtained from a basic "wavelet" function $\psi(t)$, a is binary scaling and b is dyadic translation, $\overline{\psi(\frac{t-b}{a})}$ is complex conjugate.

$$X(a,b) = |a|^{-\frac{1}{2}} \int_{-\infty}^{+\infty} f(t)\overline{\psi\left(\frac{t-b}{a}\right)} \quad (1)$$

$$\begin{cases} h(n) = \frac{1}{\sqrt{2}} \langle \phi(t), \phi(2t-n) \rangle \\ g(n) = \frac{1}{\sqrt{2}} \langle \varphi(t), \phi(2t-n) \rangle = (-1)^n h(1-n) \end{cases} \quad (2)$$

Discretization of scaling and translation are $a_m = a_0^m$, $b_n = nb_0 a_0^m$, where m, n are integers. For dyadic transform, $a_0 = 2, b_0 = 1$, the discrete wavelets $\psi_{m,n}(t) = 2^{-m/2}\psi(2^{-m}t - n)$. The discrete wavelet transform can be described as Eq. (2), where $h(n)$ and $g(n)$ are scaling and wavelet filters; $\phi(t), \varphi(t)$ are scaling function and wavelet function respectively. The signal was convolved with $h(n)$ and $g(n)$ to get wavelet coefficients.

1.2. Wavelet Packet Transform

The wavelet packet transform (WPT) is a generalization multiresolution analysis method extended by CWT[7]. The functions W_n, $n = 2m$ or $2m + 1$, $m = 0, 1, ...$, defined as Eq. (3), with $W_0 = \phi(t)$, $W_1 = \varphi(t)$, $\phi(t), \varphi(t)$ are scaling function and wavelet function respectively.

$$\begin{cases} W_{2m} = 2 \sum_{n=0}^{2N-1} h(n) W_m(2t - n) \\ W_{2m+1} = 2 \sum_{n=0}^{2N-1} g(n) W_m(2t - n) \end{cases} \quad (3)$$

In an orthogonal case, the wavelet packet atoms are given as $W_{j,m,n} = 2^{-j/2} \sum_{n=0}^{2N-1} W_m(2^{-j}t - n)$, where j, n denote the scale parameter and time localization respectively. While giving a fix j, function $W_{j,m,n}$ analyses the signal around the position $2^j \cdot n$ at the scale 2^j. The transform in this case splits both the approximation and detail coefficients into finer components, constructing a wavelet packet tree. Therefore, it can obtain the whole frequency band information rather than the low frequency information as in WT.

1.3. Wavelet De-noising

The wavelet threshold de-noising is performed in this paper. The basic idea is that the useful signal wavelet coefficients are thought larger than the noise wavelet coefficients. A threshold parameter is subsequently defined to sparse the useful information and eliminate the noise in wavelet domain. There are two kinds of threshold, namely hard and soft, described as Eq. (4), (5) respectively. Both thresholds are evaluated to get the best results.

$$f(t_i) = \begin{cases} 0, & |f(t_i)| < T \\ h_i(t), & |f(t_i)| > T \end{cases} \quad (4) \quad f(t_i) = \begin{cases} 0, & |f(t_i)| < T \\ f(t_i) - T, & |f(t_i)| > T \end{cases} \quad (5)$$

The procedure of de-noising can be described as three steps. The signals are first transformed to wavelet domain, obtaining the detail and approximation coefficients. Then, identify the noise components and performing thresholding to shrink coefficients. Finally, reconstruct the signal by using the processed coefficients. In this paper, the HS is analyzed by using the wavelet toolbox provided in Matlab®[8]. It is critical that various parameters, including mother wavelet, decomposition level, threshold type, threshold selection rule, threshold's rescaling methods need to be optimally selected. Studies from literature [9], [10] show that the mother wavelets, namely 'db10', 'db20', 'coif5' and 'dmey' are more suitable wavelets for HS signal de-noising. Thus, these four wavelets are evaluated in this work. Based on the tests performed on different HS signals, we select the decomposition level of 10 and soft thresholding. We also note that the threshold selection, 'rigrsure', and threshold's rescaling methods, 'sln', achieve a better performance.

In WTP de-noising, the default global threshold and threshold type will be got first. The function, 'wpdencmp' is used to de-noise the HS signal, in which the mother wavelet is the same as WT. Tests show that decomposition level of three provides a good performance.

1.4. *Empirical Mode Decomposition De-noising*

EMD, proposed by N. E. Huang[11] is an adaptive time-space analysis method for processing the non-stationary and non-linear signals. EMD is a fully data-driven decomposition method that requires no prior knowledge of the signal.

Inspired by WTD, Y. Kopsinis et al.[6] developed a de-noising method by combining the concept of threshold based on the EMD. In this method, the signal is decomposed into a series of oscillatory low-high frequency (AM-FM) signal called intrinsic mode functions (IMFs) through iterative procedure. It relies on the finding[12] that the noise only IMF energy can be approximately estimated by $E_i = \rho E_i^2/\beta, i = 2,3,4 ...$, where E_i is the energy of the ith order IMF, $\rho = 0.719$ and $\beta = 2.01$. The IMF which contains a lot of noise can be discerned and discarded while signal reconstruction.

According to the character of different order IMF, different thresholds for the ith order IMF are required and can be described as $T_i = 0.7\sqrt{E_i \ln N}$, where N is the length of input signal. Unlike threshold process in WTD, the threshold directly applied to the IMF can cause some problems[6]. The concept of interval threshold was proposed to infer noise and useful signal in any interval of two zero crossing.[6] These thresholds can be described using Eq. (6), (7), where $h_i(t)$ represents i^{th} order IMF, $h_i(z_j)$ is the zero crossing point, and h_{ext} is the extremum in the interval. Eq. (8) is soft thresholding with smooth function.

$$h_i(t) = \begin{cases} 0, & |h_{ext}| < T_i \\ h_i(t), & |h_{ext}| > T_i \end{cases} \quad (6) \quad h_i(t) = \begin{cases} 0, & |h_{ext}| < T_i \\ h_i(t)\dfrac{h_{ext} - T_i}{h_{ext}}, & |h_{ext}| > T_i \end{cases} \quad (7)$$

The performance of EMD depend on many parameters, such as sampling, interpolation, border effects and the stopping criteria. In this paper, we have adapted the details and setting proposed by G. Rilling et al.[13] Tests show that an iteration of 10 times is sufficient for a good performance. Likewise, hard thresholding provides a better performance than the soft one. The iterative EMD interval-thresholding (EMD-IIT) applied the idea of translation invariant and clear iterative EMD interval-thresholding (EMD-CIIT) applied in the low noise situation were studied. Besides factors mentioned above, the noise altering method and the number of ensemble can affect the performance of EMD-IIT and EMD-CIIT. Tests show that combining of the random permutations noise altering method and 12 times ensemble are the best setting.

2. Result and Discussion

2.1. *Heart Sound Data Acquisition and Pre-Processing*

For evaluating the above methods, HS data are obtained from three different databases. The data from Michigan HS database (MHSD)[14] are artificially synthesized HS data and the other two from PhysioNet HS database (PNHD)[15] and Pascal HS database (PCHD)[16] are the real-patient data, include normal and abnormal records. All data are sampled at 44.1 kHz with variance of the absolute amplitude at a value of 1. PNHD data already downsampled to 2 kHz. First, all data without the PNHD are downsampled to 2 kHz and normalized for maintaining the signal amplitude at a range of -1 to 1. Then, white Gaussian noise is added to the original signal as $s(n) = f(n) + \sigma e(n)$, where $f(n)$ is the original signal and $e(n)$ is noise, σ is the noise strength (standard deviation) and set as 1. $n = 1, 2, \ldots, N$, where N is the length of signal.

2.2. *Quantitative Evaluation*

Assuming the noise in original signal is much lower than the added noise, the quantitative assessment is carried on by the signal-to-noise ratio (SNR) and root mean square error (RMSE) using Eq. (8) and (9), where $\bar{s}(n)$ is denoised signal and N is the length of signal.

$$SNR = 10 log_{10} \frac{\sum_{n=1}^{N} s(n)^2}{\sum_{n=1}^{N}[s(n) - \bar{s}(n)]^2} \quad (8) \quad RMSE = \sqrt{\frac{\sum_{n=1}^{N}[s(n) - \bar{s}(n)]^2}{N}} \cdot 100\% \quad (9)$$

2.3. *Result and Analysis*

Table 1 lists the SNR and RMSE of the result of a test data "a0001.wav" of PNHD. It shows that these de-noising methods could effectively reduce the white noise. WTD with discrete Meyer wavelet provides the best performance. The signal mixed with -10 dB and 10dB white Gaussian noise are tested but not listed here. We note that when the noise level is increased, the performance of noise reduction gets worse. Yet, the WTD with 'dmey' wavelet still provides the best performance.

Table 1. The SNR and RMSE of test data "a0001.wav" of PNHD.

Method	WTD				WPTD				EMD		
	dmey	db10	db20	coif5	dmey	db10	db20	coif5	IT	IIT	CIIT
SNR	11.3	11.1	11	11.2	8.9	8.7	8.8	8.8	8.3	10.5	10.3
RMSE	2.8	2.9	3.0	2.8	3.8	3.9	3.8	3.8	4	3.2	3.2

All data of MHSD are tested and statistical analysis of quantitative results based on SNR are listed in Table 2. As none of the tests show WPTD and EMD-IT provide the best result, they are not listed in Table 2, 3. The result affirms that these de-noising methods are effective to white noise de-noising. From the table, there are 10 items of totally 23 items in MHSD that shows WTD with 'dmey' wavelet is better. 2 items and 11 items confirm that performances of EMD-IIT and EMD-CIIT are good respectively. Totally 50 items are tested in the PCHD. 21 of 31 items in the 'Atraining_normal' set and 17 of 19 items in the 'Atraining_extrahls' set show that the WTD is suitable for most of data.

Table 2. The result of data mixed with 0dB white noise in Michigan and Pascal Heartsound database.

Method		MHSD(23)		PCHD			
				Atraining_normal(31)		Atraining_extrahls(19)	
		N*	P**	N	P	N	P
WTD	dmey	10	43.48%	10	32.26%	11	57.90%
	db10	0	0	1	3.23%	1	5.26%
	db20	0	0	0	0	0	0
	coif5	0	0	10	32.26%	5	26.32%
EMD	IIT	2	8.7%	4	12.90%	1	5.26%
	CIIT	11	47.82%	6	19.35%	1	5.26%

* - number, ** - percentage.

All the training data totally 3241 items of six databases in PNHD are tested and results are shown in Table 3. It clearly shows that the WTD provide a better performance for most data. Though the EMD-IIT and EMD-CIIT get a better result in some cases, the SNR of EMD and WTD do not show much difference.

The signal mixed with -10dB and 10dB white Gaussian noise are tested. The performance gets worse as the noise level increases, but the result is still similar to the cases adding 0dB noise. Though the WTD provides a better performance for most data, to select the best resemble mother wavelet is a tricky issue, as the waveform of HS varies from one to another. Moreover, the HS signal mixed with various noise makes it even more difficult to select the wavelet. In some data,

WTD with 'dmey' wavelet is better than other wavelet. However, for some data, the results are reversed. The WTD with 'dmey' wavelet provides a better performance to a certain extent. Based on all the results, WPTD provides the worst performance. Perhaps the default global threshold is not a suitable choice. As for EMD, the difficult thing lies on how to determine the noise in each IMF. EMD de-noise the HS by discarding the IMF that contains much noise and applying a general threshold to the rest IMFs. As the IMF is discarded, some useful information are lost as well.

Table 3. The result of data mixed with 0dB white noise in PhysioNet database.

Method		a(409)		b(490)		c(31)		d(55)		e(2142)		f(114)	
		N*	P**	N	P	N	P	N	P	N	P	N	P
WTD	dmey	294	71.9%	24	4.89%	25	80.7%	19	34.5%	1105	51.6%	15	13.2%
	db10	16	3.91%	24	4.89%	0	0	6	10.9%	71	3.3%	0	0
	db20	1	0.24%	33	6.73%	1	3.22%	0	0	7	0.3%	0	0
	coif5	68	16.6%	143	29.2%	4	12.9%	17	31%	325	15.1%	98	85.92%
EMD	IIT	8	1.95%	61	12.4%	0	0	3	5.45%	203	8%	0	0
	CIIT	22	5.37%	205	41.8%	1	3.22%	10	18.2%	431	20.1%	1	0.88%

* - number, ** - percentage.

3. Conclusion

This paper studies several signal de-noising methods (WT, WPT, and EMD). Threshold noise reduction methods are performed and assessed by SNR and RMSE. Results of data from MHSD, PCHD, and PNHD show that these de-noising methods can be helpful to attenuate noise from real HS signals collected from patients. The WTD provides a better performance for most data, especially with 'dmey' wavelet. Yet, the selection of wavelet still is an unsolved problem. The WPTD seems to be inferior to WTD and EMD methods while applying the default global threshold. The EMD provides a good result in some cases though it lacks mathematical foundation. All the tests show SNR of the WTD differs little from EMD-IIT and EMD-CIIT even in the cases where EMD provides the best performance. As a result, WTD is much suitable for HS de-noising in most cases. Noise is inevitable in signal processing, this paper provides a reference to select a noise reduction method.

Acknowledgments

This research was supported by Research Committee of the University of Macau under grant no. MYRG2016-00157-AMSV and by the Science and Technology Development Fund of Macau (FDCT) under grant no. 088/2016/A2.

References

1. G.D. Clifford et al., "Recent advances in heart sound analysis." *Physiol Meas.* **38** (2017): E10.
2. Y.W. Bai et al., "The embedded digital stethoscope uses the adaptive noise cancellation filter and the type I Chebyshev IIR bandpass filter to reduce the noise of the heart sound." *Healthcom*, pp. 278-281, (2015).
3. T. Tosanguan et al., "Modified spectral subtraction for de-noising heart sounds: Interference suppression via spectral comparison." *Biomedical Circuits and Systems Conference*, pp.29-32, (2008).
4. C.T. Chao et al., "On the construction of an electronic stethoscope with real-time heart sound de-noising feature." *TSP*, (2012).
5. D. L. Donoho, "De-noising by soft-thresholding." *IEEE T Inform Theory.* **41**, 613 (1995).
6. Y. Kopsinis et al., "Development of EMD-based denoising methods inspired by wavelet thresholding." *IEEE T Signal Proces.* **57**, 1351 (2009).
7. Daubechies, I. (1992). Ten lectures on wavelets (Vol. 61). *Siam*.
8. M. Misiti et al., "Wavelet toolbox." *The MathWorks Inc.*, Natick, MA 15, p. 21 (1996).
9. Zeng, Kehan, Jun Huang, and Mingchui Dong. "White Gaussian noise energy estimation and wavelet multi-threshold de-noising for heart sound signals." *Circ Syst Signal Pr.* **33**.9 (2014): 2987-3002.
10. Ali, M. N., El-Dahshan, E. S. A., & Yahia, A. H. (2017). Denoising of Heart Sound Signals Using Discrete Wavelet Transform. *Circ Syst Signal Pr.* **36**(11), 4482-4497.
11. N. E. Huang et al., "The empirical mode decomposition and the Hilbert spectrum for nonlinear and non-stationary time series analysis." *Philos T R Soc A*. (1971).
12. P. Flandrin et al., "EMD equivalent filter banks, from interpretation to applications." *Hilbert–Huang transform and its applications.* p99-116. (2014).
13. G. Rilling et al., "On empirical mode decomposition and its algorithms." *IEEE-EURASIP workshop on nonlinear signal and image processing.* Vol. 3. NSIP-03, Grado (I), (2003).
14. Heart Sound & Murmur Library University of Michigan [Online]. Available: http://www.med.umich.edu/lrc/psbopen/repo/primerheartsound/primerheartsound.html.
15. P. Bentley et al., [Online]. Available: www.prterjbenley.com/heartchallenge/index.html.
16. Liu, Chengyu et al., "An open access database for the evaluation of heart sound algorithms." *Physiol Meas.* **37**, p. 2181 (2016).

Weak boundary value problem: Fuzzy partition in Galerkin method

L. Nguyen, I. Perfilieva and M. Holčapek

University of Ostrava, Institute for Research and Applications of Fuzzy Modelling, NSC IT4Innovations, 30. dubna 22, 701 03 Ostrava 1, Czech Republic Linh.Nguyen@osu.cz, Irina.Perfilieva@osu.cz, Michal.Holcapek@osu.cz
http://irafm.osu.cz

The contribution proposes a new methodology to the construction of an approximate weak solution to the boundary valued problem (BVP). It uses the framework of the Galerkin method and formulates the approximate solution in terms of the inverse fuzzy transform with respect to a uniform fuzzy partition. We propose technical details of the corresponding approximate solution in the form of a system of linear equation.

Keywords: Boundary valued problem; fuzzy partition; weak formulation of BVP; weak solution of BVP; Hilbert space; Sobolev space; generalized derivatives.

1. Introduction

The description of the laws of physics or economics for space and time dependent problems are usually expressed in terms of partial differential equations (PDEs). Due to a high complexity of such problems their solution is seldom expressed analytically. Instead, PDEs are transformed into easier forms that can be solved approximately. The approximate forms are results of subsampling of an initial domain or embedding the whole problem into an abstract algebraic environment. In this direction, the class of Galerkin methods[3] was proposed[1,2] and widely used. One particular case of the Galerkin-type methods known as a finite element method (FEM) belongs to the most popular numerical methods for PDEs, see also.[4,5]

There are two main ideas of the Galerkin-type methods: (i) searching an approximate solution of a differential equation in the form of a linear combination of the so called basic functions; (ii) construction of the so called weak problem (based on a certain bilinear form) and the corresponding to it weak solution. Therefore on the theoretical level, the difference between Galerkin-type methods consists in the selection of a space of basic functions and a bilinear form that replaces the given differential equation.

It is worth noticing that the similar idea has been already used in the context of fuzzy control where Takagi and Sugeno proposed an approximation of a control strategy by a system of fuzzy IF-THEN rules. In their model,[6] basic functions were fuzzy sets used in the antecedents of the rules. A generalization of this idea has been proposed by Perfilieva in the form of the (ordinary) fuzzy transform.[7] Later on, the concept of a higher degree fuzzy transform[8] generalizes a linear combination of basic functions to the case where coefficients are replaced by functions. The F-transform method became very popular in the applications to image and signal processing. A different approach to the computation of coefficients in the ordinary fuzzy transform has been proposed by Crouzet[9] under the name of fuzzy projection. Comparing the approaches proposed by Melenk and Babuška in,[4] Perfilieva in[8] and Crouzet in,[9] one can recognize similarities, if omitting certain assumptions on functions satisfying the conditions of a partition of unity and a fuzzy partition. As a consequence, all the mentioned theories use different approximation spaces with different requirements on the quality of approximation.

The aim of this paper is to propose a generalization of Galerkin methods based on fuzzy partitions in approximation spaces that are considered in the theory of fuzzy transforms. For a better presentation of our ideas, we restrict ourselves to a two-points boundary problem with homogeneous Dirichlet conditions:

$$-(p(x)u'(x))' + q(x)u(x) = f(x), \quad x \in (a,b), \tag{1}$$

$$u(a) = 0, \tag{2}$$

$$u(b) = 0, \tag{3}$$

where $p \neq 0$, q, f are given functions.

The paper is structured as follows. The following section is devoted to the concept of fuzzy partition, which is used in our method. The third and main section provides the Galerkin method based on this type of fuzzy partition.

2. Fuzzy Partition

In this paper, we restrict our analysis to a uniform fuzzy partition of a closed interval of real line, which is defined as a family of fuzzy sets that are determined by a specific generating function.[10] In contrast to the standard definition, we assume that each generating function has the continuous first derivate and vanishes at the boundary $\{-1, 1\}$, i.e., it belongs to $C_0^1([-1, 1])$.

Definition 2.1. A function $K : [-1,1] \to [0,1]$ is said to be a *generating function* of a fuzzy partition, if $K \in C_0^1[-1,1]$, and it is even, non-increasing on $[0,1]$ and everywhere positive except at boundary points where it vanishes.

A basic example of a generating function frequently used in fuzzy transform theory is the raised cosine function.

Example 2.1. The function $K^{rc} : \mathbb{R} \to [0,1]$ defined by

$$K^{rc}(x) = \begin{cases} \frac{1}{2}(1+\cos(\pi x)), & -1 \le t \le 1; \\ 0, & \text{otherwise,} \end{cases}$$

for any $x \in \mathbb{R}$, is called the *raised cosine* generating functions.

Note that the triangle generating function belongs to $W^{1,2}(\mathbb{R})$, because it has the so-called generalized derivatives (see Subsection 3.1).

Definition 2.2. Let K be a generating function, let h be a positive real constant, and let $x_0 \in \mathbb{R}$. For any $k \in \mathbb{Z}$, let

$$A_k(x) = K\left(\frac{x - x_0 - c_k}{h}\right),$$

where $c_k = kh$. The family $\mathcal{A} = \{A_k \mid k \in \mathbb{Z}\}$ is said to be *a uniform fuzzy partition of the real line determined by the triplet* (K, h, x_0) if

$$\sum_{k \in \mathbb{Z}} A_k(x) = 1, \quad x \in \mathbb{R}.$$

The parameters h and x_0 are called the *bandwidth* and the *central node* of the fuzzy partition $\mathcal{A}$, respectively. Furthermore, A_k and c_k are called the *k-th basic function* and the *k-th node* of $\mathcal{A}$, respectively.

In what follows, we always consider a uniform fuzzy partitions of a close interval $[a,b]$, denoted by $\mathcal{A}_N = \{A_0, A_1, \ldots, A_N\}$, where $x_0 = a$ and $h = (b-a)/N$. For the sake of simplicity, we omit the reference to the parameters of the triplet (K, h, x_0) and simply say that $\mathcal{A}_N$ is a uniform fuzzy partition of $[a,b]$ determined by a generating function K.

3. Galerkin Method Based on a Fuzzy Partition

In this section, we discuss the notion of a *weak solution* and propose a new method how it can be obtained.

3.1. Weak solution of BVP

In order to guarantee a unique solution of (1)–(3) functions p, q, f should be sufficiently smooth. This significantly restricts the class of BVPs, so that many problems in physics and other empirical sciences cannot be even considered. To extend the range of BVPs, the notion of a weak solution has been proposed. The main idea consists in replacing equation (1) by the following integral equation

$$\int_a^b [-(p(x)u'(x))' + q(x)u(x)]v(x)dx = \int_a^b f(x)v(x)dx, \quad (4)$$

where v is any element from a set V of the so called *test* functions, and the smoothness of a solution u can be reduced to the condition of differentiability of $(p(x)u'(x))$. The goal is to select a set of test functions V such that in the particular case, where functions p, q, f are smooth enough and a classical solution exists, both solutions coincide.

Assume that the set V of test functions on $[a, b]$ is a Sobolev space $W^{1,2}(a, b)$ such that

- $W^{1,2}(a, b) \subseteq L^2(a, b)$,
- for each $v \in W^{1,2}(a, b)$, there exists $g \in L^2(a, b)$ such that $v(x) = \int_a^x g(t)dt + C$ (g is a *generalized derivative* of v, $g = v'$),
- $\|v\|_{1,2} = \left[\int_a^b (|v(t)|^2 + |v'(t)|^2)dt\right]^{1/2}$.

Sobolev space $W^{1,2}(a, b)$ is a Hilbert space with respect to

$$(v, g) = \int_a^b (v(t)g(t) + v'(t)g'(t))dt.$$

Moreover, functions in $W^{1,2}(a, b)$ are continuous on $[a, b]$.[11] Therefore, we can specify the following subspace

$$W_0^{1,2}(a, b) = \{v \in W^{1,2}(a, b) \mid v(a) = v(b) = 0\}.$$

The following definition introduces the notion of a weak solution to the BVP given by (1)–(3).

Definition 3.1. Function $u : [a, b] \to \mathbb{R}$ is a weak solution to (1)–(3), if $u \in W_0^{1,2}(a, b)$ and equation (4) holds true for every $v \in W_0^{1,2}(a, b)$.

3.2. Existence of a weak solution

Let bilinear form $\mathscr{A} : W_0^{1,2}(a,b) \times W_0^{1,2}(a,b) \to \mathbb{R}$ and linear functional $\mathscr{L} : W_0^{1,2}(a,b) \to \mathbb{R}$ be determined by

$$\mathscr{A}(u,v) = \int_a^b p(x)u'(x)v'(x)dx + \int_a^b q(x)u(x)v(x)dx,$$

$$\mathscr{L}(v) = \int_a^b f(x)v(x)dx.$$

Then, $u \in W_0^{1,2}(a,b)$ is a weak solution to (1)–(3), if

$$\mathscr{A}(u,v) = \mathscr{L}(v), \quad \text{for any } v \in W_0^{1,2}(a,b). \tag{5}$$

Below, we analyze the existence of a weak solution of the initial BVP in terms of a solution to (5). We additionally assume that $p \in W^{1,2}(a,b)$ and $q, f \in C^0[a,b]$ are such that for all $x \in [a,b]$, $0 < p_L \leq p(x) \leq p_R$ and $0 \leq q(x) \leq q_R$. We will use the Lax-Milgram lemma (see,[12]) and therefore, verify its assumptions.

By the Jensen and Hölder inequalities and the fact that a function is Lebesgue integrable if and only if it is Lebesgue integrable in the absolute value, we obtain

$$|\mathscr{A}(u,v)| \leq \max(p_R, q_R) \cdot (\|u' \cdot v'\|_1 + \|u \cdot v\|_1) \leq \max(p_R, q_R) \cdot (\|u'\|_2 \cdot \|v'\|_2 + \|u\|_2 \cdot \|v\|_2) \leq 2 \cdot \max(p_R, q_R) \cdot \|u\|_{1,2} \cdot \|v\|_{1,2},$$

where $\|\cdot\|_1$ and $\|\cdot\|_2$ denotes the L_1-norm and L_2-norm, respectively. Moreover, by the Rayleigh-Ritz inequality,[12] we have that for any $u \in W_0^{1,2}(a,b)$, the following holds:

$$\|u\|_2^2 \leq \frac{(b-a)^2}{\pi^2} \|u'\|_2^2 \tag{6}$$

Thus, we obtain

$$\mathscr{A}(u,u) = \int_a^b p(x)(u'(x))^2 dx + \int_a^b q(x)(u(x))^2 dx \geq p_L \cdot \|u'\|_2^2$$
$$\geq \frac{p_L}{2} \left[\frac{\pi^2}{(b-a)^2} \cdot \|u\|_2^2 + \|u'\|_2^2 \right] \geq \frac{p_L}{2} \cdot \min\left(\frac{\pi^2}{(b-a)^2}, 1\right) \cdot \|u\|_{1,2}^2.$$

By the Lax-Milgram lemma, there exists a unique $u_0 \in W_0^{1,2}(a,b)$, that makes (5) valid. By this, $u_0 \in W_0^{1,2}(a,b)$ solves the BVP given by (1)–(3) as well. Below in section 3.4, we propose a modification of Galerkin method that solves (5) numerically.

3.3. Polynomial approximation based on a higher degree fuzzy transform

Let $\mathbb{P}_m$ denote the linear space of polynomials up to degree m. The following lemma shows that every function with continuous derivatives up to degree $m+1$ can be approximated by the inverse m-degree fuzzy transform with polynomial coefficients.

Theorem 3.1. *Let f be a function on $[a,b]$ such that $f \in C^{m+1}[a,b]$, $m \geq 1$. Let $\mathcal{A}_N = \{A_0, A_1, \ldots, A_N\}$ be a uniform fuzzy partition of $[a,b]$ with bandwidth $h = \frac{b-a}{N}$. Then, there exists a set $\{v_0, \ldots, v_N\}$ of polynomials from $\mathbb{P}_m$, such that the inverse m-degree fuzzy transform $\hat{f}_N^m$ of $\{v_0, \ldots, v_N\}$, given by*

$$\hat{f}_N^m(x) = \sum_{k=0}^{N} v_k(x) A_k(x), \qquad (7)$$

approximates f with the following error estimate

$$\|f - \hat{f}_N^m\|_{1,2} = O(h^m).$$

Having in mind that for every $m \geq 1$, a function from $W^{m,2}(a,b)$ can be approximated (in the norm of $W^{m,2}(a,b)$) by functions from $C^m[a,b]$ with any given precision, we can apply Theorem 3.1 to functions from Sobolev spaces. This will justify the proposed below Galerkin method for BVP (1)–(3) with the second order differential equation.

3.4. Galerkin method based on fuzzy partition

Let $\mathcal{A}_N = \{A_0, A_1, \ldots, A_N\}$ be a uniform fuzzy partition of the interval $[a,b]$ such that $A_j \in W^{1,2}(a,b)$ for any $j = 0, 1, \ldots, N$. Let $m \geq 1$ be a natural number. Let

$$\mathcal{D}_m(\mathcal{A}_N) = \left\{ v(x) = \sum_{k=0}^{N} v_k(x) A_k(x) \mid v(0) = v(1) = 0, v_k \in \mathbb{P}_m, k = \overline{0, N} \right\}.$$

It is easy to see that $\mathcal{D}_m(\mathcal{A}_N)$ is a subspace of the Hilbert space $W_0^{1,2}(a,b)$[a] with the basis $\mathcal{B}$ given by $\mathcal{B} = \mathcal{B}_0 \cup \mathcal{B}_1 \cup \ldots \cup \mathcal{B}_N$, where

$$\mathcal{B}_0 = \left\{ (x - c_0)^j A_0(x) \mid j = \overline{1, m} \right\},$$
$$\mathcal{B}_k = \left\{ (x - c_k)^j A_k(x) \mid j = \overline{0, m} \right\}, \quad k = \overline{1, N-1},$$
$$\mathcal{B}_N = \left\{ (x - c_N)^j A_N(x) \mid j = \overline{1, m} \right\}.$$

---

[a] A finite dimensional linear subspace of a Hilbert space is a Hilbert space.

Below, we find a solution on $\mathcal{D}_m(\mathcal{A}_N)$ of the following problem

$$\mathscr{A}(u,v) = \mathscr{L}(v), \quad \text{for any } v \in \mathcal{D}_m(\mathcal{A}_N). \tag{8}$$

On the basis of the above reasoning, this problem has a unique solution u_N on $\mathcal{D}_m(\mathcal{A}_N)$. This solution is the desired *Galerkin approximation* of the weak solution denoted above by u_0.

Below, we give some technical details for the computation of u_N. Let

$$\phi_{0,j} = (x - c_0)^j A_0(x), \quad j = \overline{1,m}$$
$$\phi_{N,j} = (x - c_N)^j A_N(x), \quad j = \overline{1,m}$$
$$\phi_{k,j} = (x - c_k)^j A_k(x), \quad j = \overline{0,m}, k = \overline{1, N-1}.$$

We find a solution u_N of equation (8) in the following form

$$u_N(x) = \sum_{j=1}^{m} \alpha_{0,j} \phi_{0,j} + \sum_{k=1}^{N-1} \sum_{j=0}^{m} \alpha_{k,j} \phi_{k,j} + \sum_{j=1}^{m} \alpha_{N,j} \phi_{N,j}.$$

It follows that we have to solve the following system of $(m+1)(N-1) + 2m$ linear equations:

$$\sum_{j=1}^{m} \alpha_{\ell,j} \mathscr{A}(\phi_{0,j}, \phi_{\ell,i}) + \sum_{k=1}^{N-1} \sum_{j=0}^{m} \alpha_{k,j} \mathscr{A}(\phi_{k,j}, \phi_{\ell,i})$$
$$+ \sum_{j=1}^{m} \alpha_{N,j} \mathscr{A}(\phi_{N,j}, \phi_{\ell,i}) = \mathscr{L}(\phi_{\ell,i}), \quad i = \overline{1,m}, (\ell \in \{0, N\})$$

$$\sum_{j=1}^{m} \alpha_{0,j} \mathscr{A}(\phi_{0,j}, \phi_{\ell,i}) + \sum_{k=1}^{N-1} \sum_{j=0}^{m} \alpha_{k,j} \mathscr{A}(\phi_{k,j}, \phi_{\ell,i})$$
$$+ \sum_{j=1}^{m} \alpha_{N,j} \mathscr{A}(\phi_{N,j}, \phi_{\ell,i}) = \mathscr{L}(\phi_{\ell,i}), \quad i = \overline{0,m}, \left(\ell = \overline{1, N-1}\right).$$

The above given system is solvable, because its matrix of coefficients is symmetric and regular (the latter fact follows from the linear independence of functions in $\mathcal{B}$) and the weak solution exists and unique. Finally, on the basis of Theorem 3.1 we have confirmed the convergence of the sequence of Galerkin approximations $\{u_N\}_N$ to u_0.

4. Conclusions

A new methodology to the construction of an approximate weak solution to the boundary valued problem (BVP) is proposed. The purpose is to extend the scope of solvable BVP problems by introducing the notion of a weak solution. The known classical approach to the construction of a numerical approximate solution is based on the Galerkin method. We analyzed this method and proposed its modification based on the concept of a fuzzy partition. We formulated the approximate solution in terms of the inverse fuzzy transform with respect to a uniform fuzzy partition. We proposed technical details of the corresponding approximate solution in the form of a system of linear equation and discussed its solvability.

Acknowledgments

This work has been partially supported by the project "LQ1602 IT4Innovations excellence in science" and by the Grant Agency of the Czech Republic (project No. 16-09541S).

References

1. K. Rektorys, *Variational methods in mathematics, science and engineering* (R. Reidel Publishing Company, Dordrecht–Holland, 2012).
2. S. C. Brenner and L. R. Scott, *The Mathematical Theory of Finite Element Methods* (Springer Science & Business Media, 2013).
3. C. A. J. Fletcher, *Computational Galerkin Methods* (Springer Science & Business Media, 2012).
4. J. Melenk and I. Babuška, The partition of unity finite element method: Basic theory and applications, *Computer Methods in Applied Mechanics and Engineering* **139**, 289–314 (1996).
5. S. Li and W. K. Liu, *Meshfree Particle Methods* (Springer Science & Business Media, 2007).
6. T. Takagi and M. Sugeno, Fuzzy identification of systems and its application to modeling and control, *IEEE Transactions on Systems, Man, and Cybernetics: Systems*, **15(1)**, 116–132 (1985).
7. I. Perfilieva, Fuzzy transforms: Theory and applications, *Fuzzy sets syst.* **157(8)**, 993–1023 (2006).
8. I. Perfilieva, M. Daňková and B. Bede, Towards a higher degree F-transform, *Fuzzy Sets Syst.* **180**, 3–19 (2011).
9. J.-F. Crouzet, Fuzzy projection versus inverse fuzzy transform as sampling/interpolation schemes, *Fuzzy Sets Syst.* **193**, 108–121 (2012).
10. M. Holčapek, I. Perfilieva, V. Novák and V. Kreinovich, Necessary and sufficient conditions for generalized uniform fuzzy partitions *Fuzzy Sets Syst.* **277**, 97–121 (2015).
11. V. G. Maz'ja, *Sobolev spaces* (Springer-Verlag, New York, 1985).
12. A. Canada, P. Drabek, A. Fonda, Handbook of differential equations. Ordinary differential equations, Elsevier, 2004.
13. G. Meinardus and G. Merz, Praktische Mathematik I, II. Mannheim, B. I.-Wissenschaftsverlag 1979, 1982.

Part 13

Soft Computing and Data Analytic in Fashion Design and Textile Production

Intelligent application of data fusion in garment manufacturing under the thinking of "Internet plus"[*]

Liu Cui[1], Hong Dai[1] and Kai Xuan Liu[1,2]

[1]School of Apparel and Art Design, Xi'an Polytechnic University, Xi'an 710048, China
[2]Shaanxi Key Laboratory of Intelligent Clothing Design,
Xi'an Polytechnic University, Xi'an 710048, China

> With the development of the "Internet" action plan, it is an inevitable trend to realize the intelligent transformation and upgrading of modern garment manufacturing industry. Through analyzing the New advantages of Intelligent Industry Development of "Internet plus clothing". Based on the "Internet plus" and data fusion model, establishment the database of human body model, garment type database, material database. And with the help of "Internet" thinking reveals the clothing in the intelligent manufacturing process using hardware, software and data fusion realizing the intelligent of design and development; intelligent of production process; terminal product intelligent; design requirements of service intelligence. Application of science and technology information technology, realization of intelligent transfer between processes, further improve the intelligent level of garment manufacturing.

1. Introduction

In 2015, in the Government work report, the Government first proposed an "action plan for the development of the 'Internet plus', promoting Mobile Internet; cloud computing; big Data; the Internet of things combination with modern manufacturing industry, promote the development of electronic commerce and the healthy development of industrial Internet [1]." Garment and textile industry as the traditional Pillar Industry in China, developing intelligent manufacturing in China's garment industry has become an inevitable trend, its a highly cross-fusion of manufacturing technology, digital and intelligent technology. On this basis, the collected data will be analyzed again, get a bigger and deeper solution, the deep integration of big data and Internet technology. Comprehensively enhance the intelligent level of clothing research, production, management and service. Upgrading the traditional manufacturing systems to the stage of flexibility, Intelligence and Information. Establishment of innovative, efficient and intelligent garment textile industry system [2].

---

[*]This paper was financially supported by China National Endowment for the Arts.

2. New Advantages of "Internet Garment" in Intelligent Industry Development

With the help of a new generation of information technology and the integration of textile industry, "Internet plus clothing" is the main line of development. Combined with continuous scientific and technological innovation, promoting intelligent clothing manufacturing under the new modern information technology, use of automated production equipment, intelligent production and digital management system, taking the Internet as the innovation power to realize the intelligent production mode. Through intelligent technology to open up various information systems, real-time data and information sharing, avoid information exchange is not smooth. In product design, production process and supply chain management control to promote the construction of intelligent factory, and striving to achieve seamless connection in all aspects of garment manufacturing and intelligent management, accurate grasp of production information and data [3]. In the "Internet plus clothing" environment, actively cultivate a new mode of production, creating a new situation of data fusion, and promote the integration of mobile internet, cloud computing; large data; Internet of things and garment manufacturing. Construction of flexible production adapted to modern individualized customization and integration of low cost e-commerce platform system, complement the short board of traditional development mode, improve production efficiency, diversification of sales methods and quality of service. Based on the new Internet technology platform, pushing forward the change of new manufacturing mode of intelligent garment, so that the intelligent production is widely applied in the clothing industry [4].

3. Data Fusion in Garment Manufacturing

3.1. *Establishment of human body model database*

"Internet plus clothing" is a new trend of the development of garment industry. How to make use of Internet to realize the personalized customization of clothing, the collection of human data is a difficult point. The mannequin database includes the general mannequin database and the special human body model database, for example the hunchback; the convex belly; the high shoulders, the wide shoulder and so on, which are the key parts of the human body. Using 3D human scanning technology, through the combination of mobile terminals and the Internet, we can accurate obtain multiple human data in the fastest time. It not only improves the efficiency of data collection, but also satisfies the customization of special body. Through cloud computing, data analysis operations, create the same human model as a real person. Big data technology is used to collect, screen and find out the corresponding rules, so as to quickly obtain similar human body model in personality customization. The establishment of mannequin database can not only provide abundant choices for personalized customization of clothing, but also output 1:1 platform by 3D printing technology, which is fast and convenient. What is more effective is the establishment of online 3D virtual fitting system, the enhancement of human body model database, the combination of garment specification design and cloud computing, which makes the matching between clothing and body more accurate and accurate for the generation of later style layout. And realizes the interaction between designer and consumers, in order to modify in time, adjust the relevant data, and improve the fitting degree and precision of clothing and human body. Fusion of Human Model data Storage system and 3D Virtual fitting system, make intelligent clothing customization, digital production become a reality. It breaks through the limitation of traditional manual measurement body, and promote the garment manufacturing to develop in the direction of high automation and intelligence.

3.2. *Establishment of garment type database*

The traditional garment CAD plate making system cannot meet the fast changing rhythm of clothing, especially in the customization industry. Once the style changes or the size of ready-made garments changes, it is necessary to make separate versions. The contradiction between data resources and personalized design restricts the efficiency of garment customization. The Establishment of Garment Type Database on the basis of massive Database, make use of computer technology, communication technology, industry technology and intelligent control technology, collection and integration of data resources such as outline; shape; style; body shape; sex and so on, the styles include collars; pockets; sleeves, etc. They can be selected according to their own preferences, then input relevant personal key data, realize intelligent design and fast matching, and the system will automatically generate a template to meet customer requirements, and provide customers with personalized and diversified products in a convenient and efficient production mode. The personalized of Qingdao red collar clothing covers about 3000000 customers, with more than 100 trillion personalized data, 1000 trillion design combinations, and 100 trillion styles of combination data. The original "three point line" volume method collects 24 data from 19 parts of the body and inputs into the red-collar personalized customization platform. The system will model the data and form the data pattern that belongs to the customer in one minute [5]. The real implementation of the clothing Internet plus, personalized data, intelligent manufacturing production mode. Customers can not only personally experience and issue orders for clothing customization, but also can produce data according to customer requirements, and win customers on the basis of ensuring quality and service. Increasing the driving force for the sustainable development of enterprises.

3.3. Establishment of material database

Material database includes fabric, accessories and so on. The existence of fabric database meets the different needs of customers. They include fabric varieties; warp and weft density; warp and weft yarn count; organizational structure; washing method; composition content; style type and sample display data parameters, forming thousands of possibilities. A combination of thousands of possibilities to meet consumers' individual needs for one person. The auxiliary materials database includes clothing decoration and functional properties, such as zipper; button; lining; material; decoration; sewing thread; trademark; packing bag and so on [6]. With the help of 3D data, the consumer opens the page to carry on the fabric, the auxiliary material choice, such as embroidering, zipper, pattern, button and so on. Local combination, automatic seamless stitching, fabric texture concavity and convex effect, the accuracy of the fabric details show good effect, through three-dimensional virtual fitting matching effect. Through software tools, customers design independently in large databases, finding the matching according to customer needs, realizing the "private customization" requirements of customers, and opening the market of big data in an all-round way. Facilitate artificial intelligence design and potential customer consumption area mining and prediction.

4. Intelligent Application of Data Fusion in Garment Manufacturing

4.1. Intelligent research and development of garment design

Under the influence of the industrial internet background and based on the traditional garment manufacturing process, through the development of a large number of data acquisition and design software, a clothing intelligent design platform has been developed, which has become the development trend of modern manufacturing industry. Garment design and development intelligent under the use of computer technology to display the effect of product design. For example, the Internet Knitting CAD system (KDS 1.0) adopts the B/S architecture, which has the functions of pattern design, process design, fabric simulation, virtual display, etc. A knitting product database is built. The user can design the knitted fabric directly through the intelligent terminal at any time.

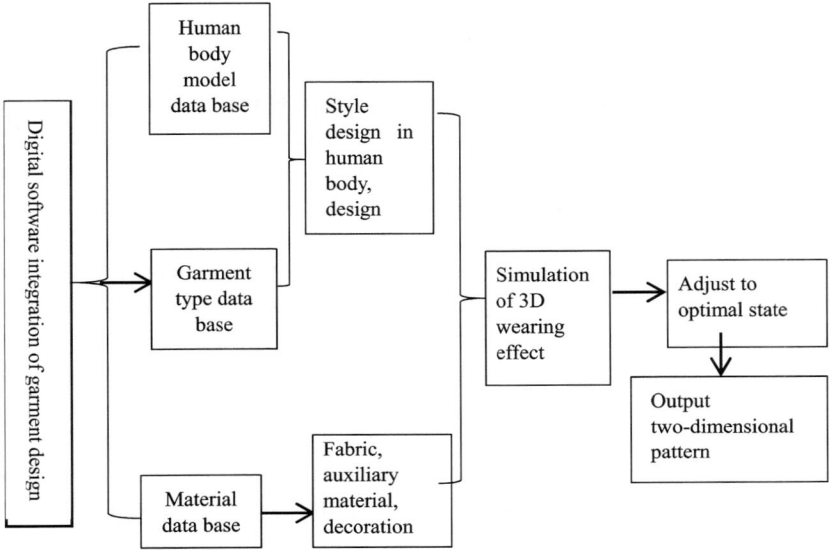

Fig. 1. Intelligent module of garment design and development.

The system combines cloud storage, knitting fabric simulation, virtual display and 3D simulation, etc. Through big data related technology, analysis of all stored information, intelligent design and optimization of knitted products [7]. Three-dimensional design in the clothing design shows the simulation three-dimensional effect, only through the clothing CAD system, has certain difficulty to realize the exaggeration modelling, the style change, the designer's creative design. Generally, the design of clothing effect diagram and style can be realized through PS, CorelDRAW and other design software. It can not only improve design efficiency, but also saves the cost of product development. The clothing design system suitable for designers is developed by synthetically designing the software such as CADX PSN CorelDraw and so on. The human body model database, garment pattern database and material database are integrated into the garment design system by computer technology. Database large capacity, wide range of choice, and software replication, cutting, deletion and other free combination design, can easily pattern, layout, fabric filling and replacement. Designers can also design on the scanned human body directly. They can directly see the fabric, color, structure and styling of the clothes they design. Modify and adjust the virtual image on the computer. The proportion of the finished design is accurate, detail is clear, and modelling is diverse. Through the computer technology and the big database may achieve the very meticulous effect.

4.2. Intelligent garment production process

With the intelligent means of industrialization, the clothing design completed by big data system was intelligent, and the machine plate making became a reality. Plate making is an important step in garment design and production, which not only achieves the efficiency of industrialization, but also provides the basis for the intellect of garment production technology. Build intelligent workshop, introduce intelligent equipment, design intelligent production line, design process according to different orders, make rational use of intelligent equipment. Such as the introduction of RFID (Radio Frequency Identification) technology as a form of data flow. After the CAD department collects the human body data for the customer, and complete big data plate, decomposed into the clothing of each part of the production process data, transmitted to the data platform of Kurt, personalized tailoring. The cut cloth is tagged into "hanging system". After transferring between the 298 processes, the RFID card maker will input all the data into an electronic label. After that, the tag like the ID card will follow the clothing that corresponds to it and go through the whole production process. Workers in each processing process receive an assigned piece of clothing. He first swipes the credit card reading and completes specific operations such as tailoring, spiking, and embroidery according to the instructions translated from the code [8]. Under the operation of intelligent production process, the production efficiency of workers in each process is monitored through the existing data platform, which is allocated to the most suitable process for workers. The data of the order, material and garment are linked together to ensure the continuous supply of materials and almost zero stock. Huge information and data chain to drive the assembly line to become the core assets of the intelligent clothing production workshop. With the innovation of the new generation of revolutionary technology, the characteristics of intelligent garment production such as self-adaptation, self-communication, self-storage, self-diagnosis and self-repair make it possible for the garment manufacturing industry from "intelligent" production to "no-man" production [9].

4.3. Intelligent clothing terminal products

With the popularization of intelligent network, more and more scientific products are being applied to clothing. Using computer information technology, textile technology and microelectronics technology. Realizing the intelligent garment terminal products has become the ultimate goal of garment manufacturing development. The intelligent clothing is mainly reflected in leisure sports, safety protection; medical care; fashion trends and so on. In 2015, the total sales volume of global smart wearable devices reached 1.16 billion, with sales revenue of 181.67 billion. Smart clothing capable of components has become a breakthrough

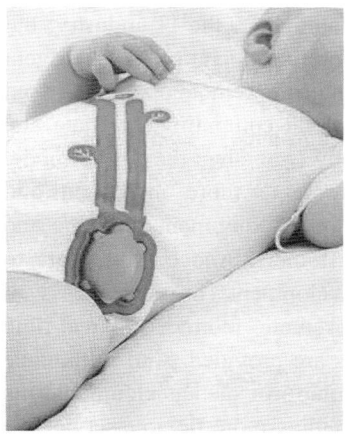 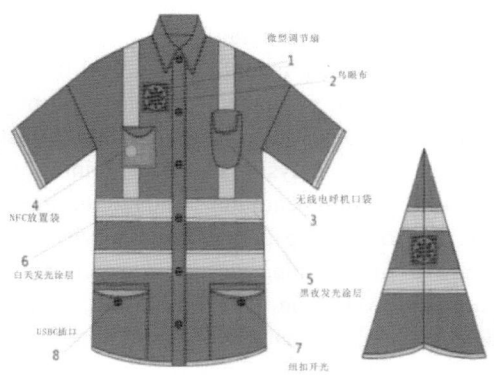

Fig. 2. Children's clothes.　　　　　　Fig. 3. Intelligent sanitation suit.

smart wearable technology, widely used in sports tracking (step number, route, energy consumption, posture) and physical sign monitoring (sleep, heartbeat, breathing, body temperature, blood pressure, body surface pressure) [10]. MIMO Jumpsuit can transmit the baby's body temperature, sleep and other physiological conditions, such as sensors, conductive fibers, like fibers woven into ordinary fabrics, so that through the network to parents' mobile phones, can be timely data processing. And the design of intelligent safety sanitation clothing, sanitation workers as a special group, summer work in high temperature, easy heat stroke; because of the nature of the work in the road vulnerable to injury, even death. Working alone for a long time may lead to physical and mental exhaustion and loneliness, so safety warning signs, heat dissipation needs and entertainment functions are urgent problems that need to be addressed [11]. In 2016, smart clothing has become an upsurge, and the development trend of intelligent clothing products cannot be underestimated in the future.

4.4. Intellectualization of services

In the heat wave of industry 4.0, clothing industry with the help of intelligent upgrading production and processing, which also brings challenges to the service model, in view of different intelligent service mode, professional service team to support. This is one of the reasons for the differentiation of competitive advantage. Lectra is a global leader in integrated technology solutions for CAD/CAM. Its technology is widely used in the fashion apparel industry, with software and services accounting for 46% of its revenue in 2016. Already outnumbering other projects. In 2017, Lectra decided to start offering SAAS models to users. The SAAS model is already familiar in the IT industry, but rarely developed in manufacturing [12]. Collaborative operation of intelligent devices in garment manufacturing, real-time monitoring, remote operation based on Internet and computer technology, through cloud computing, big data and other technologies for product tracking. Constantly strengthen the SAAS model through data analysis and data mining to promote more intelligent and networked equipment, and optimize the integration of equipment, software and services to develop personalized services to customers. And communicate to improve the clothing design research and development. In the process of garment production and processing, through remote monitoring, real-time detection, failure maintenance and other services are provided. And in the face of user experience feedback is a remote service platform such as cloud analysis that can be collected from the collected data, providing more flexible service for users [13].

5. Conclusion

Under the new economic situation, the development of intelligent manufacturing under the thinking of "Internet plus" promotes the level of information, automation and intelligence of our garment manufacturing industry, and improves the competitive advantage and advanced level of our country's garment manufacturing industry. Liberating labor force is not only the development trend of manufacturing industry in the future, but also an important measure to build a powerful manufacturing country in China. We must firmly grasp the new round of technological revolution and industrial transformation, focus on developing intelligent products and equipment, optimize the design and development mode, intellect, and promote the intellect of production process, and comprehensively enhance the intellect level of design, production, management, marketing and service. To this end, focusing on intelligent manufacturing, data acquisition is carried out on the basis of automation, and the collected data is accumulated to form a huge database under the means of information technology. In the data-driven way, the clothing can be intelligently analyzed according to the information of the database, and the correct decision is made. Promoting information manufacturing is the only way for clothing to advance towards intelligent manufacturing. With the support of big data, we should master consumer psychology, predict the trend of market discovery, and reshape the advantages of our garment manufacturing industry.

Acknowledgments

This paper was financially supported by China National Endowment for the Arts.

References

1. "Internet plus" Costume customization — Qingdao Red Collar's practice of mass customization based on the combination of two processes [J]. China Quality: 2016(04): 20–24.
2. Wang F. The first garment 4.0 factory will be put into production — An interview with Yin Zhiyong, chairman of Shanghai heying electromechanical technology Co., Ltd. [J]. Research on Textile Science: 2016(09): 52.
3. Chen M. Y. Research on intelligent manufacturing in Dong Guan clothing industry to meet the era of "industrial 4.0" [J]. Progress in Textile Science and Technology: 2017(12): 52–53+59.
4. Gao D. K. Suggestions on promoting Intelligent upgrading of Textile and Garment Industry [J]. Chinese Production: 2017(03): 54–55.
5. Wu Y. S., Sheng Y., Cai N. Research on Mass Intelligence customization based on Internet — The case of Qingdao Red Collar dress and Foshan visan furniture [J]. China Industrial Economy: 2016(04): 127–143.
6. Yi L. L. Integration mode and requirements of artificial Intelligence and Garment Design [J]. Wool Spinning Technology: 2017, 45(10): 81–85.
7. Jiang G. M., Gao Z., Gao Z. Y. Research progress of knitting intelligent manufacturing [J]. Journal of Textiles: 2017, 38(10): 177–183.
8. Zhao Y. P. Cott's Mode Innovation: data-driven mass garment personality customization [J]. Decorate: 2017(01): 26–30.
9. Mi L. C. Analysis of the current situation and trend of garment intelligent manufacturing by using the intelligent turn to reshape the new advantages [J]. Textile Clothing Weekly: 2016(34): 54–55.
10. Wu D. X. Research on the design of jogging sneakers based on the principle of human kinematics [D]. Hangzhou: Institutes of Technology of Zhejiang: 2013: 15–16.
11. Ren X. F., Shen L., Ning Y. N. An intelligent sanitation clothing design and research of [J]. Packaging Engineering, 2017, 38 (14): 164–168.
12. Zheng Y. New service mode in intelligent manufacturing [J]. IT Economic World: 2017(09): 52–53.
13. Yan Y. F. Solar energy utilization technology and its application [J]. Solar Energy Journal: 2012, 33(1): 47–56.

Garment fit evaluation using artificial intellegence technology

Kaixuan Liu, Hong Dai, Yanbo Ji, Yongmei Deng, Yongchi Xu, Yue Wang

School of Apparel and Art Design, Xi'an Polytechnic University, Xian 710048, China
Shaanxi Key Laboratory of Intelligent Clothing Design, Xi'an Polytechnic University Xian 710048, China

Xueqing Zhao, Tao Xue

School of Computer Science, Xi'an Polytechnic University, Xian 710048, China
Shaanxi Key Laboratory of Intelligent Clothing Design, Xi'an Polytechnic University Xian 710048, China

The traditional evaluation methods of garment fitting must be conducted with participation of customers. It is not suit to evaluate garment fit online. In this paper, we proposed a garment fit prediction model based on artificial neural network. The inputs of the model are different parts' clothing pressures; and the output of the model is one of three fit conditions: tight, fit and loose. In order to structure the proposed model, some measuring points were mapped on the garment patterns, and then the virtual garment pressures data were collected according to the measuring points by CLO 3D Modelist software. Finally, there are a total of sixty-nine garment pattern samples were involved model building, in which sixty are for training the proposed model and the rest are for testing the model. Test results of the garment fit prediction model shows that the prediction accuracy of the model reaches to 72.7%. If we increase the number of learning data, the perdition accuracy will be further improved.

1. Introduction

Garment fit has always been one of the major difficult problems for garment designers and consumers [1, 2]. In the traditional garment production process, fashion designers, pattern makers and clothing technologists need to communicate with each other for modifying garment pattern, and make garments again and again until the garment fit well. This process is quite time-consuming and tedious; and cost much labor power and material resources. In order to solve these issues, some virtual try-on software, such as, Clo 3D, Lectra 3D Prototype, OptiTex, V-Stitcher 3D, were developed to evaluate the garment fit. However, so far these virtual try-on software still cannot evaluate garment fit accurately. The evaluation of clothing fit is a complex process; it involves garment patterns, ease allowance, fabric properties, etc. Therefore, it is difficult to build a model to predict garment fit.

Along with rapid development of E-commerce, purchasing clothing online becomes more and more popular. As the online shopping can carry out real try-on, the evaluation of garment fit is becoming more imperative. Artificial intelligence technology is widely applied to many industries over last the decade. However, this technology is rarely used in the garment industry. Garments are labor-intensive and low-tech products. It is difficult to integrate these advanced technologies into the fashion design process. Zeng et al. proposed to evaluate garment and fabric products using fuzzy logic, artificial intelligence and sensory evaluation techniques [3-7]. Hollies, N.R. et al. proposed a human perception analysis approach to clothing comfort [8]. However, their approaches are not suit to evaluate virtual try-on. Because, one or more persons are required to participate in their evaluation experiments. It is impossible for shopping online.

As we known, the garment pattern, fabric mechanical properties, garment ease allowance [9], etc. are all influence the garment fit. No matter which parameter is changed, the clothing pressures change accordingly. A garment with whatever patterns, fabric and ease allowance, the fit condition of the garment can be ultimately reflected by garment clothing pressure. Thus, clothing pressure is an important index reflects the condition of garment fit. Moreover, clothing production enterprises have plenty of garment patterns data as the businesses grow. These data are very valuable for data mining and knowledge discovery. However, there are few garment companies carry out this work; and this result in a waste of resources.

In this context, we presented a garment fit prediction model based on Back Propagation (BP) artificial neural network. The inputs of the model are clothing pressures on different body parts; and the outputs of the model are the condition of garment fit (fit, loose, or tight).

2. General Schemes

The general scheme of the mentioned garment fit prediction model is described in Fig. 1. Its basic functions are given below.

Firstly, we have three databases of pants' patterns: DB_1, DB_2 and DB_3. A the patterns in the DB_1 are tight for a person; all the patterns in the DB_2 are fit for the person; and all the patterns in the DB_3 are loose for the person. The garment pressures on all the patterns in these three databases are measured. The learning data are collected for training the proposed model.

Secondly, a new sample, we do not know what the fit condition it is, is measured garment pressures. And then, we input the new sample's garment pressures into the fitting prediction model. The result of garment fit prediction can be acquired from the output of the proposed model finally.

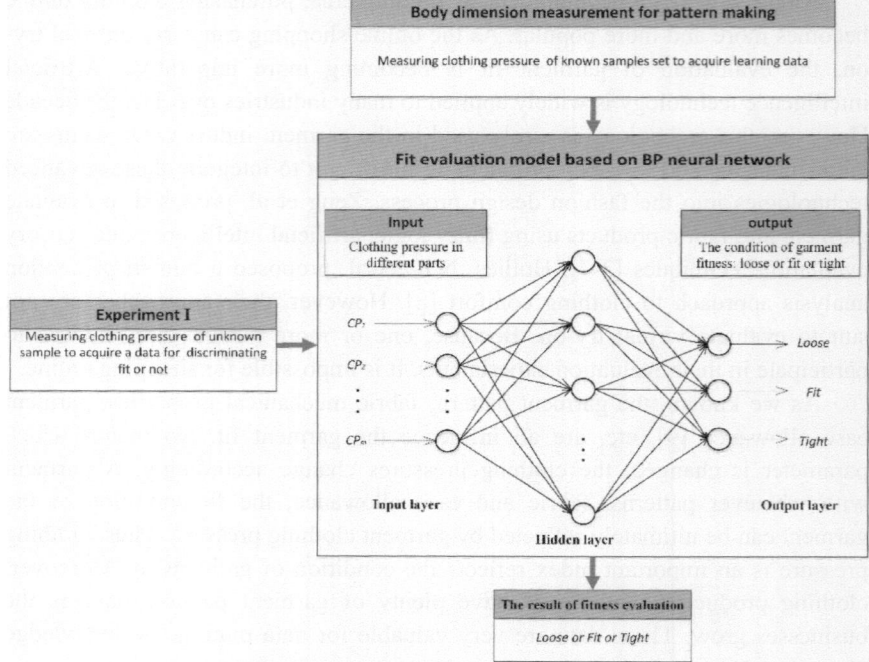

Figure 1. General schemes of fit prediction model building.

3. Experiment

In order to collect a garment pressure data for training the model, we proposed a novel method to measure clothing pressures. The method was carried out as follows:

Firstly, measuring points F_1, F_2... F_{15} were equally mapped on front piece of garment pattern; and measuring points B_1, B_2... B_5 were equally mapped on back piece of garment pattern (Figure 2(a)). As the part of human body blow knee has little influence on garment fit, we did not arrange measuring points in this part. Moreover, the human body is symmetric; therefore, we only arranged the measuring points on half of the body.

Secondly, a garment's patterns were tried on an avatar in the virtual environment by CLO 3D Modelist software (Figure 2(b)).

Finally, clothing pressures were measured according to the measuring points (Figure 2(c)). The collected data were divided into two parts: one was used for model training; the other was used for model testing.

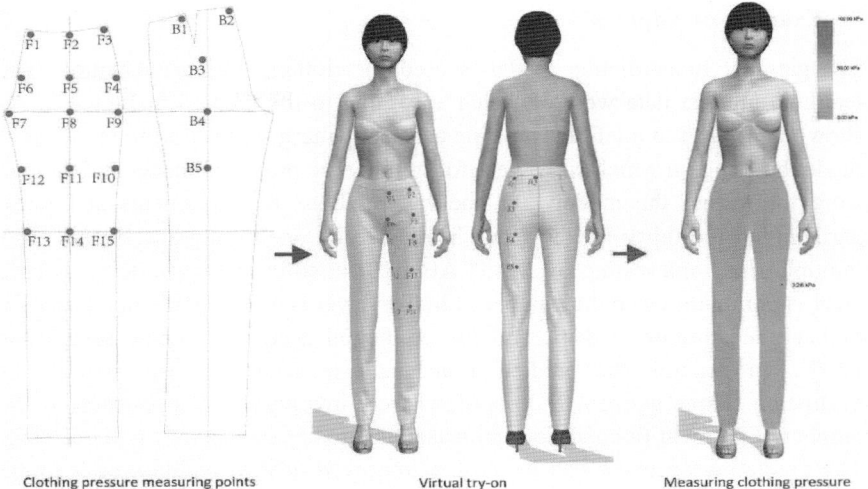

Figure 2. Clothing pressures measurement method.

4. Modeling of the Relation Between Clothing Pressure and Garment Fitting

Let CP = $(CP_1, CP_2 \ldots CP_m)^T$ be a set of m garment clothing pressures were measured by CLO 3D Modelist software. According to the proposed model in Figure 1, CP is the input data of the garment fit prediction model for training. Moreover, we set m = 20 in this research; it stand for the 20 measuring points, as shown in Figure 2.

Let GFC = $(GFC_1, GFC_2 \ldots GFC_n)^T$ be a set of n garment fit conditions. According to the proposed model in Figure 1, GFC is the output data of the garment fit prediction model for training. Moreover, we set n = 3 in this research; it stands for three garment fit conditions, there are tight, loose and fit respectively.

Let X = $(x_1, x_2 \ldots x_m)^T$ be a set of m garment clothing pressures of a new sample. We do not know whether this sample is fit. X as the input data are inputted into the model to predict garment fitting.

As three-layer BP neural network using simple nonlinear transfer functions can approximate any nonlinear functions with any precision, we adopted the BP neural network to predict garment fit condition. The S-tangent function was selected for the neural transferring function in the middle hidden layer. The S-logarithmic function was selected for the output layer. Moreover, the number of the hidden layer nodes influences the accuracy of prediction. A learning data is needed to determine the number of hidden layer in a practical application.

5. Example of Application

The garment fit prediction model is used for clothing fitting evaluation. The learning and test data were collected according to the method in Figure 2. As shown in Table 1, a total of sixty-nine garment pattern samples were involved in model building, in which sixty were for training the proposed model and the rest were for testing the model; and the fit garments, tight garments and loose garments of the thirty learning data samples each one-third respectively. After multiple trials and testing by MATLAB R2014a software, the training sample error is minimum when the number of hidden layer is 4. The prediction accuracy of back substitution is 84%, and the prediction accuracy of new samples is 72.7%. In general, the prediction accuracy is acceptable for garment fit evaluation. Three garment fit conditions were involved in this research, if the number of garment fit conditions increase, the prediction accuracy will decrease. The prediction accuracy can be further improved at the condition of learning data increase.

Table 1. Prediction accuracy and sample distribution.

	Fit condition	Fit	PA
Number of learning data samples	Loose	20	
	Fit	20	84%
	Tight	20	
Number of test data samples	Loose	3	
	Fit	3	72.7%
	Tight	3	

Note: NCP is Number of Correct Prediction; PA is Prediction Accuracy.

6. Conclusion

In this paper, we proposed a garment fitting prediction model based on artificial neural network. All the experiments in this research were conducted in a virtual environment. Compared traditional garment fit evaluation methods, the advantages of our approach is without participation of customers. Thus, the model can be applied to evaluate garment fit online. Test results of the proposed model shows that the prediction accuracy of back substitution of the prediction model is 84%, and the prediction accuracy of new samples is 72.7%.

Moreover, clothing production enterprises have plenty of garment patterns data as the businesses grow. These data are very valuable for data mining and knowledge discovery. If we use garment enterprises' data to train the proposed model, the perdition accuracy will be further improved.

Acknowledgments

This paper was financially supported by China National Endowment for the Arts.

References

1. Fan, J., W. Yu, and L. Hunter, *Clothing appearance and fit: Science and technology.* 2004, Cambridge, UK: Woodhead publishing Limited.
2. Liu, K. et al., *Fit evaluation of virtual garment try-on by learning from digital pressure data.* Knowledge-Based System, 2017. **133**: p. 174-182.
3. Zeng, X. and L. Koehl, *Representation of the subjective evaluation of the fabric hand using fuzzy techniques.* International Journal of Intelligent Systems, 2003. **18**(3): p. 355-366.
4. Zeng, X., Y. Ding, and L. Koehl, *A 2-tuple fuzzy linguistic model for sensory fabric hand evaluation*, in *Intelligent Sensory Evaluation*, D. Ruan and X. Zeng, Editors. 2004, Springer Berlin Heidelberg: Berlin, Heidelberg. p. 217-234.
5. Zeng, X. and Z. Liu, *A learning automata based algorithm for optimization of continuous complex functions.* Information Sciences, 2005. **174**(3-4): p. 165-175.
6. Zeng, X., D. Ruan, and L. Koehl, *Intelligent sensory evaluation: Concepts, implementations, and applications.* Mathematics and Computers in Simulation, 2008. **77**(5-6): p. 443-452.
7. Zeng, X. et al., *An intelligent recommender system for personalized fashion design.* in *IFSA World Congress and NAFIPS Annual Meeting (IFSA/NAFIPS), 2013 Joint.* 2013. IEEE.
8. Hollies, N.R. et al., *A human perception analysis approach to clothing comfort.* Textile Research Journal, 1979. **49**(10): p. 557-564.
9. Thomassey, S. and P. Bruniaux, *A template of ease allowance for garments based on a 3D reverse methodology.* International Journal of Industrial Ergonomics, 2013. **43**(5): p. 406-416.

A two staged forecasting scheme considering the constraints of sales forecasting in the fashion industry

R. Maleku Shrestha

Darmstadt University of Applied Sciences
64295 Darmstadt, Germany
rohan.shrestha@stud.h-da.de

G. Craparotta

Universitá di Torino
10124 Turin, Italy
giuseppe.craparotta@unito.it

S. Thomassey

ENSAIT, GEMTEX Laboratoire de Génie et Matériaux Textiles
59056, Roubaix cedex 1, France
sebastien.thomassey@ensait.fr

R. Moore

Darmstadt University of Applied Sciences, Fachbereich Informatik
64295 Darmstadt, Germany

Sales forecasting is one of the most cost effective methods to enable companies to make informed business decisions. In this paper, a two stage forecasting system is studied. A long-term forecasting system is combined with a short-term forecasting model. The proposed scheme first calculates long-term forecasts based on clustering and classification models while adjusting the effects of discounts and seasonality. These forecasts are then used as a basis for a second, short-term forecasting phase.

Keywords: Sales forecasting; fashion industry; clustering; decision trees; extreme learning machine; price effect.

1. Introduction

Sales forecasting is crucial for the fashion industry where the lead time is very long compared to the lifespan of the products. This paper presents a two-staged forecasting system. The first stage is a clustering and classification based Long Term Forecasting (LTF) system. Similar method has already been applied in multiple research.[1-3] Limited amount of historical data and availability of descriptive criterion makes this scheme favourable. Then the Short Term Forecasting (STF) is calculated with an Extreme Learning Machine (ELM) and a Feedforward Artificial Neural Network (F-ANN) using the result from the LTF.

This paper aims to improve the forecasting scheme proposed by Thomassey and Brahmadeep[4] by utilizing the historical sales data and re-using the descriptive features. In this previous work,[4] a two-stage sales forecasting system is developed to perform long-term and short-term forecasts for fast fashion products. A simulation of a store replenishment model, which is inspired from a method implemented in a famous fast fashion brand, is conducted on real data with different forecast scenarios, including the developed system. The results demonstrate that the proposed two-stage forecasting system outperforms the other scenarios in term residual inventories and total sales.

The research also focuses on adjusting the effects of discounts and seasonality on sales.

2. Constraints of Sales Forecasting in the Fashion Industry

In the fashion industry, it is commonly known that consumer demands are very volatile.[5,6] Listed below are few characteristics which should be taken into account:

(1) **Forecast Horizon:** The methods and models used to calculate the forecast for different horizons should be chosen in accordance to the forecast horizon.[7] For this paper two horizons are planned; a long term horizon and a short term horizon.
(2) **Price:** Principle of price elasticity of demand (PED) is used to show the responsiveness of the quantity demanded of a good or service to a change in its price.[8]
(3) **Seasonality:** In the fashion industry, some items are very sensitive to the seasonal variation, such as swim wears or t-shirts. Other items such as underpants are not impacted. Thus, according to the sensitivity of the considered item, the seasonality should be more or less integrated into the forecasting system for clothing sales.[7]
(4) **New products:** Fashion companies, and more especially fast fashion companies, renew their products at each collection or season. Consequently, no historical sales data of the product are available to perform the long term forecast.

3. Proposed Forecasting Scheme

3.1. *Long-term forecasting model*

The aim of the Long Term Forecast (LTF) model is to forecast the sales profile of a new product for the next season from its descriptive attributes (category, lifespan, price,...) and the sales of historical products. LTP is very challenging in the fashion market. Companies commonly rely on LTF based on average sales profiles by category. Consequently, the main expectations of the proposed model consist in giving a good overview of the sales profiles (more accurate than a average profile) and a reliable baseline for the short term forecast. The proposed LTF process is divided into 4 steps:

(1) **Step 1 — Data preparation:** Adjusting the effect of price using equation (1), the new sales volume is calculated.

$$s2 = s1(1 + e_{\langle p \rangle}(p2 - p1)/p1) \tag{1}$$

$s1$ and $s2$ are sales before and after the discount respectively, and $p1$ and $p2$ are price before and after the discount. $e_{\langle p \rangle} = -1.2$ is the coefficient of PED. Suitable $e_{\langle p \rangle}$ is chosen by repeating LTF multiple times for different values of $e_{\langle p \rangle}$ using 10-fold cross validation.

Final sales is calculated by adjusting the effect of seasonality using equation (2).

$$s3 = s2 \cdot \psi \tag{2}$$

where, s3 is sales after adjusting the effect of seasonality and ψ is pre-calculated seasonality factors.

(2) **Step 2 — Clustering of the sales profile:** A k-means clustering procedure is used to define the prototypes of sales (mean of the sales profiles of each cluster) from the historical sales.

(3) **Step 3 — Decision tree training from clustering results:** A C5.0 decision tree is trained to associate the attributes of the historical items with their prototypes/clusters determined in step 2. The optimum number of clusters is determined by repeating k-Means algorithm and the decision tree training for different number of clusters using 10-fold cross-validation, from 2 to $\sqrt{N}$, where N is the number of samples in the data set.[9]

(4) **Step 4 — Decision tree for the classification of future items:** The decision tree, previously trained, is used to assigns each new

product to one prototype of sales from its attributes. This prototype of sales becomes the sales forecast of this new product.

3.2. Short-term forecasting model

The aim of the STF is to update the LTF of the next week from real sales of the last two weeks. This kind of forecast is very useful for fashion companies to manage the replenishment of stores. STF is implemented using two different algorithms: Extreme Learning Machine (ELM) and traditional Feedforward Artificial Neural Network (F-ANN). The neural networks are particularly suitable for their learning and generalization capabilities on limited amount of data (short lifespan of products). To perform the STF of the coming week $(t+1)$, the inputs of the neural networks are: LTF (denormalized) of weeks $t-1$ to $t+1$, gross pocket price of weeks $t-1$ to $t+1$, the real sales of weeks $t-1$ and t, category, beginning_period and life_span of the item. Sigmoid function is used as the activation function of the hidden neurons. The number of neurons in hidden layer is crucial for the performance of the neural networks. Thus the ELM and F-ANN are trained for different number of hidden neurons (1 to 400) for each $week = t$. Fig. 1 illustrates the obtained RMSE on training and test datasets for weeks 9 and 29. The best number of neurons in the hidden layer is obtained for the smallest RMSE on the test datasets.

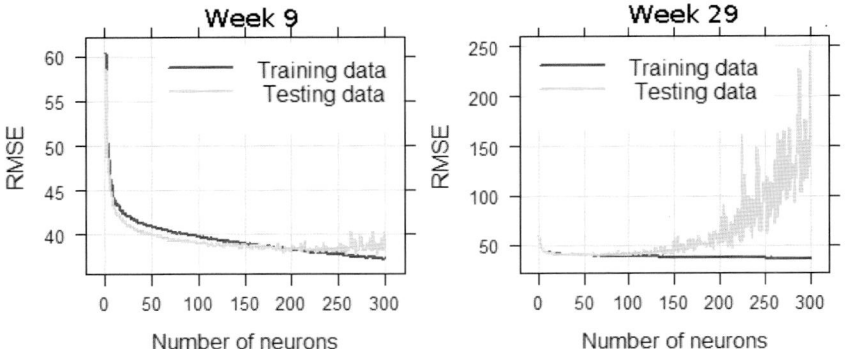

Fig. 1. RMSE according to the number of neurons in the hidden layer.

4. Empirical Study

4.1. *Data description*

Sales data from 2015 (historical data) and 2016 (future data) from an apparel store is used to evaluate the proposed forecasting scheme. There are 557 items in historical data and 695 items in future data. The Data has six descriptive criterion: ***brand***, ***category***, ***life_span*** (in weeks), ***beginning_period*** (calender week, the item was introduced to the store), ***initial_price*** (price before discount) and ***gross_pocket_price*** (price after discount).

4.2. *Empirical results*

4.2.1. *Long-term forecasting results*

Optimum number of clusters obtained from 3.1 is 7. Selecting larger number of clusters performed better at generating prototypes, but the classifier failed at classifying the items into their respective clusters. Hence, the chosen prototypes were a trade off between the classification and clustering error. Fig. 2 presents two example profile clusters and their associated prototypes. It clearly appears that the prototypes give a general overview of the sale profiles of the cluster but not a precise forecast.

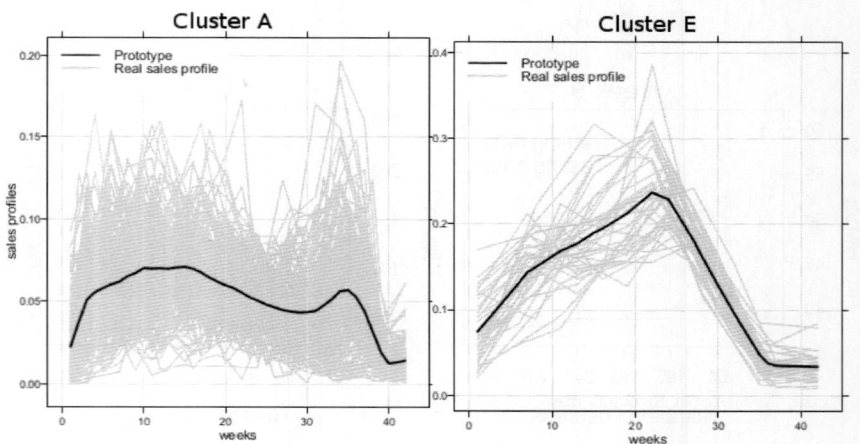

Fig. 2. Two examples of profile clusters and their associated prototypes.

4.2.2. Decision tree results

After the training process, the decision tree is composed of 73 nodes and 23 leaves. It was observed that the gross_pocket_price and initial_price were less influential than life_span and beginning_period; possibly because of the use of normalized sales instead of the real sales to create the clusters. The trained classifier is used to classify 695 future items into their respective sales prototypes (profile clusters).

4.2.3. Short-term forecast results

The STF is calculated using the ELM and the F-ANN utilizing the result from LTF. Fig. 3 illustrate the results of the forecasting scheme. It clearly appears that the LTF gives a baseline forecast and ELM and F-ANN adapt the forecast from the real sales.

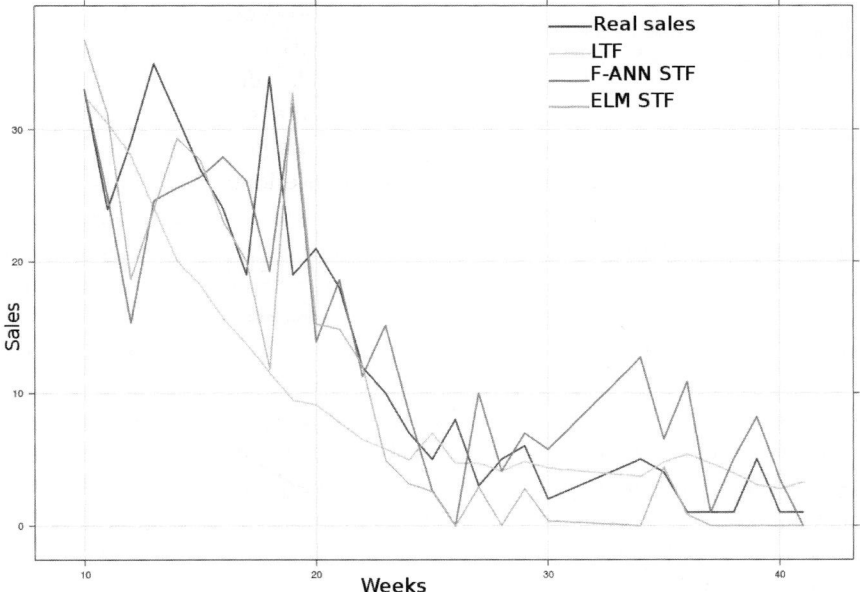

Fig. 3. Long-term forecast and short-term forecast of an item with minimum RMSE.

Table 1 shows the error in LTF and STF on future items. LTF is very challenging for fashion items and the RMSE and MAPE could appear to be very high. However, these results are still more accurate than the average profiles by category commonly used in fashion companies. STF using ELM improves the accuracy of the forecast and decreases the average RMSE by 40.00%. Similarly STF using F-ANN reduces RMSE by 1.44 compared to STF using ELM. Using F-ANN for STF decreases the overall average RMSE by 43.21% compared to LTF, resulting in more accurate forecast.

Table 1. RMSE and MPSE of LTF and STF on future items.

	Forecast	RMSE	MPSE
1	Long Term Forecast (K-means + C5.0)	45.38	187.12
2	Short Term Forecast (ELM)	27.21	90.37
3	Short Term Forecast (F-ANN)	25.77	82.74

The STF using ELM improved the accuracy of the forecast for 88% of the items (614 items out of 695) and F-ANN based forecast improved the accuracy of 92% of the items (642 items) compared to LTF. F-ANN based system improved the RMSE of 28 items that ELM based system failed to. This indicates that F-ANN based system is more relevant than ELM based STF especially for forecasting items with longer life span.

5. Conclusion

Achieving a highly accurate forecast is hard in fashion industry mainly due to many real world constraints. To enhance the forecasting accuracy, this study implemented a two stage forecasting system, the first stage demonstrates a long-term forecasting system while adjusting the effect of discount and seasonality on sales, which is then combined with a short-term forecasting model in order to increase the overall accuracy of the prediction.

Using the available descriptive attributes a LTF was calculated, LTF helps to estimate the quantity of items a company should purchase and also gives a rough prediction of the first weeks of sales. The LTF model used in this paper failed to achieve high accuracy sales forecast of future items but gives a good baseline for the STF. Some of the possible reasons behind this could be the lack of sufficient and proper descriptive attributes. Indeed, the availability and reliability of proper descriptive attributes are very crucial for accurate sales forecast,[4] and the retailers should pay more attention in setting up their information systems to store and easily extract the data.

As for the overall outcome, it is clear from the empirical results that the proposed STF scheme succeeded to improve the forecasting accuracy of more than 92% of the items compared to the LTF with an overall RMSE of 25.77. Using ELM for STF forecast did improve the accuracy but the traditional F-ANN based forecast outperformed the ELM based forecast especially for the items with the longer lifespan. The main concern for the F-ANN could be the run-time. But if the data used is not immensely large, the traditional F-ANN is advisable. Additionally, it is strongly advised to consider the effects of price and seasonality when performing a sales forecast. This study validates the concept that a clustering based long-term forecast followed by a neural net based short-term forecast can be used to improve the forecast accuracy in the fashion market. Besides, the emergence of the big data era opens new opportunities to implement deep learning techniques for fashion sales forecasting and more accurate models can be developed in the next years.

Acknowledgments

The authors gratefully acknowledge 'Evo Pricing' for supporting this research, the region "Haut de France" and the Graisyhm for their fundings.

References

1. S. Thomassey and A. Fiordaliso, *Decision Support Systems* **42**, 408 (2006).
2. S. Thomassey and M. Happiette, *Applied Soft Computing* **7**, 1177 (2007).
3. B. Hugueney, Representations symboliques de longues series temporelles (symbolic representations of long temporal series), PhD thesis, University Paris 62003.
4. Brahmadeep and S. Thomassey, Intelligent demand forecasting systems for fast fashion, in *Information Systems for the Fashion and Apparel Industry*, ed. T.-M. Choi, Woodhead Publishing Series in Textiles (Woodhead Publishing, 2016) pp. 145–161.
5. T.-M. Choi, *International Journal of Production Economics* **106**, 146 (Mar 2007).
6. A. Şen, *International Journal of Production Economics* **114**, 571 (2008).
7. S. Thomassey, *Sales Forecasting in Apparel and Fashion Industry: A Review*, in *Intelligent Fashion Forecasting Systems: Models and Applications*, eds. T.-M. Choi, C.-L. Hui and Y. Yu (Springer, Berlin, Heidelberg, 2014), Berlin, Heidelberg, pp. 9–27.
8. P. L. Anderson, R. D. McLellan, J. P. Overton and G. L. Wolfram, *McKinac Center for Public Policy. Accessed October* **13**, p. 2010 (1997).
9. J. Vesanto and E. Alhoniemi, *IEEE Transactions on Neural Networks* **11**, 586 (May 2000).

A collaborative platform with negotiation mechanism for make-to-order textile supply chain: A study based on multi-agent simulation

Ke Ma

GEMTEX (GEnie des Matériaux TEXtile), ENSAIT (Ecole Nationale Supérieure des Arts et Industries Textiles), 59100 Roubaix, FRANCE
Department of Textile and Clothing Engineering, Soochow University
Suzhou 215021, China
Department of Business Administration and Textile Management, University of Borås
50190 Borås, Sweden

Sebastien Thomassey

GEMTEX (GEnie des Matériaux TEXtile), ENSAIT (Ecole Nationale Supérieure des Arts et Industries Textiles), 59100 Roubaix, France

Xianyi Zeng

GEMTEX (GEnie des Matériaux TEXtile), ENSAIT (Ecole Nationale Supérieure des Arts et Industries Textiles), 59100 Roubaix, France

As the increasing trend of mass customization and small-series production in textile supply chain, make-to-stock strategy was not feasible in textile production anymore. Make-to-order strategy was widely applied in current textile SC. Although make-to-order textile supply chain has many advantages, there were still many defects in its traditional structure. Therefore, we proposed a collaborative platform with a novel negotiation mechanism to enhance supply chain collaboration among textile companies, so that to optimize make-to-order supply chain structure. Agent-based simulation technology was utilized in this study to realize the collaborative platform and corresponding negotiation mechanism. It was compared to the traditional make-to-order textile supply chain structure in simulation experiment. Based on simulation results, the advantages of the proposed structure were demonstrated. It can help improve multiple supply chain performances.

1. Introduction

Due to the increasing demand for customization and personalization in today's fashion industry, mass customization and small-series production play a more and more vital role in current textile supply chain (SC). Make-to-stock strategy is not appropriate anymore in textile product manufacturing, as it is almost impossible to forecast the demand of customers nowadays. Many textile companies implement make-to-order (MTO) strategy in production, which means that companies only conduct production activities after receiving an order from customers [1]. For instance, a fabric manufacturer only produces fabric specifically for a particular order rather than to use in-stock produced fabric. A common traditional MTO textile SC structure is shown in Figure 1. MTO strategy can help companies avoid inventory of semi or final products and reduce the risk of sales forecasting. However, there are still many drawbacks in such traditional MTO SC structure, e.g. high cost [2], long lead time [3] and unsmooth production flow. Therefore, we conducted this study to optimize MTO textile SC structure.

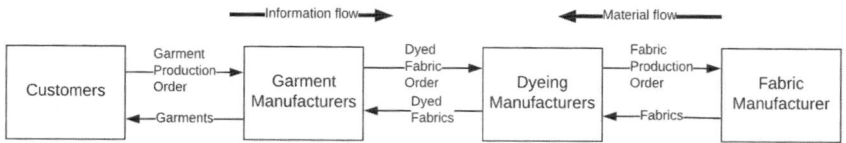

Figure 1. Traditional make-to-order textile supply chain structure.

Inter-organizational supply chain collaboration (SCC), which means "two or more autonomous firms working jointly to plan and execute supply chain operations" [4], has been widely applied in SC practice and has been discussed in many SC research, e.g. [5], [6]. It becomes an important topic and approach for SC optimization and structure upgrade. However, it is relatively less implemented in practice in textile industry due to its intense competitive environment. Only a few studies addressed textile SCC in previous research [7]–[9]. Therefore, to deal with the high competitive situation in textile industry, traditional SCC method may not an optimal option. An effective mechanism is demanded to enhance SCC among textile companies while to maintain their individual decision-making rights. In this study, we propose a novel negotiation mechanism to meet this demand for MTO textile SCC. We also develop a collaborative platform (CP) to carry on the negotiation mechanism. They are introduced in following sections.

2. Collaborative Platform with Negotiation Mechanism for Textile Supply Chain

The CP with corresponding negotiation mechanism is introduced in this section. As shown in Figure 2, the collaborative MTO textile SC structure within the CP contains same information flow and material flow as in the traditional MTO textile SC. However, companies on the CP has an option to negotiate with other companies through CP, which makes the collaborative MTO SC structure

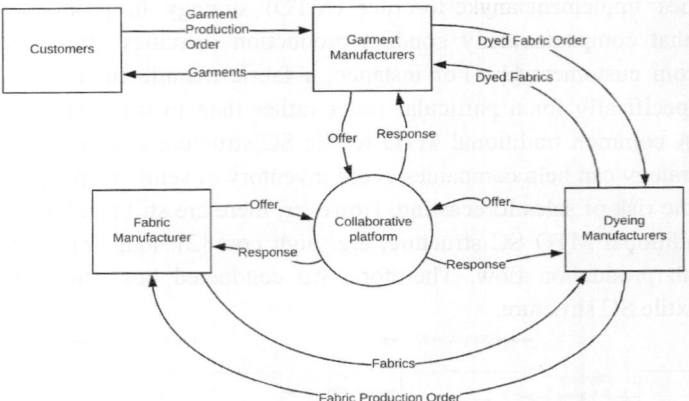

Figure 2. Make-to-order textile supply chain structure within the collaborative platform.

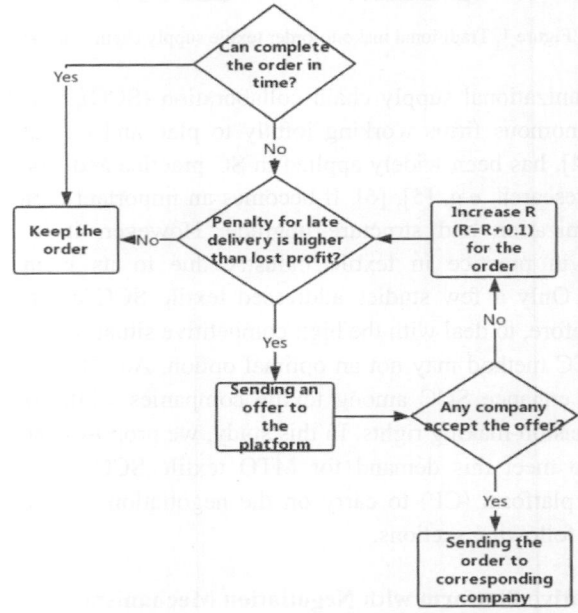

Figure 3. General decision-making process of a company on the collaborative platform.

distinct to the traditional one. The general decision-making process of a company with the negotiation process on the CP is illustrated in Figure 3. When a company cannot complete an order in time, it can send a production substitution offer with relevant data (production type, quantity and price etc.) to the CP after evaluating the penalty for late delivery and profit lost by sharing the order to another company. The price is determined by the ratio (R) of total revenue offered to another company for the production of the shared order. If any other company accepts the offer, the order is then shared to corresponding company for production. If there is no other company accept the offer, the company can increase the price of the offer by increasing R for current order and re-evaluate the penalty and profit lost. If it is still worthy for the company to share the order, it would send a new production substitution offer to the CP. This negotiation process is terminated until any other company accept the offer or R reaches over than 100%. Such negotiation mechanism can not only boost SCC by taking full advantage of resource among companies but also guarantee the individual benefit of companies by maintaining their own decision and negotiation rights.

3. Methodology

3.1. *Multi-agent simulation model*

It is difficult to experiment the CP and proposed negotiation mechanism in real-world system. Considering the high stochastic nature of SC process and relationships among companies in our proposed collaborative SC structure, traditional analytical tools or mathematical modelling methods are not applicable either. Therefore, to realize the SC process and mutual relationships among companies on CP with corresponding negotiation mechanism, agent-based simulation (ABS) technology was employed for experiment. ABS "involves building a model consisting of 'agents', who are able to interact with each other and programmed to be proactive and autonomous as well as to perceive its environment" [10]. It has emerged as a preferred tool for SC research in recent years [11] and also has been implemented in many textile SC researches, such as [12], [13].

Agents are the core components in ABS. Based on the MTO textile SC structure and our proposed CP introduced in section two, we defined three main agent types in this study, viz. (1) customer agent type, (2) supplier agent type, and (3) collaborative platform agent type. Customer agent type stands for customers in the SC structure, it is mainly responsible for order generation. Garment production orders are placed by customer agent along with order type, order size and demanded lead time. In supplier agent type, the general order decision-making process (keep the order or share the order, accept the shared order or reject the shared order), production process and corresponding resources (machines and operators) are defined. The CP agent type makes the novel negotiation mechanism applicable among companies (supplier agent type). It receives production substitution offers and sends them to all relevant companies one by one. It is also responsible for giving response to the company who places the offer.

3.2. Evaluation criteria

To comprehensively evaluate the new collaborative SC structure and compare it to the traditional one, it is necessary to consider SC performance from different angles. Therefore, five KPIs (Key performance indicators) were defined in this study as follows, which are discussed by previous studies regarding SCC in textile industry [6], [11] and generally can cover all common and important aspects for evaluating SC performance.

1. Unit cost (UC_i) is the average cost for each product unit of company i. We considered inventory cost, production cost, ordering cost and penalty for late delivery in the definition of cost, which is explains as follows.

$$TC_i = \sum_{n=1}^{N} \sum_{m=1}^{M} (PC_n \times Y_n + WC \times TT + OC \times NO_n + max(0, (LT_{nm} - RLT_{nm}) \times p)) \quad (1)$$

where TC_i is the total cost of company i, PC_n is raw material and processing cost per product n, Y_n is total yield of product n, WC is the warehouse cost per month, TT is total working time, OC is ordering cost per order, NO_n is the number of order placed, LT_{nm} is lead time for order m of product n, RLT_{nm} is required lead time for order m of product n, p is penalty for late delivery per day. Therefore,

$$UC_i = TC_i / \sum_{n=1}^{N} Y_n \quad (2)$$

2. Average lead time (ALT_i) is the average duration in company i from receiving an order to completing the order.

3. Resource utilization (RU_i) is to evaluate the utilization of resource (operators and machines) of company i. It can reflect the effectiveness and efficiency of a company.

$$RU_i = OT/TT \quad (3)$$

where OT is the effective machine operating time.

4. Delayed order percentage (DOP_i) is to evaluate the ratio of delayed order. It has influence on customer satisfaction.

$$DOP_i = NCO/TO \quad (4)$$

where *NCO* is the total number of not-on-time completed order, *TO* is the total number of completed order.

5. Yield (Y_i) is the total number of products produced in company i in the given period of time.

4. Simulation Results

A four-echelon MTO textile SC was simulated, including customer echelon, garment manufacturer echelon, dyeing manufacturer echelon and fabric manufacturer echelon. There are 50 companies in each SC echelon in the ABS model. We collected historical order data and production data from a garment company based on traditional MTO SC model located in Jiangsu Province, China as input parameters for the ABS model. To simplify our simulation model, only one type of production, viz. the production of shirt, was considered. That is to say, customer agent type only place the shirt production order. We also interviewed professionals in garment industry for necessary data for the ABS model, e.g. production time, resources needed etc. (All input parameters are available from the corresponding author on request). The simulation model was run for a duration of one year based on traditional SC structure and collaborative SC structure respectively. We compared the output of simulation (e.g. total yield) based on traditional SC structure and data collected from real industry, the simulation output is close to real data. Moreover, our ABS model was developed step by step as introduced in section 3.1. Therefore, our ABS model is representative and validated to our case. Finally, required data were extracted from simulation output for the calculation of five predefined KPIs, the final results are presented in Figure 4. They are the mean value of 50 garment manufacturers in our ABS model, thus it can represent the performance of the whole SC.

In general, the collaborative SC structure with proposed CP and negotiation mechanism performed much better than the traditional SC structure in all checked SC aspects. There are significant improvements in unit cost (a decrease of 50.9%), delayed order percentage (a decrease of 51.8%) and average lead time (a decrease of 36.8%). The improvements regarding total yield and resource utilization are relatively slight, with an increase of 3.1% and an increase of 2.6% respectively.

5. Conclusion

To meet the increasing trend of mass customization and small-series production, make-to-stock strategy is not feasible for textile SC anymore. More and more textile companies covert to make-to-order strategy in production. However,

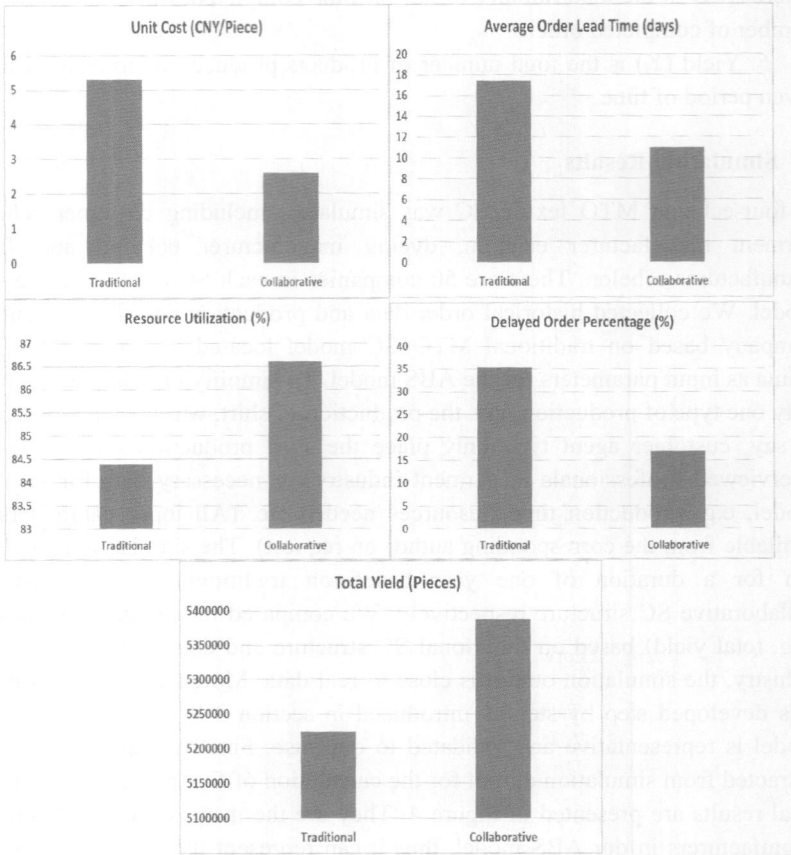

Figure 4. SC performance comparison between traditional structure and collaborative structure.

there are still many drawbacks in traditional MTO textile SC. In this study, we proposed a novel negotiation mechanism to optimize traditional MTO textile SC structure. A collaborative platform was also developed to operate such mechanism. The CP with corresponding negotiation mechanism is for enhancing SCC among textile companies while maintaining their individual decision-making right, so that to optimize the whole MTO textile SC. A multi-agent simulation model was built to realize and experiment the novel collaborative SC structure by comparing to traditional MTO SC structure. Five SC KPIs were defined and examined. Based on simulation results, the collaborative SC structure can improve all the five KPIs, which shows advantages of such innovative SC structure. It is demonstrated that the proposed negotiation mechanism can optimize MTO textile SC.

This study aims at textile industry, however, the proposed negotiation mechanism may be implemented in other similar industries or be discussed for SC research generally. Although our proposed collaborative structure with negotiation mechanism are demonstrated to be optimized on the basis of simulation results, it is interesting to examine such negotiation mechanism under distinct conditions, viz. different fashion seasons (different workloads), different applied SC echelons and different company quantities/scales. This mechanism may have distinct effects on SC performance. It is also interesting to compare such decentralized SC structure to a SC centralized optimized structure. They are all expected to be explored in future research.

Acknowledgments

This work is supported by the joint doctorate programme "Sustainable Management and Design for Textiles" which is funded by the European Commission's Erasmus Mundus programme.

References

1. F. Meisel and C. Bierwirth, Int. J. Prod. Res., vol. 52, no. 22, pp. 6590–6607, Nov. 2014.
2. K. Morikawa, K. Takahashi, and D. Hirotani, Int. J. Prod. Econ., vol. 147, no. PART A, pp. 30–37, Jan. 2014.
3. G. Ioannou and S. Dimitriou, Int. J. Prod. Econ., vol. 139, no. 2, pp. 551–563, Oct. 2012.
4. M. Cao and Q. Y. Zhang, "Supply chain collaboration: Impact on collaborative advantage and firm performance," J. Oper. Manag., vol. 29, no. 3, pp. 163–180, 2011.
5. L. Chen, X. Zhao, O. Tang, L. Price, S. Zhang, and W. Zhu, Int. J. Prod. Econ., 2017.
6. U. Ramanathan, Int. J. Oper. Prod. Manag., vol. 32, no. 5–6, pp. 676–695, 2012.
7. K. Ma, L. Wang, and Y. Chen, Sustainability, vol. 10, no. 1, p. 52, Dec. 2017.
8. T. J. Xiao and J. A. Jin, Prod. Plan. Control, vol. 22, no. 3, pp. 257–268, 2011.
9. T. C. Kuo, C. W. Hsu, S. H. Huang, and D. C. Gong, Int. J. Comput. Integr. Manuf., vol. 27, no. 3, pp. 266–280, 2014.
10. B. Ponte, E. Sierra, D. de la Fuente, and J. Lozano, Comput. Oper. Res., vol. 78, no. September 2016, pp. 335–348, 2017.
11. M. Tarimoradi, M. H. F. Zarandi, H. Zaman, and I. B. Turksan, J. Intell. Manuf., vol. 28, no. 7, pp. 1551–1579, 2017.
12. K. Ma, L. Wang, and Y. Chen, Sustainability, vol. 10, no. 1, 2018.
13. A. Pan and T. M. Choi, Ann. Oper. Res., vol. 242, no. 2, pp. 529–557, 2016.

A fuzzy signal processing approach integrated into an intelligent garment for online fetal movement monitoring

Xin Zhao, Xianyi Zeng, Guillaume Tartare, Ludovic Koehl

GEMTEX, ENSAIT, 2 allée Louise et Victor Champier
Roubaix, 59056, France

Julien De Jonckheere

INSERM, CIC-IT 1403, Maison de Régionale de la Recherche Clinique, CHRU de Lille
Lille, 59000, France

Fetal movement (FetMov) detection is one of the most important elements for fetal well-being monitoring. In this paper, we propose an online automatic fetal movement detection system based on accelerometers and the adaptive neuro-fuzzy inference system (ANFIS), which has been integrated into an intelligent garment worn by the pregnant woman. With the help of this system, fetal movement can be quickly estimated from the sensors of the garment and further classified so that relevant fetal well-being information can be extracted. This system includes the following components: 1) data acquisition; 2) signal pre-processing and feature extraction; 3) fuzzy inference system for fetal movement estimation; 4) updating of the adaptive fuzzy model by integrating new signals.

Keywords: Fetal movement detection; intelligent garment; feature extraction, ANFIS.

1. Introduction

In recent years, researchers and obstetricians paid great attention to fetal movement signal analysis for fetal well-being monitoring. Fetal movement is an important marker of fetal well-being, absence or reduction of fetal movement could be a sign of fetal death [1]. Ultrasonography, regarded as gold standard of fetal movement detection, is widely used for fetal well-being monitoring in clinical applications [2]. However, as ultrasound is expensive and only available at hospital, it is impossible for pregnant women to use it for an online and long term monitoring. Maternal perception on fetal movement is another alternative, which is easy to implement, but the drawback of this approach is in its imprecision and subjectivity. During pregnant women's normal daily activities, the sensitivity of maternal perception to fetal movement is only about 30% [3].

In this paper, a long-term fetal movement monitoring system by using accelerometers and fuzzy theory is proposed. This system has been integrated into an intelligent garment (wearable system) ensuring close contact between wearer's abdomen and acceleration sensors of the garment. Besides, long term monitoring would lead to an estimate of long term sleep-wake cycles of a fetus, which could provide an interesting key factor of well-being that could not be implemented through regular diagnosis. For simplicity, the design of the intelligent garment integrating the sensors is not discussed here and we only focus on the part of signal processing. In the literature, acceleration sensors have been proved to be relevant to lone-term fetal movement monitoring by capturing vibrations on the maternal abdomen wall [4]. Evidence begins to accumulate as studies to fetal movement detection are undertaken using accelerometers with high sensitivity, which allows us to capture all kinds of signal details with regard to system's sampling frequency. Since fetal movement signals have different signal energy distribution and different frequency components than other artifacts, it is possible to separate them from the recorded data by using amplitude detection and signal decomposition techniques in frequency space. Accelerometer sensors for fetal movement monitoring have other advantages such as low cost, low weight and high sensitivity. Furthermore, compared to ultrasound, accelerometer sensors mounted into an intelligent garment are passive and non-invasive. All these features make accelerometers an ideal tool for online fetal movement signal collection.

Based on the data collected from accelerometer sensors, many researchers involved in studies related to fetal movement detection and classification issues, and they have developed various features which could help distinguish suspected fetal movement from other signals. However, detection approach that simply based on a threshold value or some simple statistical features of the collected data could not achieve a promising outcome. Some articles describe detection methods based on a Time-Frequency (TF) signal processing approach [4]. Other classification approaches are grounded on machine learning such as Artificial Neural Network (ANN) or Genetic Algorithm (GA). These approaches could lead to an outstanding outcome, but the drawback is their high computational complexity and iterative learning process, which limit their ability in clinical applications and online well-being monitoring.

In this article, a novel fuzzy theory-based approach is proposed for fetal movement detection based on the signals measured by the accelerometers mounted into an intelligent garment. Without requirements on definition of strict thresholds and boundaries for signals and accurate mathematical models, the proposed fuzzy approach is considered as flexible, robust and capable of being adapted to all scenarios of fetal movement, which vary with each individual. Fuzzy systems have advantages of simulating human behaviors and dealing with large uncertain situations in real world. Thus, a fuzzy inference system might be an alternative for signal pattern recognition and classification. On the other hand, with the help of fuzzy theory, the signal detection system can deliver a series of simple fuzzy "IF-THEN" rules and a related inference procedure, thus the computational burden could be highly reduced. Fuzzy classification approach is suitable to ensure low energy consumption and thus a more autonomous wearable system, making long-term monitoring possible. The right balance has to be found between local (on wearable system) computation and data flow through the air transmission. Furthermore, adaptive neuro-fuzzy inference system (ANFIS) is used in this study [5]. Combining the adaptive capability of neural networks and the interpretability of fuzzy logic, ANFIS can automatically adjust membership functions and other parameters of a fuzzy model when integrating new measured signals, making the whole detection system more adaptive to various individuals and scenarios.

2. Methodology

2.1. *System Overview*

The structure of the proposed fetal movement detection system is described in Figure 1. The proposed system mainly consists of four parts: parameters training unit, signal preprocessing unit, inference and decision unit and database unit, respectively. In the parameters training unit, an ANFIS is used to update and optimize the parameters of the fuzzy model for maximizing the system's performance and robustness. In the preprocessing unit, a signal measured from an acceleration sensor is firstly filtered by using a band pass filter to eliminate noises corresponding to low and high frequencies. The filtered signal is then segmented using a moving window with 4 seconds in length and 50% window overlapping. Baseline drift correction is performed to eliminate tendency items, which can be done by applying the polynomial least-squares fitting method. Then statistics features of each signal segment are calculated and served as input vector of the fuzzy model for classification. The fuzzy model located in the inference and decision unit is used for estimating the possibility of fetal movement on each signal segment. A final output is given after a comprehensive analysis of all the consecutive signal segments along the signal time axis.

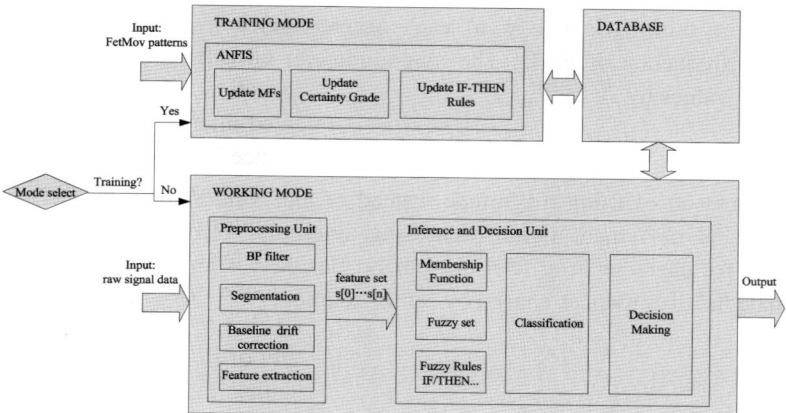

Figure 1. Structure of the fuzzy-based fetal movement detection system.

2.2. Data Acquisition

3-axis acceleration sensors was used in this study (MMA8451Q from NXP Semiconductors). During measurements, the subject was demanded to sit down or take a semi-recumbent position and stay still, the reason for this is that fetal movement detected by accelerometers often have weak amplitude which can be easily interfered by other noisy waveform.

In order to collect fetal movement related information as much as possible, four accelerometers were mounted into the intelligent garment: two placed on the upper part and two others on the lower part of the mother's abdomen. During the experiment, the subject was also asked to hold one push button to record maternal perception. This record could give us an additional reference validating the presence of the fetal movement.

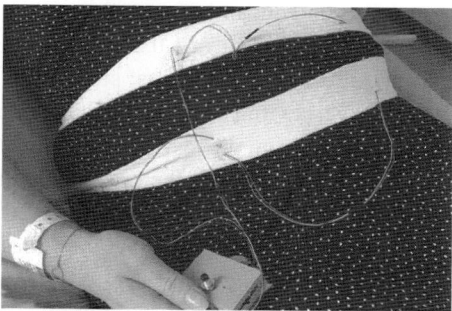

Figure 2. The intelligent garment with sensors and one push button for maternal perception.

All the three axis values of each accelerometer were collected with a sampling frequency of 60Hz. The magnitude can be calculated based on the three axis values by

$$g[n] = \sqrt{x^2[n] + y^2[n] + z^2[n]} \tag{1}$$

The reason of using magnitude instead of axis values for data analysis is that, in fetal movement signal processing, amplitude is more important than direction for characterizing fetal movement. Besides, using magnitude can also eliminate the disturbances caused by sensor rotation.

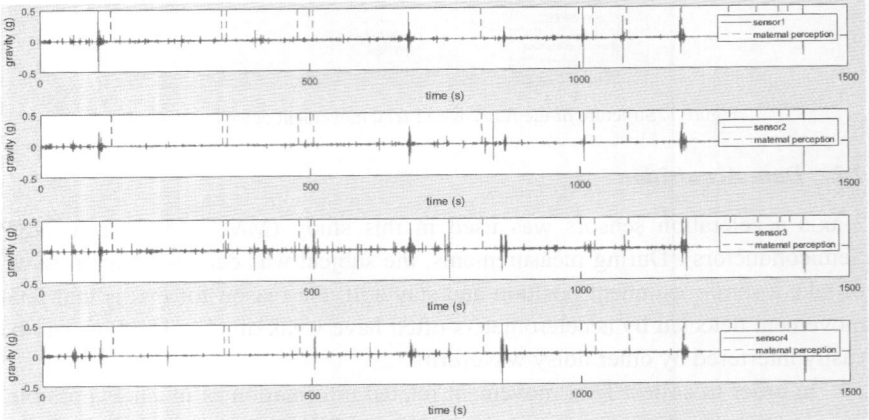

Figure 3. Fetal movement signals captured by 4 accelerometers, along with the maternal perception.

2.3. *Signal Processing*

In order to eliminate low frequencies caused by mother's breathing and other high frequency noises caused by contact between the wearer's skin and sensors, a band-pass filter with cutoff frequencies of 0.2Hz to 20Hz was applied to the magnitude values. After filtering, signal was segmented into epochs of 4 seconds with 50% window overlapping between adjacent epochs.

To our clinical observations in the experiments with different pregnant women, we got signals (see Figure 3) with the presence of picks and vibrations in a very short duration when a fetal movement occurred. The amplitude of variation is rather big (if any wearer's physical activity does not exist). In order to boost the classification accuracy, epochs that had a maximum amplitude greater than an upper threshold (0.8g) or a minimum value below a lower threshold (0.05g) were labeled as artifacts and excluded during the classification stage. Furthermore, if any one of the four sensor amplitude values exceeded the upper threshold, then signals of all the four sensors during this timestamp were labeled as artifacts.

On the other hand, the DWT (Discrete Wavelet Transform) can be thought of as a powerful tool for time-frequency analysis of non-stationary signals [6]. The DWT allows us to decompose a signal into frequency sub-bands. In the present study, statistics features of both signal amplitude in time space and wavelet coefficients in each sub-band were used to represent their distribution properties. The following statistical features were selected as inputs of the fuzzy model:

1. Mean of the absolute value of the amplitude
2. Standard deviation of amplitude values
3. Mean of the absolute values of the coefficients in each sub-band of the wavelet coefficients
4. Standard deviation of the coefficients in each sub-band of the wavelet coefficients

Table 1 shows the feature sets of one fetal movement signal segment and one noise signal segment. From Table 1 we can see that the extracted features of the two segments are different from each other, therefore, they can be used for distinguishing fetal movement signals from other artifacts.

Table 1. Extracted features of one fetal movement signal segment and one noise signal segment.

		Amplitude	D1	D2	D3	A3
FetMov	Abs Mean	0.0130	0.0157	0.0123	0.0126	0.0180
	Std	0.0228	0.0225	0.0213	0.0213	0.0269
Noise	Abs Mean	0.0024	0.0027	0.0022	0.0027	0.0028
	Std	0.0032	0.0034	0.0027	0.0036	0.0035

In real applications, we can define a relevant threshold T so that we can identify each fetal movement epoch, then combine all the neighboring epochs together to get one entire fetal movement zone if the durations between these epochs are smaller than the predefined threshold T.

3. Analysis of Experimental Results

Two raw signals of about 20 min were prepared for the experiment. They were collected from different subjects to maintain the specificity of the data. Totally 1824 patterns extracted from the first signal were used for training the ANFIS model. These patterns were labeled as 1 (FetMov) or 2 (NonFetMov) by an expert clinician with reference to the maternal perception. Selected features were extracted from these segments (patterns) and the statistical distribution of each feature was calculated and used as initial settings of the corresponding membership function in ANFIS. Then the system was trained with the SCG (Scaled Conjugate Gradient) algorithm [7] when these labeled patterns were used as input.

After training, the performance of the proposed system was examined using the second signal. Totally 2820 epochs (4 channels each with 705 epochs) were extracted from the signal. Features extracted from these epochs were used as input to fuzzy inference system for classification. The classification results were compared with the target outputs given by an expert clinician with reference to the maternal perception. Then the performance of the proposed system was evaluated. The classification accuracy was 88.51% (from 705 epochs, 624 were classified correctly). The classification result is shown graphically in Figure 4.

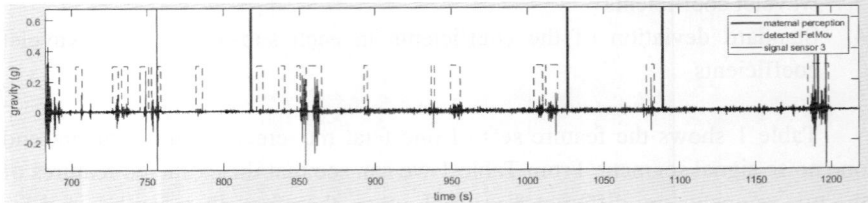

Figure 4. The classification result of the proposed system.

4. Conclusion

In this paper, we presented an ANFIS based fetal movement estimation system using four accelerometers, which has been mounted into an intelligent garment, giving real advantages to the user and making long-term fetal movement measurement possible. Accelerometers have been proved to be suitable for biomedical signal data collection. Besides, by comparing with other feature extraction techniques, we consider that a fuzzy model is appropriate for dealing with signals containing uncertainty, high volatility and diversity. Also, fuzzy systems are favorable for biomedical signal processing and classification due to their capacity of human knowledge integration and robustness. The proposed fuzzy inference system would be a good alternative for fetal movement signal processing.

The interest of using ANFIS model lies in the fact that the parameters of its membership functions can be progressively learnt with integration of new signals. In this context, the propose model is an open-source system, which will become more and more efficient and reliable with applications.

Future work aims to collect more fetal movement patterns for training the system to improve the performance of the system. A large quantity of training data would be required to boost the performance of the proposed system due to the fact that fetal movement signals vary from each individual. Future work also involves a deep analysis in different frequency components of the fetal movement signals. Besides, extra accelerometers can be added, placing to mother's thigh. In fact, during the measurement, the mother's movement can also be detected by accelerometers placed on her abdomen, so there may be an interest to add one additional sensor detecting maternal movement to obtain normal reference values for artifact suppression.

Acknowledgments

This research was supported by the project IOTFetMov ANR-14-CE24-0035-01 of the French National Research Agency (ANR). The Authors would also like to thank Mr. Hubert Ostyn, knitting workshop technician at GEMTEX laboratory, for his help in offering the resources in running the program.

References

1. De Vries, J.I., G.H. Visser, and H.F. Prechtl, *The emergence of fetal behaviour. I. Qualitative aspects.* Early human development, 1982. **7**(4): p. 301-322.
2. Manning, F.A., L.D. Platt, and L. Sipos, *Antepartum fetal evaluation: development of a fetal biophysical profile.* American Journal of Obstetrics and Gynecology, 1980. **136**(6): p. 787-795.
3. Hijazi, Z.R., S.E. Callan, and C.E. East, *Maternal perception of foetal movement compared with movement detected by real-time ultrasound: An exploratory study.* Australian and New Zealand Journal of Obstetrics and Gynaecology, 2010. **50**(2): p. 144-147.
4. Boashash, B. et al., *Passive detection of accelerometer-recorded fetal movements using a time–frequency signal processing approach.* Digital Signal Processing, 2014. **25**: p. 134-155.
5. Jang, J.S., *ANFIS: adaptive-network-based fuzzy inference system.* IEEE transactions on systems, man, and cybernetics, 1993. **23**(3): p. 665-685.
6. Daubechies, I., *The wavelet transform, time-frequency localization and signal analysis.* IEEE transactions on information theory, 1990. **36**(5): p. 961-1005.
7. Møller, M.F., *A scaled conjugate gradient algorithm for fast supervised learning.* Neural networks, 1993. **6**(4): p. 525-533.

Optimization of the body part recognition method of 3D human mesh based on propagation algorithm

Yu Chen[1], Xianyi Zeng[2], Zhebin Xue[3]

[1]*Clothing Institute, Shanghai University of Engineering Science*
333 Long Teng Road Shanghai, 201620, China
[2]*ENSAIT, Gemtex, 2. allée Louise et Victor Champier Roubaix, 59000, France*
[3]*School of Textiles and Clothing, Jiangnan University*
Wuxi, Jiangsu Province, 214122, P.R. China

Armhole specify is the base of base in 3D digital human mannequin measuring. Finding the key body part armpit is very important in the process of measuring and modeling 3D digital human polygon mannequin, especially these meshes are raw and incomplete. A recognition method by propagation algorithm which can distinguish body parts automatically has been applied in industrial production. However, this method also has the contradiction of recognition precision and speed. In this paper, we present a new method for optimizing the step of the "Isolation Plane" using binary search like method. Based on the optimized step, both recognition speed and accuracy are greatly increased. The effectiveness of our method has been validated by more than 2000 mannequins. It can also be applied for specifying other body.

1. Introduction

1.1. *Foreword*

Anthropometry is a very important research field in ergonomics and has a wide range of applications in the fields of garments, sports, national defense and so on. Among them, 3D mesh model technology has been widely used. 3D mesh model technology in accordance with the ISO20685 standard to better solve the human body's 3D measurement problems and modeling problems.[1]

However, since non-contact measurement data is acquired by scanning with a 3D scanner, some special problems often occur in the acquisition of scan data in some special parts of the human body. Especially in armpit, crotch and other body concave parts, obtained scan data was often incomplete, not precise enough or overlapping. Therefore, people have adopted some traditional remedies: the proportion method, the body characteristic point method, the slice method,[2] however, these are not very satisfactory. To this end, we have newly designed a 3D human body model recognition method based on propagation algorithm,[3] which basically solved the problem and has been applied in industry and achieved good effects.

2. The Characteristics of 3D Mesh and the Establishment of Recognition Method by Propagation Algorithm

2.1. *3D digital mesh model*

The 3D scanner scans and captures the human body's data and stores the collected data in the form of a point cloud model, which is then converted into a 3D body model. The 3D mesh human mannequin can be identified and read by the majority of 3D software. Figure 1 shows 3D mesh model consists of vertices, edges and facets

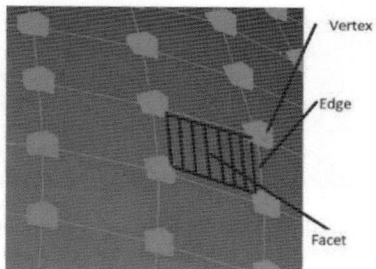

Figure 1. 3D polygon mesh diagram.

2.2. *Characteristics of mesh and propagation algorithm clothing renderings selection and processing*

Mesh model has the characteristics of the grid vertices can transfer information between each vertex. Every vertex on mesh can find his neighbors by edges. The neighbors' vertices can find his own neighbors vertices. If we mark any vertex on mesh as infected tag (called infection vertex, such as the source of infection), infection vertex will propagate infected tag to his neighbors by connected edges. New infected vertices become new source of infection, they will propagate infected tag to all neighbors' vertices. The whole process is like the spread of an infectious disease, until it infects all the vertices on the mesh.

We find that this feature of the mesh, if used well, can be used to distinguish the different regions of the mesh, and also find the location of the regional boundary. The problem is to make good use of the propagation feature of the mesh, we must establish a technology that can control its propagation, that is, stop the spread of infected mesh vertices at a certain position of the mesh. For this reason, we specially designed a kind of isolation plane, that is a non-grid surface, usually a plane, its equation in the 3D rectangular coordinate system is: AX + BY + CZ + D = 0.

The design of the isolation plane solves the constantly spreading problems that control infected vertices. Figure 2 shows schematic of the entire propagation process and propagation is blocked by isolation plane.

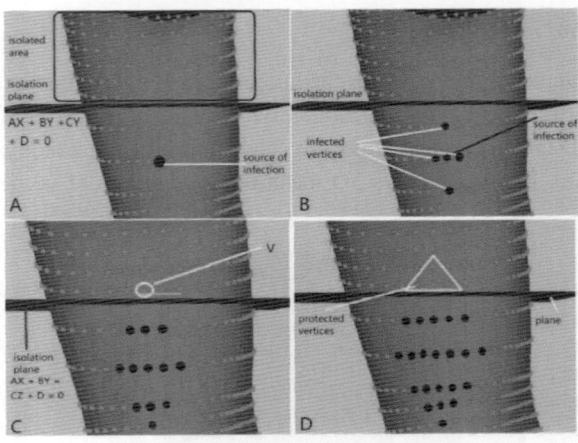

Figure 2A. The mesh vertex marked as a black dot is the source of the infection. Figure 2B. The source of the infection infected neighbor vertices, making it become to black dot. The vertices that turn into black dots transform to new sources of infection. Figure 2C. The process of infection continues, but the mesh vertex V is exempt from infection because it is on the isolation plane. Figure 2D. The vertices in the yellow triangle area are not infected because of the role of the isolation plane.

Using this feature of the mesh, we designed a propagation algorithm that can automatically identify some parts of the human body. It allows the source of infection to spread within the area under the isolation plane by setting up the source of the source and the isolation plane, then adjust the isolation plane, and check the proportion of the infected and not infected mesh vertices. By studying the rate of change of proportion, we can identify and locate some parts of the mesh. This method is named Recognition Method by Propagation Algorithm.

2.3. *The specific operation of the propagation algorithm and the core pseudo code*

In order to solve the difficulty problem of data loss or mesh overlap when we collect human body data by 3D scanner, we focus on solving the problem in the armpit, the crotch body parts which link two independent limbs. On these sites, two independent limbs are connected to each other, existing some key vertices and edges which will cause the number of infected vertices extreme growth. This key vertex and edge is the end or start point of measuring the height of the armpit and the length of the limb. By using the isolation plane to find and identify the characteristics of the body, we can find the scope, location, and size of the limbs of the armpit and arm.

In the case of the arm, the basic method is to separate the 3D human mesh into two separate parts of the arm and body by setting up the isolation plane. Continuously move the isolation plane with a certain length of step, expand the area of the quarantine area until the joints of the two parts of the arm and body, that is, two separate limbs are connected by mesh edges. So we find the height of the armpit, and we can find the size and scope of the arm.[3]

2.4. *Initial effect and problem of efficiency*

The Recognition Method by Propagation Algorithm was applied in the professional 3D human measurement modeling software. The software deals with nearly 10000 3D human mesh and can automatically and correctly find the position of the armpit. This method has been dealt with successfully in the 3D human mesh whose armpit part mesh was incomplete or overlap.

But the results show that the precision of measurement recognition must be improved by narrowing the step of the isolation plane and narrowing the step of the isolation plane will affect the speed of measurement recognition. How to solve this problem? We carried out the following studies.

3. The Improvement and Optimization of the Propagation Algorithm

We can see that the step length of the isolation plane is a very important factor.

Table 1. Results of calculation by using different step (the 3D human mesh has 6,170 vertices, 12,169 facets, the actual height of armpit is 140.54cm).

Step length	1cm	2cm
Time to calculate	4862ms	2313ms
Armpit height	139.7cm	138.7cm

We made two improvements in terms of finding the actual position of the armpit quickly and accurately.

In order to narrow the recognition area, we add 3 isolation plane.
1. Up isolation plane: normal direction is +Y, the initial position is the height of the neck.
2. Low isolation plane: normal direction is -Y, the initial position is the height of the crotch.

3. Center isolation plane: normal direction is -X, the initial position is the body center line in front view.

This can remove vertices on the other half of the body, vertices above neck and vertices below crotch. Thereby, the number of vertices that need to be calculated and identified can be reduced, thus improving the efficiency. The original isolation plane is called main isolation plane, as shown in Figure 3.

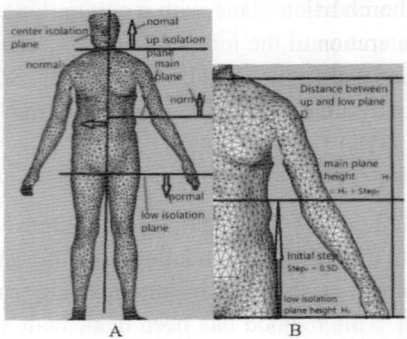

Figure 3. A: add up, low and center isolation plane, B: valid area isolated by three isolation plane.

Using method similar to the binary search method to dynamically adjust the search range,[4] so that the algorithm complexity can be reduced from O (n) to O (logn).

The specific approach to improving the design is as follows:

Set a feedback function, int SIG (float height), which uses the propagation algorithm to calculate whether the arm part and the body part are connected to each other or not at the specified height, if they are connected returns 1, else return -1, means they are not connected.

The initial step ($step_0$) takes half of the distance between the up and low isolation plane. The sign is positive, means isolation plane moving upwards. Otherwise, the sign is negative and moving plane downward. The initial height H_1 of the main isolation plane is the height of the low plane H_0 plus $step_0$, the middle of up and low isolation plane. The following step calculation formula:

$$step_{n+1} = 0.5*SIG\ (H_n) * SIG\ (H_{n+1})\ *step_n \tag{1}$$

Formula for calculating height:

$$H_{n+1} = H_n + step_n \tag{2}$$

After that, we multiply the feedback function of the last iteration H_n and the current iteration H_{n+1}, If the return value of SIG function of the two iterations are the same, both are +1 or -1 (the product is 1, indicating that the height of the isolation plane in the last iteration and this iteration are all above or below the actual armpit height), then the next step will be along the direction of the previous step forward half of previous step. Else the return value of SIG function of the two iterations are different, (the product is -1, indicates the actual height of the armpit is between the height of isolation plane in last iteration and height of isolation plane in this iteration), then the next step will be along opposite direction of the previous step, march half of previous step.

For example: in iteration n, at the height H_n, SIG function returns -1 (Arm and body are not connected), but in iteration n+1, at height H_{n+1}, SIG function returns +1 (Arm and body are connected). Based on the above formula we derive:

step $_{n+1}$ = 0.5*SIG (H_n) * SIG (H_{n+1}) *step$_n$ =>
step $_{n+1}$ = 0.5*(-1)*(+1)*step$_n$ =>
step $_{n+1}$ = -0.5*step$_n$

The height of such a new isolation plane is H_{n+2} = H_{n+1} + step$_{n+1}$ = H_{n+1} - 0.5*step$_n$ = H_n + step$_n$ -0.5*step$_n$ = H_n + 0.5*step$_n$. In order to reduce the calculation of propagation algorithm, the height of the up and low isolation plane will be updated, the update rules are as follows:

If SIG (H_{n+1}) return 1 then update up isolation plane, set height as H_{n+1},

Else SIG (H_{n+1}) return -1 then update low isolation plane, set height as H H_{n+1}. The calculation process is shown in Figure. 4A

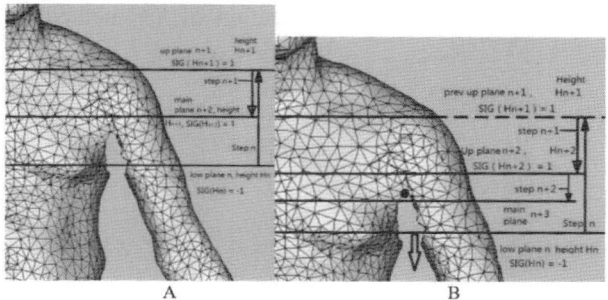

Figure 4. A: iteration n+2, B: iteration n+3.

Call SIG (H_{n+2}), if it return 1 (arm and body are connected), armpit point is below current isolation plane, base on formula (1), we can get step$_{n+2}$ = 0.5*step$_{n+1}$, H_{n+3} = H_{n+2} + step $_{n+2}$ = H $_n$ + 0.25*step $_n$. SIG (H_{n+3}) return -1, then H_{n+3} is below armpit point, This allows more accurate positioning of the armpit point height between H_{n+2} and H_{n+3}, as shown in Figure 4B.

Repeat the above steps until step$_m$ is less than the threshold we set, usually it's 0.25cm. This will be able to more accurately find the armpit position of 3D human mesh.

The core of the pseudo-code of the algorithm is as follows:
```
while (step > threshold) {
    step = 0.5*sig0*sig1*step;
    if (sig1 > 0) upPlaneHeight = height;
    else downPlaneHeight = height;
    height += step;
    sig0 = sig1;
    sig1 = SIG(height);
}
```

4. Propagation Method Optimization Results

After optimization application of the Recognition Method by Propagation Algorithm, we did the test and the result is quite satisfactory.

To test the armpit arm as an example. The test selected two typical 3D human mesh as the case object; and used two different step lengths of 1cm and 2cm as the setting for comparative test.

Mannequin1 (M1), with incomplete mesh on armpit part, it has 6,170 vertices and 12,169 facets. Mannequin2 (M2), fine mesh with partial overlap at armpit, it has 69,409 vertices and 138,814 facets.

The following Table 2, Table 3 is a test record of the effectiveness of the use of the original algorithm and the optimization of the method of propagation algorithm.

Table 2. Comparison of two calculation methods on time efficiency.

Mannequin	Mannequin M1	Mannequin M2
Original method (Step = 1cm)	4862ms	129026ms
Original method (Step = 2cm)	2313ms	63842ms
Optimized method	442ms	3632ms

Table 3. Comparison of two methods on accuracy.

Mannequin	Mannequin M1	Mannequin M2
Original method (Step = 1cm)	139.72cm	133.47cm
Original method (Step = 2cm)	138.72cm	132.47cm
Optimized method	140.31cm	133.88cm
Actual height	140.54cm	134.07cm

The test data proves that the optimized Recognition Method by Propagation Algorithm, both for the incomplete mesh model and the fine mesh model, have the same effect and regularity. They are about 4-15 times faster in recognition speed and about 7-13 times in accuracy, greatly improving the recognition efficiency of measurement calculation and the accuracy of positioning.

5. Conclusion

The optimized 3D mesh Recognition Method by Propagation Algorithm not only preserves the functions and features of the original method, but also greatly improves the speed and accuracy of the method for measurement, identification and accurate modeling. At the same time, the optimized algorithm has been applied to the professional 3D body measurement modeling software. They are used in manufacturing applications to more accurately and quickly find the arms, armpits and crotch parts.

Further, the optimized method can also be applied to various creative, experimental and production platforms and used in the measurement and modeling of various types of models, molds and 3D printing. And, it will greatly increase the speed and accuracy of their search measurements when applied to similar search methods that use fixed steps to find the key vertices and planes.

References

1. ISO 20685, 3-D scanning methodologies for internationally compatible antoropometric databases [S].
2. Shih-Wen Hsiao and Rong-Qi Chen, A study of surface reconstruction for 3D mannequins based on feature curves. Computer-Aided Design, Volume 45, Issue 11, Pages 1426–1441(November 2013).
3. Y. Chen, Body Part Recognition Method of 3D Human Polygon Meshes Based on Diffusion Algorithm. Microcomputer Applications Vol. 2, No. 4, (2016).
4. Hüseyin Hakh and Harun Uğuza and Tayfun Çay, A new approach for automating land partitioning using binary search and Delaunay triangulation. Computers and Electronics in Agriculture, Volume 125, Pages 129–136 (July 2016).

The application of process modeling in denim manufacturing*

Zhenglei He,[†] Sebastien Thomassey and Xianyi Zeng

ENSAIT, GEMTEX – Laboratoire de Génie et Matériaux Textiles, F-59000 Lille, France

Danying Zuo and Changhai Yi

Wuhan Textile University, 1st, Av Yangguang, 430200, Wuhan, China

This paper aims at briefly overview the previous use of process modeling in denim manufacturing sectors, specifically, dyeing and finishing processes. A collection of relevant works is introduced, such as modeling the dyeing process for predicting the depth of shade, the effects of alkali reductive stripping process or cellulase washing process on color properties, and certain physical properties. A specific case in regard to modeling ozone fading denim by artificial neural networks (ANN) was studied as an example at the end, the results simply revealed that modeling process using soft computing techniques is capable to accurately predict the targeted outputs which is obviously promising and potential to make a difference in the future development of denim manufacturing.

Keywords: Process modeling; denim manufacturing; ANN; dyeing; finishing.

1. Introduction

1.1. *Denim manufacturing*

Denim jeans is one of the oldest and most popular fabrics that has been generated for nearly 150 years with the widest acceptance from different age groups over the world. Nowadays, taking advantage of the worn and vintage look achieving from denim finishing, denim is continuously "young" and irreplaceable in the fashion industry [1]. By and large, denim is known as the durable 3/1 warp faced twill cotton fabric with indigo blue dyed warp yarns and gray cotton weft yarns. Resulting from the complicated structure, denim has a rather long manufacturing process which may include but is not limited to the most typical textile processes from fiber spinning to fabric weaving or clothing sewing. While particularly, the combination of dyeing and finishing processes is the soul of denim owing to the

---

*This work is supported by China Scholarship Council (CSC, Project NO. 201708420166) and financially supported by GRAISyHM et Région Hauts de France.
[†]Correspondent author: zhenglei.he@ensait.fr

most important feature of the vintage look created by indigo dyeing on denim warp yarns and finishing processes on denim garments.

1.2. Demands on modeling denim dyeing and finishing

Different from most of fabric dyeing process for other types of textile, denim dying only involves the warp yarns before weaving with undyed weft yarns, and the dyestuff of indigo is quiet spectacular that it needs a reduction process before dyeing as it is an insoluble (in water) solid pigment. Reduction is converting indigo into the water-soluble form known as leuco-indigo, a soluble form of indigo solution. It would be used to color the denim warp yarns in terms of multiple dipping and air oxidation processes to achieve the indigo blue after the restoration of the original form of indigo.

Denim finishing aims at giving a considerable value-added price to denim products by obtaining faded or other trendy aesthetic, promoting sensory quality and comfortability etc. It is also called denim washing due to the fact that most of the processes are realized by washing treatment, such as stonewash, chlorine bleaching or soften treatments and so on.

These processes, apparently, are complicated that mostly applied relying on the expertise and experience as well as a specific number of trails and errors [2]. The application is difficult to control, the process related factors and numerous possibilities of denims in multiple structures and diverse features have nonlinear and hardly-understood relationships and effects on associated product properties, which may prevent the accessibility and development from the wider use of these processes. Hence, it is necessary to systematically study the effects of and built a well-constructed model regarding denim dyeing and finishing processes taking into account of process parameters and the features of raw materials on the denim properties [3].

1.3. Process modeling techniques

Modeling techniques are plenty and well developed in recent decades. While as the fact that the correlation between the wanted inputs and outputs in manufacturing process is hard to be presented, the applicable techniques for process modeling in this field currently mostly concentrated on multiple linear or nonlinear regression models and artificial neural networks (ANN) methods as well as Fuzzy theory and a range of mathematical methods.

1.4. Organization of this paper

This paper overviewed the use of process modeling in denim dyeing and finishing and associated sectors, that mainly contains the applications and the architecture for modeling denim dyeing and finishing processes. A specific case study in modeling ozone fading reactive dyed denim using ANN would be released at the end for a broader comprehension of this work.

2. Modeling Architectures

2.1. Predicting color properties

In general, denim dyeing and the others color processes in denim finishing such as cellulose washing, laser carving, and chemical treatments are suggested to affect the color properties of denim. Meanwhile, ANN was the most popular method used to predict color properties in the area, but most of the previous studies were about trying to alternate Kubelka-Munk theory [4].

Senthkumar and Selvakumar, 2006 [5] made a difference as the beginning for modeling dyeing process to predict and achieve the expected depth of shade in dyeing process using a back propagation neural networks fed by six inputs (K/S values of undyed samples and K/S value for dyed samples after rinsing, dye fixed ratio, percentage shades, concentrations of NaCl and concentrations of Na_2CO_3). The time for primary exhaustion, and time for dye fixation were targeted as outputs with the use of Binary sigmoid activation function and an optimized structure of three hidden layers coupled with 9 neuros in each hidden layer which was optimized on the basis of experiments. 45:6 sets of data for training and texting samples were used separately and had resulted in a high performance of the 1% average error.

In a study of Balci et al., 2008 [4], the use of a Levenberg-Marquardt Algorithm based ANN was located on the alkali reductive stripping process for predicting L^* and ΔE of stripped cotton fabrics. After an optimization on the numbers of inputs and nodes and the value of mean square error (MSE) for stopping training, eight inputs such as type of the reactive dyes and type of reducing agents, owf %, original L^*, concentration of the reducing agents, caustic, process temperature and the presence of the leveling agent were given to the model with 85 nodes in one hidden layer and MSE value of 0.01 (for quieting training) to predict the L^*, whereas 2 more parameters of a^* and b^* were inputted additionally for the prediction of ΔE using 70 nodes in one hidden layer and MSE value of 0.001. The achieved R was 97%, and only 1.2 % error happened in the real use via inputting new parameters.

Cellulase is an enzyme for the degradation of cellulose materials, which has been used in denim washing process for years, and it is one of the most commonly used methods to achieve color fading effect as well as fabric softness for cotton denim. While it's hard-controlling and water-consuming etc. properties seemly are preventing it from a wider application as the industry developed rapidly. Modeling the process and predicting the color properties of K/S value and CIELab values by the inputs of treating time, temperature, pH, mechanical agitation and fabric yarn twist level using ANN, the work projected by Kan et al., 2013 [6] illustrated the potential of the wide application in denim color processes. On the basis of the parameter selecting and model structure optimization methods used in previous two studies, Kan et al., constructed a similar model and successfully verified the predicted accuracy of the model in this issue.

2.2. Predicting physical properties

Apart from color properties, the quality of denim fabric and garment also rely on their physical performance. An early well-built expert system revealed that ANN was promising and potential for modeling complex and nonlinear processes in order to predict fabric physical properties [2]. The definition of quality of fabric has been disputed for years, the satisfaction of customer in all respects is the one most accepted but combines a wide range of properties into one objective notion. In the observation of related works, the combined use of desirable function and ANN [7-9] has shown certain advantages, which may be considered in a near future to be conducted on developing the overall evaluation of denim physical properties.

3. Specific Case: Modeling Ozone Fading Denim Using ANN

Color fading is an essential finishing process for denim products, but it conventionally was achieved by chemical methods which have a high cost and water consumption, as well as a heavy burden on the environment. Instead, ozone treatment is an advanced finishing process employing ozone gas to bleach or color fade denim without water bath and consequently it causes less environmental issues. However, the complicated and nonlinear relationship between its parameters and color fading effects are still not clearly known, in other words, a model that can predict the engineering application effect is rather necessary.

According to the overviews above, it is easy to find out that ANN was the most frequently used one for predicting the denim color properties, as a result, we developed an ANN based process model in order to predict the ozone fading effect on reactive dyed denim as a specific case study.

3.1. Experimental

3.1.1. Material

Desized grey cotton (3/1 twill, 325.7g/m^2) supplied by Shunfu (Hubei, China) and the bifunctional reactive dyes (containing fluorotriazine, commercial quality) of Reactive Blue FL-RN provided by Color Root (Hubei, China) were used in the experiment. Chemicals of sodium hydroxide, hydrogen chlorine, sodium metasilicate nonahydrate, 30% hydrogen peroxide, sodium sulfate, sodium carbonate, potassium iodide (analytical grades, supplied by Sinopharm Limited, China) and OP-20 (a nonionic surfactant, chemically pure, provided by Tianjin Guangfu, China) were applied in this study.

3.1.2. Methods

In the experimental application of ozone fading dyed denim, we employed an equipment (displayed as Figure 1) to continuously feed and flowed the ozone gas (used gas would be output as exhaust) with a dosage of 137±3 mg/L·min in order to fade the dyed denim. Samples dyed by Reactive Blue FL-RN were cut to pieces and put evenly in the reactor to be treated. The fading performances at different pH1-13 and temperatures from 0-80°C with different water content (0-150%) over varied duration from 0-60 minutes were investigated.

Depending on Kubelka-Munk theory [10], it is known that the K/S reveals the color depth of samples. Consequently, we evaluated the color fading effect of ozone treatments on denim fabrics using K/S values (average of three measurements) tested by Datacolor 110 spectrophotometer (Datacolor, USA).

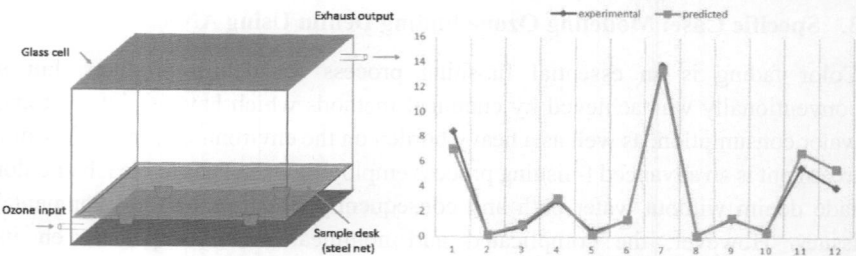

Figure 1. The reactor equipment setup.   Figure 2. Predicting accuracy on certain network.

3.1.3. Neural networks architecture

This modeling was carried out using the Neural Network Toolbox of MATLAB R2017a mathematical software. In this work, a three-layers-network was fed by four input variables of pH, temperature, water content and treating time to output the color fading performance which was characterized by K/S value, i.e. a single output variable.

A total number of 121 experimental datasets were divided into three parts for training (70%), validation (15%) and testing (15%) respectively. Sigmoid function, the most frequently used activation function, was chosen to transfer the variables to the hidden layer. Meanwhile, a linear function was used for the output transfer. An optimization of the nodes was conducted by a series of trails in terms of the network with different nodes in the hidden layer from 4-16.

3.1.4. Network training

The network was trained by scaled conjugate gradient backpropagation (*trainscg*) and levenberg-marquardt (*trainlm*) algorithm separately within 1000 epochs as the maximum, and a comparison of the results with different nodes would be introduced in the part of discussion.

3.2. Results and discussion

The evaluation of the model's predicted accuracy was presented by MSE (Mean Square Error), MAE (Mean Absolute Error) and R (Correlation coefficient) depending on the equation (1)–(3).

$$MSE = \frac{1}{n}\sum_{i=1}^{n}(e_i - p_i)^2 \qquad (1)$$

$$MAE = \frac{1}{n}\sum_{i=1}^{n}|e_i - p_i| \qquad (2)$$

$$R(e,p) = \frac{\sum_{i=1}^{n}(e_i-\bar{e})(p_i-\bar{p})}{\sqrt{\sum_{i=1}^{n}(e_i-\bar{e})^2 \cdot \sum_{i=1}^{n}(p_i-\bar{p})^2}} \qquad (3)$$

where e_i is the real experimental results, whereas p_i is the ANN predicted output.

Table 1 demonstrates the comparison of MSE, MAE and R on the network with different nodes in a single hidden layer trained with different functions of *trainscg*, *trainlm* respectively. It is noted that neural networks holding larger numbers of nodes in the hidden layer basically has higher prediction accuracy It is found additionally that the nets trained by the algorithm of *trainlm* generally performs better than the ones trained by *trainscg*, which may attribute to that the former requires more on memory to enable it to improve the fitting accuracy to be slightly higher. However, this trend is not correct to the network all the time, for instance, the nets with nodes of 4, 5, 10, and 16 were not in line with this tendency. Which indicated that related works should be careful in the selection of nodes as the overfitting happens when nodes is excessive. On the other hand, the gaps among networks without the models possessing nodes of 4, 5, 10, and 16 were not significant, so the concern should be more addressed on time-saving or other network architecture elements wile process modeling in this issue.

A contrast between a set of predicted data and real experimental data was carried out for illustrating the predicting accuracy in a 14 nodes network trained by *trainlm*, as an example, was given in the Figure 2, and the result was clear that the prediction error was close to zero.

Table 1. Comparison of MSE, MAE and R of nets with different nodes in hidden layer and trained by different functions.

		4	5	6	7	8	9	10	11	12	13	14	15	16
Train scg	MSE	3.99	2.22	3.24	2.20	1.08	0.88	0.66	1.26	1.57	1.30	1.02	1.35	1.07
	MAE	1.12	0.86	1.02	0.88	0.67	0.67	0.56	0.80	0.72	0.62	0.60	0.70	0.55
	R	0.896	0.951	0.907	0.934	0.971	0.973	0.983	0.964	0.958	0.965	0.970	0.966	0.977
Train lm	MSE	6.64	2.47	2.93	1.60	1.07	1.22	1.93	0.52	1.25	1.11	0.50	1.63	1.13
	MAE	1.05	1.02	0.93	0.70	0.59	0.70	0.79	0.39	0.60	0.63	0.47	0.68	0.52
	R	0.885	0.932	0.924	0.962	0.977	0.970	0.954	0.986	0.973	0.973	0.985	0.960	0.969

4. Conclusions

The present paper briefly introduced the denim manufacturing, especially, dyeing and finishing processes, and concluded the need of process modeling in this respect. A collection of related applications of process modeling were overviewed as well, such as modeling the dyeing process for predicting the depth of shade, the effects of alkali reductive stripping process or cellulase washing process on color properties, and certain physical performances and so on. A case study in terms of modeling ozone fading denim by ANN was illustrated at the end, it is found that modeling process using soft computing techniques is capable to accurately predict the targeted outputs which is obviously promising and potential to make a change in the future development of denim manufacturing.

Acknowledgments

Authors acknowledge the financial support from GRAISyHM et Région Hauts de France, and the first author would like to express his gratitude to China Scholarship Council for supporting this study (CSC, Project NO. 201708420166).

References

1. R. Paul, *1 - Denim and jeans: An overview*, in: *Denim*, Woodhead Publishing, 2015, pp. 1-11.
2. J. Fan, L. Hunter, *Text Res J.,* **68** 680-686(1998).
3. R. A. Jelil, X. Zeng, L. Koehl, A. Perwuelz, *Eng. Appl of A.I.*, **26** 1854(2013).
4. O. Balci, S.N. Oğulata, C. Şahin, R.T. Oğulata, *Fiber. Polym.*, **9**, 604 (2008).
5. M. Senthilkumar, N. Selvakumar, *Dyes pigm*, **68**, 89 (2006).

6. C.W. Kan, W.Y. Wong, L.J. Song, M.C. *J. Text.,* **2013**, 3 (2013).
7. H. Souid, M. Cheikhrouhou, *J. applied Sci.*, **11**, 3204, (2011).
8. M. Slah, H.T. Amine, S. Faouzi, *J. Text Inst.*, **97**, 17 (2006).
9. H. Souid, A. Babay, M. Sahnoun, *J. Comp Tech and Appl,* **3**, 356 (2012).
10. L. Yang, B. Kruse, *J. Optical Soci Ameri*, **21**, 1933 (2004).

A proposal of a rapid evaluation and analysis system of fashion design using expression analyzer and eye tracker[*]

Yu Chen[1], Xianyi Zeng[2], Xie Hong[1] and Zhebin Xue[3]

[1]*Clothing Institute, Shanghai University of Engineering Science*
333 Long Teng Road, Shanghai 201620, China
[2]*ENSAIT, Gemtex, 2. allée Louise et Victor Champier Roubaix, 59000, France*
[3]*School of Textiles and Clothing, Jiangnan University*
Wuxi, Jiangsu Province 214122, P.R. China

> Finding the most favorite garment is an important criterion in fashion design analysis. It is often taken into account in the process of fashion design expert systems. However, the existing questionnaire methods cannot provide the most favorite style, which is very important references information for new season garment fashion design decision making. They need a complex questionnaire form and some questions and answers are not very intuitive. In this paper, we propose a new method for selecting consumer wanted garment using expression analyzer and eye tracker. This two equipment can catch the expression and sight focus of consumer, which can quickly reflect the consumer's most instinctive preferences. It can also be applied on E-commerce's recommended system.

1. Introduction

1.1. Foreword

In the Internet age, fast fashion and fast spending [1] are becoming more and more mainstream. Therefore, the fashion design put forward higher requirements. In consumer-oriented fashion design, how to find an eye-catching design is a very crucial step. Because by analyzing this design can help fashion designers understand the consumer favorite style and design elements. Thus in the design of the new season clothing, the fashion designer can learn from this style and elements.

How to quickly find the consumer's favorite design, this is the primary issue, which can be collected through the questionnaire survey. A classic questionnaire is as follows:

---

[*]This work is supported by The Program for Professor of Special Appointment (Eastern Scholar) at Shanghai Institutions of Higher Learning.

Table 1. A typical questionnaire form.

age	income		
Your favorite style shape			
A: shape A	B: shape X	C: shape H	D: shape O
Your favorite color			
A: red B: blue	C: green	D: pink	E: gray

However, the answers to the questions on the questionnaire are usually closed, the answers are mostly multiple-choice questions, Even the open question, the consumer's description is also uncertain. Sometimes respondents are not always willing to take the time to do the questionnaire, some consumers may not truthfully answer the question. Another research method is choiring favorite design form a set of clothing renderings. However, this method may cause aesthetic fatigue. In this paper, we propose a new way to find the consumer's favorite clothing through the expression analyzer and eye tracker. This method judges whether the consumer is interested in the clothing by tracking the facial micro-expressions [2] and the position of the focus of the eyes when the consumer browses a large number of clothing effect figures. By analyzing the focus of the eye [3] we can know which part of the clothing to make consumers excited. In conjunction with other consumer operation actions, such as how long the picture stays and whether to look back and so on, this method can find the consumers' favorite clothes from a large number of candidate clothing samples. The following sections of this paper are arranged as follows. Section 2: the specific steps of the experiment will be introduced. Section 3: in this section, we will discuss the possible problems and feasible solutions in the implementation process.

2. Proposition of Frame of System

2.1. *Overview*

Figure 1 shows the frame of rapid evaluation system.

As shown in the figure, there are four types of main data produced by consumers when watching a clothing rendering: eye movement data, physiological data, operation action data and micro-expression data. The eye movement data is captured and processed by the eye tracker to obtain the eye movement focus diagram. The micro-expression data is captured and analyzed by the micro-expression analyzer, and the micro-expressions are classified and marked the degree of the expression. Eye movement data, micro-expression classification and degree, physiological data, and operation action data are used as input for the deep learning network. The output of the deep learning network is the consumer's degree of preference for the clothing being watched.

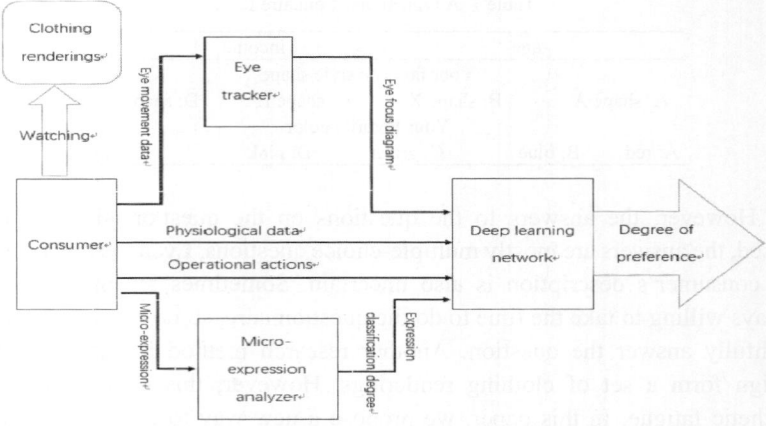

Figure 1. Frame of rapid evaluation system.

2.2. Clothing renderings selection and processing

It is recommended to use a single style of clothing, but the style can have a certain degree of change, this paper uses a young women's shirt as a research object.

1. After you have collected the clothing renderings, you need to normalize them: cut the picture to a fixed size and ratio and try to remove things that are not related to the design of the garment, such as trademarks, models' faces, etc.
2. Let the experts evaluate these clothing renderings and rank them according to their preference. Divide them into 5 levels. A, B, C, D and E. A is the favorite, E is the least favorite.
3. Clothing renderings will be viewed by the consumer according to the expert's recommendations in a certain combination order, e.g. (E, E, E, A, E), (A, A, A, E, A). With this combination of contrasts, it is easier to capture the micro-expressions and physiological data that consumers like or dislike.

2.3. Deep learning network input and output

From Figure 1, we can see that there are four types of input for deep learning networks:

1. The eye focus diagram output by the eye tracker is shown in Figure. 2. The darker (red) places have higher attention, and the lighter (green) places have less attention. The uncolored areas indicate Not concerned.

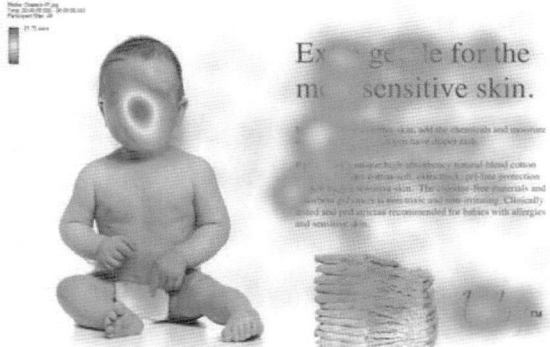

Figure 2. Eye focus diagram.

2. Physiological data includes: pulse, respiration, skin electricity, etc.
3. Operation action includes: page stop time, whether to turn over again, scale picture or not and so on.
4. Classified micro-expressions and their degree: the first expression and degree when consumers see the clothing renderings. Figure 3 shows an example of a micro-expression analyzer

Figure 3. Micro-expression analyzer and its output.

The output of the deep learning network is the consumer's preference for clothing renderings, including the overall degree of preference, the degree of preference of each design element of the garment, such as the degree of preference of the collar, the degree of preference of the sleeve, and so on.

2.4. *Deep learning network design*

Figure 4 shows a more detailed design of the deep learning network.

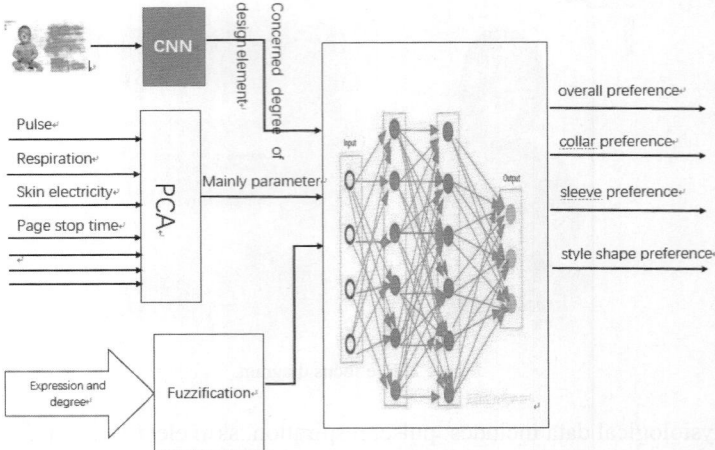

Figure 4. Detailed design of the deep learning network.

- The eye focus diagram is firstly input to a CNN convolutional neural network [4]. The output of the CNN neural network is the attention degree of individual design elements of garment.
- A series of physiological data and operation action data are processed by PCA dimension reduction to obtain low-dimensional parameter inputs.
- Classified micro-expressions and their degree will be processed by fuzzification operation. After this process, we get fuzzy vector like this {0, 0, 0.3, 1, 0.4} [5].

The above parameters will be used as inputs to neural network. The output of the neural network has been described previously.

2.5. Deep learning network training data set

Training data set is very important for neural network. This paper proposes two methods for optimizing training:
- According to the expert's suggestion, consumers watch clothing renderings sequence which is defined by expert. This will help the device capture the more obvious expressions and eye movements of consumers.
- After the consumer has finished viewing the clothing renderings, the psychologist interviews the consumer about two questions: 1, the first feeling when seeing the clothing renderings. 2, the final degree of preference of the clothing.

The above two optimization tasks can help find a more representative training set.

3. Discussion

In the last chapter, the design scheme and the basic experiment steps were put forward. This chapter puts forward some discussion about some of the problems that may be encountered in this rapid evaluation method. The problems that may be encountered are divided into two categories: one is technical and the other is about popularized.

3.1. *Technical problems*

- Normalization of micro-expressions: each consumer has a different degree of response to the received information. Some people may respond to what information is relatively flat, while others are more exaggerated, and therefore need to do a normalization operation for each consumer.
- Focus diagram: The hotspots on the diagram are not surely interest to consumers, but may be routine or disgusting.
- Clothing renderings play sequence: with different playback sequences, the consumer's micro-expression and eye movement response will have different results. This requires discussion with psychologists.

4. Future Application Proposition

This method has been further developed: through the front camera of a computer or smart phone to collect the consumer's micro-expressions and eye movement data, the pulse and breathing can be collected via a smart watch and analyzed through software. In this way, e-commerce can determine the degree of consumer preference for the product through the micro-expressions and eye-movement data of consumers when they browse their products. Of course, such application scenarios are more complex and have more interference factors. At the same time, they may also cause some legal and privacy issues.

5. Conclusion

This paper proposes a rapid evaluation method based on a micro-expression analyzer and an eye tracker, and proposes basic experimental steps. This method can quickly, directly, and less interfere get the consumer's degree of preference for clothing renderings. And the same method can also evaluate other products. If the entire rapid evaluation system can be integrated into a smart phone or other smart wearable device, e-commerce will be able to get the most direct feedback from the product, thereby improving the product or taking a new marketing strategy.

Acknowledgments

Research supported by The Program for Professor of Special Appointment (Eastern Scholar) at Shanghai Institutions of Higher Learning.

References

1. G.P. Cachon and R. Swinney, *The value of fast fashion: Quick response, enhanced design, and strategic consumer behavior*. Management science, 778-795 (2011).
2. S. Polikovsky, Y. Kameda and Y. Ohta, Facial micro-expressions recognition using high speed camera and 3D-gradient descriptor. (ICDP 2009), 2009 page 16.
3. S. Polikovsky, Y. Kameda and Y. Ohta, Eye tracker data quality: what it is and how to measure it. ISBN: 978-1-59593-982-1.
4. H. Li, Z. Lin, X. Shen and J. Brandt, A convolutional neural network cascade for face detection. Computer Vision and Pattern Recognition (pp. 5325-5334). IEEE, (2015).
5. Zadeh L. A. Fuzzy Logic [J]. Computer, 21(4):83-93, (1988).

Explorative multi-objective optimization of marketing campaigns for the fashion retail industry

H. Sundell*, T. Löfström[+] and U. Johansson[+]

Department of Information Technology, University of Borås, Sweden
hakan.sundell@hb.se

[+]*Department of Computer Science and Informatics, Jönköping University, Sweden*
{tuwe.lofstrom, ulf.johansson}@ju.se

> We show how an exploratory tool for association rule mining can be used for efficient multi-objective optimization of marketing campaigns for companies within the fashion retail industry. We have earlier designed and implemented a novel digital tool for mining of association rules from given basket data. The tool supports efficient finding of frequent itemsets over multiple hierarchies and interactive visualization of corresponding association rules together with numerical attributes. Normally when optimizing a marketing campaign, factors that cause an increased level of activation among the recipients could in fact reduce the profit, i.e., these factors need to be balanced, rather than optimized individually. Using the tool we can identify important factors that influence the search for an optimal campaign in respect to both activation and profit. We show empirical results from a real-world case-study using campaign data from a well-established company within the fashion retail industry, demonstrating how activation and profit can be simultaneously targeted, using computer-generated algorithms as well as human-controlled visualization.

Keywords: Association rules; marketing; visualization; Pareto front.

1. Introduction

A central part of industry management is wise decision-making, which in turn requires valid information. Information is needed both to understand the current situation and to predict how future changes would affect the goals, i.e., some kind of model. The task of creating a reliable predictive model is inherently complex, and will be even more difficult if aiming to improve towards several goals simultaneously.

Besides designing attractive and useful clothes, it is important for the fashion retail industry to reach prospect customers with relevant information and attractive offers. Therefore, marketing is typically done in the form of marketing campaigns where each campaign is carefully designed and personalized, based on a number of parameters. The success of the campaigns is carefully surveyed and measured on individual as well as on a total level. Based on analysis of historical data, the campaigns are then continuously re-designed and re-evaluated. The main goal of campaign optimization is to achieve as many positive responses to the campaign as possible, while achieving as high profit as possible on each order that the activated customers make. The level of how attractive each campaign is in respect of triggering customers to make an order is called *activation*, preferably measured in percentage in relation to how many prospect customers that were targeted, and the *profit* is measured in how much money that is made on each order, on average during the campaign.

In this paper, we have performed an empirical study on a well-established company within the fashion retail industry. The area of interest is optimization of the marketing campaigns, where each campaign is performed during a limited time period and the effects are continuously surveyed and measured. Each campaign is highly customized using several attributes that need to be carefully selected, to both address which subset of the prospect customers that are targeted for receiving the offer as well as how the campaign should be designed in more detail. Currently, this is done manually, where the values of the attributes are changed according to how previous campaigns have performed and using human experience. The categories of the attributes that are immediately possible to change for each campaign are as follows:

- **Campaign Type**. This is one attribute and defines how the discounts are applied.
- **Discounts**. These are several attributes that define the actual values of discounts related to the specific campaign type.
- **Add-ons**. These are several attributes of true/false type, with add-ons to the campaign offer to the customer, e.g. free delivery, or free gift.
- **Requirements**. These are several attributes of true/false type, e.g. offer limitations in time, minimal order size or minimal order value to buy.
- **Customer Type**. This is one attribute and defines how active the customer is known to be during the last period of time.

We have gathered and analyzed comprehensive data related to recently performed marketing campaigns from a well-established fashion retail company. The data set contains more than one million orders together with attribute data describing all conditions of the campaign resulting in an order. The data includes orders from several hundred uniquely designed campaigns, over a time period of two years. Some of the data has been masked or modified in order to preserve company secrets and integrity in the presentation.

In Section 2, the technique and tool used to perform the analysis is introduced. In Section 3, we explain specifically how the tool can be used for multi-objective optimization. The detailed work process of our case study aimed for multi-objective optimization is described in Section 4. In Section 5, we summarize our observations and state our conclusions.

2. Market Basket Analysis

Market basket analysis is a form of data mining typically done within the retail industry. Based on historical customer behaviour, in the form of so called *baskets*, describing which items are bought together, knowledge in the form of *association rules* can be created. Due to the extreme amount of possible combinations of all items in the baskets, the task of mining association rules is very computationally demanding, and one of the most popular and efficient algorithms is called the *Apriori algorithm*.[1]

In this paper we are investigating the possibility to apply market basket analysis within the fashion retail industry, and specifically trying to optimize the marketing campaigns toward several goals simultaneously. There are certain similarities between a market basket with varying sets of bought items and marketing campaigns with varying conditions which makes market basket analysis an interesting tool to use. To our help, we utilize a previously developed tool, that enables an exploratory work process with guidance and support from the tool in the form of extensive search as well as visualization abilities.

The tool called VISEART, which is described elsewhere,[2] is a standalone application capable of both efficient data mining of association rules and versatile exploration of the identified rules with real-time interactive visualization. The main purpose of the tool is to be a guidance support in decision making. Even though the tool only gathers historic data and does not, in a strict sense produce predictions, the analyzed data together with the versatile exploration possibilities enables simulations to be done.

3. Multi-Objective Optimization

The task of optimizing multiple objectives simultaneously can be challenging in situations where the objectives are dependent on each other. For example, increasing one objective might cause a decrease of another objective. There are multiple approaches suggested in the literature, whereas we here will focus on an approach combining algorithmic optimization and visualization.

The basis for both approaches is that we are connecting the association rules, or rather the underlying frequent itemsets, with the corresponding average values of all numerical attributes of interest for optimization, which in this case are profit and activation. The VISEART tool is capable of mining these analog values simultaneously to finding the frequent itemsets.

3.1. *Algorithmic Approach*

A situation where one objective can not be increased further without having to decrease other objectives is said to be *Pareto-optimal*. Typically, there exist multiple solutions to an optimization problem that are Pareto-optimal. The set of Pareto-optimal solutions is called the *Pareto front* and there exists several algorithms for its identification.[3] Pareto optimization can, of course, be done in more than two dimensions if necessary.

The VISEART tool can identify the Pareto front, and the candidate solutions can then be further examined manually, and the search scope be narrowed by applying constraints on minimal support etc. including the allowed interval of the objectives of main concern.

3.2. *Visualization Approach*

The Pareto front of two objectives can be visualized in a straightforward manner, by plotting the corresponding values on the x- and y-axis, and drawing connected graphical lines. As we are actually plotting association rules related to each point, each point can be individually examined for more constraints than only the two objectives of main interest. Besides illustrating the Pareto front, it can also be benefiting to plot all other association rules that technically are dominated by the Pareto front, but might offer something else in the terms of other required values of the discrete attributes, i.e., optimality of some attributes can be sacrificed on the behalf of more attractive values on the requirements for achieving those optimal values.

4. Work Process

There are several steps required before, during, and after running the VISEART tool, in order to be able to fully utilize its potential in a decision making context. In the scope of our case study with optimization of marketing campaigns, we have performed the following steps:

(1) **Data Structuring.** We identified which attributes that can serve as input respectively output, seen from a decision making point of view. The activation and profit attributes are numeric — association rules are defined for discrete values — these need therefore to be handled separately from the attributes that are to be analyzed for frequent itemsets. However, we included them in the frequent itemset generation as well by applying *binning* to create new discrete attributes, with discrete values ranging between the labels { Very low, Low, Medium, High, Very high }.

(2) **Data Cleaning.** For some of the attributes it was necessary to handle quality issues such as missing values. This is an important step for the quality of the resulting analysis.

Table 1. Example of data rows with values, that need to be transformed into baskets of virtual items so that each possible value can be uniquely identified.

Campaign Type	Discount 1	Free Delivery	Free Gift	Activation Binned
Ladder	25%	true	false	High
Ladder	40%	false	true	Very high

(3) **Basket Generation.** The marketing campaigns relational data is not directly applicable for frequent itemset generation, as this requires baskets consisting of a set of uniquely identifiable items. Thus, the various values of each attribute in the rows of data need to be transformed into a set of *virtual items*. For some attributes it might be beneficial to group them together, especially if they have true/false values. For example, the data rows in Table 1 could be transformed into $\{I_{11}, I_{21}, I_{31}, I_{44}\}$ and $\{I_{11}, I_{22}, I_{32}, I_{45}\}$ or rather just the IDs listed as $\{11, 21, 31, 44\}$ and $\{11, 22, 32, 45\}$. The actual labels associated with each virtual item and group of virtual item, needs to be stored in separate *translation tables*. See Table 2 for information regarding the generated virtual items that corresponds to the respective values from the data rows in Table 1.

(4) **Frequent Itemsets Generation.** The tool was fed with a set of transactions containing basket data together with an additional relational table with the numerical data representing the activation and profit of each order. Before the mining process was initiated, which could potentially take days for larger data sets, the suitable value for

Table 2. Example of translation tables related to the generated virtual items and the respective identified groups.

Item ID	Item Name	Group ID	Group Name
I_{11}	Ladder	G_1	Campaign Type
I_{21}	25%	G_2	Discount 1
I_{22}	40%	G_2	Discount 1
I_{31}	Free Delivery	G_3	Add-ons
I_{32}	Free Gift	G_3	Add-ons
I_{44}	High	G_4	Activation Binned
I_{45}	Very high	G_4	Activation Binned

the minimum support threshold was estimated. Given the input data size, we decided on a value of 0.1%, which resulted in appr. 750 000 identified frequent itemsets, each having a size of up to 5 items.

(5) **Association Rules Generation.** For each identified frequent itemset, the items can be combined in different ways to create several association rules. In order to reduce the total number of association rules generated, we carefully selected the appropriate subsets of attributes to be used as input and output, called antecedent and consequence in the context of association rules. In this case, we have focused on rules with the binned activation and profit attributes as output and campaign conditions as input.

(6) **Association Rules Analysis.** For each association rule, a number of metrics are computed, e.g., support, confidence and lift. These metrics can be used for evaluation, and also the additional numerical values that have been aggregated during the mining of itemsets. In order to obtain a feasible amount of rules to inspect manually, we applied various filters that ruled out groups of virtual items that appeared to have an insignificant impact on the output attributes. The first natural approach was to try to identify association rules that predict very high activation and very high profit simultaneously. However, most of the identified rules had a confidence of much less than 20%, which means that applying the conditions given in the rule, will more likely cause any other value on activation and profit than very high.

The next approach was to investigate the average values of the aggregated data from the numerical attributes, related to the corresponding association rules. The maximum values identified were 7.43% for activation and 178 for profit. Using the algorithmic approach we could identify 7 points in the Pareto front, with activation ranging from 2.13%

Table 3. Example of a Pareto front for activation and profit related to various combinations of virtual items, showing a subset of the necessary conditions.

Discount 1	Add-ons	Requirements	Other	Activation	Profit
25%	Free Delivery	Order Value	...	2.13%	178
25%	Free Delivery	Order Value	...	2.13%	173
40%	Free Delivery	Order Size	...	2.54%	147
25%	Any		...	4.64%	142
50%		Time Limit	...	6.42%	89
			...	7.43%	42
			...	7.43%	42

to 7.43%, and profit ranging from 178 to 42. See Table 3 for a brief description of the corresponding association rules related to the individual points in the Pareto front. Seen from a multi-objective optimization perspective and acknowledging how rapidly the profit decreases when activation increases, the overall optimum would be an activation of 4.64% and a profit of 142. That particular Pareto-optimal point adheres to a campaign with overall 25% discount and some kind of additions (not specified).

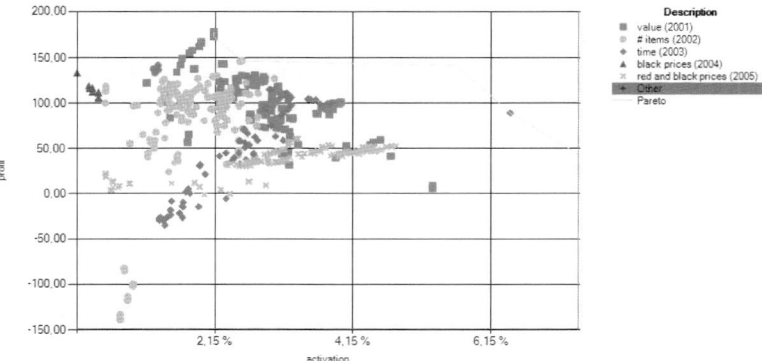

Fig. 1. Visualization of activation and profit with plotted association rules colored by type of requirements. Five different requirements are illustrated; minimal value, minimal number of items, limited time, and different types of price tags limitations.

(7) **Association Rules Visualization.** For a subset of maximum (approximately) 100 000 association rules, it can be useful to visualize the computed metrics as well as the optional numerical attributes together with the respective association rules. By visualizing the rules

in a point-based diagram with two dimensions, interesting clusters or significant characteristics can be visually identified. See Figure 1 for an example of a visualization with color-clustered results using a filter over one of the requirement item group.

Using the visualization approach and selecting various filters that graphically highlight clusters of points containing the same value for some item group, it can be identified that high level of discounts and some kind of add-ons are related to high activation. Typically, the same settings are related to low profit. However, it can be identified, that the decrease in profit could in general be significantly reduced by combining the high level of discount with some kind of requirements.

5. Concluding Remarks

We have in this paper showed how a new tool could be efficiently used for multi-objective optimization of marketing campaigns within the fashion retail industry. The two approaches combined in the tool are: 1) An algorithmic approach using Pareto front identification, 2) A visualization approach with association rules plotted along axes.

In our case study, using campaign data from a well-established company within the fashion retail industry, we selected two very important attributes for optimization: 1) Activation. This can be defined as how attractive the offer is, 2) Profit. This is dependant of several cost-related factors.

It was clearly identified that large discounts increased activation, but on the other hand could severely decrease profit. However, if combined with requirements on order size and similar constraints, relatively high values on both activation and profit could be found.

Acknowledgments

Partly funded by The Knowledge Foundation, grant nr. 20160035.

References

1. R. Agrawal and R. Srikant, Fast algorithms for mining association rules in large databases, in *Proceedings of the 20th International Conference on Very Large Data Bases*, VLDB '94 (Morgan Kaufmann Publishers Inc., San Francisco, CA, USA, 1994).
2. H. Sundell, R. König and U. Johansson, Pragmatic approach to association rule learning in real-world scenarios, in *2015 International Conference on Computational Science and Computational Intelligence (CSCI)*, Dec 2015.
3. H. T. Kung, F. Luccio and F. P. Preparata, *J. ACM* **22**, 469 (October 1975).

A data-driven approach to online fitting services

Tuwe Löfström*, Ulf Johansson*, Jenny Balkow+ and Håkan Sundell*

*Dept. of Computer Science and Informatics, Jönköping University and
Dept. of Information Technology, University of Borås, Sweden
{tuwe.lofstrom, ulf.johansson, hakan.sundell}@ju.se

+Swedish School of Textiles, University of Borås, Sweden
jenny.balkow@hb.se

Being able to accurately predict several attributes related to size is vital for services supporting online fitting. In this paper, we investigate a data-driven approach, while comparing two different supervised modeling techniques for predictive regression; standard multiple linear regression and neural networks. Using a fairly large, publicly available, data set of high quality, the main results are somewhat discouraging. Specifically, it is questionable whether key attributes like sleeve length, neck size, waist and chest can be modeled accurately enough using easily accessible input variables as sex, weight and height. This is despite the fact that several services online offer exactly this functionality. For this specific task, the results show that standard linear regression was as accurate as the potentially more powerful neural networks. Most importantly, comparing the predictions to reasonable levels for acceptable errors, it was found that an overwhelming majority of all instances had at least one attribute with an unacceptably high prediction error. In fact, if requiring that all variables are predicted with an acceptable accuracy, less than 5% of all instances met that criterion. Specifically, for females, the success rate was as low as 1.8%.

Keywords: Predictive regression; online fitting; fashion.

1. Introduction

Today, worldwide returns total more than US$ 0.5 trillion annually.[a] Furthermore, returns are especially troublesome for clothing retailers, where the return rate is over 10%, which is the highest among retail segments. For clothing retailers, this rate of returns is extremely problematic since only 48% of all returns can be resold at full value. There is, however, an even larger implication of the ever-growing number of returns. i.e., the impact

---

[a]www.marketwatch.com/story/consumers-return-6426-billion-in-goods-each-year-2015-06-18

the resulting unnecessary transports have on the environment. So, reduced returns would benefit not only e-tailers and their customers, but also the public in general. One specific problem, leading to many returns, is the fact that a customer purchasing clothes online must somehow choose a specific size, without being able to try the item of clothing on. In fact, this has lead to a behavior where a customer orders two or more identical items, of different sizes, and returns all but one.

With this in mind, the quality of the services provided for online fitting by the e-tailers in fashion, becomes very important. In this paper, we look at one specific type of fitting service, where the user inputs a few, but very important, measurements as sex, weight and height, before an algorithm predicts a number of other variables like sleeve length, neck size, waist and chest. One example of such a service is the plugin provided by the company SizeMe,[b] as used by for instance TailorStore.[c] Since the algorithm used by SizeMe is proprietary, we do not know exactly how the input data is used in order to produce the predicted sizes. Because of this, it is very important to recognize that this study in no way evaluates the specific service provided by SizeMe. Instead, we address the problem in a more general way; how can we use a data-driven approach to estimate some output variables (neck size, chest etc.) based on just a few input variables like height and weight? For this purpose we use a fairly large data set called ANSUR2, which is a public data set drawn from the 2012 U.S. Army Anthropometric Survey.[1] In the experiments, we compare using the standard technique multiple linear regression to neural networks, and we investigate the importance of adding a few readily available input measurements to the three basic variables sex, weight and height. Finally, we make a simulation where we look at the quality of the predictions, and determine whether these would be good enough if used as a guide when purchasing a shirt.

We start by providing a background for our study in Section 2, where we also give an overview of predictive modeling. In Section 3, we present the method employed, including the experiments. The main results are presented in Section 4, before we conclude and suggest some future work in Section 5.

---

[b]www.sizeme.com
[c]www.tailorstore.com/mens-body-measurements

2. Background

Garments that fit remain one of the final frontiers for a seamless online shopping experience. Companies and researchers alike struggle to find the optimum solution to help customers, which has resulted in a surge of online service providers that offer customer guidance.[2] The service provided is usually based on one of two types of logic; *self measurements* or *body scanning*. Both solutions have their inherent problems. The body scanning solutions today are based on a 3D-scanned image which becomes very accurate in measurement, but this technology is not available to all.[3] Service providers using ordinary computer cameras, such as UPcload, are emerging, but is based on 2D measures of the body and requires standing in front of a camera.[2] The other technique, which is more frequently used is based on manual body measurements. The inherent problem with this solution is twofold; first, it takes time to measure and enter data into the system which discourages the consumer and, second, previous studies show that manual measures are not very accurate.[4] The main exceptions are two measures, height and weight, that most consumers have accurate knowledge about without checking again. Thus it is very convenient to make predictions of size based on these measures only.

2.1. *Predictive modeling*

Most predictive techniques consist of a two-step process: first an *inductive step*, where a model is constructed from labeled data, and then a second, *deductive*, step where the model is applied to test instances. If the target variable is discrete, and restricted to a set of pre-defined labels, the task is called *classification*; if the target is numerical, it is instead called *regression*.

Predictive regression, consequently, maps each instance $\mathbf{x}$ to a continuous target attribute T. More technically, an algorithm uses a set of labeled training instances, each consisting of an input vector x_i and a corresponding target value t_i to learn the function $T = f(\mathbf{x}; \theta)$. During training, the parameter values θ are optimized, based on a score function. When sufficiently trained, the regression model is able to predict a value t_j, when presented with a novel (test) instance $\mathbf{x}_j$.

In this study, we compare traditional multiple linear regression to using an artificial neural network for the modeling. With multiple linear regression, the model describes the dependent variable as a linear combination of the independent variables. Artificial neural networks (ANNs) have proved to be successful in numerous data mining applications. Specifically, *multilayer perceptrons* (MLPs), i.e., feed-forward networks with one or more hidden layers using non-linear activation functions, are general function approximators; in fact MLPs with just a single hidden layer are capable of representing any continuous function.

3. Method

3.1. *The data set*

The data is divided into two sets, one for males containing 4082 instances and one for females containing 1986 instances, which include the same set of variables. The data contains about 100 body measures and some additional variables representing sex, age, position etc. Most of the measures were deemed not relevant for this study and consequently excluded. The eight variables used in the study are presented in Table 1, with descriptions taken from the documentation accompanying the data.[1] The full names used in the source are presented, with parts in brackets excluded in our text.

Table 1. Description of the data used.

Variable Name	Purpose	Description
Neck [Circumference Base]	Target	The circumference of the base of the neck encompassing the lateral neck and anterior neck landmarks
Sleeve [Outseam]	Target	The distance between the acromion landmark on the tip of the right shoulder and the stylion landmark
Chest [Circumference]	Target	The maximum circumference of the chest at the fullest part of the breast
Waist [Circumference]	Target	The horizontal circumference of the waist at the level of omphalion encompassing the waist (omphalion) landmarks
Stature	Input	The vertical distance from a standing surface to the top of the head
Weight	Input	Participant stands on the platform of the scale with weight distributed evenly on both legs
Span	Input	The distance between the tips of the middle fingers of horizontally outstretched arms
Age	Input	Age of the participant.

3.2. *Experiments*

In the first experiment, we compare using different modeling techniques (ANNs and multiple linear regression) as well as different sets of input variables. More specifically, we try the following four sets of input variables:

- **Basic**: Just the weight and the length. This represents the baseline, and is a typical choice for most online fitting services.
- **Age**: Weight, length and age. The reason for including age is that it is a variable that all users know the value for, i.e., no measurement is necessary.
- **Span**: Weight, length and span. Obviously several different variables could be used instead, but we settled for using span since it is fairly easy to measure with high precision.
- **All**: Weight, length, age and span.

Obviously we look at men and women separately. For the modeling, the tool KNIME[5] was used. After some experimentation with different parameter settings, the default settings were used for both ANNs and multiple linear regression. Min-max normalization in the interval (0,1) was used with the ANNs. The actual evaluation uses 5-fold cross-validation, i.e., all reported results are the average results from the five folds.

In the second experiment, we simulate using the predictions when ordering a shirt. This is done by, for each person in the data set, comparing the predicted value of the four key variables (Neck, Sleeve, Chest and Waist) to the true values. In order to determine whether the predictions are accurate enough, we set a threshold for every variable. It must be noted that there is no universally accepted method for setting this threshold, and the exact definition will of course impact the results significantly. For this preliminary study, we decided to base the threshold on the typical differences between two sizes, i.e., a prediction is correct if the absolute error is smaller than half the range of one size. Using this definition, we get the thresholds in Table 2 below.

Table 2. Thresholds for the variables given as maximum absolute error.

	Men	Women
Neck	10 mm	10 mm
Sleeve	5 mm	5 mm
Chest	35 mm	20 mm
Waist	30 mm	20 mm

4. Results

Starting with the results for the multiple linear regression, Table 3 below shows the accuracies for the different input variable sets.

Table 3. Accuracy using multiple linear regression.

	MAE				R^2			
Men	**Basic**	**Age**	**Span**	**All**	**Basic**	**Age**	**Span**	**All**
Neck	12.9	12.8	12.9	12.8	0.59	0.60	0.60	0.60
Sleeve	14.5	14.5	10.8	10.8	0.64	0.64	0.81	0.81
Chest	25.5	24.4	25.5	24.4	0.86	0.87	0.86	0.87
Waist	35.1	32.2	34.8	31.8	0.84	0.87	0.85	0.87
Women	**Basic**	**Age**	**Span**	**All**	**Basic**	**Age**	**Span**	**All**
Neck	10.6	10.6	10.6	10.6	0.53	0.53	0.53	0.53
Sleeve	14.2	14.2	9.6	9.6	0.63	0.63	0.83	0.83
Chest	33.9	32.7	33.7	32.6	0.73	0.75	0.74	0.75
Waist	38.4	37.4	38.3	37.4	0.76	0.77	0.76	0.77

Looking at the mean absolute errors, which of course are given in millimeters, and the R^2-values, we see that the variable age adds some information regarding Waist and Chest; both for men and women. Span, as expected, is extremely valuable for Sleeve. The best results overall are achieved when all variables are used, but to be honest the differences compared to using only weight and length must be considered marginal, with the exception of Sleeve. Comparing the results for women and men, it may be observed that the R^2-values are generally substantially higher for men; with the exception of Sleeve. Turning to the accuracy for the ANNs, we see in Table 4 below the picture is very similar.

Table 4. Accuracy using artificial neural networks.

	MAE				R^2			
Men	**Basic**	**Age**	**Span**	**All**	**Basic**	**Age**	**Span**	**All**
Neck	12.9	12.8	12.9	12.8	0.60	0.60	0.60	0.60
Sleeve	14.6	14.7	11.1	11.2	0.64	0.63	0.80	0.79
Chest	25.3	24.2	25.6	24.7	0.86	0.88	0.86	0.87
Waist	35.1	32.4	34.7	32.0	0.84	0.86	0.85	0.87
Women	**Basic**	**Age**	**Span**	**All**	**Basic**	**Age**	**Span**	**All**
Neck	10.6	10.6	10.6	10.6	0.53	0.53	0.53	0.53
Sleeve	14.3	14.3	9.9	10.1	0.63	0.63	0.82	0.81
Chest	34.0	32.8	33.9	32.7	0.73	0.75	0.74	0.75
Waist	38.6	37.6	38.8	37.6	0.76	0.77	0.76	0.77

Somewhat discouraging, when comparing the Tables 3 and 4 above, we see that there is nothing to gain from using the more advanced ANN models. If anything, the multiple linear regression models are actually marginally more accurate than the ANNs.

Turning to Experiment 2, Table 5 below shows the proportion of all instances that are deemed to be sufficiently accurate, i.e., the absolute error is smaller than the maximum absolute error, as defined in Table 2. The last column shows the proportion of all instances where all individual predictions were inside the acceptable intervals.

Table 5. Proportion of predictions that are correct, i.e., with an absolute error smaller than the threshold.

	Neck	Sleeve	Chest	Waist	Total
Men	0.476	0.227	0.730	0.503	0.045
Women	0.560	0.216	0.364	0.320	0.018

Looking first at the results for men, we see that the hardest attribute to predict with an acceptable accuracy is Sleeve. Less than one quarter of all predictions are within the acceptable interval. Waist and Neck are both also quite hard; approximately 50% of all predictions obtained a too high mean absolute error. All-in-all, less than 5% of all instances had acceptable predictions for all four output variables.

Actually, the picture for females is even worse. Here, for both Waist and (in particular) Chest, the proportions of poor predictions are much higher than for men. One key reason is of course that the acceptable intervals are smaller; but it must be noted that even a direct comparison with regard to mean absolute errors, shows that these attributes are in fact harder to predict for women. In total, only 1.8% of all females in the data set, had all four predictions inside the acceptable intervals.

While we do not claim that all instances with at least one prediction having an unacceptably high error would necessarily lead to a return, we still find the results remarkable. At the very least, it is questionable if systems based on techniques like the ones tried here really can be useful. In fact, showing predictions with a precision of 1 cm, given the magnitude of the prediction errors we observe here, might even be considered misleading.

5. Concluding Remarks

We have in this paper investigated the possibility to predict some key attributes related to fitting of shirts, based on some very basic variables like weight and height. From the experiments, we see that while the models often are fairly accurate, as seen by low mean absolute errors and high R^2 values, it remains questionable, whether the precision is of acceptable quality. For this specific data set, standard linear regression was as accurate as the more complex and potentially more powerful neural networks. Generally, the effect of adding age to the input variables was marginal, while the measure span, as expected, could lower the errors for Sleeve. Comparing the predictions to pre-defined levels for acceptable errors, it was found that less than 5% of all instances were predicted with the necessary accuracy. In fact, for females the success rate was as low as 1.8%.

Regarding future work, there are many different possibilities to pursue. It would be interesting to look for other measures that are easy to take with good precision, but with a higher predictive power. From a more technical perspective, it must be noted that both techniques used here, i.e., multiple linear regression and neural networks, are *eager* learners, meaning that they use all training instances to produce a model that is later used for the predictions. A different approach would be to use a *lazy* learner, typically some variant of k-nearest neighbors. Using k-nearest neighbors, the prediction is more specialized since it is based only on the most similar examples, i.e., here people with very similar weight, height, age etc.

Acknowledgments

Funded by The Knowledge Foundation, grant nr. 20160035.

References

1. C. C. Gordon et al., *2012 Anthropometric Survey of U.S. Army Personnel: Methods and Summary Statistics - Final Report*, tech. rep., Aemy NATICK Soldier Research Development and Engineering Center MA (2012).
2. A. Vecchi, F. Peng and M. Al-Sayegh, Size recommendations in online fashion retail: Opportunities and challenges, in *Advanced Fashion Technology and Operations Management*, (IGI Global, 2017) pp. 248–362.
3. Y.-A. Lee, M. L. Damhorst, M.-S. Lee, J. M. Kozar and P. Martin, *Clothing and Textiles Research Journal* **30**, 102 (2012).
4. P. E. Zwane, M. Sithole and L. Hunter, *International Journal of Consumer Studies* **34**, 265 (2010).
5. M. R. Berthold et al., KNIME: The Konstanz Information Miner, in *Studies in Classification, Data Analysis, and Knowledge Organization (GfKL 2007)*, (Springer, 2007).

Analysis of consumer emotions about fashion brands: An exploratory study

Chandadevi Giri[1, 2, 3], Nitin Harale[1], Sebastien Thomassey[1] and Xianyi Zeng[1]

[1]Gemtex, Ensait, 2 Allée Louise et Victor Champier, 59056 Roubaix, France
[2]University of Boras, SE-501 90 Boras, Sweden
[3]Soochow, College of Textile and Clothing Engineering, Suzhou 21506, China

Fashion products are characterized by high variability in terms of rapidly changing consumer preferences. Consumers express their emotions on social networks such as Twitter, Facebook and Instagram. The main objective of this paper is to explore Twitter data for recognizing customer sentiments about fashion brands and to analyze their overall perception towards the brands. Two brands, Zara and Levis, are considered and users' tweets related to these brands are analyzed using text mining and Naïve Bayes classifier. The results from this study suggest that social media such as Twitter can serve to be the repository of consumer sentiments and opinions. Sentiment analysis of the tweets can indicate fashion trend and thereby enable fashion brand companies to quickly respond to the ever changing consumer demands.

Keywords: Sentiment analysis; fashion industry; big data; Twitter.

1. Introduction

Fashion industry is one of those industries which are striving to exploit "Big Data" technology for the real time prediction of customer demands, their choices and sentiments about the products. Due to the advent of e-commerce, digital businesses, and the social media, consumers are rapidly voicing their preferences and opinions about the fashion products all over the internet than ever before. Due to ever-changing consumer preferences and demands, it is indispensable for the fashion brand companies to analyze the market trend in a real time. However, it poses significant challenges to the fashion designers and manufacturers to track the consumer choices and market trend as they are increasingly leaving their footprints on social media such as Twitter, Facebook, Instagram, etc. Therefore, fashion industry is flocking to "Big Data" analytics and web semantics tools to accurately forecast demands of their customers, and to prevent them from switching to other brands. Many researchers in the domain of fashion technology have argued for the extensive analysis of social media platforms to make the flawless forecasts using effective "Data Analytics" tools to spot the real time fashion trends [1]. It is in this context, this paper aims to extract the consumer sentiments from their tweets on Twitter platform.

Twitter is an online social networking and blogging service that was founded in 2006. One of the major applications of Twitter analysis has been the prediction of political election results in major countries such as USA, UK, Canada, Australia, etc. [2]. Unlike the election results, fashion industry does not depend on the fixed time market trend prediction given the rapid fluctuations in the consumer demands and choices, and therefore it is a great challenge to extract real time information from the Twitter data.

To address this challenge, this paper is an attempt to detect the real time perceptions of customers about fashion brands by analyzing Twitter data. We randomly chose only two well-known fashion brands, Zara and Levis, for the sake of small scale study. However, more number of brands could be studied for the same purpose. We applied Naïve Bayes classification method to classify emotions from the tweets as it requires only a small amount of trained data. For the computational part, we used an open-source statistical software, RStudio because of its flexible functionality for exploratory work.

The rest of the paper is structured as follows: In the second section, we present a brief overview of current approaches in the social media data analysis for fashion industry. In the section 3, we present the theoretical research framework, in which research questions, data collection and processing is described. Experimental results are discussed in the section 4, and the final section 5 presents the future scope of the study.

2. Related Literature

Ortigosa et al. [3] suggest that the companies should create their marketing campaigns based on the understanding of consumer's emotional attachments toward all the aspects of the brands and thereby provide them customized products. Vishal Vyas et al. [4] tested the performance of Rapid Miner and machine learning algorithms such as SVM, Naïve Bayes, Decision Trees by using them to derive the consumer sentiments from the tweets. Besides other prominent industries, fashion industry is investing in analytical tools and technology infrastructure required for social media analytics. However, sentiment analysis is one of the topics that remain to be the part of mainstream Big Data analytics and it warrants a further research in order for it to help today's digital businesses to survive the fierce competition. Consumer sentiment analysis using social media data could become the key part of both short and long term strategic planning of the fashion industry and it could enable them to retain and expand their customer base and make them happy and satisfied [5].

In line with the above-mentioned researches, we argue that sentiment analysis could help fashion industry to understand their customers' demands, their grievances, and product preferences in a real time to be able to gain a competitive advantage in a today's digital business era.

3. Research Framework

It is important for fashion industry to be innovative to strike the right spot for managing its brand value. Nowadays, Consumers are more brand conscious than before and their choices are changing quickly. Any outfits get outdated within a month. Thus, it is essential for the fashion industry to integrate their business with the advanced big data tools to get the real time valuable insights and exploit the opportunity of this Internet era.

As online users are increasing dramatically, our aim is to derive sentiments from the tweets of consumers focusing on fashion brands and to classify emotions and polarity of tweets.

To reach this aim, we address the following research questions;
1. Are the emotions in the tweets useful to know the popularity of fashion brands?
2. Can we identify the public opinion in the tweets about the brands?

To address our research question we choose two brands for our study 'Zara' and 'Levis'. Figure 1 depicts the research framework for this paper. The steps are followed by tweets extraction, cleaning tweets and finally doing sentiment analysis on two fashion brands.

Figure 1. Research framework.

3.1. *Data Extraction and Preprocessing*

Data is extracted from Twitter API using direct authentication with the help of "ROAuth" package in RStudio. We followed the standard method of fetching Twitter data by creating an app via which we could request API to Twitter. However, Twitter observes and strictly restricts users from accessing high load.

API, owing to which, we could fetch Tweets for only last 10 days. While fetching tweets, we used hashtags to get precise information about the brand

Zara and Levis for 10 days (from 18-02-2018 to 28-02-2018). During these 10 days, we collected 702 and 980 tweets for #Zara and #Levis respectively. As we know that the tweets data are messy and it is quite difficult to work with such cluttered data, therefore, data cleaning is the first step in text mining. Text mining was applied to remove hashtags, URLs, punctuation, retweets and whitespaces from the tweets data. As a result, the no. of tweets of the two brands is reduced to 534 and 586. Figure 2 illustrates how tweets are transformed after cleaning.

Figure 2. Tweets data before and after cleaning.

3.2. *Sentiment Analysis for Brand Zara and Levis*

Machine learning algorithms such as Naïve Bayes, SVM (Support Vector Machines), k-NN are popular techniques for opinion mining and text categorization from Text data [6]. However, we chose Naïve Bayes algorithm because it is advantageous over other methods since it requires a small amount of training data for parameter estimation. Naïve Bayes is a probabilistic classifier and it works on the principle of "Bayes' Conditional Probability Theorem." Bayes' rule is presented in Eq. (1).

$$P(Y \mid X) = \frac{P(Y \mid X)P(X)}{P(Y)} \quad (1)$$

We have a dataset T= {t_1, t_2, t_3... t_n}, which includes n tweets. For each given tweet in Twitter data "T", Naïve Bayes calculates posterior probability of classes c ∈ C, and assigns the class, $\hat{C}$, having the maximum posterior probability to the given tweet.

$$\hat{C} = argmax\ P(c|t) = argmax\ \frac{P(t|c)P(c)}{P(t)}, \text{ where } c \in C \quad (2)$$

It classifies the tweets based on the score that is calculated for the emotions present in the tweets. We trained Naïve Bayes on the dataset that is created by

[8], and it is comprised of approximately 1542 words and class labels of six emotion categories such as "joy", "sadness", "anger", "surprise", "fear," and "disgust." The score of a word "w" in a tweet "t" for a class "c" is represented as in Eq. (3).

$$\text{Score [c]} = \log \text{prior [c]} + \log \text{likelihood [w,c]} \qquad (3)$$

Where, $\log \text{prior [c]} = \log \dfrac{N_c}{N}$ and $\log \text{likelihood [w,c]} = \log \dfrac{\text{count}(w,c)+1}{\sum \text{count}(w',c)+1}$,

N = total counts of words in Dataset and N_c = counts of words from Dataset in each class c [7]. And, for each word w in L (L = Lexicon of dataset D), count (w,c) = number of occurrences of w in Tweets and count (w',c) = number of occurrences of w in L.

Emotion class having a maximum score for a given tweet out of six emotions is the best fit which can be seen in the below Figure 3. If no word in a tweet matches with the dataset [8], the best fit is assigned as "NA" which we defined as 'unknown' category.

	ANGER	DISGUST	FEAR	JOY	SADNESS	SURPRISE	BEST_FIT
5	1.46871776464786	3.09234031207392	2.06783599555953	1.02547755260094	7.34083555412328	2.78695866252273	sadness
6	7.34083555412328	3.09234031207392	2.06783599555953	1.02547755260094	1.7277074477352	2.78695866252273	anger

Figure 3. Best-fit emotion based on maximum score.

Tweets devoid of any of these emotions are then categorized as "unknown." Emotions were classified and clustered according to categories of emotions for the brands Zara and Levis and it can be visualized in Figures 4 and 5.

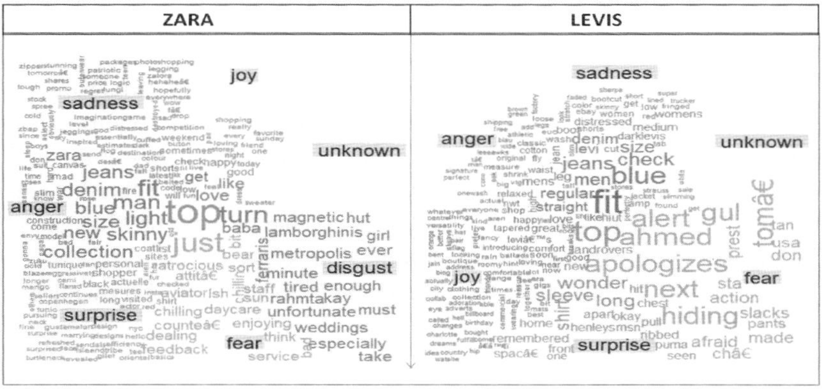

Figure 4. Word Cloud for emotions for the brand Zara and Levis.

As we can see in Figure 5 that there are significantly many tweets in the "unknown" category. Approximately 75% and 61 % of the tweets for the brand Zara and Levis respectively fall under "unknown" category. It could be attributed to the fact that the dataset we used is relatively small and the trained data entail only 1542 words. With the help of wordcloud clustering, as shown in Figure 4, we can see that words in the tweets are expressing customer sentiments for the products of the two brands. For example, as seen in Figure 4, emotion "anger" expresses certain features of brand Zara, while for Levis brand it expresses a few different features.

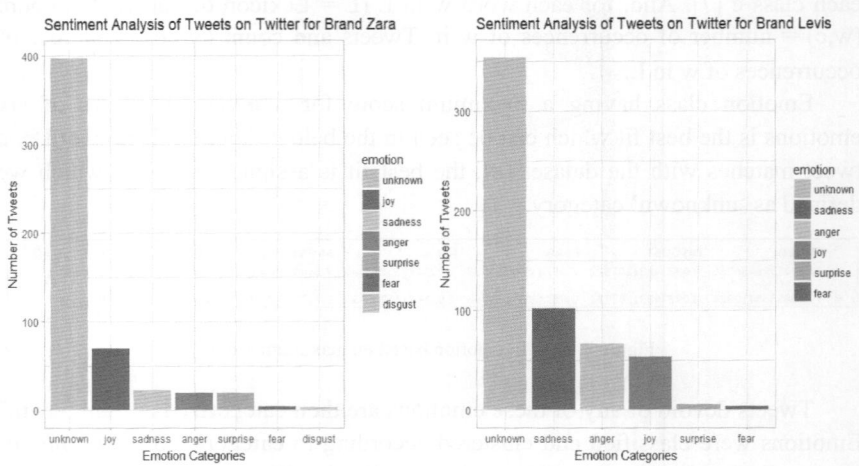

Figure 5. Sentiment Analysis of Tweets for Brand Zara and Levis.

The categories of emotions as depicted in Figure 5 can further be classified as an extension in order to discern the overall polarity of emotions such as "Positive" or "Negative" or "Neutral." as is shown in Figure 7. We used Janyce Wiebe's subjectivity lexicon [8], for the classification of polarity in the tweets. We calculated log likelihood of each tweet considering that it either belongs to "Positive," "Negative" or "Neutral" class. The value of the ratio "likelihood of Positive/likelihood of Negative" determines the class for each sentence. For example, for the brand Levis, the best fit for a polarity can be calculated as below.

if score > 1 = positive, if score < 1 = negative; if ratio score = 1 then its neutral

These categories could help designers and marketing analysts to have the clear picture of the polarity of their consumers' opinions. As we can see from the Figure 6, both brands have "Positive," "Negative," and "Neutral" tweets and their popularity is evident from the high number of positive tweets in 10 days period.

	POS	NEG	POS/NEG	BEST_FIT
5	17.9196623384892	17.1191924966825	1.04675862146897	neutral
6	17.2265151579293	0.445453222112551	38.6718836070664	positive

Figure 6. Best-fit polarity score for a tweet.

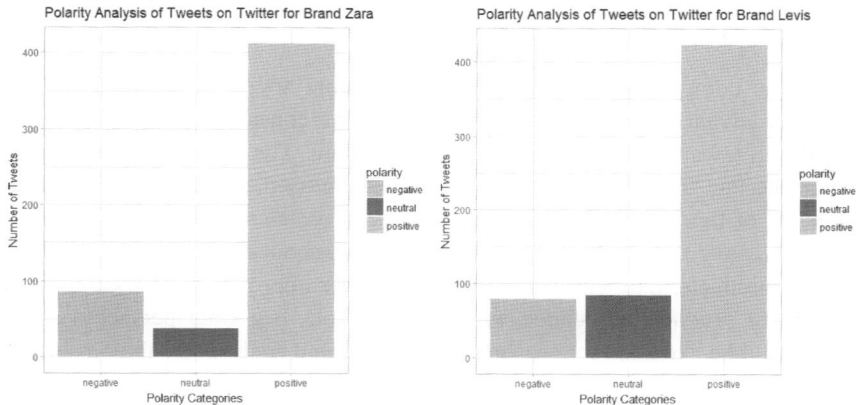

Figure 7. Polarity Analysis of Tweets on Twitter for Brand Zara and Levis.

4. Results, Discussion and Conclusion

We extracted the tweets for the brand Zara and Levis. After we applied data cleaning and text mining on tweets, the number of the tweets of Levis and Zara was reduced significantly. Sentiment analysis was performed on tweets for emotions and polarity for both the brands. The tweet emotions were clustered into seven categories, and we then analyzed the words related to each category as shown in Figure 4. It is important to note that the algorithm was unable to classify the emotions for approximately 75% tweets and 61% tweets of the brands Zara and Levis respectively. Such tweets are categorized as "unknown". It could be attributed to the problematic trained data in the sense that words included in the trained data are limited and it is not representative of all kinds of emotions. Moreover, the words devoid of expressions or the words used for certain product names, for ex. "Jeans," "Clothes", would not possibly be classified as emotions. This paper is not aimed at comparative analysis of the popularity of two brands; however, by identifying the polarity of emotions in the tweets towards "Positive," "Negative" and "Neutral," popularity of the brands could be monitored. Furthermore, the polarity classification indicates that the users are quite positive about both the brands because more than 70% tweets fall under positive category. In conclusion, sentiment analysis could be performed on the tweets and it is possible to identify consumer opinions about the fashion brands. Positive emotions could be the indicator of popularity of the brands. Sentiment analysis is the part of big data analytics and investment in such tools could enable fashion industry to improve design, production and the overall performance of their businesses.

5. Future Work

Given the complexity of sentiment analysis from social media data and the dearth of advanced research work in this area, we consider this paper to be the initial step towards the advanced research in future. The limitation of this study lies in the fact that tweets for only ten days were used for the analysis and the language of the tweets was English. Moreover, we extracted data using #Zara and #Levis and the result was for overall opinion about the brand and not about specific products. Extracting data could be further improved by using keywords specific to products associated with brands. From Big Data perspective, this work could be extended for seasonal sentiment trend analysis for famous brands if we use historical tweets spanning over a long period of time and also include different regions in the world.

Acknowledgments

We extend our sincere gratitude to the SMDTex – Sustainable Design and Management of Textiles, Erasmus Mundus Joint Doctoral commission for providing us with excellent research environment and consistent support.

References

1. Beheshti-Kashi S., Thoben K.-D. *Dynamics in Logistics Lecture Notes in Logistics* (2015).
2. Conway B.A., Kenski K., Wang D. J. *Computer-Mediated Communication 20* (2015).
3. Ortigosa A., Martín J.M., Carro R.M. *Computers in Human Behavior* (2014).
4. Vyas V., Uma V. *Procedia Computer Science* (2018).
5. Wright S., Calof J.L. *European Journal of Marketing* (2006).
6. Rezwanul M., Ali A., Rahman A. *International Journal of Advanced Computer Science and Applications* (2017).
7. Strapparava C. and Valitutti A. *Proceedings of the Fourth International Conference on Language Resources and Evaluation* (2004).
8. Riloff E., Wiebe J. *Proceedings of the 2003 conference on Empirical methods in natural language processing* (2003).

An intelligent approach to study suiting fabric formability[*]

Zhebin Xue, Lei Shen, Jianli Liu

School of Textiles and Clothing, Jiangnan University
Wuxi, Jiangsu Province, 214122, P.R. China

Yu Chen

Clothing Institute, Shanghai University of Engineering Science
Shanghai, 201620, P.R. China

In order to study the appropriateness of the samples for making good shaped men's suits, as it is called fabric formability, sensory evaluation methods have been applied to obtain panelists' assessments on the shape of the shoulder-backs. During data analysis, principal component analysis (PCA) was initially adopted to reduce the complexity of the system by extracting a small number of important mechanical properties. Then, a fuzzy neural network was developed to model the underlying relations between the samples' formability and their mechanical properties. Finally, a number of testing samples were used to verify the effectiveness of the proposed predictive model.

1. Introduction

Fabric formability is the term used to describe the appropriateness of fabric's properties to achieve the desired garment silhouette when the 2D fabric is made into 3D garment[1]. The evaluation of fabric formability for a specific garment is often conducted by a panel of experts (in tailoring or designing) based on their knowledge and experience, which is a type of sensory evaluation[2]. The reliability of the evaluation largely depends on the intuitive sense of the experts. But textile manufacturers are continuously producing new materials with properties that are different from those of conventional materials. The ever-increasing variety of materials used to make clothes has made it difficult to determine the appropriateness of a given fabric for the desired silhouette design even for very accomplished designers[3,4]. Rational systems with scientific basis must be established to help produce high quality clothes and make correct and quick response to orders[5].

In the current study, sixty-six suiting fabrics with different tactile properties

---

[*]This work is supported by the National Natural Science Foundation of China (No. 61503154).

have been selected as our experimental samples. The sample's mechanical properties have been measured by the Kawabata systems, and a group of experts have been recruited to assess the fabric formability. This assessment was performed by the examination of samples of the fabric made into a representative part of a men's suit according to standardized sensory evaluation techniques and procedures.[2] In order to reduce modeling complexity, the method of PCA has been used to select important mechanical properties for the prediction of fabric formability. On this basis, a fuzzy neural system (ANFIS) has been developed to model the relations between the extracted principal mechanical properties and the experts' sensory assessments about the fabric formability of the samples. With this system, it is possible to predict the formability of a certain fabric for the use of men's suit from its mechanical properties. Finally, a set of testing samples have been used to verify the effectiveness of the proposed predictive model.

2. Experimental

2.1. Materials

Sixty six men's suiting fabrics with different tactile properties were selected for the present study. The experimental samples were collected from fabric stores and mass manufacturers of medium-priced (price range: 500RMB–2000RMB) men's formal suits. Before the tests, all samples were labeled with random codes.

2.2. Mechanical measurement

The Kawabata Evaluation System was used to measure the mechanical properties of the 66 suiting fabrics33-36. In the current study, it is assumed that at most 10 parameters covering fabrics' tensile, shearing and bending properties are more related to the shaping behavior of a textile material. The surface properties are assumed to be less important.

2.3. Visual evaluation of fabric formability

In this section, every sample was evaluated on its formability by a panel of experts. Since the number of samples in this research is relatively large, it is not practical in terms of time, expense to make every fabric into men's suit for our study. In this situation, to find a representative part of the suit to reveal the samples' formability as truly and comprehensively as possible would be more realistic and feasible in the current study.

The representative part to be selected should meet three basic requirements: (1) the production of this part should cover the key processes to make a men's suit (including pushing, shrinking, stretching and overfeeding, etc.) to comprehensively reveal the sample's formability; (2) this part should be significant for revealing the quality of silhouette of men's suits; (3) in order to guarantee that all the samples have the same producing features to exclude as much as possible the impact of construction and tailoring techniques on the silhouette evaluation of the samples, and also for the sake of efficiency, this representative part of men's suit should not be too complicated to process.

After discussing with some experts of suit production, the "Shoulder-Back" part of a men's suit with notched lapel (shown in Fig. 1) was determined to represent the whole suit to be evaluated in the current study.

The main specifications of the "shoulder-back" are determined and illustrated in Fig. 2.

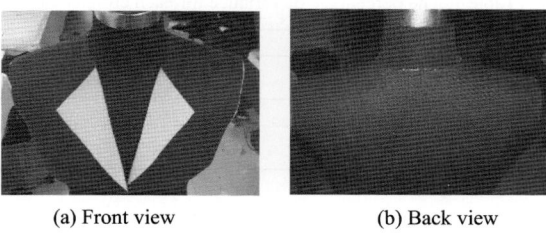

(a) Front view (b) Back view

Fig. 1. Shoulder-Back of men's suit.

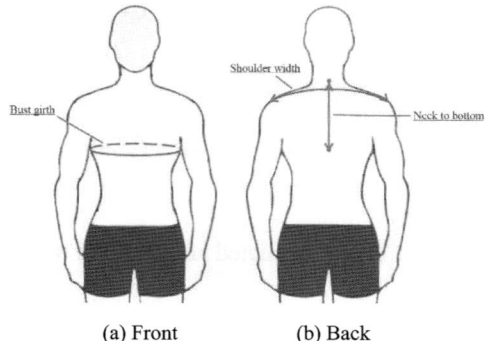

(a) Front (b) Back

Fig. 2. Illustration of major measurements of the "shoulder-back".

In the current study, five experts specialized in fashion design, textile research, and suit construction and processing were invited to participate in the formability evaluation. A training session was organized to make sure that each panelist is well aware of the experimental procedures and is capable of using pre-designed sensory techniques (including terms and scales) to evaluate the formability of the samples. A 7-point scale was used for the evaluation.

3. Results and Discussion

3.1. Feature selection

In this study, it is important to select from the original ten parameters a small number of representative ones to improve the efficiency and effectiveness of the predictive modeling in the next step. The feature selection is realized by Principle Component Analysis.

(1) Extraction of principal components

Table 1 shows the computed percentage of each component's eigenvalue to account for the total eigenvalue, which is named the percentage of variance.

Table 1. The percentage of variance for each component.

	PC1	PC2	PC3	PC4	PC5
λ_i / κ	42.164	28.853	14.133	3.964	3.550
	PC6	PC7	PC8	PC9	PC10
λ_i / κ	3.074	2.232	0.895	0.790	0.346

According to the extraction principle of PCA, the first three components are retained as the PCs (Principal Components), for their cumulative percentage of variance (85.15%) exceeds 85%. And the eigenvalues of the PCs and the corresponding component matrix are computed. According to the result, the first PC mainly composed of five parameters: B, 2HB, G, 2HG and 2HG5, respectively, which reflect the bending and shearing properties of fabrics. W, T and LT, which basically describe some basic properties of fabrics, are representative for the second PC. And the third PC with WT and RT as its major parameters embodies fabrics' tensile properties.

(2) Construction of *PC* formulas

The formula of each *PC* is constructed according to the following equations,

$$F_m = \left(a_{ij}\right)'_{p \times m} X = \left(\alpha_1, \alpha_1, \cdots, \alpha_m\right)' X \quad (1)$$
$$R\alpha_i = \lambda_i \alpha_i$$

where *m* represents the number of the *PCs*, *R* is the matrix of correlation coefficients, and λ_i and α_i correspond to the eigenvalue and the unit eigenvector of the *i*th *PC*.

Finally, the formulas of the three *PCs* are presented as follows,

$$\begin{aligned} F_1 = & 0.407ZX_1 + 0.345ZX_2 + 0.131ZX_3 + 0.007ZX_4 - 0.292ZX_5 + \\ & 0.452ZX_6 + 0.430ZX_7 + 0.447ZX_8 - 0.124ZX_9 - 0.058ZX_{10} \end{aligned} \quad (2)$$

$$F_2 = -0.228ZX_1 + 0.371ZX_2 + 0.510ZX_3 + 0.016ZX_4 - 0.163ZX_5 - \\ 0.021ZX_6 + 0.074ZX_7 + 0.044ZX_8 + 0.500ZX_9 + 0.515ZX_{10} \tag{3}$$

$$F_3 = -0.031ZX_1 + 0.057ZX_2 + 0.082ZX_3 - 0.798ZX_4 + 0.522ZX_5 + \\ 0.011ZX_6 + 0.161ZX_7 + 0.180ZX_8 + 0.135ZX_9 - 0.038ZX_{10} \tag{4}$$

where $ZX_i (i = 1, 2, \cdots, 10)$ refers to the data after standardization.

(3) Construction of the comprehensive model based on PCs

The comprehensive model of the PCs is established as follows:

$$\begin{aligned} F_G &= 0.42164F_1 + 0.28853F_2 + 0.14133F_3 \\ &= 0.1013ZX_1 + \underline{0.2642ZX_2} + \underline{0.2138ZX_3} - 0.1050ZX_4 - 0.0964ZX_5 + \\ &\underline{0.1862ZX_6} + 0.1826ZX_7 + \underline{0.2010ZX_8} + 0.1114ZX_9 + 0.1187ZX_{10} \end{aligned} \tag{5}$$

3.2. Predictive modeling of fabric formability

In this section, an ANFIS model is developed to predict the formability of men's suitings from the five principal mechanical properties. The effectiveness of the obtained model is verified at the end of the study.

(1) Data description

In the current study, the 66 suiting samples characterized in 5 major aspects are used to develop the ANFIS model. To be specific, the five properties are, Hysteresis of bending (2HB), Linearity of load (LT), Shearing rigidity (G), Hysteresis of shear force at 0.5° shear angle (2HG), and Hysteresis of shear force at 5° shear angle (2HG5), which are taken as inputs to the predictive model. The sensory evaluation results on the fabric formability of the samples are taken as the output of the model. The mechanical and sensory results on the 50 samples randomly selected from the 66 samples are used to train the ANFIS model, which are called the training set. The results of the remaining 16 samples are used to verify the effectiveness of the model, which are called the testing set.

(2) Development of the ANFIS

With the 50 sets of training data, two generalized bell-shaped major factors are employed for each of the five inputs to build the ANFIS, which leads to 32 if-then rules to be learned. All the modeling work in this study is implemented using the MATLAB programming tool. As an example, Fig. 3 shows the membership function of the mechanical variable 2HB before and after training. It is evident that the shape of the membership function has gone through significant change during the learning process.

After the parametric learning, a total of 32 if-then rules have been extracted, one of which are presented as follows (noting that input1 to 5 refer to 2HB, LT, G, 2HG, 2HG5, respectively) (see Table 2).

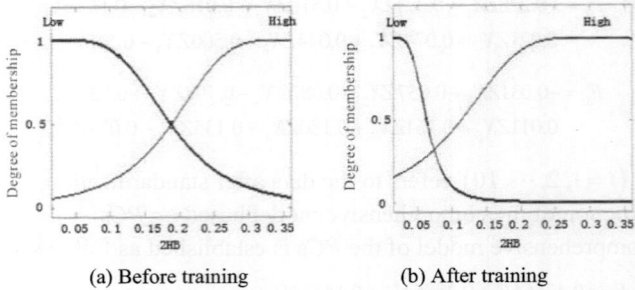

(a) Before training   (b) After training

Fig. 3. The membership function of mechanical variable 2HB before and after training.

Table 2. An example of extracted if-then rules.

Rule N.	Examples of rules
R8	If (input1 is low) and (input2 is low) and (input3 is high) and (input4 is high) and (input5 is high) then, Output = 0.1317*input1-0.1082*input2+0.4544*input3-1.659*input4-2.312*input5-0.2154

(3) Verification
a. Error testing

In order to examine the predictive performance of the obtained ANFIS model, 16 sets of testing data are loaded into the system. Table 3 shows the experimental and predictive results of the fabric formability of each sample (shoulder-back).

Table 3. Comparison between the predictive (Pre) and experimental (Ex) values.

Sample No.	Predictive	Experimental	Error	Sample No.	Predictive	Experimental	Error
1#	4.96	4.81	0.03	9#	4.67	2.26	1.07
2#	3.60	3.54	0.02	10#	5.21	4.77	0.09
3#	5.95	6.32	-0.06	11#	3.91	4.23	-0.07
4#	2.19	1.87	0.17	12#	5.89	6.91	-0.15
5#	2.80	2.30	0.22	13#	3.34	3.28	0.02
6#	5.61	5.43	0.03	14#	1.36	1.04	0.31
7#	2.59	5.01	-0.48	15#	1.88	1.56	0.21
8#	2.88	3.19	-0.10	16#	2.16	2.29	-0.06

It's not difficult to find that for most of these samples, except 9#, there exists a high level of consistence between the predictive values and the experimental ones, with the errors not exceeding 0.5.

b. Verification with linear regression test

SPSS statistics were used to carry out a linear regression test of the experimental and predictive data. Experimental variable is taken as the dependent variable, while predictive variable is the independent or predictor variable.

As the summary of the regression model, the R-square value of this model is found to be 0.947, which means that the predictive values can explain about 94.7% of the change in the experimental values.

As the ANOVA results of the model, the *p*-value (or the "sig.") of the independent variable is 0.000, which means the predictor's effect is statistically significant, or the predictive effect of our model is significant.

Table 4 shows the regression coefficient (B) and the significance level (Sig.). The significance level (or p-value) of 0.000 indicates that the predictor (the predictive values) can well predict the dependent variable (the experimental values).

Table 4. Regression coefficients.[a, b]

Model		Unstandardized Coefficients		Standardized Coefficients	t	Sig.
		B	Std. Error	Beta		
1	Predictive	0.992	0.061	0.973	16.396	.000

a. Dependent Variable: Experimental
b. Linear Regression through the Origin

3.3. *Construction of predictive chart*

Based on the data of this research, a chart (see Fig. 4) has been plotted to show the features commonly observed on the mechanical properties of high-formability suiting fabrics.

In this study, high-formability fabrics refer to the samples whose formability scores (FS) are between 5 and 7. In Fig. 10, each horizontal axis is normalized by the population mean (denoted by 0) and standard deviation (denoted by σ) of the respective parameter, i.e. 2HB, LT, G, 2HG, 2HG5 and the formability score FS. The values for high-formability suiting fabrics fall inside the "snake area" between the two solid lines. The "▼" points denote the mean values of high-formability suiting fabrics on each mechanical parameter.

This chart could be instructive to suit designers or manufacturers to judge the formability of a fabric from its mechanical properties for producing a suit of satisfactory silhouette. If for a certain fabric, the measured value(s) fall(s) outside the snake area, then it could be considered not appropriate for making men's suit. According to different situations, it could be recommended to take special controls during construction and tailoring processes or simply swift to another fabric.

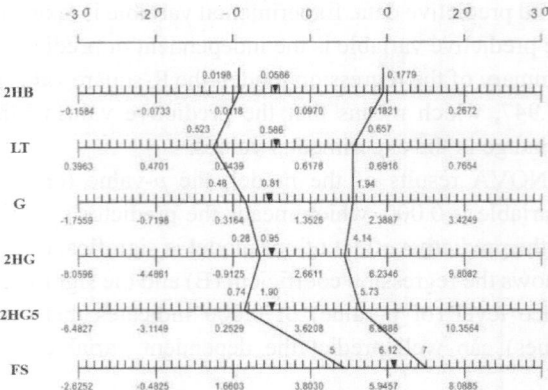

Fig. 4. Data chart for predicting high-formability suiting fabrics.

4. Conclusion

In this study, a predictive model has been developed to study the relations between fabric formability and the mechanical properties for the end-use of men's suit. Compared with conventional relational models which mainly depend on statistical methods, the proposed mathematical method is based on the cooperation of the classical tool PCA for feature selection and the intelligent method ANFIS which is endowed with the advantages of both fuzzy logic and neural network to study the complex and non-linear relations between different datasets (sensory and mechanical datasets) and to produce interpretable results. The current research was proved to be a good attempt after verification with a number of testing samples.

Acknowledgments

This work was funded by the National Natural Science Foundation of China (No. 61503154), and the Priority Academic Program Development of Jiangsu Higher Education Institutions (PAPD).

References

1. Wang F. *Property design of apparel fabrics*. Shanghai: Donghua University Press, 2002.
2. Civille G. V., Carr B. T. *Sensory evaluation techniques*. CRC Press, 2015.
3. Nayak, Rajkishore, and Rajiv Padhye, eds. Garment manufacturing technology. Elsevier, 2015.

4. El-Gamal M. A., Hassouna A., El-Newashy R. F. et al. Prediction of Garment Appearance from FAST Mechanical Properties and fabric drape [J]. Life Science Journal, 2014, 11(12).
5. Nayak R., Kanesalingam S., Wang L. et al. Artificial intelligence: technology and application in apparel manufacturing[C]//TBIS-APCC 2016. Binary Information Press, Textile Bioengineering and Informatics Society, 2016: 648-655.

Author Index

A

Abdelkhalek, Raoua; 660
Afanasyev, Alexander; 95, 1392
Afanasyeva, Tatyana; 95
Ahmad, Bilal; 117
Ahumada-Valenzuela, Omar; 987, 1018
Albayrak, Y. Esra; 1357
Alexander, Afanasyev; 95, 1392
Ali, Tayseer; 500
Almaghrabi, Fatima; 1309
Alvarez Carrillo, Pavel; 1010, 1042
Anselmo Alvarez, Pavel; 1010, 1042
Armour, Cherie; 1267
Asmuss, Svetlana; 154
Aydin, Serhat; 1050
Azad, Mohammad; 371

B

Bai, Ming Qiang; 73, 81
Balkow, Jenny; 1559
Bao, Qiong; 1341
Barbeitos, Marcos; 943
Barros, Laécio C.; 103, 937, 977
Baset, Selena; 870
Bashir, Kamal; 500, 1364
Bassanezi, Rodney; 937
Batista Florindo, João; 943
Bello-Orgaz, Gema; 591
Ben Ayed, Safa; 347
Black, Michaela; 1281
Boltürk, Eda; 1026, 1034, 1176
Bond, Raymond; 1267, 1281
Bordons, Carlos; 929

Bouaziz Mezghanni, Imen; 568
Boukhris, Imen; 660
Brigadnov, Sergey; 95
Bula, Inese; 782
Bustince, Humberto; 145
Büyüközkan, Gülçin; 1066, 1074, 1115, 1184, 1216, 1385

C

Camacho, David; 509, 591, 1083
Cao, Feng; 750, 766, 774
Cao, Yiming; 173
Carvalho, Silvia; 937
Castro, Jorge; 227
Chang, Leilei; 363
Charles, Darryl; 117
Chen, Chen; 626
Chen, Guoqing; 461, 545, 576
Chen, Hongmei; 469
Chen, Luoping; 38
Chen, Meirong; 848
Chen, Panfeng; 832
Chen, Shuwei; 725, 750
Chen, Xia; 191
Chen, Xian-Qiao; 626
Chen, Yu; 1528, 1544, 1575
Chen, Yu-Wang; 363, 389, 405
Chen, Zengqiang; 907
Chen, Zhen-Song; 1168
Cheng, Xiaochun; 585
Chikalov, Igor; 371
Choi, Kup-Sze; 1315
Cleland, Brian; 1281
Cleland, Ian; 1273, 1292, 1300
Çoban, Veysel; 1107

Craparotta, Giuseppe; 1504
Cui, Liu; 1489
Cui, Yuxin; 294

D

Dai, Hong; 1489, 1498
De Bruijn, Oscar; 405
De Jonckheere, Julien; 1520
De Tré, Guy; 198
Deng, Ping; 534, 637, 1454
Deng, Yongmei; 922, 1498
Deshpande, Ashok; 1400
Devi Sekar, Booma; 421, 1470
Diao, Hongyue; 182
Dimuro, Graçaliz Pereira; 145
Ding, Yanyan; 1208
Ding, Zhaogang; 191
Dogu, Elif; 1357
Dong, Yucheng; 191, 213
Doumouras, Craig; 355
Du, Guoping; 878, 884
Du, Shengdong; 492
Du, Zhonghe; 897
Duarte, Diogo; 439

E

Elouedi, Zied; 339, 347, 652, 660
Eray Akkaya, Ergun; 1349
Eren Akyuz, Dilek; 1349
Esmi, Estevão; 103, 943, 977
Espinilla, Macarena; 1292, 1300

F

Fahri Negüs, Ahmet; 1115
Fan, Jing; 1152
Fan, Shaocan; 1470
Fang, Xin; 302
Farooghian, F.; 617
Feng, Lifang; 848

Feng, Renyan; 832
Fernández, Javier; 145
Feyzioğlu, Orhan; 1066
Figueroa Pérez, Juan Francisco; 995, 1091
Flores Leal, Pedro; 1042
Flores, Pablo; 1425
Flores-Vidal, P. A.; 1425
Fu, Huimin; 733
Fu, Li; 109, 129

G

Galán, Ramón; 668
Gao, Jinping; 1152, 1159
Gao, Li; 1159
Gargouri, Faiez; 568
Gastelum-Chavira, Diego Alonso; 1003, 1042, 1058
Gilian, E.; 617
Giri, Chandadevi; 1567
Glauner, Patrick; 439
Gocer, Fethullah; 1066, 1385
Gómez, Daniel; 17, 1425
Gong, Xun; 492
Gonzalez, Carlos; 553
Greer, Jonathan; 1273
Grigorash, Alexander; 1267
Guan, Fei; 294
Güler, Merve; 1074, 1184
Guo, Gongde; 674, 1446
Guo, Xunhua; 545

H

Hang Pun, Sio; 1470
Harale, Nitin; 1567
Hawe, Glenn; 599
He, Xia; 878, 884
He, Xingxing; 725, 790
He, Zhenglei; 1536

Helldin, Tove; 260
Hermans, Elke; 1341
Heyburn, Rachel; 1281
Holčapek, Michal; 1478
Holmes, Danielly; 715
Hong, Long; 855, 878, 884, 891
Hou, Pingzhi; 379
Howard, Sarah; 553
Hu, Jie; 469, 477
Hu, Yunmeng; 469, 526
Huang, Darong; 379
Huang, Guofang; 959
Huang, Ming; 534
Huang, Qianqian; 446, 453
Huang, Yanyong; 453
Hula, Jan; 1440
Hurtik, Petr; 1433
Hussain, Farookh; 355
Hussain, Shahid; 371
Hussein, Ahmed; 683, 1364

I

I Vai, Mang; 1470
Ibrahim, M.; 608
İlbahar, Esra; 276
Ionova, Irina; 1392
Irvine, Naomi; 1300

J

Jaber, Noora Sabah; 683
Ji, Jianmin; 806
Ji, Yanbo; 1498
Jia, Zhen; 485, 708
Jiang, Jiang; 363
Jiang, Ming; 1245, 1254, 1333
Jiang, Nan Feng; 1446
Jiang, Yu-Ting; 24, 33
Jin, Chenxia; 308
Jin, Liuqian; 302

Johansson, Ulf; 1551, 1559
Johnston, Adrian; 12

K

Kadhim, Mustafa; 645
Kahraman, Cengiz; 276, 1034, 1099, 1107, 1176, 1192, 1199
Karasan, Ali; 1026
Karthika, Sivapathasundaram; 585
Kinghorn, Philip; 1462
Koehl, Ludovic; 1520
Koivuaho, T.; 608
Koloseni, David; 260
Kong, Mingming; 243, 268
Korobov, Alexander; 137

L

Labella, Álvaro; 227, 235
Lara-Cabrera, Raúl; 509
Larreta-Ramirez, Elsa Veronica; 1003, 1058
Laureano, Estevão; 937
Lefevre, Eric; 347, 652
Leyva López, Juan Carlos; 987, 995, 1003, 1018, 1042, 1058, 1091
Leyva Lopez, Juan; 1010
Li, Danning; 798
Li, Dingcheng; 534
Li, Fachao; 308
Li, Guo; 379
Li, Jinliang; 485, 708
Li, Mengjun; 363
Li, Tianrui; 331, 446, 453, 469, 485, 492, 500, 517, 526, 637, 683, 708, 1364
Li, Xiaodi; 534
Li, Xinzi; 182
Li, Yingfang; 790

Li, Yunlong; 492, 517
Li, Yu-Xiao; 389
Liang, Sheng; 740
Liao, Yin-Ye; 1233
Lin, Adi; 397, 413
Lin, Hua; 1323
Lin, Zhiwei; 599
Lisboa, Paulo; 10
Liu, Cheng-Gao; 1225, 1233
Liu, Dun; 446
Liu, Haozhen; 1135
Liu, Hong; 798
Liu, Jia-Lun; 626
Liu, Jianli; 1575
Liu, Jing; 252
Liu, Jun; 65, 430, 725, 750, 774, 790
Liu, Kaixuan; 922, 1489, 1498
Liu, Lan; 461
Liu, Mujiexin; 637
Liu, Pengsen; 173
Liu, Shengjiu; 517
Liu, Weifeng; 379
Liu, Xiaohong; 1127, 1208
Liu, Yangguang; 1373
Liu, Yaya; 220
Liu, Yi; 65
Liu, Yuanyuan; 674
Liu, Yunfeng; 914
Lodwick, Weldon; 937
Löfström, Tuwe; 1551, 1559
Long, Si; 1152
López Hernández, Fernado; 668
Lopez Parra, Pavel; 1010
Lopez, Dany; 553
Lu, Jie; 316, 323, 331, 355, 397, 413, 1315, 1323, 1407
Lu, Zhengqiu; 1373
Luo, Bin; 1135
Luo, Minxia; 57, 109
Luo, Shuzhen; 907

M

Ma, Baojun; 576
Ma, Fangli; 848
Ma, Feng; 626
Ma, Jun; 553, 560
Ma, Ke; 1512
Machado, Liliane; 691
Maguire, Paul; 700
Maleku Shrestha, Rohan; 1504
Manuel Escaño, Juan; 929
Martín, Alejandro; 509
Martín, Leon-Santiesteban; 987, 1018
Martínez, Luis; 137, 198, 220, 227, 235, 430
Martínez, Nuria; 17, 1425
McClean, Sally; 117, 1273
Medina, Javier; 1292
Miebach, Alessandro; 977
Miltos, Petridis; 585
Mo, Zhi-wen; 33, 46, 73, 81
Molek, Vojtech; 1440
Montero, Javier; 17
Moore, Ronald; 1504
Moraes, Ronei; 691, 715
Moshkov, Mikhail; 371
Moura, Bruno; 951
Mu, Yao; 545
Mukul, Esin; 1074, 1184
Mulvenna, Maurice; 1267, 1281
Mustafa, Kadhim; 645

N

Naderpour, Mohsen; 1407
Namvar, Anahita; 1407
Nguyen, Linh; 1478
Ni, Lei; 405

Nibouche, Omar; 700
Ning, Xinran; 733, 766, 774
Novák, Vilém; 1417
Nugent, Chris; 1292, 1300

O

O'Neill, Siobhan; 1267
Odabasi Cingi, Didem; 1349
Olaso, Pablo; 17
Onar, Sezi Çevik; 276, 1099, 1107, 1176, 1192, 1199
Öner, Mahir; 1143
Öner, S. Ceren; 1143
Orlovs, Pavels; 154
Ortega, Alfonso; 591
Oussalah, Mourad; 608, 617
Öztayşi, Başar; 276, 1099, 1107, 1143, 1176, 1192, 1199
Öztek, M. Yaman; 1115

P

Pan, Xiaodong; 89
Pan, Yu; 1159
Pang, Yongfeng; 165
Panizo Lledot, Angel; 591
Parr, Gerard; 117
Pedrycz, Witold; 3
Pei, Zheng; 243, 252, 268
Peixoto, Magda S.; 937
Peng, Bo; 526
Peng, Jia-Yin; 24
Pérez Contreras, Edgar Omar; 995, 1091
Pérez, Flávio; 943
Perfilieva, Irina; 1433, 1478
Pilla, Mauricio; 951

Q

Qiao, Xiaoshuang; 674

Qiao, Ying; 243, 252
Qin, Keyun; 220
Qin, Ya; 65
Qiu, Xiaoping; 790

R

Rainer, J. Javier; 668
Ramírez-Atencia, Cristian; 1083
Ramsey, Colette; 1267
Rankin, Deborah; 1281
Reiser, Renata; 951
Ren, Fangling; 243
Rodríguez, Rosa; 198, 220
Rodríguez-Fernández, Víctor; 1083
Rojas, Karina; 17
Romero-Serrano, Alma Montserrat; 1018
Roshany-Yamchi, Samira; 929

S

Salgado, Silvio; 103
Salunkhe, Satish; 1400
Sanchez, Daniel; 977
Santillán Hernández, Luis Carlos; 995, 1091
Schneider, Guilherme; 951
Sekar, Booma Devi; 421, 1470
Shanavas, Niloofer; 599
Shen, Lei; 1575
Shen, Yongjun; 1341
Shi, Shaodong; 959
Shu, Lan; 33
Shu, Tian-jun; 46
Siami, Mohammad; 1407
Simfukwe, Macmillan; 1364
Soares, Elaine; 691
Sofien Boutaib, Mohamed; 339
Solano-Noriega, Jesus Jaime; 987
Song, Jian; 645

Song, Siyuan; 1254
Song, Weiran; 700
Song, Yang; 323
Song, Zhenming; 897
State, Radu; 439
Stoffel, Kilian; 870
Su, Kaile; 862
Sui, Yujia; 1159
Sun, Hao; 907
Sun, Leilei; 461
Sun, Qinglin; 907
Sundell, Håkan; 1551, 1559

T

Tan, Manchun; 914
Tang, Min; 485, 708
Tao, Jin; 907
Tartare, Guillaume; 1520
Teoh, Yuen-Chun Jeremy; 1315
Thomassey, Sébastien; 1504, 1512, 1536, 1567
Torra, Vicenç; 260
Trabelsi, Asma; 652

U

Ukhanova, Maria; 1392
Ummul, F.; 608
Un Mak, Peng; 1470
Uztürk, Deniz; 1115, 1216

V

Valdez-Lafarga, Octavio; 1058
Valtchev, Petko; 439
Vianna, Rodrigo; 715
Voit, Nikolay; 95, 1392

W

Wallace, Jonathan; 1281

Wan, Hai; 806
Wan, Yan; 1152
Wang, Bin; 331
Wang, Guanjin; 1315
Wang, Hongjun; 534, 637, 645, 1454
Wang, Hui; 421, 599, 674, 700, 1300, 1446
Wang, Kewen; 824
Wang, Wei; 1323
Wang, Xiangqian; 560
Wang, Xing; 968
Wang, Xuecheng; 766
Wang, Yajing; 57
Wang, Ying-Ming; 430
Wang, Yisong; 798, 832
Wang, Yue; 1498
Wang, Zhe; 824, 862
Wang, Zhigang; 968
Wei, Qiang; 545, 576
Wen, Lian; 862
Wing, W. Y. Ng; 1300
Witheephanich, Kritchai; 929
Wu, Dianshuang; 355
Wu, Guanfeng; 848
Wu, Lixian; 109
Wu, Maonian; 840
Wu, Wannan; 907

X

Xie, Hong; 1544
Xu, Cong; 287
Xu, Desheng; 914
Xu, Dong-Ling; 287, 1309
Xu, Haiyang; 379
Xu, Jianguo; 363
Xu, Peng; 848
Xu, Weitao; 758
Xu, Xiaobin; 379

Xu, Yang; 89, 725, 733, 750, 758, 766, 774, 790, 848
Xu, Yingying; 173
Xu, Yongchi; 1498
Xuan, Junyu; 397, 1323
Xue, Tao; 922, 1498
Xue, Zhebin; 1528, 1544, 1575

Y

Yahaya, Mahama; 500
Yamin, Adenauer; 951
Yan, Li; 268
Yan, Xin-Ping; 626
Yan, Zheng; 331
Yang, Bo; 740
Yang, Jian-Bo; 287, 1309
Yang, Jie; 553, 560
Yang, Long-Hao; 430
Yang, Wei; 165
Yang, Weihua; 840
Yang, Xin; 446, 453
Yang, Xue; 33
Yang, Yan; 477, 1454
Yang, Ying; 389
Yang, Yuhang; 526
Yao, Li; 173, 182
Yatsalo, Boris; 137
Ye, Qiongwei; 576
Yen, Gary; 7
Yera, Raciel; 227
Yi, Changhai; 1536
Yin, Chengfeng; 485, 708
Yohannese, Chubato Wondaferaw; 500, 683, 1364
Yong, Longhao; 363
Yörükoğlu, Mehmet; 1050
Yu, Quan; 824
Yu, Wenyu; 205
Yuan, Chang; 1152, 1159
Yuan, Yun; 1225, 1233

Z

Zapata, Hugo; 145
Zeng, Xianyi; 1127, 1512, 1520, 1528, 1536, 1544, 1567
Zha, Quanbo; 213
Zhang, Caijuan; 1245, 1333
Zhang, Guangquan; 316, 331, 355, 397, 413, 1323
Zhang, Hu; 294
Zhang, Huarong; 57
Zhang, Li; 1462
Zhang, Mingyi; 740, 798, 840
Zhang, Shuai; 1300
Zhang, Wei; 477
Zhang, Wuyang; 733
Zhang, Xiao; 308
Zhang, Xiaobo; 1454
Zhang, Yi; 1323
Zhang, Ying; 740, 798
Zhang, Yinsheng; 814
Zhang, Zhao-Yi; 1225
Zhang, Zhen; 205
Zhao, Hailiang; 968
Zhao, Jianyang; 855, 891
Zhao, Xin; 1520
Zhao, Xueqing; 922, 1498
Zheng, Bo; 840
Zhong, Xiaomei; 725
Zhong, Xueyan; 461
Zhou, Chunliang; 1373
Zhou, Jincheng; 832
Zhou, Ningning; 959
Zhou, Si Qi; 73, 81
Zhou, Yuan; 645
Zhu, Donghua; 413
Zhu, Fujin; 413
Zhu, Huaying; 287

Zhu, Shaojun; 840
Zhu, Xinjuan; 922
Zhuang, Zhiqiang; 824
Zong, Hui; 891
Zou, Li; 173, 182
Zuo, Danying; 1536
Zuo, Hua; 316